D1524461

APPLICATION OF ACCELERATORS IN RESEARCH AND INDUSTRY

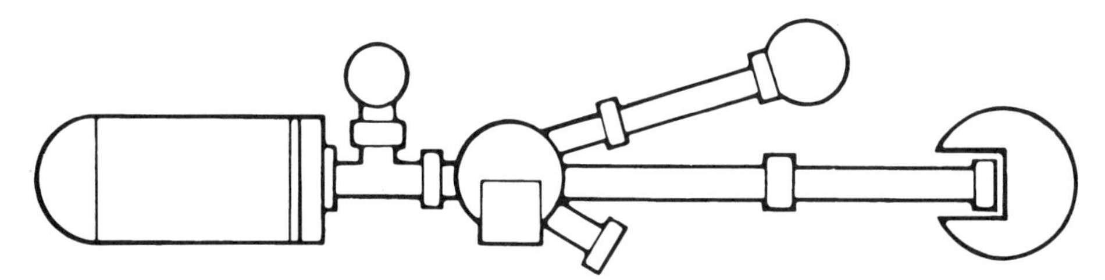

CONFERENCE ON THE APPLICATION OF ACCELERATORS IN RESEARCH AND INDUSTRY

Physics Department • University of North Texas
Denton, Texas 76203-5368

APPLICATION OF ACCELERATORS IN RESEARCH AND INDUSTRY

Proceedings of the Fourteenth International Conference

Denton, Texas November 1996

PART ONE

EDITORS

J. L. Duggan
University of North Texas, Denton

I. L. Morgan
International Isotopes Inc.

American Institute of Physics

AIP CONFERENCE
PROCEEDINGS 392

Woodbury, New York

L.C. Catalog Card No. 97-71846
ISSN 0094-243X

ISBN 1-56396-710-3 (Pt. 1)
1-56396-711-1 (Pt. 2)
1-56396-652-2 (Set)

DOE CONF- 961110

Printed in the United States of America

CONTENTS

PART ONE

Section I. Atomic and Molecular Physics

Section II. Nuclear Physics and Radioactive Ion Beam Facilities and Experiments

Section III. Positron Sources, Experiments and Theory

Section IV. Clusters, Fullerenes, Biomolecules

PART TWO

Section VI. Synchrotron Storage Rings, X-Ray Facilities and Experiments

Section VII. AMS, Microbeams, SIMS

Section VIII. Neutron Techniques for Nondestructive Analysis

Section IX. Radiography, Tomography, Neutron Tubes

Section X. Ion Implantation: Materials Modification, Semiconductors, Metalization, Focused Ion Beams

Section XI. Detectors and Spectrometers: Telescopes, Timing

Section XII. Radiation Processing, Accelerator Techniques

Section XIII. Accelerator Technology, Facilities, New Accelerators, Injectors, Ion Sources, Plasmas, Accelerator Control

Section XIV. Medical Applications: Proton Therapy, Radiation Therapy, BNCT, Isotope Production

Section XV. Accelerator Targets

Section XVI. Energy Loss and de/dx

Section XVII. Safety

PREFACE

The Fourteenth International Conference on the Application of Accelerators in Research and Industry was held on the campus of the University of North Texas (UNT) November 6–9, 1996. The major sponsors of the conference were The U.S. Department of Energy, The National Science Foundation and UNT. The conference is also a topical conference of the American Physical Society sponsored through the Division of Nuclear Physics. An industrial exhibit show composed of 47 companies that distribute components of interest to accelerator users was held in parallel to the conference for the first two days. The conference was attended by approximately 900 accelerator scientists from 47 countries. There were 496 invited papers given in 79 sessions and 260 poster papers. Of the invited papers, 27 were invited poster papers. Participants who gave invited poster papers were allowed five minutes to deliver an oral presentation during a regular session. In that oral presentation, the speaker overviewed what you would see if you visited the poster which was shown at the regular poster session which was held from 6:00 pm until 10:00 pm Friday evening, November 8th in the UNT Coliseum. A social was held during the poster session which seemed to be well received.

As was mentioned above, there were 79 four-hour sessions in the main conference. Most of the sessions had six to eight invited speakers. The conference was opened with three Plenary speakers. Gregory Norton from National Electrostatics Corporation gave a comprehensive overview of most of the uses of the 300 electrostatic accelerators that have applications spanning the field from Art History to Zoology. This talk was followed by a talk by Bill Appleton of Oak Ridge who gave an overview of the Design Project for the National Spallation Neutron Source (NSNS) which, if built, would serve the needs of the ever expanding Neutron Scattering field. The final plenary talk was given by Jerry D. Garrett, also of Oak Ridge. Jerry gave an exciting summary of Accelerated Beams of Radioactive Ions: New Research Opportunities for Nuclear Structure, Nuclear Astrophysics, and Materials Science. In this talk, he summarized the present and future facilities and applications of this exciting field.

As has been the case for most of the previous conferences in this series, Accelerator Based Atomic Physics had the most sessions. There were 18 sessions that dealt directly with atomic physics. There were also two well attended sessions on synchrotrons that contained many papers on basic Atomic Physics. In addition, there were 47 poster papers which fell under the general heading of Atomic Physics. There were eight sessions and 43 poster papers that dealt generally with the subject of Accelerator Technology. These sessions covered topics such as new accelerators, beam handling systems, ion sources, detectors, spectrometers, magnets, etc. Radioactive Beams and Nuclear Physics were the topics for eight sessions which were very well attended. There were two sessions that lasted all day Saturday, November 9th, on Positrons. There were 13 sessions and 56 posters that dealt generally with ion beam analysis. These sessions covered such topics as Rutherford Backscattering Analysis, Particle Induced X-Ray Emission, Elastic Recoil Detection and Nuclear Reaction Analysis, Accelerator Mass Spectrometry and Activation Analysis. Additional sessions and posters covered such topics as; Radiation Processing, Tomography, Single Event Upsets, Ion Implantation, Targetry for Experiments, Production of Radioisotopes for Medicine, BNCT, Radiation Safety, Detectors and Spectrometers, Energy Loss, Clusters, and Free Electron Lasers. Finally, there were three sessions on Pulsed Accelerator Applications and Non-Destructive Testing. These sessions dealt primarily with the use of neutrons for Non-Destructive Analysis.

The conference banquet was held Saturday night, November 9th, in the Silver Eagle Suites of the UNT University Union building. The banquet was followed by a concert from the award winning Two O'Clock Lab Band.

The editors would like to thank the major sponsoring agencies, namely, The United States Department of Energy, The National Science Foundation, and UNT for their continuing support of this conference series. Thanks are also due to the 47 industrial sponsors. They not only helped financially but also provided two complete days of industrial exhibits that add greatly to the total conference experience that each participant enjoyed. We are also indebted to the program and advisory committees for the excellent slate of invited speakers whose presentations were given at the conference. Our gratitude also goes out to the 79 session chairpersons who not only helped to organize many of the sessions, but also conducted the sessions at the conference. The editors now have about 500 referees that can be called on to help with the refereeing of the papers. We wish to especially thank the individuals that helped us referee papers for these proceedings. With faxes, email, and overnight mail we accomplished this task rapidly with the outstanding contribution of these referees.

We wish to thank the administrative staff, students, and professors at the University of North Texas for the monumental effort of helping us put the conference together and following through with the final publication of these proceedings. Without any question, the single person who does more work on this conference than anyone else is Barbie Stippec. As most of you know, Barbie has been with this conference almost 11 years. By now she knows this community very well. For almost each of the participants she has done something beyond the normal call of duty. She is extremely well organized, computerized, e-mailized, faxized and quite able to find and communicate with the participants no matter where they are. Barbie makes the conference work and for that we owe her a great debt. We also wish to thank Barbie's staff, Sidra Hamilton, Carly Slack, Brandy Pariso and Esperanza Sanchez, for doing an outstanding job both before and after the conference. We wish to also thank Steve Renfrow and Sameer Datar for setting up the industrial show and Alan Bigelow for planning, setting up, and conducting the poster display at the conference. We wish to thank Baonian Guo and Dan Scheffer for coordinating the Physics graduate students in the operation of the media equipment for the sessions. In addition, we extend our thanks to all of these students and our other graduate students for helping with the many other tasks associated with a conference of this magnitude. We are indebted to Dr. John Biscar who took all of the pictures that appear in the front of these proceedings.

Finally, we wish to thank the UNT administration and faculty for the support that has been given this conference series since the first time it was held at UNT in 1974. It is not easy to bring 900 visitors onto a campus during the Fall semester and provide classrooms, etc. for the meeting. Without the total support of the University, this would not be possible.

The next conference in this series will take place, as usual, on the campus of UNT, November, 1998.

J. L. Duggan
I. L. Morgan

Dedication

These Proceedings are dedicated to three "very much alive" accelerator practitioners who have pioneered the use of accelerators in the application of nuclear techniques in ion beam analysis and the design and development of high powered accelerators for radiation processing. Each of these individuals have been very important to this conference series and even more to the accelerator community. It is an honor to dedicate the proceedings to these truly outstanding individuals.

Dr. James M. Lambert

James M. (Jim) Lambert was born in Chicago, Illinois but grew up in a number of other states including Iowa and Arizona. He is truly a citizen of the U.S.A., having attended several high schools, but due to health problems was not graduated from any... He had a unique educational experience entering college (Johns Hopkins) without benefit of a high school diploma and to be consistent, went on to graduate school without benefit of a bachelors degree (it was acquired later). During his stay in Arizona, Jim met Margaret whom he followed to Maryland (Johns Hopkins). They were married while he was a graduate student and have three children. At Hopkins, Jim worked with George Owen and Leon Madansky on a variety of nuclear physics problems culminating in a thesis on nuclear polarization. Upon completing his Doctoral work, Jim spent a post doctoral year at Hopkins and two years at the University of Michigan where he became an expert on Magnet Spectrometers.

In 1964, Jim joined the faculty at Georgetown University where he began a fruitful lifetime friendship and collaboration with a Michigan colleague, Paul Treado. Jim and Paul did their research work for many years at the Naval Research Laboratory in Washington. They published an extensive series of papers on multiparticle breakup Nuclear Reactions. At the same time, they wrote a Manual of Nuclear Experiments for Canberra, Inc. This was the first composite Manual of Nuclear Experiments for undergraduate students. In 1970, they also wrote with Bill Bygrave a comprehensive book on *Accelerator Nuclear Physics*. This book contained 30 nuclear physics experiments that could be performed with small accelerators. To this day, this book is still used by universities the world over to teach Accelerator Nuclear Physics.

In 1966, Jim, Paul Treado, and Jerry Duggan designed a summer institute which was intended to teach university faculty how to use small accelerators in the teaching curriculum. It was sponsored by the National Science Foundation and held at what was then called the Oak Ridge Institute of Nuclear Studies. From this summer institute and the one that was to follow in the summer of 1967, Paul, Jim and Jerry decided to have a small conference in Oak Ridge devoted to Teaching and Research with Small Accelerators. This conference was held April 8-10, 1968. It was sponsored by the U.S. Atomic Energy Commission and the whole program was composed of 23 invited speakers. It was attended by 125 participants from four countries. This was the beginning of what is now known as the Denton Conference which usually has around 1000 participants from virtually every country that has any kind of an accelerator program. Jim has been very important to the Denton Conference. He has helped to organize all 14 of the meetings. Those of us that have helped to organize meetings understand and appreciate the work that has gone into this effort.

Because of his contributions to nuclear science, his pioneering efforts in teaching nuclear science (especially accelerator nuclear physics) and his monumental contribution to this Conference Series, these proceedings are dedicated, in part, to Jim Lambert.

William S. Rodney
Former Director of Nuclear Physics (NSF)
and Adjunct Professor at Georgetown University
Department of Physics
Washington, DC 20057

Marshall R. Cleland

Marsh Cleland was born on February 9, 1926 in the small university town of Vermillion, South Dakota. After graduating from high school in 1943, he entered the University of South Dakota (USD) to study art and music. This plan was interrupted by brief service in the U.S. Army Air Corps during World War II. He returned to USD after the atomic age had dawned and changed his major to science and mathematics. In 1947, he was graduated summa cum laude with a B.A. degree in physics. Marsh continued his studies at Washington University in St. Louis, Missouri where he was awarded a Ph.D. degree in experimental nuclear physics in 1951. In 1948, he married a classmate from USD, Rosalie McNeely. This union produced four children and six grandchildren. His first professional employment was with the Betatron Section of the National Bureau of Standards in Washington, DC, where he worked on methods to measure the energy spectra of high-energy x-rays and neutrons using large-volume liquid scintillation counters. Marsh's interest in the practical applications of nuclear physics led him to return to St. Louis in 1952 to work for the Nuclear Research and Development, Inc., a company that was started by two of his friends from Washington University to promote medical uses of radioisotopes. From these experiences, Marsh became aware of the advantages of machine generated radiation for both medical applications and for atomic and nuclear research. In 1953, he started Teleray Corporation in St. Louis to develop a high-voltage x-ray machine for cancer therapy. His initial design was based on a radio-frequency, series-coupled, Cockcroft Walton power supply. The limitations of this device were surpassed by his invention of the parallel-coupled, cascaded-rectifier system that became known as the Dynamitron accelerator. In 1958, another company, Radiation Dynamics, Inc., (RDI) was formed in Long Island, NY, to acquire Teleray's assets and to develop and manufacture Dynamitrons. By that time, the market for cancer therapy equipment was well supplied with Van de Graaff generators made by High Voltage Engineering Corp., resonant transformers made by General Electric Co. and cobalt-60 teletherapy machines made by Atomic Energy of Canada Limited and other firms. (Linacs were not yet popular). So RDI decided to focus on making accelerators for research and industrial applications.

In 1964, he designed one of the first low-energy (300 keV) ion accelerators used for ion beam modification of materials and in the early 1970s, Marsh led the joint effort by RDI and AEG Telefunken to develop a commercial high-intensity, fast-neutron generator and therapy system for the treatment of cancer. This device is still being used at the Eppendorf University Hospital in Hamburg, Germany. In RDI's early years, about 30 Dynamitron accelerators were made for research organizations. Today, the Dynamitron remains the leading industrial high-power dc accelerator in the energy range from 1.5 to 5 MeV.

In 1994, he left RDI again after completing thirty years of service. He is now the U.S. Marketing Manager for AECL Accelerators, a business unit of Atomic Energy of Canada Limited, which makes high-energy (10 MeV), high-power, microwave electron accelerators called IMPELA. His current interests are mainly environmental applications, food preservation and other applications of ionizing radiation that are facilitated by the greater penetration of higher-energy electrons.

In connection with his work, Marsh has authored or co-authored more than 60 papers on accelerator technology, radiation measurement techniques and applications of ionizing radiation. He also contributed chapters to several books on accelerators and their applications. He has participated in all but one of these accelerator conferences since their inception in 1968 and he has served as Chairman of the sessions on radiation processing for many years.

He is a member of the American Society for Testing and Materials (ASTM), American National Standards Institute (ANSI), Association for Advancement of Medical Instrumentation (AAMI), Council on Ionizing Radiation Measurements and Standards (CIRMS) and American Nuclear Society (ANS). In addition, he is a member of the American Physical Society (APS), the American Association for the Advancement of Science (AAAS) and the New York Academy of Sciences (NYAC). In 1982, Marsh received a Proclamation from the Mayor of St. Louis for professional achievement by inventing the Dynamitron accelerator. In 1987, he received an Alumni Achievement Award from the University of South Dakota and in 1992, he received an Award for Outstanding Contributions at the 8th International Meeting on Radiation Processing (IMRP) held in Beijing, China.

On a personal note, Marsh is a man who is admired and respected by all who know him. His keen wit and sense of humor make him a pleasure to be with, whether on the golf course, a skiing trip or on the way to another conference. I believe that I speak for everyone who knows him when I say that he is a kind, true, trustworthy and loyal friend. To me, he has been a mentor and a dear friend for more than 30 years. It gives me great pleasure to write this dedication in honor of Marshall R. Cleland.

J. Paul Farrell, Ph.D.
Brookhaven Technology Group, Inc.
25 East Loop Road
Stonybrook, NY 11790

Vlado Valkovic

Vlado Valkovic was born on July 19, 1939 in Draga Baska, Island of Krk, Croatia. He received his B.A. in 1961, M.A in 1963 and Ph.D in 1964 from the University of Zagreb in Nuclear Physics. In 1965 he joined the T.W. Bonner Nuclear Laboratory at the Rice University in Houston, as a research associate. From 1967 to 1970 he was chosen as head of the Nuclear Reaction Laboratory at the Ruder Boskovic Institute in Zagreb, Croatia. In 1970 he returned to the Rice University as an assistant professor. In 1971, he was promoted to associate professor and in 1975 to full professor.

During this decade, he developed analytical methods based on proton induced X-ray and gamma ray emissions. He devised a series of combined physical and chemical sample treatment techniques to substantially improve analytical detection limits. He was one of the first to use focused beams for the measurement of concentration profiles and maps. His studies of the role of the chemical elements in the biological systems had a major impact in the field of interdisciplinary sciences. He made significant contributions in the industrial applications of these methods especially to the metallurgy and coal industry.

In 1977 he gave up his full professorship at Rice University to move back to his beloved Zagreb as a Professor of Physics. There, in 1982, he founded the Laboratory of Nuclear Microanalysis and introduced locally all the necessary accelerator based analytical techniques. In 1989, he was chosen to head the Physics-Chemistry Instrumentation Laboratory of the IAEA in Seibersdorf, Austria. During this period of his life he wrote nine books, landmarks in the analytical techniques with ion beams. His description of trace elements in coal, petroleum, hair and all you need to know on nuclear microanalysis and its relation with interdisciplinary sciences, are in these books, and is a gift to all of us who work in these fields.

He served with the International Atomic Energy Agency (IAEA) until the end of 1996. During this period, the environmental pollution by radionuclides as well as by toxic elements and their pathway identification and possible impact on the environment, became his primary concern. He performed innovative work on quality assurance and control of the nuclear analytical methods. He spent a great deal of his time in developing standards, among them, air filter standards.

As an IAEA technical officer he visited numerous countries in an effort to promote accelerator based analytical technologies. Laboratories in many developing countries from all continents had the chance, with his help, to acquire modern hardware and to train their scientists and technicians in modern technologies. He also started the Accelerator Newsletter which he is continuing to publish through IAEA. Presently, back in Zagreb, he is deeply involved in the further development and application of nuclear analytical techniques and continues the study and the role of chemical elements in nature. More than 270 published papers, membership in numerous scientific societies, a wide editorial activity and a deep concern on the impact of technology in modern society completes the picture of an outstanding person in physical sciences. A person, who besides his academic qualities and achievements, has an open character and a permanent smile on his face that has earned him many friends.

In recognition of his wide and continuing contribution to accelerator based analytical techniques and to his concern on their implication to societal problems, these proceedings are in part dedicated to V. Valkovic.

T. Paradellis
Head, Laboratory for Material Analysis
Institute of Nuclear Physics
N.C.S.R. Demokritos Athens, Greece

ORGANIZATIONAL COMMITTEE

Research Session

Jerome L. Duggan, *University of North Texas,* **Chairman**
Frank T. Avignone, *University of South Carolina*
Klaus H. Bethge, *J. W. Goethe University of Frankfurt*
James W. Butler, *U.S. Department of Navy*
H. Ken Carter, *Oak Ridge Associated Universities*
Sheldon Datz, *Oak Ridge National Laboratory*
M. M. Duncan, *University of Georgia*
David L. Ederer, *Tulane University*
K. O. Groeneveld, *J. W. Goethe University of Frankfurt*
Lester D. Hulett, *Oak Ridge National Laboratory*
Brant Johnson, *Brookhaven National Laboratory*
Keith W. Jones, *Brookhaven National Laboratory*
James M. Lambert, *Georgetown University*
Joseph Macek, *University of Nebraska*
Floyd D. McDaniel, *University of North Texas*
Walter E. Meyerhof, *Stanford University*
J. Richard Mowat, *North Carolina State University*
John T. Park, *University of Missouri*
David J. Pegg, *University of Tennessee*
Friedrich Rauch, *J. W. Goethe University of Frankfurt*
John F. Reading, *Texas A&M University*
Patrick Richard, *Kansas State University*
William S. Rodney, *Georgetown University*
Claus Rolfs, *RUHR Universität*
M. Eugene Rudd, *University of Nebraska*
Stanley O. Schriber, *Los Alamos National Laboratory*
Ivan A. Sellin, *Oak Ridge National Laboratory*
Stephen Shafroth, *University of North Carolina*
W. W. Smith, *University of Connecticut*
Carroll C. Trail, *University of North Texas*
Henri Van-Rinsvelt, *University of Florida*
S. L. Varghese, *University of South Alabama*
George Vourvopoulos, *Western Kentucky University*
Richard L. Walter, *Duke University*

Industrial Session

Ira Lon Morgan, *International Isotopes Inc.,* **Chairman**
Gerald D. Alton, *Oak Ridge National Laboratory*
John R. Bayless, *Pulse Sciences, Inc.*
Joe E. Beaver, *International Isotopes Inc.*
Frank Chmara, *Peabody Scientific*
Cary N. Davids, *Argonne National Laboratory*
Charles E. Dick, *National Institute of Standards & Technology*
Barney L. Doyle, *Sandia National Laboratories*
Robert W. Hamm, *AccSys Technology, Inc.*
George M. Klody, *National Electrostatics Corporation*
Kenneth H. Purser, *Southern Cross Corporation*
James F. Ziegler, *IBM-Research*

Advisory Committee

Marshall Cleland, *AECL Accelerators*
Norman Hackerman, *Robert A. Welch Foundation*
Chris E. Kuyatt, *National Institute of Standards & Technology*
S. Thomas Picraux, *Sandia National Laboratory*
Subramanian Ram Raman, *Oak Ridge National Laboratory*

Session Chairpersons

Gerald D. Alton, *Oak Ridge National Laboratory*
Jim Arps, *Southwestern Research Institute*
Klaus H. Bethge, *Goethe-Universität*
Bert Brijs, *IMEC vzw MAP/ARS*
Joachim Burgdoerfer, *University of Tennessee*
Wei-Kan Chu, *University of Houston*
Marshall Cleland, *AECL Accelerators*
Sameer Datar, *University of North Texas*
Alan K. Edwards, *University of Georgia*
Jerry D. Garrett, *Oak Ridge National Laboratory*
David E. Golden, *University of North Texas*
Tsahi Gozani, *Science Applications Intl. Corp.*
Robert W. Hamm, *AccSys Technology, Inc.*
Yale D. Harker, *Idaho National Engineering Lab.*
Tim Z. Hossain, *AMD Incorporated*
William Dennis James, *Texas A&M University*
A. Joseph Keenan, *Texas Instruments Incorporated*
George M. Klody, *National Electrostatics Corp.*
James M. Lambert, *Georgetown University*
Gregory Lapicki, *East Carolina University*
John Lowell, *Applied Materials*
Hans J. Maier, *Universität Munchen*
Sam Matteson, *University of North Texas*
Floyd D. McDaniel, *University of North Texas*
Marcus H. Mendenhall, *Vanderbilt University*
Ira Lon Morgan, *International Isotopes Inc.*
David B. Poker, *Oak Ridge National Laboratory*
Jack Lewis Price, *Naval Surface Warfare Center*
Robert D. Rathmell, *Eaton Semiconductor Equipment*
John F. Reading, *Texas A&M University*
William S. Rodney, *Georgetown University*
David J. Schlyer, *Brookhaven National Laboratory*
Ivan A. Sellin, *Oak Ridge National Laboratory*
Stephen M. Shafroth, *University of North Carolina*
David Shiner, *University of North Texas*
Soey H. Sie, *CSIRO: Exploration Geosciences*
Willard L. Talbert, *Oak Ridge National Laboratory*
Henri A. Van Rinsvelt, *University of Florida*
Gyorgy Vizkelethy, *Idaho State University*
Richard L. Walter, *Duke University*
Rand L. Watson, *Texas A&M University*
Alex Weiss, *University of Texas at Arlington*
Roy N. West, *University of Texas at Arlington*
J. Mark Anthony, *Texas Instruments Incorporated*
Tom Aton, *Texas Instruments Incorporated*
John Biscar, *University of North Texas*
Milos S. Budnar, *University of Ljubljana*
William T. Chu, *Lawrence Berkeley Laboratory*
Sam Cipolla, *Creighton University*
John M. D'Auria, *Simon Fraser University*
F. Eugene Dunnam, *University of Florida*
James M. Feagin, *California State University*
Bruce E. Gnade, *Texas Instruments Incorporated*
Gerald Goldstein, *U.S. Dept. of Energy*
Marianne E. Hamm, *AccSys Technology, Inc.*
Cotton Hance, *Motorola, Inc.*
Richard D. Hichwa, *University of Iowa Hospitals*
Daryush Ila, *Alabama A&M University*
Keith W. Jones, *Brookhaven National Laboratory*
Joseph F. Kirchhoff, *Charles Evans and Associates*
Andreas M. Koehler, *Harvard University*
Richard C. Lanza, *Massachusetts Inst. of Technology*
Jiarui Liu, *University of Houston*
Carl J. Maggiore, *Los Alamos National Laboratory*
Daniel K. Marble, *U.S. Military Academy*
Robert May, *Thomas Jefferson Natl. Accelerator Facility*
Rahul Mehta, *University of Central Arkansas*
Gary E. Mitchell, *North Carolina State University*
Gregory A. Norton, *National Electrostatic Corp.*
Darden Powers, *Baylor University*
Carroll A. Quarles, *Texas Christian University*
Friedrich Rauch, *J.W.G. University of Frankfurt*
Patrick Richard, *Kansas State University*
Peter Rosen, *University of Texas at Arlington*
Emile A. Schweikert, *Texas A&M University*
Thomas J. Shaffner, *Texas Instruments Incorporated*
Mark H. Shapiro, *California State University*
Jefferson L. Shinpaugh, *East Carolina University*
Michael S. Smith, *Oak Ridge National Laboratory*
Vlado Valkovic, *Institute Ruder Boskovic*
S. L. Varghese, *University of South Alabama*
George Vourvopoulos, *Western Kentucky University*
H. R. J. Walters, *Queen's University of Belfast*
Duncan L. Weathers, *University of North Texas*
Stephen Wender, *Los Alamos National Laboratory*
Isao Yamada, *Kyoto University*

Fourteenth International Conference on the Application of Accelerators in Research and Industry

CAARI'96

6–9 November 1996

Industrial Exhibitors

AccSys Technology, Inc., *1177 Quarry Lane, Pleasanton, CA 94566*
Activation Technology Corporation, *2816 Janitell Road, Colorado Springs, CO 80906*
AEA Technology, *F4/135 Physics Technology, Culham, Abingdon, Oxfordshire 0X14 3DB UK*
Advanced Control Systems Corp., *Sales Office, Old Mine Rock Way, Hingham, MA 02043*
AMPTEK, Inc., *6 De Angelo Drive, Bedford, MA 01730*
Aptec Nuclear Inc., *908 Niagara Falls Blvd., Ste. 524, N. Tonawanda, NY 14120-2060*
Artep Incorporated, *6432 Ellfolk Terrace, Columbia, MD 21045*
Atlas Technologies, Inc., *2760 Washington Street, Port Townsend, WA 98368*
Balzers PFEIFFER North American, *High Vacuum Products, 8 Sagamore Park Road, Hudson, NH 12533*
Commissariat a l'Energie Atomique, *SALDUC DAM, Centre D'Etudes de Valduc, 21120 Is-Sur-Tille, France*
Computer Technology & Imaging Inc., *810 Innovation Drive, Knoxville, TN 37932*
Duniway Stockroom Corp., *Marketing Department, 1305 Space Park Way, Mountain View, CA 94043*
EG&G ORTEC, *Marketing Services, 100 Midland Road, Oak Ridge, TN 37831-0895*
Everson Electric Co., *2000 City Line Rd., Bethleham, PA 18017-2167*
Glassman High Voltage Inc., *P.O. Box 551, Route 22E, Salem Industrial Park, Whitehouse Station, NJ 08889*
GMW Associates & Danfysik, *Marketing Division, P.O. Box 2578, Redwood City, CA 94064*
Group 3 Technology Ltd., *P.O. Box 71-111, Rosebank, Auckland, New Zealand*
High Voltage Engr. Europa B.V., *Amsterdamseweg 63, 3812 RR Amersfoort, P.O. Box 99, 3800 AB Amersfoort, The Netherlands*
Huntington Laboratories, Inc., *1040 L'Avenida, Mountain View, CA 94043*
International Isotopes Inc., *523A North Elm, Denton, TX 76201*
Inverpower Controls Ltd., *835 Harrington Court, Burlington, Ontario L7N 3P3, Canada*
Ion Beam Applications SA, *Chemin du Cyclotron, 2, B-1348 Louvain-la-Neuve, Belgium*
Isotopes Products Laboratories, *1811 N. Keystone, Burbank, CA 91504*
JP Accelerator Works, *2245 47th Street, Los Alamos NM 87544-1604*
Kurt J. Lesker Company, *1515 Worthington Ave., Clairton, PA 15025*
Lake Shore Cryotronics, Inc., *Public Relations, 64 E. Walnut St., Westerville, OH 43081-2399*
LINAC Systems, *2167 N. Highway 77, Waxahachie, TX 75165*
Magnet Sales & Manufacturing Co., *Sales/Engineering/Trade Shows, 11248 Playa Court, Culver City, CA 90230-6162*
MF Physics Corporation, *5074 List Drive, Colorado Springs, CO 80919*
Micro-Analytical Research Centre, *University of Melbourne, School of Physics, Parkville 3052 Australia*
MKS Instruments, Inc., *Six Shattuck Road, Andover, MA 01810*
Motorola Inc., *Semiconductor Products Sector, 3501 Ed Bluestein Blvd., Mail Drop K-10, Austin, TX 78721-2996*
MOXTEK Soft X-Ray Technologies, *452 West 1260 North, Orem, UT 84057*
National Electrostatics Corp., *P.O. Box 620310, 7540 Graber Road, Middleton, WI 53562-0310*
Newton Scientific Inc., *7 Red Coach Lane, Winchester, MA 01890*
Northrop Grumman Aerospace Corp., *K03-14, 1111 Stewart Avenue, Bethpage, NY 11714*
Nuclear Research Corp., *125 Titus Ave., Warrington, PA 18976*
Oxford Instruments, Inc., *Nuclear Measurements Group, P.O. Box 2560, Oak Ridge, TN 37831-2560*
Peabody Scientific, *P.O. Box 2009, Peabody, MA 01960*
Princeton Scientific Corp., *P.O. Box 143, Princeton, NJ 08542*
Thomson Components & Tubes Corp., *40 G Commerce Way, P.O. Box 540, Totowa, NJ 07511*
Texas Instruments, Inc., *13510 North Central Expressway, Dallas, TX 75243*
THT Sales Company, *13438 Floyd Circle, Dallas, TX 75243*
TITAN Beta, *6780 Sierra Ct., Dublin, CA 94568*
U.S. Dept. of Energy-IPD: EG&G Mound Applied Technologies, *Stable Isotopes Group, P. O. Box 3000, Miamisburg, OH 45343-3000*
Varian Vacuum Products, *121 Hartwell Avenue, Lexington, MA 02173*
VAT Incorporated, *500 W. Cummings Park, Ste. 5450, Woburn, MA 01801*

Vector Fields, *Suite 28, 1700 N. Farnsworth Av, Aurora, IL 60505*
Whistle-Soft Inc., *168 Dos Brazos, Los Alamos, NM 87544*

MICRO ANALYTICAL

UNIWAY
Book Store

Research and industrial applications of accelerators State of the Art 1997

Jerome L. Duggan*

Department of Physics, University of North Texas, Denton, TX 76203-0368 USA

INTRODUCTION

This conference which was composed of 900 participants from 47 countries showing that research and industrial application of accelerators is a growing and vibrant area. There are, at this writing, world wide, over 350 accelerators with terminal voltages less than 20 MeV that are being used for fundamental physics or ion beam analysis and modification of materials. This paper gives a brief summary of just a few of the exciting papers that were presented at the Fourteenth International Conference on the Application of Accelerators in Research and Industry. This paper is written, for the most part, from brief summaries that were given to me after the conference by some of the session chairmen.

RADIOACTIVE ION BEAMS

Fred Becchetti of the University of Michigan gave a description of an existing radioactive beams project for astrophysics research at the Notre Dame University tandem accelerator in collaboration with the University of Michigan, and an upgrade to the system under installation, presented by M. Lee. The existing, larger-scale facilities for the Institute for Nuclear Studies (INS) at Tokyo University, at CERN in Geneva (the ISOLDE facility), the National Superconducting Cyclotron Laboratory (NSCL projectile fragmentation spectrometer) at Michigan State University, and at Oak Ridge National Laboratory (the Holifield Radioactive Ion Beam Facility, or HRIBF) were presented, with appropriate examples of in-progress or planned nuclear physics and astrophysics studies. The ISOLDE facility, as reported by Georg Bollen, has operated for a longer period of time than any other (for low-energy radioactive beams), and construction is beginning for an upgrade to provide accelerated radioactive beams. A proposed upgrade to the NSCL project was also introduced by Thomas Glasmacher. The progress in commissioning the HRIBF and associated spectrometers was described by Jerry Garrett, Jim Beene, Carl Gross and Jon Batchelder, in anticipation of imminent operation for users. The INS is also involved in a study of a future radioactive beams facility at the proposed Japan Hadron Project, as related by Ichiro Katayama of INS.

The ISAC facility at TRIUMF in Vancouver, Canada (a significant new development to supplant the existing TISOL radioactive beams mass separator) was described by Marik Dombsky, and the SPIRAL facility at GANIL in France (supplant the existing projectile fragmentation spectrometer) was presented by B. Laune, and both are under construction. These facilities, when completed and commissioned, will be world-class in their capabilities. Activities at Argonne National Laboratory were outlined by Patrick DeCrock in anticipation of a possible future radioactive beams facility at the ATLAS heavy ion accelerator, with preliminary studies emphasizing production rates for fission product radioactivities and development of a low-energy pre-accelerator.

Jerry Garrett from ORNL, informed us in his plenary talk that the 1996 Long Range Plan for Nuclear Science produced by the Nuclear Science Advisory Committee (NSAC) recommends that the next major new construction in nuclear science for the U.S. Department of Energy (DOE), after the completion of the Relativistic Heavy Ion Collider (RHIC) at Brookhaven National Laboratory, should be an ISOL type RIB facility to produce a large variety of intense RIB at energies appropriate for detailed nuclear structure and astrophysics studies.

THE NATIONAL SPALLATION NEUTRON SOURCE (NSNS)

Bill Appleton of Oak Ridge gave us an overview of the prospective design and applications of the NSNS. Figure 1 shows the accelerator system layout for the reference design chosen for the NSNS. This design project involves five DOE National Laboratories. The design is being managed by ORNL where the source will be constructed if funds become available. The anticipated completion date is in the year 2004.

The NSNS produces H^- ions that are accelerated to 1 GeV in energy in the linac accelerator, they are stripped to H^+, and injected into an accumulator ring. The ring stores

Supported in part by the Texas Advanced Research Program, the Office of Naval Research, the National Science Foundations, and the Robert A. Welch Foundation.

CP392, *Application of Accelerators in Research and Industry,* edited by J. L. Duggan and I. L. Morgan
AIP Press, New York © 1997

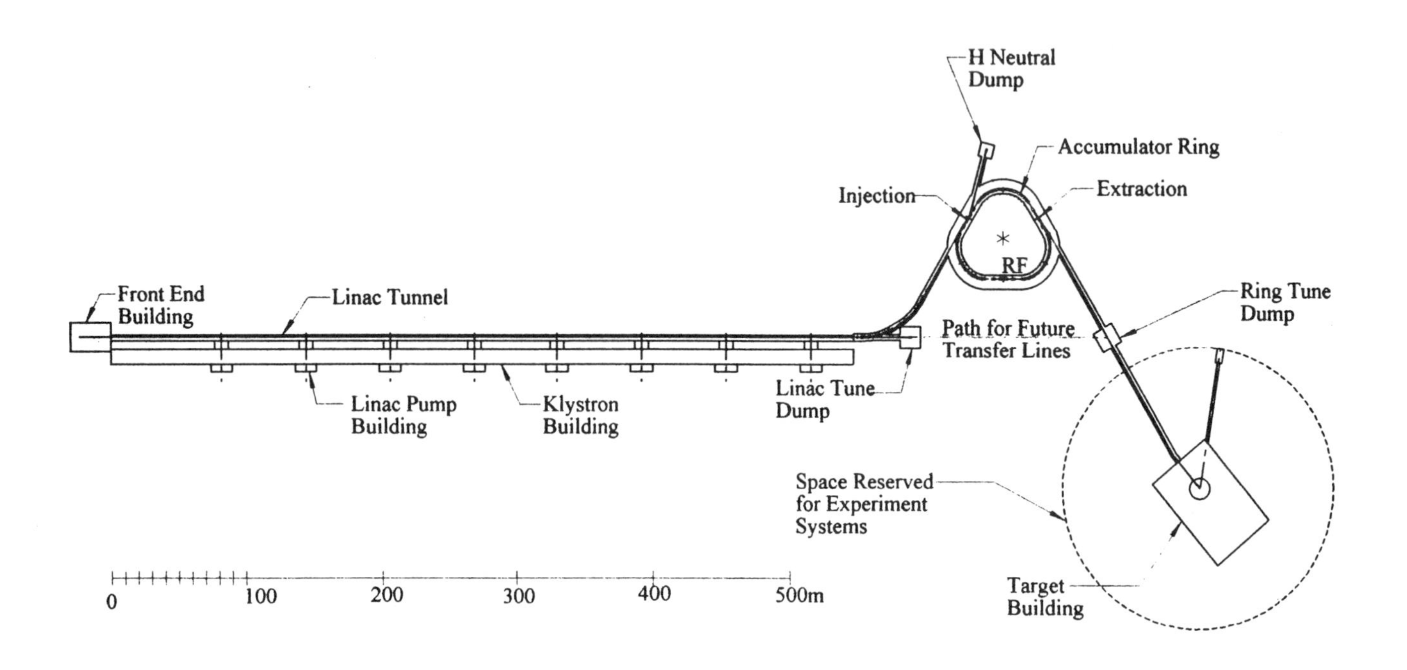

Figure 1. The NSNS Reference Design; proposed for Oak Ridge

and bunches the beam producing 1 microsecond pulses of H at a rate of 60 Hz. The proton pulses impinge on a liquid mercury target where protons are produced by the spallation nuclear reaction process. The neutrons are then moderated and guided into the experimental area where they are used for experiments. The initial beam power of NSNS will be 1 megawatt but the source is designed to be economically upgraded to 4-5 megawatts as future funds allow.

NUCLEAR PHYSICS

G.E. Mitchell, North Carolina University, reported on recent developments in nuclear instrumentation. Susan Seestrom described plans for the development of an ultra cold neutron source at the Los Alamos spallation source. The special features (advantages and disadvantages) of a pulsed source versus a reactor were emphasized. John Becker described the large scale Ge detector array (GEANIE) that has now been installed at LANSCE/WNR. Characteristics of the system (detectors plus neutron beam) were presented. Applications to science based stockpile stewardship as well as pure nuclear physics are planned. William Scott Wilburn presented an analysis of the technical problems inherent in fundamental symmetry tests, including a detailed description of expected systematic errors as well as methods to monitor and control them. Topics included residual transverse polarization moments, spin-correlated beam motion, and time-dependent detector drifts. The motivation is for the measurement of parity violation in the compound nucleus in light nuclei. Wolf Hartmut Schulte described a 4π gamma detection system that is about 90% efficient for gamma rays from 0.5 to 10 MeV. This system is ideal for the measurement of weak resonances in the proton capture reaction. The primary results reported involved weak resonances at low proton energies on silicon isotopes that could be used advantageously in depth profiling measurements. However, the system also has applications for nuclear physics and nuclear astrophysics. Vera Y. Hansper described a multi-wire proportional counter for a split-pole magnetic spectrometer. This sensitive detector system is being used in nuclear astrophysics measurements. Michal Palarczyk reported on the installation of a neutron polarimeter at IUCF. The polarimeter consists of four neutron detectors and two charged particle veto planes. This new system allows the decomposition of the (p,n) reaction into the set of spin transfer coefficients. R.W. McCullough described a high efficiency plasma source for the production of reactive atomic beams.

In the session on recent developments in nuclear physics chaired by Richard Walter, a broad spectrum of exciting topics were presented. In the few nucleon area, Adam Sarty of FSU described four massive magnetic spectrometers that can be rotated out of the horizontal reaction plane for coincidence experiments at the MIT Bates electron accelerator lab; they will be used initially to study interference terms in electron scattering from protons and deuterons. Werner Tornow of TUNL and Duke reported on the quandary between rigorous proton-deuteron calculations and data, and the implication for changes to the best models for the nucleon-nucleon interaction. T. Glasmacher of the National Superconducting Cyclotron Lab at MSU reported measurements of the energy levels for eight neutron-rich nuclei, e.g., ^{46}Ar, ^{44}S, and ^{38}Si, which were first produced with a ^{48}Ca beam incident on a Be target and then Coulomb excited in a Au foil. The enormous potential of silicon strip detectors for accumulating data in complicated multi-particle breakup reactions involving heavy-ion beams was demonstrated by Alan Wousma of ANL who exhibited data from ^{12}C + ^{12}C reactions. Phil Kerr illustrated how one can cleanly extract the tensor interaction involved in ^{6}Li reactions when the reactions are initiated with polarized ^{6}Li beams available at the FSU tandem accelerator. Lastly, Joe Hamilton showed the breadth of reactions being studied with the detector array "Gammasphere" and how one can identify spontaneous fission (SF) fragments and obtain details about the number of neutrons emitted in the SF; cases of 8 and even 9 neutrons being emitted were shown to have clean signatures for the first time.

ACCELERATOR BASED ATOMIC PHYSICS

As has been the case for the last seven conferences in this series, the largest number of papers (19 sessions) were devoted to atomic physics. Pat Richard reported that his session highlighted many aspects of the physics of highly charged ions beginning with their production in the lowest energy regime, namely in plasmas produced by the saturn pulsed-power facility at Sandia National Laboratory. Keith Wong gave this report in which he showed a 2-D x-ray spectra obtained from imaging crystals. At this stage of the work the experiments are not correctly described by available 2-D radiative MHD models. John Gillaspy described the new Electron Beam Trap (EBIT) facility at NIST which is being used to study ion-surface collisions. The work is just beginning and parties interested in the interaction of highly-charged ions with metals, insulator and semiconductor surfaces are encouraged to contact John. Mohammad Abdallah described the new "electron momentum projection spectroscopy" technique developed at Kansas State to study 2-D momentum distributions of electrons from the ionization process at low impact energies. It is found that both charge centers play an important role in producing the final electron momentum distribution and no evidence for saddle point electrons could be identified. Lokesh Tribedi presented the first DDCS measurements of ionization of atomic hydrogen by bare ions (a Kansas State-University of Nebraska collaboration). Excellent agreement is found with the CDW-EIS theory at all emission energies and angles. John Tanis of Western Michigan reported on the study of two-electron processes where one electron is captured to a bound state of the projectile and the second electron is captured to the continuum of the projectile (TI). These first measurements of this type are being pursued in an attempt to evaluate the significance of the electron-electron interaction. Steve Lundeen of Colorado State University described recent work on the electron capture from Rydberg states. He outlined a new method to describe both the initial state of the Rydberg atom and the final state in the projectile ion. This is part of a Colorado State-Kansas State collaboration to investigate these phenomena with low velocity highly charged ions.

John Reading reported that there have been many exciting advances in atomic theory and experimentation since the last Denton conference. At Reading's session Henri Bachau reported that the Bordeaux group had now developed the capability of calculating cascade processes following excitation and vacancy production in multi electron atoms. A.L. Ford described the success and failure of an Independent Event Model in excitation and ionization of helium. Joseph Macek described the development of the hidden crossing method for calculating low energy cross sections. Walter Meyerhof described an R-matrix theory for positronium formation. Amin Cassimi gave insights on multi ionizing collisions obtained with recoil spectrometry. David Church described the new RETRAP system for holding one to several thousand ions. Ivan Sellin reported new efficient techniques for recording angular correlations between auger electron and photo electrons, and auger and auger cascades, following single photo-excitation events. Gordon Berry spoke on the first observations of ion spectra observed through excitation of a single carbon k-shell electron in C60. It showed the rapid propagation of the pulse of energy from 1 atom to the whole sphere of the molecule. Uwe Becker talked about how angle resolved ionic fragmentation using coincidence techniques is a powerful tool for symmetry specific molecular spectroscopy if one takes care of the kinetic energy dependence of the fragments. Triple coincidences of the same kind provide insight into the geometry of the core excited molecules as well as their particular break up behavior. Itzik Ben-Itzak's presentation described the experimental measurement of an asymmetry in the dissociation of the electronic ground state of HD^+, with the H^+ + D(1s) dissociation limit being slightly preferred over the H(1s) + D+ limit. The measured asymmetry is in good agreement with a "single pass" Demkov theory as developed by Walter Meyerhof. Steve Shafroth reported that his session reviewed and updated selected chapters (mostly by European Authors) in a new book, *Accelerator-Based Atomic Physics Techniques and Applications* edited by S.M. Shafroth and J.C. Austin, to be published March 1997 by AIP Press. There were seven talks in the session. The one which generated the most heated discussion was given by Gianni Giardino of the University of P&M Curie in Paris, France. His subject was "Full deceleration of highly charged ions on surfaces: An insight on image acceleration." Using the techniques of high-resolution X-ray spectroscopy he, Jean Pierre Briand and coworkers observed a pronounced change in the Ka satellite spectrum when a semiconductor target such as SiH is biased so that an Ar17+ ion produced in an ECR source doesn't quite reach the surface compared to the satellite spectrum when it penetrates the surface. They show that in the case of a biased metal target such as Au, no such effect occurs because the negative image charge sucks the Ar17+ ion into the metal when it gets close to the surface. On the other hand, the SiH develops a positively charged layer at the surface because surface electrons are captured by the slowly approaching projectile. The claim is that the nearly neutral projectile ion closely approaches the surface and is repelled so as not to penetrate it. The discussion centered around the question of whether or not this was the right explanation for the change in the X-ray satellite spectrum. The session started with a most interesting talk on the present status and future prospects for ECR ion sources by Gerard Melin of the CEA, Grenoble, France. The Minimafios source has produced 1 enA of bare Ar18+ and the Caprice source 0.17 emA of U39+. Theo Zouros of the University of Crete and Institute of Electronic Structure & Lasers, Heraklion, Crete, Greece presented a stimulating talk on "Recent developments in zero degree electron spectroscopy of projectile ions." He discussed a newly developed zero degree electron spectrometer system which uses a hemispherical spectrometer and a 2-dimensional position sensitive detector with a resistive anode encoder and four element deceleration lens. This spectrometer will permit the study of low cross section processes and should lead to much new and perhaps unexpected knowledge about Auger electron processes.

Alfred Müller of Universität Giessen, Germany, brought us up to date on some of the very puzzling results obtained with several European ion storage rings, notably those at GSI, The Max-Planck-Institute für Kern Physik at Heidleberg and the Mane-Siegbahn Laboratory in Stockholm. At zero energy in the electron-ion center of mass frame exceedingly large, and much greater than predicted by theory, recombination rates were observed. In an effort to understand the rate enhancement, plasma as well as external electric field effects, were studied. Brett DePaola of Kansas State University talked about "Accelerator based Rydberg collision experiments." He gave interesting examples of how projectile Rydberg atoms or ions are prepared using beam-foil techniques. He then discussed how target Ryd berg atom production or destruction may be monitored, and ended with specific examples as well as future directions. Friedrich Aumayr of the Institut für Allgemeine Physik, Technische Universität, Wien, Austria talked about "Application of multiply charged ion sources for surface scattering studies." He brought us up to date on and compared three types of multiply charged ion sourccs, namely ECR (electron cyclotron resonance) and EBIS/EBIT (electron beam ion source/trap). The large amount of potential energy stored in a highly charged ion gives rise to new phenomena such as hollow atoms/ions when the highly charged ion approaches the surface of a metal, semiconductor or insulator. These hollow atoms/ions then emit electrons, soft x-rays, and target particles (sputtering) each of which can be studied. He presented a review of recent investigations when insulators and metal surfaces interact with highly charged ions. Hocine Merabet reported on work that he and his coworkers did at the Laboratoire de Spectroscopie Atomique, ISMRa Caen, France. They reported on work done using the techniques of zero-degree electron spectroscopy using the ECR ion source to produce Ne10+ which collided with He. They were able to show that the major contribution to stabilization ordinates from the states 3lnl' nd 4lnl' (n>4) which are populated during the collision.

In Sam Cipolla's session, Jean-Claude Dousse and Milos Budnar reported results of precision measurements on the Radiative Auger Effect (RAE) using crystal spectrometers. RAE is the simultaneous ejection of an atomic (Auger) electron and an x-ray photon which share the decay energy of an atomic excited state. It is a small effect that shows up as low-energy tailing on diagram lines, especially K_β. Dousse was able to measure RAE edge energies in medium-Z elements that compared well with Auger transition energies, and found relative RAE intensities to also compare well with relativistic Hartree-Fock theory calculations. Milos Budnar found evidence of solid-state and chemical effects in RAE from Ca to Fe, and stressed the importance of taking RAE into account when fitting X-ray spectra. This point was emphasized also by Marie-Christine Lépy who discussed all the peak-shape parameters needed to accurately describe Si(Li) x-ray spectra. In addition, one needs to be aware of the enhancement of the $K_{\beta 5}$ x-ray intensity due to 3p and 3d state mixing in transition metal elements from $20 < Z < 30$, as was described by István Török. Up to a 4% effect is seen near Cr. Gregory Lapicki reported on experiments with Fe targets that show an anisotropy in K-shell ionization using proton and C^{4+} projectiles of the same velocity. The effect is greater for heavier ions and is apparently due to double ionization involving a p-state, which is inherently anisotropic. All these findings have an impact on inner-shell ionization applications such as PIXE and XRF.

In Gregory Lapicki's session Ivo Orlić, National University of Singapore, fitted polynomials to the ratios of experimental proton-induced L-subshell cross sections that he had compiled to the ECPSSR theory. The noted discrepancies between this data base and the theory were also observed by Sam Cipolla, Creighton University, in his measurements of such cross sections by slow protons. Marek Pajek, Pedagogical University of Kielce, Poland, reported on L- and M-shell x-ray production by 0.1-2 MeV/u C, N, O ions on selected heavy target atoms and analyzed them for the direct ionization and electron capture based on the equilibrated -- rather than the incident as done in the conventional zero target thickness extrapolation -- charge state of these projectile ions. Takeshi Mukoyama, Kyoto University, Japan, reported on the results of his (with C.D. Lin) distortion approximation for inner shell ionization by charged particles while Žiga Šmit, University of Ljubljana, Slovenia, stressed the importance of renormalization in his two-state coupled channel factors multiplied by united atom cross sections as applicable in the adiabatic regime. John Reading, Texas A&M University, discussed his (with Bronk and Ford) Independent Event Model (IEM) and tried to answer the question posed by Martin and Salin i.e., is a full correlated calculation necessary for a proper calculation of the single and double ionization of He by protons and antiprotons? Reading concluded that correlations were only apparent in double ionization by protons since his IEM was successful in explaining the data for the other three processes. Using target recoil ion momentum spectroscopy, Horst Schmidt-Böcking, University of Frankfurt, Germany, measured, with his collaborators, fully differential cross sections for the proton on He transfer ionization. He presented cross sections for an intra atomic electron-electron Thomas-type scattering process so as to test the very question of intra atomic correlations. Contrary to Reading's suggestions, Schmidt-Böcking found a 7.5 dependence of these cross sections on proton velocity, in a marked difference with the conventional 11th power law that would have obtained if the electrons were uncorrelated.

NON DESTRUCTIVE TESTING AND CHARACTERIZATION

In George Vourvopoulos' session, one of the main questions was "Can nuclear techniques be used for airline security against terrorism?" This was a subject that the tragedy of TWA Flight 800 brought to the front burner again. Lee Grodzins reviewed some of the parameters that enter into the security decision making. Irrespective of the cause of the Flight 800 disaster, Grodzins said that the government and airport authorities have been sensitized again in the fight against terrorism. At this stage, however, there is no nuclear technique under development that can fulfil the security requirements for checked luggage. Although the Federal Aviation Administration is funding the development of several nuclear techniques, the only inspection system that now carries FAA certification is an x-ray imaging system. Developers of nuclear techniques, however, should not despair. Even if it is very difficult to develop a nuclear technique that can inspect 650 bags/hour, there is still need for instrumentation that can, non-intrusively and with high accuracy, inspect an individual piece of luggage for the presence of explosives.

Nuclear waste characterization is necessary for the types of nuclear waste that cannot be sampled, or are extremely radioactive. This is a problem that several countries with nuclear industry are facing. Jacques Romeyer-Dherbey from France and Ken Lamkin from Oak Ridge presented innovative techniques using an electron accelerator and a radio frequency quadrupole (RFQ) linac. Both accelerators produce neutrons that are used to interrogate a nuclear waste container for the presence of fissile material (^{235}U, ^{239}Pu, etc.).

The discussion on the detection of anti-personnel mines using nuclear techniques, Peter Trower brought up the subject of humanitarian de-mining, the clearing of the millions of mines (anti-personnel and anti-tank) that have been laid in countries such as Afghanistan, Somalia, Cambodia, Kuwait, and kill and maim innocent people every day. There are no effective techniques that can inexpensively and with 100% confidence clear an area from mines. Several governmental agencies are seeking proposals for developing de-mining techniques. The problem is there and a solution is sought. Nuclear as well as x-ray techniques show promise as being able to assist with this problem. Shall we help?

ACCELERATOR MASS SPECTROMETRY

Soey Sie reported that in his session, geophysical applications were and continue to be one of the main driving forces behind AMS development. Cosmogenic nuclides are an excellent tool for tracer and chronological applications in quaternary geology, volcanology, paleo geomorphology, paleoclimatology as demonstrated by the Purdue and Lucas Heights groups. Similar applications, including environmental concerns, drove the establishment of a new facility at Tsukuba. Anthropogenic pulses of 14C and 36Cl are now exploited more widely to trace geochemical pathways in biogenic as well as non-biogenic systems. The Chalk River group reported extensive work in the use of 36Cl to trace turnover rates in natural ecosystems. The Tokyo group reported the use of 10Be and 26Al among others to determine the sedimentation history of Lake

Baikal, the deepest fresh water body in the world. The Nagoya group reported on an elegant method to study past seismic activities including modelling to predict future occurrences, by dating pelagic sediments along fault lines. The Tucson group demonstrated the use of 14C in lunar samples to delineate the contributions of solar and galactic cosmic rays. A new direction is evident in the development of a microbeam AMS facility at CSIRO in Sydney to enable in-situ ultra trace measurement and non-cosmogenic geochronology, opening up new prospects of geological applications.

In Sameer Datar's session, the Groningen group described the operation of their AMS system. Their system is now fully automated leading to routine, almost assembly-line AMS analysis of carbon samples. The Beijing AMS facility has made significant progress and now expects to launch a major study of archaeological samples from Chinese historical sites. Jean Moran of Texas A&M University, presented a paper which describes the use of radioisotope iodine measurements to identify and trace terrestrial organic carbon in marine sediments which is of crucial importance in global carbon cycle models. Janet Sisterson's paper dealt with the measurement of excitation functions needed to interpret Cosmogenic Nuclide production in lunar rocks since AMS offers the only practical way to measuring relatively long-lived radionuclides at very low concentration levels.

CLUSTER SOLID INTERACTIONS

In Emile Schweikert's session three papers dealt with high energy clusters, fullerenes, carbon and gold clusters with total energies of up to 15 MeV. A paper by Yves LeBeyec from IPN, Orsay, France, reported new observations on nonlinear effects in secondary ion and electron emission and on the long distance (up to 100Å) spatial proximity of atomic constituents of clusters moving through a solid. As they emerge from a thin film, the charge per atom from a cluster ion is significantly smaller than the charge acquired by a single atom moving through the same thin film at the same velocity. Annie Dunlop from the Commissariat a l'Energie Atomique, France, described the damage created in various targets by MeV C_{60} and Au_5 clusters. A comparison of damage structures from GeV monoatomic and MeV cluster irradiations shows that the latter generate tracks in metals with much larger surface craters. Further experimental evidence was presented showing that the fragments of the projectile remain initially spatially correlated in their track then gradually lose constituents or dissociate into smaller size clusters. Max Döbeli from PSI, Switzerland, compared impact features (hillock formation, track diameters) in mica and PMMA from bombardment of clusters (C_{60} and Cu_{11} of up to 15 MeV) and monoatomic ions (up to 80 MeV). Possible applications of cluster impacts in lithography were suggested. Two papers dealt with cluster imprints on surfaces with projectile energies of up to 200 keV. Isao Yamada from Kyoto University discussed bombardment traces from Ar, N_2, O_2, CO_2 clusters on surfaces and reviewed the fundamentals of ion beam processing including atomic level surface smoothing and implantation with clusters. Curt Reiman from Uppsala University, presented new data on the conformation of proteins in the gas-phase revealed from their impact onto a surface. Protein folding is a key to their biological functionality. So far the proteins apomyoglobin and myoglobin have been studied. The session was rounded out with two papers discussing secondary ion mass spectrometry (SIMS), with polyatomic projectiles of a few keV to a few tens of keV in energy. Jim Delmore from Idaho National Laboratory, presented an ion gun capable of delivering nA beams of perrhenate ions. These projectiles give sensitivity enhancements of 4 to 10 over cesium ion bombardment. Perrhenate-SIMS has been applied on a range of problems from pesticides on plant leaves, to degradation products of chemical warfare agents on soil. Michael Van Stipdonk from Texas A&M University, discussed SIMS with metal oxide clusters and fullerenes. Secondary ion yields from inorganic and organic samples are significantly improved when using keV polyatomic or cluster projectiles. The increased SI yield is a benefit in applications such as sub-micron scale organic ion imaging and semiconductor dopant profiling.

Mark Shapiro's session included six invited papers (four simulation papers and two experimental papers). The four molecular dynamics simulation papers covered a range of topics related to cluster impacts with surfaces. Susan Sinnott discussed recent work in her group at the University of Kentucky using empirical, many-body, bond-order potentials to investigate chemical reactions that take place at surfaces impacted with hyperthermal organic molecules. Her work suggests that new types of thin films can be formed by these reactions. Roger Webb (University of Surrey) reported on simulations of both impacts of C-60 fullerenes with Si and graphite surfaces. He showed that C60 clusters with energies up to a few hundred eV remain relatively intact, and bounce off graphite surfaces. However, for the more rigid Si surface incoming fullerenes can either bounce or bond depending on the chemical interactions and at higher energies significant sputtering takes place. Tom Tombrello presented the results of new MD simulations of high-energy cluster impacts that included core-excitation effects. These showed that at energies greater than about 1 keV/atom there are significant inelastic effects that tend to reduce the energy of both atoms sputtered from the impacted surface, and of atoms recoiling from the incoming clusters. Mark Shapiro reported the results of simulations of very high energy (>100 keV) impacts of small gold clusters with surfaces. These showed that for open crystal orientations (100 surfaces) the atoms in the cluster can channel relatively coherently over long distances. However, for non-channeling directions very nonlinear collision cascades occur. These simulations were done on a massively parallel computer. Nobutsugu Imanishi et al. (Kyoto) reported measurements of yields of clusters from SiO2 targets bombarded with Si and Ag. They found that very large clusters could be produced by this process. Charlotte Ruelicke (Lawrence Livermore National Lab) reported on the results of molecular fragmentation experiments with highly charged ions extracted from the LLNL electron beam ion trap. She showed that short biomolecules were fragmented in highly specific patterns by this process, which should have application SIMS analysis.

RADIATION PROCESSING

In Marshall Cleland's sessions on Radiation Processing, several relatively new types of high-power industrial electron accelerators: including RPC Industries' 2 MeV, 10 kW S-band linac, Titan Scan's TB 10/15, a 10 MeV, 15 kW S-band linac, AECL's IMPELA, a 10 MeV, 50 kW L-band linac, IBA's Rhodotron, a 10 MeV, 150 kW cw vhf single-cavity recirculating-beam system and Sandia National Laboratories' Repetitive High Energy Pulsed Power 2 MeV, 300 kW induction linac were discussed. These technologies will contribute to the continuing growth of radiation processing, a field that already utilizes more than 800 electron accelerators with combined beam power ratings exceeding 50 MW. Commercial products being produced with these machines have a retail value of at least $10 billion annually. A novel method of assaying the amount of fissionable material in barrels of waste materials using pulsed high-energy bremsstrahlung from an electron accelerator was also described. New design concepts for compact high-energy electron accelerators: including a high-current linac with water-filled rf cavities and a high-current betatron for pulsed power applications and a rotating-wave single cavity system for radiation therapy were also presented. The measurements and controls used to guarantee accurate doses to irradiated products with AECL's IMPELA linac were also described. Another paper presented methods for identifying foods that have been irradiatedby means of electron spin resonance spectroscopy of long-lived free radicals. A novel means for the production and control of high-intensity plasmas was also described. This can be used for processing thin films and modifying the surfaces of materials.

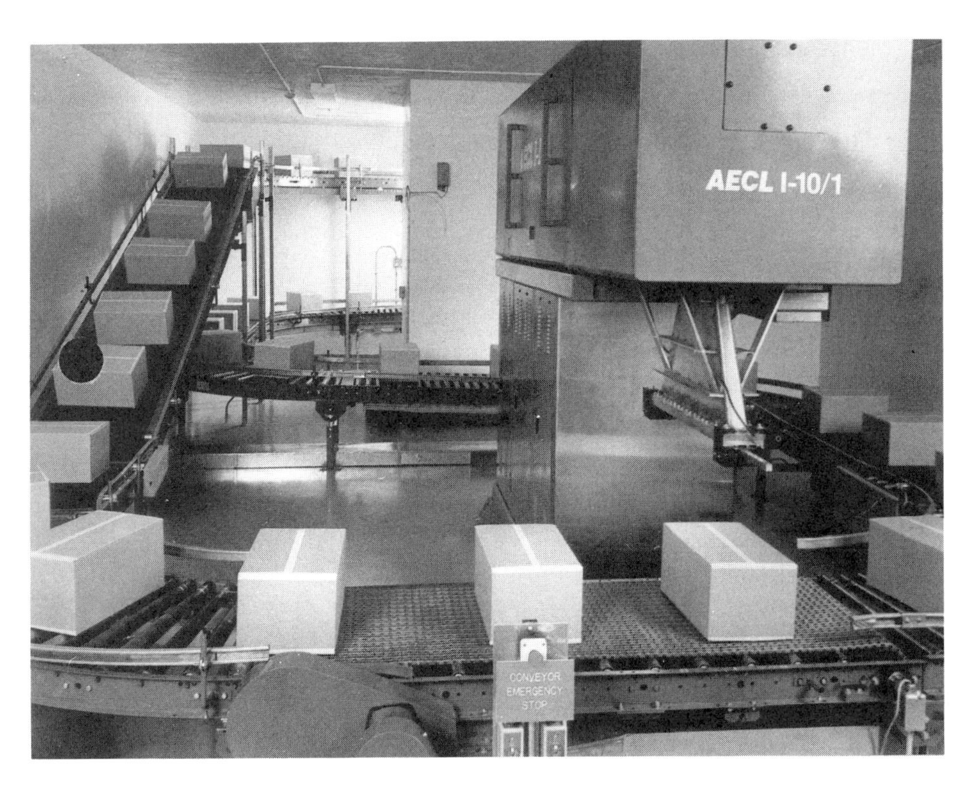

Figure 2. The AECL IMPELA which is shown sterilizing medical disposables. This is a 10 MeV L-Band Accelerator with 200 μ sec pulses and 50 kW of power.

SECTION I

ATOMIC AND MOLECULAR PHYSICS

WAVE FUNCTIONS FOR DOUBLE ELECTRON ESCAPE AND WANNIER-TYPE THRESHOLD LAWS

J. H. Macek

Department of Physics, University of Tennessee, Knoxville TN and Oak Ridge National Laboratory, OaK Ridge, TN

A representation of two-electron atomic wave functions well-adapted to computations of low-energy fragmentation states is derived. An angle-Sturmian basis is introduced. Exact wave functions are represented by sums over the angle-Sturmian functions and integrals over the index of Bessel functions. Integral representations of approximate two-electron wave functions are obtained. These representations are evaluated asymptotically and shown to give Wannier-type threshold laws for ionization of atoms by electron impact.

INTRODUCTION

The collective motion of three charged particles is fundamental to atomic dynamics when two electrons are outside of valence shells, either in doubly excited or in continuum states. In the latter case, correlated motion of two electrons produces many observable effects, of which Wannier's threshold law (1) is the most studied. Quantitative calculations in the threshold region have just begun. Among the methods that have been employed are: hyperspherical close coupling (2) convergent close coupling (3), and pseudo-state calculations (4). Crothers (5) adapted the wave functions of Refs. (6,7) to direct calculation of the ionization matrix element and thereby reproduces the Wannier threshold law. Despite the progress in extending standard methods to treat ionization, theory remains incomplete since the correlations discussed by Wannier are difficult to incorporate (8). In addition, large basis sets are required in conventional calculations, but are poorly adapted to elucidating underlying physical pictures as articulated, for example, by Fano (9).

In Fano's picture, a Schrödinger wave representing two electrons starts from a region where both electrons are close to the ionic core and propagates outward through a region, called the Coulomb zone, where the wave branches into alternative channels. Channels with increasing principal quantum number n_P are populated at successively higher values of the mean distance of both electrons from the ionic core. This process continues until a region is reached where both electrons are effectively free. The "Wannier ridge", i. e. a region in coordinate space where the two-electron potential has a local saddle point, plays a key role in this evolution. Only that part of the wave which starts on the ridge evolves into a wave representing two free electrons.

The branching off into different excitation channels is recognized in the hyperspherical adiabatic basis (10) using Demkov's construction. This construction considers that different adiabatic energy eigenvalues $\varepsilon_\mu(R)$ represent different branches of a *single* analytic function of the hyper-radius R which is single-valued on a multisheeted Rieman surface. The different sheets are connected at branch points which occur at complex $R = R_b$. The real values of the branch points equal the values of R where the outgoing wave populates the appropriate excitation channels. To incorporate this picture in the hyperspherical close coupling theory requires using large numbers of energetically allowed channel functions. Even then, the ionization region can never be rigorously represented since the part of the wave that remains on the ridge is identified only at infinite distances.

The purpose of this paper is to show how an alternate set of functions, namely the angle-Sturmian functions (11), easily incorporates the ionization component, i. e. the component that remains on the Wannier ridge as $R \to \infty$. In addition, a simple approximate expression for excitation and ionization cross sections is given. While the closed form expressions are approximate, they appear to give total ionization cross sections that are accurate to within 10 to 20%.

ADIABATIC HYPERSPHERICAL AND ANGLE-STURMIAN FUNCTIONS

Hyperspherical adiabatic bases

The hyperspherical adiabatic basis functions $\varphi(R;\Omega)$ are defined as eigenfunctions of the operator $\Lambda^2 +$

CP392, *Application of Accelerators in Research and Industry*, edited by J. L. Duggan and I. L. Morgan
AIP Press, New York © 1997

$2RC(\Omega)$ where the hyper-radius R is held fixed (10), i. e.

$$[\Lambda^2 + 2RC(\Omega)]\varphi(R;\Omega) = 2\varepsilon_\mu(R)R^2\varphi(R;\Omega). \quad (1)$$

Here Λ^2 is the grand angular momentum operator and $C(\Omega)$ is scaled potential $V(R,\Omega) = C(\Omega)/R$. The adiabatic functions are taken to be real for real R. These basis functions concentrate in the valleys of the potential V for large R where they become bound-state wave functions of one-electron atomic species (10). Accordingly, a finite number of these functions can represent excited states, but not states where both electrons are in the continuum, i.e. the hydrogen atom is ionized. Alternatively, Sturmian eigenfunctions, defined in the next section, are suitable for ionization.

The angle-Sturmian bases

The angle-Sturmian functions $S_n(\nu;\Omega)$ are defined by replacing the coefficient R of the scaled potential $2C(\Omega)$ with an eigenvalue $\rho_n(\nu)$ when the operator $\Lambda^2 + 2RC(\Omega)$ is set equal to $\nu^2 - 1/4$ to yield the eigenvalue equation

$$[\Lambda^2 + 2\rho_n(\nu)C(\Omega)]S_n(\nu;\Omega) = [\nu^2 - 1/4]S_n(\nu;\Omega). \quad (2)$$

The eigenfunctions are orthogonal with the weight $C(\Omega)$

$$\int S_{n'}(\nu;\Omega)C(\Omega)S_n(\nu;\Omega)d\Omega = \delta_{nn'} \quad (3)$$

Relation between Sturmian and adiabatic bases

It is necessary to connect the Sturmian and physical channels at large R. This is done via the connection between Sturmian and hyperspherical adiabatic functions. These in turn connect with the physical excitation channels as discussed in Ref. (10).

Since $\varepsilon(R)R^2$ is defined for all R, the equation

$$2\varepsilon(\rho)\rho^2 = \nu^2 - 1/4 \quad (4)$$

may be solved to find its roots $\rho_n(\nu)$. These are just the Sturmian eigenvalues of Eq.(2). The corresponding Sturmian eigenfunctions $S_n(\nu;\Omega)$ are, aside from normalization constants, just the adiabatic functions $\varphi(R;\Omega)$ evaluated at $R = \rho_n(\nu)$:

$$S_n(\nu;\Omega) = N(\nu)\varphi[\rho_n(\nu);\Omega], \quad (5)$$

where $N(\nu)$ is a normalization constant.

Equation (2) defines the angle-Sturmian functions and it remains to show how they are to be used to represent two-electron wave functions. In order to use these functions it is necessary to employ a suitable representation of arbitrary functions $\psi(R,\Omega)$. When the Schrödinger equation is separable in hyperspherical coordinates the solution has the form

$$\Psi(R,\Omega) = \int_c \Phi(\nu,\Omega)R^{1/2}Z_\nu(KR)\nu d\nu, \quad (6)$$

where c denotes a contour in the ν-plane that depends upon boundary conditions, $Z_\nu(KR)$ is a Bessel function and $K = \sqrt{2E}$. Remarkably, this representation also holds for any arbitrary function, thus it holds even when the Schrödinger equation is not separable. In the latter case, the function $\Phi(\nu,\Omega)$ is written as a sum over the angle-Sturmian basis functions;

$$\Phi(\nu,\Omega) = \sum_n B_n(\nu)S_n(\nu;\Omega). \quad (7)$$

Equations for the coefficients $B_n(\nu)$ are given in Ref. (11).

One-Sturmian Approximations

The simplest approximation consists of truncating the Sturmian expansion in Eq.(7) to just one term . As is usual for truncations of sums to a few terms, the one-Sturmian approximation is justified only *a posteriori*. Since only one Sturmian is used, it is convenient to omit the index n except where it is needed for clarity.

A closed-form expression for the one-Sturmian approximation to the exact wave function is given in Ref. (11). It is

$$\begin{aligned}\Psi(R,\Omega) \approx & \sqrt{\frac{2}{\pi}}\exp\left[-i\sqrt{K^2\rho(\nu_0)^2 - \nu_0^2}\right. \\ + & \left. i\nu_0 \arccos\left(\frac{\nu_0}{K\rho(\nu_0)}\right) + i\pi/4\right] \\ \times & \int_{-\infty}^{\infty} \frac{\exp\left[i\int_{\nu_0}^{\nu} \arccos\left(\frac{\nu'}{K\rho(\nu')}\right)d\nu'\right]}{\rho(\nu)\sqrt[4]{K^2 - \nu^2/\rho(\nu)^2}} \\ \times & R^{1/2}H_\nu^{(1)}(KR)S(\nu;\Omega)d(\nu^2/2).\end{aligned} \quad (8)$$

Although this function involves several approximations, we will see that it applies to a wide range of dynamical processes.

Before evaluating the integral in Eq. (8) it is useful to examine the function in detail to see how ionization channels emerge. Notice that the integral goes over all real values of ν^2, both positive and negative. When ν^2 is negative the angle-Sturmian is proportional to the corresponding adiabatic function according to Eq. (5). The integral over negative ν^2 therefore represents bound-state channels. At some value of ν^2 the eigenvalue $\rho_n(\nu)$ vanishes and the Sturmian is identical to a hyperspherical harmonic. Such harmonics represent electron motion in regions where the potential plays only a minor role (12). For ν^2 large and positive the Sturmian eigenvalue must be negative.

The meaning of negative $\rho(\nu)$ is best seen by considering the function $C(\Omega)$ in the linear model where the electrons are on opposite sides of the nucleus so that the angle between $\mathbf{r}_1$ and $\mathbf{r}_2$ equals π. Then the scaled potential becomes

$$C(\alpha) = -\frac{Z}{\sin\alpha} - \frac{Z}{\cos\alpha} + \frac{1}{\sqrt{1-\sin 2\alpha}} \tag{9}$$

When $\rho(\nu)$ is positive the angle-Sturmian function is concentrated in the regions where it represents a one-electron hydrogenic bound state function. When $\rho(\nu)$ is negative the potential effectively changes sign and the function is concentrated in the region near $\alpha \approx \pi/4$. In this region the potential is that of a harmonic oscillator in first approximation and the Sturmian function becomes a harmonic oscillator eigenstate concentrated on the Wannier ridge as postulated at the outset by Wannier. It follows that the integral over positive values of ν^2 represents ionization. In this way we see how a single Sturmian represents both the excitation and the ionization channels in a natural way. In the next section the integral in Eq. (9) is evaluated for large values of R to extract excitation and ionization matrix amplitudes.

The wave function for large R

For large R the expression Eq. (8) becomes

$$\begin{aligned}\Psi(R,\Omega) &\approx \frac{2}{\pi}\int_c \frac{1}{\rho(\nu)}\frac{1}{\sqrt[4]{K^2-\nu^2/\rho(\nu)^2}} \\ &\times \exp[i\chi(\nu)]\frac{S(\nu;\Omega)}{\sqrt[4]{K^2-\nu^2/R^2}}\nu d\nu,\end{aligned} \tag{10}$$

where $\chi(\nu)$ is a phase factor. The phase χ becomes large and the stationary phase approximation may be used to evaluate $\Psi(R;\Omega)$ asymptotically. The points of stationary phase occur at values of $\nu = \nu_\mu(R)$ given by $\rho(\nu_\mu) = R$. Defining the wave vector $K_\mu(R) = \sqrt{K^2 - 2\varepsilon_\mu(R) - 1/4R^2}$ we have for the stationary phase approximation to $\Psi(R,\Omega)$ the result, aside from an unimportant overall multiplicative constant;

$$\begin{aligned}\Psi(R,\Omega) &\approx \sum_{\text{paths}}\sum_{\mu}\frac{1}{\sqrt{K_\mu(R)}} \\ &\times \exp\left(i\int_{c_{n_*}}^{R} K_\mu(R')dR'\right)\varphi_\mu(R;\Omega).\end{aligned} \tag{11}$$

In Eq.(11), $c_{n\mu}$ denotes a contour that connects $\varphi_a(R_0;\Omega)$ at small R_0 with $\varphi_\mu(R;\Omega)$ at large R and the sum over paths goes over all such contours (11).

The probability for a transition $i \to \mu$ is just

$$P(E) = \exp\left[-2\mathrm{Im}\int_{R_i}^{R_*} K(R')dR'\right]. \tag{12}$$

Figure 1 illustrates Eq.(12) schematically. Integrating along the real axis gives the semiclassical Jost matrix element J_{11} for elastic scattering, while integrating along a path around the branch point gives the Jost matrix J_{12} for excitation. Eq. (12) follows from the relation $P(E) = |J_{12}|^2$. It must be stressed that R' of Eq.(12) is an integration variable which need not be interpreted as a physical coordinate; rather, it is identified with the Sturmian eigenvalue $\rho(\nu)$ in Eq. (2).

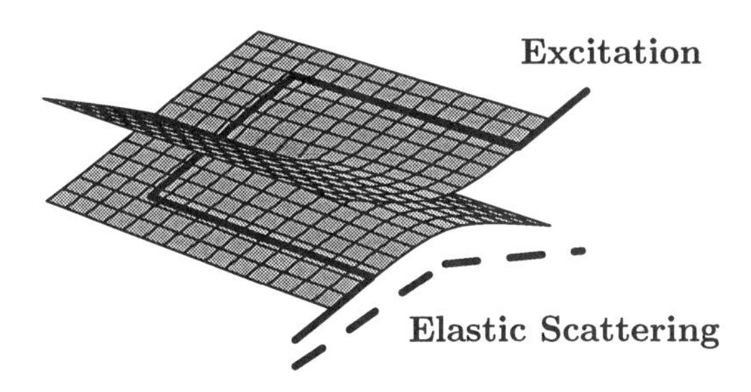

FIGURE. 1. Plot illustrating the hidden-crossing theory. Integration along the real axis describes elastic scattering and integration around the square-root branch point gives the excitation Jost matrix element.

The sum in Eq. (11) goes over both the bound state functions at large positive $\rho(\nu)$ and the harmonic oscillator states at large negative $\rho(\nu)$. These latter terms give Wannier's threshold law .

WANNIER'S THRESHOLD LAW

Large negative values of $\mathrm{Re}\rho(\nu)$ correspond to regions where the adiabatic wave function is confined to the saddle point of $C(\Omega)$ in the Wannier configuration $\mathbf{r}_1 = -\mathbf{r}_2$. Taking the hyperangle to be $\alpha = \arctan(r_2/r_1)$ and letting θ_{12} denote the angle between the position vectors, one has

$$\Lambda^2 \approx -\frac{\partial^2}{\partial\alpha^2} - 4\left(\frac{\partial^2}{\partial\theta_{12}^2} + \frac{1}{\pi-\theta_{12}}\frac{\partial}{\partial\theta_{12}}\right), \tag{13}$$

$$\begin{aligned}C(\Omega) &= -C_0 - \frac{1}{2}k_\alpha(\alpha-\pi/4)^2 \\ &- \frac{1}{2}k_\theta(\theta_{12}-\pi)^2 + \cdots,\end{aligned} \tag{14}$$

where C_0, k_α, and k_θ are constants.

Near the potential saddle and for $\mathrm{Im}\sqrt{R}$ sufficiently large, the wave functions are just those of uncoupled harmonic oscillators (11). For the lowest state of ^1S symmetry, the function $\varphi(R;\Omega)$ is

$$\begin{aligned}\varphi_{\mathrm{HO}}(R,\Omega) &\approx \exp[ia_\alpha\sqrt{R}(\alpha-\pi/4)^2] \\ &\times \exp[-a_\theta\sqrt{R}(\theta_{12}-\pi)^2],\end{aligned} \tag{15}$$

where a_α and a_θ are real constants.

On the real axis the function $\varphi_{\rm HO}(R;\Omega)$ of Eq. (15) is unbounded in α and does not satisfy appropriate boundary conditions at $\alpha = 0$ and $\pi/2$; thus it does not represent the adiabatic eigenfunction. For sufficiently large $\mathrm{Im}\sqrt{R}$, however, the function is exponentially damped and vanishingly small at the endpoints of the range of α.

The asymptotic adiabatic eigenvalues $\varepsilon_{\rm HO}(R)$ for sufficiently large $\mathrm{Im}\sqrt{R}$ are just those of uncoupled harmonic oscillators

$$\varepsilon_{\rm HO}(R) = -C_0/R - (C_{1r} + iC_1)/R^{3/2} + O(R^{-2}), \quad (16)$$

where

$$\begin{aligned} C_{1r} &= -(2n_\theta + I + 1)2^{-1/4}, \\ C_1 &= (2n_\alpha + 1)2^{-5/4}\sqrt{12Z-1}, \end{aligned} \quad (17)$$

n_α and n_θ are harmonic oscillator quantum numbers, and I is the projection of the total angular momentum L on an axis parallel to $\mathbf{r}_{12}$. Higher order terms in the series Eq.(16) represent anharmonic corrections.

For sufficiently large R_0 we may evaluate the phase integral approximately;

$$\begin{aligned} \mathrm{Im}\int_{R_b}^{\infty} K_{\rm asy}(R')dR' &\approx -C_1\sqrt{\frac{2}{C_0}}\ln\sqrt{E} \quad (18) \\ +C_1\sqrt{\frac{2}{C_0}} &\times \ln\left(\sqrt{\frac{C_0}{R_b}} + \sqrt{E + \frac{C_0}{R_b}}\right). \end{aligned}$$

Equation (18) relates to the probability for populating the state $\varphi_{\rm HO}(R;\Omega)$ at some value of R much larger than the Wannier radius $R_W = C_0/E$. To complete the computation of the cross section, a factor $P_{\rm inner}(E)$ related to the probability for reaching the point R_b must be included. In addition, the harmonic oscillator wave function analytically continued to the real axis is also present. Taking account of the three factors in the ionization probability gives the cross section for a particular partial wave L and spin S;

$$\begin{aligned} \sigma_L^{(S)} &= \frac{\pi}{2E+1}(2L+1)P_{\rm inner}(E)P_{\rm HO}(E) \quad (19) \\ &\times \int |\varphi_{\rm HO}(R_E, \Omega_E)|^2 d\Omega_E, \end{aligned}$$

where $d\Omega_E = d\alpha_E d\hat{\mathbf{k}}_1 d\hat{\mathbf{k}}_2$ and $\tan\alpha_E = k_2/k_1$. For states with $n_\alpha = 0$, integration over the five hyperangles gives $|N|^2$, the square of the normalization constant of $\phi_{\rm HO}(R_E,\Omega)$. We find $|N|^2 = R_E^{1/4}\sqrt{2a_\alpha/\pi}$, where $R_E = 4C_0/E$. The threshold cross section is then

$$\sigma \propto E^{\zeta_W^{(\rm ad)}} \quad (20)$$

which is the Wannier threshold law with the adiabatic Wannier index

$$\zeta_W^{(\rm ad)} = C_1\sqrt{\frac{2}{C_0}} - \frac{1}{4}. \quad (21)$$

For ionization of neutral atoms by electron impact $\zeta_W^{(\rm ad)} = 1.104$ which differs by 2% from the exact $\zeta_W = 1.127$.

The method given in this paper allows for essentially exact calculations of the ionization cross section (11), but its real value derives from the ability to compute higher order corrections and approximate threshold laws for a wide variety of models. For example, higher order anharmonic corrections have been shown to introduce additional nonanalytic factors of the form $\exp[-C_2\sqrt{E}]$ in threshold cross section. The constant C_2 is small for electron impact, but may not be for positron impact. This point is under study (13). As another example, the method has been applied to the Temkin-Poet model where classical calculations show that the cross section is identically zero for a range of energies just above the threshold. The quantal cross section is found to have an exponential behavior, namely $\sigma \propto \exp[-C/E^{1/6}]$ where C is a constant. The exponentially small probability is thought to be the quantal counterpart of the vanishing classical probability for a classically forbidden process.

Acknowledgments

Support by the National Science Foundation under Grant No. PHY-9222489 is gratefully acknowledged.

REFERENCES

†Managed by Lockheed Martin Energy Systems Inc. under contract No. DE-AC05-84OR21400 with the U S Department of Energy.

1. G. H. Wannier, Phys. Rev. **90**, 817 (1953).
2. D. Kato and S. Watanabe, Phys. Rev. Lett. **74**, 2443 (1995).
3. I. Bray and Stelbovics, Phys. Rev. Lett. **70**, 746 (1993).
4. J. Callaway, Phys. Rev. A **44**, 2192 (1991).
5. D. S. Crothers, J. Phys. B **19**, 463 (1986).
6. A. R. P. Rau, Phys. Rev. A **4**, 207 (1971).
7. R. Peterkop, J. Phys. B **4**, 513 (1971)
8. S. Watanabe, J. Phys. B **24**, L39 (1991).
9. U. Fano, Rep. Prog. Phys. **46**, 97 (1983).
10. J. H. Macek, J. Phys. B **1**, 831 (1968); C. D. Lin, Rept. Prog. Phys. **257**, 1 (1995).
11. J. H. Macek, S. Yu Ovchinnikov, and S. V. Pasovets, Phys. Rev. Lett. **74**, 4631 (1995); J. H. Macek, S. Yu. Ovchinnikov, Phys. Rev. **54**, 544 (1966).
12. M. Cavagnero, Phys. Rev. A **30**, 1169 (1984).
13. W. Ihra and P. F. O'Mahoney, private communication.

RESONANT-COHERENT ELECTRON EMISSION FROM CHANNELED IONS

F. J. García de Abajo

Depto. de Ciencias de la Computación e Inteligencia Artificial and Depto. de Materiales,UPV/EHU, Aptdo. 649, 20080 San Sebastián. SPAIN.

Coherent excitation of channeled ions is shown to result in the emission of bound electrons of the projectile for large enough transition energies. The energy carried by the emitted electrons is well determined, like in the case of resonant-coherent excitation, while momentum conservation leads to preferential directions of emission. The angular dependence of coherently-lost electrons is analyzed.

An ion moving under channeling conditions inside a crystal or near its surface is subject to the perturbation of both the static potential set up by the atoms ordered in the crystal lattice and the induced potential originating in the charge density fluctuations produced by the presence of the ion. The periodic components of the crystal potential may induce transitions in the internal electronic state of the ion (electron excitation and electron capture and loss) leaving the state of the target unchanged, whereas the appearance of the induced potential is connected to changes in the electronic state of the target itself.

The well-known resonant-coherent excitation[1-8] (RCE) corresponds to transitions between bound states of the ion caused by the periodicity of the crystal lattice, which is experienced by the moving ion as a periodicity in time. Focusing on a swift ion moving inside a crystal with constant velocity **v**, the crystal potential acting on a bound electron of the projectile can be expressed in terms of Fourier components labeled by reciprocal lattice vectors **g**, according to

$$V_C = \sum_{\mathbf{g}} V_{\mathbf{g}} \exp\left[i\mathbf{g}.\left(\mathbf{r}_0 + \rho + \mathbf{v}t\right)\right],$$

where ρ is the electron coordinate relative to the ion nucleus and $\mathbf{r}_0$ is the initial position of the ion. Consequently, the frequency of the perturbation attributed to harmonic **g** becomes $-\mathbf{g}.\mathbf{v}$, so that RCE with transition energy $\Delta\varepsilon$ can take place in the ion provided the condition of resonance

$$\Delta\varepsilon + \mathbf{g}.\mathbf{v} = 0 \qquad (1)$$

is fulfilled.

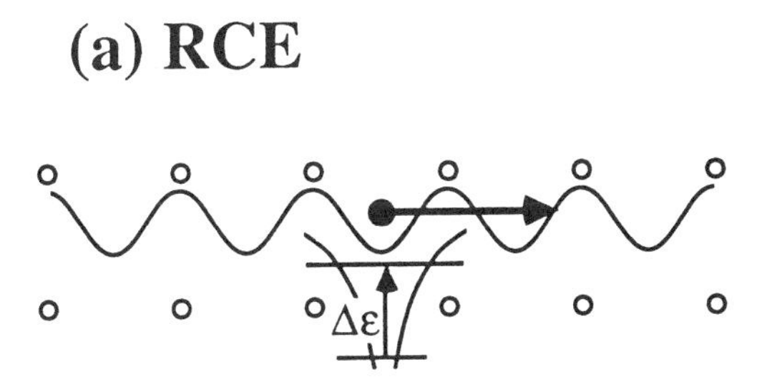

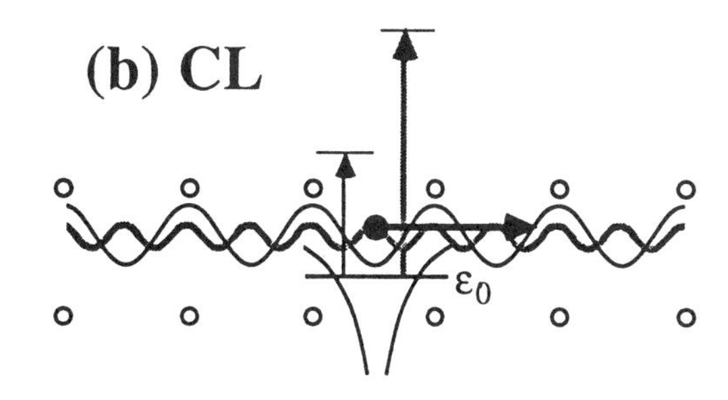

FIGURE 1. Schematic representation of **(a)** resonant-coherent excitation (RCE) and **(b)** coherent electron loss (CL) processes for a channeled ion. Open circles represent solid atoms periodically disposed along the channel. The first Fourier component of the periodic crystal potential has been represented by a black oscillatory line. The second component, of double frequency, is shown by a grey curve in the case of CL. Vertical arrows represent electronic transitions due to the corresponding Fourier components.

CP392, *Application of Accelerators in Research and Industry*, edited by J. L. Duggan and I. L. Morgan
AIP Press, New York © 1997

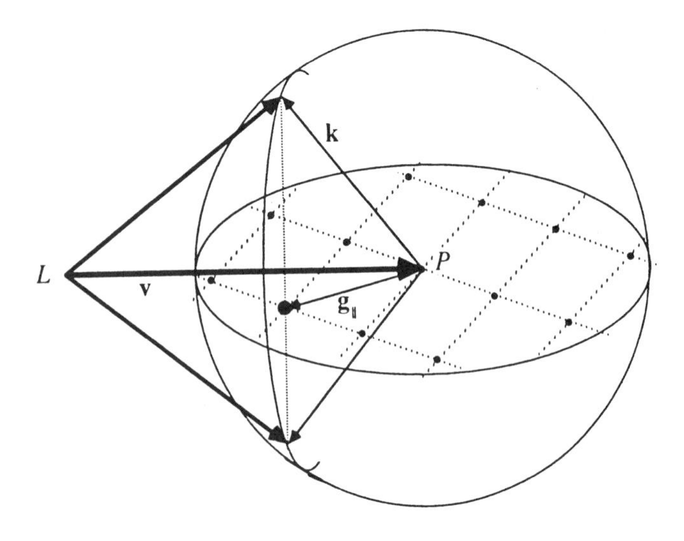

FIGURE 2. Construction of the direction of electron ejection during coherent electron loss from an ion interacting with a crystal target under planar- surface-channeling conditions. Represented is the space of momentum of final electron states, **k**. A thick arrow stands for the ion velocity. Points L and P correspond to the origin of momenta at the lab and the projectile, respectively. Reciprocal lattice sites are represented by black circles. The components of **k** parallel to the surface or plane in each case are given by $\mathbf{g}_{//}$ according to Eq. (3), i.e., the harmonic under consideration, so that the direction of ejection is determined by the projection onto the sphere of radius k [obtained from Eq. (2)] centered at the projectile, as shown in the figure.

The RCE was predicted by Okorokov,[1] who considered excitations of the nucleus rather than electronic excitations. Datz *et al.*[2] were the first to find experimental evidence in a set of elegant experiments that showed an increase in the ionized charge fraction of fixed-charge-state axially-channeled ions at the velocity of resonance, explained in terms of the shorter life-times of excited states. This effect has also been observed under planar[3] and surface[4] channeling conditions. An enhancement of the convoy-electron yield[5] and the emission of X-rays by de-excitation of resonant-coherently excited ions[6] have also been reported.

The induced part of the potential, together with the average of the crystal potential along the ion trajectory, plays the important role of mixing electronic states of the ion and shifting and splitting its energy levels. Crawford and Ritchie[7] calculated this effect, finding good agreement with experiments.

When the combination of the harmonic under consideration and the ion velocity are such that the excitation energy exceeds the ionization potential of the ion, $-\varepsilon_0$, the bound electron can be emitted to the continuum, leading to the so-called coherent loss (CL). The momentum of the excited electron is found to be, from Eq. (1),

$$k=[2(\varepsilon_0+\mathbf{g}.\mathbf{v})]^{1/2}. \qquad (2)$$

This mechanism was successfully incorporated by Sols and Flores[9] in the study of charge state fractions of ions traversing solids. Here, we are concerned with the distribution of electrons ejected via CL.

Fig. 1 schematically illustrates the RCE and CL processes for a channeled ion. The first two harmonics of the crystal potential have been considered in the latter case. The ratio of the corresponding transition energies is 2:1.

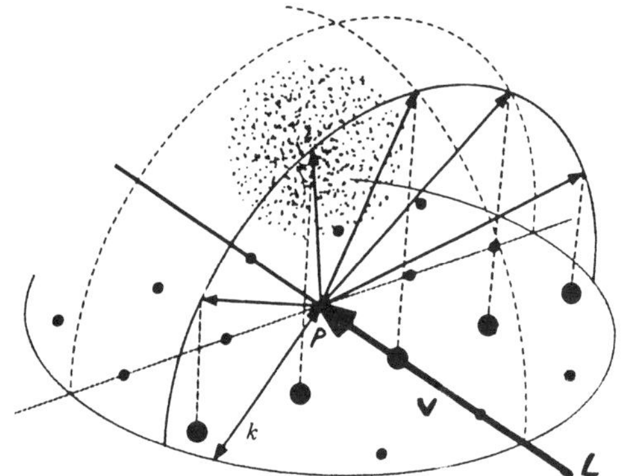

FIGURE 3. The same as Fig. 2 for surface-channeling along the direction of an axial channel within the surface. A given harmonic receives contributions from two-dimensional reciprocal lattice vectors with the same projection on the ion velocity, as shown in the figure. The dotted region on the sphere of radius k, whose size corresponds to the extension of a typical bound state in momentum space, represents the distribution of final states for the contribution of one of the two-dimensional reciprocal lattice vectors.

The energy of the emitted electron is fixed by Eq. (2). Moreover, momentum conservation leads to preferential directions of emission. Indeed, for an ion moving under surface of planar channeling conditions, as that considered in Fig. 2, the change of its momentum is approximately given by the corresponding reciprocal surface lattice vector

$$\mathbf{k}_{//} \approx \mathbf{g}_{//}, \qquad (3)$$

where the subindex $\parallel$ refers to the components parallel to the surface of planar channel in clear analogy with low-energy electron diffraction. This condition, together with Eq. (2), fixes the preferential direction of electron ejection. Now, the initial electronic state traveling with the ion can be represented by a distribution of size $\sqrt{2|\varepsilon_0|}$ centered at the ion velocity in momentum space, so that the angular spread around the preferential direction of emission may be estimated as

$$\Delta\theta = 2\sin^{-1}\left(\sqrt{2|\varepsilon_0|}/2k\right), \qquad (4)$$

so that for an axially channeled ion like that considered in Fig. 3 the distribution of ejected electron turns out to be cone-like.

A more detailed calculation supports these ideas.[10] In particular, Fig. 4 represents the angular distribution of electrons emitted in the rest frame of He^+ ions in their ground state, grazingly incident on a W(001) surface along the [100] axial channel. Figs. 4a and 4b correspond to excitation via harmonics of frequency $-\mathbf{g}.\mathbf{v} = 2\pi n v/d$ for $n = 1$ and $n = 2$, respectively, where d is the atomic spacing along the direction of the axis. The corresponding values of k are 2.5 a.u. and 4 a.u., respectively. This, together with the size of the initial state ≈ 2, gives an angular spread of 47° and 29°, respectively, according to Eq. (4), which is in qualitative agreement with the results shown in the figure.

This angular and energy confinement of the emission suggests the possibility of experimentally detecting electrons emitted via CL. This could be done under surface channeling conditions, so that the electrons do not have to travel through the solid in order to escape from the interaction region and be eventually detected. Another possibility consists in using high-energy ions, so that the electrons are emitted also with high energies and large inelastic mean free paths against the solid, preventing them from suffering further scattering processes.

In RCE experiments, high energy ions are employed to ensure that their charge state is fixed, so that the effect of the excitation on the charge state fractions are not erased by further charge changing processes. Also, frequent charge exchange would broaden the atomic state levels, leading to the consequent undesired broadening of the resonance. An alternative approach would be to consider large-life-time charge states of low-energy ions, like in the case of atomic levels lying in the band gap of an insulator, where both resonant and Auger charge transfer processes can be forbidden. In particular, recent experimental results[11] indicate that ions such as O^- in front of LiF are very stable, providing a suitable candidate for the detection of electrons emitted via CL.

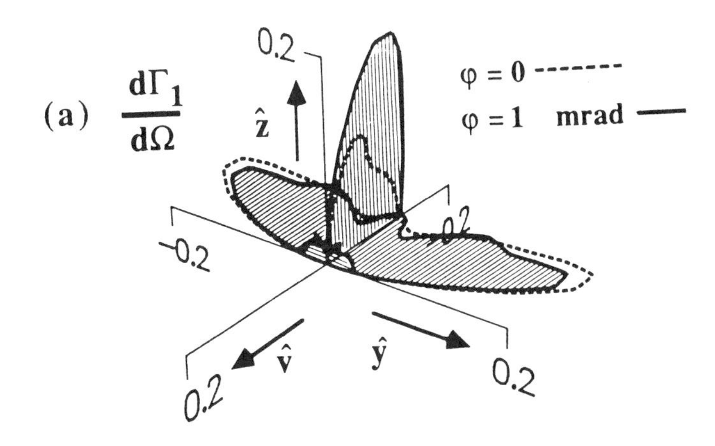

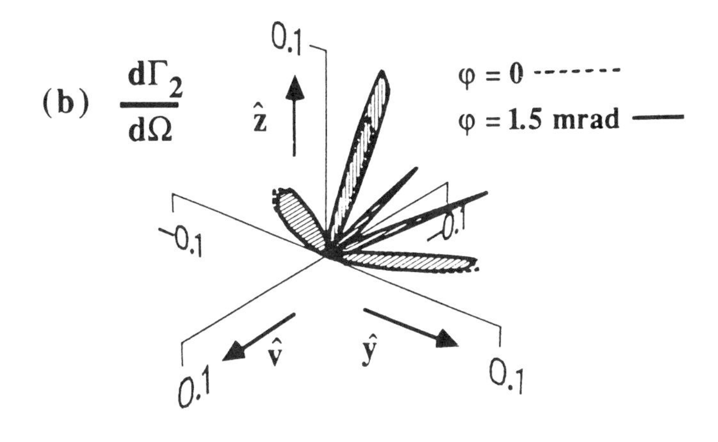

FIGURE 4. Angular dependence of the differential yield of electron emission via coherent loss (within the frame of the moving ion) for 2.5 MeV-He^+ ions incident under surface channeling conditions on a W(001) surface along the [100] direction with different angles of incidence φ, as shown in the insets. The directions of the velocity and the surface normal, z, are also indicated. (a) and (b) correspond to excitation via harmonics $n = 1$ and $n = 2$, respectively, as explained in the text.

ACKNOWLEDGMENTS

The author wants to thank V. H. Ponce and P. M. Echenique for helpful discussions and gratefully acknowledges support by the Departamento de Educación del Gobierno Vasco and the Basque Country University.

REFERENCES

1, V. V. Okorokov, JETP Lett. **2**, 111 (1965).
2, S. Datz *et al.*, Phys. Rev. Lett. **40**, 843 (1978).
3, S. Datz *et al.*, Nucl. Instrum. Methods B **170**, 15 (1980).
4, K. Kimura *et al.*, Phys. Rev. Lett. **76**, 3850 (1996).

5, K. Kimura *et al.*, Phys. Rev. Lett. **66**, 25 (1991).
6, F. Fujimoto *et al.*, Nucl. Instrum. Methods B **33**, 354 (1988).
7, O. H. Crawford and R. H. Ritchie, Phys. Rev. A **20**, 1848 (1979).
8, F. J. García de Abajo and P. M. Echenique, Phys. Rev. Lett. **76**, 1856 (1996).
9, F. Sols and F. Flores, Phys. Rev. B **30**, 4878 (1984).
10, F. J. García de Abajo, V. H. Ponce, and P. M. Echenique, Phys. Rev. Lett. **69**, 2364 (1992).
11, C. Auth, A. G. Borisov, and H. Winter, Phys. Rev. Lett. **75**, 2292 (1996).

ALIGNMENT OF EXCITED STATES PRODUCED IN ELECTRON-ION AND ION-ATOM COLLISIONS

S. R. Grabbe and C. P. Bhalla

James R. Macdonald Laboratory, Department of Physics, Kansas State University, Manhattan, KS 66506-2604

Excited states produced in electron-ion and ion-atom collisions are often aligned and the angular distribution of photons and Auger electrons resulting from the de-excitation of such states is, in general, not isotropic. We have investigated the angular distribution of photons and Auger electrons resulting from the de-excitation of collisionally aligned states of Fe^{23+} and O^{5+}. The magnetic substate cross sections for the electron impact excitation of Fe^{23+} and O^{5+} are also presented. These calculations were performed within the R-matrix formulation and the distorted wave approximation.

INTRODUCTION

When atoms (or ions) are excited by a collimated beam of electrons or protons traveling in a well defined direction, the magnetic substates of the excited atomic states are, in general, not statistically populated. The angular distribution of photons or Auger electrons emitted during the decay of these aligned states is generally nonisotropic relative to the incident beam direction (1-7).

Alignment studies of atoms and ions have a very rich history dating back to the early measurements of the angular distribution and polarization of emitted light in the '20s (3). Similarly, a series of studies by Mehlhorn and co-workers (4,5) in the '70s increased our understanding of aligned states by measuring the non-isotropic angular distribution of Auger electrons following impact ionization.

In recent years, studies of the de-excitation of highly charged collisionally aligned states have been of interest due to possible applications to plasma diagnostics (7) and the interpretation of experimental results. For example, recent measurements of electron impact excitation of helium-like and lithium-like ions on the Livermore electron-beam ion trap (EBIT) were sensitive to both the non-isotropy and polarization of the x-rays measured at 90° with respect to the electron beam direction (8). The polarization of the measured x-rays was accounted for in the data analysis by using the theoretical polarization calculations of Gullikson (9).

It is the purpose of this paper to report on the alignment of collisionally excited states of lithium-like iron and oxygen. We present magnetic substate excitation cross sections for the electron impact excitation of Fe^{23+} and O^{5+}. The angular distribution of photons emitted at $\theta = 90^o$ relative to the incident beam direction is presented for the de-excitation of the $1s^2 3p\ ^2P_{3/2}$ state of Fe^{23+}. We also compare measured and calculated differential cross sections in $O^{5+} + H_2$ collisions for the Auger electrons which result from the de-excitation of the $1s2s2p\ ^4P$ state.

RESULTS AND DISCUSSION

Lithium-like Iron

In Fig. 1, the magnetic substate excitation cross sections, σ_M, are plotted as a function of the incident electron-impact energy for the $1s^2 2s\ ^2S_{1/2} \rightarrow 1s^2 3p\ ^2P_{3/2}$ transition of Fe^{23+}.

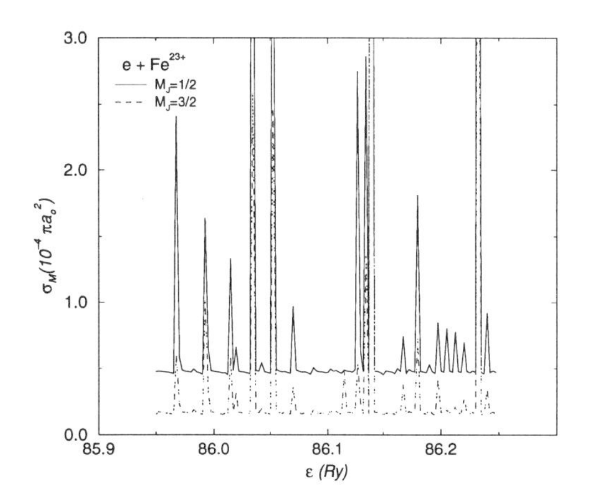

FIGURE 1. Magnetic substate excitation cross sections versus electron impact energy in Rydbergs for the transition of $1s^2 2s\ ^2S_{1/2}$ to $1s^2 3p\ ^2P_{3/2}$ in Fe^{23+}. Solid line: $|M_J| = 1/2$, dashed line: $|M_J| = 3/2$.

CP392, *Application of Accelerators in Research and Industry,* edited by J. L. Duggan and I. L. Morgan
AIP Press, New York © 1997

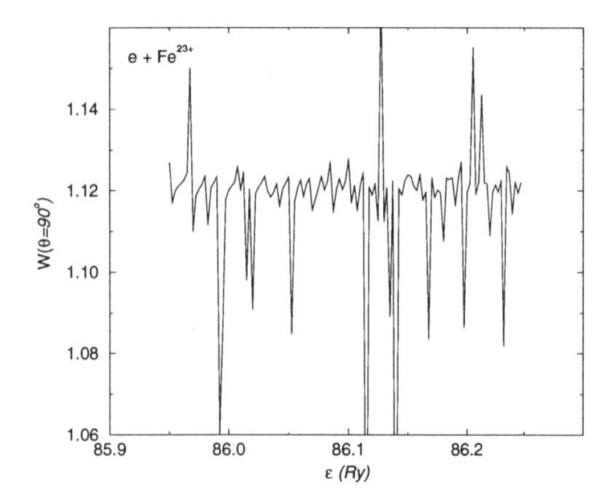

FIGURE 2. Angular distribution of photons at $\theta = 90^o$ versus electron impact energy for the transition from $1s^2 3p\ ^2P_{3/2}$ to $1s^2 2s\ ^2S_{1/2}$.

This energy region lies between the $1s^2 3p\ ^2P_{3/2}$ excitation threshold ($E_{exc} = 85.902\ Ry$) and the $1s^2 3d\ ^2D_{3/2}$ excitation threshold ($E_{exc} = 86.282\ Ry$) therefore these σ_M will not be affected by cascade contributions from higher lying states. These are the first $e + Fe^{23+}$ calculations of σ_M which account for the rich resonance structure existing at these low electron-impact energies. The calculations were performed using the R-matrix codes developed for the Opacity Project (10) and relativistic effects are included through the Breit-Pauli Hamiltonian. For the present calculations, the lowest 13 lithium-like iron bound states, up to $1s^2 4d\ ^2D_{5/2}$, have been included to represent the target in the R-matrix internal region. The R-matrix boundary radius is taken to be 3.2 a.u. and 35 continuum orbitals have been included in the expansion of the total wavefunction. The (N+1)–electron bound states that contribute to the resonance structure in the $1s^2 2s\ ^2S_{1/2} \rightarrow 1s^2 3p\ ^2P_{3/2}$ excitation can be categorized as $1s^2 nln'l'$ where $n, n' = \{2, 3, 4\}$ and $l, l' = \{0, 1, 2, 3\}$. All (N+1)–electron symmetries with even and odd parities up to total angular momentum $J = 15$ were used.

The angular distribution of photons emitted at an angle θ with respect to the incident beam direction for $Fe^{23+}(1s^2 3p\ ^2P_{3/2}) \rightarrow Fe^{23+}(1s^2 2s\ ^2S_{1/2}) + h\nu$ can be written as (11)

$$W(\theta) = 1 + \frac{1}{8}\left[9a_{3/2} + a_{1/2} - 5\right] P_2(cos\theta) \qquad (1)$$

where $P_2(cos\theta)$ is a Legendre polynomial of order 2. The $a_{|M_i|}$–coefficients are defined as

$$a_{3/2} = \frac{\sigma_{|M_i|=3/2}}{\sigma_{|M_i|=3/2} + \sigma_{|M_i|=1/2}} \qquad (2)$$

and

$$a_{1/2} = \frac{\sigma_{|M_i|=1/2}}{\sigma_{|M_i|=3/2} + \sigma_{|M_i|=1/2}}. \qquad (3)$$

Here $\sigma_{|M_i|}$ is the magnetic substate excitation cross section for $e + Fe^{23+}(1s^2 2s\ ^2S_{1/2}) \rightarrow Fe^{23+}(1s^2 3p\ ^2P_{3/2}) + e$.

The angular distribution of radiation emitted at 90° with respect to the incident beam direction for Fe^{23+} was calculated from Eq. (1) using the magnetic substate excitation cross sections plotted in Fig. 1. These results are presented in Fig. 2 as a function of the electron impact energy. The angle $\theta = 90^o$ was chosen because measured x-rays are typically detected at 90° with respect to the incident beam direction. The rich resonance structure dominating the angular distribution at $\theta = 90^o$ for these low electron-impact energies makes this and similar systems excellent candidates for future experimental studies.

Lithium-like Oxygen

Magnetic substate excitation cross sections, σ_M, were calculated for the $1s^2 2s\ ^2S_{1/2}$ to $1s2s2p\ ^4P_J$ transitions in $e + O^{5+}$ collisions. These calculations were performed within the distorted wave approximation. The results are presented in Figs. 3 and 4 as a function of the electron impact energy in threshold units ($E_{threshold} = 40.81\ Ry$).

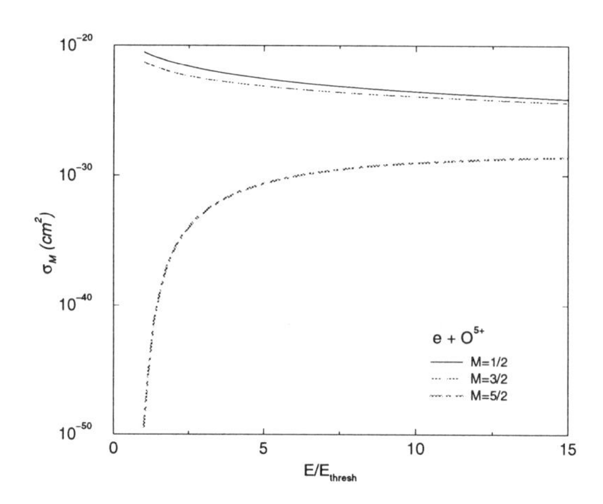

FIGURE 3. Magnetic substate excitation cross sections versus electron impact energy in threshold units for the transition of $1s^2 2s\ ^2S_{1/2}$ to $1s2s2p\ ^4P_{5/2}$ in O^{5+}. Solid line: $|M_J| = 1/2$, dashed line: $|M_J| = 3/2$, dash-dot line: $|M_J| = 5/2$.

These electron impact excitation cross sections are used to calculate the differential cross sections, $\frac{d\sigma}{d\Omega}$, for the Auger electrons which result from the de-excitation of the ^4P_J states as follows:

$$\frac{d\sigma}{d\Omega}(\epsilon, \theta) = \sum_J \left(\frac{d\sigma}{d\Omega}\right)_J = \frac{1}{4\pi}\sum_J \sigma_J(\epsilon) w_J \widehat{W}_J(\theta). \qquad (4)$$

Here σ_J is the total electron impact excitation cross section for populating the autoionizing $1s2s2p\ ^4P_J$ state of O^{5+}, w_J is the Auger branching ratio from $1s2s2p\ ^4P_J$

to $1s^2\ ^1S$, and $\widehat{W}_J(\theta)$ is the angular distribution of the outgoing Auger electron with respect to the incident beam direction (2). The angular distribution, $\widehat{W}_J(\theta)$, depends on the magnetic substate population of $O^{5+}(1s2s2p\ ^4P_J)$. The differential cross sections, $(\frac{d\sigma}{d\Omega})_J$, and, $\frac{d\sigma}{d\Omega}$, for the Auger electrons resulting from the de-excitation of the $1s2s2p\ ^4P_J$ states in $e + O^{5+}$ collisions are plotted in Figs. 5 and 6 respectively.

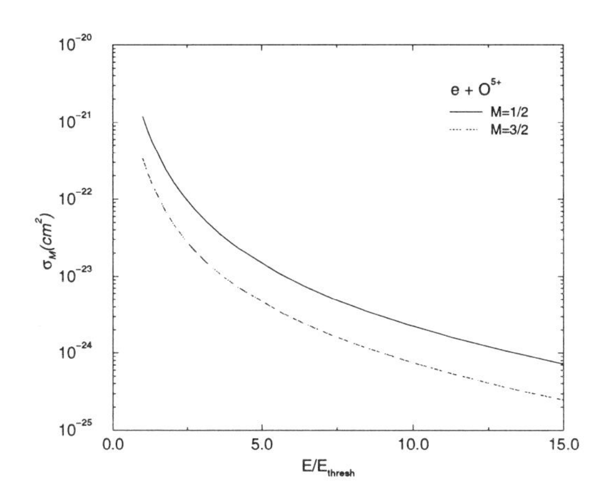

FIGURE 4. Magnetic substate excitation cross sections versus electron impact energy in threshold units for the transition of $1s^22s\ ^2S_{1/2}$ to $1s2s2p\ ^4P_{3/2}$ in O^{5+}. Solid line: $|M_J| = 1/2$, dashed line: $|M_J| = 3/2$.

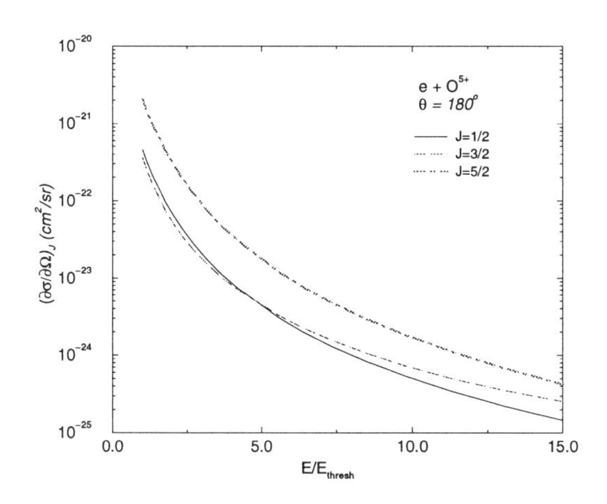

FIGURE 5. Differential cross sections for total angular momentum, J, at $\theta = 180^o$ versus electron impact energy in threshold units for $e + O^{5+}$ collisions. Solid line: $J = 1/2$, dash line: $J = 3/2$, and dash-dot line: $J = 5/2$.

In a recent experiment (12) at $\Theta_{lab} = 0^o$, the Auger electrons were measured for the following ion-atom collision

$$O^{5+}(1s^22s\ ^2S) + H_2 \rightarrow O^{5+}(1s2s2p\ ^4P) + H_2$$
$$\downarrow$$
$$O^{6+}(1s^2\ ^1S) + e_A.$$

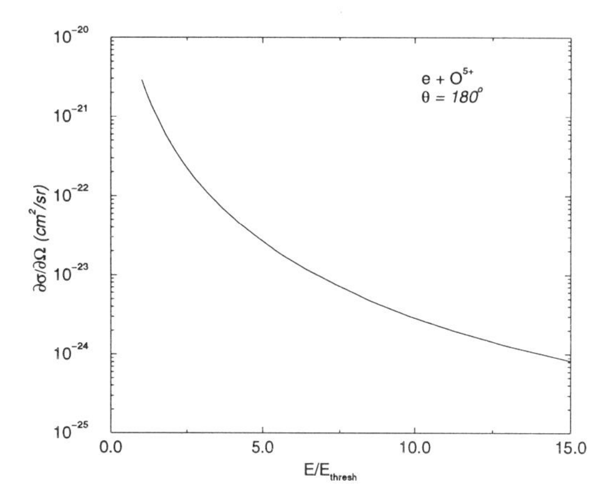

FIGURE 6. Differential cross sections summed over $J = 1/2, 3/2, 5/2$ at $\theta = 180^o$ versus electron impact energy in threshold units for $e + O^{5+}$ collisions.

In the impulse approximation, the differential cross sections, $\frac{d\sigma}{d\Omega}$, can be related to the corresponding differential cross sections, $\frac{d\Sigma}{d\Omega}$, in $O^{5+} + H_2$ collisions as follows:

$$\frac{d\Sigma}{d\Omega}(E,\theta) = \int d\epsilon \left(\left[\frac{J(Q)}{Q + V_p} \right] \frac{d\sigma}{d\Omega}(\epsilon, \theta) \right). \qquad (5)$$

Here $Q = (\sqrt{2(\epsilon + \epsilon_I)} - V_p)$, V_p is the projectile velocity in atomic units, and ϵ_I is the ionization energy of the H_2 target electrons ($\epsilon_I = 0.57\ a.u.$). $J(Q)$ is the experimentally measured Compton profile of H_2 which has been fit by least squares to a polynomial to yield the following analytic expression (13):

$$J(Q) = \frac{a_1}{[1 + (Q/\xi_1)^2]^3} + \frac{a_2}{[1 + (Q/\xi_2)^2]^4} \qquad (6)$$

where $a_1 = 1.0012$, $a_2 = 0.5383$, $\xi_1 = 0.9896$, and $\xi_2 = 1.5566$. The differential cross section in Eq. (5) is in the projectile frame of reference and can be transformed to the laboratory frame. It should be noted that an angle of $\theta = 180^o$ in the projectile frame corresponds to an angle of $\Theta_{lab} = 0^o$ in the laboratory frame.

The measured and calculated differential cross sections, $\frac{d\Sigma}{d\Omega}$, are plotted as a function of the projectile energy at $\Theta_{lab} = 0^o$ in Fig. 7. Very good agreement is found between the theory and the experimental data.

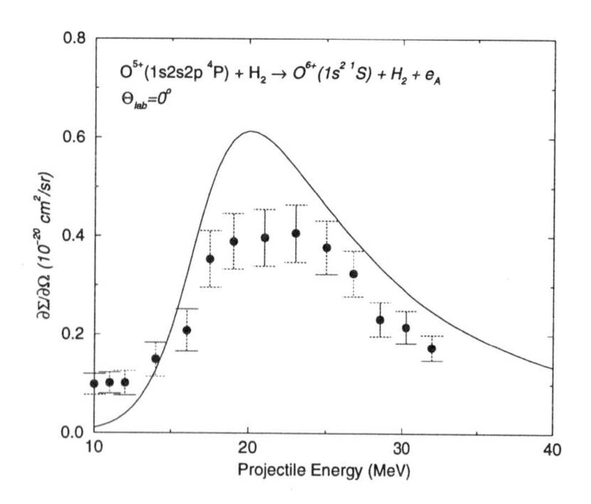

FIGURE 7. Differential cross sections at $\Theta_{lab} = 0^o$ versus projectile energy in MeV for $O^{5+} + H_2$ collisions. Solid line: calculations, filled circles: experiment.

CONCLUSIONS

The contribution of resonances to the magnetic substate excitation cross sections, and the corresponding angular distribution of the radiation from an excited state of Fe^{23+} are presented for electron impact excitation. The rich structure observed in these calculations makes collision systems of this type excellent candidates for future experimental studies. We have also demonstrated that $e + O^{5+}$ calculations can be used to interpret the production of Auger electrons in collisions of $O^{5+} + H_2$.

This work was supported by the Division of Chemical Sciences, Office of Basic Energy Sciences, Office of Energy Research, U.S. Department of Energy.

REFERENCES

1. Aberg T. and Howat G., *Handbuch Der Physik*, New York: in Springer-Verlag, 1982, vol. 31, pp. 543-555.
2. Mehlhorn W., Lectures on the Auger Effect, University of Nebraska (1969).
3. Fano U. and J.H. Macek, *Reviews of Modern Physics* **45**, 553-573 (1973), and references contained therein.
4. Cleff B. and Mehlhorn W., *J. Phys. B: At. Mol. Opt. Phys.* **7**, 593-604 (1974).
5. Döbelin E., Sandner W., and Mehlhorn W., *Phys. Lett.* **49A**, 7-15 (1974).
6. Bhalla C.P., *Phys. Rev. Lett.* **64**, 1103-1106 (1990).
7. See for example, Kawachi T. and Fujimoto T., *Phys. Rev. E* **51**, 1440-1448 (1995).
8. Wong K.L., Beiersdorfer P., Reed K.J., and Vogel D.A., *Phys. Rev. A* **51**, 1214-1220 (1995).
9. Gullikson E.M. (private communication to K.L. Wong *et al.* (8)).
10. Berrington K.A., Burke P.G., Butler K., Seaton M.J., Storey P.J., Taylor K.T., and Yu Yan, *J. Phys. B: At. Mol. Opt. Phys.* **20**, 6379-6397 (1987).
11. De Groot S.R., Tolhoek H.A., and Huiskamp W.J., *Alpha-, Beta- and Gamma-Ray Spectroscopy*, edited by K. Siegbahn, Amsterdam: North-Holland Publishing Company, 1965, pp. 1199-1208.
12. Toth G., Grabbe S., Richard P., and Bhalla C.P., submitted to *Nucl. Instr. and Meth.* (1996).
13. Lee J.S., *J. Chem. Phys.* **66**, 4908-4913 (1977).

ELECTRON DETACHMENT IN HYDROGEN NEGATIVE ION – ATOM AND MOLECULE COLLISIONS

T. J. Kvale[1], A. Sen[1,2], and D. G. Seely[3]

[1] *Department of Physics and Astronomy, University of Toledo, Toledo, Ohio 43606 USA*

[2] *Radiation Oncology Center, Bryan, Ohio 43506 USA*

[3] *Department of Physics, Albion College, Albion, Michigan 49224 USA*

This paper summarizes the recent measurements of single electron detachment (SED) and double electron detachment (DED) processes for 5–50 keV H^- ions incident on noble gases and for 10–50 keV H^- ions incident on CH_4 molecules which were conducted in this laboratory. In addition to the detachment cross section measurements, the scattered beam growth curves yielded information about other charge–changing cross sections in the hydrogen–atom (molecule) collision systems.

INTRODUCTION

Due to their relatively large cross sections, the detachment of electron(s) from negative ions in collisions with atomic gases is an important tool in understanding charge–changing ion – atom collisions. These large detachment cross sections result in significant projectile partial beams of the other hydrogen charge states (H° and H^+) which can interact with the target gas atoms. In this paper, we summarize our recently completed studies of electron detachment in H^- + noble gas [1-3] and methane collision systems [4] with emphasis on the information gained from quadratic analyses of the growth curves in the H^- collision systems in regards to the other hydrogen charge state collision systems. The apparatus used in the present measurements has been described earlier [1]. The simultaneous detection of all three scattered beams (H^-, H°, and H^+) has the distinct advantage of allowing SED and DED total cross sections to be determined absolutely *in situ* without having to rely on other experiments and/or theoretical models.

SCATTERED BEAM FRACTIONS

The charge–state composition of the incident H^- ion beam evolves as it passes through the target gas due to charge–changing collisions with the target atoms. The target thickness, π, is defined by $\pi = nl$, where n is the target gas density, and l is the distance (scattering length) the beam has traversed in the gas. For a fixed H^- ion impact energy, the beam–attenuation curve for the negative–ion current, $I_{-1}(\pi)$, and the beam–growth curves for the proton current, $I_1(\pi)$, and the neutral detector current, $S_0(\pi)$, were obtained by measuring these currents as a function of the target thickness. The target thickness ranged from $\pi = 0$ to 3×10^{14} molecules/cm^2 by changing the target gas pressure in the target cell. The target gas density, n, is obtained from the ideal gas law and the target gas pressure and temperature.

The scattered beam fraction F_i is defined to be the background subtracted flux of hydrogen ions or atoms with charge state i divided by the flux of incident H^- ions

$$F_i(\pi) \equiv \frac{[I_i(\pi) - I_i(0)]}{I_{-1}(0)} \tag{1}$$

where i indicates charge–states 0 or 1. For the neutral hydrogen atom (i = 0) beam fraction, $I_0 \equiv S_0/\gamma$, where S_0 is the measured neutral detector current and γ is the secondary emission coefficient of the neutral detector. The H° atom and the H^+ fraction growth curve data were fit to quadratic functions of target thickness:

$$F_0 = a_0 + a_1\pi + a_2\pi^2 \tag{2}$$

$$F_1 = b_0 + b_1\pi + b_2\pi^2 \tag{3}$$

where the a_i and b_i coefficients are the fit parameters in the analyses. The constant coefficients a_0 and b_0 were included as fit parameters as an experimental check on the adequacy of the background subtraction of the data prior to analysis. For both F_0 and F_1, the constant coefficients returned from the fits were within one standard deviation of zero thus giving assurance that the backgrounds were adequately subtracted in the data of Refs. 1-4.

Assuming a pure incident H^- ion beam and at sufficiently low values of π, the solutions (see, for example, Ref. 5) to the differential equations for growths of the F_0 and F_1 scattered beam fractions are:

$$F_0 = \sigma_{-10}\pi + \tfrac{1}{2}[\sigma_{-11}\sigma_{10} - \sigma_{-10}(\sigma_{-10}+\sigma_{-11}+\sigma_{01}+\sigma_{0-1})]\pi^2 \tag{4}$$

and

$$F_1 = \sigma_{-11}\pi + \tfrac{1}{2}[\sigma_{-10}\sigma_{01} - \sigma_{-11}(\sigma_{-11}+\sigma_{-10}+\sigma_{10}+\sigma_{1-1})]\pi^2 \tag{5}$$

CP392, *Application of Accelerators in Research and Industry*, edited by J. L. Duggan and I. L. Morgan
AIP Press, New York © 1997

where σ_{ij} is the total cross section for the charge–changing process $i \rightarrow j$. The linear coefficients a_1 and b_1 from the growth curve fits are the cross sections for that particular detachment process (σ_{-10} and σ_{-11}, respectively), while the quadratic coefficients a_2 and b_2 are sums of cross section pairs as shown in Eqs. 4 and 5. Thus for single detachment

$$a_1 = \sigma_{-10}\,; \qquad a_2 = \tfrac{1}{2}[\sigma_{-11}\sigma_{10} - \sigma_{-10}(\sigma_{-10}+\sigma_{-11}+\sigma_{01}+\sigma_{0-1})] \tag{6}$$

and likewise for double detachment

$$b_1 = \sigma_{-11}\,; \qquad b_2 = \tfrac{1}{2}[\sigma_{-10}\sigma_{01} - \sigma_{-11}(\sigma_{-11}+\sigma_{-10}+\sigma_{10}+\sigma_{1-1})]\,. \tag{7}$$

As can be deduced from these equations, non–zero quadratic coefficients are expected. The cross section pair adding positively to the sum in Eq. 6 or 7 ($\sigma_{-11}\sigma_{10}$ or $\sigma_{-10}\sigma_{01}$, respectively) comprising the quadratic coefficient in either case arises from the indirect, two–step process -- the scattering of H^- into the different charge state (H^+ for SED, H^o for DED) followed by a second scattering into the final, correct charge state (H^o for SED, H^+ for DED). The first cross section pair adding negatively to the sum in Eq. 6 or 7 ($\sigma_{-10}\sigma_{-10}$ or $\sigma_{-11}\sigma_{-11}$, respectively) is identical to the quadratic term in the expansion of the exponential ($\exp[-\sigma_{-10}\pi]$ or $\exp[-\sigma_{-11}\pi]$) if one assumes only the direct scattering process. The second cross section pair adding negatively to the sum ($\sigma_{-10}\sigma_{-11}$ or $\sigma_{-11}\sigma_{-10}$) arises from the decrease of the incident H^- beam due to detachment into the other charge state in the first collision and then the remaining H^- beam scattering into the correct charge state from a second collision. The remaining cross section pairs adding negatively to the sum are from scattering into the correct charge state followed by scattering out of it into the two different charge states. As such, the quadratic coefficients hold information about all three hydrogen projectile charge states incident on the target gas even though only H^- was the initial projectile beam. It is emphasized that all of these charge–changing processes listed above are inherent in the scattering of a negative hydrogen ion beam through a gas target -- even under tenuous target gas pressures. A linear fit to the growth curve data is valid only when the cross section pairs comprising the quadratic coefficients sum to zero.

RESULTS AND DISCUSSION

Figure 1 shows the measured and the expected quadratic coefficients in the scattered beam growth curves for H^- + helium collisions. Even though the agreement was the best for helium of the three noble gas targets studied in this series of experiments [2], the agreement was also reasonable for the other two noble gas target species. The cross section values were taken from the ORNL compilation data tables of recommended cross section values [6,7] to generate the dashed curves in this figure. Cross section values were not found for several processes in the case of methane, so expected quadratic coefficients could not be determined. The F_0 growth curves were found to have relatively large negative quadratic coefficients, whereas the F_1 growth curves have positive quadratic coefficients. The absolute values of the magnitudes of the quadratic coefficients in F_0 are larger than their F_1 counterparts for the collision systems studied.

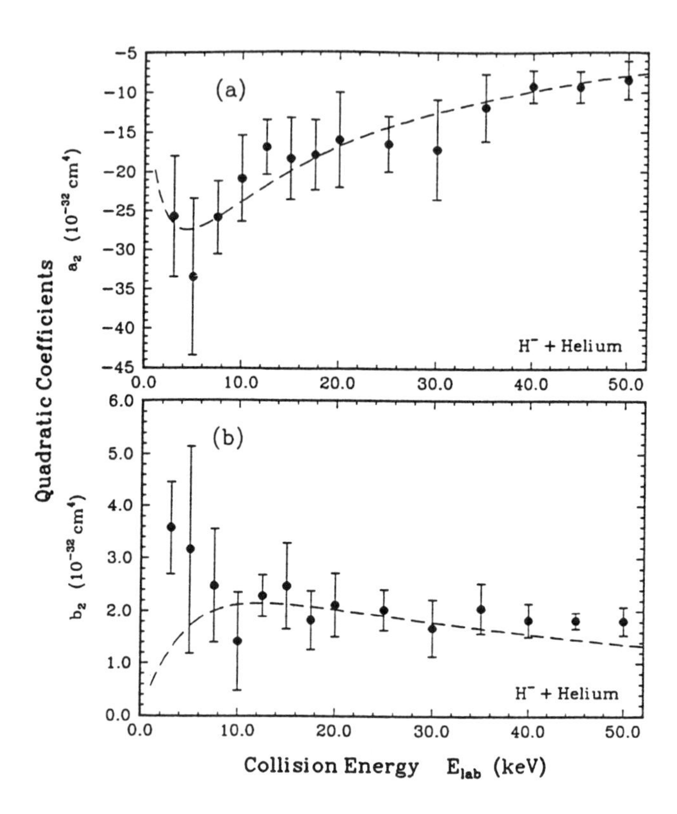

FIGURE 1. Quadratic coefficients of scattered beam growth curves in H^- + helium collisions. (a): from F_0 growth curves; (b): from F_1 growth curves. Solid circles -- measured coefficients; dashed curves -- expected coefficients generated from known cross section values.

Further information about other charge–changing scattering cross sections can be inferred from the quadratic coefficients as is seen from Eqs. 6 and 7. In order for b_2 to be positive in the case of DED, $\sigma_{-10}\sigma_{01}$ must dominate over the sum of the other cross section pairs (which includes the direct scattering process of $\sigma_{-11}\sigma_{-11}$). This being the case, one can conclude that the two–step process of SED (σ_{-10}) followed by H^o projectile ionization (σ_{01}) is dominant over the direct double electron detachment process (σ_{-11}).

For F_0 growth curves, the measured curvature is negative and the situation is reversed. The two–step process of DED (σ_{-11}) followed by electron capture (σ_{10}) by the proton beam is weaker than the other scattering processes contributing to SED. One of the cross section pairs adding negatively to the sum comprising a_2 is the square of the direct single electron detachment process ($\sigma_{-10}\sigma_{-10}$). Because this cross section (σ_{-10}) typically dominates over the other charge–changing scattering cross sections in a_2, it undoubtedly has a significant influence on the sign and the magnitude of the a_2 coefficients.

TOTAL CROSS SECTIONS

Total cross sections for single- and double-electron detachment (SED or σ_{-10} and DED or σ_{-11}, respectively) in collisions between 3– to 50–keV H^- ions and helium atoms, and collisions of 5– to 50–keV H^- ions with neon and argon atoms are shown in Figure 2 and have been reported in Refs. 1,3. Previous results have been omitted from this figure for clarity but comparisons have been made in Refs. 1,3. In general, the agreement of present cross sections with the previous measurements can be explained to a large extent by whether or not the curvature present in the scattered beam growth curves was accounted for in the analysis of the data. For sake of brevity, the reader is referred to those references for specific details (including tables reporting the present results) concerning these cross sections. In the case of methane targets, the collision energy region was from 10– to 50–keV. One notable result from that study is that the SED and DED cross sections for methane targets are very similar to that of the argon targets.

The SED and DED cross sections were determined directly from the growth of the H^0 and H^+ scattered beams, respectively, and the attenuation of the incident H^- beam. All of the necessary experimental parameters were measured, so the present cross sections reported in Refs. 1,3,4 are absolute in magnitude.

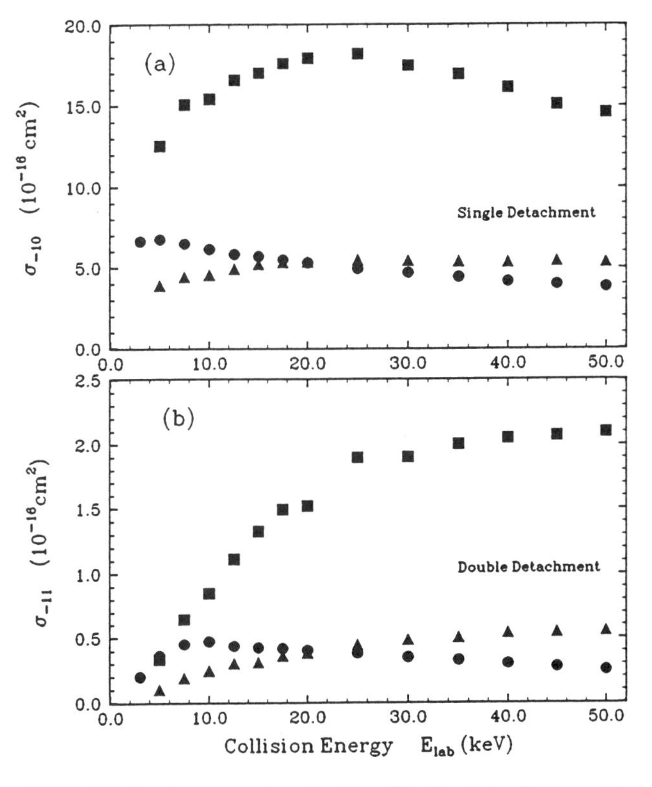

FIGURE 2. Total cross sections for H^- + (He, Ne, Ar) collisions. (a): single electron detachment; (b): double electron detachment. Circles -- helium targets; triangles -- neon targets; squares -- argon targets.

The SED and DED cross sections for helium and neon targets are of the similar magnitude, whereas the cross sections for argon targets are significantly larger than helium or neon, as shown in Figure 2. The estimated total systematic uncertainties in the present cross section values are 5.7% for σ_{-10} and 3.4% for σ_{-11}. The statistical errors in these data are typically the size of the symbols shown in Figure 2.

SUMMARY

The study of single electron detachment and double electron detachment processes in collisions between 5– to 50–keV H^- ions and atoms and/or molecules have yielded measurements of SED and DED cross sections, as well as information about some of the other charge–changing processes in the hydrogen – atom (molecule) collision systems. The growths of the H^0 atom and H^+ ion scattered beam fractions as a function of target thickness have nonlinear dependencies on target thickness which are well–described by quadratic functions of the target thickness over the range of target thicknesses used in the present experiments. The energy dependencies and magnitudes of both the SED and DED quadratic coefficients are in reasonable agreement with the expected coefficients calculated from cross sections reported in the literature for the noble gas targets.

ACKNOWLEDGEMENTS

The authors are very appreciative of the contributions to this work made by: J.S. Allen, X.D. Fang, and R. Matulioniene. This work is supported by a grant from the Division of Chemical Sciences, Office of Basic Energy Sciences, Office of Energy Research, U. S. Department of Energy.

REFERENCES

1. Kvale T.J., Allen J.S., Fang X.D., Sen A., and Matulioniene R., *Phys. Rev.* A **51**, 1351 (1995).
2. Kvale T.J., Allen J.S., Sen A., Fang X.D., and Matulioniene R., *Phys. Rev.* A **51**, 1360 (1995).
3. Allen J.S., Fang X.D., Sen A., Matulioniene R., and Kvale T.J., *Phys. Rev.* A **52**, 357 (1995).
4. Kvale T.J., Sen A., and Seely D.G., *J. Phys. B:At Mol. Opt. Phys.*, 1996 (submitted).
5. McDaniel E.W., Mitchell J.B.A:, and Rudd M.E., "*Atomic Collisions -- Heavy Particle Projectiles,*" John Wiley & Sons (1993).
6. Barnett C.F., Atomic Data for Fusion, Volume 1, "*Collisions of H, H_2, He and Li Atoms and Ions with Atoms and Molecules,*" edited by the Controlled Fusion Atomic Data Center, Publ. ORNL-6086/V1, 1990.
7. Barnett C.F., Ray J.A., Ricci E., Wilker M.I., McDaniel E.W., Thomas E.W., and Gilbody H.B., Atomic Data for Controlled Fusion Research, Volume 1, Physics Division, Oak Ridge National Laboratory, Publ. ORNL-5206/V1, 1977.

(e,2e) ON SURFACES

G. Stefani[#], R. Camilloni[&], S. Iacobucci[&]

[#] *Dipartimento di Fisica "E. Amaldi", Università di Roma Tre and Istituto Nazionale di Fisica della Materia INFM, Unita' di Roma 3 via della Vasca Navale 84, I-00146 Roma, Italy*

[&] *Istituto Metodologie Avanzate Inorganiche, CNR, Area della Ricerca di Roma, CP10, I-00016 Monterotondo and Istituto Nazionale di Fisica della Materia INFM, Unita' di Roma 3*

The application of electron-electron coincidence experiments (e,2e) to solid samples has been extensively exploited only during the past few years and the applicability to surfaces has just started to be pursued. The feasibility of (e,2e) experiments in grazing angle reflection geometry has been recently established for the first time. The possibility of using the grazing angle (e,2e) technique as a binding energy and/or momentum spectroscopy of surface states rests on the accurate knowledge of the ionization mechanism. A double collision, elastic and inelastic at large and small momentum transfer respectively, appears to be the dominant process responsible for reflection (e,2e) events. In the present paper, the results of recent (e,2e) experiments performed at energies from thresold up to 300 eV and with kinematics from normal to grazing incidence are reviewed. They allow elucidation of the (e,2e) ionization mechanism on surfaces at both grazing and normal incidence.

INTRODUCTION

The electronic structure investigation of solids is largely based upon measurements of excitation energies and transition probabilities. All of these observable quantities are determined by average characteristics of the bound state wavefunction because their expectation value comes from suitable integrals over unmeasured quantities. In spite of the large number of spectroscopies currently used to measure the aforementioned observables, none of them is capable of measuring the electron momentum density, i.e. the probability distribution for an electron to be bound with a well defined energy.

Relevance of the momentum density in understanding the electronic structure has been well established for atoms and molecules (1). To appreciate its relevance to solids, it is sufficient to consider that nowdays most calculations of the electronic structure of solids are performed within density-functional (2) theory. This approach is based on a model that describes the electron in a solid as a system of non interacting particles characterized by having spin 1/2 and the same charge density as the interacting electrons. The corresponding potential term in the Hamiltonian is an effective single particle potential which includes a correlation term. This term is only approximatively known. One of the most relevant features for an interacting electron gas is that the occupation numbers in the momentum space are no longer described by a simple step function. In the presence of correlations there is a finite probability that momentum states outside the Fermi surface are occupied. Hence, the expectation value of the energy resolved momentum density is a fingerprint more sensitive than excitation energy or transition probability to the presence of correlations in the ground state of the system. The measure of the energy resolved momentum density should provide us with a better knowledge of the correlation term for the aforementioned effective potential. Although it is very sensitive to fine details of the wavefunction, the electron momentum density is not as widely investigated as the other two observables because ionization experiments are needed to measure it, with completely determined kinematics. Coincidence experiments of this kind, in particular (e,2e) (3) and coincidence Compton scattering experiments (4), are the only available spectroscopies capable of providing both binding energies and momentum densities of the single particle states of the quantum system under study.

To perform an (e,2e) experiment, both free electrons produced by an electron impact ionization are to be analyzed in momentum and detected coincident in time, thus allowing a complete balance of energy and momentum. The unique capability of the (e,2e) technique to measure, when applied to solids, the momentum distribution of an energy selected state, as well as its dispersion in a zone of real momenta that extends all the way down to zero value, was realized (5) even before the first experiment had been performed and is not by chance that the first momentum distribution of a single orbital was measured on a carbon thin film (6). Technical difficulties, however, have hampered till very recently the application of (e,2e) spectroscopy to solids. It was not before the end of the eighties that the (e,2e) spectroscopy has been largely applied to investigate bulk properties of amorphous and crystalline solids (7).

CP392, *Application of Accelerators in Research and Industry,* edited by J. L. Duggan and I. L. Morgan
AIP Press, New York © 1997

REFLECTION (e,2e)

Even though the first application to surfaces was envisaged (8) shortly after the first (e,2e) experiment on a thin solid film (6), it has been only recently that the feasibility of such an experiment has been demonstrated. Experiments performed in the transmission mode and high energy asymmetrical kinematics must rely on the shortness of the mean free path of the low energy ejected electron in order to achieve surface sensitivity for a technique which is not inherently surface sensitive (9). Experiments performed in grazing angle reflection geometry were suggested to be surface sensitive in as much as the penetration depth of the electrons (8,10,11) does not exceed the first few layers of the sample.

The reflection (e,2e) technique consists of measuring simultaneously the energy E_0 of the incident electron, the energies E_e and E_s of the two final electrons, and the probability of them being emitted into solid angles $d\Omega_e$ and $d\Omega_s$ oriented along the directions θ_e and θ_s both on the same side of the surface as the incoming beam. At intermediate and high energies, depending on the amount of momentum transfer $\vec{K} = \vec{K}_0 - \vec{K}_s$ two approximations can be used to describe the (e,2e) process. In the dipolar limit (i.e. vanishing momentum transfer) the (e,2e) mechanism is equivalent to photoionization (12) and binding energy ($\varepsilon(\vec{q})$) spectroscopy is possible. The dipolar (e,2e) differs from photoionization in as much as the momentum associated with the transition, $\vec{q} = \vec{K}_s + \vec{K}_e - \vec{K}_0$, is not uniquely determined and can be changed by changing the geometry of the process while keeping fixed the energy balance $\varepsilon(\vec{q}) = E_0 - E_s - E_e$. In the impulsive (binary) limit, i.e. momentum transfer roughly equal to the ejected electron momentum, the spectral momentum density of the electrons bound in the target can also be measured (7). Several experiments performed on gaseous targets (13) and thin films have shown that the impulsive condition can be satisfied both in symmetric ($E_s = E_e$) and asymmetric conditions; hence both kinematics permit measurement of the momentum distribution.

In the Plane Wave Impulse Approximation (7) and within the non interacting bound particle models, it is easily verified that the five fold differential cross section can be written (with implicit antisymmetrization) (14) as:

$$\frac{d^5\sigma}{d\Omega_s d\Omega_e dE} \propto \frac{K_s K_e}{K_0} \times$$

$$\sum_t \left| \left\langle \vec{K}_e \cdot \vec{K}_s \cdot \prod_{j=1, j\neq t}^{N-1} \Psi_q(\vec{r}_j) \right| v_{0,t} \left| \vec{K}_0 \cdot \prod_{j=1}^{N} \Psi_q(\vec{r}_j) \right\rangle \right|^2$$

where E satisfies the energy balance, t and j are valence electrons indexes, $v_{0,t}$ is the Coulomb interaction between the incoming electron and the bound electron, $\Psi_q(\vec{r}_j)$ is a one electron Bloch function with energy $\varepsilon(\vec{q})$ and crystal momentum $\vec{q}' = \vec{q} + \vec{G}$, $\vec{G}$ being a reciprocal lattice vector.

The main features of the five fold differential cross section are readily understood when a simplified description of the bound states is adopted. In the quasi-free electron model, $\Psi_q(\vec{r}_j)$ can be written as linear combination of plane waves. Furthermore, in the frozen core assumption the cross section becomes

$$\frac{d^5\sigma}{d\Omega_s d\Omega_e dE} \propto \frac{K_s K_e}{K_0} \cdot \frac{1}{K^4} \cdot \sum_{\vec{G}} \left| C_{\vec{q}' - \vec{G}} \right|^2 \tag{1}$$

where the C_i's are coefficients of the ground state wave function of the linear expansion. Therefore, the five fold differential cross section factorizes in a kinematical factor depending only upon the kinematics of the collision (first two terms of (1)), times a form factor which is solely determined by the target electronic structure.

The cross section (1) has been computed assuming that the (e,2e) event is generated by a single impulsive collision where the momentum perpendicular to the solid surface necessary to reflect the final electrons in the backward hemisphere is provided by an elastic reflection of the incoming (scattered) electron with the crystal surface. In this case, and within a zero order approximation, the momentum transfer in (1) is to be replaced by a "reflected momentum transfer" $\vec{K}' = \vec{K}_0' - \vec{K}_s$, where $\vec{K}_0'$ is the specularly reflected plane wave (double step model).

The reduced penetration depth of a grazing angle reflection esperiment would make this geometry ideally suited to study molecules adsorbed on surfaces. If actually realized, such an experiment should provide a valuable insight in the chemisorption mechanisms, and also the possibility of measuring momentum distribution of spatially oriented molecules (15), in the case of physisorption.

For total energies exceeding the threshold by a few electron volts, most of the approximations for the two body interaction impulsive cross section (1) break down. To properly describe the ionization process a many body approach is needed (4).

Reflection (e,2e) measurements from surfaces have been unsuccesfully tried in a few laboratories since the late '70s. Finally, the existence of pairs of correlated electrons, escaping from surfaces under bombardment of low energy electrons at normal incidence, was demonstrated by Kirschner et al. (16). Iacobucci et al. (14) have shown the feasibility of the (e,2e) experiment in grazing angle geometry and intermediate energy. These two pioneering experiments investigate quite different kinematic regions of the (e,2e) process.

Two classes of experiments are being performed: close to threshold, whose goal is to elucidate the dynamics of generation of secondary electrons, and at grazing angle incidence intermediate energy, aimed at defining the experimental conditions for performing momentum spectroscopy on surfaces.

Normal Incidence Geometry

The first observation of correlated pairs of electrons from a solid surface (16) was performed by Kirschner et al. on a W(100) surface, using an electron beam close to normal incidence condition (energy from 10 to 100 eV). Two channeltrons detected time correlated electrons escaping from the target, using a standard coincidence counting mode (17). In two successive experiments, angular (18) and energy (19) discrimination were added as the major improvements over the first one. Among the results, the measured number of true coincidence events per primary electron increases quadratically with primary energy and an onset for production of coincidence is found, located roughly 11 eV above the Fermi energy level. The observed onset energy is explained by assuming that the pairs of electrons are generated in a single ionization impact of the primary electron, back reflected by an elastic collision with the massive crystal lattice. The two final electrons must then satisfy energy and momentum conservation and, for a quasi-free collision, the relative angle between their momentum vectors should be close to 90°. For the electrons to be able to overcome the surface barrier, both of them must have a large enough momentum component perpendicular to the surface, hence the minimum excess energy required for detecting electron pairs becomes larger than twice the work function. Further evidence for the elementary electron-electron scattering event as the dominant mechanism for electron pair generation is given by the angular and energy distribution (19) measured at normal and grazing incidence with energy selection of the two electrons. The authors have adopted a simple kinematical model that assumes the primary electron to be elastically backscattered onto the target valence electrons, which are in turn described within the free electron gas approximation. The fact that the behaviour of the cross section as a function of different quantities is reasonably well described by the simple kinematic model, argues in favour of the physical relevance of the direct collision.

Grazing incidence geometry

It seems therefore plausible that at high-intermediate energies, the first order collision model conjugated with the Plane Wave Impulse Approximation (7) can be applied to reflection (e,2e) experiments to measure the energy dispersion bands $\varepsilon(\vec{q})$ and the momentum distribution of surface states. As already mentioned, the (e,2e) event is not inherently surface sensitive and the extreme grazing geometry is designed to provide such a sensitivity. It is indeed well known that the penetration depth of an electron of few hundred of eV shrinks to a few Angstroms when it collides onto the surface at grazing angles smaller than 10°.

The experiment has been performed with an electron coincidence spectrometer (20) on a surface of highly oriented pyrolitic graphite (HOPG), with a selection of grazing angles. Energies of the ejected electron, from 8 to 21 eV, were measured and energy analyzed by a CMA mounted perpendicular to the sample surface. Hence the measured (e,2e) cross section is integrated over the solid angle subtended by the CMA. The fast scattered electrons, laying in the same plane of the primary electron beam, were detected by an hemispherical analyzer.

In order to fully exploit the potential capabilities of this technique, however, the dominant mechanism for production of the pair of ejected electrons is to be established on firm grounds. To this end, two sets of measurements were performed (21), at different grazing angles ($\theta_0 = 6.7°$ and $4.7°$), always detecting 300 eV scattered electrons correlated in time with 8 eV ejected electrons.

The mechanism responsible for generating correlated electron pairs in the grazing reflection geometry is better elucidated by measuring the (e,2e) cross section as a function of the angle θ_s. From equation (1) it can be readily seen that the angular distribution of the scattered electron is symmetric around the direction of the primary electron of the ionizing inelastic collision, i.e. around $\hat{K}_0$ or $\hat{K}_0'$ according to the validity of the single or double step model respectively. Namely, if the double scattering model is dominant, the distribution will be symmetric around the direction of the specularly reflected beam $\hat{K}_0'$, instead of the direction of the primary beam, and in the equation (1) the momentum transfer $\vec{K}$ will be replaced by the specular momentum transfer $\vec{K}' = \vec{K}_0' - \vec{K}_s$.

From the results obtained at the different grazing angles, it is evident that the five fold differential cross section is symmetric around the specular reflection direction rather than the impinging electron beam direction. Hence, the shape of the measured cross section clearly supports the double collision hypothesis. It is finally to be mentioned that the measured coincidence angular distribution was found to be very close to the corresponding one measured for the non coincident energy loss spectrum. This is good evidence for a same dominant mechanism for all of the inelastic processes at glancing angle, i.e. an elastic specular reflection plus an inelastic scattering on the valence band electrons.

The measurements performed under strict reflection geometry ($\theta_0 = \theta_s$) were aimed at measuring the binding energy spectrum. The HOPG binding energy spectrum has been measured at various reflection angles and for a selection of ejected energies. The vacuum level is the origin for the binding energy scale which has been derived from the energy conservation law. The overall energy resolution of the experiment was 1.3 eV. Under the assumptions used to derive the cross section in equation (1), the features in the binding energy spectrum are to be interpreted on the basis of the band structure of graphite. It might be speculated that the kinetic energy of the ejected electron is too low for it to be considered a plane wave. Hence the (e,2e) cross section should be determined by the joint density of states in the volume of momentum space selected by the experiment. Though this is true in principle, in practice the experiment was performed selecting a given final state and keeping it unchanged while scanning the binding energies of the initial state. This is similar to a constant final state photoemission experiment (22), hence the cross section behaviour is dominated by bound state characters. The momentum $\vec{q}$ reconstructed by the momentum conservation in the kinematics of the experiment has non vanishing components both in the parallel and perpendicular directions to the HOPG $\bar{c}$ axis. Taking into account the finite analyzer angular acceptances, the spectrometer work function and the refraction across the surface potential barrier of the sample (14), the reconstructed values of $q_{\perp}$ range from nearly the middle to the boundaries of the first Brillouin zone in the ΓMK plane (from 0.65 to 1.53 $Å^{-1}$). The $q_{//}$ component ranges between 1.67 and 2.08 $Å^{-1}$, i.e. from the bottom to the middle of the third Brillouin zone along the ΓA direction. The energy resolution of the experiment allows the resolution, in the binding energy region from 5 to 11 eV, of the ionization of the π band from the σ_3 one (23). According to the HOPG momentum density (24), in the volume of real momenta sampled by the experiment the four main bands of graphite display similar values of the momentum probability; hence they should yield similar amplitudes for the (e,2e) cross section. Consequently, based on the calculated dispersion bands (23), four main components to the (e,2e) binding energy spectrum are to be expected respectively for the π, σ_3, σ_2 and σ_1 bands. The best fit to the energy spectrum locates the position of the individually contributing peaks very close to the expected binding energy dispersion bands (23). Furthermore, the σ_2 peak is broader than the others, which is to be expected as the band σ_2 is affected by the largest energy dispersion in the region of momentum investigated. The peak corresponding to the band σ_1 is deeper than expected by about 1 eV. This discrepancy is overcome if it is assumed for its dispersion curve the one previously measured in a transmission (e,2e) experiment (24), instead of the calculated one (23).

SUMMARY AND CONCLUSIONS

Although reflection (e,2e) experiments have become feasible only during the last three years, some firm results have been already secured.

The feasibility of (e,2e) experiments in reflection geometry has been established for energies of the primary electrons from threshold up to 300 eV and for both normal and glancing angle incidence.

The mechanism that leads to formation of correlated electron pairs appears to be the same irrespective of the energy and of the kinematics chosen: namely, an elastic reflection by the crystal lattice or by the massive surface potential barrier, plus an inelastic collision with the valence band electrons.

A simple kinematical model, at low energy, and the impulse approximation, at intermediate energies, are sufficient to explain the main features of the measured cross sections.

ACKNOWLEDGMENTS

This work has been made possible by the financial support of EEC Human Capital and Mobility, Contract No. ERBCHRXCT930359 and Progetto Finalizzato Chimica Fine del CNR.

REFERENCES

1. M.A. Coplan, J.H. Moore, J.P. Doering, Rev. Mod. Phys. **66,** 98 (1994)
2. W. Kohn, L. Sham, Phys. Rev. A **190**, 1123 (1965)
3. G. Stefani, L. Avaldi, R. Camilloni, J. de Phys. IV, Colloq. **C6/3**, 1 (1993)
4. J.R. Schneider, F. Bell, Th. Tschentschez, A.J. Rollason, Rev. Sci. Instrum. **63**, 1119 (1992)
5. V.G. Neudachin, G.A. Novoskoltseva, Yu.F. Smirnov, Z. Exsper. Teo. Fiz. **55**, 1039 (1968) [Soviet. Phys. J. Exper. Ther. Phys. **28**, 540 (1969)]
6. R. Camilloni, A. Giardini Guidoni, R. Tiribelli and G.Stefani, Phys. Rev. Lett. **29,** 618 (1972)
7. I.E. McCarthy, E. Weigold, Rep. Prog. Phys. **54**, 789 (1991)
8. A. D'Andrea and R. Del Sole, Surf. Sci. **71**, 306 (1978)
9. Y.Q. Cai, M. Vos, P. Storer, A.S. Kheifets, I.E. McCarthy, E. Weigold, Phys. Rev. B **51**, 3449 (1995)
10. V.G. Levin, V.G. Neudachin and Yu.F. Smirnov, Phys. Stat. Sol. (b) **49**, 489 (1972)
11. O.M. Artomonov, Zh. Tekh. Fiz. **55**, 1190 (1985) [Sov. Phys. J. Tech. Phys. **30**, 681 (1985)]
12. A. Hamnett, W. Stoll, G. Branton, C. E. Brion and M. J. Van der Wiel, J. Phys. B **9**, 945 (1976)
13. L. Avaldi, R. Camilloni, E. Fainelli and G. Stefani, J. Phys.B: At. Mol. Phys. **20**, 4163 (1987)
14. S. Iacobucci, L. Marassi, R. Camilloni, S. Nannarone, G. Stefani, Phys. Rev. B **51**, 10252 (1995)
15. A. Giardini Guidoni, R. Tiribelli, D. Vinciguerra, R. Camilloni, G. Stefani, J. Electron. Spectrosc. Relat. Phenom. **12**, 405 (1977)
16. J. Kirschner, O. M. Artamov and A. N. Terekhov, Phys. Rev.Lett. **69**, 1711 (1992)
17. G. Stefani, L. Avaldi, R. Camilloni, New Directions in Research with Third Generation Soft X-ray Synchrotron Radiation Sources, NATO ASI E Vol. 254, *Edited by* A.F. Schlachter and F.J. Wuilleumier, Kluwer Academic Publisher 1994, p. 161-190 and references therein
18. O..M. Artamonov, M. Bode, J. Kirschner, Surf. Sci. **307-309**, 912 (1994)
19. O.M. Artamonov, S.N. Samarin, J. Kirschner, Phys. Rev. B **51**, (1995)
20. S. Iacobucci, P. Luches, L. Marassi, R. Camilloni, B. Marzilli, S. Nannarone, G. Stefani, *in* Book of Invited Papers of the XX International Conference on Photon and Electron Atomic Collisions, Whistler, July 26-August1 1995.p. 825
21. S. Iacobucci, L. Marassi, R. Camilloni, B. Marzilli, S. Nannarone, G. Stefani, J. Electron Spectrosc. Relat. Phenom. in press
22. A.R. Law, M.T. Johnson, H.P. Hughes, H.A. Padmore, J. Phys. C: Solid State Phys. **18**, L297 (1985)
23. R.C. Tatar, S. Rabii, Phys. Rev. B **25**, 4126 (1982)
24. M.Vos, P. Storer, S.A. Canney, A.S. Kheifets, I.E. McCarthy, E. Weigold, Phys. Rev. B **50**, 5635 (1994)

ENHANCEMENT EFFECTS ON ELECTRON-ION RECOMBINATION RATES

A. Müller(1), T. Bartsch(1), C. Brandau(1), H. Danared(2), D. R. DeWitt(3), J. Doerfert(4), G. H. Dunn(5,3), H. Gao(3), W. G. Graham(6), A. Hoffknecht(1), H. Lebius(3), J. Linkemann(1,7), M. S. Pindzola(8), D. W. Savin(9), S. Schippers(1), M. Schmitt(7), R. Schuch(3), D. Schwalm(7), W. Spies(1,3), O. Uwira(1), A. Wolf(7), W. Zong(3)

(1) *Institut für Kernphysik, Strahlenzentrum der Justus-Liebig-Universität, D-35392 Giessen, Germany*
(2) *Manne Siegbahn Laboratory, Stockholm University, S-10405 Stockholm, Sweden*
(3) *Department of Atomic Physics, Stockholm University, S-10405 Stockholm*
(4) *Institut für Experimentalphysik III, Ruhr-Universität Bochum, D-44780 Bochum, Germany*
(5) *Joint Institute for Laboratory Astrophysics, University of Colorado, Boulder, Colorado 80309-0440, USA*
(6) *Department of Pure and Applied Physics, The Queen's University of Belfast, Belfast BT7 1NN, Northern Ireland*
(7) *Max-Planck-Institut für Kernphysik, D-69117 Heidelberg, Germany*
(8) *Department of Physics, Auburn University, Auburn, Alabama, AL 36849, USA*
(9) *Department of Physics, University of California, Berkeley, CA 94720, USA*

Recombination of highly charged ions with electrons is studied by accelerator based merged-beams experiments. In most measurements excessively high recombination rates are observed at low relative energies which cannot be explained by ordinary radiative recombination. The enhanced rates lead to serious losses of ions during the electron cooling process in ion storage rings. Another type of rate enhancement, known from previous experiments on dielectronic recombination, is related to external fields in the electron-ion collision region. By introducing controlled electric fields in the cooler of a storage ring it has been possible for the first time to obtain quantitative results for effects of electric fields on dielectronic recombination of highly charged ions.

I. INTRODUCTION

When the first direct measurements of electron-ion recombination cross sections in colliding beams experiments became available, it was immediately noticed, that the experimental data exceeded the theoretical expectations for dielectronic recombination (DR) by large factors [1] . Theory soon provided an interpretation of the observations on the basis of Stark mixing of states in external electric fields [2]. The influence of well controlled fields on DR in a wide range of specified Rydberg states was experimentally studied for singly charged Mg^+ ions [1]. However, prior to the present work no quantitative measurements with controlled external electric fields were reported for highly charged ions.

Few years after the first direct observation of DR, again an apparent deviation of measurements from theoretical expectations was reported [3]. In merged-beams experiments at very low electron-ion center-of-mass energies, the measured recombination rates were sometimes even more than a hundred times bigger than those calculated for radiative recombination (RR). This enhancement effect grossly increases with the ion charge state. It is observed in almost all experiments, but has not yet been quantitatively explained [4]. The present short contribution presents evidence for the role of DR on recombination rate enhancement effects.

II. FIELD EFFECTS

Study of DR is now committed almost entirely to ion storage rings. The quality of the data obtained using rings is emphatically higher than with any other technique [5]. An important issue in DR, however, remains to be clarified. The early measurements on Mg^+ [6] demonstrated a clear dependence of DR cross sections for experimentally specified principle quantum numbers on known ambient electric fields in the collision region. The previous disagreement by factors 2-10 between theory and experiment [1] was resolved satisfactorily [7,8]. Nevertheless, there still remained some, possibly serious, discrepancies. Also, anomalous results in a first generation of merged beams experiments at ORNL [9] could be interpreted in terms of similar electric field mixing; though the fields in the collision region could only be estimated. Inclined beams experiments at Harvard [10] have attempted to resolve inconsistencies in the interpretation of results for Li-like ions obtained at ORNL, but the precision and scope of these experiments have left ambiguity. Effects of external electric fields on dielectronic recombination were also invoked in the interpretation of results obtained in a second generation of merged beams experiments. Examples are the studies on C^{3+} and O^{5+} [11], on N^{4+}, F^{6+} and S^{11+} [12], and on Ar^{15+} [13]. As in all other merged-beams measurements of cross sec-

CP392, *Application of Accelerators in Research and Industry*, edited by J. L. Duggan and I. L. Morgan
AIP Press, New York © 1997

tions and rates for DR of ions carried out so far, however, the external electric fields were not experimentally controlled, leaving a free parameter for theory to adjust the calculated data to the experiments.

In general, the experimental basis for a detailed understanding of field effects on DR rests almost solely on one experiment on Mg^+. In particular, field effects on the recombination of highly charged ions have not been studied in detail so far and deserve special attention.

Quantitative investigations of field effects on DR require controlled fields in the electron-ion collision region. Since an electrostatic external electric field would corrupt the energy definition of the electrons, $\vec{E}_c = \vec{v} \times \vec{B}$ motional electric fields are the best choice for those studies ($\vec{v}$ is the ion velocity and $\vec{B}$ the flux density of the magnetic field). External motional electric fields can be introduced in the cooler of a storage ring by tilting the axis of the longitudinal magnetic field that guides the electron beam. An accurately known angle between the ion beam and the magnetic field lines facilitates the precise knowledge of the size of the external electric field experienced by the ions in the collision region.

Defined transverse magnetic field components $B_\perp$ were applied to the electron-ion merging path in the cooler of the ion storage ring CRYRING in Stockholm by using the correction coils which normally serve for optimum adjustment of the electron and ion beam axes. Absolute rates for $\Delta n = 0$ DR of 10 MeV/u lithium-like Si^{11+} were measured for 9 different transverse magnetic fields $B_\perp$ between 0 and $4.2 \cdot 10^{-4}$ T corresponding to motional electric fields $E_c = vB_\perp$ between 0 and 183.1 V/cm [14]. Figure 1 shows one set of measured DR spectra taken in a single run with 5 different transverse magnetic fields. The data were taken in cycles covering one complete measurement of all these spectra at different values of $B_\perp$ using only one filling of the storage ring. Each cycle started with the injection of Si^{11+} ions into the ring from a 300 keV/u RFQ accelerator fed by the Stockholm cryogenic electron beam ion source CRYSIS. The ions were subsequently accelerated to 10 MeV/u by running the ring in synchrotron mode. After a cooling period the measurements were started, i.e. the rate of recombined ions detected with a solid state detector behind the first bending magnet downbeam from the cooler was recorded as a function of time. Shortly after the start, the cathode voltage of the cooler was switched away from cooling potential at 5.7 kV to about 6.5 kV and then slowly ramped down to 5.9 kV (in about 4 s). No additional magnetic field was applied to the cooler in this phase, so that a field-free DR spectrum (apart from the residual space-charge fields and small $\vec{v} \times \vec{B}$ fields due to the small angular spread of the ion beam) was measured in the range of the $\Delta n = 0$ resonances characterized by $e + (1s^2\,2s) \rightarrow (1s^2\,2pn\ell)$ with $n = 9, 10, \ldots$. After such a scan, covering about 25 eV in the electron-ion center-of-mass frame, the ion beam was cooled again for several seconds, the correction coils were set to produce a field $B_\perp = 1 \cdot 10^{-4}$ T and then a new scan was started. These scans were repeated with $B_\perp = 2, 3$, and $4 \cdot 10^{-4}$ T so that 5 different spectra were measured during one filling of the ring. The remaining stored ions were then dumped and the whole cycle started over and was repeated for typically 1500 times in order to obtain the level of statistics shown in Fig.1.

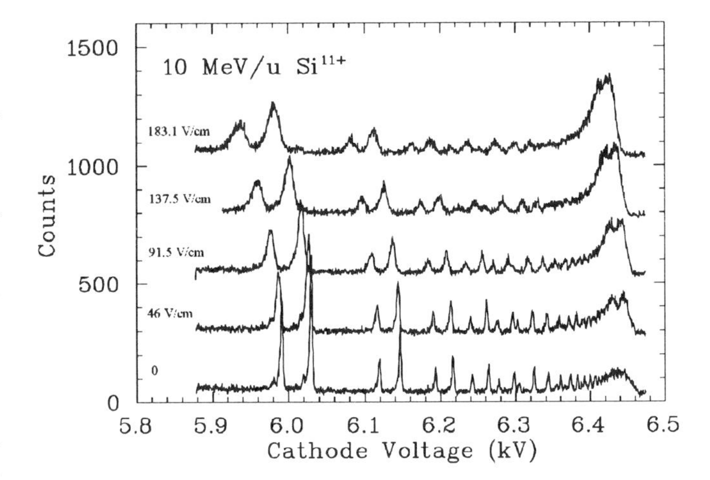

Figure 1. Measured relative rates for $\Delta n = 0$ dielectronic recombination of lithium-like Si^{11+} ions in the presence of imposed external electric fields E_c=0, 46, 91.5, 137.5, and 183.1 V/cm (from bottom to top). The spectra for increasing fields $E_c > 0$ are offset from the field-free spectrum by integer multiples of 200 counts. The abscissa is the electron laboratory acceleration voltage applied to the electron gun of the cooler at CRYRING.

The data displayed in Fig.1 are normalized to identical ion current, electron current and data-taking time so that they can be directly compared. Two features can be immediately recognized: (a) With increasing magnetic field the low-energy peaks are shifted to lower cathode voltages, i.e. to lower electron laboratory energies. The reason is that with increasing $B_\perp$ the electron beam is tilted with respect to the ion beam. With increasing angle between electron and ion beams the center-of-mass energy increases if the laboratory energy does not change. Since the resonances occur at fixed center-of-mass energies their position in the laboratory energy spectrum shifts downwards with increasing angle. (b) With increasing magnetic field the energy resolution of the measurements deteriorates. The reason for this is the increasing misalignment of the electron beam with respect to the ion beam. As a result, each ion probes an increasingly broader range of the electron-beam space charge potential distribution. However, even in the worst case, i.e. for the highest $B_\perp$, the energy spread is still of the order of only 0.4 eV for the $2p9\ell$ resonances. Thus, the relatively broad peak at the high-energy end of the spectrum is not very much influenced by the decreasing energy resolution: it is marginally broadened for the highest transverse magnetic fields used in this experiment. With this in mind, one can easily see that the size of this peak increases with the external electric field. In the peak, unresolved Rydberg states with $n > 20$ lump together and it is their contribution to the total DR rate which is subject to effects of the fields applied in this experiment. The dip on top of the peak is a result of the fine structure splitting of the $2p$ levels and the associated separated series limits of $2p_{1/2}n\ell$ and $2p_{3/2}n\ell$ Rydberg states. A detailed analysis of these experimental results with respect to the DR enhancement is presently underway.

III. RATE ENHANCEMENT AT COOLING

Before the construction of storage rings for heavy ions there was a major concern about recombination losses during the cooling by electrons. RR with its cross section diverging at zero energy was considered the beam lifetime limitation particularly for highly charged ions [15]. Regarding recombination losses due to dielectronic recombination (DR) near zero relative energy $E_{rel} = 0$ in the electron-ion collision system, a case study for Ar^{15+} ions was also carried out [16], but the results did not appear very threatening to storage and cooling of these ions.

The scene changed when unexpected high recombination rates at $E_{rel} = 0$ were observed in an experiment with 6.3 MeV/u U^{28+} ions employing a cold dense electron target at the linear accelerator UNILAC of the GSI in Darmstadt [3]. Within an energy range between 0 and roughly 1 meV the observed recombination rate dropped by a factor of 2 from about $2 \cdot 10^{-7}$ cm^{3}s^{-1}, while a realistic estimate of the RR rate yields approximately $1 \cdot 10^{-9}$ cm^{3}s^{-1} [4]. It was immediately recognized that the lifetime of a stored cooled beam of U^{28+} ions would only be seconds considering the normal density of electron coolers, and yet, this had not been widely noticed until after a series of experiments at the Low Energy Accumulator Ring LEAR at CERN in which Pb^{53+} ions were stored and cooled for preparation of further acceleration [17]. For $n_e = 4.4 \cdot 10^7$ cm^{-3} the beam lifetime was only 2 s which is intolerable for an efficient handling and acceleration of the ion beam.

While recombination rates for Pb^{53+} and its neighbouring charge states were determined only indirectly from life time measurements at LEAR, recently a special effort was made at the TSR [18] to investigate recombination of gold ions Au^{50+} which are isoelectronic with Pb^{53+}.

For the measurement of recombination spectra at low relative energies E_{rel} between electrons and ions in storage ring coolers, the effects of the cooling forces have to be considered. With fast switching between cooling- and scanning-energies together with time-resolved data taking the drag effects of the cooling forces on the ion beam can be corrected for. Figure 2 shows the recombination spectrum taken with Au^{50+} ions in the TSR. The ion energy was 3.6 MeV/u, close to the maximum possible at the accelerator facilities in Heidelberg for these heavy ions. The data are plotted on a double logarithmic scale in order to visualize the extremely sharp peak at $E_{rel} = 0$. Within approximately $4 \cdot 10^{-4}$ eV the recombination rate drops from its maximum which is as high as $1.8 \cdot 10^{-6}$ cm^3 s^{-1} to half this value. This is the highest recombination rate coefficient ever observed in merged beams experiments. The solid line in Fig.2 represents a calculation of the rate for RR based on the Bethe-Salpeter formula [19] which was modified by the corrections previously described by Andersen et al. [20]. The enhancement factor beyond the expectation of RR theory is about 60. At the electron density $n_e = 10^7$ cm^{-3} of the adiabatically expanded electron beam of the cooling device the lifetime of the circulating ion beam during electron cooling was only about 2 seconds with the settings of the present measurement.

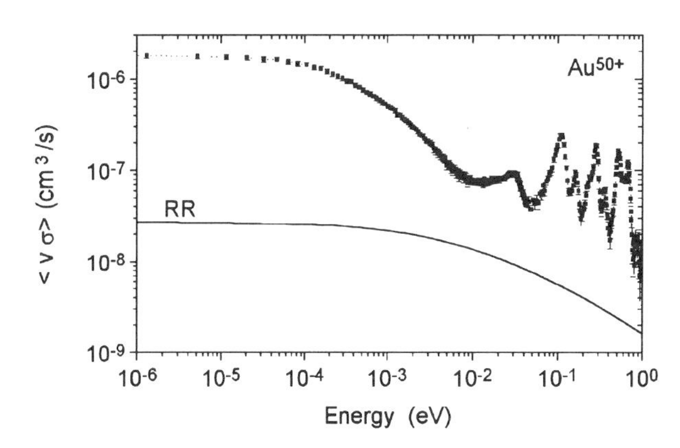

Figure 2. Recombination rates of electrons with 3.6 MeV/u Au^{50+} ions measured at the TSR as a function of the relative energy in the electron-ion center-of-mass frame. The curve denoted by RR is a calculation of the rate for radiative recombination. It agrees with the experimental rate observed at energies between 4 and 5 eV where no resonances appear to be present.

Apart from the recombination maximum at $E_{rel} = 0$ there are huge dielectronic recombination peaks. The one with the lowest energy is already at $E_{rel} = 30$ meV. The full width at half maximum of this resonance is only about 15 meV, which is the smallest width ever observed in electron collision experiments with multiply charged ions. The DR resonances are so densely spaced in energy that one can easily imagine strong resonance contributions also to the peak at zero energy. The DR resonance energy of an ion depends on the individual structure of this ion; hence, DR resonances of Au^{50+} may be fortuitously placed near $E_{rel} = 0$. One can then expect completely different recombination rates in isonuclear ions with different charge states and different electronic structure. Therefore, we also measured recombination rates of Au^{49+} and, indeed, the maximum recombination rate at $E_{rel} = 0$ is roughly a factor of 10 lower than that of Au^{50+}. DR resonances are not nearly as densely spaced as in Au^{50+} and probably are absent at $E_{rel} = 0$. Nevertheless, there is still a considerable enhancement (by about a factor of 4) of the measured rate beyond the expectations according to the theory for RR. A similar result is obtained with Au^{51+} ions. There, however, no obvious resonances are found at all at energies from 0 to 1 eV. In conclusion, excessive recombination losses during cooling in a storage ring can probably be avoided by proper choice of the ion charge state. Though, the phenomenon of enhanced recombination at low energies, also found for completely stripped ions which do not support DR, is not understood yet and needs further investigation.

IV. ACKNOWLEDGEMENTS

The authors thank the staff and the technicians of the TSR and the CRYRING for their assistance in these experiments. This work was supported by the Human Capital and Mobility Programme of the European Community, by the Gesellschaft für Schwerionenforschung (GSI), Darmstadt, by the Max-Planck-Institut für Kernphysik, Heidelberg, and by the German Ministry of Education, Science, Research and Technology (BMBF).

---

[1] Dunn, G. H., "Early dielectronic recombination measurements: singly charged ions", in *Recombination of Atomic Ions*, Graham, W. G., Fritsch, W., Hahn, Y. and Tanis, J. A., eds., NATO ASI Series B: Physics, Vol. 296, Plenum, New York, 1992, pp. 115-131

[2] Hahn, Y., "Electron-ion recombination processes: a general review", ibidem, pp. 11-29

[3] Müller, A., Schennach, S., Wagner, M., Haselbauer, J., Uwira, O., Spies, W., Jennewein, E., Becker, R., Kleinod, M., Pröbstel, U., Angert, N., Klabunde, J., Mokler, P. H., Spädtke, P., Wolf, B., Phys. Scripta **T37**, 62-65 (1991)

[4] Müller, A., Wolf, A., "Production of antihydrogen: what can we learn from electron-ion collision studies?", in *Proceedings of the International Workshop on Antimatter Gravity and Antihydrogen Spectroscopy*, Sepino, Italy, 1996, Holzscheiter, M., ed., Hyp. Int., in print

[5] Müller, A., Comments At. Mol. Phys. **32**, 143-166 (1996)

[6] Müller, A., Belić, D. S., De Paola, B. D., Djurić, N., Dunn, G. H., Mueller, D. W., Timmer, C., Phys. Rev. Lett. **56**, 127-130 (1986); and Phys. Rev. A **36**, 599-613 (1987)

[7] (6) La Gattuta, K., Nasser, I., Hahn, Y., Phys. Rev. A **33**, 2782-2785 (1986)

[8] Bottcher, C., Griffin, D. C., Pindzola, M. S., Phys. Rev. A **34**, 860-865 (1986)

[9] Dittner, P. F., Datz, S., "Early measurements of dielectronic recombination : multiply charged ions" , same volume as Ref. [1]

[10] Young, A. R., Gardner, L. D., Savin, D. W., Lafyatis, G. P., Chutijan, A., Bliman, S., Kohl, J. L., Phys. Rev. A **49**, 357-362 (1994)

[11] Andersen, L. H., Bolko, J., Kvistgaard, P., Phys. Rev. A **41**, 1293-1302 (1990)

[12] Andersen, L. H., Pan, G. Y., Schmidt, H. T., Pindzola, M. S., Badnell, N. R., Phys. Rev. A **45**, 6332-6338 (1992)

[13] Schennach, S., Müller, A., Uwira, O., Haselbauer, J., Spies, W., Frank, A., Wagner, M., Becker, R., Kleinod, M., Jennewein, E., Angert, N., Mokler, P. H., Badnell, N. R., Pindzola, M. S., Z. Phys. D **30**, 291-306 (1994)

[14] Bartsch, T., Müller, A., Spies, W., Linkemann, J., Danared, H., DeWitt, D. R., Gao, H., Zong, W., Schuch, R., Wolf, A., Dunn, G. H., "Measurement of field effects on dielectronic recombination of Si^{11+} ions", *Proceedings of the 2nd Euroconference on Atomic Physics with Stored Highly Charged Ions*, Stockholm, Sweden, 03.06.-07.06.1996, Hyp. Int., in print

[15] Beyer, H. F., Liesen, D., Guzman, O., Part. Accel. **24**, 163-175 (1989)

[16] Carlsund, C., Elander, N., Mowat, R., Griffin, D. C., Pindzola, M. S., Phys. Scripta **T22**, 243-247 (1987)

[17] Baird, S., Bosser, J., Carli, C., Chanel, M., Lefèvre, P., Ley, R., Maccaferri, R., Maury, S., Meshkov, I., Möhl, D., Molinari, G., Motsch, F., Mulder, H., Tranquille, G., and Varenne, F., Phys. Lett. B **361**, 184-187 (1995)

[18] Uwira, O., Müller, A., Linkemann, J., Bartsch, T., Brandau, C., Schmitt, M., Wolf, A., Schwalm, D., Schuch, R., Zong, W., Lebius, H., Graham, W. G., Doerfert, J., Savin, D. W., "Recombination measurements at low energies with $Au^{49+,50+,51+}$ at the TSR", *Proceedings of the 2nd Euroconference on Atomic Physics with Stored Highly Charged Ions*, Stockholm, Sweden, 03.06.-07.06.1996, Hyp. Int., in print

[19] H.Bethe and E.Salpeter, *Quantum Mechanics of One- and Two-Electron Systems*, Springer, Berlin, Heidelberg, New York, (1957)

[20] Andersen, L. H., Bolko, J., Phys. Rev. A **42**, 1184-1191 (1990)

THREE-DIMENSIONAL MOMENTUM DISTRIBUTION OF LOW ENERGY ELECTRONS EJECTED BY FAST PROJECTILES

A. Landers*, C.L. Cocke, M.A. Abdallah, and C. Dilley

James R. Macdonald Laboratory, Department of Physics, Kansas State University, Manhattan, KS 66506-2604

An imaging technique for electron spectroscopy has been developed which measures three dimensional momentum distributions for continuum electrons. Two-dimensions are obtained using position information from a channel plate detetector; for the third dimension, electron time of flight measurements are made possible by using the KSU LINAC to time focus projectile ion bunches. Bunch widths of less than 400 psec have been observed. This time of flight combined with position information yields full three-dimensional momenta of ejected electrons. Results for collisions between 30 MeV carbon ions and targets of helium and hydrogen gas are presented.

INTRODUCTION

The ionization of neutral targets by fast charged projectiles is a process for which there exists an extensive body of work, both experimental and theoretical. The dominant method used to study ejected electrons has been "traditional" electron spectroscopy, in which doubly differential cross sections in energy and angle are measured (1). However, in most cases the final charge state of the target atom is not observed, and distinction between single and multiple ionization is not possible. Other techniques have been employed to observe the target recoil in conjunction with the projectile without the ejected electron to study the ionization process (2). More recently, measurements of the final recoil state *and* two dimensions of the ejected electron momentum have been made (3). The technique described in this work enables the measurement of both the final recoil charge state and the three-dimensional momentum of ejected electrons.

Comparison between our results and those of traditional spectroscopy allows us to normalize our measured cross sections and provides a check of our method. In order to accomplish this, we have converted our measurements to cross sections differential in angle which are directly comparable to results of other groups. We also compare our data directly with cross sections differential in longitudinal momentum that can be calculated from traditional spectroscopy.

In this paper we present a detailed explanation of the experimental technique and some results, including comparison between our results and those of Tribedi *et al.* (1,4). Distributions of momenta for electrons in coincidence with singly ionized targets (He^+ and H_2^+) in all three cartesian dimensions are presented along with correlation between transverse and longitudinal momenta. Results for double ionization of helium and hydrogen targets were also measured, but will be published at a later date.

*Corresponding author eMail: allenl@phys.ksu.edu.

EXPERIMENTAL METHOD

A bunched beam of C^{6+} projectiles was accelerated to 30 MeV by the J.R. Macdonald EN Tandem Van de Graaff Accelerator and time focused by the KSU LINAC. The width of the bunches was measured to be less than .4 nsec. This beam was then tightly collimated, and directed into a target region of diffuse gas. Ejected electrons and recoil ions produced in the target region were extracted and accelerated toward position sensitive microchannel plate detectors by a uniform electric field (between 30 and 400 V/cm) oriented perpendicular to the beam (see Fig. 1).

The time-of-flight (TOF) of the electrons was measured relative to a start signal synchronized with the beam bunch, while the TOF of the recoil ions was measured relative to the electron time signal. Because the recoil TOF was much larger than the electron TOF, the final recoil charge states were clearly distinguished in the TAC spectrum.

The electron momentum was determined in the following way. The target recoils were born with low energies, then accelerated in the electric field of the spectrometer, and traveled in nearly straight line trajectories to the recoil detector, pinpointing the origin of the ionization event. The electrons were accelerated by the same field, but in the opposite direction toward another

CP392, *Application of Accelerators in Research and Industry*, edited by J. L. Duggan and I. L. Morgan
AIP Press, New York © 1997

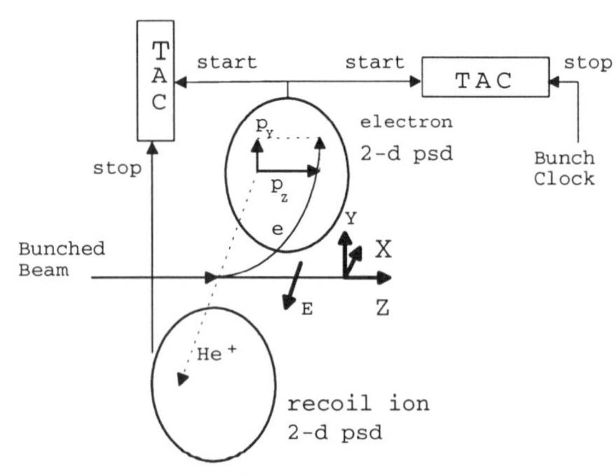

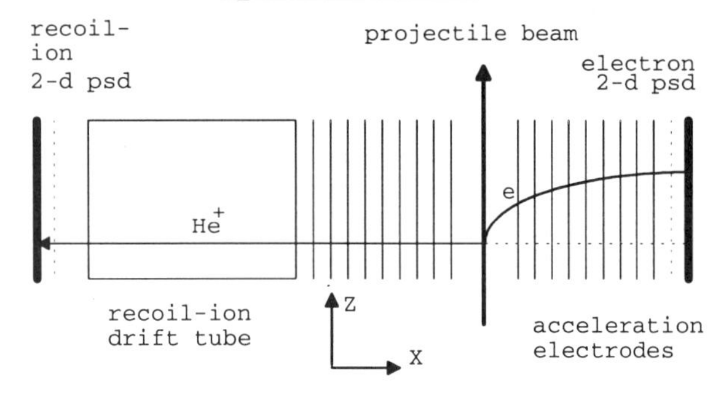

FIGURE 1. Experimental Setup

detector. The recoil position (birth-place of the electron) along the beam axis (Z) was then subtracted from the electron Z-position. (Fig. 1) Due to the symmetry of the collision system, the corresponding Y-position of the electron was measured from the center of its distribution. This is done because the resolution of the recoil detector is greater than the width of the collimated beam. The resulting (Y, Z) position in conjunction with the electron time of flight is used to determine the (Y, Z) momentum by the following equations:

$$P_Y = \frac{mY}{t} \qquad P_Z = \frac{mZ}{t} \tag{1}$$

where m and t are the electron mass and TOF respectively. The value of P_Z also requires knowledge of the relative location of the centers of the electron and recoil detectors. Because the measurement is very sensitive to this information, the following procedure was employed. P_Z was determined as described above. The resulting distribution was compared directly with data derived from traditional electron spectra by Tribedi *et al.* A shift in the absolute position of the distribution was made, while the relative values remained the same.

The electron momentum along the X-axis perpendicular to the detectors was determined solely from its TOF, the spectrometer dimensions and the electric field. They are related by the following equation:

$$P_X = \frac{md}{t} - \frac{eEt}{2} \tag{2}$$

where E is the electric field, t is the electron TOF, and d is the distance from the collision to the electron detector. The quantities m and e are the mass and charge of the electron respectively.

The resolution of the electron momentum depends on several factors, the most important being the position resolution of the detectors and the resolution for the electron TOF. The resolution of the measurement of Y- and Z-momenta is dominated by the position information of the detectors. The size of this effect was determined to be at most .03 a.u., much smaller than the widths of the distributions. From cylindrical symmetry the X- and Y-momentum distributions should be the same, yet the distribution in the X- direction is clearly broader. This is due to the resolution in the TOF measurement which is limited by the width of the projectile bunch and electronics. A resolution of .9 nsec allows for this difference.

One advantage of this technique is the ability to distinguish the final charge state of the target recoil ion. In the case of multiply charged recoils more than one electron will be accelerated towards the electron detector, and if multiple electrons are detected, the position information is lost. Because of the efficiency of the detector, however, the probability of detecting only one electron can be made much greater than that for detecting multiple electrons by placing transmission limiting grids in front of the electron detector. In addition, when multiple electrons *are* detected, the amplitudes of the detector signals are proportionaly larger. These large signals may be filtered out during data analysis.

RESULTS

Momentum distributions in all three dimensions are shown in Figs. 2 and 3 for electrons in coincidence with He^+ and H_2^+ recoil ions respectively. Included is the comparison between the Z-momentum distributions we measured and that of traditional spectroscopy by Tribedi *et al.* (4). To get absolute values for these measurements, we have normalized to the data in this peak. Also, a shift of -.24 a.u.in the absolute Z-momentum was necessary for the two methods to agree. Note, however, that the widths of the distributions for the independent methods agree remarkably well. In Figs. 2d and 3d we present a correlation between longitudinal (Z) and transverse ($P_R = (P_x^2 + P_y^2)^{1/2}$) momenta.

In Fig. 4 we present a comparison of the angular distribution for both methods, again for electrons corresponding to singly ionized targets. Althogh there is qualitative agreement, there is an apparent deficit in the forward and backward scattering angles for the helium data. This is primarily due to the inflated X-momentum distribution.

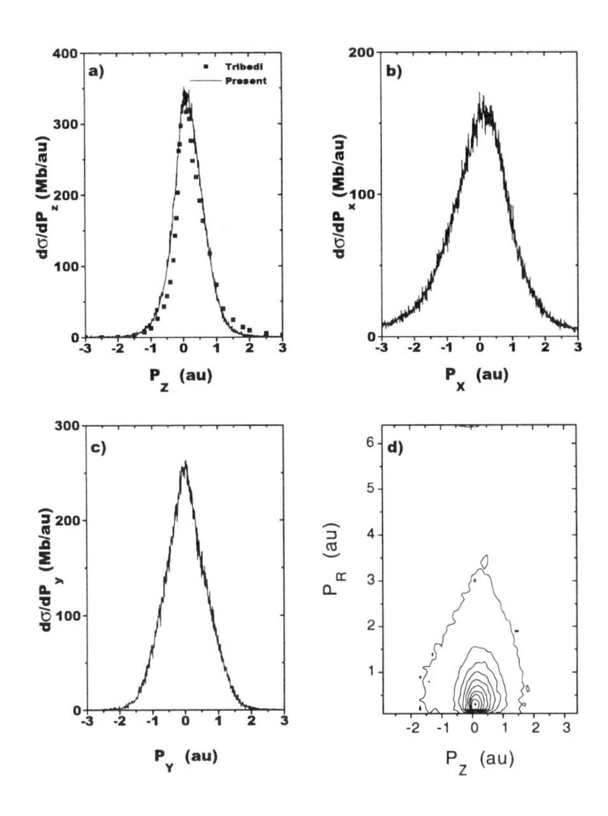

FIGURE 2. Ejected electron momenta for helium target.

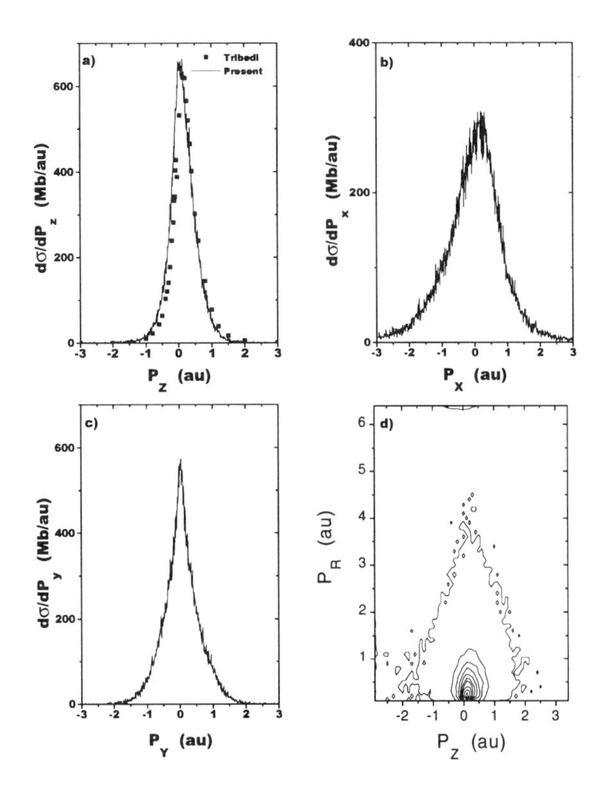

FIGURE 3. Ejected electron momenta for hydrogen target.

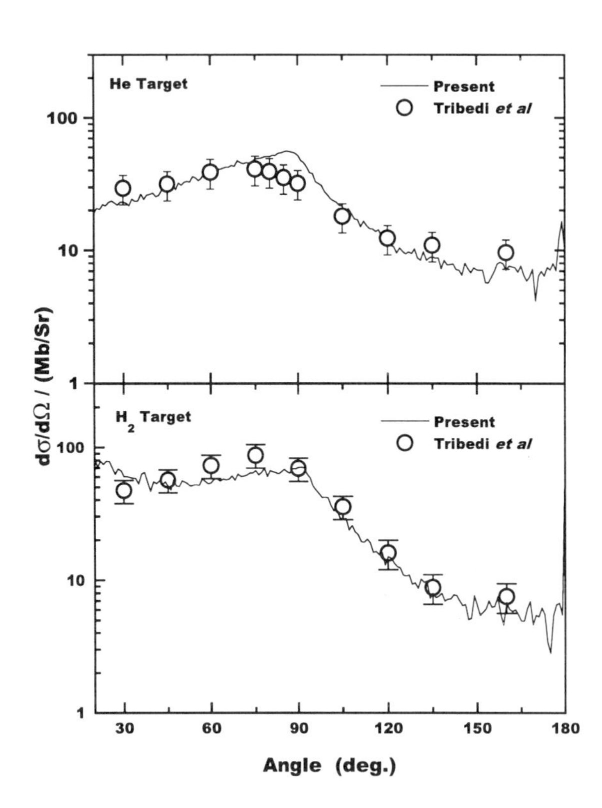

FIGURE 4. Cross sections differential in angle for 30 MeV C^{6+} projectiles on helium and hydrogen targets.

In summary, we have developed a new technique which allows us to measure the complete momentum of ejected electrons in conjunction with the final charge state of the target recoil. Our results agree favorably with that of traditional electron spectroscopy, especially in the width of the longitudinal momentum distribution. Future plans include the study of electrons from multiply ionized targets and correlations between electron momenta and target molecule orientation.

ACKNOWLEDGMENT

Authors would like to thank P. Richard and Lokesh C. Tribedi for several helpful discussions. This work was supported by the Division of Chemical Sciences, Office of Basic Energy Sciences, Office of Energy Research, U.S. Department of Energy.

1. Lokesh C. Tribedi, P. Richard, D. Ling, Y.D. Wang, C.D. Lin, R. Moshammer, G.W. Kerby III, M.W. Gealy, and M.E. Rudd *Phys. Rev. A* **54**, 2154-2160 (1996), and references contained therein.

2. W. Wu, C.L. Cocke, J.P. Giese, F. Melchert, M.L.A. Raphaelian, and M. Stockli *Phys. Rev. Lett.* **75**, 1054-1057 (1995).

3. S.D. Kravis, M. Abdallah, C.L. Cocke, C.D. Lin, M. Stockli, B. Walch, Y.D. Wang *Phys. Rev. A* **54**, 1394-1403 (1996)

4. Trebedi, Lokesh C. (Private communications, to be published).

EXPERIMENTAL DETERMINATION OF ORBITAL AND SPIN CONTRIBUTIONS TO DENSITY MATRICES OF EXCITED ATOMIC STATES FORMED IN TRANSFER-EXCITATION

O. Yenen, D. H. Jaecks and B. W. Moudry

Department of Physics, University of Nebraska-Lincoln, Lincoln NE 68588-0111

For a quantum system, the density matrix contains all the statistical information that can be obtained from measurements. For excited atomic states, the coefficients in a spherical basis expansion of the density matrix, up to rank two, are proportional to Alignment and Orientation Parameters given by Fano and Macek[1] and are called the state multipoles or multipole moments. They can be experimentally determined from the measurements of the Stokes parameters of the emitted photon. To obtain maximum possible information, one needs to break the axial symmetry of the experiment by specifying a reflection plane determined through a coincidence measurement. For LS coupled atomic states, if one can spectroscopically resolve J-multiplets, and if the process is spin dependent, information about the contributions to the state multipoles due to orbital and spin angular momenta can be extracted. In the process, it is also possible to determine the octupole moments of the excited state due to orbital and spin angular momenta. An illustrative example from transfer-excitation reactions in He^{+}-Ar collisions at low keV energies will be presented. For coupling schemes other than LS coupling, using a similar decoupling one can determine the contributions due to appropriate angular momenta. The method is general enough that one can apply it to ion, atom, electron or photon projectiles incident on atomic targets.

INTRODUCTION

We use the density matrix formulation, introduced by Von Neumann in 1927, whenever we have to deal not only with pure states but also with a mixture of independently prepared states $|\Psi_n>$ with statistical weights w_n. In that sense, the density matrix is more general than the wave function of the system. It contains all the statistical information that can be obtained from measurements for both mixtures and pure states. The expectation value of any operator can be obtained from the density matrix using (1)

$$\langle Q \rangle = \frac{tr(\rho Q)}{tr\rho}$$

When working with angular momentum eigenstates, the expansion of the density matrix in a spherical basis brings out the rotational symmetry properties of the system under study.

$$\rho = \sum_{KQ} \rho_{KQ}(JJ')T_{KQ}(JJ')$$

where $T_{KQ}(JJ')$ is the spherical tensor operator basis set defined by (2)

$$T_{KQ}(JJ') = \sum_{MM'} (-1)^{J-M} \langle J' M J \text{-} \mathrm{M} \mid \mathrm{KQ} \rangle \mid J' M' \rangle \langle JM \mid$$

Specially for excited atomic states, this expansion is particularly useful, since the expansion coefficients $\rho_{KQ}(JJ')$ are identical to the expansion of the electric and magnetic radiation fields in terms of 2^{p} poles of the charge, current, and spin magnetic moment density distributions of the excited atomic state which causes the radiation. These expansion coefficients $\rho_{KQ}(JJ')$ are called the multipole moments or state multipoles.

Furthermore, the state multipoles up to rank two are related (2) to the Fano-Macek alignment and orientation coefficients (4) which themselves are determined from the measurement of Stokes parameters. To obtain maximum possible information it is necessary to break the axial symmetry of the experiment. This is accomplished by defining a reflection plane through a coincidence measurement.

For LS coupled systems, we will describe below a method to extract information about the contributions of the total charge circulation and total spin of the electrons to the state multipoles provided that the excitation process is spin-dependent. We will present the method by applying it to transfer-excitation collisions. He^{+} - Ar collisions at low keV energies will serve as an illustrative example.

METHOD

For the two-electron process of transfer-excitation process

$$He^{+} + Ar \rightarrow He + Ar^{+}\left(3p^{4}[^{1}D]4p,\ ^{2}F^{0}_{7/2,5/2}\right)$$

CP392, *Application of Accelerators in Research and Industry*, edited by J. L. Duggan and I. L. Morgan
AIP Press, New York © 1997

we measure the normalized Stokes parameters P_1, P_2, P_3 for the radiation perpendicular to the collision plane, as well as the linear polarization P_4 for radiation emitted in the collision plane in coincidence with the scattered neutralized He. The coincidence measurement ensures the breaking of the axial symmetry by defining the reflection plane to be the plane determined by the momenta of the incoming He^+ and the scattered He. Using the formalism provided by Fano-Macek (4), we determine the alignment and orientation parameters:

$$A_o^{col}(J)=\frac{\langle 3J_z^2-J^2\rangle}{J(J+1)}=\frac{-2[P_1(2P_4+1)+P_4]}{h^{(2)}[P_1P_4-P_1-P_4-3]},$$

$$A_{1+}^{col}(J)=\frac{\langle J_xJ_z+J_zJ_x\rangle}{J(J+1)}=\frac{-2P_2[P_4+1]}{h^2[P_1P_4-P_1-P_4-3]},$$

$$A_{2+}^{col}(J)=\frac{\langle J_x^2-J_y^2\rangle}{J(J+1)}=\frac{2P(P_1-P_4)}{h^2[P_1P_4-P_1-P_4-3]},$$

$$O_{1-}^{col}(J)=\frac{\langle J_y\rangle}{J(J+1)}=\frac{2P_3[P_4+1]}{h^{(1)}[P_1P_4-P_1-P_4-3]},$$

where $h^{(k)}$ are the ratio of two 6-j symbols (5).

In the next step, the alignment and orientation parameters are related to the state multipoles (2)

$$A_o^{col}(J)=\sqrt{\frac{(2J+3)(2J-1)}{5J(J+1)}}\frac{\rho_{20}(J)}{\rho_{00}(J)}$$

$$A_{1+}^{col}(J)=\sqrt{\frac{2(2J+3)(2J-1)}{15J(J+1)}}\frac{\rho_{21}(J)}{\rho_{00}(J)}$$

$$A_{2+}^{col}(J)=\sqrt{\frac{2(2J+3)(2J-1)}{15J(J+1)}}\frac{\rho_{22}(J)}{\rho_{00}(J)}$$

$$O_{1-}^{col}(J)=\sqrt{\frac{2}{3J(J+1)}}\frac{\rho_{11}(J)}{\rho_{00}(J)}$$

where $\rho_{00}(J)$ is a normalization factor. From these equations we determine the ratio $\rho_{ij}(J)/\rho_{00}(J)$.

If the excited state which emits the radiation is well described by LS coupling, which is the case for the $^2F_{7/2,5/2}$ states of Ar^+ (6), then the spherical tensors $T_{KQ}(JJ')$ can be expressed as direct products of spherical tensors of the total angular momentum L and total spin angular momentum S. Assuming that the state multipoles in the uncoupled tensor basis (LS basis) can be given by the product of orbital and spin factors, one can obtain the L and S decoupling equation (7)

$$\rho_{kq}(JJ)=\sum_{k_Lq_Lk_Sq_S}(2J+1)\sqrt{(2k_L+1)(2k_s+1)}\begin{Bmatrix}L & S & J\\ L & S & J\\ k_L & k_S & k\end{Bmatrix}\times\langle k_Lq_Lk_Sq_S\mid kq\rangle\ \rho_{k_Lq_L}(LL)\rho_{k_Sq_S}(SS)$$

The coincidence data determine the left side of the above equation for k=2,1. Since S=1/2, from the coupling rules, we must have $k_S\leq 1$. Since we couple a tensor of rank $k_S=1$ to another tensor of rank k_L to get a tensor of rank k=2, or 1, the coupling rules dictates the rank k_L to be 3, 2, or 1. From the Stokes parameter measurements of two spectroscopically resolved lines of the multiplet, we obtain eight equations from the above expansion. The unknowns are $\rho_{11}(LL)$, $\rho_{20}(LL)$, $\rho_{21}(LL)$, $\rho_{22}(LL)$, $\rho_{31}(LL)$, $\rho_{32}(LL)$, $\rho_{33}(LL)$, and $\rho_{11}(SS)$. The eight equations with these eight unknowns form a set of non linear equations. The numerical solution of these equations (7) determine the values of the contributions of orbital and spin angular momenta to the Fano-Macek alignment and orientation parameters. The nuclear motion is treated within the impact parameter approximation, thus associating the reduced angles with an impact parameter (7).

Figs. 1 and 2 show the contributions to the orientation parameter due to orbital and spin angular momenta as a function of impact parameter. The error bars in figures 1 and 2 are obtained from the propogation of statistical uncertainties of the coincidence data.

Although we used a transfer-excitation reaction to illustrate the method, the formalism should still be applicable to other spin-dependent excitation processes. Experiments at the Advanced Light Source are underway to test the method when the radiation comes from satellite states formed after photoionization.

When the excited state is formed by a spin-independent process, i.e., $\rho_{11}(SS)=0$, the two sets of decoupling equations for the fine structure multiplets are linearly dependent, and a set of measurements for a single J is enough to determine the state multipoles in the L representation.

The method we described is also applicable to coupling schemes other than LS coupling. In this case, the state multipoles in appropriate angular momentum representations are obtained from the decoupling equations.

In figure 1, the orientation coefficient due to the total orbital angular momentum L has a larger magnitude than the value that can be obtained from a 4p electron alone. This suggests that the core electrons contribute to the orientation parameters: they circulate in the same direction as the 4p electron, resulting in a collective motion. This collective motion is impact parameter dependent. In figure 2, the excitation process is slightly spin-dependent at certain impact parameters. This, in turn, implies that at these impact parameters the spins are not isotropically distributed showing deviations from the Percival-Seaton hypothesis (8).

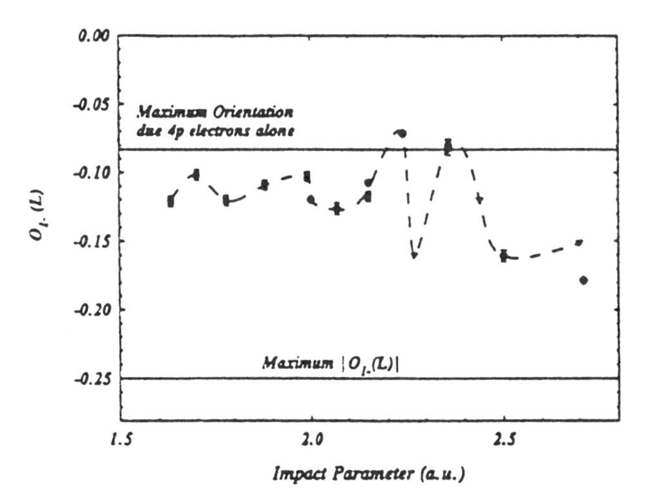

FIG. 1. The orientation parameter in the L representation O1–(L), as a function of impact parameter. This parameter is proportional to the magnetic dipole moment of the excited state produced by the orbital circulation of the charge. Filled squares are from 2 keV, filled circles from 1 keV, and filled triangles from 0.675 keV coincidence data. The dashed line is a cubic spline interpolation to guide the eye.

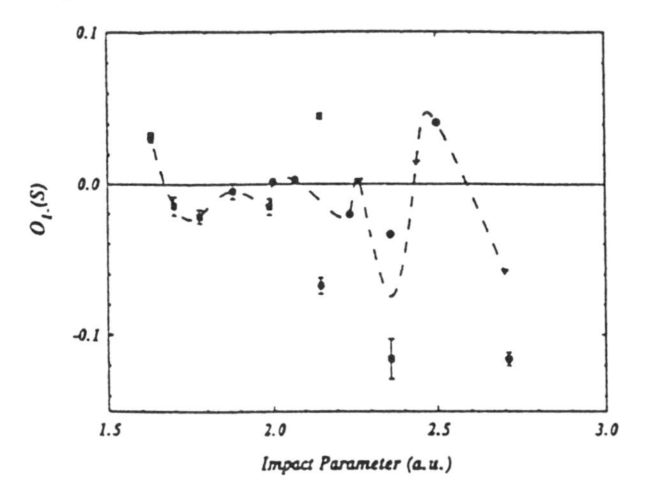

FIG. 2. The contribution to the orientation parameter due to the total spin of the excited state as a function of impact parameter. This parameter is proportional to the magnetic dipole moment of the excited state due to its total spin S. Filled squares are from 2 keV, filled circles from 1 keV, and filled triangles from 0.675 keV coincidence data. The dashed line is a cubic spline interpolation to guide the eye.

ACKNOWLEDGMENTS

We gratefully acknowledge the support of this work by the National Science Foundation under Grant No. PHY-9120213. We also thank to Professor J.H. Macek for many insightful discussions.

REFERENCES

1. Roman, P., *Advanced Quantum Mechanics*, New York, Addison-Wesley, 1966, ch. 1, p. 93.
2. Blum, K., *Density Matrix Theory and Applications*, New York, Plenum, 1981.
3. Blatt, J.M., and Weiskopf, V.F., *Theoretical Nuclear Physics*, New York, Wiley, 1952
4. U. Fano and J. Macek, *Rev. Mod. Phys.* **45**, 553 (1973)
5. B. W. Moudry, O. Yenen, and D.H. Jaecks, *Z Phys.* D **30**, 199 (1994)
6. Minhagen, L., *Ark. Fys.* **25**, 203 (1977)
7. B.W. Moudry, O. Yenen, D.H. Jaecks, and J.H. Macek, *Phys Rev.* A **54**, Nov. 1996.
8. I.C. Percival and M.J. Seaton, *Philos. Trans. R. Soc. London, Ser. A* **25**, 113 (1958).

DOUBLE ELECTRON CAPTURE IN Ne^{10+}+He COLLISIONS IN THE SUB-keV ENERGY RANGE.

J.-Y. Chesnel, A. Spieler, M. Grether, and N. Stolterfoht

Hahn-Meitner Institut GmbH, Bereich Festkörperphysik, Glienickerstrasse 100, D-14109 Berlin, Germany

H. Merabet,* C. Bedouet, F. Frémont, X. Husson, and D. Lecler

Laboratoire de Spectroscopie Atomique-ISMRA, 6 Boulevard Maréchal Juin, F-14050 Caen Cedex, France

We studied mechanisms for double electron capture in slow Ne^{10+} + He collisions using the method of high-resolution Auger-electron spectroscopy. Projectile energies as low as 500 eV were investigated. At such very low energies the production of the configurations $3\ell n\ell'$ ($n \geq 6$) of non-equivalent electrons is found to become dominant. The comparison of the experimental data with model calculations indicates that the non-equivalent electron states $3\ell n\ell'$ ($n \geq 6$) are created by dielectronic mechanisms due to electron-electron interaction. At sub-keV energies the major contribution to double electron capture is due to dynamic electron correlation.

INTRODUCTION

During the last decade much interest was focused on double electron capture in slow collisions of multiply-charged ions with neutral atomic targets (1-6). Two-electron transfer results in the production of configurations $n\ell n'\ell'$ involving doubly-excited states of the projectile. Extensive experimental and theoretical results were obtained in the so-called low-energy range of multicharged ion-atom collisions (1-6). For the example of Ne^{10+} ions, the field of low projectile energies concerns energies of typically a few ten keV to a few hundred keV (7,8).

Only little is known about multicharged ions at energies lower than 10 keV. In many fields such as astrophysics, there is a considerable need of data in this energy range. Moreover, multicharged ions at energies lower than 1 keV have practically not been explored up to now. This is due to difficulties encountered in the production of such very slow ions. In particular, although recent advances, made in the field of the Electron Cyclotron Resonance (ECR) ion sources, allow the production of ions at energies higher than a few ten keV, specific techniques are required for producing ion beams of energies lower than 10 keV (9).

In the present work, we focus the attention on double electron capture in ion-atom collisions involving sub-keV projectile energies. In the range of higher energies of several ten keV two kinds of double-capture mechanisms are invoked (6). These mechanisms are referred to as *monoelectronic* and *dielectronic* processes. Monoelectronic processes due to nucleus-electron interaction produce two-step electron transfers. Furthermore, mutual interactions of two electrons cause dielectronic processes. The latter processes lead primarily to configurations of non-equivalent electrons ($n' >> n$), where one electron is transferred to a deeper lying level, while the second electron is excited to a higher (Rydberg) lying level (3).

During the collision, molecular orbitals (MO) are formed by adiabatic interaction of a few atomic orbitals. The interaction of the atomic orbitals causes avoided-crossings between the MO energy curves, where non-adiabatic charge-transfer transitions can take place. New phenomena for the double capture mechanisms are expected in the sub-keV energy range. In this latter range, the potential energy becomes significantly larger than the kinetic energy of the incident ion. For instance, the projectile velocity is at least 10 times smaller than the mean velocity of the target electrons. Hence, the electron orbitals evolve in an extremely adiabatic manner. Because of this extreme adiabaticity, non-adiabatic transitions are expected to become weak. If a significant number of transitions diminish at sub-keV energies, an increased selectivity in the production of double-capture states is likely to occur. In this context, considering both monoelectronic and dielectronic processes, the question arises of which kind of processes is favoured by such a selectivity.

In the present work, we investigate the system Ne^{10+} + He at projectile energies lower than 10 keV to study collisions under extreme adiabatic conditions. Specific effort is made to perform measurements in the sub-keV range. It is noted that the present measurements extend our previous results obtained at higher energies above a few ten keV (7,8). First, we describe the experimental method. Then, experimental results for

*Present address: Department of Physics/220, University of Nevada, Reno NV 89557-0058 USA.

CP392, *Application of Accelerators in Research and Industry*, edited by J. L. Duggan and I. L. Morgan
AIP Press, New York © 1997

double capture cross sections are given. Model calculations of cross sections are also presented. The experimental and calculation results are compared from the perspective of electron-correlation effects in double electron capture.

EXPERIMENTAL METHOD

The experiments were carried out at the 14 GHz Electron Cyclotron Resonance Ion Sources at the Grand Accélérateur National d'Ions Lourds (GANIL) in Caen and at Hahn-Meitner Institut (HMI) in Berlin. The beam energies of a few keV as well as the sub-keV energies were achieved in Berlin. Ne^{10+} beams of higher energies ranging from 30 to 250 keV were produced in Caen. At both accelerators we used Auger-electron-spectroscopy apparatus (10,11) developed at HMI.

Ions of Ne^{10+} extracted from the ion source were magnetically analyzed and directed into the scattering chamber. For the beam energies lower than 10 keV, specific techniques were used at HMI since the deflection of the ions due to the Coulombic repulsion increases with decreasing beam energy. A high negative voltage of about 10-20 kV was set to the beam line which was insulated from the ion source and the scattering chamber. This high voltage ensured an efficient beam transport because the ion energy inside the beam line was larger than 100 keV. Before entering the chamber, the beam was decelerated by means of an electrostatic lens system.

In the center of the scattering chamber an effusive gas jet was installed providing an atomic beam target of helium. Auger electrons created after the collision were energy analyzed using a tandem electron spectrometer (10,11). A constant energy resolution of 2 eV was achieved by decelerating the electrons in the region between the analyzers, to 40 eV.

For beam energies larger than 5 keV, the average target pressure was estimated to be 10^{-4} mbar. The residual pressure in the chamber was 10^{-5} mbar, while the base pressure was below 5×10^{-7} mbar. At lower projectile energies, the target pressure was reduced to obtain a residual pressure as low as 5×10^{-6} mbar. These pressures were sufficiently low to maintain essentially single-collision conditions. The fraction of charge states other than the primary one present in the incident beam was estimated to be about 15%.

RESULTS AND DISCUSSION

Figure 1 shows typical L-Auger spectra measured at 150, 10 and 1 keV. The 150-keV spectrum has been shown before (7). The observation angle is 40° with respect to the incident beam direction. The observed peaks originate from doubly-excited states due to the configurations $3\ell n\ell'$ $(n\geq4)$ and $4\ell4\ell'$, which decay to the $2\ell\varepsilon\ell'$ continuum by means of Auger transitions. The peak centered at 180 eV is due to the configurations $4\ell4\ell'$ and the series limit of the configurations $3\ell n\ell'$ $(n\geq9)$. In the electron-energy range from 20 to 50 eV, M-Auger electrons from the configurations $4\ell n\ell'$ $(n=5\text{-}7)$ have also been observed. The $4\ell7\ell'$ Auger emission is barely visible in the spectra.

In contrast with the previous results obtained at 150 keV (7), the L-Auger spectra obtained at few-keV and sub-keV energies (Fig. 1) show that the most prominent lines originate from the non-equivalent electron configurations $3\ell n\ell'$ with n=6 to 8. At few-keV energies Auger emission from the configurations $3\ell n\ell'$ with n=4 and 5 gives rise to low-intensity lines.

Total cross sections were evaluated from the measured electron spectra (Fig. 1). The method was described previously (10) so that only a few details are given here. The Auger spectra were integrated to determine single-differential cross sections $d\sigma^a_{n,n'}/d\Omega$ for Auger emission attributed to a given complex n,n'. Total double-capture cross sections $\sigma_{n,n'}$ were evaluated by integration of $d\sigma^a_{n,n'}/d\Omega$ over the electron-emission angle and by dividing the results by the corresponding average Auger yield $a_{n,n'}$ calculated theoretically (12). The absolute

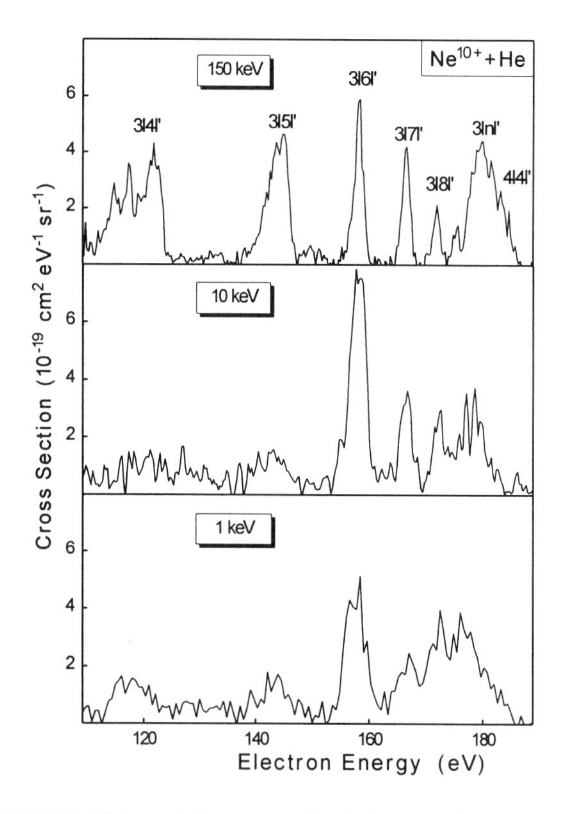

FIGURE 1. High resolution spectra of L-shell Auger electrons produced in Ne^{10+} + He collisions at projectile energies of 1, 10 and 150 keV. Each peak corresponds to the Auger decay of states associated with a configuration $3\ell n\ell'$ (n=4-9). The peak centered at 180 eV corresponds to the limit of the $3\ell n\ell'$ series and to the configurations $4\ell4\ell'$ which decay to the $2\ell\varepsilon\ell'$ configurations.

TABLE 1. Total cross sections $\sigma_{n,n'}$ for producing the configurations $3\ell n\ell'$ and $4\ell n\ell'$ $(n\geq 4)$ in 0.5 to 250-keV Ne^{10+} + He collisions. The experimental uncertainties are about 20% (see text). The average Auger yields $a_{n,n'}$, which were previously calculated (12), are presented in the last row. The projectile velocities corresponding to the different collision energies are given in the second column.

Projectile energy (keV)	Projectile velocity (*a.u.*)	$\sigma_{n,n'}$ $(10^{-17}$ cm$^2)$								
		$3\ell 4\ell'$	$3\ell 5\ell'$	$3\ell 6\ell'$	$3\ell 7\ell'$	$3\ell 8\ell'$	$3\ell n\ell'(n\geq 9)$	$4\ell 4\ell'$	$4\ell 5\ell'$	$4\ell 6\ell'$
0.5	0.030	-	-	2.73	1.20	1.70	2.2	1.2	-	-
0.7	0.036	< 1.2	< 1.8	3.96	1.41	1.80	2.2	1.2	< 2.2	< 0.8
1	0.04	1.22	1.88	5.14	1.70	1.81	2.2	1.4	3.19	0.89
5	0.10	1.64	2.17	6.12	1.94	1.79	2.2	1.4	3.50	0.96
10	0.14	1.66	1.97	6.28	2.36	1.69	1.9	1.6	4.00	1.10
30	0.24	2.60	1.99	4.86	2.48	1.21	2.0	2.4	4.14	1.67
50	0.30	2.56	2.36	4.04	2.47	1.21	2.1	3.0	4.60	2.07
70	0.36	3.62	2.78	2.49	1.93	1.07	1.7	4.2	6.76	2.23
100	0.43	4.02	3.84	2.22	2.02	1.20	1.8	5.6	9.11	2.48
150	0.53	5.43	4.46	2.52	1.82	0.96	1.6	6.1	10.76	2.97
200	0.61	6.25	7.24	2.60	0.86	0.72	0.8	5.7	11.19	3.66
250	0.68	7.24	8.10	1.91	0.61	0.43	0.6	5.5	12.15	4.04
$a_{n,n'}$		0.75	0.67	0.60	0.56	0.52	0.45	0.9[a]	0.91	0.79

[a] See Refs. (4,13).

uncertainties for the cross sections are about 30% and the relative uncertainties with respect to a variation of the emission angle are 20%.

Table I and Figure 2 give the total double capture cross sections obtained for Ne^{10+} + He collisions at energies from 0.5 to 250 keV, i.e. at projectile velocities from 0.03 to 0.68 *a.u.* The comparison of the cross sections exhibits specific features of the energy dependence of double electron capture in Ne^{10+} + He collisions (Fig. 2). The cross sections associated with the non-equivalent electron configurations $3\ell n\ell'$ $(n\geq 6)$ increase strongly when the collision energy decreases from 250 to 10 keV. Moreover, double electron capture into these configurations becomes dominant in the few-keV and sub-keV energy ranges.

To discuss the present results, it should be recalled that for the energy of 150 keV the non-equivalent electron configurations $3\ell n\ell'$ $(n\geq 6)$ were previously shown to be uniquely created by dielectronic processes (7). We expect that this finding can be extended to both few-keV and sub-keV energy ranges. On the other hand, model calculations in Ref. (7) suggested that the equivalent electron configurations $3\ell n\ell'$ and $4\ell n\ell'$ $(n=4,5)$ are primarily produced by monoelectronic processes.

For a qualitative understanding of the present experimental results, the molecular orbitals formed during the collision are considered with respect to the double capture mechanisms. Figure 3 shows examples of approximate MO diagrams for the two-state system $(Ne + He)^{10+}$. In Fig. 3.a. the diagram is relevant for the coupling between the incident and the 4ℓ-single-capture channels, while Fig. 3.b corresponds to the incident and the $3\ell 6\ell'$-double-capture channels. The MO curves were evaluated by diagonalizing the electronic Hamiltonian for-

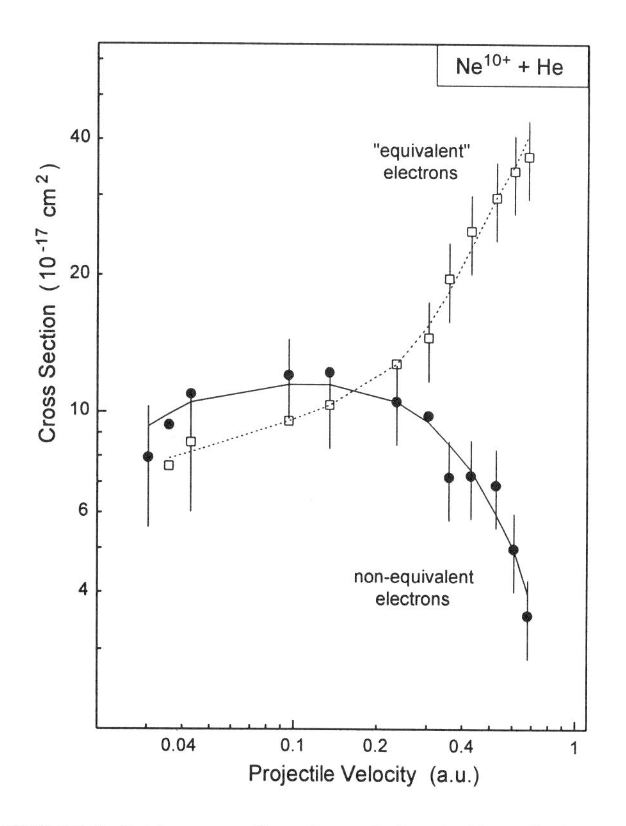

FIGURE 2. Total cross sections for producing doubly-excited states $n\ell n'\ell'$ in Ne^{10+} + He collisions at impact energies ranging from 0.5 and 250 keV (i.e. at projectile velocities from 0.03 to 0.7 *a.u.*). The square symbols represent the double capture cross sections for populating the (near)-equivalent electron configurations $3\ell n\ell'$ $(n=4-5)$ and $4\ell n\ell'$ $(n=4-6)$. The cross sections associated with the non-equivalent electron configurations $3\ell n\ell'$ $(n\geq 6)$ (circle symbols) are found to increase with decreasing energy.

med by the potential curves and the coupling matrix elements given in Ref. (7). Realistic values for the monoelectronic and dielectronic matrix elements were obtained from the formulas of Olson and Salop (14) and Stolterfoht (15), respectively.

At internuclear distances of about 4 *a.u.*, the interaction of the chosen states causes a repulsion of the MO energy curves so that an avoided crossing occurs, where non-adiabatic transitions can take place. At the monoelectronic avoided crossing (Fig. 3.a) the minimum distance between the MO curves is a factor of ten larger than the corresponding energy difference at the dielectronic crossing (Fig. 3.b). This is consistent with the fact that dielectronic interactions are at least ten times smaller than monoelectronic interactions (7).

Since dynamic couplings decrease with decreasing collision energy, the probability for transitions at strongly avoided crossings (see, e.g., Fig. 3.a.) becomes weak in the sub-keV range. Hence, the dielectronic process is likely to gain importance with respect to monoelectronic transitions. At very low energies, our experiments clearly confirm this expectation based on adiabaticity consideration.

Summarising, we found that the dielectronic processes producing the non-equivalent electron states $3\ell n\ell'$ ($n \geq 6$) are dominant at few-keV and sub-keV energies (Fig. 2).

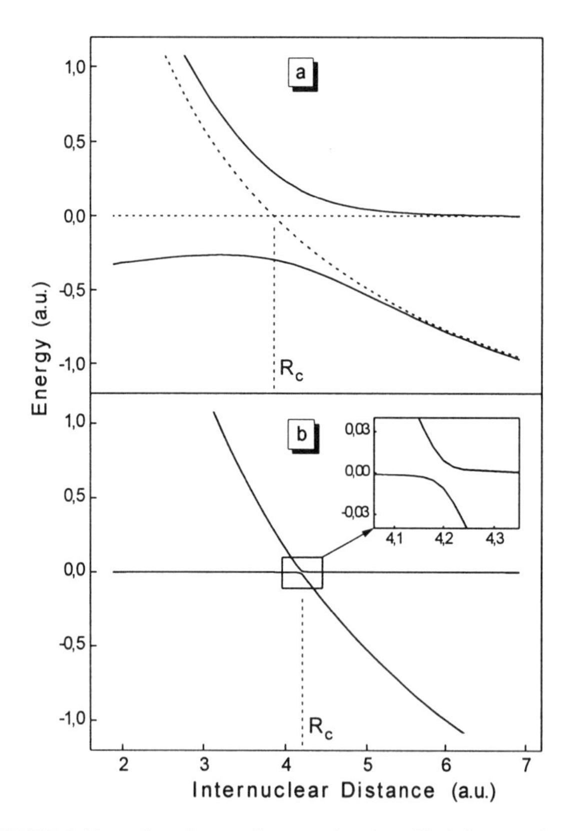

FIGURE 3. Examples of approximate molecular orbital diagrams for the two-state system (Ne + He)$^{10+}$. In (a) the diagram is relevant for the coupling between the incident and the 4ℓ-single-capture channels, while (b) corresponds to the incident and the $3\ell 6\ell'$-double-capture channels.

We note the surprising fact that dielectronic processes exceed in importance the monoelectronic transitions. Moreover, the cross sections associated with the states $3\ell n\ell'$ ($n \geq 6$) are found to increase significantly with decreasing energy (Fig. 2), whereas the population of (near)-equivalent electron states by monoelectronic processes decreases. As suggested in Fig. 3, dielectronic interactions are noticeably smaller than the monoelectronic interactions. In this context, for dielectronic transitions the criterion of extreme adiabaticity of the collision becomes valid at lower projectile velocities than in the case of monoelectronic transitions. Hence, the reduction of the dielectronic transitions is likely to occur at lower energies. To provide a quantitative understanding of charge-transfer dynamics in very slow multiply charged ion-atom collisions, further work (16) will be done using the Landau-Zener model.

ACKNOWLEDGMENTS

We are much indebted to the staffs of the ECR sources in Berlin and Caen for their generous assistance. This work was supported by the "Programme d'Actions Intégrées Franco-Allemand PROCOPE" under the contract number 95060.

REFERENCES

1. Bordenave-Montesquieu A., Benoit-Cattin P., Gleizes A., Marrakchi A. I., Dousson S., and Hitz D., *J. Phys. B* **17**, L223-L227 (1984).
2. Stolterfoht N., Havener C. C., Phaneuf R.A., Swenson J. K., Shafroth S. M., and Meyer F. W., *Phys.Rev. Lett.* **57**, 74-77 (1986).
3. Stolterfoht N., *Physica Scripta* **42**, 192-204 (1990).
4. Van der Hart H. W., Vaeck N., and Hansen J. E., *J. Phys. B* **27**, 3489-3514 (1994).
5. Bachau H., Roncin P., and Harel C., *J. Phys. B* **25**, L109-L115 (1992).
6. Stolterfoht N., *Physica Scripta* **T51**, 39-46 (1994).
7. Fremont F., Merabet H., Chesnel J.-Y., Husson X., Lepoutre A., Lecler D., and Stolterfoht N., *Phys. Rev. A* **50**, 3117-3123 (1994).
8. Chesnel J.-Y., Merabet H., Frémont F., Cremer G., Husson X., Lecler D., Rieger G., Spieler A., Grether M., and Stolterfoht N., *Phys. Rev. A* **53**, 4198-4204 (1996).
9. Martin B., Grether M., Köhrbrück R., Stettner U.,and Waldmann H., *KVI Report No. 996* (unpublished).
10. Stolterfoht N., *Z. Phys.* **248**, 81-91 (1971); **248**, 92 (1971).
11. Itoh A., Schneider T., Schiwietz G., Roller Z., Platten H., Nolte G., Schneider D., and Stolterfoht N., *J. Phys. B* **16**, 3965-3971 (1983).
12. Merabet H., Cremer G., Fremont F., Chesnel J.-Y., and Stolterfoht N., *Phys. Rev. A* **54**, 372-378 (1996).
13. Van der Hart H. W., Vaeck N., and Hansen J. E., (Private Communication).
14. Olson R. E. and Salop A., *Phys. Rev. A* **14**, 579-585 (1976).
15. Stolterfoht N., *Physica Scripta* **T46**, 22-33, 1993.
16. Chesnel J.-Y. *et al.*, *Phys. Rev. A*, to be published; Chesnel J.-Y., *PhD. Thesis*, University of Caen, France, 1996.

TECHNIQUE FOR STUDIES OF STATE - SELECTIVE ELECTRON CAPTURE IN COLLISIONS INVOLVING SLOW STATE-PREPARED MULTIPLY CHARGED IONS

J.B. Greenwood, D.Burns, R.W. McCullough, J. Geddes and H.B. Gilbody

Department of Pure and Applied Physics, The Queen's University of Belfast, Belfast, United Kingdom.

The intense beams from an ECR ion source/accelerator have been utilised in a new apparatus for studies of state-selective electron capture by multiply charged ions in well defined initial states by double translational energy spectroscopy (DTES). This avoids the ambiguities in previous TES studies carried out with beams containing metastable ions. The effectiveness of DTES is demonstrated by studies of one-electron capture at 4 keV by pure beams of ^1S ground state and ^3P metastable C^{2+} ions in collisions with H_2, He and Ar.

INTRODUCTION

The well established technique of translational energy spectroscopy (TES) has been extensively used both in this and in other laboratories (see review[1]) to provide detailed information on electron capture processes of the type

$$X^{q+} + Y \rightarrow X^{(q-1)+}(n,l) + Y^{+}(n',l') \qquad (1)$$

At velocities v < 1 au , electron capture into excited states n,l can take place very selectively through pseudo-crossings of the potential energy curves describing the initial and final molecular systems. In TES, measurements of the difference ΔT between the kinetic energy of the primary X^{+} and forward scattered excited $X^{(q-1)+}(n,l)$ product ions allow identification and an assessment of the relative importance of the excited product channels since, under the usual conditions of the experiment (see [2]), $\Delta T \approx \Delta E$ where ΔE is the energy defect for the channel. However, the results of many TES experiments carried out with primary beams containing unknown fractions of metastable as well as ground state ions, have shown that collision channels associated with metastable primary ions are often dominant. A detailed analysis of the observed energy change spectra is precluded unless the metastable content of the beam is known. This depends strongly on the ion source used and the conditions of operation. In any case, collision channels associated with ground and metastable primary ions cannot always be satisfactorily distinguished with the available energy resolution.

For accurate modelling of both astrophysical and fusion plasmas, it is now recognised that detailed information on collision processes involving ions in both ground state and metastable states is important for many partially ionized species. In order to overcome the problems inherent in the TES approach we have now successfully developed a new apparatus which employs double translational energy spectroscopy (DTES), a technique first demonstrated by Huber et al [3]. In this approach, TES is used to select in well defined states ions formed by one-electron capture in a suitable gas target. These selected ions are then used in a second stage of TES to identify and determine the relative importance of the collision product channels arising from the target gas of interest. In this work, we have demonstrated the effectiveness of our DTES apparatus, by measurements of state-selective electron capture at 4keV by pure beams of $C^{2+}(1s^2 2s^2)$ ^1S ground state and pure beams of $C^{2+}(1s^2 2s2p)$ ^3P metastable ions in collisions with H_2 , He and Ar.

EXPERIMENTAL APPROACH

Our new double translational spectrometer was developed from a TES apparatus used in many previous measurements in this laboratory [1] and many of the components including the electrostatic lens sytems, the electrostatic hemispherical analysers and the ion detection and counting sytem were the same as used previously. Only an outline of the main features of the new apparatus will be given here.

An intense beam of C^{3+} ions from an ECR ion source/accelerator, after momentum analysis, was focused and decelerated by an electrostatic lens system prior to analysis by an electrostatic analyser where it emerged with an energy of 40q eV where q = 3. The emergent beam with

CP392, *Application of Accelerators in Research and Industry,* edited by J. L. Duggan and I. L. Morgan
AIP Press, New York © 1997

an energy spread of about 1 eV FWHM , after acceleration and focusing then passed through the first target gas cell containing helium at a pressure set to maximise the forward scattered C^{2+} ions formed by one-electron capture. These C^{2+} ions at an energy determined by potentials applied to a lens system, were passed through a second electrostatic analyser with a pass energy set at 60(q - 1) eV. This arrangement provided a TES energy change spectrum, the well resolved products from which could be selected to provide a source of C^{2+} ions in either the ^1S ground or ^3P metastable states. Actually, the $C^{2+}(1s^22s2p)^1P$ excited products [4] which decay rapidly to the ground state were selected as the source of C^{2+} ^1S ions.

In this way, a beam of either pure ground state ^1S or pure metastable ^3P C^{2+} ions was prepared and, after focusing and acceleration to the required energy, passed through a cell containing the target gas of interest at a pressure low enough to ensure single collisions. The forward scattered C^+ products formed by one-electron capture in this cell then passed through a third electrostatic analyser, which was used in conjunction with a computer controlled position sensitive counting system [5], to provide an energy change spectrum for the process of interest. In the present work the energy scales on these spectra were determined by reference to our previous TES measurements for one-electron capture in C^{2+} - Ar collisions [4], well established energy level tabulations [6, 7] and, in the case of H_2 , the potential energy curves of Sharp [8].

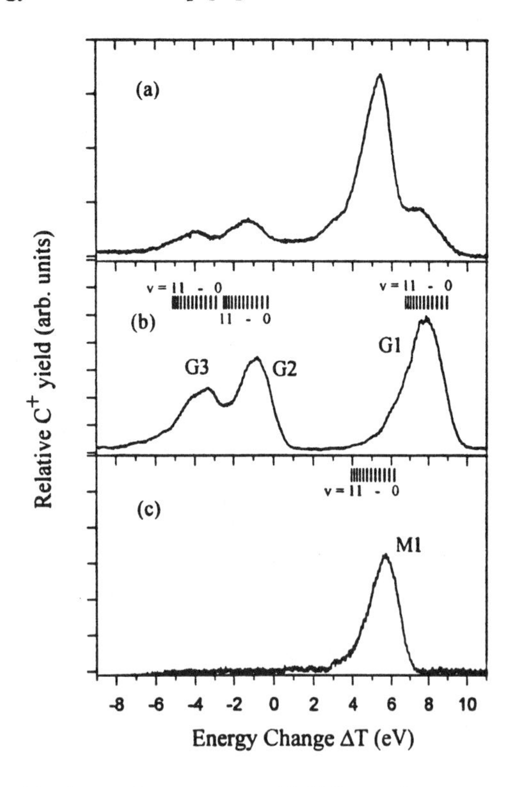

FIGURE 1. Energy change spectra at 4 keV for one-electron capture from H_2 by (a) C^{2+} ions of unknown metastable content derived directly from our ECR ion source, (b) pure ^1S ground state C^{2+} ions and (c) pure metastable ^3P C^{2+} ions. Note that different yield scales are used in each spectrum.

RESULTS AND DISCUSSION

One-electron capture in C^{2+} - H_2 collisions

Figure 1 shows energy change spectra obtained for 4 keV C^{2+} ions in collisions with H_2 . The spectrum in figure1a, which was obtained using a C^{2+} beam of unknown metastable composition directly from our ECR ion source, exhibits the same main features previously observed in TES measurements of Unterreiter et al [9]. Figures 1b and 1c show energy change spectra obtained using DTES with pure ^1S ground state and pure ^3P metastable C^{2+} ion beams. For ground state primary ions, the product channels G1, G2 and G3 correspond to C^+ $(2s^22p)$ ^2P, $C^+(2s2p^2)$ ^2D and $C^+(2s2p^2)^2S$ formation respectively of which the exothermic G1 channel is clearly dominant. All three channels involve non-dissociative electron capture with the formation of ground state H_2^+ $X\,^2\Sigma_g^+$ ions in vibrationally excited states v = 0 to 11. In the case of metastable primary ions, only the exothermic channel M1 is significant and this also involves the formation of vibrationally excited H_2^+ $X\,^2\Sigma_g^+$ ions.

The asymmetry evident in the G and M peaks in figure 1 reflects the influence of the Franck-Condon factors for $H_2X \rightarrow H_2^+X$ transitions on the probability for electron capture into a particular vibrational level. The respective Franck-Condon factors for v = 0, 1, 2, 3, 4, 5, 6, 7, 8, 9, 10 and 11 are known [10] to be 0.087, 0.156, 0.171, 0.152, 0.119, 0.088, 0.062, 0.043, 0.030, 0.021, 0.014 and 0.010 respectively. It is interesting to note that, with the exception of peak G1, the positions of the maxima in the peaks correspond roughly to v = 2 where the Franck-Condon factor is at a maximum.

The clean separation of the ground and metastable primary collision channels evident in figure 1 allows us to confirm the absence of the spin -forbidden product channel C^+ $(2s2p^2)^4P$ + H_2^+ $X\,^2\Sigma_g^+$ for which ΔE = 3.6 - 1.4 eV in the ground state spectrum.

One-electron capture in C^{2+} - He collisions

In figure 2a the TES energy change spectrum obtained using a 4 keV C^{2+} ion beam of unknown metastable content derived directly from our ECR ion source may be compared with the energy change spectra in figures 2b and 2c obtained with the pure ^1S ground state and pure ^3P metastable C^{2+} ions using DTES. In the ground state spectrum, the exothermic peaks G1 and G2 both arise from collisions with impurity species (believed to be mainly N_2 and O_2) in the collision region and account for about 29% of the total electron capture signal. This is surprising since the helium target gas pressure was up to 100 times larger than the background pressure and it indicates that the cross section for one-electron capture by ^1S ground state ions in the impurity gases is very much larger than in helium. In the case of the ^3P metastable ions, this is clearly not the case since only product channels corresponding to electron capture in helium are observed.

For ground state C^{2+} ions where there are no exothermic channels, the peaks G3 and G4 correspond to the formation of $C^{+}(1s^{2}2s^{2}2p)$ ^{2}P and $C^{+}(1s^{2}2s2p^{2})$ ^{2}D products respectively together with $He^{+}(1s)$ ^{2}S ground state ions. The G3 channel accounts for about 67% of the total C^{+} signal (including the impurity contribution) while the much smaller contribution from the more endothermic G4 channel involves core electron rearragement. In the case of electron capture by metastable ^{3}P ions, the dominant exothermic M1 channel (like the G3 channel) results in the formation of $C^{+}(1s^{2}2s^{2}2p)$ ^{2}P product ions. The minor peak M2 corresponding to the much smaller energy defect $\Delta E = 0.95$, corresponds to C^{+} $(1s^{2}2s2p^{2})$ ^{4}P formation.

The present DTES measurements for C^{2+} - He collisions avoid the ambiguities in interpretation inherent in some previous measurements carried out with beams of unknown metastable content. For example, the unexplained peak in the energy change spectrum obtained by Lee et al [12] at 2.8 eV can now be identified with the impurity background peak G2 observed in the present work.

One-electron capture in C^{2+}- Ar collisions

Although there have been a number of previous studies of one-electron capture in C^{2+} - Ar collisions [4, 9, 11], the present DTES studies with pure beams of ^{1}S and ^{3}P C^{2+} ions allow a clean identification and an assessment of the relative importance of the product channels for each species. In figure 3 the energy change spectrum observed with C^{2+} ions obtained directly from the ECR ion source can be seen to contain unresolvable contributions which are

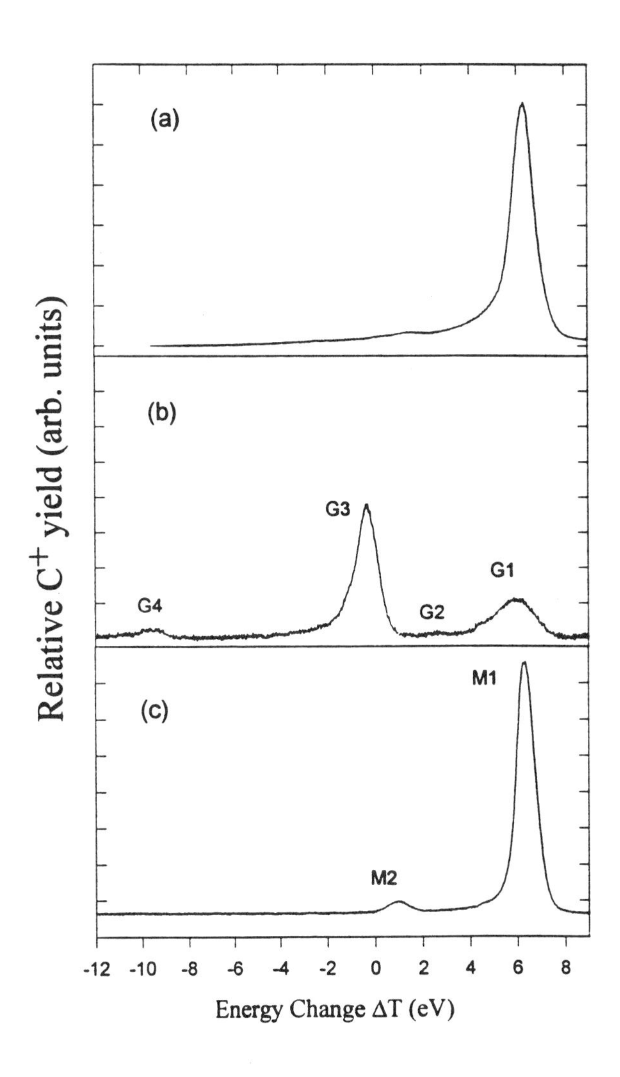

FIGURE 2. Energy change spectra at 4 keV for one-electron capture from He by (a) C^{2+} ions of unknown metastable content derived directly from our ECR ion source, (b) pure ^{1}S ground state C^{2+} ions and (c) pure metastable ^{3}P C^{2+} ions. Note that different yield scales are used in each spectrum.

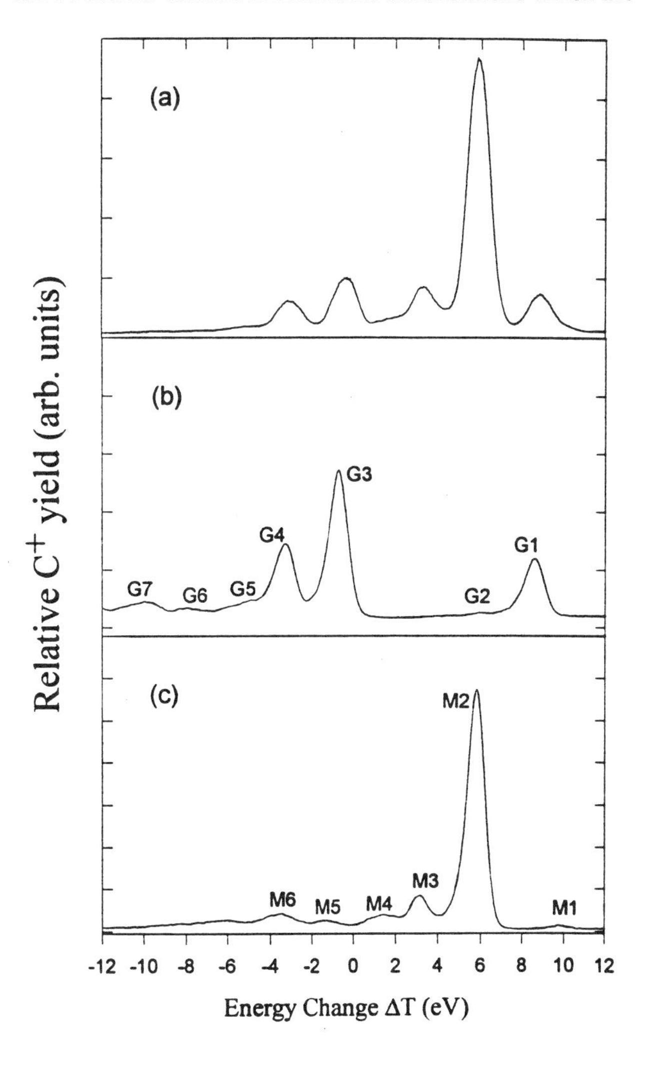

FIGURE 3. Energy change spectra at 4 keV for one-electron capture from Ar by (a) C^{2+} ions of unknown metastable content derived directly from our ECR ion source, (b) pure ^{1}S ground state C^{2+} ions and (c) pure metastable ^{3}P C^{2+} ions. Note that different yield scales are used in each spectrum.

clearly separated in the DTES spectra for the ground and metastable ions. The latter exhibit similar features on the endothermic side with the peaks in the metastable spectrum shifted relative to those in the ground state spectrum by the +6.5 eV corresponding to the energy separation between the ^1S and ^3P states. This indicates that the same final product states are dominant in this region. However, a number of closely spaced channels can contribute to some of the observed peaks making a precise assessment difficult.

For example, the peaks G7 and M6 could contain contributions from the product channels

(a) $C^+ (2s2p^2)^4P + Ar^+(3s3p^6)^2S$
(b) $C^+ (2p^3)^2D + Ar^+ (3s^23p^5)^2P$
(c) $C^+ (2s^22p)^2P + Ar^+ (3s^23p^43d)^2D$
(d) $C^+ (2s^22p)^2P + Ar^+ (3s^23p^43d)^2F$
(e) $C^+ (2s^22p)^2P + Ar^+ (3s^23p^44s')^2D$

while G6 and M5 could involve

(f) $C^+ (2s^22p)^2P + Ar^+ (3s^23p^44s)^4P$
(g) $C^+ (2s^22p)^2P + Ar^+ (3s^23p^43d)^4D$
(h) $C^+ (2s^23p)^2P + Ar^+ (3s^23p^5)^2P$

and G5 and M4 the following channels

(i) $C^+ (2s2p^2)^2S + Ar^+ (3s^23p^5)^2P$
(j) $C^+ (2s^22p)^2P + Ar^+ (3s3p^6)^2S$

For G5 and M4, channel (i) is likely to be the main contributor while, for G6 and M5, since both (f) and (g) are spin forbidden for ground state primary ions, (h) is likely to be the main contributor. In the case of peak G7, channel (a) is spin forbidden while (b) would involve capture of an electron into the 2p state of C^{2+} together with double excitation of the core $2s^2$ electrons to $2p^2$. The other three channels (c), (d) and (e) involve the capture of a 3p electron from Ar with promotion of another 3p electron to 3d or 4s. As in the case of the C^{2+} - He data, the small peak G2 in figure 3 arises from collisions with background impurities.

CONCLUSIONS

The DTES measurements presented here demonstated the considerable potential of this approach for providing detailed information on state-selective electron capture by multiply charged ions in both ground and selected metastable states. This approach overcomes the ambiguities in interpretation of TES data obtained previously using beams containing unknown fractions of metastable species. Our apparatus does not yet allow absolute measurements of electron capture cross sections with state-prepared beams. However, collision product channels for a well defined primary species can be identified unambiguously and their relative importance assessed.

ACKNOWLEDGEMENTS

This work is part of a large program supported by the UK. Engineering and Physical Sciences Research Council.

REFERENCES

1. Gilbody, H. B. *Advances in Atomic, Molecular and Optical Physics*, Vol.33, eds. B Bederson and H. Walther, New York:Academic, 1994, pp. 149-181.
2. McCullough, R. W., Wilkie, F. G. and Gilbody, H. B., *J. Phys. B: At. Mol. Opt. Phys.* **17** 1373-1382 (1984).
3. Huber, B. A., Kahlert, H. J. and Wiesemann, K. *J., Phys. B: At. Mol. Opt. Opt. Phys.* **17**, 2883-2895 (1984).
4. Lennon ,M., McCullough, R. W. and Gilbody, H. B., *J. Phys. B: At. Mol. Phys.* **16** 2191-2204 (1983).
5. Wilkie, F. G., McCullough, R. W. and Gilbody, H. B., *J. Phys. B. At. Mol. Opt. Phys.* **19** 239-251 (1986).
6. Bashkin, S., and Stoner, J. O. Jr., *Atomic Energy Levels and Grotrian Diagrams* Amsterdam: North Holland (1978).
7. Kelly, R. L,*ORNL Report* 5922, Oak Ridge National Laboratory, USA (1992).
8. Sharp, T.E., Atomic Data **2**, 119-169 (1972).
9. Unterreiter, E., Schweinzer, J. and Winter, H. *J. Phys. B. At. Mol. Opt. Phys.* **24** 1003-1016 (1991).
10. Dunn ,G..H., 1966, J. Chem Phys. **44**, 2592-2594 (!966)
11. Lee, A. R., Wilkins, A. C. R, Enos, C. S. and Brenton, A. G. *Int. J. Mass. Spectrom. Ion Processes* **134**, 213-220 (1994).

ELECTRON IMPACT DISSOCIATION OF COLD CH^+: CROSS SECTIONS AND BRANCHING RATIOS

Z. Amitay and D. Zajfman
Department of Particle Physics, Weizmann Institute of Science, Rehovot, 76100, Israel

P. Forck*, U. Hechtfischer, M. Grieser, D. Habs, D. Schwalm, and A. Wolf
Max-Planck-Institut für Kernphysik and
Physikalisches Institut der Universität Heidelberg, D-69029 Heidelberg, Germany

The experimental cross sections for dissociative recombination (DR) and dissociative excitation (DE) of ground-state CH^+ ions, as measured using the heavy-ion storage ring technique, are presented. Measurements of the branching ratios for the different final DR states using two-dimensional fragment imaging technique are also presented. In the low energy region, the DR cross section exhibits several narrow resonances, which are attributed to indirect recombination via core-excited Rydberg states, and the branching ratio measurement shows that the most important final state is the $H(1s)+C(^1D)$ asymptote.

INTRODUCTION

The electron impact dissociation of molecular ions is a process of great significance in a wide variety of plasma environments, and consequently in many fields of physics and chemistry, such as interstellar chemistry, atmospheric chemistry and plasma physics [1]. At low electron energies (i.e. when the energy is below the dissociation energy of the molecular ion) the most important process is dissociative recombination (DR). In this process, the electron is captured by the molecular ion in a dissociative doubly excited (neutral) state which then dissociates into neutral fragments. Once the energy of the free electron is above the dissociation energy of the molecular ions, another process, called dissociative excitation (DE) is energetically allowed. In this process usually the free electron excites the molecular ion to an ionic, singly excited dissociating state, yielding (for a diatomic molecular ion) one charged and one neutral fragment; however, also the formation of a neutral, doubly excited state which then autoionizes and dissociates can result in DE. Both DE and DR are strongly influenced by the electronic structure of the molecular target. This structure is responsible for the various resonances or thresholds which can be found in the respective cross sections.

The interest in DR first arose more than 50 years ago out of the need to understand the properties of the Earth ionosphere [2,3] and since then has been studied extensively both theoretically and experimentally. On the other hand, much less attention has been paid to the DE process, and only a limited set of calculations and experiments have been carried out. In both cases, the main theoretical problem lies in the complexity of the processes, which often involve excited Rydberg states of the neutral molecules and can proceed through a variety of channels and mechanisms. Experimental measurements of DR and, to a less extent, of DE have been made using a number of different techniques. Most of them, however, have inherent uncertainties in the initial vibrational excitation of the molecular ions produced in the ion source [4]; this sets a serious limitation on the comparison between theoretical calculations and experimental results, as the two processes are extremely sensitive to the initial vibrational population of the molecular ion.

During the past few years, substantial progress in DR and DE investigation has been achieved with the introduction of new types of experiments using heavy-ion storage rings. In a storage ring, molecular ions can be stored for a time which is long enough to allow complete vibrational deexcitation via spontaneous radiative transitions. The cold molecular ions are then merged with an intense electron beam, and the outgoing neutral fragments can be detected ahead of the interaction region. Using this technique, both absolute cross section and branching ratios to the different final electronic states have been measured [5,6].

In this paper, we present experimental absolute cross sections for the DR and DE of ground-state CH^+, as well as a measurement of the branching ratios for the population of the final electronic states which are produced in the DR process. For CH^+, the DR process can be depicted as:

$$CH^+(v=0) + e^-(E) \rightarrow C(nl) + H(n'l') \,, \qquad (1)$$

and the DE process as:

$$CH^+(v=0) + e^-(E) \rightarrow \left\{ \begin{array}{l} C + H^+ \quad \text{or} \\ C^+ + H \end{array} \right. \qquad (2)$$

where v denotes the initial vibrational quantum state of CH^+, E is the electron energy, and nl and $n'l'$ are the principal and angular quantum numbers for the C and H fragments, respectively. The DR of CH^+ has been a subject of a wide range of theoretical studies [6], the most recent and complete one having been done by Takagi [7] using the multichannel quantum defect theory. On the

CP392, *Application of Accelerators in Research and Industry,* edited by J. L. Duggan and I. L. Morgan
AIP Press, New York © 1997

experimental side, the DR cross section has been measured by Mul *et al.* [8] using the merged beams method; however, it is not clear whether the CH^+ ions were fully relaxed in this arrangement. An extensive study of the DR process for CH^+ has recently been carried out by us using the heavy-ion storage ring technique [6]. Here, new additional data from this experiment, regarding the DE process, are presented and aspects of more general interest to the study of different molecular dissociation processes, including both DE and DR, are discussed.

EXPERIMENTAL METHOD

Cross Section Measurement

The experimental method has already been described in detail [6] and only a short description will be given here. The CH^+ beam was produced by a tandem accelerator and injected at an energy of 7.2 MeV into the heavy-ion Test Storage Ring (TSR) at the Max-Planck-Institut für Kernphysik, Heidelberg, Germany [9]. After each injection, typically 10^6 particles circulated in the ring in a (ring-averaged) vacuum of 7×10^{-11} mbar with a beam lifetime of 10 s, allowing statistically significant measurements up to 25 s after the injection. At each turn, the ion beam was merged with the 5.0-cm-diameter electron beam of the electron cooler over a length of nominally 1.5 m at an electron density of 1.7×10^6 cm^{-3}. The electron beam had transverse temperature of $kT_\perp = 17$ meV, and longitudinal temperature of $kT_\parallel \approx 0.5$ meV.

The DR measurements were performed by detecting the neutral fragments C and H produced in the interaction region, using a solid state detector mounted straight ahead of the cooler at a distance of ≈ 6 m. The DE channel was measured using a Channeltron detector located inside the closed ion orbit behind the bending magnet following the interaction region, and was used to detect charged C^+ fragments. Thus, only the $C^+ + H^0$ channel in the DE cross section was measured in the present experiment.

Both DR and DE cross sections are given after corrections for background due to collisions with the residual gas in the storage ring. The detailed procedure used to extract the absolute values can be found elsewhere [6].

Final States and Branching Ratio Measurement

In order to measure branching ratios in the DR of molecular ions, a 2-D imaging technique was developed. The two-dimensional detection system was installed downstream of the removable solid state detector mentioned above and consisted of an 80-mm diameter Chevron Micro-Channel-Plate (MCP) coupled to a phosphor screen located 1 mm behind it. Each impact on the MCP produced a light spot on the phosphor screen, whose image was digitized at a rate of 25 frames per second using a CCD camera, coupled to a fast frame grabber device [10]. On analysis, the positions of the light spots were determined using a peak finding procedure, and for each frame that contained two hits their relative distance D on the detector plane was deduced. The detector was operated in a trigger mode, that is, after each event with usually one or two fragments hitting the detector it was switched off within less that 1 μs until the camera started its next frame. The position resolution of the detector was ≈ 100 μm and the minimum distance which could be distinguished between two hits was ≈ 2 mm.

From the distribution $P(D)$ of the projected distances D between fragments, as measured on the detector surface, the final kinetic energy release $E_{k,n}$ in the center-of-mass frame of reference can be extracted. The value of $E_{k,n}$ by itself is a signature of the final quantum states of the fragments, as

$$E_{k,n} = E - E_n \tag{3}$$

where E is the electron energy and E_n the fragments' internal energy relative to the ground state of the molecular ion.

The distribution $P(D)$ for the production of a single quantum state and assuming an isotropic distribution for the internuclear axes is given by [5,6]

$$P_{n,iso}(D) = (\arccos x_2 - \arccos x_1)/L\delta_n \tag{4}$$

where $x_{1,2} = \min(1, D/s_1 s_2 \delta_n)$, $L = s_2 - s_1 = 1.4$ m is the interaction region length, s_1=5765 mm and s_2=7165 mm are the distances from each edge of the electron cooler to the detector, and

$$\delta_n = \frac{(m_C + m_H)}{\sqrt{m_C m_H}} \sqrt{\frac{E_{k,n}}{E_i}}\,, \tag{5}$$

with m_C and m_H being the fragment masses and E_i the laboratory ion beam energy.

In general, the projected distance distribution $P_n(D)$ for a single final state n is obtained by an average over the longitudinal extension of the interaction region and over the molecular orientation according to the distribution of the dissociation angle. Since the internuclear axes of the stored diatomic molecular ions are randomly oriented, this distribution reflects the angular dependence of the DR cross section to this final state. For an isotropic distribution (Eq. 4), this spectrum rises from 0 at $D = 0$ up to a maximum at $D = s_1\delta_n$ and then drops to 0 at $D = s_2\delta_n$. A complete formalism for handling different angular distributions, although limited to some degree of anisotropy, is given in Ref. [6].

The total projected distance spectrum is given by summing over the different final states with the (normalized) coefficients b_n as the branching ratios:

$$P(D) = \sum_n b_n P_n(D)\,. \tag{6}$$

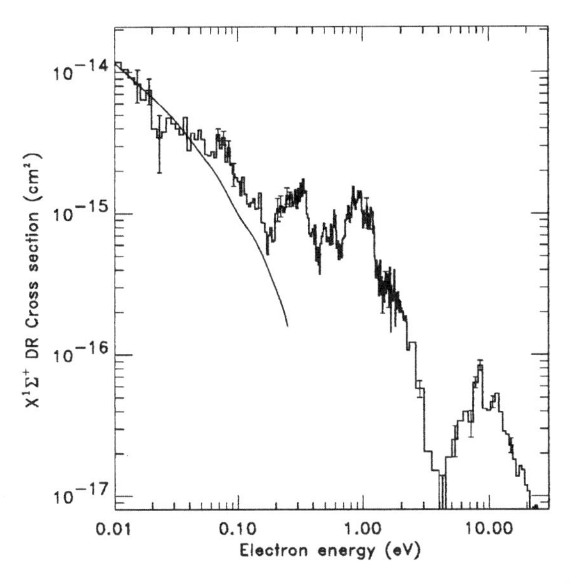

FIGURE 1. Experimental DR cross section of the ground $X^1\Sigma^+(v = 0)$ state. Also shown the theoretical calculation by Takagi [7], convoluted with the experimental energy resolution (smooth line). The estimated systematic error of the cross section scale is ±50%.

RESULTS AND DISCUSSIONS

Electron Impact Dissociation Cross Sections

Figure 1 shows the experimental absolute DR cross section for vibrationally relaxed CH^+ ions as a function of the center-of-mass electron energy as obtained from the average relative velocity of electrons and ions. Also shown is the theoretical calculation performed by Takagi [7] up to $E = 0.3$ eV. The calculation has been convoluted with the present experimental energy resolution. The absolute values of the measured and calculated cross sections agree very well with each other at the lowest energies, but diverge above $E = 0.17$ eV. Also, the low energy resonances observed in the experiment are not reproduced by the theory. A possible explanation for these resonances is related to an indirect DR process, in which the electron is first captured in one of the Rydberg states of the neutral molecule having as a core a low-lying electronically excited, bound state of CH^+. The recombination is then completed by a (pre-)dissociation along one of the potential curves coupled to these Rydberg states. Such mechanism, which we name a "core-excited indirect DR process" was already discussed in connection with our previous measurements of the DR of CD^+ [11] and of OH^+ [12]. At higher energy (above E=4 eV) two prominent peaks are present in the cross section. They are also shown in Fig. 2(a) in linear scales. These resonances are attributed to DR processes occuring via direct transitions to doubly excited dissociative Rydberg states of the neutral CH molecule such as $d^3\Pi(n \geq 3)$, $c^3\Sigma^+(n \geq 3)$, $3^1\Sigma^+(n \geq 3)$ or higher states. More details can be found elsewhere [6].

The DE cross section for the channel $CH^+ + e^- \rightarrow C^+ + H^0 + e^-$ is shown in Fig. 2(b). In a simple picture, DE is due to direct electron impact excitation to an ionic dissociative potential curve; the minimum energy needed for breakup in this case would be the energy difference between the ground vibrational state $X^1\Sigma^+(v = 0)$ and the first ionic dissociative state leading to the asymptote $C^+ + H^0$, namely the $c^3\Sigma^+$ state [6]. Taking into account the Franck-Condon overlap of the nuclear wavefunctions, a threshold of about 9.7 eV is expected. However, as can be seen in Fig. 2(b), the observed DE cross section rises already at ≈ 4 eV. We interpret the "below threshold" DE events between ≈ 4 and 9.7 eV as originating from autoionization of those doubly excited dissociative Rydberg states of the neutral CH molecule which are also responsible for the high energy resonances in the DR cross section starting around 4 eV (see Fig. 2(a) and the discussion above). Thus, the DE and DR processes in this region appear to be closely related to each other, the ratio between the cross section being determined by the probability of autoionization during the dissociation of the doubly excited molecular Rydberg state in which the electron is first captured. In the present case, all five lower excited states of CH^+, including the electronic ground state $X^1\Sigma^+$, correlate to the $C^+ + H^0$ asymptote, and as such can play a role in this specific DE channel (see Ref. [6] for relevant potential curves of CH^+).

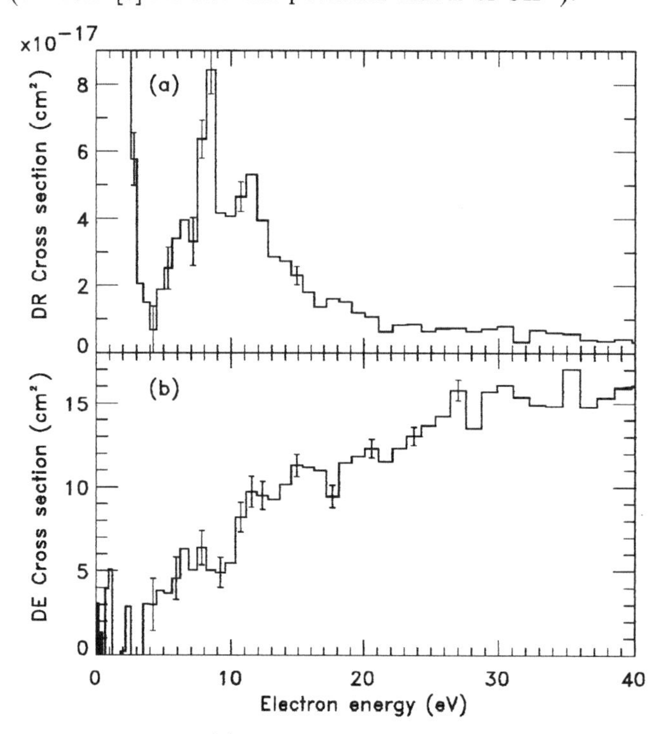

FIGURE 2. (a) High energy region of the DR cross section of the $X^1\Sigma^+(v = 0)$ state. (b) Measured DE cross section for the channel $CH^+ + e^- \rightarrow C^+ + H^0 + e^-$. The estimated systematic error of the cross section scales is ±50%.

A comparison of the measured cross sections in Fig. 2(a,b) indeed indicates the presence of structure in the DE cross section in regions where the DR cross section exhibits resonances. Within the above interpretation, the first rise of the DE cross section between 4 and $\approx$ 10 eV can be assigned to the capture of an electron in the $c^3\Sigma^+(nl)$ and $d^3\Pi(nl)$ Rydberg states, followed by autoionization. The shoulder at $\approx$ 10 eV could then indicate the threshold of direct DE through the ionic $c^3\Sigma^+$ state, while capture followed by autoionization of the neutral $3^1\Sigma^+(nl)$, and $2^1\Pi(nl)$ manifolds continues to contribute to the DR cross section. The observed onset of DE near E = 4 eV is in good agreement with the dissociation energy of CH^+ (D_0=4.08 eV). The ratio of the absolute values of the DR and DE cross sections gives valuable information about the probability of autoionization during the dissociation of the Rydberg states involved.

Branching Ratios

As an example of our measurements [6] of branching ratios in the DR process we present here results obtained at matched average velocities of the electron and the ion beam, that is, for electrons hitting the CH^+ molecule at random directions with an average kinetic energy of $kT_\perp$ = 17 meV. Results for higher relative energies, spread over the entire range of the cross section measurement shown in Fig. 1, can be found in Ref. [6]. According to the calculation by Takagi [7] the only dissociative state involved in DR at near-zero energy and up to E =0.3 eV is $2^2\Pi$, which crosses the $X^1\Sigma^+$ state close to the left turning point of the low vibrational level. This state correlates to the separate atom limit H(1s)+C(^{1}D), for which a kinetic energy release of $E_{k,^1D}$ = 5.92 eV is expected.

Figure 3 shows the projected distance spectrum $P(D)$ measured at near-zero electron energy using the two dimensional detector setup described above. The smooth line represents a fit based on Eqs. 4–6, using the known kinetic energy releases for the various energetically accessible asymptotic limits. As can be seen, the data can only be described by a superposition of *two* asymptotic final states: H(1s)+C(^{1}D) ($E_{k,^1D}$ = 5.92 eV) with a branching ratio of (79±10)% and H(1s)+C(^{1}S) ($E_{k,^1S}$ = 4.50 eV) with (21±10)%. Our result thus shows that, although the H(1s)+C(^{1}D) asymptote directly correlating to $2^2\Pi$ is indeed the dominant recombination channel, the H(1s)+C(^{1}S) channel is also significantly populated. A possible explanation for this observation could be the occurrence, during the dissociation process along the $2^2\Pi$ curve, of a transition to the $2^2\Sigma^+$ state, which crosses $2^2\Pi$ at 1.75Å and correlates to H(1s)+C(^{1}S). Thus, spin-orbit and rotational coupling between the $2^2\Pi$ and the $2^2\Sigma^+$ states would be responsible for the production of C(^{1}S) in about 1/5 of the recombination events.

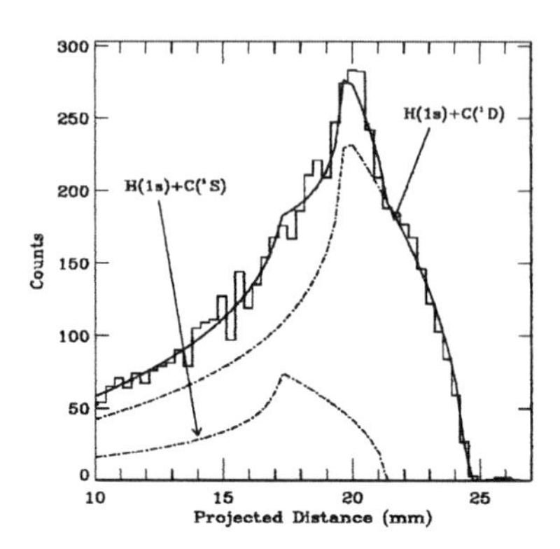

FIGURE 3. Projected distance spectrum as measured for DR with near-zero energy electrons. The solid line is a least-squares fit to the distribution, as explained in the text, using the components indicated by the dot-dashed lines.

To conclude, we have shown that the combined use of the merged-beams technique in a heavy-ion storage ring and of two-dimensional fragment imaging opens up the possibility to study electron-impact dissociation of vibrationally relaxed molecular ions in great detail. A future extension of this technique to full 3-D imaging, as well as recombination studies for well defined (non-zero) vibrational states [13], are on the way.

This work has been partially funded by the German Federal Minister for Education, Science, Research and Technology (BMBF) under Contract No. 06HD562I(3), by the German Israel Foundation (GIF) under Contract No. I-0452-200.07/95, and by a research grant from the Philip M. Klutznick fund for research.

* Present Address: Gesellschaft für Schwerionenforschung (GSI), D-64220 Darmstadt, Germany

[1] *Dissociative Recombination: Theory, Experiment and Applications*, edited by J. B. A. Mitchell and S. L. Guberman (World Scientific, Singapore, 1988).

[2] J. Kaplan, *Phys. Rev.* **38**, 1048 (1931).

[3] D. R. Bates, and H. S. W. Massey, *Proc. Roy. Soc. (Lond.)* **192**, 1 (1947).

[4] H. Hus *et al., Phys. Rev. A* **38**, 658 (1988).

[5] D. Zajfman *et al.*, Phys. Rev. Lett. **75**, 814 (1995).

[6] Z. Amitay *et al., Phys. Rev. A* **54**, 4032 (1996), and references therein.

[7] H. Takagi, N. Kosugi, and M. Le Dourneuf, *J. Phys. B* **24**, 711 (1991).

[8] P. M. Mul *et al., J. Phys. B* **14**, 1353 (1981).

[9] D. Habs *et al., Nucl. Instrum. Meth.* **B43**, 390 (1989).

[10] D. Kella *et al., Nucl. Instrum. Meth.* **A329**, 440 (1993).

[11] P. Forck *et al., Phys. Rev. Lett.* **72**, 2002 (1994).

[12] Z. Amitay *et al., Phys. Rev. A* **53**, R644 (1996).

[13] D. Zajfman and Z. Amitay, *Phys. Rev. A* **52**, 839 (1995).

CROSSED BEAMS MEASUREMENTS OF ELECTRON IMPACT DISSOCIATIVE EXCITATION OF MOLECULAR IONS

Nada Djurić, Yang-Soo Chung, and Gordon H. Dunn

JILA, National Institute of Standards and Technology and the University of Colorado, Boulder, Colorado 80309-0440

ABSTRACT

A crossed beams technique developed at JILA especially to detect and measure light fragment ions from electron impact dissociation is described. Preliminary results of absolute cross sections for obtaining D^+ fragments from dissociation of deuterated molecules CD^+, CD_2^+, and CD_3^+ are presented.

INTRODUCTION

The electron impact dissociation of molecular ions is an important process in the astrophysics of stars and galaxies, atmospheric science, plasma physics, and chemical reactions. For molecular studies, one must deal with a broad range of processes: ionization (bound and dissociative), dissociative recombination, and excitation (bound and dissociative), and at the same time recognize the possible importance of internal energies (especially vibrational) of the target ions.

We restrict our consideration to the dissociative processes leading to at least one ionic fragment in combination with other charged and neutral fragments. Thus, the processes of interest may be represented,

$$e + AB^{+} \begin{array}{l} \rightarrow A^{+} + B^{+} + 2e \\ \rightarrow A^{+} + B + e \\ \rightarrow A^{+} + B^{-} \end{array} , \qquad (1)$$

and are referred to respectively as dissociative ionization (DI), dissociative excitation (DE), and ion pair (IP) formation. The field has shown little activity since the first results on H_2^+ (1,2,3,4). Only a few experimental cross sections have been reported including some for homonuclear ions (1,2,3,5,6), and a few for heteronuclear species CO^+ (7), HeH^+ (8), D_3^+ (9), H_3^+ (10), H_3O^+ (D_3O^+) (11), and CD_4^+ (12) where "heavy" dissociation products were detected, e.g., C or O from CO^+, OH^+ from H_3O^+, or C^+ from CD_4^+. Early studies on vibrationally hot H_2^+(1,2,3,4) revealed very interesting results. It was concluded from the synergistic interplay of theory and experiment that there is dramatic dependence of the cross section for DE on the vibrational population distribution of H_2^+. As the interaction energy decreases, the contribution from the higher vibrational levels becomes increasingly important. For electron energies above 3 eV, the cross section for DE is larger than that for DR. Until now, no serious efforts have been made to investigate these effects for other molecular ions.

There is renewed interest in experimental DE studies, primarily due to the increased demands for cross sections for modeling of plasma generators for etching and deposition (13), and for modeling of edge plasmas of fusion reactors (14). At the same time, there have been numerous technological improvements in ion sources, vacuum systems, and sensitive detectors that provide better experimental conditions. The new merged beams technique that employs heavy ion storage rings has opened up new possibilities in dissociation studies of vibrationally relaxed molecular ions. The method has been used extensively during recent years for studies of dissociative recombination of molecular ions. One can expect similar progress in DE studies as broader use is made of storage rings for such studies. Despite all these experimental improvements there is only one recently published result on DE, and that is on HD^+ (15).

The greatest difficulties in performing DE experiments are due to the fact that dissociation products are born with broad kinetic energy and angular distributions. This leads to a problem in collecting and quantitatively measuring the ion products. At the same time, a large number of different fragments with a variety of masses is formed even from a simple hydrocarbon like CH_4^+. Since every dissociation process is characterized by unique energy and angular distributions, it is almost impossible to have a relatively simple apparatus that will provide 100% collection and identification of each fragment ion. Our approach to the

CP392, *Application of Accelerators in Research and Industry,* edited by J. L. Duggan and I. L. Morgan
AIP Press, New York © 1997

problem mentioned above was to set up an apparatus that would be suited to measure light fragment ions.

EXPERIMENT

The JILA crossed beams apparatus (16,17) has been modified and a new detector chamber and a new dispersion and detection system has been designed, as shown schematically in Fig. 1. This apparatus is uniquely designed to detect light fragment ions from a dissociation process. This is difficult because in the dissociative processes the light fragment—to conserve momentum—gets almost all of the released kinetic energy. This sends the fragments off at large angles, making them difficult to collect.

Since a detailed description of the JILA crossed beams experiment has been presented previously (16,17), only a few details concerning the detector chamber and procedure for obtaining the experimental data are presented below.

The primary ion beam is extracted from a Colutron ion source (18) and accelerated and transported by 7 kV potential. Source conditions were empirically optimized for maximum beam intensity. Using a mixture of CD_4:He of roughly 1:1 gave us stable and long lasting ion source conditions. The target ions produced this way may be vibrationally and rotationally hot and we can not easily control or calculate the distribution of vibrational states. This beam is mass analyzed by a sector magnet and transported through different stages of collimation and differential pumping before entering the collision chamber (10^{-7} Pa). A magnetically confined electron beam (19) intersects the ion beam at 90°. The electron beam is chopped at 1000 Hz, and the detector is gated to record counts with electrons on and electrons off in order to separate the true signal produced in the electron-ion collisions from background events. A scanning slit probe is located in the center of the collision box and is rotated to measure spatial profiles of either the electron or the ion beam. Two pairs of deflector plates are mounted

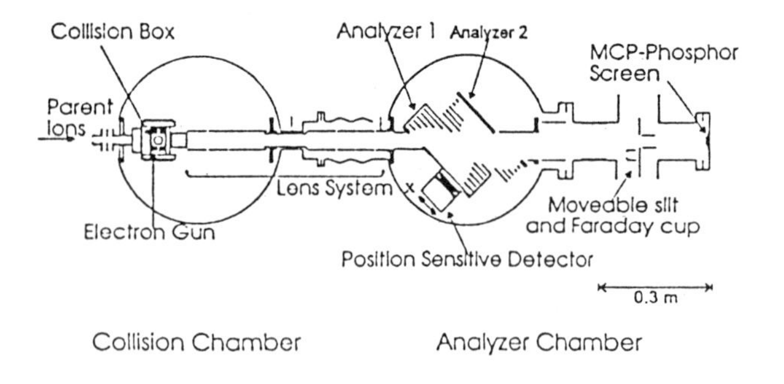

FIGURE 1. Crossed electron-ion beams interaction and fragment ion analysis chamber.

before and after the collision box to compensate for the ion beam deflection in the 0.006 T magnetic field of the electron gun. After the collision, the target ions and fragment ions enter the cylindrically symmetric lens system designed to collect, transport, and accelerate the fragment ions of interest into the analyzer chamber. Two 45° electrostatic analyzers are located in the analyzer chamber. The first analyzer separates selected fragment ions from other products and from the primary beam. It deflects the selected ion products onto the position sensitive detector (PSD) with sensitive diameter of 40 mm. The detector is mounted on a linear motion feedthrough with a travel of 50 mm (x-direction at the PSD). The primary ions will be deflected somewhat by the first analyzer, and the second one is used to redirect them out of the analyzer chamber towards the electrically isolated smaller chamber. This ancillary chamber is also used as a large ion collector. It contains three sets of horizontal deflectors, an inline movable slit, and a microchannel plate followed by a fluorescent screen, which are used as diagnostic tools.The detector is made wide and open to accommodate all light fragment ions with broad energy and angular distributions. This makes it vulnerable to backgrounds of different origin. To determine the background origin, we employed a time-of-flight technique. The arrival time spectra, acquired by the multichannel analyzer, consists of three peaks. The first two peaks were identified as coming from the dissociation of the parent ion, most likely in a collision with the surface in the vicinity of the collision box. The third peak is coming from photons produced by the parent beam hitting any surface inside the isolated ion collector chamber and traveling back directly or via reflection to the PSD. The background coming from photons was almost eliminated by coating the small chamber and all surfaces capable of reflecting photons towards the PSD with aerodag (contains graphite and isopropil alcohol).

Extensive ion trajectory modeling using SIMION (20) was carried out to design and investigate the capability of the apparatus to collect the fragment ions under investigation. The model calculations were performed for simple deuterated ions (CD_3^+, CD_2^+, CD^+) traveling at a fixed incident velocity (corresponding to 7 keV) breaking up into the light fragments D^+, D_2^+, and D_3^+ along with unspecified other fragments. The excess kinetic energy in our simulation was chosen to be not more than 10 eV per fragment ion. The relevant broad ranges of transverse and longitudinal energies were inserted into the modeling, and conditions determined whereby the desired particles would be fully collected. Both the transport and analyzer parameters vary as the target ion changes.

For the initial experiments, we measured the cross section for D^+ formation from electron-impact dissociation of CD_3^+, CD_2^+ and CD^+ from the threshold to 70 eV.

The signal distribution on the PSD of D^+ fragment ions from $CD_3^+$is shown in Fig. 2. It is similar to that for other

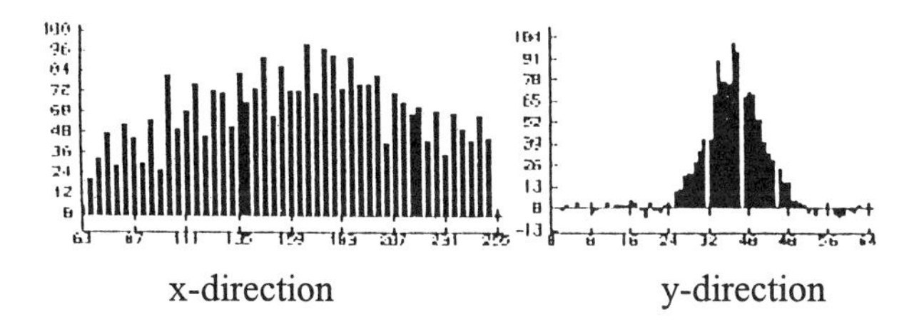

FIGURE 2. Signal distribution in the x and y direction (for the 256 x 64 array) on the PSD.

studied target ions. The D^+ distribution at the PSD is found to be relatively narrow in the y-direction, while in the x-direction (direction of PSD movement) it is broader and it can become so great that significantly less than 100% of the fragments are detected (at that PSD position). In such a case, moving the PSD to different positions is necessary, and a procedure for summing the signal at different PSD positions has to be applied. As one changes target ions from CD_3^+ towards the lighter target ions, the D^+ distribution on the PSD in the x-direction becomes narrower, while the signal distribution in the y-direction does not change much. This is to be expected, since only the width of the signal in the longitudinal direction (corrresponds to the x-direction on the PSD) is affected by the energy spread of the fragment ions.

A typical procedure of data taking (for the fragment ion of interest) begins with setting the PSD position and all voltages upstream from the collision region according to SIMION trajectory modeling. The primary ion beam is then tuned to achieve minimum background, and the form factor is measured at the chosen electron energy. Collection of data in the two channels proceeds until adequate statistical uncertainties are reached. The position of the PSD (in the x-direction) is then changed (usually 10mm) and the collection of data repeated. The signal detected at the new PSD position is a combination of part of the signal measured at the previous PSD position ("overlapped" signal) and the signal that was missed at the previous position. Changing the PSD position is continued as long as detectable signal is collected (usually three positions covering the linear motion of 50-60 mm are enough to collect all signal ions). Final data at a given interaction energy are then obtained by summing the total signal at one PSD position and by adding the "missing" signal for that position as measured at other PSD positions. The electron energy is then changed and all procedures repeated.

The efficiency of the PSD was measured in a separate experiment by directing a very small current of D^+ ions alternatively onto the PSD and into the ion collector. The number of counts detected by the PSD is then converted into a current and compared with the current measured in the ion collector and as read by a vibrating-reed electrometer.

The preliminary cross section results above 25 eV for electron impact dissociation of D^+ from CD^+, CD_2^+, and CD_3^+ are observed to have similar energy dependencies. Thus, above 25 eV the cross sections for the studied target ions remain nearly constant ($1.7x10^{16}$ cm^2) to 70 eV, the highest electron energy at which the cross sections were studied. The cross section for production of D^+ from CD^+ displays the lowest threshold at around 5 eV and is up to the maximum by around 20 eV. The cross section for D^+ from CD_2^+ has a threshold at 10 eV and reaches a maximum at around 25 eV, while the cross section for D^+ from dissociation of CD_3^+ has a threshold around 14 eV reaching a maximum value at 30 eV. The minimum energy (21) for obtaining C and H^+ (from the ground state of the CH^+) is about 6 eV. For obtaining CH and H^+ (from CH_2^+) the minimum energy is about 7.5 eV (22). The preliminary character of the data and the lack of molecular data for CD_2^+ and CD_3^+ restrict us from any conclusion about the level of vibrational state population of the targets as well as about the possiblility that some indirect mechanisms could be involved in the D^+ production.

In this brief report, we have discussed a technique to measure DE. It provides detection of light fragment ions from heavy and probably vibrationally and rotationally hot target ions. The knowledge of the cross sections for vibrationally hot ions is important for applications, and also for comparison with results for cold ions. Two observation should be mentioned: (1) D^+ fragment ions are formed with wide kinetic energy distributions; and (2) going from CD^+ towards the heavier target ions, the kinetic energy distribution of D^+ becomes broader. It is our hope that these measurements, as well as the results from the storage rings, will stimulate some theoretical analysis of DE.

ACKNOWLEDGMENTS

This work was supported in part by the Office of Fusion Energy of the Department of Energy, Contract No. DE-A105-86ER53237 with the National Institute of Standards and Technology.

REFERENCES

1. Dunn, G. H., and Van Zyl, B., and Zare, R. N., *Phys.Rev. Lett.*, **15**, 610-12 (1065); Dunn, G. H., Van Zyl, B., *Phys. Rev.* **154**, 40-51 (1967).
2. Dance, D. F., Harrison, M. F. A., Rundel, R. D., and Smith. A. C. H., *Proc. Phys. Soc.* **92**, 577-88 (1967).
3. Peart, B., and Dolder, K. T., *J. Phys. B (London)* **4**, 1496-05 (1971); **5**, 1554-58 (1972).
4. Peek, J. M., *Phys. Rev.* **154**, 52-56 (1967); Peek, J. M., and Green, T. A., *Phys. Rev.* **183**, 202-212 (1969).
5. Van Zyl, B., and Dunn, G. H., *Phys. Rev.* **163**, 43-45 (1967).

6. Hus, H., Yousif, F., Noren, C., Sen, A., and Mitchell, J. B. A., *Phys. Rev. A* **60**, 1006-09 (1988); Noren, C., Yousif, F. B., and Mitchell, J. B. A., *J. Chem. Soc., Faraday Trans.* **2**, 1697-793 (1989); Yousif, F. B., and Mitchell, J. B. A., *Z. Phys. D.* **34**, 195-97 (1995).
7. Mitchell, J. B. A., and Hus, H., *J. Phys. B (London)* **18**, 547-55 (1985).
8. Yousif, F. B., and Mitchell, J. B. A., *Phys. Rev. A*, **40**, 4318-21 (1989); Yousif, F. B., Mitchell, J. B. A., Rogelstad, M., Le Paddelec, A., Canosa, A., and Chibisov, M. I., *Phys. Rev. A* **49**, 4610-15 (1994).
9. Van der Donc, P., Yousif, F. B., and Mitchell, J. B. A., *Phys. Rev. A* **43**, 5971-74 (1991).
10. Yousif, F. B., Van der Donk, P. J. T., Orakzai, M., amd Mitchell, J. B. A., *Phys. Rev. A* **44**, 5653-58 (1991).
11. Schulz, P. A., Gregory, D. C., Meyer, F. W., and Phaneuf, R. A., *J. Chem. Phys.* **85**, 3386-94 (1986).
12. Gregory, D. C., and Tawara, H., "Dissociation and Ionization of CD_4^+ by electron impact," *Abstracts of Contributed Papers* (Eds: Dalgarno, A., Freund, R. S., Lubell, M. S., and Lucatorto, T. B.), XVI ICPEAC, New York, 1989, p. 352.
13. Flamm, D. L., "Introduction to plasma chemistry," in *Plasma Etching - An Introduction* Eds: Manos, D., and Flamm, D. L.: Boston, Academic Press, 1989, ch. 2, pp 91-183.
14. Dunn, G. H., *Nucl. Fus. Supl. At. Plasma Mater. Interaction Data Fusion* **2**, 25-39 (1992).
15. Forck, P., Grieser, M., Habs, D., Lampert, A., Repnow, R., Schwalm, D., Wolf, A., and Zaifman, D., *Nucl. Instr. and Meth. B* **79**, 273-75 (1993).
16. Rogers, W.T., Stefani, G., Camillioni, R., Dunn, G. H., Msezane, A. Z., and Henry. R. J. W., *Phys. Rev. A* **25**, 737-48 (1982).
17. Djurić, N., Bell, E. W., Daniel, E,. and Dunn, G. H., *Phys. Rev. A* **46**, 270-74 (1992).
18. Menzinger, M., and Wählin. L., *Rev. Sci. Instrum.* **45**, 538-44 (1974).
19. Taylor, P. O., Dolder, K. T., Kauppila, W. E., and Dunn, G. H., *Rev. Sci. Instrum.* **45**, 538-44 (1974).
20. SIMION 3D, Version 6.0, David A. Dahl, Idaho National Engineering Laboratory.
21. Lorquet, A. J., Lorquet, J. C., Wankenne, H., and Momigny, J., *J. Chem. Phys.* **55**, 4053-61 (1971).
22. Schuette, G. F. and Gentry, R., *J. Chem. Phys.* **78**, 1777-85 (1983).

ANGULAR DISTRIBUTION AND SPIN POLARIZATION OF MOLECULAR AUGER PROCESSES

B. Lohmann, S. Bonhoff, K. Bonhoff, J. Lehmann, K. Blum

Universität Münster, Institut für Theoretische Physik I, Wilhelm-Klemm-Str. 9, D-48149 Münster, Germany

The general theory for angular distribution and spin polarization of molecular Auger electrons emitted from freely rotating diatomic molecules is discussed within the framework of a two-step model. Assuming electron impact ionization then, in contrast to a primary photoionization process, the number of independent parameters is no longer restricted by dipole selection rules. Different spin polarization states of the primary electron beam will be considered. The physical importance of a coherent excitation process is discussed within a simple example. Numerical results for the anisotropy parameters characterizing the dynamics of the Auger emission process are discussed for HF.

INTRODUCTION

Until today, most of the experimental and theoretical studies of molecular Auger processes concentrated on determining the energy and transition probability of a given Auger line (1-6). For atoms, the past years showed that more refined information can be obtained by investigating angle and spin resolved processes (7-10). General formulas for the angular distribution and spin polarization of molecular Auger electrons following photoabsorption or photoionization have already been developed (11-13). Recently, experimental results for angle-resolved molecular Auger processes have been obtained (14,15). On the other hand, we investigated the angular distribution of Auger electrons emitted by diatomic molecules in the gas phase after electron impact ionization (16).

In the present paper, the theory for the angular distribution will be discussed and further extended by including the case of a spin polarized Auger emission. Although the general structure is similar to the treatments mentioned above, there are some important differences.

For a primary photoionization process, the number of independent parameters is limited by the dipole selection rules. They no longer apply for the case of electron impact ionization and thus additional parameters need to be determined within an experiment. Different states of polarization of the ionizing electron beam are discussed.

The molecular ensemble M^+ will be parameterized in terms of "order parameters" which reflect the anisotropy produced during the primary ionization.

The importance of the coherences produced in the primary ionization process will be considered within a simple example. A coherent ionization requires the introduction of additional parameters, even in the case of photoionization as initial process, which are therefore necessary to correctly interpret the experimental data. A detailed discussion of these parameters has recently been given by Bonhoff *et al* (16).

The main focus of the paper is on diatomics. However, the theory can be applied to polyatomics in a similar way. A discussion of the angular distribution of Auger electrons emitted from non-linear polyatomic molecules may be found in the contribution by Lehmann *et al* (17) where results for the H_2O molecule are presented.

THEORY

General Considerations

The Auger process can be described by the well-observed (18) two-step model

$$\mathrm{M} + e \longrightarrow \mathrm{M}^+ + e + e' \tag{1a}$$

$$\searrow \mathrm{M}^{++} + e_{\mathrm{Auger}} \tag{1b}$$

where the initial vacancy has been created by electron impact. The following assumptions have been made: we restrict ourselves to diatomic molecules, the rotation of the molecules between the primary ionization and the subsequent Auger decay can be neglected (13), the emitted electrons e and e' of the primary process are not detected, the molecules M are unpolarized and initially in Σ states and freely rotating; i.e. the initial axis distribution is isotropic.

Using a space-fixed coordinate system XYZ, with the incoming electron beam axis as quantization axis, and a molecule-fixed coordinate system xyz, with the molecular axis $\mathbf{n} \parallel z$, the Auger process can be illustrated as in Fig. 1. The three Euler angles $(\alpha\beta\gamma)$ re-

CP392, *Application of Accelerators in Research and Industry*, edited by J. L. Duggan and I. L. Morgan
AIP Press, New York

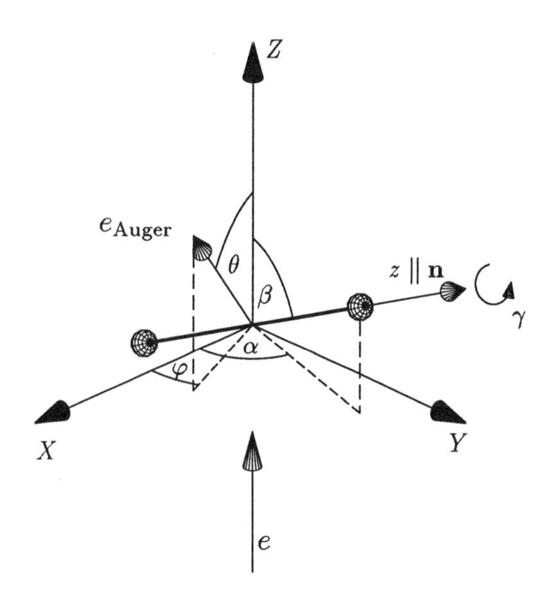

FIGURE 1. Direction of Auger emission, molecular (**n**), and space-fixed coordinate frame.

late the two systems. For diatomic molecules the angle γ can be omitted. The direction of the emitted Auger electrons will be characterized by the angles $(\varphi\theta\chi)$ with respect to the XYZ system. Generally, the angle χ can be chosen as zero.

Auger emission

Generally, applying statistical tensor methods, the state multipoles $\langle t_{kq}\rangle$ describing the emitted Auger e-lectrons can be written as

$$I\,\langle t_{kq}\rangle = \sum_{\substack{KQ'\\ \Omega'\Omega}} \langle \mathcal{D}^{(K)*}_{Q',\Omega-\Omega'}(\Omega'\Omega)\rangle \times A_{Kkq}(\Omega'\Omega)\; \mathcal{D}^{(K)}_{Q'q}(\varphi\theta\chi)\,, \tag{2}$$

where Ω denotes the total angular momentum of the molecules.

The intensity I of the emitted Auger electrons and the components of the spin polarization vector **P**, with respect to the Auger system, can be obtained from the state multipoles as

$$I = \sqrt{2}\; I\,\langle t_{00}\rangle\,, \tag{3a}$$

$$I\,P_x = I\,(\langle t_{1-1}\rangle - \langle t_{11}\rangle)\,, \tag{3b}$$

$$I\,P_y = i\,I\,(\langle t_{1-1}\rangle + \langle t_{11}\rangle)\,, \tag{3c}$$

$$I\,P_z = \sqrt{2}\; I\,\langle t_{10}\rangle\,. \tag{3d}$$

The expansion coefficients or "order parameters" are defined by

$$\left\langle \mathcal{D}^{(K)*}_{Q'Q}(\Omega'\Omega)\right\rangle = \int_0^{2\pi} d\alpha \int_0^{\pi} d\beta \sin\beta \int_0^{2\pi} d\gamma \times \langle \Omega'\mathbf{n}|\rho|\Omega\mathbf{n}\rangle\; \mathcal{D}^{(K)}_{Q'Q}(\alpha\beta\gamma)^*\,. \tag{4}$$

Their properties will be discussed in the contribution by Lehmann *et al* (17). Here, we only point out that the order parameters are determined by the spin polarization of the incoming electron beam via

$$Q' = m_{s_0} - m'_{s_0}\,, \tag{5}$$

and by the transition matrix elements of the primary ionization process. Since for diatiomcs the third Euler angle γ can be chosen arbitrarily, we get

$$Q = \Omega - \Omega'\,. \tag{6}$$

The anisotropy parameters $A_{Kkq}(\Omega'\Omega)$ are defined in analogy to the atomic case and characterize the dynamics and geometry of the Auger decay

$$\begin{aligned} A_{Kkq}(\Omega'\Omega) = & \sum_{\substack{\Omega_f L l l'\\ m m' m_s m'_s}} \frac{1}{4\pi\, k_0}\,(-1)^{m'+m'_s-q-\frac{1}{2}}\, i^{l-l'} \\ & \times \sqrt{(2l+1)(2l'+1)(2k+1)}\,(2K+1)\,(2L+1) \\ & \times \langle \Omega_f l' m' m'_s|\mathbf{T}|\Omega'\rangle\, \langle \Omega_f l m m_s|\mathbf{T}|\Omega\rangle^* \\ & \times \begin{pmatrix} l' & l & L\\ 0 & 0 & 0\end{pmatrix} \begin{pmatrix} l' & l & L\\ m' & -m & m-m'\end{pmatrix} \\ & \times \begin{pmatrix} \frac{1}{2} & \frac{1}{2} & k\\ m'_s & -m_s & m_s-m'_s\end{pmatrix} \begin{pmatrix} k & L & K\\ -q & 0 & q\end{pmatrix} \\ & \times \begin{pmatrix} k & L & K\\ m'_s-m_s & m'-m & \Omega-\Omega'\end{pmatrix}. \end{aligned} \tag{7}$$

The partial wave quantum numbers l, m of the Auger electron as well as its spin quantum number m_s refer to the molecular system.

The angular dependency of intensity and spin polarization is expressed in terms of rotation matrix elements $\mathcal{D}^{(K)}_{Q'q}(\varphi\theta\chi)$. The quantum numbers Q' and q are connected with the degree of spin polarization of the incoming beam and the emitted Auger electrons, respectively.

Each rotation matrix element has the same rank K and quantum number Q' as its associated order parameter. This shows, that the anisotropy of intensity and spin polarization is a direct consequence of the anisotropy in the molecular ensemble M^+, and the anisotropy parameters $A_{Kkq}(\Omega'\Omega)$ can be interpreted as the different weight factors for the state multipoles $I\,\langle t_{kq}\rangle$; e.g. the intensity I of the Auger electrons is weighted by the anisotropy parameters A_{K00}. Note, that, besides the usual angular momentum coupling rules, the sum over K is not restricted.

Unpolarized electron beam

Assuming an unpolarized electron beam for the primary ionization, from eq. (5) we have $Q' = 0$. The angular distribution becomes independent on the angle φ because of the axially symmetric distribution of the molecular ensemble M^+. Thus, the intensity may be written as

$$I(\theta) = \sum_{K\Omega'\Omega} \left\langle \mathcal{D}^{(K)*}_{0\,\Omega-\Omega'}(\Omega'\Omega)\right\rangle A_K(\Omega'\Omega)\, P_K(\cos\theta)\,. \tag{8}$$

This is the same expression as derived by Bonhoff *et al* (16), where, using our definition of eq. (7) the anisotropy parameters have been abbreviated as

$$A_K(\Omega'\Omega) = \sqrt{2}\, A_{K00}(\Omega'\Omega)\,. \tag{9}$$

The angular dependence is given by the Legendre polynomials $P_K(\cos\theta)$. The factors describing the primary ionization and the Auger transition are clearly separated. The coherences produced in the primary process are taken implicitly into account by the order parameters. They are missing in other theories (12,13).

The system is invariant under reflection in the reaction plane defined by the incoming beam axis and the direction of Auger emission. Thus, the polarization vector $\mathbf{P}$ must be perpendicular to the reaction plane. Choosing the coordinate frame of the Auger system in such a way that $\chi = 0$, the x- and z-component of the spin polarization vector vanish

$$P_x = P_z = 0\,. \tag{10}$$

I.e. only the y-component can be expected to be nonzero. It may be expressed as

$$IP_y = \sum_{K\Omega\Omega'} 2i \left\langle \mathcal{D}^{(K)*}_{0\,\Omega-\Omega'}(\Omega'\Omega)\right\rangle A_{K11}(\Omega'\Omega) d^{(K)}_{0\,1}(\theta)\,, \tag{11}$$

where $d^{(K)}_{Q'\,q}(\theta)$ denotes the reduced rotation matrices.

Transversely polarized electron beam

Using a transversely polarized electron beam, the Auger intensity shows a dependence on the azimuth φ which is caused by the spin-orbit interaction

$$I(\theta\varphi) = I(\theta)_{unpol} - B(\theta)\sin\varphi\,. \tag{12}$$

Here, I_{unpol} is given by eq. (8) and the function $B(\theta)$ is defined as

$$B(\theta) = \sum_{K\Omega\Omega'} 2i \left\langle \mathcal{D}^{(K)*}_{1\,\Omega-\Omega'}(\Omega'\Omega)\right\rangle A_K(\Omega'\Omega)\, d^{(K)}_{1\,0}(\theta)\,. \tag{13}$$

Thus, measuring the intensity of the Auger electrons in the reaction plane, we obtain the same result as for the unpolarized case.

Neglect of spin-orbit interaction

$B(\theta)$ can be expected as small for light molecules since the spin-orbit interaction is weak. Without proof we note that $B(\theta) = 0$ for a vanishing spin-orbit interaction. Thus, the intensity of the Auger electrons becomes independent on the polarization of the electron beam,

$$I(\theta\varphi) = I(\theta)_{unpol}\,. \tag{14}$$

Further, using an unpolarized electron beam and neglecting the spin-orbit interaction, we always end up with unpolarized Auger electrons,

$$\mathbf{P} = 0\,. \tag{15}$$

Primary ionization and coherence

We will now consider the physical importance of the order parameters $\left\langle \mathcal{D}^{(K)*}_{Q'Q}\right\rangle$. In the following, the spin is neglected for brevity. Denoting the angular momentum by Λ, we assume that states with different $|\Lambda|$ are energetically well separated, i.e. they overlap incoherently. The states $\Lambda = \pm|\Lambda|$ are, however, very nearly degenerate and will be coherently excited. Restricting ourselves to the case $|\Lambda| = 1$, we introduce the so-called "directed" electronic orbitals (19),

$$|\Pi_x, \mathbf{n}\rangle = -\frac{1}{\sqrt{2}}\left[|\Lambda = 1, \mathbf{n}\rangle - |\Lambda = -1, \mathbf{n}\rangle\right]\,, \tag{16a}$$

$$|\Pi_y, \mathbf{n}\rangle = \frac{i}{\sqrt{2}}\left[|\Lambda = 1, \mathbf{n}\rangle + |\Lambda = -1, \mathbf{n}\rangle\right]\,. \tag{16b}$$

The corresponding electronic lobes are directed along the molecular x-and y-axis for the Π_x- and Π_y-states, respectively. Within this basis set and using the fact that the plane containing the molecular axis $\mathbf{n}$ and the Z-axis is a symmetry plane, interference terms between the two states vanish, i.e. $\langle \Pi_x, \mathbf{n}|\rho|\Pi_y, \mathbf{n}\rangle = 0$.

The order parameters $\left\langle \mathcal{D}^{(K)*}_{00}\right\rangle$ characterize the distribution of the molecular axes. In order to see the meaning of the interference terms $\left\langle \mathcal{D}^{(K)*}_{0\Lambda-\Lambda'}\right\rangle$ we consider the angular distribution of Auger electrons where only terms with $K = 0, 2$ contribute to equation (8); e.g. photoionization of homo-nuclear diatomics.

Assuming an initial vacancy in a pure Π_x-orbital, we obtain, for instance, for the angular distribution of the Auger electrons

$$\begin{aligned} I(\theta) &= 2\left\langle \mathcal{D}^{(0)*}_{0\,0}(1,1)\right\rangle A_0(1,1) \\ &+2\left\langle \mathcal{D}^{(2)*}_{0\,0}(1,1)\right\rangle A_2(1,1)\, P_2(\cos\theta) \\ &+2\left\langle \mathcal{D}^{(2)*}_{0\,2}(-1,1)\right\rangle A_2(-1,1)\, P_2(\cos\theta)\,. \end{aligned} \tag{17}$$

For the opposite case of a pure Π_y-orbital we obtain the same expression except that $\left\langle \mathcal{D}^{(2)*}_{02}\right\rangle$ changes its sign. The two orbitals $|\Pi_x\rangle$ and $|\Pi_y\rangle$ differ only in their direction in space. Thus, the order parameters $\left\langle \mathcal{D}^{(2)*}_{02}\right\rangle$ can be seen as a direct measure for the influence of the spatial orbital orientation on the Auger emission.

NUMERICAL VALUES FOR THE ANISOTROPY PARAMETERS

Numerical Methods

The calculation of the $\mathbf{T}$ matrix elements by means of the Greens operator formalism is an expansion of the computation of Auger spectra by Schimmelpfennig *et al* (4). The molecular wave functions are optimized in a SCF-CI calculation. In contrast to the method described ibid. a one-center expansion of the basis set is used to obtain the phases of the transition matrix

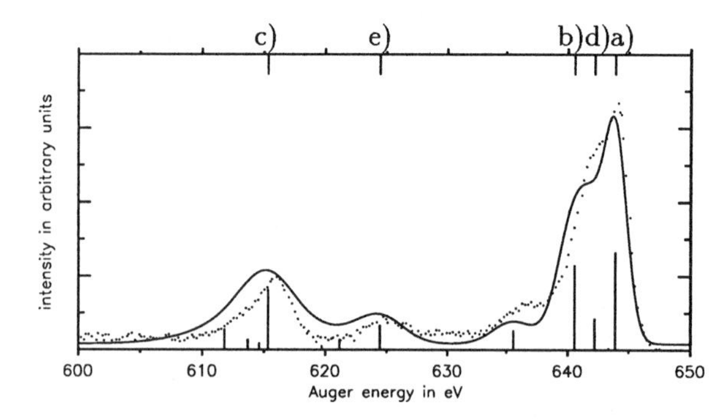

FIGURE 2. Auger spectrum of HF, our calculations: —, experiment by Shaw and Thomas (20): ···, labels a)-e) refer to the calculated anisotropy parameters (see text below)

elements, which are necessary for the calculation of the anisotropy parameters.

Calculations for HF

As an example for diatomic molecules we consider HF. Choosing the $^2\Sigma^+$ $(1\sigma^{-1})$ as initial state of the molecules M^+ we calculated the energies and intensities for all substantial Auger transitions. The resulting spectrum is shown in Fig. 2.

The anisotropy parameters were calculated for the final states with the largest total intensity, i.e. $^1\Delta\,(1\pi^{-2})$ (a), $^1\Pi(3\sigma^{-1}1\pi^{-1})$ (b), $^1\Pi(2\sigma^{-1}1\pi^{-1})$ (c), $^1\Sigma^+(1\pi^{-2})$ (d), $^3\Pi(2\sigma^{-1}1\pi^{-1})$ (e). They are given in table 1. As can be seen large values of anisotropy parameters can be expected for all Auger transitions considered. In contrast to the atomic case of angular distribution, where only even-rank anisotropy parameters α_K can occur (21), odd-rank anisotropy parameters can also be nonzero for diatomic molecules. Generally, the magnitude of the anisotropy parameters $A_K(\Omega'\Omega)$ decrease with increasing rank. A detailed discussion of the results for HF will be published elsewhere (22).

TABLE 1. Anisotropy parameter $A_K(\frac{1}{2},\frac{1}{2})$ in 10^{-4} a.u.

K	a) $^1\Delta$	b) $^1\Pi$	c) $^1\Pi$	d) $^1\Sigma^+$	e) $^3\Pi$
0	1.3252	1.1433	0.8147	0.4264	0.3183
1	0.0000	-0.3622	0.2142	-0.2113	-0.0084
2	-1.8932	0.7525	-0.7846	1.2156	-0.3182
3	0.0000	0.3625	-0.2139	-0.2111	0.0084
4	0.5679	-1.8953	-0.0302	0.8686	-0.0001
5	0.0000	0.0001	-0.0003	0.0078	0.0000
6	0.0000	-0.0004	0.0001	-0.0015	0.0001
7	0.0000	-0.0004	0.0000	0.0064	0.0000
8	0.0000	0.0000	0.0000	0.0000	0.0000

ACKNOWLEDGMENTS

The numerical calculations of the transition matrix elements have been carried out in collaboration with Drs. B. Schimmelpfennig and B. Nestmann, and Prof. S.D. Peyerimhoff, University of Bonn, Germany. B.L. acknowledges a Habilitation grant and assistance with travel grants by the Deutsche Forschungsgemeinschaft (DFG). This work has been supported by Sonderforschungsbereich (SFB) 216 "Polarization and Correlation in Atomic Collision Complexes" of the DFG.

REFERENCES

1. Aksela, S., Sairanen, O. P., Aksela, H., Bancroft, G. M., Tan, K. H., *Phys. Rev. A* **37**, 2934 (1988)
2. Carravetta, V., Ågren, H., *Phys. Rev. A* **35**, 1022 (1987)
3. Ferrett, T. A., Piancastelli, M. N., Lindle, D. W., Heimann, P. A., Shirley, D. A., *Phys. Rev. A* **38**, 701 (1988)
4. Larkins, F. P., *J. Electr. Spectr. & Rel. Phen.* **51**, 115 (1990)
5. Schimmelpfennig, B., Nestmann, B., Peyerimhoff, S. D., *J. Electr. Spectr. & Rel. Phen.* **74** 173 (1995)
6. Zähringer, K., Meyer, H. D., Cederbaum, L. S., Tarantelli, F., Sgamellotti, A., *Chem. Phys. Lett.* **206**, 247 (1993)
7. Lohmann, B., Hergenhahn, U., Kabachnik, N. M., *J. Phys. B: At. Mol. Opt. Phys.* **26**, 3327 (1993)
8. Kuntze, R., Salzmann, M., Böwering, N., Heinzmann, U., Ivanov V. K., Kabachnik, N. M., *Phys. Rev. A* **50**, 489 (1994)
9. Müller, N., David, R., Snell, G., Kuntze, R., Drescher, M., Böwering, N., Stoppmanns, P., Yu, S. W., Heinzmann, U., Viefhaus, J., Hergenhahn, U., Becker, U., *J. Electr. Spectr. & Rel. Phen.* **72**, 187 (1995)
10. Lohmann, B., Fritzsche, S., Larkins, F. P., *J. Phys. B: At. Mol. Opt. Phys.* **29**, 3327 (1996)
11. Chandra, N., *Phys. Rev. A* **40**, 752 (1989)
12. Chandra, N., Chakraborty, M., *J. Chem. Phys.***97**, 236 (1992)
13. Dill, D., Swanson, J. R., Wallace, S., Dehmer, J. L., *Phys. Rev. Lett.* **45**, 1393 (1980)
14. Hemmers, O., Heiser, F., Eiben, J., Becker, U., *Phys. Rev. Lett.* **71**, 987 (1993)
15. Zheng, Q., Edwards, A. K., Wood, R. M., Mangan, M. A., *Phys. Rev. A* **52**, 3940 (1995)
16. Bonhoff, K., Nahrup, S., Lohmann, B., Blum, K., *J. Chem. Phys.* **104**, 7921 (1996)
17. Lehmann, J., Bonhoff, K., Bonhoff, S., Lohmann, B., Blum, K., "Angular Distribution of Molecular Auger Electrons", in *Conf. Proc. of the XIVth CAARI'96*, to be published (1996)
18. Mehlhorn, W., X-Ray and inner shell processes, ed.: Carlson, T. A., Krause, M. O., Manson, S. T., New York, *AIP Conf. Proc.* **215**, 465 (1990)
19. Wöste, G., Fullerton, C., Blum, K., Thompson, D., *J. Phys. B: At. Mol. Opt. Phys.* **27**, 2625 (1994)
20. Shaw Jr., R. W., Thomas, T. D., *Phys. Rev. A* **11**, 1491 (1975)
21. Berezhko, E. G., Kabachnik, N. M., *J. Phys. B: At. Mol. Opt. Phys.* **10**, 2467 (1977)
22. Bonhoff, S., Bonhoff, K., Schimmelpfennig, B., Nestmann, B., in progress (1996)

ANGULAR DISTRIBUTION OF MOLECULAR AUGER ELECTRONS

J. Lehmann, K. Bonhoff, S. Bonhoff, B. Lohmann, K. Blum

Universität Münster, Institut für Theoretische Physik I, Wilhelm-Klemm-Str. 9, 48149 Münster, Germany

We present a general theory for the angular distribution of molecular Auger electrons that can be applied to both diatomic and polyatomic molecules. The Auger process is described in the two-step model. By choosing a convenient parameterization we obtain a formula for the angular distribution which clearly separates the effects of the two steps. The geometry and dynamics of the primary ionization are described by a set of order parameters. The information available on the dynamics of the Auger transition is contained in a set of anisotropy parameters, which have been calculated for HF and H_2O.

INTRODUCTION

Most of the experimental and theoretical studies of molecular Auger processes have concentrated on determining the energy and probability of a given Auger transition (1-6). Additional information can be obtained by investigating the angle-resolved process. General formulas for the angular distribution of molecular Auger electrons following photoabsorption or photoionization have already been developed (7-9). Recently we have developed a general theory for the angular distribution of Auger electrons emitted by molecules in the gas phase after ionization by electron impact (10, 11). Although the general structure is similar to the treatments mentioned above, there are some important differences.

- In case of photoionization as initial process the number of independent parameters is limited by the dipole selection rules. No such restrictions exist in case of ionization by electron impact and additional parameters are necessary for a correct description.
- We introduce a parameterization of the molecular ensemble M^+ in terms of "order parameters", which reflect the anisotropy produced in the primary ionization. Furthermore this set of parameters allows to present the final result in a compact and transparent form which is particularly convenient for analysing experimental data.
- Of particular interest are the coherences produced in the primary ionization process. They require the introduction of additional parameters even in the case of photoionization as initial process and are necessary for any future numerical calculation. A detailed discussion of the importance of these parameters has recently been given by Bonhoff *et al* (10) for diatomic molecules. However, for non-linear molecules, these coherence parameters have not been taken properly into account and the theories mentioned above must further be developed (11).

GENERAL THEORY

Basic Framework

We discuss the Auger process of diatomic and polyatomic molecules in the two step model (12)

$$\mathrm{M} + e \longrightarrow \mathrm{M}^+ + e + e' \qquad (1a)$$

$$\searrow \mathrm{M}^{++} + e_{\mathrm{Auger}} \qquad (1b)$$

under the following assumptions:

a) the rotation of the molecules between the primary ionization and the subsequent Auger decay can be neglected (9),

b) the initial molecules and electrons are unpolarized and the final spin states are not detected,

c) the emitted electrons e and e' of the primary process are not detected,

d) initially the molecules M are freely rotating and the axis distribution is isotropic.

The process will be described in a space-fixed coordinate system XYZ where Z coincides with the incoming beam axis. A second right-handed coordinate system xyz will be chosen to be rigidly connected with the molecular framework. The orientation of the molecular system with respect to the space-fixed system will be specified by the three Euler angles $\alpha\beta\gamma$. The direction of the emitted Auger electrons will be characterized by polar and azimuth angles θ and ϕ with respect to the XYZ system. The angles are illustrated in Fig. 1 for a diatomic molecule.

Primary process

The initial molecules are assumed to be in their electronic and vibrational groundstate. The states of the molecules M^+ can in general be degenerated and will therefore be *coherently* excited (13). For a correct description of these coherences we have to apply density

CP392, *Application of Accelerators in Research and Industry*, edited by J. L. Duggan and I. L. Morgan
AIP Press, New York

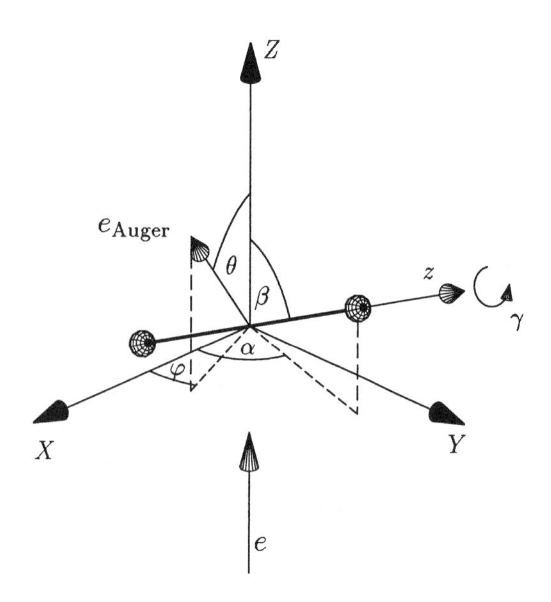

FIGURE 1. Illustration of angles definitions

matrix methods. We denote the density matrix describing the molecules M^+ immediately after the ionization process by $\boldsymbol{\rho}$ and its elements by $\langle \Psi' \omega | \boldsymbol{\rho} | \Psi \omega \rangle$, where Ψ collectively denotes the quantum numbers describing the molecular state. We normalize in such a way that the diagonal elements with $\Psi' = \Psi$ are equal to the probability density of finding a molecule in the state Ψ with an axis orientation specified by $\omega = \alpha\beta\gamma$. The off-diagonal elements characterize the coherence between states $\Psi \neq \Psi'$ produced in the primary process. The elements of the density matrix $\boldsymbol{\rho}$ are functions of $(\alpha\beta\gamma)$ and can therefore be expanded in terms of the rotation matrix elements $D^{(K)}_{Q'Q}(\alpha\beta\gamma)$ (conventions of Zare (14)):

$$\langle \Psi' \omega | \boldsymbol{\rho} | \Psi \omega \rangle = \sum_{KQ'Q} \frac{2K+1}{8\pi^2} \langle D^{(K)*}_{Q'Q}(\Psi', \Psi) \rangle \, D^{(K)}_{Q'Q}(\alpha\beta\gamma) \,. \quad (2)$$

The expansion coefficients $\langle D^{(K)*}_{Q'Q}(\Psi', \Psi) \rangle$ ("*order parameters*") are defined by

$$\langle D^{(K)*}_{Q'Q}(\Psi', \Psi) \rangle = \int_0^{2\pi} d\alpha \int_0^{\pi} d\beta \sin\beta \int_0^{2\pi} d\gamma \times \langle \Psi' \omega | \boldsymbol{\rho} | \Psi \omega \rangle \, D^{(K)*}_{Q'Q}(\alpha\beta\gamma). \quad (3)$$

Discussion of the order parameters

From the assumptions made above it is clear that the ensemble M^+ is axially symmetric around Z and that any plane through Z is a symmetry plane of the system. Hence

$a)$ $\langle D^{(K)*}_{Q'Q}(\Psi', \Psi) \rangle = \langle D^{(K)*}_{0Q}(\Psi', \Psi) \rangle$,

$b)$ $\langle D^{(K)*}_{0-Q}(\Psi', \Psi) \rangle^* = (-1)^Q \langle D^{(K)*}_{0Q}(\Psi, \Psi') \rangle$,

$c)$ $\langle D^{(K)*}_{0Q}(\Psi', \Psi) \rangle = \langle D^{(K)*}_{0Q}(\Psi, \Psi') \rangle^*$.

In order to illuminate further the geometrical meaning of the order parameters we will consider primary ionization by photon absorption:

$$\gamma + M \longrightarrow M^+ + e \,. \quad (4)$$

Because of dipole selection rules only order parameters with $K \leq 2$ contribute to eq. (2). Assuming linearly polarized light and choosing the direction of the electric field vector as Z-axis, it has been shown that no axis-orientation can be produced and that only order parameters with $K = 0$ and $K = 2$ are non-vanishing (13, 15). Furthermore we restrict to molecules of non-degenerate point groups ($\Rightarrow \Psi' = \Psi$) possessing a twofold rotation axis which gives the requirement:

$$\langle D^{(K)*}_{0Q}(\Psi) \rangle = (-1)^Q \langle D^{(K)*}_{0Q}(\Psi) \rangle \,. \quad (5)$$

Consequently we are left with two independent real parameters and expansion (2) reduces to

$$\langle \Psi \omega | \boldsymbol{\rho} | \Psi \omega \rangle = \frac{1}{8\pi^2} \left[1 + 5 \left\langle D^{(2)\,*}_{00}(\Psi) \right\rangle \frac{1}{2} (3\cos^2\beta - 1) + 5 \left\langle D^{(2)\,*}_{02}(\Psi) \right\rangle 2 \sqrt{\frac{3}{8}} \sin^2\beta \cos 2\gamma \right] \quad (6)$$

with $\left\langle D^{(2)\,*}_{00} \right\rangle = \langle \frac{1}{2}(3\cos^2 zZ - 1) \rangle$

and $\left\langle D^{(2)\,*}_{02} \right\rangle = \sqrt{\frac{3}{8}} \left[\langle \cos^2 xZ \rangle - \langle \cos^2 yZ \rangle \right]$

where $\cos nZ$ denotes the cosine of the angle between the molecular n-axis ($n = x, y, z$) and Z.

Because of $\cos^2 xZ + \cos^2 yZ + \cos^2 zZ = 1$ the parameters are not independent. Following essentially the formalism outlined in ref. (15) this can be visualized by constructing the "*orientation triangle*" shown in Fig. 2. Any possible axis distribution characterized by definite values of the two order parameters can be represented by points located within this triangle. For example the isotropic distribution is represented by the point $\langle D^{(2)\,*}_{00} \rangle = \langle D^{(2)\,*}_{02} \rangle = 0$ and all points on the line $\langle D^{(2)\,*}_{00} \rangle = \sqrt{6} \cdot \langle D^{(2)\,*}_{02} \rangle + 1$, which limits the triangle to the left, correspond to distributions where all molecular x-axes are perpendicular to Z. Further examples are given in (11).

Secondary process

The intensity $I(\theta, \varphi)$ of the emitted Auger electrons is given by (10, 11)

$$I(\theta, \varphi) = \sum_{K,Q} \sum_{\Psi, \Psi'} \left\langle D^{(K)\,*}_{0Q}(\Psi', \Psi) \right\rangle \times A_{KQ}(\Psi', \Psi) \, P_K(\cos\theta) \quad (7)$$

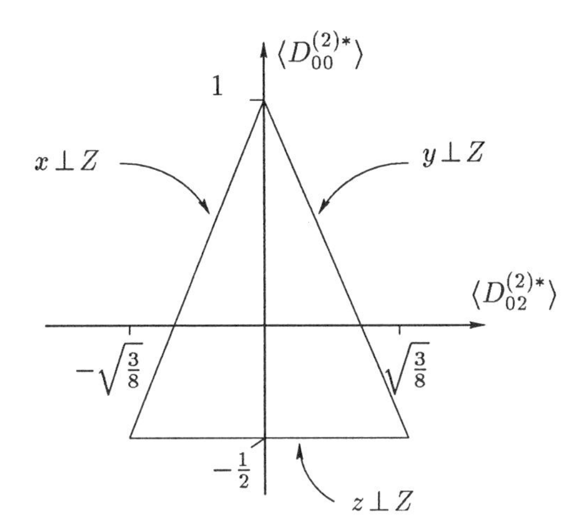

FIGURE 2. Orientation triangle

where the "*anisotropy parameters*" $A_{KQ}(\Psi', \Psi)$ are given by the expression

$$A_{KQ}(\Psi', \Psi) = \sum_{\Psi_f, m_s} \sum_{\substack{lm \\ l'm'}} \frac{1}{4\pi\, k}\, i^{l-l'}\, (-1)^m$$
$$\times\, (2K+1)\sqrt{(2l+1)(2l'+1)}$$
$$\times\, \langle \Psi_f, l'\, m', m_s | \mathbf{T} | \Psi' \rangle\, \langle \Psi_f, l\, m, m_s | \mathbf{T} | \Psi \rangle^*$$
$$\times \begin{pmatrix} l' & l & K \\ 0 & 0 & 0 \end{pmatrix} \begin{pmatrix} l' & l & K \\ m' & -m & Q \end{pmatrix}. \qquad (8)$$

Ψ_f denotes the quantum numbers of the final molecular states. The partial wave quantum numbers l, m of the Auger electron as well as its spin quantum number m_s refer to the molecular system. Thus the anisotropy parameters do not depend on the molecular orientation. The angular distribution is expressed in terms of the Legendre polynomials $P_K(\cos\theta)$. The factors describing the primary ionization and the factors describing the Auger transition are clearly separated. The anisotropy parameters $A_{KQ}(\Psi', \Psi)$ describe the geometry and dynamics of the Auger transition. The intensity is independent of φ because of the axially symmetric distribution of the molecular ensemble M^+. The coherences produced in the primary process are taken implicitly into account by the order parameters $\langle D_{0Q}^{(K)\,*}(\Psi', \Psi)\rangle$. Their importance for a correct description of the angular distribution has been pointed out in refs. (10, 11).

Diatomic molecules

Conventionally the molecular states are characterized by the projection Ω of the total angular momentum on the molecular z-axis. The Euler angle γ refers to rotation around the molecular axis. From $|\Omega\, \alpha\beta\gamma\rangle = e^{-i\Omega\gamma}|\Omega\, \alpha\beta 0\rangle$ we find $\langle D_{0Q}^{(K)\,*}(\Omega', \Omega)\rangle \sim \delta_{Q,\Omega-\Omega'}$. Hence eqs. (7, 8) can be considerably simplified. We refuse to write fully the explicit formulas. They can be found in refs. (10, 16). A detailed discussion of the significance of the coherence terms can be found ibidem. In addition ref. (16) expands the theory by taking the spin polarization of the Auger electrons into account.

Molecules of C_{2v} *symmetry*

The molecular states are classified according to the irreducible representations of the point group C_{2v}, i.e. $|\Psi\rangle = |\Gamma, S\, M_s\rangle$ with $\Gamma = A_1, A_2, B_1, B_2$.

The partial waves of the Auger electron can be symmetrized to exactly these representations by means of well-known projection operator techniques (18). In particular we find:

$$\begin{aligned}
&1)\ m = 0 &: |l\,0\rangle &= |A_1^{(l,0)}\rangle \\
&2)\ m > 0 \text{ even} &: |l\,m\rangle &= \tfrac{1}{\sqrt{2}}\left(|A_1^{(l,m)}\rangle + |A_2^{(l,m)}\rangle\right) \\
&3)\ m < 0 \text{ even} &: |l\,m\rangle &= \tfrac{1}{\sqrt{2}}\left(|A_1^{(l,|m|)}\rangle - |A_2^{(l,|m|)}\rangle\right) \\
&4)\ m > 0 \text{ odd} &: |l\,m\rangle &= \tfrac{1}{\sqrt{2}}\left(|B_1^{(l,m)}\rangle + |B_2^{(l,m)}\rangle\right) \\
&5)\ m < 0 \text{ odd} &: |l\,m\rangle &= \tfrac{1}{\sqrt{2}}\left(|B_1^{(l,|m|)}\rangle - |B_2^{(l,|m|)}\rangle\right)
\end{aligned}$$

Using these symmetry-adapted combinations, the orthogonality theorems of representation theory (18) lead to the selection rules for the **T**-matrix elements given in table 1.

TABLE 1. Selection Rules for the **T**-Matrix Elements in C_{2v} Symmetry

$\langle \Gamma_f, lm \vert \mathbf{T} \vert \Gamma\rangle = 0$ if		$\vert\Gamma_f\rangle$ $\vert A_1\rangle$	$\vert A_2\rangle$	$\vert B_1\rangle$	$\vert B_2\rangle$
$\vert\Gamma\rangle$	$\vert A_1\rangle$	m odd	m= 0 or odd	m even	m even
	$\vert A_2\rangle$	m= 0 or odd	m odd	m even	m even
	$\vert B_1\rangle$	m even	m even	m odd	m= 0 or odd
	$\vert B_2\rangle$	m even	m even	m= 0 or odd	m odd

Hence $m + m'$ must be even in eq. (8) and consequently all anisotropy parameters A_{KQ} with Q odd vanish if the molecules possess C_{2v} symmetry.

NUMERICAL VALUES FOR THE ANISOTROPY PARAMETERS

Numerical Methods

The calculation of the **T** matrix elements by means of the Greens operator formalism is an expansion of the computation of the Auger spectra, published by Schimmelpfennig *et al* (2). The molecular wave functions are optimized in a SCF-CI calculation. In contrast to the method described ibid. an one-center expansion of the basis set is used to obtain the phases of the transition matrix elements, which are necessary for the calculation of the anisotropy parameters.

Calculations for HF

Exemplarily for diatomic molecules we consider HF. The calculated spectrum as well as the anisotropy parameters for the most significant Auger transitions are given in (16).

Calculations for H_2O

As an example for polyatomic molecules we consider H_2O. Choosing the $^2A_1\,(1a_1^{-1})$ state as initial state of the molecules M^+ we calculated the energies and intensities for all significant Auger transitions. The resulting spectrum is shown in Fig. 3.

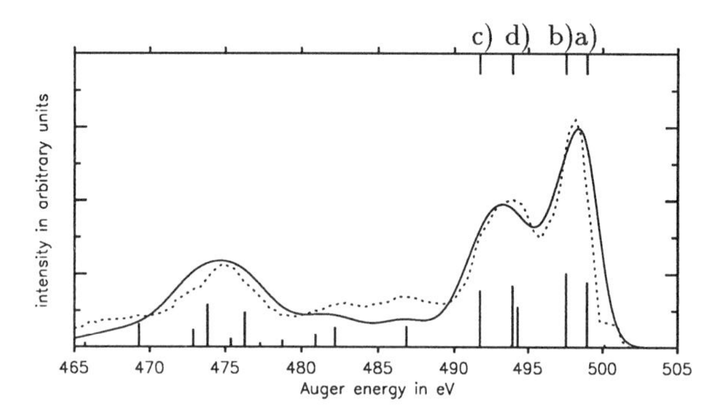

FIGURE 3. Auger spectrum for H_2O, our calculations: —, experiment by Siegbahn *et al* (4): ···, labels a)-d) refer to the calculated anisotropy parameters (see text below)

The anisotropy parameters have been calculated for the final singlet states with the largest total intensity, i. e. $^1A_1\,(1b_1^{-2})$ (a), $^1B_1\,(3a_1^{-1}, 1b_1^{-1})$ (b), $^1B_2\,(3a_1^{-1}, 1b_2^{-1})$ (c), $^1A_2\,(1b_2^{-1}, 1b_1^{-1})$ (d).

TABLE 2. Anisotropy Parameters $A_{KQ}(\Gamma, M_s = +\frac{1}{2})$ in 10^{-4} a.u. for H_2O

(K,Q)	a) $^1A_1$	b) $^1B_1$	c) $^1B_2$	d) $^1A_2$
(0,0)	0.4434	0.5035	0.3860	0.4202
(1,0)	-0.0569	-0.1262	-0.1253	0.0002
(2,0)	-0.0734	0.3049	0.2085	-0.6004
(2,±2)	0.3290	0.4517	-0.3514	-0.0000
(3,0)	-0.0752	0.1262	0.1258	-0.0003
(3,±2)	0.0448	-0.1154	0.1162	-0.0002
(4,0)	0.5089	-0.8084	-0.5947	0.1801
(4,±2)	-0.4824	0.6391	-0.4702	0.0000
(4,±4)	0.5017	0.0000	0.0000	-0.7535
(5,0)	-0.0025	-0.0001	0.0014	0.0001
(5,±2)	-0.0012	-0.0000	0.0007	0.0002
(5,±4)	0.0035	-0.0002	-0.0029	-0.0004
(6,0)	0.0009	-0.0000	-0.0000	0.0000
(6,±2)	0.0009	-0.0000	0.0001	-0.0000
(6,±4)	-0.0001	0.0001	0.0004	-0.0000
(6,±6)	0.0005	-0.0000	-0.0002	0.0000

H_2O belongs to the point group C_{2v} and consequently the conclusions of the previous section hold true. Actually all calculated anisotropy parameters with Q odd vanish and are therefore not listed in table 2. In general we can see that the magnitude of the anisotropy parameters decreases for increasing K due to the fact that the transition matrix elements diminish for large values of l.

ACKNOWLEDGMENTS

The numerical calculations of the transition matrix elements have been carried out in collaboration with Dr. B. Schimmelpfennig, Dr. B. Nestmann, and Prof.Dr. S.D. Peyerimhoff, University of Bonn. This work has been supported by the Deutsche Forschungsgemeinschaft (DFG) in Sonderforschungsbereich 216 "Polarization and Correlation in Atomic Collision Complexes". B.L. is thankful to the DFG for a Habilitation grant.

REFERENCES

1. Aksela, S., Sairanen, O. P., Aksela, H., Bancroft, G. M., Tan, K. H., *Phys. Rev. A* **37**, 2934 (1988)
2. Schimmelpfennig, B., Nestmann, B. M., Peyerimhoff. S. D., *J. Electr. Spectr. Relat. Phen.* **74**, 173 (1995)
3. Zähringer, K., Meyer, H. D., Cederbaum, L. S., Tarantelli, F., Sgamellotti, A., *Chem. Phys. Lett.* **206**, 247 (1993)
4. Siegbahn, H., Asplund, L., Kelfve, P., *Chem. Phys. Lett.* **35**, 330 (1975)
5. Carravetta, V., Ågren, H., *Phys. Rev. A* **35**, 1022 (1987)
6. Økland, M. T., Fægri Jr., K., Manne, R., *Chem. Phys. Lett.* **40**, 185 (1976)
7. Chandra, N., *Phys. Rev. A* **40**, 752 (1989)
8. Chandra, N., Chakraborty, M., *J. Chem. Phys.* **97**, 236 (1992)
9. Dill, D., Swanson, J. R., Wallace, S., Dehmer, J. L., *Phys. Rev. Lett.* **45**, 1393 (1980)
10. Bonhoff, K., Nahrup, S., Lohmann, B., Blum, K., *J. Chem. Phys.* **104**, 7921 (1996)
11. Lehmann, J., Blum, K., *J. Phys. B*, to be published (1996)
12. Mehlhorn, W., X-Ray and inner shell processes, in *AIP Conf. Proc. 215*, New York, (1990)
13. Blum, K., *Density Matrix Theory and Applications*, 2nd ed., to be published, New York: Plenum Press (1996)
14. Zare, R. N., *Angular Momentum*, New York: Wiley (1988)
15. Michl, J., Thulstrup, E. W., *Spectroscopy with polarized light*, New York: VCH Publ. (1986)
16. Lohmann, B., Bonhoff, S., Bonhoff, K., Lehmann, J., Blum, K., "Angular Distribution and Spin Polarization of Molecular Auger Processes", in *Conf. Proc. of the 14th CAARI'96*, to be published (1997)
17. Kettle, S. F. A., *Symmetry and Structure*, Chichester: Wiley (1995)
18. Tinkham, M., *Group Theory and Quantum Mechanics*, New York: McGraw-Hill (1964)

AUGER-ELECTRON SPECTROSCOPY IN SLOW HIGHLY CHARGED ION-ATOM COLLISIONS : DOUBLE CAPTURE AND STABILIZATION

H. Merabet
Department of Physics, University of Nevada, Reno, NV 89557 USA

J.-Y. Chesnel, F. Frémont, G. Cremer, C. Bedouet, X. Husson and D. Lecler
Laboratoire de Spectroscopie Atomique, ISMRa, Bd Maréchal Juin, F-14050 Caen cedex France

A. Spieler, M. Grether and N. Stolterfoht
Hahn- Meitner-Institut GmbH, Bereich Festkorperphysik, Glienickerstrasse 100, D-14109 Berlin, Germany

The method of Auger electron spectroscopy was used to measure the cross sections for Auger decay of the configurations $3\ell n\ell'$ and $4\ell n\ell'$ $(n \geq 4)$ created in the collision system Ne^{10+} + He at impact energies ranging from 10 keV to 250 keV. Auger yields were calculated to convert the experimental cross sections for Auger-electron emission into the corresponding cross sections for double electron capture. The different contributions to stabilization were extracted from the experimental double-capture cross sections and calculated fluorescence yields. For all energies, radiative stabilization is found to be about 0.3 when referred to the total double capture. At energies lower than 100 keV, the major contribution to the stabilization originates from the states $3\ell n\ell'$ $(n \geq 6)$ which are populated by the collisional dielectronic process of autoexcitation. A small contribution is found to be due to the configurations $3\ell n\ell' (n > 9)$ created by dielectronic process phenomena in the postcollisional and asymptotic regions (≈ 0.04).

INTRODUCTION

The interaction of slow, highly charged ions with neutral atomic targets has attracted a great deal of interest during the last decade. In particular, increasing attention has been devoted to double-electron capture (1 -5). Such a process leads to doubly excited states which decay either by Auger electron emission or by photon emission. Therefore, for the method of Auger spectroscopy, we have to take into account the competing decay when determining absolute cross sections for double electron capture. On the other hand, the photon emission gives rise to the radiative stabilization of the electrons on the projectile ions.

Generally, Auger decay is dominant when the configurations of equivalent or near-equivalent electrons $n\ell n'\ell'$ $(n \approx n')$ are populated during the collision. Hence, the related fluorescence yields are rather weak. This is the case for the configurations $3\ell 3\ell'$ and $3\ell 4\ell'$ of C^{4+} (4). The situation is quite different for non-equivalent electron configurations which include a core electron and a highly excited Rydberg electron. For example this is the case for the configurations $2\ell n\ell'$ in C^{4+}, in which the radiative decay branch is found to be as large as 80 % for $n = 7$ (5). Therefore, photon emission becomes the most important decay process for the configurations of non-equivalent electrons.

Recently, several experiments have been performed to study the radiative stabilization mechanisms (6 ,7). Roncin *et al.* (6) quoted that the fluorescence yields are relatively large for equivalent electron configurations in many systems. To explain such large contributions to radiative stabilization, Bachau *et al.* (7) have proposed a postcollisional autoexcitation process, referred to as auto transfer to Rydberg states (ATR), where the two-electron transition occurs at one atomic center. However, these authors found that the experimental values for radiative stabilization are generally higher than those expected from their calculations.

To explain the high amount of radiative stabilization, other dielectronic processes have to be taken into account as shown by Vaeck and Hansen (8), Merabet *et al.* (9) and Chesnel *et al.* (10). Indeed, various dielectronic mechanisms are responsible for the transfer to high Rydberg states. In addition to the collisional autoexcitation process (AE), atomic configuration interaction (CI) of equivalent electrons and nonequivalent electrons may enhance stabilization. More information about the mechanisms can be found in the work of Chesnel *et al.* (10).

In the present paper, the different contributions to radiative stabilization in the collision system Ne^{10+} + He at impact energies ranging from 10 keV to 250 keV, are discussed. Furthermore, the attempt is made to give a comparative study of the role of the dielectronic processes in the stabilization. In this work, experimental cross sections are combined with theoretical results in order to

CP392, *Application of Accelerators in Research and Industry*, edited by J. L. Duggan and I. L. Morgan
AIP Press, New York © 1997

evaluate the radiative stabilization ratio. Moreover, the contributions of the radiative stabilization mechanisms are determined at different collision energies.

RESULTS AND DISCUSSION

The measurements at the projectile energies between 30 keV and 250 keV were carried out at the 14-GHz Electron Cyclotron Resonance Ions Sources at the Grand Accélérateur d'Ions Lourds (GANIL) in Caen. The measurement with low beam energy (i. e. 10 keV) was performed at the Hahn-Meitner Institut (HMI) in Berlin. The experimental set up and the method to evaluate cross sections have been developed previously (11). Auger electrons produced in the collision were detected at several angles up to 140° with respect to the incident-beam direction.

Figure 1 exhibits typical L-shell Auger spectra at the projectile energies of 10, 150 and 250 keV, and at the observation angle of 40°. The most prominent lines are attributed to the configurations $3\ell n\ell'$, which are well separated from $n=4$ to 9. The line group centered at 180 eV can be attributed to the limit of the $3\ell n\ell'$ Rydberg series and to the configurations $4\ell 4\ell'$, decaying to the $2\ell\varepsilon\ell'$ configurations by mean of Auger transitions. In addition, the configurations $4\ell n\ell' (n=5-7)$ are observed in the electron-energy range from 20 to 50 eV which is not included in Fig. 1.

The average L-Auger and M-Auger yields $\xi_{n,n'}$ for the populated configurations $3\ell n\ell' (n=4-9)$ and $4\ell n\ell' (n=4-7)$ were calculated by means of the Hartree-Fock code by Cowan (12). Transition rates for radiative and nonradiative decays were used to calculate the individual Auger yields $\xi(n\ell n'\ell'\gamma\ J_\gamma)$ (9,13). Then, the average Auger yields for a given n and n' were obtained using the expression

$$\xi_{n,n'} = \sum_{\ell,\ell',J,\gamma} Q_{n,n'}(\ell,\ell',\gamma,J_\gamma)\,\xi(n\ell n'\ell'\gamma\ J_\gamma) \quad (1)$$

where $Q_{n,n'}(\ell,\ell',\gamma,J_\gamma)$ is the probability for the production of the singlet state $\left| n\ell n\ell'\gamma\ J_\gamma \right\rangle$ for given n and n' with the normalization $\sum_{\ell,\ell',\gamma} Q_{n,n'}(\ell,\ell',\gamma,J_\gamma)=1$. A simple model was adopted in which this probability is factorized (14),

$$Q_{n,n'}(\ell,\ell',\gamma,J_\gamma) = q_n(\ell)\,q_{n'}(\ell')\,p(J_\gamma)\,s(\gamma), \quad (2)$$

where $q_n(\ell)$, $q_{n'}(\ell')$ and $p(J_\gamma)$ are the occupation probabilities associated with the quantum numbers ℓ, ℓ' and J_γ, respectively, and $s(\gamma)$ is the squared coefficient of the singlet component of the intermediate coupling state γ. The probability $p(J_\gamma)$ was set to be proportional to $2J_\gamma+1$. The probabilities $q_n(\ell)$ and $q_{n'}(\ell')$ were first, estimated using the model by Burgdörfer *et al.* (15) in the case of $150-\text{keV}\ Ne^{10+}+He$ collisions. Then, various distributions $q_n(\ell)$ and $q_{n'}(\ell')$ (see Fig. 2) were used in order to test the variation of the average Auger yields with respect to the collision energy. For example, in accordance with experimental results of Meyer *et al.* (16), the population of high-angular momentum states was included in the probability $q_{n'}(\ell')$. It is important to note that even when extreme hypotheses are chosen for the probabilities $q_n(\ell)$ and $q_{n'}(\ell')$ as shown in Fig. 2, no significant variation is found for the average Auger yields $\xi_{n,n'}$. Hence, these latter are quite insensitive to the distributions $q_n(\ell)$ and $q_{n'}(\ell')$.

Now we focus our attention on the determination of different contributions to radiative stabilization. First, the calculated average L-Auger and M-Auger yields were used to determine the total double-capture cross sections $\sigma_{n,n'}$ by dividing the Auger emission cross sections by the related average Auger yields $\xi_{n,n'}$ (Table 1). Then, the different contributions to radiative stabilization are extracted using the experimental cross sections :

$$y^X_{Mono} = \sigma_{n,n'} \cdot \omega_{n,n'} / \sigma_{tot} \quad (3)$$

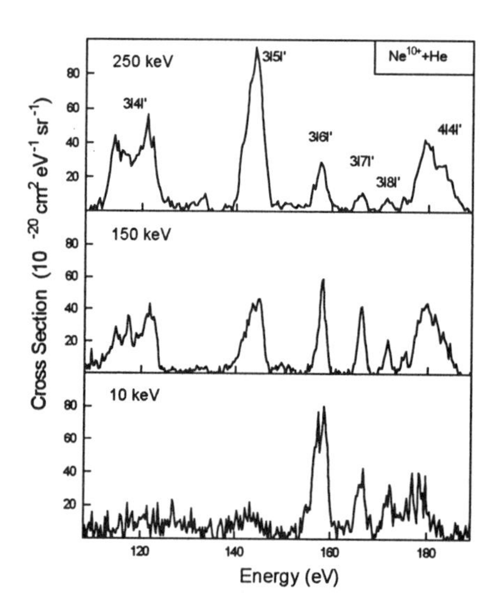

FIGURE 1 : High resolution spectra of L-shell Auger electrons produced in $Ne^{10+}+He$ collisions at projectile energies of 10 keV, 150 keV and 250 keV. Each peak corresponds to the Auger decay of states associated with a configuration $3\ell n\ell'$ (n=4-9). The peak centered at 180 eV corresponds to the limit of the $3\ell n\ell'$ series and to the configurations $4\ell 4\ell'$ which decay to the $2\ell\varepsilon\ell'$ configurations.

$$y^X_{AE} = \sigma_{3,n} . \omega_{3,n} / \sigma_{tot} \quad (4.a)$$

$$y^X_{CI} = \sigma_{4,4} . \tau . \omega_{3,n>9} / \sigma_{tot} \quad (4.b)$$

where the quantity y^X_{Mono} refers to the contribution which follows from the decay of the near-equivalent electron configurations $3\ell n\ell'$ (n=4-5) and $4\ell n\ell'$ (n=4-6) mainly

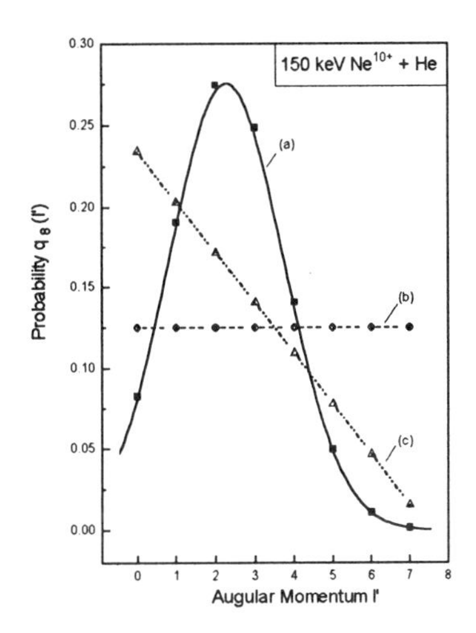

FIGURE 2 : Different distributions $q_{n'}(\ell')$ used in the calculation of the average Auger yields. The example of n'= 8 is shown. The label "a" refers to the model of Burgdörfer *et al.*(15). The labels "b" and "c" denote a uniform and linear decreasing distributions $q_8(\ell')$, respectively.

produced by monoelectronic processes. The contribution originating from the decay of the non-equivalent electron configurations $3\ell n\ell'$ (n=6 to 9) created by the dielectronic autoexcitation process AE is referred to as y^X_{AE}. The quantity y^X_{CI} is the contribution to stabilization due to photon emission from the high Rydberg states $3\ell n\ell'$ produced in the post-collisional and asymptotic regions.

In the expressions (3) and (4), σ_{tot} is the total cross section including all double capture states, $\omega_{n,n'}$ is the average fluorescence yield associated with a given configuration $n\ell n'\ell'$ summed over ℓ and ℓ', and the quantity τ is the total fraction of the initial $4\ell 4\ell'$ population which dilutes into the nonequivalent electron configurations $3\ell n\ell'$ by configuration interactions (CI). Recently, a fraction $\tau \approx 0.6$ has been found by Sánchez and Bachau (17,18). Equation (3) refers to the uncorrelated two-electron transfers and Equations (4) to the dielectronic processes (10). It is pointed out here that the radiative stabilization is given by the sum of the individual contributions as

$$S= y^X_{Mono} + y^X_{AE} + y^X_{CI} \quad (5)$$

Table 2 gives the results obtained for the different contributions to radiative stabilization for impact energies of 10 keV, 150 keV and 250 keV. The radiative stabilization value is found to be about 0.3 for all the collision energies (Fig. 3). In the case of 250 keV Ne^{10+} + He collisions, the decay of equivalent electron configurations provides the largest contribution to radiative stabilization. The contribution associated with the equivalent electrons is still significant at 10 keV ($\approx$0.1) (Table 2). On the other hand, at impact energies below 150 keV, a significant contribution to radiative stabilization originates from the configurations $3\ell n\ell'$ ($n \geq 6$) created by the collisional process of autoexcitation. This contribution increases strongly with decreasing projectile energy so as to become the dominant contribution to stabilization (0.24) at 10 keV. However, the decay of the configurations $3\ell n\ell'$($n > 9$) due to postcollisional and asymptotic CI processes gives rise to the smallest contribution to stabilization for the different

TABLE 1. Total Auger-electron emission cross sections $\sigma^a_{n,n'}$ and total double-electron capture cross sections $\sigma_{n,n'} = \sigma^a_{n,n'} / \xi_{n,n'}$ obtained for the configurations of doubly-excited states $3\ell n\ell'$ and $4\ell n\ell'$ ($n \geq 4$) produced in 10-keV, 150 keV and 250-keV Ne^{10+} + He collisions. The experimental uncertainties are ± 20% (10). The corresponding Auger yields $\xi_{n,n'}$ and fluorescence yields $\omega_{n,n'}$, which were previously calculated (13) are also given. The method used to evaluate $\sigma_{4,4}$ is given in Ref. (10).

Configurations	$\xi_{n,n'}$	$\omega_{n,n'}$	$\sigma^a_{n,n'}$ (10^{-17} cm^2)			$\sigma_{n,n'}$ (10^{-17} cm^2)		
$n\ell n'\ell'$			10 keV	150 keV	250 keV	10 keV	150 keV	250 keV
$3\ell 4\ell'$	0.75	0.25	1.25	4.07	5.43	1.66	5.43	7.24
$3\ell 5\ell'$	0.67	0.33	1.32	2.99	5.43	1.97	4.46	8.10
$3\ell 6\ell'$	0.60	0.40	3.77	1.50	1.15	6.28	2.52	1.91
$3\ell 7\ell'$	0.56	0.44	1.32	1.02	0.34	2.36	1.82	0.61
$3\ell 8\ell'$	0.52	0.48	0.88	0.50	0.22	1.69	0.96	0.43
$3\ell 9\ell'$	0.47	0.53	0.44	0.33	0.17	0.93	0.71	0.26
$3\ell n\ell'(n > 9)$	0.45	0.55	0.44	0.41	0.18	0.97	0.92	0.30
$4\ell 4\ell'$	0.90(19)	0.10 (19)	-	-	-	1.57	6.05	5.50
$4\ell 5\ell'$	0.91	0.09	3.64	9.80	11.06	4.00	10.76	12.15
$4\ell 6\ell'$	0.79	0.21	0.87	2.35	3.19	1.10	2.97	4.04

TABLE 2. Contributions to radiative stabilization of doubly-excited Ne^{8+} ions originating from the 10-keV, 150 keV and 250-keV Ne^{10+} + He collisions. These contributions y^X are due to the decay of the doubly-excited states $n\ell n'\ell'$ which are produced by the different mechanisms. In the first column, the label "Mono" refers to mechanisms involving monoelectronic processes and, AE and CI refer to dielectronic processes. In the last row, the fraction S of ions Ne^{8+} which stabilize radiatively with respect to the total number of doubly-excited ions Ne^{8+} is given.

Mechanisms	Configurations	Contributions to stabilization y^X		
	$n\ell n'\ell'$	10 keV	150 keV	250 keV
Mono	$3\ell n\ell'$ ($n = 4-5$) and $4\ell n\ell'$ ($n = 4-6$)	0.08	0.14	0.20
AE	$3\ell n\ell'$ ($n \geq 6$)	0.24	0.09	0.03
CI	$4\ell 4\ell \rightarrow 3\ell n\ell'$ ($n > 9$)	0.02	0.05	0.05
Stabilization S		0.34	0.29	0.28

collision energies. This is in clear disagreement with the conclusion by Roncin *et al.* (6) who claimed that the postcollisional CI provides the major contribution to stabilization in the 10 keV-Ne^{10+} + He system.

In conclusion, our analysis shows that the major contribution to stabilization of the system Ne^{10+} + He is due to the decay of the states $3\ell n\ell'$ ($n \geq 4$) configurations which are populated during the collision (Fig. 3). On the contrary, the contribution of the postcollisional and asymptotic processes (CI) is found to be negligible for the whole range of collision energies. This can be related to the fact that the cross sections for producing the configurations $4\ell 4\ell'$ are rather small in comparison with the corresponding cross sections for $4\ell 5\ell'$ and $3\ell n\ell'$. Furthermore, for the determination of the average fluorescence yields, we found that, within the uncertainti-es of our calculations (9,13), the occupation probabilities (see Eq. 2) do not depend on the collision energy. Therefore, the main conclusions of the present study are not affected by the uncertainties of the fluorescence and Auger yields.

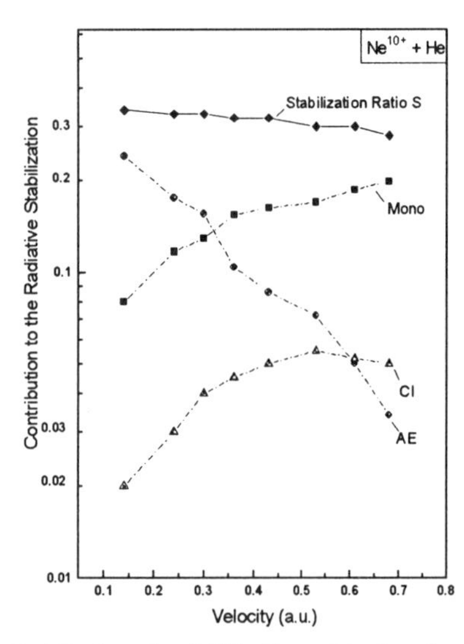

FIGURE 3 : Contributions to stabilization originating from radiative decay of doubly-excited states produced by monoelectronic and dielectronic processes in Ne^{10+} + He system as a function of the projectile velocity. The label "Mono" denotes the contribution associated with the uncorrelated processes dominantly responsible for the production of equivalent electron configurations $3\ell n\ell'$ (n=4-5) and $4\ell n\ell'$ (n=4-6) during the collision. The labels AE and CI refer to the dielectronic processes responsible for the creation of non-equivalent electron configurations $3\ell n\ell'$ in the collisional region, post-collisional region and asymptotic region, respectively.

ACKNOWLEDGMENTS

We gratefully acknowledge the staffs of the ECR sources in Berlin and Caen for their generous assistance. We are indebted to Drs. J. E. Hansen and H. Bachau for stimulating discussions. We thank Drs. R. Ali and R. Bruch for helpful comments. H. Merabet acknowledges the travel support of the Nevada NSF EPSCoR program.

REFERENCES

1 . Bordenave-Monterquieu, A., Benoit-Cattin, P., Gleizes, A., Dousson, S., and Hiltz ,D., *J. Phys. B* **18**, L195 (1985).
2 . Stolterfoht, N., Havener, C.C., Phaneuf, R. A., Swenson, J. K., Shafroth, S. M., and Meyer, F.W., *Phys. Rev. Lett.* **57**, 74 (1986).
3 . Mack, M., and Niehaus, A., *Nucl. Instr. Meth. B* **23**, 116 (1987).
4 . Van der Hart, H. W., and Hansen, J. E., *J. Phys. B* **26**, 641 (1993).
5 . Stolterfoht, N., Sommer, K., Swenson, J. K., Havener, C. C., and Meyer, F. W., *Phys.Rev. A* **42**, 5396 (1990).
6 . Roncin, P., Gaboriaud, M. N., and Barat, M., *Europhys. Lett.* **16**, 551 (1991) ; Roncin *et al., Proc. 18th International Conference of the Physics of Electronic and Atomic Collisions*, (Aarhus 1993).
7 . Bachau, H., Roncin, P. and Harel, C., *J. Phys. B* **25**, L109 (1992).
8 . Vaeck, N., and Hansen, J. E., *J. Phys. B* **26**, 2977 (1993).
9 . Merabet, H., Frémont, F., Chesnel, J.-Y., Cremer, G., Husson, X., Lecler, D., Lepoutre, A., Rieger, G. and Stolterfoht, N., *Nucl. Instr. and Meth. B* **99**, 75 (1995) ; Merabet, H., *Ph.D thesis, Universite de Caen, France* (1996).
10 . Chesnel, J.-Y., Merabet, H., Frémont, F., Cremer, G., Husson, X., Lecler, D., Rieger, G., Spieler, A., Grether, M., and Stolterfoht, N., *Phys. Rev. A* **53**, 4198 (1996).
11 . Stolterfoht, N., *Z. Phys.* **248**, 81 (1971); **248**, 92 (1971).
12 . Cowan, R. D., *The Theory of Atomic Structure and Spectra*, (University of California Press, Berkeley, 1981).
13 . Merabet, H., Cremer, G., Frémont, F., Chesnel, J.-Y., and Stolterfoht, N., *Phys. Rev. A* **54**, 372 (1996).
14 . Stolterfoht, N., Sommer, K., Swenson, J. K., Havener, C. C. and Meyer, F. W., *Phys. Rev. A* **42**, 5396 (1990).
15 . Burgdörfer, J., Morgenstern, R. and Niehaus, A., *J. Phys. B* **19**, L507 (1987).
16 . Meyer, F. W., Griffin, D. C., Havener, C. C., Huq, M. S., Phaneuf, R. A., Swenson, J. K. and Stolterfoht, N., *Phys. Rev. Lett.* **60**, 1821 (1988).
17 . Sánchez, I., and Bachau, H., *J. Phys. B.* **28**, 795 (1995).
18 . Bachau, H., private communication (1996).
19 . Hansen, J. E., Vaeck, N., and Van der Hart, H. W., private comminucation (1996).

CALCULATION OF SADDLE-POINT ELECTRON DISTRIBUTIONS IN ION-ATOM COLLISIONS

Marko Horbatsch

Department of Physics, York University, Toronto, Ontario M3J 1P3, Canada

The time-dependent Schrödinger equation for the electronic motion in ion-atom collisions is solved numerically in a discretized coordinate space representation. The wavefunction is analyzed for its continuum contributions by a projection/ Fourier transform technique. The scaling behaviour of saddle-point electron distributions with the target charge parameter is investigated.

Recently progress has been made in recoil-ion momentum spectroscopy to obtain differential information about ionized electron momenta transverse and longitudinal to the beam axis in coincidence with the transverse recoil ion momentum [1]. These data are obtained e.g. for proton-helium collisions at low to intermediate energies, and should be directly comparable to calculated electron distributions for nuclear trajectories with fixed impact parameter.

Classical trajectory (CTMC) calculations describe the experimental data with moderate success [1]. While they can show that the collisions are violent for small impact parameter, and that a majority of ionized electrons appear in a velocity region around the saddle velocity (the velocity with which the saddle point in the classical two-centre potential moves), they fail to account for some salient features in the experimental electron distributions. In particular, experiment reveals a significant variation in the emission in the transverse direction as a collision parameter is changed. Electrons can emerge either on the same side as the recoil or opposite to it (with the reverse situation for the projectile), and the behaviour appears to vary dramatically as one varies the collision energy.

Nonperturbative quantal calculations of the phenomenon were reported for the (experimentally less interesting) case of zero impact parameter in the p-H system [2]. We have embarked in a novel program to solve the time-dependent Schrödinger equation (TDSE) on a mesh, and to come up with a method to analyze the three-dimensional wavefunction after the collision [3].

Our findings were that three collision mechanisms can be operating at the same time: *(i)* population of target continuum states (which may deexcite by radiative capture not included in the calculation); *(ii)* capture to the continuum (which may also combine radiatively); and a saddle-type mechanism resulting in electrons with small nonzero transverse momenta (moving either towards or away from the projectile), and longitudinal momenta in the vicinity of the saddle-point velocity.

To explore the comparison with the CTMC calculations which are dominated by the interaction potential we extend in this contribution the proton-hydrogen study to a range of projectile and target charge values. This is to clarify to what extent one is justified in talking about saddle-point electrons, i.e., to understand better whether indeed the region where a classical particle would be field-free has a dominating effect on the electron.

Based on one-dimensional quantum mechanical model calculations [3,4] we can conclude that the saddle region itself cannot be too important. The wavefunction describing the ionized electrons is spread out in coordinate space with a position-dependent local velocity that ranges in the lab frame from zero near the target to values around the projectile velocity v_{P} near the projectile. Ahead of the projectile it exceeds v_{P}, e.g., to describe binary-encounter electrons. The modulus of the wavefunction has a small curvature so that the spreading of the wavepacket is proceeding with growing uncertainty and a time-constant momentum distribution. In the one-dimensional model calculations no particular enhancement is found in the probability density near the saddle-point velocity.

CP392, *Application of Accelerators in Research and Industry*, edited by J. L. Duggan and I. L. Morgan
AIP Press, New York © 1997

The three-dimensional results, however, are in dramatic contrast in this respect. It appears as if the additional transverse degree of freedom permits the electronic wavefunction to be pronounced in its longitudinal momentum dependence. The question that arises in this context is the importance of the quantum pressure term in the Madelung formulation of quantum mechancis. In this formulation the electronic motion can be thought of as being the result of an ensemble evolution according to fluid dynamics (as opposed to phase space dynamics as in CTMC) with the classical potential acting on the particles as well as the density-dependent quantum potential

$$V_q(\mathbf{r}, t) = \frac{1}{2} \frac{\nabla^2 \sqrt{\rho(\mathbf{r}, t)}}{\sqrt{\rho(\mathbf{r}, t)}} \quad (1)$$

This potential is responsible for the spreading of free wavepackets, such as the Gaussian or Lorentzian packets. For an asymptotic electron momentum distribution to emerge at the end of a TD calculation V_q has to become negligible. It is evident that this can be achieved in three dimensions in more ways than in the 1D situation (where the second derivative has to vanish).

Results from CTMC calculations suggest that ionized electrons are to be found at velocities and locations where they are field-free, i.e., near the trajectory in phase space of the equilibrium point in the classical potential. The latter scales in a characteristic way with the projectile and target charges Z_p and Z_t respectively. The saddle velocity is defined as

$$v_s = \frac{v_p}{1 + \sqrt{Z_p/Z_t}} \quad (2)$$

In this paper we analyze the quantum mechanical results in order to test the saddle-point hypothesis.

From a theoretical perspective one is interested in the question of the relevant scale parameters. The considerations based on the classical saddle point alone assume that the location of the maximum in the longitudinal direction depends on the charge values Z_p and Z_t only through the equilibrium point in the two-centre force. Other dependencies may arise, however, both in the quantum and in the semiclassical picture. Simple examples are the binding energies of the electron with respect to either charge, which scale as Z_p^2 and Z_t^2 respectively. Strictly quantal phenomena such as tunneling depend on the entire correlation diagram (in a symmetric system such as p-H on the evolution of the energy level difference between the gerade and ungerade states with the internuclear separation). Thus, it is not clear a priori whether one should expect a simple scaling of the results with either nuclear charge.

The investigation of the distribution of momenta as a function of projectile and target charges is also of interest for the case of many-electron atoms. In a simplistic model of two-electron atoms one can consider spherical charge distributions bound by effective charge values Z_t. Thus, simple hydrogen-like calculations can provide hints as to what to expect, e.g., in single and double ionization of helium atoms.

We solve the TDSE for the electronic motion with a prescribed straight-line nuclear trajectory $\mathbf{R}(t)$. In atomic units (au) it reads

$$i\frac{\partial}{\partial t}\psi(\mathbf{r}, t) = [-\frac{1}{2}\nabla^2 - \frac{Z_t}{r} - \frac{Z_p}{|\mathbf{r} - \mathbf{R}(t)|}]\psi(\mathbf{r}, t) \quad (3)$$

The wavefunction is discretized on a mesh that covers a box of 20 a_0^3 (Bohr radii) in the lab frame with good resolution and then becomes coarser at large distances. This technique which is based on a mapping of the Cartesian coordinates by a tangent transformation allows us to avoid reflections of probability density from artificial boundaries [5]. On the other hand it forces us to analyze the wavefunction soon after the collision - in practice at a final separation of the nuclei of about 20 a_0.

The wavefunction is analyzed in the following way. Bound-state contributions are projected out from the numerical wavefunction so that a state vector $|\chi\rangle$ is produced whose norm equals the ionization probability for the given impact parameter b. Next the mesh representation of this state, i.e., $\langle \mathbf{r}_\nu | \chi \rangle$ ($\nu = \{\nu_x, \nu_y, \nu_z\}$) is interpolated to a uniform spatial mesh and then Fourier transformed to momentum space. This discrete Fourier transform $\langle \mathbf{p}_\mu | \chi \rangle$ is interpreted as the probability amplitude for an electron to emerge ionized with given Cartesian momentum vector components. A projection onto the scattering plane (summation over all possible momenta in the z direction) results in a distribution as shown in fig. 1 for the case of $b = 1$ au p-H(1s) collisions at 15 keV ($v_p = 0.77$ au). The contours interpolate linearly between zero and the maximum value. The impact parameter was chosen to represent a range from which the total ionization cross section receives a major contribution at this energy.

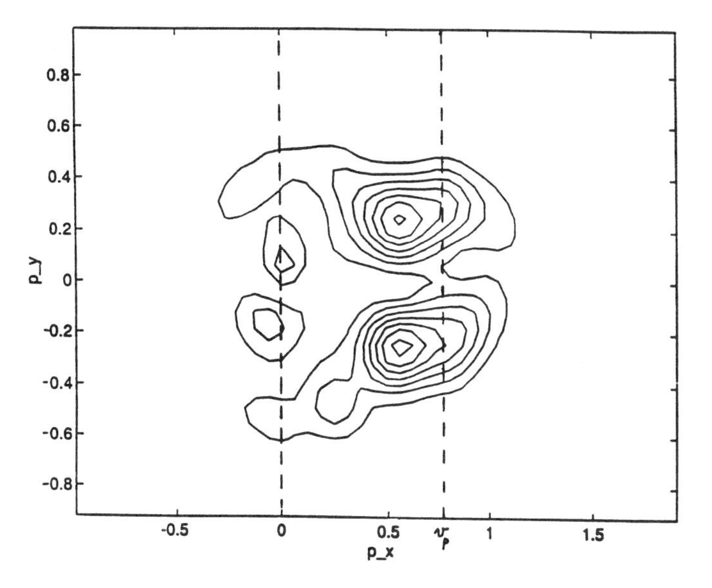

FIGURE 1. Linear contour diagram of the electron emission probability as a function of longitudinal momentum p_x and transverse momentum p_y in the scattering plane defined by the impact parameter. Upper half-plane: motion towards the projectile, lower half-plane: opposite from projectile. $b = 1$ au, $v_p = 0.77$ au (15 keV); $Z_p = 1$, $Z_t = 1$. The total ionization probability amounts to 0.17

One of the difficulties with the technique becomes apparent when one realizes typical momenta in bound states (for projectile or target). The distinction of the rest frames associated with the target, projectile or the saddle region is smaller than the typical distribution of momenta in bound states (except for the higher Rydberg states with small binding energies). It is necessary to perform a subtraction of all bound states with occupation probabilites larger than a few percent of the total ionization probability in order to avoid spurious behaviour in the contour diagram. Of course, the bound-state contributions from the numerous shells and subshells vary as the charge parameters change: a well-known example is the resonant capture to excited levels in He^{++}-H(1s) collisions.

One would like to vary the charges over a wide range. However, care must be taken to ensure that the discretized version of the SE is capable to represent the relevant states. Obviously, only a finite number of the Rydberg levels exists on any mesh, and one has to make a compromise between a reasonably accurate representation of the ground state and the total number of bound states included. Therefore, we performed at first a variation of the charge parameters on a very modest scale.

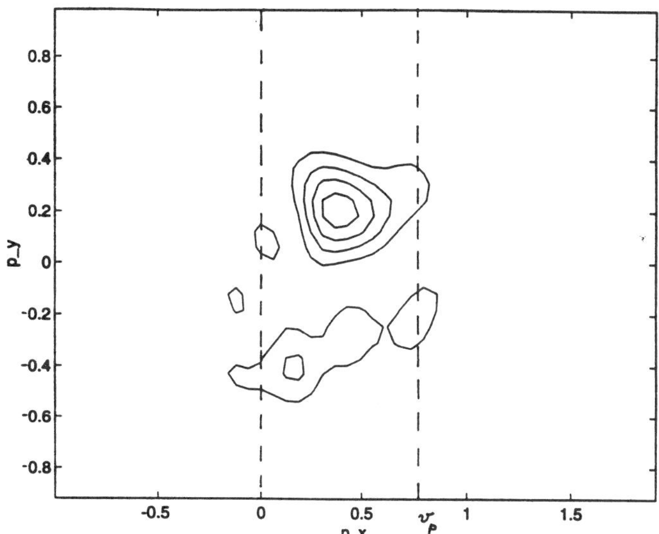

FIGURE 2. Same as in fig. 1 but for an increased target charge, i.e., $b = 1$ au, $v_p = 0.77$ au (15 keV); $Z_p = 1$, $Z_t = 1.4$. The total ionization probability amounts to 0.08

To illustrate the qualitative change in behaviour as the target charge is varied, fig. 2 provides the analogue of fig. 1, but for the charge-asymmetric case $Z_p = 1$, $Z_t = 1.4$ (which could serve as a naive calculation of single ionization of helium). It can be seen that the position of the maximum in the longitudinal direction has shifted towards the point $0.5v_p$. This is in agreement with the experimental findings of ref. [1]. We also note that the distribution is no longer symmetric in the transverse direction in the scattering plane, and that the total ionization probability represented in fig. 2 is about one-half of the value for the p-H(1s) case.

We have extracted the position of the maximum in the longitudinal direction as a function of the target charge Z_t. Fig. 3 displays the results of the position of the maximum in the longitudinal direction from several calculations and compares it with the saddle-point scaling prediction. It is evident that the results are in strong contrast with each other. This calls into question the idea that ionized electrons exist predominantly in the force-free zone, i.e., near the equilibrium point.

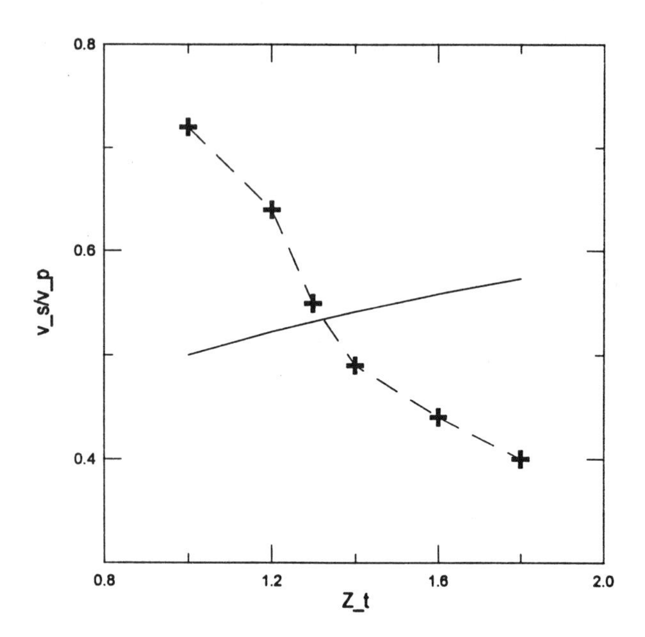

FIGURE 3. Variation of the peak value of longitudinal momentum in 15 keV, $b = 1$ au proton - hydrogen-like atom collisions expressed as the ratio v_s/v_p as a function of the target charge Z_t (crosses connected by dashed line). Solid curve: prediciton of eq. (2).

One can conclude that the numerical integration of the TDSE on a mesh can provide interesting insights into the problem. A quantity of interest would be a graph of the probabilty density for the ionized electrons in coordinate space in order to corroborate the finding that the saddle point itself in the classical potential does not play a major role.

Further work will include the investigation of the dependence on the projectile charge (there are some interesting experimental findings on that problem [6]). Furthermore it is of interest to investigate other initial states than the hydrogenic ground state. The use of effective potentials that incorporate screening, i.e., are not simply Coulombic, as well as different intial states than the ground state will permit an extension to study many-electron systems in an independent electron model.

ACKNOWLEDGMENT

I thank J. Ullrich, R. Dörner, C.L. Cocke, as well as H.J. Lüdde and R.M. Dreizler for discussions. The financial support by NSERC of Canada is greatly appreciated.

REFERENCES

1. Dörner R., *et al* , Phys. Rev. Lett., in press (1996).
2. Ovchinnikov S., Macek J., Phys. Rev. Lett. **75**, 2474 (1995).
3. Horbatsch M., Phys. Lett. **A**, in press (1996).
4. Chassid M., Horbatsch M., J. Phys. **B 28**, L 621 (1995).
5. Horbatsch M., Phys Rev. **A 44**, R5346 (1991).
6. Kravis S.D., *et al* , Phys. Rev. **A 54**, 1394 (1996).

MEASUREMENTS OF SINGLE ELECTRON CAPTURE CROSS SECTIONS IN COLLISIONS OF Kr^+ WITH Xe

H. Martínez, C. Cisneros, I. Dominguez and I. Alvarez

Instituto de Física, Laboratorio de Cuernavaca, UNAM, Apartado Postal 48-3, 62191, Cuernavaca, Morelos, México.

Absolute differential and total cross sections for single electron capture were measured from Kr^+ ions on Xe in the energy range of 1.0 to 5.0 keV. The absolute total cross section shows an oscillatory behaviour. The total cross section is compared with other available measurements and with a semiempirical model. The results are in good agreement with that model as well as the values measured at low energies.

INTRODUCTION

Electron capture processes of Kr^+ ions in collisions with Xe are of great importance in the fields of thermonuclear fusion research and astrophysics. There has been a continuous interest in energy-transfer reactions between rare gas metastable atoms and rare gas ions because of their application to the pumping of rare gas ion lasers (1, 2). Most experiments concerned with the measurement of single electron capture cross sections for Kr^+ (3) have been performed at energies below 350 eV. On the other hand, data have not been published on Kr^+-Xe collisions in the low keV region. This results have in some sense motivated the present study. In particular, the reason for choosing this system for study was that there are not sets of measurements above 1.0 keV.

We report absolute measurements of the differential and total cross sections of single electron capture for Kr^+ collisions with Xe atoms. In addition, we present the results of a calculation based on the Olson (4) theoretical analyses for single electron capture. The energy range of the present study is 1.0 to 5.0 keV. This energy range includes the low-energy region where a quasimolecular picture of the collision is appropriate, as well as the region where molecular level crossings associated with the evolution of potential states of the collision system play a critical part in the dynamics of the collision.

EXPERIMENT

Details of the experimental arrangement have been given previously (5, 6), so only a brief survey of the experimental setup will be given here. A schematic diagram of the apparatus is shown in fig 1. Kr^+ ions were formed in a colutron-type ion source and accelerated in the energy range of 1.0 to 5.0 keV. The selected ion beam was velocity analyzed by a Wien filter, passed through cylindrical plates to deviate it by 10°, and through a series of collimators before it entered the gas target cell. This cell consists of a cylinder of 2.5 cm in length and diameter, with a 1-mm entrance aperture, and a 2-mm wide, 6-mm long exit aperture. All apertures and slits had knife edges. The target cell was located at the center of a rotatable, computer controlled vacuum chamber that moved the whole detector assembly which was located 47 cm away from the target cell. A precision stepping motor ensured a high repeatability in the positioning of the chamber over a large series of measurements. The detector chamber housed a Harrower-type parallel plate electrostatic analyzer, located at 45^0 with respect to the incoming beam direction, with two channel electron multipliers (CEMs). The Kr^0 atoms formed by electron capture passed straight the analyzer through a 1-cm orifice on its rear plate, and impinged on a CEM so that the neutral counting rate could be measured. Separation of charged particles occurred inside the analyzer, which was set to detect the Kr^+ species with a second CEM. This flux was used as a measure of the stability of the beam during the experiment. To measure the angular distributions, a 0.36-mm diameter pinhole was located at the entrance of the analyzer. Path lengths and apertures gave an overall angular resolution of the system of 0.1°. The target thickness was $\approx 10^{13}$ atoms/cm^2 in order to ensure a single collision regime. Absolute gas pressures in the cell were measured by a capacitance manometer. The absolute differential cross section was calculated from the relation:

$$\frac{d\sigma}{d\Omega} = \frac{I(\theta)}{nLI_0} \qquad (1)$$

where I_0 is the number of Kr^+ ions incident per second on the target, n, the number of Xe atoms per unit volume, L,

CP392, *Application of Accelerators in Research and Industry,* edited by J. L. Duggan and I. L. Morgan
AIP Press, New York © 1997

the effective length of the scattering chamber, and $I(\theta)$, the number of Kr° counts per unit solid angle per second detected at laboratory angle (θ) with respect to the incident beam direction. The total cross section was derived by integrating the differential cross section over the solid angle $d\Omega$:

$$\sigma = 2\pi \int_0^{\pi} \frac{d\sigma}{d\Omega} \sin(\theta) d\theta \qquad (2)$$

The Kr^{+} beam intensity was measured before and after each angular scan. Measurements not agreeing to within 5% were discarded. Angular distributions were measured on both sides of the forward direction to assure they were symmetric. The estimated rms error is 15%, while the cross sections were reproducible to within 10% from day to day.

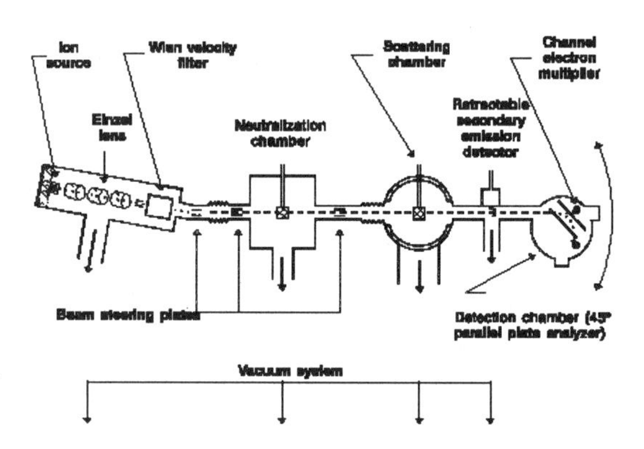

FIGURE 1. Schematic diagram of the apparatus.

RESULTS AND DISCUSSION

In order to get a direct comparison of data obtained at different values of the energy, the reduced variable $\tau = E_0\theta$ was used, where E_0 is the impact energy and θ the scattering angle. The reduced cross section is the quantity $\rho(\tau) = (d\sigma/d\Omega)\theta\sin(\theta)$. Reduced differential cross sections for single electron capture of Kr^{+} ions in Xe for several values of the incident energy are presented in Fig. 2. $\rho(\tau)$ has a rather similar behaviour at all energies. It increases rapidly with τ up to a maximum which is located near 1.5 keV-deg. This maximum becomes sharper as the energy decreases and is followed, in some cases, by very slight oscillations which confirm the existence of a curve crossing; it then decreases uniformly. An approximate value of the internuclear distance R_c corresponding to this crossing can be determined from the value of τ, by using an exponentially shielded Coulomb potential (7). This gives a value of $R_c = 1.6\ a_0$. Our results of the total cross sections for single electron capture in krypton are shown in Fig. 3. In order to obtain a preliminary understanding of the trend of the total cross sections, the charge transfer cross section was discussed in terms of the model for near-resonant charge transfer proposed by Olson (4). Since no theoretical potentials for $KrXe^{+}$ have been published so far, the total single electron cross sections are then calculated using the following parameters: the crossing distance $R_c = 1.6\ a_0$ was taken from the experimental reduced differential cross sections; the coupling matrix element $H_{12} = 0.007$ a.u. was calculated through the expression: $H_{12}(R_c) = R^{*} \exp(0.86R^{*})$, where $R^{*} = (\alpha+\gamma)R_c/2$. We used $\alpha^2/2 = 12.127$ eV as the effective ionization potential of the target and $\gamma^2/2 = 13.999$ eV as the ground state electron affinity (8). $|\Delta V'| = 4.26$ a.u.a_0^{-1} , which was fitted until the same value of the experimental cross section at 3 keV was obtained, together with the universal reduced cross section of Olson (4). The results of this calculation are shown in Fig. 3. Although the Olson model calculations (4) are not expected to be highly reliable, the calculation of σ_{10} is seen to agree (see Fig. 3) in shape and magnitude with the measurement over the entire energy range of the present study. In this case, it is clear that the calculations by the Olson model reproduce the energy dependence of the cross section well.

Plotted in Figure 4 are the total cross sections together with the reported values by Maier (3) at low energies. Here it can be seen that the low energy data seem to fit well into the present measurements. The shape of the cross section (the increment of the total cross section at keV-energies followed by a structure in the cross section at low energy) suggests that two distinct mechanisms are important in determining the single electron capture process in Xe. A complete explanation of the two mechanisms involved in the formation of Kr^{0} requires a careful study of the molecular potential energy curves of the collision system and a determination of the strength of the coupling matrix elements associated with avoided crossing of the molecular states during the collision. It is interesting to speculate that the cross section can be expressed as a function of the energy in the entire energy range. It was found that the cross section could be determined from the relationship

$$\sigma(E) = a + b\sqrt{E} + \frac{c}{E\sqrt{E}} + \frac{d}{E^2} \qquad (3)$$

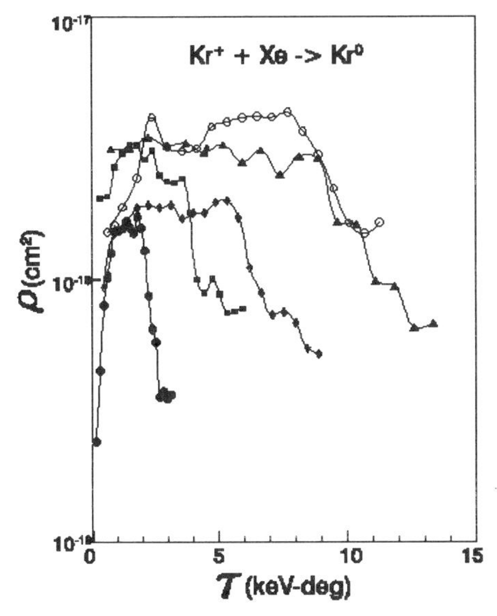

FIGURE 2. Experimental reduced differential cross sections for single electron capture of Kr^+ ions in Xe. (●) 1.0 keV; (■) 2.0 keV; (♦) 3.0 keV; (O) 4.0 keV; (▲) 5.0 keV.

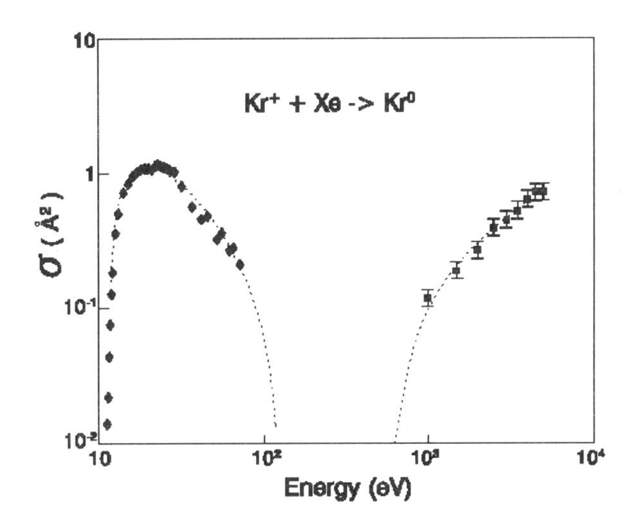

FIGURE 4. Total cross sections for single electron capture of Kr^+ ions in Xe. (●) present measurements; (♦) from Ref. 3; (----) fitting using both Maier and Present data.

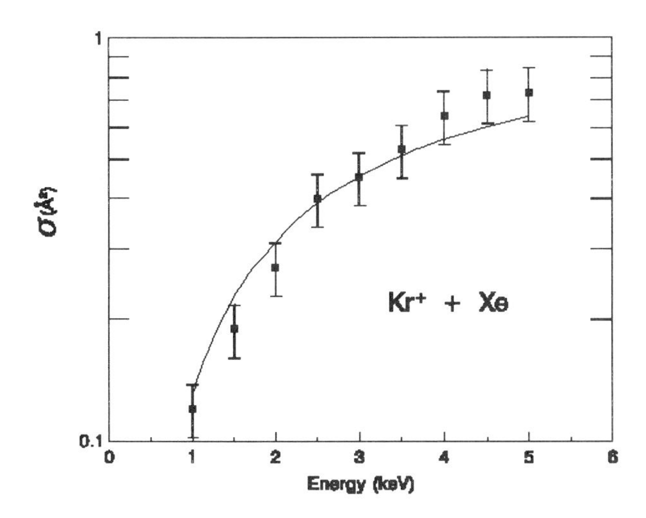

FIGURE 3. Total cross sections for single electron capture of Kr^+ ions in Xe. (●) present measurements; (__) semiempirical calculation of Olson [4].

which represents a minimum square fit using both Maier and present data and is shown as a dotted curve in Fig. 4 (a, b, c and d are constants). The agreement between the "curve fitting" and the data is quite good. This "fitting curve" presents a minimum of $\sigma_{10} \approx 0$ Å^2 about 250 eV, which should be interesting to confirm experimentally.

The results of the present work can be summarized as follows:

a) Differential and total cross sections for single electron capture of Kr^+ on Xe are reported.

b) The reduced differential cross sections present a maximum at a value of $\tau \approx 1.5$ keV-deg, which suggests the intervening of a curve crossing at $R_c = 1.6\ a_0$. It this appeared desirable that a detailed theoretical analysis be carried out to confirm this critical transition region around $R_c = 1.6\ a_c$.

c) The total cross sections for single electron capture agree reasonably well with previous experimental work (3).

d) The shape and magnitude of the present total cross sections for single electron capture are in good agreement with the semiempirical model of Olson (4) over the entire energy range of the present study.

ACKNOWLEDGMENTS

We are grateful to B. E. Fuentes for helpful suggestions and comments. The authors wish to thank Fis. P. G. Reyes for his technical assistance. This work was supported by the DGAPA-IN109196.

REFERENCES

1. Tsuji, M., Kaneko, N., Furusawa, M., Muraoka,, T., and Nishimura, Y., *J. Chem. Phys.* **98**, 8565 (1993).
2. Tsuji, M., Kaneko, N., and Nishimura, Y., *J. Chem. Phys.* **99**, 4539 (1993).
3. Maier II, W. B., *J. Chem. Phys.* **69**, 3077 (1978).
4. Olson, R. E., *Phys. Rev.* **A2** , 121 (1970).
5. Martínez, H., Fuentes, B. E., Alvarez, I.; Cisneros, C. and. de Urquijo, J., *Chem. Phys.* **190**, 139 (1995).
6. Martínez, H., Alvarez, I., Cisneros C., and de Urquijo, J., *Nuclear Instrum. and Methods* **B82**, 389 (1993).
7. Smith, F. T., Marchi, R. P., and Dedrick, K. G., *Phys. Rev.* **150**, 79 (1966); Smith, F. T., Marchi, R. P., Aberth W. and Lorents, D. C., *Phys. Rev.* **161**, 131 (1967).
8. McDaniel, E. W., *atomic collisions (Wiley, New York, 1989) p. 652.*

ANGLE AND VELOCITY VARIATION OF ELECTRON CAPTURE FROM A STARK STATE

K.B. MacAdam, D.M. Homan, O.P. Makarov and O.P. Sorokina

Department of Physics and Astronomy, University of Kentucky, Lexington KY 40502, USA

Total electron capture by K^+ or Cs^+ from a Na Rydberg $n = 24$ extremal Stark state, aligned with a weak external electric field, has been measured as a function of the angle between the atoms' electric dipole moment and the projectile beam in the reduced-velocity range $\tilde{v} = 0.15$ to 1.05. A novel "Stark barrel" is described that produces the arbitrarily directed uniform electric field, smoothly switchable in both magnitude and direction in under 2 μs. The results illustrate the importance of spatial effects as opposed to velocity matching for a full understanding of electron capture near $\tilde{v} = 1$.

INTRODUCTION

Electron capture from a Stark state, that is, from an atomic eigenstate whose electronic wavefunction is distorted by an electric field, offers a new variety of insights and phenomena in an otherwise well-studied subject. The usual targets for gas-phase electron-capture experiments are closed- or open-shell ground-state atoms or molecules; aligned or oriented alkalis in the first excited state (1); or Rydberg states selectively excited by lasers (2, 3, 4) or nonselectively excited by collisional means (5). Measurements of capture from Rydberg states have offered the opportunity for study of Coulomb three-body rearrangement under circumstances where classical theories are competitive with quantal ones or, indeed, where only classical calculations are tractable (6, 7). The juxtaposition of classical insights with empirical outcomes has shed new light on ion-atom collisions that are important in plasma and astrophysical contexts but more importantly has advanced both quantum and classical mechanics in the Correspondence Principle regime.

High-n states in the linear Stark effect are mixtures, generally field-dependent, of the angular momentum states $|\ell m\rangle$. They have non-zero electric dipole moments $\vec{\mu}_e = -e\langle\vec{r}\rangle$, where $\langle\vec{r}\rangle$ is the displacement of the center of negative charge $-e$ from the point-like core consisting of the nucleus and unexcited electrons, having net charge $+e$. We have excited the extremal "blue" (highest energy) $m = 0$ Stark state of the $n = 24$ manifold of Na as a target for capture by singly charged K^+ and Cs^+ ions traveling at reduced velocities $\tilde{v} = v_{\text{ion}}/v_{\text{Bohr}}$ in the range $\tilde{v} = 0.15$ to 1.05. $v_{\text{Bohr}} = 1/n$ a.u. is the Bohr-orbital velocity for principal quantum number n and conveniently marks the divide between perturbatively fast and adiabatically slow collisions for any n (8).

We have prepared the Na(n=24, m=0) top-Stark-state target ("Na(24*top*)") inside a novel device that we call a "Stark barrel," which can provide a uniform, variable electric field over the target region whose magnitude *and direction* can be switched smoothly on a sub-μs time scale. This allows laser excitation of the target in a standard "Stark field" 150 V/cm, aligned with the major axis of the apparatus and parallel to laser polarizations, followed by a rapid but adiabatic reduction of magnitude to the "barrel field" 1-2 V/cm and rotation of the field axis through an arbitrary angle in a horizontal plane. The Stark field is strong enough to allow full resolution of adjacent Stark levels, and the barrel field is weak enough to cause no significant sideways deflection of the incident ion beam while still providing an orienting field for the Rydberg atom. We have made two types of measurement:

1. Comparisons of total capture cross sections for the extreme cases, electron density extending upstream toward the oncoming beam and directly downstream, hiding from the approaching ion behind the core. We refer to this as the **fore-and-aft ratio for total capture** from the Stark state.
2. Variations of the total capture cross section with the angle of the directed Stark state over the range 0-360° with respect to the momentum of the incident ion.

Because of the electric-field control, these collision studies differ from all others previously performed (9). Targets prepared in angular momentum eigenstates in a region free of electric fields have zero mean electric-dipole moment. In a semiclassical picture of an ensemble of Keplerian electronic orbits, as many aphelion vectors have positive projections as have negative ones

CP392, *Application of Accelerators in Research and Industry*, edited by J. L. Duggan and I. L. Morgan
AIP Press, New York © 1997

along any given direction in space. There is little correspondence between the quantum-mechanical angular wavefunctions and elliptical orbits upon which our classical visualizations are based. Much has been made of the role of velocity matching in electron capture from laser-excited targets (10): There is an accentuation of capture for cases in which the target electron moves with speed and direction similar to that of the projectile at some phase of the initial orbit. But contributions of velocity-matching orbits to capture must be *equal* for upstream- and downstream-directed Stark states. Thus any departure of a fore-and-aft capture ratio from exactly *one* is a signature of mechanisms other than velocity matching, *e.g.* mechanisms that depend on the *spatial distribution* of electron probability. We have observed striking effects of this nature with velocity variations that appear to separate multiple-swap contributions predicted by classical theory (11). Full-angle studies show a considerable variety of behaviors that evolve throughout the velocity region we have examined.

STARK BARREL AND EXPERIMENTAL METHOD

The Stark barrel is an array of sixteen molybdenum rods, each 1.59 mm in diameter, equally spaced around a $D = 20$ mm-diameter cylinder in the manner of staves around a barrel (Fig. 1). Computer-controlled voltages are applied to each stave to create an electric field inside the cylinder by superposition. With potentials $V(\theta_0; i) = V_0 \cos(\theta_i - \theta_0 + \pi)$ applied to staves located at angles θ_i around the barrel, the Dirichlet boundary conditions of a uniform field $E = 2V_0/D$ at angle θ_0 are approximated over a large fraction of the interior of the barrel. The axis of the barrel is vertical, and horizontal atom and ion beams pass through the barrel between the staves, crossing paths at the center. The thermal (200°C) Na atomic beam is excited stepwise by nanosecond dye lasers, aimed antiparallel to the Na beam (yellow, 589 nm) and vertically downward parallel to the axis of the barrel (blue, 411 nm), respectively, so that a small cloud of Na Rydberg atoms is formed in the beam a few mm to the side of the intersection point with the ion beam. In 5-10 μs the excited atoms drift across the K^+ or Cs^+ beam, which is accelerated to 50-2000 eV. Charge transfer collisions that occur at the beam intersection result in fast neutral K or Cs Rydberg atoms moving with the unreacted ions. They pass out between the staves of the barrel toward a detector that is sensitive only to capture products in field-ionizable Rydberg states ($n \gtrsim 15$) (12). From a nanoampere ion beam two or fewer counts are registered in a delayed 10-μs counting window following each laser shot at 20 Hz repetition rate. The unreacted Na Rydberg atoms eventually (after 30-50 μs) pass out through the staves into a pulsed field-ionization detector, which provides a target-normalization signal.

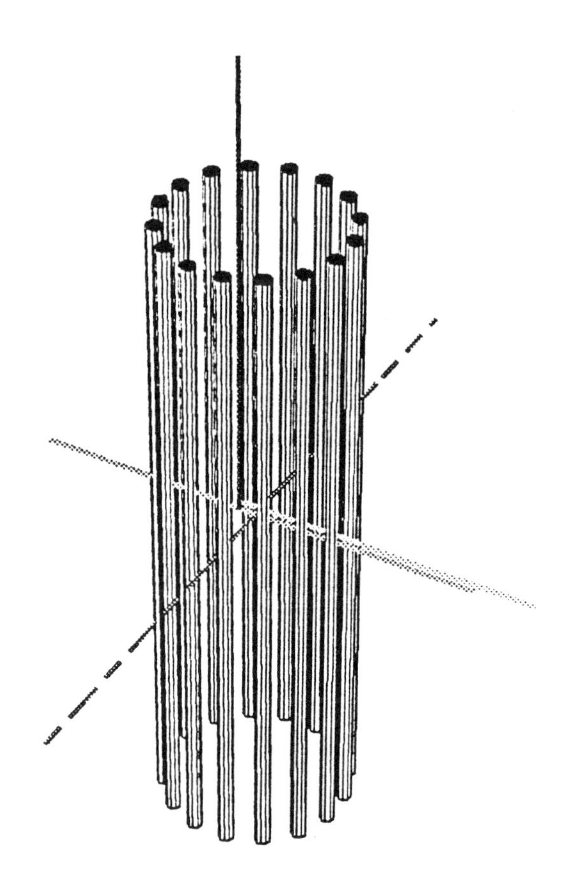

FIGURE 1. Stark barrel and beams: Ion beam, front to back. Atomic beam, left to right. Yellow laser, right to center. Blue laser, top to center.

Before the lasers fire, the electric field stands at about 150 V/cm and is aligned along (or opposite to) the ion-beam direction, $\theta_0 = 0°$ (or 180°). The yellow and blue lasers, linearly polarized parallel to the ion beam, excite a stepwise $\Delta m = 0$ transition through the 3^2P, $j = 3/2$ intermediate state to the next-to-top state of the $n = 24$ linear manifold, shown by an open circle in Fig. 2. A wavelength scan of the manifold at 0.06 cm^{-1} linewidth exhibits a pattern of well-resolved peaks in the uniform Stark field. The staves are switched immediately after the laser flash to a different set of voltages calculated by the computer for a desired V_0 and θ_0 and delivered by digital-to-analog converters. The electric field, not guaranteed to be uniform during the transition between Stark and barrel fields, adiabatically switches the Rydberg atoms down to the highest parabolic state of the $n = 24$, $m = 0$ Stark manifold, which may be considered the linear state of $n = 24$. The electronic wavefunction is strongly concentrated along the field direction at angle θ_0 outward from the core: $|\psi|^2$ has a cone-like profile that opens at a full angle 30° along the field with the nucleus at the apex. The stave potentials settle within 2 μs while atoms are drifting at approximately 1 mm/μs toward their intersection with the projectile beam. The 1-2 V/cm barrel field maintains the direction of the linear state at angle θ_0 against possible perturbation by smaller stray fields and has only a small effect on the ion beam, an effect which can be removed in normalization. The ambient magnetic field is nulled to zero.

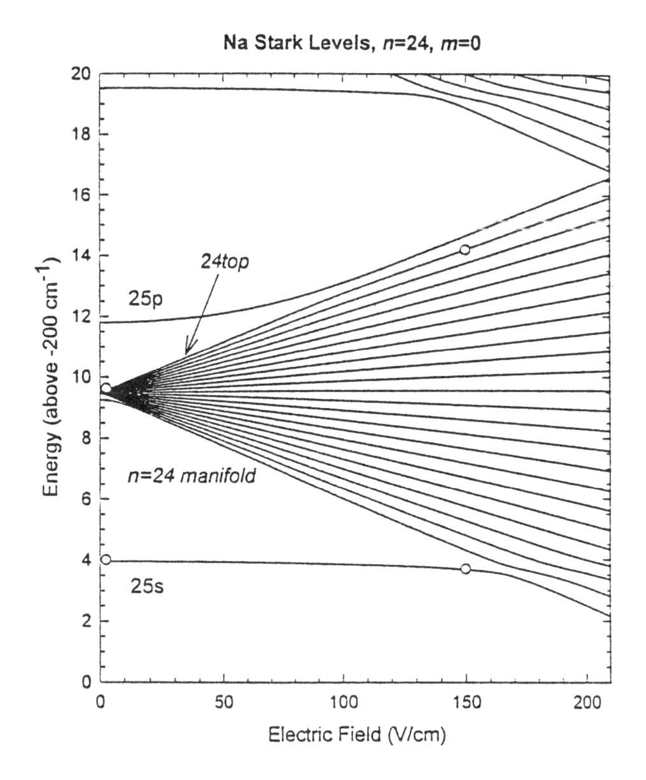

FIGURE 2. Na Stark energy levels near n=24. The 24*top* state is excited at 150 V/cm, then switched adiabatically down to 1-2 V/cm. Measurements of capture from the 25*s* state are interspersed for normalization.

In this manner, the Na(24*top*) state is presented to ion impact at fixed angles that may be set anywhere in the range 0-360°. By comparing capture data taken with the Stark field "up" ($\theta_0 = 180°$) and "down" ($\theta_0 = 0°$) we have confirmed that the Rydberg Stark states adiabatically rotate from the Stark to the barrel fields even through angles as great as ±170°. That is, the initial-angle-dependence of capture is the same for Stark-to-barrel evolutions $180° \rightarrow \theta_0$ and $0° \rightarrow \theta_0$ except for a small range of angles θ_0 for which the field vector passes too close to zero and the adiabatic criterion is violated. Adiabaticity is easily satisfied for all other cases. Fore-and-aft capture ratios are measured without any field rotation by reversing the Stark as well as the barrel fields.

Angle data are taken by alternately measuring capture from Na(24*top*) and Na(25*s*), since the *s* state at 1-2 V/cm is isotropic to collisions. Thus a small systematic effect of transverse deflection of the ion beam in the barrel field is removed: The overlap of ions with Rydberg atoms and the corresponding capture count rate are affected in the same way for 24*top* and 25*s* states, and the variation of the ratio of counts with angle may be attributed entirely to properties of the 24*top* directed Stark state. The lasers are retuned between the two states every few minutes under computer control.

Fore-and-aft capture ratios are also measured as ratios of 24*top* and 25*s* for each of the two field directions, and the ratio-of-ratios yields the result attributable to the oppositely directed 24*top* states alone. The 25*s* state functions as a transfer standard removing possible effects of ion-beam focussing as the beam enters the barrel in field-up or field-down configurations.

RESULTS

Figure 3 shows the results of angle-of-approach studies at three velocities.

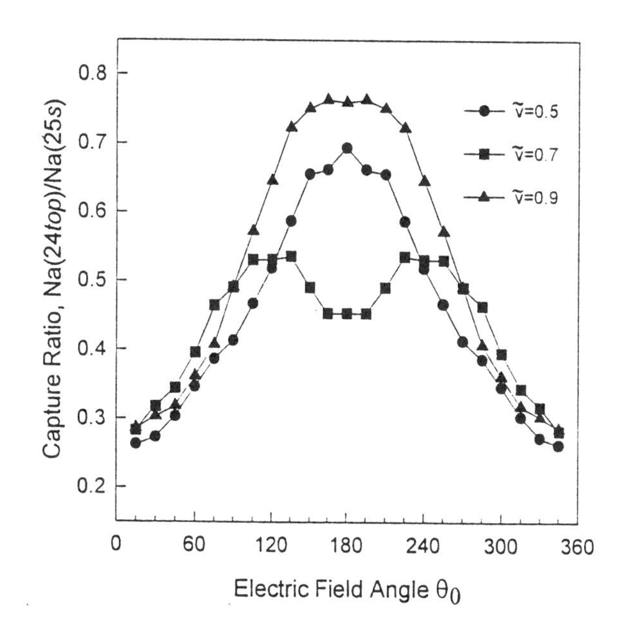

FIGURE 3. Relative capture cross sections in the Stark barrel. At $\theta_0 = 180°$ the target electron lies on the upstream side of the target atom.

The shapes evolve smoothly throughout the entire velocity range. (After we checked experimentally that the curves were even functions of θ_0 about 180° within counting statistics as expected, the data were symmetrized for display.) The side lobes at $\tilde{v} = 0.7$ are visible at lower velocities, where they lie closer to 0° and 360°, and they coalesce to form the flat-topped maximum at $\tilde{v} = 0.9$.

Figure 4 shows the fore-and-aft ratio of total capture, that is, the ratio of capture cross sections for upstream- and downstream-directed Stark states. The minimum of this ratio is associated with the dominance of the velocity-matching lobes near $\tilde{v} = 0.7$. The shoulder at $\tilde{v} = 0.5$ is due to three-swap capture (13), and the peak near $\tilde{v} = 0.3$ results from five- and higher-swap captures. The multiple swaps depend strongly on the presence of the electron on the upstream "exposed" side of the target atom during the ion's approach.

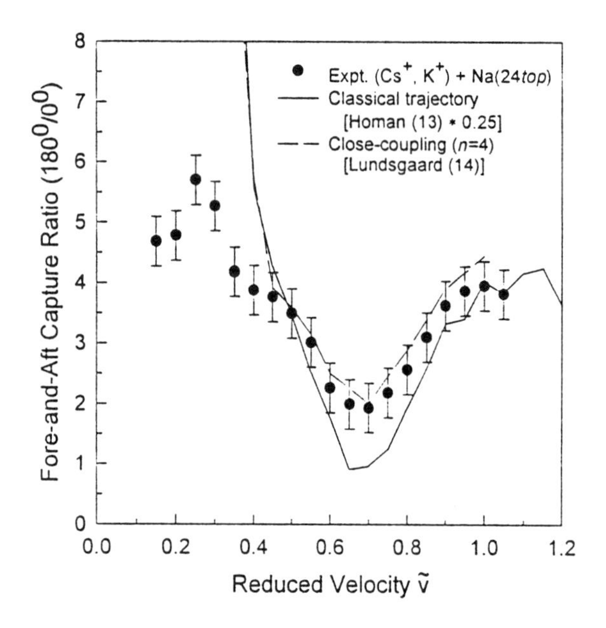

FIGURE 4. Velocity dependence of the fore-and-aft total capture ratio.

DISCUSSION

Homan (13) has carried out classical-trajectory Monte Carlo (CTMC) calculations for comparison with the experiment. The extremal Stark state is represented by an initial highly eccentric Keplerian orbit ($\epsilon = 0.994$) whose aphelion vector is directed at angle θ_0. The capture cross section may be uniquely partitioned into one-swap, three-swap, ..., (odd)-swap contributions (11). The CTMC calculations agree well with the profiles of Fig. 3 and identify the side lobes as the result of one-swap, velocity-matching capture. Three-swap capture dominates at $\tilde{v} = 0.5$ (see also (2)) and is strongly concentrated around $\theta_0 = 180°$. The velocity dependence of capture at $\theta_0 = 0°$ is nearly featureless, small and dominated by one-swap. The structure of the fore-and-aft ratio (Fig. 4) is almost entirely due to the interplay of one- and (higher)-swap processes at $\theta_0 = 180°$. Lundsgaard (14) has performed corresponding quantal close-coupling calculations for the extremal parabolic state of $n = 4$. His results also agree qualitatively with the angular features in Fig. 3 and are in excellent agreement with the fore-and-aft ratio for $\tilde{v} \geq 0.5$.

We have established unambiguously in this work that the electron's *location*, as well as its vectorial velocity distribution, is of paramount significance for electron capture at and below the matching velocity.

ACKNOWLEDGEMENTS

This work was supported in part by the Research Corporation. Graduate fellowship support was received from the U.S. Dept. of Education GAANN (DMH) and from the U. Kentucky Graduate School (OPM and OPS). We thank J.-C. Aguilar, W.L. Fuqua III, J.L. Horn and G. Porter for help with the apparatus, and M.F.V. Lundsgaard for close-coupling calculations.

REFERENCES

1. Lauritsen, J.H.V. *et al.*, *J. Phys. B* **29**, 1093-1100 (1996).
2. MacAdam, K.B. *et al.*, *Phys. Rev. Lett.* **75**, 1723-6 (1995).
3. DePaola, B.D. *et al.*, *Phys. Rev. A* **52**, 2136-40 (1995).
4. Ehrenreich, T. *et al.*, *J. Phys. B* **27**, L383-9 (1994).
5. Koch, P.M. and Bayfield, J.E., *Phys. Rev. Lett.* **34**, 448-51 (1975).
6. Pascale, J., Olson, R.E. and Reinhold, C.O., *Phys. Rev. A* **42**, 5305-14 (1990).
7. Homan, D.M., Cavagnero, M.J. and Harmin, D.A., *Phys. Rev. A* **51**, 2075-84 (1995).
8. Bransden, B.H. and McDowell, M.R.C., *Charge Exchange and the Theory of Ion-Atom Collisions*, New York: Oxford Univ. Press, 1992.
9. But see the fascinating comments in Condon, E.U. and Shortley, G.H., *Theory of Atomic Spectra*, New York: Cambridge Univ. Press, 1964, p. 401. I am grateful to P.M. Koch for this reference.
10. Kohring, G.A., Wetmore, A.E. and Olson, R.E., *Phys. Rev. A* **28**, 2526-28 (1983).
11. Homan, D.M., Cavagnero, M.J. and Harmin, D.A., *Phys. Rev. A* **50**, R1965-8 (1994).
12. Hansen, S.B., Gray, L.G., Horsdal-Pedersen, E. and MacAdam, K.B., *J. Phys. B* **24**, L315-19 (1991), and corrigendum **24**, 4475 (1991).
13. Homan, D.M., private communication.
14. Lundsgaard, M.F.V., Toshima, N. and Lin, C.D., *J. Phys. B* **29**, 1045-62 (1996), and private communication.

ELECTRON-ION IONIZATION CROSS SECTIONS FROM ELECTRON-LOSS BY ATOMIC HYDROGEN: A CASE STUDY.[1]

M. M. Sant'Anna, W. S. Melo[2], A. C. F. Santos, G. M. Sigaud, E. C. Montenegro[3], W. E. Meyerhof[4] and M. B. Shah[5]

Departamento de Física, Pontifícia Universidade Católica do Rio de Janeiro Caixa Postal 38071, Rio de Janeiro, RJ 22452-970, Brazil.

In the last decade a great advance has been obtained in the understanding of mechanisms leading to the electron loss process by multiply-charged ions. In particular, the electron-electron contribution to electron loss has been extensively studied and its connection with the electron-ion ionization process has been established, both theoretically and experimentally. In this paper we use recent experimental results obtained by Sant'Anna et al. from C^{3+} collisions on atomic Hydrogen to obtain the electron ionization cross section of C^{3+} ions. A comparison with results obtained directly from crossed-beams experiments indicates that electron-loss collisions can give a simple and reliable alternative way to obtain electron ionization cross sections of multiply-charged ions.

INTRODUCTION

During the last ten years, a great theoretical and experimental effort has been made out to establish the equivalence between the electron-electron contribution to electron loss and electron impact ionization of multiply-charged ions. This equivalence connects two important branches of atomic collisions. On the one hand, the balance between electron capture and electron loss controls the charge-state of an ion, which is one of the main parameters determining the properties of several processes such as, target ionization, energy loss, sputtering yield, desorption yield, etc. On the other hand, electron-ion ionizing collisions are one of the most predominant processes in high-temperature plasmas, which can be studied in the laboratory only through elaborate crossed-beams experiments. In view of the broad range of phenomena covered by these two processes, it would be extremely useful to have a firm connection between these two branches of atomic collision physics.

The electron loss of a projectile in the intermediate-to-high velocity regime by a neutral target is due to two competing mechanisms. The more intuitive one is due to the screened nuclear interaction of the target atom with the active projectile electron. In this case, the projectile is ionized while the target remains in the ground state. The second possibility is the electron-electron interaction between one of the target electrons and the active projectile electron. In this case, both projectile and target are usually ionized and the target nucleus does not participate actively in the process. These two mechanisms have been called screening and antiscreening modes, respectively [1].

An important alternative view of the antiscreening mode was proposed by Anholt in 1986 [2] and justified in a recent paper by Montenegro and Zouros [3]. Anholt suggested that the electron-electron contribution to electron loss can be simulated as an ionizing collision between a free electron and the projectile ion. This idea was developed almost simultaneously by Zouros et al. [4] in the study of projectile excitation, by considering that, from the projectile point of view, the interaction with the target electrons can be seen as an interaction with a beam of electrons modulated by the Compton profile corresponding to the state of the electron in the target atom. Several authors pursued this idea, identifying various signatures of the screening and antiscreening mechanisms, such as the presence of a threshold associated to the electron-ion collision [5], the impact parameter dependence of these two modes [6-8], as well as performing their experimental separation [9]. This last goal was recently achieved in a clear way through the recoil-ion momentum spectroscopy technique by Dörner et al. [10] and Wu et al. [11]. These studies show that the analogy between the antiscreening mode for electron loss and electron-ion ionizing collisions is strongly supported by the experiment. In fact, the use of electron-ion cross sections in the analysis of the electron loss data shows an excellent agreement between theory and experiment [12,13].

---

[1]Work supported in part by CNPq, FINEP, CAPES, MCT and FAPERJ.
[2]Present address: Instituto de Física, Universidade Federal Fluminense, Brazil.
[3]Corresponding Author.
[4]Permanent address: Department of Physics, Stanford University, USA.
[5]Permanent address: Department of Physics, The Queen's University of Belfast, Northern Ireland.

These findings allow us to look at the above-mentioned analogy in the reverse way, i.e., to obtain electron-ion data from projectile electron (singles) loss experiments. From the experimental point of view, this is an advantageous alternative because singles loss experiments are usually simpler than crossed-beams experiments. The analysis of this possibility is the main purpose of this paper. To this end, we use two general features in the collision regime where antiscreening is important. First, we note that the onset for the antiscreening occurs approximately at projectile energies where the screening contribution is maximum. Consequently, for the energy range where the antiscreening is important, the screening mode is in the high-energy regime, where simple pertubative approaches, such as the Plane Wave Born Approximation (PWBA), give a good description of this mechanism. Second, in this intermediate-to-high velocity regime, electron capture becomes small. Higher-order processes, such as transfer-loss, can then be neglected. Finally, our analysis is carried out using atomic Hydrogen targets. The use of atomic Hydrogen gives not only a sharp Compton profile [14], making this target an excellent candidate to simulate a free electron beam, as well as eliminating undesirable two-electron processes which occur if two-electron targets, such as molecular Hydrogen or Helium, are used. Furthermore, atomic Hydrogen is the target with the smallest charge possible which can be used to induce projectile ionization. This is an important point to justify the use of first-order perturbation theory in the analysis which follows.

PROJECTILE ELECTRON LOSS ON ATOMIC HYDROGEN

The electron loss cross section on atomic Hydrogen, for a projectile which has one active electron, can be written as:

$$\sigma_{\text{loss}} = 2\pi\int_0^\infty \text{bdbP}_S(1-P_C(b)) + 2\pi\int_0^\infty \text{bdbP}_A(b) \qquad (1)$$

where $P_S(b)$ and $P_A(b)$ are the electron loss probabilities as a function of the impact parameter, b, for the screening and antiscreening modes, respectively, and $P_C(b)$ is the electron capture probability.

The above equation has three terms. The first corresponds to the screening cross section, σ_S. The second corresponds to a two-step process in which the projectile active electron is ionized by a nucleus-electron interaction while the target electron is simultaneously captured. This transfer-loss process, with a corresponding cross section σ_{TL}, does not change the projectile charge state and, as a consequence, contributes subtractively to the total electron loss cross section. The third term corresponds to the antiscreening cross section, σ_A. Equation 1 can then be rewritten as:

$$\sigma_{\text{loss}} = \sigma_S - \sigma_{TL} + \sigma_A \qquad (2)$$

The relative importance of the three terms in Eq. 2 depends on the particularities of the collision system. For example, if the projectile is an one-electron, highly charged ion, such as O^{7+}, there is no region where the three processes occur simultaneously. At low velocities, where transfer-loss is important, the antiscreening process is energetically not allowed. For velocities above the antiscreening threshold, transfer-loss is negligible. If the binding energy of the active projectile electron decreases, as for example for C^{3+} ions, the onset velocity for antiscreening also decreases and there is a small range of velocities where all the three above-mentioned processes are important. This situation always happens if the projectile charge state is still smaller, as for C^{+}, for example. Our present analysis is restricted to cases where the overlap between the three processes is either negligible or small. C^{3+} projectiles fulfill this condition. In order to proceed with the analysis of the C^{3+} case, it is useful to divide Eq. 2 by σ_S and define the ratio R between the total electron loss cross section and the screening contribution. Then we have:

$$R = \frac{\sigma_{\text{loss}}}{\sigma_S} = 1 - \frac{\sigma_{TL}}{\sigma_S} + \frac{\sigma_A}{\sigma_S} \qquad (3)$$

At low velocities, below the antiscreening threshold, antiscreening does not contribute and R is smaller than unity due to the negative contribution of the transfer-loss. At high velocities, above the antiscreening threshold, transfer-loss is negligible and R is larger than unity due to the antiscreening contribution. Thus, the deviation of

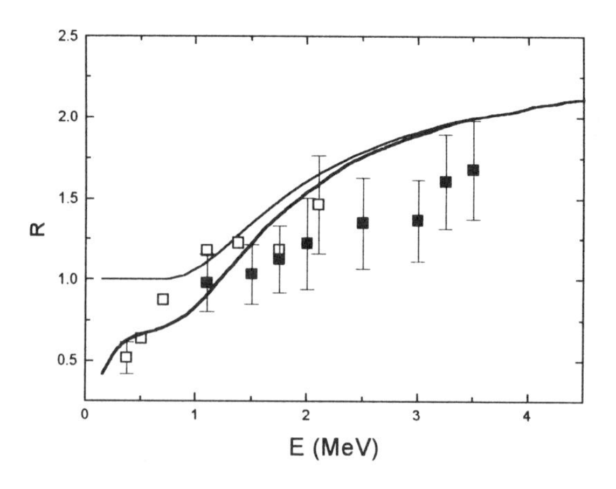

FIGURE 1. Ratio of total to screening cross sections for the electron loss of C^{3+} projectiles impinging upon H targets as a function of the projectile energy. Experimental results (■, Ref. [16]; □, Ref [15]) are compared with PWBA calculations with (thick solid line) and without (thin solid line) the transfer-loss contribution (Ref [17]).

the ratio R from unity gives a good indication about which process contributes more, apart from screening. Figure 1 shows the experimental data of Goffe et al. [15] and Sant'Anna et al. [16], divided by σ_S calculated from Ref.[17], for electron loss of C^{3+} on atomic Hydrogen. The thin solid line shows the PWBA results for the screening and antiscreening processes [17], without including transfer-loss. The thick solid line is the PWBA calculation for σ_S and σ_A, together with SCA calculation for transfer-loss [17]. It can be seen that inclusion of transfer-loss gives good agreement with the experiment for the low energy-range of the data. It can also be seen that transfer-loss becomes essentially negligible for projectile energies above 1.5 MeV, approximately where the antiscreening contribution begins to be significant. Around 4.0 MeV, the antiscreening contribution is almost the same as the screening one.

COMPARISON WITH ELECTRON-ION IONIZATION CROSS SECTIONS

As mentioned above, for energies slightly above the antiscreening threshold, the transfer-loss process can be neglected. Under these circumstances the ratio R can be approximated as:

$$R \approx 1 + \frac{\sigma_A}{\sigma_S} \tag{4}$$

If the antiscreening process can be viewed, in the projectile frame, as a collision between a beam of electrons (associated with the bound electron of the

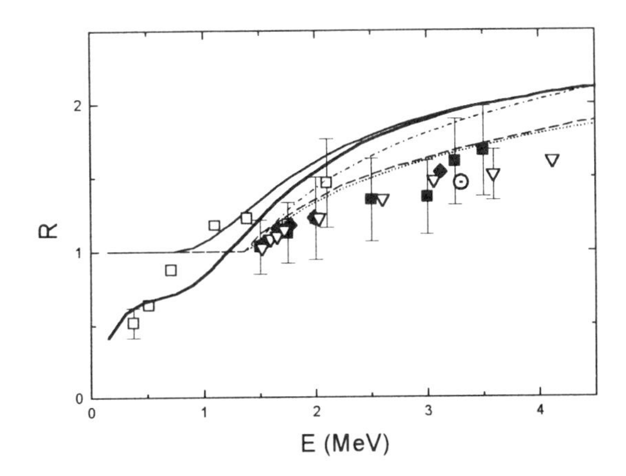

FIGURE 2. Values of R using antiscreening (H^0 + C^{3+}), eq. 4, or ionization (e^- + C^{3+}), eq.5, cross sections as a function of the projectile energy. Experimental results for C^{3+} loss (■, Ref. [16]; □, Ref [15]) and ionization of C^{3+} (V, Ref. [18]; ◆Ref. [19]; ⊙, Ref. [21]) are compared with PWBA calculations for loss with (thick solid line) and without (thin solid line) the transfer-loss contribution (Ref [17]) and calculations for ionization by electron impact (dashed line, Ref. [24]; dashed-dotted line, Ref. [23]; dotted line, Ref.[25]).

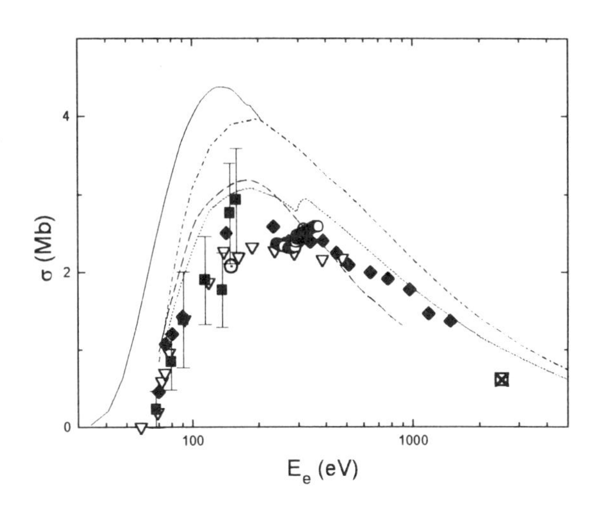

FIGURE 3. Ionization cross section by electron impact (e^- + C^{3+}) and σ_{ee} from eq.(6) for H^0 + C^{3+}, as a function of the projectile energy. Experimental results for C^{3+} loss (■, Ref. [16]) and ionization of C^{3+} (V, Ref. [18]; ◆, Ref. [19], O, Ref. [20]; ⊙, Ref. [21]; ⊠ Ref. [22]) are compared with PWBA calculations for loss without transfer-loss contribution (solid line, Ref [17]) and calculations for ionization by electron impact (dashed line, Ref. [24]; dashed-dotted line, Ref. [23]; dotted line, Ref.[25]).

atomic Hydrogen target) impinging on the projectile, Eq. (4) can be further approximated as:

$$R \approx 1 + \frac{\sigma_{ee}}{\sigma_S} \tag{5}$$

where σ_{ee} is the electron ionization cross section of C^{3+} ions. Within this approximation, we can compare the data obtained from electron loss experiments with the electron ionization results obtained from crossed-beams experiments. Figure 2 compares the loss results of Fig. 1 with the electron ionization measurements of Crandall et al. [18], Crandall et al. [19], Müller et al. [20], Kunze [21] and Donets [22]. Theoretical results for electron impact ionization of Lotz [23], Bray [24] and Hu [25] are also shown. The energy associated with the electron ionization data corresponds to that of carbon ions with the same velocity. It can be seen from this figure that the agreement between the electron loss data and the crossed-beam data lies within the experimental errors. Figure 3 illustrates our analysis from the electron ionization point of view. We plot the electron results together with the electron loss experimental results for projectile energies greater than 1.5 MeV. The cross section for electron ionization is obtained from the loss experiments through the relation:

$$\sigma_{ee} = \sigma_{loss} - \sigma_S \tag{6}$$

where σ_S is the PWBA calculation for the screening mode, as in Eq. (2). The corresponding electron velocity is the same as the ion velocity. It can be seen from Fig. 3

that the electron-loss and the electron-impact data are in good agreement with each other. It should be noted however, that the absolute values obtained from eq. (6) are strongly dependent on the uncertainties associated to both the experimental and theoretical cross-sections. The measured electron-loss cross sections are particularly sensitive to these errors when atomic Hydrogen is used as target.

CONCLUSIONS

In this work we analyzed a method which is alternative to crossed-beams measurements to determine ionization cross section of multiply-charged ions by electron impact. The relationship between electron-impact ionization and the antiscreening mode for electron loss, which has been well established in the last years for high-velocity, multiply-charged ions, is extended to the intermediate-velocity regime. In this regime, the transfer-loss process plays an important role in the analysis of the electron-loss measurements and can obscure a direct relationship between electron-impact ionization and antiscreening. Our analysis show that, for collision systems where transfer-loss and antiscreening have a small overlap at intermediate velocities, it is possible to obtain reliable cross sections for electron impact ionization from singles (non-coincidence) electron-loss measurements. The accuracy of the proposed method relies essentially in the quality of the cross section calculations for the screening mode and the errors associated to the determination of absolute cross sections in atomic Hydrogen targets.

REFERENCES

1. Montenegro, E.C., Meyerhof, W.E., and McGuire, J.H., *Advances in Atomic and Molecular Physics* **34**, 249-300 (1994).
2. Anholt, R., Phys. Lett. **114A**, 126 (1994).
3. Montenegro, E.C., and Zouros, T.J.M., *Phys. Rev.* **A50**, 3186 (1994).
4. Zouros, T.J.M., Lee, D.H., and Richard, P., *Phys. Rev. Lett.* **62** , 2281 (1989).
5. Feinberg, B., Meyerhof, W.E., Belkacem, A., Alonso, J.R., Blumenfeld, L., Dillard, E.D., Gould, H., Guardala, N., Krebs, G.F., McMahan, M.A., Rhoades-Brown, M.E., Rude, B.S., Schweppe, J., Spooner, D.W., Street, K., Thieberger, P., and Wegner, W., *Phys. Rev.* **A44** , 1712 (1991).
6. Montenegro, E.C., and Meyerhof, W.E., *Phys. Rev.* **A44**, 7229-7233, (1991).
7. Montenegro, E.C., and Meyerhof, W.E., *Phys. Rev.* **A46**, 5506-5512 (1992).
8. Montenegro, E.C., Belkacem, A., Spooner, D.W., Meyerhof, W.E., and Shah, M.B., *Phys. Rev.* **A47**, 1045-1051 (1993).
9. Montenegro, E.C., Melo, W.S., Meyerhof, W.E., and Pinho, A.G., *Phys. Rev.Lett.* **69**, 3033 (1992).
10. Dörner, R., Mergel, V., Ali, R., Buck, U., Cocke, C.L., Froschauer, K., Jagutzki, O., Lencinas, S., Meyerhof, W.E., Nüttgens, S., Olson, R.E., Schmidt-Böcking, H., Spielberger, L., Tökesi, K., Ullrich, J., Unverzagt, M., and Wu, W., *Phys. Rev. Lett.***72**, 3166 (1994).
11. Wu, W., Wong, K.L., Ali, R., Chen, C.Y., Cocke, C.L., Frohne, V., Giese, J.P., Raphaelian, M., Walch, B., Dörner, R., Mergel, V., Schmidt-Böcking, H., and Meyerhof, W.E., *Phys. Rev. Lett.* **72**, 3170 (1994).
12. Wu, W., *Thesis*. Kansas State University, 1994, unpublished.
13. Cocke, C.L., and Montenegro, E.C., *Comm. Atom. Mol. Phys.* **32**, 131-141 (1996).
14. Zouros, T.J.M., *Comm. Atom. Mol. Phys.* **32**, 291 (1996).
15. Goffe, T.V., Shah, M.B., and Gilbody, H.B., *J. Phys. B: At. Mol. Opt. Phys.*, **12**, 3763-3773 (1979).
16. Sant'Anna, M.M., Melo, W.S., Santos, A.C.F., Sigaud, G.M., Montenegro, E.C., Meyerhof, W.E. and Shah, M.B., (to be published).
17. Montenegro, E.C., and Meyerhof, W.E., *Phys. Rev.* **A43**, 2289-2293 (1991).
18. Crandall, D.H., Phaneuf, R.A., and Taylor, P.O., *Phys. Rev.* **A18**, 1911-1924 (1991).
19. Crandall, D.H., Phaneuf, R.A., Hasselquist, B.E., and Gregory, D.C., *J. Phys. B: At. Mol. Opt. Phys.*, **12**, L249-L256 (1979).
20. Müller, A, Hofman, G, Tinshert, and Salzborn, E., *Phys. Rev. Lett.* **61**, 1352-1355 (1988).
21. Kunze, H.J., *Phys. Rev.* **A21**, 937 (1971).
22. Donets, E.D., IEEE Trans. Nucl. Sci. **NS23**, 897 (1976).
23. Lotz, W., *Z Phys.*, **216**, 241 (1968).
24. Bray, I., *J. Phys. B: At. Mol. Opt. Phys.*, **28**, L247-L254 (1995).
25. Hu, W., Chen, C., Fang, D., Wang, Y., Lu, F., and Yang, F., *J. Phys. B: At. Mol. Opt. Phys.*, **29**, 2887-2895 (1996).

HYPERFINE TRANSITIONS IN GROUND STATE HYDROGEN-LIKE $^{165}Ho^{66+}$ AND $^{185,187}Re^{74+}$

J. R. Crespo López-Urrutia, P. Beiersdorfer, D. W. Savin, and K. Widmann

Lawrence Livermore National Laboratory, Livermore, CA 94550, USA

Spontaneous line emission due to the hyperfine splitting of the ground state of highly charged hydrogen-like ions excited by electron collisions was measured using an Electron Beam Ion Trap. The F=4 to F=3 transition of the 1s $^2S_{1/2}$ configuration of $^{165}Ho^{66+}$ was identified at (5726.4 ± 1.5) Å. The F=3 to F=2 transition for the two isotopes $^{185,187}Re^{74+}$ were found at (4512 ± 2) Å and (4557 ± 2) Å. We infer the nuclear dipole magnetic moment of ^{165}Ho to be 4.1267(11) n.m., with five times higher accuracy than previous measurements. For $^{185,187}Re$ we determine 3.153(2) n.m. and 3.184(2) n.m., respectively, in disagreement with a tabulated NMR measurement.

INTRODUCTION

The large hyperfine splitting (HFS) (several eV's) of the ground state of a highly charged hydrogen-like ion is sensitive to the nuclear magnetic moment, to nuclear size effects and to QED radiative corrections. Only sparse data are available; the spontaneous 1s hyperfine transition in H, D and He^+ has been observed in astrophysical sources, but only few laboratory measurements have been carried out (H, D, T, He^+) due to the extremely long lifetime of the upper hyperfine levels (1,2) in neutrals and low-charged ions. Laser pumping has been recently applied to H-like Bi^{82+} in a heavy-ion storage ring (3) resulting in the first measurement of the hyperfine splitting in a multiply charged ion.

We have measured the F=4 to F=3 hyperfine transition of the 1s ground level of H-like $^{165}Ho^{66+}$ using passive emission spectroscopy (4), and also the F=3 to F=2 transition in the two rhenium isotopes ^{185}Re and ^{187}Re. We use the results to determine their respective nuclear magnetic moments.

EXPERIMENT

Hydrogen-like holmium and rhenium ions are produced and stored in a high-energy electron beam ion trap (SuperEBIT) by an electron beam of variable energy axially compressed by a high magnetic field. The ion trap is monitored by two germanium detectors to determine the ionic species present and the charge balance. At a beam energy of 132 keV and 285 mA beam current, the He-like Ho^{65+} ion was the most abundant species in the trap (40 %), followed by Li-like Ho^{64+} (25%). The concentration of H-like Ho^{66+} was 6 %, that of bare Ho^{67+} around 0.5%. Lower charge states made up the rest. For the rhenium experiment, the beam energy was 163 keV.

Two main processes are expected to populate the upper hyperfine level of the H-like ion in our trap. The most important is collisional ionization of He-like ions, were one of the two 1s electrons is removed leaving a H-like ion behind with similar probability for population in any of the two HFS states. The second process is radiative recombination of beam electrons with bare ions, also populating both HFS levels nearly evenly. Collisional excitation requiring a spin flip is far less probable than these two mechanisms and can be neglected. With these processes, the observed count rate can be explained.

We use an optical prism spectrograph equipped with a cryogenic CCD camera to detect the spontaneous emission from the hyperfine transitions in those H-like ions. Due to the extreme low number of emitting ions (thousands), and the low excitation rate ($<10\ s^{-1}$), great care was necessary taken to separate the signals on the order of few hundred photons per hour from much higher levels of background. The thermal and atomic line background are eliminated by subtracting from every spectrum taken with hydrogen-like ions in the trap another one taken without any hydrogen-like. Cosmic rays are detected by the CCD during the long exposure times. Their contribution is largely reduced by appropriate software discriminator level during the data reduction. To obtain spectra from the two-dimensional images, the pixel counts on the CCD detector are integrated along one dimension. Many days of observation were necessary to obtain appropriate signal-to-noise ratios for the wavelength determination.

The holmium result, after addition of 18 background corrected spectra (36 hours of data) is displayed in Fig 1. A single feature at (5726.4 ± 1.5) Å with a peak height 10 times larger than the standard deviation of the background the line and a FWHM of 15 Å appeared. The total number of counts above the background over the line profile was around 4000. We attribute this feature to the hyperfine transition in Ho^{66+}. To exclude the possibility that the line

CP392, *Application of Accelerators in Research and Industry*, edited by J. L. Duggan and I. L. Morgan
AIP Press, New York © 1997.

was emitted by a lower charge state, the experiment was repeated, but making sure that no H-like Ho^{66+} could be produced, while

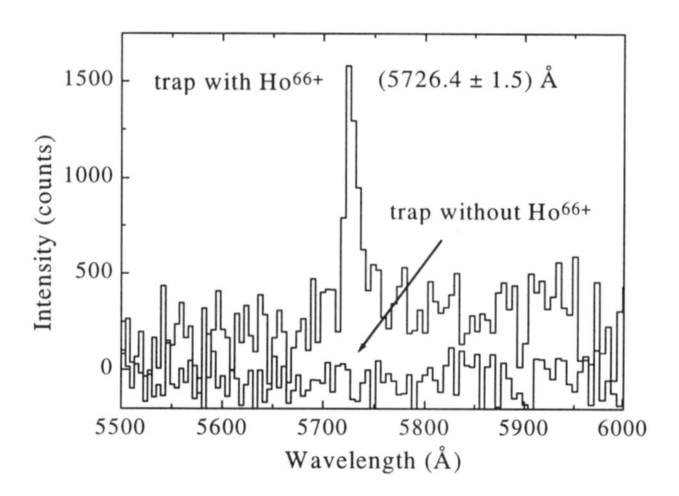

FIGURE 1. The F=4 to F=3 hyperfine transition in hydrogen-like Ho^{66+} and its baseline.

keeping the other charge states in the trap more or less unchanged. Again, spectra with and without holmium were subtracted from each other. The resulting spectra displayed no indication of any line at any position throughout the range of observation. The rhenium measurement was performed in the same way, and delivered two features at (4512 ± 2) Å and (4557 ± 2) Å, which are identified as the HFS ground state transitions of the two natural isotopes of that element. Their relative intensities are consistent with the isotopic abundances for natural rhenium.

DISCUSSION

The energy difference Δ between two neighboring levels in a H-like ion is given by Shabaev (5) as

$$\Delta = \frac{\alpha^4 Z^3}{n^3} \cdot \frac{\mu_I}{I} \cdot \frac{m_e}{m_p} \frac{(I+j)\cdot m_e c^2}{j\cdot(j+1)\cdot(2l+1)} \left[A\cdot(1-\delta)\cdot(1-\varepsilon)+\kappa_{rad}\right]$$

with following terms: α, fine-structure constant; Z, nuclear charge; n, j, l: principal, total moment and orbital moment electron quantum numbers; μ_I, nuclear magnetic moment; m_e, m_p: electron, proton mass; I, nuclear spin; A, relativistic factor; δ, nuclear charge distribution correction; ε, nuclear magnetization distribution (Bohr-Weisskopf) correction; κ_{rad}, QED radiative corrections. The two dominant QED corrections have recently been calculated (6, 7). Based on the calculations of the self energy correction by Persson et al. (6), we estimate for holmium (Z=67) a -19.3 meV shift. The vacuum polarization was given by Schneider et al. in (7). From their work we estimate this contribution to be 9.4 meV. The total net QED correction therefore is -9.9 meV. Using the above formula with the mentioned QED corrections, we obtain a nuclear magnetic moment of 4.1267(11) n.m., in good agreement with the latest published value from the Shirley table of isotopes (8) of 4.132(5). For $^{185,\,187}Re$, a QED net shift of -14.0 meV as given in (6,7) leads to nuclear magnetic moments of 3.153 n.m. and 3.184 n.m. respectively. This results disagrees with values (3.1871 n.m. and 3.1921 n.m.) tabulated in (8) where a standard diamagnetic correction was applied to the original NMR data of Alder (9) (3.1437 n.m. and 3.1759 n.m.) measured in a liquid solution of $NaReO_4$. Chemical shifts of the NMR frequency, which were not accounted for, and the fact that the electronic structure of the rhenium in that experiment was not that of a free atom, but a chemically bound Re^{7+} i.e., Re(VII), making the diamagnetic correction inaccurate, can however explain the difference.

The HFS of the ground state in H-like ions is sensitive to contributions from the nuclear magnetic moment, nuclear size effects and QED. The small QED contributions can be predicted with high accuracy; the nuclear size corrections as well, to a lesser extent. The absolute size of the nuclear magnetic moment can therefore be determined from the 1s hyperfine splitting. This method avoids both the uncertainty in the theoretical atomic diamagnetic correction, which affects atomic beam resonance measurements, and chemical frequency shifts, for the values obtained through NMR measurements. Nuclear size effects are at present the largest source of systematic error.

ACKNOWLEDGMENTS

This work was in part supported by the Office of Basic Energy Sciences and was performed under the auspices of the U.S.D.O.E. by Lawrence Livermore National Laboratory under contract # W-7405-ENG-48.

REFERENCES

1. Essen, L, Donaldson, R. W., Bangham, M. J., and Hope, E. G., Nature, **229**, 110 (1971).
2. Gould, R. J., Astrophys. Jour., **423**, 522 (1994).
3. Klaft, I., et al. Phys. Rev. Lett. **73**, N18, 2425 (1994).
4. Crespo López-Urrutia, J. R., Beiersdorfer, P., Savin, D. W., and Widman, K., Phys. Rev. Lett. **77**, 826 (1996).
5. Shabaev, V. M., J. Phys. B: At. Mol. Opt. Phys. **27**, 5825 (1994).
6. Persson, H. et al, Phys. Rev. Lett. **76**, 1433 (1996)
7. Schneider, S. M., Greiner, W., and Soff, G., Phys. Rev. **A50**, 118 (1994).
8. Firestone R. et al. in *Table of Isotopes*, ed V. S. Shirley, 8th ed.; New York: Wiley,1996, Appendix E.
9. Alder, F., and Yu, F. C., Phys. Rev. **82**, p. L105 (1951).

PHOTON AND ELECTRON EMISSION EXPERIMENTS AT THE UNIVERSITY OF NEVADA, RENO MULTICHARGED ION FACILITY

R. A. Phaneuf, R. Bruch, X. Da, V. Golovkina, I. Golovkin, V. Kantsyrev, H. Li, R. Rejoub, A. Shlyaptseva and J. K. Swenson

Department of Physics, University of Nevada, Reno, Reno NV 89557-0058

D. Hitz

CEN Grenoble, DRFMC/PSI, 85X, 38041 Grenoble CEDEX, France

V. V. Afrosimov and A. P. Shergin

A.F. Ioffe Physical-Technical Institute, 194021 St. Petersburg, Russia

The multicharged ion research facility at the University of Nevada, Reno is based on a 14.4-GHz Caprice ECR ion source. Using this ion-beam facility, high-resolution measurements have been made of extreme ultraviolet (EUV) photon emission produced in low-energy single collisions of C^{4+}, C^{5+} and O^{6+} ions with helium atoms. The observed EUV photon emission spectra result from single and double electron-capture collisions which selectively populate excited states of the multiply charged projectile ion, and from single electron capture accompanied by 1s-2p excitation of the He target atom. The sensitivity of the EUV measurements was enhanced by approximately one order of magnitude using a new type of glass-capillary-converter which guides photons emitted along a horizontal ion beam segment to the vertical entrance slit of a 2.2-m grazing-incidence monochromator. A high-sensitivity cylindrical-mirror electron energy analyzer has also been used to study electron emission resulting from low-energy collisions of hydrogen-like C^{5+} ions with helium atoms. The objective of these experiments was to investigate the role of multiple-electron transitions in such collisions.

INTRODUCTION

Processes involving single-electron transitions produced in slow collisions of multiply charged ions with neutral atoms and molecules generally have large cross sections and have been extensively studied, both experimentally and theoretically. Such collisions are relatively well understood within the framework of an independent-electron model, and may be quantitatively described in terms of transitions at avoided crossings of the adiabatic interatomic potential energy curves [1,2]. However, the situation is rather different for collisions involving multiple-electron transitions, whose theoretical description must in general go beyond the independent-electron approximation. Consequently, the role of electron-electron correlations in such processes has been a topic of considerable discussion and the motivation for a number of experiments during the last several years [3-5]. A central theme of these discussions has been whether multiple electron transitions such as double electron capture take place simultaneously at an avoided crossing at a single internuclear distance, or via a two-step mechanism involving successive single-electron transitions at avoided curve-crossings occurring at different internuclear distances. While the existing body of experimental evidence for double electron capture is rather extensive, it cannot be considered definitive, favoring the former interpretation in some cases, and the latter in others. While the general features and scalings of the cross sections for a wide range of collision systems may be predicted remarkably well by the extended classical over-barrier model of Niehaus [6], a rigorous quantal theoretical treatment of double electron capture is not yet available.

A common objective of the experiments reported here was to investigate collision processes in which single electron capture by a multiply charged ion from a neutral atom or molecule is accompanied by an excitation of either the projectile ion core or the target atom. In such "transfer-excitation" collisions, electron-electron correlations are expected to be of considerable importance, and quantitative theoretical studies have been attempted only recently by Fritsch [7] for He^{2+}, Be^{4+} and C^{6+} collisions with He. These calculations involve the consideration of a very large

CP392, *Application of Accelerators in Research and Industry*, edited by J. L. Duggan and I. L. Morgan
AIP Press, New York © 1997.

increasing collision energy below 1 atomic unit.

Since the cross sections for such multielectron processes are relatively small, their experimental signatures may easily be masked by other processes, and high sensitivity and/or high spectral resolution may be required to guarantee unambiguous measurements. Extreme care must also be exercised to ensure that such measurements on multi-electron transitions are performed under single-collision conditions.

Absolute cross sections for single electron capture by He-like C^{4+} and N^{5+} ions from He atoms accompanied by 1s-2p excitation of He were first reported by Dijkkamp et al. [8] and more recently by Hoekstra et al. [9] via photon-emission measurements in the vacuum ultraviolet region at a wavelength of 30.4 nm. The N^{5+} + He measurements of Hoekstra et al. indicated an anomalous energy dependence which was accounted for by a modification of the "dynamic" classical over-barrier model to include energy sharing between the two participating electrons. Since double electron capture by multiply charged ions from neutral atoms and molecules populates primarily doubly excited levels of the projectile ion, high-resolution Auger-electron spectroscopy is a powerful technique for detailed investigation of such processes, particularly for a low-Z ion where the Auger stabilization channel is significant. As a result, Auger electron emission due to the dominant double electron capture channels in slow multicharged ion-neutral collisions has been extensively studied with high energy resolution [11-13], and has provided detailed information about these channels.

The present work is an investigation of low-energy collisions of multiply charged ions with helium atoms using both high-sensitivity Auger-electron spectroscopy and high-resolution EUV photon spectroscopy. The common objective is a search for evidence of low-probability transitions, not only in the projectile ion but also in the target, with the hope of shedding new light on the role of correlated two-electron transitions.

EXPERIMENTS AND RESULTS

The experiments reported here were conducted at the multicharged ion research facility at the University of Nevada, Reno. The heart of this facility, whose layout is shown in Fig. 1, is a 14.4 GHz Caprice electron-cyclotron-resonance (ECR) multicharged ion source [10]. This facility first became operational for atomic collisions research in January, 1995.

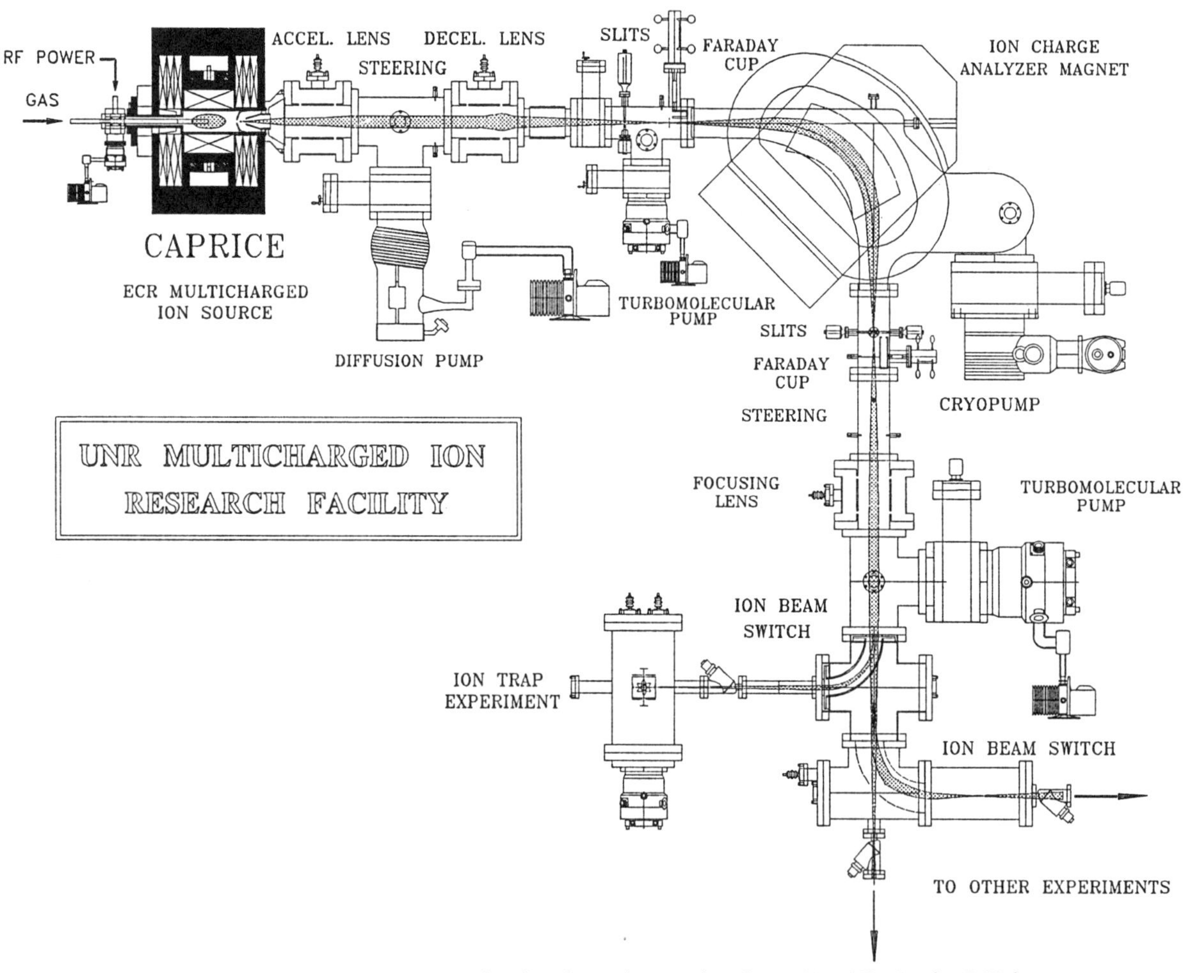

FIGURE 1. Layout of multicharged ion research at the University of Nevada, Reno based on a 14.4-GHz Caprice ECR ion source.

High-Sensitivity Electron Spectroscopic Measurements

The high-sensitivity energy analyzer used for the ejected-electron measurements was supplied by the Ioffe Institute. It is a two-stage cylindrical-mirror analyzer whose outer cylindrical electrodes consist of a high-transmission grid to suppress the detection of secondary electrons. As configured for the present experiments, the instrument has a dynamic range of $\sim 10^6$ and an energy resolution of 4%.

Measured electron energy spectra produced in collisions of C^{5+} ions with He atoms at collision energies ranging from 25 to 100 keV and an electron ejection angle of 125^0 are presented in Fig. 2. Electron spectra measured for such systems with two active electrons are dominated by the decay of projectile states. The structure observed at electron energies below 100 eV is due to LMx and Lxx decays of lithium-like configurations resulting from double-electron capture into $n \geq 3$ states of the projectile. The two prominent peaks near 250 eV are due to the decay of 1s2l2l' and 1s2lnl' series of autoionizing states. These dominant structures have been measured previously with high electron energy resolution [11-13]. The present data indicate almost no collision energy dependence in this range for these KLL and KLx transitions. The additional broad features at higher electron energies are of particular interest. Electrons ejected at energies between approximately 300 eV and the ionization potential of C^{4+} at 392 eV are attributed to double-electron capture into Rydberg levels, and are associated with Kxx Auger transitions. Electron capture to these levels is endoergic, which accounts for the observed increase of this feature with collision energy. Electrons ejected with energies between 392 eV and the ionization potential of C^{5+} at 490 eV are attributed to single electron capture accompanied by excitation of the 1s electron on the C^{5+} projectile. As expected, the cross section for this endoergic process shows a strong increase with collision energy. Corresponding measurements for H_2 and N_2 targets and for N^{6+} projectile ions reveal similar features.

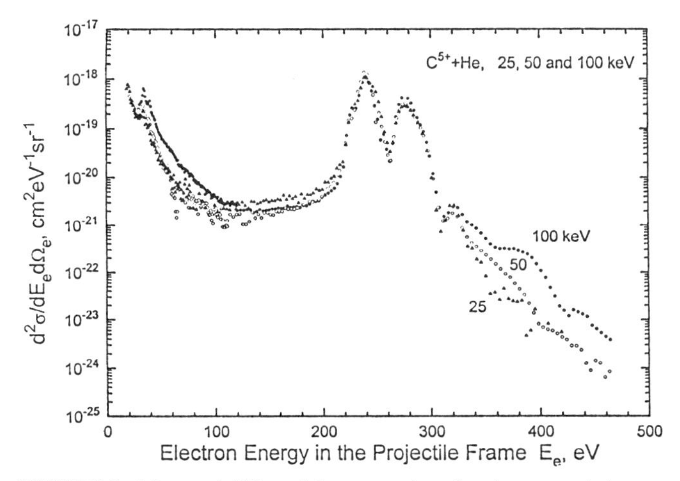

FIGURE 2. Measured differential cross sections for electron emission at an angle of 125 degrees due to collisions of C^{5+} ions with He atoms at collision energies of 25, 50 and 100 keV.

High-Resolution EUV Spectroscopic Measurements

A 2.2-meter grazing incidence vacuum ultraviolet monochromator fitted with a 600 line/mm grating was used for the high-resolution measurements of EUV spectra produced in low-energy collisions of multicharged ions with helium atoms. The sensitivity of the EUV measurements was increased by approximately one order of magnitude using a new type of glass-capillary-converter device [14] which guides photons emitted along a horizontal ion beam segment in a differentially pumped gas target cell to the vertical entrance slit of the grazing-incidence monochromator. This device consists of a bundle of tiny hollow quartz capillaries which transfer the EUV light with high efficiency by reflection at grazing angles. More details about this device and its applications are available in a paper by Bruch and Kantsyrev at this conference.

The primary objective of the present experiment was to apply high-resolution EUV spectroscopy to the investigation of single-electron capture by a multiply charged ion from a helium target atom, accompanied by 1s-2p excitation of the remaining helium target electron. As noted earlier, absolute measurements of the He^+ (2p-1s) line at 30.38 nm have been reported by the FOM group for C^{4+} + He and N^{5+} + He collisions [8,9]. To apply these absolute data to normalize the sensitivity of the present apparatus, high-resolution measurements were made for C^{4+} + He collisions at an energy of 40 keV.

Similar high-resolution EUV measurements in the vicinity of the 30.38 nm line were carried out for C^{5+} + He collisions at a collision energy of 50 keV. The EUV spectrum measured with resolution of 60 is presented in Figure 3. The He^+ line is evident in this spectral region, with a measured emission cross section 3.8 times larger than for C^{4+} + He at 40 keV. From these measurements, the cross section for single electron capture accompanied by target excitation in C^{5+} + He collisions is estimated to be $(4 \pm 1) \times 10^{-17}$ cm^2 at 50 keV.

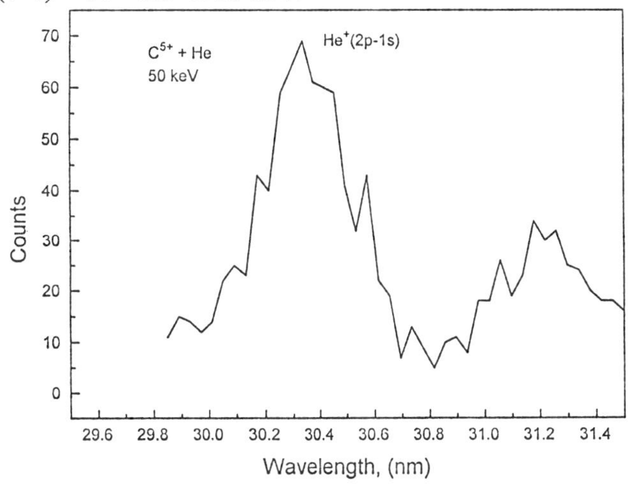

FIGURE 3. Measured high-resolution EUV spectrum in the vicinity of the He^+ (2p-1s) line at 30.38 nm, produced in collisions of C^{5+} ions with He atoms at 50 keV. The experimental resolution is approximately 60.

Similar measurements were also made for the O^{6+} + He collision system at an energy of 60 keV with a spectral resolution of approximately 200. In this case, for which the measurements are presented in Fig. 4, the measured spectrum is rich in partially resolved lines that are extremely close to the 30.38 nm He^+ line, making it nearly impossible to extract a relative cross section for the capture-plus-target-excitation process. It was found in this case that several of the observed lines are not tabulated in the compilation by Kelly [15]. As a result, extensive calculations were made using the atomic structure code of Cowan [16], in which 63 configurations were simultaneously considered for relevant transitions in lithium-like and beryllium-like oxygen. The results of these calculations are presented in Table 1, where the numbering of lines is the same as that used in Fig. 4. In this case, it is likely that a number of transitions originating from the $1s^23p$, 1s2s3s and 1s2s3d levels of O^{5+}, and the $1s^22s4d$ levels of O^{4+} appear in second order in the observed spectrum. It has been shown that 4% of the O^{6+} ions produced by an ECR ion source are in the metastable $1s2s^3S$ state [17], so the line assignments in Table 1 for the observed spectrum are plausible.

TABLE 1. Calculated Wavelengths for transitions in O^{5+} and O^{4+}.

Line	Expt. λ (nm)	Theory λ (nm)	Ion	Transition
1.	29.79			
2.	28.84			
3.	29.89	14.944	O^{5+}	$1s^23p\ ^2P_{3/2}$ - $1s^22s\ ^2S_{1/2}$
4.	29.89	14.948	O^{5+}	$1s^23p\ ^2P_{1/2}$ - $1s^22s\ ^2S_{1/2}$
4.	29.92	14.956	O^{5+}	$1s2s3d\ ^2D_{3/2}$ - $1s2s2p\ ^2P_{1/2}$
4.	29.92	14.959	O^{5+}	$1s2s3d\ ^2D_{5/2}$ - $1s2s2p\ ^2P_{3/2}$
4.	29.92	14.961	O^{5+}	$1s2s3d\ ^2D_{3/2}$ - $1s2s2p\ ^2P_{3/2}$
5.	29.97			
6.	30.07			
7.	30.18	15.072	O^{5+}	$1s2s3s\ ^4S_{3/2}$ - $1s2s2p\ ^4P_{1/2}$
7.	30.18	15.079	O^{5+}	$1s2s3s\ ^4S_{3/2}$ - $1s2s2p\ ^4P_{3/2}$
7.	30.18	15.089	O^{5+}	$1s2s3s\ ^4S_{3/2}$ - $1s2s2p\ ^4P_{5/2}$
8.	30.20	15.102	O^{4+}	$1s^22s4d\ ^3D_1$ - $1s^22s2p\ ^3P_0$
8.	30.20	15.110	O^{4+}	$1s^22s4d\ ^3D_2$ - $1s^22s2p\ ^3P_1$
8.	30.20	15.110	O^{4+}	$1s^22s4d\ ^3D_1$ - $1s^22s2p\ ^3P_1$
8.	30.20	15.117	O^{4+}	$1s^22s4d\ ^3D_3$ - $1s^22s2p\ ^3P_2$
8.	30.20	15.117	O^{4+}	$1s^22s4d\ ^3D_3$ - $1s^22s2p\ ^3P_2$
9.	30.36	30.378	He^+	$2p\ ^2P_{3/2}$ - $1s\ ^2S_{1/2}$
9.	30.36	30.379	He^+	$2p\ ^2P_{1/2}$ - $1s\ ^2S_{1/2}$
10.	30.44	15.206	O^{5+}	$1s2s3d\ ^2D_{3/2}$ - $1s2s2p\ ^2P_{1/2}$
10.	30.44	15.214	O^{5+}	$1s2s3d\ ^2D_{5/2}$ - $1s2s2p\ ^2P_{3/2}$

ACKNOWLEDGMENT

Support of this research in part by DOE and NSF Nevada Chemical Physics EPSCoR and in part by Lawrence Livermore National Laboratory is gratefully acknowledged.

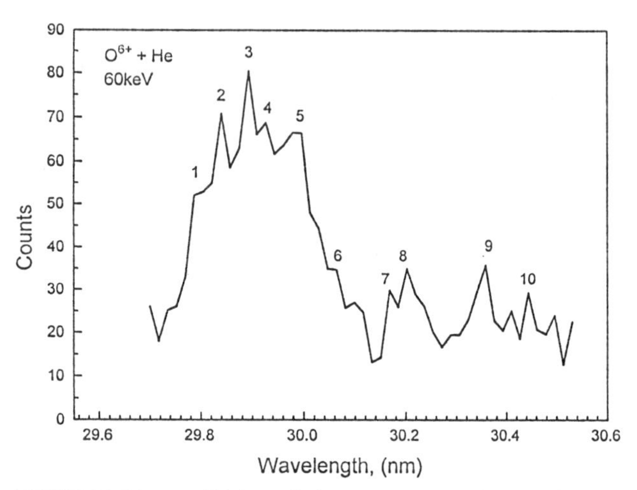

FIGURE 4. Measured high-resolution EUV spectrum in the vicinity of the He^+ (2p-1s) line at 30.38 nm, produced in collisions of O^{6+} ions with He atoms at 60 keV. The experimental resolution is approximately 200.

REFERENCES

1. R.K. Janev and H. Winter, Phys. Rep. **117**, 265 (1985).
2. H.B. Gilbody, p. 143 in *Advances in Atomic and Molecular Physics*, Vol. 22, D. Bates and B. Bederson, eds., Academic Press (1986).
3. M. Barat and P. Roncin, J. Phys. B **25**, 2205 (1992).
4. R. Ali, C.L. Cocke, M.L.A. Raphaellian, and M. Stöckli, J. Phys. B **26**, L177 (1993).
5. S. Martin, J. Bernard, Li Chen, A. Denis and J. Désesquelles, Phys. Rev. A **52** 1218 (1995)
6. A. Niehaus, J. Phys. B **19**, 2925 (1986).
7. W. Fritsch, Nucl. Instrum. Meth. Phys. Res. **B98**, 246 (1995); Phys. Lett. A **192**, 369 (1994).
8. D. Dijkkamp, D. Čirič, E. Vlieg, A. Boer, and F.J. de Heer, J. Phys. B **18**, 4763 (1985).
9. R. Hoekstra, J.P.M. Beijers, F.J. deHeer and R. Morgenstern, Z. Phys. D **25**, 209 (1993).
10. D. Hitz, P. Ludwig, G. Melin and M. Pontonnier, Nucl. Instrum. Meth. Phys. Res. **B98**, 517 (1995).
11. A. Bordenave-Montesquieu, P. Benoit-Cattin, A. Gleizes, A. I. Marrakchi, S. Dousson and D. Hitz, J. Phys. B **17**, L223 (1984).
12. R. Mann, Phys. Rev. A **35**, 4988 (1987).
13. N. Stolterfoht, C.C. Havener, R.A. Phaneuf, J.K. Swenson, S, Shafroth and F.W. Meyer, Nucl. Instrum. meth. **B27**, 584 (1987).
14. V. Kantsyrev, R. Bruch, M. Bailey and A. Shlyaptseva, Appl. Phys. Lett. **66**, 3567 (1995).
15 R.D. Cowan, *The Theory of Atomic Structure and Spectra*, Univ. of California Press, 1981.
16. R.L. Kelly, *Atomic and Ionic Spectrum Lines Below 2000 Angstroms*, Report ORNL-5922, Oak Ridge National Laboratory (1982)
17. F.W. Meyer, C.C. Havener, K.J. Snowdon, S.H. Overbury, D.M. Zehner and W. Heiland, Phys. Rev. A **35**, 3176 (1987).

PRODUCTION AND COLLISIONS OF NEUTRAL BEAMS IN THE 100 keV TO 6 MeV RANGE

J.M. Sanders and S.L. Varghese

Dept. of Physics, University of South of Alabama, Mobile, AL 36688

S. Datz, E. Deveney, H.F. Krause, J.L. Shinpaugh, and C.R. Vane

Physics Division, Oak Ridge National Laboratory, Oak Ridge, TN 37831

R.D. DuBois

Dept. of Physics, University of Missouri-Rolla, Rolla, MO 65401

High energy beams of neutral atoms can be produced by using electron capture to neutralize an accelerated ion beam. Useful beams of neutral atoms with Z=1 to 6 and squared-velocities up to 0.5 MeV/u have been produced in this manner. The neutral beams may then be used to study a wide variety of processes: electron capture, electron loss from the projectile, ionization of the target, and multiple-electron processes involving both collision partners. Systematics of these processes as functions of energy, projectile charge state, and projectile Z will be discussed.

INTRODUCTION

Collisions involving charged projectiles have been studied in great detail over the last seventy years, and many important interactions between projectile ions and target atoms have been investigated. High energy collisions involving neutral projectiles have received far less attention, but there are several reasons why investigation of collisions of high energy neutral projectiles would be of interest. Neutral beams provide cross-sections relevant to neutral-beam heating of plasmas. Using neutral projectiles allows measurement of cross sections for processes where an atomic target would be cumbersome, e.g. a useful C^0 beam may be more readily reliably obtained than a C^0 target. In addition to the more practical reasons, there are some more fundamental advantages to neutral-neutral collisions. For example, electron-electron interactions are maximized, since such a collision contains the greatest number of electrons while the nuclear charges are screened. Finally, since neither collision partner is charged, there are no strong long-range interactions in neutral-neutral collisions.

Collisions involving neutral projectiles have been extensively studied at low (less than 10 keV) energies (1), but collisions at high energies have received relatively less attention. For heavy projectiles at higher energy, C^0 and O^0 projectiles at 10 to 65 keV in collisions with noble gases and H_2, N_2, and O_2 were studied by Fogel, Ankudinov, and Pilpenka (2); Dmitriev *et al.* investigated C^0, N^0, and O^0 projectiles at 35 keV colliding with Ne, H_2, and Ar (3); Hird *et al.* (4,5,6) performed experiments using neutral beams of the halogens with energies ranging from 20 to 110 keV; and Nakai and Sataka studied C^0 projectiles on He at 25 keV/u to 125 keV/u (7). In each of these experiments, total cross sections for electron capture and for electron loss by the projectile were measured. Our recent work at Oak Ridge National Laboratory has extended the energy range up to 500 keV/u, and we have also begun investigating the capture and loss cross-sections coincident with particular final charge-states of the target atom. In work currently underway at the University of South Alabama, these coincidence measurements are being extended to the lower energy range (5 to 10 keV/u).

PRODUCTION OF NEUTRAL BEAMS

A very common method for producing neutral beams is by charge transfer to accelerated singly-charged ions from a gaseous target. In the present experiment, we have used both Ar and N_2 as the neutralizing gas, and we found that the two give very similar degrees of neutralization.

If we restrict our consideration to single capture and loss processes, then the fraction ϕ_0 of the incident beam which emerges in the neutral charge state from the neutralizing cell is given (to second order in the areal target

CP392, *Application of Accelerators in Research and Industry*, edited by J. L. Duggan and I. L. Morgan
AIP Press, New York © 1997.

thickness x) by

$$\phi_0 = \sigma_{10}x - \frac{1}{2}\sigma_{10}\sigma_{01}x^2 - \frac{1}{2}\sigma_{10}^2x^2 \quad (1)$$

where σ_{10} is the capture cross section for the singly-charged ion and σ_{01} is the loss cross section for the neutral atom. Eq. (1) will be maximized when

$$x_m = \frac{1}{\sigma_{01} + \sigma_{10}}, \quad (2)$$

and if, as is frequently the case, the loss cross section is much larger than the capture cross section, then

$$x_m \approx \frac{1}{\sigma_{01}}. \quad (3)$$

When the target thickess has this maximizing value, the neutral fraction is

$$\phi_0 \approx \frac{1}{2}\frac{\sigma_{10}}{\sigma_{01}}. \quad (4)$$

As an example, if 0.1 keV/u C^{1+} projectiles were to be neutralized by He, Eqs. (3) and (4) indicate that the maximum neutral fraction would occur when $x = 3.5 \times 10^{-15}$ atom/cm^2, and neutrals would comprise about 8% of the beam.

This method of neutralizing a positive beam by electron capture has some drawbacks. As Eq. (4) shows, the neutral yield is determined by the ratio of capture cross section to the loss cross section, and this ratio will generally decrease rapidly with increasing beam energy, so neutralization by capture becomes progressively more difficult as the beam energy is increased. In particular, for projectiles with 100 keV/u or greater, σ_{10} scales roughly as E^{-4} while σ_{01} is proportional to $\ln E/E$, so the neutral fraction

$$\phi_0 \propto \frac{1}{E^3 \ln E}. \quad (5)$$

Increasing the beam energy from 100 keV/u to 500 keV/u would, therefore, reduce the expected yield of neutral beam by a factor of more than 125. Clearly, the electron capture method for producing neutral beams will have limited usefulness for energies much greater than 500 keV/u.

The second drawback of this method for producing neutral beams is that the electron can be captured into long-lived metastable states. Therefore, the beam will not consist of one well-defined ground state, but a mixture of ground-state and metastable state atoms. While some processes of interest may be relatively insensitive to the initial state of the projectile, some, such as electron loss may have quite different cross sections for the ground state versus a metastable state. Hird *et al.* (4,5,6) investigated an alternate method of producing a neutral beam by electron detachment from a negative ion. Neutral beams formed in this fashion should have only a very small metastable component. The beams produced by Hird *et al.* were for the halogens F^0, Cl^0, and I^0. The results for electron loss cross sections for Hird's F^0 beam could be compared to the earlier results of Fogel *et al.* (8) who had produced their beam using the capture method. For He and Ne targets, the two methods gave very similar results, but for Kr and Xe targets the Hird's loss cross sections were a factor of 1.5 to 2 higher. Thus for He and Ne targets, it would appear that any metastable fraction in the incident beam does not significantly affect the cross sections.

EXPERIMENT

In the present experiment, the beams were made by neutralizing a singly-charged, positive beam by electron capture as discussed above. A schematic of the apparatus is given in Fig. (1).

Beams of singly-charged ions of the desired energy and species were obtained from the accelerator and directed into the neutralizing cell. Upon emerging from the neutralizing cell, the beam passed through a magnetic field which swept away the remaining charged beam particles. The neutral beam particles then entered the target (or recoil) cell which contained the target gas. In the present experiment, He was used as the target. The beam emerging from the target cell passed through an electrostatic deflector which analyzed the various charge states in the beam, and finally the beam particles were detected by a position-sensitive detector. The target cell had a small electric field perpendicular to the beam direction which served to extract the recoil ions and direct them onto a detector. The time-of-flight of the recoil ions provided the charge state of each recoil ion.

For each beam, a measurement of total charge-changing cross sections was performed by varying the pressure in the target gas cell. The relevant cross section was then proportional to the slope of the fraction of the charge changed beam as a function of pressure. In addition, however, the number of projectiles in each charge state coincident with recoil charge states were also counted. In this fashion, cross sections could be obtained which specified final charge state of the target (9).

RESULTS AND CONCLUSIONS

Examples of cross sections of neutral beams colliding with He are given in Figs. (2) and (3). In Fig. 2, cross sections for electron capture, transfer ionization, single ionization of He, and double ionization of He are shown. In Fig. (3), the cross sections are for ionization of He

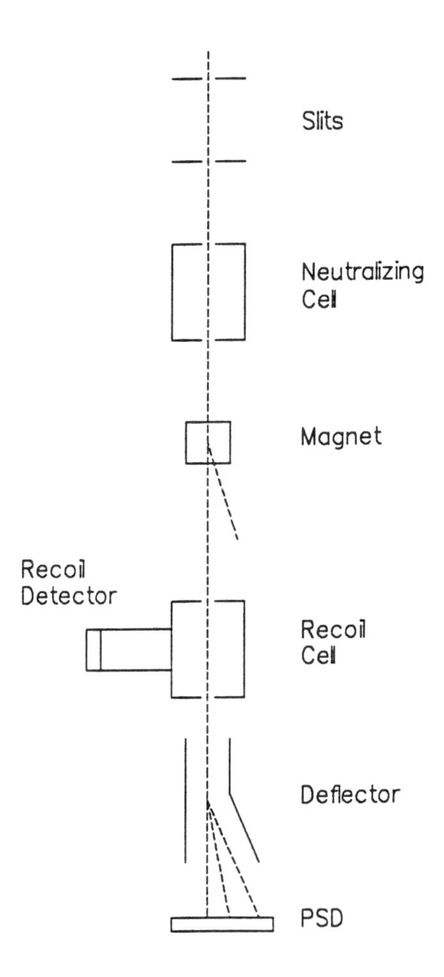

FIGURE 1: Schematic of the experimental apparatus. The dotted line represents the path taken by the beam particles. At the magnet following the neutralizing cell, unwanted ions are swept away. The beam is charge-state analyzed by an electrostatic deflector following collisions with the target gas in the recoil cell.

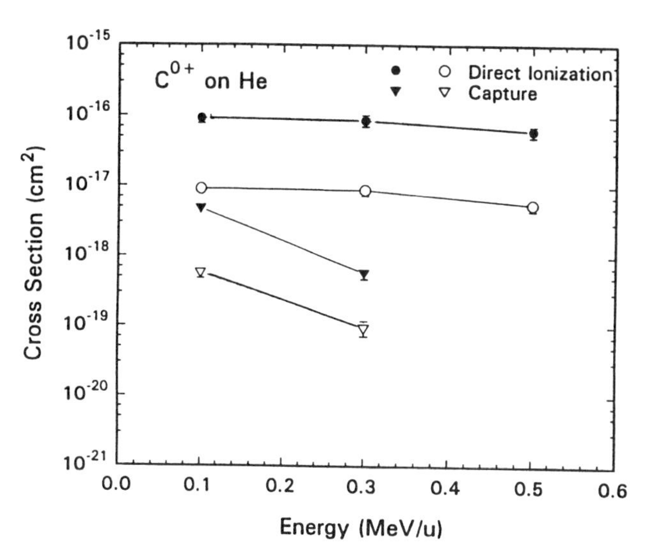

FIGURE 2: Cross sections for single capture, transfer ionization, single ionization of He, and double ionization of He in collisions of 100 keV/u C^0 with He. The solid symbols represent those processes which leave a He^{1+} recoil, while the open symbols are for processes which leave a He^{2+}. The lines are drawn to guide the eye.

in addition to different degrees of electron loss from the projectile. A noticable feature of these cross sections is that the cross sections for processes which leave an He^{2+} recoil maintain roughly a constant ratio to the corresponding processes which leave a He^{2+}. The ratio, which is approximately an order of magnitude, agrees with the ratio between single and double ionization. We may conclude therefore that these collisions can be treated in an independent electron approximation where one simply multiplies the probability for each process independently.

In conclusion, we have found that useable neutral beams of heavy projectiles up to Z=6 and at squared-velocities up to 0.5 MeV/u can be produced by electron capture to singly-charged ions. Although some fraction of the neutral atoms may be in metastable, many cross sections of interest seem to be relatively insensitive to the exact initial state of the projectile. We have additionally found that multiple loss coincident with ionization of He can be understood within an independent electron model.

ACKNOWLEDGEMENTS

Part of this research was supported by the USDOE, Office of Basic Energy Sciences, Division of Chemical Sciences, under Contract No. DE-AC05-96OR-22464 with Lockheed Martin Energy Research Corp. and by appointments to the USDOE Laboratory Cooperative Postgraduate Research Training Program administered by ORISE.

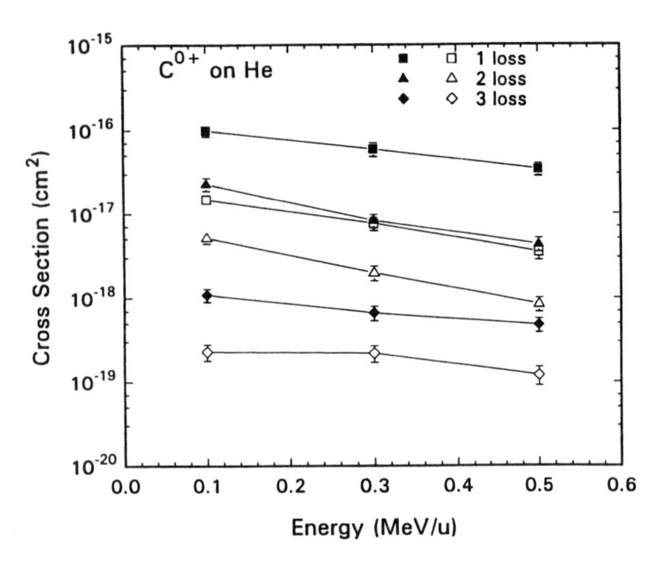

FIGURE 3: Cross sections for projectile electron loss in coincidence with single and double ionization of He in collisions of 100 keV/u C^0 with He. The solid symbols represent those processes which leave a He^{1+} recoil, while the open symbols are for processes which leave a He^{2+}. The lines are drawn to guide the eye.

REFERENCES

1. McDaniel, E.W., Mitchell, J.B.A., and Rudd, M.E., *Atomic Collisions: Heavy Particle Projectiles*, New York: John Wiley & Sons, 1993.
2. Fogel, I.M., Ankudinov, V.A., and Pilipenko, D.V., *Soviet Physics: JETP* **8**, 601-605 (1958).
3. Dmitriev, I.S., Nikolaev, V.S., Teplova, Y.A., Popov, B.M., and Vinogradova, L.I., *Soviet Physics: JETP* **23**, 832-837 (1966).
4. Hird, B., Rahman, F., and Orakzai, M.W., *Can. J. Phys.* **66**, 972-977 (1988).
5. Hird, B., Rahman, F., and Orakzai, M.W., *Phys. Rev. A* **37**, 4620-4624 (1988).
6. Hird, B., Orakzai, M.W.,and Rahman, F., *Phys. Rev. A* **39**, 5010-5013 (1989).
7. Nakai, Y. and Sataka, M., *J. Phys. B: At. Mol. Opt. Phys.* **24**, L89-L91 (1991).
8. Fogel, Y.M., Ankudinov, V.A., and Pilipenko, D.V., *Soviet Physics: JETP* **11**, 18 (1960).
9. Shinpaugh, J.L., Sanders, J.M., Hall, J.M., Lee, D.H., Schmidt-Böcking, H., Tipping, T.N., Zouros, T.J.M., and Richard, P., *Phys. Rev. A* **45**, 2922-2928 (1992).

MEASUREMENTS OF THE DIFFERENTIAL CROSS SECTIONS FOR K-SHELL IONIZATION AND BREMSSTAHLUNG WITH A MULTI-PARAMETER COINCIDENCE SPECTROMETER

Vinod Ambrose* and C. A. Quarles

Physics Department,TCU, Fort Worth TX 76129

A multi-parameter coincidence spectrometer (MPCS) coupled to a 300 kV electron accelerator has been developed for the study of inelastic electron-atom scattering cross sections. As a first application, the spectrometer has employed three detectors: one Si(Li) electron detector at 45°, one Ge photon (x-ray) detector in the electron scattering plane at -45° and a second Ge photon (x-ray) detector at 90° to the scattering plane. The doubly differential cross section (DDCS) for K-shell ionization and the triply differential cross section (TDCS) for bremsstrahlung have been measured at 100 and 140 keV for targets of Cu, Ag and TbF_3. The K-shell DDCS exhibit the increase in cross section at higher scattered electron energy seen in earlier work. The initial results at the two different photon angles suggest that there may be some angular dependence in the K x-ray emission angle observed which indicate that the data should be interpreted as triply differential in the K x-ray emission angle as well as the electron angle and energy. The TDCS for bremsstrahlung is in fair agreement with the Bethe-Heitler formula. While the present data do not extend to low enough scattered electron energy (high enough radiated photon energy) to be compared with the recent calculation of Shaffer and Pratt, it is hoped that further data will provide overlap for a few points and a test of the theoretical calculations. Future plans to measure the above cross sections at different angles and energies, as well as plans for new studies of double bremsstrahlung and triple coincidence processes with the new MPCS will be discussed.

INTRODUCTION

An electron inelastically scattering from an atom can result in X-ray emission, either by the process of bremsstrahlung, or by inner-shell ionization or excitation. In the first process, the electron is accelerated by the Coulomb field of the screened nucleus or the orbital electrons. This acceleration can result in the emission of the X-ray with an energy that can range up to the incident electron energy. In the second process, the atom is excited when a bound electron is either ejected from the atom or raised to a higher energy state. The subsequent return of the ion (atom) to the ground state can result in the emission of characteristic X-rays.

These methods of X-ray production find many applications in industry. Bremsstrahlung is the method by which X-rays are produced in X-ray tubes. In various electron microscopy techniques, samples are examined by observing the X-ray spectrum from electrons incident on the sample. It is therefore important to study these cross sections at a fundamental level.

In the present work, a Multi-Parameter Coincidence Spectrometer (MPCS) has been developed as a versatile tool for measuring these cross sections. With the MPCS these cross sections can be measured differential in the emitted photon's and scattered electron's energy and direction. The MPCS is coupled to a 300 keV electron accelerator, and thin-film targets of approximately 50 $\mu g/cm^2$ are used. The MPCS has been described in a preliminary way previously (1,2), and a more detailed description of the hardware and software is being submitted for publication separately.

There are two major advantages of the MPCS. First, it is capable of detecting two-fold coincidences between up to five electron and/or photon detectors simultaneously. Second, the measurements are energy dispersive in both photon and electron energy. These two improvements result in a dramatic increase in data collection rates. It is now possible to measure triply differential cross sections (TDCS) for bremsstrahlung as a function of emitted photon energy, and doubly differential inner-shell ionization cross sections (DDCS) as function of outgoing electron energy simultaneously. Cross sections at several emission angles can also be measured in a single run.

THE MPCS APPARATUS

As a first application the MPCS has been set up with three detectors. The electron detector is a cooled Si(Li) detector with a 0.3 mil Be window placed at 45° to the incident electron beam. The photon detectors are high purity germanium detectors with 1 to 3 mil Be windows, separated from the target chamber vacuum by 5 mil Mylar windows. These windows act as electron stoppers, but are thin enough to efficiently transmit photons above about 5 keV. One photon detector is placed at -45° to the beam, and the other at 90° in a non-coplanar arrangement. All three detectors are collimated to reduce efficiency loss edge effects, but to keep the solid angles as large as possible. Large solid angles facilitate testing of a coincidence system, although at the expense of good angular discrimination. Use of a small 3-inch target chamber also permits larger solid angles. The coincidence electronics is a combination of NIM and CAMAC hardware. The system control, data acquisition and data analysis software has been written in Visual Basic and runs on a PC computer under Windows 3.1.

CP392, *Application of Accelerators in Research and Industry*, edited by J. L. Duggan and I. L. Morgan
AIP Press, New York

Test case of K-L cascade for Tb

As an example of the MPCS, the raw data for coincidences between two photon detectors for 100 keV electrons incident on a TbF_3 target are shown in figure 1. The time spectrum is to the extreme left. The time scale is from -400ns to +400ns. A window within which real coincidences are expected to occur is marked. Left of the time spectrum and on the top are the energy spectra of the two detectors. Bottom right is the two-dimensional histogram of coincidences as function of detector energies. The density of points in this plot is approximately indicative of the number of coincidences. An energy window (box) used for cross section extraction is shown. The corresponding location of the window in the energy spectra is also shown. The run parameters, and the counts within the time and energy windows are shown in the upper left. The K-L characteristic X-ray

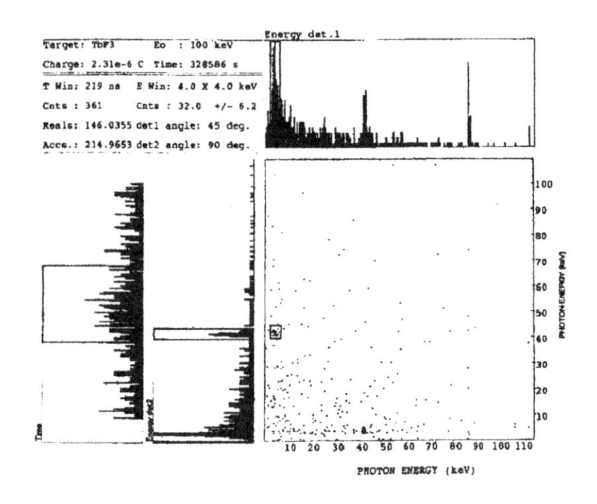

FIGURE 1. Raw data spectrum for TbF_3. The energy box is centered on one of the K-L cascade coincidence regions at (7 keV, 43 keV).

cascade is seen as a cluster of points at (7 keV, 43 keV) and at (43 keV, 7 keV). The K-shell ionization cross section calculated from these cascade coincidences is 8.8± 1.2 barns and agrees well with 8.1±1.2 from theory (3). This measurement with a precision of about 15% serves as a calibration of the MPCS for photon-photon coincidence cross sections.

RESULTS FOR 100 AND 140 KEV FOR CU, AG, AND TBF_3

Figure 2 shows the raw data for electron-photon coincidences for 100 keV electrons incident on Ag. The vertical band from the K X-rays can be seen at 22-25 keV. Also visible is the diagonal bremsstrahlung band which corresponds to the energy conservation requirement that the sum of the scattered electron and emitted photon energy be equal to the incident electron energy.

The Si(Li) detector efficiency response for electrons is different from its well-understood response to x-rays. The mono-energetic elastically scattered electron peak is accompanied by a low-energy tail. This is due to scattering with the free electrons within the detector crystal, some of which leave the active area causing a net lower energy to be deposited. The intensity of this tail diminishes with increasing electron energy, and is not a problem as long as enough electron energy channels (corresponding to 5 to 10 keV) are summed. The energy loss in the 0.3 mil Be window and Si detector deadlayer can be corrected by using values for electron energy loss tabulated by Berger and Seltzer (4). Electrons below an energy of ~25 keV do not penetrate the window. In the cross sections presented below, correction has been made for electron energy loss in the window and detector dead layer.

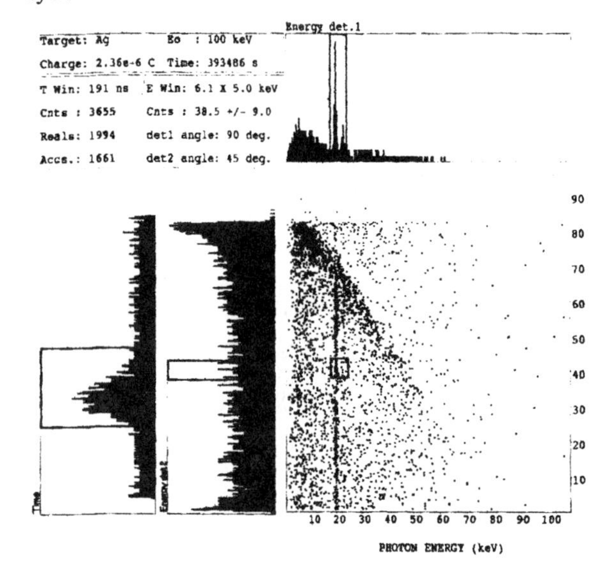

FIGURE 2. Raw spectrum for Ag. The energy box is centered on the K x-rays at an electron energy of 40 keV to compute the DDCS for K-shell ionization.

DDCSK and K x-ray isotropy

Preliminary values for the DDCS for K-shell ionization at 140 keV for Ag calculated from the vertical band is shown in figure 3 for 45° and 90° photon emission in comparison with previous data. The data agree well with the results of Genz (5) and Quarles and Faulk (6).

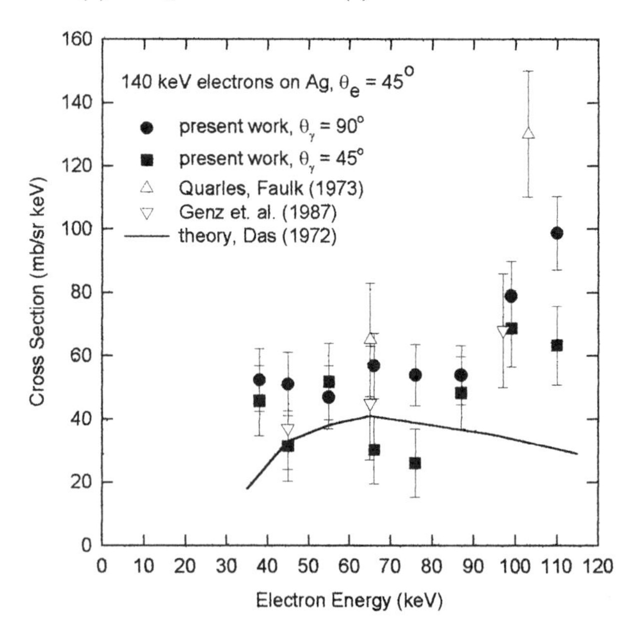

FIGURE 3. K-shell DDCS for 140 keV electrons on Ag.

Results for 100 keV on Cu and Ag are shown in Figure 4. It is clear that the present data confirm the trend of the K-shell DDCS to increase as the detected electron energy increases toward the maximum of the incident energy minus the K-shell binding energy. This is in disagreement with the theoretical calculations for the angle of 45°. It should be pointed out, however, that in the present data no correction has been made for the angular resolution which is about ± 5°.

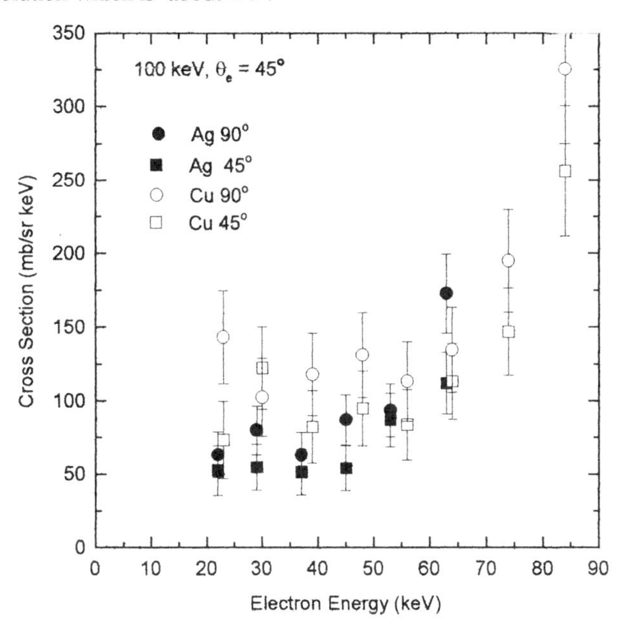

FIGURE 4. K-shell DDCS for 100 keV electrons on Cu and Ag.

The theoretical calculation of Das(7) using Dirac plane waves for the incident and scattered electron, non-relativistic hydrogenic wave functions for the bound electron and Coulomb wave functions for the ejected electron is also shown in Fig. 3. The theoretical problem has been considered most recently by Sud and Moattar (8) substituting Darwin wave functions for the K-shell atomic and ejected electrons. The results are not very sensitive to the wave functions used for the K-shell or ejected electron or to the inclusion of spin flip or exchange. While the order of magnitude of the experimental results are given rather well for the intermediate electron energy (~70 keV), the theory does not predict the observed increase of the cross section with increasing detected electron energy. Perhaps this is due to the use of Dirac plane waves for the incident and scattered electron, a common assumption of all versions of the theory. Calculations are not yet available for the 100 keV case shown in Fig. 4.

Alternatively, the data could be interpreted as a triply differential cross section for the emission of the K x-ray into a particular angle. However, the dominant ionization process is generally assumed to be isotropic and to proceed by a two-step process in which the K x-ray is emitted by the ion at a time sufficiently long after the ionization that there is no memory of the incident electron direction. This has been the assumption of the current theories discussed above. While the data at 90° is about one standard deviation higher than the data at 45°, we do not consider this to be significant at this time since there are several systematic errors related to the determination of the detector solid angles which are of this order. With more precise measurement of the detector solid angles, it should be possible to look for an anisotropy in the K x-ray emission which may arise from the polarization bremsstrahlung process.

TDCS for Bremsstrahlung

Figures 5 and 6 show the first results obtained with the MPCS for the bremsstrahlung TDCS at 45° and 90° photon emission for 140 keV on Ag and for 100 keV on Cu, Ag and TbF_3. The cross sections are scaled by $1/Z^2$ and compared with the Bethe-Heitler formula, including the Elwert factor. At 140 keV, the agreement with Bethe-Heitler is better at 90° where the cross section is larger than at 45°. For 100 keV, the agreement with Bethe Heitler is somewhat better for Cu and Ag than for Tb as one may expect, and again, the agreement is better, especially with regard to energy dependence for the 90° data where the cross section is larger. Of course, it is hoped that the data can be compared with the recent partial wave calculations of Shaffer and Pratt presented at this conference (9). It is also hoped that future data planned at 200keV can provide a larger region of overlap between experiment and theory by extending the range of measurable photon energy to a larger fraction of the incident energy. The experiment, with the present electron detector, is able to detect electrons down to about 40-50 keV. At 100 keV incident energy this corresponds to a maximum radiated fraction of energy of about 0.5-0.6. At 200 keV incident energy, however, the radiated fraction would extend up to 0.75-0.8.

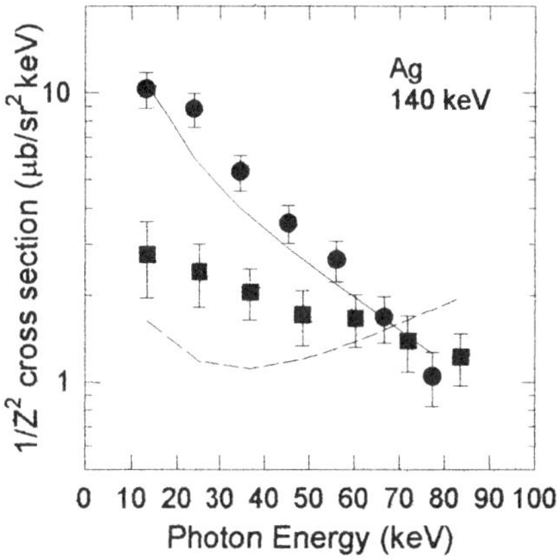

FIGURE 5. TDCS for bremsstrahlung for 140 keV electrons on Ag for θ_e = 45°. The circles are for θ_γ=90°, the squares are for θ_γ= 45°. The solid line is the Bethe-Heitler formula for 90°, the dashed line is for 45°.

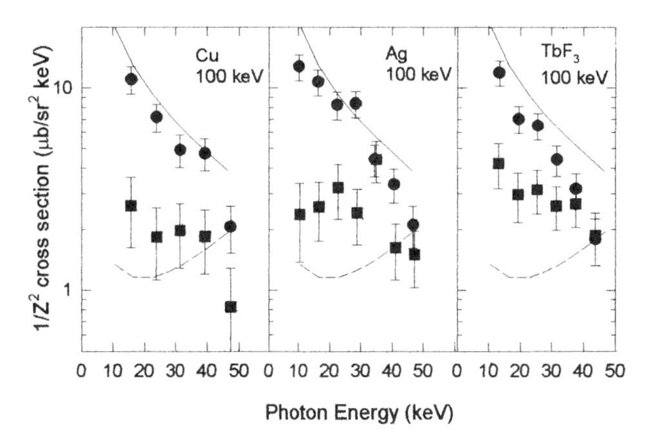

FIGURE 6. Same as Fig. 5 except for 100 keV electrons on Cu, Ag, and TbF_3.

FUTURE PROSPECTS

We are satisfied that the measurements presented here demonstrate the capability, both in accuracy and precision, of the MPCS. We have the hardware, the data acquisition, and data analysis software debugged and working well. We can anticipate some minor refinements in operation, but mainly we plan to add one or two additional detectors to bring the number up to the design level of five detectors. We will design and construct additional target chambers to investigate other angles of interest. The results presented here are the first from the MPCS, but they are not the best we can do. Several refinements are planned to reduce further the usual systematic errors in a cross section measurement. We can improve on the measurement of detector solid angles, which has been done to date using a radioactive source placed at the position of the target. One interesting refinement, which is possible when three or more detectors are used, is the normalization to the accidental rate if the source of the singles counting rate in each detector is well understood as it is in this experiment. While target thickness remains the largest source of error in these experiments, it is possible to focus on relative measurements or ratios among detectors which can significantly reduce the errors involved and provide good tests of theory.

The versatility of the MPCS permits us to focus on several problems. First, we want to investigate the K-shell ionization cross section further as a function of both electron scattering angle and x-ray emission angle. The object is to test the range of validity of the theoretical model which already seems to break down at higher detected electron energy and larger scattering angle. We also want to investigate the isotropy of the emission of the K x-ray. We can begin a study of L shell ionization, although the L-shell cross section behavior is considerably more complicated to unravel, and there are no theoretical calculations for the L shell available yet. (We have not analyzed the Tb L-shell data obtained in the present experiment).

For bremsstrahlung, we intend to select carefully and with some input from the theoretical side a variety of energies and angles to provide a test of the latest theory at several points. In order to have a larger range of photon energy to compare with theroy, we may want to raise the beam energy to 200 keV. While seemingly straightforward to do this, we anticipate more significant shielding and coincidence cross-talk problems between detectors as the beam energy is increased. Additional shielding or cross-talk minimization between detectors may necessitate a larger target chamber reducing solid angles and increasing run times required to achieve the desired level of statistics.

Finally, there is continuing interest in double bremsstrahlung. Of particular interest is the discrepancy observed in the two-photon coincidence measurement when one photon is at the characteristic K x-ray energy of the target and the second photon is a continuum bremsstrahlung photon.(10). A possible model for the observed enhancement has been discussed at this conference by Kukushkin(11). The MPCS is capable of measurement of double bremsstrahlung cross sections. Indeed, photon-photon coincidences were obtained while the present data were being collected. (See Fig. 1, for example). However, because the cross section for double bremsstrahlung is several orders of magnitude lower than the cross sections reported here, much longer run times are required to obtain any statistical significance. While we could take data on the ionization or one-photon bremsstrahlung processes simultaneously with double bremsstrahlung, it is probably better to focus on the process of interest and try to optimize the conditions for the measurement of the much smaller cross section.

We also anticipate looking for triple coincidences. This could be $(ee\gamma)$ for K-shell ionization or radiative ionization (bremsstrahlung + ionization). Or it could be $(e\gamma\gamma)$ in the case of double bremsstrahlung, where we would detect both photons in coincidence with the scattered electron. Such a fully differential cross section is well within the capability of the MPCS as currently set up. While we did detect several triples during the course of the initial runs presented here, we did not have sufficient triple coincidence events to merit analysis. The data analysis program needs to be upgraded to include triple coincidence event analysis, such as the inclusion of energy and timing cuts to distinguish real from accidental coincidences. Also, perhaps it is not so well known, but the measurement of triple coincidences requires a careful selection of beam rate to optimize the statistical error in the real minus accidental correction. For double coincidences, it is always optimum to run at the highest beam rate consistent with an acceptable pile-up or deadtime correction, although few coincidence experiments do this consistently. For triples, on the other hand, running at the highest rate is not optimum. So, detection of triples with optimum statistics requires tuning the MPCS to the optimum counting rate, which is not typically the best for studying doubles.

ACKNOWLEDGEMENTS

We appreciate the support of the Welch Foundation during the early stages of this research, and the continuing support of the TCU Research Fund. We thank Michael Murdock for much help, especially with the target chambers.

REFERENCES

* Now at Digital Equipment Corp., 77 Reed Rd., HLO2-3/J9, Hudson MA 01749-2895

1. Jingai Liu and C. A. Quarles, *Nucl. Instr. Meth. Phys. Res.* **B79**, 825(1993).
2. Vinod Ambrose and C. A. Quarles, *Nucl. Instr. Meth. Phys. Res.* **B99,** 170 (1995).
3. M. Gryzinski, *Phys. Rev. A* **138** 395; C.J. Powell, *Rev. Mod. Phys.* **48**, 33(1976).
4. M. J. Berger and S. M. Seltzer, "Tables of Energy Losses and Ranges of Electrons and Positrons," NASA (1964).
5. H. Genz, "Determination of the K Shell Momentum Distribution of Cu and Ag using Relativistic Electrons," AIP Press (1990).
6. C. A. Quarles and J. D. Faulk, *Phys. Rev. Letters* **31**, 859 (1973).
7. J. N. Das, *Nuovo Cimento* **12B**, 197 (1972); J. N. Das, *J. Phys. B: At. Mol. Phys.* **7**, 923 (1974); J. N. Das and A. N. Konar, *J. Phys. B: At. Mol. Phys.* **7**, 2417(1974) ; J. N. Das and S. Chakrobarty, *Phys. Rev. A* **32** , 176(1985).
8. K. K. Sud and S. Moattar, *J. Phys. B: At. Mol. Opt. Phys.* **23**, 2363 (1990).
9. C. D. Shaffer and R. H. Pratt, paper in this conference proceedings.
10. D. L. Kahler, Jingai Liu, and C. A. Quarles, *Phys. Rev. A* **45**, R7663 (1992).
11. A. B. Kukushkin, paper in this conference proceedings.

INNER-SHELL IONIZATION OF ATOMS BY CHARGED PARTICLES IN THE DISTORTION APPROXIMATION

T. Mukoyama* **and C.-D. Lin†**

*Institute for Chemical Research, Kyoto University, Uji, Kyoto 611, Japan

†Department of Physics, Kansas State University, Manhattan, Kansas 66506, USA

The inner-shell ionization cross sections by charged-particle impact have been calculated in the distortion approximation. A finite-basis-set variational approach is used to solve the Schrödinger or Dirac equation for atomic potentials. The obtained atomic orbitals and pseudostates are employed to calculate the inner-shell ionization cross sections. The results are compared with the experimental data and other theoretical calculations.

INTRODUCTION

In order to calculate inner-shell ionization cross sections by charged particles, the first-order theories, such as the plane-wave Born approximation (PWBA) (1) and the semiclassical approximation (SCA) (2), have been successfully applied. However, for low-energy and/or highly charged projectiles the discrepancy between these simple theories and experimental results become significant. Although several correction methods have been suggested (3), it is important to develop more elaborate theoretical models to calculate the ionization cross sections taking into account various effects. In the ionization process, the final electron is in the continuum state. However, it is not easy to calculate transition matrix elements with continuum wave functions. This fact makes it difficult to treat the ionization process in higher-order approximations than the first-order Born theories.

Recently we have shown (4) that the finite-basis-set expansion method is useful to calculate the inner-shell ionization cross sections by heavy charged particles. In this approach, the target electron wave function is expanded in terms of square-integrable (L^2) basis functions and *pseudostates*, i.e. discretized states with positive energy eigenvalues, obtained by diagonalizing the atomic Hamiltonian, are used instead of final continuum states. The main advantage of this method is its simplicity because all wave functions both for bound and continuum states are expressed in terms of the same basis functions. This fact is very useful when we go beyond the first-order Born approximation.

Using the pseudostates, the calculations of the ionization cross sections are made with the distortion approximation. This approximation was first developed by Bates (5) for calculations of the $1s$–$2s$ excitation cross sections of hydrogen atoms by protons and α particles. Since then many applications have been reported for excitation and charge-transfer processes. The calculations of K-shell ionization cross sections by charged particles have been made in the distortion approximation and the results are in good agreement with the experimental data.

In the low-energy region where the distortion approximation is useful, the electronic relativistic effects also play an important role (5). In order to take into consideration the relativistic effects, we have extended the finite-basis-set expansion method for the relativistic case and calculated the K-shell ionization cross sections in the distortion approximation.

METHOD

Nonrelativistic case

The Schrödinger equation for an electron in a potential $V(r)$ is written by

$$\boldsymbol{H}\phi = E\phi , \tag{1}$$

where the Hamiltonian $\boldsymbol{H}$ is

$$\boldsymbol{H} = -\frac{1}{2}\nabla^2 + V(r) , \tag{2}$$

where r is the radial distance of the electron from the nucleus. We use atomic units throughout the present work.

We expand the eigenfunction of the Hamiltonian $\boldsymbol{H}$ in terms of a set of L^2 basis functions $\chi_i(r)$

$$\phi_{nlm}(\boldsymbol{r}) = \sum_i^N c_i\,\chi_i(r)\,Y_{lm}(\hat{\boldsymbol{r}}) , \tag{3}$$

where c_i is the expansion coefficient, n, l, m are the principal, orbital, and magnetic quantum numbers, $Y_{lm}(\hat{\boldsymbol{r}})$ is the spherical harmonics, $\hat{\boldsymbol{r}}$ is the unit vector in the direction of the position vector $\boldsymbol{r}$.

CP392, *Application of Accelerators in Research and Industry*, edited by J. L. Duggan and I. L. Morgan
AIP Press, New York © 1997.

In the present work, Slater-type orbitals (STO's) are used as basis functions:

$$\chi_i(r) = N_i\, r^{n_i-1} \exp(-\zeta_i r) , \qquad (4)$$

with

$$N_i = \frac{(2\zeta_i)^{n_i+1/2}}{\Gamma(2n_i+1)^{1/2}} .$$

Here n_i is an integer and ζ_i is the orbital exponent of the STO, and $\Gamma(x)$ is the gamma function.

Using Eq. (3), the Hamiltonian in Eq. (2) is diagonalized and the energy eigenvalues and eigenfunctions are obtained. In addition to the eigenvalues corresponding to the bound atomic states, there exist positive discrete energy eigenvalues. These pseudostates are used to calculate the inner-shell ionization cross sections instead of continuum states.

Relativistic case

In the relativistic case, $\boldsymbol{H}$ in Eq. (1) is replaced by the Dirac Hamiltonian

$$\boldsymbol{H} = c\boldsymbol{\alpha} \cdot \boldsymbol{p} + c^2 \beta' + V(r) . \qquad (5)$$

Here c is the velocity of light, $\boldsymbol{p}$ is the momentum of the electron, and $\boldsymbol{\alpha}$ and β' are two-component matrices.

The eigenfunction of Eq. (5) with the principal quantum number n, the Dirac quantum number κ, and the magnetic quantum number m is expressed in the form

$$\phi_{n\kappa m}(\boldsymbol{r}) = \begin{pmatrix} G_{n\kappa}(r)\, \chi_\kappa^m(\hat{\boldsymbol{r}}) \\ \mathrm{i}F_{n\kappa}(r)\, \chi_{-\kappa}^m(\hat{\boldsymbol{r}}) \end{pmatrix} , \qquad (6)$$

where $\chi_\kappa^m(\hat{\boldsymbol{r}})$ is the spin-angular function and $G_{n\kappa}(r)$ and $F_{n\kappa}(r)$ are the large and small components of the radial wave function, respectively.

We expand the radial wave function in terms of STO's

$$G_{n\kappa}(r) = \sum_p c_{\kappa p}\, f_{\kappa p}(r) ,$$

$$F_{n\kappa}(r) = \sum_p d_{\kappa p}\, f_{\kappa p}(r) ,$$

where $f_{\kappa p}(r)$ is the STO with non-integer principal quantum number

$$f_{\kappa p}(r) = (2\zeta_{\kappa p})^{n'_{\kappa p}+1/2} |\Gamma(2n'_{\kappa p}+1)|^{-1/2} r^{n'_{\kappa p}-1} \exp(-\zeta_{\kappa p} r), \qquad (7)$$

$$n'_{\kappa p} = n_{\kappa p} + (\kappa^2 - \alpha^2 Z^2)^{1/2} - |\kappa| \qquad n_{\kappa p} = 1, 2, \ldots , \qquad (8)$$

and $\zeta_{\kappa p}$ is the orbital exponent, Z is the atomic number of the target atom, α the fine structure constant, and $c_{\kappa p}$ and $d_{\kappa p}$ are the expansion coefficients.

Using this basis set, the Dirac Hamiltonian, Eq. (5), is diagonalized to obtain energy eigenvalues and eigenfunctions. For a hydrogenic potential, Drake and Goldman (7) pointed out that stable solutions can be obtained by the use of the STO basis set, but there is a spurious root for $\kappa > 0$. We have shown (8) that, despite of this spurious root, the present method can give good approximate energy eigenvalues and atomic wave functions for realistic relativistic potential.

Distortion approximation

In the distortion approximation (5), the ionization probability as a function of impact parameter b is given by

$$P(b) = \sum_j \left| \int_{-\infty}^{\infty} V_{j0}(t) \exp[-\mathrm{i}\eta(t)]\, dt \right|^2 , \qquad (9)$$

where the phase factor is defined as

$$\eta(t) = \int_{-\infty}^{t} \{[\epsilon_j + V_{jj}(t')] - [\epsilon_0 + V_{00}(t')]\}\, dt' , \qquad (10)$$

and the subscript 0 and j denote the initial state and a pseudostate, respectively. The matrix element is expressed as

$$V_{jk}(t) = -\int d\boldsymbol{r}\, \phi_j^*(\boldsymbol{r}) \frac{Z_1}{|\boldsymbol{r} - \boldsymbol{R}|} \phi_k(\boldsymbol{r}) , \qquad (11)$$

where Z_1 is the charge of the projectile, $\boldsymbol{R}$ is the internuclear distance between the projectile and the target, and ϵ and ϕ are the energy eigenvalue and the eigenfunction of $\boldsymbol{H}$:

$$\boldsymbol{H}\, \phi_k = \epsilon_k\, \phi_k . \qquad (12)$$

When $\eta(t)$ in Eq. (10) is replaced by $(\epsilon_j - \epsilon_0)t$, the ionization probability in the first Born approximation is obtained. This corresponds to the SCA by the use of pseudostates.

RESULTS AND DISCUSSION

In the nonrelativistic case, we used the Hartree-Fock-Slater (HFS) potential (9) in Eq. (2). The obtained energy eigenvalues for the bound states are in good agreement with the original HFS values. It is also found that the pseudostates wave functions reproduce well the shape of the continuum wave functions with the same kinetic energy in the low-energy region and for small radial distances. The wave functions for atomic bound states and pseudostates thus obtained are used to calculate the ionization cross sections by charged-particle impact.

In Table 1, the K-shell ionization cross sections for protons on copper are compared with other theoretical models and the experimental results. The Born cross sections are obtained by the present method using pseudostates with a straight-line trajectory. These values agree well with the PWBA cross sections (10), but systematically overestimates the experimental results (11–13) and the reference data (14). On the other hand, bility.

Table 1: Comparison of K-shell ionization cross sections for protons on Cu (barns).

Source	Energy (MeV) 0.5	1.0	2.0	Ref.
Born	2.93	22.3	111	
PWBA	2.88	22.3	113	
DA	1.44	14.9	90.1	
PWBA-BC	1.41	14.3	85.8	
Experiment	1.134±0.227	14.545±0.909	89.1±6.6	11
	1.72±1.7	16.0±0.5	96.2±2.9	12
	1.58±0.19			13
Reference	1.618	15.84	99.14	14

the present values with the Coulomb trajectory in the distortion approximation (DA) are in good agreement with the PWBA values corrected for the binding-energy and Coulomb-deflection effects (PWBA-BC) (15). The distortion approximation gives slightly smaller values than the experimental results. This discrepancy can be ascribed to the electronic relativistic effects, which is neglected in the nonrelativistic calculations.

Figure 1 shows the comparison of the present results with the experimental data (16) for the K-shell ionization cross sections of He by Li^{3+} particles. It can be seen that the distortion approximation is in good agreement with the experimental cross sections.

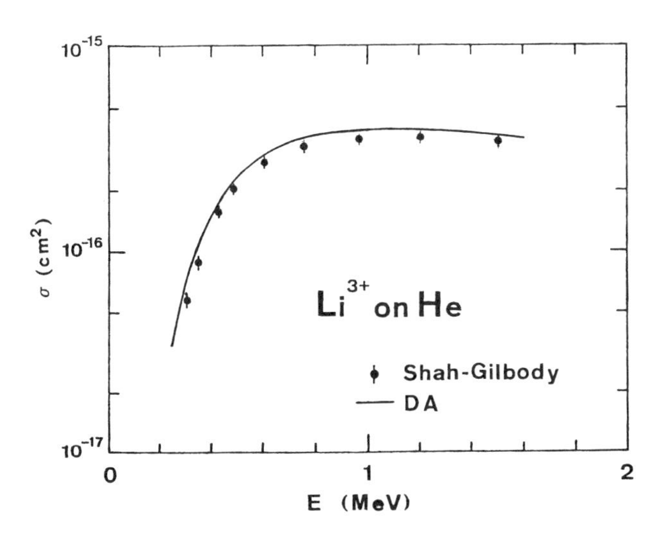

Figure 1. K-shell ionization cross sections of He by Li^{3+}.

In the relativistic case, the calculation of the K-shell ionization cross sections were made with the Dirac-Fock-Slater (DFS) potential (17). As pointed out previously (8), there is a spurious root for $\kappa > 0$, but we simply exclude this state from later calculations because this causes no trouble to predict the transition probability.

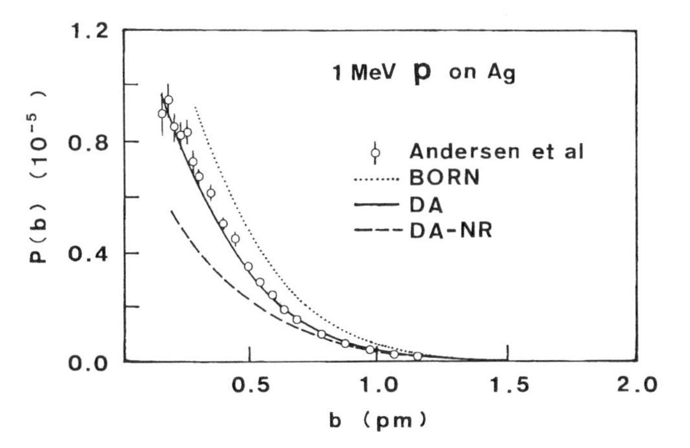

Figure 2. K-shell ionization probabilities for 1-MeV protons on silver.

Figure 2 shows the K-shell ionization probabilities for 1-MeV protons on silver as a function of impact parameter. The experimental data are taken from Andersen *et al.* (18). The solid curve indicates the distortion approximation using pseudostates (DA), the dotted curve represents the first Born approximation with same pseudostates (BORN). For comparison, the results in the distortion approximation in the nonrelativistic pseudostates are shown by the dashed curve (DA-NR). All the calculations were made using the same parameters for STO's, n and ζ.

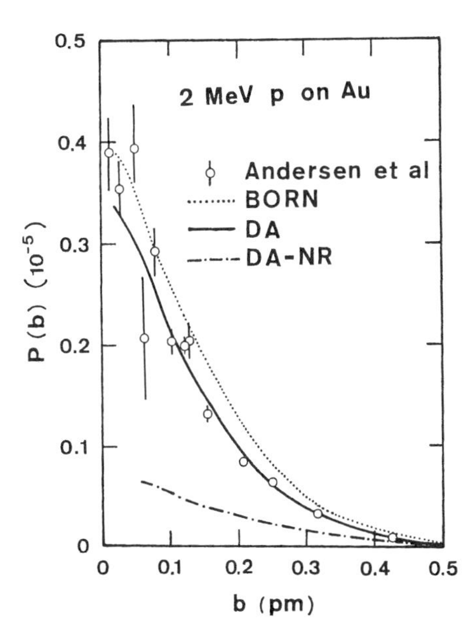

Figure 3. K-shell ionization probabilities for 2-MeV protons on gold.

The similar comparison for 2-MeV protons on gold is shown in Fig. 3. The experimental data are also taken from Andersen *et al.* (18). In this figure, the nonrelativistic results in the distortion approximation are indicated by the dot-dashed curve. The relativistic effect is larger for Au than for Ag. It is clear from both figures that the distortion approximation with relativistic pseudostates is in good agreement with the experimental data. On the other hand, the Born approximation overestimates the ionization probabilities, while the nonrelativistic calculations underestimate them.

CONCLUSION

We have calculated the K-shell ionization cross sections in ion-atom collisions in the distortion approximation. For this purpose, we employed the finite-basis-set expansion approach. Using the STO's as basis functions, the atomic states and the pseudostates were obtained by diagonalizing the atomic Hamiltonian with the HFS potential. The pseudostate wave functions were used to represent the continuum wave functions for the calculations of the ionization cross sections. We calculated the K-shell ionization cross sections for protons on copper and for Li^{3+} on helium. The results are in good agreement with the experimental data and other theoretical values.

We have extended this method to the relativistic case. The eigenfunctions of the Dirac Hamiltonian were expanded in terms of the STO's with non-integer principal quantum numbers. Stable solutions for the Dirac equation with the DFS potential were obtained. Using the relativistic pseudostates, the K-shell ionization probabilities were calculated in the distortion approximation for protons on silver and gold as a function of impact parameter. It is found that the electronic relativistic effects play an important role in these cases and the distortion approximation with relativistic pseudostates is in good agreement with the experimental results.

In the present work, we have demonstrated that by the use of pseudostates the distortion approximation can be easily applied to calculate the inner-shell ionization processes in ion-atom collisions. The calculated cross sections are in better agreement with the experimental values than the first-order theories. The extension to the relativistic case can also be made easily. The present results indicate the validity and the usefulness of the distortion approximation for calculations of inner-shell ionization cross sections in ion-atom collisions.

ACKNOWLEDGMENTS

This work is supported in part by the U.S. Department of Energy, Office of Basic Energy Sciences, Division of Chemical Sciences. Travel support under the US-Japan (NSF/JSPS) cooperative research program is also acknowledged.

REFERENCES

1. Merzbacher, E., and Lewis, H. W., *Handbuch der Physik*, Vol. 34, ed. by S. Flügge, Berlin: Springer, 1958, p. 166.
2. Bang, J., and Hansteen, J.M., *K. Dan. Vidensk. Selsl. Mat. Fys. Medd.* **31**, No. 13 (1959).
3. Mukoyama, T., *Intern. J. PIXE* **1**, 209 (1991).
4. Mukoyama, T., and Lin, C.-D., *Phys. Rev. A* **40**, 6886 (1989).
5. Bates, D. R., *Atomic and Molecular Processes*, ed. by D. R. Bates, New York: Academic, 1962, p. 549.
6. Mukoyama, T., and Sarkadi, L., *Phys. Rev. A* **28**, 1303 (1983).
7. Drake, G. W. F., and Goldman, S. P., *Phys. Rev. A* **23**, 2039 (1981).
8. Mukoyama, T., and Lin, C.-D., *Phys. Rev. A* **35**, 4942 (1987).
9. Herman, F., and Skillman, S, *Atomic Structure Calculations*, Englewood Cliffs, NJ: Prentice-Hall, 1963.
10. Choi, B.-H., Merzbacher, E., and Khandelwal, G. S., *Atomic Data* **5**, 291 (1973).
11. Lear, R., and Gray, T. J., *Phys. Rev. A* **8**, 2469 (1973).
12. Lægsgaard, E., Andersen, J. U., and Høgedal, F., *Nucl. Instr. and Meth.* **169**, 293 (1980).
13. Sera, K., Ishii, K., Kamiya, M., Kuwako, A., and Morita, S., *Phys. Rev. A* **21**, 1412 (1980).
14. Paul, H., and Sacher, J., *At. Data Nucl. Data Tables* **42**, 109 (1989).
15. Brandt, B., and Lapicki, G, *Phys. Rev. A* **20**, 465 (1979).
16. Shah, M. B., and Gilbody, H. B., *J. Phys. B: At. Mol. Phys.* **18**, 899 (1985).
17. Liberman, A., Cromer, D. T., and Waber, J. T., *Comput. Phys. Commun.* **2**, 107 (1971).
18. Andersen, J. U., Lægsgaard, E., and Lund, M., *Nucl. Instr. and Meth.* **192**, 79 (1982).

Pure ionization in collisions of fast H^+ and He^{2+} ions with gallium atoms

K. Lozhkin, C.J. Patton, P. McCartney, M. Sant'anna, M.B. Shah, J. Geddes and H.B. Gilbody.

Department of Pure and Applied Physics, The Queen's University of Belfast, Belfast, United Kingdom.

A crossed beam technique incorporating TOF analysis and coincidence counting of the collision products has been used to study pure ionization of ground state gallium atoms in collisions with H^+ and He^{2+} ions. Cross sections for processes leading to the formation of up to four-fold ionized gallium have been determined within the energy range 35 - 1440 keV amu^{-1}. Calculations based on an independent electron model of ionization, which has provided a good description of our previous measurements in Fe and Cu, is shown to fit our measured cross sections in Ga very satisfactorily even though our simple model does not allow for possible contributions from Auger ionization.

INTRODUCTION

Detailed information on the various collision processes leading to the multiple ionization of heavy metal atoms is necessary for the accurate modelling of both astrophysical and fusion plasmas. In previous work in this laboratory, [1, 2, 3] we have used a crossed beam technique incorporating time-of-flight analysis and coincidence counting of the collision products to study electron capture and ionization in collisions involving H^+ and He^{2+} ions with Fe and Cu in the energy range 35 - 1440 keV amu^{-1}. In these measurements, a specially developed oven source [4] was used to provide thermal energy beams of ground state $(3p)^6(3d)^6(4s)^2$ $^5D_4$ Fe and $(3p)^6(3d)^{10}(4s)$ $^2S_{1/2}$ Cu atoms. The process of transfer ionization (in which electron capture takes place simultaneously with ionization) was found to lead to up to seven fold ionized products while pure ionization, which became dominant at the higher energies, resulted in up to four fold ionized products. It was found that the measured cross sections could be satisfactorily described by an independent electron model approximation which assumed that electron removal takes place primarily from the 4s and 3d subshells.

In the present work, we have used the same experimental approach to study Ga^{q+} formation in collisions of H^+ and He^{2+} ions with ground state $(3d)^{10}(4s)^2(4p)$ $^2P_{1/2}$ Ga atoms within the energy range 35 - 1500 keV amu^{-1}. Over most of this energy range, pure ionization rather than electron capture is dominant and cross sections ${}_{10}\sigma_{1q}$ for the process

$$H^+ + Ga \rightarrow H^+ + Ga^{q+} + qe \qquad (1)$$

in which q = 1 - 4 and ${}_{20}\sigma_{2q}$ for the process

$$He^{2+} + Ga \rightarrow He^{2+} + Ga^{q+} + qe \qquad (2)$$

in which q = 1 - 3 have been determined. The extent to which these measured cross sections can be satisfactorily described in terms of the simple independent electron model used previously for Fe and Cu has also been investigated.

EXPERIMENTAL APPROACH

The apparatus, measurement and normalization procedure was similar to that used previously in this laboratory [5, 6] and only the essential features need be outlined here.

A primary beam of momentum analysed H^+ or He^{2+} ions of the required energy was arranged to intersect (at right angles) the thermal energy beam of Ga atoms in a high vacuum region. The slow Ga^{q+} ions and electrons formed in the crossed beam region were extracted with high efficiency by a transverse electric field applied between two high transparency grids and, after a further stage of acceleration, were separately counted by particle multipliers. Ga^{q+} ions in any particular charge state q were selectively identified and distinguished from background gas product ions by time-of-flight analysis.

It was necessary to distinguish the Ga^{q+} ions from the required pure ionization processes from those arising from electron capture processes of the type

$$H^+ + Ga \rightarrow H + Ga^{q+} + (q-1)e \qquad (3)$$

where q = 1 corresponds to simple charge transfer and q > 1 corresponds to transfer ionization. The Ga^{q+} ions arising from one-electron capture collisions could be identified by counting them in coincidence with the fast H atoms (or He^+ ions) arising from the same events which were recorded, after charge analysis by electrostatic deflection by a third particle multiplier located beyond the beam

CP392, *Application of Accelerators in Research and Industry,* edited by J. L. Duggan and I. L. Morgan
AIP Press, New York

intersection region. The Ga^{q+} ions arising from both transfer ionization and pure ionization were identified by counting them in coincidence with the electrons arising from the same events.

A careful analysis of the Ga^{q+}- fast atom/ion and the Ga^{q+} - electron coincidence spectra in the way described previously (see for example [3]) allowed determination of the separate pure ionization cross sections $_{10}\sigma_{1q}$ for 80 - 1400 keV amu^{-1} H^+ ions where q = 1 - 4 and $_{20}\sigma_{2q}$ for 43 - 300 keV amu^{-1} He^{2+} ions where q = 1 - 3. Our measured relative cross sections were normalized to our recently measured cross sections [7] for multiple ionization of Ga by electron impact. This procedure, which has been described in detail previously [6], involved the substitution of the primary ion beam by a beam of electrons while the target beam conditions remained unchanged.

RESULTS AND DISCUSSION

Cross sections $_{10}\sigma_{1q}$ for pure ionization of gallium by H^+ impact for q = 1 - 4 are shown in figure 1. The cross sections $_{10}\sigma_{11}$ for single ionization are very large and still rising with decreasing energy at our low energy limit. The cross sections $_{10}\sigma_{12}$, $_{10}\sigma_{13}$ and $_{10}\sigma_{14}$ all pass through flat maxima within the range considered and decrease by less than an order of magnitude as q increases.

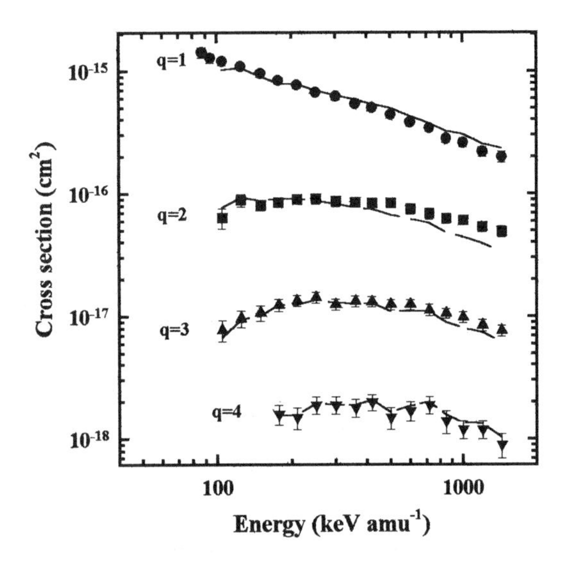

FIGURE 1. Measured cross sections for pure ionization $_{10}\sigma_{1q}$ (shown as symbols) in collisions of H^+ ions with ground state Ga atoms leading to Ga^{q+} formation compared with fits (shown as curves) based on an independent electron model of ionization .

The corresponding cross sections $_{20}\sigma_{2q}$ for pure ionization in He^{2+}-Ga collisions are shown in figure 2. Again the cross section $_{20}\sigma_{21}$ for Ga^+ formation is very large and still rising with decreasing energy at our low energy limit. The pure ionization cross sections $_{20}\sigma_{22}$ and $_{20}\sigma_{23}$ for Ga^{2+} and Ga^{3+} formation, which do not extend to as high an energy as the corresponding data for H^+, can be seen to be still rising with increasing energy within the present range.

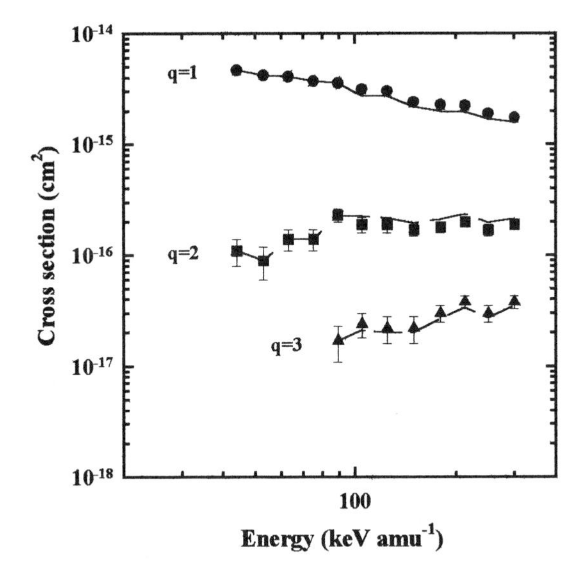

FIGURE 2. Measured cross sections for pure ionization $_{20}\sigma_{2q}$ (shown as symbols) in collisions of He^{2+} ions with ground state Ga atoms leading to Ga^{q+} formation compared with fits (shown as curves) based on an independent electron model of ionization .

It is interesting to try to describe the present cross sections for pure ionization in terms of the same simple model based on an independent electron description of multiple ionization [9] that we used in our studies of the multiple ionization of Fe and Cu [3]. We assume that the probability P for the removal of an electron from a particular subshell in the process of ionization can be approximated by the expression (see [9])

$$P(b) = P(0)\exp(-b/R) \tag{4}$$

where b is the impact parameter and P(0) and R are constants for a particular subshell. As in our previous work, we make the simple assumption that pure ionization primarily involves only the outermost subshells. For gallium these are the 4p, 4s and 3d subshells which contain 13 electrons. We also assume that the individual probabilities P(0) are the same for each of these subshells. The cross sections for pure ionization resulting in the removal of q electrons from a total of N may then be described by the expression

$$\sigma = 2\pi \int_0^\infty \binom{N}{q} P(b)^q (1 - P(b))^{(N-q)}\, b\, db \tag{5}$$

where $\sigma = {}_{10}\sigma_{1q}$ or $_{20}\sigma_{2q}$ and $\binom{N}{q}$ is the binomial coefficient.

Values of ${}_{10}\sigma_{1q}$ or ${}_{20}\sigma_{2q}$ for pure ionization by H^+ and He^{2+} ions predicted by equation (5) have been fitted to our measured values using a weighted least-squares fit as in our previous work [3]. The result of this fit for H^+ impact (figure 1) can be seen to be very satisfactory. In the case of He^{2+} impact where the data are confined to lower velocities than for H^+ impact, the agreement (see figure 2) between the two sets of curves is also very satisfactory. The values of both P(0) and R derived from the fitting procedure (figures 3 and 4) for both H^+ and He^{2+} impact exhibit an energy dependence very similar to that observed in our previous measurements for Fe and Cu [3].

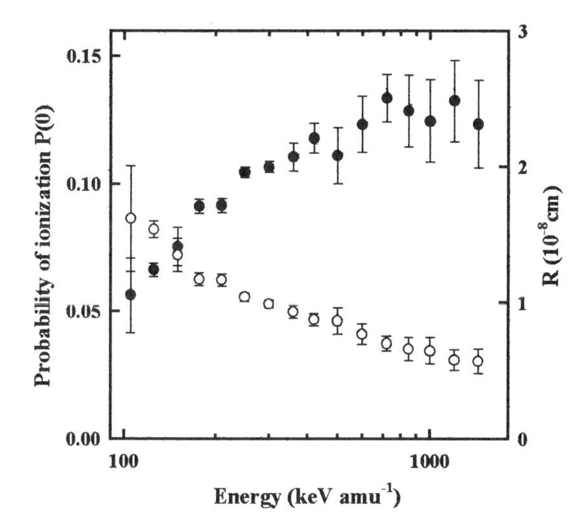

FIGURE 3. Plot showing the energy dependence of the fitting parameters P(0) (shown as ●) and R (shown as O) in the expression for the ionization probability P(b) = P(0)exp(-b/R) for the ionization of Ga by H^+ ions.

The success of our simple model in the present case is noteworthy particularly since no allowance has been made for the contributions from Auger ionization processes which are expected to be significant for Ga. For example, it is well known from studies of the electron impact ionization of Ga (see [10]) that removal of a 3d electron can result in many states which decay very effectively by autoionization leading to Ga^{2+} formation.

CONCLUSIONS

In this work we have studied, for the first time, pure ionization in collisions of H^+ and He^{2+} ions with ground state Ga atoms at energies within the range 35 - 1400 keV amu^{-1}. Cross sections for the formation of up to four fold ionized gallium have been determined. The same independent electron model, previously applied successfully to describe studies of pure ionization of Fe and Cu, has been fitted very satisfactorily to the present cross sections for pure ionization in spite of the fact that our simple model makes no allowance for contributions from Auger processes which are possible in gallium.

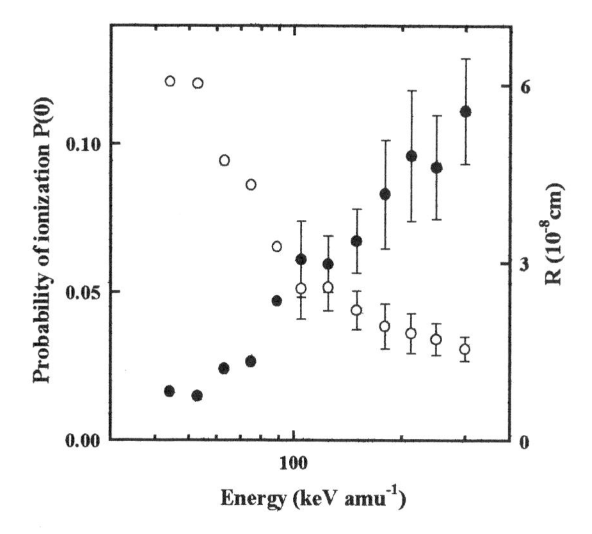

FIGURE 4. Plot showing the energy dependence of the fitting parameters P(0) (shown as ●) and R (shown as O) in the expression for the ionization probability P(b) = P(0)exp(-b/R) for the ionization of Ga by He^{2+} ions.

ACKNOWLEDGEMENTS

This research was supported by the U.K. Engineering and Physical Sciences Research Council. Two of us (C.J.P and P.McC) received Research Studentships from the Department of Education, Northern Ireland while the visit of M.S. from the Pontificia Universidade Católica do Rio de Janeiro was assisted by I A E S T E .

REFERENCES

1. Patton C.J., Bolorizadeh M.A., Shah M.B., Geddes J. and Gilbody H.B. *J. Phys. B: At. Mol. Opt. Phys.* **27**, 3695 -3706 (1994).
2. Shah M.B., Patton C.J., Bolorizadeh M.A., Geddes J. and Gilbody H.B. *J. Phys.B: At.Mol.Opt.Phys.* **28**, 1821-1833 (1995).
3. Patton C.J., Shah M.B., Bolorizadeh M.A., Geddes J. and Gilbody H.B. *J.Phys.B: At.Mol.Opt.Phys.* **28**, 3889-3899 (1995).
4. Shah M.B., Bolorizadeh M.A, Patton C.J and Gilbody H.B. Meas. Sci. and Technol. 7, 709- 711 (1996).
5. Shah M.B. and Gilbody H.B. *J.Phys.B: At Mol. Phys.* **14**, **2361-2377** (1981).
6. Shah M.B., McCallion P., Itoh Y. and Gilbody H.B. *J.Phys.B. At. Mol. Opt. Phys.* **25**, 3693 -3708 (1992).
7. Bolorizadeh M.A., Patton C.J., Shah M.B. and Gilbody H.B. *J.Phys.B: At. Mol. Opt. Phys.* **27**, 175-183 (1994).
8. McGuire J.H. *Advances in Atomic, Molecular and Optical Physics*, Vol. 29 , eds. D.R. Bates and B.Bederson (New York:Academic) .217-323 (1991).
9. Matsuo T., Tonuma T., Kumagai H. and Tawara H.H. Phys. Rev. A **50**, 1178 - 1183 (1994).
10. Vainshtein L.A., Golovach D.G., Ochkur V.I., Rakovskii V.I., Rumyantsev N.M. and Shustryakov V.M. Sov. Phys. - JETP **66**, 36-39 (1987).

HIGH RESOLUTION STUDY OF Kα HYPERSATELLITES SPECTRUM OF $_{42}$Mo ATOMS INDUCED BY 17 MeV/u ^{16}O BEAM

P.Rymuza[1], D.Chmielewska[1], T.Ludziejewski[1], Z.Sujkowski[1], D. Castella[2], D.Corminboeuf[2], J.-Cl.Dousse[2], B.Galley[2], Ch.Herren[2], J.Hoszowska[2], J.Kern[2], M.Polasik[3], M.Pajek[4]

[1]Soltan Institute for Nuclear Studies, 05-400 Swierk, Poland
[2]Physics Department, University of Fribourg, CH-1700 Fribourg, Switzerland
[3]Faculty of Chemistry, Nicholas Copernicus University, 87-100 Torun, Poland
[4]Institute of Physics, Pedagogical University, 25-509 Kielce, Poland

The Kα hypersatellite X-ray spectrum of $_{42}$Mo target bombarded with 17 MeV/u ^{16}O ions was measured with a high-resolution transmission bent crystal spectrometer. Well resolved lines of L-satellites and K-hypersatellites of $K\alpha_1$ and $K\alpha_2$ transitions were observed. Such data give valuable and rich information for testing the approximation used for the description of the multielectron system. The $K^h\alpha_1$ hypersatellite, treated as spin-flip $^1S_0 \rightarrow {}^3P_1$ transition, is forbidden in pure L-S coupling. In that case the relative yields of the Kα hypersatellites reflect directly the degree of intermediate coupling in the atom. Moreover, energies of the $K^h\alpha_{1,2}L^{0,1}$ transitions and the $I(K^h\alpha_1L^{0,1})/I(K^h\alpha_2L^{0,1})$ intensity ratios are much more influenced by the Breit interaction than the diagram lines. The discussion of these problems is illustrated with the measured Kα hypersatellite spectrum of $_{42}$Mo.

INTRODUCTION

Measurements of K X-ray spectra with resolution comparable to the natural width of the emission lines provide rich and valuable information about the structure of singly and multiply ionized atoms. Satellites appearing in such spectra arise as a consequence of the reduced screening of the nuclear charge due to additional holes in inner electron shells. If the initial state of an atom is doubly ionized in the K-shell, the radiative $K^{-2} \rightarrow K^{-1}L^{-1}$ decay is characterized by the emission of so-called hypersatellite X-ray $K^h\alpha$. A measurement of the energies of the X-ray transitions between multivacancy states offers a possibility to test the atomic structure models under unique conditions. Moreover, since the $K\alpha_1$ hypersatellite transition is forbidden in the pure L-S coupling scheme, the relative yields of the $K^h\alpha_1$ and $K^h\alpha_2$ hypersatellites strongly differ from those of the corresponding diagram lines. Thus the measured $K^h\alpha_1$ / $K^h\alpha_2$ intensity ratios reflect directly the influence of mixing between j-j and L-S coupling schemes which occurs for medium Z atoms. They are further influenced by the Breit interaction [1].

EXPERIMENT

The measurement was carried out at the PSI variable energy cyclotron in Villigen, Switzerland, using a 277 MeV ^{16}O beam. The X-ray were measured with a high resolution crystal spectrometer in the modified Du Mond slit

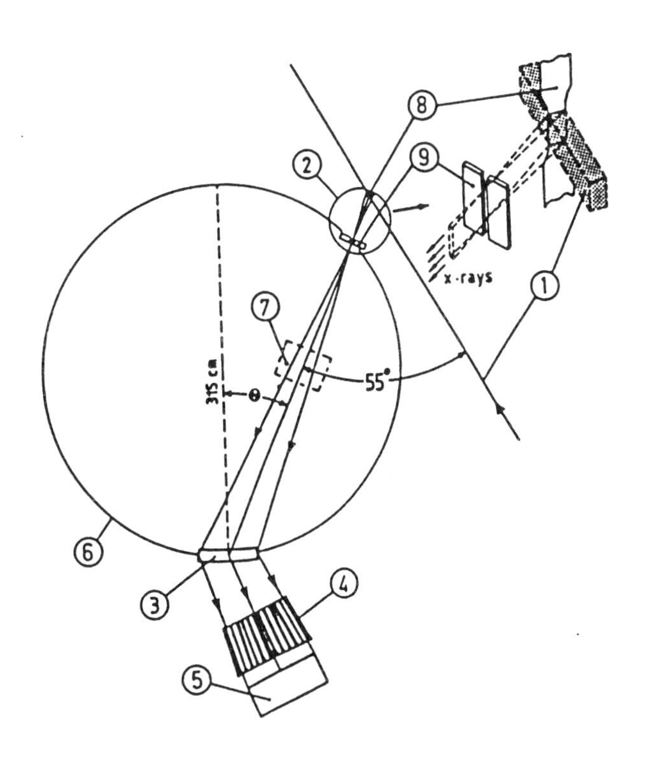

FIGURE 1. Schematic diagram of the Du Mond slit geometry used in this experiment (not to scale): (1) beam, (2) target chamber, (3) crystal, (4) collimator, (5) detector, (6) focal circle, (7) monitor detector, (8) target and (9) tantalum slit. Θ represents the Bragg angle.

CP392, *Application of Accelerators in Research and Industry,* edited by J. L. Duggan and I. L. Morgan
AIP Press, New York © 1997.

geometry [2]. A self-supporting metallic target of natural $_{42}$Mo with thickness of 9.26 mg/cm^2 was used. Schematic drawing of the experimental set-up is given in Fig.1. The energy calibration was performed by measuring the $K\alpha_1$ diagram line of $_{42}$Mo from an X-ray tube.

DATA ANALYSIS

The $_{42}$Mo Kα X -ray spectrum is shown in Fig.2. The L satellites of the diagram lines as well as of the hypersatellite lines are well resolved in the spectrum. The presence of the M satellites results in a slight broadening and an energy shift of the lines. The spectrum was analyzed by means of a least-square fit program using a single Voigt function resulting from the folding of the Lorentzian and Gaussian line shapes. The Lorentzian represents the natural X-ray line shape, whereas the Gaussian takes into account the instrumental response of the spectrometer and the multiple ionization in the outer shell (for more details of this method see [3,4]). The natural line width for the diagram lines and their satellites was taken from table [5], valid for singly ionized atoms. The width of the Lorentzian for hypersatellites was adjusted. The energy, height and width of the Gaussian for each peak was fitted, except for the $K\alpha_2L^2$, $K^h\alpha_2L^2$ and very weak $K\alpha_1L^4$ lines. The energies

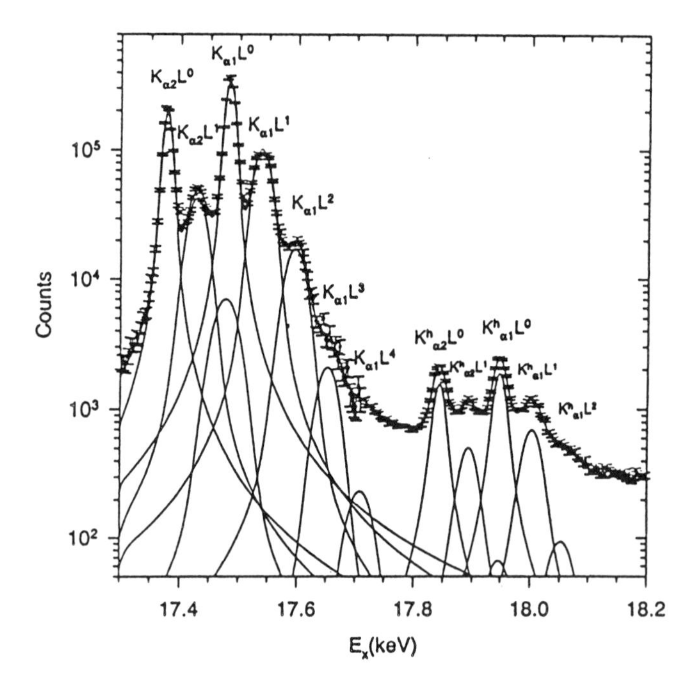

FIGURE 2. Crystal spectrometer Kα X-ray spectrum of molybdenum induced by 17 MeV/u ^{16}O ions. The energy region of diagram lines (below 17.7 keV) was scanned with 50 times shorter measuring time. In order to present in common plot the full Kα X-ray spectrum the counts and errors in that region were multiplied by factor 50.

and widths of those lines are obtained from theoretical simulations using the MCDF Grasp code [6]. The intensities of $K\alpha_2L^2$ and $K^h\alpha_2L^2$ lines were deduced from those of the $K\alpha_1L^2$ and $K^h\alpha_1L^2$ transitions.

TABLE 1. Calculated average energy shift of the diagram and the hypersatellite transitions due to additional one M-shell hole using MCDF method.

	Energy shift (eV)	
	diagram	**hypersatellite**
$K\alpha_2M^1$	2.24	2.17
$K\alpha_1M^1$	2.84	2.87

The observed broadening and the energy shifts of $K\alpha_2L^0$ $K\alpha_1L^0$ relative to the diagram line is caused by additional M-shell ionization. The collision which results in removing one or two K shell electrons can be treated as a central one in the M-shell scale [3]. Therefore the average number of the M-shell holes at the moment of the emission of the satellite and hypersatellite X-ray, n_M, are approximately the same (from the observed energy shifts n_M is estimated to 0.74).

Moreover from a calculation using MCDF code [6] (see Tab. 1) one may learn that average shifts of satellites and hypersatellites due to an additional M-shell hole are almost the same (differences less than 0.1 eV). Therefore the observed energy shifts $E(K^h\alpha L^0)$ - $E(K\alpha L^0)$ with a good approximation can be compared with the theoretical results obtained for atoms which are only singly or doubly ionized in K-shell in initial state. The advantage of using heavy ions rather than X-rays to induce hypersatellite transitions is the significant improvement in peak to background ratio. These ratios are ~$3x10^{-2}$, ~$4x10^{-1}$ and ~4 in the spectra induced by X-rays [7], He ions [8] and O beams, respectively.

RESULTS AND DISCUSSION

It was shown [1] that the Breit interaction, i.e. the interaction between any two electrons due to the exchange of the single transverse photon, substantially affects the multiplet splitting of double vacancy configurations in atom. The largest influence of this interaction is for the double K vacancy configuration. Tab.2 presents a comparison between the observed energy shifts of the $K^h\alpha$ and MCDF calculation with and without the Breit term.

Another test of theoretical description is the measurement of $I(K^h\alpha_1)/I(K^h\alpha_2)$ intensity ratio. For singly ionized atoms the $I(K\alpha_1)/I(K\alpha_2)$ are close to two and their variation with the atomic number is small, but for the hypersatellite transitions this ratio varies strongly with Z. Indeed, the $K^h\alpha_1$ spin-flip transition ($^1S_0 \rightarrow {}^3P_1$) is forbidden in light atoms

TABLE 2. Measured energy shift in eV of the $K^h\alpha_{1,2}$ hypersatellites with respect to the $K\alpha_{1,2}$ diagram lines compared to the calculations using MCDF method.

	Experiment	MCDF	
		with Breit	without Breit
$K\alpha_2$	465.9 ± 0.4	464.92	438.63
$K\alpha_1$	465.5 ± 0.4	465.00	441.57

where the LS coupling is dominant. For that reason, the relative yields $I(K\alpha_1)/I(K\alpha_2)$ are nearly zero. In heavy atoms, where the *jj* coupling is predominant, the $K\alpha_1$ transitions are no longer hindered and the intensity ratio $I(K^h\alpha_1)/I(K^h\alpha_2)$ tends to two as in the case of the diagram transitions. For medium Z atoms this ratio represents thus a sensitive tool to check the intermediate coupling model. Also the intensity ratio of $K\alpha_1$ to $K\alpha_2$ is influenced of by the Breit interaction much more strongly for the hypersatellites transitions than for diagram ones. For example for $_{42}$Mo atom the $I(K\alpha_1)/I(K\alpha_2)$ intensity ratio is 1.9040 and 1.9036 calculated with and without Breit term, respectively. The influence of the Breit interaction for the $I(K^h\alpha_1)/I(K^h\alpha_2)$ intensity ratio in comparison with our experimental results for hypersatellites is presented in Tab.3.

TABLE 3. Comparison of the hypersatellite intensity ratio $I(K^h\alpha_1)/I(K^h\alpha_2)$ between our experimental result and calculation using MCDF method. The theoretical results were corrected for additional M shell vacancies. The correction is small and is equal to ~1.5% .

Experiment	MCDF	
	with Breit	without Breit
1.23 ± 0.06	1.161	1.092

In conclusion it was shown that MCDF calculation including the proper Breit term reasonably describes the measured energy shifts and intensity ratio of $K\alpha$ hypersatellites transitions.

ACKNOWLEDGMENTS

This work was supported in part by the Polish Committee for Scientific Research (KBN) Grants No. 2 P302 119 07 and No. 2 P03B 007 11. The authors are grateful to Dr. Stammbach and Dr. Schmelzbach and the cyclotron crew for good beam conditions.

REFERENCES

1. Chen M.H. et al., *Phys.Rev.* **A25**, 391 (1982)
2. Perny B. et al., *Nucl.Instr.Meth.* **A267**, 120 (1988)
3. Rymuza P. et al., *Z.Phys.* **D14**, 37 (1989)
4. Carlen M. et al., *Phys.Rev.* **A46**, 3893 (1992)
5. Salem S.I., Lee P.L., *ADNDT* **18**, 233 (1976)
6. Polasik M., *Phys.Rev.* **A40**, 4361 (1989)
7. Salem S.I., *Phys. Rev.* **A21**, 858 (1980)
8. Boschung B. et al., *Phys.Rev.* **A51**, 3650 (1995)

L SUBSHELL X-RAY PRODUCTION CROSS SECTIONS FOR Pd, Ag, and Sn BY 75-300 keV PROTONS

Sam J. Cipolla, Michael J. Dolezal, and Larry O. Casazza

Physics Department, Creighton University, Omaha NE 68178

L subshell x-ray production cross sections have been measured for 75-300 keV protons impacting pure elemental thick targets of Pd, Ag, and Sn. The results are compared with the ECPSSR theory.

INTRODUCTION

The ionization of the K shell of atoms struck by accelerated protons and other light ions has been shown to be described well by the ECPSSR(1) formulation. This theory predicts ionization cross sections by means of a perturbed stationary state modification of the plane wave Born approximation, with corrections made for binding and polarization effects, Coulomb deflection, and relativity. Much work currently is involved with testing ECPSSR in describing L-shell ionization. X-ray detectors of improved resolution and efficiency have made it possible to investigate lighter elements than before. In particular, there are very limited data available for elements lighter than Z=55, especially in regards to subshell cross sections. A review of experimental results by Orlic(2) indicates that the ECPSSR theory generally underestimates cross section measurements, with the discrepancy increasing for lighter elements and for low incident proton velocities. This paper presents measurements of L subshell x-ray production cross sections for selected elemental targets with $Z \leq 50$ struck by 75-300 keV protons.

EXPERIMENTAL METHODS

A high-resolution 30mm^{2}x3mm Si(Li) detector equipped with an ultra-thin boron nitride window was used to measure L x-ray spectra for low energy protons striking thick foil targets of Pd, Ag, and Sn. Additionally, one or two 6-μm aluminized Mylar absorbers were used to cut down M x-ray detection. The detector was calibrated for efficiency using a model-based formula presented earlier(3). The protons were obtained from a 350-kV Cockcroft-Walton accelerator equipped with beam optics to steer and focus the beam, a mass analyzer magnet, and a biased 3-mm beam collimator. The targets were arranged on vertical ladder so that a different spot was used in each measurement. A biased screen surrounded the ladder to suppress secondary electrons. Beam currents were kept low (< 2% dead time) to minimize pile-up.

All prominent L x-ray intensities were obtained by fitting the x-ray spectrum with a quadratic background and gaussians. From the peak energies, the peak centroids and widths were determined during the spectrum analyses by fits to model functions for these parameters.

X-ray production cross sections, $\sigma_x(E)$, for each transition were obtained through thick target analysis of the x-ray yields, according to(4),

$$\sigma_x(E) = \frac{1}{N\epsilon}\left[S(E)\frac{dY}{dE} + \frac{\mu}{\rho}Y(E)\right] \quad (1)$$

where $Y(E) = N_x/N_p$, with N_x = peak x-ray count and N_p is the number of protons hitting the target, is the x-ray yield, S(E) is the target stopping power(5), μ/ρ is the target mass absorption coefficient(6), N is the target atom density, ε is the detection efficiency, and dY/dE was determined from fitting the yields to $Y(E) = A(E-C)^B$, where A, B, and C are the parameters of the fit. These were then used to derive the L x-ray production cross sections for each sub-shell, according to,

$$\sigma^x_{L_i} = \sigma_x \frac{\Gamma_{L_i}}{\Gamma_x} \quad (2)$$

where Γ_x and Γ_{Li} are the x-ray transition rates for a transition (peak) to the L_i(i=1,2,3) sub-shell and the total rate for the sub-shell, respectively, using Scofield's values as interpolated by Campbell and Wang(7). The L1, L2, and L3 cross sections were derived from the L_α, $L_{\gamma 2,3}$, and $L_{\beta 1}$ peaks, respectively.

CP392, *Application of Accelerators in Research and Industry,* edited by J. L. Duggan and I. L. Morgan
AIP Press, New York © 1997.

TABLE 1. Measured L subshell x-ray production cross sections(barns)[a].

E	L_ℓ	L_α	$L_{\beta 1}$	$L_{\beta 3,4}$	$L_{\beta 2,15}$	$L_{\gamma 1}$	$L_{\gamma 2,3}$
				Pd			
75	1.72-3±1.9-4	4.05-2±4.3-3	1.34-2±1.4-3	4.95-3±1.2-3	3.96-3±4.3-4	1.22-3±1.3-4	9.74-4±1.3-4
100	1.03-2±1.1-3	2.33-1±2.4-2	7.90-2±8.3-3	2.09-2±4.6-3	2.21-2±2.4-3	7.49-3±7.9-4	4.00-3±4.9-4
125	2.90-2±3.1-3	6.56-1±6.9-2	2.28-1±2.4-2	4.37-2±1.1-2	6.04-2±6.5-3	2.17-2±2.3-3	8.51-3±1.1-4
150	6.04-1±6.5-2	1.37±0.14	4.86-1±5.1-2	7.21-2±1.9-2	1.23-1±1.3-2	4.64-2±4.9-3	1.45-2±1.8-3
175	1.07-1±1.1-2	2.41±0.25	8.70-1±9.1-2	1.06-1±3.0-2	2.11-1±2.2-2	8.21-2±8.6-3	2.13-2±2.6-3
200	1.68-1±1.8-2	3.84±0.40	1.40±0.15	1.35-1±2.7-2	3.29-1±3.5-2	1.33-1±1.4-2	2.94-2±3.6-3
225	2.47-1±2.6-2	5.65±0.59	2.09±0.22	1.84-1±5.0-2	4.74-1±5.0-2	1.98-1±2.1-2	3.85-2±4.5-3
250	3.46-1±3.7-2	7.91±0.82	2.96±0.31	2.11-1±4.8-2	6.55-1±7.0-2	2.81-1±2.9-2	4.77-2±5.5-3
275	4.63-1±4.9-2	10.60±1.10	3.99±0.42	2.77-1±7.2-2	8.61-1±9.1-2	3.79-1±4.0-2	5.70-2±6.5-3
300	5.99-1±6.3-2	13.87±1.44	5.27±0.55	2.93-1±5.2-2	1.11±0.12	5.11-1±5.3-2	6.76-2±7.6-3
				Ag			
E	L_ℓ	L_α	$L_{\beta 1}$	$L_{\beta 3,4}$	$L_{\beta 2,15}$	$L_{\gamma 1}$	$L_{\gamma 2,3}$
75	1.05-3±1.3-4	2.17-2±2.6-3	7.22-3±8.6-4	2.83-3±4.0-4	2.34-3±2.7-4	6.66-4±7.5-5	6.10-4±8.4-5
100	6.37-3±7.4-4	1.40-1±1.6-2	4.69-2±5.4-3	1.51-2±2.0-3	1.46-2±1.6-3	4.61-3±5.1-4	3.11-3±3.9-4
125	1.93-2±2.3-3	4.27-1±5.1-2	1.46-1±1.7-2	3.71-2±5.1-3	4.39-2±4.9-3	1.47-2±1.6-3	7.13-3±9.1-4
150	4.21-2±4.8-3	9.34-1±1.1-1	3.24-1±3.7-2	6.79-2±8.8-3	9.55-2±1.0-2	3.27-2±3.6-3	1.22-2±1.5-3
175	7.95-2±9.5-3	1.741±0.206	6.09-1±7.1-2	1.10-1±1.6-2	1.76-1±2.0-2	6.28-2±6.9-3	1.93-2±2.4-3
200	1.31-1±1.5-2	2.884±0.338	1.01±0.12	1.57-1±2.1-2	2.89-1±3.2-2	1.06-1±1.2-2	2.72-2±3.4-3
225	2.01-1±2.3-2	4.340±0.507	1.55±0.18	2.16-1±3.0-2	4.35-1±4.8-2	1.62-1±1.8-2	3.53-2±4.3-3
250	2.88-1±3.3-2	6.205±0.719	2.23±0.26	2.75-1±3.6-2	6.21-1±6.8-2	2.38-1±2.6-2	4.52-2±5.4-3
275	3.98-1±4.5-2	8.498±0.978	3.08±0.35	3.46-1±4.5-2	8.48-1±9.3-2	3.34-1±3.6-2	5.46-2±6.3-3
300	5.28-1±6.0-2	11.25±1.28	4.09±0.46	4.17-1±5.2-2	1.12±0.12	4.46-1±4.8-2	6.43-2±7.3-3
				Sn			
E	L_ℓ	L_α	$L_{\beta 1}$	$L_{\beta 3,4}$	$L_{\beta 2,15}$	$L_{\gamma 1}$	$L_{\gamma 2,3}$
100	1.81-3±4.5-4	4.49-2±1.1-2	1.59-2±3.8-3	1.13-2±2.9-3	6.10-3±1.5-3	2.13-3±5.1-4	2.39-3±6.4-4
125	7.69-3±1.7-3	1.95-1±4.2-2	6.80-2±1.5-2	3.51-2±8.5-3	2.52-2±5.5-3	8.94-3±1.9-3	7.18-3±1.7-3
150	1.88-2±3.9-3	4.65-1±9.4-2	1.66-1±3.4-2	6.78-2±1.6-2	6.11-2±1.3-2	2.17-2±4.3-3	1.34-2±3.0-3
175	3.69-2±9.9-3	8.99-1±2.3-1	3.22-1±8.5-2	1.10-1±3.5-2	1.18-1±3.1-2	4.30-2±1.1-2	2.22-2±6.3-3
200	5.84-2±1.1-2	1.43±0.27	5.17-1±9.8-2	1.52-1±3.2-2	1.90-1±3.6-2	6.74-2±1.2-2	2.86-2±5.9-3
225	9.35-2±2.4-2	2.23±0.58	8.05-1±2.1-1	2.11-1±6.3-2	2.95-1±7.7-2	1.10-1±2.6-2	4.26-2±8.4-3
250	1.25-1±2.3-2	2.99±0.56	1.09±0.20	2.54-1±5.1-2	4.04-1±7.6-2	1.44-1±2.5-2	4.60-2±8.3-3
275	1.84-1±4.6-2	4.28±1.09	1.55±0.40	3.33-1±9.3-2	5.71-1±1.5-1	2.20-1±5.0-2	6.69-2±1.5-2
300	2.21-1±4.1-2	5.22±0.99	1.92±0.39	3.70-1±7.1-2	7.10-1±1.3-1	2.58-1±4.4-2	6.64-2±1.1-2

[a]Values of $N \times 10^{-n}$ are given as N-n.

RESULTS

The x-ray production cross sections for the transitions that were analyzed in the measured spectra are given in table 1. Although the L_η peaks were resolved from L_α in all the spectra, these results are not given in table 1 for brevity.

The measured sub-shell x-ray production cross sections determined from eqn.(2) were compared with calculations from the ECPSSR theory using the PC program ISICS(8). Krause's(9) values of fluorescence yields and Coster-Kronig parameters were employed. The measured x-ray production subshell cross sections are compared with the ECPSSR calculations in Fig. 1-3.

Also of interest are ratios of subshell x-ray production cross sections for each element. Figures 4-7 present different ratios in comparison with corresponding ratios using ECPSSR calculations.

Standard error propagation procedures were employed throughout. Typical uncertainties were: efficiency, 10%; mass absorption coefficient, 5%; stopping power, 5%. Yield uncertainties depended on the intensity of the peaks.

DISCUSSIONS AND CONCLUSION

Figures 1-3 show that the ECPSSR theory generally underestimates the experimental results, which is in accordance with other measurements at low energies. We could find no other measurements comparable to ours with which to make comparisons for these elements and the proton energy range used. The trend of the measurements matches well with the trend of the ECPSSR values in all cases.

The cross section ratios shown in Fig. 4-7 are sensitive somewhat to the nodal structure of the electronic wave functions, especially the L3/L1 ratio (and L1/L2), as noted by Shingal, et al.(10). They point out the utility of the energy dependence of these ratios in testing the binding energy correction of ECPSSR. Another advantage of using ratios here is that uncertainties due to the detector efficiency cancel out. The L1/L2 ratios are all smaller than the ECPSSR values, with the results for Sn being the most discrepant. However, all the measurements follow the ECPSSR trend, with the tendency towards agreement with ECPSSR as the projectile energy increases. The measured L3/L1 ratios also follow the same trend as ECPSSR ratios, and the results are consistently very close to ECPSSR values, with the Sn results being systematically larger than ECPSSR. For L3/L2, the measurements show almost no change with proton energy, which agrees with ECPSSR for Sn, but is a departure for Pd and Ag from ECPPSR which predicts a slow increase as the proton energy decreases. Finally, the L3/(L1+L2) ratios follow the ECPSSR trend, with the Pd and Ag measurement falling short of the ECPSSR values, and the Sn measurements exceeding the ECPSSR values at low energies and coinciding at the higher energies.

Of further interest are the transition intensity ratios, L_ℓ/L_α, $L_{\gamma 1}/L_{\beta 1}$, and $L_{\gamma 2,3}/L_{\beta 3,4}$, relating to the L3, L2, and L1 subshells, respectively. From table 1, the ratios, averaged all proton

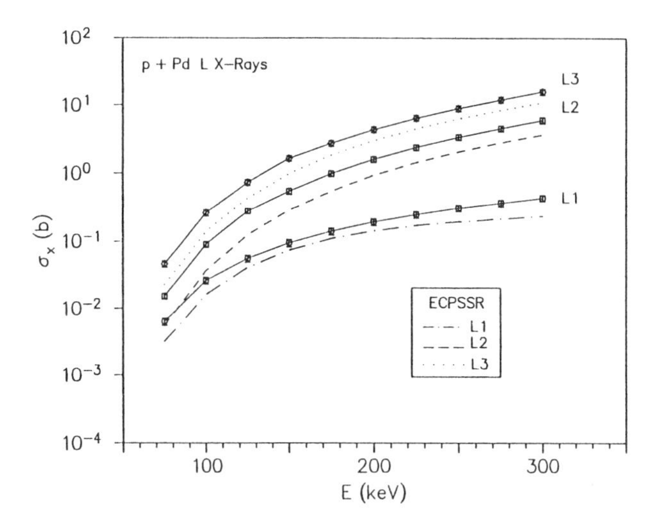

FIGURE 1. Comparison of Pd cross section measurements with ECPSSR theory.

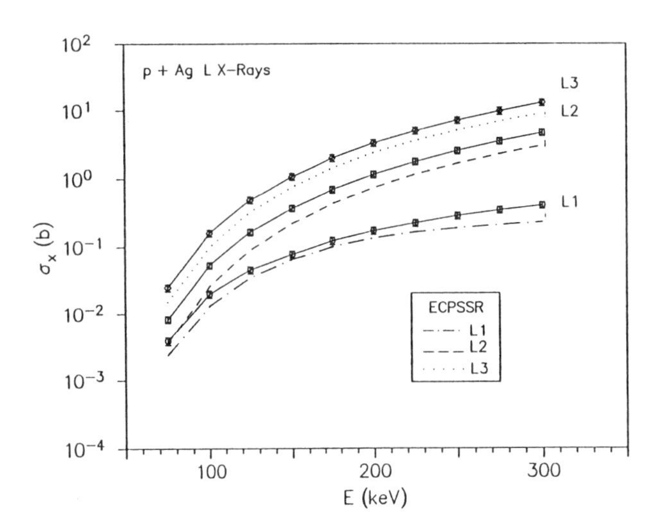

FIGURE 2. Comparison of Ag cross section measurements with ECPSSR theory.

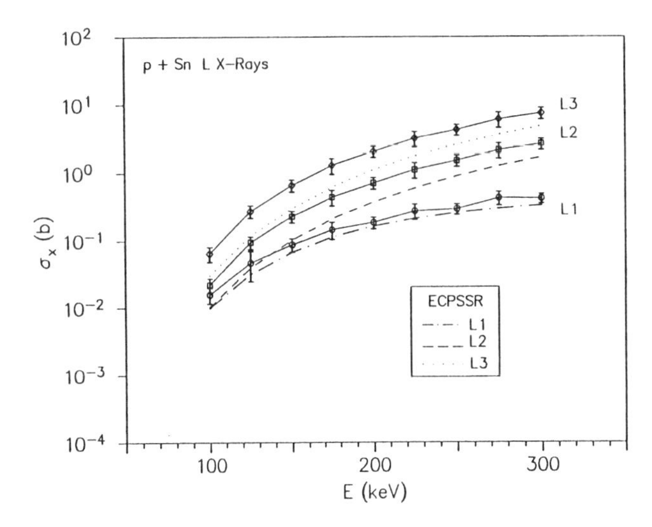

FIGURE 3. Comparison of Sn cross section measurements with ECPSSR theory.

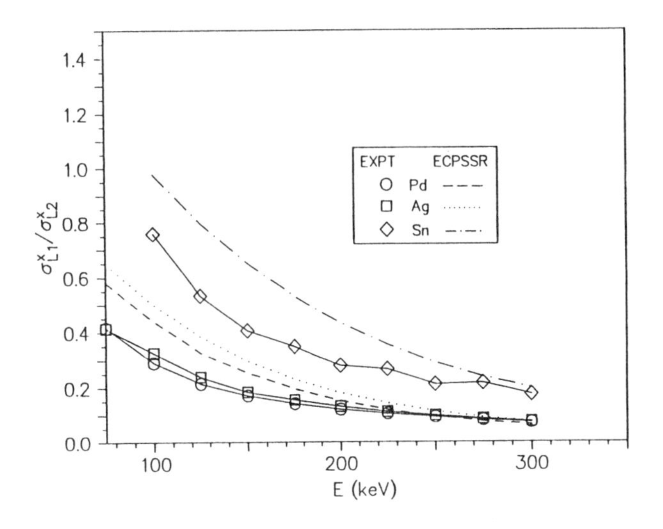

FIGURE 4. Comparison of measured L1/L2 cross section ratio with ECPSSR values.

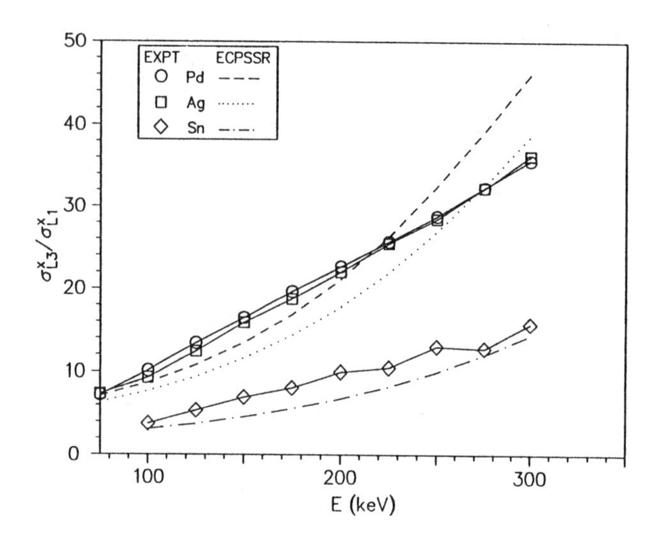

FIGURE 5. Comparison of measured L3/L1 cross section ratios with ECPSSR values.

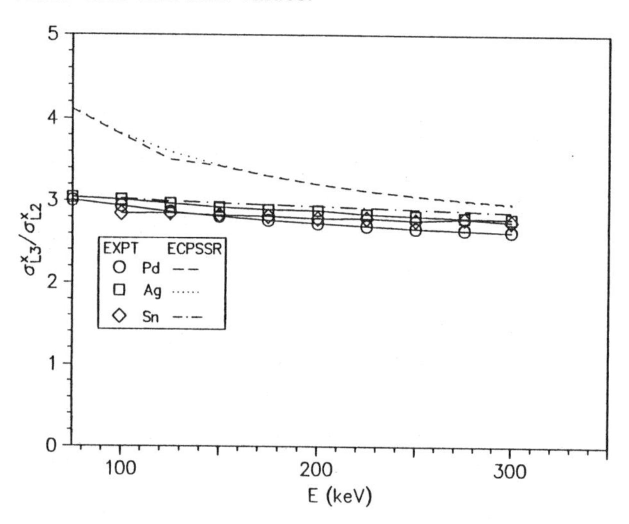

FIGURE 6. Comparison of measured L3/L2 cross section ratios with ECPSSR values.

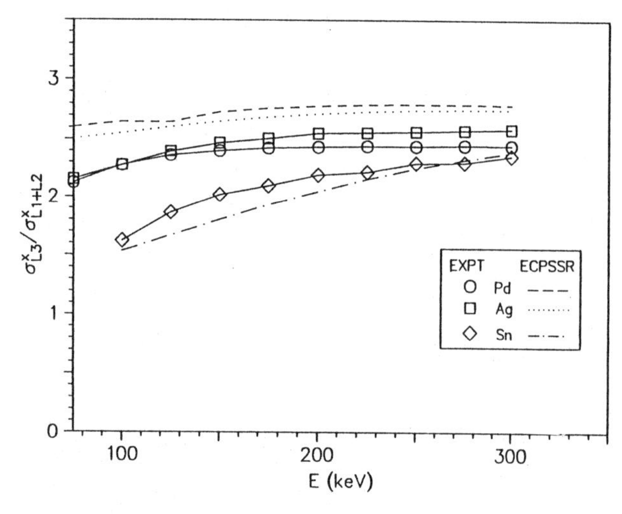

FIGURE 7. Comparison of measured L3/(L1+L2) cross section ratios with ECPSSR values.

energies, are: L_ℓ/L_α = 0.0439±0.0025 (0.037 1), $L_{\gamma1}/L_{\beta1}$= 0.0953±0.0050 (0.106), $L_{\gamma2,3}/L_{\beta3,4}$ = 0.209±0.021 (0.182) for Pd; L_ℓ/L_α = 0.0459±0.0029 (0.372), $L_{\gamma1}/L_{\beta1}$ = 0.104±0.006 (0.106), $L_{\gamma2,3}/L_{\beta3,4}$ = 0.171±0.011 (0.188) for Ag; L_ℓ/L_α = 0.0413±0.0048 (0.375), $L_{\gamma1}/L_{\beta1}$ = 0.135±0.015 (0.134), $L_{\gamma2,3}/L_{\beta3,4}$ = 0.195±0.024 (0.207) for Sn. The quantities in parentheses above are the corresponding transition rate ratios interpolated from Scofield's work(7). Our results are slightly higher than theory for L_ℓ/L_α, and in very good agreement with theory for $L_{\gamma1}/L_{\beta1}$ and $L_{\gamma2,3}/L_{\beta3,4}$.

In summary, our results generally confirm that ECPSSR x-ray production cross sections underpredict measured values for all three subshells, but are seen to follow the same trend with proton energy. Subshell cross section ratios vary in comparison with ECPSSR, but also generally exhibit the same trend. No definite conclusions can be reached from these subshell cross section ratios on the effectiveness of the ECPSSR binding energy correction. The various peak intensity ratios are mostly in good agreement with Scofield's theory calcualtions.

ACKNOWLEDGEMENTS

This work was supported by a grant from the Research Corporation and was conducted at the University of Nebraska at Lincoln.

REFERENCES

1. Brandt, W., and Lapicki, G., *Phys. Rev.* A **23**, 1717 (1981).
2. Orlic, I., Sow, C.H., and Tang, S.M., *At. Dat. Nucl. Dat. Tables* **56**, 159-210 (1994).
3. Gallagher, W. and Cipolla, S., *Nucl. Instr. & Meth.* **122**, 405 (1974).
4. Merzbacher, E., and Lewis, H., *Handbuch der Physik*, vol. **34,** Berlin: Springer Press, 1958, p. 166.
5. Ziegler, J., "Transport of Ions in Matter," ver. 91.07, IBM, Yorktown Hts., July 26, 1991.
6. Berger, M.J. and Hubbell, J.H., "XCOM: Photon Cross Section on a Personal Computer," NBS Report NBSIR 87-3595, Gaithersburg, Md., 1987.
7. Campbell, J. and Wang, J.-X., *At. Dat. Nucl. Dat. Tables* **43**, 281-291 (1989).
8. Liu, Z. and Cipolla, S., *Comp. Phys. Comm.* **97**, 315-332 (1996).
9. Krause, M.O., *J. Phys. Chem. Ref. Data* **8,** 307 (1979).
10. Shingal, R., Malhli, N.B., Gray, T.J., *J. Phys.* B **25**, 2055-2063 (1992).

L SHELL IONIZATION BY SLOW PROTONS

Sam J. Cipolla

Physics Department, Creighton University, Omaha NE 68178

L subshell and total x-ray production cross sections have been measured for 75-300 keV protons impacting pure elemental thick targets ranging from Zr through Yb. The results are compared with the ECPSSR(1) theory.

INTRODUCTION

The ECPSSR(1) theory does a satisfactory job of describing K-shell ionization by proton impact. Much attention now is being devoted to the study of L-shell ionization. A review of experimental results by Orlic(2) indicates that ECPSSR generally underestimates cross section measurements, with the discrepancy increasing for lighter elements and for low incident proton velocities. These are also the areas where more data are needed to better test ECPSSR. This paper presents measurements of L subshell x-ray production cross sections for elemental targets between Zr and Yb struck by 75-300 keV protons.

EXPERIMENTAL METHODS

A high-resolution 30mm^2x3mm Si(Li) detector equipped with an ultra-thin boron nitride window was used to measure L x-ray spectra for low energy protons striking thick foil targets of Zr, Nb, Mo, Pd, Ag, Sn, Ce, Gd, Er, and Yb. Additionally, one or two 6-μm aluminized Mylar absorbers were also used to cut down M x-ray detection. The detector was calibrated for efficiency using a model-based formula presented earlier(3). The protons were obtained from a 350-kV Cockcroft-Walton accelerator equipped with beam optics to steer and focus the beam, a mass analyzer magnet, and a biased 3-mm beam collimator. The targets were arranged on vertical ladder so that a different spot was used in each measurement. A biased screen surrounded the ladder to suppress secondary electrons. Beam currents were kept low (< 2% dead time) to minimize pile-up. See Fig. 1.

All prominent L x-ray intensities were obtained by fitting the x-ray spectrum with a quadratic background and gaussians for which the centroids and widths were determined by model functions.

X-ray production cross sections, $\sigma_x(E)$, for each transition were obtained through thick target analysis of the x-ray yields, according to(4),

$$\sigma_x(E) = \frac{1}{N\epsilon}\left[S(E)\frac{dY}{dE} + \frac{\mu}{\rho}Y(E)\right] \quad (1)$$

where $Y(E) = N_x/N_p$, with N_x = peak x-ray count and N_p is the number of protons hitting the target, is the x-ray yield, S(E) is the target stopping power(5), μ/ρ is the target mass absorption coefficient(6), N is the target atom density, ε is the detection efficiency, and dY/dE was determined from fitting the yields to $Y(E) = A(E-C)^B$. These were then used to derive the L x-ray production cross sections for each sub-shell, according to,

$$\sigma^x_{L_i} = \sigma_x \frac{\Gamma_{L_i}}{\Gamma_x} \quad (2)$$

where Γ_x and Γ_{Li} are the x-ray transition rates for a transition (peak) to the L_i(i=1,2,3) subshell and the total rate for the subshell, respectively, using Scofield's values as interpolated by Campbell and Wang(7). In all measurements, the L1 and

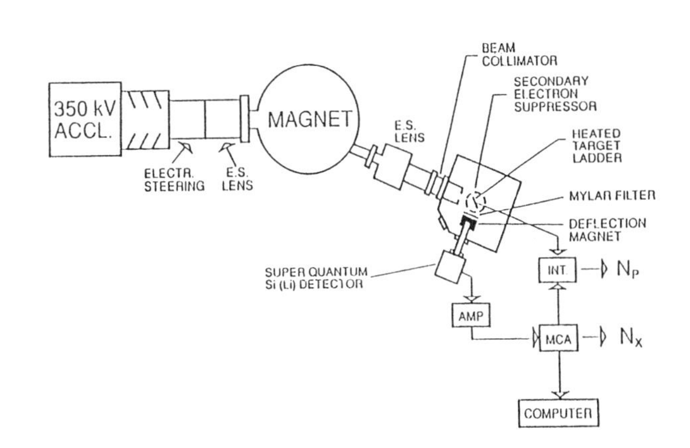

FIGURE 1. Experimental System.

CP392, *Application of Accelerators in Research and Industry,* edited by J. L. Duggan and I. L. Morgan
AIP Press, New York © 1997.

L3 cross sections were derived from the L_{α} and $L_{\gamma2,3}$ peaks, respectively. Except for Ce (where $L_{\gamma1}$ was used as $L_{\beta1,4}$ could not be resolved), the L2 cross section was derived from the $L_{\beta1}$ peak.

RESULTS

The measured subshell x-ray production cross sections were compared with calculations from the ECPSSR theory using the PC program ISICS(8). Krause's(9) values of fluorescence yields and Coster-Kronig parameters were employed. Experimental results are given in table 1. The ratios $\sigma_{expt}/\sigma_{ECPSSR}$ as a function of the reduced velocity parameter ξ_s^r for each sub-shell are presented in figures 2-4, where ξ_s^r is defined as,

$$\xi_s^r = \sqrt{m_s^r}\frac{2v_1}{\theta_s v_{2s}} \quad (3)$$

and v_1 is the proton speed, θ_s is the reduced target-atom s-subshell binding energy, v_{2s} is the s-subshell electron speed, m_s^r is the s-subshell relativity correction from ECPSSR theory. The cross section ratios L_ℓ/L_{α} (L3), $L_{\gamma1}/L_{\beta1}$ (L2), and. $L_{\gamma2,3}/L_{\beta3,4}$ (L1), averaged over the proton energies, are presented in table 2, where they are compared with corresponding transition rate ratios from ref(7).

DISCUSSIONS AND CONCLUSION

Figures 2-4 show that the ECPSSR theory generally underestimates the experimental results, which is in accordance with other measurements at low energies. Exceptions here are Ce, which shows the opposite trend, and Gd, Er, and Yb, where $\sigma_{expt}/\sigma_{ECPSSR}$ is less than unity (ECPSSR over-predicts) for ξ_s^r between 0.2 and 0.3. For L3, and to a lesser extent L2 where the data are more scattered, the trend is toward agreement with ECPSSR with increasing ξ_s^r. The L1 results seem to oscillate, with a minimum around ξ_s^r = 0.4.

From table 2, the ratios L_ℓ/L_{α} follow the general trend of the theory, and show good agreement for the high-Z targets, but the results exceed theory progressively more as Z_2 decreases. An anomaly is the low value for Zr. The ratios $L_{\gamma1}/L_{\beta1}$ also follow the trend of the theory, with good agreement, except for low values for Ce and Gd, and a high value for Nb. The trend is likewise for $L_{\gamma2,3}/L_{\beta3,4}$. Ratios are given only for cases where the peaks were cleanly resolved in the spectra.

The compilation of Orlic, et al.(2), shows that very few L subshell measurements in this energy region have been reported. Most of these extend only down to 250 keV, except for the 100-200 keV measurements of Harrigan, et al.(10) for protons on Gd, and the 200-300 keV protons on Er measurements of Braziewicz, et al.(11). Harrigan, et al., are in good agreement

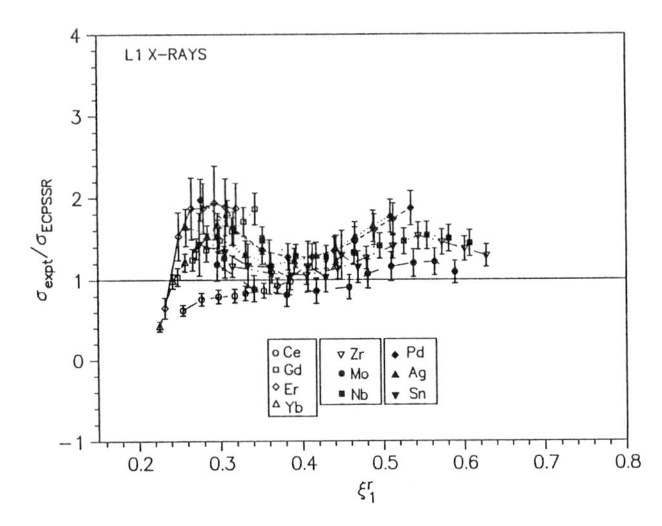

FIGURE 2. Comparison of L1 cross section measurements with ECPSSR theory.

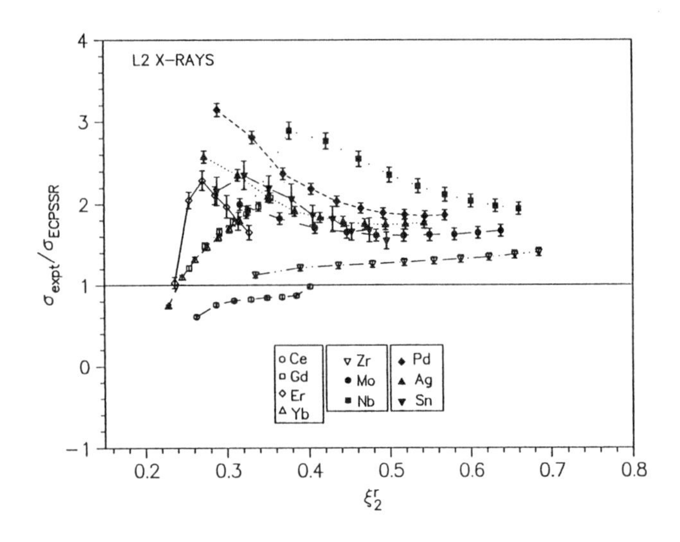

FIGURE 3. Comparison of L2 cross section measurements with ECPSSR theory.

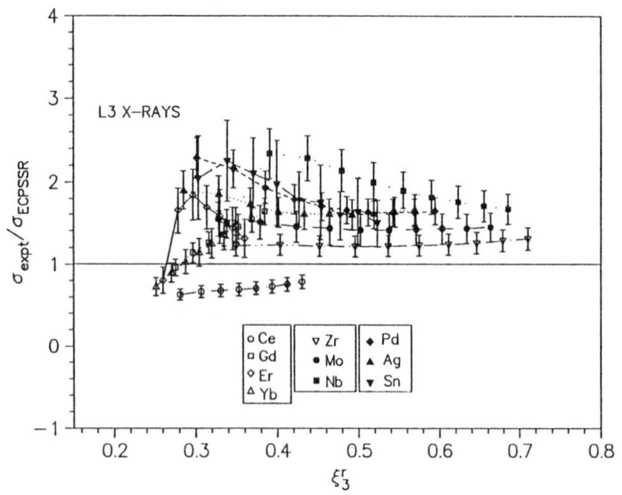

FIGURE 4. Comparison of L3 cross section measurements with ECPSSR theory.

TABLE 1. Measured L subshell x-ray production cross sections(barns)[a].

E		L1	L2	L3		L1	L2	L3
75	Zr	2.16-2±2.4-3	7.92-2±8.1-3	2.53-1±2.6-2	Nb	2.18-2±2.5-3	7.56-2±8.9-3	2.16-1±2.6-2
100		6.03-2±6.5-3	3.57-1±3.6-2	1.06±0.11		8.35-2±9.4-3	5.32-1±6.6-2	1.49±0.19
125		1.08-1±1.2-2	9.28-1±9.5-2	2.63±0.27		1.40-1±1.5-2	1.39±0.16	3.77±0.44
150		1.63-1±1.8-2	1.85±0.19	5.12±0.52		2.05-1±2.2-2	2.66±0.31	7.02±0.82
175		2.25-1±2.4-2	3.18±0.32	8.62±0.88		2.69-1±2.9-2	4.30±0.50	11.15±1.30
200		2.89-1±3.1-2	4.92±0.50	13.13±1.34		3.34-1±3.6-2	6.33±0.74	16.19±1.89
225		3.63-1±3.9-2	7.11±0.726	18.71±1.91		3.95-1±4.2-2	8.74±1.01	22.02±2.55
250		4.16-1±4.4-2	9.75±0.995	25.46±2.60		4.80-1±5.1-2	11.51±1.32	28.78±3.33
275		4.98-1±5.3-2	12.96±1.32	33.60±3.43		5.67-1±6.1-2	14.62±1.69	36.32±4.17
300		6.16-1±6.6-2	16.66±1.70	42.70±4.36		6.88-1±7.5-2	18.19±2.07	44.69±5.10
75	Mo	1.47-2±2.5-3	5.28-2±7.4-3	1.54-1±2.1-2	Pd	6.38-3±8.4-4	1.53-2±1.6-3	4.61-2±4.9-3
100		3.91-2±7.1-3	2.40-1±3.4-2	7.00-1±1.0-1		2.62-2±3.2-3	9.02-2±1.0-2	2.65-1±2.8-3
125		7.30-2±1.0-3	6.40-1±8.7-2	1.82±0.23		5.58-2±7.0-3	2.61-1±2.7-2	7.47-1±7.8-2
150		1.18-1±2.2-2	1.32±0.20	3.65±0.51		9.47-2±1.2-3	5.55-1±5.9-2	1.56±0.16
175		1.62-1±2.7-2	2.29±0.33	6.19±0.85		1.40-1±1.7-2	9.93-1±1.0-1	2.74±0.29
200		2.27-1±3.7-2	3.62±0.53	9.59±1.31		1.93-1±2.3-2	1.60±0.17	4.37±0.46
225		2.79-1±4.2-2	5.32±0.75	13.87±1.86		2.50-1±3.0-2	2.38±0.25	6.43±0.67
250		3.27-1±4.6-2	7.37±0.99	18.97±2.46		3.13-1±3.6-2	3.38±0.35	9.00±0.94
275		3.86-1±6.2-2	9.84±1.28	24.83±3.10		3.74-1±4.3-2	4.56±0.48	12.07±1.25
300		4.20-1±5.4-2	12.73±1.61	31.71±3.86		4.43-1±5.0-2	6.01±0.63	15.78±1.64
75	Ag	3.89-3±5.4-4	8.25-3±1.0-3	2.49-2±3.0-3				
100		1.99-2±2.5-3	5.36-2±6.1-3	1.60-1±1.9-2	Sn	1.38-2±3.7-3	1.78-2±4.3-3	5.01-2±1.2-2
125		4.55-2±5.8-3	1.66-1±2.0-2	4.91-1±5.8-2		4.14-2±2.0-2	7.60-2±1.6-2	2.14-1±4.6-2
150		7.77-2±9.5-3	3.70-1±4.2-2	1.07±0.12		7.70-2±1.7-2	1.85-1±3.8-2	5.19-1±1.0-1
175		1.23-1±1.6-2	6.96-1±8.1-2	2.00±0.24		1.28-1±3.7-2	3.60-1±1.0-1	1.00±0.26
200		1.73-1±2.2-2	1.16±0.14	3.31±0.39		1.65-1±3.4-2	5.78-1±1.1-1	1.59±0.30
225		2.25-1±2.7-2	1.77±0.20	4.99±0.58		2.45-1±6.4-2	9.00-1±2.4-1	2.49±0.64
250		2.88-1±3.4-2	2.55±0.29	7.13±0.83		2.65-1±4.8-2	1.22±0.23	3.34±0.62
275		3.49-1±4.0-2	3.51±0.40	9.77±1.12		3.86-1±8.8-2	1.73±0.44	4.78±1.21
300		4.10-1±4.7-2	4.67±0.53	12.92±1.48		3.83-1±6.4-2	2.14±0.41	5.83±1.11
125	Ce	2.02-3±2.2-4	1.84-3±2.1-4	6.63-3±7.1-4				
150		7.50-3±8.0-4	7.92-3±8.9-4	2.37-2±2.6-3	Gd	2.12-3±2.2-4	2.22-3±2.4-4	6.64-3±6.9-4
175		1.71-2±1.9-3	2.12-2±2.4-3	6.02-2±6.7-3		6.87-3±7.2-4	8.16-3±8.9-4	2.32-2±2.4-3
200		3.10-2±3.3-3	4.42-2±4.9-3	1.25-1±1.4-2		1.58-2±1.6-3	2.11-2±2.3-3	5.90-2±6.1-3
225		5.03-2±5.4-3	8.09-2±9.1-3	2.27-1±2.5-2		3.01-2±3.2-3	4.46-2±4.9-3	1.24-1±1.3-2
250		7.39-2±7.9-3	1.32-1±1.5-2	3.77-1±4.2-2		5.08-2±5.3-3	8.21-2±9.0-3	2.31-1±2.4-2
275		1.04-1±1.1-2	2.04-1±2.3-2	5.88-1±6.5-2		7.86-2±8.2-3	1.38-2±1.5-2	3.90-1±4.0-2
300		1.37-1±1.5-1	2.94-1±3.2-2	8.68-1±9.6-2		1.16-1±1.2-2	2.18-1±2.4-2	6.18-1±6.4-2
150	Er	4.87-4±9.3-5	6.34-4±1.7-4	2.04-3±4.1-4	Yb	1.63-4±2.5-5	2.17-4±2.6-5	9.14-4±1.26-4
175		3.37-3±6.3-4	4.05-3±7.8-4	1.31-2±2.2-3		1.31-3±1.2-4	1.20-3±1.3-4	4.06-3±5.3-4
200		9.30-3±1.9-3	1.10-2±2.3-3	3.52-2±5.9-3		4.00-3±3.6-4	3.87-3±4.3-4	1.23-2±1.8-3
225		1.82-2±3.0-3	2.20-2±4.3-3	6.90-2±1.1-2		9.19-3±8.5-4	9.50-3±1.1-3	2.97-2±4.3-3
250		3.19-2±7.4-3	3.78-2±1.06-2	1.18-1±2.5-3		1.77-2±1.6-3	1.95-2±2.2-3	6.12-2±8.6-3
275		4.72-2±8.7-3	5.72-2±1.36-2	1.76-1±3.2-2		3.05-2±2.7-3	3.56-2±3.9-3	1.13-1±1.6-2
300		6.62-2±1.1-2	8.10-2±1.77-2	2.48-1±4.3-2		4.85-2±4.3-3	6.00-2±6.7-3	1.95-1±2.7-2

[a]Values of $Nx10^{-n}$ are given as N-n.

TABLE 2. Measured Intensity Ratios Compared with Theory.

	L_ℓ/L_α Expt.	Theor.	$L_{\gamma1}/L_{\beta1}$ Expt.	Theor.	$L_{\gamma23}/L_{\beta34}$ Expt.	Theor.
Zr	0.0218±0.0011	0.0379				
Nb	0.0510±0.0031	0.0375	0.0431±0.0025	0.0310		
Mo	0.0493±0.0033	0.0372	0.0475±0.0033	0.0421		
Pd	0.0439±0.0025	0.0371	0.0953±0.0050	0.106	0.209±0.021	0.182
Ag	0.0459±0.0029	0.0372	0.104±0.006	0.106	0.171±0.011	0.188
Sn	0.0413±0.0048	0.0375	0.135±0.015	0.134	0.195±0.024	0.207
Ce	0.0415±0.0022	0.0392	0.129±0.007	0.178		
Gd	0.0424±0.0018	0.0414	0.162±0.009	0.186	0.273±0.018	0.251
Er	0.0469±0.0041	0.0434	0.182±0.021	0.187	0.261±0.024	0.252
Yb	0.0502±0.0037	0.0445	0.198±0.012	0.189	0.277±0.018	0.253

with our measurements at higher energies for the L2 and L3 shells, but are more than twice our values for the L1 shell. Braziewicz, et al., report cross sections that are much smaller than ours for all subshells, except at 300 keV, where they are larger. They note that ECPSSR over-estimates their results for $\xi_s^r \leq 0.35$, which we see also for Er and other rare earths.

To summarize, these results generally confirm that, for $\xi_s^r <$ 0.7, ECPSSR x-ray production cross sections under-predict measured values for all three subshells, with a few exceptions: ECPSSR completely over-predicts the Ce data, and over-predicts for Gd, Er and Yb below approximately $\xi_s^r = 0.3$. The data are most scattered for the L2 subshell due to the difficulty of resolving the $L_{\beta1}$ peak in the spectral fits. The L1 subshell results are better determined and show an oscillatory trend. The subshell ratios, L_ℓ/L_α, $L_{\gamma1}/L_{\beta1}$, and $L_{\gamma2,3}/L_{\beta3,4}$, follow the general trend of the corresponding ratios of transition rates interpolated from Scofield's work(7).

This work along with others points to a possible shortcoming in the ECPSSR theory for describing L subshell ionization by slow protons for a range of atomic targets. Smit(12) has pointed out that ionization cross section calculations at low projectile speeds are extremely sensitive to projectile kinematics, for which ECPSSR may give results that differ by a factor of two. It is also possible that enhancement effects in metallic solids could add to measured cross sections(13) and cross section ratios(14). Clearly, more data are needed to further fill in the gaps in order to note trends, especially for the lighter elements. The ECPPSR formalism is attractive because of its utility, so efforts should continue to clarify the source of discrepancies with experiment.

ACKNOWLEDGEMENTS

This work was supported by a Cottrell College Science grant from the Research Corporation and was conducted at the University of Nebraska at Lincoln.

REFERENCES

1. Brandt, W., and Lapicki, G., *Phys. Rev.* A **23**, 1717-1729 (1981).
2. Orlic, I., Sow, C.H., and Tang, S.M., *At. Dat. Nucl. Dat. Tables* **56**, 159-210 (1994).
3. Gallagher, W. and Cipolla, S., *Nucl. Instr. Meth.* **122**, 405-414 (1974).
4. Merzbacher, E., and Lewis, H., *Handbuch der Physik*, vol. **34,** Berlin: Springer Press, 1958, p. 166.
5. Ziegler, J., "Transport of Ions in Matter," ver. 91.07, IBM, Yorktown Hts., July 26, 1991.
6. Berger, M.J. and Hubbell, J.H., "XCOM: Photon Cross Section on a Personal Computer," NBS Report NBSIR 87-3595, Gaithersburg, Md., 1987.
7. Campbell, J. and Wang, J.-X., *A T. Dat. Nucl. Dat. Tables* **43**, 281-291 (1989).
8. Liu, Z. and Cipolla, S., *Comp. Phys. Comm.* **97**, 315-332 (1996).
9. Krause, M., J. Phys. Chem. Ref. Data **8**, 307 (1979).
10. Harrigan, M.F., Spicer, B.M., and Cohen, D.D., Aust. J. Phys., **37**, 475-486 (1984).
11. Braziewicz, J., Braziewicz, E., Pajek, M.,Czyewski, T.,Glowacka, L., Jaskola, M., Kobzev, A.P., Kauer, T., Trautmann, D., and Kretschmer, W., J. Phys. B **24**, 1669-1682 (1991).
12. Smit, Z., Phys. Rev. A**53**, 4145-4150, 1996.
13. Hoszowska, J. and Dousse, J-Cl., J. Phys. B**29**, 1641-1653, 1996.
14. Petukhov, V.P., Romanovsky, E.A., and Kerkow, H., Nucl. Instr. Meth. **B109/110**, 19-22, 1996.

M-SHELL IONIZATION OF ATOMS BY C, N, AND O IONS

M. Pajek[1], J. Braziewicz[1], J. Semaniak[1], T. Czyżewski[2], L. Głowacka[2], M. Jaskóła[2] M. Haller[3], R. Karschnick[3], W. Kretschmer[3], A. P. Kobzev[4], D. Trautmann[5] and G. Lapicki[6]

1) Institute of Physics, Pedagogical University, 25-509 Kielce, Poland
2) Soltan Institute for Nuclear Studies, 05-400 Otwock-Świerk, Poland
3) Physikalisches Institut, Universität Erlangen-Nürnberg, D-91058 Erlangen, Germany
4) Joint Institute for Nuclear Research, Dubna, Russia
5) Institute of Physics, University of Basel, CH-4056 Basel, Switzerland
6) Department of Physics, East Carolina University, Greenville, NC 27858, USA

M-shell ionization in selected heavy atoms (Au, Bi, Th and U) by energetic C^{q+}, N^{q+} and O^{q+} ions of different charge states (q=1-6) has been studied in the energy range 0.1-2 MeV/amu. The measurements were performed using target thicknesses allowing ion charge equilibration in the target. Derived equilibrium M-shell ionization cross sections are compared with the theoretical predictions based on the semiclassical (SCA) and the PWBA approximations for direct ionization and the OBK approximation for the electron capture, as well as the ECPSSR theory including the corrections for higher-order effects. Substantial contribution of the electron capture caused by the ion charge equilibration is observed for high energies. The influence of the multiple ionization in M-, N- and O-shells on measured cross sections is discussed.

INTRODUCTION

Inner shell ionization by charged particles was studied intensively in last years, mainly, due to the importance of the particle-induced x-ray emission (PIXE) method in analytical studies. On the other hand, available experimental ionization cross sections forced the development of the theoretical models to describe this fundamental process. Presently, the existing data for K-, L- and M-shell ionization by light ions can be reproduced quite well by the theoretical models based on the plane-wave Born approximation (PWBA) or the semiclassical approximation (SCA), when the corrections for higher-order effects are included in these approaches.

For heavy ions, perturbing strongly initial electronic state, the first-order theories cannot describe the data so well. In this case substantial discrepancies were reported, mostly for M- and L-shells (1,2). The present work summarizes our systematic studies of M-shell ionization in heavy atoms by C, N and O ions. However, it is focused on a presentation of the experimental method, data analysis and the results for selected $O^{q+} \rightarrow$ Au system, showing all typical features found for other systems. We note here that a part of our results was published earlier (3), and the recent results for M-shell ionization by heavy ions reported by other authors can be found in Ref. (4-6).

Present results, covering energy range 0.1-2 MeV/amu, are compared with the predictions of the ECPSSR theory describing both the direct ionization (DI) and the electron capture (EC), and with the SCA calculations for direct ionization, performed in the separated (SCA-SA) and the united (SCA-UA) atoms limit. In these comparisons, the effect of the ion charge equilibration in the target is taken into account by calculating the "equilibrium" ionization cross sections.

EXPERIMENTS AND METHOD

The measurements of M-shell ionization in Au, Bi, Th and U by C^{q+}, N^{q+} and O^{q+} ions were performed at two accelerators: the Van de Graaff machine at Joint Institute for Nuclear Research in Dubna for low energies (1-4 MeV); and the tandem accelerator at University of Erlangen for higher energies (4-40 MeV).The ionization cross sections were obtained from M-x-ray yields measured with carefully efficiency calibrated (7) Si(Li) and HPGe detectors. The L-x-ray spectra were also measured to estimate the multiple ionization effects.The absolute values of the cross sections were obtained by normalizing the x-ray yields to the numbers of elastically scattered projectiles detected by Si surface barrier detector. Further details, concerning both the experiments and data analysis, can be found elsewhere (2). In the present work, two important effects occuring for heavy ion impact, namely, the ion *charge equilibration* in the target and the *multiple ionization* effects, are discussed in more details.

For heavy ions a contribution of the electron capture, which depends on the ion charge state q, becomes

CP392, *Application of Accelerators in Research and Industry,* edited by J. L. Duggan and I. L. Morgan
AIP Press, New York © 1997.

comparable with the direct ionization. Consequently, in comparison of experimental results with the theoretical predictions the ion charge state in the target has to be known. In experiments using the targets of intermediate thicknesses, here 20-40 μg/cm^2, the equilibrium charge state distribution is reached, so essentially, the *equilibrium* x-ray and ionization cross sections are measured. Such equilibrium ionization cross sections have to be compared with the theoretical cross sections σ_{eq} calculated for the equilibrium charge state distribution (8):

$$\sigma_{eq} = \sum_{q} F(q) \cdot \sigma(q) \tag{1}$$

The equilibrium ion charge fractions F(q) are presently well known and can be found in Ref. (9). We note here that this approach is an alternative to the well known method (8) of measuring the ionization cross sections σ(q) for a given charge state q, in which vanishingly thin targets (below 1 μg/cm^2) are used.

Another important aspect in measurements of ionization cross sections for heavy ions is the *multiple ionization* effect, which complicates a comparison of experimental results with the theoretical prediction. In this case the atomic parameters (fluorescence and Coster-Kronig yields) should be known for the multi-vacancy configurations, but practically, only their single-vacancy values are available. In order to correct for this effect the energy shifts of L-x-ray lines can be measured to estimate the probabilities of ionization of M-, N, and O-shells, which in turn allow one to correct approximately (10) the atomic parameters for the multiple ionization effect. The data, if corrected in this way for the multiple ionization effects, are accurate, typically, within 20-25%. Such measurements of M-shell ionization cross sections are accurate enough to discuss the prediction of different theoretical approaches describing the M-shell ionization by heavy ion impact.

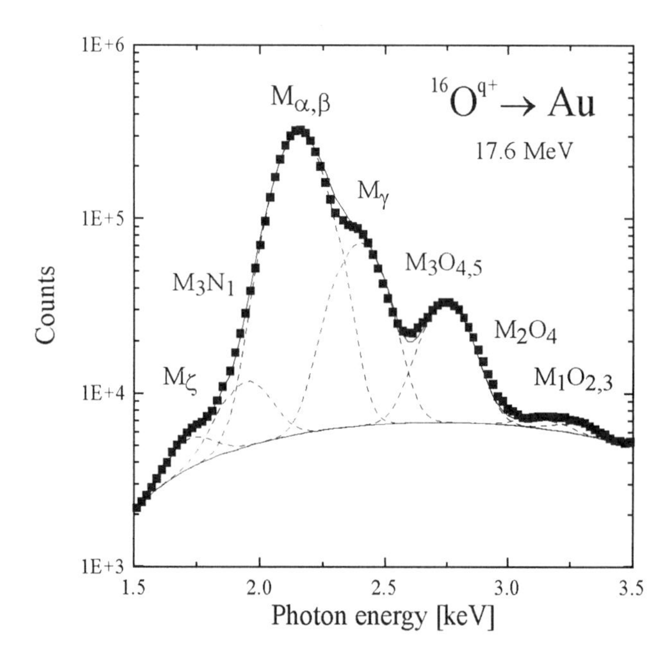

FIGURE 1. Typical M-x-ray spectrum measured for gold target bombarded by 17.6-MeV $^{16}O^{q+}$ ions.

RESULTS

M-shell ionization

M-shell ionization cross sections for Au, Bi, Th and U by C^{q+}, N^{q+} and O^{q+} ions of energy 0.1-2 MeV/amu were measured at equilibrium ion charge state condition. In measured x-ray spectra M_ζ, M_3M_1, $M_{\alpha,\beta}$, M_γ, $M_3O_{4,5}$, M_2O_4 and $M_1O_{2,3}$ transitions were analysed (see Fig.1) to derive the total M-shell x-ray cross sections, which were further converted to the M-shell ionization cross sections using the effective M-shell fluorescence yields from Ref.(1). Measured M-shell ionization cross sections were compared with the predictions of the equilibrium ionization cross section (see Eq.1), in which the cross section for a given charge state q consisted of contribution of DI and charge state dependent EC process, i.e.

$$\sigma(q) = \sigma_{DI} + \sigma_{EC}(q) \tag{2}$$

The theoretical cross sections for direct ionization within the semiclassical approximation (SCA) were performed using hyperbolic trajectories and relativistic hydrogenic wave functions (11). The binding effect was simulated in calculations by two extreme cases of separated (SA) and united (UA) atom limits. In the ECPSSR theory (12), based on the PWBA for direct ionization and the Oppenheimer-Brinkman-Kramers approximation modified by Nikolaev (OBKN) for the electron capture, the corrections for binding-polarization, Coulomb deflection, relativistic and energy-loss effects are included. Generally, the ECPSSR theory predicts dominating contribution of the EC process for fully stripped and H-like ions (see Fig.2), which occur in large fractions in equilibrium distribution for higher energies, e.g. 30% and 50%, respectively, for 2 MeV/amu oxygen ions (4).

In Fig. 3 measured equilibrium M-shell ionization cross sections for $O^{q+} \rightarrow$ Au are compared with the theoretical predictions calculated using Eqs.1 and 2. The present data agree within 10% with the result reported by Andrews et al. (4) for 2 MeV/amu, after converting their data to the

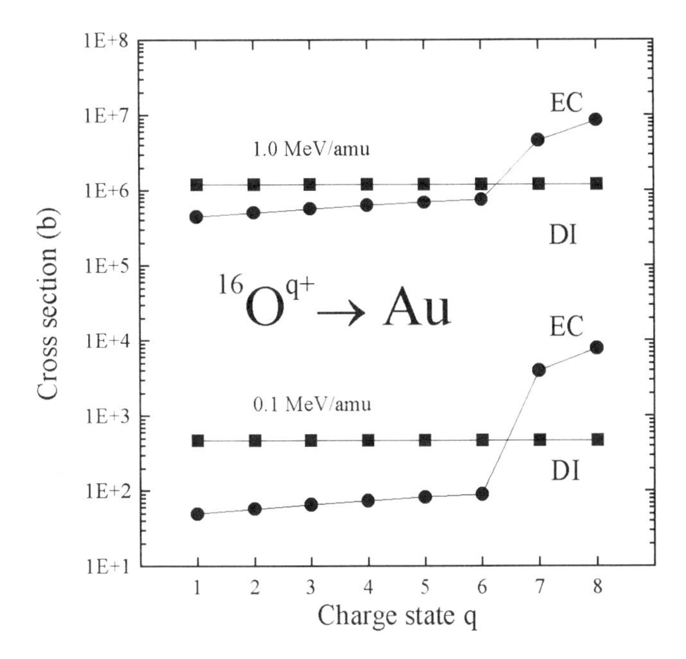

FIGURE 2. Theoretical M-shell cross sections for direct ionization (DI) and electron capture (EC) for $O^{q+} \rightarrow$ Au calculated according to the ECPSSR theory for 0.1 and 1.0 MeV/amu.

equilibrium cross section. From comparison of different theoretical predictions of M-shell ionization by oxygen ions the following conclusions can be drawn: (i) the SCA calculations in the separated atom limits (SA) predict quite satisfactory the data below 1.5 MeV/amu, but for higher energies the data are higher than calculations, which do not include, however, the EC contribution; (ii) the ECPSSR theory slightly overestimates the data above 1 MeV/amu, but fails evidently for low energies; (iii) it is worth noting that the PWBA describes the data quite well over a whole energy range. These observations suggest some defficiency of the ECPSSR theory in the low energy regime and, on the other hand, the importance of the electron capture contribution for high energies.

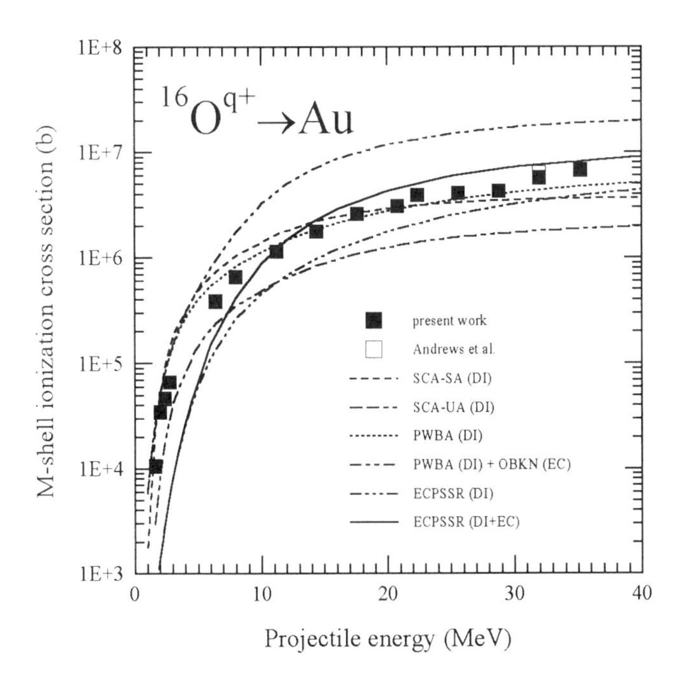

FIGURE 3. Measured M-shell ionization cross setions for $O^{q+} \rightarrow$ Au compared with the predictions of the SCA-SA (-UA) calculations, the PWBA and OBKN approximations and the ECPSSR theory for, as indicated on the figure, direct ionization (DI) and electron capture (EC) processes.

Multiple ionization effects

For the same systems as studied in the M-shell case, the L-x-ray spectra were measured in the energy 0.1-2 MeV/amu. Here the individual x-ray transitions were resolved, and consequently L-subshell cross sections were derived. However, in the present paper the L-subshell ionization cross sections will not be discussed, referring the reader to our ealier publications on this topic (2). On the other hand, we show here that analysis of measured L-x-ray spectra allows one to estimate the ionization probabilities per electron, p, for M-, N- and O-shells, and thus estimate an influence of the multiple ionization effect on measured M-shell cross sections.

The $L_\ell(M_1 \rightarrow L_3)$ and $L_\alpha(M_{4,5} \rightarrow L_3)$ transitions are well separated for heavy atoms and their energies can be determined quite accurately (± 5 eV). We found that the energy shifts for these transitions are about 25 eV, indicating small multiple ionization effects in the M-shell ($p_M \approx 0.05$). The separation of individual transition in Lγ group is more complicated (see Fig.4) due to small energy difference between the peaks and, additionally, their higher energy shifts. Careful analysis of the energy shifts in Lγ group gives the estimates of the ionization probabilities for N- and O-shells. The result of such analysis for 35.2-MeV $O^{q+} \rightarrow$ Au is shown in Fig.4. Here the peaks are shifted, independent of the ion energy, of about 100 eV for $L\gamma_{1,2,3,5}$ $(N_{4,3,2,1} \rightarrow L_{2,1})$, 200 eV for $L\gamma_{4,4'}$ $(O_{3,2} \rightarrow L_1)$ and 320 eV for $L\gamma_6$ $(O_4 \rightarrow L_2)$ transitions. Using these results, and relating (10) the observed energy shifts for N→L transitions with N-shell ionization probability per electron, one arrives at the estimation of ionization probability $p_N \approx 0.15\text{-}0.20$. Observed energy shifts for O→L transitions indicate that the ionization probability for O-shell should not exceed (13) a value found for the N-shell, i.e $p_O \leq p_N$. These estimates of the ionization probabilities for M-, N- and O-shell allow us to conclude, using simple relation (10)

$$\overline{\omega}_M = \omega_M / (1 - p_N(1 - \omega_M)) \qquad (3)$$

accounting for a change of the fluorescence yields, that the effect of the multiple ionization in outer shells does not increase the M-shell fluorescence yields by more than about 25%. This result shows that the multiple ionization effect can thus reduce the measured M-shell ionization cross sections only by less than 25%, i.e. by an amount, which does not change our earlier conclusions.

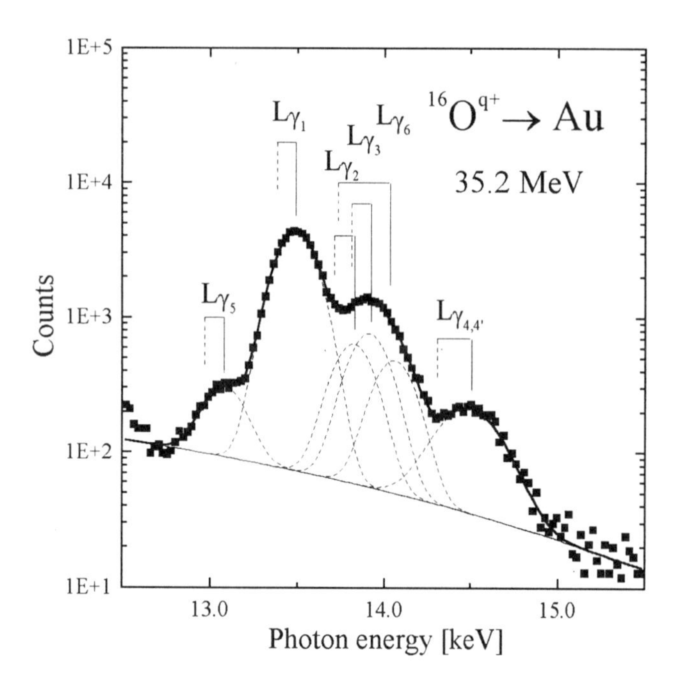

FIGURE 4. Measured L_γ x-ray spectrum in gold for 35.2-MeV O^{6+} ion impact showing resolved structure of individual transitions used to estimate the ionization probabilities.

CONCLUSIONS

M-shell ionization cross sections measured for equilibriated ion charge state distributions can be directly compared with the theoretical predictions by weighting the cross sections by known equilibrium charge state fractions. Both SCA and ECPSSR predictions do not reproduce satisfactorily the data in a whole energy range studied. The electron capture process dominates the direct ionization for high energies due to the ion charge equilibration in the target. A magnitude of the effect of multiple ionization can be estimated by studying the energy shifts of L-x-ray transitions.

ACKNOWLEDGMENTS

The authors acknowledge the support provided by the German WTZ Grant No. N-88-94 and the Polish Committee for Scientific Research.

REFERENCES

1. Pajek, M., Kobzev, A.P., Sandrik, R., Skrynik, A.V., Ilkhamov, R.A., Khusmurodov, S.H., and Lapicki, G., Phys. Rev. A**42**, 261 (1990).
2. Semaniak, J., Braziewicz, J., Pajek, M., Czyżewski, T., Głowacka, L., Jaskóła, M., Haller, M., Karschnick, R., Kretschmer, W., Halabuka, Z., and Trautmann, D., Phys. Rev. A**52**, 1125 (1995).
3. Czyżewski, T., Głowacka, L., Jaskóła, M., Semaniak, J., Braziewicz, J., Pajek, M., Haller, M., Karschnick, R., Kretschmer, W., Kobzev, A.P., and Trautmann, D., Nucl. Instrum. and Methods, B**109/110**, 52 (1996).
4. Andrews, M.C., McDaniel, F.D., Duggan, J.L., Miller, P.D., Pepmiller P.L., Krause, H.F., Rosseel, T.M., Rayburn, L.A., Mehta, R., and Lapicki, G., Phys. Rev. A**36,** 3699 (1987).
5. Yu, Y.C., Sun, H.L., Duggan, J.L., McDaniel, F.D., Yin, J.Y.,and Lapicki, G., Phys. Rev. A**52**, 3836 (1995).
6. Carlen, M.W. , Boschung, B., Dousse, J.-Cl., Halabuka, Z., Hoszowska, J., Kern, J., Rheme, Ch., Polasik, M., Rymuza, P. and Sujkowski, Z., Phys. Rev. A**49**, 2524, (1994).
7. Pajek, M., Kobzev, A.P., Sandrik, R., Ilkhamov, R.A., and Khusmurodov, S.H., Nucl. Instrum. and Methods, B42, 346 (1989).
8. Gardner, R.K., Gray, T.J., Richard, P., Schmiedekamp, C., Jamison, K.A., and Hall, J. M., Phys. Rev. A**15**, 2202 (1977).
9. Shima, K., Mikumo, T., and Tawara, H., Atomic Data and Nuclear Data Tables, **34**, 357 (1986); **51**, 173 (992).
10. Lapicki, G., Mehta, R., Duggan, J.L., Kocur, P.M., Price, J.L., and McDaniel, F.D., Phys. Rev. A**34**, 3813 (1986).
11. Trautmann, D. and Kauer, T., Nucl. Instrum. and Methods, B**42**, 426 (1989).
12. Brandt, W., and Lapicki, G., Phys. Rev. A**20**, 465 (1979); A**23**, 1717 (1981).
13. Ito, S., Shoji, M., Madea, N., Katono, R., Mukoyama, T., Ono, R., and Nakayama, Y., J. Phys. B**20**, L597 (1987).

EXPERIMENTAL STUDIES OF HIGHLY-CHARGED ION COLLISIONS WITH RYDBERG ATOMS

S.R. Lundeen and C.W. Fehrenbach
Colorado State University, Ft. Collins, CO 80523

Charge transfer collisions between both singly and multiply charged ions and Rydberg atoms have been studied using a dense Rydberg target, constructed by CW laser excitation of a thermal Rb beam. Total charge transfer cross sections were measured over a range of velocities for a few ion charges up to Q=40. A Doppler-tuned CO2 laser was used to selectively detect neutral Rydberg products in a particular n,L level in the studies with singly charged ions, and promises to become a valuable tool in studies of multiply charged ions.

I. INTRODUCTION

Charge transfer collisions between ions and highly excited atoms are both of scientific and of practical interest. The two most significant features of such collisions are 1) large cross sections and 2) quasi-resonant capture. The cross sections are large since, in order to capture the highly excited electron, a slow ion of charge Q need only approach to within the distance $R = Q^{1/2}n^2a_0$, at which the electric field of the ion equals the field binding the electron in the atom. The capture is quasi-resonant in the sense that the final binding energy of the captured electron is approximately equal to it's initial binding energy in the Rydberg atom. Both of these feature assume increasing importance as the charge of the ion increases. The higher the ion charge, the more likely it is to capture an electron, and the larger is the excitation of the captured electron in the product ion. For example, a slow fully stripped oxygen ion captures the electron from a hydrogen atom in the n=10 level with a cross section of about 80,000 a_0^2, and usually results in a Rydberg state of O^{7+} bound by about 0.23 eV, 870 eV above the ground state of O^{7+}. If the resulting ion relaxes by photon emission, this can become an efficient cooling mechanism for a hot plasma containing both excited hydrogen atoms and highly charged impurity ions.

In addition to it's practical importance, the ion Rydberg charge transfer process is also scientifically intriguing because it is a qualitatively rich phenomena where no successful quantum mechanical calculations exist, and where, to date, all predictions are based on classical models. This, in spite of the fact that the process is usually assumed to be equivalent to a three body problem, i.e. a fully stripped ion incident on an excited hydrogen atom.

II. PREVIOUS EXPERIMENTAL STUDIES

Experimental studies of ion–Rydberg atom charge transfer collisions were pioneered by MacAdam and collaborators at the Univ. of Kentucky (1). In a series of experiments, a beam of low energy singly charged ions crossed a thermal Na atomic beam excited by two pulsed dyel lasers to levels near $n_T = 28$. The captured electrons were analyzed with a Stark Ionization detector, which infers the principal quantum number of the Rydberg products by measuring the field strength at which they reionize . These experiments confirmed the large total cross section and the quasi-resonant nature of the capture, the two main features of the collisions, and stimulated substantial advances in the classical theories. The Stark Ionization measurements on the product Rydberg states could be interpreted as a map of population vs. n. These plots resembled the predictions of Classical Trajectory Monte Carlo (CTMC) calculations (2).

Recently, the Stark Ionization method was also applied to study charge transfer collisions of highly charged ions (3). A thermal Rb beam, excited by a single UV laser to the 17P state, was crossed with a Kr^{8+} beam, and the product Kr^{7+} Rydberg states were analyzed in a Stark Ionization detector. The initial results of this experiment showed rather broad n distributions of the product ions, qualitatively similar to the predictions of CTMC theory.

In both of these Stark Ionization experiemnts, close comparison of the measured n distributions with theoretical calculations show some clear differences. It is uncertain, however, if these differences indicate difficulties in the theory, or whether the experimental data have been properly interpreted. The basic problem is that different Stark levels of the same principal quantum number ionize at fields which vary by as much as a factor of three. Consequently, an n-distribution can be derived from experimental data only after making assumptions regarding the L and m distributions. These assumptions cannot be independently tested.

III. RESIS DETECTION OF RYDBERG LEVELS

Thus, while the Stark Ionization method of detection of Rydberg states has been extremely useful in past studies,

CP392, *Application of Accelerators in Research and Industry,* edited by J. L. Duggan and I. L. Morgan
AIP Press, New York © 1997.

it suffers from being only moderately selective in discriminating between many possible Rydberg levels. There is an alternative method of detecting Rydberg states which is much more successful in isolating an individual n,L,m state. This method selectively ionizes particular Rydberg levels in a two-step process consisting of 1) selective laser excitation to a highly excited discrete level and 2) Stark Ionization of the highly excited level with collection of the resulting ion. Increased selectivity is obtained from the first step of the process where the high frequency resolution of the laser can select an individual n,L state to be excited and ionized, to the exclusion of all others. This two step detection method, sometimes referred to as Resonant Excitation Stark Ionization Spectroscopy (RESIS), offers a much more discriminating, though still highly efficient detector of Rydberg states. While it is, in principle, widely applicable, it has so far been implemented only using a Doppler-tuned CO_2 laser for the excitation step. This limits the method to detection of n=9 or 10 Rydberg levels, whose binding energies are just slightly greater than the 0.12 eV energy of a CO_2 laser photon. While this is clearly a significant limitation in comparison to Stark Ionization, it is balanced by the much higher selectivity of the RESIS technique.

IV. RYDBERG TARGET STUDIES WITH SINGLY-CHARGED IONS

We have used the RESIS technique for several years now, mostly for spectroscopic studies of atomic and molecular Rydberg states (4,5). Since the use of a CO_2 laser limits these studies to n=9 and 10 Rydberg levels in a fast beam, we decided, several years ago, to build a dense Rydberg target which could be used as a charge exchange target in forming these fast beams. Our hope was to produce large populations concentrated in the n = 9 and 10 levels.

We first built such a Rydberg target in 1993 (6). It consists of a thermal Rb beam excited by three CW lasers to the nF state, where $7 < n < 14$. Figure 1 shows the excitation scheme for the target. The first laser is a single mode diode laser at 780 nm, tuned to the $5S_{1/2}$ – $5P_{3/2}$ resonance transition in Rb. The second laser is a color center laser at 1529 nm, which excites the $5P_{3/2}$ – $4D_{5/2}$ transition. Finally, a Ti:Sapphire laser excites the $4D_{5/2}$ state to one of the F states with n between 7 and 13. One of the most interesting aspects of the operation of this target is that it radiates stongly in the blue-green. This was completely unexpected since the spontaneous decay of any of the nF levels to visible radiation is expected to be small. In the case of the 10F target, the blue-green radiation occurred at a wavelength of 508 nm, the wavelength of 11D - 5P spontaneous decay. The 11D state, which lies just 75 cm^{-1} below the 10F level, was clearly not being pumped, yet it was radiating strongly. Eventually, we understood this to be due to the natural formation of a mirrorless maser (7).

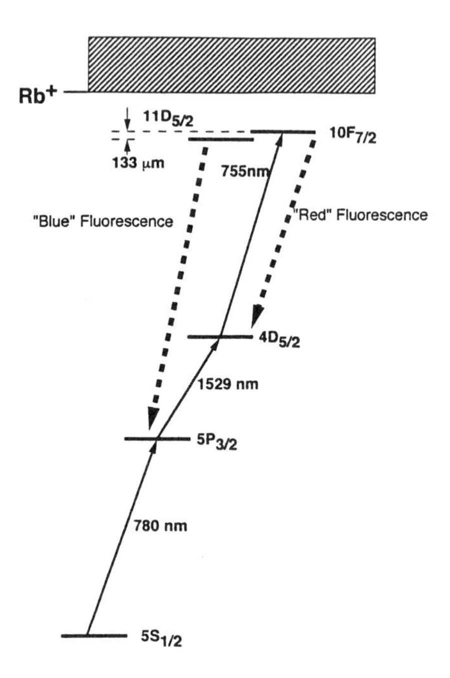

FIG. 1. Level diagram of Rb "10F" Rydberg target. The 10F level is excited in a thermal Rb beam by three single frequency CW lasers. The excited population is rapidly shared between the 10F and 11D levels by mirrorless maser action. Fluorescence radiation from both levels is monitored as a measure of target thickness.

The density of 10F atoms was sufficiently high that gain existed on the 75 cm^{-1} 10F–11D transition. This gain was sufficient to amplify spontaneous or black-body radiation to the point of saturating the 10F–11D transition. In this way, the population originally excited to the 10F level was rapidly and efficiently shared between the 10F and 11D levels. Thus the "10F" target was, in reality, a nearly equal mixture of 10F and 11D states. In operation, we routinely monitored the radiation emitted from both levels, the "red" (10F–4D) radiation from the F level, and the "blue" (11D–5P) radiation from the D level. For F populations well above the threshold for maser action, these two levels were approximately proportional, and their sum was used as a monitor of target density. Similar radiation (at appropriately different wavelengths) is observed for the other nF targets in the range $7 < n < 13$. From measurements of the intensity of the fluorescence from the 10F target, we determined the total number of excited atoms to be about 2×10^{10} atoms in a spherical volume of diameter about 2 mm, giving a target thickess of about 1×10^{10} cm^{-1}.

As an initial demonstration of the operation of this target in combination with the RESIS method of Rydberg detection, we studied charge transfer collisions by 11 KeV S^+ ions on the Rydberg target (6). Figure 2 shows the apparatus used for this study. The S^+ ions, at v/c = .00086, captured electrons from the Rydberg target and proceeded down the beamline. Those in n=9

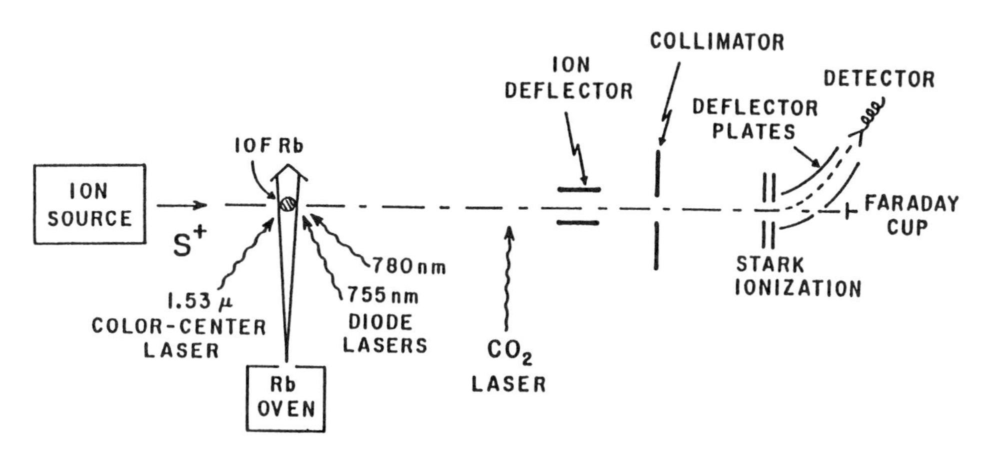

FIG. 2. Schematic diagram of the apparatus used to demonstrate use of the Rydberg target/RESIS method of studying charge transfer collisions. A S^+ beam passes through the Rb Rydberg target where it is partially neutralized to form sulfur Rydberg levels. Those in n=9 or 10 states can be detected by the RESIS method. They are selectively excited to a higher level (9 to 17, 10 to 30) which is subsequently Stark ionized. The Doppler-tuned CO_2 laser excites only one selected n,L level.

or 10 states were excited by the Doppler-tuned laser to an upper state which could be easily Stark ionized. The current obtained in the Stark Ionizer could then be plotted as a function of the variable angle between the fast S beam and the fixed frequency CO_2 laser , or equivalently the Doppler-shifted laser frequency. Such a spectrum, plotted in Fig. 3, clearly resolves the separate RESIS peaks due to n=10 levels with L = 4,5,6,7,8, and 9. By calculating the excitation efficiency of the CO_2 laser, and accounting for the differing spontaneous decay rates of the several L levels, the heights of the well-resolved signals could be used to infer the initial populations of each 10L level. This analysis clearly showed that the L levels were populated non-statistically, something that had been previously predicted by CTMC calculations, but never so directly observed.

Since only n = 9 or 10 levels are easily detected with the RESIS technique, it is not possible to directly map out the n-distribution of excited levels resulting from charge exchange with this method . However, an indirect test of predictions can be made by measuring both the RESIS signal from a particular n,L level and the total charge transfer beam. Their ratio reveals the fraction of the total product beam in the particular state being detected. Then, as n_T is varied, this ratio should change in a predictable way as the detected level is either favored or disfavored in the capture process.

We illustrate this in Fig. 4, which shows CTMC predictions (8) for the ratio of 10G population to total capture for He^+ ions of 0.1 a.u. velocity incident on Rydberg targets of various n_T's. The ratio shows a peak near $n_T = 8$ or 9, and falling off for either higher or lower values of n_T. It should be possible to test these predictions quite unambiguously with the RESIS technique. This will provide definitive tests of the final state distributions predicted by any of several theories of the charge capture process.

V. STUDIES WITH MULTIPLY-CHARGED IONS

The thickness of our Rydberg target is sufficient to neutralize about 0.5% of a singly-charged beam at matching velocity. Since the cross section for charge transfer increases linearly with Q, this should be sufficient to cause very substantial charge changing in collisions with multiply charged ions.

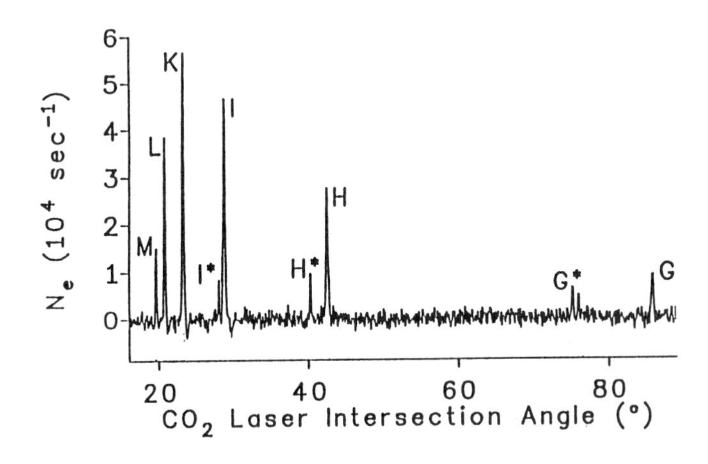

FIG. 3. RESIS spectrum showing 10-30 transitions in sulfur after charge transfer in the 10F Rydberg target, plotted as a function of the intersection angle of the beam and CO_2 laser, measured from anti-parallel. The S beam velocity is 0.00086 c, and the 10R(18) line of the CO2 laser is used (ν = 974.621cm^{-1}). Excitation of L=4,5,6,7,8, & 9 levels in n=10 are all well resolved. The strongest lines represent transitions from 10L to 30(L+1) levels, such as 10H–30I. The lines labeled with asterisks represent transitions to 30(L-1) levels, such as 10H–30G.

During the summer of 1994, we transported the Rydberg target to the J.R. Macdonald Laboratory at Kansas State University, where the CRYEBIS ion source provides a convenient source of slow highly charged beams. There we measured the relative charge transfer cross section in collisions with the 10F Rydberg target for a range of ion charges and velocities. They showed the predicted flat behavior at low velocities, followed by a steep drop above a critical velocity which increased slowly with ion charge. By measurements of the flourescence intensity and the target geometry, we were able to infer the absolute target thickness, and hence the absolute charge transfer cross sections, with a precision of 30%. All of these measurements were in reasonable agreement with the predictions of classical theories. A preliminary report of the results of this study has appeared (9) and a more complete report is in preparation (10).

RESIS detection of the products of collision by multiply charged ions with the Rydberg target appears to be quite feasible. The key observation is that the most probable states formed in such charge transfer collisions have about the same binding energy, regardless of the ion charge. Thus, they are at about the correct energy for RESIS detection using a Doppler -tuned CO_2 laser, even though they may have a large principal quantum number. For example, in capture by fully stripped oxygen ions on an n=10 target, the most probable binding energy is predicted to be about 0.23 eV, which corresponds to n=62 in hydrogenic oxygen. There are a number of transitions which could be induced by a CO_2 laser to detect the n=62 level or several nearby levels, for example the 62–95 transition at 1049 cm^{-1}. The matrix elements for these transitions are reduced only by a factor of Q below those of comparable transitions in neutral Rydberg levels, making the transitions observable with our existing lasers. As well, the electric field necessary to Stark ionize a level with principal quantum number n and core charge Q scales like Q^3/n^4, and therefore the n=95 level of hydrogenic oxygen should ionize at approximately the same fields as n=20 in hydrogen. Thus, there appear to be no substantial barriers to applying the RESIS detection technique to moderately charged Rydberg atoms formed by charge transfer with the Rydberg target. In fact, we have already observed, in a preliminary study, one such transition in hydrogenic carbon, the 47 - 73 transition (12). However, in this experiment, the measured signal sizes were not yet sufficiently reproducible to permit a reliable determination of, for instance, the capture fraction into n=43. Improvements in the detector design (13) should help to alleviate these problems in future experiments.

ACKNOWLEDGMENTS

This work was supported by the Division of Chemical Sciences, Office of Basic Energy Sciences, Office of Basic Energy Research, U.S. Department of Energy.

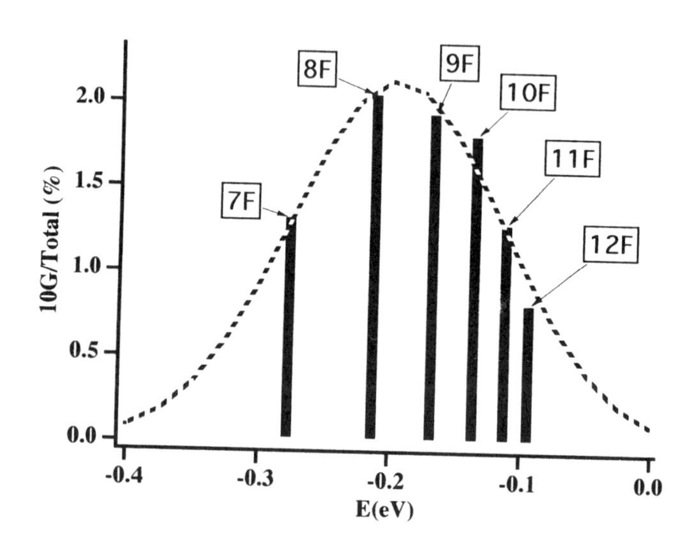

FIG. 4. Predicted capture fractions into the 10 G level by singly charged ions at v = 0.1 a.u. incident on six different Rydberg targets. The vertical axis is the predicted ratio of 10G population to total neutral product. The horizontal axis is the binding energy of the six targets, from 7F to 12F. The dashed line is a gaussian fit, drawn to guide the eye.

REFERENCES

1. K.B. MacAdam, in *Atomic Physics 12*, Proceedings of the 12th International Conference on Atomic Physics, eds. Jens C. Zorn and Robert R. Lewis, AIP Conference Proceedings 233, New York 1991, pp. 310-324
2. J. Pascale, R.E. Olson, and C.O. Reinhold, *Phys. Rev.* **A 42**, 5305 (1990)
3. A. Pesnelle, et. al., *Phys. Rev. Lett.*, **74** 4169 (1995)
4. E.A. Hessels et. al., *Phys. Rev.* **A 46** 2622 (1992)
5. Z.W. Fu, E.A. Hessels, and S.R. Lundeen, *Phys. Rev.* **A 46**, R5313 (1993)
6. F.J. Deck, E.A. Hessels, and S.R. Lundeen, *Phys. Rev.* **A 48**, 4400 (1993)
7. C.W. Fehrenbach, S.R. Lundeen, and O.L. Weaver, *Phys. Rev.* **A 51**, R910 (1995)
8. B.D. DePaola, private communication
9. B.D. DePaola, et. al., *Phys. Rev.* **A 52**, 2136 (1995)
10. M.-T. Huang, PhD thesis, Kansas State University (in preparation)
11. C.W. Fehrenbach and S.R. Lundeen, *Bull. Am. Phys. Soc.*, **40**, 1335 (1995)
12. D.S. Fisher, C.W. Fehrenbach and S.R. Lundeen, *Bull. Am. Phys. Soc.*, **41** 1120 (1996)

COLLISIONS AND SPECTROSCOPY OF LOW-ENERGY HIGHLY-CHARGED IONS USING AN ION TRAP

D. A. Church[a], J. Steiger[b], G. Weinberg[a], B. R. Beck[b], J. McDonald[b], L. Gruber[b], and D. Schneider[b]

[a]*Physics Dept., Texas A&M University, College Station, TX 77843-4242*
[b]*Lawrence Livermore National Laboratory, Box 808, Livermore, CA 94550*

ABSTRACT

Electron transfer from H_2 to highly-charged Xe^{q+} (q = 35, 43 - 46) and Th^{q+} (q = 73 -80) ions at center-of-mass energies near 6 eV has been studied using the RETRAP system at Lawrence Livermore National Laboratory. The ions were produced in the Electron Beam Ion Trap and retrapped in the Penning ion trap. Initial cross section data are in reasonable accord with the predictions of the absorbing sphere model, and true double capture is found to be about 25% of the total. The development with time of the charge of a single ion undergoing collisions has been observed non-destructively. Certain spectroscopic measurements are planned, following cooling of the stored ions to cryogenic temperatures.

INTRODUCTION

Ion traps have been used extensively in recent years to laser-cool certain singly-charged ions for purposes of precision laser-and microwave-spectroscopy, studies of one-component plasmas and ion crystals, precision mass spectroscopy, and much other basic physics. Progress has been summarized at the Nobel Symposium on Trapped Charged Particles and Related Fundamental Physics (1). Stored ion research on multiply-charged ions is considerably less extensive, but includes the use of multi-charged stored ions as targets for inner-shell photoionization using synchrotron radiation (2), lifetime measurements on metastable ion states (1,3), precision mass spectroscopy (1), and electron transfer collision studies of highly-charged ions with H_2 (4-6).

The expanding utilization of the Electron Beam Ion Trap (EBIT) (7), and its extension Super-EBIT to produce high charge high-Z ions, plus the ability to extract at low energy (8) and transport (9,10) these ions are the driving forces for further rapid developments in this area of research. High-charge high-Z ions extracted from EBIT have been re-captured into a cryogenic Penning ion trap system called RETRAP at Lawrence Livermore National Laboratory, where they have been stored for tens of seconds under benign ultra-high vacuum conditions. Electron transfer collision measurements on e.g. Th^{79+} with H_2 at mean center-of-mass energies near 6 eV have been completed (4-6) and are discussed below. Further, the stored ions have been cooled by the method of resistive damping using a resonant circuit, by tuning the ion axial oscillation frequency to resonance. The amplified induced voltage on this circuit has been used to monitor the ions, and recently has been used to observe the effects of laser-cooling of Be^+ ions, and through collisional coupling Be^{2+} ions, in a new RETRAP ion trap configuration. The combination of resistive damping plus collisional energy exchange via collisions with a cloud of laser-cooled Be^+ ions promises to reduce the energy of the highly-charged ions to the point of crystallization under the combined interaction of their mutual repulsion plus the confining fields of the trap.

The motivation for these advances derives from investigations of nature under extreme conditions. White dwarf stars are predicted to have interiors composed of crystallized fully-stripped ions (11). Studies of the crystallization of binary mixtures of stripped ions is planned in RETRAP. Precision laser spectroscopy of highly-charged ions is best carried out on a sample with minimal Doppler shifts and external interactions, confined in a small region of space. Studies of the g-factor of a single electron bound to a high-Z ion provides information on relativistic and QED interactions that scale as powers of $(Z\alpha)$, a parameter that can be 0.6 or higher, leading to slower convergence of the perturbation expansion. The hyperfine structure transition of several high-Z one-electron ions lies in the laser range, allowing precision measurements of an electron-nucleus interaction which can be compared with theory with relative ease. Beryllium-like ions are highly metastable in the ground state of the triplet system, $^3P_0$. Fine-structure transitions to the nearby $^3P_1$ level lie in the laser range for several values of Z, permitting interesting fine-structure interval and lifetime measurements which test relativity and QED in atomic theory.

COLLISION MEASUREMENTS

To quantitatively investigate the time scales of ion storage with charge state q, electron transfer collisions with the common residual gas H_2 in a cryogenic vacuum were undertaken using Ar^{11+}, $Xe^{35+,43\text{-}46+}$, and $Th^{73\text{-}80+}$ ions at center-of-mass energies near 6 eV. A direct measurement of the H_2 density at the trap site was not feasible. However, the cross section σ_{11} for electron

CP392, *Application of Accelerators in Research and Industry*, edited by J. L. Duggan and I. L. Morgan
AIP Press, New York © 1997.

transfer from H_2 to Ar^{11+} had been independently measured vs. energy in a beam measurement (12), so this result along with the measured charge-state storage time constant τ_{11} for the ions in RETRAP allowed a calculation of the target H_2 density according to the equation $n(H_2) = (\tau_{11}\sigma_{11}v_{11})^{-1}$, where the speed v_{11} was calculated from the mean ion energy. The rate coefficients $k_q = (n(H_2)\,\tau_q)^{-1} = \sigma_q v_q$ for the other ion charge states were then calculated using their measured storage constants τ_q. The mean cross sections σ_q were obtained with estimates of v_q. The results were compared with the absorbing sphere model (13), which is valid for these charge states and energies (5,6). The data were in approximate agreement with this theory. True double capture, in which both captured electrons remain on the ion, was found to occur about 25% of the time at high q.

The charge state storage time constants were measured as follows. The "compensation electrodes" of the cylinder trap (see Fig. 1) were connected by a high-Q inductor to form a parallel tuned circuit at a frequency of 1.21 MHz. By appropriately adjusting the relative potentials on the trap electrodes (5,6), the ions oscillated axially at a well-defined frequency ω_z near this value. Their frequencies ω_z (proportional to $(q/m)^{1/2}$) were periodically swept through tuned-circuit resonance by a voltage ramp. The ions efficiently induced charge on the compensation electrodes, resulting at resonance in an induced voltage signal large compared to the noise contributions from the tuned circuit and preamplifier near 4K temperature. Since this method is non-destructive, the same ion or ions could be monitored sequentially in the pre-determined storage intervals. As the ion charge state decreased due to electron transfer, the ion resonance signal moved to higher ramp voltages. It was found that energy transferred in the collisions was insufficient to remove the highly-charged ions from the trap.

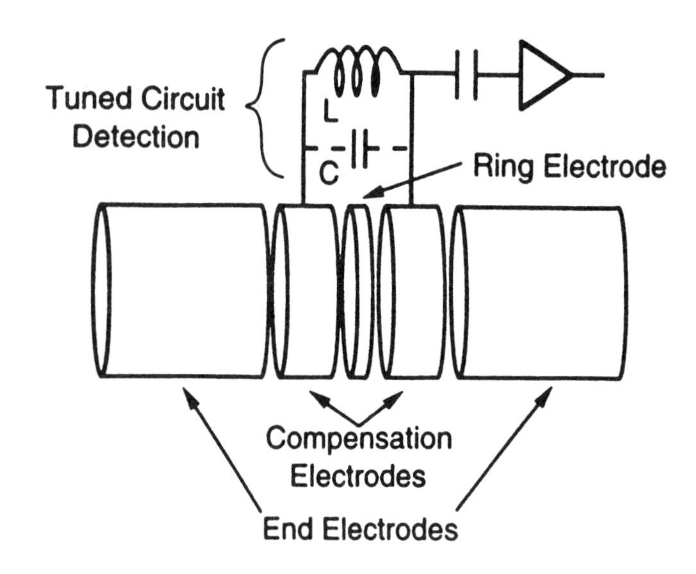

FIGURE 1. Cylinder Penning ion trap, with electrodes identified. The detection circuit, trap, and pre-amplifier were all cooled by the liquid helium dewar of the superconducting solenoids. The magnetic field is parallel to the trap symmetry axis.

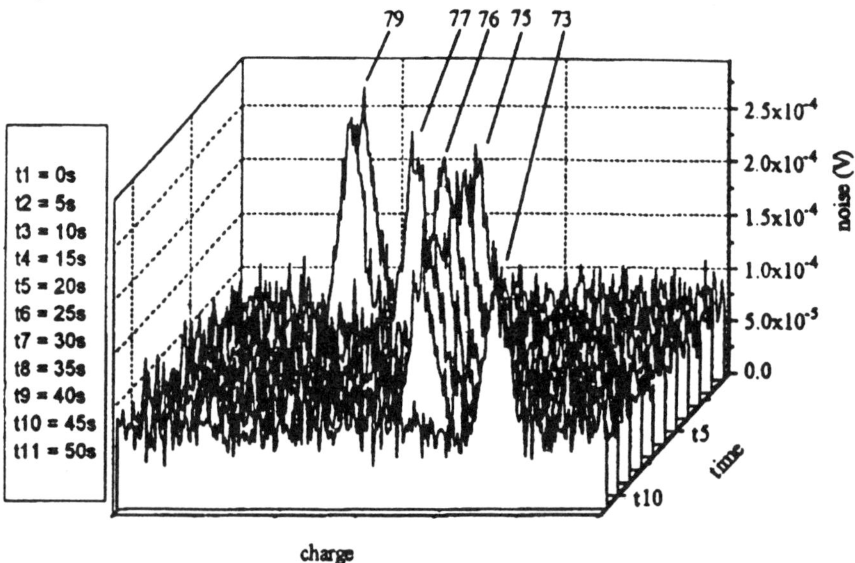

FIGURE 2. The signal due to a single Th ion, initially with q = 79+, obtained at 5 s intervals. As the ion captures one or two electrons in random collisions with H_2, the charge of the ion decreases. Small decreases in signal amplitude also occur due to ion energy loss during detection.

Fig. 2 shows the signal from a single Th ion, initially with q = 79+, as a function of storage time. At random times, the ion captures one or two electrons in collisions with H_2, reducing its charge state at the detection intervals so that q = 73+ at the end of the measurement. Signals were followed from a number of histories such as this, and analyzed assuming that the charge transfer rate was the same for each charge state, a reasonable approximation for high q and small Δq.

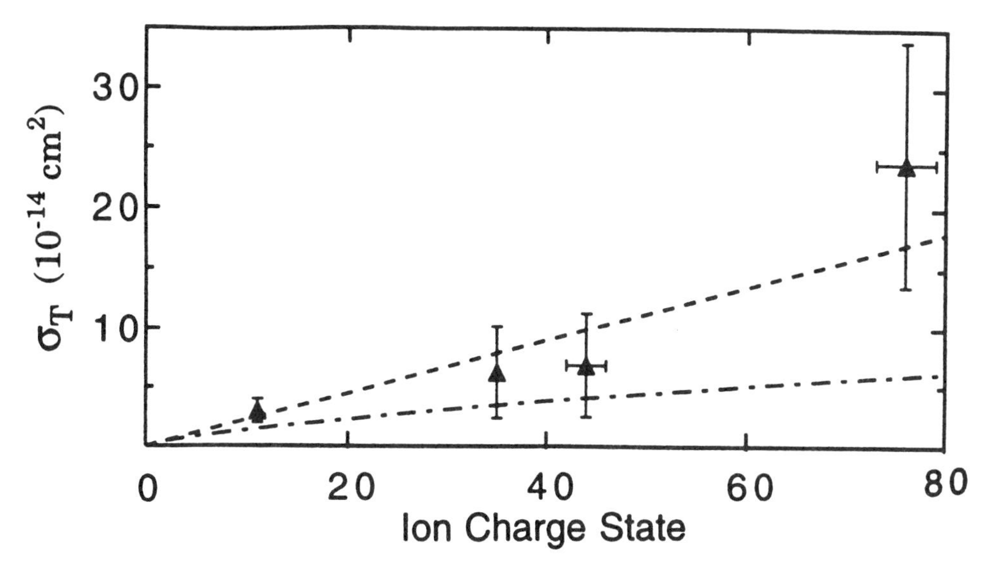

FIGURE 3. The measured cross sections for electron transfer, some averaged over 10% ranges of charge, are plotted vs. charge state. The dashed line is a linear fit to the data, and the dot-dashed line is a plot of the prediction of the absorbing sphere model.

The probabilities in each measurement interval t for 0, 1, or 2 steps in charge can be written

$$P_0 = \exp(-\lambda_1 t)$$
$$P_1 = (\lambda_s t)\, P_0$$
$$P_2 = (\lambda_D t + (\lambda_s t)^2/2)\, P_0$$

where $\lambda_1 = \lambda_s + \lambda_D$ is the sum of the single (λ_s) and true double (λ_D) charge transfer rates. The quadratic term in P_2 corrects for two single-electron transfers in the interval t. The mean of the results yielded $\lambda_D/\lambda_1 = 0.21$ (4). All of the data, including much data from several simultaneously-stored ions, were separately analyzed by an analog method described in refs. (5) and (6). Results for the cross sections σ_q derived from all of the data are plotted vs. q in Fig. 3. Some data are lumped into 10% ranges of q for this plot. A calculation of the absorbing sphere prediction (13) is also plotted.

SPECTROSCOPY

A new Penning trap system has been designed and installed in RETRAP, to permit separated laser-cooling of Be^+ ions and resistive cooling of highly-charged high-Z ions. These pre-cooled ion clouds will then be merged for sympathetic Coulomb-collisional cooling of the highly-charged ions by the laser-cooled Be^+ ions. The new trap system is diagrammed in Fig. 4. It consists of four traps: two rudimentary Penning traps at the ends, and two optimally-designed (14), compensated hyperbolic-electrode traps at the center. Each hyperbolic trap is constructed with a tuned circuit, consisting of an external inductor connected between the two end electrodes of the trap. The circuit is used to measure voltages induced by the axial motion of the confined ions and to cool the ion center-of-mass motion. The ring electrode of each trap is split into four quadrants, to aid in cyclotron resonance excitation and detection. The ring electrode of the lower hyperbolic trap is fenestrated to pass laser-beams and collect fluorescence from the stored ions. The dewar and thermal shields of the superconducting solenoids which create the magnetic field for the trap have radial tubes to windows in the vacuum shell to pass this light. The four ion traps are separated by control electrodes which facilitate ion capture and transfer. The highly-charged ions will be stored in the upper hyperbolic trap and resistively cooled by the tuned circuit (time constant < 1 s), while Be^+ ions are laser-cooled in the lower hyperbolic trap. The ions will then be merged in the lower hyperbolic trap for further cooling and detection. Ions in the outer traps can be moved to the central traps.

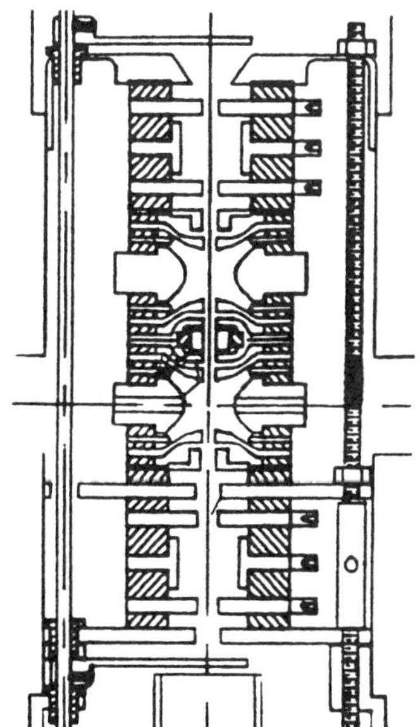

FIGURE 4. System of four Penning traps in RETRAP. The outer two traps are rudimentary; the inner two have hyperbolical electrode surfaces and are equipped with circuits for axial and radial excitation and detection of the ions. The third trap from the top has apertures for optical excitation and detection of the ions.

Be^+ ions are laser-cooled by 313 nm radiation from a tunable, frequency-doubled dye laser. For cooling, the wavelength is tuned slightly above that of the cycling transition 2s $^2S_{1/2}(m_J = 1/2, m_I = 3/2)$ - 2p $^2P_{3/2}(m_J = 3/2, m_I = 3/2)$ in the magnetic field B near 4 T. The ions are optically pumped to the lower level by near-resonance radiation (15).

A mixture of 10^4 to 10^5 total Be^+ and Be^{2+} ions produced in a pulse from a Metal Vapor Vacuum Arc (MEVVA) has been trapped in the lower hyperbolic trap, with the Be^+ axial frequency tuned to resonance with the tuned circuit in a well with axial depth near 225 V. The mean squared voltage signal induced by the ions (proportional to ion number and mean energy) dropped slowly (times of several hundred s) due to resistive cooling, which has a nominal one-dimensional time constant $T_Z \cong 100$ s for Be^+. When the ion temperature had dropped by a factor near 100, tuning the laser wavelength above (or below) the cycling resonance cooled (or heated) the ions, as evidenced by short term decreases or increases in the Be^+ tuned circuit signal. The fluorescence from the ions at 313 nm was also imaged using a cooled CCD detector. At sufficiently low ion temperatures, a dark region appeared at the center of the fluorescing ion cloud, currently interpreted as the start of centrifugal separation (16) of the fluorescing Be^+ ions and the unexcited Be^{2+} ions. Verification of this interpretation will constitute a first demonstration of the basis for the planned spectroscopy on cooled highly-charged ions.

Certain cold, highly-charged ions will be detected by observing the fluorescence of "forbidden" fine-structure or hyperfine structure transitions excited using a second tunable dye laser. Level lifetimes may be in the range of 0.1 - 10 ms. Consequently the weak fluorescence can be detected essentially background-free by chopping or pulsing the exciting lasers. Estimates of signal rate indicate that only a few stored ions will be sufficient. For hydrogen-like ions with a hyperfine splitting in the optical region, the Zeeman effect of the hyperfine structure occurs in the weak-field appoximation in the trapping field B. The current calculation of the bound-electron g-factor (17) includes relativistic and radiative corrections to order $(Z\alpha)^2$, $(Z\alpha)^2(m/M)^2$, $\alpha(Z\alpha)^2$, and $\alpha(Z\alpha)^2(m/M)^2$. Further theoretical effort is underway. Because $(Z\alpha)$ can be 0.6 or higher, the corrections to the g-factor can exceed 10%. The ground state hyperfine-structure of H-like Bi has been recently measured in a storage-ring at GSI, along with the lifetime of the hyperfine transition (18). Neither result agrees with current ab initio calculations. The lifetime disagreement is about 15%. Measurements of this type with significantly smaller excitation linewidths should be feasible in RETRAP, since the Doppler widths will be considerably reduced.

ACKNOWLEDGMENTS

This research was supported in part by the U.S. Department of Energy under contract No. W-7405-ENG-48 with LLNL. G.W. and D.A.C. were supported in part by the Robert A. Welch Foundation. The authors thank E. Magee and D. Nelson for outstanding technical support.

REFERENCES

1. Bergstrom, I., Carlberg, C. and Schuch, R., *Trapped Charged Particles and Related Fundamental Physics*, Singapore, World Scientific, 1995.
2. Kravis, S.D., Church, D.A., Johnson, B.M., Meron, M., Jones, K.W., Levin, J.C., Sellin, I.A., Azuma, Y., Mansour, N., Berry, H.G. and Druetta, M., *Phys. Rev.* **A45**, 6379 (1992).
3. Church, D.A., *Physics Reports* **228**, 254 (1993).
4. Steiger, J., Church, D.A., Weinberg, G., Beck, B., McDonald, J., and Schneider, D., *Hyperfine Interactions* (in press).
5. Beck, B.R., Steiger, J., Weinberg, G., Church, D.A., McDonald, J., and Schneider, D., *Physical Review Letters* **77**, 1735 (1996).
6. Weinberg, G., Steiger, J., Beck, B., Church, D.A., McDonald, J. and Schneider, D., (to be submitted to Phys. Rev. A.).
7. Levine, M.A., Marrs, R.E., Henderson, J.R., Knapp, D.A., Schneider, M.B., *Physical Scripta* **T22**, 157 (1988).
8. Schneider, D., DeWitt, D., Clark, M.W., Schuch, R., Cocke, C.L., Schmieder, R., Reed, K.J., Chen, M.H., Marrs, R.E., Levine, M., and Fortner, R., *Phys. Rev.* **A42**, 3889 (1990).
9. Schneider, D., Church, D.A., Weinberg, G., Steiger, J., Beck, B., McDonald, J., Magee, E., and Knapp, D., *Rev. Sci. Instrum.* **65**, 3472 (1994).
10. Pikin, A., Morgan, C.A., Bell, E.W., Ratliff, L.P., Church, D.A., and Gillaspy, J.D., *Rev. Sci. Instrum.* **67**, 2528 (1996).
11. Chabrier, G., Ashcroft, N.W., and DeWitt, H.D., *Nature* **360**, 48 (1992).
12. Kravis, S.D., Saitoh, H., Okuno, K., Soejima, K., Kimura, M., Shimamura, I., Awaya, Y., Kaneko, Y., Oura, M., and Shimakura, N., *Phys. Rev.* **A52**, 1206 (1995).
13. Olson, R.F. and Salop, A.. *Phys. Rev.* **A14**, 579 (1976).
14. Gabrielse, G., *Phys. Rev.* **A27**, 2277 (1983); see also Ghosh, P.K., *Ion Traps*, Oxford Press, 1995.
15. Brewer, L.R.., Prestage, J.D., Bollinger, J.J., Itano, W.M., Larson, D.J., and Wineland, D.J., *Phys. Rev.* **A38**, 859 (1988).
16. O'Neil, T.M., *Phys. Fluids* **24**, 1447 (1981).
17. Grotch, H. and Hegstrom, R.A., *Phys. Rev.* **A4**, 59 (1971).
18. Klaft, J., Borneis, S., Engel, T., Fricke, B., Grieser, R., Huber, G., Kuhl, T., Marx, D., Neumann, R., Schroeder, S., Seelig, P., and Voelker, L., *Phys. Rev. Letters* **73**, 2425 (1994).

PROJECTILE CHARGE DEPENDENCE OF IONIZATION AND DISSOCIATION OF CO IN FAST COLLISIONS

Vidhya Krishnamurthi*, I. Ben-Itzhak†, and K.D. Carnes

James R. Macdonald Laboratory, Department of Physics,
Kansas State University, Manhattan, KS 66506-2604
**Present address: Etec Systems Inc., 26460 Corporate Ave., Hayward CA 94545*

Experiments have been carried out to study how changes in the interaction strength (defined as q/vb) of a fast ion-molecule collision affect the ionization and dissociation of the molecular target, in this case CO. The coincidence time-of-flight technique was used for collisions at fixed velocity (energy of 1 MeV/amu). The interaction strength was changed by varying the charge of the projectile ion. The cross sections for single and multiple ionization of CO increase rapidly for small q, approximately as q^{2n} (where n is the number of ionized electrons), and more slowly for larger values of q. A rather simple theoretical model based on the independent electron approximation and perturbation theory is in good agreement with the data [1]. The dissociation patterns of the transient CO^{Q+} molecular ions also exhibit a dependence on the projectile charge which is qualitatively explained by the same model.

INTRODUCTION

Interactions between fast, charged projectiles and molecules result in the formation of transient molecular ions which eventually either stabilize and maintain their molecular nature or dissociate into fragment ions or atoms. In this paper we report the results of a systematic study of the effect of changes in the "interaction strength" (defined as q/vb) on ionization and dissociation of these molecular ions. The interaction strength was varied by changing the charge of 1 Mev/amu projectiles. The results obtained experimentally have been compared with those given by a simple theoretical model for multiple ionization based on perturbation theory within the independent electron approximation [2].

EXPERIMENTAL METHOD

A detailed description of the experimental method used to carry out these measurements is given in the papers by Ben-Itzhak *et al.* [3,4]. In short, a bunched beam of X^{q+} ions is accelerated to an energy of 1 MeV/amu in the J.R. Macdonald tandem Van de Graff accelerator and then charge separated by a 90° analyzing magnet. These ions then interact with the CO molecules in the collision region, under single collision conditions. The recoil ions formed in the interaction region are extracted and accelerated by the fields between the parallel meshes that define the interaction region. These ions then drift through a field-free region and are detected by a microchannel plate detector at the end of the drift region. The time of flight of all these recoil ions is recorded relative to a timing signal synchronized with the narrow beam bunches, thus maintaining the correlation of all charged fragments of a molecule. The single ion and ion-pair yields are then converted to relative cross sections and the kinetic energy released in each disoociation channel is evaluated from the time difference between the ion-pairs. For further experimental details of the aparatus and data reduction see Krishnamurthi *et al.* [1].

†Corresponding author eMail: ibi@phys.ksu.edu.

RESULTS AND DISCUSSION

Ionization

The model for multiple ionization, as suggested by Ben-Itzhak *et al.* [2], is based on perturbation theory within the independent electron approximation. Given a target with N equivalent electrons out of which n are ionized during a single collision with a projectile of charge q, the cross section for the process is taken to be

$$\sigma_n = 2\pi \int_0^\infty b\, db\, \binom{N}{n} P^n(b)[1-P(b)]^{N-n} \qquad (1)$$

where $P(b)$ is given, under the first Born Approximation as $P(b) = q^2 P_H(b)$. $P_H(b)$ is the ionization probability of the active electron of the same target caused by a proton at the same velocity. For large values of q, $P(b)$ becomes larger than 1. To avoid violation of unitarity,

CP392, *Application of Accelerators in Research and Industry,* edited by J. L. Duggan and I. L. Morgan
AIP Press, New York © 1997.

$P(b)$ is replaced by $P'(b) = 1 - e^{-P(b)}$ following the suggestion by Sidorovitch *et al.* [5]. The model is still valid since the value of the probability function $P(b)$ is negligible at impact parameter values where perturbation theory fails, i.e. its contribution to the ionization cross section of the many electron target at small impact parameters is negligible [1,2].

The values for the active-electron ionization probabilities for proton impact calculated using hydrogenic wave functions within the semiclassical Coulomb approximation (SCA) have been tabulated by Hansteen *et al.* [6], using the scaling law $P(q, Z, v; b) = q^2/Z^2 P(1, 1, v/Z; Zb)$ for targets with $Z > 1$. We, however, had to solve the problems of (i) defining the effective Z of our target molecule and (ii) approximating CO by hydrogenic wavefunctions, before we could use the SCA values for $P_H(b)$. The latter problem was solved by using a $2p$ hydrogenic wavefunction in our calculations since this best describes the spatial charge cloud distribution of CO. The effective Z of the valence electrons of CO was determined by using it as a fitting parameter to obtain the best agreement between the calculated and experimentally measured values of the single ionization cross section of CO caused by 1 MeV proton impact. This value of Z was then used to determine all the other single-, double- and triple-ionization cross sections caused by 1MeV/amu X^{q+} impact.

Comparisons of the calculated and observed values of single-, double- and triple-ionization cross sections for CO, caused by 1 MeV/amu X^{q+} ions, are shown in Fig. 1. The overall agreement between experiment and theory seems to be rather good. We know that by definition, $P(b)$ is quadratic in q. However, the figure indicates that the cross sections are almost linear in q for large q values. This can be explained in terms of the "window effect". In Fig. 2 we have plotted the single ionization probability function (defined in Eq. 1), as a function of b, for different projectile charge states. For a given projectile charge, single ionization takes place over a certain range of impact parameter values, peaking at a certain value. Double ionization occurs at a different range of impact parameter values, shifted towards lower impact parameter values as compared to single ionization. This is further repeated for triple ionization, the range moving down to still lower values of impact parameters. This suggests that the lower values of impact parameters do not contribute towards a given ionization as the projectile charge increases.

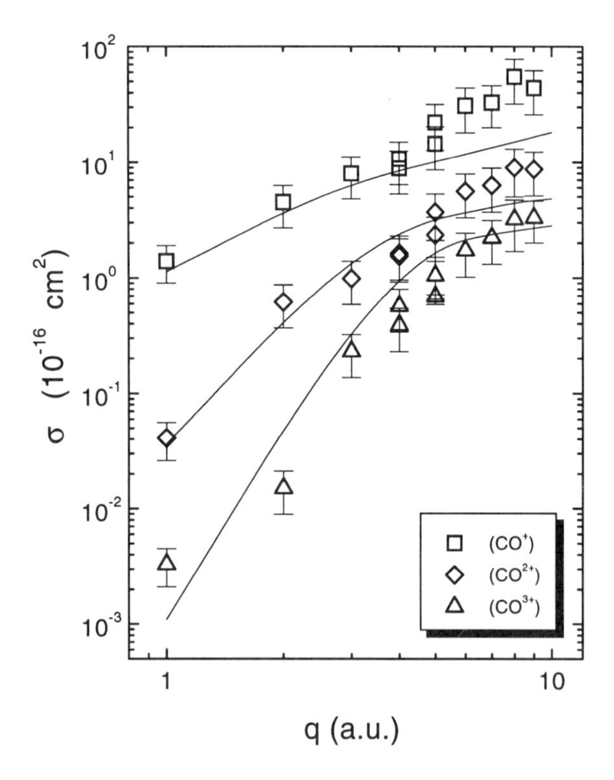

FIGURE 1. Comparison between model calculations (solid curves) and the measured single-, double- and triple- ionization cross sections of CO in 1 MeV/amu collisions.

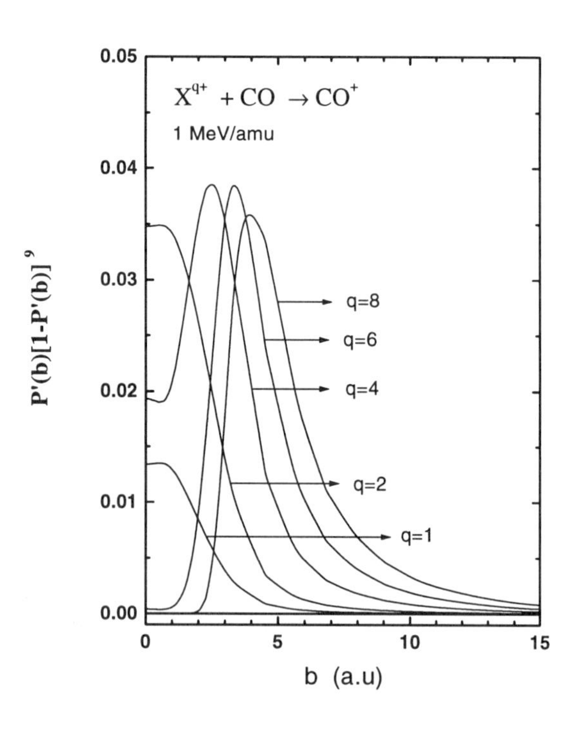

FIGURE 2. Plot of the probability function for single ionization as a function of the impact parameter b for different values of q.

The rate of increase of ionization cross sections is reduced from the q^2 dependence expected if the whole range of impact parameters were to contribute to the cross section. Thus, even though the probability within first order perturbation theory scales as q^2, it is clear from the figure that the shift of the impact parameter range towards higher b and the exclusion of contributions from the range where b is small slows down the increase of the ionization cross section.

Dissociation

If the transient molecular ion is in an excited state which is dissociative in nature, the ion will fragment. One would expect the probability of finding the transient ion in a given excited state to scale as some function of q^2, just as in the case of ionization that has been discussed above. Given the "window effect", however, it is not clear how the distribution of states should scale with q. We have carried out the exercise of plotting the average interaction strength defined as

$$\left\langle \frac{q}{vb} \right\rangle_n = \frac{\int_0^\infty b\, db\, P'^n(b)\, [1-P'(b)]^{N-n}(q/vb)}{\int_0^\infty b\, db\, P'^n(b)[1-P'(b)]^{N-n}}, \quad (2)$$

as a function of q (see Fig. 6 of Ref. [1]). That figure indicates that although the average interaction strength varies very rapidly with q while q is small, it varies more slowly for larger values of q. This suggests that for small values of q the excited state distribution of the transient molecular ions varies very rapidly with q, but it slows down for larger values of q. Since molecules dissociate, measuring the branching ratios or the kinetic energy released in the dissociation should give us some insight into this shift in excited state distribution.

We have plotted the variation of the branching ratios for CO^+, CO^{2+} and CO^{3+} with projectile charge in Fig. 3. In all cases, the most dominant dissociation channel, which is also the energetically lowest state of the transient molecualar ion, decreases with the projectile charge while the others increase. This variation, however, is slow and it flattens out at large values of q. These observations are in qualitative agreement with the predictions of our model, that the interaction strength and therefore the excited state distribution should vary slowly with q. The next figure (Fig. 4) shows the average value of the kinetic energy release upon dissociation of CO^{2+} into C^++O^+ as a function of the projectile charge. This first moment of the distribution shifts to higher energy values with increasing q in an approximately linear fashion $\overline{KER} = (14.3 \pm 0.5) + (0.6 \pm 0.1)q$. The reason for this approximately linear dependence is not clear yet. This variation is slow, but it does not seem to flatten out for large q values like the previously discussed quantities.

We have shown from the previous discussions that a rather simple model describes the dependence of single-, double- and triple-ionization as a function of projectile

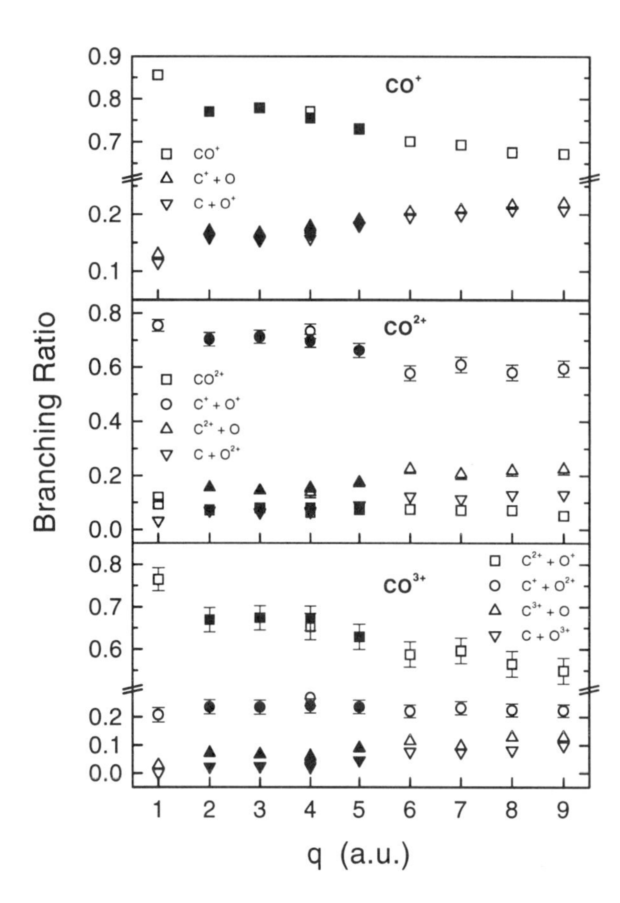

FIGURE 3. Branching ratios of CO^+, CO^{2+}, and CO^{3+} as a function of the projectile charge q. (Open symbols: H^+ and $F^{(4-9)+}$; full symbols: $B^{(2-5)+}$.

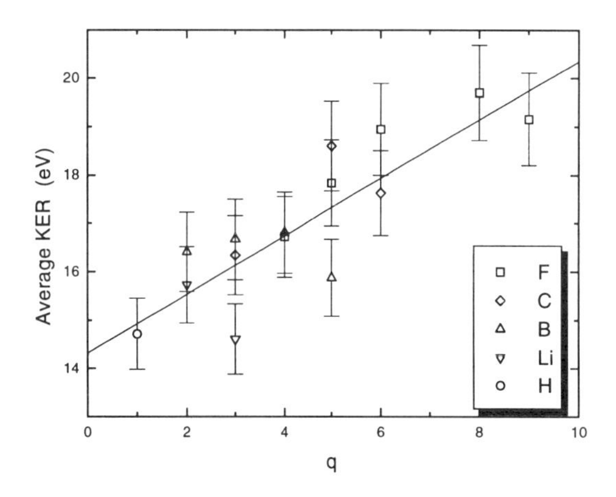

FIGURE 4. Average KER as a function of q.

charge rather well. The model suggests that the rate of change of ionization should be slow (some function of q) even though $P'(b)$ is a function of q^2. This is explained using the "window" picture, that as q increases, the ionization probability function shifts towards higher impact parameters, thereby reducing the change in the interaction strength. The dissociation studies indicate that the distribution of transient molecular ion excited states shifts slowly with q and flattens out for large values of the projectile charge. This is also consistent with the slow variation predicted by the model.

ACKNOWLEDGMENT

This work was supported by the Division of Chemical Sciences, Office of Basic Energy Sciences, Office of Energy Research, U.S. Department of Energy.

---

[1] Krishnamurthi, Vidhya, Ben-Itzhak, I., and Carnes, K.D., *J. Phys. B* **29**, 287 (1996) and references within.
[2] Ben-Itzhak, I., Gray, T.J., Legg, J.C., and McGuire, J.H., *Phys. Rev. A* **38**, 3685 (1988).
[3] Ben-Itzhak, I., Ginther, S.G., and Carnes, K.D., *Nucl. Instrum. and Meth. B* **66**, 401 (1992).
[4] Ben-Itzhak, I., Ginther, S.G., and Carnes, K.D., *Phys. Rev. A* **47**, 2827 (1993).
[5] Sidorovitch, V.A., Nikolev, V.S., and MacGuire, J.H., *Phys. Rev. A* **31** 2193 (1977).
[6] Hansteen, J. H., Johansen, O.M., and Kocbach, L., *At. Data Nucl. Data Tables* **15**, 305 (1975).

ABOVE AND BELOW SURFACE INTERACTIONS OF HIGHLY CHARGED IONS ON METALS, INSULATORS OR SEMICONDUCTORS

J.-P. Briand

Laboratoire de Physique Atomique et Nucléaire—Institut du Radium, Université P&M Curie 4 Place Jussieu F-75252 Paris Cedex 05. Unité associée au CNRS ERS 112

We summarize in this paper recent experiments which demonstrate, for the first time, that different kinds of hollow atoms are formed during the interaction of slow highly charged ions **above** and **below** metal, insulator and semiconductor surfaces. These results show that, in both cases, the conduction or valence character of the most weakly bound target electrons plays an essential role in the interaction. The ions are accelerated above metal surfaces (contact) and backscattered on insulators by the holes formed during the capture of valence electrons. Below the surface the feeding of the hollow atoms strongly depends on the metal or insulator character of the target and is found to be much faster in metals than in insulators.

INTRODUCTION

When a slow highly charged ion approaches or penetrates a surface many electrons can be captured in excited states leading to the formation of hollow atoms (1) (2). Above a surface the capture process feeds, at large distances, the Rydberg states of the ion. These hollow atoms are referred as "Rydberg hollow atoms". Below or at surface, when the collision holds at closer distances (with contact), the electrons are captured in lower excited states e.g. in the M and N shells for argon (while the K and L shells remain empty); these hollow atoms are named "surface hollow atoms". The capture process is similar to that of a highly charged ion approaching an isolated atom. In this case single capture processes (σ_1) are dominant because the removal of a second, more bound electron is more difficult than for the first one ($\sigma_1>\sigma_2>\sigma_3$...). In ion—surface collisions, a metal is an infinite reservoir of weakly bound electrons, each having the same energy, and a great number of them may be captured in excited states of the ion in a single event.

When capture processes occur at large distances (i.e., larger than about an atomic radius: "no contact") like in ion—isolated atom collisions, or in collisional processes far above a surface, these events are successfully explained in some well suited overbarrier models. The distances of capture (and geometrical cross sections for isolated atoms), as well as the principal quantum number where it occurs, are then satisfactorily explained by simply knowing the binding energy of the target electrons.

At closer distances (e.g. below the surface), these capture processes have been first described in highly charged ions—surface collisions in the framework of the Fano-Lichten electron promotion model (3), or in Landau-Zener models (4) involving innershells.

Both models (barrier and electron promotion) played an important, conceptual, and practical role in the first days of the study of the interaction of highly charged ions with surfaces. They nevertheless suffer genuine diseases. For quantum mechanical reasons (5), for instance, it is not clear how, far above a surface, are captured the removed electrons. Most authors assume that some of the electrons are step by step captured (following an arbitrary sequence law), the others going into the continuum (assuming an arbitrary branching ratio). Below the surface, the electron promotion, or the innershell overbarrier models imply binary collisions, while a very highly charged ion interacts continuously with many atoms. These models have to be reformalized to overcome these difficulties.

In the last six years the main problem brought forward by all the groups working in the field was to determine whether the Auger electrons or the X rays emitted in flight came from ions below or above the surface. It was first recognized (3) that, at keV/q energies, the observed interactions mainly held below the surface. Many groups have recently decelerated the ions and successfully observed the "outside" interactions.

The most exciting query today is to know whether the ion surface interaction depends or not on the metal, semiconductor, or insulator character of the surface (6). This question obviously stands for above surface interactions where the captured electrons are the most weakly bound conduction or valence electrons, but also for below or at surface interactions, as demonstrated below, where they play a major role in the filling of the M and N shells of the ions.

This paper deals with this subject. We will present in Section I the experimental procedures. In Section II we will review some recent experiments which demonstrate, for the first time, that different kinds of hollow atoms are formed above metal, insulator and semiconductor surfaces, and which show that the conduction or valence character of the most weakly bound electrons plays an essential role in the interaction. In Section III we will present some preliminary results which also show that different kinds of hollow atoms are formed below metal and insulator surfaces. These results are tentatively explained via Auger neutralization processes.

1. Experimental procedures

These experiments were carried out in various places where ECR sources produce Ar^{17+} ions (CEN at Grenoble, GANIL at Caen, AECR at Berkeley). The technique was

CP392, *Application of Accelerators in Research and Industry*, edited by J. L. Duggan and I. L. Morgan
AIP Press, New York © 1997.

either to study the K X rays emitted in flight by the ions, at high resolution, with a crystal spectrometer, or to compare the K and L X rays of the ions with the target X rays, with a low energy SiLi detector.

The principle of the experiments was based on the atomic clock property of the hollow atoms (1) (2). This property comes from the step by step filling of the eight L holes of the ion. In both cases (Rydberg or surface hollow atoms) (7) the L shell is filled by the M (N) electrons through a sequence of eight, very fast (few 10^{-16} s) LMM Auger transitions, in a way similar to the filling of the lower reservoir of a sandglass. Consequently there is a competition between the filling of the K hole(s), by the emission of the Kα X ray we observe, and that of the eight holes of the L shell. The Kα X ray is then emitted in presence of a certain number x of L spectator electrons and displays a characteristic array of (up to eight) KL^x satellites which can be resolved in crystal spectrometry. When the filling of the L shell is slow (compared to that of the K hole), like it happens for Rydberg hollow atoms formed above the surface, the distribution of these KL^x satellites peaks on the KL^1 or KL^2 lines. When it is faster than that of the K hole (surface hollow atoms), the L shell is quickly filled and the K X ray is emitted in presence of more L electrons (KL^x satellites with large x values).

2. Above surface interactions

The 11 keV/q Ar^{17+} ions produced by the ECR sources of CEN at Grenoble or GANIL at Caen were, at normal incidence, decelerated, in a single gap, down to a tunable energy range of 0<E<14 eV/q, and sent onto three different surfaces: polycrystalline gold, SiO_2 and SiH (111 plane of Si passivated by one monolayer of hydrogen). The first experimental results, recently published (8), can be summarized as follows.

(i) Very different K X-ray spectra (Fig. 1) are observed during the interaction of the same Ar^{17+} ions at 1 eV/q kinetic energy on metal (Au) and semiconductor (SiH) surfaces, which have roughly the same work function. With the silicon target one observes the KL^x satellite spectrum (KL^1>KL^2>KL^3) which is expected (9) for the decay in vacuum of hollow atoms formed above a surface (i.e., before any contact).

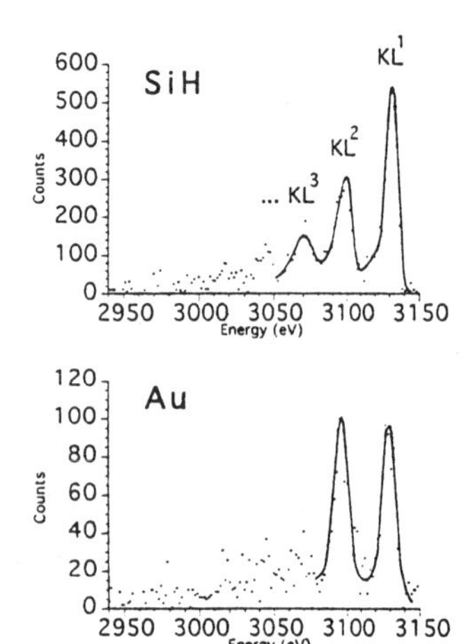

FIGURE 1. K X-ray spectra of 1 eV/q Ar^{17+} ions on SiH and Au surfaces.

In contrast one observes on gold targets an X-ray spectrum peaked on the KL^2 satellite which corresponds to the sum of the "outside" X-ray spectrum mentioned above, plus a signal of an "at surface" interaction. When an ion approaches a surface, at distances below 1 Å, the electrons are captured, as experimentally demonstrated, in the M and N shells of the ion instead of the Rydberg states. One or two electrons are then quickly captured in the M shell of the ion at the time of the contact. When e.g. two "more" electrons are captured, a fast ($\tau \sim 10^{-15}$ s or less) LMM Auger transition may occur, accelerating the still operative filling of the L shell (one more L electron) which increases significantly the intensity of the KL^2 satellite and gives a signature of the contact of the ion (the LMM Auger transition at such low velocities —few 10^4 ms^{-1}— holds along ~0.1 Å of the trajectory of the ion, i.e., just where the capture process has occured).

(ii) In both cases, the exact energy of the X-ray lines indicates that, in the KL^1 state, the ion has a mean number of e.g. two M electrons (KL^1M^{2+}... state), which is in agreement with the expected triangular population of these hollow atoms (ions) (9) and is thus highly ionized ($q \geq 10$) (1). This result clearly means that the reionization processes (Auger cascade and resonant ionization) are slightly faster than the capture processes.

In more recent experiments we have studied, in the energy range of 0<E<14 eV/q, the evolution of the X-ray spectra, for both targets, as a function of the initial kinetic energy of the incoming ions. The most interesting result obtained is that the relative intensity of the KL^x lines does not change in the considered range for gold, whereas that of the KL^{2-3} satellites dramatically increases for SiH with increasing kinetic energies of the ion. Above E=14 eV/q both spectra tend to be similar.

The invariance of the relative intensities of the KL^x satellites on the gold target can be easily explained by the acceleration of the ion by its own image (10). According to the presently accepted models, the considered ion starts capturing electrons at z_0~20 Å above the surface. Above z_0 the ion is accelerated when approaching the surface and gains about 3-4 eV/q of kinetic energy. Below z_0 the "extra" energy gain of the ion is smaller owing to its partial neutralization, but is still substantial according to our results (see above) which show that the ion has still a high charge below z_0. The ions then drop irremediably on the metal surface. At very low initial kinetic energies the kinematics of the ion is then governed by its image acceleration and not by its initial kinetic energy (E<12 eV/q).

On silicon the continuous decrease of the relative intensities of the KL^2 and KL^3 satellites with the initial kinetic energy of the ion down to nearly zero, clearly demonstrates that the kinematics of the ions does depend on the initial conditions and that the image acceleration vanishes or is overcome. These findings can be explained by the insulator character of the target: in silicon the captured electrons come from the valence band instead of the conduction band and some positive holes appear at the surface [up to about 3 times the initial charge of the ion according to the measured number of ejected electrons (11)]. These numerous positive charges, twice as close to the ion as the image, easily overcome the image acceleration. The ion still having a charge rather large, as discussed above, is backscattered above the surface, like on a trampoline, but without any contact, in agreement with the above observation of a pure "outside" spectrum. When the energy of the ions decreases to nearly zero the relative intensity is more peaked on the KL^1 satellite. This general trend, in the framework of the atomic clock property of the hollow atoms, means that the L shell is slower filled at lower energies, and consequently that the capture processes becomes slower than the reionization processes or holds in higher n states. These findings are consistent with a shorter penetration of the ion at lower energies in the capture area (below z_0). The asymptotic limit of this evolution has recently been observed in Ar^{17+} ion collisions with C_{60} (12), a case where the ion, after having captured several electrons, quickly escapes the capture area and loses all its electrons but one in Auger cascades (no refeeding), and exhibits a pure KL^1 X-ray spectrum. The penetration of the ions in the capture area below 20 Å is then controlled by tuning their kinetic energy between 0 and ~12 eV/q.

In a last series of experiments we have studied the X-ray spectra emitted by the same ions at 1 keV/q kinetic energy and grazing incidence, $1<\theta<6°$, corresponding to the same normal kinetic energies, as in the previous experiments ($0<E<12$ eV/q). On SiH and Au surfaces we observed a full invariance of the shape of the X-ray spectra with the incidence angle, which means that the image acceleration is still operative at grazing incidence for insulators or semiconductors as it is for metals. On silicon the positive charges (holes) appearing on the surface being, at grazing incidence, "left behind", there is no overcoming of the image effect, which is in perfect agreement with the above statement.

In conclusion it can be said that the great difference between the configurations of hollow atoms formed above metals, semiconductors (or insulators) having the same work function comes from the changes induced in the kinematics (approach) of the ion by the structure of the lattice of the target (i.e., the presence or not of conduction electrons or holes), and is independent of the nature (atomic structure) of the target atoms. This opens the way to many applications for the study of semiconductors or insulators with slow highly charged ions, or the control of their movements.

3. Below surface interactions

We present Fig. 2 the K X-ray spectra observed with the same Ar^{17+} ions at 10 keV/q on gold and silicon targets, at normal incidence, i.e., when most of the X rays are emitted by the ion below the surface.

One clearly sees on this figure dramatic changes in the relative intensity of the eight KL^x lines which show that different kinds of hollow atoms are also formed below metal and insulator surfaces. The general trend of the evolution of the eight KL^x lines relative intensity, observed on many different targets, is that the spectra for metals are peaked on the satellites corresponding to a large number of L spectator electrons (KL^8 and KL^7 mainly). In the framework of the atomic clock property of the hollow atoms (1) (2) this means that the L shell of the ions is more quickly populated in metals than in insulators. The L shell being filled by the M and N electrons (7) through the fast LMM (LMN...) Auger transitions, these findings mean, in turn, that the M and N shells of the ions are faster filled in metals than in insulators (by conduction electrons). Some additional weaker and non monotonic changes in the relative intensities of the KL^X satellites have also been found in various insulators or semiconductors.

These results have to be compared to those obtained in experiments already presented (13) which clearly indicate that the main interaction processes below a surface is the Auger neutralization.

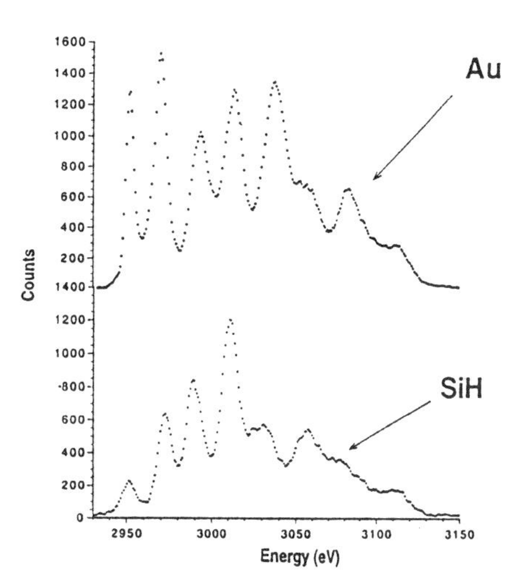

FIGURE 2. K L^x-ray spectra of Ar^{17+} ions impinging on SiH and Au surfaces.

In our first experiment in 1990 (1) when we found that Ar^{17+} ions are fed, below a surface, in the M and N shells leaving the K and L ones empty (hollow atoms), we tentatively explained this selective filling by an energy resonance between the M and N shells of the ion and e.g. the K shell of C and O surface contaminants.

More recently we studied with a low energy SiLi detector the characteristic X-ray spectra of argon ions of different initial charge states (from 8+ up to 17+) on different targets (C, SiH or SiO_2). In the case of carbon targets, for instance, we did not observe any C K X ray for Ar^{9+} and Ar^{12+} ions, which means that the captured electrons do not come from the K shell of the target but only from the L one. For Ar^{16+} ions, a clear C Kα peak is observed but its intensity is not large enough to explain the whole electron capture processes observed. The main capture process is then the capture of C L electrons. Below a surface the capture process of weakly bound L electrons through a Landau-Zener mechanism is not possible, the distance of interaction being larger than the interatomic distances in the lattice. Nearby a highly charged ion the bound character of a L electron of C is also questionable and does not make possible any electron promotion process. Some other mechanisms must thus be found to explain the observed transfer. The most likely process would be the Auger neutralization process, first introduced by Hagstrum (14) to describe the close interaction (contact) of singly charged ions on surfaces. The velocity of the ions being comparable, or smaller than that of the "L" or valence (V) electrons, the ion may be quasibound in the solid. These target electrons may then participate in Auger MVV transitions filling the M shell. Inside a solid the argon ions having no discrete state above n=5 (dissolution into the continuum due to the external screening), the highest ion levels to be filled in this collision are consequently the M and N shells. The Auger transitions being more probable for the ejected electrons of the lowest energy, this may also explain the high selectivity of the population into the M and N shells of argon.

These experiments show that some Auger neutralization type processes must also be taken into account beyond the more conventional transfer mechanism of e.g. K electrons in some specific cases (resonant). This mechanism may then explain that the filling of the M (N) shells of the ion is faster in metals than in insulators, the density of states being much larger in metals than in insulators.

CONCLUSION

One can say that below and outside a surface the hollow atoms are, in most cases, formed by capture of the most weakly bound electrons of the target. The conduction or valence character of these electrons has been found to play, in both cases, a very important role during the interaction.

Above a metal surface where the captured electrons come from the conduction band, the collective response of the free electrons leads to the formation of an image charge which irremediably attracts the ion towards the surface (the ion touches the surface). One then observes, in addition to the resonant neutralization processes which hold above the surface, another very specific interaction at the surface.

Above an insulator or a semiconductor the captured electrons come from the valence band and the image acceleration is overcome by the appearance of positive holes. The ion instead of touching the surface is backscattered and one only observes an outside resonant neutralization process which holds along a period of time depending on the ion initial kinetic energy.

Below the surface, the capture of these weakly bound electrons proceeds via an Auger neutralization process which transfers, either the conduction electrons (in metal), or the valence ones (insulators) into the excited states of the ion. The neutralization rate is found much larger in metals than in semiconductors owing to the larger density of states in metals.

Above metal or insulator surfaces low energy ions may experience at grazing incidence a specular reflection. In this case the ions feel the repulsive electric field of the nuclei of the top layer atoms. They must then touch or, penetrate, at least slightly, the surface. Highly charged ions have been found to be very quickly neutralized when they are specularly reflected (15): at surface they capture a certain number of electrons in M or N shell (for e.g. Ar) and may continue to do so in higher n states on their way back.

Above an insulator or a semiconductor surface, at normal incidence, the ions start capturing electrons at a distance z_0 (e.g. 20 Å) and are backscattered well before touching the surface. They escape the capture area with an certain number of electrons in excited states. These excited electrons are further lost during the Auger cascades. Their charge and energy depend on how deep they penetrate the capture area (or how long they stay in), and thus only on the initial kinetic energy of the ions. One can then control the approach of the ions, between e.g. 0 and 20 Å, just by tuning their velocity in an energy range of 0-~12 eV/q (trampoline effect).

ACKNOWLEDGEMENTS

The experiments were carried out with L. de Billy, G. Borsoni, B. d'Etat-Ban, G. Giardino, V. Le Roux and S. Thuriez of our laboratory, in collaboration with M. Froment of the Laboratoire de Physique des Liquides et Electrochimie (Université P&M Curie and CNRS Paris), M. Eddrief and C. Sébenne of the Laboratoire de Physique des Solides (Université P&M Curie and CNRS Paris), D. Schneider and S. Bardin of LLNL Livermore, M. Prior, H. Khemliche, J. Jin and Z. Xie of LBNL Berkeley. I would like to express our gratitude to G. Melin and A. Brenac of CEN Grenoble, C. Lyneis of LBNL Berkeley, and D. Lecler and J.-P. Grandin of GANIL, who welcomed us at their ECR sources, for their constant and efficient help during the runs.

REFERENCES

1. Briand, J.-P., de Billy, L., Charles, P., Essabaa, S., Briand, P., Geller, R., Desclaux, J.-P., Bliman S., and Ristori , C., *Phys. Rev. Lett.* **65**, 159-162 (1990).
2. Briand J.-P., *Comments in At. Mol. Phys.* **32** (1996).
3. Meyer, F.W., Havener, C.C., Overbury, S.H., Reeds, K.J., Snowdon, K.J., and Zehner, D.M., *J. Phys.* (Paris) Coll. **50**, C1, 263-275 (1989).
4. Stolterfoht, N., Arnau, A., Grether, M., Köhrbrück, R., Spieler, A., *Phys. Rev. A* **52**, 445-456 (1995).
5. Briand, J.-P., "Study of the Interaction of Highly Charged Ions with Metallic Surfaces" in *Proceedings of the Fifteenth International Conference on X-Ray and Inner-Shell Processes.*, Knoxville, Tennessee, July 9-13, 1990. Editors Carlson, T.A., Krause, M.O., Manson, S.T., *AIP Conference Proceedings 215.* New York: AIP, 1990, pp.513-523.
6. Morgenstern, R., and Briand, J.-P., Invited talks at the 9th General Conference of the European Physical Society, Florence, September 14-17, 1993. Unpublished.
7. Briand, J.-P., Giardino, G., Borsoni, G., Froment, M., Eddrief, M., Sébenne, C., Bardin, S., Schneider, D., Jin, J., Khemliche, H., Xie, Z., and Prior, M., *Phys. Rev. A* **54**, Oct. (1996).
8. Briand, J.-P., Thuriez, S., Giardino, G., Borsoni, G., Froment, M., Eddrief, M., *Phys. Rev. Lett.* **77**, 1452-1455 (1996).
9. d'Etat-Ban, B., Briand, J.-P., Ban, G., de Billy, L., Desclaux, J.-P., Briand, P., *Phys. Rev. A* **48**, 1098-1106 (1993).
10. Winter, H., Europhys. Lett. **18**, 207-212 (1992).
11. Aumayr, F., Winter, HP., *Comments in At. Mol. Phys.* **29**, 275-303 (1994).
12. Briand, J.-P., de Billy, L., Jin, J., Khemliche, H., Prior, M., Xie, Z., and Nectoux, M., *Phys. Rev. A* **53**, 2925-2928 (1996).
13. Briand, J.-P., et al. See contributed papers in Book of Abstracts of the Nineteenth International Conference on the Physics of Electronic and Atomic Collisions, Whistler, BC, Canada. July 26-August 1, 1995. Edited by Mitchell, J.B.A., McConkey, J.W., Brion, C.E., Vol. 2, pp.714, 715, 716, 724, and 733.
14. Hagstrum, H.D., *Phys. Rev.* **96**, 336-365 (1954).
15. Schneider, D., Briere, M., *Phys. Scripta* **53**, 228-242 (1996).

MEASUREMENT OF THE 1s2s $^1S_0$ - 1s2p $^3P_{0,1}$ TRANSITIONS IN HELIUMLIKE NITROGEN BY FAST-BEAM LASER SPECTROSCOPY

E.G. Myers, J.K. Thompson, D.J.H. Howie, E.P. Gavathas and N.R. Claussen

Department of Physics, Florida State University, Tallahassee, Florida 32306

J.D. Silver

Clarendon Laboratory, University of Oxford, Oxford OX1 3PU, U.K.

For a large range of Z the intercombination $1s2s\ ^1S_0 - 1s2p\ ^3P_1$ interval in heliumlike ions is accessible to spectroscopy with infra-red lasers. Using a CO_2 laser and a nitrogen ion beam from a Van de Graaff accelerator, together with careful energy calibration, we obtained 986.321(7) cm^{-1} for the N^{5+}, $2^1S_0 - 2^3P_1$ transition. This result is a severe test of two-electron Lamb Shift calculations. In $^{14}N^{5+}$ there is also sufficient hyperfine mixing of the 2^3P_0 and 2^3P_1 levels to permit spectroscopy of the otherwise totally forbidden $2^1S_0 - 2^3P_0$ interval. By careful choice of CO_2 laser line it was possible to observe the $2^1S_0 - 2^3P_0$ and the $2^1S_0 - 2^3P_{1,F=2}$ transitions at similar ion beam energies, yielding a precise result for the $2^3P_1 - 2^3P_0$ fine-structure interval of 8.6709(10) cm^{-1}. This result provides a test of current calculations of QED contributions of order $\alpha^7 m_e c^2$, which are required for obtaining a new value of the finestructure constant from the finestructure of helium.

INTRODUCTION

Helium and heliumlike ions are the prototypical many electron system and provide an important testing ground for relativistic quantum mechanics. The energy separations of the $n = 2$ levels, shown in fig. 1, are particularly sensitive to interesting relativistic and quantum-electrodynamic (QED) effects. Recent theoretical developments have included relativistic many-body perturbation theory [1,2] and relativistic configuration interaction theory [3,4] which obtain all "structure", but not all QED contributions to $O(Z\alpha)^4$ atomic units (a.u.). Using highly precise non-relativistic wave-functions and operators derived from QED, Drake and his collaborators aim to extend calculations of the 1s2p ^3P_J finestructure splittings in helium to $O(\alpha^5)$ a.u. [5–7], and hence by comparison with experiment [8,9], obtain a new "atomic physics" value for the finestructure constant. There is also considerable interest, e.g. see [4,10–13], in developing new methods and improving the precision of two-electron Lamb Shift calculations.

Because of the different Z scalings of the various theoretical contributions to the energy spacings of the $n = 2$ levels of two-electron ions, it is important to have precision measurements at low, intermediate, and high Z. Recent experimental work has concentrated on the $1s2s\ ^3S_1 - 1s2p\ ^3P_J$ transitions, and the results have been compared with theory, *e.g.* in ref. [2]. Precision laser measurements have been carried out up to Z=5 [12,14,15]. However for higher Z these intervals lie in the vacuum ultraviolet and measurements have been carried out in emission using concave grating spectrometers *e.g.* see [16,17]. At the precision currently obtainable with these techniques there is generally good agreement with existing theory.

The general objective of our program is to extend the precision of laser techniques to higher Z heliumlike ions. We make use of the fact that the intercombination $1s2s\ ^1S_0 - 1s2p\ ^3P_1$ interval and the $1s2p\ ^3P_J -{}^3P_{J'}$ fine-structure splittings are in the infra-red for various Z. Although these transitions are first-order forbidden, they are accessible to spectroscopy with medium-to-high power infra-red lasers and measurements are practical using the doppler-tuned fast-beam laser technique. A further advantage of measuring infra-red rather than ultraviolet splittings is that a given fractional precision (which is often limited by the Doppler effect), translates into a higher absolute precision for testing the theory.

CP392, *Application of Accelerators in Research and Industry*, edited by J. L. Duggan and I. L. Morgan
AIP Press, New York © 1997.

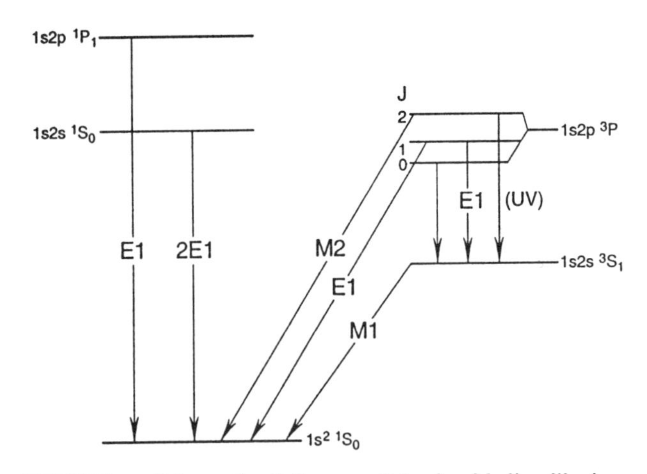

FIGURE 1. Schematic of the $n = 2$ levels of heliumlike ions showing their principal decay modes.

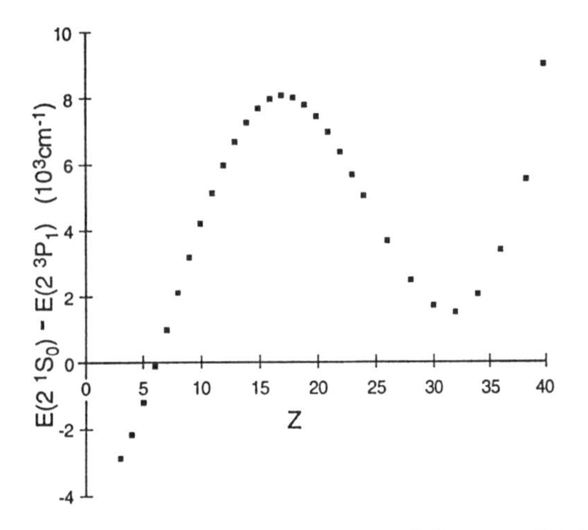

FIGURE 2. Z dependence of the $1s2s\ ^1S_0 - 1s2p\ ^3P_1$ interval in heliumlike ions.

MEASUREMENT OF THE 1s2s $^1S_0$ - 1s2p $^3P_1$ INTERVAL

As was noted some time ago [18], the intercombination $2^1S_0 - 2^3P_1$ interval remains in the infra-red for a large range of Z, see fig. 2, and although first-order forbidden, is accessible to precision laser spectroscopy. As a comparatively small splitting between an S and a P state, this interval is very sensitive to the QED electron self-energy.

We have demonstrated the utility of this approach to precision spectroscopy in helium-like ions by a fast-beam laser resonance measurement on N^{5+} using a CO_2 laser [19], see fig. 3. A 6 MeV N^{5+} beam was obtained by foil stripping a N^+ beam obtained from a specially constructed ion source in the terminal of the Florida State University FN tandem [20]. The beam was energy analysed in a 90 degree analysing magnet and then merged with a 200 W beam from a grating tuned

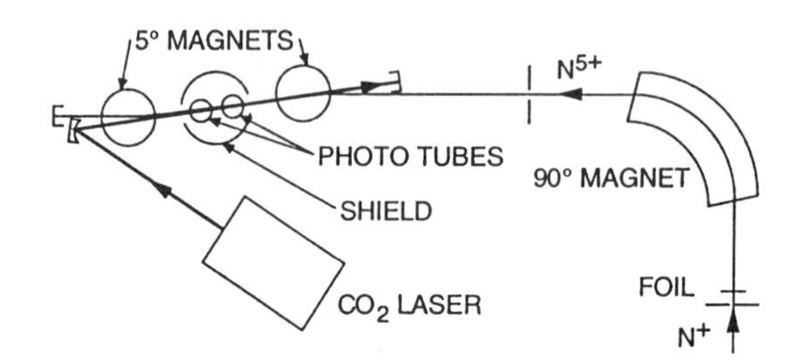

FIGURE 3. Simplified schematic of the experimental arrangement.

	E (cm^{-1})
This experiment [19]	986.321(7)
Drake [10]	986.579
Plante, Johnson, and Sapirstein [2]	984.7
Cheng et al. [4,21]	993.6

TABLE 1. Our result for the 1s2s $^1S_0$ – 1s2p $^3P_1$ interval in $^{14}N^{5+}$ without hyperfine interaction, compared with recent theory.

CO_2 laser. Transitions induced from the metastable 2^1S_0 level to the ns-lived 2^3P_1 level were detected via the subsequent UV transitions to 2^3S_1, using a pair of photo-multiplier tubes. The resonances were scanned by varying the beam velocity so as to vary the doppler shift. A survey scan of the three induced transitions (three because of hyperfine structure) is shown in fig. 4.

By a careful energy calibration of the 90 degree analysing magnet involving the well known $^1H(^{15}N, \alpha\gamma)^{12}C$ resonance, as well as a time-of-flight system, we were able to achieve a beam energy calibration to better than 3 keV and a final wavenumber precision of 7 ppm. Our result for the 1s2p $^1S_0$ – 1s2s $^3P_1$ interval in N^{5+} is compared with current theory in table 1. The result clearly differentiates between the recent, but less precise relativistic calculations [2,3], and the 'unified' result of Drake [10]. The remaining discrepancy (35 experimental standard deviations) provides a stimulus for better inclusion of correlation effects in the relativistic calculations, and for a better treatment of the Bethe-logarithm in the calculation of the two-electron self-energy.

HYPERFINE-INDUCED TRANSITION AND FINESTRUCTURE MEASUREMENT

Although quite rare, hyperfine induced transitions are of considerable interest in both atomic physics and

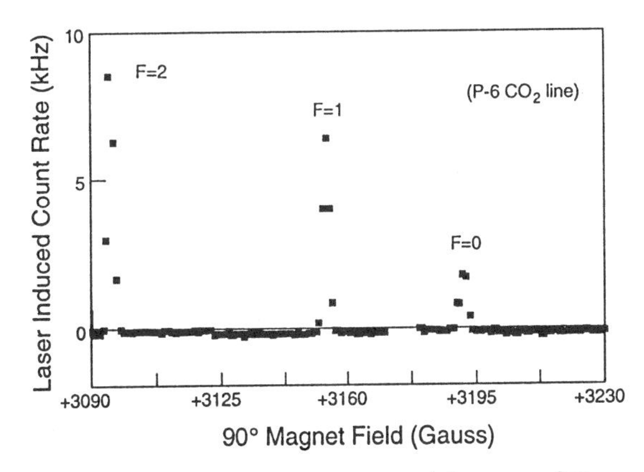

FIGURE 4. Survey scan of the $1s2s\ ^1S_0 - 1s2p\ ^3P_{1,F}$ transitions in $^{14}N^{5+}$.

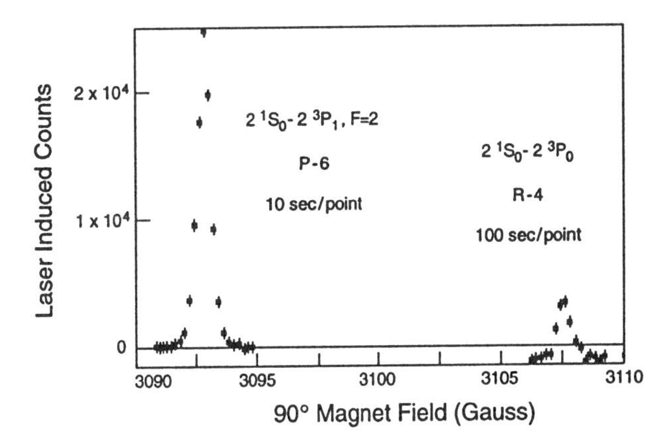

FIGURE 5. Scan showing the $1s2s\ ^1S_0 - 1s2p\ ^3P_{1,F=2}$ transition and the hyperfine induced $1s2s\ ^1S_0 - 1s2p\ ^3P_0$ transition.

astrophysics, e.g. see [22] and references therein. In atoms or ions with zero nuclear spin, single-photon electromagnetic transitions in which $J = 0 \rightarrow J' = 0$, *e.g.* transitions of the type $1sns\ ^1S_0 - 1sn'p\ ^3P_0$ in helium-like ions, are absolutely forbidden. However, the hyperfine interaction, by mixing levels of different total electronic angular momentum J, in our case the $2\ ^3P_0$ and $2\ ^3P_1$ levels, can produce observable transition probabilities [23].

Following the observation of the laser induced $2^1S_0 - 2^3P_1$ resonances, it was realized, because of the excellent signal-to-noise, and because in N^{5+} the $J = 0$ to $J = 1$ finestructure is unusually small, leading to relatively large hyperfine mixing, that it would be possible to induce and observe the $2^1S_0 - 2^3P_0$ transition as well. Further, by measuring the difference (in beam velocity) between this resonance and a $2^1S_0 - 2^3P_1$ hyperfine component, a value for the $2^3P_0 - 2^3P_1$ interval would be obtained in which a large fraction of the systematic uncertainty due to the beam velocity would cancel out. The systematic uncertainty due to the beam velocity measurement would be further reduced by changing the laser line so that the resonances occurred as close in beam velocity (or energy) as possible.

We therefore scanned the relatively strong $2\ ^1S_0 - 2\ ^3P_{1,F=2}$ transition which occurs near 6.15 MeV with the $P6$, 956.1850 cm^{-1} [24], CO_2 laser line. We then changed laser line to $R4$, 964.7690 cm^{-1}, and searched for the the $2\ ^1S_0 - 2\ ^3P_0$ transition at a beam energy approximately 75 keV higher. After various improvements were made to the interaction chamber and the experimental procedure to ensure efficient overlap of the laser and ion beams, the very weak hyperfine induced transition was observed with enough signal-to-noise for a precise measurement [25], see fig.5. The relative strengths of the two resonances is consistent with estimates based on the hyperfine mixing coefficients of ref. [23]. Using the $P6$ line the two hyperfine intervals of the 2^3P_1 level were also measured.

	ΔE_{01} (cm^{-1})
This experiment [25]	8.6709(10)
Yan and Drake [5]	8.68213(2) [a]
Zhang, Yan and Drake [7]	8.686(20) [b]
Chen, Cheng, and Johnson [3]	8.67(2)
Plante, Johnson, and Sapirstein [2]	8.73(2)

TABLE 2. Our results for the $^{14}N^{5+}$, $2\ ^3P_0 - 2\ ^3P_1$ finestructure interval compared with recent theory.

[a] Error is from computational uncertainty.
[b] Error is from estimate of uncalculated terms.

Our result for the $2^3P_0 - 2^3P_1$ finestructure interval in $^{14}N^{5+}$ is compared with theory in table 2. It is consistent with the order α^4 a.u. calculations of ref. [5]. In contrast to helium, the recently calculated order $\alpha^5 \log \alpha$ contribution [7] slightly worsens the agreement for this interval. Hence our result will provide a strong test of the remaining order α^5 a.u. contributions, which are currently being calculated [6]. When the equivalent calculations for helium are combined with recent high precision finestructure measurements [8,9], it should be possible to extract a new value for α. This should help resolve the current three-way discrepancy between the values for α obtained from the quantized Hall effect, the ac Josephson effect and from "$g - 2$", as discussed in ref. [5].

ACKNOWLEDGEMENTS

This work was supported in part by the National Science Foundation.

REFERENCES

1. W.R. Johnson and J. Sapirstein, Phys. Rev. A **46**, R2197 (1992).
2. D.R. Plante, W.R. Johnson, and J. Sapirstein, Phys. Rev. A **49**, 3519 (1994).
3. M.H. Chen, K.T. Cheng, and W.R. Johnson, Phys. Rev. A **47**, 3692 (1993).
4. K.T. Cheng, M.H. Chen, W.R. Johnson, and J. Sapirstein, Phys. Rev. A **50**, 247 (1994).
5. Z.-C. Yan and G.W.F. Drake, Phys. Rev. Lett. **74**, 4791 (1995).
6. T. Zhang, Phys. Rev. A **53**, 3896 (1996).
7. T. Zhang, Z.-C. Yan and G.W.F. Drake, Phys. Rev. Lett. **77**, 1715 (1996).
8. D. Shiner, R. Dixson, and P. Zhao, Phys. Rev. Lett. **72**, 1802 (1994); R. Dixson and D. Shiner, Bull. Am. Phys. Soc. **39**, 1059 (1994).
9. G. Gabrielse, private communication (1996).
10. G.W.F. Drake, Can. J. Phys. **66**, 586 (1988).
11. G.W.F. Drake, I.B. Khriplovich, A.I. Milstein, and A.S. Yelkhovsky, Phys. Rev. A **48**, R15, (1993).
12. E. Riis, A.G. Sinclair, O. Poulsen, G.W.F. Drake, W.R.C. Rowley, and A.P. Levick, Phys. Rev. A**49**, 207 (1994).
13. H. Persson, S. Salomonson, P. Sunnergren, and I. Lindgren, Phys. Rev. Lett. **76**, 204 (1966).
14. T.J. Scholl, R. Cameron, S.D. Rosner, L. Zhang, R.A. Holt, C.J. Sansonetti, and J.D. Gillaspy, Phys. Rev. Lett. **71**, 2188 (1993).
15. T.P. Dinneen, N. Berrah-Mansour, H.G. Berry, L. Young, and R.C. Pardo, Phys. Rev. Lett. **66**, 2859 (1991).
16. K.W. Kukla, A.E. Livingston, J. Suleiman, H.G. Berry, R.W. Dunford, D.S. Gemmell, E.P. Kanter, S. Cheng, and L.J. Curtis, Phys. Rev. A **51**, 1905 (1995).
17. D.J.H. Howie, J.D. Silver and E.G. Myers, J. Phys. B**29**, 927 (1996).
18. E.G. Myers, Nucl. Instrum. Methods, B**9**, 662 (1985).
19. E.G. Myers, J.K.Thompson, E.P. Gavathas, N.R. Claussen, J.D. Silver and D.J.H. Howie, Phys. Rev. Lett. **75**, 3637, (1995).
20. E.G. Myers, J.K. Thompson, P.A. Allen, P.E. Barber, G.A. Brown, V.S. Griffin, B.G. Schmidt, and S. Trimble, Nucl. Instr. and Methods Phys. Res., A**372**, 280 (1996).
21. Because Cheng *et al.* [4] give results only for even Z we have obtained a value for $Z = 7$ by interpolating the differences between their results and those of ref. [2] for $Z = 6$ and $Z = 8$. The large difference between this result and the other calculations is mainly due to their different values for the QED contributions. If one combines an interpolation of their relativistic energies with Drake's QED correction [10], one obtains 985.9 cm^{-1}, in better agreement with experiment.
22. A. Aboussaid, M.R. Godefroid, P. Jonsson, and C. Froese Fischer, Phys. Rev. A **51**, 2031 1995.
23. P.J. Mohr, in *Beam-Foil Spectroscopy*, ed. I.A. Sellin and D.J. Pegg (Plenum, New York, 1976), vol.1., p.97.
24. F.R. Petersen, E.C. Beaty, and C.R. Pollock, J. Molec. Spectosc.**102**, 112 (1983).
25. E.G. Myers, D.J.H. Howie, J.K. Thompson and J.D. Silver, Phys. Rev. Lett., **76**, 4899 (1996).

SEMICLASSICAL FRAGMENTATION INTO CHARGED PARTICLES

Jan M. Rost

Fakultät für Physik, Universität Freiburg, D–79104 Freiburg, Germany

The development of semiclassical approximations is briefly reviewed. A semiclassical S–matrix theory is described for inelastic scattering of atoms with electrons or photons. As an example ionization of hydrogen by electron impact resonance formation below the ionization threshold are discussed.

INTRODUCTION

Semiclassical approximations have served two rather different purposes: first, to explore the classical limit of quantum mechanics and to gain more insight into the nature of quantum phenomena. Second, to develop a theory that provides reliable approximations for cases in which it is not possible to solve the full problem quantum mechanically. To the first category belongs the formulation of the semiclassical propagator initiated by van Vleck in 1928 [1] and completed by Maslov [2] and Gutzwiller [3]. Within the second category the WKB–approach has been most successful in different areas of physics. The essentially one–dimensional theory has led to useful results in times where we lacked the computer power to treat complicated, more dimensional problems. For scattering problems a logical application has been the calculation of WKB–phase shifts for elastic scattering, pioneered by Ford and Wheeler [4].

With the improvement of the computers, even the most general semiclassical formulation by van Vleck has become computationally feasible and thus conceptually interesting again. Miller, Marcus and others showed in the seventies that semiclassical approximations for the Green's function itself can lead to remarkable results for reactive scattering in molecular complexes, whose dynamics can not be characterized by short wave lengths [5]. In this paper a semiclassical formulation of inelastic electron–atom scattering derived from the path integral representation of the S–matrix will be sketched.

SCALING IN THE CLASSICAL ATOM

A classical N–electron atom has some remarkable properties that are important for the semiclassical description of the scattering process. They do not depend on the choice of the coordinate system but are particularly simple to derive in hyperspherical coordinates. The hyperradius $\mathcal{R}^2 = \sum_i r_i^2$, composed of all electron–nucleus distances, measures the overall extension of the atom. All other coordinates transform to a space of $3N-4$ angles Ω on the hypersphere with radius $\mathcal{R}$. We scale the coordinates, momenta and the hamiltonian H itself with the energy

$$\begin{aligned} \mathcal{R}' &= E\mathcal{R} \\ \mathcal{P}' &= E^{-1/2}\mathcal{P} \\ H' &= E^{-1}H \equiv 1. \end{aligned} \tag{1}$$

The angles and the corresponding generalized momenta are dimensionless and therefore not affected by the scaling. The scaled hamiltonian reads

$$H' \equiv 1 = \frac{\mathcal{P}'^2}{2} + \frac{\Lambda'^2(\Omega)}{2\mathcal{R}'^2} + \frac{C(\Omega)}{\mathcal{R}'} \tag{2}$$

where $\Lambda(\Omega)$ is the grand angular momentum operator, which contains the dependence upon all the momenta in the angles Ω while $C(\Omega)$ can be viewed as an angle dependent generalized charge whose exact form is not important for the present context. For Eq. (2) one can derive the following properties [6]:

(i) As a function of time the hyperradius $\mathcal{R}(t)$ has one extremum that is a minimum. In fact this can be shown for arbitrary masses of the $N+1$ particles.

(ii) In the limit $E \to 0$ the dynamics for any (preserved) total angular momentum L in a two–electron atom is governed by the same effective hamiltonian of $L = 0$. The angular moment has the dimension of action [length]× [momentum]. From Eq. (1) follows that $L' = L\sqrt{E}$. Hence, for any finite L the effective angular momentum is zero in the limit $E \to 0$. Note also that the action scales as $S = S'/\sqrt{E}$ where S' does not depend on the energy E.

(iii) The angle $\theta_{12} = 180°$ between the electrons in a two–electron atom is a fixed point of the classical equations of motion, i.e. if $\theta_{12}(t_0) = 180°$ and $\dot{\theta}_{12}(t_0) = 0$ then $\theta_{12}(t) = \theta_{12}(t_0)$ for all times t.

With (i) – (iii) we can justify a considerable simplification for electron–atom scattering. (The atom is treated as a core with one active electron). The reason for only using $L = 0$ provided by (ii) is very different from elastic scattering with short range potentials where a partial wave analysis shows that only $l = 0$ survives for $E \to 0$. In Coulomb scattering, all angular momenta contribute with a priori unknown weight but the dynamics for each L is determined in the limit $E \to 0$ by the same hamiltonian as for the S–wave. Furthermore, because of (iii), for the total cross section it is sufficient to calculate the process at fixed $\theta_{12} = 180°$. Hence, the problem reduces from 12 phase space variables to 4, the two electron–nucleus distances r_i and the conjugate momenta p_i.

THE SCATTERING AMPLITUDE

In the "one–dimensional world" (only one coordinate, the distance r_i describes each electron) the cross section reduces to a probability directly proportional to the S–matrix. Clearly, the ability to calculate the absolute cross section is lost in this approximation. However, most experiments gain their absolute measurements by comparison with a known value in some limit of the cross section. We can either normalize the theoretical curve to the same limit or directly to the relevant experiment at one data point.

The scattering amplitude reads [7]

$$S_{\epsilon,\epsilon^-}(E) = \sum_j \sqrt{\mathcal{P}_j(E,\epsilon,\epsilon^-)} \exp\{\frac{i\Phi_j}{\hbar} - \frac{i\nu_j\pi}{2}\}, \quad (3)$$

with the classical probability for the j–th trajectory

$$\mathcal{P}_j(E,\epsilon,\epsilon^-) = \frac{1}{R}\left|\frac{\partial\epsilon}{\partial r^-}\right|^{-1}_{\epsilon^-}. \quad (4)$$

The normalization R will be discussed in the next section, r^- is the initial position of the projectile. The sum runs over all classical trajectories j that take the projectile from energy ϵ^- to ϵ during the collision. Each trajectory accumulates a phase, which is defined by the classical action $\Phi = \int q_1 dp_1 + \int q_2 dp_2$ and a contribution of $\nu\pi/2$ from caustics along the trajectory [3,5]. The true initial state is given as a product of the bound state of the target electron with binding energy ϵ_B and the free projectile state with energy $\epsilon^- = E + \epsilon_B$.

ELECTRON IMPACT IONIZATION

For pure Coulomb interactions (no perturbation by a core potential) one can see from property (i) that only a *single* trajectory fulfills these boundary conditions simultaneously. This is illustrated in Fig. 1 where the function $\epsilon(r^-)/E$ is plotted for electron–hydrogen scattering. The final energy of the projectile can can be negative for initial conditions $r^- \in I(1)$ (then the projectile has become bound in exchange for the target electron). If $\epsilon > E$ we have an excitation process ($r^- \in I(3)$, the target electron is still bound after the collision) and for $0 < \epsilon < E$ ($r^- \in I(2)$) the system has fragmented into the original projectile electron, the target electron and the ionized target. Each point of the function represents a trajectory that connects an initial condition r^- with a given final energy ϵ. Since the function is monotonic the connection between final and initial conditions is uniquely established by a single trajectory. This is certainly a remarkable property: it reduces the sum of Eq. (3) to one term. Hence, the *semiclassical* analysis of the threshold dynamics under the simplifications made possible through the properties $(i) - (iii)$ reveals an essentially *classical* behavior [7].

Classically we can calculate the ionization probability

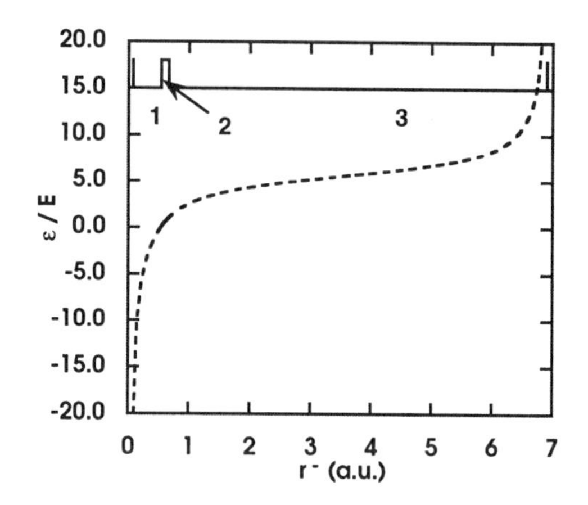

Figure 1: Energy dependence $\epsilon/|E|$ of the projectile on its initial position 1000 a.u.+r^-, normalized to the total energy $E = 0.1$a.u.. The inverse derivative of this function enters Eq. (4). Intervals $I(i)$ correspond to different events, see text.

as

$$\int_0^E \mathcal{P}(E,\epsilon,\epsilon^-)d\epsilon = \frac{1}{R}\int_0^E d\epsilon\frac{dr^-}{d\epsilon} = \frac{\Delta r^-(2)}{R}. \quad (5)$$

The normalization is given by all possible processes,

$$R = \int_{-\infty}^{+\infty} d\epsilon\frac{dr^-}{d\epsilon} = \Delta r^-(1) + \Delta r^-(2) + \Delta r^-(3). \quad (6)$$

The ionization probability of Eq. (5) describes a problem of distinguishable Coulomb particles such as positron–hydrogen scattering [8]. For electron–hydrogen scattering the ionization probability must be modified due to the Pauli–principle which requires the differential scattering amplitude to be symmetric with respect to electron exchange. This could lead to interferences between the two contributions. However, the action is symmetric under electron exchange in the final channel, $\Phi(\epsilon,\epsilon^-,E) = \Phi(E-\epsilon,\epsilon^-,E)$ so that the scattering probability

$$P_\epsilon^\pm(E) = \left[\mathcal{P}(E,\epsilon,\epsilon^-)^{\frac{1}{2}} \pm \mathcal{P}(E,E-\epsilon,\epsilon^-)^{\frac{1}{2}}\right]^2 \quad (7)$$

is independent of the phase factor in Eq.(3). The "+" sign in Eq. (7) corresponds strictly speaking to the ${}^1S^e$ partial wave while the "−" sign represents ${}^3S^e$ symmetry. The total ionization probability $P^\pm(E)$ is now obtained by integrating explicitly over the energy ϵ of one electron,

$$P^\pm(E) = \int_0^{E/2} P_\epsilon^\pm(E)d\epsilon. \quad (8)$$

$P^+(E)$ is in excellent agreement with all data points of the threshold experiment [10] up to excess energies as high as 8 eV (Fig.2) [7]. Written in the form

$$P(E) \propto E^{\alpha(E)} \quad (9)$$

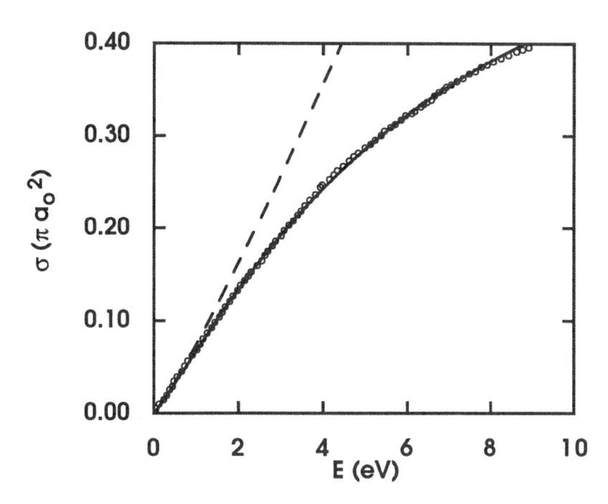

Figure 2: The total ionization cross section for electron impact on $H(1s)$. The experimental data points are taken from [10]. The calculated cross section (solid) has been normalized to the experimental data at 5.84 eV. The dashed line is the Wannier cross section $\sigma(E) = \sigma_0 E^{1.127}$.

the exponent is $\alpha = 1.127$ in the limit $E \to 0$ as analytically derived by Wannier [9]. However, as can be seen from Fig.2, the Wannier cross section and the (semi)–classical cross section differ for finite E. Note that the semiclassical S–matrix approach is capable of providing a cross section with an absolute energy scale. Only recently another theoretical approach succeeded to reproduce this dependence correctly [11].

CHAOTIC SCATTERING

If the energy ϵ^- of the incident electron beam is lowered the total energy remains negative and no ionization can take place. What happens with the initial conditions which lead to ionization for $E > 0$? The deflection function (Fig. 3) can answer this question. Comparing Fig. 1 with Fig. 3 one sees that *ionizing* orbits get replaced by *chaotic scattering* orbits [12]. Chaos is generated by trajectories which hit the potential boundary since there is not enough energy for double–escape of the electrons. Hence both electrons bounce many times into the nucleus. Thereby, they suffer a time delay compared to electrons following 'direct' trajectories from the intervals 1 and 3. The time delay is the signature of a resonance [13]. What had been ionization above threshold, described by a regular behavior of interval 2 in Fig. 1, becomes resonance formation below threshold, characterized by the irregular scattering in interval 2 of Fig. 3a. Although it might seem hopeless to deal with a fractal object like the interval 2 in Fig. 3 we can form the analogy to the total classical ionization cross section. Close to threshold it is justified to neglect all phases since according to the scaling (see (ii)) they oscillate rapidly as $\exp(-iS'/\sqrt(|E|))$. As in Eq.(5) for ionization we integrate the fractal initial conditions from interval 2 over all electron energies ϵ. The quantity

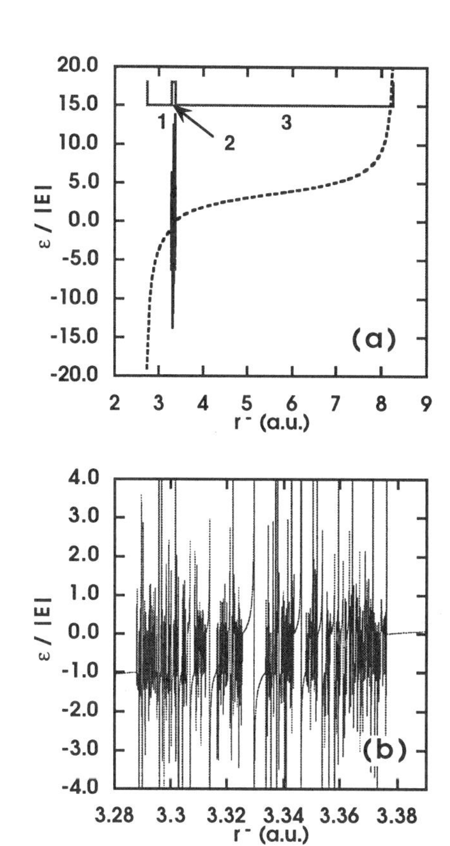

Figure 3: Classical deflection function as in Figure 1, but for $E = -0.1$ a.u. with interval 2 for chaotic scattering. Part (b) shows a detail of part (a) highlighting the fractal structure.

we will obtain is the probability for resonance scattering, that is, the probability that the electron suffered a time delay in the scattering event and, for a short time, an (excited) 3–body complex had been formed. For each ϵ there is not only one but an infinite set of trajectories $\{i\}$ contributing to the sum of Eq. (3). However, to each index i belongs a continuous branch of trajectories with all energies ϵ. This branch yields upon integration over ϵ a small subinterval $\delta_i(r^-)$. The sum over i recovers then the entire interval $\Delta r^-(2) = \sum_i \delta_i$ of chaotic scattering, which replaces the corresponding interval in Eq. (5):

$$P_{frac}(E) = \frac{1}{R}\sum_i \int_{-\infty}^{\infty} \left|\frac{\partial \epsilon}{\partial r_i^-}\right|_{\epsilon^-}^{-1} d\epsilon = \sum_i \frac{\delta_i r^-}{R} = \frac{\Delta r^-(2)}{R}. \tag{10}$$

Thus, the classical probability for resonance formation below threshold is in exact analogy to the total classical ionization cross section above threshold. Together the probabilities represent 3–body events where both electrons have to participate in the scattering process simultaneously [15]. Figure 4 shows the "3–body" cross section across the ionization threshold. The full circles are calculated according to Eq. (10) while the solid line is a fit with

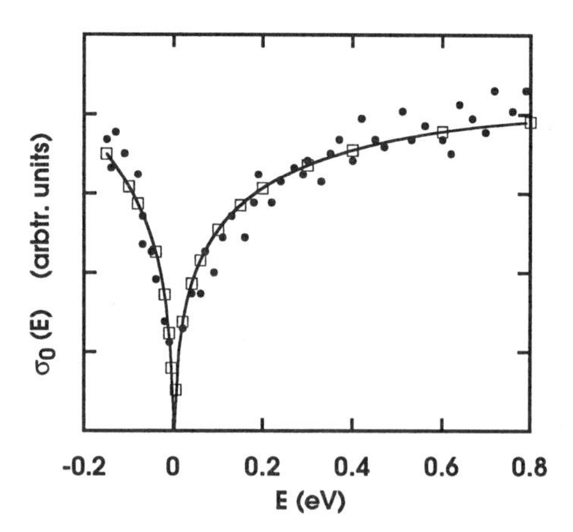

Figure 4: Total cross section for electron impact on atoms close to the fragmentation threshold. The black circles are the theory for a hydrogen target $H(1s)$ (Eq. (10)). The solid line is a fit of the theory according to Eq. (11). The open squares are experimental data for a He(1s^2) target [14]. Shown is the cross section $\sigma_0(E) = \sigma(E)/E$ to emphasize the dip at $E = 0$.

the function

$$\sigma_{3-body}(E) \propto |E|^{1.127}(1 + a|E|^{\frac{1}{2}} + bE). \qquad (11)$$

For $E > 0$ Eq. (11) reflects the result of Eq. (9) that the ionization cross section close to threshold is proportional to the 1.127th power of E. Noting from Fig. 4 and Eq. (11) that this holds also below threshold we can conclude that $|E|^{1.127}$ originates in the 3–body amplitude of the S–matrix and should be independent of the process through which it was activated. The residual dynamics, essentially the density of states, depend on the initial state and the excitation as a whole (by photon or particle impact etc.). However, the residual part varies slowly with energy. Hence, we can take it into account with a Taylor expansion about $E = 0$ in Eq. (11).

In their threshold experiment (electron impact on He) Cvejanović and Read have probed positive as well as negative energies [14]. Within the experimental accuracy the data coincide with our "3–body" cross section although it is for hydrogen and not for helium (Fig. 4). However, one core electron participates only as a spectator forming a He^+ core which is equivalent to H^+ for the non–core penetrating final states considered here.

SUMMARY AND OUTLOOK

In this contribution some special properties of the classical three–body Coulomb system have been described. They lead in the context of a semiclassical S–matrix theory to a surprisingly simple description of the ionization of atoms. Even more interesting is the picture which emerges from the semiclassical perspective for the two–electron spectrum as a function of excitation energy below the ionization threshold: for low energies resonances are formed. Semiclassically, they are generated by a coherent superposition of many orbits where the phases are substantial for the resonance spectrum [16]. Very close to threshold the phases fluctuate randomly, the resonances disappear and the inelastic cross section may be described essentially classically. In between, phase differences (and this means in the semiclassical context differences in the actions of orbits) play a role to first order which leads to the semiclassical version of Ericson–fluctuations in the inelastic cross section [17]. Work to confirm this interpretation quantitatively and to investigate its consequences is in progress.

ACKNOWLEDGMENT

Financial support by the DFG within SFB 276 and from the Gerhard-Hess-Programm is gratefully acknowledged.

REFERENCES

1. J. M. van Vleck, Proc. Acad. Nat. Sci. USA **14**, 178 (1928).
2. V. P. Maslov and M. V. Fedoriuk, *Semi-Classical Approximations in Quantum Mechanics* (Boston, Reidel, 1981).
3. M. C. Gutzwiller, *Chaos in Classical and Quantum Mechanics* (Springer, New York, 1990).
4. K. W. Ford and J. A. Wheeler, Ann. Phys. N.Y. **7**. 259 (1959).
5. W. H. Miller, Adv. Chem. Phys. **56**, 38 (1974) and references therein.
6. J. M. Rost, J.Phys.B**28**, 3003 (1995).
7. J. M. Rost, Phys.Rev.Lett. **72**, 1998 (1994).
8. J. M. Rost and E. J. Heller, Phys.Rev. **A49**, R4289 (1994).
9. G. H. Wannier, Phys.Rev. **90**, 817 (1953).
10. J. W. McGowan and E. M. Clarke, Phys.Rev. **167**, 43 (1968).
11. J. H. Macek and S. Yu. Ovchinnikov, Phys. Rev. Lett. **74** 4631 (1995).
12. J. M. Rost, J. Phys. B**27**, 5923 (1994).
13. B. Eckhardt, CHAOS **3** 613, (1993).
14. S. Cvejanović and F. H. Read, J. Phys. B **7**, 1841 (1974). In this experiment the yield of very low energy electrons is measured. They are interpreted as products of a "three–body" process as discussed in the text.
15. J. M. Rost and D. Wintgen, Europhys. Lett. **35**, 19 (1996).
16. D. Wintgen, K. Richter, G. Tanner, CHAOS **2**, 19 (1992) and references therein.
17. T. Ericson, Phys. Rev. Lett. **5**, 430 (1960), R. Blümel and W. P. Reinhardt, *Directions in Chaos, vol. 4, Quantum Non-Integrability*, ed. D.H. Feng and J.-M. Yuan, p 245 (World Scientific, Singapore, 1992).

RESONANT COHERENT EXCITATION OF SURFACE CHANNELED B^{4+} IONS

K. Kimura, H. Ida, M. Fritz and M. Mannami

Department of Engineering Physics and Mechanics, Kyoto University, Kyoto 606-01, Japan

Charge-state distributions of surface channeled B ions are measured, when 4.5 ~ 6.5-MeV B^{3+} ions are incident on a SnTe(001) surface along a [100] axis at glancing angles 2 ~ 6 mrad. Small reduction of B^{4+} fraction is observed at ~ 5.5 MeV, which corresponds to the energy of the resonant coherent excitation (RCE) of the [100] surface channeled B^{4+} ion from n = 1 to n = 2 states. The observed reduction is attributed to the ionization of the B^{4+} ions excited by the RCE. The RCE is also observed even when the ions are incident on the surface at azimuthal angles 100 ~ 300 mrad off the [100] direction, i.e. even under surface planar channeling conditions. The observed resonance energy is shifted to higher energy from that observed under the [100] surface channeling condition. The observed shift was explained by taking account of the 2D periodicity of the crystal plane.

INTRODUCTION

An axial channeled ion collides with crystal atoms periodically. These collisions cause a periodic perturbation. If the excitation energy of the ion coincides with the frequency or one of its higher harmonics, resonant coherent excitation (RCE) may occur (1). The condition of the RCE for the axial channeled ion is given by

$$2\pi\hbar\frac{kv}{d} = \Delta E, \tag{1}$$

where k is an integer, v is the ion velocity, d is the atomic spacing along the atomic row and ΔE is the excitation energy. The RCE was first observed by Datz *et al* (2) through the reduction of the charge state fraction of the incident ion. In recent studies, enhancement of projectile x-ray emission (3,4) and of convoy electron emission (5) due to RCE were also observed.

The RCE can occur not only inside a crystal but also on a crystal surface (6). Theoretical studies revealed some interesting features of the RCE of surface channeled ions (7 - 10). Creation of a strongly poralized excited state was predicted by Burgdörfer *et al* (7). An advantage of surface channeling over transmission geometry was suggested for obtaining stimulated emission of soft x-rays from ions (8). A new spectroscopy for analyzing bound states of projectile ions in the vicinity of the surface was proposed (9). Garcia de Abajo *et al* studied the resonant coherently induced electron loss to the continuum for the case of surface channeling (10). Recently, we have observed the RCE of surface channeled ions through the change of the charge-state distribution (CSD) of the surface channeled ions (11). In the present paper, we report the recent observation of the RCE of surface axial channeled ions (11) together with results of new measurements under surface planar channeling conditions.

EXPERIMENTAL

A single crystal of SnTe(001) was prepared by epitaxial growth in situ by vacuum evaporation on a cleaved surface of KCl at 250°C in a UHV chamber. The crystal was mounted on a five-axis precision goniometer. As SnTe has a NaCl-type crystal structure, Sn atoms and Te atoms are alternately arranged along a [100] direction with a spacing d = 6.0 a.u.

Beams of 4.6 - 6.5 MeV B^{3+} ions from the 1.7 MV Tandetron accelerator of Kyoto University were collimated by a series of apertures to less than 0.1 × 0.1 mm^2 and to a divergence angle less than 0.3 mrad. The beams were incident on the crystal at glancing angles θ_i of 2 - 6 mrad measured from the surface plane. The ions scattered from the crystal surface at a scattering angle θ_s were selected by a movable aperture (ϕ = 0.2 mm and the acceptance angle ± 0.25 mrad). The ions passing through the aperture were energy analyzed by a 90° sector magnetic spectrometer. Microchannel plates located in the focal plane of the spectrometer served as position-sensitive detector (length 55 mm, resolution 0.13 mm) for the energy-analyzed ions. Thus the energy spectrum of the scattered ions could be measured without sweeping magnetic field. The energy window of the spectrometer was about 9 % and the energy resolution was better than 0.1 % including the energy spread of the incident beam.

The CSD of the scattered ions was measured as follows: The magnetic field of the spectrometer was changed periodically so that the ions of each charge state could reach the detector alternately. The energy spectrum of q+ ions was registered in the qth memory group of a multichannel analyzer. This was done to reduce the error

CP392, *Application of Accelerators in Research and Industry,* edited by J. L. Duggan and I. L. Morgan
AIP Press, New York © 1997.

caused by the fluctuation of the incident beam. In the present energy region, the CSD of the scattered ions depends weakly on the incident charge state (12, 13) and the B^{4+} fraction is dominant (larger than 50 %) irrespective of the incident charge state. This means that there is no frozen charge state which is desirable for the observation of the RCE.

RESULTS AND DISCUSSIONS

Figure 1 shows an example of the observed energy spectrum of the B^{4+} ions scattered at θ_s = 8 mrad when 5.5-MeV B^{3+} ions were incident along the [100] axis at a glancing angle θ_i = 4 mrad. There are a few peaks at ~ 5.4, ~ 5.34 and ~ 5.26 MeV. In our previous study on the energy loss of surface channeled ions, we have shown that these peaks correspond to the particular ion trajectories (14). The corresponding ion trajectories projected on the plane perpendicular to the atomic string are shown in the inset. The peak of the highest energy corresponds to the on-string trajectory labeled "A" which has the shortest path near the surface. Here, we concentrate on this trajectory. For the ion that follows this trajectory, the probability for the occurrence of the RCE is expected to be large because of the shortest distance of the closest approach to the atomic string. As the electron loss probability for the excited ion is much larger than that for the ion in the ground state, a change in the CSD of the channeled ions should occur at the resonance energy, as had been observed in the transmission experiment (2).

Figure 2 shows the ratio of the B^{4+} fraction to the B^{5+} fraction observed at the specular angle ($\theta_s = 2\theta_i$) when B^{3+} ions were incident at θ_i = 4 mrad along the [100] axis. The statistical error is smaller than the symbols. There is a

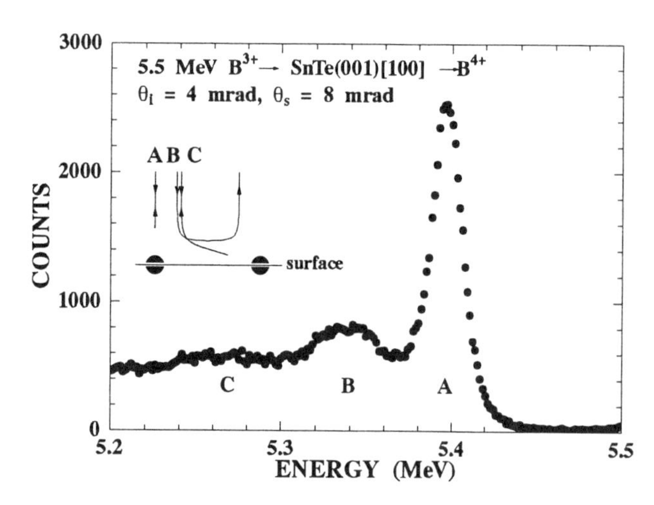

FIGURE 1. Energy spectrum of the [100] surface channeled B^{4+} ions observed at θ_s = 8 mrad when 5.5-MeV B^{3+} ions are incident on a SnTe(001) at θ_i = 4 mrad. There are several well defined peaks that correspond to the specific trajectories shown by the inset, where closed circles show the atomic strings. The highest energy peak corresponds to the trajectory "A", the second and third peaks correspond to "B" and "C".

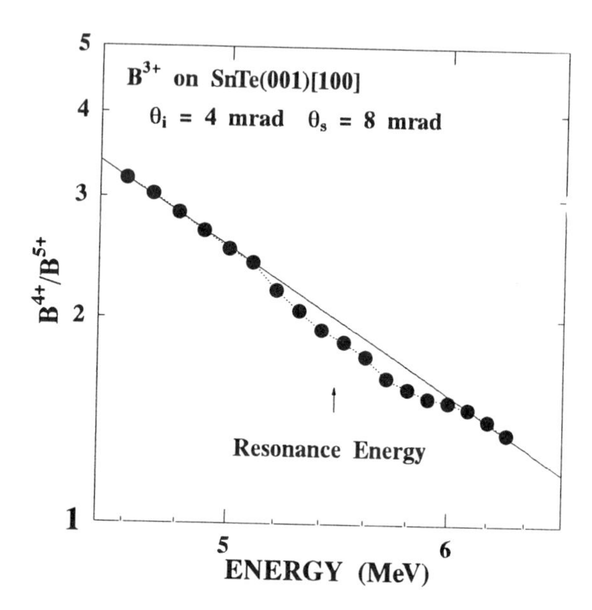

FIGURE 2. Ratio of the B^{4+} fraction to the B^{5+} fraction observed at the specular angle when B^{3+} ions are incident at θ_i = 4 mrad on a SnTe(001) surface along the [100] axis. The lines are drawn to guide the eye. The statistical error is smaller than the symbols.

small reduction of the B^{4+} fraction around 5.5 MeV. In order to see clearly the observed reduction, the general trend of the energy dependence (shown by a solid line in Fig. 2) is subtracted. The background-subtracted result is shown in Fig. 3 together with the results for other incident angles. The reduction of the B^{4+} fraction can be seen around 5.5 MeV for various incident angles. This energy agrees with the calculated resonance energy (5.46 MeV) for the RCE of the B^{4+} ion from n = 1 to n = 2 with the second harmonic (k = 2). Although the observed B^{4+} reduction is very small, this good agreement indicates that the observed B^{4+} reduction is caused by the RCE. The small effect of RCE on the CSD can be attributed to the large charge exchange probabilities at the surface. The subsequent charge exchange processes almost erase the B^{4+} reduction caused by

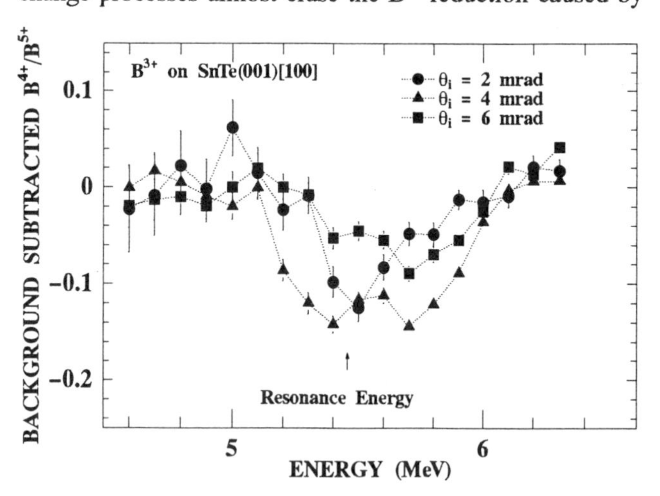

FIGURE 3. Background-subtracted ratio observed at the specular angle when B^{3+} ions are incident at various incident angles on a SnTe(001) surface along the [100] axis.

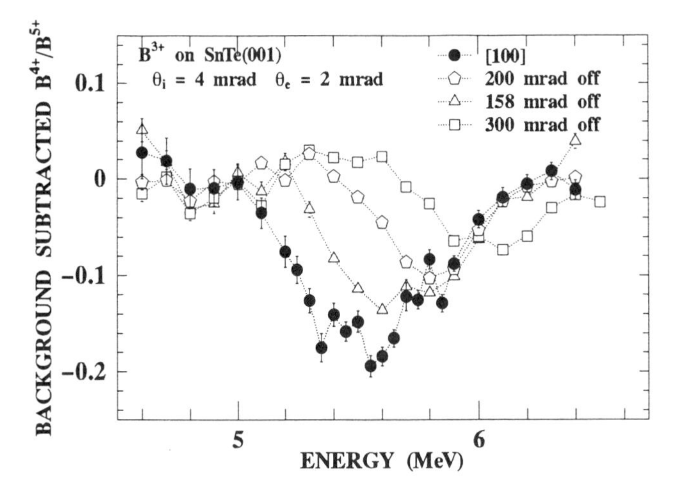

FIGURE 4. Background-subtracted ratio B^{4+}/B^{5+} observed at θ_s = 6 mrad when B^{3+} ions are incident at θ_i = 4 mrad with various azimuthal angles on a SnTe(001) surface.

the RCE (11).

The CSD of surface channeled ions was measured at off-specular conditions. Figure 4 shows the background-subtracted ratio of B^{4+}/B^{5+} observed at θ_s = 6 mrad when B^{3+} ions were incident at θ_i = 4 mrad along the [100] axis (closed circle). The observed B^{4+} reduction is slightly enhanced as compared with the results at specular angle (see Fig. 3). This can be attributed to the effect of surface steps. Some ions are scattered at down steps as shown by the inset in Fig. 5. After passing over the down step, the charge exchange probabilities are so small that the CSD at the point passing over the down step is preserved. These ions appear at scattering angles smaller than the specular angle and the energy losses of these ions are smaller than that of the ion specularly reflected from a flat surface because the path lengths of theses ions is smaller in the surface region (13).

Figure 5 shows the energy spectrum of B^{4+} ions scattered at θ_s = 6 mrad when 5.75–MeV B^{3+} ions were incident at θ_i = 4 mrad. There is a shoulder at the high energy side of the peak "A" (~ 5.7 MeV). This shoulder corresponds to the ions scattered at down steps (13). Thus the subsequent charge exchange processes are partly suppressed by observing the CSD at $\theta_s < 2\theta_i$ and the larger reduction of B^{4+} fraction can be seen.

Figure 4 also shows the results observed under surface planar channeling conditions, i.e. at various azimuthal angles ϕ = 100 ~ 300 mrad from the [100] axis (open symbols). The reduction of B^{4+} fraction is also seen for the surface planar channeling. The observed resonance energy increases with increasing azimuthal angle. The condition of the RCE for planar channeling is given by (15)

$$2\pi\hbar \frac{k \cos\phi + l \sin\phi}{d} v = \Delta E \qquad (2)$$

where, k and l are integers. This equation takes account of 2D periodicity of the crystal plane and is also valid for surface planar channeling.

The azimuthal angle dependence of the observed resonance energy is shown in Fig. 6. The solid curve shows the calculated result with k = 2 and l = 0. Although the experimental resonance energy is slightly larger than the calculated result, the agreement is reasonably good. If the energy loss (~ 100 keV) of the surface channeled ion is taking into account, the agreement becomes much better. Thus the observed result can be explained by taking account of the 2D periodicity of the surface.

The observed reduction of the B^{4+} fraction decreases with increasing azimuthal angle (see Fig. 4). However, this does

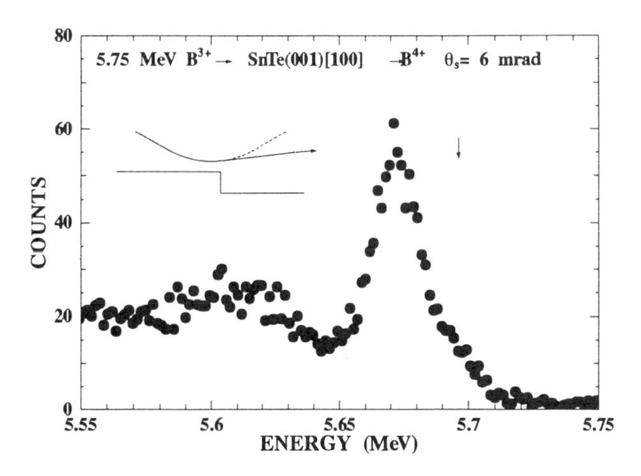

FIGURE 5. Energy spectrum of the [100] surface channeled B^{4+} ions observed at θ_s = 6 mrad when 5.5–MeV B^{3+} ions are incident on a SnTe(001) at θ_i = 4 mrad. Peak "A" has a shoulder at the high energy side, which corresponds to the ions scattered at down steps. The inset shows a possible ion trajectory of these ions.

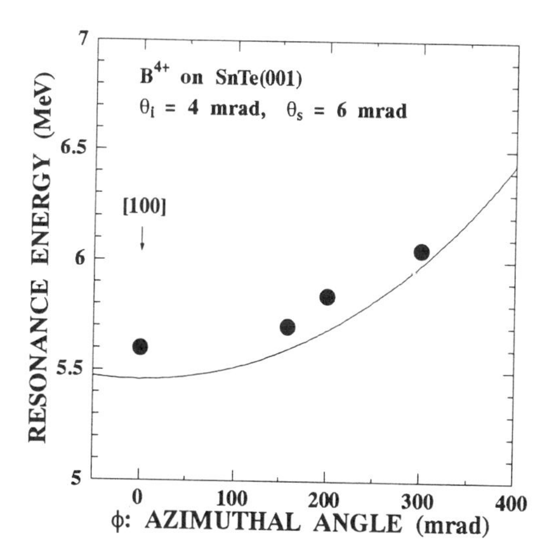

FIGURE 6. Observed resonance energy as a function of the azimuthal angle. The solid curve shows the calculated result with eq. (2).

not imply that the RCE probability decreases with azimuthal angle. The RCE probability can be estimated by the ratio $\Delta F(4+)/F(4+)$, where $\Delta F(4+)$ is the reduction of B^{4+} fraction due to the RCE and $F(4+)$ is the B^{4+} fraction at the resonance energy. With increasing azimuthal angle, the resonance energy increases and so $F(4+)$ decreases as can be seen in Fig. 2. Thus, both $\Delta F(4+)$ and $F(4+)$ decrease with azimuthal angle. As a result, the estimated RCE probability is found to be almost independent of the azimuthal angle.

CONCLUSION

Resonant coherent excitation of surface channeled B^{4+} ions on a SnTe(001) surface is observed through reduction of B^{4+} fraction. The observed B^{4+} reduction is small because of large charge exchange probabilities in the vicinity of the surface. It is shown that these charge exchange processes can be partly suppressed by utilizing surface steps. The reduction of the B^{4+} fraction is also observed under surface planar channeling conditions. The observed resonance energy can be explained by taking account of the 2D periodicity of the crystal surface.

ACKNOWLEDGMENTS

We would like to thank the members of the Department of Nuclear Engineering of Kyoto University for the use of the Tandetron accelerator. This work was supported in part by a Grant-in-Aid for Scientific Research from the Ministry of Education, Science and Culture.

REFERENCES

1. Okorokov, V.V., Pis'ma Zh. Eksp. Teor. Fiz. **2**, 175 (1965) [JETP Lett. **2**, 111 (1965)].
2. Datz, S., Moak, C.D., Crawford, O.H., Krause, H.F., Dittner, P.F., Gomez del Campo, J., Biggerstaff, J.A., Miller, P.D., Hvelplund, P., and Knudsen, H., Phys. Rev. Lett. **40**, 843–847 (1978).
3. Fujimoto, F., Komaki, K., Ootuka, A., Vilalta, E., Iwata, Y., Hirao, Y., Hasegawa, T., Sekiguchi, M., Mizobuchi, A., Hattori, T., and Kimura, K., Nucl. Instrum. Methods Phys. Res., Sect. B **33**, 354–357 (1988).
4. Datz, S., Dittner, P.F., Gomez del Campo, J., Kimura, K., Krause, K.F., Rosseel, T.M., Vane, C.R., Iwata, Y., Komaki, K., Yamazaki, Y, Fujimoto, F, and Honda, Y., Radiat. Eff. Def. Solids **117**, 73–77 (1991).
5. Kimura, K., Gibbons, J.P., Elston, S.B., Biedermann, C., DeSerio, R., Keller, N., Levin, J.C., Breining, M., Burgdörfer, J., and Sellin, I.A., Phys. Rev. Lett. **66**, 25–28 (1991).
6. Kupfer, E., Gabriel, H., and Burgdörfer, J., Z. Phys. A **300**, 35–41 (1981).
7. Burgdörfer, J., Gabriel, H., and Kupfer, E., Nucl. Instrum. Methods **194** 337–340 (1982).
8. Elci, A., Phys. Rev. Lett. **53**, 1696–1699 (1984).
9. Kawai, R, and Kawai, M., Surf. Sci. **195**, 535–556 (1988).
10. Garcia de Abajo, F.J., Ponce, V.H. and Echenique, P.M., Phys. Rev. Lett. **69**, 2364–2367 (1992).
11. Kimura, K., Ida, H., Fritz, M., and Mannami, M., Phys. Rev. Lett. **76**, 3850–3853 (1996).
12. Narumi, K., Fujii, Y., Toba, K., Kimura, K., and Mannami, M., Nucl. Instrum. Methods Phys. Res., Sect. B **100**, 1–9 (1995).
13. Fritz, M., Kimura, K., Kuroda, H., and Mannami, M., Phys. Rev. A **54** (1996).
14. Fujii, Y., Kishine, K., Narumi, K., Kimura, K., and Mannami, M., Phys. Rev. A **47**, 2047–2054 (1993).
15. Datz, S., Moak, C.D., Crawford, O.H., Krause, H.F., Dittner, P.F., Gomez del Campo, J., Biggerstaff, J.A., Knudsen, H., and Hvelplund, P., Nucl. Instrum. Methods, **170**, 15–20 (1980).

FORGOTTEN (?) X-RAY INTENSITY ENHANCEMENT IN SOLIDS AT LINES RELATED TO NOT COMPLETELY FILLED SHELLS

I. Török*, J. Pálinkás*#, M. Budnar^, M. Kavcic^, A. Mühleisen^, and J. Kawai$

**ATOMKI, Institute of Nuclear Research of H.A.S., Debrecen, Pf 51. H-4001, Hungary*
#Dept. of Experimental Physics, Kossuth University, Debrecen, Pf. 105. H-4001, Hungary
^J. Stefan Institute, University of Ljubljana, Jamova 39, p. p. 3000, SI-1001 Ljubljana, Slovenia
$Dept. of Materials Science and Engineering, Kyoto University, Sakyo-ku, Kyoto 606-01, Japan

Compared to relative intensity values calculated for free atoms, some x-ray lines in solids show a 1-2 order of magnitude enhancement in low Z regions. These lines correspond mainly to quadrupole transitions at higher Z, but when the subshell involved is not completely filled, significant dipole component can occur, which is of higher probability. This fact was known in the 30-ies already, but it was almost forgotten, and revealed once more recently. There are guessings why is it not widely known: (i) the early results failed in assigning the transitions, (ii) the results were published in not too well known journals. The paper shows the enhancement of $K\beta_5$ line around Z=24 (Cr), using experimental values measured in metals. Other lines are mentioned, where similar behaviour of the relative intensity was indicated. The significance of the effect for various x-ray analytical methods is presented: The present result is useful for the improvement of the accuracy of trace elemental analysis of 3d transition metals using e. g. PIXE or total reflection x-ray fluorescence (TXRF).

INTRODUCTION

"My grandmother knew much more about household tricks (e. g. conservation of foods, etc.), than my mother, and my mother knew much more about such things, than my wife. The newer and newer generations got many things prepared ready, so they do not need the knowledge and skills, mastered by their parents. So many useful tricks are going to be forgotten. A similar effect exists also in physics. The newer and newer instruments, e. g., make obsolete the skills of the fathers and mothers of physics. In Paris there is yet existing the electroscope used in the early radioactivity measurements by M. Curie, who could measure with a certain accuracy. F. Joliot could measure with the same instrument only with an order of magnitude worse. The professor in the 70-ies could do it also with another order of magnitude worse. Sorry, sometimes also basic knowledge could fall out from the collective memory of physicists." (I. T.)

Compared to relative intensity values calculated for free atoms e.g. by Scofield (1), some x-ray lines in solids show a 1-2 order of magnitude enhancement in low atomic number (Z) regions. These lines at high Z are mainly quadrupole transitions, but when the subshell involved is not completely filled, significant dipole component can occur, which is of higher transition probability.

As it looks the enhancement effect for the $K\beta_5$ line was first measured in the late twenties of our century (2). In the paper we discuss this enhancement, we try to find the reasons, why it could be almost forgotten, and mention other x-ray lines, where this effect was observed, and the consequencies of this effect in basic physics and in analytical applications.

THE ENHANCEMENT OF THE $K\beta_5$ LINE AROUND Z=24 (CHROMIUM)

Early measurements

The low intensity $K\beta_5$ line in the K x-ray spectrum is often regarded as a K-$M_{IV,V}$ transition. For free atoms and for Z>30 elements the $K\beta_5$ line can be described as a quadrupole (E2) transition. For elements between Z values of 17 to 29 in solids or in compounds a significant dipole component may appear due to solid state or chemical effects. This causes an enhancement of the intensity of the $K\beta_5$ line and gives a peak at Z=24 (Cr) in the $K\beta_5/K\alpha_1$ relative intensity as a function of the atomic number. This enhancement is discussed in the book of Agarwal(3) and Blokhin(4). In both edition of Agarwal's book a factor of 10^{-3} is missing from the figure which can lead to the conclusion that the vertical axis is calibrated in arbitrary units.

CP392, *Application of Accelerators in Research and Industry*, edited by J. L. Duggan and I. L. Morgan
AIP Press, New York © 1997.

Tracing the references of the Agarwal book, we found the paper of H.-T. Meyer (2), as the oldest one on this subject. There are at least two indications why the effect had been omitted: (i) The Wiss. Veröff. Siemens-Konzern is not perhaps a well known journal today; (ii) The author writes about the $K\beta_2$ line, and only at the end of the paper one can find a remark, that in the intensity of the $K\beta_2$ line at elements with Z<29(Cu), also the outer M-levels can have a role. This means, that the assignment of the transitions was ambiguous in that time. (From the mentioned books of Blokhin (p. 62) and Agarwal (p. 100 in the 2nd Ed.) one can learn, that the $K\beta_5$ line with increasing atomic number appears first in solids at Z=17(Cl), and in free atoms at Z=21(Sc), and the level becomes filled at Z=29(Cu). (Similarly for $K\beta_2$ the respective values are Z=30, Z=31, Z=36.) It means, that in the paper where first was reported this Z dependence of the $K\beta_5$ line, this line is not mentioned!

On the other hand the $K\beta_5$ line is a rather weak one, which might be also the source of its neglecting. Furthermore near to the $K\beta_5$ line there is the K absorption edge, what makes difficult to evaluate the spectra in this region. This can be the additional reason for not revealing it in the x-ray spectra. Anyhow, probably in the former Soviet Union (SU), there were further measurements, which were tabulated in an x-ray table collection by Blokhin and Svejcer (5). But as before the World War II. the results from the SU were not available for a wide western public, it could happen, that in 1984 (when ref. (5) was published) such statements could appear (6), that there are Z areas where experimental values are not available: There are no measured values of the $K\beta_5$ (K-M_4,M_5) for elements with Z<63, and no measured values for $K\beta_2$ (K-N_2,N_3) for elements with atomic number Z<36. The results of Meyer were obtained by photo-plate - densitometer methods, which is now regarded as unreliable, but his values fit well into values from the modern measurements.

Originally we thought, that the curve given in the Agarwal book is a theoretical one (because of the lack of experimental points), but we could not find a theory giving this result, so now perhaps it is better to suppose, that this is a curve guiding the eye through the early experimental data.

Compilation of relative intensities

Experimental values of this enhanced intensity are given in e. g. the x-ray tables of Blokhin and Svejcer (5). Many other tables, however, give the values of Scofield (1), calculated for free atoms, and the difference can reach a factor of 70. We recently compiled the available experimental $K\beta_5/K\alpha_1$ x-ray line intensity ratio values (7, 8), published from the late twenties to 1994, and adding some values of our own measurements or derived graphically by us from spectra published in literature, and they are shown in Fig. 2. of ref. (8) as a function of the atomic number of the emitting atoms.

In Fig. 1. of this paper we give the Z dependence of the $K\beta_5/K\beta_{1,3}$ x-ray intensity ratio also as a function of the atomic number of the emitting element. These values are related to the $K\beta_5/K\alpha_1$ x-ray line intensity ratio through the $K\beta_{1,3}/K\alpha_1$, which itself is Z dependent (1).

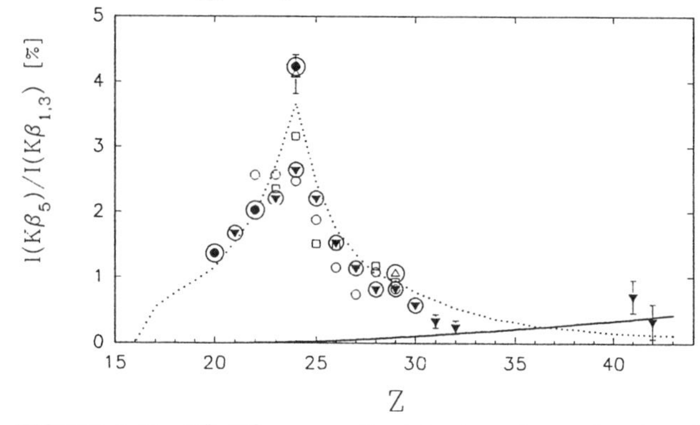

FIGURE 1. The $K\beta_5/K\beta_{1,3}$ x-ray line intensity ratio as a function of the atomic number of the emitting element. Symbols: experimental data of different origin (sources are given in ref. (8), encircled full dots: our own measurements; encircled other symbols: derived graphically by us from spectra published in literature), curves: full line: theoretical values of Scofield(1), dotted line: Agarwal(3).

Recent measurements

As it turned out, since the experimental values for the intensity of the $K\beta_5$ line and those for the radiative Auger emission lines for medium Z elements are rather uncertain in the literature, the $K\beta$ spectra of Ca, Ti, and Cr were measured (9). From these spectra, as well as from the repeated measurements performed in Ljubljana, using 1.2 MeV protons from a Van de Graaff generator to excite the metallic Ca, Ti, Cr, and Fe samples, the $K\beta_5$ line intensities were extracted. A flat crystal spectrometer equipped with position sensitive detector was used to measure the x-ray spectrum. The target was seen also by a Si(Li) detector simultaneously. With this set up good quality spectra could be obtained despite the low intensity of the $K\beta_5$ line. The evaluation of the spectra is in progress, using the EWA code (10). A sample spectrum is shown in Fig. 2.

The subject is a living one: Measurements of V and Cr $K\beta_5$ line intensities were performed recently at the Kyoto University (11), where Tochio et al. used x-ray tube for excitation of the samples. They investigated the intensity enhancement depending on the chemical surrounding of the atoms. They found a difference between tetragonal and octahedral crystal systems. The origin of this difference is that p and d orbitals mix each other in the tetragonal symmetry, where as they do not mix in the octahedral symmetry (12).

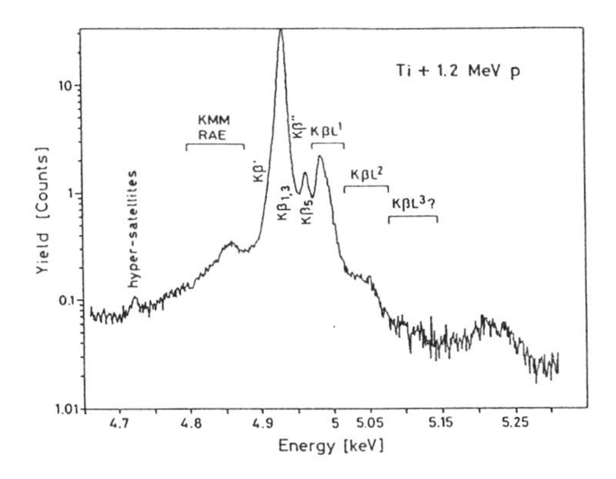

FIGURE 2. The Kβ region of Ti x-ray spectrum, obtained by 1.2 MeV proton bombardment.

OTHER LINES SHOWING ENHANCEMENT EFFECT

Already Blokhin (4) and Agarwal (3) mentioned, that lines other than the $K\beta_5$ can show similar behaviour, when the subshell involved is not fully filled. Recently in works from a group in Fribourg experimental detection of such effect was reported. They measured medium Z elements, like Zr, Mo, Pd, when they observed enhancement of the normally quadrupole $K\beta_4$ line. For Mo this enhancement is of a factor of about 40 comparable to the enhancement of the $K\beta_5$ x-ray around chromium. They also observed similar effect in the L lines (13, 14).

LINE INTENSITY ENHANCEMENT AND ACCURACY OF X-RAY ANALYTICAL METHODS

The enhanced intensity of the $K\beta_5$ x-ray can reach up to 4% of the $K\beta_{1,3}$ line (Fig. 1.), and one has to consider this enhancement in the energy dispersive x-ray spectrometry. If the K x-ray spectra are fitted with two peaks (Kα and Kβ), only a modest fit can be obtained (e. g. 15). Using additional peaks of relative intensities of a few percent, at the energies where the low intensity $K\beta_5$ or radiative Auger emissions should appear, the fit can be considerably improved (16). Also the multiple ionization $K\beta L^i$ satellites can influence the picture, mainly at ion excitation. The involvement into the spectrum evaluation of such weak, but in certain cases of several percent intensity lines can improve even the measurements of the Kβ/Kα ratios. For improving PIXE data base it seems to be important to measure the Kβ region with crystal spectrometers, and determine the accurate energies and intensities of the low intensity components. These values can be used in the evaluation of x-ray spectra obtained by semiconductor detectors.

CONCLUSION

The enhancement of the Kβ5 x-ray line intensity around Cr (3d transition metals), and similar effects at other lines, at other Z regions, must be considered in spectra evaluations in basic physics and in analytical methods like PIXE, to avoid significant systematic errors.

ACKNOWLEDGEMENTS

This work was supported by OTKA, grant No. 3011 and No. TO 16636, furthermore by the Hungarian-Slovenian Intergovernmental S&T Cooperation Project No. SLO-9/95.

REFRENCES

1. Scofield, J. H., *Phys. Rev.* **179**, 9-16 (1969); *Phys. Rev. A* **9,** 1041-1049 (1974); *Atomic Data and Nuclear Data Tables* **14**, 121-137 (1974).
2. Meyer, H.-T., *Wiss. Veröff. Siemens-Konzern* **7**, 108-162 (1929).
3. Agarwal, B. K., *X-Ray Spectroscopy. An Introduction.* Springer Series in Optical Sciences, Berlin:Springer-Verlag, Vol. 15, 1st Ed., Berlin: Springer, 1979, p. 116, Fig. 2.32; 2nd Ed., Berlin: Springer, 1991, p. 100, Fig. 2.32.
4. Blokhin, M. A., *Fizika rentgenovskih luchei,* Moscow: Gosudarstvennoe izdatel'stvo tekhniko-teoreticheskoi literatury, 1953. Fig. 24. on p. 64.
5. Blokhin, M. A., Svejcer, I. G., *Rentgenospektralny spravochnik,* Moscow: Nauka, 1982, Table 12.1, p. 70. (In Russian).
6. Salem, S. I., and Scott, B. L., *Phys. Lett.* **103 A,** 321-322 (1984).
7. Török, I., Papp, T., Pálinkás, J., Budnar, M., Mühleisen, A., Kawai, J., Campbell, J. L., "Solid state effects in the relative intensity of the $K\beta_5$ x-ray line around Z=24(chromium)," in *XIX. ICPEAC, Scientific program and abstracts of contributed papers,* Whistler, British Columbia, Canada, 26. July-1. Aug. 1995. Eds. J. B. A. Mitchell et al. Vol. 2. p.732.
8. Török, I., Papp, T., Pálinkás, J., Budnar, M., Mühleisen, A., Kawai, J., Campbell, J. L., *Nucl. Instr. Meth. B* **114,** 9-14 (1996).
9. Budnar, M., Mühleisen, A., Hribar, M., Janzekovic, H., Ravnikar, M., Smit, Z., and Zitnik, M., *Nucl. Instr. Meth. B* **63,** 377-383 (1992)
10. Végh, J., *Thesis*, Debrecen (1990).
11. Mukoyama, T., private communication.
12. Kawai, J., *J. Mater. Sci. Lett.* **11**, 1096-1098 (1992).
13. Ludziejewski, T., Rymuza, P., and Sujkowski, Z.; Boschung, B., Dousse, J.-Cl., Halabuka, B., Herren, Ch., Hoszowska, J., Kern, J. and Rhême, Ch.; Polasik, M., *Phys. Rev. A* **52,** 2791-2803 (1995).
14. Hoszowska, J., and Dousse, J-Cl., *J. Phys. B: At. Mol. Opt. Phys.* **29,** 1641-1653 (1996).
15. Lépy, M. C., Plegnard, J., and Morel, J., *Nucl. Instr. Meth. A* **339,** 241-247 (1994).
16. Johansson, A. E., Campbell, J. L., Malmquist, K. G., *Particle induced X-ray Emission Spectroscopy (PIXE),* New York: J. Wiley & Sons, 1995, p.19.

Precision Angle Resolved Autoionization Resonances in Ar and Ne

N. Berrah[1], B. Langer[1], T. W. Gorczyca[1], R. Wehlitz[2], A. Farhat[1]. J. D. Bozek[3],

[1]Physics Department, Western Michigan University, Kalamazoo, MI 49008.
[2]University of Tennessee, Physics Department, Knoxville, TN 37996-1200.
[3]Lawrence Berkeley Laboratory, Advanced Light Source Mail Stop 2-400, Berkeley CA 94720.

We have used angle-resolved photoelectron spectroscopy and improved R-matrix calculations to study the Ar $3s^23p^6 \rightarrow 3s3p^6$ np (n=4-16) and Ne $2s^22p^6 \rightarrow 2s2p^6$ np (n=3-5) autoionization resonances, and the Ne $2s^22p^6 \rightarrow 2p^43s3p$ doubly excited resonance. The photoelectron angular distribution parameters β have been parameterized according to the prescription of Kabachnik and Sazhina (1) and have been compared with our R-matrix calculations and those of Taylor (2). We have also analyzed the cross section shape for the resonances, which agree well with previous experiments done for the lower resonances.

INTRODUCTION

Theoretical work has shown that the electron angular distribution (3) and the shape of the autoionization resonances (4) are crucial to the understanding of certain types of electron-electron correlation. Autoionization resonances in Ne (Ar) result from the decay of the excited discrete state Ne^* $2s2p^6$ np (Ar* $3s3p^6$ np) into the continuum state Ne^+ $2s^22p^5$ + e^- (ks,kd) (Ar^+ $3s^23p^5$ + e^- (ks,kd)). Since the continuum can also be reached by direct photoionization, both paths add coherently, giving rise to interferences that produce the characteristic Beutler-Fano line shape (5). Determinations of the absorption spectrum of Ne (6) and the angular distribution anisotropy parameter β have been reported (7-9) with low resolution and with no detailed analysis. Detailed measurements of the shape of the Ar autoionization series have been conducted previously (10-13). Also, angular distributions for the first three autoionization resonances have been measured (14-16).

In this work, we report on quantitative angle-resolved electron spectrometry studies of (a) the Ne $2s^2 2p^6 \rightarrow 2s2p^6$ np (n=3-5) autoionizing resonances and the $2s^22p^6 \rightarrow 2p^43s3p$ doubly excited resonance, (b) the Ar $3s^23p^6 \rightarrow 3s3p^6$ np (n=4-9) autoionization resonances and (c) extended R-matrix calculations of the angular-distribution parameters for both Ne and Ar measurements. Our results are compared with previous theoretical work by Taylor (2).

EXPERIMENTAL PROCEDURE.

Two components of instrumentation were needed in these photoemission measurements: high-resolution and high-brightness beamline and an angle-resolved technique to measure the angular distribution of the emitted photoelectrons. The experiment was performed in two-bunch operation at the Advanced Light Source (ALS) at Lawrence Berkeley Laboratory. The experiment was carried out with two (three in the case of the Ne experiment) time-of-flight (TOF) spectrometers (17) to record spectra simultaneously (while the photon beam was scanned) and to allow *both* partial-cross section and angular-distribution measurements. The two TOFs are mounted perpendicularly to the incoming photon beam. Details of the experimental apparatus are described in another work (18). The monochromator bandpass was 19 meV in the resonance region of Ne and 6 meV in the case of Ar, and the scans were normalized to the incident photon flux. We determined the degree of linear polarization P_1 of the monochromatic light in the resonance region to be P_1=0.99. The tilt of the polarization plane with respect to the experimental system was measured to be zero.

DATA ANALYSIS AND R-MATRIX CALCULATIONS

We have used a parameterization of the variation in β over autoionizing resonances introduced by Kabachnik and Sazhina (1). It is based on the Fano (19) parameterization for the total photoionization cross section given by Taylor (2) where A, B, C, and ϵ are listed in ref.1 and where the quantities X,Y, Z are considered as free parameters in the fit to the β data. β is expressed as follows:

$$\beta = \frac{X\epsilon^2 + Y\epsilon + Z}{A\epsilon^2 + B\epsilon + C} . \tag{1}$$

We used a sophisticated configuration interaction (CI) basis to perform R-matrix calculations to compute total cross sections and angular distribution parameters. These calculations are an extension of Taylor's calculations (2). In

CP392, *Application of Accelerators in Research and Industry*, edited by J. L. Duggan and I. L. Morgan
AIP Press, New York © 1997.

the case of neon, we investigated the resonance contributions due to capture into the $2s2p^6$ np Rydberg states, and the low -lying $2s^22p^4$ 3lnl' doubly-excited autoionizing states. Thus, the required Ne^+ targets states included the $2s2p^6$ and the $2s^22p^4$ 3l (l=0,1) states in addition to the $2s^2 2p^5$ ground state. The most important correlation effect was found to be the $2p^2 \rightarrow 3\bar{d}^2$ two-particle, two hole interaction. For the similar process in Ar the doubly-excited $3s^23p^44lnl'$ resonances lie well above the region of interest, namely the energy range containing the $3s3p^6np$ Rydberg series, so they do not need to be considered. However, due to the greater number of electrons compared to Ne, the similar two particle two hole type correlations were not enough to converge the continuum wavefunction description.

RESULTS AND DISCUSSION

The case of Neon

The aim of this work is to extract the angular distribution parameters. We have done this by fitting our total cross section data using the parameterization of Kabachnik and Sazhina (1). The results for the total photoionization cross-section are shown in Fig. 1 for one double-excitation and three single-excitation autoionization resonances.

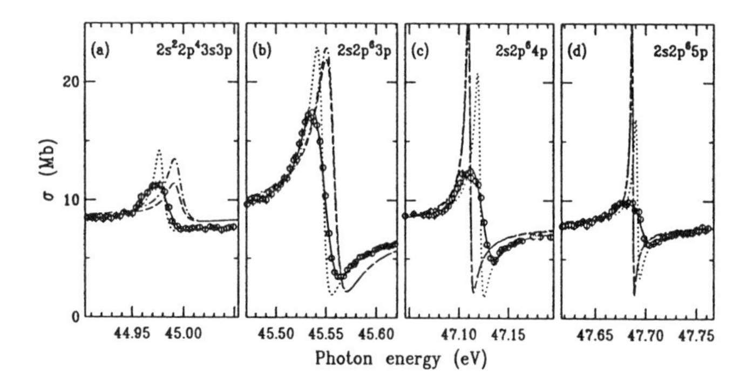

FIGURE 1. Cross section behavior of the Ne $2s^22p^6$ (a) $2s^22p^43s3p$ and (b-d) $2s2p^6np$ (n=3-5) autoionization resonances. The circles are the experimental data taken at the magic angle. The solid lines represent a Fano curve fit to our data. The dotted lines are created using the Fano parameters without bandpass; the dot-dashed and the dashed lines are our theoretical R-matrix calculations in length and velocity form, respectively.

The data are plotted along with a least square fit (represented by the solid line going through the data), as well as our R-matrix calculations. In order to take into account the monochromator bandpass, we convoluted the line shape calculated on the basis of a Fano shape with the slit function before comparing it to the normalized data. The slit function was a normalized Gaussian curve with FWHM equal to the measured bandpass. The fitting parameters, the width Γ, the line profile q and the correlation coefficient ρ^2 for the photoionization cross section in the region of the first 4 resonances were extracted and can be found in Berrah et al. (20)

The pronounced behavior of the angular distribution parameter β as a function of the photon energy is shown in Fig. 2 for the $2s2p^6$ (^2S) np resonances, the strong $2p^43s3p$ resonances and the weak $2p^4nln'l'$ resonances.

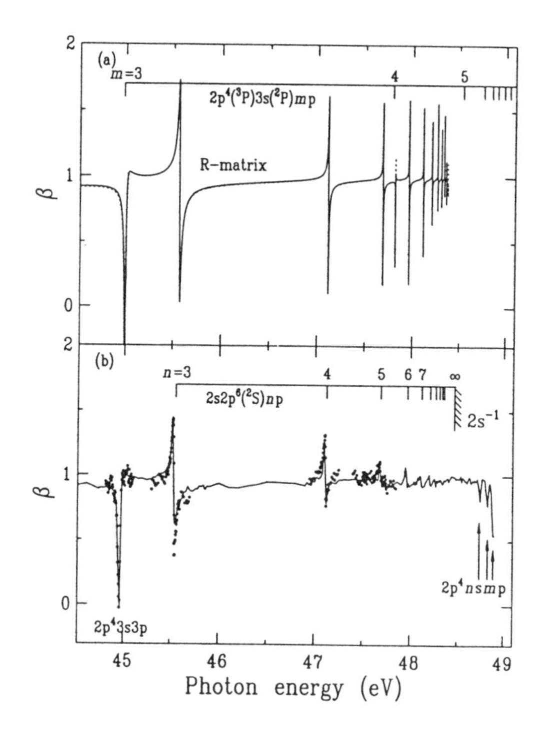

FIGURE 2. (a) Theoretical calculations and (b) experimental results of the angular distribution anisotropy parameter β for the Ne 2p line in the region of the autoionization resonances below the 2s threshold. In (a), the dotted line and the solid lines, which are nearly identical on this scale, show the length and the velocity forms of our calculation, respectively. In (b), the solid line represents the β parameter as obtained from our scans.

In Fig. 2, angular distribution data are shown for the $2p^2 \rightarrow 3s3p$ resonance as well as the $2s \rightarrow np$ (n=3-5) resonances and are compared with a fit curve. As can be seen from

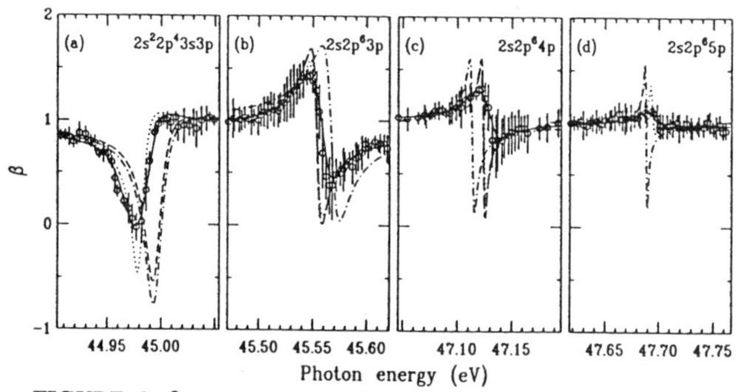

FIGURE 3. β parameter as a function of photon energy across the resonances of the Ne $2s^22p \rightarrow$(a) $2s^22p^43s3p$ and (b-d) $2s2p^6np$ (n=3-5) autoionization resonances. The circles are our experimental data; the solid lines represent a curve fit to our data using the parametrization;the dot-dashed (in (a-d)) and the short-dashed (only in (a)) lines are our theoretical R-matrix calculations in length and velocity forms, respectively; the long dashed curves in (b) and (c) which is on top of our deconvoluted results (dotted lines) are taken from ref (2) and shifted to the resonance position.

Fig. 2, the parameterization of Kabachnik and Sazhina (1) gives a very good fit to the experimental angular distribution data. The angular distribution is also compared in Fig. 3 with the never experimentally tested R-matrix results of Taylor (2) and our R-matrix results.

From the figure we note that the entire Rydberg series $2s2p^6np$ generated by our calculations lines up well with the experimental values, indicating that our computed quantum defect μ is quite reasonable. Also, the calculated widths of these resonances, the degree of interference with the background, and the shape of the background itself all show good agreement with experiment. We also note good agreement with Taylor's (2) calculation which we shifted to match the data since he didn't give the energy positions of the lines. In all cases, the dotted lines are deconvoluted fit curves. As can be seen from the figure, agreement between the data and both theories is excellent. The parameters X,Y and Z produced by our fit and our calculations are listed in Table 1 along with the results from Taylor (2) available only for n=3-4.

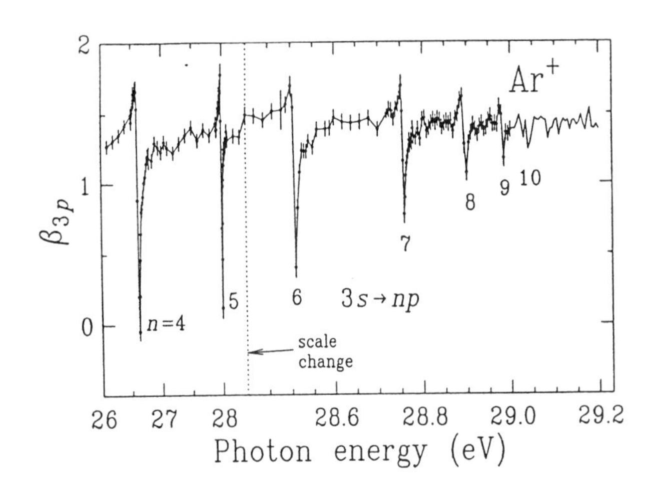

FIGURE 4. Angular distribution anisotropy parameter ß measurements of the Ar $3s^23p^6 \rightarrow 3s3p^6$ np autoionization resonances

TABLE 1: Angular distribution profile parameters for the Ne $2s^22p^6 \rightarrow 2s^22p^43s3p$ and $2s2p^6np$ (n=3-5) autoionization resonances compared with theoretical calculations.

	Method	$2s^22p^43s3p$			$2s2p^63p$		
		X	Y	Z	X	Y	Z
Taylor (1977)	CI (L)				-0.35	1.23	-1.27
	CI (V)				-0.26	0.96	-0.84
present results	theo. (L)	-0.28	-0.14	0.35	-0.28	1.06	-1.02
	theo. (V)	-0.28	-0.14	0.21	-0.28	1.12	-1.17
	exp.	-0.291(1)	-0.22(1)	0.23(3)	-0.285(1)	1.08(6)	-1.06(4)
	Method	$2s2p^64p$			$2s2p^65p$		
		X	Y	Z	X	Y	Z
Taylor (1977)	CI (L)	-0.33	1.33	-1.37			
	CI (V)	-0.29	1.04	-0.93			
present results	theo. (L)	-0.27	1.14	-1.23	-0.27	1.16	-1.32
	theo. (V)	-0.27	1.12	-0.97	-0.27	1.14	-1.28
	exp.	-0.312(1)	1.02(15)	-0.92(10)	0.301(1)	0.7(3)	-0.8(3)

The Case of Argon

We have fit both the total photoionization cross section data and the angular distribution parameter as in the neon case. The variation of ß as a function of photon energy is shown in Fig. 4. Comparison of the n=4,5 resonances with the CI calculation of Taylor (2) is shown in Fig. 5 (a),(b). It also shows the comparison of the resonances with our R-matrix calculation.

As can be seen, our calculation appears shifted to higher energies with respect to the measurements for the n=4 resonance but this shift almost vanishes for higher resonances. The parameters X, Y, and Z produced by our fit are listed in Table 2 for n=4 to n=9 along with the results from Taylor (2). From Table 2, one can note the excellent agreement between the data and the CI calculations. It is clear that the CI calculation is indeed a better model for the data because, in the case of n=4, the single configuration calculation did not even predict the correct sign of the Y parameter. We also would like to point out that our R-matrix calculations tried, unlike others, to obtain the energies position. The result is fair for the low resonances but agrees better with the measurements for the higher resonances.

CONCLUSION

We have observed the Ne 2s→np (n=3-5) and the Ne $2p^2 \rightarrow nln'l'$ resonances. The analysis of the differential cross-section resonances agree well with previous measurements (6). The angular distribution shows pronounced resonances, more pronounced than in previous measurements. Comparisons of the extracted parameters with Taylor's results (2) and with our R-matrix results are very good. We have also made detailed angle-resolved measurements of the Ar $3s^23p^6 \rightarrow 3s3p^6$ np autoionization resonances as well as R-matrix calculations. The comparison of our results with our R-matrix calculations and those of Taylor (2) is found to be excellent.

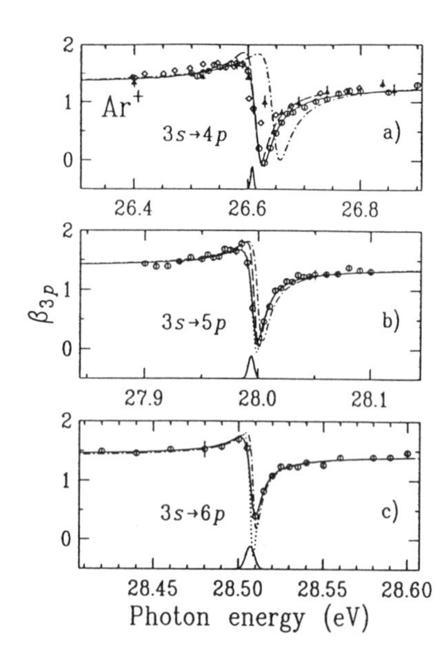

FIGURE 5. Angular distribution of the Ar $3s^23p^6$ - $3s3p^6$ np autoionization resonances for n= 4,5 and 6 fitted to Kabachnik and Sazhina (1976) model . They are compared to the calculation of Taylor (1977) for n=4 and n=5 (long dashed lines) and to our calculations (dot-dashed lines) for n=6. The dotted lines are deconvoluted fits. The triangles are data from Svensson et al (1987) and the diamonds are taken from Codling et al (1980). The Gaussian curve at the resonance position represents a monochromator bandpass of 6 meV.

REFERENCES

1. Kabachnik, N. M. and Sazhina, I. P., *J. Phys. B: At. Mol. Opt. Phys* **9**, 1681 (1976).
2. Taylor, K. T., *J. Phys. B: At. Mol.* Opt. **10**, L699 (1977).
3. Amusia, M. Ya., *Atomic photoeffect* (plenum, New York) (1990).
4. Feshbach, H., *Ann. Phys.* **19**, 287(1962).
5. Beutler, H., *Z. Phys.* **93**, 177(1935).
6. Codling, K., Madden, R. P., and Ederer, D. L., *Phys. Rev.* **155**, 26 (1967).
7. Schmidt, V., *Z. Phys. D* **2**, 275-283 (1986).
8. Becker, U. and Wehlitz, R., *J. Elect. Spectrosc. Relat. Phenom.* **67**, 341-361 (1993).
9. Hall, R. I., Dawber, G., Ellis, K., Zubek, M., Avaldi, L. and King, G. C., *J. Phys. B: At. Mol. Opt. Phys.* **24**, 4133 (1991).
10. Madden, R. P., Ederer, D. L., and Codling, K ., *Phys. Rev.* **177**, 136 (1969).
11. Baig, M. A. and Ohno, M., *Z. Phys. D* **3**, 369-373 (1986).
12. Sorensen, S. L., Åberg, T., Tulkki, J., Rachlew-Källne, E., Sundström, G. and Kirm, M. *Phys. Rev. A* **50**, 1218-1230 (1994).
13. Wu, S. L., Zhong, Z. P., Feng, R. F., Xing, S. L., Yang, B. X. and Xu, K. Z., *Phys. Rev A* **51** 4494-4500 (1995).
14. Codling, K. et al. *J. Phys. B: At. Mol. Phys.* **13** L693-L697 (1980).
15. Svensson, A., Krause, M.O., and Carlson, T. A., *J. Phys. B: At.Mol. Phys.* **20**, L271 (1987).
16. van der Meulen, P., Caldwell, C. D., Whitfield, S. B. and De Lange, C. A., *Phys. Rev. A* **46** 2468-2485 (1992).
17. Becker, U. and Shirley, D. A., *Phys. Scrip.* **T31**, 56 (1990).
18. Langer, B., Farhat, A., Nessar, B., Berrah, N., Hemmers, O. and Bozek, J., *" Resonant Raman effect in the sudy of Xe" in Proc. Fourth Workshop on Atomic Physics with hard X-Rays from High Brilliance Synchrotron Light Sources* (1996) (in press).
19. Fano, U., 1961 *Phys. Rev.* **124**, 1866 (1961).
20. Berrah, N., Langer, B., Farhat, A., "High Resolution Excitation and Photoionization using synchrotron Radiation from the Advanced Light Source", in Proceedings of the US-Indo workshop, 1996.

TABLE 2. Comparison of the fitted (solid line through the data), and calculated angular-distribution parameters for the $3s3p^6$ 4p and 5p $^1P^o$ resonances in Ar.

	Method	$3s \rightarrow 4p$			$3s \rightarrow 5p$			$3s \rightarrow 6p$		
		X	Y	Z	X	Y	Z	X	Y	Z
Taylor (1978)	SC(L)	−1.53	−2.88	−1.21						
	SC(V)	−0.81	−0.81	−0.09						
	CI (L)	-1.79	1.85	-0.44	-1.77	1.33	-0.14			
	CI (V)	-1.47	1.44	-0.31	-1.46	1.01	-0.06			
present results	theory (L)	-1.73	1.99	-0.56	-1.67	1.55	-0.41	-1.66	1.40	-0.34
	theory (V)	-1.41	1.73	-0.54	-1.36	1.35	-0.39	1.35	1.22	-0.32
	experiment	-1.567(12)	1.49(2)	-0.322(12)	-1.483(11)	1.00(2)	-0.158(13)	-1.46(2)	0.95(4)	-0.10(2)

	Method	$3s \rightarrow 7p$			$3s \rightarrow 8p$			$3s \rightarrow 9p$		
		X	Y	Z	X	Y	Z	X	Y	Z
present results	theory (L)	-1.65	1.36	-0.35	-1.64	1.22	-0.26	-1.63	1.29	-0.33
	theory (V)	-1.34	1.19	-0.33	-1.33	1.08	-0.25	-1.32	1.13	-0.31
	experiment	-1.507(10)	0.97(4)	-0.15(3)	-1.489(14)	0.79(7)	-0.13(6)	-1.49(2)	0.56(9)	-0.46(8)

ACKNOWLEDGMENTS

This work was supported by the US Department of Energy, Office of Basic Energy Science, Division of Chemical Science under contract No. DE-FG02-92ER14299. B.L. and R. W. are indebted to the Alexander von Humboldt Foundation for partial support.

ANGLE RESOLVED RESONANT RAMAN AUGER SPECTROSCOPY OF THE XE $4d \rightarrow 6p$ TRANSITION

B. Langer[1], A. Farhat[1], B. Nessar[1], N. Berrah[1], O. Hemmers[2], J. D. Bozek[3]

[1]*Physics Department, Western Michigan University, Kalamazoo, MI 49008*
[2]*Chemistry Department, University of Nevada, Las Vegas, NV 89154-4003*
[3]*Lawrence Berkeley National Laboratory, Advanced Light Source, Mail Stop 2-400, Berkeley, CA 94720*

We studied the angular distributions and decay rates of the Xe $4d_{5/2} \rightarrow 6p$ resonant Auger lines using the high resolution and high flux of undulator beamline 9.0.1 at the Advanced Light Source. The electron spectra were recorded by two time-of-flight (TOF) spectrometers with energy resolutions as low as to 43 meV. This made it possible to determine the angular distribution parameters β of almost all possible final ionic $5p^4(^3P,^1D,^1S)6p$ states. A variation of the β parameter in an energy scan over the resonance shows evidence for a possible interference of the resonant with the nonresonant path way to the $5p^4(^3P)6p(^2P_{3/2})$ final state.

INTRODUCTION

The resonant Auger decay of the Xe $4d_{5/2} \rightarrow 6p$ resonance has been studied for almost two decades since the first spectrum was reported by Eberhardt *et al.* in 1978 (1). Their work was followed by other experimental and theoretical studies (2-7). It was more than a decade after the first observation that measurements on the angular distribution were performed by Carlson *et al.* (8), who found anomalously negative β-values in the decay spectrum. Such behavior was first explained theoretically for the decay of the Ar $2p \rightarrow 4s$ resonance by Cooper (9), who applied angular momentum transfer theory, treating the resonant decay as a single-step process. Kämmerling *et al.* (10) and Becker (11) compared resonant Auger and normal Auger angular distributions experimentally and theoretically treating the decay as a two step process with a spectator electron. These experimental studies were limited by the low resolution of the photon sources as well as of the electron spectrometers, making it difficult to compare the results with the various theoretical calculations (9,10,12-15).

Recently, however, the development of new synchrotron sources and high resolution monochromators in combination with high-resolution electron spectrometers has made it possible to study the energy positions and intensities of the peaks in the Xe $4d_{5/2}6p \rightarrow 5p^46p$ decay spectrum with a resolution better than the natural linewidth (106 meV (16)) of the $4d$ inner shell hole by utilizing the Auger resonant Raman effect (17,18).

A xenon atom in the ground state has a filled $4d$ inner shell as well as filled $5s$ and $5p$ valence shells. After absorbing a photon with sufficient energy, an electron may be excited out of the $4d$ shell (actually we studied the $4d_{5/2}$ component) into an unoccupied Rydberg state (here $6p$), which then can decay into one of several possible final ionic states. Figure 1 schematically shows the excitation and

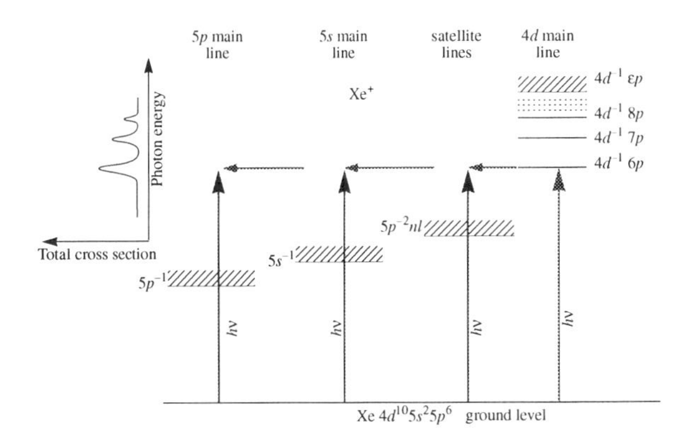

FIGURE 1. Schematic picture of the Xe $4d \rightarrow 6p$ autoionization resonances.

decay into the $5p^{-1}$ and $5s^{-1}$ final ionic states (main lines) as well as the decay into the $5p^{-2}nl$ states (satellite lines). Note, that all of these final ionic states can also be reached via direct photoionization which could lead to interference effects. The cross section will only be little affected due to the relative large strength of the resonance compared to the nonresonant photoionization. However, the angular distribution parameter β is a much more sensitive probe for such interferences since it includes also the phases of the outgoing electron waves.

The angular distribution photo- and Auger electrons is, in general, not spherically symmetric. For linearly polarized photons and randomly oriented targets the differential cross section $d\sigma/d\Omega$ in the dipole approximation can be expressed

CP392, *Application of Accelerators in Research and Industry*, edited by J. L. Duggan and I. L. Morgan
AIP Press, New York © 1997.

by a 2nd order Legendre polynomial (19). If the linear polarization of the incident light is less than 100% the differential cross section can be described by (20):

$$\frac{d\sigma}{d\Omega} = \frac{\sigma}{4\pi}\left[1 + \frac{\beta}{4}(1 + 3P_1\cos 2\Theta)\right] \qquad (1)$$

where P_1 is the degree of linear polarization (here: $P_1 = 0.991(2)$) and Θ is the angle between the polarization and the direction of the emitted electrons. Thus, only a single parameter β is necessary to describe the angular distribution. $\beta = 0$ describes a spherically symmetric distribution of the emitted electrons, whereas positive β values stand for emission preferably in the direction of the electric field vector of the incident light. Negative β values describe emission patterns where the electrons are emitted preferably perpendicular to the electric field vector.

In order to determine the angular distributions experimentally it is, in general, sufficient to measure the differential cross sections at two different angles, although more angles will always give more reliable results.

EXPERIMENTAL SETUP

Our electron spectra were recorded using time of flight (TOF) electron analyzers. Figure 2 shows one of the analyzers. Electrons created in the source volume fly straight to the micro channel plates (MCPs) where they are detected. In order to measure the flight time of the electrons we have to use a pulsed light source i. e. the synchrotron ring has to be operated in timing mode where there are only few bunches (here: two) of electrons in the ring. Electrons of all kinetic energies are detected simultaneously which allows measurements independent of the fluctuations of the target density or the photon flux. The kinetic energy resolution of our analyzers is about 1% of the kinetic electron energy. In order to improve the energy resolution we can apply a retarding voltage at the retarding cage and the inner nose.Two of these TOF analyzers were mounted to a rotatable chamber. In normal position one analyzer takes spectra at the magic angle $\Theta_m = 54.7°$ and the other analyzer records spectra at 0° with respect to the electric field vector of the synchrotron light. This setup is similar to the one used by the group of U. Becker in Berlin (21,22) which itself was an improvement of the original TOF analyzers developed in the group of D. A. Shirley in Berkeley (23)

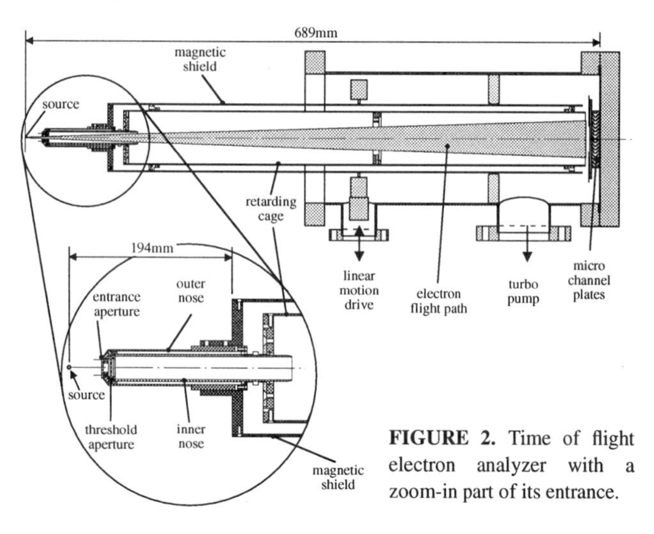

FIGURE 2. Time of flight electron analyzer with a zoom-in part of its entrance.

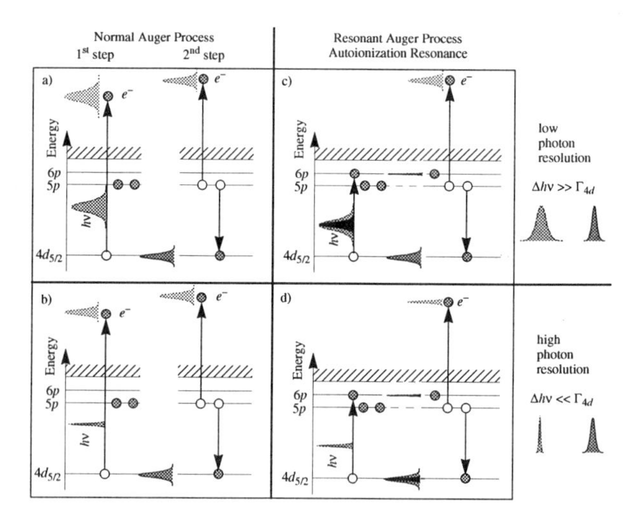

FIGURE 3. Comparison of normal (a,b) and resonant (c,d) Auger processes with low (a,c) and high (b,d) photon resolution. The Gaussian curves represent the natural line widths, the photon bandpass, and the line width of the outgoing electrons, respectively.

The experiment was performed at the Advanced Light Source (ALS) synchrotron radiation facility of the Lawrence Berkeley National Laboratory (LBNL) using double bunch operation. Xenon atoms were ionized by monochromatic synchrotron radiation from an 8-cm, 55-period undulator and spherical grating monochromator on beamline 9.0.1. We used a 925 lines/mm grating with a 100 µm entrance slit and a 60 µm exit slit providing us with a photon resolution of about 15 meV at the energy of the Xe $4d_{5/2} \rightarrow 6p$ resonance at 65.11 eV.

NORMAL AUGER VS. RESONANT AUGER/AUTOIONIZATION RESONANCE

In most cases the normal Auger decay can be considered as a two step process (Fig. 3a,b). In the first step an inner shell hole is created. In the second step the inner shell hole is filled by a valence electron and the gained energy is transferred to another valence electron which leaves the ion. In contrast to the simultaneous double ionization (shakeoff) where the energy of the two outgoing electrons is shared arbitrarily, the normal Auger process leads to discrete lines in an electron spectrum.

In the resonant Auger process, which can also be referred to as an autoionization resonance, the inner shell hole is created by resonantly exciting an inner shell electron into an unoccupied Rydberg level. The decay of this resonance can

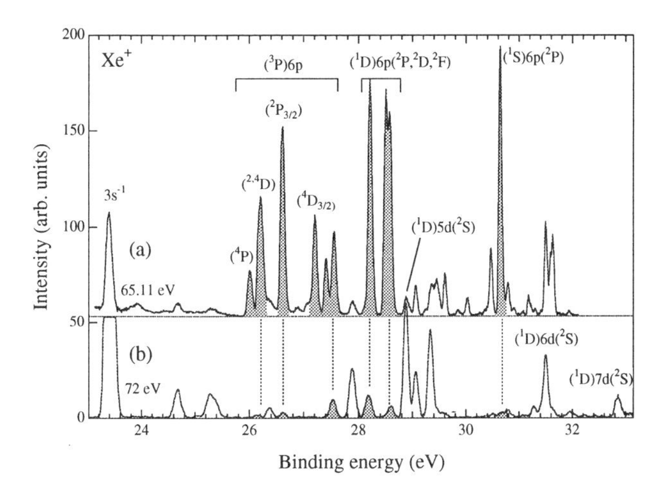

FIGURE 4. (a) Xenon satellite spectrum taken on top of the $4d_{5/2}\rightarrow 6p$ resonance with 30 V retarding. (b) The same spectrum well above the resonances at 72 eV (from Ref. (27)).

either involve the excited $6p$ electron (participator decay) or leave the $6p$ electron unaffected (Fig. 3c and d, spectator decay). In both cases the final ionic state is singly charged. The participator decay leaves the ion in a final state with one hole in the outer shell which is the state of a main line whereas the spectator decay results in a double hole in the outer shell and an excited electron which is the final ionic state of a satellite line. In this paper we focus on the spectator decay which is dominant in the Xe $4p \rightarrow 6p$ case (6).

Figure 3 shows schematically the energy distribution of electrons resulting from normal (Fig. 3a and b) and from resonant Auger processes (Fig. 3c and d) when photons with different energy resolutions are used to respectively ionize and excite the Xe atom. In the upper part (Fig. 3a and c) we have a low photon energy resolution with a bandpass much larger than the natural linewidth of the inner shell hole. In contrast, the photon bandpass in the lower part (Fig. 3b and d) is much smaller than the inner shell linewidth. The linewidth of the outgoing electrons in the cases shown in Fig. 3a,b, and c is limited by the natural life time width of the inner shell hole due to relaxation respectively the photon bandpass. In the resonant case with small photon bandpass (Fig. 3d) there is no possibility to relax before the decay due to energy conservation. Thus, the narrow linewidth is carried away by the outgoing electron. The outgoing electron then shows dispersion when the photon energy is changed. Again, the direct photoionization is also possible. In contrast to the case in Fig. 3c, here the resonant decay lines and the direct photolines are indistinguishable.

The process described in Fig. 3d has been referred to as Auger resonant Raman scattering (24,25) since it has the following features (26): i) linear dispersion of the outgoing electron, ii) intensity enhancement when tuning the photon energy, and iii) linewidth solely depends on the photon bandpass (narrowing). Note, in order to take advantage of the narrow linewidth, the electron spectrometer must have a comparable resolving power.

RESULTS

Figure 4 compares photoelectron spectra taken at the magic angle "on" the Xe $4d_{5/2}\rightarrow 6p$ resonance with spectra taken well above that resonance at $h\nu$ = 72 eV (27). Note, that the resonantly enhanced satellite lines differ from those most prominent $(^1D)nd(^2S)$ lines in the nonresonant spectrum. Nevertheless, there is some nonresonant intensity for the $(L)6p(L')$ lines which are grayed in Fig. 4. The relative intensities and β parameters for these lines on resonance can be found in Ref. (28).

In order to map out the resonance, we took several photoelectron spectra in the resonance region. Figure 5 shows the part of the spectra in the kinetic energy range from 36.0 to 37.3 eV. The linear dispersion of the outgoing electron can clearly be seen since the individual spectra are only shifted vertically. Figure 5 also shows that there are lines which are almost unaffected by the resonance whereas the lines in the center show clearly the resonant enhancement. These spectra, taken under the magic angle, were simultaneously recorded with spectra under 0° with respect to the polarization of the synchrotron light. Figure 6 shows the results of the resonance profiles in the intensity as well as in the angular distribution for the $(^3P)6p(^2P_{3/2})$ line and two central line groups from Fig. 5. The intensities for these selected lines as a function of the incident photon energy show Lorentzian profiles sitting on top of a small non-resonant background (Fig. 6b). The width of the profile are in good agreement with recently published data (16). The β parameters (Fig. 6a) shows strong fluctuations in the low energy range which might be artifacts caused by instable beam conditions at that time. For the other energies, the angular distributions of the line groups $(^1D)6p(^2P_{1/2},{}^2D)$ and $(^1D)6p(^2F_{7/2},{}^2P_{3/2})$ are essentially flat or slowly varying as can be expected since the resonant intensities are much stronger than the non-resonant satellite intensities. The β parameter for the $(^3P)6p(^2P_{3/2})$ line, however, shows a step (arrow in Fig. 6a) in the vicinity of the maximum of the res-

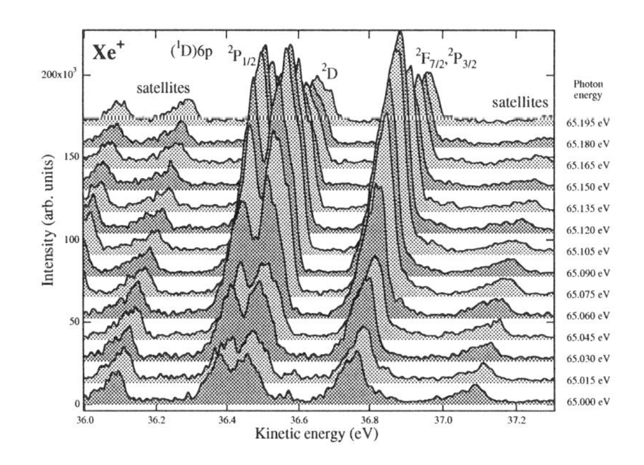

FIGURE 5. Scan over the Xe $4d_{5/2}\rightarrow 6p$ resonance. The spectra were recorded using a 32 V retardation potential

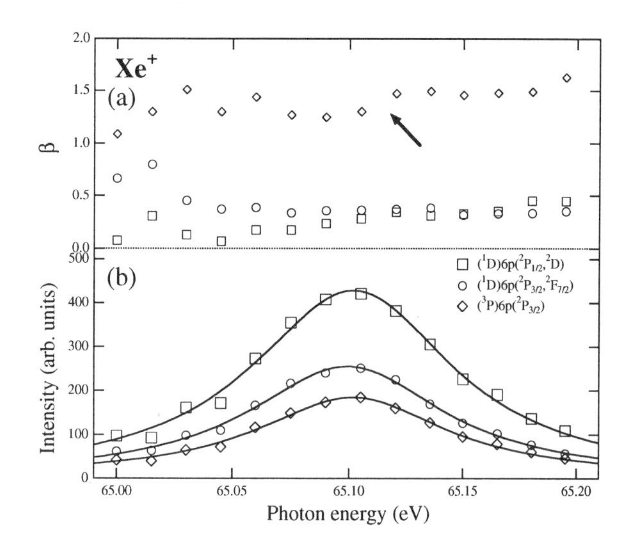

FIGURE 6. Resonance profiles of (a) the β parameters and (b) the intensities of different lines or line groups. The solid curves in (b) are Lorentzian profiles fitted to our data. The linewidth was measured to be 105 meV which is good agreement with recent published measurement for the natural linewidth of the Xe $4d_{5/2} \to 6p$ resonance (16).

onance which might indicate an interference of the resonant with the non-resonant path way to the same final ionic state.

CONCLUSIONS

Time-of-flight electron spectroscopy in combination with tunable synchrotron radiation provided by a 3rd generation storage ring has been used to study the angular distributions of the Xe spectator lines following Xe $4d_{5/2}{\to}6p$ excitation. The high resolving power of the monochromator as well as of our electron analyzers allowed us to determine the β parameters of almost all possible final ionic $5p^4(^3P,^1D,^1S)6p$ states with many of them being much closer than the natural linewidth of the $4d$ inner shell hole. At least for the $(^3P)6p(^2P_{3/2})$ line, the resonance profile of the angular distribution parameter indicates an interference of the resonant with the non-resonant path way to the same final ionic state.

ACKNOWLEDGMENTS

We wish to thank the ALS for providing an excellent source of photons. BL is indebted to the Alexander von Humboldt Foundation for partial financial support in the frame of the Feodor Lynen program. This work was supported by the US Department of Energy, Office of Basic Energy Science, Division of Chemical Science, under contract No. DE-FG02-92ER14299.

REFERENCES

1. Eberhardt, W., Kalkhoffen, G., and Kunz, C., *Phys. Rev. Lett.* **41**, 156-159 (1978).
2. Schmidt, V., Krummacher, S., Wuilleumier, F., and Dhez, P., *Phys. Rev. A* **24**, 1803-1811 (1981).
3. Aksela, H., Aksela, S., Bancroft, G. M., Tan, K. H., and Pulkkinen, H., *Phys. Rev. A* **33**, 3867-3875 (1986).
4. Becker, U., Prescher, T., Schmidt, E., Sonntag, B., and Wetzel, H.-E., *Phys. Rev. A* **33**, 3891-3899 (1986).
5. Heimann, P. A., Lindle, D. W., Ferret, T. A., Liu, S. H., Medhurst, L. J., Piancastelli, M. N., Shirley, D. A., Becker, U., Kerkhoff, H.-G., Langer, B., Szostak, D., and Wehlitz, R., *J. Phys. B: At. Mol. Phys.* **20**, 5005 (1987).
6. Becker, U., Szostak, D., Kupsch, M., Kerkhoff, H.-G., Langer, B., and Wehlitz, R., *J. Phys. B: At. Mol. Phys.* **22**, 749-762 (1989).
7. Caldwell, C. D., in "Alignment in low-energy photoionization" in *Proceedings of the 15th International Conference on X-Ray and Inner-Shell Processes*, edited by Carlson, T. A., Krause, M. O., and Manson, S. T., (American Institute of Physics, Knoxville, 1990), pp. 685-695.
8. Carlson, T. A., Mullins, D. R., Beall, C. E., Yates, B. W., Taylor, J. W., Lindle, D. W., and Grimm, F. A., *Phys. Rev. A* **39**, 1170-1185 (1989).
9. Cooper, J. W., *Phys. Rev. A* **39**, 3714-3716 (1989).
10. Kämmerling, B., Krässig, B., and Schmidt, V., *J. Phys. B: At. Mol. Phys.* **23**, 4487-4503 (1990).
11. Becker, U., "Synchrotron radiation experiments on atoms and molecules" in *Proceedings of the 16th International Conference on the Physics of Electronic and Atomic Collisions*, edited by Dalgarno, A., Freund, R. S., Koch, P. M., Lubell, M. S. and Lucatorto, T. B. (American Institute of Physics, New York, 1990), pp. 162-175.
12. U. Hergenhahn, N. M. Kabachnik, and B. Lohmann, *J. Phys. B: At. Mol. Opt.* **24**, 4750 (1991).
13. Hergenhahn, U., Lohmann, B., Kabachnik, N. M., and Becker, U., *J. Phys. B: At. Mol. Opt.* **26**, L117-L121 (1993).
14. Chen, M. H., *Phys. Rev. A* **47**, 3733-3738 (1993).
15. Tulkki, J., Aksela, H., and Kabachnik, N. M., *Phys. Rev. A* **50**, 2366-2375 (1994).
16. Masui, S., Shigemasa, E., Yagishita, A., and Sellin, I. A., *J. Phys. B: At. Mol. Opt.* **28**, 4529-4536 (1995).
17. H. Aksela, S. Aksela, O.-P. Sairanen, A. Kivimäki, A. Naves de Brito, E. Nõmmiste, J. Tulkki, S. Svenson, A. Ausmees, and S. J. Osborne, *Phys. Rev. A* **49**, R4269 (1994).
18. H. Aksela, O.-P. Sairanen, S. Aksela, A. Kivimäki, A. Naves de Brito, E. Nõmmiste, J. Tulkki, S. Svenson, S. J. Osborne, and A. Ausmees, *Phys. Rev. A* **51**, 1291 (1995).
19. A. F. Starace, in *Corpuscles and Radiation in Matter* I, edited by W. Mehlhorn (Springer-Verlag, Berlin, 1982), pp. 1.
20. Derenbach and V. Schmidt, *J. Phys. B: At. Mol. Phys.* **17**, 83 (1984).
21. U. Becker, D. Szostak, H.-G. Kerkhoff, M. Kupsch, B. Langer, R. Wehlitz, A. Yagishita, and T. Hayaishi, *Phys. Rev. A* **39**, 3902 (1989).
22. U. Becker and R. Wehlitz, *Phys. Scr.* **41**, 127 (1992).
23. M. G. White, R. A. Rosenberg, G. Gabor, E. D. Poliakoff, G. Thornton, S. H. Southworth, and D. A. Shirley, *Rev. Sci. Instrum.* **50**, 1268 (1979).
24. Brown, G. S., Chen, M. H., and Crasemann, B., *Phys. Rev. Lett.* **45**, 1937 (1980).
25. Crasemann, B.,"Atomic and molecular physics with synchrotron radiation" in *Proceedings of the XVII International Conference on the Physics of Electronic and Atomic Collisions*, edited by I. E. McCarthy, W. R. MacGilliray and M. C. Standish (American Institute of Physics, Brisbane, Australia, 1992), pp. 69.
26. Åberg, T., "Radiative and radiationless resonant raman scattering by synchrotron radiation", in *Proceedings of the Conference on Atomic Physics at High Brilliance Synchrotron Sources*, edited by Berry, G., Cowan, P., and Gemmel, D. (Argonne National Laboratory, Argonne, 1994), pp. 167.
27. Whitfield, S. B., Langer, B., Viefhaus, J., Wehlitz, R., Berrah, N., Mahler, W., and Becker, U., *J. Phys. B: At. Mol. Phys.* **27**, L359 (1994).
28. Langer, B., Berrah, N., Farhat, A., Hemmers, O., and Bozek, J. D., *Phys. Rev. A* **53**, R1946 (1996).

NEGATIVE ION PRODUCTION IN SMALL ANGLE SCATTERING OF HIGHLY CHARGED IONS FROM THE (0001) SURFACE OF HIGHLY ORIENTED PYROLYTIC GRAPHITE

M. Reaves, Q.C. Kessel, E. Pollack, and W.W. Smith

Department of Physics and The Institute of Materials Science, The University of Connecticut, Storrs, CT 06269

M.A. Briere* and D.H. Schneider

Lawrence Livermore National Laboratory, Livermore CA 94550

Highly charged N, O, F, and S ions, chosen for their electron affinities, were extracted from the Lawrence Livermore National Laboratory's EBIT II. After collimation , these ions struck a target of highly oriented pyrolytic graphite (HOPG) at an incident angle of 1.0 degree. Those ions scattered by 2.35 degrees (1.35 degree with respect to the surface) were charge state analyzed and the predominant charge state fractions were determined. As might be expected, there is a tendency for the fraction of negative ions to increase with increasing electron affinity; however, the negative ion yield is also strongly dependent on the ion velocity. For example, for sulfur the negative ion yields measured range from 0.13 to 0.23 of the scattered ions while for fluorine the range was 0.35 to 0.40. A pronounced velocity dependence found for the S^- ions is described well by a Saha-Langmuir-type equation.

INTRODUCTION

The availability of highly charged ions with relatively low kinetic energies from electron cyclotron resonance (ECR) and electron beam ion trap (EBIT) ion sources has led to a number of unique experiments which provide new insight into ion-surface interactions. In 1990 Briand and co-workers provided evidence for the creation of "hollow atoms" during collisions of Ar^{17+} ions with a solid silver target, at normal incidence (1). They observed x-ray lines from highly excited states of argon with a high resolution crystal spectrometer and concluded that large numbers of electrons from the surface were being captured into the outer shells of the Ar ions, hence the term "hollow atoms." This picture has been refined as a result of theoretical work and a number of additional experiments (2-8). Burgdörfer and co-workers (2) present a picture in which the highly charged ion (HCI) approaches the conducting surface causing the surface electrons to congregate near the point of impact (forming the classical image charge). The image charge accelerates the ion toward the surface and as the ion gets closer, a large number of these electrons are drawn from the surface to the approaching ion. In the Burgdörfer classical over-the-barrier model, the image interaction decreases the height of the potential barrier between the HCI and the surface. At a critical distance the conduction band electrons from the Fermi level of the metal cross over the barrier, primarily into the outer shells of the incoming ion. Capture occurs preferentially to these outer levels because their energies are more likely to be in resonance with the electrons in the Fermi level. In this model, as relaxation takes place, x rays, such as those observed by Briand (1) and others (3-4) may be emitted while additional electron captures take place; however, although the ion may be neutralized, most of the electrons are loosely bound and stripped off when the ion enters the surface. Electrons emitted from the projectile both before and after entering the surface have been observed by Meyer and co-workers (5).

As Burgdorfer and co-workers note, because of the image charge acceleration of the incident ion towards the surface, the time for neutralization and relaxation of the atom or ion is limited. This problem can be partially overcome by using glancing angle collisions, in which the ions do not penetrate the surface, instead of using normal incidence. This is done in the present work and the result is a high degree of neutralization and the creation of a substantial number of negative ions for those elements which have positive

CP392, *Application of Accelerators in Research and Industry,* edited by J. L. Duggan and I. L. Morgan
AIP Press, New York © 1997.

electron affinities, *i.e.* for those atoms which have bound states capable of attracting an extra electron. Preliminary results of this experiment, utilizing S, O, and F ions were presented in 1994 (6), and Folkerts and co-workers have used glancing angle collisions to investigate the fast neutralization and charge equilibration of O^{q+} in grazing incidence collisions with surfaces of Au crystals (7). They also observed the creation of O^- ions and the fractions of all the charge states they observed agree well with calculations by Burgdörfer and co-workers (2). It is interesting to note that the model used for their calculation predicts nearly a 100 percent of the O to be in the negative ions state in the vicinity of the surface. If this is true, then the large fraction of neutral atoms leaving the interaction region must become neutral either through resonant ionization or autoionization of excited negative configurations on the outgoing path. In this context, it is interesting to note that we observed no doubly negative ions, nor did we observe negative N ions, which exist only in an excited state (a negative electron affinity equal to -0.07 eV)(9).

EXPERIMENT AND RESULTS

The EBIT and its external beam line for extracted ions are described elsewhere (10). In brief, HCIs are extracted from the plasma, accelerated to kinetic energies of the extracted charge state times the accelerating voltage, 7 kV, in the present experiment. The ions are then magnetically analyzed and directed onto the target. The target, the C(0001) plane of HOPG is placed at an angle so that its normal is 89 degrees with respect to the ion beam, i.e., the angle of incidence of the beam is 1 degree. Slits parallel to the plane define a scattering angle which was chosen to be 2.35 degrees with respect to the incident beam and this corresponds to an exit angle of 1.35 degrees for the scattered ions, measured with respect to the surface. Calculations using the Marlowe program (11) show such ions rarely penetrate the surface (12). The narrow angular distributions observed in this experiment also imply that most of the ions scatter as a result of specular reflection. The scattered ions and atoms pass through an electrostatic field to spatially separate the charge states and then strike a position-sensitive channel plate detector. A second channel plate detector is mounted parallel to the target, and each scattering event generates enough photons and secondary electrons to trigger it. These signals are used to gate the charge state detector, thus eliminating most of the background noise as well as signals from unwanted scattering events.

The target, highly oriented pyrolytic graphite (HOPG), was freshly cleaved within one hour of its mounting in the target chamber and the chamber's evacuation; it was heated by a quartz lamp before each experiment. Measurements of the profile of the cleaved surfaces show atomically flat terraces of tens of microns dimensions. During the experiments the vacuums were maintained at 10^{-8} Torr or better. Table 1 presents the results for a number of ions and energies together with the electron affinities for the corresponding negative ions. The negative ion percentages are plotted versus their electron affinities in Fig. 1. Also shown is a vertical bar representing the range of negative ion yields for glancing angle collisions measured by Folkerts and coworkers (7) for O ions undergoing surface channeling on a gold crystal; for these collisions the negative ion yields do not depend on the incident charge state. Figure 2 is a plot of the negative ion yields for the various charge states of S used in the present experiment versus the incident ion velocity, v, which for these collisions is a close approximation to the parallel velocity.

TABLE 1. Electron Affinities and Experimental Charge state Yields.

Incident Ion	Ion velocity	Electron Affinity[a]	X^-	X^0	X^+	X^{2+}
N^{6+}	0.24 a.u.	-0.07 eV	0.00	0.91	0.09	0.00
O^{7+}	0.35 a.u.	1.46 eV	0.25	0.63	0.11	0.01
F^{7+}	0.32 a.u.	3.04 eV	0.40	0.51	0.08	0.01
F^{9+}	0.37 a.u.		0.35	0.55	0.10	0.00
S^{7+}	0.25 a.u.	2.08 e.V	0.23	0.60	0.15	0.02
S^{9+}	0.28 a.u.		0.18	0.58	0.20	0.04
S^{13+}	0.34 a.u.		0.15	0.54	0.26	0.05
S^{15+}	0.36 a.u.		0.13	0.56	0.27	0.04

[a]From Ref. 8, the values represent the differences in total energies of the normal states of the atoms and their negative ions.

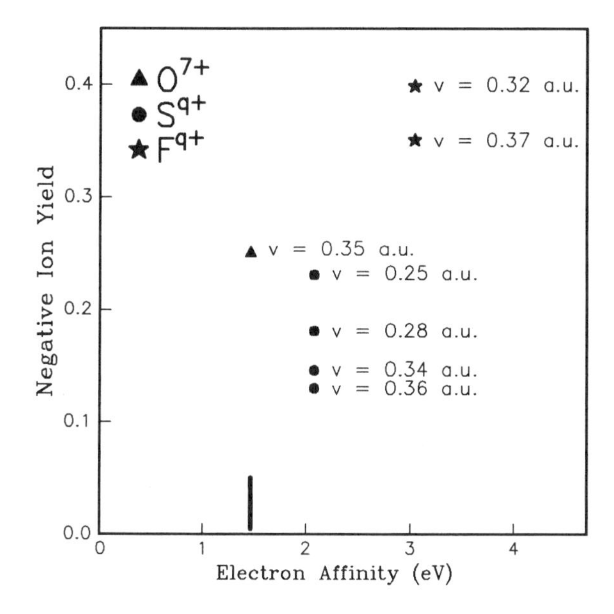

Figure 1. Negative ion percentages plotted versus electron affinity. The vertical bar for O^- production is from the work of Folkerts and coworkers (7) (7) and represents values for velocities from 0.1 to 0.5 a.u. The two sets of O data provide a strong indication that factors other than the electron affinity of the incident ion are important in determining the negative ion yield.

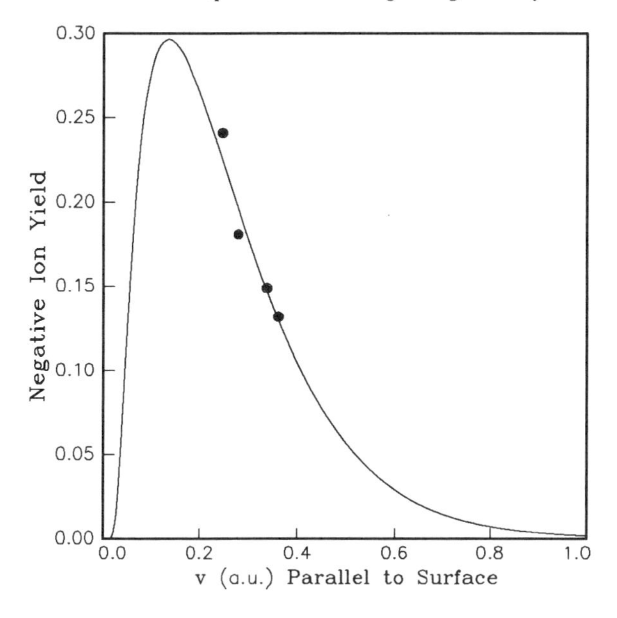

Figure 2. Negative ion percentages for S^- production plotted as a function of velocity. The statistical counting errors in the data do not exceed the size of the data points The solid curve is fit to the data following the procedure of Winter (14).

ANALYSIS OF THE DATA

Figure 2 shows the negative ion yield plotted versus the projectile velocity. The solid curve in Fig. 2 is, based on the Saha-Langmuir equation (13) and generated by the formulation given by Winter (14). According to Winter P^-, the yield for conversion of neutral atoms to negative ions, is given by

$$P^- \approx 1/\{1 + (g^-/g^+)\exp[(E_g(y_S) + v^2/2)/v_C v)]\}.$$

In general, E_g is the energy difference, expressed in atomic units, between the Fermi level and the atomic level, e.g., at infinite separation, E_g equals the work function minus the electron affinity. As the ion approaches the surface, E_g becomes smaller since the interaction with the surface draws the atomic level closer, in energy, to the Fermi level. The factors g^+ and g^- are introduced to take into account the spin degree of freedom and the degeneracy of the atomic level. For S the ratio g^-/g^+ has been taken to be 1/3. The energy gap $E_g(y_S)$ in the equation refers to its value at the distance y_S from the image plane defined in eq. (9) of Winter's article (14) as the atomic "freezing distance." For $y < y_S$, transition rates are larger and for $y > y_S$, the probability for transitions becomes negligible. The quantity v_C is a parameter used to fit the equation to the experimental results. To evaluate $E_g(y_S)$ and v_C, it is convenient to define

$$f(v) = (1/v_C)v^2/2 + E_g/v_C.$$

Figure 3 shows f(v) plotted versus $v^2/2$ for the S data.

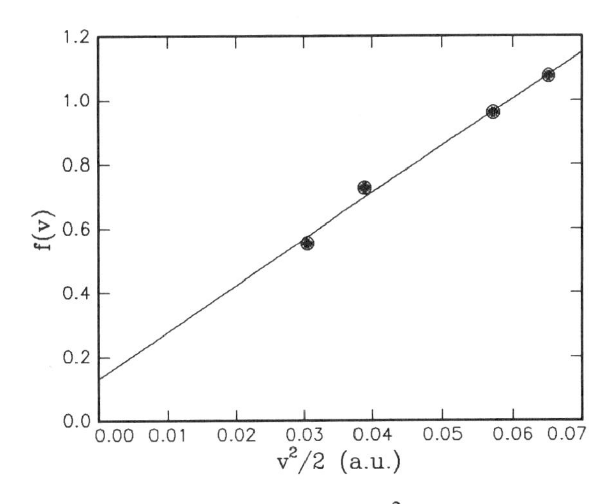

Figure 3. The function f(v) plotted versus $v^2/2$. The statistical counting errors in the data do not exceed the size of the data points.

The straight line in Fig. 3 and the fit of the curve to the data in Fig. 2 show that the data are consistent with Winter's description of the neutralization and creation of negative ions in these glancing angle collisions (14). In this model the image interaction of the ion with the conducting surface increases the affinity energies of negative ions. As a consequence, charge transfer from the surface to the ions can take place via resonant one-electron tunneling and this process should show a well defined dependence upon the component of velocity parallel to the surface. The shape and amplitude of the fitted curve in Fig. 2 are sensitive to the work function of the conducting solid, the electron affinity of the ion and the parallel velocity. In particular, it is sensitive to the energy gap between the Fermi-level and the atomic level, and this gap changes along the trajectory as the electron affinity increases. If this increase is sufficient to bring the atomic level into resonance with the Fermi level, electron transfer to an already neutralized atom is likely to occur.

The least squares fit shown in Fig. 3 gives a value of 0.25 eV for the energy gap at the distance of closest approach to the image plane, $E_g(y_s)$. Taking the photoelectric work function of C to be 4.81 eV and the electron affinity of S to be 2.08 eV, we have an energy gap of 2.73 eV at infinite separation. The value of 0.25 eV for the energy gap at the distance of closest approach implies an increase in the effective electron affinity to 4.81 - 0.25 = 4.56 eV or a change in electron affinity of 2.48 eV. This, in turn, implies a distance of closest approach to the image plane of about -2.7 a.u. (Ref. 14, eq. 1). As the image plane is likely to lie about 4 a.u. above the top layer of surface atoms, the apparent height for resonance transfer is approximately 1 to 2 a.u. or 0.5 to 1 Å above the surface.

ACKNOWLEDGMENTS

The authors are pleased to acknowledge very helpful conversations with Dr. Fred Meyer of Oak Ridge National Laboratory and Professor Yukap Hahn of the University of Connecticut. This research was performed under the auspices of the US Dept. of Energy by Lawrence Livermore National Laboratory under contract No. W-7405-ENG-48. This research was also supported the University of Connecticut Research Foundation and for one of us, (QCK), the Research Corporation.

REFERENCES

*Present address: Department of Physics, University of Rhode Island, Kingston, R.I. 02881-0817.

1. Briand, J.P., de Billy, L., Charles, P., Essabaa, S., Briand, P., Geller, R., Desclaux, J.P., Bliman, S., and Ristori, C., *Phys. Rev. Lett.* **65**, 159-162 (1990).
2. Burgdörfer, J., Reinhold, C., and Meyer, F., *Nucl. Instr. Meth. in Phys. Res. B* **98**, 415-419 (1995).
3. Andra, J.J., Simionovici, A., Lamy, T., Brenac, A., Lamboley, G., Pesnelle, A., Andriamonje, A., Fleury, A., Bonnefoy, M., Chassevent, M., and Bonnet, J.J., "Interaction of multi charged ions at low velocities with surfaces and solids," in *Electronic and Atomic Collisions: Invited Papers of the XVII ICPEAC, Brisbane,* 1991, pp.89-104
4. Clark, M.W., Schneider, D.H., Deveney, E.F., Kessel, Q.C., Pollack, E., and Smith, W.W, *Nucl. Instr. Meth. in Phys. Res. B* **79**, 183-185 (1993).
5. Meyer, F.W., Overbury, S.H., Havener, C.C., Zeijlmans van Emmichoven, P.A., Burgdörfer, J, and Zehner, D.M., *Phys. Rev. A* **44**, 7214-7228 (1991).
6. Briere, M.A., Reaves, M, Kessel, Q.C., Pollack, E., and Smith, W.W., "Negative ion production from shallow angle scattering of slow, highly charged ions by highly oriented pyrolytic graphite, C(0001)," in *Seventh International Conference on the Physics of Highly Charged Ions, Book of Abstracts*, Ed., Aumayr, F., and Winter, H.P., (1994) Vienna, p. Mo52. The results of this research are being submitted to the University of Connecticut as a portion of the Ph.D. thesis of M.R. Reaves.
7. Folkerts, L., Schippers, S., Zehner, D.M. and Meyer, F.W., *Phys. Rev. Lett.* **74**, 2204-2207, **75**, 983 (1995).
8. Schneider, D.H.G., and Briere, M.A., *Physica Scripta* **55**, 228-242 (1996).
9. McDaniel, E.W., Mitchell, J.B.A., and Rudd, E.R., *Atomic Collisions: Heavy Particle Projectiles*, New York: John Wiley & Sons, Inc., 1993, ch. 6, pp.425-432.
10. Schneider, D., DeWitt, D., Clark, M.W., Schuch, R., Cocke, C.L., Schmieder, R., Reed, K.J., Chen, M.H., Marrs, R.E., Levine, M., and Fortner, R., *Phys. Rev. A* **42**, 3889-3895 (1990).
11. Robinson, M.T., and Torrens, I.M., *Phys. Rev. B* **9**, 5008-5024 (1974).
12. Private Communication, Dr. Christina Höfner.
13. McDaniel, E.W., *Collision Phenomena in Ionized Gases*, New York: John Wiley & Sons, Inc., 1964, Ch. 13, pp. 684-685.
14. Winter, H., *Comments At. Mol. Phys.* **26**, 287-320 (1991).

THEORETICAL STUDIES OF PHOTOIONIZATION OF HELIUM

Jian Zhi Tang [a], **Isao Shimamura** [b], and **Joachim Burgdörfer** [a]

[a] *Department of Physics, University of Tennessee, Knoxville, TN37996-1200 and Oak Ridge National Laboratory, Oak Ridge, TN37831-6377*

[b] *The Institute of Physical and Chemical Research (RIKEN), Wako, Saitama 351-01, Japan*

Motivated by recent high-resolution atomic photoionization spectroscopy using synchrotron radiation, a theoretical framework the hyperspherical close-coupling method, has been developed to unravel novel aspects of strong electron-electron correlations in multiply excited states. The theory leads to excellent agreement with measured spectra featuring rich structures due to two-electron excited states. Furthermore, we apply the method to the energy region above the double ionization threshold. The obtained double photoionization cross sections are compared with other theoretical results and with the most recent experiments.

1 INTRODUCTION

With the advent of intense photon sources and the newly developed high-resolution spectroscopic techniques, it has become possible to investigate the dynamics of strong electron-electron correlations in atomic spectra in great detail. Photoionization of He attracts considerable attention, both experimentally and theoretically. High-resolution measurements have revealed rich structures in the spectra of total (TCS) and partial (PCS) photoionization cross sections and in the angular distributions of photoelectrons [1,2] for energies below the double photoionization threshold. Theoretically, the accurate computation of the double excitation to high-lying doubly excited states of He is a nontrival problem and demands both considerable computational and conceptual efforts. One of most important theoretical difficulties lies in the fact that either pseudostates and/or trial wavefunctions are used in most computational methods for expanding the total wavefunctions of the He atom. Unphysical resonances result from using pseudostates while for trial wavefunctions the variational parameters have to be determined separately at each energy. Practically, it is extremely difficult to choose pseudostates and/or variational parameters that are suitable in the region where the cross section displays rich double-electron excitation structures. Many theoretical methods developed recently produce results that agree very well with each other and with experiments on the TCS and the resonance parameters for doubly excited states converging to low-n thresholds. However, there are significant discrepancies among theoretical results [3,4] on the photoelectron angular distributions or the differential cross sections (DCS), and hence on the asymmetry parameter often denoted by β.

CP392, *Application of Accelerators in Research and Industry,* edited by J. L. Duggan and I. L. Morgan
AIP Press, New York © 1997

When the photon energy exceeds the double ionization threshold, either one or both of the two electrons in He may be ionized. The ratio of the double to single photoionization cross sections is a parameter characteristic for electron-electron correlations and is commonly measured by experiments. The values of the ratio in the low-energy region has long been controversal until very recently [5–7]. New measurements using the cold target recoil ion momentum spectroscopy (COLTRIMS) technique [5] appear to yield improved results for the ratio of double to single photoionization cross sections. They display significant differences to most of the previous theoretical and experimental data. In the present paper, we report on a theoretical study of DCS for double excitation to high-lying doubly excited states and double photoionization in the low-energy region of He.

2 FORMULATION

Only doubly excited states of the $^1P^o$ symmetry are populated in the photoionization of He in the ground state. These states can interact with several continua resulting in autoionization,

$$\mathrm{He}(1s^2) + h\nu \longrightarrow \mathrm{He}^{**} \longrightarrow \mathrm{He}^+(nl) + e \quad (1)$$

The partial differential cross section for photoionization process in Eq. 1 by linearly polarized radiation may be written as,

$$\frac{d\sigma_{nl}}{d\Omega}(\theta) = \frac{\sigma_{nl}}{4\pi}[1 + \beta_{nl}P_2(\cos\theta)], \quad (2)$$

where θ is the angle between the direction of photoelectron ejection and the polarization axis and σ_{nl} is the partial cross section for the production of $\mathrm{He}^+(nl)$. β_{nl} is the asymmetry parameter which measures the anisotropy in the angular distribution, and P_2 is the second-order Legendre polynomial

$$P_2(\cos\theta) = \frac{3\cos^2\theta - 1}{2}. \quad (3)$$

Theoretically, β and σ are obtained from the calculated dipole transition amplitudes between the initial and final states. σ depends only on the absolute values of the amplitudes, while β depends also on the relative phases between the amplitudes. This implies that β contains more information on the process of photoionization than does σ. For example, weak resonance structures in the spectra of σ may be enhanced in β due to the interference of the amplitudes.

The dipole transition amplitudes between the initial ground state of He and the final continuum states involving doubly excited states or double continuum states are calculated by the hyperspherical close-coupling (HSCC) method [8]. In the hyperspherical coordinates, the independent particle radial distances of two electrons, r_1 and r_2, are replaced by a pair of collective variables, i.e., the hyperradius $R = \sqrt{r_1^2 + r_2^2}$ measuring the size of the electron pair and the hyperangle $\alpha = \tan^{-1}(r_2/r_1)$ describing the degree of electron-electron radial correlation. The total wave function of He is expanded in terms of hyperspherical basis functions for both the initial bound state and the final continuum state. This expansion is found to converge rapidly. The Schrödinger equation is cast into close-coupling equations and these equations are solved from $R = 0$ to a large distance R_m. At R_m, the wave function is connected smoothly to an outer-region wave function expressed in independent-electron coordinates. At each R between 0 and R_m, diabatic basis functions are set up by solving the Schrödinger equation in the hyperangle space. Unlike variational trial wavefunctions, the diabatic basis functions contain no adjustable parameters. This method has proved to be quite powerful in reproducing the experimental TCS and PCS measured with high resolution [9,10] and has greatly improved the accuracy over approaches based on conventional independent-electron coordinates.

3 RESULTS AND DISCUSSION

There exist many Rydberg series converging to each ionization threshold $He^+(n)$ for high n. Doubly excited states below the threshold become narrower as n becomes larger. The detection of these high-lying doubly excited states becomes possible only with the advent of bright, narrow-bandpass photon sources. Very recently, a photoionization experiment of He could extract the PCS for all the energetically accessible decay paths between the n =4 and 5 thresholds [2]. With an energy resolution as high as 12 meV, the modulations of the overlapping resonances in DCS and β could be investigated. The theoretical results from the HSCC method are found to agree very well with the experimental data for total and partial cross sections of all the possible decay channels even on an absolute scale. Those for β agree well with experiment in the details of the resonance structures, but differ slightly in the absolute value. Fig. 1(a) shows the calculated $\beta_{n=3}$ between the n = 4 and 5 threshold and which should be compared with the experimental data [2] displayed in Fig. 1(b). The excellent agreement with the experiment in the details of the resonance structures is demonstrated in Fig. 1(c) where we have shifted the experimental β value by 0.08. The origin of the difference is not yet fully understood.

To calculate the double photoionization cross section in the low energy region above the double ionization threshold, the electron -electron correlations in both the initial and final state wavefunctions have to be treated very accurately. Previously, various computational methods have given different results which strongly

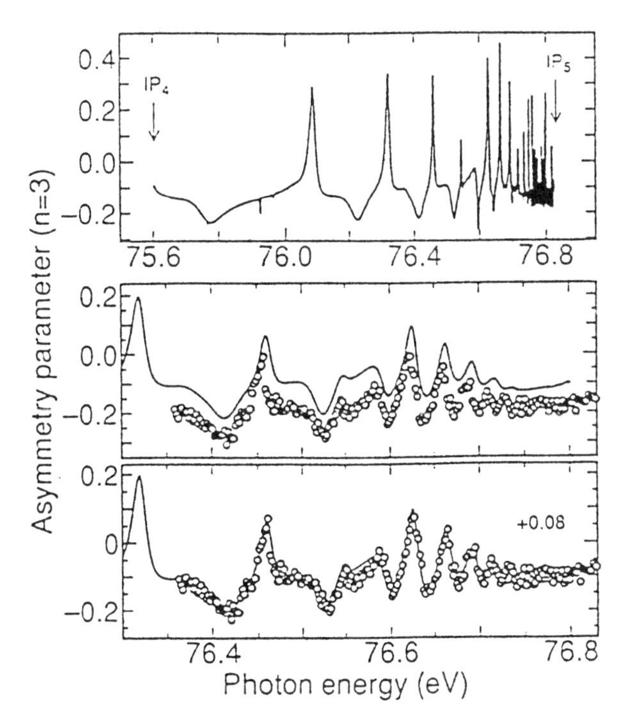

Fig. 1. (a)The calculated asymmetry parameter β_3 for the decay channel $He^+(n = 3)$. (b) Comparison of β_3 in (a) convoluted with a resolution of 12 meV (solid curve) with experimental data (circles). (c) Same as (b) but with the experimental data shifted up by 0.08.

depend on the gauge of the dipole transition operator [11,12]. Unlike the double photoionization problem at high energies, there is no established argument to determine which gauge is most reliable in the low-energy region.

We have extended the HSCC method to study both single and double photoionizations. We find the length form and the acceleration form agree very well with each other [13]. The calculated ratios for of the cross sections for double to single photoionization are smaller than most experimental data measured before 1994. However, very recent experimental measurements give smaller values for the ratio in the low-energy region and agree very well with our results [5,7]. Fig. 2 shows the calculated acceleration-form ratio with the most recent and reliable experimental data [5,7].

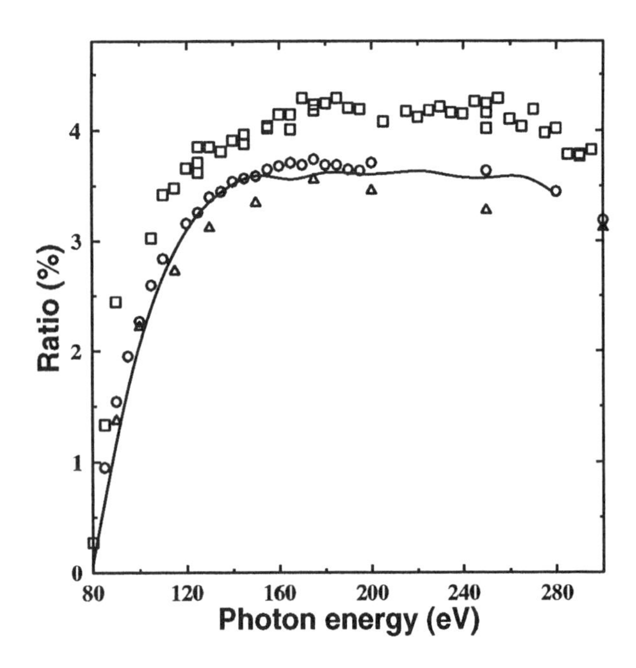

Fig. 2. Comparison of the calculated ratio (solid curve) with experimental data of Ref. [5] (triangles), Ref. [6] (squares), and Ref. [7] (circles).

4 CONCLUSION

The hyperspherical close-coupling method has been used to study the dynamics of photoionization process of He. The results are in good agreement with measured differential cross sections in the energy region involving high-lying doubly excited states and the double photoionization cross sections in the energy region above the double ionization threshold.

References

[1] M. Zubek et al., *J. Phys.* B **24**, L337 (1991).

[2] A. Menzel et al., *Phys. Rev. Lett.* **75**, 1479 (1995).

[3] M. A. Hayes and M. P. Scott, *J. Phys.* B **21**, 1499 (1988).

[4] I. Sanchez and F. Martin, *Phys. Rev.* A **48**, 1243 (1993).

[5] R. Dörner et al., *Phys. Rev. Lett.* **76**, 2654 (1996).

[6] J. C. Levin et al., *Phys. Rev. Lett.* **76**, 1220 (1996). A more recent measurement gives smaller ratios in the low-energy region.

[7] W. Stolte and J. A. R. Samson (private communication).

[8] J.-Z. Tang, S. Watanabe, and M. Matsuzawa, *Phys. Rev.* A **46**, 2437 (1992).

[9] J.-Z. Tang et al., *Phys. Rev. Lett.* **69**, 1633 (1992).

[10] J.-Z. Tang and I. Shimamura, *Phys. Rev.* **50** , 1321 (1994).

[11] K. Hino et al., *Phys. Rev.* A **48**, 1271 (1993).

[12] K. W. Meyer and C. H. Green, *Phys. Rev.* A **50**, R3573 (1994).

[13] J.-Z. Tang and I. Shimamura, *Phys. Rev.* A **52** , R3413 (1995).

Double Photoionization of Helium

Kurt W. Meyer, Chris H. Greene, and Brett D. Esry
Department of Physics and JILA,
University of Colorado, Boulder, CO 80309-0440
(September 27, 1996)

The eigenchannel R-matrix method is combined with a finite element basis set to calculate double photoionization cross sections of helium for photon energies in the range 80-180 eV. These cross sections are calculated in the length, velocity, and acceleration gauge forms. The gauge discrepancies among the cross sections have been greatly reduced in comparison to previous results calculated with a hydrogenic basis. The ratio of double to single photoionization, a key parameter in characterizing electron correlations, is also presented and compared to recent theoretical and experimental values.

I. INTRODUCTION

Accurate quantitative descriptions of two-electron escape processes in the low and intermediate energy regions have only recently been developed. One such process which has received a great deal of attention, both theoretically and experimentally, is double photoionization. Of particular interest is the ratio of double to single photoionization in helium, as this ratio characterizes the importance of electron correlations. (To first order, double photoionization can not occur if electron-electron interactions are absent.) Recent experimental measurements of this ratio by Dörner *et. al* [1] and Levine *et. al* [2] over the intermediate energy range indicate a maximum value of roughly 4%. Although the profiles of this ratio are similar for the two sets of measurements, they differ in magnitude by 10-20%. Recent theoretical methods used in calculating helium double photoionization cross sections include the convergent close-coupling calculations of Kheifets and Bray [3], the hyperspherical close-coupling calculations of Tang and Shimamura [4], and the eigenchannel R-matrix calculations of Meyer and Greene [5], among others. Although good agreement was achieved between two different gauge forms for the former two calculations, the profiles of the ratio for the two methods differ greatly. The results from the eigenchannel R-matrix method are characterized by significant discrepancies between various gauge form calculations, and so can provide only a rough estimate of the ratio in the intermediate energy range.

II. METHOD

The eigenchannel R-matrix method is adopted to calculate helium double photoionization. This method has been very successful in describing single electron escape processes [6,7]. A previous application of this method to double photoionization resulted in cross sections that were slightly lower than the experimental data available at that time for helium. Recent experimental data, however, has been closer to the range predicted by that first theoretical calculation using this technique. Although the eigenchannel R-matrix method seems capable of describing two-electron escape processes accurately, some problems arose in its application to helium double photoionization. A discrepancy between the various gauge forms used in calculating dipole matrix elements indicated inaccuracies in the initial and/or final state wave functions. In particular, the discrepancy between the velocity and acceleration forms was roughly 15%, and the length form gave results which were nearly an order of magnitude larger than the results obtained for other gauge forms. Interestingly, the convergent close-coupling method of Kheifets and Bray [3] gave unreasonable results in the length form, much like our R-matrix calculations. However, the hyperspherical close-coupling method of Tang and Shimamura [4] was able to obtain good agreement between length and acceleration forms.

The eigenchannel R-matrix method [8] and its application to helium double photoionization [5] have been described extensively elsewhere, and will not be discussed here. The difference between our previous application of this method and the current approach involves our choice for the variational basis set. Instead of forming a two electron basis set that consists of products of one electron hydrogenic orbitals, a finite element basis set is adopted for the present study. Our hope is that the finite element method will be able to represent the initial and final state wave functions more accurately, and thereby reduce the discrepancy between gauge forms. The usefulness of a finite element approach for calculating accurate bound state wave functions has been previously illustrated [9,10]. This method has also been adopted for calculating accurate bound state excitation cross sections in elastic scattering [11]. The finite element set used in our approach consists of a set of six fifth order Hermite interpolating polynomials, as described elsewhere [11, 12].

In our previous approach, we solved for the radial part of the one-electron hydrogenic orbitals $\phi_k(r)$ subject to "box" boundary conditions, that is, boundary conditions imposed at the surface of the reaction volume. Orbitals that vanished at the outer radius of the R-matrix box were referred to as "closed" orbitals, while "open" orbitals referred to orbitals which were nonzero at the box radius. We now wish to expand these hydrogenic orbitals

CP392, *Application of Accelerators in Research and Industry*, edited by J. L. Duggan and I. L. Morgan
AIP Press, New York © 1997

with a finite element basis set:

$$\phi_k(r) = \sum_{n=1}^{N}\sum_{i=1}^{6} C_i^n u_i(x_n), \tag{1}$$

where n refers to the sector number, and i refers to one of the six Hermite interpolating polynomials. These six polynomials are defined by imposing six "boundary" conditions: the polynomial value and its first derivative at the two endpoints and the center of the sector. Each of the six polynomials will have only one nonzero boundary condition, with this boundary condition chosen to be unity (see Fig. 1). Also, $u_i(x_n)$ is defined to be nonzero only in sector n, and x_n is only defined over the interval [-1,1]. The transformation to the physical sector $[r_i, r_{i+1}]$ is then given by

$$r = a_i x_i + b, \tag{2}$$

where

$$a_i = \frac{r_{i+1} - r_i}{2}, b_i = \frac{r_{i+1} + r_i}{2}. \tag{3}$$

Although there appears to be $6N$ independent coefficients in the above expression, by imposing boundary conditions at the origin and the box radius r_0, and also by imposing the constraint that the wave function and its first derivative be continuous across sector boundaries, the number of independent coefficients is reduced to $4N$.

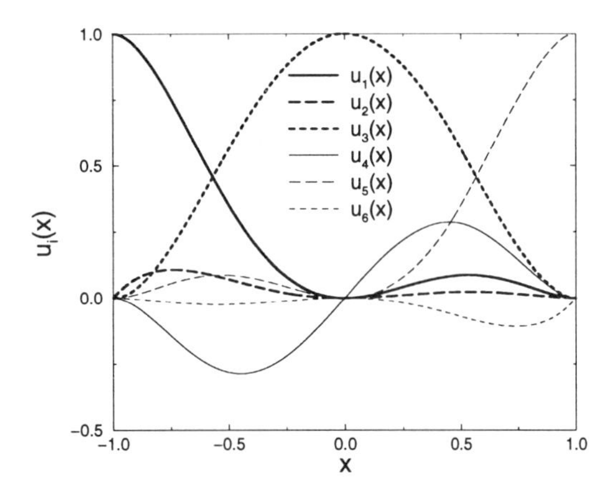

FIG. 1. The set of six Hermite interpolating polynomials adopted as the finite element basis set in this study.

As in our earlier R-matrix calculations, a two-electron wave function is formed by taking products of one electron orbitals:

$$\psi(\vec{r}_1, \vec{r}_2) =$$
$$\mathcal{A}\sum_{n=1}^{N}\sum_{m=1}^{M}\sum_{i=1}^{6}\sum_{j=1}^{6} C_{ij}^{nm} u_i(x_n) u_j(y_m) \mathcal{Y}_{l_1 l_2}^{LM}(\hat{r}_1, \hat{r}_2), \tag{4}$$

where $\mathcal{A}$ is the antisymmetrization operator. Our two-dimensional sectors are chosen to be square, so that $u_i(x_n)$ and $u_j(y_m)$ are of identical form. Although there are 36 coefficients for each sector, enforcing continuity of the wave function and first derivatives across sector boundaries reduces the number of independent coefficients to 16. To obtain the ground state wave function, we impose the boundary condition that the wave function vanish on the surface of the reaction volume. This is easy to implement with a Hermite interpolating polynomial basis set since only one of the six polynomials is nonzero on the outer boundary. Simply setting the corresponding coefficient of this wave function to zero ensures that the wave function vanishes at the box radius.

In order to express the final state correctly, our expansion must include basis functions which are nonvanishing on the surface of the reaction volume in order to allow for ionization. We retain the idea from our previous approach of including closed-open two electron configurations consisting of a product of a closed and an open one-electron hydrogenic orbital. However, the open hydrogenic orbital is replaced by a single Hermite interpolating polynomial in the outermost sector which is nonzero at the outer boundary (denoted by u_5). In other words, we add basis functions of the form

$$\mathcal{A}\sum_{m=1}^{M}\sum_{j=1}^{6} C_{5j}^{Nm} u_5(x_N) u_j(y_m) \mathcal{Y}_{l_1 l_2}^{LM}(\hat{r}_1, \hat{r}_2) \tag{5}$$

to our expansion in Eq. 4. Retaining the closed-open type of two-electron configurations allows us to use the same criterion to distinguish between single and double photoionization. Namely, if the energy of our closed orbital representing the "inner" electron is negative, then we consider it to remain bound (single ionization), while a positive closed orbital energy is considered ionized (double ionization). This criterion along with a box averaging technique was used to obtain single and double photoionization cross sections in our original study. An improvement over this approach has been developed since then. In actuality each closed orbital will contribute to both the bound state excitation and the ionization cross section. The contribution of each closed orbital to ionization of the "inner" electron can be determined by projecting the wave function onto a continuum hydrogenic wave function. This frame transformation technique [13] was used to obtain the results shown here.

Upon first glance, it would appear that the finite element basis set has a disadvantage of requiring the solution of a larger number of expansion coefficients than would be required for a hydrogenic basis as used previously, thereby limiting the number of sectors which could be used. However, since the local basis functions are only defined to be nonzero for a given sector, the Hamiltonian and overlap matrices are sparse. Efficient numerical routines are implemented to take advantage of these sparse, symmetric matrices. A Lanczos algorithm [14] is used to obtain the helium ground state wave function and energy. The final state is obtained by solving the usual generalized eigenvalue problem

$$\underline{\Gamma} C = b \underline{\Lambda} C, \tag{6}$$

where

$$\underline{\Gamma} = 2[E\underline{S} - \underline{H} - \underline{L}], \tag{7}$$

$\underline{S}$ is the overlap matrix, $\underline{H}$ is the Hamiltonian matrix, $\underline{L}$ is the Bloch operator, and $\underline{\Lambda}$ is the surface overlap matrix. The matrices are partitioned into closed and open basis sets

$$\begin{bmatrix} \underline{\Gamma}^{cc} & \underline{\Gamma}^{co} \\ \underline{\Gamma}^{oc} & \underline{\Gamma}^{oo} \end{bmatrix} \begin{bmatrix} C^c \\ C^o \end{bmatrix} = b \begin{bmatrix} 0 & 0 \\ 0 & \underline{\Lambda}^{oo} \end{bmatrix} \begin{bmatrix} C^c \\ C^o \end{bmatrix}. \tag{8}$$

This leads to the reduced generalized eigenvalue problem:

$$\underline{\Omega}^{oo} C^o = b \underline{\Lambda}^{oo} C^o, \tag{9}$$

where

$$\underline{\Omega}^{oo} = \underline{\Gamma}^{oo} - \underline{\Gamma}^{oc} (\underline{\Gamma}^{cc})^{-1} \underline{\Gamma}^{co}. \tag{10}$$

The streamlined eigenchannel treatment involves a transformation which diagonalizes the Hamiltonian matrix. For a closed two-electron basis set which is orthogonal, the closed portion of the overlap matrix is diagonal, so that $\underline{\Gamma}^{cc}$ is diagonal after the Hamiltonian is transformed and therefore trivial to invert. Since the transformation only needs to be performed once, this provides a great computational improvement over directly solving for the generalized eigenvalue problem at each desired energy.

For our purposes, application of the streamlined transformation of Ref. [8] is impractical owing to the large dimension of the Hamiltonian and overlap matrices. Rather than solve the original generalized eigenvalue problem, we would like to solve the reduced eigenvalue problem, given by Eq. 9. Note that $\underline{\Omega}^{oo}$ involves the inverse of $\underline{\Gamma}^{cc}$, which is large but sparse. Instead of directly calculating the inverse of this matrix, we use a linear algebraic routine to solve the sparse system of equations

$$\underline{\Gamma}^{cc} \underline{Q}^{co} = \underline{\Gamma}^{co}. \tag{11}$$

The $\underline{\Omega}$ matrix is then given by

$$\underline{\Omega}^{oo} = \underline{\Gamma}^{oo} - \underline{\Gamma}^{oc} \underline{Q}^{co}. \tag{12}$$

For our finite element application, the open-open part of the surface overlap matrix is diagonal, and in fact equal to the identity matrix. This reduces the generalized eigenvalue problem given in Eq. 9 to the ordinary eigenvalue problem

$$\underline{\Omega}^{oo} C^o = b C^o \tag{13}$$

The greatest computational effort required to obtain results with a finite element basis involves solving for Eq. 11. A generalized minimum residual iterative method with an incomplete LU factorization for preconditioning is adopted from the sparse linear algebra package (SLAP) [15] to perform this task. The dimension of $\underline{\Gamma}^{cc}$ in this study is 6912, and about 5% of its elements are nonzero. The amount of computational effort required to solve Eq. 11 increases with energy. This is due in part to a larger number of iterations required to solve for each right hand side, and in part due to an increase in the number of right hand sides (the number of closed-open two electron configurations included). Although the results presented here are limited to energies no higher than 100 eV above the double ionization threshold, it is anticipated that further improvements will allow us to perform calculations at higher energies more efficiently.

III. RESULTS

The results presented here were obtained by using three partial waves in the initial state and four partial waves in the final state, with a reaction volume radius of 10 a.u.. Adding more partial waves did not significantly change our results. Twelve sectors were used for each dimension. The calculated ground state energy of -2.9028 a.u. for helium improved significantly over the previous value of -2.8953 a.u. obtained in our original R-matrix calculations with a hydrogenic basis. (The "exact" calculated value [16] of the helium ground state energy is -2.9037 a.u. in infinite mass units.)

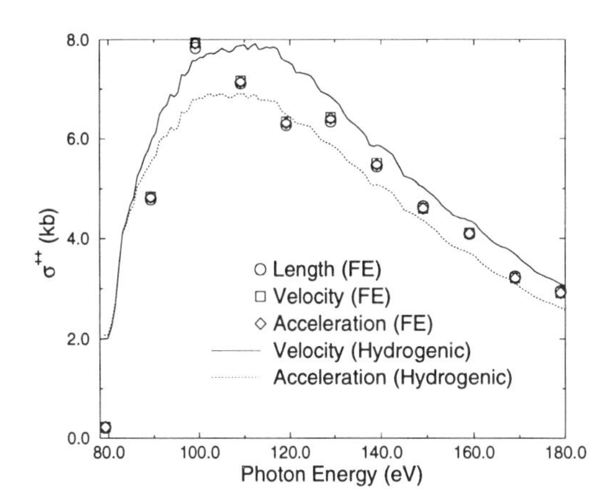

FIG. 2. Comparison of double photoionization cross sections calculated with eigenchannel R-matrix for a finite element (FE) basis, denoted by the open symbols, and a hydrogenic basis, denoted by the solid and dotted lines.

Figure 2 compares our present calculations of the double photoionization cross sections to our original R-matrix calculations. The most obvious difference in the two calculations is the improved agreement between the gauge forms. (The original length gauge form calculations are not shown, as they are roughly an order of magnitude larger than the other gauge form calculations.) Kheifets and Bray [3] had suggested that our previous gauge discrepancy could be improved if we used a more accurate helium ground state. We have not yet determined whether our gauge agreement is due more to an improvement in the initial state or the final state. Although our finite element results are consistent with our previous results for 50 eV and higher, there appears to be some fluctuations in our cross sections at lower energies. These fluctuations may be due to the presence of pseudoresonances. In addition to the frame transformation a box averaging technique may be required to obtain a smoother and more accurate cross section profile.

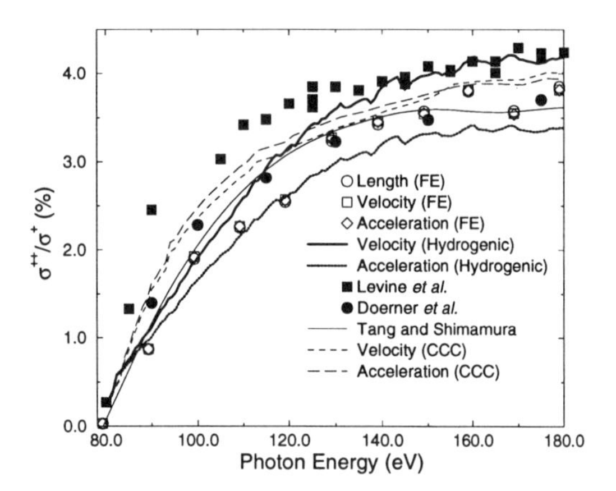

FIG. 3. Comparison of ratio of double to single photoionization with other theoretical and experimental values. Our present results are denoted by the open symbols, the experimental values are denoted by filled symbols, and the lines denote various theoretical calculations. The convergent close-coupling calculations of Kheifets and Bray are denoted by CCC.

Figure 3 compares our present results to some of the more recent theoretical and experimental values. Fluctuations of ratio values in our finite element calculations are observed even at higher energies. It is difficult to determine whether our results are more consistent with those of Kheifets and Bray or Tang and Shimamura due to these fluctuations. Our results are in good agreement with the experimental values of Dörner for energies above 50 eV. Levin's experimental values are slightly higher than our results over the entire energy range shown in Figure 3. Further calculations will be required to understand and resolve the nature of the fluctuations in our data.

IV. CONCLUSION

Much effort has been expended in trying to understand the origin of the discrepancy between various gauge forms in our previous eigenchannel R-matrix treatment of helium double photoionization. A finite element basis set was adopted in an attempt to improve the accuracy of our initial and final state wave functions. Double ionization cross sections calculated with this basis set show a dramatic improvement in agreement between the length, velocity, and acceleration gauge forms. Furthermore, these results are consistent with the (broader) range of values calculated previously with a hydrogenic basis. Because of the large computational effort required to perform calculations with our method, we have not yet had the opportunity to explore our method for several box sizes or a larger energy range. Further exploration will be required before we can fully access the usefulness of our method. However, our preliminary results are promising. The use of a finite element basis set in conjunction with the eigenchannel R-matrix method gives a more stable theoretical description, and presumably a more accurate description of the double continuum process as well.

ACKNOWLEDGMENTS

We thank J. Bohn for discussions related to this study. This work was supported in part by the National Science Foundation.

---

[1] R. Dörner *et. al*, Phys. Rev. Lett. **76**, 2654 (1996).
[2] J.C. Levin, G.B. Armen, and I.A. Sellin, Phys. Rev. Lett. **76**, 1220 (1996).
[3] A. Kheifets and I. Bray, Phys. Rev. A **54**, R995 (1996).
[4] J.Z. Tang and I. Shimamura, Phys. Rev. A **52**, R3413 (1995).
[5] K.W. Meyer and C.H. Greene, Phys. Rev. A **50**, R3573 (1994).
[6] C.H. Greene and M. Aymar, Phys. Rev. A **44**, 1773 (1991).
[7] F. Robicheaux and C.H. Greene, Phys. Rev. A **46**, 3821 (1992).
[8] C.H. Greene and L. Kim, Phys. Rev. A **38**, 5953 (1988).
[9] M. Braun, W. Schweizer, and H. Herold, Phys. Rev. A **48**, 1916 (1993).
[10] J. Ackermann and J. Shertzer, Phys. Rev. A **54**, 365 (1996).
[11] J. Shertzer and J. Botero, Phys. Rev. A **49**, 3673 (1994).
[12] J.P. Burke, C.H. Greene, and B.D. Esry, Phys. Rev. A (in press).
[13] K.W. Meyer, C.H. Greene, and I. Bray, Phys. Rev. A **52** 1334 (1995).
[14] M.T. Jones and M.L. Patrick, Applied Numerical Mathematics, **12**, 377 (1993).
[15] M.K. Seager and A. Greenbaum, netlib online documentation.
[16] E. Lindroth, Phys. Rev. A **49**, 4473 (1994).

DYNAMICS OF THE DOUBLE IONIZATION PROCESS FROM e,(3-1)e AND (e,3e) EXPERIMENTS

A. Lahmam-Bennani, A. Duguet

Laboratoire des Collisions Atomiques et Moléculaires (Unité associée au CNRS N°281)
Université de Paris-Sud, Bât. 351, 91405 Orsay Cedex, France

The recent advent and extension of multianalysis and multidetection systems have allowed spectacular developments in the study of electron impact double ionization (DI) using double or triple coincidence techniques, yielding the most detailed insight into the fundamentals of the DI process. Illustrative results from both e,(3-1)e and (e,3e) experiments are discussed. It is shown that a suitable choice of the kinematical parameters allows a particular DI mechanism to be favoured, and hence to be identified via its signature in the measured angular distributions.

INTRODUCTION

The study of the double ionization (DI) process under charged particle impact has been the subject of many experimental and theoretical investigations for about thirty years. Spectacular advances in coincidence detection techniques have made it recently possible to perform the so-called (e,3e) experiments (1). Such experiments yield highly differential cross sections that contain the most detailed information about the fundamentals of the DI process. A major aim of these measurements is to study the various interactions between all particles involved in the collision and to determine their influence on the angular distribution of the DI products. This allows for a detailed understanding of the role of electron-electron correlations, both in the initial and in the final states, and may enable disentangling between various possible DI mechanisms.

The complete DI experiment (Fig. 1) is characterized by an incident electron labeled 0 and three outgoing electrons, labeled a for the fast scattered one and b and c for the two ejected ones (though they are of course undistinguishable). To fully determine the kinematics, one needs to measure all three final electrons energies and angles and detect them in a triple coincidence, an (e,3e) experiment. The measured quantity is then a five-fold differential cross section (5DCS), $d^5\sigma/dE_a dE_b d\Omega_a d\Omega_b d\Omega_c$, analogous to the corresponding single ionization (e,2e) triple differential cross section (2).

If one excludes Auger processes, the conservation laws for energy and momentum read

$$E_t+E_0=E_a+E_b+E_c+E_I^{2+}$$

$$\mathbf{q_t}+\mathbf{k_0} = \mathbf{k_a}+\mathbf{k_b}+\mathbf{k_c}+\mathbf{q_r}$$

where E_i and $\mathbf{k_i}$ (i=0,a,b,c) are the incoming and outgoing electron energies and momenta, $\mathbf{K}=\mathbf{k_0}-\mathbf{k_a}$ is the total momentum transfer to the target, and $\mathbf{q_t}$ and E_t are the initial target momentum and energy. In practice, thermal atoms in their ground state -usually rare gases- are used.

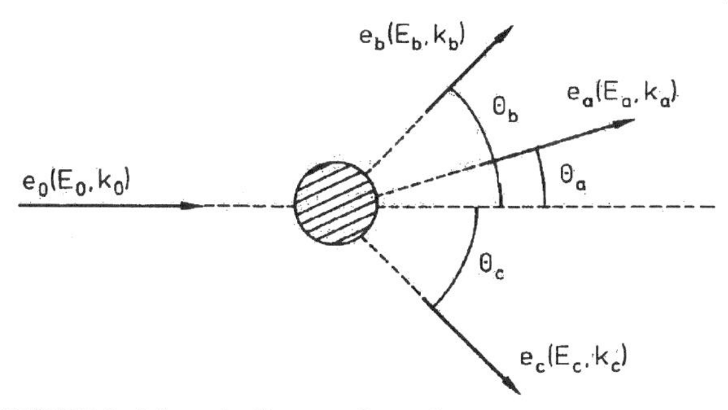

FIGURE 1. Schematic diagram of a coplanar (e,3e) experiment.

Hence, E_t and q_t are both ~0. The quantity $\mathbf{q_r}$ is the ion recoil momentum, and E_I^{2+} is the DI energy of the target, leaving the doubly charged residual ion in a particular final state. This latter quantity is known from the measurement of the other four energies, hence one knows precisely from which orbital(s) the two ejected electrons have been knocked out.

EXPERIMENTAL

The first (e,3e) spectrometer was described by Lahmam-Bennani and co-workers (1,3). Briefly, a few kiloelectronvolt electron beam (e_0) crossed a target gas beam at right angles. The scattered (e_a) and ejected (e_b and e_c) electrons were analyzed in a 127° cylindrical analyzer and two twin hemispherical analyzers, respectively. Conventional electronics were used to process the signals, and two time-to-amplitude converters were used to register coincident a-b and a-c events. An AND gate permitted only triple coincidence events to be recorded. This system suffered from two severe limitations. First, angle θ_c was fixed, while θ_b could only be varied over a limited range. Second, and most important, the three electrons were detected at three preselected angles. This "monodetection" system led to very long accumulation times, several days

CP392, *Application of Accelerators in Research and Industry,* edited by J. L. Duggan and I. L. Morgan
AIP Press, New York © 1997

for each single data point, or up to one month for one single angular distribution.

For these reasons, a second generation spectrometer was built, including multiangle (and potentially multienergy) detection of the two slow electrons. The only other existing (e,3e) spectrometer was recently described by Ford et al. (4). The design philosophy and the objectives are quite different from those described above. Our system is based on a double toroidal analyzer (5), each half of which collects electrons e_b and e_c, respectively, in almost the totality of the collision plane ($\mathbf{k_0}$, $\mathbf{k_a}$). The angular information contained in this plane, i.e. the ejection angles θ_b and θ_c is preserved upon arrival on the position sensitive detectors. By measuring the correlation between all possible pairs of ejected electrons and the fast scattered electron e_a detected at fixed angle θ_a, it is possible to simultaneously perform ≈200 experiments analogous to the previous (e,3e) ones.

Figure 2 shows a triple coincidence time spectrum. One may distinguish

(i) the uniform accidental contribution, due to fully uncorrelated a,b, and c events,

(ii) the three semi-accidental contributions, due to a pair of correlated electrons, the third one being random (the so-called ac and ab walls along the *x-y* axis, respectively, and the bc wall along the diagonal), and

(iii) at their intersection, the triple coincidence peak. To obtain the desired triple coincidence signal, one has to subtract from the intensity in the peak region the contribution of each of these four random coincidences, see (6) for more details.

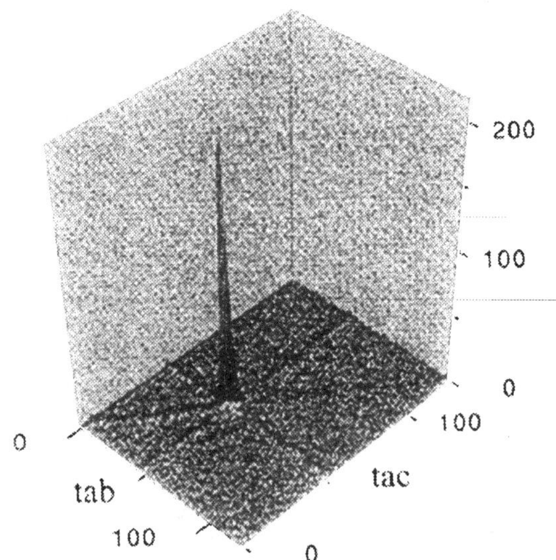

FIGURE 2. Triple coincidence time spectrum. Two identical time-to-amplitude converters are simultaneously started by an a-event, and stopped by a b-event (t_{ab} axis) or a c-event (t_{ac} axis).

A major difficulty of these (e,3e) experiments is due to the very low coincidence counting rates, typically < 20 counts/hr. This figure cannot be improved by increasing the target density, n, or the incident beam intensity, I. Indeed, the fully accidental coincidences contribution in Fig. 2 increases as $(nI)^3$, while the signal is directly proportional to (nI). The consequence is that there exists an optimum value $(nI)_{opt}$, which minimizes (6) the percentage statistical uncertainty for a given accumulation time. The quantity $(nI)_{opt}$ is usually rather small, 1-10 nA, hence the small triple coincidence rate, which results in very long accumulation times. For this reason, it is also very interesting to measure double coincidences between pairs of the three outgoing electrons, an e,(3-1)e experiment. Of course, the price is that information is lost about the ion final state.

RESULTS

In this section, two selected results will be given to illustrate how it is possible in these highly differential experiments to select the kinematics such that a particular DI mechanism is favoured, and hence can be identified via its signature in the measured angular distributions. Both results are concerned with the DI of argon.

The first result was obtained (7) in an "incomplete" (e,3e) experiment, so called e,(3-1)e experiment, where in the present case the fast a electron is not detected, yielding a four-fold differential cross section (4DCS). Very useful information about the DI process is still obtained, without the difficulties of a triple coincidence.

El Marji et al (7) measured, at 5.5 keV impact energy, the angular correlation between the two "atomic" electrons knocked out in DI of Ar, at fairly unequal energy sharing of the available excess energy. The results are shown in Fig. 3a and 3b. The angular distribution of the fastest ejected electron is found to be always peaked at ~80° (the 'binary' peak direction), and is independent, both in shape and in magnitude of the direction of observation of the other electron. In contrast, the slowest electron is found to be almost isotropically distributed, with an intensity which depends on the direction of emission of the other electron. These observations were predicted by Lahmam-Bennani et al (8). They can be interpreted by attributing the DI process, under these kinematical conditions, to a single interaction mechanism, the so-called shake-off (SO) mechanism (9), where the incident electron collides with a single target electron, which is ejected with a high kinetic energy in a binary (e,2e)-like process, hence the peaked distribution. Ejection of the second, slow electron occurs due to the electronic relaxation of the target caused by the sudden change in the effective charge seen by this electron. Hence the isotropic distribution. El Marji et al's conclusion is fully consistent with the prediction made by McGuire (9) that at the impact energy of this experiment, the SO mechanism should be the dominant one.

The second result (10) that will be discussed here is a 'complete' (e,3e) experiment, again for DI of Ar, but where now the two ejected electrons equally share the excess energy, and are rather slow, 10 eV each. A sample of the results is shown in Figure 4. One ejected electron is detected at a fixed angle along the direction **-K**, i.e. opposite to the momentum transfer, and the other one is mapped in the opposite half plane.

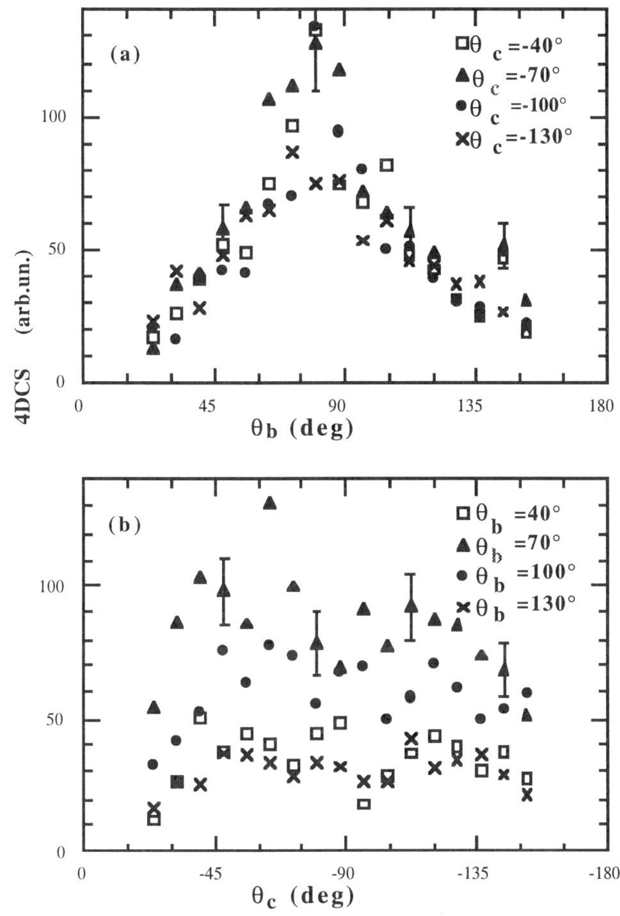

FIGURE 3. The relative four-fold differential cross section for double ejection from Ar at E_0=5500 eV. (a) The slow electron with energy E_c=17.5 eV is detected at fixed θ_c angle, and the fast one with E_b=96 eV is mapped versus θ_b in the opposite half-plane. (b) Same, with the fast electron angle, θ_b, fixed, and the slow one, θ_c, variable.

The first observation is that the measured distribution is not peaked in the **+K** direction, the maximum being shifted some 50 degrees backwards. That is, the two electrons do not preferentially emerge back-to-back, as one may reasonably expect from the Coulomb repulsion in the final state of these two slow electrons. More important, the **K** direction is clearly not a symmetry axis for the measured distribution. In DI, similar to (e,2e) single ionization, such a symmetry around **K** is inherent (11) to any first order model - e.g. the first Born approximation (FBA) - where only single interaction mechanisms such as the SO are included. Therefore, the symmetry breaking in Fig 4 is a clear indication that second (or higher) order effects must play an important role in the collision process. In fact, it is possible to show (12) from these data and many similar ones, that the angular distribution is made up by the sum of two contributions, as tentatively drawn with broken lines in Fig. 4, the first order contribution centered around **K**, and the second order contribution which is the dominant one and is shifted backwards. This contribution is attributed (12) to a two step (TS) mechanism where the incident electron suffers two successive collisions with two 'atomic' electrons, resulting in their quasi-simultaneous double ejection.

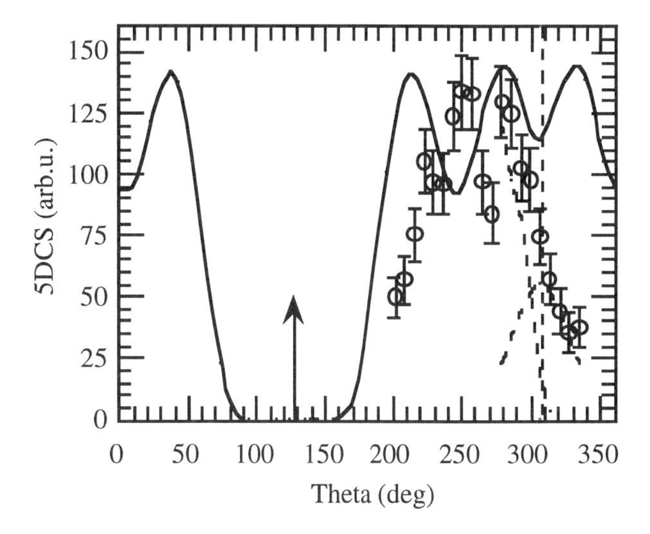

FIGURE 4. The relative (e,3e) cross section for double ionization of argon, from (10). Kinematical parametrers are E_0=5563 eV, E_a=5500 eV, E_b=E_c=10 eV. One ejected electron is detected along the **-K** direction, shown by the arrow, while the second one is mapped in the opposite half plane. The **+K** direction is shown by the vertical dashed line. Error bars are one standard deviation statistical error. The full curve is a first Born calculation (13), arbitrarily normalized to the experiments. The dashed curves are tentative deconvolution of the experimental results into first and second order contributions, see text.

The data of Fig 4 are also compared with theoretical predictions (13) performed within the first Born approximation and using an approximate BBK (14) wavefunction for the final state description. Clearly, such first order theory, which only includes the SO contribution, fails to describe the observations, hence supporting the above conclusion that the TS mechanism plays here a dominant role.

When combined together, the results of these two experiments seem to indicate that under the given kinematics (5.5 keV impact energy) :

(i) at unequal energy sharing between the two ejected electrons, the DI process is due to a SO mechanism, whereas

(ii) at equal energy sharing, the TS mechanism is the dominant one.

Item (ii) is in apparent contradiction with the above cited prediction by McGuire (9) only. Indeed, the energy distribution function among the two electrons is well known (15) to be of a U-type, where the most significant part of the differential cross section corresponds to asymmetric sharings. When integrated, it is this part which largely dominates the total cross section, and which is found both in the e,(3-1)e experiments and in McGuire's predictions to be due to a SO process. The part of the differential cross section close to the equal energy sharing corresponds to a marginal fraction of the total cross section. The (e,3e) experiment attributes it to a TS process, but this fraction is small enough so that the calculated total cross section is not sensitive to its effect. In fact, one may speculate that if Mc Guire's calculations were repeated for differential cross sections at equal energy sharing, one would likely find a different behaviour as a function of impact energy, where the TS process would dominate the SO at ~5 keV.

CONCLUSION

Two illustrative results have been presented for the DI of argon, showing the sensitivity of these highly differential experiments to the collision dynamics. In these examples, a suitable choice of the kinematics yields angular distributions which are selectively dominated either by a first order or a second order interaction process.

ACKNOWLEDGMENTS

This work was partially supported by the European Community Contract No. CHRX-CT93-0350.

REFERENCES

1. Lahmam-Bennani, A., Dupré, C., and Duguet A., *Phys.Rev.Lett.* **63**, 1582 (1989)
2. Ehrhardt, H., Jung, K., Knoth, G., and Schlemmer, P., *Z.Phys.D,* **1**, 3 (1986)
3. Lahmam-Bennani, A., *J.Phys.B : At.Mol.Opt.Phys.*, **24**, 2401 (1991)
4. Ford, M.J., Doering, J.P., Moore, J.H., and Coplan, M.A., *Rev.Sci.Instrum.*, **66**, 3137 (1995)
5. Leckey, R.C.G., and Riley, J.D., *Appl.Surf.Sci.,* **22/23**, 196 (1985)
6. Dupré, C., Lahmam-Bennani, A., and Duguet, A., *Meas.Sci.Technol.*, **2**, 327 (1991)
7. El Marji, B., Lahmam-Bennani, A., Duguet, A., and Reddish, T.J., *J.Phys.B : At.Mol.Opt.Phys.*, **29**, L157-161 (1996)
8. Lahmam-Bennani, A., Ehrhardt, H., Dupré, C., and Duguet, A., *J.Phys. B : At.Mol.Opt.Phys.,* **24**, 3645-53 (1997)
9. McGuire, J.H., *Phys.Rev.Lett.,* **49**, 1153 (1982)
10. El Marji, B., Schröter, C.D., Duguet, A., Lahmam-Bennani, A., Lecas, M., and Spielberger, L., *to be published* (1997)
11. Joulakian, B., and Dal Cappello, C., *Phys.Rev.A*, **47**, 3788 (1993)
12. Lahmam-Bennani, A., Duguet, A., and El Marji, B., "Double ionization mechanisms from triple coincidence experiments" in *Proceedings of the EuroConference on Coincidence Studies of Electron and Photon Impact Ionization,* Belfast, Sept 5-7, 1996, pp20
13. Dal Cappello, C., and Joulakian, B., *to be published* (1997)
14. Brauner, M., Briggs, J., and Klar, H., *J.Phys.B : At.Mol.Opt.Phys.*, **22**, 2265 (1989)
15. Duguet, A., Dupré, C., and Lahmam-Bennani, A., *J.Phys.B : At.Mol.Opt.Phys.*, **24**, 675 (1991)

APPLICATION OF GLASS - CAPILLARY CONVERTERS TO HIGH RESOLUTION EUV SPECTROMETRY : FIRST RESULTS FOR ION - ATOM COLLISIONS

V.Kantsyrev, R.Bruch, R.Phaneuf, V.Golovkina
Physics Department, University of Nevada Reno,Reno,NV89557,USA

D. Schneider
N-Division, Lawrence Livermore National Laboratory, Livermore, CA94550, USA

V.Leroux
University of Paris, Paris, France

We have applied a new more efficient diagnostic method for studies of extreme ultraviolet (EUV) radiation produced by multicharged ions interacting with He gas. Following single and double electron capture and one electron capture plus target-ion excitation in He^{2+} + He collisions the subsequent emitted EUV photons are analyzed with a high-resolution 2.2 m grazing incidence monochromator in conjunction with a new type of glass capillary converter (GCC), specifically designed for the EUV wavelength region (10.0nm< λ <100.0nm). With this new technique a flux density enhancement of spectral line radiation on the entrance window of the monochromator of about 10 has been achieved over a distance of 60 cm. By further optimizing this method an enhancement of the flux density of EUV radiation of the order of 20 to 30 is expected.

INTRODUCTION

In the study of radiative stabilization in highly charged ion-atom collisions experiments [1], one of the main problems is: (i) the analysis and detection of small intensity extreme ultraviolet (EUV), soft X-ray (SXR) and X-ray signals and (ii) the capability to image an extended ion beam source area or volume onto a detection system with a specific cross sectional area. Therefore we have developed and tested for the first time new type of glass capillary converters (GCCs) for the EUV region to increase the spectral sensitivity of our short wavelength spectrometry instrumentation [2,3]. Such devices can concentrate, guide and focus short wavelength radiation from a source onto a detection system, and at the same time optimize the shape of the cross section of the photon beam for specific configuration of the aperture of the analyzing system. In contrast to frequently used optical systems such as grazing incidence toroidal mirrors or multilayer mirrors (MLM) to enhance and focus weak optical signals in the EUV region, GCCs can be easily adapted to more complex imaging problems. These wide band optical systems are cost effective, relatively simple and have the potential to enhance the flux density of radiation by a factor of several hundred [4]. Such GCCs consist of a bundle of capillaries, and the guiding properties are based on the effect of almost total external reflection of radiation [5] from inner smooth surfaces of these capillaries at grazing incidence (from several arc sec to 10° -15°). The basic ideas regarding capillary optics stem from the 1950's [6], and first pioneering experiments with single capillary optical devices were provided in the 1970's [7-10]. One of the first fields of application of simple capillary optics was concerned with bio-medical X-ray microscopy [11] and X-ray structural analysis [12]. At the end of the 1980's and at the beginning 1990's a rapid advancement of this new technology took place leading to more sophisticated multicapillary devices, i.e. concentrators [13-15], and converters [16]. Such GCCs were used in particular for solid state physics experiments [17-18] with guiding and focusing of soft x-ray radiation from a table-top Z-pinch plasma source to a small focus spot (1-2 mm) with a flux density of soft x-rays to 10^6 W/ cm^2. Another application was concerned with X-ray diffraction studies [19].

CP392, *Application of Accelerators in Research and Industry,* edited by J. L. Duggan and I. L. Morgan
AIP Press, New York © 1997

EXPERIMENTAL CONDITIONS

In this work we have designed and tested new GCC devices for studies of EUV photons emitted from multicharged ion beams interacting with He gas [3]. This diagnostic technique is based on the development of the first GCC device specifically constructed for the high-resolution EUV spectrometry of ion-atom collisions [3]. In Fig. 1 we show a schematic layout of our EUV high resolution, high sensitivity 2.2 m grazing-incidence monochromator with GCC. This experimental apparatus is an integral part of the University of Nevada, Reno Electron Cyclotron Resonance (ECR) multicharged ion facility [1]. Our experimental apparatus consists of a target chamber (see Fig. 1) with a differentially pumped gas cell, a collimator system and quadrupole lens for focusing and steering of the ion beam (1-5 μA) and a Faraday cup. The target cell is operated at a gas pressure of about 0.2 - 5 mTorr. The gas pressure in the target cell is controlled via a capacitance manometer.

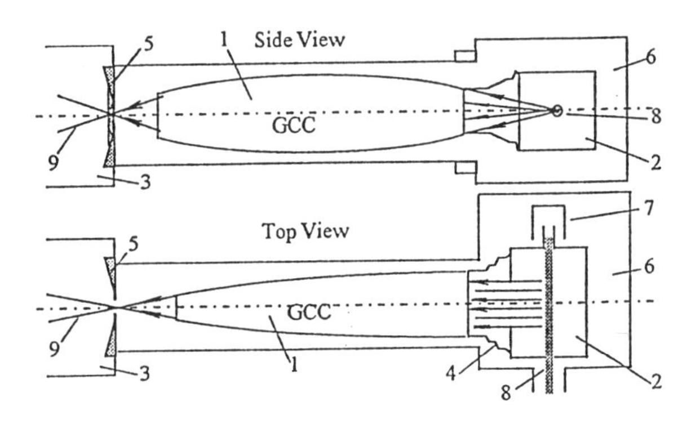

FIGURE 1. Schematic diagram of coupling EUV monochromator to the collision cell with complex EUV-GCC device. 1. GCC, 2. Gas cell, 3. Grazing incidence monochromator, 4. Flexible housing, 5. Monochromator entrance slit, 6. Target chamber, 7. Faraday cup, 8. Ion beam, 9. EUV radiation

The EUV photons emitted by excited projectile and / or target ions exit the target cell window at 90° with respect to the ion beam and are collected, guided and focused by our new GCC device onto the entrance slit of the 2.2m grazing incidence monochromator. Selected photons are detected by a channel electron multiplier (CEM). Data acquisition and control is performed with a PC-CAMAC computer system, where the photon intensity is normalized with respect to the amount of charge collected in the Faraday cup.

Specifically for this investigation in the EUV spectral region a new type of GCC device with maximum efficiency in the wavelength range 25.0 - 60.0 nm has been constructed [3]. In our experiment the GCC system images a horizontal (cylindrical) ion beam segment ("line"-shape source with an effective diameter 2mm) and focuses the radiation onto the vertical entrance slit of the monochromator ("line"-shape focus as rectangular area 0.8 *10 mm). The distance from the gas-cell to the entrance slit of the monochromator is 60 cm. We note that the GCC consists of 170 quartz monocapillaries with an inner diameter of 390μm and outer diameter of 590 μm. The monolithic aluminum body of the GCC is 480 mm long and 57 mm high. The capillaries are fixed inside this body by several metal support patterns. For the GCC entrance a rectangular cross section of 24x12 mm^2 has been adopted, which corresponds to the exit window of the gas cell. We further note that the GCC exit dimensions are 3x21 mm^2. Another important design characteristic is concerned to the effective photon transmitting area, divided by the total illuminated area. For the entrance and the exit of the GCC these ratios are 7% and 40%, respectively. The specific construction of the GCC ensures that the diffraction grating of the spectrometer is illuminated over a homogeneous area, defined by the grating mask and entrance slit. In order to reduce the outstreaming of gas, the GCC shell is directly connected to the gas cell via a flexible housing. For alignment purposes the GCC optics is mounted on a (X, Y, Z) positioner in a vacuum chamber and can be moved along three mutually perpendicular directions. The intrinsic focal distances of the GCC at the entrance and exit plane are 50 and 60mm, respectively (see Fig.1). The efficiency ν of concentration of the GCC is approximatly 10^{-2} (ν is the efficiency of concentration of radiation, defined as the ratio of the emitted photons reaching the rear focal area of the GCC per unit time compared to the total number of photons from the source area per unit time emitted into the solid angle 4π).

In addition we have tested our system using as prototype test cases EUV spectra arising from N^{5+} + He, O^{6+} +He and Ar^{8+} + He collisions. For the experimental determination of the coefficient of enhancement η of the GCC (η is defined as the ratio of flux density on the monochromator entrance slit with and without GCC) we used strong well resolved spectral lines. Using this method we obtained as a lower limit η=4.7 for λ=15.0 nm, η=10 for λ=30.4 nm, η=5 for λ=50.8 nm and η=1.7 for λ=104 nm. Our theoretical estimates of η predict a maximum enhancement of 20 to 30 in the spectral region between 30.0 and 60.0 nm for such a "line source - line focus" GCC compared with experimental values of

η = 100 in the 0.8 -1.0 nm spectral range for "point source - point focus" GCC system [4].

We have determined also the effective resolution of our detection system with the GCC optics [20,21]. Using 40 to 200 μm monochromator entrance slits, a wavelength resolution of 500 to 100 has been achieved. The observed Doppler broadening with the GCC from projectile spectral lines is about 5 to 10 %. This broadening results from a 5° to 7° angular acceptance angle of the capillaries in our set - up.

Our experimental results clearly show that this new spectroscopic technique can be used successfully to investigate electron correlation effects in transfer-excitation and ionization-excitation process of He by multicharged ions. In Fig.2 we have displayed a high resolution spectrum produced in 20 keV He^{2+} + He collisions. As may be seen, the transitions He^{+} (np)-> 1s with up to n = 5 are clearly resolved.

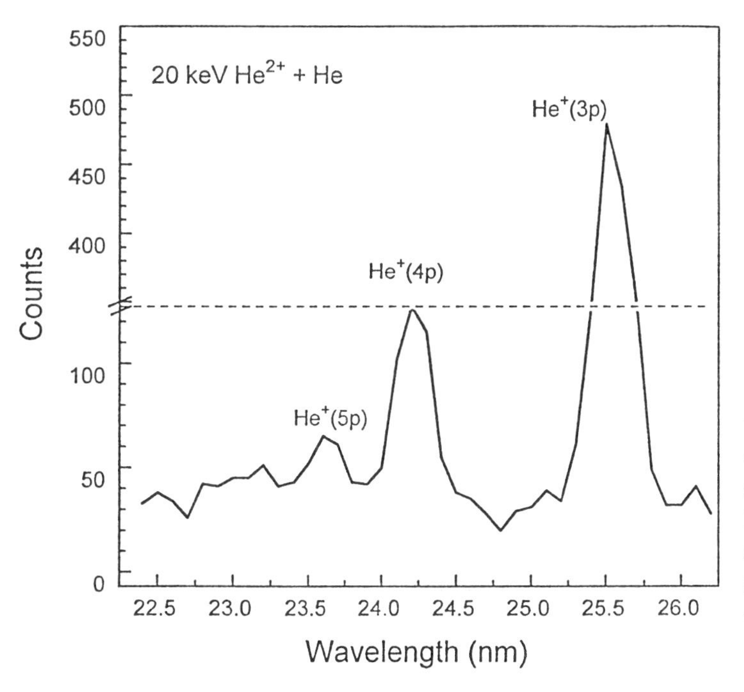

FIGURE 2. Characteristic high resolution EUV emission spectra showing the HeII 3p->1s, 4p->1s, and 5p->1s radiative decays following 20 keV He^{2+} + He single collisions. The observed lines arise mainly from one-electron capture and target-ion excitation.

The signal statistics in Fig.2 have been dramatically improved compared to other measurements without the GCC [3, 20]. The enchanced sensitivity and improved resolution allows us to perform the first detailed studies on transfer-excitation in slow ion-helium collisions [22] with state selective observation of the exited He^{+} (np) states with n ≥ 3. The He^{2+} + He collision system is one of the most fundamental test cases where the role of one and two electron processes can be studied. In this work we have been particularly interested in the population of target as well as projectile He^{+}(np) states, which can not be distinguished under 90 degree observation. Such states may be produced by the following processes:

One-Electron Capture

He^{2+}(P)+He($1s^2$, T)->He^{+}(np, P)+He^{+}(1s, T) ; (1)

Capture Plus Simultaneous Excitation of the Target Ion

He^{2+}(P)+He($1s^2$, T)->He^{+}(1s, T)+He^{+}(np, P) ; (2)

Single Ionization and Simultaneous Excitation of the Target Ion

He^{2+}(P)+He($1s^2$,T)->He^{2+}(np,T)+e^-. (3)

Here P and T represent the projectile ion and target atom respectively. From our data we have deduced He^{+}(np) emission cross sections normalized to the 2p cross section of Folkerts et al. [23]. Our 3p and 4p results at 20 keV impact energy (5 keV amu^{-1}) are in excellent agreement with the data from the Netherlands group. For He^{+}(5p) we report here the first cross section, namely σ (5p) = $0.45x10^{-18}$ cm^2 .

CONCLUSION

In conclusion, we have demonstrated for the first time that complex GCC optics can dramatically enhance collisional EUV spectra arising from multicharged ion-atom collisions. Our new method is suitable not only for high-resolution collisional spectroscopy but also for accurate cross section measurements related to multielectron transfer processes at intermediate and low velocities.

ACKNOWLEDGMENT

This study has been supported by the NAS/CAST Program, ACSPECT Corp., Reno, Nevada and in part by LLNL and AVS.

REFERENCES

1. R. Phaneuf, "Advanced sources of highly charged ions", *in Experimental Methods in the Physical Sciences,* (F. B. Dunning and R. G. Hullet, Ed), Acad. Press, **29A**, 169-188 (1995).

2. V. Kantsyrev, R. Bruch, M. Bailey, A. Shlyaptseva, *Appl. Phys. Lett.*, **66**, N26, 3567 (1995).
3. V. Kantsyrev, R. Bruch, Rev. Sci. Instr.(in press).
4. V. Kantsyrev, N. Mingalev, O. Petrukhin, S. Pikuz, W. Romanova, T. Shelkovenko, A. Shlyaptseva, A. Faenov, Quantum electronics, **23**,12, 1181 (1993).
5. A. Compton, *Phil. Mag.*, **45**, 1121 (1923).
6. R. Pound, G. Rebka, *Phys. Rev. Let.*, **3**, 439 (1959).
7. P. Mallozi, H. Epstein, R. Yung, D. Applebaum, B. Fairand, W. Gallangher, R. Uecker, M. Muckerheide, *J. Appl. Phys.*, **45**, 1891 (1974).
8. O. Ananyin, Yu. Bykovsky, V. Kantsyrev, Yu.Kozyrev, Inventor Certificate No 520863, filled 15 October 1974 (USSR), *Bull. Izobret. USSR*, **11**, 229 (1979).
9. W. Vetterling, R. Pound, *J. Opt. Soc. Am.*, **66**, N10, 1048 (1976).
10. V. Arkad'ev, M. Kumakhov, Poverkhn (Fiz. Khim. Mekh) 10, 25 (1986).
11. Yu. Bykovsky, V. Kantsyrev, B. Komarov, Yu Kozyrev, N Permyakov, P. Pleshanov, *Archive of Pathology (USSR)*. 12, 78 (1978).
12. A. Rindby, *Nucl. Instr. Meth.*, **A249**, 536 (1986).
13. V. Arkad'ev, A. Kolomiitsev, M. Kumakhov, I Ponomarev, I. Knodeev, Yu. Chertov, I. Shakhparonov, *Sov., Phys., Usp.*, **32**, 271 (1989).
14. O. Ananyin, Yu. Bykovsky, A. Zhuravlev, V. Znamensky, V. Kantsyrev, S. Frolov, *Sov. Tech Phys. Lett.*, **16**, N1, 65 (1990).
15. V. Kantsyrev, A. Kologrivov, K. Kopytok, A. Shlyaptseva, *Proc. 12 Intern. Conf. on Coherent and Nonlinear Optics* (KINO'91), **1**, 103 (Leningrad, 1991).
16. V. Kantsyrev, K. Kopytok, A. Shlyaptseva, *"Dense Z - pinches". Third Intern. Conf.* (London, 1993), *AIP Conf. Proc.* N299, 612.
17. V. Kantsyrev, A. Kologrivov, K. Kopytok, A Zyabnev, S. Kraevsky, A. Shlyaptseva, *Proc.II Intern. Conf. "Physics of Glass Solid State"*(Riga, Latvia, 1991),73.
18. O. Ananyin, Yu. Bykovsky, V. Kantsyrev, A. Kologrivov, K. Kopytok, A. Shlyaptseva, M. Yakovlev, *Sov. Tech. Phys. Let.*, **18**, N5, 331 (1992).
19. V. Kovantsev, J. Pant, V. Pantojas, N. Nazaryan, T. Hayes, P. Persaus, *Appl. Phys. Let.*, **62**, N23, 2905 (1993).
20. R.Bruch, V.Kantsyrev, V.Golovkina, A.Shlyaptseva, I.Golovkin, R.Phaneuf, D.Schneider, Rev.Sci. Instr. (in press).
21. R.Bruch, V.Kantsyrev, R.Phaneuf, V.Golovkina, A.Shlyapseva, I.Golovkin, D.Shneider, Phys. Scr.(in press).
22. W.Fritsch,. Phys. Lett. A. **192**, 369 (1994).
23. H.O Folkerts, F.W.Bliek, L.Meng, R.Morgenstern, M. van Hellermann, H.P.Summers, R.Hoekstra, J.Phys.B, **27**, 3475 (1994).

PHOTO-DOUBLE IONIZATION AND THE ROLE OF INDIRECT PROCESSES IN MOLECULES

G.C. King, D.B. Thompson, G. Dawber, and N. Gulley

Department of Physics and Astronomy, University of Manchester, Manchester M13 9PL, UK

A novel threshold-photoelectron photoelectron coincidence technique has been used to study the role of indirect processes in molecular photo-double ionization. For the case of the molecule CO, photo-double ionization is observed below the adiabatic double ionization threshold and is interpreted in terms of the excitation of intermediate singly-charged ion states.

INTRODUCTION

There has been significant progress in the study of photo-double ionization processes in recent years, both experimentally (1-3) and theoretically (4,5). For atomic photo-double ionization the rare gases have been studied in greatest detail with the main goal of investigating the role of electron correlations in the three-body Coulomb breakup. So far most of the attention has focused on the direct ionization process. Attention is now, however, turning to indirect photo-double ionization mechanisms. These mechanisms may play a dominant role in the photo-double ionization of heavier rare gases, via creation of highly excited, singly-charged positive ion (cation) states (6). Also of increasing interest is the study of photo-double ionization in molecular systems. Very little experimental information exists about these targets, but their multi-centered nature is expected to produce new effects. For example, dissociation processes now become involved.

Adiabatic photo-double ionization of a molecule corresponds to absorption of an energetic photon with the immediate ejection of two electrons. The resulting doubly-charged positive ion (dication) is always unstable against dissociation into two cation fragments (however, a binding force produced by exchange interactions between the remaining electrons may create a local potential minimum capable of supporting vibrational motion). Cation pair production can however also occur at photon energies below a molecule's double ionization threshold (here, *double ionization threshold* refers to the adiabatic formation energy of the lowest dication state); this was first observed in photoion-photoion-coincidence (PIPICO) studies of water (7). Clearly, the two cations cannot result from adiabatic photo-double ionization, but must be due to other indirect double ionization mechanisms. Here we describe a novel threshold-photoelectron photoelectron coincidence (TPEPECO) technique to investigate these mechanisms below the double ionization threshold of CO.

Apparatus and Experimental Method

The main elements of the experimental apparatus are a photon beam, an effusive gas beam, and an electron coincidence spectrometer consisting of two electron energy analyzers. Tunable energy photon beam was provided by the Daresbury Laboratory Synchrotron Radiation Source via a toroidal grating monochromator and entered the apparatus through a 1 mm diameter glass capillary where it intersected the gas emanating from a hollow needle. The coincidence spectrometer is described in detail elsewhere (8). It is based on a pair of 127° cylindrical deflector analyzers. One analyzer is operated in the threshold mode by being tuned to collect zero-energy electrons using the penetrating field technique (9). The other analyzer is operated in a conventional mode to detect finite-energy electrons.

The apparatus was used to detect coincidences between the zero-energy electrons detected by one analyzer and the finite-energy electrons detected by the other. The study includes two types of measurement. In the first type, the photon energy was held at a constant energy while the collection energy of the conventional analyzer was scanned in uniform steps. In the second type, the conventional analyzer was set to detect electrons at a constant energy, E_E ,and then the photon energy was scanned in uniform steps. In all measurements, the threshold analyzer operated with an energy resolution of 20 meV, whereas the resolution of the conventional analyzer was 50 meV; the photon beam energy spread was 100 meV. All spectra have been corrected for variations in photon intensity with energy by normalization to a photodiode output, acquired simultaneously. Also, spectra of the first type have been

CP392, *Application of Accelerators in Research and Industry,* edited by J. L. Duggan and I. L. Morgan
AIP Press, New York © 1997

corrected for variations in analyzer transmission as a function of electron energy; this transmission function was reasonably flat with a maximum at 1.5 eV. To calibrate the photon energy scale, threshold-photoelectron spectra of argon were collected in the region of the Ar^+ $3s^{-1}$ (^+S) line at 29.24 eV (10). The energy calibration of the conventional analyzer was then determined from the position at which the same argon state appeared in photoelectron spectra acquired by it. These calibration procedures led to an overall uncertainty in the energy scales of ±30 meV.

Results and Discussion

We have recently collected TPEPECO spectra of CO. Figure 1 shows four constant photon energy coincidence spectra of the first type, where the collection energy of the conventional analyzer was scanned from 0 to 2 eV. The photon energies for the bottom three spectra all lie below the CO double ionization threshold energy of 41.25 eV (11), while the photon energy for the top spectrum lies above it. The bottom three spectra appear nearly identical; in each a prominent structure arises at the same kinetic energies. Since the photon energies of the bottom three spectra all lie below the double ionization threshold, the signal appearing in them cannot originate from adiabatic double ionization.

A TPEPECO spectrum of the second type, in which the collection energy of the conventional analyzer is fixed at E_E = 0.5 eV, is shown in Fig. 2. This TPEPECO spectrum follows the evolution with photon energy of the 0.5 eV peak of Fig. 1 from below the lowest CO dissociation limit to well above the double ionization threshold. In this TPEPECO spectrum, the coincidence signal appears as a broad band with superimposed fine structure. We note that above 41.74 eV (0.5 eV above the double ionization threshold) the superimposed fine structures can all be identified as unresolved lines due to overlapping vibrational sequences of dication states; when shifted by 0.5 eV, the peak positions match those seen in threshold-photoelectrons coincidence (TPEsCO) spectra of CO (11,12). Thus, all the peak positions in the top spectrum of Fig. 1 have counterparts here. The spectrum also shows structure below the double ionization threshold, confirming the existence of double ionization processes below this energy.

In photoion photoion coincidence (PIPICO) studies of CO, Lablanquie et al (13) observed an onset for C^++O^+ pair production at a photon energy of 38.4 eV, i.e. below the double ionization threshold. To explain this observation, the authors suggested the existence of a two-step double ionization mechanism. This two-step mechanism, also proposed by Winkoun et al (7), involves the creation and

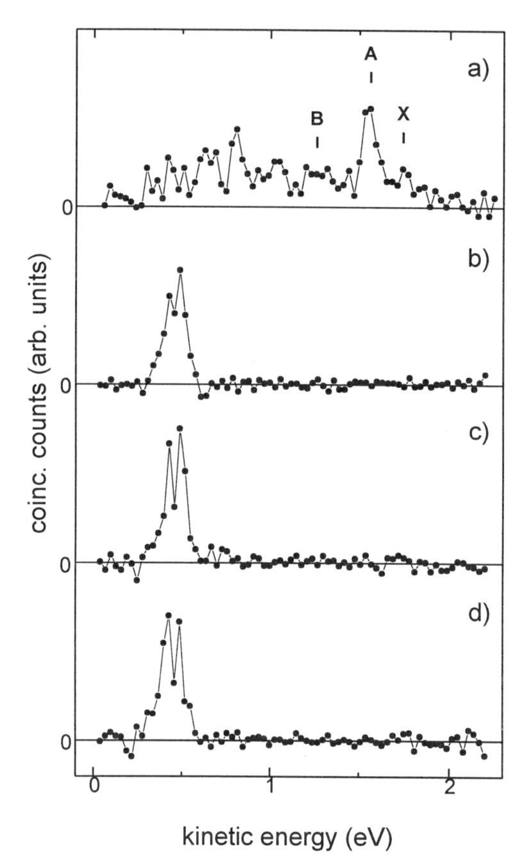

Figure 1. TPEPECO spectra of CO collected at the following constant photon energies: a) 43 eV b) 41 eV c) 40 eV d) 38 eV. In the top spectrum the X, A, and B mark the lowest vibrational levels of the first three CO dication states, as determined from Ref. (11).

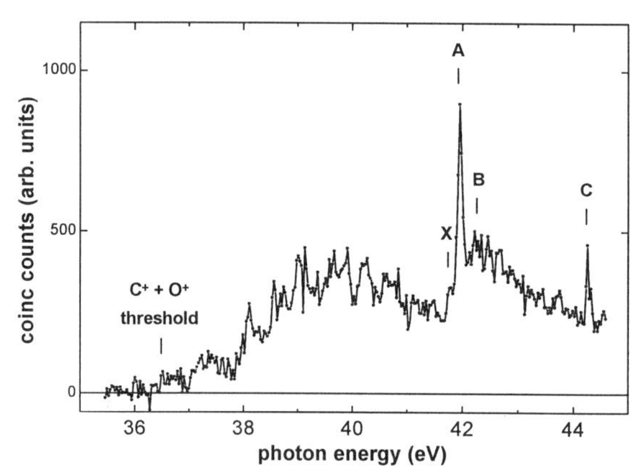

Figure 2. TPEPECO spectrum of CO showing the yield of coincidences between zero-energy and 0.5 eV photoelectrons. The X, A, B and C mark the lowest vibrational-level excitation-energies of the first four CO dication states, as determined in Ref. (11).

subsequent electronic decay of a "precursor" state corresponding to an excited state of the singly-charged ion; $CO + h\nu \rightarrow (CO^+)^* + e^- \rightarrow CO^{2+} + 2e^- \rightarrow C^+ + O^+ + 2e^-$. The precursor ion state can be repulsive, so that electronic decay may occur along a continuum of internuclear separations. As proposed by Becker et al (14), it may happen that electronic decay occurs after complete dissociation of the molecular ion, via autoionization of an atomic fragment; for example, $CO + h\nu \rightarrow (CO^+)^* + e^- \rightarrow C^+ + O^* + e^- \rightarrow C^+ + O^+ + 2e^-$. Such a process was first identified in the photo-double ionization by He IIα radiation of molecular oxygen *above* its double ionization threshold (15).

An analysis of measured kinetic energy release distributions of C^++O^+ at photon energies close to the double ionization threshold (16) suggests that the associated electrons must have kinetic energies very close to zero. That these two electrons cannot both have zero kinetic energy has been shown by TPEsCO spectroscopy (11,12). Support for the existence of the precursor ion states appears in recent inner-valence photoelectron spectra of CO (17). These spectra show a single band, superimposed with dense fine structure, stretching across the entire binding energy region between 36 and 42 eV. The accompanying calculations associate this broad band with repulsive ion states produced by ionization of the 3σ inner-valence shell of CO. The superimposed fine structure is attributed to single ion Rydberg series converging to vibrational sequences of dication states

In the bottom three spectra of Fig. 1, the signal occurs at constant kinetic energy regardless of photon energy; this is indicative of an Auger or autoionization process. Molecular autoionization is expected to produce electrons exhibiting a wide, slowly varying distribution of kinetic energies (14). In contrast, in the bottom three coincidence spectra the signal appears to be composed of two narrow lines. In fact, these energies correspond to kinetic energies associated with well-known atomic oxygen autoionization transitions (18,19). Thus, the signal appears to originate from complete dissociation of the molecular ion followed by autoionization of an atomic oxygen fragment. We note that a high degree of final state selectivity exists in these spectra; in photoelectron spectra collected at comparable photon energies (14) many more lines, also associated with atomic oxygen autoionization transitions, appear at kinetic energies below 2 eV. In contrast with the bottom three spectra, the top spectrum in Fig. 1 exhibits a coincidence signal that is continuous across the entire kinetic energy range, implying the presence of different ionization processes. As the photon energy lies above the double ionization threshold, this result is not unexpected. The energy positions of peaks in the spectrum, subtracted from the photon energy, match those seen in TPEsCO spectra (11,12) that are associated with the population of overlapping vibrational levels of several dication states.

In the TPEPECO spectrum of Fig. 2, the coincidence signal appears as a broad band with superimposed fine structure. In the TPEPECO spectrum of Fig. 3, the coincidence signal extends to about 36.5 eV, much lower than the onset for C^++O^+ production observed by Lablanquie et al (13). Indeed, the onset appears at the low est possible photon energy, i.e. the energy limit for dissociation into C^++O^+, lying at 35.99 eV (20,21), summed with the electron kinetic energies detected in this experiment. With the assumption that we are detecting coincidences due to precursor ion states whose formation is accompanied by ejection of a near-zero-energy electron, we can interpret the TPEPECO spectrum as a representation of the spectroscopy of the precursor ion states.

Summary

The present TPEPECO measurements locate a coincidence signal from electrons associated with cation pair production in CO at photon energies below the adiabatic double ionization threshold. The variation of the signal as a function of electron kinetic energy suggests that C^++O^+ production proceeds by complete dissociation of a excited molecular ion followed by autoionization of the atomic oxygen fragment. The signal reveals an onset at a photon energy much lower than the previously measured onset for C^++O^+ production. The structure of this signal is attributed to the precursor ion states involved in this process. However, with the current lack of theoretical information about the evolution of inner valence single ion states of CO, we can interpret this structure only qualitatively. The present measurements also demonstrate the potential of the TPEPECO technique to study the spectroscopy and evolution of precursor states involved in double ionization processes.

References

1. Schwarzkopf O., Krassig B., Elminger J., and Schmidt V., *Phys. Rev. Lett.* **70**, 3008 (1993).
2. Huetz A., Lablanquie P., Andric L., Selles P. and Mazeau J., *J. Phys. B: At. Mo. Opt. Phys.* **27**, L13 (1994).
3. Dawber G., Avaldi L., McConkey A. G., Rojas H., MacDonald M. A., and King G. C., *J. Phys. B: At. Mo. Opt. Phys.* **28**, L271 (1995).
4. Maulbetsch F. and Briggs J. S., *J. Phys. B: At. Mo. Opt. Phys.* **26**, 1679 (1993).
5. Maulbetsch F. and Briggs J. S., *J. Phys. B: At. Mo. Opt. Phys.* **27**, 4095 (1995).
6. Dawber G., Avaldi L., Zubek M., McConkey A.G., Rojas H., Gulley N., MacDonald M. A., King G. C., and Hall R. I., *Can. J. Phys* (in press).
7. Winkoun D., Dujardin G., Hellner L., and Besnard M. J., *J. Phys. B: At. Mo. Opt. Phys.* **21**, 1385 (1988).
8. Hall R. I., McConkey A. G., Ellis K., Dawber G., Avaldi L., MacDonald M. A., and King G. C., *Meas. Sci. Tech.* **3**, 316 (1992).
9. Cvejanovic S. and Read F. H., *J. Phys. B: At. Mo. Opt. Phys.* **7**, 1180 (1974).
10. Hall R. I., Avaldi L., Dawber G., Rutter P. M., MacDonald M. A., and King G. C., *J. Phys. B: At. Mo. Opt. Phys.* **22**, 3205 (1989).

11. Hochlaf M., Hall R. I., Penent F., Kjeldsen H., Lablanquie P., Lavollee M., Eland J. D. H., *Chem. Phys.* (in press).
12. Dawber G., McConkey A. G., Avaldi L., MacDonald M. A., King G. C., and R. I. Hall, *J. Phys. B: At. Mo. Opt.Phys.* **27**, 2191 (1994).
13. Lablanquie P., Delwiche J., Hubin-Franskin M.-J., Nenner I., Morin P., Ito K., Eland J. D .H., Robbe J.-M., Gandara G., Fournier J., and Fournier P. G., *Phys. Rev. A* **40**, 5673 (1989).
14. Becker U., Hemmers O., Langer B., Menzel A., Wehlitz R., and Peatman W. B., *Phys. Rev. A* **45**, R1295 (1992).
15. Price S. D. and Eland J.H.D., *J. Phys. B: At. Mo. Opt. Phys.* **24**, 4379 (1991).
16. Masuoka T., *J. Chem. Phys.* **101,** 322 (1994).
17. Baltzer P., Lundqvist M., Wannberg B., Karlsson L., Larsson M., Hayes M. A., West J. B., Siggel M. R. F., Parr A.C., and Dehmer J. L., *J. Phys. B: At. Mo. Opt. Phys.* **27**, 4915 (1994).
18. Wills A. A., Cafolla A. A., and Comer J., *J. Phys. B: At. Mo. Opt. Phys.* **24**, 3989 (1991).
19. Cermak V. and Sramek J., *J. Elec. Spec. and Rel. Phen.* **2**, 97 (1973).
20. Moore C. E., *Atomic Energy Levels* **1**, Washington D.C.: U.S. GPO, 1958.
21. Huber K. P. and Herzberg G., *Molecular Spectra and Molecular Structure IV. Spectroscopic Constants of Diatomic Molecules,* 2nd ed., New York: Van Nostrand Reinhold, 1979.

DOUBLE PHOTOIONIZATION IN ATOMS

L. AVALDI

IMAI-CNR, Area della Ricerca di Roma, CP10, 00016 Monterotondo, Italy

The study of double photoionization is of fundamental interest because the process is entirely due to relaxation and electron correlation effects and it allows one to tackle directly the still unsolved three-body Coulomb problem. The results of recent electron-electron coincidence experiments, which are the most suited tool to investigate double photoionization, are presented and discussed.

INTRODUCTION

Double photoionization (DPI) occurs when a single, energetic photon is absorbed by an atom resulting in the emission of two electrons. The process is of fundamental interest because it is entirely due to relaxation and correlation effects. DPI results in a doubly charged ion and two escaping electrons, which are subject to long range Coulomb forces. In the case of helium, where a bare nucleus is produced and no processes other than direct double ionization occur, DPI represents the most suited process to study the still unsolved three-body Coulomb problem. Several kind of informations on DPI can be obtained by the measurements of the total cross section σ^{2+} or the partial cross sections $d\sigma/dE_i$ and $d^2\sigma/dE_i d\Omega_i$, where E_i (i=1,2) is the kinetic energy of one of the two electrons. However only the measurement of the triple differential cross section $d^3\sigma/dE_1 d\Omega_1 d\Omega_2$, TDCS, enables the complete determination of the energy and angular pattern of this break-up reaction. Such a measurement implies experiments where the two photoelectrons are detected in coincidence, after energy and angular selection, and provides the deepest "insight" into the process. After a brief description of a typical experimental set-up, the main achievements in the DPI studies of rare gases by coincidence experiments are reviewed.

EXPERIMENTAL

Double ionisation potentials of rare gases are located above 30eV, thus a photon source in the VUV region is needed. Apart from the first measurements [1,2] performed with a HeII lamp, the most suited source for these studies is monochromatized and tunable synchrotron radiation.
A typical apparatus (Fig.1) comprises two electrostatic analyzers which detect the two electrons produced by the interaction of the incident light with the target gas. The signals from the two analyzers are processed by pulse electronics units and finally fed into a time-to-amplitude converter (TAC). The fact that the coincidences arising from electrons produced from the double photoionisation event are correlated in time, enables one to separate them from accidental coincidences which are uncorrelated. In the experiments one of the analyzers, analyzer 1 in Fig.1, is kept fixed while the other one is rotated. The experiments are classified according to how the two photoelectrons share the excess energy $E=h\nu-IP^{2+}=E_1+E_2$, where IP^{2+} is the double ionisation potential. In an equal sharing experiment $E_1=E_2=E/2$, while in an unequal sharing experiment $E_1\neq E_2$ and usually one of the photoelectrons is much faster than the other one. The low value of the DPI cross section and the limited beam-time at synchrotron radiation laboratories have hampered these studies and such experiments became feasible only recently [3]. Since that time, considering that a complete mapping of the TDCS needs a large set of measurements in different kinematic regions, new experimental approaches have been developed. These new set-ups either explore simultaneously the TDCS over the full space with an unique analyzer characterized by a large solid angle [4-6] or by a set of analyzers independently controlled and rotatable [7].

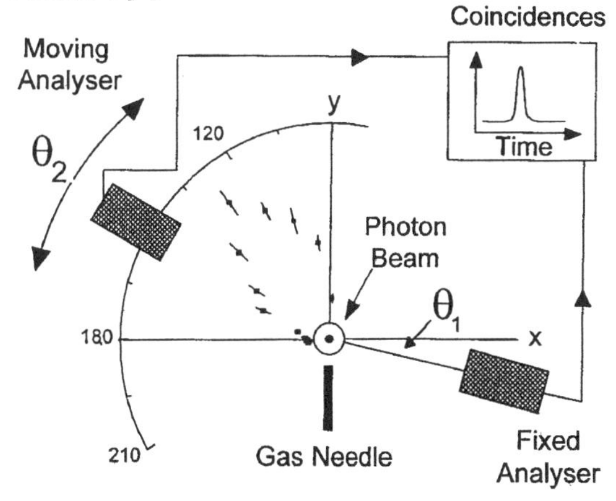

FIGURE 1 Schematic of a coincidence apparatus

RESULTS

TDCS of helium

Despite the fact that He represents the most favourable system for DPI studies the first experimental data were not obtained until 1993 [8], due to the low value of σ^{2+}. Since then few other experiments have been reported [9-13].

CP392, *Application of Accelerators in Research and Industry,* edited by J. L. Duggan and I. L. Morgan
AIP Press, New York © 1997

Finally in 1995 [14] a set of data at E=20eV, in equal sharing condition, has been put on the absolute scale. Absolute TDCS's enable a more detailed comparison with theoretical models that sometime predict similar shapes, but sizes that may differ by orders of magnitude. In Fig.2 some examples of the TDCS measured for both sharing conditions are shown. This figure displays the main features of the TDCS patterns when hν is varied from the threshold region up to 130eV.

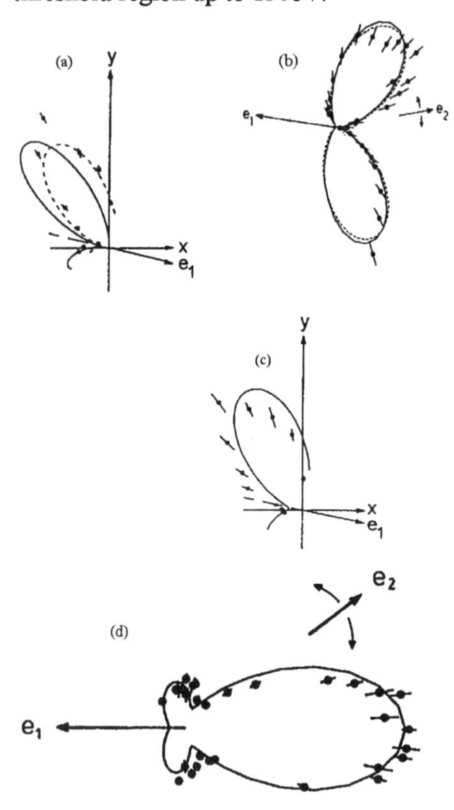

FIGURE 2 He TDCS for equal sharing conditions at E=0.6 (a) [11] and 20eV (b) [8] and in unequal sharing conditions E_1=1.85, E_2=0.15(c) [11] and E_1=0.5, E_2=47.9 eV(d) [10].

The main features can be summarized as follows :

- equal energy sharing experiments (Fig. 2a,b)

the TDCS's display a node at both ϑ_{12}=180° and 0° and two lobes almost symmetric about the direction of the fixed analyzer independent of E.

- unequal energy sharing experiments (Fig. 2c,d)

a dramatic change in shape (Fig.2d) is observed at the highest excess energy, with most of the intensity in the direction opposite to the direction of the slow electron. For almost the same ratio E_1/E_2, but at 2eV excess energy (Fig. 2c) no appreciable change in shape is observed moving from the equal to the extreme unequal energy sharing.

The trend shown in Fig.2 has been confirmed by an extensive set of measurements from the Paris group[12].

DPI experiments in a coplanar geometry can be described by placing the z axis along the photon direction and x and y along the polarization ellipse axis. Then assuming complete linear polarization of the incident light the TDCS can be written [15]:

$$\mathrm{TDCS}_{\varepsilon} = \left| \begin{array}{l} a^{g}(E_1,E_2,\vartheta_{12})(\cos\vartheta_1 + \cos\vartheta_2) + \\ a^{u}(E_1,E_2,\vartheta_{12})(\cos\vartheta_1 - \cos\vartheta_2) \end{array} \right|^2 \quad (1)$$

where ε=x is the direction of the polarization of the incident radiation, $\vartheta_1=(\mathbf{e},\mathbf{k}_1)$, $\vartheta_2=(\mathbf{e},\mathbf{k}_2)$ and the relative angle $\vartheta_{12}=(\mathbf{k}_1,\mathbf{k}_2)$. Equation (1) disentangles the geometrical properties of the TDCS from the dynamical ones, which are included in the complex amplitudes a^g and a^u, respectively symmetric and antisymmetric in E_1 and E_2. The effects of the electron-electron and electron-residual ion interactions are included in the ϑ_{12} and E dependence of the amplitudes. In the case of equal sharing equation (1) reduces to

$$\mathrm{TDCS}_{\varepsilon} = \left|a^{g}(E/2,E/2,\vartheta_{12})\right|^2 (\cos\vartheta_1 + \cos\vartheta_2)^2 \quad (2)$$

and all the physical information is in the correlation factor

$$C(E,\vartheta_{12}) = \left|a^{g}(E/2,E/2,\vartheta_{12})\right|^2$$

The two nodes can be easily understood from equation (2) : the vanishing of the geometrical term results in the node at $\vartheta_{12}=\pi$, while the electron-electron repulsion prevents the emission of the two photoelectrons in the same direction. Several models have been proposed to describe DPI. On the one hand, by revisiting the Wannier model Huetz et al.[15] have predicted the shapes of the TDCS by analyzing the spatial symmetries of the electron-pair wavefunction. Kasansky and Ostrovsky [16] instead have proposed an extended Wannier model which accounts for the boundary conditions at the border of the reaction zone and allows one to calculate numerically the time evolution of the two electrons in the free zone. On the other hand, other authors [17-20] have performed ab-initio calculations. The differences between the proposed models stem either from the description of the two electron wavefunctions or from the way of calculating the scattering amplitude. Coulomb wave functions with angle- and velocity-dependent effective charges have been used by LeRouzo and Dal Capello[17], Proulx and Shakeshaft [19] and Pont and Shakeshaft [20], while Maulbetsch and Briggs [18] made use of the 3-C wavefunctions where all the interactions in the final state are treated on equal footing. Then Maulbetsch and Briggs[18], as well as LeRouzo and Dal Capello[17], evaluated the dipole matrix element directly, while Proulx and Shakeshaft [19] and Pont and Shakeshaft [20] in their 2SC model used a flux formula.For the sake of simplicity, in Fig. 2 the equal sharing data have been compared with the Wannier model and the 3-C model. In the Wannier model the correlation factor has been represented by a gaussian function

$$C(E,\vartheta_{12}) = \exp\left[-4\ln 2(180^{\circ} - \vartheta_{12})^2 / \vartheta_{1/2}^{2}\right], \vartheta_{1/2} = \vartheta_0 E^{1/4}$$

By using ϑ_0 as a free fit parameter a satisfactory representation of the data (dashed curve in Fig.2a,b) is obtained when ϑ_0 assume the values of 77±3.5 and 43±3 deg eV$^{-1/4}$. Several values have been proposed for ϑ_0, ranging from 66.7 to 103° [16], however a recent analysis [16] on the available experimental data predicts a decrease of ϑ_0 with increasing excess energy. Good agreement with the data is also obtained by the 3C (full line in Fig.2) and the 2SC (not shown in the figure) models.

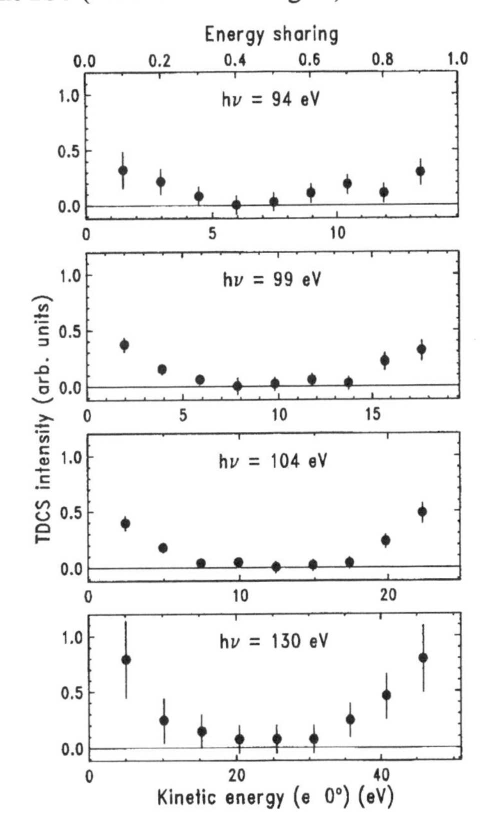

FIGURE 3 TDCS for DPI of He versus E_1. The emission directions are along the direction of the linear polarization at ϑ_{12}=180°.

However when the absolute TDCS at E=20eV are considered the 2SC results are in remarkable agreement with the experiments, while the 3C calculations depending on the adopted gauge can overestimate the experiment by a factor of ten [21].

The unequal sharing kinematics is the most severe test for the models. However, some of the experimental observations can be explained in terms of equation(1). From (1) it is evident that only the a^u amplitude is responsible for the filling of the node at $\vartheta_{12}=\pi$ (Fig. 2d). On the other hand the fact that at the lowest excess energy the node still exists indicates that close to threshold the a^g amplitude is dominant all over the angular range. These observations have been confirmed by an experiment, where a coincidence time-of-flight spectrometer has been used to collect the photoelectron pairs emitted at $\vartheta_{12}=\pi$ ($\vartheta_1=0,\vartheta_2=\pi$) [13]. The measurements, Fig.3, show that near threshold the TDCS does not vary appreciably with the energy sharing, while vice versa a rapid variation is observed moving away from the equal sharing condition at higher hν. Ab-initio calculations reproduce satisfactorily the data at larger E's (Fig.2d) while some discrepancies are observed approaching the threshold (Fig.2c).

TDCS of the heavier rare gases

The process of DPI in He is fully described by one electron-pair wavefunction, namely the ^1P_o one, and the shape of the TDCS pattern is subject to severe constraints. In the case of the heavier rare gases, where the doubly charged ion ground state is np^4, the ^3P, ^1D and ^1S configurations are possible and therefore different electron-pair wavefunctions exist [15]. The first investigation in the equal sharing kinematics of the DPI of Ne, with the atom left in $2s^22p^4$ $^1S^e$ state has shown a surprising pattern [4,21]. In this case conservation rules impose the ^1P_o symmetry for the outgoing electron pair, the same symmetry as for He. Therefore a two lobe pattern was expected. Surprisingly the TDCS displayed a "butterfly" shape with two small additional lobes. The physical explanation of the extra lobes in an independent particle model relies on the fact that the photoelectrons ejected from a p-orbital have two continua, namely εsεp and εdεp, accessible. A description with a factor accounting for this and a correlation factor similar to $C(E,\vartheta_{12})$ reproduces satisfactorily the observed features [4,21]. Another explanation [16] considers the two active electrons in a p^2 $^1S^e$ initial state which contains a $\cos\vartheta_{12}$ factor due to the property of the spherical harmonics. It is this factor that leads to the extra-nodes in the TDCS.

DPI in the heavier rare-gases may also occur via a two step mechanism with either an inner shell ionization followed by Auger decay or the formation of a satellite state of the singly charged ion and its subsequent autoionization. The indirect processes alter the magnitude as well as the shape of the TDCS as shown in a recent extensive study in the region of the $Ne^+2s2p^4(^3P^o)3p$ $^2S^e$ resonance by the Paris and Freibourg group [23]. A further aspect has to be considered when studying indirect DPI processes. If the energy of the incident photon beam is such that the photoelectron has the same energy as the Auger/autoionizing electron, it is not possible anymore to "label" and to distinguish the two electrons. The two-step model fails and the process must be treated as a *one-step* process showing resonance character. The antisymmetry of the final-state wavefunction and the property of the Coulomb operator lead to two amplitudes where both the electrons may play the role of the photoelectron/Auger electron [24]. These two amplitudes result in remarkable interference effects in the angular correlation pattern [23,25]. The strong dependence of this effect on the ratio of the experimental energy resolution and the natural width of the state has limited these studies to a few favorable cases. The advent of third generation synchrotron radiation sources and new monochromators which provide high fluxes at resolving power of about

10000, throughout the range of interest, will boost these investigations.

Threshold photoelectron(s) coincidence technique (TPEsCO) and the spectroscopy of doubly charged ions

Compared with the other experimental techniques the threshold photoelectron(s) coincidence technique, TPEsCO, [26] has unique characteristics: high efficiency for electrons with energy ≤10 meV and high resolution. In close proximity to threshold, double photoionisation yields two near zero energy electrons which can be detected by the two analyzers, using the penetrating field technique [27]. A true coincidence signal is then a signature of a double ionisation event and a TPEsCO spectrum is the double ionisation analog of a threshold photoelectron spectrum in single ionisation. TPEsCO spectra [26], measured for all the rare gases in the energy region which includes the $np^4(^3P, ^1D$ and $^1S)$ states and the $nsnp^5(^3P$ and $^1P)$ states, with the exception of xenon, have shown that all the possible ionic states are accessible. Their relative intensity follows the hierarchy of "favouredness" proposed by Huetz et al. [14] in their extension of the Wannier theory. Moreover TPEsCO allows the investigation of "satellite" states of the doubly charged ion, where two electrons are ejected and a third one is promoted to an unoccupied orbital. Fig.4 shows the Ar^{2+} spectrum above 62eV. The observed features have been assigned as $Ar^{2+}(3s^23p^3nl)$ satellite states converging to Ar^{3+}. These satellite states are expected to provide a fertile ground for the investigation of novel electron correlation effects because they involve a three-electron excitation process.

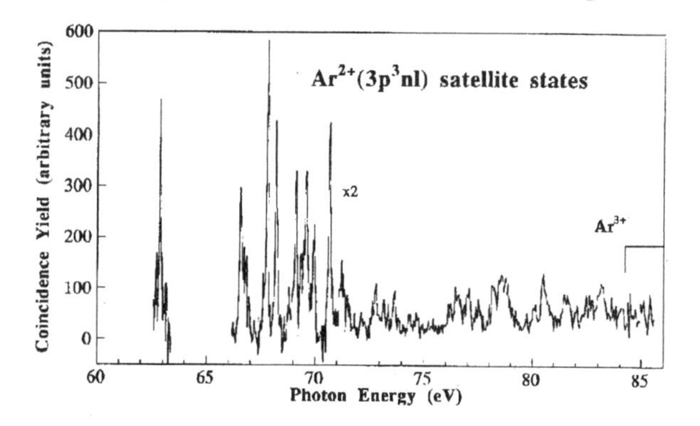

FIGURE 4 TPEsCO spectrum of Ar^{2+} satellite states.

CONCLUSIONS

A brief review of the physical mechanisms which lead to double ionisation of rare gases has been given. The interest in the field has grown quite rapidly in the last few years and already the measured TDCS's have shed light on several dynamical aspects and on the role of electron-electron correlations in their different facets.The necessity of a large amount of experimental information to produce a full picture of DPI is challenging experimentalists to build more and more sophisticated multicoincidence set-ups. As far as the theoretical aspects are concerned, although ab-initio calculations have shown to be the better suited tool to understand the finest detail of DPI, a general parametric representation of the TDCS, like the one available for the angular distribution of single photoionization, would be highly desirable.

Finally the advent of the third generation radiation sources is expected to enable DPI studies at higher excess energy. In this regime it will be possible to verify whether TDCS's provide direct information on the electron-electron correlation in the initial states, as predicted in the first proposal of these experiments [28].

ACKNOWLEDGMENTS

It is a great pleasure to thank G.C. King, U. Becker, R.I. Hall, G. Dawber and J. Viefhaus, whom I had the privilege to work with in the last few years. Work partially supported by the NATO C.R.G.960141

REFERENCES

1. Mazeau J., Selles P., Waymel D. and Huetz. A, Phys. Rev. Lett. **67**, 820 (1991)
2. Waymel D., Andric L., Mazeu J, Selles P. and Huetz A., J. Phys. **B26**, L123 (1993)
3. Krässig B., Schwarzkopf O. and Schmidt V., J. Phys. **B26**, 2589 (1993)
4. Huetz A., Andric L., Selles P, Jean A., Lablanquie P. and Mazeau J in "Proc. 19th Int. Conf. Physics of Electronic and Atomic Collisions", (New York:AIP Press),p139 (1996)
5. Dörner et al..Phys. Rev. Lett. 77, 1024(1996)
6. Cvejanovic S, Bagley GW and Reddish TJ J. Phys. **B27**, 5661 (1994)
7. Wehlitz R., Viefhaus J., Wieliczek K., Langer B., Whitfield S.B. and Becker U., Nucl. Instr. Meth. **B90**, 257 (1995)
8. Schwarzkopf O., Krässig B., Elminger J. and Schmidt V., Phys. Rev. Lett. **70**, 3008 (1993)
9. Huetz A., Lablanquie P., Andric L., Selles P. and Mazeau J., J. Phys. **B27**, L13 (1994)
10. Schwarzkopf O., Krässig B., Schmidt V., Maulbetsch F. and Briggs J., J. Phys. **B27**, L347 (1994)
11. Dawber G., Avaldi L., McConkey A.G., Rojas H., MacDonald M.A. and King G.C., J. Phys. **B28**, L271 (1995)
12. Lablanquie P., Mazeau J., Andric L., Selles P. and Huetz A., Phys. Rev. Lett. **74**, 2192 (1995)
13. Viefhaus J et al.,J. Phys. **B29**, L729 (1996)
14. Schwarzkopf O. and Schmidt V., J. Phys. **B28**, 2847 (1995)
15. Huetz A., Selles P., Waymel D. and Mazeau J., J. Phys. **B24**, 1917 (1991)
16. Kazansky and Ostrovsky V.N., J. Phys. **B27**, 447 (1994); J. Phys. **B28**, 1453 (1995);Phys. Rev. **A51**, 3712 (1995); Phys. Rev. **A51**, 3698 (1995)
17. LeRouzo H. and DalCapello C., Phys. Rev. **A43**, 318 (1991)
18. Maulbetsch F. and Briggs J.S., J. Phys. **B26**, 1679 (1993);J. Phys. **B26**, L647 (1993); J. Phys. **B27**, 4095 (1994); J. Phys. **B28**, 551 (1995)
19. Proulx D. and Shakeshaft R., Phys. Rev. **A48**, R875 (1993)
20. Pont M. and Shakeshaft R., Phys. Rev. **A51**, R2676 (1995)
21. Maulbetsch F, Pont M, Briggs J.S. and Shakeshaft R., J. Phys. **B28**, L341 (1995)
22. Schaphorst S.J., Krässig B., Schwarzkopf O., Scherer N. and Schmidt V., J. Phys. **B28**, L233 (1995)
23. Schaphorst S.J. et al.J. Phys. **B29**, 1901 (1996)
24. Vegh L. and Macek J.H., Phys. Rev. **A50**, 4031 (1994)
25. Schwarzkopf O. and Schmidt V.,J. Phys. **B29**, 3023 (1996)
26. King G.C. et al. J. Electron Spectrosc. Relat. Phenom. **76**, 253 (1995)
27. Hall R.I. et al. Meas. Technol. **3**, 316 (1992)
28. Levin V.G., Neudatchin V.G., Pavlitchenkov A.V., Smirnov Yu. F., J. Phys. **B17**, 1525 (1984)

CHARGE DEPENDENCE OF ONE AND TWO ELECTRON PROCESSES IN COLLISIONS BETWEEN HYDROGEN MOLECULES AND FAST PROJECTILES

E. Wells, I. Ben-Itzhak*, K.D. Carnes, and Vidhya Krishnamurthi†

James R. Macdonald Laboratory, Department of Physics,
Kansas State University, Manhattan, KS 66506-2604
†Present address: Etec Systems Inc., 26460 Corporate Ave., Hayward CA 94545.

The ratio of double to single ionization (DI/SI) as well as the ratio of ionization-excitation to single ionization (IE/SI) in hydrogen molecules was studied by examining the effect of the projectile charge on these processes. The DI/SI and IE/SI ratios were measured using the coincidence time of flight technique at a fixed velocity (corresponding to 1 MeV/amu) over a range of projectile charge states (q = 1-9,14,20). The results for a highly charged Cu^{20+} projectile, for example, indicate that the DI/SI and IE/SI ratios are 11% and 45%, respectively, a large increase from the ratios of 0.13% and 1.95%, respectively, for H^+ projectiles. The DI/SI ratio increases much more rapidly with projectile charge than the IE/SI ratio. In addition, the results show that the rate of increase of both these ratios decreases for highly charged projectiles.

INTRODUCTION

The ratio of double to single ionization (DI/SI) of helium has been of great interest in recent years (see for example the review of Knudsen and Reading [1]). The constant ratio reached at the high velocity limit indicates that double ionization is caused by electron-electron interactions after the ionization of one electron by the incoming projectile. At lower impact velocities, double ionization is dominated by the direct interactions between the projectile and both target electrons. Between these extremes, interference between the two mechanisms can occur [2]. Similar behavior is expected for other two-electron processes such as ionization-excitation. Determining the IE rate of a He target is difficult, however, since one must measure the emitted photons. Similar studies of DI and IE have been conducted on the hydrogen molecule which is also a two-electron target and has the additional advantage that when ionization-excitation from the ground state of the neutral molecule occurs, it is immediately followed by dissociation into $H^+ + H(nl)$. This is due to the fact that all of the potential energy curves of the excited electronic states of H_2^+ are dissociative in the Franck-Condon region. Single ionization, however, leads to one of two possible results: (i) H_2^+ in bound vibrational states of the $1s\sigma$ electronic ground state, and (ii) H_2^+ in the vibrational continuum of the same electronic state which immediately dissociates. The first process is the predominant one, with the ratio of the ground-state dissociation process to the non-dissociative single ionization process having been measured to be approximately 1.45% for H_2 [3].

Most of the previous studies of the DI/SI ratio for the hydrogen molecule have concentrated on the dependence on the sign and mass of singly charged projectiles. For example, Edwards *et al.* [4] measured the $H^+ + H^+$ yield produced by intermediate velocity electron and proton impact, using a back-to-back detection system. The $H_2{}^+$ yield, however, had to be measured using a different apparatus, leading to some uncertainty in the absolute magnitude of their data, as they have discussed in a later publication [5]. Kossmann *et al.* [6] have reported measurements of double and single ionization cross sections of H_2 due to fast electron impact. Ben-Itzhak *et al.* [7] used the coincidence time-of-flight technique to measure the velocity dependence of DI/SI for fast proton impact. Using the same technique, Ben-Itzhak *et al.* [3] measured the ratio of ionization-excitation to single ionization (IE/SI) of H_2 by fast proton impact.

There have been, however, comparatively few experiments investigating the double ionization and ionization-excitation of a hydrogen molecule as a function of projectile charge. Shah and Gilbody [8,9] have measured the ratio of dissociative to non-dissociative ionization of H_2 by H^+, He^{2+}, and Li^{3+} projectiles. Krishnakumar *et al.* [10] reported results for IE/SI for fast (3.3-7.0 MeV/amu) C^{6+}, O^{8+}, and S^{16+} projectiles where the measured values have been corrected for dissociative ionization of the $H_2{}^+$ electronic ground state and contributions from double ionization were

*Corresponding author eMail: ibi@phys.ksu.edu.

CP392, *Application of Accelerators in Research and Industry*, edited by J. L. Duggan and I. L. Morgan
AIP Press, New York © 1997

considered negligible. Cheng *et al.* [11] measured cross sections for single ionization, double ionization, and ionization-excitation of D_2 by 0.5-1.25 MeV/amu O^{8+} projectiles. In this study, we extend their work by measuring the IE/SI and DI/SI ratios for projectiles ranging in charge from $q = 1$ to $q = 20$ at a fixed velocity corresponding to 1 MeV/amu. Our results are compared to the previous measurements discussed above.

EXPERIMENTAL METHOD

A bunched beam of (H^+, $Li^{(2-3)+}$, $F^{(4-9)+}$, $Cu^{14+,20+}$) projectiles was accelerated to 1 MeV/amu by the J.R. Macdonald EN Tandem van de Graaff accelerator, collimated, and directed into a target cell containing a thin hydrogen gas target. Recoil ions produced in the target cell were extracted and accelerated toward a Z-stack microchannel plate detector by a strong uniform electric field (typically 1200 V/cm) of a time-of-flight spectrometer (see Fig. 1).

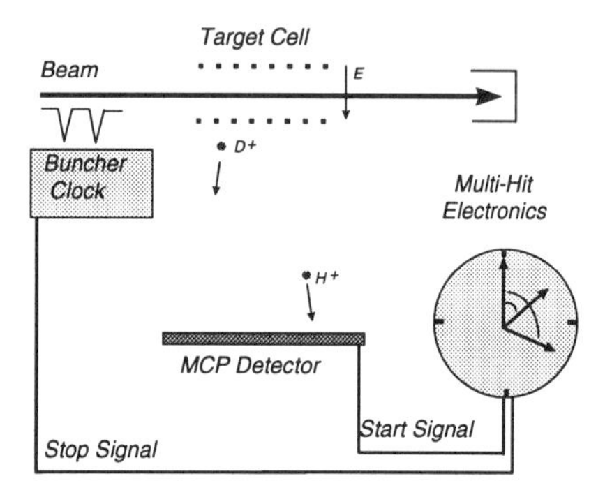

FIGURE 1. Conceptual view of the experimental setup.

The time-of-flight of the different ions was recorded by a multi-stop system relative to a common start signal synchronized with the beam bunch [12]. Typical one- and two-dimensional spectra are shown in Fig. 2.

The Cu^{q+} beams, however, could not be bunched properly due to the mass of the projectile. In this case a dc beam was used, and a signal from a scintillator, serving as a projectile detector, was used as the common stop (see Ref. [13] for further details). For the dc beam case, a simulation was used to correct for random stops, i.e. projectiles that preceded the true projectile by less than the time difference between the recoil-ion and the true projectile.

To improve the accuracy of the measurement of the double ionization channel we used the deuterium hydride isotope of hydrogen. For this isotope of the hydrogen molecule the flight time of the two fragments is significantly different even when using the strong extraction field needed to collect all fragments. Explicitly, the time of flight of D^+ was about 100 ns longer than that of the H^+ fragment, much longer than the time spread caused by the kinetic energy released in the dissociation, for the experimental conditions used. The mass difference between the HD and H_2 isotopes is not expected to affect the electronic transitions during the collision as they are the same within the Born-Oppenheimer approximation. In our data analysis we have assumed that the double ionization of hydrogen is isotropic as was shown for example by Edwards *et al.* [14] and Yousif *et al.* [15]. However, if this process is not isotropic it will have only a minor effect on our results because we have angular discrimination effects on less than 10% of the target length. Thus the contribution from most of the target length is independent of the angular distribution of the fragments.

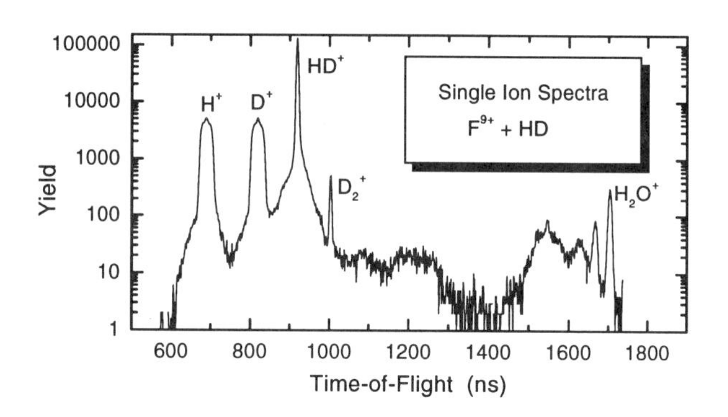

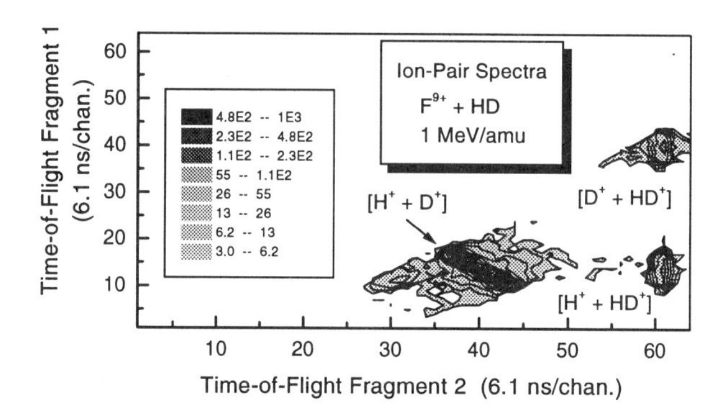

FIGURE 2. Top: Time of flight spectrum of single recoil ions created by 1 MeV/amu F^{9+} impact on HD. Bottom: Density plot of ion pairs from the same collision system.

The ionization-excitation of hydrogen molecules can be determined from the contribution of single H^+ and single D^+ fragments because all the excited states of HD^+ are dissociative within the Franck-Condon region. These fragments typically have a few eV (> 2 eV) of kinetic energy, and thus can be separated from the very slow (< 0.5 eV) H^+ and D^+ fragments from the dissociation of the electronic ground state of HD^+, which takes place if the vertical transition populates its vibrational continuum. The contribution of the ground-state dissociation channel was measured using a weak extraction field as described in detail by Ben-Itzhak *et al.* [3]. Briefly, when a weak extraction field is used most of the fast H^+ or D^+ fragments from IE miss the detector, while those which still hit the detector are shifted to much smaller or larger times. The fragments from the ground-state dissociation channel have a kinetic energy that is typically < 0.5 eV, and reach the detector despite the weak extraction field. For high projectile charge, however, it became difficult to evaluate the ground-state dissociation contribution due to the increased amount of fast fragments from the IE contribution. In these cases, the value of ground-state dissociation by fast proton impact was used (assuming that ground-state dissociation is due to the overlap between the initial and final vibrational wave functions, and thus is independent of q). Once the contribution of slow H^+ and D^+ fragments was determined, it was subtracted from the total yield of these fragments measured with a strong extraction field in order to determine the IE channel.

The ratio DI/SI is evaluated by comparing the number of $[H^+ + D^+]$ ion-pair events, i.e. H^+ in coincidence with D^+, to the number of HD^+ recoil ions. Random ion-pairs must be subtracted from the DI yield. These are events in which both a D^+ and a H^+ recoil ion are recorded for the same beam bunch, but they do not originate from the same molecule. This channel can be estimated by using the purely random $[H^+ + HD^+]$ and $[D^+ + HD^+]$ ion-pair channels, which must come from double collisions in the same beam bunch [16]. The ratio IE/SI is evaluated by comparing the sum of H^+ and D^+ single events to the number of HD^+. To this ratio one has to perform two corrections: (i) "Lost fragments" from ion-pair events must be subtracted from the IE yield. These are double ionization events in which only one of the two recoil ions was detected because the detection efficiency is smaller than one. This correction for lost fragments is discussed in detail in our earlier publication [16]. (ii) The ground-state dissociation, i.e. the yield of slow H^+ and D^+ fragments, as described above, has to be subtracted from the IE channel and added to the SI channel.

RESULTS AND DISCUSSION

The measured ratios IE/SI and DI/SI are shown in Fig. 3 as a function of projectile charge, for q = 1-20 at a fixed collision velocity corresponding to 1 MeV/amu.

The DI/SI ratio begins to increase rapidly as a function of the projectile charge from an initial value of 0.13% for proton impact, but slows its increase somewhat for $q > 4$. For Cu^{20+}, the value of the DI/SI ratio reaches 11% (see Fig. 3). While the contribution of double ionization to the dissociative ionization channel is negligible for proton impact, it is clearly not negligible for more highly charged projectiles. Furthermore, the contribution of DI to the H^+ yield when a H_2 homonuclear molecule is used as a target might be even larger because the detection efficiency of at least one fragment out of two might be close to twice the detection efficiency of one fragment out of one. The IE/SI ratio shows a similar rapid increase with projectile charge, increasing from 1.95% for proton impact to about 25% for F^{9+} projectiles. The rate of increase then slows somewhat, reaching a value of 45% for Cu^{20+}.

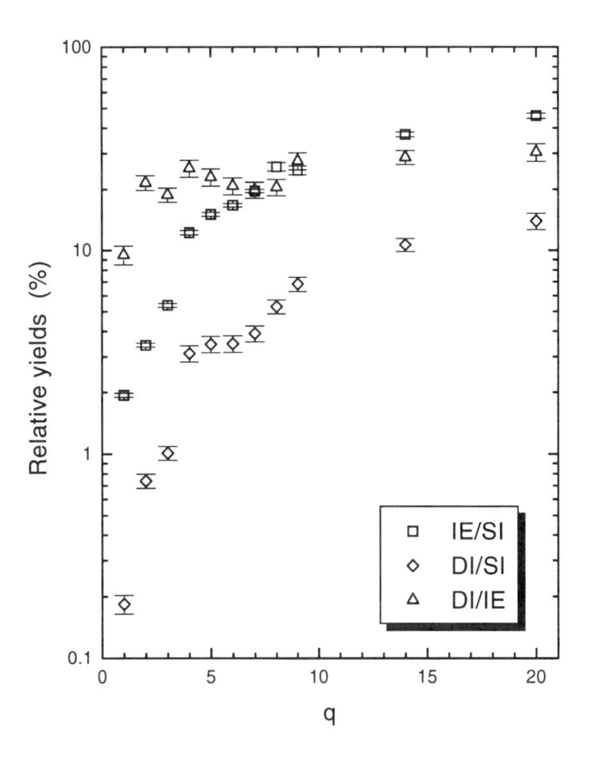

FIGURE 3. The relative yield of SI/IE, IE/SI, and DI/IE in HD as a function of the projectile charge.

The ratio of DI/IE also increases with projectile charge, approximately linearly for $q > 2$ as can be seen in Fig. 3. This increase of DI relative to IE indicates that while for fast proton impact DI can be considered a negligible contribution to dissociative ionization in comparison to IE and ground-state dissociation, this is not the case for highly charged ions. For the latter, DI represents a sizable portion of the total dissociative ionization.

The ratio of dissociative to non-dissociative ionization has been measured by other groups [8–11], and a scaling law suggested by Knudsen *et al.* [17] for He double ionization has been adapted by Krishnakumar [10]. This scaling law has the form:

$$R = A + B\ q^2\ /\ v_p^2\ ln(9.18 v_p) = A + B\ q^2\ /\ V^* \quad (1)$$

where q is the projectile charge, v_p is the projectile velocity in atomic units, V^* has been defined as equal to $v_p{}^2 ln(9.18 v_p)$, and the constants have been found by fitting the available data to be: A = 0.0277 and B = 0.407 [10]. We have compared our data to that of the groups cited above in Fig. 4 and generally found it to be in good agreement with previous results and the scaling law. In order to compare our results to the other measurements of R, we have added our separate measurements in the following manner:

$$R = IE/SI + \alpha(DI/SI) \quad (2)$$

where α is the ratio of the probability of detecting at least one out of two recoil ions to the probability of detecting one recoil ion. For our experimental setup, this ratio was found to be 1.625. There are some deviations between our data and previous results which might be explained by screening effects in some of our projectiles. The other measurements of R were done using bare projectiles (H^+, He^{2+}, Li^{3+}, C^{6+}, O^{8+}, S^{16+}).

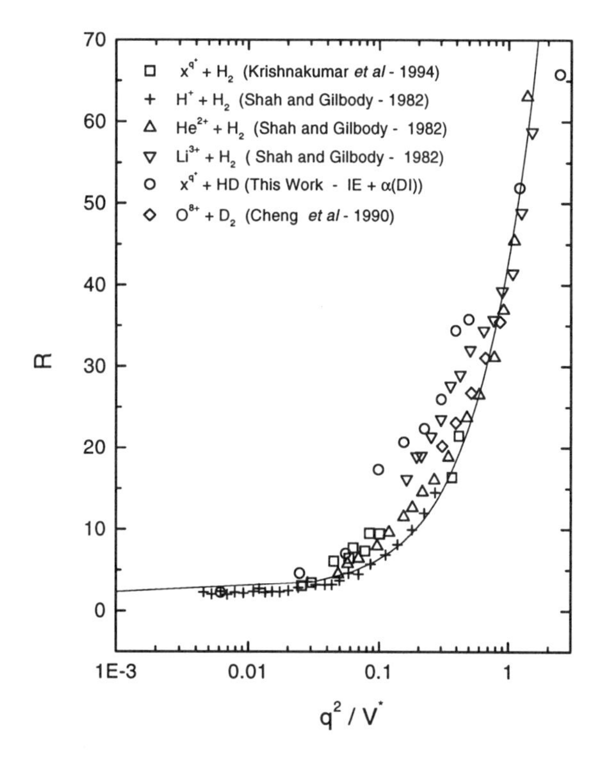

FIGURE 4. The ratio, R, of dissociative to non-dissociative ionization of hydrogen molecules as a function of q^2/V^*. The solid line represents the scaling law used by Krishnakumar.

In summary, we have measured the IE/SI and DI/SI ratios in hydrogen molecules as a function of projectile charge from $q = 1$ to 20 at a fixed velocity (1 MeV/amu) using the coincidence time-of-flight technique. The contribution from dissociation of the electronic ground state of molecular hydrogen was taken into account, and the IE/SI and DI/SI ratios were measured simultaneously. The contribution of double ionization to dissociative ionization is negligible for proton impact, but as projectile charge increases, it rapidly becomes significant. For a highly charged Cu^{20+} projectile, the DI/IE ratio is about 30%. The rate of increase of both the IE/SI and DI/SI ratios with q decreases for higher projectile charge. Further work is required to better understand this trend.

ACKNOWLEDGMENT

This work was supported by the Division of Chemical Sciences, Office of Basic Energy Sciences, Office of Energy Research, U.S. Department of Energy.

[1] Knudsen, H., and Reading, J.F., *Phys. Reports* **212**, 107 (1992).

[2] McGuire, J.H., *Phys. Rev. Lett.* **49**, 1153 (1982).

[3] Ben-Itzhak, I., Krishnamurthi, Vidhya, Carnes, K.D., Aliabadi, H., Knudsen, H., Mikkelsen, U., and Esry, B.D., *J. Phys. B* **29**, L21 (1996).

[4] Edwards, A.K., Wood, R.M., Beard, A.S., and Ezell, R.L., *Phys. Rev. A* **37**, 3697 (1988).

[5] Edwards, A.K., Wood, R.M., Beard, A.S., and Ezell, R.L., *Phys. Rev A* **42**, 1367 (1990).

[6] Kossmann, H., Schwarzkopf, O., and Schmidt, V., *J. Phys. B* **23**, 301 (1990).

[7] Ben-Itzhak, I., Krishnamurthi, Vidhya, Carnes, K.D., Aliabadi, H., Knudsen, H., and Mikkelsen, U., *Nucl. Instr. and Meth. B* **99**, 104 (1995).

[8] Shah, M.B., and Gilbody, H.B., *J. Phys. B* **15** 3441, (1982).

[9] Shah, M.B., and Gilbody, H.B., *J. Phys. B* **22** 3983, (1989).

[10] Krishnakumar, E., Bapat, Bhas, Rajgara, F.A., Krishnamurthy, M., *J. Phys. B* **27**, L777 (1994).

[11] Cheng S., Cocke C.L., Kamber E.Y., Hsu C.C., Varghese S.L., *Phys. Rev. A* **42**, 214 (1990).

[12] Ben-Itzhak, I., Ginther, S.G., and Carnes, K.D., *Nucl. Instrum. and Meth. B* **66**, 401 (1992).

[13] Ben-Itzhak, I., Carnes, K.D., and DePaola, B.D., Rev. Sci. Instrum. **63** (1992) 5780.

[14] Edwards, A.K., Wood R.M., and Ezell, R.L., *Phys. Rev. A* **31** 99 (1985).

[15] Yousif F.D., Lindsay B.G., and Latimer, C.J., *J. Phys. B* **21** 4157 (1988).

[16] Ben-Itzhak, I., Ginther, S.G., and Carnes, K.D., *Phys. Rev. A* **47** 2827 (1993).

[17] Knudsen H., Anderson L.H., Hvelplund, P., Astner, G., Cederquist, H., Danared, H., Liljeby, L., and Rensfelt, K-G., *J. Phys. B* **17** 3545 (1984).

Large Scale Molecular Dynamics Simulation of a Surface Coulomb Explosion.

Hai-Ping Cheng
Department of Physics & QTP
University of Florida
Gainesville Fl 32611

and

J. D. Gillaspy
Physics Laboratory
National Institute of Standards and Technology
Gaithersburg, MD 20899

Highly charged ions colliding with a solid have been predicted to produce localized Coulomb explosions on the surface. We have modeled the explosion using a large scale molecular dynamics simulation. Our results show the temporal evolution of three different types of craters which are formed when an incident highly charged ion produces 100 singly-charged Si atoms in various initial distributions on the surface. The total number of ejected particles ranges from 245 to 317 and appears to be determined by the initial shape of the ionized region rather than simply by the initial repulsive energy restored in the charged region. Contrary to intuition, a long and thin cylindrical distribution is the most efficient pattern for ejecting particles. In all three cases, the number of ejected neutral particles is much greater then the number of ejected ions (6-10 times as many atoms as ions). The angular distribution of ejected particles is also analyzed.

INTRODUCTION

Recent developments in highly charged ion-surface bombardment experiments (reviewed in [1]) have generated a major need for theoretical modeling and understanding at the atomic level. In some of the experiments, when a slow, highly charged ion (HCI) collides with a non-metallic surface, a hillock or crater of nanometer-size can be generated and directly imaged with atomic force microscopy [1-2]. Several goups have speculated that research with HCIs may lead to important practical applications in the fabrication of nanostructures on surfaces[1-7].

The experimental investigations have shown that the creation of surface features is sustained down to very low kinetic energies, at which point most of the energy is deposited in the form of internal potential energy of the ion [2]. A Coulomb explosion was postulated as a mechanism for the formation of such craters as early as 1979 [8]. Since then, considerable effort by a number of groups [1-15] has been put into studying the electronic transitions and ionization processes during HCI-surface bombardment.

Explicit descriptions and depictions of the surface Coulomb explosion dynamics [4, 5, 7, 9, 15] have been largely phenomenological in character, however . Since the lattice dynamics governs the formation of craters, understanding the Coulomb explosion at the atomic level is a first step towards a more complete and fundamental understanding of the overall process.

Our previous paper [16], involving approximately 300 surface charges, was the first that provided a microscopic description of Coulomb explosion based on a large-scale molecular dynamics simulation. We demonstrated the explosion process on a silicon surface with specified model initial conditions. Shockwave propagation, ultra-rapid evaporation of surface atoms, and the formation of craters were analyzed in detail, and atomic-scale visualizations of the explosion process were presented.

In this work, we carry out additional simulations involving fewer initial surface charges and we focus on a comparative study of the influence of the initial shapes of the charged region on the final crater shape, as well as on the angular distribution of emitted particles.

SIMULATIONS

Numerical solutions of the Newtonian equations of motion are obtained via Gear's predict-corrector algorithm [17] for an N particle system. Each

CP392, *Application of Accelerators in Research and Industry,* edited by J. L. Duggan and I. L. Morgan
AIP Press, New York © 1997

of the 34560 particles interacts with each of the other particles. We use the Tersoff potential [18] for Si-Si interactions, 1/r Coulomb potential for Si^+-Si^+ interactions and a 12-4 potential of the following form for Si-Si^+ interactions:

$$V(r_{ij}) = \varepsilon\left[\left(\frac{\sigma}{r_{ij}}\right)^{12} - \left(\frac{\sigma}{r_{ij}}\right)^{4}\right] \qquad (1)$$

The parameters ε and σ are obtained as described previously [16] by combining first-principle calculations and experimental data.

The substrate is represented by a slab that is periodically repeated in x and y directions. In the positive z direction, we use free boundary condition for the surface. We use a few static layers at the bottom of the substrate to stabilize the systems. By controlling the temperature of the layer(s) between dynamical layer and static layers, we simulate a infinite substrate in -z direction. To properly treat the system we are interested in, the periodic boundary conditions in x, y directions are not applied to ions.

At t=0, 100 Si ions are embedded in the Si surface. The shape of the three initial charge distributions we choose to study are hemispherical, flat disc, and long thin cylindrical. The initial temperature of the substrate in all three cases is very low, about 10 K. The temperature of the static layers is fixed during the explosion processes to provide a heat bath for the dynamical layers above. We use time interval, Δt = 0.4 fs, to achieve energy conservation during the processes. The simulations are carried out for 1.6 ps.

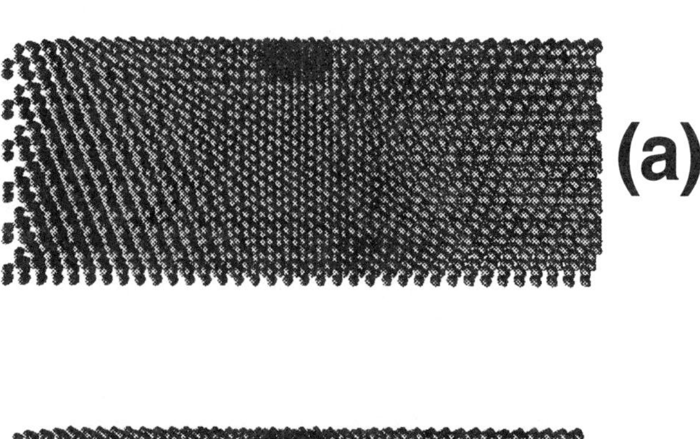

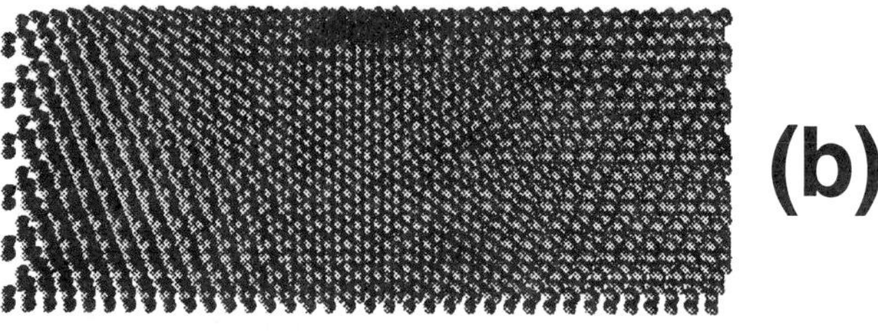

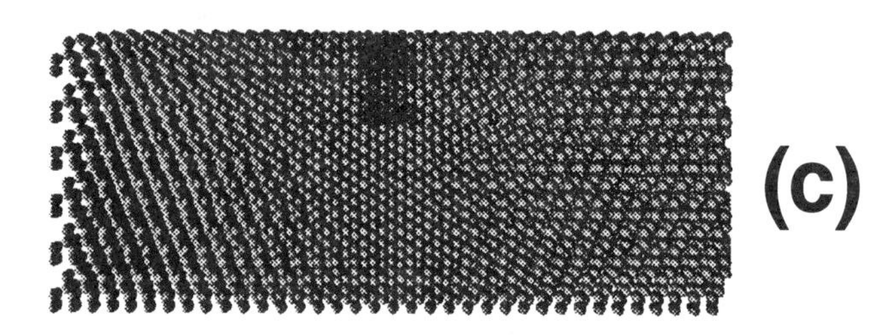

FIGURE 1. Molecular dynamics simulation of a Coulomb explosion in silicon at t=0.0 for initial charged region shaped as (a) hemisphere, (b) disc, and (c) cylinder. Ions are indicated with a daker shade than the atoms.

RESULTS

Three initial shapes of the charged region is shown in figure 1. The positions of the atoms and ions at 160 fsec after the inditial condition is illustrated in figure 2. At 1.6 ps, we collect particles that are more than 0.6 nm (12 a.u.) above the surface layer for each of the three simulations. The total numbers of ejected particles are 245, 245, and 317 for (a) hemisphere, (b) flat disc, and (c) long cylinder. The number of ejected ions of these three situations are 31, 35, and 24 respectively. At a first glance, the relative numbers of ejected ions is intuititive. Since (b) has the most ions on the surface layer, it is relatively easier for the ions to be pushed out. For the same reason, case (c) has the least number of ejected ions. Nevertheless, the total numbers of ejected particles are not what we would expect since they do not increase or decrease monotonically as a function of total initial energies. The repulsive Coulomb energies stored in the charged region for the three shapes (a)-(c) at t=0 are, 9.61 keV, 8.95 keV and 9.33 keV respectively. By summing up the amount of energy dissipated into the substrate, we find that case (c) is higher than the other two cases. The higher efficiency of this geometry for depositing energy into the lattice explains the increased evaporation of surrounding lattice atoms.

From the simulation data, angular distribution of the total ejected particles can be obtained. Fig.3 displays the distribution of azimuthal angles of three simulations. All of the three histograms have a similar broad distribution over 0-100 degrees. On the other hand, the distribution patterns in the ejected ions of the three situations are quite different from each other (see Fig.4). The cylindrical shape has a relatively uniform distribution.

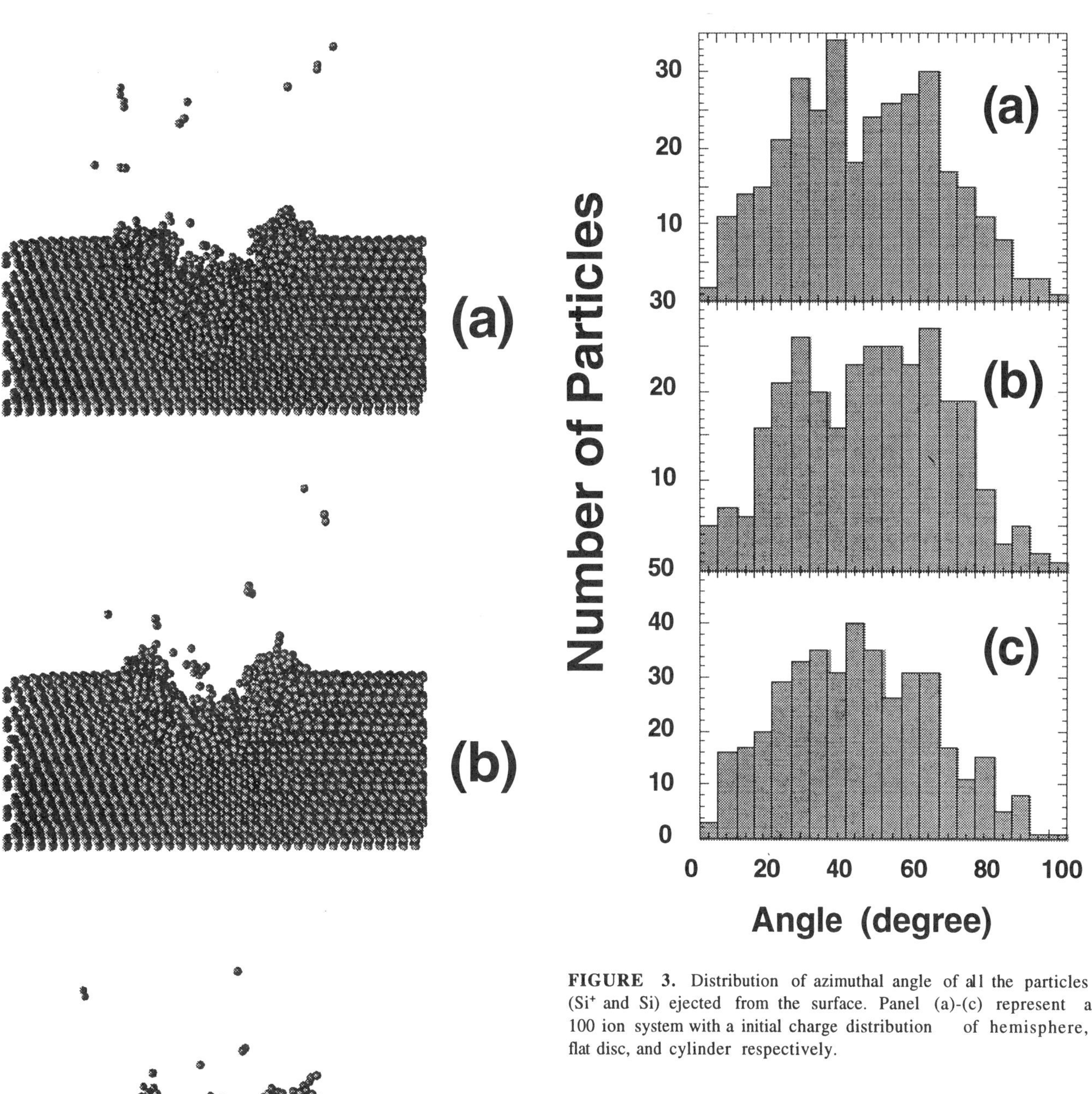

FIGURE 2. Molecular dynamics simulation of a Coulomb explosion in silicon at t=0.160 for initial charged region shaped as (a) hemisphere, (b) disc, and (c) cylinder.

FIGURE 3. Distribution of azimuthal angle of all the particles (Si^+ and Si) ejected from the surface. Panel (a)-(c) represent a 100 ion system with a initial charge distribution of hemisphere, flat disc, and cylinder respectively.

CONCLUSIONS

Our simulation of Coulomb explosions on a Si surface suggest that t≈100 fs is an important time scale. In comparing this work to our previous work, we find that for all the systems under investigation, such as systems with 100, 265, and 365 ions, 10^2 fs is the time for shockwaves to collapse and for potential energies to drop to about e^{-1} of their initial values. The numbers of ejected particles depend on the energy that is dissipated into the substrate. This is the mechanism for the initial

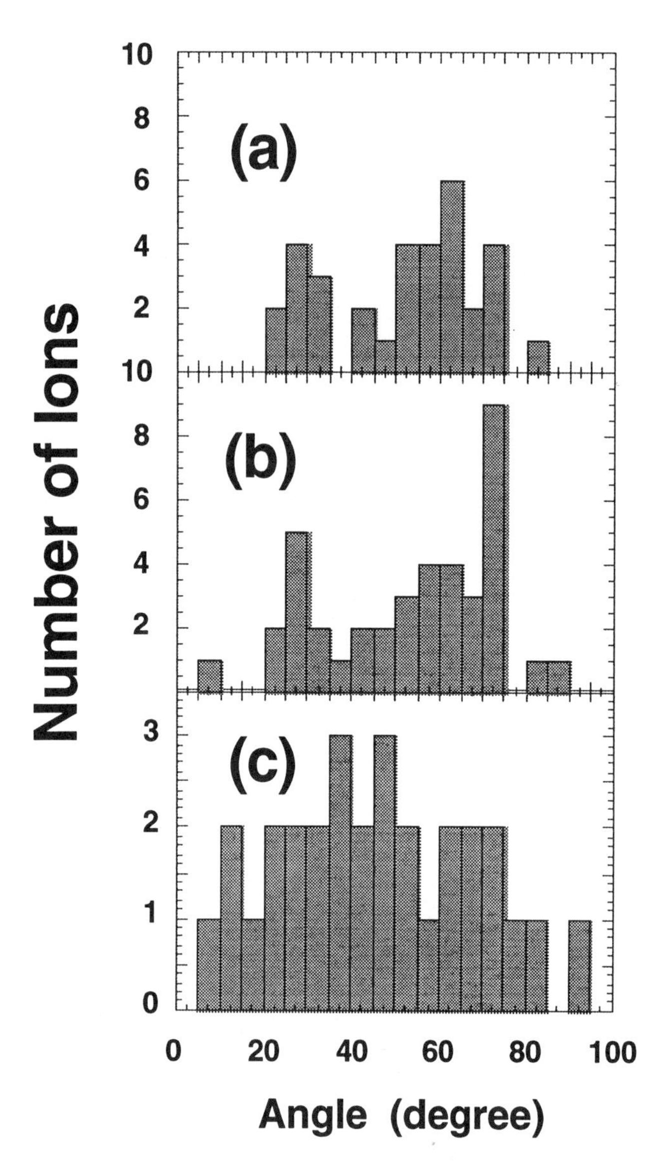

FIGURE 4. Distribution of azimuthal angle of all the ions (Si^+ only) ejected from the surface. Panel (a)-(c) represent a 100 ion system with a initial charge distribution of hemisphere, flat disc, and cylinder respectively. Notice the stronger shape dependence in fig.4 than in fig.3

shape dependence of the total ejected particles. Finally, we conclude that the fraction of energy that goes into the substrate decreases as the number of ions increases from 100 to 365.

REFERENCES

1. D.H. Schneider, M.A. Briere, *Physica Scripta*, **53**, 228 (1996).
2. D.C. Parks, R. Bastasz, R.W. Schmieder, and M. Stöckli, *J. Vac. Sci. Technology*. **B13**, 941(1995).
3. R.W. Schmieder and R.J. Bastasz, *Proc. VIth Intl. Conf. Physics Highly Charged Ions*, AIP Conf. Proc. **274**, 675 (1992).
4. Itabashi, 1995: N. Itabashi, K. Mochiji, H. Shimizu, S. Ohtani, Y. Kato, H. Tanuma, N. Kobayashi, *Jpn. J. Appl. Phys.*, **34**, 6861 (1995).
5. 11. R. Morgenstern and J. Das, *Europhysics News*, **25**, no. 1, 3 (1994)
6. J.D. Gillaspy, Y. Agilitskiy, E.W. Bell, C.M. Brown, C.T. Chantler, R.D. Daslettes, U. Feldman, L.T. Hudson, J.M. Laming, E.S. Meyer, C.A. Morgan, A.L. Pikin, J.R. Roberts, L.P. Ratliff, F.G. Serpa, J. Surga, and E. Takas, *Physica Scripta*, **T59**, 392 (1995).
7. I. Hughes, *Physics World*, **8**, no. 4, 43 (1995);
8. I.S. Bitenskii, M.N. Murakhmetov and E.S. Parilis, *Sov. Phys. Tech. Phys*. **24**, 618(1979).
9. D. Schneider, M.A. Briere, T. Schenkle, EBIT, N-Division, Livermore annual report (1994).
10. P. Apell, J. Phys. B, **21**, 2665 (1988); N. Vaeck and J. E. Hansen, *J. Phys. B*, **28**, 3523 (1995);
11. F. Aumayr, H.. Kurz, D. Schneider, M.A. Briere, J.W. McDonald, C.E. Cunningham, and H.P. Winter, *Phys. Rev. Lett.*, **71**, 1943 (1993); P. Varga, *Comments At. Mol. Phys*., **23**, 111 (1989).
12. J.P. Briand, B. d'Etat-Ban, D. Schneider, M.A. Briere, V. Decaux, J.W. McDonald, and S. Bardin, *Phys. Rev. A*, **53**, 2194 (1996).
13. J. Burgdorfer, C. Reinhold, and F. Meyer, *Nucl. Instrum. and Methods in Phys. Res.*, B, **98**, 415 (1995).
14 M. Grether, A. Spielder, R. Kohrbruck, and N. Stolterfoht, *Phys. Rev. A*, **52**, 426 (1995).
15. E. Parilis, Caltech preprint.
16. Hai-Ping Cheng and J.D. Gillaspy, Phys. Rev. **B**, Jan. 15 (1996).
17. M.P. Allen and D.J. Tildesley, *Computer Simulations of Liquids* (Clarendon, Oxford, United Kingdom, 1987).
18. J. Tersoff, *Phys. Rev. B* **39**, 5566 (1988).

COMPARISON OF SINGLE-IONIZATION OF HYDROGEN MOLECULES TO THAT OF HELIUM IN COLLISIONS WITH FAST PROTONS

E. Wells, D. Studanski*, I. Ben-Itzhak† and K.D. Carnes

James R. Macdonald Laboratory, Department of Physics,
Kansas State University, Manhattan, KS 66506-2604
**Permanent address: Department of Physics, Drake University, Des Moines, IA 50311*

The ratio of single ionization (SI) of hydrogen to that of helium was measured using the time of flight technique for 1-12 MeV proton impact ionization. A gas target composed of a 1:1 mixture of D_2 and ^{3}He was used to determine the ratio of single ionization of these two-electron targets relative to each other. Using the known double- to single-ionization ratios for both of these targets, we have also evaluated the ratio of their double ionization relative to each other. Preliminary results show that the ratio $D_2^+/^3He^+$ is 1.79 ± 0.05. In contrast, double ionization of these two targets is approximately equal. It is suggested that single ionization of hydrogen molecules is more likely due to its smaller binding energy while the stronger electron-electron interaction in helium compensates for the smaller probability of proton impact ionization and leads to roughly equal double ionization of both targets.

INTRODUCTION

The ratio of double- to single-ionization of helium has been of great interest in recent years (see for example the review of Knudsen and Reading [1]). Similar studies have been conducted on the hydrogen molecule which is also a two electron target. We have adopted a different approach in which we have tried to compare the single ionization by fast proton impact of these two-electron targets to each other. A similar comparison for double-ionization has also been carried out. This project was motivated by our previous work [2] in which it was suggested that single ionization of hydrogen molecules by fast proton impact is twice as likely as single ionization of helium in similar collisions while double ionization is roughly equal. However, the comparison was done using total single ionization cross sections from a few sources [3–5] in addition to the ratios of double- to single-ionization of each target, and the former are typically not known as precisely as the latter. Our goal in this experiment was to make a direct comparison between single- and double-ionization of these two-electron targets relative to each other. From this comparison we hoped to find the main target properties causing similarities and/or differences between the ionization of hydrogen molecules and helium atoms, targets which differ in electron binding energy, wave functions, potential symmetry *etc.* The comparison was done for 1 to 12 MeV proton impact, where collision velocities are high enough such that double ionization is dominated by electron-electron interaction, i.e. double ionization caused by the interaction between the projectile and each target electron is negligible.

†Corresponding author eMail: ibi@phys.ksu.edu.

EXPERIMENTAL METHOD

A beam of protons was accelerated to 1-12 MeV by the J.R. Macdonald EN Tandem and was directed into a target cell that contained a one to one gas mixture of D_2 and ^{3}He. Ions produced in the target cell were extracted and accelerated toward a Z-stack microchannel plate detector by strong uniform electric fields of the time of flight spectrometer shown in Fig. 1 (see Ref [6] for further details). The beam current was monitored by a Faraday cup down stream from the target cell in order to evaluate total cross sections. The proton beam was bunched to ~1 ns pulses 5.3 μsec apart, and the timing signals correlated with these beam bunches were used to stop the time to amplitude converter which was started by the timing signal from the recoil-ion detector. A typical time of flight spectrum of single recoil ions is shown in Fig. 2.

The main peaks in this spectrum are the $^3He^+$ and D_2^+ associated with single ionization. The main contamination to the gas target is water, and it does not overlap with the peaks of interest in this work except

CP392, *Application of Accelerators in Research and Industry,* edited by J. L. Duggan and I. L. Morgan
AIP Press, New York © 1997

for a small contribution of H_2^+ which has the same m/q as the D^+ peak. This was the reason for choosing D_2 over H_2 as a hydrogen target. The $^3He^+$ and $^3He^{2+}$ are clearly separated from all other peaks except for a negligible amount of HD^+ which overlaps with the $^3He^+$. HD is less than 0.1% relative to D_2 and thus can be ignored relative to the main $^3He^+$. We insured that all recoil ions were detected with the same efficiency by accelerating them to a few keV energy and by using a sufficiently low discrimination level for their signals. We have also checked that the ratio of He^{2+}/He^+ for 4 MeV proton impact agrees with the measurement of Knudsen *et al.* [7]. All measurements were carried out at single collision conditions which were tested by a standard pressure dependence.

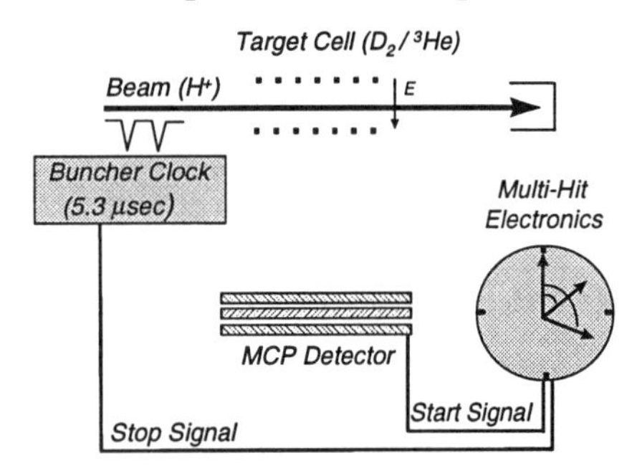

FIGURE 1. Experimental setup.

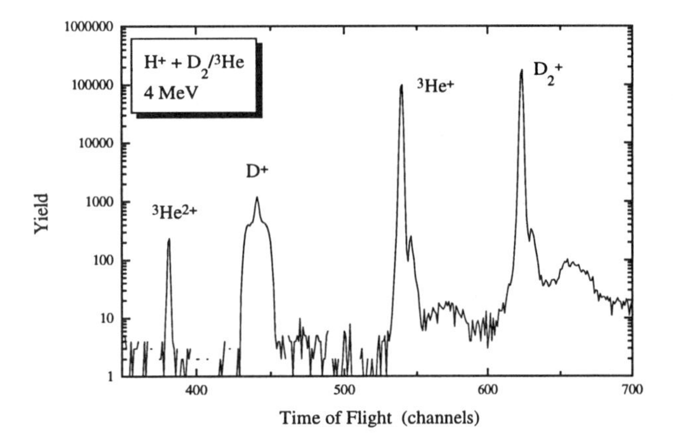

FIGURE 2. Time of flight spectrum of single recoil ions created by 4 MeV proton impact on a 1:1 gas mixture of D_2 and ^{3}He.

RESULTS AND DISCUSSION

The ratio of single ionization of D_2 and ^{3}He was measured for 1 to 12 MeV proton impact as described in the previous section. This ratio is evaluated from the single recoil ion time of flight spectrum. One must be aware, however, that some of the He^+ is left in excited states, a process commonly named ionization-excitation (IE). In contrast, all the excited electronic states of D_2^+ populated by vertical transitions (i.e. within the Franck-Condon region) dissociate rapidly into D^++D(nlm). Thus, in order to compare the same processes for both targets it is necessary to either subtract the ionization-excitation contribution from the measured He^+ yield or add this channel to the measured yield of D_2^+. As we have no idea what fraction of He^+ is in its excited electronic states the first option is not possible. On the other hand, the ionization-excitation channel which leads to D^++D formation is directly measured and it is relatively easy to add it to the true single ionization. When performing this correction one has to take into account two things: (i) Double ionization producing D^++D^+ ion pairs is also contributing to the D^+ peak in the singles spectrum. This channel has a higher detection efficiency because it is sufficient to detect one out of the two recoil ions (which hit the detector too close in time to be distinguished from a single D^+ hit). Double ionization of hydrogen molecules is very small for the collisions under study, about 0.13 % of the D_2^+ formation [2]. (ii) Some (about 0.5%) of the singly ionized D_2^+ in its electronic ground state also dissociates because it is in the vibrational continuum and it should be added to the D_2^+ yield [8]. The contributions of ionization-excitation and ground state dissociation are independent of the collision energy over the energy range investigated, and they add up to approximately 2.5%, 3.0%, and 3.5% for D_2, HD, and H_2, respectively [2,8], relative to the singly charged molecular ion yield. In order to simplify data analysis we have just multiplied the D_2^+ yield by 1.025. The corrected yield then includes all single ionization, nondissociative as well as dissociative, and the ionization-excitation channel, and thus can be compared with the He^+ yield which includes the same electronic processes. We have therefore calculated the ratio of single ionization of hydrogen molecules to that of helium atoms as

$$R_1 = 1.025\ A(D_2^+)\ /\ A(^3He^+)\ , \tag{1}$$

where A is the area under the respective peak in the TOF spectrum (Fig. 2). This ratio was calculated for all measured target pressures for which single collision conditions were valid and the average was used as the result while the standard deviation was used as the best estimate of the error. This ratio, which is equal to the ratio of $\sigma^+(D_2)/\sigma^+(He)$, is shown in Fig. 3 as a function of the proton energy. It can be seen that this ratio is constant over the energy range studied, and its value is $R_1 = 1.79 \pm 0.05$.

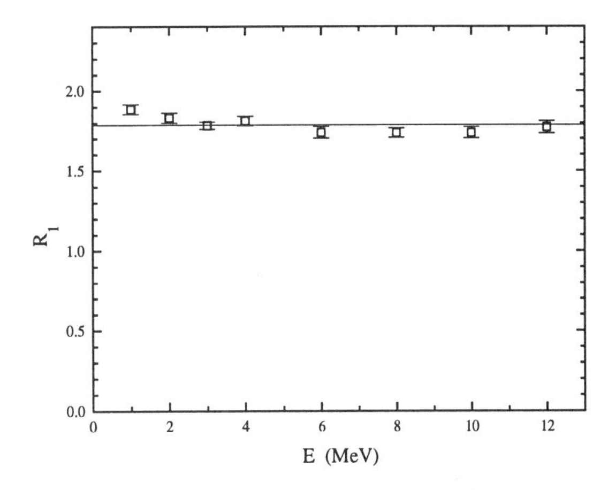

FIGURE 3. The ratio, R_1, of single ionization and ionization excitation of D_2 to that of He as a function of the proton energy.

Recently, Wu *et al.* [9] reported the following scaling law for single ionization by slow highly charged ions,

$$\sigma_s^+ = \sigma^+ I^\alpha / q \qquad v_s = v / I^{1/2} q^{1/4} \ , \tag{2}$$

where σ_s^+ and v_s are the scaled cross section and scaled collision velocity, respectively, I is the ionization potential (i.e. binding energy of the electron), q is the projectile charge, and α is a fitting parameter which was found by them to be 1.3 ± 0.1. Even though the collisions in our study are at the other velocity extreme, i.e. far above the ionization threshold, we have decided to try and probe the validity of the scaling with the ionization potential. The ratio R_1 of the single ionization of hydrogen molecules to helium atoms should scale like

$$R_1 = \frac{\sigma^+(D_2)}{\sigma^+(He)} = [\frac{I(He)}{I(D_2)}]^\alpha \ , \tag{3}$$

where I(He)=24.5876 eV and I(D_2)=15.4666 eV. Our data shown in Fig. 3 fits best to a value of $\alpha = 1.25 \pm 0.06$ which is within the range of values suggested by Wu *et al.* [9]. This suggests that the scaling with ionization potential is not limited to the near threshold region. However, *the origin of the value of* $\alpha = 1.25$ *is still to be explained.* Furthermore, this analysis suggests that *the difference between the single ionization of hydrogen molecules and helium atoms could stem from the difference in their binding energies.*

The ratio of double ionization of hydrogen molecules to helium atoms can also be evaluated from our results as follows, even if not with the same level of accuracy,

$$R_2 = \frac{\sigma^{++}(D_2)}{\sigma^{++}(He)} = R_1 \times \frac{R(D_2)}{R(He)} \ , \tag{4}$$

where $R(D_2)$ and $R(He)$ are the double- to single-ionization ratios of each target, respectively. The ratio $R(He)$ can be determined from our measurements directly by dividing the area of the $^3He^{2+}$ by that of the $^3He^+$ peak. However, the double ionization cross section for the collisions under study is so small that our measurements were not as precise as previous ones due to limited statistics. To improve the calculation of R_2 defined above we have used the double to single-ionization ratio $R(He)$ reported by Andersen *et al.* [10]. The need to interpolate and extrapolate had a minor effect on the accuracy because the ratio He^{2+}/He^+ is quite constant from 2 MeV and higher. We also used the ratio of double- to single-ionization of hydrogen molecules previously reported by this group [2]. The ratio of double ionization of hydrogen molecules relative to that of helium atoms is shown as a function of proton impact energy in Fig. 4. This ratio is 1.03 ± 0.10, and it is quite flat over the energy range measured, similar to the ratio of single ionization of these targets. Double ionization of hydrogen molecules seems to be as likely as double ionization of helium in these fast collisions in contrast with the fact that it was more likely to singly ionize the hydrogen, i.e. $R_2 \sim 1$ while R_1 is about 1.8.

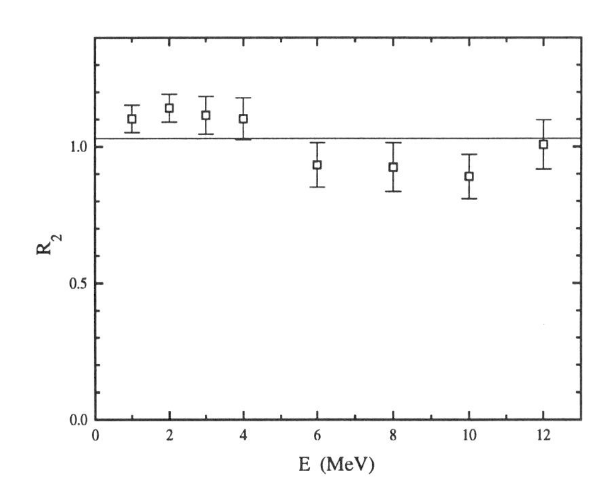

FIGURE 4. The ratio, R_2, of double ionization of D_2 to that of He as a function of the proton energy.

For the high velocity collisions under study, which are well described by perturbation theory, double ionization by two direct interactions between the projectile and the target electrons is negligible. Double ionization proceeds mostly by a "two-step" mechanism: first the projectile ionizes one electron and then the other electron is ionized by the interaction between the electrons. In order to have the same likelihood for double ionization of these two-electron targets, there must be a stronger probability to ionize the second electron by electron-electron interactions in helium than in hydrogen in order to compensate for the higher probability to ionize the first electron of hydrogen by the projectile. Qualitatively, one should expect the electron-electron interaction to be larger in helium than in a hydrogen molecule because the average electron-electron distance

$\overline{r_{12}}$ is smaller in helium, about 1.4 a.u. in helium [11] and 2.4 a.u. in hydrogen [12]. However, the question of *whether this difference is sufficient to explain the fact that double ionization is equal for helium atoms and hydrogen molecules* calls for a quantitative theoretical analysis.

To summarize, the ratio of single ionization of hydrogen molecules was measured relative to that of helium atoms for fast proton impact. It is more probable to singly ionize the hydrogen molecule by about a factor of 1.8. This result is in accord with the scaling law for single ionization in slow collisions suggested by Wu *et al.* [9] thus extending scaling with ionization potential to the high velocity regime. However, it is an open question as to *why the single ionization should scale with the ionization potential to a power of 1.25.* In contrast, the ratio of double ionization of hydrogen molecules to that of helium atoms is approximately one. It is suggested that the stronger electron-electron interaction in the helium target caused by the smaller average distance between the target electrons compensates for the higher probability for ionizing the first electron in hydrogen. The question *of whether the ratio of these two steps will approximately cancel each other and lead to approximately equal double ionization cross sections* calls for quantitative theoretical work.

ACKNOWLEDGMENT

This work was supported by the Division of Chemical Sciences, Office of Basic Energy Sciences, Office of Energy Research, U.S. Department of Energy. D.S. was also supported by NSF grant, "Undergraduate Summer Research in Experimental and Theoretical Physics".

---

[1] Knudsen, H. and Reading, J.F., *Phys. Reports* **212**, 107 (1992).

[2] Ben-Itzhak, I., Krishnamurthi, Vidhya, Carnes, K.D., Aliabadi, H., Knudsen, H., and Mikkelsen, U., *Nucl. Instr. and Meth. B* **99**, 104 (1995).

[3] Shah, M.B., and Gilbody, H.B., *J. Phys. B* **15**, 3441 (1982).

[4] Shah, M.B., and Gilbody, H.B., *J. Phys. B* **18**, 899 (1985).

[5] Edwards, A.K., Woods, R.M., Beard, A.S., Ezell, R.L., *Phys. Rev. A* **37**, 3697 (1988).

[6] Ben-Itzhak, I., Ginther, S.G., and Carnes, K.D., *Nucl. Instrum. and Meth. B* **66**, 401 (1992).

[7] Knudsen, H., Andersen, L.H., Hvelplund, P., Astner, G., Cederquist, H., Danared, H., Liljeby, L., and Rensfelt, K.-G., *J. Phys. B* **17**, 3545 (1984).

[8] Ben-Itzhak, I., Krishnamurthi, Vidhya, Carnes, K.D., Aliabadi, H., Knudsen, H., Mikkelsen, U., and Esry, B.D., *J. Phys. B* **29**, L21 (1996).

[9] Wu, W., Deveney, E.F., Datz, S., Desai, D.D., Krause, H.F., Sanders, J.M., Vane, C.R., Cocke, C.L., and Giese, J.P., *Phys. Rev. A* **53**, 2367 (1996).

[10] Andersen, L.H., Hvelplund, P., Knudsen, H., Møller, S.P., Sørensen, A.H., Elsener, K., Rensfelt, K.-G., and Uggerhøj, E., *Phys. Rev. A* **36**, 3612 (1987).

[11] Coulson, C.A. and Neilson, A.H., *Proc. Roy. Soc.* **78**, 831 (1961).

[12] Barnett, M.P., Briss, F.W., and Coulson, C.A., *Mol. Phys.* **1**, 44 (1958).

ASPECTS OF (e,2e) : 6 DIMENSIONAL INTEGRAL METHOD

S. P. Lucey†, J. Rasch†, Colm T. Whelan†, H. R. J. Walters‡

†Department of Applied Mathematics and Theoretical Physics,
University of Cambridge, Silver Street, Cambridge, CB3 9EW, England
‡Department of Applied Mathematics and Theoretical Physics,
David Bates Building, Queen's University of Belfast, Belfast, N.I. BT7 1NN

An outline of the 6 Dimensional Integral Method, (6DIME), is presented. The application of 6DIME to $(e, 2e)$ and $(\gamma, 2e)$ is discussed. Examples of calculations using DS3C wavefunction, (Berakdar and Briggs (1994)) are made for $(e, 2e)$ on Hydrogen using 6DIME. Results are compared to the DWBA.

INTRODUCTION

In several areas of atomic physics and in particular in the fields known as $(e, 2e)$, $(\gamma, 2e)$, the transition matrix one needs to evaluate can be written in terms of an integral dependent upon the spatial coordinates of the two final state electrons. An $(e, 2e)$, $((\gamma, 2e))$, event is one in which an electron, (photon) collides with a target, inducing the ionisation of an electron from the target. The two final state continuum electrons are then detected in coincidence providing one with a knowledge of the full kinematics of the system. In order to describe the numerous experiments in this field, (Whelan et. al. [2] and references cited therein), one must attempt to solve the time independent Schrödinger equation,

$$(H_T - E_T)\,\Psi^{\pm} = 0, \tag{1}$$

where, E_T, is the total energy and H_T is the total Hamiltonian of the system. Both the initial and final states of the collision are a many body problem and hence it is not possible to explicitly write down the exact form of the total wavefunction of the system, $\Psi^{\pm}$. The different methods for approximating, $\Psi^{\pm}$, have been reviewed by Whelan et. al. [2].

Typically, when calculating a triple differential cross section, (TDCS), of an (e,2e) process one has to calculate highly oscillatory functions in a 6 dimensional coordinate space. To calculate a 6 dimensional integral is extremely difficult, thus one usually attempts to reduce the 6 dimensional integral to one of fewer dimensions, (see Lewis [9] and Brauner et. al. [11]). This reduction of calculable dimensions, however, can generally only be carried out if either the initial state or the final state of the system is of a separable form. Hence a constraint is enforced upon the sophistication of approximations that one can use in either the initial of final state.

In this paper we would like to outline a computational method that has been devised to overcome these difficulties. The computer code carries out a 6 dimensional integration and hence allows for the first time the use of both good initial and good final state wavefunctions in the calculation $(e, 2e)$ and $(\gamma, 2e)$ processes. The method, 6DIME, is general and can be applied to all such 6 dimensional problems.

As an example of the accuracy of 6DIME, a very difficult 6 dimensional $(e, 2e)$ calculation on Hydrogen will be presented. Firstly, the necessary theory of $(e, 2e)$ on Hydrogen will be explained, secondly, the 6DIME will be outlined, thirdly, the actual 6 dimensional calculation will be written down and finally the results of this method will be compared to experiment and other theory. Atomic units are used throughout.

GENERAL THEORY

The TDCS for an $(e, 2e)$ event on a neutral hydrogen target can be written,

$$\frac{d^3\sigma}{d\Omega_f d\Omega_s dE_s} = (2\pi)^4 \frac{k_f k_s}{k_0} \mid T_{ji} \mid^2, \tag{2}$$

where (Ω_s, E_s, k_s), (Ω_f, E_f, k_f), are the solid angles, energies and momenta of the two detected electrons, which will always be subscripted s and f in this work, k_0 is the momentum of the incident electron. If one sums over initial, i, and averages over all final, j, spin states then the spin averaged transition matrix can be written as,

$$\mid T_{ji} \mid^2 = \frac{1}{4}\left(|f + g|^2 + 3|f - g|^2\right). \tag{3}$$

CP392, *Application of Accelerators in Research and Industry,* edited by J. L. Duggan and I. L. Morgan
AIP Press, New York © 1997

The direct, f, and exchange, g, ionisation amplitudes can be written in a variety of ways depending upon how one specifies the total wavefunction of the system, (Goldberger & Watson [7] and Whelan et. al. [8]). In the prior formalism,

$$f = \langle \Psi_j^-(\mathbf{r}_s, \mathbf{r}_f)|V_i|\phi_i(\mathbf{r}_s, \mathbf{r}_f)\rangle, \quad (4)$$

$$g = \langle \Psi_j^-(\mathbf{r}_f, \mathbf{r}_s)|V_i|\phi_i(\mathbf{r}_s, \mathbf{r}_f)\rangle, \quad (5)$$

where, Ψ_j^-, is the final state wavefunction that is an approximation to the solution of equation (1), with incoming wave boundary conditions. $\phi_i(\mathbf{r}_s, \mathbf{r}_f)$, is the asymptotic initial state wavefunction and V_i is that part of the total Hamiltonian undiagonalised by the initial state, ϕ_i. The total Hamiltonian for this system can be written,

$$H_T = -\frac{1}{2}\nabla^2_{\mathrm{r_s}} - \frac{1}{2}\nabla^2_{\mathrm{r_f}} - \frac{1}{\mathrm{r_s}} - \frac{1}{\mathrm{r_f}} + \frac{1}{|\mathbf{r}_s - \mathbf{r}_f|}, \quad (6)$$

The asymptotic initial state is represented by the product of an incoming plane wave with the ground state wavefunction of Hydrogen, ϕ_H,

$$\phi_i(\mathbf{r}_s, \mathbf{r}_f) = \frac{e^{i\mathbf{k}_0 \cdot \mathbf{r}_f}}{(2\pi)^{3/2}} \phi_H(\mathbf{r}_s). \quad (7)$$

Since, ϕ_i is the eigensolution of $H_i\phi_i = E_i\phi_i$, one can define the potential, V_i, as $V_i = H_T - H_i$, hence,

$$V_i = -\frac{1}{\mathrm{r_f}} + \frac{1}{|\mathbf{r}_s - \mathbf{r}_f|}. \quad (8)$$

6 DIMENSIONAL INTEGRAL METHOD

A general 6 dimensional integral over the spatial coordinates of two electrons can be written,

$$J(\mathbf{r}_s, \mathbf{r}_f) = \int \mathrm{d^3 r_f} \ \mathrm{f}(\mathbf{r}_\mathrm{f}) \int \mathrm{d^3 r_s} \ \mathrm{V}(\mathbf{r}_\mathrm{s}, \mathbf{r}_\mathrm{f}) \mathrm{g}(\mathbf{r}_\mathrm{s}), \quad (9)$$

where $f(\mathbf{r}_f)$, $g(\mathbf{r}_s)$, are complex functions that one generally finds in problems in scattering or ionisation theory, $V(\mathbf{r}_s, \mathbf{r}_f)$, is a function of both integration variables that usually contains terms such as equation (8). In Rasch et. al. [10], a computationally efficient and accurate method for calculating integrals such as equation (9) was presented. The outline of the method is as follows.

A good method for numerically evaluating integrals of the form equation (9) must have three essential components. Firstly, since the outer integral is looping over the inner integral several times, the inner integral must be very accurate and very fast. The method we have used for this is the one detailed in, Lagaris and Papageorgiou [1], the authors initially adopt a spherical coordinate system for the inner integral, $I(\mathbf{r}_s)$,

$$I(\mathbf{r}_s) = \int \mathrm{d^3 r_s} \ \mathrm{V}(\mathbf{r}_\mathrm{s}, \mathbf{r}_\mathrm{f}) \mathrm{g}(\mathbf{r}_\mathrm{s}), \quad (10)$$

$$I(\mathbf{r}_s) = \int_0^{R_s} \mathrm{dr_s} \ \mathrm{r_s^2} \int_0^{\pi} \mathrm{d}\theta_\mathrm{s} \ \sin(\theta_\mathrm{s}) \int_0^{2\pi} \mathrm{d}\phi_\mathrm{s} \ \mathrm{V}(\mathbf{r}_\mathrm{s}, \mathbf{r}_\mathrm{f}) \mathrm{g}(\mathbf{r}_\mathrm{s}), \quad (11)$$

where instead of integrating the r_s component from $0 \rightarrow \infty$ we have cutoff this integral at some large value, R_s. The integration variable, ϕ_s is then linearly transformed, such that,

$$\phi = \phi_s - \phi_f \quad (12)$$

Futhermore, ϕ is transformed to $\mathrm{r_{sf}} = |\mathbf{r}_\mathrm{s} - \mathbf{r}_\mathrm{f}|$,

$$\mathrm{d}\phi = \frac{1}{\sin(\theta_\mathrm{s})\sin(\theta_\mathrm{f})\sin(\phi)} \frac{\mathrm{r_{sf}}}{\mathrm{r_s r_f}} \mathrm{dr_{sf}}, \quad (13)$$

using equation (13) it is possible to make a further transformation on $\mathrm{r_{sf}}$,

$$\mathrm{r_{sf}} = \frac{1}{2}(\mathrm{b} - \mathrm{a})\mathrm{t} + \frac{1}{2}(\mathrm{b} + \mathrm{a}). \quad (14)$$

Hence, the inner integral is now,

$$I(\mathbf{r}_s) = \frac{4}{\mathrm{r_f}} \int_0^{R_s} \mathrm{dr_s} \ \mathrm{r_s} \int_0^{\pi} \mathrm{d}\theta_\mathrm{s} \ \sin(\theta_\mathrm{s}) \int_{-1}^{1} \frac{\mathrm{dt}}{\sqrt{1-\mathrm{t}^2}} \frac{\mathrm{r_{sf}} \mathrm{V}(\mathbf{r}_\mathrm{s}, \mathbf{r}_\mathrm{f}) \mathrm{g}(\mathbf{r}_\mathrm{s})}{\sqrt{(b-a)t + 3b + a}\sqrt{(b-a)t + b + 3a}}, \quad (15)$$

where,

$$a^2 = \mathrm{r_s^2} + \mathrm{r_f^2} - 2\mathrm{r_s r_f} \cos(\theta_\mathrm{f} - \theta_\mathrm{s}),$$
$$b^2 = \mathrm{r_s^2} + \mathrm{r_f^2} - 2\mathrm{r_s r_f} \cos(\theta_\mathrm{f} + \theta_\mathrm{s}). \quad (16)$$

Secondly, the integration method overcome the problem of integrating the singularity caused by potential functions of the form equation (8). After this series of transformations, however, the singularity now occurs at the end points of the inner integration in equation (15). By using a Gauss-Chebychev integration for the t variable the singular term, $1/\sqrt{1-t^2}$, in equation (15) is included in the integration weights and hence can be dealt with in a computationally efficient manner.

Thirdly, if the outer integrand, $f(\mathbf{r}_f)$ of equation (9), contains exponentially decreasing functions, as is the case in $(\gamma, 2e)$, then one can employ standard integration techniques, (Press et. al. [12]), in its evaluation. However, in $(e, 2e)$ $f(\mathbf{r}_f)$ is an undamped oscillating function and thus it is very difficult to achieve a convergent result for the whole integral, $J(\mathbf{r}_s, \mathbf{r}_f)$. In the results presented below we have multiplied, $f(\mathbf{r}_f)$ by a minimal damping function,

$$J(\mathbf{r}_s, \mathbf{r}_f) = \int \mathrm{d^3 r_f} \ \mathrm{f}(\mathbf{r}_\mathrm{f}) \mathrm{e}^{-\lambda \mathrm{r_f}} \mathrm{I}(\mathbf{r}_\mathrm{f}), \quad (17)$$

where $\lambda \leq 0.001$. This parameter has the effect of ensuring a convergent result to the integration. In tests the addition of this parameter was shown to produce $\approx 0.1\%$ difference between analytical and calculated results.

APPLICATIONS TO (e,2e) ON H

In order to demonstrate the versatility of this method we shall use the final state wavefunction, known as the Dynamically Screened 3 Coulomb wavefunction, (DS3C), devised by Berakdar and Briggs [3]. This wavefunction, Ψ^-_{DS3C}, is the eigensolution of an approximate Hamiltonian, H_{DS3C}, and

$$\begin{aligned}\Psi^-_{DS3C} &= N e^{i\mathbf{k}_s.\mathbf{r}_s} e^{i\mathbf{k}_f.\mathbf{r}_f}\ {}_1F_1\left(i\beta_s, 1, -i(k_s r_s + \mathbf{k}_s.\mathbf{r}_s)\right) \\ &\quad {}_1F_1\left(i\beta_f, 1, -i(k_f r_f + \mathbf{k}_f.\mathbf{r}_f)\right) \\ &\quad {}_1F_1\left(i\beta_{sf}, 1, -i(k_{sf} r_{sf} + \mathbf{k}_{sf}.\mathbf{r}_{sf})\right), \qquad (18)\end{aligned}$$

where

$$N = \frac{e^{-\frac{(\beta_s+\beta_f+\beta_{sf})}{2}}}{(2\pi)^3}\Gamma(1-i\beta_s)\Gamma(1-i\beta_f)\Gamma(1-i\beta_{sf}), \qquad (19)$$

The remainder of the total Hamiltonian $H_{rem} = H - H_{DS3C}$ unoperated upon by DS3C is minimised by an appropriate choice of Sommerfeld parameters in the hypergeometric functions, Berakdar and Briggs [3]. From this formulation one obtains,

$$\beta_s = -\frac{4Z - \sin(\Theta)}{4k}, \quad \beta_f = -\frac{4Z - \sin(\Theta)}{4k},$$
$$\beta_{sf} = -\frac{1 - \sin^2(\Theta)}{2k\ \sin(\Theta)} \qquad (20)$$

where, Z, is the nuclear charge of the residual target and Θ, is half the angle between the detected electrons. The direct ionisation amplitude for this process is thus,

$$\begin{aligned}f(\mathbf{r}_s, \mathbf{r}_f) &= \int d^3 r_f\ e^{i\mathbf{q}.\mathbf{r}_f}\ {}_1F_1\left(i\beta_f, 1, -i(k_f r_f + \mathbf{k}_f.\mathbf{r}_f)\right) \\ &\quad \int d^3 r_s\ e^{-i\mathbf{k}_s.\mathbf{r}_s} e^{-r_s}\left(\frac{1}{r_{sf}} - \frac{1}{r_f}\right) \\ &\quad {}_1F_1\left(i\beta_s, 1, -i(k_s r_s + \mathbf{k}_s.\mathbf{r}_s)\right) \qquad (21) \\ &\quad {}_1F_1\left(i\beta_{sf}, 1, -i(k_{sf} r_{sf} + \mathbf{k}_{sf}.\mathbf{r}_{sf})\right),\end{aligned}$$

where, $\mathbf{q} = \mathbf{k}_0 - \mathbf{k}_f$, is the momentum transfered to the target in the collision. The combination of the three oscillating hyper-geometric coordinates and the complex exponential in the outer integration make this a very challenging integral to calculate. The results obtained using the 6DIME for equation (22) have been checked using a different computer code based on the method of, Brauner et. al. [11]). The agreement between the two has been found to be no worse than 2%. Each point on the TDCS, (Figures 1a) ,b) ,c)) took approximately 1 hour to calculate on a Dec-Alpha work station.

In the following section the results of Berakdar & Briggs [3] will be compared to, the Distorted Wave Born Approximation, (DWBA) and experimental results detailed in Röder et. al. [6].

RESULTS

The DWBA is described in detail elsewhere, (Whelan et. al. [13], Rasch [5]), here we only outline its main features. The distorted waves in the initial channel are generated in the static exchange potential of the atom and the distorted waves for the outgoing electrons are generated in the field of the ion. Final channel 3 body effects are included via the Gamov factor, (Botero & Macek [4], Whelan et. al. [13] and Röder et. al. [6]), and initial channel distortion effects via the polarisation potential of Whelan et. al. [13]. The advantage of this simple model is that initial and final channel 3 body effects can be switched on and off at will.

Figures 1a), b), c), are TDCS on Hydrogen in an Energy sharing geometry at an incident electron energy, $E_0 = 27.2eV$. In Figure 1a), the DS3C does extremely well in producing the binary peak of the experimental data and predicts the height of the recoil peak very well. The DWBA is able to produce the recoil lobe successfully though it fails to predict the magnitude of the binary lobe. It is interesting to note in Figure 1a) that the difference between the DWBA with and without polarisation is very small. In Figures 1b), c), one can see that the DS3C approach no longer describes the binary or recoil peak correctly, failing to reproduce the size of the binary peak in Figures 1b) and 1c) and falsely predicting the position of experiment in the recoil peak. In these two figures the DWBA with polarisation does however predict the size and position of both peaks extremely well. One can also note in Figures 1b), c), that the influence of the polarisation potential, is very large. This statement is emphasised by looking at the binary peak in Figure 1c), both the DS3C and DWBA approaches predict similar results but with the wrong size.

From these figures, one is able to conclude that the DS3C approach which ignores all initial channel polarisation effects fails to agree with experiment in cases where these effects appear to be strong. The DWBA approach agrees well with experiment, as in Figures 1b), c), however, in Figure 1a) it is unable to reproduce the binary peak. From this, one can conclude that in order to have a theory that agrees with experiment in different geometries, it is necessary to improve these approximations. Up until now, this was not possible because one could not calculate, for example, the DS3C approach with a sophisticated initial state. This is particularly true when one

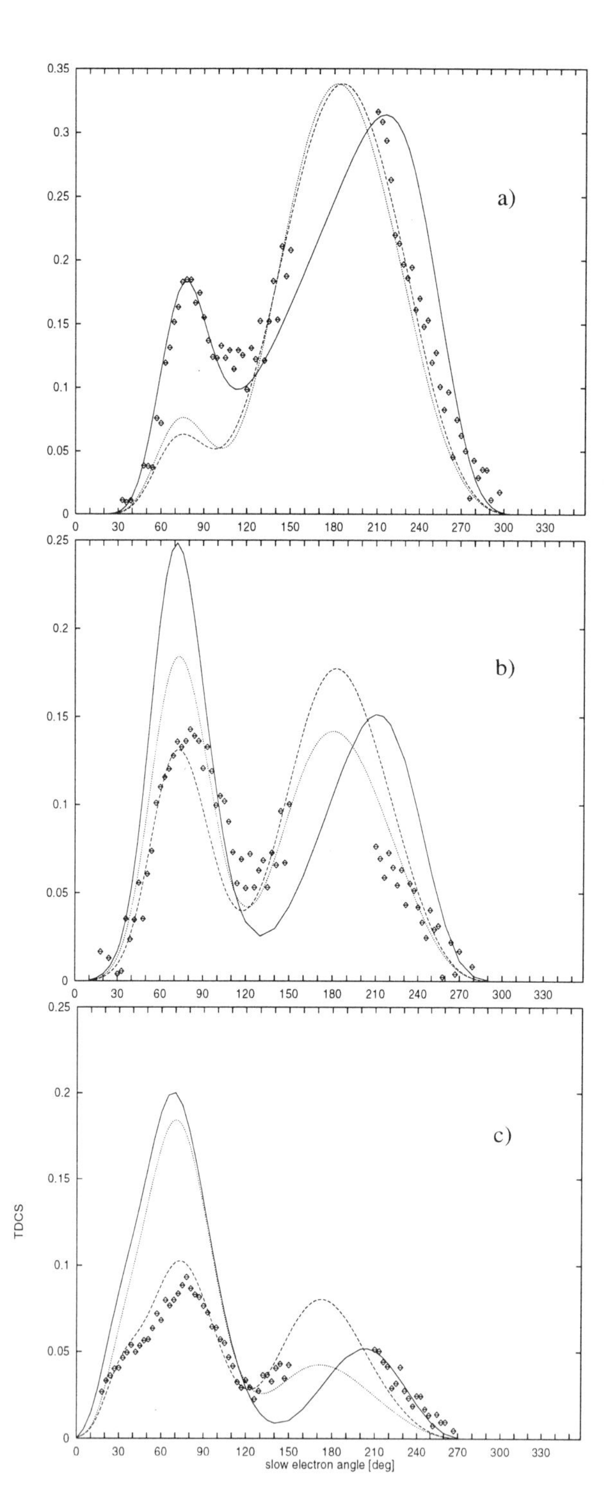

Figure 1: TDCS for $(e, 2e)$ on Hydrogen. Incident electron energy, $E_0 = 27.2eV$, detected electron energy, $E_s = E_f = 6.8eV$. The angle of the 'fast electron' detector is fixed at; a) 15°, b) 30°, c) 45°, and the 'slow electron' is rotated around the ionisation plane. The results are taken from Röder et. al. [6]. In Figures a), b), c), the solid line is the DS3C approximation, the dashed line is the DWBA approximation with polarisation potential and the dotted line is the DWBA without polarisation potential.

carries out $(e, 2e)$ calculations on Helium or other heavier targets in which initial state correlation plays a more important role than for the case of Hydrogen.

Using the 6DIME, it should now be possible, in the future, to use a more sophisticated combination of initial and final states to hopefully improve our theoretical predictions of experiment.

ACKNOWLEDGMENTS

We are grateful to Professors J. H. Macek and Y. Popov for valuable discussions. We gratefully acknowledge support from the Royal Society, the EC(HCM CHCRX CT93 0350), NATO(CRG 950 407) and P.P.A.R.C. studentships.

References

[1] Lagaris I. E. and Papageorgiou D. G. *Comp. Phys. Comun.*, 76:80, 1993.

[2] Whelan Colm T. Walters H. R. J. Lahmam-Bennani A. Ehrhardt H. *(e, 2e) and related processes.* Kluwer Academic Press: NATO ASI Series, 1993.

[3] Berakdar J and Briggs J S. *Phys. Rev. Lett.*, 72(24):3799–82, 1994.

[4] Botero J. and Macek J. H. *Phys. Rev. Lett.*, 68:576–9, 1992.

[5] Rasch J. PhD thesis, University of Cambridge, 1996.

[6] Röder J. Rasch J. Jung K. Whelan Colm T. Ehrhardt H. Allan R. J. and Walters H. R. J. "On the role of Coulomb 3 body effects in low energy impact ionisation of H(1s)". *Phys. Rev. A..*, 62:1, 1996.

[7] Goldberger M. L. and Watson K. M. *Collision Theory.* Structure of Matter Series. John Wiley & Sons, Inc., London, thrid printing edition, 1967.

[8] Whelan Colm T. Walters H. R. J. Hannsen J. Dreizler R. M. *Aust. J. Phys.*, 44:39–58, 1991.

[9] Lewis R. R. *Phys. Rev.*, 102(2):537–544, 1956.

[10] Whelan Colm T. Rasch J., Lucey S. P. submitted to Computer Physics Communications, 1996.

[11] Brauner M. Briggs J. S. and Klar H. *Journal of Physics B: Atomic, Molecular, Optical Processes*, 22:2265–2287, 1989.

[12] Press W. H. Tuekolsky S. A. Vetterling W. T. and Flannery B. P. *Numerical Recipes in Fortran.* Cambridge University Press, 1992.

[13] Walters H. R. J. Whelan Colm T., Allan R. J. *Journal de Physique*, 3(C6):39, 1993.

RECENT RECOIL ION MOMENTUM SPECTROSCOPY EXPERIMENTS AT KSU

M. Abdallah, C.L. Cocke, S. Kravis[1], E.C. Montenegro[2], R. Moshammer[3], L. Saleh, J. Ullrich[4], S.L. Varghese[5], W. Wolff[6] and H. Wolf[6]

J.R. Macdonald Laboratory, Physics Dept., Kansas State University, Manhattan, KS 66506-2604
Present addresses: [1]NOVA R&D,1525 Third St., Riverside, CA 92507
[2]Pontificia Univ. Catolica do Rio de Janeiro, C.P. 38071, Rio de Janeiro 22452-970 Brazil
[3]U. Frankfurt, IKF, August Euler St. 6, 60486 Frankfurt/Main, Germany; [4]GSI Darmstadt, Postfach 11 05 52, 64220, Darmstadt, Germany; [5]Physics Department, University South Alabama, Mobile, AL
[6]Instituto de Fisica, Universidade Federal do Rio de Janeiro, Rio de Janeiro, Brazil

Recoil momentum spectroscopy is used to study collisions involving both fast and slow projectiles on He targets. Experiments have been performed on electron capture and loss from fast ions from the KSU LINAC and slow ions from the KSU CRYEBIS using a supersonic jets with a momentum resolution below 0.5 au. Using fast ions, the final states populated in electron capture from He by 10 MeV F^{8+} have been resolved with a Q-value resolution of 18 eV, sufficient to separate final channels in which the He^+ ion is left excited from those in which He^+ is left in its ground state. With slow ions, electron capture from He by slow bare Ne ions has been studied. A few recent results are discussed.

INTRODUCTION

The use of recoil momentum spectroscopy to obtain kinematic information on the products of ion-atom collisions is becoming widespread. The idea was inherent is several experiments in the sixties [1] and was revived in the late eighties, starting with the pioneering work of Ullrich et al. with room temperature targets [2]. With the development by the Frankfurt group of recoil momentum spectroscopy with cold targets (COLTRIMS) [for a recent review, see Ref. 3], this approach to determining final state kinematics has now become a truly high-resolution tool for probing ion-atom and photon-atom collisions. In this paper we describe some recent ion-atom collisions results from KSU obtained using recoil momentum spectroscopy .

High Velocity Collisions

Review of Principle

In a two-body collision, the momentum received by the recoil ion p_z is uniquely related to the electronic energy release in the collision, Q, by $p_z = Q/v - nv/2$, where v is the projectile velocity and n the number of electrons transferred from target to projectile. Thus the recoil momentum spectrum becomes a final-state Q spectrum for the cases of pure electron capture or pure excitation (no electrons left in the continuum). For high velocity and highly charged projectiles incident on He, this channel is generally not the strongest one, since single and even double ionization is a very important channel, and transfer ionization generally is at least as large as pure single capture (hereafter referred to simply as single capture). Nevertheless the examination of the single capture channel provides some insights into the collision process.

Apparatus

The experimental apparatus presently used in the fast-ion experiments at KSU is shown in Fig. 1. A supersonic gas jet of He, precooled to liquid nitrogen temperature, is skimmed and directed to cross the beam at right angles. A very uniform electric field of strength typically 1-2 V/cm collects the recoil He ions and directs them onto the face of a position sensitive channel-plate detector located 35 cm from the jet. The target density achieved is typically 10^{11} atoms/cm^2 and the momentum spread transverse to the jet is less than 0.15 a.u. The three dimensional momentum vector of the recoil ion is reconstructed from the position with which it hits on the detector and from its flight time to the detector. The latter is measured by timing on the projectile, which is charge-state analyzed before detection. A set of helmholtz coils surrounding the whole apparatus is used to shift the recoil "picture" onto the detector surface, in a way similar to the scan operation in a TV tube, since the recoil momentum can be easily so large as to throw the recoil off the detector surface.

CP392, *Application of Accelerators in Research and Industry*, edited by J. L. Duggan and I. L. Morgan
AIP Press, New York © 1997

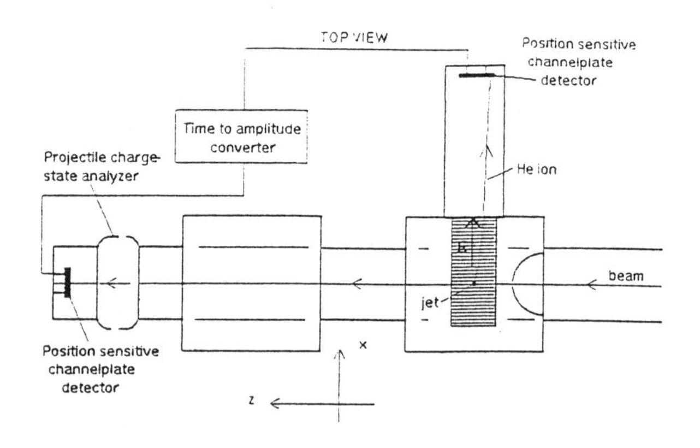

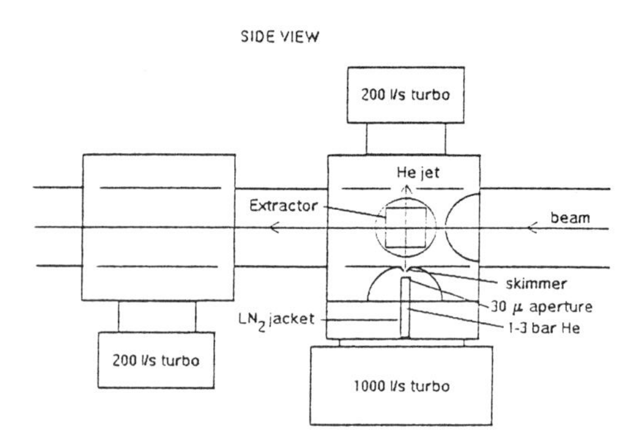

FIG. 1. Schematic of apparatus used for COLTRIMS at KSU LINAC.

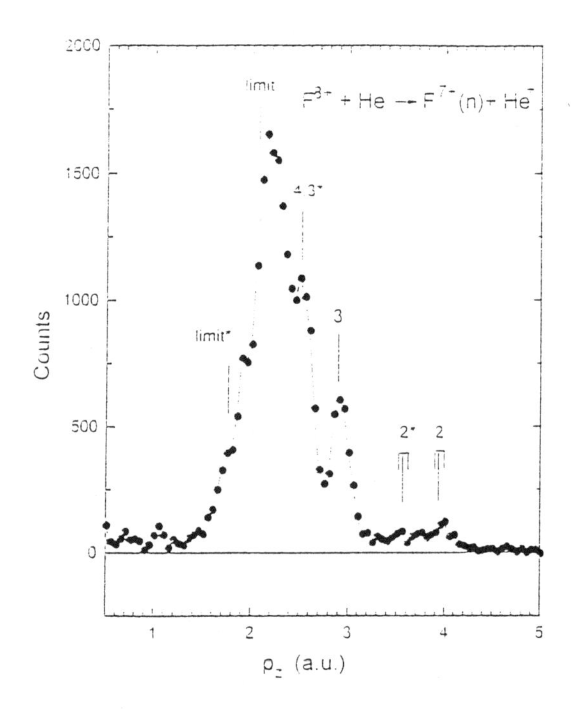

FIG. 2. Longitidunal momentum spectrum for 10 MeV F^{8+}+He producing F^{7+}+He^{+}. The final n of the captured electron is indicated; the * denotes the expected location of capture acompanied by excitation of the He+ ion to its series limit.

Results

Single electron capture and transfer ionization for F^{+8} on He at 10 MeV has been studied with the apparatus. At this energy, the single capture cross section, while still much smaller than that for single ionization, is sufficiently large that random coincidences from the single ionization channel do not overwhelm the single capture channel. Fig. 2 shows the Q spectrum for the single capture channel, shown as the longitidunal momentum transfer spectrum of the recoil He^{+} ions detected in coincidence with F^{+7} ions. Randoms have been subtracted. The resolution in p_z is 0.15 a.u., which, at this collision velocity (4.58 a.u.) corresponds to an energy resolution of 18 eV. To acheive an equivalent resolution by measuring the projectile energy gain would require a fractional resolution of about one part in 10^{6}, clearly not possible. The final states populated in capture at this energy can be identified from this spectrum. In this channel, capture to K and L shells of the F is almost negligible. The main capture is to n=3 and higher, and indeed, the process is dominated by capture to states sufficiently near the series limit that they cannot be resolved with the present apparatus. This conclusion is qualitatively what one would expect from an OBK calculation, but is quite different from the conclusions reached earlier on the basis of Auger electron spectroscopy [4], which showed much stronger L shell capture. Interpretation of the strengths of Auger lines in terms of initial populations is always complicated by the necessity to take cascade rearrangements into account, and this problem was recognized as a major one in the earlier work. The present direct Q value measurements do not suffer from this.

The momentum distributions can also sliced in p_z so as to show $p_r = \sqrt{(p_x^2 + p_y^2)}$ spectra, the transverse momentum spectra. If one assumes a universal relationship between impact parameter (b) and recoil transverse momentum (a pure Coulomb deflection function would be $p_r = 2Z_1Z_2/bv$, where Z_1 and Z_2 are the core charges), one can deduce a b dependence for the capture process to each final state, which can then be compared to a theoretical calculation done in an impact paramater format. Such a distribution is shown in Fig. 3, for various final state captures, where we have taken $Z_1 = 8$, $Z_2 = 1$. We note, however, that such a universal deflection function is probably not justified, and this whole transformation is very problematic and an interesting object of study in itself.

Low Velocity Collisions

Energy gain spectroscopy and angular distribution measurements on low energy capture by highly charged ions have been heavily studied over the past decade [see Refs. 5 and 6 for reviews]. The energy gain of the projectile determines the final state into which the capture goes, and the angular distribution provides

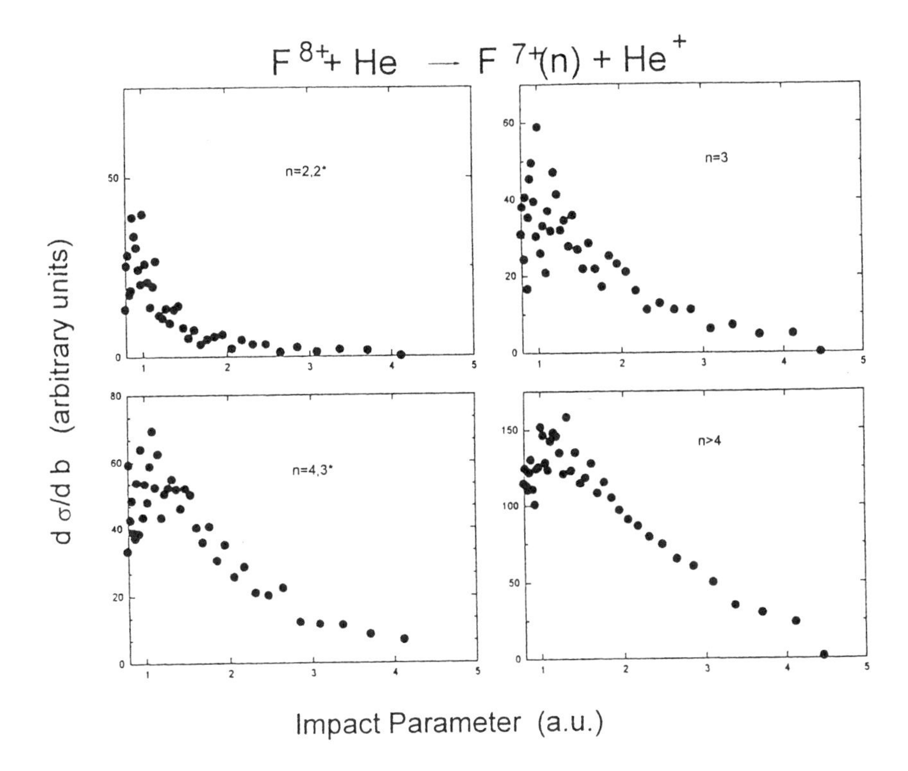

FIG 3. Impact parameter dependence for capture to various slices of the data of Fig. 2 for 10 MeV F^{8+} on He, deduced from transverse recoil momentum spectra.

information on the trajectory in "molecular orbital" space which the system follows. We have recently installed a COLTRIMS [3] system at the KSU EBIS. The apparatus is very similar in principle to that shown in Fig. 1; essentially the identical system is described in Ref. 3.

We show here some first results on electron capture from He by Ne^{+10} at a velocity of 0.65 au (53 keV incident energy). This system has also been studied in Refs. 7 and 8, and our results are in good agreement with those. In Fig. 4 are shown the longitidunal momentum spectra for single capture (SC), double capture with radiative stabilization (DC) and transfer ionization (TI), separated by detecting the charge state analyzed projectile in coincidence with the charge state analyzed He ion. The SC channel shows clearly resolved capture to n=4,5 and 6. The expected capture from the classical barrier model [9] is n=5, in good agreement with the results shown. In the double capture channel, the stronger channel, TI, corresponds to capture to (n,n')=(4,n'), where n'=4,5,6. These states, are relatively symmetric in principal quantum number, and are characterized by small fluorescence yields, decaying mainly by autoionization. However, the (n,n') = (3,n), with n=5,6,.... series, with relatively asymmetric principal quantum numbers, lies in the same region of the "favored" Q window [5,6] and is also populated. This series is expected to have larger fluorescence yields, becoming progressively larger as n' becomes larger and progressing to unity as n' goes to infinity. Thus near the series limit, any population of this series (which is coincident in Q value with the symmetric (4,4) states) will be reflected by enhanced radiative stabilization. This is seen clearly in Fig. 4, which shows strong DC near the (3,n') series limit. As n' becomes smaller, the fluorescence yield of this series becomes smaller, and this series shows up in the TI channel instead. The angular distributions for each of these final states are in the data, and can be extracted to provide information on the mechanism(s) by which each final state is populated.

In addition to the study of capture, this apparatus is also being used to determine the impact parameter dependence for single ionization of He by slow, highly charged projectiles. The mechanism for this process has been the center of considerable debate in recent years [10-13], with the first impact parameter dependence of the continuum electron spectra for proton bombardment having recently been reported [14].

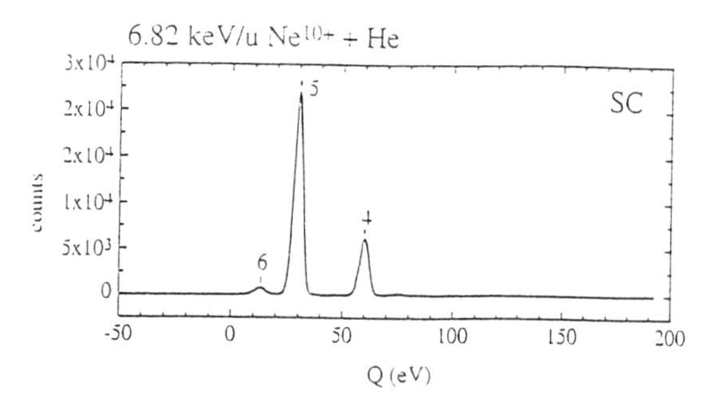

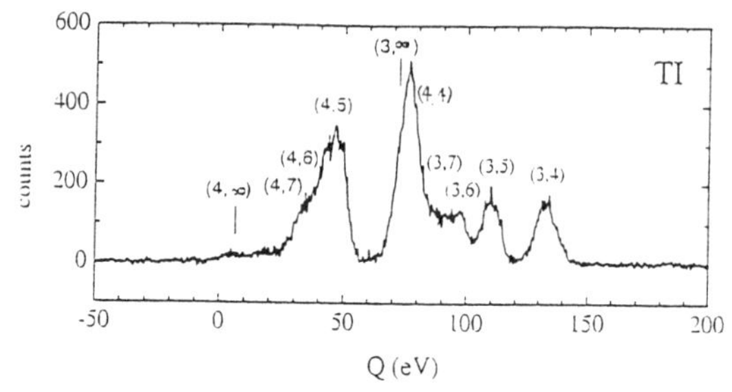

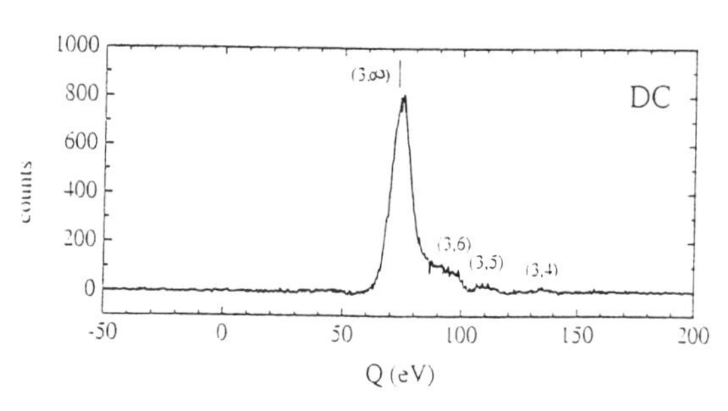

Fig. 4. Longitidunal momentum spectra, converted to Q spectra, for 6.82 keV/amu Ne^{10+} on He. Top: Single capture (Ne^{+9}+He^{+}); with n of captured electron indicated. Middle: transfer ionization (Ne^{+9}+He^{+2}), with (n,n′) of the captured electrons indicated. Bottom: double capture (Ne^{+8}+He^{+2}).

ACKNOWLEDGMENTS

This work was supported by the Division of Chemical Sciences, Office of Basic Energy Sciences, Office of Energy Research, U.S. Department of Energy.

1. For a review, see B. Fastrup, in "Methods of Experimental Physics: Atomic Physics Accelerators," ed. P. Richard, Vol. 17, p. 149 (Academic Press, New York) 1980.

2. J. Ullrich, H. Schmidt-Boecking and C. Kelbch, *Nucl. Inst. Meth. A* **268**, 216 (1988); R.E. Olson, J. Ullrich and H. Schmidt-Boecking, *Phys. Rev. A* **39**, 5572 (1989).

3. J. Ullrich *et al.*, *Comm. At. Mol. Phys.* **30**, 285 (1994).

4. J. Newcomb *et al.*, Ph.D. Thesis, Kansas State University (unpublished, 1982).

5. M. Barat and P. Roncin, *J. Phys. B* **25**, 2205 (1992).

6. C.L. Cocke, in "Review of Fund. Proc. and Appl.of Atoms and Ions," ed.C.D. Lin, p. 111 (World Scientific, Singapore, 1993).

7. A. Cassimi *et al.*, *Phys. Rev. Lett.* **76**, 3679 (1996).

8. P. Roncin and M. Barat, Eighth Intl. Conf. Phys. Highly Charged Ions, contributed abstracts,125 (unpublished, 1996).

9. A. Niehaus, *J. Phys. B* **19**, 2925 (1986).

10. M. Pieksma *et al.*, *Phys. Rev. Lett.* **73**, 46 (1994).

11. S.Yu. Ovchinnikov and J.H. Macek, *Phys.Rev. Lett.* **75**, 2474 (1995).

12. S. Kravis *et al.*, *Phys. Rev. A* **54**, 1394 (1996) and ref. cited therein.

13. W. Wu *et al.*, Phys. Rev. Lett. **75**,1054 (1995).

14. R. Doerner *et al.*, *Phys. Rev. Lett.* (to appear,1996)

STUDIES OF PHOTOABSORPTION AND COMPTON SCATTERING USING COLD TARGET RECOIL ION MOMENTUM SPECTROSCOPY

L. Spielberger, O. Jagutzki, R. Dörner, V. Mergel, U. Meyer, Kh. Khayyat, T. Vogt, M. Achler, H. Schmidt-Böcking,
Institut f. Kernphysik, August-Euler-Str. 6, D-60486 Frankfurt, Germany
J. Ullrich, M. Unverzagt,
Gesellschaft f. Schwerionenforschung, D-64291 Darmstadt, Germany
B. Krässig, M. Jung, E.P. Kanter, D.S. Gemmell,
Physics Division, Argonne National Lab., Argonne IL 60439, USA
M.H. Prior, H. Khemliche,
Lawrence Berkeley National Lab., Berkeley CA 94720, USA
C.L. Cocke
Dept. of Physics, Kansas State University, Manhattan KS 66506, USA

We present experimental values of $R = \sigma^{++}/\sigma^{+}$ for photoabsorption at photon energies ranging from 90 eV to 400 eV and for Compton scattering at an energy of 58 keV. The photoabsorption values are significantly lower than previously published ones. The technique of Cold Target Recoil Ion Momentum Spectroscopy (COLTRIMS), that was used in these experiments, permits a detailed control of the experimental conditions either in measuring the recoil ion momentum distribution (*momentum mode*) or in imaging the target volume (*monitor mode*).

INTRODUCTION

The investigation of dynamical electron–electron correlation effects in multi-electron systems is one of the central goals of nowadays atomic physics. Most of these studies concentrate on the Helium atom, the simplest multi-electron system. Kinematical complete experiments on the double ionization, as (γ,2e) after photoabsorption [1, 2, 3] or (e,3e) after electron impact ionization [4, 5] yield the most detailed information on the dynamics of the ionization process and the initial correlations in the bound target atom. In the case of classical electron spectroscopy, these experiments suffer from a small coincidence solid angle in the range of $10^{-4} - 10^{-6}$ of 4π. Therefore, this type of experiments is extremely time consuming. For the Helium atom, they could only be realized for some selected arrangements of outgoing electron energies and angles in the case of photoabsorption at low photon energies [1, 6, 7]. In the case of electron impact ionization they could still not be performed [8]. The situation is getting even worse with decreasing cross sections at high photon energies or large momentum transfers for charged particle impact, respectively, where the results become to be mostly sensitive on correlation effects in the bound target atom.

Therefore, the investigation of the ratio of total cross sections for double to single ionization, $R = \sigma^{++}/\sigma^{+}$, which was historically the first quantity to be investigated in the field of correlation effects, is still essential. Here, much more efficient ion collection techniques can be used. In comparison with theoretical calculations, R probes the understanding of the dynamics in many body systems. Of special importance is R at high photon energies. It is reported to reach a constant asymptotic high energy value, R^{∞}, which is mostly probing the internal electron–electron correlations of the Helium atom [9].

At low photon energies from double ionization threshold up to several hundred eV the absorption of the photon is the only process leading to ionization. Several experimental investigations in this energy range culminated in recommended experimental values of $R_{\rm ph}$ [10, 11]. The recent advent of modern synchrotron radiation machines made higher photon energies above some keV accessible. In this regime the second ionization process, Compton scattering of the photon, becomes important. The cross sections for single ionization from photoabsorption and Compton scattering of Helium are equal at about 6 keV [12], Compton scattering dominates at higher energies. In this regime the double ionization cross section is in the range of 10^{-26} cm^2 and the available photon flux is decreasing. Therefore, only few experimental data points

CP392, *Application of Accelerators in Research and Industry*, edited by J. L. Duggan and I. L. Morgan
AIP Press, New York © 1997

up to 20 keV were existing [11]. A constant asymptotic value, R^{∞}, is expected to exist for both processes.

In this paper we present experimental values of the ratio of double to single ionization of Helium obtained with the method of Cold Target Recoil Ion Momentum Spectroscopy (COLTRIMS) [13] in the low energy regime from 90 eV to 400 eV for photoabsorption, R_{ph} [14], and at the highest energy reported so far, at 58 keV for Compton scattering, R_{C} [15]. Due to the intrinsic differences in momentum space between both processes COLTRIMS allowed the first experimental separation of photoabsorption and Compton scattering at about 8 keV [16]. Here the predicted asymptotic value for photoabsorption, $R^{\infty}_{\text{ph}} = 1.67\%$ could be confirmed. In this paper, we focus on the strongly increased experimental sensitivity of COLTRIMS with respect to classical time-of-flight techniques either in measuring the recoil ion momentum distribution for photoabsorption (*momentum mode*) or in a 3-dimensional monitoring of the target volume for Compton scattering (*monitor mode*). These techniques can also be applied to the investigation of charged paticle impact ionization.

RESULTS

Our data are shown in figure 1. The low energy pho-

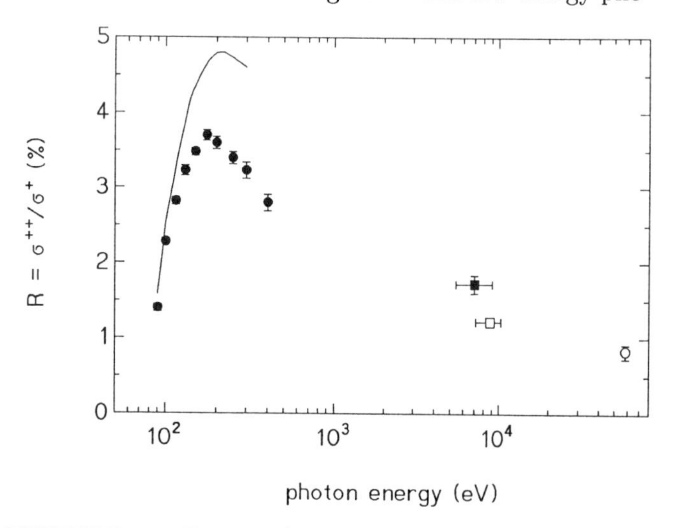

FIGURE 1. Ratios of double to single ionization for photoabsorption (full symbols) and Compton scattering (open symbols). The previously published data for photoabsorption are represented by the recommended values of [10] (full curve). Our results: full circles: [14], open circle: [15], squares: [16].

toabsorption values are significantly below the previously published results, which are represented by the recommended values from [10]. Our photoabsorption data show a close agreement with the calculations by Tang and Shimamura [17] and by Pont and Shakeshaft [18]. Our high energy Compton scattering result is also lower than the value published by Wehlitz *et al.* of $R_{\text{C}} = (1.25 \pm 0.3)\%$, however their significantly larger error bar almost overlaps with or data point.

It is obvious from the deviation of the COLTRIMS data from the earlier work as well as from the large scatter between different experiments combined in the recommended values, that it is most cruical for such experiments to exclude systematical errors. COLTRIMS offers a detailed control of the experimental parameters and allows to check for all the different sources of systematical errors in a measurement of the ratio of total cross sections R that have been discussed in the literature so far. We first list those possible problems before we discuss how COLTRIMS allows to control them.

1. Contributions of lower energetic stray light resulting in double ionization with a different value of R. Especially at photon energies in the keV regime with the ionization cross section for photoabsorption in the $E_{\gamma}^{-3.5}$ scaling being by orders of magnitude smaller than at the threshold [12], even small impurities can cause large systematical errors on R. Photons with an energy below the double ionization threshold would lead to a reduction of the apparent R.

2. Ionization by low energetic secondary electrons produced by photons impinging on solids in the experimental setup. The effect on R is similar to the one due to point 1).

3. A different detection efficiency of doubly and singly charged Helium ions on the ion detector, most probable increasing the apparent R.

4. Secondary collisions of the recoiling ions with residual gas resulting in charge exchange, decreasing the apparent R.

5. A different detection solid angle for different charge states, which would result in a higher apparent R.

Due to the inherent difference between the outgoing recoil ion momentum distributions for photoabsorption and Compton scattering [16], the respective techniques yielding this experimental control within COLTRIMS are different. In the following, they shall be discussed separately.

Photoabsorption: The spectrometer setup, which is shown in figure 2, consists of the following basic elements: The well localized reaction volume that is created by the overlap region of the photon beam and a supersonic Helium gas jet, and a precisely controlled electrostatic field that extracts all ions onto a position sensitive detector. If the extraction field is "weak" (in the order of several V/cm) the detected position and time-of-flight are broadened due to the recoil ion momentum. The respective deviations from the mean values give the information on the ion momentum vector (*momentum mode*).

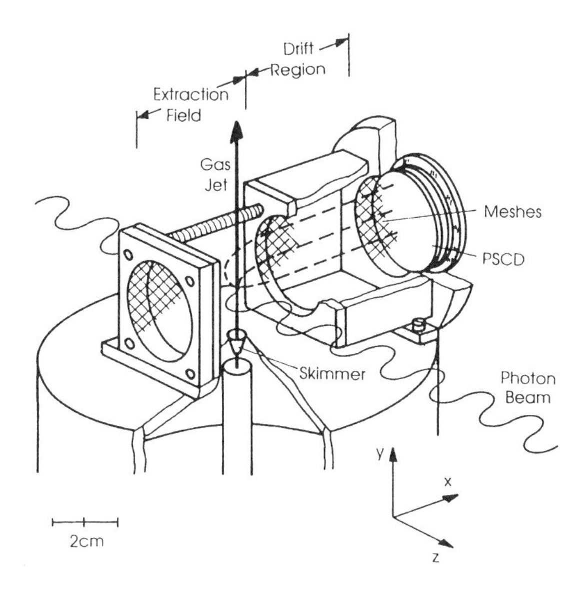

FIGURE 2. Sketch of the recoil ion momentum spectrometer.

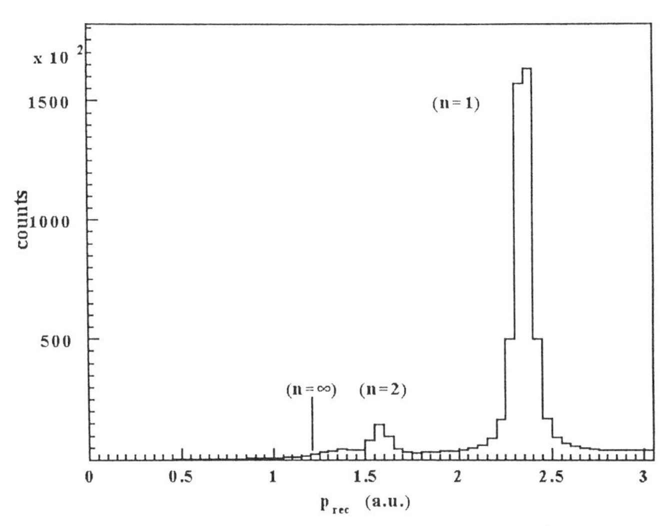

FIGURE 3. Momentum distribution of He^+ ions produced by 100 eV photons, integrated over all emission angles. No background was subtracted. (from [14])

For single ionization induced by the absorption of low energy photons, the outgoing recoil ion momentum vector mirrors the one of the photoelectron:

$$\vec{p}_{\mathrm{He}^+} = -\vec{p}_{\mathrm{e}^-} , \qquad (1)$$

as the photon momentum k can be neglected and the incoming momentum of the Helium atom is small due to low internal temperature of the supersonic gas jet. With the ionization potential of Helium from the ground state without excitation of the remaining electron of 24.6 eV one obtains $|\vec{p}_{\mathrm{He}^+}| = |\vec{p}_{\mathrm{e}^-}| = 2.35$ a.u. for $E_\gamma = 100$ eV and 24.2 a.u. for $E_\gamma = 8$ keV. As $|\vec{p}_{\mathrm{He}^+}|$ is measured, the energy of the photon, that was absorbed in the ionization process is measured for each ionization event. The He^+ momentum distribution from absorption of 100 eV photons is shown in figure 3. Besides the groundstate ionization, ionization with excitation of the remainig electron can clearly be seen. No absorption of contaminating lower energetic photons *within the photon beam* (see point 1) listed above) is observed.

The data of figure 3 are integrated over all recoil ion emission angles. They show a dipolar angular distribution [16, 3, 14]. Ions from ionization induced by stray light or secondary particles would not be restricted to the well localized target volume and would be visible as background in figure 3 or in the dipole emission pattern. No such contamination, according to the above listed points 1) and 2) can be seen in both distributions.

For each ionization event, the detector pulse height was additionally recorded which therefore can be checked during the experiment and in the off-line data analysis. The effect of any threshold to discriminate against noise can thus be controlled in the off-line data analysis. The pulse height distribution was found to be identical for both Helium charge states (see figure 2 in [14]). Therefore, a difference in the detection efficiency according to point 3) can be excluded within the given error bars.

Charge exchange in secondary collisions of the recoiling ions with the residual gas (point 4.) is strongly unlikely due to the well localized gas jet target with a background pressure of a few 10^{-7} mbar. In addition such collisions would strongly change the measured momenta of the ions and thus be visible in figure 3.

In the *momentum mode* the ion extraction field is strong enough to project all ions onto the detector and both helium charge states have the same detection solid angle. This excludes errors according point 5).

Compton scattering can be regarded as a scattering of the photon from an atomic electron with the nucleus remaining as a spectator. The outgoing recoil ion momentum is therefore close to the one of the atomic nucleus in the atom, which is the compensated bound electron momentum distribution [16, 15] with the FWHM of about 1 a.u.. If the details of this momentum distribution are not of interest, the ion extraction voltage can be increased until the momentum-broadening does not affect the recoil ion TOF and position and both quantities reduce to an image of the localized target volume in three dimensions (*monitor mode*). With the help of software windows in the off-line analysis the correct target size and TOF window can be selected and one has the ability to check for sources of background ions. This is demonstrated in figure 4. The background that was strongly suppressed in the "gated" spectrum shown in figure 4 was detector dark counts, that are randomly distributed in time and position. Due to the limited photon flux at the photon energy of 58 keV their rate was roughly a factor 30 higher than the ion rate. No ionic background arising from other other sources as the target volume, according to the above point

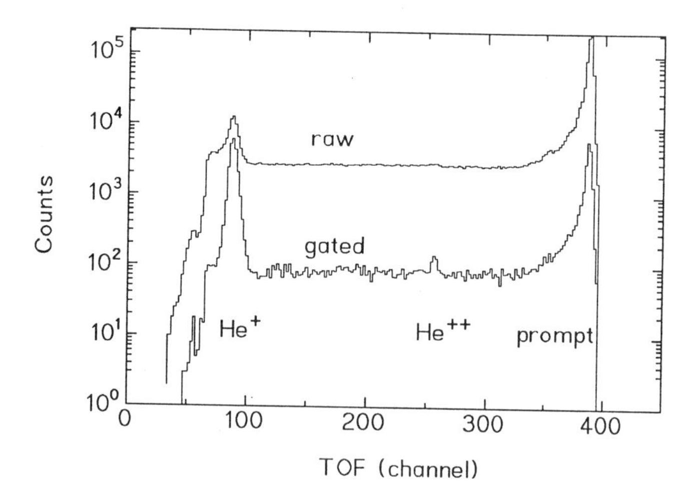

FIGURE 4. Ion TOF spectrum for Compton scattering at 58 keV. The spectrum on top is the experimental raw spectrum, the one labeled with "gated" was obtained in the off-line analysis. Here, only events with a position in the target area are selected.

1) and 2) could be found.

The remaining sources of possible error are avoided or can be checked like in the photoabsorption case.

DISCUSSION

With the above described capability of controlling the experimental conditions COLTRIMS provides a much more sensitive control of possible sources of systematic errors than all previous experiments in this field, that measured only the time-of-flight of ions produced in an extended effusive gas target. A recent consistency check of the data on $R_{\rm ph}$ for photoabsorption with the well established ratio for charged particle impact, $R_{\rm cp}$, strongly supports this claim [3]. Furthermore, our values of $R_{\rm ph}$ in the low energy regime are in excellent agreement with the most recent calculations by Tang and Shimamura [17] and by Pont and Shakeshaft [18]. Our low energy data are also in good agreement with new results by Stohlte and Samson [19] unpublished so far. A further set of data which is in between the recommended data and the new COLTRIMS values has been published by Levin *et al.* [11]. These data have been taken by detecting the ion charge state only. The reason for the disagreement is not obvious to us.

Our value of $R_{\rm C} = (0.84^{+.08}_{-.11})\%$ at 58 keV for Compton scattering is in agreement with the asymptotic high energy ratio of 0.8% [20, 21, 22] but is in contrast with another prediction of $R_{\rm C}^{\infty} = 1.67\%$ being the same value as the asymptotic value for photoabsorption [23].

Beyond its powerful abilities in determing R, COLTRIMS in the *momentum mode* is well suited for measuring highly differential cross sections. Completed by a position sensitive electron detector the above described spectrometer based on the ion collecting technique represents an extremely efficient tool for coincidence measurements [24, 25, 3]. Recently we have obtained fully differential cross sections for all angles and energies of the photoelectrons close to threshold [3].

ACKNOWLEDGEMENTS

We are grateful for the financial support from the DFG, BMBF, EU, Alexander v. Humboldt Stiftung and the DOE. We received indispensable help in building the supersonic gas jet from U. Buck. We are grateful for the support of Universität Frankfurt for the transportation of the experimental setup to the ESRF in Grenoble, in particular to Ch. Kazamias and W. Schäfer.

REFERENCES

1. O. Schwarzkopf *et al.* *Phys. Rev. Lett.*, **70**, 3008 (1993)
2. A. Huetz *et al.* *J. Phys.*, **B27**, L13 (1994)
3. R. Dörner *et al.* *Phys. Rev. Lett.*, **77**, 1024, 1996
4. A. Lahmam-Bennani *et al.* *J. Phys*, **B25**, 2873 (1992)
5. M. J. Ford *et al.* *Phys. Rev.*, **A51**, 418 (1995)
6. O. Schwarzkopf *et al.* *J. Phys.*, **B27**, L347, (1994)
7. P. Lablanquie *et al.* *Phys. Rev. Lett.*, **74**, 2192, (1995)
8. B. El Marji *et al.* *J. Phys*, **B28**, L733, (1995)
9. J.H. McGuire *et al.* *J. Phys.*, **B28**, 913 (1995) and references therein.
10. J.M. Bizau and F.J. Wuilleumier. *J. Electron Spectrosc. & Relat. Phenom.*, **71**, 205, (1995)
11. J. C. Levin, G. B. Armen, and I. A. Sellin. *Phys. Rev. Lett.*, **76**, 1220 (1996)
12. L.R. Andersson and J. Burgdörfer. *Phys. Rev. Lett.*, **71**, 50 (1993)
13. J. Ullrich *et al.* *Comm. At. Mol. Phys.*, **30**, 285 (1994)
14. R. Dörner *et al.* *Phys. Rev. Lett.*, **76**, 2654 (1996)
15. L. Spielberger *et al.* *Phys. Rev. Lett.*, **76**, 4685 (1996)
16. L. Spielberger *et al.* *Phys. Rev. Lett.*, **74**, 4615 (1995)
17. J.Z. Tang and I. Shimamura. *Phys. Rev.*, **A52**, 1 (1995)
18. M. Pont and R. Shakeshaft. *Phys. Rev.*, **A51**, 494 (1995)
19. W.C. Stohlte and J.A.R. Samson. priv. communication.
20. L.R. Andersson and J. Burgdörfer. *Phys. Rev.*, **A50**, R2810 (1994)
21. T. Surić *et al.* *Phys. Rev. Lett.*, **73**, 790 (1994)
22. M.A. Kornberg and J.E. Miraglia. *Phys. Rev.*, **A53**, R3709 (1996)
23. M. Ya Amusia and A.I. Mikhailov. *J. Phys.*, **B28**, 1723 (1995)
24. R. Moshammer *et al.* *Phys. Rev. Lett*, **73**, 3371, (1994)
25. R. Dörner *et al.* *Phys. Rev. Lett*, (1996), accepted for publication.

CHANGE IN FLUORESCENCE YIELDS OF F^- IONS AT MULTIPLY IONIZED STATES

T. Yamamoto[a], T. Takenaga[a], and M. Uda[a,b]

[a]Department of Materials Science and Engineering, Waseda University
3-4-1 Ohkubo, Shinjuku-ku, 169 Tokyo, Japan
[b]Laboratory for Materials Science and Technology, Waseda University
2-8-26 Nishiwaseda, Shinjuku-ku, 169 Tokyo, Japan

Chemical bond effects in F Kα K^1L^1 satellite intensities are explained quantitatively for alkali-metal fluorides. Change in fluorescence yields for multiply ionized states is caused by different resonant orbital rearrangement (ROR) probabilities, and plays one of the most important roles in these chemical bond effects.

INTRODUCTION

Particle Induced X-ray Emission (PIXE) has widely been used as a tool for trace elemental analysis for a quarter of a century. It has also been extensively utilized for studying chemical bond effects which are reflected in fine structures of PIXE spectra caused by transitions of the valence electrons. Multiply ionized satellite intensities of F Kα PIXE spectra emitted from a series of fluorides (1-3) are especially sensitive to change in chemical environments. However, intensity change in such satellite lines has not yet fully been explained theoretically. We have developed a method of calculating ionization cross sections for electrons on molecular orbitals in condensed matter (4) using the semi classical approximation (SCA) (5). The intensities of F Kα K^1L^1 multiply ionized satellite spectra emitted from a series of a rutile type fluorides were estimated by this method (4). In these we assumed orbital rearrangement after inner shell ionization but before X-ray emission occurs by reducing the binding energies (negative) of all the orbitals concerned but with no change in numbers of electrons on the orbitals. However, the above mentioned orbital rearrangement is not straightforward during inner shell ionization of F ions in alkali-metal fluorides. Change or chemical bond effect in fluorescence yields for the multiply ionized state is caused by difference in degree of the orbital rearrangement between the highest occupied molecular orbital at K^0L^0 and the lowest unoccupied molecular orbital at K^1L^0. This kind of rearrangement was defined as resonant orbital rearrangement (ROR) by the present authors (6). We have tried to account for change in the fluorescence yields for the satellite X-rays together with change in multiple ionization cross sections due to formation of chemical bonding states. This allows the chemical bond effect on the intensities of the K^1L^1 multiply ionized satellite lines(X1) in F Kα X-rays emitted from a series of alkali-metal fluorides to be explained quantitatively.

MULTIPLE IONIZATION CROSS SECTION

The K^1L^n multiple ionization probability is expressed using the binomial equation (4), if the electrons on the L shell are constituents of the molecular orbitals in condensed matter

$$P^M_{K1Ln} = \binom{2}{1}[P_K(b)]^1[1 - P_K(b)]^{2-1}$$
$$\times \sum_{n_1,n_2,\cdots,n_k} \binom{e_1}{n_1}\left[\left(C_L^{(1)}\right)^2 P_L(b)\right]^{n_1}\left[1 - \left(C_L^{(1)}\right)^2 P_L(b)\right]^{e_1-n_1}$$
$$\times \binom{e_2}{n_2}\left[\left(C_L^{(2)}\right)^2 P_L(b)\right]^{n_2}\left[1 - \left(C_L^{(2)}\right)^2 P_L(b)\right]^{e_2-n_2}$$
$$\vdots$$
$$\times \binom{e_k}{n_k}\left[\left(C_L^{(k)}\right)^2 P_L(b)\right]^{n_k}\left[1 - \left(C_L^{(k)}\right)^2 P_L(b)\right]^{e_k-n_k} \tag{1}$$

where $\binom{e_k}{n_k}$ denotes the binomial coefficient, e_k the number of electrons in the ground state, n_k the number of electron vacancies in each molecular orbital, with $n_1 + n_2 + \cdots + n_k = n$, C the partial expansion coefficient, and P_K and P_L the ionization probabilities of the K and L shell electrons calculated for an isolated atom using the semi classical approximation (SCA) (5), respectively. Then the K^1L^n multiple ionization cross section for condensed matter is expressed as

$$\sigma^M_{K1Ln}(E) = 2\pi\int_0^{r_0} bdbP^M_{K1Ln}(b) \cdot \tag{2}$$

RESULTS AND DISCUSSIONS

Observed F Kα spectra induced by 5.5MeV He^+ ions emitted from thin films of alkali-metal fluorides of Benka et

CP392, *Application of Accelerators in Research and Industry*, edited by J. L. Duggan and I. L. Morgan
AIP Press, New York © 1997

al. (3) are shown in Fig. 1. These spectra are composed of the diagram lines (X0) emitted from the K^1L^0 initial states and the satellite lines (X1 and X2) from the K^1L^1 and K^1L^2 multiply ionized initial states, respectively, which were deconvoluted here by the least squares employing the Lorentzian distribution function, where the line widths are 2.3, 3.8 and 4.0eV for X0, X1 and X2, respectively. Resulting relative intensities (X1/X0) are summarized on Table 1.

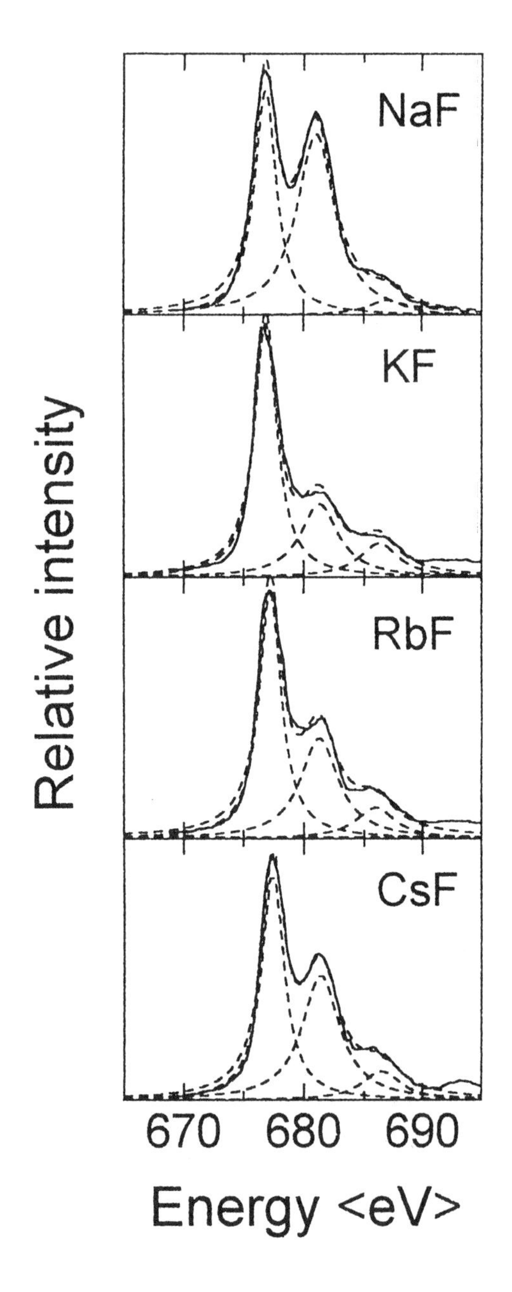

Figure 1. Observed F Kα spectra (3) emitted from alkali-metal fluorides by 5.5 MeV He^+ impacts (solid line), and each component of them deconvoluted using Lorentzian distribution function (dashed line).

TABLE 1. Comparison of relative intensities of X1/X0 between experiments and calculated multiple ionization cross sections.

	Relative intensity	
	Observed	Cross section
NaF	0.81	0.22
KF	0.29	0.11
RbF	0.42	0.11
CsF	0.55	0.11

In order to explain dramatic change in satellite intensities of X1 lines emitted from K^1L^1 doubly ionized states, the multiple ionization cross sections for the K^1L^0 and K^1L^1 states were estimated with the aid of eq. (5), where the partial expansion coefficients were calculated using the DV-Xα molecular orbital (MO) method (7). In this MO calculations, a $(M_6F)^{5+}$ cluster and the potential well of $9a_0/10$ a.u. and -3.0 Hartree in width and depth, respectively, were adopted, where M is a representative of Na, K, Rb and Cs, and a_0 is the atomic distance. Resulting relative multiple ionization cross sections (X1/X0) are also shown in Table 1. However, the observed relative intensities (X1/X0) have not yet fully been explained by the calculated relative cross sections.

Let us consider the excitation processes to form the K^1L^1 state, which is schematically shown in Fig. 2. A ground state (K^0L^0) goes to K^1L^0, K^1L^1 and K^1L^n states due to the direct coulomb interaction potentials acting between incident ions and orbital electrons, where $n \geqq 2$. Parts of the ionized states, however, go up to higher ionized states through the shake process. Then, the multiple ionization cross sections for K^1L^0 and K^1L^1 states, σ_{K1L0} and σ_{K1L1}, are expressed as

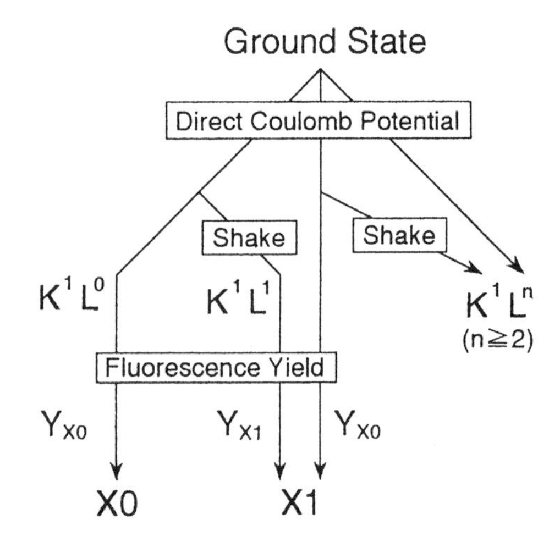

FIGURE 2. Schematic diagram of excitation process in the case of ion impacts.

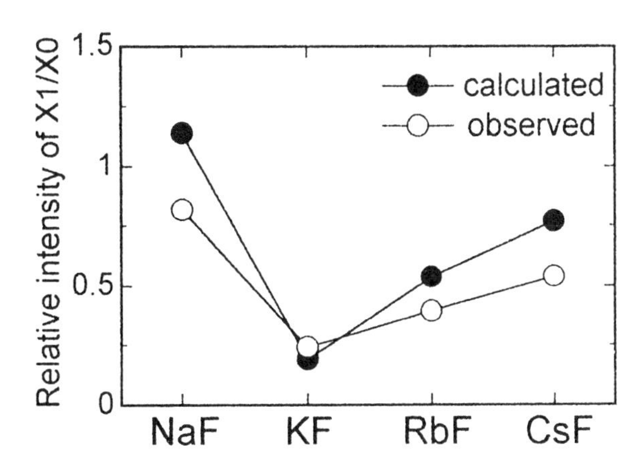

Figure 3. Comparison of relative intensities of X1/X0 in ion induced X-ray spectra between calculations and experiments.

$$\sigma_{K1L0} = \sigma^{DC}_{K1L0}(1 - P_{shake}) \quad (3)$$

and

$$\sigma_{K1L1} = \sigma^{DC}_{K1L1}(1 - P_{shake}) + \sigma^{DC}_{K1L0} P_{shake} \quad (4)$$

where the shake probabilities from K^1L^0 to K^1L^1 and from K^1L^1 to K^1L^2 are assumed to be the same. Here σ^{DC}_{K1L0} and σ^{DC}_{K1L1} denote the multiple ionization cross sections to produce the K^1L^0 and K^1L^1 states by the direct coulomb interaction, respectively, and P_{shake} is the shake probability for the isolated F atom which was calculated by Mukoyama and Taniguchi (8), i.e. 0.1845 for F atom.

Each ionized state emits X-ray by the amount expected from the fluorescence yield. The fluorescence yield for the isolated F atom from Krause (10), i.e. Y_{X0}=0.013, was adopted here for the diagram line of X1, because the numbers of electrons on the F L shells in all the alkali-metal fluorides are almost the same, i.e. 7.5～8. However, a striking change in fluorescence yields for the X1 line in fluorescent X-ray spectra was found, when K^1L^0 state goes to K^1L^1 state through the ROR process. Then, the fluorescence yield Y_{X1} estimated from the observed F Kα fluorescent X-ray and F KVV Auger spectra (6) was employed in the ionization mode of $K^1L^0 \rightarrow K^1L^1$ through the shake process. However, a part of the K^1L^1 state which was produced by the direct coulomb interaction emits the amount of X-ray expected from the fluorescence yield for the isolated atom, i.e. 0.013, because ROR does not occur in this case. The total X-ray production cross sections for X0 and X1 lines, σ_{X0} and σ_{X1} can be expressed as

$$\sigma_{X0} = Y_{X0}\sigma^{DC}_{K1L0}(1 - P_{shake}) \quad (5)$$

and

$$\sigma_{X1} = Y_{X0}\sigma^{DC}_{K1L1}(1 - P_{shake}) + Y_{X1}\sigma^{DC}_{K1L0} P_{shake} \quad (6)$$

Calculated relative intensities (X1/X0) using eq. (5) and (6) are compared with the observed ones in Fig. 3. Change in fluorescence yields for the satellite X1 lines in the F Kα fluorescent X-ray spectra is successfully explained by introducing ROR even for the ion induced F Kα satellite spectra emitted from the alkali-metal fluorides. As the same theoretical treatment has been successfully applied to fluorescent X-rays, it is concluded that the new concept of ROR can universally be used for every kind of ionization process.

REFERENCES

1. Uda, M., Endo, H., Maeda, K., Sasa, Y., Kumagai, H. and Tonuma, T., *Phys. Rev. Lett.* **42** 1257-1260 (1979).
2. Deconninck, G., and Van Den Broek, S., *Proc. Int. Conf. X-ray Processes and Inner-Shell Ionization*, 881- (1980).
3. Benka, O., Watson, R. L., and Kenefick, R. A., *Phys. Rev. Lett.* **47** 1202-1205 (1980).
4. Mochizuki, M., Yamamoto, T., Nagashima, S., and Uda, M., *Nucl. Instr.and Meth.* **B109/110** 31-38 (1996).
5. Hansteen, J. M., and Mosebekk, O. P., *Nucl. Phys.* **A 201** 541-560 (1973).
6. Uda, M., Yamamoto, T., and Takenaga, T., submitted to *Adv. Quantum Chem.*
7. Adachi, H., Tsukada, M., and Satoko, C., *J. Phys. Soc. Jpn.* **49** 875-883 (1978).
8. Mukoyama, T., and Taniguchi, T., *Phys. Rev.* **A36** 693-698 (1987).
9. Benka, O., and Uda, M., *Phys. Rev. Lett.* **56** 1667-1670 (1986).
10. Krause, M. O., *J. Phys. Chem. Ref. Data* **8** 307-322 (1979).

DIPOLE-FORBIDDEN TRANSITIONS AND CLOSE-COUPLING IN ATOMIC COLLISIONS

M.I. SYRKIN

Maritime College, State University of New York, Bronx, NY 10465

Recent experimental and theoretical studies indicate that the role of nondipole transitions in atomic collisions with charged projectiles was substantially underestimated in the past. The mechanism of dipole-forbidden transitions appears to be especially nontrivial in slow collisions wherein it manifests via multistep chains of virtual dipole transitions between closely coupled levels and significantly outperforms direct one-step dipole transitions. The results are important for an analysis of beam experiments and the kinetics of atomic populations in plasmas.

INTRODUCTION

Collisionally induced atomic collisions differ from those caused by radiative interactions in that they do not discriminate over the transitions with the orbital momentum change greater than unity, i.e. so- called dipole-forbidden transitions. In fact, the experimental and theoretical studies performed in the last decade with highly excited (Rydberg) atoms, wherein an active electron behaves mostly as in a hydrogen-like atom, clearly indicated an important role of nondipole transitions in both fast and slow collisions with charged projectiles. These transitions lead to readily observable experimental effects and result in significant implications for an analysis of beam experiments and the radiative-collisional kinetics of atoms in plasmas. This trend is also expected to persist (although in rather disguised form) for the lower part of an atomic spectrum. The aim of the paper is to review the current status of the matter and emphasize implications for applications. In what follows we assume the hydrogen-like structure of a target atom with an active electron having principal and orbital quantum numbers n and l, respectively. The length and velocity atomic units are designated by a_0 and v_0. The velocity of an atomic electron is given by $v_n = v_0/n$.

FAST COLLISIONS

If a projectile is sufficiently fast compared to the orbital velocity of an atomic electron and the energy transfer to the atom is small (i.e. n-change $\Delta n \sim 1$) then the transition amplitude is small and meets the perturbation theory requirements. The cross sections obtained in the first order of the perturbation theory constitute what is known as a standard Born approximation (see, for instance, (1-3)). In that case each transition with fixed l-change Δl is a one-step transition and is induced solely by the corresponding term in the multipole expansion of an atom-projectile interaction, the dipole transition $\Delta l=1$ overwhelmingly dominating over all others. The dipole approximation then is obtained by retaining dipole transitions only with the cross-section given by the Bethe-Born formula and various modifications thereof. For decades the latter result has been a common tool of the atomic collisions theory. In the meantime for large energy change $\Delta n > 1$ the difference between dipole and nondipole transitions becomes less pronounced. Although each dipole-forbidden transition is still noticeably inferior to the dipole-allowed one, but taken collectively they become quite competitive with dipole transitions as the collision velocity reduces. This means that the total depopulation cross section (i.e. the cross section of leaving the initial nl-level for all other levels) is substantially contributed by nondipole transitions. A direct indication of that fact can be found in the early numerical data of Omidvar (4) on the l-resolved transitions between relatively low levels and on the nl-n'l' form-factors between Rydberg levels by Flannery and McCann (5). An analytic treatment of dipole versus nondipole contributions to the l-averaged $n \rightarrow n'$ cross sections has been given by Beigman (see (2) and (6)). In fact, the contribution of dipole-forbidden transitions to the total depopulation in fast collisions typically varies around 20-30%, but can readily grow up to 50% as the velocity reduces (long before the perturbation theory fails). An analytical approach to dipole-forbidden transitions based on the comparison of exact quantum and semiclassical treatments enables transparent closed-form formulas for individual and total nondipole cross sections (Syrkin, (7)):

$$\sigma = 8\pi a_0^2\ (nn')^2\ (v/v_0)^{-2}\,\Omega_{nl\text{-}n'l'}\ /\ (2l+1)$$

$$\Omega_{nl\text{-}n'l'} = \Delta n^{-2}\,\Delta l^{-2}\min(\Delta n^{-1}, \Delta l^{-1})\,(2\Delta l+1)(2l'+1)\times \qquad (1)$$

$$V(ll'\Delta l\,|\,000)\ K(\Delta n, \Delta l, \varepsilon),\ \ \varepsilon = (1 - l^2/n^2)^{1/2}$$

where V is a 3j-symbol, $K(\Delta n, \Delta l, \varepsilon)$ is a slow function of its arguments and varies around ~ 0.015. According to the formula (1) the total nondipole contribution to the l-averaged $n \rightarrow n'$- transition is

CP392, *Application of Accelerators in Research and Industry,* edited by J. L. Duggan and I. L. Morgan
AIP Press, New York © 1997

$$\sigma^{nd} = 8\pi a_0^2\ (nn')^2\ (v/v_0)^{-2}\ K'\ f(\Delta n)/(\Delta n)^3 \qquad (2)$$

where $f(\Delta n) = 2[\ln \Delta n + 1/(1 + 1/\Delta n)]$, $K' \approx 0.017$. Recently Ivanovsky , Yanev and Solov'ev (8) have published nl-n'l' cross sections based on purely classical mechanics, the analytic expressions obtained in the high energy limit being different from the formula (1). Authors assert that their results should be valid down to the maximum of the cross section and even below, although no comparison with quantal calculations has been presented.

The role of nondipole interactions looks even more impressive for the collisions with spin exchange. In particular, dipole transitions are no longer singled out and are inferior to dipole-forbidden transitions even taken individually (Syrkin, (9)). For example, transitions $n,l \rightarrow n'l'$ with $n=10$, $n'=15$ and $\Delta l > 1$ contribute more than 80% to the averaged cross section $n \rightarrow n'$. This appears quite justified considering that an exchange scattering occurs predominantly in close encounters with large energy transfer, which in turn is caused by short range nondipole interactions. The cross sections may be obtained in the semiclassical version of the Ochkur approximation, which is a suitable first-order modification of the Born-Oppenheimer approximation for spin exchange (Syrkin, (9)):

$$\sigma = 8\pi a_0^2\ (nn')^{-2}\ M\ (v/v_0)^{-6}\ \Omega_{nl-n'l'}\ /\ (2l + 1)$$

$$\Omega_{nl-n'l'} = \Delta n\text{-}1/2\ (2l+1)(2l'+1)\ B \times \qquad (3)$$

$$[(l + l')^{2/3} - |\,l - l'\,|^{2/3}]\ /\ [P + C]$$

$$P = \min(l, l'), \quad \Delta n>0$$

where $M = (2s + 1)/2(2s'+ 1)$, s and s' being spins of the atom and the atomic core after collision, $B = 0.0638$, $C = 0$ if $\Delta n = 1$, and $C = [P/(P + 5)]^2$ otherwise.

SLOW COLLISIONS

At intermediate and low velocities the scheme of transitions undergoes radical change. At intermediate velocities an atom-projectile interaction is not weak and is no longer a mere perturbation. Instead it couples together a broad variety of levels forcing an electron to visit many states in one single act of collisions. Since the major term in the atom-projectile interaction is a dipole coupling, the whole process can be viewed as a chain of virtual dipole transitions. These virtual transitions can be adequately described by a close coupling within the so-called "m-averaged" version of the semiclassical impact parameter method (Beigman and Syrkin, (10)). The latter approach is based on two major assumptions. First, since the dipole interaction is a long range potential the projectiles can be viewed as moving along classical trajectories. Second, the averaging over magnetic quantum numbers m (whose role is minor in this context) allows to reduce considerably a computational complexity while preserving major features of the process. Such an approach leads to the close-coupling system of the Schrodinger equations for amplitudes:

$$i\hbar\ da_{nl}(\rho,t)/dt = \Sigma_{n'}\ \Sigma_{l'=1\pm1}\ V_{nl-n'l'}(\rho,v,t)\times \qquad (4)$$

$$\exp(-i\omega_{nl-n'l'}\,t)\ a_{n'l'}(\rho,t)$$

where ρ is an impact parameter and $V(\rho,v,t)$ is a dipole potential averaged over magnetic quantum numbers (for details see (10)). The solution of Eq.(4) gives the probabilities of dipole chain transitions and at low velocities results in large Δl, thus considerably outperforming one-step dipole $\Delta l = 1$ transitions. It is in this sense that the dipole interaction causes dipole-forbidden transitions and the total depopulation is therefore dominated by nondipole transitions $\Delta l > 1$ in contrast to fast collisions. The situation, however, proves somewhat more involved in adiabatic collisions (Syrkin, (11)). The competition between dipole-allowed and dipole-forbidden transitions in slow collisions entails the nonmonotonic behavior of the dipole fractional depopulation (DFD) (i.e. the fraction of the dipole-allowed transitions in the total depopulation cross section). Namely, the velocity dependence of the DFD is the curve with a minimum and the evolution from direct transitions in fast collisions to the chain of virtual transitions at intermediate velocities (which is common for any multilevel system) represents only a downhill part of the curve preceding the minimum. As the projectile velocity drops below the velocity of an atomic electron (the so-called "matching" velocity v_n) the actual balance between dipole-allowed and dipole-forbidden transitions depends on the type of a spectrum, or, more precisely, on the energy level splitting. From the analysis of the multilevel close-coupling system (4) it follows, in particular, that the minimum of the DFD is attained, roughly speaking, around the velocity $v_m \sim n^3 (\Delta E/Ry) v_n$, where ΔE is the energy spacing adjoining the depopulated level caused by either a quantum defect or a relevant level width or splitting, whichever is greater. Therefore, if the level separation ΔE is small (as for large l-levels) then the minimum v_m shifts into the deep adiabatic region and the dipole-forbidden depopulation prevails practically over the entire low velocity region. This situation takes place, in particular, in the depopulation of d-levels in the Rydberg atoms of Na and was tested in detail in the series of the experiments by MacAdam with co-workers (see (12) and references therein) The results are in a good accord (Fig. 1a) with the predictions of the semiclassical close-coupling theory with "m-averaging" (10,11) and with the recent purely classical calculations of Kazansky and Ostrovsky (13). In contrast, for large energy gaps (as for strongly nonhydrogenic s- and p-levels in alkalis) the minimum is reached readily below the matching velocity, beyond which point the role of dipole transitions increases again. The measurements of the s-levels depopulation performed recently by Rolfes *et. al.* (14) and by Irby *et. al.* (15) also revealed a close correspondence with the dipole-forbidden mechanism down to the matching velocity $v \sim v_m$. According to the close-coupling theory (11) for lower velocities (Fig. 1b) the DFD slowly revives. The experimental demonstration of that prediction would be of a special interest.

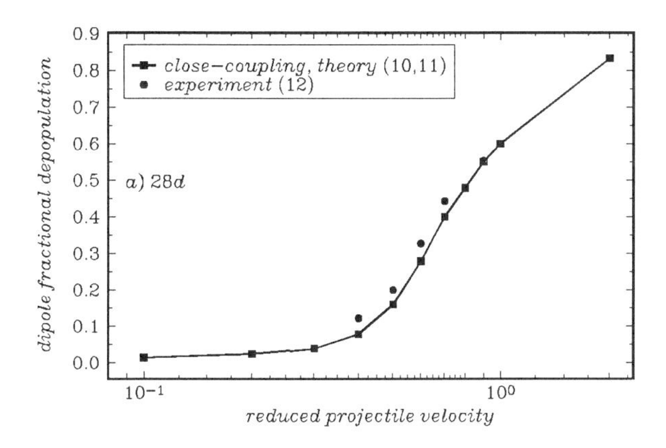

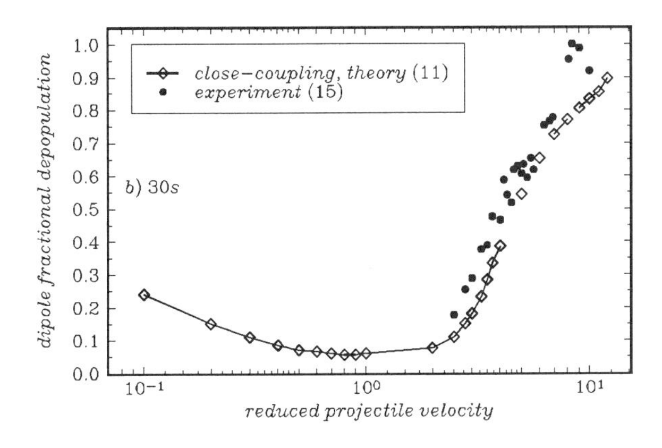

FIGURE 1. Fractional dipole depopulation in Na: a) 28d and b) 30s. Reduced projectile velocity defined as v/v_n.

When extending the above theory to the lower levels of alkalis or more complex atoms one does not foresee changes other than in energy spacings and nonhydrogenic corrections to matrix elements. These certainly do not affect the qualitative aspect of the competition between dipole and nondipole transitions. In fact, calculations show that the cross sections obtained for high n remain valid down to n = 4 or even n = 3 within the factor of ~ 2. A certain experimental support to this viewpoint comes also from the recent collisional experiments with metastable levels of He by Lagus *et. al.* (16). However, a systematic quantitative analysis of close-coupling in lower levels of complex atoms is quite complicated by the abundance of competing channels and intricacy of the spectrum and requires more experimental and theoretical efforts.

CONCLUSIONS

As shown above nondipole transitions represent an efficient alternative to dipole transitions in the redistribution of atomic populations (especially in slow collisions) and lead to obvious implications for atomic kinetics. In particular, in view of the substantial computational complications coupled with the resolution over orbital quantum numbers the majority of calculations in plasma kinetics as a matter of practice was performed under the assumption of the statistical equilibrium over l-sublevels - with surprisingly good results. However, at low temperatures and densities the equilibrium condition for relatively low levels seems dubious if not invalid at all. The issue is considerably clarified by an account for dipole-forbidden transitions which sharply increase the average angular momentum transferred to an atom and intensify the collisional l-mixing thereby placing the thermal equilibrium assumption on more reliable grounds.

Overall, unlike the widespread misconception dipole-forbidden transitions play an important role in collisions with charged projectiles and as such should be given an adequate consideration whenever necessary.

ACKNOWLEDGMENTS

I am grateful to K.B. MacAdam for fruitful collaboration and to I.L. Beigman for enlightening contributions at the early stage of this work. The remarks on various aspects of the problem by I.I.Fabrikant, M.R.Flannery, A.B.Kazansky, V.I.Ochkur are appreciated. I thank P.Levy and Maritime College Joint Fund for support.

REFERENCES

1. Mott, N.F. and Massey, H.S.W., *Theory of Atomic Collisions*, London: Oxford University Press, 1987.
2. Sobel'man, I.I., Vainshtein, L.A. and Yukov, E.A., *Excitation of Atoms and Broadening of Spectral Lines*, Berlin: Springer, 1981.
3. Flannery, M.R., in *Rydberg States of Atoms and Molecules*, edited by Stebbings, R.F. and Dunning, F.B., New York: Cambridge University Press, 1983, ch. 11.
4. Omidvar, K., *Phys. Rev.* **188**, 140-151 (1966).
5. Flannery, M.R. and McCann, K.J., *J. Phys. B* **12**, 427-445 (1979).
6. Beigman, I.L. and Lebedev, V.S., *Phys. Rep.* **250**, 95-238 (1995).
7. Syrkin, M.I., *Lebedev Institute Reports* **6**, 6-10 (1989).
8. Ivanovsky, J., Yanev, R.K. and Solov'ev, E.A., *J.Phys. B* **28**, 4799-4809 (1995).
9. Syrkin, M.I., *Phys. Rev. A* **50**, 2284-2291 (1994); Syrkin, M.I., *Phys. Rev. A* **52**, 847-850 (1995).
10. Beigman, I.L. and Syrkin, M.I., *Zh. Eksp. Teor. Fiz.* **89**, 400-406 (1989) (*Sov.Phys. JETP* **62**, 226- 232 (1985)).
11. Syrkin, M.I., *Phys. Rev. A* **53**, 825-830 (1996).
12. Sun, X. and MacAdam, K.B., *Phys. Rev. A* **47**, 3913- 3922 (1993).
13. Kazansky, A.B. and Ostrovsky, V.N., *Phys. Rev. A* **52**, R1811-R1814 (1995).
14. Rolfes, R.G., Irby V.D., Makarov, O.P. Dickinson, R.C. and MacAdam, K.B. , *J.Phys. B* **27**, 1167- 1174(1994).
15. Irby, V.D , Rolfes, R.G., Makarov O.P., MacAdam K.B. and Syrkin, M.I., *Phys.Rev. A* **52**, 3809-3815 (1995).
16. Lagus, M.A., Boffard, J.B., Anderson, L.V. and Lin C.C., *Phys. Rev. A* **53**, 1505-1518 (1996).

DOUBLE EXCITATION OF HELIUM PRODUCED BY FAST ION IMPACT

L.S. Pibida, R. Wehlitz, R. Minniti, and I.A. Sellin.

The University of Tennessee, Department of Physics, Knoxville, TN 37996-1200, USA and Oak Ridge National Laboratory, Oak Ridge, TN 37831-6377, USA.

We have measured the electron emission yield for autoinization from the doubly excited states $2p^2(^1D)$ and $2s2p(^1P)$ of helium produced by C^{q+} (q=4,5,6) ions at a reduced energy of 1.67 MeV/u. The observation angles were between 47.7° and 132.3° with respect to the ion beam direction. The angle and charge state dependence for these two states are discussed.

INTRODUCTION

During recent years there has been increased interest in studying electron-correlation effects in atomic collisions. For example, these effects have been studied in the doubly excited states of helium by Pedersen *et al.*[1] and Giese *et al.*[2,3] Using 20 MeV ion impact we present data in a wide angular range at a resolution not accomplished before, allowing us to better study these effects. One of the simplest models used to describe the double-excitation process in detail is the independent-electron model (IEM). The IEM assumes that each target electron interacts independently with the projectile [8,9]. For two-electron excitation this is known as a second-order process [5] (TS-2) and does not include any correlation between the electrons. The total cross section for this process is expected to vary as q^4, where q is the projectile charge state. There are also first-order mechanisms [2,5] (TS-1) which are used to describe the double excitation process, and are roughly divided into two groups [2]: static and dynamic correlation of the target electrons. The static mechanism assumes that the projectile interacts with only one target electron. The second electron is excited due to a final state rearrangement of the target produced by a static correlation of the electrons, i.e. a "shake up" process. The dynamic mechanism also assumes that the projectile interacts with only one electron, but now this electron can for example collide with the other one resulting in a two-step process. The total cross section for these two first-order mechanisms is expected to vary as q^2. A quantum interference between the first- and second-order mechanisms will give a term in the total cross section that is proportional to q^3.

EXPERIMENTAL SETUP

The C^{q+} (q = 4, 5, 6) ions were provided by the 7.0 MV EN Tandem Van de Graaff Accelerator at Oak Ridge National Laboratory. The ions having a reduced energy of 1.67 MeV/u were entering the experimental chamber, interacting with the helium atoms in an approximately 1 mm diameter diffusive gas jet crossing the ion beam and collected in a Faraday Cup for normalization purposes. The beam was collimated by two apertures of 1 mm and 0.75 mm diameters respectively, located 2 m apart. To provide a differential pumping stage a 5 mm diameter differential pumping hole was placed between the experimental chamber and the rest of the beam line. The electrons emitted from the interaction region were energy and angle analyzed by a cylindrical mirror analyzer (CMA) that has a position sensitive detector (PSD) mounted 17 mm down stream of its exit aperture. The CMA transmission is such that the image space is a one to one replica of the object space with unit magnification, except aberrations . The gas needle as well as the ion beam was positioned at the focal point of the CMA, as shown in Fig. 1.

The CMA was operated in its high resolution mode, so the electrons passed through the analyzer with an energy of 5 eV and the energy resolution in this case was of 0.1 eV. We measured the electrons emitted between 47.7° and 132.3° with respect to the ion beam direction. The experimental chamber was covered inside with a CO-NETIC AA foil so that the electrons were minimally affected by external magnetic fields. The background pressure of the chamber was 2.7×10^{-8} mbar. With the high purity He gas in the chamber the pressure was 6.6×10^{-6} mbar. The electrons were position detected by the PSD, their signal was amplified and analyzed by an acquisition system specially designed for this kind of application. The acquisition system consisted of three custom made CAMAC modules, a Digital to Analog Converter (DAC), a CAMAC Crate Controller and a program that interacts with these modules through a PC. The CMA was set to scan from 32 eV to 42 eV in order to measure the

CP392, *Application of Accelerators in Research and Industry,* edited by J. L. Duggan and I. L. Morgan
AIP Press, New York © 1997

electron emission yield from the doubly excited states $2s^2(^1S)$, $2p^2(^1D)$, $2s2p(^1P)$ and $2p^2(^1S)$ in helium.

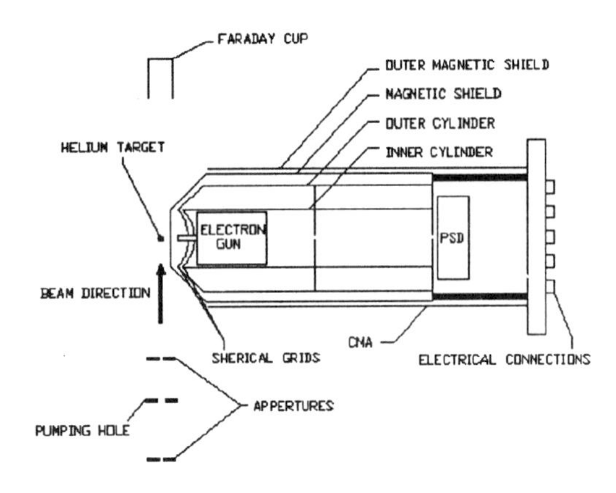

FIGURE 1. Schematic drawing of the apparatus used in the experiment. Not to scale.

DATA ANALYSIS AND RESULTS

Peaks corresponding to the $2p^2(^1D)$ and $2s2p(^1P)$ states were least-squares fitted to extract the Shore parameters [11] from the spectra shown in Fig. 2. The theoretical line shape used in the fit was[5]

$$\frac{d\sigma^2}{dEd\Omega} = C(E,\theta) + \Sigma \frac{A_j(\theta)\varepsilon_j + B_j(\theta)}{\varepsilon_j + 1} \qquad (1)$$

where E is the electron kinetic energy, θ is the electron emission angle with respect to the ion beam direction, $\varepsilon_j = 2\frac{E-E_{Rj}}{\Gamma_j}$, E_{Rj} is the energy of the j resonance, Γ_j is the natural width, and A_j and B_j are the Shore parameters of the j resonance. $C(E,\theta) = D(\theta) + L(\theta) \times (E - E_{R1})$ is the contribution to the double differential cross section from the direct ionization process, $D(\theta)$ and $L(\theta)$ are arbitrary functions of θ. The single differential cross section for each resonance is obtained by integrating equation (1) over the electron energy

$$\frac{d\sigma}{d\Omega} = C(\theta) + \frac{1}{2}\pi\Gamma_j B_j(\theta) \qquad (2)$$

where $C(\theta) = \int C(E,\theta)dE$ is the contribution from the direct ionization. The line shape and intensity of the peaks were obtained from the fitting using equation (1). The natural width of the peaks were fixed in the fit with values equal to 0.072 eV for the $2p^2(^1D)$ state and 0.041 eV for the $2s2p(^1P)$ state[2,7]. The fitting routine includes the spectrometer function that was earlier determined to be approximately Gaussian with a full width half maximum of 0.15 eV. The width used here was determined from the measured width of the Xe Auger lines $N_4O_{23}O_{23}$ ($^1S_0$) and $N_5O_{23}O_{23}$ ($^1S_0$). The energy spectrum of the Xe Auger lines $N_{45}O_{23}O_{23}$ with electron energies $27 \leq E_e \leq 37$ eV was used to obtain the energy calibration for the CMA. The images obtained in the PSD for the Xe Auger lines $N_4O_{23}O_{23}$ ($^1S_0$) and $N_5O_{23}O_{23}$ ($^1S_0$), which are known to give an isotropic angular distribution, were used to obtain the angular detection efficiency of the multichannel plates (MCP). The measured peak position for the $2p^2(^1D)$ state is 35.27 ± 0.04 eV and the $2s2p(^1P)$ state is 35.48 ± 0.04 eV which agree within the error bars with the values from references [2,6,7], so the energy difference between these states is 0.21 ± 0.06 eV.

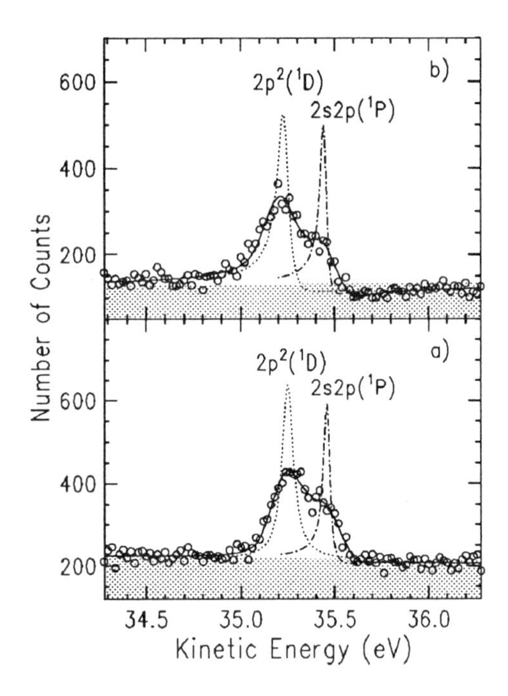

FIGURE 2. The fit by Fano profiles of the electron spectra at a) 63° and b) 116° produced by C^{4+} ions. The dotted line is the profile of the $2p^2$ resonance, the dashed-dotted line is the 2s2p resonance and the solid line is the contribution of both lines convoluted with the resolution function of the spectrometer. The shaded area corresponds to direct ionization.

From the Shore parameters we were able to obtain the single differential cross section as a function of the emission angle using equation (2) which is shown in Fig. 3. The θ dependence for the single differential cross section was obtained by fitting the data points shown in Fig. 3 with a function $F(\theta)$, where

$$F(\theta) = \alpha + \beta\cos\theta + \gamma\frac{1}{2}(3\cos^2\theta - 1) \qquad (3)$$

so that a curve for the single differential cross section for angles between 0° to 180° was obtained and is shown in

Fig. 4 for the $2p^2(^1D)$ state. The α, β, and γ parameters obtained for the $2p^2(^1D)$ state are the following (in units of 10^{-20} cm^2/sr):

C^{4+}

$\alpha = 3.96$ $\quad\beta = 0.26$ $\quad\gamma = 4.78$

C^{5+}

$\alpha = 7.66$ $\quad\beta = 0.29$ $\quad\gamma = 9.72$

C^{6+}

$\alpha = 9.88$ $\quad\beta = 1.65$ $\quad\gamma = 11.45$

The total cross section is then obtained by integrating $F(\theta)$ in the entire angular range as

$$\sigma_{t_j} = \int F(\theta) d\Omega. \tag{4}$$

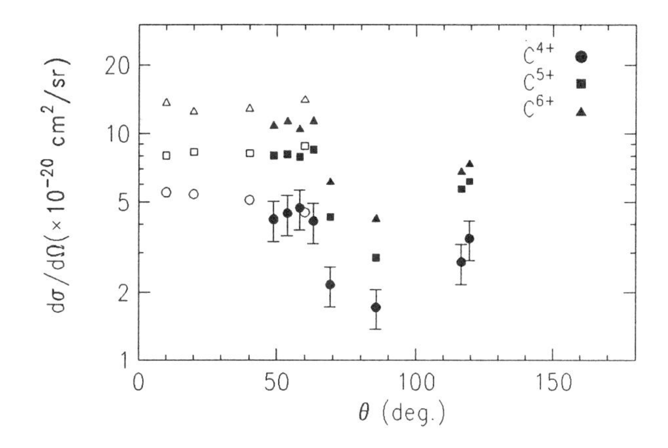

FIGURE 3. The differential emission cross section for autoionization from the $2p^2(^1P)$ state as a function of the electron emission angle after excitation by carbon ions. The solid symbols are the present data and the open symbols represent the data of Giese *et al.*[2]

From what was discussed in the introduction the total cross section can be parametrized by

$$\sigma_t(q) = aq^2 + bq^3 + cq^4 \tag{5}$$

In a good approximation a simpler parametrization has been used to describe the q dependence of the total double-excitation cross section [2]. So we were able to obtain the total cross section for each resonance as a function of the incoming ion charge state q by the parametrization

$$\sigma_{t_j} = Aq^B \tag{6}$$

which gives for the $2p^2(^1D)$ state $A = 2.25 \times 10^{-20}$ cm^2 and $B = 2.27 \pm 0.18$. For the $2s2p(^1P)$ state the A and B are: $A = 1.72 \times 10^{-20}$ cm^2 and $B = 2.02 \pm 0.19$, see Fig. 5. The data from Giese *et al.*[2] shown in this figure was obtained by assuming that the measured peak could be fitted by two theoretical peaks representing these two states. From the present results we are tempted to conclude that first-order processes are dominant for both states for $4 \le q \le 6$. These results are in good agreement with previous results[2] and calculations[5].

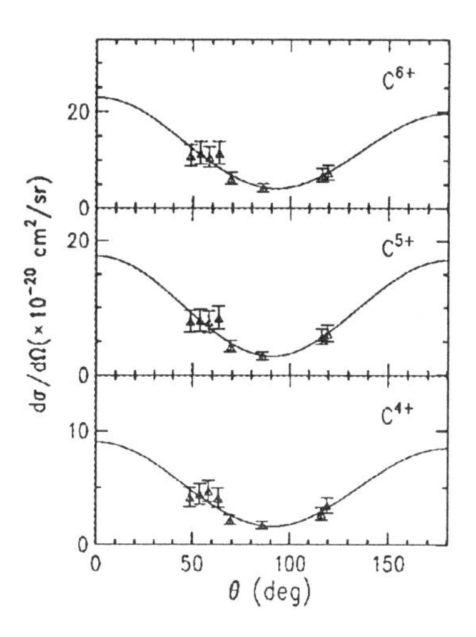

FIGURE 4. Single differential cross section from the $2p^2(^1D)$ state as a function of the emission angle for different incoming C^{q+} ions. The solid line is the fit curve of the data using the function $F(\theta)$.

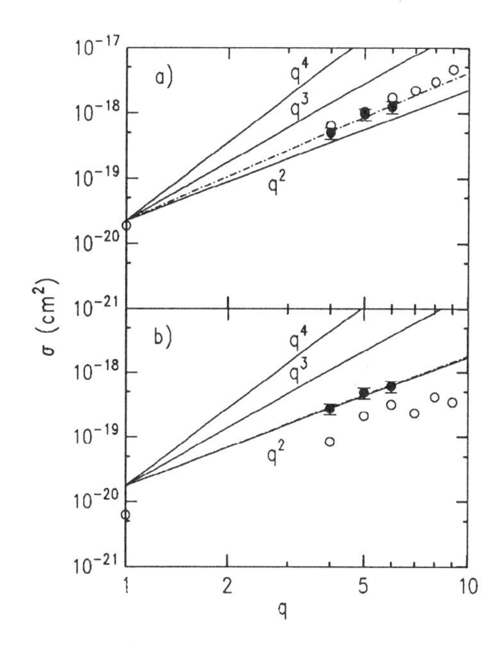

FIGURE 5. a) The total emission cross section from the $2p^2(^1D)$ state as a function of the projectile charge state. b) same as a) but from the $2s2p(^1P)$ state. The solid symbols are the present data, the open symbols are Giese *et al.*[2] data, the dashed line is the fit curve of the data using $\sigma \approx aq^b$, the solid lines indicate the slopes of q^2, q^3 and q^4 dependence.

From equation (2) we were able to obtain the single differential cross section for the $2p^2(^1D)$ and $2s2p(^1P)$ states. We found that the differential cross section for the $2p^2(^1D)$ state is approximately 1.85 ± 0.20 times bigger than for the $2s2p(^1P)$ for all angles $48.8^0 \leq \theta \leq 119.5^0$ for C^{4+} ions and approximately 2.03 ± 0.22 times bigger for C^{5+} and C^{6+} ions. The ratio of the total double excitation cross section between the $2p^2(^1D)$ and the $2s2p(^1P)$ states is 1.97 ± 0.25 for q=4,5,6.

CONCLUSIONS

In conclusion we can say that the $2p^2(^1D)$ and the $2s2p(^1P)$ states are primarily populated by the first-order mechanisms, for projectile charge states q=4,5,6. For projectiles with higher incident charge state theory[5] predicts that second-order mechanisms become more important for the $2p^2(^1D)$ state, which is subject of further investigations. The behavior of the angular dependence of the single differential cross section of the $2p^2(^1D)$ state is similar for all three charge states. Theoretical calculations of the single differential cross section of the $2p^2(^1D)$ state at our projectile velocity are scarce, some calculations are available for the double differential cross section of the $2s^2(^1S)$ state [4,10]. We can also say that the differential cross section of the $2p^2(^1D)$ is approximately 1.85 times bigger than for the $2s2p(^1P)$ in the measured angular range for C^{4+} and approximately 2.03 times bigger for C^{5+} and C^{6+} ions. The ratio of the total double excitation cross section between the $2p^2(^1D)$ and the $2s2p(^1P)$ states is approximately equal to 1.97 for q=4,5,6. Theoretical calculations by Fritsch and Lin [5] for C^{6+} at 1.5 MeV/u give a ratio of 0.96, which is about a factor of 2 difference between the measured and the calculated value.

ACKNOWLEDGMENTS

This work was supported by NSF and U.S. DOE, Office of Basic Energy Science, Division of Chemical Science, under contract number DE-AC05-96OR22464 with Lookheed Martin Energy System Research Cooperation.

R. Wehlitz, acknowledges partly financial support by a Feodor-Lynen-Fellowship of the A.V. Humboldt Foundation.

The participation of I. Sellin was sponsored by the U.S. Department of Energy, Office of Basic Energy Science, Division of Chemical Science, under Contract No. DE-AC05-96OR22464 with Lockheed Martin Energy Research Corporation.

REFERENCES

1. Pedersen J.O.P. and Hvelplund P., *Phys. Rev. Lett.* **62** 2373-2376 (1989).
2. Giese J.P., Schulz M., Swenson J.K., Schone H., Benhenni M., Varghese S.L., Vane C.R., Dittner P.F., Shafroht S.M. and Datz S., *Phys. Rev. A* **42** 1231-1244 (1990).
3. Giese J.P. and Pedersen J.O.P., *Nucl. Instrum. Methods B* **56/57** 176-179 (1991).
4. Martin F. and Salin A., *J. Phys. B: At. Mol. Opt. Phys.* **28** 2159-2171 (1995).
5. Fritsh W. and Lin C.D., *Phys. Rev. A* **41** 4776-4782 (1990).
6. Arcuni P.W. and Schneider D., *Phys. Rev. A* **36** 3059-3070 (1987).
7. Shearer-Izumi W. , *Atm. Data Nucl. Data Tables* **20** 531-561 (1977).
8. McGuire J. H., *Phys. Rev. A* **36**, 1114 (1987).
9. Stolterfoht N., *Comm. At. Mol. Phys.*, July (1988).
10. Martin F. and Salin A., *J. Phys. B: At. Mol. Phys.* **28** 639-652 (1995).
11. Shore B. W., *Phys. Rev.* **171** 43-54 (1968)

LYMAN AND BALMER ALPHA EMISSION IN COLLISION OF IONS WITH HYDROGEN

J. Geddes

The Queen's University of Belfast, Belfast BT7 1NN, UK

A detailed knowledge of processes involving collisions of hydrogen and helium ions with atomic and molecular hydrogen is of continuing interest. Processes that result in the formation of H(n=2) and H(n=3) excited products are of special importance. Apart from the fundamental interest, some of these processes are relevant to controlled thermonuclear research and the physics of proton aurorae in planetary atmospheres. In spite of the relative simplicity of such collision systems, the validity of the numerous theoretical descriptions of particular collision processes is still far from clear. We are studying these processes through observation of the Lyman and Balmer alpha emissions from the excited products formed in single collisions.

INTRODUCTION

Reliable measured cross sections for the excitation of atomic hydrogen by H^+, He^+ and He^{2+} are essential for an assessment of the range of validity of the different theoretical descriptions of such collisions. Although a large variety of theoretical treatments of excitation have been developed (cf (1)), in contrast there have been a relatively small number of experiments (cf (2)) due to the inherent difficulties of providing a well defined target of atomic hydrogen. Many of the excitation processes are directly relevant to the heating, modelling and diagnostics of magnetically confined plasmas in fusion energy research and the understanding of proton aurorae in planetary atmospheres.

In the laboratory at Belfast excitation of the n=2 and n=3 states of hydrogen has been investigated for collisions of H^+, He^+ and He^{2+} within the energy range 3-100 kev amu^{-1}. For the H(n=2) state, measurements of the excitation cross sections for 2p and 2s state formation have been based on observation of the Lyman alpha radiation emitted both from the spontaneous decay of H(2p) and the electric field induced decay of H(2s). In separate experiments Balmer alpha emission has been observed from decay of the n=3 state to n=2. In terms of cross sections $\sigma(3l)$ for excitation of the fine-structure states of H(n=3), the Balmer alpha emission cross section $\sigma(H_\alpha) = \sigma(3s) + 0.12\sigma(3p) + \sigma(3d)$ where the factor 0.12 is the branching ratio for 3p decay to the 2s state.

EXPERIMENTAL APPROACH

A momentum analysed H^+, He^+ or $^3He^{2+}$ beam of the required energy intersected (at right angles) a thermal energy beam of ~ 70% dissociated hydrogen. The crossed beam intersection region was maintained at a pressure not exceeding 3×10^{-6} Torr.

The hydrogen beam was derived from a 2.45 GHz microwave discharge sources developed in the Belfast laboratory (3,4). The Lyman and Balmer alpha emission directly from the sources was found to give an unacceptably high background signal in the detectors. This background radiation was minimised either by using a light baffle made from teflon in the exit canal of the atom source or by transporting the hydrogen through a long (400 mm) bent teflon tube from the discharge source to the interaction region. Although the source is capable of providing high intensity beams with dissociation fractions D greater than 0.9, the presence of the light baffle reduced D to ~ 0.7. The dissociation fraction was determined using a quadrupole mass spectrometer. A small sample of the beam was passed into a differentially pumped chamber where it was modulated by a mechanical chopper and the modulated H_2^+ signal $S(H_2^+)$ from the mass spectrometer was recorded. The dissociation fraction D of the hydrogen beam is given by

$$D = 1 - S_o(H_2^+) / S_f(H_2^+)$$

where $S_o(H_2^+)$and $S_f(H_2^+)$ are the H_2^+ signals with the discharge on and off respectively. Checks have been made to verify that the atom beam is at room temperature (4).

The Lyman alpha detector consisted of an 18 stage EMI 9642 electron multiplier fitted with a LiF window and a shield to limit the angular acceptance. The transmission characteristics ensure that Lyman alpha radiation (121.6 nm) is the only hydrogen atom transition to be recorded.

The Balmer alpha detector comprised a lens to form a parallel beam, an interference filter and a cooled photomultiplier EMI 9658QB. The 4 period filter centred at 656.4 nm had a 0.3 nm FWHM and a 10^{-5} reject ratio.

CP392, *Application of Accelerators in Research and Industry,* edited by J. L. Duggan and I. L. Morgan
AIP Press, New York © 1997

When H^+ projectile ions were used, this filter accepted Balmer alpha radiation from excited target H but rejected Doppler-shifted Balmer alpha radiation arising from electron capture by the fast H^+ projectile into the n=3 state.

The H(2p) and H(n=3) atoms, that are formed by excitation, decay spontaneously within the field of view of the detectors. In the measurement of the associated cross sections the radiation detector was set at 54.7° with respect to the incident ion beam thereby obviating the need for polarisation corrections. H(2s) atoms formed in the beam interaction region were detected at a point 1.5 cm beyond the crossing region and along the thermal beam direction where an electric field of 15 V cm^{-1} was applied. The Stark mixing which resulted in Lyman alpha emission was sufficient to quench the thermal energy H(2s) within the field of view of a Lyman alpha detector. H(2s) resulting from dissociative excitation of molecular H_2 was sufficiently energetic that only minimal quenching occurred in the detector's field of view. In order to facilitate signal detection the ion beam was modulated by electrostatic deflection prior to the interaction region and the resulting modulated signal from the radiation detectors was recorded by a narrow band lock-in amplifier and phase sensitive detector.

It was necessary to take account of the contribution to the Lyman or Balmer alpha signal from both the background gas and the undissociated H_2 molecules in the beam. The background contribution was accessed by turning the atom beam off and using a separate gas feed to raise the background pressure to the same value as that recorded when the atom beam was present. Under constant total mass flow conditions it can be shown (5) that the cross sections σ_a and σ_m for radiation arising from atoms and molecules respectively in the thermal energy beam are related by the expression

$$\sigma_a/\sigma_m = (S_o/S_f - (1-D))/\sqrt{2}\,D$$

where S_o and S_f are signals with the discharge on and off.

All cross sections have been normalised to earlier measurements of Lyman or Balmer alpha emission cross sections in the Belfast laboratory for collisions of H^+ with either H or H_2 (5, 6).

RESULTS

Cross sections measured by the methods described earlier are now compared with the results of other experimental and theoretical investigators.

H^+ - H Collisions

Cross sections for the excitation of H(2p) by H^+ have been measured by a number of laboratories within the energy range 2-800 keV and there is excellent agreement with theory (cf (2)).

Figure 1 presents the measured cross sections of the Belfast laboratory (5,7) and Chong and Fite (8) for H(2s) formation by H^+ collisions. This reaction channel is approximately an order of magnitude weaker than the 2p excitation channel. In contrast to the larger 2p product channel, none of the calculations (Figure 1) provide a good fit to the experimental data for H(2s) formation. Most of the two centres calculations that use large basis sets show oscillatory structure which is not observed experimentally. The general belief is that the larger the basis sets the better the results of calculations. This idea was recently challenged by Slim and Ermolaev (9). These authors have compared a semiclassical impact parameter coupled channel calculation that uses a basis set with only one projectile and 50 target states (curve SE51) and a similar calculation with a basis set where 37 states are centred on both the target and projectile (curve SE74). Their results show that oscillatory structure arises in the latter case. Kuang and Lin (10) (not shown) have recently investigated this problem in detail by examining the convergence of the two centre atomic orbital close coupling method within the impact parameter approximation. They conclude that spurious oscillations are due to the simultaneous use of short range pseudo-continuum states on both projectile and target centres.

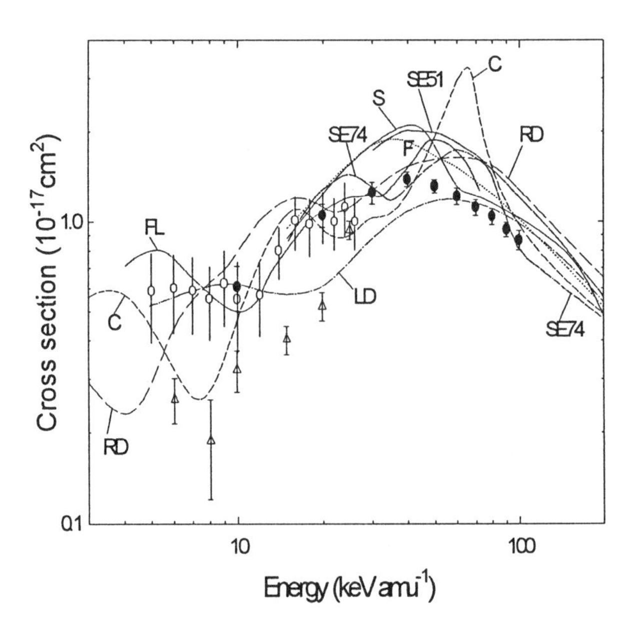

FIGURE 1. Cross sections for 2s excitation of H atoms in H^+ - H collisions. Experiment: •, Higgins et al (7). o, Morgan et al (5). Δ, Chong and Fite (8). Theory: FL, Fritsch and Lin (1), two-centre AO expansion. C, Cheshire et al (27), seven state close coupling. RD, Rapp and Dinwiddie (28), seven state close coupling. S, Shakeshaft (29), two centre expansion. LD, Lüdde and Dreizler (30), two-centre optical model. F, Ford et al (31), single-centre expansion. SE51, Slim and Ermolaev (9), two centre AO 51 state expansion. SE74, Slim and Ermolaev (9), two centre AO 74 state expansion.

Balmer alpha emission cross sections of target excitation in H^+ - H collisions are presented in Figure 2. The experimental values of the Belfast group (11) are significantly larger than those reported by the Giessen group (2) even if a 15% reduction (11) is applied for the cascade contribution The former cross sections were obtained by a direct observation of the Balmer alpha radiation while the later were derived indirectly from the expression $\sigma(H_\alpha) = \sigma(n=3) - 0.88\sigma(3p)$ using the cross section $\sigma(n=3)$ for H(n=3) production (12) and their own

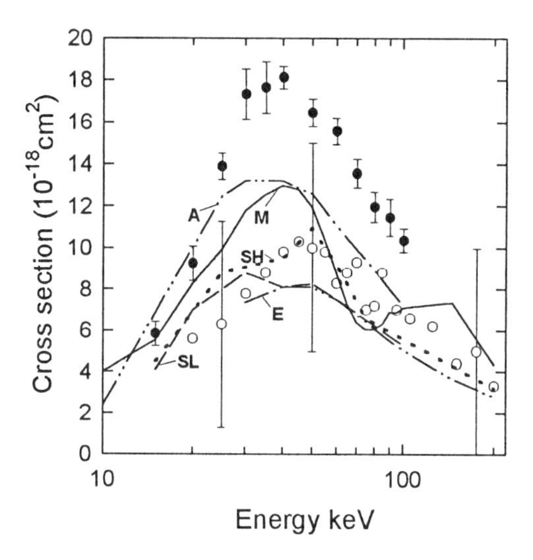

FIGURE 2. Cross sections for Balmer alpha emission in H^+ - H collisions. Experiment: •, Donnelly et al (11). o, Detleffsen et al (2). Theory: M, McLaughlin et al triple centre close coupling. A, Ast et al (26) one centre optical potential. SH, Shakeshaft (29) two centre symmetric state impact parameter coupled state, SL, Slim (14) two centre symmetric basis coupled channel. E, Ermolaev (13) impact parameter close coupled one projectile state.

measured value of $\sigma(3p)$. This discrepancy in the experimental data motivated the more recent calculation by Ermolaev (13), Slim (14) and McLaughlin et al (15). Further work is required to resolve the discrepancy

He^+ - H Collisions

Cross sections for the excitation of H(2p) in collisions of He^+ with H are shown in Figure 3. Until recently there has been little experimental or theoretical effort on this interesting two electron system. The experimental results from Belfast (16, 17) and Giessen (2, 18) are shown. Comparison of these cross section with those for equi-velocity H^+ - H collisions (2) shows that both have similar structure with a minimum at ~11 keV amu^{-1}. This structure arises due to interference between capture and excitation channels where there is a negligibly small difference in the energy defects. In this case the product channels are He(1s, 2p) + H^+ and He^+(1s) + H(2p). The calculations of Kuang et al (19) and Ermolaev et al (20) using atomic orbital basis sets have an energy dependence similar to the experimental values. The molecular orbital calculations of Errea et al (21) are not expected to be accurate at energies where ionisation becomes significant.

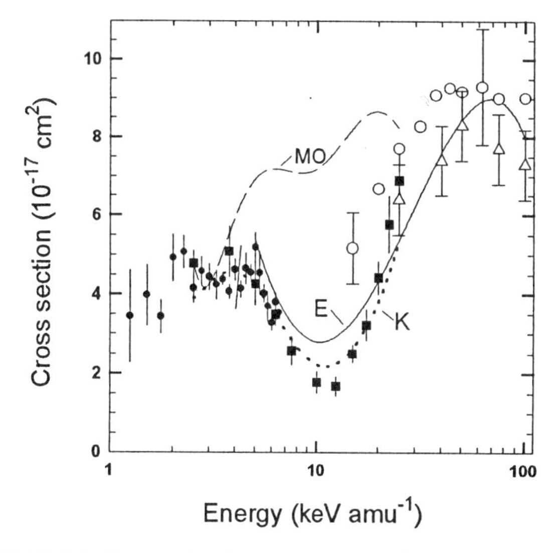

FIGURE 3. Cross sections for 2p excitation in He^+ - H collisions. Experiment: •, McKee et al (16). ■, Geddes et al (17). o, Detleffsen et al (2). Δ, Schartner et al (18). Theory: MO, Errea et al (21) MO expansion. E, Ermolaev (20) pseudo-one electron AO basis. K, Kuang et al (19) 2 centre AO method.

The measured cross sections of Geddes et al (17) for H(2s) formation in He^+ - H collisions increase in magnitude as the energy increases from 2 keV amu^{-1} and reach a plateau extending from 16 to 30 keV amu^{-1}. Agreement with the calculations of Kuang et al (19) is reasonable while Errea et al (21) and Ermolaev et al (20) do not predict the observed energy dependence.

Cross sections for Balmer alpha emission in He^+ - H collisions are shown in Figure 4. The experimental values of Donnelly et al (11) exhibit structure at approximately

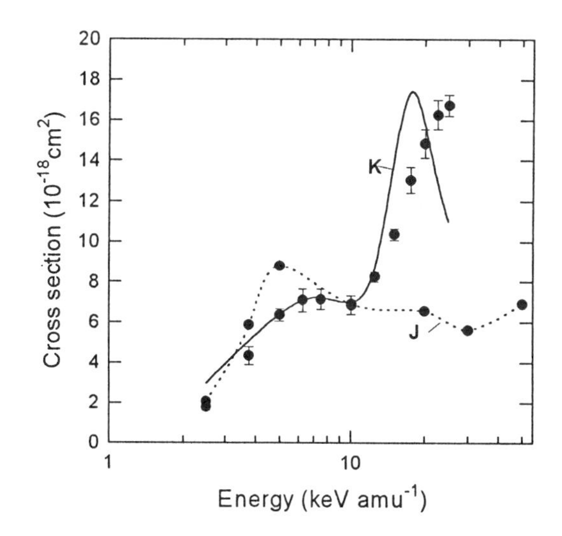

FIGURE 4. Cross sections for Balmer alpha emission in He^+ - H collisions. Experiment: •, Donnelly et al (11). Theory: K, CEng et al (19) 2 centre atomic orbital method. J, Jackson et al (22) 2 centre atomic orbital method

the same energy (~11 keV amu^{-1}) as that observed in 2p formation. Again, this could be due to interference between capture and excitation channels (9). The calculations of Kuang et al (19) predict the observed structure. The calculated values of Jackson et al (22) are close to those of Ermolaev et al (20) but in poor agreement with experiment.

He^{2+} + H Collisions

In figure 5 cross sections for 2p state excitation in He^{2+} - H collisions measured by Higgins et al (23), Hughes et al (24) and Detleffsen et al (2) are presented. The structure in the calculated cross sections of Fritsch et al (25) between 2 - 8 keV amu^{-1} and 15-40 keV amu^{-1} have been confirmed experimentally by Higgins et al (23) and Hughes et al (24).

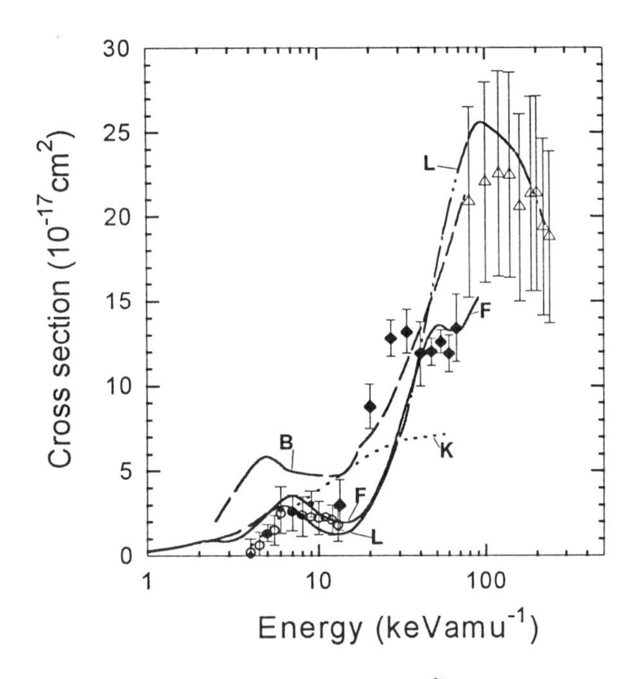

FIGURE 5. Cross sections for 2p excitation in He^{2+} - H collisions.

Experiment: •, Higgins et al (23). ♦, Hughes et al (24). Δ, Detleffsen et al (2). o, Hoekstra and Beijers (cited in(2)). Theory: F, Fritsch et al (25). B, Bransden and Noble (32). K, Krstic and Janev (33). L, Lundsgaard and Nielsen (34).

Measured cross sections for H(2s) formation in the energy range 30-70 keV amu^{-1} are in good accord with the semiclassical close coupling calculations of Fritsch et al (25).

The measured cross sections of Donnelly et al (11) for Balmer alpha emission in He^{2+} - H collisions in the energy range 15-70 keV amu^{-1} increase in value as the energy increases from 15 keV amu^{-1} and reach a plateau above 50 keV amu^{-1}. These measurements are in excellent accord with the calculations of Ast et al (26) where a two centre atomic orbital expansion was used in a close coupled set of equations in the local optical potential approximation. There is also good agreement with the close coupling calculations of Fritsch et al (25)

ACKNOWLEDGMENTS

This work forms part of a programme supported by the UK Engineering and Physical Sciences Research Council. The author would like to thank Prof. H. B. Gilbody and Drs. A. Donnelly, D. P. Higgins, M. P. Hughes and R. W. McCullough for their contribution to the work.

REFERENCES

1. Fritsch, W. and Lin, C. D., *Physics Reports* **202**, 1-97 (1991)
2. Detleffsen, D., Anton, M., Werner, A. and Schartner, K-H, *J. Phys. B.*, **27**, 4195-4213 (1994)
3. McCullough, R. W., Geddes, J., Donnelly, A., Liehr, M., Hughes, M. P. and Gilbody, *Meas. Sci Technol.* **4**, 79-82, (1993)
4. Donnelly, A., Hughes, M. P., Geddes, J. and Gilbody, H. B., *Meas. Sci Technol.*, **3**, 528-32 (1992)
5. Morgan, T. J., Geddes, J. and Gilbody, H. B. *J Phys B*, **6**, 2118-38 (1973)
6. Williams, I.D., Geddes, J. and Gilbody, H.B. *J Phys B* **15**, 1377- 89 (1982)
7. Higgins, D. P., Geddes, J. and Gilbody, H. B., *J. Phys. B.* **29**, 1219-24 (1996)
8. Chong, Y. P. and Fite, W. L., *Phys. Rev.* **A16**, 933-42 (1977)
9. Slim, H. A. and Errmolaev, A. M. , *J. Phys. B.* **27**, L203-9 (1994)
10. Kuang, J. and Lin, C. D., *J. Phys. B.* **29**, 1207-18 (1996)
11. Donnelly, A., Geddes, J. and Gilbody, H. B., *J Phys B* **24**, 165-72 (1991).
12. Park, J. T., Aldag, J E., George, J. M. and Peacher, J. L., *Phys. Rev.* **A14**, 608-14 (1996)
13. Ermolaev, A. M., *J. Phys.* B. **4**, L495-99 (1991)
14. Slim, H. A. *J Phys. B.* 26, L743-46 (1993)
15. McLaughlin, B. M., Winter, T. G. and McCann, J. F., *J. Phys. B.* submitted for publication
16. McKee, J. D. A., Sheridan, J. R., Geddes, J. and Gilbody, H. B., *J. Phys. B.* **10**, 1679-86 (1977)
17. Geddes, J., Donnelly, A., Hughes M. P., Higgins, D. P. and Gilbody, H. B., *J. Phys. B.* **27**, 3037 - 43 (1994)
18. Schartner, K. H., Detleffsen, D. and Somer, D., *Phys. Lett..* **136A**, 55-58 (1989)
19. Kuang, J., Chen, Z. and Lin, C. D., *J. Phys. B.* **28**, 2173 - 79 (1995)
20 Ermolaev, A. M., Jackson< D., Shimakura, N. and Watanabe, T., J. Phys. B., **27**, 4991-5009 (1994)
21 Errea, L. F., Mendez, L. and Riera, A., *Z. Phys. D.* **14**, 229-36 (1989)
22. Jackson, L. F., Slim, H. A., Bransden, B. H. and Flower, D. R., *J. Phys. B.* **25**, L127-30 (1992)
23. Higgins, D. P., Geddes, J. and Gilbody, H. B., *J. Nucl. Instr. Methods Phys. Research* **B103**, 120-22 (1995)
24. Hughes, M. P., Geddes, J. and Gilbody, H. B., *J. Phys. B.* **27**, 1143-50 (1994)
25 Fritsch, W., Shingal, R. and Lin, C. D., *Phys. Rev.* **A44**, 5686-92 (199125
26. Ast, H., Lüdde, H. J. and Dreizler, R. M., *J. Phys. B.* **21**, 4143-56 (1989) and private communication
27.Cheshire, I.. M., Gallaher, D. F. and Taylor, A. J., *J. Phys. B.* **3**, 813-32 (1970)
28. Rapp, D. and Dinwidde, D., *J. Chem. Phys.* **57**, 4919-27 (1972)
29. Shakeshaft, R., *Phys. Rev.* **18**, 1930-34 (1978)
30. Lüdde, H. J., and Dreizler, R. M., *J. Phys. B.* **22**, 3243-54 (1989)
31. Ford, A. L., Reading, J. F. and Hall, K. A., *J. Phys. B.* **26**, 4537-51 (1993)
32. Bransden, B. H. and Noble, C. J., *J. Phys.* B. **14**, 1849-56 (1981)
33. Krstic, P. S., and Janev, R. K., *Phys. Rev.* **A47**, 3894-3912 (1993)
34. Lundsgaard, M. F. V., and Nielsen, S. E., Z. Phys. D **34**, 97-105 (1995)

THE DISSOCIATIVE RECOMBINATION OF COLD POLYATOMIC MOLECULAR IONS MEASURED AT A STORAGE RING

O. Heber*, L. H. Andersen+, D. Kella+, H. B. Pedersen+, L. Vejby-Christensen+, and D. Zajfman*

+ *Institute of Physics and Astronomy, University of Aarhus, DK-8000 Aarhus C, Denmark*
**Department of Particle Physics, Weizmann Institute of Science, Rehovot, 76100, Israel*

Dissociative Recombination (DR) is a major process in planetary ionospheres and interstellar clouds. The branching ratio in the dissociative recombination of the astrophysically important species H_3O^+, H_2O^+ and CH_3^+ with low energy electrons is presented. The molecular ions were stored and vibrationally cooled in the ASTRID storage ring of Aarhus Denmark. The recombination was measured by merging the ions with the electron cooler. All the branching ratios were measured directly for the first time. The three body dissociation was found to be a major channel, a fact that was neglected previously in common interstellar chemical models.

INTRODUCTION

In interstellar media and in planetary ionospheres where the temperature and density are low, collisions between neutral atoms or molecules have often a small reaction probability. On the other hand, reactions where ions are involved are much faster, and are among the most important processes in the physics of interstellar clouds. The dissociative recombination (DR) of molecular ions with free electron is one of these processes which are relevant to the understanding of the formation of interstellar matter [1]. In DR, a molecular ion capture a free electron in a dissociating (doubly) excited state, leading to the production of fast neutral fragments in various configuration. A crucial condition in laboratory studies of DR processes, is the molecular ion internal temperature. The DR process is known to be extremely sensitive to the initial vibrational state of the molecular ion. Storage rings have the advantage that molecular ions can be stored for seconds, a time which is usually long enough to cool the ions vibrational degrees of freedom to the ground state. In the last few years many new experiments have been done using stored molecular ions in storage rings [2]. The resulting new detailed information for the DR of simple diatomic molecules such as HD^+ and HeH^+ [2], have also prompted the theoretical study and modeling of the processes involved. Much more challenging and not less important is the DR process for polyatomic molecules which play a major role in the interstellar matter reaction chain. The DR of polyatomic molecules leads to several products of neutral molecules and atoms. In this report, we present results for the DR branching ratio of H_3O^+, H_2O^+ and $CH_3{}^+$. The energetically allowed DR channels for these molecules are as follows:

$$H_3O^+ + e^- \rightarrow \begin{cases} H_2O + H & 6.4\,eV \quad N_a \\ OH + H_2 & 5.7\,eV \quad N_b \\ OH + H + H & 1.3\,eV \quad N_c \\ O + H_2 + H & 1.4\,eV \quad N_d, \end{cases} \qquad (1)$$

$$H_2O^+ + e^- \rightarrow \begin{cases} OH + H & 7.4\,eV \quad N_a \\ O + H_2 & 7.6\,eV \quad N_b \\ O + H + H & 3.1\,eV \quad N_c, \end{cases} \qquad (2)$$

$$CH_3^+ + e^- \rightarrow \begin{cases} CH_2 + H & 5.0\,eV \quad N_a \\ CH + H_2 & 5.2\,eV \quad N_b \\ CH + H + H & 0.7\,eV \quad N_c \\ C + H_2 + H & 1.6\,eV \quad N_d, \end{cases} \qquad (3)$$

The energy releases written above are for the formation of ground state products, starting from ground state molecular ions and zero kinetic energy electrons. The N_i are the normalized branching ratios.

EXPERIMENT

The experiment was carried out at the heavy ion storage ring ASTRID in Aarhus, Denmark. The ions were produced in an electron-impact ion source. H_2O^+ ions were produced from water vapor, H_3O^+ from a mixture of water vapor and H_2, and $CH_3{}^+$ from methane. The ions were injected to the ring at an energy of 150 keV and accelerated by an RF system to 5-6 MeV. After the acceleration stage the ions were merged with an electron beam at the same velocity where the DR reaction takes place. The produced neutral were collected and measured by an energy-sensitive semiconductor detector with an active surface area of 40 by 60 mm^2 located about 6 m from the interaction region and outside the storage path of the ions in the ring. A DR signal on the detector is characterized by a single peak in the energy spectrum proportional to the full energy of the molecules (see Fig. 1). Since all the neutral fragments hit the detector practically at the same time, no branching ratio can be deduced from this full energy signal. The background signals that are produced by collisions between the ions and the residual gas, is subtracted by chopping the electron beam.

CP392, *Application of Accelerators in Research and Industry,* edited by J. L. Duggan and I. L. Morgan
AIP Press, New York © 1997

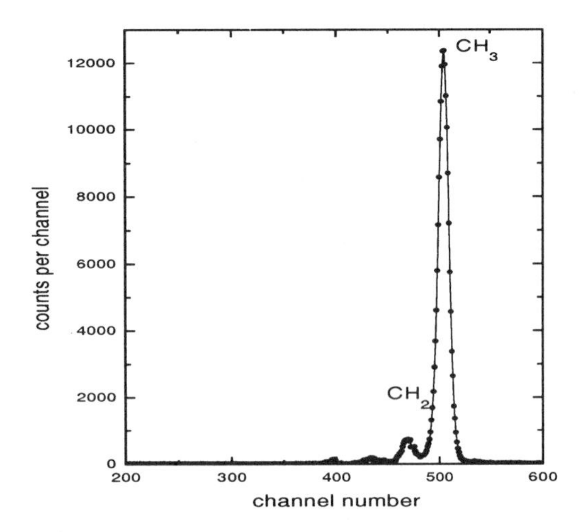

FIGURE 1. Energy spectrum for the DR of CH_3^+. The small peak annotated CH_2 is due to hydrogen atoms which have missed the detector surface.

The different branching channels in each full energy peak are determined by using the grid method [3]. In this method, a grid with a known transmission is inserted in front of the detector. A single particle that travel towards the grid can either pass through it with a probability T or be stopped by the grid with a probability 1-T. Similarly two particles from the same molecule can both pass the grid with a probability T^2. The probability for each one of these particles to pass through the grid while the other do not pass, is T(1-T). The probability that neither of them pass through the grid is $(1\text{-}T)^2$. The same type of arguments can be applied for three or more fragments of a molecule. The energy spectrum measured by the detector with the grid is thus characterized by additional lower energy peaks, as only some of the particles are detected, while the others are stopped by the grids. For example the branching ratios for the DR of CH_3^+, as depicted in Eq. 3 can be solved using the following set of equations:

$$\begin{pmatrix} N(C+3H) \\ N(C+2H) \\ N(C+H) \\ N(C) \\ N(3H) \\ N(2H) \\ N(H) \end{pmatrix} = \tag{4}$$

$$\begin{pmatrix} T^2 & T^2 & T^3 & T^3 \\ T(1-T) & 0 & 2T^2(1-T) & T^2(1-T) \\ 0 & T(1-T) & T(1-T)^2 & T^2(1-T) \\ 0 & 0 & 0 & T(1-T)^2 \\ 0 & 0 & 0 & T^2(1-T) \\ 0 & T(1-T) & T^2(1-T) & T(1-T)^2 \\ T(1-T) & 0 & 2T(1-T)^2 & T(1-T)^2 \end{pmatrix} \begin{pmatrix} n_a \\ n_b \\ n_c \\ n_d \end{pmatrix}$$

Where N(X) are the integral counts under each peak in the energy spectrum and n_i are the required (non-normalized) branching ratio. Solving this (over-determined) set of equations for n_i yield the desired branching ratio. The grids used in the experiments had a nominal transmission of 70% and 36%. As will be discussed in the next section, the actual grid transmissions were measured during the experiment, and found to be slightly lower. A more detailed description can be found elsewhere [4], including the DR and dissociative excitation cross sections.

RESULTS AND DISCUSSION

The DR energy spectra after transmission through the 70% grid, are presented in Figs. 2 to 4 for H_3O^+, H_2O^+ and CH_3^+ respectively. The peaks, after background subtraction, were fitted with a Gaussian function, and the net integrals were inserted in the equations sets and solved numerically.

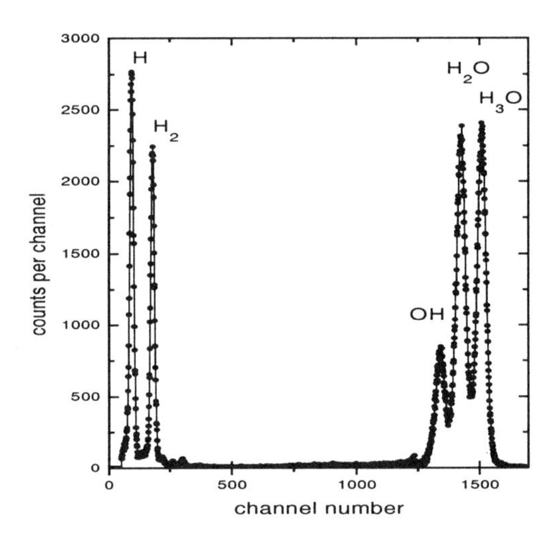

FIGURE 2. Energy spectrum for the DR of H_3O^+ with the 70% grid.

The grid transmission parameter was measured with heavy particles (F^+ beam) and light particles (H^+ beam). The transmissions were found to be T=0.285±0.02 and T=0.635±0.02 for the F^+ ions and 0.24±0.04 and 0.58±0.07 for the proton beam. Using these measured grid transmissions for heavy and light particles gave consistent branching ratios for the two different measurement sets (the two grids). As an additional test, the grid transmission constants were used also as free parameters in the equations sets, yielding similar branching ratios and the same transmission constants as measured within the experimental errors. Trying to solve Eq. 4 with only one transmission probability value was impossible, as it produced negative value for some of the branching ratios. The reason for the different transmission constants for light and heavy particles is not clear at present, and simple edge scattering models were found inadequate to explain these differences.

The results for the branching ratios of the DR of H_3O^+ are:

$$\begin{array}{lll} N_a & (H_2O+H) & = 0.33 \pm 0.17 \\ N_b & (OH+H_2) & = 0.18 \pm 0.09 \\ N_c & (OH+H+H) & = 0.48 \pm 0.24 \\ N_d & (O+H+H_2) & = 0.01 \pm 0.08. \end{array} \quad (5)$$

The most important branching ratio is the water fraction. This result, together with interstellar chemical models, makes it possible to calculate the correct amount of water present during star formations [5]. As water is believed to be the main mediator for heat exchange between the gravitational collapse energy release and radiation loss, the above number is of high importance for these models. Also, it clearly shows that most of the oxygen in interstellar clouds is probably stored in form of water , and not under oxygen molecular form. This is evident as the branching ratio for the production of O (channel N_d) is negligible.

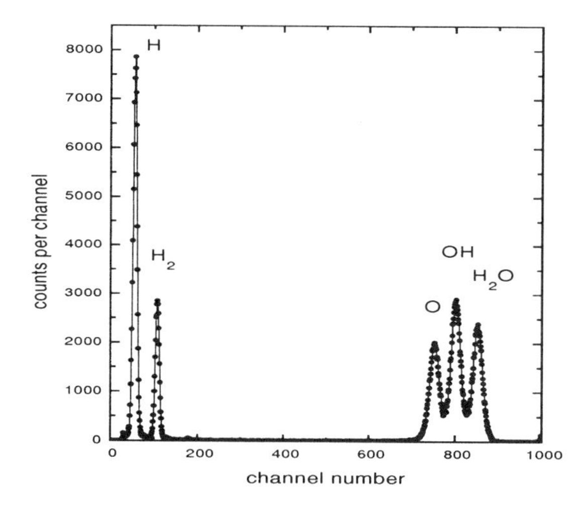

FIGURE 3. Energy spectrum for the DR of H_2O^+ with the 70% grid.

The results for the branching ratios of H_2O^+ are:

$$\begin{array}{lll} N_a & (OH+H) & = 0.22 \pm 0.10 \\ N_b & (O+H_2) & = 0.10 \pm 0.10 \\ N_c & (O+H+H) & = 0.68 \pm 0.20 \end{array} \quad (6)$$

The dominant channel here is the three body break-up (channel N_c), in contradiction to the original model of Bates [6] which, predicted mainly that the release of single hydrogen atom would be the dominant process in the DR of these species.

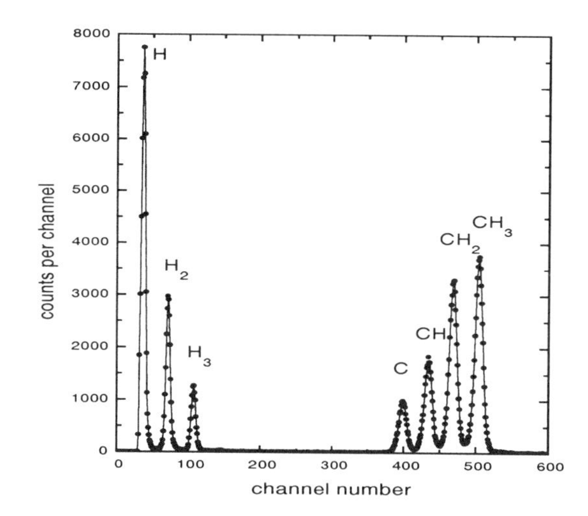

FIGURE 4. Energy spectrum for the DR of CH_3^+ with the 70% grid.

The results for the branching ratios of $CH_3{}^+$ are:

$$\begin{array}{lll} N_a & (CH_2+H) & = 0.40 \pm 0.10 \\ N_b & (CH+H_2) & = 0.14 \pm 0.10 \\ N_c & (CH+H+H) & = 0.16 \pm 0.15 \\ N_d & (C+H+H_2) & = 0.30 \pm 0.08. \end{array} \quad (7)$$

Here again the three body dissociation is strong. The 30% carbon production channel (N_d) may affect the interstellar clouds models which have a problem of predicting the large observed abundance of atomic C [7].

An important result from the branching ratios measured here is the large and sometimes dominant fraction of three body dissociation. In common interstellar models [7], the three body dissociation channels have usually been neglected. Clearly, more theoretical calculations are needed in order to understand the reason for this predominance.

SUMMARY

The DR of H_3O^+, H_2O^+ and $CH_3{}^+$ polyatomic molecular ions was measured with cooled ions in their ground vibrational states. All the branching ratios have been directly measured. The water production in the DR of H_3O^+ was found to be 33%, a value of high interest in interstellar clouds modelling. The three body dissociation channels are major routes in the DR for all the ions measured so far. It is hoped that the above results will promped new theoretical models, which will lead to a more basic understanding of the fragmentation pattern. This is an important issue as interstellar cloud modellers need litteraly thousands of these branching ratios.

ACKNOWLEDGMENTS

This work has been supported by the Israel Science Foundation and by the Danish National Research Foundation through the Aarhus Center for Advanced Physics (ACAP). We thank the ASTRID staff for support and help during the experiment.

---

[1] Herbst, E., *Astrophys. J.* **222**, 508 (1978).
[2] Zajfman, D., *et al. Dissociative Recombination Theory Experiment and Applications III:* World Scientific 1996.
[3] Heber, O,. *et al. J. Phys. B* **18**, L201 (1985).
[4] Vejby-Christensen, L., *et al.* submitted for publication.
[5] Millar, T. J., *et al. Astron. Astrophys.* **194**,250 (1988).
[6] Bates, D. R., *J. Phys. B* **24**, 3267 (1991).
[7] Sternberg, A., and Dalgarno, A., *Ap. J. Suppl.* **99**, 565 (1995).

IONIZATION OF ATOMS BY BARE ION PROJECTILES

Lokesh C. Tribedi[1†]

[1] *J.R. Macdonald Laboratory, Department of Physics, Kansas State University, Manhattan, Kansas 66506-2604*

The double differential cross sections (DDCS) for low energy electron emission can provide stringent tests to the theoretical models for ionization in ion-atom collision. The two-center effects and the post collision interactions play a major role in ionization by highly charged, high Z projectiles. We'll review the recent developments in this field and describe our efforts to study the energy and angular distributions of the low energy electrons emitted in ion-atom ionization.
PACS Number: 34.50.Fa

Ionization is an important inelastic process in fast ion-atom collisions and has been the subject of theoretical and experimental studies for a long time. Most of these studies involve the measurements of the total or single differential ionization cross sections using light ion projectiles like H^+ or He^+. The doubly differential cross sections (DDCS) in ejected electron energy and angle can provide detailed information on the ionization dynamics. Recently Rudd and co-workers have measured the energy and angular distributions of the electron DDCS of atomic and molecular hydrogen by low energy (20-114 keV) protons and He^+ [1,2]. Stolterfoht and co-workers [3] have reported earlier, the DDCS measurements for helium bombarded by protons at energies between 300 keV to 5 MeV. The first Born approximation (FBA) could reproduce quite well the data for 5 MeV protons. The FBA also gives reasonable agreement for 1 MeV protons except some discrepancies for forward angles. At lower energies the FBA failed to explain the data both for protons and helium ions [1,2]. Suárez et al. [4] have investigated the low energy electron emission in ionization of Ne by low energy H^+ and $^3He^{2+}$.

Compared to ionization by low-charged projectiles, ionization mechanisms involving highly charged ions are more complicated. The low energy electrons play a dominant role in the ionization DDCS although they are difficult to detect. Because of the high charge of the projectile, the ejected electron spectra are strongly influenced by the two-center (projectile and target nuclei) Coulomb potential and by the post-collision interactions. Such effects can not be adequately described by the first Born approximation even at relatively high projectile velocities. At intermediate to high collision energies, the theoretical method commonly employed to incorporate the two-center effect is the CDW-EIS (continuum distorted wave-eikonal initial state) approximation of Crothers and McCann [5] (see also the review by Fainstein et al. [6]). In this approximation the ionized electron is assumed to be influenced by the long-range Coulomb field of both the target and the projectile. The wave functions used in this model satisfy the correct asymptotic boundary conditions of the Coulombic three body system. On the other hand, the plane wave Born approximation often fails to provide accurate cross sections for ionization by highly charged ions (HCI) projectiles. However, there are not many measurements using HCI as projectiles. Only a few measurements have been reported for bare ions [7–9]. Pedersen et al. [7] and Stolterfoht et al. [8] have reported the DDCS measurements for ionization of He by a variety of bare ion projectiles. Most recently Tribedi et al. [9] measured the details of energy and angular distributions of low energy electrons emitted in ionization of H_2 by energetic bare carbon ions. These measurements clearly show that the two-center mechanism plays a major role in the emission of low energy electrons.

We describe here, in brief, the experimental technique used at Kansas State University (KSU) for measuring the electron DDCS. Bare carbon ions of different energies were obtained from the tandem van-de-Graaff accelerator of the J. R. Macdonald Laboratory, of the Physics Department at KSU. The energy and charge state selected highly collimated beam was made to collide with H_2 or He target gas flooded inside the chamber. The gas pressure was kept low (0.1 mT) in order to minimize the scattering of the low energy electrons emitted in the ionization of the target. At present the experiment is being carried out using atomic hydrogen target [1]. A hemispherical electrostatic analyzer [1] with energy resolution about 5% was used. The spectrometer could be rotated between 15° and 160° and the electrons were detected at different angles. Extreme care was taken to detect the lowest energy ($\leq$ 1eV) electrons. For normalization of the present data we have measured the electron DDCS for 1.5 MeV proton on He for which the absolute cross sections are known [10].

CP392, *Application of Accelerators in Research and Industry,* edited by J. L. Duggan and I. L. Morgan
AIP Press, New York © 1997

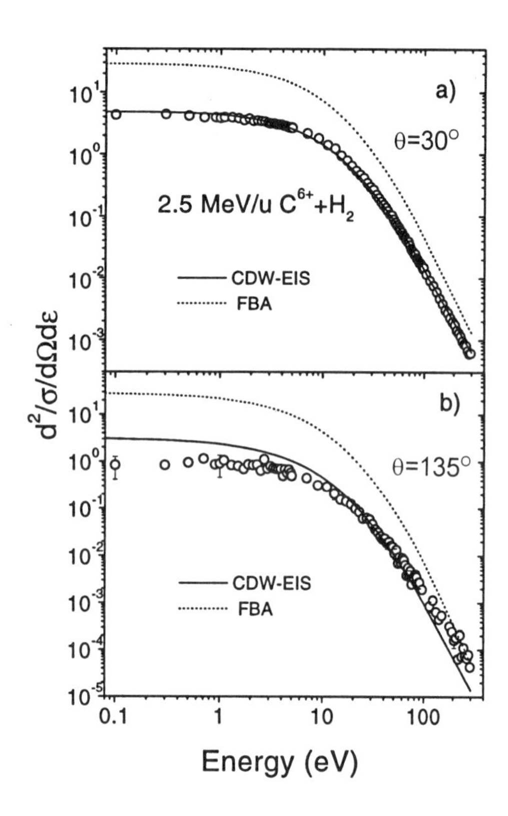

FIGURE 1. Double differential cross sections of electron emission for (a) $\theta = 30°$ and (b) $\theta = 135°$ for 2.5 MeV/u C^{6+}+H_2.

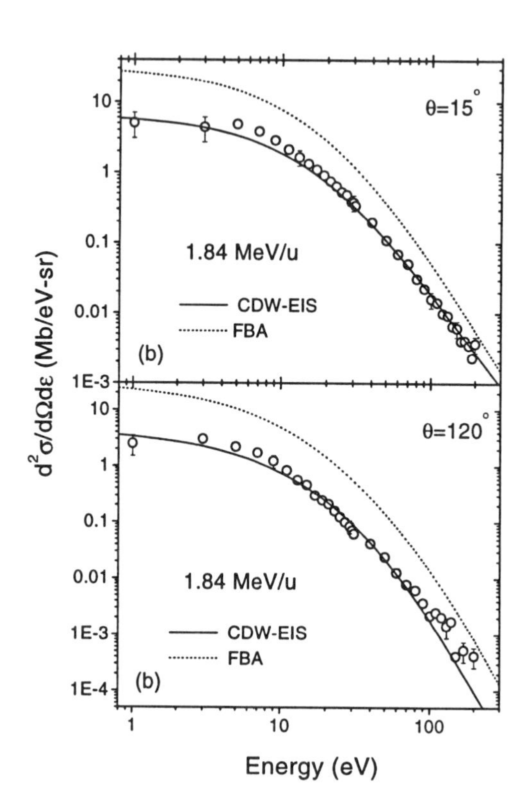

FIGURE 2. Double differential cross sections of electron emission for $\theta = 15°$ and $\theta = 120°$ for 1.84 MeV/u C^{6+}+H_2.

To demonstrate the two-center effect and post collision interaction we give some examples on electron DDCS spectra measured recently by the KSU group. Electron DDCS spectra for 2.5 MeV/u C^{6+}+H_2 are shown in Fig. 1 for two different electron emission angles (θ) i.e. 30° and 135° (taken from Ref. [9]). Very similar data for 1.84 MeV/u C^{6+}+H_2 are shown in Fig. 2 for $\theta = 15°$ and 120° [11]. It is clear that for both beam energies there are large discrepancies between the data and FBA calculations (dotted lines). The largest discrepancy occurs for low-energy electrons of both the forward and backward angles. It clearly shows that the FBA is not adequate to describe soft electron emission. The FBA results come closer to the data for electrons with energy above 100 eV. The low energy electrons are strongly influenced by the post collision interaction at all emission angles, as seen by the poor agreement with the theory. Furthermore, these observations show that the two-center mechanism of ionization and the post collision effects are more important in heavy-ion impact ionization. An excellent agreement is found between the data and the CDW-EIS theory for both the energies and for all angles. It shows that the CDW-EIS theory is adequate to describe the two-center nature of the ionization process. However, in case of backward angle there are some discrepancies at lowest and the highest electron energies for beam energy 2.5 MeV/u.

The FBA and CDW-EIS models are compared with the angular distributions at fixed electron energies in Fig. 3 for a beam energy of 2.5 MeV/u. The distributions peak around 75°. As described before the FBA calculations tend to agree with the data above 100 eV. For the highest electron energy (295 eV) the FBA reproduces the distribution quite well except for some discrepancies at forward angles. This indicates that the post collision interaction has substantial influence on higher energy electrons emitted in the forward directions. The CDW-EIS reproduces the distributions quite well. However, there are some discrepancies between data and calculations of the large backward angles for the lowest and highest energy electrons which are not understood at this stage.

Our theoretical treatment is based on the independent electron model which ignores electron-electron interaction. Furthermore, we simplify the molecular hydrogen target as an effective one-electron hydrogenic target with charge $Z_{eff} = 1.064$, where $Z_{eff}^2/2$ gives rise to the ionization potential of H_2. As demonstrated by Gulyás et al. [12], improved agreement with the DDCS at backward angles could be made if Hartree-Fock wave functions are used in both the initial and final states. This model has provided a good agreement with the data of Stolterfoht et al. [8] for ionization of He target. The present author and co-workers are extending the electron DDCS measurements to collisions involving atomic hydrogen as target.

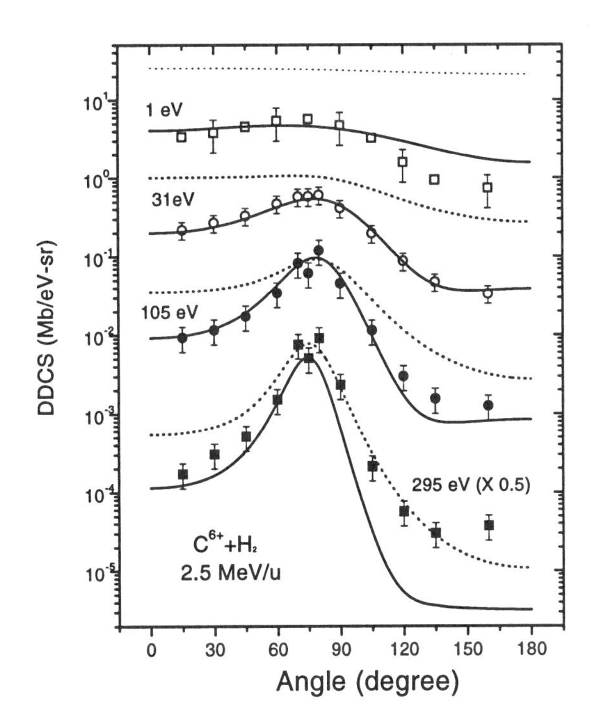

FIGURE 3. Angular distributions of electron DDCS for different electron energies. The solid (dotted) line is the CDW-EIS (FBA) calculation.

ACKNOWLEDGMENT

This work is carried out in collaboration with P. Richard, Y.D. Wang, D. Ling, C.D. Lin and M.E. Rudd. We would like to thank T.J.M. Zouros for reading this manuscript carefully and correcting it. This work is supported by the Division of Chemical Sciences, Office of Basic Energy Sciences, Office of Energy Research, U.S. Department of Energy.

†On leave from Tata Institute of Fundamental Research, Homi Bhabha Road, Bombay-400005, India.

[1] Gealy, M.W., Kerby III, G.W., Hsu, Y.-Y., and Rudd, M.E., *Phys. Rev. A***51**, 2247 (1995).

[2] Hsu, Y.-Y., Gealy, M.W., Kerby III, G.W., and Rudd, M.E., *Phys. Rev. A***53**, 297 (1996).

[3] Manson, S.T., Toburen, L.H., Stolterfoht, N., *Phys. Rev. A***12**, 60 (1975).

[4] Suárez, S., Garibotti, C., Bernardi, G., Focke, P., and Meckbach, W., *Phys. Rev. A*.**48**, 4339 (1993).

[5] Crothers. D.S.F., and McCann, J.F., *J. Phys. B***16**, 3229 (1983).

[6] Fainstein, P.D., Ponce, V.H., and Rivarola, R.D., *J. Phys. B***24**, 3091 (1991).

[7] Pedersen, J.O.P., Hvelplund, P., Petersen, A., and Fainstein, P., *J. Phys. B***24**, 4001 (1991).

[8] Stolterfoht, N., Platten, H., Schiwietz, G., Schneider, D., Gulyás, L., Fainstein, P.D., and Salin, A., *Phys. Rev A***52** 3796 (1995).

[9] Tribedi, L.C., Richard, R., Wang, Y.D., Ling, D., Lin, C.D., Moshammer, R., Kerby, G.W., and Rudd, M.E., *Phys. Rev. A***54**, 2154 (1996).

[10] Rudd, M.E., Toburen, L.H., and Stolterfoht, N., *Nuclear Data Tables* **18**, 413 (1976).

[11] Tribedi, L.C., *et al.* (to be published).

[12] Gulyás, L., Fainstein, P.D., and Salin, A., *J. Phys. B***28**, 245 (1995).

DEPTH PROFILES OF K X-RAY EMISSION FROM HIGHLY STRIPPED ARGON IONS TRAVERSING SOLID MEDIA

R. L. Watson, V. Horvat, and J. M. Blackadar

Cyclotron Institute and Department of Chemistry, Texas A&M University, College Station, TX 77843

K x-rays emitted at 90° from an Ar beam incident on thick targets of Be, C, NaF, and KCl at an energy of 6-MeV/u were measured with a curved crystal spectrometer. The observed (Doppler shifted) average energies of the H-like Ar 2p(^{2}P) to 1s(^{2}S) and the He-like Ar 1s2p(^{1}P) to 1s^2(^{1}S) transitions provided a means of testing the projectile energy dependence of theoretical initial state population fractions. Good agreement was found for all of the targets except Be. These population fractions were used to calculate depth profiles for emission and detection of the x-rays.

INTRODUCTION

Recent attempts at developing *ab initio* models for calculating the charge distributions of ions traveling in matter have recognized the need to express the rate equations in terms of cross sections for populating the various quantum states involved in the charge equilibration process (1-3). While current theoretical methods are unable to provide the level of accuracy required to treat ions containing large numbers of electrons, considerable success has been realized in the prediction of charge distributions for fast ions in which only a restricted number of quantum states make significant contributions. Work by Rozet et al. (3,4), for example, has led to the availability of a program (ETACHA) that can handle ions with up to 28 electrons distributed over all subshells with principal quantum number $n < 4$.

Since the ability of such models to accurately predict charge distributions rests on the reliability of the theoretical quantum state population fractions, it is of interest to devise experimental ways of testing these quantities. One method that recently has been used to examine the energy dependence of the 2p population fraction for H-like Ar ions involves measuring the average Doppler shift of x-rays emitted as the ions slow down in thick solid targets (5). Because the kinetic energy of the ion is directly conveyed by the Doppler shift of the emitted x-ray, both the average x-ray energy and the x-ray line profile observed for a single incident projectile energy are sensitive to the entire excitation function of the initial state population fraction.

Previous measurements focused on the 2p population fraction of H-like Ar at an incident beam energy of 15-MeV/u (5). In the present work, a 6-MeV/u Ar beam has been employed to test the projectile energy dependence of the 1s2p population fraction in He-like Ar ions and to further examine the behavior of the H-like Ar 2p population fraction.

EXPERIMENTAL METHODS AND ANALYSIS

Details of the experimental methods are identical to those described in reference 5 and so only a brief summary will be presented here. The spectra of K x-rays produced by a beam of 6-MeV/u Ar ions incident on thick solid targets of Be, C, NaF, and KCl were recorded using a curved crystal spectrometer with its focal circle plane oriented perpendicular to the beam axis. The spectrum of x-rays observed with the Be target is shown in Fig. 1. The most prominent x-ray transitions are labeled by their initial-state electron configurations. The two transitions of interest in the present work - 1s2p(^{1}P) to 1s^2(^{1}S) and 2p(^{2}P) to 1s(^{2}S) - are labeled 1s2p and 2p, respectively. It should be noted that the detection efficiency of the proportional counter used in these measurements was a factor of 2.5 higher for x-rays above the Ar absorption edge at 3202.9 eV than for those below. Determination of the exact spectrometer observation angle was accomplished by measuring the Doppler shifts of K x-rays from the 1s2p(^{1}P) He-like and 2p(^{2}P) H-like initial states, excited by passing the Ar beam through a thin (2.3 mg/cm^2) Al target foil.

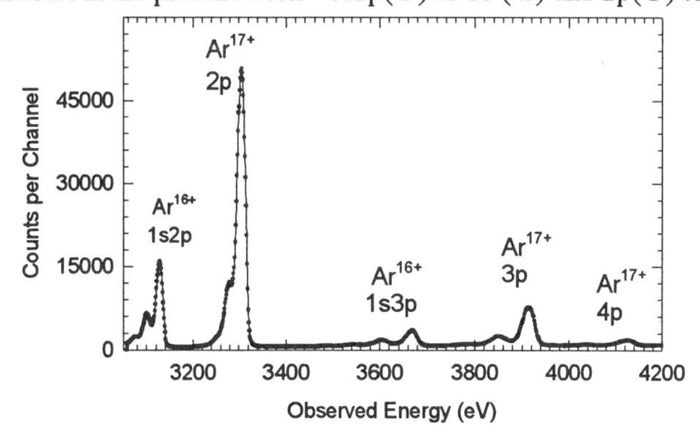

FIGURE 1. Spectrum of K x-rays from Ar ions traveling in a thick Be target at an incident energy of 6-MeV/u.

If the initial state population fraction as a function of projectile energy $f_i(\varepsilon)$ (hereafter referred to as the population fraction excitation function) is known, the rate of x-ray emission per incident ion as a function of projectile energy $R_e(\varepsilon)$ may be calculated from (5)

$$R_e(\varepsilon) = \frac{1}{N_p}\frac{dN_x}{d\varepsilon} = \frac{M\lambda f_i(\varepsilon)}{1.389\times10^{12}\rho\sqrt{\varepsilon}\,S}, \qquad (1)$$

where N_p is the number of incident ions, N_x is the number of emitted x-rays, M is the projectile mass (amu), λ is the transition rate constant (s^{-1}), ρ is the target density (g/cm^3), ε is the

CP392, *Application of Accelerators in Research and Industry,* edited by J. L. Duggan and I. L. Morgan
AIP Press, New York © 1997

projectile energy (MeV/amu), and S is the stopping power ($MeV/mg/cm^2$). Correcting this rate for absorption in the target yields $R_d(\varepsilon)$, the rate of detectable x-ray emission in the direction of the spectrometer per incident ion;

$$R_d(\varepsilon) = R_e(\varepsilon)\, e^{-\mu x(\varepsilon)}, \qquad (2)$$

where μ is the x-ray mass absorption coefficient and x is the x-ray path length in the target. These rates are transformed to corresponding rates as a function of target thickness, $R_{e\ or\ d}(x)$ and as a function of x-ray energy , $R_{e\ or\ d}(E_x)$ upon multiplication of equations (1) and (2) by $d\varepsilon/dx$ and $d\varepsilon/dE_x$, respectively. Finally, the average x-ray energy may be obtained by performing the following integrations

$$\langle E_x \rangle = \frac{\int_0^{\varepsilon_i} E_x R_d(E_x)\, dE_x}{\int_0^{\varepsilon_i} R_d(E_x)\, dE_x} \qquad (3)$$

and compared with the experimental value.

RESULTS AND DISCUSSION

Excitation functions for the population fractions of He-like Ar 1s2p and H-like Ar 2p states in the various targets investigated were calculated using program ETACHA and they are shown in Fig. 2. It may be seen that the two population fractions behave quite differently as a function of projectile energy. These theoretical population fractions were used to compute the "expected" average x-ray energies via equation (3). Values of the required stopping powers and x-ray absorption coefficients were taken from references (6) and (7), respectively. Other data needed were the rest frame x-ray energies - 3321.4 eV for the H-like Ar transition [refs. (8) and (9)] and 3139.7 eV for the He-like Ar transition [ref. (9)], and the transition rate constants -

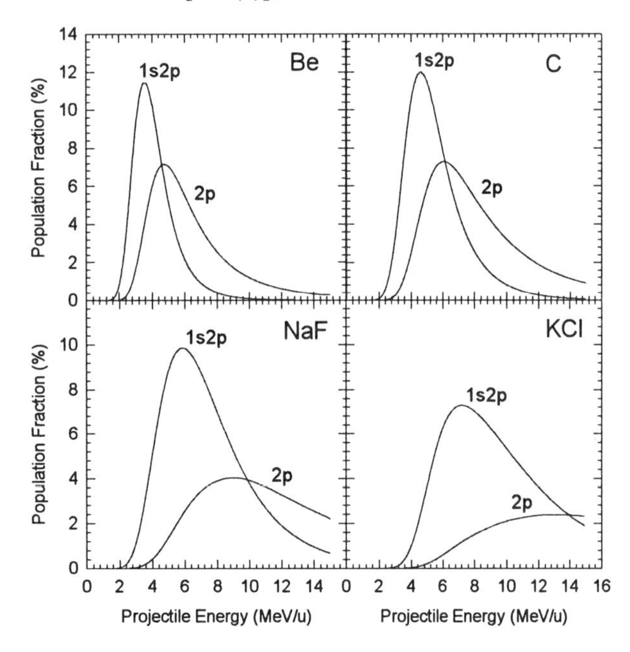

FIGURE 2. Population fractions for the H-like Ar 2p and He-like Ar 1s2p states over the projectile energy range of 0 to 15 MeV/u.

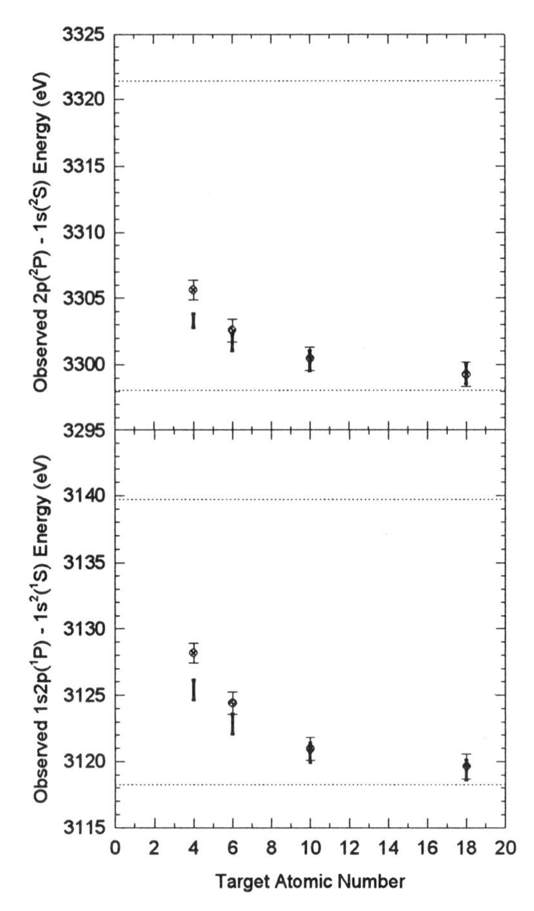

FIGURE 3. Comparison of the measured average x-ray energies (circles) with those predicted by the model calculations (vertical bars). The dashed lines indicate the expected x-ray energies for Ar ions at rest (top of each frame) and for Ar ions having the full incident energy (bottom of each frame).

$6.58 \times 10^{13}\ s^{-1}$ for the H-like Ar transition [ref. (10)] and $1.07 \times 10^{14}\ s^{-1}$ for the He-like Ar transition [ref. (11)]. The average x-ray energies observed in the various targets are shown in Fig. 3, where they are compared with the calculated average x-ray energies. The same general trends are displayed by both sets of data. Specifically, it is apparent that the agreement between theory and experiment is excellent for the KCl and NaF targets, good for the C target, and only fair for the Be target. This is in contrast to the results obtained previously for the H-like ^2P to ^2S transition using a 15-MeV/u Ar beam, where the agreement between theory and experiment was found to be excellent for the Be target, fair for the C target, very poor for the NaF target and good for the KCl target (5). The data points for Be and C are most sensitive to the shapes of the population fraction excitation functions because of the relatively low degree of x-ray absorption in these targets. The x-ray absorption coefficients in NaF and KCl, on the other hand, are so large that x-rays emitted from the interior regions of the targets are highly absorbed. The effect of absorption is readily apparent in the calculated depth profiles presented in Fig. 4.

Considering the results for the Be target of both the 6-MeV/u and the 15-MeV/u studies, it may be concluded that the actual 1s2p and 2p excitation functions are probably somewhat broader than predicted by the ETACHA program, especially in the low

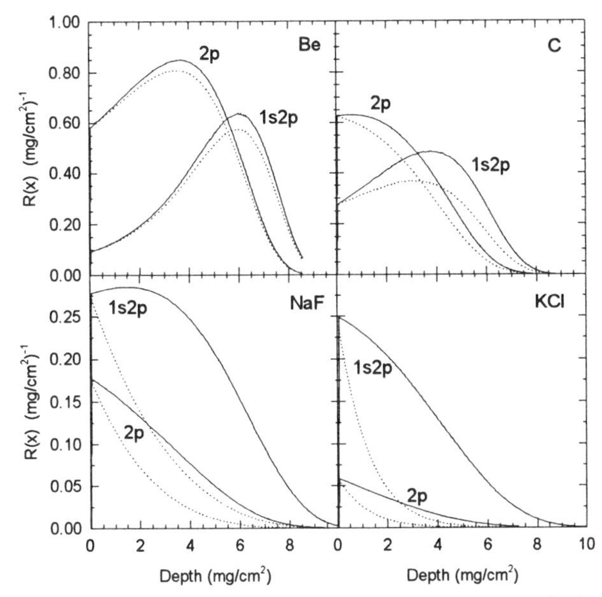

FIGURE 4. Dependence of the x-ray emission rates on depth in the target for an incident energy of 6-MeV/u. The solid curves show the total emission rates $R_e(x)$, while the dotted curves show the detectable emission rates $R_d(x)$.

projectile energy region. In the case of the C target, the combined results indicate the theoretical excitation functions need to be shifted down in energy to produce better agreement with the measured energies. The NaF results provide an interesting puzzle. At 15-MeV/u, where the average x-ray energy is least sensitive to the population fraction excitation function, the measured x-ray energy corresponds to a much higher average projectile velocity than predicted, while at 6-MeV/u, the measured and calculated x-ray energies are in excellent agreement. In attempting to account for this discrepancy, one is led to speculate that the stopping powers for Ar in NaF may be underestimated at the higher projectile energies.

ACKNOWLEDGMENTS

We thank J. P. Rozet for providing a copy of his program ETACHA and instructing us in its use. This work was supported by the Robert A. Welch Foundation.

REFERENCES

1. Anholt, R. and Meyerhof, W. E., *Phys. Rev.* A 33, 1556 (1986).
2. Betz, H. D., Höpler, R., Schramm, R., and Oswald, W., *Nucl. Instrum. Methods* B 33, 185 (1988).
3. Rozet, J. P., Chetioui, A., Piquemal, P., Vernhet, D., Wohrer, K., Stéphan, C., and Tassan-Got, L., *J. Phys.* B 22, 33 (1989).
4. Rozet, J. P., Stéphan, C., and Vernhet, D., *Nucl. Instrum. Methods* B 107, 67 (1996).
5. Watson, R. L., Horvat, V., and Blackadar, J. M., *Phys. Rev.* A (in press).
6. Hubert, F., Bimbot, R., and Gauvin, H., *At. Data and Nucl. Data Tables* 46, 1 (1990).
7. Henke, B. L., Lee, P., Tanaka, T. J., Shimabukuro, R. L., and Fujikawa, B. K., *At. Data and Nucl. Data Tables* 27, 1 (1982).
8. Garcia, J. D., and Mack, J. E., *J. Opt. Soc. Am.* 55, 654 (1965).
9. Briand, J. P, Mossé, J. P., Indelicato, P., Chevallier, P., Girard-Vernhet, D., Chetioui, A., Ramos, M. T., and Desclaux, J. P., *Phys. Rev.* A 28, 1413 (1983).
10. Bethe, H. A. and Salpeter, E. E., *Quantum Mechanics of One- and Two-Electron Atoms*, New York: Academic Press, 1957.
11. Lin, C. D., Johnson, W. R., and Dalgarno, A., *Phys. Rev.* A15, 154 (1977).

INTENSITY ESTIMATION OF AN F Kα K^1L^1 SATELLITE PRODUCED BY ION IMPACTS BY USE OF THE MO SCHEME

M. Mochizuki[a], T. Yamamoto[a] and M. Uda[a,b]

[a] *Department of Materials Science and Engineering, Waseda University 3-4-1 Ohkubo , Shinjyuku-ku, Tokyo 169, Japan*
[b] *Laboratory for Materials Science and Technology, Waseda University 2-8-26 Nishiwaseda, Shinjyuku-ku, Tokyo 169, Japan*

To explain the chemical bonding effect appeared in X-ray satellite spectra, an attempt has been made to deconvolute the observed spectra and to calculate the relative intensities of them in view of the Molecular-Orbital(MO) scheme. Relative intensities of F K^1L^1/K^1L^0 for MgF_2 and NiF_2 produced by 2MeV He^+ and 84MeV N^{4+} impacts were estimated here successfully based on the Semi-Classical Approximation.

INTRODUCTION

Significant chemical bond effects have been displayed in the intensity distributions of X-ray satellite spectra produced by ion impacts, which were emitted from multiply ionized initial states (1-6). The structures of the satellites are sensitive to change in chemical environments around the atom of interest. Then the study of intensity distributions in the satellites has been used for the chemical state analysis of solid materials (7-11).

Chemical bond effects is one of the most important causes to introduce change in ionization cross sections and fluorescence yields. In our previous report(11) calculations of the ionization cross sections for electrons on the molecular orbitals (MO) were performed in view of the Semi Classical Approximation, SCA(12), which has successfully been used to estimate the ionization cross sections of electrons on the atomic orbitals(13).

In alkali fluorides F2p orbitals have almost atomic nature though the partial expansion coefficients of F2p becomes a little smaller than 1(1). Then a F K^1L^0 X-ray line of the alkali fluoride with the rock salt type of structure, which is emitted from one K-shell vacancy state without any vacancies on the L-shell, constitutes only one line even in the molecule. On the other hand the covalency of F^- ions in fluorides with the rutile type of structure is larger than that of alkali fluorides at ground states(1), and then the F2p orbitals should contribute to form several MOs in the rutile type of fluorides. This requests us to treat the K^1L^0 line being composed of several lines instead of one. In this paper we have attempted to estimate observed F K X-ray intensities with the aid of the MO concept which can lead to deconvolute observed spectra into components.

THEORETICAL

We have expanded the ionization cross section for an isolated atom to that for an atom embedded in a molecule or a condensed matter (11). The multiple ionization probability of one K and n L electrons, K^1L^n, for the molecule (M) in which valence electrons are composed, at least in part, of the F L shell electrons is expressed as

$$
\begin{aligned}
P^M_{K^1L^n}(b) &= \binom{2}{1} \cdot \left[P_K(b)\right]^1 \cdot \left[1 - P_K(b)\right]^{2-1} \\
&\times \sum_{n_1 \cdot n_2 \cdots n_k} \binom{e_1}{n_1} \cdot \left[\left(C_L^{(1)}\right)^2 \cdot P_L(b)\right]^{n_1} \cdot \left[1 - \left(C_L^{(1)}\right)^2 \cdot P_L(b)\right]^{e_1 - n_1} \\
&\times \binom{e_2}{n_2} \cdot \left[\left(C_L^{(2)}\right)^2 \cdot P_L(b)\right]^{n_2} \cdot \left[1 - \left(C_L^{(2)}\right)^2 \cdot P_L(b)\right]^{e_2 - n_2} \\
&\vdots \\
&\vdots \\
&\times \binom{e_k}{n_k} \cdot \left[\left(C_L^{(k)}\right)^2 \cdot P_L(b)\right]^{n_k} \cdot \left[1 - \left(C_L^{(k)}\right)^2 \cdot P_L(b)\right]^{e_k - n_k}
\end{aligned}
\qquad (1)
$$

where $C_L^{(k)}$ denotes the partial expansion coefficients of F L shell wave function, $\binom{e_k}{n_k}$ the binomial coefficient, $P_K(b)$ and $P_L(b)$ the atomic ionization probabilities of K and L shells, respectively, for the isolated F atom, and e_k and n_k the number of electrons and vacancies in the k-th MO with $n_1+n_2+....+n_k=n$. Then the multiple ionization cross section of one K and n L electrons, K^1L^n is described by

$$
\sigma^M_{K^1L^n}(E) = 2\pi \int_0^\infty b db \cdot P^M_{K^1L^n}(b) \qquad (2)
$$

For estimating observed X-ray intensities precisely in the case of thick target, the energy loss of the projectile ion through the target and the self absorption of the X-rays produced by the projectile ion in the target should also be taken into account. X-ray yields from an element Z can now be expressed for a K^1L^n initial state (14) as

$$Y_{K^1L^n} \propto \omega_{K^1L^n} \cdot \int_{E_i}^{E_0} \frac{\sigma^M_{K^1L^n}(E) \cdot T^M_{K^1L^n}(E)}{S^M(E)} dE \qquad (3)$$

where $\omega_{K^1L^n}$ is the fluorescence yield, E_i the incident energy of the projectile, E_0 the final energy of the projectile, $T^M_{K^1L^n}(E)$ the transmission rate of the X-rays with the energy E through the target, $S^M(E)$ the matrix stopping power of the projectile with energy E in the target, and $\sigma^M_{K^1L^n}(E)$ the K^1L^n ionization cross section expressed by eq.(2). From eq.(3) we get the relative intensity of the X-rays emitted from the K^1L^1 and K^1L^0 initial states where the valence orbitals are composed, at least in part, of the F L orbital.

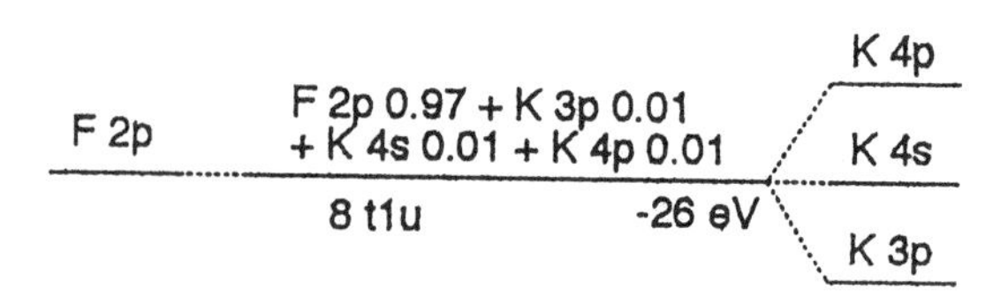

FIGURE 1. A molecular orbital diagram for KF (a representative alkali fluoride) at the K^0L^0 or ground state. The molecular orbital is composed of atomic orbitals of F2p, K3p, K4s and K4p. Here the orbital components are expressed by the sum of the squares of the partial expansion coefficients and the overlap populations for the above orbitals.

PROCESSING OF EXPERIMENTAL DATA AND CALCULATION OF IONIZATION CROSS SECTION IN THE MO FRAME

Electronic structures of the fluorides were calculated by employing the DV-Xα molecular orbital calculation method(15). Here a $(FM_3)^{5+}$ cluster with the C_{2v} symmetry was used where M is a representative of Mg and Ni. The orbital components related to the F L shell for MgF_2 and NiF_2 are summarized on Table 1.

In case of alkali fluorides, the F 2p orbitals are almost degenerated in one orbital, as shown in Fig.1. On the other hand in NiF_2 the covalency of the F^- ion is much larger than that of alkali fluorides, leading to F L orbital splitting.

A schematic diagram of molecular orbitals for valence shells belonging to an A_1 symmetry block of NiF_2 at the K^0L^0 or ground state is shown in Fig.2(a) as an example of a series of MO's. F2p orbitals are bonded with 3d, 4s, and 4p orbitals of Ni in appreciable amounts. The same is also true for B_1 and B_2 symmetry blocks. This suggests that MO's with F2p components are not degenerated but spread out in a wide energy range. Then profile of the K^1L^0 line should be written by summation of each MO line whose intensity is proportional to the partial coefficient of F 2p.

TABLE 1. The orbital components for molecular orbitals which involve F L shell orbitals in MgF_2 and NiF_2 at the ground state. Here the orbital components are expressed by the sum of the squares of the partial expansion coefficients and the overlap populations.

Compounds	MO	Orbital components
MgF_2	$9a_1$	F2p 0.99
	$10a_1$	F2p 0.96 + Mg3s 0.02 + Mg3p 0.01
	$6b_1$	F2p 0.97 + Mg3s 0.02 + Mg3p 0.01
	$3b_2$	F2p 0.99 + Mg3p 0.01
NiF_2	$14a_1$	F2p 0.98 + Ni4s 0.01+ Ni4p 0.01
	$15a_1$	F2p 0.85 + Ni3d 0.10 + Ni4s 0.03+ Ni4p 0.02
	$18a_1$	F2p 0.03 + Ni3d 0.96 + Ni4s 0.01
	$10b_1$	F2p 0.86 + Ni3d 0.10 + Ni4s 0.02+ Ni4p 0.01
	$13b_1$	F2p 0.07 + Ni3d 0.92 + Ni4s 0.01
	$5b_2$	F2p 0.91 + Ni3d 0.08 + Ni4p 0.01
	$7b_2$	F2p 0.06 + Ni3d 0.94

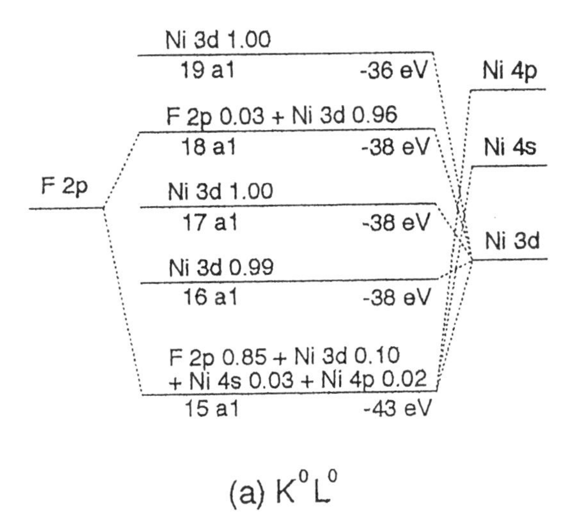

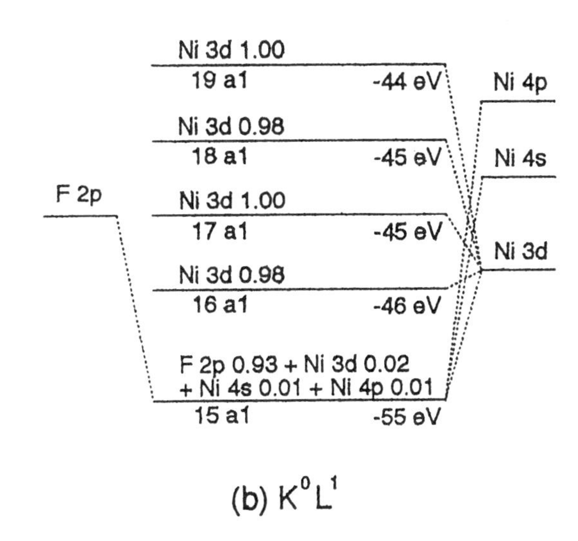

FIGURE 2. A molecular orbital diagram for NiF_2 at the K^0L^0 or ground state (a) and K^0L^1 state (b), which belongs to an A_1 symmetry block made of atomic orbitals of F2p, Ni3d, Ni4s and Ni4p. Here the orbital components are expressed by the sum of the squares of the partial expansion coefficients and the overlap populations for the above orbitals.

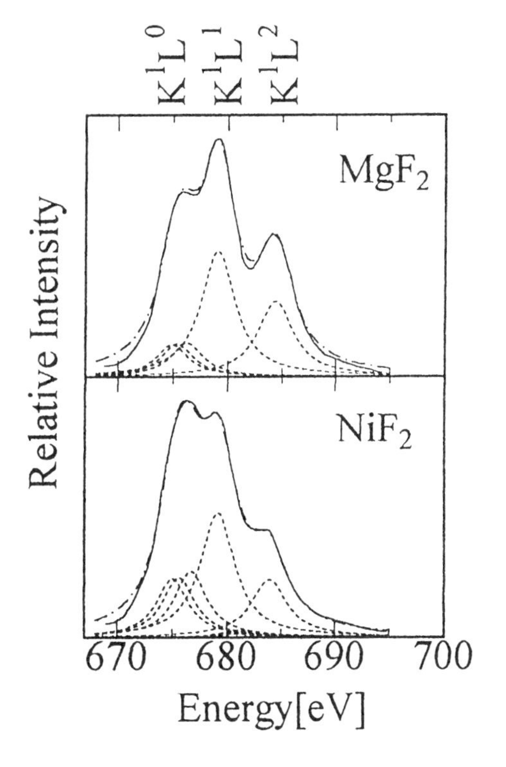

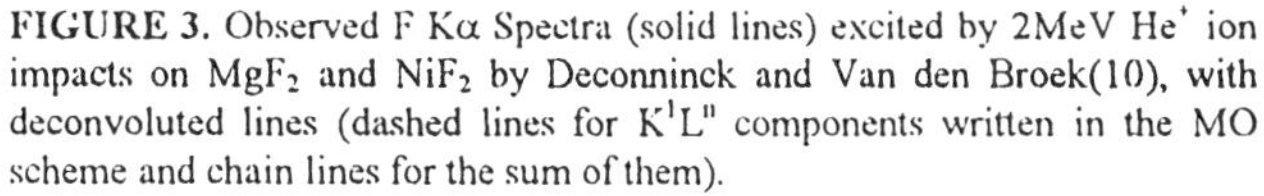
FIGURE 3. Observed F Kα Spectra (solid lines) excited by 2MeV He^+ ion impacts on MgF_2 and NiF_2 by Deconninck and Van den Broek(10), with deconvoluted lines (dashed lines for K^1L^n components written in the MO scheme and chain lines for the sum of them).

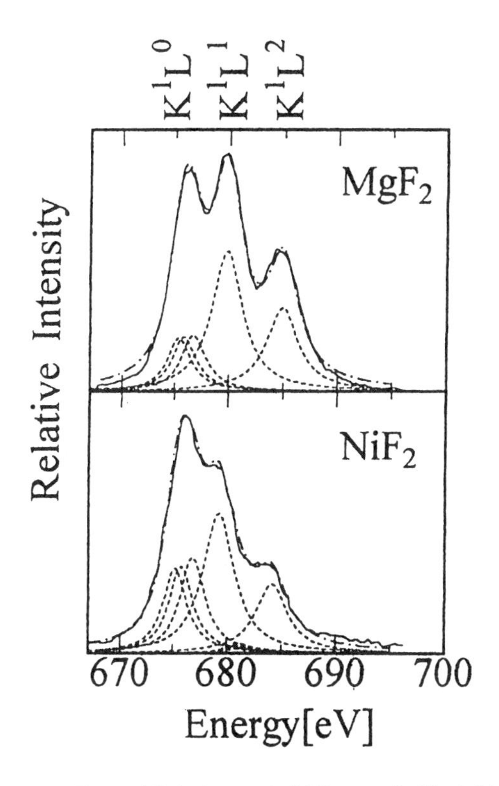

FIGURE 4. Observed F Kα Spectra (solid lines) excited by 84MeV N^{4+} ion impacts on MgF_2 and NiF_2 by M. Uda et. al. (3) , with deconvoluted lines (dashed lines for K^1L^n components written in the MO scheme and chain lines for the sum of them).

TABLE 2. Comparison of observed and calculated relative intensities of F K^1L^1 / K^1L^0.

Compounds	2MeV He^+		84MeV N^{4+}	
	Obs.	Calc.	Obs.	Calc.
MgF_2	1.41	1.24	1.03	1.01
NiF_2	0.66	1.11	0.59	0.93

At the K^0L^0 ground state of NiF_2, F2p orbitals are mixed with those of Ni in two MOs, i.e. $15a_1$ and $18a_1$ as shown in Fig.2(a). On the other hand at the ionized states of K^0L^1, K^0L^2,···, which are considered to be hypothetical initial states for producing K^1L^1, K^1L^2,··· ionized states, the F 2p orbitals in NiF_2 are localized on the central F^- ion by use of only one MO, i.e. $15a_1$ as shown in Fig.2(b). These facts lead to form several X-ray lines for K^1L^0 but only one line for K^1L^1, K^1L^2,··· transitions.

The observed F K^1L^0 line was deconvoluted into each MO component using the Lorentzian distribution function. F Kα satellite spectra observed by Deconninck and Van den Broek under 2MeV He^+ excitation conditions (10) are reproduced in Fig.3 with deconvoluted components here calculated. Widths of K^1L^0, K^1L^1 and K^1L^2 lines were selected as 3.6, 3.8 and 4.0 eV, respectively. F Kα satellite spectra observed by Uda et. al. under 84 MeV N^{4+} excitation conditions (3) are shown in Fig.4 together with deconvoluted components here calculated. Line widths for the K^1L^0, K^1L^1 and K^1L^2 were selected as 2.8, 3.4 and 3.6 eV, respectively. Observed relative intensities of K^1L^1 / K^1L^0 are summarized on Table 2.

The K^1L^1 / K^1L^0 relative intensities were estimated with the aid of eq.(3) which are summarized on Table 2. Coincidence between the observed and calculated relative intensities is fair but not satisfactory yet. This is because only the direct Coulomb ionization process was considered in this paper, which acts between the projectile and the target atom's electrons. Change in fluorescence yields, and degree in the shake off and resonant orbital rearrangement (ROR) (16) as a function of chemical environments should be considered for advanced calculations, which are in due course.

REFERENCES

1. Uda, M., Endo, H., Maeda, K., Awaya, Y., Kobayashi, M., Sasa, Y., Kumagai, H., and Tonuma, T., *Phys. Rev. Lett.* **42** 1257-1260 (1979).
2. Uda, M., Endo, H., Maeda, K., Sasa, Y., and Kobayashi, M., *Z. Phys.* **A300** 1-6 (1981).
3. Uda, M., Benka, O., Fuwa, K., Maeda, K., and Sasa, Y., *Nucl. Instr. and Meth.* **B22** 5-12 (1987).
4. Benka, O., and Uda, M., *Phys. Rev. Lett.* **56** 54-57 (1986).
5. Benka, O., Watson, R. L., and Kenefick, R. A., *Phys. Rev. Lett.* **47** 1202-1205 (1980).
6. Watson, R. L., Leeper, A. K., Sonobe, B. J., Chino, T., and Jenson, F. E., *Phys. Rev.* **A15** 914-925 (1977).
7. Raman, S., and Vane, C. R., *Nucl. Instr. and Meth.* **B3** 71-77 (1984).
8. Kauffman, R. L., Jamison, K. A., Gray, T. J., and Richard, P., *Phys. Rev.Lett* **36** 1074-1077 (1976).
9. Demarest, J. A., and Watson, R. L., *Phys. Rev.* **A17** 1302-1313 (1978).
10. Deconninck, G., and Van Den Broek, S., *IEEE Trans. Nucl. Sci.* **NS28** 1404-1406 (1981).
11. Mochizuki, M., Yamamoto, T., Nagashima, S., and Uda, M., *Nucl. Instr. and Meth.* **B109/110** 31-38(1996).
12. Hansteen, J. M., and Mosebekk, O. P., *Nucl.Phys.* **A201** 541-560 (1973).
13. Mukoyama, T. and Ito, S., *ICR Report* **91-2** (1991)
14. Johansson, S. A. E., and Campbell, J. L., *PIXE, A Novel Techniqe for Elemental Analysis* : John Wiley and Sons, 1988, pp. 96-.
15. Adachi, H., Tsukada, T., and Satiko, C., *J. Phys. Soc. Jpn.* **49** 875-883 (1978).
16. Uda, M., Yamamoto, T., and Takenaga, T., *Advanced Quantum Chemistry*, in press.

BREMSSTRAHLUNG FROM IONS

N. B. AVDONINA, R. H. PRATT

Department of Physics and Astronomy, University of Pittsburgh, Pittsburgh PA 15260, USA

We show that bremsstrahlung cross sections for ions can be obtained with the good accuracy from Coulombic cross sections and neutral atom cross sections, using an interpolation procedure. The new analytical prescriptions for both the Coulombic and the neutral atom cross sections are given.

Introduction

Electron bremsstrahlung is one of most important mechanisms of the X ray production. However, the calculation of the bremsstrahlung cross sections from many-electron ions and atoms usually is a very time consuming problem, even for a single spectrum of an electron of some specific initial energy, particularly for targets consisting heavy elements.

One of the ways of simplification of the problem of calculation of bremsstrahlung in a wide energy region is to present simple analytical expressions for the cross sections. Such attempts were made in [1,2] for the range 1 - 500 keV. The parametrization was based on an ionization factor approach [1], with which the ionic bremsstrahlung cross sections are accurately obtained using interpolation between the two limiting cases of the neutral atom and the completely stripped ion (pure Coulomb case). The results were compared with the numerical results obtained using a partial-wave and multipole expansion of the relativistic matrix elements in the Dirac-Slater central potential, and a general accuracy of 20% was established. However, at electron energies T_i > 200 keV the accuracy of the results for ions became much worse in a wide region of hard photon bremsstrahlung spectra. because the parametrization results for both the pure Coulomb case and the neutral atom case became unreliable. The accuracy obtained generally deteriorated to worse than 50%, particularly for heavy elements.

We have improved the results of papers [1,2] by analyzing and modifying the usual Elwert-Born expressions for high energy electron Coulombic bremsstrahlung (CBS) and bremsstrahlung from neutral atoms (NBS), in this way expanding the area of their applicability to higher electron energies. We also introduce an empirical higher Born correction, independent of the fraction of energy radiated and linear in incident energy. Additional partial-wave data for the bremsstrahlung cross sections from many different ions of Al, Fe, Mo, W, and U have been obtained, partly in order to verify the accuracy of the proposed analytical prescription scheme, partly in order to extend the very restricted data base of numerical results. By comparison with this data a 10% accuracy of the present parametrization throughout the spectrum, for all elements of the periodic table, in the energy range 0.02 - 2.00 MeV, is established.

The Pure Coulomb case

In the energy region 1 - 500 keV, discussed in [2], the prescriptions for bremsstrahlung of photons of energy k from the pure Coulombic potential were to use either (1) the classical or (2) the Elwert-Born approximations. The choice depended on the value of the energy parameters $\nu_i = Z\alpha/p_i$ and $\tau = k/T_i$, where $p_i = \sqrt{2T_i}$ ("natural units" with $\hbar$=m=c=1 are being used, Z is the nuclear charge, α = 1/137 is the fine structure constant). If $\nu_i\ (1-\tau) > 0.7$ the classical approach was recommended and used for high Z elements. In the soft photon region of the CBS (small $\mu = k\nu_i/2T_i$) the classical expression nuclear charge, α = 1/137 is the fine structure constant). If $\nu_i\ (1-\tau) > 0.7$ the classical approach was recommended and used for high Z elements. In the soft photon region of the CBS (small $\mu = k\nu_i/2T_i$) the classical expression was for the scaled bremsstrahlung cross section $\sigma(k) = (kd\sigma/dk)(\beta_i/Z)^2$ (β_i is the initial velocity of the scattering electron) was approximated as [3]

$$\sigma_{cl}(k)=16/3\alpha^3(1+\mu\pi)\ln(2/\gamma\,\pi)+O(\mu^2). \qquad (1)$$

The corresponding approximate expression used for the hard photon region of the bremsstrahlung spectra from heavy nuclei (large μ) was

$$\sigma_{cl}(k)=(16\pi/3\sqrt{3})\alpha^3[1+d_1\mu^{-2/3}+d_2\mu^{-4/3}+d_3\mu^2+O(\mu^{8/3})], \qquad (2)$$

where d_1 = 0.217747, d_2 = -0.0131214, and d_3 = -0.0057, Euler's constant γ = 1.78. Both expressions in the region of their applicability give results with an accuracy better than 5% for all elements considered.

If $\nu_i(1-\tau) < 0.7$ the prescriptions of the previous approach give us a choice between nonrelativistic and relativistic Born formulas. Both these formula involve multiplication of Born results by Elwert factors [4]:

$$F_{rel} = \beta_i[1 - \exp(-2\pi Z\alpha/\beta_i)]/\beta_f[1 - \exp(-2\pi Z\alpha/\beta_f)]. \qquad (3)$$

for the relativistic Bethe-Heitler formula with $\beta_{i(f)} = [1 - 1/(1+T_{i(f)})^2]^{1/2}$, and

$$F_{nrel} = p_i[1 - \exp(-2\pi Z\alpha/p_i)]/p_f[1 - \exp(-2\pi Z\alpha/p_f)], \qquad (4)$$

CP392, *Application of Accelerators in Research and Industry,* edited by J. L. Duggan and I. L. Morgan
AIP Press, New York © 1997.

Table 1. Coulombic Bremsstrahlung $\sigma(Z = 92)$ in mb

note $\sigma_{Kramers} = 5.61$ mb

		k/T$_i$						
		0.3	0.4	0.5	0.6	0.7	0.8	0.9
$T_i = 200$ keV								
	PW	8.83	7.99	7.36	6.84	6.41	6.03	5.69
	σ_c	8.88	7.89	7.13	6.52	6.00	5.55	5.16
$T_i = 500$ keV								
	PW	8.23	7.16	6.37	5.73	5.18	4.70	3.83
	σ_c	7.86	6.80	5.98	5.33	4.78	4.32	3.91
$T_i = 800$ keV								
	PW	7.74	6.73	5.88	5.19	4.60	4.13	3.64
	σ_c	7.41	6.35	5.53	4.88	4.35	3.90	3.53
$T_i = 1000$ keV								
	PW	7.47	6.50	5.68	5.00	4.38	3.86	3.39
	σ_c	7.16	6.12	5.32	4.68	4.17	3.76	3.40

for nonrelativistic Born approximation. Note that in the Elwert-Bethe-Heitler (EBH) formula relativistic kinematics were used, while in nonrelativistic Elwert-Born (EB) formula $p_{i(f)}$ are nonrelativistic momenta of incoming and outgoing electrons.

At 500 kev these predictions fail in the tip region for heavy elements. We now propose a relativistic modification of the Elwert factor based on our comparison of EBH and EB approximations, and numerical partial wave (PW) data for bremsstrahlung cross sections. Our prescription is to use only the Bethe-Heitler formula but with the modified Elwert factor in the form (2)

$$F_{mod} = p_f[1 - \exp(-2\pi Z\alpha/p_i)]/p_i\,[1 - \exp(-2\pi Z\alpha/p_f)], \quad (5)$$

but with relativistic $p_{i(f)} = \sqrt{T_{i(f)}(2 + T_{i(f)})}$.

The accuracy of the new prescription is much better than both the Elwert-Bethe-Heitler results and Elwert-Born results. However, for heavy nuclei in many cases it is still worse than 10%. To improve our prescriptions and to reach 10% accuracy throughout the nonclassical energy region we have made an additional higher Born correction of order $(Z\alpha)^2$ to the bremsstrahlung cross sections $\sigma_c(k)$, multiplying them by the empirical factor

$$C(T_i,Z)=1+1/4(Z\alpha)^2(2-T_i). \quad (6)$$

Taking into account the correction (6) and the modification of the Elwert factor (5) we have for the bremsstrahlung cross section

$$\sigma_c = C(T_i,Z)\sigma_{BH}F_{mod}, \quad (7)$$

where σ_{BH} is the Bethe-Heitler cross section.

Comparison of the results σ_c of approximation (7) with the partial-wave calculations shows that they agree very well in the energy range 2 keV - 2 MeV for all elements of the periodic table, if (7) is used for $\nu_i \leq 1$ (see Table 1). For $\nu_i = Z\alpha/\beta_i > 1$, starting from the soft photon region, we should use the classical approximation (1) with a switch to the expression (2) when it crosses (1). If the classical approximation curve crosses σ_c with increasing k/T_i another switch to σ_c from (7) generally gives the desired 10% level accuracy.

The Neutral Atom Case.

In the completely screened case (neutral atoms) the bremsstrahlung cross sections were presented for all elements of the periodic table, based on numerical calculations [5] in the partial-wave approximation. In [2] analytical prescriptions for the neutral atom bremsstrahlung cross sections were based on the exact nonrelativistic Born result for electron scattering in the field of the Yukawa potential $V(r) = - Z\alpha \exp(-Q_s r)/r$ [6] :

$$\sigma_B(k)=(8\alpha^3/3)\{Q_{min}^2/(Q_{min}^2+ Q_s^2) - Q_{max}^2/(Q_{max}^2+ Q_s^2)+\ln[(Q_{max}^2+Q_s^2)/(Q_{min}^2+ Q_s^2)]\}, \quad (8)$$

where $Q_{max(min)} = p \pm p_f$, $p_{i(f)} = \sqrt{2T_{i(f)}}$ are the initial (final) momenta of the scattering electron.

Table 2. Bremsstrahlung from ions (in mb),

note $\sigma_{Kramers} = 5.61$ mb.

			k/Ti 0.2	0.4	0.6	0.8
$\sigma(Z = 26)$	Zi =8	pw	7.64	5.32	3.69	2.47
		σ_i	7.57	5.19	3.65	2.54
	Zi =14	pw	7.65	5.33	3.70	2.47
		σ_i	7.58	5.19	3.65	2.54
	Zi =20	pw	7.73	5.36	3.71	2.48
		σ_i	7.62	5.20	3.66	2.54
$\sigma(Z = 92)$	Zi =9	pw	9.62	6.60	5.34	4.41
		σ_i	9.89	6.60	5.25	4.28
	Zi =24	pw	9.62	6.60	5.34	4.41
		σ_i	9.89	6.60	5.25	4.28
	Zi =46	pw	9.68	6.76	5.44	4.49
		σ_i	10.08	6.64	5.26	4.29
	Zi =74	pw	10.07	6.76	5.44	4.49
		σ_i	10.75	6.71	5.29	4.30
	Zi =80	pw	10.45	6.81	5.48	4.53
		σ_i	11.08	6.73	5.30	4.30

The expression (8), with the screening constant $Q_S = 0.016Z^{0.35}$, multiplied by Elwert factor F_{nrel} in the form (4) works with good accuracy for most atoms. However, it fails for light atoms in the soft photon region, where prescriptions (8) gives too small results (sometimes 35% less than the PW results). This situation can be improved by changing the screening constant. We use the fact that, the smaller is T_i and/or k/T_i, the more important is the screening effect. Thus we can find Q_S, fitting $\sigma_B(k)$ from Eq. (8) to the partial-wave results for neutral atoms from [5] at $k/T_i = 0$, for T_i small enough that the parameter $\nu_i = Z\alpha/\sqrt{2T_i} > 1$ (the region where classical approximation for the Coulombic bremsstrahlung is valid). The expression (8), with the screening constant $Q_S = 0.225 + 0.205\,Z - 0.00086\,Z^2$ obtained in this way, gives accurate results in the classical region.

The prescription (8) multiplied by Elwert factor F_{nrel}, suggested in [2], also fails, exept in the soft photon regions, for all elements in the hard photon region, if the parameter $\nu_i = Z\alpha/\sqrt{2T_i} < 1$. The bigger is Z, the wider is the region where the prescription (8) does not work with a good enough accuracy. In the hard photon region it becomes too large by a factor of two. The main reason for the disagreement of thepredicted screening results with the results of calculations in the partial-wave approximation [5] is that formula (8) does not take into account relativistic effects. For $\nu_i < 1$ the incident electron energy exceeds the K-shell ionization threshold $[T_i > (Z\alpha)^2/2]$, and bremsstrahlung from a neutral atom should be close to the pure relativistic Coulombic bremsstrahlung from the nuclei, i.e. very different from Eq. (8). The screening effect should be taken into account only in the soft photon part of the spectrum where a nonrelativistic treatment, which can include screening, becomes better (nearly exact in the Coulomb case). Keeping this information in mind we used in our calculation the modified formula

$$\sigma_n = \sigma_B - (16\alpha^3/3)\ln(Q_{max}/Q_{min}) + \sigma_c \tag{9}$$

with σ_n from Eq. (6). Nonrelativistic kinematics was used for $Q_{max(min)}$. The third term is our corrected prescription for the pure Coulombic case, Eq. (7). In the hard photon region the two first terms cancel and we have the pure Coulombic result. It is interesting to note that in the soft photon region the two last terms partly cancel, and using prescription (9) we have results even better than we had in the pure nonrelativistic screened case using Eq. (6).

Ions

For bremsstrahlung from ions we exploit the idea of interpolating as a function of degree of ionicity Z_i/Z between neutral atom and nuclear point Coulomb cases. Due to the monotonic behavior of the ionic bremsstrahlung cross sections with the degree of ionicity, they can be fairly accurately obtained using the cross sections for the neutral atom σ_n and for the completely stripped ion σ_c:

$$\sigma_i = \sigma_n + I(k, T_i, Z_i/Z, Z)(\sigma_c - \sigma_n), \qquad (10)$$

where Z is the nuclear charge, and Z_i is the ionic charge. As in [1,2] the ionization factor was used in the simplest form $I = (Z_i/Z)^3$.

We are obtained additional partial-wave data for Al, Fe, Mo, W, and U ions to verify the accuracy of the proposed prescription scheme. In the Tabl. 2 we present the comparison of our new analytical calculations with the partial-wave results for ions of Iron and Uranium. The accuracy of our calculations is better than 10% for all cases considered.

1. C. M. Lee, R. H. Pratt and H. K. Tseng, Phys. Rev. **A16** (1977) 2169.
2. I. J. Feng and R. H. Pratt, Internal Report, PITT-266 (1981).
3. V. Florescu and A. Costescu, Revue Roumaine de Physicue **23**, 131 (1978).
4. G. Elwert, Ann. Physik **34**, 178 (1939).
5. R. H. Pratt, H. K. Tseng, C. M. Lee, L. Kissel, C. MacCallum and M. Riley Atomic Data and Nuclear Data Tables **20**, 175 (1977); Erratum **20**, 477 (1977).
6. H. A. Bethe and E. E. Salpeter, Quantum Mechanics of One- and Two-Electron Atoms, (Academic Press, Inc., New York, 1957).

COMPARISON BETWEEN MEASUREMENT AND MONTE CARLO SIMULATION OF A BREMSSTRAHLUNG SOURCE

N. Stritt [a], J. Jolie [a], Th. Materna [a], W. Mondelaers [b]

[a] *Institut de physique, Université de Fribourg, Pérolles, CH-1700 Fribourg, Switzerland*
[b] *Universiteit Gent, Proeftuinstraat 86, B-9000 Ghent, Belgium*

10 MeV electrons delivered by the 15 MeV linear electron accelerator of the University of Ghent are used to irradiate a tantalum/graphite bremsstrahlung source. Since earlier work showed that the theory given by Monte Carlo simulations of the source using the codes EPCOT and EGS4 overpredict the spectrum's intensity at low energy, a new direct measurement of the bremsstrahlung source was conducted with a germanium detector at different angles. Another germanium detector was used which improved the quality of the gamma-ray spectrum and similar but better experimental set-up geometries were constructed. In order to make a clear comparison between simulations and direct measurements, the Monte Carlo calculations included not only precise simulations of the sources but also of the germanium detector response. All these improvements were done to understand the disagreement between the Monte Carlo simulations and the first measurement.

INTRODUCTION

In an earlier work [1], the determination of a full energy spectrum of a bremsstrahlung source was obtained as a function of angle with a germanium detector. The spectra were compared to Monte Carlo simulation using a FORTRAN code named EPCOT [2]. However, the measured and the calculated spectra showed some discrepancies at low energy (E<1MeV). In order to study this difference and to see if these discrepancies arise at other angles, further measurements were done in the same configuration as the previous experiment, but with another germanium detector. The experimental set-up, including the detector shielding, was also improved. The use of a smaller germanium detector (70 cm^3) with a better energy resolution and a lower efficiency improved the quality of the spectra. In particular a lower efficiency reduced the counting rate in the detector and minimized the Ge detector dead time effects. The same procedure mentioned in ref. [1] using the EGS4 [3] code is applied to include the detector response in order to compare the experimental and simulated spectra.

THE EXPERIMENTAL SET-UP

The bremsstrahlung source consists mainly of a 1 mm tantalum plate surrounded by graphite. A schematic geometry of the source is shown in Fig. 1. The source has a cylindrical symmetry along the beam axis. For this experiment, the linear accelerator delivered electrons with an energy of 10 MeV, with an intensity in the order of a few μA and with a pulse repetition rate of 4000 Hz. The photons produced are measured with a 70 cm^3 Ge detector at three different angles (1.5°, 4.5°, 7.5°) relative to the direction of the electron beam at a distance of 11 meters from the source. The selection of the photon beam was realized with two collimators. A lead collimator with an aperture of 2 mm cut the height of the photon beam. A 12 cm thick tungsten collimator with an aperture of 1 mm was installed before the detector and set the angular resolution of the measurement at ±0.005°.

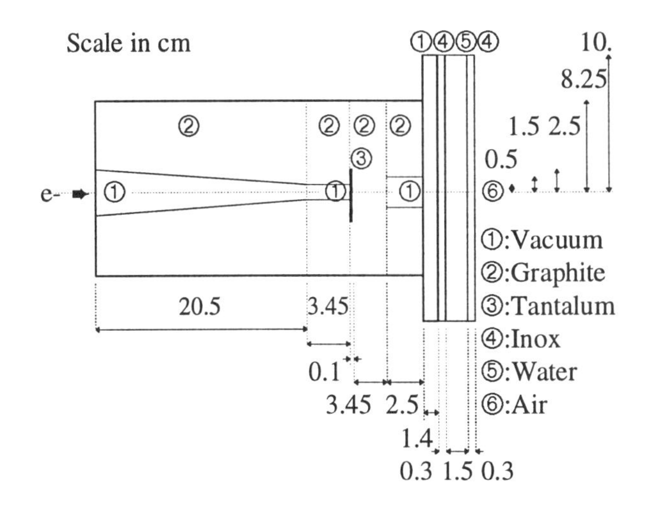

FIGURE 1. Schematic geometry of the bremsstrahlung source installed in Ghent and used in the EPCOT simulation.

A more complete description of the experimental set-up can be found in ref. [1]. Energy and efficiency calibration of the Ge detector were realized with radioactive sources of ^{56}Co, ^{152}Eu and ^{226}Ra. As it was not possible with sources to calibrate the detector higher than the most energetic radioactive gamma-ray lines (^{56}Co: 3547 keV), the electronics was set to cut all events having an energy slightly above this highest energy. As the energy and efficiency calibration could not be experimentally checked by other means, we preferred not to extrapolate the calibrations to energies well above 4 MeV. The shape of the electron beam was determined in a previous work [4] by visual control by means of closed

CP392, *Application of Accelerators in Research and Industry,* edited by J. L. Duggan and I. L. Morgan
AIP Press, New York © 1997.

TV-circuit images of BeO targets inserted in the accelerator transport channel, and located proximal to the actual beam stop. The incident electron beam direction was determined before measurements with an ionization chamber.

RESULTS OF THE EXPERIMENT

The result of the spectra in the Ge detector at angles (1.5°, 4.5°, 7.5°) are shown in fig. 2. In order to have a clearer comparison, a binning of 25 keV was realized on the spectra. Because the electron intensity can not be known with a sufficient precision at this low current (higher current cause dead time problems in the detector), an absolute photon fluence is not possible. Therefore, the photon intensity is normalized to the total number of gamma ($N_{tot,gamma}$) in the spectra. The acquisition time for all spectra was 10800 seconds. A background measurement realized by closing the collimator in front of the detector was subtracted. The mean energy of the different measurements is summarized in table 1. and is calculated by

$$\overline{E} = \frac{1}{N_{tot}} \int_{E_0}^{E_{max}} I(E) E dE \,, \qquad (1)$$

with N_{tot} the total number of photons between E_0 and E_{max}. E_{max} was set by the electronic to 4115 keV. Problems with the subtraction of the background arise for the 511 keV line. The intensity of the 511 keV line in the background measurement was higher than in the source itself. When the bremsstrahlung photon beam is stopped in the tungsten collimator it creates by the pair creation and annihilation process many 511 keV gamma rays which are then detected in the Ge detector. If no element is placed in between the source and the detector, the electrons or positrons emitted during the stopping process in the detector itself as well as the 511 keV gamma rays will all contribute to the detector signal, whereas when a beamstop is placed only the 511 keV gamma rays reach the detector. Therefore the intensity of the 511 keV line is increased when the tungsten collimator is closed in front of the Ge detector. After the subtraction of the background the too low intensity of the spectrum at this energy (see Fig. 2.) affects the mean energy. To remove this anomaly, the channels containing the 511 keV line are left out to calculate the mean energy given in table 1. As it can be seen in Fig. 2. and Table 1, the spectrum at 1.5° is harder than the one at 4.5° and 7.5°. The average photon energy decreases with increasing angle, mainly because the primary electrons, which creates the bremstrahlung photons, have undergone higher energy loss due to multiple scattering than those propagating in the forward direction. The photon spectrum at 4.5° and 7.5° are very similar.

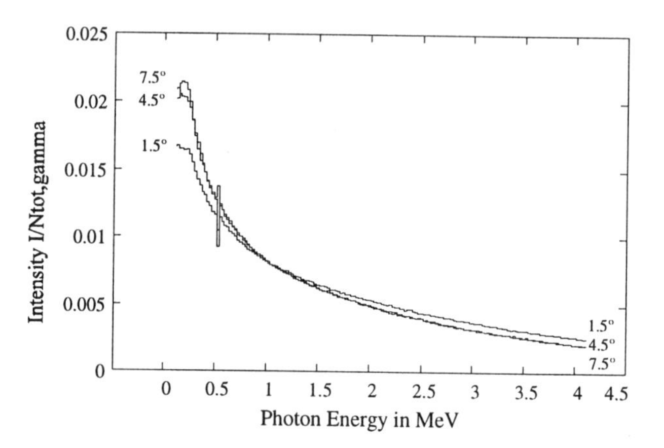

FIGURE 2. Photon spectrum in the Ge detector of the bremsstrahlung source at angles 1.5°, 4.5°, 7.5°.

TABLE 1. Mean energy of the bremsstrahlung source measured by the 70 cm^3 Ge detector.

Angle	Mean Energy in MeV from E_0 =100 keV to E_{max} = 4115 keV
1.5°	1.51 ± .02
4.5°	1.39 ± .02
7.5°	1.38 ± .02

MONTE CARLO SIMULATION OF THE SOURCE

The bremsstrahlung simulation were realized with the FORTRAN code EPCOT [2]. This program uses the bremsstrahlung cross section derived by Selzer and Berger [5]. These data are a combination of the accurate results given by Pratt *et al.* [6] who obtained the electron-nucleus cross section for 1 keV to 2 MeV incident electron, on the basis of numerical phase-shift calculation, and of the cross section obtained by Davies *et al.* [7] and Olsen [8] for high energy (>50 MeV). The angular distribution of the bremsstrahlung photons is sampled from a gaussian distribution having a mean square angle of m_oc^2/E_e, where m_oc^2 is the rest mass of the electron and E_e is the incident electron total energy. In the simulation of the bremsstrahlung process, the electron scattering is negligible compared to the multiple scattering process. The energy threshold was set to 100 keV for electrons and positrons and to 10 keV for photons. The geometry of the source has been simulated as shown in Fig 1. The photon scoring is achieved by counting the number of photons falling in a ring, separated from the source by 11 meters of air, of inner radius R1 and outer radius R2, corresponding to the different angles. The angular resolution was set to ±0.250°. The simulations were performed with a mono-energetic incident electron beam modeled from visual measurements as having a Gaussian radial spread of σ_0 = 0.35 cm. The number of histories per run was equal to 2 x 10^8. The calculated photon fluence at angles 1.5°, 4.5°, 7.5° and 15° is given in Fig. 3.

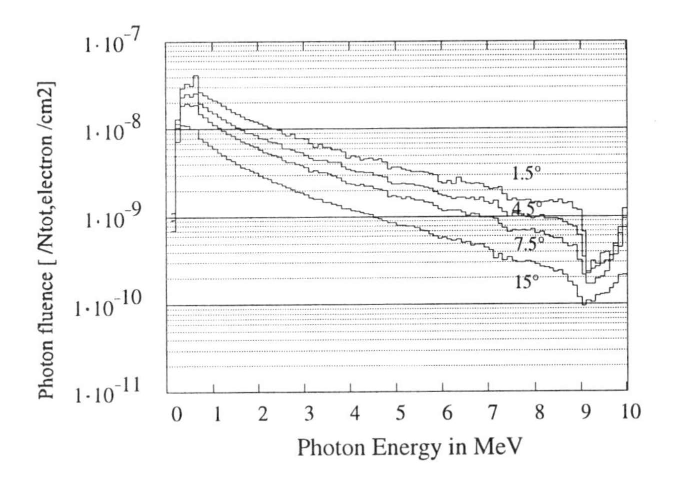

FIGURE 3. Absolute photon fluence calculated by the EPCOT simulation at angles 1.5°, 4.5°, 7.5° and 15°.

The same technique mentioned in ref. [1] was used to include the Ge detector response in order to compare the measurement and the simulation. The computer code EGS4 [3] calculated the detector response function of 100 gamma rays of energy ranging from 0.1 to 10 MeV with an energy resolution of 100 keV. A convolution of the simulated bremsstrahlung spectra with these detector responses was than realized. Fig. 4. shows the bremsstrahlung spectra at 7.5° before and after the convolution technique normalized to the total number of photons in the spectrum ($N_{tot,gamma}$). The mean energy of the bremsstrahlung spectra calculated by the simulation before and after the convolution are given in table 2. The uncertainties quoted in this table are statistical errors.

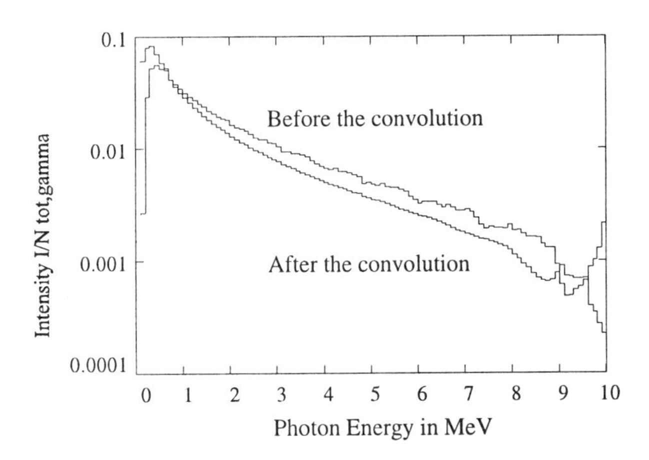

FIGURE 4. Photon fluence normalized to $N_{tot,gamma}$ calculated by EPCOT before and after the convolution with the by EGS4 calculated detector response at 7.5°.

COMPARISON BETWEEN SIMULATION AND MEASUREMENT

Fig. 5 shows a comparison between the convoluted simulation and the measured spectra at the three angles. The ratio between experiments and simulations varies from 0.581 to 1.455 for the 1.5° angle, from 0.700 to 1.256 for the 4.5° angle, and from 0.652 to 1.308 for the 7.5° angle. A systematic deviation between theory and measurement occurs which tends to increase as the angle goes down. The mean energy from 100 keV to 4115 keV measured and calculated is shown in Fig. 6. The biggest difference between simulation and measurement appears at 1.5°. At this angle, the measured mean energy is 1.29 times higher than the simulated one.

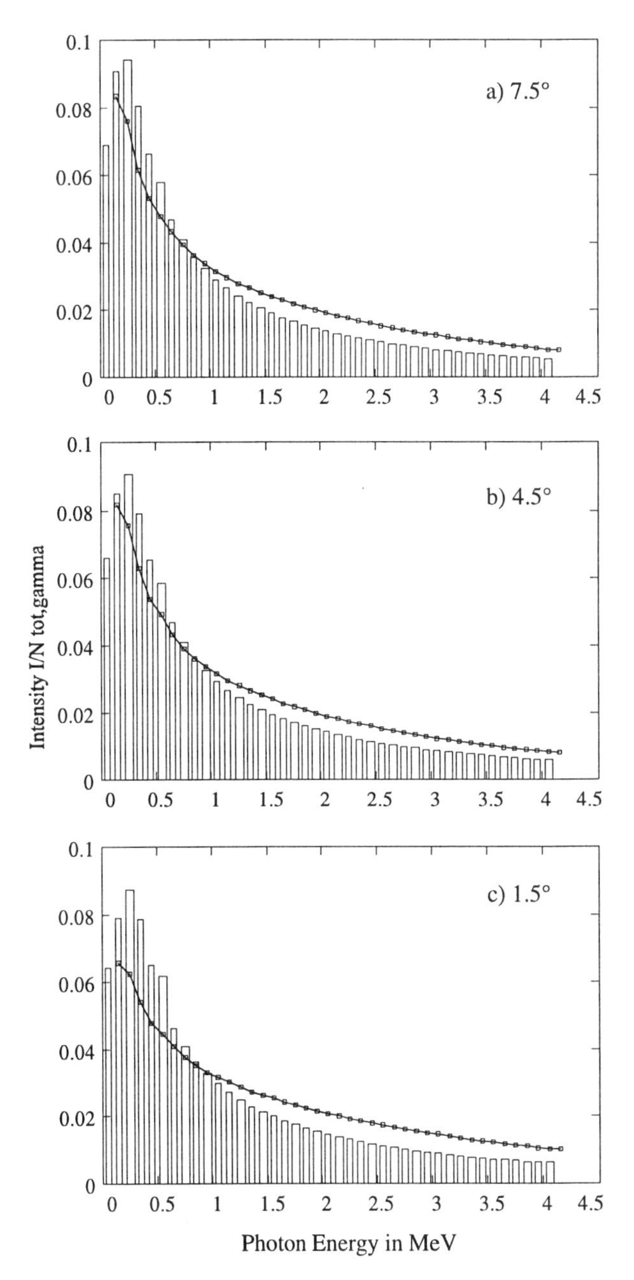

FIGURE 5. Comparison between the convoluted simulation and the measurement of the source at a) 1.5°, b) 4.5° and c) 7.5° (histogram: simulation; line and squares: measurements).

TABLE 2. Mean energy of the simulated bremsstrahlung source from 0 to 10 MeV.

Angle [Angular resolution: ±0.250°]	Mean energy before the convolution [MeV]	Mean energy after the convolution [MeV]
1.5°	2.44 ± .02	1.84
4.5°	2.36 ± .01	1.78
7.5°	2.242 ± .008	1.68
15.0°	2.067 ± .002	1.55

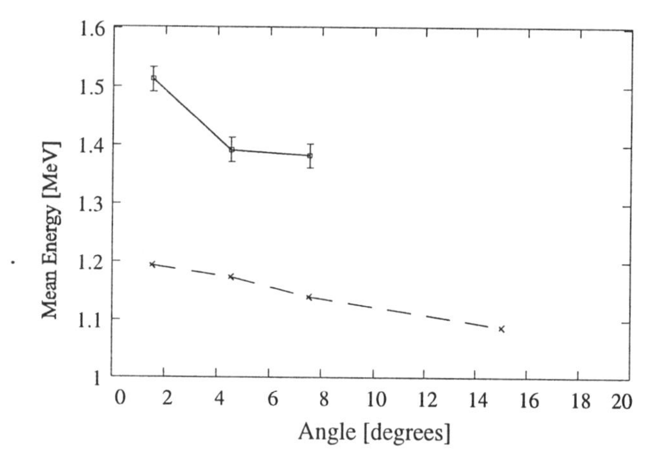

FIGURE 6. Comparison between the mean energy from 100 keV to 4115 keV of the convoluted simulation and the measurement of the source at 1.5°, 4.5°, 7.5°, 15° for the simulation only. (dashed line and cross: convoluted simulations; full line and error bars: measurements).

CONCLUSION

A comparison between simulated and measured spectra was realized for three different angles. A similar discrepancy as observed in ref. [1] persists in the present analysis. In the comparison of the energy spectra as well as the mean energy, the simulation results tend to give too much of low energy photons. A more realistic angular distribution of the created bremsstrahlung photons can be found in a new version of the EGS4 [9] code which includes a distribution calculated by Schiff. According to ref. [10], the difference between the Schiff and the default version of EGS4 angular distribution is not negligible, especially at small angles. This could be an explanation of the discrepancies observed between the measurement and the EPCOT simulation as the EPCOT bremsstrahlung angular distribution is close to the default version of EGS4. A new version of the EPCOT code will be developed including this new distribution and new simulations will be performed.

In conclusion, two sets of measurements (in ref. [1] and in this work) using two different detectors were realized. They both give similar results which tend to confirm the validity of the measurements.

ACKNOWLEDGMENT

This work was supported by the Swiss National Science Foundation and by the Nationaal Fonds voor Wetenschappelijk Onderzoek Belgium.

REFERENCES

1. Stritt N., Bertschy M., Jolie J., Mondelaers W., Nucl. Instr. Meth. B113 (1996) 150.
2. Van Laere K., Ph.D. Thesis, Rijksuniversiteit Ghent (1991), unpublished.
3. Nelson W. R., Hirayama H. and Regulla D. F., The EGS4 code System, SLAC Publication 285.
4. Van Laere K. and Mondelaers W., Int. J. Appl. Radiat. Isot. in Press.
5. Selzer S. M. and Berger M. J., At. Nucl. Dat. Tabl. 35 (1986) 345.
6. Prat R. H., Tseng H. K., Lee C. M., Kissel L., MacCallum C. and Riley M., At. Nucl. Dat. Tabl. 20 (1977) 477.
7. Davies H., Bethe D. A. and Maximon L. C., Phys. Rev. 93 (1954) 778.
8. Olsen H., Phys. Rev. 99 (1955) 1335.
9. Bielajew A. F., R., Mohan R. and Chui C. S., National Research Council of Canada internal Rep. PIRS-0203 (1989).
10. Faddegon B. A., Ross C. K. and Rogers D. W., Med. Phys. 18 (4) (1991) 727.

HIGH EFFICIENCY X-RAY BREMSSTRAHLUNG SOURCES

V.K. Grishin, B.S. Ishkhanov, S.P. Likhachev, D.A. Rodionov, and V.I. Shvedunov

Institute of Nuclear Physics, Moscow State University, Moscow 119899 Russia

We propose to generate X-rays by having an electron beam in a magnetic field cross thin foil radiators multiple times. We report that our simulation and analytic results agree and indicate that the X-ray yield will be at least doubled over that from a single, equivalently thick, target.

INTRODUCTION

X-rays, widely used in many fundamental and applied fields [2], are usually generated by accelerated electrons passing through a heavy metal radiator. Unfortunately, the efficiency of the conversion of electron energy to that of X-rays is at most ~3% for 1 MeV and ~8% for 5 MeV electrons [3,4]. As the radiator thickness increases, the number of generated X-rays increases while their survival decreases, so that the optimal thickness for heavy metal radiators like W, Au, and Pt is usually chosen to be equal to about half the electron mean free path [3,5]. Consequently, less than a third of the produced X-rays emerge from the radiator as seen in Fig. 1. It is to liberate a larger percentage of these produced X-rays that motivates our work described here.

We propose to increase the X-ray yield [1] by replacing a single heavy metal target by multiple thin foils which are emersed in a magnetic field into which the electron beam is introduced through a magnetically shielded channel. These electrons then traverse the foils thereby considerably increasing the yield over a similar thickness single target.

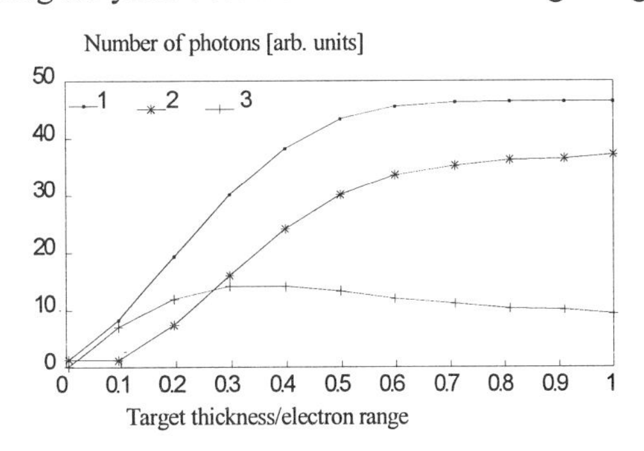

Figure 1. X-rays with depth: (1) total produced; (2) absorbed; and (3) surviving.

The X-ray yield for N crossings of foils with D thickness (total thickness, L = ND) and N >> 1 is

$$Y = \frac{A}{\mu}(1 - e^{-\mu x}) \quad , \tag{1}$$

where A is the average foil X-ray generation efficiency, μ is the foil X-ray absorption coefficient usually averaged to the spectrum [3,5,6], and x is L for a single target and D for foils. For a target, $\mu L >> 1$ while for a foil, $\mu D < 1$, so $\exp(-\mu L) << 1$ and $\exp(-\mu D) \leq 1$. The foil-to-target X-ray yield ratio for N foil crossings to one target traversal is

$$\frac{Y_{ND}}{Y_L} \approx \mu ND = \mu L >> 1 \quad . \tag{2}$$

Our multiple crossing X-ray generator can be realized in several ways depending on the accelerator parameters and the application. The simplest scheme for irradiating extended objects has a transverse magnetic field located at the accelerator beam exit. The magnetic field bends the electrons through a small angle and the foils are placed at intervals on their trajectory so that the X-rays from each foil emerge in the beam direction as seen in Fig. 2b.

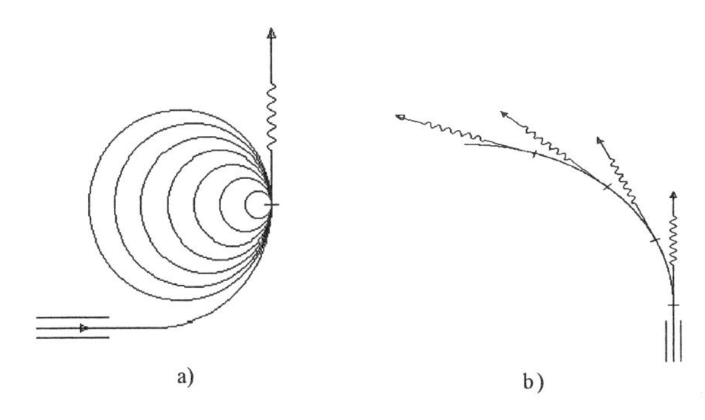

Figure 2. X-ray sources to produce (a) pencil and (b) fan beams.

CP392, *Application of Accelerators in Research and Industry*, edited by J. L. Duggan and I. L. Morgan
AIP Press, New York © 1997.

The scheme shown in Fig. 2a produces a narrow directed X-ray beam. Here electrons multiply cross a single foil, their return to the foil being assured by the magnetic field. The foil thickness is chosen so that first turn electrons do not interfere with electrons exiting the accelerator. For electrons with energies between 200 keV and 5 MeV, magnetic fields of less than 100 G to 2 kGs accomplish this while reducing the orbits by ~1 cm when ~10 % of the initial electron energy is lost. The mean free path of 5 MeV electrons in tungsten is 3.04 g/cm^2, so in a 0.3 g/cm^2 thick foil these electrons will lose 0.5 MeV. Therefore, electrons can ideally make up to 10 passes through the foil, increasing the X-ray yield between two and four times over that from a single pass through an equally thick target [1].

SIMULATION RESULTS

To verify our analytical yield estimates [1], we Monte-Carlo simulated [7] our two X-ray generating schemes in runs with 10,000 electrons bombarding tungsten foils whose thickness was 0.1 of their mean free path. X-rays were registered that were within a 50° cone whose axis was the electron momentum. These simulations agreed well with our analytic results.

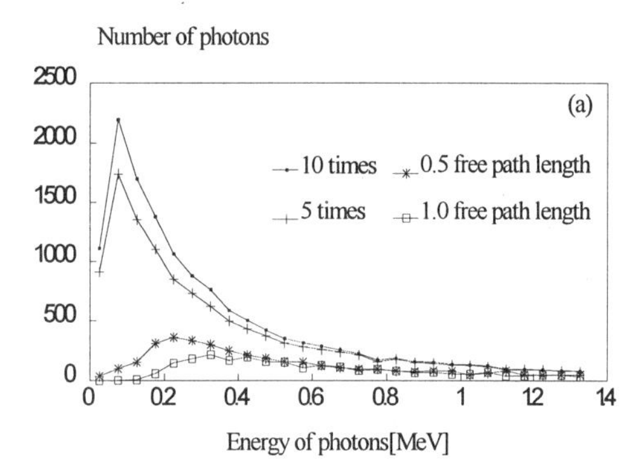

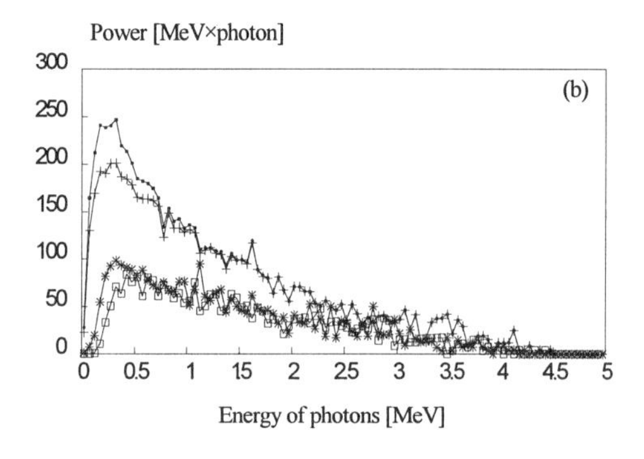

Figure 3. X-ray (a) spectra and (b) power spectra for 5 MeV electrons.

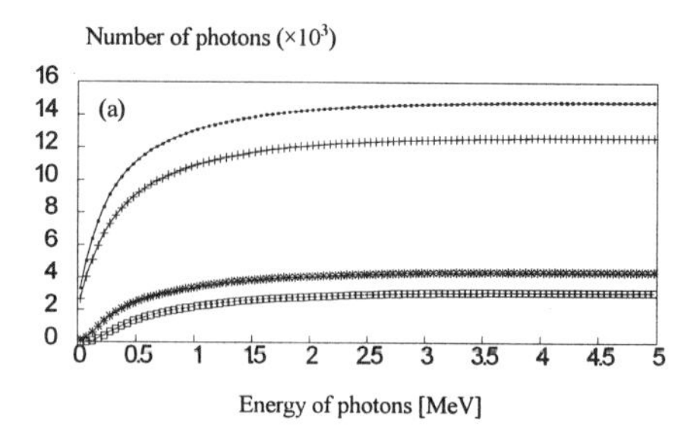

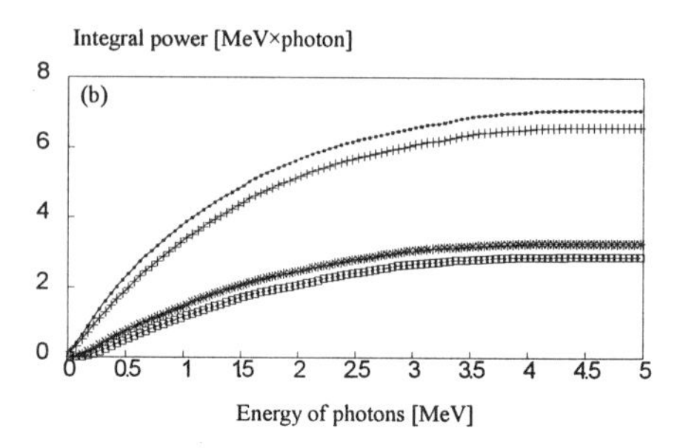

Figure 4. X-ray energy for 5 MeV electrons with integrated (a) number of X-rays and (b) radiated X-ray power.

The foil crossing X-ray production efficiencies for 5 and 10 MeV electrons are seen with X-ray energy in Figs. 3a and 6, the X-ray power spectra in Fig. 3b, and the integral X-ray power spectra in Figs. 4 and 5. Here the multiple crossing foil X-ray yield is four times greater than the single crossing equal thickness target yield. The integral X-ray power is more than twice that of the single target crossing power. Note that the absorption of soft X-rays is disproportionally reduced which benefits our scheme in this energy region.

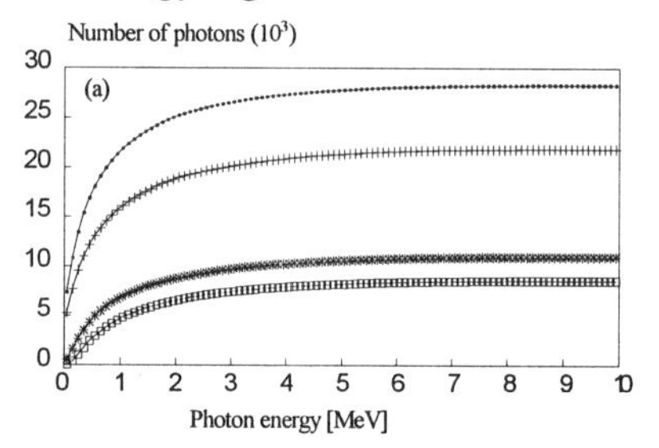

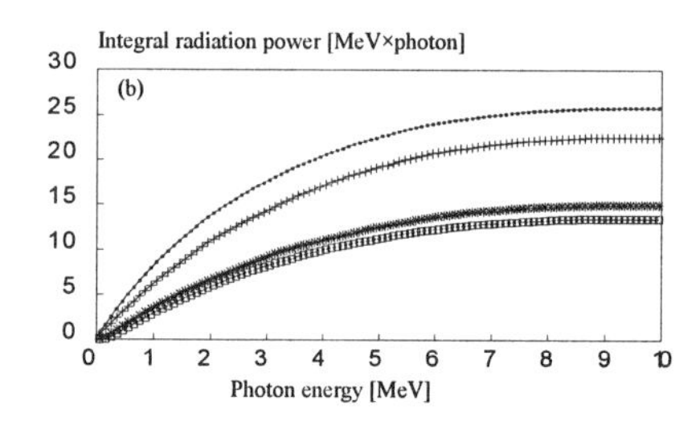

Figure 5. X-ray energy for 10 MeV electrons with integrated (a) number of X-rays and (b) radiated X-ray power.

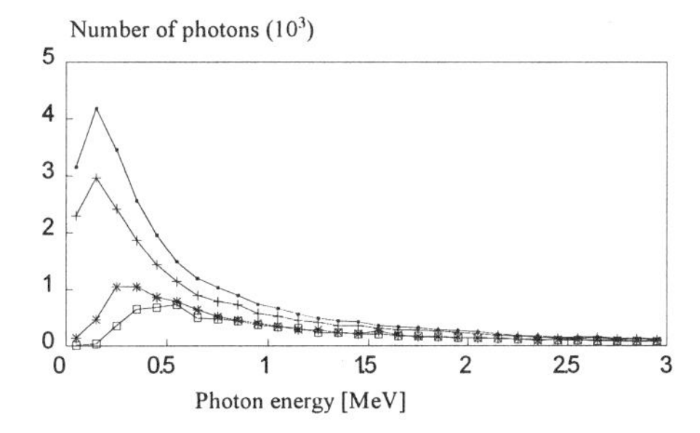

Figure 6. X-ray spectra for 10 MeV electrons.

Although the agreement between our simulations and analytic predictions is good, there are important questions which must be solved to implement these schemes in hardware. For example, electrons will have considerable angular scattering in the foils and must be strongly focused vertically, so a modest strength inhomogeneous transverse magnetic field will be needed. To solve this and other engineering problems, we are currently developing codes that simulate electron motion in magnetic fields and X-ray generation and heat dissipation in foils. However, we have shown that we can substantially increase the efficiency of bremsstrahlung sources and the practical development of such sources is well within the current capabilities of the technology.

ACKNOWLEDGMENT

We thank W.P. Trower for fruitful discussions.

REFERENCES

1. V.K. Grishin, B.S. Ishkhanov, and V.I. Shvedunov, Vestnic Moscovskogo Universiteta, **37** (1996) 83 (in Russian).
2. B.S. Ishkhanov, and I.M. Kapitonov, Moscow, Izdatelstvo Moscovskogo Universiteta (MSU Publications, Moscow, 1979) p. 215 (in Russian).
3. M.J. Berger and S.M. Seltzer, Phys. Rev. **C2** (1970) 621.
4. D.J.S. Findlay, Nucl. Instrum. Meth. **B50** (1990) 314.
5. M.J. Berger and S.M. Seltzer, NASA SP-3012 (1964).
6. E. Storm and H. Israel, LANL (1987).
7. R. Brun, F. Bruyant, M. Maire, A.C. McPherson, and P. Zanarini, **GEANT 3.17** User's Guide CERN, DD/EE/95-1 (1987).

RECENT EXPERIMENTAL WORK ON BREMSSTRAHLUNG IN KEV-ENERGY RANGE

R. SHANKER AND S.K. GOEL

Atomic Physics Laboratory, Physics Department, Banaras Hindu University, Varanasi-221 005, India.

Experimental investigations on the thick-target bremsstrahlung (TTB) spectra are presented for 7.0-keV electrons incident on Ag and Au targets. The relative shape and the absolute intensity distributions of the bremsstrahlung photons have been measured and compared with the thin- as well as with the thick-target theories. Agreement between experiment and theory is satisfactory. The discrepancy observed between the two is suggested to arise mainly from the solid-state effects.

INTRODUCTION

The experimental investigations of electron bremsstrahlung process have been made in the past for both thin and thick targets with electrons having energy in the range of keV to MeV (1,2). The low keV energy spectra of the bremsstrahlung radiation was studied by a few workers in the last decades for free atoms and molecules with atomic numbers Z in the range Z=2-92 (3-5). The interest in the bremsstrahlung spectra produced by the keV-electrons incident on the thick targets has grown in recent years because of their importance in understanding the basic processes of penetration of electrons through matter. The practical interest lies in the widespread commercial use of x-rays, electron beams, electron bursts and in upper atmosphere. The thick-target bemsstrahlung (TTB) spectra produced in the keV energy range were measured by Chervenak etal. (6) and Ambrose etal. (7). These authors have compared their TTB results with different models and have suggested the need to develop a comprehensive theory for accounting the solid state effects.

The present work was undertaken to study the atomic field bremsstrahlung process by keV-energy electrons using a newly developed experimental setup in the atomic physics laboratory. Bremsstrahlung intensity distributions from keV-electrons incident on the solid targets have been measured and compared with the existing theories for thin- as well as for thick- targets. The discrepancies existing between experiment and theory are pointed out.

EXPERIMENTAL PROCEDURE

A monoenergetic beam of 4.0-8.0 keV electrons was obtained from a home-built electron gun and was made to strike the Ag and Au targets having thicknesses in the range 150 μgm / cm^2 -200 μgm /cm^2 at 45 to their normals and the resulting bremsstrahlung radiation transmitted through a 6- micron thick hostaphan foil was recorded by a Si (Li) dectector (FWHM = 250 eV at 5.9 keV) positioned at 90° to the incident beam direction. The targets were thick enough to arrest incident electrons but thin enough to transmit a substantial fraction of photons. The incident beam current was monitored on the target itself. Fig. 1 shows a schematic diagram

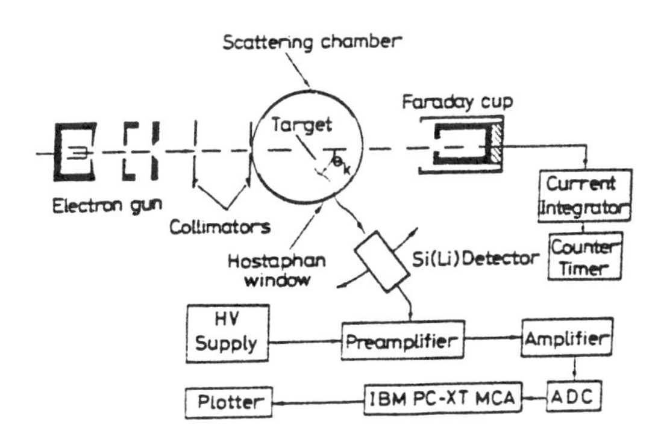

Fig. 1 Schematic diagram of the experimental setup.

of the experimentalsetup. The procedure for determining the efficiency of Si (Li) detector and that for the calibration of photon energy spectra, the background subtraction and for data collection etc. have been described in recent papers (8,9). The beam current used in recording the spectra was kept below 3.0 nA. About 20 to 30 min. times were found to be sufficient for obtaining the data of less than 10% counting statistics. The number of bremsstrahlung photons N_b (k) recorded by a detector within an energy window of Δk , and a solid angle $\Delta\Omega$ at an angle θ_k to the incident beam direction is given by

$$N_b(k) = N_e N_t \, (d^2\sigma/dkd\Omega)\Delta k\Delta\Omega \, \epsilon(k) \qquad \text{-------(1)}$$

where N_e and N_t are the numbers of electrons incident on the target and that of target atoms per cm^2 respectively, ϵ(k) is the detector's efficiency at photon energy k and $d^2\sigma$ /dkdΩ is the energy and angualr distribution of the photons or the doubly differential cross section (DDCS).

CP392, *Application of Accelerators in Research and Industry*, edited by J. L. Duggan and I. L. Morgan
AIP Press, New York © 1997.

BREMSSTRAHLUNG SPECTRA AND ABSOLUTE INTENSITY

The intensity of TTB photons on an absolute scale can be obtained by using the expression,

$$I_{Eo,k} = \frac{1.602x10^{-3}x4\pi\ N_k\ ergs}{N_q \Delta k \Delta\Omega \epsilon(k)\ sec\ mA\ keV} \quad \text{----------(2)}$$

where N_k and N_q are the numbers of photons in energy window Δk at k and that of charge (in nano-Coulomb) on the target. The solid angle subtended by the detector at the target was $\Delta\Omega/4\pi = 1.73x10^{-4}$ sr while the photon energy window was taken to be Δk=0.25 keV.

The uncertainties in incident beam energy, solid angle determination, detector's efficiency calibration, charge collection and in counting statistics were estimated to be 1%, 17%, 5%, 10% and 3% respectively. Hence, the total uncertainty in determination of the absolute intensity of the TTB spectra, when combined in quadrature amounted to within 21%.

RELATIVE SHAPE OF THE BREMSSTRAHLUNG-PHOTON ENERGY SPECTRUM

Theoretical calculations (10) predict slowly varying changes in the bremsstrahlung energy spectrum as a function of incident energyβ (β is the velocity of incident electrons in units of the speed of light) and of photon energy k, in addition to a strong dependence of Z^2. Measurements of absolute DDCS often bear large uncertainty due to which the above mentioned variations are not found to be very sensitive. However, by taking the ratio of the DDCS for two targets at the same incident energy and same angle of photon detection θ_k and $\Delta\Omega$, the experimental uncertainty can be reduced to a few percent. For example, we have measured (see, Fig.1)the ratio of the DDCS of the bremsstrahlung spectrum produced from 7.0-keV electrons incident on thin Ag and Au targets (see, Ref. 9,11). The agreement between experiment and theory for the relative shape of the photon energy spectrum was found to be excellent within the estimated 5% uncertainty in the theory and less than 4% in the experiment; however, the thick target effects were found to introduce a discrepancy between experiment and theory of about 20%. This discrepancy has been suggested to arise due to energy loss, backscattering of incident electrons, straggling and x-ray absorption in the target. These effects, as a matter of fact are not considered in the development of the thin target theory (10). We have cross checked the experimental DDCS ratio for Ag and Au targets, for example, at k/T = 0.8; the corresponding values are found to scale within 19% with (Z_{Ag}/Z_{Au}) and within 42% with $(Z_{Ag}/Z_{Au})^2$. This analysis suggests that the data are influenced with the thick-target effects which follow a Z-dependence in contrast to the thin targets for which the data are expected to follow a Z^2-dependence.

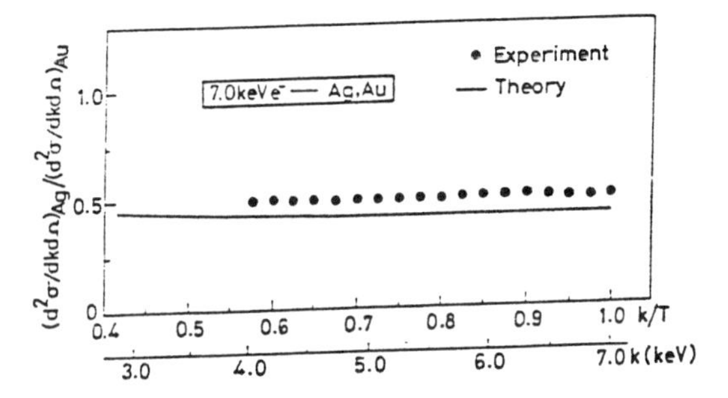

Fig.2 : Ratio of the double differential cross-sectoin of bremsstrahlung photons for 7.0-keV electrons on Ag and Au vs the fraction of photon energy radiated (k/T).

THE THICK-TARGET BREMSSTRAHLUNG INTENSITY DISTRIBUTION

The thick-target bremsstrahlung intensity distributions for 7.0-keV electrons incident on Ag & Au targets have been measured (12) and calculated by using a modified Kramers-Kulenkampff-Dyson (KKD) formula :

$$I_{Eo,k} = CZ(E_0\text{-}k)\ R_{Eo,k}\ f_{Eo,k,\alpha}\ ergs/(sec\ mA\ keV\ sr) \quad \text{--------(3)}$$

where $I_{E0,k}$ is the thick-target bremsstrahlung emission intensity, C is the Kramers constant, Z is atomic number of the target atom, $f_{Eo,k,\alpha}$ is the photon attenuation correction and $R_{Eo,k}$ is the electron backscattering correction.

Further, a semiemperical formula by Storm (13) has also been used to calculate the intensity distribution of the bremsstrahlung spectra produced in the above collision system. The experimental thick-target bremsstrahlung spectra were put on an absolute scale following the expression given by Equation (2). The Storm formula is given by

$$I_{Eo,k} = \frac{[11/4\pi Z(E_o\text{-}k)(1\text{-}e^{-3k/EI})]\ f_{Eo,k,\alpha}.ergs}{[(k/E_o)^{1/3}(1\text{-}e^{-Eo/EI})](sec\ mA\ keV\ sr)} \quad \text{------(4)}$$

where the term in large parentheses is an approximate to the Born approximation thin formula (14) with electron energy- and backscattering losses included and further multiplied by a term for the photon attenuation correction. E_o and E_I are the incident electron energy and the average ionization potential of L-and M-shells of Ag and Au inkeV. The predictions of the modified KKD and that of semiemperical formula are shown in Fig. 3, wherein a thick-target bremsstrahlung intensity distribution for 7.0-keV electrons on Au are plotted as a function of the photon energy radiated in the collision.

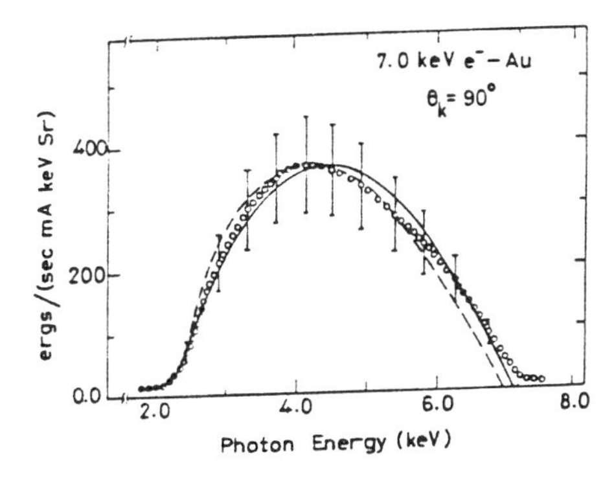

Fig.3. Absolute thick-target bremsstrahlung intensity distribution vs photon energy for 7.0-keV electrons incident on Au. o: present experimental data; — : modified KKD formula; ---- : Storm's semiemperical formula.

The predictions are found to describe the behaviour of the data for the measured spectra within 21%. The error bars shown represent the systematic uncertainty in the measurements. There is a satisfactory agreement between experiment and thick-target theories (i.e., KKD and semiemperical) after we took various thick-target effects, namely, electron energy loss, electron back scattering, photon attenuation in the target and inherent filtration of the photons in the vacuum chamber window, air, berillium window, gold and dead layers of the detector, into account. The largest contribution of the uncertainty comes from the solid angle determination and from charge collection uncertainties.

CONCLUSIONS

An experimental setup for studying the electron bremsstrahlung process with low keV-electron incident on semi thick self supported targets has been developed. The ratio of the DDCS of the bremsstrahlung spectrum from 7.0-keV electrons on Ag and Au targets has been measured and compared with the best theory available today. The agreement between experiment and theory for the relative shape of the photon energy spectrum is found to be excellent within the estimated experimental uncertainty of less than 4%; however, about 20% discrepancy is observed between them due to the thick-target effects . Further, the absolute intensity distributions of the bremsstrahlung photons has been calculated for the measured spectra and have been compared with the modified KKD model and that with a semiemperical formula. The agreement between experiment and model predictions is satisfactory within the systematic uncertainty of the measurements.

ACKNOWLEDGMENTS

Authors wish to thank Prof. Dr. R. Hippler and Prof. C. A. Quarles for extending their help and taking interest in this work. The work was supported by the Department of Science and Technology (DST), New Delhi under Project No. SP/S2/K-37/89.

REFERENCES

1. Motz, J.W.and Placious, R.C.,Phys.Rev.**109**,235(1958).
2. Quarles, C.A.and Heroy, D.B., Phys.Rev. **A24,**48(1981).
3. Aydinol, M., Hippler, R., McGregor, I. and Kleinpoppen, H., J.Phys.B**13,**989(1980).
4. Hippler, R. Saeed,K.,McGregor,I. and Keinpoppen, H., Phys.Rev.Lett. **46,**1622(1981).
5. Semann, M.and Quarles, C., Phys. Rev**A26,**3152(1982).
6. Chervenak, J.G. and Liuzzi, A, Phys.Rev.**A12,**26(1975).
7. Ambrose, R. Kahler, D.L., Lehtihet,H.E.andQuarles, C. Nucl.Instr.Meth. Phys. Res. **B56/57**,327(1991).
8. Goel, S.K., Singh, M.J.and Shanker,R.,Pramana:J.Phys., **45**,291(1995).
9. Goel, S.K., Singh, M.J. and Shanker,R.,Phys.Rev. **A52,** 2453(1995).
10. Kissel, L., Quarles, C.A. andPratt,R.,At.DataNucl.Data Tables, **28**,381(1983).
11. Goel,S.K.,Ph.D.Thesis, Banaras Hindu University, Varanasi, India
12. Goel, S.K. and Shanker, R.,Phys. Rev. **A54,**.(1996).
13. Storm, E., Phys.Rev. **A5,**2328 (1972).
14. Koch,H.W.andMotz,J.W.,Rev.Mod.Phys.**11**,920(1959).

PROPERTIES OF INTENSE QUASI-MONOCHROMATIC X RAYS PRODUCED BY RESONANT TRANSITION RADIATION

T. Awata[‡†], K. Yajima[‡], T. Tanaka[‡], M. Imai[‡], A. Itoh[‡], N. Imanishi[‡], K. Yoshida[§], K. Nakayama[¶], and A. P. Potylitsin[*]

[‡]*Kyoto University, Kyoto 606-01, Japan,* [§]*Hiroshima University, Hiroshima 727, Japan,* [¶]*Toshiba Co., Kanagawa 210, Japan,* [*]*Nuclear Physics Institute, Tomsk 634041, Russia*

To develop the possibilities of resonant transition radiation (RTR) as an intense quasi-monochromatic X-ray source, we have measured energy spectra and angular distributions of the RTR X rays emitted from Kapton- and silicon-foil stacks by varying the number of foils, foil thickness, and spacing between adjacent foils. The intensity of RTR X rays increases nonlinearly with the number of foils in case of 12.5-μm thick Kapton- and silicon-foil stacks. The energy of the RTR X rays can be tuned by changing only the thickness of the foil and will be defined well when the X rays are sliced at a fixed emission angle. The obtained brilliance of 1.4×10^{12} photons/(s·mrad^2·mm^2·0.1% b.w.·mA) for the peak energy of 6.7-keV X rays is comparable to those of synchrotron radiation (SR) emitted by using bending magnets in GeV-electron facilities.

INTRODUCTION

In recent years, high brilliant X-ray beams are being widely used in various fields, and it is demanded to develop and design an inexpensive intense X-ray source (1-3). For this purpose we have started to study whether transition radiation can be applied to a quasi-monochromatic X-ray source. Transition radiation (TR) is emitted when a charged particle crosses boundaries of media having different dielectric constants (4). When a relativistic electron passes through many thin foils placed periodically in a vacuum, the TR X rays emitted from the respective boundaries interfere with one another and become a quasi-monochromatic beam. The interfered TR is called resonant transition radiation (RTR). RTR is classified into two types (5, 6). One is the resonance between TR emitted from the front and back surfaces of each foil, and the other is the resonance between different foils. The former is called intra-foil resonance and the latter inter-foil resonance.

In this paper, we report experimental results which have been performed using a 1-GeV electron beam passing through Kapton- and silicon-foil stacks. We have measured energy spectra and angular distributions of the RTR X rays as a function of thickness and spacing, and then compared them with calculated results. A feasibility of the RTR X rays as an alternative X-ray source will be discussed.

RESONANT TRANSITION RADIATION

When a relativistic electron of a velocity v passes through N foils placed periodically in a vacuum with the same thickness l_1 and spacing l_2, the intensity of emitted RTR per photon energy per solid angle is given by (7)

$$\frac{d^2P}{d\omega d\Omega}=\frac{\alpha\omega\sin^2\theta}{16\pi^2c^2}(Z_1-Z_2)^2F_{1\,\mathrm{foil}}F_{N\,\mathrm{foils}}, \tag{1}$$

where P is the number of photons, θ is an emission angle, ω is an angular frequency of photon, and α is the fine structure constant. Z_i (i=1,2) are the formation lengths of the media expressed by

$$Z_i=\frac{4c\beta}{\omega\left\{\left(\frac{1}{\gamma}\right)^2+\left(\frac{\omega_i}{\omega}\right)^2+\theta^2\right\}}, \tag{2}$$

where β is v/c, c is the velocity of light, and γ is the Lorenz factor of the electron. The plasma frequencies of the respective media ω_i are obtained by the following equation:

$$\omega_i=\sqrt{\frac{4\pi Ne^2}{m_e}}, \tag{3}$$

where N is the electron density in the media and m_e is the electron rest mass. $F_{1\mathrm{foil}}$ and $F_{N\mathrm{foils}}$ are the intra- and inter-foil resonance factors, respectively, expressed by

$$F_{1\,\mathrm{foil}}=1+e^{-\mu_1 l_1}-2e^{-\mu_1 l_1/2}\cos\frac{2l_1}{Z_1}, \tag{4}$$

†And JSPS Research Fellow.

CP392, *Application of Accelerators in Research and Industry,* edited by J. L. Duggan and I. L. Morgan
AIP Press, New York © 1997.

$$F_{N\text{ foils}} = \frac{1+e^{-N\sigma} - 2e^{-N\sigma/2}\cos 2NX}{1+e^{-\sigma} - 2e^{-\sigma/2}\cos 2X}, \quad (5)$$

where $\sigma = \mu_1 l_1 + \mu_2 l_2$, $X = l_1/Z_1 + l_2/Z_2$, and $\mu_{1,2}$ are the X-ray absorption coefficients of the media.

EXPERIMENTAL

The experiment was performed using the 1.3-GeV electron synchrotron at the Institute of Nuclear Study, the University of Tokyo (INS-ES). The experimental setup is shown in Fig.1. The 1-GeV electron beam extracted from INS-ES was made to cross a thin-foil stack in a target chamber, and was swept away by a bending magnet. The average electron beam current was monitored using an ionization chamber located behind the bending magnet, and was kept at a given value between 0.2 pA and 20 pA, depending on the measurement. The size of the electron beam was about 3 mm in diameter at the position of the thin-foil stack. The RTR X rays emitted from the thin-foil stack were detected by an X-ray crystal spectrometer equipped with a LiF(200) crystal and having an energy resolution of about 3.3% at 8.1 keV of the Cu-K X ray. The background originating from Bremsstralung was monitored by inserting an aluminum shutter absorbing the RTR X rays. The angular distributions of the RTR X rays were measured by moving an X(horizontal)- and a Y(vertical)-slit across the beam axis in front of the crystal spectrometer. The slit has a 2 mm square (0.4 mrad×0.4 mrad) aperture, and was moved by a step of 0.4 mrad. The thin-foil stacks used in the experiment were made of 12.5-μm thick and 50-μm thick Kapton as well as 15-μm thick silicon.

RESULTS AND DISCUSSION

Total energy spectra were taken for the RTR X rays emitted in an emission-angle range of -3 to 3 mrad, and were corrected for detection efficiencies. Figure 2 shows examples of the measured spectra for the one- and eight-foil stacks of 12.5-μm thick Kapton with a spacing of 100 μm. The relative statistical errors of the yields were typically less than 1% including the contribution of the background. The energy spectra calculated from Eq. (1) with a Monte Carlo method using the same parameters as used in the experiment are also shown in Fig. 2. The calculation takes into account the respective sizes of the source and the detector of 3 mm in diameter and 30 mm × 30 mm wide as well as the self absorption of X rays by the stacks (8), and the values were multiplied by a factor of 6 to fit the experimental yields of the one foil. The energy spectra show broad single peaks at 6.7 keV for both targets and are reproduced well by the calculation. The peak intensity of the 6.7-keV X rays for the eight-foil stack is about 8.4 times

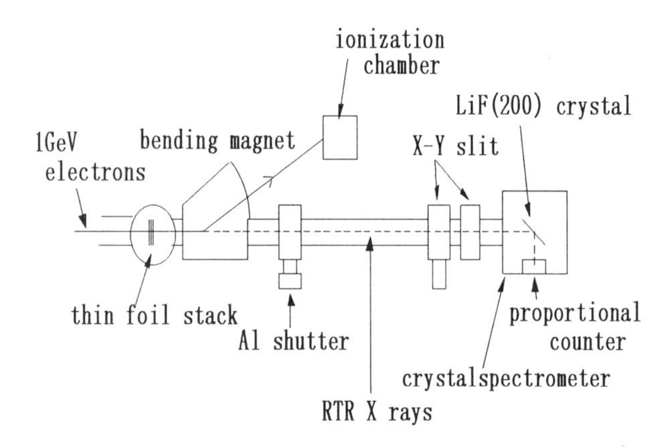

FIGURE 1. Schematic diagram of the experimental setup. After crossing a thin-foil stack, the 1-GeV electron beam was bent by the magnet into the ionization chamber. X rays emitted from the thin-foil stack were detected by the X-ray crystal spectrometer. The aluminum shutter was used to evaluate background.

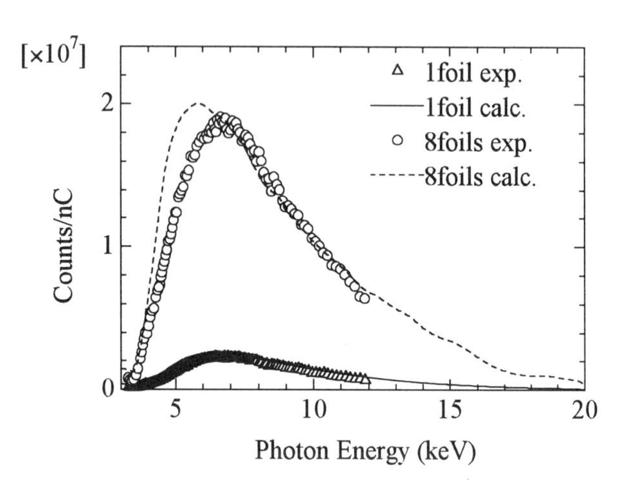

FIGURE 2. Measured and calculated energy spectra for the one- and eight-foil stacks made of 12.5-μm thick Kapton with a spacing of 100 μm. Relative statistical errors are typically less than 1%. The calculated values were multiplied by a factor of 6.

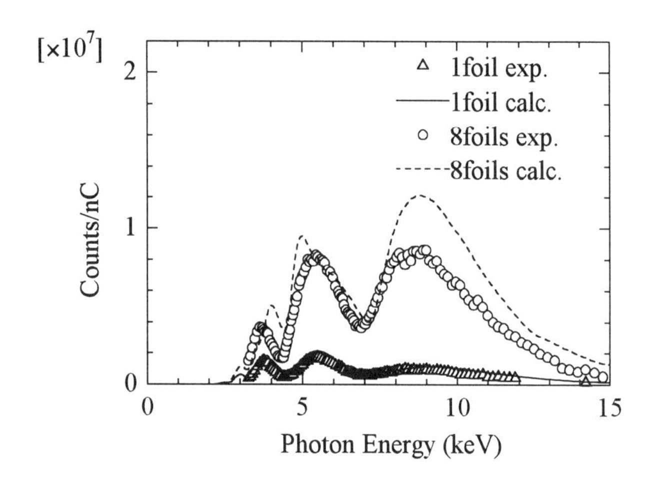

FIGURE 3. Same as Fig.2 for the 50-μm thick Kapton stacks with a spacing of 200 μm. Relative statistical errors are typically less than 1%. The calculated values were multiplied by a factor of 5.

as high as that of the one foil, and much higher than the sum of the absorption-corrected intensities of X rays emitted from the individual foils.

The nonlinear dependence is also depicted in Fig. 3 which shows the measured and calculated energy spectra for one- and eight-foil stacks of 50-μm thick Kapton with a spacing of 200 μm. The energy spectra show three peaks at 3.8, 5.5, and 8.7 keV for both targets. The intensity of the 8.7-keV X rays for the eight-foil stack nonlinearly increases with increasing the number of foils. However, being caused by the strong self absorption of low-energy X rays, the intensity of the 5.5-keV X rays is lower than the sum of the one foil's.

Figure 4 shows the energy spectra for one- and two-foil stacks of 15-μm thick silicon with a spacing of 100 μm along with the calculated results. The energy spectra have two peaks at 4.3 and 10 keV, and the intensity of the 10-keV peak for the two-foil stack is higher than twice the intensity of the one-foil stack. The stack composed of foils with low X-ray absorption coefficients like beryllium foils will show a more prominent nonlinear dependence of the RTR-X-ray intensity on the number N of foils in the stack, as expected from Eq. (5).

It was shown in Figs. 2-4 that the energy of the RTR X rays can be changed by varying foil thickness. This feasibility comes from the intra-foil resonance. When the absorption by the foil stack itself is ignored, Eq. (4) is reduced to the following simple form:

$$F_{1\text{foil}} = 4\sin^2\left(\frac{l_1}{Z_1}\right). \quad (6)$$

Then, the condition of the intra-foil resonance is

$$\frac{l_1}{Z_1} = \frac{2n-1}{2}\pi, \quad (7)$$

where n is a positive integer, and is called the order of the intra-foil resonance. Substituting Eq. (2) into Eq. (7) under the condition of $v \approx c$, one obtains the following equation:

$$\omega = \frac{\omega_1^{\,2} l_1}{2c}\frac{1}{(2n-1)\pi}. \quad (8)$$

Then, the peak energy relates to the foil thickness l_1, and can be calculated using Eq. (8). The experimental and calculated results are shown in Table 1, and a good agreement was obtained between them. Thus, the energy of X rays can be tuned over a wide range of energy by changing only foil-material and thickness.

Figure 5 shows the angular distributions of the RTR X rays measured at X-ray energies of 3.7 and 5.5 keV for the

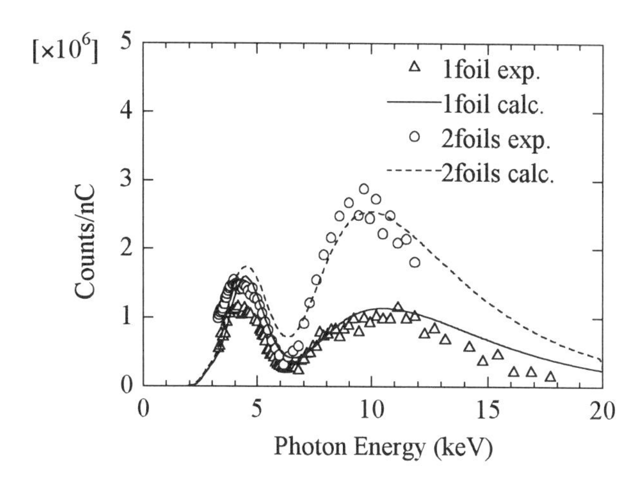

FIGURE 4. Same as Fig.2 for the 15-μm thick silicon stacks with a spacing of 100 μm. Relative statistical errors are typically less than 6%. The calculated values were multiplied by a factor of 5.5.

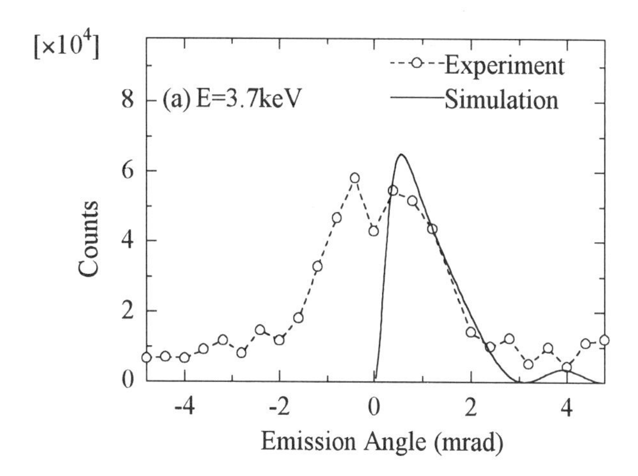

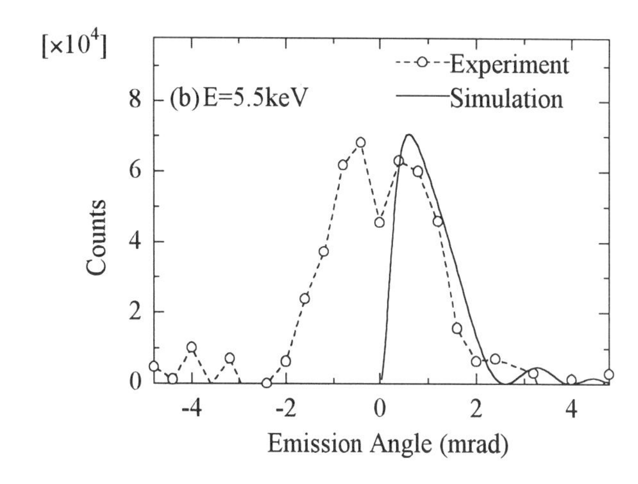

FIGURE 5. Measured and calculated angular distributions at X ray energies of (a) 3.7 keV (b) 5.5 keV for the one-foil stack made of 50-μm thick Kapton.

TABLE 1. Comparison between the Measured and Calculated Peak Energies of the RTR X rays

	n=1		*n*=2		*n*=3	
Foil	**Measured (keV)**	**Calculated (keV)**	**Measured (keV)**	**Calculated (keV)**	**Measured (keV)**	**Calculated (keV)**
12.5-μm thick Kapton	6.7	6.20	-	2.00	-	-
50-μm thick Kapton	-	25.8	8.6	8.7	5.5	5.15
15-μm thick silicon	10	11.7	4.3	3.88	-	2.33

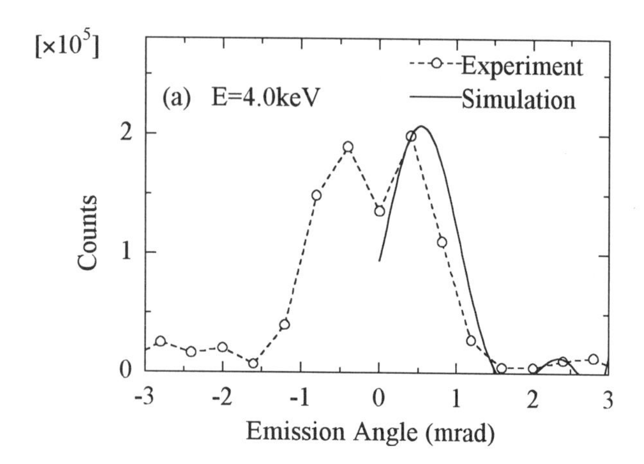

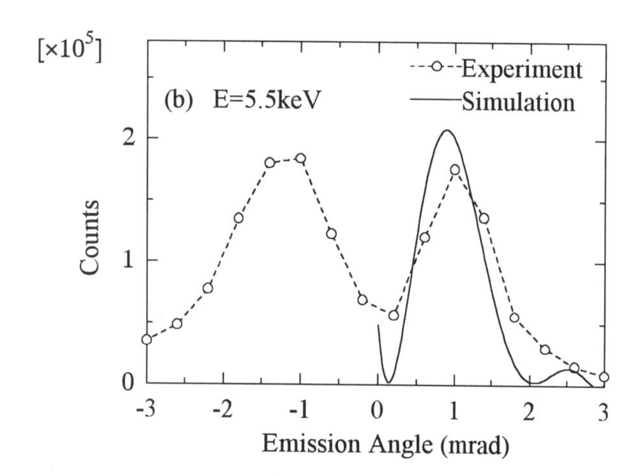

FIGURE 6. Measured and calculated angular distributions at X-ray energies of (a) 4.0 keV (b) 5.5 keV for the eight-foil stack made of 12.5-μm thick Kapton with the spacing of 100 μm.

one 50-μm thick Kapton foil along with the calculated results. Figure 6 presents the results for the eight-foil stack of 12.5-μm thick Kapton with the spacing of 100 μm. The calculation considers the effect of the beam size. The agreement between the measured and the corresponding calculated results is fairly good for all of the conditions. By comparing Figs. 5 and 6, one knows that the angular distribution depends on the X-ray energy in the case of the multiple-foil stack but it does not change irrespective of the X-ray energy for the one foil. That is, the resonance effect in the multi-foil stack modifies the emission angle of the RTR X rays depending on their energy. Then, one can produce the RTR X rays defined well in energy resolution when they are sliced at a fixed emission angle.

One can estimate the feasible brilliance of RTR X rays on the basis of the measured intensities of the RTR X rays for the eight-foil stack composed of 12.5-μm Kapton with the spacing of 100 μm as shown in Fig. 2. The estimated brilliance for the 6.7-keV X rays is about 1.4×10^{12} photons/(s·mrad^2·mm^2·0.1%b.w.·mA). This brilliance is comparable to that of synchrotron radiation (SR) emitted by using a bending magnet in a GeV-electron facility (9).

From the above result, we conclude that the RTR has a possibility to be used as an intense quasi-monochromatic X-ray source, especially under a pulse mode operation.

ACKNOWLEDGMENTS

The authors would like to thank the staffs of INS-ES, the University of Tokyo for the operation of the electron synchrotron. This work was supported in part by a Grant-in-Aid Contract No.06554012 from the Ministry of Education, Science, Sports, and Culture.

REFERENCES

1. Rullhusen, P., "Novel X-ray sources produced by electron beams," in *Proceedings of the Conference on Third European Particle Accelerator(EPAC92)*, 1992, **1**, pp. 240-244.
2. Chu, A. N., Piestrup, T. W., Barbee, Jr., and Pantell, R. H., *J. Appl. Phys.* **51**, 1290-1293 (1980).
3. Cesareo, R., Hanson, A. L., Gigante, G. E., Pedraza, L. J., and Mahtaboally, S.Q.G., *Phys. Rep.* **213**, 117-178 (1992).
4. Ter-Mikaelian, M. L., *High-Energy Electromagnetic Processed in Condensed Media*: Wiley-Interscience, 1972.
5. Piestrup, M. A., Boyers, D. G., Li, Q., Moran, J., Buskirt, F. R., Maruyama, X. K., Neibours,R., Robinson, R. M., and Snyder, D. L., *IEEE Trans. Nucl. Sci.* NS-**35**, 464-469 (1988).
6. Tanaka, T., Awata, T., Itoh, A., Imanishi, N.,Yamakawa, T., Oyamada, M., Urasawa, S., and Nakazato, T., *Nucl. Instr. and Meth.* B **93**, 21-25 (1994).
7. Cherry, M. L., Hartmann, G., Müller, D., and Prince, T. A., *Phys. Rev.* D **10,** 3594-3607 (1974).
8. Henke, B. L., Gullinkson, E. M., and Davis, J. C., *At. Data Nucl. Data Table* **54,** 181-342 (1993)
9. Kitamura, H., private communication (1996).

ION SCATTERING OFF MAGNETIC SURFACES

M. Dirska, J. Manske, G. Lubinski, M. Schleberger
Universität Osnabrück, FB Physik, D-49069 Osnabrück, Germany

R. Hoekstra
KVI, Atomic Physics, Zernikelaan 25, NL-9747 Groningen, The Netherlands

A. Närmann
Universidad del País Vasco, Facultad de Química, Departamento de Física de Materiales, Apto. 1072, E-20080 San Sebastián, Spain

Most of the ions scattered off a metal surface at low energies ($\approx$ 10 keV) under grazing incidence conditions are neutralized. A few of these particles are neutralized into excited states which subsequently decay by light emission. In the case of magnetic surfaces the information about the spin of the electrons involved in the neutralization process might be accessed via the polarization of the emitted light. The method is very well suited to investigate the properties of magnetic surfaces and multilayers. A UHV-setup that serves this purpose is presented as well as some recent results obtained for magnetized Fe(110) surfaces probed with He ions.

I. INTRODUCTION

The investigation of surface magnetism is of high technical relevance in information science where an ongoing trend to magnetic recording devices involving structures still decreasing in size can be observed. Thus, the need for suitable techniques to study the magnetic properties of surfaces and layer systems is growing rapidly.

There are two types of different methods used for probing magnetic surfaces: methods that involve electron beams at low energies which have the disadvantage that they are extremely sensitive to external magnetic fields, and methods using photons incident on the surface (MOKE, SMOKE) where the surface sensitivity is limited by the wavelength of the light.

In our method we use low energy ions to probe surface magnetism. This technique is superior in that it provides monolayer sensitivity by properly choosing the kinetic energy of the ions and the scattering geometry. Furthermore, ions are much less affected by external magnetic fields. For a description of the historical development see, e.g., [1].

The main idea of our technique is the following: A beam of singly charged He ions with a kinetic energy in the low energy range (up to 15 keV) is grazingly scattered off a magnetic Fe(110) surface. This ensures extreme surface sensitivity [2]. Most of the incoming ions are neutralized into the ground state by an Auger process [3–5]. There is also the possibility of resonant electron capture into an excited metastable state. The subsequent decay leads to the emission of polarized light. This polarization is related to the the total angular momentum $\vec{J} = \vec{L} + \vec{S}$ of the state. It is well known that the anisotropic population of the magnetic substates gives rise to the emission of polarized light (see, e.g., [1]).

We have studied the following transitions: He $1s2p^3P$–$1s3d^3D$(λ=587.6nm) and He $1s3d^1D$–$1s2p^1P$ (λ=667.8nm). The triplet transition is sensitive to the magnetization of the surface, since electrons with different spin polarization will populate different states $(L+1, L, L-1)$. The singlet transition is not sensitive to the magnetization, since all electrons regardless of their spin will populate the degenerated state $J = L$.

In our experiment we measure the Stokes parameters, which describe the polarization of the emitted light [7]. The third Stokes parameter (the S parameter) represents the circular polarization of the emitted light. Using the optical convention for the relation between left– and right–handed circularly polarized light and the relative Stokes parameter S/I, one gets:

$$\frac{S}{I} = \frac{I_R - I_L}{I} \tag{1}$$

where I is the total intensity (polarized and unpolarized light) and I_R and I_L are the intensities of the right– and left–handed polarized light, respectively. The sign of S/I represents the sense of rotation of the electric field vector.

For magnetized targets it is very useful to introduce the following quantity:

$$\Delta S/I = S_+/I - S_-/I \tag{2}$$

where S_+/I is the circular polarization for magnetization of the target parallel to the z–axis (see Fig. 1) and S_-/I is that for antiparallel magnetization.

In order to obtain the direction of the spin of the captured electrons one has to determine the circular polarization of the emitted light in the triplet case. Therefore,

CP392, *Application of Accelerators in Research and Industry*, edited by J. L. Duggan and I. L. Morgan
AIP Press, New York © 1997.

it is the change in the circular polarization of the light emitted along the z-axis that is the tool to probe the surface magnetization.

II. EXPERIMENT

Our experiments were performed in the UHV setup PLEASE (**P**olarized **L**ight **E**mission **A**fter **S**cattering **E**vents). The base pressure in the UHV chamber is in the 10^{-10}mbar range.

Fig. 1 shows the main part of our experimental setup. A beam of singly-charged and mass-selected He ions in the low energy range is focused onto the Fe(110) target. Typical beam currents range from $0.05 - 2\mu$A depending on the beam energy. The crystal is mounted on a yoke made of soft iron. Around the yoke, an insulated silver wire is coiled. A magnetization parallel or antiparallel to the z–axis can be achieved simply by selecting the appropriate direction of the current through the wire.

The sample can be rotated by 90°. This allows the magnetization of the crystal along the magnetically easy or hard axis.

The emitted light is detected perpendicular to the incoming beam. The detector, which is capable of determining the polarization of the light, is mounted outside the UHV chamber. It consists of a rotating retarder, an interference filter to select the wavelength, and a fixed polarizer. The intensity of the emitted light is measured as a function of the relative angle between the retarder and the polarizer. From the angle-dependent intensity, all three Stokes parameters can be determined [6,7]. It takes only a few minutes to analyze the polarization properties of the light. A more detailed description of our experimental setup and procedures is presented in [8].

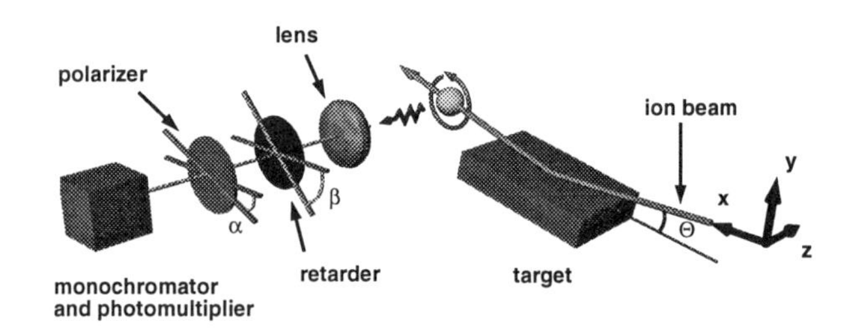

FIG. 1. Experimental setup

III. RESULTS

In Fig. 2 we show the effect of reversing the direction of the surface magnetization on the degree of circular polarization (triplet transition). Obviously there is a strong magnetization effect. The difference between the two relative Stokes parameters is $\Delta S/I \approx 27\%$, independent of the primary energy of the ion beam.

As an additional check we present in Fig. 3 for the singlet transition the relative Stokes parameter S/I for both directions of magnetization as a function of the primary energy. As expected, there is no effect due to the spin polarization of the captured electrons. The absolute value of the degree of circular polarization coincides with the value for the unmagnetized target in the triplet case [1].

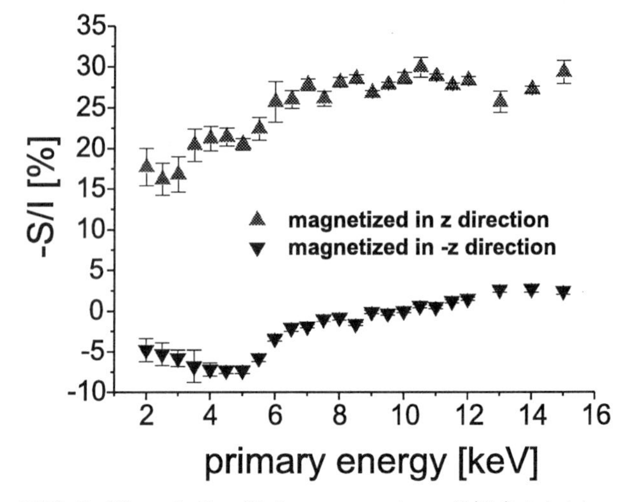

FIG. 2. The relative Stokes parameter $-S/I$ (triplet transition) as a function of the primary energy of the He^+ beam. The Fe(110) target is magnetized parallel and antiparallel to the z–axis, respectively.

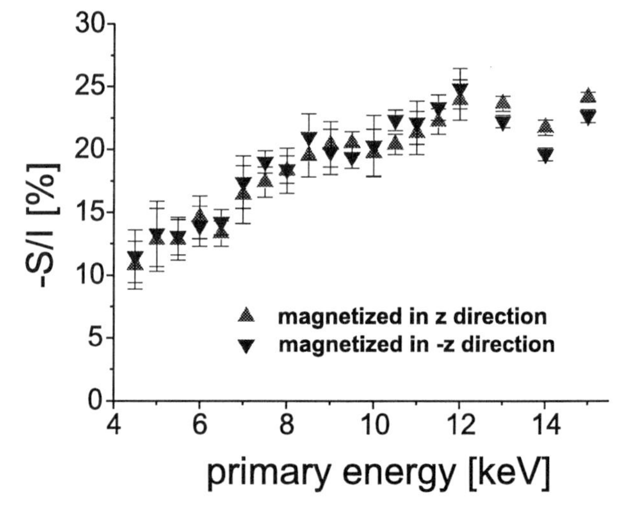

FIG. 3. The relative Stokes parameter $-S/I$ (singlet transition) as a function of the primary energy of the He^+ beam. The Fe(110) target is magnetized parallel and antiparallel to the z–axis, respectively.

Note that for the results presented in Figs. 2 and 3 our sample was magnetized along the easy axis of magnetization. The error bars indicate the statistical error of the photon counting rate and the uncertainty of the fitting procedure used to calculate the Stokes parameters. In

Fig. 4 we show the difference in the degree of polarization obtained after magnetizing the crystal along the easy and hard direction of magnetization, respectively. Obviously there is an easily measurable difference between the two types of magnetization. Depending on whether the direction of magnetization is along the magnetically soft or hard direction, there is a significant or only a marginal effect on the circular polarization, respectively. An effort to directly relate the magnetization of the surface to the polarization signal is currently under way.

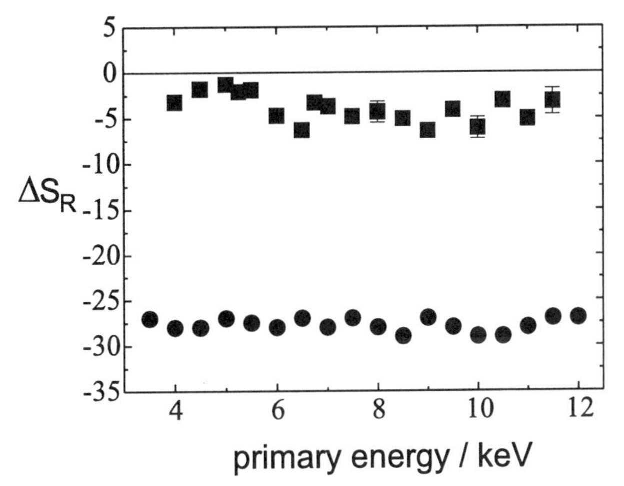

FIG. 4. The difference in circular polarization, $\Delta S_R = \Delta S/I$ as a function of primary energy for magnetization along the magnetically soft (circles) and hard (squares) direction.

IV. CONCLUSIONS AND OUTLOOK

We have shown that a low-energy ion beam can be used to study surface magnetism. With the tools described above we are now able to study magnetic layer systems with monolayer sensitivity. For example, one can follow the transition from two– to three–dimensional magnetism: For a Co monolayer on Cu the magnetization is oriented perpendicular to the surface and changes to in–plane magnetization with increasing number of Co layers. With the method described here we can follow the magnetization of the topmost atomic layer during the growth process. Another example is the coupling of magnetic layers through non–magnetic layers (for example Co/Cu/Co). Experiments to investigate these types of systems are being prepared.

V. ACKNOWLEDGEMENTS

We would like to thank W. Heiland for many helpful discussions and support of the project. Financial support from the Deutsche Forschungsgemeinschaft for M.S. and from the Universität Osnabrück for J.M. is gratefully acknowledged. The stay at the Universität Osnabrück of R.H. is sponsored by the EC HCM network program, Grant No. ERBCHRXCT930103. A. N. acknowledges financial support from the EC, Grant No. ERBFMBICT950038.

[1] J. Manske, M. Dirska, M. Schleberger, A. Närmann, and R. Hoekstra. *J. Magn. Magn. Mater.*, 1996. submitted

[2] C. Höfner, A. Närmann, and W. Heiland. *Nucl. Instrum. Methods B*, 72:227-233, 1992.

[3] H. D. Hagstrum. In N. H. Tolk, J. C. Tully, W. Heiland, and C. W. White, editors, *Inelastic Ion-Surface Collisions*, page 1. Academic Press, New York, 1977.

[4] M. Alducin, A. Arnau, and P. M. Echenique. *Nucl. Instrum. Methods B*, 67:157-159, 1992

[5] N. Lorente, R. Monreal, and M. Alducin. *Phys. Rev A*, 49:4716-4725, 1994

[6] D. Clarke and J. F. Grainger. *Polarized Light and Optical Measurement*, Vol. 35 of *Int. Series of Monographs in Nat. Philosophy.* Pergamon Press, Oxford, New York, 1971.

[7] M. Born and E. Wolf *Principles of Optics.* Pergamon Press, New York, 1970.

[8] M. Dirska, J. Manske, G. Lubinski, M. Schleberger, A. Närmann, and R. Hoekstra. 1996. to be published

DECAY MODES OF MULTIPLY EXCITED IONS THE CASE OF TRIPLY EXCITED STATES

H. BACHAU

Laboratoire des Collisions Atomiques, C.P.T.M.B. (URA 1537 du CNRS)
351, Cours de la Libération, 33405 France

We report on calculations of triply excited states of Li and Ne^{7+}. After a brief presentation of the method, we will discuss the post collisional evolution of these states in the context of triple electron capture in collision of multicharged ions on atom at few KeV. The possibility of radiative and autoionizing decays is discussed in view of recent experimental observations.

INTRODUCTION

In the course of our ongoing work on the assessment of radiative stabilisation and autoionization processes occurring after ion-atom collisions, we have been led to examine the de-excitation cascades for multiply excited states of ions. Depending on the relative position of the I^{q+} multiply excited states and $I^{(q+i)+}$ thresholds, the cascade process may involve autoionization (ejection of one or more electrons in the continua) and/or radiative stabilization. The simplest case is the collision Ne^{10+}+He which has received a considerable attention from both experimental and theoretical sides (1,2,3,4,5). In the collision energy range 10-150 KeV, Ne^{8+}(3,n), (4,4) and (4,5) states are populated through double electron capture. We focus here on Ne^{8+}(4,4) states which lie in the region of the Ne^{9+}(N=3) threshold. The post-collisional interaction between (4,4) and (3,n) Rydberg states results on the possibility for (4,4) states to transfer to (3,n) series which may decay radiatively or autoionize (5). Very recently, evidences of radiative decay have been reported from two different coincidence experiments where triply excited states of C^{3+} and Ne^{7+} have been observed (6). This raise the question of the properties of triply excited ions which has not yet received as much attention as the doubly excited state case. We briefly present two approaches (details and references can be found in (7)) to investigate triply excited states and we discuss the cases of Li(2,2,2) and Ne^{7+}(4,n,n') states.

THEORETICAL APPROACHES

The Space Partition Approach

In principle, the Feshbach formalism and the definition of the projection operators can be extended to three (or more) electron systems. The rigorous construction of the projection operators becomes cumbersome when more than two electrons are considered. Nevertheless, it is possible to build projection operators such that the wave-functions are represented in the zeroth order of the perturbation theory (where only the two first terms of the energy Z-expansion are correctly evaluated). For example, the resonant part of the (3,3,3) autoionizing states is build on the basis of antisymmetrized configurations $\{n_1l_1,n_2l_2,n_3l_3\}$ with $n_1,n_2,n_3>2$. Practically, we build a projection operator $Q=q_1q_2q_3$ where q_i excludes the electron i from 1s, 2s and 2p orbitals. As a matter of fact, this method is well adapted for the case of multicharged ions where intrashell correlations dominate. The difficulty is that there exist in general an infinite number of open channels where two-electrons are ejected. Nevertheless, we take advantage that the (3,3,3) states mainly autoionize to the (2,3,k) continua, this first step being followed by the ejection of a second electron or a radiative decay to (1s,nl,k) state. The above approximation considerably simplifies the problem of the determination of the lifetime of resonances. Although calculations are much more involved than in the case of two electron systems, we have calculated positions and widths of N^{4+}(3,3,3) autoionizing states (7) and we found a good agreement with other calculations.

The Stabilization Method

We focus here on an approach of the stabilization method which has been recently proposed by Mandelshtam, Ravuri and Taylor (8) and successfully used in various contexts. This method consists in diagonalizing the total Hamiltonian for various values of the box size R_{max}. A diagram is obtained by plotting the eigenvalues as a function of the box size. The density of states is evaluated from the corresponding histogram in the energy region of interest. Peaks in the density diagram are associated to resonances whose parameters (positions and total widths) can be extracted through a fitting procedure. The advantage of this approach is that no restriction in the configuration interaction is used, except, of course, the unavoidable one due to the limited size of the basis. In order to avoid to recalculate two-electron matrix elements for each box size, we introduce a potential barrier in the Hamiltonian:

$$H = \mathrm{H}+\Omega(r_1)+\Omega(r_2)+\Omega(r_3)$$

where, for each electron i, the barrier is defined by $\Omega(r_i)=0$ for $0\le r_i\le L$, $L_{\min}\le L\le R_{\max}$ and

CP392, *Application of Accelerators in Research and Industry*, edited by J. L. Duggan and I. L. Morgan
AIP Press, New York © 1997.

$\Omega(r_i) = W$ for $L \le r_i \le R_{max}$. L being varied instead of R_{max}, one needs *one* evaluation of the two-electron matrix elements at R_{max}. Only one-electron matrix element needs to be recalculated for each L value and a considerable computation time is saved.

Numerical Approach

Antisymmetrized three-electron configurations are defined on the basis of hydrogenic one-electron orbitals. The radially dependent part of these orbitals are expanded on B-splines functions defined from $r_i = 0$ to $r_i = R_{max}$. Therefore one obtains a discretized representation of the atomic spectrum. The lower eigenstates represent bound states while the ones lying above the threshold represent the continuum.

RESULTS AND DISCUSSION

We have shown (7) that the two theoretical approaches give similar results for N^{4+}(3,3,3). The advantage of the space-partition approach is that it requires one diagonalization in each channel. However, this method is restricted to heavy ions while the stabilization approach can be applied to lighter systems (provided that the basis set is large enough). Fig. 1 shows a density diagram for the $2s^22p$ $^2P^o$ state of lithium. We also show on the graph the position and autoionizing width values obtained from a fit, they are close to recent experimental results (9).

We examine now the case of triplet excited states of multicharged ions. In a recent contribution, J. Bernard et al (6) have reported experimental evidence of fully stabilized (true) triple electron capture. A possible mechanism was suggested where C^{3+}(2,n,n') and Ne^{7+}(3,n,n') are populated and (partly) decay radiatively. The ions have been produced during triple electron capture collision of C^{6+} and Ne^{10+} on Ar at few Kev.

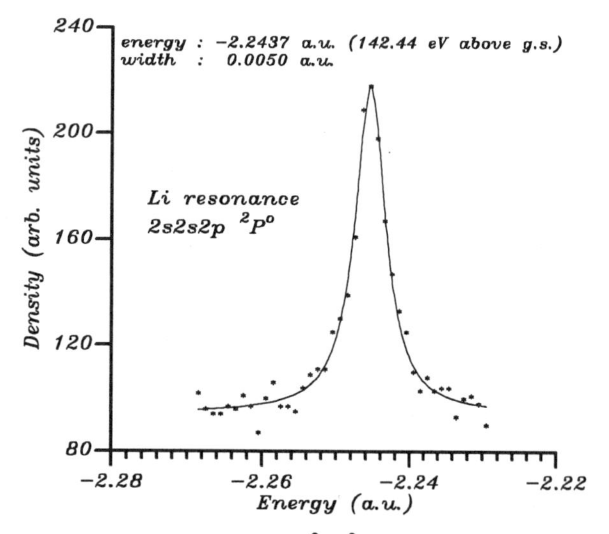

FIGURE 1. Density of the lowest $2s^22p$ $^2P^o$ triply excited state of lithium. The full line represents a lorentzian fit whose parameters are shown on the figure. The Li ground state (g.s.) energy is taken at -7.478678 a.u..

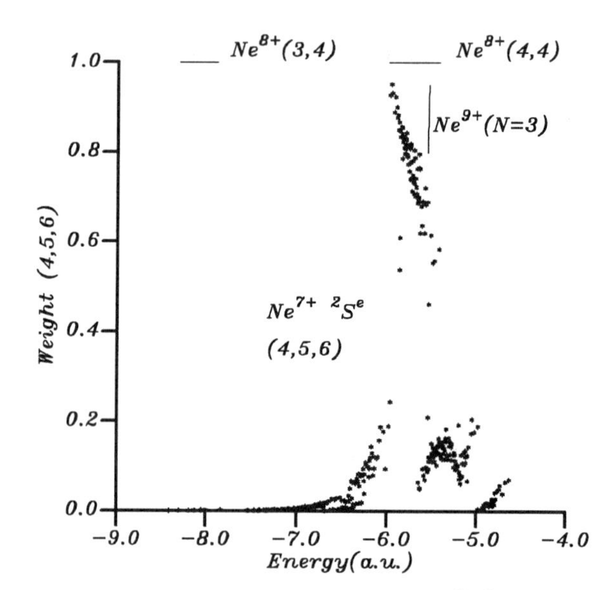

FIGURE 2. Weight of the (4,5,6) components of Ne^{7+} $^2S^e$. We show the energy range for Ne^{8+}(3,4) and (4,4) series (horizontal lines) and the vertical line represents the position of the Ne^{9+}(N=3) Threshold.

In the case of Ne^{7+}, the (4,5,6) states are initially populated (6) and the (3,n,n') states are populated in the way out of the collision through transfer to Rydberg states (ATR). Using the space-partition approach, we have calculated the position and component of (4,5,6) series. There is a strong configuration interaction with other series (e.g. with (4,5,5)) and we show in Figure 2 the total weight of the (4,5,6) component, it is worth noting that the (4,5,6) states lie in the region of the Ne^{8+}(4,4) states and of the Ne^{9+}(N=3) threshold. Therefore, besides the autoionization channels, there exists a possibility for the (4,5,6) states to transfer to (4,4,n) series, which can either autoionize or undergo a second transfer to (3,n,n') series. This support the analysis proposed in (6). There is now evidences that the cascade process occurring after multiple electron capture involves autoionization and radiative stabilization. The quantitative evaluation of the relative importance of these processes is the goal of future investigations.

REFERENCES

1. Roncin, P., Gaboriaud, M.N., Szilagyi, Z., and Barat, M., *AIP Conference Proceedings 295 (ICPEAC XVIII)*, 1993, pp. 537-546
2. Bordenave-Montesquieu, A., Moretto-Capelle, P., Gonzalez, A., Benhenni, M., Bachau, H. and Sanchez, I, *J. Phys. B* 27, 4243-61 (1994)
3. Fremont, F., Merabet, H., Chesnel, J.Y., Husson, X., Lepoutre, A., Lecler, D., Rieger, G. and Stolterfhot, N., Phys. Rev. A 50, 3117-23 (1994)
4. Martin, S., Bernard, J., Chen, Li, Denis, A. and Desesquelles, J., *Phys. Rev. A* 52, 1218-23 (1995)
5. Sanchez, I. and Bachau, H., *J. Phys. B* 28, 795-806 (1995)
6. Bernard, J., Chen, Li, Denis, A., Desesquelles, J., Martin, S., Roncin, P. and Barat, M., "Double Rydberg States in Stabilised Triple Capture in Bare Ions (Z=6-10) on Rare Gas Target Collisions", Presented at the HCI96 Conference on, Omiya, Saitama, Japan, September 23-26 1996
7. Bachau, H., *J. Phys. B.* 29, 4365-79 (1996)
8. Mandelshtam, V.A., Ravuri, T.R. and Taylor H.S., *Phys. Rev. Lett.* 70, 1932-35 (1993)
9. Kiernan, L.M., Lee, M-K., Sonntag, B.F., Sladeczek, P., Zimmermann, P., Kennedy, E.T., Mosnier J-P. and Costello, J.T., *J. Phys. B* 28, L161-68 (1995)

CHARACTERISATION AND APPLICATIONS OF A NEW REACTIVE ATOM BEAM SOURCE

R. W. McCullough

Department of Pure and Applied Physics, The Queen's University of Belfast, Belfast, United Kingdom

A compact high efficiency 2.45 GHz microwave plasma source for the production of reactive atom beams has been developed for accelerator target studies and the treatment, modification and growth of new materials. The source employs twin slotted line radiators for efficient microwave coupling into the plasma tube. An axial magnetic field provided by NdFeB permanent magnets is used for plasma confinement and to enable electron cyclotron heating of the plasma. The source has been used to produce highly dissociated beams of hydrogen, oxygen, chlorine and nitrogen. Fluxes of up to $4x10^{18}$, $7x10^{15}$, and $4x10^{15}$ atoms s^{-1} have been measured at the exit of the source for H, O and Cl respectively. In the case of nitrogen the source has achieved a dissociation efficiency of up to 67% compared with the few per cent reported previously for other sources. The source is currently being used for thin film diamond etching and growth experiments and to provide atomic targets for ion beam collision studies.

INTRODUCTION

Studies of collision processes involving unstable reactive atomic targets are frequently carried out using a crossed beam configuration where a thermal energy reactive atom beam is crossed by an ion beam from an accelerator. Such applications require well characterised, stable, high intensity atom beams derived from pure gases. In this paper the progress made in our laboratory (1)-(4) to develop simple microwave discharge sources for these studies is reviewed.

DESCRIPTION OF THE SOURCES

Two different sources have been built and characterised. In each case the microwave power is coupled to the plasma using slotted line radiators. One source has an axial magnetic field applied.

Source without magnetic field

A schematic diagram of the source is shown in Fig. 1. A cylindrical pyrex or quartz discharge tube T, 150 mm long, outside diameter 26 mm and wall thickness 1.7 mm, is clamped between two slotted line radiators R1 and R2 of the type described by Lisitano et al (5). Gas is introduced through a 10 mm diameter tube at one end while the atom beam emerges from the other end through an exit canal C of diameter ranging from 1 to 1.7 mm and length ranging from 2 to 20 mm. R1 and R2 were machined from 1.5 mm thick half cylinders of copper and mounted between two water cooled flanges F1 and F2. Four long slots of length 54 mm and width 2 mm with short interconnecting slots are cut into each half cylinder and fed with 2.45 GHz microwave power by N type coaxial connectors C1 and C2. The input power was normally in the range 30-175 W and fed from a single power supply via two coaxial cables.

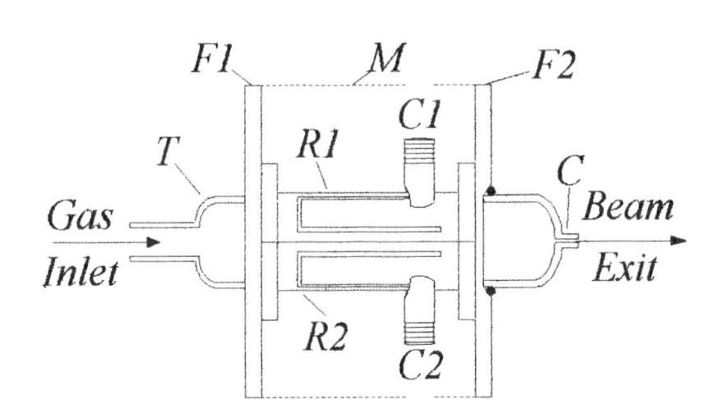

FIGURE 1. Schematic diagram of the source without a magnetic field

The reflected power was no more than 15% with the discharge running. Wall temperatures of the discharge tube T did not exceed 70°C. A cylindrical radiation screen M was placed at a distance $\lambda/4$ from the radiators R1 and R2. The discharge tubes were cleaned with concentrated orthophosphoric acid and rinsed in distilled water prior to assembly of the source.

Source with magnetic field

A schematic diagram of this source is shown in Fig. 2. The discharge tube, radiators and microwave supplies are

CP392, *Application of Accelerators in Research and Industry,* edited by J. L. Duggan and I. L. Morgan
AIP Press, New York © 1997.

as described above. An axial magnetic field is provided by an assembly of NdFeB ring magnets separated by aluminium alloy rings. An inner copper shield protects the magnets from microwave heating effects.

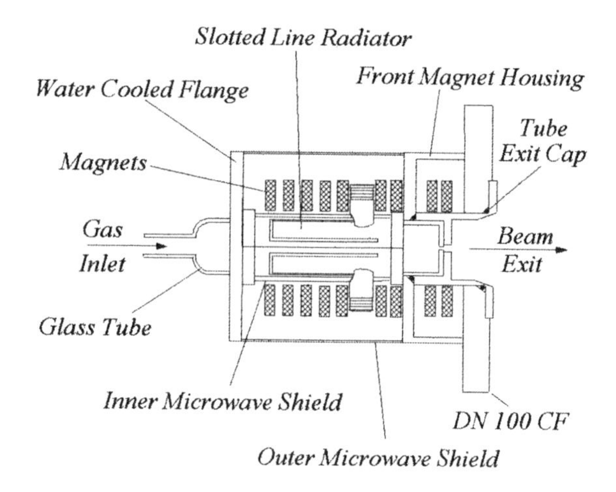

FIGURE 2. Schematic diagram of the source with a magnetic field.

SOURCE CHARACTERISATION

The procedures for measurement of the dissociation fractions, atom fluxes and beam densities have been described in detail previously (6) and are briefly summarised here. The beam emerging from the source passed between electrostatic sweep plates placed to remove a small low energy (<60 eV) charged particle component. It was then mechanically chopped at 180 Hz and a small fraction passed into a second differentially pumped chamber where it was sampled with a quadrupole mass spectrometer whose output was directed to a phase sensitive lock-in amplifier system. The dissociation fraction was obtained from the relation

$$D = (S_1 - S_2) / S_1$$

Where S_1 is the N_2^+ molecular ion signal with the discharge off and S_2 the corresponding signal with the discharge on. The atom flux was obtained from measurements of the rate of change of pressure in a resevoir of known volume feeding the source.

Performance of the source without magnetic field

In Figs. 3a, 3b and 3c we show the observed dependence of the dissociation fraction D and the atom flux (in sccm) on discharge tube gas pressure for hydrogen, oxygen and chlorine respectively operating with microwave input power of 150 W and exit canal 20 mm long and 1 mm diameter. The beam flux was found to depend approximately on the cubed power of the canal diameter and inversely on its length. Measured dissociation fractions increased only slightly at higher input power and decreased by 25% at 30 W input power. Similar values of D were obtained when the beam exit canal was varied in length from 2 to 20 mm and in internal diameter from 1 to 1.7 mm.

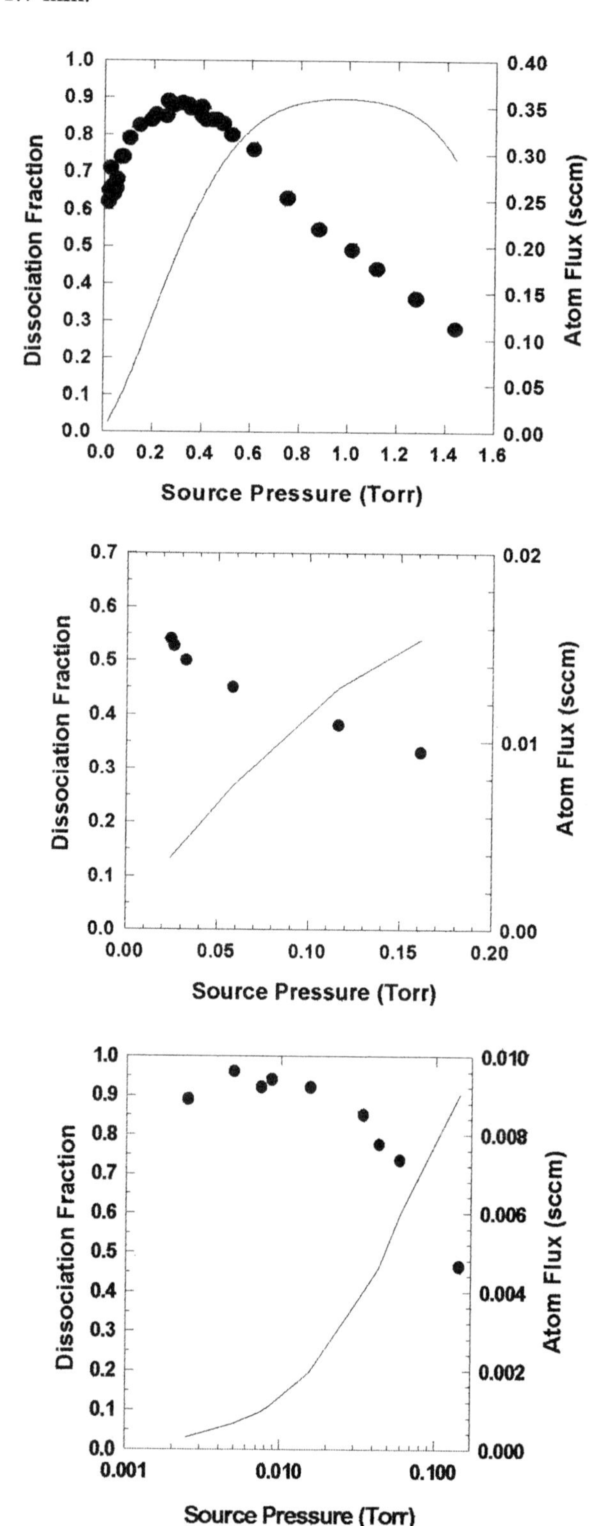

FIGURE 3. Dissociation fraction and atom flux versus source gas pressure for (a) hydrogen (b) oxygen and (c) chlorine. • Dissociation fraction D, — Estimated atom beam flux (To convert from sccm to atoms cm^{-2} s^{-1} multiply by $4.5x10^{17}$.)

This source configuration without an applied magnetic field gave values of D for nitrogen of less than 0.04.

Performance of the source with magnetic field

In a preliminary investigation an axial magnetic field was generated by placing a solenoid coil over the complete source structure. The variation of the dissociation fraction D with magnetic field is shown in Fig. 4 for nitrogen gas at a pressure of 3×10^{-3} Torr and 100W of microwave power. Similar behaviour was observed with oxygen. The origin of the large and rapid variations in D are associated with electron cyclotron resonance (ECR) heating of the plasma and the presence of plasma oscillations (3).

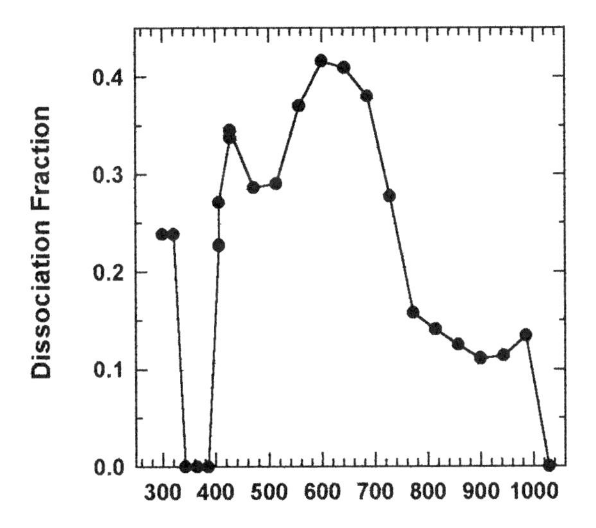

FIGURE 4. The dependence of dissociation fraction with applied magnetic field for nitrogen.

Coupling of the microwave power to the plasma occurs at the electron cyclotron resonance (ECR) condition where the angluar frequency ω of the applied field satisfies the relation $\omega = eB_c/m$ where B_c is the applied magnetic field and e and m are the charge and mass of the electron respectively. For 2.45 GHz radiation B_c = 875 G. In addition to the fundamental angular frequency harmonics are also generated (7) so that, in our case, microwave absorption also occurs at magnetic field values for which 2.45 GHz corresponds to a harmonic of the fundamental frequency. In fig. 4 the dissociation fraction D peaks at magnetic field values of 438 G and 292 G corresponding to cyclotron resonance at the first and second harmonics respectively.

The compact permanent magnet structure, used in the present source, has been designed to give a magnetic field configuration that maximises the value of D. Figure. 5 shows the observed dependence of D and atom beam flux on gas pressure for nitrogen with a microwave input power of 150 W and an exit canal 1.7 mm diameter and 4 mm long. When fed with oxygen the dissociation fractions were comparable in value with those shown in Fig. 5 for nitrogen. In the case of hydrogen the source performance was similar with and without a magnetic field.

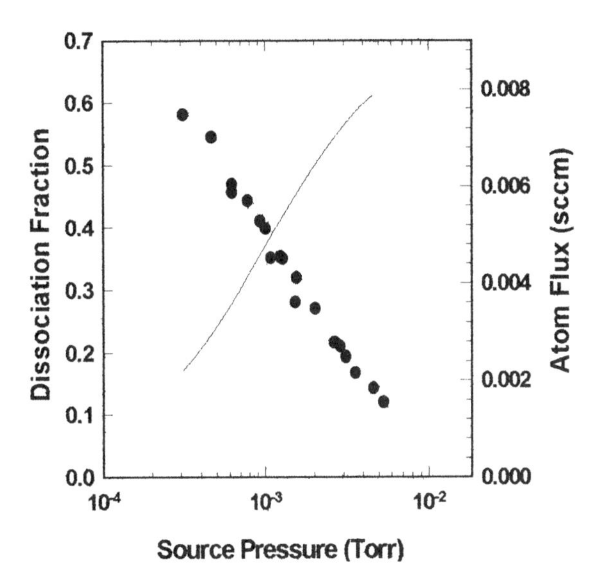

FIGURE 5. The variation of dissociation fraction and atom flux on source pressure for nitrogen, • Dissociation fraction D, — Estimated atom beam flux

APPLICATIONS

The source has been successfully used in crossed beam ion atom collision experiments to produce highly dissociated targets of atomic hydrogen. Cross sections for the formation of H(2s) in collisions of 10 - 100 keV H^+ with H have been measured recently in our laboratories (8) using the source without a magnetic field in a crossed beam configuration. In another application the feasability of using the source for studies of state selective electron capture by multiply charged ions in atomic targets is being investigated using the technique of translational energy spectroscopy (9). Figure 6 shows a schematic diagram of the experimental arrangement of the accelerator target region. In this arrangement the distance between the end of the source exit and the ion beam can be adjusted. The source is electrically isolated from the chamber in which it is mounted. High transparency meshed grids are mounted on either side of the crossed beam interaction region and, together with the main source body, are maintained at a different voltage with respect to the chamber. This voltage labelled collision region arrangement enables ions which have undergone electron capture collisions in the labelled region, where the target is predominantly atomic, to be distinguished from those outside this region where the gas is predominantly molecular.

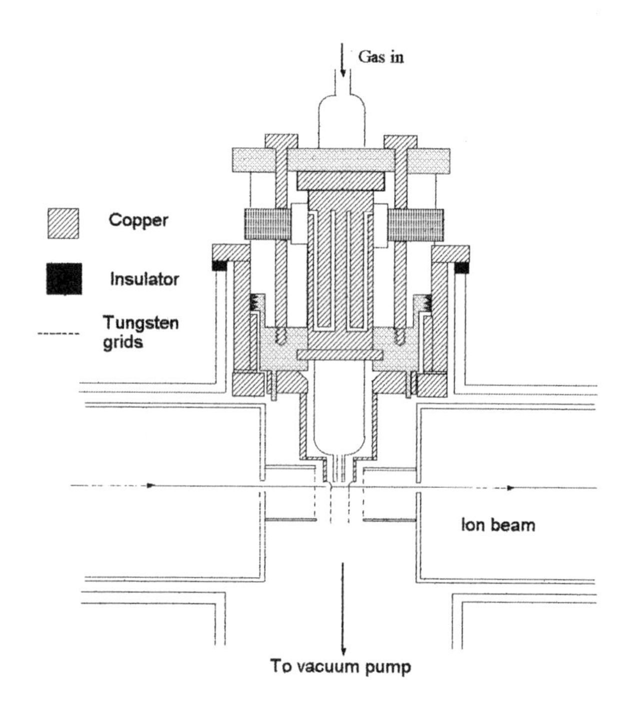

FIGURE 6. Schematic diagram of atom source without a magnetic field shown in crossed beam configuration.

CONCLUSIONS

We have shown that highly dissociated, high intensity thermal energy beams of hydrogen, oxygen and chlorine atoms can be generated from a plasma that is sustained in a simple glass discharge tube by microwave radiation from special slotted line radiators. Highly dissociated beams of nitrogen were produced when an axial magnetic field was applied.

The feasability of using the source for accelerator target studies has been demonstrated for atomic hydrogen and work is in progress to extend these studies to an atomic nitrogen target.

ACKNOWLEDGEMENTS

The work described in this paper was carried out in collaboration with C M Donnelly, J Geddes and H B Gilbody, The Queen's University of Belfast, UK: J M Woolsey, University of Stirling, UK; M Schlapp, Argonne National Laboratory, USA and E Salzborn, University of Giessen, Germany. The work was supported by the UK Engineering and Physical Sciences Research Council and The British Council Academic Research Collaboration (ARC) Programme.

REFERENCES

1. McCullough, R.W., Geddes, J., Donnelly, A., Liehr, M., Hughes, M.P. and Gilbody, H. B., *Meas. Sci. Technol.* **4**, 79-82. (1993)
2. McCullough, R.W., Geddes, J., Donnelly, A., Liehr, M. and Gilbody, H.B., *Nucl. Instrum. Methods Phys. Res.*, **B79**, 708-710. (1993)
3. Geddes, J., McCullough, R.W., Higgins, D.P., Woolsey, J.M. and Gilbody, H.B., *Plasma Sources Sci. Technol.* **3**, 58-60. (1994)
4. Higgins, D. P., McCullough, R. W., Geddes, J., Schlapp, M., Woolsey, J., Salzborn, E. and Gilbody H. B., *Nucl. Instrum. and Methods Phys. Res.* **B103**, 508-510 (1995)
5. Lisitano, G., Ellis, R.A., Hooke, W.M. and Stix, T.H., *Rev. Sci. Inst.* **39**, 295-297 (1968)
6. Donnelly, A., Hughes, M.P., Geddes, J. and Gilbody, H.B., *Meas. Sci. Technol.* **3**, 528-532. (1992)
7. Stix, T. H., *Theory of Plasma waves*, New York: McGraw - Hill, 1962.
8. Higgins, D. P., Geddes, J. and Gilbody, H. B., *J Phys B: At. Mol. Opt. Phys.* **29**, 1219-1224 (1996)
9. McCullough, R. W., McLaughlin T. K., Hodgkinson J. M and Gilbody H. B., *Nucl. Instrum. Methods Phys. Res.* **B98**, 199-203 (1995)

SECTION II

NUCLEAR PHYSICS AND RADIOACTIVE ION BEAM FACILITIES AND EXPERIMENTS

PROTON CAPTURE RESONANCE STUDIES

G. E. Mitchell,[1] E. G. Bilpuch,[2] C. R. Bybee,[1] J. M. Cox,[3] L. M. Fittje,[3] M. A. Labonte,[1] E. F. Moore,[1] J. D. Shriner,[1] J. F. Shriner, Jr.,[3] G. A. Vavrina,[1] and P. M. Wallace[2]

[1]*North Carolina State University, Raleigh, North Carolina 27695*
and Triangle Universities Nuclear Laboratory, Durham, North Carolina 27708
[2]*Duke University, Durham, North Carolina 27708*
and Triangle Universities Nuclear Laboratory, Durham, North Carolina 27708
[3]*Tennessee Technological University, Cookeville, Tennessee 38505*
and Triangle Universities Nuclear Laboratory, Durham, North Carolina 27708

The fluctuation properties of quantum systems now are used as a signature of quantum chaos. The analyses require data of extremely high quality. The ^{29}Si(p,γ) reaction is being used to establish a complete level scheme of ^{30}P to study chaos and isospin breaking in this nuclide. Determination of the angular momentum J, the parity π, and the isospin T from resonance capture data is considered. Special emphasis is placed on the capture angular distributions and on a geometric description of these angular distributions.

INTRODUCTION AND MOTIVATION

Bohigas *et al.* (1) published a seminal paper in 1984, conjecturing that the quantum analogs of classically chaotic time-reversal-invariant systems show level fluctuations consistent with the Gaussian Orthogonal Ensemble (GOE) of random matrix theory. It is now generally accepted that analogs of classically chaotic systems show GOE statistics and that analogs of classically integrable systems exhibit Poisson statistics.

Tests of energy eigenvalue distributions for comparison with the GOE require data of extremely high quality: data must be complete (few or no missing levels) and pure (few or no levels with misassigned quantum numbers), or the energy eigenvalue distributions can be severely biased (2). Analyses of high quality proton and neutron resonance data (3,4) show agreement with GOE and suggest chaotic behavior above the separation energies. At very low energies, an analysis that combines data from different nuclides shows behavior which is GOE for light masses and moves toward the Poisson distribution for heavier masses (5). Only one nuclide – ^{26}Al – has data of sufficient quality and quantity that extend from the ground state well into the resonance region (6,7). An analysis of the fluctuation properties of ^{26}Al showed behavior between GOE and Poisson (8). These data also offered the first experimental test of how a broken symmetry (in this case, isospin) affects eigenvalue fluctuations. The data were consistent with predictions by Dyson (9) and Pandey (10) that even a small amount of symmetry breaking would have the same effect on the fluctuations as the complete absence of the symmetry. These results prompted several new theoretical studies specifically examining isospin breaking (11-14).

However, limited statistics precluded a definitive conclusion. In order to obtain additional data suitable for studying the effects of isospin breaking on statistical properties, we are establishing a complete level scheme for ^{30}P up to $E_x \approx 8$ MeV. The key reaction in determining the complete spectroscopy for ^{30}P is the ^{29}Si(p,γ) reaction. In this paper we revisit the utilization of the resonance capture reaction for determination of the relevant quantum numbers. The first step was to measure excitation functions. Since the capture reaction is so sensitive that even very weak states can be observed, this measurement is crucial to ensure completeness above the separation energy. Excitation functions in the range E_p = 1.0 – 3.3 MeV (15,16) identified 17 previously unknown resonances in ^{30}P. For each of the 47 resonances in the energy range E_p = 1.0 – 2.5 MeV, detailed γ-ray spectra have been measured; these measurements are described in Sect. 2. For most resonances, angular distributions are necessary to assign all of the quantum numbers; a discussion of the analysis of angular distributions of primary γ-rays for this spin $I = 1/2$ target is given in Sect. 3. The use of angular distributions of secondary γ-rays is illustrated in Sect. 4, and a summary is presented in the final section.

CP392, *Application of Accelerators in Research and Industry*, edited by J. L. Duggan and I. L. Morgan
AIP Press, New York © 1997.

Table 1. Branching ratios for the $E_x = 7921.8$ keV state and previous results.

E_f (keV)	J^π;T	Branching Ratio (%)	Previous Values (%)[a]
709.02	1^+;0	0.3 ± 0.1	-
1454.67	2^+;0	0.5 ± 0.1	-
1973.62	3^+;0	10.8 ± 0.9	11
2539.03	3^+;0	16 ± 1	20
2839.9	3^+;0	51 ± 3	51
3928.9	3^+;0	1.3 ± 0.2	-
4182.65	2^+;1	0.7 ± 0.1	-
4298.1	4^+;0	3.5 ± 0.3	3
4343.6	5^+;0	15 ± 1	15
4736.4	3^+;0	1.1 ± 0.2	-

[a]Ref. (19)

FIXED-DETECTOR MEASUREMENTS

Once the resonances were identified, the next step was to measure γ-ray branching ratios. The KN accelerator and high resolution system of the TUNL High Resolution Laboratory provided beam energy resolution of about 220 eV FWHM, allowing resolution of close-lying resonances. The γ-rays were detected with a pair of 60% efficient HPGe detectors, one of which employed a large BGO Compton-suppressor (17). The suppressed detector was located at 55° and the unsuppressed detector at 90°. Previous branching ratio measurements for ^{29}Si(p,γ) resonances had been performed by Reinecke *et al.* (18) in the energy range $E_p = 0.3 - 2.3$ MeV and by Cameron (19) in the energy range $E_p = 2.3 - 3.3$ MeV. The branching ratios were remeasured because greater sensitivity was expected due to improved beam energy resolution and detection efficiency. As an example, the branching ratios for the $E_x = 7921.8$ keV state ($E_p = 2.4077$ MeV) are listed in Table 1 and compared with previous values. In general our results are consistent with previous measurements but show much greater sensitivity to weak branches.

Once branching ratios were determined, all observed transitions were assumed to be dipole (E1, M1), quadrupole (E2, M2), or electric octupole (E3). Since $J^\pi;T$ values are already known for most states below $E_x \approx 5$ MeV, this assumption limited the $J^\pi;T$ values for the resonance. For the $E_x = 7921.8$ keV state, this step led to possible assignments of $J^\pi = 2^-, 3^+, 3^-$, or 4^-. Since the γ-ray strength $S = (2J+1)\Gamma_\gamma/\Gamma_{\text{total}}$ is known for this resonance (15), a lower limit can be obtained for Γ_γ. This limit can be converted to lower limits on the partial γ-ray widths to each final state. These partial widths can be compared with Endt's recommended upper limits (RUL's) for strengths in this mass region (20). For all assignments except $J^\pi;T = 3^+;0$ at least one of the primary transitions significantly exceeds its RUL. Thus the $E_x = 7921.8$ keV state is assigned $3^+;0$. However, a unique $J^\pi;T$ assignment from fixed detector measurements is the exception, not the rule.

^{29}Si(p,γ) ANGULAR DISTRIBUTIONS

If the capture spectrum does not provide a unique assignment of the resonance quantum numbers, angular distributions are the next step. For isolated resonances that conserve parity, the angular distribution of a primary γ-ray can be written

$$W(\theta) = A_0[1 + a_2P_2(\theta) + a_4P_4(\theta) + \ldots], \qquad (1)$$

where A_0 is an overall normalization factor and P_k is a Legendre polynomial. Because ^{29}Si has a nonzero spin, the coefficients a_k for any primary γ-ray depend not only on the γ-ray mixing ratio δ_γ, but also on the proton mixing ratio δ_p. In this particular case, the target spin/parity are $1/2^+$ and natural parity compound states can have channel spin mixing in the entrance channel while unnatural parity compound states can have orbital angular momentum mixing in the entrance channel. The practical effect of having both entrance channel and exit channel mixing is to complicate enormously the fitting procedure. Since all primary γ-rays share a common entrance channel, the angular distributions for all primary transitions should be fit simultaneously. If N distributions are measured for a given resonance, that implies N overall normalization constants A_0, a single proton mixing ratio δ_p, and up to N γ-ray mixing ratios δ_γ. Since the fit is nonlinear (the mixing ratios occur quadratically), even a relatively small value of N can lead to a fitting space of high dimension.

Although a detailed fit is necessary to determine δ_γ, our primary goal is the assignment of $J^\pi;T$ to each level. Often the limits on J^π can be improved by comparing the experimental coefficients a_k to their allowed region in the coefficient space. Possible J^π values for which the experimental coefficients are significantly outside their allowed range of values can be eliminated without the need for a detailed fit. As a simple example, consider a compound state of 1^- and a final state of 2^+. The entrance channel has orbital angular momentum $\ell = 1$ and channel spins $s = 0$ or 1. The exit channel is assumed to be pure E1. The proton mixing ratio is the ratio of the reduced width amplitudes $\gamma_{s\ell}$ (for $s = 0$ and 1) in the entrance channel. Only the a_2 coefficient is non-zero, and it can be expressed simply in terms of δ_p. The possible range of a_2 is $-0.10 \leq a_2 \leq 0.05$. Any transition which does not have a_2 within or near this range and higher a_k's near zero cannot represent a $1^- \rightarrow 2^+$ transition.

In general, more coefficients may be non-zero; the most common case in this reaction has both a_2 and a_4 with possible non-zero values. Consider a compound state with $J^\pi = 2^+$ and a final state with $J^\pi = 1^+$.

Table 2. Expected a_2 and a_4 values and the experimental coefficients for the secondary 2724-keV $\rightarrow$ 0 transition following the primary transition from the 7759-keV state.

	a_2	a_4
$J^\pi = 3^+$	0.59 ± 0.03	0.10 ± 0.01
$J^\pi = 2^+$	-0.09 ± 0.14	-0.03 ± 0.02
Exp.	0.50 ± 0.02	0.16 ± 0.04

The entrance channel has $\ell = 2$ and $s = 0$ or 1; the exit channel can be either M1 or E2. The expressions for a_2 and a_4 are very complicated. (Both a_2 and a_4 are quadratic in both of the mixing ratios.) The allowed values define a region in a_2-a_4 space. In the simplest cases with both a_2 and a_4 allowed, the allowed values lie along a line segment or along an ellipse. In more complicated cases, such as this one, the allowed values lie inside a two-dimensional region of the space. It is interesting to note that often there is another excluded region inside the main allowed region. Allowed regions for four different combinations of initial and final states, including this example, are shown in Fig. 1. In this case, for a fixed value of δ_p the allowed values of a_2 and a_4 lie on an ellipse. A different value of δ_p produces a different ellipse. The allowed region shown comprises the union of these ellipses as δ_p is allowed to vary over all possible real values.

As an application of this approach, consider the angular distributions measured at the $E_x = 7759$ keV state ($E_p = 2.2381$ MeV). Although Reinecke *et al.* (18) had previously assigned this state $J = 2$, our fixed detector measurements allowed $J^\pi = 1^+, 2^+, 2^-$, or 3^+. Angular distributions were measured for seven primary transitions, four of which decayed to final states with $J^\pi = 3^+$. In Fig. 2, the experimental values of a_2 and a_4 are plotted for each of these four transitions and the allowed regions for each of the four possible J^π assignments to the resonance state. For both 1^+ and 2^-, at least one transition has experimental coefficients well outside the allowed region. A complete least-squares analysis is consistent with this result, excluding the 1^+ and 2^- possibilities for the $E_x = 7759$ keV state. The least squares analysis also shows that both 2^+ and 3^+ are allowed. The γ-ray mixing ratios extracted for each case can be converted to partial γ-ray widths and compared to Endt's RUL's. This comparison indicates that the isospin is $T = 0$ if $J^\pi = 2^+$ and is $T = 1$ if $J^\pi = 3^+$.

ANGULAR DISTRIBUTIONS OF SECONDARY GAMMA-RAYS

For those cases (such as the $E_x = 7759$ keV state discussed in Sect. 3), where the fixed-detector measurements and the primary γ-ray angular distributions do not yield a unique $J^\pi;T$ assignment, additional information is required. One possible source of such information is the angular distribution of the secondary γ-rays. As an illustration, consider the transition from the 2724-keV state to the ground state following the primary transition from the 7759-keV state to the 2724-keV state. The 2724-keV state has $J^\pi = 2^+$, and the ground state has $J^\pi = 1^+$; thus the 7759 $\rightarrow$ 2724 $\rightarrow$ 0 cascade corresponds to either $3^+ \rightarrow 2^+ \rightarrow 1^+$ or $2^+ \rightarrow 2^+ \rightarrow 1^+$. For each of the two possible assignments to the resonance state, the primary γ-ray angular distributions provided the proton mixing ratio and γ-ray mixing ratio for the primary γ-ray. The γ-ray mixing ratio for this secondary transition has already been determined as $\delta_\gamma = -3.0 \pm 0.4$ (21). The expected a_2 and a_4 values for each case are listed in Table 2 along with the experimental values. The experimental results are consistent only with $J^\pi = 3^+$. Since the only isospin assignment consistent with $J^\pi = 3^+$ was $T = 1$, the value of $J^\pi;T$ is uniquely determined for this state.

SUMMARY

Measurements are in progress to establish a complete level scheme for ^{30}P by studying the ^{29}Si(p, γ) reaction. Fixed detector measurements have been performed for each of 47 resonances in the energy range $E_p = 1.0$ – 2.5 MeV. Branching ratios have been determined, and the partial γ-ray widths compared to Endt's RUL's in order to limit $J^\pi;T$ possibilities. Angular distribution measurements are being performed for states where a unique assignment has not yet been obtained. A geometric method of analysis that utilizes the allowed values of coefficients to limit possible J^π assignments has proven quite useful. The angular distributions of secondary transitions are also valuable in assigning quantum numbers.

ACKNOWLEDGMENTS

We wish to thank C. A. Grossmann and L. K. McLean for their assistance in data collection. This work was supported in part by the U. S. Department of Energy, Office of High Energy and Nuclear Physics, under grants No. DE-FG05-88ER40441, DE-FG05-91ER40619, and DE-FG02-96ER40990.

REFERENCES

1. Bohigas, O., Giannoni, M. J., and Schmit, C., *Phys. Rev. Lett.* **52**, 1 (1984).
2. Shriner, J. F. Jr., and Mitchell, G. E., *Z. Phys. A* **342**, 53 (1992).
3. Haq, R. U., Pandey, A., and Bohigas, O., *Phys. Rev. Lett.* **48**, 1086 (1982).
4. Bohigas, O., Haq, R. U., and Pandey, A., *Phys. Rev. Lett.* **54**, 1645 (1985).

5. Shriner, J. F. Jr., Mitchell, G. E., and von Egidy, T., *Z. Phys. A* **338**, 309 (1991).
6. Endt, P. M., de Wit, P., and Alderliesten, C., *Nucl. Phys. A* **459**, 61 (1986).
7. Endt, P. M., de Wit, P., and Alderliesten, C., *Nucl. Phys. A* **476**, 333 (1988).
8. Shriner, J. F. Jr., *et al.*, *Z. Phys. A* **335**, 393 (1990).
9. Dyson, F. J., *J. Math. Phys.* **3**, 1191 (1962).
10. Pandey, A., *Ann. Phys.* **134**, 110 (1981).
11. Guhr, T., and Weidenmüller, H. A., *Ann. Phys.* **199**, 412 (1990).
12. Paar, V., *et al.*, *Phys. Lett. B* **271**, 1 (1991).
13. Ormand, W. E., and Broglia, R. A., *Phys. Rev. C* **46**, 1710 (1992).
14. Hussein, M. S., and Pato, M. P., *Phys. Rev. C* **47**, 2401 (1993).
15. Frankle, S. C., *et al.*, *Phys. Rev. C* **45**, 2746 (1991).
16. Vavrina, G. A., Ph.D. Thesis, North Carolina State University, 1996.
17. Shriner, J. F., Jr., *et al.*, *Nucl. Instrum. Methods B* **99**, 641 (1995).
18. Reinecke, J. P. L., *et al.*, *Nucl. Phys. A* **435**, 333 (1985).
19. Cameron, J. A., *Phys. Rev. C* **47**, 1498 (1993).
20. Endt, P. M., *At. Data Nucl. Data Tables* **55**, 171 (1993).
21. Endt, P. M., *Nucl. Phys. A* **521**, 1 (1990).

Figure 1. Allowed regions of the angular distribution coefficients a_2 and a_4 for four combinations of initial and final state J^π. See text for additional details.

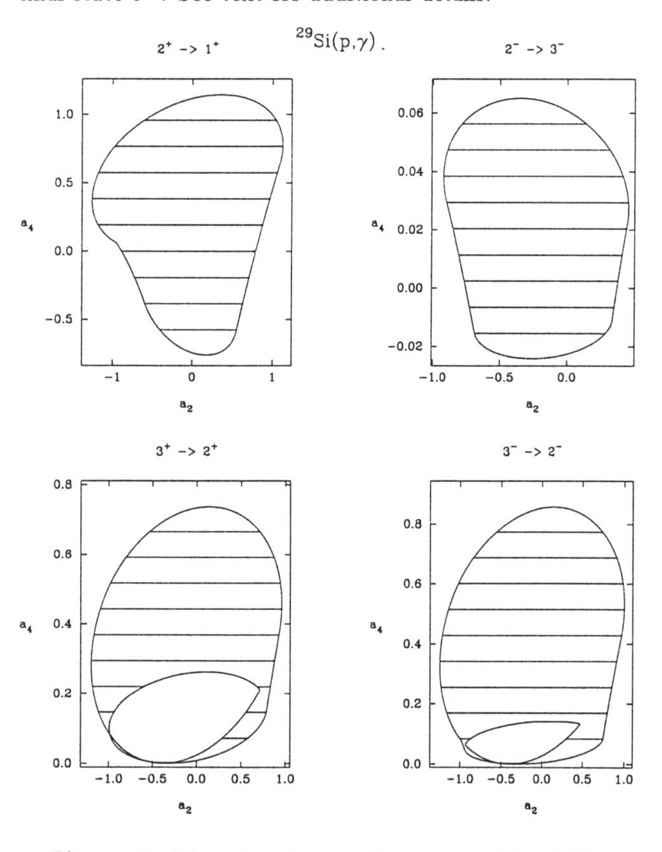

Figure 2. Allowed regions and experimental coefficients for four primary transitions from the 7759-keV resonance to 3^+ final states. The four plots show the four possible J^π assignments for the compound state determined from our fixed detector measurements. The allowed regions in the left two plots are line segments along the a_4 axis and are denoted by the broad black lines.

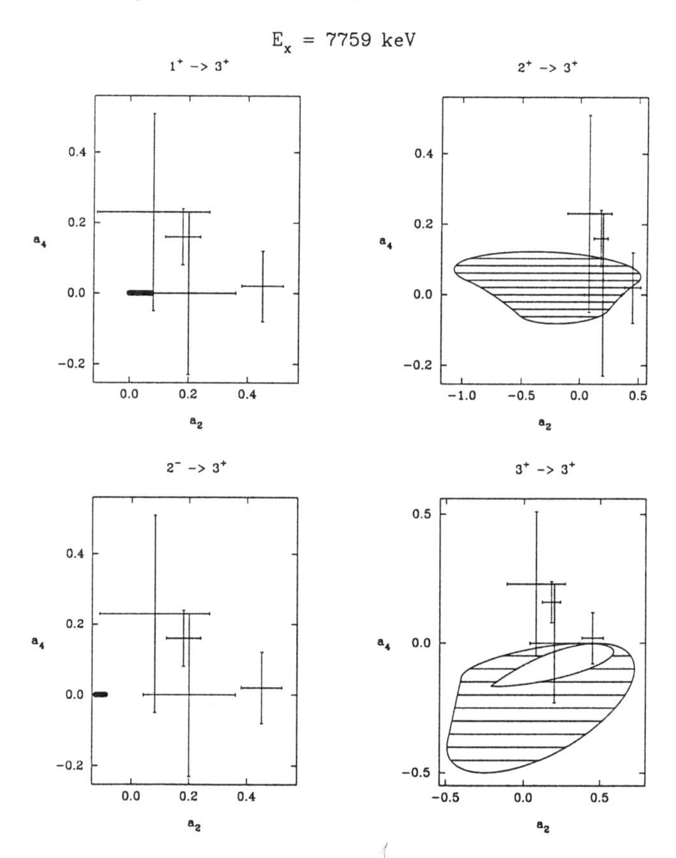

EXPERIMENTAL APPROACH TO A DIRECT STUDY OF THE NUCLEAR REACTION H(^{7}Be,γ)^{8}B

U. Greife[1], K. Brand[3], L. Campajola[2], A. D'Onofrio[4], L. Gialanella[1,2], E. Huttel[5], R. Kubat[5], G. Oliviero[2], H. Rebel[6], V. Roca[2], C. Rolfs[1], M. Romano[2], M. Romoli[2], S. Schmidt[1], W.H. Schulte[1], F. Strieder[1], F. Terrasi[7], H.-P. Trautvetter[1], D. Zahnow[1]

[1]) *Institut für Experimentalphysik III, Ruhruniversität Bochum, Germany*
[2]) *Dipartimento di Scienze Fisiche, Università Federico II, Napoli and INFN, Napoli, Italia*
[3]) *Dynamitron-Tandem Laboratorium, Ruhr-Universität Bochum, Germany*
[4]) *Dipartimento di Fisica Teorica SMSA, Università di Salerno and INFN, Napoli, Italy*
[5]) *Forschungszentrum Karlsruhe, Hauptabteilung Zyklotron, Germany*
[6]) *Forschungszentrum Karlsruhe, Institut für Kernphysik, Germany*
[7]) *Facoltà di Scienze Ambientali, Seconda Università di Napoli, Caserta and INFN, Napoli, Italy*

Nuclear fusion reactions play a key role in the understanding of energy production, neutrino emission and nucleosynthesis of the elements in stars. The direct measurement of the cross section of these reactions at the relevant energies is usually hampered by cosmic radiation, beam induced background and/or the radioactivity of the nuclei involved. In order to suppress these background contributions one can either place an experiment underground [1,2] or exploit coincedence techniques between the reaction products. The major aspect in this contribution is the efficient detection of the recoil nucleus in a direct capture reaction.

Radiative capture reactions A(x,γ)B are among the most important reactions for the formation of the elements. They are usually studied in the laboratory by detecting the emitted γ-rays. If the capture cross section is small, one of the nuclei involved is radioactive and/or competing reactions produce a high γ-ray background, even measurements with high-resolution Ge detectors have reached their limitations. It has been shown [3] that the direct detection of the recoiling nucleus B can greatly improve the experimental sensitivity. In summer of 1994 the NABONA project was initiated to combine the two fields of nuclear astrophysics and accelerator mass spectrometry with the aim to determine reaction rates of radiative capture reactions important for nuclear astrophysics.

The absolute cross section $\sigma(E)$ of the reaction ^{7}Be(p,γ)^{8}B influences sensitivly the calculated flux of high energy solar neutrinos and must therefore be known with adequate precision. [4 and references therein]. Using a radioactive ^{7}Be target ($T_{1/2}$ = 53.29 d) the $\sigma(E)$ data were derived from the β-delayed α-decay of ^{8}B. The work of several investigators led to a fairly consistent picture of the energy dependence of $\sigma(E)$ - or equivalently of the astrophysical S(E) factor - but not on the absolute value: the extrapolated absolute S(0) factor ranges from 16 to 45 eV b [5 and references therein]. The discrepancy is most likely to be found in the complicated target stoichiometry of the ^{7}Be target (produced via hot chemistry).

It is the aim of a project at the 3 MV TTT-3 tandem accelerator in Naples (Fig. 1) to provide an improved $\sigma(E)$ value in the nonresonant region, i.e. at $E_{c.m.}$ = 1.0 MeV [3]. The reaction is studied in inverse kinematics, p(^{7}Be,γ)^{8}B, i.e. a radioactive ^{7}Be ion beam of E_{lab} = 8.0 MeV is guided into a windowless gas target system filled with H_2 gas (pressure p(H_2) = 5.0 mbar), thus avoiding the above problems of target stoichiometry. As a novel technique the ^{8}B residual nuclides are detected directly in an efficient recoil separator. Alternatively this setup also allows to implant the ^{8}B recoils in a rotating wheel and to detect their subsequent β-delayed α-activity. Since the elastic scattering yield is observed concurrently with the ^{8}B yield, $\sigma(E)$ is related ultimately to the Rutherford scattering cross section ("relative measurement"). Due to the low capture cross section (about 0.5 μb) and the low target density (about 10^{19} atoms/cm^2) a ^{7}Be beam intensity of about 100 ppA is needed in order to achieve sufficient statistical accuracy in a finite time. Moreover, a high purity of the beam, and in particular the absence of isobaric

CP392, *Application of Accelerators in Research and Industry*, edited by J. L. Duggan and I. L. Morgan
AIP Press, New York © 1997.

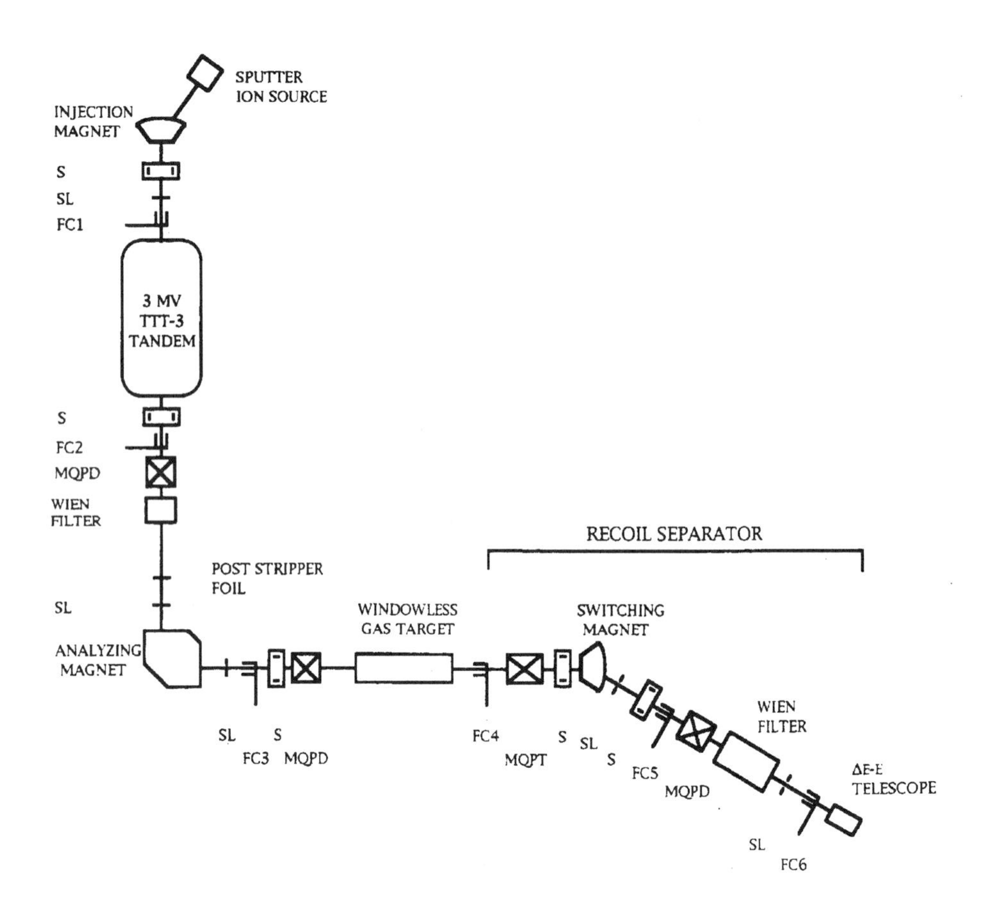

Figure 1: Schematic diagramm of the 3 MV TTT-3 tandem accelerator at Naples, the beam transport system, the windowless gas target, and the recoil separator (S = X-Y steerers, SL = slits, FC = Faraday cup, MQPD = magnetic quadrupole doublet, MQPT = magnetic quadrupole triplet).

contaminants (for a unique analysis of the p+^{7}Be elastic scattering yields), is needed. The setup including the recoil separator was tested [6] with the H(^{12}C,γ)^{13}N radiative capture reaction and the result was in excellent agreement with preyious work. Furthermore, the setup of the rotating wheel was involved [7] in recent cross section measurements of the d(^{7}Li,p)^{8}Li reaction near the E=0.61 MeV resonance. The ^{7}Be nuclides were produced in a Li_2O matrix via the ^{7}Li(p,n)^{7}Be reaction using a 13 MeV proton beam (about 10 μA) from the KIZ-cyclotron at Karlsruhe. Up to now, after a proton bombardment of 230 hours (over a time period of 6 weeks) a total ^{7}Be activity of 6.0 GBq (corresponding to about 4×10^{16} ^{7}Be nuclides) was produced.

In the sputter source, the ^{7}Be nuclides in the Li_2O matrix were extracted in form of a $^7BeO^-$ molecular ion beam. Setting the 35° injection magnet to mass-23 ions, this beam was accompanied by an intense $^7LiO^-$ molecular beam. Both beams were focused by two gridded lenses and accelerated to the terminal voltage U = 2.42 MV of the tandem. After stripping in a 3 μg/cm^2 thick C foil, the 8.0 MeV ions of $^7Be^{3+}$ and $^7Li^{3+}$ emerging from the accelerator were focused by a magnetic quadrupole doublet on the object slits of a 90° analysing magnet and this double focusing magnet focused the beam on the image slits. Inserting a post-stripper C foil near the object slits, fully stripped $^7Be^{4+}$ ions were produced. These $^7Be^{4+}$ ions were selected by the analysing magnet, while the accompanying intense $^7Li^{3+}$ ions were filtered. It was found that another mass-23 ion beam, i. e. $^{23}Na^-$ ions, was also extracted from the sputter source leading to similar rigidity as $^7Be^{4+}$. This and other possible contaminant beams were finally suppressed using a Wien filter before the analysing magnet (Fig. 1). The purity of the resulting $^7Be^{4+}$ beam was tested by guiding the beam through the gas target and the recoil separator (with its Wien filter switched off) into the ΔE-E ionisation chamber for particle identification. For this purpose the beam intensity was reduced by closing the low-energy slits. The resulting identification matrix (Fig. 2a) demonstrates the high cleanliness of the ^{7}Be ion beam.

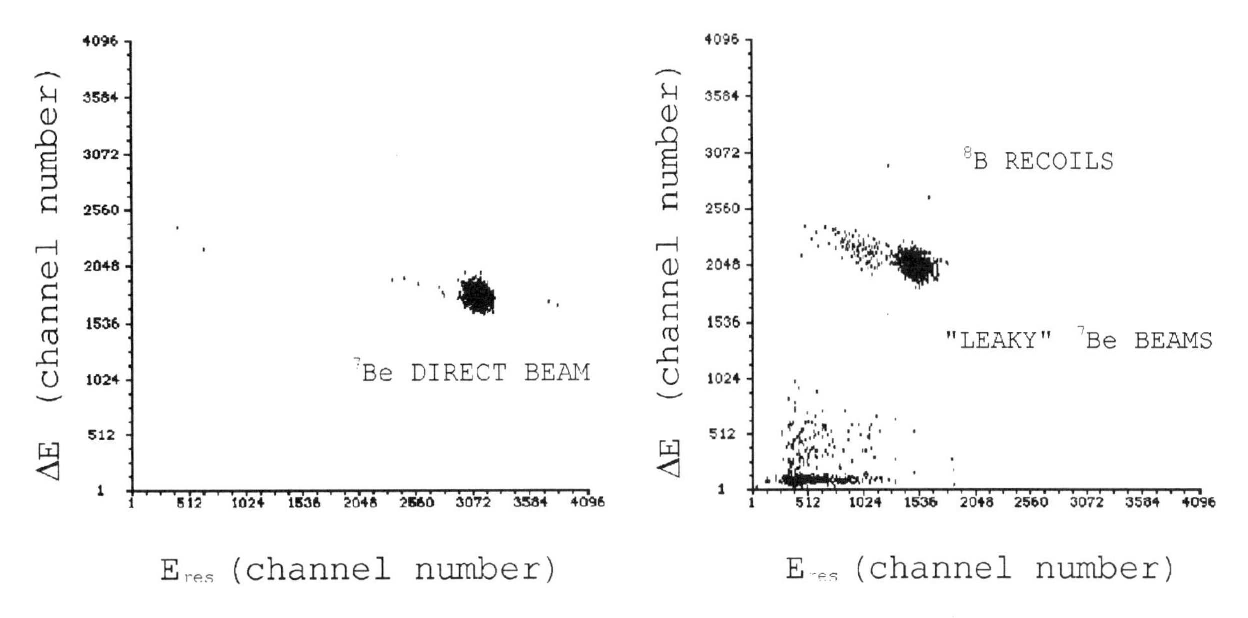

Figure 2: Two-dimensional density plot of the $\Delta E - E$ telescope measured with the recoil separator tuned (a) to the 8 MeV $^7Be^{4+}$ beam and (b) to the $^8B^{5+}$ recoils.

The analysed $^7Be^{4+}$ current amounted up to 70 pA and lasted for about 48 hours. With a 50 % transmission through the gas target system a time-averaged $^7Be^{4+}$ current of about 18 pA was available in the target zone. The ^8B residual nuclides from $p(^7Be,\gamma)^8B$ at E_{lab} = 8 MeV were produced in the H_2 gas target and guided through the recoil separator with a 100 % efficiency. For the above ^7Be current, $p(H_2)$ = 5.0 mbar, and σ = 0.5 μb one expects a ^8B count rate of about 1 event per 3 hours, consistent with observation (Fig. 2b): the two events per 5 hours are clearly resolved from the counts due to the "leaky" ^7Be particles [6], where the ^7Be beam was suppressed in the recoil separator by a factor 5×10^{-10}. Several improvements are possible, such as higher activation of the cathodes (e.g. 20 GBq) and higher tandem transmission, leading to a tenfold increase in the ^8B counting rate.

In summary, a 8.0 MeV ^7Be radioactive ion beam has been produced with a tandem accelerator having sufficient purity and intensity for the measurement of the $p(^7Be,\gamma)^8B$ capture cross section. In the next months the beam intensity has to be improved in order to obtain a high counting rate with both detection setups. The next aim of the NABONA collaboration is the determination of the cross section for the reaction $^{12}C(\alpha,\gamma)^{16}O$ in inverse kinematics. The experimental setup will consist of a pointlike gas jet target, a large acceptance recoil spectrometer and a detector array for the measurement of γ-recoil coincidences. It is foreseen to install the setup after this experiment at one of the now emerging radioactive beam facilities, where the advantages of this approach are even more apparent. Furthermore, the experience gained in this development project could be an important input for the RIB laboratories which will install a dedicated recoil mass spectrometer for nuclear astrophysics.

REFERENCES

1. U. Greife, C. Arpesella, C.A. Barnes, F. Bartolucci, E. Bellotti, C. Broggini, P. Corvisiero, G. Fiorentini, A. Fubini, G. Gervino, F. Gorris, C. Gustavino, M. Junker, R.W. Kavanagh, A. Lanza, G. Mezzorani, P. Prati, P. Quarati, W.S. Rodney, C. Rolfs, W.H. Schulte, H.P. Trautvetter and D. Zahnow, Nucl. Instr. Meth. **A350** (1994) 327
2. M. Junker, C. Arpesella, C. Broggini, P. Corvisiero, G. Fiorentini, A. Fubini, G. Gervino, U. Greife, C. Gustavino, J. Lambert, P. Prati, W.S. Rodney, C. Rolfs, H.P. Trautvetter, D. Zahnow and S. Zavatarelli, this proceedings
3. M.S. Smith, C. Rolfs and C.A. Barnes, Nucl. Instr. Meth. **A306** (1991) 233
4. A. Dar and G. Shaviv, to be published in ApJ. **468** (1996)
5. B.W. Filippone, Ann. Rev. Nucl. Part. Sci. **36** (1986) 717
6. L. Gialanella, F. Strieder, K. Brand, L. Campajola, A. D'Onofrio, U. Greife, E. Huttel, F. Petrazzolo, V. Roca, C. Rolfs, M. Romano, M. Romoli, S. Schmidt, W.H. Schulte, F. Terrasi, H.P. Trautvetter and D. Zahnow, Nucl. Instr. Meth. **A376** (1996) 174
7. F. Strieder, L. Gialanella, U. Greife, C. Rolfs, S. Schmidt, W.H. Schulte, H.P. Trautvetter, D. Zahnow, F. Terrasi, L. Campajola, A. D'Onofrio, V. Roca, M. Romano and M. Romoli, Z. Phys. **A355** (1996) 209

Interactions obtained from precision polarized ^{6}Li scattering experiments

P.L. Kerr[a], P.V. Green, K.W. Kemper, A.J. Mendez[c], E.G. Myers, E.L. Reber[b] and B.G. Schmidt

Physics Department, Florida State University, Tallahassee, Florida 32306-3016

Detailed analyzing power angular distribution measurements allow the contributions from the different interaction terms responsible for producing them to be isolated. The high quality data needed for these studies can now be taken because of the development of an intense laser pumped alkali ion source with close to maximum polarization and greatly improved reliability.

Early speculation [1] of the magnitude of spin effects in heavy-ion scattering and reactions predicted them to be quite small. This result was based on folding model calculations that used proton spin-orbit potentials. Measurements of large spin-flip probabilities [2], analyzing powers with polarized alkali beams [3] and projectile fragment polarizations [4] suggest that the early ideas are incomplete when dealing with spin effects in nuclei rather than nucleons. In fact, virtual excitation of the projectile or fragment [5] during the scattering or reaction process gives spin dependent effects that are much larger than those arising from explicit spin-orbit effects.

The projectile-target relative angular momenta are quite a bit larger in heavy-ion scattering than in proton or deuteron scattering which results in highly structured scattering angular distributions and analyzing powers. The rapidly changing analyzing powers mean that measurements must be made in small angle steps with good angular resolution. For example, as can be seen in Fig. 1, the ^{6}Li + ^{12}C elastic scattering analyzing power iT_{11} changes from +.04 to -.12 in two degrees. Because large analyzing powers usually occur at angles where the cross sections are small, ion sources are needed that provide robust beam currents and high polarization if the analyzing power angular distributions are to be mapped out in sufficient detail to extract the physics from them.

The Florida State University laser pumped polarized Li ion source has increased in reliability, intensity and beam polarization so that it is now possible to tackle experiments where the cross section of the scattering or reaction being measured is as small as 5μb/sr. The ion source will operate for a week, delivering on target currents of 150nA and beam polarizations t_{10} and t_{20} of 1.1 and -1.1, respectively. The polarizations measured at the source are 1.15 and -1.20, showing that the polarization losses are minimal. The lifetime of the source is governed by the blockage of the oven exit collimators which are used to define the atomic beam before the laser pumping region.

A summary of the analyzing powers measured to date at FSU is: iT_{11}, T_{20}, T_{21} and T_{22} for elastic and inelastic scattering of ^{6}Li + ^{4}He, ^{9}Be, and ^{12}C, ^{9}Be(^{6}Li,^{6}He), ^{12}C(^{6}Li,α) and ^{12}C(^{6}Li,d) at energies close to 30 and 50 MeV, and iT_{11} and $^TT_{20}$ for ^{12}C(^{6}Li,^{3}He). In addition, numerous beam polarization monitoring scatterings have been established along with beam polarization calibration standards for es-

CP392, *Application of Accelerators in Research and Industry*, edited by J. L. Duggan and I. L. Morgan
AIP Press, New York © 1997.

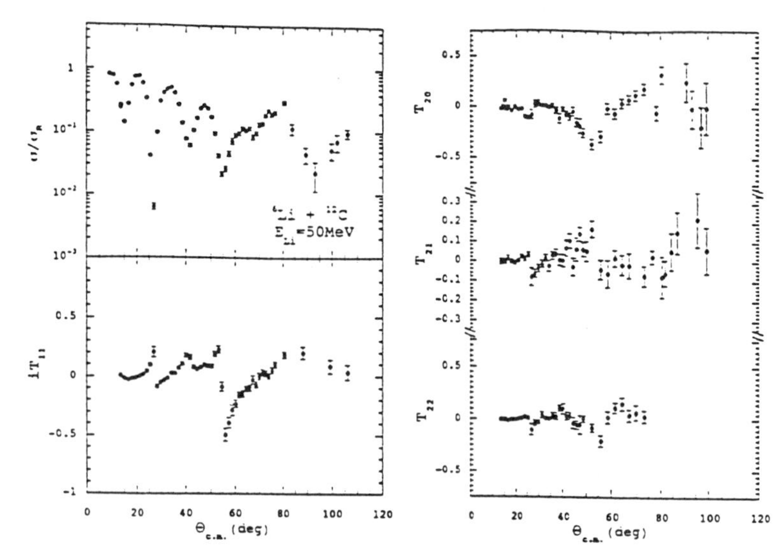

FIGURE 1. The experimental data for ^{6}Li + ^{12}C elastic scattering at 50 MeV. No calculations are shown so that the rapid changes in the analyzing powers can be clearly seen.

tablishing the on-target absolute beam polarization.

The first analyzing power [3] measured with either a polarized ^{6}Li or ^{7}Li beam was of the ^{6}Li elastic scattering vector analyzing power (VAP) at Heidelberg. These data were understood as arising from an explicit spin-orbit interaction. Folding model calculations [6] predicted that ^{6}Li and ^{7}Li would produce the same sign VAP. Naturally, the measured [7] VAP's were of the opposite sign. The accepted explanation for the cause of the VAP's [5] is that they arise from virtual excitation of the projectile to its excited states. Now, tensor analyzing powers (TAP) have also been measured. For ^{7}Li, they arise from re-orientation of its very large ground state quadrupole moment [8]. Since the quadrupole moment of ^{6}Li is close to zero there must be different effects that produce the TAPs when compared with ^{7}Li.

The approach we have taken to the understanding of the origin of the ^{6}Li analyzing powers is to produce the real interaction potential with the double folding model assuming the M3Y effective nucleon-nucleon interaction. We have assumed a Woods-Saxon imaginary interaction. Since we have measured inelastic cross sections for both projectile and target excitation for ^{6}Li + ^{12}C scattering, we have concentrated our analysis on this system. We have carried out coupled channels analyses of this system with the magnitude of the target and projectile excitations fixed by matching the measured inelastic cross sections. By then turning on and off different couplings and interaction terms, one can learn about the origin of the different analyzing powers.

Originally, most of the experimental effort went to measuring vector analyzing powers probably because they carried considerable information in deuteron scattering, which has spin 1 as does ^{6}Li. Now that we have been able to analyze complete sets of analyzing powers for ^{6}Li we see that the origin of the VAP is perhaps the hardest to understand because it arises from interferences between many different contributions. The forward angle structure in the analyzing power T_{21} is in fact the easiest to understand since neither virtual projectile excitation nor the spin-orbit interaction gives any contribution. A tensor interaction can reproduce T_{21} quite well. Further calculation shows that the ground state re-orientation is responsible for producing T_{21} and the magnitude of this analyzing power is able to set the re-orientation strength. The other tensor analyzing powers, T_{20} and T_{22}, arise from projectile virtual excitation and from g.s. re-orientation. The spin-orbit term gives very little contribution to these APs in contrast with deuteron scattering, where the second order spin-orbit terms give rise to T_{20} and T_{22}. The analyzing power iT_{11} is the most difficult to describe because all terms give contributions to it. However, even here it is possible to learn about the

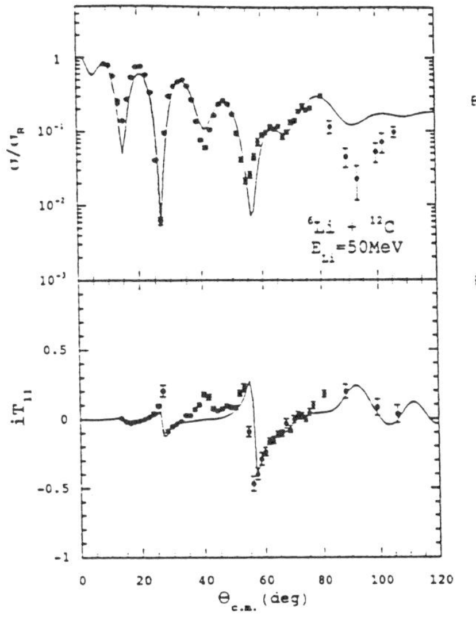

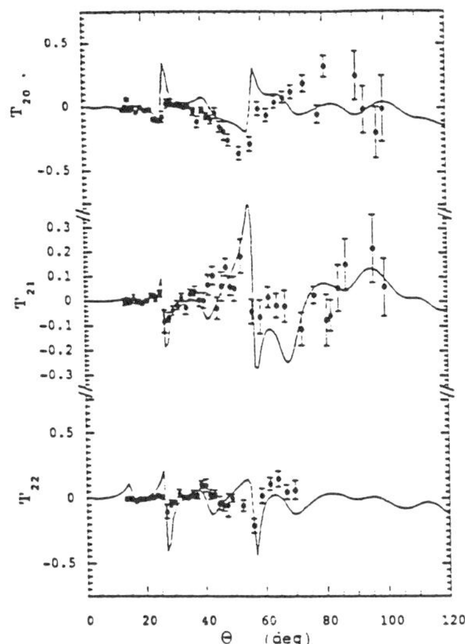

FIGURE 2. The Solid line is the result of a coupled channels calculation that includes, spin-orbit, virtual excitation, and ground state re-orientation terms. The analyzing power T_{21} arises solely from the ^{6}Li g.s. re-orientation.

underlying interaction because different angular ranges are sensitive to different interaction terms. Luckily, target excitation [9] does not give a significant contribution to the APs so that the extracted physics is independent of the target.

The major failure in our understanding of polarized ^{6}Li scattering now centers on target excitation. The experimental cross sections for both the 4.44 MeV, 2^+ and 9.64 MeV, 3^- states in ^{12}C are well reproduced by calculation, but the measured analyzing powers are much larger than any calculations have been able to produce to date.

Our understanding of heavy-ion spin effects has made a complete circle from the first days of its study 20 years ago. The initial folding model estimates of the size of the spin-orbit interaction are in fact correct. They are small. The analyzing powers provide a detailed look at how the projectile structure and the excitation of the projectile influence heavy-ion scattering. It is possible to extract this information only because ^{6}Li has a small g.s. quadrupole moment, and the spin-orbit interaction is weak. These two facts make ^{6}Li a special tool for increasing our understanding of heavy-ion scattering and reactions.

ACKNOWLEDGEMENTS

This work was supported in part by the National Science Foundation and the State of Florida.

[a] LANL, Los Alamos, N.M

[b] INEL, Idaho Falls, ID

[c] NEC Corp., Middleton, WI

REFERENCES

[1] H. Amakawa and K.-I. Kubo, Nucl. Phys. **A266** (1976) 521.

[2] W. Dünnweber *et al.*, Phys. Rev. Lett. **43** (1979) 1642.

[3] W. Weiss *et al.*, Phys. Lett. **61B** (1976) 237.

[4] K. Sugimoto *et al.*, Phys. Rev. Lett. **39** (1977) 323.

[5] H. Nishioka *et al.*, Phys. Rev. Lett. **48** (1982) 1795; H. Ohnishi *et al.*, Nucl. Phys. **A415** (1984) 271; F. Petrovich *et al.*, Nucl. Phys. **A425** (1984) 609.

[6] F. Petrovich *et al.*, Phys. Rev. C **17** (1978) 1642.

[7] G. Tungate *et. al.*, Phys. Lett. **98B** (1981) 347.

[8] W. Dreves *et al.*, Phys. Lett. **78B** (1978) 36.

[9] P.V. Green *et al.*, Phys. Rev. C **53** (1996) 2862; P.L. Kerr *et al.*, Phys. Rev. C **54** (1996) in press.

PION PRODUCTION FROM PROTON-NUCLEUS COLLISIONS: RECENT RESULTS AT THE MOSCOW MESON FACTORY

A. Badalà, R. Barbera, F. Librizzi, F.Marzo, D. Nicotra, A. Palmeri, G. S. Pappalardo, F. Riggi, G. Russo, R. Turrisi
Istituto Nazionale di Fisica Nucleare, Sezione di Catania, and Dipartimento di Fisica dell'Università di Catania, Catania, Italy

V.Aseev, Yu. Gavrilov, F.Guber, M.Golubeva, T.Karavicheva, A.Kurepin, K.Shileev, V.Tiflov
Institute for Nuclear Research, Russian Academy of Sciences, Moscow, Russia

Recent results on pion production from proton-nucleus collisions at the Moscow Meson Factory are presented and compared with previous data. Two main topics are discussed: the pion production below the free NN threshold and the long-standing problem of the *anomalous* production of π^+ near 350 MeV, for which a definite evidence of a narrow resonance has been obtained.

I. INTRODUCTION

The production of pions in heavy-ion collisions at bombarding energies near the absolute nucleon-nucleon (NN) threshold is a complex phenomenon, which involves several aspects of the heavy-ion dynamics. Recent exclusive results obtained in this energy domain have been discussed with reference to stopping effects of the projectile (1), as well as to pion reabsorption and shadowing from the residual nuclear matter (2). The interpretation is however complicated due to the complex structure of both projectile and target nuclei. Furthermore, for heavy-ion reactions at energies around 100 MeV/nucleon, the available energy of the interacting system is high compared to the energy carried out by the outgoing pion and the corresponding phase-space for the residual system is very large. In case of proton-nucleus collisions at energies near or below the NN threshold, the restriction of the final phase space allows one to study the importance of cooperative effects versus the incoherent nucleon-nucleon mechanism. Moreover the knowledge of the initial stage of the collision allows one to go into the details of the elementary processes involved, which are not smeared out so dramatically as in heavy-ion collisions. A large experimental and theoretical effort has been spent, however, only in the case of proton-nucleus collisions leaving the residual nucleus in a well defined state near the kinematical limit, mainly for spectroscopic studies (3). Less attention has been devoted to the continuum part of the pion spectra and to the related production mechanisms. Only few data exist on charged pion production from p+A collisions at energies below the free NN threshold ($\simeq$ 280 MeV). These have been mainly obtained at IUCF (4) and Orsay (5,6). The theoretical treatment of such a process needs also to be improved in several respects, including the role of secondary collisions and the pion reabsorption/rescattering in the nuclear matter.

This is a report concerned with new experimental results obtained at the Moscow Meson Factory (MMF), by the magnetic spectrometer CLAMSUD (7), in the framework of a joint INFN (Italy)- INR (Russia) collaboration. A comparison to previous results obtained at 201 MeV with the Orsay Synchrocyclotron (6) is also reported.

II. EXPERIMENTAL TECHNIQUE

The experiments discussed here have been carried out at the end of the linac tunnel of the MMF accelerator. This was used to provide proton beams with energies ranging between 200 and 400 MeV at a reduced intensity, suitable for the measurements. Charged pions were detected by the CLAMSUD spectrometer (7). This is a compact dipole magnet with non parallel poles, which is able to deflect pions and protons with momenta up to 250 MeV/c. For the experiments discussed here, the CLAMSUD spectrometer was equipped with a focal plane detector which includes two multiwire X-Y drift chambers for the reconstruction of trajectories and two planes of segmented scintillators for energy loss (ΔE) and time-of-flight (TOF) measurements. An additional hodoscope of scintillators was located on the entrance side of the spectrometer, to provide a signal to start the time-of-flight measurement. The solid angle of the spectrometer was about 12 msr. Charged pions were identified by their ΔE and TOF. Due to the compactness of CLAMSUD, pion flight paths were limited to about 2 m, causing approximately only 20 % of pions to decay in flight for 60 MeV pions. A detailed description of the experimental set-up and its installation at the MMF has already been reported (8). Here we will focus mainly on some new results which have been obtained by this experimental set-up. Two main topics are discussed, namely the pion production below the NN threshold and the so-called *anomalous* production of pions near 350 MeV.

CP392, *Application of Accelerators in Research and Industry*, edited by J. L. Duggan and I. L. Morgan
AIP Press, New York © 1997.

III. RESULTS

A. Pion production below the NN threshold

Previous investigations on pion production from p+A collisions at subthreshold energies (6) have been concerned with the systematics of total cross sections as a function of the bombarding energy and target mass number, with the energy dependence of the angular behaviour, and with the π^+ to π^- ratio. The angular distributions were found to be strongly dependent on the pion kinetic energy with a forward peaking for the most energetic pions. The origin of this bump is not yet fully understood. Positive pions show a larger forward contribution. The ratio between positive to negative pion yields depends upon the proton incident energy, reaching its maximum value (corresponding to the elementary NN pion production cross section ratio) near the NN threshold.

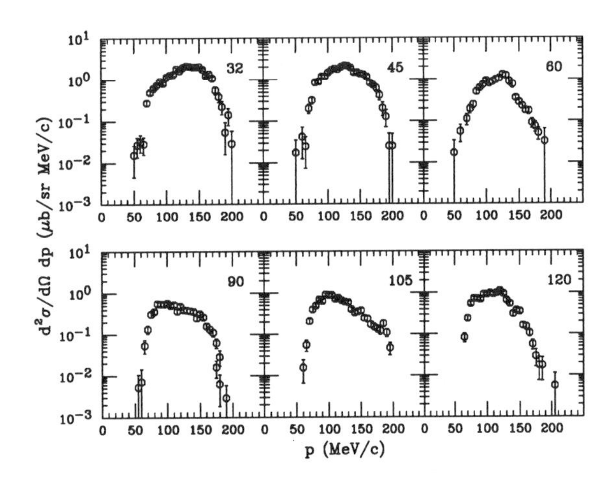

FIGURE 1. Momentum spectra of π^+ measured at 248 MeV in the reaction p+Cu. The detection angle in the lab system is shown for each spectrum.

While at bombarding energies well above the NN threshold, the picture considering the collision as a superposition of independent elementary NN interactions is usually adopted, cooperative effects could in principle become important near the threshold and the pion yield could be strongly affected by the Fermi motion and Pauli blocking. Reliable calculations are, however, needed to evaluate the importance of these effects and the main features of the pion production close to the threshold. New results which have been obtained at the MMF, concerning the production of charged pions from p+Cu at 248 MeV were compared to previous results at 201 MeV and to model calculations. Momentum spectra were measured at several angles. As an example, fig. 1 shows the absolute differential cross sections of positive pions. The absolute values of the negative pion yields are approximately a factor of 10 lower than the corresponding values for π^+.

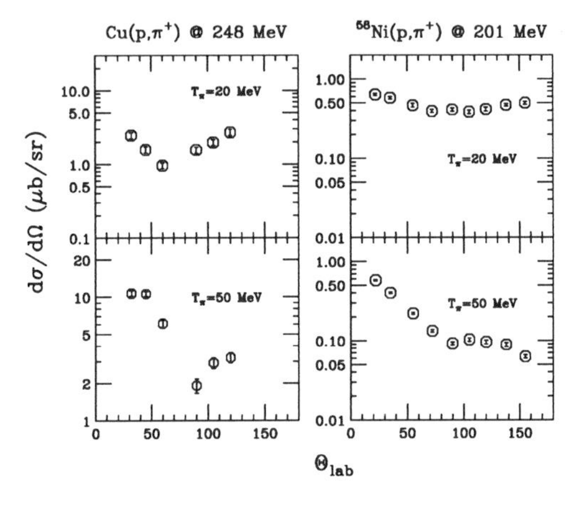

FIGURE 2. Angular distributions of low- and high-energy π^+ from p+^{58}Ni at 201 MeV (right) and p+Cu at 248 MeV (left)

The integrated angular distribution of π^+ is peaked at forward angles and shows a slight enhancement at backward angles. Angular distributions of low and high energy pions are shown in fig. 2 for the p+Cu system at 248 MeV, together with previous data from p+^{58}Ni at 201 MeV. As it is seen from fig. 2, the data at 248 MeV show a phenomenology which is similar to the data already measured at 201 MeV on a system with similar mass. The forward peak is associated to the most energetic pions, whereas low energy pions are more isotropic. A consistent theoretical description based on a multiple scattering model (9) is in progress to try to reproduce both the results at 201 MeV and at 248 MeV. This model incorporates the effect of the pion reabsorption and rescattering in the nuclear matter. A preliminary comparison of the data to the model predictions shows that the inclusion of secondary collisions is essential, especially to understand the relatively large yield observed at backward angles, whereas the forward yield is dominated by first-chance collisions. Further work is under way along this line.

B. Anomalous production of π^+ near 350 MeV

A relevant part of the experimental program at the Moscow Meson Factory has been devoted to the anomalous production of positive pions observed in the p+Cu reaction around 350 MeV. After the first observation by Krasnov et al. (10), several experiments have tried to confirm the existence of a narrow resonance in pion production near 350 MeV (11,12,13). Contradictory results have been obtained so far. The origin of this anomalous production, if any, may have several explanations.

The first proposed explanation was related to the possible excitation of the $^3F_3$ dybarion resonance (M=2220 MeV) almost at rest, with subsequent three body decay π^++p+n (10). It was also suggested that the enhanced production of low energy pions near 350 MeV could be due to the formation of a resonant nuclear state near 350 MeV excitation energy with a small width, which decays predominantly by emission of two pions. This interpretation was based on the experimental observation that the anomaly was seen only for pion energies below about 70 MeV, together with the fact that in such a process on a heavy nucleus, the maximum energy of a pion would be 350 MeV - 2 $m_\pi \simeq$ 70 MeV. Alternative possibilities were discussed in terms of an apparent transparency of nuclei to low energy pions or to bound pion Cooper pairs in nuclei.

The enhancement of the cross section, which is seen particularly for low energy pions, is however very small and the expected width of this bump is only 5 MeV, which makes this experiment difficult to perform.

A definite evidence has been obtained now for such a phenomenon in the Cu(p,π^+) near 350 MeV (14). For this experiment, the beam energy was varied by means of a special tuning procedure of the linac with about 2 MeV steps. A time-of-flight measurement with two detectors placed along the linac channel was carried out to get the absolute value of the beam energy. This is known to a precision less than 1 MeV, whereas the energy spread was estimated to be about ± 1 MeV. Particular attention was devoted to avoid any systematical error due to the experimental set-up or to beam focussing conditions. Most of the measurements were repeated twice, in different runs, and the results were confirmed.

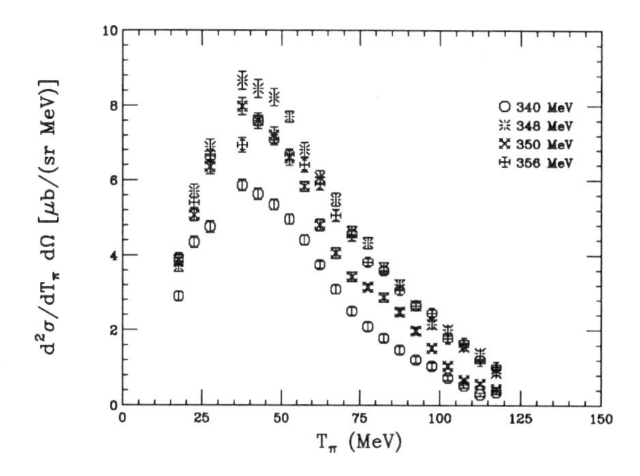

FIGURE 3. Energy spectra of π^+ from p+Cu reaction at several proton bombarding energies around the resonance.

Measurements of π^+ from p+Cu reaction were done at 90° in the lab system, for several proton energies, from 340 to 364 MeV, in small energy steps. For each bombarding energy the π^+ momentum spectrum was measured from the threshold up to the maximum momentum which can be reached by the CLAMSUD spectrometer. Figure 3 shows some of the spectra obtained at several incident energies. As it can be seen, there is a definite change in the shape and absolute value of the cross section near 348-350 MeV. Several methods of analysis were used to extract the excitation functions of selected regions of the measured spectra, their slopes and the ratio between the low energy and high energy pion yields. If the ratio between the low energy and high energy sides of each spectrum (fig.4, left) is plotted against E_p, a bump is observed near 350 MeV, with a width of the order of 5 MeV. The statistical error bars for this set of data are in the order of 3-4 %. The discrepancy of the measured yields from the average regular behaviour is in the order of 25 % for the chosen momentum regions. Since each spectrum is obtained by a superposition of different (five) field settings of the spectrometer, systematic errors due to relative normalization could be a possible source of problems within each spectrum. To exclude this possible source of systematic errors, the ratio between the low energy and the high energy sides of a portion of spectrum, measured at a given field setting, was studied as a function of the incident energy. This spectrum is obtained for a region of pion momentum of approximately ±15% wide with respect to the central momentum. This corresponds to a constant efficiency as a function of the momentum.

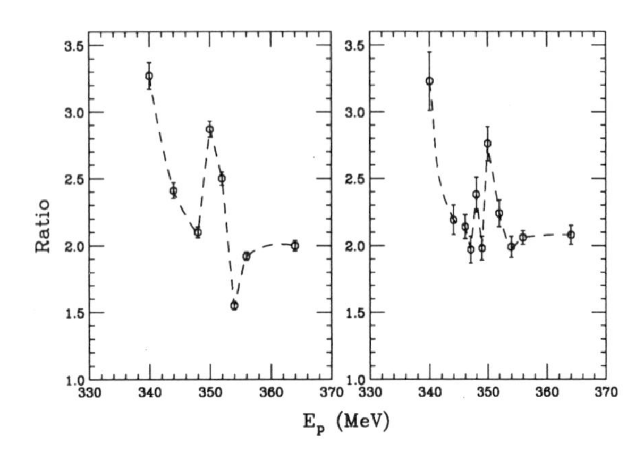

FIGURE 4. Ratio of low energy to high energy pion yield as a function of the bombarding energy. The integrated cross section between 30 and 60 MeV was divided by the corresponding value obtained for the region between 75 and 120 MeV (left). The low energy (75-97 MeV) and high energy (97-120 MeV) sides of the same portion of the spectrum were considered to build up the ratio (right).

Also in this case a bump was observed, consistent with that extracted from the analysis of the overall shape of the spectrum. As an example, fig. 4(right) shows the result for the ratio between the cross section in the region 75-97 MeV and that in the region 97-120 MeV, extracted from a single field setting.

In conclusion, the already observed structure is thus confirmed and an estimation of the width of this reasonance gives a value which is comparable to that estimated before (11). Additional data have been measured and some of the already reported measurements have been repeated to get unambiguously evidence for such a process. Further work is presently in progress to analyze all the data in more detail with several method of analysis.

REFERENCES

1. Badalà A., et al., *Phys. Rev.* **C53**,1782-1791(1996) and references therein.
2. Badalà A., et al., *Phys. Rev.* **C** , in press.
3. Hoistad B., *Advances in Nuclear Physics* **11**,135-178(1979).
4. Throwe T.G., et al., *Phys. Rev.* **C35**,1083-1088(1987) and references therein.
5. Bimbot L., et al., *Nucl.Phys.* **A440**,636-646(1985).
6. Badalà A., et al., *Phys. Rev.* **C46**,604-615(1992).
7. Anzalone A., et al., *Nucl. Instr. and Methods in Phys. Res.* **A308**,533-538(1991).
8. Badalà A., et al., *Nucl. Instr. and Methods in Phys. Res.* **B99**,657-660(1995).
9. Sibirtsev A., *INFN Report* **BE-96/5**(1996).
10. Krasnov V.A., et al., *Phys.Lett.* **108B**,11-14(1982).
11. Julien J., et al., *Phys.Lett.* **142B**,340-343(1984).
12. Julien J. et al., *Z.Physik* **A347**,181-184(1994).
13. Yen Y., et al., *Phys.Lett.* **269B**,59-62(1991).
14. Aseev A., et al., *Phys. Lett.* **B**, submitted.

TECHNIQUES FOR PRECISION MEASUREMENTS OF PARITY VIOLATING ASYMMETRIES

W.S. Wilburn,[1] G.E. Mitchell,[2] N.R. Roberson,[1] and J.F. Shriner, Jr.[3]

[1]*Duke University, Durham, North Carolina 27708*
and Triangle Universities Nuclear Laboratory, Durham, North Carolina 27708
[2]*North Carolina State University, Raleigh, North Carolina 27695*
and Triangle Universities Nuclear Laboratory, Durham, North Carolina 27708
[3]*Tennessee Technological University, Cookeville, Tennessee 38505*
and Triangle Universities Nuclear Laboratory, Durham, North Carolina 27708

Recent parity non-conservation (PNC) measurements in nuclei use the compound nucleus as a laboratory for the study of symmetry breaking, with the symmetry breaking matrix elements treated as random variables. We plan to measure parity-violating longitudinal analyzing powers A_z in $(\vec{p},\alpha)$ scattering on resonances in $A \sim 30$ nuclei. Since the resonances are often narrow, the proton beam energy must be carefully controlled. In addition, since the effects are expected to be small ($A_z < 10^{-3}$), systematic errors must be carefully controlled. Expected sources of systematic errors will be discussed as well as techniques for monitoring and controlling them. These include residual transverse polarization moments, spin-correlated beam motion, and time-dependent detector drifts.

INTRODUCTION

Recent parity non-conservation (PNC) measurements in nuclei use the compound nucleus as a laboratory for the study of symmetry breaking, with the symmetry breaking matrix elements treated as random variables. This approach is made appealing by the observation of very large parity violation in neutron resonances (1). One key issue is the mass dependence of the effective nucleon-nucleus weak interaction. Since the neutron measurements are not feasible for light nuclei, we have considered charged-particle reactions (2–4). In particular, we are designing an experiment to measure longitudinal analyzing powers $A_z(\theta)$ in $(\vec{p},\alpha)$ reactions with target nuclei in the $A \sim 30$ mass region. This region is chosen because it is very different from the neutron studies ($A \sim 100$ and 230) and because high-resolution resonance data exist for five nuclei.

Measurements of A_z will be performed with a longitudinally polarized beam of protons incident on a thin target supported by a carbon foil. The α-particles produced in the target will be counted in a large solid angle detector at backward angles, while four small surface barrier detectors at forward angles will be used to monitor residual transverse beam polarization via elastic scattering from the carbon target backing. Figure 1 is a schematic drawing of the apparatus.

A_z measurements at the 1×10^{-4} level require that both the statistical and systematic uncertainties are reduced to the few times 10^{-5} level. This level of statistical uncertainty will be achieved with the large solid angle detector and with electronics capable of handling the high counting rates (up to 1 MHz) from the detector. The systematic uncertainties are minimized by several means. First, the azimuthal symmetry of the detector, which covers essentially 360° in ϕ, will greatly reduce the sensitivity of the apparatus to many systematic effects, including the leading-order contributions from beam misalignment and transverse polarization components. Second, the proton beam properties, such as position, angle, energy, etc., will be monitored and controlled as much as is feasible. Finally, the proton polarization and energy are varied in a manner designed to cancel systematic effects.

SOURCES OF SYSTEMATIC ERRORS

Most systematic effects are eliminated by the periodic reversal of the proton beam helicity. Comparing count rates for opposite helicity states, for example removes the need for absolute knowledge of flux, target thickness, detector solid angles, etc. In addition, the azimuthal symmetry of the detectors further suppresses systematic effects. For example, a transverse polarization component might cause an increase in count rate in the detectors on the right side of the target, due to

CP392, *Application of Accelerators in Research and Industry*, edited by J. L. Duggan and I. L. Morgan
AIP Press, New York © 1997.

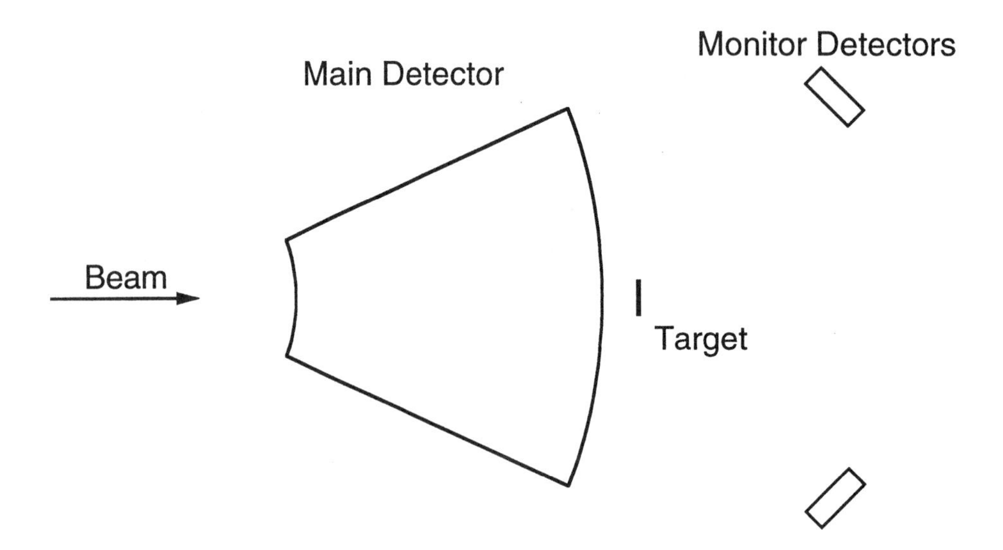

FIGURE 1. Proposed experimental configuration showing the relative positions of the target, main detector, and the four monitor detectors. This view is from the top looking down. The side view is the same.

the parity allowed analyzing power A_y. The left detectors would, however, see a corresponding decrease in count rate and averaging the measurements obtained from the two would cancel the false asymmetry, to the extent that the detectors are identical. Systematic effects that are correlated with either proton helicity or detector position, however, must be considered. Also, time-dependent effects are not completely removed by these simple methods and require more careful treatment. Because the count-rate asymmetries are small, rate-dependent effects such as dead time and pile up do not produce significant systematic asymmetries.

Correlated Errors

These effects can be understood by considering four detectors, located at the same polar angle θ, but different azimuthal angles ϕ, corresponding to the directions up, down, left, and right. The count rate N_α in an individual detector is given by

$$N_\alpha = N_0 x \sigma (1 + \vec{P} \cdot \vec{A}) \Omega, \tag{1}$$

where N_0 is the incident proton flux, x is the target thickness, σ is the cross section, and Ω is the solid angle subtended by the detector. $\vec{P}$ and $\vec{A}$ are vectors describing the polarization of the beam and the analyzing power of the reaction, respectively.

Ideally, the quantities in Eq. 1 are the same for all four detectors, with the exception of $\vec{A}$ which depends upon ϕ. However, the proton fluxes and polarizations in general are not equal for both helicities, the differences being described by the quantities δN_0, δP_x, δP_y, and δP_z. In addition, since the beam position on target may be helicity dependent, a helicity-dependent target thickness difference δx is included. Detector-correlated differences arise if opposing detectors have different solid angles, the differences given as $\delta_x \Omega$ between the left and right detectors, and $\delta_y \Omega$ between the up and down detectors. In addition, since θ may differ between detectors, we also consider $\delta_x \sigma$ and $\delta_y \sigma$, and δA_x and δA_y. The count rate in each detector can then be written

$$N_{L\pm} = N\left(1 \pm \frac{\delta N_0}{N_0}\right)\left(1 \pm \frac{\delta x}{x}\right)\left(1 + \frac{\delta_x \sigma}{\sigma}\right)\left(1 + \frac{\delta_x \Omega}{\Omega}\right)\left[1 \pm P_y A_y \left(1 + \frac{\delta A_y}{A_y}\right)\left(1 \pm \frac{\delta P_y}{P_y}\right) \pm P_z A_z \left(1 \pm \frac{\delta P_z}{P_z}\right)\right], \tag{2}$$

$$N_{R\pm} = N\left(1 \pm \frac{\delta N_0}{N_0}\right)\left(1 \pm \frac{\delta x}{x}\right)\left(1 - \frac{\delta_x \sigma}{\sigma}\right)\left(1 - \frac{\delta_x \Omega}{\Omega}\right)\left[1 \mp P_y A_y \left(1 - \frac{\delta A_y}{A_y}\right)\left(1 \pm \frac{\delta P_y}{P_y}\right) \pm P_z A_z \left(1 \pm \frac{\delta P_z}{P_z}\right)\right], \tag{3}$$

$$N_{U\pm} = N\left(1 \pm \frac{\delta N_0}{N_0}\right)\left(1 \pm \frac{\delta x}{x}\right)\left(1 + \frac{\delta_y \sigma}{\sigma}\right)\left(1 + \frac{\delta_y \Omega}{\Omega}\right)\left[1 \pm P_x A_x \left(1 + \frac{\delta A_x}{A_x}\right)\left(1 \pm \frac{\delta P_x}{P_x}\right) \pm P_z A_z \left(1 \pm \frac{\delta P_z}{P_z}\right)\right], \tag{4}$$

$$N_{D\pm} = N\left(1 \pm \frac{\delta N_0}{N_0}\right)\left(1 \pm \frac{\delta x}{x}\right)\left(1 - \frac{\delta_y \sigma}{\sigma}\right)\left(1 - \frac{\delta_y \Omega}{\Omega}\right)\left[1 \mp P_x A_x \left(1 - \frac{\delta A_x}{A_x}\right)\left(1 \pm \frac{\delta P_x}{P_x}\right) \pm P_z A_z \left(1 \pm \frac{\delta P_z}{P_z}\right)\right], \tag{5}$$

where $\pm$ refers to the beam helicity and $N = N_0 x \sigma \Omega$ is the count rate for unpolarized beam with no systematic variations. For each detector an asymmetry ε is formed:

$$\varepsilon = \frac{N_+ - N_-}{N_+ + N_-}. \qquad (6)$$

Averaging the asymmetries for all four detectors and neglecting terms higher than first order in the variations, we obtain the measured asymmetry at angle θ

$$\varepsilon(\theta) = P_z A_z + (P_x\, \delta A_x + P_y\, \delta A_y)/2 + \frac{\delta \mathcal{L}}{\mathcal{L}}, \qquad (7)$$

where $\mathcal{L} = N_0 x$ is the luminosity. As noted in the introduction, most of the systematic effects are cancelled by the azimuthal symmetry, leaving only three terms in the expression.

The first term in Eq. 7 is the parity violating asymmetry. The last term is the helicity-dependent luminosity difference and is easily removed by monitoring the count-rate asymmetry of the elastically scattered protons, where the parity violating term is suppressed. The middle term arises from the product of two systematic effects and is more problematic. This term consists of the products of the residual transverse polarizations P_x and P_y with the pairwise differences in analyzing powers δA_x and δA_y, all of which vanish in the ideal case. This term can be reduced to a negligible level by simultaneously reducing the transverse polarization and the analyzing power differences to the 10^{-3} level, making the products less than 1×10^{-5}.

Time-Dependent Errors

Time-dependent errors arise from drifts in experimental parameters such as detector gains, incident flux, and beam position on target. The primary technique for reducing the false asymmetries generated by these changes is fast spin reversal, at a rate of 10 Hz. This allows measurements to be made in each helicity state very close together in time, before significant drift occurs. Periodic changes, such as in beam intensity or polarization, will be averaged to zero if the frequency of the variation is less than 10 Hz. Previous measurements have shown that a 10 Hz spin-reversal frequency is sufficient to reduce such effects to the 10^{-6} level (5). In addition, by carefully choosing the sequence of spin reversal, the effects of any remaining drifts can be further reduced.

The importance of the spin reversal sequence can be illustrated by considering a linear detector gain drift with a positive slope (increasing with time). If the proton helicity is simply alternated between the two states $\rightarrow\leftarrow\rightarrow\leftarrow\ldots$, the second state will always have more counts, leading to a false asymmetry. However, if the four-step sequence $\rightarrow\leftarrow\leftarrow\rightarrow$ is chosen instead, the effect is cancelled. Carrying the process one step further gives the eight-step sequence $\rightarrow\leftarrow\leftarrow\rightarrow\leftarrow\rightarrow\rightarrow\leftarrow$, which cancels time-ordered drifts to quadratic order (6). This sequence is chosen for the proton helicity reversal.

Transverse Polarization Moments

In the previous discussion of the effects of transverse polarization, we implicitly assumed that the polarization was constant at all points in the beam. However, Simonius *et al.* have shown (7) that higher-order transverse polarization moments can contribute a false asymmetry even if the average transverse polarization is zero. For example, if the right half of the proton beam has a positive P_y and the left side has a negative P_y, then the false asymmetry seen by the left and right detectors will have the same sign, since the left detector sees more counts from the left side of the beam, and similarly for the right detector. Thus, the effects do not cancel when the results from the two detectors are averaged.

Fortunately, these effects are typically much smaller than 10^{-5} (8,9) and should not be significant in our case. This conjecture will be verified by directly measuring the first moments of transverse polarization with specially constructed carbon analyzer foils that are thick on one half and thin on the other. By measuring asymmetries with mirror-image pairs of analyzer foils, for example one foil thick on the right half, and the other foil thick on the left half, both x and y moments can be determined.

ESTIMATING AND MONITORING SYSTEMATIC EFFECTS

Limits on false asymmetries arising from systematic effects are determined by first measuring the sensitivities of the observed asymmetry to the effects and then measuring or estimating the size of the effect. For example, the sensitivity to transverse polarization is found by measuring the asymmetry produced by a purely transverse beam. The size of the transverse polarization during data collection is then monitored by observing the left-right scattering asymmetry from the ^{12}C target backing. Multiplying the observed transverse polarization by the sensitivity gives the false asymmetry. This procedure is repeated for each effect, including spin-correlated beam position, angle, intensity, and energy variations.

In spite of these efforts, it is possible for an unexpected false asymmetry to be present. As a final check, the energy of the proton beam will be ramped over the resonance of interest. This is achieved by applying a voltage to the target rod. The voltage is increased in small steps, typically 10 V every 800 ms, for approximately 1000 steps. At the end of the cycle, the voltage is reset and the cycle repeated. Since the energy dependence of the parity violating asymmetry is determined by the known resonance parameters, a false asymmetry will appear as a background and can be subtracted.

SUMMARY

A_z measurements at the 1×10^{-4} level are required for studies of parity violation in $(\vec{p}, \alpha)$ reactions on $A \sim 30$ nuclei. Reducing systematic uncertainties below this level requires a number of precision techniques. The azimuthal symmetry of the apparatus is maximized to reduce the sensitivity to many systematic effects, including the leading-order contributions from beam misalignment and transverse polarization components. Sensitivities to known systematic variations are determined and these variations are monitored as much as is feasible. Finally, the proton polarization and energy are varied in a manner designed to cancel systematic effects.

ACKNOWLEDGMENTS

This research was supported in part by the U.S. Department of Energy, Office of High Energy and Nuclear Physics, under Grants No. DE-FG05-91-ER40619, DE-FG05-88-ER40441, and DE-FG05-87-ER40353. The authors wish to thank D.P. Balamuth for informative discussions.

REFERENCES

1. J.D. Bowman *et al.*, Ann. Rev. Nucl. Part. Sci. **43**, 829 (1993).
2. J.F. Shriner, Jr. and G.E. Mitchell, Phys. Rev. C **49**, R616 (1994).
3. G.E. Mitchell and J.F. Shriner, Jr., Nucl. Instrum. Methods B **99**, 305 (1995).
4. G.E. Mitchell and J.F. Shriner, Jr., Phys. Rev. C **54**, 371 (1996).
5. P.R. Huffman *et al.*, Phys. Rev. Lett. **76**, 4681 (1996).
6. N.R. Roberson *et al.*, Nucl. Instrum. Methods A **326**, 549 (1993).
7. M. Simonius, R. Henneck, C. Jacquemart, and J. Lang, Nucl. Instrum. Methods **177**, 471 (1980).
8. R. Balzer *et al.*, Phys. Rev. C **30**, 1409 (1984).
9. V.J. Zeps *et al.*, Phys. Rev. C **51**, 1494 (1995).

RESONANCE NEUTRON CAPTURE STUDIES

F. Corvi[1], K. Athanassopulos[1], H. Beer[2], P. Mutti[1], H. Postma[3] and L. Zanini[1]

1. *Institute for Reference Materials and Measurements, Geel, Belgium*
2. *Forschungszentrum, Karlsruhe, Germany*
3. *Delft University of Technology, Delft, The Netherlands*

Neutron capture studies taking place at the newly refurbished GELINA pulsed neutron facility are presented. These investigations are of two types: first, the total neutron capture cross section is measured with high resolution in the keV region for isotopes which are particularly interesting for the study of s-process nucleosynthesis. From the obtained resonance parameters and/or capture areas, one can derive the stellar capture rate as a function of the temperature kT. Results have recently been obtained for the bottlenecks ^{138}Ba and ^{208}Pb, and for the s-only isotope ^{136}Ba.

A second type of experiment consists of measuring with Ge-detectors primary and secondary gamma-rays from neutron capture in single resonances in an effort to derive their spins and parities. These values are important for the interpretation of the results of parity-non-conservation measurements performed at LANSCE (LANL) by the TRIPLE collaboration. Results have just been obtained for the isotope ^{109}Ag.

INTRODUCTION

The research activity presented here has been performed at the GELINA pulsed neutron source facility situated at the EC Institute for Reference Materials and Measurements of Geel, Belgium.

The facility is based on an s-band linear accelerator operated typically to provide electron bursts of 100 MeV average energy and 1 ns width at a repetition frequency of 800 Hz and an average beam current of about 70 μA. With these parameters, fast neutrons are produced by bremsstrahlung γ-rays inside a rotary U-target at an average rate of ≃ 3×10^{13} n/s. They are subsequently moderated in two 4 cm thick water slabs in order to provide a white neutron spectrum going from thermal energy up to several MeV.

The main goal of the research studies at GELINA is to satisfy neutron data requests for the nuclear energy programme, namely for both thermal and fast reactors and associated fuel cycles. However, some research of a more fundamental character is allowed when carried out in collaboration with, or even better when stimulated by outside Institutes or Universities. A powerful help in providing the needed manpower has been the EC programme on training and mobility of young scientists. In the following, two different types of research are described: fast neutron capture measurements for nuclear astrophysics and spin and parity assignments of neutron resonances in support of parity violation (PV) experiments.

NEUTRON CAPTURE CROSS SECTIONS OF ASTROPHYSICAL INTEREST

The present measurements, being concerned with capture in stable isotopes, contribute to the study of the s-process of stellar nucleosynthesis which is mainly responsible for the build up of elements heavier than iron. The issue here is to provide values of the Maxwellian-averaged capture (MAC) cross section as a function of stellar temperature kT in the range from 5 to 100 keV. Since most, if not all, of the relevant data have already been produced in the last 20-30 years, the idea is to concentrate on a few cases which provide a deeper insight into the s-process and for which a greater accuracy (typically ± 2%) is required. Also, an important reason for repeating some measurements is that they were performed with a lower energy limit of 3 to 5 keV which is too high to determine accurately the MAC cross section at kT=8-12 keV, which is the range of stellar temperatures favoured by the most recent stellar models.

In general the nuclear transmutation flow of the synthesis is governed by the slowest conversion rates which correspond to the smallest MAC cross

CP392, *Application of Accelerators in Research and Industry,* edited by J. L. Duggan and I. L. Morgan
AIP Press, New York © 1997.

sections. These nuclides are the so-called bottleneck isotopes, i.e., those with magic neutron numbers of 50, 82 and 126. The most important of them are ^{88}Sr, ^{138}Ba and ^{208}Pb, respectively. Their role appears clearly from the approximate expression $N_A \sim [\sigma_A + \tau_o^{-1}]^{-1}$ where N_A and σ_A are the abundance and the MAC cross section, respectively, of a stable isotope of mass A, and τ_o is the average neutron exposure. When $\sigma_A \gg \tau_o^{-1}$, as in most cases, no information can be derived on the exposure. On the other hand, when $\sigma_A \leq \tau_o^{-1}$, as for the bottlenecks, the abundances are sensitive to τ_o. In the last 3-4 years we have remeasured two of these bottlenecks, namely ^{138}Ba and ^{208}Pb. Details of the measurements and results are given in refs.1,2.

The measurements were performed using the time-of-flight technique and so-called total energy detectors, i.e., rather small C_6D_6-based liquid scintillators whose amplitude pulses were appropriately weighted in order to make the efficiency proportional only to the total γ-energy emitted. In the case of ^{138}Ba, the capture areas of 138 resonances were determined in the energy range 0.5-200 keV. The derived MAC cross section is compared in Fig. 1 with the ORELA (3) data and with a Karlsruhe (4) activation data point at 25 keV. While both curves agree with this last value, they diverge at lower temperature, our data suggesting a much steeper rise with decreasing kT. The reason for the discrepancy is that we found two new strong p-wave levels below 3 keV.

In a second exercise, the MAC cross section of ^{208}Pb was derived by measuring the capture areas of 12 resonances observed below 400 keV. In this case, however, the compound capture cross section is so small that direct capture cannot be neglected. This second component, mainly consisting of p-wave capture, was calculated by H. Oberhummer et al. (5). The two contributions, are separately plotted *vs* kT in Fig. 2: note that their sum, also plotted, is in excellent agreement with an activation measurement which is sensitive to both compound and direct capture (6). From Fig. 2 we also note that the σ-curve derived from the data of Macklin et al. (7) is about a factor of two larger than ours: we ascribe this discrepancy to an insufficient correction for the prompt neutron scattering background in resonances with large values of the scattering-to- capture ratio.

Finally we have recently measured neutron capture in the s-only isotope ^{136}Ba (8): this nuclide, as well as the other barium isotopes, has recently received much attention because s-process calculations yield a barium overproduction of at least 20% of the known solar abundance.

The MAC cross section values, derived from the capture areas of 184 resonances observed by us below 60 keV, are listed in Table 1 together with the corresponding values recently obtained in Karlsruhe (9) and Oak Ridge (10) : the agreement is excellent particularly with the Karlsruhe data. The fact that this agreement has been obtained with two very different detection techniques (total energy and total absorption detectors, respectively) is the best demonstration that very accurate values can be obtained in most cases.

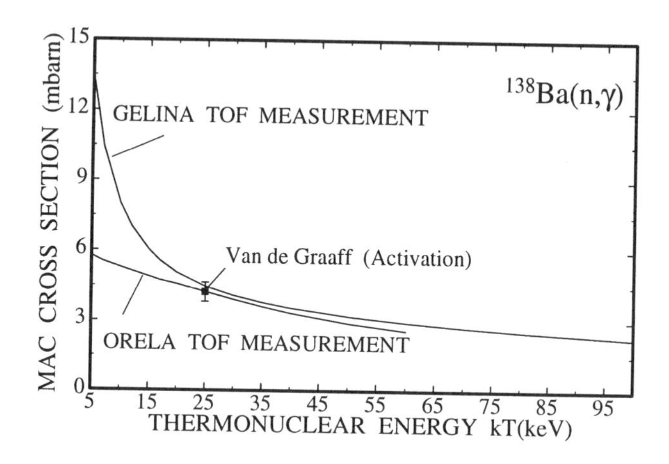

FIGURE 1. The MAC cross section of ^{138}Ba.

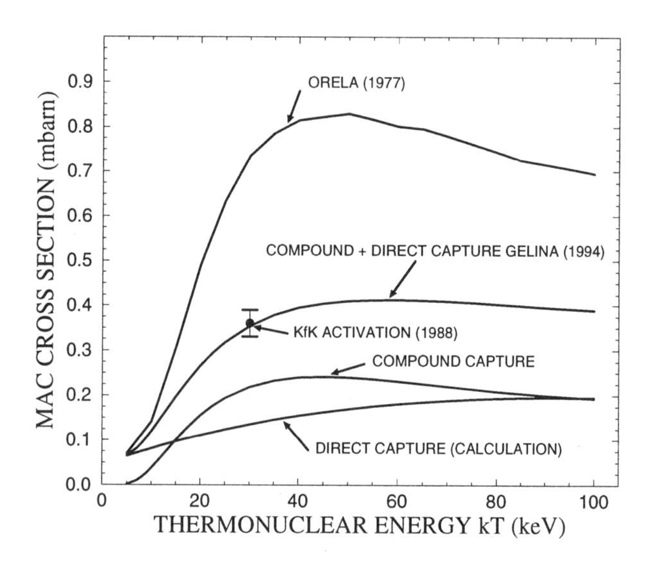

FIGURE 2. The MAC cross section of ^{208}Pb.

TABLE 1. The MAC cross section of ^{136}Ba

kT (keV)	$<\sigma V>/V_T$ (mb) Present Work	Voss et al.[9]	Koehler et al.[10]
5	166.3 ± 4.9		186.0 ± 6.1
8	124.2 ± 4.5		139.4 ± 4.5
10	108.6 ± 4.0	109.7 ± 4.6	122.0 ± 3.9
12	97.7 ± 3.8	99.3 ± 3.9	
15	86.2 ± 3.6		94.6 ± 3.0
20	74.1 ± 3.3	74.9 ± 2.7	79.2 ± 2.5
25	66.2 ± 3.1	66.5 ± 2.3	69.2 ± 2.2
30	60.4 ± 2.9	60.6 ± 2.1	62.0 ± 2.0

SPIN AND PARITY ASSIGNMENTS OF NEUTRON RESONANCES

Extensive investigations on the parity violation (PV) of neutron resonances have been carried out in recent years by the TRIPLE collaboration at the LANSCE facility in the Los Alamos National Laboratory (11,12). After having successfully detected PV-effects in several p-wave resonances of ^{238}U and ^{232}Th, the TRIPLE collaboration has more recently concentrated its attention on the region of the 3p peak of the neutron strength function. Isotopes such as ^{106}Pd, ^{108}Pd, ^{107}Ag, ^{109}Ag, ^{113}Cd and ^{115}In have been studied.

Knowledge of the spins of the relevant resonances are of great help in the analysis of the PV-data; in particular this information becomes essential in the case of non-zero spin targets for which not only the spins of the p-wave resonances but also those of the s-wave resonances are required. For this reason we have extended our spin determination program to the target nucleus ^{109}Ag after having successfully completed the ^{238}U and ^{113}Cd case (13,14).

PV effects originate from the mixing of compound nuclear states of same spin but opposite parity. In the case of the odd-even target nucleus ^{109}Ag, with spin and parity $I^\pi=1/2^-$, the p-resonances with $J^\pi=0^+$ and 1^+ can mix with $J^\pi=0^-$, and 1^- s-waves, respectively. On the other hand $J^\pi=2^+$ p-waves cannot show any PV effect.

We have applied, as in previous cases, the low-level population method of spin assignment which exploits the fact that the population of the low-lying excited states reached by radiative neutron capture depends significantly on the initial spin.

This work has recently been completed and has already been presented in some detail (15). In the following only a summary of the results is given.

For assigning the spins of s-wave levels, we have calculated the intensity ratios of the two secondary γ-rays at 338.9 and 350.1 keV for 55 resonances in the neutron energy range 10-1400 eV.

The ratios are plotted *versus* neutron energy in the upper part of Fig. 3: there is a clear splitting into two groups, which we associate with the two possible spin values. In this way we could assign $J^\pi=0^-$ to twelve and $J^\pi=1^-$ to forty-three resonances.

In the case of p-levels, we calculated the intensity ratios between the doublet at 235.7-237.1 keV and the peak at 191.5 keV for twenty-three resonances: the results, given in the lower part of Fig. 3, show clearly a split into three groups corresponding to the three possible spin values.

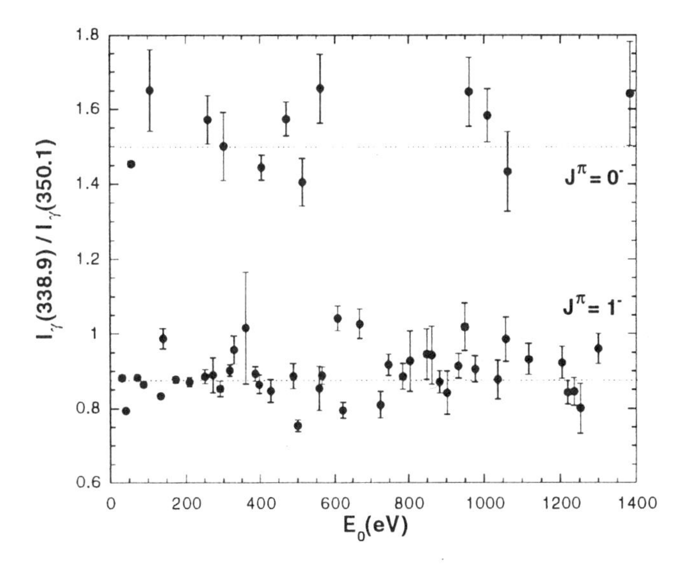

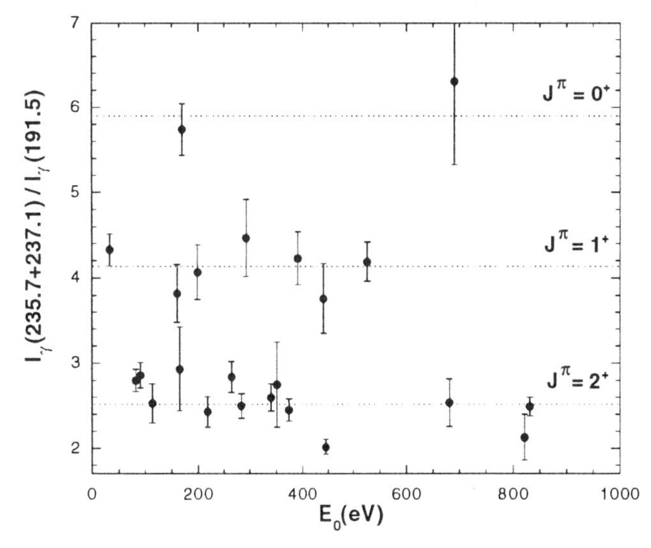

FIGURE 3. Intensity ratios between indicated gamma lines plotted *versus* the energy of 55 s-wave (upper) and 23 p-wave resonances (lower).

In Table 2 we list our spin assignments for the four l=1 levels exhibiting a PV effect, according to the TRIPLE results (16). The P_i/σ_i values in the second column are the preliminary ratios between the longitudinal asymmetries and their standard deviations, and as such they quantify the statistical significance of the effect. In the 4th and 5th column the nearest s-wave levels and their spins are given. We notice that, for the three most significant results, we assign J=1 to both the l=1 and the nearest l=0 levels, in keeping with expectations. On the other hand we disagree for the E_o = 91.5 eV resonance since $J^{\Pi} = 2^+$ states cannot present any PV effect.

TABLE 2. Spin values of both the p-wave resonances exhibiting statistically significant PV effects and the nearest s-wave levels.

p-wave resonances			s-wave resonances	
E_o (eV)	P_i/σ_i	J^{Π}	E_o (eV)	J^{Π}
32.7	42.4	1^+	30.4	1^-
91.5	3.6	2^+	87.7	1^-
199.0	4.3	1^+	209.6	1^-
293.3	8.6	1^+	291.0	1^-

In the interpretation of these spin assignment experiments, we are sometimes confronted with the problem of resonance parity determination. When data are not available from the literature, the usual procedure is to apply the Bayes theorem on conditional probability. However, when the probabilities of a level being l=0 or l=1 are comparable, this method gives an ambiguous answer. In an effort to gather additional evidence in favour of one l-value, we have closely examined the low energy γ-ray spectra and have found that there is a general tendency to populate more those ^{110}Ag bound levels of parity opposite to that of the initial state. Thus an intensity ratio between two transitions de-exciting two states of opposite parity should give an indication on the l-value of the resonance.

In Fig. 4 are plotted, for all resonances with J=1, the intensity ratios of the line at 198.7 keV, de-exciting a 2^+ state, and of the doublet at 235.7-237.1 keV, both de-exciting negative parity states. The result is striking: there is a clear-cut division between the two groups of s- and p-wave levels and the magnitude of the effect is comparable to the one related to the spin dependence.

A similar picture is obtained for the sample of J=0 resonances. We believe that this unexpected result is reported for the first time in the literature.

We have exploited the effect for assigning the parity of the resonances at 106.3, 169.8 and 360.4 eV, and for checking all other values. Besides this application, however, the effect deserves to be studied on its own merit since it can yield information on the mechanism of radiative capture. We intend therefore to pursue these investigations in two directions: first, by searching for the effect in other nuclides, second by trying to reproduce the effect in simulations of the gamma decay.

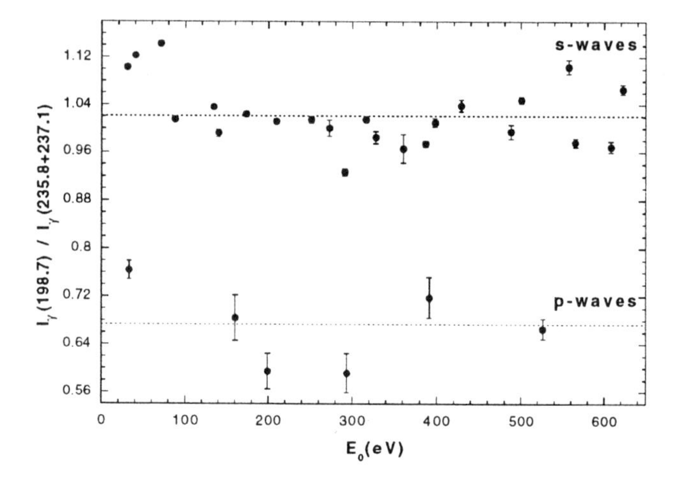

FIGURE 4. Intensity ratios for J=1 resonances.

REFERENCES

1. Beer, H. et al., *Capture Gamma Ray Spectroscopy* ed. J. Kern, World Scientific, Singapore, 1994, p. 698.
2. Corvi, F. et al., *Nuclei in the Cosmos III*, ed. M. Busso et al., AIP Press, New York, p. 165, 1995
3. de L. Musgrove, A.R. et al., *Aust. J. Phys.* **32**, p. 213, 1979
4. Beer, H. et al., *Phys. Rev.* **C21**, 534, 1980
5. Oberhummer, H. et al., *Surveys in Geophys*, in print.
6. Ratzel, U., KfK Karlsruhe, report, unpublished, 1988.
7. Macklin, R.L. et al., *Ap. J.* **217**, 222, 1977.
8. Mutti, P. et al., *Nuclei in the Cosmos IV*, ed. M. Wiescher, *Nucl. Phys. A* (Suppl.) in print.
9. Voss, F. et al., *Phys. Rev.* **C52**, 1102, 1995.
10. Koehler, P.E. et al., *Phys. Rev.* **C54**, 1463, 1996.
11. Zhu, X. et al., *Phys. Rev.* **C46**, 768, 1992.
12. Frankle, C.M. et al., *Phys. Rev.* **C46**, 778, 1992.
13. Gunsing, F. et al., *Capture Gamma-Ray Spectroscopy*, ed. J. Kern, World Scientific, 1994, 707.
14. Corvi, F. et al., *Nuclear Data for Science and Technolgoy*, ed. J.K. Dickens, ORNL, 1994, p. 201.
15. Zanini, L. et al., *Proc. 9th. Intern Symp. on Capture Gamma-Ray Spectroscopy*, Budapest, 8-12 Oct. 1996.
16. Mitchell, G.E., *private communication*, 1996.

Trapped-Ion Based Technique for Measuring the Nuclear Radii of Highly-Charged Radioactive Isotopes

S. R. Elliott
Department of Physics, University of Washington, Seattle, WA 98195

P. Beiersdorfer, and M. H. Chen
Department of Physics and Space Technology, Lawrence Livermore National Laboratory, Livermore, CA 94550

We present measurements to isolate the variation of nuclear effects in the x-ray transitions of few-electron heavy ions. Using a technique to produce and trap radioactive ions we measured the energy difference between $2s_{1/2}$ - $2p_{3/2}$ transitions in Li-, Be-, B-, and C-like ^{233}U, ^{235}U and ^{238}U. Due to the simplified atomic structure of few electron ions, the data are readily intrepreted in terms of the variation in the mean nuclear radius. A value $\delta\langle r^2\rangle^{233,238}$ = -0.432 ± 0.043 fm^2 is found, which differs from earlier measurements based on different techniques. A value of $\delta\langle r^2\rangle^{235,238}$ = -0.250 ± 0.032 fm^2 is also found.

INTRODUCTION

The charge distribution variation, parameterized in terms of the change in mean square radius, $\delta\langle r^2\rangle$, has been inferred in the high-Z region from neutral-atom optical isotope shift studies (1) and muonic-atom x rays (2) whereas electron scattering has provided a value for $\langle r^2\rangle^{1/2}$ for ^{238}U only (3). Neutral-atom optical and Kα x-ray transition isotope shifts are sensitive to the same nuclear parameters. However, the Kα results are easier to interpret in terms of $\delta\langle r^2\rangle$. There is a large theoretical uncertainty in the optical-transition specific mass shift which makes it difficult to deduce nuclear parameters from those measurements. For example, the best uranium $\delta\langle r^2\rangle^{233,238}$ value (defined as $(r^2_{RMS})^{233}$ - $(r^2_{RMS})^{238}$ where r_{RMS} is the root mean square radius) results come from Ref. (4). Because the configuration interaction is so difficult to calculate, they used Kα data (5) to calibrate the optical $\Delta E^{233,238}$ in terms of $\delta\langle r^2\rangle^{233,238}$. Thus, even though laser spectroscopy gave a very precise value of ΔE, the uncertainty in $\delta\langle r^2\rangle$ is determined by the less precise Kα data. In muonic atoms the muon wavefunction extends much deeper into the nucleus than would that of a bound electron. Therefore higher moments in the nuclear charge density distribution ($\delta\langle r^4\rangle$, $\delta\langle r^6\rangle$, etc.) affect ΔE. In contrast, isotope shifts involving electronic transitions are primarily sensitive to the first moment, $\delta\langle r^2\rangle$. Furthermore, vacuum polarization and nuclear polarization effects contribute a much larger share to muonic energy levels than for electronic levels. Thus, although muonic atom measurements have provided data on a number of high-Z isotopes, they are complimentary to the atomic transition data. Interestingly, these earlier measurements of $\delta\langle r^2\rangle^{233,238}$ have produced discrepant results, i.e. (-0.520 ± 0.081 fm^2) (6) and (-0.383 ± 0.044 fm^2) (4,5).

We have previously demonstrated (7) the advantages of trapping few-electron, very-high-Z radioactive ions in an electron beam ion trap (EBIT) for studies of $\delta\langle r^2\rangle$. Using precision x-ray spectroscopy and exploiting the simplified electronic structure of few-electron ions, we have isolated the nuclear effects and inferred the isotopic variation of the nuclear charge distribution between ^{233}U, ^{235}U and ^{238}U.

EXPERIMENT

Our technique for determining $\delta\langle r^2\rangle$ is based on precise Doppler-shift-free measurements of the n=2 to n=2 x-ray transitions in nearly bare ions of the isotopes in question. The transitions studied are the electric dipole, $2s_{1/2}$-$2p_{3/2}$ transitions in the Li-like ion, Be-like ion, B-like ion, and C-like ion. The specific transitions studied and their previously measured energies in ^{238}U are given in Ref. (8). Because the measurements are for transitions in an inner shell, the electron wavefunction overlap, especially that of the 2s electron, with the nucleus is large. It is thus an excellent probe of the nuclear charge distribution resulting in a relatively large energy shift (ΔE) as different isotopes are measured. Compared to muonic atoms, however, the overlap is modest and nuclear polarization corrections are essentially avoided. Moreover, the atomic physics of few-electron ions is tractable, and deducing $\delta\langle r^2\rangle$ from ΔE is relatively simple. Most importantly, it is not complicated by large specific mass shift corrections necessary in neutral ions (9). In other words, in our measurement the coulomb shift (δE_{coul}), which is directly related to $\delta\langle r^2\rangle$, is by far the dominant contribution to ΔE, and other atomic or nuclear contributions are minimal. A further benefit of our technique is that the energy of the Δn=0 transitions studied falls within a range where high-precision crystal spectroscopy is easily employed.

The measurements were done at the high-energy electron beam ion trap (SuperEBIT) at Lawrence Livermore National Laboratory (10). An electron beam ionizes, excites, and radially traps the ions. The ions are trapped axially by

CP392, *Application of Accelerators in Research and Industry*, edited by J. L. Duggan and I. L. Morgan
AIP Press, New York © 1997.

potential differences between three co-linear cylindrical electrodes through which the beam passes. Low-charge ions, injected into the trap, are ionized to high charge states by successive collisions with beam electrons.

The ^{233}U and ^{235}U ions were provided by a thin wire platinum probe with a plated tip placed near the electron beam (11). This probe provides a continuous source of ions for the trap. The total mass of plated ^{233}U (^{235}U) was only 100 ngm (20 mgm) and the ^{233}U (^{235}U) was isotopically enriched to 99.92% (99.77%). The ^{238}U ions were provided by a metal vapor vacuum arc source (MEVVA) (12) using a ^{238}U cathode depleted in ^{235}U weighing 14 gm.

The ions are studied by their characteristic x rays observed through ports in the cryogenic vessels surrounding the trap. The $2s_{1/2}$-$2p_{3/2}$ electric dipole transitions, situated near 4.5 keV, were analyzed in a high-resolution von Hámos-type curved-crystal spectrometer (13). The spectrometer uses a 120 x 50 x 0.25 mm^3 LiF(200) crystal (2d = 4.027Å) bent to a 75-cm radius of curvature. X rays are recorded with a gas-filled position sensitive proportional counter. The energy resolution of the setup was 1.1 eV FWHM. The x-ray spectra of ^{233}U and ^{235}U were, each in turn, compared with that from ^{238}U.

Measuring $\Delta E^{A,238}$ requires knowledge of the dispersion of the spectrometer but an absolute calibration is unnecessary. To determine the dispersion, we employed the ^{238}U-transition energy measurements of Ref. (8). The dispersion uncertainty from this procedure is 0.4% which results in a 1-meV uncertainty in $\Delta E^{A,238}$. Because we are measuring energy differences between nearby lines, many systematic errors, such as detector non-linearities, cancel permitting very precise measurements.

Data collection alternated between spectra. By interleaving the spectra, we could monitor and correct for any possible electronic gain shifts. The uncertainty associated with electronic gain drifts is approximately 5 meV.

ANALYSIS

Tables 1 and 2 summarize ΔE for each transition measured. The contributions from systematic errors to the overall uncertainty of each ΔE value are small and are summarized below. Line fits to these $\Delta E^{233,238}$ ($\Delta E^{235,238}$) data results in a slope, 32 ± 36 meV/charge (15 ± 27 meV/charge), which are consistent with zero; that is, ΔE is nearly independent of the charge state. This finding is confirmed in a theoretical study of the effect of electron correlations on the transition energies. We calculated $\Delta E^{233,238}$ for the four ionization stages using a multi-configuration Dirac-Fock (MCDF) (14) code and found differences no larger than 11 meV, affirming the small size of electron correlations. We performed a second calculation of the $\Delta E^{233,238}$ for the Li-like and Be-like transitions using a relativistic configuration interaction (RCI) code. The RCI calculations were done by increasing the basis set until convergence was achieved (15). The results agreed within 0.1 meV with those from the MCDF calculations, affirming the predictive power of our calculations for ΔE and providing a uncertainty of less than 1 meV in the calculated size of the isotopic variation in the electron correlations.

Table 1: A summary of the δE_{coul} and the deduced $\delta\langle r^2\rangle^{233,238}$ values for each charge state. The uncertainties listed are entirely statistical.

Key	$\Delta E^{233,238}$ (meV)	δE_{coul} (meV)	$\delta\langle r^2\rangle^{233,238}$ (fm^2)
Li	256 ± 118	260 ± 118	-0.338 ± 0.153
Be	300 ± 61	304 ± 61	-0.409 ± 0.081
B-1,2	320 ± 52	324 ± 52	-0.428 ± 0.068
C	362 ± 62	366 ± 62	-0.488 ± 0.083

Table 2: A summary of the δE_{coul} and the deduced $\delta\langle r^2\rangle^{235,238}$ values for each charge state. The uncertainties listed are entirely statistical.

Key	$\Delta E^{235,238}$ (meV)	δE_{coul} (meV)	$\delta\langle r^2\rangle^{235,238}$ (fm^2)
Li	63 ± 81	65 ± 81	-0.085 ± 0.106
Be	209 ± 42	211 ± 42	-0.284 ± 0.056
B-1,2	200 ± 38	202 ± 38	-0.267 ± 0.050
C	175 ± 52	177 ± 52	-0.236 ± 0.070

In order to infer δE_{coul} and thus $\delta\langle r^2\rangle$ from ΔE, we need to estimate the isotopic variation of the specific mass shift, of the QED terms, and of the nuclear polarization (9). The advantage of our technique is that all these terms are small with correspondingly small uncertainties. The specific mass shift, also called the mass polarization contribution, has been calculated for the Li-like U^{89+} ion (16,17) and is similar for all ionization states under consideration here. It is found to be of the order of 50 meV with a theoretical uncertainty of 100% due to presently ignored terms of order $(Z\alpha)^2$. We estimate the isotopic variation of this value to be on the order of 1% (the mass difference between ^{233}U and ^{238}U) or less than 1 meV.

The estimate of the QED self energy contribution to these energy levels is about 57 eV (18). The finite nuclear size correction to this value is about 800 meV for ^{238}U. The dependence of this value on isotope can be estimated from Ref. (18) to be 8 meV (4 meV) for ^{233}U (^{235}U). The QED vacuum polarization contribution to the energy levels also has a nuclear size correction. The vacuum polarization contributes about -14 eV and the nuclear size correction is about -760 meV (17). Note that this value is almost equal and of opposite sign of the self energy contribution. Thus not only is the isotopic dependence of these two effects small, but they tend to cancel and can be ignored in the analysis of these data.

Nuclear polarization, or nuclear polarizability, calculations have been done for the 1s, 2s, and 2p levels in H-like U^{91+} ions for the even-A isotopes (19). These calculations show a modest isotopic dependence which must be taken into consideration in our data. It would be preferable if calculations existed also for the odd-A isotopes. We are forced to extrapolate the values of the even-A results to that for ^{233}U and ^{235}U. (Note that Refs. (4, 6) indicate that any even-odd staggering in this isotopic region is small compared to present experimental precision.) Since the entire correction for the singly-excited $2s_{1/2}$ - $2p_{3/2}$ transitions measured in this work comes from the 2s shell, the values calculated for the H-like U^{91+} $2s_{1/2}$ level accurately approximates that of all the charge states considered here. The nuclear polarization contribution difference between ^{233}U (^{235}U) and ^{238}U is 4 meV (2 meV). The authors of Ref. (19) estimate the uncertainty in their calculations of the absolute size of the nuclear polarization contribution to be ±25%. Thus we take the difference value also to be uncertain by 25%, or ±1 meV. Eliminating the nuclear polarization contribution to ΔE yields a final values for δE_{coul} for each charge state as summarized in Tables 1 and 2.

THE NUCLEAR SIZE

The MCDF (14) calculations use a nuclear charge density function, ρ(r), described by the two-parameter Fermi distribution

$$\rho(r) = \rho_o/(1 + e^{[r-\mu]/\tau})$$

where r is the radius, μ is the half-density radius, and τ is the skin thickness. The resultant energy level determinations, however, are not sensitive to actual charge distribution provided that the associated root mean square radius (r_{RMS}) is reproduced. We calculated δE_{coul} for each charge state for 22 values of μ between 7.03811 and 7.14395 fm, holding τ constant, and computed the corresponding r_{RMS}. The results of these calculations provide $\delta\langle r^2\rangle^{A,238}$ as a function of δE_{coul} for each ionization stage. The origin is defined as the values for ^{238}U (μ = 7.13753 fm and τ = 0.52339 fm) which correspond to a two-parameter Fermi distribution with r_{RMS} = 5.8610 fm. This r_{RMS} is equal to the value one derives from a four-parameter deformed-Fermi distribution using the parameters given in Ref. (6).

To deduce $\delta\langle r^2\rangle^{A,238}$ from δE_{coul} using the curves described above, we did a quadratic interpolation between the calculated points for each charge state. The results for A=233, listed in Table 1, were then averaged and we find $\delta\langle r^2\rangle^{233,238}$ = -0.432 fm^2 with a statistical uncertainty of 0.042 fm^2. This procedure of deducing $\delta\langle r^2\rangle$ for each charge state seperately and then averaging, ensures proper treatment of the electron correlation contribution. Figure 1 shows $\delta\langle r^2\rangle^{A,238}$ as a function of charge state. The systematic uncertainty in δE_{coul} (5 meV) translates into a systematic uncertainty in $\delta\langle r^2\rangle^{233,238}$ of 0.006 fm^2. Adding the uncertainties in quadrature, the final result is $\delta\langle r^2\rangle^{233,238}$ = -0.432 ± 0.043 fm^2. (Note that this value differs by 5% from that of Ref. (7). The entire difference is attributed to an error in the nuclear polarization contribution by Plunien *et al.* (19) corrected in the subsequently published erratum (20).) This result can be compared with that of previous studies; -0.383 ± 0.044 fm^2 (4, 5) and -0.520 ± 0.081 fm^2 (6). The present measurement thus favors neither of the earlier measurements. The weighted mean of all measurements is -0.422 ± 0.029 fm^2. All three experiments are consistent with this mean value to within 1 to 2 standard deviations.

The results for A=235, listed in Table 2, were then averaged and we find $\delta\langle r^2\rangle^{235,238}$ = -0.250 fm^2 with a statistical uncertainty of 0.031 fm^2. The systematic uncertainty in δE_{coul} (5 meV) translates into a systematic uncertainty in $\delta\langle r^2\rangle^{235,238}$ of 0.006 fm^2. Adding the uncertainties in quadrature, the final result is $\delta\langle r^2\rangle^{235,238}$ = -0.250 ± 0.032 fm^2.

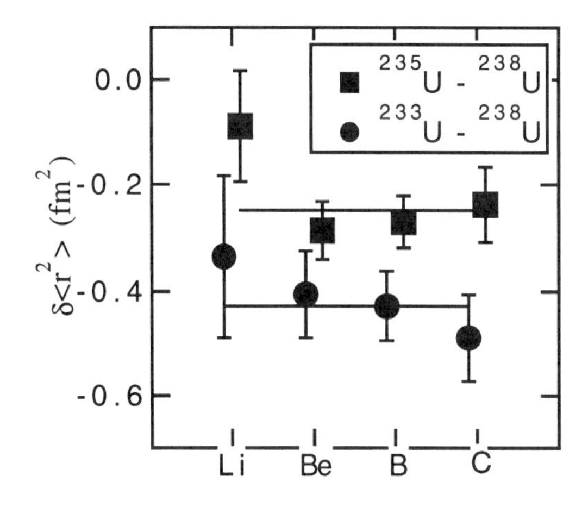

Figure 1: The measured $\delta\langle r^2\rangle^{A,238}$ as a function of ion charge state. The solid lines represent the weighted average of the 4 transitions.

DISCUSSION

Ref. (4) used the Kα value for $\delta\langle r^2\rangle^{233,238}$ (5) to calibrate isotopic frequency shifts (δν) in terms of $\delta\langle r^2\rangle^{A,238}$. We performed a similar procedure using the two $\delta\langle r^2\rangle$ values presented in this letter to calibrate the δν values. The deduced constant to convert $\delta\nu^{233,238}$ to $\delta\langle r^2\rangle^{233,238}$ is -3.27±0.24 x 10^{-5} fm^2/MHz. The deduced constant to convert $\delta\nu^{235,238}$ to $\delta\langle r^2\rangle^{235,238}$ is -2.95±0.38 x 10^{-5} fm^2/MHz. The weighted average of these two constants is -3.16±0.20 x 10^{-5} fm^2/MHz. Using two calibration points results in a lower uncertainty as compared to the previous determination based on a single Kα measurement. The value of $\delta\langle r^2\rangle^{233,238}$ determined in this way is -0.416 ± 0.028 fm^2. Figure 2 shows a summary of all the U $\delta\langle r^2\rangle$ measurements under discussion here.

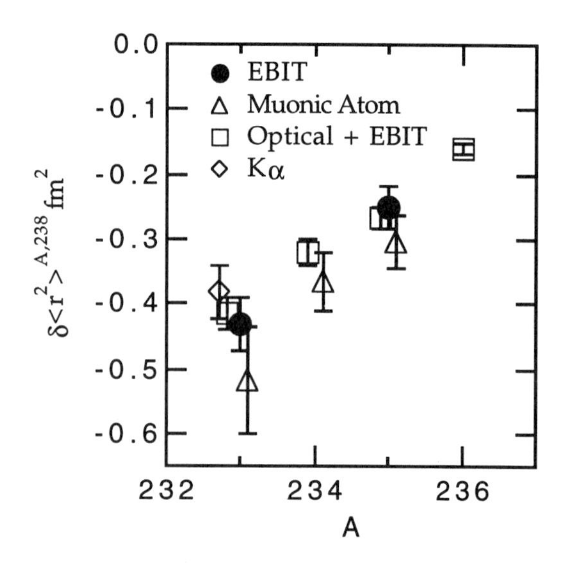

Figure 2: A summary of the measurements of $\delta\langle r^2\rangle^{A,238}$ in the U system. The optical data was normalized as described in the text using the current results as the normalization.

In summary, we have performed high-precision spectroscopy on radioactive few-electron highly charged uranium ions trapped in an EBIT. These measurements in radioactive ions were made possible by the application of an ion source which permits the use of material in nanogram samples. The isotopic energy shift between ^{235}U and ^{238}U was measured for $2s_{1/2}$ - $2p_{3/2}$ transitions in U^{86+} through U^{89+} ions. In addition, we reevaluated the previous EBIT $\delta\langle r^2\rangle^{233,238}$ value in light of a recent erratum concerning the contribution of nuclear polarization. These precise isotope shift measurements have been made in electronic transitions that are strongly afected by quantum electrodynamics, effectively measureing the Lamb shift for various isotopes of U. The simple atomic structure of these few electron ions permitted the relatively easy interpretation of these energy shifts in terms of nuclear radii changes. The current results are statistically limited.

ACKNOWLEDGMENTS

We are grateful for the technical expertise and assistance of E. W. Magee and R. Lougheed. This work was performed under the auspices of the U. S. Department of Energy at Lawrence National Laboratory under Contract No. W-7405-Eng-48.

REFERENCES

1. P. Aufmuth, K. Heilig, and A. Steudel, At. Data Nucl. Data Tab. **37**, 455 (1987).
2. R. Engfer, H. Schneuwly, J. L. Vuilleumier, H. K. Walter, and A. Zehnder, At. Data Nucl. Data Tab. **14**, 509 (1974).
3. H. de Vries, C. W. de Jager, C. de Vries, At. Data Nucl. Data Tab. **36**, 495 (1987).
4. A. Anastassov, Yu. P. Gangrsky, K. P. Marinova, B. N. Markov, B. K. Kul'djanov, and S. G. Zemlyanoi, Hyper. Inter. **74**, 31 (1992).
5. R. T. Brockmeier, F. Boehm, and E. N. Hatch, **15**, 132 (1965).
6. J. D. Zumbro, E. B. Shera, Y. Tanaka, C. E. Bemis, Jr., R. A. Naumann, M. V. Hoen, W. Reuter, and R. M. Steffen, Phys. Rev. Lett. **53**, 1888 (1984).
7. S. R. Elliott, P. Beiersdorfer, and M. H. Chen, Phys. Rev. Lett. **76**, 1031 (1996).
8. P. Beiersdorfer, D. Knapp, R. E. Marrs, S. R. Elliott, and M. H. Chen, Phys. Rev. Lett. **71**, 3939 (1993).
9. W. H. King, Isotope Shifts in Atomic Spectra (Plenum Press, New York, 1984).
10. D. A. Knapp, R. E. Marrs, S. R. Elliott, E. W. Magee, and R. Zasadzinski, Nucl. Instr. and Meth. **A334**, 305 (1993).
11. S. R. Elliott, and R. E. Marrs, Nucl. Instr. and Meth. **B100**, 529 (1995).
12. I. G. Brown, J. E. Galvin, R. A. MacGill, and R. T. Wright, Appl. Phys. Lett. **49**, 1019 (1986).
13. P. Beiersdorfer, R. E. Marrs, J. R. Henderson, D. A. Knapp, M. A. Levine, D. B. Platt, M. B. Schneider, D. A. Vogel, and K. L. Wong, Rev. Sci. Instr. **61**, 2338 (1990).
14. I. P. Grant, B. J. McKenzie, P. H. Norrington, D. F. Mayers, and N. C. Pyper, Comput. Phys. Commun. **21**, 207 (1980); and B. J. Mckenzie, I. P. Grant, and P. H. Norrington, Comput. Phys. Commun. **21**, 233 (1980).
15. M. H. Chen, K-T. Cheng, and W. R. Johnson, Phys. Rev. **A47**, 3692 (1993).
16. S. A. Blundell, W. R. Johnson, and J. Sapirstein, Phys. Rev. **A41**, 1698 (1990).
17. S. A. Blundell, Phys. Rev. **A46**, 3762 (1992).
18. P. J. Mohr and G. Soff Phys. Rev. Lett. **70**, 158 (1993).
19. Guenter Plunien, Berndt Müller, Walter Greiner, and Gerhard Soff, Phys. Rev. **A43**, 5853 (1991); and Guenter Plunien and Gerhard Soff, Phys. Rev. **A51**, 1119 (1995).
20. Guenter Plunien and Gerhard Soff, Phys. Rev. **A53**, 4614 (1996).

NEW BINARY AND TERNARY SPONTANEOUS MODES FOR ^{252}Cf AND NEW BAND STRUCTURES WITH GAMMASPHERE

J.H. Hamilton[1], A.V. Ramayya[1], G.M. Ter-Akopian[1,2,3], J.K. Hwang[1], J.Kormicki[1], B.R.S. Babu[1], A. Sandulescu[1,3,4], A. Florescu[1,3,4], W. Greiner[1,3,5], Yu. Ts. Oganessian[2], A.V. Daniel[1,2], S.J. Zhu[1,3,6], M.G. Wang[6], T. Ginter[1], J.K. Deng[1], W.C. Ma[7], G.S. Popeko[2], Q.H. Lu[1], E. Jones[1], R. Dodder[1], P. Gore[1], J.O. Rasmussen[8], S. Asztalos[8], I.Y. Lee[8], S.Y. Chu[8], K.E. Gregorich[8], A.O. Macchiavelli[8], M.F. Mohar[8], S. Prussin[9],M.A. Stoyer[10], R.W. Lougheed[10], K.J. Moody[10], J.F. Wild[10], J.D. Cole[11], R. Aryaeinejad[11], Y.X. Dardenne[11], M.W. Drigert[11], K. Butler-Moore[11]

[1]*Vanderbilt University, Physics Department, Nashville, TN 37235*
[2]*Joint Institute for Nuclear Research, Dubna 141980, Russia*
[3]*Joint Institute for Heavy Ion Research, Oak Ridge, TN 37831*
[4]*Institute of Atomic Physics, Bucharest, P.O. box MG-6, Romania*
[5]*Institute für Theoretische Physik, der Universitat Frankfurt/Main, D-60054 Germany*
[6]*Tsinghua University, Physics Department, Beijing, P.R. China*
[7]*Mississippi State University, Physics Department, Mississippi State, MS 39762*
[8]*Lawrence Berkeley National Laboratory, Berkeley, CA 94720*
[9]*Univeristy of California/Berkeley, Berkeley, CA 94720*
[10]*Lawrence Livermore National Laboratory, Livermore, CA 94550*
[11]*Idaho National Engineering Laboratory, Idaho Falls, ID 83415*

Prompt γ-γ-γ coincidence studies following spontaneous fission of ^{252}Cf were carried out at Gammasphere first with 36 and later with 72 Ge detectors. Yields and neutron multiplicities were measured directly for Sr-Nd, Zr-Ce Mo-Ba, Ru-Xe, and Pd-Te correlated pairs. A new hot fission mode was discovered going via only ^{108}Mo-^{144}Ba, ^{107}Mo-^{145}Ba, and/or ^{106}Mo-^{146}Ba pairs where one or more of ^{144,145,146}Ba are hyperdeformed at scission with 3:1 axis ratio. For the first time correlated pairs involved in cold ternary fission were observed with the third particle ^{4}He, possibly ^{6}He, and ^{10}Be. The $^{142-146}$Ba data provide evidence for the predicted disappearance of stable octupole deformation at high spins. New level structures and isotopes include new octupole deformations, identical bands and other structures.

EXPERIMENTAL DESCRIPTION

Studies of the spontaneous fission of actinide elements with new large detector arrays provide new opportunities to probe the fission process and structures of neutron-rich nuclei. The spontaneous fission (SF) of a source of ^{252}Cf was studied in December 1995 with 72 large Compton suppressed Ge detectors in the Gammasphere array at Lawrence Berkeley National Laboratory. Events in which three or more γ-rays were detected in coincidences were recorded. A γ-γ-γ cube data set was built from the data. These data allow one to gate on one known gamma ray in each of the partner fragments or two gamma rays in one partner and look for all the other partners corresponding to different neutron channels. By doing this for several different partner γ-rays in both fragments and in each fragment, one can uniquely identify the partners and obtain the relative yields of each of the various partners in coincidence with a given light or heavy fragment. In this way we have uniquely obtained the yields for correlated fragment pairs and via the missing mass and charge compared to ^{252}Cf identified ternary fission modes and finally investigated the level schemes of many neutron-rich nuclei.

ULTRA HOT FISSION MODE

The first direct measurements of yields and neutron multiplicities of correlated fragment pairs of Zr-Ce and Mo-Ba in the spontaneous fission (SF) of ^{252}Cf were recently reported (1,2). We have repeated these measurements with the 1995 Gammasphere data. The 8-10 neutron multiplicities in Mo-Ba pairs are considerably enhanced compared to those for gross (total) neutron multiplicities (3) and other pairs (see Fig. 1). From both the new and older data, we find that there are two fission modes in the breakup into Mo-Ba: first, the normal mode with large total kinetic energy (TKE) and broad mass distribution of primary fragments, and second a new ultrahot mode in which the TKE is much lower and the mass distribution is limited to one or more of three pairs, ^{108}Mo-^{144}Ba, ^{107}Mo-^{145}Ba, ^{106}Mo-^{146}Ba where at scission at least one or all of ^{144,145,146}Ba are

CP392, *Application of Accelerators in Research and Industry*, edited by J. L. Duggan and I. L. Morgan
AIP Press, New York © 1997.

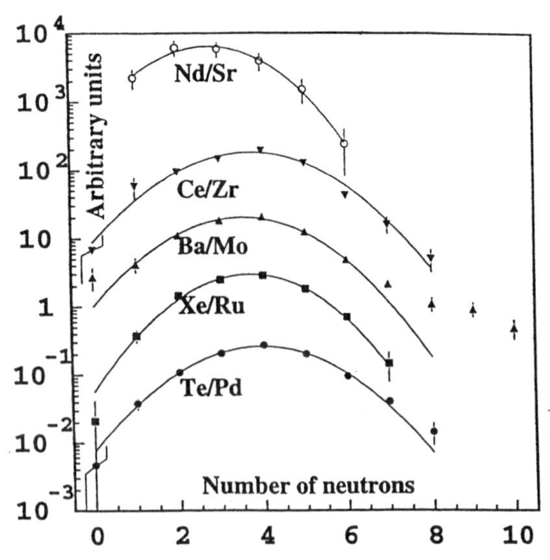

Figure 1. Multiplicity distribution of prompt neutrons for ^{252}Cf.

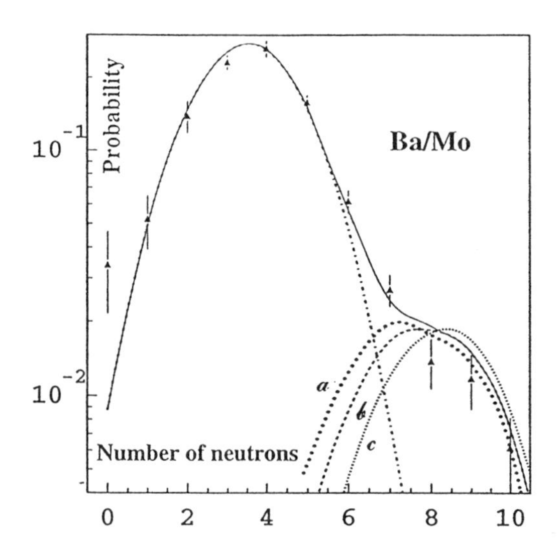

Figure 2. Fit for Mode 2: 146/106 (a), 145/107 (b), and 144/108 (c).

hyperdeformed (2).

The yields of correlated fragment pairs and neutron multiplicities shown in Fig. 1 originate as a result of the de-excitation processes of primary fission fragments and carry information about the mass and excitation energy distributions of the primary fission fragments. A least square best fit of the calculated yields $Y_i^{calc}(A'_L, A'_H)$ to the pattern of the experimental yields after neutron evaporation $Y_i(A'_L, A'_H)$ revealed that the mass and excitation-energy-distributions of primary fission fragments are fitted by a single Gaussian distribution for Zr-Ce and Ru-Xe, but a satisfactory fit can be obtained for the Mo-Ba data only when two distinct fission modes contribute to the formation of the primary Mo-Ba fission fragments. In fact, good fits to the data were obtained for the first fission mode and a new ultra hot second mode in which only one Mo-Ba primary fragment pair, ^{108}Mo-^{144}Ba or ^{107}Mo-^{145}Ba, or ^{146}Mo-^{146}B, contributes. For each of these primary fragment pairs, essentially the same value of <TKE> = 153 ± 3 was found for the second fission mode as a result of the unfolding procedure. For Mode 1 <TKE> = 189 ± 1 MeV independent of which of the three pairs contribute to Mode 2.

The ^{106}Mo-^{146}Ba primary pair gives the best fit to the 7-10 neutron multiplicities (see Fig. 2). Mode 1 looks like a familiar fission mode of ^{252}Cf in the respect that its principal characteristics (<TKE>, σ_{TKE}, $\bar{A}_H$, σ_{A_H}) and excitation energies of the primary fragments are like those known before (3). The new Mode 2 appears quite peculiar because of its low TKE value, low Coulomb barrier, and very hot Ba excitation. For the Mo-Ba division of ^{252}Cf, one and the same fragments, ^{144}Ba, ^{145}Ba and/or ^{146}Ba, appear either with a "standard" or enormously high excitation energy in the first and second fission modes, respectively. This implies that ^{144}Ba, ^{145}Ba and/or ^{146}Ba are found in two states which are remarkable for their very different deformations at scission. From their excitation energies, one can estimate the ratios of the axes at scission: a/b = 2.8, 3.0 and 3.2 for ^{144}Ba, ^{145}Ba, and ^{146}Ba, respectively. This is the first evidence for the long-predicted hyperdeformed shape (4) in nuclei.

Recently, Donangelo et al. (5) repeated the analysis of the yields and reached the same conclusion that there is a new Mode 2 and that at scission one or more of ^{144,145,146}Ba is hyperdeformed with most likely ^{146}Ba. They further carried out an analysis in which it is argued that the structure of the system at its scission point is that of a three cluster system, ^{106}Mo-^{14}C-^{132}Sn, which predominantly breaks into a Mo-Ba split. In this case, the high excitation of the ^{146}Ba fragment results from the coalescing of ^{14}C and ^{132}Sn.

The first "normal" fission mode has features typical of the bulk of fission events of ^{252}Cf, whereas the second abnormal mode revealed here for the first time provides evidence for a hyperdeformed state or states in ^{144,145,146}Ba with ≈3:1 axis ratio.

COLD BINARY AND ^{4,6}He, ^{10}Be TERNARY FISSION

The spontaneous cold decay of heavy nuclei is now a widely observed phenomenon, including the emission of α particles and heavier clusters up to ^{34}Si (6) and the cold (neutronless) fission of many actinide nuclei into fragments with masses ranging from ≈70 to ≈160 (7,8) confirming theoretical predictions for such cold decays to the ground states of both final fragments (9). We made the first direct observation of cold (neutronless) fragmentations in the spontaneous fission of ^{252}Cf (8) and more recently extended that work with Gammasphere (10).

Spontaneous fragmentations involving more than two final nuclei are very rare processes. Ternary fragmentations observed in the spontaneous or thermal neutron-induced fission of heavy nuclei are usually related to the emission of a light third fragment, the most probable being an α particle (11). It is very important to establish theoretically and experimentally that cold (neutronless) ternary fragmentations and the pairs associated with each similar to the binary ones already observed to exist. From our new data, the experimental yields and the correlated pairs associated with

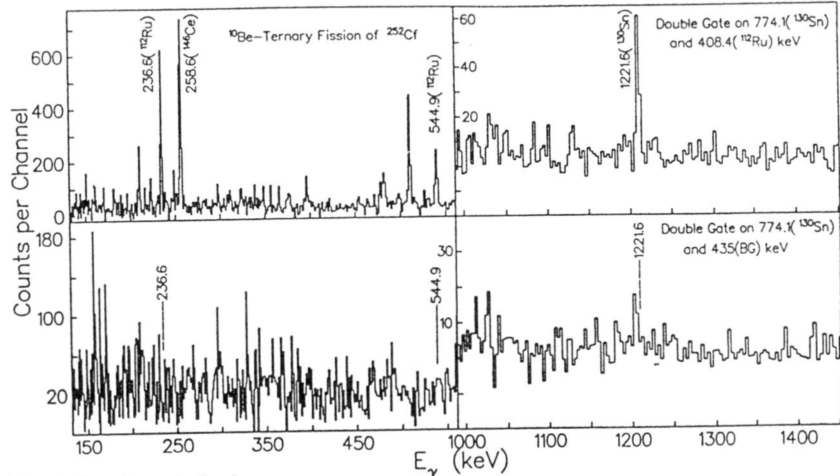

Figure 3. Gamma spectra double gated as shown in the figure.

^{4}He, (^{6}He or α2n) and ^{10}Be ternary fission of ^{252}Cf were detected.

To search for events corresponding to neutronless ternary fission from ^{252}Cf, the secondary fragments were correlated by observing the γ-rays emitted by each of them. The yields of all the partner nuclei that can involve from 0 to 10 evaporated neutrons in binary fragmentations along with neutronless ternary fragmentations were extracted. To obtain the neutronless ternary yields per 100 fission events, the binary yields were normalized to the summed yield for the partner given by Wahl (12). As a cross-check on the ternary events, gates were set on transitions in each of the two fragments correlated with the light ternary fragment. An example of the data are shown in Fig. 3 where double gates on the $4^+\rightarrow 2^+$ transitions in ^{130}Sn and ^{112}Ru and on one $4^+\rightarrow 2^+$ and a background are shown. The missing N and Z between ^{130}Sn and ^{112}Ru correspond to ^{10}Be. Similar data were obtained for ^{4,6}He and other ^{10}Be channels. A total of 16, 15, and 10 ternary cold neutronless channels were observed for ^{4,6}He and ^{10}Be ternary fission, respectively, and tentatively several for ^{14}C ternary fission. The Coulomb barrier in these very neutron-rich nuclei preclude prompt charged particles being emitted by one fragment. Unfortunately, we cannot distinguish between the ^{6}He and α,2n fission.

Theoretical yields were obtained by calculating the penetrabilities through a double-folded potential barrier with trajectories calculated as a function of time by using the M3Y nucleon-nucleon force between the deformed fragments with the third light particle being spherical. Satisfactory agreement was obtained between the relative experimental and theoretical yields for cold binary and ^{4,6}He and ^{10}Be ternary fission. These data (13) provide the first direct evidence for the fragment pairs with which ternary fission occurs and so provide exciting new insights for a more microscopic understanding of fission.

NEW INSIGHTS INTO NUCLEAR STRUCTURES

Static Octupole Deformation

The concept of spontaneous intrinsic symmetry-breaking in many body systems has a long history in nuclear physics going back to A. Bohr (14). Strong octupole correlations occur when a pair of single-particle orbitals with $\Delta N = 1$, $\Delta\ell = 3$, and $\Delta j = 3$ with large octupole interaction matrix elements lie close to each other as occurs just above the closed shells, for example, for N = 88-90 Ba-Ce nuclei associated with the $\nu f_{7/2}$-$\nu i_{13/2}$, and for Z = 56-58 the $\pi d_{5/2}$-$\pi h_{11/2}$ orbitals. Opposite parity intertwined bands connected by enhanced E1 transitions were discovered in $^{144}_{56}Ba_{88}$ (15) and $^{146}_{58}Ce_{88}$ (16) to confirm the theoretical predictions (17) of the importance of stable octupole deformation in this region, but such were not reported in ^{142}Ba (15) where it was also predicted. With Gammasphere, new high spin states up to 16^+ and 19^- in $^{142\text{-}146}$Ba and in $^{143\text{-}147}$Ba were obtained from SF of ^{252}Cf (for example Fig. 4, refs. 18,19). The new N = 86, ^{142}Ba and ^{144}Ce levels now exhibit similar intertwined positive- and negative-parity bands connected by enhanced E1 transitions to clear up the previous discrepancy between theoretically-predicted (17) static octupole deformation for ^{142}Ba and ^{144}Ce and experiment (15,16).

Cranked shell model calculations predict that stable octupole deformation increases with increasing spin for the Ba-Ce nuclei around N = 88 but vanish at higher spins (20). A shape transition towards $\beta_3 = 0$ is expected at frequencies above 0.3 MeV after the alignment of the $\nu i_{13/2}$ and $\pi h_{11/2}$ pairs (20). Indeed, now the ground band is observed to be crossed above spin 10^+ in ^{146}Ba and while intertwined transitions to 16^+, 15^- are seen in ^{144}Ba, no intertwined connecting transitions are seen above 10^+ in ^{146}Ba with our 1995 data (Fig. 4). These results (18,19) provided the first confirmation in any mass region of the theoretical prediction (20) of the vanishing of stable β_3 above $\hbar\omega = 0.3$ MeV. The electric dipole moments (D_0) are larger in ^{142}Ba than in ^{144}Ba and drop dramatically in ^{146}Ba as predicted (21).

Identical Bands

Another area of much interest is bands with identical moments of inertia. Identical ground state bands are observed in neutron-rich ^{98,100}Sr, ^{108,110}Ru, ^{144,146}Ba, $^{156\text{-}160}$Sm and, for the first time, octupole states in ^{144,146}Ba. These identical bands have quite different patterns as functions of N and Z to present new challenges for theory. The levels of $^{156,158}_{62}\mathrm{Sm}_{94,96}$ and recently discovered $^{160}_{62}\mathrm{Sm}_{98}$ (19,22) are all characteristic of well-deformed shapes with a small, smooth decrease in the transition energies and moments of inertia for every state from 2^+ to 14^+ as A increases. However, it is surprising to find that these three successive nuclei have such very similar E_γ, J_1 and J_2 moments of inertia. Indeed, E_γ, J_1 and J_2 have constant differences 2.5-4.1% for 156-158 and , even more identical, 3.0-3.6% for 158-160 at all spins. These are a new type of identical bands with constant differences in J_1 and J_2.

ACKNOWLEDGMENTS

Work at Vanderbilt and INEL supported by U.S. DOE grant and contract DE-FG05-88ER40407 and DE-AC07-76ID01570, respectively, and at the Joint Inst. for Nuclear Res. by grant 94-02-05584-a of the Russian Fed. Foundation of Basic Sciences. The Joint Inst. for Heavy Ion Res. is supported by its members, U. TN, Vanderbilt U. and U.S. DOE through contract DE-FG05-87ER40361 with the U. of TN. Work at Tsinghua U. supported by the Nat'l. Natural Science Foundation of China. Work at LBNL and LLNL supported by U.S. DOE grants DE-AC03-76SF00098 and W-7405-ENG48. A. Sandulescu and A. Florescu partially supported by Twinning Pro. of Nat'l. Res. Council. W. Greiner supported by joint program with NSF.

REFERENCES

1. Ter-Akopian, G. M., et al., Phys. Rev. Lett. **73**, 73 (1994).
2. Ter-Akopian, G. M., et al., Phys. Rev. Lett. **77**, 32 (1996).
3. Wild, J., et al., Phys. Rev. **C41**, 640 (1990).
4. Dudek, J., et al., Phys. Lett. **B211**, 252 (1988).
5. Donangelo, R., et al., J. Mod. Phys. Letts. (submitted).
6. P.B. Price, et al., Nucl. Phys. **502**, 41C (1989).
7. Hulet, E. R., et al., Phys. Rev. Lett. **56**, 313 (1986).
8. Hamilton, J. H., et al., J. Phys. G: Nucl. Part. Phys. **20**, L85 (1994).
9. Sandulescu, A. and Greiner, W., Rep. Prog. Phys. **55**, 1423 (1992).
10. Sandulescu, A., et al., Phys. Rev. **C54**, 258 (1996).
11. Vandenbosch, R. and Huizenga, J., *Nuclear Fission,* C. Wagemans, ed., Boca Raton: CRC Press (1991).
12. Wahl, A. C., At. Data Nucl. Data Tables **39**, 1 (1988).
13. Ramayya, A.V., et al., *in* Third Int. Conf. on Dynamical Aspects of Nuclear Fission (8/ 30-9/ 4, 1996), submitted.
14. Bohr, A., Mat. Fys. Dan. Vid. Selsk **26**, No. 14 (1952).
15. Phillips, W. R., et al., Phys. Rev. Lett. **57**, 3257 (1986).
16. Phillips, W. R., et al., Phys. Lett. **B212**, 402 (1988).
17. Nazarewicz, W., et al., Phys. Rev. Lett. **52**, 1272 (1984).
18. Zhu, S. J., et al., Phys. Lett. **B357**, 273 (1995).
19. Hamilton, J. H., et al., *Progress in Particles and Nuclear Physics,* Proc. Erice Conf. (Sept. 1996), to be published.
20. Nazarewicz, W. and Tabor, S., Phys. Rev. **C45**, 2226 (1992).
21. Butler, P. A. and Nazarewicz, W., Nucl. Phys. **A533**, 249 (1991).
22. Zhu, S. J., et al., J. Phys. G: Nucl. Part. **21**, L57 (1995).

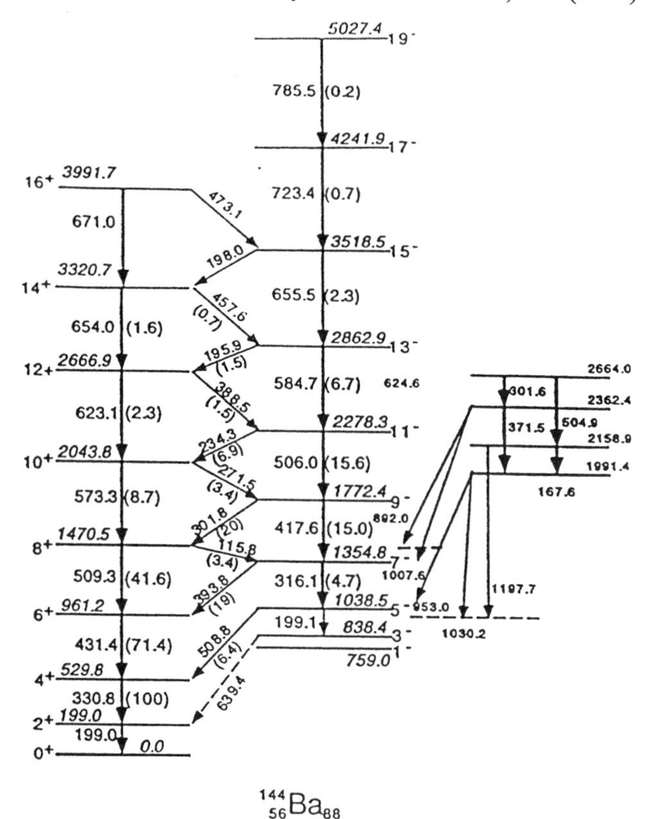

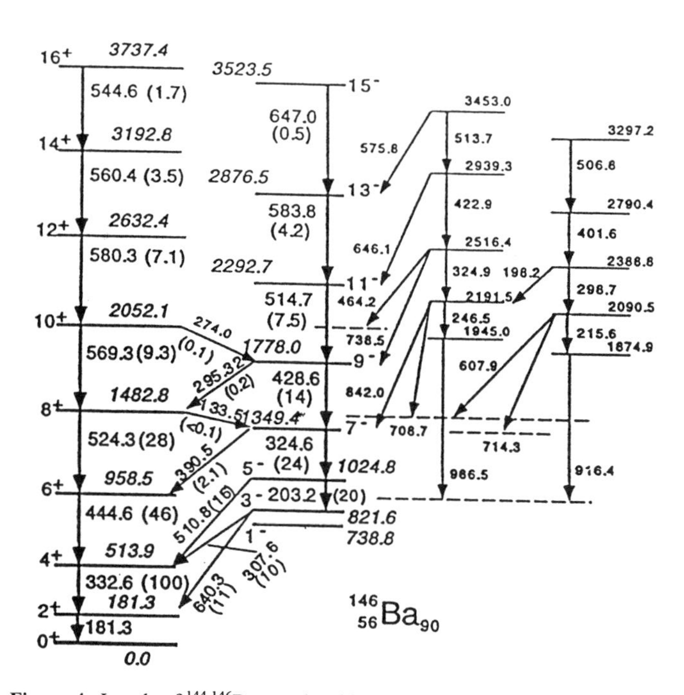

Figure 4. Levels of ^{144,146}Ba populated in spontaneous fission.

LOW-ENERGY NUCLEAR RESONANCES IN (p,γ)-REACTIONS ON SILICON ISOTOPES*

W.H. Schulte [a], H.v. Bebber [a], M. Mehrhoff [a], H.W. Becker [a], M. Berheide [a], L. Borucki [a], J. Domke [a], F. Gorris [a], Ch. Grunwald [a], S. Kubsky [a], N. Piel [a], C. Rolfs [a], and G.E. Mitchell [b]

[a] *Institut für Experimentalphysik III, Ruhr-Universität Bochum, 44780 Bochum, Germany*
[b] *North Carolina State University, Raleigh, NC 27695, USA and*
Triangle Universities Nuclear Laboratory, Durham, NC 27708, USA

A high efficiency γ-ray detection system based on a 12" x 12" NaI(Tl) bore hole detector has been used to measure the strength of low-energy narrow resonances in proton-induced capture reactions. The strengths of the E_p = 324 keV and the E_p = 416 keV resonances in ^{29}Si(p,γ)^{30}P and of the E_p = 620 keV resonance in ^{30}Si(p,γ)^{31}P have been measured. The applicability of these resonances for high resolution depth profiling in materials analysis is discussed.

INTRODUCTION

Resonances in proton induced capture reactions on different isotopes have found applications in materials science, especially to perform investigations on the depth distributions of light elements in thin films (1). Due to the superior depth resolution, special emphasis has recently been placed on the application of resonances at projectile energies below 0.5 MeV (2). It has been demonstrated that narrow, low-energy resonances can be applied successfully (2,3). The use of such resonances benefits from the large stopping power for protons at these energies, which results in an increased depth resolution. Furthermore, the reaction products of these resonances often can be detected with a high signal-noise ratio. Since the applications of many resonances in material analysis suffer from their low reaction cross section, an efficient detection system is needed to keep the ion doses moderate and to obtain precise depth profiles.

In the special case of depth profiling of silicon, proton induced resonances are present at projectile energies below 1 MeV for all silicon isotopes. From the viewpoint of high resolution depth profiling and isotopic tracing experiments, the E_p = 324 keV and the E_p = 417 keV resonances in ^{29}Si(p,γ)^{30}P and the E_p = 620 keV resonance in ^{30}Si(p,γ)^{31}P are of special interest. Although these resonances are frequently used for calibration purposes, e.g. to convert relative resonance strength to absolute values by normalization to these resonances, and although some of them have already been applied in materials science investigations (4-9), there remain uncertainties about resonance parameters.

This is particularly true for the 620-keV resonance in ^{30}Si(p,γ)^{31}P, where serious discrepancies exist between the reported resonance strength measurements (10,11). For the 324-keV resonance in ^{29}Si(p,γ)^{30}P the resonance strength has only been determined with large uncertainty (12,13).

It should be noted that knowledge of the parameters of the low-energy proton induced resonances on silicon isotopes is also of interest in the field of nuclear astrophysics, in order to determine reaction rates in a stellar environment (10).

The present paper focuses on the application of a large 12" x 12" NaI(Tl) spectrometer for the determination of the strengths of low energy proton induced resonances in silicon isotopes. Here we concentrate on the measurement of the strengths of the 324-keV and the 417-keV resonances in ^{29}Si(p,γ)^{30}P and of the 620-keV resonance in ^{30}Si(p,γ)^{31}P.

EXPERIMENTAL ARRANGEMENTS

A high-efficiency γ-ray spectrometer has been installed at the Dynamitron Tandem Laboratory (DTL) at the Ruhr-Universität Bochum (14). The spectrometer has mainly been used for DIGME (Deuterium Induced Gamma Ray Emission) analysis and nuclear-resonance depth profiling experiments.

The experimental set-up is shown schematically in Fig. 1, including a 12" x 12" NaI(Tl) bore hole detector. Since the target is positioned in the center of the detector the solid angle is nearly 4π. Due to the high efficiency various γ-rays resulting from a cascade decay of a resonant state are effectively summed and angular distribution effects can be neglected. The measured values of the total efficiency, as well as the photo peak efficiency, are in excellent agreement with computer

* Supported by NATO grant CRG950635 and U.S. Department of Energy grant no. DE-FG05-88ER90991.

CP392, *Application of Accelerators in Research and Industry*, edited by J. L. Duggan and I. L. Morgan
AIP Press, New York © 1997.

simulations using the GEANT code (15). Extensive simulations and analyses have also been performed to determine the effect of cascade decays on the γ-spectra (16).

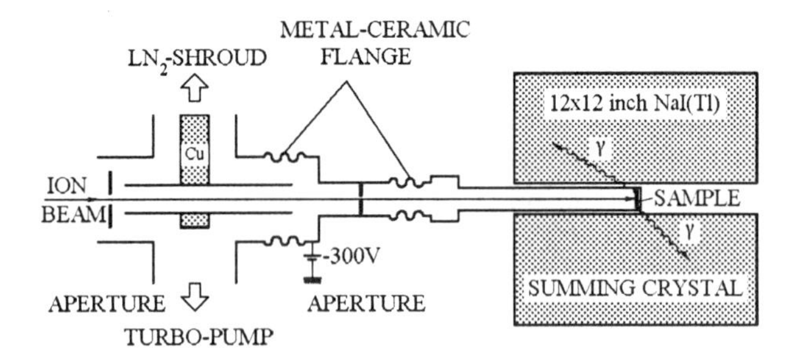

FIGURE 1. Schematic drawing of the experimental set-up including a 12" x 12" NaI(Tl) bore hole detector.

Figure 2 shows the total efficiency of the detector as a function of γ-energy for monoenergetic γ-rays. The energy resolution of the detector is $\Delta E/E \approx 2$ % at a γ-energy of 10 MeV (16).

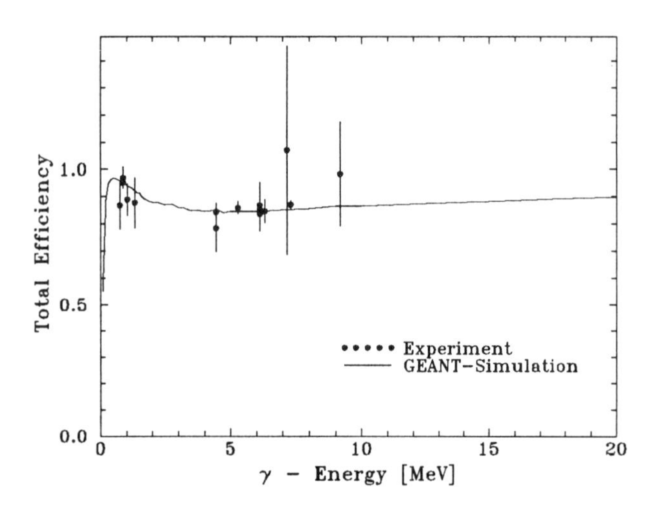

FIGURE 2. Total efficiency of the NaI(Tl)-detector for monoenergetic γ-rays as a function of γ-energy.

The target system, designed in UHV technology, is evacuated by a magnetically coupled 340 l/s turbo molecular pump. A cooling trap consisting of a copper pipe at LN_2-temperature prevents hydrocarbon deposition on the target during the measurements. A pressure of ~ 5 × 10^{-8} mbar near the sample is achieved within one hour of pumping. The sample is placed near the end of a stainless steel vacuum pipe reaching into the centre of the 35 mm diameter bore hole of the γ-detector. The target holder was designed to minimize absorption of γ-rays. Alternatively air cooling and liquid nitrogen cooling can be applied to the target holder.

Thin silicon-oxide films thermally grown on silicon substrates were used as targets. These films are known to be amorphous. Complementary measurements were performed using Si-wafers as target material; in this case there were possible channeling effects. In order to reduce these channeling effects we irradiated the targets with a high proton dose of more than $10^{17}/cm^2$ before measuring the resonance yield. All targets were of natural isotopic composition.

Proton beams provided by the 450-kV electrostatic accelerator as well as by the 4-MV Tandem accelerator of the DTL were used with intensities of 1-2 μA .

A standard Faraday cup arrangement including a calibrated current integrator was used to measure the ion dose. Alternatively a spinning wire dosemeter mounted in front of the set-up allowed precise and sample independent ion dose determination at a current range between 10 pA and 10 μA (17).

MEASUREMENTS AND RESULTS

The strength ωγ of a narrow resonance can be calculated from the step in the thick target γ-ray yield by (18):

$$\omega\gamma = \frac{2}{\lambda^2}\frac{M}{m_p + M}\varepsilon_{eff}\frac{N_\gamma}{N_p \eta_{det}} \tag{1}$$

with λ the de Broglie wavelength of the incident projectile, M the target nucleus mass, m_p the projectile mass, ε_{eff} the effective stopping power, N_γ the number of resonant γ-rays, N_p the number of projectiles, and η_{det} the detector efficiency.

Assuming the validity of Bragg's-rule, ε_{eff} is given by

$$\varepsilon_{eff} = \varepsilon_a + \frac{1}{N_a}\sum_i N_i \varepsilon_i \tag{2}$$

with ε_a and N_a the stopping power and the number of the active target atoms, and ε_i and N_i the stopping power and the number of the inactive target atoms (18).

The "SiO_2" stoichiometry of the silicon oxide targets and the film thickness were determined by ellipsometry. For both the silicon wafer and the silicon oxide targets, the stopping power ε_{eff} was determined using the TRIM95 computer code (19). The uncertainties of these values are estimated to be ±10 %.

A γ-spectrum obtained from a silicon target at a projectile energy of 420-keV is shown in Fig. 3. A large fraction of the resonant counting rate is in the full energy peak corresponding to the excitation energy of the level of the ^{30}P compound nucleus. Due to the large volume of the detector significant contributions to the background counting rate arise from cosmic rays and natural radioactivity.

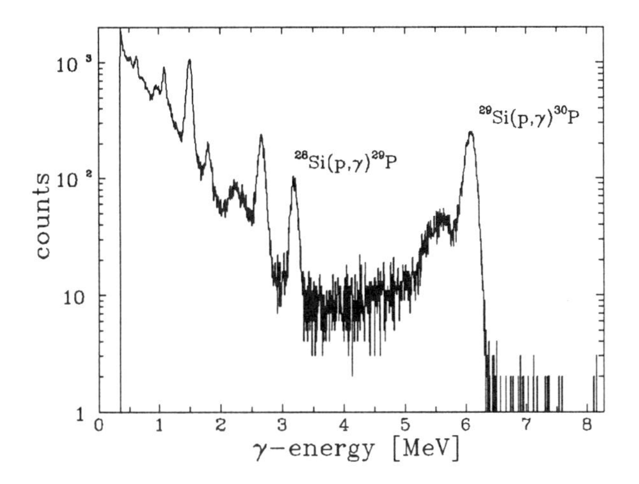

FIGURE 3. γ-spectrum obtained from a Si-target at a projectile energy of 420 keV.

For the extraction of the resonance strength we used the total resonant counting rate of the NaI(Tl)-detector, i.e. the counting rate of the entire γ-spectrum after subtraction of the background counting rate and the contributions of lower energy resonances. To take into account the counting rate at low γ-energies, $E_\gamma \leq 1$ MeV, which is obscured by intense background counting rate and detector noise, we extrapolated the γ-spectra according to the shape obtained from GEANT simulations to $E_\gamma = 0$ MeV.

The level in ^{31}P which is excited by the 620-keV resonance in ^{30}Si(p,γ)^{31}P (E_X = 7.898 MeV) leads to isotropic γ-emission and decays 95±1 % directly to the ground state (10). Thus a nearly monoenergetic γ-spectrum is produced. The resonant counting rate from the SiO_2 targets is $Y = N_\gamma/N_p = (1.52\pm0.05) \times 10^{-11}$. Using the efficiency of the NaI(Tl)-detector of $\eta_{det} = 85.5\pm2.5$ % and $\varepsilon_{eff} = 862 \times 10^{-15}$ eV·cm^2 we obtain $\omega\gamma = 2.09\pm0.23$ eV. From pure silicon targets ($(Y = 3.64\pm0.11) \times 10^{-11}$ and $\varepsilon_{eff} = 357 \times 10^{-15}$ eV·cm^2) we obtain $\omega\gamma = 2.07\pm0.23$ eV.

The decay of the excited state in ^{30}P at E_X = 6.002 MeV excited by the 417-keV resonance in ^{29}Si(p,γ)^{30}P predominantly produces γ-rays with energies of about 5.3 MeV (10). The resonant counting rate from the SiO_2 targets was $Y = (3.00\pm0.09) \times 10^{-12}$. Using the efficiency of the NaI(Tl)-detector of $\eta_{det} = 85.0\pm2.5$ % and ε_{eff} = 731 × 10^{-15} eV·cm^2 we obtain $\omega\gamma = 0.230\pm0.025$ eV. Using the pure silicon targets ($Y = (7.72\pm0.22) \times 10^{-12}$ and $\varepsilon_{eff} = 296\times 10^{-15}$ eV·cm^2) we obtain $\omega\gamma = 0.245\pm$ 0.027 eV.

The γ-ray decay scheme of the resonance at 324 keV in ^{29}Si(p,γ)^{30}P (E_X = 5.92 MeV) is dominated by the decay to the ground state (90 %), with about 10 % of the decay to the second excited state in ^{30}P, thus producing 5.21 MeV and 0.71 MeV γ-rays. The resonant counting rate from the SiO_2 targets is $Y = (4.13\pm0.16) \times 10^{-13}$. With $\eta_{det} = 85.0\pm2.5$ % and $\varepsilon_{eff} = 846 \times 10^{-15}$ eV·cm^2 we obtain $\omega\gamma = 0.030\pm0.003$ eV.

Table 1 compares the resonance strengths from previous and present work.

TABLE 1. Strength ωγ of proton induced resonances on silicon isotopes at low projectile energies

Reaction	E_p [keV]	ωγ [eV] previous	ωγ [eV] present
^{29}Si(p,γ)^{30}P	324	0.015±0.005 [a] 0.027±0.005 [b]	0.030±0.003[e]
^{29}Si(p,γ)^{30}P	417	0.260±0.025 [c]	0.230±0.025 [e] 0.245±0.027[f]
^{30}Si(p,γ)^{31}P	620	2.50±0.15 [c] 1.94±0.13[d]	2.09±0.23 [e] 2.07±0.23[f]

[a] Ref. 12
[b] Ref. 13, corrected with ωγ = 0.260±0.025 eV of the E_p = 417 keV resonance (see also 12)
[c] Ref. 10
[d] Ref. 11
[e] SiO_2-targets
[f] Si-targets

DISCUSSION

The values for resonances strengths obtained from SiO_2-targets are in excellent agreement with those obtained from Si-targets. No influence of the effect of channeling was observed, indicating an amorphous structure of the Si-target after the pretreatment with a high ion dose.

The resonance strengths obtained can be compared with reported values (table 1).

In the case of the 324-keV resonance our result is in good agreement with the result obtained by Harris et al. (13), after normalizing their measurement to the more recently measured strength of the 417-keV resonance. Our result disagrees with the result published by Reinecke et al. (12).

We find good agreement with the strength for the 417-keV resonance obtained by Riihonen et al. (10).

For the 620-keV resonance we find fair agreement with the measurement of Riihonen et al. (10). Our result agrees very well with that obtained by Paine and Sargood (11).

The 620-keV resonance in $^{30}Si(p,\gamma)^{31}P$ is the strongest of the resonances considered here; thus it would be favorable for depth profiling applications. Its width $\Gamma_R = 68\pm9$ eV would allow a depth resolution for near surface regions somewhat above 1 nm, which could be improved by tilting the sample with respect to the incident ion beam. Experimental verification of this prediction may be hampered by the required ion beam energy resolution and the effect of thermal Doppler broadening (which amounts to about 100 eV at room temperature). Furthermore, background radiation arising from the 499-keV resonance in $^{30}Si(p,\gamma)^{31}P$ ($\omega\gamma \approx 0.12$ eV (20)) and the low energy resonances in other silicon isotopes must be considered. The 620-keV resonance has been applied by Pruppers et al. for profiling thin layers of hydrogenated amorphous silicon (5) and by Pelloie et al. to study silicon mobility (8).

The $\omega\gamma$ obtained for the 417-keV resonance in $^{29}Si(p,\gamma)^{30}P$ ($\Gamma_R \leq 100$ eV) is a factor of 8 weaker compared to $\omega\gamma$ of the 620-keV resonance in $^{30}Si(p,\gamma)^{31}P$, but the attainable depth resolution is better. This is due to the smaller Doppler broadening at the lower projectile energy and to the larger stopping power. Furthermore this resonance can be measured using low energy implantation accelerators that can provide an excellent energy resolution in the eV range (2). Thus a depth resolution of ≤ 1 nm may be possible, making this resonance favorable for applications, where the samples are sufficiently stable under radiation. For example Antilla and Hirvonen studied silicon diffusion in aluminum using this resonance as a probe (4).

The resonance at 324-keV in $^{29}Si(p,\gamma)^{30}P$ ($\Gamma_R \leq 30$ eV (2)) in principle even allows a better depth resolution. It has been applied for investigations on silicon mobility during the growth of thin dielectric films (6,7). The low resonance strength restricts applications of this resonance to isotopic tracing experiments with highly enriched ^{29}Si or to samples that are highly stable under ion beam radiation.

SUMMARY

The resonance strengths of low energy resonances in proton induced capture reactions of silicon isotopes at low projectile energies have been determined using a high efficiency NaI(Tl) detection system. Due to the excellent efficiency calibration an accuracy of better than ± 5 % may be achieved. But presently the accuracy is limited by the uncertainty in the stopping power.

REFERENCES

1. Amsel, G., and Maurel, B., *Nucl. Instr. Meth.* **218**, 183-196 (1983).
2. Schulte, W.-H., Ebbing, H., Becker, H.W., Berheide, M., Buschmann, M., Angulo, C., Rolfs, C., Amsel, G. Trimaille, I., Battistig, G., Mitchell, G.E., and Schweitzer, J.S., *Vacuum* **44**, 185-189 (1993).
3. Battistig, G., Amsel, G., d' Artemare, E., and Vickridge, I., *Nucl. Instr. Meth.* **B66**, 1-10 (1992).
4. Antilla, A., and Hirvonen, J., *Thin Solid Films* **33**, L13-L14 (1976).
5. Pruppers, M.J.M., Zijderhand, F., Maessen, K.M.H., Bezemer, J., Habraken, F.H.P.M., and van der Weg, W.F., *Nucl. Instr. Meth.* **B15**, 512-515 (1986).
6. Rolfs, C., and Baumvol, I.J.R., *Z. Phys.* **A353**, 127-140 (1996).
7. Baumvol, I.J.R., Borucki, L., Chaumont, J., Ganem, J.-J., Kaytasov, O., Piel, N., Rigo, S., Schulte, W.H., Stedile, F.C., and Trimaille, I., *Nucl. Instr. Meth.* **B** (in press).
8. Pelloie, B., Perrière, J., Siejka, J., Debenest, P., Straboni, A., and Vuillermoz, B., *J. Appl. Phys.* **63**, 2620-2627 (1988).
9. Vickridge, I., *Thesis*, Université Paris 7, 1990.
10. Riihonen, M., Keinonen, J., and Antilla, A., *Nucl. Phys.* **A313**, 251-268 (1979).
11. Paine, B.M., and Sargood, D.G., *Nucl. Phys.* **A331**, 389-400 (1979).
12. Reinecke, J.P.L., Waanders, F.B., Oberholzer, P., Janse van Rensburg, P.J.C., Cilliers, J.A., Smit, J.J.A., Meyer, M.A., and Endt, P.M., *Nucl. Phys.* **A435**, 333-351 (1985).
13. Harris, G.I., Hyder, A.K., and Walinga, J., *Phys. Rev.* **187**, 1413-1444 (1969).
14. Piel, N., Schulte, W.H., Berheide, M., Becker, H.W., Borucki, L., Grama, C., Mehrhoff, M., and Rolfs, C., *Nucl. Instr. Meth.* **B** (in press).
15. *GEANT-Detector Description and Simulation Tool*, Program Library W5013, CERN, Geneve, 1994.
16. Mehrhoff, M., Borucki, L., Becker, H.W., Berheide, M., Piel, N., Rolfs, C., and Schulte, W.H., to be published.
17. Piel, N., Berheide, M., Polaczyk, Ch., Rolfs, C., Schulte, W.H., *Nucl. Instr. and Meth.* **A349**, 18-26 (1994).
18. Rolfs, C.E., and Rodney, W.S., *Cauldrons in the Cosmos*, The University of Chicago Press, Chicago, 1988.
19. Ziegler, J.F., *TRIM Computer Program - Version 95.06* (1995).
20. Endt, P.M., *Nucl Phys.* **A521**, 1-830 (1990).

FINE STRUCTURE IN THE α DECAY OF ^{192}Po

J. Wauters, C.R. Bingham, W. Reviol and B.E. Zimmerman

University of Tennessee, Knoxville, TN 37996 USA

A.N. Andreyev, N. Bijnens, M. Huyse and P. Van Duppen

Katholieke Universiteit Leuven, Celestijnenlaan 200D, B-3001 Leuven, Belgium

I. Ahmad, D.J. Blumenthal, C.N. Davids, R.V.F. Janssens and D. Seweryniak

Argonne National Laboratory, Argonne IL 60439 USA

J.C. Batchelder

Louisiana State University, Baton Rouge, LA 70803 USA

L.F. Conticchio and W.B. Walters

University of Maryland, College Park, MD 20742 USA

X.S. Chen and P.F. Mantica

Oak Ridge National Laboratory, Oak Ridge, TN 37831 USA

B.C. Busse

Oregon State University, Corvallis, OR 97331 USA

L.T. Brown

Vanderbilt University, Nashville, TN 37235 USA

H. Penttila

University of Jyvaskyla, Jyvaskyla, FIN 40531, Finland

Fine structure in the α decay of ^{192}Po has been studied in the reaction of ^{36}Ar on ^{160}Dy at 176 MeV. Evaporation residues were selected in-flight using the Argonne Mass Analyzer and implanted into a double sided silicon strip detector. The correlation technique between implants and subsequent decays was used to observe fine structure in the α decay of ^{192}Po leading to the identification of an excited 0^+ state in ^{188}Pb at 571(31) keV. The half-life of ^{192}Po has been determined to be 33.2(14) ms. The observation of a low-lying 0^+ state is discussed in terms of proton particle-hole pair excitations across the Z=82 shell gap. The small hindrance factor to the excited 0^+ state relative to the ground state supports the picture of shape coexistence in light even-even Po isotopes.

CP392, *Application of Accelerators in Research and Industry,* edited by J. L. Duggan and I. L. Morgan
AIP Press, New York © 1997.

Shell-model intruder states are often associated with the phenomenon of shape coexistence in the region of singly-closed shells (1). In the case of even-even isotopes, these states appear as low-lying 0^+ states. In the even-even Pb nuclei, the 0^+ bandhead of a slightly oblate band can be followed from near N=126 down to N=108 (^{190}Pb): the parabolic behavior of its excitation energy is well understood within the framework of 2-particle-2-hole proton excitations interacting with the valence neutrons (1). In the very neutron-deficient isotopes ^{186}Pb and ^{188}Pb, the oblate band is no longer observed but a prolate-deformed intruder band has dropped down in energy near where the oblate band is expected (2,3). Due to the decay out of the band, no 0^+ bandhead was found using in-beam spectroscopy and therefore α and β decay spectroscopy has to be used.

The production of ^{192}Po was achieved by using the ^{160}Dy(^{36}Ar,4n)^{192}Po reaction. Beams of ^{36}Ar ions (2.3 10^{10} ions/s) were accelerated by the Argonne Accelerator Facility ATLAS to an energy of 176 MeV and were used to bombard a 510 μg/cm^2 66% enriched ^{160}Dy target. Recoiling ions were passed through the Fragment Mass Analyzer (FMA) (4) to separate them from the primary beam and disperse them in mass/charge (A/Q) at the focal plane. A parallel grid avalanche counter (PGAC) was used at the focal plane to provide horizontal and vertical positions of ions resulting in A/q, timing and energy loss information. After passing through the PGAC, the ions were implanted in a double-sided silicon strip detector (DSSD) (5) located 40 cm behind the PGAC, where their subsequent charged-particle decays were detected. The DSSD had a thickness of 65 μm, a total area of 16x16 mm^2 divided in 48 horizontal and 48 vertical strips. Behind the DSSD a 1 mm thick, 300 mm^2 Si detector was placed at 5 mm, yielding an efficiency for detecting electrons emitted from the DSSD of 26%. Implantation and decay events were defined by a coincidence, resp. anticoincidence between the PGAC and DSSD. Decay events opened the gate on the Si detector in order to record coincident electron radiation. Due to the segmentation of the DSSD detector (2304 pixels), it was possible to obtain time- and position correlations between the implantation of an ion and its subsequent decay.

Figure 1a shows the total α spectrum recorded in the DSSD. The 7.167 MeV α line of ^{192}Po (5) is the most energetic line in the spectrum but not the dominating one as reactions on target impurities (heavier Dy isotopes) leading to heavier Po isotopes have much higher cross-sections. The half-life of ^{192}Po, 34(3) ms (6), is improved to 33.2(14) ms. The spectrum of decays that follow the implantation of a A=192 product within 100 ms is shown in Fig. 1b, within 100-200 ms (Fig. 1c) and 200-300 ms (Fig. 1d). The spectrum in Fig. 1b is dominated by the 7.167 MeV α line of ^{192}Po: α particles completely stopped in the DSSD give the full enery peak while α particles escaping the DSSD leave only partially their energy giving counts between 0.5 and 7.167 MeV. On the basis of the comparison of TRIM calculations with the experimental spectrum, it was concluded that the count rate in the region between 6.5 and 7.0 MeV due to escape alpha's is only 0.08 counts/channel.

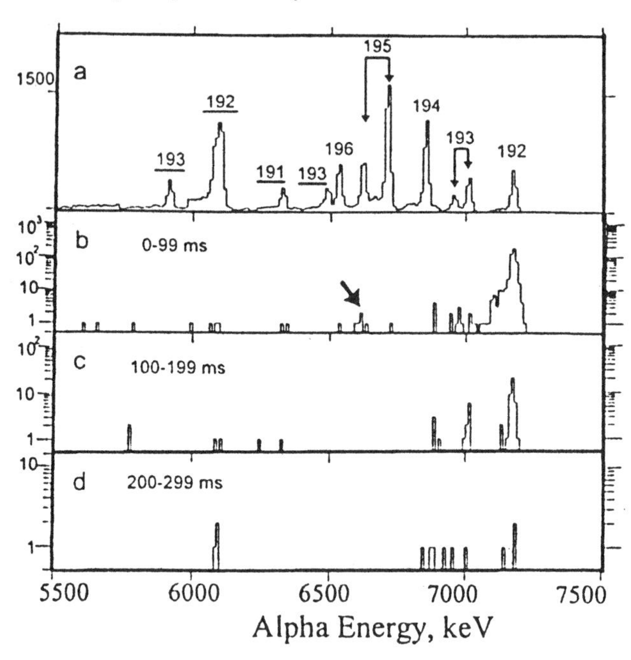

Fig.1. Part of the DSSD decay spectra: a) total α spectrum, the mass number of the different Po and Bi (underlined) α lines are given; b-d) A=192 mass-gated α spectra with different time gates as mentioned in the figures.

Figure 1b clearly shows a cluster of 5 α counts. The escape background alpha's would be 0.4 counts. The half-life behavior of this line, with an energy of 6.61(3) MeV, is in agreement with a ^{192}Po assignment (in a recent decay study of ^{192}Po at Jyvaskyla an α line with a similar energy has been observed (7)). None of the 5 counts give a α-e coincidence but another decay event with the same energy, 50 ms after an implant gives a prompt coincidence with an electron of 470 keV. Unfortunately, due to an electronic problem, no mass identification was obtained for this event. However, the probability of a chance coincidence was estimated to be 2×10^{-4}. From all these arguments, we assign the line at 6.61(3) MeV to fine structure in the α decay of ^{192}Po feeding a level at 571(31) keV in ^{188}Pb. The low hindrance factor (HF) of this transition, relative to the 7.167 MeV ground-state α line, 0.9(4), is a strong fingerprint for an s-wave transition, giving 0^+ for the spin and parity of the 571 keV state in ^{188}Pb. In order to further determine the HF to greater accuracy, the α decay of ^{192}Po was recently remeasured in collaboration with Rutgers University, using the reaction $^{164}Er(^{32}S,4n)^{192}Po$ (8). Preliminary analysis confirm the earlier obtained results and would give HF=0.7(3).

By correlating subsequent α decays within the same DSSD pixel, it is possible to deduce α-branching ratios. For the ^{194}Po-^{190}Pb decay chain, we obtained 5(2) 10^{-3} for the α-branching ratio of ^{190}Pb, in agreement with the literature value 4.0(4) 10^{-3} (8), and for ^{192}Po-^{188}Pb, we obtained a value of 0.085(13) for ^{188}Pb, a more precise result than the literature value of 0.03-0.1 (9). Our improved α-branching ratio of ^{188}Pb now firmly ends a long standing discussion on the persistence of the Z=82 shell closure near midshell manifested in the reduced α-decay width systematics (for a discussion see (9)).

The energy of the oblate 0^+ intruder states in the even-even Pb nuclei has been calculated by Heese et al. (2) from the positions of the intruder bandheads in the neighboring odd Tl and Bi isotopes, using the prescription of Ref. (10), and by applying the N_pN_n scheme of Brenner et al. (11). The results are shown in Table 1 together with the experimental excitation energies for $^{188-196}Pb$, and the extrapolation of the rotational bandstructure assuming a modified rigid rotor model ("Rot" in Table 1). Clearly, the experimental excitation energy for ^{188}Pb is close to the calculations based on the oblate intruder state, but deviates from the rotational band extrapolation.

Table 1. Comparison of calculated excitation energies of excited 0^+ states with the experimental values (this work and ref. to other work in (2) except for the Bi value for ^{188}Pb (12)). The energies are given in keV.

Pb	Exp.	Tl & Bi	N_pN_n	Rot.
196	1143	1143	1134	
194	931	926	935	
192	769	766	772	
190	658	683	645	
188	571	677	530	720
186				570

The HF of the 6.61 MeV α line of ^{192}Po fits nicely in the systematics of the s-wave hindrance factors of the heavier Po isotopes towards oblate intruder states in Pb. The evolution from 2.8 (^{198}Po), 2.5 (^{196}Po), 1.1 (^{194}Po) to 0.9 (^{192}Po) can be understood as due to the increased mixing between the proton 4p-2h intruder state and the ground state in the Po parent as discussed in Ref. (13,14) resulting in a reduction in α-decay strength to the ground state. This is also evidenced by the behavior of the Po reduced α widths as, with decreasing neutron number, they tend to saturate at smaller values than expected from reduced width systematics in this region (13,14).

REFERENCES

1. Wood, J. L., *et al., Phys. Rep.* **215**, 101 (1992).
2. Heese, J., *et al., Phys. Lett.* **B302**, 390 (1993).
3. Baxter, A. M., *et al., Phys. Rev.* **C48**, R2140 (1993).
4. Davids, C. N., *et al., Nucl. Instr. & Meth.* **B70**, 358 (1992).
5. Wauters, J., *et al., Phys. Rev.* **C47**, 1447 (1993)
6. Leino, M. E., *et al., Phys. Rev.* **C24**, 2370 (1981).
7. Page, R., private communication.
8. Cizewski, J., Wauters, J., *et al., ATLAS proposals 587 and 466-3,* 1996
9. Wauters, J., *et al., Z. Phys.* **A342**, 227 (1992).
10. Van Duppen, P., *et al., Phys. Rev.* **C35**, 1861 (1987).
11. Brenner, D. S., *et al., Phys. Lett.* **B293**, 282 (1992).
12. Wauters, J., *et al.,* to be published.
13. Wauters, J., *et al., Phys. Rev. Lett.* **72**, 1329 (1994).
14. Bijnens, N., *et al., Phys. Rev. Lett.* **74**, 3939 (1995).

Angle-Corrected Doppler-Shift Attenuation Analysis*

S.L. Tabor, R.A. Kaye, and G.N. Sylvan

Physics Department, Florida State University, Tallahassee, Florida 32306

If all nuclei from a reaction recoil from a thin target at essentially the same velocity, it is easy to correct for the variation of γ-ray energy with emission angle due to the Doppler shift to combine data from different detectors and improve the statistical accuracy. However, the situation is more complex with thick-target experiments in which γ emission occurs over a wide range of velocities. Now charged-particle detector arrays allow the determination of the exact recoil angle and velocity from fusion-evaporation reactions on an event-by-event basis. Corrections for the variations in the Doppler shift due to different angles between each detector and each recoil event and for the variations in the recoil velocities would provide better statistical accuracy and more accurate line shapes for analysis using the Doppler-shift attenuation method to infer mean lifetimes of excited states. A technique to make these corrections has been developed using simulated line shapes.

INTRODUCTION

The measurement of the mean lifetimes of nuclear levels determines the strengths of electromagnetic transitions between states and provides valuable information concerning nuclear shapes, degrees of collectivity, magnetic moments, and other nuclear structure properties. Such measurements can be improved substantially if the Doppler-shifted γ-ray line shapes can be combined for detectors at different angles and for events with different recoil angles and velocities. Techniques for adjusting line shapes to different angles and velocities have been tested using Doppler-shift simulations.

Doppler-Shift Attenuation Method

A technique for measuring the mean lifetimes of nuclear levels in the femtosecond range by comparing their rate of decay with the rate at which the nuclei slow down in the target and backing is called the Doppler-shift attenuation method (DSAM). The velocity at which each decay occurs can be inferred from the Doppler shift observed in γ detectors placed at forward or backward angles. The mean lifetime of the level can be determined from either the centroid or the detailed shape of the observed Doppler-broadened line shape. In the latter case the entire line shape is usually compared with a theoretical line shape obtained by simulating the slowing down and decay processes with a computer code. The hypothetical mean lifetime used in the simulation is varied to obtain the best agreement with the experimental line shape.

An example of a simulated line shape which would be observed at an angle of 30° is shown in Fig. 1, with (solid curve) and without (dashed curve) taking into account the effects of the finite detector resolution of 3.1 keV full width at half maximum. The line shapes simulated the decay of a state with a mean lifetime of 0.4 ps starting from an initial velocity of 3.3% of the

CP392, *Application of Accelerators in Research and Industry*, edited by J. L. Duggan and I. L. Morgan
AIP Press, New York © 1997.

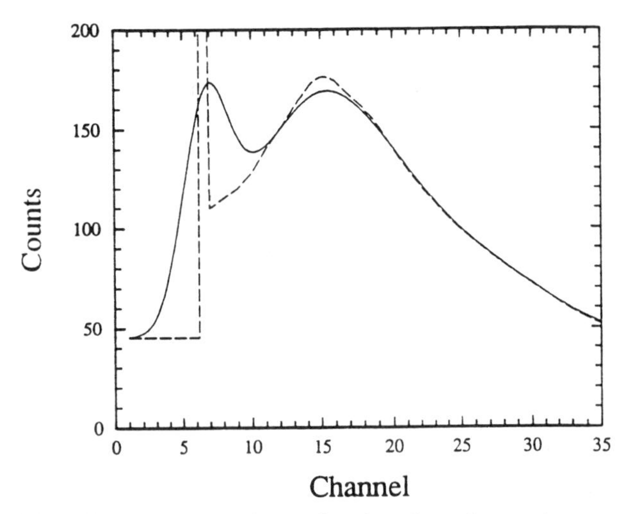

FIGURE 1. A simulated forward angle Doppler-shifted line shape comparing the effects of including (solid curve) or not including (dashed curve) the finite resolution of the γ detector.

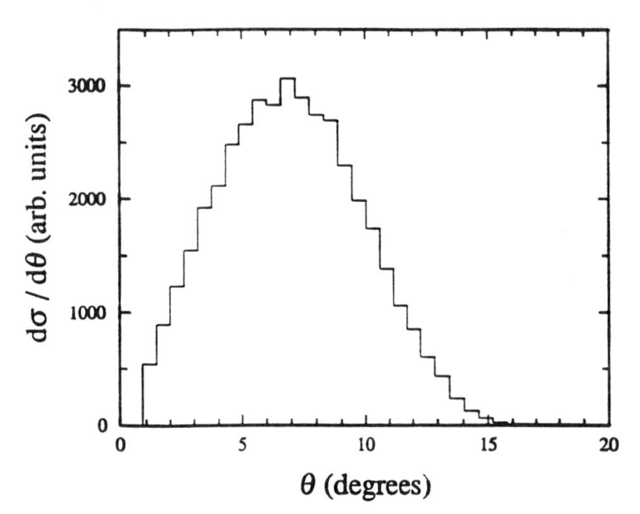

FIGURE 2. Angular distribution of ^{80}Sr reaction products from the ^{58}Ni(^{28}Si,$\alpha 2p$)^{80}Sr reaction at 130 MeV.

speed of light. The dispersion in the spectrum is 0.8 keV per channel.

Heavy-Ion Reactions

In the application of the DSAM to high-spin states, the nuclei of interest are typically produced in heavy-ion fusion-evaporation reactions with an initial recoil velocity of 2% to 4% of the speed of light. Although the compound nuclei produced in the fusion step of the reaction have a constant velocity along the beam direction, the evaporation of protons, neutrons, and α particles in the second step leads to a variation in the speed and direction of the final reaction products. Arrays of charged-particle detectors with full-sphere coverage such as the Microball [1] make it possible to reconstruct the exact recoil direction of each event, except for the small effect of any undetected neutrons. An example of the distribution of evaporation residues in the polar angle θ is shown in Fig. 2 for the ^{58}Ni(^{28}Si,$\alpha 2p$)^{80}Sr reaction at 130 MeV using the Microball and the Gammasphere γ-detector array. The distribution peaks at about 7° and extends out to 15°.

ANGLE CORRECTIONS

The lowest order formula for the Doppler shift suggests a method of shifting spectra to a different angle of observation. The γ energy after Doppler shifting E is given by

$$E = E_0[1 + \beta(t)\cos(\theta)], \quad (1)$$

where E_0 is the unshifted γ-ray energy, θ is the angle between the direction of motion and the γ detector, and $\beta(t)$ is the velocity of the nucleus relative to that of light at the time of emission t. The line shape is the ensemble of all such shifts integrated over the decay time t and weighted according to the mean lifetime of the state. The function $\beta(t)$ decreases from the initial recoil velocity to zero as the recoiling ion slows down in the target and backing. It is independent of the angle of observation.

At first the problem of angle transformation seems simple since for each event one needs only to scale the amount by which a γ ray differs from the unshifted value by the ratio of the cosines of the angles. *I.e.*,

$$\Delta E_{\rm new} = \Delta E_{\rm old}\frac{\cos(\theta_{\rm new})}{\cos(\theta_{\rm old})}, \quad (2)$$

where $\Delta E = E - E_0$.

Finite Resolution

The complicating fact is the finite resolution of the γ detector, as illustrated in Fig. 1. In particular, some of the events corresponding to emission at rest (for which no angle correction should be applied) are mixed up with some of the events corresponding to emission from moving nuclei (for which the angle correction should be applied).

As an approximate solution to this problem we have used a Fermi function to provide a smooth cutoff of the angle transformation:

$$\Delta E_{\text{new}} = \Delta E_{\text{old}}[R + F(\Delta E_{\text{old}})(1 - R)], \quad (3)$$

where R is the ratio of cosines

$$R = \frac{\cos(\theta_{\text{new}})}{\cos(\theta_{\text{old}})}, \quad (4)$$

and $F(x)$ is the Fermi function

$$F(x) = \frac{1}{1 + \exp[(x - x_0)/a]}. \quad (5)$$

The parameter x_0 is the center of the Fermi function at which $F(x) = 0.5$ and a is the "diffuseness" which determines how quickly the function switches between its limiting values of zero for $x << x_0$ and one for $x >> x_0$.

The effect of this angle transformation can be seen in Fig. 3. The top panel of this figure shows simulated line shapes calculated for observation angles of 20° (dashed curve) and 30° (solid curve). In the lower panel the dashed curve has been transformed from an observation angle of 20° to 30° using Eq. 3. The Fermi parameters used are $x_0 = 2$ and $a = 1$ in units of channel number relative to the position of the stopped peak at channel 6.3. The transformation is quite good, but not perfect because of the effects of finite resolution. Some variation of the Fermi function parameters x_0 and a may be needed when the lifetime varies substantially from this intermediate value of 0.4 ps.

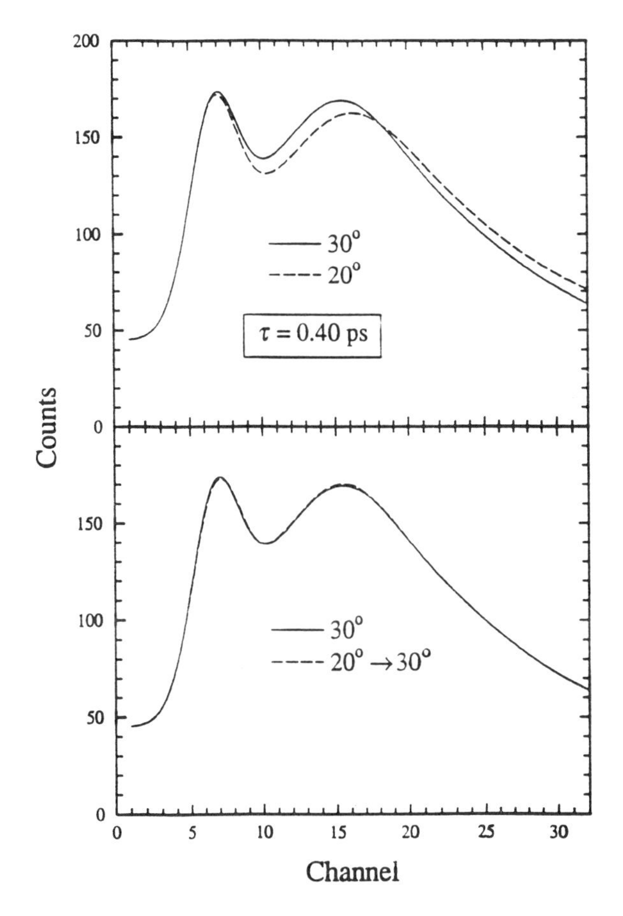

FIGURE 3. Calculated DSAM line shapes at 20° and 30° (top panel). In the lower panel the 20° curve (dashed line) has been transformed to an observation angle of 30°.

VELOCITY CORRECTIONS

The evaporation of light particles from the compound nucleus leads to a range of recoil velocities as well as angles. The magnitude of the recoil velocity can also be determined on an event-by-event basis by kinematic reconstruction if all the evaporated particles are detected in a full-sphere detector array. However, the problem of transforming events corresponding to different recoil velocities to some common velocity is more complicated than the corresponding angle transformation. In this case different recoil velocities correspond to different decay-weighted distributions of the function $\beta(t)$.

Therefore energy transformations suffer from the complexity of $\beta(t)$ as well as the problem of finite γ detector resolution. An obvious approximation is to scale the function $\beta(t)$ linearly with the initial recoil velocity. The line shape can then be transformed using Eq. 3 with a different scaling ratio R based on

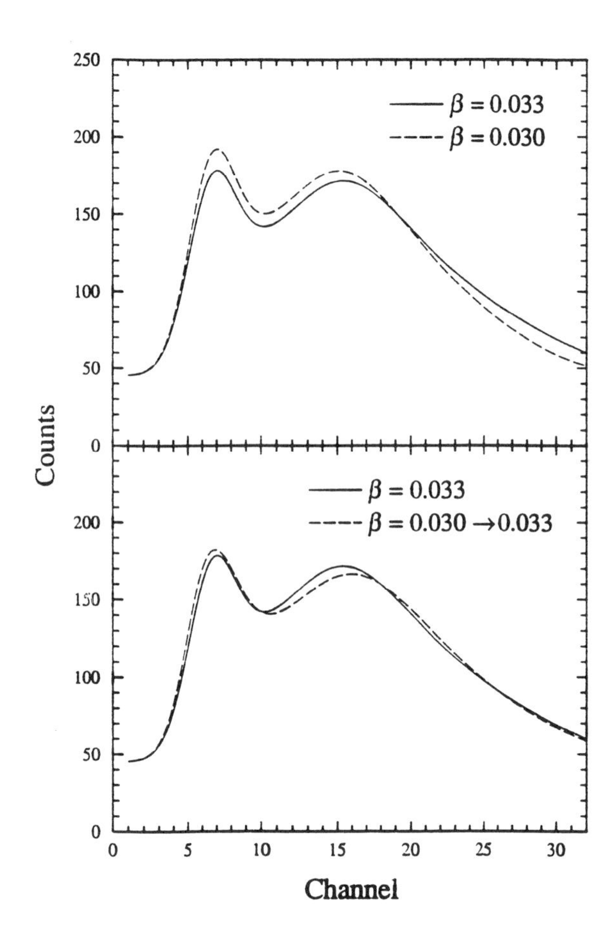

FIGURE 4. Calculated DSAM line shapes for initial recoil velocities of $\beta = 0.033$ and $\beta = 0.030$ (top panel). In the lower panel the $\beta = 0.030$ curve (dashed line) has been transformed to an initial velocity of $\beta = 0.033$.

$$R = \frac{\beta_0(\text{new})}{\beta_0(\text{old})}, \tag{6}$$

where $\beta_0(\text{old})$ is the kinematically reconstructed initial recoil velocity for a particular event and $\beta_0(\text{new})$ is a common initial recoil velocity to which all events should be transformed.

The results of this approximate transformation from a simulation calculated for an initial recoil velocity of $\beta = 0.030$ to $\beta = 0.033$ are shown in Fig. 4. Although the curves are more similar after the transformation (lower panel) than before (upper panel), this approximate transformation does not work as well as the angle transformation shown in Fig. 3.

It is possible to devise an almost exact energy transformation for the data based simply on a shift of the dashed curve in the upper panel of Fig. 4 to the position of the solid curve. The limitation of this technique is that the transformation will depend somewhat on the lifetime of the state. Therefore, initial lifetime estimates will be needed before all the events are transformed to a common recoil velocity and somewhat different transformations will be needed for different states.

SUMMARY

A Doppler-shifted line shape observed at one angle can be transformed rather well to correspond to a different angle of observation. The transformation is based on the ratio of the cosines of the angles and must be smoothly cut off where the stopped peak joins the Doppler-shifted component of the line shape. This transformation is useful to combine the data from γ detectors placed at somewhat different angles in an array. It can also be used to compensate for variations in the recoil angle if the kinematics of each event can be reconstructed from information about the energies and angles of emission of all the recoil particles.

Kinematic reconstruction also leads to a knowledge of the variation of recoil velocity in the evaporation process. A simple velocity compensation technique based on the ratios of the initial recoil velocities does not work as well as the angle correction because of the complex dependence of the function $\beta(t)$ on the decay history and on the non-linear deceleration process. A somewhat lifetime-dependent correction can be applied, based on the simulated line shapes.

REFERENCES

* Supported in part by the National Science Foundation.

1. D.G. Sarantites *et al.* (to be published).

HADRON-INDUCED REACTIONS: FROM BASIC RESEARCH TO NEW TECHNOLOGICAL APPLICATIONS

Henry H.K. Tang

IBM Microelectronics, Semiconductor Research and Development Center, East Fishkill Laboratory, Hopewell Junction, NY 12533

J. L. Romero

Department of Physics, University of California, Davis, CA 95616

We focus on nucleon- and pion-induced reactions on light elements, over the energy range from MeVs to GeVs. We discuss their significance in several important technological applications: 1) single event upsets (SEUs) in microelectronic systems, and 2) particle-beam radiology. We describe the work of nuclear modeling at IBM (NUSPA model) which is applied to simulations of cosmic-ray-induced SEUs at the device and circuit levels. We dicuss some current modeling issues. The modeling of recoil spectra of 80 MeV proton on Si is stringently tested by a recent experiment at the National Superconducting Cyclotron Laboratory of Michigan State University, using reverse kinematics. We also report new simulation results of thick targets irradiated by high-energy protons (50-250 MeV). They show significant secondary radiation due to nuclear fragments, a basic problem, which up till now, has not been systematically tackled. We discuss the implications of these results on future SEU studies and radiation-related problems.

INTRODUCTION

Nucleon- and pion-induced reactions over various energy regions are recognized to be important probes with which nuclear dynamics and structure can be studied. In recent years, several important radiation- related areas have been identified: 1) single event upsets (SEUs) in microelectronic systems, and 2) particle-beam radiology. Also many important space radiation problems are related to SEU and radiation effects in biological systems. (1) They all require data or modeling of hadron-induced reactions as crucial inputs.

SINGLE EVENT UPSETS AND NUCLEAR MODELING

For many years, it has been known in the microelectronics industry that single event upsets impose a serious product reliability issue. In the last ten years or so, there has been much research activities on SEU problems.

By now, it is recognized that a significant component of SEU is of nuclear origin. For example, high-energy cosmic-ray particles like nucleons and pions are a major source of SEUs for many products of current and future technologies. These particles interact with Si and O nuclei in the device and produce secondary fragments like hydrogen ions, helium ions, and recoil nuclei, which in turn produce spurious electron-hole pairs along their tracks. (2) These electron-hole pairs, if produced in the neighborhood of a pn junction, can be further separated by the strong fields. Carriers of appropriate sign can be swept to a nearby contact. If the collected charge exceeds a certain threshold value (critical charge), the memory state of the device can be flipped. (2)

A key parameter of SEU problems is the energy deposited by the secondary charged fragments. For Si, each MeV deposited is associated with electron-hole pairs of 44.5 fC. To develop SEU modeling capabilites at the level of device and circuit, it is essential to have a detailed model for the distributions of the secondary fragments.

Serious SEU modeling efforts at IBM were started around 1986. Nuclear modeling for SEU work was undertaken by the first author. The result was the development of a nuclear spallation model (NUSPA) (2-4). In the early years of this development, focus was on applications at sea level where the cosmic ray spectra are dominated by nucleons. A critical review of nucleon-induced reactions and comparison of NUSPA with a large body of nuclear data (mainly inclusive spectra of light ions) was reported in Ref. 3. In the past few years, in anticipation of future applications at high altitudes where pion fluxes are large, the model was extended to include pion-nucleus reactions.

The basic formulation of the NUSPA model is given in Ref. 3. Here we summarize the main features. The model assumes that the nuclear reaction is a two-step process. The first stage of the reaction is an intranuclear cascade of quasi-free nucleon-nucleon or pion-nucleon scatterings. In the case of nucleon reaction, Pauli blocking and Fermi motion of target nucleons are taken into account. The second stage of the reaction is a statistical decay process of an excited compound nucleus formed at the end of the cascade stage. The computational scheme of NUSPA is based on Monte Carlo techniques. The model computes absolute cross sections and it simulates all exclusive reaction channels on an equal footing. Also the model satisfies a class of fundamental sum rules which set up rigorous constraints on the

CP392, *Application of Accelerators in Research and Industry,* edited by J. L. Duggan and I. L. Morgan
AIP Press, New York © 1997.

reaction cross section and cross sections of all the exit channels. Table 1 of Ref. 3 summarizes the database which is used to check the model.

In parallel to modeling work, a collaboration between IBM and the UC Davis experimental group went on for a number of years. Neutron reactions at energies below 50 MeV were studied, and the data obtained at UC Davis were used to verify the reaction model (5-7).

Since the kinetic energies carried by heavy recoil nuclei from spallation reactions are distributed over short distances, this constitutes a significant source of secondary radiation. Up till now, due to the intrinsic difficulties of conventional experiments with stationary targets, data of recoil spectra are almost non-existent. In the last two years, the second author developed an experimental program of measuring recoil spectra using reverse kinematics. Preliminary results of a recent experiment at the National Superconducting Laboratory at Michigan State University for the reaction Si28(HI,A)x using a Si beam of 80 MeV/nucleon indicates that the recoil spectra are fairly well predicted by the NUSPA model (8-9). As new technologies push for devices with smaller sizes and critical charges, SEU issues will be more critical for designers. More accurate SEU models will be needed and recoil data will be very useful in complementing the existing data of light ions.

SECONDARY RADIATION EFFECTS IN THICK TARGETS

In the case of thick targets irradiated by high-energy protons, the role of secondary nuclear fragments is much more subtle than in the case of SEUs in devices. The fundamental principle of proton therapy is based on electronic stopping power (electronic energy loss per unit path). For proton, it is small near the target surface, but is large in the Bragg region near the end of proton range. Since the nuclear cross sections are much smaller than the cross sections associated with electronic stopping power, these high-order nuclear processes are usually neglected in dosimetry calculations.

With the advent of new proton facilities and demands for new dosimetry calibrations, it is important to address the question of energies from secondary fragments and to quantify their radiation effects. To address this problem, we did Monte Carlo simulations of water irradiated by high energy protons. The calculations are based on nuclear cross sections computed from the NUSPA model. We have come to the conclusion that the energies associated with secondary radiation may not be as small as is usually believed. The detailed analysis will be reported in a forthcoming paper (10). Here, we highlight some of the major results.

We compute, the mean kinetic energies of charged fragments from Monte Carlo-generated events of p + H_2O, as function of target depth. We also compute the total kinetic energies of the charge fragments distributed over the entire target volume. From the standpoint of energy deposition, the nuclear fragments fall into two categories.

The first class consists of the heavier fragments of recoil nuclei and helium ions. They have typical mean kinetic energies of a few MeVs, which are distributed over small volumes due to short particle ranges.

The second class consists of secondary hydrogen ions (protons, deuterons and tritons). Their energy spectra are broad, and some of these secondary light ions have long ranges. Hence, the energy deposition of H ions is non-local in the sense that the energy is distributed over a large volume, in contrast to the local nature of energy deposition of recoils and He.

If we take a small volume (but large enough such that quantities like stopping power vary smoothly), the ratio of the mean total kinetic energies of secondary charged fragments (secondary radiation energy) to the electronic energy loss of the primary proton in the volume can be shown to be

$$\rho\ ((E_1+E_2)\ \sigma_R + E_R\ \sigma_{EL} + E_p\ \sigma_{pp})\ /\ (dE/dx)_e.$$

Here ρ is the number density of water molecules; σ_R is the p + O reaction cross section; σ_{EL} is the elastic cross section of p + O; σ_{pp} is the cross section of free pp scattering; E_1 is the mean total kinetic energy of recoils and secondary H; E_2 is the mean total kinetic energy of secondary H ions; E_R is the mean recoil energy of O in elastic p + O scattring; E_p is the mean kinetic energy of primary proton in the volume in question; and $(dE/dx)_e$ is the electronic stopping power of the primary proton .

Note that σ_{pp} is well known; σ_{EL} is calculated from optical model, from which E_R is readily derived; σ_R, E_1 and E_2 are computed from the NUSPA model.

Figure 1 shows a plot of this ratio as a function of target depth for beam energies of 250, 200, 150 and 100 MeV. The solid curves correspond to energies from H ions and the dotted curves correspond to energies from recoils and He. Plotted in this way, these curves are very close to being linear with respect to target depth. It is significant that for beam energies from 250 MeV down to 150 MeV, from target surface down to about half of the proton range, the secondary energy produced at each point can be larger than 10% of the electronic energy loss.

Table 1 summarizes the ratio of total radiation energy to electron energy loss for beam energies of 250, 200, 150, 100 and 50 MeV. Whereas the effects of secondary fragments near the end of the proton range are negligible, there is a fairly large distance over which the secondary energies are significant.

SUMMARY AND CONCLUSION

Much of the work on hadron-induced reactions discussed in this paper was originally motivated by reliability concerns of electronic systems affected by SEUs. It turns out that results of these reactions have a much wider range of applications. One potentially fruitful development is the systematic recoil measurements of light elements by exploiting reverse kinematics. The data acquired will be of great value for many radiation-related applications. Also our simulation results seem to indicate that energies due to secondary nuclear fragments in thick targets are not always negligible. All these suggest that hadron-induced reactions will continue to play a significant role in basic and applied work in the future.

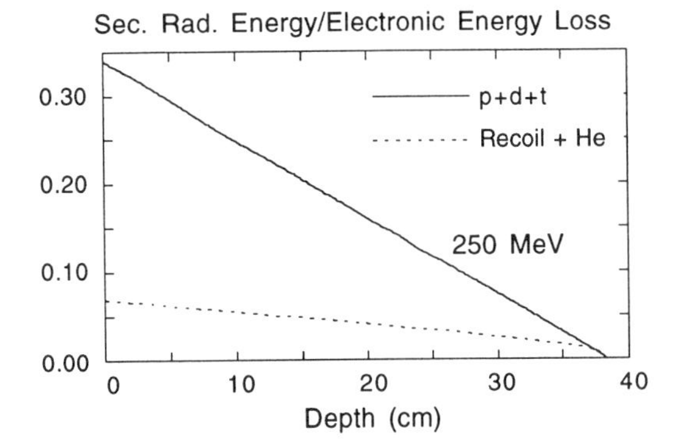

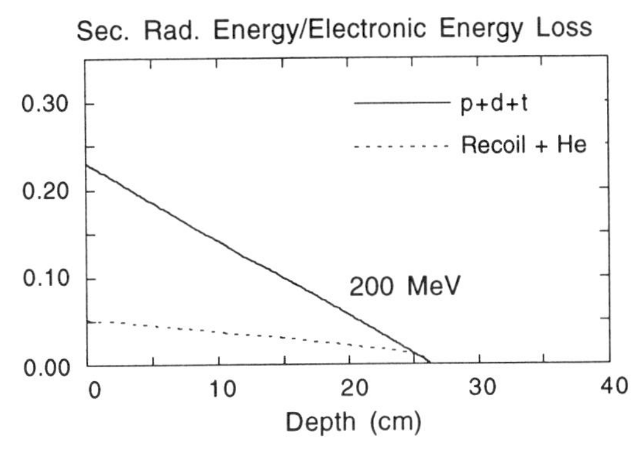

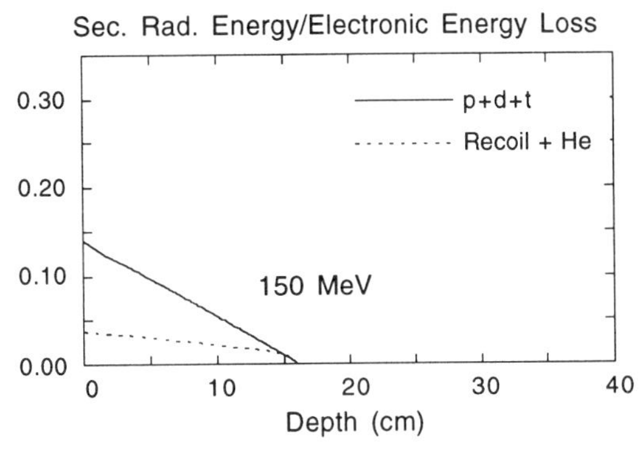

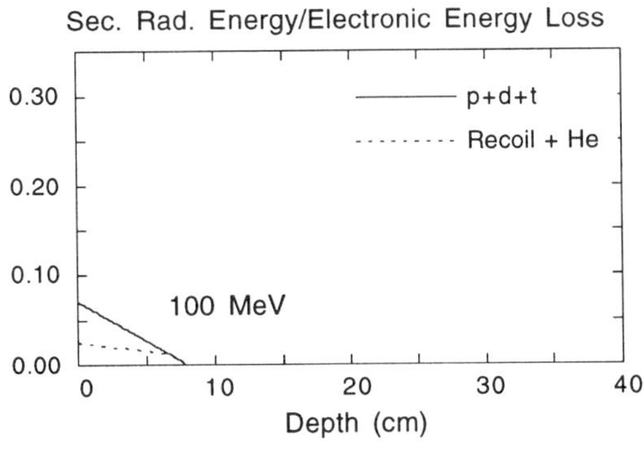

FIGURE 1 Ratio of secondary radiation energy to electron energy loss of the primary proton as a function of target depth for beam energies of 250, 200, 150 and 100 MeV.

TABLE 1. Ratio of the total secondary radiation energy to the electronic energy loss of the primary proton.

Beam Energy (MeV)	Recoil + He / E.E.L.	p+d+t / E.E.L.
50.	0.0079	0.0086
100	0.0137	0.0264
150	0.0195	0.0519
200	0.0258	0.0847
250	0.0327	0.1245

REFERENCES

1 J. W. Wilson, L. W. Townsend, W. Schimmerling, G. S. Khandelwal, F. Khan, J. E. Nealy, F. A. Cucinotta, L. C. Simonsen, J. L. Shinn, and J. W. Norbury. NASA Reference Publication 1256 (1991).

2. H. H. K. Tang, IBM J. Res. and Dev. 40(1), 91-108 (1996). See also other articles in January issue of IBM J. Res. Dev.40(1) (1996).

3 H. H. K. Tang, G. R. Srinivasan and N. Azziz, Phys. Rev. C42, 1598-1622 (1990).

4 H. H. K. Tang, Nuclear Spallation Model/Codes System (NUSPA), unpublished.

5 P. S. Rezentes, Ph.D. Thesis, University of California, Davis, 1993.

6. P. S. Rezentes, C. M. Castaneda, J. L. Romero, H. H. K. Tang, T. A. Cahill, and J. R. Drummond, UC Davis Preprint (1995), and to be published.

7 E. L. Hjort, F. P. Brady, J. R. Drummond, B. McEachern, J. H. Osborne, J. L. Romero, and D. S. Sorenson, Phys. Rev. C53, 237-242 (1996).

8 J. L. Romero, F. P. Brady, D. A. Cebra, J. Chance, J. Kintner, J. H. Osborne, G. Acevedo-Bolton, D. J. Morrissey, M. Fauerbach, P. Pfaff, C. F. Powell, B. M. Sherrill, and H. H. K. Tang,. BAPS 41, 987 (1996).

9 J. L. Romero, H. H. K. Tang, F. P. Brady, D. A. Cebra, J. Chance, J. Kintner, J. H. Osborne, D. J. Morrissey, M. Fauerbach, P. Pfaff, C. F. Powell, B. M. Sherrill, contributed paper of this conference.

10 H. H. K. Tang, J. L. Romero, F. P. Brady, and S. Catto, IBM SRDC Preprint (1996), and to be published.

IS THE TWO-NUCLEON/THREE NUCLEON PARADIGM AN ILLUSION?

W. Tornow

Department of Physics, Duke University, Durham, North Carolina 27708
and Triangle Universities Nuclear Laboratory, Duke Station, Durham, North Carolina 27708

H. Witała

Institute of Physics, Jagellonian University, PL-30059 Cracow, Poland

An important breakthrough recently achieved by a group of theoreticians at Pisa makes it now possible to compare the wealth of accurate proton-deuteron scattering data at energies below E_p=3.3 MeV to three-nucleon calculations that include, for the first time, the Coulomb interaction in a rigorous way. Previous comparisons of theoretical predictions and experimental data were restricted to neutron-deuteron scattering where the data base is quite small. The calculated proton-deuteron results for the observables $A_y(\theta)$ and $iT_{11}(\theta)$ deviate considerably from experimental data. Using the Pisa results as starting values in a phase-shift analysis, one finds that only two parameters (out of 62) have to be modified slightly in order to obtain a good description of all proton-deuteron data. However, these modifications are inconsistent with both nucleon-nucleon data and the nucleon-nucleon potential models used in the calculations. Possible explanations of this observation will be discussed.

INTRODUCTION

Figure 1 shows neutron-deuteron (n-d) analyzing power $A_y(\theta)$ data in the neutron energy range from 5 to 8.5 MeV in comparison to rigorous three-nucleon (3N) calculations (1) using the Paris nucleon-nucleon (NN) potential model (2). Here, "rigorous" refers to the 1% computational accuracy achieved in the solutions of the Faddeev 3N equations for the continuum. The surprisingly large difference between data and calculations observed for the magnitude of $A_y(\theta)$ is called the "$A_y(\theta)$-puzzle". Since its discovery a few years ago, the n-d $A_y(\theta)$-puzzle remains the most elusive problem in low-energy three-nucleon (3N) elastic scattering (3). Although several groups have performed extensive studies with modern nucleon-nucleon (NN) potential models, including various types of three-nucleon forces (3NFs), it turned out to be impossible to account for the 25-30% discrepancy between calculations and experimental data. Sensitivity studies have clearly shown that $A_y(\theta)$ in n-d elastic scattering is governed by a complicated interplay between the total angular momentum L=1 NN ^3P_j interactions (1). As has been known since quite some time, the ^3P_j interactions are also responsible for the NN $A_y(\theta)$, and conversely, they have been determined mainly from phase-shift analyses of NN $A_y(\theta)$ data.

Witała and Glöckle (4) studied the on-shell aspect of the $A_y(\theta)$ puzzle. They found, using a trial and error approach, a combination of ^3P_j NN interactions that describes both the available n-d and NN $A_y(\theta)$ and cross-section $\sigma(\theta)$ data below 50 MeV. In addition, the proton-deuteron (p-d) $A_y(\theta)$ and $\sigma(\theta)$ data are also well described, although the calculations did not include the

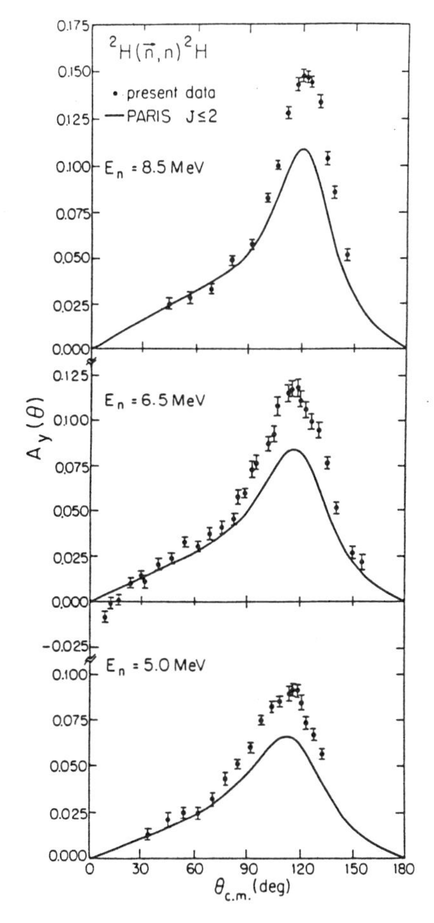

FIGURE 1. Neutron-deuteron elastic analyzing power $A_y(\theta)$ data at E_n=5.0, 6.5, and 8.5 MeV in comparison to rigorous Faddeev calculations using the Paris NN potential.

CP392, *Application of Accelerators in Research and Industry,* edited by J. L. Duggan and I. L. Morgan
AIP Press, New York © 1997

Coulomb interaction. However, the ^3P_j interactions obtained by Witała and Glöckle exhibit an unexplained large breaking of charge independence. In addition, the sign of the charge-independence breaking (CIB) is inconsistent with theoretical expectations based on the meson-exchange theory of the NN interaction. The unexpected finding of Witała and Glöckle casts doubt on the validity of the paradigm that the comparison of 3N data and 3N calculations provides a powerful tool for testing and determining the NN interaction used in the calculations. In the following we will refer to this paradigm as the "2N/3N paradigm".

After more than thirty years of intensive work the 3N scattering problem is now solvable with the Coulomb interaction taken into account in a rigorous way (5). Unfortunately, the solution of the charged 3N scattering problem is restricted to energies below the deuteron breakup threshold (i.e., E_p=3.3 MeV in the case of p-d scattering or E_d=6.6 MeV in the case of d-p scattering.) Nevertheless, this new theoretical development not only allows for a much more stringent test of the 2N/3N paradigm than is possible with n-d data, but it also provides some important additional information. First, it confirmed speculations that an $A_y(\theta)$ puzzle exists also for $\vec{p}$-d elastic scattering (6) (see Fig. 2b, solid curve). Second, it showed that a similar problem exists with respect to $iT_{11}(\theta)$ in $\vec{d}$-p scattering (7) (see Fig. 2c, solid curve).

II. PROTON-DEUTERON PHASE-SHIFT ANALYSIS

For p-d and d-p scattering extremely accurate sets of observables exist at $E_{c.m.}$=2.0 MeV (E_p=3.0 MeV and E_d=6.0 MeV (8, 9)) and $E_{c.m.}$=1.67 MeV (E_p=2.5 MeV and E_d=5.0 MeV (9)): $\sigma(\theta)$, $A_y(\theta)$, $iT_{11}(\theta)$, $T_{20}(\theta)$, $T_{21}(\theta)$, and $T_{22}(\theta)$. Starting from the Pisa phase-shift calculations (6) at $E_{c.m.}$=2.0 MeV based on the Argonne AV18 NN potential (10) + Urbana 3NF (11), a phase-shift search was performed using the high-accuracy data of the

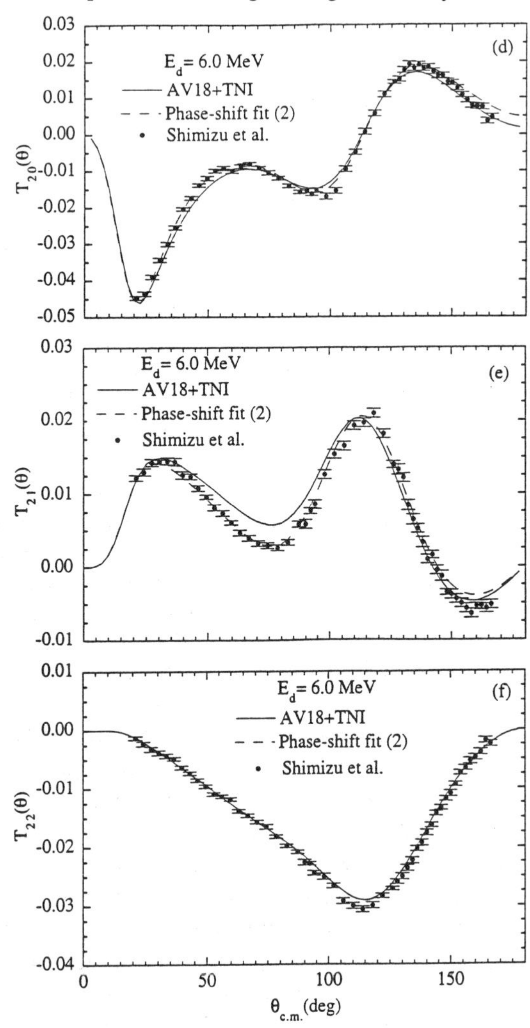

FIGURE 2. Comparison of rigorous 3N calculations (solid curves) and experimental data for p-d and d-p scattering observables at $E_{c.m.}$=2 MeV. The dashed curves were calculated from the phase shifts given by AV18 + 3NF with the exception of $^4P_{1/2}$ and $\varepsilon_{3/2^-}$, which were replaced by the values obtained from a 2-parameter phase-shift search.

Kyushu group (9). The main difference between the "experimental" and theoretical phase shifts was found for the $^4P_{1/2}$ phase shift and the $^2P_{3/2}$ - $^4P_{3/2}$ mixing parameter $\varepsilon_{3/2^-}$. Therefore, we searched only on these two parameters, i.e., a total of 60 parameters were left unchanged. We had to lower $^4P_{1/2}$ from 22.3° to 21.54° and to raise $\varepsilon_{3/2^-}$ from 2.23° to 2.47°. The dashed curves shown in Fig. 2 represent the results of our 2-parameter phase-shift fit. Clearly, all observables are described very well. We multiplied the n-d $^4P_{1/2}$ and $\varepsilon_{3/2^-}$ phase-shift parameters by the factors found for p-d scattering at E_p=3.0 MeV and calculated the n-d $A_y(\theta)$ at E_n=3 and 5 MeV. As can be seen from Fig. 3 (dashed curves) the data are well described by out modified phase shifts. As has been shown by the Bochum-Cracow group (13), a strong correlation exists between the n-d ^4P_J and the NN ^3P_j phase shifts: A change of the $^3P_0$ NN interaction influences the n-d $^4P_{1/2}$ phase shift. We expect such a relation to exist also between the $^4P_{1/2}$ p-d phase shift and the $^3P_0$ NN interaction. However, in the theoretical approach used by the Pisa group to incorporate the Coulomb interaction in 3N calculations, this relation cannot be calculated. Therefore, using the information given in Ref. (6) about the ratio of the p-d and n-d $^4P_{1/2}$ phase shifts at E_n=3 MeV, the p-d $^4P_{1/2}$ phase shift obtained from our phase-shift search was converted into a n-d $^4P_{1/2}$ phase shift. Subsequently, we used the relation established by the Bochum-Cracow group between $^4P_{1/2}$ and $^3P_0$ and determined the scale factor λ_{eff}=0.91 with which the AV18 $^3P_0$ NN interaction had to be multiplied in order to reproduce our new n-d $^4P_{1/2}$ phase shift. It should be pointed out that the new NN $^3P_0$ interaction did not modify the mixing parameter $\varepsilon_{3/2^-}$. In fact, it turned out to be impossible to find a single NN interaction parameter or any combination of particular NN interaction parameters (for example $^3P_1$ and $^3P_2$ NN force parameters) that can account for the 10% larger value obtained for $\varepsilon_{3/2^-}$ in the p-d phase-shift search and simultaneously describes the NN data. Finally, employing the relation $\lambda_{eff}=2/3\lambda_{pp}+1/3\lambda_{np}$ with λ_{pp}=0.91 and λ_{np}=0.91 new $^3P_0$ p-p and n-p phase shifts were obtained. It turned out that these phase shifts overpredict the n-p $A_y(\theta)$ and underpredict the p-p $A_y(\theta)$ by 50% and 10%, respectively. Unfortunately, experimental n-p and p-p data are not available at E_n=3 MeV. Assuming the "phase-shift data", to which the AV18 NN potential is fitted, to be accurate at the 5% level, we must conclude that our solution λ_{pp}=0.91 and λ_{np}=0.91 is incorrect. In fact, it turned out that the NN data can be described only with $\lambda_{pp}\neq\lambda_{np}$, i.e., one has to introduce a considerably large breaking of charge independence, as was done already by Witała and Glöckle. The parameter λ_{np} must be close to 1.0 in order to describe the n-p $A_y(\theta)$ data. This implies $\lambda_{pp}\approx$0.87, and this value is in fact in good agreement with p-p $A_y(\theta)$ data.

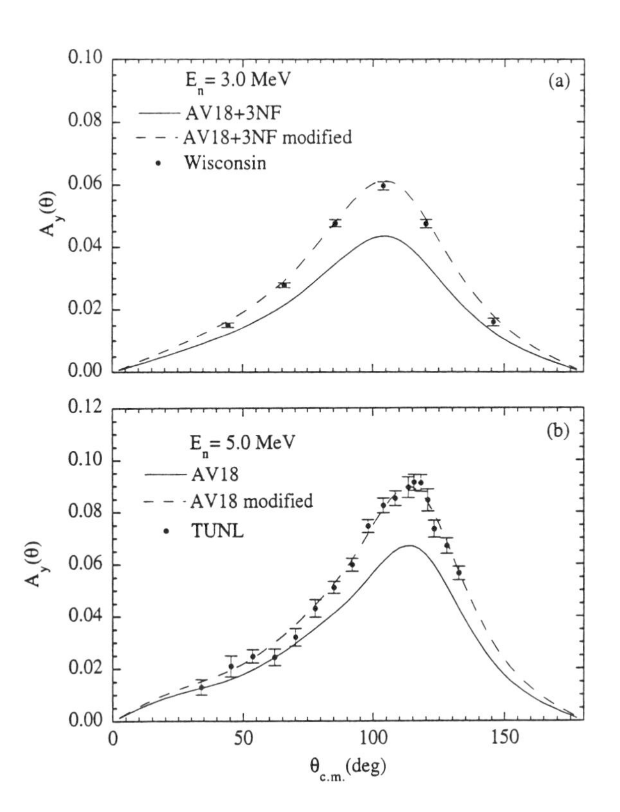

FIGURE 3. Comparison of rigorous 3N calculations (solid curves) and experimental n-d analyzing power data at E_n=3.0 (12) and 5.0 MeV (1). The dashed curves were calculated from phase shifts generated from the original potentials except for $^4P_{1/2}$ and $\varepsilon_{3/2^-}$ where we used Coulomb corrected values obtained from fits to p-d data at E_p=3.0 MeV.

DISCUSSION

We are confronted with the expected observation that the 2N/3N paradigm yields a satisfactory description of both 2N and 3N data. However, the completely unexpected finding is the fact that the associated ^3P_j NN interactions are in severe contradiction to theoretical expectations. In addition, the $\varepsilon_{3/2^-}$ mixing parameter cannot be affected by any reasonable modification of the NN interactions, i.e., modifications that describe the NN data within their associated uncertainties.

Of course, the 2N/3N paradigm is based on the assumption that the 3N system is governed by the same "free" NN interactions as is the NN system, i.e., 3NF effects do not play a significant role. Considering the $^4P_{1/2}$ wave, we notice that this phase shift is associated with a peripheral wave. Therefore, 3NF effects must be very small. In fact, this conjecture is in perfect agreement with theoretical calculations. In 3N scattering, only the $^2S_{1/2}$ phase shift is affected by the traditional short-range 3NFs. However, this phase shift has hardly any influence on $A_y(\theta)$ and $iT_{11}(\theta)$.

One may argue that the 2-parameter phase-shift search described above is too restrictive and that our result is an artifact caused by our approach. In order to check on this conjecture, we searched on five specific parameters: $^4P_{1/2}$, $^4P_{3/2}$, $^4P_{5/2}$, $\varepsilon_{1/2^-}$, $\varepsilon_{3/2^-}$. It was found that these parameters are related to the NN $^3P_0$, $^3P_1$, and $^3P_2$ interactions, which in turn are largely responsible for the NN $A_y(\theta)$ and 3N $A_y(\theta)$ and $iT_{11}(\theta)$. It turned out that our earlier finding is very stable with respect to the specific search procedure. In fact, even when we searched on all 62 parameters (up to angular momentum L=4) we always observed that $^4P_{1/2}$ has to be smaller and $\varepsilon_{3/2^-}$ has to be larger by about the same amount as found in the 2-parameter search.

In any attempt to rescue the 2N/3N paradigm one has to keep in mind that the n-d $A_y(\theta)$ is about a factor of 10 larger than the n-p $A_y(\theta)$ at low energies (E_n<25 MeV), thus providing a greatly enhanced sensitivity to the ^3P_j NN interactions. It should also be noted that the p-p $A_y(\theta)$ is governed by the Coulomb interaction. In fact, the p-d $A_y(\theta)$ at low energies is almost two orders of magnitude larger than the p-p $A_y(\theta)$. Therefore, n-d and p-d data provide a magnifying glass for studying the ^3P_j NN interactions.

We performed sensitivity studies in the NN system, i.e., we modified the ^3P_j phase shifts and investigated their influence on all NN observables (see also Ref. (14)). It turned out that in the energy range of interest for low-energy 3N calculations, i.e., energies up to the Fermi energy, basically only $A_y(\theta)$ is sensitive to modifications of the ^3P_j phase shifts. Furthermore, according to our experience, the three ^3P_j phase shifts cannot be determined in a unique way: a band of solutions exists that almost equally well describes the low-energy NN $A_y(\theta)$ data. This observation raises the question about the accuracy of the extrapolation procedure used in global NN phase-shift analyses to obtain low-energy phase shifts from high-energy data where several observables are sensitive to the ^3P_j NN interactions.

In order to support our suspicion we consider again E_n=3 MeV. Using a trial and error approach we found that the phase-shift combination $0.976\times{}^3P_0$, $0.912\times{}^3P_1$, and $1.16\times{}^3P_2$, describes all calculated NN observables almost as well as does the original phase-shift combination $^3P_0$, $^3P_1$, and $^3P_2$ obtained from the AV18 NN potential. Here, "almost as well" refers to the fact that differences between the modified and original descriptions of up to 2% were tolerated. This solution uses exactly the same small CIB and charge-symmetry breaking as embedded in the original AV18 NN potential model. At E_n=10 and 25 MeV, our modification factors are not valid anymore. In fact, at these energies we observed differences of up to 7% and 13% for n-p and p-p scattering observables, respectively, i.e., our modified NN phase shifts do not describe the NN "phase-shift data" accurately. Nevertheless, if one multiplies the AV18 interactions with the λ_{eff} factors associated with $0.976\times{}^3P_0$, $0.912\times{}^3P_1$, and $1.16\times{}^3P_2$, one finds that the n-d $A_y(\theta)$-puzzle is reduced by a factor of two. Of course, this surprising result may just be accidental. On the other hand, it clearly demonstrates what has to be done: on the phase-shift level one has to determine smoothly varying, energy dependent multiplication factors for the AV18 $^3P_0$, $^3P_1$, and $^3P_2$ phase shifts in the energy range from 0 to about 50 MeV. These new phase shifts must connect smoothly to the existing (assumed "correct") high-energy phase shifts. On the potential level, $^3P_0$, $^3P_1$, and $^3P_2$ NN interactions have to be constructed that reproduce the new ^3P_j phase shifts. It may be necessary to readjust slightly other NN interaction components in order to describe the "recommended" NN data base to which the original AV18 potential was fitted. Finally, this new potential model must be used in 3N calculations and the results must be compared to n-d and p-d data. Depending on the results, several iterations of the entire procedure may be required. Only then will it be possible to answer the question whether the 2N/3N paradigm is really an illusion, or whether nuclear physics has not found yet the "correct" ^3P_j NN interactions at low energies.

ACKNOWLEDGMENTS

We acknowledge Prof. L.D. Knutson for providing us with a phase-shift search code, and Dr. A. Kievsky from the Pisa group for performing rigorous proton-deuteron calculations. This work was supported in part by the U.S. Department of Energy, Office of High Energy and Nuclear Physics, under grant no. DEFG05-91-ER40619 and by the Maria Sklodowska-Curie II Fund under grant no. MEN/NSF-94-161. Some of the numerical calculations have been performed on the Cray Y-MP of the North Carolina Supercomputing Center at Research Triangle Park, North Carolina, and on the Cray Y-MP of the Höchstleistungsrechenzentrum in Jülich, Germany.

REFERENCES

1. Tornow, W., *et al.*, Phys. Lett. **B257** 273-277 (1991).
2. Lacombe, H., *et al.*, Phys. Rev. **C21** 861-873 (1980).
3. Witała, H., Hüber, D., Glöckle, W., Phys. Rev. **C49** R14-R16 (1994) and references therein.
4. Witała, H., and Glöckle, W., Nucl. Phys. **A528** 48-72 (1991).
5. Kievsky, A., Viviani, M., and Rosati, S., Nuclear Physics **A577** 511-527 (1994).
6. Kievsky, A., Viviani, M., and Rosati, S., Phys. Rev. **C52** R15-19 (1995).
7. Kievsky, A., Rosati, S., Tornow, W., and Viviani, M., Nucl Phys. A (1996), in press.
8. Knutson, L.D., Lamm, L.O., McAninch, J.E., Phys. Rev. Let. **71** 3762-3765 (1993).
9. Shimizu, S., *et al.*, Phys. Rev. **C52** 1193-1202 (1995).
10. Wiringa, R.B., Stoks, V.G.J., and Schiavilla, R., Phys. Rev. **C51** 38-51 (1995).
11. Pudliner, B.S., *et al.*, Phys. Rev. Lett. **74** 4396-4399 (1995).
12. McAninch, J.E., Lamm, L.O., and Haeberli, W., Phys. Rev. **C50** 589-601 (1994).
13. Hüber, D., Golak, J., Witała, H., Glöckle, W., and Kamada, H., Few-Body Systems **19** 175-193 (1995).
14. Tornow, W., and Walter, R.L., Few-Body Systems **8** 11-21 (1990).

MOMENT MEASUREMENTS OF EXOTIC NUCLEI

P. F. Mantica[a,b,*], R. W. Ibbotson[b], D. Anthony[b], M. Fauerbach[b], D. J. Morrissey[a,b], C. F. Powell[b], J. Rikovska[c,d], M. Steiner[b], N. J. Stone[c], W. B. Walters[d]

[a]*Department of Chemistry, Michigan State University, East Lansing, Michigan 48824, USA*
[b]*National Superconducting Cyclotron Laboratory, Michigan State University, East Lansing, Michigan 48824, USA*
[c]*Department of Physics, Oxford University, Oxford OX1 3PU, UK*
[d]*Department of Chemistry and Biochemistry, University of Maryland, College Park, Maryland 20742, USA*

A program directed towards the measurement of ground state moments in nuclei far from stability has been initiated at the National Superconducting Cyclotron Laboratory at Michigan State University. Spin-polarized fragments are produced at small angles relative to the primary beam axis by intermediate-energy heavy-ion reactions. The fragments are collected and analyzed using the A1200 fragment separator. The β-NMR technique is then used to detect the resonance frequency of the spin-polarized exotic fragments. The results of the first experiments using the β-NMR system recently installed at the NSCL will be reported, including measurements of the polarization of ^{12}B fragments following the reaction of ^{18}O at 80 MeV/A on a Nb target.

INTRODUCTION

Measurements of the ground state moments of exotic nuclei provide important information on the extent to which single-particle and/or collective features dominate the low energy structure of these nuclei. Such measurements, however, are difficult for nuclei far off the line of stability due to the short half-lives of these species. The pioneering experiments by Asahi *et al.* (1), in which they measured the polarization of secondary fragments produced off the central beam axis following intermediate-energy heavy ion reactions, now provide a unique opportunity to measure ground state moments in a variety of light, exotic nuclei. A β-NMR technique can be applied to measure the nuclear hyperfine splitting resonance curve corresponding to the appropriate nuclear moment, where, for the case of a dipole interaction only in an applied magnetic field B, the peak in the resonance curve (given as the Larmour frequency, ν_L) is directly related to the nuclear g factor through the relation $h\nu_L = gB$.

To date, ground state magnetic moments have been measured using the β-NMR technique following intermediate-energy heavy-ion collisions for ^{14,15,17}B and ^{17}N (2,3), ^{9}C and ^{13}O (4), ^{21}F (5), and ^{43}Ti (6). Electric quadrupole moments have also been measured for spin-polarized fragments produced in projectile fragmentation, most recently for ^{14,15}B (7). Through the implementation of a β-NMR apparatus following the A1200 fragment separator at the National Superconducting Cyclotron Laboratory (NSCL) at Michigan State University, we plan to take advantage of the high quality beams of exotic nuclei presently available at the NSCL to make precise measurements of nuclear magnetic and quadrupole moments for nuclei having Z less than ≈ 40, and with half-lives less than a few seconds. This contribution describes the initial experiments used to verify the polarization of fragments following projectile fragmentation and to test the new β-NMR system.

EXPERIMENTAL TECHNIQUE

The β-NMR system used in the initial measurements at the NSCL was similar in many respects to that described by Asahi *et al.* (8). The system resided between the pole faces of a large dipole magnet (pole gap of 10.2 cm), which provided the nuclear Zeeman splitting and the directional holding field for the spin-polarized secondary beams. The detectors consisted of two β telescopes, located at 0° and 180° with respect to the direction of the holding field of the dipole magnet. The telescopes were each composed of a 4.4 cm x 4.4 cm x 3 mm thick ΔE plastic scintillator and a 5.1 cm x 5.1 cm x 25 mm thick total energy plastic scintillator. Each scintillator detector was coupled to a long (> 56 cm) acrylic light guide to place the photomultiplier tube of each telescope element beyond the fringe field of the dipole magnet. The telescopes were placed 9 mm from the catcher foil, and covered approximately 33% of the 4π solid angle.

The catcher foil for the experiments reported here was a 2.5 cm x 2.5 cm x 250 μm thick Pt foil, annealled at 630° C for 10 hours in air. The foil was mounted between the two β telescopes, and tilted at an angle of 45° relative to the beam axis of the A1200. Surrounding the catcher foil was a set of radiofrequency (RF) coils, which provided the oscillating magnetic field for the resonance measurement. The RF coils were two 30-turn loops (diameter 2.3 cm) of 28 AWG magnet wire, arranged in a Helmholtz-like configuration, with a separation distance of 3.0 cm. The coil inductance was measured to be 77 μH. The RF coils were configured as part of an RCL circuit, which also included a 50 Ω resistor and a variable capacitor to provide the maximum alternating magnetic field by matching the impedance of the circuit to the output impedance (50 Ω) of the RF source.

A secondary beam of ^{12}B ($T_{1/2}$ = 20 ms, $I^{\pi} = 1^{+}$, Q_{β} = 13 MeV) was produced using a primary beam of ^{18}O at 80 MeV/A incident on a 216 mg/cm^2 ^{93}Nb target. The secondary fragments were selected by the A1200 fragment separator (9). Two steering magnets located upstream of the A1200 target position allowed the collection of fragments in the range of +3° to -3° relative to the primary beam axis. The full angular acceptance of fragments in the deflection plane at the target position was approximately 1°. Fragments were also accepted within the range of 1% of the chosen central momentum of the A1200 as defined by momentum slits placed at the first momentum-dispersed image of the device. Beam identification was accomplished at the A1200 focal plane using the energy loss of the fragments measured in a 300 μm Si PIN detector and the fragment time-of-flight (TOF) referenced to the K1200 Cyclotron radiofrequency. A second 300 μm Si PIN was located behind the catcher foil position of the β-NMR apparatus to allow for redundant fragment identification when the catcher foil was removed, again using energy loss and TOF information.

Two data acquisition techniques were employed during the ^{12}B polarization experiments. The first involved pulsing the primary ^{18}O beam from the cyclotron. During beam-on cycles, ^{12}B fragments at 35 MeV/A were directed from the A1200, energy degraded to $\approx$ 13 MeV/A using Al degrader foils, and implanted into the Pt catcher foil for 20 ms. During the beam-off cycles, which were 40 ms in duration, the RF coils surrounding the catcher foil were energized for the entire beam-off period every other cycle, and β spectra were collected for the ^{12}B spin-polarized fragments. The RF-on spectra were then normalized to the spectra collected during the RF-off condition in order to correct for possible changes in the position of the beam at the catcher foil. Valid β events required signals in both the ΔE and E detectors of a single β telescope within a time period of 100 ns.

In the second data acquisition mode the ^{12}B activity was collected continuously for a given run, with the RF coils energized during the entire run period. This method was employed to test the feasibility of performing, more efficiently, moment measurements for fragments having long (> 10 s) decay half-lifes and spin-lattice relaxation times. Normalization of the β spectra was accomplished by continuously collecting the ^{12}B activity with no signal supplied to the RF coils. This second, or batch, technique had not been employed in the previous β-NMR experiments described in the literature; however, as will be discussed in the following section, the observed magnitude of the destruction of the spin polarization was comparable to that measured using the more conventional beam-on/beam-off acquisition method.

RESULTS AND DISCUSSION

To test the proper operation of the β-NMR system, and also to confirm the observed polarization of fragments following intermediate-energy heavy-ion collisions, we completed two measurements of the polarization of ^{12}B fragments. The first measurement was the reproduction of the resonance curve for ^{12}B. Using a static field of 0.124 T, an incident beam angle of +3° and selecting fragments on the peak of the momentum yield curve, the frequency range from 930–970 kHz was scanned, using a frequency modulation of $\pm$ 10 kHz. During this measurement, the batch implantion method described above was employed. The resulting resonance curve is shown in Fig. 1. Fitting this curve to a Lorentzian peak shape, we have extracted a linewidth of 7 kHz, and a Larmour frequency of 947(1) kHz. From the measured Larmour frequency, we deduce a value of 1.002(2) μ_N for the ground state

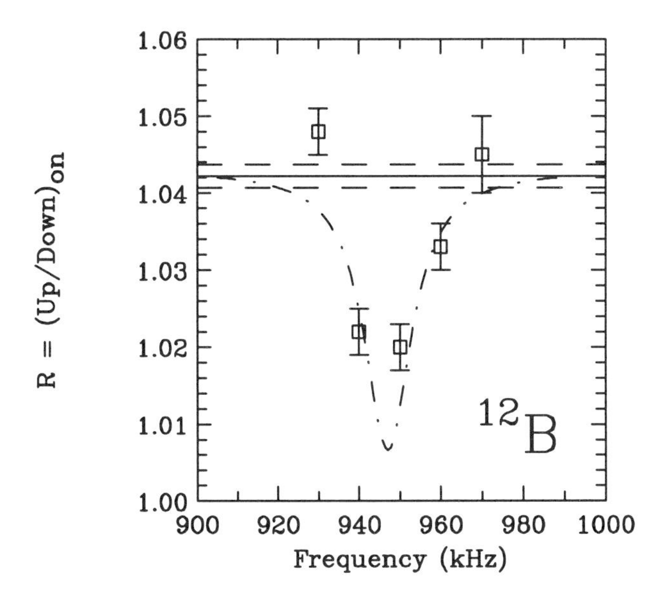

FIGURE 1. Resonance curve for ^{12}B using the batch implantation technique. The frequency modulation employed was ± 10 kHz. The full line is the average Up/Down ratio for all ^{12}B runs when no RF signal was applied to the implanted sample (RF-off). The dashed lines indicate the error attributed to the averaged RF-off ratio. The dot-dash line is a Lorentzian fit to the data, using a peak centroid of 947 kHz and a linewidth of 7 kHz.

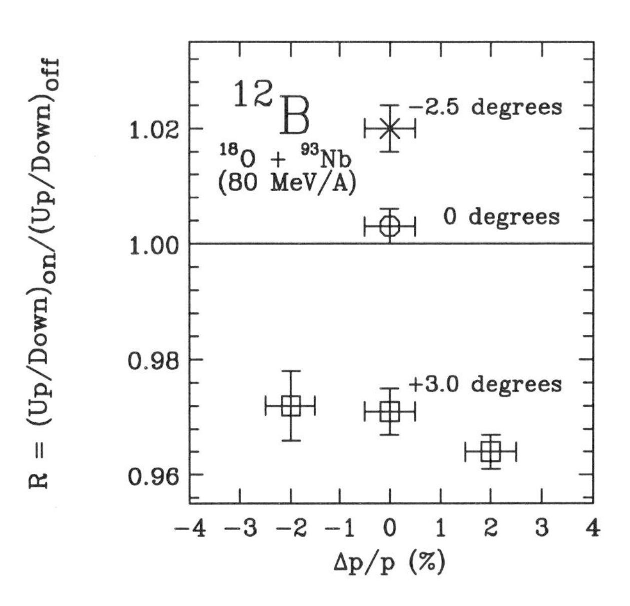

FIGURE 2. Dependence of the observed NMR effect (in %) on the fragment momenta for ^{12}B fragments measured for positive beam deflection (□), negative beam deflection (×), and normal incidence (○) following the reaction ^{18}O + ^{93}Nb at 80 MeV/A. For these measurements, the RF was set to the resonance peak, with a frequency modulation of ± 30 kHz. The value $\Delta p/p$ gives the deviation of the momentum of the ^{12}B fragments from the peak of the momentum yield curve, and the error bars on these values indicate the momentum acceptance of the A1200.

magnetic moment of ^{12}B, which agrees well with the adopted value of $+1.00306(^{+15}_{-14})$ μ_N given in Ref. (10).

We also explored the dependence of the observed polarization of the ^{12}B fragments on the longitudinal fragment momentum distribution and on the angle of the emitted fragments. For these measurements, we used the more conventional beam-on/beam-off data acquisition method. The results for ^{12}B fragments are shown in Fig. 2. One can observe that the polarization of ^{12}B fragments has little dependence on fragment momentum in the range from -2% $\leq \Delta p/p \leq$ +2% with the incidence angle of the primary ^{18}O beam at +3.0 degrees to the axis of the A1200, in agreement with the results of Okuno *et al.* (11). In the previous work on the reaction of ^{15}N on ^{93}Nb at 67.3 MeV/A, only a small dependence of the ^{13}B fragment polarization was observed over the same range of fragment momenta. Our result also corroborates the observations of Okuno *et al.* that intermediate mass targets produce non-zero values for the polarization of fragments at the peak in the momentum yield curve (11).

As for the dependence of the ^{12}B polarization on the incident beam angle, we have observed a change in the sign of the polarization of the ^{12}B fragmentswith a change in the direction of the incident angle of the primary ^{18}O beam to -2.5 degrees. This result supports the hypothesis (11) that the mean deflection angle for the projectile-like fragments, and hence the dominance of near- or far-side trajectories, determines the sign of the observed polarization. A null result for an ^{18}O beam at normal incidence was also observed. These results support the preliminary results from GSI obtained for ^{37}K fragments produced following high-energy (500 MeV/A) fragmentation (12).

An additional result from these measurements is that the magnitude of the peak of the resonance curve for ^{12}B shown in Fig. 1, which again was measured using the batch implantation method, is similar to that observed using the more conventional beam-on/beam-off data acquisition method (see Fig. 2). This result suggests that the polarization of the ^{12}B fragments is not disturbed by the application of an alternating RF field during fragment implantation. Although the batch implantation technique may provide for more efficient data collection, especially for long half-life species, it may not be as good as the conventional acquisition technique for cancelling systematic errors attributed to changes in beam position on the catcher foil.

ACKNOWLEDGEMENTS

We would like to thank K. Johnson (MSU Chemistry Department) and J. Yurkon (MSU/NSCL) for their technical assistance, and K. Matsuta (Osaka University) for enlightening discussions on the applications and experiences of the Osaka group using the β-NMR technique. This work was supported in part by the National Science Foundation under Contract No. PHY-95-28844.

REFERENCES

1. Asahi, K. *et al.*, Phys. Lett. B **251**, 488-492 (1990).
2. Okuno, H. *et al.*, Phys. Lett. B **354**, 41-45 (1995).
3. Ueno, H. *et al.*, Phys. Rev. C **53**, 2142-2151 (1996).
4. Matsuta, K. *et al.*, Hyperfine Int. **97/98**, 519-526 (1996).
5. Okuno, H. *et al.*, RIKEN Accel. Prog. Rep. **26**, 28 (1992).
6. Matsuta, K. *et al.*, Hyperfine Int. **78**, 123-126 (1993).
7. Izumi, H. *et al.*, Phys. Lett. B **366**, 51-55 (1996).
8. Asahi, K. *et al.*, Nucl. Instr. and Meth. **220**, 389-398 (1984).
9. Sherrill, B. *et al.*, Nucl. Instr. and Meth. B **56/57**, 1106-1110 (1991).
10. Raghavan, P. At. Data Nucl. Data Tables **42**, 189-291 (1989).
11. Okuno, H. *et al.*, Phys. Lett. B **335**, 29-34 (1994).
12. Schmidt-Ott, W.-D. *et al.*, "Observation of spin polarization of projectile fragments at 500 Mev/u," in *Nachrichten, GSI 12-95*, 1995, pp. 8-9.

LABORATORY FOR UNDERGROUND NUCLEAR ASTROPHYSICS (LUNA): FIRST RESULTS OF THE MEASUREMENTS OF THE REACTION $^3He(^3He,2p)^4He$ AT SOLAR ENERGIES

M. Junker[1,8], C. Arpesella[1], E. Bellotti[2], C. Broggini[3], P. Corvisiero[4], G. Fiorentini[5], A. Fubini[6], G. Gervino[7], U. Greife[8], C. Gustavino[1], J. Lambert[9], P. Prati[4], W.S. Rodney[9], C. Rolfs[8], H.P. Trautvetter[8], D. Zahnow[8], S. Zavatarelli[4]

[1]) *Laboratori Nazionali del Gran Sasso, S.S. 17bis, km 18.910, I-67010 Assergi(AQ)*
[2]) *Universitá di Milano, Dipartimento di Fisica and INFN, Milano*
[3]) *INFN, Padova*
[4]) *Universitá di Genova, Dipartimento di Fisica and INFN, Genova*
[5]) *Universitá di Ferrara, Dipartimento di Fisica and INFN, Ferrara*
[6]) *ENEA, Frascati and INFN, Torino*
[7]) *Politecnico di Torino, Dipartimento di Fisica and INFN, Torino*
[8]) *Institut für Experimentalphysik III, Ruhr-Universität Bochum*
[9]) *Georgetown University, Washington*

A compact 50 kV accelerator facility has been installed in the underground laboratory of the Laboratori Nazionali del Gran Sasso (LNGS) in Italy in the framework of the LUNA—project. Because of the specially designed accelerator setup and the suppression of the cosmic ray background in the underground laboratory the sensitivity of the set up gives the possibility to measure the cross sections of nuclear reactions at energies far below the coulomb barrier, where cross sections are as low as a few fbarn. Currently the reaction $^3He(^3He,2p)^4He$, which is important for the Solar Neutrino Puzzle, is being investigated at energies relevant in our sun (E=15–27 keV). These are the first direct measurements of a nuclear cross section covering the whole thermal energy range of the solar plasma. This contribution reports on the measurements covering the energy range between E= 20.7 keV and 25 keV.

The nuclear reactions of the pp-chain (Fig. 1) play a key role in the understanding of energy production, nucleosynthesis and neutrino emission of the elements in stars and especially in our sun [1]. A comparison of the observed solar neutrino fluxes measured by the experiments GALLEX/SAGE, HOMESTAKE and KAMIOKANDE provides to date no unique picture of the microscopic processes in the sun [2 and references therein]. A solution of this so called "solar neutrino puzzle" can possibly be found in the areas of neutrino physics, solar physics (models) or nuclear physics. In view of the important conclusions on non-standard physics, which might be derived from the results of the present and future solar neutrino experiments, it is essential to determine the neutrino source power of the sun more reliably.

The reaction $^3He(^3He,2p)^4He$ is one of the major sources of uncertainties for the calculation of the neutrino source power. It has been studied previously [3 and references therein] down to about E= 25 keV, but there remains the possibility of a narrow resonance at lower energies that could enhance the rate of path I at the expense of the alternative paths of the pp chain that produce the high-energies neutrinos (E_ν > 0.8 MeV). Such a resonance level in ^6Be has been sought without

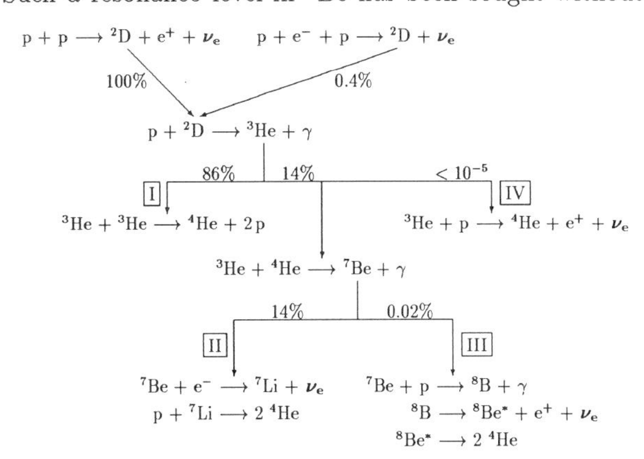

FIGURE 1: The pp–chain

CP392, *Application of Accelerators in Research and Industry*, edited by J. L. Duggan and I. L. Morgan
AIP Press, New York © 1997

success by various indirect routes, and it is also not predicted by most nuclear-structure theories. However, the existence of this hypothetical resonance can be positively dismissed only by direct measurements over the full range of the solar Gamow peak (E_t = 16 to 25 keV), which represents the relevant thermal energy region in the sun.

Due to the Coulomb Barrier E_c involved in the nuclear fusion reactions, the cross section $\sigma(E)$ of a nuclear reaction drops nearly exponentially at energies $E < E_c$, leading to a low-energy limit E_l of the feasible $\sigma(E)$ measurements in a laboratory at the earth surface, mainly due to the influence of cosmic rays in the detectors [1,2]. Since E_l is far above E_t, the high energy data have to be extrapolated down to E_t. As usual in physics, extrapolation of data into the "unknown" can lead onto "icy ground". Although experimental techniques have improved over the years to extend $\sigma(E)$ measurements to lower energies, it has not yet been possible to measure $\sigma(E)$ within the thermal energy region in stars.

The low-energy studies of thermonuclear reactions in a laboratory at the earth's surface are hampered predominantly by the effects of cosmic rays in the detectors [1,2]. Passive shielding around the detectors provides a reduction of gammas and neutrons from the environment, but it produces at the same time an increase of gammas and neutrons due to the cosmic-ray interactions in the shielding itself. A 4π active shielding can only partially reduce the problem of cosmic-ray activation. The best solution is to install an accelerator facility in a laboratory deep underground [5].

The worldwide first underground accelerator facility has been installed at the Laboratori Nazionali del Gran Sasso (LNGS) in Italy, based on a 50 keV accelerator. This pilot project is called LUNA and has been supported since 1992 by INFN, BMBF, DAAD-VIGONI and NSF/NATO. The major aim of this project is to measure $\sigma(E)$ of ^{3}He(^{3}He,2p)^{4}He over the full range of the solar Gamow peak, where the cross section is as low as 8 pbarn at E= 25 keV and about 20 fbarn at E= 17 keV.

The 50 kV accelerator facility as well as the gas target and the detector setup is described in detail in [6]. A schematic design of the facility is shown in Fig. 2. The accelerator consists of a duoplasmatron ion source (energy spread lower than 20 eV) mounted on a high voltage terminal which can be biased with a high voltage of maximum 50 kV (ripple less than $5\cdot10^{-4}$, longterm stabilty better $1\cdot10^{-4}$). Since the extraction voltage is identical to the acceleration voltage, the extraction electrode can be moved in situ relative to the outlet of the ion source,

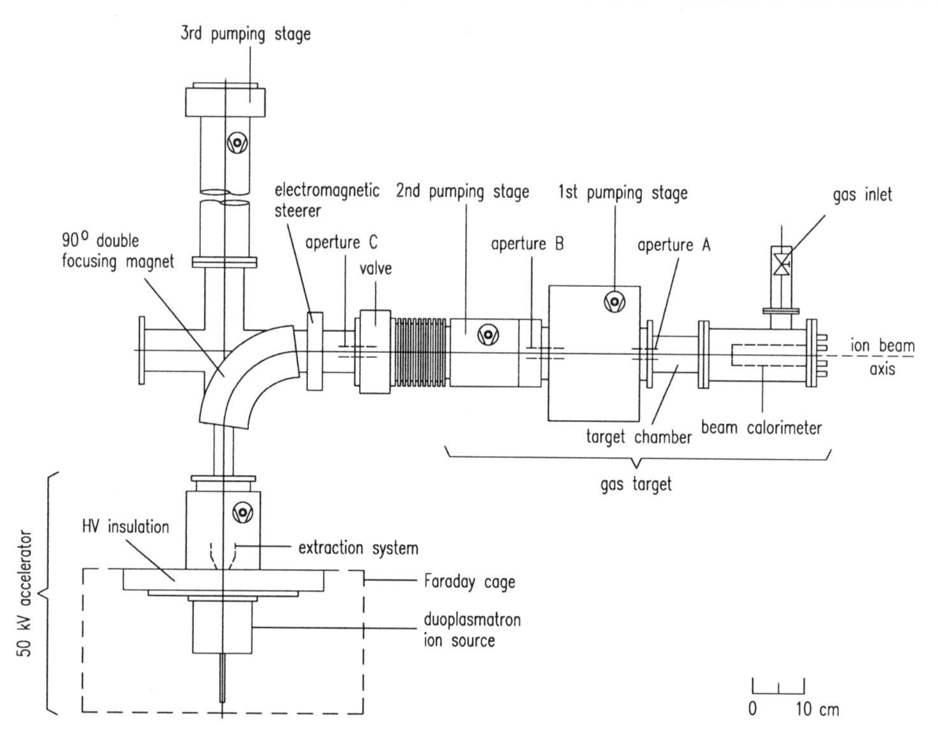

FIGURE 2: Schematic design of the 50 kV accelerator and the gas target setup installed by the LUNA–Collaboration at the LNGS.

maintaining the same electric field and thus beam currents when varying the beam energy. The beam transport is provided by a double-focusing 90° analyzing magnet and a electromagnetic steerer. The shim angles of the analyzing magnet are adjustabile between 25° and 40° degrees, to allow focussing/defocussing of the beam.

The beam intensity is determined by a beam calorimeter with a constant temperature gradient.

The target is provided by a windowless target chamber filled with 0.3 mbar of ^{3}He. The vacuum system has been designed to reach a pressure of $8\cdot10^{-6}$ mbar near the analyzing magnet and allowing gas recirculation for the target gas as well as for the ^{3}He out of the ion source.

The apparatus is equipped with an interlock system which allows to run the system without an operator on site. The duty time of the accelerator in this running conditions is about 90% with the weekly servicing time being 8h. As the typical beam current in the target area obtained down to E_{lab}= 40 keV is of about 400 μA of ^{3}He, a weekly charge of 200 Cb can be accumulated on the target.

In order to design the detection setup for the measurements of the ^{3}He(^{3}He,2p)^{4}He reaction and to understand the resulting spectra for quantitative analyses, a Monte Carlo program was written. Various quantitative tests of the Monte Carlo predictions have been carried out successfully [7]. Based on the results of the Monte–Carlo simulation four ΔE-E telescopes were placed (in a rectangular target chamber) around the beam at a distance of 2.5 and 3.5 cm from its axis (Fig 3). Each telescope consists of transmission surface barrier silicon detectors with a 0.25 μm thick Al layer deposited on both sides of the detectors. The ΔE and E detectors both have an active (square) area of 2500 mm^2; the ΔE (and E) detector has a thickness of 140 μm (and 1000 μm). A mylar foil and an Al foil (each of 1.5 μm thickness) are placed in front of each telescope. They stop the intense elastic scattering yield and shield the detectors from beam-induced light; the detectors are maintained permanently at low temperature (about -20°C) using a liquid recirculating cooling system. The ^{4}He ejectiles from ^{3}He(^{3}He,2p)^{4}He ($E_{^4\text{He}}$ ~ 0 to 4.3 MeV) are stopped in the ΔE detectors, while the ejected protons (E_p ~ 0 to 10.7 MeV) leave signals in both the ΔE and E detectors of a given telescope (coincidence requirement). Standard NIM electronics are used in connection with the four telescopes. The signals are handled and stored using a CAMAC multiparametric system. A pulser is permanently used in all detectors to check for dead time and electronic stability. The acquisition system also stores concurrent information on experimental parameters (such as ion beam current and gas pressure in the target chamber) via CAMAC scalers.

The accepted events have to occure in coincidences in the 2 detectors of a detector telescope. This coincidence requirement of each telescope significantly minimizes events due to the intrinsic radioactivity of the detectors themselves and of surrounding materials. In addition the events had to lay in a closed region of the ΔE-E plane (Fig. 4), whose borders are first tracked by Monte Carlo simulations and then fixed in order to cut electronic noise. A further noise reduction was optained accepting only events occuring in one single E–detector; events which triggered more than one E–detector were rejected.

The chosen region together with the anticoincidence requirement led to an absolute detection efficiency of 8.5 ± 0.2% as determined by the Monte Carlo program. In the ΔE-E energy region of events from ^{3}He(^{3}He,2p)^{4}He (Fig. 4), no background events have ever been observed at LNGS since the installation of the equipment (january 1994) during several "no beam" and/or "no target" background measurements (up to two months running) as well as during a 10 day run

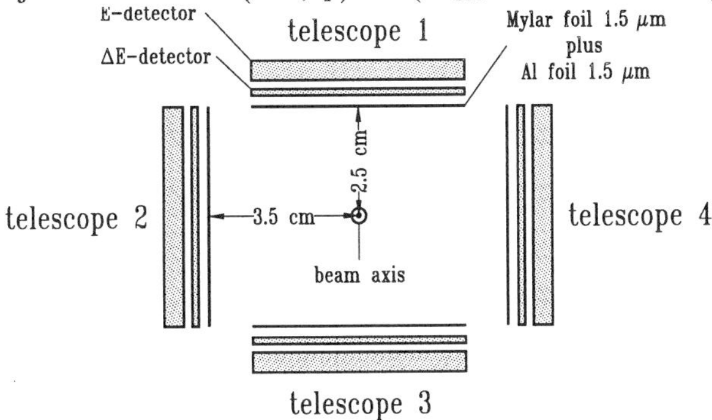

FIGURE 3: Schematic design of the detector setup

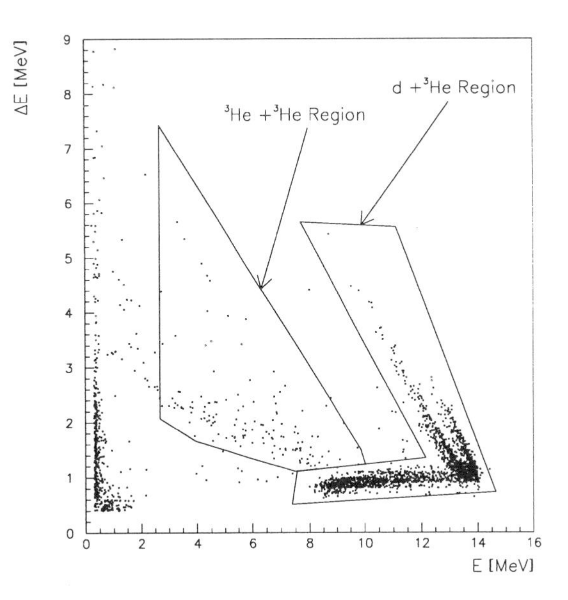

FIGURE 4: ΔE–E plot for one of the telescopes at E = 24.7 keV: the ^{3}He+^{3}He and d+^{3}He regions are shown; note the beam-induced noise at the left edge of the spectrum.

with a ^{4}He beam and a ^{4}He target gas. At the Bochum laboratory (earth surface) a background rate of about 10 events/day was found.

A second region is indicated in Fig. 4 corresponding to protons produced by the ^{3}He(d,p)^{4}He or D(^{3}He,p)^{4}He contaminant reactions due to deuterium contamination in the beam (as HD$^+$ molecules of mass 3) or in the target gas. This reaction has a cross section one millionfold higher than that of ^{3}He(^{3}He,2p)^{4}He at E_{lab} = 40 keV, mainly due to the barrier ratio, E_c(d+^{3}He)< E_c(^{3}He+^{3}He), and thus extremely small deuterium contaminations can lead to sizeable event rates. Although the selected regions in the spectra of the telescopes allow for a clear separation of the events from both reactions (Fig.4), a few protons from the contaminant reactions can hit the detectors near the edges of their active volumes loosing only a fraction of their energy and thus leading to a background rate in the ΔE-E region of the ^{3}He(^{3}He,2p)^{4}He events. The probability of such events was investigated by the Monte Carlo program as well as by direct measurements using projectiles with $Z/A = 0.5$ (selected by the 90° analyzing magnet). The ratio between the background events in the ^{3}He–^{3}He region and those in the clearly separated d-^{3}He region (= monitor) turned out to be (0.40 ± 0.04)%. During the experiments the deuterium contamination d/^{3}He ranged between $5{\cdot}10^{-8}$ and $5{\cdot}10^{-6}$(mainly in the ion beam).

Fig. 5 summarizes the ^{3}He(^{3}He,2p)^{4}He results obtained until 1. Aug. 1996 with the LUNA setup at the 50 kV underground accelerator facility (LNGS), with the lowest counting rate of 3 events per day at E= 20.7 keV. At this energy about 1000 Cb of ^{3}He$^+$ have been accumulated on the target. The data obtained at higher energies (450 kV accelerator in Bochum) with the LUNA setup [6] are also included for completeness. Previous data obtained [3] at E= 25 to 343 keV are also shown in Fig. 5. The LUNA data have been obtained at energies within the solar Gamow peak, i.e. below the 21 keV center of this peak, and represent the first measurement of an important fusion cross section in the thermal energy region; they also demonstrate the usefulness of an underground accelerator facility for such noval aims. The new data give no evidence for the existence of the hypothetical resonance, at least down to E = 20.7 keV. The observed energy dependence $S(E)$ was parameterized by the expressions [2,8]

$$S_b(E) = S(0) + S'(0)E + \frac{1}{2}S''(0)E^2 \quad (1)$$

$$S_s(E) = S_b(E)\exp(\pi\eta U_e/E), \quad (2)$$

where $S(0)$, $S'(0)$, $S''(0)$, and U_e are fit parameters. A fit to equation (1) at $E \geq 100$ keV leads to $S(0) = 5.1 \pm 0.1$ MeVb, $S'(0) = -2.6 \pm 0.7$ MeV and $S''(0) = 2 \pm 1$ MeV/b, which in turn gives at lower energies U_e = 290 ± 40 eV with a reduced χ^2 value of 0.70 using equation (2). The resulting curves $S_b(E)$ and $S_s(E)$ are shown in Fig. 5 as dashed and solid curves, respectively. The deduced screening potential is consistent with the calculated value of 240 eV (adiabatic limit). As a consequence the new data remove the apparent absence of the effects of electron screening in ^{3}He(^{3}He,2p)^{4}He. However, it should be noted that the low-energy data appear not to follow a simple exponential behaviour: improved data at E = 25 to 50 keV are thus highly desirable. Aside from this aim the LUNA project will extend the measurements down to E = 17 keV requiring an additional running time of about 1 year. Definite conclusions with respect to the expected solar neutrino fluxes have to await the results of these experiments.

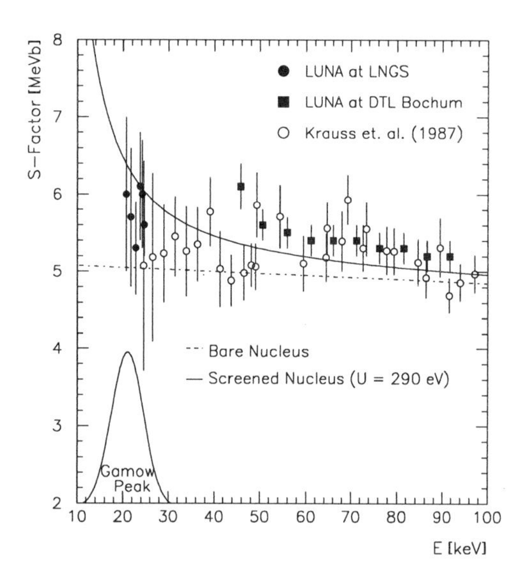

FIGURE 5: The $S(E)$-factor of the reaction ^{3}He(^{3}He,2p)^{4}He: data measured by the LUNA collaboration (filled-in points) and by previous work (circles = [3]). The Gamow peak is shown in arbitrary units.

REFERENCES

1. Rolfs,C. and Rodney,W.S., *Cauldrons in the Cosmos*, Chicago and London: University of Chicago Press (1988)
2. Bahcall,J.N. and Pinsonneault,M.H., *Rev.Mod.Phys.* **64**, 885 (1992)
3. Krauss,A. et al.,*Nucl.Phys.A* **467**, 273 (1987)
4. Fowler,W.A., *Rev.Mod.Phys.* **56**, 149 (1984)
5. Fiorentini,G., Kavanagh,R.W. and Rolfs,C., *Z.Phys.A* **350**, 289 (1995)
6. Greife,U. et al., *Nucl.Instr.Meth.A*, **350**, 327 (1994)
7. Arpesella,C. et al., *Nucl.Instr.Meth.A* **360**, 607 (1995)
8. Assenbaum,H.J., Langanke,K. and Rolfs,C., *Z.Phys.A* **327**, 461 (1987)

INVESTIGATION OF SHORT-LIVED PT AND PB α EMITTERS NEAR THE PROTON DRIP LINE

C. R. Bingham, J. Wauters, and B. E. Zimmerman

University of Tennessee, Knoxville, TN 37996 USA

K. S. Toth

Oak Ridge National Laboratory, Oak Ridge, Tennessee 37831 USA

J. C. Batchelder and E. F. Zganjar

Louisiana State University, Baton Rouge, LA 70803 USA

D. J. Blumenthal, C. N. Davids, D. J. Henderson, and D. Seweryniak

Argonne National Laboratory, Argonne, IL 60439 USA

L. T. Brown

Vanderbilt University, Nashville, TN 37235 USA

B. C. Busse

Oregon State University, Corvallis, OR 97331 USA

L. F. Conticchio and W. B. Walters

University of Maryland, College Park, MD 20742 USA

T. Davinson, R. J. Irvine, and P. J. Woods

Edinburgh University, Edinburgh, EH9 3JZ UK

In a series of experiments at the Argonne ATLAS Accelerator Facility, several α emitters near the proton drip line were produced with fusion evaporation reactions, separated from the beam and dispersed in M/Q with a recoil mass spectrometer, and implanted and studied in a double-sided silicon strip detector. In ^{78}Kr bombardments of ^{92}Mo and ^{96}Ru, the new isotopes ^{166}Pt and ^{167}Pt were identified *via* their α-decay properties and more accurate half-lives were measured for ^{168}Pt and ^{170}Pt. The light isotopes of lead, ^{180}Pb, ^{182}Pb, and ^{184}Pb were produced in Mo bombardments of Zr target nuclei. The α-decay energies and half-lives of the new isotopes are as follows: 1) ^{166}Pt, E_α = 7110(15) keV, $T_{1/2}$ = 0.3(1) ms; and 2) ^{167}Pt, E_α = 6988(10) keV, $T_{1/2}$ = 0.7(2) ms. Also, the half-life of ^{168}Pt, which was previously unknown, was determined to be 2.0(4) ms and that of ^{170}Pt was observed to be 14.7(5) ms. The tentative α-decay energies and half-lives of the even Pb isotopes are: 1) ^{184}Pb, E_α = 6625(10) keV, $T_{1/2}$ = 500(25) ms; 2) ^{182}Pb, E_α = 6895(10) keV, $T_{1/2}$ = 62(5) ms; and 3) ^{180}Pb, E_α = 7250(15) keV, $T_{1/2}$ = $5.8^{+2.8}_{-1.4}$ ms. The α-decay rates for these Pt and Pb nuclides are compared with earlier measurements and systematic trends of the reduced widths with neutron number are discussed.

CP392, *Application of Accelerators in Research and Industry*, edited by J. L. Duggan and I. L. Morgan
AIP Press, New York © 1997

INTRODUCTION

The study of nuclei near the proton drip line presents a number of experimental challenges. First of all, the production of these very radioactive nuclei can be accomplished only with particle accelerators under very selective conditions, usually with a very small cross section. Secondly, the greater production cross sections for nuclei nearer stability results in a very high background level, thus requiring special techniques to observe the relatively small fraction of the desired product. A spectrometer to separate the various activities according to their mass/charge ratios is a useful tool to purify the sources. The development of high efficiency charged particle detectors makes detection of relatively few decays by proton or alpha emission practical. In this paper we describe, briefly, an experimental setup at Argonne National Laboratory to make these studies and report on some of the α-decay results of nuclei near the proton drip line produced with that facility.

The emission of α particles is a decay mode that can be observed sensitively with charged particle detectors. The energy of the observed α gives a direct measurement of the mass difference of the α emitter and its daughter nucleus and provides a sensitive test of mass formulas. The α-decay rate depends on two factors, one which expresses the probability of the α penetrating the combined Coulomb and centrifugal barrier, and another which depends on the probability of formation of the α within the nucleus, and thus, the nuclear structure of the parent and daughter. Within a formalism developed by Rasmussen (1), the decay constant λ is expressed as $\delta^2 P/h$, where P is the probability of penetrating the barrier and δ^2 is called the reduced width. A study of the systematic variation in δ^2 with proton and neutron numbers reveals changes in nuclear structure. In particular, closed neutron and proton shells have a huge effect on the reduced widths. It is of interest to study these reduced widths in the region near Z and N = 82.

EXPERIMENTAL METHOD

The Pt α emitters studied here were produced by bombarding a ^{92}Mo (>97% enrichment) metal foil (0.565 mg/cm2) thick) with a 5 particle nA beam of ^{78}Kr beams from the ATLAS accelerator facility at Argonne National Laboratory. Incident energies of 357 and 384 MeV were used to emphasize A = 167 and A = 166 isotopes, respectively. The light Pb nuclides were produced by bombardment of metal foils of several Zr isotopes with a 422-MeV ^{92}Mo beam. The recoiling products were separated from the incident beam and separated by their mass/charge ratio (A/Q) by use of the Fragment Mass Analyzer (FMA) (2) and, after passing through a parallel grid avalanche detector (PGAC) located at the FMA focal plane, were implanted in a Double-sided Silicon Strip Detector (DSSD). Signals from the PGAC were used to identify implantation events in the DSSD and to determine A/Q for the implant. For the Pt experiments, the DSSD was of the same size (1.6 x 1.6 cm), granularity (48 x 48 strips), and thickness (≈65 μm) as the ones recently developed (3) for use in proton and α-decay studies. The DSSD used for the Pb experiments was of the same thickness but had an area of 4 x 4 cm and a granularity of 40 x 40. The decay-energy signals from the different strips were gain matched by use of a mixed ^{244}Cm and ^{240}Pu α source. Final energy calibrations for implanted α emitters were provided by using known α lines of Hf, Yb, W, Ta, Lu, Os, Hg, and Pb.

Implant and decay events in each pixel location of the DSSD were time stamped with signals from a continuously running clock. Both position and time correlations between individual implants and their subsequent decay events were observed. Half-lives of the implanted parent nuclei could be determined from the differences between implantation and first decay times within the same pixel. Also, in similar fashion the half-lives of the subsequent decay products can also be determined if they decay by proton or α decay. The method of maximum likelihood was used to obtain half-lives and uncertainties.

EXPERIMENTAL RESULTS

The α spectra obtained by bombarding ^{92}Mo with 357-MeV ^{78}Kr under various gating conditions are shown in Fig. 1. Parts (a) and (b) show spectra gated with A = 168 and A = 167 recoils, respectively. Parts (c) and (d) show the same spectra with additional requirements, namely, correlation with subsequent α decay in the same pixel of ^{164}Os and ^{163}Os, respectively. These spectra demonstrate unambiguously the assignments of the 6832 and 6988 keV peaks to ^{168}Pt and ^{167}Pt, respectively. A similar set of spectra taken at the bombarding energy of 384 MeV reveals an α line in the A = 166 spectrum which is correlated with the α decay of the ^{162}Os daughter at an energy of 7110 keV. This permits the assignment of this peak to the new isotope ^{166}Pt. Similarly, by bombarding ^{96}Ru with 420-MeV ^{78}Kr, we have produced ^{170}Pt and observed its decay via a 6550-keV α particle.

The α spectra obtained by bombarding ^{90}Zr with 422-MeV ^{92}Mo under several different gating conditions are shown in Fig. 2. Part (a) shows the α particles which were observed within 10 ms after an implant was incident on the same pixiel. The broad peak near 7.2 MeV belongs to the decay of ^{179}Tl and to ^{180}Pb which was just

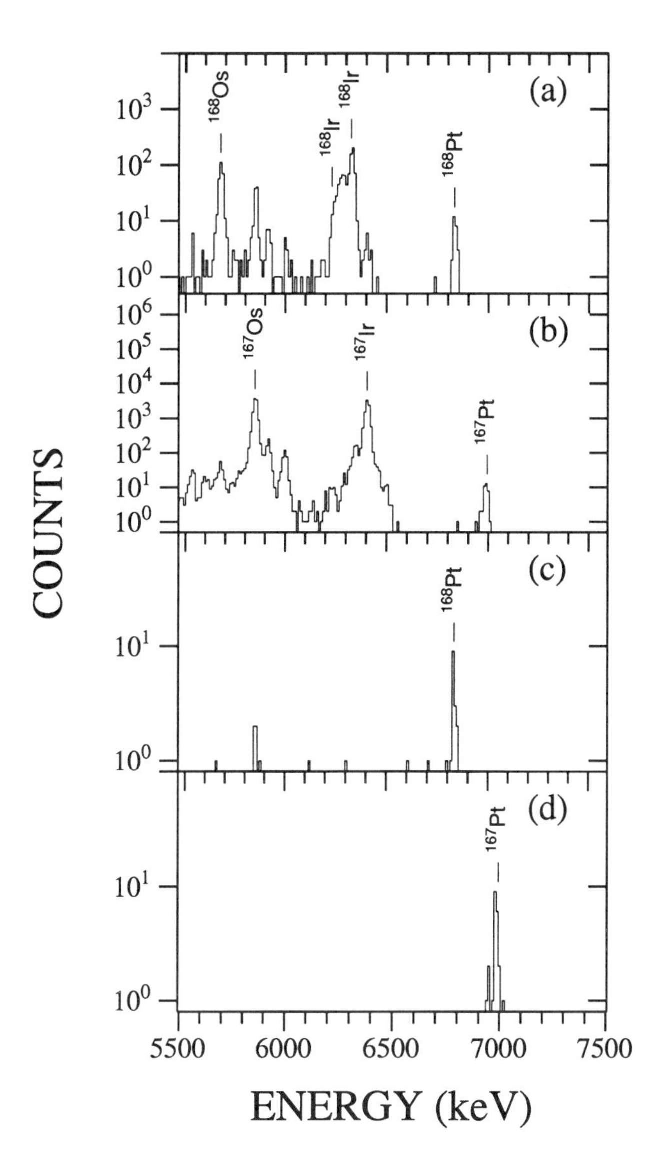

Fig. 1. High-energy portion (5.5-7.5 MeV) of the α spectrum resulting from bombardment of ^{92}Mo with 357-MeV ^{78}Kr displayed with various gates: (a) A = 168 gate, (b) A = 167 gate, (c) A = 168 correlated with the α decay of ^{164}Os, and (d) A = 167 correlated with ^{163}Os α decay.

recently identified [4]. Parts (b) and (c) show the resulting spectra when additional gates on A = 179 and A = 180, respectively, are included. Part (d) shows the result when an additional requirement that an α particle was emitted later by ^{176}Hg (6750-keV) in the same pixel is placed on the spectrum in part (c). The results of these gated spectra permit an unequivocal assignment of the 7250-keV line to the decay of ^{180}Pb.

The half-lives of all the α peaks observed were determined through fitting the distribution of time differences between the implant and the decay with an

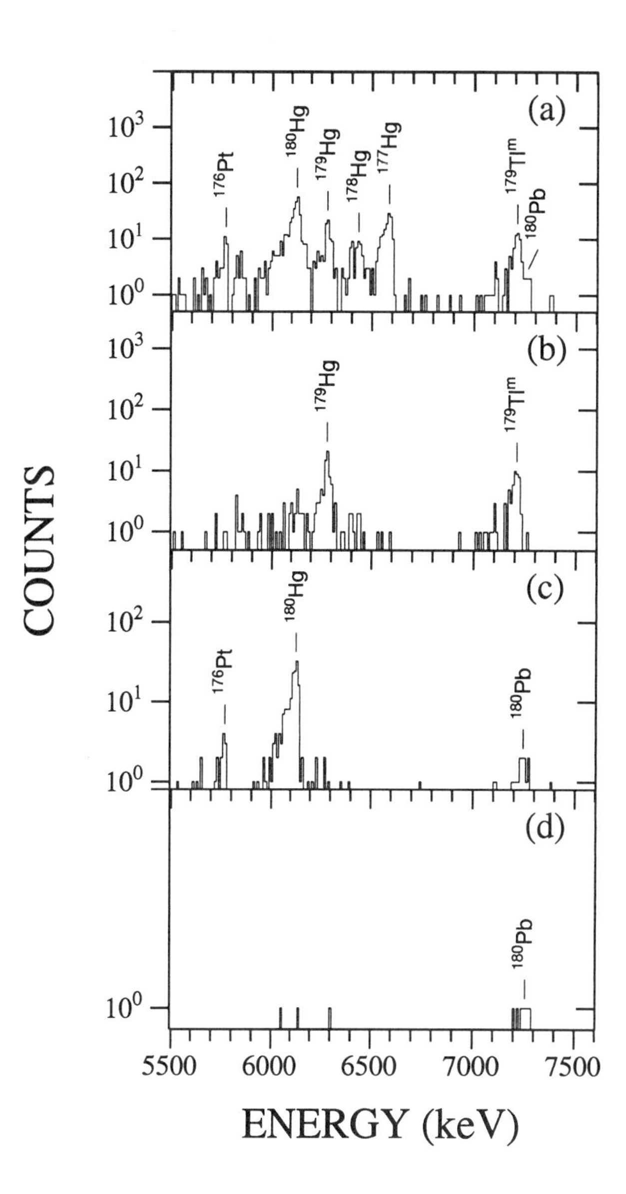

Fig. 2. High-energy portion of the α spectrum resulting from bombardment of ^{90}Zr with 422-MeV ^{92}Mo displayed with various gates: (a) decay occurred within 10 ms of implant in same pixel, (b) same with A = 179, (c) same with A = 180, (d) A = 180 correlated with ^{176}Hg α decay.

exponentially decaying function, utilizing the method of maximum likelihood for the weakly-produced activities. A summary of the observed α energies and half-lives observed in the present experiments are given in Table I along with some results from previous experiments (4-7). Where earlier results were available, the current energies and half-lives generally agree with the previous results, though error bars are significantly less in the present results.

Table I. Decay energies, half-lives, and reduced widths of proton-rich platinum and lead α emitters.

	E_α(keV)	$T_{1/2}$(ms)	δ^2(keV)	Ref.
^{166}Pt	7110(15)	0.3(1)	88^{+58}_{-29}	Present
^{167}Pt	6988(10)	0.7(2)	88^{+41}_{-24}	Present
^{168}Pt	6832(10)	2.0(4)	96^{+14}_{-26}	Present
	6824(20)			(7)
^{170}Pt	6550(6)	14.7(5)	114^{+10}_{-9}	Present
	6545(8)	6^{+5}_{-3}	290^{+180}_{-140}	(7)
^{180}Pb	7250(15)	$5.8^{+2.8}_{-1.4}$	40^{+47}_{-11}	Present
	7230(40)	4^{+4}_{-2}	68^{+115}_{-43}	(4)
^{182}Pb	6895(10)	62(5)	54^{+10}_{-8}	Present
	6919(15)	50^{+40}_{-35}	55^{+152}_{-28}	(5)
^{184}Pb	6625(10)	500(25)	58^{+10}_{-9}	Present
	6632(10)	550(60)	49^{+11}_{-8}	(6)

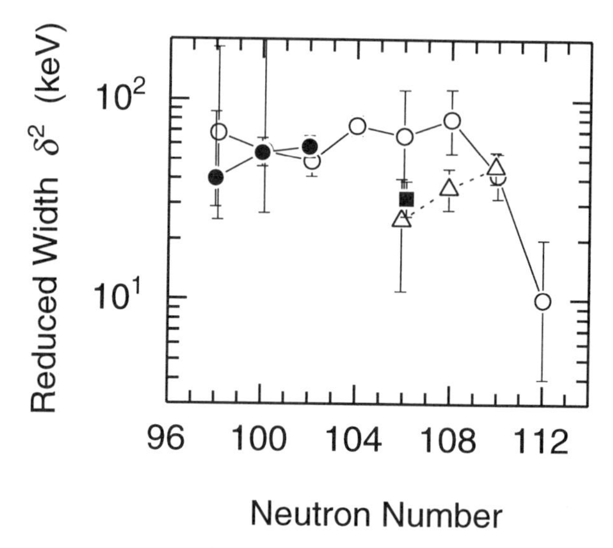

Fig. 3. Present reduced widths (solid circles) in comparison with those calculated from E_α, $T_{1/2}$, and B_α of earlier measurements tabulated in Ref. (4-7, 9-11).

DISCUSSION AND CONCLUSIONS

The α-decay rates of Pt and Pb isotopes listed in Table I were examined within the formalism of Rasmussen (1) wherein the α reduced width δ^2 is defined as λh/P, where λ is the α decay constant and P is the α-particle penetrability factor. Calculated widths are given in Table I where error bars were determined by utilizing the experimental uncertainties on E_α and $T_{1/2}$. It was assumed that the α-decay branch is 100% for each of the nuclides treated. The reduced widths for ^{166}Pt , ^{168}Pt, and ^{170}Pt lead to a smooth trend with neutron number N for the widths of the light even-even Pt nuclei. There is a slight downward trend as one approaches the N = 82 neutron shell as shown graphiclly in a recent publication [8]. The δ^2 for the new isotope ^{167}Pt is close to the values of its even-A neighbors, which indicates that it is an unhindered transition. This implies that the spin and nucleon configuration for ^{167}Pt and ^{163}Os are the same. The reduced widths for Pb isotopes are of particular interest since they have a closed proton shell. The reduced widths from the present results are compared with those resulting for heavier Pb isotopes in Fig. 3. The open circles, resulting from one set of measurements, tend to follow the trend of the present measurements. The open triangles from another set of measurements and the solid square from another recent FMA measurement (11) are somewhat lower suggesting a dip in δ^2 at N = 106. If this dip is real it may result because the ground state of ^{184}Hg is at the point of maximum shape mixing in the mercuries, thus reducing its nuclear similarity with spherical ^{188}Pb.

ACKNOWLEDGEMENTS

Nuclear physics research at The University of Tennessee, The University of Maryland, Vanderbilt University, and Louisiana State University is supported by the U. S. Department of Energy through Contracts No. DE-FG02-96ER40983, DE-FG05-88ER40418, DE-FG05-88ER40407, and DE-FG05-84ER40159, respectively. Research partially sponsored by Oak Ridge National Laboratory, managed by Lockheed Martin Energy Research Corporation for the U. S. Department of Energy under Contract No. DE-AC05-96OR22464. Argonne National Laboratory is operated by the University of Chicago for the U. S. Deparatment of Energy under Contract No. W-31-109-Eng-38. P.J.W. and T. D. would like to thank NATO for support. R.J.I. would like to thank EPSRC for financial support.

REFERENCES

1. Rasmussen, J. O., *Phys. Rev.* **113**, 1593 -1598 (1959).
2. Davids, C. N., *et al., Nucl. Instrum. Methods* B **70**, 358-361 (1992).
3. Sellin, P. J., *et al., Nucl. Instrum. Methods* A **311**, 217-223 (1992).
4. Toth, K. S., *et al., Nucl. Phys.* A **355**, 225-226 (1996).
5. Toth, K. S., *et al., Phys. Rev. C* **35**, 2330-2332 (1987).
6. Schrewe, U. J., *et al., Phys. Lett.* **91B**, 46-48 (1980).
7. Hofmann S., *et al., Z. Phys. A* **299**, 281-282 (1981).
8. Bingham, C. R., *et al., Phys. Rev. C* **54,** R20-R23 (1996).
9. Toth, K. S., *et al., Phys. Rev. Lett.* **53**, 1623-1626 (1984).
10. Wauters, J., *et al., Phys. Rev. C* **47**, 1447-1453 (1993).
11. N. Bijnens, *et al., Z. Phys. A,* (in press).

POLARIZED EPITHERMAL NEUTRON STUDIES OF MAGNETIC DOMAINS

V. P. Alfimenkov, A. N. Chernikov, L. Lason+, Yu. D. Mareev, V. V. Novitsky, L. B. Pikelner, V. R. Skoy and M. I. Tsulaya
Frank Laboratory of Neutron Physics, Joint Institute for Nuclear Research, 141980 Dubna, Russian Federation

C. R. Gould and D. G. Haase
Physics Department, North Carolina State University, Raleigh, NC 27695-8202, USA and the Triangle Universities Nuclear Laboratory, Durham, NC 27708-0308

N. R. Roberson
Physics Department, Duke University, Durham, NC 27708-0305, USA and the Triangle Universities Nuclear Laboratory, Durham, NC 27708-0308

The average size and shape of magnetic domains in a material can be determined from the precession of polarized neutrons traversing the material. Epithermal neutrons (0.5 eV < E_n < 100 eV), which precess more slowly than thermals, effectively probe the internal structure of samples that are thick or have large domains or large internal fields. Such epithermal neutron measurements require a neutron polarizer and analyzer based on cryogenically polarized spin filters. We discuss the measurement at JINR, Dubna, of magnetic domains in a 2.0 cm. diam. crystal of holmium using 1.7 to 59 eV neutrons polarized by a dynamically polarized proton target and analyzed with a statically polarized dysprosium target.

INTRODUCTION

It is possible to measure the average size and shape of magnetic domains in solids by measuring the depolarization of thermal neutrons.(1) However thermal neutrons are too strongly depolarized in large internal magnetic fields or in thick samples. We demonstrate here how magnetic domains in the interior of such samples can be characterized using epithermal neutrons. We have measured the average size and shape of magnetic domains in a 2.0 cm. diam single crystal cylinder of ferromagnetic holmium using 1.7 to 59 eV polarized neutrons.

The initial polarization f_{no} of a beam of neutrons traversing a ferromagnetic material is reduced to a value f_n by the precession of the neutron magnetic moments in the local fields of the individual domains. A classical neutron of energy E passing through a holmium domain of length δ precesses an angle

$$\Delta\Theta = (\gamma_n)\, B_\perp \frac{\delta}{\sqrt{\frac{2E}{m_n}}} \qquad \text{(Eq. 1)}$$

where γ_n is the neutron gyromagnetic ratio, m_n is the mass of the neutron and $B_\perp$ is the component of the local field perpendicular to the neutron polarization axis. For example, a 59 eV neutron would precess one radian in 880 micrometers in a saturated holmium domain ($B_{internal}$ = 0.66 T).

The polarization ratio f_n/f_{no} in a sample of randomly oriented domains is found to be proportional to $\exp(-2D'_{random}t)$ where t is the sample thickness. If the precession in any one domain is small the depolarization parameter D'_{random} is inversely proportional to the neutron energy:

$$D'_{random} = \frac{\gamma_n^2 \left\langle B_\perp^2\, \delta^2 \right\rangle_{avg}\, n}{8\left(\frac{E_n}{m_n}\right)} \qquad \text{(Eq. 2)}$$

where n is the number of domains per unit length.(2,3)

In a sample in which the domain magnetization directions are highly correlated, e.g. a single crystal, the polarization ratio becomes dependent upon the relative orientation of the preferred domain magnetization direction and the neutron polarization axis. Another depolarization parameter

$$D'_{bulk} = \frac{\gamma_n^2 B_{int}^2}{16\, v_n^2}\, \delta \qquad \text{(Eq. 3)}$$

where v_n is the neutron velocity and B_{int} the total internal field, is used to describe the depolarization due to the distribution of these correlated domains (4).

EXPERIMENT

For this experiment neutrons from the IBR-30 pulsed source are polarized by passage through a cryogenic dynamically polarized proton target. The neutron polarization is about 0.6 and the neutron flux exiting the target is 300,000 $E^{-0.9}$ neutrons/second-eV. The neutrons proceed through an adiabatic spin flipper and enter a drift tube in which there is a guide field of 0.04 T (5).

CP392, *Application of Accelerators in Research and Industry*, edited by J. L. Duggan and I. L. Morgan
AIP Press, New York © 1997

The holmium target is located in a cryostat at 4.2 Kelvin in the neutron guide field. The geometry of the target and neutron beam is shown in Figure 1. The angle between the spin and the *c* axis was varied in ten steps: θ = -15° to 180° by a shaft into the cryostat. Upon exiting the holmium target the neutron beam is guided to a cryogenic statically polarized natural Dy target. The Dy target consists of a 1.4 mm thick stack of foils cooled to 40 mK in a 1.5 T polarizing field parallel to the plane of the target plate. Resonances in Dy at 1.7, 2.7, 14, 16, 18, 36, and 59 eV have spin dependent neutron transmission(5) and act as analyzers for neutrons with these energies. The neutrons passing through the Dy analyzer are then counted by a ^{3}He ionization detector 30 m distant from the neutron source. Neutron energies are determined by the time-of-flight technique.

Neutron yields were collected for both parallel (N_+) and anti parallel (N_-) neutron and dysprosium spin orientations, flipping the spin each five minutes. An average transmission asymmetry $<\varepsilon> = (N_+ - N_-)/(N_+ + N_-)$ proportional to f_n was determined by integration over the time of flight channels corresponding to a particular Dy resonance(6). Because holmium is paramagnetic at room temperature, the ratio of the initial polarization f_{no} to the polarization f_n of the transmitted neutron beam could be calculated as $f_n/f_{no} = <\varepsilon>_{4.2\,K} / <\varepsilon>_{300K}$.

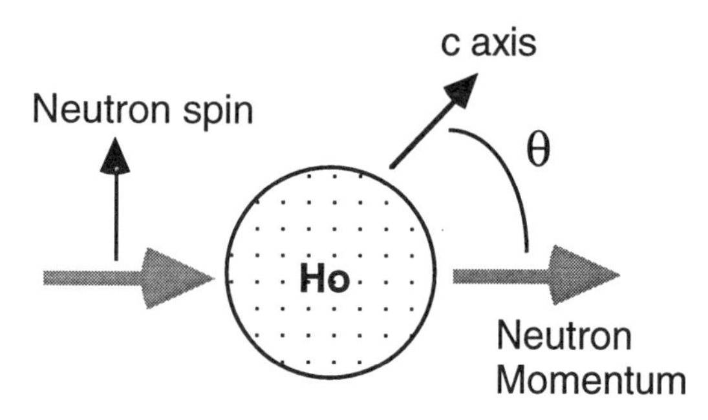

Figure 1. Experimental geometry used for this investigation. The neutron spin, neutron momentum and the *c* crystalline axis are coplanar. The cylindrical axis of the crystal is perpendicular to that plane. The neutron polarization axis is perpendicular to the neutron momentum.

EXPERIMENTAL RESULTS

The energy dependence of f_n/f_{no} is shown in Figure 2. At high energies f_n/f_{no} is proportional to $\exp(-\alpha/E_n)$ as expected and the slope α is dependent upon the orientation of the crystal. This determination of f_n/f_{no} over a wide energy range is a significant improvement over previous measurements.

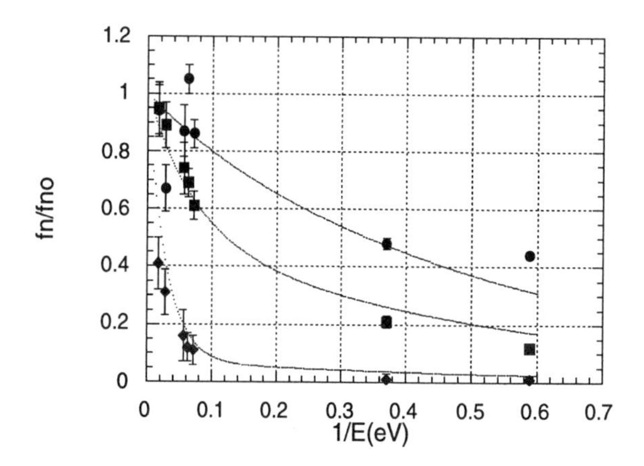

Figure 2. Energy dependence of the polarization ratio f_n/f_{no} as measured at selected orientation angles. Diamonds, $\theta = 0^{o}$; squares, $\theta = 30^{o}$; dots, $\theta = 90^{o}$. Error bars indicate statistical errors only. The dotted lines indicate transmissions calculated from the model using constants obtained from model fits.

The polarization ratio at single energies (Figure 3) shows that the neutrons are strongly depolarized when the neutron spin is nearly perpendicular to the *c* crystalline axis. (angle ~ -15° and $+165^{o}$) This implies that the domains are needle-like, or laminar, with the long dimension nearly parallel to the *c* axis. Because the 59 eV neutrons are completely depolarized at these angles the domain length in the long dimension is presumed to be greater than the 880 micron limit discussed above. The average width δ of the domains must be much smaller because the polarization ratio is relatively large at intermediate angles.

We have fit the polarization ratio results at each energy to a model of the form:

$$f_n/f_{no} = \exp(-A) \times \{\sin^2(\theta + \phi) + \cos^2(\theta + \phi) \exp(-B/|\sin(\theta + \phi)|)\}$$

where ϕ, A and B are fit parameters. This is an extension of the formalism of Ref. 4. To reduce errors the 14, 16, and 18 eV data were averaged together to one set before fitting. The 1.7 and 2.7 eV data are insensitive to the value of B. Therefore this term was not included in those fits.

The weighted average value of the angle ϕ was found to be 15.5 degrees, indicating that the long axes of the domains are shifted with respect to the *c* axis. The parameters A and B are proportional to the depolarization factors D'_{random} and D'_{bulk}, respectively. The $\exp(-A) = \exp(-2D'_{random}t)$ term is presumably due to residual random domains in the crystal and is orientationally independent.

By fitting over the 1.7 to 59 eV range it was found that $D'_{random} = 0.58/E_n(eV)$ /cm.

In the model it is explicitly assumed that the domains have an average width δ in the direction perpendicular to the orientation direction ϕ. Therefore at each crystal orientation the effective average domain length traversed by a neutron will be increased by $1/|\sin(\theta + \phi)|$. By fitting over the 1.7 to 59 eV data we find that $D'_{bulk} = 2.8/E_n(eV)$ /cm. From Eq. 3 and the average cylinder thickness t = 1.57 cm., we then find the domains have an average width of $\delta = 59 \pm 6$ microns.

Through measurement of both the energy and orientation dependence of f_n/f_{no} we have deduced the average shape and size of magnetic domains in the interior of a thick ferromagnetic sample. A three parameter model successfully describes the salient features of the depolarization in this, admittedly, basic magnetic system.

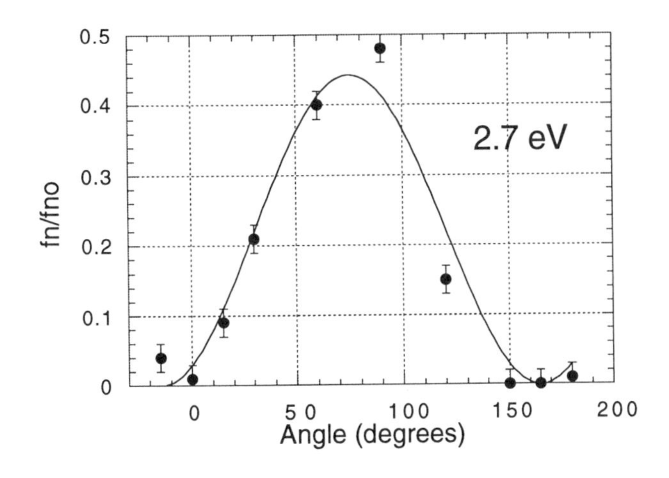

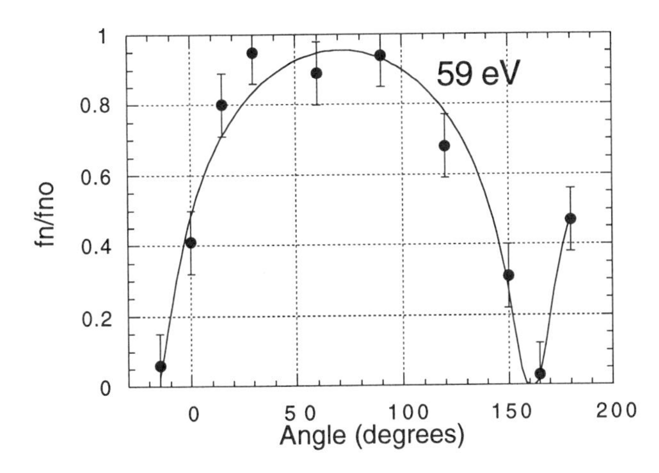

Figure 3. The measured neutron polarization transmission f_n/f_{no} as a function of the crystal orientation at two energies . The lines indicate f_n/f_{no} calculated from the model.

ACKNOWLEDGMENTS

This research was supported by Grant No. NK1300 from the International Science Foundation, by the Russian Government, and by the U.S. Department of Energy, Office of High Energy and Nuclear Physics, under Contract Nos. DE-AC05-76-ER01067 and DE-FG05-88-ER40441.

REFERENCES

+ Permanent Address: Dept. of Nuclear Physics, Lodz University, Poland

1. W. G. Williams, *Polarized Neutrons* (Clarendon, Oxford, 1988), and references therein.
2. O. Halpern and T. Holstein, Phys. Rev. 59, 960 (1941).
3. Hans Postma, H. Marshak, V. L. Sailor, F. J. Shore, and C. A. Reynolds, Phys. Rev. 126, 979 (1962).
4 D. G. Haase, J. D. Bowman, P. P. J. Delheij, C. M. Frankle, C. R. Gould, J. N. Knudson, J. E. Koster, G. E. Mitchell, S. Penttila, H. Postma, N. R. Roberson, S. J. Seestrom, S. H. Yoo, and V. W. Yuan, Phys. Rev. B46, 11290 (1992).
5. V. P. Alfimenkov, S. B. Borzakov, J. Wierzbicki, A. I. Ivanenko, Yu. D. Mareev, O. N. Ovchinnikov, L. B. Pikelner and E. I. Sharapov, Nucl. Phys. A376, 229 (1982).
6. V. P. Alfimenkov, Yu.D. Mareev, V. V. Novitsky, L. B. Pikelner, V. R. Skoy, C. R. Gould, D. G. Haase, N. R. Roberson, Nuclear Instrum. and Methods in Physics Research A352, 592 (1995).

COULOMB EXCITATION OF NEUTRON RICH LIGHT NUCLEI

T. Glasmacher[a,b], H. Scheit[a,b], B.A. Brown[a,b], J.A. Brown[a], P.D. Cottle[d], P.G. Hansen[a], R. Harkewicz[a], M. Hellström[e], R.W. Ibbotson[a], J.K. Jewell[d], K.W. Kemper[d], D.J. Morrissey[a,c], M. Steiner[a], P. Thirolf[a], and M. Thoennessen[a,b]

[a] *National Superconducting Cyclotron Laboratory, Michigan State University, East Lansing, Michigan 48824, USA*
[b] *Department of Physics and Astronomy, Michigan State University, East Lansing, Michigan 48824, USA*
[c] *Department of Chemistry, Michigan State University, East Lansing, Michigan 48824, USA*
[d] *Department of Physics, Florida State University, Tallahassee, Florida 32306, USA*
[e] *Gesellschaft für Schwerionenforschung, D-64291 Darmstadt, Germany*

We have produced beams of light neutron rich nuclei ($Z < 20$) in the A1200 fragment separator at the NSCL by fragmenting high intensity ^{48}Ca beams (E=70 and 80 MeV/nucl.) in a 379 mg/cm^2 beryllium target. The exotic beams then impinged on a secondary gold target where they were excited in the Coulomb field of the target. The nuclei de-excite from a bound state by emission of a discrete photon, thus establishing the existence and energy of this state. Using a high efficiency position sensitive photon spectrometer we were able to detect photons from eight even-even nuclei far from the valley of stability (^{44,46}Ar, ^{40,42,44}S, ^{34,36,38}Si) and to correct for the considerable Doppler shifts and Doppler broadening of the photo peaks.

INTRODUCTION

For many years Coulomb excitation has been used to study low lying states in stable nuclei. Typically the incident beam energy was chosen to be below the Coulomb barrier in order to avoid excitations of the nuclei by the strong interaction. Stable enriched targets were bombarded with a heavy beam which excites low lying states in the target through exchange of a virtual photon. A real photon is emitted when the nucleus de-excites and can be detected in a high resolution photon detector. Alternatively, a particle beam of interest can impinge on and be stopped in a heavy target. In this case the emitted photon will be Doppler shifted and Doppler broadened; however due to the generally slow beam velocity these effects are small.

For exotic nuclei far from the valley of β-stability with short lifetimes it is impossible to fabricate targets. However, beams of such nuclei with energies of several tens of MeV/nucl. are now available. If one bombards a heavy target with such exotic beams, the projectile is excited in the Coulomb and strong fields of the target and de-excites from a bound state by emission of a discrete photon. These photons establish the existence and energy of a state in the projectile. For intermediate energy beams (30-50 MeV/nucl.) the Coulomb excitation cross section can be much larger at small scattering angles than the nuclear excitation cross section. In cases where the nuclear excitation cross section is negligible, it is then not only possible to measure the excitation energy of a state, but one can also measure absolute cross sections of photon yields to determine transition matrix elements. For example, the 2^+ state of the neutron-rich nucleus ^{32}Mg has recently been studied using a beam of ^{32}Mg at 50 MeV/nucl. and a ^{208}Pb target at RIKEN (1) and ^{11,12,14}Be nuclei were investigated at GANIL (2). We have started a program at the National Superconducting Cyclotron Laboratory directed towards measurements of the energies and transition rates of first excited states in neutron rich even-even nuclei far from stability.

CP392, *Application of Accelerators in Research and Industry*, edited by J. L. Duggan and I. L. Morgan
AIP Press, New York © 1997

EXPERIMENTAL TECHNIQUE

The A1200 fragment separator at the NSCL (3) is ideally suited to provide many different beams of exotic nuclei, at energies of several tens of MeV/nucleon.

We constructed and commissioned a high efficiency photon spectrometer to identify and measure the excitation energies of bound states in even-even nuclei far from the valley of stability. The spectrometer consists of 42 position sensitive cylindrical NaI(Tl) detectors, each of which is read out by two phototubes. Information about the energies and interaction points of the incident photons can be reconstructed from the two photomultiplier tube signals by light division. The NaI(Tl) crystals are 18 cm long and 5.75 cm in diameter and arranged in three concentric rings around a 6 inch diameter beamline. To shield the array from room background photons and to provide an environment with stable temperature the detector array is shielded by a 16.5 cm thick wall of low background lead on each side. Figure 1 shows the detector setup.

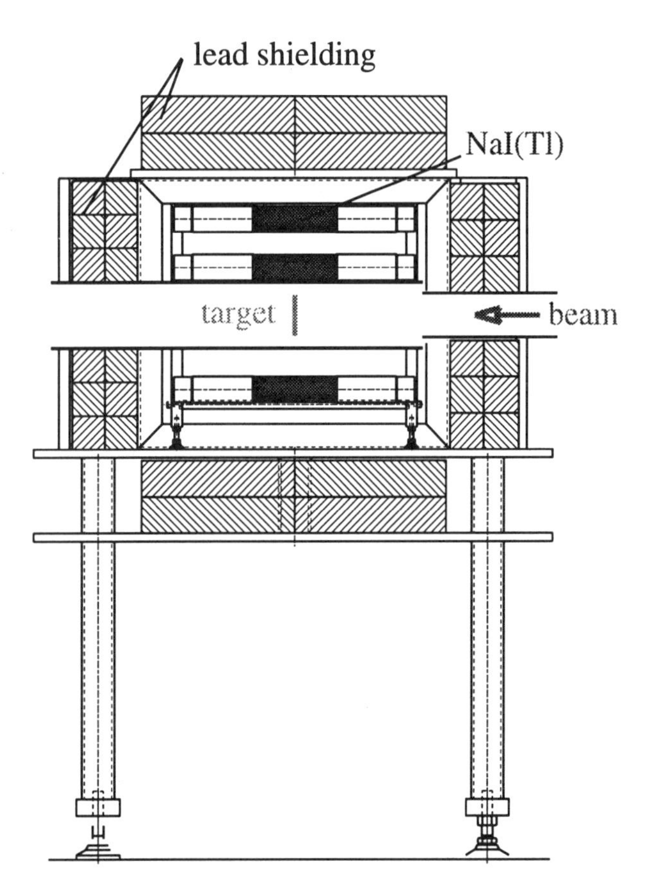

FIGURE 1. Side view of the mechanical setup of the NaI(Tl) detector array, the support structure and the lead shielding. The secondary target is located in the beamtube in the center of the array.

In a first series of experiments beams of $^{48}Ca^{12+,13+}$ and $^{40}Ar^{12+}$ with energies up to 80 MeV/nucleon and intensities as high as 20 pnA were produced with the NSCL room temperature electron cyclotron resonance (RTECR) ion source and accelerated in the K1200 cyclotron. Secondary sulfur and argon beams were obtained via the fragmentation of the primary beams in a 379 mg/cm^2 ^{9}Be primary target located at the mid-acceptance target position of the A1200 fragment separator. A measurement of the time of flight between a thin plastic scintillator located after the A1200 focal plane and a fast plastic telescope located after the secondary target was recorded for each fragment and provided positive identification of the fragment. The exotic beams were excited in a gold target, located in the center of the position sensitive NaI(Tl) detector array as indicated in Fig. 1. Photons were detected in coincidence with scattered beam particles. By setting software gates on the energy-loss - time-of-flight spectrum (shown in Fig. 2) it is possible to project out photons associated with transitions in the nuclei of interest. Photons originating from the gold target can be clearly distinguished from γ-rays associated with the exotic beams by their respective Doppler shifts.

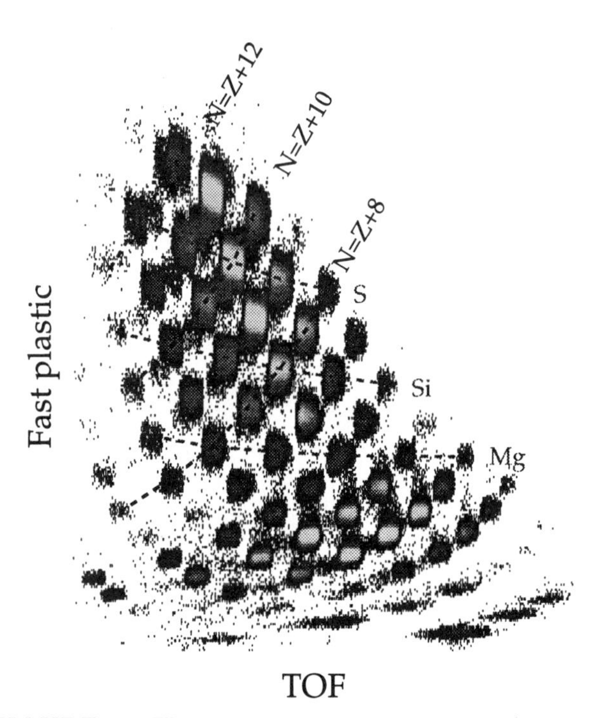

FIGURE 2. The measurement of energy-loss (in a fast plastic detector located after the secondary target) and time of flight (TOF) from the A1200 focal plane allows identification of individual isotopes on an event-by-event basis.

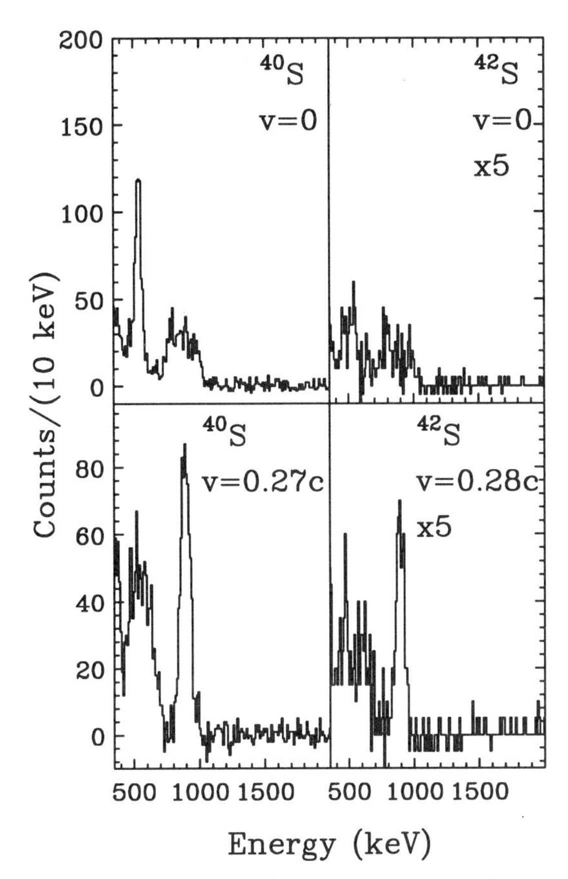

FIGURE 3. Representative energy spectra observed in coincidence with ^{40}S and ^{42}S particles. The top panels in the laboratory frame show peaks corresponding to the 547 keV ($7/2^{+}$ →g.s.) transition in the gold target while the peaks corresponding to transitions in ^{40}S and ^{42}S in the bottom panels become only visible after shifting into the projectile frame.

Figure 3 shows representative photon spectra in the laboratory frame (top panels) with the ($7/2^{+}$ →g.s.) transition in ^{197}Au clearly visible at 547 keV. Peaks corresponding to transitions in the projectile become only visible after Doppler shifting to the projectile rest frame (bottom panels). The results from this first series of experiments are being published (4).

ACKNOWLEDGEMENTS

We thank Professor Lee Sobotka (Washington University, St. Louis), who salvaged the NaI(Tl) detectors and made them available to us. This work was supported by the National Science Foundation under Grant Nos. PHY-9528844, PHY-9523974 and PHY-9403666.

REFERENCES

1. Motobayashi, T. *et al.*, Phys. Lett. B **346**, 9-14 (1995).
2. Anne, R. *et al.*, Z. Phys. A **352**, 397-401 (1995).
3. Sherrill, B. *et al.*, Nucl. Instr. and Meth. B **56/57**, 1106-1110 (1991).
4. Scheit, H. *et al.*, Phys. Rev. Lett. in press (1996).

BEHAVIOR OF INTRUDER BASED STATES IN LIGHT Bi AND Tl ISOTOPES: THE STUDY OF ^{187}Bi α DECAY

J. C. Batchelder and E. F. Zganjar,

Louisiana State University, Baton Rouge, LA 70803 U.S.A.

K. S. Toth

Oak Ridge National Laboratory, Oak Ridge, TN 37831 U.S.A.

C. R. Bingham and J. Wauters

University of Tennessee, Knoxville TN 37996 U.S.A.

C. N. Davids and D. S. Seweryniak,

Argonne National Laboratory, Argonne, IL 60439 U.S.A.

R. J. Irvine

University of Edinburgh, Edinburgh, EH9 3JZ, U.K.

W. B. Walters and L. F. Conticchio

University of Maryland, College Park, MD 20742 U.S.A.

L. T. Brown

Vanderbilt University, Nashville TN 37235 U.S.A.

Intruder state excitation energies in odd-mass nuclei just outside a closed proton shell plotted versus neutron number generally exhibit parabola-shaped curves with minima near neutron mid-shells. The Bi isotopes, however, do not seem to follow this trend. Recent experiments performed at Argonne National Laboratory have identified the previously unobserved ^{187}Bi ground state ($h_{9/2}$) to ^{183}Tl ground state $s_{1/2}$ α transition. Its energy when combined with those of two earlier known transitions, namely ^{187}Bi ($h_{9/2}$) $\rightarrow$ ^{183m}Tl ($h_{9/2}$) and ^{187m}Bi ($s_{1/2}$) $\rightarrow$ ^{183}Tl($s_{1/2}$), establishes the excitation energies of the ^{183m}Tl and ^{187m}Bi to be 620(20) keV and 110(20) keV, respectively. This value for ^{187m}Bi is 80 keV lower than the excitation energy of the same intruder level in ^{189}Bi. Implications of this result with respect to intruder-state systematics are discussed.

INTRODUCTION

In the spherical shell model, the proton $1h_{9/2}$ orbital lies above the Z = 82 closed shell while the proton $3s_{1/2}$ lies below. Any $1h_{9/2}$ configurations in Z < 82 nuclei and $3s_{1/2}$ configurations in Z > 82 nuclei are referred to as proton "intruder" states. There has been much experimental evidence to show that the excitation of these intruder states exhibit a parabolic dependence of its excitation energy with a minimum when the neutron number is midway between the major neutron shell closures. This behavior has been observed in both the Z = 50 and Z = 82 region (1,2). This is the result of the strongly attractive proton-neutron force between the

CP392, *Application of Accelerators in Research and Industry*, edited by J. L. Duggan and I. L. Morgan
AIP Press, New York © 1997

"intruder" protons and the available valence neutrons. Hence, the maximum binding energy is obtained at the mid-point of the neutron shell (3).

While the picture presented above for the $1h_{9/2}$ intruder in Z < 82 nuclei is quite convincing, it was actually the odd mass Bi isotopes that provided the first evidence for intruder states in the Z = 82 region (4) However, recent results indicate that, unlike the Tl $9/2^-$ intruder state (which reaches a minimum at N = 108), the excitation energy of the $1/2^+$ intruder state of Bi continues to drop as one proceeds from N = 108 [^{191}Bi; 242 keV] (5) to N= 106 [^{189}Bi; 190(40) keV] (6). Indeed this trend seemed to continue in ^{187}Bi as reported in Ref (7), although this was based on data with very poor statistics (8). With this in mind, we reinvestigated the α decay of ^{187}Bi.

EXPERIMENTAL

Bismuth-187 was produced via the ^{97}Mo(^{92}Mo,pn) reaction. A 1-mg/cm^2 thick target of ^{97}Mo (93% enrichment) was bombarded with 420-MeV ^{92}Mo ions accelerated at the Argonne National Laboratory ATLAS facility. The average beam current on target was ~ 2 pnA. Following production , recoils of interest were passed through a fragment mass analyzer (FMA) (9) and a gas-filled parallel grid avalanche counter (PGAC) (for mass/charge identification), and then implanted into a 60-μm thick double-sided silicon strip detector (DSSD) with 40 horizontal and 40 vertical strips. This strip arrangement results in a total of 1600 pixels. For each event in the DSSD, the time, energy, and event type (recoil or decay, depending on whether it is in coincidence with the PGAC or not) were recorded. Subsequent decays in a pixel can then be correlated with the parent allowing for nuclidic identification. This technique is known as parent-daughter correlations. Additional technical details have recently appeared in a paper (10) dealing with the proton decay of ^{185m}Bi.

RESULTS AND DISCUSSION

Figure 1(a) shows the total decay spectrum accumulated over a period of ~ 3 days with an average beam intensity of 2 pnA. One observes the well known Hg and Pb α-peaks arising from the larger reaction channels. In addition, there are events ≥ 7 MeV that we assign to ^{186}Bi (11) and ^{187}Bi.

Figure 1(b) shows the same decay spectrum correlated with a previous recoil of mass 187, and a time between decay and recoil implantation of ≤ 250 ms. There are six peaks evident in this spectrum that we assign to the α-decay of ^{187}Bi. A further constraint on the time between implantation and decay of ≤ 1.0 ms results in Fig 1(c). This figure clearly demonstrates that peaks 5 and 6 have much shorter half-lives than peaks 1-4 so that

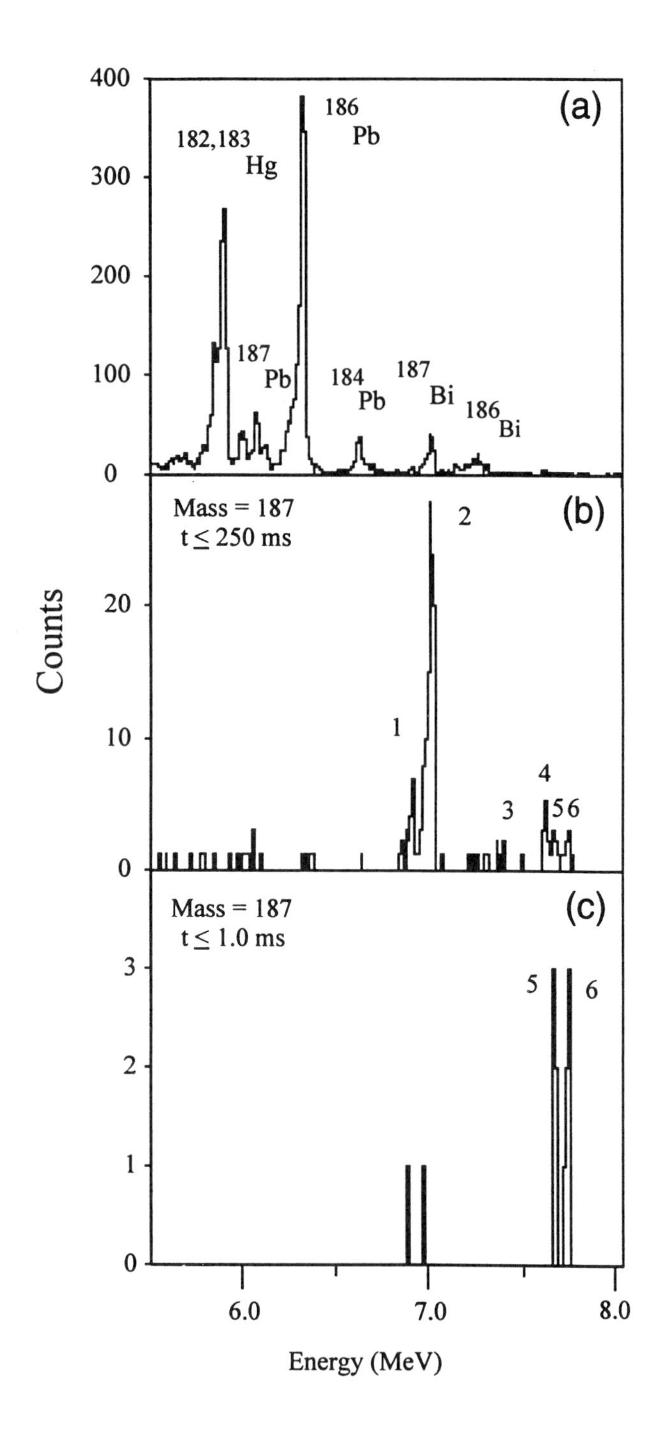

FIGURE 1. Alpha particle spectrum recorded in the DSSD during ^{92}Mo bombardments on ^{97}Mo. Part (a) shows the total decay spectrum accumulated. Part (b) is the spectrum obtained by gating on the recoils of mass = 187, and with a time between decay and recoil implantation of ≤ 250 ms. In part (c) there is a further constraint on the time to ≤ 1.0 ms. Numbered peaks are discussed in the text.

Table 1. Alpha-decay energies, half-lives, relative intensities and reduced widths in the decay of ^{187}Bi.

#	E_α (keV)	$t_{1/2}$ (ms)	$I\alpha$ (%)	$I_i \longrightarrow I_f$	δ^2 (ℓ=0) (keV)	HF
1	6905(15)	36^{+10}_{-6}	13.7	$9/2^- \longrightarrow (11/2^+)$	29(4)	2.5
2	7006(10)	32(3)	74.6	$9/2^- \longrightarrow 9/2^-$	70(9)	1
3	7379(25)	21^{+29}_{-8}	3.7	$9/2^- \longrightarrow (3/2^+)$	0.2(7)	175
4	7624(20)	25^{+9}_{-5}	8.0	$9/2^- \longrightarrow 1/2^+$	0.08(2)	875
5	7670(25)	$0.27^{+0.16}_{-0.07}$	37	$1/2^+ \longrightarrow (3/2^+, 1/2^+)$	27(10)	2.6
6	7734(20)	$0.30^{+0.12}_{-0.06}$	63	$1/2^+ \longrightarrow 1/2^+$	30(10)	2.3

as in the case of heavier odd-A Bi isotopes, there are two isomers in ^{187}Bi. Following the systematics of the heavier odd-mass Bi isotopes, we assign spins of $9/2^-$ for the ground state, and $1/2^+$ for the short-lived isomer. The relevant information on each of these six transitions is summarized in Table 1. These values compare with data from a previous report (8) of 6986(10) keV; 35(4) ms for the ground state decay, and 7583(10) keV; 0.8(6) ms.

The alpha peaks labeled 1-4 are assigned to the decay of the $9/2^-$ ground state of ^{187}Bi. All four have individual half-lives which are consistent with each other and lead to our adopted value of 32(3) ms for the $9/2^-$ ground state. In addition, peaks 1, 2 and 4 are correlated not only with A = 187 implants, but have parent-daughter correlations with ^{183}Hg 5.89-MeV α particles. These ^{183}Hg alpha decays originate from the decay sequence $^{187}\text{Bi} \xrightarrow{\alpha} {}^{183}\text{Tl} \xrightarrow{\beta+/EC} {}^{183}\text{Hg} \xrightarrow{\alpha} {}^{179}\text{Pt}$. The second step is not observed in our spectra since the thin (60μm) DSSD is not sensitive to β^+ particles, i.e., they do not leave enough energy in the detector to rise above the noise threshold. Peak #2 also has a parent-daughter correlation with the known alpha transition arising from the $9/2^-$ isomer of ^{183}Tl (12). This corresponds to an alpha branching ratio of ~ 1.5%. A previous estimate based on systematics was 3% (11).

A calculation of the alpha reduced width (13) with ℓ = 0 reveals that peaks # 1 and 2 are essentially unhindered. On the basis of systematics of the heavier odd-mass Bi isotopes, peak # 2 is clearly the $9/2^- \longrightarrow 9/2^-$ transition. Peak #1 is a transition to a state which lies 103 keV above the 9/2- isomer level. From the reduced width we can conclude that this state is probably 9/2- or 11/2+. Peaks 3 and 4 have hindrance factors of 175 and 875 respectively. These are consistent with transitions from the $9/2^-$ ground state to a 250(25) keV $3/2^+$ state and the $1/2^+$ ground state respectively. Peaks # 5 and 6 are both unhindered alpha transitions from the short-lived $1/2^+$ isomer of ^{187}Bi. These are interpreted as transitions to the ground state and

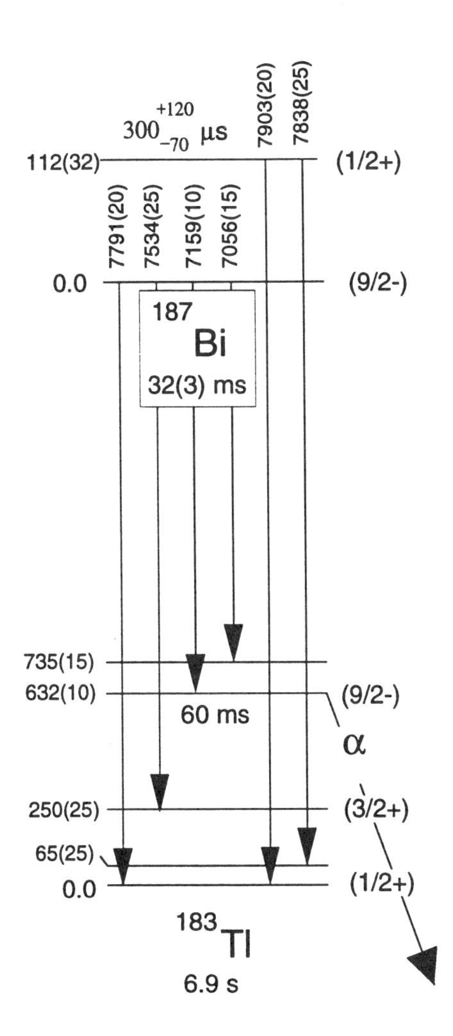

FIGURE 2. Partial α decay scheme of ^{187}Bi. All Energies shown are in keV. Note that for the ^{187}Bi α transitions, energies shown are Q_α values.

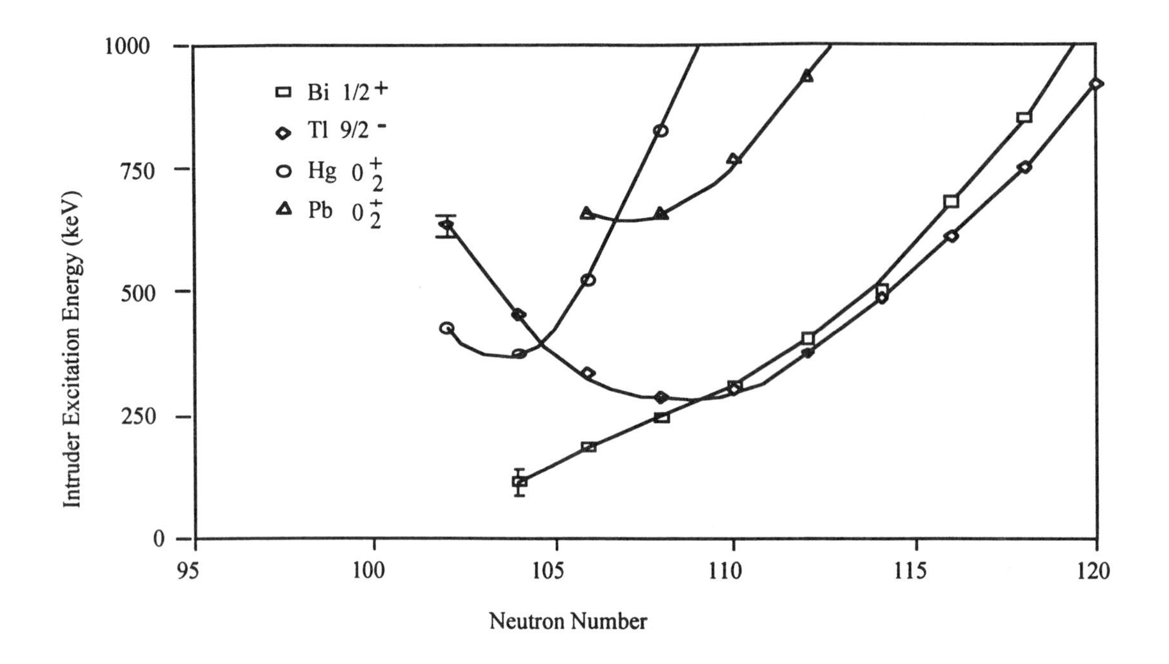

Figure 3. Plot of the intruder state excitation energies versus N for odd-mass Tl ($\pi h_{9/2}$), Bi ($\pi s_{1/2}$) and even-mass Pb and Hg isotopes.

a low-lying 1/2+ or 3/2+ state at 65(25) keV, respectively. A partial α decay scheme for ^{187}Bi is shown in figure 2. It should be noted that the alpha energies of peaks # 2, 4 and 6 establish the excitation energies of both the ^{183}Tl and ^{187}Bi isomers.

Figure 3 shows a plot of level energies of the intruder states in odd-A Tl and Bi nuclei and even-mass Pb and Hg nuclei versus neutron number, including the new values of 112 (28) keV for the $\pi s_{1/2}$ state of ^{187}Bi and 632(22) keV for the $\pi h_{9/2}$ state of ^{183}Tl. The Tl $\pi h_{9/2}$ levels fall on a parabola-shaped curve with a broad minimum at N ≈ 108. The light Bi isotopes, however continue to drop further down in both energy and neutron number.

The $\pi s_{1/2}$ intruder state in the Bi isotopes is believed to be due to transferring a proton from the $s_{1/2}$ shell to the $h_{9/2}$ shell, resulting in a two-particle one hole state (2p-1h). The further dropping of the excitation energy of this state may indicate that it is a more strongly deformed prolate structure than the corresponding Pb intruder state configuration. This could result in a high degree of mixing, where a hole mixes with the 4p-3h state similar to the 4p-2h deformed structure in the Po nuclei.

Acknowledgments

This work was supported in part by the U.S. D.O.E. under contracts # DE-FG05-84ER40159 (Louisiana State University), DE-AC05-84OR21400 (Oak Ridge National Laboratory), W-31-109-ENG-38 (Argonne National Laboratory), DE-FG05-88ER40418 (University of Maryland, and DE-FG05-87ER40361 (University of Tennessee).

References

1. Heyde, K., *et al., Phys. Rep.* **102**, 291-393 (1983).
2. Wood, J. L., *et al., Phys Rep.* **215**, 101-201 (1992).
3. Heyde, K., *et al., Nuc. Phys* **A484**, 275-294 (1988).
4. Alpsten, M. and Astven, G., *Nucl. Phys.* **A134**, 407-418 (1969)
5. Leino, M. E., *et al., Phys. Rev.* **C24**, 2370 (1981).
6. Batchelder, J. C., *et al., Phys. Rev* C **52**, 1807-1809 (1995).
7. Coenen, E., *et al., Phys. Rev. Lett.* **54**, 1783-1786 (1985).
8. J. Schneider, Gesellschaft für Schwerionenforschung, Darmstadt, Thesis Report No. GSI-84-3, 1984 (unpublished).
9. Davids, C.N., *et al., Nucl. Instr. Meth.* **B70**, 358-361 (1992).
10. Davids, C.N., et al., *Phys. Rev. Lett.* **76**, 592-595 (1996).
11. Batchelder, J. C., et al., submitted to *Z. Phys. A*.
12. Schrewe, U. J., et al., *Phys. Lett* **.91B**, 46-48 (1980).
13. Rasmussen, J. O., Phys. Rev. **113**, 1953-1598 (1959).

THE LOW ENERGY KAON PROGRAM AT THE CELSIUS STORAGE RING

A. Badalà, R. Barbera, M. Gulino, F. Librizzi, A. Mascali, D. Nicotra, A. Palmeri, G. S. Pappalardo, F. Riggi, A. C. Russo, G. Russo, A. Santoro, R. Turrisi
Istituto Nazionale di Fisica Nucleare, Sez. di Catania, and Dipartimento di Fisica dell'Università di Catania - Italy

V. Dunin, C. Ekström, G. Ericsson, B. Höistad, J. Johansson, T. Johansson, L. Westerberg, J. Zlomaczhuk
The Svedberg Laboratory - Uppsala - Sweden

The CLAMSUD spectrometer has been recently installed at the jet-target position of the CELSIUS ring at "THE SVEDBERG LABORATORY". The physical purpose is the study of kaon production at energies below the N-N threshold. Due to the low cross-section and short lifetime of kaons we increased the solid angle by means of two quadrupoles positioned at the entrance of the dipole. The experimental quality of the measurements due both to the beam characteristics and to the CLAMSUD detector will be shown.

I. INTRODUCTION

The study of particles production at energies below the absolute NN threshold is a very important field of investigation in nuclear physics. This study allows to reach information on the reaction dynamics and on the momentum distribution of nucleons in target and projectile nucleus. In the last two decades this study was essentially performed through pion production [1]. More recently the subthreshold kaon production is under investigation [2]. There are many advantages in using this "probe": a) Since kaon has a strangeness different from zero, it doesn't interact strongly with the nuclear matter and hence no distortion due to final state interaction is expected, b) The threshold energy for kaon production (T_{th}=1.58 GeV) in the free NN reaction is much higher than for pions (T_{th}=0.290 GeV), since with the kaon also a Λ particle is emitted to fulfill the strangeness conservation law. For this reason the kaon production in nucleus-nucleus collisions can supply information on the nuclear matter at higher densities: the description of the reaction dynamics by means of the equation of state of the nuclear matter is sensible to the nuclear compressibility. This is not the case for subthreshold pion production.

To compare the experimental results with the calculations arising from the solution of the nuclear equation of state, it is necessary to know accurately the cross section for the NN process and it is also useful to study kaon production induced by protons in order to study the kaon production when the nuclear matter has the normal compressibility.

In this context we started our investigation on kaon production by using a 1.2 GeV proton beam on a CH_4 gaseous target (jet-target). The beam is supplied by the storage ring CELSIUS [3]. The CLAMSUD magnetic spectrometer [4], successfully employed as a pion detector [5], is used to detect kaons. Because of the low cross-section for kaon production, two quadrupoles were put between the target position and the entrance flange of the spectrometer in order to increase its solid angle.

The CLAMSUD as well as the CELSIUS beam characteristics are discussed in the mainframe of the kaon detection in Sect. II. Sect. III is devoted to the presentation of the quality of the experimental results from the technical point of view.

II. THE CELSIUS STORAGE RING AND THE CLAMSUD SPECTROMETER

The CELSIUS storage ring, that employes beams coming from a cyclotron, has a circumference of 82 m and is able to accelerate the beam up to 1.36 GeV. The ring consists of four 90° arcs and four straight sections (see Fig. 1). Each arc is built up of 10 dipole magnets with a common coil. One straight section is used for the beam injection, the second one is used for beam diagnostics and pellet target. The third section holds the electron cooler and a RF cavity. The fourth section holds the jet-target station with its scattering chamber.

The ring may operate in static mode or in cycles. When operating in cycles, the beam may be accelerated or decelerated and the magnet excitation as well as the RF frequency are programmed to be consistent with the particle momentum. A typical period of a cycle is about 900 s and the useful time interval is about 840 s. During the useful time the beam circulates with constant energy and may be used for physics measurements. The resolution $\Delta p/p = 4 \cdot 10^{-4}$ of the beam is suitable for our purposes.

In the jet-target the gas beam is obtained by pressing a gas through a cooled nozzle, keeping the pressure-temperature conditions such that the gas is close to the transition to liquid. Then the target beam passes a set of collimator and the central part of the scattering chamber. It is finally collected in a cryogenic beam dump, that is regenerated after about 40 running hours. The intersection between the projectile beam and the jet-target beam gives the dimension of the target point. This intersection

CP392, *Application of Accelerators in Research and Industry*, edited by J. L. Duggan and I. L. Morgan
AIP Press, New York © 1997

TABLE I. CLAMSUD main characteristics in the CELSIUS configuration

Maximum central momentum	200 MeV/c
Solid angle	16 msr
Momentum acceptance	±20 %
Momentum resolution	$2.6 \cdot 10^{-3}$
Dispersion	1.16 cm/%
Magnification	0.8
Mean trajectory length	1.66 m
Reference gap	8.5 cm
Weight	20 Tons
Power	50 kW

is given roughly by a cylinder with the axis parallel to the beam axis. The section of the cylinder is roughly a circle having a diameter of 4 mm (the beam spot) and its height is about 10.5 mm (due to the gas jet profile). The thickness of the methane target is 3.9 10^{13} atoms/cm^2, that joint with the beam intensity, gives a luminosity of $\sim 10^{28}$ cm^{-2} s^{-1}. The thickness of the scattering chamber window is 1 mm of steel.

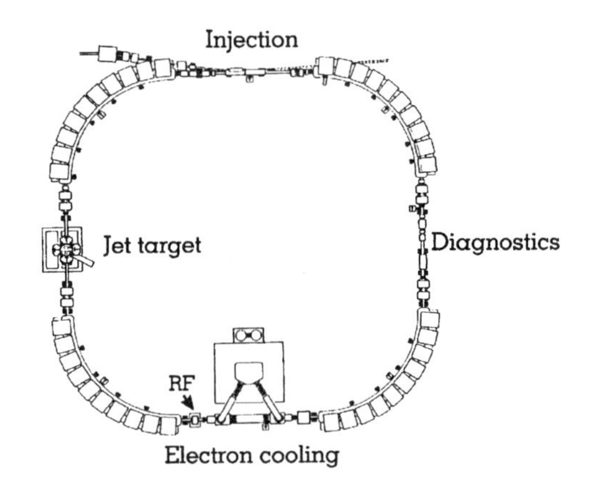

FIGURE 1. Layout of CELSIUS

The CLAMSUD spectrometer is a compact dipole magnet with non parallel poles [4]. To increase the solid angle, two quadrupoles were designed and added between the target point and the entrance flange. The main characteristics of the whole spectrometer (dipole plus the two quadrupoles) in the CELSIUS placing are shown in Table I. A sketch of the spectrometer is shown in Fig. 2.

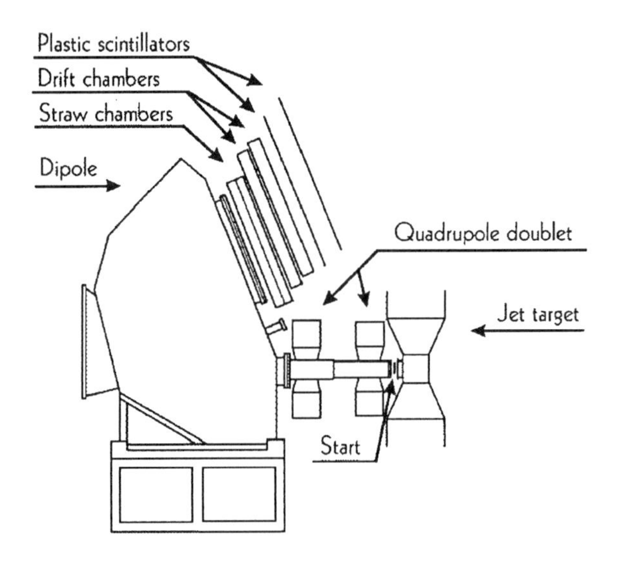

FIGURE 2. Sketch of the CLAMSUD magnetic spectrometer

The spectrometer is equipped with a focal plane detector consisting of two multiwire X-Y drift chambers and two planes of segmented plastic scintillators. Before the last run two straw chambers were also added between the exit flange and the drift chambers in order to better reconstruct the particle trajectory. A start detector is mounted near the entrance of the first quadrupole. Timing and energy loss information are provided by the two planes of scintillators, while the coordinates of particles hit, given by the two wire chambers, allow the measurement of the linear momentum of the detected particle. A single wire drift time is used for the radial coordinate, while a delay line read-out is used for the axial one.

The spatial resolution of 300 μm in the radial axis and of 1 mm in the axial one ensure an overall energy resolution of 0.2%.

Simulations have been performed to establish the spectrometer parameters such as solid angle and energy resolution [4]. The pion, kaon and proton time of flight were calculated for different values of linear momentum taking into account the overall time resolution.

III. INSTALLATION AND FIRST EXPERIMENTAL RESULTS

The CLAMSUD spectrometer was installed at CELSIUS on June '95. In November '95 a first test measurement on kaon production was performed: the distance between the entrance flange and the jet-target was 1.5 m, which resulted in a solid angle of the spectrometer of about 2.8 msr. In this test kaons were produced from proton-Carbon collision at 1.2 GeV. A CH_4 gas jet target was used, which allowed simultaneous luminosity determination via elastic proton scattering. The kinematical coincidence between two protons has been measured by means of two scintillator detectors; spurious coincidences out of the kinematics were also measured in order to subtract the background due to proton coming from Carbon. This first test show the feasibility of the measurement of kaon production with our experimental apparatus, in spite of the ratio $\sim 10^{-4}$ between kaons and pions detected in the same interval of momentum.

After the installation of the two quadrupoles, which allows a solid angle of 14 msr in the momentum interval $P_0 - 15\% \div P_0 + 15\%$ (P_0 being the central momentum), a new test measurement was performed in March '96. In the scatter plot of Fig. 3 the time of flight for two corresponding detectors in the two planes B and C is shown. The points at the center of the scatter plot have a time of flight within the limits obtained by the simulation for kaons. The kaon identification is also supported by the cross check with the energy loss spectra.

In the present experimental conditions the count rate is ~13.6 kaons per hour in the momentum interval 170 - 230 MeV/c that corresponds to an energy interval of about 22 MeV.

IV. CONCLUSIONS

Preliminary measurements have shown the feasibility of subthreshold kaon production experiments induced by protons by using the CLAMSUD spectrometer upgraded with two quadrupoles. The luminosity of the CELSIUS beam coupled with the upgraded CLAMSUD solid angle gives a reasonable production rate of kaons. Kaon identification among a strong background of pions has been shown to be possible owing to the high separation power of the time of flight.

ACKNOWLEDGMENTS

The authors are very grateful to the continuous technical assistance given by O. Byström, L. Petterson, R. Wedberg and to the CELSIUS staff for the good beam quality.

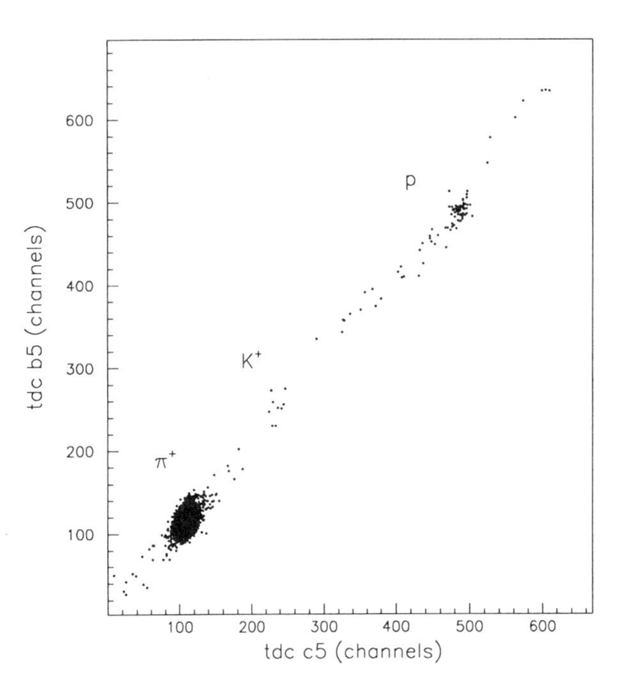

FIGURE 3. Time of flight relative to a detector in the first plane of scintillators versus the corresponding in the second plane

REFERENCES

1. Braun-Munzinger, P., Stachel, J., *Ann. Rev. Nucl. Part. Sci* **37**, 97 (1987), and references therein; Cassing, W., Metag, V., Mosel, V., Niita, K., *Phys. Rep.* **188**, 363 (1990), and references therein
2. Hartnack, C., *et al.*, *Nucl. Phys. A* **580**, 643 (1994); Sibirtsev, A., Büscher, M., *Z. Phys. A* **347**, 191 (1994).
3. *The Svedberg Laboratory Report 1987-1991.*
4. Anzalone, A., *et al.*, *Nucl. Instr. and Meth. in Phys. Res. A* **308**, 533 (1991); Badalà, A., *et al.*, *INFN Report TC* **93/07**, (1993).
5. Badalà, A., *et al.*, *Nucl. Instr. and Meth. in Phys. Res. B* **99**, 657 (1995).

Split-Target Neutronics and the MLNSC Spallation Target System

G. J. Russell, P. D. Ferguson, E. J. Pitcher and J. D. Court

Manuel Lujan, Jr., Neutron Scattering Center, Los Alamos National Laboratory, Los Alamos, New Mexico 87545

The Manuel Lujan, Jr., Neutron Scattering Center (MLNSC) at the Los Alamos National Laboratory is one of four operating Short-Pulse Spallation Sources worldwide. The MLNSC target system (composed of targets, moderators, and reflectors) was first installed in 1985. The target system employs a split tungsten spallation target with a void space in between (the flux-trap gap); this target system will be upgraded in 1998. The ability to efficiently split a spallation target allowed us to introduce the concept of flux-trap moderators and ultimately the notion of backscattering and upstream moderators. The upgraded MLNSC target system will employ both flux-trap and upstream/backscattering moderators to simultaneously service 16 neutron flight paths with high-intensity neutron beams for materials science research.

INTRODUCTION

The Manuel Lujan Jr. Neutron Scattering Center (MLNSC) at the Los Alamos National Laboratory is one of four operating Short-Pulse Spallation Sources (SPSS) worldwide. Protons from the 800-MeV Los Alamos Neutron Science Center accelerator (formerly Los Alamos Meson Physics Facility) impinge vertically downward onto the MLNSC spallation target system composed of spallation targets, moderators, and reflectors.[1] The MLNSC target system(2), which was first installed in 1985, employs a split tungsten target with a void space in between (the flux-trap gap); this target system will be upgraded in 1998. The ability to efficiently split a spallation target allowed us to introduce the concept of flux-trap moderators and ultimately the notion of backscattering and upstream moderators. Flux-trap moderators have several inherent neutronic advantages: a) all moderators are high-intensity; b) the neutron spatial distribution is fairly uniform over the moderator surface, and c) the moderators can be viewed in either transmission or backscattering geometry. We will discuss the rationale behind split targets, flux-trap and backscattering moderators, and the application of these concepts to the existing and upgraded MLNSC target systems.

Figure 1 shows the basic target-moderator geometries that have been (or will be) utilized in SPSS target systems. The most traditional geometry is a solid spallation target and wing moderators. The notion of a split spallation target was pioneered at Los Alamos, and is currently used in the MLNSC target system.

SPLIT-TARGET NEUTRONICS

An important general objective in the design of a spallation source target system is to maximize neutron production. Total neutron production (per incident particle) depends essentially on the target material, the amount of material in the incident beam, geometry, and the energy and type of the incident particles. Two projectiles are generally considered for the incident beam: protons and deuterons; however, we confine our discussion here to protons. Once neutrons are produced inside the target, they must leak from the target before they can be used. Therefore, the other crucial aspect of spallation source target design is the maximization of the leakage of low-energy neutrons from the target. The main factors controlling neutron leakage are

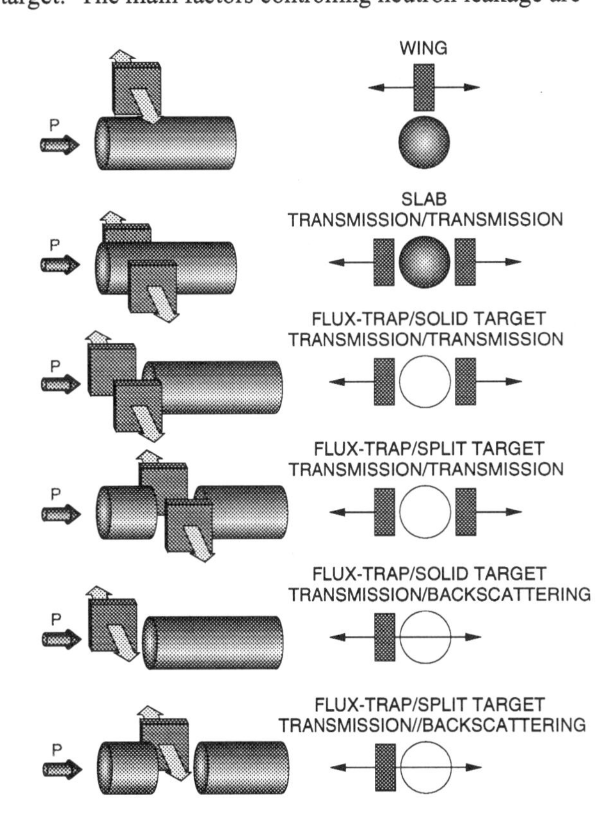

FIGURE. 1. Basic spallation target-moderator configurations.

[1] The basic notion of spallation and spallation targets is discussed in Ref. (1).

CP392, *Application of Accelerators in Research and Industry,* edited by J. L. Duggan and I. L. Morgan
AIP Press, New York © 1997

parasitic absorption in the target material and target geometry. The three materials of choice for practical neutron production (solid) targets are lead, tantalum, and tungsten. Depleted uranium has been used at the ISIS facility, and liquid mercury is under study as a target for the European Spallation Source (ESS) project.

Figure 2 shows the effect of splitting a tungsten spallation target: the neutron leakage for a split target is only about 10% less than a solid target with a target gap (flux trap) of 14 cm and a parabolic proton beam profile. Flux-trap gaps do not really affect the protons as they travel unhindered from target region to target region until a stopping length of target material finally halts their movement. Whether the stopping length of material is lumped in one solid piece or spread over a number of target segments is almost of no consequence for the primary protons.

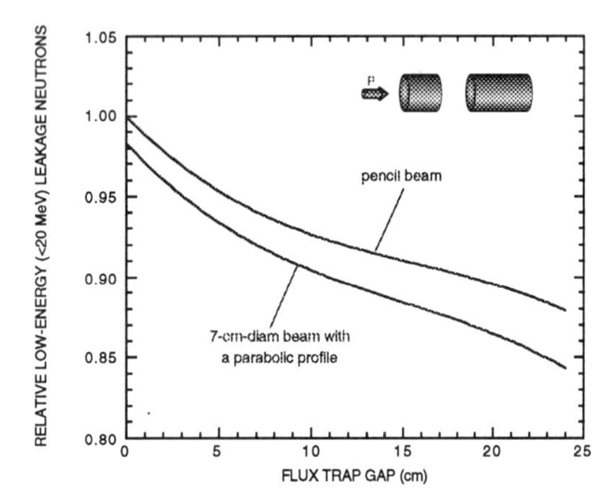

FIGURE 2. MLNSC split-target, flux-trap gap study.

The ability to efficiently split a target allowed us to introduce the concept of flux-trap moderators(3) and ultimately the concept of backscattering and upstream moderators(3). The relative performance of upstream, central, and downstream flux-trap moderators is illustrated in Fig. 3; the data shows the potential of "upstream" moderators compared to "downstream" moderators.

POISONS, DECOUPLERS, AND LINERS

For most users of a pulsed spallation-neutron source, useful neutrons can be defined as those headed in the right direction with appropriate energy at the right time. Unfortunately, spallation neutrons produced directly in the target rarely have the desired characteristics. We must, therefore, add the necessary systems and devices to the bare neutron production target in order to tailor the neutron pulse so that its characteristics are as close as possible to the users' requirements. As mentioned above, a complete target system consists not only of target(s) for the production of neutrons, but also of moderators, reflectors, and, in the case of an SPSS, poisons, decouplers, liners.

In addition to the choice of material, temperature, geometry (*e.g.*, wing versus flux-trap moderators), and the presence or absence of a reflector, moderator neutronic performance is also strongly tied to the presence or absence of poisons, decouplers, and liners. The choice of materials and thickness for these target system components is a crucial part of moderator design(3).

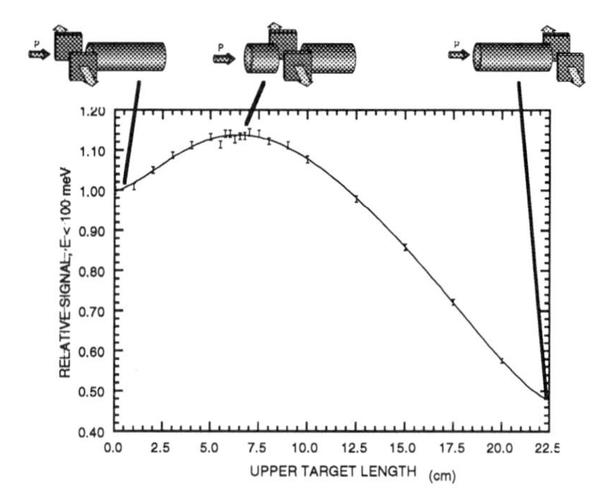

FIGURE 3. Illustration of upstream, central, and downstream flux-trap moderators viewed in transmission.

The function of poisons, decouplers and liners is to tailor the temporal and energy characteristics of the neutron pulses emitted by the moderator(4). Figure 4 shows the arrangement of poisons, decouplers and liners in the split-target, flux-trap moderator geometry.

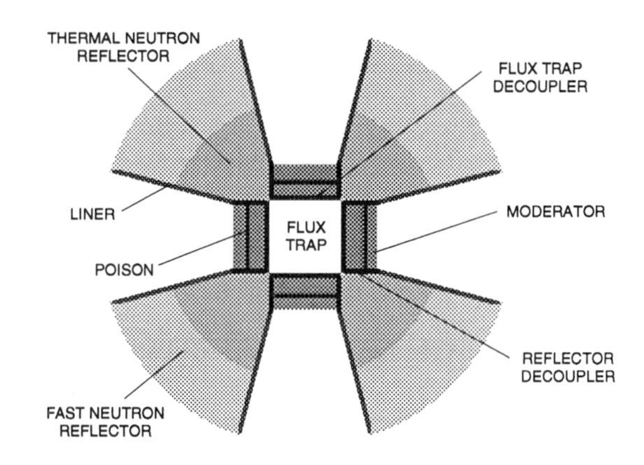

FIGURE 4. Arrangement of poisons, decouplers, and liners in a flux-trap moderator geometry. Poisons are typically oriented parallel to and positioned some distance (≈1 to 3 cm) behind the moderator viewed surfaces. The flux-trap decouplers neutronically insulate moderators from one another whereas the reflector decouplers neutronically isolate moderators from the adjoining reflector material. Liners neutronically insulate the reflector from the moderator viewed-surface.

For thermal neutrons, the poison neutronically defines that part of the moderator "viewed" by an experiment. Decouplers surround a moderator and both geometrically and neutronically isolate it from the reflector. Liners geometrically and neutronically isolate the moderator "viewed surface" from the reflector. The goal of short-pulse moderator design is to get as much useful neutron intensity from a moderator as possible with little or no attendant degradation in the neutron pulse width.

CALCULATED RESULTS

The time-averaged neutron source brightness from 5x13x13 cm liquid hydrogen flux-trap moderators and composite reflectors is illustrated in Figs. 5a and 5b. The moderator geometry is depicted in Fig. 4; the overall reflector size was 114 cm diam. by 114 cm high. The proton energy was 800 MeV, and the targets were stopping-length (22.5 cm), light-water-cooled (pure) tungsten plates with a diameter of 10 cm. The type and size of the inner reflector was varied in the calculations. The ortho/para-hydrogen mix was assumed to be 50/50 v%. We show data for four composite reflectors: beryllium/lead, graphite/lead, light-water/lead, and heavy-water/lead (lead is always the outer reflector).

For a decoupled system with an inner graphite reflector, moderator performance is essentially independent of the size of the inner reflector. For a decoupled system and light-water and heavy-water inner reflectors, moderator neutronic performance decreases when the inner reflector radius is increased. This is due to too much moderation occurring in the inner reflector with subsequent capture of neutrons in the decoupler/liner materials. For an inner reflector of beryllium, the moderator performance continually increases with increased radius of the inner reflector. Note that asymptotic neutronic performance is reached for an inner reflector radius of 30-35 cm.

For a coupled system with beryllium and liquid-deuterium reflectors, moderator performance increases as the size of the inner reflector becomes larger. Except for the very first calculated point at 15 cm radius, the moderator neutronic performance for an inner light water reflector of light water decreases with increasing inner reflector radius, reaching an asymptotic value at around 35 cm radius. The data point at 15 cm is interesting and indicates that neutronic gains may be made using light water as a premoderator for a liquid hydrogen moderator and a lead reflector. For a coupled system with an inner graphite reflector, moderator performance is essentially independent of the size of the inner reflector. For coupled moderators, the neutronic performance of the various composite reflectors reach asymptotic values at different inner reflector radii.

We show here only time-integrated data. Clearly, for short-pulse spallation source (decoupled systems) and long-pulse spallation source (coupled systems) applications, adequate time-dependent neutronic performance is imperative(5). We have time-dependent data for these calculations, but it is beyond the scope of this work to discuss the results.

We have calculated the neutronic performance of coupled light water moderators in a solid beryllium reflector. The four flux-trap moderators were in the geometry depicted in Fig. 4 (except with no poisons, decouplers, and liners). The light-water moderators were 5x13x13 cm, and the solid beryllium reflector was 200 cm diam and 200 cm high. The proton energy was 800 MeV, and the light-water cooled (pure) tungsten targets were 10 cm diam and 22.5 cm long (total equivalent tungsten length).

The results of this moderator thickness study are depicted in Fig. 6. Note that the time-integrated leakage flux peaks at a moderator thickness of about 4 cm.

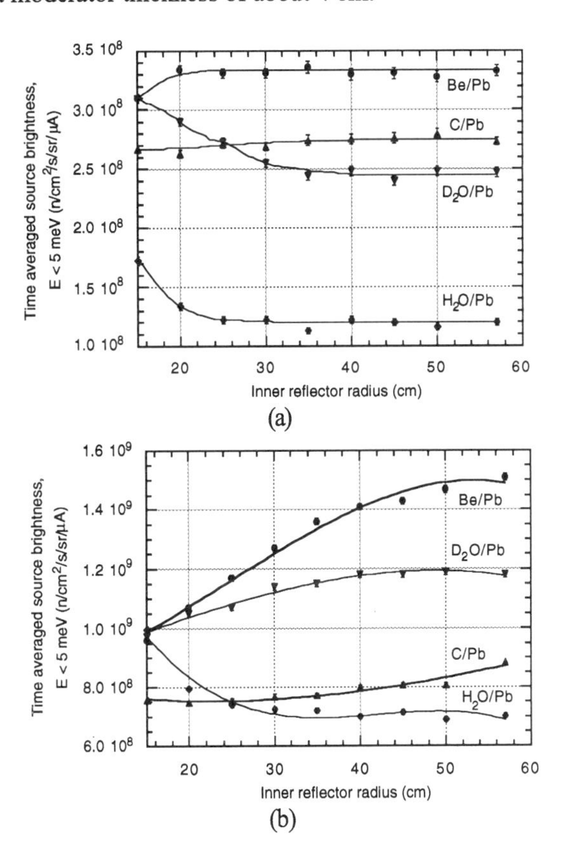

FIGURE 5. Time-averaged moderator source brightness for 5x13x13 cm liquid hydrogen flux-trap moderators (ortho/para at 50/50 v%) for decoupled (a) and coupled (b) composite-reflector systems.

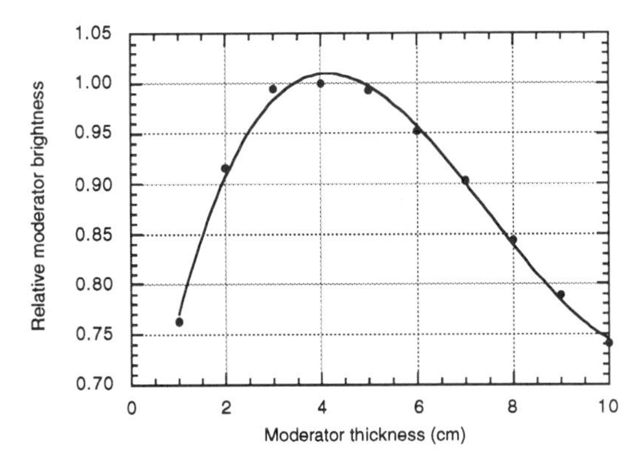

FIGURE 6. Time-averaged moderator source brightness for 5x13x13 cm coupled light-water flux-trap moderators versus moderator thickness. The solid beryllium reflector was 200 cm diam and 200 cm high.

THE MLNSC TARGET SYSTEMS

The MLNSC 800-MeV proton beam impinges vertically downward into the target system. The innovative split-target/flux-trap-moderator arrangement was introduced in 1983 to take advantage of the vertical proton beam injection scheme. This target system was implemented in 1985, using a composite beryllium-nickel reflector-shield(3). The split-target with four flux-trap moderators (viewed in transmission) as used in the original MLNSC as-built target system are depicted in Fig. 7a. Figure 7b shows the upgraded MLNSC target-moderator arrangement with the addition of two upstream backscattering moderators(6). Figure 7c shows the complete MLNSC upgraded target system (targets, moderators, and reflectors) . Calculated time averaged brightness ratios for the MLNSC as-built and the MLNSC upgraded target are given in Table 1.

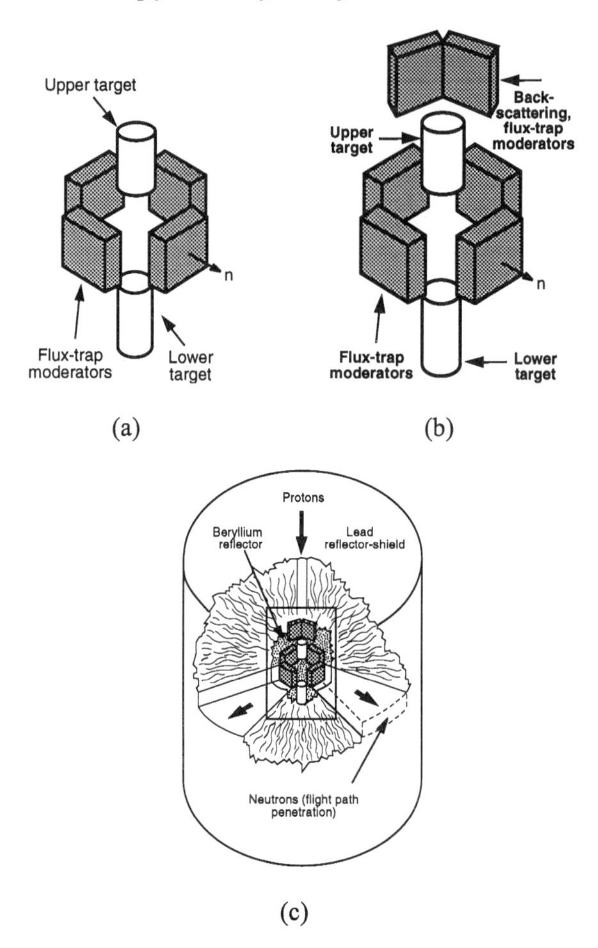

FIGURE 7. (a) the MLNSC as-built, split-target, flux-trap moderator configuration; (b) the MLNSC upgraded split-target, flux trap/backscattering moderator configuration; and (c) the MLNSC upgraded target system (targets, moderators, and reflectors).

TABLE 1. Relative time averaged performance of the MLNSC Upgrade target system. The coupled LH_2 is relative to the MLNSC as-built decoupled LH_2 and the coupled H_2O is relative to the MLNSC as-built HI H_2O (3,4,5).

	MLNSC as built	MLNSC Upgrade
decoupled LH_2	1.00	1.14
HR H_2O	1.00	1.18
HI H_2O (3,4,5)	1.00	1.10
HI H_2O (6,7,8)	1.00	1.14
coupled LH_2		5.34
coupled H_2O		6.56

DISCUSSIONS AND CONCLUSIONS

The upgraded MLNSC target system will employ both flux-trap and upstream/backscattering moderators to simultaneously service 16 neutron flight paths with high-intensity neutron beams for materials science research. We have calculated the relative neutronic performance of the as-built MLNSC moderators to the upgraded MLNSC moderators. One design goal was to add the two additional moderators and keep the relative performance of the four flux-trap moderators to within 10% of each other to account for engineering penalties as the design progress. This objective has been fulfilled.

ACKNOWLEDGMENTS

This work was supported by the U. S. Department of Energy, BES-DMZ, under contract No. W-7405-Eng-36.

REFERENCES

1. G. J. Russell, E. J. Pitcher, and L. L. Daemen, "Introduction to Spallation Physics and Spallation-Target Design," in Proceedings of the International Conference on Accelerator-Driven Transmutation Technologies and Applications, AIP Conference Proceedings 346, 1994, pp. 93-104.

2. G. J. Russell, C. D. Bowman, E. R. Whitaker, H. Robinson, and M. M. Meier, "LANSCE High-Power (200 uA) Target-Moderator-Reflector-Shield," in ICANS-VIII: Proceedings of the Eighth Meeting of the International Collaboration on Advanced Neutron Sources, Rutherford-Appleton Laboratory report RAL-85-110, 1985, pp. 272-293.

3. N. Bultman, A. Jason, E. Pitcher, G. Russell, W. Sommer, D. Weinacht, and R. Woods, "Los Alamos Next-Generation Spallation Source," Volume I, Los Alamos National Laboratory report LA-UR-95-4300 (December 1995).

4. G. J. Russell, E. J. Pitcher, and P. D. Ferguson, "Coupled Moderator Neutronics", in ICANS-XIII: Proceedings of the 13th Meeting of the International Collaboration on Advanced Neutron Sources, G. S. Bauer and R. Bercher, Eds., Paul Scherrer Institut report PSI 95-02, ISSN 1019-6447, 1995, pp. 495-509.

5. E. J. Pitcher, G. J. Russell, P. A. Seeger, and P. D. Ferguson, "Performance of Long-Pulse Source Reference Target-Moderator-Reflector Configurations," in ICANS-XIII: Proceedings of the 13th Meeting of the International Collaboration on Advanced Neutron Sources, G. S. Bauer and R. Bercher, Eds., Paul Scherrer Institut report PSI 95-02, ISSN 1019-6447, 1995, pp. 323-329.

6. P. D. Ferguson, G. J. Russell, and E. J. Pitcher, "Reference Moderator Calculated Performance for the LANCE Upgrade Project," in ICANS-XIII: Proceedings of the 13th Meeting of the International Collaboration on Advanced Neutron Sources, G. S. Bauer and R. Bercher, Eds., Paul Scherrer Institut report PSI 95-02, ISSN 1019-6447, 1995, pp. 510-517.

ZEOLITE TARGET DEVELOPMENT AT TISOL

M. Dombsky[1] and J. M. D'Auria[2]

(1) TRIUMF, 4004 Wesbrook Mall, Vancouver, B.C., Canada V6T 2A3
(2) Department of Chemistry, Simon Fraser University, Burnaby, B.C., Canada V5A 1S6

In on-line isotope separation, rapid efficient release of short-lived radioactive products from thick targets has traditionally been accomplished by the use of refractory target materials heated to elevated temperatures. At high temperatures, (~ 2000° C) the diffusion rate of the desired product from the target matrix is maximized. Unfortunately, the requirement for high temperature limits the choice of target materials to the refractory metals and ceramics, usually carbides. However, it is possible to achieve efficient release of products at significantly lower temperatures by the use of porous target materials in which the desired product has a sufficiently high diffusion rate. One such class of materials is the zeolites. The structure of zeolites is composed of aluminum and silicon oxides with exchangeable cations balancing the charge. The target composition can be altered through cation exchange. Results of on-line isotope production from zeolite targets at TRIUMF are presented.

INTRODUCTION

The goal of beam development at on-line isotope separator (ISOL) facilities has been to provide the maximum yields of the isotopic beams required for experiments. Since, in general, the desired nuclides are short-lived, optimum delivery requires maximizing the efficiency of production, release, transport and ionization components of the beam generation process. Choosing an appropriate target material involves finding a candidate that can provide both a high production cross-section for the desired product and a rapid release of the product from the target matrix.

The production rate (R) can be estimated from:

$$R = \sigma \Phi N_t \quad (s^{-1}) \qquad (1)$$

where (Φ) is the projectile flux, (N_t) the number of target nuclei and (σ) the production cross-section. Although it may not always be simple to account for scattering losses and incident beam energy degradation in thick targets (1), relative production rates can be estimated from the abundant experimental cross-sections available, and from semi-empirical formalisms such as those of Silberberg and Tsao (2).

The release properties of product/target combinations are less simple to predict. In theory, relative diffusion rates of products in the target matrix can be estimated from diffusion coefficients (D) derived from an Arrhenius relationship:

$$D = D_0 \exp\{-\Delta E/kT\} \quad (cm^2/s) \qquad (2)$$

where (D_0) is the temperature independent pre-exponential diffusion coefficient and (ΔE) is the activation energy for diffusion (3). In practice, it is difficult to find literature values of D_0 and ΔE that are specific to the product/target species being considered. Furthermore, there are no semi-empirical formulae for estimating diffusion co-efficients. Rough estimates of diffusion coefficients have been obtained by considering the self-diffusivities of the target elements as lower limits (4). However, even for self-diffusion in such apparently simple systems as oxides, various values of D_0 and ΔE may result in predictions varying by several orders of magnitude (5).

Desorption of products from the target material and other surfaces encountered during transport is similarly governed by an Arrhenius relationship:

$$\tau = \tau_0 \exp\{\Delta H_a/kT\} \quad (s) \qquad (3)$$

where (τ) is the mean sticking time of the product on a surface, (τ_0) is a constant of order 10^{-13} s, and (ΔH_a) is the adsorption enthalpy of the product on the particular surface. Estimates of sticking times can be obtained from both experimental (6) and semi-empirical (7) compilations of ΔH_a.

Even in the absence of sufficient information on diffusion and desorption behavior, it is clear from equations 2 and 3 that an increased target temperature will enhance the transport of products from a target. Traditional target strategies, employed with ISOLS, have generally involved heating the target material to as high a temperature as possible. While this is feasible for many products, it presents problems for production of beams of the first-row non-metals such as C, N and O.

PRODUCTION STRATEGIES FOR C, N AND O

The non-metals of the first row of the periodic table are capable of combining with each other to form gaseous oxides, nitrides and carbides. The volatility of such molecules enhances their release probabilities. However, the same elements also form refractory oxides, nitrides and carbides with most metals used as targets or containers for ISOL purposes. Once formed, the refractory products are unlikely to be transported to the ion source before the desired nuclide decays.

Strategies for generating carbon, nitrogen and oxygen beams have included the use of targets such as platinum in graphite, where stable Pt species are not formed at high temperature (8), or graphite with flowing nitrogen, where the ^{13}N product was released as molecular N_2 (9).

At TISOL, the emphasis has been on maximizing production of ^{16}N ($t_{1/2}$ = 7.13 s). Initial attempts concentrated on production of NO from oxide targets coupled to an electron cyclotron resonance (ECR) ion source equipped with a quartz plasma chamber. An attempt to produce N_2 from an AlN target was also made. In all cases, the targets were heated to the maximum temperatures consistent with their vapor pressure properties. The maximum ^{16}N yields, of order ~ 10^4 /s/μA, were insufficient for experimental purposes. In a change of strategy, an attempt was made to generate nitrogen from a highly porous target running at low temperature. The synthetic zeolite Linde 13X (Na^+ cation) was initially tested at a temperature of 500° C. Since then, zeolite targets have been heated up to 650° C. Sufficient ^{16}N was produced to satisfy the experimental requirements. The maximum ^{16}N yields from zeolite NaX are compared to yields from other TISOL targets in Table 1.

ZEOLITE STRUCTURE

Zeolites are natural or synthetic aluminosilicate minerals with open "cage-like" structures based on interconnected AlO_4 and SiO_4 units. Due to charge imbalance in the lattice, zeolite neutrality is maintained by one trapped cation incorporated for each AlO_4 unit. The cations can be exchanged, either partially or entirely, by generally simple chemical substitution in an aqueous medium. Traditional zeolite nomenclature designates both the cation and the structural form of a particular species, as for example, NaX, KX or CaA.

The specific structure of a particular zeolite determines the size of the cavity formed by the oxide units. Due to the variation in the "cage" cavity dimensions, different zeolites are capable of trapping guest molecules of varying dimensions while allowing smaller molecules to pass through the structure. Because of this capability, zeolites are commonly called "molecular sieves". The commonly

Table 1. Maximum yields from TISOL Targets.

Target	^{16}N Yield (/s/μAp$^+$)	Ion Beam
MgO	9.5×10^4	N^+
	3.8×10^4	N_2^+
	1.4×10^4	NO^+
CaO	2.3×10^2	N^+
	1.2×10^4	N_2^+
	3.2×10^3	NO^+
AlN	1.8×10^3	N^+
	3.9×10^3	N_2^+
	2.9×10^3	NO^+
Al_2O_3	2.2×10^2	N^+
	4.9×10^1	N_2^+
NaX	1.8×10^6	N^+
	1.1×10^6	N_2^+
	3.4×10^6	NO^+

available synthetic zeolites, usually with Na^+ cations, are Linde A and X. The effective pore sizes of zeolites A and X are ~ 4 Å and ~ 8.5 Å respectively (10). The exact open dimensions depend on the nature and position of the cations in the lattice. While the open pore size of zeolite A is on the order of the kinetic dimensions of molecules such as N_2, O_2, CO and the noble gases, the pores of zeolite X are large compared to such species as CCl_4 and SF_6 (10). To date, the Na^+, Ca^{2+}, and Tl^+ forms of zeolite X have been used as targets at TISOL. The formula of dehydrated NaX is $Na_2O \cdot Al_2O_3 \cdot 2.5SiO_2$.

ZEOLITE THERMAL STABILITY

The thermal stability of zeolites varies with structure and cation. At higher temperatures, the zeolite lattice collapses to an amorphous phase or recrystallizes into a non-zeolitic species. Estimates of the breakdown temperatures have been obtained by monitoring the changes of zeolite X-ray diffraction patterns as a function of temperature. The temperatures at which initial structure degradation is observed are given in Table 2.

Table 2. Thermal Stability of Some Zeolites (11).

Zeolite	°C	Zeolite	°C
NH_4A	120	NH_4X	80
LiA	660	LiX	680
NaA	660	NaX	660
KA	800	KX	600
MgA	620	MgX	700
CaA	540	CaX	500
LaA	740	LaX	500
BaA	90	BaX	500

RADIATION EFFECTS

Despite the possibility of structural collapse, there is some evidence that zeolites may be sufficiently resistant to radiation damage to be acceptable ISOL targets. Radiation damage to the zeolite lattice has been studied using both neutron and gamma irradiation. Rees and Williams (12) reported initial damage to zeolite X starting at 6.2×10^{17} n^0/cm2, with complete collapse of the zeolite structure at 7×10^{19} n^0/cm2. Campbell (13) successfully produced and eluted ^{142}Pr from ^{141}Pr exchanged zeolite X using 5×10^{13} $n^0/cm^2/s$. Stamires and Turkevich (14) studied γ-induced damage to zeolites X and Y. Radiation induced lattice defects were followed by esr spectroscopy of samples exposed to up to 20 megaRoentgen, using a ^{60}Co source. Induced lattice damage could be annealed by heating samples of X and Y above 350° C and 500° C, respectively.

At TISOL, a NaX target was operated at 650° C for 6 weeks under irradiation of 1-2 μA of 500 MeV protons. No degradation of the ^{16}N yield was observed during this time.

DIFFUSION IN ZEOLITES

The molecular sieving capability of zeolites, has instigated much research into their absorption and diffusion properties. Unfortunately, most of the literature on diffusion in large pore zeolites (X and Y) has been concerned with mass transport of large organic molecules with dimensions on the order of the zeolite pore size (15). A relative estimate of diffusion rates can be made by comparing the self-diffusion of Ca^{2+} in CaA and CaX with the self-diffusion observed for CaO, a more traditional higher temperature target. The self-diffusion coefficients derived from values of D_0 and ΔE using equation (2) are presented in Fig 1. The curves in Fig. 1 suggest that diffusion in zeolite X may be 10 orders of magnitude faster than in calcium oxide.

ZEOLITE TARGETS AT TISOL

Zeolite target development at TISOL has specifically centered on obtaining beams of ^{16}N, ^{9}C and ^{31}Ar. Beams of other gaseous products have been observed incidentally. For ^{16}N and ^{9}C, NaX targets were used. For ^{31}Ar, NaX zeolite was exchanged in a $CaCl_2$ solution to provide a CaX target. An approximately 57% exchanged TlX target was also prepared by treating NaX with a solution of $TlNO_3$. The intent of the TlX target was to study release of higher mass products. All zeolite targets were used in conjunction with the ECR ion source and a cooled transfer line.

Using NaX, sufficient yields of both ^{16}N and ^{9}C were obtained to satisfy experimental requirements. However, no ^{31}Ar was detected using the CaX target, although $^{32-35}$Ar were observed.

The TlX target run was plagued with vacuum problems in the TISOL target chamber and was stopped prior to completion of a thorough product survey. However, beams of ^{45,46}Ar and $^{90-92}$Kr were observed and there was evidence of multiply-charged beams of long-lived Hg products. Mercury γ-rays were observed in spectra obtained at lower mass positions. The maximum yields of species obtained using zeolite targets are presented in Table 3. Argon obtained from the NaX target is possibly produced from potassium impurities in the zeolite.

CONCLUSIONS

Zeolite targets have been successfully operated at TISOL to provide radioactive beams of carbon, nitrogen, oxygen and the noble gases. Future plans include further investigation of zeolites as possible targets for oxygen beam production. Investigation of cation exchanged forms of zeolite X, as well as other, less common zeolites are of interest.

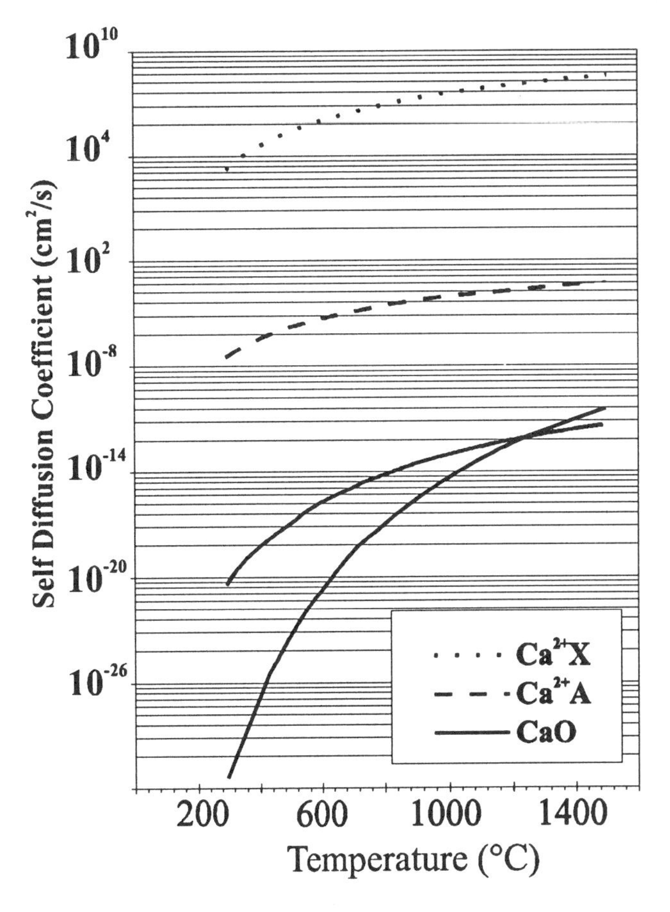

FIGURE 1. Self-diffusion coefficients for Ca^{+2} in CaA, CaX and CaO. Based on D_0 and ΔE values compiled in Ref. (16,17)

Table 3. Maximum yields from TISOL Targets.

Target	Product	Ion Beam	Yield (/s/μAp$^+$)	Target	Product	Ion Beam	Yield (/s/μAp$^+$)
Na^+X	^8He	He^+	1.4×10^3	$Ca^{2+}X$	^{32}Ar	Ar^+	5.1
	^{18}Ne	Ne^+	3.3×10^6		^{33}Ar	Ar^+	1.5×10^3
	^{19}Ne	Ne^+	3.3×10^8		^{34}Ar	Ar^+	1.4×10^5
	^{23}Ne	Ne^+	1.7×10^7		^{35}Ar	Ar^+	2.1×10^6
	^{24}Ne	Ne^+	1.4×10^6	$Tl^+_{.57}X$	^{10}C	CO^+	3.7×10^6
	^{25}Ne	Ne^+	4.3×10^3		^{15}C	A = 47	8×10^3
	^{34}Ar	Ar^+	2.9×10^4		^{16}N	A = 47	8.8×10^4
	^{35}Ar	Ar^+	1.1×10^6		^{17}N	N^+	8×10^3
	^9C	CO^+	1.0×10^4		^{18}N	NO^+	2.6×10^3
	^{10}C	C^+	1.4×10^6		^{18}Ne	Ne^+	1.5×10^6
		CO^+	1.2×10^7		^{23}Ne	Ne^+	7.5×10^4
	^{15}C	C^+	5.6×10^4		^{25}Ne	Ne^+	5.4×10^3
		CO^+	1.3×10^4		^{34}Ar	Ar^+	5.4×10^3
	^{16}N	N^+	1.8×10^6		^{35}Ar	Ar^+	6.1×10^3
		N_2^+	1.1×10^6		^{34}Ar	Ar^+	1.1×10^4
		NO^+	3.4×10^6		^{34}Ar	Ar^+	1.9×10^3
	^{17}N	N^+	1.8×10^5		^{45}Ar	Ar^+	1.1×10^4
		NO^+	2.5×10^5		^{46}Ar	Ar^+	1.2×10^3
	^{18}N	N_2^+	1.5×10^3		^{90}Kr	Kr^+	2.1×10^5
		NO^+	3.4×10^3		^{91}Kr	Kr^+	4.6×10^4
	^{14}O	O^+	5.7		^{92}Kr	Kr^+	1.1×10^3
		O_2^+	8.5×10^4		long lived Hg	?	?
		A = 32	1.3×10^4				

ACKNOWLEDGMENTS

The authors gratefully acknowledge the contributions of H. Biegenzein, D. Jones, P. Machule, H. Sprenger and M. Trinczek.

REFERENCES

1. Talbert, W. L., and Hsu, H.-H., *Nucl. Instr. And Meth.* **A362**, 229-233 (1995)
2. Silberberg, R., and Tsao, C. H., *Astroph. J. Suppl.*, 315-367 (1973)
3. Borg, R. B. and Dienes, G. J., *An Introduction to Solid State Diffusion*, San Diego: Academic Press, 1988, ch. 3, p. 60.
4. Carraz, L. C., Haldorsen, I. R., Ravn, H. L., Skarestad, M. and Westgaard, L., *Nucl. Instr. And Meth.* **148**, 217-230 (1978)
5. Wuensch. B. J., "Diffusion in Stoichiometric Close-packed Oxides" in *Proceedings of a NATO Advanced Study Institute on Mass Transport in Solids*, 1981, p.353-.376.
6. Kirchner, R., *Nucl. Instr. And Meth.* **B70**, 186-199 (1992)
7. Eichler, B., Hübener, S., and Roßbach, H., *ZfK Rossendorf Reports* **560** and **561** (1985)
8. Ravn, H. L., *Nucl. Instr. And Meth.* **B70**, 107-117 (1992)
9. Van Duppen, P., et al, *Nucl. Instr. And Meth.* **B70**, 393-397 (1992)
10. Breck, D. W., *Zeolite Molecular Sieves*, New York: John Wiley & Sons, 1974, ch. 8, p. 637.
11. Ref. 10, ch. 6, p. 495.
12. Rees, L. V. C. and Williams, J., *Trans. Farad. Soc.*, **62**, 1481 (1965)
13. Campbell, D. O., *Inorg. Nucl. Chem. Lett.*, **6**, 103-109 (1970)
14. Stamires, D. N., and Turkevich, J., *J. Amer. Chem. Soc.*, **73**, 757-761 (1964)
15. Kärger, J. and Ruthven, D. M., *Diffusion in Zeolites and Other Microporous Solids*, New York: John Wiley & Sons, 1991, ch. 12, pp. 427-466.
16. Ref. 10, ch. 7, p. 576.
17. Mrowec, S., *Defects and Diffusion in Solids, An Introduction*, New York: Elsevier/North Holland, 1980, ch. 4, p. 432.

LOW-ENERGY RADIOACTIVE BEAM EXPERIMENTS USING THE UM-UND SOLENOID RNB APPARATUS AT THE UND TANDEM: PAST, PRESENT AND FUTURE

F.D. Becchetti
Department of Physics, University of Michigan
Ann Arbor, Michigan 48109 USA

J.J. Kolata
Department of Physics, University of Notre Dame
Notre Dame, IN 46556 USA

Approximately ten years ago (1987) one of the first operational low-energy radioactive nuclear beam (RNB) facilities was put into place at the University of Notre Dame (UND) as a joint project between the University of Michigan (UM) and UND. The key elements to the success of the project were the installation of a large-bore 3.5 Tesla superconducting solenoid (supplied by UM) to collect and focus secondary radioactive beams, combined with an upgrade of the UND FN Tandem, the latter including the addition of a high-intensity sputter ion source. The resulting secondary beams (^{8}Li, ^{6}He, ^{7}Be, ^{8}B, ^{18}F^{m}, ...) are generally produced by selective, high-cross-section direct reactions. These beams are sufficiently intense (*viz.* 10^4/s to 10^8/s) to permit measurement of many low-energy reaction cross sectons of interest to nuclear astrophysics, nuclear reaction theory, and high-isospin nuclear physics. A review of past and recent RNB data obtained with this apparatus will be presented together with plans for a major upgrade using a pair of 6T solenoids (M.Y. Lee *et al.* - this conference).

INTRODUCTION

The low-energy radioactive nuclear beam (RNB) facility operated at the University of Notre Dame (UND) FN-tandem Van de Graaff as a joint project between the University of Michigan (UM) and UND was one of the first operational RNB facilities in the world (1987). Until recently (mid-1996), it still produced some of the most intense low-energy RNBs available, specifically beams of ^{6}He, ^{8}Li, ^{7}Be, ^{8}B, ^{18}F, ^{18}F^{m} (and others) with intensities $> 10^4$/s, with some beams (^{8}Li) approaching 10^9/s. These relatively high RNB intensities are generally needed to measure cross sections relevant to nuclear astrophysics (and other nuclear phenomena) since unlike projectile break up or elastic scattering, the cross sections are small (*viz.* mb vs. b). Thus, the success of the facility required the development of several high-intensity RNBs and related high-efficiency detection devices and techniques.

We'll now briefly describe some of these and then summarize the RNB experiments done with the facility (up to mid-1996) together with some of the more important discoveries made. Finally, we'll outline plans for a new, upgraded facility which will utilize two large s.c. solenoid magnets (additional information is available at UM Website http://www.physics.lsa.umich.edu/twinsol).

RNB PRODUCTION

The key developments (1985-96) were as follows:

a) development and installation of a large air-core superconducting solenoid as the secondary RNB collection and focussing device (1987–89)
b) utilization of selective, high-cross-section direct- or compound-nuclear reactions for the RNB production reactions (1987–)
c) installation of a high-intensity SNICs-type sputter ion source and rotating production target (1990)
d) improvements in the FN tandem accelerator (accelerator tubes; vacuum; strippers; time-of-flight beam pulser and sweeper; etc)

Although a rotating target mount was designed, built and installed, radiation damage still remains a problem for certain production targets, and to this day often limits the amount of primary beam which can be used (by x5 to x10). Most recently, a ^{3}He gas cell has been used for the efficient production of a ^{8}B beam via the ^{6}Li(^{3}He, n)^{8}B reaction (Table I). This cell has Havar windows 2 microns thick and has been successfully run at 1 atm of ^{3}He for a week with ca. 1 μA of 30 MeV ^{6}Li primary beam. (It has recently been tested to 10 atms.) Many future experiments will utilize this cell or a similar one.

We generally select a *direct nuclear reaction* to form the RNB of interest via a specific nuclear state having a high production cross section (*viz.* > 10 mb/sr at lab angles $\theta < 10°$). Additionally, the production reaction is done, if possible, on a high-melting-temperature target in reverse kinematics (heavy beam on a light target) so that one gains a large forward-angle enhance-

CP392, *Application of Accelerators in Research and Industry*, edited by J. L. Duggan and I. L. Morgan
AIP Press, New York © 1997

ment from the c.m. to the lab system. Unlike fragmentation reactions, which are typically used at higher energies, the use of a more selective direct reaction is much more efficient in terms of the RNB yield for a given RNB energy spread (phase space). This is important for producing useable RNBs having a good energy resolution, as an example to resolve final nuclear states which typically require δE (FWHM) $\lesssim$ 1 MeV. Also, we have exploited production-reaction selectivity to produce short-lived, excited nuclear isomer beams, specifically $^{18}F^{m}(J^{\pi} = 5^{+};\ E_{x} = 1.12$ MeV; $T_{1/2} = 163$ nsec). The mechanism for producing $^{18}F^{m}$ at low energies [via the $^{12}C(^{17}O,^{18}F^{m})$ reaction] has also proved effective at higher energies (see Tables). Details of specific RNB production reactions and their yields may be found in the references given in Tables I and II.

After RNB production at the primary-beam target, it is then desirable to collect, separate and then refocus the selected RNB as efficiently as possible. Since, even with kinematic focussing the RNB production "cone" can be fairly large at low energies, the best device to refocus the beam is one with a large, axi-symmetric acceptance. Large, superconducting solenoids appear to be well-suited for this task[1] owing to their simple yet "fast" optics *i.e.* their good "f/#" (the ratio of aperture to focal length) which is a measure of the collection efficiency.

We discovered, rather serendipitously, the advantages of using such devices as RNB collectors during a set of experiments done in 1984–86 at the ATLAS facility at Argonne National Laboratory (ANL). As part of a Ph.D. thesis project, a 3.5 T, 20 cm bore, 40 cm long air-core, superconducting solenoid (obtained surplus by UM from the DoE laser-fusion program) was installed and configured for use as a large solid angle, "zero-degree" reaction-product spectrometer[1]. This concept was based on the work of Schapira *et al.* at IPN-Orsay who had used an iron-yoked s.c. solenoid for this purpose[2].

In the ANL-ATLAS experiments, which used a ca. 100 MeV ^{18}O beam on various targets, a major limitation was pulse pile up at the solenoid focal-plane detector due to "unwanted" reaction products such as $^{16,17}O$, ^{19}F, *i.e.* stable ions, as well as the radioactive ions (RNBs) $^{15,19}O$, $^{17,18,20}F$, where the latter were often at the 10^{2} to 10^{3}/s level. At that point it was realized that the solenoid magnet could be utilized as a simple but effective RNB collector. It was therefore subsequently moved to UND for this purpose[3] as part of a joint UM-UND project (Fig. 1). Recent improvements at UND now permit operation to ca. 10 MV terminal voltage with primary beams up to $A \sim 40$, with μA's of intensity, with variable pulse width ($\delta t \sim 1$ nsec) and beam-pulse spacing.

In the course of operating the RNB apparatus, we've developed other axi-symmetric ion-optical systems[4] which we plan to utilize in future work. These include an $\vec{E} \otimes \vec{B}$ system using a radial electric field lens, and systems based on focussing and *defocussing* solenoids. In addition, funding from DoE and UM was obtained to build a large-bore 7T system for use at NSCL[5]. A dual solenoid system[6] which can utilize time-of-flight for improved RNB energy resolution was also developed.

RNB REACTIONS STUDIED, 1987–1996

A summary of the RNB reactions studied with the 3.5T apparatus is given in Table I. Appropriate references for this table and related work are given in Table II. A number of these studies were the first of their type done with an RNB, and they revealed a number of important properties. Specifically, due to the nature of the unstable projectile RNB-induced nucleon transfer reactions and fusion-evaporation reactions are often *highly* exothermic ($Q > 0$). Hence we find:

1) large cross sections to *excited* states, even at low energies (Fig. 2), due to
2) a favorable reaction "Q-window"; and hence
3) RNB n-tunneling at low energies is favored, and
4) very energetic neutrons and α particles can be produced in low-energy RNB fusion-evaporation (CN) reactions.

We also observe:

5) large BE2 $\Uparrow$ for certain RNB excitations,
6) with ^{8}Li elastic scattering and optical-model (OM) potentials similar to ^{6}Li and ^{7}Li elastic scattering (and OMs) but ^{6}He is *not* similar to ^{4}He elastic scattering and its OM (Instead, $^{6}He \approx\ ^{6}Li$, ^{7}Li and ^{8}Li elastic scattering and OM-see Fig. 3).

Also,

7) measurements of many of the ^{8}Li S-factors needed for accurate input to Big-Bang nucleosynthesis and the "missing mass" problem were done including, recently,
8) measurements of low-energy ^{8}B break-up cross sections (Fig. 4) related to the solar neutrino problem (too few solar ν's).

(While the latter measurements have helped to clarify both the "missing mass" and the solar neutrino problems, they unfortunately have not provided "solutions" to these problems.)

Finally, we have

9) demonstrated the feasibility of producing high-intensity isomeric, high-spin, excited RNBs for new types of nuclear-reaction studies using "γ-tagged" RNBs.

THE FUTURE

The 3.5T RNB apparatus has been removed from its original beam line and will be re-installed elsewhere at UND. In its place a pair of 30 cm bore 6T solenoids have been installed (funded by the US-NSF, UND and UM) and are presently being tested. This new system will provide a wider variety of RNBs including higher-mass, higher-energy (to 10 MeV/u), and higher-purity beams. (The latter will be accomplished via. a radial $\vec{E}$ electrode and/or absorber placed between the two magnets; see M. Lee *et al.*, this conference).

We also intend to use fast timing of certain RNBs to provide an energy-compensated beam. This should permit high-resolution (*viz.* $\delta E \lesssim 100$ keV) RNB measurements for Coulomb excitation, transfer reactions and other phenomena. The dual system will also permit low-background $\gamma-\gamma$ and γ-particle RNB measurements, low-energy RNB total reaction cross section measurements[7], and radioactive-ion trapping. We also plan to extend some of our previous measurements (Table I) to higher energies (and other nuclei) as well as pursue several new aspects of RNB research at both the nuclear and atomic levels.

ACKNOWLEDGEMENTS

This work has been funded by grants from the National Science Foundation, the University of Notre Dame, and the University of Michigan (Office of Vice-President for Research and the Department of Physics). We also gratefully acknowledge the post-doctoral scholars, Ph.D. students, and other UND and UM personnel who have participated in major aspects of this project, specifically W. Liu, R. Stern, R. Smith, J. Brown, D. Roberts, A. Morsad, K. Lamkin, L. Lamm, and E. Berners.

REFERENCES

1. See [St87a,b] and [StC87] in Table II
2. J.P Schapira, S. Gales, and H. Laurent, Orsay Report IPNO-PhN-7921 (1979); J.P. Schapira, F. Azaiea, S. Fortier, S. Gales, E. Hourani, J. Kumpulainen, and J.M. Maison, *Nucl. Instrum. Meth.* **224**, 337 (1984).
3. See [St87a,b,c], [StC87], [BeC88], [Ko89], and [RNB1,2,3,4] in Table II
4. See [Li86], [Li89], [O'DC94] and [O'D94] in Table I
5. F.D. Becchetti, J. Bajema, J.A. Brown, F. Brunner, H. Griffin, J.W. Jänecke, T.W. O'Donnell, R.S. Raymond, D.A. Roberts, R.S. Tickle, R.M. Ronningen, H. Laumer, N. Orr, A. Zeller, and R.E. Warner, *Nucl. Instrum. Meth.* **B79**, 326 (1993).
6. See M.Y. Lee *et al.* [BAPS **41**, 1028 (1996) and this conference] and [Be96b] in Table II.
7. D.A. Roberts, *et al.*, *Nucl. Phys.* **A588**, 247c (1995); R.E. Warner, *et al.*, *Phys. Rev.* **C52**, R1166 (1995) and *Phys. Rev C* (1996), in press.

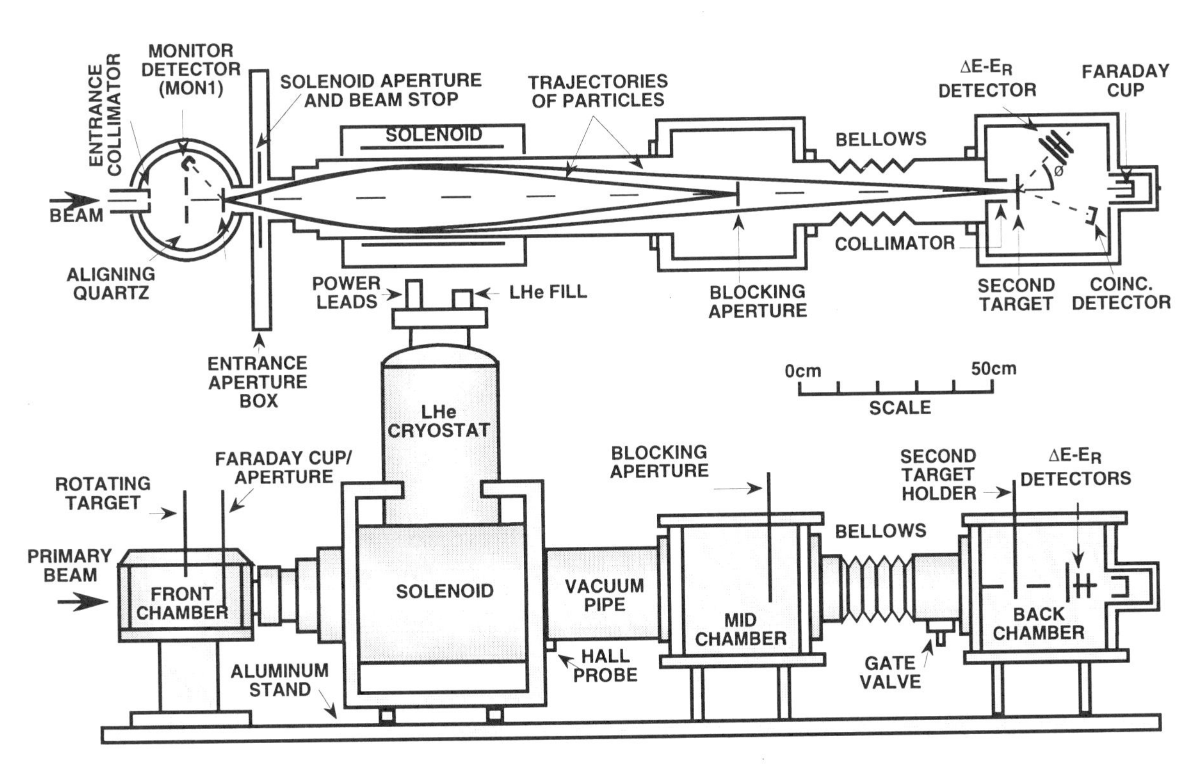

Figure 1. Top and side view of the UM-UND 3.5T RNB apparatus (1987–1996; from [Li89], Table II)

TABLE I: RNB Reactions Studied, 1987–1996

Reaction Studied	E(MeV)	Measurement	References[a)]	Comments
^{8}Li(d,t)^{7}Li	14.8	Limited $\sigma(\theta)$	BeC88, Be89, Li90, FaC91, Be93	CD_2 target; inverse kinematics; large cross section observed; relevant to Big-Bang nucleosynthesis
^{8}Li(d,t)^{7}Li	6–14.8	Excitation function	FaC91, Ba93, Ba95	Degraded ^{8}Li beam; determination of astrophysical S-factors
^{8}Li(p,α)^{5}He	14.8	$\sigma(\theta)$	Be92	Highly exothermic; large cross section; determination of astrophysical S-factors
^{8}Li(p,n)^{8}Be	14.2	$\sigma(\theta)$	Fl92, Ca93	RNB charge exchange; coincident detection $^{8}Be^* \to \alpha + \alpha$
^{8}Li(d,n)^{9}Be	6–14	$\sigma(\theta)$ and excitation function for $^{9}Be_{g.s.}$ and $^{9}Be^*$	Ba93, Ba95, La96	Utilized coincident detection $^{9}Be^* \to \alpha + \alpha + ...$
^{8}Li(α,n)^{11}B	1–14	Excitation function and limited $\sigma(\theta)$ for $^{11}B_{g.s.}$ and $^{11}B^*$	Gu95	Utilized gas-filled, multi-anode detector (MUSIC) as ^{4}He "target"; key reaction in Big-Bang nucleosynthesis; most of cross section is to *excited* states; *i.e.* much larger than g.s. $\to$ g.s. cross sections
^{6}He + ^{9}Be – Au ^{6}He + ^{12}C, Ni, Au	7–9 10	Elastic scattering "	Sm91a Wa95	Deduced LE ^{6}He optical potentials; ^{6}He scattering resembles ^{6,7,8}Li scattering rather than ^{4}He
^{12}C(^{8}Li,^{8}Li)^{12}C	14.8	Elastic and inelastic scattering, $\sigma(E,\theta)$	Sm91b	Indicates large BE2 ⇑ for $^{8}Li^*$ (E_x = 0.98 MeV); ^{8}Li + ^{12}C optical potential determined; ^{8}Li+^{12}C similar to ^{6,7}Li + ^{12}C scattering
^{8}Li + ^{2}H, Be, ^{12,13}C, N, Al, Ni	13–20	Elastic scattering; RNB fusion-evaporation; RNB transfer reactions, $\sigma(E,\theta)$	Be93	Systematics of LE ^{8}Li-induced RNB reactions; deduced ^{8}Li optical potentials; very-high-energy α's observed from (^{8}Li,α); exothermic n-transfers observed with good Q-matching characteristics; FRDWBA analysis and spectroscopic factors consistent with a direct transfer mechanism.
^{nat}Ni(^{8}Li,^{8}Li*)	14.8, 19.6	Coulomb excitation of ^{8}Li(E_x = 0.96 MeV)	Br91	Large BE2 ⇑ for $^{8}Li^*$ deduced; related to solar-neutrino "problem"
^{58}Ni(^{8}Li,^{7}Li)^{59}Ni	14.8	Neutron transfer, $\sigma(E,\theta)$	Be96	$\sigma(E,\theta)$ consistent with neutron tunneling; reaction favors excited states due to exothermic Q values (*viz.* $Q_{opt} \doteq 0$)
Ni(^{8}B,^{7}Be p)	26	^{8}B breakup at low energy	Sc96	Set limits on E2 and E1 LE capture rates for ^{7}Be(p,γ) as related to the solar neutrino problem
$^{18}F^m$ + ^{12}C, Au	70	Limited scattering	Be90, Be91 Br93, Br95	Demonstrates feasibility of producing short-lived, high-spin isomeric RNBs and scattering

a)See Table II; early versions of most work are presented in [RNB1, 2, 3, 4] of Table II

TABLE II: Bibliography for TABLE I (and Related References)

[Li86] "Focusing of Multiply-Charged Energetic Ions Using Solenoidal B and Radial E Fields," W.Z. Liu, F.D. Becchetti, and R. Stern, Rev. Sci. Instrum. **58** (1986) 220.

[St87a] "Tests of a Large Air-core Superconducting Solenoid Reaction Product Spectrometer," R.L. Stern, F.D. Becchetti, T. Casey, J.W. Jänecke, P.M. Lister, W.Z. Liu, D.G. Kovar, R.V.F. Janssens, M.F. Vineyard, W.R. Phillips and J. Kolata, Rev. Sci. Instrum. **58** (1987) 1682.

[Ko89] "A Radioactive Beam Facility using a Large Superconducting Solenoid," J.J. Kolata, A. Morsad, X.J. Kong, R. Warner, F.D. Becchetti, W.Z. Liu, D.A. Roberts, and J.W. Jänecke, Nucl. Instrum. Meth. **B40/41** (1989) 503.

[Be89] "Measurements of Discrete Nuclear Reactions Induced by a Radioactive ^{8}Li Beam," F.D. Becchetti, W.Z. Liu, D.A. Roberts, J.W. Jänecke, J.J. Kolata, A. Morsad, X.J. Kong, and R.E. Warner, Phys. Rev. C **40** (1989) R1104.

[Li89] "Large-aperture, Axially-symmetric Ion-optical Lens Systems Using New Types of Electrostatic and Magnetic Elements," W.Z. Liu and F.D. Becchetti, Rev. Sci. Instrum. **60** (1989) 1228.

[Sm90] "Production and Use of Radioactive ^{7}Be Beams," R. J. Smith, J. J. Kolata, K. Lamkin, A. Morsad, K. Ashktorab, F. D. Becchetti, J. Brown, J. W. Jänecke, W. Z. Liu and D. A. Roberts, Nucl. Instrum. Methods A **294** (1990) 26.

[Be90] "Production of an Isomeric, Excited Radioactive Nuclear Beam," F. D. Becchetti, K. Ashktorab, J. A. Brown, J. W. Jänecke, D. A. Roberts, J. van Klinken, W. Z. Liu, J. J. Kolata, K. Lamkin, R. J. Smith and R. E. Warner, Phys. Rev. C **42** (1990) R801.

[Sm91a] "Scattering of ^{6}He from ^{197}Au, ^{nat}Ti, ^{27}Al, ^{nat}C and ^{9}Be at E=8-9 MeV," R.J. Smith, J. J. Kolata, K. Lamkin, A. Morsad, K. Ashktorab, F. D. Becchetti, J. B. Brown, J. W. Jänecke, W. Z. Liu and D. A. Roberts, Phys. Rev. C **43** (1991) 761.

[Be91] "Production and use of ^{6}He, ^{7}Be, ^{8}Li, ^{12}B and Metastable Nuclear Beams,"F. D. Becchetti, J. A. Brown, K. Ashktorab, J. W. Jänecke, W. Z. Liu, D. A. Roberts, R. J. Smith, J. J. Kolata, K. Lamkin, A. Morsad and R. E. Warner, Nucl. Instrum. Methods B **56/57** (1991) 554.

[Sm91b] "Elastic and Inelastic Scattering of ^{8}Li from ^{12}C," R.J. Smith, J. J. Kolata, K. Lamkin, A. Morsad, F. D. Becchetti, J. A. Brown, W. Z. Liu, J. W. Jänecke, D. A. Roberts and R. E. Warner, Phys. Rev. C **43** (1991) 2346.

[Br91] "Coulomb Excitation of ^{8}Li," J. A. Brown, F. D. Becchetti, J. W. Jänecke, K. Ashktorab, D. A. Roberts,J. J. Kolata, R.J. Smith, K. Lamkin, and R. E. Warner, Phys. Rev. Lett. **66** (1991) 2452.

[Be92] "The Reaction Rates for $^{8}Li(p,\alpha)$ and $^{8}Li(p,\ n\ \alpha)$ and their Effect on Primordial Nucleosynthesis," F.D. Becchetti, J.A. Brown, W.Z. Liu, J.W. Jänecke, D.A. Roberts, J.J. Kolata, R.J. Smith, K. Lamkin, A. Morsad, R.E. Warner, R.N. Boyd, and J.D. Kalen, Nucl. Phys. A **550** (1992) 507.

[Fl92] "Applications of Kinematics to Nuclear Reactions Using Radioactive Ion Beams," N.R. Fletcher, Nucl. Instrum. and Meth. in Phys. Res. A **316** (1992) 143.

[Ca93] "Cross Sections for the Primordial Reaction ^{8}Li(p,n)^{8}Be(g.s.) at $E_{c.m.}$ = 1.5 MeV," D.D. Caussyn, N.R. Fletcher, K.W. Kemper, E.E. Towers, J.J. Kolata, K.L. Lamkin, R.J. Smith, F.D. Becchetti, J.A. Brown, J.W. Jänecke, D.A. Roberts and .L. Gay, Phys. Rev. C **47** (1993) 387.

[Be93] "Systematics of ^{8}Li-Induced Radioactive Beam Reactions: E = 13–20 MeV," F.D. Becchetti, W.Z. Liu, K. Ashktorab, J.F. Bajema, J.A. Brown, J.W. Jänecke, D.A. Roberts, J.J. Kolata, K.L. Lamkin, A. Morsad, R.J. Smith, and X.J. Kong, Phys. Rev. C **48** (1993) 308.

[Ba93] "^{2}H Induced Reactions on ^{8}Li and Primordial Nucleosynthesis," M.J. Balbes, M.M. Farrell, R.N. Boyd, X. Gu, M. Hencheck, J.D. Kalen, C. Mitchell, J.J. Kolata, K. Lamkin, R. Smith, R. Tighe, K. Ashktorab, F.D. Becchetti, J.B. Brown, D.A. Roberts, T.-F. Wang, D. Humphreys, G. Vourvopoulos, and M.S. Islam, Phys. Rev. Lett. **71** (1993) 3931.

[O'D94] "Mass Separator Using a Large Solenoid "Lens" with Time of Flight, and Position-Sensitive Detectors," T.W. O'Donnell, E. Aldredge, F.D. Becchetti, J.A. Brown, P. Conlan, J. Jänecke, R.S. Raymond, D.A. Roberts, R.S. Tickle, H.C. Griffin, J. Staynoff, and R. Ronningen, Nucl. Instrum. Meth. A **353** (1994) 215.

[Ba95] "Cross Sections and Reaction Rates of $d+^8Li$ Reactions Involved in Big Bang Nucleosynthesis," M.J. Balbes, M.M. Farrell, R.N. Boyd, X. Gu, M. Hencheck, J.D. Kalen, C.A. Mitchell, J.J. Kolata, K. Lamkin, R. Smith, R. Tighe, K. Ashktorab, F.D. Becchetti, J. Brown, D. Roberts, T.-F. Wang, D. Humphreys, G. Vourvopoulos, and M.S. Islam, Nucl. Phys. A **584** (1995) 315.

[Br95] "Proton Scattering from an Excited Nucleus ($^{18}F^m$, $J^\pi = 5^+$, $E_x = 1.1$ MeV) Using a γ-Ray Tagged Secondary Isomeric Nuclear Beam," J. Brown, F.D. Becchetti, J. Jänecke, D.A. Roberts, D.W. Litzenberg, T.W. O'Donnell, R.E. Warner, N.A. Orr, and R.M. Ronningen, Phys. Rev. C **51** (1995) 1312.

[Gu95] "The $^8Li(\alpha,\ n)^{11}B$ Reaction and Primordial Nucleosynthesis," X. Gu, R.N. Boyd, M.M. Farrell, J.D. Kalen, C.A. Mitchell, J.J. Kolata, K. Lamkin, K. Ashktorab, F.D. Becchetti, J. Brown, D. Roberts, K. Kimura, I. Tanihata, and K. Yoshida, Phys. Lett. B **343** (1995) 31.

[Wa95] "Elastic Scattering of 10 MeV ^6He from ^{12}C, ^{nat}Ni, and ^{197}Au," R.E. Warner, F.D. Becchetti, J.W. Jänecke, D.A. Roberts, D. Butts, C.L. Carpenter, J.M. Fetter, A. Muthukrishnan, J.J. Kolata, M. Belbot, K. Lamkin, M. Zahar, A. Galonsky, K. Ieki, and P. Zecher, Phys. Rev. C **51** (1995) 178.

[Be96] "Quasi-elastic Neutron Transfer from a Low-Energy Radioactive-Beam: $^8Li +^{58} Ni$, $E(^8Li) = 19.6$ MeV," F.D. Becchetti, J.F. Bajema, K. Ashktorab, J.A. Brown, J.W. Jänecke, D.A. Roberts, J.J. Kolata, K. Lamkin, R.J. Smith, and R.E. Warner, in [RNB3] (submitted to Phys. Rev. C).

[Sc96] "Sub-Coulomb Dissociation of ^8B," J. von Schwarzenberg, J.J. Kolata, D. Peterson, P. Santi, M. Belbot, and J.D. Hinnefeld, Phys. Rev. C **53** (1996) R2598.

Conference Proceedings-Invited and Contributed Papers (1987-96)

[StC87] "Test of A Large Air-core Superconducting Solenoid as a Nuclear Reaction Product Spectrometer and Radioactive Beam System," R.L. Stern, F.D. Becchetti, T. Casey, J.W. Jänecke, P.M. Lister, W.Z. Liu, D.G. Kovar, R.V.F. Janssens and J.J. Kolata, *5th International Conference on Nuclei Far From Stability*, Rosseau Lake, Ontario, Canada (September 14-19, 1987), p. N7.

[BeC88] "A Radioactive Beam Facility Using a Large Superconducting Solenoid," F. Becchetti *et al., International Symposium on Heavy Ion Physics and Nuclear Astrophysical Problems*, Tokyo, Japan (July 1988).

[RNB1] *Radioactive Nuclear Beams, The First International Conference* Berkeley, CA (October 1989), edited by W.D. Myers, J.M. Nitschke, and E.B. Norman, World Scientific (Singapore 1989).

[FaC91] "Study of Deuteron Induced Reactions on ^{8}Li of Interest to Primordial Nucleosynthesis", M.M. Farrell *et al., Int'l. Conf. on Unstable Nuclei in Astrophysics* (Tokyo, June 1991).

[RNB2] *Proc. Second International Conf. on Radioactive Nuclear Beams*, Th. Delbar ed., Lovain-la-Nauve, Belgium 1991; Adam Hilger (Bristol 1992).

[RNB3] *Proc. Third International Conference on Radioactive Nuclear Beams,* Michigan State Universitiy, eds. D. Morrisey *et al.*, Michigan State University, East Lansing, MI, 24-27 May 1993; Editions Frontieres (Gif-sur-Yvette Cedex, France 1993).

[O'DC94] "Mass Separator Using a Large Solenoid "Lens" with Time of Flight, and Position-Sensitive Detectors," T.W. O'Donnell, E. Aldredge, F.D. Becchetti, J.A. Brown, P. Conlan, J. Jänecke, R.S. Raymond, D.A. Roberts, R.S. Tickle, H.C. Griffin, J. Staynoff, and R. Ronningen, *Proc. 1994 Symp. on Radiation Measurements and Applications*, Ann Arbor, MI; Nucl. Instrum. Meth. A **353** (1994) 215.

[BeC96] "TwinSol: A Dual Superconducting Air-core Solenoid Apparatus for the Production and Utilization of Radioactive Nuclear Beams," F.D. Becchetti, M.Y. Lee, T.W. O'Donnell, D.A. Roberts, J.J. Kolata, and M. Wiescher, *Proceedings Fourth International Conference on Radioactive Nuclear Beams*, Omiya, Japan (June 3-7, 1996).

[RNB4] *Proc. Fourth International Conference on Radioactive Nuclear Beams*, Omiya, Japan (June 3-7, 1996), in press.

[St87b] "Design and Utilization of an Air-core Superconducting-Solenoid Nuclear-Reaction-Product Spectrometer," R.L. Stern, Ph.D. Thesis, University of Michigan, UM-NP-RS-87-1 (1987). [Includes research done at ANL-ATLAS (Argonne, IL)].

[Li90] "Production and Use of Radioactive Ion Beams for Measurements of Nuclear Reactions," W.Z. Liu, Ph.D. Thesis, University of Michigan (1990).

[Br93][b)] "Production and Use of Isomeric Beams," J.A. Brown, Ph.D. Thesis, University of Michigan (1993). [Includes research done at the NSCL (MSU-E. Lansing, MI)].

[La96] "The Study of the d+^{8}Li Reaction at Astrophysical Energies," K. Lamkin, Ph.D. Thesis, University of Notre Dame (1996).

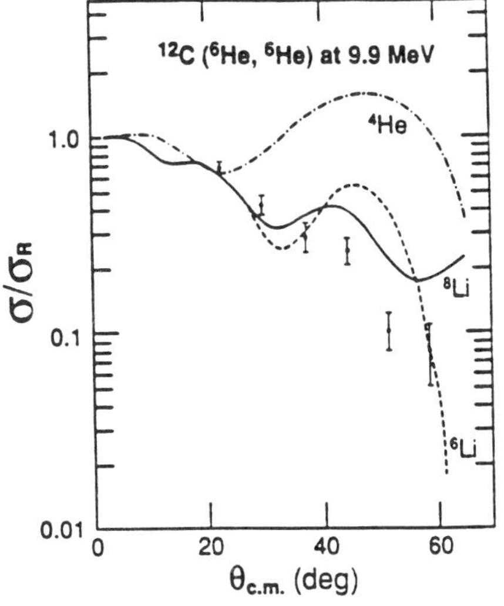

Figure 3. Elastic scattering of 10 MeV ^{6}He from ^{12}C compared with ^{4}He, ^{6}Li and ^{8}Li OM predictions ([Wa95], Table II).

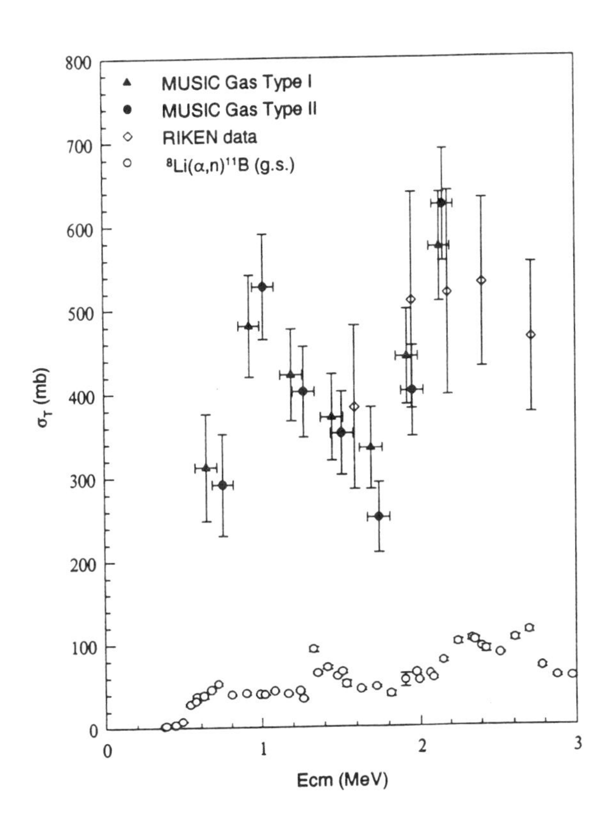

Figure 2. Cross section for ^{8}Li$(\alpha,n)^{11}$B$^* + ^{11}$B$_{g.s.}$ (top) compared with ^{8}Li$(\alpha,n)^{11}$B$_{g.s.}$ (from [Gu95], Table II).

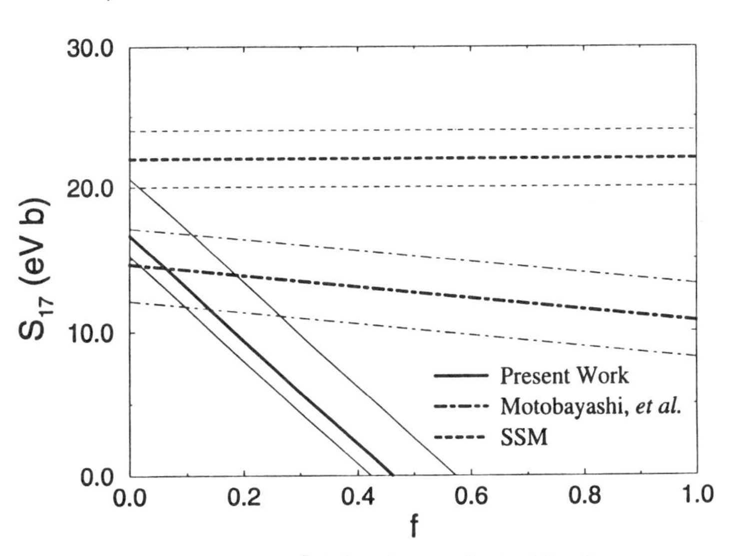

Figure 4. Plot of the ^{7}Be(p,γ) astrophysical S_{17} factor extrapolated to solar energies, as a function of the $E2$ component of the breakup yield, for the recent UND work ([Sc96], Table II), the data of T. Motobayashi *et al.*, [Phys. Rev. Lett. **73**, 2680 (1994)] and the standard solar model.

RESULTS OF THE RIST EXPERIMENT TO DESIGN A HIGH POWER TARGET FOR THE PRODUCTION OF RADIOACTIVE NUCLEAR BEAMS

C.J. Densham[a], J.R.J. Bennett[a], R.A. Burridge[a], T.A. Broome[a], P. Drumm[a], W.R. Evans[a], I.S.K. Gardner[a], M. Holding[a], G.M. McPherson[a], G.R. Murdoch[a], V. Panteleev[a], T.G. Walker[a], T.W. Aitken[b], J. Kay[b], S. Metcalf[b], H. Price[b], D.D. Warner[b], H. Ravn and the ISOLDE Group[c]

[a] *CLRC, Rutherford Appleton Laboratory, Chilton, Oxfordshire OX11 0QX, UK*
[b] *CLRC, Daresbury Laboratory, Warrington, WA4 4AD, UK*
[c] *CERN, CH-1211 Geneva 23, Switzerland*

The results of the RIST experiment, to design a high power tantalum foil target for the production of radioactive nuclear beams (RNB), are presented. The target was constructed from 6000 tantalum foil discs (25 μm thick) and spacer washers, diffusion bonded together to form a rigid structure. A target of similar geometry was successfully tested at the ISOLDE facility, CERN, and produced RNB yields and release times at least as good as a normal ISOLDE target. Tests using the ISIS proton beam demonstrated a method to overcome high voltage fluctuations induced by the pulsed proton beam. Operation in a mass separator produced beams of stable ions from a surface ioniser. Thermal tests using electron beam heating showed that the finned target surface gives an enhanced emissivity which radiates 24 kW (design value) at 2300 K, equivalent to running with the 800 MeV proton beam from ISIS at a current of 100 μA.

INTRODUCTION

Various proposals [1] have been made world-wide for a high intensity facility for accelerated radioactive beams. A standard production mechanism is by proton spallation of a thick target, but a limit on the achievable intensity lies with the technical difficulty of producing a target and ion source capable of withstanding the energy deposited by the primary proton beam. The objective of the RIST experiment [2,3] was to develop the technology required for a high power tantalum foil target and ion source. Radioactive beams were to be produced by bombardment of the target with an 800 MeV proton beam at the ISIS facility [4]. This would show if the yields were comparable with those from ISOLDE [5], CERN, at the same beam current, and whether the radioactive beams increase in proportion to the proton current up to 100μA.

KEY PROBLEM AREAS FOR TARGET DESIGN

- Power dissipation from the target at high proton beam currents
- Additional heating of target at low beam currents
- Structural integrity of the target and ion source assembly
- Voltage stability during proton beam pulses
- Fast and efficient transport of radioactive products to ion source
- High current operation of ion source

TARGET DESIGN AND MANUFACTURE

The target comprises a stack of 25 μm thick tantalum foil discs diffusion bonded together to form a rigid cylinder 40 mm in diameter. To allow the radioactive products to diffuse out, each foil is separated from its neighbour by a tantalum washer. The energy profile dE/dx generated by the proton beam decreases along the length of the target, so to achieve a uniform power deposition the thickness of the spacer washers between the individual foils is reduced along the length. A 10 mm diameter hole is laser cut into the centre of each target foil to create a channel for the effusion of the products to the ion source. The central hole also reduces the radial temperature gradients in the foils.

The target is cooled by thermal radiation to a water cooled jacket. To increase the effective emissivity of the surface, the entire length of the target is covered with 1 mm high fins spaced 0.3 mm apart [6, 7]. The finned surface is achieved by inserting larger diameter discs at regular intervals into the assembly. The target is closed at the ends

CP392, *Application of Accelerators in Research and Industry*, edited by J. L. Duggan and I. L. Morgan
AIP Press, New York © 1997

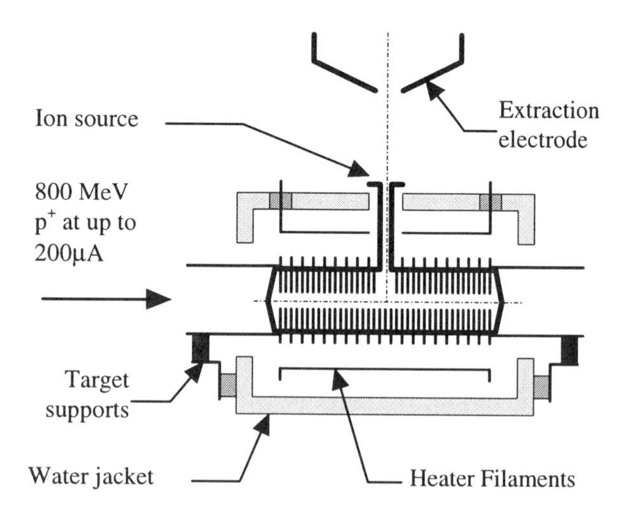

FIGURE 1. Schematic diagram of target within front end assembly.

with 0.1 mm thick end windows pressed into a conical shape to reduce the thermal stresses. The entire foil assembly is diffusion bonded together by sealing it in an evacuated nickel can and applying a 125 MPa pressure for over 15 hours at 900°C.

FRONT END ASSEMBLY

A schematic diagram of the tantalum target, tungsten ion source and water cooled copper jacket is shown in Fig. 1, and a photograph is shown in Fig. 2. Seven tungsten filaments run along the 180 mm active length of the target. These heat the target by electron bombardment during conditioning off-line, and at low proton beam currents when on-line. An ISOLDE tungsten surface ionization source is fitted into a hole at the centre of the target. Positive ions are extracted from the ion source at 30kV by an earthed extraction electrode and focussed into a beam by an electrostatic Einzel lens.

The target, water jacket and extraction optics are mounted beneath a flange and contained within a water cooled vacuum bucket. The assembly is supported beneath a 4 m high steel shield plug through which pass the radioactive beam pipe and services. The front end assembly is tilted at an angle of 10° and the beam is deflected to the vertical by an electrostatic bend just before it enters the shield plug. This ensures that there is no direct shine path for radiation from the target though the beam pipe. The front end assembly is lowered into the ISIS target station and a shielding bunker built above it. The bunker incorporates a 90° electrostatic bend, bringing the beam into the horizontal plane where it enters the isotope separator and beamline to the detector station.

OFFLINE THERMAL TESTS

The target and ion source were fully conditioned and tested off-line by electron beam heating from the filaments surrounding the target. The target voltage was raised to 10 kV and a beam of stable potassium extracted from the ion source and focussed through the separator. Accurate target temperature measurements of the inside of the ion source tube were made using an optical pyrometer. Plotted in Fig. 3 is the ‘effective’ emissivity of the target, calculated from the temperature and power dissipated during the tests. The effective emissivity was found to rise steadily with temperature to over 0.7 at around 2000°C. At this temperature the target radiated 24 kW, equivalent to intercepting around 100 μA of the 800 MeV ISIS proton beam. The data shows good agreement with the curve, which shows the effective emissivity calculated by the ANSYS® finite element program for the fin geometry. The calculation used a line of best fit for the material emissivity of 0.3 at 1200 °C rising to 0.5 at 2200 °C, from measurements made during earlier heating tests.

FIGURE 2. Photograph of the RIST target, ion source and heater filaments supported within the upper water cooled copper jacket, shown with the lower jacket removed.

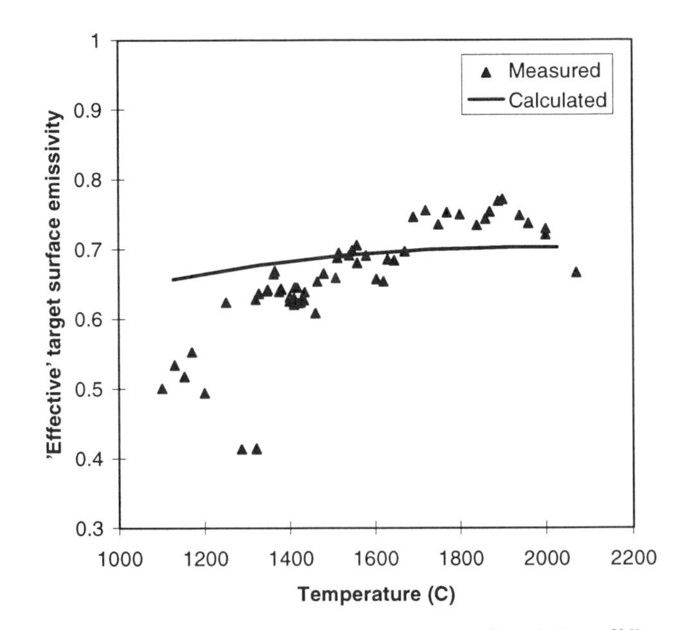

FIGURE 3. Effective emissivity of RIST target surface during off-line tests, calculated from temperature and power dissipated by target. Calculated curve result from ANSYS program using measured material emissivity values of 0.3 (1000°C) - 0.5 (2000°C).

TARGET STRUCTURAL INTEGRITY

The ISIS proton beam is circular with a parabolic distribution and at 100 μA deposits a maximum of 3.5 kW per centimetre thickness of tantalum. For a 40 mm diameter foil with a 10 mm hole in the centre cooled from the outside the temperature gradient is around 448°C. If it assumed that each target foil deforms elastically and is only constrained by its neighbours to deflect within a plane, then the internal and external tangential stresses are 100 N/mm^2 (compressive) and 117 N/mm^2 (tensile) respectively. In practice, at high temperatures plastic deformations would relieve these stresses, but it was not clear how many thermal cycles a target, in particular the diffusion bonds, could be expected to withstand.

An off-line experiment was performed to simulate these thermal stresses using electron beam heating. Three test samples were diffusion bonded together from 25 μm foils as part of the development required to manufacture a target by this method. The test samples were heated by a filament passing through the centre. The central hole was increased in diameter to allow for the more concentrated heating that results from electron bombardment compared with proton beam heating.

Each test target was able to withstand the power and temperature gradient that would be generated by a 100 μA proton beam at temperatures in excess of 2000°C for a minimum of twenty heating and cooling cycles. There was no sign of excessive distortion or failure in any of the tests, and the diffusion bonds remained intact. It appears that at these high temperatures the tantalum is in a state where all the induced stresses are relieved. By the time the target has cooled down to the level where stresses can no longer be relieved, the temperatures have equilibrated so that the gradients are low enough for the residual stresses to be withstood by the material. The conclusion from this experiment is that a RIST target is capable of withstanding many cycles in the ISIS proton beam at 100 μA.

ON-LINE RIST TARGET TESTS AT ISOLDE

Two on-line tests of the RIST style target geometry have now been performed at ISOLDE. A 25 μm thick foil target was run in December 1994 and a 0.1 mm foil target in November 1995. In both cases a standard ISOLDE 20 mm diameter target tube was packed with circular foils to reproduce as closely as possible the RIST geometry. The proton beam was de-focused to give a standard deviation of 2.72 mm. To allow the foils to intercept sufficient current, the central hole diameter was 7 mm. A hot tungsten surface ioniser was used, as for the RIST experiment. Release curves were obtained for ^{8}Li, ^{9}Li, ^{25}Na, ^{26}Na, ^{46}K, ^{80}Rb, ^{124}Cs, ^{144}Eu and ^{176}Yb.

During the 25 μm foil test a beam of natural Yb was introduced into the ion source to simulate the expected load due to rare earth atoms building up in the RIST target. At the total expected rare earth current of 200 nA, the efficiency of the ion source was not affected for the observed beams of ^{25}Na and ^{176}Yb.

The central hole in the targets meant that only around half of the beam current was intercepted, but total yields per proton intercepted were comparable with the ISOLDE standard results. In addition the response curves were faster, even in many cases for the 0.1mm foil target. It appears from this that the open channel design successfully increased the speed of effusion of isotopes from the target.

MONTE CARLO SIMULATION OF TARGET DELAY TIMES

It was necessary to make the RIST target almost double the 20mm diameter of ISOLDE targets. It was not known whether the increased target volume would cause a greater delay and consequently offset the gain from the larger primary proton current and any benefit from changes to the foil geometry. A Monte Carlo program was written [6, 8] to model the geometry and investigate the effect on the release curves of diffusion, decay, sticking delays and the time-of-flight for different masses at different temperatures.

Published data on diffusion coefficients and sticking times is scarce for the alkali and rare earth metals in tantalum [9-12]. Consequently the model was written to allow the diffusion coefficient and sticking times to be adjusted to fit the code results to the release curves for the

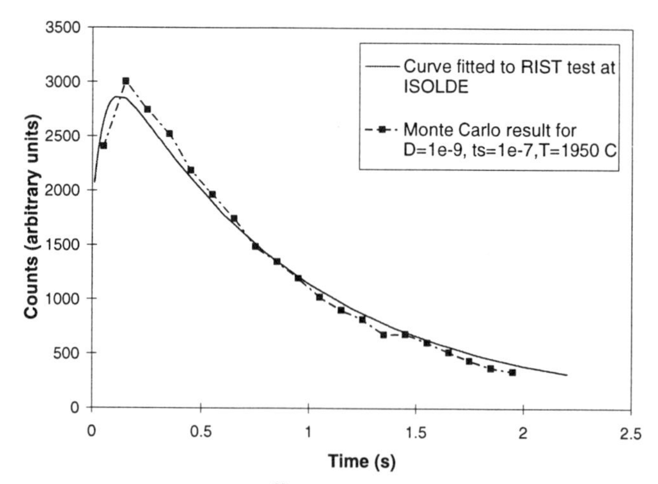

FIGURE 4. Release curve of ^{80}Rb from 25 μm foil RIST target tested at ISOLDE, plotted against Monte Carlo results with diffusion coefficient D (m^2/s), sticking time ts (s) and temperature T (°C) indicated.

different isotopes produced. Both the 25 μm foil and 0.1 mm foil targets tested at ISOLDE were modelled, and compared with the experimental results. Results for ^{80}Rb from the 25 μm foil target are shown in Fig. 4. Estimates could be made of diffusion coefficients and sticking times for many of the elements investigated [6, 8]. The sticking times agreed well with those by Kirchner [12] but diffusion coefficients were found to be typically an order of magnitude greater than those measured by Beyer et al [9]. In general, the surface sticking time appeared to be the limiting factor for most alkali metals from the thinner foil target, while for the thicker foils the main cause of delay was diffusion.

The derived diffusion coefficients and sticking times have been used to model the full-size RIST target geometry. Preliminary results show that the release times will be noticeably delayed compared with the 20 mm diameter target tested at ISOLDE, and will probably be similar to the release from the standard ISOLDE rolled foil geometry. So far the work has indicated the critical target delay factors, and it is intended to continue tests at ISOLDE in conjunction with the Monte Carlo calculations to optimise target geometries for the production of maximum yields of particular isotopes.

VOLTAGE STABILITY

The pulsed proton beam can cause voltage fluctuations of the target, which is a severe problem for a high resolution separator. The ISOLDE front end operates in air, and solves the voltage fluctuation problem by switching off the front end voltage for the duration of the pulse, however this was not thought to be a viable solution for the ISIS beam which is pulsed at up to 50Hz.

To test the effect, an experiment was performed with the ISIS proton beam at a low repetition rate. The beam was passed through the walls of concentric tubular electrodes at up to 30kV, with a vacuum surrounding the high voltage elements. The problem was found to be considerably reduced by the vacuum, however fluctuations were caused by ionisation of the insulation in the adjacent cables by the intense radiation. This effect was eliminated by the addition of a sufficiently large capacitor across the terminals.

CONCLUSIONS

It can be seen that all key problem areas in the design of a high power radioactive beam target as outlined at the start of this paper have been successfully addressed, without there being a need for a run on the ISIS proton beam. A foil geometry was designed to give sufficient power dissipation and a fast response time, and was successfully tested at ISOLDE. A vacuum tight target with an enhanced surface emissivity was diffusion bonded together and survived vigorous heating tests at the design temperature and heat load for a target subject to a 100 μA proton beam.

REFERENCES

1. See for example J.M.Nitschke, "Accelerated Radioactive Nuclear Beams: Existing and Planned Facilities", in *Proc. of 13th Int. Conference on Cyclotrons and their Applications,* Vancouver, 6-10 July 1992, pp 713-720.
2. J.R.J Bennett, R.A.Burridge, T.A.Broome, C.J.Densham et al, "The RIST Project at RAL", in *Proceedings of the 4th European Particle Accelerator Conference* 1994, World Scientific pp 1415 -1417.
3. J.R.J Bennett, R.A.Burridge, T.A.Broome, C.J.Densham et al, "Progress of the RIST Project", in *Proceedings of the 5th European Particle Accelerator Conference* 1996, to be published.
4. J. L. Finney, "ISIS: A Resource for Neutron Studies of Condensed Matter", Europhysics News, Vol. 20, 1989.
5. H. J. Kluge (Editor), "ISOLDE Users Guide", CERN 86-05, 1986.
6. C.J.Densham, DPhil thesis (in preparation).
7. J.R.J. Bennett, C.J. Densham, P.V. Drumm et al, "The Design and Development of the RIST Target", presented at the 13th International Conference on Electromagnetic Isotope Separators 1996, to be published in Nucl. Inst. & Methods B.
8. C.J.Densham, C.Thwaites and J.R.J.Bennett, "Critical Parameters for the Release Time of a RIB Target and Ion Source System", to be published in Nucl. Inst. & Methods B as for ref.[7].
9. G.J. Beyer, W.D. Fromm and A.F. Novgorodov, "Tracer Diffusion of Different Nuclear Reaction Products in Polycrystalline Tantalum", Nucl. Inst. & Meth. 146 (1977) 419-430.
10. R. Kirchner, "On the Release and Ionisation Efficiency of Catcher-Ion-Source Systems...", Nucl. Inst. & Meth. B70 (1992) 186-199.
11. A.J. Aas, "Diffusion in ISOLDE Target", thesis submitted 1994 to University of Oslo.
12. R. Kirchner, O.Klepper, D. Marx, G.E. Rathke and B. Sherrill, "An Ion Source with Storage Capability for Bunched Beam Release...", Nucl. Inst. & Meth. A247 (1986) 265-280.

THE SPIRAL RADIOACTIVE ION BEAM FACILITY AT GANIL

B. Launé
R&D Accélérateurs
Institut de Physique Nucléaire
91406 Orsay Cedex, France
and
SPIRAL Group
GANIL
BP 5027
14021 Caen Cedex, France

ABSTRACT

The interest for the study of nuclei far from stability has started at GANIL as early as 1984 with projectile fragmentation experiments using the LISE fragment separator, which has been constantly improved ever since. Another dedicated device for fragment production (SISSI) was added in 1994. These equipments allowed many pioneering experiments. In order to expand the range closer to the Coulomb barrier, it was decided to build the SPIRAL facility, based on ISOL production technique associated with a cyclotron for the post-acceleration. In parallel the THI program was initiated in order to increase the performance of GANIL as a primary beam producer: up to 6 kW of beam power. The SPIRAL facility, that is expected to produce its first post-accelerated radioactive beams in late 1998, will be described as well as the future prospects.

1. INTRODUCTION

Investigation of nuclei far from the stability has already allowed the discovery of new phenomena in the field of nuclear physics.[1] The availability of exotic nuclei is expected to open up new ways for the exploration of the nuclear structure problems as well as to provide for advances in related fields such as astrophysics or solid state physics. This is the reason why GANIL, a facility producing a large spectrum of heavy ion beams of intermediate energy (25-95 MeV/nucleon) since 1983,[2,3] has pioneered this field as early as 1984, using the projectile fragmentation method in combination with the LISE separator[4,5] and later the SPEG spectrometer.[6] The dedicated SISSI[7] device was put in operation in 1994 in order to improve both the secondary beam quality and the rate of production. It is composed of two 10 T-superconducting solenoids, the first one focussing the beam on the target, the second one collecting the products with a large angular acceptance. GANIL has consequently made available to its users competitive medium energy radioactive beams (50-95 MeV/A) in combination with state-of-the-art equipements.

However, producing secondary beams through the fragment separation method reaches its optimum when these secondary beams have a velocity similar to the primary beam's; this implies increasing losses in terms of intensity and beam quality as the beam is slowed down. For lower energy, the ISOL (Isotopic Separation On-Line), pioneered and developped at ISOLDE (CERN)[8] is more adequate. The primary beam is stopped in a thick target where the radioactive species are created through spallation and fragmentation processes. They may then diffuse out of the hot target (1800 to 2300 K), be ionized in an ion source (i. e. an Electron Resonance, Source) and be further accelerated, thus producing RIB (Radioactive Ion Beams) of high optical quality and which energy can be adjusted. It was then decided to build at GANIL a facility using the ISOL method, SPIRAL,[9-11] with the present GANIL cyclotrons as primary beam producer. The intrinsic separation of the cyclotron is a particularly useful feature, as demonstrated in the Louvain-la-Neuve Facility.[12] In parallel a program has been launched to increase the primary beam intensities (THI).[13] A target station will be developped and a cyclotron has been chosen as post-accelerator allowing final energies in the range 2-25 MeV/A, while the RIB will be made available at the source energy for material or atomic science. SPIRAL stands for *Séparateur et Postaccélérateur d'Ions Radioactifs Accélérés en Ligne.*

In the following, the SPIRAL project will be described in more details, but it should be stressed that GANIL will soon be a unique facility for the production of radioactive beams, either by fragmentation method, chemically insensitive, more suitable for very short lifetime species and in the intermediate energy range, either by the ISOL method, which produces high intensity and good quality RIB but chemically sensitive, closer to the Coulomb barrier (up to 25 MeV/A). In the case of the ISOL method, an advantage of the GANIL facility over the light ion beam technique will be the possibility to vary not only the energy of the projectile but also its nature, allowing the optimization of the production rate.

CP392, *Application of Accelerators in Research and Industry,* edited by J. L. Duggan and I. L. Morgan
AIP Press, New York © 1997

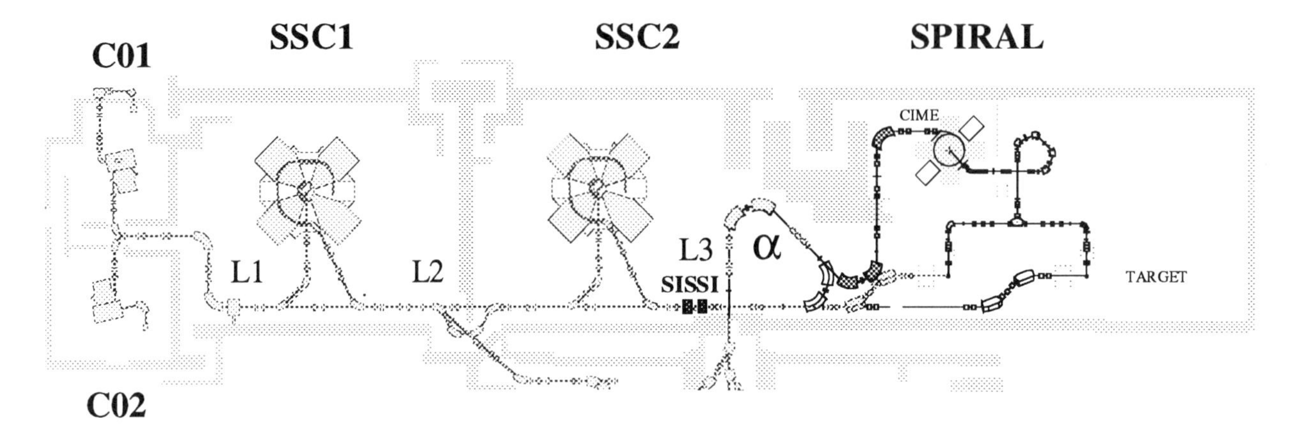

Fig. 1 Schematic view of the GANIL cyclotrons and the SPIRAL addition

2. THE SPIRAL PROJECT

The layout of the SPIRAL installation is shown in figure 1. The primary beam from the present Separated Sector Cyclotrons (SSC) may be directed down to a vault housing two target stations. The RIB's are then axially injected into the cyclotron, by means of the Low Energy beam line, providing the matching for injection and a 5.10^{-3} resolution in terms of charge-to-mass ratio for a 80π mm.mrad emittance (possibly a few 10^{-4} if the LAMS spectrometer[14] is later built). The beam is then extracted and transported to the experimental areas via a "S"-shaped spectrometer, re-using the second half of the α spectrometer, allowing beam purification through energy-loss method. That scheme makes use of all the existing GANIL experimental areas.

3. GANIL INTENSITY INCREASE PROGRAM

The present GANIL cyclotrons produce beams with a power intentionnaly limited to 400 W. The combination of an ECR4 (Electron Cyclotron Resonance Ion Source) operating at 14.5 GHz and an injection stage at 100 kV makes possible the acceleration of higher intensity beams.[15] The goal of the THI operation *(Transport de Haute Intensité)* is to boost the intensity by a factor of 15 for species up to argon (2 10^{13} pps) and hence cope with 6 kW beam power. To reach it, i) the beam quality must be improved, ii) the losses must be controlled, iii) the lifetime of the stripper foils must be extended. For that purpose, a rebuncher (R2) has been installed between the two SSC's, new diagnostics and sensors are being added and a movable stripper foil system is being tested in order to first accelerate high power beams by the end of 1996 and be ready for routine production for SPIRAL in 1998.

4. TARGET AND SOURCE

An extended development program has been set up for the production of intense multicharged RIB's.[16] Two main axes of research are persued: external target for the production of gases, mainly noble, where the target is linked by a transfer tube to an ECR source; the second one uses an internal target and is more devoted to condensable elements.

4.1. External Target

This has already been successfully tested at the SIRa test bench, installed on one of the present GANIL experimental station, and revealed good yields for the production of gases (see table 1). A large effort has also been made on the target design: it should not only withstand the beampower, but also have favorable micro- and macroscopic structure to release rapidly the radioactive products. As the heavy ions have a narrow Bragg peak, the power is better distributed using a conical target.[17] This carbon target has been successfully tested on a 6 kW proton beam at Louvain-la-Neuve. The experimental behavior has been found to agree with the simulation. This target, associated with a simplified NanoganII ECR source with permanent magnets will initially equip the first station.

4.2. Internal Target

For the production of condensable species, the target-source assembly must be as compact as possible and kept at highest possible temperature. Permanent magnets for the source are very attractive but it was shown that they could not stand the high neutron flux. A new design is being tested, SHyPIE, where the permanent magnets are located out of the 0-90° area. Another promising solution, proposed at Grenoble for PIAFE[18] is the use of a very simple 1+ source, followed by an ECR source for charge state breeding.

RIB (charge state and life time)	primary beam	yield (pps) with present intensities	projected SPIRAL intensities (pps)	projected SPIRAL intensities (pA)
^{77}Kr (10+) 74.4 ms		6.1×10^6	9.2×10^7	150
^{76}Kr (10+) 14.8 h		3.6×10^6	5.3×10^7	85
^{75}Kr (10+) 4.3 ms	^{78}Kr, 73 MeV/A	1.8×10^5	2.7×10^6	4
^{74}Kr (9+) 11.5 ms	35 pnA	7.0×10^4	1.1×10^6	1.5
^{73}Kr (9+) 27.0 s	(2.2×10^{11} pps)	2.5×10^3	3.8×10^4	0.05
^{72}Kr (9+) 17.2 s		2.6×10^2	3.9×10^3	0.005
^{35}Ar (8+) 1.77 s		6.2×10^7	4.7×10^8	600
^{34}Ar (7+) 844.5 ms	^{36}Ar, 96 MeV/A	1.4×10^6	1.1×10^7	10
^{33}Ar (8+) 173 ms	115 pnA	1.8×10^4	1.4×10^5	0.2
^{32}Ar (8+) 98 ms	(7.2×10^{11} pps)	1.0×10^2	7.8×10^2	0.001
^{19}Ne (5+) 17.22 s	^{20}Ne, 96 MeV/A	5.0×10^7	3.8×10^8	300
^{18}Ne (5+) 1.67 s	208 pnA	3.1×10^6	2.3×10^7	20
^{17}Ne (5+) 109 ms	(1.3×10^{12} pps)	3.1×10^4	2.3×10^5	0.2
^{8}He (1+) 119 ms	^{13}C, 75 MeV/A 421 pnA (2.6×10^{12} pps)	2.1×10^5	1.5×10^6	0.2

Table 1 Measured yields with present GANIL intensities and projected SPIRAL yields assuming 6 kW primary beam and a 50% transmission inside the CIME cyclotron

5. CYLOTRON

CIME is a 4-sector, 2-dee, K265 room-temperature compact cyclotron. The working diagram is shown on figure 2. The average magnetic field extends between 0.75 and 1.56 T, the extraction radius is 1.5 m. The radiofrequency ranges from 9.6 to 14.5 MHz and possible harmonic modes are 2, 3, or 4, while the possibility of running harmonic 5 (down to 1.7 MeV/A) using the same central region, is being investigated. The beam is axially injected with a maximum voltage of 34 kV. A Mueller inflector (34 mm radius) is used for harmonic 2 and 3, while a spiral inflector has been selected for harmonic 4 and 5 (45 mm radius). Pumping is accomplished by two large cryopanels associated with turbomolecular pumps. The cryopower is transported to the panels via heat pipes, a technique developed for the AGOR superconducting cyclotron.[19] Extraction is accomplished by a double electrostatic deflector and two passive magnetic channels. The magnet[20] and the vacuum chamber have been built and assembled. The first maps are being measured and have already shown a good agreement with the TOSCA[21] 3D-magnetic calculations and the correct identity of the four sectors. First acceleration of stable beams is expected in 1997.

6. TUNING SPECIFICITIES

The radioactive elements will be accelerated in a wide range of intensities (down to a few pps) and the wanted beam will be in most cases mixed with other species. Special tuning method and dedicated diagnostics tools had to be developed.[22] The 5.10^{-3} resolution of the analyzing part of the injection beam line is not sufficient to clean the beam from all the contaminants and isomeric or isobars will not even be separated by the cyclotron. Direct tuning seems difficult if not impossible, so the beam line and the cyclotron will be pretuned with an analog beam, close to the selected one in terms of charge-to-mass ratio. This pretuning can be made in parallell with the tuning of the primary beam, thus saving time. Afterwards, a shift is operated: in frequency, in magnetic field or by a combination of both. Nevertheless, diagnostics (efficient down to a few pps) are required: in the cyclotron a plastic scintillator mounted in the radial probe, scanning the acceleration, has been selected. The light is taken out of the magnetic field region by an optical fiber to a photomultiplier. The time resolution is good and it is rather cheap and robust. This scheme can be completed by silicon detectors that allow the identification ($E.\Delta E$). A test of a scintillator has been successfully performed in the present GANIL Separated Sector cyclotron and showed that isochronism can be precisely optimized with current down to 10^{2-3} pps. A identification station, based on radioactive decay, will also be installed prior to injection.

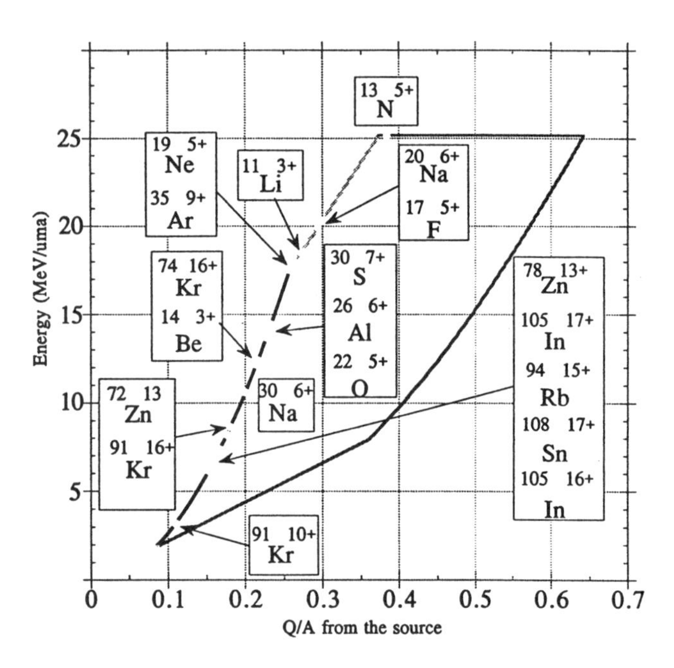

Fig. 2 Working diagram of the CIME cyclotron

7. CONCLUSIONS

Funding is provided by CNRS/IN2P3, CEA/DSM and the Lower Normandy Region. Many French laboratories contribute to the construction and/or the R&D effort. SPIRAL will become a major European user facility. It relies on two international advice commitees: a technical and a scientific one. The definition of the associated instrumentation is underway and a letter-of-intent procedure will soon be launched for the first experiments, that should start at the end of 1998.

8. AKNOWLEDGEMENTS

It is a pleasure to thank A. C. Mueller (IPN Orsay), M.-P. Bourgarel, M. Lieuvin, and A. C. C. Villari (GANIL).

9. REFERENCES

1) A. C. Mueller, B. M. Sherrill, Ann. Rev. Part. Sci. 43 (1993) 529
2) A. Joubert, Status Report on the GANIL Facility, 10th International Conference on Cyclotrons and their Applications, East Lansing, 1984
3) M. Bajard et al. GANIL Status Report, Fourth European Particle Accelerator Conference, London, 1994
4) R. Anne, D. Bazin, A. C. Mueller et al., NIM A257 (1987) 215
5) R. Anne, and A. C. Mueller, NIM B70 (1992) 276
6) L. Bianchi, B. Fernandez, J. Gastebois, A. Gilibert, W. Mittig, J. Barrette, SPEG, an Energy Losss Spectrometer for GANIL, NIM A276 (1989) 509
7) A. Joubert et al., 1991 Particle Accelerator Conference, IEEE Vol.1 (1991) 594
8) E. Hagebo et al., NIM B70 (1992) 165
9) The SPIRAL Radioactive Facility, GANIL R 94 02, May 1994
10) Radioactive Ion Beams at SPIRAL, A. C. C. Villari et al., Nuclear Physics A588 267c-272c
11) M. Lieuvin, the SPIRAL Radioactive Ion Beam Facility, 14th International Conference on Cyclotrons and their Applications, Cape Town, October 1995
12) M. Loiselet et al., in Radioactive Nuclear Beams, D. J. Morrissey Ed. Editions Frontières, Gif-sur-Yvette (1993), p. 179
13) E. Baron, Upgrading the GANIL Facilities for High-Intensity Heavy Ion beams (T.H.I. Project), 14th International Conference on Cyclotrons and their Applications, Cape Town, October 1995
14) A. Chabert, Ch. Ricaud, B. Bru, Principle of a new kind of Large Acceptance Mass Separator, NIM A351 (1994) 371
15) Ch. Ricaud et al., Commissioning of the New High Intensity Axial Injection System for GANIL, 13th International Conference on Cyclotrons and their Applications, Vancouver, July 1992
16) A. C. C. Villari et al., News from the SPIRAL Project at GANIL, Fourth International Conference on Radioactive Nuclear Beams, OMIYA, Japan, June 1996
17) P. Foury et al., to be submitted to NIM (1996)
18) C. Tamburella et al., PIAFE Int. Rep. ISN 96-15 (1996)
19) S. Buhler, A. Horbowa, the AGOR Cryopumps, 12th International Conference on Cyclotrons and their Applications, Berlin, 1989
20) M. Duval, M.-P. Bourgarel, F. Ripouteau, New Compact Cyclotron Design for SPIRAL, GANIL S 95 01
21) TOSCA, Vector Fields Ltd, Oxford, England
22) B. Launé, the Diagnostics System for the SPIRAL R.I.B. Facility, 14th International Conference on Cyclotrons and their Applications, Cape Town, October 1995

HIGH POWER TARGET APPROACHES FOR INTENSE RADIOACTIVE ION BEAM FACILITIES

W. L. Talbert[1,2,3], T. A. Hodges[2], H.-H. Hsu[4] and M. M. Fikani[1]

[1] *Amparo Corporation, Santa Fe, NM 87504*
[2] *TRIUMF, Vancouver, B.C., Canada V6T 2A3*
[3] *Oak Ridge National Laboratory, Oak Ridge, TN 37830*
[4] *Los Alamos National Laboratory, Los Alamos, NM 87545*

Development of conceptual approaches for targets to produce intense radioactive ion beams is needed in anticipation of activity for a next-generation, intense ISOL-type radioactive beams facility, strongly recommended in the NSAC 1995 Long Range Plan for Nuclear Science. The production of isotopes in vapor form for subsequent mass separation and acceleration will depend on the ability to control target temperature profiles within the target resulting from interactions of the intense production beams with the target material. A number of earlier studies have identified promising approaches which need, however, to be carefully analyzed for specific target systems. A survey will be made of these earlier concepts employing various cooling techniques, including imposition of thermal barriers between the target materials and cooling systems. Some results of preliminary analyses are summarized.

INTRODUCTION

Over the past 25 years, thick targets have been used for the production of short-lived radioisotopes with incident high-energy (≤1 GeV) proton beams. For such targets, the requirement for fast release of the produced activities leads to operation of the targets at elevated temperatures in order to facilitate rapid and copious release of the activities.

At the incident beam currents presently in use (≤1-2 μA), the desired operating temperatures are achieved through external heating of the target. This results in uniform heating of the target material, and has the additional benefit of providing controllable operating temperatures. The so-called thick target systems developed to operate at the ISOLDE facility[1] represent the basis for the robust targets required at intense radioactive ion beam facilities employing primary proton beams with intensities up to 100 μA.

At high production beam intensities, the internal heating of the target material by the production beam interactions becomes significant or even dominant, in contrast to the requirement to supply external heating for targets operated with low-intensity beams. This internal heating requires that target systems involving forced cooling be developed, even for targets that normally operate at high temperatures.

In order for heat to be removed, yet maintain the target material at elevated temperatures, the target system must employ a thermal impedance between the target and coolant, except for target systems that can be operated under radiant cooling conditions. Also, an internally heated target using a normal incident beam intensity profile is heated more intensely in the central region, and temperature uniformity within the target material depends strongly upon the heat transport within the target itself.

In 1991, Talbert, Hsu and Prenger[2] addressed the energy deposition in high-power targets using a comprehensive charged-particle transport code,[3] and assessed cooling approaches that might be used. In particular, the use of an adjustable thermal impedance between the target and cooling system was suggested to provide control of the target operating conditions.

In this work, preliminary analyses have been made for possible thermal barriers in three types of targets; a high-temperature target (Ta), a moderate-temperature target (Ti) and a low-temperature target (molten Pb). The analyses will need validation when high-intensity proton beams are available, and they are considered merely to be guidelines for follow-on engineering of real targets.

TARGET ISSUES

The target design issues include: (1) a determination of the internal deposition of energy in the target material by the production beam; (2) specification of the target operating parameters such as operating temperature and target material characteristics; (3) the transfer of heat from the target material to the cooling system; and (4) the effects of the non-uniform beam energy deposition, both radially and longitudinally. Issues associated with possible chemical selectivity of available radioactive species associated with release, transport and ionization are not addressed in this work.

Choice of target systems

To provide a generalized approach, target systems were chosen to illustrate a range of thermal conductance requirements from the above issues of (1) the heat generated by beam energy deposition in the target materials, (2) the tolerance of these materials to temperature, and (3) knowledge of the heat transfer properties of the target materials. The high-temperature (~2200°C), moderate-power target system considered was a metallic tantalum

CP392, *Application of Accelerators in Research and Industry*, edited by J. L. Duggan and I. L. Morgan
AIP Press, New York © 1997

target (density of 8.8 g/cm^3 or 0.6 normal density). Metallic titanium (density of 2.7 g/cm^3 or 0.6 normal density) was chosen as an example of a low-power, moderate temperature (~1600°C) target. A high-power, low-temperature (≤800°C) target is the molten lead system (density of ~9.3 g/cm^3). For the metallic targets, it is assumed that the targets consist of stacked foils, hence the less-than-full densities.

The target chamber concept used is of cylindrical geometry, with target length 6 cm (arbitrarily chosen to illustrate the concept), target radius of 7 mm, a target container "case" of 0.5-mm thickness in ideal thermal contact with a thermal barrier (of small thickness, determined in the analyses) and a graphite sleeve of 1.75-cm nominal thickness coupled to the cooling jacket of inner radius 2.5 cm (the sleeve was assumed to have a thin SiC coating on the inside surface to inhibit any chemical reaction with the thermal barrier material). Mechanical stresses arising from thermal gradients were included in the analyses, where appropriate. In view of the conceptual nature of this study, and to simplify the analyses, radiative thermal losses from the ends of the target (or within the target) were not considered.

THERMAL BARRIERS AND ANALYSES

The choice of thermal barrier was dictated by the temperature gradient to be maintained, and thus the target operating temperature was controlled by the design of the thermal barrier. Three different barriers were considered, in order of increasing thermal conductivity: (1) contact thermal resistance, which has a rich literature and, so long as elastic deformations are not exceeded, is amenable to quantitative description;[4] (2) use of refractory materials, such as zirconia (ZrO_2) and carbon-bonded carbon fibers (CBCF) treated at 2400°C to stabilize the thermal conductivity;[5,6] and (3) porous metal thermal barriers, which exhibit a universal correlation between thermal and mechanical properties.[7]

While the use of a thermal barrier appears to be a trivial approach, the choice of suitable materials requires considerable analysis.

The numerical analyses utilized the LAHET code system[3] to predict beam energy deposition rates and distributions. For the thermal and stress analyses, the ANSYS code[8] was employed. Both codes have been widely benchmarked.

BEAM ENERGY DEPOSITION

The beam energy depositions in the target assemblies all showed a characteristic profile that spread out progressively down the axis of the target for an incident central (Gaussian) profile beam of 2.0-mm Gaussian parameter, resulting in a flatter profile at the exit end of the target than at the entrance. The energy deposition rates are indicated for the tantalum target in Fig. 1. The data are shown as an axial function at various mean radii, with the innermost region having the highest energy deposition rate, and spreading of the beam as it propagates down the target is apparent. Such spreading, with accompanying central energy deposition attenuation, was very evident in the higher-density targets. Note that actual targets will generally be longer than the 6-cm length used in this study, and that the energy deposition will decrease along the length more than depicted here.[9]

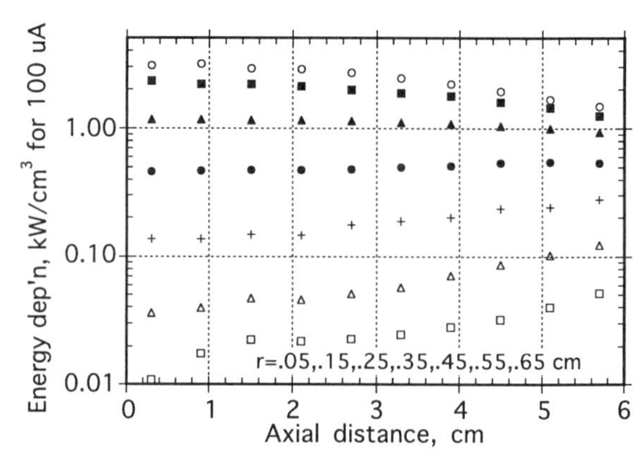

FIGURE 1. Energy deposition rate for tantalum target, from 500-MeV proton irradiation.

The total energy deposition rates for a 100-µA, 500-MeV proton beam were, for the respective target systems, 7.07 kW (Ta), 3.77 kW (Ti) and 10.18 kW (molten Pb). The energy deposition profiles were used as source terms for the finite element analyses of thermal energy transport of the targets through candidate thermal barriers, consistent with the imposed target operating temperature limits.

THERMAL PERFORMANCE

The challenge presented in the thermal analyses was to select an appropriate thermal barrier to allow transport of the heat out of the target region, while maintaining the desired target operating temperature. In the metallic targets, because of the limited radial heat transfer within the target material itself (despite their relatively high conductivities), it was determined that an annular beam (rather than a central, Gaussian-shaped beam) formed by fast circular sweep with a 5-mm radius of a tightly-focused beam (Gaussian parameter of 0.2 mm) provided a satisfactory thermal gradient within the target material, with the added benefit that the central portions of the assumed stacked metallic foils could be removed to facilitate release of the produced radioactivities.

In the three cases, two types of thermal barrier were tried. The results are summarized, case by case, below.

Tantalum target

The first approach was to use two simple contact thermal resistances between the tantalum target container and the SiC coating on the graphite sleeve, and between the graphite sleeve and the cooling jacket. With a central Gaussian profile beam, the temperature gradient within the target material was quite large, of the order of 900°C, shown in Fig. 2. Use of an annular beam reduced this radial thermal gradient to less than 200°C, and resulted in a maximum operating temperature of about 2200°C. Thermal stresses, however, on the inner thermal barrier required that either the tantalum target case be increased in thickness or that an initial gap of about 0.1 mm be incorporated in the design. A single thermal contact resistance was inadequate to sustain the required thermal gradient of nearly 1800°C.

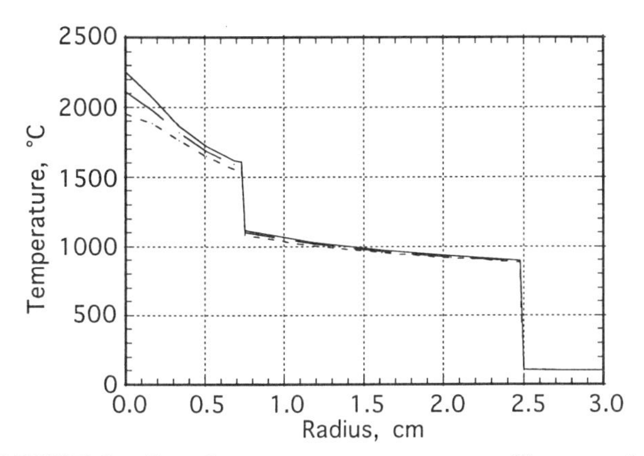

FIGURE 2. Tantalum target temperature profiles at axial distances of 0, 3, 6 cm for two contact barriers.

The second approach was to use a CBCF thermal barrier between the tantalum target container and the graphite sleeve (with the outer graphite-cooling jacket joint being shrunk-fit to provide good thermal contact). In this case, a pyrolytic graphite sleeve was employed, with the high thermal conductivity planes being normal to the axis of the target. For an annular beam, a thickness of CBCF of 0.4 mm provided a target operating temperature of 2200°C. This approach is very attractive in that the thermal stresses between the tantalum target case and the graphite sleeve are cushioned by the CBCF layer, which has been shown to be stable under such stresses.[10]

Titanium target

The first barrier considered was that of single barrier of plasma-stabilized ZrO_2, and the operating temperature of the titanium target could be held below 1600°C with a barrier thickness of 0.31 mm, for a central Gaussian beam profile. For this case, however, a target radial thermal gradient of nearly 500°C resulted.

Consideration of a CBCF single thermal barrier with a pyrolitic graphite sleeve and an annular beam resulted in acceptable operational conditions for a thermal barrier thickness of 0.32 mm.

A third thermal barrier of porous metallic titanium was considered for an annular beam and pyrolytic graphite sleeve, and a 50% porous barrier of 1.1-mm thickness provided the proper operating conditions. The resulting temperature profiles are illustrated in Fig. 3. For this case, a pressure of approximately 1.5 MPa was imposed on the thermal barrier. A more porous barrier would be thinner, in approximate proportion to the porosity.

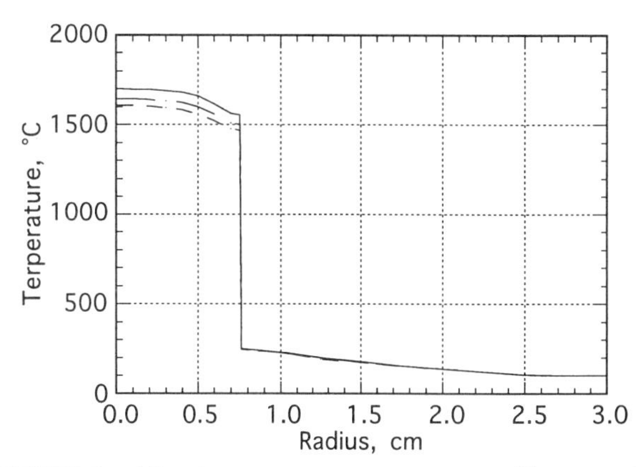

FIGURE 3. Titanium target temperature profiles at axial distances of 0,3,6 cm for a 1.1-mm porous titanium barrier and normal graphite sleeve.

Molten lead target

The lead target analysis proved to be quite difficult, given the very large energy deposition density and the requirement that convection within the target material be considered (the linear thermal conductivity of molten lead is quite poor). While the ANSYS code does have provision for convective heat transfer, the application to the small dimensions of the target required very small zoning which could not be accommodated in the calculation. Hence, an effective thermal conductivity was developed on the basis of experiences with molten lead targets elsewhere[11] which simulated the combined effects of conductive and convective heat transfer. The resulting effect was to provide a radial gradient of about 60°C in the target material, and a 30°C gradient across the tantalum target container, resulting in a thermal barrier gradient of 575°C (the gradient through the graphite sleeve was 135°C).

Analysis of the thermal barrier requirement using CBCF was not encouraging, with a barrier thickness of 0.045 mm resulting.

A 50% porous stainless steel barrier analysis showed that for a thickness of 0.11 mm the appropriate gradient was established, and a pressure of 1.4 MPa resulted. The thickness could be made larger for a less porous barrier choice.

The low operating temperature of the molten lead target allows for a wide choice of possible porous metal barrier materials. This study simply illustrates that a single material (stainless steel) exists which meets the requirement.

CONCLUSIONS

This study of a high-power target concept used for production of radioactive beams at high-energy proton accelerators resulted in evaluation, through simulations, of various thermal barrier materials for application to a range of target types. These simulations were made for a suite of target types chosen to encompass the expected range of operating parameters (temperature and energy deposition density) for possible targets.

Of the thermal barrier materials considered in this study, CBCF appears to have the greatest potential use, because of its "adjustability" through thermal pre-treatment and good mechanical tolerance to thermal stresses within the target assembly. The fact that the use of CBCF as a thermal barrier satisfied most of the cases considered in this study (tantalum and titanium targets, and with central as well as annular incident beams) highlights the versatility of this material. In most cases, the CBCF thickness required was nominal (a few tenths of a mm), serving also as a "cushion" to thermal stresses between the target components. However, in the case of the lead target (in which a central beam only was used), the CBCF thickness was marginally low.

The zirconia barrier material was also useful in this study and thicknesses required were quite reasonable for barrier gradient conditions less rigorous than for CBCF (the barrier maximum temperature should not exceed 1600°C). While small thicknesses of zirconia appear feasible to apply with plasma spray coating, the hardness of the material may make target components susceptible to damage with the thermal stresses imposed, even for moderate thermal gradients.

Porous metal barriers are attractive for cases where the beam heating rates are very high and the target temperature tolerance is low. It is not easy to simulate the operational characteristics of porous metal thermal barriers because of the requirement to couple thermal stress and thermal flow calculations. However, as a class, porous metals offer a potentially higher thermal conductivity than do refractory compounds or CBCF, and are useful in special cases where the thermal loads are high combined with low temperature tolerances. The thermal properties of porous metals are also variable by adjusting porosity.

The thermal consequences arising from the use of an annular beam was particularly evident for the tantalum and titanium targets considered. Such a configuration for the incident beam, while not easy, seems to be feasible at existing high-energy accelerators with intense beams, and is highly desired to control the thermal gradients within the target materials.

Two important high-power target issues remain to be evaluated. First, many target materials in use, or proposed for use, are of the form of powder or compressed powder. Because of the importance of internal heat transport within the target itself, subsequent studies should identify techniques to enhance the target material conductivity, possibly by admixing graphite fibers or boron nitride grains in the target material (with partial abundances up to perhaps 30% of the target composition). It follows that such dilution of the target material will decrease the radioactive beam production rate, but improved target temperature profiles may partially compensate for such production rate losses.

Second, the parameter spaces of porous metal barriers (metal choice and porosity) need to be more comprehensively characterized to enable consideration of such thermal barrier materials over a wider range of target operating conditions than was the case for this study.

The important conclusions from this study are (1) the use of thermal barriers has been shown to be a versatile approach, with a variety of useful barrier materials available, and (2) the target examples treated here span the apparent ranges of parameters for target heating and target temperature tolerance, and the general target concept studied appears to be suitable for application in future target designs.

ACKNOWLEDGMENTS

This work benefited greatly, especially in defining the target systems of interest, from discussions with Prof. John D'Auria of Simon Fraser University and Drs. Jack Beveridge, Ian Thorson and Marik Dombsky of TRIUMF, the Canadian National Meson Facility. Discussions with Dr. Richard Prael of Los Alamos National Laboratory and Dr. Pierre Grand of Amparo Corporation provided guidance in LAHET calculation interpretations. Dr. William Miller of HYTEC, Inc. provided insight to the porous metal thermal barrier analyses. One of us (WLT) expresses gratitude to the TRIUMF organization, and especially to Dr. Erich Vogt, for partial support during the initial phase of this work, and to the U.S. Department of Energy, who supported the bulk of this work through Grant No. DOE-FG05-94ER81704.

REFERENCES

1. See, for example, *The ISOLDE Users' Guide*, Ed. H.-J. Kluge, CERN report 86-03 (1986).
2. W. L. Talbert, H.-H. Hsu and F. C. Prenger, *Nucl. Instrum. Methods* **B70**, 175 (1992).
3. R. E. Prael and H. Lichtenstein, *User Guide to LCS: The LAHET Code System*, Los Alamos National Laboratory report LALP 91-51 (1991).
4. Benjamin Gebhart, *Heat Conduction and Mass Diffusion*, New York, McGraw Hill, 1993, Sections 2.1.4, 7.2 and 7.3.
5. H. E. Eaton, J. R. Linsey and R. B. Dinwiddie, *Proceedings of the 22nd International Conference on Thermal Conductivity*, Ed. T. W. Tong, Lancaster, Technomic Publishing, 1994, p. 289.
6. R. B. Dinwiddie, Oak Ridge National Laboratory Metals and Ceramics Division letter report 1101-43-94, 1994.
7. R. G. Miller and L. S. Fletcher, *AIAA Progress in Astronautics and Aeronautics: Thermophysics, Vol. 39,* Ed. M. Yovanovich, 1975, p. 81.
8. *ANSYS Users' Guide for Revision 5.0*, Swanson Analysis Systems, Inc., Houston, PA (1993).
9. W. L. Talbert and H.-H. Hsu, *Nucl. Instrum. Methods* **A362**, 229 (1995).
10. C. E. Weaver, private communication (May, 1996).
11. T. A. Hodges, private communication (January, 1995).

Ionization efficiency and effusive delay time characterization of high temperature target-ion sources for RIB generation

R.F. Welton[a], G.D. Alton[b], B. Cui[c] and S.N. Murray[b]

[b] *Oak Ridge National Laboratory, P.O. Box 2008, Oak Ridge, TN 3783-6368, USA*
[a] *Oak Ridge Institute of Science and Engineering, Oak Ridge, TN 3783*
[c] *China Institute fo Atomic Energy, Beijing, China*

Ion sources for radioactive ion beam (RIB) generation must efficiently ionize short-lived-radioactive nuclei released from on-line targets with minimal delay times. Delay times attributable to interactions between chemically active species and surfaces of the vapor transport system which are long compared to the half-life of the desired radioactive atom and/or low ionization efficiency of the target/ion source (TIS) will result in a severe reduction of the RIB intensity available for research. We have developed complementary off-line techniques for directly measuring both effusive delay times and ionization efficiencies for chemically active species in high temperature TISs using only the stable complements of the radioactive element of interest. Equipment, designed and developed for these measurements, include: a high-temperature Ta valve; a differentially cooled injection nozzle; and a gaseous flow measurement and control system. These techniques are employed in a systematic investigation of fluorine transport and ionization in an electron-beam-plasma target/ion source (EBPTIS) designed for initial use at the Holifield Radioactive Ion Beam Facility (HRIBF).

INTRODUCTION

Off-line techniques have been developed for ionization efficiency and effusive delay time measurements for chemically reactive species in high temperature target/ion sources (TISs). Molecular feed materials, containing a large fraction of the reactive atom of interest, are introduced into the TIS at well defined flow rates and time distributions. Feed materials are selected that are gaseous at the temperatures (20-300 C) of the flow control system used to inject the material into the TIS and to fully dissociate at the operating temperatures of the TIS (1000-2000 C). Since the feed material is introduced at a controlled rate from an external injection system rather than through diffusion from solid target material, as in on-line process, effusive delay times in the TIS can be delineated from of the diffusion release times associated with the target material.

Interest in generating radioactive ion beams (RIBs) of ^{17}F and ^{18}F for astrophysics research [1] at the Holifield Radioactive Ion Beam Facility (HRIBF) [2] and the use of the electron beam plasma target/ion source (EBPTIS) for their generation was the motivating factor for studying the effusive flow characteristics of fluorine and fluoride compounds formed in this type of ion source. The TIS used in these studies is very similar to sources described previously [3]. When employed in on-line RIB generation, nuclear reaction products, created by high energy light ion bombardment of thick target materials are first diffused from the target material located inside a target material reservoir and then transported to the ionization chamber of a TIS through a vapor transport tube where they undergo electron impact ionization. For the HRIBF TIS used in these experiments, the target material reservoir and transport tube are resistively heated to temperatures in excess of 1000 C. Ion beams extracted from the ionization chamber are accelerated to energies of 20-60 keV. All measurements were made first with the target material reservoir empty and then with fiberous Al_2O_3 target material with a nominal fiber diameter of 3μm [4]. Ionization efficiencies for Xe during operation were typically 15 % , falling to 6.5% with Al_2O_3 present due to the increased vapor pressure of the SiO_2 contaminants in the target material.

FLOW MEASUREMENT AND CONTROL APPARATUS

During RIB generation, the flow rates of the radioactive species are typically 10^8 to 10^{13} atoms/s, which, for most cases, is insufficient to form a monolayer on the interior surfaces of the TIS. Therefore, studies must be made in a flow-rate regime comparable to those that occur in typical on-line operations for RIB generation. Toward this end, we have developed a simple, static-vacuum, flow measurement system ,capable of directly measuring gaseous flows down to ~10^{12} particles/s and well defined, reproducible feed material flow-rates down to ~10^{10} particles/s. Molecular gases were introduced into the TIS from a feed material reservoir, maintained at pressures

CP392, *Application of Accelerators in Research and Industry,* edited by J. L. Duggan and I. L. Morgan
AIP Press, New York © 1997

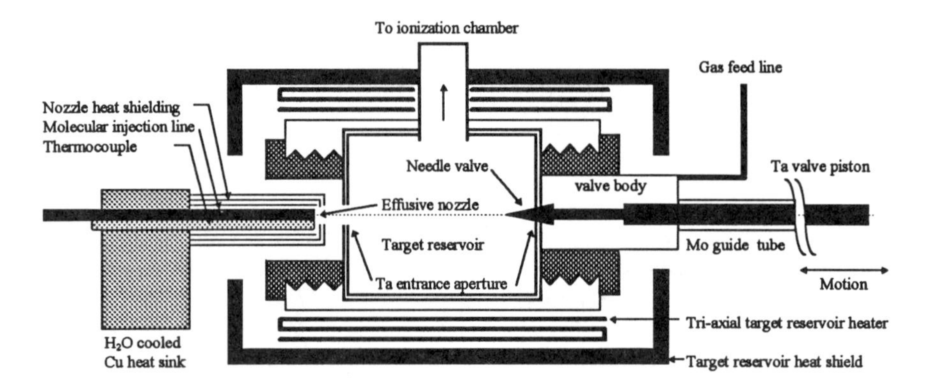

FIGURE 1. A schematic illustration of the high temperature valve assembly used in measuring th effusive flow delay times (Right) and the sample injection system used for the introduction of intact molecules into the TISS for the determination of ionization efficiencies (Left).

between 1 and 1000 Torr. The reservoir provided a static pressure for controlling the feed-rates of the material in question through a set of calibrated and variable leaks; the feed material is then directed into a small flow measurement volume, maintained at 1-100mTorr, or fed directly into the TIS by means of a two-way valve. The measurement volume is independently evacuated and can be isolated to time the rate-of-rise in pressure due to gas flow through the leak valves. If j_{mv}, $N_{mv}(t)$, $n_{mv}(t)$, $P_{mv}(t)$, T_{mv} are, respectively; the feed material leak rate, number of molecules, number density, pressure and temperature in the flow measurement volume, then

$$j_{mv} = \frac{dN_{mv}}{dt} = V_{mv}\frac{dn_{mv}}{dt} = \frac{V_{mv}}{kT_{mv}} \cdot \frac{dP_{mv}}{dt}. \quad [1]$$

Capacitance manometers are used to measure pressures in the feed material reservoir and the flow measurement volume. The background out-gassing rate of the measurement volume was typically less than 10^{12} molecules/s.

The magnitudes of feed material flow-rates from the flow-rate measurement system were confirmed with a calibrated Xe leak supplied by VTI Scientific [5] and with a He leak detector, calibrated against two standard leaks. Measured flow-rates always varied less than 10% through a given leak and agreed, typically, with calibrated leak rates to within ~1%. The system was also tested by flowing Ne, Ar, Kr, SF_6, CF_4 and CCl_2F_2 through the calibrated leak and each species was seen to obey the theoretically predicted $M^{-1/2}$ flow dependence to within a few percent. In addition, a linear relationship was observed between the feed material reservoir pressure and ion current for each of the species in the flow regime studied. After measuring the flow rate of the desired species in the 10^{12} to 10^{13} molecules/s range, the reservoir pressure was reduced to provide scaled flow-rates of the species of interest as low as 10^{10} molecules/s.

EFFUSIVE DELAY TIME MEASUREMENTS

The mean effusive delay time for a given species within the TIS can be determined by either rapidly establishing or interrupting the flow into the TIS while monitoring the time dependence of the mass selected ion current. A high temperature valve, close coupled to the target material reservoir, is employed to minimize contributions of particle delay from the injection feed line to the measured delay functions of the TIS. The valve must have a minimal influence on the target material reservoir temperature when either opened or closed to avoid sources or sinks of sample particles. The valve assembly is schematically depicted in Fig. 1. The valve seat occurs between a carbonized tapered Ta needle and a thin Ta plate. The needle and supporting shaft are thermally isolated from the external environment and maintained at the target reservoir temperature by radiant heat. The valve assembly is enclosed in the heat shielding of the TIS where W-Re thermocouples continuously monitor the temperatures of the target reservoir, valve body, valve needle and Mo guide tube. By displacing the piston, the conductance between the valve body and target reservoir can be varied over several orders of magnitude.

The delay function is defined as the time response of the mass selected ion current, $I(t)$, to changes in the flux of the particles entering the TIS, $j(t)$. If $j(t)$ is small enough so that ionization conditions are not perturbed, the total number of sample particles in the TIS, $N(t)$, is proportional to $I(t)$ [6]. Furthermore, if $j(t)$ can be rapidly changed with respect to the delay times of the species within the TIS, the mean particle delay time can be determined from the measurement. By applying formalism from the kinetic theory of gases to the dynamic particle balance within the TIS, under molecular flow conditions, we obtain,

$$I(t) \propto N(t) = e^{-\alpha t} \int_{-\infty}^{t} j(t') e^{\alpha t'} dt' \quad where \ \alpha = 1/\tau. \quad [2]$$

Here τ is the mean delay time for the sample particle in the TIS. Eq. 2 is readily solved for the case where $j(t)$ is rapidly reduced from j_1 to j_0 at time t_1

$$N(t) = \frac{j_1 - j_0}{\alpha} e^{-\alpha(t-t_1)} + \frac{j_0}{\alpha} + k \quad [3]$$

where k~0 if the valve closes rapidly with respect to τ. Similarly if $j(t)$ is rapidly raised from j_0 to j_1 at time t_1 the number of particles is given by [7]

$$N(t) = \frac{j_1}{\alpha} - \frac{j_1 - j_0}{\alpha} e^{-\alpha(t-t_1)} - k. \quad [4]$$

Effusive delay time measurements were made by recording $I(t)$ for the sample species and the background Ta current during each valve opening and closing. Initial tests with Xe show excellent agreement between the measured delay function and the delay function predicted from evacuation of the TIS volume, assuming conductance limited flow through the extraction opening (τ=volume/conductance). Fluorine delay functions were studied by introducing SF_6 to create a source of atomic F within the valve body which was maintained at temperature above of 1200 C to allow complete thermal dissociation of this molecule [8]. In addition to F^+, the mass spectrum of the extracted beams also contained AlF^+ and BeF_2^+ with no SF_x^+ present. The metal fluorides, likely result from the interaction of atomic fluorine with impurities which were always present in the TIS over ~1000 hours of operation in spite of several TIS cleanings and replacement of the feed system.

An example of the measured delay function characteristic of AlF is shown in Fig. 2; The solid curves are fits to the data derived by use of Eqs. 3 and 4. The delay times were studied as a function of SF_6 flow into the TIS; a very clear flow independent regime was found for equivalent F flow-rates of less than ~2 μA. These results are shown in Fig. 3. The addition of the Al_2O_3 target material to the target reservoir had little effect on the observed delay times. In addition to SF_6, CCl_2F_2 and

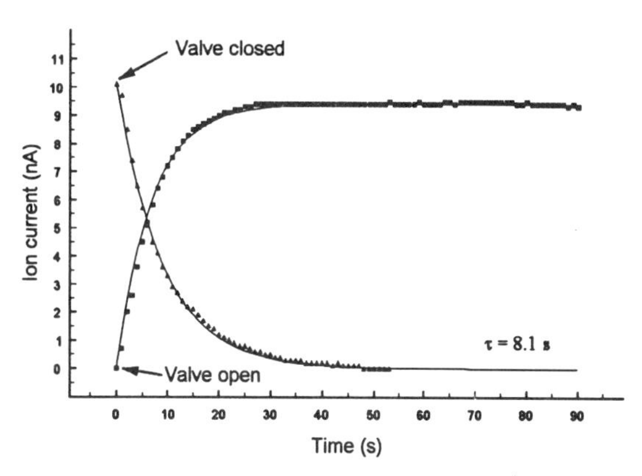

FIGURE 2. Example delay time measurement for AlF^+ from the TIS with Eq. 3 and 4 fitted to the data to determine the effusive delay time.

CF_4 were also employed as means for supplying atomic F. The delay times for F^+, AlF^+ and BeF_2^+ from CCl_2F_2 and SF_6 were nearly identical while the delays for F^+ and CF_x^+ from CF_4 were considerably shorter. This can be explained on the basis of complete molecular dissociation of both SF_6 and CCl_2F_2 in the TIS while CF_4 remains intact at the TIS operating temperatures [8].

IONIZATION EFFICIENCY MEASUREMENTS

Experience shows that physically attaching the gas feed line to the high temperature TIS results in significant conductive heat transfer to the feed line and consequently causes thermal dissociation of SF_6 in relatively cool regions of the line (650-1200 C). This causes substantial and indeterminal losses of throughput due to chemical reaction of the reactive F feed material with the walls of the feed tube. To circumvent this problem, we designed, developed and characterized an effusive-flow nozzle, separated from the high temperature target material reservoir by a gap of 4.2 mm, which operates at temperatures below the threshold for molecular dissociation of SF_6 (650 C). The nozzle permits injection of known fractions of intact molecular feed gases into the TIS. The SF_6 molecules will fully dissociate at the temperatures of the TIS (1200-1400 C) [8]. Ionization efficiencies of F are determined by comparing the mass analyzed ion currents with the total flux of F delivered to the TIS by the SF_6 gas, after correcting for losses to the vacuum system due to the injection gap. A schematic representation of the water cooled injection system is shown in Fig. 1.

Fractional losses of particles across the injection gap into the surrounding vacuum system were determined for two limiting cases: (i) particles which pass through the injection gap in the vacuum system are lost (maximum loss scenario) and (ii) molecules have a maximum

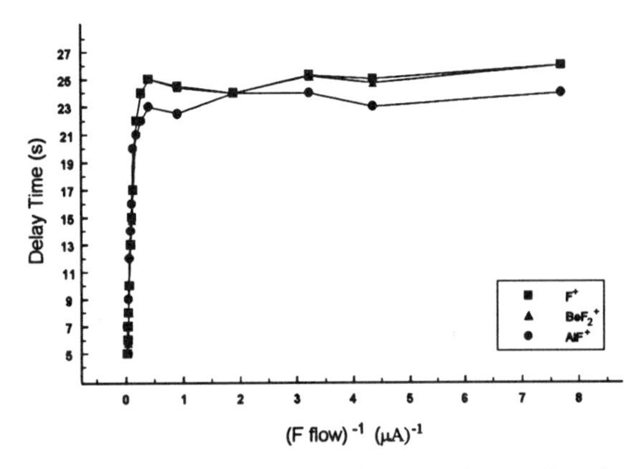

FIGURE 3. Fluorine/fluoride delay times plotted versus the reciprocal of the equivalent F flow rate introduced into the TIS using SF_6 as the carrier species at a target reservoir temperature of 1525 C.

probability of reentering the TIS through the injection aperture since no adsorption losses are assumed (minimum loss scenario). The maximum fractional loss, case (i), can be determined either by use of Monte Carlo effusive flow simulations of the feed-line/TIS injection system [6] or by employing analytical approximations derived from the kinetic theory of gases under Knudsen flow conditions [9]. The minimum fractional loss, case (ii), can be determined from experiment by introducing an equivalent mass and flow-rate of noble gas (no dissociation/adsorption losses) into the TIS and monitoring the ion current yielded with the injection gap open and then with the gap sealed. Once the two limiting extremes of the fractional loss are determined, the ionization efficiency can be calculated for each case from the measured ion currents. If the dimensions and placement of the nozzle are chosen correctly, the uncertainty in the ionization efficiency measurement can be made reasonably small. For these experiments, the minimum possible transmission of particles into the TIS from the feed system (maximal loss) was calculated and found to vary from 3 to 13%, depending on the size of the entrance aperture used. The maximum possible transmission (minimal loss) was measured using noble gases and also found to vary from 10 to 40 % depending on the size of the entrance aperture.

Ionization efficiencies were determined from the mass selected ion currents of F and the fluorides extracted from the EPBTIS over equivalent F flow-rates from 1 to 10 μA, correcting for particle transmission across the injection gap. The ionization efficiencies were found to be: F: 1.6 to 0.8%; AlF: 1.8 to 0.9%; and BeF_2: 1.4 to 0.7%, respectively. The addition of Al_2O_3 target material had the effect of reducing both the Xe and all fluoride efficiencies by a factor of ~2.3 ostensibly due to a persistent vapor load on the TIS.

DISCUSSION

The measured delay functions for both fluorides and Xe can be represented theoretically using Eqs. 3 and 4. The measured mean delay times for evacuation of Xe from the TIS agrees with the theoretical values. Examination of the measured fluoride delay times versus flow-rate clearly shows an independent flow-rate regime below 2 μA of F flow and a linear dependence on $(\text{flow rate})^{-1}$ at feed rates above 2 μA. This behavior is predicted from the adsorption kinetics of a surface which has reached saturation at 2 μA of flow [10]. The delay times for F and BeF_2 are identical thus, BeF_2 appears to be the dominant transport agent for the F through the TIS. The measured AlF and BeF_2 delay times of 25s are in approximate agreement with the delay time expected from adsorption enthalpy data (3.2, 3.1 eV, respectively), mean number of wall collisions within the TIS (χ~1000) and the estimated effective temperature of the TIS [11].

The ionization efficiency measurement technique detailed in this report is also able to measure efficiencies for reactive F at flow rates which have been shown, from delay time studies, to preclude surface saturation effects and are therefore in a flow-rate regime comparable to those for RIB generation. Because many molecules decompose at temperatures available in RIB TISs, the techniques and apparatus used to measure delay times and ionization efficiencies of fluorine are applicable to other RIB TIS systems and for other RIB species. These methods provide direct measurements of delay times and ionization efficiencies for candidate RIB species independent of the release characteristics of the target allowing delineation of effusive and diffusive flow processes when compared with on-line studies.

REFERENCES

1. M.S. Smith, NIMB 99 (1995) 349
2. D.K. Olsen, Nucl. Instr. Meth. A328 (1993) 303
3. G.D. Alton, Rev. Sci. Instrum. 65 (1994) 2012
4. Alumina fiber supplied by RATH Preformance fibers, Wilmington, MA. USA
5. Vacuum Technology Incorperated, Oak Ridge, TN 37832 USA
6. R.F. Welton, G.D. Alton, A. Piotrowski, S.N. Murray, Rev. Sci. Instrum. 67 (1996) 1670
7. R. Kirchner, Nucl. Instrum. Methods B70 (1992) 186
8. Calculations were made by using The Thermocalc which is a product of the Royal Institute of Technology Stockholm, Sweden and HSC Software from Outokumpu Research Oy, Pori, Finland
9. A. Roth, 'Vacuum Technology', Elsevier, New York (1990)
10. V. Ponec, Z. Knor and S. Cerny, 'Adsorption on solids', Butterworth & Co, London (1974)
11. G. Bolbach, J.C. Blais, and A. Marilier, Sur. Sci. 90 (1979) 65

FIRST ON-LINE RESULTS FOR As AND F BEAMS FROM HRIBF TARGET / ION SOURCES

D. W. Stracener[a], H. K. Carter[b], J. Kormicki[b,†], J. B. Breitenbach[b,‡], J. C. Blackmon[c], M. S. Smith[d] and D. W. Bardayan[e]

[a] *Joint Institute for Heavy Ion Research, Oak Ridge, TN 37831*
[b] *Oak Ridge Institute for Science and Education, Oak Ridge, TN 37831*
[c] *The University of North Carolina at Chapel Hill, Chapel Hill, NC 27599*
[d] *Oak Ridge National Laboratory, Oak Ridge, TN 37831*
[e] *A. W. Wright Nuclear Structure Laboratory, Yale University, New Haven, CT 06511*

The first on-line tests of the ion sources to provide radioactive ion beams of ^{69,70}As and ^{17,18}F for the Holifield Radioactive Ion Beam Facility (HRIBF) have been performed using the UNISOR facility at HRIBF. The target/ion source is an electron beam plasma (EBP) source similar to the ISOLDE design. The measured efficiencies for ^{69}As and ^{70}As were 0.5 ± 0.2% and 0.8 ± 0.3%, respectively. The arsenic hold-up time in the tested target/ion source was 3.6 ± 0.3 hours as measured with ^{72}As at a target temperature of 1300° C. The measured efficiencies for ^{17}F and ^{18}F were 0.0052 ± 0.0008% and 0.06 ± 0.02%, respectively. The source hold-up time for fluorine was measured with Al^{18}F since 88% of the observed radioactive fluorine was found in this molecule. The Al^{18}F hold-up time was 16.4 ± 0.8 minutes at a target temperature of 1470° C.

INTRODUCTION

The recently completed Holifield Radioactive Ion Beam Facility (HRIBF) is designed to provide energetic radioactive ion beams for nuclear physics and nuclear astrophysics research. To produce these beams, light ions from the K=100 Oak Ridge Isochronous Cyclotron (ORIC) will impact a thick target in the ion source of an isotope separator. The extracted radioactive ions, after mass analysis and charge exchange, are injected into the 25 MV tandem accelerator. One of the most challenging aspects of this project is the performance of the target/ion source of the isotope separator. This paper presents results of on-line tests of the initial version of the ion source used to produce arsenic beams. Also, preliminary results for fluorine beams from a similar source are described.

EXPERIMENTAL RESULTS

The UNISOR (1) separator on-line to the Holifield tandem accelerator provides a unique capability (2) to investigate target/ion source performance. Proton and deuteron beams of energy up to 40 MeV are used to produce the desired radioisotopes directly in the target/ion source under investigation. The mass analyzed ions are counted using traditional on-line nuclear spectroscopy techniques with a moving tape system and a γ-ray detector. Alternatively, in experiments prior to those reported here, heavy ion implantation experiments, similar to those performed by Kirchner (3), were used to measure release times and efficiencies of various beam/target combinations (2) in preparation for designing and constructing the target/ion source used here.

Target/Ion Source

The target/ion source used in these studies is a modified version of the general design (4) for the HRIBF facility which is similar to the CERN/ISOLDE design (5). Figure 1 shows the electron beam plasma ion source as used in these experiments. The following modifications were made specifically for a liquid germanium target which will be used to produce radioactive arsenic ion beams. Since the vapor pressure of germanium is 10^{-4} torr at 1100° C, it was expected that this would be the maximum operating temperature for the target. With this in mind, the design of the target heater was simplified to a single-pass current path, with an attached heat shield. The cathode current connection was moved from midway along the cathode transfer tube to the rear-most point. This ensures that the transfer line will be at the highest possible temperature which will help to reduce the sticking time (6) of arsenic on the tantalum surfaces. Several layers of 0.025 mm thick tantalum around the transfer line provide additional heat shielding. A carbon target holder, with a 1 mm thick entrance window, is used since tantalum reacts strongly with germanium. The germanium target, designed to stop 40 MeV protons, is 4 mm thick by 9 mm diameter.

CP392, *Application of Accelerators in Research and Industry*, edited by J. L. Duggan and I. L. Morgan
AIP Press, New York © 1997

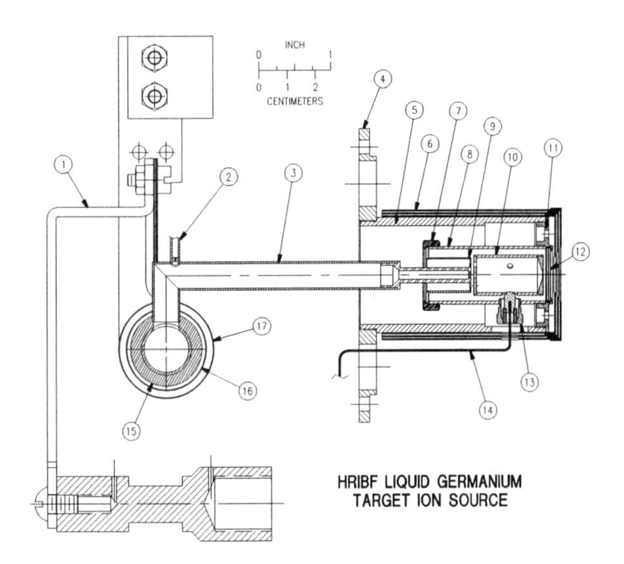

FIGURE 1. Axial cut of the HRIBF liquid germanium target/ion source. Part numbers and construction material are indicated: 1. cathode current lead (Ta), 2. gas (Xe) transfer line (Ta), 3. transfer line (Ta), 4. head flange (C), 5. outer tube (C), 6. heat shields (Mo or Ta), 7. cathode support nut (Mo), 8. anode support tube (Ta), 9. cathode (Ta), 10. anode tube (Ta), 11. end flange (Mo), 12. anode heat shields (Mo), 13. anode support assembly (Ta, BeO, Mo), 14. anode wire (Mo), 15. target holder (C for Ge, Ta for Al_2O_3), 16. target heater (Ta), 17. heater heat shield (Ta).

Several copies of this target/ion source have been constructed and tested off-line to determine typical operating parameters. The single pass target heater has operated reliably over a wide range of target temperatures starting at 900° C, with no heater current, up to 1570° C with a heater current of 480 A. Moving the cathode connection to the rear of the transfer tube enables the entire line to be operated at temperatures in excess of 1700° C. This was measured at the coolest point which is next to the cathode current connection, with a current of about 350 A.

The other operational parameter which is important is the target/ion source efficiency. Typical ion source efficiency for Xe is 10-15% with a cathode current of 350 A and anode current of 200 mA at 150 volts. Under these conditions the germanium target is around 900° C without any target heater current. When the target temperature is raised the vapor pressure of the target material at some point becomes so high that the efficiency of the ion source is decreased. For the germanium target/ion source we observed no significant increase in efficiency up to the highest temperature measured, namely, 1300° C. Subsequent modifications allow the target to be run at much higher temperatures. We are presently trying to determine the highest temperature at which the germanium target may be held without a loss in ion source efficiency. For on-line experiments the Xe gas inlet is restricted to 0.3 mm in order to reduce the flow of reaction products into the gas line. With this arrangement the on-line Xe efficiency is reduced to 3%.

On-Line Arsenic Experiments

The performance of the ion source for the production of arsenic beams was tested using (p,n) and (p,2n) reactions on a natural 99.999% pure germanium target mounted in the ion source. The extracted arsenic ions were deposited onto the tape transport system for 10 minutes and then moved to the detector station and the activity of the sample was counted for 8 minutes. The count rates shown in Fig. 2 are net counts in the indicated γ-ray transitions. These data and the calculated production rates (7) are used to determine the absolute efficiency for the target/ion source. The efficiency of the target/ion source at 1270° C was 0.8 ± 0.3 % for ^{70}As and 0.5 ± 0.2 % for ^{69}As. These data also clearly show a hold-up time in the target/ion source and an improvement in yield as the target temperature is increased.

Since the growth curves seem to be dominated by the half-life of the respective isotopes, the hold-up time as implied from Fig. 2 must be longer than the half-life of either ^{69}As ($t_{½}$ = 15.2 min) or ^{70}As ($t_{½}$ = 52.6 min). In order to measure the hold-up time in the target/ion source it was necessary to use an arsenic isotope with a longer half-life. After proton irradiation for several hours enough ^{72}As ($t_{½}$ = 26.0 hours) activity is built up in the target to enable such a measurement. The proton beam is stopped and the rate of ions released from the source is monitored as a function of time. This is shown in Fig. 3 for ^{72}As. Since the radioactive half-life in this case is much longer than the apparent hold-up time, we say that the hold-up time is 3.6 ± 0.3 hours at 1300° C.

Target temperature is an important parameter for any target/ion source. Figure 4 shows the equilibrium yield for ^{69}As and ^{70}As as a function of the inverse of the target temperature. The data show an exponential increase in the yield with increasing temperature and

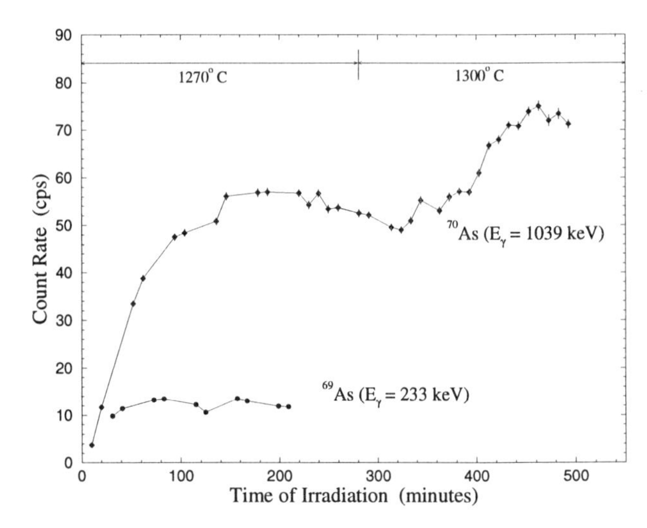

Figure 2. Release of ^{69}As and ^{70}As. The net count rate of a characteristic γ-ray from each isotope is plotted. No corrections for detector efficiency or branching ratios have been made. These data were taken at the two indicated target temperatures.

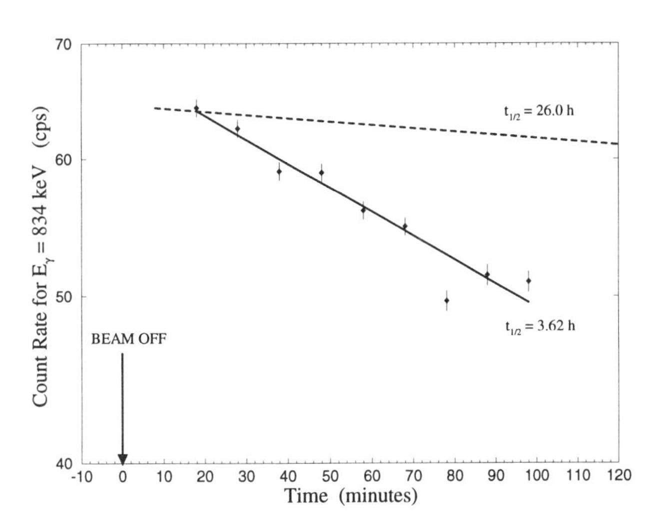

FIGURE 3. Release of ^{72}As after proton bombardment is stopped. The net count rate for the 834 keV γ-ray from ^{72}As is plotted versus the time with the solid line being an exponential fit to the data. The dashed line represents the half-life of ^{72}As.

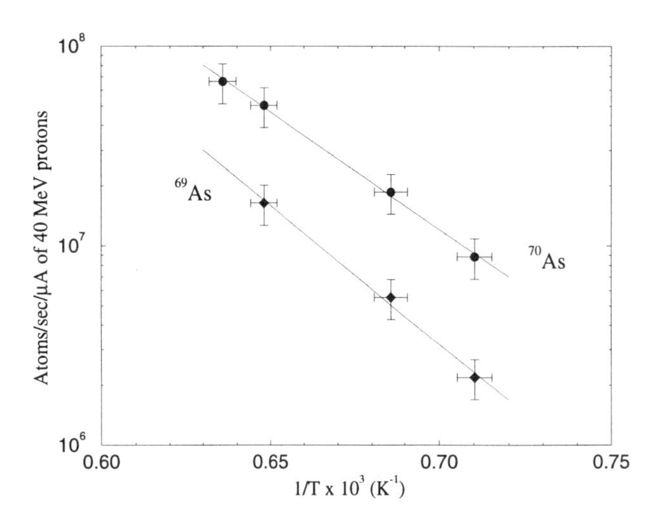

FIGURE 4. The normalized yield of ^{69}As and ^{70}As as a function of the inverse target temperature. The lines are exponential fits to the data. The counts are corrected for detector efficiency and branching ratio and are normalized to a proton beam intensity of 1 μA on a ^{70}Ge target.

indicate that the target can be operated at much higher temperatures. The ion source efficiency was constant over the range of the data shown. The data in Fig. 4 are corrected for radioactive decay, detector efficiency and branching ratio and are normalized to 1 μA of protons on an enriched ^{70}Ge target.

Off-Line Fluorine Release Measurements

Refractory oxides have been explored as a promising category of target materials for production of radioactive fluorine beams. The most encouraging results so far have been obtained with fibrous Al_2O_3. The material (8) consists of 3 μm fibers of Al_2O_3 with a SiO_2 binder (2% by weight) and a density of 0.12 g/cm^3. This Al_2O_3 was tested in an activation and off-line release measurement to determine the fluorine release efficiency. The target material was irradiated with a low-intensity proton beam, which produced ^{18}F ($t_{1/2}$ = 110 minutes) from the $^{18}O(p,n)^{18}F$ reaction. The γ-rays emitted from the target were measured with a high-purity germanium detector to determine the quantity of ^{18}F in the target. The Al_2O_3 target was then heated in a vacuum oven, and the ^{18}F activity was measured again to determine the fractional release. Activation and release measurements performed with Al_2O_3 in a crystalline form (ρ=3.7 g/cm^3) showed no significant release. However, measurements performed with the fibrous Al_2O_3 showed near complete release for temperatures greater than 1400 °C. The only radiation emitted by the decay of ^{18}F is a 511-keV gamma ray, so the number of 511-keV gamma rays was measured as a function of time to determine the half-life of the activity.

The results of one measurement in which the fibrous Al_2O_3 material was heated to a maximum temperature of 1700 °C is shown in Fig. 5. The slope of the decay of the activity measured before heating is in agreement with the known half-life of ^{18}F. Following heating in the vacuum oven, only a few percent of the expected activity was observed to remain in the target material. Furthermore, the half-life of the remaining activity was found to be 3.6 hours. This activity is not from the decay of ^{18}F, but probably from the decay of an isotope produced by reactions on a contaminant in the Al_2O_3. We conclude that greater than 95% of the ^{18}F is released after heating to 1700 °C. Subsequent tests showed similar results with releases in excess of 80% for targets heated to T > 1400 °C. For targets heated to T < 1200 °C, no significant release was measured.

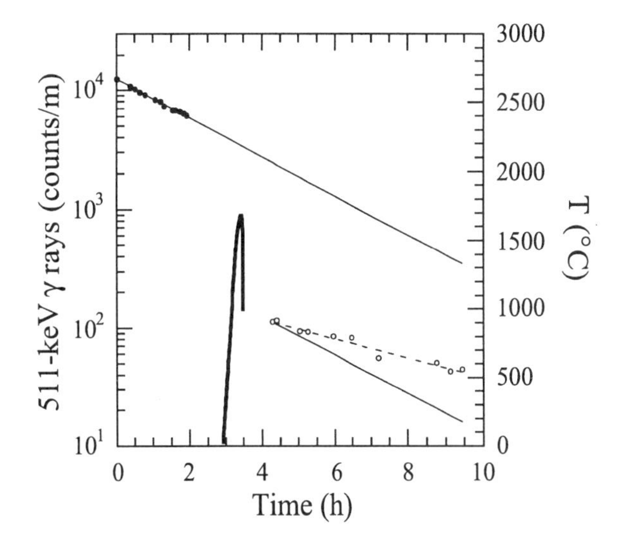

FIGURE 5. Activation and release measurement of ^{18}F from a fibrous Al_2O_3 target material. The filled circles were measured prior to heating, and the open circles measured after heating. The heavy line is the temperature of the target during heating. The thin solid lines show the expected activity for a 110 minute half-life. The dashed line is an exponential fit to the data after heating which gives a half-life of 3.6 hours.

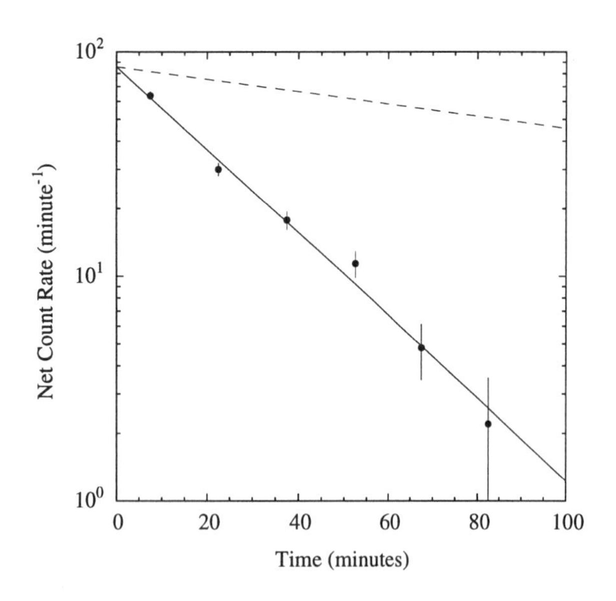

FIGURE 6. Release of $Al^{18}F$ after proton bombardment is stopped. The net count rate for the annihilation radiation is plotted versus the time with the solid line being an exponential fit to the data. The dashed line represents the half-life of ^{18}F.

On-Line Fluorine Experiments

The performance of a target/ion source for the production of $^{17,18}F$ was also tested. The only changes in the target/ion source from that described above were the target holder and target material. The target material was the Al_2O_3 fibers and the target holder was made of tantalum and welded to the transfer tube. The reactions used were $^{16}O(d,n)^{17}F$ ($t_{½}$ = 64.5 seconds) and $^{18}O(p,n)^{18}F$ ($t_{½}$ = 110 minutes).

Due to the extreme reactivity of atomic fluorine, the fluorine isotopes are transported in molecular form with 88% of the observed radioactive fluorine found in mass 44 (45) corresponding to $Al^{17}F$ ($Al^{18}F$). This is based on measurements of emitted γ-rays at other masses which could correspond to the following molecules (% yield): F(3), HF(2) and SiF(7). In addition, the following masses (molecules) were checked and negligible amounts of radioactive fluorine were observed: 28(BeF), 42(NaF), 47(BeF_2), 58(KF) and 62(AlOF).

At a target temperature of 1470° C, the efficiency of this source for ^{17}F (found in the $Al^{17}F$ molecule) was 0.0052 ± 0.0008%, and the efficiency for ^{18}F was 0.06 ± 0.02%. The production rates were calculated from the measured cross sections (9) and tabulated stopping powers (10). The uncertainty in the efficiency is primarily due to the estimated uncertainty in the proton beam current on the target. The hold-up time was measured at 1470° C using the longer lived ^{18}F isotope. The results of this measurement are shown in Fig. 6. By fitting the activity with an exponential, the hold-up time was determined to be 16.4 ± 0.8 minutes. We are currently exploring the effects of source parameters, such as temperature and surface passivation, on efficiencies and hold-up times.

ACKNOWLEDGMENTS

The authors wish to acknowledge the contribution of G. D. Alton who suggested and obtained the Al_2O_3 fiber material. We also appreciate the many helpful discussions with J. R. Beene, S. Ichikawa, H. Ravn, S. Sundell, W. L. Talbert and R. F. Welton. We are also particularly grateful for the contribution to the arsenic experiments by M. J. Brinkman. The technical drawings were made by G. D. Mills and C. A. Reed. Operation, upgrades and maintenance of the separator were ably performed by A. H. Poland.

Oak Ridge Institute for Science and Education is managed by Oak Ridge Associated Universities for the U.S. Department of Energy under contract number DE-AC05-76OR00033. Oak Ridge National Laboratory is managed by Lockheed Martin Energy Research Corp. for the U.S. Department of Energy under contract number DE-AC05-96OR22464.

REFERENCES

† Also Physics Department, Vanderbilt University, Nashville, TN 37235, USA.

‡ Present address: Zur Steinhelle 31, Dertingen, Germany.

1. Carter, H. K., Mantica, P. F., Kormicki, J., Reed, C. A., Poland, A. H., Croft, W. L., and Zganjar, E. F., "UNISOR Separator Upgrade", Physics Div. Prog. Rpt. ORNL-6842, Martin Marietta Energy Systems, Inc. Oak Ridge National Laboratory, September 30, 1994.
2. Carter, H. K., Kormicki, J., Breitenbach, J., Ichikawa, S., Mantica, P. F., Alton, G. D., and Dellwo, J., "On-Line Ion Source/Target Release Time and Efficiency Experiments", Physics Div. Prog. Rpt. ORNL-6842, Martin Marietta Energy Systems, Inc. Oak Ridge National Laboratory, September 30, 1994.
3. Kirchner, R., Nucl. Instr. and Meth. in Phys. Res. **B70**, 186-199 (1992).
4. Alton, G. D., Dellwo, J., Murray S. N., and Reed. C. A., "Target Ion Source Development", and following articles, Physics Div. Prog. Rpt. ORNL-6842, Martin Marietta Energy Systems, Inc. Oak Ridge National Laboratory, September 30, 1994.
5. Ravn, H. L., Nucl. Instr. and Meth. in Phys. Res. **B70**, 107-117 (1992).
6. Kirchner, R., Nucl. Instr. and Meth. in Phys. Res. **B26**, 204-218 (1987).
7. Gomez del Campo, J., private communication using a statistical model code, LILITA, 1996.
8. Alcen®, alumina fiber supplied by RATH Performance Fibers, Wilmington, DE, USA.
9. Gruhle, W., Schmidt, W., and Burgmer, W., Nucl. Phys. **A186**, 257-263 (1972).
10. Anderson, H. H., and Ziegler, J. F., *Hydrogen Stopping Powers and Ranges in All Elements*, New York: Pergamon Press, 1977.

TWINSOL: A DUAL SUPERCONDUCTING SOLENOID SYSTEM FOR LOW-ENERGY RADIOACTIVE NUCLEAR BEAM RESEARCH*

M.Y. Lee, F.D. Becchetti, J.M. Holmes[a], T.W. O'Donnell, M.A. Ratajczak, D.A. Roberts, J.A. Zimmerman[b]

Department of Physics, The University of Michigan, Ann Arbor, MI 48109-1120 USA

J.J. Kolata, L.O. Lamm, J. von Schwarzenberg, M. Wiescher

Department of Physics, University of Notre Dame, Notre Dame, IN 46556 USA

A unique type of apparatus is currently under construction as part of an upgrade to the radioactive ion beam facility at the University of Notre Dame Nuclear Structure Laboratory. The device will consist of a pair of large in-line superconducting solenoids ($B_0 = 6$ tesla, bore = 30 cm) which will be used to produce, collect, transport, focus and analyze both stable and radioactive nuclear beams. This apparatus in conjunction with the recently upgraded accelerators at Notre Dame is especially well suited for the production and utilization of intense (*viz.* > 10^6/sec), low-energy (1 - 10 MeV/u), stable and radioactive nuclear beams relevant to the study of reactions involved in astrophysical processes. These improvements will allow for the production of radioactive beams of greater intensity, higher purity and at both higher and lower energies than previously available at this facility. The first phase of construction and results of initial tests will be reported.

INTRODUCTION

In the last decade many nuclear physics laboratories have successfully utilized various methods to produce and use radioactive nuclear beams (RNBs). A number of facilities have recently either upgraded their RNB capabilities or announced plans to do so.[1] Although the majority of these facilities are using or developing Isotopic Separation On-Line (ISOL) methods of RNB production, a few facilities, like the UM-UND RNB apparatus at the University of Notre Dame Nuclear Structure Laboratory (UND), have produced usable secondary radioactive beams by collecting reaction products from the bombardment of a primary beam upon specially selected targets. In contrast with other labs such as NSCL and GANIL that use a similar method (projectile fragmentation), but at much higher kinetic energies, UND is in a unique position to study low-energy nuclear reactions, especially those that are relevant to astrophysical processes.

A new device, "TwinSol", is currently being installed as a joint UM-UND project funded by NSF, UND and UM to extend the successful RNB research program at UND which began nine years ago.[2] This apparatus is an adaptation of the original concept of a dual solenoid spectrometer at IPN-Orsay, France, by J.P. Schapira *et al.*[3,4] The design and development of TwinSol is based on the previous experiences of our group in using superconducting solenoids to produce and utilize RNBs[5,6,7,8,9] which included the installations of a 3.5 T magnet at UND[10,11,12] and a 7 T solenoid at NSCL.[13,14] Thus TwinSol represents the third generation of superconducting solenoidal RNB research systems developed by our group.

DESIGN

TwinSol is built around two, in-line, large-bore, simple (no yoke or end compensation) custom designed superconducting solenoids spaced approximately 4 meters apart in the West Target Room experimental area of the UND Nuclear Structure Lab.[c] Primary beams are accelerated from a SNICS source through a 10 MV FN-Tandem Van de Graaff. The basic setup is shown in Fig. 1.

Each solenoid is rated for a maximum central field strength of 6 tesla at a current of 100 amperes with an axial field integral of $\int B dl = 3.8$ T m. The cryostats are ca. 98 cm in length with a 30 cm diameter warm bore which is slightly larger than the outer diameter of the beamline pipe. Since the magnets are mounted on rails, they can slide along the support structure independently of the other beamline components provided they do not come into contact with chambers or vacuum tees. In most instances the centerline of the first solenoid will be about 1.1 m from the production target, whereas the second solenoid position will vary depending upon the specific experiment and mode of operation. Typically the large magnetic forces between the two solenoids will prevent us from placing them closer than about 2 m apart from each other.

---

* Additional information may be found on the World Wide Web. The HTML document may be accessed at: http://www.physics.lsa.umich.edu/twinsol.
a) Permanent address: Physics Department, University of Texas, Austin, TX 78712.
b) Permanent address: Department of Chemistry, University of Michigan, Ann Arbor, MI 48109-1055.
c) The solenoids were manufactured by Cryomagnetics, Inc., 1006 Alvin Weinberg, Dr. Oak Ridge, TN 37830.

CP392, *Application of Accelerators in Research and Industry*, edited by J. L. Duggan and I. L. Morgan
AIP Press, New York © 1997

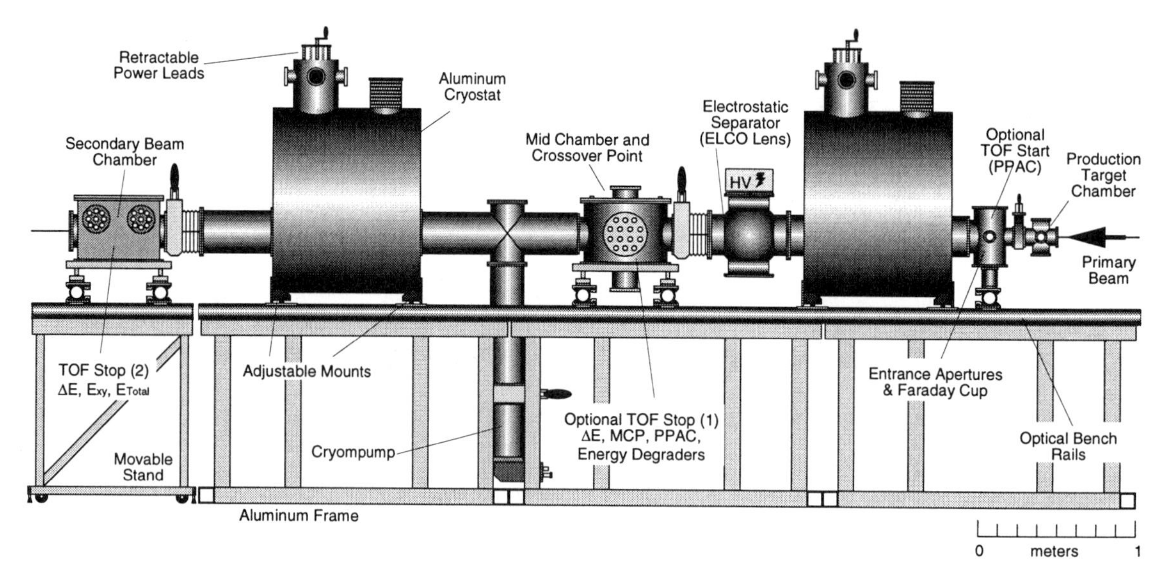

FIGURE 1. Scale illustration of the University of Michigan - University of Notre Dame double solenoid ion-optical RNB system, TwinSol.

Each magnet consists of ten concentrically wound coils of multi-filament NbTi wire that spliced together form a solenoid of length ca. 60 cm with inner and outer diameters of ca. 18 cm and 21 cm respectively. The coils are aligned to ± 1 mm and suspended inside the aluminum cryostat by a series of thin fiberglass and stainless steel supports that are designed to optimize the balance between physical strength characteristics and thermal insulation properties. The axes of the coils are offset below the axes of the cryostats such that 70% of the liquid helium (LHe) contained in the 290-liter capacity cryostat will be above the coils (see Fig. 2). In fact, the bottom of the inner wall of the cryostat is less than 2.25 cm away from the bottom of the outside edge of the coil. The LHe cryostat is insulated with superinsulation, an 80° K liquid nitrogen shield and an inner helium vapor-cooled 20° K radiation shield.

The offset-bore design, the special retractable power lead/persistent heater switch system, and the unique supports allow for a low helium loss rate cryostat. Initial tests with the first solenoid demonstrated a LHe boil off rate of less than 0.17 liters per hour with leads retracted. Thus, optimistically we can run for about 40 days with a persistent current in the magnets without refilling helium. This is essential as there is no LHe liquifier or recovery system available and therefore LHe costs must be minimized.

The large-bore all-aluminum cryostats provide a relatively lightweight system that unlike stainless steel is also less susceptible to long-lived neutron activation. Wherever possible, we have also used aluminum components for the various cryostat and beamline subsystems. Besides the possibility of activation however, neutrons may also cause the decay of the magnetic field in a persistent current superconductor. Recently, experimenters at GANIL reported significant neutron heating in their SISSI RNB production system.[15] Although UND operates at very much lower energies and hence lower neutron fluxes than GANIL, whether or not TwinSol will experience similar difficulties remains to be seen (provisions are available for running in non-persistent mode). One of the advantages of the combination of large-bore cryostat and small production target chamber is that the geometry allows for tight neutron shielding around the production target combined with a suitable setback, thus offering greater radiation protection for the coil.

The production chamber is a custom aluminum ISO-100 4-way cross separated from the ISO-200 aperture and Faraday cup chamber by a gate valve. In addition to conventional production targets, a rotating target and a newly developed gas target cell (1 - 10 atm) which was successfully implemented in a recent RNB experiment[16] are available. Behind the first solenoid is an ISO-250 4-way cross that will serve as the housing for a radial electrostatic field electrode (ELCO lens)[17] that is being developed to further select reaction products exiting from the first solenoid (see below). The secondary beam chamber will be mounted on a movable stand. For certain experiments that require especially low background, e.g. RNB γ-spectroscopy, the secondary chamber can be placed in the next room, behind a 1.3 m thick concrete wall.

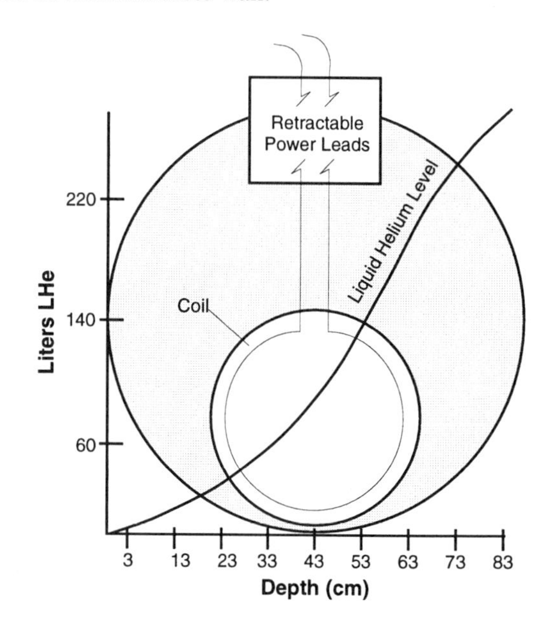

FIGURE 2. Offset bore and retractable power lead design help minimize the LHe loss rate.

MODES OF OPERATION

There are two basic configurations available as shown in Fig. 3. The "parallel" flight mode is reserved for studying only the most magnetically rigid (high-mass, low charge-state) ions. The "crossover" mode offers several advantages over the parallel mode. First, because of the small area of the crossover point, better radiation shielding is available in the space between the two magnets.

Second, there are several useful devices which can be placed at the crossover point. Small area (thus low capacitance) detectors at the crossover focus will allow for fast timing of particles. Such a timing detector can be thin (*viz.* 10 μm) silicon, a scintillator foil, or foil and microchannel plate. When detectors become count rate limited for high-intensity RNBs, then the intrinsic primary beam pulsing properties of the tandem accelerator will be used for timing. Beam pulses with good timing ($\delta t \sim 1$ ns) coupled with a typical flight time of 200 ns over the ca. 6 m flight path implies that 1% energy resolution is attainable, e.g. 200 keV for a 20 MeV RNB. This is sufficient to resolve most nuclear states.

Other possibilities include placing a position sensitive detector (PSD) near the crossover point to provide ray tracing information. Prototype work with the UM 7 T solenoid at NSCL has shown that a PPAC, even in a very high magnetic field, is usable as both an active "aperture" PSD and a timing detector.(13) In another configuration, unwanted reaction products that have the same rigidity as the secondary beam of interest and come to a focus at the crossover point may be filtered out by using a combination of apertures and "mid-plane absorbers" (energy loss foils) in conjunction with the second solenoid. In some instances the detector at the crossover point may also serve as the absorber. With the older 3.5 T solenoid RNB setup, ^{8}Li beams (10^8 particles per sec.) of 80% purity were attainable at the focal plane. Using absorbers between the two solenoids in the new setup should provide up to 95% beam purity after the second magnet. In general, however, particle identification from time-of-flight information and the use of position sensitive detectors as active apertures will allow us to "separate" out the impurities during the data analysis.(13) These methods will be tested once the second solenoid is operational (late 1996).

Notice in Fig. 3 that the magnetic fields of the two solenoids are shown as parallel. An anti-parallel field configuration where the two magnets are opposing each other will still focus the beam as it passes through the two solenoids. Since solenoids are approximately double focusing - they focus particles as they enter the bore and then also focus as the particles leave - the difference in the modes is that anti-parallel field modes (due to the superposition of opposing fields in the central region) will have slightly reduced bending powers, hence longer crossover focal lengths, compared to parallel field modes (this has been verified using orbit-simulation codes). However, when certain detectors (PMTs/MCPs) are present between the two magnets, we may deliberately choose the anti-parallel configuration in order to reduce the field in the central region. Another device that is currently being developed is a high voltage electrostatic filter to be used to separate secondary beams. This consists of two concentric, cylindrically-shaped electrodes placed in between

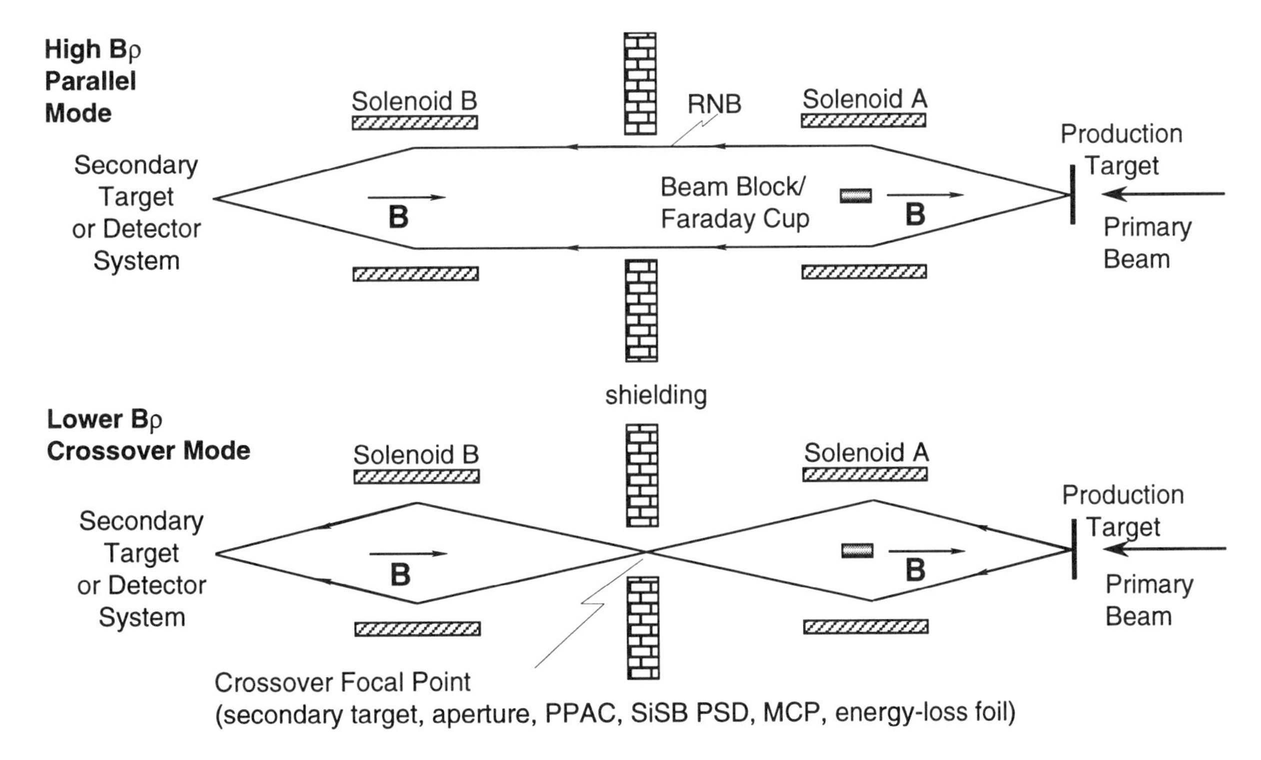

FIGURE 3. Parallel and crossover modes of operation.

the two magnets. Simulations show that commercially available, modest (a few tens of kV) power supplies will sufficiently separate the sort of low-energy RNBs that we will be producing. Figure 4 shows trajectories for a scenario of two helium isotopes with the same rigidity at several different exit angles from the production target. The background $^4He^{++}$ would typically overwhelm the $^6He^{++}$ of interest at the focal point. Initially, both follow identical paths through the solenoid but are separated by the electrostatic filter. The mean change in focal lengths for this particular simulation is 50 cm. These calculations indicate that the effects of the radial electric field are most useful only if a small range of production angles ($\delta\theta \sim 2°$) is selected or ray-tracing is utilized. This method is therefore best suited only with certain RNBs and solenoid modes.

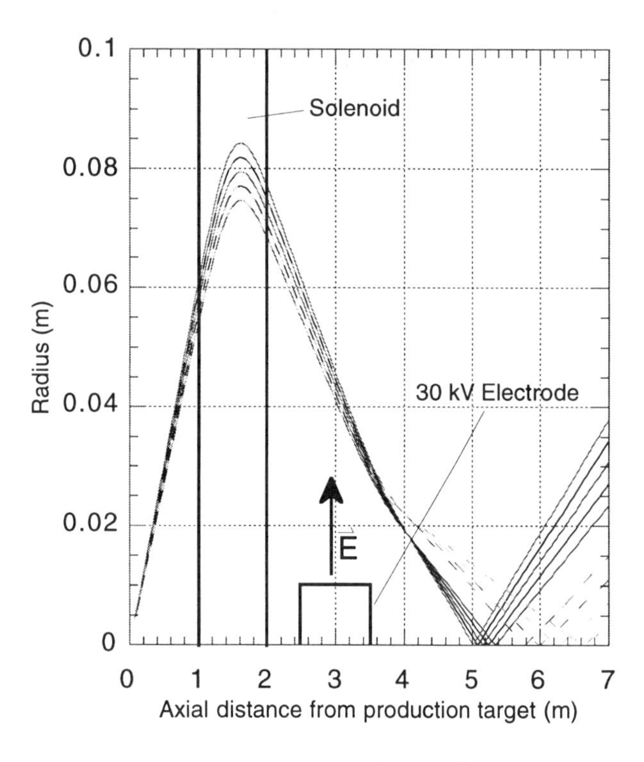

FIGURE 4. Calculated trajectories of ^{4}He and ^{6}He at 30 MeV and 20 MeV respectively with exit angles between 3.0° and 3.5°. The inner electrode has radius 1 cm, length 1 m, and electric potential at 30 kV. The grounded outer electrode is off scale in this plot (radius = 0.15 m). Typical potentials will range from 30 to 80 kV.

DISCUSSION AND CONCLUSIONS

Although TwinSol is a major part of the upgrade to the UND Nuclear Structure Laboratory, other improvements at UND include the recent installation of a small KN Van de Graaff (for low-energy stable beam astrophysics research), a new helium ion source and on-going work with the FN-Tandem which has successfully delivered beams at a 10.4 MV terminal voltage. As such, TwinSol will allow for experiments with RNBs at higher energies (E/A $\rightarrow$ 8 MeV/u) than previously studied at UND with the older, single UM 3.5 T solenoid (E/A < 3 MeV/u).

TwinSol will be especially well suited to the study of reactions involving both stable and radioactive light heavy ions (A < 40) near the Coulomb barrier where there is unique and interesting physics. We anticipate that TwinSol will be fully operational by late 1996. As of October 1996, the first half of the apparatus, up to the mid chamber shown in Fig. 1, has been put in place. Updated information may be found on our WWW page.

ACKNOWLEDGMENTS

TwinSol is funded jointly by the University of Michigan, University of Notre Dame and the NSF through the Academic Research Infrastructure program, grant PHY-9512199.

REFERENCES

1. *Proc. Fourth International Conference on Radioactive Nuclear Beams*, Omiya, Japan, June 4-7, 1996, Amsterdam, North-Holland, in press.
2. Becchetti, F.D., "LOW-ENERGY RADIOACTIVE BEAM EXPERIMENTS USING THE UM-UND SOLENOID RNB APPARATUS AT THE UND TANDEM: PAST, PRESENT AND FUTURE," presented at 14th Int'l Conf. on the Applications of Accelerators in Research and Industry, Denton, TX, Nov. 6-9, 1996, (this conference).
3. Schapira, J.P., Gales, S., and Laurent, H., "UN SPECTROMETRE CONSTITUE D'UN OU DE DEUX SOLENOIDES CRYOGENIQUES COAXIAUX," IPN-Orsay Report IPN0-PhN-7921, 1979.
4. Schapira, J.P., Azaiez, F., Fortier, S., Gales, S. Hourani, E., Kumpulainen, J., Maison, J.M., *Nucl. Instrum. Methods*, **224**, 337-46, 1984.
5. Kolata, J.J., Morsad, A., Kong, X.J., Warner, R.E., Becchetti, F.D., Liu, W.Z., Roberts, D.A., Jänecke, J.W., *Nucl. Instrum. Methods*, **B40/41**, 503-6, 1989.
6. Becchetti, F.D., Liu, W.Z., Roberts, D.A., Jänecke, J.W., Kolata, J.J., Morsad, A., Kong, X.J., Warner, R.E., *Phys. Rev. C*, **40**, R1104, 1989.
7. Smith, R.J. Kolata, J.J., Lamkin, K., Morsad, A., Ashktorab, K., Becchetti, F.D., Brown, J., Jänecke, J.W., Liu, W.Z., Roberts, D.A., *Nucl. Instrum. Methods*, **A294**, 26, 1990.
8. Becchetti, F.D., Brown, J.A., Ashktorab, K., Jänecke, J.W., Liu, W.Z., Roberts, D.A., Smith, R.J., Kolata, J.J., Lamkin, K., Morsad, A., Warner, R.E., *Nucl. Instrum. Methods*, **B56/57**, 554-8, 1991.
9. Becchetti, F.D., Liu, W.Z., Ashktorab, K., Bajema, J.F., Brown, J.A., Jänecke, J.W., Roberts, D.A., Kolata, J.J., Lamkin, K.L., Morsad, A., Smith, R.J., Kong, X.J., Warner, R.E., *Phys. Rev. C*, **48** , 308-18, 1993.
10. Stern, R.L., Becchetti, F.D., Casey, T., Jänecke, J.W., Lister, P.M., Liu, W.Z., Kovar, D.G., Janssens, R.V.F., Vineyard, M.F., Phillips, W.R., Kolata, J., *Rev. Sci. Instrum.*, **58**, 1682, 1987.
11. Stern, R.L. "Design and Utilization of an Air-Core Superconducting-Solenoid Nuclear-Reaction-Product Spectrometer," Ph.D. Thesis, University of Michigan, UM-NP-RS-87-1, 1987
12. Liu, W.Z., "Production and use of Radioactive Ion Beams for Measurements of Nuclear Reactions," Ph.D. Thesis, University of Michigan, 1990.
13. O'Donnell, T.W., Aldredge, E., Beccheti, F.D., Brown, J.A., Conlan, P., Jänecke, J., Raymond, R.S., Roberts, D.A., Tickle, R.S., Griffin, H.C., Staynoff, J., Ronningen, R., *Nucl. Instrum. Methods*, **A353**, 215-6, 1994.
14. O'Donnell, T.W., Ph.D. Thesis, University of Michigan, in progress, 1997.
15. Savalle, A., Baelde, J.L., Baron, E., Berthe, C., Gillet, J., Grunbrerg, C., Jamet, C., Lemaitre, M., Moscatello, M.H., Ozille, M., "THE SISSI FACILITY AT GANIL," GANIL Report A96.02, 1996.
16. von Schwarzenberg, J., Kolata, J.J., Peterson, D., Santi, P., Belbot, M., Hinnefeld, J.D., *Phys. Rev. C*, **53**, R2598, 1996.
17. Liu, W.Z., Becchetti, F.D., Stern, R., *Rev. Sci. Instrum.*, **58**, 220-2, 1987, and references cited there.

INITIAL RESULTS OF THE COMMISSIONING OF THE HRIBF RECOIL MASS SPECTROMETER

C. J. Gross

Physics Division, Oak Ridge National Laboratory, Oak Ridge, TN 37831
and
Oak Ridge Institute for Science and Education, Oak Ridge, TN 37831

T. N. Ginter

Department of Physics and Astronomy, Vanderbilt University, Nashville, TN 37235

Y. A. Akovali, M. J. Brinkman, J. W. Johnson, J. Mas, J. W. McConnell, W. T. Milner, and D. Shapira

Physics Division, Oak Ridge National Laboratory, Oak Ridge, TN 37831

A. N. James

Department of Physics, University of Liverpool, Liverpool L69 3BX, UK
and
Joint Institute for Heavy Ion Research, Oak Ridge, TN 37831

The recoil mass spectrometer at the Holifield Radioactive Ion Beam Facility is currently undergoing commissioning tests. This new spectrometer is designed to transmit ions with rigidities of K = 100 resulting from fusion-evaporation reactions using inverse-kinematics. The device consists of two sections: a momentum separator to provide beam rejection and a mass separator for product identification. Using normal-kinematic and symmetric reactions, the commissioning tests have shown that the A/Q acceptance is almost ± 5%, the energy acceptance is approximately ± 12%, and there has been little, if any, primary beam observed on the focal plane. Commissioning tests are presently underway with reactions using inverse-kinematics.

A recoil mass spectrometer (RMS) is currently undergoing commissioning tests at the Holifield Radioactive Ion Beam Facility (HRIBF). Similar devices, which are often used to identify exotic nuclei resulting from compound nuclear reactions, are located at many laboratories throughout the world [1-4]. The RMS at the HRIBF represents a new generation of spectrometers [5] by combining a momentum separator for primary beam rejection and a mass separator for product identification.

A. COMPONENTS OF THE RMS

A schematic of the K = 100 RMS is shown in Fig. 1 and has been described in-depth in Ref. [5-7]. In summary, it consists of three magnetic dipoles (D), seven quadrupoles (Q), two sextuples (S), and two electric dipoles (E). The physical properties of each element may be found in Ref. [5]. The first two quadrupoles gather the recoiling nuclei and determine the momentum focus which is located inside Q3. The first magnetic dipole separates the recoiling nuclei from the beam based on momentum. The primary beam, originating from a tandem accelerator, has a well-defined momentum and should be focused spacially according to its charge-state distribution. The recoils have a large momentum distribution caused by the evaporation process and will thus, fill the available space. Small rods called "fingers" may be inserted through the split poles of Q3 to intercept the primary beam at its focus and still have minimal impact on the overall transport efficiency for recoil products. These "fingers" are intended for use with inverse-kinematic reactions where beam particles and recoils have similar rigidities and hence, would all be accepted by the RMS, these "fingers" could provide beam rejection factors of 10^{13}.

The second half of the momentum separator, D2-Q4-Q5, provides a second focus which serves as the object for the mass separator portion of the spectrometer. This position, called the achromat, is energy independent when Q3 is correctly adjusted. The two sextuples provide second-order corrections in the final focus.

The mass separator section contains a magnetic dipole between two electric dipoles. The electric elements separate the recoils as a function of kinetic energy and charge (E/Q) and the magnetic dipole separates as a function of momentum and charge (P/Q). The net result are recoils separated as a function of mass and charge (A/Q). The magnitude of the mass dispersion is determined by the last pair of quadrupoles, Q6 and Q7. Together with Q4 and Q5, the position of the final focal plane is determined.

CP392, *Application of Accelerators in Research and Industry*, edited by J. L. Duggan and I. L. Morgan
AIP Press, New York © 1997

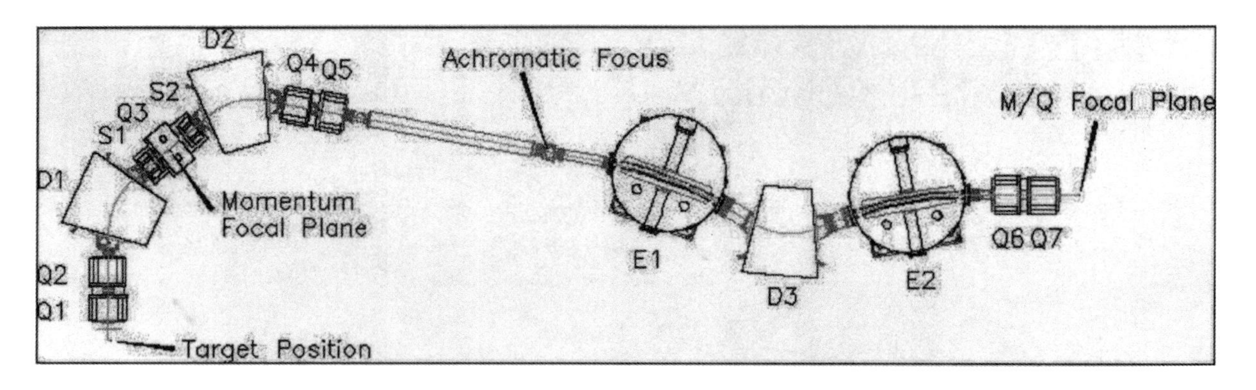

FIGURE 1 The layout of the RMS. The fingers are located at the momentum focal plane located inside Q3.

The performance parameters of the spectrometer, such as the size of the final image and the mass dispersion, often depend only upon specific groups of elements of the RMS. These elements may be adjusted together without affecting the field values of the other elements. Thus, several "knobs" have been incorporated into the control system so that users may adjust these parameters according to the requirements of the experiment.

B. DESIGN PERFORMANCE PARAMETERS

Various acceptances govern the performance of the RMS and its efficiency to detect recoil products. The RMS is a zero degree device and has an asymmetric, overall solid-angle acceptance. The horizontal acceptance is ± 30 mrad and the vertical acceptance is ± 115 mrad and is determined by the positions of Q1 and Q2. The target-to-Q1 distance is 75 cm. However, other apertures in the system affect the effective overall solid angle which ranges between 10- and 15 msr. The primary beam spot on target should be a vertical line 0.5 mm in width and 2 mm in length. The energy acceptance is determined, in part, by the length of the plates of the electric dipoles and is estimated to be some ± 15% at the base. The A/Q acceptance, reflecting the various apertures throughout the flight path is near ± 5%. Overall mass resolution (M/ΔM) has been calculated to be 540 although through collimation, it is estimated that this value may approach 1000 (including software corrections).

C. COMMISSIONING DETECTORS

The mainstay detector of the RMS is the position-sensitive avalanche counter (PSAC) located at the final focal plane of the RMS. This detector has an active area of 36 cm by 10 cm and is filled with isobutane gas. The arrangement of planes with respect to the beam is cathode, horizontal sensing plane (electrically split in half), anode, and vertical sensing plane. The anode and cathode have 20 μm gold-plated tungsten wires spaced 1 mm apart. The position sensing planes have the same wires which are spaced 2 mm apart and are connected to delay-lines using 2 ns taps per wire. The overall position resolution attainable with this device is approximately 2 mm. The total PSAC efficiency (defined as those events which cause an anode signal) is better than 90%.

Behind the focal plane, several detector systems may be used. Double-sided silicon-strip detectors (see contribution by K. S. Toth, *et al.*), an ionization chamber, and a moving tape collector have been constructed and will undergo commissioning tests this year. At the target position, a germanium detector array consisting of segmented-Clover detectors (4 Ge detectors in a single housing) and conventional Ge detectors will be used. In the present tests, only four conventional Ge detectors have been used in conjunction with the PSAC to confirm mass identification.

D. INITIAL PERFORMANCE

An alpha source and four reactions have been used to commission the RMS and these are listed in Table 1.

TABLE 1. Reactions used in the initial commissioning tests of the RMS.

Beam	Energy (MeV)	Target	Thickness ($\mu g/cm^2$)	Goal
^{4}He	5.8			initial parameters
^{32}S	120	^{58}Ni	300	mass identification
^{58}Ni	250	^{98}Mo	900	strip detectors[1]
^{58}Ni	250	^{92}Mo	500	strip detectors[1]
^{58}Ni	220	^{60}Ni	300	mass separation

[1]Attempts to identify ground state alpha and proton decays from the reaction products using double-sided silicon-strip detectors. See the contribution to these proceedings by K. S. Toth, *et al.* for information about these tests.

The highest mass resolution achieved so far has been M/ΔM = 450 using the Ni+Ni reaction and is shown in Fig. 2. The energy acceptance of the spectrometer has been measured to be at least ± 12.5%. This value is smaller than similar devices (± 20%) due to the longer plates of the electrostatic dipoles. The A/Q acceptance has been measured to be in excess of 4.5%. A total reaction efficiency (not corrected for dead time) has been estimated to be 1-2%. Because this efficiency is for all

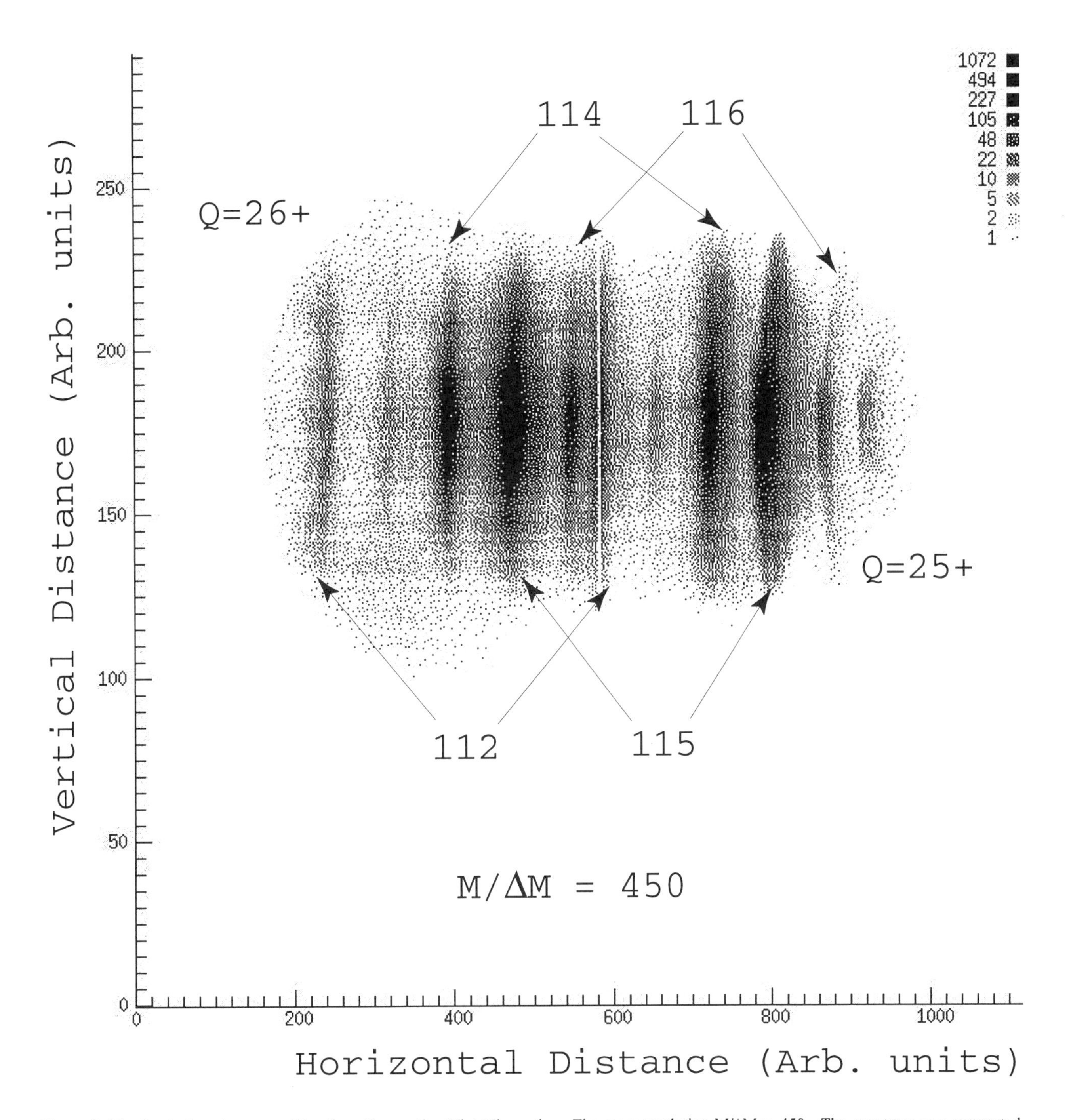

Figure 2 The focal plane image resulting from the reaction Ni + Ni reaction. The mass resolution M/ΔM = 450. The spectrum was generated from the PSAC requiring that the sum of each delay-line signal falls within a relatively small gate and the entire range is shown. This spectrum has no other gating conditions. Notice the absence of beam in this spectrum.

fusion-evaporation recoils, individual channel efficiency will vary with pure nucleon evaporation channels being more efficient than alpha evaporation channels. These efficiencies will increase when inverse-kinematic reactions are used.

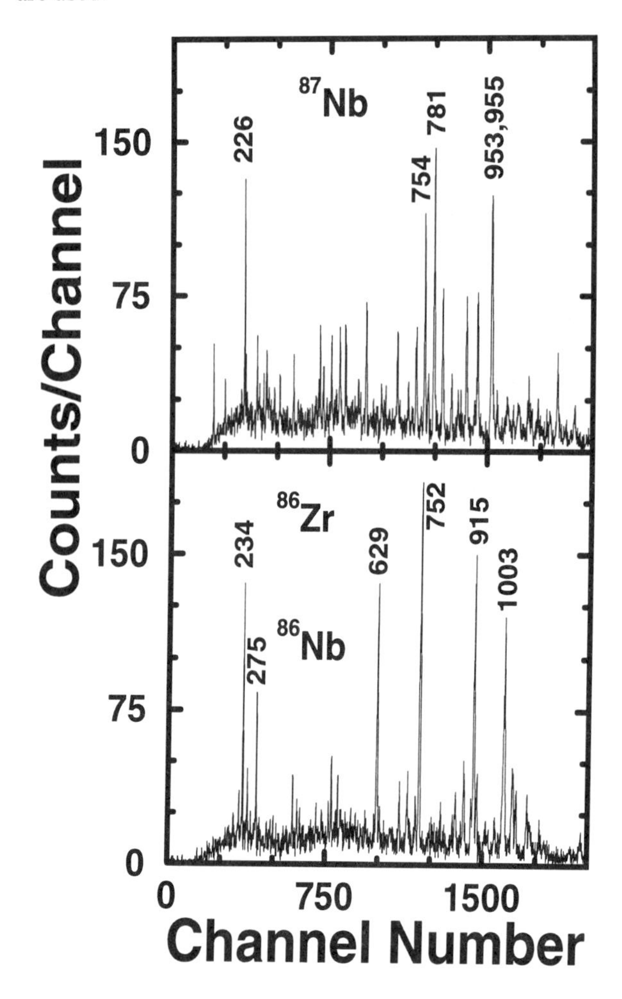

Figure 3 Mass 86 and 87 gated γ ray spectra from the S + Ni reaction. All indicated transitions in the upper spectrum correspond to transitions observed [8] in 87Nb. The indicated transitions in the lower spectrum correspond to known transitions in 86Nb [9] (275keV) and 86Zr [10] (all others).

One persistent feature of the commissioning tests so far, has been the absence of beam on the focal plane. The momentum separator is clearly preventing scattered beam from reaching the PSAC allowing high primary beam currents to be used. Most tests have been conducted with 15 pnA, or more, yet the PSAC has always run at less than 1 kHz.

Mass 86 and 87 γ ray spectra are shown in Fig. 3. These data were taken with the S+Ni reaction and demonstrate the channel selection sensitivity of the RMS. Only those known transitions in ^{86}Nb, ^{86}Zr, and ^{87}Nb can be observed in their respective mass-gated spectrum.

Further commissioning tests will be continuing throughout this year and in 1997. Inverse-kinematic reactions and the successful use of the "finger" system will constitute a large portion of our future efforts. Our upgraded Ge array will become available in 1997, as well as, a 4π charged particle ball located at the target position. Neutron detectors will also augment this in-beam spectroscopy system. The ionization chamber and tape collector plus pair spectrometer plus Clover Ge system will also be tested in the coming months. As radioactive ion beams become available for nuclear structure studies, the RMS stands ready to investigate nuclei far from stability.

E. ACKNOWLEDGMENTS

We would like to acknowledge the work performed by P. F. Mantica, J. J. Das, and R. L. Auble on the installation of the RMS. We acknowledge support from the SERS program administered by Oak Ridge Associated Universities (J. M.) and the University-DOE Laboratory Cooperative Graduate Student Research Participation Program administered by the Oak Ridge Institute for Science and Education (T. N. G.). Research sponsored by the Oak Ridge National Laboratory, managed by Lockheed Martin Energy Research Corporation for the U.S. Department of Energy under contract number DE-AC05-96OR22464. This work was also supported by the U.S. DOE under contract number DE-AC05-76OR00033 (Oak Ridge Institute for Science and Education).

F. REFERENCES

[1] Cormier, T. M., *et al.*, Nucl. Instrum. Methods **212**, 185 (1983).

[2] James, A. N., *et al.*, Nucl. Instrum. Methods in Phys. Res. **A267**, 144 (1988).

[3] Spolaore, S., *et al.*, Nucl. Instrum. Methods **A238**, 381 (1985).

[4] Davids, C. N., *et al.*, Nucl. Instrum. Methods in Phys. Res. **B70**, 358 (1992).

[5] Cole, J. D., *et al.*, Nucl. Instrum. Methods in Phys. Res. **B70**, 343 (1992).

[6] Cole, J. D., *et al.*, *in Exotic Nuclear Spectroscopy*, edited by C. McHarris, (Plenum Press, New York, 1990), p 11.

[7] Mantica, P. F., Nucl. Instrum. Methods in Phys. Res. **B99**, 338 (1995).

[8] Jungclaus, A., et al., Z. Phys. **A340**, 125 (1991).

[9] Gross, C. J., et al., Nucl. Phys. **A535**, 203 (1991).

[10] Chowdhury, P., et al., Phys. Rev. Lett. 67, 2950 (1991)

INVESTIGATING PROTON EMITTERS AT THE LIMITS OF STABILITY WITH RADIOACTIVE BEAMS FROM THE OAK RIDGE FACILITY

K. S. Toth

Oak Ridge National Laboratory, Oak Ridge, TN 37831 USA

J. C. Batchelder and E. F. Zganjar

Louisiana State University, Baton Rouge, LA 70803 USA

C. R. Bingham and J. Wauters

University of Tennessee, Knoxville, TN 37996 USA

T. Davinson, J. A. MacKenzie, and P. J. Woods

University of Edinburgh, Edinburgh, EH9 3JZ UK

By using beams from the Holifield Radioactive Ion Beam Facility at the Oak Ridge National Laboratory it should be possible to identify many new ground-state proton emitters in the mass region from Sn to Pb. In these investigations nuclei produced in fusion-evaporation reactions will be separated from the incident ions and dispersed in mass/charge with a recoil mass separator and then implanted into a double-sided Si strip detector for the study of proton (and α-particle) radioactivity. This paper summarizes data presently extant on proton emitters and then focuses on tests and initial experiments that will be carried out with stable beams and with radioactive ions as they are developed at the Oak Ridge facility.

BACKGROUND

Because of the repulsive Coulomb force the proton drip line is located much closer to the valley of stability than the drip line on the neutron-rich side. Not only is the proton drip line more accessible, but even if a nucleus is proton unbound, the Coulomb barrier slows down proton emission. One can therefore obtain structure information for nuclei existing *beyond* the drip line, something that cannot be done on the neutron-rich side where decay takes place essentially instantaneously if the last neutron is unbound.

In proton decay an energetically unbound proton emerges from the nucleus by tunneling through the Coulomb and centrifugal barriers. Because only a single nucleon is involved, this process is simpler to describe than α decay since there is no particle preformation factor to consider. In addition, the decay width is much more sensitive to the shape and size of the centrifugal barrier (and therefore ℓ) than in the case of α-particle emission. Thus, measurements of proton-decay half-lives and energies can provide spectroscopic information. For example, the use of ℓ = 2 and ℓ = 5 potentials for the emission of a 1-MeV proton in the rare earth region results in two barrier penetration times that differ by a factor of 100.

Ground-state proton decay as a factor in determining the limit to nuclear existence had long been discussed in theoretical papers, but it was not observed until 1981 when ^{151}Lu was reported (1) to be a proton emitter. Within a few years other isotopes, ^{109}I, ^{113}Cs, and ^{147}Tm (and possibly ^{150}Lu) were found to decay by proton emission [see the summary in Ref. (2)]. Further searches proved not to be fruitful until 1992 when the Daresbury Recoil Separator was used, in conjunction with the newly-developed double-sided strip Si detectors (3), to observe proton decay from ^{156}Ta and ^{160} Re (4), ^{146}Tm (5), ^{150}Lu (6), and ^{112}Cs (7). The development of a more sensitive technique was necessitated by the fact that while the first set of emitters (2) were produced in (^{58}Ni, p2n) reactions induced on a variety of targets with cross sections of about 50 μb, the ones found at Daresbury, were produced in (^{58}Ni, p3n) reactions with cross sections down in the few μb range.

This same technique has now been used at Argonne National Laboratory during the past few years where a

CP392, *Application of Accelerators in Research and Industry*, edited by J. L. Duggan and I. L. Morgan
AIP Press, New York © 1997

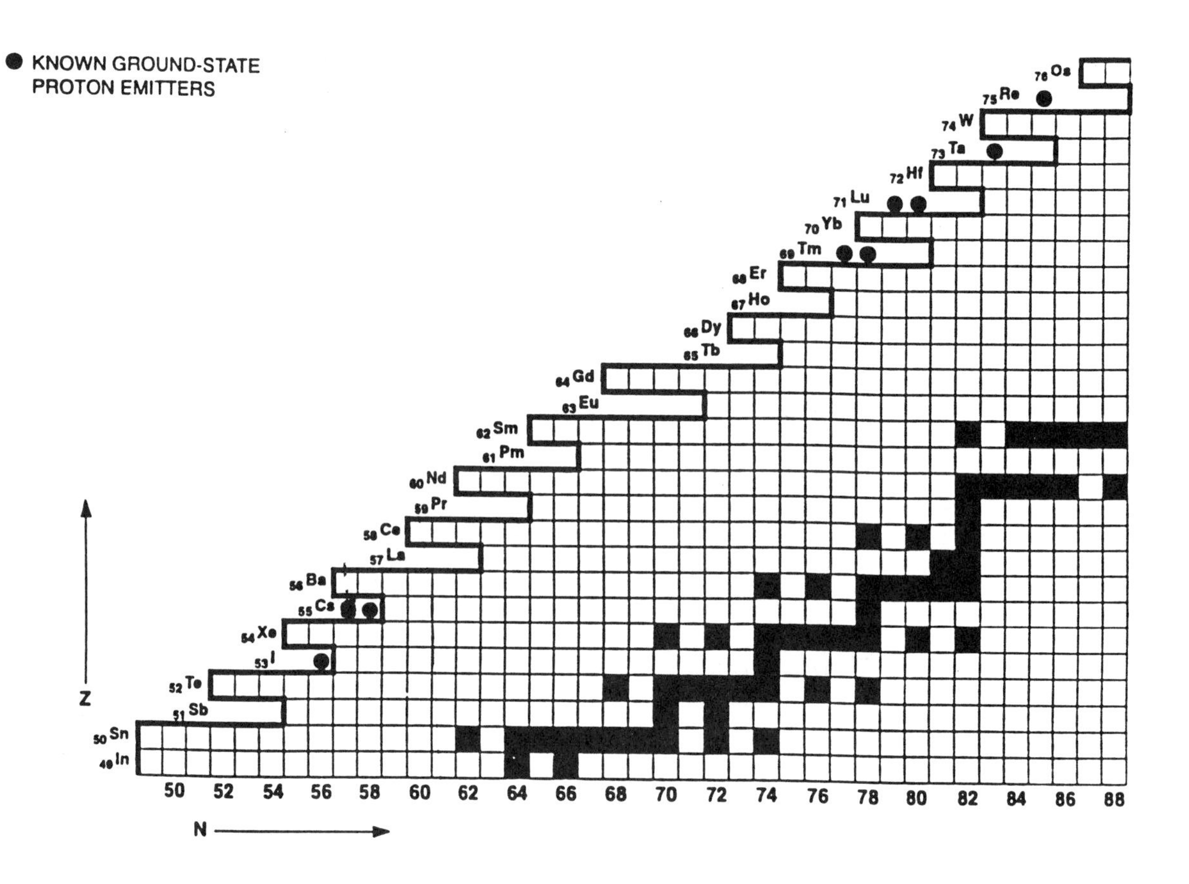

FIGURE 1. Portion of the nuclidic chart which shows one predicted (10) proton drip line (heavy border on the left side of the diagram) and the nine isotopes in this mass region that are known to emit protons from their ground states.

group of investigators (including the present authors) has studied ground-state proton decay in elements above rhenium utilizing strip detectors and a fragment mass analyzer (8). Several new cases of proton emission have been identified; that of $^{185}Bi^{m}$ has been described in a recent publication (9).

The location of known emitters (2,4-7) up to rhenium is shown in Fig. 1 where one predicted (10) drip line is indicated by the heavy border on the left side of the diagram. For the emitters near N = 82, lifetimes calculated with a one-body barrier penetration model and shell-model spin and parity assignments based on systematics of single-proton levels have been found to agree with experimental values. However, for ^{112}Cs, ^{113}Cs, and ^{109}I experimental rates are slower than calculated half-lives by factors of between 10 and 100. These hindrances are thought to be due to the fact that the three nuclei are transitional rather than spherical in shape. Bugrov and Kadmenskii (11) have recently examined these nuclei and have addressed theoretically the question of deformation effects on proton-decay half-lives. Neutron-deficient isotopes with masses between ^{113}Cs and ^{147}Tm have been predicted to be well-deformed and their proton decay rates should provide even more stringent tests for theoretical calculations. Experimental efforts to produce these emitters have not been fruitful [see e.g. Ref. (12)] partly as a result of a dearth of suitable target and beam combinations.

PROPOSED EXPERIMENTS

At the Oak Ridge National Laboratory we are mounting a research program to search for and investigate new cases of proton emission in the mass region between Cs and Tm. The attempts will be made by utilizing strip detectors, a recoil mass spectrometer (13), and beams from the Holifield Radioactive Ion Beam Facility (14).

Figure 2 shows compound systems in the In - Os mass region that could be produced with three of the radioactive beams (^{33}Cl, ^{58}Cu, and ^{69}As) considered for early delivery incident on extremely neutron-deficient even-even targets. One sees that as far as getting beyond the drip line ^{58}Cu does best overall with ^{69}As better for a

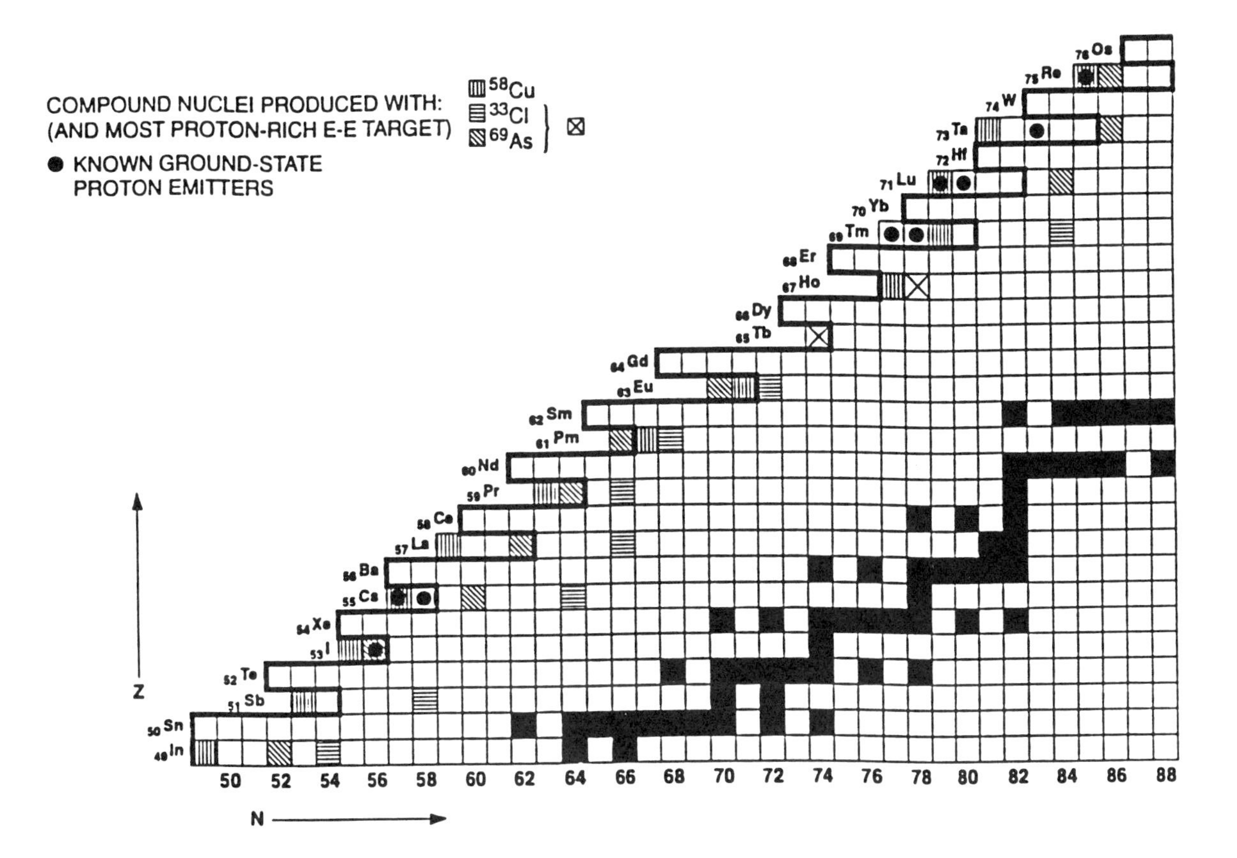

FIGURE 2. Portion of the nuclidic chart showing compound nuclei that could be formed with stable neutron-deficient even-even targets and radioactive beams of ^{33}Cl, ^{58}Cu, and ^{69}As. The two squares marked with an "X" indicate nuclei that can be formed with both ^{33}Cl and ^{69}As. Known proton emitters and one predicted (10) proton drip line (heavy border on the left side of the diagram) are also shown.

few elements and ^{33}Cl clearly not as well. Obviously (xn) evaporation products would be channels of choice but here at the drip line their cross sections are extremely low and one has to rely on reactions wherein charged particles are evaporated together with neutrons. The (pxn) channels lead to even-Z elements and, except for the heavier nuclei in Fig. 2, not beyond the drip line. However, the (αxn) channels, produce very neutron-deficient odd-Z isotopes and for many of these elements the (αn) and (α2n) channels should get us far enough to observe proton radioactivity in the region intermediate between ^{113}Cs and ^{147}Tm.

Development of intense radioactive beams at the Oak Ridge Facility has proceeded. Recently, a low intensity beam of ^{70}As was produced and accelerated through the facility's 25-MV tandem. Along with these developments, tests with stable beams from the tandem accelerator have been used to commission the recoil mass spectrometer and its ancillary experimental equipment. In a recent test, a double-sided Si strip detector of thickness 60 μm, area 4 x 4 cm, and having 40 orthogonal strips on the front and rear was placed close to the focal plane of the recoil spectrometer. A 0.5 mg/cm^2 thick foil enriched in ^{92}Mo was bombarded with ^{58}Ni and products with A = 147 were then implanted into the detector. Figure 3 shows the spectrum recorded and one sees protons that were emitted from the ^{147}Tm 0.6-s ground state and its 360-μs isomer. Further tests to fine tune the strip detector electronics have been scheduled for the near future.

During these searches for new proton emitters most of the nuclei encountered will primarily EC/β^+ decay. With this in mind, we will place a thick Si detector directly behind the strip detector and Ge detectors outside the strip detector chamber to record β-delayed proton and γ-ray spectra, respectively. For half-lives > 0.5s, we plan to use a tape collection system to extract and transport nuclides to shielded areas where their radioactivities can be studied with large Ge detectors in close geometry. The nuclides will be collected at the focal plane from either neighboring mass positions or adjacent charge states of the masses being studied with the strip detector.

In conclusion we note that the mass-tagging capabilities of the recoil separator can be used in conjunction with a Ge array at the target position. This combination permits the simultaneous study of level properties by in-beam measurements of nuclei whose decays are being investigated with detectors at the focal plane. Indeed, for in-beam studies that require additional filtering, correlation of events at the target position with decays of particular recoil products can provide isotopic identification.

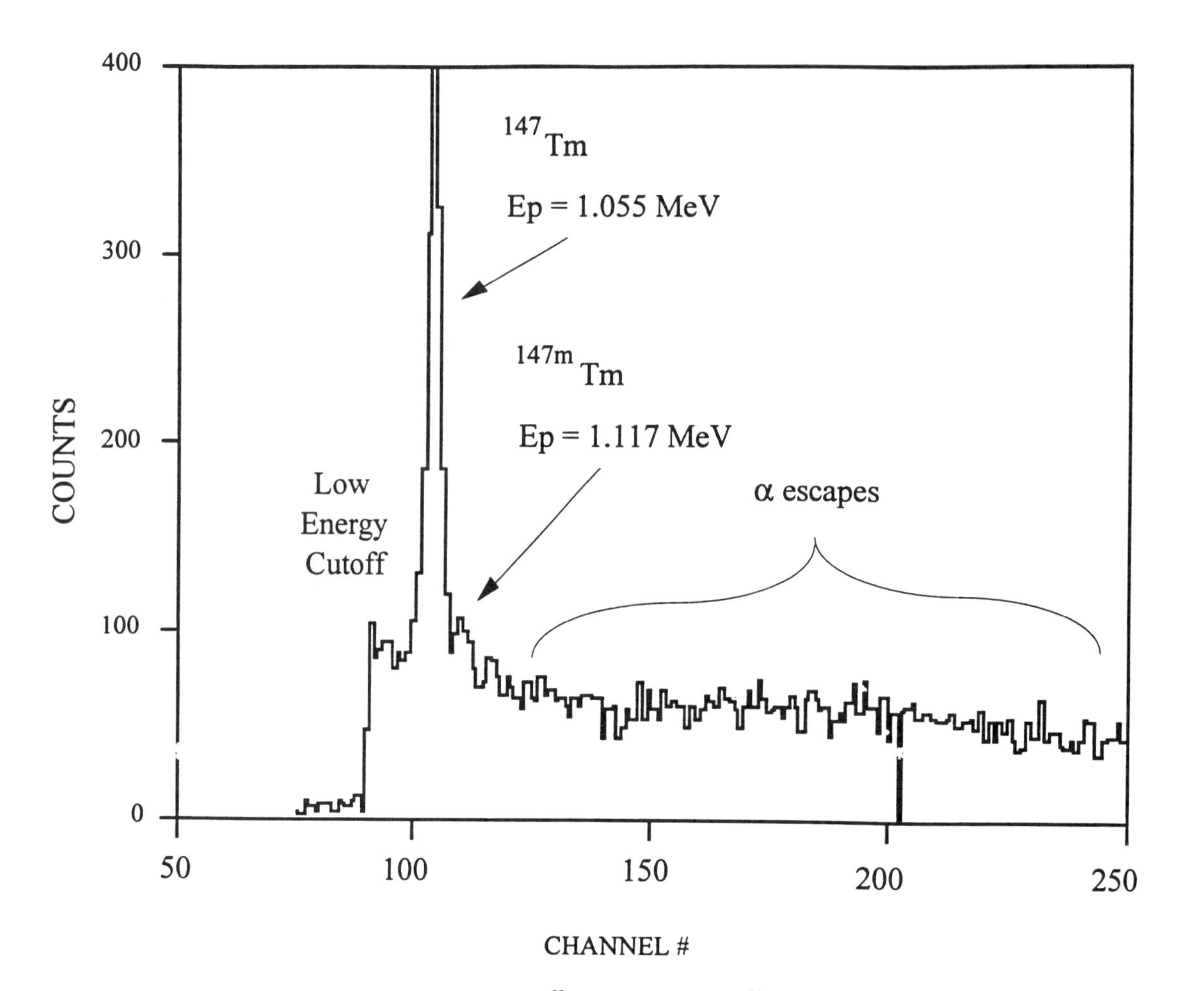

FIGURE 3. Spectrum recorded in a double-sided Si strip detector during ^{58}Ni bombardments of ^{92}Mo. The detector was positioned at the focal plane of the recoil mass spectrometer to accept A = 147 products. Protons from the ^{147}Tm 0.6-s ground state and the isotope's 360-μs isomer are seen in the spectrum together with escape α particles emitted by nuclides produced in reactions on heavier Mo isotopes present in the target.

ACKNOWLEDGMENTS

Research sponsored by the Oak Ridge National Laboratory, managed by Lockheed Martin Energy Research Corporation for the U.S. Department of Energy under contract number DE-AC05-96OR22464. This work was also supported by the U.S. DOE under contract numbers DE-FG05-84ER40159 (Louisiana State University) and DE-FG05-87ER40361 (University of Tennessee).

REFERENCES

1. Hofmann, S., *et al.*, *Z. Phys.* **A305**, 111-123 (1982).
2. Hofmann, S., *et al.*, "Experiments on Proton Radioactivity," in *Proceedings of the 7th International Conference on Atomic Masses and Fundamental Constants*, Darmstadt-Seeheim (1984), pp. 184-195.
3. Sellin, P. J., *et al.*, *Nucl. Instrum. Methods Phys. Res.* **A311**, 217-223 (1992).
4. Page, R. D., *et al.*, *Phys. Rev. Lett.* **68**, 1287-1290 ((1992).
5. Livingston, K., *et al.*, *Phys. Lett.* **B312**, 46-48 (1993).
6. Sellin, P. J., *et al.*, *Phys. Rev.* **C 47**, 1993-1942 (1993).
7. Page, R. D., *et al. Phys. Rev. Lett.* **72**, 1798-1801 (1994).
8. Davids, C. N., *et al.*, *Nucl. Instrum. Methods Phys. Res.* **B70**, 358-361 (1992).
9. Davids, C. N., *et al.*, *Phys. Rev. Lett.* **76**, 592-595 (1996).
10. Comay, E., Kelson, I., and Zidon, A., *At Data Nucl. Data Tables* **39**, 235-240 (1988).
11. Bugrov, V. P. and Kadmenskii, S. G., *Sov. J. Nucl. Phys.* **49**, 967-972 (1989).
12. Livingston, K., *et al.*, *Phys. Rev. C* **48**, 3113-3114 (1993).
13. Gross, C. J., *et al.*, "Initial Results of the Commissioning of the HRIBF Recoil Mass Spectrometer," in *Proceedings of the 14th International Conference on the Application of Accelerators in Research and Industry*, November 6-9, 1996, Denton, Texas, U.S.A.
14. Garrett, J. D.,.Alton,G. D., Baktash, C., Olsen, D. K., and Toth, K. S., *Nucl. Physics* **A557**, 701c-714c (1993).

NUCLEAR STRUCTURE WITH RADIOACTIVE BEAMS AND THE ISOSPIN LABORATORY (ISL) INITIATIVE

R. F. Casten

WNSL, Yale University, New Haven, Connecticut 06520-8124

A selection of recent developments in nuclear structure physics are briefly discussed in the context of plans for an advanced, next-generation ISOL-based radioactive beam facility in North America. These developments focus on several important areas, such as the structure of N=Z nuclei, the robustness or fragility of magic numbers and the underlying shell structure of nuclei, the effects of the valence p-n interaction, and the discovery of simple, compact, nearly universal correlations of nuclear observables. Brief comments are made on recent advances in experimental instrumentation and techniques that will allow meaningful nuclear structure information to be extracted from experiments with beam intensities that can be many orders of magnitude weaker than we are accustomed to.

INTRODUCTION

An exciting era in nuclear physics is being ushered in with the advent of new accelerators capable of producing beams of exotic nuclei (Radioactive Nuclear Beams or RNBs) far from the valley of stability. Powerful facilities utilizing the projectile fragmentation technique already exist and/or are being upgraded, and first generation ISOL-machines are also operating. In the recent DOE/NSF Long Range Plan for Nuclear Science the top priority for new construction projects is a twin recommendation for an upgrade to the MSU facility and the development of a cost effective plan for construction (after RHIC) of an advanced, next-generation ISOL type facility with capabilities similar to those envisioned in the IsoSpin Laboratory (ISL) concept(1). The advent of these facilities will open up whole new horizons in nuclear physics and give us access to regions of nuclei where there is every likelihood to encounter phenomena unlike anything found to date near stability. Already such ideas as halo nuclei have become major areas of active research and give a taste of the prospects to come.

NUCLEAR PHYSICS IN EXOTIC NUCLEI

In the rest of the paper, we discuss some research areas in nuclear structure that will be of particular interest in RNB studies. Among the plethora of new ideas for RNB studies, we will focus on just a few that typify the opportunities at hand.

Robustness of Shell Structure

The microscopic underpinnings of nuclear physics lie in the concepts of shell structure and magic numbers as embodied in the Shell Model. The shell gaps at nucleon numbers 2,8,20,50,82, and 126 are renowned benchmarks along the nuclear chart. Nearly all models of nuclear structure focus on the valence or near-valence space. Yet, since at least as far back as the early 1970's there have been hints of the possible fragility of structure. Proton sub-shell gaps at Z = 40 and 64, for certain neutron numbers only, were discovered(2-5), the former in early experiments on neutron-rich nuclei near A = 100. More recently, the foundations of structure in the idea of immutable magic numbers were more severely shaken by the studies (6) of ^{32}Mg which, though it has N = 20, is not only non-magic in character but probably deformed.

The disappearance of benchmark shell gaps or the appearance of new ones has taken on additional credibility recently due to noteworthy calculations(7) of Dobaczewski, Nazarewicz and their colleagues, who studied the approach to the neutron drip line (e.g., in Sn). In near neutron drip line nuclei the outer realms consist of low density, diffuse, weakly bound, spatially extended, nearly pure neutron matter (neutron skin or halo). Combined with the effects of nucleon scattering into the (nearby) continuum, such a structure seems unlikely to be able to support the sharp contours of the traditional shell model potential (Woods-Saxon). The calculations of ref. 7 suggest a melting of shell structure, a major re-ordering of single particle level sequences and, if magic numbers retain validity at all, they are likely to be quite different from the standard ones. (e.g., N = 70, 112 instead of 82 and 126). Such a scenario is contrasted with the usual shell sequence in Fig. 1. Note that the usual $\Delta j = 1$ sequence of normal parity orbits is replaced by a sequence with $\Delta j = 2$.

Since the evolution and manifestations of collectivity in nuclei depend on shell structure, radical changes in

CP392, *Application of Accelerators in Research and Industry*, edited by J. L. Duggan and I. L. Morgan
AIP Press, New York © 1997

that structure can lead to collective modes and their phenomenology quite different than anything known near stability. This area of research is already active in very light nuclei where the exotic properties of nuclides such as the halo nucleus ^{11}Li are well known and intensely studied. A great opportunity awaits when new facilities with ISL-like capabilities will allow us to approach the drip line in heavier mass regions.

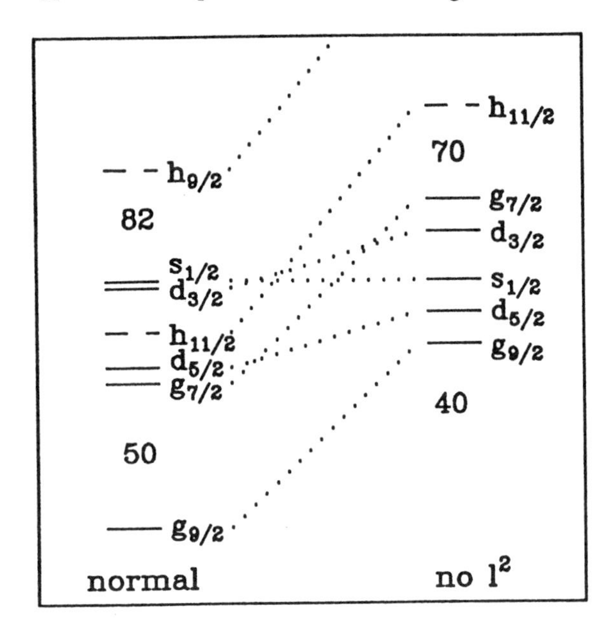

FIGURE 1. Normal and exotic shell structure.

N = Z Nuclei

Nuclei with N=Z are unique in exhibiting anomalously enhanced T=0 pn interactions. This shows up in a number of observable features, most notably in certain double differences of binding energies (8). These double differences, constructed from neighboring even-even nuclei, give a measure of the average pn interaction strength of the last proton with the last neutron. They show a smooth phenomenology from the lightest nuclei to the actinides with the sole exception of N=Z nuclei where there are huge singularities corresponding to enhanced interaction strengths. The strengths of the observed spikes have a characteristic variation across a shell. Data are only available, however, up to about A ~ 60. With RNBs, though, it may be possible to extend our empirical information even to A ~ 80 and beyond and thereby to test ideas of pn interactions in a mass region where Coulomb effects are larger, where shells can contain more particles, and where pairing efforts can scatter particles over several orbits.

Another feature of N = Z nuclei — which also reflects enhanced pn interactions — is that of anomalous shapes. Even ^{80}Zr, which might have been thought of as doubly magic at N = Z = 40, is probably deformed (and possibly axially asymmetric as well). The new capability to extend our knowledge of N = Z nuclei towards, and to, ^{100}Sn offers a superb way of studying the pn interaction in heavy nuclei. One goal is to investigate the structure of ^{100}Sn, presumably a doubly magic nucleus, and the heaviest stable N = Z nucleus.

If the T = 0 interaction leads to T = 0 pn pairing in 0^+ states, other exotic effects, especially at high-spin, such as diminished Coriolis effects and delayed backbending, could result. Note that N = Z nuclei are absolutely unique: there is only one set of N = Z nuclei yet unstudied and we are on the threshold of access to them. ISOL facilities such as HRIBF will start us on the path to studies of new N = Z nuclei and next generation ISOL facilities should allow detailed spectroscopy of the heaviest stable N = Z nuclei.

Correlations of Nuclear Observables

Recently, remarkable correlations of nuclear properties have been observed (9,10) that point to a striking and unexpected simplicity in the heretofore thought complexity of nuclear structural evolution. These correlations concern the most basic nuclear observables such as (for even-even nuclei), single nucleon separation energies, $E(2^+_1)$, $R_{4/2} = E(4^+_1)/ E(2^+_1)$, and $B(E2:2^+_1 \rightarrow 0^+_1)$ values. They relate such observables either to extrinsic quantities such as the valence nucleon product N_pN_n or to other collective observables themselves.

An example of the latter is shown in Fig. 2 where $E(4^+_1)$ is plotted against $E(2^+_1)$ for all even-even nuclei in the Z = 50-82, N = 82-126 shell. A striking tri-linear correlation appears, in contrast to a jumble of points if $E(4^+_1)$ were plotted, for example, against N, Z, or A. Moreover, the correlation splits nicely into three linear segments with slopes of 3.33 (rotor nuclei), 2.00 (anharmonic vibrators) and 1.00 (pre-collective nuclei).

Such correlations are relevant to RNB studies for several reasons. First, they provide easy ways to predict (often by interpolation not extrapolation) the properties of new nuclei far from stability and hence provide guidelines to the design of experiments. Secondly, since they require only the easiest-to-obtain data they provide a means of assessing the structure of newly measured nuclei where, with RNBs, it might well be that these are the only observables measurable with weak beam intensities. Thirdly, of course, by providing compact trajectories they provide new paradigms of behavior and therefore act as a magnifying glass that brings deviant behavior to light. This has been shown recently(11,12) in studies of intruder states in Hg and of 2-hole nuclei (Cd, Hg, N = 80). Such deviations are critical since they will provide a means of testing whether the correlations persist in new regions. In fact, in many ways, one of the most interesting questions in RNB studies of nuclear structure is whether such correlations apply only to known nuclei near stability or to all nuclei. In the first case, one then needs to ask what is special about such nuclei. In the second case, the question is what very general features of shell structure and interactions are at work. Given the above discussion of the fragility of magic numbers and of anomalies in N = Z nuclei, it is quite possible that these correlations will in fact break down in exotic regions. If they do, this fact will give instantaneous hints of radical underlying changes in shell structure. If they do not it will provide new, simpler, and more efficient signatures of structure, based on data that are easily observable even with beam intensities of 10^5 p/sec or less.

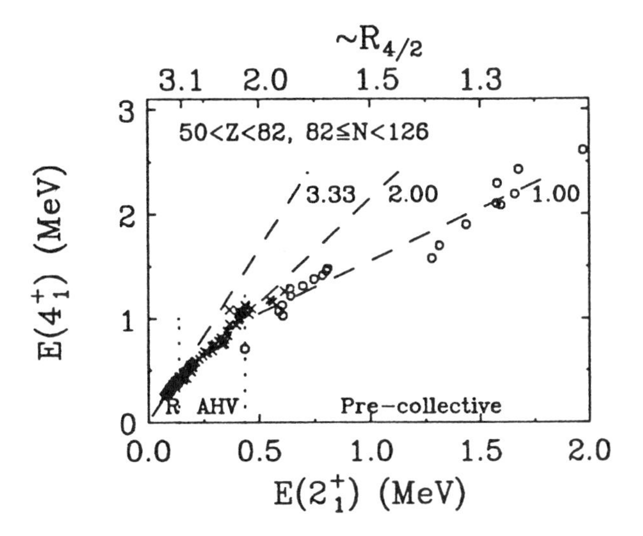

FIGURE 2. Correlations of $E(4^+_1)$ with $E(2^+_1)$. Based on Ref. 10.

EXPERIMENTAL APPROACHES

There are, of course, many approaches to the experimental study of exotic nuclei. They all, however, face two key issues -- low beam intensities (often lower by 6 or more orders of magnitude compared to our experience with stable beams) and high background levels (from the radioactivity of the beam nuclei and their daughters). There are several techniques designed to obviate these difficulties. One is the use of inverse kinematics, as illustrated in Fig. 3. Instead of the usual situation of a light projectile bombarding a heavy target, with the scattered particles emitted in all angles, a massive projectile (an RNB nucleus) strikes a light target. From the kinematics, all the reaction products go forward in a narrow cone. For example, for elastic scattering of ^{12}C on Os nuclei, the largest kinematically allowed scattering angle is 3.6° in the lab. Inverse kinematics has at least two important advantages. First, all the radioactive debris exits the reaction area and cannot scatter at large angles to strike chamber walls or detectors, producing background. Secondly, if one uses a detector system at forward angles to catch the reaction products, relatively small solid angle devices can essentially observe the total cross section rather than a small piece of the differential cross section.

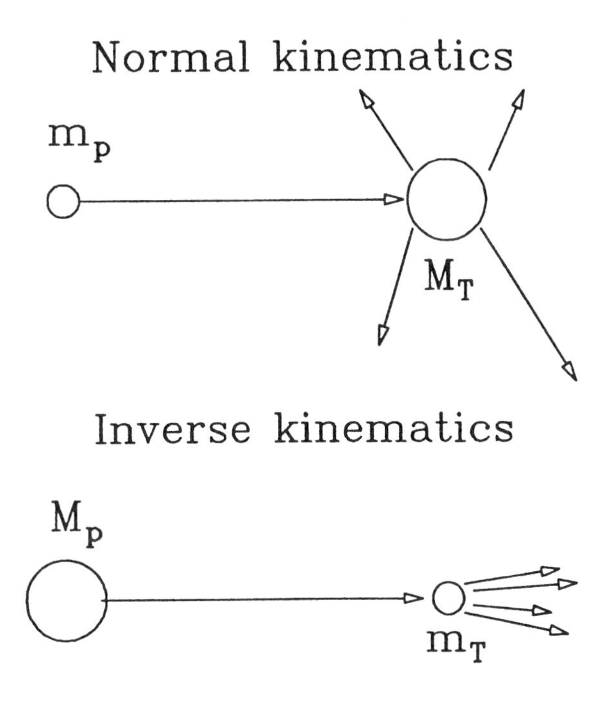

FIGURE 3. Normal and inverse kinematics.

One outstanding exploitation of this idea is the use of fragment mass analysers or recoil separators to identify various reaction fragments, for example, in a heavy ion induced reaction. These devices are essential in many RNB experiments since there can be many reaction channels and often the most interesting nuclei — those farthest from stability — will be produced with very weak cross sections. Such focusing and identification systems can be used to tag specific channels in coincidence with γ-ray spectra from modern Ge arrays, leading to lower γ-ray backgrounds and higher sensitivities. They can also isolate individual reaction products and focus them on secondary targets for further reactions if the cross sections are sufficient. Devices at ANL and HRIBF are fine examples of this genre.

As an example of the use of inverse kinematics, our

Yale-BNL-Clark group has developed an instrument(13) for measuring Coulomb excitation cross sections [B(E2: $0^+_1 \rightarrow 2^+_1$) values] for excitation of the first 2^+ state in even-even nuclei with RNBs. The device is illustrated in Fig. 4. The heavy RNB passes along the central axis of a through-well NaI (Tl) detectors and strikes a target (e.g., ^{12}C) at its mid-point. The scattered nuclei (Coulomb excited or in their ground states) exit the back of the detector and can be counted downstream in a Faraday cup or with a scintillator.

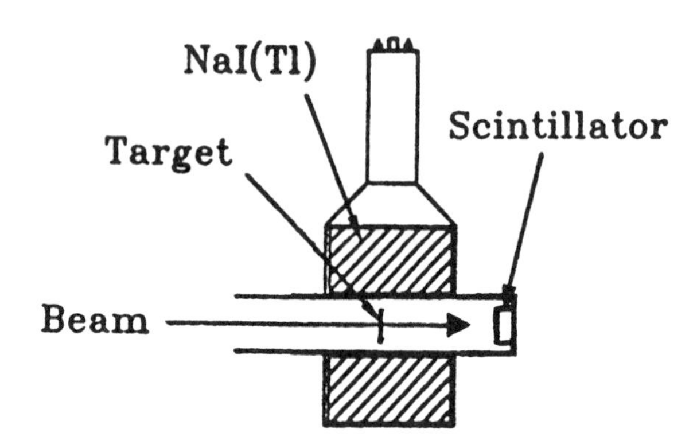

FIGURE 4. Coulomb excitation instrument for use in inverse kinematics.

The detector has 80-90% solid angle (the exact solid angle depends somewhat on the 2^+ level lifetime) for detection of the de-excitation γ-rays. This instrument has been tested with stable beams at Yale [Os on C] and with RNBs at MSU [^{108}Ru on C]. These tests, plus simulations, indicate that a typical B(E2: $0^+_1 \rightarrow 2^+_1$) value for a heavy collective nucleus can be measured in about a 1-day run with an RNB of only 10^4 particles/sec. It will therefore be ideal for survey experiments on a wide variety of new nuclei made accessible with RNBs at future ISOL facilities with capabilities envisioned in the ISL concept.

ACKNOWLEDGEMENTS

Work supported under U.S. Department of Energy Grant DE-FG02-91ER40609.

REFERENCES

1. The North American Steering Committee For the IsoSpin Laboratory, "The Isospin Laboratory", LALP 91-51, 1991.
2. E. Chiefetz et al., *Phys. Rev. Lett.* **25**, 38 (1970).
3. R. F. Casten et al., *Nucl. Phys.* **A184**, 357 (1972).
4. M. Ogawa et al., *Phys. Rev. Lett.* **47**, 289 (1978).
5. R. F. Casten et al., *Phys. Rev. Lett* **47**, 1433 (1981).
6. D. Guillemaud-Mueller et al., *Nucl. Phys.* **A246**, 37 (1984); T. Motobayashi et al., *Phys. Lett.* **B346**, 9 (1995).
7. J. Dobaczewski et al., *Phys. Rev. Lett.* **72**, 981 (1994).
8. J. Y. Zhang, R. F. Casten and D. S. Brenner, *Phys. Lett.*, **B227**, 1 (1989).
9. R. F. Casten, *Phys. Lett.* **152B**, 145 (1985).
10. N. V. Zamfir, R. F. Casten, and D. S. Brenner, *Phys. Rev. Lett.* **72,** 3480 (1994).
11. R. F. Casten and N. V. Zamfir, *J. Phys. G.*, (in press).
12. R. V. Jolos et al., to be published.
13. C. Barton et al., *Nucl. Instr. and Meth.*, to be published.

DIFFUSION PROCESSES AND RIB TARGETS

Paul Shewmon

Materials Science. & Engineering, Ohio State Univ., Columbus OH 43220

The effectiveness of RIB targets is strongly influenced by the diffusion controlled release of product isotopes. Equations are given relating the release probability of the product to its half life, diffusion coefficient, D, and target geometry. D for the product isotope can very profoundly. In metal targets it depends primarily on relative atomic size of the product and target atoms. In covalent compounds, it is determined by the degree to which the product will bond to the target atoms. Exceptionally fast diffusion/release will be observed for isotopes in covalent compounds which do not chemically bond in the lattice, and for atoms which occupy interstitial positions in metals and compounds.

Radiation damage produced in the target by ion bombardment will markedly increase D for atoms which occupy normal lattice sites in metals and compounds, but will have relatively little effect on isotopes that occupy interstitial positions.

Isotopes may be produced uniformly throughout a thin target. But before they can be studied, they must escape from the solid, and the only way this can happen is by solid-state diffusion. Diffusion occurs by well established laws which indicate clearly how the geometry influences isotope escape, but this is the simple part of the problem. The difficult question is what the diffusion coefficient of coefficient will be for an alien isotope created in a solid target. Our treatment of that question will take make up the larger part of the paper and consist, at best, of informed guesses and speculation.

DIFFUSION EQUATIONS

The first equation used to describe diffusion is a flux equation, with the flux, J, proportional to the concentration gradient. The constant of proportionality is called the diffusion coefficient, D. So, in one dimension,

$$J = -D(\partial c/\partial x)_t. \qquad (1)$$

The second diffusion equation describes the change in concentration in a given volume. It is the divergence of the flux. Taking D to be a constant, gives [1],

$$(\partial c/\partial t)_x = D(\partial^2 c/\partial x^2)_t. \qquad (2)$$

The dimensions of D are $(\text{length})^2/\text{time}$, usually cm^2/s. Eqn. (2) ignores the production rate of the isotope and considers only its diffusion once produced.

Examination of solutions to the diffusion equation show they always depend on the dimensionless variable $x/\sqrt{Dt}$.[2] For example, a plane of tracer initially at some fixed value of x in a thick sample will, after diffusing for a time t, spread out to have a half-width of $2\sqrt{Dt}$. Or, consider the production of an isotope uniformly in a thin sheet (target) of thickness h, with the isotope diffusing to the surface where it escapes. If one suddenly stops producing the isotope, the average concentration in the sheet will drop to 30% of its initial value in a relaxation time of $\tau = h^2/\pi^2 D$. (If the isotope is diffusing out of a sphere of diameter d, the relaxation time is $\tau = d^2/4\pi^2 D$.) If the isotope being produced in the sheet has a half life λ which is less than τ, much of the isotope will decay before it has time to diffuse out of the target. Or, if you have a thick target, the chances are only 1 in 10 that an isotope produced at a depth of $x=2.3\sqrt{D\lambda}$ will escape from the free surface before it decays. Chances are 50:50 for an isotope starting at $x=0.69\sqrt{D\lambda}$.[3] It should be clear from these examples that the diffusivity D and the dimension like h are important quantities in operating an effective target.

FACTORS DETERMINING D

Metals – The diffusion coefficient for an atom is its jump frequency, Γ, times its jump distance, α, squared. In 3 dimensions, $D=(1/6)\Gamma\alpha^2$.[ref.1,ch2] The jump frequency varies exponentially with temperature, and also depends on defect concentrations. For example, in metals the host atoms always move by a vacancy mechanism, that is an atom moves only when it can jump into a neighboring site which has become vacant. Thus Γ is the number of nearest neighbors, z, times the fraction of sites that are vacant, N_v, times the jump frequency into a neighboring vacant site, w. Thus, $\Gamma = zwN_v$, or $D = \gamma a_o^2 N_v w$, where a_o is the lattice constant, and γ a geometric constant which often equals one. The jump frequency w increases exponentially with temperature, $w = \nu \exp(-H_m/kT)$ where ν is the vibration frequency of the jumping atom. If there is no radiation hitting the metal, N_v has its equilibrium value for each temperature and $N_v = \exp(-H_v/kT)$.

The equation for D for self-diffusion by a vacancy mechanism can be written as $D = D_v N_v$, or

CP392, *Application of Accelerators in Research and Industry*, edited by J. L. Duggan and I. L. Morgan
AIP Press, New York

$$D = D_o \exp((-H_m - H_v)/kT) = D_o \exp(-Q_{SD}/kT). \quad (3)$$

For metals one often finds that

$0.1 < D_o < 10 cm^2/s$ Zener's Rule

and there is a good theoretical basis for this, developed about 40 years ago by Zener. Also, for metals

$Q_{SD}/T_m \cong 0.0015 eV/°K$ (fcc and bcc metals)

is a reasonable approximation.

Substitutional alloying elements obey roughly the same rules, i.e. they diffuse using the same vacancies and with values of D that differ by less than a factor of 10 from that of the solvent atoms in an alloy of the same composition. Smaller impurities, which occupy interstitial sites, diffuse much faster than atoms that sit on lattice sites, often with a Q that is less than half of Q_{SD}. (Hydrogen and Deuterium diffuse with a Q that is less than $0.1Q_{SD}$). D_o again follows Zener's rule. Fig. 1 shows a comparison of D for iron, carbon and hydrogen atoms in iron between room temperature and 1000°K. Hydrogen diffuses most rapidly and has the smallest activation energy (slope in Fig. 1). Carbon, still an interstitial atom, but a much larger one, is intermediate both in D and in Q. There is extensive data on diffusion in most metals and alloys. Good summaries can be found in [4] and [5].

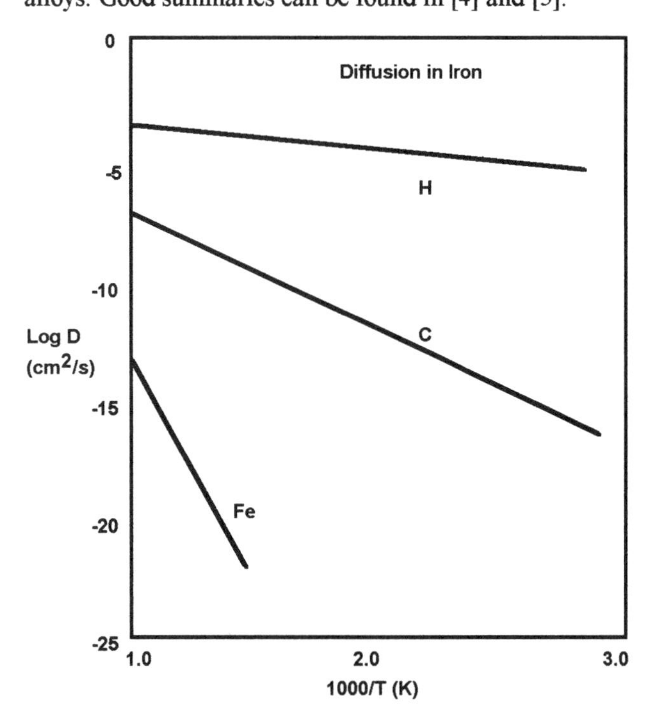

FIGURE 1 - D for hydrogen, carbon and iron atoms in α-iron between 1000°K and room temp.[Ref. 1, ch. 3]

Irradiation of a metal with high energy ions (>0.1 MeV) create mostly heat thru electronic excitation. However, some of the energy goes into elastic nuclear collisions that knock atoms off their lattice sites with enough energy to produce many more displaced atoms before they come to rest. This results in the production of many interstitials and vacancies. Both of these defects increase D of the lattice atoms. The equation for the radiation enhanced self-diffusion coefficient now becomes

$$D_{RED} = D_v N_v + D_i N_i \quad (4)$$

where D_i and N_i are the diffusion coefficient and atom fraction of host atoms diffusing as interstitials. Since vacancies and interstitials are produced in equal numbers and both diffuse out to sinks, it turns out that $D_v N_v = D_i N_i$.[6] At low temperatures ($T<0.5T_m$) the radiation induced increase in D can be many orders of magnitude because the radiation induced concentrations of defects, N_i and N_v, are so much higher than the low equilibrium (thermally generated) concentrations at these temperatures. At higher temperatures (perhaps $T>0.7T_m$) the percentage increase becomes smaller because the radiation generated defect concentration is larger and the radiation induced defects anneal out faster.

The radiation induced increase in D for smaller atoms, that normally rest on interstitial sites, would be expected to be much smaller than that for substitutional atoms. Since the interstitial atoms do not need defects to move, they will not show a marked increase in D with irradiation. The local heating will speed up their jumps, but they may tend to be trapped by vacancies and dislocations created by irradiation. So the net effect is a limited percentage increase in D for the interstitial impurities.

Non-Metals – There has been much less work on diffusion in non-metals. The most stable non-metals are those with strong covalent bonds (MgO, Al_2O_3, graphite, SiO_2, BN, etc.). A satisfactory surrogate for these might be silicon which diffusion has been studied because of its use as a semiconductor. The atomic spacing of metals is determined largely by the interaction of the closed/complete electron shells around the nucleus. Thus the structures can be thought of as being made up by packed spheres, and the energy required to put a host atom in an interstitial position is much greater than that required to form a vacancy. In covalent solids the interatomic spacing is much greater than the radius of the filled electron shells. As a result the energy of pushing a silicon atom into an interstitial position is much closer to that for forming a vacancy in silicon than it is in a metal. Figure 2 shows a plot of log(D) vs. 1/T for silicon, and many solutes therein. Note the following: D is the lowest and slope (activation energy, Q/k) highest for pure silicon. Group III and V elements as well as oxygen, that bond reasonably with silicon diffuse faster than silicon atoms. Elements which have much larger ion cores, but bond weakly with silicon, e.g. Fe, Cu, Ni have a much lower Q and diffuse 10 to 20 orders of magnitude faster than does silicon, depending on the temperature.

Impurities diffuse rapidly in silicon when they move as interstitials. The lattice is relatively open, so an interstitial Cu atom can jump quite rapidly. At equilibrium the Cu atom may well find a Si vacancy and drop into it. Thus most of the copper atoms in a crystal of silicon reside on normal lattice sites. However, Cu interstitials jump so much more frequently Cu atoms on lattice sites that the diffusion rate of Cu is controlled/determined by the small fraction of atoms in interstitial sites.

The same sort of behavior should prevail in compounds like Al_2O_3, BN, and graphite. These compounds all have very high melting points, and self diffusion is so slow it is very difficult to measure. However, Na and Li ions diffuse so fast thru alumina that beta-alumina is used as a solid electrolyte in sodium-sulfur batteries operating at a few hundred degrees Centigrade. That is, the diffusion coefficient of (interstitial) Na and Li ions in beta-alumina is many orders of magnitude faster than Cu in silicon (see Fig. 2) even though aluminum and oxygen move more slowly in Al_2O_3 than Si does in silicon. Also, He (from air) diffuses thru quartz glass at room temperature at a rate that limits the ultimate vacuum attainable in ultra high vacuum systems made of glass. In conclusion then, atoms that do not bond with the host atoms in a covalent compound can diffuse rapidly thru the compound at temperatures far below the melting point of the compound.

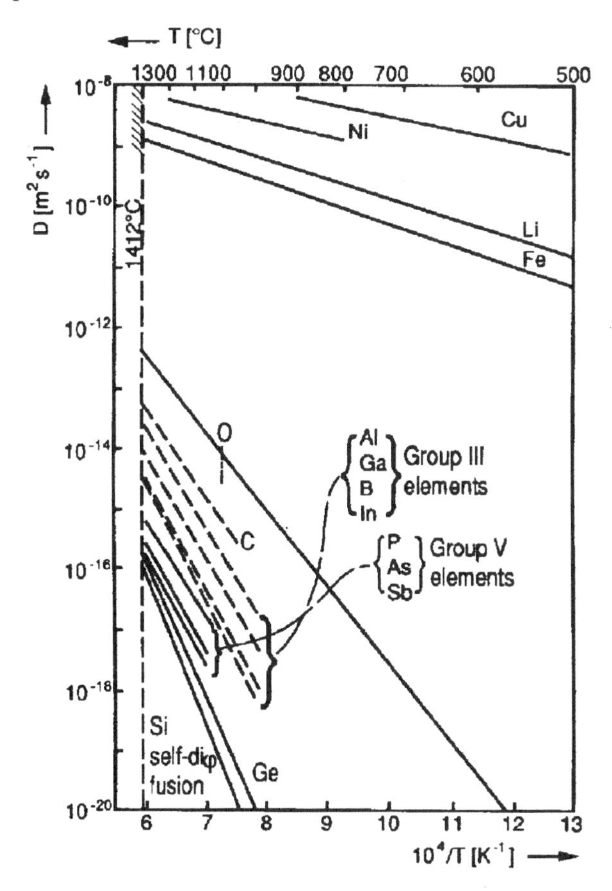

FIGURE 2 - D for various types of elements in silicon. [From Shewmon, ref.[1], ch.6]

Radiation of covalent, or ionic, compounds should increase self diffusion appreciably by raising the vacancy and interstitial concentrations.[7,8] The effect of radiation should be negligible for fast diffusing elements that do not covalently bond in the lattice and are already mobile interstitials, like our examples of He in silica glass, or Na in β-Al_2O_3. For solute that bind weakly on lattice sites, like Cu in silicon, the enhanced concentration of radiation produced interstitial host atoms, e.g. Si in Si, will push more of the solute atoms into interstitial sites by reactions of the sort $Si_I + Cu_{Si} \rightarrow Si_{Si} + Cu_I$, where the subscript 'Si' indicates an atom on a lattice site and 'I' indicates the atom occupies an interstitial site.

DIFFUSION AT DEFECTS AND INTERFACES

Along internal defects like dislocations and grain boundaries atoms diffuse with a lower activation energy, and thus faster than they do thru the lattice. Thus, these defects could help in the release of isotopes from targets. However, only a small fraction of the atoms in a sample are at such defects so their contribution to overall diffusion is usually only observed at lower temperatures where lattice diffusion is very slow. Also, radiation will enhance lattice diffusion more than it along defects because radiation raises the concentration of defects in the lattice more than it does at grain boundaries or dislocations. The only time dislocation and grain boundary effects might be important is if you are forced to work at a temperatures which are a small fraction of the melting point of the target material.

REFERENCES

1. Shewmon, P.G., *Diffusion in Solids, 2nd ed.*, Warrendale, PA: TMS, 1987, ch. 1.

2. A summary of solutions to the diffusion equation can be found in Carslaw, H., Jaeger, J., *Conduction of Heat in Solids,* Oxford: Clarendon Press, 1959. See also, Crank, J., *Mathematics of Diffusion, 2nd ed.*, Oxford: Clarendon Press, 1975

3. More completely, the fraction of atoms produced at a depth x which decay before they escape is: $1\text{-}\exp(-x/\sqrt{D\lambda})$.

4. Smithells, *Metals Ref. Book, 6th ed.*, London: Butterworth 1983, ch. 13.

5. Landolt-Boerstein, New Series, vol. III/26, *Diffusion in Solid Metals & Alloys*, Berlin: Springer Verlag, 1990.

6. Averback, R.S., et al., *Atomic Transport in Irradiated Solids*, Matl. Res. Soc. Sym. Proc. v.311, Phase Transformations in Thin Films, 1993

7. Pells, G.P., *Radiation Effect in Ceramic Insulators*, Matl.Res. Soc. Sym. Proc. v.373, Microstructure of Irradiated Materials, 1995

8. Van Sambeek, Andrew, et al., *Radiation Enhanced Diffusion in MgO,* Matl. Res. Soc. Sym. Proc. v.373, Microstructure of Irradiated Materials, 1995.

THE REX-ISOLDE PROJECT

O. Kester, D. Habs, K. Rudolph
Sektion Physik, LMU München, D-85748 Garching, Germany
G. Hinderer, E. Nolte
TU München, D-85748 Garching, Germany
G. Bollen, H. Raimbault-Hartmann, H. Ravn
CERN, CH-1211 Geneva 23, Switzerland
L. Liljeby, K.G. Rensfelt
Manne Siegbahn Institute of Physics, S-10405 Stockholm, Sweden
D. Schwalm, R. von Hahn, R. Repnow
Max-Planck-Institut für Kernphysik, D-69029 Heidelberg, Germany
A. Schempp
J.W.Goethe-Universität Frankfurt, D-60325 Frankfurt am Main, Germany
U. Ratzinger
Gesellschaft für Schwerionenforschung, D-64220 Darmstadt, Germany

and the REX-ISOLDE collaboration

Abstract

REX-ISOLDE is a pilot experiment at ISOLDE/CERN to study the structure of neutron-rich nuclei (N=20, N=28) with post-accelerated radioactive ion beams (1). Therefore radioactive ions with charge state 1^+, which are delivered by the online mass separator ISOLDE, are accelerated up to 2.2 MeV/u by means of a new concept. The radioactive ions are first accumulated in a Penning trap, then charge breeded to a charge-to-mass ratio of 1/4.5 in an Electron Beam Ion Source (EBIS) and finally accelerated. The LINAC consists of three components, namely a Radio Frequency Quadrupole (RFQ) accelerator, which accelerates the ions from 5 to 300 keV/u, an interdigital H-type structure (IH) with a final energy between 1.1 and 1.2 MeV/u and three seven gap resonators, which allow to vary the final energy between 0.8 and 2.2 MeV/u. Pulsing of the EBIS high voltage platform allows deceleration of the ions from the trap, which leads to a rather low injection energy of the RFQ. Thus a more efficient, adiabatic bunching and better output emittances than in the proposal design of the RFQ are the results. The IH-structure is now similar to that of the GSI HLI-structure (2), because of the increased exit energy in comparison to (1). An overview of the project, emittances of beam dynamics calculations for the design of the accelerator components and the present state of the accelerator are presented.

INTRODUCTION

The physics with accelerated radioactive nuclear beams (RNB) represents one of the frontiers of nuclear physics, which explains the build up of numerous RNB-facilities all over the world. Many discussions at several conferences (3-5) have shown that RNBs are new tools to study nuclei far off stability. The efficient production and acceleration of the RNB's is the crucial task. REX-ISOLDE (**R**adioactive Beam **EX**periment at **ISOLDE**) makes use of the availability of radioisotopes from 68 elements from the online mass separator at the new PS-ISOLDE (6). The production of radioactive ions and the yields special for REX-ISOLDE are described in several papers (1), (7).

REX-ISOLDE is a pilot experiment aiming at two main goals. One of the aims is the demonstration of a new concept to bunch, charge-breed and post-accelerate single charged, low energetic ions in an efficient way. The other aim is to study the structure of very neutron-rich Na, Mg, K and Ca isotopes in the vicinity of the closed neutron shells N = 20 and N = 28 by Coulomb excitation and neutron transfer reactions with a highly efficient γ- and particle-detector array MINIBALL. The experiment dwells on established techniques, but represents a new way of combination of these structures.

MOTIVATION OF REX-ISOLDE

In the first experiment the accelerated ions are used to study the dynamic properties and to examine the shapes of very neutron rich nuclei close to semimagic shells. Due to a rather low final energy of 2.2 MeV/u the first experiment is limited to rather light nuclei ($A \leq 50$) to reach the Coulomb barrier. In constrast to the structure of stable nuclei, which are thoroughly investigated experimentally and theoretically, the structure of nuclei close to the neutron dripline is not deduced yet. REX-ISOLDE provides a test of the shell model over a wide range in isospin.

In experiments with relativistic Coulomb excitation, there a strong deformation for the 2^+-state of ^{32}Mg has been measured (8). The ground state of ^{32}Mg is interpreted as a strongly-deformed neutron intruder state. REX-ISOLDE will show that $^{30-33}$Na and ^{31}Mg exhibit a strong ground state deformation too. The first measurements will perform a detailed nuclear level spectroscopy to get more insight into the different collective and single-particle states of that region in the nuclear

CP392, *Application of Accelerators in Research and Industry*, edited by J. L. Duggan and I. L. Morgan
AIP Press, New York © 1997

chart. The expected difference in the deformation of the neutron and proton distribution in very neutron rich nuclei will be examined via neutron transfer reactions. These reactions are sensitive to the distribution of neutrons in the nuclear skin.

THE ACCELERATION CONCEPT

The basic concept of REX-ISOLDE is to inject the ions from the separator continously into a Penning trap, where they are accumulated and cooled. After 20 ms, bunches will be transferred to an electron beam ion source (EBIS). After charge breeding (10ms) to a charge-to-mass ratio of 1/4.5 the ions are injected into a radio frequency quadrupole (RFQ) accelerator via a mass separator which is similar to the well known Nier-spectrometer (9). The accelerator consists of a RFQ, an interdigital H-type (IH) structure and three seven-gap (7-gap) resonators. The 7-gap resonators and the IH-Structure will allow an energy variation between 0.8 and 2.2 MeV/u to meet the experimental requirements. The detector set-up is described in (7). Figure 1 shows the present state of the experimental set-up.

The accumulator

Due to the proposed charge-breeding and acceleration scheme, accumulation of the cw-beam from the ISOL is required. Therefore the ions are injected into a large Penning trap, where they are slowed down by collisions with the atoms of a buffer gas. The energy loss of the ions during a single oscillation through the trap has to be as large as the initial energy spread of the ISOLDE ions to reach efficiencies of about 100%. A special side-band cooling technique of the magnetron motion in the trap (10) will be used to separate the radioactive ions from unwanted species. A report of the present status of the trap will be presented in (11).

The charge breeder

In contrast to a plasma ion source, an EBIS uses monoenergetic electrons from an electron gun focused by a strong magnetic field to produce highly charged ions (12). In an EBIS the ions are confined radially by the potential depression of the negative space charge of the electrons, while the longitudinal confinement is performed by potential barriers, which are established by cylindrical electrodes surrounding the beam. Trapped low-charged ions will undergo stepwise ionisation via electron impact until they are extracted by rise of a potential gradient along the axis of the electron beam. The parameters of the REX-EBIS are: A current density of 200 A/cm^2, a beam current of 0.5 A, a 1.5 m Solenoid providing a trap length of 0.8 m with a tip field of 2 T and a beam energy between 5 and 10 kV. Concerning these parameters the charge to mass ratio of 1/4.5 is reached for Na in 6 ms and for K in 8 ms. To get a high efficiency in charge breeding, the overlap of injected ions and the electron beam has to be as good as possible. Hence a rather low extraction emittance of 3 π mm mrad of the Penning trap is required for the injection of the radioactive ions into the EBIS. While the voltage of the trap platform is fixed to 60 kV to decelerate the ions from ISOLDE, the platform of the EBIS can be pulsed in the ionisation cycle. Thus the 5 keV/u injection energy can be provided. Higher efficient bunching and lower extraction emittances of the ions in the RFQ are advantages of the pulsed EBIS platform as well.

The mass separator

As the intensity of the radioactive ions is much smaller than the intensity of ions from residual gas, a mass separator is required. The mass separator consists of an electrostatic 90^0 cylinder deflector and a 90^0 magnetic bender build up in a vertical S-shape (Fig. 1).

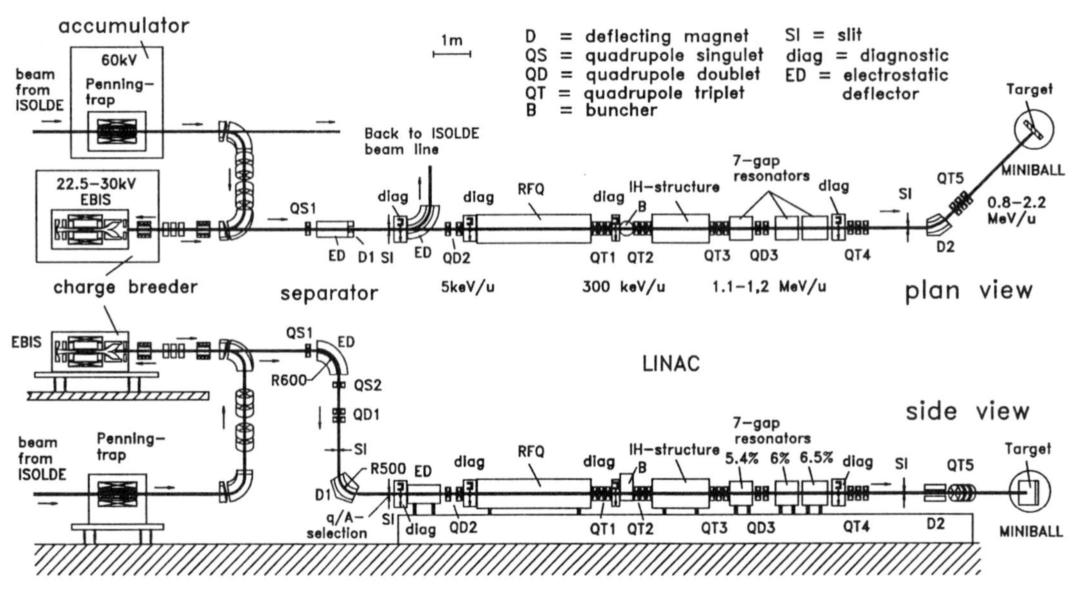

FIGURE 1. Schematic of REX-ISOLDE

Due to the potential depression of the electron beam, the extracted ions will have an energy spread, which limits the q/A-resolution of a magnetic separator to 200. To get rid of this limit a first design of a separator similar to the Nier-spectrometer was carried out. The electrostatic deflector separates the ions according to their energies irrespective of their masses. The focal plane is projected to a plane between the doublet QD1 and the magnetic bender. There the energetic width of the ion beam can be restricted. The correct charge-to-mass ratio is selected in the focal plane of the bending magnet. In case of a proper design the energy dispersion is compensated by the energy dispersion of the magnetic field (13). The calculated q/A-resolution of the separator concerning an extraction emittance of the EBIS of 3 π mm mrad is about 760. Higher emittances will decrease the resolution, because of larger entrance angles of the ions and higher order optic effects of the electrostatic deflector.

The accelerator

The linear accelerator of REX-ISOLDE consists of a 3m 4-rod RFQ, a matching section, a 20 gap IH-structure and three 7-gap resonators. All resonators operate at 101.28 MHz with a duty factor of 10% due to bunching and charge-breeding. The macrostructure of the accelerated ions will have a typical bunch width of 100 μs and a pulse distance of 20 ms. The microstructure will have a pulse width depending on the final energy between 1.9ns at 2.2 MeV/u and 13 ns at 0.8 MeV/u. The time between the pulses will be 10 ns.

4-rod-RFQ

The 4-rod-RFQ accelerates the radioactive ions with a charge-to-mass ratio larger than 1/4.5 from 5 keV/u to 300 keV/u. This kind of RFQ is a well tested structure (14) which accelerates low energy ion beams efficiently. The rf quadrupole field provides transverse focusing for the low energy ions while a modulation of the four rods performs smooth bunching of the injected dc-beam and acceleration. Similar RFQs to the REX-RFQ are used at the GSI HLI-LINAC (15) and at the Heidelberg high-current injector (16). The design of the modulation of the rods has been optimized for the specific charge-to-mass ratio of 1/4.5. The electrode voltage has been reduced to 42 kV in comparison to the first design to reduce the rf-power requirement. The final design parameters along the beam axis calculated with PARMTEQ are shown in Fig.2. For the design of the REX-RFQ the assumed normalized injection emittances have been $\epsilon_{x,y} = 0.61\ \pi$ mm mrad. The final transverse emittances are: $\epsilon_{x,x'} = 0.68\ \pi$ mm mrad and $\epsilon_{y,y'} = 0.67\ \pi$ mm mrad assuming 95% transmission. The resulting emittance growth is only 11%. In longitudinal direction the phase spread and the energy spread are: $\Delta\varphi = \pm 14^0$ and $\Delta W/W = \pm 1.5\%$. Thus the longitudinal emittance of $\epsilon_{long} = 1.89$ keV/u*ns is very small.

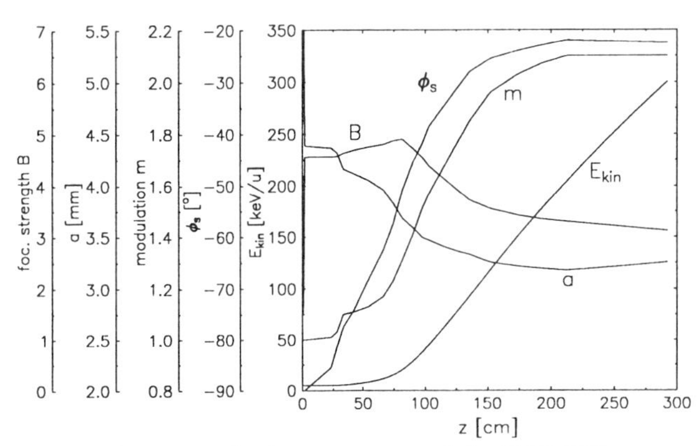

FIGURE 2. Design parameters of the REX-RFQ

At present calculations with MAFIA have performed to evaluate the field distribution along the axis. These calculations will show how to size the tuning plates to obtain good field flatness. From calculations and from similar 4-rod-RFQs the shunt impedance is estimated to be $R_p = 100$ kΩm. Thus the peak-power consumption will be 53 kW. The production of the rods and of the stems should be finished at the end of 1996. The RFQ-tank will be completed at the beginning of 1997. After copper plating of tank and rods the RFQ will first be built at the Munich tandem laboratory for tuning and high power tests.

matching section

To match the beam into the acceptances of the IH-structure a section consisting of two magnetic quadrupole triplet lenses and a rebuncher is required. This section is similar to those of the GSI HLI-accelerator and of the CERN lead-LINAC. The first triplet lens focuses the beam through the rebuncher and produces a waist for diagnosis. The rebuncher is a two gap $\lambda/4$-coaxial resonator with a peak voltage of 28.5 kV. The second triplet lens matches the beam to the transverse acceptance of the IH-structure. The aperture of the lenses is 2 cm. The design of the matching section is finished and parts will be ordered soon.

IH-structure

The IH-structure accelerates the ions from 0.3 MeV/u to an extraction energy between 1.1 and 1.2 MeV/u. In comparison to (1), the energy shift of the IH-Structure is increased by 0.4 MeV/u. An enhancement of the final energy of the IH-structure is required, because of the increased final energy at the target. The variation of the final energy of the IH-structure can be achieved by adjusting the gap voltage distribution via capacitive plungers. The lower final energy of the IH-structure is important for deceleration of the ions down to 0.8 MeV/u. Otherwise the deceleration from 1.2 MeV/u down to 0.8 MeV/u through the 7-gap resonators would perfom a non-acceptable phase spread of the ions at the target which could not be reduced by a rebuncher without losses.

The IH-struture (17) for REX-ISOLDE is a very efficient drift-tube structure which can be scaled easily from similar structures like the GSI HLI-IH-structure or 'tank 1' from the IH-structure of the lead LINAC at CERN (2, 18). The used beam dynamics concept of the 'Combined Zero Degree Structure' (18) was developed and applied for both IH-DTL's mentioned above and resulted in a very reliable beam operation and in the reach of the specified beam parameters (19). After a first acceleration section with a 0^0-synchronous particle the ions are focused in transverse direction by an inner-tank triplet lens. Just behind the triplet the ions are rebunched by a -30^0-synchronous particle section (3 gaps), followed by a second acceleration section. The whole structure has 20 gaps and a total length of 1.5 m and is shown in Fig.3. The resonator tank is cut into an upper and a lower shell. The drift tubes and the magnetic flux inductors are mounted on the center frame.

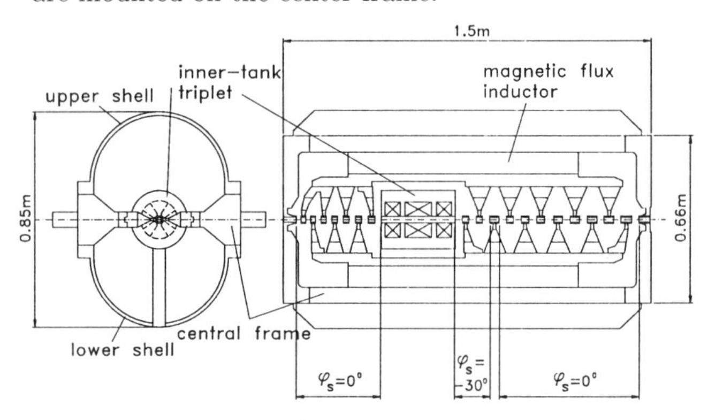

FIGURE 3. Schematic of the REX-IH-structure

For beam dynamics calculations and drift tube design of the IH-structure normalized injection emittances of $\epsilon_{x,x'}$ = 0.59 π mm mrad, $\epsilon_{y,y'}$ = 0.56 π mm mrad and ϵ_{long} = 2.31 keV/u*ns have been assumed. The accepted energy/phase spread are $\pm 10^0/\pm 3.4\%$ respectively. The final emittances of the present design are $\epsilon_{x,x'}$ = 0.59 π mm mrad, $\epsilon_{y,y'}$ = 0.63 π mm mrad and ϵ_{long} = 2.65 keV/u*ns. The final energy/phase spread are $\pm 7^0/\pm 1.2\%$. Assuming a shunt impedance of 300 MΩ/m, the total peak power consumption will be 36 kW. Field distributions have been studied by MAFIA calculations. In order to obtain a flat gap voltage distribution the magnetic flux inductors have to be adjusted properly. The drift tube structure will be completed in the beginning of 1997 and the construction of the tank is just finished.

7-gap resonators

The last section of the accelerator consists of three 7-gap resonators similar to those built for the high-current injector at the MPI für Kernphysik in Heidelberg (20). These special types of split ring resonators are designed and optimized for synchronous particle velocities of β_s = 5.4%, 6.0% and 6.6%. The total resonator voltage is about 1.7 MV. The required peak power will be 80 kW. The resonator has a single resonance structure, which consists of a copper half shell and three arms attached to both sides of the shell (21). The output of the IH-structure is matched with a triplet lens to the first 7-gap resonator. Between first and second resonator there is an additional doublet for transverse focusing. Detailed beam dynamics calculations have been performed, showing normalized acceptances of 1.2 π mm mrad for the x,x'-plane and 3 π mm mrad for the y,y'-plane. The phase and energy spread of the injected ions have been assumed to $\pm 10^0$ and $\pm$ 1.3% respectively. Presently, bead perturbation measurements with down scaled models are carried out and the power resonators are manufactured. MAFIA calculations have been performed to evaluate the gap voltage distribution, which show a good agreement with model measurements.

REFERENCES

1. "Radioactive beam Experiment at ISOLDE: Coulomb Excitation and Neutron Transfer Reactions of Exotic Nuclei", *Proposal to the ISOLDE committee*, CERN-ISC94-25
2. Ratzinger, U., "A low Beta RF Linac-Structure of the IH-Type with Improved Radial Acceptance", in *Proc. of the LINAC88*, Williamsburg, CEBAF-89-001, 185-187 (1988)
3. Myers, W. D., Nitschke, J. M., and Norman, E. B., *Proc. of the RNB1-1989*, Berkeley (California), World Scientific, Singapore 1990
4. Delnar, Th., *Proc. of the RNB2-1991*, Louvain-la-Neuve (Belgium), Adam Hilger, Bristol 1992
5. Morrissey, D. J., *Proc. of the RNB3-1993*, East Lansing (Michigan), Edition Frontieres, France, 1993
6. Albrow, M., *et al.*, "The ISOLDE Facility at the PS-Booster", CERN-PSCC-89-29
7. Habs, D., *et al.*, "The REX-ISOLDE Project", in *Proc. of the RNB4-1996*, Omiya, Japan, 1996, *to be published*
8. Motobayashi, T., *et al., Phys. Lett. B* **346**, 9 (1995)
9. Nier, A. O., Roberts, T. R., and Franklin, F. J., *Phys. Rev.* **75**, 346 (1949)
10. Bollen, G., *et al., Nucl. Inst. and Meth.* A **368**, 675 (1996)
11. Bollen, G., *these proceedings*
12. Donets, E. D., and Ilyushchenko, V. I., JINR R7-4124, 1968
13. Aston, F. W., *Philos. Mag. 38*, 709 (1919)
14. Schempp, A., *et al., Nucl. Inst. and Meth.* **B10/11**, 831 (1985)
15. Friedrich, J., *et al.*, "Properties of the GSI HLI-RFQ Structure", in *Proc. of the IEEE-PAC91*, San Francisco (California), 3044 (1991)
16. Kleffner, C.-M., *et al.*, in *Proc. of the EPAC92*, 1340 (1992)
17. Nolte, E., *et al., Nucl. Inst. and Meth.* **158**, 311 (1979)
18. Ratzinger, U., "The IH-Structure and its Capability to Accelerate High Current Beams", in *Proc. of the IEEE-PAC91*, San Francisco, 567-571 (1991)
19. Lasheras, N. C., Crescenti, M., and Vretenar, M., "Commissioning and Experience in Stripping, Filtering and Measuring the 4.2 MeV/u Lead Ion Beam at CERN LINAC3", in *Proc. of the LINAC96*, Geneva, Switzerland, *to be published*
20. von Hahn, R., *et al., Nucl. Inst. and Meth.* **A328**, 270-274 (1993)
21. von Hahn, R., *et al.*, in *Proc. of the EPAC92*, 1313 (1992)

NOVEL TARGETS FOR THE PRODUCTION OF THE SECONDARY RADIOACTIVE BEAMS: ^7Be and ^{15}O *

M. Gai, and R.H. France III
Dept. of Physics, U46, University of Connecticut, 2152 Hillside Rd.,
Storrs, CT 06269-3046
and
Ch. Barue, M. Cogneau, Th. Delbar, P. Leleux, M. Loiselet, C. Michotte, G. Ryckewaert
Universite Catholique de Louvain, B-1348 Louvain-La-Neuve, Belgium
and
M. Gaelens
Inst. voor Kern-en Stralinsfysique, Katholique Universiteit Leuven, Leuven, Belgium
and
S. Zeisler
Dept. of Radiochemistry, German Cancer Res. center, Heidelberg, Germany

January 14, 1997

Abstract

In this paper we describe recent advances in the production of RNBs of ^7Be and ^{15}O with possible application for the production of ^{17}F. The production of ^7Be relies on chemical separation method of ^7Be from a massive LiF production target, and for the ^{15}O we used targets fabricated at the Jet Process Corporation (JPC), New Haven, CT, using the proprietary Jet Vapor Deposition (JVD)^{TM} method.

1 Introduction

The production of Radioactive Nuclear Beams (RNB) is of high priority for studies in Nuclear Physics, Nuclear Astrophysics as well as other applications in basic and applied sciences. Of particular importance are beams to be used in studies in Nuclear Astrophysics as well as beams of chemically active elements such as oxygen and fluorine. For example, a beam of ^7Be is essential for progress in studies of the Standard Solar Model [1] by the measurement of the $^7Be(p,\gamma)^8B$ reaction rate. And beams of ^{15}O are crucial for studies of the breakout from the hot-CNO cycle into the rp-process predicted to occur in Nova bursts. This breakout mainly occurs through the reaction $^{15}O(\alpha,\gamma)^{19}Ne$ [2], predicted to have a slow rate. These studies require very intense secondary RNBs. The production of $C^{15}O$ also appears to be of importance for medical applications and PET-cameras [3]. In the past, several groups developed techniques to extract ^7Be from metallic Li targets, and we have applied a similar chemical extraction technique for separation of ^7Be out of a large (approximately 20 grams) LiF target bombarded with 200 μA proton beams of 30 MeV. The production of secondary RNBs of chemically active elements is hampered by the low yield for release from the hot production target. JPC have used the Jet Vapor Deposition (JVD)^{TM} method [4] to produce a prototypical target with 350 double-layers, with each

*Work Supported by a grant from NATO for US-Belgium Collaborative Research.

CP392, *Application of Accelerators in Research and Industry*, edited by J. L. Duggan and I. L. Morgan
AIP Press, New York © 1997

double layer constituting of $Ti^{15}N$ backed by carbon (1,000 Å each). These prototypical targets were bombarded by 55 μA proton beams of 7.6 MeV. All experiments were performed at Louvain-La-Neuve (LLN) using proton beams extracted from CYCLONE 30.

2 The Production of ^7Be Beam

We have measured the production of ^7Be in a LiF thick target using 30 MeV proton beams, to be $4.3(6) \times 10^{-4} I_0$, with I_0 the primary beam intensity. The thick target yield for 30 MeV proton beam on a metallic Li target can be estimated by integrating the known [5] $^7Li(p,n)$ cross section $= 3.5 \times 10^{-3} I_0$. The reduction by approximately a factor of 10 for our LiF target is due to target stoechiometry and geometry (angle of tilt with respect to beam) as the LiF melts during bombardment and aggregates at the bottom of the cylindrical carbon container including the target material. This production rate however, appears reasonable and should allow for achieving the design goal of an accelerated beam intensity equal to 10^9 $^7Be/sec$ on target. With an assumed ^7Be ionization efficiency of 10%, and accelaration and transmition of a few percent. For example for this production rate a measurement at one energy (E_{cm}= 0.5, 0.75, 0.9, or 1.1 MeV) that would last for approximately two days each, would also require two days of bombardment per one energy for preparing the ^7Be source. Rough measurements of activity in target container chamber apparently show that very little ^7Be leaves the hot target to stick to the neighboring walls during bombardment.

Due to the large amount of Li present in our 20 gram LiF target, special care is needed for chemically separating the ^7Be from ^7Li to yield the ratio of $^7Be/^7Li = 1$ in the target. Note that for such a ratio the LLN cyclotron can easily produce a relatively pure ^7Be beam due to the mass ratio $M(^7Be)/M(^7Li) = 1.000132$. We performed a test of Ion Exchange Chromatography using a small amount from an old LiF target bombarded for several weeks (approximately a year after bombardment). The target indeed contained the expected amount of activity when corrected for half-life of ^7Be.

After dissolution of the target in concentrated HNO_3 (or in in a mixture of $HNO_3 - HCl$) and evaporation to dryness, the residue was dissolved in 20 ml of H_2O containing a small quantity of Fe_2NO_3 (100 μg of Fe). The solution was treated by NH_4OH to co-precipitate iron and beryllium as hydroxide. After centrifugation, precipitate was washed two or three times with 2 ml of H_2O and dissolved in HCl(12N). By transfer on ion exchange resin (Merck M5080, with column Aldrich flexcolumn, i.d. of 7mm, length of 100 mm) ^7Be was isolated in 10 ml of HCl (12N). The solution was evaporated to dryness and dissolved in 100 μL of water. Using this Ion Exchange method the ^7Be extraction yield was estimated to be better than 70%; with an estimated loss of ^7Be (in the solvent) of 10% or less. We estimate this method reduces 25 grams of 7LiF to a few micrograms, thus yielding a ratio close to 1:1 of separated ^7Be to ^7Li for the design sample to be placed in the ion source, either in a form of $^7BeCl_2$ (evaporation/ECR ion source) or plated onto a sputtering plate (sputtering/ECR source).

3 The Production of Oxygen RNBs

For the production of RNBs of chemically active elements such as oxygen (and possibly fluorine) we are using targets fabricated by the Jet Process Coprporation (JPC), New Haven, CT. The targets are produced by JPC using their proprietary Jet Vapor Deposition (JVD)^{TM} method [4]. Two kinds of targets are considered. First, a recoil escape multilayered targets, with the RNBs produced in one thin layer and recoil to the adjacent layer where they are stopped and produce volatile gaseous molecule such as the $C^{15}O$. Second, are highly permiable targets composed of atomic clusters of aluminum oxyde. Such a powder target contains approximately 40% void which allows for a rapid diffusion out of the target and transfer to the ion source. Atomic cluster powder targets could possibly be considered for the production of ^{17}F. We plan to test such targets at the HRIBF facility at Oak Ridge during December 1996.

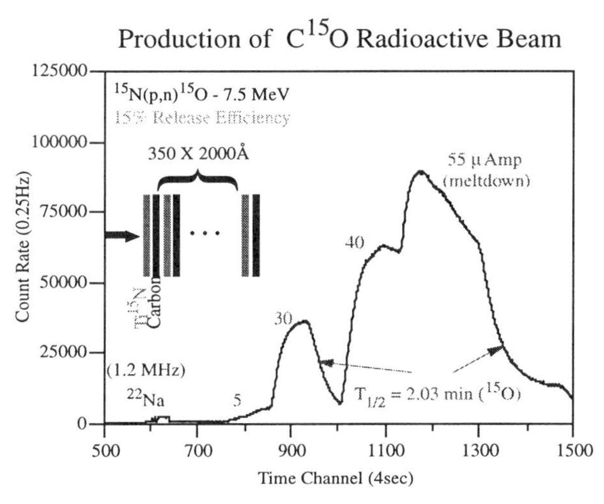

Fig. 1: Measured production and extraction efficiency of $C^{15}O$.

The recoil release multilayered target is designed using Monte Carlo techniques [7], so that the recoiling ^{15}O nuclei are stopped in the carbon layers. The stopped ^{15}O nuclei produce $C^{15}O$ molecules that are readily extracted from the hot foil. The $C^{15}O$ molecules were collected in a cryogenic getter for counting of the β^+ decay. Using this setup at LLN we measured a production of 5×10^8 $^{15}O/sec$ for the thin $(350 \times 2,000\mathring{A} = 70\mu m)$ prototypical target, with an extraction efficiency of approximately 15%, limited by target meltdown. We project for the design goal target with approximately 10,000 double layers (2 mm thick), a production of $C^{15}O$ secondary RNB of the order of 10^{11} $/sec$, which is at least a factor of 100 better than could be obtained with the use of the $^{12}C(\alpha,n)^{15}O$ reaction. We continue our effort to improve target resilience and better extraction efficiency, by using different nitrites as well as allowing for water cooling of the thick target. Based on our experience with the prototypical thin tharget we do not expect problems with the production of the thicker foils with approximately 10,000 layers (2 mm). And the production of a test multi-layered aluminum nitrides target is planned at LLN in February 1997.

References

[1] N. Hata and P. Langacker; Phys. Rev. **D50**(1994)632, ibid **D52**(1995)420.

[2] R.K. Wallace and S.E. Woosley; ApJ. Suppl. **45**(1981)389.

[3] M. Gai, J.-Z. Zhang, B.L. Halpern, S.R. Boorse, J.J. Schmitt, Th. Delbar, P. Leleux, M. Loiselet, C. Michotte; 6th Workshop on Targetry and Target Chemistry, Vancouver, August 17-20, 1995.

[4] J.J. Schmitt and B.L. Halpern; US patent #4,788,082 (11/29/88).

[5] R.R. Borchers and C.H. Poppe; Phys. Rev. **129**(1963)2679.

[6] B.W. Filippone and M. Wahlgren; Nucl.Inst.Meth. **A243**(1986)41.

[7] J.F. Ziegler and J.P. Biersack, Code TRIM92, unpublished.

DESORPTION KINETICS AND RIB SOURCES

S.D. Kevan

Physics Department, University of Oregon, Eugene, OR. 97403

The various factors that effect the desorption of atoms and molecular fragments from surfaces and their impact on the efficiency of radioactive ion beam (RIB) sources will be discussed. The important role played by entropy in determining both the low- and high-coverage desorption rate will be emphasized. Also, the possibility of accomplishing the desorption step using a pulsed laser, followed by resonance ionization of the desorbed species will be proposed as a low duty-cycle but highly selective approach to the RIB design problem. Finally, various design factors such as the incident and exit beam size and temporal character and the structure of the target will be summarized.

INTRODUCTION

Following production of a trace component of a desired radioactive species in a solid target, a radioactive ion beam (RIB) can be produced through a series of kinetic processes: diffusion to the surface of the target, desorption into the vapor phase, and transport into an ionization region. To achieve the highest possible RIB intensity, each of these processes must be controlled since their relevant time constants must be made as short as possible compared to the half life of the often short-lived radioactive species. In the present paper, I will focus on the desorption kinetic step. It can be argued that this is often the most problematic kinetic event in RIB sources since chemical bonds actually need to be broken. At a minimum, this implies rather large activation energies and thus relatively slow rates. At worst, the desorption event can be so slow that either the species remains on the surface for many half lives or, at higher temperature where the entropy of dissolution starts to dominate, the species migrates into the bulk of the target. My focus is thus to define the extent to which desorption kinetic events can be controlled and/or modified to produce a desired outcome.

THERMAL DESORPTION KINETICS

In current RIB designs, the radionuclide desorbs spontaneously from the surface due to thermal fluctuations. The process is said to activated since there is obviously an energy barrier to desorption. In most cases of interest, the reverse step - adsorption - is not activated. This leads to a conceptually simplified 1D model. The adsorbed atom or molecule resides in a well typically $E_0 \sim$ 1-3 eV deep, and the interaction potential energy increases smoothly and monotonically as the atoms or molecules move away from the surface. This model is surely oversimplified since it largely neglects motion parallel to the surface. However, the diffusion barriers parallel to the surface are nearly always smaller than the desorption barrier, so surface diffusion kinetics are always fast on the time scale for desorption. This leads to a useful and nearly always correct 'quasiequilibrium approximation'.[1-3] This approximation is used to evaluate the desorption rate in terms of mostly equilibrium quantities. It holds that the desorption rate under non-equilibrium conditions is equal to the equilibrium desorption rate. The justification is that the events that maintain equilibrium are fast compared to desorption, so a desorbing particle loses memory of how it was adsorbed - whether from an equilibrium vapor phase or, e.g., by diffusion from the bulk. In this circumstance, we can equate the equilibrium rate of adsorption from the kinetic theory of gases to the equilibrium, first-order desorption rate. The desorption rate constant then becomes:

$$k_d = \frac{SA_o q_{int} k_B T}{N_o h \Lambda^2} \frac{\lambda(\theta,T)}{\theta} \tag{1}$$

where T is the surface temperature, S is the sticking coefficient, q_{int} is the internal partition function of the gas phase molecule, $\Lambda = (h^2/2\pi M k_B T)^{1/2}$ is the thermal wavelength, A_o is the area of the surface having N_o adsorption sites, and $\lambda(\theta,T) = \exp[\mu(\theta,T)/k_B T]$ is the equilibrium activity as a function of surface coverage, θ, and temperature T. At low surface density, the chemical potential $\mu(\theta,T)$ is just given by E_0, the binding energy of the molecule to the surface, and the desorption rate constant is

$$k_d = A\exp\left(-E_o/k_B T\right) \tag{2}$$

CP392, *Application of Accelerators in Research and Industry*, edited by J. L. Duggan and I. L. Morgan
AIP Press, New York © 1997

This equation has the same form as that predicted from Transition State Theories, where the prefactor includes partition functions and a dynamical crossing probability (S, in our case). This is not surprising since the underlying approximations are quite similar. The dominant effect of temperature is found in the exponential. All current and planned RIB sources operate at high temperature, partly to achieve large desorption rates. The factor $SA_o/N_o\Lambda^2$ is near unity, so the prefactor is approximately given by

$$A \sim \frac{q_{\text{int}}}{q_{ads}} \frac{k_B T}{h} \tag{3}$$

where q_{ads} is the partition of a particle adsorbed at a single site. The factor k_BT/h is of order 10^{13}, and this value is commonly referred to as a 'normal' desorption prefactor. In atomic systems, the desorption prefactor is often close to this value since the partition functions refer to relatively high energy internal degrees of freedom which are frozen at the operating temperature. In molecular systems, the internal degrees of freedom include molecular vibrations and rotations that occur at low energy. Indeed, gas phase rotational partition functions are typically of order 10 per degree of freedom. Since the rotational partition functions multiply, the prefactor in molecular systems can be as large as 10^{15} - 10^{16}. This increased prefactor for molecular versus atomic desorption systems is an artifact of entropy. The adsorbed molecule is normally constrained to lie near a particular orientation on the surface. The free molecule is free to rotate and consequently has a larger internal entropy. Our first important observation is thus that, for equal adsorption energy E_o, the desorption rate for molecules and molecular fragments will be 2-3 orders of magnitude larger than for atoms. Other things being equal, one ought to search for molecular desorption species in RIB sources.

This important role played by the entropy associated with internal degrees of freedom has another, more subtle implication. Consider the following thought experiment. Situate an ensemble of molecules adsorbed on a surface at a temperature where the desorption rate is negligibly small. Now raise the temperature so that the desorption rate is very large. One would think that the molecules would desorb, but, of course, there is another kinetic path available: dissolution into the bulk. In some instances, particularly in the case of relatively light adsorbed atoms, the activation energy for dissolution is smaller than for desorption. The reason for this is simply that in diffusion, the energy of the activated complex is often reduced by the ability to form a second bond before the first is broken, while in desorption a bond simply must be broken. By the same reasoning as above, however, the prefactor for dissolution is normally much lower than for desorption (even for atoms) since the entropy gain upon desorption is larger than for dissolution. The prefactor can be considered to be the rate at which the system tries to surmount the activation energy barrier. This means that the dominant kinetic process - desorption vs. dissolution - is determined by how quickly the temperature is raised. When the temperature is raised quickly, the larger attempt frequency for desorption will favor desorption. When it is raised slowly, the barrier always dominates and dissolution can be favored. Our second observation is thus that it might be useful to use the temporal evolution of the RIB target temperature to optimize the desired product.

Finally, it is useful to discuss briefly the coverage-dependence of the desorption rate. This is a complicated subject that has received quite a bit of effort, but we will simplify matters manifestly. Specifically, we return to Eq. (1) and write a virial expansion for the chemical potential, retaining only the first term in the expansion. The result is[4,5]

$$k_d = \frac{SA_o}{N_o} \frac{q_{\text{int}}}{q_{ads}} \frac{k_B T}{h} \exp\left(-\tfrac{E_o}{k_B T} + 2B_2\theta\right) \tag{4}$$

This equation is only strictly valid in the zero-coverage limit, but we nonetheless see an exponential coverage dependence governed by the second virial coefficient, $B_2(T)$. Recall that $B_2(T)$ normally adopts a particular form, independent of system and statistical model: it is negative at low T, passes through zero at moderate T, and becomes almost constant at high T. The constancy at high temperature is associated with the highly repulsive potential barrier when molecules approach too closely - atoms and molecules occupy space. This strong coverage dependence of the desorption rate can be understood in terms of the configurational entropy. As saturation is approached, the configurational entropy becomes small since there are few accessible configurations due to the hard-wall repulsion. The system gains entropy rapidly by desorbing molecules. The important observation is that the desorption rate typically increases, sometimes dramatically, as a function of coverage. It might be useful to consider RIB sources for which the surface plays the role of a reservoir that fills almost to capacity in steady state so that desorption occurs at high coverage.

RIB SOURCE BASED ON PULSED LASER DESORPTION AND IONIZATION

The observations from the previous section emphasize the role played by entropy in desorption. The conclusions are that 1) use of molecules and molecular fragments might be advantageous, 2) rapid increases in target temperature favors desorption over dissolution, and 3) operation at finite surface coverage tends further to favor rapid desorption. Current RIB sources do use molecular fragments in some instances, though it is not clear whether this results from consideration of the entropic factor discussed above. At present, I know of no RIB source that has attempted to program the temporal behavior of the target temperature or to use the surface as a reservoir. How might one accomplish these in a real design?

A logical place to start is to use a pulsed laser to desorb the adsorbed atoms and molecules. There is quite a bit of work at present to try to use lasers to guide surface chemical reactions - 'surface photochemistry'.[6] However, in many instances, directing a pulsed laser at a surface simply raises the surface temperature for a few microseconds, and the induced surface chemistry is actually thermal in nature.[7] Thus, it has been shown that pulsed lasers can be used, in some though not all instances, to thermally desorb atoms and molecules on a microsecond time scale, a scale determined by the thermal properties of the substrate. With proper tuning, this can be quantitative. For example, stable molecules like CO and NO have been 100% desorbed with a pulsed laser. Atoms that are strongly bound to surfaces - carbon is a prime example - are difficult to desorb by any thermal technique. Examples also exist where desorption is favored over other kinetic routes on these short time scales. The methoxy fragment CH_3O- can be quantitatively removed by laser desorption, but it tends to decompose to constituent atoms upon slower heating.[7]

A major drawback to pulsed laser desorption is the rather poor duty cycle for essentially all pulsed lasers. A q-switched Nd:YAG laser, for example, operates at only a few tens of Hz with a pulse duration of tens of nanoseconds. In order for pulsed laser desorption to be practical for RIB sources, this duty cycle problem will need to be overcome. Once again, the obvious solution to this problem is to utilize a pulse laser to ionize the desorbing products. The quiescent temperature of the target would be held high enough that bulk diffusion to the surface would be fairly rapid and favored thermodynamically, but low enough that desorption would be slow. In this case, the desired radionuclides will segregate to the surface while the desorption laser is off. They will be desorbed quickly and nearly quantitatively and then ionized very close to the surface, directly after desorption.

Such a scheme might work well for atomic species, since the resonance enhanced multiphoton ionization (REMPI) is quite efficient and exquisitely selective in this case.[8] Indeed, the selectivity could obviate the need for mass filtering after the RIB source. REMPI is somewhat less useful in the case of molecules, since these will generally desorb into a wide range of rotational-vibrational levels and a single laser wavelength is correspondingly of lower overall efficiency. In this case, one would need to use a higher power laser and do non-resonant multiphoton ionization, a process that is reasonably efficient though not selective.

GEOMETRICAL CONSIDERATIONS AND TRANSPORT

The Liouville theorem does not apply to RIB sources since they are not conservative. A relatively bright source of low mass MeV-energy ions is directed at the target, and a thermal source of the desired radionuclide results. Existing RIB sources transport the neutral beam over a distance of typically 10 cm to the ionization region. The efficiency of this process could possibly the improved by moving the ionization region in closer proximity to the target. This could possibly be implemented with the pulse laser scheme discussed in the previous section. Specifically, ions incident from the cyclotron focused into a spot of order 1 mm provide a spatially well-defined region where the desired radionuclides are produced. Lateral diffusion in the target will be small, since the deposition depth is on the micron length-scale. Thus most of the desired radionuclides will accumulate on a small spot on the surface of the target. The desorption laser could be focused onto the same spot, and the ionization laser could be directed parallel to but ~ 1mm above the surface plane. A similar scheme has been implemented to detect secondary neutrals desorbing from surfaces.[9] With this geometry, the RIB source will nearly preserve the optical brightness of the incident ion beam, thereby improving overall efficiency.

ACKNOWLEDGEMENTS

This work was supported in part by the US Department of Energy under grant DE-FG06-86ER45275.

REFERENCES

1. Iche, G. and Nozières, P., *J. Physique* **37**, 1313-1323 (1976).
2. Ibach, H., Erley, W. and Wagner, H., *Surface Sci.* **92**, 29-42 (1980).
3. Leuthauser, U., *Z. Phys.* B **37**, 65-67 (1980).
4. Peterson, L.D., and Kevan, S.D., *Phys. Rev. Lett.* **65**, 2563-2566 (1990).
5. Kevan, S.D., Wei, D.H., and Skelton, D. C., *CRC Crit. Rev. Surface Chem.* **3**, 77-140 (1994).
6. Zimmermann, F. M., and Ho, W., *Surface Sci. Repts.* **22**, 129-247 (1995).
7. Hall, R.B., and Bares, S. J., edited by R.B. Hall and D.B. Ellis, *Chemical and Structure and Interfaces*, Deerfield Beach, Fl.: VCH 1986, p. 85.
8. Pfab, J., edited by Clark, R.J.H., and. Hester, R. E., *Spectroscopy in Environmental Science*, New York: Wiley 1995.
9. Kobrin, P. H., Schick, G. A., Baxter, J. P., and Winograd, N., *Rev. Sci. Inst.* **57**, 1354-1362 (1986).

ADVANCED TARGET CONCEPTS FOR RIB GENERATION

G. D. Alton
Oak Ridge National Laboratory, P. O. Box 2008, Oak Ridge, Tennessee 37831-6368

In this report, we describe highly permeable composite target matrices that simultaneously incorporate the short diffusion lengths, high permeabilities, and heat removal properties neccessary to effect maximum diffusion release rates of short-lived, radioactive species as required for efficient radioactive ion beam (RIB) generation in nuclear physics and astrophysics research programs. The RIB species are generated by either fusion or fission nuclear reactions between high energy ^{1}H, ^{2}H, ^{3}He or ^{4}He ion beams and specific nuclei which make up the target material. The target materials may be used directly as small diameter particulates coated or uncoated with Re or Ir to minimize adsorption following diffusion release and eliminate sintering of the particulates at elevated temperatures; plated onto both sides of thin disks of C, for example; or plated, in thin layers, onto low density, Ir or Re coated carbon- bonded-carbon-fiber (CBCF) or reticulated-carbon-fiber (RCF) to form sponge-like composite target matrices; or in other cases, where applicable, the target material of interest can be grown in crystalline fibrous form and fabricated in woven mats of the target material to form a highly permeable fibrous structure.

INTRODUCTION

Many of the reactions fundamentally important in nuclear physics, and astrophysics are inaccessible to experimental study using stable/stable beam/target combinations and therefore can only be studied with accelerated radioactive ion beams (RIBs). The availability of RIBs offers unique opportunities to further our knowledge about the structure of the nucleus, the stellar processes which power the universe and the nucleosynthesis burn cycles responsible for heavy element formation. As a consequence of world-wide interest in the potential benefits of using RIBs, facilities have been built, funded for construction or proposed for construction in Asia, Europe, and North America [1]; these include the Holifield Radioactive Ion Beam Facility at the Oak Ridge National Laboratory (ORNL) [2]. These facilities are based on the use of the well-known on-line isotope separator (ISOL) technique in which radioactive nuclei are produced by nuclear reactions in selectively chosen target materials by high-energy proton, deuteron, He, heavy ion beams or neutrons. The ISOL technique has been utilized to produce low energy ion beams from over 600 isotopes [3]. In order to successfully perform nuclear physics and astrophysics research with radioactive ion beams (RIBs), intensities ranging from 10^{5} to 10^{12} particles/s must be delivered to the experimental station. While these intensities appear modest, they are not easily achievable because of the limited production rates of the species and their lifetimes in relation to the times required for diffusion from solid state targets and effusion into the ion source. The production rates in the target are set by practical limits on the primary beam intensity in terms of the maximum permissible, on-target, power density which can be used without compromising the efficiency of the ion source or the physical integrity of the target, and the reaction cross sections for producing the species of interest. The radioactive product beam intensity is determined by the rate at which the particular species can be diffused from the target, the average residence time of the radioactive species on surfaces between the target and the ionization chamber of the source in relation to their lifetimes, the ionization efficiency of the ion source, and losses in beam transport between the ion source and experimental station. The upper temperature at which the target/ion source can be operated without deleteriously affecting the ionization efficiency is set by the vapor pressure characteristics of the target material [4, 5]. Because of these factors, careful attention must be given in the selection of the most appropriate target material and in the design of composite target/heat-sink systems which incorporate the short diffusion lengths and high permeability attributes required for prompt and efficient diffusion release of the species of interest at controlled temperatures.

In this report, we briefly review present efforts at the HRIBF to design advanced highly permeable composite targets required for effecting fast and efficient diffusion release of the radioactive species of interest from the target material.

PROCESSES WHICH AFFECT RELEASE TIMES

The principal means whereby short half-life radioactive species are lost between initial formation and utilization are associated with diffusion and surface adsorption processes where the delay times are long with respect to the life-time of the species in question for forming RIBs. The diffusion and surface

CP392, *Application of Accelerators in Research and Industry*, edited by J. L. Duggan and I. L. Morgan
AIP Press, New York © 1997

adsorption/desorption processes depend exponentially on the operational temperature of the target/ion source. Because of the strong temperature dependence of the adsorption process, the materials of construction of the target chamber, vapor transport tube and ionization chamber of the source can significantly affect the residence times of the radioactive species through the magnitudes of the enthalpies of adsorption on the surfaces of materials that the particle comes in contact with following diffusion release and transport to the ionization chamber of the source. The diffusion and surface adsorption processes are briefly described below.

Diffusion Theory

The time and temperature-dependent release of a nuclear reaction product species, embedded in a chemically dissimilar target material, implies the presence of a binary diffusion mechanism which underlies the release process. Whenever there is a concentration gradient of impurity atoms or vacancies in a solid material, the atoms or vacancies will move through the solid until equilibrium is reached. The net flux, J, of either the atoms or the vacancies is related to the gradient of concentration, ∇n, by Fick's first equation given by:

$$J = -D\nabla n \tag{1}$$

where D is the diffusion coefficient.

The time-dependent form of Eq. 1 is known as Fick's second equation. The three-dimensional form of this equation, which allows for the creation of particles S(x,y,z,t) as well as the loss of particles E(x,y,z,t), can be expressed in a Cartesian coordinate representation as follows:

$$\frac{\partial n}{\partial t} = D\left(\frac{\partial^2 n}{\partial x^2} + \frac{\partial^2 n}{\partial y^2} + \frac{\partial^2 n}{\partial z^2}\right) + S(x,y,z,t) - E(x,y,z,t) \tag{2}$$

where D is assumed to be independent of concentration.

The target geometries that will be used or are envisioned for use at the HRIBF include: planar, cylindrical and spherical geometries. For example, the time-dependent form of Eq. 2 appropriate for planar targets is given by

$$\frac{\partial n}{\partial t} = D\frac{\partial^2 n}{\partial x^2} + S(x,t) - E(x,t) \tag{3}$$

The equation, appropriate for diffusion from cylindrical geometry targets, where the diffusion process is assumed to move only in the r direction, can be expressed in the following form

$$\frac{\partial n}{\partial t} = \frac{D}{r}\frac{\partial}{\partial r}\left(r\frac{\partial n}{\partial r}\right) + S(r,t) - E(r,t) \tag{4}$$

While for spherical-geometry targets, where the diffusion process is assumed also to be solely in the radial direction, Fick's second equation takes the form:

$$\frac{\partial n}{\partial t} = \frac{D}{r^2}\frac{\partial}{\partial r}\left(r^2\frac{\partial n}{\partial r}\right) + S(r,t) - E(r,t)\ . \tag{5}$$

Solutions to the respective time dependent forms of Equations 3, 4, and 5, appropriate for the particle target material geometry, can be found either by separation of variables, the use of Lapace or Fourier transformation techniques, or by standard numerical computational techniques or in combination.

We assume that the particles are uniformly distributed during production of the radioactive species. For the more general case, the distribution function S must be chosen to represent the actual distribution of the radioactive species within the target material. For a uniform distribution of particles such as assumed in the production of radioactive species with a primary ion beam of intensity I and charge Z, S is given by

$$S(x,t) = \sigma n I \ell / ZeV \tag{6}$$

where e is the charge on the electron, n the number of interaction nuclei per unit volume, ℓ is the length of the target material, σ is the cross section for production of the species of interest, and V is the volume of the irradiated sample.

For production of radioactive species with half-life $\tau_{1/2}$, E is given by

$$E(x,t) = n\lambda \text{ where } \lambda = 0.693/\tau_{1/2} \tag{7}$$

For solids, the diffusion process is dependent on the activation energy H_A required to move the atoms or vacancies from site to site and the temperature T according to:

$$D = D_0 \exp(-H_A/kT) \tag{8}$$

where D_0 is the intrinsic diffusion coefficient of the atom within the particular crystal matrix. D is related to the vibrational frequency and lattice parameters of the particular atom and crystal and k is Boltzmann's constant. H_A can be extricated from experimental data by measuring the dependence of D on target temperature T.

Figure 1 displays a computational simulation of the release of ^{58}Cu and ^{63}Cu from thin layers (5 μm) of Ni metal electroplated on Ir coated (1 μm) C fibers of diameter 6 μm as computed with DICTRA [6].

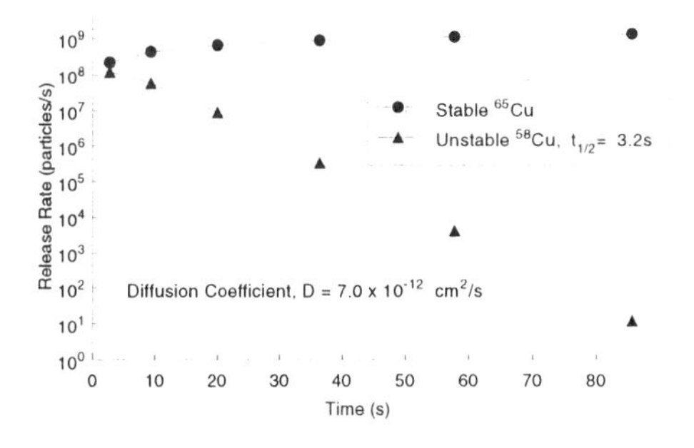

Fig. 1. Simulation of the diffusion release of ^{58}Cu from an annular deposit of Ni onto Ir coated, RVCF, similiar to the composite target illustrated in Fig. 4 (Ni coating: 5 μm; Ir coating: 1 μm; C diameter: 5 μm). The target is assumed to operate at 1000°C. Diffusion is assumed to move radially outward due to a reflecting boundary condition imposed at the interface between the Ni deposit and the Ir coating.

Theory of Adsorption

Time delays, associated with adsorption processes, that are excessively long in relation to the life-time of the radioactive species can result in significant losses of beam intensity in an ISOL facility. The residence time of a particle on a surface is given by the Frenkel equation:

$$\tau = \tau_0 \exp\left[H_{ad}/kT\right] \tag{9}$$

where H_{ad} is the heat of adsorption or enthalpy required to evaporate the atom or molecule from the surface, k is Boltzmann's constant, T is the absolute temperature, and τ_0 is the time required for a single lattice vibration ($\sim 10^{-13} - 10^{-15}$ s). The heat of adsorption increases with increasing chemistry between the adsorbed atom and the surface where the adsorption takes place. This value varies widely depending upon the adsorbent/adsorbate combination.

The desorption rate of atoms per unit area dN/dt in thermal equilibrium with a surface at temperature T is given by

$$\frac{dN}{dt} = \frac{S(T)NkT}{h} \exp\left[(\Delta ST - H_{ad})/kT\right] \tag{10}$$

where S(T) is the temperature-dependent probability that the particle will stick to the surface (sticking coefficient), N is the number of atoms adsorbed per unit area, h is Planck's constant and ΔS is the change in entropy of the adsorbed particle.

Since it is desired to minimize the residence times of atoms/molecules on surfaces in the target/ion source, the choice of the materials of construction for the vapor transport system of the source is extremely important. Coating the inner surfaces of the vapor transport tube (usually made of Ta or W) with a chemically inert material is expected to significantly reduce residence times and, therefore, the noble metals, iridium or rhenium, have been recommended for this use [7, 8].

NEW TARGETS AND TARGET CONCEPTS

New target concepts are presently under development at the HRIBF that, in principle, can be applied in the design and fabrication of targets with short diffusion lengths and high permeability properties required if present RIB facilities are to be successful in meeting the intensity and species needs of their respective research programs [9]. These new technologies include the choice of low-density target matrixes, such as carbon-bonded-carbon fiber (CBCF) or reticulated vitreous carbon fiber (RVCF), which can be used as the thermal transport structure for the removal of heat from the target material and as the plating matrix for the target material itself. Since short-lived particles must swiftly diffuse from the target material, the diffusion lengths must be short (thin target materials) and the target temperature must be as high as practicable. Techniques are presently available that can be used to uniformly deposit specified thicknesses of the material in question onto the support matrix of choice. At this point in time, several target coating schemes are being used or are under consideration for this purpose, they include: 1) chemical vapor deposition (CVD); 2) physical vapor deposition (PVD); electrolytic deposition (ED) and 3) sol-gel coating. In addition, cataphoretic or electrophoretic coating techniques are also under consideration as viable means for coating target matrix support structures with specified thicknesses of material.

Fibrous target materials.

In a few cases, fibrous materials with small fiber diameters are available. Figure 2 shows photographs of

Fig. 2. Photograph of an Al_2O_3 fibrous matrix (fiber diameter: 3 μm) which has been successfully used for the low-intensity, on-line production and release of ^{17,18}F as a means of evaluating the material for potential RIB generation.

Al_2O_3 fibrous materials. The Al_2O_3 fibrous target material has been successfully used to produce and release $^{17,18}F$ for potential use in the HRIBF research program [10]. As noted, the target matrix is highly permeable and, therefore, the release efficiency should be high.

Composite Target Matrix Designs

The objective of the target development program is to develop thin but highly permeable target matrix/heat-sink systems that, in combination, will permit operation of the target matrix at the maximum temperature allowable during production of radioactive species.

Reticulated-vitreous-carbon fibers (RVCF) and carbon-bonded-carbon (CBCF) fibers offer generic matrices for coating target materials. Figure 3 is a photograph of a carbon-bonded-carbon fiber (CBCF) fiber target matrix (fiber diameter: 6 µm) precoated with ~ 1 µm of Re followed by chemical vapor deposition (CVD) of 5 µm of SiC. The composite target matrix is highly permeable and will be tested for the release of short-lived P and S isotopes. Figure 4 displays electroplated Ni/RVCF composite target for future evaluation as a possible target material for use in generating ^{58}Cu. The choice of the particular target matrix system will depend on the availability of techniques for uniformly coating the target material onto the composite target matrix at the thicknesses required for efficient and fast diffusion release and the results of thermal transport studies [9].

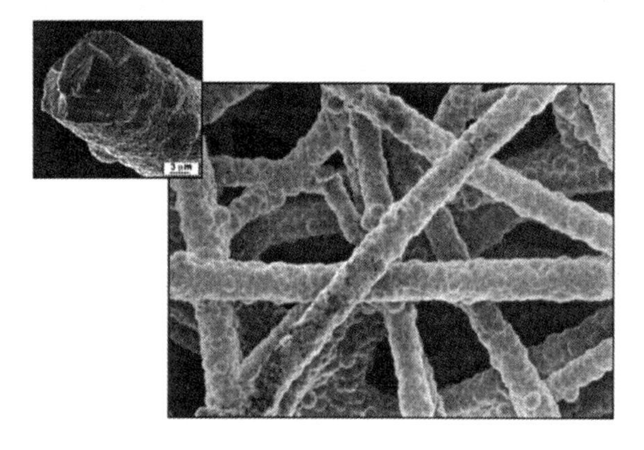

Fig. 3. Photograph of a carbon-bonded-carbon fiber (CBCF) matrix (fiber diameter: 6 µm) precoated with ~ 1 µm of Re followed by chemical vapor deposition (CVD) of 5 µm of SiC. The composite target matrix is highly permeable and will be tested for the release of short-lived P and S isotopes.

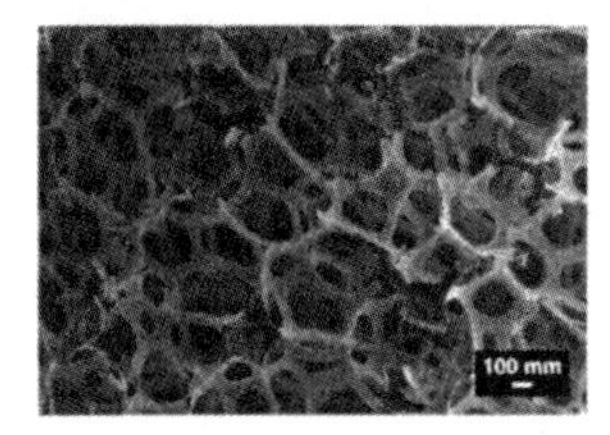

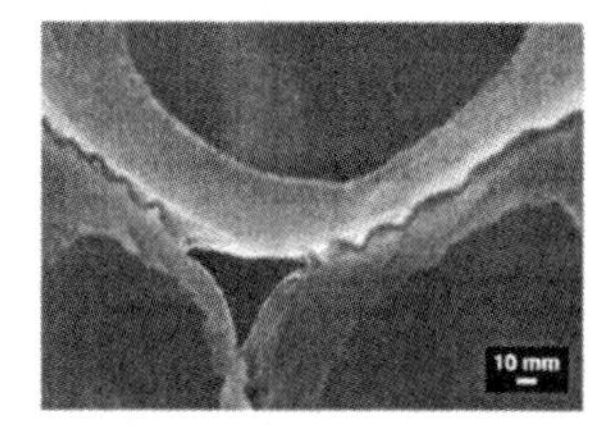

Fig. 4. Photograph of reticulated-vitreous-carbon-fiber (RVCF) matrix (fiber diameter: 54 µm) electroplated with ~ 3 µm Ni for potential use in generating ^{58}Cu.

The efficiency of the ionization process is strongly dependent on the pressure in the ionization chamber of the source and thus the vapor pressure of the target material; the vapor pressure of the target material is, in turn, strongly dependent on the target temperature; therefore, control and maintenance of a specific temperature, below a critical value, in the target holder/target matrix region of the target/heat sink assembly is critically important for successful and efficient generation of high intensity RIBs. We are presently designing a heat/sink system for mounting target matrices for control of target temperatures [9]. To achieve the primary objective of the project, the target/heat-sink assembly design will be optimized so that the particular target composite matrix can be maintained at a prescribed temperature which is limited by the maximum vapor pressure of the target material that can be tolerated before compromising the ion source efficiency and/or destroying the target through sublimation/vaporization processes. Research sponsored by the Oak Ridge National Laboratory, managed by Lockheed Martin Energy Research Corporation for the U.S. Department of Energy under contract Number DE-AC05-96OR22464.

REFERENCES

[1] D. K. Olsen, Nucl. Instr. and Meth. **A328**, 303 (1993).

[2] A Proposal for Physics with Exotic Beams at the Holifield Heavy Ion Research Facility, eds. J. D. Garrett and D. K. Olsen, Physics Division, ORNL, March 1991, unpublished.

[3] H. L. Ravn in ISOLDE User's Guide, Ed. H. -J. Kluge, CERN 86-05 (1986).

[4] R. Kirchner and E. Roeckl, Nucl. Instr. and Meth. **133**, 187 (1976).

[5] P. Van Duppen, P. Decrock, M. Huyse, and R. Kirchner, Rev. Sci. Instr. **63**, 2381 (1992).

[6] DICTRA is a database diffusion code which is a product of the Royal Institute of Technology, Stockholm, Sweden.

[7] G. D. Alton, H. K. Carter, I. Y. Lee, C. M. Jones, J. Kormicki, and D. K. Olsen, Nucl. Instr. and Meth. **B66**, 492 (1992).

[8] G. D. Alton and J. Dellwo, Nucl Inst. and Meth. **A382**, (1996) 225.

[9] G. D. Alton, J. W. Middleton, and J. R. Beene, to be published in Nucl. Instr. and Meth..

[10] D. Stracener, H. K. Carter, J. B. Breitenbach, J. Kormicki, J. C. Blackmon, M. S. Smith and D. W. Bardayan, these proceedings.

STATUS OF THE RNB FACILITY AT INS

I.Katayama, T.Nomura, S.Arai, Y.Arakaki, Y.Hashimoto, A.Imanishi, S.C.Jeong, T.Katayama, H.Kawakami, S.Kubono, T.Miyachi, H.Miyatake, K.Niki, M.Okada, M.Oyaizu, Y.Shirakabe, P.Strasser, Y.Takeda, J.Tanaka, M.H.Tanaka, E.Tojyo, M.Tomizawa, M.Wada, S.Kato[a], T.Shinozuka[b], and H.Wollnik[c]

Institute for Nuclear Study, University of Tokyo (INS), 3-2-1 Midorichou, Tanashi, Tokyo, 188 Japan

[a] *Physics Department, Yamagata University, Yamagata, 990 Japan*

[b] *Cyclotron and Radioisotope Center, Tohoku University, Aoba, Sendai, 980-77 Japan*

[c] *Department of Physics, Gissen University, Gissen, Germany*

A Radioactive beam facility using a thick target, an ISOL and heavy ion linacs will soon come into operation at INS, University of Tokyo. The status of the project together with experimental programs are reported.

INTRODUCTION

At INS, Univ. of Tokyo, an ISOL based radioactive nuclear beam (RNB) facility is now, after 4 years construction, under final tuning. The ISOL based RNB has better energy resolution and higher brightness compared with a heavy ion projectile fragmentation based RNB. With a proton beam from the K=67 SF cyclotron, we can expect a reasonably intense RNB, i.e., up to 10^{10} atoms/sec for nuclei located near the stability line. The INS RNB facility is a prototype for the Exotic Nuclear Beam Arena (E-Arena) of the Japanese Hadron Project (JHP) [1]. There are several crucial techniques to be developed, such as a highly efficient ion source, a high resolution isotope separator, a versatile secondary heavy ion accelerator, and a technique for handling highly radioactive materials and high radiation levels. In this paper, we report the status of developments of these techniques, except for high-radiation handling. The status of two experimental projects associated with the facility, i.e., nuclear astrophysics and laser-ion-trap, is also reported.

LAYOUT OF THE FACILITY

Figure 1 shows the layout of the present facility. The radioactive nuclides are produced in a thick target with the cyclotron beam. Using an ion source of proper type for the element of interest, the nuclides are ionized and extracted for isotope separation. The separated beam, after deceleration to 2 keV/amu to match the injection

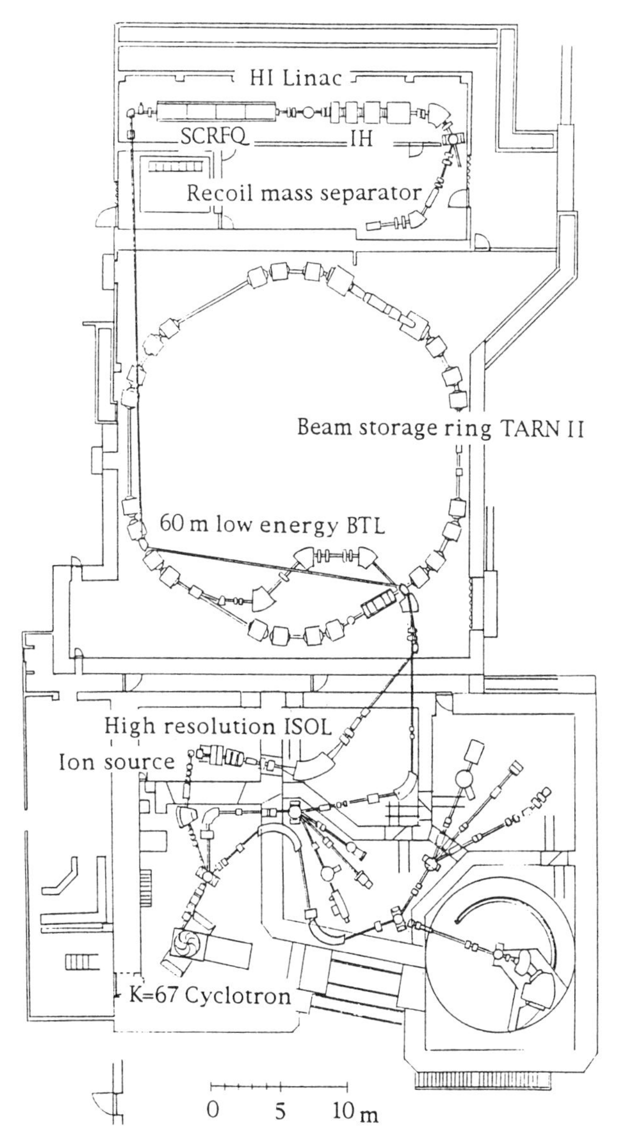

FIGURE 1. Layout of the INS Radioactive Nuclear Beam Facility and the cyclotron.

CP392, *Application of Accelerators in Research and Industry*, edited by J. L. Duggan and I. L. Morgan
AIP Press, New York © 1997

energy of the Split Core RFQ (SCRFQ) linac, is transported through the 60m low energy beam transport line (60m BTL) to the linac. The linacs are of two types, i.e., SCRFQ and Interdigital H (IH). The SCRFQ linac, operated at 25.5MHz, accelerates ions with q/m of more than 1/30 to 170 keV/amu, while the IH linac, operated at 51 MHz, accepts ions with q/m of 1/10. In order to fulfill these requirements, a carbon stripper and a buncher are installed between the two linacs. The output energy of the IH linac is variable from the injection energy of 170 keV/amu to 1.05 MeV/amu. The IH linac beam is transported through a bending magnet and a switching magnet to one of three target stations. For the moment, only one beam line, with 60 degree bending angle, is to be used for nuclear astrophysics experiments using a recoil mass separator.

ION SOURCE

Three different types of ion sources have been developed to ionize various elements. The surface ionization and FEBIAD types, which are mounted in a standardized vessel like one used at ISOLDE, employ a short transfer tube at high temperature. The FEBIAD type with this structure is efficient for ionizing metallic elements. An ECR type, which is just under on-line test, has a long, curved transfer tube at room temperature (as well as sometimes at liq. N_2 temperature) as shown in Fig. 2 [2], and is used to ionize gaseous elements. This structure, which enables a decoupled target system, is effective for reducing gas loading in the ECR and contaminants. To accomodate these different ion sources, the first focus element, consisting of a quadrupole doublet, is made movable along the beam line.

Ion bunching, is often used to give on-line diagonastics of the ionization efficiency [3].

HIGH RESOLUTION ISOL SYSTEM

The ISOL consists of double stage separator. The first stage has a mass dispersion of 170mm while that of the second stage is 3800mm. All the components are electrically isolated from ground potential. The schematic diagram of the potential distribution of the system is shown in Fig.3. This scheme has the following features:

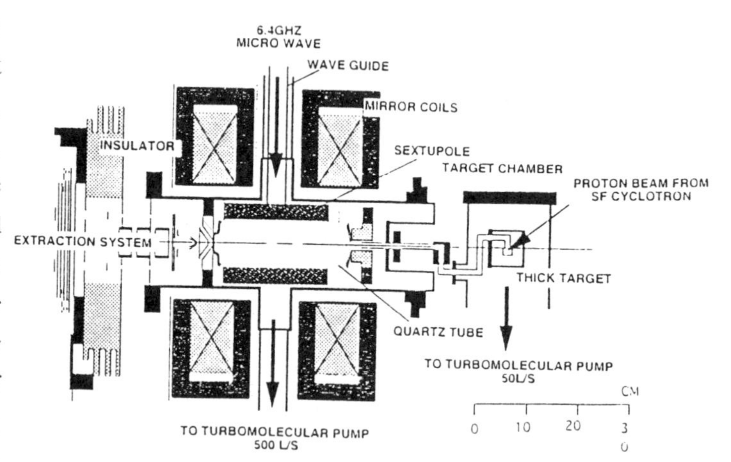

FIGURE 2 . Schematic view of the ECR source with a long waved transfer tube [2].

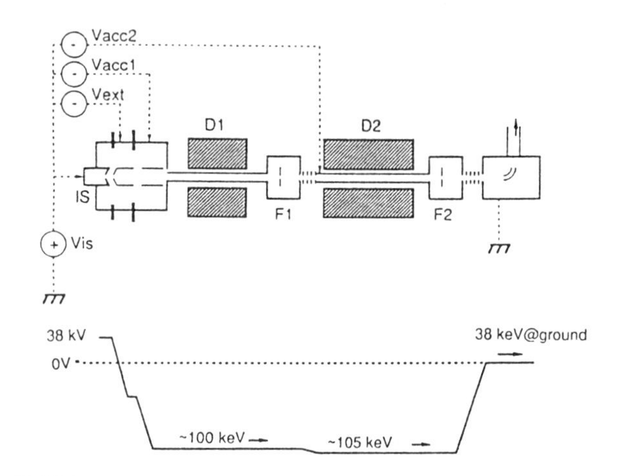

FIGURE 3 . Potential distribution of the present electrically floating ISOL [4].

1) The ion source potential is chosen to match the ion energy to the specification of the linac, i.e., 2 keV/amu.

2) The potentials of the 1st and the 2nd stage are at the maximum tolerable voltage to give a better mass separation at the final focusing point.

3) A difference in potentials between the 1st and 2nd stage can be used to remove contaminant ions at the final focusing point.

The design mass resolutions of the 1st and 2nd stages are about 300 and 9000, respectively, in the case of an ideal beam emittance of ± 0.2mm and $\pm$ 20mrad. In the off-line test of the ISOL, a mass resolution of 3300 with a limited emittance of about 8πmm·mr was achieved for mass 20 isobars [4]. The use of electric elements is avoided as much as possible in order to keep the space charge compensation in an intense low energy beam. The ISOL has already been tested on-line for 2 years with the three kinds of ion-sources previously mentioned. Radioactive nuclides of ^{39}K, ^{20}Na, ^{18}Ne and ^{19}Ne have been separated. Tests of the electrically floating capability of the ISOL have started recently.

60M BEAM TRANSPORT LINE

Since the linac was installed in a room converted from the old cyclotron vault, the ISOL beam had to be transported through a beam storage ring room (see Fig. 1.). The 60m BTL was designed using only electric ion optical elements [4]. The beam envelope calculated by the program GIOS is shown in Fig. 4. The ion optical system of the 60m BTL comprises 7 deflectors and 134 quadrupole singlets. Emittance matching quadrupoles are also placed in the first injection part, and in the entrance and exit of each deflector unit. The first two deflectors lift up the beam from the ISOL level (1.2m) to 2.3m height, and the last two deflectors bring the beam back to the linac level of 1.2 m. Six sets of symmetric quadrupole lenses, each having a bore diameter of 36 mm and a length of 100 mm, connect the 7 deflectors. A set of quadrupoles is aligned as a periodic array of alternating focusing and defocusing lenses with an interval distance of 40 cm. The acceptance of such a periodic lens system is proportional to the square of the diameter of the lenses, and inversely proportional to the distance of the interval. The phase advance of each half unit was chosen to be 77 degrees to provide maximum acceptance. The design acceptance of the 60m BTL is 170 πmm·mr. Each straight section is equipped with one or two beam diagnastic devices consisting of a cross hair beam profile monitor, a slit, and a beam stopper. Seven local CPUs are used to control about 100 power supplies and 10 diagnastic devices.

HEAVY ION LINACS

The 1st linac (SCRFQ) has been steadily tested for the last two years [5]. The second linac (IH) has been recently successfully tested achieving its design specification [6]. Figure 5 shows the energy tunability of the IH linac beam measured with the 1st dipole magnet as a magnetic analyser.

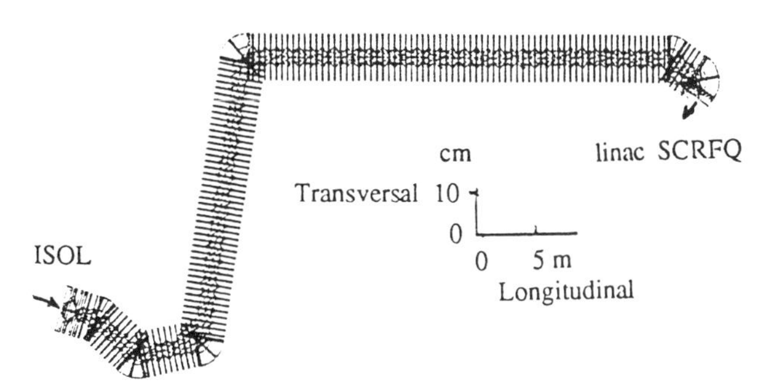

FIGURE 4 . Beam envelope of the 60m Beam Transport Line [4].

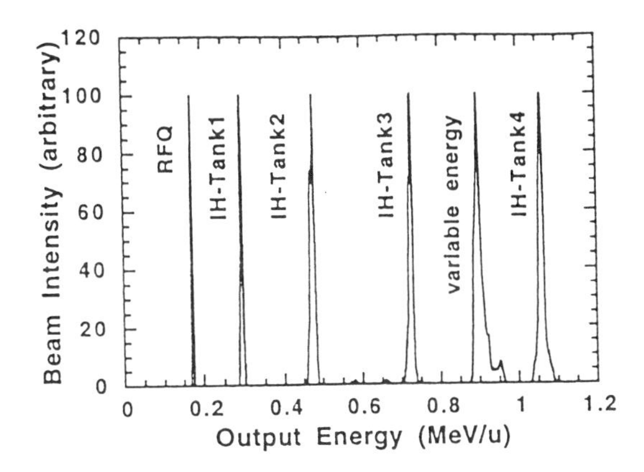

FIGURE 5 . Energy profiles of the IH beam, which were measured with different IH cavity excitations.[6].

EXPERIMENTAL APPARATUS

Nuclear Astrophysics

A recoil mass separator (RMS) and a low-background, high efficiency NaI(Tl) gamma ray detector are prepared for nuclear astrophysics study [7]. The RMS shown in Fig. 6 has a mass resolving power of about 60, and a beam spread acceptance of 5%, which allows the use of a beam charge stripper between the two linacs. The RMS is used to reduce the gamma ray background for capture gamma measurements. A windowless gas target with differential pumping and a blow-in geometry is also prepared. There are several candidates for unstable nuclei close to the stable nuclear region from the interest of nuclear astrophysics point of view. Among them, $^{15}O(\alpha,\gamma)^{19}Ne(p,\gamma)^{20}Na$ and $^{14}O(\alpha,p)^{17}F(p,\gamma)^{18}Ne(\alpha,p)^{21}Na$ are considered of first importance since they are associated with the ignition mechanism of explosive hydrogen burning.

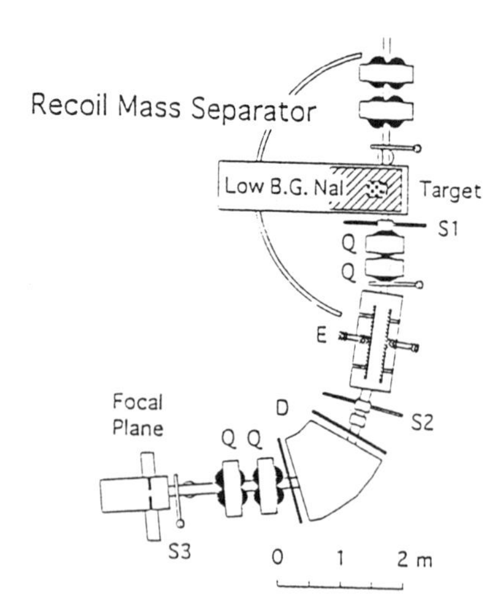

FIGURE 6 . Layout of the recoil mass separator and the low - background gamma detector for nuclear astrophysics experiments [7].

Laser Ion Trap

An ion trap with a laser for unstable nuclear spectroscopy has been installed in the ISOL room. As a first step, measurements on nuclear magnetic moments and their distribution in the unstable nuclei ^{7,11}Be were taken using a laser-microwave double resonance method [8]. A laser cooled ion crystal consisting of a few stable $^{9}Be^{+}$ ions was observed, and the ground state hyperfine splitting of $^{9}Be^{+}$ was measured with a precision of 10^{-5}. A search for a resonance of $^{7}Be^{+}$ is in progress. Also, a technique for trapping of ions transported from the isotope separator is under development. Recently, a linear trap comprising a sextupole ion guide (SPIG) and three ring electrodes outside of the SPIG (see Fig. 7), shows promise for efficient ion trapping [9].

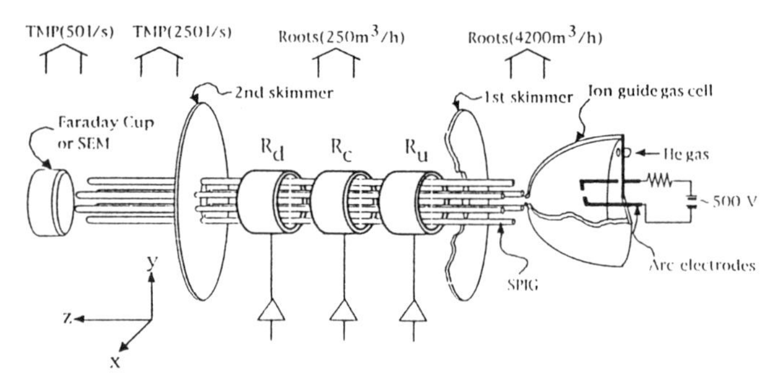

FIGURE 7 . A schematic drawing of the SPIG-trap for trapping ions from the ion guide [9].

CONCLUSION

INS and KEK, the National Laboratory for High Energy Physics, are going to unite in order to start a new organization in April 1997. This new National Laboratory aims at promoting the JHP. In the E-Arena, it is planned to use a 10μA proton beam from the 3 GeV rapid synchrotron booster for the production of exotic nuclei. Almost all the instruments presently used for RNB at INS will be transferred to the E-Arena. We will add a 3rd IH linac by which the final energy of the RNB can be increased to 6.5 MeV/amu. Details of the new layout are presently under intensive discussion.

REFERENCES

1. JHP Plan Working Group, *JHP*, INS Pamphlet(1995)
2. Jeong S.C., Oyaizu M., Kawakami H., Shirakabe Y. and Nomura T.,"Beam-bunching in an ECR ion source by the pulsed gating -potential method", INS-Rep-1107 (to be published in Nucl. Instr. Meth.).
3. Nomura T., Shirakabe Y., Ikeda N. and Shinozuka T., *Nucl Instr. Meth.* **B93**, 492-498 (1994).
4. Wada M., Katayama I., Tanaka J., Strasser P., Jeong S.C., Yahata K., Tomizawa Y., Nomura T., Fujioka M. and Wollnik H., "High resolution mass separator and 60m beam transport line for the radioactive nuclear beam facility at INS", presented at the Conference on EMIS13, Bad Durkheim, Germany, September 23-27, 1996.
5. Arai S., Imanishi A., Niki K., Okada M., Takeda Y., Tojyo E., and Tokuda N.,"Construction and beam tests of a 25.5-MHz split coaxial RFQ for radioactive nuclei" INS-Rep-1152 (submitted to *Nucl.Instr.Meth.*).
6. Arai S., Araki Y., Hashimoto Y., ,Imanishi A., Katayama T., Masuda T, Niki K., Okada M., Takeda Y., Tojyo E., Tomizawa M., Yoshida K., and Yoshizawa M., "Beam test results of the INS RFQ/IH linacs", presented at the 18th International Linac Conference, Geneva, Switzerland, August 26-30, 1996.
7. Kubono S. et al., "New ISOL-based radioactive nuclear beam facility at INS", presented at the Conference on RNB4, Omiya, Japan, June 3-8, 1996.
8. Wada M., ,K.Okada K., Wang H., Enders K., Kurth F., Nakamura T., S.Fujitaka S., Tanaka J., Kawakami H., Ohtani S., and Katayama I.,"Laser-microwave spectroscopy of the hyperfine structure of ^{9}Be for the investigation of unstable Be isotopes", presented at the Conference on STORI'96, Bernkastel-Kues, Germany, September 30-October 4, 1996.
9. Fujitaka S., Wada M., ,H.Wang H., Tanaka J., Kawakami H., Katayama I., Ogino K., Katsuragawa H., Nakamura T., Okada K., and Ohtani S., "Accumulation of unstable nuclear ions for ion trap experiments", presented at the Conference on EMIS13, Bad Durkheim, Germany, September 23-27,1996.

PRESENT AND FUTURE RADIOACTIVE NUCLEAR BEAM DEVELOPMENTS AT ARGONNE

P. Decrock

Physics Division, Argonne National Laboratory, Argonne, Illinois 60439

A scheme for building an ISOL-based radioactive nuclear beam facility at the Argonne Physics Division, is currently evaluated. The feasibility and efficiency of the different steps in the proposed production- and acceleration cycles are being tested. At the Dynamitron Facility of the ANL Physics Division, stripping yields of Kr, Xe and Pb beams in a windowless gas cell have been measured and the study of fission of ^{238}U induced by fast neutrons from the $^{9}Be(d,n)$ reaction is in progress. Different aspects of the post-acceleration procedure are currently being investigated. In parallel with this work, energetic radioactive beams such as ^{17}F, ^{18}F and ^{56}Ni have recently been developed at Argonne using the present ATLAS facility.

INTRODUCTION

Within the last decade a growing interest in the production and use of energetic radioactive nuclear beams (RNB) has come forth (1,2). Experiments at first generation RNB facilities have demonstrated the capability of studying new research frontiers in the fields of nuclear physics and nuclear astrophysics. In order to meet the challenging goals of future experimental programs, a concept for a second generation facility, providing high intensity RNB and covering a broad range of nuclei, is highly desired. A novel scheme for producing intense, high-quality RNB based on the two accelerator method has been proposed by a group in the Physics Division at Argonne. Details of this plan, including research possibilities, can be found in a working paper entitled "Concept for an advanced Exotic Beam Facility Based on Atlas" (3). A brief overview of the basic concept is presented in the first section of this paper. At present, R&D related to different aspects of the proposed production- and acceleration methods are in progress. The present status of these RNB developments at Argonne are discussed in the following section of this paper.

THE FACILITY CONCEPT

The proposed method for producing energetic RNB at Argonne, is based on the ISOL (Isotope Separator On-Line) approach, i.e. post-acceleration of low-energy ISOL beams. The different elements involved in the production process are shown schematically in Fig.1. The proposed driver accelerator is a 215-MV drift tube linac which can deliver several light ions (^{1}H,^{2}H,^{3}He,^{4}He,...) with high beam intensities. The addition of a 30-MV RFQ/Linac injector for q/m=1/6 would also permit the acceleration of heavier beams such as ^{18}O and ^{36}Ar up to 100 MeV per nucleon. With this variety of primary beams, several different nuclear reaction mechanisms can be exploited (4). One interesting mechanism is fission of ^{238}U using 0.1 pmA of 100 MeV neutrons from the breakup of 200 MeV deuterons (5). The separation between the breakup target and the thick uranium production target can solve problems related to the high beam power. After release from the target, the produced nuclei will be ionized and mass separated using standard ISOL techniques and an isobar separator, delivering 100 kV 1^{+} beams to an RFQ

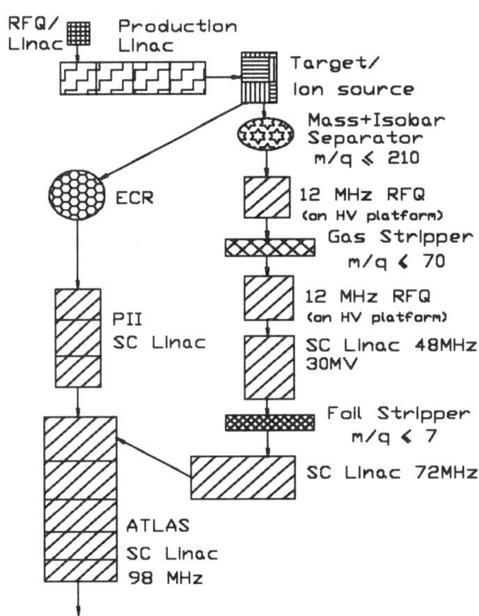

FIGURE 1. Block diagram of the Argonne RNB-Facility concept.

The post-acceleration of the low energy ISOL beams is based on the existing superconducting ATLAS linacs. However, four new acceleration subsystems are required to pre-accelerate the initial low q/m beams to energies of 0.5-1.0 MeV per nucleon, before injection into the existing ATLAS linacs and the further acceleration from 1 to 15 MeV per nucleon, can be achieved. These four new systems consist of a low-frequency 12 MHz RFQ, positioned on a 300 kV high voltage platform, followed by a second 12 MHz RFQ and two new sections of superconducting linacs optimized for low q/m

CP392, *Application of Accelerators in Research and Industry*, edited by J. L. Duggan and I. L. Morgan
AIP Press, New York © 1997

beams. In this acceleration scenario, the RNB's have to be stripped twice: a first time, gas stripping from 1^+ to 2^+ will occur just off the RFQ high voltage platform and a second time foil stripping to q/m=0.15, just in front of the third matching pre-accelerator linac section, is necessary. The energy range of this acceleration method is quite flexible, i.e. RNB's can be selected at different stages in the acceleration process (i.e. for research in atom or ion traps without acceleration, selection after the pre-acceleration at 0.5-1.0 MeV/nucleon for nuclear astrophysics experiments and at 1-10 MeV per nucleon after acceleration by ATLAS).

For noble gases (e.g. Kr and Xe isotopes) or volatile compounds, an alternative and efficient scheme for producing RNB's at ATLAS can be conceived, as is also shown in Fig 1. In this case, transport of the radioactivity from the shielded target to an ECR ion source, positioned on a 350 kV platform, can be accomplished by using a cold transfer line, which also enables a certain isobaric selectivity. The now highly charged ion beams from the ECR ion source can then directly be injected in the existing SC positive ion injector of ATLAS (PII) (6).

PRESENT RNB DEVELOPMENTS AT ARGONNE

Stripping in a gas target

As described in the previous section, gas stripping of the low energy 1^+ ISOL beams will be performed at about 8 keV per nucleon, just before injection into the second RFQ. Stripping from 1^+ to 2^+ and from 1^+ to 3^+ for masses larger than 70 and 140, respectively, is necessary in order to match the velocity requirements of the pre-accelerator. Since, at these low energies, few experimental data are available, a systematic study of the stripping efficiency and the multiple scattering of N^{1+}, Kr^{1+}, Xe^{1+} and Pb^{1+} beams in He and N_2 gas targets, has been performed. The beams were delivered by the 5.0 MV Dynamitron accelerator at the ANL Physics Division. An overview of the results is given below and the details of these experiments will be presented elsewhere (7).

The 1^+ beams from the Dynamitron were first analyzed by two bending magnets and then further collimated using different circular apertures along the beam line, yielding a beam with an angular divergence < 0.1 mrad. This beam was then focused on a 10 cm long gas cell having an entrance aperture of 2 mm in diameter and a vertical aperture of 2 mm wide and 6 mm high as exit. The different charge states were analyzed using a parallel plate deflector and the charge state fractions were measured by using a movable silicon particle detector, positioned 5.4 m downstream with respect to the gas target. The best results were obtained with He as a stripper gas. High efficiencies of 40%, 49% and 50% were obtained for stripping Kr^{1+} (0.8 MeV), Xe^{1+} (1.0 MeV) and Pb^{1+} (1.0 MeV), respectively, into the 2^+ charge state. An efficiency of 33% for stripping a 1.0 MeV Pb beam from 1^+ to 3^+ in a He target, has been measured. Figure 2 shows the results of the charge state fractions, obtained with a 1.0 MeV Xe and Pb beam, versus the He target thickness.

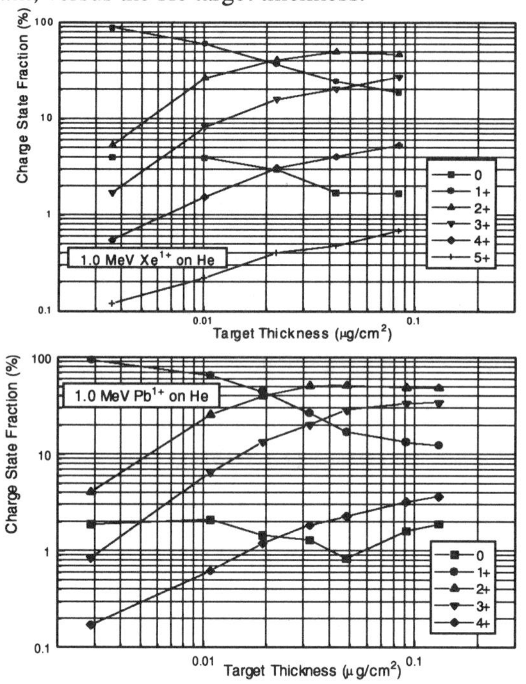

FIGURE 2. Measured charge state fractions for Xe^{1+} and Pb^{1+} (1.0 MeV) beams after passing a He gas target.

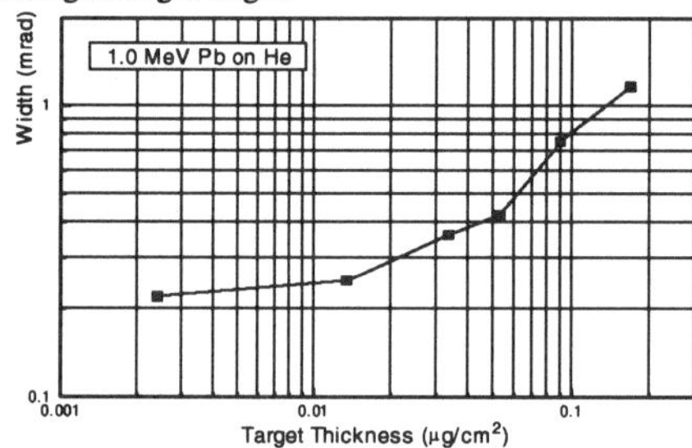

FIGURE 3. Measured full width at half maximum (in mrad) of a 1.0 MeV Pb^{1+} beam after passing a He gas target.

By installing a small rectangular slit (width = 1.6 mm) in front of the silicon detector a scan of the beam profile with an angular resolution of 0.2 mrad could be performed and information about multiple scattering in the gas target could be obtained. In Fig. 3 the full width at half maximum (FWHM) of the angular distribution of a Pb beam after passing a He target (including all charge states) is plotted as a function of the target thickness. At a He target thickness of 0.1 $\mu g/cm^2$, which is a thickness adequate for obtaining high 2^+ and 3^+ stripping yields (Fig. 2.), the effects of multiple scattering are small (FWHM $\approx$ 1 mrad).

Target concept for the production of fission fragments

At the 5.0 MV Dynamitron Facility of the Argonne Physics Division, a test area for production- and ion source studies has been constructed. A shielded cave has been build, as shown in Fig 4, which enables the use of high intensity deuteron beams for production studies. As explained above, the fission of ^{238}U by secondary neutrons is a promising reaction mechanism for the production of a variety of neutron rich isotopes. This reaction mechanism can be studied at the Dynamitron facility. A 4.8 MeV deuteron beam will be used to generate neutrons from the $^{9}Be(d,n)^{10}B$ reaction which will then be used for the irradiation of an ^{238}U target. Absolute fission yields and release studies from the target material will be performed.

FIGURE 4. View of the shielded cave (during construction) at the Argonne Dynamitron Facililty, which will be used for fission studies of ^{238}U, target release measurements and on-line ion source developments.

A calculation of the production yields of neutron rich Kr, Xe and Sn isotopes, obtainable with this specific production method, has been made. The results of this production calculation and the geometry of the target system are shown in Fig 5. These calculations include the experimentally determined neutron energy distribution (at 0°) from the $^{9}Be(d,n)^{10}B$ reaction (8) and the fission cross sections and the fractional yields are taken from Ref. (9) and (10), respectively. These results show that with a 100 pμA deuteron beam, high yields for these isotopes can be obtained and that systematic studies of the production and the release properties of fission products from a ^{238}U target matrix are possible. Irradiations will start in the fall of 1996.

Ion Source Developments

In order to produce the 1^{+} ISOL beams, we will build upon the vast realm of experience existing at ISOL facilities such as CERN-ISOLDE or GSI Darmstadt (11,12). Unfortunately, no universal ion source exists, specific ion sources have to be optimized for specific elements or groups of elements (13).

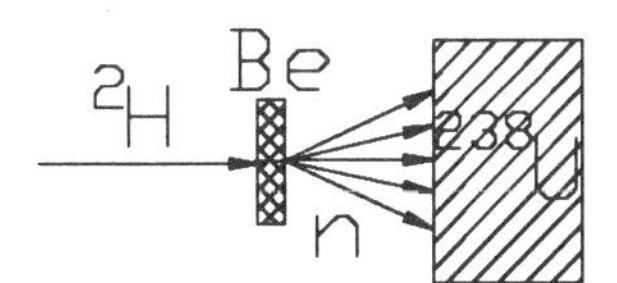

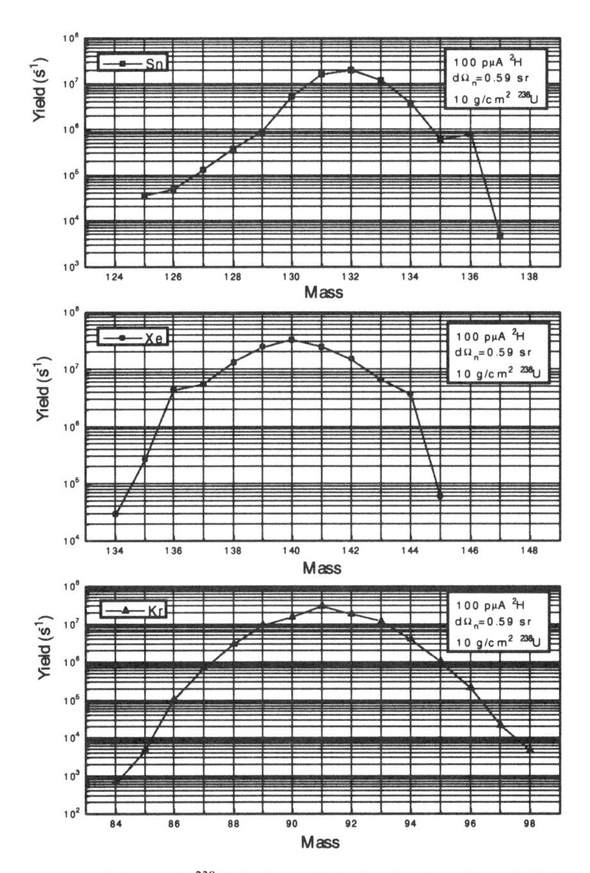

FIGURE 5. Calculated ^{238}U fission yields for the Sn, Xe and Kr isotopes. Fission is induced by fast neutrons from the $^{9}Be(d,n)$ reaction. Yields are calculated for a 4.8 MeV deuteron beam (intensity=100 pμA) and only the neutrons emitted within a solid angle of 0.59 sr are taken into account. Isotopes with half-lives in the seconds to minutes range can be studied.

A test bench for ion sources is now available at ATLAS and, as a first step, a Gill-Piotrowski-type high temperature plasma ion source has been installed (14). In the future other types of plasma sources might be developed and installed at the test stand in order to gain experience in this particular ion-source technology and to elaborate more the specifications, such as ionization efficiency, emittance and energy spread, of these sources.

As mentioned above, an alternative way of producing RNB's at ATLAS, is by coupling the production target with the present positive ion injector by means of an efficient high-charge-state ECR ion source. On-line ECR ion sources for low charge states have been used successfully since several years at the Louvain Facility (15,16) and the TISOL facility (17). Efforts to develop on-line ECR ion sources for efficient production of high charge states at Ganil (18) look very promising. Possible losses due to the transport time between target and the ion source, can be compensated by the gain in overall acceleration efficiency, mainly due the fact that with this method at least the first gas stripping is not required. Of course, the target-ion source distance should be kept as short as possible. This method of ionizing the radioactive atoms in an ECR ion source, followed by post acceleration with a SC linac, will be tested using ^{18}F ($T_{1/2}$=109 m) atoms. A proposal has been submitted at ATLAS to produce ^{18}F at the Dynamitron facility and post accelerate the ^{18}F atoms with the positive ion injector of ATLAS (19). Fluorine and neon efficiency measurements and the fluorine release time out of the ATLAS ECR ion source have been performed. The production of ^{18}F at the Dynamitron will start in the fall of 1996. The beam will be used for nuclear astrophysical measurements (19). This ^{18}F beam development is a continuation of a former program at ATLAS with radioactive ^{18}F beams (see below).

An interesting progress could be the combination of using 1^+ ISOL beams and an ECR ion source playing the role of a charge-state amplifier. First results with a high-charge-state breeder, developed for the PIAFE project, show that 1^+ ISOL beams can be transformed into higher charge states with relatively high efficiencies (20). Optimization and generalization of such a device could be a helpful solution for coupling ISOL beams with the ATLAS positive ion injector.

Acceleration Developments

Different aspects of the post-acceleration method are currently being examined (21,22). A low-charge-state CW RFQ prototype is under construction (23).

Present Radioactive Beams at ATLAS

Besides this specific R&D for a future RNB facility, first generation radioactive beams are presently being developed at ATLAS. Large amounts of ^{18}F were produced by the ^{18}O(p,n)^{18}F reaction, using the medical radioisotope production cyclotron from the University of Wisconsin (24). With these samples, a radioactive ^{18}F beam has been generated at Argonne using the ATLAS Tandem accelerator and nuclear physics experiments have been performed (24,25). The development of a ^{56}Ni beam at ATLAS is currently in progress (26). In addition, a ^{17}F beam has been developed using an in-flight technique; by bombarding a hydrogen gas target with a 83 MeV ^{17}O beam, a ^{17}F beam was produced (p,n reaction) and further transported to the experimental area (27).

SUMMARY

An ISOL-based facility for the production of energetic RNB at Argonne has been proposed. The post acceleration of the beams is based on Superconducting Linacs, a technology which has been developed successfully during the last decade at Argonne. At present, R&D related to important parts of the production process, such as primary production targets, ion sources, gas targets for beam stripping, and different other issues related to the development of a RNB facility, is in progress at the Argonne Physics Division. In parallel with these developments first generation radioactive beams are being developed and used for research at ATLAS.

This research was supported by the US DOE Nuclear Physics Division under contract W-31-109-ENG-38.

REFERENCES

1. *Proceedings of the Third International Conference on Radioactive Nuclear Beams*, Gif-sur-Yvette, Editions Frontieres (Edited by D.J. Morrissey),1993
2. Mueller A.C., *"Improvements of Present Radioactive Beam Facilities and New Projects,"* presented at the International Conference on Exotic Nuclei and Atomic Masses, Arles, France, June 19-23, 1995
3. *"Concept for an advanced Exotic Beam Facility Based on Atlas"*, Physics Division, Argonne National Laboratory, (1995)
4. Nolen J.A., *Rev. Sci. Instrum.* **67**, 935-937 (1996)
5. Nolen J.A., *"A Target Concept for Intense Radioactive Beams in the ^{132}Sn Region"*, presented at the Third International Conference on Radioactive Nuclear Beams, East-Lansing, MI, May 24-27, 1993
6. Bollinger L.M. et al., *Nucl. Instr. Meth.* **B79**, 753-757 (1993)
7. Decrock P., Kanter E., Nolen J.A., *Nucl. Instr. Meth.*, to be published
8. Meadows J.W., *Nucl. Instr. Meth.* **A324**, 239-246 (1993)
9. *Neutron Cross Sections vol. 2*, San Diego: Academic Press, 1988, p 717
10. Crouch E.A.C., *At. Nucl. Data Tables* **19**, 417-532 (1977)
11. Ravn H.L., *Nucl. Instr. Meth.* **B70**, 107 (1992)
12. Kirchner R., *Nucl.. Instr. Meth.* **B70**, 56 (1992)
13. Ravn H.L. et al., *Nucl. Instr. Meth.* **B88**, 441-461 (1994)
14. Gill R.L and Piotrowski A., *Nucl. Instr. Meth.* **A234**, 213-217 (1985)
15. Loiselet M. et al, *"Aspects of the Louvain-la-Neuve results and projects,"* presented at the Third International Conference on Radioactive Nuclear Beams, East-Lansing, MI, May 24-27, 1993
16. Decrock P. et al, *Nucl. Instr. Meth.* **B58**, 252-259 (1991)
17. Buchmann L. et al, *Nucl. Instr. Meth.* **A295**, 291 (1990)
18. *"The Spiral Radioactive Ion Beam facility"*, GANIL, int. report (1994)
19. Decrock P., *"Measurement of the $^{18}F(p,\gamma)/^{18}F(p,\alpha)$ branching ratio using a ^{18}F beam post accelerated by the positive ion injector of ATLAS"*, ATLAS proposal, Argonne Nat. Lab., (1996)
20. Geller R., Tamburella C. and Belmont J.L., *Rev. Sci. Instrum.* **67**, 1281-1285 (1996)
21. Shepard K.W. and Kim J.W., *"A low-charge-state injector Linac for ATLAS"*, presented at the 1995 Particle Accelerator Conference, Dallas, Texas, May 1-5, 1995
22. Kim J.W., Shepard K.W. and Nolen J.A., *"A High Gradient Superconducting Quadrupole for a Low Charge State Ion Linac,"* presented at the 1995 Particle Accelerator Conference, Dallas, Texas, May 1-5, 1995
23. Shepard K.W. and Sellyey W.C., "A *Low-frequency RFQ for a Low-Charge-State Injector for ATLAS,"* presented at the XVIII International Linac Conference, Geneva, Switzerland, August 26-30, 1996
24. Roberts A.D. et al, *Nucl. Instr. Meth.* **B103**, 523-528 (1995)
25. Rehm K.E. et al, *Phys. Rev C* **53**, 1950-1954 (1996)
26. Rehm K.E., *"Study of the Single-Particle States around N,Z=28 with a Radioactive ^{56}Ni Beam"*, ATLAS proposal, Argonne Nat. Lab. (1996)
27. Harss B. et al, *"The First Production and Transport of Radioactive ^{17}F at ATLAS for Research"*, presented at the XVIII International Linac Conference, Geneva, Switzerland, August 26-30, 1996

NUCLEAR ASTROPHYSICS AT THE HOLIFIELD RADIOACTIVE ION BEAM FACILITY

Jeff C. Blackmon

Oak Ridge National Laboratory, PO Box 2008, Oak Ridge, TN 37831
and
The University of North Carolina at Chapel Hill, Chapel Hill, NC 27599-3255

Reactions involving radioactive nuclei play an important role in explosive stellar events such as novae, supernovae, and X-ray bursts. The development of accelerated, proton-rich radioactive ion beams provides a tool for directly studying many of the reactions that fuel explosive hydrogen burning. The experimental nuclear astrophysics program at the Holifield Radioactive Ion Beam Facility at Oak Ridge National Laboratory is centered on absolute cross section measurements of these reactions with radioactive ion beams. Beams of ^{17}F and ^{18}F, important nuclei in the hot-CNO cycle, are currently under development at HRIBF. Progress in the production of intense radioactive fluorine beams is reported. The Daresbury Recoil Separator (DRS) has been installed at HRIBF as the primary experimental station for nuclear astrophysics experiments. The DRS will be used to measure reactions in inverse kinematics with the techniques of direct recoil detection, delayed-activity recoil detection, and recoil-gamma coincidence measurements. The first astrophysics experiments to be performed at HRIBF, and the application of the recoil separator in these measurements, are discussed.

INTRODUCTION

The rates of nuclear reactions are important parameters in many astrophysical models. In explosive stellar events, the temperatures and densities may be so extreme that reactions occur on time scales as short as seconds. Under such conditions, reactions involving radioactive nuclei play a key role in energy generation and nucleosynthesis. The rates of reactions involving radioactive nuclei are essential input for models of explosive events, but there exists relatively little experimental information on these reactions (1). The primary objective of the nuclear astrophysics program at the Holifield Radioactive Ion Beam Facility (HRIBF) at Oak Ridge National Laboratory (ORNL) is the direct measurement of reactions involving radioactive nuclei that are important for astrophysics (2).

Radioactive ion beams are produced at HRIBF by an ISOL-type target/ion source (3-5). Intense light-ion beams from the Oak Ridge Isochronous Cyclotron (ORIC) pass through a high-temperature target. Reaction products diffuse out of the target material and pass through a transfer tube to an ion source, where they are ionized and extracted. The radioactive ion beam is then accelerated in the 25-MV tandem accelerator, so either a negative beam is extracted directly or a positive beam is extracted and charge exchanged. Two stages of mass analysis before injection into the tandem provide a mass resolution of 1 part in 20,000.

Special attention must be given in an experiment with a radioactive ion beam to maximize the detection signal-to-noise because of the relatively low beam intensities and high backgrounds. A versatile experimental station for astrophysics research is currently under construction at HRIBF centered around the Daresbury Recoil Separator. We discuss the installation of the separator and supporting equipment, and its planned application in the first experiments.

Most astrophysical reaction rates are dominated by contributions from low energy resonances. The spectroscopic properties of each resonance (resonance energy, spin, parity, total and partial widths) determine its contribution to the reaction rate (6). Beams produced by the ISOL technique are well-suited for the study of these resonance properties because intense beams can be produced at low energy with excellent mass and energy resolution. However, the target material must be carefully selected to maximize production of the species of interest, while at the same time allowing for fast diffusion of the reaction products out of the target. Likewise, the ion source must also be tailored to provide a high efficiency for the species of interest. A significant amount of development must be performed for each beam species (4).

Beams of radioactive fluorine are currently one focus of development at HRIBF. The $^{17}F(p,\gamma)^{18}Ne$, $^{18}F(p,\alpha)^{15}O$, and $^{18}F(p,\gamma)^{19}Ne$ reactions are all important in the hot-CNO cycle, the sequence of reactions that fuel nova explosions (1). Two recent measurements using a radioactive ^{18}F beam have determined the strength of an important resonance in the $^{18}F(p,\alpha)^{15}O$ reaction corresponding to a state at E_x = 7.063 in the compound nucleus ^{19}Ne (7,8). These measurements have accurately determined the $^{18}F(p,\alpha)^{15}O$ reaction rate for temperatures greater than 5×10^8 K. The experimentally determined rate differs by as much as a factor of 10 from previous estimates based on incomplete spectroscopic information. These new results emphasize the need for direct measurements of important reaction rates using radioactive ion beams. The rates of the $^{17}F(p,\gamma)^{18}Ne$ and $^{18}F(p,\gamma)^{19}Ne$ reactions remain uncertain, and the spectroscopic properties of lower energy states in ^{19}Ne must be measured to accurately determine the $^{18}F(p,\alpha)^{15}O$ reaction rate for $T<5\times10^8$ K, temperatures characteristic of most novae. We discuss recent progress that has been made at HRIBF in the development of radioactive fluorine beams, and initial measurements planned with these beams.

We are also planning a measurement of the $^7Be(p,\gamma)^8B$ reaction. The next generation of neutrino detection experiments, which are coming on-line in the next few years, will provide greatly improved measurements of the solar neutrino energy spectrum. The uncertainties in the predicted solar neutrino fluxes need to be reduced for comparison to these experiments. The largest nuclear physics uncertainty in the flux of high-energy neutrinos comes from uncertainty in the $^7Be(p,\gamma)^8B$ reaction rate (9). Measurements using a radioactive ^7Be target have established the energy dependence of the cross section, but there is some uncertainty regarding the overall normalization (10). Our plans for a measurement of the $^7Be(p,\gamma)^8B$ cross section in inverse kinematics using a radioactive ^7Be beam are also discussed.

THE DARESBURY RECOIL SEPARATOR

Recoil detection has been shown to be particularly advantageous for the measurement of capture reactions in inverse kinematics (11). An experimental station for astrophysics experiments is being constructed at HRIBF with the Daresbury Recoil Separator (DRS) (12) as its core. The DRS separates recoiling reaction products from the primary beam in two 1.3-m-long ExB velocity filters and a 50° dipole bending magnet, which provides a q/m focus. Three sets of quadrupole-triplet magnets focus the recoils, and two sextupole magnets remove higher-order aberrations. Because of the strong focusing of the recoil particles at forward angles, the DRS has effectively a 100% recoil detection efficiency for capture reactions. The DRS is also well-suited for these measurements due to its long velocity filters. In capture reactions, and particularly in (p,γ) capture reactions, the recoiling reaction products have nearly the same momentum as the beam particles. For example, in the $^{17}F(p,\gamma)^{18}Ne$ reaction, the momentum difference between the ^{17}F and ^{18}Ne is only 1%, but the separation between the two will be greater than 2 cm at the exit of the first velocity filter owing to the 5% difference in the velocities.

The DRS was transferred to ORNL in the fall of 1994, and the physical installation of the separator is complete. All of the elements have been assembled, aligned, and tested. The magnetic fields are measured and regulated using hall probes, and the magnetic field resolution has been measured to be 1 part in 10^4. The velocity filters' high voltage plates (±300 kV) are currently being conditioned. All elements have also been interfaced to a VME front-end processor that is independent of the data acquisition system. We are currently developing a program to allow control and regulation of the DRS elements through a user-friendly graphical interface.

An ion optical solution to optimize the performance of the DRS was developed. We have increased the distance between the two velocity filters slightly from the original configuration and have installed a new set of slits between the two velocity filters. These new slits will allow rejection of a large fraction of the primary beam far upstream of the focal plane. In addition, any particles that scatter from these slits will be rejected by the second velocity filter.

The focal plane detector system for the DRS was developed and tested in stable beam measurements at Yale University. The basis of the system is a position-sensitive carbon-foil microchannel plate detector and ΔE-E gas ionization counter. The ion counter was moved 2 m downstream to allow for additional focal plane elements. A second microchannel plate detector (timing signal only) is available for use in an upstream focal-plane chamber for time-of-flight measurements. Optics calculations indicate that a reasonably tight, parallel focus of the recoils is possible through the 2-m focal plane length. A moving tape system, which may be used for delayed-activity detection, is also planned. A new gas-handling system for the ion chamber is currently being constructed.

Two new target chambers have been constructed for the DRS at the University of North Carolina at Chapel Hill. One chamber is designed for studying capture reactions using a foil target. This chamber allows external gamma ray detectors, e.g. a BaF_2 detector array, to be placed close to the target for maximum efficiency in recoil-gamma coincidence measurements. A second larger chamber was constructed to house an annular array of silicon surface barrier detectors based on the Louvain-Edinburgh Detector Array (LEDA) design (13,14). The array may be positioned up to 20 cm from the target, subtending $\Delta\theta\approx20$-30° in 16 radial segments. Detection of light ions in the array is advantageous in transfer reactions where the efficiency for recoil detection is low. For example, the efficiency of the DRS for recoil detection in the $^{18}F(p,\alpha)^{15}O$ reaction is only about 1%, but the silicon detector array has a 40% efficiency for the detection of alpha particles. Both chambers mount on the same

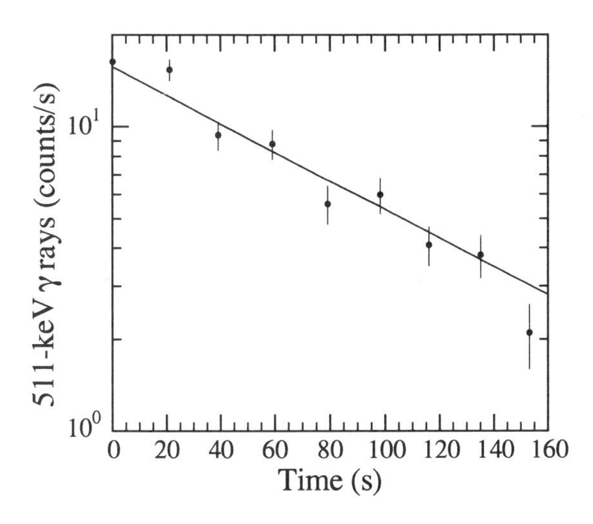

FIGURE 1. Measured activity following the implant of mass 44 ($Al^{17}F$). The line shows the expected decay from the 64.5 s half-life of ^{17}F.

aligning base and can be interchanged with relative ease. Commissioning of the DRS with stable beams will begin once the installation of the target chambers and focal plane detectors is complete.

^{17}F AND ^{18}F BEAMS

Beams of ^{18}F and ^{17}F are currently being produced in tests at HRIBF using an electron beam plasma target/ion source and fibrous Al_2O_3 target material. A high yield of either ^{18}F or ^{17}F is produced from the $^{18}O(p,n)^{18}F$ or $^{16}O(d,n)^{17}F$ reactions on Al_2O_3. The fibrous material contains short path lengths for diffusion, and sustains high temperatures without sintering. Beams of positive ions are extracted from the ion source, mass analyzed using the UNISOR isotope separator, and implanted onto a movable tape. The radioactive fluorine current is then determined from the measured yield of 511-keV gamma rays. The activity is measured as a function of time to confirm that the half-life of the activity corresponds to that of ^{17}F or ^{18}F.

Because of the extreme reactivity of atomic fluorine, it is likely the observed fluorine is transported in molecular form. Most (95%) of the ^{17}F intensity we observe is at mass 44, and most of the ^{18}F at mass 45. Figure 1 shows the measured activity of 511-keV gamma rays following the implantation of mass 44 (with a deuteron driver beam). We conclude that fluorine which is released from the target promptly is released primarily as AlF. At a target temperature of 1500 °C, our preliminary measurements of the source efficiency are $(5\pm2) \times 10^{-5}$ for $Al^{17}F$ and $(6\pm2) \times 10^{-4}$ for $Al^{18}F$, with a Xe efficiency of 1.5%. The uncertainties include estimates of the systematic uncertainties, and the largest uncertainty is in the hydrogen driver beam current.

The hold-up time in the source was determined by turning off the proton driver beam and measuring the current of $Al^{18}F$ as a function of time. At a target temperature of 1500 °C the hold-up time was determined to be 16.4 ± 0.8 m for AlF. This agrees with the value of 13 ± 6 m which we estimate from the ratio of the efficiencies for $Al^{17}F$ and $Al^{18}F$. We have also observed that the hold-up time decreases sharply with increasing target temperature. We are currently studying the efficiency of the source as a function of operating parameters, such as target temperature. Further details of the fluorine beam development may be found in Stracener (15).

The $^{17}F(p,\gamma)^{18}Ne$ reaction rate is believed to be dominated by an unconfirmed resonance corresponding to an excited state in ^{18}Ne at $E_x = 4.561$ MeV (16). Comparison to the analog states in ^{18}O indicates that this state may be a 3^+ state, making it a strong s-wave resonance. We are planning a study of $^{17}F(p,p)^{17}F$ scattering cross section in the energy region of the 4.561 MeV state using a radioactive ^{17}F beam and a thin (5 $\mu g/cm^2$) CH_2 target. The scattered protons will be detected in the LEDA array. Given the present source efficiency for $Al^{17}F$, a beam current greater than 10^5 $^{17}F/s$ on target is expected. We will measure an excitation function (at 10-15 different energies). By summing the azimuthal and some radial segmentation of the LEDA array, less than 5% statistics can be achieved in one day per energy with a beam current of 10^5 $^{17}F/s$. If the presence of the resonance is confirmed, we will utilize the full radial segmentation of the LEDA array to measure the angular distribution of the scattered protons to determine the spin and parity of the resonance.

A substantial increase in the ^{17}F beam current is required for a direct measurement of the $^{17}F(p,\gamma)^{18}Ne$ reaction. With a beam current of 10^8 $^{17}F/s$, a count rate of 2 $^{18}Ne/hr$ is expected in the focal plane of the DRS. Therefore, more than a hundredfold increase in the ^{17}F current is needed to make a direct measurement of the $^{17}F(p,\gamma)^{18}Ne$ reaction realistic. A ^{17}F beam may also be used to study the $^{14}O(\alpha,p)^{17}F$ reaction by measuring the inverse reaction $^{17}F(p,\alpha)^{14}O$. The rate of the $^{14}O(\alpha,p)^{17}F$ reaction is particularly important in determining the amount of "leakage" of material out of the hot-CNO cycle to higher masses (1). A measurement of the $^{17}F(p,\alpha)^{14}O$ reaction would be performed using the silicon detector array. A count rate of 4 counts/hr is expected in the detector array with a beam current of 10^8 $^{17}F/s$. A measurement of the $^{17}F(p,\alpha)^{14}O$ reaction also requires a substantial increase in beam intensity from that achieved thus far. A significant increase in fluorine current should be obtainable with the current source design. For example, the overall (Xe) efficiency of the particular source used in these tests was about a factor of 6 less than that of other sources of the same design. We are also investigating design changes that could result in substantial improvement in the efficiency for AlF. of the $^7Be(p,\gamma)^8B$ reaction using a radioactive ^7Be beam.

THE $^7Be(p,\gamma)^8B$ REACTION

Six measurements of the $^7Be(p,\gamma)^8B$ reaction have been reported. The four most precise of these measurements differ in overall normalization by about 30%. The difference is most likely due to systematic uncertainties, for example in the ^7Be target density. We are planning a measurement of the $^7Be(p,\gamma)^8B$ cross section in inverse kinematics using a ^7Be beam. The ^7Be beam will be produced by a sputter ion source as BeO, and dissociated in the terminal of the tandem. A second, high-energy stripper foil will produce a high efficiency of $^7Be^{4+}$, so that the tandem analyzing magnet can be used to help eliminate Li contaminants from the beam.

A thin CH_2 target (3×10^{18} atoms/cm^2) will be used, and the cross section will be measured by detecting recoiling ^8B nuclei in the Daresbury Recoil Separator. The product of beam current and target density will by monitored by measuring elastic scattering from the target. Tests of the ΔE-E gas ionization counter show it has sufficient resolution to distinguish ^7Be and ^8B. With a ^7Be beam current of 100 pA, we expect 55 counts/day at a center-of-mass energy of 1 MeV. Therefore, a run of 8 days duration is required to achieve 5% statistics. We expect 15 counts/day at a center-of-mass energy of 0.4 MeV. A somewhat longer run will be required at this lower energy, but a thicker target could reduce the required beam time. While this approach also has systematic difficulties, it is a different technique from that used previously. A similar experiment is currently being conducted by the Napoli-Bochum (NABONA) collaboration using a windowless hydrogen gas cell (17,18).

CONCLUSION

The astrophysics experimental station at HRIBF will allow for a high detection efficiency for a variety of techniques, such as direct recoil detection, recoil-gamma coincidence measurements, and charged particle detection in a silicon detector array. Stable beam commissioning will begin once the installation of the target chamber and focal plane are complete.

Our initial tests with fibrous Al_2O_3 as a target material for the production of ^{17}F and ^{18}F beams are encouraging. We have produced a ^{17}F beam with intensity sufficient for a measurement of $^{17}F(p,p)^{17}F$ scattering cross sections. An improvement in efficiency by about a factor of several hundred would result in sufficient current for a measurement of the $^{17}F(p,\gamma)^{18}Ne$ reaction and the $^{14}O(\alpha,p)^{17}F$ reaction by measurement of the $^{17}F(p,\alpha)^{14}O$ inverse reaction. We are also developing a measurement

ACKNOWLEDGMENTS

The development of radioactive fluorine beams is being conducted at ORNL by D. W. Bardayan, J. C. Blackmon, H. K. Carter, J. Kormicki, A. H. Poland, M. S. Smith, D. W. Stracener, and other staff at HRIBF. The Daresbury Recoil Separator is being installed by D. W. Bardayan, J. C. Blackmon, A. E. Champagne, P. D. Parker, D. E. Pierce, M. S. Smith, and other members of the astrophysics collaboration at HRIBF. Oak Ridge National Laboratory is managed by Lockheed Martin Energy Research Corporation for the U. S. Department of Energy under Contract No. DE-AC05-96OR22464.

REFERENCES

1. Champagne, A. E., and Wiescher, M., *Annu. Rev. Nucl. Part. Sci.* **42**, 39-76 (1992).
2. Smith, M. S., *Nucl. Inst. Meth. Phys. Res.* **B99**, 349-353 (1995).
3. Garrett, J. D., Alton, G. D., Baktash, C., Olsen, D. K., and Toth, K. S., *Nucl. Phys.* **A557**, 701c-714c (1993).
4. Alton, G. D., Haynes, D. L., Mills, G. D., and Olsen, D. K., *Nucl. Inst. Meth. Phys. Res.* **A328**, 325-329 (1993).
5. Ravn, H. L., *Nucl. Inst. Meth. Phys. Res.* **B70**, 107-117 (1992).
6. Rolfs, C. E., and Rodney, W. S., *Cauldrons in the Cosmos*, Chicago: U. of Chicago Press, 1988.
7. Rehm, K. E., *et al., Phys. Rev. C* **53**, 1950-1954 (1996).
8. Coszach, R., *et al.*, Phys. Lett. **B353**, 184-188 (1995).
9. Bahcall, J. N., and Pinsonneault, M. H., *Rev. Mod. Phys.* **64**, 885 (1992).
10. Filippone, B. W., *Ann. Rev. Nucl. Part. Sci.* **36**, 717 (1986).
11. Smith, M. S., Rolfs, C. E., and Barnes, C. A., Nucl. Inst. Meth. Phys. Res. **A306**, 233- 239 (1991).
12. James, A. N., Morrison, T. P., Ying, K. L., Connell, K. A., Price, H. G., and Simpson, J., *Nucl. Inst. Meth. Phys. Res.* **A267**, 144-152 (1988).
13. Bain, C., Ph. D. Dissertation, U. of Edinburgh, 1996, pp. vi-xxix.
14. Sellin, P. J., *et al., Nucl. Inst. Meth. Phys. Res.* **A311**, 217-223 (1992).
15. Stracener, D. W., Carter, H. K., Kormicki, J., Poland, A. H., Smith, M. S., Blackmon, J. C., Bardayan, D. W., "Studies on the Production and Release of F and As Isotopes from the HRIBF Target/Ion Source," in *Proceedings of the 14th International Conference on the Application of Accelerators in Research and Industry*, 1996.
16. Garcia, A., Adelberger, E. G., Magnus, P. V., Markoff, D. M., Swartz, K. B., Smith, M. S., Hahn, K. I., Bateman, N., and Parker, P. D., *Phys. Rev. C* **43**, 2012-2019 (1991).
17. Gialanella, L., *et al.*, preprint, accepted for publication in *Nucl. Inst. Meth. Phys. Res. A.*
18. Campajola, L., *et al.*, preprint, accepted for publication in *Zeit. fur Phys A.*

SECTION III

POSITRON SOURCES, EXPERIMENTS AND THEORY

PRODUCTION OF AN INTENSE SLOW POSITRON BEAM BY USING A COMPACT CYCLOTRON AND ITS APPLICATIONS

M. Hirose*

Laboratory for Quantum Equipment Technology, Sumitomo Heavy Industries, Ltd.
2-1-1 Yato-cho, Tanashi-city, Tokyo 188, Japan

An intense slow positron beam has been produced for the first time by using a compact proton cyclotron. A slow positron beam intensity of about 2 x 10^6 e^+/s has been achieved using a proton current of 30 μA. In the near future beam intensities of the order of 10^7-10^9 e^+/s are expected from this system. In order to apply the slow positron beam from the compact cyclotron to the analyses for material surfaces, two applications are under development. One is variable-energy positron lifetime spectroscopy and the other is polarized positron beam application.

INTRODUCTION

Slow positron beam spectroscopies are extremely useful to the analyses for material surfaces. For example, Reflection High-Energy Positron Diffraction (RHEPD) makes it possible to analyze the structure of the topmost atomic layer (1). Positron Re-emission Microscopy (PRM) is capable of visualization of lattice defects near the surface (2). Positron annihilation induced Auger Electron Spectroscopy (PAES) is used to characterize the elemental content of the topmost atomic layer (3). Some of the other new spectroscopies using slow positron beams for the material surface analyses will be detailed later.

In order to realize such new spectroscopies, slow positron beams have been developed in many laboratories in recent years. In such cases two main methods have been developed in order to produce positrons. One uses radioisotopes (RI) and the other makes use of an electron linear accelerator (LINAC). Generally the former is compact and inexpensive, but the intensity of the slow positron beam is not so high. On the other hand, the latter method can produce an intense slow positron beam, but its cost is very high and the system is normally very large.

We have succeeded in developing a compact and inexpensive intense slow positron beam production system using a compact cyclotron by producing and using activity-saturated radioisotopes (4). Used in an on-line mode, an intense positron source has been produced by making use of the $^{27}Al(p,n)^{27}Si$ reaction. A slow positron beam intensity of about 2 x 10^6 e^+/s has been achieved using a proton current of 30 μA. In the near future beam intensities of the order of 10^7-10^9 e^+/s are expected from this system. A comparison of primary sources for the production of a slow positron beam is given in Table 1.

In order to apply the slow positron beam from the compact cyclotron to the research for material surfaces, two attractive applications are under development. One is variable-energy positron lifetime spectroscopy and the other is polarized positron beam application. The details of these applications are also described.

SLOW POSITRON BEAM PRODUCTION

A schematic cross section of the slow positron beam generator using a compact cyclotron is shown in Figure 1. As shown in Fig. 1, protons from a compact cyclotron irradiates an aluminum target and the nuclear reaction $^{27}Al(p,n)^{27}Si$ follows in it. The β^+ rays emitted from the ^{27}Si are partly converted into slow positrons at the tungsten moderator.

On-Line Use of the Cyclotron

In general, there are two methods with which to use a slow positron beam created by the cyclotron produced ion beam irradiation of a target material. One is off-line mode (i.e. using the slow positron beam after the completion of the irradiation) and the other is on-line mode (i.e. using the slow positron beam during the irradiation).

The off-line mode has the merit of avoiding damage induced in the moderator by the irradiation, but the intensity of the slow positron beam decreases according to the half-life of the radioisotopes. Therefore, radioisotopes with a short half-life, and which in general have a high radioactivity, cannot be used in this mode. When using radioisotopes with a long half-life, it may take much time to start and stop the slow positron beam. The inconvenience of this mode is thus immediately apparent.

On the other hand, the on-line mode has the merit that radioisotopes with a short half-life can be used. However, there had been a problem of a decrease in the slow positron beam intensity due to damage induced in the moderator by the irradiation.

In order to avoid this problem, we have invented a

CP392, *Application of Accelerators in Research and Industry*, edited by J. L. Duggan and I. L. Morgan
AIP Press, New York © 1997

TABLE 1. A Comparison of Primary Sources for the Production of a Slow Positron Beam

Primary source	Production process	Slow e^+ beam intensity	Feature
Long half-life RI	β^+ decay	$10^{4\text{-}6}$ e^+/s	Compact Inexpensive Low intensity DC beam Polarized beam
Electron LINAC	Pair production	$10^{7\text{-}8}$ e^+/s	Very large Expensive High intensity Pulsed beam
Compact cyclotron	β^+ decay	$10^{7\text{-}9}$ e^+/s	Compact Inexpensive High intensity DC beam Polarized beam

"transmission type target system" (as shown in Fig. 1) instead of a "reflection type target system" (5). This system eliminates damage induced in the moderator by the irradiation and makes it possible to produce an intense slow positron beam continuously by on-line use of the cyclotron.

Aluminum Target

The requirements for the target materials in which a nuclear reaction occurs are as follows: 1) High yield of the β^+ decay radioisotopes. 2) High percentage of the β^+ decay branching of the radioisotopes. 3) High maximum energy of the β^+ rays emitted from the radioisotopes. 4) A short half-life. 5) High thermal conductivity.

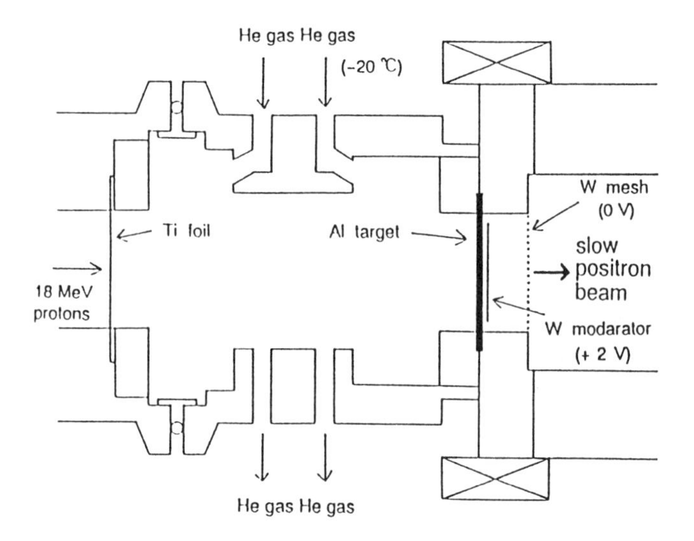

FIGURE 1. Schematic cross section of the slow positron beam generator using a compact cyclotron.

We have chosen "aluminum" as the target material which best meets the above mentioned requirements for a "proton" incident ion beam. The reaction $^{27}Al(p,n)^{27}Si$ has a threshold energy of 5.8 MeV. The percentage of the β^+ decay branching of the ^{27}Si is 100%. The ^{27}Si has a half-life of 4.1 s. The maximum energy of the β^+ rays emitted from the ^{27}Si is 3.85 MeV. The maximum range of 18 MeV protons in the aluminum target is about 1.8 mm. Hence the aluminum target should be 1.8 mm or more thick for the transmission type target system. When the moderator is set behind an aluminum target of this thickness, direct damage which would be induced in the moderator by proton irradiation is avoided. The saturated radioactivity of the ^{27}Si for a proton current of 1 μA is evaluated at 8.5 GBq, and the radioactivity is distributed widely from the proton injected surface to a depth of about 1.5 mm.

Fig. 2 shows the calculated intensity of the slow positron beam which is generated by the β^+ rays emitted from the same surface as the proton injected surface (reflection type) and from the opposite surface to that (transmission type). The ordinate units refer to the slow positron beam intensity that is obtained from each area per 0.1 mm of depth. The sum of these values gives the real slow positron beam intensity for a proton current of 1 μA. The slow positron beam intensities are $5.4 \times 10^5 \times \alpha$ e^+/s (reflection type) and 4.3×10^5 e^+/s (transmission type), respectively. α represents the decrease factor due to the damage induced in the moderator by the irradiation. When consideration of the damage induced in the moderator by the irradiation is taken into account, the slow positron beam intensity obtained for the transmission type target system is rather better than that for the reflection type.

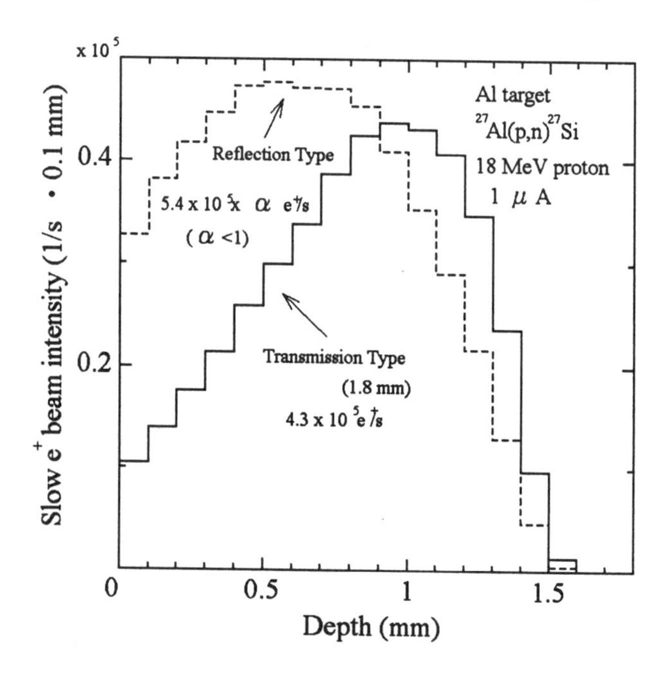

FIGURE 2. Calculated intensities of the slow positron beam formed by the $^{27}Al(p,n)^{27}Si$ reaction for a proton current of 1 μA. In the reflection type target system (dashed line), α represents the decrease factor due to the damage induced in the moderator by proton irradiation. In the transmission type target system (solid line), the thickness of the aluminum target is 1.8 mm.

Slow Positron Beam Intensity

The calculated intensities of the slow positron beam obtained by using a compact cyclotron are summarized in Table 2. As described in the previous section , the intensity of the slow positron beam is 4.3 x 10^5 e^+/s when a proton beam with an energy of 18 MeV and a current of 1 μA irradiates an aluminum target with a thickness of 1.8 mm and the slow positron beam is extracted using the transmission type target system at a conversion efficiency of 10^{-4}. Based on this value an intensity of 3.0 x 10^7 e^+/s should be achieved using the compact cyclotron that Sumitomo Heavy Industries, Ltd. manufactures (commercial name: CYPRIS; maximum proton beam current: 70 μA). In addition, if the proton beam current and the conversion efficiency are improved up to 1 mA and 10^{-3} respectively in the future, it might be possible to produce a super intense slow positron beam giving the order of 10^9 e^+/s.

At any rate, considering that a penning trap for making a DC positron beam is not required and a compact cyclotron is not so expensive, the fact that there is the possibility of obtaining an intense slow positron beam with the order of 10^7 e^+/s using the present compact cyclotron is very attractive.

Experimental

This study has been performed by using the compact cyclotron (CYPRIS MODEL-370 manufactured by Sumitomo Heavy Industries, Ltd.) of SHI Examination and Inspection, Ltd. CYPRIS MODEL-370 can accelerate protons and deuterons up to 18 and 10 MeV, respectively, and it can also accelerate He ions. The maximum beam current is 70 μA. The approximate size of the compact cyclotron is only 5 m long, 4 m wide and 3 m high.

In this study, protons with an energy of 18 MeV and an aluminum target with a thickness of 2.0 mm have been applied in the transmission type target system. The slow positron beam is generated from the 25 μm thick tungsten moderator set behind the aluminum target and is transported to the detector using a magnetic field produced by solenoid coils. The detector is composed of a microchannel plate (MCP) and a phosphorescent plate. The intensity of the slow positron beam is determined by stopping all the positrons and measuring the annihilation γ-rays with a Ge semiconductor detector.

TABLE 2. The Calculated Intensities of the Slow Positron Beam Obtained by Using a Compact Cyclotron

Moderator conversion efficiency	Proton current 1 μA	70 μA	1 mA
10^{-5}	4.3 x 10^4 e^+/s	3.0 x 10^6 e^+/s	4.3 x 10^7 e^+/s
10^{-4}	4.3 x 10^5 e^+/s	3.0 x 10^7 e^+/s	4.3 x 10^8 e^+/s
10^{-3}	4.3 x 10^6 e^+/s	3.0 x 10^8 e^+/s	4.3 x 10^9 e^+/s

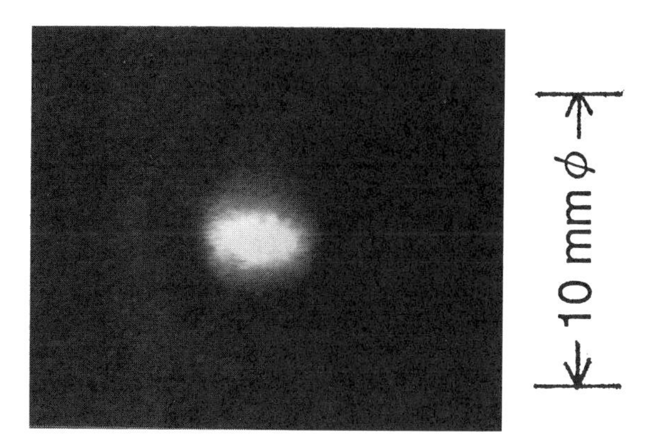

FIGURE 3. The slow positron beam profile observed at the center of the MCP by using the compact cyclotron.

Results

The slow positron beam profile obtained by using the compact cyclotron is shown in Fig. 3.

A slow positron beam intensity of about 2 x 10^6 e^+/s (as measured with a Ge semiconductor detector) has been achieved using a proton current of 30 μA. This value should be compared to the theoretical maximum intensity of 1.1 x 10^7 e^+/s. The intensity is expected to be improved in the future by refinement of various adjustments.

APPLICATIONS

Variable-Energy Positron Lifetime Spectroscopy

Variable-energy positron lifetime spectroscopy is a very effective method to inspect lattice defects non-destructively. In order to realize this spectroscopy, a pulsed slow positron beam with a pulse width of about 150 ps is required. A new positron pulsing system has been developed for the first time using an induction cavity to bunch the DC slow positron beam produced using the compact cyclotron (6).

The most suitable electric potential change according to time to supply for the slow positron beam buncher is given as

$$V(t) = -\frac{mL^2}{2et^2} + E_0 \qquad (1)$$

where t (<0) s is time when positrons pass the bunching gap and all positrons are time converged on the target at $t = 0$, E_0 eV is the initial positron energy, L m is the distance from the gap to the target. As shown in Equation (1), the most suitable electric potential change according to time to supply for the slow positron beam buncher $V(t)$ is t^{-2} function, not sine function.

In general, sine function is applied in order to bunch charged particles (7,8). However, in the case of bunching for low energy particles such as slow positrons, applying sine function is not efficient. That is the reason why we apply an

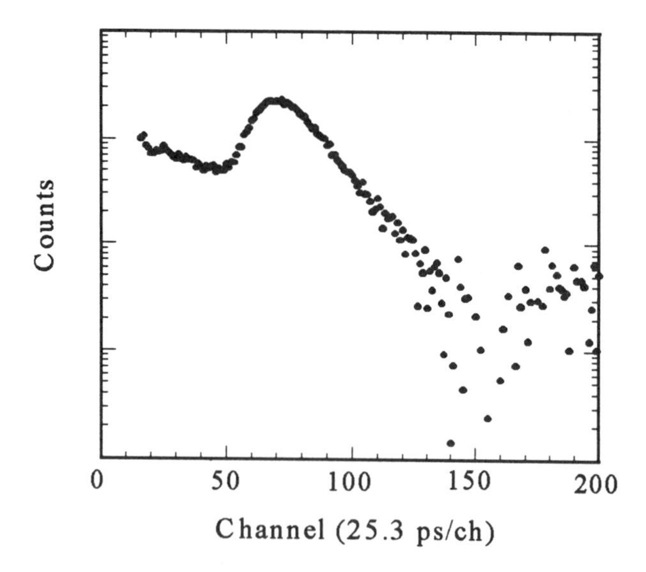

FIGURE 4. Time spectrum of the preliminary slow positron beam bunching test. The pulse width of the bunched positrons is about 300 ps.

induction cavity to slow positron beam bunching and do not apply a usual rf cavity.

The result of a preliminary slow positron beam bunching test using an induction cavity is shown in Fig. 4. In the test $V(t)$ changed from +70 V to -70 V during time of 5 ns under the condition of $L=$ 1.0 m and E_0 = 850 eV. The acquired pulse width of the bunched positrons is about 300 ps. The tuning for bunching was not sufficient in the preliminary experiment. A fine tuning of $V(t)$, E_0 and L for achieving a pulse width of 150 ps is in progress.

Polarized Slow Positron Beam

It is well known that positrons emitted from β^+ decay radioisotopes are longitudinally polarized with the helicity of v/c, where v and c being velocities of the positron and light, respectively. Hence the slow positron beam produced by using a compact cyclotron is supposed to be polarized, because the positrons are originally emitted from the β^+ decay radioisotopes ^{27}Si.

In order to confirm that the slow positron beam is polarized, a special polarimeter which applies magnetic quenching of ortho-positronium has been developed. By using this polarimeter, the slow positron beam polarization has successfully been measured as 33.4$\pm$5.3% (9). This value is consistent with the estimated polarization of 38.1$\pm$1.4% based on the Monte Carlo simulations.

CONCLUSION

An intense slow positron beam has been produced for the first time by applying a compact cyclotron. The compact cyclotron is much smaller and much more inexpensive than the electron LINAC that is used for producing an intense slow positron beam at present. In the near future, the compact cyclotron is almost sure to take the place of the electron LINAC for its outstanding ability to produce an intense slow positron beam economically.

A new positron pulsing system has been developed for the first time by applying an induction cavity. In a preliminary slow positron beam bunching test, a pulse width of about 300 ps has been achieved. In the near future variable-energy positron lifetime spectroscopy with the slow positron beam pulse width of 150 ps by applying an induction cavity will be realized.

A polarized slow positron beam has successfully been produced. The slow positron beam polarization has been measured as about 33% using a special designed polarimeter by means of magnetic quenching of ortho-positronium. The polarized slow positron beam will be applied to the research for magnetic material surfaces, biomaterials and so on.

ACKNOWLEDGMENTS

Professor H. Kobayashi of National Laboratory for High Energy Physics (KEK) and Mr. T. Kobayashi of Japan Synchrotron Radiation Research Institute are gratefully acknowledged for giving support in the design of the slow positron beam buncher using an induction cavity. We would like to thank Professor Y. Ito of The University of Tokyo and his collaborators for their useful discussions. This work has been performed by the assistance of SHI Examination and Inspection, Ltd.

*Research performed in collaboration with, at Sumitomo Heavy Industries, Ltd.: T. Nakajyo, M. Washio, at Tokyo Metropolitan University: M. Chiba, R. Hamatsu, T. Hirose, H. Iijima, M. Irako, N. Kawasaki, T. Kumita, T. Matsumoto, J. Yang, and at National Laboratory for High Energy Physics: Y. Kurihara, T. Omori and Y. Takeuchi.

REFERENCES

1. Ito, Y., Hirose, M., Kanazawa, I., Sueoka, O., and Takamura, S., *Materials Science Forum* **105-110**, 1893-1896 (1992).
2. House, J. V., and Rich, A., *Phys. Rev. Lett.* **61**, 488-491 (1988).
3. Weiss, A., Mayer, R., Jibaly, M., Lei, C., Mehl, D., and Lynn, K. G., *Phys. Rev. Lett.* **61**, 2245-2248 (1988).
4. Hirose, M., Washio, M., and Takahashi, K., *Appl. Surf. Sci.* **85**, 111-117 (1995).
5. Stein, T. S., Kauppila, W. E., and Roellig, L. O., *Rev. Sci. Instrum.* **45**, 951-953 (1974).
6. Hirose, M., Nakajyo, T., and Washio, M., *Appl. Surf. Sci.* (submitted).
7. Schodlbauer, D., Sperr, P., Kogel, G., and Triftshauser, W., *Nucl. Instrum. Methods* **B34**, 258-268 (1988).
8. Suzuki, R., Mikado, T., Chiwaki, M., Ohgaki, H., and Yamazaki, T., *Appl. Surf. Sci.* **85**, 87-91 (1995).
9. Kumita, T., Chiba, M., Hamatsu, R., Hirose, M., Hirose, T., Iijima, H., Irako, M., Kawasaki, N., Kurihara, Y., Matsumoto, T., Nakabushi, H., Omori, T., Takeuchi, Y., Washio, M., and Yang, J., *Appl. Surf. Sci.* (submitted).

POSITRON BEAM LIFETIME SPECTROSCOPY AT LAWRENCE LIVERMORE NATIONAL LABORATORY

Richard H. Howell[a], Thomas E. Cowan[a], Jay H. Hartley[a] and Philip A. Sterne[a,b]

[a] Lawrence Livermore National Laboratory, Livermore CA 94550
[b] Physics Dept., University of California, Davis CA 95616

Defect analysis is needed for samples ranging in thickness from thin films to large engineering parts. We are meeting that need with two positron beam lifetime spectrometers: one on a 3 MeV electrostatic accelerator and the second on our high current linac beam. The high energy beam spectrometer performs positron lifetime analysis on thick sample specimens which can be encapsulated for containment or for *in situ* measurements in controlled environments. At our high current beam we are developing a low energy, microscopically focused, pulsed positron beam to enable positron annihilation lifetime spectroscopy for defect specific, 3-dimensional maps with sub-micron location resolution. The data from these instruments is interpreted with the aid of first principles calculations of defect specific positron lifetimes.

INTRODUCTION

Our understanding of many important mechanical and electrical properties is limited by our knowledge of the microstructure and defects. There are many techniques that can identify the species and size of impurities. There are few however that can detect open volume defects such as atomic vacancies or voids, fig. 1. All positron annihilation spectrographic methods are particularly sensitive to determining low concentrations of the class of defects including vacancies, voids and negatively charged defects. This sensitivity stems from the attractive interaction between these defects and positrons which often binds the positron to the defect site. Using positron annihilation lifetime spectroscopy determines the size or charge of defects in metals, semiconductors and molecular or organic compounds.

Positron annihilation lifetime spectroscopy measures the electron density at the annihilation site. Since the electron density is sensitive to the defect volume there is a distinct correlation between the positron lifetime and defect size and condition. In metals and some compounds the correlation between lifetime and defect size can be calculated from first principles with sufficiently high accuracy to differentiate between major defect classes such as vacancies, vacancy clusters and voids. In the best case the positron lifetime is sensitive to the vacancy volume change associated with a dislocation strain field.

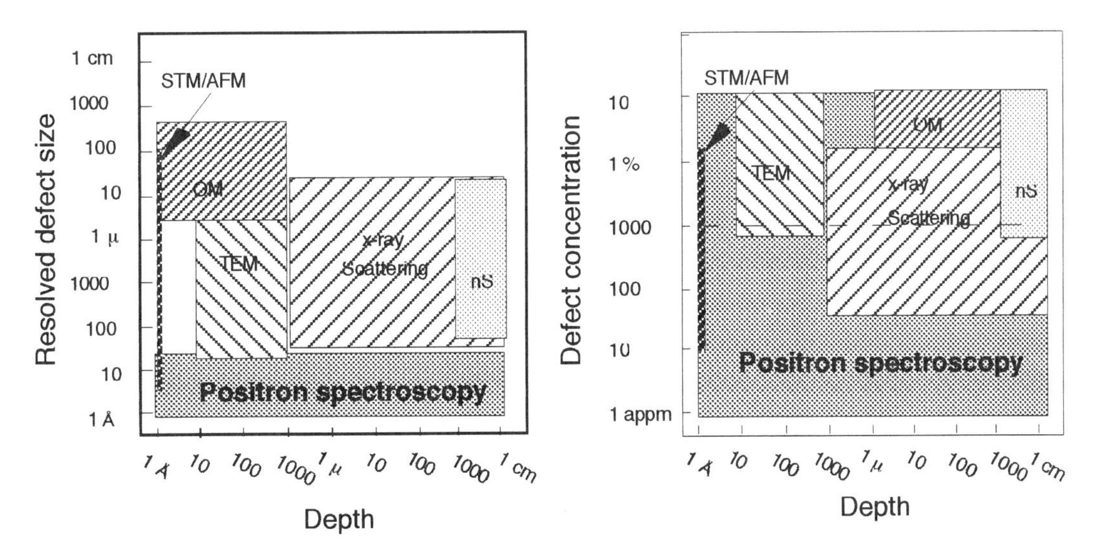

Figure 1. Positron annihilation lifetime analysis fills a special niche for techniques for general vacancy defect analysis, optical microscopy, OM, neutron scattering, nS, transmission electron microscopy, TEM, scanning tunneling microscopy, STM and atomic force microscopy, AFM, and x-ray scattering. Positron techniques are highly sensitive and can resolve the size of atomic vacancies at any depth in a sample.

CP392, *Application of Accelerators in Research and Industry*, edited by J. L. Duggan and I. L. Morgan
AIP Press, New York © 1997

LABORATORY CAPABILITIES

Positron annihilation spectroscopies have been traditionally performed by placing the sample in proximity to a radioactive source. This "source-sample" technique has been used at Livermore to measure electron momenta to determine Fermi surfaces through positron annihilation angular correlation and defect properties by positron annihilation lifetime spectroscopy.

Significant advantages can be achieved by performing positron measurements with a monoenergetic beam. There is a relaxation of the sample geometry so that engineering samples can be measured. Better definition of the measurement volume is obtained by controlling the energy and direction of the positrons during implantation in the sample. For high energies the positron beam can even pass through thin windows before implanting into the sample allowing *in situ* measurements in controlled environments. In positron annihilation lifetime spectroscopy higher data rates are achieved by a positron beam due to the nearly 100% efficiency in determining the implantation time. Also in a well controlled geometry only annihilation events from the sample will contribute to the data which eliminates systematic contributions from the positron source and simplifies the data analysis.

High Energy Beam

There are both low and high energy positron beams available at LLNL. Our high energy beam is derived from a 100 mCi ^{22}Na source moderated by a 2 micron thick tungsten single crystal foil positioned in the terminal of a 3 MeV Pelletron electrostatic accelerator, see fig. 2. Both moderated positrons and positrons directly emitted from the source are captured and accelerated. This beam contains a current of $5 \cdot 10^{5}$ $e^{+}s^{-1}$ captured directly from the radioactive source at 3 MeV. A similar current of moderated positrons can be obtained. To perform positron lifetime spectroscopy the time of implantation is determined for each positron as it passes through a 2 mm plastic scintillator. Annihilation gamma-rays of the implanted positrons are detected by a BaF_2 detector. The annihilation lifetime is calculated from the time difference in the two detectors with a system resolution of ~250 ps. Positrons leave the implantation detector with a broad energy distribution, an average energy of 2.6 MeV and a 1 cm effective beam diameter at a sample placed 4 cm down stream. At this energy positrons implant from mm to cm into a sample depending on the sample density. Positrons that scatter away from the sample are rejected by an anticoincidence detector. In the pictured close coupled geometry a small, < 1 %, contribution from annihilations in the detector system is found in measured spectra and 2000 counts per second can be achieved. The annihilation detector can also be run in coincidence with a second detector to constrain all events to annihilation in the sample. This geometry reduces the counting rates to less than 1000 counts per second This system is modeled on the one used successfully at Stuttgart Germany[1]

We are using positron annihilation lifetime spectroscopy with the high energy beam to determine aging effects in carbon fiber resin composites. We are measuring the changes in the hole volume brought on by accelerated aging of the composites at elevated temperature and in hostile atmospheres. Data describing changes in the hole volume are correlated with infrared spectroscopy and mechanical tests to provide a complete description of the changes during aging. We are also determining in radiation damage in metals from induced radiation from high energy accelerator beams and from self irradiation in radioactive species. Interpretation of lifetimes found in the irradiated alloys is aided by first principles calculations of the bulk and defect lifetimes. We are also using the high energy beam to determine changes in polymeric and molecular solids brought on by irradiation or chemical degradation.

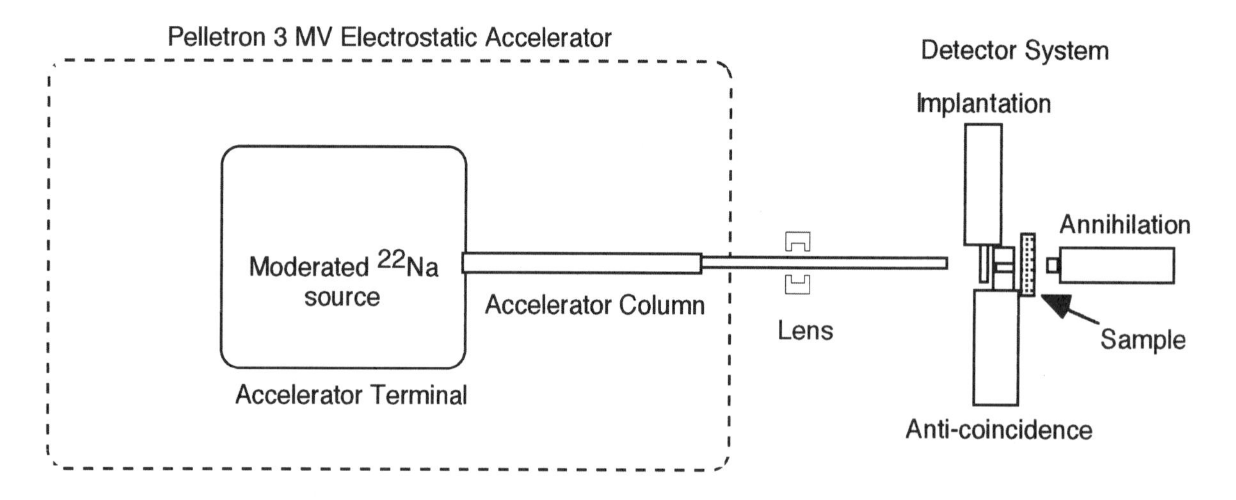

Figure 2. Schematic diagram of the critical components of the high energy beam positron annihilation lifetime system.

3-D Scanning Positron Microprobe

We are developing a new capability at our high current low energy positron beam located at the LLNL 100 MeV electron linac. By controlling the implantation location and energy of a pulsed, microscopically focused positron beam we will perform positron annihilation lifetime spectroscopy to determine defect concentration and size over spatial volumes of 0.025 μm^3. Our high current positron beam has recently been upgraded to 10^{10} e^+s^{-1}. These high initial currents will enable us to perform rapid analysis and 3 dimensional maps of defect characteristics on a practical time scale.

3-Dimensional mapping of defects requires several separate qualities for the positron beam. 3-D spatial sensitivity requires a highly focused, < 1 micron final spot size, and variable energy 1-50 keV positron energy. Maximum implantation depths for these energies range from 1 to 10 microns depending on sample density. The location of buried features can be located with high precision by sweeping the energy and location of the positron beam in small steps. The size and concentration of the defect species at the end of the positron range are then determined by positron annihilation lifetime spectroscopy measurements correlated to defect size through validated theoretical calculations. To perform positron annihilation lifetime spectroscopy requires pulsing the focused beam with a pulse duration less than 100 ps. To obtain high counting rates our positron pulsing rate will be 15 MHz and multiple detectors will be used for the annihilation gamma rays.

Many of the beam characteristics described above have been obtained in separate systems [2-4] and a design for an efficient, low source strength microprobe has been reported [5]. The LLNL microprobe design integrates new features and those of the published designs to achieve a beam system with improved efficiency and pulse duration. A schematic of the elements of this system is shown in Fig. 3.

In an optical system the initial conditions of the beam entering a pulsing or focusing system limits quality of the system output. In moderated positron systems positrons are expelled into the vacuum by the negative workfunction of a moderator material. Initial conditions are set by the spot size and the distribution of directions and energies of positrons emitted from the material used to thermalize the initial high energy positron source. In focusing systems the overall brightness of the beam limits the final conditions for a lens system. This limitation has been defeated in positron optical transport systems by resetting the initial conditions by implanting and thermalizing the positron beam in one or more intermediate steps. At each thermalization step the history of energy and dispersive distributions is lost and new initial conditions are established. This brightness enhancement technique produces highly focused beams with high transmission efficiency. The initial energy spread of the beam is the critical parameter for pulsing systems. The time compression of the pulse is limited only by time of flight spreading from the energy distribution.

The initial linac beam consists of bright pulses of high intensity and short duration. The first step of the 3 dimensional scanning pulsed positron microprobe will be to trade the short time duration for narrow energy. The initial high current linac beam has an energy width of 4 eV in 3 μs pulses at 300 Hz. This beam is transported in a magnetic field to a penning trap pulse stretcher which is the first element of the microprobe (see

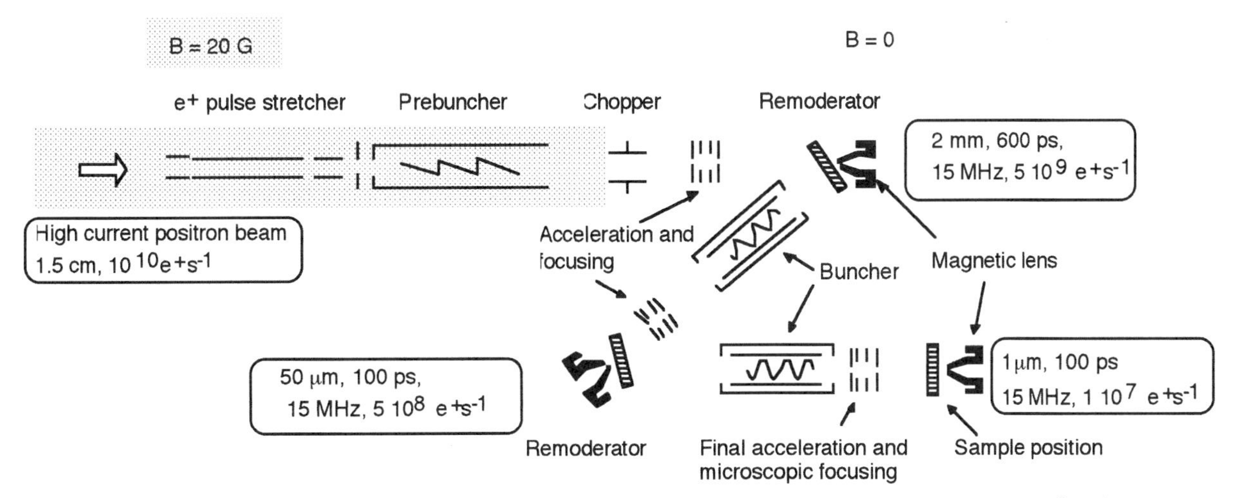

Figure 3 Schematic of the main elements of the 3-D pulsed positron microbeam system. The final beam of more than 10^7 e^+s^{-1} will be less than 1 micron diameter and 0.5 to 50 keV. Beam position will be scanned in the lateral dimensions by magnetic coils and scanned in depth by energy variation to provide a 3-dimensional map of defect concentrations and size.

Fig. 3). We will trap the positron pulses and slowly raise the trap voltage so that the highest energy positrons will spill over a fixed voltage barrier. The stretcher will convert the 4 eV, 3 μs beam pulse into a ~20 meV, 300 ms beam pulse with average energy set by the endcap voltage. This provides a significantly reduced energy distribution for input to the first pulsing electrodes and makes it practical to benefit from an ideal pulsing shape generated by a fast programmable Tektronix AWG2041 waveform generator to obtain high efficiency, sub nanosecond pulses in a single stage. This pulsed beam will then be accelerated to 15 keV and extracted from the magnetic field of the penning trap. After extraction the beam will be decelerated and focused on the first brightness enhancement moderator by a combination of electrostatic, and single pole magnetic lenses. The beam incident on the first remoderator is expected to be 2 mm in diameter and have $5 \cdot 10^9$ e^+s^{-1} in 600 ps pulses at 15 MHz. The penning trap stretcher, first pulsing section and magnetic filed extraction are now fabricated and will be available for test in the near future.

A second stage of focusing and pulsing includes a resonant buncher driven by a Rohde & Schwartz SMT02 sine wave generator that has been phase and frequency locked to the waveform input of the first pulsing section. This stage completes the pulse compression to 100 ps and focuses the beam on a second remoderator which serves as the input to the microprobe acceleration column. The beam at this point is expected to have 50 micron diameter and $5 \cdot 10^8$ e^+s^{-1}. Following the second remoderator is a final stage similar to published designs [6] with variable acceleration and focusing to a one micron beam spot to reach the final beam characteristics of $1 \cdot 10^7$ e^+s^{-1} , 1-50 keV, 15 MHz, 100 ps.

When completed we anticipate using the 3D scanning defect microprobe to study defect distributions in metals and alloys, in inhomogenous systems such as semiconductor devices, and in thin film and composite polymeric materials. With the microprobe we will determine the spatial defect distributions in 3 dimensions near microscopic features such as grain boundaries, cracks, precipitates, and buried interfaces.

ACKNOWLEDGEMENTS

We wish to thank Drs. W. Triftshäuser, G. Kögel, P. Sperr, and D.T. Suzuki for their valuable conversations on beam pulsing and focusing. This work was performed under the auspices of the US Department of Energy by LLNL under contract No. W-7405-ENG-48.

REFERENCES

1. W. Bauer, J. Briggmann, H.-D. Carstanjen, S. H. Connell, W. Decker, J. Diehl, K. Maier, J. Major, H. E. Schaefer, A. Seeger, H. Stoll and E. Widmann, Nucl. Inst. and Meth. B50 , 300 (1990)

2. P. Willutzki, J. Störmer, G. Kögel, P.Sperr, D. T. Britton, R. Steindl and W. Triftshäuser, Materials Science Forum 175-178, 237 (1995)

3. R. Suzuki, T. Mikado, M. Chigaki, H. Ohgaki, and T. Yamazaki Applied Surface Science 85, 87 (1995)

4. G.R. Brandes, K. F. Canter, T. N. Horsky and P.H. Lippel and A. P. Mills, Rev. Sci. Instrum. 59, 228 (1988)

5. A. Zecca, R. S. Brusa, M. P. Duarte-Naia, G. PP Karwasz, J. Paridaens, A. Piazza, G. Kögel, P. Sperr, D. T. Britton, K. Uhlmann, P. Willutzki and W. Triftshäuser, Europhys. Lett. 29, 617 (1995)

6. K. Uhlmann, D. T. Britton and G. Kögel, Meas. Sci. Technol. 6,932 (1995)

EXPERIMENTAL INVESTIGATION OF SLOW POSITRON BEAMS PRODUCED FROM ACCELERATOR–GENERATED ^{13}N

B.J. Hughey, R.E. Shefer, and R.E. Klinkowstein,

Newton Scientific, Inc., Cambridge, MA 02141

K.F. Canter

Department of Physics, Brandeis University, Waltham, MA 02254

Positron beams are powerful diagnostic probes for the nondestructive evaluation of metals, semiconductors, and polymers. An impediment to the further development and commercialization of positron-based diagnostics is the present lack of suitable sources of intense (>10^7 e^+/sec) slow positron beams. A slow positron source is being developed which uses a compact deuteron accelerator to produce the positron emitter ^{13}N via the ^{12}C(d,n)^{13}N reaction. This paper describes the design of a target and moderator holder to be used in experiments to determine the practically achievable slow positron yield from accelerator produced ^{13}N using an existing 4 MeV tandem accelerator. A W(100) foil is used to moderate the positrons, which are then guided through a bent solenoid to an annihilation target. The effects of neutron irradiation on W(100) moderator efficiency has been measured and found to be unimportant. Experiments are underway to measure the slow positron flux for a range of deuteron beam currents and bombarding energies between 1.5 and 4 MeV.

INTRODUCTION

The recent explosion of positron beam applications in basic and applied research (1–3) has created a need for both multi-user high flux (>10^9 e^+/sec) facilities as well as a need to increase the relatively low fluxes (10^5-10^6 e^+/sec) commonly available to individual investigators. A compact accelerator based positron source offers a convenient, cost-effective alternative which can provide moderate to intense flux slow positron beams (10^6 –10^8 e^+/sec). When coupled with high efficiency moderators, compact accelerator sources can potentially reach into the 10^8–10^9 e^+/sec range, and may rival the deliverable beam brightness offered by many high flux beam production schemes presently under consideration.

Nitrogen-13 is a short-lived ($\tau_{½}$ = 9.97 min.) radioisotope which decays entirely by positron emission (β^+, 1.19 MeV, 100%). ^{13}N can be produced at high yields by bombardment of ^{12}C with relatively low energy (1–4 MeV) deuteron beams (4). The feasibility of using the ^{12}C(d,n)^{13}N reaction to produce slow positron beams was recently demonstrated at low beam power at Brookhaven National Laboratory (5). These experiments indicate that slow positron fluxes in excess of 10^7 e^+/sec should be achievable with a 1.5 MeV, 1 mA deuteron beam, and that fluxes in excess of 10^8 could be obtained with a 4 MeV, 2 mA beam. The issues important to scaling the low power Brookhaven results to the production of intense positron beams include: (1) achievable W(100) moderator efficiency when coupled with an accelerator target, (2) survivability of carbon targets under high power deuteron beam bombardment, (3) neutron damage, if any, to the moderator during ^{13}N production, and (4) the effect of positron absorption in the ^{12}C target on achievable slow positron flux. In this paper we describe simulations and preliminary results of experiments designed to quantitatively determine the maximum slow positron flux that can be realistically obtained from accelerator-produced ^{13}N.

^{13}N-BASED SLOW POSITRON SOURCE

A high flux positron beam can be produced using a low-energy (1.5 MeV) deuteron accelerator capable of delivering milliampere currents. A highly efficienct accelerator technology with these capabilities has been developed by Newton Scientific, Inc. (NSI) (6). Such an accelerator-based positron system can be made quite compact, and thus suitable for installation in a small laboratory. For example, the 1.5 MeV accelerator is 1.8 m in length and 0.8 m in diameter. The system under development incorporates a target consisting of two

CP392, *Application of Accelerators in Research and Industry*, edited by J. L. Duggan and I. L. Morgan
AIP Press, New York © 1997

graphite disks brazed into a copper cylinder on opposite sides. Cooling will be provided by water flowing along the axis of the cylinder. In this way, one target region can be switched through 180° to face a W(100) foil moderator while the other is irradiated to replenish the ^{13}N activity. By switching the target with a time interval short compared to the ^{13}N half-life, a nearly continuous β^+ flux can be extracted. In the limit of rapid switching of the target, the maximum available activity will be one half of the ^{13}N saturated yield.

Effective positron yield from target

The fast positron flux emitted from the carbon target face will depend on the deuteron penetration depth, the ^{13}N production rate as a function of depth, and the attenuation of the fast positrons as a function of depth. As a first approximation to the effect of positron attenuation in the carbon target, we have calculated the "effective ^{13}N yield" as a function of deuteron beam energy. The deuteron energy as a function of depth in the target was calculated using stopping cross-sections generated by the ion transport code TRIM (7). The ^{13}N production rate as a function of depth was then obtained using the calculated deuteron energy vs. depth profile and literature values for the yield of the $^{12}C(d,n)^{13}N$ reaction (4,8). The number of fast positrons escaping the target from each layer at depth x_n was then determined by integrating the positron track through the material as a function of angle to the surface normal, θ, at which the positron
was emitted. The equation used was:

$$P_n = \int_0^1 \exp\left(-\frac{x_n/\cos\theta}{x_o}\right) d\cos\theta, \qquad (1)$$

where x_o is the attenuation length of the ^{13}N positrons in carbon, given from Ref. (9) to be approximately 70 mg/cm^2. (The factor of ½ due to the fact that only half the emitted positrons are directed out of the target is omitted because it is included in the conventional definition of moderation efficiency as the number of slow positrons after the moderator divided by the β^+ activity of the radioactive material.) Figure 1 shows the "effective ^{13}N yield" (including positron attenuation) as well as the thick target yield as a function of incident deuteron energy. As expected, the increased deuteron range at higher energy leads to a decrease in the fraction of positrons escaping from the target surface.

Effect of neutron irradiation on moderator foil

Neutron damage to the tungsten moderator is of concern because thermalized positrons will become trapped in radiation-induced defects in the metal lattice, resulting in attenuation of the slow positron

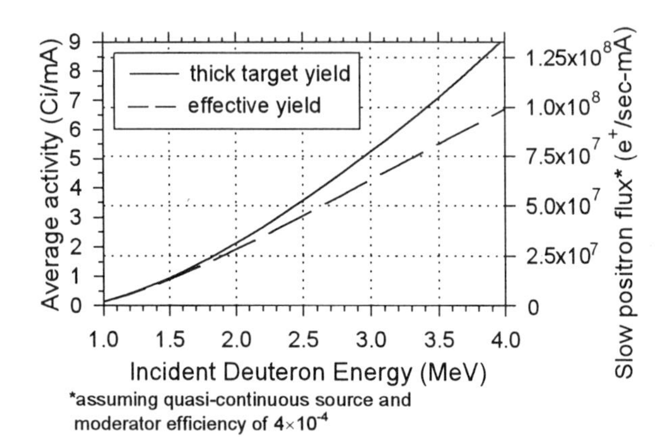

FIGURE 1. Calculated effect of positron attenuation in the carbon target on effective yield of the $^{12}C(d,n)^{13}N$ reaction as a function of energy. The solid curve is a fit to the data of Refs. (4,8), and the dashed curve includes the effect of β^+ attenuation in the target.

flux. The $^{12}C(d,n)^{13}N$ reaction is endothermic, producing a neutron spectrum with an endpoint energy approximately 0.3 MeV less than the incident deuteron beam energy. The neutron fluence at the moderator is a function of the ^{13}N production rate, distance from target to moderator, and irradiation time. As an example, irradiation of a carbon target for 500 hours at an average ^{13}N yield of 1 Ci will produce a total fast neutron fluence of 1.7×10^{15} n/cm^2 at a moderator foil placed 2.5 cm from the target.

The effect of neutron damage on moderator efficiency had not previously been directly measured. It is possible to estimate the order of magnitude of this effect using data on neutron damage to reactor wall materials (10) combined with the diffusion properties of thermalized positrons in metals (1). The probability of positron trapping in a lattice monovacancy after neutron irradiation to a fluence of 2×10^{15} n/cm^2 is estimated to be negligible (5×10^{-3}). The definitive test is to irradiate a tungsten moderator foil with neutrons and compare its performance to that observed prior to irradiation. This was accomplished by exposing a W(100) moderator to neutrons from the Massachusetts Institute of Technology Research Reactor (MITRR).

The moderation efficiency prior to neutron irradiation was measured in an ion pumped solenoidal positron beam system at Brandeis University which incorporates a 15 mCi ^{22}Na source. A W(100) foil (11) was subjected to three electron bombardment annealing cycles in a separate chamber, then transported in air to the positron beam system. Slow positrons were accelerated to 30 eV and transported through a 60 cm, 60° bent solenoid to an annihilation detector. The slow positron flux was measured to be 6.1×10^3 e$^+$/sec.

The foil was then irradiated with neutrons in the 2PH1 port of MITRR for 47 min. Since neutron damage to metals is primarily caused by fast neutrons (10), the moderator foil was covered with a layer of cadmium absorber to eliminate the large thermal neutron component present in the reactor spectrum. The fast neutron flux (>3 keV) in this port is approximately $(4.5\pm2.5)\times10^{11}$ n/cm^2-sec. This irradiation therefore corresponded to a fast neutron dose of $(1.3\pm0.7)\times10^{15}$ n/cm^2, or the equivalent of 380±200 hours in the 1 Ci ^{13}N system described above. Space constraints in the reactor port required that the moderator foil be removed from its holder, covered directly with cadmium foil, and then inserted in a small plastic vial. Even in the absence of neutron damage to the foil, some degradation in foil performance is expected under these conditions due to defects introduced by handling of the single crystal foil.

Following irradiation, the foil was re-inserted in the holder, and the slow positron flux was measured with no further processing to be 3.7×10^3 e$^+$/sec. The observed 40% decrease in moderator efficiency can probably be attributed primarily to mechanical damage introduced by the severe handling of the foil required for insertion into the reactor port. After the foil was annealed in situ, the slow positron flux increased to 8.1×10^3 e$^+$/sec, confirming the expected restoration of moderator efficiency after in situ annealing. These experiments show that long term exposure of the moderator foil to neutrons in an ^{13}N-based positron source will not be a substantial problem, requiring at most in situ annealing of the foil every few hundred hours.

TARGET AND MODERATOR DESIGN

Experiments to determine the realistically achievable slow positron flux as a function of ^{13}N activity produced in the target for a variety of beam energies and currents are presently underway. The deuteron beam will be obtained from a NSI high current tandem accelerator installed in the Nuclear Engineering Department at the Massachusetts Institute of Technology (6). The accelerator will be operated at energies in the range 1.5–4 MeV. Figure 2 is a photograph of the experimental system. The slow positron flux will be measured following transport through a 90° bent solenoid by monitoring with a NaI detector the 511 keV annihilation gamma radiation produced when the beam strikes an annihilation target. Varying the bias on the grid immediately after the moderator from approximately –30 V to +30 V will allow separation of the signal due to the slow positron beam from background due to fast positrons annihilating elsewhere in the system.

Figure 3 is a cross-sectional drawing of the accelerator target and positron moderator holder, and Fig. 4 is a photograph of the holder with the target rod installed. The holder has been designed to locate the moderator foil as close to the target as practical (<0.7 mm). During irradiation, the target is rotated to face the deuteron beam. Immediately following irradiation, the target will be rotated 180° to face the moderator foil. In

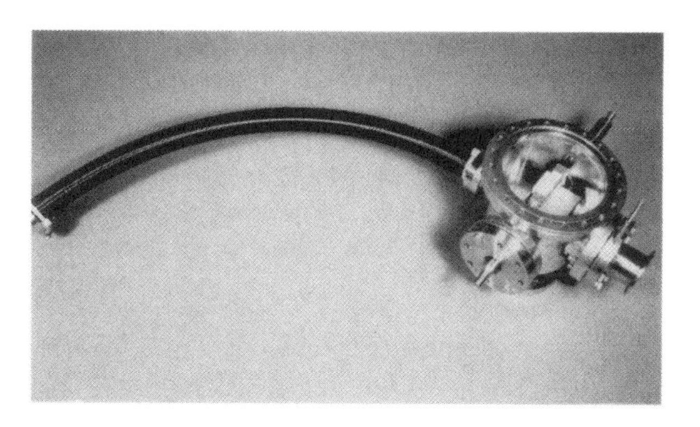

FIGURE 2. Photograph of target chamber and bent solenoid. The deuteron beam enters the chamber through port at the right, and the water cooled target rod passes through the chamber at right angles to the ion beam. Moderated positrons are guided with external coils (not shown) to the 90° bent solenoid extending to the left.

situ annealing will be accomplished by rotating the moderator holder up to the position shown, and bombarding the foil with electrons from a retractable filament positioned as shown.

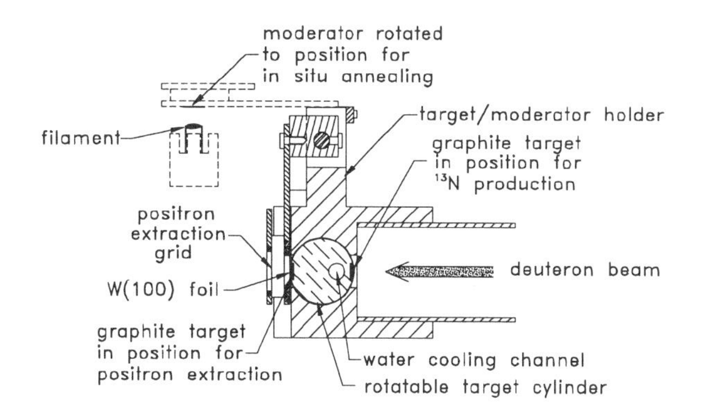

FIGURE 3. Cross-sectional drawing of target/moderator holder.

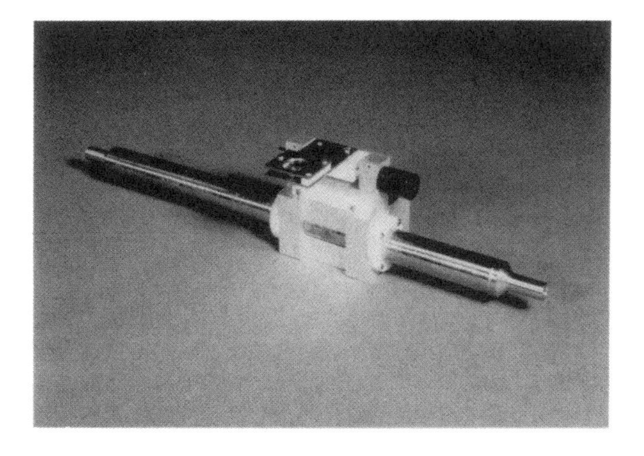

FIGURE 4. Photograph of target/moderator holder. The graphite target is shown rotated to face the moderator, and the moderator holder is shown in the raised position for in situ annealing. During irradiation, the target is rotated 180° to face the deuteron beam.

Graphite target design

Generation of intense slow positron beams (10^7 –10^8 e^+/sec) will require a graphite target capable of withstanding bombardment with deuteron beams of 1 kW (10^7 e^+/sec at 1.5 MeV, see Fig. 1) to 4 kW (10^8 e^+/sec at 4 MeV). Presently available tungsten moderator foils (11) have a diameter of 0.9 cm, and thus a maximum beam diameter of about 0.7 cm is desirable in order to obtain high moderation efficiency. The minimum beam power density required for the generation of an intense slow positron beam is thus approximately 3–10 kW/cm^2. Targets capable of withstanding even higher power densities are desirable for applications requiring high beam brightness, since the beam brightness is inversely related to the square of the positron source diameter (i.e. the deuteron beam diameter).

Previous researchers have demonstrated that high velocity linear water flow across the back surface of a metallic target will allow removal of beam power densities up to about 10 kW/cm^2 (12,13). They accomplished this by providing a thin rectangular channel behind the target in order to obtain water flow of 400–1300 cm/sec. We have designed the prototype accelerator target with a simplified version of this geometry: a coolant channel 0.64 cm in diameter is drilled through a copper target cylinder to which is brazed a 0.6 cm diameter × 0.1 cm thick piece of high thermal conductivity AGKSP graphite (14). The back surface of the graphite is located 0.2 cm from the outer edge of the water cooling channel. We expect to obtain a flow rate of at least 400 cm/sec with this design coupled to an available source of chilled water at 2 gpm. The maximum heat flux which can be removed from this target will be determined for deuteron beam energies between 1.5 and 4 MeV.

Discussion

We plan to perform experiments to determine the efficiency of slow positron beam generation using the target and moderator geometry described above. Single crystal W(100) transmission moderators exhibit a wide range of moderation efficiencies, since the moderation efficiency depends on the geometry of source encapsulation, distance from source to moderator foil, relative sizes of source and foil, and foil pre-treatment methods, such as annealing or oxygen treatment. An accelerator-based source has different mechanical, thermal, and vacuum characterisitcs and constraints than a radioisotope source such as ^{22}Na. For this reason, it is important to experimentally determine the slow positron production efficiency obtainable using accelerator-produced ^{13}N. We will accomplish this by measuring the slow positron flux obtained with a W(100) moderator foil as a function of deuteron beam energy and current. These measurements will be compared with the predicted flux shown in Fig. 1. The results of these measurements will allow us to determine the accelerator parameters required (beam energy and current) for an intense ^{13}N-based positron beam, and to optimize the target/moderator holder design for maximum slow positron production efficiency.

ACKNOWLEDGMENTS

The assistance of A. Krupyshev, J. O'Neal, and R. Xie in the measurements of moderator efficiency is gratefully acknowledged. This work is supported by National Science Foundation Grant No. DMI-9561398.

REFERENCES

1. Schultz, P.J. and Lynn, K.G., *Rev. Mod. Phys.* **60**, 701 (1988).
2. See articles in *Appl. Surf. Sci.* **85** (1995).
3. See articles in *Slow Positron Beam Techniques for Solids and Surfaces*, Ottewitte, E. and Weiss, A.M., Eds., New York: AIP Press, 1994.
4. Shefer, R.E., Hughey, B.J., Klinkowstein, R.E., Welch, M.J., and Dence, C.S., *Nucl. Med. Biol.* **21**, 977 (1994).
5. Xie, R., Petkov, M., Becker, D., Canter, K., Jacobsen, F.M., Lynn, K.G., Mills, R., and Roellig, L.O., *Nucl. Instrum. Meth. Phys. Res.* **B93**, 98 (1994).
6. Shefer, R.E., Klinkowstein, R.E. and Yanch, J.C. in *Proc. Fifth Int'l. Conf. on Appl. of Nucl. Tech.*, June, 1996, to be published by SPIE.
7. TRIM-91, J.F. Ziegler, IBM Research Center, Yorktown, NY 10598
8. Jaszczak, R.J., Macklin, R.L., and Gibbons, J.H., *Phys. Rev.* **181**, 1428 (1969).
9. Knoll, G.F., *Radiation Detection and Measurement, 2nd ed.*, New York: John Wiley & Sons, 1989, p. 48.
10. English, C.A., *J. Nucl. Mat.* **108&109**, 104–123 (1982).
11. J. Chevalier, University of Aarhus, Aarhus, Denmark.
12. Hanley, P.R., Haberl, A.W., and Taylor, A. *IEEE Trans. Nuc. Sci.*, June, 1967, 933
13. Seiler, R.F., Cleland, M.R., and Wegner, H.E., *IEEE Trans. Nuc. Sci.*, June, 1967, 943
14. Carbone of America, Bay City, MI

AN INTENSE POSITRON BEAM USING A LARGE AREA ^{64}Cu SOURCE

F.M. Jacobsen[a], A. Koymen[a], A.H. Weiss[a], S. Ovunc[a], E. Srinivasan[a], S. Goktepeli[b], K. Unlu[b] and B.W. Wehring[b]

a) Department of Physics, University of Texas at Arlington, Arlington, TX 76019-0059
b) Nuclear Engineering Teaching Lab., University of Texas at Austin, Austin TX 78712

In this paper we describe the University of Texas at Arlington and the University of Texas at Austin approach to an intense positron, e$^+$, beam. An approximately 50 Ci ^{64}Cu source distributed over 1000 cm^2 combined with a solid Kr moderator/remoderator is expected to yield a sub cm diameter e$^+$ beam with an intensity in the range of 10^8 to 10^9 e$^+$/sec.

INTRODUCTION

Standard university positron, e$^+$, beams use β^+ emitting isotopes such as ^{22}Na and ^{58}Co to supply fast (0 - 500 keV) positrons. A small fraction (10^{-4} -10^{-3}) of the β^+ particles is moderated to a few eV by use of either a well annealed few thousand Å thick tungsten (1) (or Ni, Cu) film or a Rare Gas Solid, RGS, moderator (2). The e$^+$ affinity to tungsten is about -3 eV and, therefore, a stopped e$^+$ that diffuses to the tungsten surface within its lifetime may be ejected into the vacuum with a kinetic energy of 3 eV. The RGS moderators work differently in that their e$^+$ affinities are positive. The driving force of RGS moderators is hot diffusion once the e$^+$ has slowed down below the threshold for positronium formation leaving only the weak phonon interaction available for further slowing down. By using 100 mCi of ^{22}Na (about the maximum source strength commercially available) combined with a RGS moderator an e$^+$ intensity of about 5 10^6 e$^+$/sec may result (1-3).

For many e$^+$ techniques such as Low Energy Positron (Positronium) Diffraction (4), the Positron Reemission Microscope (5), two dimensional Angular Correlation of gamma Rays (6) and the formation of e$^+$ micro beams for three dimensional defect mapping of semiconductors, polymers and metals an intensity of about 10^8 - 10^9 e$^+$/sec or greater is required.

Also many fundamental experiments within the field of atomic physics will greatly benefit from the availability of intense low energy positron beam. This include differential scattering measurements (7), formation of antihydrogen (8), Bose condensation (9) and Ps liquid studies (10).

Basically, two options are available to increase significantly the e$^+$ beam intensity over that of a standard e$^+$ beam. One is to increase the number of fast positrons available for moderation and another options is to improve the efficiency of the moderation procedure. While the latter route to intense e$^+$ beam is the elegant one (see ref. 11) it is the first method that is pursued by most groups around the world and it involves the use of high current and high energy (100's MeV) e$^-$ LINACs for the production of fast positrons via pair production (12), production of strong ^{64}Cu sources in high neutron flux reactors (13) and other techniques (14-15).

The most intense low energy e$^+$ beam is currently the LINAC beam at Lawrence Livermore National Laboratory, LLNL, (16) that produces 10^{10} e$^+$/sec whereas other LINAC beams yield intensities in the range of 10^7 - 10^8 e$^+$/sec (17). Presently, the strongest e$^+$ beam based on radioactive isotopes is that at Brookhaven National Laboratory, BNL, yielding about 10^8 e$^+$/sec using ^{64}Cu.

Our approach to an intense low energy e$^+$ beam is to use the Nuclear Engineering Teaching Laboratory, NETL, reactor at Austin to produce ^{64}Cu. The NETL reactor is a low neutron flux reactor and to offset this disadvantage our scheme necessitates the use of a large area ^{64}Cu source. The main challenge is then to reduce the beam diameter to a usable size with a minimum loss in e$^+$ intensity.

GENERAL CONSIDERATIONS

The e$^+$ beam at Austin will be electrostatically guided and will deliver low energy positrons to user beam lines in the energy range from a few eV to 50 keV at an intensity greater than 10^8 e$^+$/sec. Initially, the e$^+$ beam size at the target will be about 1 mm. Later, however, a brightness enhancement stage will be included whereby the beam diameter at the target may be reduced to the μm range. The e$^+$ beam is situated outside the reactor and the ^{64}Cu source is transported from the core of the reactor to

CP392, *Application of Accelerators in Research and Industry*, edited by J. L. Duggan and I. L. Morgan
AIP Press, New York © 1997

the source position of the beam under vacuum. The vacuum of the beam port into the reactor and e^+ beam vacuum is separated by a gate valve. The advantage of this procedure is that the ^{64}Cu source can be removed from the e^+ beam apparatus in a matter of minutes providing quick access for maintenance and other purposes. Furthermore, in comparison to schemes in which the β^+ source region is located at the reactor core (14), our beam design does not differ much from that of conventional e^+ beams based on a ^{22}Na source.

THE ^{64}Cu SOURCE

In a thin target approximation we can calculate the density of activated ^{64}Cu, n_{64}, as

$$n_{64} = n_{63}\sigma_n I_n(1 - \exp(-t/\tau))\,\tau \qquad (1)$$

with n_{63} being the density of ^{63}Cu (abundance 69%), σ_n is absorption cross-section for thermal neutrons ($4.5\ 10^{-24}$ cm^2), I_n is the neutron flux ($2\ 10^{12}$ n/(cm^{2}sec)), τ is the mean lifetime of ^{64}Cu (18.5 hours) and t corresponds to the irradiation time. If we take t = 6 hours we get the following value for the β^+ activity

$$A_{\beta+} = 0.8\ \mathrm{Ci/cm^3} \qquad (2)$$

where we have factored in the branching ratio for β^+ emission. The attenuation length, l, of the β^+ particles in the Cu source is 0.025 mm (18). If the thickness of the Cu source is greater than say $3l$ then the activity available for low energy e^+ production is simply given as

$$I_{\beta+} = A_{\beta+}l = 2\ \mathrm{mCi/cm^2} = 7\ 10^7\ \beta^+\ \mathrm{cm^{-2}sec^{-1}} \quad (3)$$

whereas the total activity is greater than 0.04 Ci/cm^2. By assuming a moderation efficiency of $2\ 10^{-3}$ (see below) we obtain the intensity of low energy e^+ as

$$I_{e+} = 1.5\ 10^5\ e^+\mathrm{cm^{-2}sec^{-1}} \qquad (4)$$

so to obtain 10^8 e^+/sec a source area of about 1000 cm^2 is required. The source will consist of a 10 by 10 cm^2 base copper plate onto which is mounted 400 copper cylinders with a height of 1 cm and a diameter of 0.5 cm (see Figure 1). The total surface area is 1300 cm^2. To produce the low energy e^+ the ^{64}Cu is cooled to 15 K and coated with a few µm of solid Kr and $I_{e+} = 2\ 10^8$ e^+/sec should result.

The aspect ratio, R (length/diameter), of the cylinders is that found optimum for magnetically guided e^+ beams (19). However, as the present beam will be electrostatically guided it is possible that R should be larger whereby the surface area can be further increased for a fixed source volume (10x10x1 cm^3). From the point of view of the source strength, R should be as large as possible. However, for sufficiently large R some of the low energy e^+ will not escape from the source due to their many interactions with the surface of the solid Kr whereby annihilation may occur.

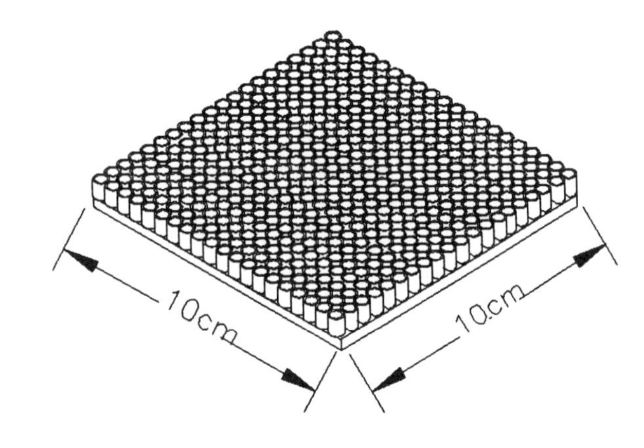

FIGURE 1. ^{64}Cu source consists of 400 cylinders mounted onto 10 by 10 cm^2 copper plate. The total surface area is 1300 cm^2.

SOURCE - MODERATOR CONFIGURATION

To enable transport of the low energy e^+ beam its size has to be reduced considerably. The traditional way of reducing the diameter of an e^+ beam is through the so-called remoderation technique (20) in which the e^+ beam is accelerated to about 5 kV and then strongly focused onto a thousand Å tungsten film. About 30% of the e^+ are re-emitted on the opposite side of the film (21) with a diameter equal to that of the incident beam but with an energy of only a few eV. The efficiency of this method to brightness enhance e^+ beams is well documented in the literature (5).

We have, however, opted for a potentially more efficient (though untested) method of reducing the initial large area of our e^+ beam. The idea is to combine the moderation and remoderation into one stage. Figure 2 shows the ^{64}Cu source plate as forming one face of a box with the remaining faces being made out of tungsten. The entire box is cooled to 15 K and then coated with solid Kr. The face of the box opposite to the ^{64}Cu source contains a hole in it with a diameter of 0.5 - 1 cm and it is positrons that emerge from this exit aperture that form the e^+ beam.

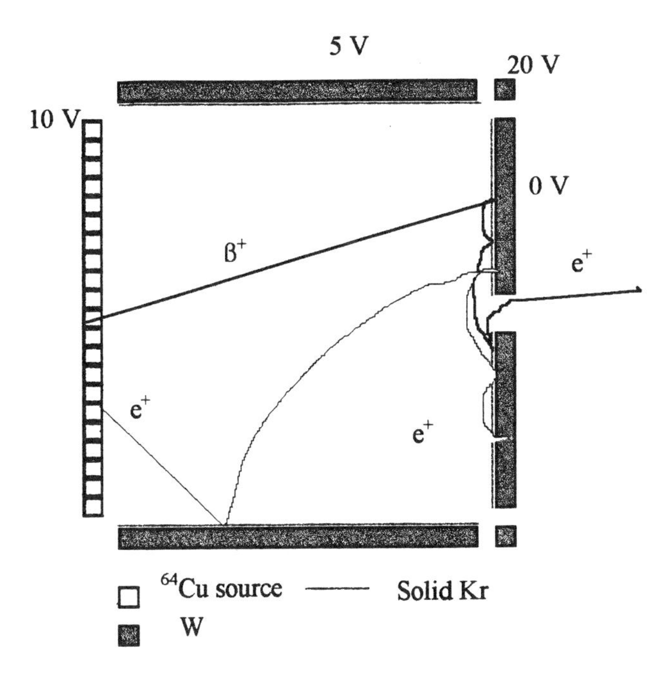

FIGURE 2. This figure illustrates our combined moderation and remoderation procedure for forming an intense e^+ beam with diameter of about 1 cm using a large area ^{64}Cu source. Note that the entire unit is cooled to 15 K and all the surfaces are coated with solid Kr that acts as a moderator as well as a remoderator.

The low energy e^+'s are emitted from the solid Kr with a kinetic energy of about a few eV (22) and when they interact with the solid Kr on the other faces of the box there is a high probability of the e^+ being reemitted into the vacuum. It has been shown that the reemission probability is at least 90% (23). If no electric fields exist inside the box then the probability for an e^+ to exit the box would be about 10%. However, by applying potentials to the various faces of the box as shown in Fig. 2 electric field components pointing toward the beam axes are created whereby the exit probability is increased to about 50%. The potentials shown in Fig. 2 may not be the optimum ones but they are of the right magnitude.

A further advantage of the arrangement of Fig. 2 is that none of the β^+ particles can escape the box without having first interacted with the solid Kr moderator. This effect should result in a higher primary moderation efficiency (19).

The further transport of the e^+ beam to the target chamber is accomplished by using standard positron transport optics (24).

Conclusion

We have described a method of producing an intense low energy e^+ beam using a large area ^{64}Cu source. With an activation time of the ^{64}Cu much longer than the mentioned 6 hours an e^+ intensity approaching the 10^9 e^+/sec level becomes possible.

ACKNOWLEDGEMENT

This work was supported by the State of Texas, Advanced Research Program.

REFERENCES

1. Gramsch E., Throwe J. and Lynn K.G., Appl. Phys. Lett. **51**, 492 (1987).
2. Gullikson E.M. and Mills Jr. A.P.. Phys. Rev. Lett **57**, 376 (1986).
3. Lynn K.G., private communication.
4. Chen X.M., Canter K.F., Duke C.B., Paton A., Lessor D.L. and Ford W.K., Phys. Rev. B **48**, 2400 (1993).
5. Brandes G.R. Canter K.F., Mills Jr. A.P., Phys. Rev. Lett. **61**, 492 (1988) and Van House J. and Rich A., Phys. Rev. Lett. **61**, 488 (1988).
6. Chen D.M., Berko S., Canter K.F., Lynn K.G. and West R.N., Phys. Rev. Lett. **58**, 921 (1987).
7. See Hyperfine Interaction **89**, (1994).
8. Haarsma L.H., Abdullah K. and Gabrielse. Phys. Rev. Lett. **75**, 806 (1995).
9. Platzman P.M. and Mills Jr. A.P., Phys. Rev. B **49**, 454 (1994).
10. Mills Jr, A.P., Science **218**, 335 (1982).
11. Gerola D., Waeber W.B. and Shi M., Nucl. Instr. Meth. A **364**, 33 (1995).
12. Howell R,H., Alvarez R.A. and Stanel M., Appl. Phys. Lett. **40**, 751 (1982).
13. Lynn K.G., private communication.
14. Triftshauser G., Kogel G. and Triftshauser W., Matr. Sci. Forum **175-178**, 221 (1995).
15. Xie R., Petkov M., Becker D., Canter K., Jacobsen F.M., Lynn K.G., Mills R. and Roellig L.O., Nucl. Instr. Meth. B **93**, 98 (1994).
16. Howell R.H., private communication.
17. "Intense Positron Beams", Eds. Ottowitte E.H.and Kells W.P. (World Scientific, Singapore, 1988) and Appl. Surf. Sci **85**, 87-182 (1995).
18. Linderoth S., Hansen H.E., Nielsen B., and Petersen K., Appl. Phys. A **33**, 25 (1988).
19. Khatri R., Charlton M., Sferlazzo P., Lynn K.G., Mills Jr. A.P. and Roellig L.O., Appl. Phys. Lett. **57**, 2374 (1990).
20. Mills Jr. A.P., Appl. Phys. **23**, 189 (1980).
21. Chen D.M., Lynn K.G., Pareja R. and Nielsen B., Phys. Rev. B **34**, 4123 and Jacobsen F.M., Charlton N., Chevallier J Deutch B.I. and Poulsen M.R., J. Appl. Phys. **67**,. 575 (1990).
22. Vasimathi D., Amarendra G., Canter K.F. and Mills Jr. A.P., Appl. Surf. Sci. **85**, 154 (1995).
23. Petkov M., private communication.
24. Canter K.F. *Positron Studies of Solids, Surfaces and Atoms*, Singapore, World Scientific, 1986, 102.

POSITRON EMITTERS FOR IN VIVO PLANT STUDIES

Y. Fares, J. D. Goeschl, C. E. Magnuson, C. J. McKinney, R. L. Musser and B. R. STRAIN

Phytotron and Botany Department, Duke University, Durham, NC 27706, USA

The use of short-lived positron emitter isotopes in studying the dynamics of biological systems provides an indepth understanding of the regulating functions of the system, that is otherwise unattinable. When we coupled such studies with tracer kinetics models, and a system approach of data analysis, in vivo simaltaneous processes and their interactions are understood. The techniques applied, results of their applications and system analysis of data are reported.

INTRODUCTION

In plants, questions of whether photosynthesis or translocation control carbon dioxide partioning and hence productivity, is of great agronomic importance because many efforts are being directed at selecting plant varieties with high rates of photosynthesis via genetic engineering and / or selective breeding. Effects of environmental factors such as water stress, elevated CO_2 or temperature on carbon assimilation and allocation in plants, and the interaction of these processes are not well understood and cannot be predicted with any degree of confidence even though they have been studied extensively (1).

Absorption o f nitrate by roots provide the predominant source of N_2 for the growth and yield of most crop species, yet the internal factors regulating its uptake from the soil solution, and initial stages in its subsequent metabolism are poorly understood (2).

These deficiencies in our understanding of the dynamics of the growth regulating factors result in part from the limitation in the experimental techniques and the lack of dynamic models that describe these processes (3). In view of the above discussion, we have, in the last 20 years adopted the following philosophy in our program for in vivo plant research:
i) developed mathematical models to describe the dynamic behavior of the intact biological system. ii) used systems approach in the analysis of the data of the modeled dynamic behavior, iii) used short-lived positron emitters analogues of the natural constituents of biological tissue, namely C-11 and N-13, to experimentally validate the mathematical models.

SYSTEMS APPROACH IN THE ANALYSIS OF PLANTS BEHAVIOR

The prime importance of adobting a systems approach inthe behavior of plant systems is that all the factors are considered simaltaneously. Feedback mechanism, for example, is one of the most important mechanisms ooccuring in dynamic biological systems, enabling the appropriate function to be maintained. such as what regulates carbon partioning in plants; is it photosynthesis or translocation? (1 & 4).

Conceptually, in the plant carbon is considered moving through a series of compartments with characteristic kinetics. When methods of compartmental analysis are applied, time-dependent as well as steady state parameters are obtained. Parameters, such as carbon exchange rates, relative pool sizes, turnover rates, export rates and translocation speeds can be calculated and compared to those predicted by detailed mechanistic models (1,5,6).

Since most other measurements require destructive sampling at various intervals, much of the short- term interactive information is lost, and the impact of the experimental conditions and procedures on the plant's behavior may not be known. Real-time nondestructive techniques for measuring many of the dynamic parameters of carbon flow in plants are needed (2). The use of $11\text{-}CO_2$ makes these measurements, in vivo, possible, where it is unnecessary to know either the biochemical pathway or the constituents of each compartment in order to analyze the data (5,6). This approach proved extremely valuable in biomedical studies (7).

EXPERIMENTAL VALIDATION OF MODELS USING SHORT-LIVED POSITRON EMITTERS

Specifically, it was our desire to develop a method to follow the uptake, transport, and sink assimilation of CO_2 in intact plants as affected by: (a) environmental factors such as temperature, light intensity, water stress, and mineral nutrition, and (b) by biological factors such as efficiency of the transport system and sinks, and their hormonal control. Furthermore, the method(s) must produce information of sufficient detail that the

CP392, *Application of Accelerators in Research and Industry*, edited by J. L. Duggan and I. L. Morgan
AIP Press, New York © 1997

mechanistic aspects of detailed biophysical models-e.g. leaf photosynthesis, phloem translocation and assimilation - are verified.

The possible use of C-11 was first suggested in 1930 -see (3and references therein). As tracer C-11 has several advantages. First it decays by positron emission followed by positron-electron annihilation with the emission of two gamma rays at 180^0. These rays have sufficient energy (0.511 MeV) to penetrate several cm of tissue, and be detected in vivo and in time coincidence. This makes it possible to localize the source, and to reduce undesired background activity. Secondly the short half-life of C-11 (20.3min) makes it possible to perform several experiments on the same set of plants under the same environmental conditions. Thirdly the half-life is comparable with the turnover time of the photosynthetic pool and velocity of transport, making possible dynamic measurements that cannot be done with long-lived tracers. The same rational is applied to N-13, although it was first used in soil denitrification studies. The fact that nitrogen -like cabon-is a natural constituent of living tissue and the environment, ultimately compelled researchers to use its isotopes in whole plant studies (3,8).

Theory

Examination of mathematical models of carbonallocation in plants indicates that much needed information could be obtained by 11-CO_2 labeling methods (10, 11). Emphasis is placed on the significance of bringing the system to steady state conditions. Often the analytical treatment of data, i.e., mathematical techniques of tracer kinetics, is used with the implicit assumption that the steady state exists. The continuous infusion of tracer till the system is brought to steady state with respect to the tracer, proved extremely valuable in biomedical studies (7).

Figure 1 shows a special profile of the tracer input into the leaf, I (t) , which is very close to an ideal step input function (or as we call it, extended square wave, ESW) of height h, starting at $t = t_0$ and stopping at time $t = \Theta$ as depicted in Fig. 2a. If $t_0 = 0$, then

$$I(t) = h\,F(t - \Theta) \quad (1)$$

Where O is the time for tracer input termination. If we consider the leaf as a linear system of n compartments Fig. 2b , then the change in activity in the ith compartment with time is given by;

$$dX(i)/dt = f_{ii} X_i + \sum_{j \neq i} f_{ij} X_j + I_i ; \quad n=1,2...n \quad (2)$$

where I_i (t) is the step input function of 11-CO_2 into compartment i, $f_{ii} = -[f_{0i} + \sum_{j \neq i} f_{ji}]$, and $(-f_{0i} X_i)$ is the rate of loss from compartment i, e.g. via respiration , decay of C-11, etc. $f_{ij} X_j$ is the rate of transfer of material from compartment j into i and $f_{ji} X_i$ is the rate ofexchange from the ith to the jth compartment. e.g. loading of photo-synthates into the phloem (12).

At time t_0 we start introducing a tracer amount of 11-CO_2 into one or more compartments of steady state system (the leaf) continuously at a constant level, Fig. 3. We assume there is complete mixing of labeled and unlabelled material in each compartment, and this mixing is rapid in comparison to the rates of transfer of material between compartments, and that the amount of tracer added to any compartment is negligible.

Let N_i be the total amount of labeled material in the ith compartment at time (t) measured in the same units as X_i and let k_i be the conversion factor from mass units to radioactivity units, then the specific activity is $s_i = k_i N_i / X_i$. By steady state assumptions f_{ij} and f_{ji} are all constant and $I(t)\, s_i/k_i = I'(t)$ is constant. Since $N_i' = s_i X_i / k_i$ therefore the change in the amount of radioactivity in the ith pool is given by equation 3, (12)

$$dN_i/ dt = f_{ii} N_i + \sum_{j \neq i} f_{ij} N_j + I_i'(t) \quad (3)$$

The rate of change in the total activity in the leaf is then given by

$$dN_i/dt = dN_1/dt + dN_2/dt + ... + dN_n/dt \quad (4)$$

Theoretically for any system of strongly connected compartments represented by equation (4) , in which the input is a step function, its time dependent response should be made of three segments, i) the initial buildup , followed by ii) the steady state segment, then iii) the washout curve of accumulated material after isotope input termination (8,1). This is exactly what we observed experimentally as illustrated by Fig. 3.

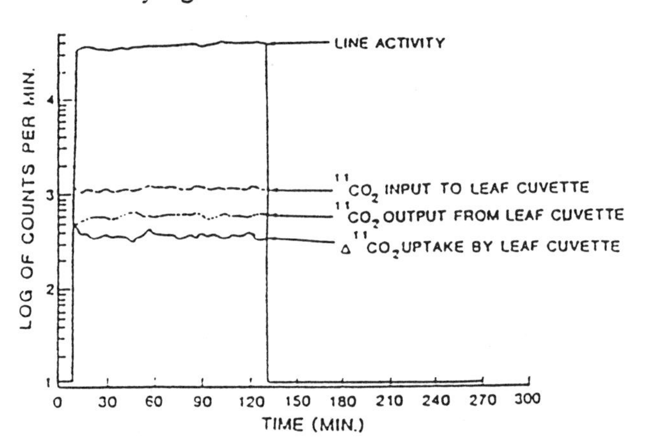

Figure 1. This illustrates the difference between the input and output activity to the leaf . Leaf uptake remains practically constant,close to a step function.

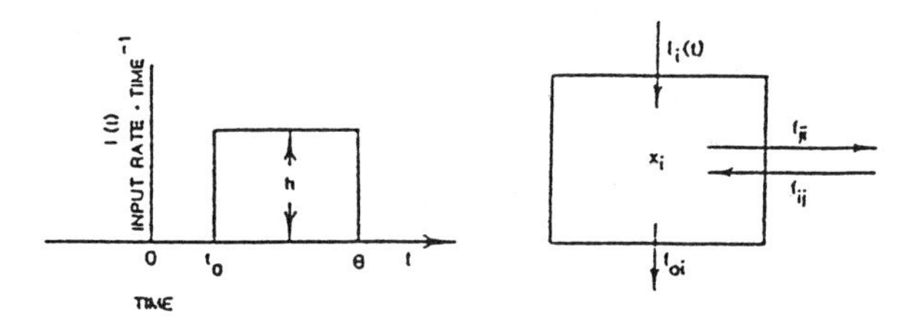

Figure 2a. Step input function I(t)of height h.
Figure 2b. N-Compartment system.

Experimental

The details of experimental set up and the innovative technique employed are described in (1, 8) and illustrated in Fig. 4 . A mixture of labeled 11-CO_2 and 12-CO_2 was diluted with free air and the final mixture contained 400 ppm cm^{-3} 12-CO_2 with specific activity of 6.7×10^9 Bq mol^{-1}. All system parameters were continuously monitored for stability and calibrated before and after each experiment. The CsF detectors had an i.d. of 15 mm and were arranged at a distance from the plant stem that gave a minimum spatial resolution of 6 mm along the plant stem.

Translocation of Photosynthates

By placing pairs of detectors alongside the petiole and stem of the plant under study as described in references (1,8),we were able to obtain information such as profile #5 in Fig.3, which represents the shape of the photosynthate efflux or export profile in the phloem as measured at the petiole. Profiles #6,#7 and #8 show what happens to this profile as the photosynthates move down the stem towards the roots. Obviously, the shape of the profile has already been set by the loading mechanism of the leaf. If we consider profile #5 in Fig.3 as input into that segment of the stem and #6 as output some distance away from it, then we have both spatial and temporal relationships of the translocating photosynthates, Fig.5. The relationship between input and output is sometimes called the transfer function h(t). This quantity is of great importance, not only because it tells us what happened to the translocates as they travel the distance between input and output, but it is a measure of the transit time or speed of translocation since it is the probability density function of transit times (5,8,12,13).

REPRESNTATIVE STUDIES WITH C-11

Experimental Tests Of The Munch-Horowitz Theory Of Phloem Transport

Because of the limited space, we will present one example of data of experiments conducted with 11-CO_2 ESW,which provide the appropriate parameters to test the theoretical predictions of Munch-Horowitz (M- H) model (14,15) (results with N-13 will be presented elsewhere).

In all the experiments that we conduct with ESW , besides all the biological and environmental factors such as light intensity, temperature, humidity, respiration etc., we measured the following parameters : average activity count rate in the line above the ESW, which corresponds to the average activity in the step function input (ACL), overall leaf input count rate (ACI), and the average leaf output activity (ACO) over the ESW period, total flow rate through the input cuvette to the leaf (TF), the input-output CO_2 mixture to the leaf, the net photosynthetic rate (CER), and specific activity of the input mixture to the leaf (SA). We also calculate the rate of accumulation of C-11 in the leaf at steady state (RE), the rate of storage in the leaf at steady state (RS), the turnover time (TE) of the export pool in min, the ratio of the rate of export of 11-C to that of assimilation (RE/RT) which we call % E, the relative export pool size, E. pool in Bq cm^{-2} and the fractional exchange constant of the export pool Lam E (Table 1).

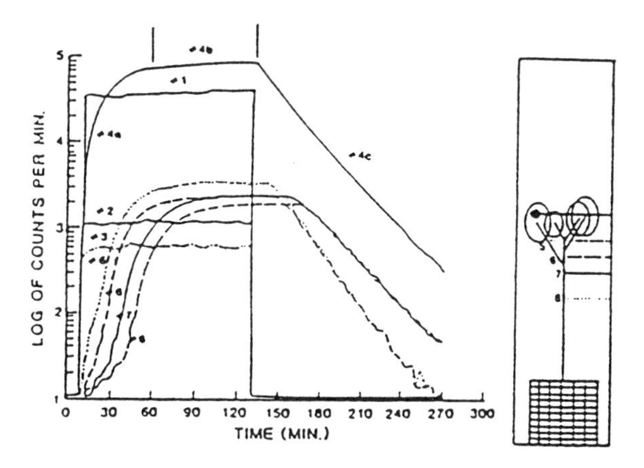

Figure 3. Real Time display of the activity in various parts of the C-11 system, i.e. ESW. Profiles #1, #2, and #3 show activities flowing in the line from the accelerator, at the entrance to the leaf cuvette, and the exit from the leaf cuvette respectively. Activity in the labeled area portion of the leaf (dark area on plant diagram, inset right) is shown by traces #4a, #4b, and #4c which indicate the buildup, steady state, and washout behavior respectively. Profiles #5, #6, #7,and #8 show activity at points along the petiole, and stem as indicated on the plant diagram.

Response To CO_2 Concentration

A. theophrasti plants were tested on the morning of subsequent days at mean CO_2 concentrations of 240, 350, and 675 cm^3 m^{-3}. The effects of the resulting increased loading rates on the speed and concentration of labeled photosynthates in the phloem, normalized to 350 cm^3m^{-3}, are represented in Figure(5) by the solid and dashed lines respectively.Increasing CO_2 from 240 to 350 cm^3m^{-3}led to a 40% increase in loading rate, a 12% increase in speed and 33% increase in relative concentration. Increasing the loading rate from 350 to 675 cm^3 cm^{-3} caused a 21% increase in loading, a 33% increase in relative concentration, but an 11% decrease in transport speeds. Several plant species were tested, and the general patterns were always maintained with some species variations.

As predicted by the M-H model, increased loading rates resulted in increased levels of C-11 activity (proportional to concentration) of translocated solutes in the phloem of all specimens tested, e.g. cotton, Fig 6 . Increased loading also resulted in increased transport speed, except in D. decumbens where transport speed decreased and A. theophrasti where speed increased, then decreased as loading rates increased. These unusual patterns are also consistent with predictions of the model if one assumes unloading to be limited by enzyme-like kinetics (14,15).

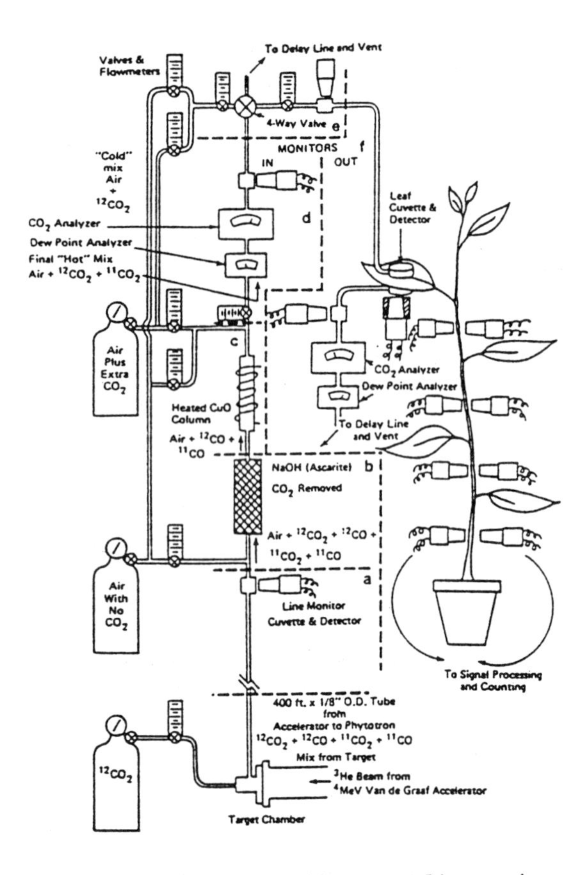

Figure 4. The experimental layout. (a) Line monitor, (b) Removal of target material, (c) Oxidation of 11-CO, (d) In line monitor, CO_2 analyzer and dew point analyzer, (e) Final monitor, (f) Out monitor, leaf cuvette and plant and detector assembly in the controlled environment chamber (CEC).

Transport speed and concentration were affected equally by diurnally increased loading rates in A.theophrasti while speed was increased proportionately more than concentration in I. batata, E. crus-galli and G. hirstum. Based on the predictions of M-H theory(16) these responses are suggestive that the latter three sets of plants were sink-limited than A. theophrasti plants Figure (6. a &b).

Conversely, concentration was increased substantially more than speed in P. plicatum, which has very sluggish growth, especially root growth, in comparison to other bunch grass. Our initial interpretation of transport responses in P. plicatum is that they are consistent with those predicted by linear unloading model with low unloading conductance (Table I).

DISCUSSIONS AND CONCLUSION

Munch-Horwitz (M-H) model of phloem transport was used to predict how a hypothetical sieve tube might respond to different rates of solute loading (16). One important prediction was that increased loading rates always led to increased solute concentration along the entire axis of the sieve tube. Increased loading was also predicted to cause increased transport speed, except where hypothetical unloading mechanism had saturable kinetics. In the latter case tranport speed increased until the loading rates approached the maximum (V_{max}), then the speed decreased while concentration rose sharply. Although the mechanism(s) of unloading is presently open to question , recent experimental studies suggest a combination of linear and saturable kinetics in the uptake of sucrose and other assimilates by sink tissue, see (14,15,16 & references therein). The relative contribution of each mechanism can vary. Thus, one may expect any one of the patterns predicted by the model depending on the species and physiological condition of the test plants. In our studies, we observed instances where loading rates were altered by endogenous, physiological changes, e.g. diurnal, or by environmental factors, e.g. CO_2 concentration. We measured the speed of transport which is proportional to translocate concentrations in the phloem at high and low loading rates in six species of plants. In most cases speed also increased, however, in two cases speed was lower and tracer activity was much higher at the higher loading rate. All the responses are consistent with the M-H theory of phloem loading when using the unloading equation of Goeschl & Magnuson (16),e.g. the latter two cases are consistent with the assumption that the unloading rate was limited by a process with saturable kinetics, i.e. enzyme-like.

The non-destructive application of short-lived C-11(and N-13) in stydying the dynamics of the intact plant , when coupled with mathematical models describing the plant's behavior and systems approach of data analysis, has enabled us to better understand the interactions of plant functions under various environmental and physiological factors. This type of information, i.e. the measurement of parameters of the dynamics of a plant are of extreme agro-economical and scientific value. A host of ecologically, agriculturally and genetically important questions can be answered using this technique.

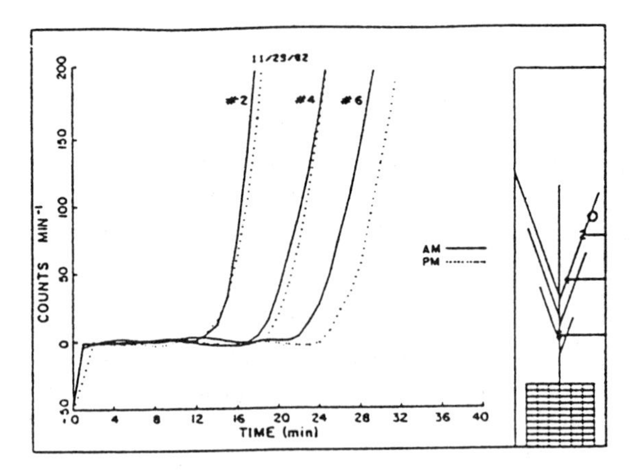

Figure 5. Tracer front patterns at three selected positions (#2, #4, #6) on the same E. crus-galli plant (shown on plant diagram, inset right). These times and known distances were used to calculate the mean speed of transport.

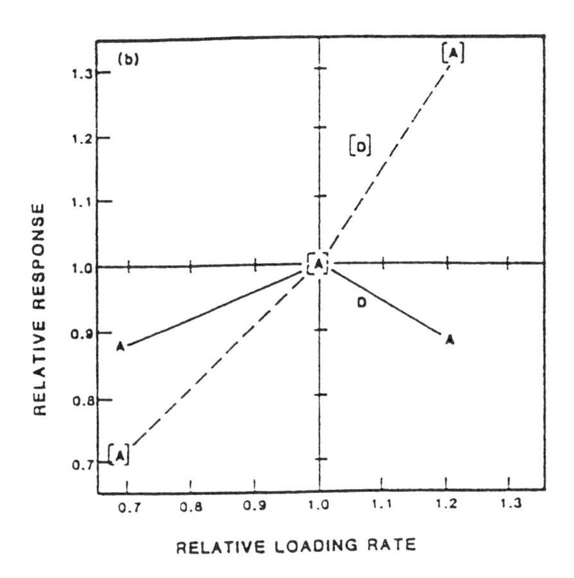

Figure 6 A. Effects of different loading rates respectively on relative C-11activity and speed in D. decumbeus ([D] & D) resulting from diurnal changes, and A. theophrasti (solid and dashed lines) resulting from changes in the ambient CO_2 concentrations.

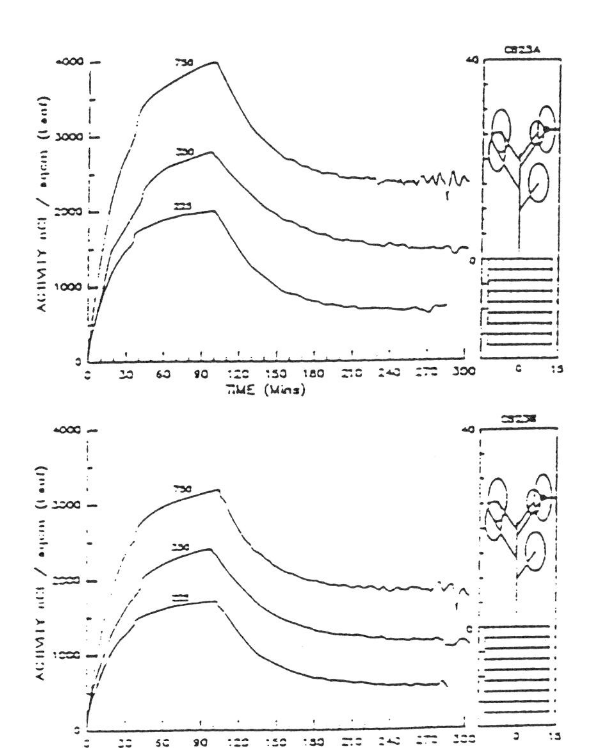

Figure 6 B. Real-time C-11 tracer kinetic data in the leaves of two cotton plants (A & B) during and following a 100 min ESW labeling with $11\text{-}CO_2/CO_2$ at three different concentrations of 225, 350, and 750 uL L^{-1}. The labeling was done at the same time of the day on three consecutive days.

TABLE I.

PLANT	LOADING RATE nmol cm^{-2} s^{-1}		STORAGE nmol cm^{-2} s^{-1}		CER nmol cm^{-2} s^{-1}		STOM.RES. s cm^{-1}	
	High	Low	HLR	LLR	HRL	LLR	(HLR	LLR)
Abutilon								
(AM/PM)	1.32	1.14	.47	.63	1.79	1.77	1.41	1.55
(675/350)	1.59	1.32	1.03	.47	2.62	1.79	1.89	1.41
Digitaria	1.46	1.38	.60	.62	2.06	2.00	2.85	2.77
Echinochloa	2.35	1.97	.68	.86	3.03	2.83	1.13	1.14
Gossipium	1.30	1.09	.56	.72	1.86	1.81	1.28	1.47
Impomoea	0.91	0.70	.53	.68	1.45	1.38	0.75	0.69
Paspalum	1.38	1.00	.23	.59	1.61	1.59	1.8	1.8

Table I. Physiological parameters at theHigh Loading Rates (HLR) compared to those corresponding to Low Loading Rate.

ACKNOWLEDGMENT

This work was supported by U.S. Department of Energy, Office of Carbon Dioxide and Climate Research (DOE101-81ER60012), and DOE, SBIR # $DEACO_2$-86ER80379. We also acknowldge NSF Biological Research Resources program (DEB 80-21312) for support of Duke University Phytotron.

REFERENCES

1. Neals, T. F., Incol, L. D., *Bot. Rev.* **10**, 107-120 (1968).
2. Wray, J. L. *Plant, Cell and Environment* **11**, 369- 376 (1988).
3. Fares, Y., et al., *International Journal of Applied Radiation and Isotopes* **29**, 431-441 (1978).
4. Gifford, R. M., and Evans, L. T., *Ann. Rev. Plant Physiol.* **30**, 485-495 *(1981)*.
5. Moorby, J. and Jerman, P. D., *Planta (Berlin)* **122**, 155-160 (1975).
6. Geiger, D. R., *Botanical Gazzet* **140**, 241-248 (1979).
7. Yamamoto, Y. L. et al., Phelps, M. E. and Heiss, W. D., eds., *Positron Emission Tomography of The Brain*, (Springer,Berlin, Heidelberg,New York,1983) 78-86.
8. Magnuson, C. E. et al., *Radiat Environ Biophys* **21**, 51-56 (1982).
9. Hole, J. D., Emran, A. M., Fares, Y., and Drew, M. C., *Plant Physiol.* **93**, 642-647 (1990).
10. Thornley, J. H. M., *Models in Plant Physiology*. Academic Press, New York (1976).
11. DEMichele, D. W., Sharpe, P. J. H. and Goeschl, J. D., *Towards The Engineering of Photosynthesis. CRC Critical Revies in Bioengineering*, 1979.
12. Jacquez, J. A., *Compartmental Analysis in Biology and Medicine*. An Arbor, The University of Michigan Press, 1985.
13. Fares, Y. et al., *J. Radioana. Nucl. Chem.* **24**, vol **1**,105-122 *(1988)*.
14. Magnuson, C. E., Goeschl, J. D. and Fares, Y., *Plant, Cell and Environment* **9**, 103-109(1986).
15. Goeschl, J. D. and Magnuson, C. E., *Plant, Cell and Environment* **9**, 95--102 (1986).
16. Goeschl, J. D., Magnuson, C. E., DE Michele, D. W. and Sharpe, P. J. H. *Pl. Physiol.* **58**, 556-562 (1976).

---

* Send correspondence to: Y. Fares, 1118 Neal Pickett, College Station, Tx 77840 U.S.A.

THE HIDDEN CROSSING THEORY APPLIED TO POSITRONIUM FORMATION

S. J. Ward,[†] J. H. Macek[*] and S. Yu. Ovchinnikov[*]

[†]*Department of Physics, University of North Texas, Denton, Texas 76203*
[*]*Department of Physics and Astronomy, University of Tennessee, Knoxville, Tennessee 37996-1501 and Oak Ridge National Laboratory, Post Office Box 2008, Oak Ridge, Tennessee 37831*

We have applied the hidden crossing theory to compute S- and P-wave positronium formation cross sections for positron-hydrogen collisions in the Ore gap. This calculation, which is the first application of the hidden crossing theory to positron collisions, provides a physical explanation of why the S-wave cross section is so small.

I. INTRODUCTION

The recent development in the theoretical treatment of positron collisions has been promoted by the experimental advances made possible by the development of slow monoenergetic positron beams of increasingly high intensity. In 1992 the Bielefeld group reported the first measurements of the cross section for positronium (Ps) formation in positron-hydrogen collisions (1,2). In 1996 the Wayne-State group also measured this cross section using a different experimental technique (3). Further experimental investigation of Ps-formation in positron-hydrogen collisions is underway by the UCL group (4). Ps-formation in positron-hydrogen collisions is one of the simplest fundamental three-body processes. This process is of interest in astrophysics because 511keV gamma rays, arising from positron annihilation, have been observed from solar flares and from the center of the galaxy (5,6). It has been known for some time that the S-wave partial contribution to the Ps-formation cross section for positron-hydrogen collisions in the Ore gap region is very small (7-9). (The Ore gap is the energy region between the Ps-formation threshold and the first excitation level of the target atom.) However, the physical explanation of why the S-wave cross section is so small was not known. An important virtue of the hidden crossing theory is that it is well adapted to elucidating underlying physical pictures. The hidden crossing theory has been applied primarily to inelastic processes in ion-atom collisions and successfully describes these processes (10-12). Using the hyperspherical representation, the hidden crossing theory has very recently been applied to, and successfully describes, electron-impact ionization of hydrogen (12-14). We have applied the hidden crossing theory to Ps-formation in positron-hydrogen collisions in the Ore gap, and computed the S- and P-wave cross sections. This calculation, which is the first application of the hidden crossing theory to positron collisions, provides a physical explanation of why the S-wave cross section is so small. We compare the hidden crossing cross section to accurate Kohn variational (7-9,15,16), hyperspherical (17, 18) and close coupling cross sections (19) to determine the reliability of the theory in describing Ps-formation.

Section II outlines the hidden crossing theory, section III presents the results and discussion, and section VI provides a summary. Atomic units are used throughout unless explicitly stated.

II. THEORY

The hidden crossing theory has it origins in the work of Landau (20) on transitions when the motion is quasiclassical. It has very recently been shown to be the lowest level of approximation to a new theory (12). In this new theory the exact wave function of three charged particles is represented by a sum over the angle-Sturmian functions and an integral over the index of Bessel functions. Replacing the sum of angle-Sturmian functions by a single angle-Sturmian function and evaluating the approximate integral for large hyper-radius R leads to the hidden crossing theory. The advantage of this formation of the hidden crossing theory (12), compared to the conventional formulation of the hidden crossing theory for ion-atom collisions (11), is that it does not employ a semiclassical approximation. In fact it provides the fundamental basis for extending the hidden crossing theory to electron impact and to the correlated motion of three charged particles of arbitrary mass and charge. The key quantity in the hidden crossing theory is a function $\varepsilon(R)$. The importance of this function was recognized by Demkov (21) who remarked that adiabatic eigenvalues $\varepsilon_\mu(R)$ correspond to different branches of the same function $\varepsilon(R)$ which is single valued on a multisheeted Riemann surface. Surfaces corresponding to different eigenvalues are connected at branch points. In ion-atom collisions it is customary to take the reac-

CP392, *Application of Accelerators in Research and Industry,* edited by J. L. Duggan and I. L. Morgan
AIP Press, New York © 1997

tion coordinate R as the internuclear distance. However, to describe the correlated motion of three charged particles of arbitrary mass and charge it is appropriate to take R as the hyper-radius (12). The branch points on the Riemann surface are square root branch points. This means that the appropriate Riemann surface can be constructed by plotting $\Re[\varepsilon(R)]$ versus R. Surfaces corresponding to different sheets are joined at square root branch points and associated branch cuts. The computation of $\varepsilon(R)$ for complex R is the first part of the hidden crossing calculation.

For Ps-formation in positron-hydrogen collisions it is necessary to use the adiabatic approximation in the hyperspherical representation. The coupled partial differential eigenvalue equations ($hf = (2\varepsilon(R)R^2 + 1/4)f$) in the hyperspherical representation are derived by Zhou and Lin (17) and also given by Passovets et. al. (14). The quantities f and h are given by Eqs. (4) and (5) of Passovets et. al. (14). The eigenvalue equations are written in terms of the hyperangle $\alpha (= \tan^{-1} r_2/r_1)$ and the angle $\theta_{12} (= \cos^{-1} \hat{\mathbf{r}}_1 \cdot \hat{\mathbf{r}}_2)$, where r_1 (r_2) is the distance of the positron (electron) with respect to the proton which we take to be infinitely massive. We solved these eigenvalue equations to compute $\varepsilon(R)$ as a function of R using the basis-spline method and codes developed by Bottcher and co-workers (22,23). We employed fifth-order polynomials for each interval in the range of the coordinate α and the coordinate θ_{12}. We used a non-uniform distribution of collocation points that we chose for the scaled potential $C(\alpha, \theta_{12})$ (14,17). For the S-wave cross section we used 32 collocation points for the α coordinate and 30 for the θ_{12} coordinate whereas for the P-wave cross section we used 26 collocation points for both the α and θ_{12} coordinates. For the integration in the complex plane we used a mesh spacing of 0.025 for the S-wave and of 0.05 for the P-wave.

The second part of the hidden crossing calculation for Ps-formation in the Ore gap is to compute the S-matrix element S_{12}. The expression for this S-matrix in terms of the elements of the Jost matrix J is (12)

$$S_{12} = \sum_{a=1}^{2} [(J^-)^{-1}]_{1,a} \, J^+_{a,2} \,. \tag{1}$$

The term with $a = 1$ in Eq. (1) is given by integrating inward from a point R_i on the first sheet of the Riemann surface (the $e^+ - H(1s)$ sheet) to the classical turning point R_1^t using the negative branch of the wave vector $K_1(R)$, then outward using the positive branch of $K_1(R)$, going clockwise around the branch point R_b to get to a point R_f on the second sheet (the $p + Ps(1s)$ sheet). The term with $a = 2$ corresponds to integrating inward from R_i using the negative branch of K_1, going counterclockwise around the branch point to the turning point of the second sheet R_2^t, and outward using the positive branch of K_2 to the point R_f. The Ps-formation matrix element S_{12} is the *coherent* sum of the contribution corresponding to the two paths. The wave vectors are defined according to

$$K_{1,2}^2 = K^2 - 2\varepsilon_{1,2}(R) - \frac{1/4}{R^2} \tag{2}$$

where K^2 is twice the total energy E of the $e^+ - H$ system (12).

From Eq. (1), $|S_{12}|^2$ can be expressed in the form

$$|S_{12}|^2 = 4P \sin^2 \Delta_{12} \,. \tag{3}$$

where P is the single-passage probability given by

$$P = \exp\left[-2\left|\Im \int_c K(R)\, dR\right|\right] \tag{4}$$

and Δ_{12} is the phase given by

$$\Delta_{12} = \left|\Re \int_c K(R)\, dR\right| . \tag{5}$$

The contour c is from R_1^t, around R_b, to R_2^t. As is standard procedure in computing $|S_{12}|^2$, we multiplied the *r.h.s.* of Eq. (3) by $(1-P)$. The $(1-P)$ term comes from the unitarity of the Jost matrix. Its justification can be found in a number of places, see for instance Demkov and Osherov (24) and Nikitin and Umanskii (25). Finally, the partial-wave (L) Ps-formation cross section in the Ore gap is obtained using

$$\sigma_{12}^{(L)} = \frac{(2L+1)}{(2E+1)} |S_{12}|^2 \quad (\pi a_o^2). \tag{6}$$

III. RESULTS AND DISCUSSION

1. S-wave

The phase Δ_{12}, defined by Eq. (5), for S-wave scattering is close to π ($\Delta_{12} \sim 3.2$) which means that the two paths that lead to Ps-formation (Paths $a = 1$ and 2 of Eq. (1)) interfere almost completely destructively. Hence, the S-wave cross section is very small and highly sensitive to the accuracy with which the phase is determined. The S-wave cross section for Ps-formation within the Ore gap is shown in table 1 and compared with Kohn variational (7-9), hyperspherical (17,18) and 18-state close coupling (cc) (19) results. The Kohn variational results agree with the exact results to within 10% (8) and are considered to be the most accurate in this region. In table 1 we also show the cross section computed by adding an ad-hoc factor $\frac{1/4}{R^2}$ to the *r.h.s.* of Eq. (2). (In the table we refer to this calculation as modified hidden crossing.) Even though this factor is ad-hoc, it shows the factor required to obtain a cross section in reasonable agreement with the Kohn variational results. It also shows the extreme sensitivity of the S-wave cross section to a small change in the phase—a change of 2.7–2.8% in the phase causes the cross section to increase by a factor 13–14. We believe that the S-wave results have converged to within 2% with respect to the number of

TABLE 1. S-wave Ps-formation cross section in the Ore gap (πa_o^2)

k	Hidden Crossing	Modified Hidden Crossing	Variational (7-9)	Hyperspherical (17,18)	18-state cc (19)
0.71	0.00031	0.0042	0.0041	0.00397	0.0033
0.75	0.00041	0.0055	0.0044	0.00427	0.0030
0.80	0.00045	0.0062	0.0049	0.00483	0.0041
0.85	0.00047	0.0066	0.0058	0.00557	0.0047

TABLE 2. P-wave Ps-formation cross section in the Ore gap (πa_o^2)

k	Hidden Crossing	Variational (15,16)	18-state cc (19)
0.75	0.364	0.365	0.376
0.80	0.548	0.482	0.485
0.85	0.693	0.561	0.573

collocation points, and to 9% with respect to the number of integration points. Interestingly, for $k = 0.71$, 0.75 and 0.80 *a.u.*, S-wave Ps-formation occurs via tunneling—the value of R_2^t is larger than $\Re(R_b)$.

1. P-wave

Preliminary P-wave Ps-formation cross sections are shown in table 2 and compared with Kohn variational (15,16), and close coupling (cc) (19) results. Our results agree with the Kohn variational results within 0.3, 14 and 24% for $k = 0.75, 0.80$ and 0.85 *a.u.*, respectively. As for the S-wave case, we also computed this cross section by adding the ad-hoc factor $\frac{1/4}{R^2}$ to the *r.h.s.* of Eq. (2). However, this factor decreases the P-wave cross section by less than 5%. We believe that our preliminary P-wave results have converged to within 3% with respect to the number of collocation points, and to within 0.4% with respect to the number of integration points. For P-wave scattering, Ps-formation occurs via tunneling for the entire Ore gap.

IV. SUMMARY

Using the hidden crossing theory we have computed the S- and P-wave Ps-formation cross sections for positron-hydrogen collisions in the Ore gap.
These calculations are the first application of the hidden crossing theory to positron collisions. The hidden crossing theory has provided a physical explanation of why the S-wave cross section is so small; namely, the two paths leading to Ps-formation destructively interfere. The P-wave cross section is in reasonable agreement with the accurate Kohn variational results.

ACKNOWLEDGMENTS

J. H. M. gratefully acknowledges support by the National Science Foundation under Grant No. PHY-9222489. S. J. W. gratefully acknowledges prior support by the National Science Foundation under Grant No. PHY-9213900.

REFERENCES

1. Sperber, W., Becker, D., Lynn, K. G., Raith, W., Schwab, A., Sinapius., G., Spicher, G., and Weber, M. *Phys. Rev. Lett.* **68**, 3690-3693 (1992).
2. Weber, M., Holmann, A., Raith, W., Sperber, W., Jacobsen., F., and Lynn, K. G., *Hyperfine Interact.* **89**, 221-242 (1994).
3. Zhou., S., Li., H., Kauppila., W. E., Kwan, C. K., and Stein, T. S., Accepted for publication in *Phys. Rev. A* (1996).
4. Laricchia, G., *Private Communication*, (1996).
5. Massey, H. S. W., *Can. J. Phys.* **60**, 461-470 (1982).
6. Leventhal, M., and Brown, B. L., in *Proceedings of the Third International Workshop on Positron (Electron) -Gas Scattering*, (Editors: Kauppila, W. E., Stein, T. S., and Wadehra), Sinapore: World Scientific, 1986, pp. 140-151.
7. Humberston, J. W., *Can. J. Phys.* **60**, 591-596 (1982).
8. Humberston, J. W., *J. Phys. B* **17**, 2353-2361 (1984).
9. Humberston, J. W., *Adv. At. Mol. Phys.* **22**, 1-36 (1986).
10. Ovchinnikov, S. Yu., and Solov'ev, E. A., *Comments At. Mol Phys.* **22** 69-85 (1988).
11. Solov'ev, E. A., *Usp. Fiz. Nauk* **157** 437-476 (1989) [*Sov. Phys. Usp.* **32** 228-250 (1989)].
12. Macek, J. H. and Ovchinnikov, S. Yu., *Phys. Rev. A* **54** 544-560 (1996).
13. Macek, J. H., and Ovchinnikov, S. Yu., *Phys. Rev. Lett.* **74** 4631-4634 (1995).
14. Passovets, S. V., Macek, J. H., and Ovchinnikov, S. Yu., in *Proceedings of Invited Talks of the XIX International Conference on the Physics of Electronic and Atomic Collisions, AIP Conference Proceedings* **360**, (edited by Dubé, L. J., Mitchell, J. B. A., McKonkey, J. W., and Brion, C. E.), Woodbury New York: AIP Press, 1995 pp. 347-355.
15. Brown, C. J., and Humberston, J. W., *J. Phys. B* **17** L423-L426 (1984).
16. Brown, C. J., and Humberston, J. W., *J. Phys. B* **18** L401-L406 (1985).

17. Zhou, Y., and Lin, C. D., *J. Phys. B* **27** 5065-5081 (1994).
18. Zhou, Y., and Lin, C. D., *J. Phys. B* **28** 4907-4925 (1995).
19. McAlinden, M. T., Kernoghan, A. A., and Walters, H. R. J., *Hyperfine Interact* **89** 161-194 (1994).
20. Landau, L. D., and Lifshitz, E. M., *Quantum Mechanics: Non-Relativistic Theory*, Oxford: Pergam Press, 1981 ch. 7, pp. 164-196.
21. Demkov, Yu.,in *Proceedings of Invited Talks of the XI International Conference on the Physics of Electronic and Atomic Collisions*, Leningrad, 1967, (edited by Flaks, I. P., and Solov'ev, E. A.) (Joint Insitute for Laboratory Astrophysics Boulder, CO, 1968), pp. 186.
22. Bottcher, C., *Adv. At. Mol. Phys.* **25** 303-322 (1989).
23. Wells, J., Oberacker, V. E., Umar, A. S., Bottcher, C., Strayer, M. R., Wu., J. S. and Plunier, G., *Phys. Rev. A* **45**, 6296-6312 (1992).
24. Demkov, Yu. N., and Osherov, V. I., *Zh. Eksp. Teor. Fiz.* **53** 1589-1599 (1967) [*Sov. Phys. JETP* **26** 916-921 (1968)].
25. Nikitin, E. E., and Umanskii, S. Yu., *Theory of Slow Atomic Collisions*, Berlin, Heidelberg, New York, Tokyo: Springer-Verlag, 1984.

MICROSTRUCTURAL CHARACTERIZATION OF EARTH AND SPACE PROCESSED POLYMERS WITH POSITRONS

Jag J. Singh
NASA Langley Research Center, Hampton, VA 23681

Positrons provide a versatile probe for monitoring microstructural features of molecular solids. In this paper, we report on positron lifetime measurements in two different types of polymers. The first group comprises polyacrylates processed on earth and in space. The second group includes fully-compatible and totally-incompatible Semi-Interpenetrating polymer networks of thermosetting and thermoplastic polyimides. On the basis of lifetime measurements, it is concluded that free volumes are a direct reflection of physical/electromagnetic properties of the host polymers.

INTRODUCTION

Positron annihilation spectroscopies have been used to infer host polymer properties for a number of years (1). Our earlier studies of epoxies (2), polyamides (3) and polyimides (4) have indicated that positronium (Ps) atoms form readily in epoxies and polyamides, but do not seem to form in polyimides studied. These conclusions were drawn as a result of the observations that lifetime spectra did not have a significant longlife component in the case of polyimides. Of course, one could argue that Ps atoms could have formed and been quenched so fast that they could not be distinguished from localized positron annihilations. However, the experimentally observed relative intensities of the short and intermediate lifetime components in polyimides argue against the formation of Ps atoms in them (5). We will review here the positron behavior in two different categories of polymers: (i). Contact Lens Polymers (Epoxies), and: (ii). Semi-Interpenetrating Polymer Networks (S-IPN) of polyimides. The contact lens samples studied included earth- and space- processed rods. The S-IPN samples included fully- compatible and totally-incompatible thermoset and thermoplastic constituents.

EXPERIMENTAL PROCEDURES

Material Preparation

Contact lens samples were synthesized from methyl methacrylate and varying amounts of cross-linking ethylene glycol di-methacrylate. The polymerization process was initiated at 40 ^{0}C by using thermal catalysts. The S-IPN samples were synthesized from three different types of polyimides. LaRC^{TM} - RP46 / LaRC^{TM} - IA polymers where the thermosetting and the thermoplastic components were fully compatible and LaRC^{TM} - RP46 / LaRC^{TM} - SI polymers where the two components were totally-incompatible. The relative concentration of the two components varied from 100 : 0 to 0 : 100 percent by weight.

Positron Lifetime Measurements

Positron lifetime measurements were made using a standard fast - fast coincidence measurement system. A 25 μC Na^{22} positron source, deposited on a 2.54 μm thick kepton foil folded on itself, was sandwiched between 2.54 cm x 2.54 cm x 0.25 cm test coupons and the spectra were accumulated for 24 hours. This counting period produced a total of more than 10^5 counts in the peak and over 10^6 in each spectrum. The time resolution of the lifetime system was 265 picoseconds. All measurements were made at room temperature and in dry samples. The lifetime spectra were analyzed with PATFIT program (6).

EXPERIMENTAL RESULTS

Contact Lens Samples

The positron lifetime spectra in these materials exhibited 3-clear components. The longest lifetime components (τ_3) are associated with the quenching of orthopositronium (O-Ps) atoms. These lifetimes are quantitatively related (7) to the dimensions of the microvoids where O-Ps atoms form and annihilate. The free volume fractions (f) have been calculated using the relation: $f = CI_3Vf_3$, where C has been determined by equating f in N4a samples with its saturation moisture

CP392, *Application of Accelerators in Research and Industry,* edited by J. L. Duggan and I. L. Morgan
AIP Press, New York © 1997

content. This is acceptable since H_2O molecules can not enter these materials chemically. The results are summarized in Tables I and II. Also listed in the table are the densities of various samples. The densities of various samples are consistent with their free volume fractions. The designations Top/Bottom refer to the stoppered end and the opposite end, respectively, of the glass tube in which the sample rods were polymerized.

TABLE I. Summary of Orthopositronium Lifetimes and Associated Parameters of the Earth - Processed Contact Lens samples. (The designations Top/Bottom refer to the stoppered end and the opposite end, respectively, of the glass tube in which the sample rods were polymerized.)

Sample Designation		Top/Bottom	τ_3(ps)/I_3(%)	f (%)	ρ (gm/cc) (± 0.0020)
N4a	{Two percent cross-	Top	2487 ± 13/ 20.8 ± 0.1	1.40 ± 0.02	1.1686
N4a	linking agent}	Bottom	2498 ± 11/ 21.3 ± 0.1	1.45 ± 0.03	1.1689
N4b	{Five percent cross-	Top	2122 ± 20/ 23.2 ± 0.3	1.17 ± 0.04	1.1836
N4b	linking agent}	Bottom	2283 ± 14/ 22.4 ± 0.2	1.30 ± 0.03	1.1796

TABLE II. Summary of Orthopositronium Lifetimes and Associated Parameters of Space-Processed Contact Lens Samples. (The designations Top/Bottom refer to the stoppered end and the opposite end, respectively, of the glass tube in which the sample rods were polymerized.)

Sample Designation		Top/Bottom	τ_3(ps)/I_3(%)	f (%)	ρ (gm/cc) (± 0.0020)
N4a	{Two percent cross-	Top	2498 ± 14/ 21.9 ± 0.2	1.49 ± 0.03	1.1685
N4a	linking agent}	Bottom	2498 ± 14/ 22.2 ± 0.2	1.51 ± 0.03	1.1686
N4b	{Five percent cross-	Top	2046 ± 14/ 23.1 ± 0.3	1.09 ± 0.03	1.1836
N4b	linking agent}	Bottom	2183 ± 13/ 22.7 ± 0.3	1.21 ± 0.03	1.1820

S-IPN Samples

The lifetime spectra in these samples were also analyzed into 3-components. However, the intensities of the third components (τ_3) in all S-IPN samples were very low (< 1 %) and these were, therefore, attributed to the background. The intermediate lifetime components (τ_2) were the strongest in all S-IPN samples. We attribute these components to the trapped positron annihilations rather than fast-quenched O-Ps atoms. Just like the O-Ps annihilation lifetimes, the trapped positron annihilation lifetimes can be quantitatively related (8) to the trap dimensions. The lifetime results, free volume fractions, densities and dielectric constants of S-IPN samples are summarized in Tables III and IV. The free volume fractions (f) have been calculated by equating f in LaRC™ - IA with its saturation moisture content. This is justifiable since hydration of this thermoplastic polyimide is miniscule. It is instructive to note distinct differences between the properties of the compatible and incompatible S-IPN samples.

DISCUSSION

Contact lens rods polymerized in microgravity environment should be homogeneous in composition and density, compared with 1-g processed materials, where the bottom samples may be more compacted than the samples from the top section of the rods. However, the data summarized in Tables I and II show no differences in the free volume and densities of the two types of contact lens samples. A closer investigation of the exact geometrical orientations of the processing chambers revealed that the polymerization cylinders in the ground laboratory and the space shuttle bay were in horizontal planes as opposed to the vertical orientations that the experimental plans called for. Under these circumstances, we should have expected what we actually observed! We do, however, find large differences in the values of free volume fractions and the densities between top and bottom sections of the 5 % cross-linked rods polymerized, both in the space and the ground laboratories, while no such differences are noted in the

TABLE III. Summary of Positron Lifetime and Associated Parameters in Fully-Compatible S-IPN Samples (8).

Sample Composition LaRC^{TM}- RP46 : LaRC^{TM} - IA	τ_2 (ps) / I_2 (%)	f (%)	ρ (gm/cc)	ε (10 GHz) (± 2%)
0 : 100	488 ± 3/80.0 ± 2.0	2.81 ± 0.08	1.3402 ± 0.0029	2.82
25 : 75	422 ± 5/79.3 ± 3.0	2.58 ± 0.12	1.3815 ± 0.0028	3.01
35 : 65	399 ± 4/78.9 ± 2.9	2.48 ± 0.10	1.3668 ± 0.0020	3.11
50 : 50	394 ± 1/79.4 ± 0.1	2.48 ± 0.09	1.3713 ± 0.0041	3.14
65 : 35	400 ± 2/76.3 ± 1.4	2.42 ± 0.07	1.3629 ± 0.0032	3.14
75 : 25	399 ± 2/79.6 ± 1.8	2.50 ± 0.07	1.3546 ± 0.0036	3.13
100 : 0	400 ± 4/81.4 ± 2.2	2.56 ± 0.08	1.3572 ± 0.0020	3.13

TABLE IV. Summary of Positron Lifetimes and Associated Parameters in Totally-Incompatible S-IPN Samples.

Sample Composition LaRC^{TM}- RP46 : LaRC^{TM} - SI	τ_2 (Ps) / I_2 (%)	f (%)	ρ(gm/cc) (± 0.0020)	ε (10 GHz) (± 2%)
0 : 100	400 ± 2/93.2 ± 0.4	2.48 ± 0.12	1.3727	3.13
25 : 75	394 ± 1/93.4 ± 0.5	2.47 ± 0.12	1.3696	3.13
35 : 65	397 ± 2/95.0 ± 1.0	2.52 ± 0.14	1.3630	3.12
50 : 50	395 ± 1/92.3 ± 0.4	2.44 ± 0.11	1.3667	3.13
65 : 35	396 ± 2/93.1± 0.8	2.47 ± 0.13	1.3620	3.13
75 : 25	398 ± 2/93.2 ± 0.3	2.48 ± 0.11	1.3693	3.13
100 : 0	396 ± 1/94.7 ± 0.6	2.51 ± 0.08	1.3789	3.12

case of 2 % cross-linked rods. This is probably due to some chemical reaction between the cross-linking agent and the silicone rubber stopper.

For the compatible LaRC^{TM} - RP46/LaRC^{TM} - IA samples, it is found that the free volume appears to go through a minimum at 50:50 composition. The densities and dielectric constants also follow a consistent trend as seen in the data summarized in Table V and illustrated in Figures 1 and 2. It is noted that both the density and the dielectric constant go through a maximum at about 50:50 composition where the electrostatic interaction between the two overlapping chains is expected to be maximum.

For the incompatible LaRC^{TM} - RP46/ LaRC^{TM} - SI samples, on the other hand, no such trends were observed. The stiffness of the LaRC^{TM} - SI chain precluded all electrostatic effects observed for the more compliant LaRC^{TM} - IA member (9).

CONCLUDING REMARKS

Positrons are ideal probes for monitoring microstructural properties of polymers. We have observed direct correlations between the atomic scale free volume

TABLE V. Comparison Between the Experimental and Computed Values of the Densities and Dielectric Constants of Fully-Compatible S - IPN Samples (9).

Sample Composition	Density (ρ), gm/cc		Dielectric Constant (ε) at 10 GHz	
LaRC™- RP46 : LaRC™- IA	Experimental	Computed (*)	Experimental	Computed (**)
0 : 100	1.3402 ± 0.0029	1.3402 ± 0.0029	2.82 ± 0.06	3.10 ± 0.03
25 : 75	1.3815 ± 0.0028	1.3558 ± 0.0031	3.01 ± 0.06	3.11 ± 0.03
35 : 65	1.3668 ± 0.0020	1.3620 ± 0.0036	3.11 ± 0.06	3.12 ± 0.03
50 : 50	1.3713 ± 0.0041	1.3714 ± 0.0045	3.14 ± 0.06	3.12 ± 0.03
65 : 35	1.3629 ± 0.0032	1.3672 ± 0.0034	3.14 ± 0.06	3.13 ± 0.03
75 : 25	1.3546 ± 0.0036	1.3643 ± 0.0026	3.13 ± 0.06	3.12 ± 0.03
100 : 0	1.3572 ± 0.0020	1.3572 ± 0.0020	3.13 ± 0.06	3.11 ± 0.03

(*) ρ (computed) = $w_1\rho_1 + w_2\rho_2 + (1/2)\alpha\beta(\rho_1+\rho_2)$, where $\alpha = 0.0168 \pm 0.0031$ and β = Molecular chain overlap parameter.
(**) ε (computed) = $\{(1-f)/(\varepsilon_R) + (f)/(\varepsilon_{Air})\}^{-1}$, where $\varepsilon_R = 3.30 \pm 0.03$.

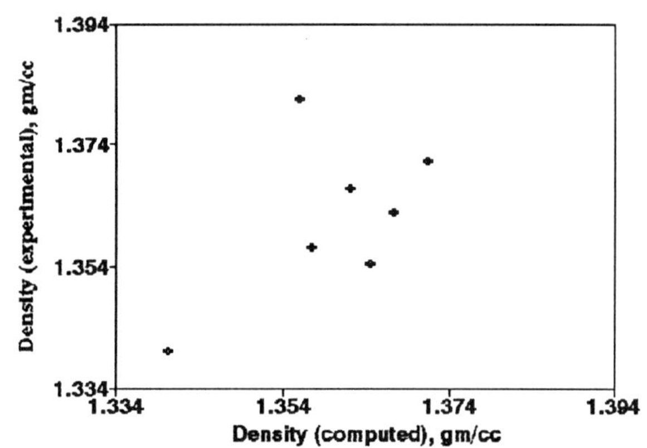

Figure 1. Comparision Between the Experimental and the Computed Values of Densities of Fully-Compatible S-IPN Samples.

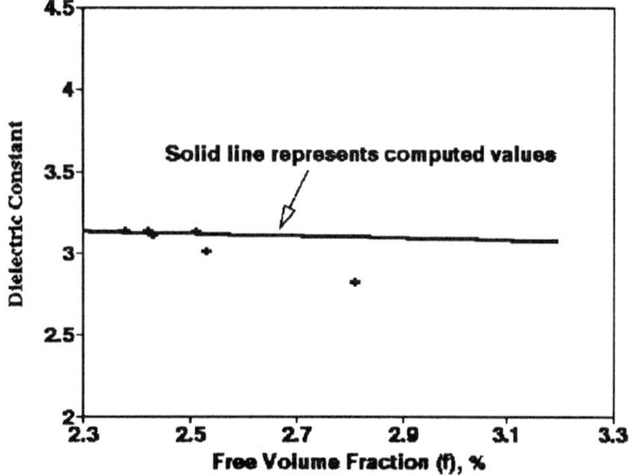

Figure 2. Dielectric Constant vs Free Volume Fraction in Fully-Compatible S-IPN Samples.

holes and macroscopic mechanical and electromagnetic properties of the polymeric materials - not only qualitatively but also quantitatively.

ACKNOWLEDGEMENTS

I would like to thank Drs. Billy T. Upchurch and Ruth H. Patter of this Center for providing the contact lens and the S-IPN samples, respectively.

REFERENCES

1. Jean, Y.C.: Characterizing Free Volumes and Holes in Polymers by Positron Annihilation Spectroscopy, in Proceedings of NATO Advanced Research Workshop on Advances with Positron Spectroscopy of Solids and Surfaces, held at Varenna, Italy on July 16-17, 1993, pages 563-580.
2. Singh, J.J.; Stoakley, D.M.; Holt, W.H.; Mock, W.M.; and Teter, J.P.: Effect of Transition Metal Ions on Positron Annihilation Characteristics in Epoxies, *Nucl. Instr. and Methods,* **26(b)**, 598(1987).
3. Singh, J.J.; St. Clair, T.L.; Holt, W.H.; and Mock, W.M.: Moisture Dependence of Positron Annihilation Spectra in Nylon-6, *Nucl. Instr. and Methods,* **221**, 427(1984).
4. Singh, J.J.; Eftekhari, A.; and St. Clair, T.L.: A Low Energy Positron Flux Generator for Microstructural Characterization of Thin Polymer Films, in AIP Conference Proceedings #303, edited by E. Ottewitte and A.H. Weiss (AIP Press, New York, 1994) pp. 516 - 525.
5. West, R.N.: Positron Studies of Condensed Matter, *Advances in Physics,* **22**, 263(1973).
6. Kirkegaard, P.; Pedersen, N.J.; and Eldrup, E.: PATFIT-88 (RISO-M-2740), 1988.
7. Nakanishi, H.; Wang, S.J.; and Jean, Y.C.: in *Positron Annihilation Studies of Fluids,* edited by S.C. Sharma (World Scientific, Singapore, 1988) p.292.
8. Deng, Q.; Sundar, C.S. and Jean, Y.C.: Pressure Dependence of Free Volume Hole Properties in an Epoxy Polymer, *J. Phys. Chem,* **96(1)**, 492(1992).
9. Singh, J.J.; Pater, R.H.; and Eftekhari,, A.: Microstructural Characterization of Semi- Interpenetrating Polymer Networks by Positron Lifetime Spectroscopy, NASA Technical Paper # 3617, 1996.

SECTION IV

CLUSTERS, FULLERENES, BIOMOLECULES

PROSPECTS OF MATERIALS PROCESSING BY GAS CLUSTER ION BEAMS

Isao YAMADA

Ion Beam Engineering Experimental Laboratory, Kyoto University
Sakyo, Kyoto 606-01, Japan

A number of new technological advancements are strongly related to ion beam equipment development. The author has proposed gas cluster ion beam processing which can produce unusual very low energy and extremely high density bombardment effects associated with cluster ion - solid interactions. This paper reviews status of cluster ion beam processing technologies such as shallow ion implantation, high yield sputtering and smoothing and low temperature thin film formation.

INTRODUCTION

Ion beam processes had already been suggested in 1885 when J.J. Thomson discovered an ion beam by repeating a Canal Ray experiment performed originally by Goldstein [1]. The construction and operation of present ion beam equipment is essentially similar to that of Thomson. Active research and development of ion beam processing was started in the early 1960s. Over recent decades ion implantation has been aggressively adopted in production processes for leading edge integrated circuit fabrication. In recent implantation technology for very shallow junction formation, ion implantation at an energy lower than 1 keV is proposed. Very high energy implantation from 200 keV to 3 MeV is also starting to be used for device production [2]. Large area ion implantation has become of recent interest relative to an expanding demand to fabricate flat panel display devices. A non-mass analyzed system is being successfully applied in forming TFT elements, an unusual process for semiconductor device fabrication. This processing will be one of a large field of promising new applications in addition to LSI production.

Due to the progress of ion beam technology, remarkable applications now exist in fields other than the electronic industries, for examples, precise surface machining of very hard materials and surface modification for creation of unusual materials by ion implantation. A large scale Japanese national project has made a major contribution in this area. High current metal ion beam equipment, a high current sheet ion beam system, very low energy ion beam equipment, high brightness focused ion beam system, ion beam deposition system for high quality thin film formation over large area and very high energy ion implantation equipment have been developed [3].

In addition to traditional research and development of ion beam processing, quite new and unique procedures have been proposed. One is a technique called Plasma Immersion Ion Implantation (PIII) which was originated by Conrad of the University of Wisconsin [4]. The principle of this method is to use ions accelerated through a negative sheath formed in a plasma discharge region. This makes it possible to implant ions into a surface of complicated configuration. This process has been widely applied as a surface modification method for various large metal parts, for example, automobile crank shafts, gears etc. Semiconductor doping by plasma immersion implantation developed by N. Cheung of the University California, Berkeley [5], is also successful even though the process is performed without mass analysis. The most important breakthrough of these technologies is the implantation of ions by means of a plasma which is one of classical methods of surface processing.

The author has proposed "Gas cluster ion beam processing" which can be used to produce unusual effects associated with cluster ion-solid interactions: very low energy and extremely high density bombardment effects [6]. Since the discovery and subsequent development of ion beam processing, beams of atomic or molecular ions have been used. In cluster ion beam technology, beams consisting of a few hundreds to thousands of atoms generated from various kinds of gas materials are employed. Multi-collisions during the impact of accelerated cluster ions upon the substrate surfaces produce fundamentally low energy bombarding effects in a range of a few eV to hundreds of eV per atom at very high density. These bombarding characteristics can be applied to shallow ion implantation, high yield sputtering and smoothing, surface cleaning and low temperature thin film formation. Cluster ion beam procedures are now expanding into new industrial fields which are presently limited by available atomic and molecular ion beam processes [7,8].

GAS CLUSTER ION BEAM TECHNOLOGY

Since the discovery of ion beams more than 100 years ago, beams of atomic or molecular ions have been used for materials processing [1]. While many of the techniques for monomer ion implantation beam transport and wafer handling can be directly applied to cluster-beam implantation systems, the materials aspects of the process

CP392, *Application of Accelerators in Research and Industry*, edited by J. L. Duggan and I. L. Morgan
AIP Press, New York © 1997

are significantly different. Cluster ion beams consisting of a few hundreds to thousands of atoms have been generated from various kinds of gas materials. Multi-collisions during the impact of accelerated cluster ions upon the substrate surfaces produce fundamentally low energy bombarding effects in a range of a few eV to hundreds of eV per atom at very high density. These bombarding characteristics can be applied to shallow ion implantation, high yield sputtering and smoothing, surface cleaning and low temperature thin film formation. This concept has been nurtured for a long time by the present author who started a systematic experimental study in 1988 under funding from the ministry of education and JRDC (Japanese Research and Development Cooperation).

Cluster ions are fundamentally different from monomer ions. First, the kinetic energy of an ionized cluster is shared to each constituent atom so that very low energy processes can be achieved. Second, the cluster ions have extremely low charge to mass ratios, and third, collective motions of the cluster atoms during impact play dominant roles in the process kinetics. These characteristics lead to

Table 1 Characteristics of cluster ion beam processes

PROCESS REQUIREMENT	*CHARACTERISTICS OF CLUSTER ION BEAM RECESSING*
	IMPLANTATION
Shallow implantation	Very low energy per constituent atom (~1/N)
High dose implantation	High material transport per charge (~x N)
Low charge accumulation	Low charge per implanted atom (~1/N)
	SPUTTERING
High rate sputtering	One or two orders higher (>N)
Surface smoothing	Strong lateral sputtering
Low sputtering energy	Low energy sputtering due to multi-collision effects
	THIN FILM FORMATION
High quality film formation	Low energy bombardment 0.1 - 1eV (~1/N)
High deposition rate	Current times cluster size (~N)
Very smooth surface	Control of nucleation and growth processes
Compound formation	High reactivity

many new and unique opportunities for surface modification processing. Examples include very high yield sputtering (sputter yields more than two orders of magnitude higher than those produced by monomer ions), very shallow atomic penetration during implantation (less than 50 nm in Si at 150 keV), and strong chemical interactions at low temperatures (very thin oxide layers formed at room temperature).

In Table 1, various basic requirements for present ion beam processes, which are being expanded into areas of semiconductor devices, optical devices and other nano-scale materials modifications, and characteristics of cluster ion beam processes are listed. In Figure 1, cluster ion beam processes and their characteristics are summarized.

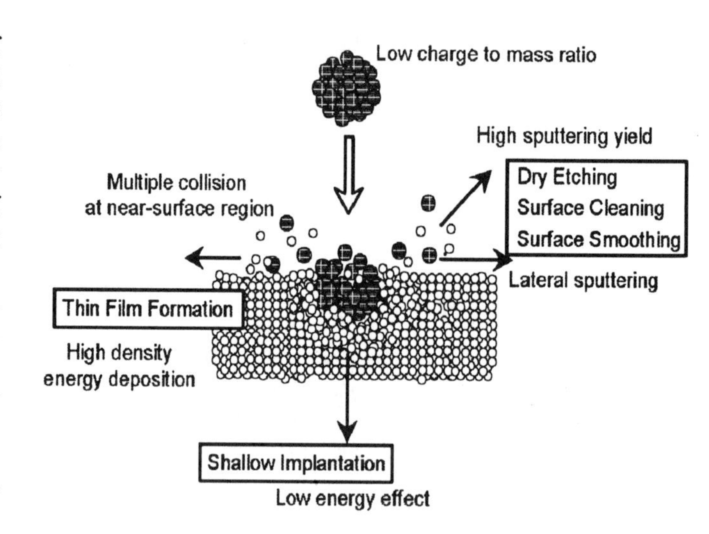

Fig. 1 Characteristics of cluster ion beam processes

Three different cluster ion beam systems which can accelerate cluster ions up to 10 kV, 30 kV and 200 kV have been constructed. The 10 kV and 30 kV gas cluster ion beam systems can be used for surface processing, mainly for sputtering, cleaning and surface modifications. The 200 kV equipment can be used for shallow ion implantation.

Figure 2 shows a schematic diagram of the 30 kV gas-cluster ion beam equipment. Cluster ion beams from gaseous materials such as Ar and CO_2, are generated by expanding the gases through a Laval nozzle into a high-vacuum chamber after collimation by a skimmer. Production of cluster beams from O_2, N_2 and SF_6 has also been studied recently.

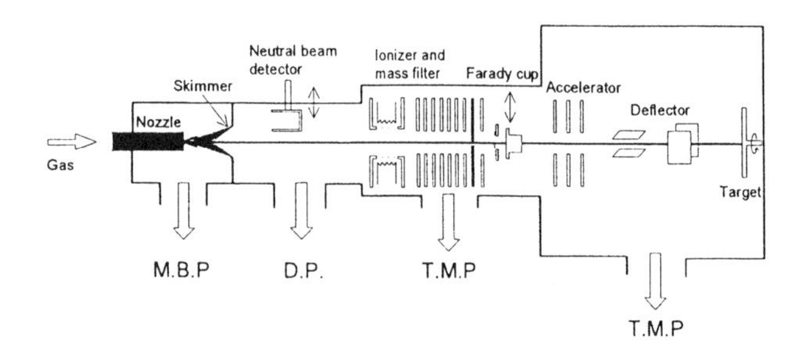

Fig. 2 Schematic illustration of 30kV Gas cluster ion beam system

Ionization is performed by electron bombardment. Mass analysis in the 30 kV gas cluster ion beam equipment is accomplished by an electrostatic retarding potential method (deceleration) based upon the different kinetic energies of clusters of different masses. Mass-selected clusters with sizes up to 5000 atoms have been

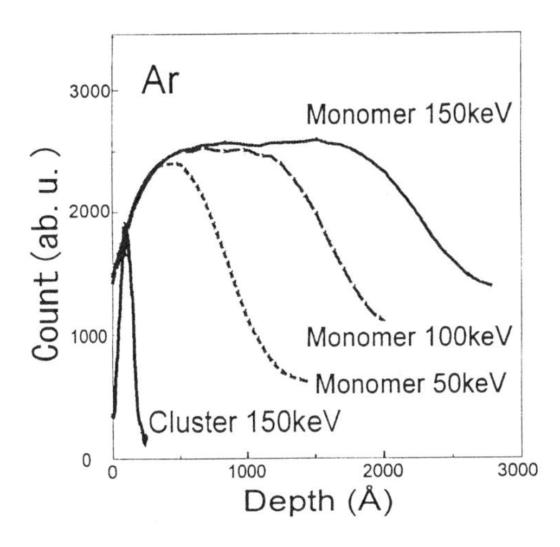

Fig. 3 Channeling spectra of ion bombarded Si(100) surfaces (average cluster size is 3000 atoms/cluster) and monomer ions accelerated at 50,100 and 150keV.

electrostatically accelerated with voltages in the range of 30 kV. Monomer ions can be effectively suppressed by applying a low extraction voltage and adjusting the electrostatic lenses. The highest cluster ion beam current obtained has been about 1 μA which corresponds to an Ar atom flux of approximately $2x10^{16}$ atoms/s/cm^2.

Shallow ion implantation

Since a few data is available for cluster ion-solid surface interactions[8], fundamental studies related to the bombardment have to be done. In order to establish a quantitative relationship between the stopping distribution of cluster atoms and the cluster size/energy characteristics, ion implantation by decaborane ($B_{10}H_{14}$) has been studied [7]. Decaborane, which represents a cluster containing 10 boron atoms, was ionized by electron bombardment and accelerated to 20 keV to impact upon a Si (100) substrate at room temperature. For comparison, atomic boron ions were also implanted into a similar Si(100) substrate at 2 keV. The range distribution of the 10 boron atom clusters implanted at 20 keV is essentially identical to the distribution from monomer ions implanted at 2keV. The results confirm that the effective implantation energy is equivalent to the value obtained by dividing the given energy by cluster size (N).

Bombarding effects have been studied by RBS and channeling methods. Figure 3 shows the channeling spectra of Si (100) surfaces bombarded with 150 keV Ar clusters and with 150, 100 and 50 keV Ar monomer ions. Doses of cluster and monomer ions were $2.5x10^{13}$ and $1x10^{15}$ ions/cm^2, respectively. The Ar cluster ion beam current was 5 nA.

By assuming that the cluster size in Figure 3 is 3000 atoms, the average energy per constituent atom of the cluster is estimated to be approximately 50 eV. The data of Figure 3 shows that the thickness of the damaged layer caused by cluster ion implantation was less than 250Å. This depth of damage is much less than the depth produced by 150 keV monomer ions, but it is greater than the predicted maximum penetration range of a 50 eV ion. In order to penetrate to a depth of 250 Å, an energy of almost 10 keV would be required. According to molecular dynamics (MD) simulations, this discrepancy can be explained by shock wave effects. Shock waves produced by cluster ion impact are capable of generating lattice damage at depths substantially greater than the range of the cluster atoms themselves.[9].

Sputtering and smoothing

Surface atom sputtering characteristics associated with cluster ion bombardment are very different from those of monomer ions [7,8]. Monomer ion bombardment produces rough surfaces when the surfaces are bombarded at normal incidence. However, it has been found that the smoothest surface resulting from cluster ion bombardment is obtained when the cluster ions are incident from the normal direction. Figure 4 shows AFM images of as-deposited and ion-bombarded vacuum-deposited Cu film surfaces. 20keV Ar cluster ions of size 3000 to a dose of 8×10^{15} ions/cm^2 and monomer Ar ions to a dose of 1.2×10^{16} ions/cm^2 were used for bombardment.

Original surface (Ra=5.8 nm)

Ar cluster ion bombardment (Ra=1.3 nm)
Dose: of 8×10^{15} ions/cm^2 at 20 keV

Fig.4 AFM images of Cu film surfaces before and after sputtering.(The scales are the same, lateral size is $1\mu m \times 1\mu m$)

The surface roughness (Ra) after cluster ion bombardment was reduced to 1.3 nm from original roughness of 5.8 nm. In the case of monomer ion bombardment, the Ra value was 4.9 nm after the dose of

1.2×10^{16} ions/cm^2. The dependence of the sputtering yields upon the atomic number of the target material has been investigated.

Figure 5 shows the dependence of sputtering yield on atomic number for bombardment by 20 keV Ar_{3000} cluster ions. The sputtering yield is much higher than that of atomic ions, by about one to two orders of magnitude. With the same equipment, Ar monomer ions have also been produced, and the same targets have been sputtered. The monomer results indicate normal dependence of the sputtering yield on the atomic number of the target materials. Figure 4 also indicates reactive sputtering yields of Si , W and Cu due to SF_6 ion beam bombardment shown in comparison with the yields due to Ar cluster beam sputtering. Sputtering yields by SF_6 are considerably higher than those by the Ar cluster ion bombardment.

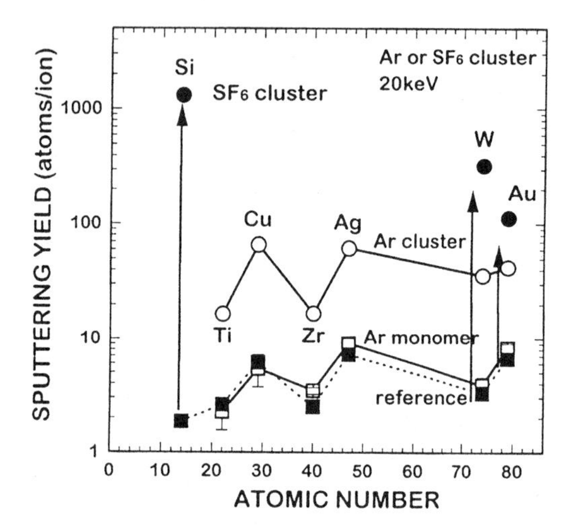

Fig. 5 Dependence upon atomic number of sputtering yields due to 20 keV-Ar cluster ions and monomer ions. Reactive sputtering yield of Si, W and Cu due to SF_6 cluster ion beam bombardment is also shown.

Thin film formation

High density bombardment with low energy cluster ions can modify thin substrate surface layers. Ti and Si substrates have been bombarded with 20 keV O_2 and CO_2 cluster ions. Stoichiometric TiO_2 and SiO_2 layers could be formed at room temperature on the substrate surfaces [7]. SiO_2 films of 11 nm thickness were formed by irradiation by 7 keV O_2 cluster ions to a dose of 5×10^{15} ions/cm^2. In the case of monomer ion oxidization, SiO_2 films thicker than 6 nm were not obtained. These results indicate that gas cluster ions have strong chemical reactivity. Fourier transform infrared FT-IR spectra of silicon surfaces irradiated with CO_2 gas cluster ions at room temperature shows that stoichiometric SiO_2 films can be formed. The results of XPS and TEM analysis of film thickness were consistent. By conventional thermal oxidation, high temperature treatment (>700°C) is necessary to obtain such a thick SiO_2 layer

MOS capacitors were used to evaluate the damage caused by gas cluster ion irradiation [10]. 18 nm thick gate oxide films thermally grown at 900 °C were directly irradiated with 5 keV- Ar cluster and monomer ions. Samples were annealed at 400 °C in N_2 for 30 minutes after the irradiation. Degradation of the I-V characteristics was observed in both oxides before annealing. However, in the case of irradiation with cluster ions, the I-V characteristics of the annealed samples coincided with those of the unirradiated samples, which meant that the damage was recovered by the annealing. On the other hand, the damage caused by monomer ion irradiation was not recovered. These results suggests that the radiation damage caused by cluster ions is much less than that by monomer ions and low damage processing should be possible.

SUMMARY

Emerging technologies of ion beam processing have been reviewed. High current metal ions, wide sheet beams, very high energy beams at very high currents, and very fine intense beams will all provide new opportunities to develop new materials. With unusual bombardment effects by cluster ion beams, new areas of materials processing such as shallow implantation, low damage and high rate sputtering, thin film formation and surface cleaning are expected to be applied in various industries.

REFERENCES

[1] I.Yamada, Electronics (Japanese, Ohm Sya Publishing Co.) Dec.(1994) and Jan., Feb., Apr. and Nov. (1995).

[2] IIT'96 Extended Abstract.

[3] I.Yamada, Mat. Res. Soc. Proc. Vol.268, (Materials Research Society, 1992) p.261.

[4] J.R. Conrad, J.Radtke, R.A. Dodd, and F.J.Worzala, J.Appl. Phys., 62 (1987)4591

[5] N.W.Cheung, Nuclear Instrument and Methods, B55 (1991) 811.

[6] I. Yamada, Proc. 14th Symp. on Ion Sources and Ion-Assisted Technology, Tokyo, 1991 (The Ion Engineering Society of Japan, Tokyo) p.227.

[7] Recent reports, I.Yamada and J.Matsuo, 1996 MRS spring Meeting April 19, 1996 San Francisco. in Print. and I.Yamada, J.Matsuo, Z.Insepov, D.Takeuchi, M.Aakizuki and N.Toyoda, J.Vac. Sci. Technol. A.14 (3) May/Jun (1996) 781.

[8] M.G.Blain, S.Della-Negra, H.Joret, Y.Le Beyec and E.A. Schweikert, Phys. Rev. Lett. 63 (1989) 1625

[9] Z.Insepov and I.Yamada, Nuclear Instrument and Methods, B99 (1995) 248.

[10] M.Akizuki, J.Matsuo, H.Harada, S.Ogasawara, A.Doi, and I.Yamada, Extended Abstracts of the 1995 International Conf. on Solid State Devices and Materials, Osaka, 1996, pp.31-33 (Japan Society of Applied Physics).

THE SPUTTERING EFFECTS OF CLUSTER ION BEAMS

N.Toyoda, J.Matsuo and I.Yamada

Ion Beam Engineering Experimental Laboratory, Kyoto University, Sakyo, Kyoto 606-01, JAPAN

It has been observed that a cluster ion which contains several thousands of atoms produces unique sputtering effects. It has a high yield and a strong smoothing effect for various materials such as CVD diamond films. In this work, the sputtering yield, surface smoothing and angular distribution of the sputtered atoms have been measured, and the different sputtering mechanisms of cluster ions have been revealed. Cu films were irradiated with Ar cluster ion beams accelerated up to 20keV, at several incident angles. From the energy and incident angle dependence of the sputtering yield, it was seen that the energy density is responsible for the sputtering. The Cu surfaces irradiated with Ar cluster ions were observed by Atomic Force Microscope. The surface roughness was smallest at normal incidence, and the surface roughness increased with the incident angle. A ripple structure was observed at incident angle of 60°. The angular distribution of the sputtered atoms displayed an under-cosine shape at normal incidence, and many sputtered atoms were distributed in a lateral direction. A single trace of an Ar cluster ion impact measured by STM confirmed this result.

INTRODUCTION

A cluster consists of two to several thousands of atoms or molecules loosely combined with each other by attractive intermolecular forces. If a cluster ion, containing several thousands of atoms, is accelerated to an energy of a few tens of keV, each constituent atom has an energy of only a few eV. It is difficult to obtain beams of such low energy monomer ions because of dispersion due to space-charge effects. In addition, a cluster ion beam can transport larger quantities of atoms than a monomer ion beam for the same ion current.

In energetic cluster-ion bombardment, several thousands of atoms collide with target atoms within a radius of a few tens of Å in a few pico-seconds. Consequently, the impact processes of cluster ions are quite different from those of monomer ions. There are multiple collisions near to the surface and a high density of energy deposition into the near surface.

We have previously reported experimental results of the unique physical sputtering effects of gas cluster ion beams, such as surface smoothing, cleaning, thin film formation at low temperature, shallow implantation and so on[1-6].

In this work, we report the energy and incident angle dependence of the sputtering yield, the surface smoothing effects with Ar cluster ion beams and the angular distribution of the sputtered atoms.

EXPERIMENTS

The experimental apparatus used in this study is shown in figure 1. This 30keV gas cluster ion beam equipment consists of four parts; source chamber, differential pumping chamber, ionizing chamber and target chamber.

Neutral clusters are generated with the cooling during an adiabatic expansion. High pressure (2000 ~ 4000 Torr) gases expand through a Laval nozzle (0.1mm diameter) into a source chamber pumped down to 0.01 Torr by a mechanical booster pump. When the cluster beam collides with the residual gas in the source chamber, a shock wave called a Mach disk appears in front of the beam and causes a decrease of the beam intensity[7]. Therefore, a skimmer (0.3mm in diameter) to collimate the cluster beam has to be placed upstream of the Mach disk.

Subsequently, the neutral cluster beam is ionized by electron bombardment and extracted with a very low extraction voltage (V_{ext}), to reduce the fraction of monomer

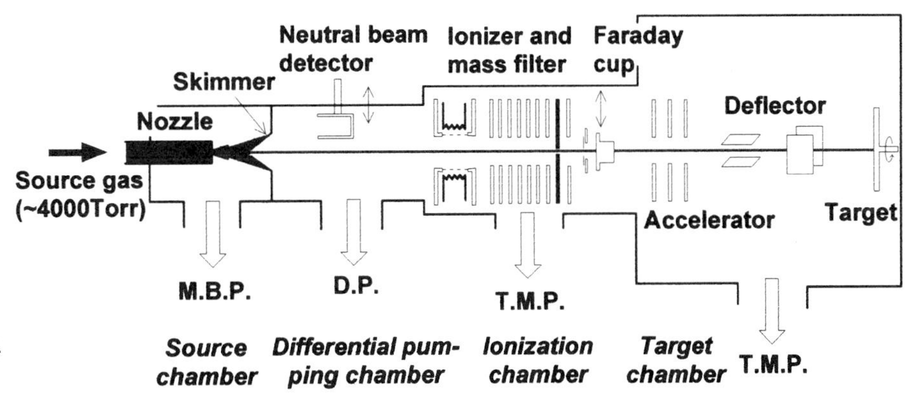

FIGURE 1. Schematic diagram of the 30keV gas cluster ion beam equipment.

CP392, *Application of Accelerators in Research and Industry*, edited by J. L. Duggan and I. L. Morgan
AIP Press, New York © 1997

ions by means of space charge effects. Mass separation of clusters is performed in the ionization chamber using an electrostatic lens system which was designed to utilize inherent chromatic aberration.

Mass selected cluster ions are accelerated up to 30keV and scanned by deflectors in the target chamber to obtain uniform irradiation. The cluster ion current density is 2.8μA/cm^2 with an energy of 20keV.

Cu and Ag films deposited on a Si substrate were irradiated with Ar cluster ion beams with energies from 10 to 25keV and at several incident angles. The average cluster size was about 3000, measured by the retarding potential method. Subsequently, sputtered depth and surface morphology were measured by contact profilometer and AFM(Atomic Force Microscope).

Cu films were irradiated with Ar cluster ion beams, and sputtered particles were collected by several Si substrates located on a semicircular holder at regular intervals. The angular distribution of sputtered atoms was obtained from the Cu deposited on each Si substrate using a RBS measurement. The trace of a single Ar cluster ion impact was measured by STM(Scanning Tunneling Microscope).

RESULTS AND DISCUSSIONS

Energy and incident angle dependence of the sputtering yield

Figure 2 shows the energy dependence of the sputtering yield of Cu and Ag with Ar cluster ions. The average cluster size is about 3000, and the targets were irradiated at normal incidence. The sputtering yield is

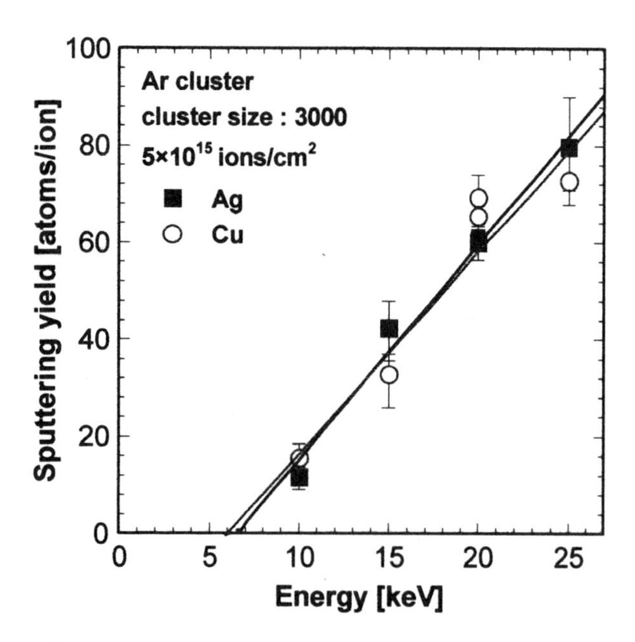

FIGURE 2. The energy dependence of the sputtering yield of Cu and Ag by Ar cluster ion beams. The sputtering yield increases linearly with the ion energy, and the threshold energy for sputtering of both metals is 6keV.

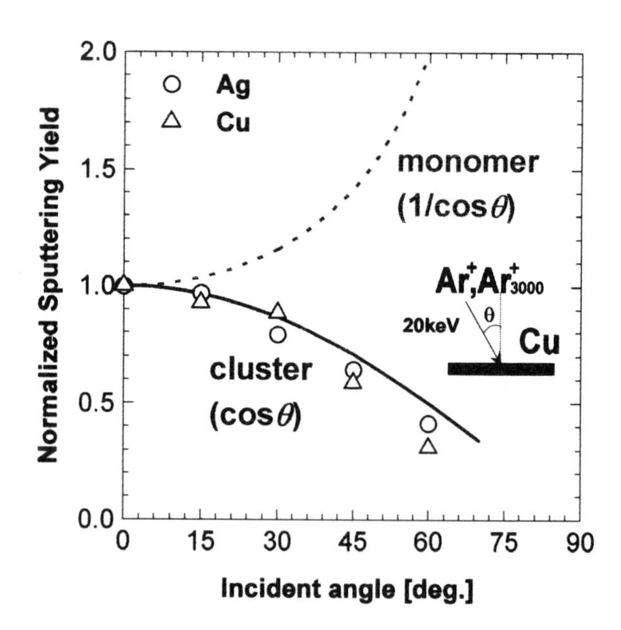

FIGURE 3. The incident angle dependence of the sputtering yield of Cu and Ag by Ar cluster ion beams. The sputtering yield of both metals decrease with the incident angle according to cosθ.

defined as an average number of atoms sputtered by one cluster ion (atoms/ion).

The sputtering yields of Cu and Ag with cluster ions increase linearly with the ion energy. The threshold energy of the sputtering for both Cu and Ag is 6keV, which corresponds to an energy of 2eV per constituent atoms of the cluster. In the case of an Ar monomer ion, the threshold energy reported for Ag and Cu is 15 and 17eV, respectively[8]. These energies are almost one order of magnitude higher than the energies of the constituent atoms of the clusters. This indicates that the sputtering mechanism associated with cluster ions is totally different from that of monomer ions.

Figure 3 shows the incident angle dependence of the normalized sputtering yield of Cu and Ag with Ar cluster ions. The ion energy was fixed at 20keV. The incident angle (θ) is the angle from the surface normal.

The sputtering yield of both metals decreases with increase of the incident angle according to cosθ. However, according to the collision cascade theory[9], that of monomer ions increases ($\cos^{-n}\theta$, n=1~2). This difference is caused by the change in sputtering mechanism. Each constituent atom of a cluster ion has an energy of only a few eV. The energy of the cluster ion is deposited directly to the surface layers of the target, and atoms in the surface layer are sputtered as a consequence of the multiple collision with cluster and target atoms. Also, as shown figure 2, the sputtering yield at normal incidence increases linearly with the energy. From these results, the sputtering yields of cluster ions are proportional to the energy density, which decreases in proportion to cosθ.

Smoothing effect of cluster ion beams

When a cluster ion containing a few thousand atoms bombards a target, multiple collisions between incident atoms and target atoms occur near the surface, and many atoms are sputtered from the surface in lateral directions. This phenomenon is predicted by molecular dynamic simulations[10], and is called "lateral sputtering".

Figure 4 shows the ion dose dependence of the average surface roughness (R_a) of Cu surfaces irradiated with Ar cluster ions with an energy of 20keV. The surface roughnesses were measured by Atomic Force Microscope (AFM).

The initial surface, with a R_a of 59Å, is dramatically smoothened by Ar cluster ion beams, and R_a decreases as the ion dose increases. Finally a very flat surface (Ra=12Å) was obtained. No roughening mechanism is observed in cluster ion irradiation.

Figure 5 shows AFM images of the morphology of Cu surfaces irradiated with Ar cluster ions at several incident angles. The ion energy and ion dose were 20keV and 5×10^{15} ions/cm^2. Scan area of the AFM was 1μm×1μm.

The average roughness of the initial surface was 59Å. However, the average roughness at normal incidence was 10Å, and R_a increased with the incident angle. In the case of 60°, a ripple morphology whose wave vector is parallel to the incident direction is observed, which has been reported in the case of heavy monomer ion irradiation at oblique incidence[11]. From figure 5, the smoothing effect of cluster ion beams is most effective at normal incidence.

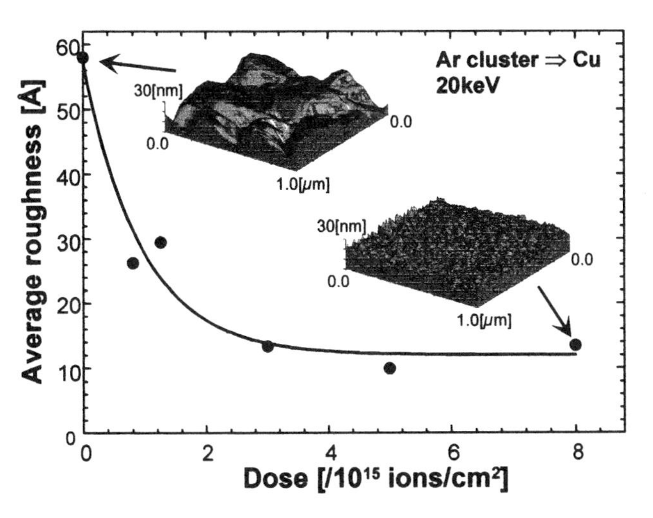

FIGURE 4. The dose dependence of the surface roughness of Cu irradiated with Ar cluster ion beams. The initial surface roughness (Ra=59Å) decrease with the ion dose.

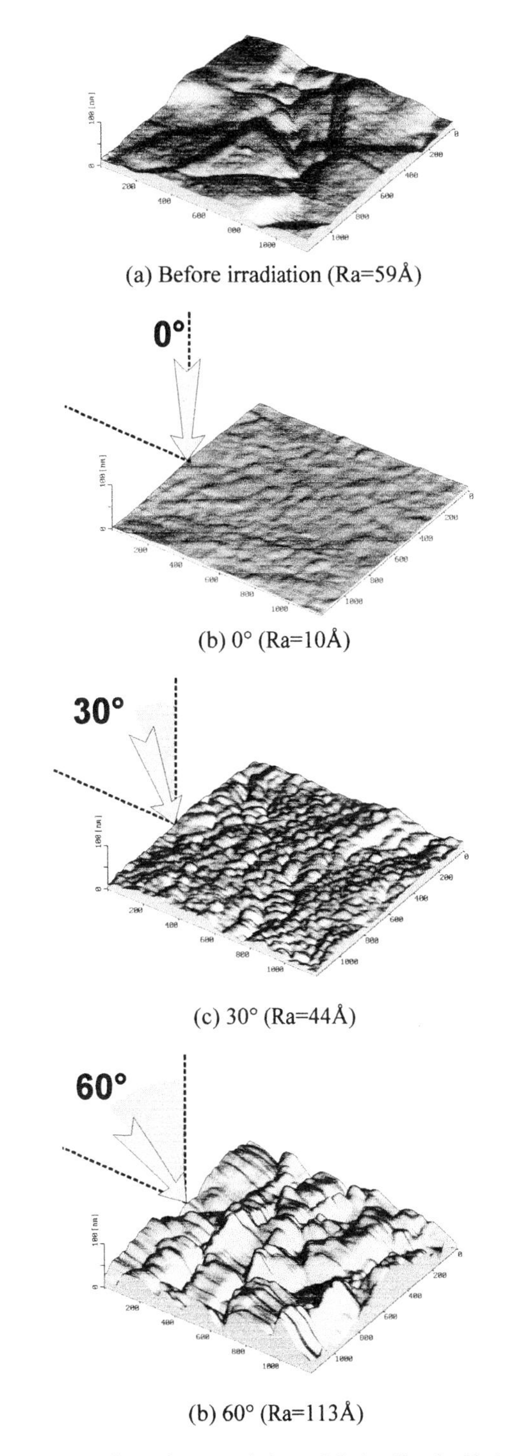

(a) Before irradiation (Ra=59Å)

(b) 0° (Ra=10Å)

(c) 30° (Ra=44Å)

(b) 60° (Ra=113Å)

FIGURE 5. The surface morphology of Cu irradiated with Ar cluster ion at several incident angles. The surface roughness is most improved at normal incidence. At the incident angle of 60°, ripple structure whose wave vector is parallel to the incident direction is observed.

Angular distribution of sputtered atoms

In order to study the smoothing mechanisms of cluster ions, information about the angular distribution of sputtered atoms is very important.

Cu films were irradiated with 20keV Ar cluster ion beams collimated to less than 1.5mm diameter by an aperture above the target. Sputtered Cu atoms were collected by several Si substrates located on a semicircular holder at a distance of 50mm from the target. Subsequently, the Cu densities on the Si substrates were measured by RBS measurements.

Figure 6 shows the angular distributions of the atoms sputtered by Ar cluster and monomer ions at normal incidence. The angular distribution with monomer ions displays a cosine distribution which is predicted theoretically from the cascade collision theory for high energy bombardment[9]. However, the angular distribution with cluster ions displays an under-cosine shape, which is similar to that of low-energy monomer ion bombardment, and many atoms are sputtered to the lateral directions. This result indicates the " lateral sputtering " effect of gas cluster ion impacts.

A single track of a gas cluster ion impact gives information about the angular distribution of sputtered atoms. Figure 7 shows a STM image of a single track of an Ar cluster ion impact on a Highly-Oriented Pyrolitic Graphite (HOPG) surface. The ion energy was 150keV and the average cluster size was 100 atoms/cluster. A crater structure is observed with a diameter of 200Å. The diameter of the cluster ion is about 20Å. This track indicates the local high-density energy deposition by cluster ion impact. Many atoms sputtered from the crater cavity redeposit isotropically around the crater edge. This result shows good agreement with the observations presented in figure 6.

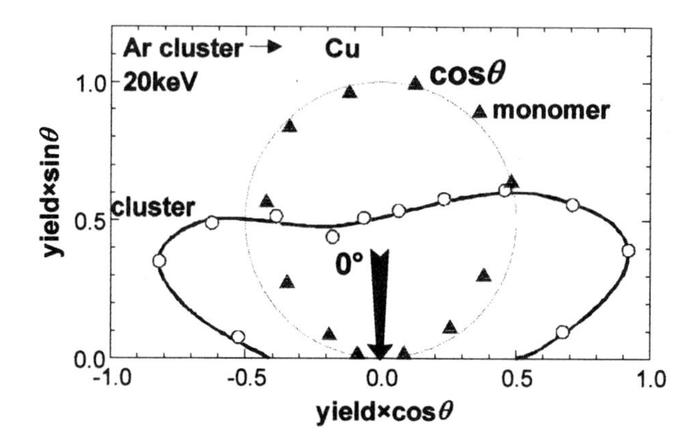

FIGURE 6. The angular distribution of sputtered atoms by Ar monomer and cluster ion beams at normal incidence. The angular distribution of monomer ion displays cosine distribution, however, that of cluster ion displays under-cosine distribution with many sputtered atoms distributed in lateral directions.

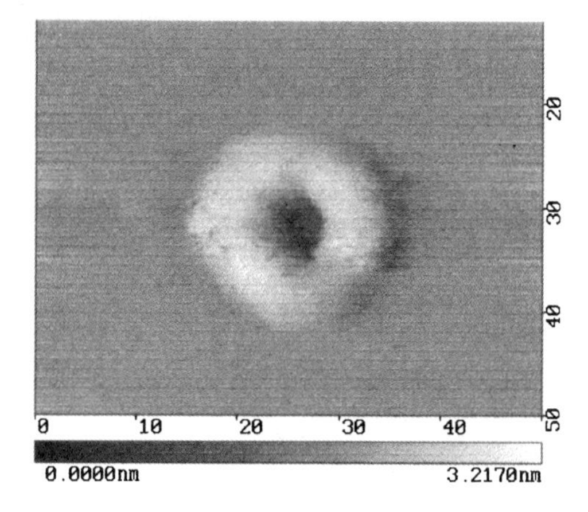

FIGURE 7. The STM image of a single track of Ar cluster ion impact at normal incidence. Crater structure is formed, and the sputtered atoms from the hole are redeposited isotropically around the crater.

CONCLUSIONS

The sputtering effects of Ar cluster ion beams have been studied. The sputtering yield of an Ar cluster ion increases linearly with the energy, and the threshold energy for sputtering of Cu and Ag is 6keV. This energy is only 2eV for each cluster constituent atom.

The incident angle dependence of the sputtering yield is proportional to $\cos\theta$, attributed to energy density changes. From cascade collision theory, the angular distribution of sputtering by monomer ions is proportional to $\cos^{-n}\theta$. These results indicate that the sputtering process for cluster ions is quite different from that of monomer ions.

In the case of cluster ion bombardment, the surface roughness decreases with ion dose. An initial average roughness of 59Å is improved to 12Å by Ar cluster ion bombardment. The surface is smoothest after irradiation at normal incidence. A ripple structure is observed at an incident angle of 60°.

From the angular distribution of the sputtered atoms, it has been seen that many sputtered atoms are distributed in a lateral direction. In the monomer ion case, the angular distribution shows a cosine distribution, which is predicted theoretically from the cascade collision theory. A single trace of a gas cluster ion impact indicates the high-density energy deposition given to a local area by a cluster ion impact.

From these results, the sputtering by gas cluster ions occurs from the very near surface of the target, and the sputtering mechanism is quite different from that of monomer ions.

REFERENCES

1. I.Yamada, W.L.Brown, J.A.Northby and M.Sosnowski, Nucl. Instr. and Meth. B, **79**, 223(1993).
2. I.Yamada, J.Matuso, Z.Insepov and M.Akizuki, Nucl. Instr. and Meth. B, **106**, 165(1995).
3. T.Yamaguchi, J.Matsuo, M.Akizuki, C.E.Ascheron, G.H.Takaoka and I.Yamada, Nucl. Instr. and Meth. B, **99**,237(1995).
4. M.Akizuki, J.Matsuo, I.Yamada, M.Harada, S.Ogasawara and A.Doi, Jpn. J. of Phys., **35**, Pt.1,No.2B(1996).
5. N.Toyoda, H.Kitani, J.Matsuo and I.Yamada, Nucl. Instr. and Meth. B, to be published.
6. H.Kitani, N.Toyoda, J.Matsuo and I.Yamada, Nucl. Instr. and Meth. B, to be published.
7. H.W.Liepmann and F.S.Sherman, *Rarefied Gasdynamics*, **II**, 1181(1979).
8. K.Kanaya, K.Houjyou, K.Koga and F.Toki, Jpn. J. Appl. Phys., **12**, 1297(1973).
9. P.Sigmund, Phys. Rev., **184**, 383(1969).
10. Z.Insepov, M.Sosnowski, G.H.Takaoka and I.Yamada, Mater. Res. Soc. Symp. Proc.,**316**, 999(1994).
11. T.M.Mayer, E.Chason and A.J.Howard, J. Appl. Phys., **76**, 1633(1994).

USE OF EMPIRICAL MANY-BODY POTENTIALS IN THE SIMULATION OF CLUSTER IMPACTS ON SURFACES

Lifeng Qi and Susan B. Sinnott*

Department of Chemical and Materials Engineering
University of Kentucky, Lexington, KY 40506

Molecular dynamics simulations have been performed to investigate the chemical and structural processes which occur when molecular clusters of acetylene and ethylene impact a non-rigid, hydrogen-terminated diamond (111) surface at two incident hyperthermal kinetic energies. The atoms are characterized by a realistic many-body empirical potential for hydrocarbon systems. The goal is to explore the various chemical and mechanical processes which take place within the molecular clusters and between the clusters and the non-rigid surface on impact. Important processes observed during the simulations include surface deformation, cluster fragmentation, adsorption (tethering) of individual cluster molecules on the surface, and polymerization of the cluster molecules.

INTRODUCTION

In the last decade, interactions between clusters and solid surfaces have attracted much attention in the surface science and materials engineering communities because of their importance in many crucial applications, such as surface cleaning, catalysis, thin film growth, and the formation of novel chemical products (1-4). Simulation studies not only help guide experimental researchers, but are becoming an increasingly powerful tool in the interpretation of many experimental observations (5). There have been numerous atomistic simulations of cluster-surface collisions (6-12), but few studies have dealt with resulting chemical reactions *within* clusters, especially polymerization reactions within organic molecular clusters. There is therefore much that is not known about the chemistry that occurs during such collisions, despite the fact that beams of organic molecules and hydrogen are the starting materials in such important process as the chemical vapor deposition of diamond films (13), and unsaturated organic molecules are monomers of many important commercial polymers.

In this paper, we employed molecular dynamics simulations to investigate the impact of organic molecular clusters with a non-rigid hydrogen terminated diamond (111) surface. We examine the effects of varying the incident kinetic energy of the clusters and the bond saturation of the carbon atoms within the molecular clusters.

COMPUTATIONAL DETAILS

The computational approach is molecular dynamics simulations, where Newton's equations of motion are integrated with a third-order Nordsieck predictor corrector (14) using a time step of 0.2 fs. The forces on the individual atoms are derived from a reactive empirical bond order (REBO) hydrocarbon potential that realistically describes covalent bonding within both the clusters and the surface (15,16). Originally developed to examine the chemical vapor deposition of diamond thin films (15), it has been successfully used to study a variety of surface science processes such as surface sputtering (17), indentation (18), tribochemistry (19), and patterning (20). The cohesive van der Waals interactions between molecules in the clusters are characterized with a long-range Lennard-Jones potential. The combined expression for the binding energy of the clusters and surface is therefore:

$$E_b = \sum_{i}\sum_{j<i} [\, V_r(r_{ij}) - B_{ij}\, V_a(r_{ij}) + V_{vdw}(r_{ij})] \qquad (1)$$

where E_b is the binding energy, r_{ij} is the distance between atoms i and j, V_r is a pair-additive term that takes into account the interatomic core-core repulsive interactions, V_a is a pair-additive term that models the attractive interaction due to the valence electrons, and B_{ij} is a many-body empirical bond-order term that modulates valance electron densities (19,20) and depends on atomic coordination and angles (21). V_{vdw} is a Lennard-Jones potential that is only non-zero after the short-ranged REBO potential goes to zero.

The three-dimension molecular clusters were constructed and allowed to fully equilibrate at 5 K prior to impact. Each cluster contained 64 molecules, or 256 atoms in the acetylene cluster and 384 atoms in the ethylene cluster. The non-rigid, hydrogen terminated diamond (111) surface was also equilibrated at 5 K and consisted of 3136 atoms in 12 carbon

* Corresponding Author

CP392, *Application of Accelerators in Research and Industry,* edited by J. L. Duggan and I. L. Morgan
AIP Press, New York © 1997

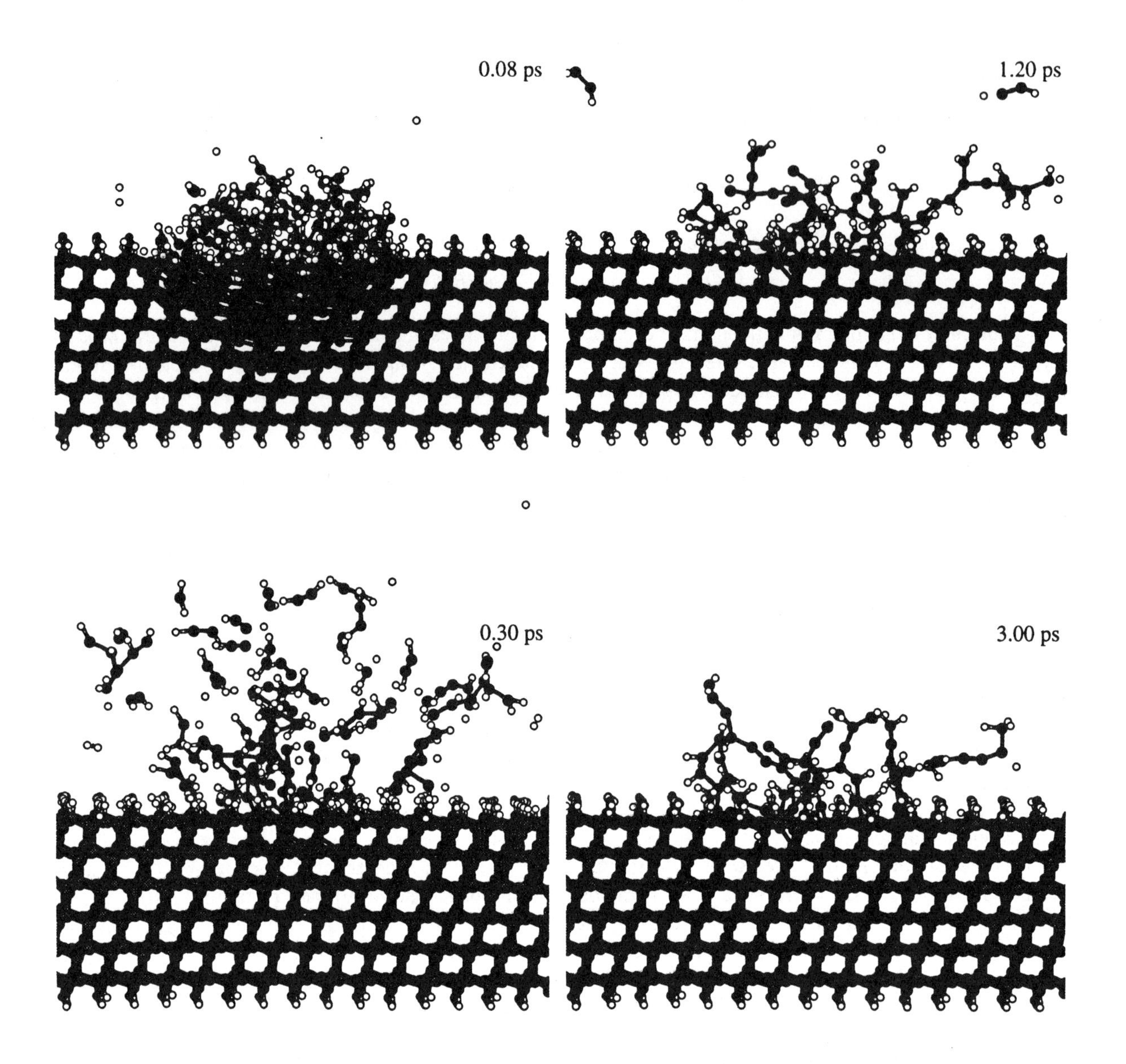

Figure 1. Snapshots from a representative simulation where an acetylene cluster is impacting a hydrogen terminated diamond (111) surface. The cluster contains 64 molecules and has an incident velocity of 12 km/s. The white spheres are hydrogen atoms and the gray spheres are carbon atoms.

layers, terminated by hydrogen layers on the top and bottom to satisfy the coordination requirements of the carbon atoms. The low equilibration temperature was used to minimize complications from thermal motion on the results of the cluster-surface collisions. The two surface layers (one layer of carbon and one layer of hydrogen) furthest from the cluster were held rigid. Moving towards the cluster, the next 3 carbon layers had Langevin forces applied to maintain the surface at a constant temperature of 5 K and mimic the heat dissipation properties of a real diamond surface. The remaining atoms in the surface, and all of the atoms in the cluster, were allowed to evolve in time according to Newtons equations of motion with no additional constraints.

RESULTS AND DISCUSSIONS

First, we examined the effect of incident kinetic energy on the impact of identical molecular clusters. Figure 1 shows selected snapshots from the collision of an acetylene cluster with the diamond (111) surface where the cluster has a velocity normal to the surface of 12 km/s (i.e., an external kinetic energy of 19.43 eV per molecule). At the beginning of the simulation, around time = 0.04 ps, the molecules in the cluster do not interpenetrate each other, or the intermixing within the cluster is small . At time = 0.08 ps, the cluster

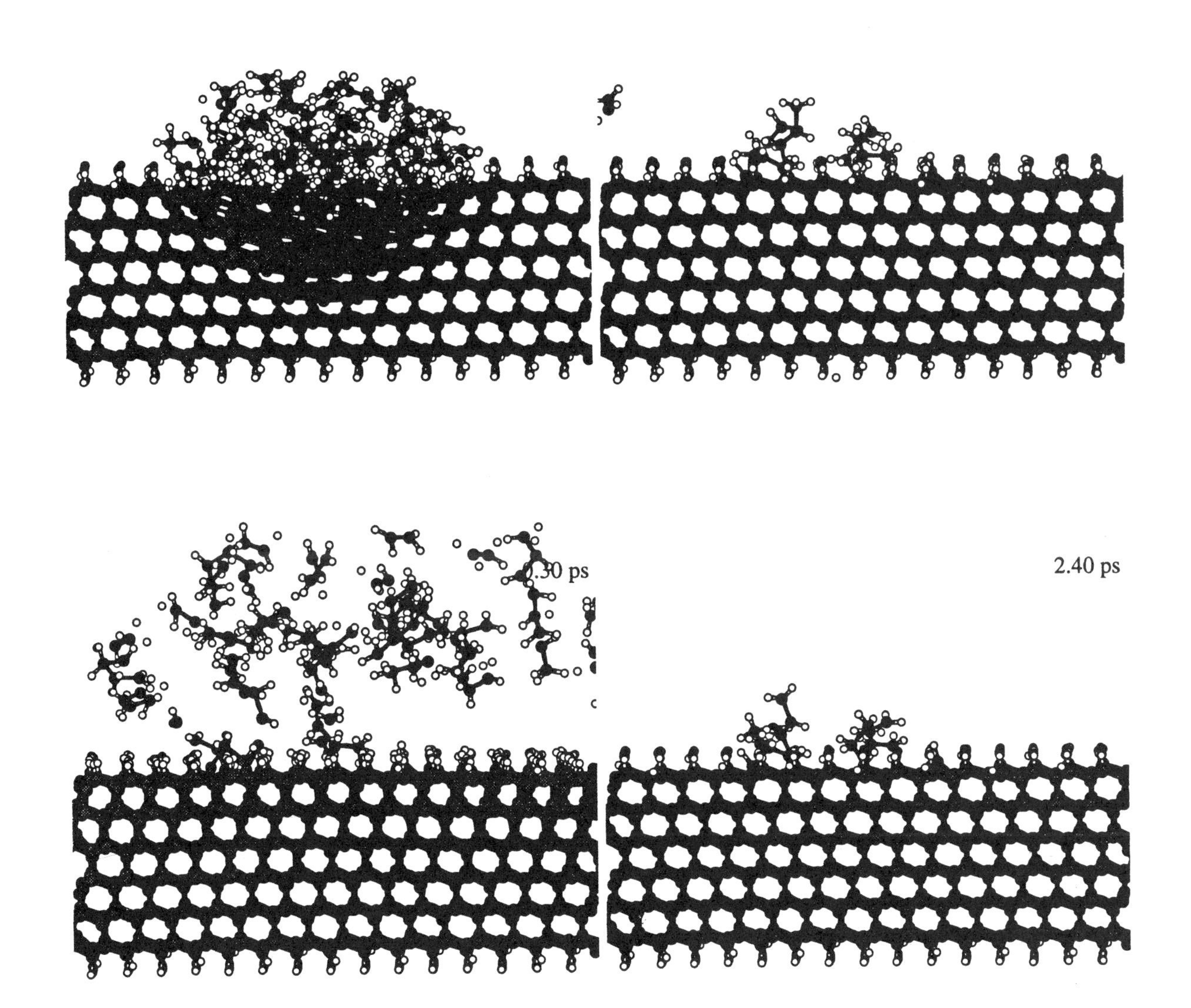

Figure 2. Snapshots from a representative simulation where an ethylene cluster is impacting a hydrogen terminated diamond (111) surface. The cluster contains 64 molecules and has an incident velocity of 12 km/s. The white spheres are hydrogen atoms and the gray spheres are carbon atoms.

has made full contact with the surface, which has deformed elastically to accommodate the force of impact. As the cluster and surface decompress, however, between time = 0.30 ps and time = 1.20 ps, a significant number of polymerization reactions are observed.

Approximately 76% of the carbon atoms in the cluster are involved in some form of polymerization, with about 37% of the polymer chains sticking to the surface. The products of the polymerization reactions range from dimers to chains that include up to 12 carbon atoms. Although most of them have a linear skeletal structure, branched and even network polymer products are also observed.

We find that this phenomena is strongly dependent on the collision energy. In simulations where the clusters have an incident velocity of 8 km/s (or an external incident kinetic energy of 8.63 eV/molecule), mostly nonreactive, elastic collisions are observed. The molecular cluster breaks apart as a result of impact with the surface. Following the collision with the surface, the molecules backscatter away, with only one acetylene molecule chemisorbing to the surface.

The second effect to be examined is that of carbon bond saturation within the molecular clusters. To facilitate this comparison, clusters of ethylene are given incident velocities of 12 km/s, which equals an external kinetic energy of 20.92 eV/molecule. Polymerization reactions between the molecules in the cluster are again observed (Fig. 2). Again, both linear and branched polymers result from the collision, with the longest polymer chain consisting of 8 carbon atoms. About 61 % of the ethylene carbon atoms participate in some sort of polymerization reaction, with about 2% of the polymer chains sticking to the surface after impact.

When the ethylene cluster is allowed to impact the surface at a velocity of 8 km/s, the effect is very similar to the acetylene collision at this velocity. Following impact, the cluster is fragmented and only one ethylene molecule remains chemisorbed on the diamond (111) surface.

As expected, each cluster-surface impact involves significant energy exchange between the cluster and the surface. At the point of impact, the cluster's external kinetic energy decreases dramatically as it is converted into internal kinetic energy and potential energy. In each simulation, the system potential energy reaches a maximum at the point of maximum surface deformation, and then decreases to a point near the initial potential energy.

CONCLUSIONS

We have investigated the impact of organic molecular clusters with a non-rigid, diamond (111) surface at two hyperthermal, incident velocities using molecular dynamics simulations. Mostly nonreactive collisions result at cluster incident kinetic energies that are less than the total bond energy of a single molecule within the cluster. In these cases, the main result of the cluster-surface collision is cluster fragmentation as almost all the molecules backscatter off the surface. In contrast, at incident kinetic energies that are greater than or equal to the total bond energy of a single molecule within the cluster, significant polymerization of the cluster molecules is observed, with only 24-39% of the molecules backscattering off the surface without reacting.

As the bond saturation of the carbon atoms within the molecular clusters increases, the degree of polymerization as a result of impact decreases significantly. The bond saturation also has an affect on the percentage of molecules adsorbing to the surface as a result of impact. The collision of the more reactive acetylene cluster resulted in 37% of the molecules chemisorbing to the surface either individually or as part of a polymer chain that chemisorbed, while only 2% of the ethylene molecules remained connected to the surface. These chemisorbed molecules/chains may be viewed as the initial stages of thin film growth, where the exact nature and morphology of the film is unclear. They may also be viewed as very short "tethered chains", a well-known method of strengthening interfaces (22).

These simulations thus qualitatively predict that impacting molecular organic clusters against a surface at hyperthermal velocities is a viable way of creating short polymer chains, tethering these chains to a surface, and possibly constructing a thin film. Many issues remain to be addressed, such as the manner in which the results vary as a function of surface composition and cluster size. Work is currently underway to address these issues.

ACKNOWLEDGMENTS

The authors are grateful for the support of the Office of Naval Research under grant number N00014-95-1-1183.

REFERENCES

1. Hsieh, H., Averback, R. S. Sellers, H; and Flynn, C. P., *Phys. Rev. B* **45**, 4417, 1992.
2. Yamada, Y., Appl. *Surf. Sci.* **43**, 23, 1989; Yamada, Y., *Nuc. Inst. and Meth.* **B45**, 707, 1990.
3. Tsukuda, T., Yasumatsu, H., Sugai, T., Terasaki, A., Nagata, T., and Kondow, T., *J. Phys. Chem.* **99**, 6367, 1995.
4. Gaps, J., and Kreig, G. J., *J. Chem. Phys.* **61**, 4037, 1974.
5. Halicioglu, T., and Bauschlicher Jr., C. W., *Rep. Prog. Phys.* **51**, 883, 1988.
6. Weistein, J. D., Fisher, R. T., Vasanawala, S., Shapiro, M. H., and Tombrello, T. A., *Nuc. Inst. and Meth.* **B88**, 74, 1994.
7. Shapiro, M. H., Tosheff, G. A., and Tombrello, T. A., *Nuc. Inst. and Meth.*, **B88**, 81, 1994.
8. Haberland, H., Insepov, Z., and Moseler, *Phys. Rev. B* **51**, 11061, 1994.
9. Cleveland, C. L., and Landman, U., *Sience* **257**, 355, 1992.
10. Xu, G. Q., Holland, R. J., Bernasek, S. L., and Tully, J. C., *J. Phys. Chem.* **90**, 3831, 1989.
11. Nakamura, M., Tsukada, M., and Aono, M., *Surf. Sci.* **283**, 1993.
12. Yarbrough, W.A., and Messier, R., *Science* **247**, 688, 1990.
13. Allen, M. P. and Tildesley, D. J., *Computer Simulation of Liquids*, 1987, ch.3, pp 82-83.
14. Gear, C. W., *Numerical Initial Value Problems in Ordinary Differential Equations*, New Jersey: Prentice-Hall, Englewood Cliffs, 1971.
15. Brenner, D. W., *Phys. Rev. B* **42**, 9458, 1990.
16. Brenner, D. W., Harrision, J. A., and Sinnott, S. B. (unpublished)
17. Taylor, R. S., and Garrison, B. J., *J. Am. Chem. Soc.* **116**, 4465, 1994.
18. Harrison, J. A., White, C. T., Colton, R. J., and Brenner, D. W., *Surf. Sci.* **271**, 57, 1992.
19. Harrison, J. A., White, C. T., Colton, R. J., and Brenner, D. W., *Mat. Res. Soc. Bull.* **18**, 50, 1993.
20. Sinnott, S. B., Colton, R. J., White, C. T., and Brenner, D. W., *Surf. Sci.* **316**, L1055, 1994.
21. Tersoff, J., *Phys. Rev. B.* **37**, 6991, 1988.
22. Smith, J. W., Kramer, E. J., and Mills, P. J., *J. Poly. Sci. B* **32**, 1731, 1994.

NON-LINEAR EFFECTS IN HIGH ENERGY CLUSTER ION IMPLANTATION

D.Takeuchi, T.Aoki, J.Matsuo and I.Yamada
Ion Beam Engineering Experimental Lab., Kyoto University, Sakyo, Kyoto, JAPAN

Cluster ion implantation realizes ultra-shallow (<50nm) junction formation. Cluster ions can provide a low-energy and high-current beam, because each constituent atom in the cluster, with sizes of a few thousands, has an energy of only a few eV. Non-linear effects of high energy cluster ion implantation caused by multiple-collisions and high-density energy deposition within a local surface region, have been investigated. Decaborane ($B_{10}H_{14}$), C_{60} and large Ar cluster ions with sizes of from 100 to 3000 were implanted into Si and sapphire substrates at energies up to 300keV. The number of disordered substrate atoms per atom in a cluster ion was much larger than the damage done by a monomer ion with the same velocity. Single traces of cluster ions on solid surfaces were observed by STM and AFM to investigate the mechanism of the non-linear effects. Those of large Ar cluster ions looked like crater shapes where the crater diameter was proportional to one-third of the cluster ion energy.

INTRODUCTION

Ion beams are being applied for surface modification of materials in a variety of different ways: ion implantation, ion beam mixing, sputtering, and particle or cluster beam assisted deposition. Fundamental to all of these processes is the deposition of a large amount of energy, generally some keV, in a localized area. This can lead to the production of defects, atomic mixing, crystal disorder and in some cases, amorphization.

On the other hand, the interaction of energetic cluster ions with solid surfaces has different impact process results than monomer ions. This process is of interest because a large amount of energy can be deposited in a small region, entirely amorphizing the surface layer (1). In this paper, non-linear effects of high energy cluster ion implantation caused by multiple-collisions and high-density energy deposition within a local surface region have been investigated.

EXPERIMENTAL

A representative 200 keV gas cluster implantation system developed at Kyoto University (2) consists of a cluster generator, an electron bombardment ionizer, an E×B mass-filter (Wien-filter), an acceleration tube and a target chamber, as shown in Fig.1. Adiabatic expansion of high-pressure gas through a nozzle is utilized for the formation of a high-intensity gas cluster beam (3). The size distributions of Ar cluster ions are analyzed by the E×B mass-filter over a range of source pressures (2). The size distribution can be observed within the region of 95 to 110 in the case of 100 atom clusters, of 400 to 600 in the case of 500 atom clusters and of 1000 to 5000 in the case of 3000 atom clusters at a source pressure of 4 atm. Single traces of cluster ions on solid surfaces were observed by STM and AFM to investigate the mechanism of the non-linear effects. Large Au grains on mica (4) and sapphire(0001) surfaces were used. Before irradiation, the large grain gold films were prepared by vapor deposition and annealing in UHV and the sapphire substrate was annealed in air by a furnace in order to prepare a wide, atomically-flat surface.

The fullerene ions were produced in a hot hollow cathode ion source in a conventional ion implantor. Fullerene powder was evaporated in an oven close to the anode of the ion source, and then led to the ionizing part. Electron bombardment is used for ionization. The fullerene ions were extracted at 10 kV and then mass-separated by a 90° sector magnet. After mass-separation, the C_{60} fullerene ions were accelerated to energies up to 380 keV at the rear of the 90° sector magnet. Subsequently, the fullerene ion beam was swept, to irradiate the target. The ion currents were measured by a Faraday cup in the target chamber.

The bombarding effects of Ar clusters and fullerene ions on sapphire(0001) surfaces were investigated by RBS channeling measurements. We used a 2 MeV $^4He^+$ analyzing beam for these measurements. Sapphire(0001) surfaces were irradiated with size-selected Ar clusters or fullerene ions at energies of up to 300 keV. Sufficient monomer ion suppression could be achieved using mass-filters. Since secondary electrons and ions disturb accurate measurements of ion beam current, a special Faraday cup was installed in the irradiation chamber for these experiments (2).

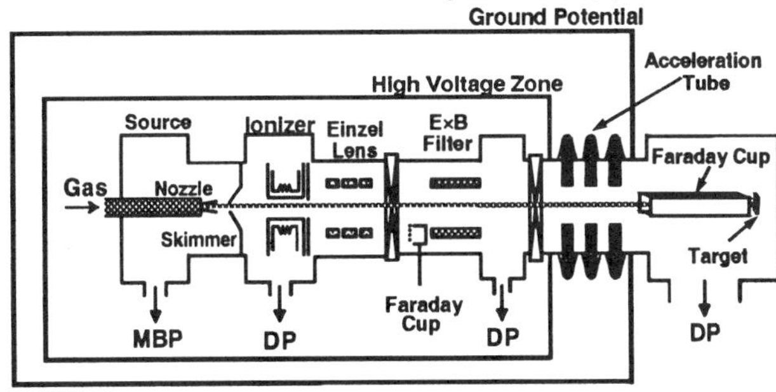

FIGURE 1. Schematic diagram of 200keV gas cluster ion implanter.

CP392, *Application of Accelerators in Research and Industry*, edited by J. L. Duggan and I. L. Morgan
AIP Press, New York © 1997

Si(100) substrates were also irradiated by C_{60} fullerene singly charged ions with ion dose up to 6×10^{13} ions/cm^2. The total number of irradiating carbon atoms was as high as 3.6×10^{15} atoms/cm^2, because one ion transports 60 carbon atoms. The cluster energy is varied between 70 and 300 keV (i.e. 1.2 and 5.0 keV per carbon atom in C_{60} fullerene). During irradiation, Si substrates were kept at room temperature and the pressure in the target chamber was below 1×10^{-6} Torr.

$B_{10}H_{14}$ sublimates easily at room temperature. Sublimation gas from $B_{10}H_{14}$ was ionized by electron bombardment and extracted for a $B_{10}H_{14}$ ion beam. $B_{10}H_{14}$ ions were implanted into Si(100) at an energy of 20 keV, and damage induced in the Si was investigated by RBS channeling.

RESULTS AND DISCUSSIONS

Observation of Ion Traces

Figure 2 shows a STM image of an ion trace on a gold grain due to the bombardment by 150 keV Ar cluster ions with a size of 3000 atoms. An AFM image of a trace on a sapphire(0001) surface, due also to a 100-atom cluster was also observed. The former has a crater shape, while the latter has a hill shape. All STM images of Ar cluster ion traces on Au grains or HOPG surfaces had a crater shape. It is likely that AFM resolution is one order smaller than STM.

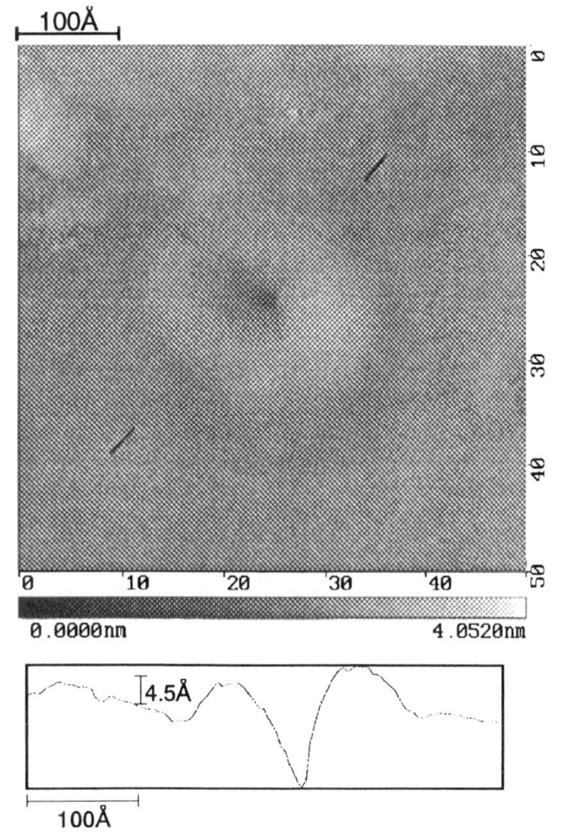

FIGURE 2. A STM image of a trace of Ar cluster ion bombardment on an Au surface. It has a crater-like shape with 200Å diameter. The diameter has proved to be proportional to cubic root of acceleration voltage.

Therefore, the Ar cluster ion trace on sapphire probably also has a crater shape. This result is quite different from that of monomer ion case.

The diameter of these traces was about 200 Å when the energy of the cluster ions was 150 keV. The diameter observed by AFM and STM proved to be proportional to the cubic root of the acceleration voltage of the cluster ions. This indicates that the total energy of cluster ions is isotropically distributed into the substrate. Molecular Dynamics simulation confirms that hemispherical damage regions are formed at the moment of impact of an energetic ion cluster beam (5).

Non-linear Effects of Cluster Ion Bombardment

RBS channeling measurements have been used to study damage on sapphire surfaces irradiated by Ar cluster ions at various energies and ion doses. $B_{10}H_{14}$ and fullerene ions damage to sapphire and Si substrates has also been studied at various energies and ion doses.

Figure 3 shows the number of disordered atoms for a variety of ion doses with three different cluster sizes at 150 keV. The total number of disordered atoms was obtained by integration of the surface-peak in the channeling spectra (6). The density of disordered atoms on the unirradiated surface, due to surface distortion and intrinsic vacancies, is shown as a background in Fig.3. The number of disordered atoms due to cluster bombardment increases with ion dose and finally reaches saturation. This suggests that the modified layer was totally amorphized. A slight cluster size dependence was observed. In the case of clusters with the sizes of 500 and 3000, the saturation values of disordered atoms were about 1.8×10^{16} and 2.0×10^{16} aluminum atoms/cm^2, respectively. This corresponds to a depth of about 40Å.

In order to compare with the monomer Ar ion irradiation case, the number of disordered atoms were predicted using TRIM(TRansport of Ions in Matter) calculations (7). TRIM

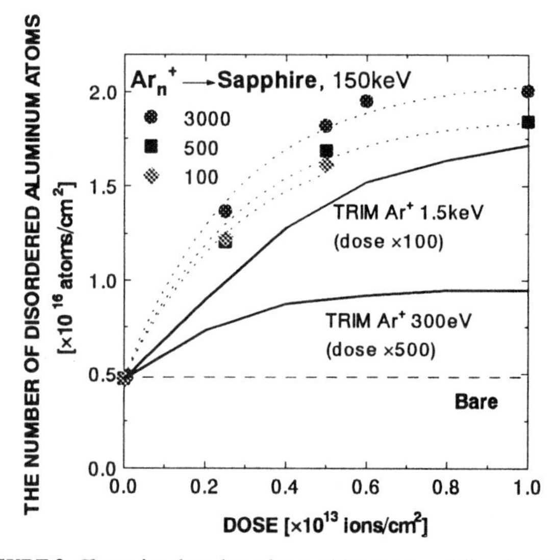

FIGURE 3. Cluster ion dose dependence of the number of disordered atoms for various cluster sizes. The values were calculated from the area of surface peaks of RBS channeling spectra.

can be used to estimate the probability of vacancy production per ion. It can also estimate vacancy distribution versus depth. From the concentration of aluminum in sapphire, the number of aluminum atoms per 1Å layer depth is 4.7×10^{14} atoms/cm^2/Å. The probability of vacancy production given by the TRIM calculation multiplied by the ion dose predicts the number of vacancies in the layer. We assume that the layer was completely amorphized when the calculated number of vacancies exceeded 4.7×10^{14} atoms/cm^2/Å. The number of disordered atoms in each 1Å layer no longer increases once the layer becomes completely amorphized. The total number of disordered aluminum atoms was found by integration over the number calculated with this assumption. The results of the calculations are also depicted in Fig.3. The disordered atom number appears to saturate for monomer implant also.

The case of 500 atom Ar clusters accelerated through 150 kV, and bombarded with a dose of 1×10^{13} ions/cm^2, can be considered similar to the case of 300 eV Ar ions implanted to a dose of 5×10^{15} ions/cm^2. The case of 100 atom Ar clusters accelerated through 150 kV and implanted at a dose of 1×10^{13} ions/cm^2, can be compared to the prediction for the case of 1.5 keV Ar ions implanted with 1×10^{15} ions/cm^2. The case of 3000 atom Ar clusters accelerated through 150 kV and implanted at a dose of 1×10^{13} ions/cm^2 should be compared to the case of 50 eV Ar ions implanted with 3×10^{16} ions/cm^2, but the TRIM simulation is not suitable for such a low energy regime. All the results in Fig.4 include the background disorder value.

In the case of TRIM calculations, the number of disordered atoms drastically decreases with decreasing Ar ion energy, because of the reduction in the ion projected range. In comparing TRIM calculations with experimental results, the energy of constituent atoms has a larger influence on the TRIM calculations. In the case of 50 eV, it is impossible for Ar monomer ions to produce large damage

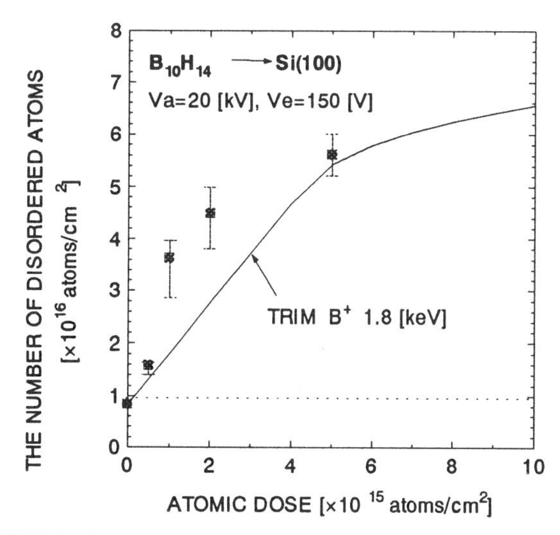

FIGURE 4. The relation between the number of disordered Si atoms and atomic dose of B. $B_{10}H_{14}$ ions were accelerated to 20 kV, for a similar B SIMS profile to 2 keV B monomer ion implantation.

on a substrate. Even with such a low velocity in the cluster irradiation case, the thickness of the damaged layer was about 40Å, which was higher than the value calculated by TRIM. Therefore, cluster ion bombardment has a quite different mechanism of damage formation than monomer ion bombardment even at the same velocity.

In the region of ion dose less than 2×10^{13} ions/cm^2, the slope of each line corresponds to the number of disordered atoms caused by a single cluster ion. The slope is larger than in the case of a monomer ion. Therefore, an impact by a cluster ion produces a much larger displacement than a monomer ion does.

The results of $B_{10}H_{14}$ ion irradiation on Si(100) is shown in Fig.4. Figure 4 shows the dependence of the number of disordered Si atoms on B atomic dose. This figure also shows the non-linear effect of cluster ions in the low dose region, while the saturation number of disordered atoms of $B_{10}H_{14}$ ion irradiation is almost the same as that caused by monomer B ions. The depth distribution of the B atoms implanted as $B_{10}H_{14}$ ions at the energy of 20 keV was checked by SIMS, and it was quite similar to the profile from B ions implanted at the energy of 2 keV. In addition, this similarity in damage characteristics was seen when comparing C_{60} impacts and monomer carbon ones. These damage characteristics are different from the ones caused by large Ar cluster bombardment (8). These ions are what are called small clusters. At the same velocity, it seems that the damage layer thickness in the case of small cluster ions such as $B_{10}H_{14}$, C_{60} and Ar_{100}, is the same as that in the case of monomer ions, while the damage layer thickness in the case of large cluster ions such as Ar_{500} and Ar_{3000}, is thicker than in the case of monomer ions. Samples implanted with $B_{10}H_{14}$ and B were annealed to look at damage recovery. Superior crystalline quality was recovered in $B_{10}H_{14}$ sample which leads us to believe, amorphization may be more complete in $B_{10}H_{14}$ sample.

In order to estimate the non-linear effects of cluster ion bombardment, a Molecular Dynamics (MD) simulation was carried out (9). Figure 5 shows the dependence of the number of disordered Si atoms on B cluster size calculated by MD. The larger the cluster is, the higher the number of disordered atoms until the number seems to saturate for larger size. This result shows good agreement with the conclusion that large cluster ions produce a larger amount of defects, as shown in the lower dose region of Fig.3 and Fig.4.

The dependence of the penetration depth of ions on Ar cluster size is shown in Fig.6. The large filled circles indicate MD results and the small white squares are TRIM results, which represent monomer-like impinging. There is a crossover point around the size of 30 which indicates the threshold between monomer and cluster. This result is confirmed in Fig.3, where the saturation damage number is similar for the different cluster sizes (dashed line).

According to Fig.3 and Fig.4, $B_{10}H_{14}$ and Ar_{100} ions produced amorphous layers faster than monomer ions, while the thickness of the damaged layers was the same. However, larger cluster ions such as Ar_{500} and Ar_{3000} show much larger damage than monomer Ar ions. This result shows that the threshold between monomer and cluster is at least several tens, and this corresponds to Fig.6. We could not find any difference in the number of disordered aluminum atoms caused by each size of Ar cluster ions at the same total energy in Fig.3. This is reasonable, as shown in Fig.6.

As described above, damage formation by cluster ions is quite different from monomer ion damage. A monomer ion forms defects in the substrate by cascade collisions. On the other hand, a cluster ion forms a completely amorphized local area in the surface region, as indicated by STM and AFM observation and Molecular Dynamics simulations.

CONCLUSIONS

Features of cluster ion bombardment on solid surfaces have been studied by implantation of various sizes and energies of Ar clusters, fullerene ions and $B_{10}H_{14}$. Non-linear effects of cluster ion impact have been seen in damage formation on solid-surfaces. The total number of disordered atoms depends on the total kinetic energy of cluster ions, and disorder caused by a cluster ion is much larger than that caused by a monomer ion. This difference is correlated to the one in STM and AFM images of single ion traces. The crater-like shape of cluster ion traces is the result of high-density energy deposition during cluster ion impact. In addition, the larger clusters, which have more than several tens of constituent atoms, penetrate deeper than monomer and small clusters, even at the same kinetic energy as shown in MD simulations.

REFERENCES

1 D.Takeuchi, N.Shimada, J.Matsuo and I.Yamada, *Nucl. Instr. and Meth. in Phys. Res.* **B**, to be published.
2 D.Takeuchi, J.Matsuo, A.Kitai and I.Yamada, *Mater. Sci. & Eng. A*, MSA217/218, 74-77, 1996
3 I.Yamada and G.H.Takaoka, *Jpn. J. Appl. Phys.*, **32**, 2121-2141, 1993
4 V.M.Hallmark, S.Chiang, J.F.Rabolt, J.D.Swalen and R.J.Wilson, *Phys. Rev. Lett.*, **59**, 2879-2882, 1987
5 Z.Insepov and I.Yamada, *Nucl. Instr. and Meth. in Phys. Res.* **B 99**, 248-252, 1995.
6 M.Yoshimoto, T.Maeda, T.Ohnishi and H.Koinuma, *Appl. Phys. Lett.*, **67** **(18)**, 2615-2617, 1995
7 J.P.Biersack and L.G.Haggmark, *Nucl. Instr. and Meth. in Phys. Res.* **174**, 257, 1980.
8 D.Takeuchi, K.Fukushima, J.Matsuo and I.Yamada, *Nucl. Instr. and Meth. in Phys. Res.* **B**, to be published.
9 T.Aoki, J.Matsuo, Z.Insepov and I.Yamada, *Nucl. Instr. and Meth. in Phys. Res.* **B**, to be published.

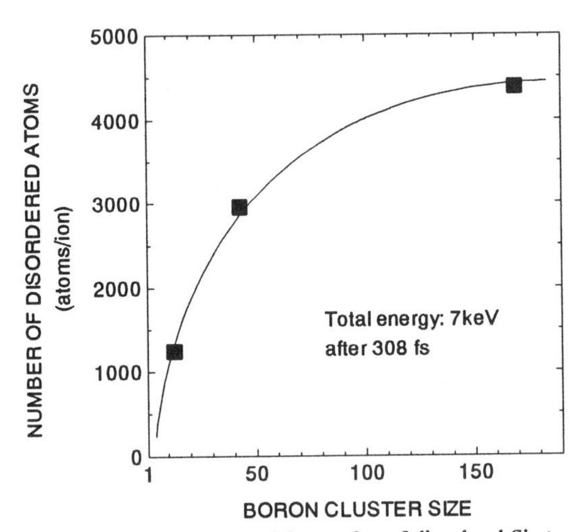

FIGURE 5. The dependence of the number of disordered Si atoms on B cluster size calculated by MD. The number of disorder represents at the time after 308 fs of cluster ion impact, which seems to decrease along with time. The larger the cluster is, the more the number of disordered atoms increase, while the number seems to saturate for larger size.

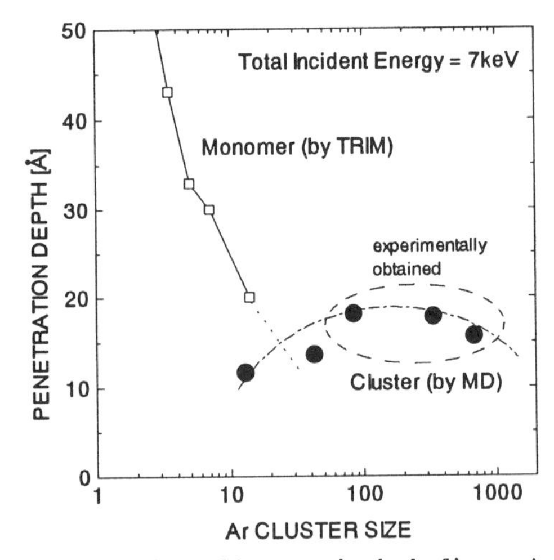

FIGURE 6. The dependence of the penetration depth of ions on Ar cluster size. The large filled circle indicates MD results, while small white square does TRIM results, which represents monomer-like impinging. There is a crossover point around the size of 30 which indicates the threshold between monomer and cluster. Dashed circle around the size from 100 to 1000 shows cluster sizes used in Fig.3.

PLASMA CLUSTER BEAM SOURCES FOR THIN FILM DEPOSITION

P. Milani and P. Piseri

INFM-Dipartimento di Fisica, Universita' di Milano, Via Celoria 16, 20133 Milano, Italy

The deposition of ionized and neutral cluster beams has been proposed as an interesting technique for the growth of thin films and nanostructures. Clusters are aggregates of atoms or molecules ranging from tens up to several thousands of units. The control of composition, size and kinetic energies is of central importance for the synthesis of thin films with tailored properties. Several methods are currently used for the production of cluster beams. Here a review will be given with particular attention to pulsed plasma sources and mass spectrometric characterization techniques. An example of deposition of nanocrystalline carbon thin films will be presented.

INTRODUCTION

The study of free and supported clusters has gained increasing interest for the synthesis of an entire new class of nanostructured materials (1). A necessary requisite to the realization of cluster-assembled materials with tailored properties is the possibility of controlling and characterizing cluster formation, size and deposition parameters.

The application of molecular beam techniques to the study of free clusters represented a fundamental breakthrough for the understanding of these systems (2) and offered many possibilities of application in the domain of surface processing (3), shallow implantation (3), ULSI fabrication (4), nanostructured material synthesis (5).

Deposition of thin films by cluster beams was originally proposed by Takagi and Yamada at Kyoto University (6). Ionized cluster beam deposition (ICBD) was presented as an alternative to the ion-assisted deposition technique (7). A metal vapour is expanded from a crucible designed to obtain a supersonic expansion. Atoms and eventually clusters formed in the expansion are subsequently ionized by electron impact and accelerated to few keV towards a substrate. Since the kinetic energy of the cluster is shared among its constituents, it is possible to transfer a considerable amount of energy while keeping the bombarding effects confined to the first monolayers. This enhances the formation of nucleation sites and adatom diffusion thus improving the density and the crystallinity of the film (7). Cluster beams can represent an alternative to ion beams when low ion beam energies and high fluxes are requested.

ICBD has been applied to a large variety of materials (6) obtaining high quality thin films. However systematic mass spectrometric studies showed that the pure expansion techniques were not efficient for cluster production and that clusters may not change significantly the film growth processes (7).

These results pointed out that mass spectrometry is a necessary diagnostic tool for cluster beam deposition and that the development of new sources with a better control of neutral and ion cluster production is an essential prerequisite.

The relative facility of producing clusters of gaseous and volatile species stimulated the development of supersonic cluster beams as tools for surface reactive etching, smoothing, oxidation and shallow implantation (3). Recently the development of new supersonic sources capable of working with refractory materials has opened new perspectives to thin film deposition (8, 9, 10, 5). In this contribution we will discuss the operation principles of an apparatus based on these sources and the possible applications.

CLUSTER SOURCES AND MOLECULAR BEAM EQUIPMENT

Clusters can be formed by the condensation of a vapour expanding out of a nozzle into a vacuum. Expanding jets are widely used to produce supersonic beams of atomic and molecular clusters (11). By varying the expansion conditions the mass distribution and velocities of the clusters can be controlled. Vapour of solid materials can be formed in a joule-heated oven, however vapour pressures attainable with this method are often too low to achieve an efficient condensation. In this case seeded beams can be used: the sample material is evaporated in a cold buffer gas which favors supersaturation and condensation (11). Gas aggregation and seeded supersonic sources are very efficient for the production of neutral clusters; ionized aggregates are obtained with relatively high efficiency by electron impact ionization downstream from the expansion.

Mass distributions of clusters in the beam are characterized by mass spectrometric techniques (2). In particular, time-of-flight mass spectrometry (TOF/MS) can be conveniently used. The operating principle is very simple: ions with the same kinetic energies but with different masses must have different velocities. If they are injected into a drift tube the ions will separate according to their masses. TOF/MS can detect ionized particles: in the case of gas aggregation sources the clusters must be ionized by photons or electrons. Electrons are traditionally used to achieve high flux of ions, however electron impact ionization produces a severe fragmentation of the aggregates thus changing the original mass distribution. This makes it very difficult to establish a correlation between the source conditions and the cluster mass distribution.

The most important parameters for cluster beams for surface processing are ion cluster intensity and kinetic energy: the ionized particles must be accelerated and focused with high spatial resolution. This approach is suitable for the formation of thin films with a high crystalline character, good adhesion and in general without any memory of the parent clusters (10). A fine control on the cluster mass distribution is not crucial.

A different and, in a certain sense, opposite approach is necessary for the synthesis of nanocrystalline thin films which exhibit interesting physical-chemical properties (1). In this case clusters must retain their individuality and the deposition of neutral and/or ionized particles should be performed with a careful control of the mass distribution and impact energy. These parameters have a strong influence on the particle diffusion and coalescence after the deposition.

The synthesis of nanocrystalline films of semiconductors, carbon or oxides requires the development of sources capable of providing stable and intense beams for these refractory materials with the possibility of tuning the mass distribution charge, kinetic energy and temperature of the aggregates.

The introduction of plasma cluster sources (PCS) allowed the extension of cluster beam techniques to refractory materials and enabled researchers to partially meet these requirements. In PCS the target material is vaporized with a high-power pulsed laser (laser vaporization source or LVS) (12) or by a pulsed electric discharge which erodes the electrode material (pulsed arc cluster ion source or PACIS) (9). The produced plasma is mixed with a buffer gas pulse delivered by a valve and then expanded. A schematic representation of a typical PCS is shown in figure 1. These sources have several advantages:

- pulsed regime: reduced gas load and interfaciability with UHV deposition apparatus.
- compact design and high flexibility.
- efficient production of ionized clusters.
- control on cluster temperature.

The major drawback is represented by the low duty cycle compared to continuous supersonic sources.

Several modifications have been introduced in order to improve PCS from the point of view of intensity and stability. In particular the use of a chamber where the plasma thermalizes before expansion drastically increased the mass range and the intensity attainable (8) (fig.1).

The presence of the cavity strongly modifies the expansion dynamics of the cluster beam and widens the pulse duration. In many cases the theory developed for the description of gas pulse evolution in this kind of sources is not completely valid (13). The volume of the chamber can influence the mass distribution but also the charge of the particles, as long residence times cause the neutralization of the charged particles.

It has been frequently assumed that ion and neutral clusters produced by PCS have the same mass distribution (14). This assumption is not justified in general. It should be reiterated that gas dynamics in PCS are very complex and, up to now, poorly characterized. Cluster formation takes place under non-equilibrium conditions and the presence of a thermalization cavity does not imply that neutral and ion clusters spend the same amount of time in the source before the expansion. In the case of the PACIS source, we have verified that neutral and ion clusters have different mass distributions (see below).

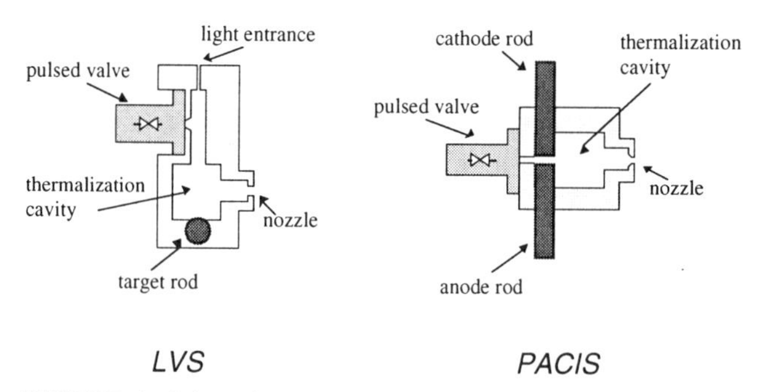

FIGURE 1. Schematic representation of Laser Vaporization Source (LVS) and Pulsed Arc Cluster Ion Source (PACIS).

PCS can be interfaced with deposition apparatus as shown in fig.2. As we have already underlined a good diagnostic of cluster beams is necessary. Figure 2 is a schematic representation of the apparatus used for the deposition of nanocrystalline carbon films (see the next section). The size distribution of clusters can be monitored by a TOF/MS. Ion optics can separate neutral from ion clusters and control the kinetic energy of the latter. A chopper allows the characterization of the supersonic expansion. Cluster deposition rates and fluxes can be monitored by Faraday cup or a quartz micro balance. Standard surface and thin film characterization techniques can be added. This apparatus has a modular structure which allows interfacing with other UHV facilities.

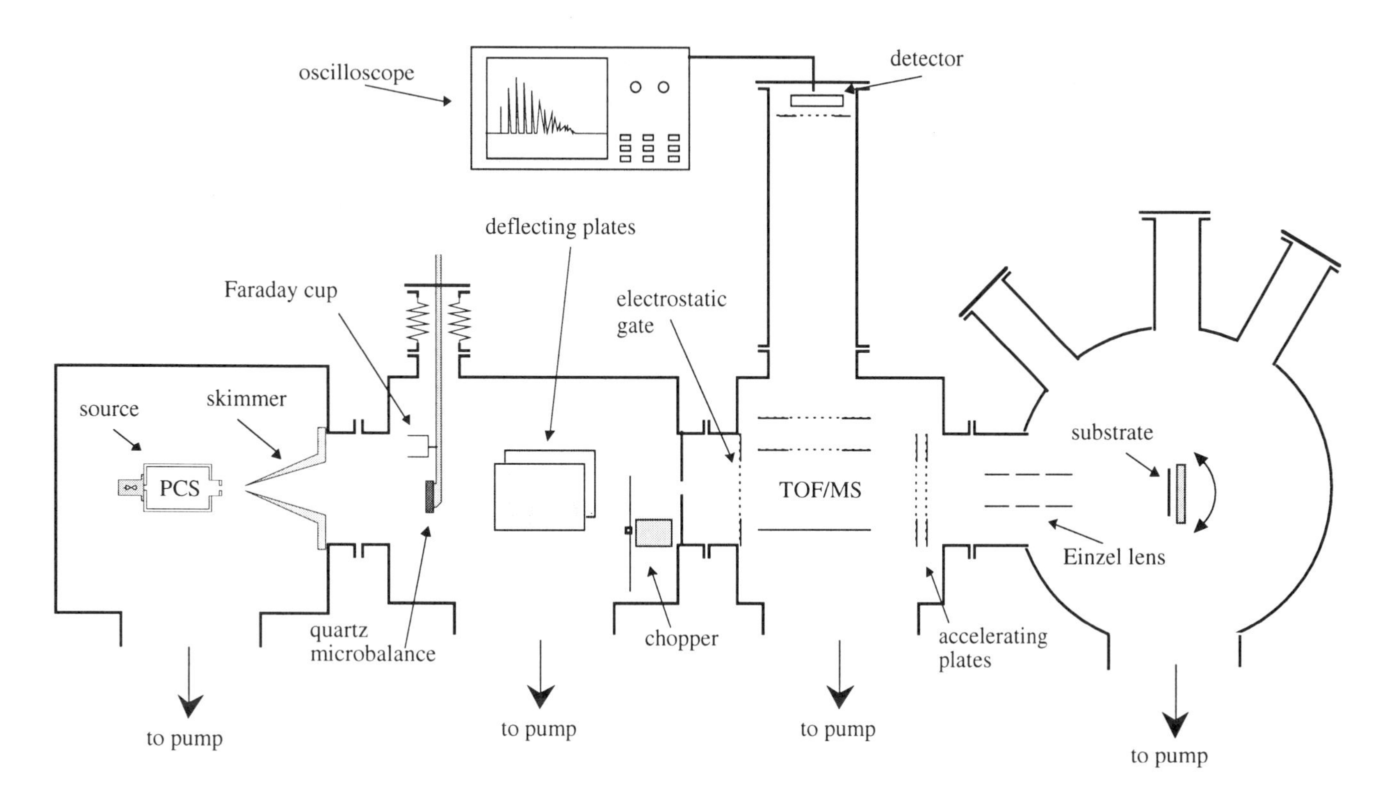

FIGURE 2. Schematic diagram of a typical deposition apparatus. It is composed of four differentially pumped chambers. From the left: the source chamber, a beam manipulation/characterization chamber, the TOF/MS chamber and finally the deposition one.

SOURCE CHARACTERIZATION AND THIN FILM DEPOSITION

The study of carbon clusters in molecular beams has been a very fertile field in the last few years. Fullerene discovery (15) has stimulated a renewed interest and suggested the possibility of using carbon cluster beams as a synthetic route towards new carbon-based materials (14).

Carbon ions produced by laser ablation or arc discharge have been extensively used to grow thin films and coatings. The electronic and structural properties of the film can be controlled by varying the carbon ions' kinetic energy and the substrate temperature during the deposition (16).

Carbon clusters can be used to control the hybridization of the film. Clusters of different mass exhibit different structural and electronic properties: small C_n clusters (n~10-20) should exhibit an sp^3 hybridization whereas clusters with hundreds of components should be sp^2. Controlling the mass distribution and the kinetic energies of carbon clusters should allow the synthesis of films with tailored electronic properties (band-gap) and morphology ranging from low density networks, hollow zeolite-like structures, to high-density diamond-like films (14).

Low energy deposition of neutral carbon clusters has been reported (14) and the dependence of the final material on the precursor clusters has been inferred. However, a systematic characterization of the cluster beam was not performed. Moreover the reactivity and coalescence of small carbon clusters is very poorly known. This renders the interpretation of the thin film structure very difficult.

We have undertaken carbon film deposition with cluster beams produced by a PACIS source. Compared to LVS, PACIS provides higher flux of neutral and ionized clusters. Our source has been modified compared to the original design (9). The thermalization cavity and the nozzle has been designed to favour plasma thermalization. A characterization of the expansion shows that large mass clusters are efficiently produced as in the case of gas aggregation sources.

The carbon cluster beam have been characterized to find the content of different cluster sizes and charge states. By growth rate measurements with a quartz microbalance, the overall intensity has been estimated to be in the range $3\cdot10^{14}$ - $2\cdot10^{15}$ s^{-1} sr^{-1} depending on the discharge voltage. A fraction on the order of ten percent of the total average flux is due to anions, while cations are about 2%.

Looking at the time evolution of the cluster intensities, we find a substantially different behaviour for cluster ions when compared to neutrals. Pulse duration is in the range of 10 - 20 ms for neutral clusters, whereas it is much shorter for the ions (70 ± 10 µs for cations, 50 ± 15 µs for anions). This implies that instantaneous intensities of the cluster ion beam must be higher than that of the neutrals. Using a Faraday cup we have measured fluxes of $3.5\cdot10^{15}$ s^{-1} sr^{-1} for cations and $1.4\cdot10^{17}$ s^{-1} sr^{-1} for anions. Neutrals, ionized with a laser, have been characterized with time-of-flight techniques.

The mass range of cluster ions (cations) is also substantially different from that of neutrals, i.e. 0 - 1200 atoms/cluster (with the maximum of mass distribution at 350 atoms/cluster) against 0 - 3000 atoms/cluster (maximum at 750 atoms/cluster). This is probably due to a trade-off between cluster growth and charge neutralization, that is, to grow big a cluster needs long residence time inside the source, but this also means a bigger chance to be neutralized.

Cluster velocity for neutrals is in the range 1400 - 1800 m/s depending on exit time from the source, with most of the clusters near the lower boundary of this range. The typical kinetic energy of a medium-size cluster is thus about 0.3 keV. The velocity of cluster ions is spread over the range 1200 - 1900 m/s, the distribution being peaked at 1700 m/s. A typical mass spectrum of neutral carbon clusters is shown in fig.3.

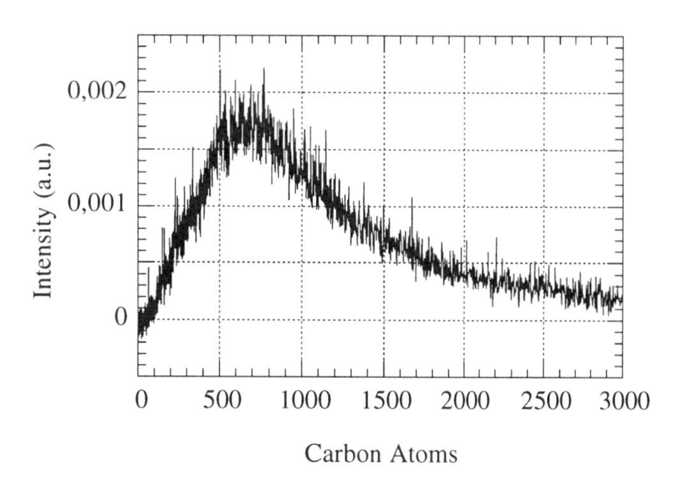

FIGURE 3. A typical mass distribution of carbon clusters synthetized with a PACIS source. To detect them with the MS, clusters have been ionized using the fourth harmonic of a Nd:YAG laser (266 nm).

Thin films have been deposited on Silicon surfaces (100) and on metallic substrates (Cu, Mo). Neutral and ion clusters have been deposited without observing charging effects. As suggested by the mass spectra reported in fig. 3, fullerene-like sp^2 clusters are predominant. The films have been characterized by Scanning Electron Microscopy and Raman spectroscopy. SEM micrographs show the formation of nanocrystalline structures with a mean diameter of few tens of nanometers. Raman spectra are typical of a-C with a high sp^2 character (17). The substrate temperature seems to have a strong influence on the structure of the film. Depositions performed with the substrate at 250 °C showed a tendency to graphitization less marked than if the annealing is performed after the deposition. a-C films deposited with conventional techniques show the same behavior for considerably higher temperatures (18).

CONCLUSIONS

The development of PCS sources has shown new opportunities for creating nanostructures materials with structural architecture based on atomic and molecular clusters. The possibility of controlling the properties of cluster-based materials will be enhanced by the ability of controlling cluster production and deposition. Efforts should be directed towards the characterization of the mechanisms underlying cluster formation and coalescence. Cluster beam techniques are a powerful tool to achieve this goal and they may provide even more exciting opportunities in the future.

ACKNOWLEDGMENTS

This work has been supported by INFN, Centro Innovazione Lariano, CNR and INFM.

REFERENCES

1. Siegel, R.W., *Mater. Sci. Eng. B* **19**, 37-43 (1993)
2. de Heer, W.A., *Rev. Mod. Phys.* **65**, 611-676 (1993)
3. Yamada, I., *Nucl. Instr. and Meth. B* **112**, 242-247 (1996)
4. Yamada, I., and Matsuo, J., "Gas cluster ion beam processing for ULSI fabrication", presented at the 1996 MRS Spring Meeting, San Francisco, CA, April 8-12, 1996
5. Barborini, E., Piseri, P., Milani, P., and Iannotta, S., *Materials Science Forum*, in press
6. Yamada, I., and Takagi, T., *Thin Solid Films* **80**, 105-115 (1981)
7. Brown, W.L., et al., *Nucl. Instr. and Meth. B* **59/60**, 182-189 (1991)
8. Milani, P., and de Heer, W.A., *Rev. Sci. Instrum.* **61**, 1835-1839 (1990)
9. Siekmann, H.R., Luder, C., Fachrmann, J., Lutz, H.O., and Meiwes-Broer, K.H., *Z. Phys D* **20**, 417-420 (1991)
10. Haberland, H., Karrais, M., Mall, M., and Turner, Y., *J. Vac. Sci. Technol. A* **10**, 3266-3271 (1992)
11. Hagena, O.F., *Rev. Sci. Instrum.* **63**, 2374-2379 (1992)
12. Michalopoulos, D.L., Geusic, M.E., Hansen, S,G,, Powers, D.E., and Smalley, R.E., *J. Phys. Chem.* **86**, 3914-3917 (1982)
13. Gentry, W.R., "Low energy pulsed beam sources" in *Atomic and Molecular Beam Methods*, New York: Oxfors University Press, 1988
14. Melinon, P., et al., *Int. J. Mod. Phys. B* **9**, 339-397 (1995)
15. Kroto, H.W., Heath, J.R., O'Brien, S.C., Curl, R.F., and Smalley, R.E., *Nature* **318**, 162-164 (1985)
16. Lossy, R., Pappas, D.L., Roy, R.A., Cuomo, J.J., and Sura V.K., *Appl. Phys. Lett.* **61**, 171-173 (1992)
17. Schwan, J., Ulrich, S., Batori, V., Ehrhardt, H., and Silva, S.R.P., *J. Appl. Phys.* **80**, 440-447 (1996)
18. Cho, N.H., et al., *J. Appl. Phys.* **71**, 2243-2248 (1992)

CLUSTER ION ASSISTED THIN FILM FORMATION

Jiro Matsuo, Makoto Akizuki and Isao Yamada

Ion Beam Engineering Experimental Laboratory, Kyoto University
Sakyo, Kyoto 606-01, Japan

A new thin films formation technique using gas-cluster ion beams is demonstrated. Oxygen cluster ion beams are utilized to grow oxide films. We have found that high quality SiO_2 films of up to 11 nm thick can be formed on Si substrate surfaces at room temperature by direct oxidation using O_2 and CO_2 cluster ion beams at acceleration voltages of less than 10 keV. O_2 cluster ions were also used to irradiate Si substrates covered with SiO_2 films, during the evaporation of Pb atoms. Increasing the acceleration voltages, the O_2 cluster ion bombardment enhanced the oxidation and surface smoothing. At the acceleration voltages above 5 kV, the polycrystalline lead oxide films were oriented with their (111) direction parallel to the surface. A significant smoothing effect was observed, even if the acceleration voltage was as low as 1 kV. An average surface roughness of 0.9 nm was obtained at 7 kV. These results indicate that O_2 cluster ions strongly enhanced the oxidation and the smoothness without causing irradiation damage.

INTRODUCTION

Gas cluster ion beams have been applied recently as a new technique for surface modification, providing shallow implantation, high-rate sputtering, surface smoothing, cleaning and film formation as a consequence of the unique irradiation effects.[1-10] The average energy per constituent atom of a 10 keV cluster ion containing 1000 atoms is only 10 eV. As a result, the interactions between the cluster and substrate atoms occur in near-surface regions. Moreover, cluster ions can deposit their energy with a high density within a very localized surface region. Therefore, the irradiation of low-energy gas cluster ions can be expected to enhance the chemical reactions on the substrate surface.

There is a strong requirement to provide low temperature processing for semiconductor and optical device fabrication. For example, the fabrication processes must be carried out at a lower temperature to prevent the degradation of device performance, due to impurity redistribution and the formation of stress-induced defects. Cluster ion beam processing is one of the candidates for use in low temperature processing.

Oxidation is a key issue for semiconductor devices and many kinds of low temperature oxidation techniques, such as plasma oxidation at 600 °C[11] and ion-assisted oxidation at 450 °C[12], have been intensively studied. Room temperature oxidation by a low-energy oxygen ion beam has also been reported.[13]

In this paper, O_2 cluster ion-assisted deposition was examined to form not only SiO_2 but also PbO_x films. Lead oxide is an important material for thin lead-based ferroelectric films used in nonvolatile memory devices.[14] It is important to form these films at low temperature because lead and lead oxide are very volatile materials. The irradiation effects of cluster ions on the oxidation, crystallinity and surface morphology were investigated.

EXPERIMENTS

Figure 1 shows a schematic diagram of the low-energy gas cluster ion beam equipment for film formation. The technique for the generation of gas cluster ion beams and for the cluster-size separation has been described previously[3,6] Cluster beams from gas sources such as O_2, CO_2 and Ar gas can be formed by adiabatic expansion through a Laval nozzle into a vacuum chamber. The clusters were ionized by electron bombardment. The size selection of the cluster ions was performed using a lens system.

High current O_2 cluster ion beams, of a current density of 100 nA/cm^2, were obtained by diluting O_2 gas with He gas and cooling the inlet gases.[9] O_2 cluster ions were accelerated up to 7 kV. The minimum and mean sizes of O_2 cluster ions used in this experiment were 250 and 2000 atoms, respectively. The maximum and average energy per constituent atom were less than 14 eV and 1.75 eV, respectively.

Oxidation of a silicon substrate was performed using O_2 or CO_2 cluster ions. N-type (100) 2-3 Ωcm silicon wafers were used as the substrates. Al gate MOS capacitors of 1.7×10^{-4} cm in area were fabricated on the wafer to examine the electrical characteristics. The thickness of SiO_2 films formed by the cluster ion irradiation was evaluated by X ray photoelectron spectroscopy (XPS).

CP392, *Application of Accelerators in Research and Industry*, edited by J. L. Duggan and I. L. Morgan
AIP Press, New York © 1997

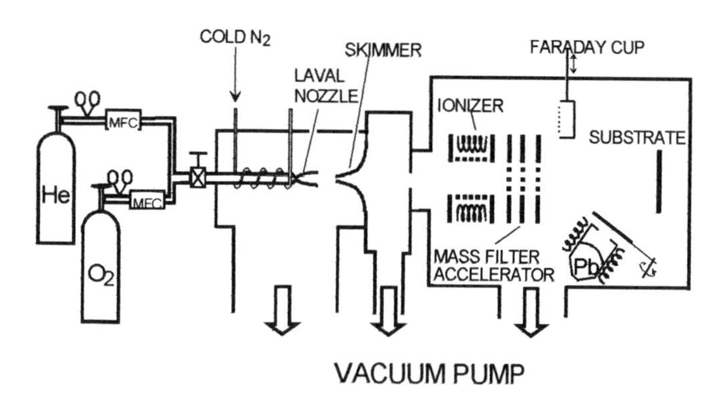

FIGURE 1 Schematic diagram of the low energy gas cluster ion beam apparatus for cluster ion-assisted deposition.

The lead oxide films were grown onto SiO_2 films, formed on (100) silicon wafers, by the deposition of the lead concurrently with O_2 cluster ion irradiation. The lead was evaporated from a carbon crucible with a resistance heater. The deposition rate was 2.4 nm/min at a crucible temperature of 580 °C and varied from 2.4 to 0.7 nm/min. The substrate temperature during the deposition was 65°C, which was caused by the radiation from the crucible. The thickness of the films was approximately 80 nm for all samples.

The thickness of the films was measured by contact profiler. The composition was measured by auger electron spectroscopy (AES) and the surface smoothness was evaluated by atomic force microscopy (AFM). X-ray diffraction (XRD) and reflection high energy electron diffraction (RHEED) were used to evaluate the crystalline structures.

RESULTS AND DISCUSSIONS

Oxidation by O_2 cluster ion beams

Low-temperature oxidation of silicon was performed by O_2 and CO_2 cluster ion irradiation. Figure 2 shows the dependence of SiO_2 film thickness on the ion dose, with various acceleration voltages (Va) of O_2 and CO_2 cluster ions. The sample temperature was kept at room temperature during cluster irradiation. The thickness of the SiO_2 tends to saturate at a dose above $5x10^{15}$ ions/cm^2. The saturated value of the thickness increases with the cluster ion energy and layers formed with O_2 cluster ions were thicker than those by CO_2 for the same acceleration voltage. This result indicates that chemical reactivity of O_2 cluster ions is much stronger than that of CO_2. 11 nm SiO_2 films were formed with 7 keV O_2 cluster ion irradiation at a dose of $5x10^{15}$ ions/cm^2. At this acceleration voltage, the maximum energy per constituent molecule was 28 eV. The irradiated dose of $5x10^{15}$ ions/cm^2 corresponded to $1x10^{19}$ molecules/cm^2, because the mean cluster size was 2000.

It has been reported that a silicon surface is oxidized by 40 - 200 eV O_2 monomer ion irradiation at room temperature when using a Kaufman-type source. 5nm SiO_2 films have been formed by irradiation with a composite argon-oxygen ion beam with the energy of 60 eV at a dose of 10^{18} ions/cm^2. The oxide thickness reached a limiting value of 4-6 nm at a dose of $3x10^{17}$ ions/cm^2, even if at higher monomer oxygen energy. This is caused by the balance between oxidation rate and sputtering rate.[13)] Thicker SiO_2 films (>11 nm) could be obtained by O_2 cluster ion irradiation. Even though the maximum energy per molecule in the clusters is less than 28 eV, thicker oxide films were formed by oxygen cluster ion irradiation. These results reveal that the oxidation mechanism using cluster ion irradiation is of a different form than that of monomer ion irradiation.

According to molecular dynamics simulation, cluster ion bombardment creates a very high-temperature and high-pressure region near the substrate surface.[15,16)] Enhancement

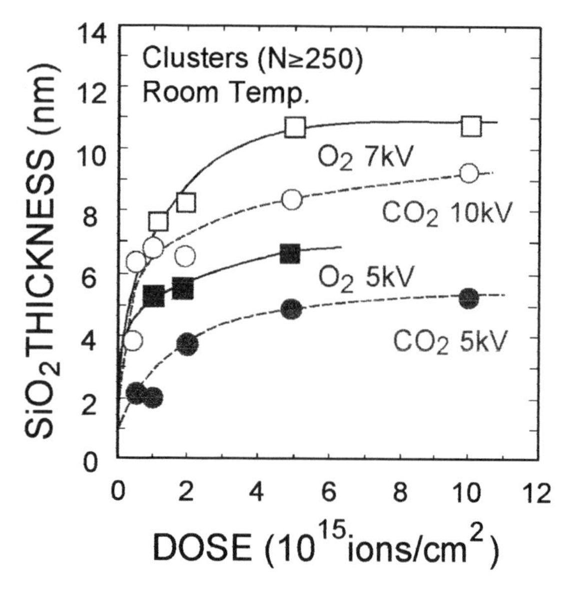

FIGURE 2 Dose dependence of the thickness of SiO_2 films formed by O_2 and CO_2 cluster ion irradiation.

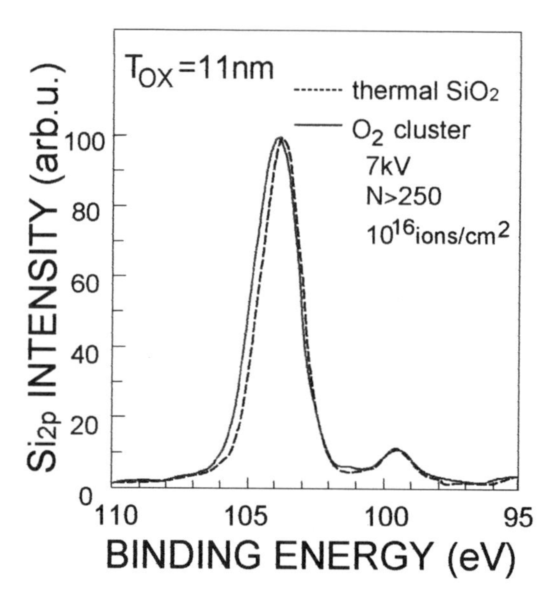

FIGURE 3 Si 2p lines of the X ray photoelectron spectra for the 11 nm thick SiO_2 films formed by irradiation with 7 keV O_2 cluster ions and by thermal oxidation at 900 °C.

of the oxidation process by O_2 cluster ion irradiation can be explained by the generation of excited oxygen and silicon atoms around the impact region caused by the high-density energy deposition. The activation energies for O_2 cluster ion oxidation were 0.018 and 0.020 eV at doses of $1x10^{15}$ ions/cm^2 and $5x10^{15}$ ions/cm^2, respectively. These activation energies are two orders of magnitude lower than the activation energy for the linear rate constant in the case of thermal oxidation.[17)] The activation energy for cluster ion oxidation is close to that for monomer ion oxidation (0.025 eV[12)] and 0.007 eV [13)]). These results indicate that thermal process is not dominant in O_2 cluster ion oxidation and ion bombardment enhances the oxidation. The cluster ion dose is 2-3 orders of magnitude lower than that of monomer ions. This is important for minimizing the degradation of semiconductor devices by charge-up effects.

Figure 3 shows the XPS spectra of Si2p for the 11 nm thick SiO_2 films formed by 7 keV O_2 cluster ion irradiation at a dose of $1x10^{16}$ ions/cm^2 and by thermal oxidation at 900°C. The two spectra are quite similar, indicating that these two SiO_2 films had similar structures. According to a cross-sectional TEM observation, the interfaces of SiO_2/Si formed by both O_2 and CO_2 cluster ion irradiation were very smooth. Quite good C-V characteristics were obtained for a MOS capacitor, whose gate oxide was grown to 6.5 nm in thickness by irradiation with 5 keV O_2 cluster ions at room temperature. The dose was $5x10^{15}$ ions/cm^2. High-quality SiO_2 films could be formed at room temperature.

Cluster assisted deposition of PbO_x

Figure 4 shows the XRD patterns of samples deposited with O_2 cluster ion irradiation as a parameter of the acceleration voltage (Va). The lead films deposited without oxygen clusters were polycrystals with a (111) preferred orientation, which was consistent with the previous reported results.[18)] Increasing the acceleration voltage of oxygen cluster ions, the Pb (111) peak disappeared with the appearance of a PbO (111) peak in the XRD patterns. The lead oxide films had (111) orientations when grown at an acceleration voltage above 5 kV. According to the RHEED observation, polycrystal PbO films were obtained.

The lead oxide films with strongest (111) preferred orientations were grown by decreasing the deposition rate from 2.4 to 0.7 nm/min. These results clearly demonstrate that the bombardment of O_2 cluster ions can enhance the crystallization of the films during the deposition. It has been reported that the randomly oriented lead oxide films are formed by reactive ion beam sputtering deposition[19)] and that lead oxide films with a (100) preferred orientation were formed by the oxidation of Pb (111) films.[20)] The crystalline structures of lead oxide films, deposited with O_2 cluster ion irradiation are different from those formed by other techniques.

Increasing the acceleration voltages of O_2 cluster ions, not only enhances the crystallization but also improvessurface smoothness. Figure 5 shows the AFM images of the PbO_x films deposited with 7 keV O_2 cluster ions (a), with 0.15 keV O_2 cluster ions (b). Dramatic improvement of surface morphology is clearly observed in these AFM images. The

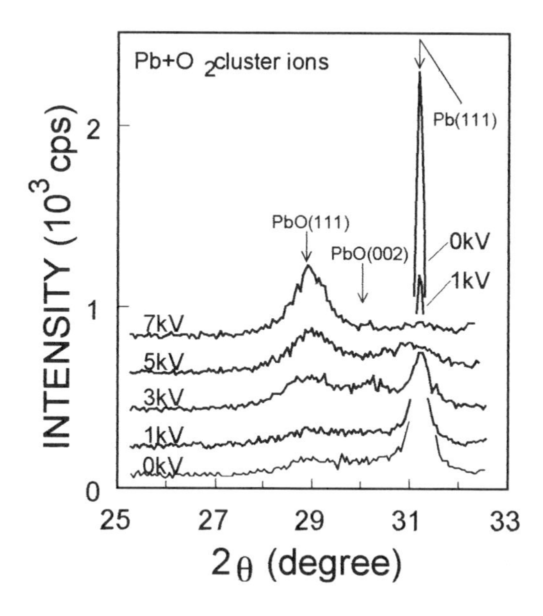

FIGURE 4 XRD patterns of the lead oxide films deposited with O_2 cluster ion irradiation as a function of the acceleration voltage.

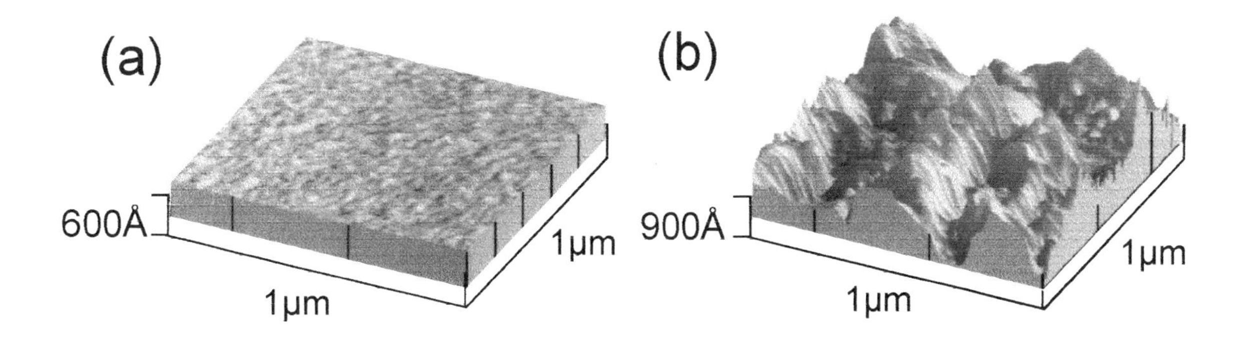

FIGURE 5 AFM images of lead oxide films deposited with 7 keV O_2 cluster ion irradiation (a), with 0.15 keV(b).

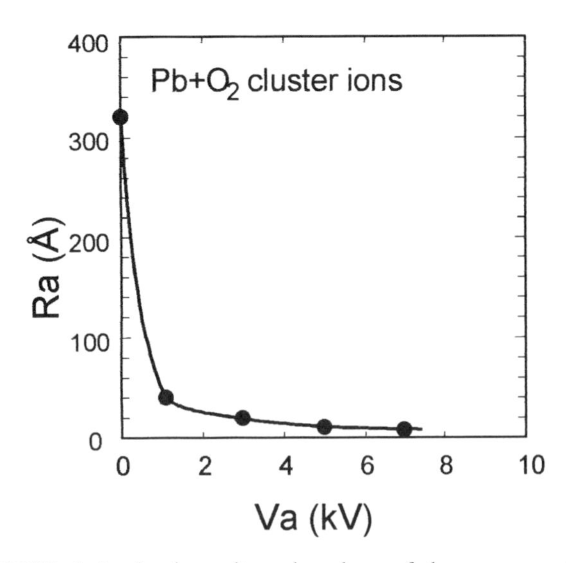

FIGURE 6 Acceleration voltage dependence of the average surface roughness (Ra) of lead oxide films deposited with O_2 cluster ion irradiation.

average surface roughness(Ra) of films deposited with 7keV and 0.15keV O_2 cluster irradiation were 0.9 nm and 32 nm, respectively. The Ra value of the lead films deposited without oxygen was 35 nm, which is close to the value for the films deposited with 0.15keV O_2 cluster irradiation. For comparison, PbO_x films were deposited with an O_2 partial pressure of $2x10^{-4}$ Torr. Although the XD patterns of the films deposited with an O_2 partial pressure showed the PbO_x crystalline structure, the Ra of the film was 5.6 nm, which is about five times larger than the value of the films deposited with 7keV O_2 cluster irradiation. Furthermore, in the case of irradiation by 3 keV O_2 cluster ions, a Ra value of 2.2 nm was obtained, even though the film included the Pb crystalline structure. These results indicate that the surface smoothness can be improved by O_2 cluster ion irradiation.

Figure 6 shows the acceleration voltage dependence of the average surface roughness (Ra) of lead oxide films deposited with O_2 cluster ion irradiation. The Ra was reduced dramatically with increasing acceleration voltage. Although films deposited with 1keV O_2 cluster ion irradiation had (111) preferred orientation crystalline structures similar to Pb films deposited with no oxidation, the films produced using cluster irradiation had significantly lower Ra. The Ra of the films deposited with 5 keV O_2 cluster ion irradiation had the same Ra of 1.1 nm over the range of deposition rate from 0.7 to 2.4 nm/min. No deposition rate dependence of Ra was found, in spite of the increasing of the intensity of the PbO (111) peak as the deposition rate decreased. The surface smoothness is not attributable to the crystalline structure of the films.

A significant smoothing effect of gas cluster ion beams has been reported on various kinds of surfaces, such as Pt, Cu, polycrystalline Si, SiO_2 and Si_3N_4 films.[1-10] This smoothing effect can be explained by lateral sputtering, which has been demonstrated both experimentally and theoretically.[2,15] However, the smoothing effect was observed even at an acceleration voltage of 1keV. Another smoothing mechanism may be responsible for smoothing at such low energy.

CONCLUSIONS

High-quality SiO_2 films of up to 11 nm thick were formed at room temperature on Si surfaces by O_2 cluster ion beam irradiation. A strong enhancement of the oxidation was found using O_2 cluster ion irradiation. This could be the result of high-density energy deposition at near surface regions by gas cluster ion irradiation. PbO_x films were also grown on SiO_2 films by O_2 cluster ion-assisted deposition. The irradiation of the energetic O_2 cluster ions enhanced the oxidation, the crystallization and smoothness of lead oxide films. The polycrystalline lead oxide films formed with O_2 cluster ion irradiation were preferentially oriented with their (111) direction parallel to the surface at acceleration voltages above 5 kV. Significantly smoothed PbO_x film surfaces were obtained with O_2 cluster ion irradiation even if the acceleration voltages were less than 1 kV. An average surface roughness of 0.9 nm was obtained at the acceleration voltage of 7 kV.

REFERENCES

1. I. Yamada, W.L. Brown, J.A. Northby and M. Sosnowski, Nucl. Instr. and Meth. B, **79**, 223(1993)
2. I. Yamada, and J. Matsuo, J.Vac. Sci. Technol. (in press)
3. J. Matsuo, H, Abe, G.H. Takaoka and I. Yamada, Nucl. Instr. and Meth., **B99** (1995) 244.
4. R. Beuhler and L. Friedman, Chem.Rev., **86** (1986) 521
5. J. Matsuo, M. Akizuki, J.A. Northby, G.H. Takaoka and I. Yamada, Proc. of ICSSPIC-7, in :Surf. Rev. Lett. in press.
6. O. F. Hagena, Rev. Sci. Instrum. **63** (1992)2374
7. I. Yamada, J. Matsuo, Z. Insepov and M. Akizuki, Nucl. Inst. and Meth., B106 (1996) 165.
8. M. Akizuki, J. Matsuo, I. Yamada, M. Harada, S. Ogasawara and A. Doi, Nucl. Instr. and Meth. B (1996), in press.
9. M. Akizuki, J. Matsuo, S. Ogasawara, M. Harada, A. Doi and I. Yamada, Jpn. J. Appl. Phys. 35 (1996) 1450.
10. M. Akizuki, J. Matsuo, M. Harada, S. Ogasawara, A. Doi, K. Yoneda, T. Yamaguchi, G. H. Takaoka, C. E. Ascheron and I. Yamada: Nucl. Instrum. & Methods, B99 (1995) 229.
11. Kimura, E. Murakami, T. Warabisako, H. Sunami and T. Tokuyama: IEEE Electron Device Lett., vol. EDL-7 (1986) 38.
12. Kawai, N. Konishi, J. Watanabe and T. Ohmi: Appl. Phys. Lett. 64 (1994) 2223.
13. S. Todorov and E. R. Fossum: J. Vac. Soc. Technol. B6 (1988) 466.
14. J. F. Scott and C. A. Araujo, Science 246 (1989) 1400.
15. Insepov and I. Yamada: Nucl. Instrum. & Methods, B99 (1995) 248.
16. T. Aoki
17. E. Deal and A. S. Grove: J.Appl.Phys. 36 (1965) 3770.
18. M. Murakami, J. Angelillo, H. C. W. Huang, A. Segmuller and C. J. Kircher, Thin Solid Films, 60 (1979) 1.
19. J. M. Vandenberg, S. Nakahara and A. F. Hebard, J. Vac. Sci., Technol., 18 (1981) 268.
20. J. W. Matthews, C. J. Kircher and R. E. Drake, Thin Solid Films, 42 (1977) 69.

REACTIVE ACCELERATED CLUSTER EROSION (RACE) FOR MICROMACHINING

J. Gspann

Universität Karlsruhe und Forschungszentrum Karlsruhe, Institut für Mikrostrukturtechnik, Postfach 3640, D-76021 Karlsruhe, Germany

Accelerated ionized cluster beams are used for micromachining of bulk diamond, CVD diamond films, single-crystalline silicon, or Pyrex glass, among others. Beams of clusters of CO_2 or of SF_6 with about 1000 molecules per unit charge are accelerated to up 120 KeV kinetic energy for mask projective surface bombardment. Patterning is achieved via physical as well as chemical surface erosion: reactive accelerated cluster erosion (RACE). Very smooth eroded surfaces result for bulk natural diamond, silicon, metals and glass. Polycrystalline, strongly faceted CVD diamond films are effectively planarized. Submicrometer structures with adjustable wall inclination can be generated. Surface melting seems to govern the cluster impact induced nanomodifications.

INTRODUCTION

Acceleration of singly ionized clusters of 1000 molecules of CO_2 yields cluster velocities of about 20 km/s which is higher than the speed of sound in any material. Consequently, impacts of such high speed clusters onto whatever target lead to shock wave phenomena, superficial melting, vaporization, and debris ejection /1/. At the immediate vicinity of the locus of impact, very high temperatures of several thousand Kelvin have been calculated to persist for times of the order of a picosecond /2,3/, eventually providing also core-excitations /4/. The resulting plasma enables the involved materials to react chemically even if the cluster material is originally rather inert, as in the case of CO_2 or SF_6. With volatile reaction products, an additional way of erosion is opened up. Otherwise, a surface modification will result. Prolonged bombardment of a target by accelerated cluster beams has been used for surface micromachining /5/. Including the aspect of the chemical transformation, the method is now called RACE: reactive accelerated cluster erosion /6/.

EXPERIMENTAL

The clusters are generated by adiabatic nozzle expansion of CO_2 or SF_6 gas, the converging-diverging nozzle having 0.1 mm throat diameter, 10° angle of initial divergence, and 28 mm length of the divergent part (Figure 1). Most of the expanding gas is frozen onto cryopanels attached to a liquid nitrogen bath cryostat. The core of the expanding nozzle flow is transferred to high vacuum via two skimming orifices. Electrons of 150 eV energy partly ionize the cluster beam which may then be focussed by up to 10 kV potential negative with respect to the acceleration potential. [7]

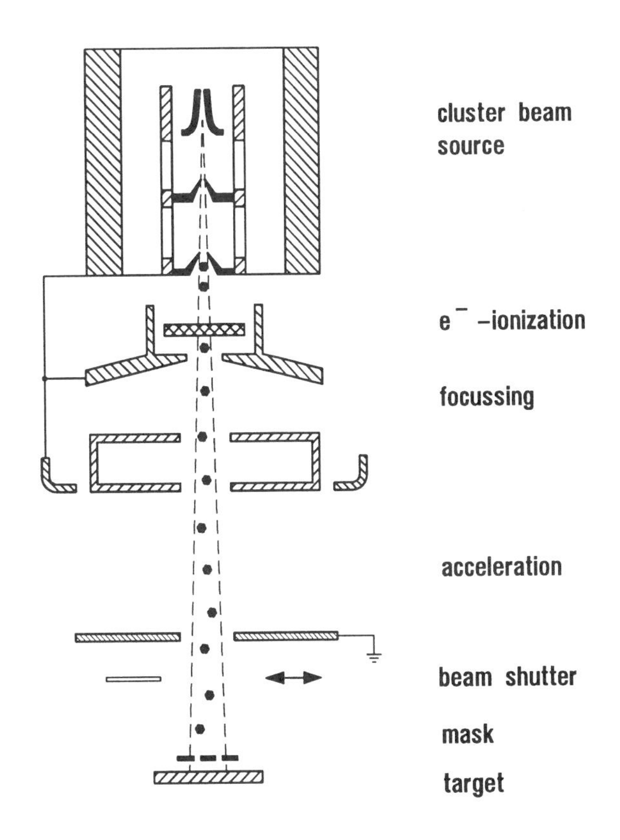

FIGURE 1: Schematic view of the setup for reactive accelerated cluster erosion (RACE).

The cluster mean sizes can be measured by using a dedicated time-of-flight spectrometer in place of the target [8]. The nozzle feed pressure is chosen to provide clusters of a suitable size, e.g. 1000 molecules of CO_2 per unit charge, in order to ensure high speed impacts. At the

CP392, *Application of Accelerators in Research and Industry*, edited by J. L. Duggan and I. L. Morgan
AIP Press, New York © 1997

chosen 100 kV acceleration voltage the ionized clusters impinge on the target with about 20 km/s speed.

The experimental set-up described earlier [7] is now equipped with a rapid beam shutter which allows times of exposure as short as 0.5 ms. A manually operated additional beam flag serves to protect the rapid shutter from prolonged cluster beam erosion. Target, mask and beam shutters are all kept at ground potetial. The cluster beam source can be operated at high potential of either polarity. In the present case, a positive potential of 100 kV is used to accelerate positive cluster ions towards the ground electrode which they pass via a 100 mm diameter central orifice.

Most often, a microstructured nickel foil of 8 μm thickness, which is generated via e-beam lithography and nickel galvanoforming, serves as a stencil mask. Its proximity distance from the target is of the order of 50 μm, as obtained by optical microscope inspection.

After exposure to the accelerated ionized cluster beam, the targets are transferred through ambient atmosphere without particular precautions to either a Scanning Electron Microscope or a Digital Instruments Nanoscope III. They are investigated by atomic force microscopy in the so-called contact mode using silicon nitride cantilevers, or in the tapping mode with Si cantilevers.

RESULTS

Glass and Quartz

Figure 2 shows hexagonal blind holes eroded by impact of carbondioxide clusters of about 1000 molecules into pyrex glass using a nickel mask at about 50 μm distance from the target surface. The bottom planes are at least as smooth as the original glass surface. Inspite of the large proximity distance, steep sidewalls are achieved due to the rather parallel cluster trajectories. Some vertical striations may be seen in the sidewalls of the blind holes which at least in part result from corresponding structures of the mask.

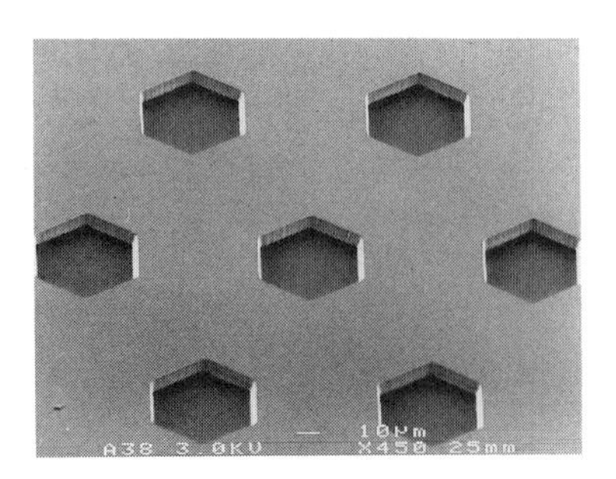

FIGURE 2: Blind holes eroded into pyrex glass by carbondioxide clusters.

Figure 3 shows a similar blind hole eroded into polycrystalline quartz. Obviously, the process of cluster erosion leads to a smoothening of the eroded surface which is reminiscent of melting. In the sidewalls, no vertical but horizontal striations may be recognized. This type of striations is also observed regularly with metal targets, such as gold or copper. These striations are found to remain horizontal also if the target is tilted against the vertical direction. Hence, they may be due to a thin layer of fluid which freezes onto the sidewalls.

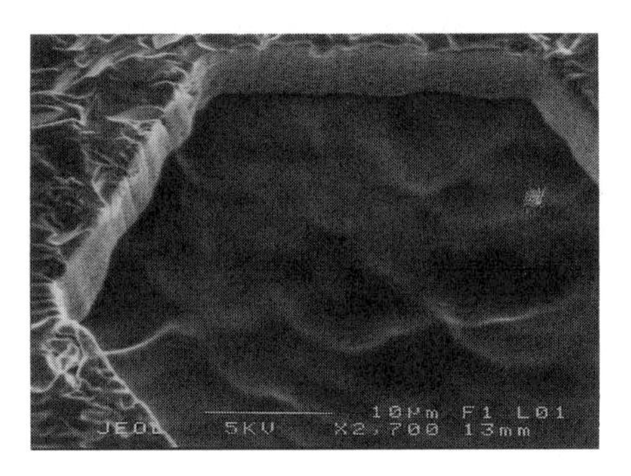

FIGURE 3: Blind hole eroded into polycrystalline quartz by carbondioxide clusters.

Natural Diamond

Figure 4 shows hexagonal blind holes eroded by accelerated CO_2 clusters into natural diamond. The upper right corner of the hexagon in Figure 4 is tilted and enlarged in Figure 5 to show the smoothness of the diamond bottom plane. This hole bottom is as transparent as the original diamond, at least under visual inspection. The granular appearance of the top surface of Figure 5 is mostly due to ejected debris.

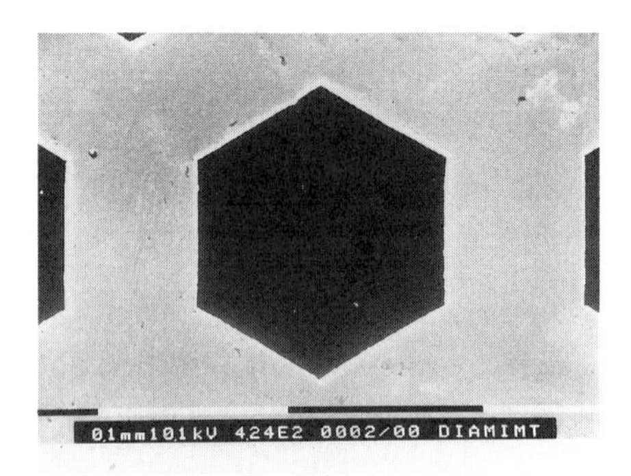

FIGURE 4: Hexagonal blind holes eroded into natural diamond by CO_2 clusters.

FIGURE 5: Upper right corner of the hexagon in Fig. 4, enlarged and tilted to show the very smooth bottom plane.

CVD diamond planarization

The results shown in Figures 3 to 5 motivated the use of cluster erosion also for planarizing polycrystalline artificial diamond films /9/. Figure 6 gives an example of a part of a hexagon eroded into diamond film which has been deposited by chemical vapor deposition (CVD).

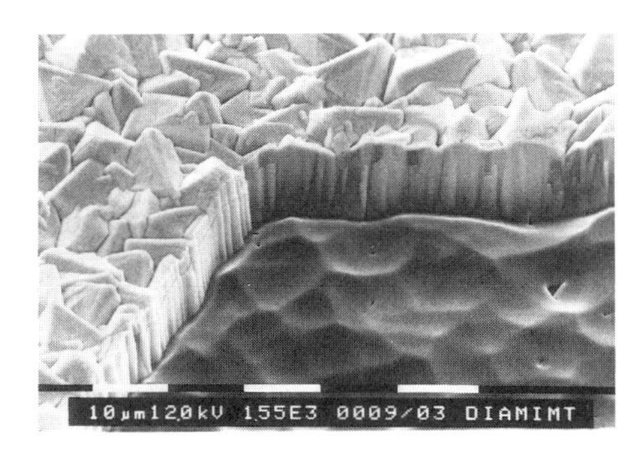

FIGURE 6: Artificial polycrystalline diamond film partially planarized by accelerated CO_2 cluster bombardment.

The very pronounced smoothing achieved is similar to the results obtained with quartz (Figure 3), again indicating some kind of superficial fluidization.

Atomic force microscopy

Finally, the effect of single cluster impacts has been studied by atomic force microscopy of a silicon surface bombarded for very short times /10/. The main result is the observation of very flat hillocks of about 1 nm height instead of craters of some 10 nm depth, as expected in view of the results of more macroscopic impacts /1/. Obviously, craters do not persist due to too slow solidification so that a rather perfect recovery takes place which in turn explains the smoothness of the eroded surfaces.

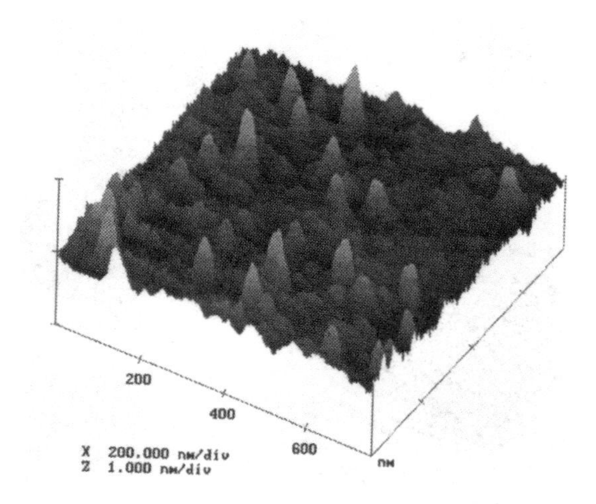

FIGURE 7: Isolated cluster impact induced hillocks on polished silicon as obtained by atomic force microscopy

CONCLUSIONS

The reactive accelerated cluster erosion allows to micromachine, to planarize and to pattern a wide variety of materials. All the energy applied for acceleration is harnessed for the micromachining process unlike, e.g., in the lithography with high-energy phonons. Freedom of lateral pattern design and very smooth eroded surfaces characterize the method. The observed crater recovery is thought to remove the crater dimensions as a limiting factor for the achievable spatial resolution.

REFERENCES

/1/ J. Gspann, From Clusters to Crystals, P. Jena et al. eds., Kluwer Acad. Publ. 1992, 1115

/2/ U. Landman and C.L. Cleveland, Science 257, (1992) 355

/3/ Z. Insepov, M. Sosnowski and I. Yamada, Laser and Ion Beam Modification of Materials, I. Yamada et al. eds., Elsevier Science B.V. (1994) 111

/4/ M.H. Shapiro and T.A. Tombrello, Phys. Rev. Lett. 68 (1992) 1613

/5/ P.R.W. Henkes and R. Klingelhöfer, J. de Physique 50 C2 (1989) 159

/6/ J. Gspann, Nucl. Instr. and Methods B 112 (1996) 86

/7/ J. Gspann, Surface Review and Letters 3 (1996) 897

/8/ J. Gspann and H. Vollmar, J. Chem. Phys. 73 (1980) 1657

/9/ J. Gspann, Sensors and Actuators A 51 (1995) 37

/10/ P. v. Blanckenhagen, A. Gruber and J. Gspann, Nuclear Instr. and Methods B (in press)

CLUSTER-ION FORMATION BY HEAVY-ION-INDUCED ELECTRONIC SPUTTERING

N. Imanishi, S. Kyoh, K. Takakuwa, M. Umezawa, Y. Akahane, M. Imai, and A. Itoh

Department of Nuclear Engineering, Kyoto University, Sakyo, Kyoto 606-01, Japan

Cluster-ion yields have been measured for a SiO_2 target bombarded by Si ions at energies of 1.4 to 5.3 MeV. Positively-charged cluster ions were observed up to a mass-to-charge ratio of 850 u/e and were classified into three series of singly-charged $Si(SiO_2)_p$, $SiO(SiO_2)_p$, and $SiO_2(SiO_2)_p$ ions, where p denotes an integer. The yields were analyzed with a power function of the electronic energy loss, and the obtained exponent varies systematically with cluster size. The yield of clusters composed of a few atoms was found to depend on the intensity of Si ions. This result is attributable to secondary reactions in a selvage region in vacuum adjacent to the target surface.

INTRODUCTION

Electronic sputtering in a MeV-energy region is one of the most interesting and applicable phenomena in ion-solid interactions (1 and references cited therein). In the past decade great progress has been made on the sputtering of frozen gases and organic molecules. Several models have been proposed to interpret the nonlinear dependence of the secondary-ion yield on the electronic stopping power (2-6). A pressure-pulse model based on a molecular-dynamics calculation gains success especially for the desorption of large organic molecules induced by fast heavy ions (5).

So far, systematic experimental studies on the yields of secondary ions for tightly-bound chemical compounds have been scarce except for alkali halides which have been a matter of considerable concern from a theoretical point of view. It has just started to study effects of ion and electron bombardment on SiO_2, which has attracted much interest because of its technological importance to silicon devices. For examples, Jacobsson and Holmen measured the sputtering yield in O_2, CO, and N_2 ambient for several ions with energies of 150-300 keV (7). Sugden *et al.* observed the variation of sputtering yield with ion species, charge state, and incident angle in a MeV-energy range (8). Electron stimulated desorption of ions was studied by Petravic *et al.* (9). Data are, however, far from systematic to draw any picture of sputtering mechanism.

In the present study, we have measured in more detail the yields of individual secondary ions produced from a silicon-oxide target bombarded by Si ions at impact energies of 1.4 to 5.3 MeV, where the electronic stopping power is dominant (10). Thus, any role of the electronic collision will be revealed by the production of secondary ions containing large-sized clusters emitted from the tightly-bound chemical compounds.

EXPERIMENTAL

Measurements of secondary cluster ions were carried out using a method described in detail previously (11) and briefly described below. A Si ion beam from the Kyoto University 1.7-MV tandem Cockcroft-Walton accelerator was collimated to a spot of 1 mm in diameter and incident on a 400-nm thick SiO_2 target epitaxially grown on a silicon wafer. The secondary atomic and cluster ions sputtered from the target were extracted through a 1-mm-diam aperture positioned at 5 mm apart from the target. They were, then, accelerated and focused with an einzel lens onto a 6-mm-diam entrance slit of a magnetic mass-analyzer and were finally detected with a channel electron multiplier (Ceratron) (12) through a 1×5-mm^2 slit, resulting in a mass resolution of 1.6%. Mass spectra of secondary ions were measured for Si projectiles in the energy and current regions of 1.4 - 5.3 MeV and 0.02 - 10 particle nA (abbreviated nA hereafter), respectively. The obtained mass spectra were reproducible and hardly depended on the charges of the incident ions.

RESULTS

An example of the obtained mass spectra of secondary-ion species is shown in Fig.1. Dominant species found were $Si_nO_m^+$ species with m being equal to $2n$, $2n$-1, and $2n$-2, but oxygen rich species of $m \geqq 2n+1$ were not observed. The secondary ions could be fully observed at a normal extracting voltage down to 0 V, but were substantially suppressed by applying a reverse voltage even lower than 1 V. Therefore, the target was certainly free from the electrical charging-up, and the spectra were taken without applying the extracting voltage throughout the experiment.

CP392, *Application of Accelerators in Research and Industry*, edited by J. L. Duggan and I. L. Morgan
AIP Press, New York © 1997

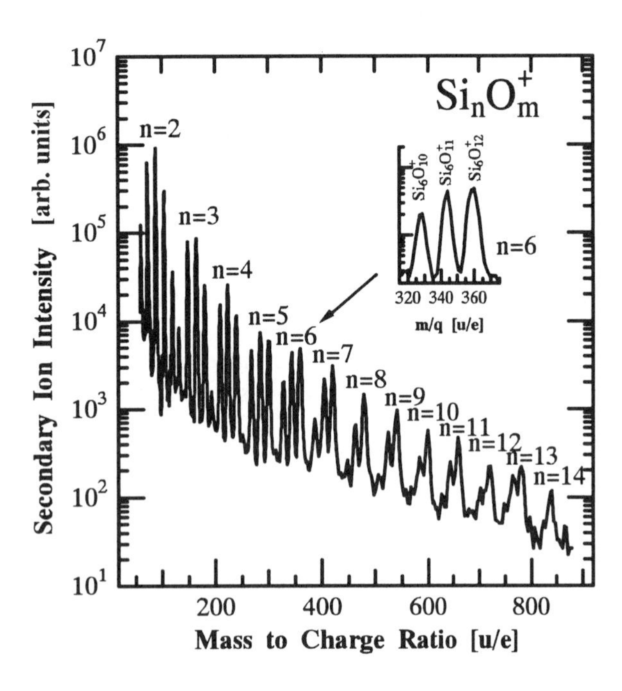

FIGURE 1. An example of the obtained mass spectra of secondary ions for a SiO_2 target bombarded by 4.0-MeV Si ions. Dominant species found were $Si_nO_m^+$ species with m being equal to $2n$, $2n$-1, and $2n$-2 (see the inset).

Figure 2 shows the secondary-ion yields as a function of the Si incident energy. In the present energy region, the electronic stopping power is dominant and increases with increasing incident energy and vise versa for the nuclear stopping power (10). The steep increase of the yields therefore reflects the increasing tendency of the electronic stopping power of the Si-SiO_2 projectile-target system.

Figure 3 shows typical examples of the dependence of the yields on incident current. Data were taken from 0.02 nA to 10 nA for the incident Si ions at 3.0 MeV. The yields of rather simple ions such as Si^+, Si_2^+, SiO^+, and Si_2O^+ increased with increasing beam current. On the other hand those of complex cluster ions and oxygen ions were found to be rather independent of the beam current. A similar current dependence was found in sputtering and/or desorption induced by electrons (13).

DISCUSSION

So far, mechanisms based on the electronic energy deposition have been proposed to explain secondary-particle emission under high-energy ion bombardment. However, different models predict different dependences on the electronic stopping power and the detailed mechanism is still an open problem. Most of the models try to explain the total yield of the secondary particles and do not step into the explanation of the yields of individual components.

In the present experiment it was found that the relative secondary-ion yield Y is given by the following simple power function of the electronic stopping power S_e [eV/Å]

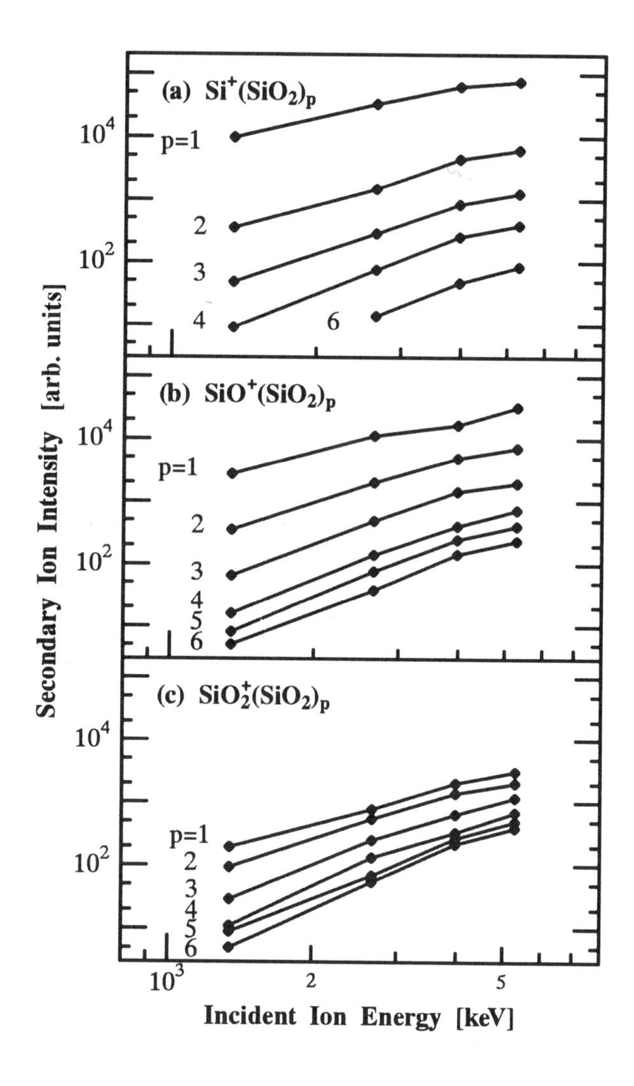

FIGURE 2. Incident-energy dependence of the secondary-ion yields for the SiO_2 target bombarded by Si ions: (a) $Si^+(SiO_2)_p$, (b) $SiO^+(SiO_2)_p$, (c) $SiO_2^+(SiO_2)_p$.

and a size parameter P

$$Y = 10^{6.7} \times F^{P^{0.38}} \tag{1}$$

$$F = \left(\frac{S_e}{10^{3.4}}\right)^{2.4}, \tag{2}$$

where the size parameter P is defined as a number of atoms in the cluster divided by 3 and represents that the cluster ions are formed by SiO_2 molecules and SiO_2 constituents. In Fig. 4, the reduced logarithmic yields $\log(Y/10^{6.7})/\log F$ of all the observed cluster ions for the different incident energies are plotted as a function of P for the Si-SiO_2 system. The factor F depends on the electronic stopping power in the form of power function with an exponent of 2.4, as described in Eq. (2). As can be seen from Eq. (1) the factor F relates to the yield of the main component SiO_2 molecule and thus to the total sputtering yield.

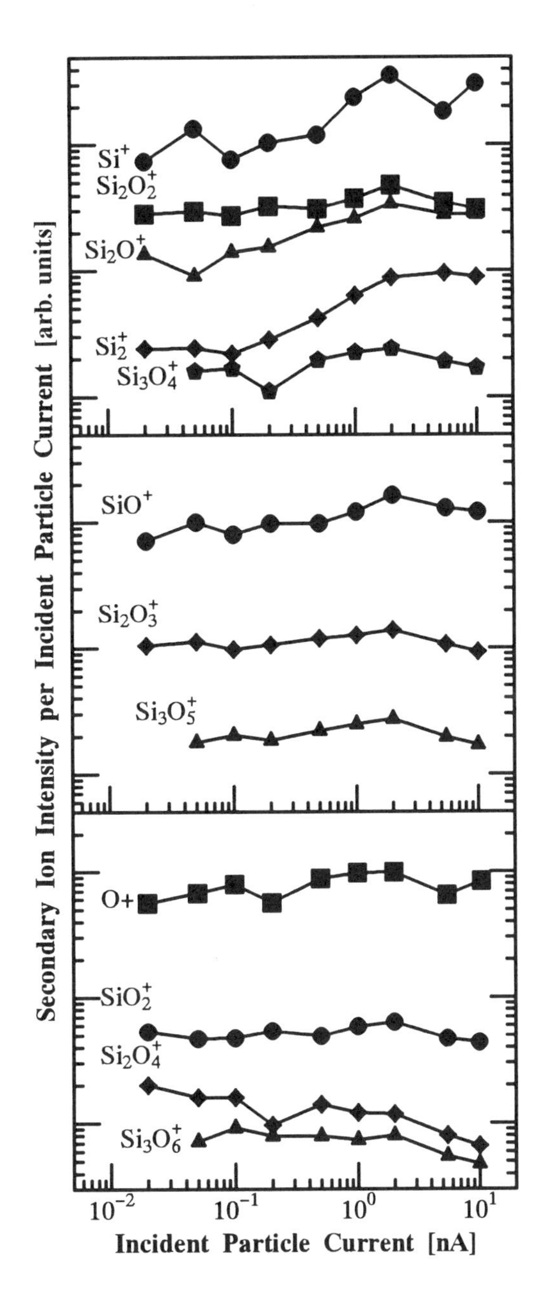

FIGURE 3. Incident-current dependence of the secondary-ion yields for the SiO_2 target bombarded by 3-MeV Si ions.

In previous experiments, it was found that the total sputtering yield is proportional to the electronic stopping power at a low energy deposition and has a nonlinear dependence at high energy deposition. The nonlinear dependence was predicted by several models such as the shock wave (2), the Coulomb repulsion (3), the thermal spike (4), and the pressure pulse models (5). The respective exponents of the power function of the electronic stopping power are predicted to be 1.5, 2, 2, and 3. In the present Si-SiO_2 case, the exponent of 2.4 was obtained from the experimental yields, and this value is not inconsistent with the exponents predicted by the models except for the shock wave model.

The relative yields of all of the observed cluster ions were

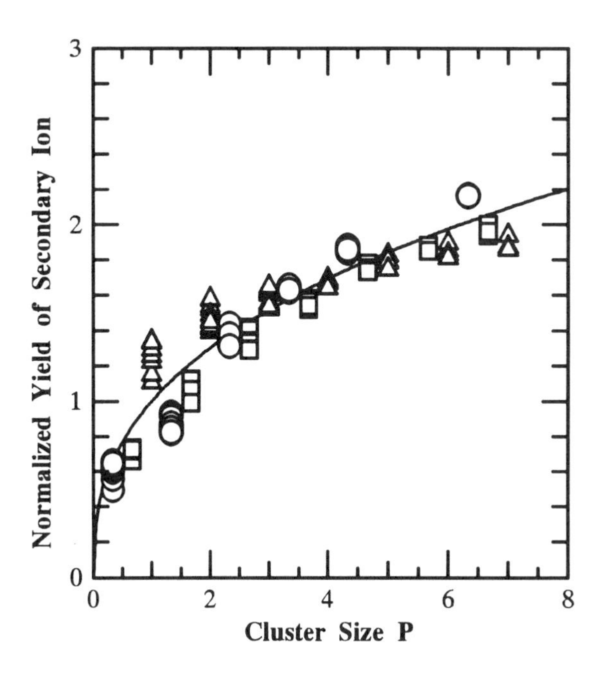

FIGURE 4. The logarithmic yields of the observed ions are normalized by a factor of log$F$, and plotted as a function of size parameter P of the secondary ions (see Eqs. (1) and (2) in detail). The circles, squares, and triangles represent the $Si^+(SiO_2)_p$, $SiO^+(SiO_2)_p$, and $SiO_2^+(SiO_2)_p$ ions, respectively.

found to be approximately proportional to the power function F with the exponent $P^{0.38}$, as shown in Eq. (1) and Fig. 4. When it is assumed that the cluster ions are formed by the coagulation of SiO_2 molecules in a vacuum, the exponent should not be $P^{0.34}$ but P because of the incoherent nature of sputtering. In addition, the density of SiO_2 molecules in a selvage region in vacuum just outside the target surface is estimated to be less than 10^9/cm^3/nA by referring to sputtering yields for the Cl-SiO_2, F-UF_4, and Cl-UF_4 projectile-target systems in an energy region of 0.05-2 MeV/u (8,14). This estimated density is not enough to coagulate to form large clusters composed of up to 14 molecules. When the observed cluster ions are assumed to be formed by decomposition of very large metastable clusters after sputtering, intense massive parents should be observed but there is no indication that any clusters composed of a particular number of SiO_2 play an essential role in the observed spectra shown in Fig. 1. Another possibility is that the clusters are formed in the near-surface region due to the multiple breaking of bonds directly by the incident ions and indirectly by the shower of secondary electrons as proposed and named an ion-track model by Hedin *et al.* (6). This collective effect was supported by the observation that the sputtering process does not occur in the case of the silicon oxide layer thinner than a critical thickness (8,14). In the present case, an energy of 100-300 eV/Å is deposited to the near-surface region and 60% of the deposit energy, that is, 60-180 eV/Å, is consumed in the excitation process of bonding electrons. This energy is sufficient to break a large number of covalent bonds (4.7-8.2 eV/bond) at the same time (15). On the basis of the ion-track model, the cluster-ion yield is described by a power function of S_e with an exponent depending on the size of the emitted ions. In the present experiment, the exponent depends on the size parameter P as given by $P^{0.38}$. Thus, the experimental result qualitatively supports the ion-track model.

A gradual increase of the yield with increasing incident current was observed for relatively small cluster ions at the 0.1-10 nA range, where the probability for a track region formed by a projectile to be hit directly by a subsequent projectile is very low because its radius is narrower than about 2 nm (16) and its life time is 10^{-11}-10^{-10} s. The dislocation of atoms in the track region may induce memory effects. The dislocation is more pronounced in the collision of low-energy ions than that of MeV ions, because the former induces more violent collisions through nuclear collisions than the latter. However, the current dependence of the yields have not been observed in the nA region for the low energy ions (17). In addition, SiO_2 is lost rather than one element preferentially (8). The secondary-ion yields at the respective beam currents were reproducible irrespective of the history of the ion bombardment. Therefore, the current dependence cannot be explained by the atomic processes happening in solid.

The importance of the interaction in the selvage region in vacuum just outside the target surface was pointed out in previous reports (1). Based on the finding of the dependence of ion yield on incident electron current, Walkup *et al.* claimed that the positive ions produced by electron bombardment of alkali halides are formed by gas-phase ionization of neutral atoms and molecules (13). In sputtering, the number of neutrals ejected is generally much larger than the number of ions produced (1). Then, it is plausible that the atomic and small cluster ions are additionally formed through the secondary process that the neutral clusters floating in the selvage region are dissociated into ions via collisions with incident ions and the accompanying secondary electrons.

CONCLUSIONS

The secondary-ion yields have been measured for the SiO_2 target bombarded by Si projectiles at energies between 1.4 and 5.3 MeV, where the electronic collision becomes more dominant than the nuclear collision in energy deposition. The observed cluster ions are classified into three series of singly-charged $Si(SiO_2)_p$, $SiO(SiO_2)_p$ and $SiO_2(SiO_2)_p$ ions, where p denotes an integer. All the yields increase with increasing incident energy and are characterized by a power function of the electronic energy loss, whose exponent varies with cluster size. From the fact we can conclude that the cluster ions are produced directly through the collective process in the solid. In addition, an incident-current dependence of the yield was found for some small cluster ions composed of a few atoms beyond a certain value of incident current. Therefore, in the case of the small ions, the secondary process between the ejected neutrals and incident ions in the selvage region outside the target surface plays a substantial role.

ACKNOWLEDGMENTS

This work was done with the Experimental System for Ion Beam Analysis at Kyoto University. We thank K. Yoshida and K. Norizawa for their useful advice and technical support during the experiments. It has been supported in part by a Grant-in-Aid for Scientific Research from the Ministry of Education, Science, Sports, and Culture of Japan.

REFERENCES

1. Sundqvist, B. U. R., ed. Behrisch, R., and Wittmaack, K., *Sputtering by Particle Bombardment III*, Berlin, Heidelberg: Springer, 1991, p.257, and refs. cited therein.
2. Bitensky, I. S., and Parilis, E. S., *Nucl. Instr. Meth.* B **21**, 26 (1987).
3. Johnson, R. E., and Brown, W. L., *Nucl. Instr. Meth.* **209/210** 469 (1983).
4. Johnson, R. E., *Inst. J. Mass Spectrum. Ion Process*, **78**, 357 (1978).
5. Fenyo, D., Sundqvist, B. U. R., Karlsson, B. R., and Johnson, R. E., *Phys. Rev.* B **42**, 1895 (1990).
6. Hedin, A., Hakansson, P., Sundqvist, B. U. R., and Johnson, R. E., *Phys. Rev.* B **31**, 1780 (1985).
7. Jacobsson, H., and Holmen, G., *Nucl. Instr. Meth.* B **82**, 291 (1993).
8. Sugden, S. Sofield, C. J., and Murrell, M. P., *Nucl. Instr. Meth.* B **67**, 569 (1992).
9. Petravic, M., Williams, J. S., and Wong, W. C., *Nucl. Instr. Meth.* B **78**, 333 (1993).
10. Ziegler, J. F., Handbook of Stopping Cross Section for Energetic Ions in All Elements (Pergamon, New York, 1980); Also Biersack, J. P., and Ziegler, J. F., TRIM code.
11. Kyoh, S., Takakuwa, K., Sakura, M., Umezawa, M., Itoh, A., and Imanishi, N., *Phys. Rev.* A **51**, 554 (1995).
12. Murata Mgf. Co., Ltd. in Japan.
13. Walkup, R. E., Avouris, Ph., and Ghosh, A. P., *Phys. Rev.* B **36**, 4577 (1987).
14. Seiberling, L.E., Meins, C.K., Cooper, B.H., Griffith, J.E., Mendenhall, M.H., and Tombrello, T. A., *Nucl. Instr. Meth.* **198**, 17 (1982); Meins, C. K., Griffith, J. E., Qiu, Y., Mendenhall, M. H., Seiberling, L. E., and Tombrello, T. A., *Radiat. Eff.* **71**, 13 (1983).
15. Malherbe, J. B., Hofmann, S., and Sanz, J. M., *Appl. Surf. Sci.* **27**, 355 (1986).
16. Tombrello, T. A., *Nucl. Instr. Meth.* B **94**, 424 (1994).
17. Postawa, Z., Kolodziej, J., Czuba, P., Piatkowski, P., Szymonski, M., Bielanska, E., Camra, J., Ciach, T., Faryna, M., and Rakowska, A., *Nucl. Instr. Meth.* B **78**, 314 (1993).

EXPERIMENTS ON THE FRAGMENTATION OF C_{60} MOLECULES

H. Gordon Berry

Department of Physics, University of Notre Dame, Notre Dame IN 46556

Ion-fragment yields from the fullerene molecule C_{60} in vapor form have been reported for two different collision systems: the impact of high-energy, highly charged xenon projectiles, and the impact of x-ray photons at energies close to the 1s K-edge of carbon in the range of 270 to 320 eV. We compare the results of these two experiments in terms of the fragmentation processes taking place, and also report on a further experiment of observing fluorescence following the impact of fast heavy ions. One goal of this last experiment was to observe the giant dipole(plasmon) resonance of the C_{60} molecule in emission and help in interpreting the two earlier experiments.

INTRODUCTION

The general problems of the nature of clusters and their dynamics lie between those of atoms and molecules on the one hand, and with solids on the other. The former have become accessible to detailed quantum mechanical calculations which take into account the full wavefunctions of at least all the valence electrons and, where necessary can include inner-shell electrons of individual atoms. The C_{60} cluster of carbon atoms is a prime example of a cluster whose atoms are all on the surface, as pointed out in the early work on fullerenes.(1)

Their quasispherical cage structures are amongst the most stable large molecular structures known. Hence, it is of interest to study in detail their break-up mechanisms, following the input of various amounts of energy.

Our initial experiments(2) were aimed at studying the fragment products following extremely violent collisions induced by fast highly-charged heavy ion projectiles. Following the surprising results of these experiments, and the preliminary interpretations, two further experiments were performed which comprise the main focus of this brief report: The first consisted of depositing a very well defined amount of energy in the C_{60} fullerene by bombardment with x-rays close to the energy needed to liberate one K-shell electron from one of the carbon atoms of the molecule, and again study the ion fragment spectrum(3); the second, was the observation of fluorescence from the C_{60} molecule, following heavy-ion impact(4), a similar collision system to the experiments of ref. 2.

The heavy-ion collision experiments were carried at Argonne National Laboratory by the Argonne atomic physics group, the authors of the papers in references 2 and 4. The X-ray experiments utilized the MAX I synchrotron storage ring in Lund, Sweden, the experiments being performed by the Finnish collaboration headed by S. Aksela, and who appear as authors of the publication of ref. 3. We describe the three different collision experiments below.

ION FRAGMENTATION FOLLOWING HEAVY ION IMPACT ON C_{60}

The arrangement for heavy-ion collision experiments is shown in Fig. 1. The C_{60} vapor target was produced in an oven heated to approximately 490 degrees C, and was bombarded with a pulsed beam of highly charged xenon ions of energies in the range of 420 to 625 MeV.

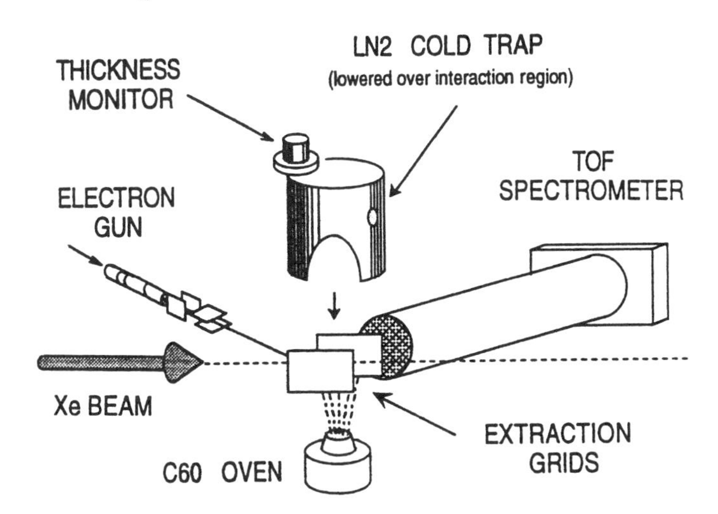

Figure 1. Experimental layout for ion fragment measurements from Xe ion collisions with C_{60}.

The beam was produced at ATLAS, the Argonne superconducting accelerator, and most experiments utilized 625 MeV Xe^{35+}. The resulting ion fragments were extracted

CP392, *Application of Accelerators in Research and Industry*, edited by J. L. Duggan and I. L. Morgan
AIP Press, New York © 1997

and counted in an ion time-of -flight mass spectrometer. Up to 8 multiple coincidences could be recorded in the flight times of up to 8 microseconds between successive pulses of the ion beam. Further experimental details are given in reference 2.

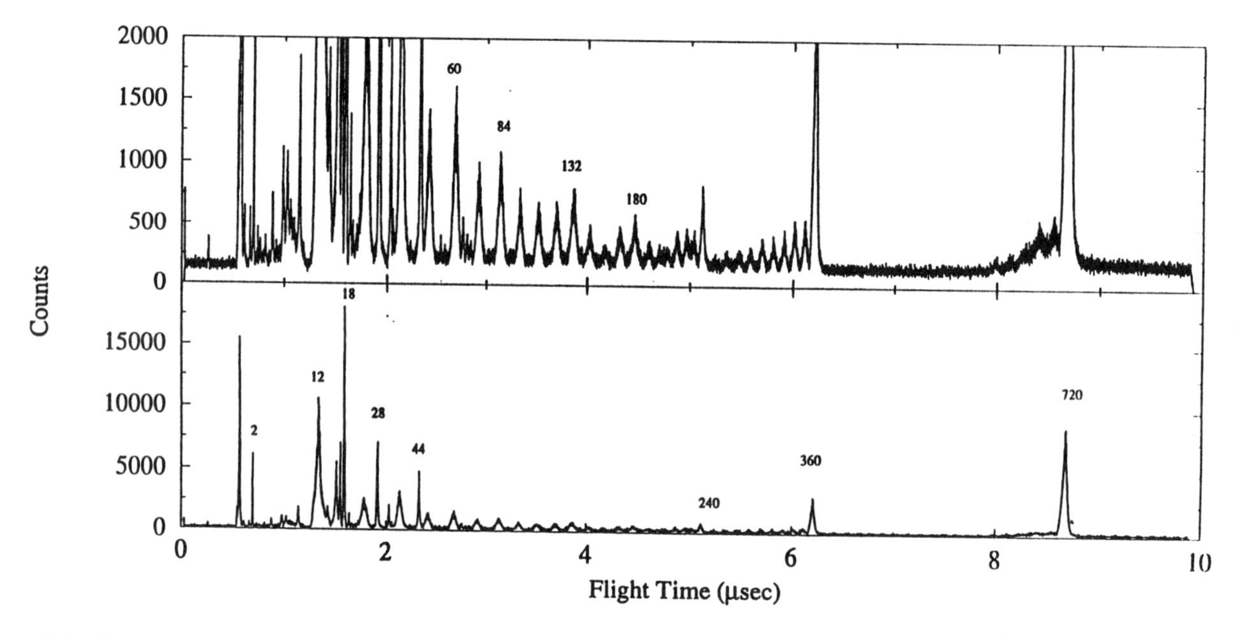

Figure 2. Ion fragment time-of-flight spectrum following xenon ion collisions with C_{60}.

General observations of this spectrum show positively charge carbon fragments from the single atom up to and including C_{60}. Figure 3 shows more clearly that multiply-charged C_{60} ions are seen, plus the even-numbered carbon masses of 58 down to down to 48, for singly-, doubly, and probably three-times ionized ions. The peaks are all narrow, indicating no time delays in ejection, and small amounts of kinetic energy.

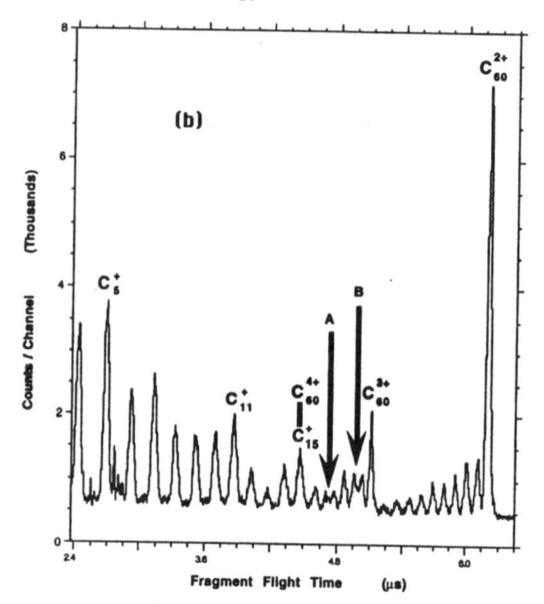

Figure 3. Charged fragments in the range of M/Q from C_5^+ to C_{60}^{2+}.

Further details on the dependence of the yields on ion-beam energy, incident xenon charge state, and on the coincidence measurements can be found in ref. 2. We focus here on the proposed mechanisms for the modes of excitation of the C_{60} molecule. For large impact parameters, the dominant mode of excitation is through a giant dipole resonance (GDR) that was initially predicted(5). And then measured (6)(7) in ultraviolet photoexcitation and dissociation of the fullerene molecule. This resonance has an energy of about 20 eV, with a full halfwidth of about 10. The theoretical model treats a "jellium-like" solid shell, with the inner electrons and ion centers smeared out to create a spherical shell of radius 3.55Å and a thickness of about 0.2Å, in which the 240 valence electrons, 180 σ electrons and 60 more weakly bound π electrons, are confined. The latter lie mainly on the inner and outer rims of the shell, and all 240 electrons participate in the collective excitations.

The model described in (2) also provides a prediction of the distribution of ion fragment masses, which is compared with the experimental results in Fig. 4. A principal input into the model is the energy deposition through the collision with the xenon ion. Thus, the model can be utilized for different processes for depositing energy into the fullerene cage. In particular, it can be applied to the impact of X-ray photons.

X-RAY IMPACT ON C_{60}

The experimental setup of Figure 1 was used, eith the xenon ions being replaced by a beam of photons of energies 280 to 313 eV from the MAX-I storage ring in Lund,

Sweden. The oven was of similar construction and also operated at a temperature of close to 500 degrees C. Further experimental details of the X-ray experiment are given in (3).

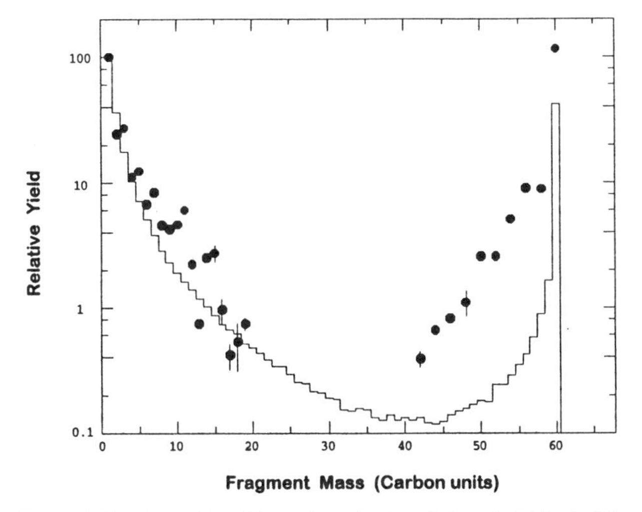

Figure 4. Experimental (points) and theoretical predictions (solid line) of the distribution of ion fragment masses, following heavy-ion impact.

The goal of the experiment was to selectively excite electrons of an individual or localized carbon site, to different bound, unoccupied valence levels and to the continuum, and to compare the dissociation pathways as a function of energy. The ionic states created in core-hole decay processes are different for resonant excitation below and ionization above the ionization threshold. The low fluorescence yield means that resonant excitation is accompanied by the resonant Auger process leaving the molecule singly ionized. Photoionization leads to a doubly ionized molecule through the normal Auger process. In both cases a fast Auger electron is emitted with the released transition energy but can lose part of it in multielectron processes (shakeup, plasmon excitations) to the remaining ionic system. Electrons and/or fragments are then emitted, giving rise to multiply charged ions or to different fragment ions. The interaction between the fast electrons emitted in resonant or normal Auger processes and the rest of the molecule will be very similar, because differences in the kinetic energies are only a few eV at 270 eV. However, slow photoelectrons with varying kinetic energies can cause enhanced fragmentation, following possible excitation of a giant dipole/plasmon resonance.

Figure 5 shows time of flight ion fragment spectra made at six different incident photon enenergies. The inset of the figure indicates the total ion yield spectrum, and is similar to previously published spectra taken from solid C_{60} (8), C_{60} in a solid frozen gas matrix (8b) and vapor phase absorption (8c).

The sharpest features *A*, *B*, and *C* in the spectrum are due to the excitation of the electron to the $5t_{1u}$, $2t_{29}$, $5t_{2u}$ and $4a_g$ orbitals of the π^* character. The ionization threshold at 290 eV is followed by strong σ^* resonances up to around 310 eV Terminello *et al.*(8).

TOF ion spectra (Fig. 5) were collected for incident photon energies of the resonance peaks *A*, *B*, and *C* and above the C-ionization threshold at 6 photon energies of 292-313 eV.

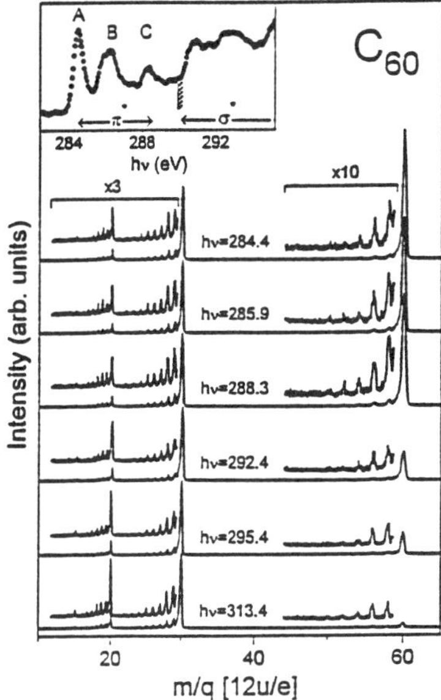

Figure 5. A set of TOF spectra at the indicated incident photon energies. The inset shows the total ion yield spectrum.

The spectra show rich structure revealing peaks due to multiply charged C_{60}^{q+} (q=1 to 4) ions and associated mass-loss peaks. Some general features of the spectra are obvious. At the π^* resonances the C_{60}^{+} lines dominate but the relative intensities of the C_{60}^{2+} peaks are rather high even for these resonant excitations. Note that in these resonant cases, the resonant Auger decay process leaves the molecule initially in a singly ionized state. The high intensity of the doubly-charged ions indicates that second-step Auger transitions are energetically possible for a large fraction of the final states (40%) populated in the course of the resonant Auger process. Multiply charged ions can also be produced by shakeoff processes during the resonant Auger transition, but this is not believed to be a significant channel.

Other details explaining these fragment ion mass spectra are discussed in ref. 3.

FLUORESCENCE YIELDS FOLLOWING HEAVY-ION IMPACT WITH C_{60}

The experimental setup was similar to that used in Fig. 1, with the TOF analyzer being replaced by a 0.2-m vacuum monochromator, equipped with a channelplate detector in its exit focal-plane to collect the wavelength-dispersed photons. 5-15 pnA beams of multicharged xenon with mean charge of 40 bombard the C_{60} vapor target.Further experimental details are given in ref. 4.

If the giant dipole (plasmon) resonance is observed in emission, it should be roughly Gaussian in shape, centered at 20 eV, and with a width of 5.4 eV. Its wavelength distribution is then centered at 620Å, with a full-width of 330Å, but skewed to give a long tail at higher wavelengths. Fig. 6 shows an emitted photon spectrum obtained within this spectral region. The region is clearly rich in well-defined transitions, which dominate the expected giant dipole resonance emission.

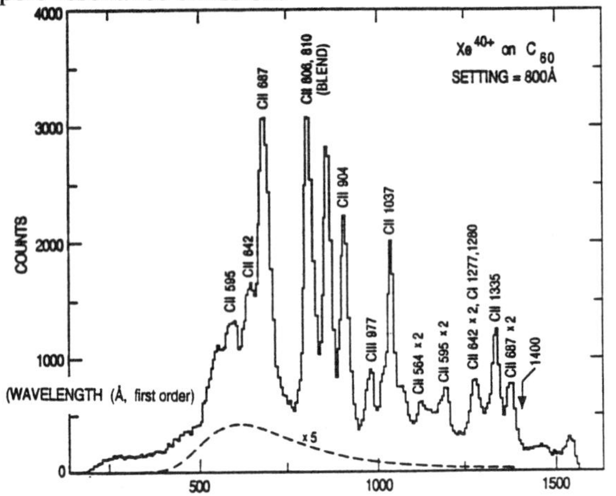

Figure 6. VUV spectrum from C_{60} following collisions with 625 MeV Xe^{+40} ions.

All lines in this spectrum can be associated with well-known transitions in neutral and ionized carbon atoms C I. C II, and C III. The dashed curve in the figure indicates the expected profile of the dipole resonance.

The strength of the plasmon profile takes into account a careful calculation of the efficiency of the detection system, and of the number density of C_{60} molecules, an estimate of the number of xenon ion projectiles, and an estimate of the branching ratio for fluorescence of the dipole itself. Thus, we used a cross-section for excitation of the resonance of 387 $Å^2$, a photon emission branch of $5x10^{-5}$, an efficiency for detection of 10^{-4}, and a target thicknes of $3x10^{11}$ cm^2.

In Fig. 7 we show that the time decay of the fluorescence is rapid. We have been able to fit these decays to multi-exponentials which yield lifetimes similar but slightly longer that the atomic lifetimes of their upper states.

The principal conclusion of these experiments is that we are unable to verify the presence of the plasmon resonance as a main channel for the excitation energy in these experiments. However, it remains the most likely energy exchange pathway.

References

1. H.W. Kroto, J. R. Heath, S. C. O'Brien, R. E. Curl, and R. E. Smalley, Nature (London) **318**, 162 (1985).
2. LeBrun, T., Berry, H. G., Cheng S., Dunford, R. W., Esbensen, H.,Gemmell, D. S., Kanter, E. P., and Bauer, W., Phys. Rev. Lett. **72**, 3965 (1994).
3. Aksela, S., Nõmmiste, E., Jauhiainen, J., Kukk, E., Karvonen, J., Berry, H. G., Sorensen, S. L., and Aksela, H., Phys. Rev. Lett. **75**, 2112 (1995).
4. Ali, R., Berry, H.G., Dunford, R.W., Gemmell, D.S., Kanter, E.P., LeBrun, T., Reichenbach, H.M., and Young, L. J. Phs. B. to be published (1996)
5. Bertsch, G.F., Bulgac. A., Tomanek, D., and Wang, Y., Phys. Rev. Lett. **67**, 2690 (1991).
6. Hertel, I.V., Steger, H., de Vries, J., Weisser, B., Menzel, C., Kamke, B., and Kamke, W., Phys. Rev. Lett. **68**, 784 (1992).
7. Yoo, R. K.,Ruscic, B., and Berkowitz, J., J. Chem. Phys. **96**, 911 (1992).
8. (a) Terminello, L. J., et al, Chem. Phys. Lett. **182**, 491 (1991). (b)Wästberg, et al, Phys. Rev. B **50**, 13031 (1994). (c) Krummacher, S., Biermann, M., Neeb, M., Liebsch, A., and Eberhardt, W., Phys. Rev. B **48**, 8424 (1993).

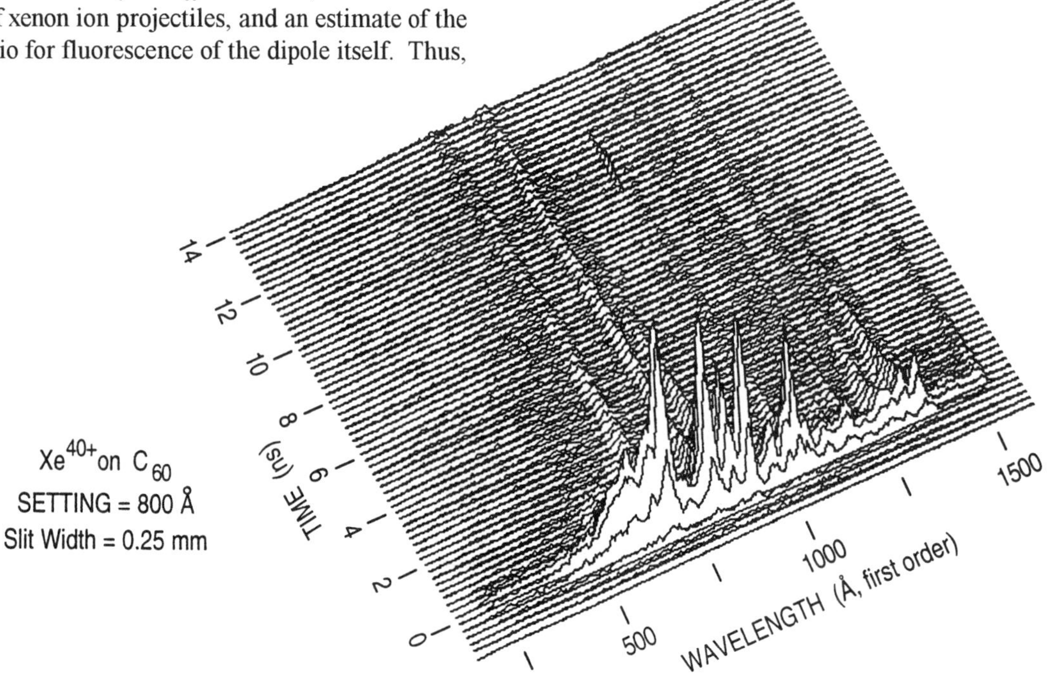

Figure 7. Spectrum of wavelength versus time.

SIMULATION OF ENERGETIC PARTICLE-SURFACE IMPACTS INVOLVING FULLERENES

Roger Smith, Steve Hobday[1] and Roger Webb[2]

[1]*Loughborough University, Leicestershire LE11 3TU, U.K.*
[2]*University of Surrey, Guildford GU2 5XH, U.K.*

The impacts of C_{60} fullerenes with graphite surfaces and the impact of Ar ions with fullerite are studied by means of molecular dynamics simulations. For the case of the C_{60} impacts on graphite, surface waves spread from the point of impact and the molecule bounces off the surface at energies up to a few hundred eV. At higher energies the molecule embeds within the surface and can form a crater. The size of the craters have been calculated and shown to be in agreement with experimental results The impact of energetic ions with fullerite causes sputtering by two mechanisms. First, individual C atoms are ejected over a short times scale of a few hundred femtoseconds. Secondly molecules are desorbed from the surface over timescales of a few picoseconds. The desorption of up to 5 C_{60}'s can cause large craters to form on the surface. Movies of these phenomena illustrate the physical processes involved.

INTRODUCTION

The use of energetic ions and clusters as tools for surface and material modifications is becoming widespread. Here we report on some simulation results involving fullerenes. We examine the energetic interaction of a fullerene particle with a graphite lattice and also the interaction of an energetic single Ar ion with the fullerite lattice.
The simulations were motivated by experiments undertaken to examine the effects of the particle bombardment on graphite and fullerite crystals. The results of these experiments are briefly outlined in the next section.

EXPERIMENTAL RESULTS

Experiments have been conducted of fullerene impacts on graphite [1] in which a HOPG target was irradiated with C_{60} molecules in the energy range 500 eV to 23 keV. The surface after irradiation was imaged by the STM. Hillocks of between 30-50 Å in diameter and 2-3 Å in height, surrounding a crater caused by the impact, were observed. The penetration depth (i.e. depth of the crater) was found to scale with $E^{0.5}$ where E is the energy of the incoming molecule. This penetration depth was measured by an oxidation technique that selectively etched away layers that contain defects.
Experiments have also been conducted on Ar ion-bombarded fullerite films [2]. Raman spectra indicate a transition to amorphous carbon and AFM pictures of the surface show that large damage occurs even for small ion doses.

SIMULATION DETAILS

The Molecular Dynamics simulations were carried out using the many-body potential formalisms of Tersoff [3] and Brenner [4]. For the fullerene impacts on graphite the Tersoff potential was used but for the Ar impacts on fullerite, a many-body potential modified by the inclusion of long-ranged terms was used. This is because the forces which bind the individual carbon atoms together are covalent whereas the C_{60} 's are held together in the lattice essentially by long-ranged Van der Waals-like forces. To model the short-ranged covalent forces, the Brenner potential [4] was used, with the repulsive two-body term splined to the ZBL screened Coulomb potential [5] at close particle separation. A hybrid potential developed previously by the authors [6] which combines the long-range and short-range forces was used to model the crystal binding. The pair part of the potential is long-ranged and cuts off at a distance of 11 Å. The potentials fits the cohesive energy and lattice spacing of the fcc bulk fullerite structure. No other properties of the bulk C_{60} structure were used in the fitting process. The binding energy of the C_{60} molecules in the fullerite structure (~2 eV per molecule) is much less than the binding energy of the individual C atoms within the C_{60} structure itself (~ 7 eV per atom).
For the C_{60} impacts on graphite, a target of 46,720 atoms in 8 layers was used with free boundaries and the initial temperature of the lattice was at 0 K. For the Ar impacts on fullerite most simulations were conducted on a target which consisted of 4800 atoms, 80 molecules arranged as 16 C_{60} molecules in five layers with the surface plane being the {111} orientation. Some simulations were also carried out on a larger lattice consisting of 36 atoms in five layers. The molecules were randomly oriented within the fullerite structure. Periodic boundary conditions were applied to the sides of the target. Atoms within 2 Å of the target base were held fixed, the other atoms in the target being free to move. These were given kinetic and potential energy at the start of the simulation equivalent to a temperature of 300 K. Equilibration for 0.5 ps was followed by the impact of the single incoming ion for a further 5 ps. The direction of the argon ion was chosen to be normally incident to the C_{60} film. A separate simulation for each impact location was conducted with the incident Ar at energies of 300 eV, 600 eV and 1 keV. After each trajectory had run its course, a new fcc lattice was constructed with a different orientation of the fullerene molecules.

SIMULATION RESULTS

For the C_{60} bombardment of graphite there is good agreement between the experimental results and the simulations. Figure

CP392, *Application of Accelerators in Research and Industry,* edited by J. L. Duggan and I. L. Morgan
AIP Press, New York © 1997

1 shows examples of typical damage induced in the graphite by energetic impacts at 3 and 6 keV. The deepest penetration depth is 4 and 6 monolayers respectively at these energies. At 23 keV the simulation results show a penetration depth of 10 monolayers. The measured experimental results always seem to be one monolayer less than the simulation result. The most likely explanation for this is that over a longer time than the simulation, the deepest layer might well anneal and not become etched. The predicted hillock heights around the crater edges are also in good agreement with experiment. However the width of these rims is about 15% larger in the simulation than the experiment. This disagreement would improve if the simulation was to be run for a longer time on a bigger lattice.

It was not found necessary to generate statistics from a large number of different impact points on the target surface. Most trajectories were found to be qualitatively similar.

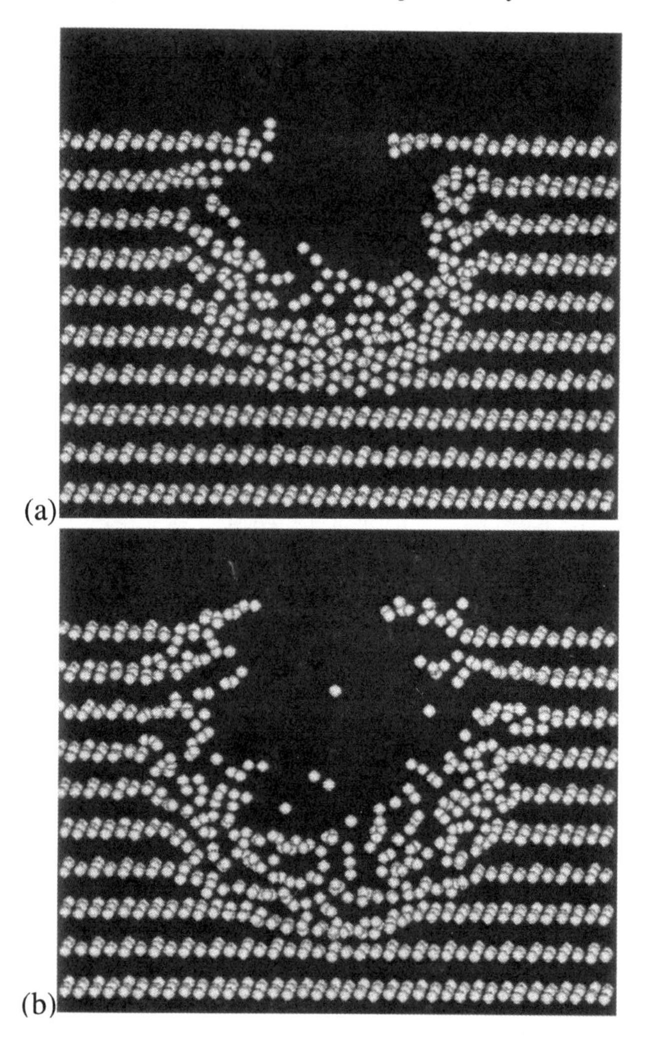

FIGURE 1. Cross-sections of the crater formed on graphite by (a) a 3 keV C_{60} molecule incident normally on a graphite surface 500 fs after impact (b) a 6 keV impact.

In contrast, for Ar impact on fullerite a sequence of trajectories had to be run to generate good statistics and to understand the collisional processes involved. Figure 2 is a set of pictures depicting the temporal progression of one trajectory of a 600 eV impacting Ar atom after the initial equilibration period with the fcc fullerite lattice. No individual C atoms are sputtered. Instead the cage of the surface C_{60} molecule into which the Ar passes is peeled open and although it begins to reform as time elapses, sufficient momentum has been imparted to destroy the cage causing local amorphisation of the surface. In addition a single C_{60} molecule desorbs from the crystal. Figure 3 gives examples of snapshots from two other trajectories. One where an individual C atom is emitted. The other shows the desorption of 5 surface molecules. Examination of statistics from a number of such trajectories shows that particles can be emitted from the surface in two ways. First individual carbon atoms can be ejected usually as a direct result of a collision with the incoming argon ion or primary recoil. These usually appear within the first 0.5 ps following impact. Secondly larger particles can be emitted from the surface. These are mainly C_{60} molecules which are desorbed as a result of energy being transferred to surface molecules from the ion. This phenomenon occurs over the period of about 5 ps. The individual C atom yield increases with increasing ion energy over the range studied (300eV -1 keV) whereas the yield of C_{60}'s decreases. At 1 keV, an average of 0.36 individual (i.e. not part of a cluster) C atoms are ejected per impact. Calculations for graphite [7] show that this yield would be obtained from a graphite surface at an incidence angle of about 35°. The surface C_{60} 's are weakly bound to the lattice and are easily desorbed by the addition of energy to the crystal. The calculations show that up to 5 C_{60} 's could be emitted as a result of the impact. Molecules ejected from near the edges of the surface could be overestimated. Nevertheless the results show an interesting trend. At the lower ion energy of 300 eV where the energy is deposited closer to the surface, more emission of C_{60} 's occurs than at the higher energy of 1 keV where the ion has more chance to travel further and deposit its energy deeper. The desorbed C_{60} 's also do not appear to have any directional dependence. This is in contrast to sputtering of atoms from an fcc lattice which show a distinct correlation due to channeling and blocking by surface atoms. In the case of bombardment of fullerite, the prediction of a yield of up to five individual C_{60}'s from the surface means that pits an order of magnitude larger than those caused by impact on a single crystal should be observed. So far however the resolution of our AFM measurements has been insufficient to observe this prediction.

Generally, however, the combination of the experimental results from the STM and AFM and the results of the atomistic computer simulations helps throw new light on atomic scale damage produced by energetic particle or cluster beams.

ACKNOWLEDGMENT

Steven Hobday was in receipt of a studentship from the EPSRC which enabled part of this work to be carried out. We would also like to thank M. Kappes, and G. Brauchle, A. Richter and U. Gibson for useful discussions.

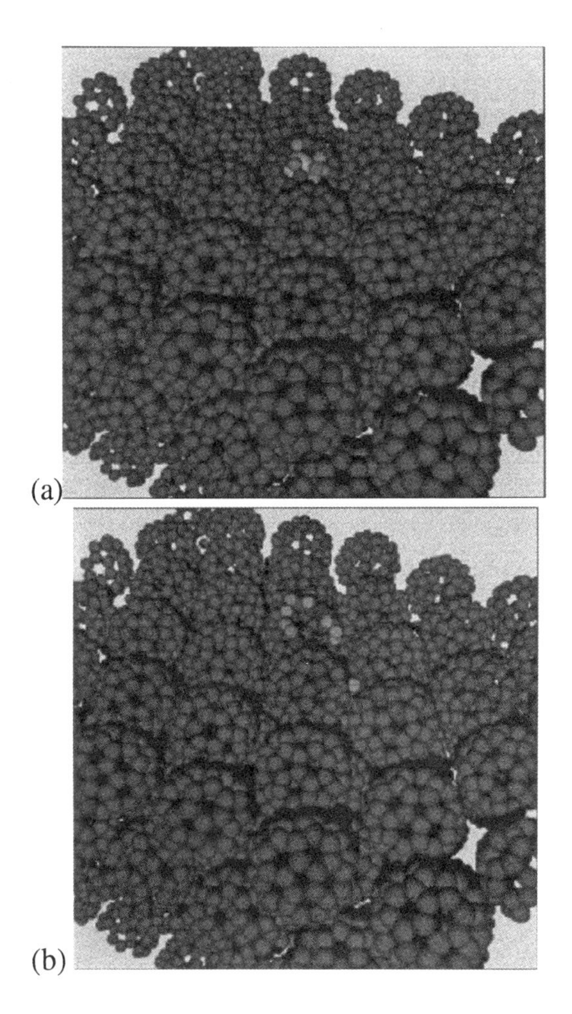

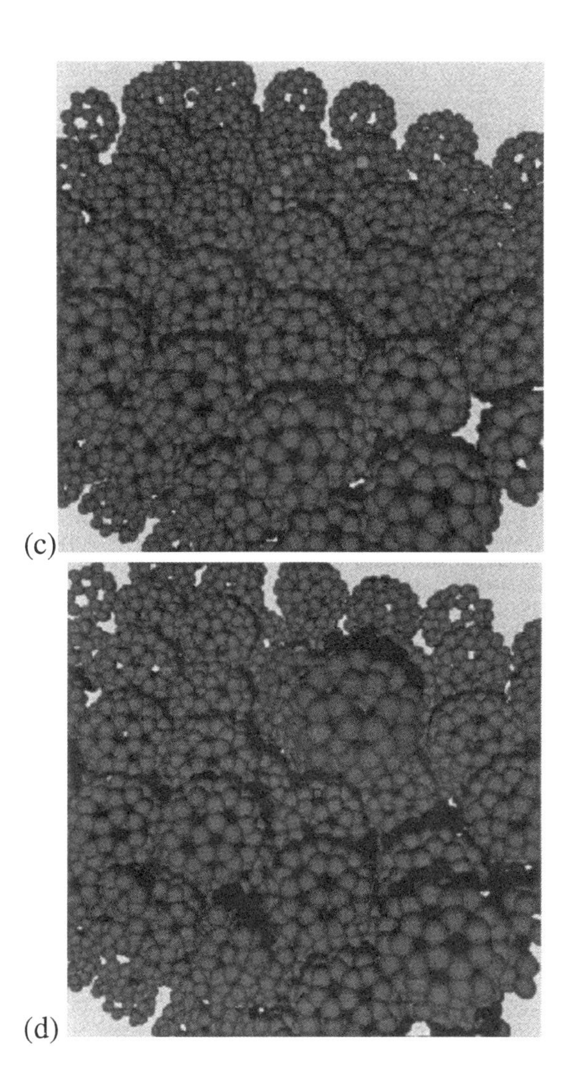

FIGURE 2. An example of a 600 eV Ar impact which causes a C_{60} molecule to desorb. On impact the Ar ion peels open a surface C_{60} molecule destroying it. Energy is imparted to the lattice which causes a nearby C_{60} to desorb. (a) 10fs after impact, (b) 25 fs, (c) 200 fs, (d) 5 ps after impact..

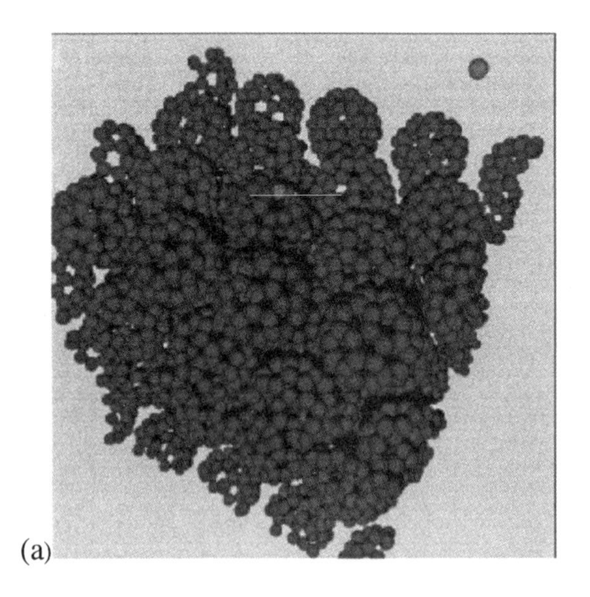

(a)

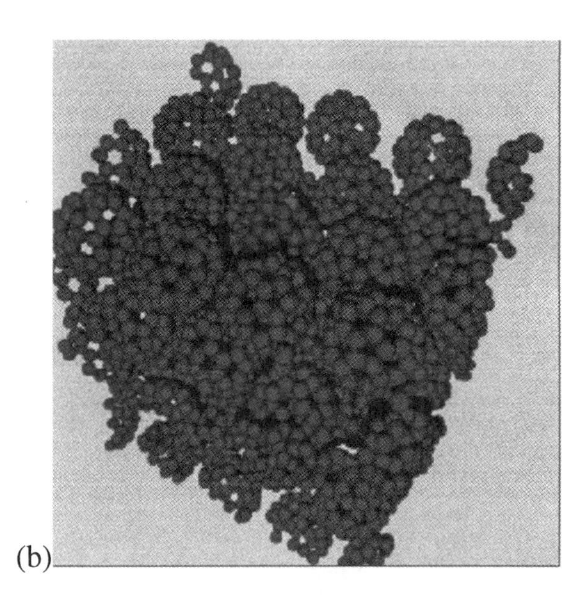

(b)

(c)

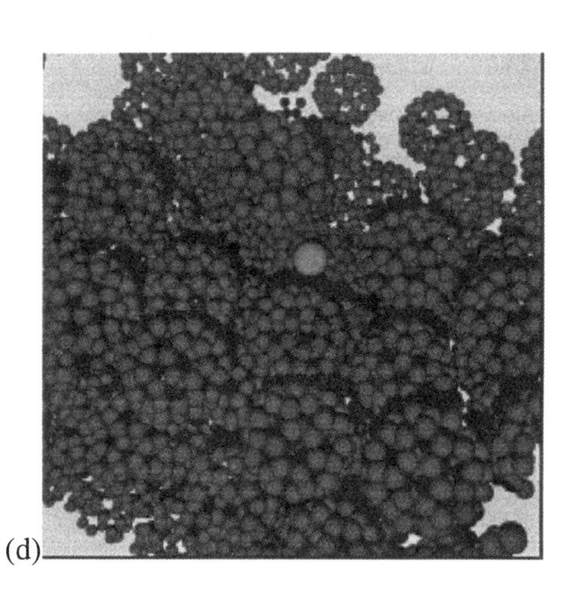

(d)

FIGURE 3 Examples of (a) and (b) a 1 keV Ar impact causing the ejection of a single C atom but with the surface molecules remaining intact, (c) and (d) an impact at 600 eV where the Ar atom diffuses out of the lattice (the large sphere shown in d) causing 5 surface molecules to be desorbed. The time frames are (a) 200fs (b) 1ps (c) 125 fs (d) 5ps.

REFERENCES

1. Webb, R.P., Kerford, M., Kappes, M. and Brauchle, G., *A comparison between fullerene and single ion impacts on graphite*, Proc. EMRS, 1996 in press.
2. Hobday, S., Smith, R., Richter A. and Gibson, U., Rad. Eff. Defects. in Sol. 1996 in press.
3. Tersoff, J. , Phys. Rev. Lett. **61**, 2879 (1988).
4. Brenner, D. W., Phys. Rev. B **42**, 9458 (1990); Phys. Rev. B **46**, 1948 (1992).
5. Ziegler, J.F., Biersack, J.P. and Littmark, U., *The stopping and range of Ions in Solids*, Pergamon, New York, 1985 p 41.
6. Smith, R. and Beardmore, K. M., Thin Solid Films **272**, 255 (1996).
7. Smith, R. and Webb, R. P., Nucl. Instrum. and Meth. B **59/60**, 1378 (1991).

FRAGMENTATION OF BIOMOLECULES USING SLOW HIGHLY CHARGED IONS

Christiane Ruehlicke[1‡], Dieter Schneider[1], Robert DuBois[2], Rodney Balhorn[3]

[1]Physics and Space Technology, [3]Biology and Biotechnology, Lawrence Livermore Natl. Lab., Livermore, CA 94550, [2] Department of Physics, University of Missouri-Rolla, Rolla, MO 65401

We present first results of biomolecular fragmentation studies with slow highly charged ions (HCI). A thin layer of the tripeptide RVA was deposited on gold targets and irradiated with slow (few 100 keV) ions, e.g. Xe^{50+} and Xe^{15+}, extracted from the LLNL EBIT (electron beam ion trap). The secondary ions released upon ion impact were mass analyzed via Time-Of-Flight Secondary-Ion-Mass-Spectrometry (TOF-SIMS). The results show a strong dependence of the positive and negative ion yields on the charge state of the incident ion. We also found that incident ions with high charge states cause the ejection of fragments with a wide mass range as well as the intact molecule (345 amu). The underlying mechanisms are not yet understood but electron depletion of the target due to the high incident charge is likely to cause a variety of fragmentation processes.

INTRODUCTION

Interactions between slow (~ 2keV/amu) highly charged ions (HCI) with biomolecules are being investigated. Currently the emphasis of the studies is on exploring the basic mechanisms of the interaction, e.g. breakup, in particular in polypeptides. These experiments might reveal a new method of degrading biomolecules. A potential application could be the fragmentation of proteins, which is used in peptide mapping and sequencing. The availability of proteases, which are commonly used for protein fragmentation, is very limited, therefore there is a demand for additional fragmentation techniques.

Studying the impact of HCI upon biomolecules also is an extension of studies of HCI solid surface interaction, which have shown that HCI impact causes the emission of a large number of electrons from the surface (1,2), enabling a variety of secondary processes to take place.

The ions used in these studies are produced by an electron beam ion trap (EBIT) (3,4) and charge states up to 50+ have been used. The high potential energy (few 100 keV) enhances the effect of electronic interactions with solids or molecules as compared to the nuclear interactions that occur in the collisions. The potential energies of the ions are comparable to their kinetic energy, giving rise to new effects.

EXPERIMENTAL

The target, the polypeptide RVA, which consists of a sequence of the aminoacids Arginine, Valine and Alanine, was synthesized using a PS3 Peptide Synthesizer (Rainin Instruments) and purified by reversed-phase high performance liquid chromatography. Aliquots of the peptide (50 µl of a 10 mg/ml solution in water) were deposited on flat gold disks and allowed to dry for 15 h in the presence of a desiccant. The targets were clamped onto a sample ladder and mounted in the experimental area within one day. TOF-SIMS was performed in a high vacuum (10^{-10} Torr) chamber equipped for surface analysis at the end of the ion extraction beamline on EBIT. Xe gas was injected via a ballistic jet into the trap region of EBIT and Xe ions with charge states ranging from 15+ to 50+ were extracted at kinetic energies of 700 keV or 105 keV resp.

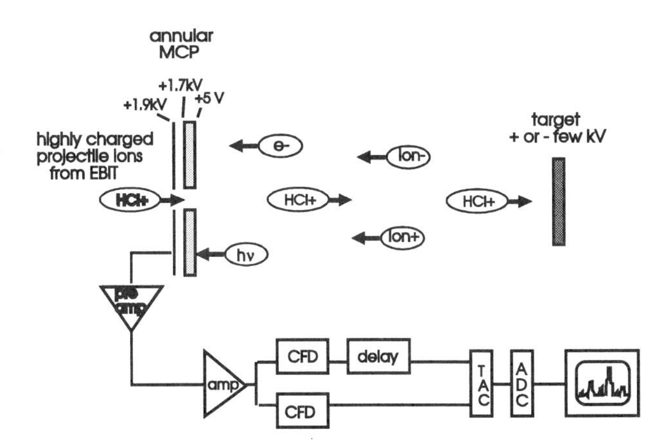

FIGURE 1. The Time-Of-Flight system used in our experiments is shown in this figure. Secondary ions and electrons ejected from the target are accelerated towards the microchannelplate detector. The target is biased positive or negative depending on which spectra are taken. Start and stop pulses are determined due to their pulse heights and arrival time at the detector.

CP392, *Application of Accelerators in Research and Industry,* edited by J. L. Duggan and I. L. Morgan

The fluxes were ca. 1000 ions /s, where the Xe^{15+} was extracted in pulses of a few ms length at 2 Hz and the Xe^{50+} was extracted in a continuous mode. The TOF spectra were obtained using a TOF-SIMS spectrometer (Fig. 1) (5). Secondary ions are accelerated between the target and a channelplate detector at voltages of a few kV. For negative ions the start signal was taken from electrons emitted from the sample upon individual ion impact, while protons and in rare cases H_2^+ provided the start signal in the positive case. The stop signals were given by the secondary ions. The flight time of the secondary ions increases with charge and decreases with the mass at the same time. The positive secondary ion yield depends on the probability of proton emission, which might follow a different charge state dependence than the larger mass secondary ions. However, this affects the secondary ion yield only if there is no proton ejected after impact of a primary ion.

RESULTS AND DISCUSSION

Examples of TOF-SIMS spectra are shown in Fig. 2. For both the negative and the positive secondary ion spectra the yields are much higher for Xe^{50+} being the primary ion compared to Xe^{15+} . This general trend has also been observed with incident ions of other charge states. The Xe^{50+} spectra show a wide mass range of secondary ion peaks, many of which are below the background intensity in the corresponding Xe^{15+} spectra. This demonstrates a higher sputtering efficiency for ions with high charge states. The large numbers of very low mass ions, e.g. H, C, and O compounds, mostly stem from contaminant on the surface rather than from the peptides and are common in these spectra regardless of the target material. While these compounds as well as atomic ions of alkali metals and halides used during sample preparation dominate the spectra up to masses of ca. 100, most larger mass fragments are unique to the peptide sample and appear to be molecular fragments. Both the positive and the negative spectra show a contribution at the mass of the intact molecule in the Xe^{50+} spectra, which indicates that it is also possible to lift intact molecules with both negative and positive excess charge from a solid surface. The occurrence of even higher masses suggests that the intact molecule combines with other fragments, possibly Na. Since we can not distinguish between different charge states with the current setup, we assume all fragments to be singly charged. In addition to these peaks and a wide distribution of fragments in both the positive and negative spectra the negative spectrum also shows some distinctive peaks in the mass range between 110 and 250 amu. The Xe^{15+} spectra both show fragments up to ca. 100 amu, higher mass peaks are not removed efficiently enough to be distinguished from the background counts on the spectra. This is true also for the peak corresponding to the intact molecule, whose mass position is marked in the positive spectrum.

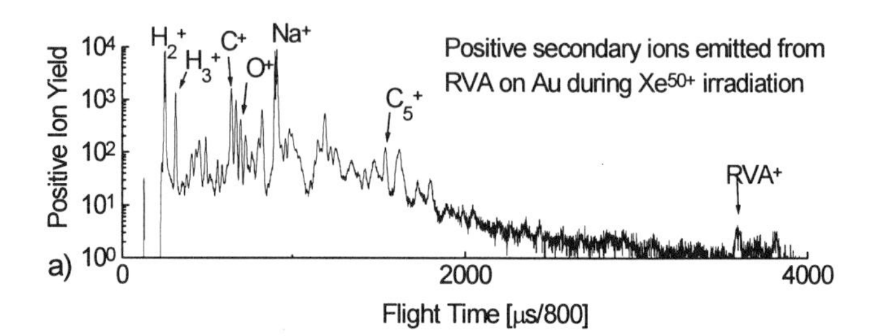

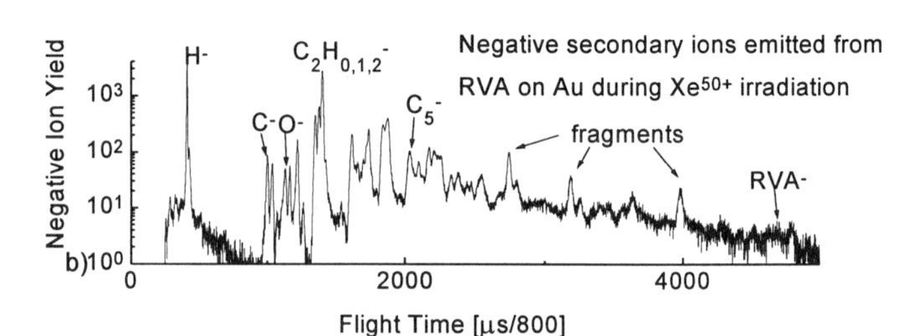

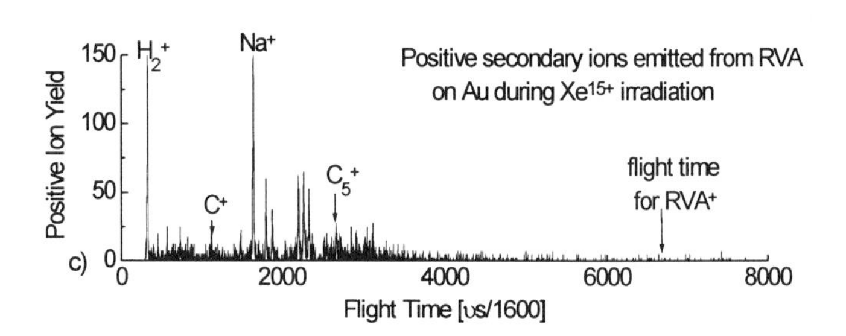

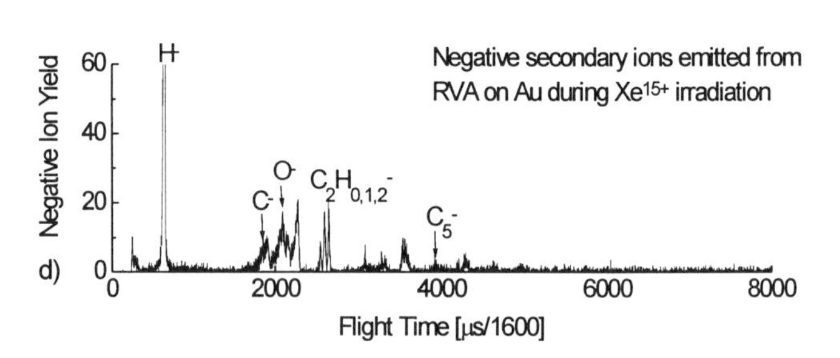

FIGURE 2. TOF-SIMS spectra showing the negative and positive secondary ion yields for projectile ions Xe^{50+} and Xe^{15+} impact on the tripeptide RVA deposited on a Au substrate. The yields, which were normalized to 10^6 incident projectile ions, are plotted as a function of flight times of the secondary ions. A linear background was subtracted from the spectra. A few characteristic peaks are marked in all spectra, in c) we indicate the position where the RVA^+ peak would be expected, while d) actually shows a shorter mass range than the other spectra.

The underlying mechanisms for breakup and ablation are not yet understood but it is assumed that different processes occur, which originate in the large electric field induced by the high ion charge. This high field could cause the removal of binding electrons, therefore causing the formation and ejection of molecular fragments.

The electron depletion of the solid substrate by the HCI is also expected to be a factor in the ablation and ejection process, where weakening of the binding between the molecule and the surface precedes the desorption.

While the structure, binding and adhesion of the polypeptide molecules differs significantly from the solid surfaces, some of the responses to HCI impact are similar, e.g. the higher secondary ion yields and occurrence of high mass clusters due to HCI impact have been observed in sputtering solid surfaces with HCI as well (5).

CONCLUSIONS

First results of fragmentation studies of biological molecules by HCI are presented. The TOF-SIMS spectra of the tripeptide RVA show the ablation of the intact molecule from a gold surface due to HCI impact as well as the fragmentation into a series of smaller fragments. The sputtering efficiency was shown to increase with increasing projectile ion charge and negative and positive secondary ion spectra were found to show different fragment yields. Some parallels can be drawn to interactions that occur between solid surfaces and HCI but the underlying mechanisms are not yet understood.

ACKNOWLEDGMENT

This work has been performed under the auspices of the U.S. Department of Energy by Lawrence Livermore National Laboratory under Contract No. W-7405-ENG-48.

‡ Christiane Ruehlicke is affiliated with Universität Bielefeld (Bielefeld, Germany) through Rainer Hippler.

REFERENCES

1. McDonald, J. W., Schneider, D., Clark, M. W. and Dewitt, D. Phys. Rev. Lett. **68**, 2297 (1992)
2. Kurz, H., Aumayr, F., Winter, H. P., Schneider, D., Briere M. and A., McDonald, J. W., Phys. Rev., **A49**, 4693 (1994)
3. Levine, M. A., Marrs, R. E., Henderson, J. R., Knapp, D. A., and Schneider, M. B., Physica Scripta **T22**, 17 (1988)
4. Schneider, D. H., et al., Phys. Rev. **A44**, 3119 (1991)
5. Schneider, D. H. G. and Briere, M. A., Physica Scripta **53**, 228 (1996)

SECTION V

PIXE, RBS, CHANNELING, NRA, ACTIVATION, ERD, MICROPROBES, RESONANCE IONIZATION SPECTROSCOPY

THE ROLE OF ION BEAM BASED ANALYSES IN GLOBAL CLIMATE RESEARCH

Thomas A. Cahill and Juan Zubillaga*

Departments of Atmospheric Sciences, LAWR, Physics, and the Air Quality Group, Crocker Nuclear Laboratory, University of California, Davis 95616, U.S.A.

Ion beam based analytical techniques are beginning to have a significant role in global climate research because of the growing importance of atmospheric aerosols in global climate models (GCM). The aerosol component of GCMs is not adequate to match the complexity of the observed data, opening the way for ion beam techniques to guide the modelers into a greater degree of accuracy. The well established role of PIXE in size-compositional profiles of aerosols has been extended by use of proton scattering techniques (PESA; PESA-C) for hydrogen analogs and chemical states of sulfates, and AMS for ^{14}C measurements of organic aerosols and, potentially, CO_2 itself. The global span of PIXE programs has allowed the beginning of global maps of continental aerosols of climatic importance, a development fostered by the development of small, portable samplers designed to use PIXE based ion milliprobe analyses.

INTRODUCTION

It is my experience that if an atmospheric physicist or chemist can buy a device that sits on a bench and gives the information he or she desires on atmospheric constituents, they will never use ion beams in their investigations. This is true even if the information they gather may be inferior to that given by ion beam analysis, until they start to become non-competitive with ion beam based analytical techniques. Only then will they, with obvious reluctance, include ion beam techniques in their work.

It is thus not surprising, therefore, that the development of ion beam based analytical techniques in atmospheric physics and chemistry has come largely from accelerator physicists who are comfortable with such techniques and have access to the expensive hardware required to do such work. However, it almost automatically makes the accelerator physicist an "outsider" in the discipline of atmospheric science, further hindering the wide application of ion beam based techniques in areas such as global climate research. It is almost as though we, as accelerator physicists, found that only a single accelerator in a foreign national laboratory could accomplish our research. We would use it if we must, but the barriers would be high and our enthusiasm low. We would probably spend serious effort finding some way to avoid having to use the foreign facility and do the work at our home or a nearby university. Logistical problems are not the sole reason for this attitude. Local resources also allow for greater interaction, changes in facilities, specialized equipment, and flexibility that advances our research interests.

Thus, it is not surprising to find that accelerator based analytical techniques are used by atmospheric scientists and global climate researchers only when they deliver information that is otherwise very hard to gather by other means. This problem does not exist in analyzing gaseous components of the atmosphere, as these can be studied by a variety of optical and chemical methods, especially with the rapid advances in the use of tunable lasers. Atmospheric aerosols, however, pose an entirely more intractable problem for standard methods. Aerosols have the additional obstacles of particulate size and shape, very low concentrations in the atmosphere requiring separation from the predominant gases, and enormous chemical complexity.

The complexity of atmospheric aerosols was one of the reasons that emphasis was placed on gaseous compounds in global climate research until just a few years ago. Vast amounts of effort was put into understanding the effect of the "greenhouse gases" (in order of importance, H_2O, CO_2, CH_4, ...) on global climate in the global climate models (GCMs). However, no amount of "tweaking" could avoid the unpleasant fact that the predicted temperature rise of about 1.5 C was two or three times greater than the observe temperature rise. This discrepancy was seized by some policy makers as cause to reject the whole idea of significant climatic impacts of human activities, leading to opposition to controls on emissions of greenhouse gases.

A byproduct of the 1991 Kuwaiti oil fires was a renewed interested in atmospheric aerosols as a factor in

CP392, *Application of Accelerators in Research and Industry*, edited by J. L. Duggan and I. L. Morgan
AIP Press, New York © 1997

the global climate. The theory of a "nuclear winter" was tested by these fires, and found to be seriously lacking. But the calculation did show that anthropogenic aerosols could be a factor in climate, leading to a cooling that, could partially counteract the heating effects of CO_2. Suddenly, aerosols were very important, and the lack of information on aerosols on a global scale became an immediate hindrance to quantitative calculations of human impacts on global climate.

Our primary focus at UC Davis was in atmospheric visibility monitoring, first in California (1973-1978), and later in national parks, monuments, and wilderness areas throughout the U.S. (1977-present). The original EPA and NPS effort grew into the IMPROVE* program, presently operating about 72 sites in the US (1). This program measures atmospheric aerosols by size and composition for comparison to visibility, identifies sources of haze, and in some cases results in abatement of these sources. Nevertheless, the same measurements of visibility, and of the fine particles that cause it are also the key measurements for understanding the effects of aerosols on the Earth's radiation budget. For visibility, we look horizontally. For climate, we look vertically.

Our role in global climate research was modest before 1991. We had done some original work at Mauna Loa Observatory on transported Asian dust (2) and airborne sampling of Arctic Haze (3). Our earlier work in Arctic haze lead to a request that we operate on the NOAA aircraft over Kuwait, as well as at ground sites within the oil fields (4; 5). It is also perhaps instructive that over half of all the compositional data on Kuwaiti aerosols was gathered by ion beam based methods.

From the Kuwaiti work, and the papers on climatic effects of aerosols (6), we realized that our work in visibility was relevant to studies of aerosol effects on global climate. All one had to do to see this fact was to compare the predictions of the GCMs for aerosol like sulfates with our IMPROVE measurements. We performed an analysis of the seasonal variation of aerosols as predicted by the global models, and compared it to the data of IMPROVE aerosol samplers at Shenandoah National Park. Shenandoah National Park represents the center of the eastern US haze blanket, that has been estimated by modelers to produce a local cooling effect greater in magnitude than the heating due to increasing CO_2. The models had roughly predicted the average annual sulfates (7). However, the seasonal results were both quantitatively and qualitatively different (Figure 1). Clearly, the models needed better data which our ion beam based IMPROVE network could provide.

The key factor that makes model calculations so difficult, and in this case so inexact, is the extreme sensitivity of optical scattering to particle size and composition. Particles in the critical size range, on the order of 0.5 to 1 μm in diameter, scatter light 100 times more efficiently than those between 0.05 and 0.1 μm in diameter. Further, hygroscopic particles such as sulfates and nitrates can, by picking up water, grow into the critical size range, which other particles, such as hydrophobic organic particles and soils, will not. Thus, one must know the size and composition of particles in the size rage between 0.05 and about 2.5 μm in diameter to understand aerosol effects on visibility and global climate. Additionally, the question of optically absorbing particles, which behave yet differently, remains. This complexity opens the door for ion beam based methods.

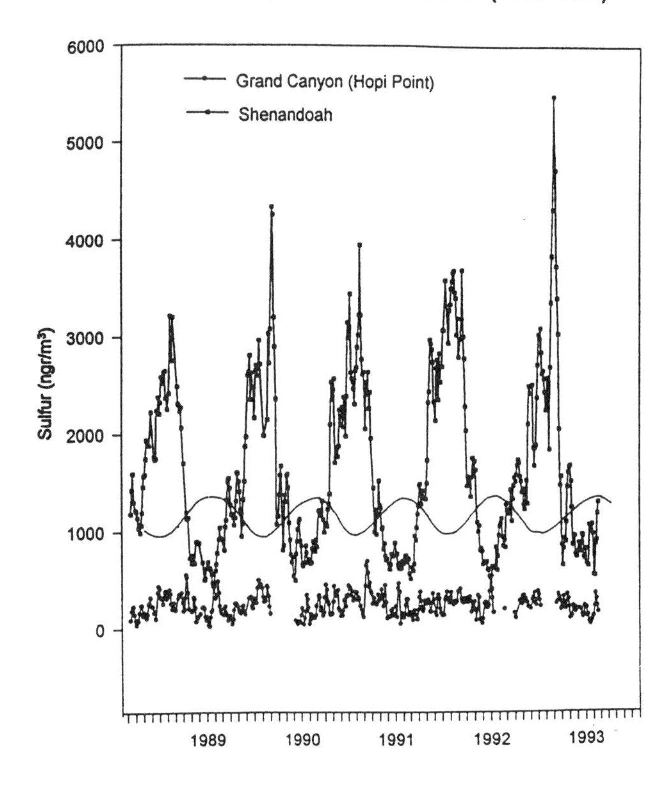

FIGURE 1. Seasonal variation of fine sulfate aerosols at Shenandoah National Park from IMPROVE as compared to typical results of global climate models (solid line).

ION BEAM BASED AEROSOL MEASUREMENTS

What are the ways in which ion beam based techniques can assist global climate modelers? In what ways are ion beam techniques currently in use in these areas? There are a number of applications in use, and I will do little more than list some of them. The list continues to grow.

1. Spatially and temporally resolved measurements of fine aerosols

This work was pioneered by Jack Winchester of Florida State, and later by Swedish and Belgium scientists using PIXE. Since aerosol modeling does not yet work well, researchers require measurements made at many places and times, an effort well matched to the aerosol groups around the world (8). We are attempting to gather this diverse work into a single global map of fine sulfates, a map repeatedly and urgently requested by global modelers (9). Note that one does not require expensive equipment to do this work, yet it is vitally important. The use of continuously collecting samplers such as the Florida State "streaker" greatly eases the problem of sample collection over extended periods of time. We have encouraged many aerosol groups in the past decade, "loaning" equipment and performing analyses and quality assurance checks.

2. Size profiles of fine aerosols

Size profiles are an area in which PIXE has almost become a dominant force. The logic is straightforward. To collect size resolved particles, one usually uses an inertial impactor. But in order to collect sizes accurately, one cannot allow a thick layer of particles to build up. If one has only a few monolayers of 1 μm diameter particles, one has only a few 10s to 100s of μg/cm^2 of mass, far too little for most chemical methods. PIXE, however, needs small masses to be quantitative, and hence its importance in this area (10).

3. Complementary ion beam techniques

A very important aspect of ion beam based analytical systems is the ability to simultaneously use atomic and nuclear techniques to extend the amount of information available. We have used ^{1}H (p, p)^{1}H scattering at 4.5 MeV proton elastic scattering analysis (PESA), to measure hydrogen. This technique allows one to measure a surrogate of organic matter, generally the most important component of fine mass in remote areas. It also allows measurement of the chemical states of sulfates by simply comparing the quantities of hydrogen and sulfur atoms. Sulfuric acid has 2 hydrogen atoms per sulfur, while ammonium sulfate has 8. An example of a PIXE plus PESA analysis is given below for size segregated aerosols at Shenandoah National Park, Virginia, in September, 1991.(Figure 2) (11) Note that for particles between 0.34 and 1.15 μm in diameter, the plot of 4 times H versus S shows a big disagreement on some days, indicating periods of almost pure sulfuric acid in the midst of a predominantly ammonium sulfate aerosol.

4. AMS measurement of ^{14}C content of aerosols

We are starting to see the use of AMS to measure the ^{14}C content of atmospheric aerosols. This measurement is critical to separate fossil fuel organic particles, which have no ^{14}C, from plant derived and natural organic sources, including wood smoke, which have modern levels of ^{14}C. The problem is the very tiny amount of carbon collected in the typical air sampler, and the need for hundreds of micrograms of carbon for an AMS reading. However, for non-AMS based ^{14}C measurements, which typically require 10 times more mass than AMS the problem is worse. The problem is reduced if one has a CO_2 based gaseous ion source, then the efficiency of transfer from the aerosol into the accelerator is enhanced by at least a factor of 10, greatly aiding in the collection problem and opening the way for vastly increased use of AMS in organic analysis.

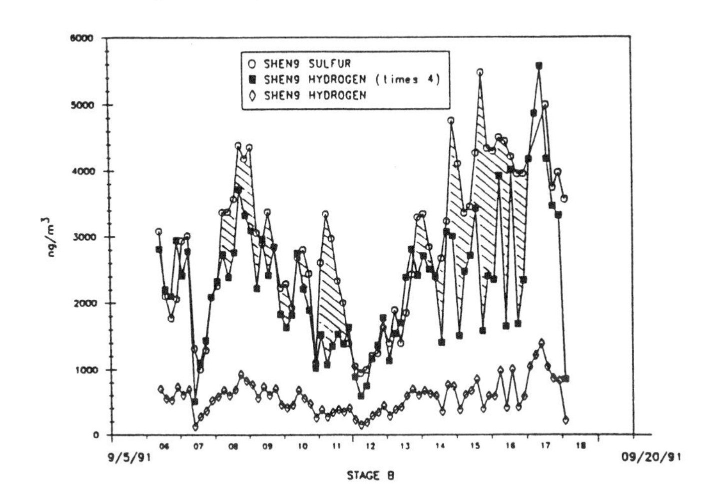

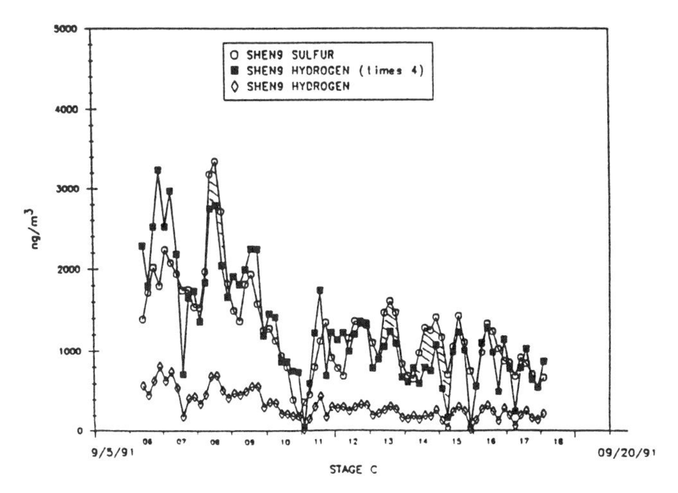

FIGURE 2. Sulfur, hydrogen and hydrogen times 4 for Shenandoah National Park, September, 1991. The top graph is for particles 0.34 μm < D_p < 1.15 μm, the optically efficient aerosols. The bottom is for particles 0.07 μm < D_p <0.34 μm, too small to efficiently scatter light. If hydrogen times 4 equals sulfur, the results are consistent with ammonium sulfate. The cross hatched areas are more acidic, indicating almost pure sulfuric acid in some periods.

5. Gaseous artifacts in quartz air filters

Most modern methods of analysis of organic aerosols require collection on a rather thick quartz air filter. This type of filter, can collect significant gaseous artifacts. We have modified a technique utilized in 1970 to analyze steel for hydrogen content to do hydrogen depth profiles in quartz filters (12). In this technique, the coincident protons from ^{1}H (p, p)^{1}H are both measured and their energies summed. The energy loss is sensitive to whether the scattering was on the front surface or at depth. We have called this technique C-PESA (C for coincidence). Again, the use of energetic ions gives the analysts advantages that simply can not be matched by non ion beam techniques, even at much greater cost. We have found that a good fraction of what others had been calling organic particles was actually organic gaseous artifact (Figure 3). This finding has a direct influence on the predicted role of trees in causing haze. We find than natural vegetation causes only about 10% of the source of haze at places such as Great Smoky Mountains National Park each summer.

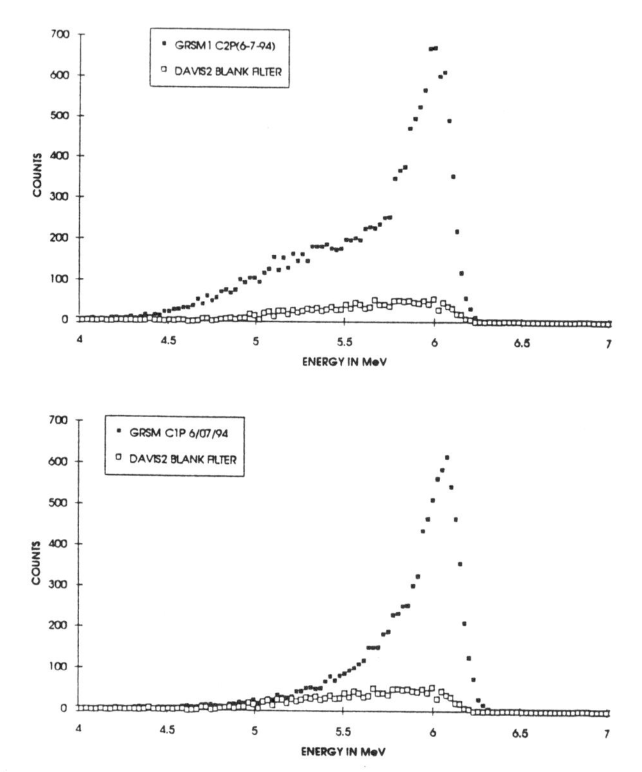

FIGURE 3. C-PESA spectra of two 6 mg/cm^2 quartz air filters from Shenandoah National Park summer, 1993. The protons from the surface have energies about 6 MeV, those from the back surface, roughly 4.5 MeV. The bottom spectrum shows a correct collection of particles on the surface, while the top spectrum shows massive hydrogen (organic?) artifact throughout the filter.

6. Portable air samplers and proton milliprobes

The ability of ion beam methods, and especially PIXE, to utilize only small amounts of mass in their analysis allows for the development and deployment of new, smaller air samplers. Samplers such as the aforementioned streaker (13) and the inertial impactor of rotating drum design (DRUM samplers) (14; 15) are becoming steadily smaller and more portable. The key factor here is that ion beam techniques are sensitive only to areal density ($\mu g/cm^2$), not mass. By focusing the ion beams to small areas (proton milliprobes), the required total collected mass can also be made very small. Samplers can thus be battery or solar powered, making measurements easier at remote sites, on airplanes, balloons, etc.

7. Single particle analysis by proton microprobes

The use of proton microprobes is not as promising in some ways. The key particles for optical scattering are in the 0.5 to 1 μm diameter range, which pushes the limits of microprobes. Nevertheless, important work is possible.

8. Synchrotron based XRF measurements

The last area that has begun to be used in global climate aerosol work is synchrotron based XRF (S-XRF) measurements. The great sensitivity of this tool for transition and heavy metals was utilized on aerosols in our work over Kuwait (4), together with PIXE and PESA. Note that that work also used laser beams to measure optical absorption and a microbalance to measure mass, two crucial and non-ion beam measurements that help one to achieve closure on mass balance. The S-XRF was the only method to achieve the sensitivity necessary to measure the nickel and vanadium components of the aerosol cloud, which allowed us to trace individual well fields and later to measure the same components using ground based inertial impactors of the rotating drum design (5). Due to the infrequent access to the facility, and the cost and difficulty of moving an analytical team to the Stanford SSRL site for a week at a time, we reserve these measurements for exceptional situations when higher sensitivity must be achieved. It is, however, a powerful tool that complements the other ion beam techniques since it can use exactly the same filters used by other techniques.

This last point should be emphasized. The fact that most ion beam techniques are non destructive allows repeated measurements from the same filters. Repeated analyses deliver information that would otherwise require a whole suite of different air samplers running together in the field, and several different laboratories, each performing specialized types of analyses. This scenario would pose serious logistical and cost constraints for key measurements at remote sites, seriously restricting acquisition of required data for viable global modeling of optically important aerosols. Once all the ion beam based measurements are completed, the sample can then be sacrificed for traditional, destructive analyses, such as ion chromatography (IC). In tests of the IMPROVE system, there was very good agreement between the PIXE and IC analysis from Teflon filters. This last point may end up being one of the most important in the wider use of ion beam techniques, since there is not enough money to form and staff a large number of fully instrumented sites around the world. We hope and anticipate that ion beam techniques will become a better established part of programs such as NOAA, ALERT in Canada, and the UN's WMO Global Atmospheric Watch Program. This process that will be greatly aided by the efforts of accelerator laboratories to make their capabilities known and used by aerosol scientists at their home universities and institutes.

---

*The IMPROVE (Interagency Monitoring of Protected Visual Environments) fine aerosol monitoring network is a cooperative program funded by the U.S. National Park Service, Environmental Protection Agency, U.S. Forest Service, Fish and Wildlife Service, and the Bureau of Land Management.

ACKNOWLEDGEMENTS

None of this work could behave been done without the continuing support of the scientists of the Air Quality Group. One of us (JZ) acknowledges the hospitality of the Department of Atmospheric Sciences, Department of Land, Air and Water Resources, during his sabbatical stay at Davis.

---

*Present address: Depto. Fisica de la Materia Condensada, Facultad de la Ciencias, Universidad del Pais Basca, APDO - 644, Bilbao, 48010, SPAIN.

REFERENCES

1. Malm, W.C., Sisler, J.F., Huffman, D., Eldred, R.A. and Cahill, T.A. Spatial and seasonal trends in particle concentration and optical extinction in the United States. *Journal of Geophysical Re-search,* VOL. 99, No. D1, 1347-1370, January 20, 1994.
2. Braaten, D.A. and T.A. Cahill. Size and composition of Asian dust transported to Hawaii. *Atmospheric Environment* (Great Britain). 20:1105-1109 (1986).
3. Cahill, Thomas A. and Robert A. Eldred. Elemental composition of arctic particulate matter. *Geophysical Letters.* 11:413-416 (1984).
4. Cahill, Thomas A., Kent Wilkinson, and Russ Schnell. Composition analyses of size-resolved aerosol samples taken from aircraft downwind of Kuwait, Spring, 1991. *Journal of Geophysical Research.* Vol. 97, No. D13, Paper no. 92JD01373, Pp. 14513914520, September 20 (1992).
5. Reid, Jeffrey S., Thomas A. Cahill, and Michael R. Dunlap. Geometric/aerodynamic equivalent diameter ratios of ash aggregate aerosols collected in burning Kuwaiti well fields. *Atmospheric Environment* (1994), Vol.28, No. 13, pp. 2227-2234.
6. Charlson, S.E. Schwartz, J.M. Hales, R.D. Cess, J.A. Coakley Jr., J.E. Hansens and D. J. Hofmann, *Science* 255 (1992) 423.
7. Engardt and H. Rodhe, *Geophys. Res. Lett.* 20 (1993) 117.
8. Cahill, Thomas A., Javier Miranda, and Roberto Morales. Survey of PIXE programs - 1991. *International Journal of PIXE.* S. Morita, Editor-in-Chief. Vol. 1, No. 4, Pp. 297-310 (1991).
9. Cahill, T.A. Climate Forcing by Anthropogenic Aerosols: The Role of PIXE *Nuclear Instruments and Methods in Physics Research B: Bean Interactions with Materials and Atoms,* 109/110 (1996) 402-406.
10. Cahill, Thomas A. and Paul Wakabayashi. Compositional analysis of size-segregated aerosol samples. Chapter in the ACS book *Measurement Challenges in Atmospheric Chemistry.* Leonard Newman, Editor. Chapter 7, Pp. 211-228 (1993).
11. Thomas A. Cahill, Wakabayashi, Paul H. and James, Teresa A. Chemical states of sulfate at Shenandoah National Park during summer, 1991. *Nuclear Instruments and Methods in Physics Research B: Bean Interactions with Materials and Atoms,* 109/110 (1996) 542-547.
12. Cahill, T. A., Casteneda, C.M., Romero, J.L., King, R.B. In Press, *Nucl. Instr. & Methods (1996).*
13. Annegarn, H.J., T.A. Cahill, JPF Sellschop, and A. Zucchiatti. Time sequence particulate sampling and nuclear analysis. Adriatico Research Conference on Aerosols. Trieste, Italy, July 22-25, 1986. *Physica Scripta.* 37:282-290 (1988).
14. Cahill, Thomas A., Patrick J. Feeney, and Robert A. Eldred. Size-time composition profile of aerosols using the drum sampler. Fourth International PIXE Conference. Tallahassee, FL, June 9-13, 1986. *Nuclear Instruments and Methods in Physics Research.* B22:344-348 (1987).
15. Raabe, Otto G., David A. Braaten, Richard L. Axelbaum, Stephen V. Teague, and Thomas A. Cahill. Calibration Studies of the DRUM Impactor. *Journal of Aerosol Science.* 19.2:183-195 (1988).

MATRIX EFFECTS IN PIXE ELEMENTAL ANALYSIS OF THICK CALCULI TARGETS.

Wojciech M. Kwiatek[1], Janusz Lekki[1], Tomasz Nowak[1], Erazm M. Dutkiewicz[1], and Czeslawa Paluszkiewicz[2]

1. Henryk Niewodniczański Institute of Nuclear Physics, 31-342 Cracow, Poland

2. Regional Laboratory, Jagiellonian University, 31-059 Cracow, Poland

The PIXE technique for Trace Element Analysis have been applied to the studies of mineral deposits such as kidney stones in human organism. The calculi mainly composed of phosphates, oxalates and uric acid were extracted during surgical operations and were measured at the proton beam as thick targets. Trace elements studies of such samples are influenced by the thick targets matrix effects and by the sample composition changes caused by energy deposition in the target due to the proton beam irradiation. These both difficulties are especially pronounced in the case of the biological samples. In this paper the procedure dealing with the above problems is described, basing on calculations with the use of principal formula for the detected X-ray yields and two complementary techniques for PIXE experiments such as Fourier Transform InfraRed Spectroscopy (FTIR) and Elastic Back Scattering (EBS). A rough estimation of sample chemical composition was achieved by means of the FTIR analysis, which also may serve as a tool for local sample temperature estimation during beam irradiation. Composition of major target elements, needed for beam stopping and X-rays attenuation calculations were determined using the EBS technique applied simultaneously with PIXE. The above approach was used to estimate elemental contents of several samples. Comparison between traditionally calculated and improved results is presented.

INTRODUCTION

The aim of this study was to estimate the magnitude of the corrections needed to be incorporated to the trace element concentration calculations with an external standard technique for thick biological samples. As an example of biological samples the kidney stones were selected. The samples were obtained during the surgical operations done at the Urological Clinic at the Collegium Medicum of the Jagiellonian University.

PIXE measurements of biological samples that are composed of an organic matrix are suffering from several difficulties. For thick targets, to avoid matrix effects like X-rays attenuation and depth dependent X-ray production cross sections, an external standard technique is traditionally applied. However, the irradiating proton beam deposits a significant energy on the sample and therefore the sample matrix may change its elemental composition and the concentration level of trace elements can be significantly altered [1,2]. From our experience those effects are important particularly for relatively heavy matrices and light traces.

The external standard technique in PIXE assumes that the standard is composed from the similar matrix as the measured sample. This assumption is sometimes hardly acceptable and can be corrected only if the real sample composition is known. Backscattered protons spectra detection performed simultaneously with PIXE brings such complementary information about sample matrix and allows for correction. Fortunately, backscattering protons cross sections for main biological elements as carbon or oxygen exceed several times cross sections calculated using the Rutherford formula for standard 2-3 MeV accelerator beam energy [3,4,5] and therefore the possibility of light, biological sample composition determination is significantly enhanced.

As a second complementary technique, FTIR gives the rough information for sample chemical composition and may bring information about local sample temperature during ion beam irradiation.

EXPERIMENTAL

Targets for the analysis from samples and a standard were prepared as thick targets in form of pellets of 10 mm in diameter and the thickness of 1 mm. X-ray spectra were detected by a Si(Li) detector with an energy resolution of 165 eV for the 5.89 keV X-rays (Mn K-alpha

CP392, *Application of Accelerators in Research and Industry*, edited by J. L. Duggan and I. L. Morgan
AIP Press, New York © 1997

line). Backscattered protons spectra were collected using a surface-barrier detector characterised by the energy resolution of 18 keV for an Am-241 source. PIXE and EBS spectra were registered simultaneously by the multiplexing ORTEC 919 Multi-Channel Buffer controlled by an IBM PC computer [6]. For FTIR analysis the samples were prepared according to the KBr technique [7]. Those analyses were performed at the Regional Laboratory of the Jagiellonian University in Cracow, Poland, using the BioRad spectrometer with a 2 cm^{-1} resolution. The spectra were measured in the wave-number region from 400 cm^{-1} to 4000 cm^{-1}.

RESULTS

Using FTIR technique the kidney calculi samples were classified as:

- phosphates including:
  - calcium phosphate (apatite),
  - magnesium ammonium phosphate hexahydrate (struvite),
- oxalates including:
  - calcium oxalate monohydrate (whewellite),
  - calcium oxalate dehydrate (weddellite), and
- uric acid type of stones and their mixtures.

Because kidney stone matrices are mostly composed of calcium compounds [8] the IAEA H-5 (animal bone) was chosen as an external standard for PIXE elemental analysis. The IAEA H-5 standard was determined by FTIR analysis as a mixture of calcium phosphate as amorphous carbonate apatite and proteins. Its matrix and trace element content changes due to annealing was presented in our earlier publication [2]. Figure 1 shows a comparison of a FTIR spectrum of this standard and one of studied samples. Comparing those spectra one can see similarities in bonds positions. The bands of the wave number region 500-650 cm^{-1} correspond to the existence of PO_4^{3-} group as well as the bands in the region of 900 - 1100 cm^{-1}. The bands of 1420, 1450 cm^{-1} correspond to the existence of Co_3^{2-} ions. The broad bands about 3400 cm^{-1} indicates the existence of H_2O molecule. In addition the bands 1480 - 1780 cm^{-1} and 2800 - 3000 cm^{-1} are characteristic for protein bands.

For presentation in this paper four elements were chosen: Ca, Fe, Zn, and Sr. The X-ray energies of those elements cover large extent of the PIXE spectrum. According to the IAEA Report [9] the H-5 standard contains 21,2% of Ca, 79 ppm of Fe, 89 ppm of Zn and 96 ppm of Sr.

The basis of the computational procedure for X-ray yields was the principal formula (1):

$$Y = N_t(\Omega / 4\pi)\, \varepsilon \omega \exp(-\mu_a x_a) \cdot \int_0^X N_p(E_p)\, \sigma(E_p) \exp(-\mu x)\, (dE_p / dx)\, dx \qquad (1)$$

where N_t = number of atoms in the target within the beam volume, $\Omega/4\pi$ = detector solid angle, ε = detector efficiency, ω = fluorescence yield, μ_a = linear attenuation coefficient for Al, x_a = Al foil thickness, X = proton range in a target (mass units, g/cm^2), N_p = number of irradiating protons, E_p = energy of irradiating protons, σ = ionisation cross section, μ = mass attenuation coefficient for the target matrix, x = depth (mass units, g/cm^2).

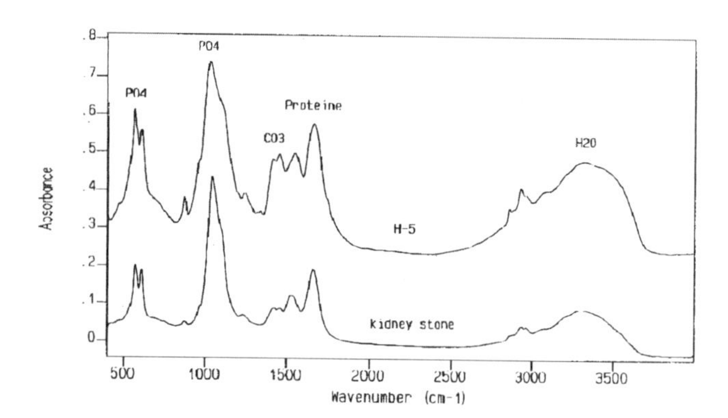

FIGURE 1. FTIR spectrum of the IAEA H-5 standard (a) and a spectrum of kidney stone - calcium phosphate type (as carbonate apatite and proteins).

In the integral part of the above formula the constant attenuation factor μ as well as variable dE_p/dx (the stopping power derivative as a function of sample depth x) require the knowledge of sample elemental composition. Sample composition was determined using the results of EBS spectra analysis. The EBS spectra were evaluated using our EBS-ERD software [10] which is applicable for uniform, thick targets. This restriction of the computer code is sometimes harmful, as the target composition and elemental depth distribution may be changed during the beam irradiation. However, especially for heavier elements the reasonable agreement may be achieved and target matrix composition may be determined. X-ray data as attenuation factors, fluorescence yields, X-ray excitation cross sections were taken from the literature [11,12]. The effect of secondary X-ray emission was estimated to be not significant. Proton beam stopping powers for compound targets were calculated using Ziegler's code TRIM [13].

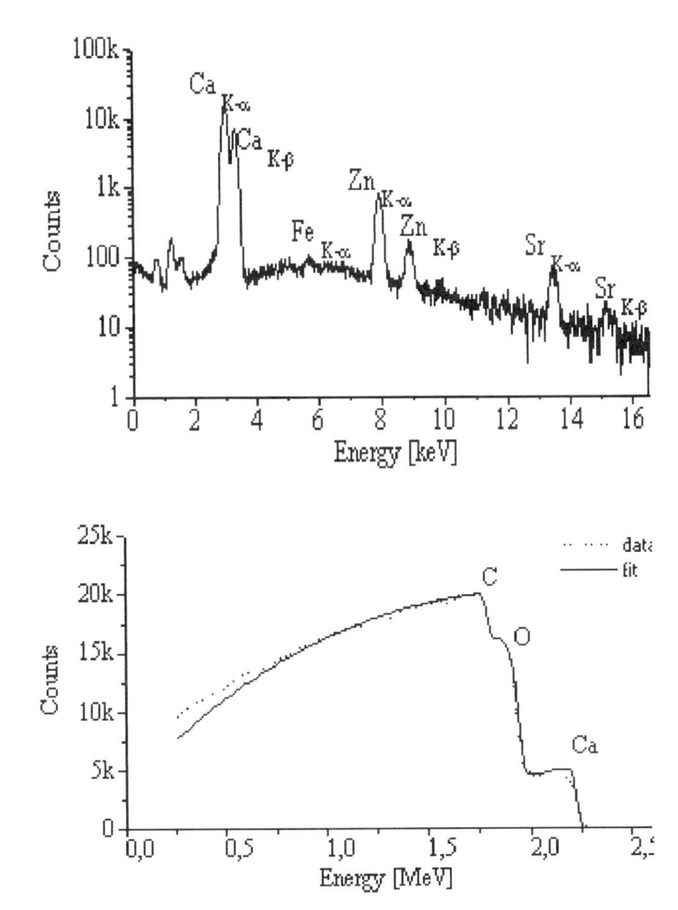

FIGURE 2. PIXE (top) and EBS (bottom) spectra of apatite type kidney stone sample. Only elements of interest are marked.

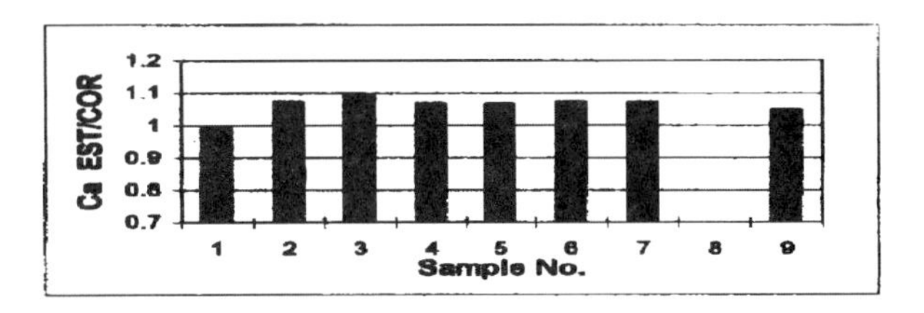

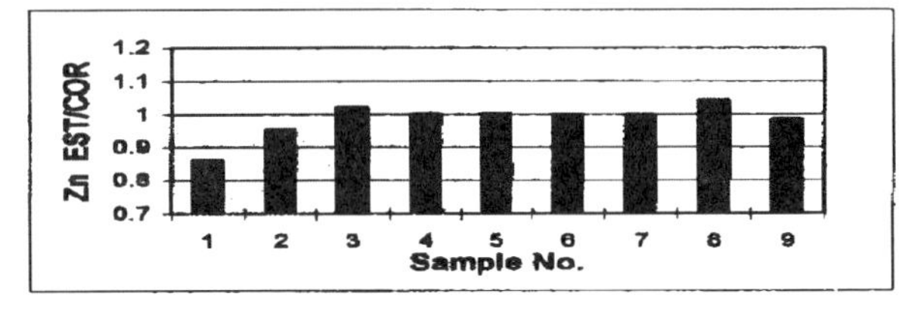

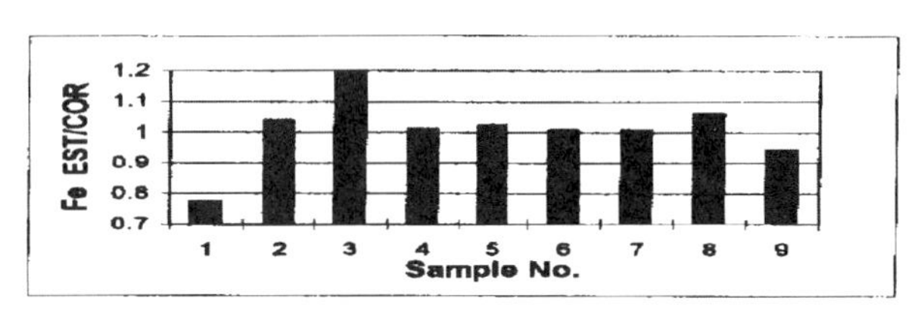

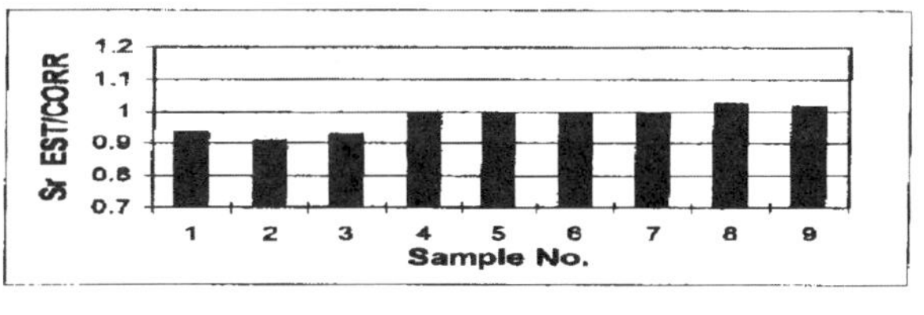

FIGURE 3. Plot of the ratio between EST and COR values.

Figure 2 shows an example of complementary PIXE and EBS spectra of the apatite type kidney stone sample.

Table 1 summarise the results obtained for all samples measured. Absolute concentrations of elements determined according to the external standard technique with the use of the IAEA H-5 standard are shown. The table presents also concentrations of the same four elements rescaled using results of principal formula calculations basing on sample composition obtained with EBS.

Figure 3 shows plots of the ratio between EST and COR values that provides information about significance of the correction.

As one can see the correction factors depend on sample type. In all cases the corrections for Ca were up to 10 % while for heavier elements the corrections were neglectable for whewellite and weddellite type of kidney stones.

This proves that IAEA H-5 standard was perfectly chosen as an external standard for trace element analysis in whewellite cases. For the other kidney stones samples the corrections were in the range of 5% to 20 %.

As is seen, the most significant changes were observed in the samples composed of the organic matrices as samples 1-3, and of struvit matrix as sample 8. Those samples contain Ca on the level below 15%.

TABLE 1. Concentrations of Ca, Fe, Zn, and Sr in kidney stones samples determined using external standard technique (columns with EST labels) and corrected using principal formula calculations (columns with COR labels). Values for Fe, Zn, and Sr are given in ppm. Estimated uncertainty is in the range of 5–10%. Shortcuts: ND = not detected, uric = uric acid, wew. = whewellite, and wed. = weddellite.

Sample No & Type	Ca EST	Ca COR	Fe EST	Fe COR	Zn EST	Zn COR	Sr EST	Sr COR
1 uric+wew.	9.4%	9.5%	14	18	200	235	115	120
2 uric+wew.	11.2%	10.5%	45	43	59	61	28	31
3 uric+wew.	4.4%	4.1%	4.9	4.1	5.6	5.5	17	19
4 wew.	27.8%	26.1%	217	215	330	330	31	31
5 wew.	30.6%	28.7%	15	14	30	30	72	73
6 wew.	29.5%	27.5%	7.5	7.4	19	19	25	25
7 wed.	26.5%	24.7%	24	23	24	24	34	34
8 struvit	ND	ND	41	39	95	91	150	146
9 apatite	19.6%	18.6%	13	14	585	595	270	267

SUMMARY

Results presented in this paper show that the presence of the matrix effects is significant in PIXE analysis of thick target composed of mixed matrix containing organic structures. This effect should be taken into consideration. Simple, traditional interpretation of X-ray spectra should be improved using complementary techniques. Additional information about target composition gathered simultaneously with PIXE data may also serve to minimise errors stemming from matrix changes during beam heating of the sample. This work shows the possibility of matrix correction in trace element analysis due to selection of the standard.

ACKNOWLEDGMENTS

Many thanks are given to R. Hajduk, the VdG group: S. Łazarski, Cz. Sarnecki, and Z. Szklarz for their help during measurements, and to L. Glebowa for sample preparation. Thanks are also given to Dr. M. Galka for providing of the kidney stones samples. This work has been supported by the State Committee for Scientific Research (KBN), Poland, Grant No. 2P03B04508.

REFERENCES

[1] M. Cholewa, G. Bench, B.J. Kirby, and G.J.K. Legge, *Nucl. Instr. and Meth.* **B54**, 101 (1991).
[2] W.M. Kwiatek, J. Lekki, C. Paluszkiewicz, and N. Preikschas, *Nucl. Instr. and Meth.* **B64**, 512-516 (1992).
[3] E.Rauhala, Nucl. Instr. and Meth. **B12**, 447 (1985).
[4] M.Luomajarvi, Nucl. Instr. and Meth. **B9**, 255 (1985).
[5] W.M.Wilson et al. Nuclear Phys. **A245**, 262 (1975) .
[6] E. Rokita, A. Wróbel, W.M. Kwiatek, E. Dutkiewicz, *Nucl. Instr. and Meth.* **B109/110**, 109-112 (1996).
[7] S.E. Wiberley, The encyclopedia of Spectroscopy, ed. L. Clarc (Reinhold, New York, Chapman and Hall, London, Waverly, Baltimore, 1960) 569-574.
[8] H.J. Schneider, Urolithiasis, Handbook of Urology Vil 17/I Springer-Verlag, Berlin-Heidelberg-New York-Tokio 1995.
[9] IAEA Report **TECDOC-880** Survey of reference materials.
[10] S. Kopta, J. Lekki, and B. Rajchel, *Journal of Radioanalytical Chemistry* Vol.172 No.1, 3-17 (1993).
[11] E.C. Montenegro, G.B. Baptista, and P.W.E.P. Duarte, *Atomic Data and Nuclear Data Tables*, **22**, 131-177 (1978).
[12] D.D. Cohen and M. Harrigan, *Atomic Data and Nuclear Data Tables*, **33**, 255-343 (1985).
[13] J.F. Ziegler, J.P.Biersack, and U.Littmark, The Stopping and Ranges of Ions in Solids, Vol.1, Pergamon Press (1985).

STUDY OF THE ELEMENTAL LATERAL DISTRIBUTION IN FIBER OPTIC CONDUCTORS

G. Bernasconi, A. Tajani, M. Dargie and V. Valkovic

International Atomic Energy Agency, Agency's Laboratories, A-2444 Seibersdorf, Austria

Fiber optic conductors have recently acquired an important role in high speed communications applications ranging from local area computers network to intercontinental data transmission. The elemental composition, the distribution and the dimensions of the different layers in the optical fiber affect its data transmission characteristics as well as signal losses and therefore should be carefully controlled. In this paper a study of the radial concentration profile of elements in different glass fibers is performed by using tube excited X-ray microanalysis (μ-XRF) and micro-PIXE (μ-PIXE) techniques.

INTRODUCTION

Recent technological advances in materials science, particularly in the fiber optics field, have revolutionized world wide communications. The availability of high purity materials for construction and doping, in addition to improved non-linear optics techniques, lead to the development of fibers capable of sending high rates of information at distances of tens and even hundreds of kilometers without need of intermediate amplification devices. For optical fibers the effective bandwidth is proportional to the inverse of the propagation distance whereas for electrical (coaxial) conductors is proportional to the inverse of the square of the distance. Due to this fact optical fibers have better capability of sending information at long distances than coaxial cables. Other applications of optical fibers include illumination and imaging of inaccessible places and recently IR - visible - UV spectrometry in harsh environments (e.g. high temperatures or pressures) (1). Optical fibers are cylindrical wave guides which conduct electromagnetic radiation mainly in the wavelength range from the UV to the infrared although no single fiber can cover the whole range. Fibers are made of plastic, compound glass, silica-based glass and fluoride based glass. Plastic is used for optical wavelengths, silica is used for UV - near infrared region, and fluoride fibers extend further into the infrared. Fibers consist of two or more concentric cylinders as illustrated in Fig. 1.

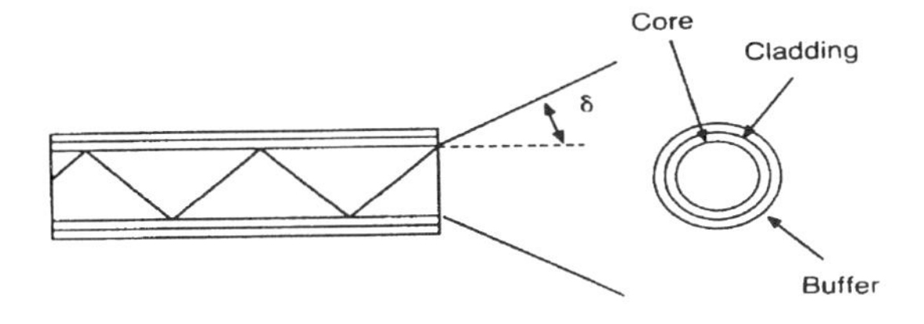

FIGURE 1. Typical fiber optic structure.

The efficient optical transmission is realized by total reflection in the interface between the core and the cladding therefore the core must have a higher refraction index than the cladding. This is usually achieved by different chemical composition between core and cladding (doping). Doping elements commonly used include Ge and Er and have to be added in high concentrations in order to achieve the desired optical properties (2). The choice of materials and fabrication techniques for optical fibers has to consider mechanical properties as well as light propagation; traction resistance and bending capabilities are important and usually specified by the manufacturers. There are three basic types of fibers (3): multimode step index, multimode graded index and single mode (Fig. 2). For the first type the core size is 50 μm or higher (up to 1000 μm); this type is mostly used to transmit light for illumination or in bundles to explore inaccessible places (e.g. inside human body or machines, endoscopes). Graded index multimode fibers have a core size from 50 μm to 100 μm; in the core material the refraction index decreases steadily with increasing radius keeping the photons propagating in the fiber inside its core. The last fiber type has core sizes of 5-10 μm which support only a single mode, very advantageous for

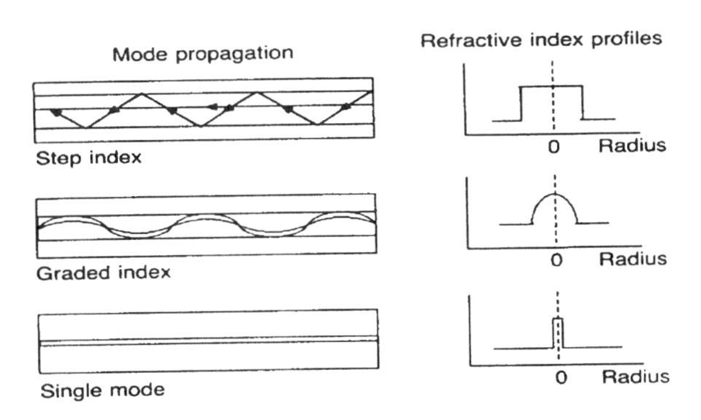

FIGURE 2. Propagation modes and refraction index of different fiber types.

CP392, *Application of Accelerators in Research and Industry*, edited by J. L. Duggan and I. L. Morgan
AIP Press, New York © 1997

telecommunications applications (due to the fact that the path length through the fiber is the same for all photons).

The microanalytical study of optical fibers measures radial variations in the fiber composition which create the necessary refraction index profiles. Microanalysis studies may lead to the determination of irregularities and imperfections that degrade the quality of fiber, helping to find their origin and improving the production process. On the other hand good quality fibers, which have large lengths of uniform cores, might be used as standard reference material for lateral elemental mapping and quantitative microanalysis. In this paper lateral elemental maps of various fibers have been performed using both μ-XRF and μ-PIXE techniques.

EXPERIMENTAL

For this work, a "Laboratory Scale" μ-XRF set up developed at the IAEA Laboratories in Seibersdorf (Austria) and a tandem accelerator μ-PIXE device located at the Ruder Boskovic Institute in Zagreb (Croatia) were used.

The laboratory scale μ-XRF set up, which has been described in detail (4, 5), uses a Mo-anode fine focus X-ray tube with 2 kW power, with a Siemens Kristalloflex high voltage supply. The tube is coupled to a glass capillary with ellipsoidal profile (6) from X-ray Capillary Optics (XCO) Sweden. The capillary outlet has a diameter of about 10 μm giving a practical beam size at the sample surface (1 mm distance from the tip) of 15~20 μm. The sample is mounted in air. The characteristic X-rays produced by the irradiated samples are measured by a Si(Li) X-ray Link detector whose resolution is 165 eV at 5.9 keV. Samples are placed on a sample holder mounted over a rotation stage and three translation stages for three dimensional positioning that are all computer controlled. Each coordinate can be set with an accuracy of 1 μm (100 μrad for the rotation). A computer program was written to take care of the sample movement and spectrum acquisition, allowing for long, unattended scans of samples. The program allows also to gather an optical image of the sample through a microscope and TV camera and select the region of the sample to be analyzed. The program measures each point of the selected region in sequence during a preset time using a predefined number of steps. The resultant spectrum can be optionally saved for each point and/or the intensities of up to 10 ROIs can be stored in the map file together with the position and live measuring time for each point (7). If the individual spectra have been saved it is possible to analyze them in batch mode and generate a map from the results. Another program makes 2-d maps based on information of up to 3 elements simultaneously where the normalized intensity from an analytical line of each element contributes to a different color. The colors resultant from the combination identify the regions in the sample with different elemental composition. Alternatively the program can generate false color and gray scale images. The proton microprobe uses a proton beam coming from a 3 MeV Van de Graff tandem accelerator. The beam is focused by a quadrupole magnets from Oxford (England) and steered by 2 separate magnetic coils. The size of the spot at the focus is about 2÷4 μm and the beam can be steered by controlling the current in the coils. A computer controls a set of ramp generators and amplifiers which sweep the beam in a rectangle on the sample with a maximum area of about 1 mm^2. The exact area can be adjusted by changing the gain of the amplifiers and a subregion can be selected by the computer software. The sample is placed in a vacuum chamber which contains a Si(Li) X-ray detector (Link) and a detector for charged particles (for Rutherford Backscatter measurements). During the measurement, the proton beam scans the sample and the data is collected using a multiparameter analyzer (MPA) which is triggered by the pulses coming from the detector. The position of the beam and the energy of the pulses are digitalized simultaneously by 3 analog to digital converters and the values are stored in the MPA memory. The computer program makes a map using this data that contains position in steps (0-127) for X and Y coordinate and counts for each ROI. A pulse-height spectrum containing the energies of the pulses from the whole scanned area can also be generated. This spectrum is useful to select the different ROIs which will be used to select the counts for the elemental maps. In the μ-XRF case, each position of the sample is measured for a relatively long time and a pulse height spectrum from each point is produced. This allows to store the areas of each ROI with optional background subtraction or alternatively perform sophisticated spectrum analysis. In the μ-PIXE case the proton beam, sweeping the selected rectangle, passes many times over each point. Events, containing energy and position, are classified according to the current ROIs list and added to the map of the correspondent ROI. Therefore in this case background subtraction or other kind of spectrum analysis are not easily performed.

Sample Preparation

Table 1 lists the fiber optic samples, obtained from manufacturers or vendors, together with the specifications provided. The optical fiber samples were cut in segments of 20 mm length. Each segment was inserted in a hole drilled in the center of a small plastic cylinder (10 X 10 mm) in order to hold it vertical. The plastic cylinder with the fiber was centered inside a cylindrical

TABLE 1. Analyzed fibers

Fiber Code (this paper)	Manufacturer code	Manufacturer / Vendor	Type	Core and cladding diameters (declared by manufacturer).
G50	01-G50/VNJH-D L027	Huber+Suhner AG	Gradient	50 / 125 μm
H200	01-H200/VNJH-D L027	Huber+Suhner AG	Step	Max = 600 μm
125D	50/125 D	Hirshmann	Gradient	-
BT	-	British Telecom	Monomode	5.6 /125 μm
BRUGG	-	Brugg Co.	Gradient	-
159/125	-	Siecor	Step	-

polyethylene vessel (10 mm internal diameter and 20 mm height) and embedded by filling the vessel with Araldite resin (mixture CY 212). The embedding material was hardened at 60 °C for 48 hours and then the top 10 mm, containing the embedded fiber, was cut in 1 mm slices using a low speed diamond saw. Each slice was carefully polished to a flat surface using silicon carbide paper and a polishing turntable.

MEASUREMENTS AND RESULTS

The samples were first measured using μ-XRF with a Mo-anode tube excitation operated at 40 kV and 20 mA. These measurements show that all the fibers, except 2, have a core doped with Ge. These 2 fibers (H200 and 159/125) did not show any particular element but the shape of the core could be inferred from the coherent to incoherent scatter ratio indicating that they are probably of the step index type. For the BRUGG fiber, measured with Mo-anode excitation, the effect of the penetration of the primary and fluorescent X-rays can be observed for the high Z elements (Ge and Zn). This fiber has 6 cores and 2 of them are surrounded by a thin Zn layer (the reason is not known to the authors). The penetration of X-rays makes this layer look like a tube in the μ-XRF plot (Fig. 3-A). The maps of these elements show a shadow in a particular direction due to the fact that the fiber is not perfectly perpendicular to the sample surface .

The μ-PIXE results show the effects of better resolution (beam size of 2~4 μm) and lower penetration of the proton beam compared to the X-ray beam. The previously mentioned shadows, observed in the case of μ-XRF, are not observed for μ-PIXE (Fig. 3-B). In the μ-PIXE measurement all the Ge cores as well as the claddings, identified by their Si content, are well seen. The sample 125D was mapped with a Cr-anode tube excitation in order to asses the ability of μ-XRF to map low Z elements. Figure 4-A displays the map of Ge-Kα line excited by the continuum of the tube passing through the capillary, whereas Fig. 4-B displays the same plot performed with μ-PIXE. Figure 5 shows the Si intensity for the same fiber, μ-XRF case (5-A) and μ-PIXE (5-B). In Fig. 5-A shadow effects, due to the very low penetration of the Si characteristic emission, are observed. These effects are not so evident for μ-PIXE (Fig 5-B) because in this case the information depth is significantly lower.

A plot of the intensity of different elements as function of the radial position for the fiber 125D shows dimensions of the different layers that compose the fibers (Figs. 6 and 7). According to the μ-XRF measurements the diameter of the core (full width at half maximum FWHM) is 50 μm and that of the cladding is 120 μm. The actual dimensions are probably less, due to the X-ray beam diameter (15~20 μm at the sample position). The measured profile is a convolution of the actual concentration profile and that of the X-ray beam.

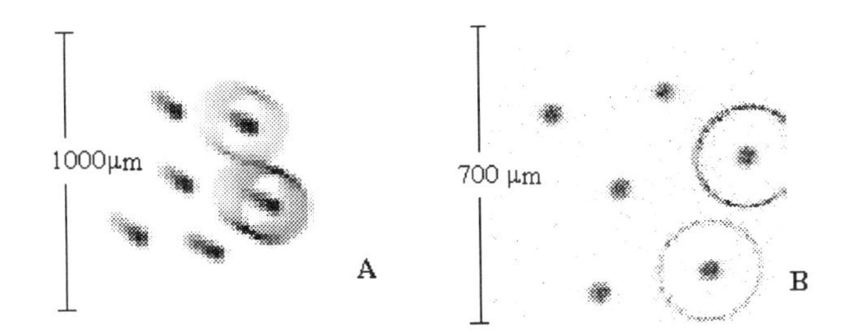

FIGURE 3. BRUGG Fiber. Zn (2 circles) and Ge (6 cores) intensities plotted together as gray levels. Darker gray corresponds to higher intensity. μ-XRF (3-A) and μ-PIXE (3-B) maps.

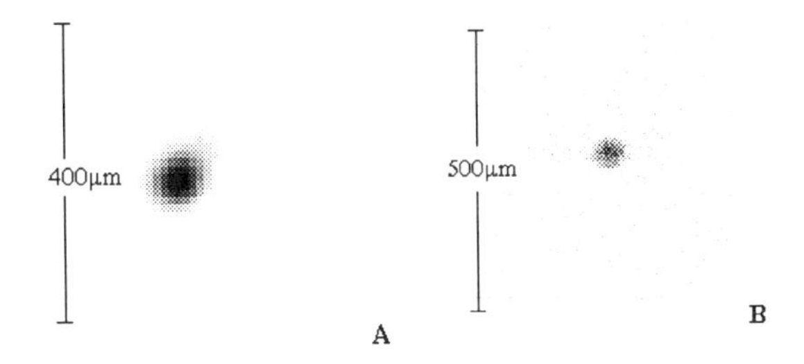

FIGURE 4. 125D Fiber core. Ge intensity plotted as gray level. Darker gray corresponds to higher intensity. μ-XRF (4-A) and μ-PIXE (4-B) maps.

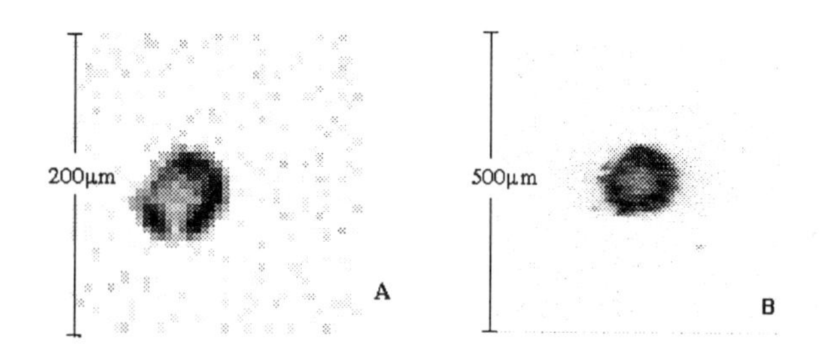

FIGURE 5. 125 D Fiber core and cladding. Si intensity plotted as gray levels. µ-XRF (5-A) and µ-PIXE (5-B) maps.

From the µ-PIXE measurements, lower values of the diameters are inferred, consistent with the small particle beam diameter and therefore higher resolution. However the scale of the map has a bigger uncertainty than in the case of µ-XRF. The fibers that contain Ge have lower Si intensity in the core than in the cladding probably due to the high concentration of Ge in the core which both displaces Si and reduces the Si intensity (even if the concentration is constant) with its high absorption coefficient The concentration of Ge (in the form of GeO_2) is reported to be in the range of 10% (8).

A single mode fiber (BT) has also been analyzed with both methods. This fiber is reported to have a Ge doped core of only 5.6 µm diameter. Although this core was well detected in the µ-XRF experiments (67 cps maximum intensity for Ge-Kα) it could not be detected in the µ-PIXE measurements.

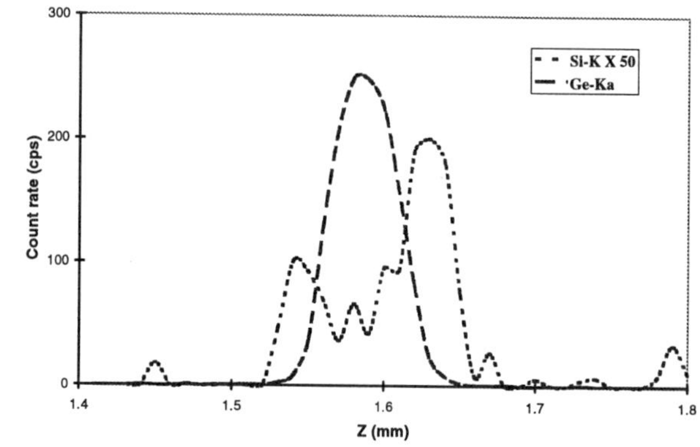

FIGURE 6. 125 D Fiber, vertical cut of core and cladding taken from µ-XRF map. Note that Si intensity is multiplied by 50 due to the low sensitivity of µ-XRF for Si.

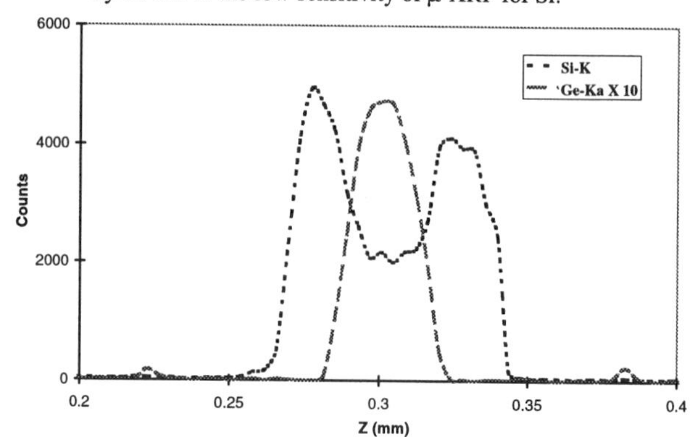

FIGURE 7. 125 D Fiber, vertical cut of core and cladding taken from µ-PIXE map. Note that Ge intensity is multiplied by 10 due to the low sensitivity of µ-PIXE for Ge.

DISCUSSIONS AND CONCLUSIONS

The results demonstrate the suitability of both methods for analysis of optical fibers and the advantages and limitations of each of them. The main advantages of µ-XRF are the low construction and maintenance cost of the set up compared to a suitable accelerator together with the micro beam line and the higher sensitivity for high Z elements. The main disadvantages are that much longer measurement time is required to map the same area and the sensitivity for low Z elements (e.g. Si or Al) is much lower. The long measurement time is somewhat compensated by the possibility to run long unattended experiments. µ-PIXE has the advantage of smaller beam size and therefore better spatial resolution.

PIXE also has the potential to use other analytical techniques as Rutherford Backscatter and perhaps PIGE (Proton Induced Gamma Emissions). For example RBS spectra of the core of one fiber indicated also low Z elements like N and O. These methods (RBS and PIGE) can be used for local analysis but have not been implemented yet for elemental mapping.

ACKNOWLEDGMENT

The authors are thankful to the Institute Ruder Boskovic, Zagreb, Croatia for their kind cooperation in performing all the PIXE measurements.

REFERENCES

1. Tran, C. D., and Gao, G. H., *Anal Chem* **68**, 2264-2269 (1996).
2. Allen, W. B., *Fiber Optics Theory and Practice*, New York: Plenum Press, 1973.
3. Soliman, F. A. S., *Appl. Radiat. Isot.* **47 No 2**, 175-183 (1996).
4. Bernasconi, G., Haselberger, N., Markowicz, A., and Valkovic, V., *Nucl. Instr, and Meth.* **B 86**, 333-338 (1994).
5. *Microanalysis Data Acquisition and Control Program*, Computer Manual Series No 9, International Atomic Energy Agency, 1996, ch. 4, pp. 17-19.
6. Attaelmanan, A., Voglis, P., Rindby, A., Larsson, S., and Engstrom, P., *Rev. Sci. Instrum.* **66 (1)** , 24-27 (1995).
7. *Microanalysis Data Acquisition and Control Program*, Computer Manual Series No 9, International Atomic Energy Agency, 1996, ch. 1, p. 9, ch 5, pp. 25-27.
8. Tamburrini, M., Department for Non Linear Optics, FUB, Rome, Personal Communication describing the structure of optical fibers for long distance communications (1996).

Background due to energy-loss of γ-rays in SSD in PIXE induced XRF

M. UDA[a,b], K. MORITO[a], H. MATSUI[a], T. KOTANI[a], M. NAKAMURA[c] and H. ISE[c]

[a]*Department of Materials Science and Engineering Waseda University,*
[b]*Labolatory for Materials Science and Technology Waseda University,*
[c]*Tokyo Metropolitan Isotope Research Center*

PIXE induced XRF spectra induced by Co Kα primary X-rays had a rather high background in steel analysis, which was excited with 2.6 MeV H^+. This was caused by continuous X-rays produced by energy loss in Si(Li) of γ-rays emitted through nuclear reactions between accelerated H^+ beam, and a Co foil used as a primary target and an exit nozzle for the H^+ beam. The background was reduced significantly by lowering H^+ energies from 2.6 to 2.0 MeV, and by replacing an Al nozzle by a Cu nozzle. Resulting S / B ratios and M.D.L.s were much improved.

INTRODUCTION

PIXE induced XRF [1-3] has been effectively used to analyze trace amounts of impurities whose atomic numbers are smaller than that of a matrix element by a factor of 1～4, where PIXE and XRF mean Particle Induced X-ray Emission and X-Ray Fluorescence, respectively. Then Mn, Cr, V and Ti of 100 parts per million (ppm) or less in steel have successfully been analyzed [4,5]. However, the PIXE induced XRF spectra induced by Co Kα primary X-rays had a rather high background in steel analysis.

The background at the energy region of Mn Kα (～6 keV) in the steel analysis is composed of low energy tails of Rayleigh, Compton and Raman scattered Co Kα and Fe Kα, and continuous X-Rays produced by energy loss of γ-rays in Si(Li) which are emitted from a Co foil for Co Kα primary X-ray source and from a nozzle for proton beam extraction from vacuum to air atmosphere. Among these the Rayleigh, Compton and Raman scattered Co Kα can not be reduced because these X-rays are induced by incidence of the Co Kα primary X-ray on a specimen to be analyzed. However, the scattered Fe Kα and γ-rays can be reduced by adopting appropriate experimental conditions.

Fe Kα emissions are originated from two sources: 1) an Fe foil used for reducing Co Kβ emitted from the primary target made of the Co foil and 2) the steel to be analyzed. The contribution from the former source can be reduced to practically zero by using Co foil thicker than the proton range in the foil, leading to avoidance of direct excitation of the Fe foil by the proton. The second source does also not contribute to emitted Fe Kα by absorbing appreciable amounts of the Co Kβ X-ray through the Fe foil.

γ-rays caused by the nuclear reaction between accelerated protons and the Co foil used for the primary X-ray source, ^{59}Co (p,nγ) ^{59}Ni, can be reduced by lowering accelerating voltage of the protons to an amount which can not excite the most intense γ-ray in the nuclear reaction i.e. 343 keV γ-ray. γ-rays emitted from the beam nozzle can also be reduced by employing a low cross section material for nuclear reactions concerned.

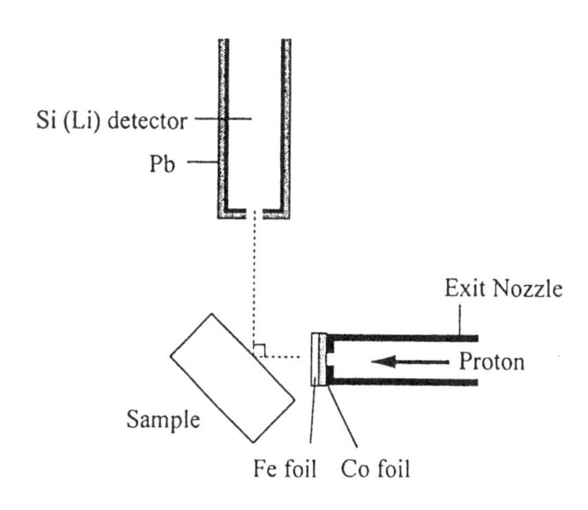

FIG.1. Layout of the PIXE induced XRF system.

CP392, *Application of Accelerators in Research and Industry,* edited by J. L. Duggan and I. L. Morgan
AIP Press, New York © 1997

EXPERIMENTS AND RESULTS

Experimental setup for PIXE induced XRF is shown in FIG.1 schematically where HV-4117 (High Voltage Engineering Europe) was used for proton incidence on the Co and Fe foils to produce Co Kα X-rays. A PIXE induced XRF spectrum is shown in FIG.2 taken from steel with 64 ppm Mn by irradiating Co Kα produced by 2.6 MeV H^+ through (20 μm Co + 20 μm Fe) sandwiched foils. The background at the Mn Kα energy region is composed of low energy tails of Rayleigh, Compton and Raman scattered Co Kα and Fe Kα, and continuous X-Rays produced by energy loss of γ-rays in Si(Li). The γ-rays are originated from nuclear reactions between H^+, and Co used for the Co Kα source and Al used for the exit nozzle, and also between neutrons and Al. Here neutrons might be produced by a nuclear reaction between H^+ and C used for inner wall coating. γ-ray spectra measured with a Ge solid state detector (SSD) placed perpendicular to the Si(Li) detector are shown in FIG.3 where the Al nozzle with a 3mm pinhole in diameter covered with the sandwiched Co and Fe foils was used, whose inner wall was coated with enough thick graphite to stop proton beams. γ-rays originated from ^{59}Co (p,nγ) ^{59}Ni, ^{27}Al (p,p'γ) ^{27}Mg or ^{27}Al (n,pγ) ^{27}Al, and ^{12}C (p,nγ) ^{12}N can clearly be seen in FIG.4(a) together with a β-decay γ-ray (511keV). All the assignments of the γ-rays are given in TABLE 1.

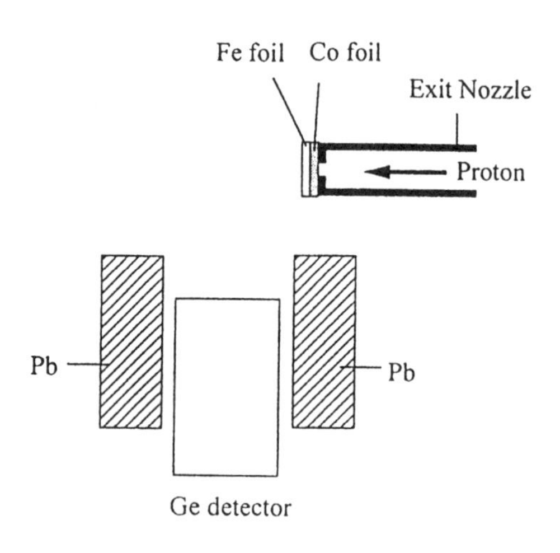

FIG.3. Layout of γ-ray measurement emitted from the primary target and the exit nozzle.

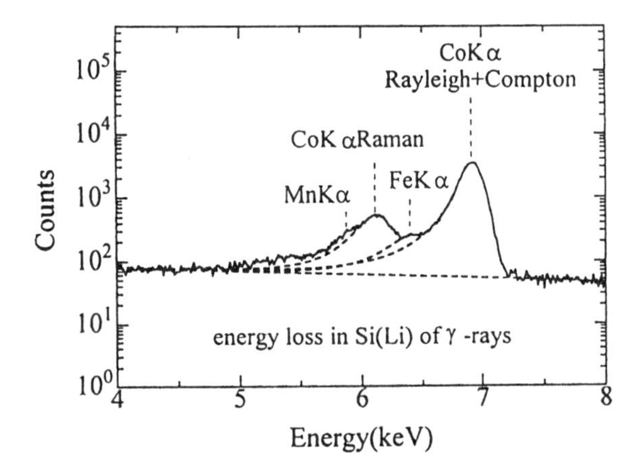

FIG.2. PIXE induced XRF spectrum taken from the steel with 64 ppm Mn. Here combined foils of 20 μm Co and 20 μm Fe were irradiated with 2.6 MeV H^+ to emit Co Kα. The background at the energy range of Mn Kα is decomposed to components concerned.

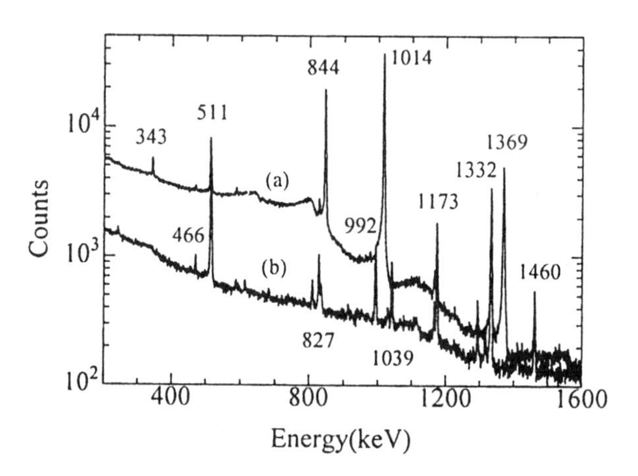

FIG.4. γ-rays spectra emitted from nuclear reactions of 2.6 MeV protons with the Co foil and the exit nozzle made of Al (a), and of 2.0MeV protons with the Co foil and the exit nozzle made of Cu (b).

TABLE 1. γ-ray assignment

γ-ray energy	nuclear reaction	
keV	2.6 MeV Proton impacts	2.0 MeV Proton impacts
343	^{59}Co(p, nγ)^{59}Ni	-
466	-	^{59}Co(p, γ)^{60}Ni
511	β-decay	β-decay
827	-	^{63}Cu(p,αγ)^{60}Ni
844	^{27}Al(p,p' γ)^{27}Mg, ^{27}Al(n,p γ)^{27}Mg	-
992	-	^{63}Cu(p,γ)^{64}Zn
1014	^{27}Al(p,p' γ)^{27}Mg, ^{27}Al(n,p γ)^{27}Mg	-
1039	-	^{65}Cu(p,γ)^{66}Zn
1173	^{59}Co(p, γ)^{60}Ni	^{59}Co(p, γ)^{60}Ni
1332	^{59}Co(p, γ)^{60}Ni	^{59}Co(p, γ)^{60}Ni
1369	^{27}Al(n,α γ)^{24}Na	-
1460	^{40}K	^{40}K

To reduce these γ-rays, 2.0 MeV H^+ was employed to irradiate the Co foil, which could not excite the most intense γ-ray produced by ^{59}Co (p,nγ) ^{59}Ni, i.e. 343 keV γ-ray, whose threshold energy is 2.28 MeV. As the range of proton in the Co foil is estimated to be 17.6 μm, thickness of the Co foil was selected to be 20 μm commercially available, which is thick enough not to excite the Fe foil by protons directly. Thickness of the Fe foil placed downstream was chosen as 20 μm which gives a small enough ratio of Co Kβ / Co Kα, i.e. 1.3×10^{-3}. γ-ray spectra taken under the above experimental conditions are shown in Fig.4(b) where the Cu nozzle with a 3mm pinhole in diameter without graphite coating was used for proton exit. Significant reduction in γ-rays due to nuclear reactions between the Co foil or the Al nozzle and proton can be seen on both characteristic γ-rays and continua originated from energy loss of them.

Under the experimental conditions mentioned above, PIXE induced XRF spectra were measured for high purity (99.998 wt.%) Fe and steel with 64 ppm Mn, which are shown in FIG.5(a) and (b). By subtracting the spectra for the high purity Fe from those for the Mn bearing steel, much improved values of Signal/Background (S/B) ratio and Minimum Detection Limit (M.D.L.) were obtained, as shown in TABLE 2, where data acquisition time was chosen so as to get the same count rates of Raman scattered Co Kα for both experiments.

The authors wish to thank Dr. S.Nagashima for his helpful discussions.

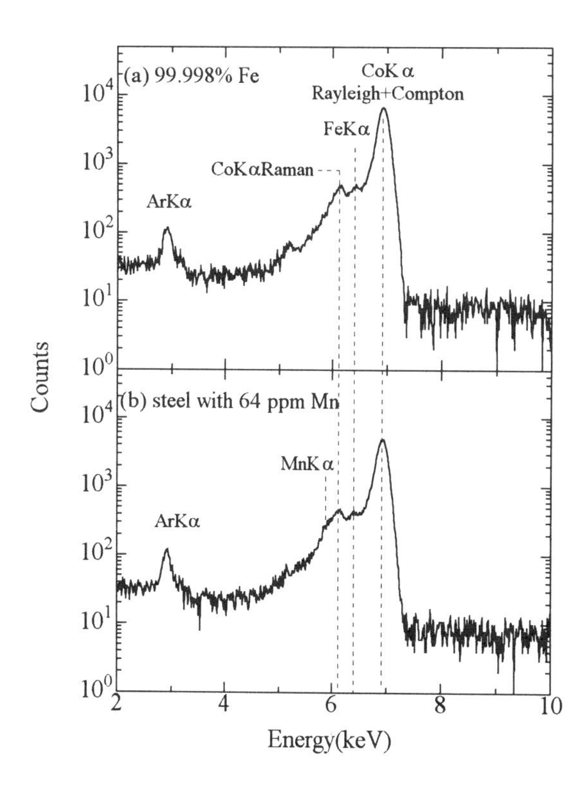

FIG.5. PIXE induced XRF spectra taken from high purity (99.998%) Fe (a) and steel with 64 ppm Mn (b), which were excited with Co Kα X-rays emitted from Co 20 μm and Fe 20 μm foils by 2.0 MeV proton impacts.

TABLE 2. Signal, background and their ratio, and M.D.L. for the steel with 64 ppm Mn by the use of PIXE induced XRF

H^+ Energy(MeV)	2.0	2.6
Signal(counts)	8.4×10	6.7×10
Background(counts)	2.9×10^3	3.6×10^3
S/B ratio	0.29	0.19
M.D.L.	12	17

REFERENCES

1. Tsang-Lang Lin, Chien-Shu Luo and Jen-Chang Chou, Nucl. Instr. and Meth. **151** (1978) 439.
2. Le Huong Quynh, I. Demeter and Z. Szokefalvi-Nagy, Nucl. Instr. and Meth. B **49** (1990) 566.
3. N. B. Kim, H. J. Woo, D. K. Kim, K. Y. Lee and K. S. Park, Nucl. Instr. and Meth. B **75** (1993) 379.
4. K.Morito, S.Nagashima, M.Kato, T.Kotani, M.Uda, Nucl. Instr. and Meth. B **109/110**, (1996) 569.
5. S.Nagashima, T.Kotani, M.Kato, K.Morito, H.Matsui, M.Uda, Nucl. Instr. and Meth. B **109/110**, (1996) 243.

TRACE ELEMENTAL ANALYSIS OF CORAL AND BONE SAMPLES

Amy J. Kastner[†], Chandrika Narayan, Michael O'Connor, Gunter Kegel and Arthur Mittler

Radiation Laboratory, University of Massachusetts Lowell, Lowell, MA 01854.
[†] Susquehanna University, Selinsgrove, PA 17870

Biomaterials have been developed for various medical applications. Strength factors as well as the deterioration rate of the biomaterials in the human body must be considered. Extensive research covers the use of hydroxyapatite, a coralline substance for biomedical applications. Coral, bone and hydroxyapatite were examined using external beam Proton Induced X-ray Emission (PIXE), Neutron Activation Analysis (NAA) and X-ray Diffraction (XRD) to determine the similarities among these materials.

INTRODUCTION

For several years, the use of biomaterials in various medical applications has been studied. Bone repair, prosthetic manufacturing, joint replacement and oral surgery are just a few examples of areas where biomaterials are used. Many design factors go into developing a material that, when incorporated into the human body, will not cause infection and be rejected due to incompatibility with the tissues of the body.

Research has been conducted to find methods of repairing bone that has been subjected to injury or reconstruction surgery. It has been found that the chemical constitution and the macro structure of coral are quite similar to that of bone allowing the bone to grow into the coral matrix. Hence the use of coral in these types of surgeries is beneficial. The shearing strength of natural coral, however, is low and other agents must be added to coral to obtain a stronger material. Thus, Holmes *et. al.* [1] studied the coralline hydroxyapatite (HA) material obtained by a hydrothermal exchange process which converts the calcium carbonate structure of coral into pure HA. Investigations of HA ceramics as biomaterials for bone grafts were performed by Akino [2]. Implantation of the coralline HA and the bone ingrowth were studied by Martin *et al.* [3]. Cook *et al.* investigated HA coated materials for hip replacement [4], and for bone ingrowth and interface attachment strength [5]. Constantz of Norain Corp., Cupertino, CA proposed the use of a new biomaterial (called Norain SRS - "skeletal repair system") while studying the integral structure of coral and its resemblance to bone. With his co-workers [6], he developed a process for the *in situ* formation of the mineral phase of bone, inorganic calcium and phosphate sources to form a paste that could be surgically implanted by injection.

In the past PIXE has been used in biomedical applications to examine biopsy samples [7], synthesized drugs in AIDS research [8], contamination levels in estuarine sediments [9], *etc*. This paper investigates similarities among coral, bone and HA using various analytical tools. This will assist in judging the compatibility of biomaterials plus coral with natural bone.

EXPERIMENT

Three analytical tools, external beam PIXE, NAA and XRD were employed in the comparative study of coral, bone and hydroxyapatite material.

Proton Induced X-ray Emission (PIXE)

At the Lowell Radiation Laboratory, a 5.5 MV Van de Graaff accelerator is used to accelerate protons which penetrate a thin window at the end of the beam line reaching the sample to perform external PIXE. The experimental set up was the same as in Ref. [10]. The incident proton energy was chosen to be 2 MeV with a beam spot size of about $1 mm^2$, and the resolution of the x-ray Ge-detector (HPGe) was ~ 190 eV. An x-ray spectrum of Fe was obtained using a pure Fe-foil. The centroid of the K_α-peak in this spectrum can be determined with an accuracy of about 100 eV. Hence this peak provides adequate accuracy to calibrate the x-ray energy scale.

Neutron Activation Analysis (NAA)

Coral, bone and HA samples were packed in polyethylene containers and irradiated at 1 MW for 10

CP392, *Application of Accelerators in Research and Industry*, edited by J. L. Duggan and I. L. Morgan
AIP Press, New York © 1997

minutes using the nuclear reactor of the Radiation Laboratory. The neutron fluence was about 10^{12} neutrons/cm^2-sec. Variations in neutron fluence were accounted for by using an internal gold monitor. The gold monitor is a dilute (202 μg/mL) solution of gold in HCl. About 50 μL of this solution was added to each sample vial including the standard, and the vials were dried overnight at ~ 50°C.

Gamma-ray spectra were analyzed with Canberra Accuspec software [11] and the elemental concentrations in the samples were determined with reference to a separate standard. The standard was a liquid standard containing known quantities of Ca, Mn, and other elements.

X-ray Diffraction (XRD)

In order to obtain a quantitative phase analysis, XRD was performed with a Philips diffractometer at the UML Center for Advanced Materials. X-rays from a Cu-source (λ = 1.5405 A°) impinged on a powder sample. An angular scan between 10° and 90° resulted in a diffraction pattern from which the chemical phase of the sample was obtained by reference to the International Center for Diffraction Data (ICDD) files [12].

RESULTS AND DISCUSSION

The PIXE pulse height spectra of coral, bone, and HA are shown in Figs. 1, 2, and 3 respectively. A peak due to atmospheric Ar can be seen in all these spectra, it is an artifact of the external beam. A comparison of the spectra indicates the presence of small amounts of S, Mn and Ni in addition to the major peaks for Ca. Furthermore the phosphorus peak is quite prominent in bone and HA spectra while missing from the spectrum for coral, suggesting that the missing phosphorus content in coral can be provided by HA in the mixing process prior to bone implantation. Since no standards were used, relative concentration ratios of different elements with reference to Ca ware evaluated by determining the area under the K-peak for each element.

Analysis of the gamma spectrum from the NAA indicated that in each sample there were peaks associated with Mn^{56} (at 846.8 keV and 1810.9 keV) which confirmed the presence of Mn as observed from the PIXE. The peaks due to P or S were not observed since the half-lives of the corresponding isotopes are only a few seconds. Using the number of counts in the Au peak in both the standard and the samples, the amount of Mn and Ca was evaluated in ppm.

From XRD analysis, the chemical phase composition

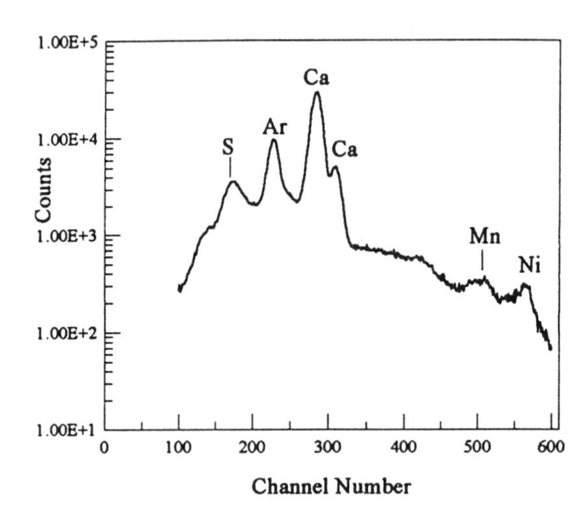

FIGURE 1. PIXE spectrum of coral sample.

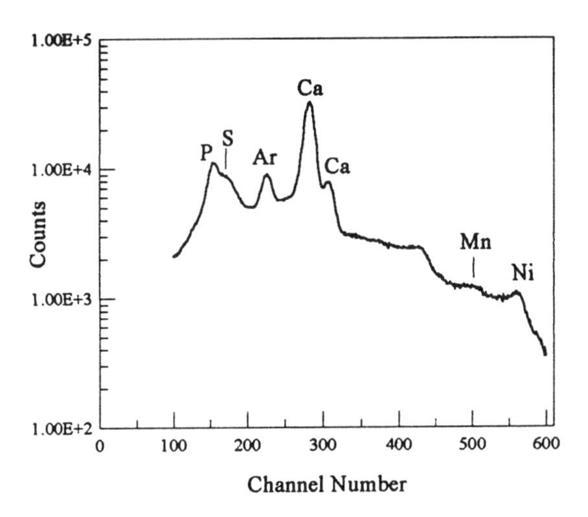

FIGURE 2. PIXE spectrum of bone sample.

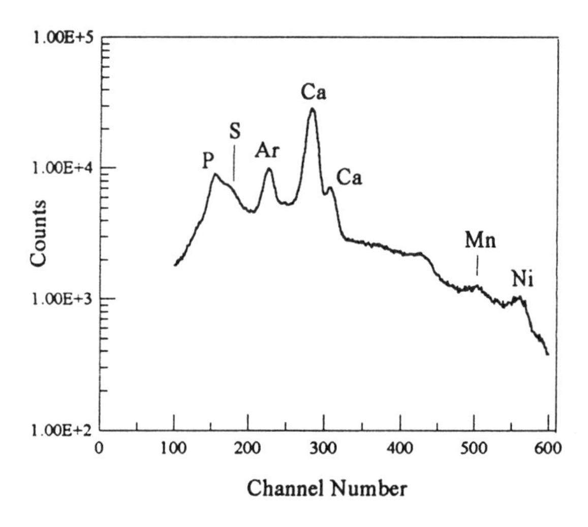

FIGURE 3. PIXE spectrum of hydroxyapatite material.

TABLE 1. Comparison of the results obtained by different techniques.

Sample	PIXE Relative concentration ratio for S, P, Mn and Ni with ref. to Ca				NAA		XRD
	P/Ca	S/Ca	Mn/Ca	Ni/Ca	Ca (ppm)	Mn(ppm)	
Coral	----	0.204	0.042	0.031	736160	7.02	$CaCO_3$ Aragonite
Bone	0.287	0.279	0.066	0.033	876720	14.6	$(Ca)_3(PO)_2\ xH_2O$ Ca- Phosphate Hydrate
Hydroxyapatite	0.319	0.221	0.073	0.045	364535	1.12	$(Ca)_5(PO_4)_3(OH)$ Ca- Hydroxide Phosphate

of the three samples were determined. The diffraction patterns indicated that each sample is made up of only one phase. Due to the detection limits of this method (3%), the presence of Mn or Ni was not evident.

Results of the above measurements are summarized in Table 1. From the PIXE analysis the relative concentration ratio obtained for each element in the three samples are compatible even though better estimates could be done using standards. Further, the above analyses indicate that the elemental constituents obtained from PIXE and NAA agree with each other. The phase compositions obtained from XRD for coral and bone agree with their chemical compounds, while the phase composition of HA agrees with the one given by the chemical supplier. The presence of Mn and Ni in all the samples might be due to the sample handling method and could not be substantiated further.

CONCLUSION

The coral alone is not strong enough and the biomaterial alone does not have the ability to bond to bone efficiently. Hence a combination of the two is optimal for bone replacement surgeries. In this paper external beam PIXE, NAA and XRD techniques were used to confirm the chemical compatibility of coral-hydroxyapatite mixture and natural bone.

ACKNOWLEDGEMENTS

The authors would like to thank the members of the accelerator group and the nuclear reactor group at the Radiation Laboratory for their friendly assistance. In particular we would like to thank Dr. Bettenhausen, Mary Montesalvo, Jerry Falo, Dave Medich and Tony Honnellio for sample irradiation and NAA and Tom Regan for providing the coral samples. Thanks goes to Mike Downey of Center of Advanced Materials for his assistance in XRD. American International Chemical, Inc., Natick, MA are acknowledged for the supply of hydroxyapatite material. Dr. R. Foster and Shirley Parasovich are thanked for their help in supplying information on replacement surgery with coral and HA. One of the authors (A.K.) was supported by the National Science Foundation funding for the Research Experiences for Undergraduates (REU) in physics.

REFERENCES

1. Holmes, R., Mooney, V., Bucholz, R., and Tencer, A., *Clinical Orthopaedics and Related Research*, **188**, 252-262 (1984).
2. Akino, M., *Hokkaidi Journal of Med. Sci.*, **66 (4)**, 468-481 (1991).
3. Martin, R.B., Chapman, M.W., Sharkey, N.A., Zissimos, S.L., Bay, B., and Shors, E.C., *Biomaterials*, **14 (5)**, 341-348 (1993).
4. Cook, S.D., Enis, J., Armstrong, D., and Lisecki, E., *Dental Clinics of North America*, **36 (1)**, 235-255 (1992).
5. Cook, S.D., Thomas, K.A., Dalton, J.E., Volkman, T.K., Whilecloud, T.S., and Kay, J.F., *Journal of Biomaterials Research*, **26**, 989- 1001 (1992).
6. Constantz, B.R., Ison, I.C., Fulmer, M. T., Poser, R.D., Smith, S.T., VanWagoner, M., Ross, J., Goldstein, S.A., Jupiter, and J.B., Rosenthal, D.I., *Science*, **267**, 1796-1798 (1995).
7. Valkoic, V., Bernasconi, G., Haselberger, N., Makarewicz, M., Ogris, R., Moschini. G., Bogdanovic, I., Jaksic, M., and Valkovic. O., Nucl. Instr. and Meth., **B 75**, 155-159 (1993).
8. Cholewa, M., Legge, G.J.F., Weigold, H., Holan, G., and Birch, C., Nucl. Instr. and Meth., **B 77**, 282-286 (1993).
9. Al-Jundi, J., Mamas, C.J.V., Sokhi, R.S., and Earwaker, L.G., Nucl. Instr. and Meth., **B 79**, 568-570 (1993).
10. Narayan, C., O'Connor M., Kegel, G.H.R., Johnson, R., Salmons, C., and White, C., Nucl. Instr. and Meth., **B 118**, 396-399 (1996).
11. Canberra Industries, Inc., Meriden, CT, U.S.A.
12. International Center for Diffraction Data, Swarthmore, PA, U.S.A.

TOWARDS INTELLIGENT SPECTRUM ANALYZING SYSTEM FOR INDUSTRY-ORIENTED PIXE

S. Iwasaki, K. Murozono, K. Ishii, and M. Kitamura

Department of Quantum Science and Energy Engineering, Tohoku University, Sendai 980-77, Japan

Outstanding features of a new method called 'pattern analysis method' are its analysis speed and easy handling, therefore enabling us to automatize the PIXE spectrum analysis. We address the following tasks within the framework of the pattern analysis method towards realization of more intelligent system for the industrial PIXE: adoption of more detailed models for all the reference X-ray spectra of element, of continuum from the Bremsstrahlung process and of the summing effect; real time inference of the content elements based on the iterative Bayes' theorem, which could be effective in the selection of candidate elements and monitoring the status of the spectrum accumulation; use of the suitable X-ray absorber with accurate data for penetration rate, being indispensable to make PAM analysis more reliable; and development of iterative method for thick sample PIXE where various non-linear effects become significant.

INTRODUCTION

Industrial applications of PIXE have not been exploited as expected partly due to the difficulty in the analysis of PIXE spectra. The conventional peak analysis has necessitated analysts to be highly experienced, because of the complexity of PIXE spectra and existence of frequent overlaps of peaks to be separated, and of the ambiguity in the estimation of the continuous background.

The present authors have developed a new method called 'pattern analysis method',[1,2] hereafter PAM. Outstanding features of this method are its analysis speed and easy handling; therefore it could enable us to automatize the process of the spectrum analysis, although there are some problems to be solved in order to apply the method in an actual system.

In this paper, we address the remained tasks of this approach, and also the prospects of the intelligent pattern analysis system applied to future industrial purposes.

PATTERN ANALYSIS METHOD AND ITS IMPROVEMENT

Pattern Analysis Method

The function of PAM is the linear transformation of a vector x , a measured spectrum of a sample, in the spectra space spanned by x_is each of which is the reference spectrum of the single element i (a continuum component due to the Bremsstrahung processes included), into another space of the elemental concentrations, y_i as $y=X^+x$. Here, X^+ is the Moore-Penrose pseudo-inverse matrix of X which is composed of x_is. X^+ is easily obtained by several mathematical algorithms such as the neural network[1], and the singular value decomposition[2], etc. Thus, our task in the analysis is only to multiply the known matrix X^+ by the measured spectrum x. In this approach, each component x_i is considered as a fundamental pattern of the measured spectra.

Fig.1 shows an example of the performance of PAM for various liver samples, which have beam measured under the same experimental condition. All measured spectra for a bovine liver, two mouse livers and two rat livers are successfully reproduced with a set of common reference spectra as the results of PAM,[2] though the simplest single Gaussian peak function model has been adopted. Prior the PAM analysis, the candidate content elements among the all possible elements of the atomic numbers from 11 to 82 in the samples and the shape of the continuum reference spectrum were determined from the preliminary analysis of the spectrum for the first bovine liver. The candidate content elements were chosen by the peak search with a priori knowledge on the liver samples. As the continuous background shape, we took a combination of a polynomial and two exponential functions (one is for the higher energy tail, and the other for gamma rays), as done in the SAPIX code.[3] The fitting was performed in an interactive way; we adjusted the model parameters of the background until an acceptable agreement was achieved by observing the fitted result.

Improvement of Reference Spectra Model

Peak Function Model

First, we should adopt more realistic peak function model, i.e., a single Gaussian accompanies by an exponential tail and a plateau with an escape peak as suggested in the literature.[4] A revised model calculation including above components was generally in good agreement with the experimental data, especially in the lower energy region. However, due to the ambiguity of the parameter values given in the literature for the above model, the fitting was not improved as expected for some cases. Therefore, further searches for some model parameters may be necessary using

CP392, *Application of Accelerators in Research and Industry*, edited by J. L. Duggan and I. L. Morgan
AIP Press, New York © 1997

the generalized least-squares method[5] for the multi-elemental standard samples.

Bremsstrahung Components Model

The above mentioned model for the continuum background also introduces uncertainty in the results because there are large ambiguities in the selection of fitting region

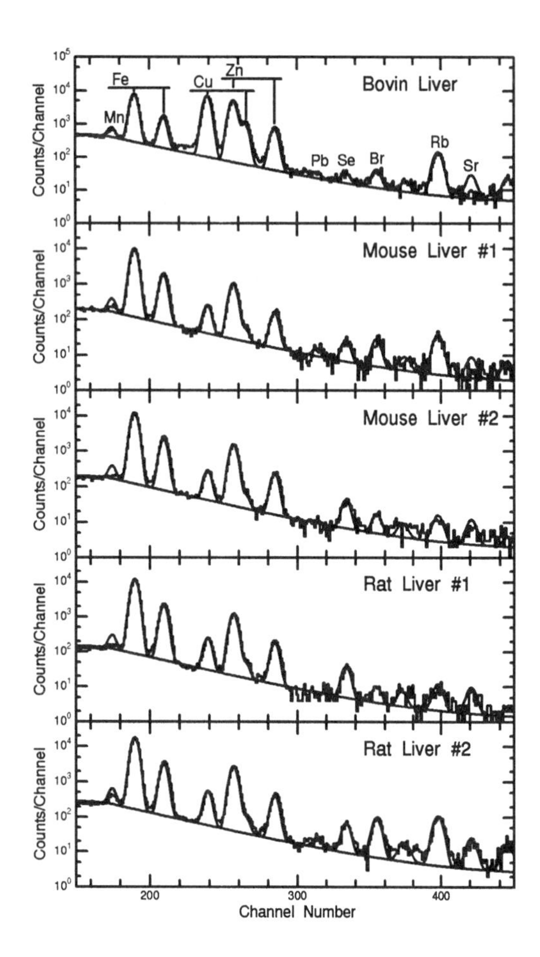

in the spectrum, and of the fitting points which are usually

FIGURE 1. PIXE spectra from samples of a bovine liver, two mouse livers and two rat livers, compared with the results of the pattern analysis method.

chosen at some extremal points between peaks. There is no physical base for the fitting functions and region, and the degree of fitting is easily changed by the level of the experiences of the analysts and status of the object spectra.

According to Ishii and Morita,[6] the continuous background is understood mainly as the sum of various Bremsstralung processes, such as atomic Bremsstrahung, secondary electron Bremsstralung, and others. They gave a comprehensive formulation for them. A theoretical model calculation based on this formulation reasonably reproduced the continuous spectrum from a thin Mylar film in shape and amplitude. This theoretical approach could reduce the ambiguity in the estimation of the continuous components.

Sum Coincidence Continuum

The possibility exists of two photons entering the detector within a sufficiently short time interval that they are recorded as a single event. The outcome is a continuum above each peak in the spectrum, culminating in a pile up peak at energy of the sum of the peak energies. In the framework of PAM, we should include the inevitable summing components of the prominent peaks into the reference spectra. The spectral shape of the summing continuum can be estimated using a statistical model.

Evaluation of Penetrability of X ray Absorber

In PAM, the correct evaluation of the penetration rate of the absorber used is indispensable, because the shape of the reference spectra varies due to the penetrability. This effect should be taken into account in the preparation phase of the reference spectra in the system.[2] Sera, et al.[7] pointed out that, e.g., in the case of Mylar absorber, the model calculation based on the data base values for the photoelectric cross sections gave fairly reliable estimation, but, the experimental penetrability data derived from their proposed procedure was more desirable because the procedure was easy, adaptable for any type of absorber, and free from the geometrical uncertainty in the calculation.

TOWARDS INTELLIGENT SYSTEM

Automatic Selection of Candidate Elements

The pre-selection of the candidate elements as mentioned above is necessary in PAM to avoid the ill-conditioned results: spurious elements or elements with negative concentration values.[2] Although the selection using the knowledge of the samples is not so large drawback of the PAM method, it is desirable to find an easier method without or with minimum process. This could be solved by a new approach to the element inference based on the Bayes' theorem. This approach was first developed for the neutron spectrum unfolding.[8]

In this approach, we repeatedly estimate the plausibility or appropriateness of the content elements (i.e., the relative concentration of the elements) from the outcome of the pulse height data. Actually, we accumulate the a posteriori probability for each possible element given by the Bayes formula every time of the event (one count) of the X-ray pulse height data taking. The accumulated posterior probabilities were renormalized (the sum of them equals to unity) and used as the new a priori probabilities for the estimation the next event. We start the estimation at the first event with the prior of the equal probability for every element in the sample, i.e., with the uniform prior distribution. The posterior probabilities gradually change as increase the number of events, and those of the probable elements in the sample approach to the approximate true relative concentration values, whilst those for the almost all improbable elements decrease to infinitely small values. Sometime, a small number of exceptional elements whose references spectra are quite similar to those of the true content elements remain. These elements can be removed from the candidate lists using the knowledge of the samples.

Thus, the list of candidate elements with the approximate concentration values would be given by this method. The intensities of the Bremsstrahlung's and others continua, relative to those of the reference ones are also inferred by the same way. This method was applied to a PIXE spectrum of a stainless steel type 316 (SUS316) sample, giving a consistent result with other analysis methods.[9]

As mentioned above, the estimation is made on the detection event basis during the measurement. In addition to the candidate element selection, such (quasi-) real-time mode analysis presenting approximate results would be quite efficient and useful for the monitoring the status of the data accumulation especially in the industrial application.

Optimization of X-Ray Absorber

Sera, et al[10] recently proposed the method where they use various types of absorbers aggressively in order to suppress a few prominent peaks. This method improved the signal to background (pile up events) ratio of the spectra, considerably, and enabled them to detect many heavier trace elements especially in the mineral samples. This idea can be applied to other types of samples. We have, actually, used a special absorber for the example case shown in Fig.1, which was a pierced Mylar absorber of 1000 μm in thickness called funny filter. The strong K-X ray peaks from a major element potassium in the liver samples were effectively suppressed, while the lower energy X-rays from the lighter elements survived through the small aperture of the filter. This type of absorber was first used to suppress an ill-condition appeared in the case of a serum sample with no absorber; the result of the indium contained as the internal standard was overestimated due to the strong interference of the potassium K-X ray peaks with the indium L-X ray peaks and partly due to the inadequacy of such simple peak shape model. The suppression of the indium L-X rays relative to its K-X ray peaks solved the ill-condition by increasing the orthogonality between the reference spectra of the potassium and indium.[2]

Future PIXE system should equip a mechanism to choose and set a specific absorber among various type of absorbers to optimize the data taking condition. It would not be difficult to build an intelligent system for the selection of the optimum absorber using the combination of the previously mentioned quasi-real-time analysis method and a small knowledge-base system.

PAM for Thick Sample Cases

If we aim to apply PIXE in the industrial field, we should consider the thick target cases where the various non-linear effects will occur, such as, the energy loss of the incident protons, absorption of the emitted X-rays in the sample, and secondary fluorescence effect in the sample. The matrix correction for the yield of the trace elements become significant in this case. In the case of the full stop thick sample, a thick sample of SUS316, we tested an iterative method to reach a set of correct reference spectra in thick sample, starting from a set of reference spectra in each pure elemental thick target. The latter spectra are sometime quite different with those of the thin target cases, and can be estimated from the model calculation considering above effects. The final obtained reference spectra could interpret the measured thick SUS316 spectrum well, and the result of the elemental concentrations was consistent with the case of the thin sample case.[11]

The above approach was in an ideal case (uniform and smooth surface) of the thick samples. Usually we have no information on the surface of the sample and inside of the matrix. It is known that the useful information can be obtained from the observation of proton spectrum back scattered from the sample surface like Rutherford back scattering (RBS). This kind of information is of particular importance if the sample surface contains lights elements whose X-rays could not observed by usual X-ray detector. Detailed analysis of the observed proton spectra we can deduce more useful information on the matrices.

CONCLUSIONS

In this paper, we discussed the status of the PAM system for the industrial application of PIXE, and prospects towards intelligent PAM system.

The present paper is summarized as follows:

- More detailed models for the reference spectrum of element, of the continuum from the Bremsstrahlung process and of the sum effect are necessary.
- Quasi-real time analysis based on the iterative Bayes' inference could be effective in the selection of candidate elements and monitoring the status of the spectrum accumulation. This method could also be a basic and indispensable technique for the intelligent system, meaning an adaptable system to the change of the status of the data

collection and to various types of samples.

- Use of suitable X-ray absorber is efficient to make the PAM analysis more reliable. The X-ray absorber selection guide and measurement method of the X-ray penetration rate in absorbers have almost been developed.

- Concerning the application of PAM to thick samples, the iterative method is promising although the status of the development is preliminary. Suitable combination of the proton spectrum measurement to PIXE could be useful.

REFERENCES

1. Iwasaki, S., Fukuda, H., Yoshizaki, K., Kitamura, M. and Ishii, K., "Application of Neural-Network Technique to Analysis of PIXE Spectra", *Internat. J. PIXE* **4**, 131-136 (1994).
2. Murozono, K., Iwasaki, S., Inoue, J., Ishii, K., Kitamura, M., Sera, K., and Futatsugawa, S., "System of Pattern Analysis in PIXE Spectra," presented at the 2nd International Symposium on Bio-PIXE, Beijing, Aug. 19-22, 1996.
3. Sera, K., Yanagisawa, T., Tsunoda, H., Futatsugawa, S., Hatakeyama, S., Saitoh, Y., Suzuki, S., and Orihara, H., "The Takizawa PIXE Facility Combined with a Baby Cyclotron for Positron Nuclear Medicine", *Internat. J. PIXE*, **2**, No. 1, 47-55 (1992).
4. Campbell, J.L., Millman, B.M.., Maxwell, J.A., Perujo, A., and Teesdale, W.J., "Analytic Fitting of Mono-energetic Peaks from Si(Li) X-ray Spectrometers," *Nucl. Instr. Meth.*, **B9**,71-79 (1985).
5. Smith, D.L., "*Probability, Statistics, and Data Uncertainties in Nuclear Science and Technology,*" OECD/NEA, Nuclear Data Committee Series, Vol.4, American Nuclear Society, Inc. (LaGrange Park, Illinois (1991) p.219.
6. Ishii, K. and Morita, S., "Continuous Backgrounds in PIXE", *Internat. J. PIXE* **1**, 1- 30 (1990).
7. Sera, K., Futatsugawa, S., Hatakeyama, S., and Saitou, S., "Determination of Physical Quantities for PIXE by means of PIXE 1. Absorption curve," *Internat. J. PIXE* **4**, 165-179 (1994).
8. Iwasaki, S., "A New Approach for Unfolding Problems Based only on the Bayes' Theorem," presented at the ninth International Symposium on Reactor Dosimetry, Prague, Czech, Sep. 2-6, 1996.
9. Iwasaki, S., Murozono, K., Suzaki, K., and Kitamura, M., "A New Method for the Spectrum Analysis Based on the Bayes' Theorem -Application to X-ray Spectrum Analysis-," The 11th Workshop on the Radiation Detectors and Their Uses, Feb. 5-7, 1997, KEK, Tsukuba, Japan.
10. Sera, K., and Futatsugawa, S., "Effects of X-ray Absorbers Designed for Some Samples in PIXE Analysis," *Internat. J. PIXE* , **5**, 2&3, 181-193 (1995).
11. Iwasaki, S., Ishii, K. Yoshizaki, K., Fukuda, H., Murozono, K., Inoue, J., Kitamura, M., Yokota, H., Iwata, Y., Orihara, H., "Vertical Beam in-air PIXE System at CYRIC"', *ibid.*, 163-173.

MICRO-PIXE STUDIES OF *Lupinus angustifolius* L. AFTER TREATMENT OF SEEDS WITH MOLYBDENUM

W.J. Przybylowicz[*], **J. Mesjasz-Przybylowicz**

Van de Graaff Group, National Accelerator Centre, Faure 7131, South Africa

K. Wouters, K. Vlassak

Laboratory for Soil Fertility and Soil Biology, Katholieke Universiteit Leuven, B-3001 Heverlee, Belgium

N.J.J. Combrink

Department of Agronomy and Pastures, University of Stellenbosch, Matieland 7602, South Africa

An example of nuclear microprobe application in agriculture is presented. The NAC nuclear microprobe was used to determine quantitative elemental distribution of major, minor and trace elements in *Lupinus angustifolius* L. (Leguminosae) after treatment of seeds with molybdenum. Experiments were performed in order to establish safe concentration levels and sources of Mo in seed treatments. Elemental distributions in Mo-treated plants and in the non-treated control plants were studied in order to explain how Mo causes toxicity. Some specific regions of Mo and other main and trace elements enrichment were identified.

INTRODUCTION

Molybdenum (Mo) is a co-factor for a few enzymes, amongst which nitrogenase and nitrate reductase are the most important. *Nitrate reductase* occurs in all plants. It converts nitrate (NO_3^-) into nitrite (NO_2^-), which then can be further reduced to ammonium-nitrogen for proteins. *Nitrogenase* is the enzyme required for *nitrogen fixation* and effective nodulation in Legumes (1). In general soils contain sufficient Mo to sustain normal plant growth, but it is less readily available at low pH (2, 3). There are a host of means to ensure its availability in such circumstances, the most popular of which is lime application. However, it is costly and a much more economical method is to directly apply Mo onto the seeds before sowing. In high concentrations toxic symptoms may occur. At present, the recommendation for Mo application on all legumes in the Western Cape region (South Africa) is 100 g/ha MoO_3. The sowing rate for *Lupinus angustifolius* L. (Leguminosae) is about 60 kg seed per ha. This implies that 1.7 g MoO_3 per kg of seed should be applied in order to reach the recommended amount of Mo per hectare.

To better understand how Mo causes toxicity, elemental distributions were studied in the Mo-treated plants and compared with those found in the non-treated (control) plants.

* on leave from the Faculty of Physics & Nuclear Techniques, Academy of Mining & Metallurgy, Cracow, Poland

MATERIAL AND METHODS

Seeds of *Lupinus angustifolius* L. were subjected to Mo at four different rates (0.59, 2.95, 5.91 and 11.82 g of MoO_3 per kg of seed) and these and control seeds were allowed to germinate. Three weeks after sowing, cross sections of leaf tips and stems for selected plants were prepared. The sections were cut with a razor blade and transferred to liquid isopentane, which was cooled by liquid nitrogen, and immediately freeze-dried. They were mounted on a separate target frames, using thin 1.5% Formvar layers as support, and carbon coated.

The analyses were performed with 3.0 MeV protons, using the nuclear microprobe of the National Accelerator Centre (NAC) at Faure, South Africa (4). A beam spot size of 4 μm was used and the proton current was 200 - 400 pA. Elemental maps were generated using Proton Induced X-ray Emission (PIXE). A Si(Li) X-ray detector (manufactured by PGT) with a 30 mm^2 active area and 8 μm Be window was positioned at 135°, about 23 mm from the target. An additional 100 μm Kapton™ filter was used to shield the detector from backscattered protons. Point analyses in a few selected regions were obtained using simultaneously PIXE and proton BackScattering (BS). Backscattered protons were detected with an annular Si surface barrier detector 100 μm thick, positioned at 176° . The Dynamic Analysis (a rapid matrix transform method) (5) was used for the production of *true elemental images* of Mo as well as other main and trace elements. The composition of cellulose ($C_6H_{10}O_5$) was used to

CP392, *Application of Accelerators in Research and Industry*, edited by J. L. Duggan and I. L. Morgan
AIP Press, New York © 1997

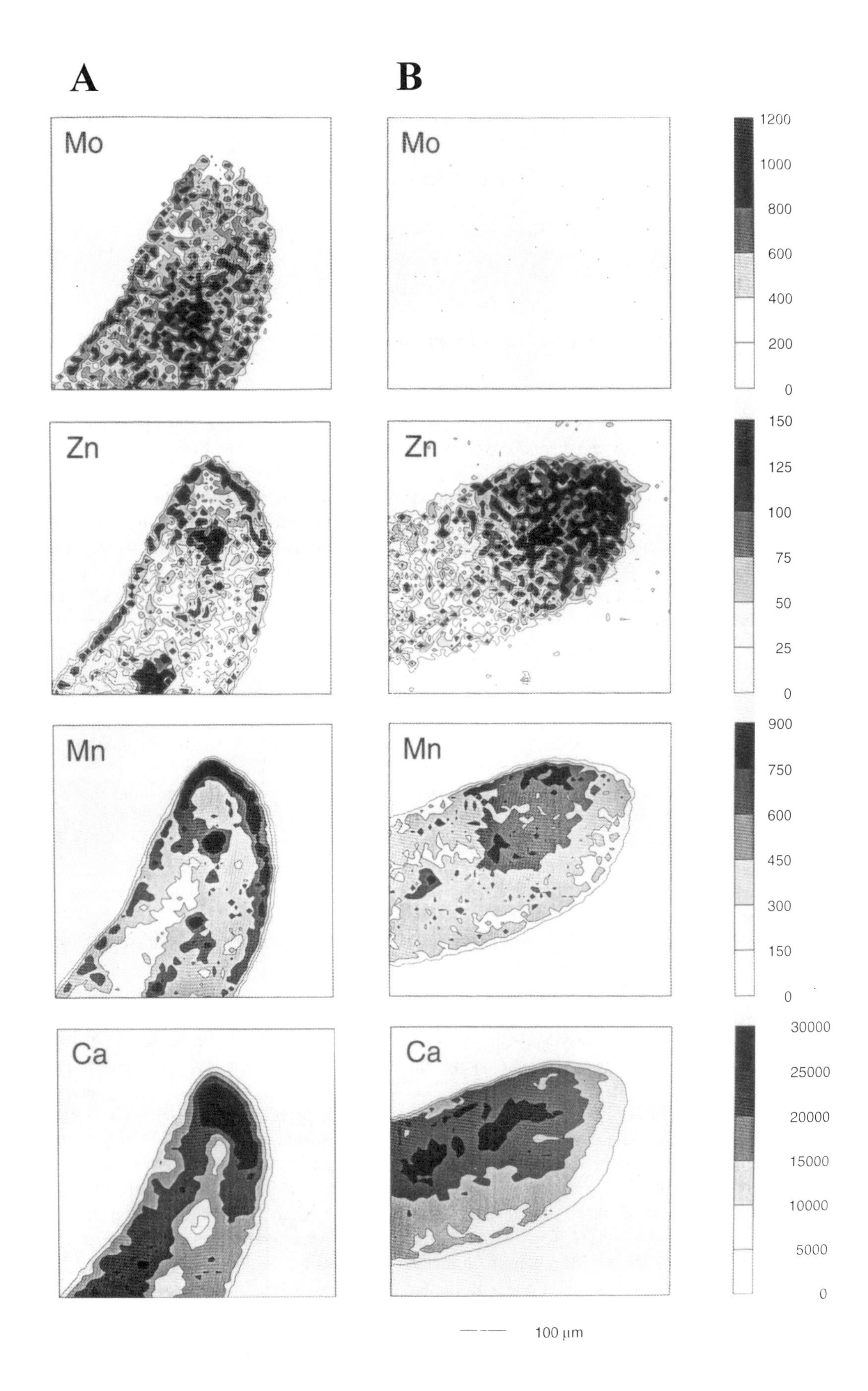

FIGURE 1. Quantitative maps of Mo, Zn, Mn and Ca from the leaf tip cross-sections of: (A) Mo-treated plant; (B) control plant. Scale of intensity in μg/g dry mass. Scan size 600 μm x 600 μm, total accumulated charge 7.51 μC for (A) and 4.03 μC for (B).

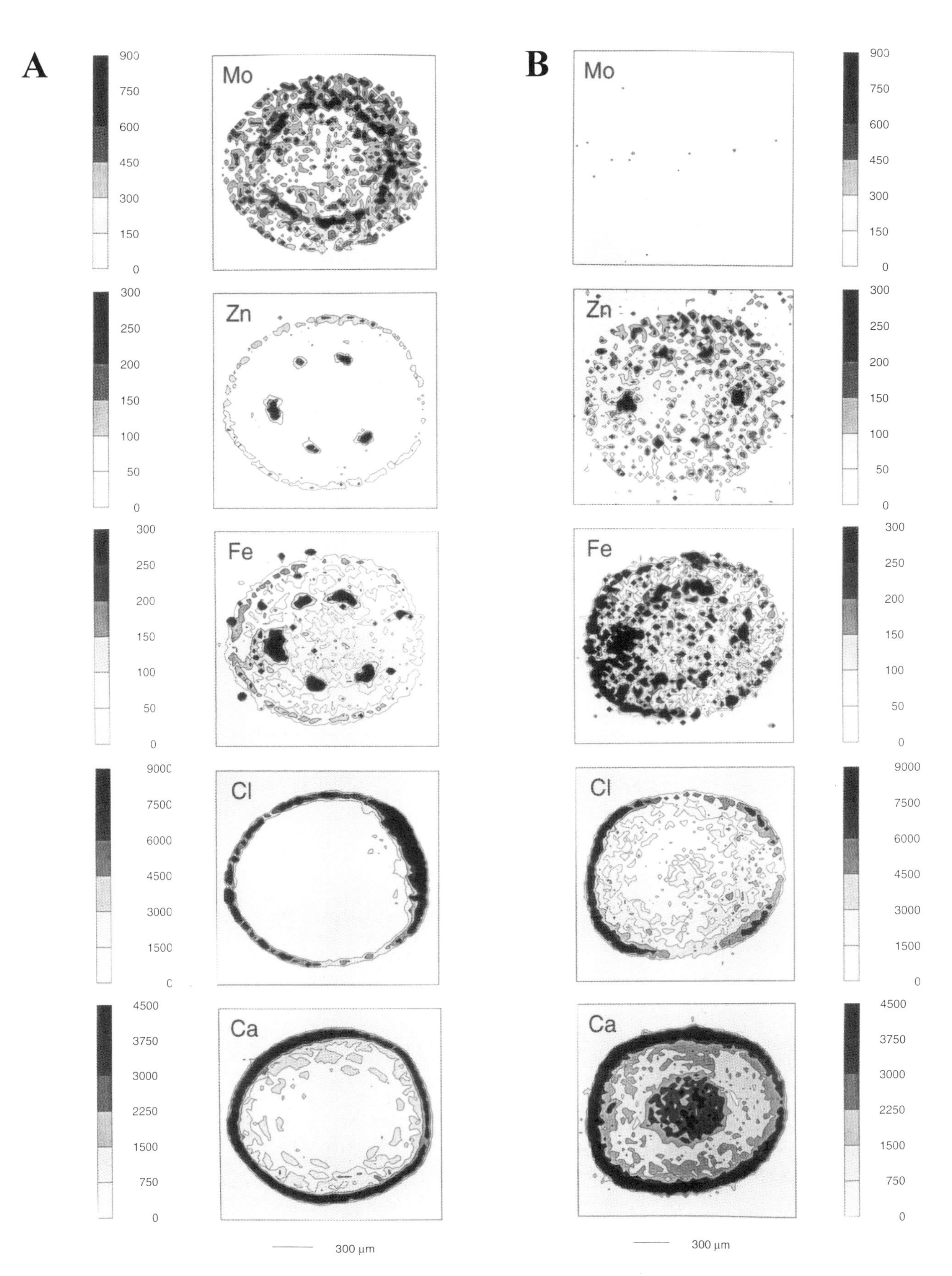

FIGURE 2. Quantitative maps of Mo, Zn, Fe, Cl and Ca from the stem cross-sections of Mo-treated and control plant. (A): Mo-treated plant, scan size 1.6 mm x 1.6 mm, total accumulated charge 8.16 µC; (B): control plant, scan size 1.8 mm x 1.8 mm, total accumulated charge 0.98 µC. Scale of intensity in µg/g dry mass.

approximate the matrix. It was earlier proved that the error due to this simplification does not exceed 10 % for elements with $22 < Z < 56$ (6), which is acceptable in this application. All measured samples could be treated as "infinitely thick" (the proton beam was completely stopped by the sample). Scanned regions were divided into 64 x 64 pixels. Elemental maps are presented using the Interactive Data Language (IDL) package (7). The maps are contours linking pixels with similar values.

RESULTS

Molybdenum distribution

Only plants that received 11.82 g of MoO_3 per kg of seed (the highest dose administered in this experiment) are discussed here. Some differences between them and the control plants were clearly observed. In the treated plants, Mo accumulates in certain regions. In the leaf Mo was mostly found in veins, and some veins are preferred to the others. This is visible in Fig. 1A which shows Mo enrichment mainly in the central vein, whereas in the remaining two veins and in the other parts of the leaf the concentration is similar. In the stem the phloem is favoured, while the central part of the stem shows depletion (Fig. 2A). In the control plants Mo was below the limit of detection (50 ppm for the leaf tip and 120 ppm for the stem). The few recognisable spots on Mo maps of the control plants (Figs. 1B and 2B) are most probably minimal misleading contributions from background.

Impact on other elements

Elemental maps of other elements show that the Mo treatment clearly influenced their distribution. The most interesting results were found for Zn, Fe, Mn, Ca and Cl. In the control plants Zn, Mn and Fe are localised in the leaf tips (Fig. 1B). In the Mo-treated plants the distribution of these elements is altered. Zn and Mn are enriched in veins, but only in those which do not show Mo enrichment. Ca is depleted in all three veins visible in Fig. 1A. In the stem, Ca and Cl have the highest concentrations in the outer cell layers (epidermis). In contrast, they show significant depletion in the central cylinder of the Mo-treated plants in comparison with the non-treated, control plants. Fe, Zn and also Mn are concentrated in regions depleted in Mo, mainly in the xylem and also in the epidermis (Fig. 2A). The reason for the inverse distribution of Zn, Fe, Mn and Mo is not known yet, but earlier studies performed on other plants (8, 9) showed similar elemental relations.

CONCLUSIONS

The sensitivity of the nuclear microprobe was sufficient to allow studies of Mo distribution at typical rates of MoO_3 administration in the seeds of *Lupinus angustifolius* L. In addition, the impact of Mo on the distribution of other elements was clearly shown. The behaviour of some of the essential elements (Ca and Cl) as well as of important co-enzymes (Zn, Mn and Fe) is visibly affected.

From studies using ICP-AES (3) it was found that the highest administered dose (i.e. 11.82 g of MoO_3 per kg of seed) was above the toxicity level for Mo fed with this fodder and eventually also for the plant itself. Present studies of elemental distributions suggest that the toxicity might not be directly caused by the Mo excess, but rather by its negative interaction with other elements. Elemental interactions are well known in plant nutrition research, but the conducted studies of elemental distribution gave direct and precise information on these interactions, helping in identification of elements on which Mo administration had the highest impact, and in creating a two-dimensional picture of such an impact.

Further studies will concentrate on setting up experiments with selection of nutrient levels of other elements (Zn, Mn, Fe) to find how they can influence the Mo distribution.

ACKNOWLEDGEMENTS

The authors wish to thank M.J. Renan for language improvements; S. Hendricks for the preparation of the photographs and K.A. Springhorn for technical assistance.

REFERENCES

1. Marschner, H., *Mineral Nutrition of Higher Plants*. London: Academic Press, 1995, pp. 231-255 and pp. 369-379.
2. Cheng, B.T., and Ouellette, G.J., 1973. Molybdenum as a plant nutrient. *Commonwealth Bureau of Soils: Soils & Fertilizers* **36**, 207-214 (1973).
3. Wouters, K., Molybdeen-zaadbehandelingen bij *Lupinus angustifolius*. M.Sc. Thesis, Faculteit Landbouwkundige en Toegepaste Biologische Wetenschappen, Katholieke Universiteit Leuven, Belgium, 1996.
4. Prozesky, V.M., Przybylowicz, W.J., van Achterbergh, E., Churms, C.L., Pineda, C.A., Springhorn, K.A., Pilcher, J.V., Ryan, C.G., Kritzinger, J., Schmitt, H., and Swart, T. *Nucl. Instr. and Methods* **B 104**, 36-42 (1995).
5. Ryan, C.G., Jamieson, D.N., Churms, C.L., and Pilcher, J.V. *Nucl. Instr. and Meth.* **B 104**, 157-165 (1995).
6. Pineda, C.A., and Peisach, M., *Nucl. Instr. and Meth.* **B 35**, 344-348 (1988).
7. Interactive Data Language, User's Manual. Research Systems Inc., Boulder, USA (1993).
8. Jung, Y.-K., 1979. Der Einfluss variierter Agebote von Eisen, Mangan, Kupfer, Zink, Molybdän und Bor auf Wachstum, Ertrag und Nährstoffgehalt bei Knaulgras (*Dactylis glomerata* L.) und Weissklee (*Trifolium repens* L.) in Rein- und Mischkultur. Script for the degree of "Doktor der Agrarwissenschaften". Technischen Universität Berlin, Germany, 1979.
9. Kabata-Pendias A., and Pendias H., *Trace elements in soils and plants*, Florida: CRC Press, 1984, p. 66.

Pixfit - A Spectral Analysis Program for PIXE

R. L. Coldwell[1] and H. A. Van Rinsvelt[2]

[1]*Constellation Technology, 9887 4th St. N., St. Petersburg FL, 33702*
[2]*Physics Department, University of Florida, Gainesville, FL 32611*

Spectra analysis software under development at Constellation Technology is tested for its ability to fit PIXE spectra. A modified chi-square is first minimized with respect to parameters in a "standard background". Then a less modified chi-square is minimized with respect to constants and knots in a spline background term. A peak shape, an efficiency curve, and a table of relative x-ray decay rates are used to construct a response function linear in each element, but also containing non-linear parameters giving the energy calibration, the peak widths, and corrections to the efficiency in the form of "absorption" coefficients. The resulting counts for each element divided by its standard deviation gives the statistical certainty of detection of the element.

1. Introduction

Several assumptions are made by the fitting routine. The first is the notion of robust fitting which is a systematic way of changing the error estimates on each data point to make weighted least squares fits "behave" [1,2]. The second assumption is that with the robustness decreasing the weights of the peaks, the only assumption needed to fit the background is the continuity of the function and its first and second derivatives throughout the region being fitted. The third assumption is that a few peak shape (frequently a single Gaussian) templates with variable locations, widths and heights can represent the response of the detector to the X-rays coming from the sample. The fourth assumption is that the relative strengths of the X-rays from each element are those due to electrons missing in the K shells of these elements which are given in the **Table of Radioactive Isotopes** [3].

The spectra tested were provided by W. Maenhaut. They are the same as those used in an earlier intercomparison of spectral data processing techniques [4]. The high statistics Bovine Liver spectrum (figure 1, BOLB) in their paper will be used to generate a standard background and an efficiency calibration which can be used to fit other spectra. This was tested here by making a quick fit to their Orchard Leaves ORLA spectrum. Owing to the fact that the high statistics and hence low errors associated with the very large peaks in these spectra can make the high count region dominate the chi-square, the code option was used in which the errors in each channel were taken to be the larger of 0.01 times the data or the square root of the data. This has no effect on the fit to the second spectrum.

2. Fitting Method

2.1 Background

The background of spectrum BOLB is fitted to a cubic spline

$$b(x) = \sum_{i=0}^{3} c_i x^i + \sum_{j=1}^{M} d_j \left(k_j - x\right)_+^3 \qquad (1)$$

where $(x)_+^3 = x^3 \quad x > 0 \qquad = 0 \quad x \leq 0$

although the code also has the capability of fitting to the exponential of a cubic spline.

It is important to notice that the k values, in addition to the c's and d's in equation 1 are varied in the process of minimizing the chi-squares. It has been long known that this takes much more computer time than simply using a large number of knots and allowing only the coefficients to vary [5]. Minimizing with respect to the knots is not a fast procedure, but it has the potential for giving a much better curve fit owing to the fact that it optimizes the locations of the third derivative discontinuities in the background representation. In general a fit to BOLB would begin with 10-20 background constants, the response function would be added and then in a series of iterations the code would add 10 to 20 knots and coefficients to achieve the best fit. It is not recommended to attempt a complete fit of the background before any peaks are introduced, but it is illustrative to do so. The code was set up to fit the data as background between channels 20 and 1024 with 60 constants, consisting of 4 polynomial constant c's, 28 spline coefficients d and 28 knots k. The resulting fit is shown in figure 1. The background fit does include the small peak at about channel 50, but does not excessively enter into any of the others.

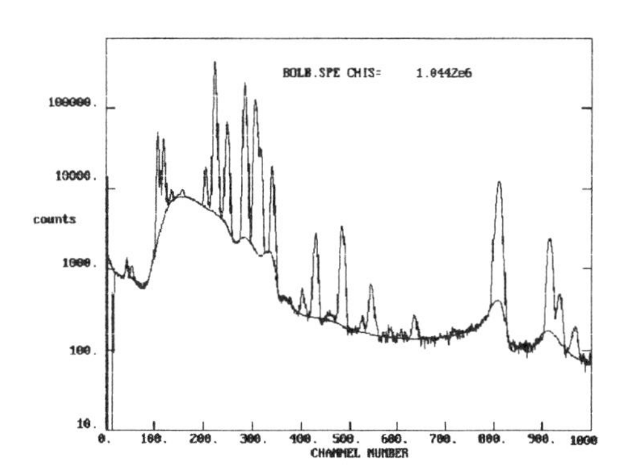

Figure 1. Fit to the background of spectrum BOLB with no peaks.

CP392, *Application of Accelerators in Research and Industry*, edited by J. L. Duggan and I. L. Morgan
AIP Press, New York © 1997

2.2 The peak shape

The code is designed to allow the user to select a large multicomponent peak that has been fitted with one shape, to remove a single component, and to fit the residual as a new peak shape. In practice this can require a few iterations to converge on the optimal peak shape. In this case, however, the bulk of the variation in observed peak shapes came simply from the separations and intensities of the $K_{\alpha 1}$, $K_{\alpha 2}$, etc. lines making up the response function for each element. All of the fits shown here were made using a Gaussian shape for the individual peaks. The non-Gaussian parts to the peaks are thus fitted as part of the background.

2.3 The response function

The elements of interest are put in a file consisting of their names. Each name refers in turn to a file consisting of the K line energies and their intensities ordered by energy. These files, along with a template for the detector's response to an individual line enable the code to form a complete response function. This response function contains constants representing the energy calibration, the full width at half maximum, and the efficiency/absorption calibration. This last can be turned on and off for individual elements. An initial crude calibration based on the silver and potassium peaks is needed to start the fit. An estimate of the full width at half maximum accurate to within an order of magnitude is also necessary to start the fit. The fit also needs three energies associated with large peaks so that the minimization can improve the initial calibration sufficiently to start the fit. With the efficiency assumed constant, the fit is most noticeably off in the potassium/calcium region owing to the rather large absorption for this energy range as shown in figure 2, and also in the silver region as shown in figure 3 due to the fact that the fit is required to be accurate in the high counting rate range (5-10 keV) and there is no allowance for the decrease in efficiency at high energies. The total number of counts in the silver peaks is incorrectly given as 47216 ± 1865. At this point the standard deviation in this number simply means that if an equivalent but random data set were fitted with a constant efficiency, the fit would yield this same obviously too low value.

It should be noted, however, that the full width at half maximum and the energy calibration have been accurately determined at this point.

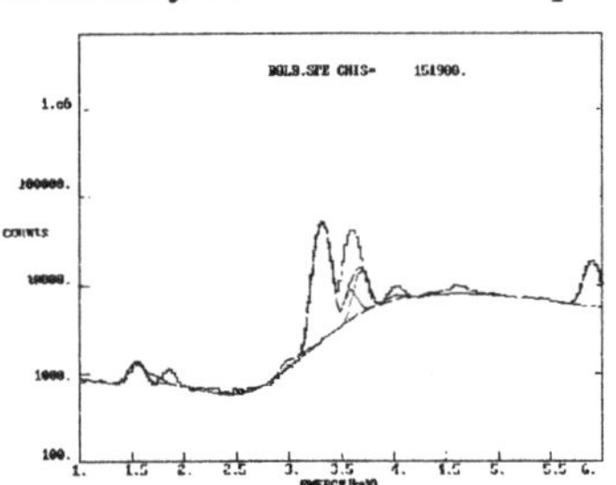

Figure 2. Fit with constant efficiency in the potassium/calcium region.

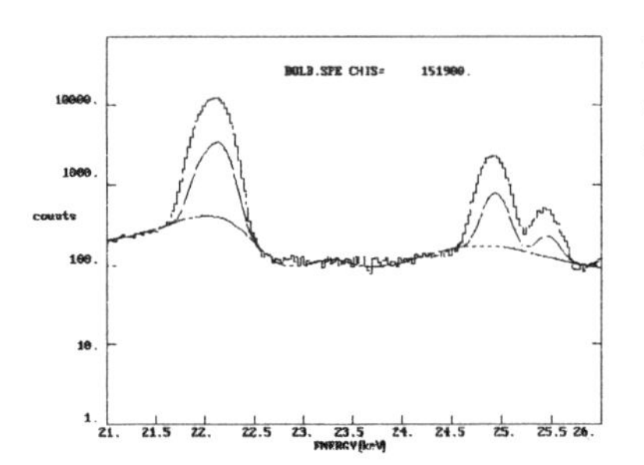

Figure 3. Fit with a constant efficiency in the silver region.

2.4 Absorption/efficiency

The absorption is fitted as the exponential of a linear term plus a two knot spline with variable knots. The efficiency has the same form but with the opposite sign. The difference is that the efficiency is not varied in the process of fitting. The first fit to the efficiency corrects the silver region, but not the potassium region. The constants as determined by this fit were then put into the efficiency file and the fit was restarted thereby finding a second linear term plus two knot spline to add to the first. Figures 4 and 5 show the fits obtained with a 9 constant efficiency in the potassium/calcium region and in the silver region respectively.

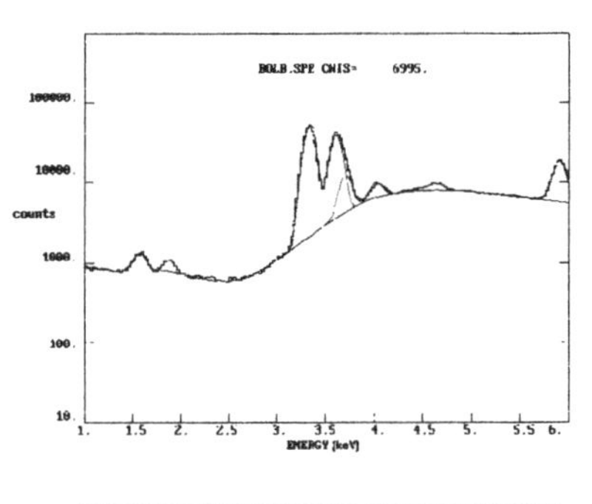

Figure 4. Fit with a 9 constant efficiency in the potassium/calcium region.

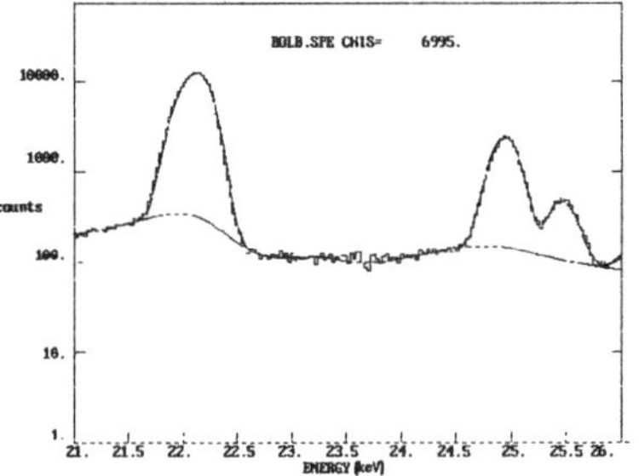

Figure 5. Fit with a 9 constant efficiency in the silver region.

The log of the efficiency is now given by a 9 constant expression:

$$\ln eff(E) = cE + \sum_{j-1}^{4} d_j \left(k_j - E\right)_+^3 \qquad (2)$$

where the limitation to a linear form at large energy values incorporates the theoretical form and gives the most accurate extrapolation. Figure 6 shows the efficiency as a function of the energy as determined by this fitting procedure. The mylar absorption edge at the low energy end and the slow fall off at high energy values are clearly visible in this plot.

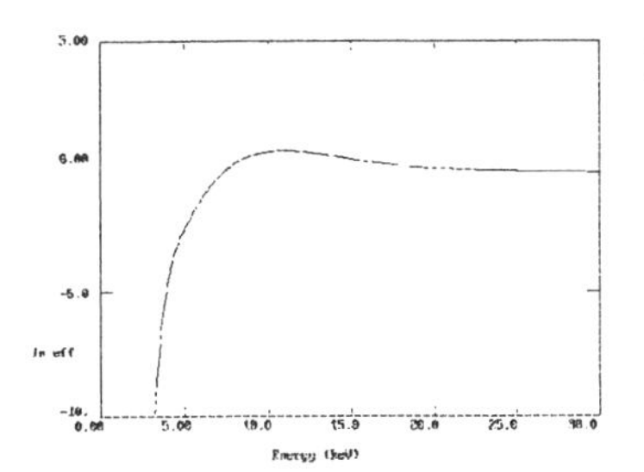

Figure 6. Efficiency as a function of energy

2.5 Determination of the standard background.

The next fit to BOLB was then made by making the above efficiency a fixed function. The rest of the fit was started from scratch in a mode where the background was fitted to 10 constants, the response function fitted, then the background with 12 constants, then the response function is refitted, then the background again until 56 background constants have been fitted. Finally all potential peaks with strengths greater than 5 times the standard deviation in the underlying background were added. The resulting background along with the energy coefficients was then placed in a standard background file. The constants in this file become a function which is fitted to the data in terms of an overall constant, energy parameters and absorption coefficients.

3. Results

3.1 The final fit to BOLB

The final fit was made by varying the parameters in the background function and adding a single overall background constant. Since the fit is "essentially" of the background to itself, the fact that a good fit is achieved is necessary but not sufficient to demonstrate that the method works. The results are the "unknown" peak list [table 1] of those peaks not contained in the response function such as other elements, escape peaks, sum peaks, etc. and the final response function values in table 2. Table 2 contains the information for each observed element, including the total counts in each element. Also shown, on the line below the information for each element, are the number of counts in the K_α line, and the ratio of the counts in the K_α line to its average value as determined by the various codes used in reference 4. The K_α areas in table 2 are all within the range of the average of the various $P/\overline{P}$ values listed in figure 9 of reference 4. The largest difference occurs on the rather small Sb peak. This is approximately a 3 sigma effect occurring at the end of the spectrum where the background is least well determined. The most significant difference is on the calcium K_α line where the value found here is 0.89 of the average and 4.5 standard deviations off. This line completely overlaps the potassium line and is in the region where the efficiency changes rapidly. This is probably the only code to fit the efficiency with the peaks, and figure 4 is such strong evidence of a good fit that it seems reasonable to conclude that the other codes may have slightly overestimated the amount of calcium in this peak.

Table 1. The "unknown" peaks in BOLB. The chi-square is that of the entire fit.

```
FIT TO BOLB.SPE
IP=     2
     8 PEAKS 1005 CHAN, CHIS=      5118.        BCHIS=
2178.      NNB,NITB=  512     1 1
 CUTOFF=    5.00
 CIS(1,2,3)     .391553593891       .268995627007E-01
-.859565413887E-07
 channel energy error fwhm err  strength error
  43.452  1.560 .003  .141 .006  3067.32  123.3
  54.636  1.861 .004  .126 .009  1765.14  108.0
 157.447  4.625 .005  .120 .013  4402.39  401.8
 524.136 14.467 .016  .196 .037  509.069  83.88
 584.622 16.088 .037  .503 .083  811.671  120.9
 859.128 23.438 .041  .547 .075  686.885  103.2
 896.209 24.130 .042  .829 .099  1877.75  206.0
1003.002 27.285 .029  .742 .065  1020.21  80.70
```

Table 2. The file BOLB.OUT with the line for each element followed by the graphical determination of the K_α line and its ratio to the average value given in reference 4. The chi-square is only over the 630 points for which the response function is non-zero.

```
 UNKFI CHI-       4857.48       NPMIN-      630
  I,          CIS(I), STATISTICAL ERROR
 1   .391553593891       .8040E-03     ENERGY
 2   .268995627007E-01 .4332E-05     ENERGY
 3 -.859565413887E-07 .4664E-08     ENERGY
 9   17.4143819820       .5255         FWHM**2
10   .828107539801E-01 .3085E-02     FWHM**2
11   .158824470676E-04 .4403E-05     FWHM**2
   ACTIVITY     ERR     COUNTS  ERR   NAME
COUNTS/ERR
12 .13217E+09 .111E+07 461783. 3880.    K  119.
                       260000        1.01
13 .12455E+07 .301E+05 66898.4 1617.    CA 41.4
                       44040         0.89
14    8579.59     139.3  95149.7 1544.    MN  61.6
                       79460         0.97
15    147187.     900.7 0.2786E07 .17E05  FE  163.
                      0.23451 E+07   1.00
16    27818.3     177.4 0.1653E07 .11E05  CU  157.
                      0.14339 E+07   1.01
17    13632.6     85.89 0.1039E07 6547.   ZN  159.
                      0.09066 E+07   1.03
18    21.6617     1.468   2681.90 181.7   SE  14.8
                        2368          0.95
19    189.835     3.930   24136.0 499.7   BR  48.3
                        21340         1.00
20    244.512     4.656   29883.4 569.1   RB  52.5
                        26380         1.06
21    20.3671     1.652   2017.26 163.7   MO  12.3
                        1740          1.16
22    2048.76     14.00   190442. 1302.   AG  146.
                        160400        1.01
23    28.0778     2.197   2195.78 171.8   SB  12.8
                        2192          1.24
24    54.2045     2.580   5001.41 238.0   GA  21.0
                        4375          0.83
25    10.5478     1.403   1227.64 163.3   AS  7.52
26    10.5690     1.221   1175.71 135.8   Y   8.66
```

3.2 The first fit to ORLA

The low statistics Orchard Leaves A spectrum of reference 4, also provided by W. Maenhaut, was used to test the code's ability to use the efficiency and background information to quickly fit a different spectrum taken under the same conditions.

The calibration file is the same as that for BOLB except that the initial calibration energies were those of Ca, Fe, and Ag. There are no background constants except for the BOLB background which is manipulated by means of absorption constants. The elements in the response function are those from BOLB which do not include all of the elements in ORLA. The elements with strength larger than 25 times the error in the underlying background were added to the fit. The resulting fit is shown in figure 7. Four large peaks (more than 25 times the standard deviations of the underlying background) were added to the BOLB response function as shown in table 3. The first two are escape peaks while the next two represent elements reliably present in ORLA but not in BOLB.

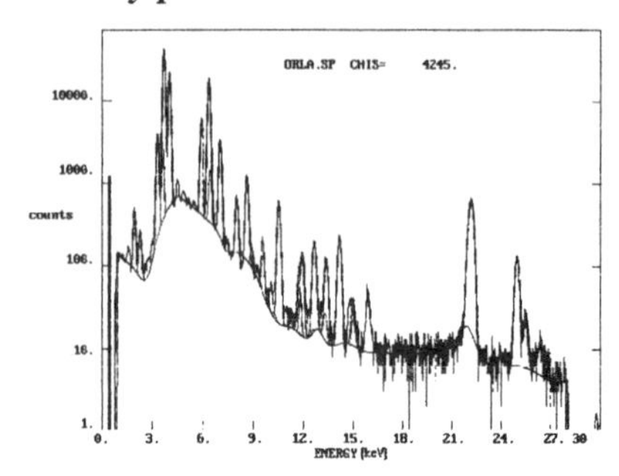

Figure 7. First fit to ORLA

Table 3. The four large additional peaks needed in addition to those in BOLB in order to fit ORLA.

FIT TO ORLA.SP
IP= 2
4 PEAKS 1004 CHAN, CHIS= 4245. BCHIS= .0000 NNB,NITB= 512 0 0
CUTOFF= 25.00
CIS(1,2,3) .380136195521 .269940585580E-01 -.108665891755E-06

channel	energy	error	fwhm	err	strength	error
57.828	1.941	.004	.143	.008	2293.83	132.2
69.928	2.267	.007	.162	.015	1259.46	108.5
454.506	12.627	.005	.222	.011	1577.99	83.37
511.177	14.150	.004	.228	.007	1911.09	66.09

4. Conclusions.

The results show that the software Pixfit is definitively capable of fitting the data. The detailed results of the fit to BOLB show that they are "essentially" in agreement with the results of the codes tested in reference 4. The known ratios of the peaks making up the K and L X-ray lines have been used to determine an efficiency sufficiently accurate to fit the K_α and K_β lines of each element without introducing a separate parameter for each. This allows both lines to participate in determining the presence of an element, increasing the statistics by about 20% or so. The fact that the response functions automatically use all of the components of the lines made it possible to fit the above data with a single Gaussian shape (with a full width at half maximum a quadratic function of the energy) eliminating the need to construct a separate template for each element.

The background fits are the most time consuming. The quadratic radius of convergence of the chi-square becomes very small when the knots are included, and this means a lot of steps. The time required goes approximately as the 4th power of the total number of knots. Typically 4 hours or more on a IBM 486 DX2 are required for the 56 and 60 background fits to BOLB. The absorption/efficiency fits also involve moving knots, and these typically require 1/2 to 1 hour. These two separate fits were indeed required in the work described here. The fit to the response function, the addition of well separated peaks, and the modifying of the background is relatively fast. The final fit to BOLB and the initial fit to ORLA required 5 to 10 minutes.

The notion is that once a method of taking data has been determined, a data set with a large number of counts can be taken as in BOLB above. Extensive work can then be done to determine a standard background and an efficiency curve. These results combined with the files containing the K_α and K_β lines and intensities can then be used to quickly fit any spectra that are taken. Since the information used is maximized and the number of parameters is minimized with this approach, the ability to pull lines from noise is optimized.

The accuracy of PIXFIT was investigated here. Owing to its extensive use of non-linear minimization it uses much more CPU time than other standard codes. The user interface is still under development. The inputs are the spectra. a very approximate calibration, and a list of expected elements. The output is a list of total counts with error estimates for these same elements plus a list of other energies with significant counts. This is a little less than is required of codes such as AXIL and thus ultimately the code should be slower, but more accurate and easier to use.

References

1. R.L. Coldwell, Nucl. Instr. and Meth. **A242** (1986) 455
2. R.L. Coldwell and G.J. Bamford, **The Theory and Operation of Spectral Analysis Using Robfit,** AIP, New York (1991) pp. 50-58
3. E. Browne and Richard B. Firestone; V.S. Shirley, Editor **Table of Radioactive Isotopes,** John Wiley, New York (1986) pp. C-19, C-24.
4. J.L. Campbell, W. Maenhaut, E. Bombelka, E. Clayton, K. Malmqvist, J.A. Maxwell, J. Pallon, and J. Vandenhaute, Nucl. Instr. and Meth. **B14** (1986) 204
5. C. de Boor, **A Practical Guide to Splines,** Springer-Verlag, New York (1978) p.180

ANALYSIS OF ROCKY MOUNTAIN LICHENS USING PIXE: CHARACTERISTICS OF IRON AND TITANIUM

B. M. CLARK, N. F. MANGELSON

Department of Chemistry and Biochemistry, Brigham Young Univeristy, Provo, Utah, 84602

L. L. ST. CLAIR, K. T. ANDERSON

Department of Botany and Range Science, Brigham Young University, Provo, Utah, 84602

L. B. REES

Department of Physics and Astronomy, Brigham Young University, Provo Utah, 84602

Lichens have been shown to be effective biomonitors of air quality. They are currently being used to characterize background element levels and to identify air pollution effects on federally administered lands in the Rocky Mountain region of the western United States. PIXE analysis for twenty elements has been performed on over two hundred lichen specimens collected from various national forests, national monuments, and national parks in the region. This paper reports on patterns of iron and titanium accumulation in lichen tissues. Data show a strong relationship between concentrations of iron and titanium. The Fe/Ti ratios agree well with values reported in similar lichen studies; however, our values for both iron and titanium concentrations are ten times greater than other reports. A distribution function for the log of iron concentrations is distinctly bimodal. The lower concentration mode contains fruticose lichens from bark substrates and the higher concentration mode contains foliose lichens from rock substrates. High iron concentrations in fruticose lichens along the Wasatch Front suggest air pollution impact from a local steel plant.

INTRODUCTION

Over the last several decades lichens have been used extensively as biomonitors of air quality (1). Lichens are non-vascular, composite organisms consisting of a fungus and an alga or cyanobacterium which live together in a complex symbiotic relationship. Lichens, which lack a true root system, obtain most of their nutrients through wet and dry atmospheric deposition. Lichens demonstrate several growth form patterns. In this study species of the foliose and fruticose growth forms were used. Foliose lichens are dorsoventrally flattened and generally somewhat adherent to the substrate. Foliose species commonly occur on either rock or bark. Fruticose species are densely branched with a single point of attachment and generally occur on bark. Due to differences in attachment patterns, foliose species have a greater degree of contact with the substrate.

In the last fifteen years we have established lichen air quality biomonitoring baselines throughout the Intermountain Area. Most of our work has been in federally administered wilderness areas and national parks and monuments. One interesting observation from our work has been the relationship between iron and titanium concentrations in lichen thalli. Other researchers, particularly Takala, *et al.* (2) and Nieboer, *et al.*(3), have also reported on iron and titanium concentrations in lichens. While these other studies used different analytical methods, their data is comparable to ours. Nieboer, *et al.*(4) have suggested that due to the relative consistency of iron/titanium ratios it should be possible to accurately predict accumulation of iron or titanium by lichens from air pollution sources.

EXPERIMENTAL

A total of 204 lichens from fourteen sites in Idaho, Utah, Montana, Wyoming, Colorado, and Arizona were sampled. Approximately 70% of the specimens were from the following list: *Letharia vulpina* (22%), *Xanthoparmelia cumberlandia* (19%), *Rhizoplaca melanophthalma* (10%), *Usnea hirta* (9%), *Umbilicaria vellea*(6%), and *Xanthoria polycarpa*(5%). Due to unique species distribution patterns, not all species were present at each site. Lichen samples were detached from the substrate, transported to Brigham Young University, and prepared as described by Williams, *et al.*(5). A minor deviation from this procedure was used for small samples. Cutting up the sample in a blender

CP392, *Application of Accelerators in Research and Industry*, edited by J. L. Duggan and I. L. Morgan
AIP Press, New York © 1997

TABLE 1. Average Fe/Ti ratio for each site with standard deviation and number of lichens sampled.

Site[a]	Mean Fe/Ti	s	Fe range (ppm)	N	Site	Mean Fe/Ti	s	Fe range (ppm)	N
Salmon N. F.	7.55	2.0	47.8-6880	53	High Uinta W. A.	6.61	.85	1050-4000	7
Sawtooth N. F.	8.30	3.1	190-2200	6	Clearwater N. F.	8.08	2.0	83.5-7440	10
Manti-Lasal N. F.	8.14	1.1	1950-4570	6	Nez Perce N. F.	6.25	1.3	90-3310	8
Cedar Breaks N. M.	6.78	.93	2400-7900	4	Fort Bowie N. H. S.	6.89	.45	7190-26000	2
Anaconda-Pintler W. A.	6.10	1.6	170-8700	17	Uinta N. F.	9.51	1.4	1570-16800	22
Capitol Reef N. P.	6.15	1.5	3860-5740	3	San Juan N. F.	7.26	1.1	292-13100	39
Bridger-Teton W. A.	7.62	1.8	500-8600	7	Chiricahua N. M.	7.77	.74	95.8-11700	19

[a]Abbreviations: N. F. (National Forest), N. M. (National Monument), N. P. (National Park), W. A. (Wilderness Area), N.H.S. (National Historic Site).

was deemed unnecessary. Instead, the sample was placed directly into a teflon vessel, and homogenized in a Mikro-Dismembrator. Approximately 1 milligram of powdered lichen sample was secured to a thin polycarbonate film with polystyrene dissolved in toluene. Samples approximately 1 mg/cm^2 thick were then irradiated with a beam of 2.25 MeV protons. Data analysis procedures are also described in Williams, *et al.*(5).

DISCUSSION

The Fe/Ti ratio for a set of samples was calculated as

$$Fe/Ti = \sum(Fe/Ti)_i/N$$

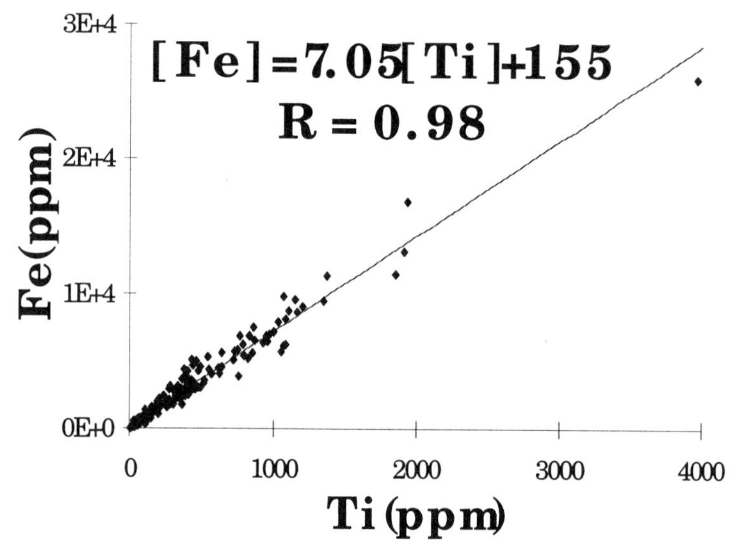

Figure 1. Iron concentration versus titanium concentration for 204 lichens.

where $(Fe/Ti)_i$ is the ratio for a single sample in the set and N is the number of samples. Figure 1 plots iron concentration versus titanium concentration for all 204 lichens along with a linear regression line through the data. The equation for the line and the R value are also given. The correlation is remarkably strong, even at higher concentrations. This is because iron and titanium oxides are thought to form under similar geological conditions. The slope of 7.05 is quite consistent with the average lithosphere ratio of Fe/Ti = 6.5 (6). A similar study in New Brunswick, Canada by Nieboer, *et al.*(3) reported Fe/Ti ratios for *Cladonia* spp. of 8.6 with an iron range from 242 to 1292 ppm. A study in Finland by Takala *et al.*(2) report Fe/Ti ratios of 7.46 and 6.12 for *Hypogymnia physodes* and *Pseudevernia furfuracea* with iron ranges of 450 to 2580 and 700 to 1720 ppm respectively. The iron range for our study was 47.8 to 26,000 ppm. The ratios in these studies are consistent with ratios in this study; however, both studies cite ranges for Fe and Ti which are more than an order of magnitude less than our samples.

The mean Fe/Ti ratios for this study were calculated for each site (Table 1). Some variance between sites is expected, but most ratios fall between 6 and 8.5. One exception is the samples from the Uinta National Forest. This forest includes much of the of the Wasatch Mountains (commonly known as the Wasatch Front) which extend in part from the north to the south end of Utah Valley, a distance of approximately 50 miles. The Fe/Ti ratio from this area (9.51) is the highest of any of our sites in the Rocky Mountains. If this site is set aside ratios for all other sites agree within their standard deviations.

TABLE 2. Average Fe/Ti ratio for 6 species.

Genus/Species	Mean Fe/Ti	s	N
Xanthoparmelia cumberlandia	7.35	0.90	39
Letharia vulpina	6.31	1.4	44
Rhizoplaca melanophthalma	7.20	1.2	21
Usnea hirta	7.67	1.1	18
Umbilicaria vellea	8.96	1.9	13
Xanthoria polycarpa	10.0	1.3	11

Fe/Ti ratios for the six most commonly sampled species are reported in Table 2. All agree within standard deviations except for *Xanthoria polycarpa*, which has a high value of 10.0. All samples for this species are from the Uinta National Forest and were collected along the Wasatch Front. The two most frequently collected lichens are *Letharia vulpina* (a fruticose lichen) and *Xanthoparmelia cumberlandia* (foliose lichen). *Letharia vulpina* was collected primarily in north central Idaho and Montana, while *Xanthoparmelia cumberlandia* was collected extensively from sites throughout the Intermountain area.

The average Fe/Ti ratio for foliose lichens was 7.81 with a standard deviation of 2.1 while fruticose lichens had a Fe/Ti ratio of 7.46 with standard deviation of 2.3. These data suggest that there is not a significant difference in Fe/Ti ratios for the two growth forms when averaged across all samples.

The distribution functions of the log of iron concentrations were plotted for fruticose and foliose lichens and is shown in Fig. 2. Instead of a single distribution, two separate peaks occur. These distributions patterns show that foliose lichens have much higher iron concentrations (10x) than fruticose lichens. Separation between the two distributions occurs at 1250 ppm iron. The mean and standard deviation of the data on the logarithmic scale for both iron and titanium are given in Table 3. One possible explanation for iron concentration differences between growth forms may be the fact that the foliose species (mostly *Xanthoparmelia cumberlandia*) used in this study occur exclusively on rock substrates; which may result in greater access to and uptake of soil and ultra-fine rock debris. As may be inferred from the strong correlation between iron and titanium, the titanium distribution follows a similar bimodal pattern.

TABLE 3. Log[Fe] and log[Ti] distribution data.

Growth Form	log [Fe] Mean	s	log [Ti] Mean	s
Fruticose	2.75	0.44	1.90	0.39
Foliose	3.50	0.33	2.62	0.35

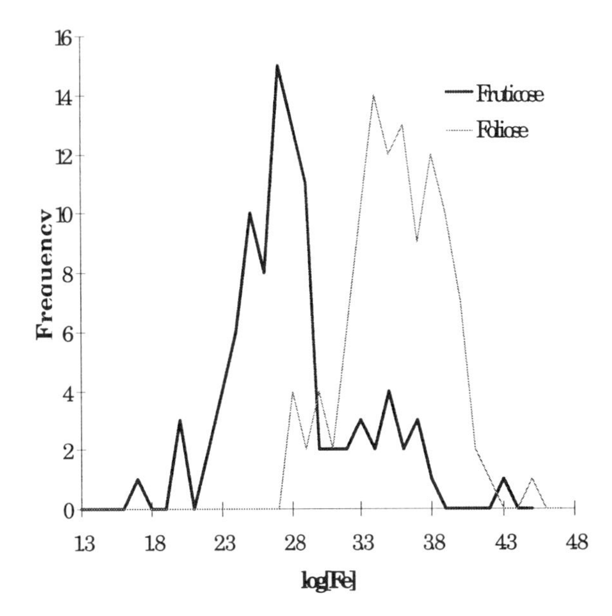

Figure 2. Log [Fe] distribution for fruticose and foliose lichens.

One area of particular interest occurs for 3.2<log[Fe]<3.7. The fruticose distribution follows a descending trend between 2.7 and 3.2, but increases again between 3.2 and 3.7. A total of sixteen fruticose lichens are represented by the data from 3.2 to 37. Close examination of the data in the region of interest shows that two of the samples were *Xanthoria elegans* from the Manti-Lasal National Forest which were collected from rock substrate. In addition, all eleven *Xanthoria polycarpa* lichens are in this region. These thirteen lichens account for the anomaly in fruticose species distribution.

As shown in Table 2, the Fe/Ti ratio is much higher for *Xanthoria polycarpa* than other species. Explanations for this pattern could be biological or environmental. No samples of this species were taken at any other site, so comparison on the basis of site is not possible. Some species of lichens are known to have an affinity for certain elements (*Dermatocarpon miniatum* for sulfur (5)), but data verifying selective binding of iron by *Xanthoria polycarpa* is incomplete.

Another possible explanation may be related to the presence of a steel mill in Utah county and the close proximity of the the sample sites to this facility. High concentrations of iron oxide have been reported in the air of Utah Valley. Another interesting trend is that the Fe/Ti ratios for foliose lichens in the same area seem normal. This fact suggests that airborne iron particles may accumulate to a greater degree in lichens which grow above ground level. Furthermore, as mentioned earlier, foliose lichens on rock substrate may be getting a larger fraction of their iron and titanium from the substrate. The eleven foliose lichens sampled in the same location have log[Fe] values ranging from 3.38 to 3.80, well within the general pattern for foliose species.

CONCLUSIONS

Elemental analysis of lichen tissues show a remarkably consistent relationship between iron and titanium. Our data show that the distribution functions of the log of iron concentrations for fruticose lichens is smaller than for foliose lichens. These relationships may be useful in identifying air pollution-related impact in the Rocky Mountain region.

ACKNOWLEDGMENTS

We wish to recognize the U. S. Forest Service and Brigham Young University for their support of this project.

REFERENCES

1. Ahmadjian, V., *The Lichen Symbiosis*, New York: John Wiley & Sons, Inc, 1993, pp 142-146.
2. Takala, K., Olkkonen, H., Salminen, R., *Environmental Pollution* **84**, 131-138 (1994).
3. Nieboer, E., Richardson, D. H. S., Boileau, L. J. R., Beckett, P. J., Lavoie, P., Padovan, D., *Environmental Pollution* (Series B) **4**, 181-192 (1982).
4 Nieboer, E., Richardson, D. H. S., Tomassini, F. D., *The Bryologist* **81**, 226-246 (1978).
5. Williams, R. N., Casellas, R. C., Mangelson, N. F., Rees, L. B., St. Clair, L. L., Schaalje, G. B., Swalberg, K. D.,*Nucl. Instr. Methods B* **109/110**, 336-340 (1996).
6. Poldervaart, A., Green, J., *Geotimes* **3** (No. 8), 25-27, (1959).

PIXE FOR THIN FILM ANALYSIS

M. H. Tabacniks[1,2], A. J. Kellock[1] and J. E.E. Baglin[1]

[1]IBM Almaden Research Center, San Jose CA 95120, [2] University of Sao Paulo, Brazil.

Abstract

PIXE (Particle Induced X-ray emission) is a highly sensitive, element specific analytical technique for surfaces and thin films. A new PIXE spectrometer has been installed and calibrated at the IBM Almaden Research Center (ARC) Ion Beam Laboratory. Its performance has been optimized for elements in the region 15<Z<35, specifically addressing the elements of greatest interest for magnetic storage technology and research. The system and its performance are fully described. One significant advantage is gained by correlated analysis using both PIXE and RBS on one sample. The application of such analysis for a spin-valve-like multilayered thin film is discussed.

Introduction

The PIXE technique is based upon x-ray spectrometry where instead of using primary x-rays a proton (or helium ion) beam is used to eject inner-shell electrons from atoms in a sample. When the resulting vacancies are filled spontaneously by outer-shell electrons, characteristic x-rays are emitted. The transitions filling the innermost shell are called K x-rays, those filling the L shell are L x-rays and so on. Due to the L sub-shell structure, the L-series transitions are more complex than the K-series. Nevertheless a superposition of two elemental spectra can readily be resolved using spectrum fitting routines based on chi-square minimization . Since the PIXE technique is well established, the reader is referred to several reviews (1,2,3) and a very useful book (4) that summarize the available knowledge with several examples of applications in a practical and clear text.

PIXE formulation for thin film analysis

For a thin film PIXE analysis run, the number of x-rays detected, N_x, is proportional to the quantity t_z (atoms/cm^2) of species Z present in the target, and the collected charge Q, (4,5):

$$N_x(Z) = \frac{\Omega}{4\pi} \sigma_x \varepsilon \, T \, t_Z \frac{Q}{qe} \qquad (1)$$

In equation 1, Ω is the solid angle subtended at the sample by the detector, σ_X is the characteristic x-ray line production cross section, ε is the detector efficiency, T is the transmittance of the x-ray absorbers placed between sample and detector, q is the incident particle charge state, e the electron's charge, and t_z the total number of atoms/cm^2 for the sample in question.

Equation 1 can be rewritten defining an effective x-ray yield per (atom/cm^2), per unit charge of incident ions, $Y_{eff}(Z) = (\Omega/4\pi)\sigma_x \varepsilon \, T(1/qe)$ which is a function of the experimental parameters, atomic number and selected x-ray line (e.g., Kα, Kβ, Lα ..):

$$N_x(Z) = Y_{eff}(Z) \, t_Z \, Q \qquad (2)$$

PIXE-ARC setup

PIXE and RBS (Rutherford Backscattering Spectrometry (6)) are both multi-elemental ion beam based analytical techniques. At ARC, PIXE and RBS share the same experimental setup that includes a 3 MeV single ended Pelletron accelerator connected to a high vacuum analysis chamber. Operating incident beam currents for PIXE are usually some tens of nanoamperes, similar to those of RBS/FRS analysis. At ARC, the sample size can extend up to 7cm x 7cm, being limited mainly by the target holder dimensions. The actual analyzed area is defined by the ion beam and can be set from 0.002 to 0.1 cm^2 or sample rastering can spread the analysis over 7x7cm^2. To adapt the existing setup used for RBS/FRS analysis at ARC, a Si(Li) detector (Kevex, 146eV FWHM @ MnKα, 5μm Be window) and an anti-scattering beam collimator were installed and operational conditions optimized. As a result, PIXE and RBS/FRS analysis can be done on the same sample spot by just switching beams, if necessary, from H^+ to He^+.

The ARC-PIXE system was optimized to detect and quantify any element with atomic number Z>15, having its best sensitivity for elements in the range 22<Z<40 and Z>60, at about 10^{14} at/cm^2. The design parameters of the PIXE-ARC system were optimized for the analysis of thin films deposited on common substrates that include C (UDAC - Ultra Dense Amorphous Carbon), Si, SiO_2, Al_2O_3. In order to avoid counting saturation from the Al or

Si in the substrates, a 130μm thick Mylar absorber was installed in front of the x-ray detector, limiting PIXE sensitivity to elements with atomic number Z>15. RBS/FRS analysis has good ability to distinguish and detect the elements with Z<15, and may thus provide information in this range not seen by PIXE. Thus, the combination of the RBS/FRS and PIXE techniques enables the detection of any element in the periodic table with a single experimental setup. PIXE discriminates neighbor elements that cannot be clearly separated by the RBS technique alone as in the case of NiFe alloys for example, while RBS/FRS handles the low-Z elements beyond the PIXE limit down to hydrogen. However, only the RBS/FRS technique is able to depth profile the elemental concentrations.

Calibration

Experimental calibration of a PIXE setup can be done by evaluating Y_{eff} (Equation 2) with a set of elemental standards. A "working curve" for all elements may be obtained by fitting a theoretical curve describing Y_{eff} while optimizing the experimental input parameters.

PIXE-ARC calibration was done using thin (500Å) elemental reference films vapor deposited on UDAC substrates. The advantages of using UDAC for thin film analysis have been demonstrated previously (7). Each film thickness was measured in at/cm^2 using both H^+ and He^+ RBS analysis. In both cases the analytical Rutherford scattering cross section is valid, thus providing an absolute reference for sample thickness calibration. PIXE effective yield curves for K and L lines were fitted using the theoretical description of Y_{eff} and parameters of the experimental setup.

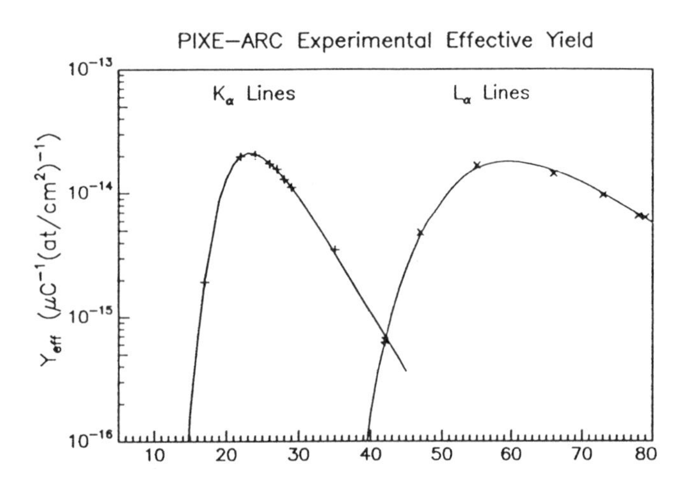

Figure 1. PIXE-ARC experimental effective yields and calibration curve as a function of atomic number. (2.3MeV H^+, 130μm Mylar absorber)

Figure 1 shows the experimental data and fitted curves for Kα lines (Z<45) and Lα lines (Z>40). The average standard deviation was 4.5% for the K lines and 3.8% for the L lines. These curves were calculated using a polynomial parameterization (8) of ECPSSR cross sections (9), theoretical values for the x-ray emission rates (10) and x-ray absorption coefficients (11). Because of the limited energy range of the x-ray Si(Li) detector, detection of Kα lines is used for Z<45 while Lα lines are used for elements with Z>40. The beam energy was set to 2.3MeV either for H^+ or He^+, matching the energy normally used for RBS analysis. At this energy, a 50μm thick Be foil is needed to stop the scattered H^+ particles from entering the Si(Li) detector, while only 5μm Be is enough to stop any scattered He^+ ions. When using Si, quartz or sapphire substrates, a 130μm mylar foil is used to block the scattered particles in addition to absorbing the Si and Al x-rays from the substrate. To avoid the need for a stopping foil, thus avoiding any extra x-ray absorption, a He^+ beam with energy lower than 1.05 MeV should be used.

Minimum Detection Limit

The minimum detection limits for PIXE analysis of thin elemental films on carbon and quartz substrates as a function of atomic number are shown in Figure 2. The minimum detection limit is the lowest amount that may be positively detected against background using the 3σ criterion. According to this criterion the minimum detection limit, MDL, was calculated using the equation (12):

$$MDL(counts) = 3 \left\{ \sqrt{\sum_{FWHM} PkBg} \right\} \qquad (3)$$

where $\sum PkBg$ sums the background counts delimited by the FWHM of the corresponding Kα or Lα peak. Analytical conditions were: 2.3MeV H^+ beam and 30μC of collected charge. Values obtained here are in good agreement with previously published PIXE sensitivities (4).

RBS - PIXE comparison

PIXE and RBS techniques show similar, and also remarkably complementary features. Both techniques are multielemental and non destructive, they share the same setup and type of electronics, they use similar beam energies and beam currents and they require similar analysis time and they have comparable sensitivities.

To reveal their complementary nature, three common characteristics ought to be compared:

- the cross sections, which ultimately determine how fast data can be collected to an acceptable statistical significance
- the elemental resolution, which represents the power to distinguish the signals from neighbor elements
- the signal to background or noise ratio, which ultimately defines the minimum detection limits.

While cross section is an intrinsic physical property, resolution and noise depend on the particular experimental conditions and thus are capable of being optimized. Under the experimental conditions used at ARC, PIXE or RBS data acquisition takes less than 15 minutes per sample. The PIXE x-ray production cross section for 2.3 MeV protons is roughly a function of $(1/Z)^8$ while for a fixed detection geometry, the RBS cross section is a function of Z^2. But, as outlined above, yield is not the only determining factor in choosing between PIXE and RBS. In fact, both PIXE and RBS should often be used concurrently to improve detection and quantification of the elemental composition and depth profile of thin films.

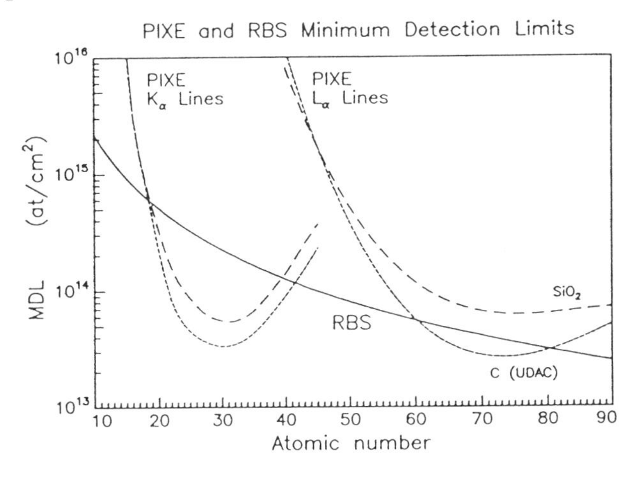

Figure 2. Minimum Detection Limits for PIXE analysis (2.3 MeV H^+) of a thin film on carbon and quartz substrates compared to the calculated MDL of 2.3MeV He^+ RBS at 170°. Accumulated charges were 30μC and at least 10 counts in a peak were required for the RBS MDL.

A measure of the resolving power of each analytical system can be obtained by calculating the ratio, *r*, of the energy difference of the signals of two immediate neighbor elements to the detector resolution, for a very thin film:

$$r(Z) = \frac{E(Z) - E(Z-1)}{FWHM} \qquad (4)$$

where Z is the atomic number, E(Z) is the energy of the detected signal (back scattered particle or x-ray) for element Z, and FWHM is the full width at half maximum of the detector response (particle or x-ray). As defined, r>1 means good element discrimination, while for r<1 the signals of neighbor elements may overlap, thus reducing specificity. For RBS, neighbor element resolution is a function of the kinematic factor $K=E(Z)/E_0$, where E_0 is the energy of the incident beam. At 170°, K behaves approximately as $(1/A)^2$ where A is the atomic mass number. For PIXE data, and according to Moseley's law, the K or L line energy is a function of Z^2. The FWHM for a Si(Li) detector can be calculated as $FWHM = \sqrt{noise^2 + 2.35 E f}$ where E is the x-ray energy and f is the Fano factor (4). Combining both factors, the discrimination power of PIXE is expected to grow linearly with Z and stabilize to an almost constant value. Figure 3 compares the discrimination power for PIXE and RBS as described above.

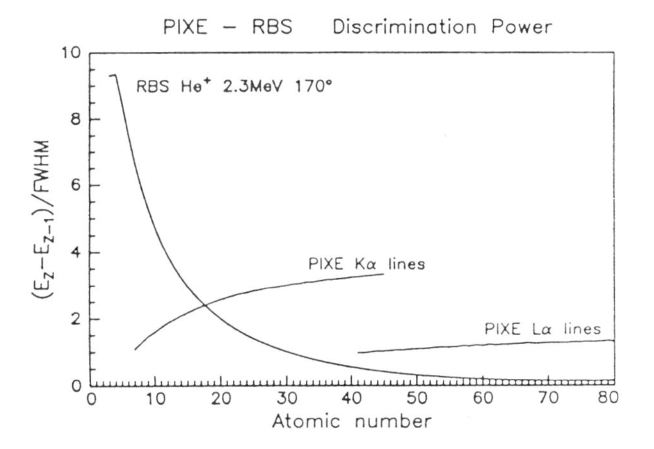

Figure 3. Discrimination power for 2.3MeV H^+ PIXE and 2.3MeV He^+ RBS setup at 170°. PIXE and RBS detector FWHM values are respectively 170 eV @ 5.9keV and 20 keV.

RBS is an absolute and depth sensitive multielemental technique. PIXE is more sensitive to elements with Z<40 while RBS is extremely sensitive to high Z elements. RBS resolving power is higher than 1 only for Z<30 while PIXE resolving power is >1 for all Z. It should be noted that resolving power is rather a figure of merit, not a real limitation. It is only important when a sample contains a series of neighbor elements (e.g. magnetic alloy films) in which case PIXE and RBS data from the same spot can be used to discriminate and quantify the interfering elements. PIXE can be used to determine total elemental abundances, that can then be used to assist or constrain the RBS depth profile fitting for consistent but overlapping components. The minimum detection limits also show the complementary nature of the two techniques.

Application example

To illustrate the power of using PIXE in conjunction with RBS, a typical thin film structure was analyzed.The sample is a spin valve type magnetic sensor structure with many thin layers, some of which are binary alloys. The challenge was to determine the alloy composition, as well as the thickness of each layer. The nominal sample structure, from the top down was: 75Å Ta / 100Å FeMn / 20Å Co / 20Å Cu / 100Å $Ni_{80}Fe_{20}$ / 50Å Ta / Al_2O_3 substrate. Figure 4a shows the PIXE spectrum of this sample. There are well defined Kα peaks for Mn, Fe, Co, and Ni. The Cu Kα and the Ta Lα peaks are overlapped, but the Cu Kβ and Ta Lβ lines are well separated, allowing the measurement to be made with these weaker but measurable peaks. By measuring the peak areas for each element, its thickness and/or concentration can be calculated. Figure 4b shows the corresponding RBS spectrum for this sample. The most obvious feature is that there are only four peaks. The Ta peak is split into signals corresponding to the top and bottom Ta layers, there is a small Cu signal visible, and all the other four metals Mn, Fe, Co, Ni are lumped together into an unresolvable peak. From this spectrum the thickness of each Ta layer and the Cu layer can be determined. Since the Mn, Fe, Co, and Ni signals were not resolved, the PIXE data was used to determine the concentrations of these elements. A simulated fit to the RBS data using the PIXE measured NiFe and NiMn concentrations is shown overlaid in Figure 4b. The fit is extremely good, giving confidence in the analysis. Each element was plotted separately, so that its contribution to the total peak shape can be seen. There is excellent agreement of the results of both techniques within experimental uncertainties.

References

1. Johansson, S.A.E. and Johansson, T.B. Analytical Application of Particle Induced X-Ray Emission. *Nucl. Instr. Meth.* **137**, (1976) p.473-516.
2. Khan, M.M., and Crumpton, *CRC Crit. Rev. Anal. Chem.* **11**(1981) p.103 and p.161.
3. Mitchell I.V. and Barfoot, K.M. *Nucl. Sci. Appl.* **1** (1981) p. 99.
4. Johansson, S.A.E., and Campbell, J.L. PIXE, *A Novel Technique for Elemental Analysis*. John Wiley and Sons, 1988.
5. Tabacniks, M.H. *Calibration of the PIXE-SP System for Elemental Analysis*, Institute of Physics, University of Sao Paulo, April, 1983. (M.S. thesis, in Portuguese).
6. Chu, W., Mayer, J.W., and Nicolet, M.A. *Backscattering Spectrometry*. Academic Press. NY. 1978.
7. Tabacniks, M.H. Fajardo, F., Comedi, D., Chambouleyron, I., and Kellock, A.J. PIXE and RBS analysis of thin germanium films. *13th Int. Conf. on the Application of Accelerators. in Research and Industry*. Denton, TX. November 7-10, 1994.
8. Maxwell, J.A., Campbell, J.L. and Teesdale, W.J. The Guelph PIXE software package. *Nucl. Instr. Meth. Phys. Res.* **B43**, (1989) p. 218-30.
9. Chen, M.H. and Crasemann. *Atomic and Nucl. Data Tables* **33** (1985) p. 217.
10. Scofield, J.H. *Phys Rev.* **A9** (1974) p. 1041.
11. Berger, M.J. and Hubbell, J.H. *XCOM 1.2.*. National Bureau of Standards Report NBSIR 87-3597.
12. Currie, L.A. Limits for qualitative detection and quantitative determination. *Anal. Chem.* **40** (1968) 587.
13. RUMP/GENPLOT. Doolittle,L.R. *Nucl. Instr. Meth.* **B15**, (1986) p.227. Available from Computer Graphics Service, Lansing, NY

Acknowledgments

The authors thank Wolf Hanish, who spent his summer 1995 NSF fellowship at IBM-ARC helping to prepare the initial set of PIXE standards

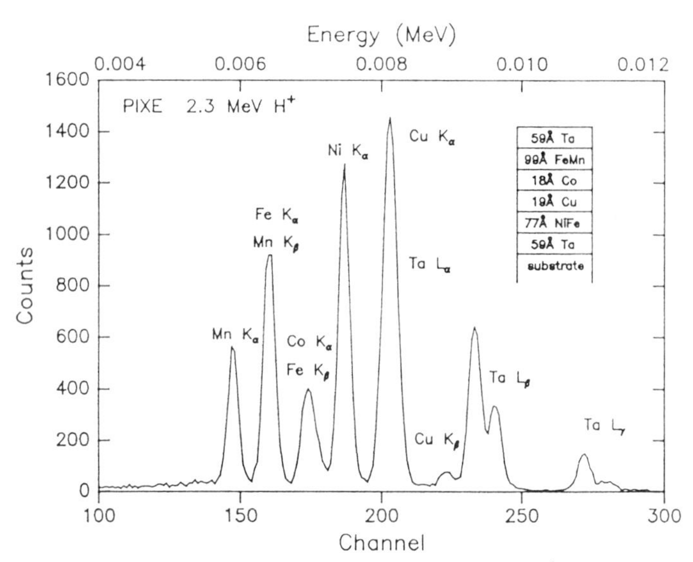

Figure 4a. PIXE spectrum of a spin-valve-like multi-layered thin film.

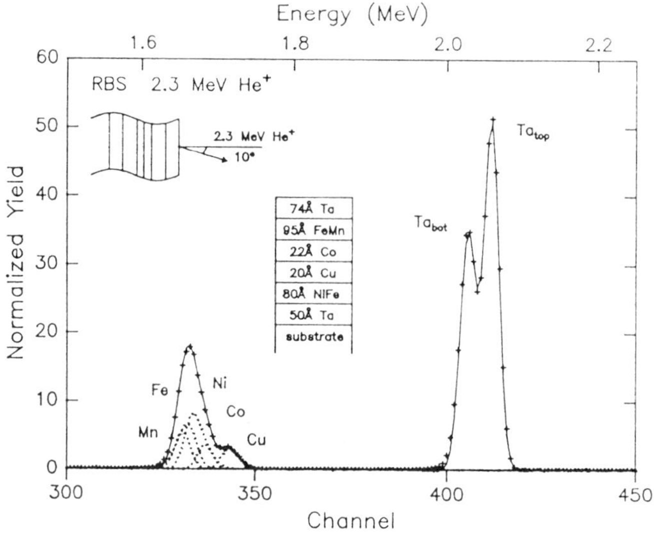

Figure 4b. RBS spectrum of a spin-valve-like multilayered thin film and the simulation obtained for best fit, using RUMP (13), constrained by elemental ratios given by PIXE.

EVIDENCE FOR ENHANCED ALUMINUM CONCENTRATION IN BRAIN TISSUE FROM ALZHEIMER'S DISEASE PATIENTS USING PIXE

M.E.Debray [1,2], A.J.Kreiner[1,2,§], M.Buhler[1], M. A.Cardona[1,2], D.Hojman[1,2,§], J.M. Kesque[1], G.Levinton[1], J.J.Menéndez[1], F.Naab[1,2], M. J.Ozafrán[1], H. Somacal[1,2], M. E. Vázquez[1], H.Grahmann[1], M.Davidson[3,§], J.Davidson[3,§], M.E.Levin[4], C.A.Mangone[4], R.L. Caccuri[4], A.Tokuda[4], A.A.Eurnekian[4], D.González[5], C.López[5] and O.E.Roses[5]

[1] *Departamento de Física, Comisión Nacional de Energía Atómica, 1429 Buenos Aires, Argentina*
[2] *Universidad Nacional de Gral. San Martín, Buenos Aires*
[3] *Departamento de Física, Facultad de Ciencias Exactas y Naturales, Universidad de Buenos Aires*
[4] *Hospital Santojanni, Buenos Aires*
[5] *Cátedra de Toxicología y Química Legal, Facultad de Farmacia y Bioquímica, Universidad de Buenos Aires*

The Particle Induced X-Ray Emission (PIXE) analytical technique with ^{16}O ion beams (18 MeV) was applied to the study of elemental composition at different brain regions of patients with a confirmed post-mortem diagnosis of Alzheimer's disease and in samples from control subjects. The results obtained in the actual study show a clear correlation between occurrence of Alzheimer's disease and the presence and increased concentration of aluminum (Al).

INTRODUCTION

Alzheimer's disease (AD) is a progressive, degenerative illness of the brain now considered as a leading cause of dementia. It affects about 10% of the population older than sixty. The cause for AD is unknown and it can only be diagnosed with certainty through post-mortem histopathological studies (1,2). The relation between aluminum (Al) and AD has a long history but nevertheless the direct evidence on elevated Al concentration in AD cerebral tissue samples is contradictory and inconclusive (1,2).

A number of microprobe (1-5) studies of Al in AD have been performed, but they have failed to resolve the controversy surrounding this issue. Specifically a study (2) using a proton microprobe, carried out in 1992, reported the absence of enhanced Al in neuritic plaque cores in AD deceased patients in samples which were not treated by the usual chemical staining techniques and the suggestion was made that earlier Al evidence could be linked to contaminated reagents. These studies, however, were conducted at a detectable limit (5) of analysis of 15 ppm. In the present work we have studied bulk matter samples from both AD and control subjects and hence no staining agents were used. Very recently and in parallel with the present investigation heavy ion (5 MeV Si^{2+}) and proton (2 MeV) microprobe samples (hippocampus) of AD patient studies (6) determined the presence of Al in cell nuclei of brain.

§ Member of CONICET, Argentina

EXPERIMENTS AND RESULTS

A heavy-ion oxygen (^{16}O) probe was utilized here (7), which has roughly a two orders of magnitude larger cross section for characteristic X-ray production than the hitherto used protons (8,9), finding restricted but clear-cut evidence of enhanced Al presence in AD samples. The cross section depends roughly on Z^2(proj.), [σ(p, 2.5MeV) $\approx$ 940 b vs. $\sigma(^{16}O$, 18MeV) $\approx$ 5.7 10^4 b] implying an increased sensitivity for equal irradiation times (or equal integrated particle current on the target).

The ^{16}O beams were provided(8) by the TANDAR Tandem Van de Graaff accelerator of the Atomic Energy Commission in Buenos Aires. The PIXE analytical technique was optimized in the 14-50 MeV ^{16}O-ion bombarding energy range by measuring the Al K X-ray production cross section. Although this production reaches its maximum at approximately 30 MeV, the measurements were carried out at 18 MeV, where the cross section is somewhat smaller, in order to minimize the background of nuclear origin, since at this energy one is below the Coulomb interaction barrier of the projectile with all light major bioelements (H, C, N and O) composing the tissue matrix. The X-rays were measured with a Si(Li) detector (2 mm crystal thickness) with a resolution of 180 eV at 5.9 keV and a 0.5 mil Be window looking into the reaction chamber through a 2.5 μm Mylar window and placed at 90° respect to the beam axis.

CP392, *Application of Accelerators in Research and Industry*, edited by J. L. Duggan and I. L. Morgan
AIP Press, New York © 1997

TABLE 1. Frecuency of ocurrence (F.O.) and average concentrations of Al in Alzheimer disease and control patients brain samples measured using PIXE with ^{16}O (18 MeV) ion beams.

Sample Description	Control [F.O.]	Al [a] [μg/g]	AD [F.O.]	Al [a] [μg/g]
Temporal white matter	1/6 [b]	1.8(1.2)	2/4	11.7(3.5)
Frontal cortex matter	0/1	—	2/3	3.6(1.4)
Frontal white matter	0/1	—	2/2	11.5(2.5)
Frontal cortical and sub-cortical matter	0/4	—	—	—
Frontoparietal cortical and sub-cortical matter	—	—	1/1	3.5(1.4)
Parietal cortical and sub-cortical matter	1/1	2.7(0.9)	1/1	5.8(1.8)
Cortical and sub-cortical left-uncus matter	—	—	0/1	—
Parietal cortical matter	1/1	1.2(0.7)	1/1	6.2(2.0)
Parietal white matter	0/1	—	1/1	3.4(1.1)
Frontoparietal white matter	—	—	1/1	[c]
Temporal white matter and cortical matter	—	—	1/1	9.2(2.8)
Temporal cortical matter	0/1	—	—	—
Positive Samples [d]	**3/16** **19%**	**1.9(0.8)**	**12/15** **80%**	**6.9(1.0)**

a) The Al concentration values in [μg/g] were determined only for thin targets.
c) No thin target available. Al present in thick sample.
b) Patient died from perforated ulcer.
d) Percentage of positive cases and corresponding average Al concentrations. The mean concentration values were taken on the positive cases only.

To enhance the transmission of Al X-rays the small (1.5 mm) gap between the chamber and the detector's Be was filled with a controlled atmosphere of He gas. The normalization was made by a Faraday cup current integration and by measuring Au L X-rays produced in a thin Au foil (~ 300 μg/cm2) placed upstream from the target with a planar Ge detector.

Samples coming from 5 different post-mortem histopathologically confirmed AD subjects and different brain locations (see Table 1) were measured and compared with identically treated samples from 5 controls (no AD). Typically 20 mg of brain tissue material were extracted by a stainless steel needle from the region of interest and "digested" in a borosilicate micro-kjeldhal, using only nitric acid. The mineralized residue was diluted with bidistilled water, and microdrops of 10 μl were pipetted on a thin electronic grade, self-supporting carbon backing of ~ 30 $\mu g/cm^2$ to totalize a solid deposit of about 1 mg/cm^2 thickness. For each sample material also a thick target was made. All samples were treated with the same reagents. Also blank targets were produced using the same backings, reagents, water and vessels but no organic matter, showing no Al above detection limits.

Figure 1. shows three PIXE X-ray spectra. The upper one corresponds to material from a control subject containing no Al above detection limits (defined (9) as three times the square root of the background area under the full-width at half maximum range in the Al peak region). The middle frame shows a spectrum corresponding to an AD thin sample (thin means in this context no significant energy loss by the projectile, typically a surface density of about or less than 1 mg/cm^2) where an Al peak is clearly visible. This spectrum corresponds to a sample of frontal white matter and contains a concentration of 12 ppm in weight (or μg/g) which translates into 15 ng of Al in 1.2 mg of original substance deposited on the backing. The detection limit for Al of this spectrum is 3 ng.. The background contribution

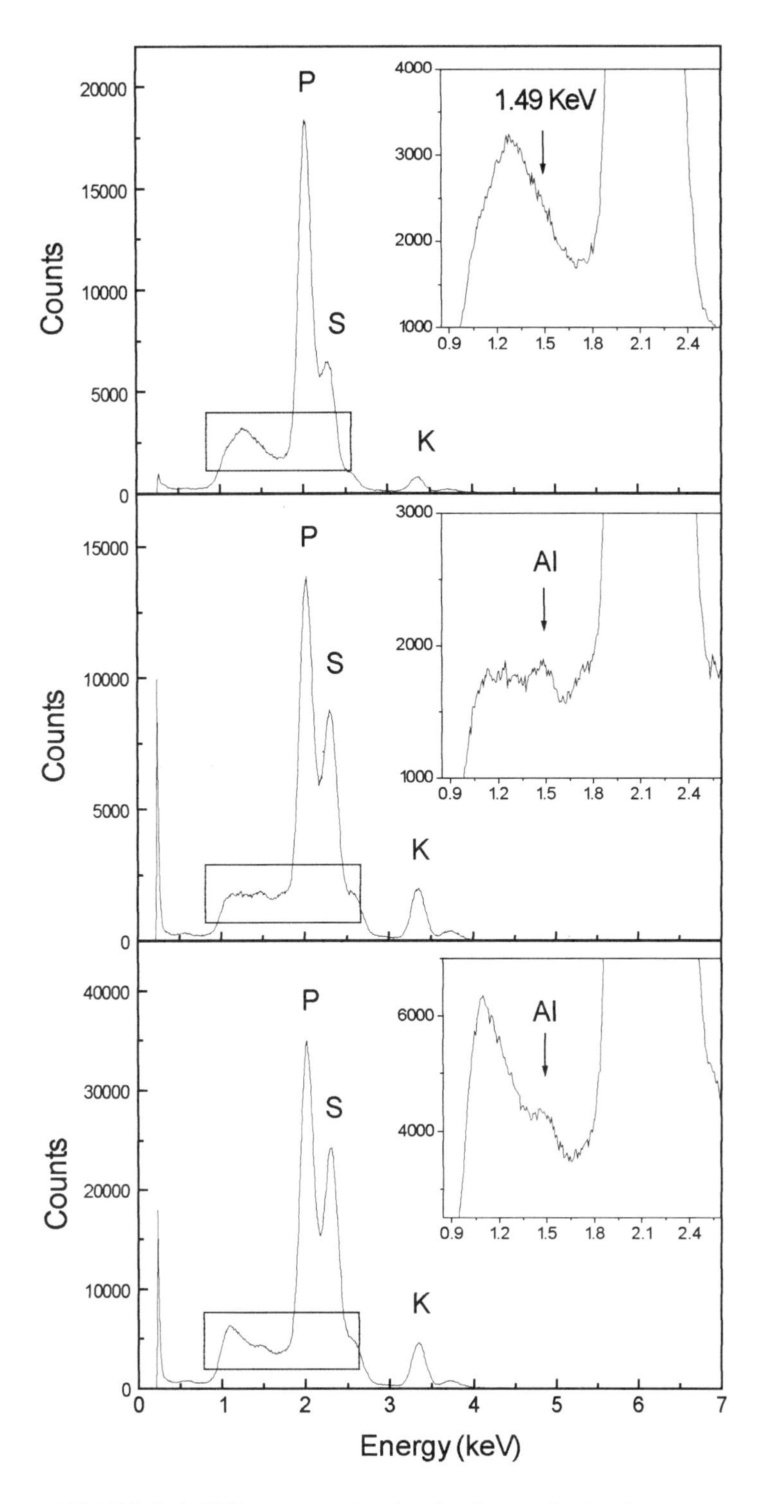

FIGURE 1. X-Ray spectra showing the absence (top) and presence (middle and bottom) of Aluminum(Al). Sulfure(S), Phosphorus(P) and Potassium(K) peaks can also be seen.

of the carbon backing to the aluminum detection limit for the 1.2 mg mineralized deposit was approximately 0.3 ppm, that is, the backing represents less than 10% of the total background at 1.5 keV and hence did not have an effect on the minimun detection limit obtained here for aluminum of brain samples. The low detection limit for Al is explained by the large X-ray production(6) with heavy ions and the high transmission of our set-up for the Al X-ray energy. The lower frame spectrum is obtained from a thick sample of the same material. No cuantitative analysis is attempted for such samples.

Table 1. shows the concentrations in μg/g (ppm) of 32 brain tissue samples prepared from histopathologically characterized AD's and control patients classified by the brain sampling location. These data show clearly, within the scope of this limited set, that the presence of Al, both in frequency and concentration (80% and 6.9 ppm), is greatly enhanced in AD patients as compared to control subjects (19% and 1.9 ppm). It should be mentioned that one of the control patient which shows Al presence (in temporal white matter died from stomach ulcer having been treated with Al-containing drugs. Another interesting observation is a tendency to higher Al concentration in white matter samples.

The measurements on samples from well characterized cases of AD contradict previous results indicating simultaneous presence of Al and Si in form of alumino-silicates. No correlations between them were found in the analyzed spectra.

CONCLUSIONS

Concluding one may say that the PIXE technique with ^{16}O beams of 18 MeV, combined with bulk material digestion and preconcentration is capable of detecting trace amounts of Al (and other elements) in the ppm range in brain tissue samples. In the restricted set of data presented here a clear correlation between presence of Al (both in frequency and concentration) and existence of AD is apparent. The results presented in this work are the first of a more complete study which aims at constructing a cerebral AD aluminum concentration map.

REFERENCES

1. Markesbery W.R. and Ehmann W.D., Brain trace elements in Alzheimer's Disease in *"Alzheimer Disease"*, (Terry R.D., Katzman R. and Bick K.L. eds., Raven Press, New York, 353-369(1994)).
2. Landsberg J.P, McDonald B. and Watt F., *Nature* **360**, 65-68(1992).
3. Landsberg J.P., McDonald B.,Roberts J.M., Grime G.W. and Watt F., *Nuc.Instr. and Meth.* **B54** 180-185(1991).
4. Pinheiro T. ,Taper U.S.A.S., Sturesson K. and Brun A., *Nuc.Instr. and Meth.* **B54** 186-190(1991).
5. Watt F. and Grime G.W., in *"Particle -Induced X-ray Emission Spectrometry"*, eds. Johansson S.A.E., Campbell J.L. and Malmquist K.G., *Chemical Analysis*, Vol. **133**, J.Wiley (1995) p.135
6. Yumoto S. et al., Nucl. *Instr. and Meth.* **B109**, 362-367(1996)
7. Debray M.E. el al., XI International Conference "Alzheimer's Disease and Related Disorders", Buenos Aires, Argentina, September 12-16 (1995), "PIXE experimental investigation on samples of Alzheimer's Disease brain tissue", *Revista Neurológica Argentina* Vol. 20 Suplemento 1, 20(1995)
8. Ozafrán M.J. et al., *Nucl. Instr. and Methods* **B99**, 384-386(1995).
9. Johansson S.A.E. and Campbell J.L., *"PIXE: A Novel Technique for Elemental Analysis"* (Wiley, New York, 1988) and references therein.

DYNAMIC MAPPING ANALYSIS OF RECURRENT CALCIUM-RICH KIDNEY STONES SEQUENTIALLY EXCRETED FROM A SINGLE PATIENT

C.A. Pineda[1,3✉] A.L. Rodgers[2], V.M. Prozesky[3] and W.J. Przybylowicz[3*]

[1] *Groote Schuur Hospital, Observatory, 7925, South Africa*
[2] *Department of Chemistry, University of Cape Town, South Africa*
[✉3] *Van de Graaff Group, National Accelerator Centre, P.O. Box 72, Faure, 7131, South Africa*
**On leave from the Faculty of Physics and Nucl. Techniques, The University of Mining and Metallurgy, Cracow, Poland*

The technique of Dynamic Analysis (DA) has been applied to obtain information on the distribution of minor components and trace elements (TE) in a series of calcium oxalate (CaOx) human kidney stones by nuclear microprobe (NMP) determination. In the present study microanalysis was further expanded to include information on Na, Mg, Al, Si, P, S and Cl. Interest was also focused on determining levels of directional variability in elemental concentrations of Ca, minor components and TE throughout selected micro-regions of single stones by DA.

INTRODUCTION

Preliminary investigations on Ca-rich kidney stones have suggested that analysis by proton microprobe could further expand the interpretations of the mechanisms and role of TE in the build-up of Ca-rich concretions. As a continuation to this work particularly from subjects with recurrent urolithiasis (1,2) the present study extends the work to the analysis of elements of lower atomic mass than Ca.

In the context of calcium oxalate stones, particular interest has been paid recently to the effect of minor and trace element aggregates as promoters or inhibitors in the three stages of stone formation (ie. nucleation, growth and aggregation). Stone formation is thought to be caused, in part, by the high supersaturation of urine with respect to the stone forming minerals including calcium oxalate in its various forms of hydration: calcium oxalate monohydrate (COM), dihydrate (COD) and tri-hydrate (COT) (3). Formation begins with the deposition of biopolymers and/or protein molecules, which provide the framework for subsequent mineral deposition.

The aim of this work was to characterise the levels of minor and TE in CaOx concretions and in particular to focus attention on the interrelationship of different kinds of spatial elemental distributions with time of excretion and type of stone in a series of CaOx kidney stones sequentially excreted from a single patient. The model presented herewith suggests that TE and minor components are deposited in the bulk of the concretion according to a particular phase in the nucleation process.

MATERIAL AND METHODS

Samples preparation

A series of seven CaOx urinary stones of different sizes, excreted by a male patient (present age 45 years) over a period of 11 years, at different intervals, was collected for analysis. For simplicity they were numbered 1, 2, 3, 4, 5, 6 and 7 in chronological order. The masses of the stones varied from 5 to 50 mg and the size varied from about 0.5 mm to 3-4 mm in diameter. No particular treatment for cleaning of the stones was implemented other than simple washing in methanol to remove possible blood constituents.

Each stone was embedded in Tensol Cement, a dichloromethane mixture supplied by Messrs African Explosive and Chemical Industry, Wadeville, South Africa, to act as a binder and fitted in a perspex holder. The holder was then positioned on a specially designed mounting ladder for irradiation. Each sample was then cut with a titanium knife to produce a relatively flat surface adequate for mapping by the proton beam. Since Ca-rich materials are insulators, all samples were coated with a thin layer of carbon to prevent overcharging by secondary electrons and to keep the bremsstrahlung background at reasonable levels.

Irradiation and measurement

Major components (C, O, Ca) were detected by 2 MeV alpha-RBS spectrometry, minor components were analysed by PIXE with 1 MeV protons and elemental maps of trace

CP392, *Application of Accelerators in Research and Industry*, edited by J. L. Duggan and I. L. Morgan
AIP Press, New York © 1997

elements were determined by PIXE with 3 MeV protons using the nuclear microprobe at the Faure Van de Graaff accelerator. A detailed description of the component parts of the nuclear microprobe system have been described previously (4,5). The target was supported on an xyz manipulator with micrometer position adjustments, and viewed by an optical microscope at 45^0, to set up the beam size, focus and target position. A Link Pentafet (model 6648) X-ray detector with 80 mm^2 active area and 8 µm beryllium window was positioned at 135^0.

For irradiation with 1 MeV protons a closed geometry was used by placing the X-ray detector at about 22 mm from the target surface. A 100 µm mylar absorber was used to attenuate the high intensity low energy pulses. For 3 MeV protons the detector was positioned at 37.5 mm from the target and a 80 µm Al absorber was used to allow the detection of Ca K X-ray signals. This procedure allowed the simultaneous measurements of Ca as well as trace metals (Z>20), concentration levels of which ranged up to about 200 $\mu g.g^{-1}$. The beam current absorbed by the targets (which were infinitely thick for both the 1 and 3 MeV protons) was extracted at the specimen ladder. To prevent any possible damage or modification of the target surface during irradiation, proton beam currents of the order of 100 to 500 pA were used. In addition a negative potential of 1500 volts was applied to a Cu ring situated in front of the target to suppress the secondary electrons generated in the process. Beam size spots of about 5-10 micrometers were used.

The microprobe was interfaced to a CAMAC data acquisition and analysis computer systems which include the data analysis software package XSYS (6) used in multiparameter mode and the Geo-PIXE package (7). These constitute the main components of the DA system (8), which uses a spectral decomposition transform that closely approximates the time-consuming non-linear least-square method. The DA was performed in live time to obtain continuously updated quantitative elemental maps as data accumulated.

In order to extract additional information about directional growth from elemental maps obtained by DA the resulting 64-dimensional data matrix for each map was *compressed* into a vector by calculating the *mean value* of rows or columns. This compressed mean showed a relatively smoothed trend for most elemental maps particularly for Ca, indicating directions of gradual changes in elemental concentration.

RESULTS

Concentration levels of TE and minor components

Elements detected using 1 MeV protons were Na, Mg, Al, Si, P, S and Cl. Elements detected with 3 MeV other than Ca were Fe, Ni, Cu, Zn, Br and Sr. Transition elements such as Cd and Sn were also observed. Rare earths La, Ce and Nd (not previously reported in the literature) as well as heavy metals such as Tl, Pb and Bi were observed in small concentrations. Because of the relatively high standard absolute error and low levels of concentration, rare earths and heavy metals (with the exception of Pb) as well as the transition elements were not considered in the analysis of the data, although they could play a role in stone nucleation and growth. Concentration ranges and mean values at micro-region level for stones 2 and 4 showed similar mean values and are presented in Table 1 including the standard deviation.

Table 1. Elemental concentrations in $\mu g.g^{-1}$ for stones 2 and 4 as obtained by point and DA analyses in random selected scanned micro-regions. Concentrations levels agreed fairly well with previous reported work (9).

Z	STONE 2		STONE 4	
	CONCEN RANGE	MEAN	CONCEN RANGE	MEAN
Na	0.3-1%	0.6(0.1)%	0.1-1%	0.5(0.3)%
Mg	400-4000	2200(1300)	300-3000	1600(1000)
Al	200-2000	1300(800)	200-9800	6300(4200)
Si	900-1800	1200(800)	200-1000	700(760)
P	1200-5400	3200(1400)	700-7000	3600(2000)
S	0.1-1%	0.5(0.4)%	0.08-0.4%	0.24(0.3)%
Cl	300-2700	1500(1200)	950-9000	4200(3700)
Ca	12 - 80%	37(19)%	11 - 56%	28(11)%
Mn	1.9 -15	6.2(3.2)	3.0 - 8.1	4.7(11.5)
Fe	2.5 - 40	15(8.4)	6.2 - 65	24(15)
Cu	1.0 - 4.0	2.2(0.7)	1.1 - 3.2	2.2(0.6)
Zn	1.0 - 15	8.3(4.6)	2.0 - 16	11(4.0)
As	1.3 - 3.6	2.3(0.9)	1.1 - 3.3	2.0(0.7)
Se	1.8 - 4.6	2.8(0.7)	1.5 - 4.3	1.9(0.9)
Br	2.5 - 37	14(10)	3.0 - 31	11(7.5)
Sr	29 - 132	56(23)	30 - 96	50(16)
Pb	2.8 - 17	8.5(4.5)	4.0 - 13	8.7(2.1)

Since matrix correction factors (MCF) in thick target PIXE have been shown to be relatively independent of matrix composition (for 1 as well as for 3 MeV protons (10)), the exact matrix composition of each stone was approximated to that of hydroxyapatite (HAPA). On the other hand, since the information obtained by DA could be considered in a relative manner, these MCF effects are not particularly critical for elemental mapping of trace elements such as those investigated in the present study. Furthermore, the compressed mappings discussed below are still valid as a mean value of elemental profiles equivalent to a single general linear scan (see figure 3). The greatest variability on TE content throughout the whole body of the stone corresponds to Fe, Zn, Br and Sr.

Parallel determination by RBS indicated the presence of C, O and Ca as major components, with possible empirical formulae corresponding to those of COM and COD. The first two and the last three stones in the series had the former structure while stone 3 and 4 had the latter one. From the determined concentration ranges of P (see table 1) it's possible that P could be considered as a major component as was indicated by destructive analysis of one of the stones with a value of 30 % for the phosphate. From the most common matrices which occur in kidney stones

(11) it appears that this level of phosphorus corresponds to that of HAPA. It is therefore likely that the stones consist of CaOx, HAPA and/or brushite (12).

in all stones and which is regarded as a promoter of kidney stone formation (3).

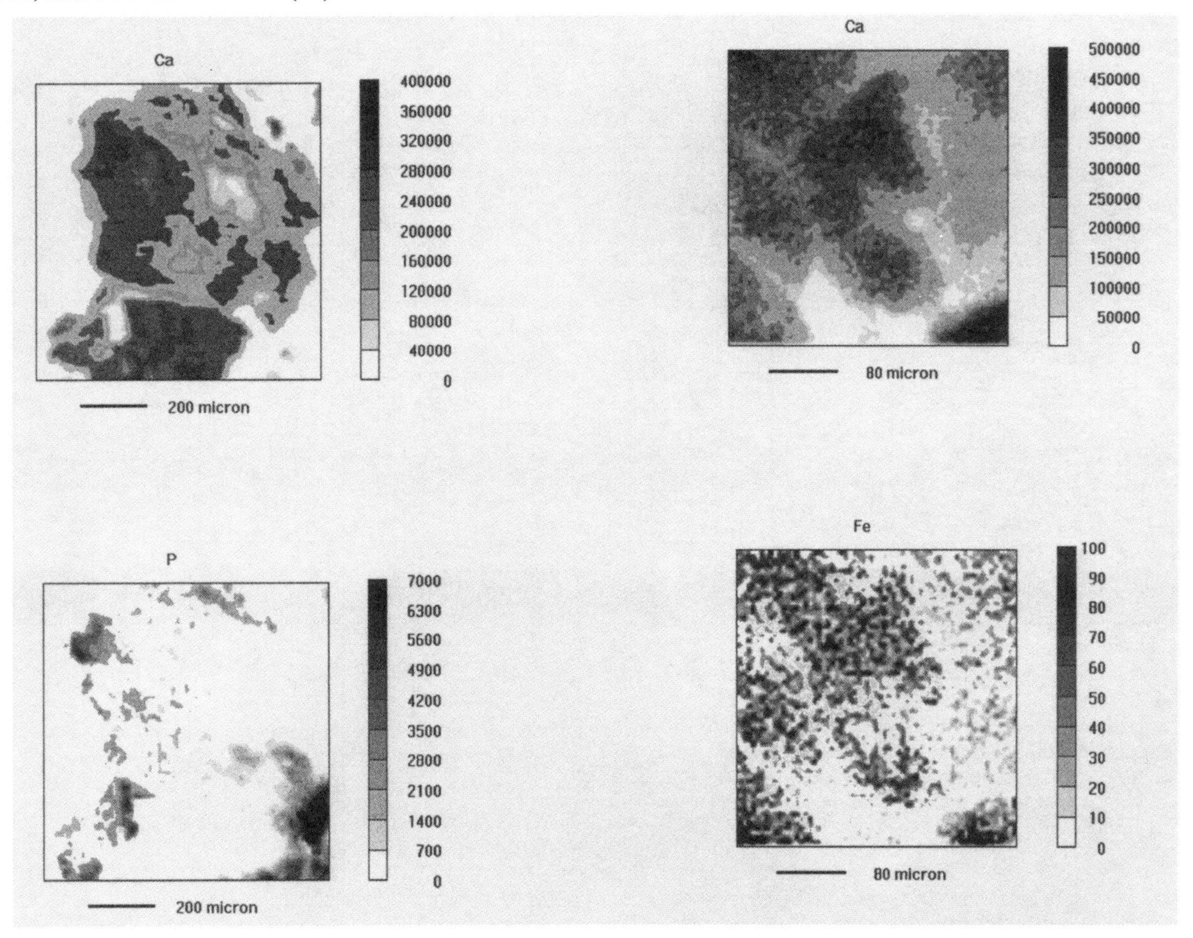

Figure 1: Contour maps made by the Interactive Data Language (IDL) package (13), for stone 7 showing the variability on metal distribution for Ca and P irradiated with a beam of 1 MeV protons scanned over an area of 1050 x 1050 μm^2, divided into 64 x 64 pixels. Accumulated charge was 1 μC. The scales shown on the right side of every map are in $\mu g.g^{-1}$ and represent quantitative elemental concentrations found by DA.

Figure 2. As fig.1 but for irradiation of stone 4 with 3 MeV protons. Scanned area was 400 x 400 μm^2 and collected charge 33 μC.

Dynamic mapping of Ca, minor components and trace elements

Figure 1 shows maps of Ca and P for stone 7. Analyses by both energies showed that in general the distribution of Ca was non-uniform, with *pockets* of widely varying concentrations being scattered randomly throughout the scan area (see figures 1,2). Since Ca concentrations ranged in general from 5-50% it is assumed that the composition of the matrix changes in certain micro-regions. For pure COM, the atomic concentration of Ca corresponds to about 31%. Thus, there are some regions in the structure of calcium oxalate stones where Ca is obviously enriched. The random deposition of Ca (which is likely to be in the form of CaOx (1)) might be indicative of variations in the presence and activity of the organic matrix which is present

DISCUSSION

Comparison of elemental mapping distributions

The correlation of different elements in the mappings with Ca suggests that such elements are bound to compounds which are Ca-rich or to proteins related to the CaOx structure (see figure 2). Perhaps these elements have the same chemistry or they are trapped by special mechanisms. Attempts have been made to study this type of mechanism but have not resulted in firm conclusions (3).

If the spatial distribution and directional growth of Ca is related to the particular type of stone phase at micro-region level, then TE which have been found to correlate with Ca could be bound to sites within the CaOx structure. Those elements which show a negative spatial correlation with Ca may not particularly bind to CaOx matrix sites but could have been deposited as isolated ion-entities in the stone matrix. Examples of these are As, Se ,Pb and in particular P (see figure 1). It is reasonable to speculate that the consistent and in most cases uniform presence of minor components and trace elements might suggest that they play a role in crystal (and hence stone) initiation. It is also likely that they might be components of the organic matrix which is present throughout the stone. On the other hand, the presence of Cu and Fe could arise from blood contamination while Zn and Sr might simply occur in all stones as a result of normal renal filtration processes. In general it was found that Ca and Sr were positively correlated to a great extent, as may be expected from their chemical nature. Correlation could also possibly arise as a result of the type of mechanism by which Sr is filtrated in the urine and bound to the CaOx matrix.

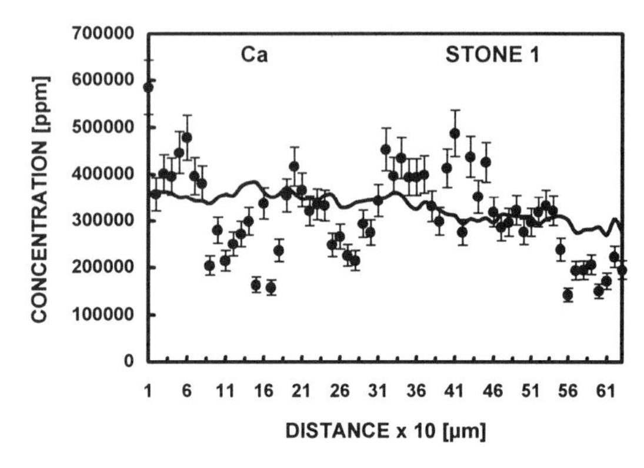

Figure 3. Plot of the mean value of the rows (line) of the Ca map for stone 1 in a 620 x 620 μm^2 micro-region indicating a directional gradual decrease in Ca deposition, which is related to stone growth. For comparison data of a pre-selected linear traverse (or single row) is plotted (dots) showing a greater degree of scatter.

Directional variability of components and stone growth

The information presented in Figure 3 on the mean value of the rows of the Ca map is an important tool for the visualisation of hidden information that otherwise could not be established. This is particularly relevant for studies of time-related deposition of layers in crystal structure growth such as CaOx human kidney stones. This work forms part of a first attempt to treat data from elemental mapping as mean values in a preferential direction. It is envisaged that this technique will also be useful in other fields of bio-medical research where similar time-dependent phenomena are involved.

REFERENCES

1. Pineda, C.A., and Peisach, M., *Nucl. Instr. and Meth.* **B85**, 896-900 (1994).
2. Pineda, C.A., Rodgers, A.L., Prozesky, V.M., and Przybylowicz, W.J., *Nucl. Instr. and Meth.* **B104,** 351-355 (1995).
3. Campbell, A.A., Fryxell, G.E., Graff, G.L., Rieke, P.C., and Tarasevich, B.J., *Scan. Micros.* 7, 423- 429 (1993).
4. Tapper, U.A.S., McMurray, W.R., Ackermann, G.F., de Villiers, G., Fourie, D., Groenewald, P.J., Kritzinger, J., Pineda, C.A., Schmitt, H. Springhorn, K., and Swart, T., *Nucl. Instr. and Meth.* **B77**, 17-24 (1993).
5. Prozesky, V.M., Przybylowicz, W.J., Van Achterbergh, E., Churms, C.L., Pineda, C.A., Springhorn, K.A., Pilcher, J.V., Ryan, C.G., Kritzinger, J., Schmitt, H., and Swart, T., *Nucl. Instr. and Meth.* **B104**, 36-42 (1995).
6. Anonymous, Indiana University Cyclotron Facility Scientific and Technical Report, 1985, pp165.
7. Ryan, C.G., Cousens, D.R., Sie, S.H., Griffin, W.L., Suter, G.F., and Clayton, E., *Nucl. Instr. and Meth.* **B47**, 55-71 (1990).
8. Ryan, C.G., Jamieson, D.N., Churms, C.L., and Pilcher, J.V., *Nucl. Instr. and Meth.* **B104**, 157-165 (1995).
9. Levinson, A.A., Nosal, M., Davidman, M., Prien,SR., E.L., Prien,JR., E.L., and Stevenson, R.G., Trace elements in kidney stones from three areas in the United States. *Invest.Urol.* **15**, 270-274 (1978).
10. Pineda C.A., Ph.D Dissertation, "Thick Target PIXE analysis", University of Cape Town, 1993.
11. Pougnet M.A.B., Peisach, M., and Rodgers, A.L., *Nucl. Instr. and Meth.* **B35**, 472-471 (1988).
12. Kwiatek, W.M., Lekki, J., Paluszkiewicz, C., and Preikschas, N., Application of FTIR, PIXE and EBS for trace element analysis in biological samples. *Nucl. Instr. and Meth.* **B64**, 512-516 (1992).
13. Anonymous, Interactive Data Language, User's Manual, Research System Inc., 777 29th Street, Boulder, USA, 1993.

ION BEAM ANALYSIS OF ANCIENT HUMAN BONE

St. Jankuhn, T. Butz, R.-H. Flagmeyer, T. Reinert, J. Vogt

Fakultät für Physik und Geowissenschaften, Universität Leipzig, D-04103 Leipzig, Germany

J. Hammerl, R. Protsch von Zieten, M. Wolf

Institut der Anthropologie, Universität Frankfurt a.M., D-60323 Frankfurt a.M., Germany

H. Baumann, K. Bethge, I. Symietz

Institut für Kernphysik, Universität Frankfurt a.M., D-60486 Frankfurt a.M., Germany

Proton backscattering spectrometry, Rutherford backscattering spectrometry, proton induced X-ray emission, and proton induced Gamma-ray emission as well as X-ray fluorescence analysis were applied to analyse major and trace elements of bones of two Merowingian populations (8th century AD). Hereof, together with quantitative digital radiography which gives the bone mineral density, we expect to obtain more information about our ancestors. As a region of special interest the femur neck (Collum femoris) was investigated to determine the bone mineral content and the Ca/P-ratio. In order to estimate the influence of the burial environment radial distributions of element concentrations of cross sections of the femur shaft (Corpus femoris) were recorded by lateral-resolved PIXE and PIGE.

INTRODUCTION

The calcification and remodelling processes of bone are complex and only partially understood so far. For some years, modern medicine uses quantitative digital radiography (QDR) to determine the bone mineral density (BMD). This technique has been established as early diagnosis of bone loss and metabolic bone diseases as, e.g., in the case of osteoporosis. The BMD value reflects the bone mineral content (BMC) but gives no evidence about all chemical elements involved in the metabolism of the bone. Ion beam analytical methods, on the other hand, are well suited for the determination of element concentrations in biological samples because these methods are fast, accurate, and allow a non-destructive multi-elemental analysis; see e.g. (3, 4, 5). The combination of proton backscattering spectrometry (PBS), Rutherford backscattering spectrometry (RBS), proton induced X-ray emission (PIXE), and proton induced Gamma-ray emission (PIGE) gives a complete overview about major as well as trace elements of bone samples.

In the present study we report on first investigations of archaeological human bones belonging to two Merowingian populations. The final aim is to reveal relations of anthropological interest such as between the biological development (e.g. effects of menopause on bone loss), health status (osteoporotical changes), and living conditions (dietary intake) of our ancestors.

EXPERIMENTAL

The femora studied derive from skeletons excavated from former Merowingian burial places near Bockenheim and Edesheim, Germany. The findings were dated about 750 ± 60 AD with the ^{14}C-method. Bones were examined in two steps: (i) Examination of age and sex of the mortal remains and determination of the BMD in the original femur with an intact proximal region (Fig. 1①) using QDR (2) and (ii) measuring the elemental concentrations of specially prepared bone samples by ion beam methods and X-ray fluorescence analysis (XRF). The latter requires cutting out and pelletizing of bone pieces from the femoral neck (Fig. 1②) including the Ward's triangle (Fig. 1③). Also, cross sections were prepared

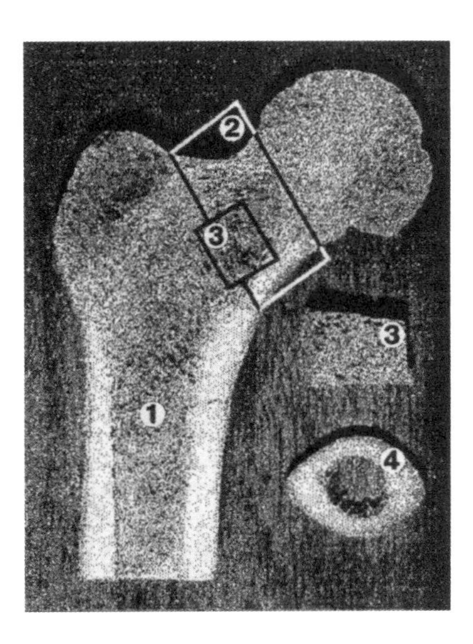

Figure 1. Cut through the proximal part of a femur (for explanations see text).

CP392, *Application of Accelerators in Research and Industry*, edited by J. L. Duggan and I. L. Morgan
AIP Press, New York © 1997

from the femoral shaft (Fig. 1④). For details of sample preparation, see (7).

A collimated beam of 1.7 MeV protons (1 mm beam spot diameter at a current of 1 nA) from the Leipzig 2 MV van de Graaff accelerator was used for PBS, PIXE, and PIGE measurements. The *simultaneous* recording of the spectra of all these ion beam methods guarantees (i) measurements on the identical region of interest, (ii) the same charge collection for each spectrum, and (iii) because of the time saving a reduction of changes in the elemental composition of the pelletized or cutted samples due to irradiation.

The backscattered ions used in PBS are detected by means of a Canberra PIPS detector with an active area of 50 mm^2 and an energy resolution of $\Delta E_p = 10$ keV. The detector is placed at a distance of 59 mm at an angle of $\Theta = 170°$ with respect to the ion beam direction.

For PIXE analysis, a Si(Li) detector made by RÖNTEC — 9.6 mm^2 active area, $\Delta E_X = 145$ eV at $E_X = 5.9$ keV — is arranged at a distance of 47 mm from the sample at $\Theta = 141.2°$ backward angle. A pierced "funny" filter with a hole of 0.5 mm diameter is placed in front of the detector to reduce the intensity of the lines of the major elements P and Ca as well as pile-up effects.

The PIGE setup allows the spectrometry of γ-rays from nuclear reactions. It consists of an ORTEC 148 cm^3 HighPurity Ge detector in 90° geometry with $\Delta E_\gamma = 1.75$ keV at $E_\gamma = 1332$ keV and the associated electronic equipment for routine analysis. For the determination of the element concentrations the following reactions were selected: $^{19}F(p,p'\gamma)^{19}F$ for fluorine at a γ-ray energy $E_\gamma = 110$ keV, $^{23}Na(p,p'\gamma)^{23}Na$ for sodium (E_γ=439 keV), and $^{31}P(p,\gamma)^{32}S$ for phosphorous (E_γ=2237 keV).

RBS measurements were carried out with 2 MeV $^4He^+$ ions from the Frankfurt a. M. 7 MV van de Graaff accelerator. For further explanations, see (7).

Besides the ion beam methods the bone pellets were investigated by XRF using the commercial system SPECTRO X-LAB. The primary X-rays were produced by a tube with a rhodium source. Five spectra per sample were accumulated under different excitation conditions, e.g. several polarizers (B_4C, Al_2O_3), secondary targets (Cd, Co), and filters (Pd, Ta, Fe) were used. The emitted characteristic X-rays were detected by a Si(Li) detector with an active area of 10 mm^2 and an $\Delta E_X < 150$ eV at $E_X = 5.9$ keV.

RESULTS AND DISCUSSION

In order to illustrate the results of the bone investigations typical spectra of the measurements are shown in Figs. 2–5.

The PBS spectra (cf. Fig. 2) were evaluated using the differentiation method described in (9). From this we get the major elemental concentrations (C, O, P, Ca). The $^4He^+$-RBS measurements provide the O, P, Ca contents, too, but not the C concentration. For the analysis of the collected data in standardless PIXE experiments the computer codes EDR288 (RÖNTEC) and YIELD (1) are used (Fig. 3). The evaluation of the PIGE measurements (Fig. 4) uses CaF_2, NaCl, and GaP as standards to determine the concentration of light elements F, Na, and P. Figure 5 shows the X-ray yield of a bone sample from the femoral neck region obtained by XRF. As recognizable from Figs. 2–5 the major elements P and Ca can

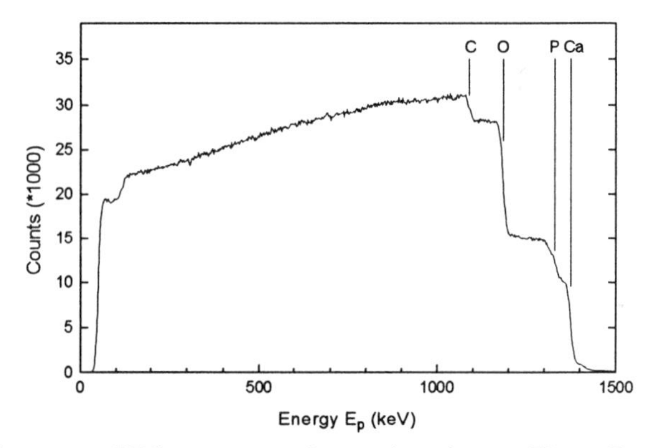

Figure 2. PBS spectrum of spongious bone. The collected charge amounts to $Q = 50\mu C$.

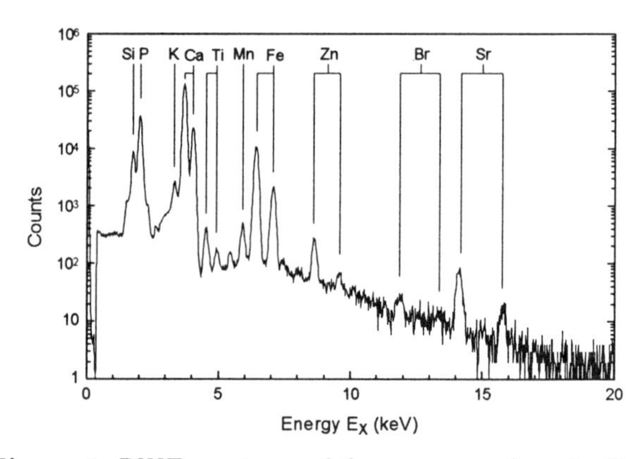

Figure 3. PIXE spectrum of the same sample as in Fig. 2.

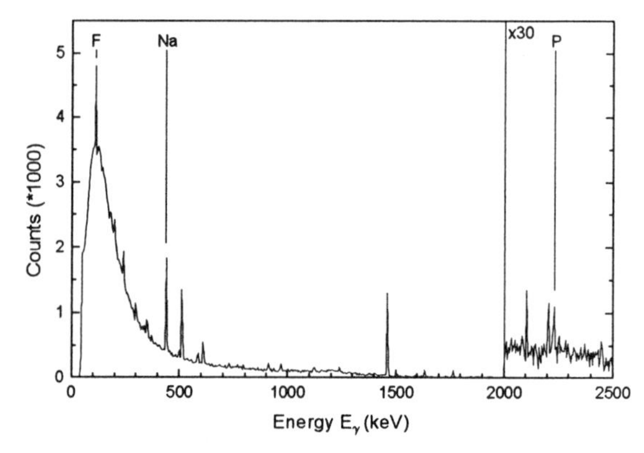

Figure 4. Typical PIGE spectrum of a dense bone sample.

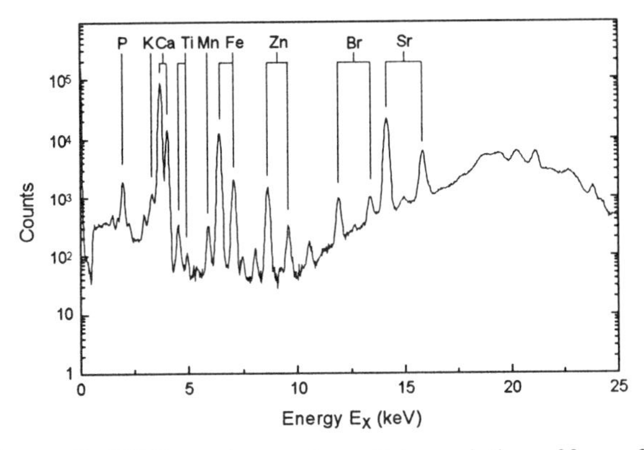

Figure 5. XRF spectrum of a pellet consisting of bone from the neck zone including the Ward's triangle measured with X-rays polarized by B_4C for 700 s.

be determined almost by all methods, whereas for trace element analysis only PIXE, PIGE, and XRF are applicable. In Table 1 typical experimental results ((8), error is about ±15%) are listed together with reference values.

BMD values of all samples are within the range of twice the standard deviation of the reference database of white U.S. Americans (in particular, cf. (2)) but the mean value is higher than the reference average value. This is in agreement with the measured Ca and P concentrations which markely exceed the reference values of recent bone (Table 1). Interesting to note that, moreover, the Ca/P ratio (2.13) also exceeds that of the reference man (2.0) (6) but equals that of Calciumhydroxyapatite (2.15, $Ca_{10}(PO_4)_6(OH)_2$) as the main fraction of the inorganic part of bones. Possible reasons are *post-mortem* changes of the elemental content (diagenesis) and/or the status of the biological development of the Merowingians. Indications for the first assumption are provided by lateral-resolved measurements of bone cross-sections using ion beam analysis and will be a point for further investigations.

Table 1. Typical BMD value in [g·cm^{-2}] of the femoral neck of a male Merowingian individuum of age group adultus II (age 31...40) and average elemental concentrations in [μg·g^{-1}] in comparison with recent bone.

	Merowingian	Today	(Refs.)
BMD	1.01	0.96	(2)
Elements			
P [mg·g^{-1}]	137	100	(6)
Ca [mg·g^{-1}]	292	200	(6)
F	464	639	(3)
Na	2550	5763	(3)
Mn	44.8	-	
Fe	58	7.58	(4)
Zn	152	144	(4)
Sr	292	47.8	(4)

CONCLUSIONS

Six different analytical methods using X-rays or high energy ions as primary radiation are applied to bone samples. In this study the basic requirements and optimal conditions for a variety of techniques are described to analyse thoroughly human femora. First measurements on archaeological bones of the Merowingian period were carried out to determine their major and trace elements. Further studies will establish a relation between BMD and elemental concentrations, and estimate *post-mortem* element exchange processes.

ACKNOWLEDGMENTS

This work is supported by the Bundesministerium für Bildung, Wissenschaft, Forschung und Technologie, Germany, core program "Einsatz neuer Technologien in den Geisteswissenschaften", under grant no. 03-BU9LEI.

REFERENCES

1. Frey, H., Vogt, J., Otto, G., *J. Radioanal. Nucl. Chem.* **99** (1), 193–202 (1986).
2. Hammerl, J., *PhD Thesis*, Univ. Frankfurt a. M., 1990.
3. Hyvönen-Dabek, M., *J. Radioanal. Chem.* **63** (2), 367–378 (1981).
4. Hyvönen-Dabek, M., Räisänen, J., and Dabek, J. T., *J. Radioanal. Chem.* **63** (1), 163–175 (1981).
5. Hyvönen-Dabek, M., Riihonen, M., and Dabek, J. T., *Phys. Med. Biol.* **24** (5), 988–998 (1979).
6. ICRP Report No. 23, *Report of the Task Group on Reference Man*, Oxford:Pergamon Press, 1975.
7. Jankuhn, St., Butz, T., Flagmeyer, R.-H., Reinert, T., Vogt, J., Hammerl, J., Protsch von Zieten, R., Wolf, M., Baumann, H., Bethge, K., Symietz, I., "Osteodensitometrical Studies and Elemental Analysis of Ancient Human Bones", submitted to *Int. Series on Optics Within Life Sciences*, Vol. IV., Berlin:Springer, 1996.
8. Reinert, T., *Diploma Thesis*, Univ. Leipzig, 1996.
9. Vogt, J., Wirth, C., Zschau, H.-E., and Otto, G., *Nucl. Instr. Meth.* **B68**, 285–288 (1992).

Two stage streaker and PIXE analysis for urban aerosol studies

P. Prati [1], F. Cardoni[1], P. Formenti[1], A. Zucchiatti[1], F. Lucarelli[2], P.A. Mandò[2] and E. Cereda[3]

1: Dipartimento di Fisica and INFN - Via Dodecaneso 33, 16146 - Genova (IT)
2: Dipartimento di Fisica and INFN - L.go E. Fermi 25, 16146 - Firenze (IT)
3: C.I.S.E. - Via Reggio Emilia 39, 20090 - Segrate - MI (IT)

A two-stage streaker sampler was exposed to collect size-fractionated aerosol samples in Genova (Italy) in the spring of 1995. Samples were analysed by Particle Induced X-ray Emission (PIXE) at the Van Der Graaff facility of the INFN of Firenze. Hourly concentrations of elements from Na to Pb were measured both in the fine (<2.5 μm) and coarse (2.5-10 μm) fraction. Data were analysed by applying the Time Series and Principal Component (PCA) Analysis and the Source Profiles by Unique Ratios (SPUR) technique. Even with a short sampling possible it was possible, exploiting the hourly time resolution and the particulate size discrimination of the sampler, to deduce a picture of the aerosol composition in one of the most polluted districts of Genova. In particular, five major aerosol sources were singled out: their profiles were obtained using a minimisation procedure.

INTRODUCTION

For the needs of a research program addressed at the characterization of the aerosol particulate in Genova, we have built an aerosol sampler based on a two stage, one week, streaker commercially supplied by PIXE International Corporation*

The use of streakers, in conjunction with the PIXE technique for aerosol studies, is well developed due to the possibility to follow the time evolution of the particulate both for monitoring purposes and for the goal of sources location (1,2). Although the separation of the fine and coarse particulate can add very useful information, the two-stage version of these devices has not been used so extensively and there is therefore a lack of knowledge about the behaviour of the sampler in separating the two aerosol fractions.

During a sampling campaign in Genova (3) we performed specific tests to asses:

- the efficacy of separation of the two particulate fractions in the impaction and filter stages of the streaker sampler;
- the amount of possible corrections to introduce during the PIXE analysis of the two stages.

The two goals have been achieved both with a SEM analysis of the deposition, which gave us informations on the particles size, and with a PIXE irradiation. These approaches are complementary and give togheter a rather clear picture of the sampler performance. In particular the SEM study allows a measurement of the size distribution of the particles in the two stages of the streaker while with a PIXE scanning it is possible a precise determination of the thickness of each element exceeding the Minimun Detection Limit (MDL) with good spatial resolution.

After these tests, we used the sampler for a week in the district of Multedo,one of the most polluted of Genova. Among the most important harbours of the Mediterranean Sea, with almost 700,000 inhabitants and several different industries, Genova is located in the north-west of Italy. The urban atmosphere is monitored with a public network of 15 air quality measuring stations. Stations are generally equipped for: Total Suspended Particulate (TSP) sampling over 24h intervals, hourly averages of pollutant gases (CO, SO_2) and hourly average standard meteorological parameters.In spite of the concern about poor air quality, the complete lack of information about elemental hourly concentrations allowed only a partial characterisation of the urban aerosol.

The amount and quality of the information provided by the use of our sampler in conjunction with PIXE analysis, is such that it was possible to reconstruct a first picture of the aerosol contributors in the area with only one week of sampling.

SAMPLER DESCRIPTION

Impactors are selective on the particles aerodynamic diameter (D_{ae}) which is related to the geometric average diameter (D_p) by the relation:

$$Dae = Dp \sqrt{\rho p} \qquad (1)$$

being ρ_p the density of the particle.

Our streaker sampler has both impaction and filter stages and it is designed so that, at the flow of 1.0 l/min, the Effective Cut Diameter (ECD) of the impaction

* PIXE International Corporation, P. O. BOX 7744 Tallhassee, FL 32316

CP392, *Application of Accelerators in Research and Industry*, edited by J. L. Duggan and I. L. Morgan
AIP Press, New York © 1997

stage, defined as the aerodynamic diameter at which 50% of the particles of that diameter impact, is of 2.5 μm. Furthermore the geometry of the nozzle and the flow rate are such to provide a rather sharp efficiency curve (4). Particles with aerodynamic diameter greater than 10 μm are stopped on a pre-impactor: a porous disk coated with APIEZON-L. Air flows continously on a kapton film (0.008 mm thick coated with paraffin). The fine particulate is finally collected on a 0.4 μm Nuclepore filter which has an efficiency very close to one for any particle diameter. Therefore the two stages essentially correspond to particles with $D_{ae} < 2.5$ μm (fine stage) and particles with 2.5 μm $< D_{ae} <$ 10 μm (coarse stage). The impaction plate and the filter are mounted on a cartridge wich rotates, around a common axis in about a week.

During the sampling period in Genova, where a set point of 0.80 ln/min was fixed,the average volumetric flow, obtained by correcting the mass flow for pressure and temperature variations (recorded hourly), was of 0.84 ± 0.01 l/min, which should corresponds to a 5% increase of the ECD. The Nuclepore filters and Kapton films used for these measurements have then been analysed with two complementary techniques.

ANALYSIS OF DEPOSITION

SEM investigation

We used a kapton film and a Nuclepore filter exposed with a flow of about 0.8 ln/min for the analysis of the particulate diameters distribution by the Scanning Electron Microscope (SEM) of C.I.S.E. in Milano. Some sectors of both stages have been coated with a thin gold layer and then exposed to a 20 keV electron beam. Secondary electrons, produced by the interaction of the beam with the deposited particulate, have been detected to distinguish the particle morphology and, in particular, their size. In order to verify the cut-off efficiency we have counted the particles in three typical regions:

1) the kapton film inside the impaction streak (487 particles in 67 fields);
2) the kapton film outside the streak (133 particles in 31 fields);
3) the Nuclepore filter (very high particle density).

It must be noted that our analysis was limited to the particles dimensions: their aerodynamic diameter is of course a function of the density which we did not measure. The results of the analysis can be summarised as follows:

a) The coarse stage (diameter > 2.5 μm) is completely collected by the Kapton film inside the streak.

b) On the Kapton film small particles can be observed as well in the streak but their mass (if we assume a constant density for all particles) is about 0.1% of the total. In other words, under protons irradiations, the X-ray yield from the Kapton would be dominated by the larger particulate. The small particles distribution is more or less uniform inside and outside the streak: actually we deduced a 1.2 ratio between the population in the two regions.

c) Counting the small particles (diameter < 1 mm) in both the Kapton and Nuclepore streaks sets an upper limit of 10% for small particles retained by the impaction stage and not reaching the Nuclepore filter. This contamination cannot be considered as a background because it is not present on the non exposed sectors of the film. A more precise evaluation of this loss of particles is forbidden by the difficulty of counting exactly the particles deposited on the Nuclepore filter.

d) The diameters of the particles trapped on the Kapton film show a two-mode log-normal distribution (fig. 1). The two modes can be separated by discriminating particles with diameters below and above 1 μm. Only the second mode (diameter > 1 μm) contributes to the mass deposition on the streak (point b) and its count median diameter (CMD) is 3.4 μm. The minimum diameter found inside this mode is 1.9 μm. The result is consistent with the nominal cut-off aerodynamic diameter of the streaker (2.5 μm for a 1.0 l/min air flow). Outside the streak the deposited particles show a single-mode distribution which corresponds to the first mode inside the streak: this is consistent with the statements at point b).

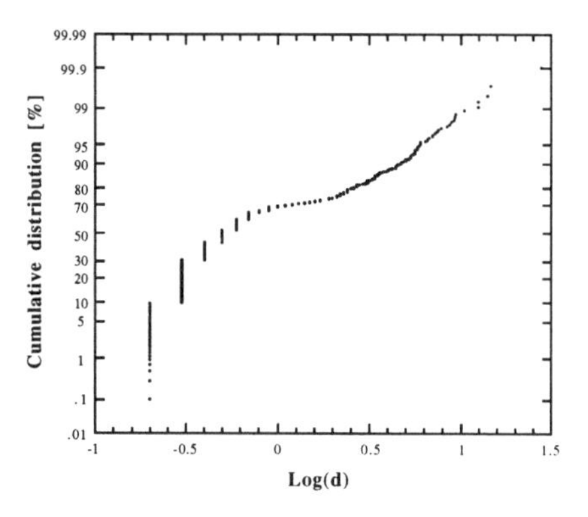

FIGURE 1: The particulate size [μm] cumulative log-normal distribution on the Kapton impaction stage exposed for a test in Genova.

PIXE analysis

A different analysis has been performed on the streaker during the PIXE irradiation of the Nuclepore and Kapton films exposed in Genova. We have bombarded the filters with 3 MeV protons at the INFN Van der Graaff accelerator in the Physics Department of the University of Florence (3,5).

We did actually detect X rays produced by the following elements: Na, Mg, Al, Si, S, Cl, K, Ca, Ti, V, Cr, Mn, Fe, Ni, Cu, Zn, Sr, Br, Pb. For our test we adopted a circular (diameter 0.5 mm) beam spot with an intensity of about 5 nA. During the analysis of a weekly deposition on the Kapton impaction stage, two points corresponding to two hours of the week (A and B in the following), where selected where a markedly different content in Na and Cl was observed. These elements are typical of the sea salt composition and are expected to be abundant mostly in the coarse stage (6). We considered their abundance an index of the aerosol composition. Because the sampling site was 50 m far from the sea, we assumed that high Na and Cl concentrations would indicate a predominantly coarse aerosol while, with low Na and Cl concentrations, we would expect more man-made pollutants, distributed in the fine stage (6). In other words the two points A and B should correspond to different aerosol compositions.There the beam was moved across the 3 mm wide streak in steps of 1 mm. We collected in this way 9 spectra per point: 3 spectra "below" the streak, 3 inside the streak and, finally, 3 "above" the streak. For each element above the MDL, we measured the thickness as a function of the "radial" position in the two points. In each case the streak profile is sharp and the thickness ratio outside and inside the streak ranges from 0.004±0.002 for Cl (pos. B) to 0.10 ± 0.05 for Fe (pos. B) with an average value of 0.06±0.04. This value, which is higher than the result of the SEM pictures observation (point b), can be due to some coarse particles collected immediately outside the streak edge and intercepted by the beam halo (beam diameter 0.5 mm). In any case, the PIXE scanning also substantially confirms that no coarse particulate is collected outside the streak in the Kapton impaction stage.

In Tab. 1 the ratios between the average thickness measured in the points A and B outside the streak in the Kapton and the corresponding thickness on the Nuclepore filter are reported. The elements quoted in the table are those above the MDL on the Nuclepore filter in both positions. The ratios range from 0.004 ± 0.003 (S, pos. A) to 0.8 ± 0.4 (Cl, pos. A) with an average value of 0.10 ± 0.07. For the case of Cl, the only element which shows a high ratio, it must be noted that its deposition thickness on the Nuclepore filter was substantially negligible during the whole sampling week while its concentration was always very high in the coarse stage. The average ratio drops to 0.05 ± 0.02 if the Cl is not considered. Since the SEM analysis showed that the fine particulate "contamination" on the Kapton film can be considered as constant, the ratios in Tab. 1 demonstrate that no significant correction to the signal of the Nuclepore filter is needed. An average 5% correction is actually of the same order of the other uncertainities introduced by the standards (5% typically), the air flow (a few per cent) and counting statistics.

TABLE 1. Measured thickness ratios between points outside the streak in the impaction stage (Kapton film) and inside the streak in the Nuclepore filter, in the two positions A and B, for the elements above the MDL. For Cu and Pb, in the position B, the outside streak thickness in the impaction stage was compatible with zero.

	Position A	Position B
Na	0.05 ± 0.03	0.14 ± 0.02
S	0.004 ± 0.003	0.005 ± 0.007
Cl	0.8 ± 0.5	0.4 ± 0.5
K	0.06 ± 0.02	0.07 ± 0.06
Ca	0.09 ± 0.06	0.04 ± 0.02
V	0.033 ± 0.006	0.02 ± 0.01
Fe	0.07 ± 0.02	0.056 ± 0.009
Ni	0.008 ± 0.005	0.008 ± 0.006
Cu	0.07 ± 0.02	-
Pb	0.005 ± 0.004	-

RESULTS

The sampler was located 5 m above ground, on the roof of one of the provincial monitoring stations, adjacent to the coast in Multedo district, overlooking a road with dense traffic. The area, one of the most industrialised in the whole city, is four kilometres from a steel smelter and six kilometres from an oil-fired power plant, both located on the coast in direction east-south-east with respect to the sampling site.We report here only the main results obtained with the one-week sampling: a complete discussion is reported elsewhere (7)

In figure 2 the elemental apportionment of the total mass measured in fine and coarse stage (fig. 2a) is shown. As expected, Genova being a coastal town, Cl and Na represent a large fraction (~50%) of the total mass of the coarse fraction. Significant contributions (10-15%) come from Fe, Ca, Si and S. In contrast, almost 50% of the fine mass is contributed by S, due to its (slowly varying) regional component. Figure 2b displays, for each element, the fine-to-coarse ratio of the collected mass. V, Ni, S, Pb typically deriving from high temperature emissions, are more abundant in the fine mode, while for Cl and Na the coarse component is prevalent. Fe and Mn, due to natural as well as anthropogenic processes, show quite balanced distributions between the two modes.

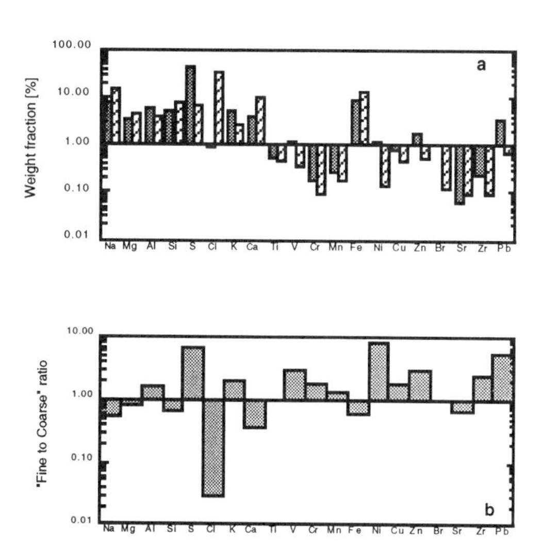

FIGURE 2. a) Elemental distribution of the total mass measured in fine (full bars) and coarse (dashed bars) stage. b) Elemental fine-to-coarse ratio of the average mass.

The study of time sequence plots allows preliminary association of element groups: this procedure is paricularly useful if it is possible to isolate the contribution of one individual source. Sources identification is possible if they are not spatially aligned, if changes in metereological conditions "switch" on or off a particular source or if one source is predominant over the others. This conditions were evident twice during the sampling week. In fig. 3.a time plots for Na and Cl in the coarse stage are shown: synchronous with a wind reversal of a few hours (from the south) was an episode, evident in the second part of the week. Since also S, Mg, Ca, K and Br concentrations increased simultaneously with Cl and Na we are confident of the isolation of a marine episode. As a second example, time variations of Fe and Cu in the fine stage are plotted in fig. 3.b. A five hours episode of very high concentrations is evident in the last days of the week: for some elements (Fe, Cu, Zn, Mn and Cl) there is a correlated increase of one or two order of magnitude.

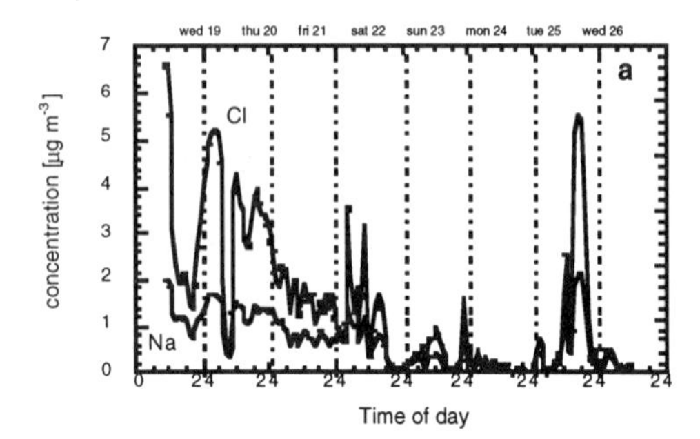

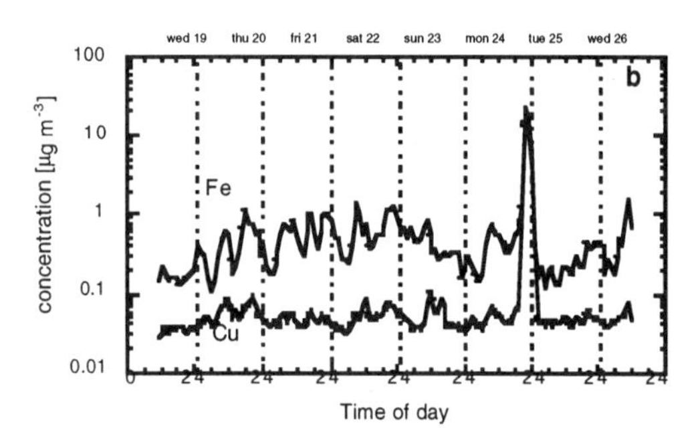

FIGURE 3. a) Time variation of Na and Cl in the coarse fraction b) time variation of Fe and Cu in the fine fraction.

We used PCA to identify the major aerosol sources which were: marine salt, urban dust, traffic, the oil-fired power plant and the big steel smelter.

Sources profiles were then extracted by applying the SPUR technique (8), where particular events are selected according to the unique elemental ratios and the differing time constants of the emitting processes. In Table 2 the profiles for the 5 major sources are reported. Due to the shortness of the sampling period we could only extract common profiles for the fine and coarse stage of the particulate. Further sampling campaigns are planned in different sites in the town and for longer periods with the goal of a complete characterisation of the Genova aerosol in the fine and coarse mass stages.

TABLE 2. Source profiles (percent of weight) extracted from the SPUR analysis for five sources The errors, not reported to save space, are on the last digit of each number.

	Marine	Dust	Traffic	P. Plant	Smelter
Na	.21	.09	.06	.14	.12
Mg	.05	.06	.06	.031	.033
Al	.014	.083	.11	.052	.050
Si	.036	.146	.09	.037	.06
S	.06	.08	.13	.51	.11
Cl	.53	.12	.021	.012	.19
K	.012	.05	.06	.038	.09
Ca	.043	.14	.07	.024	.07
Ti		.010	.016		
V		.011	.019	.033	.016
Mn					.012
Fe	.045	.19	.20	.06	.18
Ni			.010	.03	.011
Cu		.013	.024	.016	.018
Zn		.016	.022	.013	.02
Pb		.03	.12	.014	.03

CONCLUSIONS

We have analized the behaviour of a commercial streaker sampler which is part of the instrumentation used for an aerosol study campaign in the town of Genova. We were interested in particular in the streaker reliability in discriminating the two particulate fractions, fine and coarse, that characterise any aerosol. A SEM investigation and a PIXE scanning of the Kapton films and Nuclepore filters mounted on the streaker gave consistent numerical results which show the reliability of the equipment

The first results, obtained with a one-week sampling in a very polluted district of the town of Genova, confirm the utility of this approach for urban aerosols studies.

REFERENCES

1 Johansson S. A. E. and Campbell J. L. *PIXE: a Novel technique for elemental analysis*, John Wiley & Sons,1988, Chapter 12, 200-215

2 Annegarn H. J., Cahill T. A., Sellschop J. P. F., Zucchiatti A. , *Phys. Scripta* 37(1988)282-290

3 Formenti P., Prati P., Zucchiatti A., Lucarelli F. and Mandò P. A., *Nucl. Instr. and Meth. B*113 (1996), 359-362

4 Marple V. A. and Liu B. Y. H. *Environmental Science and Technology*, n. 8 (1974), 648-654

5 Del Carmine S. P., Lucarelli F., Mandò P. A., . Moscheni G., Pecchioli A. and MacArthur J. D., *Nucl. Instr. and Meth.* B45 , 1990, 341-346

6 Chow J. C. , *Journ. of the Air and Waste Management Association* 1995, 325-397

7 Formenti P., Annegarn H. J., Prati P., Zucchiatti A., Lucarelli F., Mandò P. A. First study of Genova aerosol via streaker sampling and PIXE analysis, submitted to: *Atmospheric Environment: urban atmosphere.*

8 Annegarn H. J., Braga Marcazzan G. M., Cereda E., Marchionni M. and Zucchiatti A., *Atmospheric Environment*, 1992, 26A, 333-343

STUDIES OF ATMOSPHERIC AEROSOLS IN MEXICO CITY USING PIXE

J. Miranda, I. Crespo, S. González, A. López-Suárez, M.A. Morales, B. Pablo, and R. Paredes-Gutiérrez

Instituto de Física, Universidad Nacional Autónoma de México, Apartado Postal 20-364
México, D.F. 01000, México

Along the years 1993-1995 several studies of atmospheric aerosols in the Metropolitan Area of Mexico City were performed. Typically, samples were collected in morning periods (8:00 h to 14:00 h) following different protocols, using a Stacking Filter Unit of the Davis design, separating particles with sizes between 2.5 μm and 15 μm (known as coarse fraction, deposited on polycarbonate filters) and smaller than 2.5 μm (known as fine fraction, collected on Teflon or polycarbonate filters). Elemental analysis of the particulate matter deposited onto the filters was done with Proton Induced X-ray Emission (PIXE). A summary of the results for the fine fraction is presented, including mean elemental concentrations, and results of multivariate statistical analyses, such as Cluster Analysis and Principal Component Analysis. The influence of meteorological parameters to the local elemental concentrations is discussed on the basis of multivariate statistics.

INTRODUCTION

The Mexico City Metropolitan Area (MCMA) faces an important air pollution problem, needing studies to characterize the pollutants present in its atmosphere. Atmospheric aerosols are contributors to this problem, having effects on human health, acid rain, building damage, and visibility. Proton Induced X-ray Emission (PIXE) has shown to be a very useful tool to determine the elemental composition of atmospheric aerosols. The main advantages of PIXE for these analyses are discussed elsewhere (1).

In the present work, the use of PIXE for the measurement of elemental contents in fine atmospheric aerosols (sizes below 2.5 μm) in several sites in the MCMA is exemplified. Average concentrations for several sampling periods in the years 1993-1995 are given. Multivariate statistics are applied to a sample set in order to identify pollutant sources and their relative contributions to the total mass measured in the samples. Finally, the influence of meteorological parameters, such as wind speed and direction, is discussed.

EXPERIMENTAL

The samples were collected with Stacking Filter Units (SFU) of the Davis design (2). The SFU separates particles in two fractions: the coarse one is deposited onto 8 μm pore size polycarbonate filters (Nuclepore filters, Costar Corp.), with sizes ranging from 2.5 μm to 15 μm Mean Aerodynamic Diameter (MAD); the fine fraction is accumulated on 3 μm pore size Teflon filters (Teflo filters, Gelman Sciences), for particles smaller than 2.5 μm MAD, or 0.4 μm pore size polycarbonate filters. The choice of filters for the fine fraction depends upon the kind of analysis desired. Background radiation during PIXE analyses is lower for polycarbonate filters, thus improving the sensitivity of the X-ray detection system, while Teflon filters (originally hydrogen-free) can be used for the analysis of hydrogen in the aerosols using proton elastic scattering (PESA). The air flow meters in the SFU's were calibrated using an MKS 358C mass flow meter (MKS Instruments, Andover, MA, U.S.A.).

Three different sampling protocols were adopted:

1. One site in Southwestern MCMA, from 16 October to 10 December, 1993; daily from 8:00 to 14:00;
2. Three sites (Southwestern, Western, and Northeastern MCMA) simultaneously, from 19 January to 11 May, 1994; once a week from 8:00 h to 14:00;
3. One site in Southwest MCMA, from 15 August to 14 September, 1995; three samples per day (9:00 h to 15:00 h, 15:00 h to 21:00 h, 21:00 h to 9:00 h).

Weather in the first two periods is dry, while the last one occured during the rainy season. The sampling sites characteristics are described in Table 1.

Gravimetric mass of the filters before and after exposition were measured with an Ohaus 200GD microbalance (sensitivity 10 μg, resolution 10 μg). PIXE elemental analyses were carried out following standard analytical procedures, as described elsewhere (3). Typical minimum detection limits (in μg of the element per m^3 of air) are displayed in Fig. 1. The resulting X-ray spectra were

CP392, *Application of Accelerators in Research and Industry*, edited by J. L. Duggan and I. L. Morgan
AIP Press, New York © 1997

deconvoluted with the *Axil* computer code (4) to calculate elemental mass densities on the filters. From this and the air flow measurements in the SFU, elemental concentrations in the aerosols were finally computed.

TABLE 1. Characteristics of sampling sites.

Property	SW	W	NE
Latitude	19°18' N	19°24' N	19°33' N
Longitude	99°10' W	99°16' W	99°2' W
Altitude	2300 m	2500 m	2300 m
Distance to industry	2 km	5 km	2 km
Traffic density	Medium	Low	High
Type	School	Residential	Residential

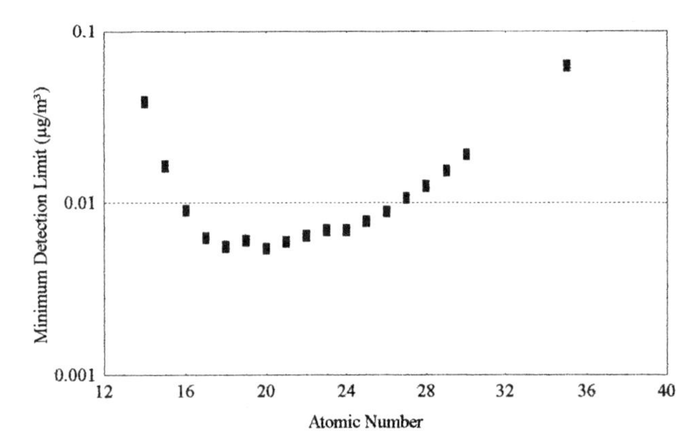

FIGURE 1. Typical minimum detection limits for fine aerosol analysis using a 2.2 MeV proton beam, and an integrated beam charge of 5 µC.

RESULTS AND DISCUSSION

A total of 18 elements were detected in the samples. Mean elemental concentrations for the 1993 period, the 1994 period at the Southwestern site, and the morning (9:00 h to 15:00 h) period in 1995 are compared in table 1, for the fine fraction. No definite pattern is found in the concentrations for these three periods. In spite of the observation of seasonal variations in other studies (5), in the present work it could not be assured. Some elements are more abundant in certain cases (Si and S in 1993 and 1994, Zn in 1995), while others do not show important variations (Ca, Ti, and Cu). This is observed in spite of the different climate during the sampling periods (dry in 1993 and 1994, rainy in 1995). Unfortunately, no meteorological data is available for 1993 and 1994, which might help in this regard. In short, no accurate statements can be made from this data, although more studies are in progress to clarify this point.

Table 3 presents data obtained for the fine fraction in the three sites during 1994. The Southwestern site has higher mass concentrations. S contents are also larger in this site, although the Western site is close to industry (a posible source of this element), and also of Zn and Pb, which are important contaminants in this area. Wind regimes may explain this behavior, as they tend to move down the Basin, carrying industrial pollutants away from the Western site.

TABLE 2. Mean concentrations in the fine fraction measured in three periods (µg·m⁻³).

Element	1993	1994	1995
Mass	50±7	72±12	33±4
Al	1.2±0.08	3.0±0.42	0.60±0.15
Si	1.9±0.11	1.8±0.15	1.1±0.17
P	0.57±0.03	0.22±0.02	NO
S	3.8±0.35	3.5±0.30	2.1±0.23
Cl	0.18±0.01	0.14±0.01	0.086±0.007
K	0.16±0.01	0.45±0.04	0.11±0.02
Ca	0.22±0.01	0.33±0.03	0.27±0.04
Ti	0.020±0.002	0.031±0.003	0.024±0.006
V	0.021±0.004	0.037±0.005	0.013±0.005
Cr	0.005±0.001	0.013±0.002	0.025±0.003
Mn	0.009±0.001	0.008±0.002	0.027±0.003
Fe	0.20±0.01	0.35±0.02	0.39±0.049
Ni	0.004±0.001	NO*	NO
Cu	0.014±0.001	0.017±0.004	0.019±0.002
Zn	0.096±0.008	0.09±0.01	0.171±0.019
Br	0.014±0.001	NO	NM**
Pb	0.081±0.006	0.12±0.01	0.083±0.010

*NO = element was not observed.

**NM = element was not measured due to Br contamination of filters.

TABLE 3. Elemental concentrations of fine fraction in three different sites in 1994 (µg·m⁻³).

Element	SW	W	NE
Mass	72±12	59±11	50±11
Al	3.0±0.42	NO*	0.91±0.19
Si	1.8±0.15	2.0±0.17	1.9±0.17
P	0.22±0.02	0.16±0.02	0.18±0.02
S	3.5±0.30	2.6±0.28	2.7±0.34
Cl	0.14±0.01	0.11±0.01	0.13±0.01
K	0.45±0.04	0.41±0.04	0.39±0.04
Ca	0.33±0.03	0.46±0.03	0.41±0.03
Ti	0.031±0.003	0.030±0.003	0.026±0.003
V	0.037±0.005	0.038±0.005	0.032±0.005
Cr	0.013±0.002	0.011±0.002	0.008±0.002
Mn	0.008±0.002	0.014±0.003	0.008±0.002
Fe	0.35±0.02	0.37±0.03	0.24±0.02
Cu	0.017±0.004	0.021±0.005	0.012±0.004
Zn	0.090±0.01	0.13±0.02	0.10±0.01
Pb	0.12±0.01	0.25±0.05	0.10±0.01

*NO = element was not observed or not reliably measured.

The observed variations of elemental concentrations along the day in the Southwestern site are completely similar to those measured by Aldape and coworkers (6), namely, higher values correspond to morning periods, decreasing in the afternoon and night. The explanation comes from increased human activities in the morning, thermal inversions, and scavenging by typical afternoon rain.

The influence of wind and relative humidity on the aerosols can be appreciated in Fig. 2. There, dendograms and wind roses for the three periods along the day are shown, as an example of Cluster Analysis. The grouping for anthropogenic and natural elements is apparent. Nevertheless, relative humidity (R.H.) normally does not correlate with any element. On the other hand, the effect of the N-S or E-W components of the wind is important. For the morning period, the wind is mainly Westerly, and the x component (Vx) correlates strongly with anthropogenic elements like Pb, Zn, Cu, and Mn. Large avenues and a few industries are located West of the sampling site, thus explaining the correlation through transport by the wind. In the afternoon and night there is a different wind regime (see the wind roses), being Northeasterly winds the dominant ones. Dendograms for these periods show a correlation among soil-derived elements (Si, K, Ca, Fe) and the y component of the wind (Vy). It means that the winds transport soil dust from the Northern zones of the MCMA.

With the complete data set for the Summer of 1995, it is feasible to carry out an Absolute Principal Components Analysis (*APCA*) (7, 8) to identify the pollutant sources and to apportion their contributions (source profiles) to the total mass quantified with PIXE. APCA identified four components in the fine fraction, and their source profiles are given in Table 4, where only profiles greater than three times the standard deviations are tabulated. Proportional contributions of each component to total mass are also shown in Table 4. The most important source is soil dust, although there is strong influence of industrial emitters of different types. Also, the traffic component (identified by Pb) is mixed with an industrial part (S, V, and Cr), probably due to wind transport, as explained above. In previous studies, soil was also known as the major contributor to fine aerosols in the Southwest of the MCMA (2).

Finally, a comparison with other large urban areas is worthwhile to evaluate the pollution levels measured in Mexico City. For this purpose, the studies carried out by Andrade *et al.* (9) in São Paulo, Brazil, and by Morales *et al.* (10) in Santiago, Chile, are appropriate, as they were performed using similar procedures, both for sampling (using SFU's and 24 h averages during the rainy season), as well as for PIXE analyses. Table 5 displays results for some representative elements.

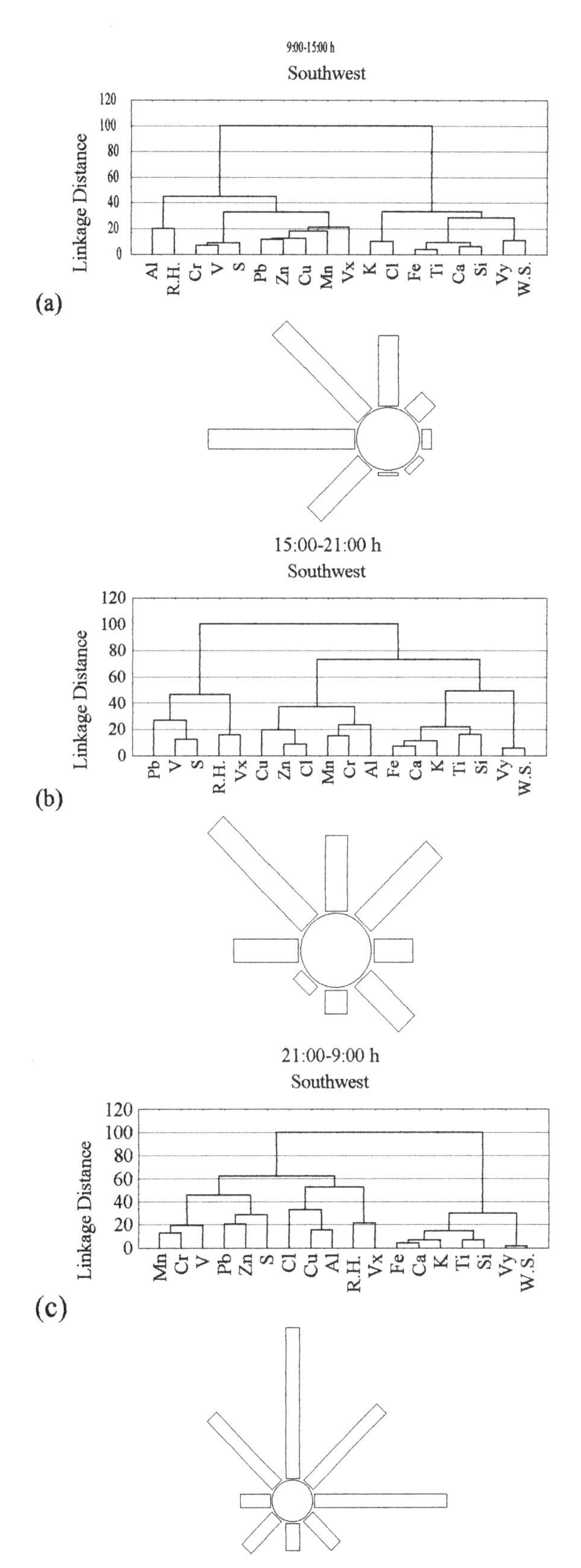

FIGURE 2. Dendograms for aerosols collected in the Summer of 1995. Wind roses for the corresponding sampling periods are also shown. (a) Morning, (b) Afternoon, (c) Night. *R.H.* = relative humidity, *Vx* = x component of wind, *Vy* = y component of wind, *W.S.* = wind speed.

TABLE 4. Source profiles (in ng·m^{-3}) for the fine fraction after a VARIMAX rotation, and contribution to total measured mass after application of APCA. Summer of 1995.

Element	Factor 1 Soil	Factor 2 Traffic-Industry	Factor 3 Cl-Zn	Factor 4 Industry
Al				420 ± 120
Si	610 ± 64			
S		780 ± 220		
Cl			35 ± 3.2	
K	35 ± 10			
Ca	130 ± 8.6			
Ti	17 ± 1.3			
V		14 ± 4		
Cr		8.1 ± 2.9		
Mn	5.8 ± 3.0			
Fe	147 ± 16			
Cu				7.0 ± 0.94
Zn			73 ± 19	
Pb		28 ± 9.2		
Contrib. to total mass	34.0 %	27.3 %	13.3 %	25.4 %

TABLE 5. Comparison of elemental concentrations (μg·m^{-3}) measured in aerosols in three large urban areas.

Element	Mexico City	São Paulo	Santiago
Year	1995	1989	1993
S	0.605	2.90	2.88
V	0.008	0.031	0.009
Fe	0.114	1.96	0.150
Zn	0.057	0.238	0.109
Pb	0.046	0.100	0.210

The figures in table 5 show that the MCMA atmosphere is cleaner than that of São Paulo and Santiago. It must be kept in mind, however, that the 1995 study in Mexico City was conducted in the Southwestern zone, which is expected to present lower elemental contents. The results from Santiago, on the other hand, correspond to a downtown site, with heavy traffic conditions. The effect of using unleaded gasoline since 1990, though, apparently has decreased the Pb contents in the aerosols in Mexico City. Further studies are under progress to firmly establish these assertions.

CONCLUSIONS

The results attained in this study are helpful in several regards. Elemental contents of atmospheric aerosols have not shown significant variations along different seasons, according to the data shown for the studies from 1993 to 1995. In spite of differences presented by some sites for certain elemental concentrations (notably, S, Zn, and Pb in the Western region of the MCMA), aerosol composition is apparently rather uniform. The effect of winds in the composition of fine aerosols in the Southwestern area seems to be transport of soil dust from the Northern zones, although morning regimes enrich the aerosols with traffic-industrial related elements from important avenues and some industry from the West. This statement is supported by the relative contributions of the most important sources, as computed from APCA, associated with soil dust and traffic-industry. On the other hand, the comparison with other large urban areas appears as if the elemental concentrations in the MCMA are lower than in Santiago and São Paulo.

ACKNOWLEDGEMENTS

Work partially supported by DGAPA-UNAM under contract IN-100493. The authors thank the support of Dr. E. Andrade and the technical assistance of K. López, J.C. Pineda, E. Pérez-Zavala, E. Santillana, and M. Galindo. Meteorological data were kindly provided by the Comisión Metropolitana para el Control y Prevención de la Contaminación Ambiental.

REFERENCES

1. Johansson, S.A.E. and Campbell, J.L., *PIXE: a Novel Technique for Elemental Analysis*, Chichester: John Wiley and Sons, 1988.
2. Cahill, T.A., Eldred R.A., Feeney P.J., Beveridge P.J., and Wilkinson L.K. *Trans. AWMA*, 213-222 (1990).
3. Miranda, J., Andrade, E., López-Suárez, A., Ledesma, R., Cahill, T.A., and Wakabayashi, P.A., *Atmospheric Environment* **30**, 3471-3479 (1996).
4. Maenhut, W. and Vandenhaute J., *Bull. Soc. Chim. Belg.* **95**, 407-418 (1986).
5. Aldape, F., Flores-M. J., Díaz R.V., Miranda J., Cahill T.A., and Morales J.R. *Int. J. PIXE* **1**, 373-380 (1991).
6. Aldape, F., Flores-M. J., Díaz R.V., Morales J.R., Cahill T.A., and Saravia L. *Int. J. PIXE* **1**, 355-371 (1991).
7. Thurston, G.D. and Spengler J.D. *Atmospheric Environment* **19**, 9-25 (1985).
8. Maenhut, W. and Cafmeyer J. *J. Trace Microprobe Techn.* **5**, 135-158 (1987).
9. Andrade, F., Orsini C., and Maenhut, W., *Atmospheric Environment* **28**, 2307-2319 (1994).
10. Morales, J.R., Dinator, M.I, and Romo, C.M., "Application of PIXE to Study the Composition of Suspended Particulate Matter in Santiago, Chile," presented at the II International Seminar on Environmental Pollution, Santiago, Chile, November 6-8, 1995.

RBS STUDIES OF SOLID-SOLID INTERFACES: TRANSITION METALS ON ALUMINUM

R. J. Smith, N. R. Shivaparan, V. Shutthanandan, and Adli A. Saleh

Physics Department, Montana State University, Bozeman, MT 59717 USA

The technique of ion backscattering and channeling has been used to study the interface stoichiometry and structure for thin transition metal films deposited on Al single crystal surfaces. Analysis of the backscattering data is based on the concepts of shadowing and blocking, and the well-known theory of Rutherford scattering. However, the analysis provides only the average stoichiometry of the interface, and can be limited by the energy resolution typical of solid state detectors. Problems in the analysis associated with near-surface dechanneling have been identified. In some cases we have used x-ray photoemission spectroscopy and computer simulations to eliminate these problems. In this paper we discuss alloy formation at interfaces of Al with Pd, Ni, Fe, and Co, and the epitaxial growth of fcc Ti and Ag on Al.

INTRODUCTION

The synthesis of layered structures on the nanometer scale has become more of a reality than a promise in recent years. This is due in part to advances in our understanding of the mechanisms active during the growth process, along with improvements in our capabilities using ultrahigh vacuum (UHV) technology. Yet, we are still lacking a comprehensive model which can be used to reliably predict the atomic structure at a metal-metal interface. To make further progress in developing such a model we need to understand how a variety of physical phenomena compete during the growth process to lower the total energy of the system, and ultimately determine the interface structure. We have focused our studies on transition-metal aluminide formation on Al single crystals because of the technological importance of the aluminides. In particular, several of the aluminides have potential applications as metallization layers on III-V semiconductors (1).

In this paper we review the utility of high-energy ion scattering (HEIS) as a tool to characterize the interface structure at solid-solid interfaces. We find that our results fall generally into three categories: 1) Overlayer growth with minimal substrate disruption; 2) Alloy formation at the interface, where HEIS accurately indicates the interface stoichiometry as verified using x-ray photoemission spectroscopy (XPS); and 3) Other situations where the stoichiometry indicated by HEIS does not agree with that determined by XPS. In this last case we have resolved some of the discrepancies between techniques by using computer simulations of the interface formation.

EXPERIMENT

High-energy ion scattering, when used in the channeling mode, provides a powerful tool to probe surface atomic structure and overlayer growth modes (2). In addition, HEIS provides a direct means for accurately measuring the overlayer coverage. In the channeling geometry the ion beam is incident along a low-index crystallographic direction of the substrate, and the energy spectrum of backscattered particles exhibits a surface peak (SP) associated with ions backscattered from the topmost layers of the solid. The SP areas are converted to areal densities of visible target atoms (atoms/cm^2) using the Rutherford scattering cross section, the solid angle subtended by the detector, and the incident ion dose. Increases in the SP area for substrate atoms may be associated with displacements of those atoms off of lattice sites, a characteristic of reconstruction or alloy formation. A decrease in the SP area of substrate atoms is associated with shadowing of those atoms by the adatoms in the overlayer. X-ray photoemission experiments are used to further characterize the chemical state of the film, using the measured binding energies, and the film morphology, using changes in emission intensity as

CP392, *Application of Accelerators in Research and Industry,* edited by J. L. Duggan and I. L. Morgan
AIP Press, New York © 1997

a function of adatom coverage. Photoelectron diffraction experiments are also used to identify localized, ordered overlayer growth.

All of the experiments were done in a UHV chamber at a base pressure of 1×10^{-10} Torr, attached to a 2 MV Van de Graaff accelerator via a differentially pumped beam line (3). The standard dose of He^+ ions for one spectrum was 1.6×10^{15} ions/cm^2. Energy analysis of the backscattered He^+ ions for HEIS was performed using a bakeable silicon detector at a scattering angle of 105°.

The Al crystals were mechanically polished and then chemically etched to remove mechanical damage. In the vacuum chamber the crystals were cleaned by Ar ion sputtering, and annealed at 500°C, until the photopeak associated with Al-oxide was completely removed from the XPS spectrum. Metals were evaporated onto the Al substrates at room temperature from resistively heated wires at typical rates of 0.5 to 0.9 monolayers per minute.

RESULTS

Overlayer Growth

Fig. 1 shows the HEIS channeling spectra of backscattered ions in the regions of the Al and Ti surface peaks for clean Al(001) and for 1.9 monolayers (ML) of Ti deposited on Al(001). One monolayer here is equal to the atomic density of the Al(001) plane, 1.22×10^{15} atoms/cm^2. From the figure it is clear that the Al SP area has decreased after Ti deposition on the surface, indicating that fewer Al atoms are visible to the incident ion beam in the presence of Ti atoms at the surface. This suggests that Al atoms are being shadowed by the Ti atoms which, because of the small shadow cone radius of approximate 0.1 Å, requires that the Ti atoms are sitting directly above Al atoms, forming an epitaxial overlayer without displacing the substrate atoms.

In Fig. 2 we plot the Al surface peak area as a function of Ti coverage on Al(110), determined by the Ti surface peak area. The yield of Al atoms per deposited Ti atom initially *decreases* up to a coverage of about 5 ML, and then begins to increase back towards its initial value. We have argued in a previous paper that these results are explained quite naturally by the formation of a Ti overlayer in an fcc structure (4). This is remarkable since Ti metal assumes a hexagonal structure at room temperature, and transforms to a bcc structure at elevated temperatures, but no equilibrium *fcc* structure is observed. The solid squares in Fig. 2 show the results of computer simulations for the scattering yield from a Ti layer which grows layer-by-layer. The good agreement with experiment supports our interpretation of an epitaxial overlayer. Similar results are observed for Ti on the Al(001) surface, although Al-Ti site exchange may occur for the first monolayer only on that surface. At a critical thickness of about 5 ML the film undergoes a structural transformation which allows Al atoms to become more visible to the incident ion beam. However, never during the experiment does the scattering yield from Al exceed the value measured for the clean surface.

This application of shadowing to locate adatoms is perhaps the trademark of HEIS as applied to solid-solid interfaces, and many similar situations have been reported. We have observed epitaxial growth and shadowing for Ag films grown on both Al(110) and Al(001). The overlayer order in that case exists for film thicknesses exceeding 30 ML, at

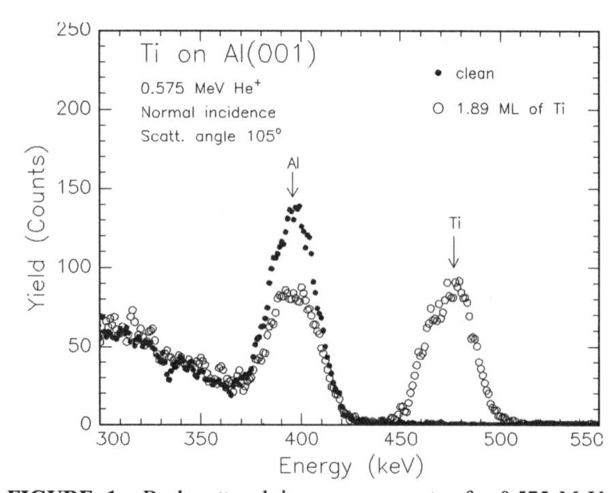

FIGURE 1. Backscattered ion energy spectra for 0.575 MeV incident He+ ions on Al(001), clean and with 1.89 ML of Ti.

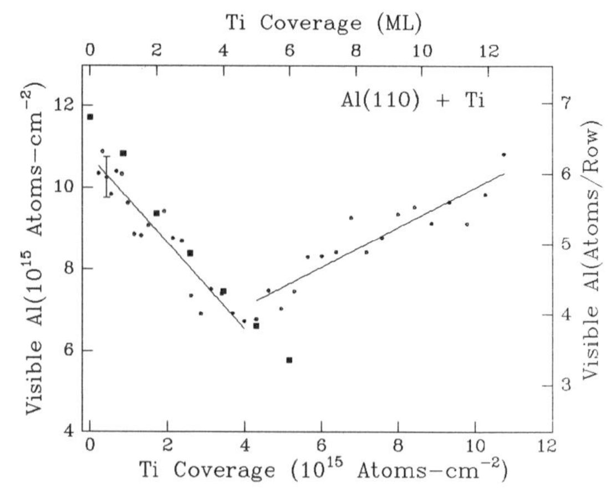

FIGURE 2. Visible Al atoms, at 0.96 MeV incident ion energy, as a function of Ti coverage on Al(110). The solid lines are linear fits to the data in the two regions indicated.

which coverage the Al surface peak has vanished completely (5). However, ordered overlayer growth need not result in shadowing. Ni films grown on W(110) surfaces assume an incommensurate structure. In this case, the SP area of the W substrate remains constant as the Ni coverage increases (6). The constant peak area indicates that W atoms are not being disturbed by the overlayer, i.e. the surface is rigid with no lateral reconstruction.

Alloy Formation

In Fig. 3 we show results for Pd deposition on Al(110) (7). The results for Al(001) are essentially the same (8). There appear to be two distinct regions of growth. First the Al yield *increases* with a slope of 1.2 Al atoms per deposited Pd atom up to about 9 ML of Pd coverage. 1 ML on Al(110) is 0.866×10^{15} at/cm^2. Beyond this coverage there is an apparent saturation state where the Al yield remains unchanged. The increase in Al yield means that more Al atoms become visible to the incident ion beam as the Pd coverage increases. That is, Al atoms are moving off of substrate lattice sites at the interface. This could be an indication of alloy formation if the atoms at the interface intermix. The stoichiometry of the mixed interface is obtained from the slope in Fig. 3, approximately 1:1 or AlPd. Of course this is only an average composition, and we have no indication that the structure is that of the AlPd compound. However, the formation energy for AlPd is quite large, which favors compound formation.

To supplement these measurements we have used XPS to look for chemical shifts which would

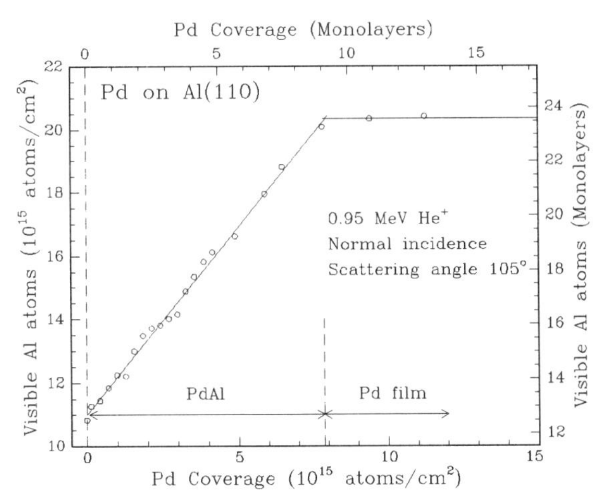

FIGURE 3. Visible Al atoms plotted as a function of Pd coverage on Al(110). The solid lines are linear fits to the data in the two regions indicated.

indicate compound formation. Measurements of the 3d core-level emission show a split pair of peaks at the binding energy of Pd metal, and a chemically shifter pair of peaks at 1.8 eV larger binding energy, attributed to AlPd (7,8). Thus, in this case there is agreement for the stoichiometry of the interface alloy, as determined using the slope of the HEIS yield, and using the binding energy as measured by XPS. In addition, XPS confirms that a metallic Pd film forms on the interface after 9 ML of coverage. The HEIS yield remains flat in this region because interdiffusion has stopped at room temperature, and no additional Al atoms are being displaced from substrate lattice sites.

Binding energy measurements are not useful for some interfaces, for example when the chemical shifts are too small to be measured, or are unknown. This is the case for Fe on Al surfaces (9). Three distinct regions of growth are seen for Fe on Al(110), with slopes for the Al yield of 3:1, 1:1 and zero, respectively. Although no chemical shift data is available to assist the analysis in this case, we have used the Fe 2p emission intensities to model the film morphology. The Fe coverages as determined using HEIS for each growth regime serve as boundary conditions in the fitting process. With this combination of XPS and HEIS measurements we have shown that, for Fe on Al surfaces, the stoichiometries determined for the interface are consistent with measurements from the two techniques. However, this success should not be taken as a guarantee that HEIS can be used to determine interface stoichiometry in all cases. There are pitfalls, as we discuss in the next section.

Special Cases: Near-surface Dechanneling

The growth of thin Ni and Co films on Al surfaces appears to be a special case where analyses by HEIS and XPS techniques yield different results for the stoichiometry of the mixed interface. In Fig. 4 we show HEIS results for Ni on Al(110). As in the case of Fe on Al(110) there are three distinct growth regimes. We have argued that these correspond to the growth of NiAl, Ni_3Al and Ni metal respectively (10). However, the slope in the first growth regime, up to 2 ML of Ni coverage, is approximately 2:1, suggesting a stoichiometry of Al_2Ni, a compound that is not found in the Al-Ni phase diagram. Although the chemical shifts measured in XPS are small, a careful lineshape analysis, including satellite photoemission structures, suggests that the initial interface is more

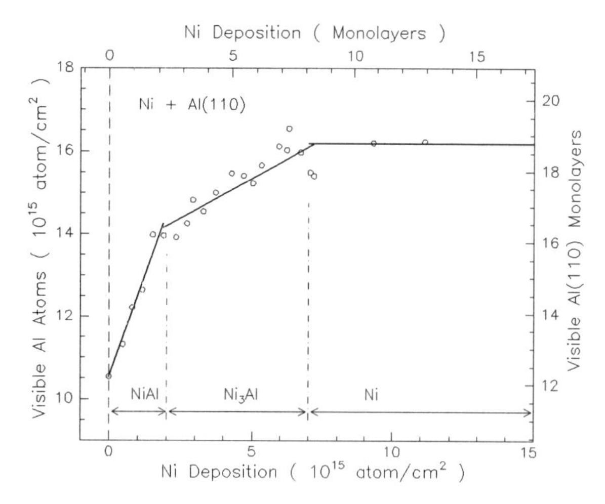

FIGURE 4. Visible Al atoms plotted as a function of Ni coverage on Al(110). The solid lines are the linear fits to the data in the three regions indicated.

characteristic of AlNi, incidentally the compound with the largest heat of formation. The apparent discrepancy was resolved in this case by using computer simulations of interface formation (11). Embedded-atom-method potentials were used in a Monte Carlo simulation to generate snapshots of the evolving interface which were in turn probed with ion scattering computer simulations. The results indicated that the initial interface formed with a 1:1 stoichiometry, i.e. AlNi, as suggested by XPS. However, there was significant *dechanneling* of the ion beam in the region of 10 to 20 layers below the interface. The depth resolution of our solid state detector is not sufficient to separate dechanneling at this depth from ion scattering at the interface. Thus, the slope of the HEIS yield is artificially high, associated with the near-surface dechanneling. We have not yet determined the extent to which such dechanneling might be present for other transition metals on Al surfaces. However, the existence of this phenomenon for Ni-Al interfaces raises a flag of caution for such interpretations of HEIS results.

We have recently measured the interface formation for Co films deposited on Al surfaces (5). Here again we find that the slopes obtained from the HEIS yields measured as a function of Co coverage suggest stoichiometries which are not confirmed by our XPS measurements. Although we do not have useful chemical shifts for the Al-Co interface, model fits to the intensity variations of the core-level emission suggest that the interface consists initially of AlCo, followed by Co metal growth. The HEIS yields suggest a more Al-rich interface compound, so we believe that surface dechanneling may again be responsible for the larger slope.

Finally, we have observed a peculiar case for the growth of Ni on Al(001) where the HEIS yield initially rises and then returns to its starting value for Ni coverages between 0 and 2 ML. We believe that this unusual behavior at low coverages may be associated with dechanneling, since photoelectron diffraction measurements indicate that Ni is forming a bcc overlayer on the Al substrate (5).

In conclusion, we have shown that there are several solid-solid interfaces where HEIS has been successfully used to determine interface structure and composition. However, as with most techniques, one must always have in mind those special situations where the blind interpretation of slopes as a measurement of interface stoichiometry may lead to mistakes. The situation may be remedied by a multitechnique investigation of the interface.

We are pleased to acknowledge the technical support of Erik Andersen and Norm Williams. This work was supported by NSF Grant No. DMR 9409205.

REFERENCES

1. Sands, T., Appl. Phys. Lett. **52**, 197-199 (1988).
2. van der Veen, J. F., Surface Science Reports **5**, 199-288 (1985).
3. Smith, R. J., Whang, C. N., Xu, M., Worthington, M., Hennesy, C., Kim, M. W., Holland, N., Rev. Sci. Instrum. **58**, 2284-2287 (1987).
4. Saleh, A. A., Shutthanandan, V., and Smith, R. J., Phys. Rev. **B49**, 4908-4914 (1994).
5. Shivaparan, N. R., and Smith, R. J., unpublished.
6. Xu, M., and Smith, R. J., J. Vac. Sci. Technol. **A9**, 1828-1832 (1991).
7. Shivaparan, N. R., Shutthanandan, V., Krasemann, V., and Smith, R. J., to appear in Surface Science (1996).
8. Shutthanandan, V., Saleh, A. A., Shivaparan, N. R., and Smith, R. J., Surface Science **350**, 11-20 (1996).
9. Shivaparan, N. R., Krasemann, V., Shutthanandan, V., and Smith, R. J., Surface Science, **365** 78-88 (1996).
10. Shutthanandan, V., Saleh, A. A., and Smith, R.J., J. Vac. Sci. Technol. **A11**, 1780-1785 (1993).
11. Shutthanandan, V., Saleh, A. A., Denier van der Gon, A. W., and Smith, R. J., Phys. Rev. **B48**, 18292-18295 (1993).

IN-SITU OBSERVATION OF REACTION BETWEEN METAL AND Si SURFACE BY LOW ENERGY RBS/CHANNELING

Masataka Hasegawa and Naoto Kobayashi

Electrotechnical Laboratory, 1-1-4 Umezono, Tsukuba, Ibaraki, 305 Japan

We have developed a low energy Rutherford backscattering spectrometry(RBS)/ion channeling measurement system for the analysis of thin films and solid surfaces with the use of several tens keV hydrogen ions, and of a time-of-flight spectrometer which was originally developed by Mendenhall and Weller. The depth resolution of our system is better than that of conventional RBS system with MeV helium ions and silicon surface barrier detectors. This RBS/ion channeling system is small in size compared to the conventional RBS/ion channeling measurement system with the use of MeV He ions, because of the small ion accelerator for several tens keV ions. The analysis of crystalline thin films which utilizes ion channeling effect can be performed with this low energy RBS/ion channeling measurement system. The *in situ* observation of the thermal reaction between iron and silicon substrate with the use of this measurement system is demonstrated. The deposited Fe (3.3ML) on Si(001) clean surface diffused into the substrate by 380℃ annealing, while on the hydrogen-terminated (dihydride) Si(001) the 480℃ annealing did not lead to the diffusion. Present results indicates that the hydrogen termination of Si(001) surface prevents the deposited Fe from diffusing into the substrate up to relatively high temperature compared to the clean surface.

INTRODUCTION

Rutherford backscattering spectrometry (RBS) and ion channeling have been a well-known analysis technique for composition, impurity distribution, and structure of thin films and solid surfaces for many years(1). Because of the typical energy resolution about 10keV of the silicon surface barrier (SSB) detector which is commonly used for the detection of scattered ions, helium or other heavier ions are commonly employed as probe ions with typical incident energies ranging from several hundreds keV to several MeV. Because of the energy resolution low energy (several tens ~ several hundreds keV) ions cannot be energy-analyzed by the SSB detector, and are not usually used for the analysis of thin films by RBS/ion channeling measurement in spite of the advantage of high sensitivity to the surface atoms due to the large scattering cross section of low energy ions. Recently Mendenhall and Weller have developed a new time-of-flight (TOF) spectrometer, and have demonstrated its performance for 100 - 500 keV helium and other heavy ions(2,3). The TOF spectrometer is able to provide better depth resolution than that of conventional MeV RBS with SSB detector.

We have developed a measurement system for *in situ* analysis of thin films and solid surfaces which are prepared in molecular beam epitaxy (MBE) apparatus by RBS/ion channeling with the use of several tens keV hydrogen ions and this TOF spectrometer. The advantages of the usage of hydrogen ions with this energy range are (A) much larger sensitivity than conventional RBS with MeV helium ions because of the larger scattering cross section, (B) smaller and simpler ion accelerator can be available compared to the usage of MeV ions, and (C) possible analysis of single crystalline thin films with the use of ion channeling effect. In this article we will report the performance of our low energy RBS/ion channeling measurement system with the use of TOF spectrometer and several tens keV hydrogen ions to the analysis of solid materials.

LOW ENERGY RBS/CHANNELING MEASUREMENT SYSTEM

CP392, *Application of Accelerators in Research and Industry,* edited by J. L. Duggan and I. L. Morgan
AIP Press, New York © 1997

In Fig.1 we present a schematic of our low energy RBS/ion channeling measurement system with a TOF spectrometer. It consists of a 30kV small ion accelerator with a microwave ion source (not shown), an ion beam transport, and the MBE chamber. Two sets of apertures which are separated by a distance 1200mm are equipped in the beam transport duct to collimate the ion beam to a divergence angle less than 1mrad for the ion channeling experiment. The base pressure in the MBE chamber is 2×10^{-10}Torr. The pressure in the chamber during the ion scattering measurement is less than 2×10^{-9}Torr. The sample is mounted on a 6-axis precision goniometer in the chamber. The dimension of whole facility is less than 5m which is much smaller than conventional RBS/channeling facility with MeV helium ions because of the small accelerator for several tens keV ions.

We have utilized the TOF ion detector which was demonstrated by Mendenhall and Weller for the energy analysis of several tens keV hydrogen ions(2,3). Two microchannel plate (MCP) detectors were used for the flight time measurement. An annular type MCP, which was called the start MCP, was installed 6mm downstream from a carbon foil coaxially on the flight path of the ion. Another MCP was at 455mm downstream from the carbon foil (called stop MCP). The carbon foil has a thickness of 1 μ g/cm^2, and is mounted on a tungsten mesh whose transmission coefficient is 94%. The ion scattered from the sample penetrates the carbon foil and emits secondary electrons which are detected by the start MCP and produce the start signal of the TOF measurement. The ion goes through the center hole of the start MCP, and generates a stop signal at the stop MCP after the flight through the length of the detector. The advantage of this TOF detection method is that we do not need the pulsing of the primary ion beam to trigger the flight time measurement. In the original scheme which was demonstrated by Mendenhall and Weller the start MCP was at the off-axis of the flight pass of the ion. We have modified the arrangement of the start MCP, the stop MCP and the carbon foil to the coaxial arrangement by adopting an annular type MCP for the generation of the start signal, as shown in Fig.1. We have an aperture of 2mm diameter just before the carbon foil, which determines the detector solid angle. In this arrangement the detector solid angle can be altered readily by changing the distance between the sample and the detector.

In Fig.2 we show the TOF spectra of 25 keV

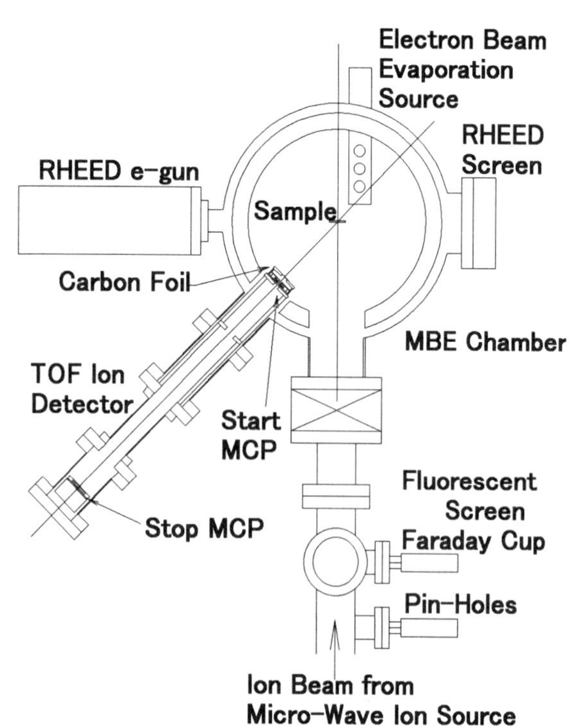

FIGURE 1. Schematic of low energy RBS/ion channeling measurement system with a time-of-flight spectrometer. MCP denotes microchannel plate detector.

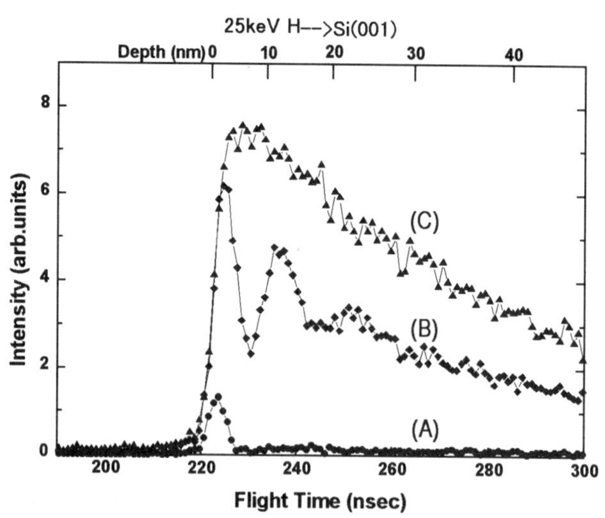

FIGURE 2. Time-of-flight spectra of hydrogen ions scattered from silicon crystal. The energy of the primary ions was 25 keV, and the detection angle was 135 degrees with respect to the direction of the primary beam. (A) The direction of the incident ion beam was along <100> axis of silicon crystal and the direction of the spectrometer was along the <110> axis of silicon substrate, so-called "double alignment geometry" (axial channeling). (B) The direction of the incident ion beam was along (110) plane (planar channeling). (C) The direction of the incident ion beam was along random orientation. The oscillatory feature which is characteristic to the planar channeling was clearly observed in curve (B).

hydrogen ions scattered from (001) substrate of silicon crystal. The energy resolution of this TOF detector for 25 keV hydrogen ions was 4.1% which corresponds to the depth resolution of 4.8nm for silicon. In the case of conventional RBS measurement with MeV helium ions and the SSB detector the depth resolution is more than 10nm for silicon. Thus the depth resolution of our system with 25 keV hydrogen ions and the TOF detector is superior to that of the conventional RBS.

The spectrum (A) in Fig.2 was observed when the direction of the incident beam was along the <100> axis perpendicular to the surface, and when the direction of the TOF spectrometer was also along the <110> axis of silicon substrate, so-called "double alignment geometry"(1). The intensity of the primary beam was about 25 nA, and the measurement duration was 15 min. The intensity of the scattered ions from the sample is reduced to the order of a few percent of that of random direction, which shows the ion channeling occurs. The large scattering cross section of low energy ions is preferable for RBS/ion channeling measurement with the double alignment condition.

Figure 2 also shows the spectrum when the direction of the incident beam is along the (110) plane of the crystal, which is called "planar channeling". The spectrum clearly shows the characteristic oscillatory shape of planar channeling(4), as shown in Fig.2 (curve B). As found from this figure our low energy RBS/ion channeling measurement system with the TOF spectrometer can be available for the analysis of thin films and solid surfaces by ion channeling effect.

IN SITU OBSERVATION OF REACTION BETWEEN Fe AND Si SUBSTRATE

We performed the *in situ* observation of the thermal reaction between Fe and Si substrate with the use of the low energy RBS/ion channeling measurement system by 25 keV hydrogen ions in the MBE chamber. Very thin Fe layer was deposited on the hydrogen terminated Si(001) substrate by an electron beam evaporator. The Si(001) substrate was hydrogenated by HF treatment(5). Subsequent thermal reaction between deposited Fe and Si substrate was caused by the direct current heating in the chamber. Figure 3 shows the spectra of the scattered ions observed when the direction of the incident ion beam was along the <100> axis of the silicon substrate, and the direction of the spectrometer was along the <110> axis of silicon substrate (double alignment). The scattered angle of the ions was 135 degrees with respect to the direction of the primary ion beam. The random spectra were also shown in the figure. The amount of the deposited Fe was measured to be 3.3 monolayers from the spectrum observed for the as-deposited film. The corrected scattering cross sections for low energy collisions were used for the estimation of the deposited amount of Fe(6). The spectra observed after the annealing of 480℃ shown in Fig.3 is almost the same as those for the as-deposited film, which shows the reaction between Fe and Si substrate has not occurred by the annealing at this temperature. After the 590℃ annealing, however, the peak of the Fe signal became smaller and wider than those for the as-deposited film, which indicates the diffusion of Fe atoms into the Si substrate. In the case of the Fe deposition on Si(001) clean surface shown in Fig.4 we observed the reaction between the deposited Fe layer and the substrate

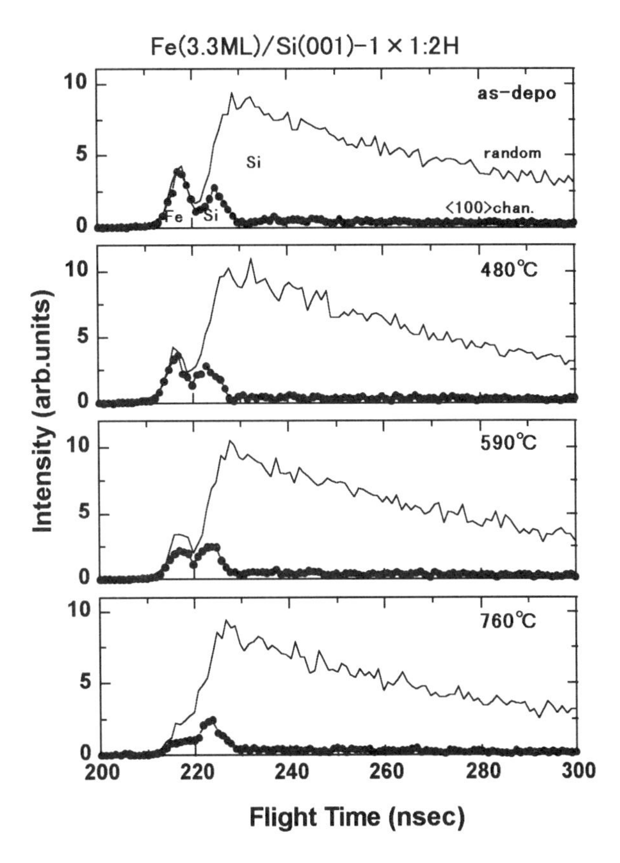

FIGURE 3. Time-of-flight spectra of hydrogen ions scattered from 3.3ML Fe deposited on Si(001)-dihydride surface before and after the annealings at 480, 590, and 700℃. The incident energy of the ion was 25keV, and the detection angle was 135 degrees with respect to the direction of the incident ion beam.

by the annealing at 380℃. Present results shows that the hydrogen termination of Si(001) surface prevents the deposited Fe from diffusing into the substrate up to relatively high temperature compared to the clean surface.

SUMMARY AND CONCLUSIONS

We have developed a low energy RBS/ion channeling measurement system for the analysis of thin films and solid surfaces with the use of several tens keV hydrogen ions and TOF spectrometer. The TOF spectrometer consists of two micro-channel-plate detectors and a thin self-supporting carbon foil. The pulsing of the primary ion beam is not necessary for this type of TOF measurement. The dimension of whole system is very compact compared to the conventional RBS/channeling measurement system with the use of MeV He ions. The energy resolution for 25keV H^+ was 4.1%, which corresponds to the depth resolution of 4.8 nm for silicon. The thermal reaction between iron and Si(001)-dihydride surface was observed with the use of this measurement system. The hydrogen termination of Si(001) surface prevents the deposited Fe from diffusing into the substrate up to relatively high temperature compared to the clean surface.

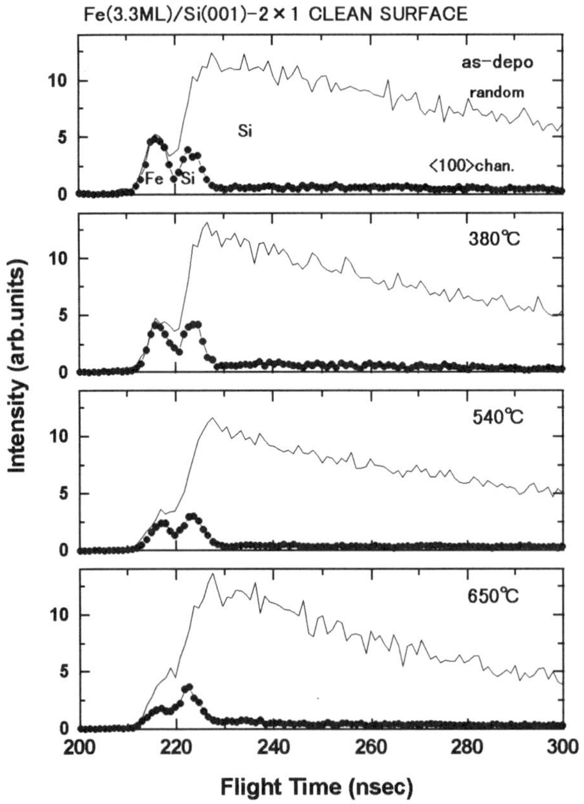

FIGURE 4. Time-of-flight spectra of hydrogen ions scattered from 3.3ML Fe deposited on Si(001)-clean surface before and after the annealings at 380, 540, and 650℃. The incident energy of the ion was 25keV, and the detection angle was 135 degrees with respect to the direction of the incident ion beam.

REFERENCES

1. For example see; Feldman, L.C., Mayer, J.W., and Picraux, S.T., *Materials Analysis by Ion Channeling*, New York: Academic Press, 1982.
   Tesmer, J.R. and Nastasi, M., *Handbook of Modern Ion Beam Materials Analysis*, Pittsburgh: Materials Research Society, 1995.
2. Mendenhall, M.H. and Weller, R.A., Nucl.Instr. and Meth.**B40/41,** 1239-1243(1989).
3. Mendenhall, M.H. and Weller, R.A., Nucl.Instr. and Meth.**B47,** 193-201(1990).
4. Gemmell, D.S., Rev.Mod.Phys.**46,** 129-227(1974).
5. Hasegawa,M.,Kobayashi,N., and Hayashi,N., Surf.Sci.**357/358**, 931-936(1996).
6. Hautala, M. and Luomajarvi, M., Rad.Eff.**45,** 159(1980).

THE USE OF MAXIMUM ENTROPY AND BAYESIAN STATISTICS IN ION-BEAM APPLICATIONS

V.M. Prozesky♦, J. Padayachee, R. Fischer*, W. von der Linden*, V. Dose* and C.G. Ryan**

Van de Graaff Group, National Accelerator Centre, P.O. Box 72, 7131 Faure, South Africa
**Surface Physics Division, Max Planck Institute for Plasma Physics, D-85748 Garching, Germany*
***CSIRO Division of Exploration and Mining, P.O. Box 136, North Ryde, NSW 2113, Australia*

An introduction to the use of maximum entropy (ME) and Bayesian methods in ion-beam applications is presented. The formalism is applied to the deconvolution of detector blurring functions in RBS and PIXE measurements. The resulting improvement in detector resolution can be as much as a factor of 5 when the detector functions are well known. The possibilities of further applications in ion-beam related work are also discussed

INTRODUCTION

The use of ion-beams for the characterization and modification of surface layers is well known and the associated techniques are often used in the studies of surfaces of a large variety of materials. The results obtained by these methods are invariably in a convoluted and imperfect form, caused by finite and complex detector response functions, as well as counting statistics. The deconvolution of such spectra, to be able to extract more information, can go a long way in improving techniques and models.

The Bayesian formalism, together with the principle of maximum entropy (ME), is a powerful technique that enables the solution of inverse problems. In the case of ill-defined problems, there are an infinte number of inverse answers, or reconstructions, that can satisfy any misfit statistic, such as a specific χ^2. In these cases the principle of ME allows the assignment of the most objective probability distributions on the basis of the imperfect knowledge available. This work shows the potential in applying these techniques in ion-beam experiments.

In techniques where counting is done by an energy dispersive detector, the resulting spectrum is a convolution of the real events with the detector response function. In general this is a convolution of the form:

$$g(y) = \int_{-\infty}^{\infty} f(x)K(x,y)dx \qquad (1)$$

where $K(x,y)$ is the detector response function. The experimenter can only measure the spectrum $g(y)$. The interest in this work is to obtain the spectrum $f(x)$, if we have adequate knowledge of $K(x,y)$. If there were no counting errors, there would only be one correct solution, but the inevitable uncertainties associated with counting, allows an infinite of answers that, when convoluted, yield similar spectra $g(y)$. The Bayesian formalism offers a solution that yields the most probable solution for $f(x)$.

THE BAYESIAN FORMALISM AND ME

Bayes theorem [1] is a formula describing the calculation of inverse quantities of a probability distribution, and was derived more than two centuries ago. If we consider the data distributions (not necessarily probability distributions) f as the physical distribution without the detector and counting influences (the reconstruction), and D as the measured spectrum respectively, and we have other relevant information X such as the detector response function for the whole range of interest, then Bayes theorem expresses the probabilities of the respective data distributions as follows:

$$p(f|DX) = \frac{p(f|X)p(D|fX)}{p(D|X)} \qquad (2)$$

In eq. (2) p is probability, the quantity of which the probability is described is before the |, and the given quantities are after the |. This means that the term $p(f|X)$ is the probability that a specific distribution f is the correct reconstruction, given that the information X is true. The probability of a specific reconstruction is therefore dependent on the probability of that distribution before the data D is taken, as well as the probability of obtaining that specific data, given that the reconstruction is correct. The

♦ Corresponding author: National Accelerator Centre, P.O. Box 72, 7131 Faure, South Africa. email: vic@nac.ac.za

CP392, *Application of Accelerators in Research and Industry*, edited by J. L. Duggan and I. L. Morgan
AIP Press, New York © 1997

term in the denominator can mostly be considered as a normalisation term, since it does not depend on the specific reconstruction, and will not be discussed in the following.

The term $p(D|fX)$ is, in counting experiments, mostly taken as a misfit term such as a χ^2 test. This term describes the closeness of the fit of the data with a reconvolution of the reconstruction f with the detector response function. In ill-defined inverse problems the minimization of χ^2 normally yields very unstable and unsatisfactory solutions. In the use of Bayes theorem, this term is offset by the 'prior' probability, as given by $p(f|X)$. This term incorporates all the prior knowledge of the reconstruction, e.g. positivity and smoothness, without any knowledge of the data D.

This probability normally has to be assigned on the basis of minimum information content, constrained by the prior information, and that is the principle of ME. The specific application of Bayes theorem to such problems [2], including the concept of adaptive kernels [3] is discussed in detail elsewhere, and will not be further discussed.

Similar concepts have been used in reconstruction of compositional depth profiles from electron probe micro-analysis data using different projectile energies [4] and the determination of depth profiles from angle-dependent X-ray photo-electron spectroscopy [5].

RUTHERFORD BACKSCATTERING (RBS) AND PARTICLE INDUCED X-RAY EMISSION (PIXE)

The two techniques of RBS and PIXE are probably the most widely used in the ion-beam analysis community, and both these techniques offer multi-elemental information, with direct depth information obtainable from RBS. The normal way of analysis of spectra obtained in these techniques is to guess the composition of the target, and to simulate the expected spectrum on the basis of that composition. In RBS the use of the RUMP [6] program is frequent, and in PIXE examples of software packages are GeoPIXE [7] and GUPIX [8]. These simulations include the convolution by the detector response function. A χ^2 type test is then used to obtain the correct composition, and in the case of RBS direct depth information. This is a satisfactory process, as smoothness in the depth distribution is guaranteed by the input from the user. Such techniques, however, can still only yield convoluted information, with information on a scale smaller than the characteristic width of the detector response function smeared out.

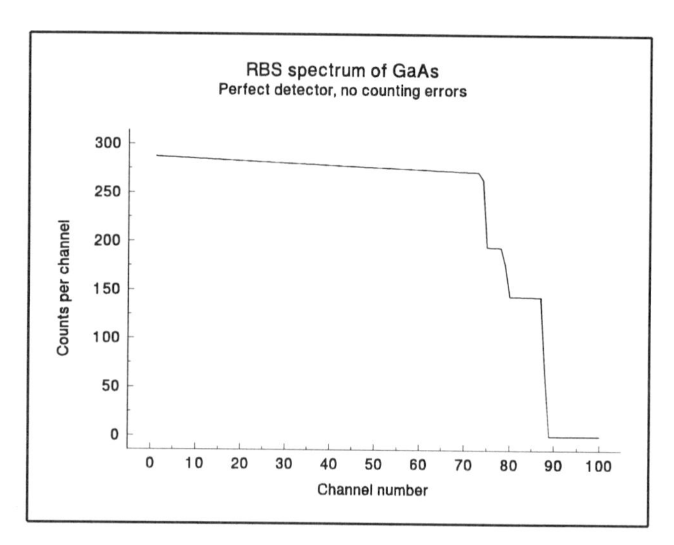

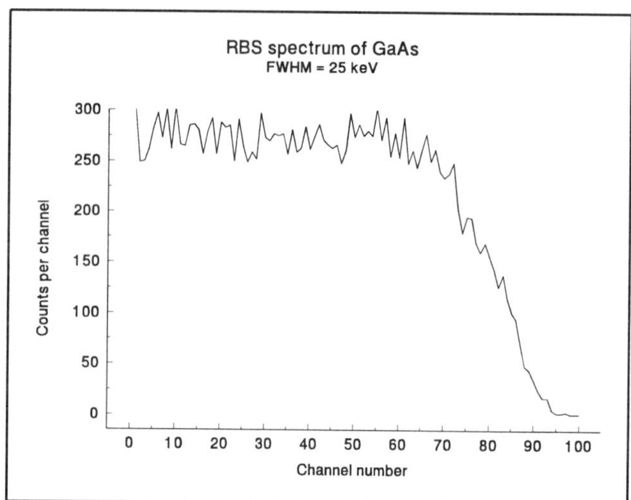

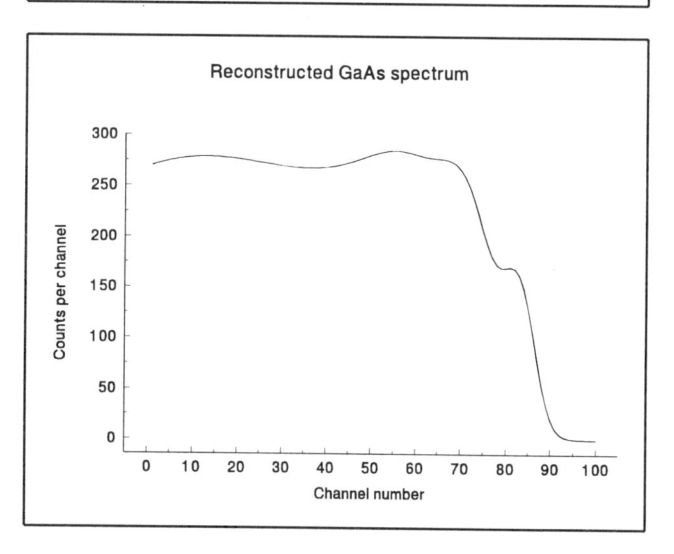

FIGURE 1. (a) A theoretical RBS spectrum of GaAs, with a perfect detector, and no counting statistical errors. (b) A corresponding experimental spectrum of GaAs, using a ^{4}He beam with an energy of 2 MeV. (c) The reconstructed spectrum that can be compared with (a) The improvement in resolution is good enough to separate Ga and As, but not to separate the isotopes of Ga, and is around a factor of 5.

In this work we present an attempt to determine the most probable reconstruction of the spectrum with the detector response function deconvoluted, based on the self-consistent Bayesian formalism. In the following analysis, the only prior knowledge assumed was that the detector response function was known, and that the resulting reconstruction was positive.

RBS and Gallium Arsenide (GaAs)

A measurement was made of thick GaAs, with an annular silicon surface barrier detector. The detector has an energy resolution of 25 keV at 5.486 MeV, and is the standard detector in a microprobe environment, where detection efficiency is important. This energy resolution has the result that the surface signals from Ga and As are convoluted in such a way that forward simulation yields a large error in determining the relative concentrations of Ga and As. In this specific case, Ga has 2 stable isotopes, namely mass 69 (60.1%) and 71(39.9%). The mass of the single stable isotope of As is 75. A theoretical spectrum of GaAs, with no detector convolution or counting errors is shown in Fig. 1(a). The steps resulting from the different nuclide masses are evident, and this will be the spectrum that is sought by deconvoluting a collected spectrum from a GaAs sample.

The detector response function was measured by collecting an RBS spectrum from a thin (1nm) film of Ir on Si, using the same beam type, energy and experimental setup as for the GaAs measurement. The thickness of the layer was such that the peak obtained from the backscattering of ^{4}He from Ir corresponded to the detector response function, with no measurable thickness contribution. The detector response function was measured for a range of incident ^{4}He energies, but was found to be constant within the precision measured, and therefore the function was considered to be independent of the measured energy of backscattered particles.

The spectrum collected in a typical setup is shown in Fig. 1(b). ^{4}He ions at 2 MeV were used as primary projectiles, and backscattered particles were detected at 176°. The reconstruction is shown in Fig. 1(c), and clearly shows the step obtained from the mass difference between the Ga and As isotopes. The improvement of the resolution is not good enough separate the two isotopes of Ga. This example clearly demonstrates both the power and limitations of the reconstruction process, and led to an enhancement of a factor 5 with respect to the experimental spectrum. Even in this case, with a detector resolution of 25 keV, this meant an effective resolution of 5 keV, better than is experimentally obtainable with a surface barrier detector. Although there is still some variation on the plateau of the spectrum, the statistical error is significantly improved, allowing improved fitting of the deconvoluted spectrum

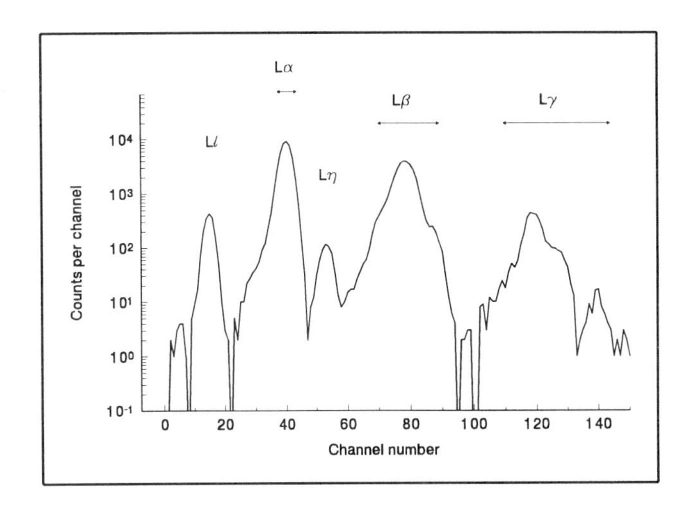

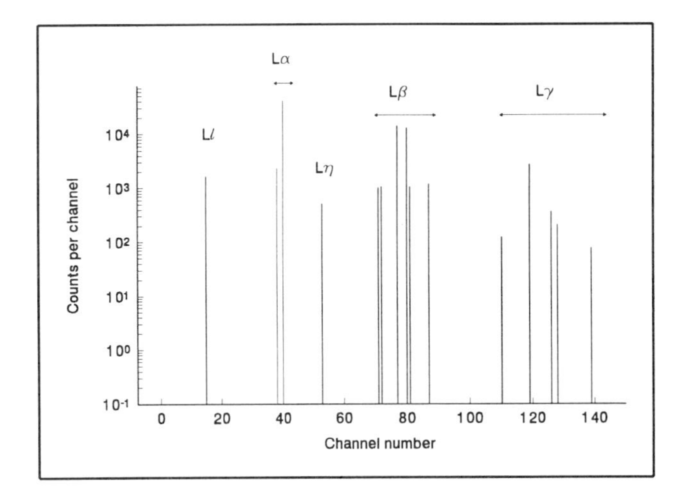

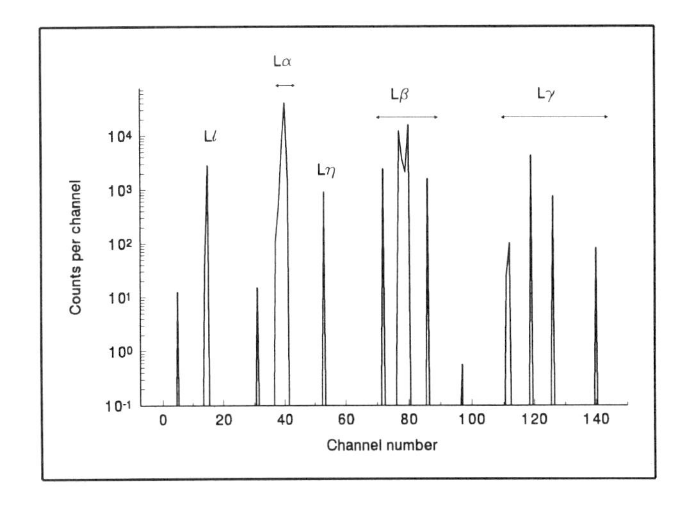

FIGURE 2. (a) Simulated PIXE data for Pt L X-rays, with gaussian noise added. (b) The original X-ray lines and intensities that were used for the simulation. (c) The reconstructed spectrum that can be compared with (b). In the above a typical detector response function was used, with an energy resolution of 200 eV at 5.898 keV, and the improvement in resolution was 4 times better than the measured spectrum.

PIXE and interfering peaks

For the study of the deconvolution of interfering peaks in PIXE, a spectrum of the Pt L lines from a thick target was simulated by convoluting the various L contributions with a typical detector response function, with the normal variation as a function of energy. Gaussian noise was then added, in accordance with counting statistics errors. This spectrum is shown in Fig. 2(a), and shows the serious overlap that occurs between various peaks. Also evident from this spectrum is the asymmetry of the peaks, with long low energy tails that worsen the overlap problem. In Fig. 2(b) the intensities and energies (channel numbers) of the peaks that were used as input are shown, with the reconstructed spectrum in Fig. 2(c). It is clear that there is significant improvement in the resolution (around 4x), and that most of the peaks are separated. Another advantage obtained is that the peaks are symmetrised, as the deconvolution process takes account of the complex shapes of the response function.

CONCLUSIONS

The Bayesian formalism, in conjunction with the use of maximum entropy, has proved to be a powerful method of obtaining information hidden through detector convolution in counting experiments in ion-beam techniques. Improvement factors of up to 5 can be obtained if the detector response function is known accurately. It is easy to show that the description of the detector response function used should contain 98% of the information of the real detector response function, to gain a factor of 5 improvement in distributions where the second moment exist [9].

Other applications in ion-beam techniques are the deconvolution of the beam spot size in nuclear microprobe mapping experiments, getting depth profiles from PIXE measurements using two detectors at different angles and ion beam tomography using a nuclear microprobe.

REFERENCES

[1] Skilling, J., Maximum Entropy and Bayesian Methods, Kluwer Academic publishers, 1989, pp. 45-52.

[2] Buck, B. and Macauly, V.A. (eds) Maximum Entropy in Action, Oxford Science Publications, Oxford University Press, Oxford, 1991.

[3] Fischer. R., Von der Linden, W., and Dose, V., Submitted at the 1996 International Workshop on maximum Entropy and Bayesian methods, Berg-en-Dal, National Kruger Park,13-18 August 1996.

[4] Smith, G.C., Park, D., and Cochonneau, O., Journal of Microscopy, Vol. 178, April 1995, pp. 48-55.

[5] Livesey, A.K., Smith, G.C., Journal of Electron Spectroscopy and Related Phenomena, 67 (1994) 439-461.

[6] Doolittle, L.R., Nucl. Instr. and Meth. B 15 (1986) 227.

[7] Ryan, C.G., Cousins, D.R., Sie, S.H., Griffin, W.L., Suter, G.F. and Clayton, E., Nucl. Instr. and Meth. B 47 (1990) 55.

[8] Maxwell, J.A., Campbell, J.L. and Teesdale, W.J., Nucl. Instr. and Meth. B 43 (1989) 218.

[9] Dose, V., Personal communication.

BSCAT - code for simulation and for analysis of the RBS/NRA spectra

Bogusław Rajchel

Institute of Nuclear Physics, 31-342 Kraków, ul. Radzikowskiego 152
Poland

In last years the beam of charged particles is frequently used for nondestructive investigations of thin surface layers. The nuclear methods such as RBS (Rutherford BackScattering Spectroscopy) and the NRA (Nuclear Reaction Analysis) have depth and mass resolution good enough for studies of the elemental composition and the structures of solid materials. To analyze and to simulate of the results collected in RBS/NRA type measurements many computer programs were created, for example RUMP[1] or DVBS[2]. Another group of the nuclear methods, called generally as PIXE (Particle Induced X-ray Emission) or PIGE (Particle Induced Gamma-ray Emission) have very good lower detection limit for determination of the concentration of elements. Unfortunattely the simple PIXE/PIGE technique can not be used for detrmination of the depth distribution of elements in target. Fortunately the simultaneous detection of backscattered particles and X-rays can be more effective way for investigatios of the composition and the structure.of thin surface layers. The BSCAT code was created for determination of solid material structures and for determination of the depth distribution of elements by using of the RBS/NRA and the PIXE/PIGE techniques. The BSCAT code can be used for simulation and for analysis of the experimental spectrum of particles backscattered on multielemental and multilayer samples. The first version of the BSCAT code was used initially for investigation of the thin carbon coatings on the Si single crystal and was presented on the ECCART'95 [3] conference. Recently BSCAT program was rebuilt and now can perform simulation and analysis of the RBS/NRA and simple PIXE spectra.

Introduction

In last years the nuclear methods such as RBS (Rutherford BackScattering Spectroscopy) and the NRA (Nuclear Reaction Analysis) methods are frequently used for nondestructive investigations of thin surface layers. Both methods have depth and mass resolution good enough for studies of the elemental composition and the structures of solid materials. The lower detection limit of the both mentioned methods is good enought for typical material investigations. Another group of the nuclear methods, called generally as PIXE (Particle Induced X-ray Emission) or PIGE (Particle Induced Gamma-ray Emission) have very good lower detection limit for determination of the concentration of elements. Unfortunattely the depth distribution of elements can not be determined by typical PIXE/PIGE technique. By all mentioned reasons, the simultaneous detection of backscattered particles and X-rays (or γ-rays) can be very effective way for investigations of the composition and the structure.of thin surface layers. To analyze and to simulate of the results collected in RBS/NRA type measurements many computer programs were created, for example RUMP[1] or DVBS[2]. The BSCAT code was created mainly for simulation of RBS/NRA experiments, for determination of solid material structures and for determination of the depth distribution of elements by using two spectra recorded by different particles detectors and by one PIXE or PIGE type spectrum. The BSCAT code can be used for simulation and for analysis of the experimental spectrum of particles back-scattered on multielemental and multilayer samples. The first version of the BSCAT code was used initially for investigation of the thin carbon coatings on the Si single crystal and was presented on the ECCART'95[1] conference. Recently BSCAT program was rebuilt and now can perform any complex analysis of the RBS/NRA and simple PIXE spectra. The BSCAT code can be used for

[1] Fourth European Conference on Accelerators in Applied Research and Technology (ECAART) Zürich, Switzerland, 1995

CP392, *Application of Accelerators in Research and Industry*, edited by J. L. Duggan and I. L. Morgan
AIP Press, New York © 1997

simulation of results in RBS/NRA type experiments.

Interaction of charged particles with solid materials

The nuclear techniques found broad applications in investigations of content material, in the study of the structure of surface layers and even in making new materials. The ion beam allows us to obtain information about the elemental composition of the surface layer in cases when no chemical method can be applied. Nuclear methods, as the RBS/NRA or the PIXE, were described with detail in many papers [4,5,6]. Now, by this reason, only fenomena, importrant for understanding of possibilities of the BSCAT code are discussed.

The impact beam of energetic ions can be described by set of following parameters: number N_0 of impact ions, energy E_0, spread of energy, atomic number Z_1 and mass M_1. The target (sample) can be divided (Fig.1) by set of layers, each with thicknes of Δx_i: Each layer can be described by: thicknes of the layer, x_{0i} - possition of the layer in sample, the atomic number Z_2 and mass M_2 . The experimental geometry can be defined by: θ - impact angle, angle of detection of scattered particles and by angle of detection of the X-rays or γ-rays. A simple trajectory of ions in thick target is shown in the Fig. 1.

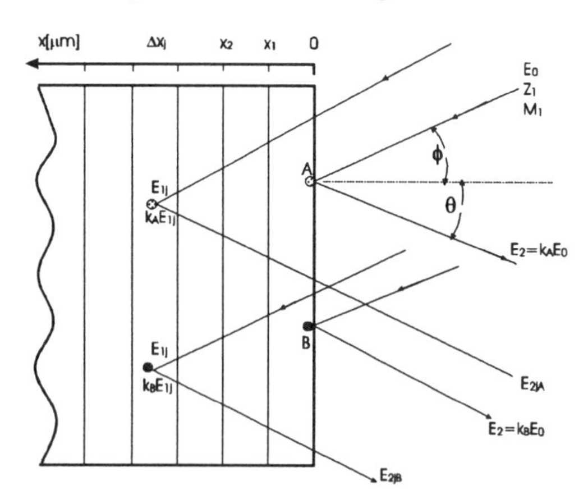

Fig.1 The trajectory of an ion in a thick amorphous target

After entering the target, the ion with an initial energy E_0 moves down and its energy decreases gradually. So at the depth **x** the ion energy is equal to E_1. Passing through a sublayer Δx at the depth x the ion is scattered at angle θ. After scattering the ion can move to the surface of the sample while still losing energy. If the ion energy is high enough the ion can leave the target with energy E_2 and will be detected by detector. So the energy spectrum of ions detected by the detector can be calculated as:

$$\frac{dN(E_2)}{dE_2} = N_0 n d\Omega \frac{d\sigma(E_1,\vartheta)}{d\Omega} \frac{dx}{dE_2} \quad (1)$$

For compound targets we have to adopt the above primary model by changing the stopping power function. In compound target, composed by l different elements, the stopping power function must be calculated as:

$$S(E) = \sum_{i=1}^{l} \omega_i S(E) \quad (2)$$

We have to handle the compound target as several independent one element targets mixed together. The ion, passing through the compound target, can be backscattered only once at the large angle θ on the atom of the one element. So, the energy spectrum particles backscatterd by compound target is:

$$\frac{dN(E_2)}{dE_2} = \sum_{i=1}^{l} \left[\frac{dN(E_2)}{dE_2}\right]_i \quad (3)$$

Using equation (3) we have to remember that the stopping power function must be calculated according to equation (2). In equation (2) the factor ω_i are the relative concentrations of the elements in compound target.

In real experiment we must remember that the energy distribution of impact particles can be described by gaussian function with an initial energy spread σE_0 determined by accelerator. The energy distribution of particles moving in solid material is modified by straggling processes. This situation is shown in. fig. 2.

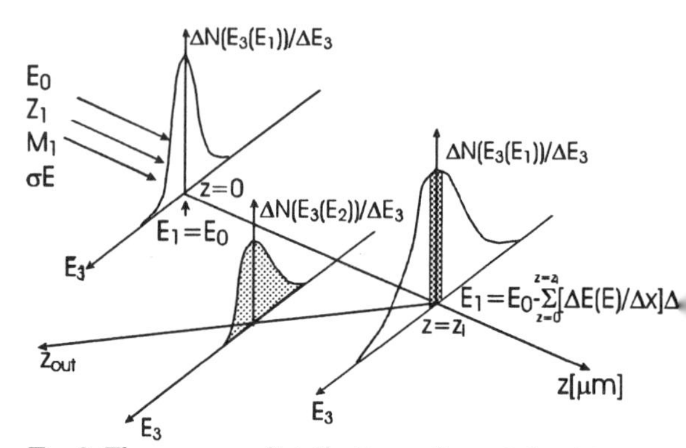

Fig.2 The energy distribution of particles in material.

In experiment we always use real particle detectors. The real detector spectrum is a convolution of the energy spectrum with a detector response function:

$$\left[\frac{dN(E_2)}{dE_2}\right]_{det} = \left[\frac{dN(E_2)}{dE_2}\right]_{t\,arg.} * g(E_2)_{det} \qquad (4)$$

In equation (4) the $g(E_2)_{det}$ is the detector response function and the real energetic ions spectrum is signed by "targ" index.

As was shown above the RBS method is a very good tool for the investigation of the multielemental targets and also to study multilayer targets. However, for light elements, we must remember that scattering cross sections are small. This situation can be improved when we work with non-Rutherford cross sections. The real cross section for scattering of particles by positive charge of nucleus can be described by the Rutherford cross section formula only when the interaction between the impact particle and the nucleus is a pure Coulomb interaction. When the energy of the impact particle is very low or when the particle is very heavy, the screening effects may be important. Also at higher ion energy and very small impact parameters there can be large departures from the Rutherford cross section, due to the interaction of the incident particle with the nucleus of the target atom. Deviations from the Rutherford due to nuclear interaction are important when the impact parameter is close to the nuclear radius. At that moment the cross section may be many times higher than the Rutherford cross section. We may use those phenomena for lowering detection limit for light elements in target. For example by using the α particles with energy bigger than 3075 keV for detection of oxygen atoms we can see that at angle 170^0 the resonance cross section is 16 times larger than the Rutherford cross section and we may use those phenomena to achieve better sensitivity for oxygen. For example the resonance phenomena may also be useful in depth profiling of the oxygen concentration in surface layer.

The particle, moving through a solid material, can induce the X-ray emission. By detection of the characteristic X-ray also the composition of materials can be determined. Description of the simulation and the analysios of the PIXE/PIGE spectra by the BSCAT code will be included in next paper.

Calculation procedure

For simulation and for analysis of the complex (multielemental and multilayer) amorphous targets a computer code was created. This code, based mainly on the formula (1-4) can simulate the X-ray spectrum and two spectra of backscattered particles at two different scattering angles. The simulation can be done for the Rutherford universal cross section formula and/or for non-Rutherford cross sections. The non-Rutherford cross section values can be entered into the code from file. All spectra can be calculated for different stopping power functions. Two different formula of the stopping power functions are included. One defined by Ziegler[7] and second by the Montenegro[8] formula. The main idea of the calculation of the spectrum of the backscattered particles is shown in fig.3.

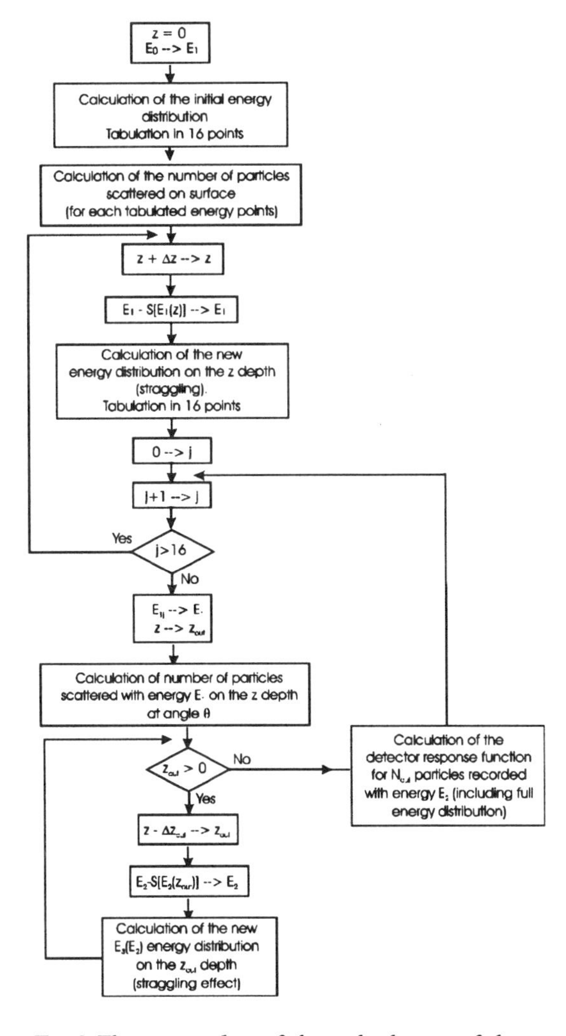

Fig.3 The main idea of the calculation of the spectrum of the backscattered particles.

Results

The BSCAT code, mainly prepared for simulation and for analyze of the spectra from RBS/NRA type experiments has been used for analysis experimental results collected for Si samples covered by multielemental thin layers such as the DLC [Diamond Like Coating] films or the TiX (X=C, N). For example, the BSCAT code was used for simulation of the spectra of protons or He^+ ions, scattered with initial energy E_0 from 1600 keV to 2200 keV, incident in a random direction on the Si single crystal covered initially by thin Ni layer and in next step covered by carbon (DLC type) film created on the surface by the Dual Beam IBAD [Ion Beam Assited Deposition] technique. The last carbon layer in final step of was doped by the ^{57}Fe atoms. The simulation was performed at 168^0 detection angle. Experimental and calculated spectra of He^+ with initial energy E_0 =1750 keV are shown in fig. 4.

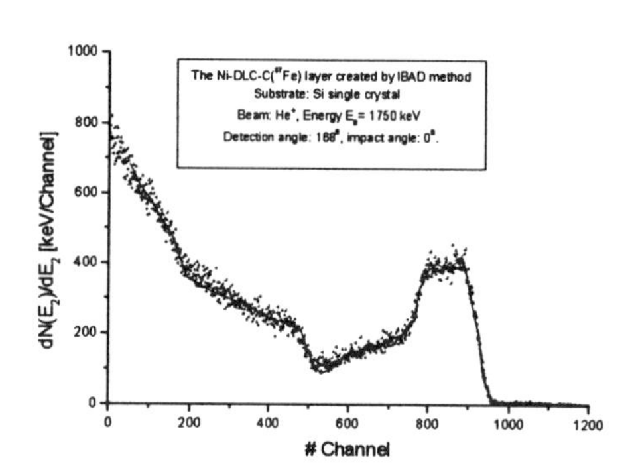

Fig.4 Two spectra of He^+ ions scattered at angle 168^0 on the Si substrate covered by thin Ni-DLC-C(^{57}Fe) film. Spectrum (+) was collected experimentally and the second spectrum (line) was calculated by the BSCAT code.

Conclusion

The BSCAT code is a powerful tool for simulation and for analysis of the complex spectra recorded in RBS/NRA/PIXE experiments. Two versions of the BSCAT code are ready: one for IBM PC type computers (requirements: i80486 or Pentium processor, MS Windows 95) and second for UNIX type computers. Unfortunately only PC version is enriched by graphics shell. The PC verision of the BSCAT code was written partially in the Microsoft Fortran and partially in Microsoft C/C++ (graphics shell). The UNIX version was written for HP and for CONVEX computers.

Acknowledgments

Work was partially sponsored by State Committee of Scientific Research Grant no. 2 P03B 11009 and Grant no. 7 T08C 031 08 and also was partially supported by the Kościuszko Foundation from USA.

Reference

1. L.R.Dolittle, NIM, B9 (1985) 344
2. V.Boháč, D.M.Shirokov, NIM, B84 (1994) 497
3. ECCART'95 - Fourth International Conference on Accelerators in Applied Research and Technology, Zürich, Switzerland, 1995
4. L.C.Feldman, J.W.Mayer, Fundamentals of Surface and Thin Film Analysis (1986)
5. W.K.Chu, J.W.Mayer, M.A.Nicolet, Backscattering Spectrometry (1978)
6. J.W.Mayer, E.Rimini, Ion Beam Handbook for Material Analysis (1977)
7. J.F.Ziegler et all. The Stopping and Ranges of Ion in Matter, (Pergamon, New York, 1985)
8. E.C.Montenegro et all. Phys. Lett. 92a, 195 (1982)

CO-EVAPORATED NOBLE METAL BEHAVIOR ON THE SI(111) SURFACE BY USING RBS

J. Yuhara and K. Morita

Department of Crystalline Materials Science, School of Engineering, Nagoya University, Furo-cho, Chikusa-ku, Nagoya, 464-01, Japan

Concentration and structure changes of monolayer single, binary, and ternary noble metal adsorbates (Au, Ag, and Cu) on the Si(111) surface by isochronal annealing at temperatures from 150 to 630°C have been studied by means of LEED-AES-RBS techniques. The Ag/Si(111)-$\sqrt{3}$x$\sqrt{3}$-Au surface changes into the 2$\sqrt{3}$x2$\sqrt{3}$-(Au,Ag) and $\sqrt{21}$x$\sqrt{21}$-(Au,Ag) structures on thermal annealing. It means that Ag on the Si(111)-$\sqrt{3}$x$\sqrt{3}$-Au structure is chemically bound to Au atoms. In the case of Cu/Si(111)-$\sqrt{3}$x$\sqrt{3}$-Ag surface, Cu dissolves into the Si bulk at 250°C, and segregates back to the surface when Ag leaves the surface on annealing at higher temperatures. It means that Cu atoms are not bound to Ag atoms on the Si(111) surface. These results indicate that there is a similarity between the surface phases of the binary noble metals on the Si(111) surface and their bulk phases. Moreover, the Ag/Si(111)-$\sqrt{3}$x$\sqrt{3}$-(Au,Cu) ternary surface changes into the 2$\sqrt{3}$x2$\sqrt{3}$-(Au,Ag,Cu) structure on thermal annealing, which has been newly found for the first time in this experiment.

INTRODUCTION

It has been reported recently that the thermal behaviors and surface structures of binary metal adsorbates on the Si(111) surface are different from those of single metal adsorbates, because the atomic interactions at the surface takes place not only between the metal adsorbates and the Si substrate but also between two different metal adsorbates [1-19]. It has been shown that the surface superstructures are closely related to the metal coverages on the Si(111) surface. It is well known that Rutherford backscattering spectrometry (RBS) is one of the most reliable and quantitative technique to measure the depth profile and composition. Therefore, we have applied the RBS technique in combination with surface sensitive probes such as LEED and AES techniques to study metal adsorbates on the Si(111) surface.

In this paper, we review the changes of concentrations and structures of single, binary, and ternary noble metal adsorbates on the Si(111) surface studied systematically by isochronal annealing. It is emphasized that the interaction between binary metal adsorbates on the Si(111) surface is similar to that in their alloys.

EXPERIMENTAL

The specimen used was a mirror-polished n-type Si(111) wafer of the resistively of 3 Ωcm, with a size of 25x3x0.5mm^3. The specimen was placed on a manipulator in a conventional UHV chamber, which was evacuated to base pressures less than 3x10^{-10}Torr. The chamber was equipped with a 4 grid optics for LEED, a double pass cylindrical mirror analyzer for AES, a water-cooled solid state detector for RBS and three evaporation sources of metals (Au, Ag, and Cu).

The specimen surface was cleaned by repeated direct current heatings, for 5min at no less than 1050°C at pressures below 5x10^{-10} Torr. After the cleaning process, a distinct 7x7 LEED pattern was observed, and the AES spectra did not show any traces of impurities such as C and O. Specimen temperatures higher than 600°C were measured with a radiation thermometer. Lower temperatures were measured with an alumel-chromel thermocouple.

Ag films sub-ML thick (1ML for Si(111) face = 7.8x10^{14} atoms/cm^2) were deposited onto the Si(111)-7x7, $\sqrt{3}$x$\sqrt{3}$-Au, "5x5"-Cu, and $\sqrt{3}$x$\sqrt{3}$-(Au,Cu) surfaces at room temperature. Single, binary, and ternary metal adsorbates on the Si(111) surface were isochronally annealed for 15min at temperatures from 150°C to 630°C; the surface structures and coverages were measured by means of LEED, AES, and RBS techniques at several stages of the annealing. The time to reach the annealing temperature was about 3min.

The amount of deposited metal atoms was measured by RBS with an accuracy of about ± 0.05ML for Au and Ag, and ± 0.1ML for Cu using a 1.5MeV He$^+$ ion beam. The effect of He$^+$ beam damages on the thermal behavir of the adsorbates and its surface coverages was carefully tested by comparing RBS spectra taken both at irradiated position and non-irradiated position until annealings at several temperatures. No visible change of the thermal behavir and the surface coverages due to beam damage was observed.

EXPERIMENTAL RESULTS

Single films of Au, Ag, and Cu, about 0.6~0.8ML thick, deposited on the Si(111)-7x7 surface were isochronally annealed at temperatures from 200°C to 630°C. The surface structures and the coverages determined by LEED, AES, and RBS at each annealing stage are

CP392, *Application of Accelerators in Research and Industry,* edited by J. L. Duggan and I. L. Morgan
AIP Press, New York © 1997

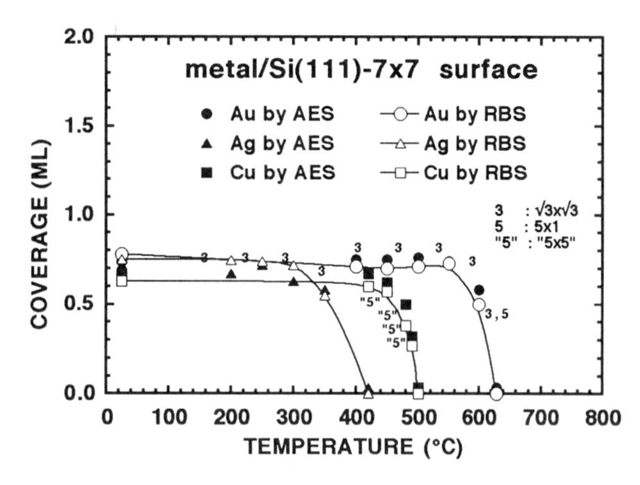

Figure 1 Coverage changes versus temperature measured by AES and RBS on the Si(111) surface for Au, Ag, and Cu.

shown in Fig.1. One notices in Fig.1 that the decay curves of Au, Ag and Cu measured by RBS are almost the same as those measured by AES. This demonstrates that all the metal adsorbates are located at the topmost Si(111) surface.

The Ag/Si(111)-$\sqrt{3}$x$\sqrt{3}$-Au surface was isochronally annealed to examine the interaction of Ag with Au atoms. The Ag and Au coverages measured by both AES and RBS are displayed as a function of annealing temperatures in Fig.2. The Au coverage does not change at any stage of the annealing, while Ag coverage gradually decreases to zero and the LEED patterns showed the structures of 2$\sqrt{3}$x2$\sqrt{3}$-(Au,Ag) ($\theta_{Ag}\geq0.6$ML), $\sqrt{21}$x$\sqrt{21}$R(±10.89°)-(Au,Ag) ($0<\theta_{Ag}\leq0.6$ML), and $\sqrt{3}$x$\sqrt{3}$-Au ($\theta_{Ag}\approx0$ML).

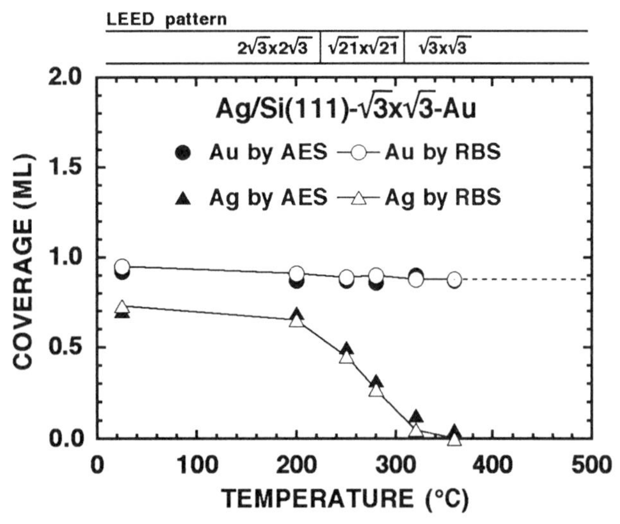

Figure 2 Coverage changes versus temperature measured by AES and RBS on the Ag/Si(111)-$\sqrt{3}$x$\sqrt{3}$-Au surface. The LEED patterns observed are shown at the top of the figure.

The Ag/Si(111)-"5x5"-Cu surface was isochronally annealed to examine the interaction of Ag with Cu atoms. The Ag and Cu coverages are displayed as a function of temperatures in Fig.3. Clearly the Cu coverage decreases to zero at a temperature of 250°C and eventually recovers at a temperature of 400°C. At 350°C, the Cu concentration in the Si bulk measured by RBS decreased to the detection limit, which corresponds to a Cu/Si concentration ratio of 0.02. This result indicates that the Cu atoms migrate into the bulk beyond the RBS detection limit that is ~3000Å.

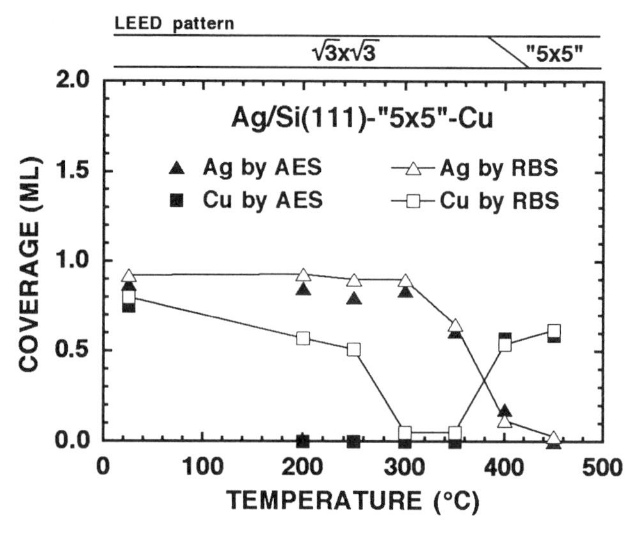

Figure 3 Same as Fig.2 but for Ag/Si(111)-"5X5"-Cu surface.

The Ag/Si(111)-$\sqrt{3}$x$\sqrt{3}$-(Au,Cu) surface was isochronally annealed to examine the effect of Ag to Au and Cu atoms. The typical RBS spectra obtained in this experiment are shown in Fig.4. The LEED observations showed patterns of $\sqrt{3}$x$\sqrt{3}$ at RT ≤ T ≤ 200°C, 2$\sqrt{3}$x2$\sqrt{3}$ at 250°C ≤ T ≤ 350°C and $\sqrt{3}$x$\sqrt{3}$ at T = 400°C. Then, the surface coverages of the Ag/Si(111)-$\sqrt{3}$x$\sqrt{3}$-(Au,Cu) surface measured both by AES and RBS are shown as a function of annealing temperatures in Fig.5. The Au coverage does not change at all in the temperature range of the present experiment. The Ag coverage gradually starts to decrease at 200°C. The Cu coverage decreases to almost zero around 300°C, and eventually recovers at 400°C.

DISCUSSION

It is seen from comparison between Fig.1 and 2 that Ag on the Si(111)-$\sqrt{3}$x$\sqrt{3}$-Au surface is thermally less stable than Ag at the Si(111)-$\sqrt{3}$x$\sqrt{3}$-Ag surface. Even when the Ag coverage decreases down to zero, the Au coverage almost keeps the initial level. This result indicates that the Au thermal stability is not influenced by co-existence of Ag. The surface structure at the Ag/Si(111)-$\sqrt{3}$x$\sqrt{3}$-Au surface continues to change into 2$\sqrt{3}$x2$\sqrt{3}$-(Au,Ag) ($\theta_{Ag}\geq0.6$ML), through $\sqrt{21}$x$\sqrt{21}$-(Au,Ag) ($\theta_{Ag}\leq0.6$ML), to $\sqrt{3}$x$\sqrt{3}$-Au ($\theta_{Ag}\approx0$ML) on the annealing. This result indicates that Ag atoms are bound to Au atoms at the Si(111)-$\sqrt{3}$x$\sqrt{3}$-Ag surface. Since 2$\sqrt{3}$x2$\sqrt{3}$ and $\sqrt{21}$x$\sqrt{21}$ spots include $\sqrt{3}$x$\sqrt{3}$ spots, the 2$\sqrt{3}$x2$\sqrt{3}$ and $\sqrt{21}$x$\sqrt{21}$-(Au,Ag) structures may come from the rearrangement of Ag atoms deposited on the Si(111)-$\sqrt{3}$x$\sqrt{3}$-Au surface. It was found that there is no equilibrium in the coverage ratio of Au to Ag. Therefore, it is concluded that Au-Ag adsorbates on the Si(111) surface form the 2D-alloy solid solution.

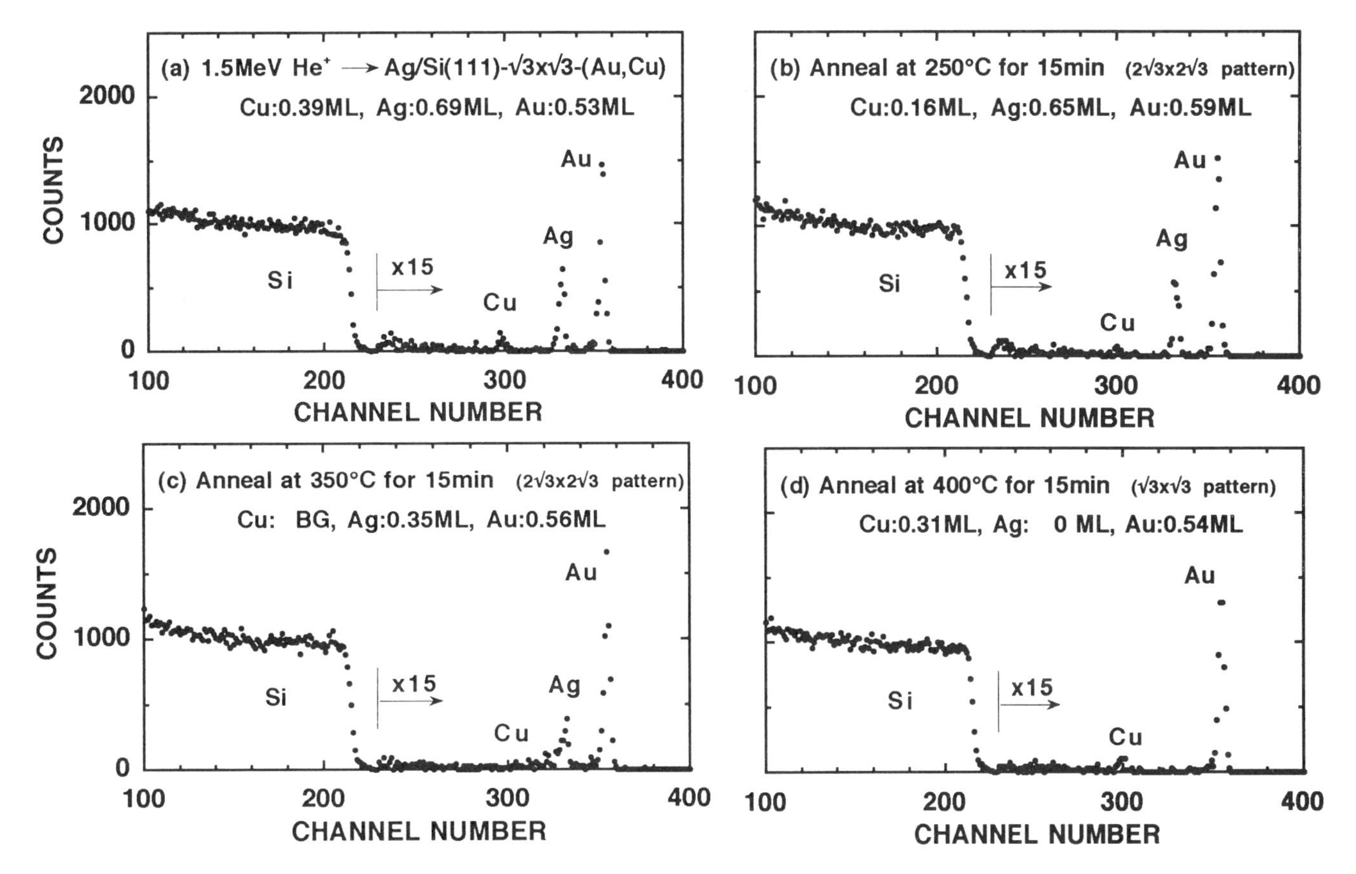

Figure 4 RBS spectra of a 1.5MeV He^+ ion beam from a Si(111)-$\sqrt{3}$x$\sqrt{3}$-(Au,Cu) surface on which an Ag layer of 0.69ML was deposited (a), subsequently annealed at 250°C for 15min (b), at 350°C for 15min (c), and at 400°C for 15min (d).

It has been reported by Ichimiya *et al.* using RHEED and STM that the Si(111)-$\sqrt{21}$x$\sqrt{21}$-(Au,Ag) was formed by deposition of gold atoms about 0.14ML thick on the Si(111)-$\sqrt{3}$x$\sqrt{3}$-Ag surface at room temperature [4,12]. Nogami *et al.* showed the similar result obtained by LEED and STM [11]. However, the Au coverage in the $\sqrt{21}$x$\sqrt{21}$-(Au,Ag) structure in the present study is considerably larger than that obtained by them. Gavriljuk *et al.* showed using LEED-AES techniques that the Si(111)-$\sqrt{21}$x$\sqrt{21}$-(Au,Ag) was formed by codeposition of Au and Ag atoms on the Si(111) surface at room temperature and post-annealing at 450°C [8]. The coverages of Au and Ag deposited are about 1ML and 0.2ML, respectively, which is in good agreement with ours. These results may suggest that there are two different $\sqrt{21}$x$\sqrt{21}$-(Au,Ag) structures formed at RT and elevated temperatures. Moreover, it seems to indicate that the $\sqrt{21}$x$\sqrt{21}$-(Au,Ag) structures have two combination of the Au and Ag coverages, which are Au-rich and Ag-rich.

Cu atoms on the Si(111)-$\sqrt{3}$x$\sqrt{3}$-Ag surface start to dissolve into the Si bulk upon annealing at a temperature of 150°C, but the Cu coverage on the clean Si(111)-7x7 surface does not decrease for the annealing up to 450°C, as shown in Fig.3. These facts indicate that the Cu thermal stability is strongly influenced by the co-existence of Ag atoms. On the contrary, the Ag thermal stability on the Si(111) surface does not change at all, even if Cu atoms co-exist. Further, the Cu atoms, which had once dissolved into the bulk, segregate back to the topmost surface, when Ag atoms leave the surface upon annealing at temperatures above 400°C. We found from the RBS spectrum that about 80% of the Cu atoms, which had dissolved into the bulk, segregated back to the surface. This behavior indicates that the location of Cu atom in the quasi5x5 structure at the surface is energetically more favorable than the solute site in the bulk. From these results, it is concluded that Ag and Cu atoms are separated from each other on the Si(111) surface. It is well known that solute Ag or Cu atoms are strongly segregated at the surface of dilute Ag-Cu alloy systems. Therefore, the present experimental data on Ag-Cu binary adsorbates on the Si(111) surface reflects the bulk properties of Ag-Cu alloy.

It has been found in the case of (Au,Cu) /Si(111) that an ordered 2D alloy of Au_4Cu is formed at the Si(111) surface on annealing at lower than 200°C [6]. This fact also indicates that there is a similarity between the ordered phase of the Au_4Cu on the Si(111) surface and the ordered phase of Au-Cu alloys. Therefore, it is concluded that the interaction between the binary metal adsorbates arrayed two-dimensionally on the Si(111) surface is similar to that in three-dimensional alloys.

It is also seen from Figs.4 and 5 that the Au coverage at the Ag/Si(111)-$\sqrt{3}$x$\sqrt{3}$-(Au,Cu) surface did not change at any stage of the annealing, which is also similar to that for the binary metal adsorbates. Therefore, it is concluded that Si-Au bonding is the most stable among Au-Ag-Cu ternary adsorbates against thermal heating and is not reduced by co-existence of other adsorbates on the Si(111) surface.

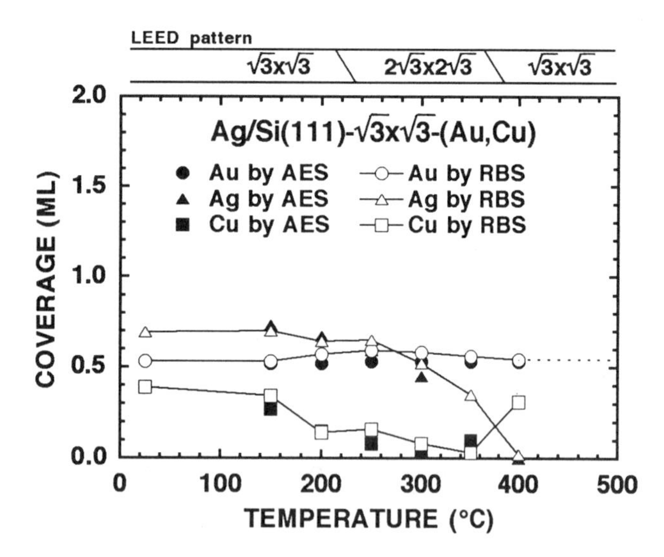

Figure 5 Same as Fig.2 but for Ag/Si(111)-√3x√3-(Au,Cu) surface.

The Ag coverage started to decrease at 200°C and reached 0ML on annealing at 400°C , as shown in Figs.4 and 5. It has been reported by the present authors that Ag on the Si(111)-√3x√3-Au surface decreases to 0ML on annealing below 350°C [7]. Thus, thermal stability of Ag on the Si(111)-√3x√3-(Au,Cu) surface became higher than that on the Si(111)-√3x√3-Au surface. Moreover, the LEED pattern of 2√3x2√3 persists instead of √21x√21 at the Si(111)-√3x√3-(Au,Ag) surface, even when the Ag coverage decreased to 0.35ML. It has been shown that the Ag/Si(111)-5x1-Au surface produces a √3x√3-(Au,Ag) and that the Ag/Si(111)-√3x√3-Au surface produces a 2√3x2√3-(Au,Ag) ($\theta_{Ag}\geq$0.6ML) and a √21x√21-(Au,Ag) ($\theta_{Ag}\leq$0.6ML) structure on thermal annealing, respectively [7]. Therefore, it is concluded that the obtained structure of the Si(111)-2√3x2√3 surface is constructed from Au, Ag, and Cu, but not from Au and Ag. The 2√3x2√3-(Au,Ag,Cu) structure has been newly produced for the first time in this experiment.

The coverages of Au, Ag, and Cu of the 2√3x2√3-(Au,Ag,Cu) structure were 0.52ML, 0.35ML, and 0.10ML, respectively, on annealing at 350°C, at which the prominently bright 2√3x2√3 LEED spots were observed. So far, it has been shown in the binary systems that Cu is not bound to Ag, and Ag and Cu are repulsive from each other. The present structure of the 2√3x2√3 indicates that Ag atoms are only locally bound to Au atoms at the Si(111)-√3x√3-(Au,Cu) surface. It also indicates that Ag/Si(111)-√3x√3-(Au,Cu) surface can not construct √21x√21 because of long range repulsive interaction with Cu atoms at the surface but construct a smaller period-structure of 2√3x2√3.

From a point of view of the device technology, the effects of impurities on the thermal stability of metal adsorbates at the Si surface are very important, because the life of electronic devices is affected drastically. The experimental results in the present study indicate that even if Au is the most stable as the contact material among Au, Ag, and Cu, impurities such as Cu in Au would be easily dissolved into the Si bulk. Therefore, the purity of conducting metal wire will be very important for formation of the thermally stable interface.

SUMMARY

The changes of the atomic structures and concentrations of single and binary metal adsorbates on the Si(111) surface induced by isochronal annealing at temperatures from 150°C to 630°C have been studied by LEED, AES and RBS. It is shown that Ag atoms are bound to Au atoms and are not bound to Cu atoms on the Si(111) surface and there is a similarity between the surface phases of the binary noble metals on the Si(111) surface and their bulk phases. The 2√3x2√3-(Au,Ag,Cu) ternary structure is newly produced for the first time in this experiment.

REFERENCES

1. H. Daimon, C. Chung, S. Ino, and Y. Watanabe, Surf. Sci. **235** (1990) 142.
2. I. Homma, Y. Tanishiro, and K. Yagi, Surf. Sci. **242** (1991) 81.
3. I. Homma, Y. Tanishiro, and K. Yagi, The Structure of Surfaces III, **Vol.24** (1991) 610.
4. A. Ichimiya, H. Nomura, and Y. Horio, Proc. Fifth Topical Meeting on Crystal Growth Mechanism, (1992).
5. O. V. Bekhtereva, B. K. Churusov, and V. G. Lifshits, Surf. Sci. Lett. **273** (1992) L449.
6. M. Sasaki, J. Yuhara, M. Inoue, and K. Morita, Surf. Sci. **283** (1993) 327.
7. J. Yuhara, M. Inoue, and K. Morita, J. Vac. Sci. Technol. **A11** (1993) 2714.
8. Y. L. Gavriljuk, V. G. Lifshits, and N. Enebish, Surf. Sci. **297** (1993) 345.
9. T. Yamanaka, A. Endo, and S. Ino, Surf. Sci. **294** (1993) 53.
10. J. Yuhara, R. Ishigami, and K. Morita, Control of Semiconductor Interfaces, ed. by I. Ohdomari et al, Elsevier Sci. B.V. (1994) pp 399-404.
11. J. Nogami, K. J. Wan, and X. F. Lin, Surf. Sci. **306** (1994) 81.
12. A. Ichimiya, H. Nomura, Y. Horio, T. Sato, T. Sueyoshi, and M. Iwatsuki, Surf. Rev. Lett. **1** (1994) 1.
13. J. Yuhara, R. Ishigami, and K. Morita, Surf. Sci. **326** (1995) 133.
14. J. Yuhara, R. Ishigami, D. Ishikawa, and K. Morita, Surf. Sci. **328** (1995) 269.
15. D. Ishikawa, J. Yuhara, R. Ishigami, K. Soda, and K. Morita, Surf. Sci. **356** (1996) 59.
16. D. Ishikawa, J. Yuhara, R. Ishigami, K. Soda, and K. Morita, Surf. Sci. **357/358** (1996) 432.
17. D. Ishikawa, J. Yuhara, R. Ishigami, K. Soda, and K. Morita, Surf. Sci. **357/358** (1996) 966.
18. J. Yuhara, R. Ishigami, D. Ishikawa, and K. Morita, Appl. Surf. Sci. (1996) in press.
19. D. Ishikawa, J. Yuhara, K. Soda, and K. Morita, Surf. Sci. (1996) in press.

DETERMINATION OF THE DENSITY OF SPUTTERED THIN FILMS BY RUTHERFORD BACKSCATTERING SPECTROSCOPY AND LOW ANGLE X-RAY DIFFRACTION

D.V.Dimitrov, G.C.Hadjipanayis

Department of Physics and Astronomy, University of Delaware, Newark, DE 19716

C.P.Swann

Bartol Research Institute, Newark, DE 19716

Rutherford backscattering spectroscopy (RBS) and low angle X-ray diffraction (LAXRD) were used to determine the density of thin Ta, Ag, Mo and Fe films as compared to the bulk value. An exact formula for the positions of the diffraction peaks as a function of the thickness of the film and its index of refraction was derived. A computer program utilizing this formula was written and used to do a least square fit to the low angle diffraction spectra and to obtain the thickness of the films. The densities of the films were found to be in the range between 84 and 98 % of the bulk value. A detailed analysis of conditions that need to be fulfilled in order for the method to give reliable results is also presented.

INTRODUCTION

Knowing the average density of a thin metallic film is often an important issue, in particular if one studies its magnetic and resistivity properties. In our recent studies [1] on thin Fe films it was found that the saturation magnetization of the films was about 10 % lower than the bulk value. Similar result were also obtained in other studies [2-3]. Different reasons aiming to explain the effect were proposed including, an existence of dead magnetic layer on the bottom of the film [2], magnetic disorder in the film, or decreased density of the film as compared to the bulk value [3]. The last explanation seems very appealing because the sputtering process, which is used to fabricate the films, is a highly nonequilibrium process which might lead to a large number of defects and voids in the film. The main goal of this work was to give the last idea a thorough experimental examination. RBS and low angle X-ray diffraction were used because they are very accurate and the quantities they provide (total number of atoms per unit area and thickness of the film) are all one needs to calculate the average density of a thin film.

SAMPLE PREPARATION AND CHARACTERIZATION

Thin Ta, Ag , Mo, and Fe films were prepared by dc magnetron sputtering on water cooled glass strips and Si wafers. The base pressure of the system was 2×10^{-7} Torr. The sputtering was done in 5 mTorr Ar atmosphere with a flow of 50 cc/min. He^{+} ions with 2 MeV energy, produced by Van de Graaff accelerator, were used to collect RBS spectra. Films with two different thicknesses were prepared to study eventual size dependence of the density. The theoretical fits of RBS data and the number of atoms projected over unit surface area of the film were done using the software package RUMP2 [4]. Low angle and high angle X-ray diffraction were taken with a commercial Philips X-ray diffractometer. The magnetic measurements of the Fe films were done using SQUID magnetometer.

THEORETICAL

In this section a formula for the positions of the X-ray diffraction maxima as a function of the thickness of the film and the index of refraction of the film is derived. Also a relationship between the thickness of the film, the number of atoms per unit area, and the density of the material is obtained. Figure 1 shows a simplified geometry of the experiment. The incident beam (1) is partially reflected from the top surface (2) and partially refracted (3). The refracted beam 3 is partially reflected at the bottom of the film and then partially refracted back into the air (4). The beams (2) and (4) interfere and if the conditions for a constructive interference is fulfilled a maximum is observed. In this case there is a complete analogy between the X-ray diffraction from a metallic film and the diffraction of light from thin dielectric film. The condition for a maximum is :

$$\Delta L = m\lambda . \tag{1}$$

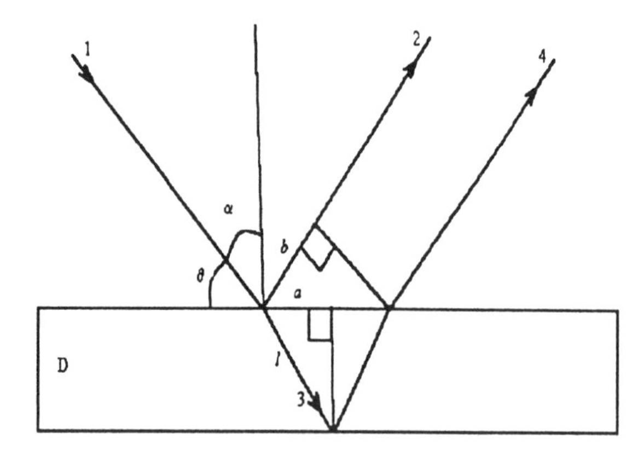

FIGURE 1. A diagram of the low angle diffraction geometry

where ΔL is the difference in the optical path between ray 2 and ray 4, m is the order of the maximum (an integer number), and λ is the wavelength of the X-ray. Assuming that the index of refraction for X-rays in the media of the film is ν we can write the condition for the maximum as follows:

$$2\nu l - b = m\lambda. \qquad (2)$$

After some straightforward geometry and algebra the following relation is obtained

$$\nu^2 - \sin^2\alpha = (m\lambda/2D)^2. \qquad (3)$$

At this point one can take into account that the index of refraction of X-rays is very close to 1 and slightly smaller, $\nu = 1 - \varepsilon$, where ε is of the order of 10^{-6} for most of the materials. Substituting for n and ignoring ε^2 as negligible relative to other terms we obtain:

$$\sin^2\theta = (m\lambda/2D)^2 + 2\varepsilon \qquad (4)$$

A computer program doing a least square fit between the equation 4 and the positions of the diffraction maxima, as measured by the X-ray spectra was written. The thickness of the film (D) was obtained as one of the fitting parameters.

By fitting RBS spectra of a film the total number of atoms per unit area (N) seen by the He^+ beam is obtained. It is straightforward to be shown that the average density (ρ) of the film is given by the formula (where μ is the mass of the atom) :

$$\rho = N\mu / D \qquad (5)$$

RESULTS AND DISCUSSION

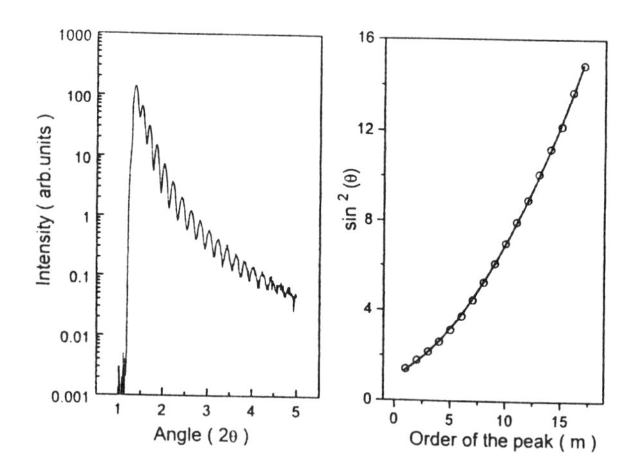

FIGURE 2. Low angle X-ray diffraction spectrum and its fit of a 426 Å thick Ta film on glass

Figure 2 shows a low angle X-ray spectrum of the Ta sample (thickness 426 Å) and a least square fit of the positions of the peaks. Figure 3 shows the RBS spectrum and the theoretical fit of the same sample.

The low angle X-ray spectra for films with heavier atoms (Ta, Ag) had higher intensities and more peaks, but even in the case of the film with the lightest element (Fe) there were at least 7 diffraction peaks which were enough to obtain reliable fits. The results of the density measurements are summarized in Table 1 below. It can be seen that in general the average density of thicker films is closer to the bulk value. The only exception are the Fe films for which there is not a significant change. To find out why the difference in Fe films detailed microstructural studies are required.

The above mentioned analysis is only true if the crystallographic structure and the lattice parameter of the grains of the film are the same as in bulk form.

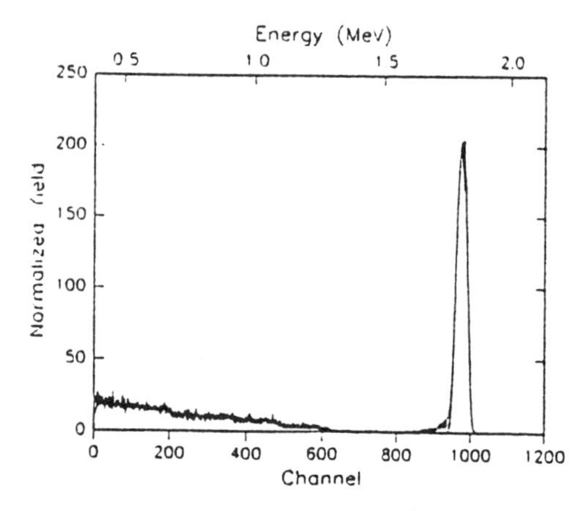

FIGURE 3. RBS spectrum and its fit of a 426 Å thick Ta film on glass. The theoretical curve matches perfectly the experimental data everywhere. The only difference is seen at the trailing edges of both the Ta and glass peaks

TABLE 1. Summary of the densities versus thickness of Ta, Ag, Mo and Fe films

Ta		Ag		Mo		Fe	
D(Å)	ρ(%)	D(Å)	ρ(%)	D(Å)	ρ(%)	D(Å)	ρ(%)
232	94	206	95	165	84	188	91
426	98	435	98	325	87	397	90

Using the method in case that the above mentioned requirements are not satisfied, one may obtain results on the film density, which look unphysical or wrong. An example is a thin Mo film grown on Si wafer instead on glass.

Using the above mentioned method straightforwardly it was found that the density of the film is 105 % as compared to the bulk value. The explanation of this discrepancy comes when one studies the high angle X-ray spectrum of that sample and compares it to the one of a Mo film deposited on a glass substrate Fig. 4. In the case of a Mo film on a glass substrate it is seen that the crystallographic structure and the lattice parameter are close to that of bulk which is obviously not the case when Mo was grown on Si wafer. The difference between Si and glass substrate is that Si wafers are crystalline in contrast to glass which is amorphous. It is a well known fact that when one grows a film on a high quality crystalline surface, as is the case of Si wafers, a film with different crystallographic structure or different lattice parameter than that in bulk can be fabricated. All films which were deposited on glass showed same crystallographic structure and lattice parameter as the ones in bulk. This fact is a natural consequence of the amorphous character of the glass substrates.

In summary, the average density of sputtered thin films is less than the bulk density because of defects and voids in the film. It is also dependent on thickness and elemental composition. For Fe films the average density of the film is in excellent agreement with the reduced saturation magnetization.

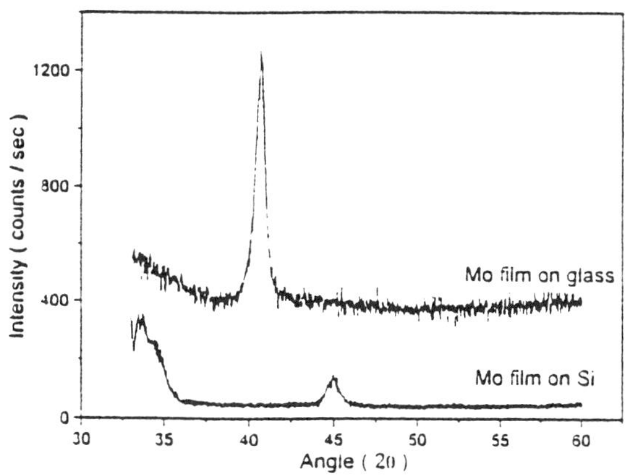

FIGURE 4. High angle X-ray diffraction spectra of Mo film on glass (top) and on Si wafer (bottom)

ACKNOWLEDGMENTS

Work supported by NSF DMR 9307676

REFERENCES

1.Dimitrov D.V., Murthy A.S., Hadjipanayis G.C., Swann C.P., "Magnetic Properties of Exchange Coupled Fe/Fe-O Bilayers", *Proceedings of the 40th Annual Conference on Magnetism and Magnetic Materials*, Philadelphia, November 7-12 (1995)

2.A.Layadi, J.O.Artman, B.O.Hall, R.A.Hoffman, C.L.Jensen, D.Chakrabarti, and D.A.Sanders, "FMR in evaporated single and multilayer thin films", *Journal of Applied Physics* **64**, 5760 (1988)

3.Y.K.Kim and M.Oliveria, "Magnetic properties of sputtered Fe thin films: Processing and thickness dependence", *Journal of Applied Physics* **74** (2), 1233 (1993)

4.Doolitttle L.R., *Proceedings of the 7th Internaional Conference on Ion Beam Analysis*, Berlin, Germany

A METHOD OF DETERMINING CHANNELING PARAMETERS IN BACKSCATTERING GEOMETRY

M. Kokkoris[a,1], S. Kossionides[a,1], T. Paradellis[a], Ch. Zarkadas[a], E. N. Gazis[b], C.T. Papadopoulos[b], R. Vlastou[b], X. Aslanoglou[c]

[a] *N. C. S. R. "Demokritos", Institute of Nuclear Physics, Tandem Accelerator, Laboratory for Material Analysis, Ag. Paraskevi 153 10, Athens, Greece.*
[b] *N. T. U. A., Department of Physics, Athens 157 80, Greece.*
[c] *University of Ioannina, Department of Physics, Ioannina, Greece.*
[1] *Authors to whom all correspondence should be addressed.*

The energy loss of channeled protons in silicon has been measured in the past in the transmission geometry and was found to be approximately half of the normal loss, thus confirming the equipartition rule. Other measurements however, concerning different crystals (e.g. Ge), deviated from this theory. In the backscattering geometry, the most successful corresponding attempts combined RBS with the nuclear resonance phenomenon. Nevertheless, they involved either considerable additions to the standard goniometer setup commonly used, or tedious Monte-Carlo calculations, thus limiting their applicability. In the present work, a method for the determination of the energy loss and dechanneling probabilities of axially channeled protons in silicon [100], in the energy range E_p = 1.7-2.6 MeV, is presented. It is carried out *in situ*, using the same experimental setup and beam properties (size, divergence) with the ones present in the actual analysis of a sample. The results obtained are in good agreement with already existing values in literature.

INTRODUCTION

Several experiments have been carried out so far to establish the energy loss of channeled particles. The first measurements of this sort were carried out on thin silicon targets in the transmission geometry (1) and the experimental energy loss was found to be approximately half of the "normal" loss (incurred in a randomly oriented crystal), and so to confirm the equipartition rule (2). Nevertheless, other experiments carried out with different crystals (such as Ge), using the same geometry, deviated from this theory (3). In all the above mentioned measurements, the determined quantity was the average energy loss of the so called "best channeled" particles. Such particles move far from the lattice nuclei in the region of low electron density and thus their energy losses are substantially less than the "poorly channeled" ones.

Recently, another experimental design has been used, in which the energy spectra of particles backscattered by a thick single crystal are registered. These measurements provided reliable numbers for practical purposes but they involved either considerable additions to the standard goniometer setup commonly used, or tedious Monte-Carlo calculations, thus limiting their applicability. In these experiments the energy spectra measured with a crystallographic direction aligned to the incident beam are formed by particles dechanneling at different depths. The most successful of the above mentioned attempts were the ones using the nuclear resonance phenomenon, combining the RBS data with those of the ^{27}Al(p, γ) reaction (5), the ^{19}F(p, α) reaction (6), or the ^{28}Si(p, γ) one (7).

In the present work a method of analyzing RBS spectra is presented in the energy range of proton resonances in ^{28}Si, allowing an estimation of the energy losses of channeled protons *in situ*, using the same experimental setup and beam properties (size, divergence) with the ones present in the actual analysis of the sample. The obvious advantage in this case is that since the channeling parameters are very sensitive, depending on the quality of the target, the energy and the collimation of the beam, the ambient temperature and the very nature of channeling (axial or planar), the method can be customized to provide reliable results in any experimental setup and it can be extended to other projectile-target combinations.

EXPERIMENTAL SETUP

The experimental setup at N.C.S.R. "Demokritos" includes a goniometer system (RBS-400 by Charles Evans and Associates) which permits experiments for backscattering spectroscopy in the case of oriented or non-oriented crystalline targets (12).

The proton beam was well collimated and the beam spot dimensions remained almost unchanged during the measurements (1.5 x 1.5 mm^2 with a variation of 10% at the

CP392, *Application of Accelerators in Research and Industry*, edited by J. L. Duggan and I. L. Morgan
AIP Press, New York © 1997

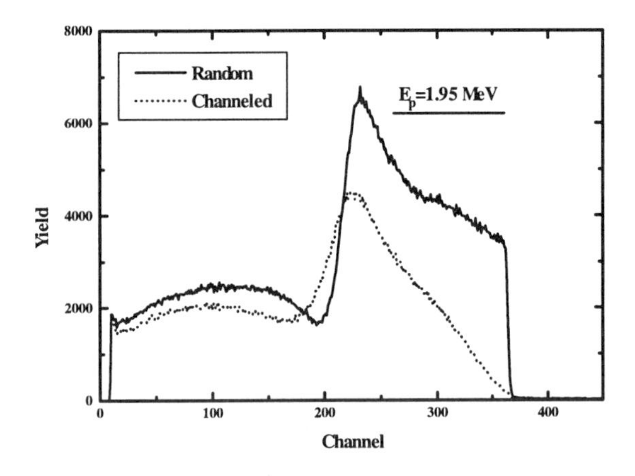

FIGURE 1. Random and channeling spectra of Si [100] at Ep = 1.95 MeV. The change of the resonance shape and the shift of the minimum are clearly visible.

most). The crystal was of average quality, showing a surface χ_{min} of approximately 5% for the analyzed energy range.

PRINCIPLE OF THE METHOD

The cross section for elastic scattering of protons shows two strong resonances in the energy range into consideration (E_p = 1.7-2.6 MeV) at E_p = 2.09 MeV and at E_p = 1.67 MeV with corresponding natural widths of 52.0 ± 0.8 and 15.6 ± 0.6 keV (8). In fig. 1, a typical backscattering spectrum is shown where in the "random" case the interference pattern between nuclear and Rutherford scattering can be observed (7, 8). The interference minima, which will be the focus of the present work, occur at E_p = 2.07 MeV and E_p = 1.62 MeV respectively and appear shifted towards lower energies in the channeling geometry. The motion of a proton originally channeled, as shown in fig. 2, can be described as follows: The proton moves within the low electron density channel for a length ℓ losing energy at a rate $S(E)_{ch}$. The energy of the proton of initial energy E_o at the point of dechanneling (A) can be described in general by the following equation:

$$E_\ell = E_o - \int_0^\ell S(E')_{ch}\,dr \qquad (1)$$

where $E_\ell \leq E' \leq E_o$.

Using the quantity ε defined (10) as the ratio of the average stopping power in the channel to the one in the amorphous medium:

$$\varepsilon = \frac{\bar{S}(E)_{ch}}{\bar{S}(E)_r} \qquad (2)$$

and replacing the integral with the average value $\bar{S}(E')$ the equation finally becomes:

$$E_\ell = E_o - \varepsilon \cdot \bar{S}(E')_r \cdot \ell \qquad (3)$$

Assuming an exponential decay law with depth for the intensity of the channeled beam, the dechanneling distance ℓ can be related to the decay constant λ as follows (6):

$$\ell = -\lambda \cdot \ln\left(\frac{1-R}{I_o}\right) \qquad (4)$$

where: I_o = fraction of initially channeled protons and R = ratio of the integrated channeled to random spectra over the same energy region.

After dechanneling, the particle moves on along the same direction (the small angle of dechanneling θ_1 can be ignored for all practical purposes) for a distance x, but now in a medium considered to be amorphous, until the backscattering occurs. The motion of dechanneled particles can be subsequently described by equations derived for the case of amorphous targets (9).

Thus, the depth D_{Rmin}, at which the resonance minimum is observed, can be expressed, according to the above arguments, as:

$$D_{R\,min} = x + \ell = \int_{E_\ell}^{E_{R\,min}} \frac{dE}{S(E'')} - \lambda \cdot \ln\left(\frac{1-R}{I_o}\right) \qquad (5)$$

where E_{Rmin} the energy of the interference minimum and $E_{Rmin} \leq E'' \leq E_\ell$.If we replace the integral with the average value of the stopping power $\bar{S}(E'')$, the above relations can be simplified and combined as follows:

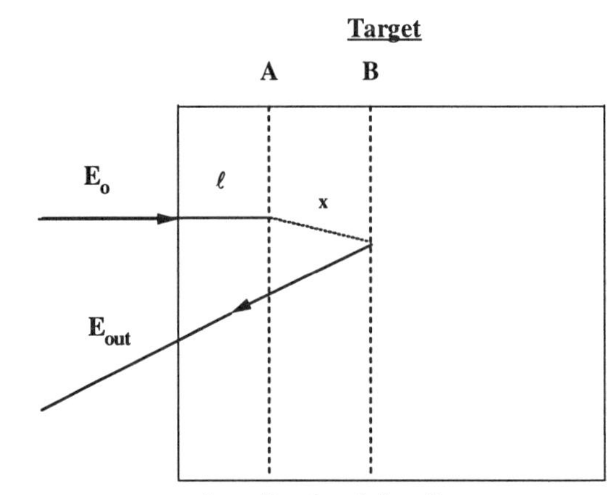

FIGURE 2. Diagram showing the scenario of the proton movement inside the silicon target.

$$D_{R\,min} = \left|\frac{E_{R\,min}-E_{\ell}}{\overline{S}(E'')}\right| - \lambda\cdot\ln\left(\frac{1-R}{I_o}\right) \Rightarrow$$

$$\Rightarrow D_{R\,min} - \frac{E_o - E_{R\,min}}{\overline{S}(E'')} = -\left(1-\varepsilon\frac{\overline{S}(E')}{\overline{S}(E'')}\right)\cdot\lambda\cdot\ln\left(\frac{1-R}{I_o}\right) \qquad (6)$$

Close to the target's surface, where $\overline{S}(E') / \overline{S}(E'') \cong 1$, the relationship is linear with slope equal to $-(1\text{-}\varepsilon)\cdot\lambda$. We can therefore use the experimental points close to the surface for an approximate determination of this product.

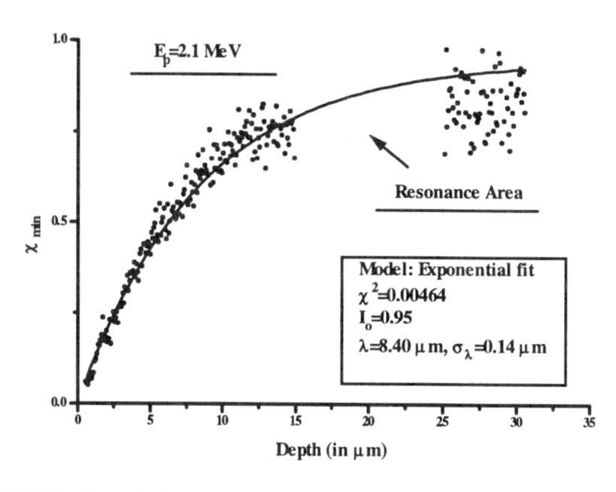

FIGURE 3. A typical spectrum at Ep = 1.7 MeV, showing χ as a function of depth (with the use of RUMP for the depth calibration of the MCA) with the appropriate exponential fit following the exclusion of the interference region where great anomalies in the ratio occur.

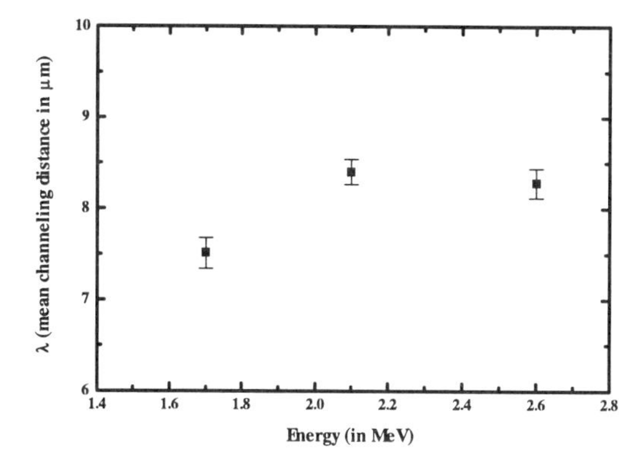

FIGURE 4. Mean channeling distance at Ep = 1.7, 2.1 and 2.6 MeV showing that λ varies insignificantly for the energy range into consideration.

RESULTS AND DISCUSSION

Dividing the aligned spectra channel by channel with the random ones, the function $\chi(x)$ is obtained. The part containing the resonance region is excluded (fig. 3) and the remaining points are fitted with an exponential function of the form: $\chi(x) = I_o(1-e^{-x/\lambda})$. For the energy range into consideration, the decay constant λ was found to vary between 7.5 ± 0.2 and 8.4 ± 0.2 μm (fig. 4), showing that the initial beam divergence was roughly constant. For the analysis we used the value: $\overline{\lambda} = 8.1 \pm 0.2$ μm.

It should be noted that the most striking difference between the random and channeled spectra is the shape of the peaks. In the random case the shape of the peak still resembles the shape of the thin target yield as presented in the literature (8). In the axial case [100], which is analyzed in the present work, the broadening is much larger. Another effect is the progressive "shallowing" of the minimum that preceded the peak in the random spectrum, which is caused by particles with a lower energy loss on their incoming trajectories. The resonant shapes are further broadened with increasing initial energy, mainly due to the energy straggling, since the shapes appear at greater depths. It is also important to note the ever increasing contributions from particles continuously dechanneling along the way till they reach the resonance energy, with the two extreme cases being particles dechanneled initially at the surface and particles dechanneling immediately before reaching the resonance energy. With the use of the mean channeling distance λ we average over the different trajectories.

If we constrain the analysis to initial energies relatively close to the energy of each resonance, meaning that the corresponding resonance shape appears at a depth of a few microns at the most, we can ignore the straggling factor. For these depths it can be assumed that since the maximum appears at the same channel –due to the strong contribution of the initially dechanneled particles of the beam, the whole information of the channeling procedure is contained in the change of the position of the minimum. The position of the resonance minimum in the random spectra can be used with a RUMP-like program to recalibrate the energy scale of the MCA into a depth scale. Thus, the position of the corresponding minimum, D_{Rmin}, from the channeled spectra can be calculated , under the assumption that the backscattered particles exit the crystal in a random direction.

The corresponding fractions R are determined by integrating over the entire range of the resonance up to 3σ, thus for 98% of the energy region where the resonant term in the total cross section is dominant. The total error in R (including statistical error of experimental data and systematic error due to the method of integration adopted) is estimated to be of the order of 10%. After the extraction of the experimental quantities involved in eq. 6, the values of the modified distance $D_{Rmin}-[(E_o-E_{Rmin}) / \overline{S}(E'')]$ versus the modified logarithm $-\ln[(1-R) / I_o]$ can be plotted as shown in fig. 5, including experimental points from both resonances. The offset of the line is very close to zero (0.09 $\pm$ 0.16), demonstrating the validity of our hypothesis near the target surface. The inclination of the line, determined with the least squares fit method assuming weighted errors in both individual parameters, was found to be 2.8 ± 0.4 μm, leading to a value of $\varepsilon = 0.64 \pm 0.07$, in good agreement with the transmission experiments. For the corresponding mean stopping powers, standard values from literature

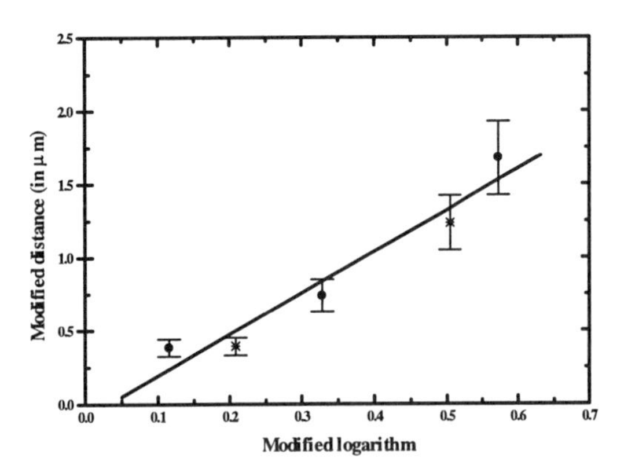

FIGURE 5. Plot of the modified distance (eq. 6) versus the modified logarithm using the experimental points at Ep = 1.7, 1.75, 1.8, 2.15, 2.2 MeV for the linear fit (• → points from the resonance at 1.67 MeV, * → points from the resonance at 2.09 MeV).

have been used (10).

CONCLUSION

This method offers a simple tool for the determination of the average channeling energy losses of protons in a silicon bulk target. The results are obtained in situ for the crystal into consideration and no complicated mathematical analysis is required. The value of ε extracted can be subsequently used in a RUMP-like program for elemental analysis in channeling geometry. It is believed that it can be applied to other projectiles, namely α-particles in silicon (11), as well as in more complicated crystals, such as Al_2O_3 and MgO.

Nevertheless, due to all the analyzed arguments, it is evident that at greater depths the simplified hypothesis cannot reproduce the experimental data, the linearity is lost, and the method used for the integration fails. A detailed simulation would have to take into account the different impact parameters, different trajectories, energy and spatial straggling, thermal effects, initial divergence of the beam as well as the total cross section for the whole spectrum. An interesting attempt presented recently (7), clearly demonstrates the difficulties and compromises in accuracy one should tolerate if such a task is undertaken. It is believed that a lot of fine and complex details need to be studied before a standard method for the analysis of any sample in channeling geometry is established.

REFERENCES

1. Appleton, B. R., Erginsoy, C., and Gibson, W. M., *Physical Review* **161**, 330-349 (1967).
2. Lindhard, J., *Mat. Fys. Medd.* **34**, n. 14, 1-64 (1965).
3. Sattler, A. R., and Dearnaley, G., *Physical Review* **161**, 244-251 (1967).
4. Rudnev, A. S.,Shyshkin, K.S., Sirotinin, E.I., and Tulinov, A.F., *Radiation Effects* **22**, 29-33 (1974).
5. Blanchin, D., Poizat, J.-C., Remillieux, J., and Sarazin, A., *NIM* **70**, 98-102 (1969).
6. Hellborg, R., *Physica Scripta* **4**, 75-82 (1971).
7. Vos, M., Boerma, D. O., and Smulders, P. J. M., *NIM* **B30**, 38-43 (1988).
8. Vorona, J., Olness, J.W., Haeberli, W., and Lewis, H.W., *Physical Review* **116**, 1563-1571 (1959).
9. Ziegler, J. F., *"The Stopping and Range of Ions in Matter"*, Pergamon Press, 1977-1985, Vol. **2-6**.
10. Gemell, D. S., *Reviews of Modern Physics* **46**, 129-227 (1974).
11. Santos, J. H. R., Grande, P.L., Boudinov, H., Behar, M., Stoll, R., Klatt, Chr., and Kalbitzer, S., *NIM* **B106**, 51-54 (1995).
12. Kokkoris, M., Huber, H., Kossionides, S., Paradellis, T., Zarkadas, Ch., Gazis, E.N., Vlastou, R., Aslanoglou, X., Assmann, W., and Karamyan, S., *presented in the 14th International Conference on the application of Accelerators in Research and Industry*, 6-9 November (1996).

ION BEAM ANALYSIS AT THE NEW CWRU-OHIO MATNET 1.7 MV TANDEM PELLETRON FACILITY†

M.A. Stan, C. A. Zorman, M. Mehregany, J. Weiss, J. Angus, and A. H. Heuer

Case Western Reserve University, Cleveland, Ohio 44106-7204

In the fall of 1995, an ion beam analysis (IBA) laboratory was established at Case Western Reserve University (CWRU) as a part of the Ohio Materials Network (MatNet). A brief overview of MatNet and the accelerator/endstation equipment will be given, followed by a sampling of IBA projects during the laboratory's first year of operation. Included are boron concentration measurements of chemical vapor deposited (CVD) diamond using the $^{11}B(p,\alpha)2\alpha$ reaction at 660 keV and the application of Rutherford backscattering spectrometry (RBS) and elastic recoil detection analysis (ERDA) for the determination of film density and hydrogen content in SiO_2 films grown on SiC.

† Supported in part by the National Science Foundation, Division of Materials Research, (Grant No. 9408720).

INTRODUCTION

The objective of this paper is to introduce our facility to the ion beam analysis (IBA) community and to present examples of how IBA was brought to bear on a sampling of materials engineering problems in the first year of the laboratory's operation.

The Ion Beam Analysis Laboratory at Case Western Reserve University (CWRU) was established in the fall of 1995 as an addition to the Ohio state-wide distributed network of materials synthesis and characterization facilities known as the Ohio Materials Network Consortium (MatNet). Currently, MatNet contains 5 major nodes, including that at CWRU, that provide user-oriented facilities and services to 14 Ohio universities with materials-related graduate programs. Additionally, the nodes service two Federal materials laboratories, NASA-Lewis Research Center, and Wright-Patterson Air Force Base, as well as local industry. At CWRU, the IBA laboratory is managed as part of the Center for the Surface Analysis of Materials (CSAM). In addition to ion beam analysis, the CSAM facility provides surface analysis and electron microscopy services to MatNet users.

The objective of the IBA Laboratory is to provide "routine" ion beam analyses, including the use of Rutherford backscattering spectrometry (RBS), ion channeling, elastic recoil detection analysis (ERDA), nuclear reaction analysis (NRA) by charged particle detection, and particle-induced X-ray emission (PIXE). In the following we will describe the laboratory instrumentation and capabilities, current and planned, followed by some applications of the IBA techniques to several on-going materials research projects.

ION BEAM EQUIPMENT

The ion source is an National Electrostatics Corporation (NEC) alphatross radio frequency (rf) source equipped with a Rb charge exchanger. It is designed for light ion production using H, He, N, and O gases. The analytical work at CWRU has required H^- and He^-, where source currents of 5 μA and 2 μA are typically used for the hydrogen and helium beams respectively.

The accelerator is an NEC model 5SDH 1.7MV tandem pelletron, equipped with a switching magnet providing beam output capability at ±15° and 0° with a mass-energy product of 12 amu-MeV. The beamline is presently configured with a single end station on the +15° port. The beam divergence with a 2 mm beam defining-apparature in the end station is less than 2 mrad. Some of the more important accelerator specifications are listed in Table 1.

TABLE 1. 5SDH Accelerator Specifications

Insulating column voltage rating	1.85 MV
Terminal voltage fluctuations	≤ 0.1 %
Singly charged ion energy range	0.6 - 3.4 MeV
Doubly charged ion energy range	to 5.1 MeV
3.4 MeV proton beam current on target†	≥ 200 nA
5.1 MeV He beam current on target†	≥ 500 nA

†- obtained with 2 mm beam defining-apparature at end station

CP392, *Application of Accelerators in Research and Industry*, edited by J. L. Duggan and I. L. Morgan
AIP Press, New York © 1997

The end station is presently equipped for RBS, ion channeling, ERDA, NRA by charged particle detection, and contains a Si(Li) detector for PIXE analysis. The Al scattering chamber (in plan view) shown in figure 1, has a diameter of 17" and is turbo-pumped. Samples are transferred to the chamber via a mechanically pumped load-lock directly above the target (T). The sample stage has approximate dimensions of 1"x 2". The scattering chamber is operated as an unsuppressed Faraday cup and is isolated from the beamline by an insulating break (IB). The beam defining apparature (BDA) is located within the collimator at the entrance to the chamber. A secondary electron shield (SES) has been added to the original equipment to block secondary electrons emitted from the target from reaching the collimator (which is electrically isolated from the rest of the chamber). The chamber is equipped with two 50 sq. mm Si surface barrier detectors. One of the detectors (SBD2) is located beneath the beamline and provides charged particle detection at a 170 ° scattering angle. A second detector (SBD1) is mounted on a moveable carousel and is typically used to collect charged particles in the forward scattering direction for ERDA. The sample manipulator is a Vacuum Generators model HPTRX. This manipulator provides both resistive heating and liquid nitrogen cooling capabilities. The manipulator provides 5 axis stepper motor controlled movement that includes two orthogonal axes of rotation for ion channeling measurement. The angular resolution of each rotation axis is of the order of 10^{-3} degrees. Sample viewing is provided by a 3" window (V) and a CCD camera. The CCD camera output at 5X magnification is displayed to a remote monitor and provides a very convenient way to locate the ion beam on the target (many materials fluoresce under ion beam irradiation). The chamber is equipped with a 10 sq. mm Si(Li) X-ray detector manufactured by Noran Instruments. The Si(Li) crystal is contained within a sealed vacuum and views the sample through a NORVAR ultra-thin window to allow for light element detection to and including B. The X-ray spectrometer has a 139 eV FWHM resolution for the Mn K_α line. Presently filters are attached to the Si(Li) detector collimator with carbon tape. Future chamber modifications will include a filter wheel and an electron flood gun.

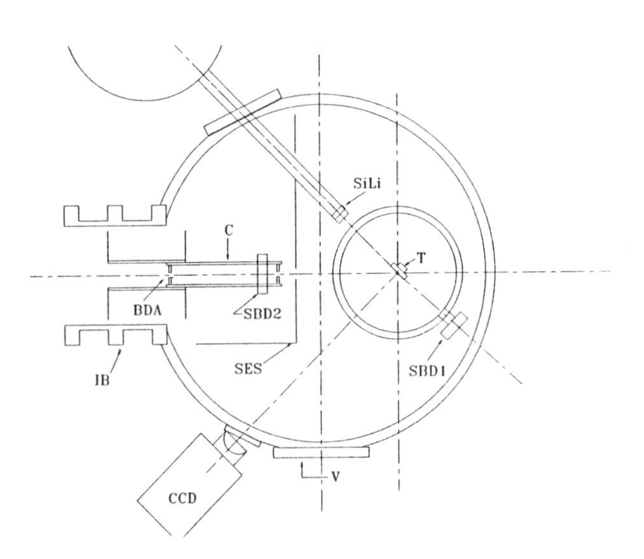

FIGURE 1. Plan view of end station vacuum chamber.

IBA IN THE LABORATORY'S FIRST YEAR

In the past year, nuclear reaction analysis has generated the most interest as measured by beam time spent on a particular technique. This is a consequence of NRA's virtually unique ability to provide non-destructive quantitative analysis of light elements, specifically B. Boron is an element not easily quantified by other techniques; its analysis is important in a variety of applications, including low pressure growth of conductive diamond films, planarization layers used in Si wafer production, and borosilicate glass manufacturing. Another area which has generated interest has been the ability of RBS to provide layer-by-layer composition, thickness, and interdiffusion analysis of optical films. RBS is generally well suited to these applications since the ion ranges are in the range of the wavelength of visible light. In these applications, IBA has a distinct advantage over competing techniques such as Auger and X-ray photo-electron depth profiling, inasmuch as the surface analysis techniques are destructive, time intensive, and suffer from spurious effects such as ion beam mixing and preferential sputtering. The IBA Laboratory has also provided ion channeling analysis to examine implant damage in semiconductor wafers.

PIXE analysis has been used only in qualitative applications to provide multi-element identification of samples whose composition is uncertain, prior to performing charged particle spectroscopy (RBS, NRA). Future enhancements to the laboratory's capabilities will include the use of the Guelph GUPIX (1) program for quantitative analysis of PIXE spectra.

There have been relatively few requests to date for hydrogen analysis by ERDA. In those cases where the hydrogen analysis was applied, issues such as the presence of surface hydrogen and beam damage of polymers resulted in difficulties with data interpretation. We present below some examples of how IBA at CWRU was successfully applied to several materials engineering problems.

BORON ANALYSIS WITH THE $^{11}B(p,\alpha)2\alpha$ REACTION

We have provided quantitative B analysis to CWRU researchers interested in solid source doping of hot-filament CVD deposited diamond films on Si. The objective of the research was to develop a model for the transport of B from a sublimating hexagonal boron nitride (h-BN) source located near the Si substrate onto which diamond was deposited.

The B doping experiments consisted of depositing CVD diamond on a series of Si substrates held at different distances from the h-BN solid source. The gas mixture, reactor temperature, and deposition time were the same for all deposition runs.

The "bulk" analysis was conducted using the large resonance (Γ=300 keV) in the $^{11}B(p,\alpha)2\alpha$ reaction at 660keV. The width of this resonance is not troublesome here, as previous secondary ion mass spectroscopy (SIMS) analysis of similar diamond films showed the B concentration to be uniform over a depth of several microns. The only interfering reaction in this application comes from ^{15}N and it has an α-particle yield 3 orders of magnitude lower than that from the B atoms (2). The interference is not expected to cause problems in the analysis, even though the diamond sees N from the solid source. This conclusion is obtained when it is realized that naturally occurring ^{15}N exists at the 0.37% level, the 3 order of magnitude difference in the α-particle yields, and recent research (3) that has shown nitrogen incorporation in diamond in an amount 2 orders of magnitude below that of B incorporation.

The α-particles were detected using a 3 msr solid angle at a 170° scattering angle with a 7.5μm Kapton foil to stop the proton recoils. A 96% dense TiB_2 pellet was used as a B containing reference standard. This material was chosen because it is conductive and can withstand the same magnitude of beam current that was applied to the diamond films. In this way, errors in the charge collection resulting from different beam profiles at different beam currents were avoided. To verify proper charge collection, the boron concentration of a piece of Pyrex glass, suitably carbon coated to minimize charging, was determined. Boron concentrations in the unknown n_x, were obtained using the following relation.

$$n_x = n_{st}[(Y_x / Q_x) S_x] / [(Y_{st} / Q_{st}) S_{st}] \quad (1)$$

The subscripts x and st refer to the sample under test and the standard (TiB_2), respectively. The α-particle yield is represented by Y, the incident proton dose by Q, and the proton stopping power in the surface energy approximation by S. The simplifying assumptions used to obtain this expression have been discussed in the literature (2). In all cases, S was obtained using the TRIM95 Monte Carlo program (4) with the theoretical densities for all materials.

The α-particle spectrum obtained from Pyrex glass indicated a boron concentration of 6.1 x 10 ^{22} atoms/cm^3. This value is high by 8.5%, but is considered to be adequate for the doping studies in the CVD diamond film growth experiments. Measurements of the α-particle yield for the diamond films were conducted with target current densities of 50 nA/mm^2 and doses in the range 200-600 μC . The well-documented high secondary electron yield of diamond films(5) was observed during the analysis. It was manifest by observing the change in the apparent beam current as the beam moved from the diamond film to the Al, target holder, during which a reduction of 20% in the apparent beam current was typically observed. This effect was dealt with by "conditioning" the diamond film for approximately 15 minutes, after which the difference in apparent beam current between the Al holder and the diamond films was reduced to less than 2%. We speculate that the reduction in the secondary electron yield under high energy ion irradiation is a consequence of graphitization of the diamond surface and accumulation of hydrocarbons in a "cracked" layer. The measured B concentration for the diamond films was in the range 1-4 x 10 ^{20} atoms/cm^3. Figure 2 shows the B incorporation in the diamond films as a function of the distance between the diamond film and the h-BN solid source. This data has been used to model the B doping as a simple one-dimensional diffusion process (6).

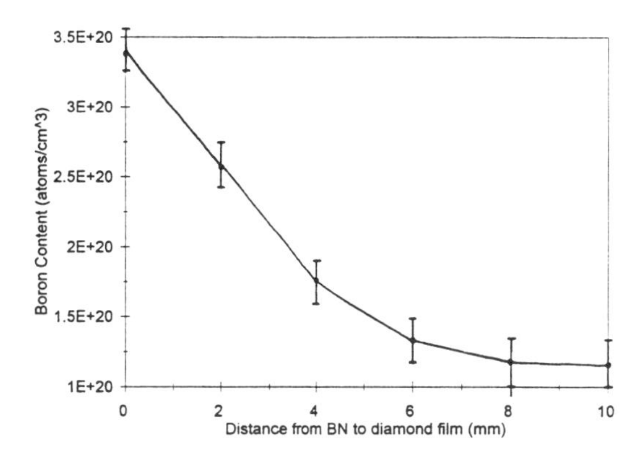

FIGURE 2. Boron content as a function of vertical distance between the h-BN solid source and the Si substrate. The solid line is a guide to the eye.

OXIDATION STUDIES OF 3C-SiC

MicroElectroMechanical Systems (MEMS) research at CWRU uses Si-based micromachining techniques to fabricate devices from structures containing 3C-SiC (the cubic SiC polytype) thin films. The micromachines require oxide layers for electrical and mechanical applications. RBS and ERDA have been employed in conjunction with scanning electron microscopy (SEM) and mechanical stylus measurements to evaluate the quality of SiO_2 layers grown on 3C-SiC using two oxidation procedures.

SiC films were oxidized with procedures designated as "dry" and "wet" oxidation. The "dry" oxidation entailed heating the SiC wafer to 1040 °C for 6 hrs. in the presence of a 6.0 slpm (standard liters per minute) O_2 flow. The "wet" oxidation procedure used a temperature of 1100 °C with an ambient produced with 5.5 slpm O_2 and 8.0 slpm$_2$ H . Following oxidation, windows were opened in the SiO_2 layers with conventional lithography and wet chemical etching. Mechanical stylus measurements were made across the step formed between the window and surrounding film to determine the oxide thicknesses. The oxide thickness grown with the "wet" oxidation procedure was 90 nm whereas that grown with the "dry" procedure was 70 nm.

RBS spectra of the films were obtained by positioning the 2.0 MeV He^+ beam on the oxide layer next to the window used for the mechanical stylus measurements. In this way, the mechanically determined thicknesses could be used in conjunction with a RUMP (7) simulation of the RBS spectra to obtain the oxide film density.

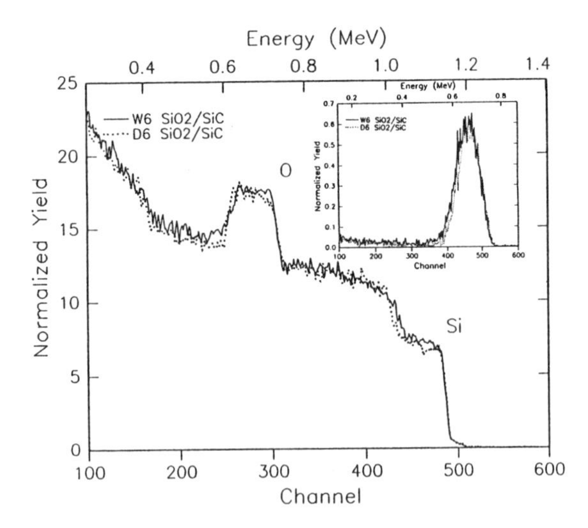

Figure 3. 2.0 MeV He^+ RBS and ERDA (inset) spectra of the "D6" and "W6" SiO_2 films grown with the "dry" and "wet" oxidation recipes, respectively.

Figure 3 shows the RBS spectra for an oxide film "W6" grown using the "wet" oxidation, and an oxide film "D6" grown using the "dry" oxidation recipe. The data were obtained with the ion beam at an inclination of 15° and a 30 μC dose. These data indicate that films are of nominally the same stoichiometry in the near surface region of the oxide. Furthermore, the low energy edge in the yield of the oxide layer of film W6 indicates the presence of a nonuniformity. This type of feature in an RBS spectrum can be attributed to at least two different mechanisms. Either the film surface is rough on the scale of the ion beam penetration, or the interface between the SiC and the SiO_2 layers is not sharp. To further elucidate the origin of the oxide nonuniformity in sample W6, SEM measurements were performed. The morphology of the D6 sample was quite smooth on the thickness scale of the oxide layer, whereas W6 contained a high density of surface defects whose height was on a scale comparable to the oxide film thickness. Consequently, the film thickness data of W6, determined by the mechanical stylus, is not accurate and cannot be used to obtain the oxide density of this sample. The mechanical film thickness measurement of sample D6 is valid, and when used in conjunction with a RUMP simulation of the data in fig. 3, yields an oxide density of 2.14 gm/cm^3, in reasonable agreement with the bulk value of 2.2 gm/cm^3.

Elastic recoil measurements of the films were performed to investigate whether any H was incorporated into the films during oxidation. The measurements were performed with 2.0 MeV He^+ ions, a 15° angle of incidence, and a scattering angle of 30°. A 7.5 μm Kapton foil was used to stop the He recoils. The scattering geometry was verified by performing ERDA on a polystyrene film whose H concentration and thickness were previously determined by NRA. The hydrogen forward recoil spectra of the two samples are shown as an inset in fig.3. These data have been fit to a model spectrum using RUMP, and are consistent with a "cracked" hydrocarbon surface layer. If the oxide films contained a detectable level of H throughout their thickness, the width of the forward spectra would be roughly twice that of the peaks shown in fig. 3. Examination of the low energy portion of the spectrum behind the surface hydrogen indicates no essential difference in the detectable H concentration of the films.

SUMMARY

An ion beam analysis laboratory was established at the CWRU node of Ohio MatNet in the fall of 1995 to provide "routine" IBA services to the participating MatNet users, neighboring Federal laboratories, and local industry. In its first year of operation, RBS, ion channeling, and charged particle NRA services were provided. Particular interest was shown in the application of RBS to optical thin films, and to the detection of boron via NRA in a variety of materials. In the upcoming year, quantitative PIXE analysis will added to the laboratory's IBA capabilities.

Boron concentration data from CVD-grown diamond films were presented. The $^{11}B(p,\alpha)2\alpha$ reaction at 660 keV provided adequate sensitivity in this application, where the B concentration was of the order of 10^{20} atoms/cm^3. Reasonable accuracy in the B concentration was obtained using a simple expression for the concentration, and a dense TiB_2 pellet as a B reference standard. Pre-conditioning of the diamond films in the ion beam was necessary to minimize spurious beam current integration, resulting from what is believed to be an initial high secondary electron yield of the diamond films.

2.0 MeV He^+ RBS and ERDA analyses were applied to the study of oxidation of 3C-SiC films. The objective was to establish the oxide film density and to determine if hydrogen was incorporated into the oxide films. The oxide produced by the "dry" recipe was uniform in thickness and had a density of 2.14 gm/cm^3 whereas the oxide grown with the "wet" recipe was non-uniform and a determination of its density was not possible. The H forward recoil spectra of the two films were essentially identical and consistent with that of a "cracked" hydrocarbon layer. Therefore the H content of the films is estimated to be less than 0.1 at %.

ACKNOWLEDGMENTS

The authors wish to thank Mr. J. Sears for SEM analysis of the oxide films, and Mr. K. Vinod for mechanical stylus measurements of the windows in the oxide films on SiC.

REFERENCES

1. Maxwell, J.A., *Nucl. Instr. and Meth.* **B95**, 407-421(1995).
2. Lappalainen, R., Raisanen, J., Antilla, A., *Nucl. Instr. and Meth.* **B9**, 55-59 (1985).
3. Samlenski, R., Haug, C., Brenn, R., Wild, C., Locher, R., and Koidl, P., *Diamond Relat. Mater.*, **3**, 947-951 (1996).
4. Ziegler, J., Biersack, J., and Littmark, U., *The Stopping and Range of Ions in Solids*, New York: Pergammon Press, (1985).
5. Mearini, G., Krainsky, I., and Dayton, J., *Surface Int. Anal.*, **21**, 138 (1994).
6. Weiss, J., *MS Thesis,* Case Western Reserve University (1996).
7. Doolittle, L.R., *Nucl. Instr. and Meth.* **B9**, 344-351 (1985).

ANALYSIS OF ALUMINUM NITRIDE THIN FILMS BY RBS AND NRA USING A DEUTERIUM BEAM

E. Andrade, J. C. Pineda, E. P. Zavala and F. Alba.

Instituto de Física, Universidad Nacional Autónoma de México, Ciudad Universitaria, Coyoacán, Apartado Postal 20-364, México, D.F., C.P 01000, México.

S. Muhl, J.A. Zapien and J. M. Mendez.

Instituto de Investigación en Materiales, Universidad Nacional Autónoma de México, Ciudad Universitaria, Coyoacán, Apartado Postal 70-360, México, D.F., 04510, México.

RBS and NRA using a low energy deuterium beam have been used in the characterization of thin films of aluminum nitride. The examination of the films prepared by DC reactive magnetron sputtering from an aluminum target and argon - nitrogen gas mixtures at various lateral distances from the center of the target, facilitated the modeling of the growing mechanism. The atoms areal density determined with the IBA measurements together with profilometry thickness measurements, gave the volumetric density of the films. X-ray diffraction measurements of the sample provide information of the AlN films structure.

INTRODUCTION.

The development of materials involves a variety of stages since the product density, purity, crystalline form, and orientation, etc., very often are determined by the fabrication process and are in their turn determinant for the material's macro properties. Additional to this is the fact that many of the details of the processes that occur during the fabrication of these new materials are frequently not well understood (1,2). These ideas have been the main basis for the research presented in this paper since aluminum nitride is of interest: it can be hard, chemically and thermally stable, it has a high thermal conductivity and its wide band gap means that it is transparent over a wide wavelength range (3). Similarly its high ultrasonic velocity, low acoustic loss and fairly large piezoelectric coupling factor signifies that it is promising for a wide range of thin film piezoelectric devices (4).

EXPERIMENTAL PROCEDURE.

In this work AlN films were prepared by DC magnetron reactive sputtering in a turbomolecular pumped vacuum system equipped with a stainless steel reaction chamber. The aluminum target was 4" diameter, ¼" thick and of 99.999% purity. The argon and nitrogen gas purity was 99.99%. The equipment was pumped down to below 2 x 10^{-6} torr prior to admitting the gases and the target was

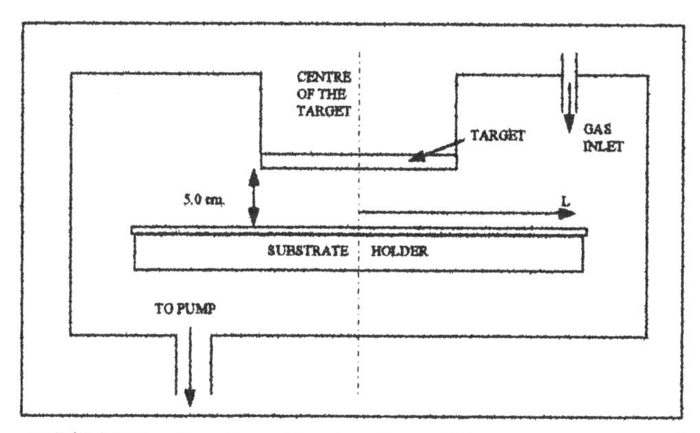

FIGURE 1. A schematic drawing of the arrangement of the sample holder and target. Illustrating the definition of the orientation of the distance L.

sputtered for 5 minutes against the shutter before the films were deposited. Pre-cleaned substrates of monocrystalline silicon, 0.5 x 0.5 cm., were used throughout at various lateral distances L (0.0, 1.3, 2.5, 3.8, 5.0, 6.5, 8.7, 10.0 and 11.5 cm), from the center of the target. The vertical distance between the substrate holder and the target was kept constant at 5 cm, see fig. 1. Samples were prepared at 250 °C, a gas pressure of 4 mTorr, total gas flow of 15 sccm which contained 70% of nitrogen and a plasma power of 300 watts.

The IBA facilities at the University of Mexico (5) based on a vertical single ended 5.5 MV Van de Graaff

CP392, *Application of Accelerators in Research and Industry*, edited by J. L. Duggan and I. L. Morgan
AIP Press, New York © 1997

accelerator were used to obtain the areal density and the composition of the AlN films. A 2 MeV ^{4}He RBS standard technique has been shown to be a satisfactory technique to analyze similar thin AlN films (6). However, ^{4}He gas was not available in our facility and instead a 960 keV deuterium beam incident perpendicular to the sample was used. At this bombardment energy the Coulomb barrier for the Si nuclei is high and the elastic cross section is about 3 orders of magnitude bigger than that of the possible ^{28}Si(d,p) and ^{28}Si(d,α) nuclear reactions (NR). A 300 µm surface barrier detector, placed at 165° laboratory angle and subtending a solid angle of 2.8 msr, was used to detect the emitted particles. Under these conditions there are various ^{14}N(d,α) and ^{14}N(d,p) nuclear reactions which are useful for analysis of the films. Both, the elastically scattered projectiles and the nuclear reaction products were detected, and for this reason no absorber was placed in front of detector. The beam current was kept sufficiently low (a few nA.) to avoid pulse pile up. A secondary electron suppressor was used for good beam integration.

A Sloan Dektak IIA profilometer was used to measure the thickness of the films and a Siemens X-ray diffractometer, model D500, using Cu Kα radiation was utilized to measure the crystallographic properties. A 1mm wide strip of silicon was used to mask the edge of the silicon substrates and in this way a step was formed between the edge of the deposit and the substrate. This step was used to measure the film thickness. These measurements together with the determination of the molecular areal density from the IBA analysis allowed us to calculate the volumetric density ρ (g/cm^3) of the samples.

EXPERIMENTAL RESULTS.

The figures 2a and b show typical energy spectra from one of the samples. Fig. 2a shows the low-energy part arising from elastically scattered particles and fig. 2b the high-energy part containing the peaks from nuclear reaction products. Using kinematic equations, Q-values, excitation energies and energy calibration, the peaks in the spectrum were identified as arising from ^{14}N(d,α)^{12}C, and ^{14}N(d,p)^{15}N. Also ^{12}C(d,p)^{13}C and ^{16}O(d,p)^{17}O peaks can be observed. They are due to surface and due to oxidation of the surface layers of the AlN. No peaks due to nuclear reactions with aluminium were observed.

The areal density (molecules/cm^2) and the composition of the AlN films was obtained from the simulation of the elastic region of the spectra (smooth curve in fig. 2a) using the well known RUMP software (7). This was aided by the thickness (~1µm) of the films since this increased the accuracy of the simulation with the software, and also because the value the deuteron beam energy chosen was such that the elastic cross section of aluminium and nitrogen can be described by the Rutherford cross section.

It was found that the areal density for the samples at distances L $\leq$ 10 cm were fairly constant and was on average about 5x10^{17} molecules/cm^2. An empirical estimation of the error in the areal density and composition of the films, by a process of trial and error, demonstrated that uncertainties are approximately ±3%. Using the value of the areal density from the RUMP fitting and the known density of AlN, 3.255 g/cm^3, a calculated film thickness of 0.99 µm is obtained, in agreement with the profilometer measurements. The areal density for the AlN films for distances L greater than 10 cms decreased to about 12% of the above value.

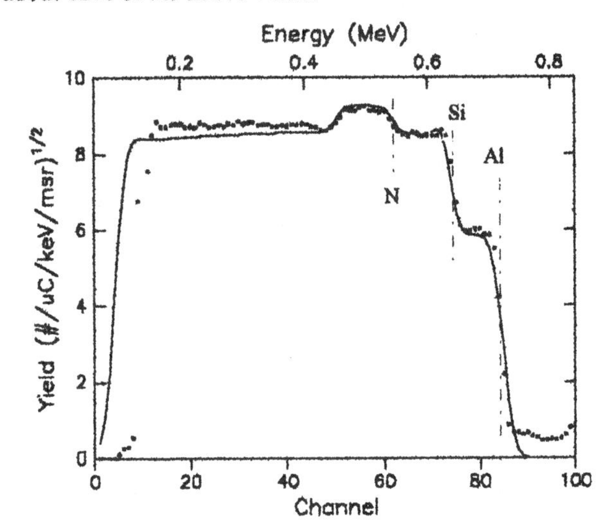

FIGURE 2a. A typical RBS spectrum, measured at 165° with 960 keV incident d ions, for an AlN film on a silicon substrate.

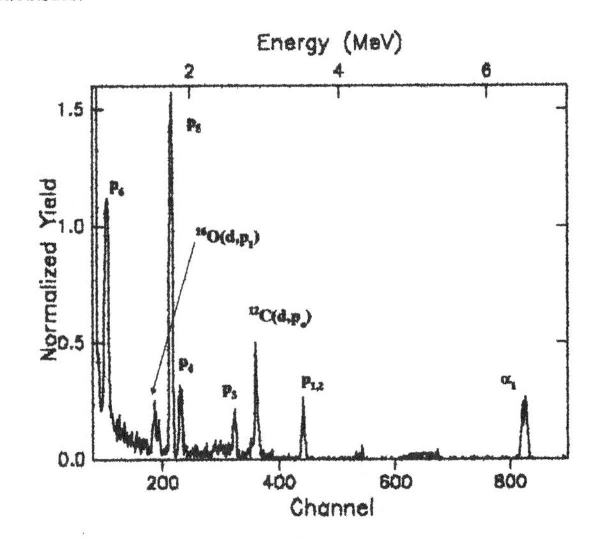

FIGURE 2b. A typical charged particle spectrum from the 960 keV d bombardment of a thin film of AlN on a silicon substrate, measured at 165°. Peaks arising from ^{12}C(d, p$_0$)^{13}C, ^{16}O(d, p$_1$)^{17}O, ^{14}N(d, α_1)^{12}C, and ^{14}N(d, p$_{1,2,3,4,5,6}$)^{15}N are labeled.

It is important to mention that the nitrogen areal density is difficult to obtain directly from the RBS spectrum since the cross section is much lower than that of aluminium, and the nitrogen counts are on top of those from the Si substrate. However, the areal density can be obtained indirectly from the lack of counts in the aluminium region. The assumption is that aluminium atoms are principally substituted by nitrogen, and that the elastic scattering cross-section is described by the Rutherford relation, the first point is supported by the comparison of the simulated spectra to the experimental data.

From each of the $^{14}N(d,p)$ and $^{14}N(d,\alpha)$ peaks in the energy spectra, the nitrogen areal density of the films can be deduced, and this gives an independent determination of the values obtained from the simulation of the RBS spectrum. The procedure involves calculations based on the area count yield of the peak, the measurement of the beam charge and the cross sections. However, the uncertainty in the values of the cross sections is a limitation. An alternative method is the use of reference materials (8) and this technique was used. A thick, high purity, stoichiometric TiN target was used as the reference material.

Comparison of the nitrogen areal density obtained with the simulation and that using the reference TiN target, resulted in a difference of about ±5 % for all the analyzed samples.

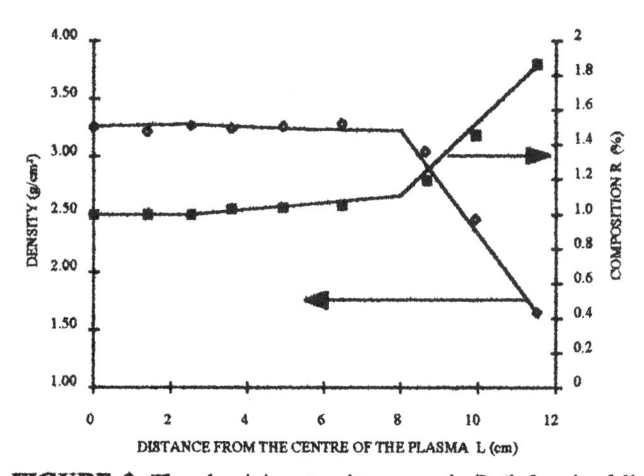

FIGURE 3. The aluminium to nitrogen ratio R (left axis, full symbols) and the calculated density (right axis, open symbols) as a function of the lateral distance from the center of the target.

The results of the compositional analysis of the AlN films, R = Al/N, and the volumetric density, calculated from the areal density divided by the measured film thickness, as a function of the lateral distance L are presented in figure 3. In a related study it was found that several film properties, optical bandgap, refractive index and dielectric strength, were constant for deposits made at L < 6.0 cm and changed almost linearly with greater distances. It is interesting to note that the aluminum to nitrogen ratio, *R*, increased from ~1 to ~2 as the distance L was varied from 0 to 11.5 cm from the center of the target.

The X-ray diffraction analysis of the films showed that the samples deposited in the central area, under the described conditions, were polycrystalline. The crystal structure was the commonly found hexagonal form typical of aluminum nitride, with the c axis perpendicular to the substrate surface (4). Samples prepared at L > 10 cm were amorphous.

Measurements of the substrate floating potential were performed in order to help explain why the properties of the deposits varied as a function of the distance from the center of the target. It was found that the potential became less negative as the distance from the center of the target increased, and hence positive ion bombardment of the film is expected to decrease with increasing L.

DISCUSSION.

This study has demonstrated various interesting aspects related to the preparation of aluminum nitride films using a DC plasma, especially in regard to the possible film formation processes. At one extreme we might consider that the compound is generated by a bombardment assisted chemical reaction on the surface of the aluminum target. Indeed studies of the deposition rate versus nitrogen concentration and those of the hysteresis involved in reactive sputtering indicate that the target surface is converted, to an unknown degree, to the nitride (9). Following this the compound would be sputtered and the AlN molecules condense to form the film. The other extreme is that aluminum may be sputtered and a combination of these atoms plus plasma generated nitrogen radicals impinge on the substrate, react and form the deposit. Obviously, combinations of these two models could also exist.

It is known that N_2^+ ions and N_2^* metastable radicals are formed within the sputtering plasma (10). In fact, the majority of such species will bombard the aluminum target and many are reflected from the target surface, with most being ejected almost perpendicular to the target surface (11). An additional source of nitrogen radicals is the fraction that are extracted from the plasma by the negative self-biasing of the substrate (12), and our measurements show that the self bias is stronger in the central region of the substrate holder. Thus, regardless of its source, nitrogen based energetic species will be incident on the substrate in a direction that is almost perpendicular to the surface. On the other hand, the angular distribution of the sputtered aluminum atoms from the target is in the form of a wide angle distribution (13). Therefore the film should become nitrogen deficient as the deposit is made further from the center of the target, since as L increases the arrival rate of the nitrogen species will decrease faster than the arrival rate of aluminium atoms. This is the observed situation for our experiments. The results of the density versus distance L show that the film prepared at 11.5 cm has a density much lower than that obtained by a composition-based estimation of the density of the deposit, 2.6 cm/gm^3. This value was determined by assuming a linear relation between the density and composition, and using the values of pure aluminium and AlN.

Similarly the variation of the floating potential with the distance, together with the observation that the films prepared at $L < 11.5$ cm were crystalline, clearly demonstrates that bombardment of the growing film by nitrogen ions is determinant for the morphology of the deposit. We can therefore conclude that the sputtering process, for a DC plasma, principally involves the ejection of aluminum atoms and nitrogen radicals bombardment of the film is determinant for the film structure and compositional formation process.

An additional interesting aspect of this work is demonstrated by the fact that some of the densities calculated from the ion beam analysis of films made at $L = 0$, but under different conditions of plasma power and nitrogen concentration, had values which would correspond to a density greater than the bulk value of aluminum nitride. A careful examination of the RBS spectra for such films showed that at the film - substrate interface there was intermixing of Al, nitrogen and Si. Now, as stated earlier, the film thickness was obtained from measurements at the edge of the film, whilst the RBS spectra was obtained in the middle area. Therefore, if etching of the substrate during deposition or diffusion of the depositing species into the substrate occurred, the RBS may have examined a thicker film than that measured at the edge and this would lead to the overestimation of the density. This situation has also been seen in similar analysis of a-C:H films deposited on silicon substrates (14). However, under the conditions used in the present work the obtained RBS spectra showed that this phenomena was not significant.

CONCLUSIONS.

Good quality aluminum nitride films have been prepared by DC reactive magnetron sputtering. The results indicate that the compound is principally formed by the arrival at the substrate of aluminum atoms and plasma generated nitrogen radicals that are independently adsorbed on the film surface and these then react to form the deposit. The morphology and crystallinity of the film depend on the nitrogen species bombardment. Finally, a combination of RBS and NRA was found to provide valuable information concerning the film composition and density.

ACKNOWLEDGMENTS

The authors wish to thank Leticia Baños for her help with the x ray diffraction analysis and Lazaro Huerta for his assistance in the operation of the accelerator. This work was supported by DGAPA-UNAM, through the project No. IN102495.

REFERENCES.

1. Messier R., *Proc. Surface Disordering: Growth, Roughening and Phase Transitions*, Les Houches, France, march 31, 1992, pp. 250-257.
2. Martin P. J., *J. Mater. Sci.*, **21**, 1-6 (1986).
3. Ed. O. Madelung, *Data in Science and Technology; Semiconductors, Group IV elements and III-V compounds*, Springer Verlag, 1991, p. 69.
4. Strite S. and Morkoç H., *J. Vac. Sci. and Technol. B*, **10**,1237-1241 (1992).
5. Andrade E., *Nucl. Inst. and Meth. Phys. Res.*, **B56/57**, 802-805 (1991).
6. Stedile F. C., Baumvol I. J. R., Schreiner W. H. and Freire Jr. F. L., *Nucl. Instrum. and Meth.*, **B79**, 501-505 (1993).
7. Doolittle L. R., *Nucl. Inst. and Meth. Phys. Res.*, **B15**, 227- 234(1986).
8. Amsel G. and Davies J., *Nucl. Instrum. and Meth.*, **218**, 177-183 (1983).
9. Morgan J. S., Bryden W. A, Kistenmacher T. J., Ecelberger S.A. and Poehler T.O., *J. Mater. Res.*, **5**, 2677-2681 (1990).
10. Harper J. M. E., Cuomo J. J. and Hentzell H. T. G., *J. Appl. Phys.*, **58** (1), 550-563 (1985).
11. Cachard A., Fillit R., Kadad I. and Pommier J. C., *Vacuum*, **41**, 1151-1153 (1990).
12. Este G. and Westwood W. D., *J. Vac. Sci. Technol.*, **A5**, 1892-1897 (1987).
13. Eds. John L. Vossen and Werner Kern, *Thin Film Processes.*, Academic Press, (1978), ch. 11-4, pp. 143-146.
14. Milne W. I., Personal Comunication.

THERMAL VIBRATION OF SURFACE ATOMS OBSERVED BY HIGH-RESOLUTION RBS

K. Nakajima, K. Kimura, and M. Mannami,

Department of Engineering Physics and Mechanics, Kyoto University, Kyoto 606-01, Japan

High-resolution RBS spectrum is measured when 300-keV He^+ ions are incident on a PbSe(100) single crystal along the [001] axis. The observed surface peak can be resolved into several peaks which correspond to successive atomic layers. This feature allows to estimate the thermal vibrational amplitude of the surface atoms. Comparison between the observed amplitude and the theoretical one calculated by means of the Debye model indicates that the thermal vibrational amplitude of the surface atoms is enhanced. The surface Debye temperature is estimated to be 113 K from the observed result while the bulk Debye temperature is 156 K.

INTRODUCTION

When a beam of fast ions is incident on a single crystal along a low index axis, the probability of the backscattering from crystal atoms is reduced (1). The energy spectrum of backscattered ions shows a peak, called "surface peak", which corresponds to the ions scattered from the atoms in the surface region. The ions of the surface peak are scattered mainly from the surface atoms because the subsurface atoms are shadowed by the surface atoms. This shadowing effect is not complete due to a possible surface reconstruction and the thermal vibration of crystal atoms. Thus the intensity of the surface peak depends on the surface atomic structure and the thermal vibration of the surface atoms. Detailed analysis of the surface-peak intensity allows to determine not only the surface atomic structure but also the dynamical property of surface atoms, *i.e.* the thermal vibrational amplitude of the surface atoms (2). It is known that the thermal vibrational amplitude is usually enhanced at the surface (3). Theoretical studies showed that the thermal vibrational amplitude is, in some cases, two times larger than the bulk value at the topmost atomic layer and it decreases very rapidly toward the bulk value with increasing depth (4). In the experimental studies, however, only an effective amplitude of surface region was estimated from the observed surface-peak intensity. If the contribution of each layer to the surface peak can be measured separately, more detailed information can be obtained, *e.g.* the amplitude for each atomic layer may be estimated.

In the present report, we observe the RBS spectrum of a PbSe(100) crystal under channeling conditions using a monolayer-resolvable high-resolution RBS system. It is demonstrated that the surface peak can be resolved into several peaks which correspond to the individual atomic layers. The observed peak yield is compared with the results of the calculation including the effects of the enhancement and the correlation of the thermal vibration at the surface.

EXPERIMENTAL

A single crystal of PbSe(100), which has an NaCl-type crystal structure, was prepared by epitaxial growth in situ by evaporation of pure PbSe on a SnTe(100) surface at 520 K in a UHV chamber. The substrate SnTe(100) crystal was also prepared in situ by epitaxial growth on a cleavage (100) surface of KCl, which was mounted on a high precision 5-axis goniometer. The surface structure of PbSe(100) surface was determined to be a bulk exposed surface using reflection high-energy electron diffraction (RHEED).

The RBS spectrum was measured using a high-resolution RBS system. The details of the high-resolution RBS system were described elsewhere (5). Briefly, a beam of 300-keV He^+ ions from the 4-MV Van de Graaff accelerator was collimated to 2.5×2.5 mm^2 and to a divergence angle less than $\pm$ 1 mrad by a series of apertures. The typical beam current was 10 nA. The ions scattered from the target crystal were energy analyzed by a 90° sector magnetic spectrometer. The acceptance angle of the spectrometer was about 5×10^{-5} sr. The energy analyzed ions were detected by a one-dimensional position sensitive detector located on the focal plane. Thus, the energy spectrum of the scattered He^+ ions were measured without sweeping the magnetic field. The energy resolution of the system was about 300 eV including the energy spread of the incident beam.

EXPERIMENTAL RESULT

Figure 1 shows an example of the observed energy spectrum of backscattered He^+ ions at an scattering angle θ_s = 94° when the beam was aligned away from any major

CP392, *Application of Accelerators in Research and Industry*, edited by J. L. Duggan and I. L. Morgan
AIP Press, New York © 1997

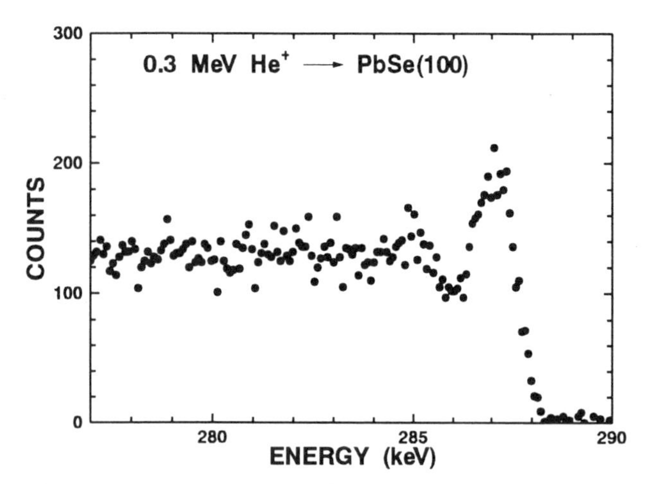

FIGURE 1. Energy spectrum of He^+ ions scattered at 94° when 300-keV He^+ ions are incident on a PbSe(100) crystal at random direction. The peaks at ~ 287 and ~ 285 keV correspond to the ions scattered from Pb atoms in the first and second atomic layers.

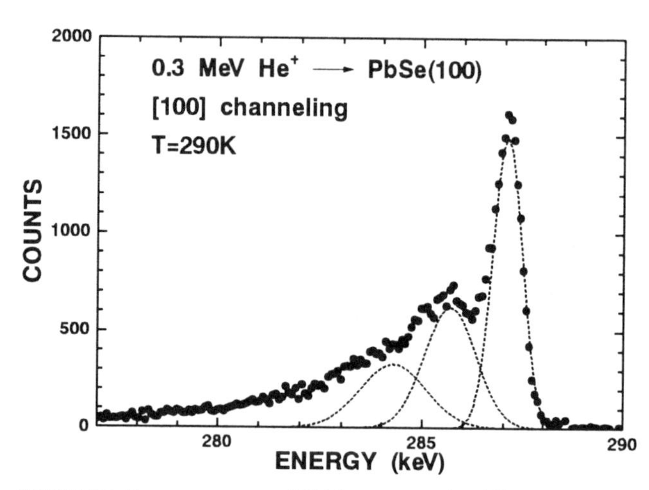

FIGURE 2. Energy spectrum of He^+ ions scattered at 94° when 300-keV He^+ ions are incident on a PbSe(100) crystal along the [100] axis. The so-called surface peaks can be resolved into several peaks which correspond to the ions scattered from Pb atoms in the successive atomic layers.

crystallographic direction. The incident and exit angles were ~ 4° and ~ 87° respectively. As the energy of the ions scattered from Se atoms are smaller than 269 keV, only the He^+ ions scattered from Pb atoms can be seen in the spectrum. There are two peaks at ~ 287 and ~ 285 keV. The peak separation agrees with the calculated energy difference of the ions scattered from adjacent atomic layer showing that these peaks corresponds to the He^+ ions scattered from the Pb atoms in the first and second atomic layers respectively. Our previous study showed that the charge state distribution of the scattered He ions is independent of the atomic layers from which the ions are scattered (6). Thus, the observed peak yield can be directly converted to the Pb concentration of each atomic layer. The observed spectrum indicates that there is no enhancement or reduction of Pb concentration at the surface region.

When the incident direction is aligned to a low index axis, the spectrum changes drastically. Although the first peak intensity does not change, the yield from subsurface layers reduces due to the shadowing effect. Figure 2 shows an example of the observed energy spectrum of backscattered He^+ ions at θ_s = 94° when 300-keV He^+ ions were incident on the PbSe(100) along the [100] axis at room temperature. In order to avoid a possible blocking effect, the exit direction was aligned away from any major crystallographic direction. The observed spectrum was deconvoluted into Gaussians as shown by dashed curves. The intensity of the second peak normalized to the first peak intensity was obtained to be 0.71.

Similar spectra were observed at various crystal temperatures. Figure 3 displays the temperature dependence of the normalized yield of the second Pb peak. The yield increases with increasing temperature.

DISCUSSION

As the PbSe(100) surface was found to be a bulk exposed surface by RHEED observation, the contribution from subsurface layers comes from the imperfect shadowing effect due to thermal vibration. Although the Monte Carlo simulation is needed to calculate the whole surface-peak yield, the contribution of the second peak can be calculated without Monte Carlo simulation. Here, we concentrate on the second peak yield.

Because the PbSe has an NaCl-type crystal structure, Pb and Se atoms are arranged alternately along the [100] direction with an interatomic distance 0.306 nm. The Pb atom in the second atomic layer is shadowed by the Se atom in the first atomic layer. The flux distribution $f(r)$ of the incident 300-keV He^+ ions around the Pb site in the second atomic layer is calculated by means of the impulse approximation for He-Se scattering and the Moliére

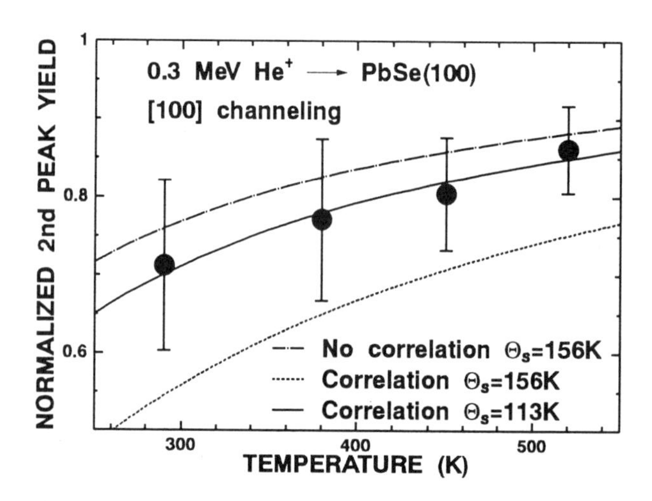

FIGURE 3. Temperature dependence of the observed normalized yield of the second Pb peak (circles). The calculated results without ether enhancement of thermal vibrational amplitude of surface atoms or the correlation of thermal vibration between the adjacent atoms (dot-dashed curve), including only the correlation effect (dashed curve), and including both the enhancement and the correlation (solid curve) are also shown.

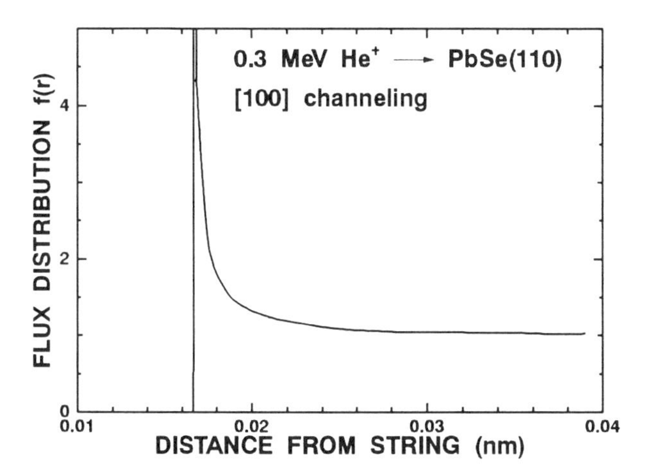

FIGURE 4. Flux distribution of [100] channeled 300-keV He^+ ions around the Pb site in the second atomic layer for a static PbSe crystal.

potential for ion-atom interaction. Figure 4 shows the calculated result for a static crystal, *i.e.* the calculation without the thermal vibration. The effect of the thermal vibration of the surface Se atom is given by a convolution with a Gaussian which describes the thermal vibration

$$F(\boldsymbol{r}) = \frac{1}{2\pi\sigma_1^2}\int_{-\infty}^{+\infty} f(\rho)\, e^{-\frac{(r-\rho)^2}{2\pi\sigma_1^2}}\, d\rho, \qquad (1)$$

where $\boldsymbol{r}$ and ρ are two dimensional vectors perpendicular to the [100] axis and σ_1 is the one dimensional root-mean-square (rms) thermal vibrational amplitude of the surface Se atom.

The normalized yield of the second Pb peak is given by

$$\begin{aligned} Y_2 &= \frac{1}{2\pi\sigma_2^2}\int_{-\infty}^{+\infty} F(\boldsymbol{r})\, e^{-\frac{r^2}{2\pi\sigma_2^2}}\, d\boldsymbol{r} \\ &= \frac{1}{2\pi(\sigma_1^2+\sigma_2^2)}\int_{-\infty}^{+\infty} f(\boldsymbol{r})\, e^{-\frac{r^2}{2\pi(\sigma_1^2+\sigma_2^2)}}\, d\boldsymbol{r}, \end{aligned} \qquad (2)$$

where σ_2 is the one dimensional rms thermal vibrational amplitude of the Pb atom in the second atomic layer. Thus the second peak yield can be calculated if σ_1 and σ_2 are given. The amplitude of the thermal vibration of the bulk crystal atoms can be calculated using the Debye model (7)

$$\sigma^2 = \frac{3\hbar^2}{Mk_B\Theta_D}\left(\frac{T^2}{\Theta_D^2}\int_0^{\Theta_D/T}\frac{x}{e^x-1}dx+\frac{1}{4}\right), \qquad (3)$$

where Θ_D is the Debye temperature, T is the crystal temperature and M is the mass number of the crystal atom. The dot-dashed curve in Fig. 3 shows the second peak yield calculated with the thermal vibration amplitude for bulk crystal. In the calculation, we employed average mass number, M = 143.1 , and Θ_D = 156 K (8). The calculated yield is slightly larger than the observed result but the agreement is rather good. However, if the enhancement of the thermal vibration of surface atoms is taken into account, the calculated yield becomes larger than that shown by the dot-dashed curve in Fig. 3 and the disagreement with the experimental result becomes larger.

This discrepancy can be solved by taking account of the correlation of the thermal vibration between the adjacent atoms. We assumed that the position distributions of the first Se atom and the second Pb atom are independent in the derivation of eq. (2). If there is a correlation, $\sigma_1^2+\sigma_2^2$ should be replaced by $\sigma_1^2+\sigma_2^2-2\sigma_1\sigma_2R_{12}$ in eq. (2), where R_{12} is the equal-time correlation between the displacement of the Se atom in the first layer and that of the Pb atoms in the second layer (9). The correlation is written as

$$R_{12} = \frac{\langle x_1x_2\rangle}{\sigma_1\sigma_2}, \qquad (4)$$

where x_1 and x_2 denote displacements of the Se and Pb atoms in the x-direction (perpendicular to the [100] axis) and $\langle\ \rangle$ indicates the time average. The correlation of adjacent atoms inside a bulk crystal can be calculated using the Debye model (10)

$$\langle x_ix_{i+1}\rangle = \frac{d^2k_BT}{2Mc^2\pi^2}\int_0^{\Theta_D/T}\left(\frac{1}{e^x-1}+\frac{1}{2}\right)\sin\left(\frac{Tq_Dd}{\Theta_D}x\right)dx, \qquad (5)$$

where q_D is the Debye cut-off wave number. Although the correlation at the surface region might be changed from the bulk value, eq. (5) is used to calculate R_{12} as the first order approximation. Including the correlation effect, the second peak yield without the enhancement of the thermal vibrational amplitude of surface atoms was calculated. The calculated result, shown by a dashed curve in Fig. 3, is smaller than the observed yield indicating that the thermal vibration is enhanced at the surface.

The effective Debye temperature Θ_s for the Se atoms of the first atomic layer was estimated as the following: Since the theoretical study showed that the enhancement of the thermal vibrational amplitude decreases rapidly toward the bulk value with increasing depth (4), we assumed that the enhancement occurs only in the first atomic layer. The thermal vibrational amplitude of the surface Se atom is calculated with eq. (3) using a surface Debye temperature Θ_s instead of Θ_D. The temperature dependence of the second Pb peak yield was calculated for various Θ_s and the results were compared with the observed one. The best fit, shown by a solid curve in Fig. 3, was obtained at Θ_s = 113 K. More detailed analysis is now in progress, which will be published elsewhere.

CONCLUSION

The energy spectrum of 300-keV He^+ ions backscattered from a PbSe(100) crystal was measured using a high-resolution spectrometer under the [100] channeling condition. The observed spectrum showed a surface peak which could be resolved into several peaks. These peaks correspond to the ions scattered from individual atomic layer. The intensity of the Pb second peak was calculated using the Debye model. Comparing the calculated result with the observed one, it was found that the thermal vibrational amplitude was enhanced at the surface. The surface Debye temperature was estimated to be 113 K while the bulk Debye temperature is 156 K.

ACKNOWLEDGEMENT

We are grateful to the member of the Department of Nuclear Engineering of Kyoto University for the use of the 4-MV Van de Graaff accelerator. This work was supported in part by a Grant-in-Aid for Scientific Research from the Japanese Ministry of ducation, Science and Culture.

REFERENCES

1. Gemmell, D.S., Rev. Mod. Phys., **46**, 129–227 (1974).
2. Feldman, L.C., Mayer, J.W., and Picraux, S.T., *Materials Analysis by Ion Channeling,* New York, Academic Press, 1982; and references therein.
3. Desjonquères, M.-C., and Spanjaard, D., *Concepts in Surface Physics,* Berlin, Springer-Verlag, 1993, ch. 4.
4. Allen, R.E., de Wette, F.W., and Rahman, A., Phys. Rev. **179**, 887–892 (1969).
5. Kimura, K. and Mannami, M., Nucl. Instrum. Methods in Phys. Res., **B113**, 270–274 (1996).
6. Kimura, K., Ohtsuka, H., and Mannami, M., Phys. Rev. Lett., **68**, 3797–3800 (1992).
7. *e.g.*, Blackman, M., *edited by S. Flügge, Encyclopedia of Physics,* Berlin, Springer, 1955, Part I, Vol. 7, p.377.
8. Lippmann, G., Kästner, P., and Wanninger, W., Phys. Status Solidi **(b) 43**, K159 (1971).
9. Jackson, D.P., and Barrett, J.H., Comp. Phys. Comm., **13** 157–166 (1977).
10. Nelson, R.S., Thompson, M.W., and Montgomery, M., Phil. Mag. **1** 1385 (1965).

CROSS SECTION FOR NON-RUTHERFORD BACKSCATTERING OF α ON ^{10}B

Jiarui Liu, O. Minayeva and Wei-Kan Chu

Texas Center for Superconductivity, University of Houston, Texas 77204-5932

Previously unreported sharp (α,α) resonance on ^{10}B has been observed and measured with good accuracy. The resonance energy is 5.085 MeV at lab angle of 165° with cross section enhancement of 16.5 relative to Rutherford cross section. Backscattering cross section on ^{10}B was measured for θ_{lab} = 165° in the 1.0 - 5.3 MeV energy range. Self-supported films of 500 Angstrom from enriched ^{10}B were used as targets. The 50% discrepancy between previous measurements in the energy range of 2.0-3.3 MeV has been resolved. The broad resonance at 4.1 MeV was also measured with better accuracy. Angular distribution at the 5.085 MeV resonance was measured for the geometric flexibility in ion beam analysis. The relative cross sections are presented in graphic and tabular forms with a typical ±2.7% uncertainty.

INTRODUCTION

Rutherford Backscattering Spectrometry (RBS) has high sensitivity to heavy elements in light matrices. Sensitivity to light elements, such as boron, carbon, nitrogen and oxygen in heavy matrices is low.[1] One of the methods to get around these limitations is to use backscattering in a higher energy range where strong departures from Rutherford scattering are observed due to nuclear interaction. Sensitivity to light elements can be enhanced at certain resonance energies by a factor of 10 to 100 relative to the classical RBS. This allows light-element detection and depth profiling in heavy matrices with better experimental error. Hence, this generalized Backscattering Spectrometry (BS) is an attractive approach, from an analytical point of view, since only a conventional RBS apparatus is required and all advantages of RBS for heavy-element analysis remain, allowing simultaneous light- and heavy-element detection and depth profiling. In BS, the scattering process is elastic and the kinematics are the same as RBS. The cross section is no longer Rutherford due to the nuclear interaction. The cross section at 170.5° increased by a factor of 22.66 at 3.034 MeV resonance and 128 at 4.265 MeV resonance for ^{16}O and ^{12}C respectively [2,3]. In our previous paper we reported the (α,α) cross section on ^{11}B with a new discovered resonance at 3.87 MeV with the cross section enhancement of a factor of 21 relative to Rutherford cross section.[4]

Boron is an important light element in semiconductors, nuclear reactors and fusion devices. There are two stable isotopes of natural boron (80.4% of ^{11}B and 19.6% of ^{10}B). The ^{11}B is the major isotope and the cross section has been published in our previous paper.[4]. The ^{10}B, although it consists of 19.6%, is an important isotope in nuclear fission and fusion technologies due to its' large thermal neutron absorption cross section. Moreover, if there is any sharp resonance for ^{10}B, it could be used for natural boron detection or for enriched ^{10}B applications due to the reasonable price of this isotope. Detection and depth profiling of boron in semiconductors have been studied for many years. Nuclear reactions, such as (p, α), (α,p), (p, γ) and (n, α) are often used for boron detection; although (α,α) BS is possible and simpler for ^{10}B measurement. There is not much cross section data for (α,α) elastic scattering in the literature, and there is only one collection of (α,α) cross section data in "*Handbook of Modern Ion Beam Materials analysis*"[5]. This data was published by McIntyre et al. for 170.5° lab angle in the energy range of 1-3.3 MeV.[6] Cross sections on ^{10}B were measured on self-supported enriched ^{10}B film with an accuracy of 7%. Previously unreported elastic scattering anomalies were observed in ^{10}B near 1.5 and 1.65 MeV. An earlier cross section on ^{10}B was published by Mo et al. in 1973 for ^{14}N energy levels study.[7] The cross sections were measured in the energy range of 2-4.3 MeV for the c.m. angles of 67.8, 80.7, 113.5, 141.8, 154,9 and 169.1 degree. The claimed error in the

CP392, *Application of Accelerators in Research and Industry,* edited by J. L. Duggan and I. L. Morgan
AIP Press, New York © 1997

absolute cross section measurements was estimated to be 3%. The comparison between the cross section at c.m. angle of 169.1° (lab angle of 162°)[7] and the cross section at lab angle of 170.5°[6] shows 50% discrepancy. There is no tabular data in both papers. The cross section was observed to be Rutherford between 1 and 1.3 MeV for both isotopes in all previous measurements.[4-8]

The objective of this work was to provide an accurate data of the (α,α) cross section for practical applications in ion beam analysis. In this paper, we present the cross sections of $^{10}B(\alpha,\alpha)^{10}B$ in the energy range of 1.0-5.3 MeV at θ_{lab} = 165° in both graphic and tabular forms. Angular distribution at the characteristic energy of 5.085 MeV was also measured for geometrical flexibility in ion beam analysis.

EXPERIMENTAL

Self-supported thin films of 500 Angstrom from 99% enriched ^{10}B were used in this measurement. Both highly enrichment and thinness helped to separate the ^{10}B signals from ^{12}C signals from contamination during sample preparation. Our measurements showed no evident interference signals from ^{12}C or ^{11}B in the energy range from 1.0 MeV to 5.3 MeV. Another essential factor for signal separations is to use a detector with good resolution. The detector used in this measurement was a partially depleted silicon detector with the energy resolution of 14 keV from Canberra.

The α-particles in the measurement were extracted from the 5SDH-2 pelletron at the Texas Center for Superconductivity, University of Houston. The details can be found elsewhere[4]. There was no measurable carbon contamination for more than 10 hrs bombardment on the sample with the 30-50 nA α-particle beam in the energy range of this measurement. The energy resolution of the detection system was 15 keV. Typical backscattering spectrum for low α–παρτιχλε energy of 1.0 MeV and resonance α -particle energy at 5.085 MeV are shown in Fig. 1. We can see from the spectrum at 1.0 MeV that the ^{10}B peak was well separated from the ^{12}C-contamination peak and there is no ^{11}B in between. The ^{12}C contamination was on both sides of the film, while the ^{16}O contamination is on the film but with non-uniform distribution. There were also Si and Ba contaminations as shown in the spectrum. The well separated ^{10}B peaks from peaks of other elements are essential factors in the cross section measurements. First, the backscattering yield can be counted without interference and the error of the relative yield measurement can be determined mainly by the statistics. Secondly at the low incident particle energy of about 1 MeV, the interaction on ^{10}B, C, O and other contamination are Rutherford so the relative contents of all elements and the energy loss in the film can be calculated for the effective energy corrections. The spectra at higher incident beam energy showed better separation of different elements and low background. The spectra were collected with ND-9900 multichannel analyzer with a ND-582 ADC with constant dead-time. The dead-time during spectra collection was less than 1% and no correction was required.

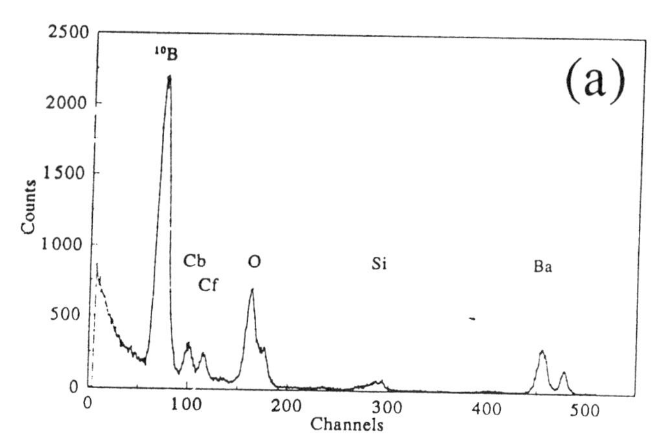

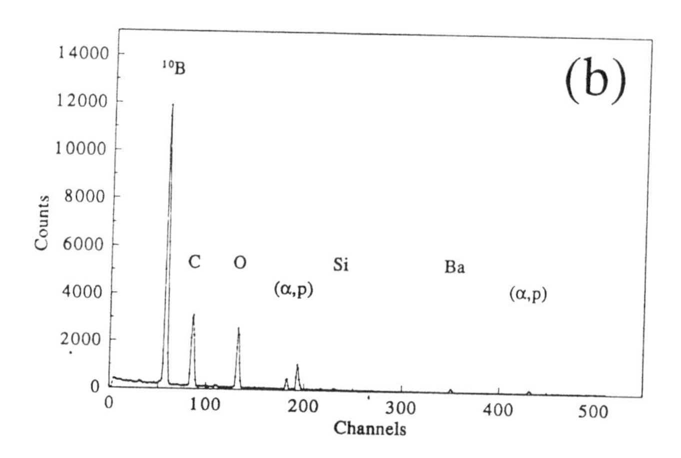

FIGURE 1. The backscattering α-particle spectrum at θ_{lab} = 165° for incident 1 MeV α-particles(a) and for resonance α-particle energy of 5.085 MeV (b). The ^{10}B backscattering peak and other elements peaks from the thin target are indicated

CROSS SECTION

The cross section calculation was performed with the same method as for ^{11}B, where the Rutherford cross section has been confirmed within ±1% in the energy range 0.98-1.33 MeV. Integrated counts under the ^{10}B peaks were extracted from 126 BS-spectra and normalized by the collected ion beam charge for cross section calculations. The cross sections were measured at $\theta_{lab} = 165° \pm 1°$.

The ratio of the cross section to the Rutherford cross section, σ/σ_R, at the same inciden particle energy of E can be calculated as:

$$\sigma/\sigma_R(E) = [A_B(E)/A_{RB}(E_R)] x [E/E_R]^2 F$$

where A_B (E) and A_{RB} (E_R) are the counts under ^{10}B peaks for the energy E and the energy of Rutherford scattering E_R, respectively. Scattering at E_R from ^{10}B was selected at a region determined to be Rutherford. F close to unity was the correction factor for the electron shell, the F-factor. The σ/σ_R ranges from 1.0 MeV to 5.3 MeV, as shown in Fig. 2.

The energy of Rutherford scattering E_R was selected around 1 MeV which is low enough to avoid cross-section deviation due to nuclear interaction. Theoretical estimate of Bozoian points a Rutherford cross-section below 1.7 MeV.[9]

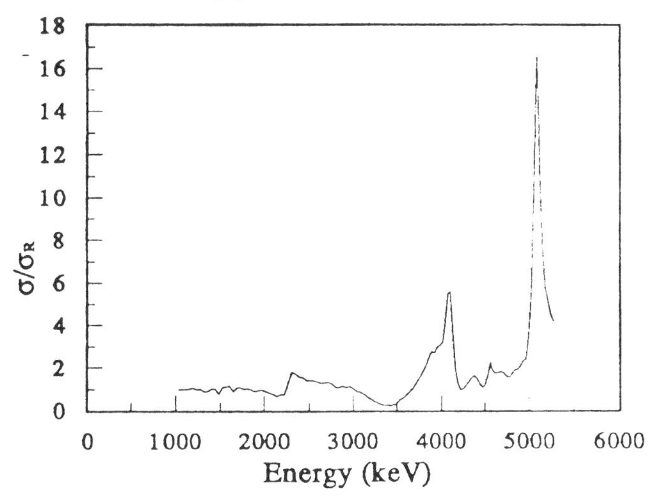

FIGURE 2. The ratio of the backscattering cross section to Rutherford cross section on ^{10}B at $\theta_{lab} = 165°$ for the α-particle energy from 1.0 MeV to 5.3 MeV.

The ratio of the cross section to the Rutherford value was close to 1 and was constant within experimental error of ±2.7% for an energy range below 1.3 MeV. The normalization cross section at 1.0 MeV as Rutherford is reliable. Another possible deviation from the Rutherford cross-section may be due to the electron shielding effect (F-factor) at low energy [10]. The estimate shows a possible 0.02% cross section deviation, which is negligible. The tabular form of the cross section is presented in Table 1 for practical use in ion beam analysis by BS at 165°. For ion beam analysis, the angular distribution at useful resonance energy of 5.085 MeV was also measured.

The cross sections in the energy range of 1 -3.3 MeV agreed with McIntyre's data well within experimental error. In the energy range between 2 MeV and 4.3 MeV, where about 50% discrepancy exists between McIntyre's (7%) and Mo's (3%) data, our measurements agreed with McIntyre's. Our measurement shows a broad resonance at 4.1 MeV with σ/σ_R =5.59. A more exciting result in this measurement is the resonance at 5.085 MeV with 16.5 times cross section enhancement relative to Rutherford cross section. This previously unreported resonance at 5.085 MeV with the half-width of about 75 keV is quite useful for non-Rutherford detection of ^{10}B.

ANGULAR DISTRIBUTION

The angular distribution was measured at selected energy of the new observed resonance at 5.085 MeV as shown in Fig. 3 for the laboratory angular range from 110° to 170°. We can see that the cross section is very sensitive to scattering angle at the resonance energies. The angular distribution is also given in Table 2 for direct use in ion beam analysis.

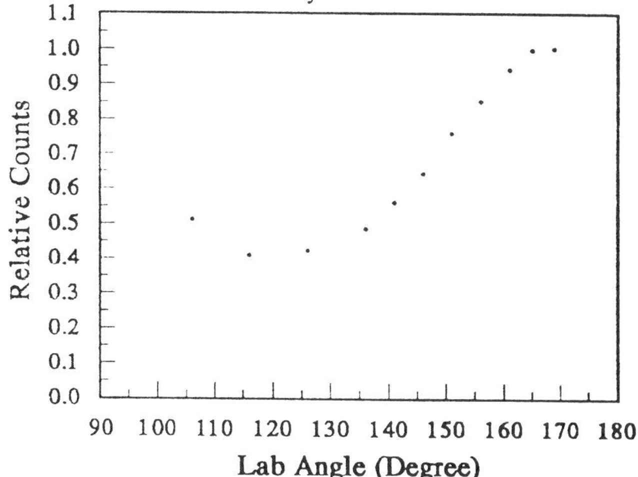

FIGURE 3. The angular distribution of (α,α) backscattering on ^{10}B at nuclear resonance energies of 5.085 MeV.

CONCLUSION

In conclusion, (α,α) cross sections on ^{10}B are presented in both graphic and tabular forms for the energy range from 1.0 to 5.3 MeV at $\theta_{lab} = 165°$. New resonance at 5.085 MeV with the σ/σ_R = 16.5 was observed. In the energy range of 1-3.3 MeV our measurement at lab angle 165° agrees well with McIntyre's data at lab angle at 170.5°, which shows that Mo's data has some problem in absolute cross section measurement. The angular distribution at new useful resonance energy is also presented as graph and table with the relative error less than 1%.

ACKNOWLEDGMENTS

This work was supported by the State of Texas through the Texas Center for Superconductivity at the University of Houston.

TABLE 1. Cross sections ratio σ/σ_R of (α,α) scattering on ^{10}B at θ_{lab} = 165° for energy range of 1.0-5.3 MeV.

Eα(keV)	σ/σR	Eα(keV)	σ/σR	Eα(keV)	σ/σR
1033	1.00	2755	1.24	4107	5.59
1074	0.98	2796	1.08	4138	3.40
1115	0.98	2837	1.08	4168	1.85
1156	1.00	2878	1.15	4199	1.21
1197	1.04	2919	1.08	4230	0.98
1238	0.98	2960	1.12	4260	1.00
1280	1.02	3001	1.02	4291	1.16
1321	0.88	3042	0.89	4322	1.33
1362	0.89	3083	0.85	4352	1.53
1402	1.00	3124	0.75	4383	1.62
1444	1.03	3165	0.62	4414	1.53
1485	0.75	3206	0.53	4445	1.26
1526	1.10	3247	0.42	4475	1.08
1567	1.11	3288	0.34	4506	1.20
1608	1.13	3329	0.27	4537	1.66
1649	0.88	3370	0.25	4567	2.23
1690	1.09	3411	0.28	4583	1.98
1731	1.06	3431	0.20	4598	1.82
1772	1.00	3452	0.28	4629	1.75
1813	1.03	3462	0.30	4659	1.79
1854	0.98	3493	0.40	4690	1.82
1895	0.90	3493	0.38	4721	1.72
1936	0.93	3523	0.46	4752	1.59
1977	0.94	3534	0.51	4782	1.56
2018	0.90	3554	0.55	4813	1.74
2059	0.82	3585	0.63	4844	1.90
2100	0.78	3615	0.75	4874	1.93
2141	0.65	3646	0.90	4905	2.10
2182	0.73	3677	1.01	4936	2.31
2223	0.73	3708	1.22	4966	2.39
2264	1.25	3738	1.43	4997	3.49
2305	1.78	3769	1.65	5028	5.60
2346	1.70	3800	1.86	5059	10.8
2387	1.57	3830	2.14	5074	15.1
2428	1.54	3861	2.45	5089	16.5
2469	1.39	3892	2.80	5105	14.3
2510	1.41	3923	2.68	5120	11.3
2551	1.38	3953	2.99	5151	7.49
2592	1.34	3984	3.04	5181	5.80
2633	1.27	4015	3.19	5212	5.04
2673	1.30	4045	4.13	5243	4.51
2714	1.32	4076	5.46	5273	4.18

TABLE 2. Angular distribution of (α,α) scattering on ^{10}B at resonance energy of 5.085 MeV. The scattering yields at different angles are normalized to 165°.

Angle(°)	Yield	Angle(°)	Yield
106	0.512	151	0.758
116	0.409	156	0.851
126	0.423	161	0.943
136	0.485	165	1.000
141	0.560	169	1.003
146	0.643		

References:

1. Chu, Wei-Kan., Mayer, James W., Nicolet, A., *Backscattering Spectrometry*. Academic Press, 1978.
2. Leavitt, J. A., McIntyre Jr, L. C., Ashbaugh, M. D., Oder, J. G., Lin, Z., and Dezvfouly-Ariomandy, B., *Nucl. Instr. and Meth* **B44**(1990)260
3. Leavitt, J. A., McIntyre Jr, L. C., Stoss, P., Oder, J. G., Ashbaugh, M. D. Dezvfouly-Ariomandy, B., Yang Z-M., and Lin, Z. *Nucl. Instr. and Meth* **B40/41**(1989)776
4. Liu, J. R., Zheng, Z. S., Chu, W. K., *Nucl. Instr. and Meth* **B108**(1996)1-6
5. Tesmer, J. R., and Nastasi, M. , *Handbook of Modern Ion Beam Materials Analysis,* MRS, 1995
6. McIntyre Jr, L. C., Leavitt, J. A., Ashbaugh, M. D., Lin, Z., and Stoner Jr. J. O., *Nucl. Instr. and Meth.* **B64**(1992)457
7. Mo, T. , and Weller, H. R., *Phys. Rev.* C8(1973)972
8. Dayras, R.A., Switkowski Z. E., and Tombrello, T. A., *Nucl. Phys.* **A261**(1976)36
9. Bozoian, M., *Nucl. Instr. and Meth.* **B82**(1993)602.
10. L'ecuyer, J. , Davies J. A., and Matsunami, N., *Nucl. Instr. and Meth.* **160**(1979)337.

RBX, COMPUTER METHODS FOR ANALYSIS AND SIMULATION OF RBS AND ERDA SPECTRA

E. Kótai

KFKI-Research Institute for Particle and Nuclear Physics, H-1525 Budapest P.o.b.49, Hungary

Since the seventies a large number of computer methods have been developed for data analysis in Rutherford backscattering spectrometry (RBS). The computer simulation method is the best way to get quantitative and extensive information from the row spectra. The main methods and approaches of spectrum synthesis are demonstrated on our RBX code. The program can use any particle-target combination and beam energy between 30 keV and 50 MeV. The calculations and fittings methods of non-Rutherford scattering cross section are included. The contribution from electronic screening, the corrected Bohr straggling, corrected stopping power in channel direction and pile-up effects are discussed. The useful techniques for extracting an accurate depth-concentration profile of impurities and crystal defects from spectra are demonstrated.

INTRODUCTION

Rutherford backscattering spectrometry (RBS) has been developed as a powerful non-destructive method to investigate the near surface region of solids. The technique is widely used for the quantitative and absolute determination of atomic concentrations and their variation with depth, the layer thickness of multielemental, multilayered films. Combining with the channeling technique it has been provided valuable information on the damage distributions and lattice location of doping atoms. Ref. (1) contains an extensive review of the backscattering spectrometry technique and presents most aspects and exhaustive bibliography on the subject.

The elastic recoil detection method (ERDA) was developed for profile analysis of light element. The few MeV He ERDA is widely used for H and D detection.

The modern computer programs for ion backscattering and ERDA data analysis should be capable of handling all kinds of spectra. The computer methods are ideally suited to account for many finer theoretical details, such as electronic screening, correction of Bohr straggling, nonlinear detector response, effects of pile-up, correction of stopping powers in channeled direction, etc.. For an accurate analysis the computer simulation method is the best way to get quantitative and extensive information from the raw spectra. The history and evaluation of computer modeling programs were published in Ref. (2, 3).

The RBX is a program (3) with an interactive graphical interface for ion scattering data analysis and simulation. This program can synthesize backscattering and ERDA spectra of multi-elemental, multi-layered films. Samples are considered to be build up with finite number of layers, each with uniform composition. The RBX code can handle the non-Rutherford cross-sections and channeling spectra. It has been successfully operating in our laboratory since May 1984.

SPECTRUM SYNTHESIS

The first step in any simulation is to describe the sample in a formal way. In the RBX code the matrix is made up of layers, in which each layer is assigned to a composition. The program makes an assumption that the composition within each layer is uniform. For the units of depth other than areal density, the density of layer must be defined. The simulated spectrum is made up of the superimposed contributions from each isotope of each sublayer in the sample. The basic method of the synthesis of RBS spectra is described in Ref. 3. In the spectrum simulation, we take into account the effect of the isotopes, effect of the non-Rutherford cross-section, the effect of the energy fluctuation due to the system resolution caused by the energy straggling and the effects of pile-up. The program is extended to simulate a channeling spectrum for damaged crystals as well.

Non-Rutherford spectra can be evaluated when tabulated or fitted scattering cross-section data are available. Fitting functions for the differential recoiled cross-sections of He ERDA are included (4, 5). For the calculation of other cross-sections the program uses an editable cross-section library. Special fitting function (Breit-Wigner, Lorentz formulas, step function etc.), energy and scattering angle dependent enhanced factor (tabulated or fitting functions) and tabulated cross-section values can be used. A separate program builds up the library. This program also converts the data measured in C.M. system to laboratory system.

The program incorporates energy straggling based on the Bohr model, as an option. The program uses the simple

CP392, *Application of Accelerators in Research and Industry*, edited by J. L. Duggan and I. L. Morgan
AIP Press, New York © 1997

asymptotic formula of Lindhard and Scharff (6) to consider the local electron density of the target atom:

$$\frac{\Gamma^2}{\Gamma^2_{Bohr}} = \begin{cases} L(\zeta)/2 & for\, \zeta \le 3 \\ 1 & for\, \zeta > 3 \end{cases} \quad (1)$$

where $L(\zeta) = 1.36\, \zeta^{1/2} - 0.016\, \zeta^{3/2}$. Here, ζ is the reduced energy variable defined by $\zeta = v^2/(Z_2 v_o^2)$ (v is the ion velocity and v_o the Bohr velocity). The program gives opportunity to modify the calculated straggling for each layers.

For the calculation of pile-up contribution a simple approximation is used (7). If the yield of the i-th channel of the synthesize spectrum is denoted by H(i) , then the pile-up contribution is given by:

$$PILEUP(i) = p\sum_{i'=1}^{i-1} H(i')\,H(i-i') \quad (2)$$

where p is the probability of the pile-up effect. The probability is proportional with the count rates. The program normalizes the calculated spectrum, so the total integrated count does not change.

Both random and channeled sample orientations can be assumed. The amount of dechanneling in the channeling spectrum is calculated for point defect, amorph clusters and extended defects (dislocations, dislocation loops, strain). The calculation of the dechanneling fraction for point defects is based on the multiple scattering model (8). This calculation is limited for single elemental crystals.

To calculate the depth profile of defects or to simulate a channeling spectrum, we need both random and channeling spectra from the same kind of perfect crystal. The normalized (channeling) yield at any depth x may be expressed as

$$\chi(x) = \chi_R(x) + [1 - \chi_R(x)]\left[\frac{N_D(x)}{N}\right] \quad (3)$$

where $\chi_R(x)$ is the random fraction of the dechanneling particles, $N_D(x)$ is the point defect distribution and N is the atomic density of the host material. The dechanneling fraction is expressed by

$$\chi_R(x) = \chi_0(x) + [1 - \chi_0(x)]\prod_i \exp\left[-\int \sigma_{Di} n_{Di}(x)dz\right] \quad (4)$$

where n_{Di} is the density of defects and σ_{Di} is the normalize dechanneling probability for each defects. This formula assumes that the dechanneling processes from different type of defects are independent. In the energy to depth calculation the stopping power ration of channeled ions to the random values must be defined for each orientation. The stopping power in a damaged crystal is calculated by:

$$\varepsilon_{oriented}(x) = \chi_R(x)\varepsilon_{random}(x) + (1 - \chi_R(x))\varepsilon_{virgin}(x) \quad (5)$$

where ε_{random} is the random stopping power, ε_{virgin} is the stopping power in the perfect crystal in the same orientation (9).

For the profile calculation we need the normalized channeling spectra for the perfect and damaged crystals and an assumption about the crystal defect type. The defect distribution is derived numerically started from the surface. The program finds n_D in the first interval, and calculates the amount of dechanneling for the next interval. The procedure is continued until the user defined depth is reached. The difference in the calculated and measured dechanneling yield behind the damaged region shows the rightness of our assumption about the defect type. The program can calculate automatically the depth profile of different crystal defect types (point defects and extended defects) in the case when the concentrations of these defects are proportional in the full region.

For the spectrum simulation of the channeling spectra of damaged crystal the program handles all defect type and distribution that is defined in depth-concentration table. The direct scattering and the dechanneling fraction are calculated and the resulting spectrum is a superposition of these two contributions.

SPECTRUM ANALYSIS

The RBX program calculates many short and approximate information about the sample (3). It displays the next data automatically using the cursor position:

1. Chemical symbol and atomic mass of the surface element.
2. Depth scale for a given element.
3. The areal density of a surface layer (at/cm^2).
4. The energy of the penetrating ions before scattering.
5. Area or integral counts of a peak.

These calculations use surface approximation. The necessary equations for such calculation are well documented and discussed (1).

In practical analyses of a simple solid (like binary compound thin films) the information is obtainable by inspecting Y(E), the yield versus energy spectrum. It is preferable to process the data and convert the energy scale into a depth scale and the yield into atomic concentration. The RBX program deconvolutes the RBS spectra by a process of surface approximation or successive simulation. The user must arrive at an inspired first estimate of the structure of the surface of the samples. The format of the layer descriptions is the same as for the spectrum synthesis. The unit of depth scale is normally the areal density (at/cm^2). For the conversion from this unit to other natural length unit, the density of the material in the surface region must be known and defined. Concentrations are expressed in atomic percentage or the ration of relative concentration of two atoms within the layer. The simulation method is based on a full but appropriate description of the sample, and uses the calculated cross section and stopping factor of each depth. In the first step the program simulates a spectrum with low and constant concentration of the given elements in all layers and calculates an approximation profile from the ratio of the yield of the experimental and synthesized data. The effect of the concentration gradient in a layer for the stopping is

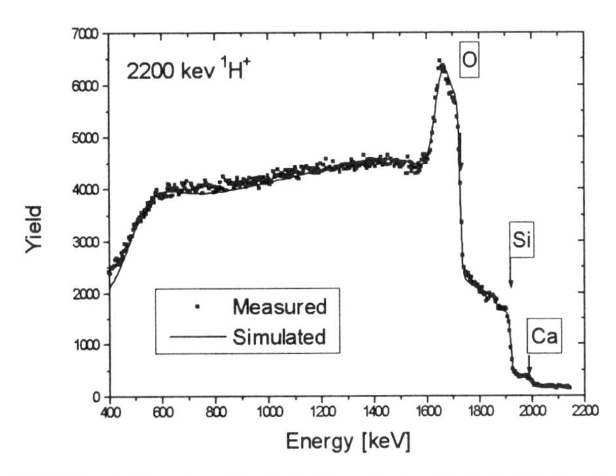

FIGURE 1. Experimental (points) and simulated (solid lines) backscattering spectra of 2200 keV $^1H^+$ on glass. The detecting angle was 165°. Five nA beam current was used for pile-up simulation.

neglected. In the second step the program simulates newly the spectra using this profile, and gets the final profile from the comparison of the experimental one (3). It is a useful and fast method, when the concentration of the given element is low or does not change rapidly.

APPLICATION

A few applications are presented to demonstrate the capabilities and limitation of the present spectrum simulation method. The measurements were carried out in the Van de Graaff laboratory of the Research Institute for Particle and Nuclear Physics. The analyzing beam was $^4He^+$ or $^1H^+$. The beam spot size was 1 x 1 mm. The beam dose was measured by a transmission Faraday cup (10). The spectrum was taken in standard RBS configuration (CORNELL geometry) with a 165° scattering angle and a 2.5 msr detector solid angle. The detector resolution was 20 keV (FWHM).

The first synthesized spectrum demonstrates the effects of non-Rutherford cross-section and pile-up (Fig.1). A glass sample was analyzed by 2.2 MeV proton beam. The $^{28}Si(p,p)^{28}Si$ reaction has two resonances below 2200 keV (11, 12). One of them at 1660 keV has a width of 55 keV, while the other one at 2080 keV has a width of 18 keV. The peak of the higher energy resonance appears near the oxygen peak, and the lower one at 600 keV. The shape of the latter peak became deformed by the energy straggling. The high energy tail is due to the pile-up effect. The program simulated the correct yield assuming 5 nA beam current.

Figure 2a. shows the channeling spectra of silicon implanted by 343 keV BF_2^+ ions and annealed at 600 °C 15 and 30 minutes. The 2 MeV $^4He^+$ beam was aligned to <100> direction. The program calculated the defect distribution in two ways: In the first case it assumes that the point defects contribute to the dechanneling only, in the second case extended defects (distortions of channels) give the dechanneling. The results are shown on Fig. 2a. The as implanted sample can describe by a 350 nm thick surface amorphous layer. After 15 minutes of annealing at 600° C the thickness of amorphous layer decreased and a significant amount of extended defects remains in the sample near the range of the implanted ions. For the spectrum synthesis I assumed that point defects dominated near the surface and extended defects in the bulk. The simulated channeling and random spectra are shown on Fig. 2b by solid lines.

Figure 3 shows the measured and simulated ERD spectra of TiH\Ni multilayers. The analyzing beam was 1700 keV $^4He^+$, the detecting angle 14°, and the sample tilting angle 85°. A 7.8μ Mylar foil was placed before the detector. Seven TiH\Ni layers were evaporated onto glass. The thickness of layers was about 7 nm. Small hydrocarbon contamination was found on the surface. The

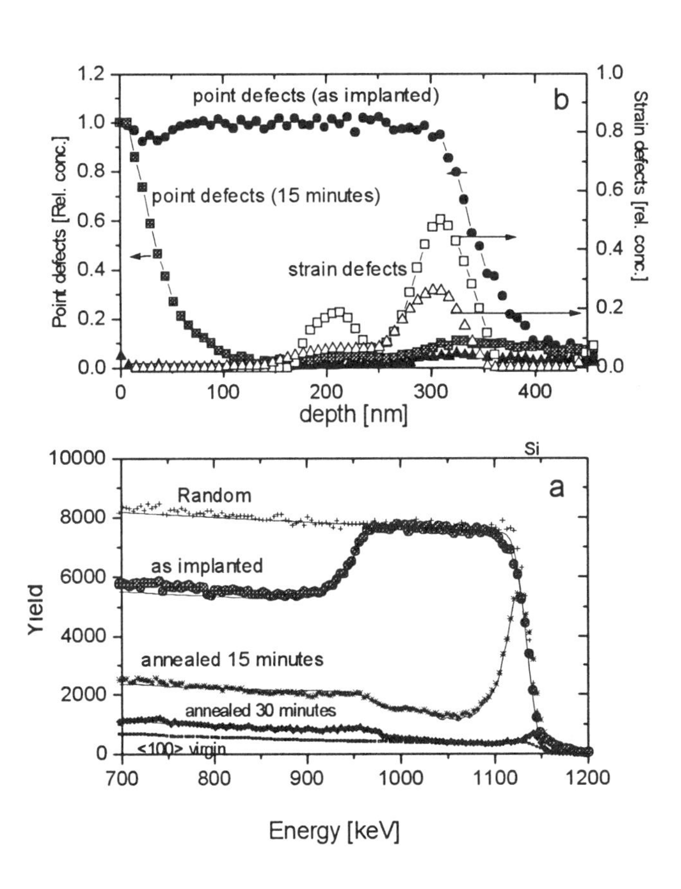

FIGURE 2(a) Observed (points) and simulated (solid curves) channeling spectra of 2000 keV $^4He^+$ ions from Si crystal implanted by 343 keV BF_2^+ molecular ions. The spectra of as implanted (circle), annealed at 600° 15 minutes (cross) and 30 minutes (triangle) are shown. The sample was oriented in <100> direction. The spectrum of a non-implanted sample is also shown for comparison (scattered line).
FIGURE 2(b) The calculated point (filled) and strain (open) defect distribution before (circle), and after annealing (square for 15 minutes, triangle for 30 minutes).

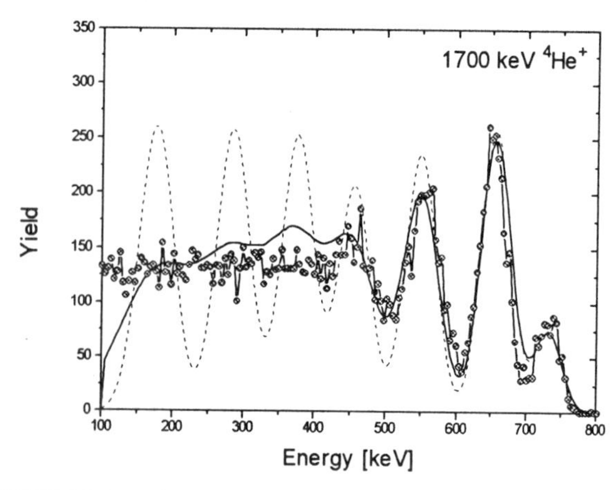

FIGURE 3. Experimental (points) and simulated (lines) ERDA spectra of 1700 keV $^4He^+$ ions on 7 x (TiH/Ni) multilayers evaporated onto glass. Small hydrocarbon contamination was found on the surface. The detecting angle was 14°. The dashed lines show the simulated spectra using pure Bohr straggling, the solid one shows the result of simulation with the correction for multiple scattering calculated by DEPTH (13) code.

dotted line shows the simulated spectra using pure Bohr straggling. This assumption is valid only near the surface, where the effect of multiple scattering is negligible. The solid line represents the result of simulation with straggling correction. The correction factors are calculated by DEPTH code (13). This program was developed by E. Szilágyi and F. Pászti to calculate the effects of multiple scattering, energy and geometry straggling on the depth resolution of RBS, ERDA and NRA methods.

SUMMARY

This paper presents an interactive nuclear backscattering analysis system used in our laboratory in the last twelve years. It is quite useful in extracting quantitative information from backscattering spectra using two methods: analyze or simulate a spectrum. The program can handle non-Rutherford scattering cross sections too. The time required to simulate a spectrum is about 1 - 20 sec depending on the complexity of the sample and the speed of the PC. The accuracy of spectrum synthesis is about 5-10 %. This program is proved to be very useful in planning a measurement, to select the best experimental parameters as well.

ACKNOWLEDGMENT

This work was supported in part by the OTKA grant T016506 of the Hungarian Academic of Science.

REFERENCES

1. Chu, W.K., Mayer, J.W., Nicolet, M.A., *Backscattering Spectrometry*, New York, Academic Press, 1978
2. Rauhala, E., *J.Appl. Phys.* **56** 3324 (1984).
3. Kótai, E. *Nucl. Instr. and Meth. in Phys. Res. B* **85** 588 (1994)
4. Szilágyi, E., Pászti, F., Manuaba, A., Hajdu, C., Kótai, E., *Nucl. Instr. and Meth. in Phys. Res. B* **43** 502. (1989).
5. Quillet, V., Abel, F., Schott, M., *Nucl. Instr. and Meth. in Phys. Res. B* **83** 47. (1993)
6. Lindhard, J. Scharff, M., *K.Dan. Vidensk. Sesk. Mat. Fys. Medd.* **28** 8 (1954)
7. Zolnai, L. Szabó, Gy., *Nucl. Instr. and Meth. in Phys. Res. B* **34** 118 (1988)
8. Felmann, L.C., Mayer, J.W., Picraux, S.T., *Material Analysis by Ion Channeling*, New York, Academic Press, 1982.
9. Kótai, E., "Measurement of the stopping powers for channeled ions in ion implanted crystals", in the *"Twelfth International Conference on Ion Beam Analysis"*, Tempe, Arizona 1995.
in print in *Nucl. Instr. and Meth. in Phys. Res. B.*
10. Pászti, F., Fried, M., Manuaba, A., Mezey, G., Kótai, E., Lohner, T., *J. Nucl. Mat.* **114** 330. (1983).
11. Vorona, J., Olness, J.W., Haerball, W. and Lewis, H.W., *Phys. Rev.* **116** 1563 (1959).
12. Rauhala, E., *Nucl. Instr. and Meth. in Phys. Res. B* **12** 447 (1985).
13. Szilágyi,E., Pászti, F., *Nucl. Instr. and Meth. in Phys. Res. B* **85** 616 (1994)

HEAVY ION BEAM BACKSCATTERING SPECTROSCOPY OF $GaAs/BaF_2/GaAs$ HETEROSTRUCTURES

M.F. Stumborg, F. Santiago, and T.K. Chu
Naval Surface Warfare Center, Dahlgren Division, Dahlgren, VA 22448

J.L. Price, N.A. Guardala, and D.J. Land
Naval Surface Warfare Center, Carderock Division, Silver Spring, MD 20903

Heavy Ion Backscattering Spectroscopy (HIBS) using 12.4 MeV Carbon ion beams was used to examine $GaAs/BaF_2/GaAs$ heterostructures deposited on GaAs substrates by Molecular Beam Epitaxy (MBE). The ability of the heavier ions to separate the backscattered signals of Arsenic and Gallium, as well as the individual isotopes of Gallium, was used to verify the 50/50 stoichiometry of the MBE grown GaAs layers. This information is not readily available using traditional *insitu* MBE diagnostic techniques. HIBS spectra were analyzed using a modification of an existing Rutherford Backscattering Spectroscopy (RBS) analysis program. The HIBS spectra were also instrumental in identifying interdiffusion of BaF_2 and GaAs layers taking place during MBE growth. This information was used to modify the MBE growth process to achieve sharper interfaces between the BaF_2 and GaAs layers.

INTRODUCTION

The goal of this study was to grow epitaxial GaAs films on BaF_2 films by Molecular Beam Epitaxy (MBE). The $BaF_2(100)$ films were grown on GaAs(100) wafers as described in a previous investigation (1). This report examines the subsequent growth of GaAs on these BaF_2 films. The MBE deposition of GaAs uses separate Ga and As effusion cell sources. This introduces the possibility that the deposited film may deviate from a 1:1 Ga-to-As ratio. Large deviations from this ratio can have adverse affects on the epitaxial quality of the film. The ability to determine the Ga-to-As ratio is therefore a very useful diagnostic aid to the MBE practitioner.

X-ray Photoelectron Spectroscopy (XPS) is commonly used in the analysis of MBE-grown materials. The extreme surface sensitivity of XPS makes it a valuable diagnostic tool, but also renders it useless when attempting to evaluate the stoichiometry of films. XPS can measure the Ga-to-As ratio only in the near surface region (< 20Å) of a film. It is well known that the surfaces of many epitaxial crystals readily reconstruct so that the stoichiometry at the surface is not the same as it is in the bulk region. This reconstruction is particularly prevalent in the III-V compound semiconductor crystals (2). An analysis technique other than XPS is needed to determine the Ga-to-As ratio throughout the GaAs layer. Heavy Ion Backscattering Spectroscopy (HIBS) is used for this purpose.

MBE GROWTH AND CRYSTAL STRUCTURE

GaAs was deposited on an MBE-grown BaF_2 film(1) in a VG Semicon V80H MBE deposition chamber at 650 °C for 2 hours. The sample was examined by *insitu* XPS after deposition. XPS indicated the presence of Ga, As, and a small amount of Ba on the sample surface. No fluorine was detected. The sample was then removed and examined by *exsitu* X-ray diffraction. Bragg diffraction peaks from (100) orientations of GaAs and BaF_2 were seen. No other orientations were present. The lattice constants calculated from these peaks were 5.65 Å for the GaAs, a value in agreement with the bulk value; and 6.30 Å for the BaF_2, which is 1.6% higher than the bulk value of 6.20 Å.

HEAVY ION BACKSCATTERING SPECTROSCOPY

The sample was loaded into the scattering chamber of an NEC model 9SDH2 Pelletron positive ion accelerator described in detail elsewhere (3). Pressure in the scattering chamber was lower than 5×10^{-7} mbar. A solid state surface barrier detector was placed at an angle of 170 ° with respect to the incident direction of the 12.4 MeV $^{12}C^{4+}$ ion beam, behind a 3/16 inch diameter entrance aperture, 8.4 inches from the sample surface. The scattering chamber was electrically isolated and grounded through a current integrator. The sample was tilted 30 ° from the incident beam direction. HIBS spectra of Counts versus

CP392, *Application of Accelerators in Research and Industry,* edited by J. L. Duggan and I. L. Morgan
AIP Press, New York © 1997

Backscattered Energy were collected by a Nuclear Data 6600 ADC/computer.

DATA ANALYSIS

The theory describing the major mechanisms by which backscattered ions lose energy (i.e. energy loss by electronic excitation of target atoms and kinematic collisions with target atoms) is very well understood. Generating a simulated spectrum composed of numbers of backscattered ions as a function of their backscattered energy is therefore a straightforward, but by no means simple, matter. This task is best handled by computer programs (4) that generate a simulated spectrum, given film parameters such as areal densities and relative concentrations of atomic constituents. The simulated spectrum is then compared to the experimental spectrum and appropriate parameters are adjusted to make the simulation agree with the experimental data. Currently available programs are capable of this analysis for standard Rutherford Backscattering Spectroscopy (RBS) where the projectile mass is much less than the target atom masses. An adaptation of such a program was used here to generate simulated spectra for HIBS data analysis. One major adaptation of the program arises from the ability of heavier projectiles to distinguish individual isotopes of atoms. For light projectiles (such as He) backscattered energy differences are often smaller than the energy resolution of the detector. In fact, backscattered energy differences between different *elements* can also be less than the detector resolution. Since isotopes are indistinguishable in RBS, a weighted average of the isotope masses is sufficient for analysis. For HIBS analysis each isotope is treated individually in the modified program.

Another consideration when moving out of the realm of Rutherford backscattering into HIBS is the effect of Coulomb screening on the backscattering cross section (5). This effect is not accounted for in the present version of the analysis program. Fortunately, this is not a major concern here because the only quantitative claims being made involve the Ga-to-As ratio. The difference of the Coulomb screening effect on the cross sections of Ga and As are relativvely insignificant, due to the similarity in the Z of these two elements.

Figure 1 shows the experimental HIBS spectrum of the MBE-grown GaAs/BaF_2/GaAs heterostructure. The peak centered at approximately 8 MeV is due to projectiles backscattered from Ba atoms in the middle layer. The GaAs/BaF_2 interface is atomically abrupt when BaF_2 is deposited on GaAs using a process developed in our laboratory(1). The asymmetric nature of the Ba peak shows that this is not the case when GaAs is deposited on BaF_2. The long tail on the high energy side suggests that Ba is present in the top GaAs layer. The counts at energies less than approximately 6400 keV are a superposition of signals due to backscattering from Ga and As atoms in the

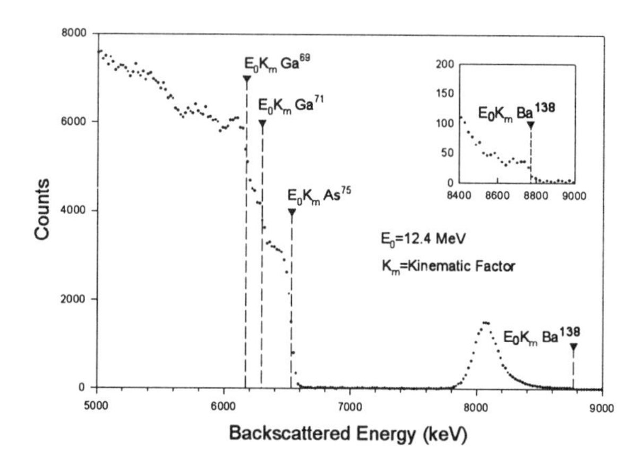

FIGURE 1. HIBS Spectrum: 12.4 MeV C^{12} incident on a GaAs/BaF_2/GaAs heterostructure and the energies of those ions after backscattering from Ga^{69}, Ga^{71}, As^{75}, and Ba^{138} atoms on the sample surface as calculated from the kinematic factors.

top layer and in the substrate. Also shown in the figure are the energies of projectiles backscattering from Ga^{69}, Ga^{71}, As^{75}, and Ba^{138} atoms that lie on the surface of the sample. Clearly, Ga, As, and a small amount of Ba are present on the sample surface. This is consistent with observations made during XPS analysis.

The clear advantage of HIBS over RBS is demonstrated in Fig. 1. The distinct features of the GaAs signal arising from the improved mass resolution make fitting simulation spectra much easier. More importantly, the Ga-to-As ratio of the top GaAs layer can be verified from the height of the As and Ga isotope signal edges. An RBS spectrum of this same sample (not shown here) does not separate the Ga isotope signals, and the separation of the Ga and As signals is barely perceptible.

The first step in analyzing the spectrum is to fit a simulation spectrum to the front edge of the GaAs surface signal. This is shown in Fig. 2. As previously mentioned, there are two significant energy loss mechanisms considered in RBS and/or HIBS. Projectiles backscattered from the surface are subject to just one of these mechanisms (kinematic collisions), making their contribution to the spectrum simpler to determine. Contributions from successively deeper layers with different atomic constituents and/or relative concentrations are then added to the simulation to build the spectrum. The simulation in Fig. 2 is of a surface layer of 1:1 GaAs with 0.25% Ba contamination. The amount of Ba contamination cannot be stated with confidence in light of the need to correct the cross sections for Coulomb screening in the analysis program. However, for the purpose of this investigation the amount of Ba contamination is not as significant as the fact that it occurs at the sample surface. The thickness of the layer was chosen to fit the dip in the experimental data at the back edge of the As peak. The thickness calculated from the simulation data is ≈ 1400Å , assuming the layer density to be that of GaAs (5.32 g/cm^3). The actual thickness is

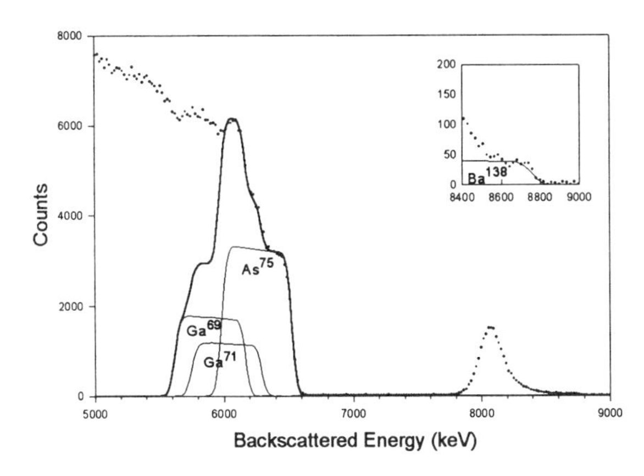

Figure 2. Simulation of a 1400Å GaAs layer containing 0.25% Ba. The GaAs signal is separated into contributions from As and the two Ga isotopes.

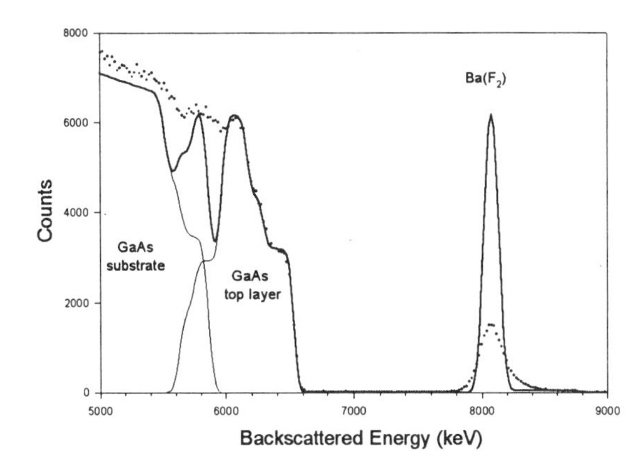

Figure 3. Simulation assuming a $GaAs/BaF_2/GaAs$ heterostructure with abrupt interfaces. The GaAs signal is separated into contributions from the GaAs layer on the sample surface and the GaAs substrate.

probably somewhat less since the Ba contamination would increase the volume density of the layer. The amount of Ba was chosen to fit the number of counts in the Ba peak at 8700 keV, as shown in the inset of Fig. 2.

The simulation is built up one layer at a time, with each layer having separate contributions from each elemental (and isotopic) species. It is therefore possible to decompose the simulation signal into its constituent parts. This is also shown in Fig. 2.[1]

The goal of this study was to grow GaAs on BaF_2 on GaAs. X-ray diffraction data strongly suggested the presence of $BaF_2(100)$ and GaAs(100) layers. The next step in the "building up" of the simulation spectrum was to add a layer of BaF_2 under the top GaAs layer, and then a GaAs substrate layer under the BaF_2 layer. A simulation spectrum following these assumptions is shown in Fig. 3. The BaF_2 thickness (assuming bulk BaF_2 volume density) of 286Å was chosen to give a simulated Ba peak with approximately the same FWHM as the experimental Ba peak. The Ga and As signals are separated into contributions from the GaAs top layer and the GaAs substrate layer.

The assumption of a $GaAs/BaF_2/GaAs$ heterostructure is not consistent with the data of Fig. 3. The layer beneath the top GaAs layer (presumably BaF_2) is strongly deficient in Ba as evidenced by the degree to which the Ba simulation peak rises above the experimental data. Note also the prominent "dips" in the simulation spectrum between 5400 keV and 6 MeV. Just as the simulation of the intermediate "BaF_2" layer over estimates the Ba concentration, it is logical to assume that some other constituents are underestimated, resulting in the dips. The only candidates for this missing material are As^{75}, Ga^{71}, and Ga^{69}. In light of the previous discussion regarding the ability of HIBS to resolve individual isotopes, it is hypothesized that the higher energy dip corresponds to As^{75} missing from the intermediate layer, and that the lower energy dip corresponds to Ga^{71} and Ga^{69} missing from the same region.

This hypothesis is further strengthened by a simple calculation. Consider a point on the simulation spectrum of Fig. 3 half way down the missing As^{75} dip, which has an energy of 5984 keV. Using the kinematic factor of As^{75}, this point corresponds to a backscattered C^{12} ion with an incident energy of 11.37 MeV. Using the kinematic factors of Ga^{71} and Ga^{69}, the same C^{12} ion will have backscattered energies of 5768 keV and 5654 keV respectively. These energies are in excellent agreement with the same point half way down the missing Ga^{71} and Ga^{69} dips. The position of these dips is consistent with a missing GaAs layer buried under enough material to slow a 12.4 MeV C^{12} ion down to 11.37 MeV.

The preceding discussion demonstrates that the simple assumption of a surface layer of GaAs on top of a second layer of BaF_2 on top of a third layer (the substrate) of GaAs is not valid. The dips and peaks of Fig. 3 suggest the necessity of removing Ba and adding Ga and As to the second layer in the simulation spectrum. The gradual tailing off of the high energy side of the Ba peak shows that the assumption of an abrupt interface between the top two layers is also invalid. In addition to removing Ba from the second layer, it will be necessary to add Ba to the top layer. Furthermore, the addition of Ba to the top layer and the removal of Ba from the second layer must be done gradually. In other words, the top two layers must be divided into thinner sublayers where the concentrations of the atomic constituents can be varied independently of all other layers.

Figure 4 is the result of a simulation after dividing each of the top two "layers" into five layers each. When the layer corresponding to the GaAs substrate is added this simulation has a total of eleven layers. The concentrations

[1] The simulation spectra shown in Figs. 2, 3, and 4 used the natural abundance of the Ga and Ba isotopes (As is mono-isotopic). There are two stable isotopes of Ga, (60.0% at 68.93 amu and 40.0% at 70.92 amu). There are seven stable isotopes of Ba, but simulation program limitations could only accommodate three isotopes. Therefore, the simulations used the two major isotopes (71.7% at 137.9 amu and 11.2% at 136.9 amu) and a weighted average of the five minor isotopes (17.1% at 135.2 amu).

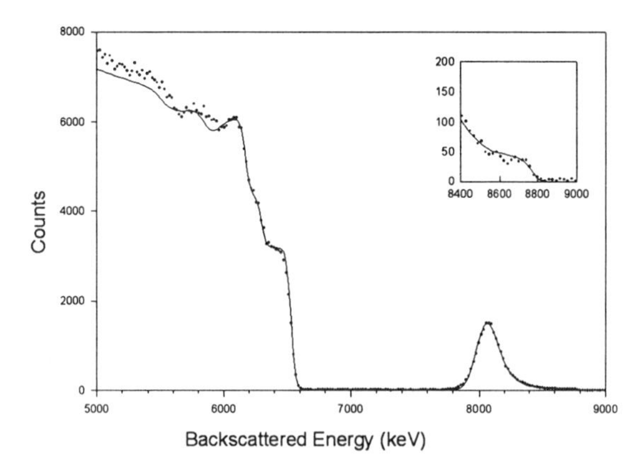

Figure 4. Simulation assuming a graded interface between the GaAs top layer and the Ba layer beneath. Data used to generate this simulation are listed in Table I.

of these layers are listed in Table I. The natural isotopic abundance was retained. All layers containing Ga and As retained a 1:1 Ga-to-As ratio. All layers containing Ba, which were originally believed to be BaF_2, did not retain a 2:1 F-to-Ba ratio. In fact, the fit to the experimental data improved as F was removed from the simulation until no F was present in any layer. Even though a F peak is not observable (due to the small mass of F with respect to the C^{12} projectile), F can still have an affect on the rest of the simulation due to its stopping power, S_e. Removing F from a layer means adding some other species. In this case all other species have a higher S_e. This higher S_e slows the projectile more than if the layer contained F. Since the scattering cross section is inversely related to the projectile energy, removing F increases the cross section for all layers below the layer from which the F is removed. The net result in the simulation of Fig. 4 is to increase the magnitude of the substrate layer signal. There is in fact a 1% increase in the magnitude of the simulation spectrum of Fig. 4 at 5 MeV over the same point on Fig. 3, where the simulation contains F.

Removing all F from the simulation spectrum in order to obtain a better fit to the experimental spectrum cannot be justified by the HIBS data alone, since F cannot be detected. An RBS spectrum of this same sample using 2.0 MeV He projectiles should show a small, but distinct F peak. This was done and no F was detected. Additionally, an XPS depth profile of this sample failed to detect F. The removal of F from the simulation was therefore justified by other methods.

CONCLUSIONS

The use of HIBS in conjunction with other analysis techniques has proved to be a very helpful diagnostic aid in the MBE deposition of $GaAs/BaF_2/GaAs$ heterostructures. HIBS analysis shows that Ga and/or As vapors are etching the BaF_2 layers in the growth chamber. Deposition parameters must therefore be changed if the goal of growing this heterostructure is to be reached. These studies are currently under way.

The enhanced mass resolution of HIBS is its greatest attribute in this study. It allows the verification of the stoichiometric ratios of the atomic constituents. This information cannot be gathered from XPS analysis, the MBE practitioner's main diagnostic tool. Furthermore, the spectral features introduced by the enhanced mass resolution serve as an aid in fitting the simulation spectrum to the experimental spectrum.

Table I. Data used in the simulation of Fig. 4. Layers 1-10 each have an areal density of 0.017 mg/cm^2. Layer 11, the GaAs substrate, is treated as infinitely thick.

Layer #	%Ba	%As	%Ga
1	0.25	49.87	49.87
2	0.30	49.85	49.85
3	0.53	49.74	49.74
4	1.29	49.36	49.36
5	1.61	49.19	49.19
6	2.02	48.99	48.99
7	4.60	47.70	47.70
8	6.45	46.78	46.78
9	12.05	43.98	43.98
10	16.52	41.74	41.74
11	0.00	50.00	50.00

ACKNOWLEDGMENTS

The authors would like to thank Mr. Patrick Cady of NSWC/Carderock for his help with the HIBS measurements. This work was supported in part by the Office of Naval Research and the NSWC/Dahlgren Division Independent Research Program.

REFERENCES

1. Stumborg, M.F., Chu, T.K., Santiago, F., Price, J.L., Guardala, N.A., and Land, D.J., *J. Vac. Sci. Technol. A*, **14(1)**, 69-79,1996.
2. Drathen, P., Ranke, W., and Jacobi, K., *Surface Science*, 77, L162-6, 1978.
3. Price, J.L., et. al., *Nucl. Instrum. Methods B*, **56/57**, 1014-1016, 1991.
4. Tesmer, J.R., and Natasi, M., *Handbook of Modern Ion Beam Materials Analysis*, Pittsburg, PA, Materials research Society, 1995, Sec 4.3.3.2.
5. L'Ecuyer, J., Davies, J.A., and Matsunami, N., *Nucl. Instrum. Methods*, **160**, 337-346, 1979.

DEVELOPMENT FROM RUTHERFORD BACKSCATTERING TO HIGH ENERGY BACKSCATTERING SPECTROMETRY

J.Y.Tang, H.S.Cheng, Z.Y.Zhou, F.J.Yang

Department of Nuclear Science, Fudan University, Shanghai 200433, China

Summarized are the experimental and theoretical studies at Fudan University in high energy backscattering spectrometry (HEBS), including the measurement of non-Rutherford backscattering cross sections and their successful applications, comments on the phenomenon of deviation from Rutherford Backscattering (RBS), and the R-matrix analysis of the elastic scattering cross sections. Advantages of the development from RBS to HEBS are emphasized and references therein are listed.

INTRODUCTION

Conventional Rutherford backscattering spectrometry (RBS) using helium ions with an energy of 1-2 MeV has been used extensively for the determination of stoichiometry, thickness, impurity concentration and depth profile in films (1). It is a non-destructive and quantitative technique. However, a disadvantage of RBS is the low detection sensitivity for light elements, particularly in the presence of a large amount of heavier elements. This limitation can be largely overcome with high energy backscattering spectrometry (HEBS) (2-4), in which the elastic scattering cross section is substantially enhanced by the nuclear interaction between the light element and helium ion with the energy higher than that used in normal RBS.

In addition to the higher detection sensitivity for light elements, advantages associated with HEBS include improved mass (depth) resolution, increased probing depth and improved accuracy in measured stoichiometric ratios. But, the inconvenience of using measured, rather than calculated (Rutherford), cross sections arises. Among other laboratories, backscattering cross sections of ^{4}He around 165^0 from a set of elements, such as C (5), N (6), O (7), Al, Si, Fe, and Co, in the energy range 2.0-9.0 MeV, F in 1.6-5.0 MeV, Na in 1.6-6.0 MeV, Mg in 2.0-8.4 MeV, Cl in 2.0-8.5 MeV, and Ni in 3.0-8.4 MeV (8,9), have been measured and evaluated at Fudan. Most of the data have been collected and published in (10).

The measurements show that the scattering remains Rutherford up to a certain energy E_{nr} until resonant structures appear, and that even in the resonance region, there may exist several energy regions where the cross sections present an enhanced and smooth variation. Either sharp resonances or slowly varying energy regions are suitable for higher energy non-Rutherford scattering analysis. Figure 1-3 shows, respectively, the measured cross sections for ^{4}He backscattering from C, N, and O in the energy range 2.0-9.0 MeV. Listed in Table 1 are the energy ranges and resonant energies suitable for HEBS analysis using C(α,α)C, N(α,α)N, and O(α,α)O reactions.

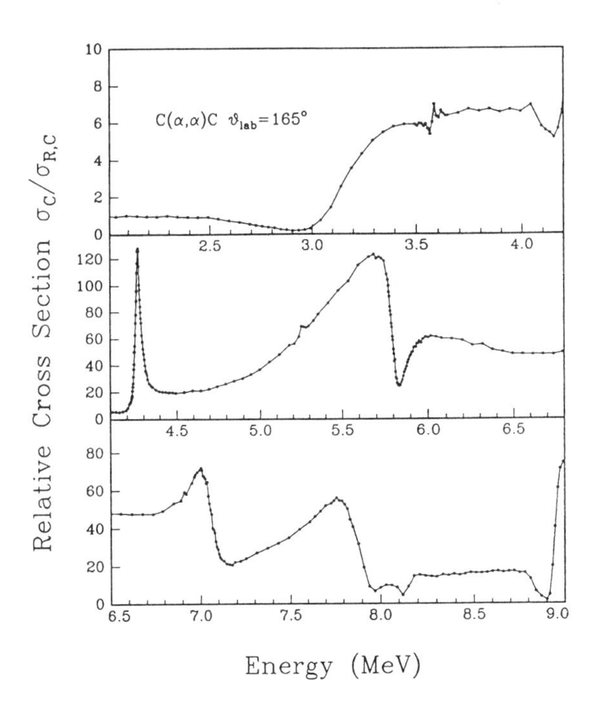

FIGURE 1. The measured cross sections for 165^0 backscattering of ^{4}He from ^{12}C in the energy range 2.0-9.0 MeV, expressed in terms of the RBS cross sections at the same energies. (5)

CP392, *Application of Accelerators in Research and Industry*, edited by J. L. Duggan and I. L. Morgan
AIP Press, New York © 1997

TABLE 1. The Energy Ranges and Resonant Energies Suitable for HEBS Analysis Using C(α,α)C, N(α,α)N, and O(α,α)O Reactions

Reaction	Energy range (MeV)	Resonant energy (MeV)
C(α,α)C (5)	3.6-4.0, 5.97-6.20, 6.4-6.8, 8.18-8.80	4.262
N(α,α)N (6)	3.15-3.55, 5.15-5.65, 5.90-6.10	3.570, 4.446, 8.650
O(α,α)O (7)	5.53-5.88, 6.0-6.2, 8.40-8.86	3.045

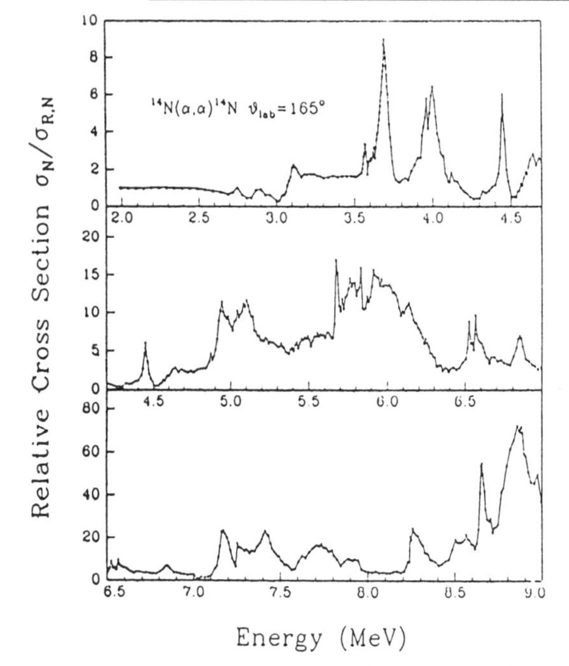

FIGURE 2. The measured cross sections for 165° backscattering of ^4He from ^{14}N in the energy range 2.0-9.0 MeV, expressed in terms of the RBS cross sections at the same energies. (6)

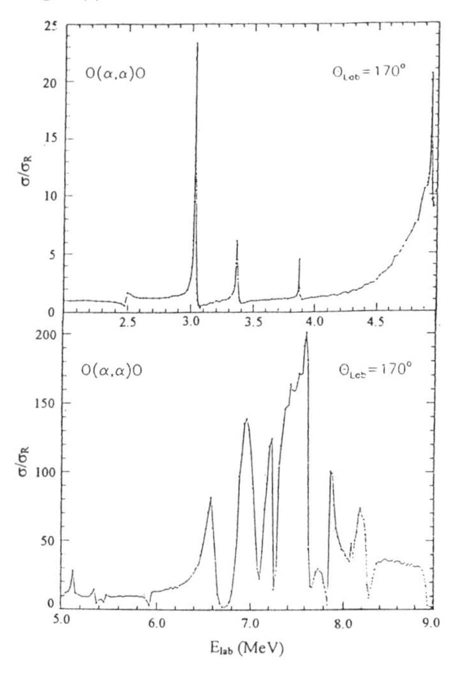

FIGURE 3. The measured cross sections for 170° backscattering of ^4He from ^{16}O in the energy range 2.0-9.0 MeV, expressed in terms of the RBS cross sections at the same energies. (7)

EXAMPLES OF SUCCESSFUL APPLICATIONS IN HEBS

HEBS of 6.2 MeV and 4.26 MeV He Ions for Analysis of Carbon and Oxygen (11)

The contaminations of C and O in the surface and interface of thin films and the depth profiles of carbides and oxides of heavier elements may not be measured by using the normal RBS methods. HEBS of 6.2 MeV helium ions was utilized for simultaneous compositional analysis of C and O.

These enhanced non-Rutherford cross sections have increased the detection sensitivity to light elements in the presence of a large amount of heavier elements. In addition, the sharp resonance for scattering of 4.26 MeV He ions by C has been utilized for carbon contamination measurement at the surface and interface, yielding an even greater sensitivity for C than the backscattering around 6.2 MeV. In the early days, the differential scattering cross sections at a scattering angle $\Theta_{lab}=165^0$ for He ions on C in the energy range 4.0-4.5 MeV and 5.7-7.0 MeV, and on O in the energy range 5.6-6.5 MeV were measured and used in calculations.

Determination of Stoichiometry of a YBaCuO Sample by Backscattering of 8.8 MeV He Ions

Non-Rutherford backscattering of 8.8 MeV He ions, which has an increased sensitivity to O by a factor of 25 over RBS (2), was utilized to measure stoichiometries in high temperature superconducting materials. Direct oxygen determination using 2 MeV He ions is unpractical because the oxygen cross section is low relative to the metals and the O peak rests on a high background. The enhanced cross sections of 8.8 MeV He ions reveals the O peak drastically, and the stoichiometry of O and metals in a YBaCuO sample was determined to be $Y_{1.2}Ba_2Cu_3O_{6.7}$ (12).

Simultaneous Analysis of Light Elements Using HEBS

A metal oxide layer, which contains elements Pt, Ru, Ni, and Fe, coated on substrate Ni was analyzed using 6.2, 7.72 and 8.79 MeV ^4He ions, with its HEBS spectra plotted in Fig.1 (13). The 2 MeV RBS spectrum provides no more information than that it is a thick sample. In the 6.2 MeV spectrum, carbon and oxygen profiles are obtained because the relative cross sections to the RBS values are as high as 58 and 12, respectively. However, the detection of nitrogen is hindered by its small cross sections in this energy region. At 7.72 MeV incident energy, it appears as an oxygen peak owing to the huge cross section bump around 7.6 MeV. Nitrogen peak cannot be seen for its relatively low cross sections in this energy region, until at 8.79 MeV where the cross section of nitrogen is enhanced by a factor of 54.

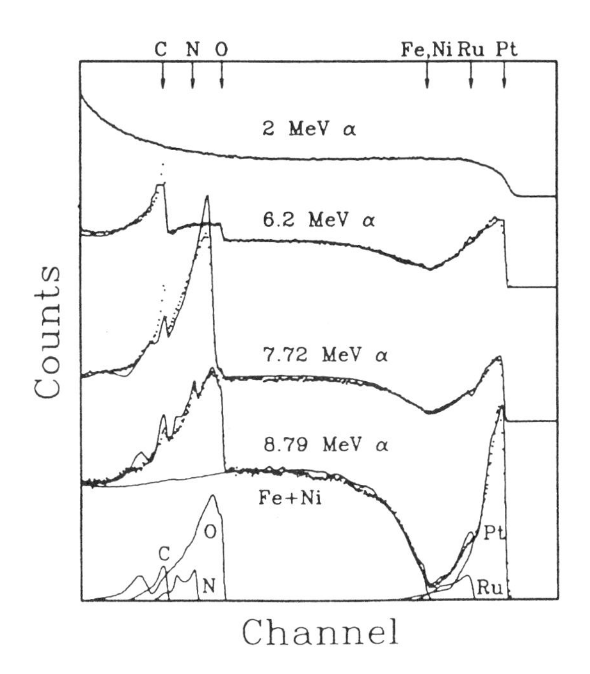

FIGURE 4. HEBS spectra of a thick coated electrode material at 6.2, 7.72 and 8.79 MeV, respectively, with a 2 MeV RBS spectrum included for comparison (13). The solid curves indicate the computer simulation results of the extended RUMP program (14).

PROBLEMS OF FURTHER DEVELOPMENT

What is the Energy E_{nr} at Which the Scattering Cross Section Begins to Deviate From Its Rutherford Value?

The phenomenon of deviation from RBS has puzzled ion beam experimenters who work in HEBS. Though cross section measurements were made in a number of laboratories (2-10), as mentioned above, no satisfactory calculations appeared. The complexity of cross section calculations in the higher energy range roots in the ever-present resonant structures. As a first step, Bozoian et al. (15) developed a simple classical model to calculate the projectile energy $E_{nr}(F)$ at which the scattering cross section deviates from its Rutherford value by fraction F. An analytical expression for $E_{nr}(F)$ has been derived and reported to have good agreements with experimental data. But after a critical examination of the formula, we found (16), through discussions and experiments, that it gives incorrect dependencies of E_{nr} on the departure factor F and the scattering angle for the ^{4}He-C and ^{4}He-O backscattering, though it is useful in roughly estimating the energy value. We asserted that these inadequacies originate from the classical approach, which is an over-simplified and phenomenal description of nature. In the meantime, Bozoian extended the model to account for isotopic variation of the incident beams of H and He (17), and to include heavier projectiles, as Li, C and O (18), on the basis of optical model calculations. A simpler expression for E_{nr} has been published in (18) and was checked by the C- and O-ion experiments of Räisänen and Rauhala (19) with reasonable accuracy. But all of these successes have not yet changed its phenomenal description of nature. The fundamental solution to the problem lies in quantum mechanics.

How to Compare the Scattering Cross Section Values Measured in Different Laboratories?

The measured cross sections, particularly for narrow resonances, often vary with the experimental conditions used, such as sample thickness, beam energy spread, and scattering angles. Therefore it is hard to compare the cross section values measured in different laboratories. Computer codes based on the differential cross section expression of R-matrix theory are needed to analyze the measured elastic scattering cross sections within a certain energy range or at several scattering angles. Level parameters of the compound nucleus, i.e., the spins and parities J^π, resonant energies, and widths of the excited levels, are identified from the analysis. These level parameters are then utilized to check either the measured values of resonant cross sections or the cross sections at certain particular angles according to the experimental requirements.

Though similar work in nuclear spectroscopy was done in the 50s and 60s, the experimental accuracy was limited by the target thickness (25-30 keV, for instance), the detector resolution (as high as ~150 keV), and the relatively simple calculating programs, in which the determination of level parameters was done by visual inspection of the excitation curves at a number of scattering angles.

From the elastic scattering cross sections for 170^0 backscattering of ^{4}He from ^{16}O in the energy range 2.0-9.0 MeV, we deduced the level parameters of the compound nucleus ^{20}N with a computer code named Multi 6 (20). The procedures and the level parameters were summarized in (21). Plotted in Fig.5 are the calculated cross sections near a resonance for samples of different energy smear, which is a parameter used in Multi 6 to account for the contributions from the beam energy spread and energy loss and straggling in

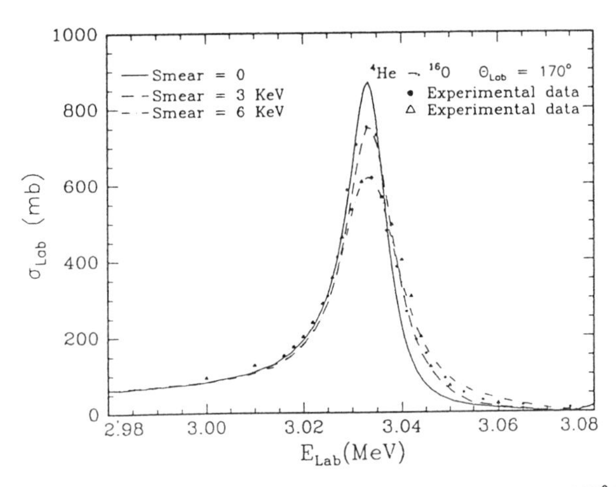

FIGURE 5. The calculated cross sections near the 3.034 resonance at 170^0 for three samples of different energy smear: 0, 3, and 6 keV. The peak value for the zero-smear sample is deduced to be 871 mb. (21)

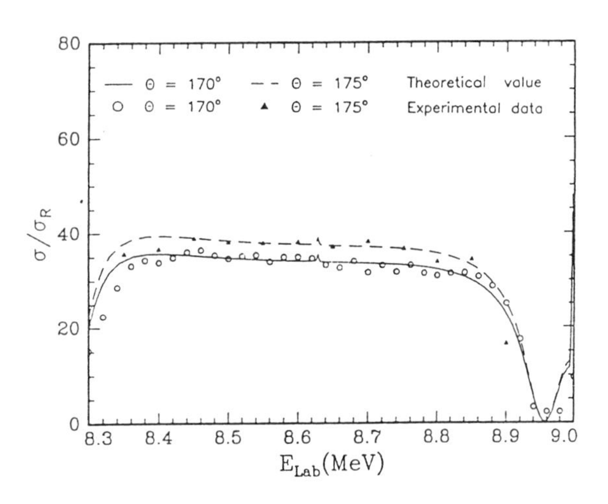

FIGURE 6. The calculated cross sections of ^{4}He from ^{16}O at 170^0 and 175^0 in the energy range 8.3-9.0 MeV. 10% increase of the cross section is seen in this smooth-varying region as the scattering angle increases by 5^0. (21)

the target. The results show that the peak cross section value decreases with increasing energy smear.

The effects of sample thickness and beam energy spread on cross section values where the cross sections present an enhanced and smooth variation are negligible. However, the cross section is dependent on the scattering angle. The larger the scattering angle, the greater the cross section (see Fig.6).

The cross sections of ^{4}He from ^{12}C in the energy range of 5.0-9.0 MeV were measured at the scattering angle of 170^0, and the measured values of Leavitt et al. (22) from 2.0-5.0 MeV at 170.5^0 were quoted as well, to deduce the level parameters of the compound nucleus ^{16}O with the Multi 6 code. As in the oxygen case, these level parameters (23) can be utilized to calculate the cross sections with different energy smears or at different scattering angles.

ACKNOWLEDGMENTS

The authors are grateful to Zhao G.Q., Shen H., and Feng Y. for their contributions. Thanks are due to Lu C.R., Yuan D.S., and Sun C.C. for providing stable beams, due to Fang D.F. for the computer code Multi 6. The financial support from the Chinese National Science Foundation is also acknowledged.

REFERENCES

1. Chu, W.K., Mayer, J.W., and Nicolet, M.,-A., *Backscattering Spectrometry*, New York, Academic Press, 1978
2. Martin, J.A., Nastasi, M., Tesmer, J.R., and Maggiore, C.J., *Appl. Phys. Lett.*, **52**, 2177 -2179 (1988)
3. Gossett, C.R., *Nucl. Inst. and Meth.*, **B40/41**, 813-816 (1989)
4. Leavitt, J.A., and McIntyre,Jr., L.C., *Nucl. Inst. and Meth.*, **B56/57**, 734-739 (1991)
5. Feng, Y., Zhou, Z.Y., Zhou, Y.Y., and Zhao, G.Q., *Nucl. Inst. and Meth.*, **B86**, 225-230 (1994)
6. Feng, Y., Zhou, Z.Y., Zhao, G.Q., and Yang, F.J., *Nucl. Inst. and Meth.*, **B94**, 11-14 (1994); *Chin. J. Nucl. Phys.*, **15**, No.4, 329-332 (1993)
7. Cheng, H.S., Shen, H., Tang, J.Y., and Yang, F.J., *Nucl. Inst. and Meth.*, **B83**, 449-453 (1993); *Acta Physica Sinica (overseas edition)*, **2**, No.9, 641-647 (1993)
8. Cheng, H.S., Li, X.Y., and Yang, F.J., *Nucl. Inst. and Meth.*, **B56/57**, 749-752 (1991)
9. Cheng, H.S., Shen, H., Yang, F.J., and Tang, J.Y., *Nucl. Inst. and Meth.*, **B85**, 47-50 (1994); *Chin, J. Nucl. Phys.*, **15**, No.4, 333-336 (1993)
10. See Appendix 7, Non-Rutherford Elastic Backscattering Cross Sections, in *"Handbook of Ion Beam Materials Analysis"*, Eds: J.R.Tesmer, M.Nastasi, MRS, 1995, pp.481-508
11. Zhou, Z.Y., Qiu, Y.X., Zhao, G.Q., Gu, X.L., Pan, L.Q., Tang, J.Y., and Yang, F.J., in *"High Energy and Heavy Ion Beams in Materials Analysis"*, Eds: Tesmer, J.R., Maggiore, C.J., Nastasi, M., Barbour, J.C., Mayer, J.W., MRS (Workshop Proceedings, Albuquerque, New Mexico, June 14-16, 1989), 1990, pp.153-164
12. Wu, S.M., Cheng, H.S., Zhang, C.T., Yao, X.W., Zhao, G.Q., Yang, F.J., and Hua, Z.Y., *Nucl. Inst. and Meth.*, **B45**, 227-229 (1990)
13. Zhou, Z.Y., Zhou, Y.Y., Zhang, Y., Xu, W.D., Zhao, G.Q.,Tang, J.Y., and Yang, F.J., *Nucl. Inst. and Meth.*, **B100**, 524-528 (1995)
14. Doolittle, L.R., in (11), pp.175-182
15. Bozoian, M., Hubbard, K.M., and Nastasi, M., *Nucl. Inst. and Meth.*, **B51**, 311-319 (1990)
16. Tang, J.Y., Sun, Y.D., Cheng, H.S., and Shen, H., *Nucl. Inst. and Meth.*, **B74**, 491-495 (1993); *Chin. J. Nucl. Phys.*, **15**, 167-171 (1993)
17. Bozoian, M., *Nucl.Inst. and Meth.*, **B58**, 127-131 (1991)
18. Bozoian, M., *Nucl. Inst. and Meth.*, **B82**, 602-603 (1993)
19. Räisänen, J., and Rauhala, E., *J. Appl. Phys.*, **77**, (4) 1762-1765 (1995)
20. Fang, D.F., *Ph.D Dissertation*, Fudan University (1988)
21. Shen, H., Cheng, H.S., Tang, J.Y., and Yang, F.J., *Nucl. Inst. and Meth.*, **B90**, 593-595 (1994)
22. Leavitt, J.A. McIntyre,Jr., L.C., Stoss, P., Oder, J.G., Ashbaugh, M.D., Dezfouly-Arjomamdy, B., Yang, Z.M., and Lin, Z.,*Nucl. Inst. and Meth.*, **B40**, 776-779 (1989)
23. Shen, H., Cheng, H.S., Tang, J.Y., and Yang, F.J., *Acta Physica Sinica*, **43**, No. 10, 1569-1575 (1994)

DETECTOR

A program for the acquisition and analysis of RBS, HFS, ERD, and PIXE Spectra

M. D. Strathman,

MeV Technology, Inc., 5150 Shadow Estates, San Jose, CA 95135.

DETECTOR / DETACQ are a suite of acquisition and analysis programs designed to operate on Microsoft 32-Bit platforms such as WinNT^{tm} and Win95^{tm}. DETECTOR, the analysis program, is capable of analyzing Rutherford Backscattering data, Hydrogen Forward Scattering data, Elastic Recoil data, and Particle Induced X-ray Emission data. In addition the program includes substantial capabilities in analyzing Lateral and Angle Resolved Images acquired from any of the above analytical techniques. DETECTOR incorporates the ability to use either five parameter or eight parameter stopping power data, screened coulomb cross sections, and an independent stopping foil description. DETECTOR allows the user to use a single sample description to fit experimental data in multiple data windows. Each data window can hold any of the above experimental data types and the theoretical fit to the experimental data, where each window has an independent calibration and excitation probe. This capability allows user to use the several different techniques (even if they were acquired at different times) and then combine them to obtain a better description of the sample which is being analyzed. DETECTOR / DETACQ are both designed with Windows^{tm} compatible graphical user interfaces. The majority of the programs are written in Microsoft Visual Basic 4.0 and therefore incorporate all of the conventional user interaction tools.

Introduction

Over the years many different computer programs for the analysis of data acquired by energetic ion beam analysis have been written.[1,2,3] As personal computers have become more powerful the analytical tools which run on these machines have also improved. Some of the early programs such as SCATT used numerous approximations to speed up the computations needed to generate the theoretical spectra. These approximations made it easy to generate interactive algorithms for interpenetrating spectra at the expense of being able to accurately model complex spectra. Programs based on slab analysis[4], such as RUMP were better at predicting subtle effects in the experimental data but were slower in execution. Today with the advent of 32 bit operating systems such as Windows95^{tm} and WindowsNT^{tm} and micro processors operating near 200 MHz, programs based on full blown slab analysis can generate theoretical spectra fast enough for interactive analysis. DETECTOR^{tm} is such an analysis program which is written in Visual Basic and has both 16 bit and 32 bit operating system versions.

Discussion

DETECTOR^{tm} is based on the concept of a VIEW or DATA FRAME where each experimental data set has an independent window or VIEW DATA. For example, multiple DATA FRAMES can be used to simultaneously display data from:

1) a normal exit (170 degree) detector,
2) a grazing exit (100 degree) detector,
3) a PIXE detector,
4) an Elastic Recoil Detector (ERD) ,
5) a normal exit detector (acquired simultaneous to ERD data), and
6) a detector with data acquired where the sample is aligned with the probe beam (channeling).

Each VIEW has its own set of calibration parameters such as:

1) beam element, energy, and current;
2) detector angle, solid angle, straggling factor, and absorber foils;
3) multichannel analyzer parameters, energy per channel, off set, and calibration; and
4) sample tilt

A single theoretical description is used to model the sample composition. This allows the user to use the data acquired from many different detectors, possibly acquired at different times with very different beam parameters, to generate a more accurate description of the sample than is possible using a single detector. Two or more VIEWs can be linked with respect to the beam current thereby allowing one

CP392, *Application of Accelerators in Research and Industry,* edited by J. L. Duggan and I. L. Morgan
AIP Press, New York © 1997

detector (such as the normal exit detector) in an ERD experiment to be used to normalize the model to correct for problems associated with beam current integration.

Theoretical Model

The theoretical model is based on the slab method of analysis. In this method, the sample is divided into layers and each layer is then subdivided into slabs that are small enough so that the scattered particles from a given slab fall into a single channel in the multichannel analyzer. The algorithm starts with the beam at the surface of the sample, the beam then traverses the first slab where the energy lost is computed using either (user selected):

1) the 5 parameter polynomial fit to the stopping power[5], or
2) the 8 parameter stopping power fit of Ziegler and Biersack[6]

the number of scattered particles (or x-rays generated) is computed using either (user selected):

1) The Rutherford[4] cross section,
2) The Screened Rutherford[7] cross section, or
3) A manually entered Ratio to Rutherford of either of the previous cross sections

the scattered energy is computed using the kinematic factor and then these particles exit the sample through all of the layers traversed along the inward path (using the selected stopping power model). Straggling and the energy width of the particles emanating from the slab are computed before the they are summed into the theoretical spectra. All of the stopping powers, scattering cross sections, and kinematic factors are computed "on the fly." That is, look up tables are not used for any of these factors. This allows the use of the calibration factors associated with each VIEW to be used when computing the theoretical spectra for the VIEW. The process is then iterated until a zero energy is obtained for the analytical particle.

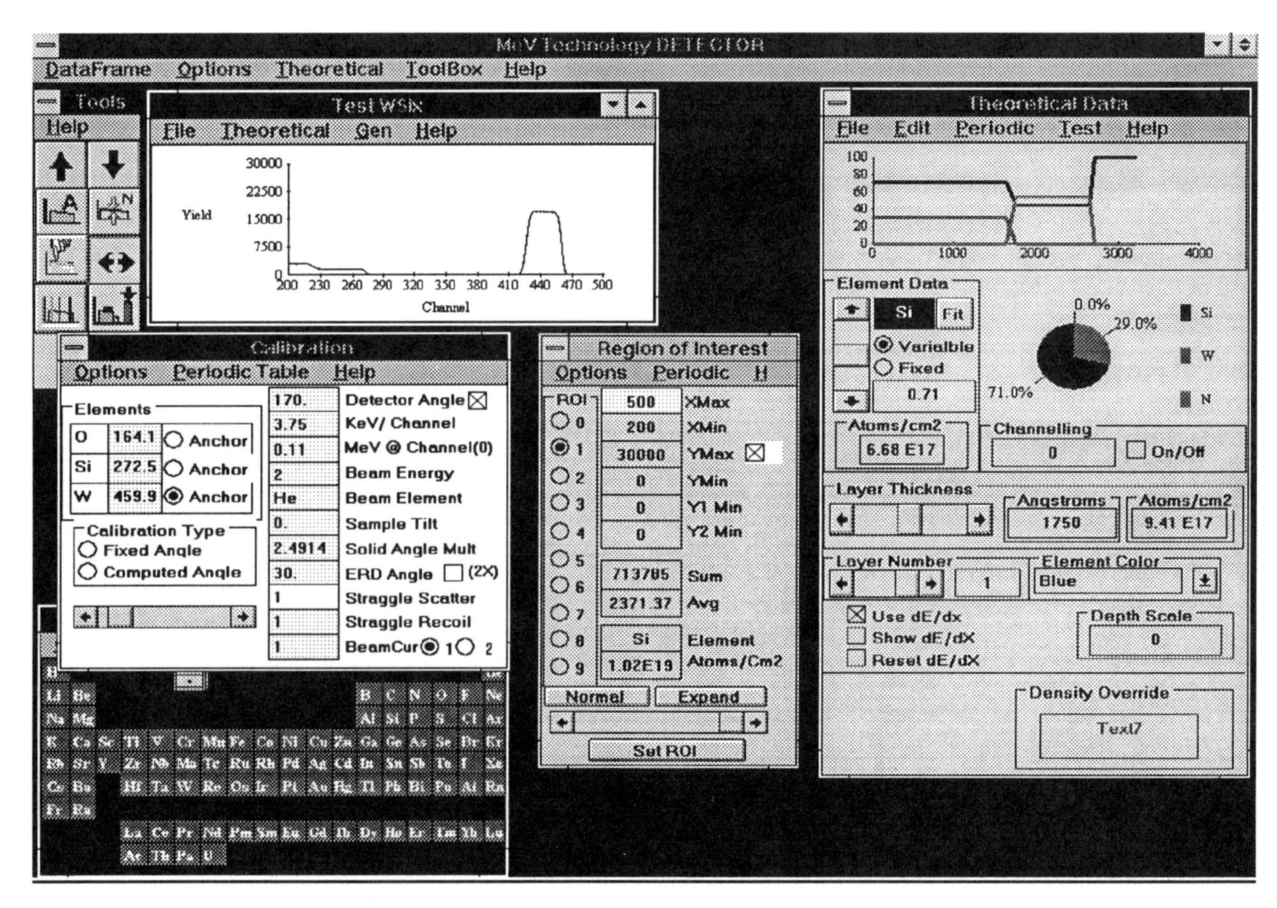

Figure #1: A screen dump of several of the DETECTOR forms

Major Forms In DETECTOR

The VIEW DATA
This is the main Data Viewing Form. All experimental data is displayed via this form. Multiple copies of this form can be loaded to allow data from different detectors or analytical runs to be used.

The Tool Box
The Tool Box holds several tools used to interact with the data including the Normalization tool, the Scaling tool, the Auto Scale tool, the Select Spectra tool, the Fill Theoretical tool, the Region of Interest (ROI) tool, and the Calibration tool.

The Layer Table
The Layer Table is the user interface to the theoretical model. This form includes all of the controls needed to input the parameters used to describe the sample composition.

The Periodic Table
This form allows the user to select an element and put it into the theoretical model, to look at the parameters associated with the element, and to use an element in the Calibration Form

The Region Of Interest Form
This form allows the user to sweep out selected regions of a given VIEW DATA form and expand the region to allow detail to better viewed. This form also allows the user to calculate the atoms/cm2 for a surface impurity. Each VIEW DATA has its own distinct set of ROI parameters.

The Calibration Form
This form allows the user to interactively calibrate a given VIEW DATA for parameters such as the beam energy, keV per channel, beam element, calibration type, and sample tilt. Each VIEW DATA has its own distinct set of Calibration parameters.

The Load Data Form
This form allows the user to look at data files and select data to be loaded into the VIEW DATA Form. This form also controls the ability to load multiple spectra into a given VIEW DATA form.

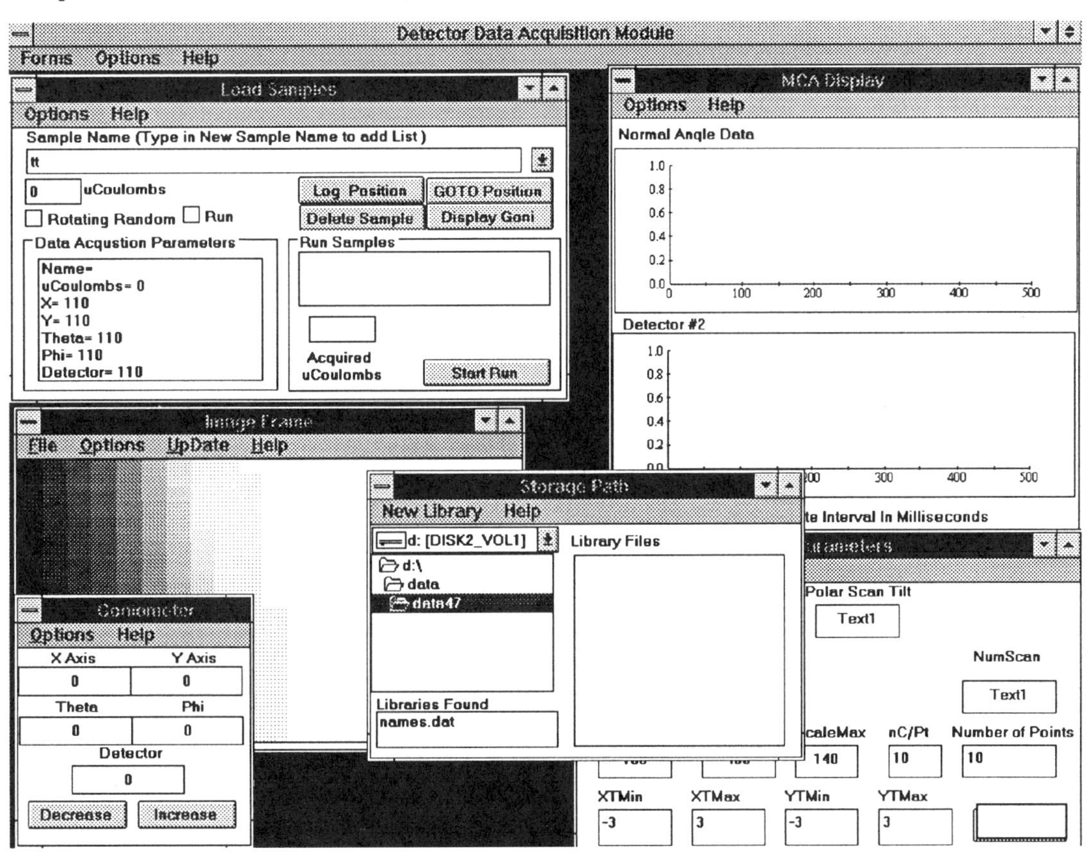

Figure #2 A standard set of forms used with DETACQ.

WINDOWS Compatibility

All of the forms used in DETECTOR conform to Windows specifications and therefore can be rearranged on the screen at the users discretion. All forms can be printed on any printer that is attached to the Windows operating platform. Additionally, the VIEW DATA and DEPTH PROFILE forms can be copied to the Windows Clipboard and then Pasted into any Windows application such as WORD or EXCEL. In addition to the depth profile plot the DEPTH PROFILE form also exports all of the parameters associated with the theoretical model to the clipboard as a text file that is easy to convert to a WORD table. All plots of data or theoretical fits are exported as graphics metafiles so that they can be rescaled in the target application and retain the maximum resolution (limited only by the output device).

Data Acquisition

Data acquisition is handled by a separate program entitled DETACQ (or DETECTOR ACQUSITION). This program includes forms for controlling the sample Goniometer, iterating the acquisition sequence through a number of samples, acquiring Lateral or Angle Resolved Images, a view of the data as it is acquired, and the storage path for the data at the end of a preset number of micro coulombs.

DETAQ currently supports three different goniometers and can accommodate both polar and line scan acquisition. The polar scan algorithm includes the ability to mark the minima in the polar scan, compute the centroid, and then drive the Goniometer to aligned position for channeling analysis.

DETACQ runs concurrently with DETECTOR on a Windows 3.1 platform thus allowing simultaneous data acquisition and analysis

Conclusion

DETECTOR/DETACQ represent a significant improvement in high throughput, fully integrated Ion Beam Analysis Acquisition and Analysis Programs for Windows based operating systems.

References

---

[1] SCATT & TOS, Copyright Charles Evans & Associates, Michael D. Strathman

[2] HYPRA, Copyright Charles Evans & Associates

[3] Larry Doolittle, "High Energy Backscattering Analysis using RUMP", pp. 175-182, High Energy and Heavy Ion Beams in Materials Analysis, Materials Research Society (1990)

[4] Chu, Mayer, and Nicolet, "Backscattering Spectrometry", Academic Press (1978)

[5] Chu, Mayer, and Nicolet, "Backscattering Spectrometry", pp. 364-365, Academic Press (1978)

[6] Tesmer and Nastasi, "Handbook of Modern Ion Beam Materials Analysis", pp. 396-397, Materials Research Society (1995)

[7] Tesmer and Nastasi, "Handbook of Modern Ion Beam Materials Analysis", p. 42, Materials Research Society (1995)

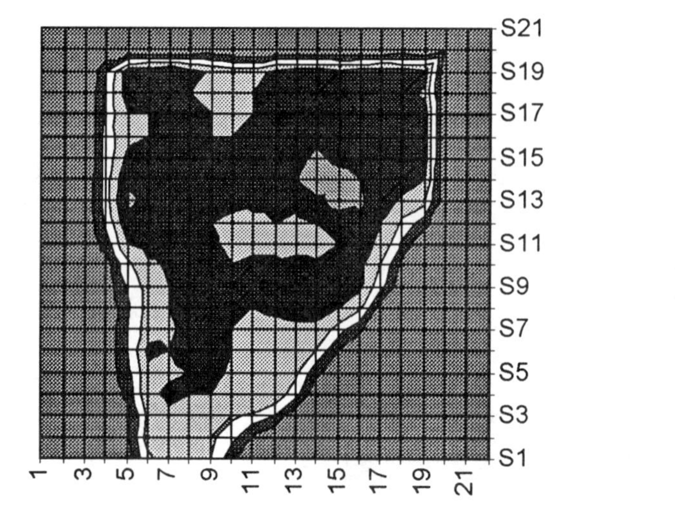

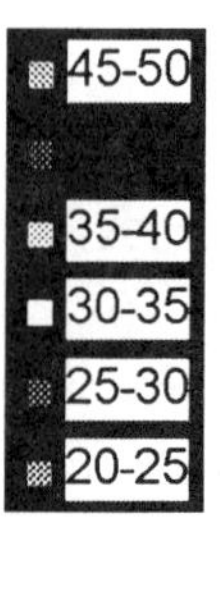

Figure #3: Lateral Image Map of the metal distribution in a silicide on ¼ Silicon Wafer, He4++, 170 Degree Backscatter RBS

INTERNAL CALIBRATION OF HYDROGEN ERD SPECTRA FROM THE FORWARD SCATTERED He SIGNAL USING A SILICON ΔE-E DETECTOR TELESCOPE

M. El Bouanani[a], P.N. Johnston[a], I.F. Bubb[a] and H.J. Whitlow[b]

(a) *Department of Applied Physics, RMIT., GPO Box 2476V, Melbourne 3001, Australia.*
(b) *Department of Nuclear Physics, Lund Institute of Technology, Box 118, S-221 00, Lund, Sweden.*

Elastic Recoil Detection Analysis (ERDA) using a He ion beam is a well established method for quantitative depth profiling of hydrogen in materials. Absolute hydrogen concentration is usually determined by direct comparison with measured hydrogen recoil spectra from standards of known hydrogen content such as polystyrene, mylar or kapton. These polymers are well known for their relative hydrogen loss under ion beam irradiation which, even if it is corrected for, adds an extra source of uncertainty. A new approach based on the use of a ΔE-E silicon detector telescope operating in coincident mode has been investigated. This allows simultaneous detection of forward scattered helium and recoiled proton for auto-normalisation thus eliminating the need for hydrogen reference samples. The use of sufficiently thin ΔE detectors provides other potential benefits such as reducing the straggling contribution in the energy resolution and increased analytical accessible depth.

INTRODUCTION

Since its development by L'Ecyuer et al. in 1975 (1) Elastic Recoil Detection Analysis (ERDA) has become a powerful tool for quantitative depth profiling of light elements in near-surface solid state materials. ERDA analysis of hydrogen using a low energy He beam was first reported by Doyle and Percy (2). One of the main advantages of using He incident ions is the minimal radiation damage of the analysed targets. This is very important for accurate measurement of hydrogen which is well known for its instability and its loss under radiation. Since then, extensive studies aimed mainly at the optimisation of the depth resolution have been undertaken (3-11). A wide range of combinations including the mass of the incident ions, kinematic considerations and different detection approaches have been used. The main detection methods are based on the use of (i) the stopping filter method which consists of placing a sufficiently thick foil to prevent the used heavy projectiles from reaching the silicon energy detector (ii) an electrostatic spectrometer (7) or an electromagnetic filter (6) (iii) particle identification methods based on telescope detection systems such as Time of Flight-Energy (8-9), ion-induced electron emission and silicon energy detectors (8) and Energy Loss-Energy (ΔE-E) silicon detectors (12). Considerable effort has been dedicated to make ERDA depth profiling of hydrogen quantitatively accurate. The determination of the non Rutherford He - H cross sections (13-15) has been until recently a subject of discrepancies which reflects the difficult problem of hydrogen loss under ion beam irradiation. In addition, reliable incident beam charge integration has to be achieved. However, quantitative depth profiles of hydrogen can be achieved by direct comparison with hydrogen recoil spectrum from polymer standards such as Mylar, Kapton or Polystyrene. Accuracy requires that the target hydrogen content has not significantly changed during ERDA measurements. The use of light incident ions such as Helium beams is then preferred in order to minimise the effect of irradiations.

Here we describe a new technique. Simultaneous detection of both the recoiled proton and forward scattered helium using a silicon ΔE-E telescope in order to eliminate the need for hydrogen standards for quantitative depth profiling. A new analysis approach for internal normalisation is described.

EXPERIMENTAL

The measurements have been performed at the RMIT 1 MV Tandem accelerator with 2.9 MeV $^4He^{++}$ beam. The experimental set up uses the well known ERDA geometry. A glancing incident beam impinged

CP392, *Application of Accelerators in Research and Industry*, edited by J. L. Duggan and I. L. Morgan
AIP Press, New York © 1997

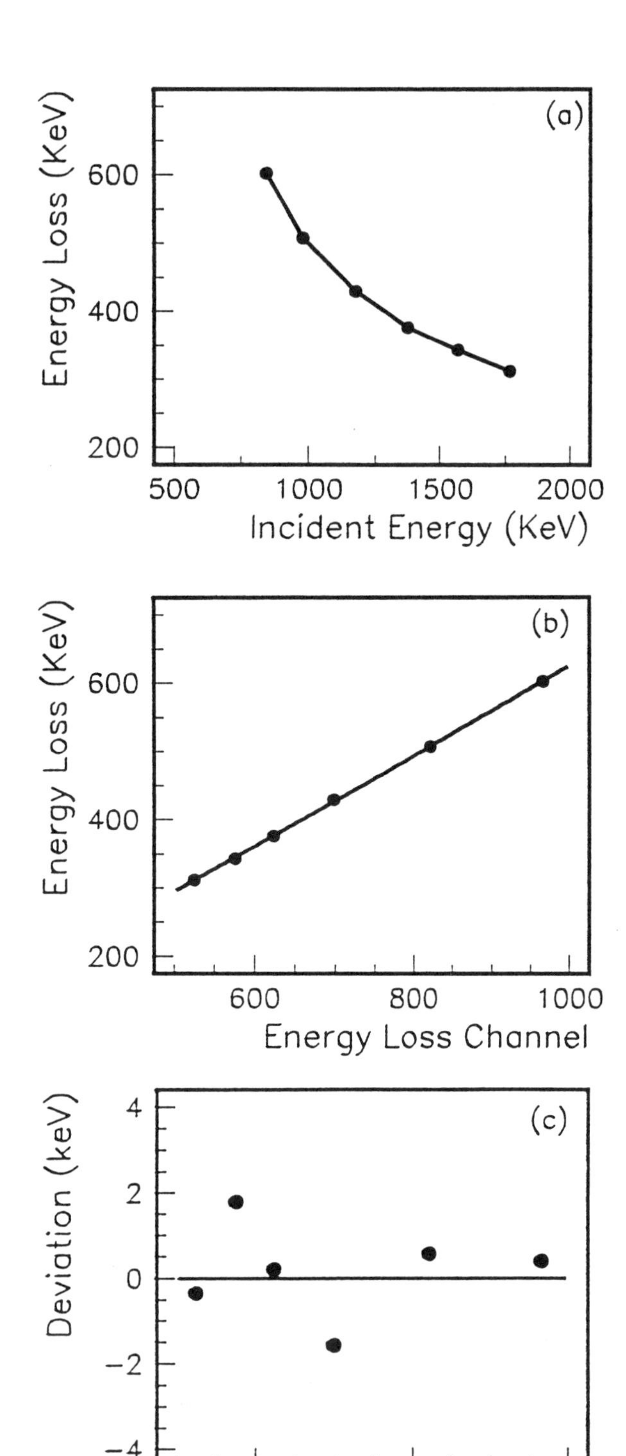

FIGURE 1. (a) Measured total energy loss in the 10 μm ΔE detector vs proton energy. (b) Energy loss response calibration of the ΔE detector. (c) The deviation of the energy loss response experimental data from the straight line fit.

at 75 degrees on the surface sample normal. The silicon ΔE-E telescope was placed at 30 degrees to the incident beam direction. A 1x5 mm^2 vertical rectangular slit collimated the ΔE-E telescope. The E detector is an ORTEC 500 μm SBD. The 10 μm p-i-n ΔE silicon detector was fabricated by SINTEF (16) as a prototype for CHICSi detector telescope array. Both E and ΔE detectors were operated in coincidence mode. The collection of the data use a CAMAC based multi-parameter pulse height analysis system controlled by a MacIntosh Quadra 900 computer and SPARROW KMAX software with event by event data storage to allow for off line analysis.

Both detectors E and ΔE were calibrated using 0.9 - 1.8 MeV protons beam scattered on a 20 nm thin gold film on silicon substrate in RBS configuration. The ΔE-E telescope was placed at a scattering angle of 150° relative to the incident proton direction. The scattered proton peak from the thin gold layer is a well defined gaussian with a FWHM of 11 keV. For each incident energy two sets of data collection were performed with and without the ΔE detector in front the E detector.This allows energy and energy loss calibration for both ΔE and E detectors. Figures 1b and 1a show the energy loss calibration and the energy loss dependence on the incident proton energies. The measured energy loss has a non linear dependence on the proton energy which reflects the expected non linear form of the stopping power. As one would expect and as shown in Fig.1b, the proton energy loss response of the ΔE detector is linearly dependent on the energy loss. The deviations of the experimental proton energy loss data from the straight line fit are presented in Fig.1c.

For 1.2 MeV protons, the energy resolutions of both detectors E and ΔE were 11 keV and 92 keV. The apparent poor resolution of the ΔE is suspected to be due to the electronic noise as a resulting from inefficient insulation of the detector ground from the vacuum chamber. For these measurements the above resolution was enough for He and H separation and the proton energy loss signal was well above the noise level.

RESULTS AND DISCUSSION

Figure 2a shows a typical two dimensional spectrum acquired using a 2.9 MeV Helium-4 beam incident on hydrogen contaminated SiN_3 sample.

The hydrogen and the forward scattered helium are both well separated. Figures 2b and 2c show helium and hydrogen energy projections. Only the high energy scattered helium ions are seen in the spectrum.

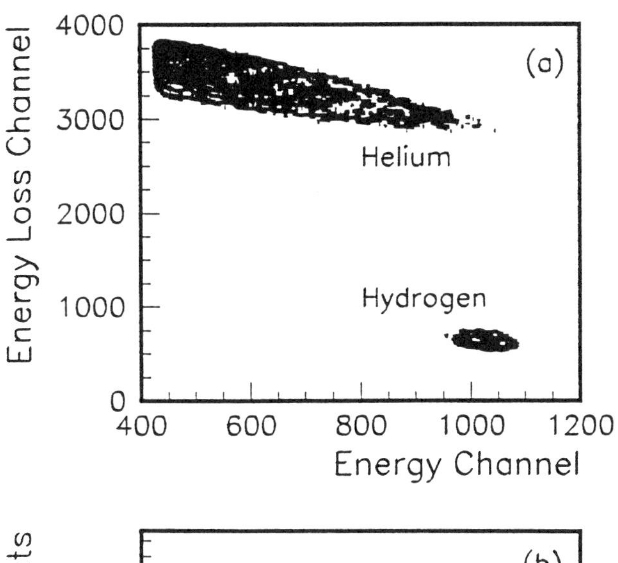

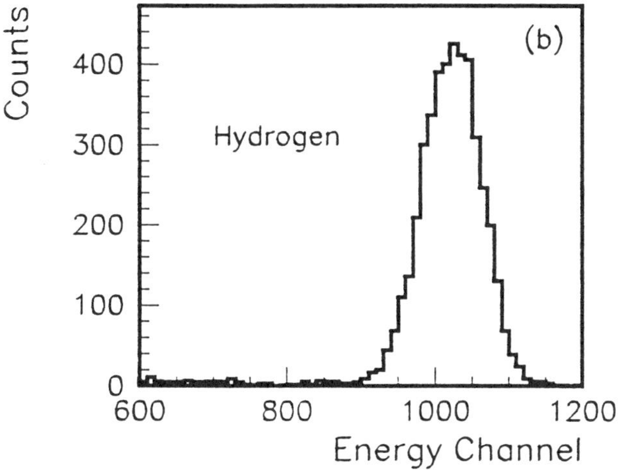

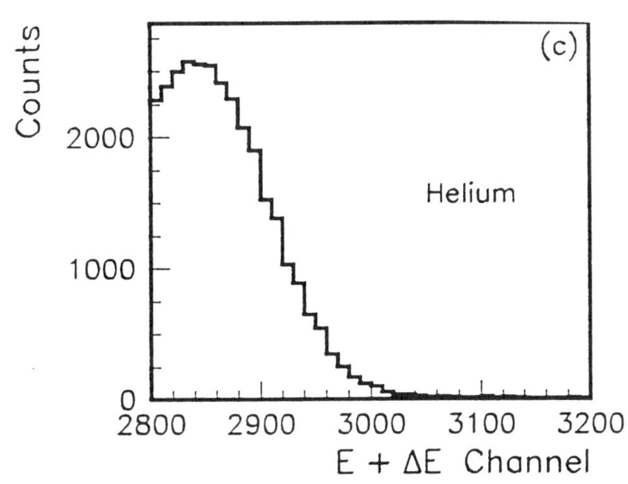

FIGURE 2. (a) ΔE-E two dimensional ERDA spectrum of hydrogen contaminated SiN_3 sample. (b) Projected energy spectrum of the hydrogen component. (c) Total energy (ΔE+E) projection of the forward scattered He component.

The choice of 2.9 MeV helium incident ions was readily available from RMIT Tandem accelerator and it was sufficient to demonstrate the advantages of the ΔE-E method. The shape of the low energy cut off in the helium spectrum is due to the straggling in the ΔE detector. The cut off is the result of incident helium ions losing all their energy in the ΔE detector.

As shown in Figure 2c, the total energy projection of the forward scattered helium was obtained by adding both E and ΔE signals which gives a much better energy resolution. This was done using the two following assumptions (i) the proton energy loss response calibration as shown in figure 1b remains linear outside the range of the measured data . (ii) The energy loss response calibration of both hydrogen and helium is the same. This assumption is known to be valid for E detectors. For ΔE detectors it should also be valid excepting the effect of the back dead layer, which is very thin. Other problems, such as pulse height defect become important only for heavy ions.

In Figure 2b the hydrogen E and ΔE signals are not added. This would worsen the 41 keV hydrogen spectrum energy resolution due to the 92 keV apparent energy resolution of the ΔE detector. The summation of both E and ΔE signals should be done only in the case where the energy resolution of the ΔE detectors is much less than their straggling contribution. This would improve the depth resolution comparing to the stopping foil technique.

Both helium and hydrogen spectra were obtained under the same experimental conditions. For samples with known composition, auto-normalisation of the forward scattered helium energy spectrum in a manner similar to Rutherford Backscattering Spectrometry (RBS) allows inter-normalisation of the hydrogen recoil spectrum. This eliminates the needs for hydrogen standards and consequently the uncertainties related to the hydrogen loss in the commonly used polymers as hydrogen references. The auto-normalisation using the forward scattered helium spectrum is a means of determining the number of incident helium particles. Since the recoils and scattered projectiles are detected in the same detector telescope the solid angles for recoil and scattered projectile detection are identical. The accuracy of the absolute hydrogen concentrations are then dependent directly and mainly on both the helium-matrix and helium-hydrogen cross section data. RUMP (17) software can be used for such analysis.

In the first instance one has to assume that the hydrogen concentration does not affect the stopping power. In fact a feed back between RUMP analysis of both the scattered helium energy and the hydrogen recoil spectra could be used to take account of the hydrogen contribution in the total stopping power. However, this would not necessarily make the depth profiling of hydrogen more accurate. The reason is that the hydrogen_contribution to the stopping power is in general smaller than the errors in the stopping power itself. As an example a hydrogen concentration as high as 30 at.% in silicon will lead to less than 5 % increase in the stopping power in pure silicon when using 2.9 MeV helium-4. The effect will be even smaller for samples containing elements heavier than silicon.

The ΔE detector is not a passive absorber and consequently its thickness is a compromise between the minimum acceptable energy loss and the electronic noise of the thin detector and associated electronics. Using TRIM (18) for 1.2 MeV protons and assuming an electronic noise of 100 keV, the thickness of the ΔE detector can be as low as 6 μm. This would result in a significant improvement in the energy resolution due to the reduction of the straggling contribution. Moreover, the analytically accessible depth would also be increased because of the lower energy cut off comparing to the stopping filter technique when using a helium probing beam.

CONCLUSION

The existence of an alternative solution to the use of hydrogen standards for extracting quantitative depth profiles of hydrogen using ERDA. have been demonstrated. The use of a ΔE-E silicon detection system with low energy helium beams permits simultaneous detection of recoiled hydrogen and forward scattered helium projectiles allowing an internal normalisation in the case of predetermined sample composition.

The use of ΔE detectors as thin as 6 μm would improve the depth resolution due to a significant reduction in the straggling contribution. Furthermore, the low energy cut-off in the recoiled hydrogen spectrum would be reduced with consequential increase in the analytically accessible depth.

ACKNOWLEDGMENT

The authors wish to thank Dr. Bo Jakobsson on behalf of CHICSi, Lund University, Sweden, for lending us the prototype ΔE detectors, and Mr Robert Short, RMIT Applied Physics, for his help on running the accelerator.

REFERENCES

1. L'Ecuyer, J., Brassard, C., Cardinal, C., Chabbal, J., Deschenes, L., Labrie, J.B., Terreault, B., Mariel, J.G. and St-Jacques, R, J. Appl. Phys. **47** (1976) 881.
2. Doyle, B.L. and Percy, P.S., Appl. Phys. Lett. **34** (1979) 811.
3. Paszti, F., Szilagyi, E and Kotai, E., Nucl. Instr. and Meth. **B 54** (1991) 507.
4. Turos, A. and Meyer, O., Nucl. Instr. and Meth. **B 4** (1984) 92.
5. Nagata S., Yamaguchi, S. and Fujino, Y., Nucl. Instr. and Meth. **B 6** (1985) 533.
6. Ross, G.G., Terreault, B., Gobeil, G., Abel, G., Coucher, C. and Veilleus G., J. Nucl. Mater. **128/129** (1978) 730.
7. Kruse, O. and Carstanjen, H.D.., Nucl. Instr. and Meth. **B 89** (1994) 191.
8. Groleau, G, Gujrathi,S.C. and Martin, J.P., Nucl..Instr. and Meth. **218** (1983) 11.
9. Thomas, J.P., Fallavier, M., Ramdane, D., Chevarier, N. and Chevarier, A., Nucl. Instr. and Meth. **218** (1983) 125.
10. Benka, O., Brandstotter, A. and Steinbauer, E., Nucl. Instr. and Meth. **B 85** (1994) 650.
11. Brice, D.K. and Doyle, B.L., Nucl. Instr. and Meth. **B 45** (1990) 265.
12. Kreissig, U., Grotzschel, R. and Behrisch, R., Nucl. Instr. and Meth. **B 85** (1994) 71.
13. Baglin, J.E.E., Kellock, A.J., Crockett, M.A.and Shih, A.H., Nucl. Instr. and Meth. **B 64** (1992) 469
14. Wang, H. and Zhai, G.Q., Nucl. Instr. and Meth. **B 34** (1988) 145.
15. Benenson, R.E., Wielunski, L.S. and Lanford, W.A., Nucl. Instr. and Meth. **B 15** (1986) 453.
16. Evensen, L., Westgaard, T., Whitlow, H.J. and Jakobsson, B., submitted to IEEE Trans. Nucl. Sci.
17. Doolittle, L.R., Nucl. Instr. and Meth. **B 15** (1986) 227.
18. Ziegler, J.F., Biersack, J.P. and Littmark, U, The stopping and ranges of ions in matter, edited by Ziegler, J.F. (Pergamon, New York) 1995.

HYDRATION DEPTH PROFILING OF OBSIDIAN SURFACES USING ^{7}LI IONS

S.R.Neve and P.H.Barker

Physics Department, The University of Auckland, Private Bag 92019, Auckland, New Zealand
sr.neve@auckland.ac.nz, ph.barker@auckland.ac.nz

Water can diffuse into the surface of volcanic and 'man-made' glass at a rate dependent on the surrounding environmental conditions, creating over time a hydration layer of the order of μm whose thickness can be measured by a variety of techniques. The 3.09MeV resonance of the nuclear reaction $^{1}H(^{7}Li,\gamma)^{8}Be$ has been employed at the University of Auckland Research Accelerator laboratory AURA2, to profile the hydration rim in the near surface region of samples of the volcanic glass obsidian. The development of this technique is presented here. Some problems encountered due to carbon contamination in the system, as well as other interfering backgrounds are discussed. The hydration rims of a selection of obsidian artefacts found in archaeological sites in New Zealand, and of artificially induced samples have been measured. We evaluate the technique as a means of providing the data for obsidian hydration dating.

INTRODUCTION

Obsidian hydration dating has been developed over the years as an alternative archaeometric technique to radiocarbon dating. It relies on the fact that the surface of a glass takes up water over a period of time, via ion exchange reactions in the matrix, forming a hydration layer in the near-surface region, whose thickness can be described by $x \approx k \cdot t^{1/2}$ (where k is the hydration rate coefficient).
An estimate of the age of an obsidian artefact can be made by determining the thickness of this layer and combining it with a knowledge of the dependence of the rate of hydration for the particular obsidian on hydrating temperature and surrounding environmental conditions. The hydration density is greatest at the surface of the sample and gradually decreases into it, reflecting the diffusion of water further into the structure. The rims formed can grow to depths of tens of μm for very old surfaces. For the obsidian artefacts found in New Zealand, which has a short history, they are small, of the order of 1μm.

Optical techniques are frequently used to measure the hydration rims. A thin section is prepared from the artefact and viewed under a microscope. However, small rims are difficult to measure and this method results in the destruction of or damage to the artefact.

A variety of ion beam techniques has been used to profile hydrogen in materials (1). In this work we utilised the 3.09MeV resonance reaction of $^{1}H(^{7}Li,\gamma)^{8}Be$ to develop a reliable and non-destructive technique for determining hydration rims in New Zealand obsidian artefacts.

EXPERIMENTAL DETAILS

The Profiling Technique

The depth profile of hydrogen was determined by varying the energy of the incident ^{7}Li beam, thus probing different depths into the sample, and detecting the characteristic γ-rays emitted (14.7 and 17.6MeV). These γ-rays were due to the de-excitation of $^{8}Be^{*}$(17.6MeV) to the 2.94MeV state and ground state respectively. The yield of γ-rays at a particular beam energy was proportional to the concentration of hydrogen at the corresponding depth in the sample. If the incident beam was at the resonance energy then the γ-rays detected were from reactions taking place on the surface of the target. If the beam energy was greater than the resonance energy, then the beam penetrated into the sample, losing energy down to 3.09MeV, at which point the resonance reaction took place. The γ-rays then detected corresponded to the hydrogen concentration at a depth in the sample. Using the stopping power compilations of Northcliffe and Schilling (2), the depth was calculated by integrating $(dE/dx)^{-1}$ of the sample from the incident beam energy down to the resonance energy. The resonance reaction at 3.09MeV is reported to have a finite energy width of 85.4 ± 3.5keV (3), which should be the full width at half maximum of the yield curve obtained from an infinitely thin layer of hydrogen. The width of the resonance limits the depth resolution obtainable by the method, and before the thickness of the hydration layer can be obtained for a particular sample, the thin target profile needs to be unfolded from its profile curve.

CP392, *Application of Accelerators in Research and Industry*, edited by J. L. Duggan and I. L. Morgan
AIP Press, New York © 1997

Experimental Set-Up

A beam of negative lithium ions was extracted from a duoplasmatron helium ion source with lithium charge exchange canal and a $^{7}Li^{2+}$ beam was accelerated to MeV energies with the AURA2 4MV folded tandem electrostatic accelerator. The ion beam was energy analysed with a 90° magnet and then passed along a length of beam line through two 8-leafed tantalum iris collimators, and into the target chamber. The beam line and target chamber were evacuated to pressures below $1x10^{-5}$torr using oil and mercury diffusion pumps. Targets to be irradiated were attached to a 27-position target ladder which was loaded into a side arm of the target chamber, at 90° to the beam. In choosing the obsidian targets to be loaded on to the target ladder, the least conchoidally fractured hydrated face was orientated for bombardment after having been wiped with AR Methanol to remove any minor surface contamination. The target ladder was moved in the chamber via a stepping motor which could be manually operated or computer controlled. A 5" x 6" NaI(Tl) detector was positioned at 90° above the targets outside the vacuum chamber to detect the reaction γ-rays.

The control of the target position and data collection was fully automated through a program purposely written for this experiment using the commercial software package Kmax (Sparrow Corporation). The program was run on a Macintosh 650 with PowerMac upgrade card, and communicated with various CAMAC modules which interfaced with the detector electronics and stepping motor controller. Changing the energy of the ^{7}Li beam from the accelerator was not via computer control and had to be performed manually, which took a matter of minutes.

Gamma-ray yield curves as a function of ^{7}Li energy were taken from approximately 2.8 to 7MeV ion energy. Beam currents ranged from 5-50nA on target (over a 3-4mm area), with dead-times generally being held below 10%. A typical yield point measurement took 10-15minutes. The beam on target was monitored during each run by using a Si(Li) detector to detect the x-rays being emitted. The potassium x-rays were used for obsidian targets and titanium x-rays for the TiH_2-Ta target. To arrive at a final value for the amount of beam which had hit the target (in arbitrary units), the x-ray yields were corrected for the energy dependence of the production cross-sections. The γ-ray yield was then normalised to this value, which produced a point on the profile curve for the particular incident energy or, equivalently, the depth in the target.

An Igor Pro (WaveMetrics Inc.) program was written to handle the data processing and unfolding of the thin target profile from the sample profile. Curve fits were first made to the two profiles, and then, instead of deconvolving the thin target curve from the sample curve, the reverse process was performed. An initial estimate of the unfolded curve was convolved with the response curve from the thin hydrogen target. Adjustments were then made to this estimate and the procedure repeated via a process of minimisation of least squares.

Various obsidian samples were analysed to investigate the ability of the technique to routinely analyse artefacts. Obsidians that had been artificially hydrated for differing amounts of time at various temperatures and in different environments (eg. soil, sand, distilled water, silica saturated water, steam) were examined, as were naturally hydrated obsidian artefacts which had been lying on or under the ground for an unknown number of years.

RESULTS

Resonance Width and Stopping Power

An very thin layer of TiH_2 was evaporated onto a substrate of tantalum and analysed to establish the intrinsic shape of the resonance. Also profiled was a freshly broken, 'inside' obsidian surface. Such a surface takes up water straight away and forms a very thin hydration layer.

From Fig. 1, the FWHM of the thin hydrogen resonance is 94 ± 5keV, and that of the obsidian thin target 100 ± 5keV, the former value agreeing quite well with that from (3).

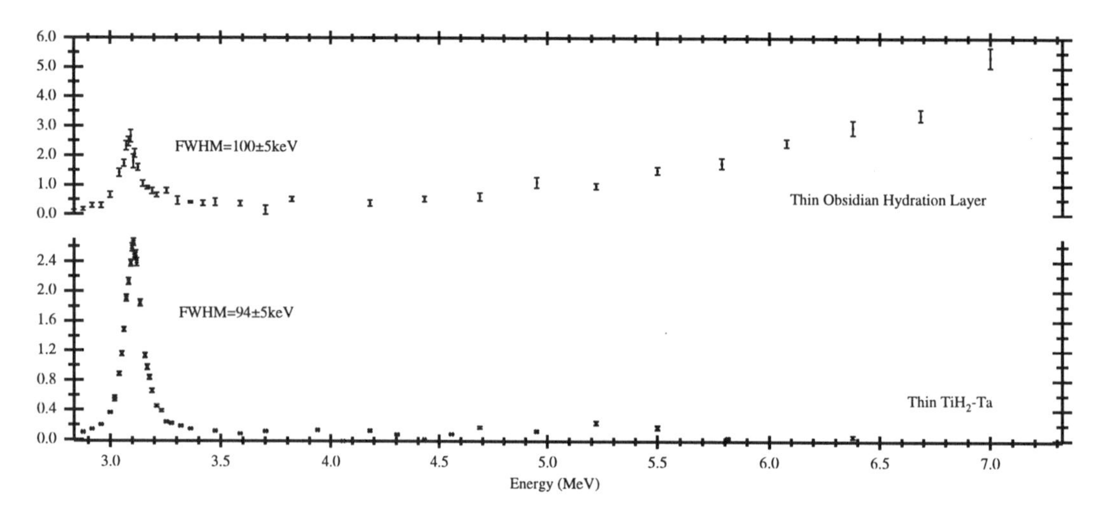

FIGURE 1. Hydrogen profile curves of a thin TiH_2-Ta target and the inside surface of a freshly broken obsidian.

The value for the width of the resonance obtained via the thin TiH_2-Ta target, was used in the unfolding procedure to arrive at a measurement of the thickness of the hydration layer. The depth resolution of the measurement was roughly ±0.08μm, which corresponded to the resonance half width but this was not the sole contributor to the ascribed error in the rim measurement. Further contributions came from the statistics of the γ-ray spectra and from the calculated stopping powers.

For each distinctive obsidian source the elemental composition (4) and density vary, and within each individual source itself there exists a chemical variability because they are not homogeneous. This leads to the value for the stopping power of the ^{7}Li beam varying between samples and even at different positions in the same sample. A typical obsidian composition and average density were used to calculate the dE/dx value which was then applied to all obsidian targets regardless of source. This introduced an error of approximately ±0.04μm to rim thicknesses of the order of 1.5μm.

Gamma-ray Spectrum Backgrounds

As the beam energy was raised beyond 3.09MeV, the yield of γ-rays of energies below 11MeV was seen to greatly increase for both the thin TiH_2-Ta and obsidian targets. This background gradually encroached into the region of interest set down for the ^{1}H(^{7}Li, γ)^{8}Be reaction γ-rays of 14.70 and 17.6MeV (see Fig. 2), resulting in a rise in the hydrogen profile curves.

We determined that this background was due to the ^{7}Li ion beam hitting the surface of objects (such as the targets and collimators) which had carbon deposits on them. The reaction that was occurring was ^{12}C(^{7}Li,α)^{15}N. This feature had also been encountered in (5). To overcome the problem we removed all sources of carbon, and altered the position of the lower energy limit to the ^{8}Be γ-ray region of interest, so that it would better discriminate against the encroaching background coming from lower energies. The carbon build-up on the iris collimators and targets still needed to be monitored during the analysis, and the background problem was largely avoided by performing the higher energy ^{7}Li runs at the beginning before the carbon had accumulated.

Elimination and control of the carbon contaminations did not entirely eliminate the background problem for the obsidian targets, whose hydrogen profiles continued to rise (Fig. 1). With the aid of a higher resolution γ-ray spectrum various coulomb excitation reactions were identified (nuclear excitation of the ^{7}Li projectile has been reported (6)), but these were not the cause of the background in the region of interest (12.5 to 18MeV). Reaction of the ^{7}Li ion beam with oxygen, the main constituent of the glass matrix, via ^{16}O(^{7}Li, γ)^{23}Na with Q=19.70MeV, produced the many high energy γ-rays, from the de-excitation of the broad resonance at ^{23}Na*(25.4MeV). This γ-ray background was obviously unavoidable. Because the elemental composition varies among samples drawn from different geographical sources we decided to measure background profiles for all the majoı sources in order to see if a single curve could be applied to all. Inside surfaces were used so that the hydration layer would not be a contributing factor to the background. The resulting profile curves and their average are shown in Fig. 3.

Up to about 7.5MeV (which equates to a depth of approximately 7.3μm) the curves were essentially the same. The average was therefore applied to all artefact target profiles regardless of their source. When an artefact was being profiled, a few points were taken at higher energies up to 7MeV, and used to subtract the appropriate proportion of the background profile. This reduced the high energy end of the artefact profile down to a fairly constant level, although not necessarily zero because the sample could have had, for example, a non-zero internal water content.

This technique for dealing with interfering backgrounds worked quite well for the obsidian artefacts that were tested (see Fig. 4 for example). However, the presence of this background limits the probe range in a target to rim thicknesses below 7μm. The hydrogen profiles of most New Zealand artefacts, however, start to tail off well below 1.5μm.

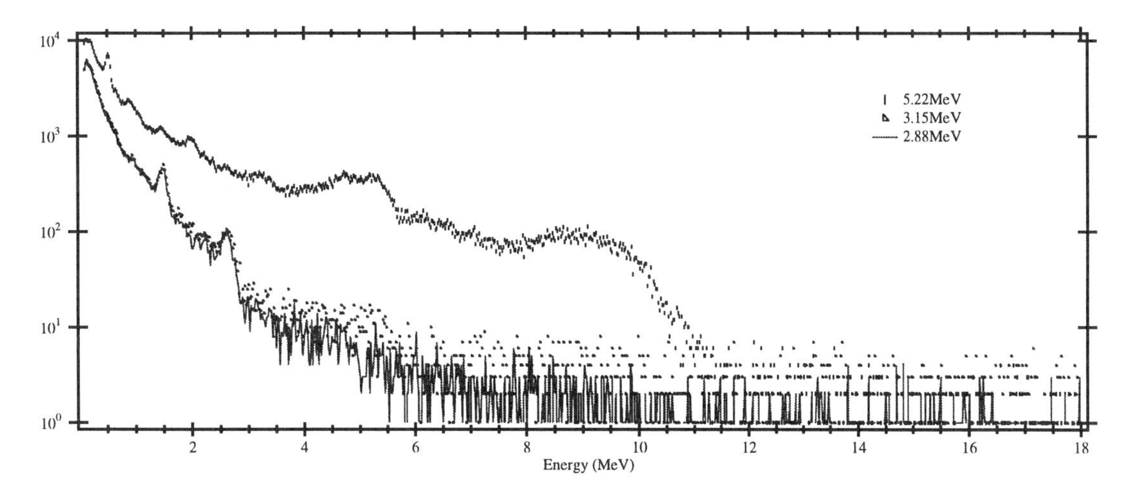

FIGURE 2. Gamma-ray spectra using three different beam energies on the TiH_2-Ta target which had a carbon deposit on its surface. The region of interest for the ^{1}H(^{7}Li, γ)^{8}Be reaction was set from approximately 12.5 to 18MeV.

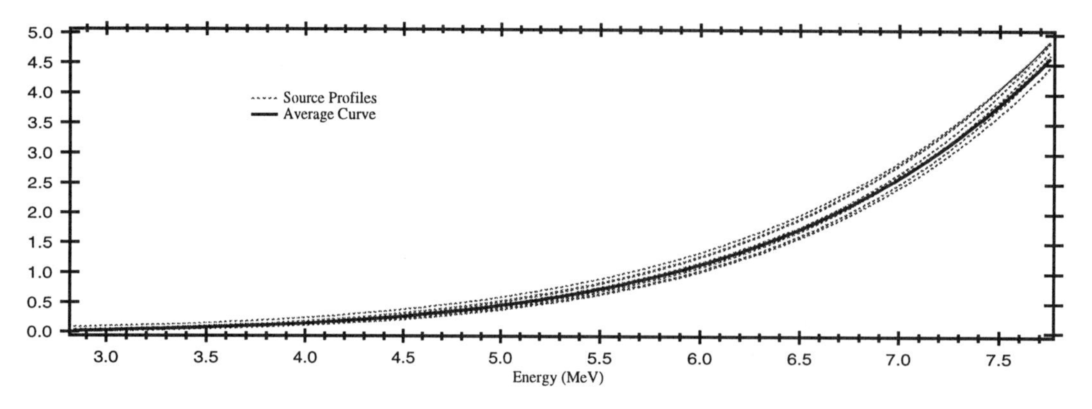

FIGURE 3. Background profile curves for 9 of the major obsidian sources in New Zealand and the average curve.

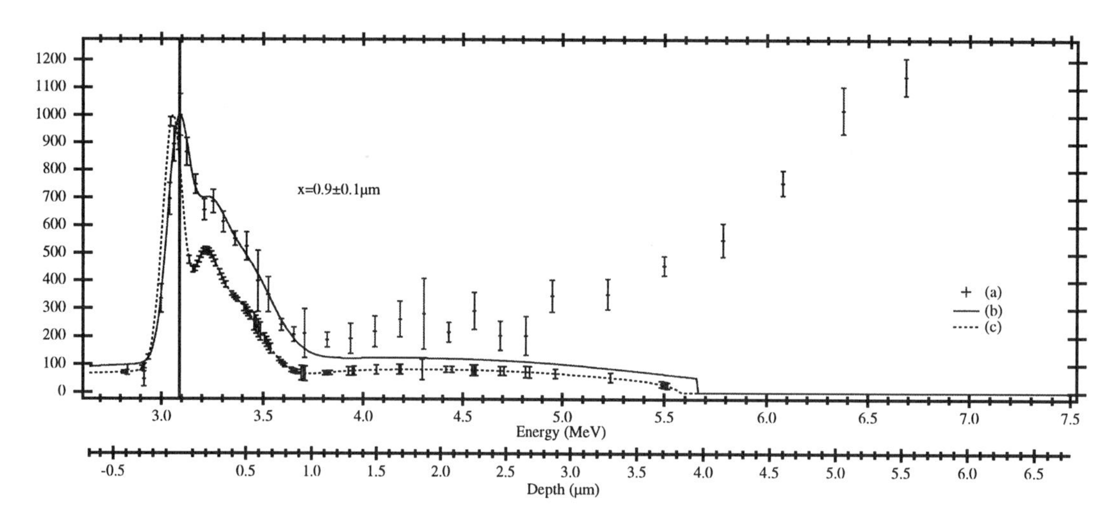

FIGURE 4. Hydrogen profile curve of a naturally hydrated obsidian target. (a) Profile of the raw data; (b) Profile after oxygen background subtraction; (c) Profile of the hydration layer.

Hydrogen mobility in the surface of the obsidians was never evident. On scanning up and down the profile curve, to fill-in or repeat any points, it was found that there was always good agreement amongst points taken at different times.

CONCLUSION

We have been able to measure the hydration rims (of less than 7μm) with an accuracy of 5-10% in artefact and artificially induced obsidians. The reliable translation of these to dates will however be problematic. Contributing uncertainties to rim measurements came from counting statistics, approximations of the stopping power value due to generalisations made about the obsidian composition and density, the irregular shape of the target surfaces and the dead-time of the data acquisition system at the higher beam energies. The beam intensity monitoring using emitted x-rays from the target worked satisfactorily. The techniques for dealing with γ-ray backgrounds were successful. Further investigations of ^{7}Li on ^{16}O reactions are on-going.

ACKNOWLEDGMENTS

Many thanks to M.J. Keeling and W.B. Wood, the technical staff of the AURA2 laboratory, for their continuing dedicated support; also to Martin Jones of the Anthropology Department at The University of Auckland for providing obsidian samples.

REFERENCES

1. Ziegler, J.F., et al., *Nuclear Instruments and Methods* **149**, 19-39 (1978).
2. Northcliffe, L.C., and Schilling, R.F., *Nuclear Data Tables* **7**, 233-463 (1970).
3. Ajzenberg-Selove, F., and Lauritsen, T., *Nuclear Physics* **A227**, 1-244 (1974).
4. Neve, S.R., Barker, P.H., Holroyd, S., Sheppard, P.J., *New Zealand Journal of Archaeology* **16**, 93-121 (1994).
5. Stauber, M.C., Padawer, G.M., Brandt, W., D'Agostino, M.D., Kamykowski, E., Young, D.A., *Proceedings of the Fourth Lunar Science Conference, Geochimica et Cosmochimica Acta* **2(4)**, 2189-2201 (1973).
6. Wang, C.W., Lin, E.K., Yu, Y.C., *Nuclear Physics* **A570**, 363c-370c (1994).

NUCLEON-INDUCED SECONDARIES: A REVIEW AND FUTURE EXPERIMENTAL DEVELOPMENTS

J. L. Romero

Department of Physics, University of California, Davis, CA 95616

Henry H.K. Tang

IBM Microelectronics, Semiconductor Research and Development Center,

East Fishkill Laboratory, Hopewell Junction, NY 12533

D. J. Morrissey, M. Fauerbach, R. Pfaff, C. F. Powell, B.M. Sherrill

National Superconducting Cyclotron Laboratory, Michigan State University

F. P. Brady, D. A. Cebra, J. Chance, J. C. Kintner, J. H. Osborne

Department of Physics, University of California, Davis, CA 95616

We discuss the experimental situation of nucleon-induced reactions. Motivated by important current technological applications such as single event upsets (SEU) in microelectronic devices and particle-beam radiotherapy, we focus on light elements and beam energies below 250 MeV. We review the data from proton- and neutron-induced reactions obtained at UC Davis and elsewhere in the last 2 decades. Despite the large compilation of data (mainly inclusive spectra of light ions), we point out the scarcity of recoil data. We emphasize the fact that recoil nuclei contribute a significant part of the secondary radiation which is a major source of SEUs in devices, and of importance for radiation-related calculations. We discuss a new program under development, in which we measure recoil spectra using reverse kinematics. We discuss results from a recent experiment of $^{1}H(^{28}Si,A)x$ at 80 MeV/nucleon performed at the NSCL at Michigan State University. The preliminary data are compared well with a cascade-statistical reaction model (NUSPA).

INTRODUCTION

In recent years, there has been a revival of interest in the study of nucleon-induced reactions as an important experimental tool in several areas of basic and applied nuclear physics. On the basic side, high-quality data from these reactions at judiciously chosen bombarding energies provide a unique means to characterize reaction mechanisms, probe nuclear structure and impose stringent tests on theoretical models. A knowledge or calculation of such cross sections is also important in other areas of basic research, such as astrophysical and cosmic-ray research where the production and/or transport of isotopes through the interstellar (H2) medium is studied and must be calculated. In applied research, it is well known that precise neutron cross sections on various elements are necessary for any reliable radiation shielding or radiation therapy calculation. In the even more complex and delicate simulations of cosmic-ray-induced soft error rates in semiconductor devices, it is imperative that the microscopic energy loss and the charge generation produced by neutron plus silicon interactions be included. (From charge symmetry and other theoretical considerations one can predict rather well neutron-induced from proton-induced cross sections at equivalent or nearby energies.)

Such calculations, to be complete, require cross sections for all channels ranging from light particle production to target fragmentation and recoil. The target fragmentation measurements are difficult since the low energy fragments become trapped in the target or have such a low energy that they cannot pass through the detection system [1].

From a theoretical standpoint, Tang et al. of I.B.M. have suggested that for a large class of problems in pure and applied nuclear research, cascade-statistical- type (C-S) models are a most appropriate analytical framework. (2) The I.B.M. theorists have recently shown that a large quantity of inclusive data of light fragments as well as heavy recoils, from targets like C, N, O, Al and Si, and in the energy range of 60 MeV - 200 MeV, can be systematically reproduced by a C-S model (2). They further argue that if one includes the medium distortions due to the target mean-field and other structure effects prominent in low-energy nuclear dynamics, the C-S model can be properly generalized and extended below 50 MeV. Thus, it can be expected that eventually the low- and medium-energy reactions can be well described within one unified model. This contention is partially supported by the work of Subramanian et al. , who found that some of the low-energy data can be explained by a C-S- cluster model.(3,4) However, in the absence of neutron and recoil data, a definitive test of the C-S model cannot be completed and theoretical progress in low-energy phenomenology remains limited.

CP392, *Application of Accelerators in Research and Industry*, edited by J. L. Duggan and I. L. Morgan
AIP Press, New York © 1997

For an extensive discussion of neutron data on C up to 100 MeV and other model calculations, see the recent evaluation by Chadwick et al. (5) The case of N and O nuclei is reviewed in ref. (6).

REVIEW OF DATA OF NUCLEON-INDUCED REACTIONS

Because of the impact of SEU on advanced microelectronic systems, the basic physics issues were addressed in a collaboration between IBM and UC Davis in the past few years. From this joint project, much work was done on the measurement and modeling of the spectra of secondary light ions from nucleon-induced reactions. It is important to point out that, up till now, data on heavy recoil spectra is almost non-existent. This is largely because in conventional proton and neutron beam experiments the heavy recoil fragments typically do not escape the target, and therefore are not detected. Yet from the standpoint of energy deposition, the recoil nucleus produced in a collision event constitutes a significant part of the total energy released and in addition produces a large energy loss density on the stopping power.

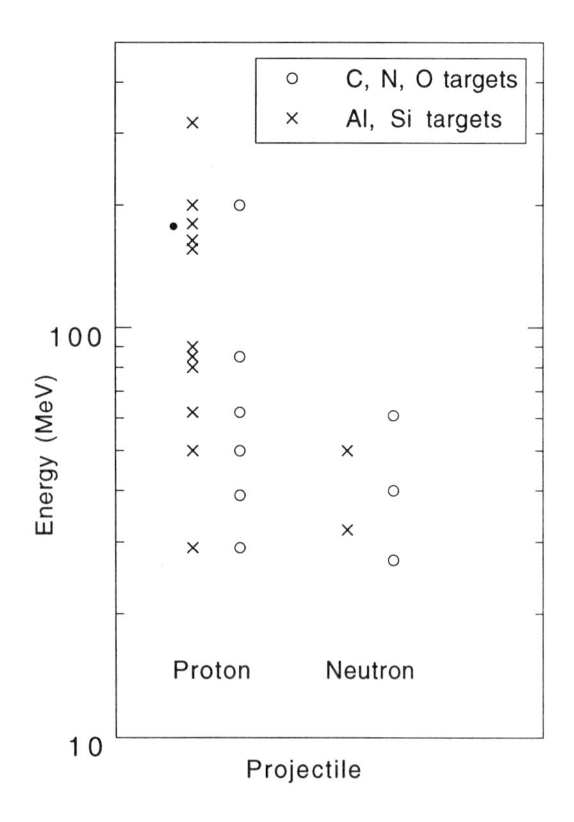

FIGURE 1 A view of the experimental data base for nucleon-induced reactions on light nuclei

As mentioned above, in all modeling endeavors it is of primary importance that one works with a comprehensive data base. For such a data base to be of practical use, it should minimally contain inclusive spectra (double differential cross sections) of light ions from the dominant channels, i.e., (p,p'x), (p,nx), (p,^{2}Hx), (p,^{3}Hx), (p,^{3}Hex), (p,^{4}Hex), (n,n'x), (n,px), (n,^{2}Hx), (n,^{3}Hx), (n,^{3}Hex), (n,^{4}Hex), and inclusive spectra of the various residual (recoil) nuclei.

Fig. 1 is based on a recent review on the status of protons- and neutron-induced reactions on light elements for energies up to 1 GeV. (7) The data in Fig. 1 is only representative, and not exhaustive. The cross and open circles correspond to meausurements of double differential inclusive spectra for n, p, d, ^{3}H, ^{3}He or ^{4}He. Only data that covers a fairly complete angular and energy region is included. A few of those measurements include light fragments up to Li. The filled circle are the measurements for heavy recoils done at the Indiana University Cyclotron Facility.(1) Several features are apparent. In the medium-energy region of 60-200 MeV, the measurements of charged ions (p, d, t, He) are reasonably complete so that meaningful comparisons can be made with theoretical predictions. Below 60 MeV, an energy range most relevant for device applications and SER phenomena, there are only a few published experiments.

The most serious drawback in the data set in Fig. 1 is the almost complete absence of measurements of recoil nuclei in the exit channels, which are needed to complement earlier measurements [3,4] of the light particle (H and He ion) spectra. Also recent neutron induced data from Louvain is now available.(7)

PROTON-INDUCED RECOILS BY REVERSE KINEMATICS

The objective of an experiment with proton beam is to obtain isotopic double differential cross sections (in angle, energy, Z and A) for the inclusive ^{28}Si (p, A)x reaction initially at a proton energy of 80 MeV and for about A≥12. Here A is the mass number of the outgoing nucleus from the fragmentation of ^{28}Si and x stands for anything else. The energy spectra for the products A are needed over nearly the complete kinematic range at a number of angles. Alternatively, this can be accomplished by using reverse kinematics on a CH_2 target (and on a C target for background).

The main characteristics and advantages of the use of reverse kinematics are well known (9): the heavy fragments tend to have relatively low energies in the projectile frame. Thus, in the lab frame the fragments tend to lie in a forward cone. For example, the fragments with 11<A<28 lie mainly in a 7° cone for the case of incident beam energy equal to 80 MeV per nucleon. Lighter fragments of higher energy in the projectile frame tend to spread to larger angles and can be measured in nucleon induced kinematic cases. Fig. 2, obtained from the NUSPA code, compares the kinematics for the direct and reverse reaction for ^{24}Mg fragments. The Mg fragments from the direct reaction Si(p,Mg x) at 80 MeV incident energy are spread out up to 180 deg. in the lab. frame and their energies are low (up to only 10 MeV) (Fig. 2a). On the other hand, the completely equivalent reverse kinematic reaction H(Si,Mg x) at 80 MeV/u produces Mg fragments which are strongly forward focused to a few degrees and their energies are high (between 1600 and 1800 MeV).

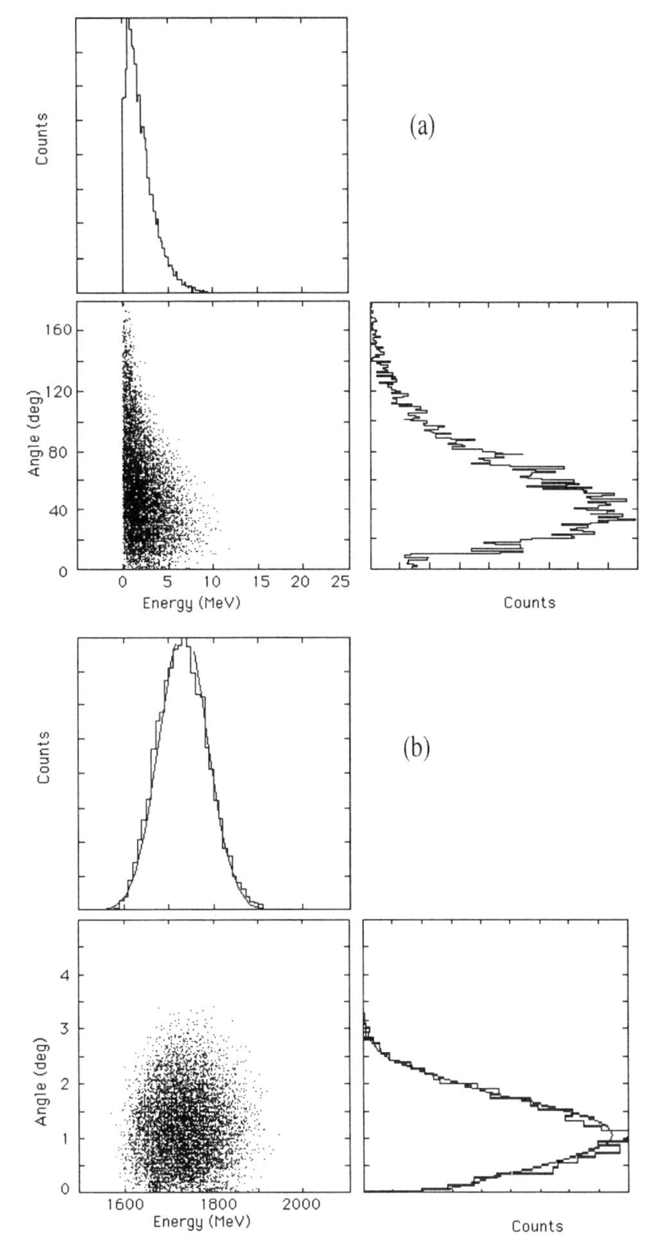

FIGURE 2 Focusing in reverse kinematics. (a) Distribution in angle and energy of Mg fragments for the inclusive reaction ^{28}Si(p, ^{24}Mg x) at 80 MeV obtained using the code NUSPA. (b) Transformed distribution for the reverse kinematic reaction ^{1}H(^{28}Si, ^{24}Mg x) at 80 MeV/u.

The first measurement of recoil spectra in this important energy range was made in our recent exploratory studies. We carried out the experiment at the National Superconducting Cyclotron Laboratory of the Michigan State University. We measured the differential cross sections of the heavy charged fragments from the reverse reaction ^{1}H(^{28}Si,A)x, using a ^{28}Si beam of 80 MeV/nucleon, and the A1200 zero degree mass spectrometer at NSCL. We reported some of the results in the General American Physical Society Meeting in May 1996. (10)

The A1200 Mass Separator has excellent mass, charge and energy resolution and it can be used at 0°.(11,12) The experimental setup takes advantage of detectors mounted in the beam line, since by its very design, the A1200 is part of the beam line. The characteristics of the setup and the location of the permanent detectors are as follows (11,12):

- Beam intensities ranging from 0.1-20 pnA (particle nanoamperes),
- A CH_2 target of 2.6 mg/cm^2. By also measuring the spectra from a C target of similar thickness, the H cross section is obtained by subtraction.
- At the Dispersive Image # 2 (refs. 11,12) two parallel plate position sensitive detectors (PPAC) allow one to measure accurately the rigidity of the fragment (momentum/Z) and also determine the angle of the fragment with respect to the beam direction.
- At the Final Achromatic Image (11,12) energy loss (dE) and time of flight detectors allow excellent fragment identification (Z and A). In this experiment, TOF was measured relative to the radiofrequency of the cyclotron. In additions, 2 PPAC's are used to further define the incident angle.
- Each run consisted of a given magnetic field setting that established the average rigidity (= momentum/Z), covering a range of rigidities of 3% (1% near the beam rigidity). Further analysis at the dispersive image # 2 allowed us to subdivide each range of rigidities into 6 regions (3 near the beam) . So the momentum resolution for each detected fragment was 0.5%. Each setting took on the average 30 minutes per target. A total of 16 setting gave a coverage of p/Z=630 to 960 MeV/c. At each rigidity setting, excellent fragment identification was achieved from Z = 4 to 14 and A = 6 to 29. So we have collected cross section information for a total of about 60 fragments at energies of 0 to the maximun allowed and angles from 0 deg. to about 80 deg. in the target frame. For the case of heavy fragments, the present setup measures about 60-70% of the total production cross section.

Only a fraction of the data has been analyzed. Fig. 3 shows preliminary results for the inclusive cross section of ^{20}Ne, ^{24}Mg, and ^{27}Al fragments produced from the reaction 80 MeV/u ^{28}Si beam on ^{1}H near 0 deg. The symbols are the experimental data points and the histogram is the theoretical prediction of the NUSPA code of H. Tang. We plot the double differential cross section against momentum per fragment charge. It should be pointed out that both the model calculations and the experimental measurements yield absolute cross sections. It is significant that the agreement in width and absolute cross section between theory and experiment is achieved without adhoc adjustments. A full analysis of the experiment including all fragments (about 60) is in progress.

CONCLUSION AND FUTURE DEVELOPMENTS

A review of data on nucleon-induced secondaries reveals an almost complete abscncc of measurements of recoil nuclei on light elements. The recoil experiment at the NSCL A1200 facility demonstrates the feasibility of reverse kinematics as an accurate means of measuring recoil spectra. This experiment opens up new possibilities. A wealth of data from light elements such as O, N and C, important for many radiation physics applications, can now be obtained by the same setup as the Si experiment described. Currently we propose to run a series of such recoil measurements in a priority according to their technological importance. The following partial list is made according to the energy range presently available at NSCL. ^{16}O (at 200 and 80 MeV/u), ^{12}C (at 200 and 100 MeV/u), ^{14}N (at 155 and 90 MeV/u), ^{40}Ca (at 55 MeV/u).

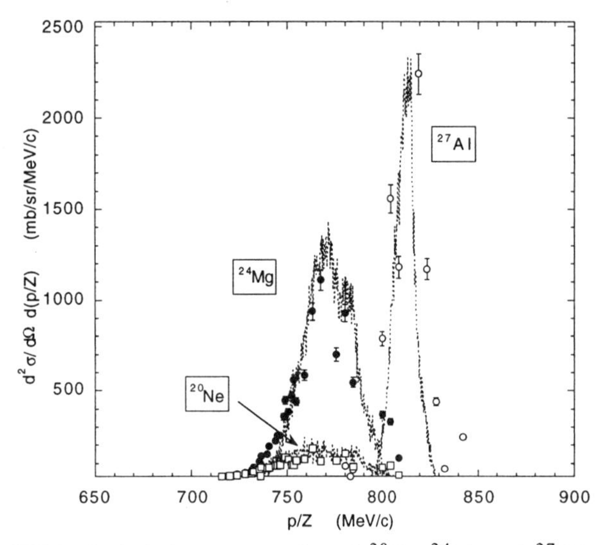

FIGURE 3. Inclusive cross section of ^{20}Ne, ^{24}Mg and ^{27}Al fragments produced from the reaction 80 MeV/u ^{28}Si beam on ^{1}H near 0 deg. The histogram is the theoretical prediction of the NUSPA code of H. Tang.

ACKNOWLEDGMENTS

We thank Gabriel Acevedo-Bolton for help on the data analysis. We acknowledge support from the NSF.

REFERENCES

1 K. Kwiatkowski , S.H. Zhou, T.E. Ward, V.E. Viola,Jr., H.Breuer, G.J. Mathews, A. Gökmen and A.C. Mignerey ,Phys. Rev. Lett. 50, 1648 (1983).

2 H.H.K.Tang, G.R. Srinivasan, and N. Azziz. Phys. Rev. C42, 1598-1622 (1990)

3 T.S. Subramanian, J. L. Romero, F. P. Brady, J.W. Watson, D.H. Fitzgerald, R. Garrett, G.A. Needham, J. L. Ullmann, C.I. Zanelli, D.J. Brenner, and R. E. Prael. Phys. Rev. C 28, 521 (1983).

4 T.S. Subramanian, J. L. Romero, F. P. Brady, D.H. Fitzgerald, R. Garrett, G.A. Needham, J. L. Ullmann, J.W. Watson, C.I. Zanelli, D.J. Brenner, and R. E. Prael. Phys. Rev. C 34, 1580 (1986).

5 M.B. Chadwick, L.J. Cox, P.G. Young, and A. Meigooni. Nucl. Sci. Eng. 123, 17 (1996)

6. M.B Chadwick and PG Young, Nucl. Sci. Eng. 123, 1 (1996)

7 H.H.K. Tang. IBM J. Res. and Dev. 40(1), 91-108 (1996).

8 I. Slypen, V. Corcalciuc, J.P. Meulders, and M. B .Chadwick, Phys. Rev. C 53, 1309 (1996).

9 W.L. Kehoe, A.C. Mignerey, A. Moroni, I. Iori, G.F. Peaslee, N. Colonna, K. Hanold, D.R. Bowman, L.G. Moretto, M.A. McMahan, J.T. Walton and G.J. Wozniak. Nucl. Instr. Meth. A311 (1992) 258. And references therein.

10 J. L. Romero, F. P. Brady, D. A. Cebra, J. Chance, J. Kintner, J.H. Osborne, G. Acevedo-Bolton, D.J. Morrissey, B.M. Sherrill, C. Powell, M. Fauerback, and H.H.K. Tang. BAPS 41, 987 (1996).

11 D. J. Morrissey, "Planning for a STANDARD A1200 Experiment", NSCL document (1993).

12 G. A. Souliotis, D. J. Morrissey, N.A. Orr, B.M. Sherrill, and J.A. Winger, Phys. Rev. C 46, 1383 (1992).

STUDY OF DEUTERON INDUCED REACTIONS ON NATURAL IRON AND COPPER AND THEIR USE FOR MONITORING BEAM PARAMETERS AND FOR THIN LAYER ACTIVATION TECHNIQUE

S. Takács[1], F. Tárkányi[1], M. Sonck[2], A. Hermanne[2] and S. Sudár[3]

[1] *Institute of Nuclear Research of the Hungarian Academy of Sciences (ATOMKI), H-4001 Debrecen, Hungary*
[2] *Vrije Universiteit Brussel (VUB), Cyclotron Department, 1090 Brussels, Belgium*
[3] *Institute of Experimental Physics, Kossuth University, H-4026 Debrecen, Hungary*

Excitation functions of deuteron induced nuclear reactions on natural iron and copper have been studied in the frame of a systematic investigation of charged particle induced nuclear reactions on metals for different applications. The excitation functions were measured up to 20 MeV deuteron energy by using stacked foil technique and activation method. The measured and the evaluated literature data showed that some reaction can be recommended for monitoring deuteron beams, and the excitation functions can be used to determine calibration curves for Thin Layer Activation Technique (TLA). Cross sections calculated by statistical model theory, STAPRE, taking into account preequilibrium effect are in reasonable agreement with the experimental results.

INTRODUCTION

In the frame of a systematic investigation of charged particle induced nuclear reactions on metals excitation functions of deuteron induced nuclear reactions were studied up to 20 MeV by using stacked foil technique and activation method. There are several applications of deuteron induced nuclear reactions (Charged Particle Activation Analysis of different substances (CPAA), wear measurement using Thin Layer Activation (TLA) technique, isotope production for medical and other purposes, monitoring deuteron beam energy and intensity) where the knowledge of the excitation function is essential. The quality of most existing experimental data, however, are far from the requirements of the mentioned applications. Therefore, the goal of this work was to investigate nuclear reactions induced by deuteron on natural copper and iron to find possible candidates among the reactions and recommend them for monitoring the performance of deuteron beams in low energy region and monitoring wear and/or corrosion of copper and/or iron containing parts of mechanical structures.

EXPERIMENT

Irradiations were carried out with the external beams of the VUB CGR 560 and the ATOMKI MGC 20E cyclotron. Stacks each containing more than ten foils of 10 μm thick Cu and stacks made of 25 μm thick Fe commercial (Goodfellow) foils were irradiated for about 1 hour with 200 nA deuteron beams of a 10, 16 and 21 MeV primary energies. The beam current was kept constant during each irradiation, and was measured in a Faraday-cup, which was equipped with a special "long" collimator and a secondary electron suppressor. The beam also was monitored via the $^{27}Al(d,x)^{24}Na$ reaction (1) placing high purity Al foils in front of each stacks. The beam current obtained by the Faraday-cup was 10% lower than determined via the monitor reaction (used values are: E_d=21 MeV σ=54 mb) (1). For calculating the cross section data we used the beam current values measured by Faraday cup. The irradiation set-up, the experimental technique, data acquisition and data evaluation were the same or similar as it was described earlier in (2). The initial energy of the particles was determined with an accuracy of ± 0.3 MeV by time of flight method at the Brussels cyclotron (3) and ± 0.2 MeV by an analysing magnet for irradiations in Debrecen. The effective "on target" energy for each foil was determined using the energy-range formula and table of Andersen and Ziegler (4). The activity of the irradiated foils was measured without chemical separation by using high resolution gamma-ray spectrometry. The detector-source distances were kept long enough (typically 20 cm) to avoid cascade and pile up losses. The decay of the activity of the samples was followed by measuring each sample several times. Taking into account the structure of the irradiated stacks no correction was necessary for the recoil effect. The cross sections were calculated for natural isotopic composition. The decay data of the investigated isotopes and the Q-values

CP392, *Application of Accelerators in Research and Industry*, edited by J. L. Duggan and I. L. Morgan
AIP Press, New York © 1997

of the contributing processes were taken from (5). The relative uncertainties in the cross section values were estimated by combining the individual relative errors in quadrature. The average error varied from 10% to 15%. The error of the energy scale was estimated taking into account the uncertainty of the primary bombarding beam energy, the effect of the beam straggling and the error in the target thickness.

RESULTS AND DISCUSSION

When bombarding natural Cu and Fe target with deuteron beam up to 20 MeV several reaction processes take place and contribute to the formation of ^{64}Cu, ^{62}Zn ^{63}Zn and ^{65}Zn isotopes on copper and ^{51}Cr, ^{52}Mn, ^{54}Mn, ^{56}Mn, ^{55}Co, ^{56}Co, ^{57}Co and ^{58}Co isotopes on iron. Regarding the decay parameters (half life, energy and intensity of the emitting gamma-lines) of the produced nuclei the processes, leading to the formation of ^{65}Zn on natural Cu, ^{56}Co and ^{57}Co on natural Fe can be important in practical applications for beam monitoring and wear measurement. These processes have reasonable high cross section to use them conveniently. Therefore, we discuss here only the formation of these three isotopes. The reactions leading to the formations of other side reactions will be published separately.

The values of the measured cross section data of the above three processes are listed in Table 1. and are presented in Figs. 1-2 in comparison with data measured by other authors earlier. Data measured on enriched targets by other authors were converted to data measured on targets with natural isotopic composition. The solid curves represent the fit of our experimental points (eye guide curves only), while the dotted lines for ^{56}Co and ^{57}Co show the result of our theoretical calculation. Cross sections calculated by statistical model theory, taking into account preequilibrium effect (STAPRE) are in reasonable agreement with the experimental results. The model and the parameters used in the calculation were the same as it was described in (10) by Sudár.

^{nat}Cu(d,x)^{65}Zn

Out of the possible nuclear reactions the ^{63}Cu(d,γ) and the ^{65}Cu(d,2n) processes are expected to contribute to the formation of ^{65}Zn using natural Cu target. The measured cross section data and the available earlier experimental results are plotted in Fig. 1. Our results are in agreement with data of Pement (6) and Okamura (7) below 10 MeV and disagree with the values of Fulmer (8) which are too low.

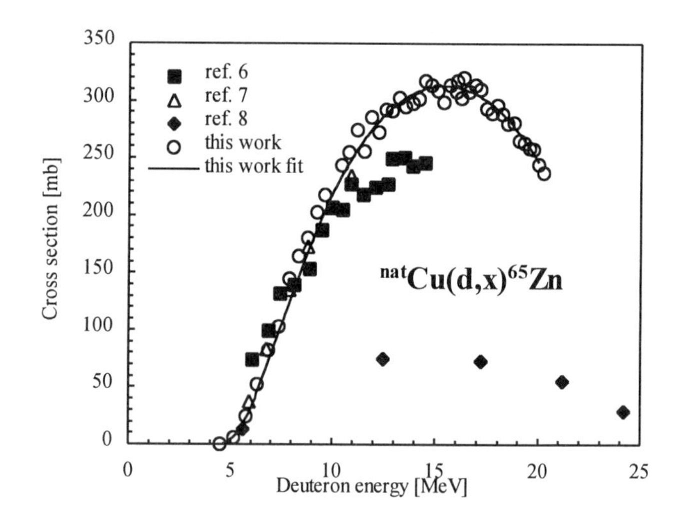

FIGURE 1: Excitation function of deuteron induced reactions on natural Cu leading to the formation of ^{65}Zn

^{nat}Fe(d,x)^{56}Co

In the investigated energy range using natural Fe target the ^{54}Fe(d,γ), ^{56}Fe(d,2n) and ^{57}Fe(d,3n) processes are expected to contribute to the formation of ^{56}Co. Taking into account the isotopic composition of the target, the Q-value of the reactions and the systematic of the excitation functions of deuteron induced reactions one can state that ^{54}Fe(d,γ) and ^{57}Fe(d,3n) reactions have only minor contribution. The main contributing process is the ^{56}Fe(d,2n) reaction.

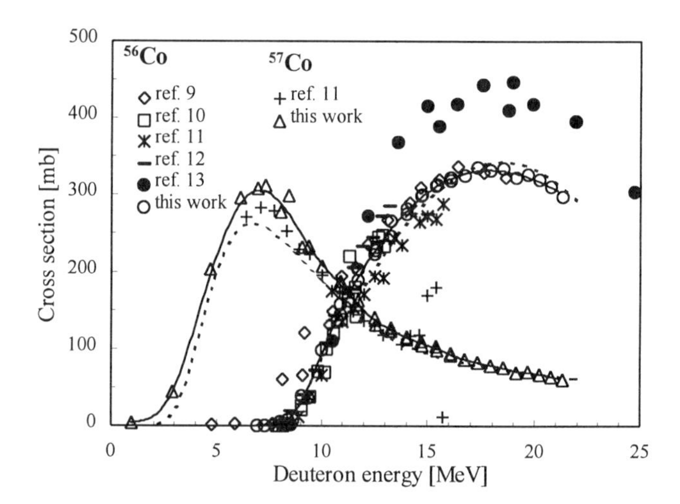

FIGURE 2: Excitation functions of deuteron induced nuclear reactions on natural Fe leading to the formation of ^{56}Co and ^{57}Co. The solid curves represent the fit of our measured points (eye guide curves) and the dotted lines show the result of STAPRE model calculation.

TABLE 1: Experimental cross section values of deuteron induced nuclear reactions on natural Cu and natural Fe.

Energy [MeV]	$^{nat}Cu(d,x)^{65}Zn$ sigma [mb]	Energy [MeV]	$^{nat}Fe(d,x)^{56}Co$ sigma [mb]	$^{nat}Fe(d,x)^{57}Co$ sigma [mb]
5.16 ± 0.47	6 ± 4	0.94 ± 0.60		5 ± 1
5.77 ± 0.43	25 ± 3	2.91 ± 0.50		44 ± 6
6.34 ± 0.39	52 ± 7	4.69 ± 0.41		203 ± 23
6.87 ± 0.35	82 ± 9	6.10 ± 0.35		296 ± 34
7.38 ± 0.32	103 ± 14	6.93 ± 0.43	0.4 ± 0.2	309 ± 35
7.87 ± 0.29	145 ± 17	7.31 ± 0.29		313 ± 36
8.34 ± 0.27	164 ± 19	8.05 ± 0.41	5.8 ± 0.8	278 ± 32
8.79 ± 0.24	180 ± 22	8.39 ± 0.24	8.5 ± 1.0	299 ± 34
9.23 ± 0.22	203 ± 24	9.07 ± 0.39	40 ± 4.5	232 ± 26
9.65 ± 0.20	217 ± 25	9.38 ± 0.20	37 ± 4.2	234 ± 28
10.43 ± 0.60	243 ± 27	10.02 ± 0.37	99 ± 11	207 ± 24
10.81 ± 0.57	255 ± 28	10.88 ± 0.55	145 ± 16	188 ± 22
11.18 ± 0.55	274 ± 30	10.90 ± 0.35	157 ± 18	182 ± 21
11.54 ± 0.52	256 ± 28	11.72 ± 0.52	188 ± 21	162 ± 19
11.90 ± 0.49	286 ± 31	11.74 ± 0.34	203 ± 23	152 ± 17
12.25 ± 0.47	273 ± 30	12.51 ± 0.49	223 ± 25	140 ± 16
12.59 ± 0.45	292 ± 32	12.54 ± 0.32	228 ± 25	131 ± 14
12.92 ± 0.43	291 ± 32	13.28 ± 0.47	247 ± 28	124 ± 14
13.25 ± 0.41	302 ± 33	13.30 ± 0.30	266 ± 30	128 ± 14
13.58 ± 0.39	295 ± 32	14.01 ± 0.45	276 ± 31	115 ± 13
13.89 ± 0.37	298 ± 33	14.03 ± 0.29	283 ± 32	113 ± 13
14.21 ± 0.35	302 ± 33	14.72 ± 0.43	300 ± 34	110 ± 13
14.51 ± 0.34	317 ± 35	14.74 ± 0.28	300 ± 34	105 ± 12
14.82 ± 0.32	314 ± 35	15.40 ± 0.41	312 ± 36	104 ± 13
15.12 ± 0.30	309 ± 34	15.42 ± 0.26	314 ± 35	99 ± 11
15.41 ± 0.29	299 ± 33	16.06 ± 0.39	319 ± 36	94 ± 11
15.70 ± 0.28	314 ± 35	16.08 ± 0.25	323 ± 36	91 ± 10
15.99 ± 0.26	308 ± 34	16.71 ± 0.37	326 ± 37	87 ± 10
16.07 ± 0.55	318 ± 35	17.33 ± 0.35	335 ± 39	82 ± 10
16.27 ± 0.25	302 ± 33	17.94 ± 0.34	332 ± 38	79 ± 9.2
16.35 ± 0.52	320 ± 35	18.53 ± 0.32	334 ± 38	76 ± 8.8
16.63 ± 0.49	308 ± 34	19.11 ± 0.30	323 ± 37	70 ± 8.3
16.91 ± 0.47	314 ± 35	19.68 ± 0.29	326 ± 37	71 ± 8.2
17.18 ± 0.45	309 ± 34	20.24 ± 0.28	320 ± 37	67 ± 7.6
17.45 ± 0.43	293 ± 32	20.78 ± 0.26	311 ± 36	65 ± 7.5
17.71 ± 0.41	289 ± 32	21.32 ± 0.25	298 ± 35	61 ± 7.0
17.98 ± 0.39	296 ± 33			
18.24 ± 0.37	288 ± 32			
18.50 ± 0.35	280 ± 31			
18.75 ± 0.34	281 ± 31			
19.01 ± 0.32	265 ± 29			
19.26 ± 0.30	263 ± 29			
19.51 ± 0.29	259 ± 28			
19.75 ± 0.28	258 ± 28			
19.99 ± 0.26	244 ± 27			
20.24 ± 0.25	237 ± 26			

The measured cross section data and the available earlier experimental results are plotted in Fig. 2. Our data are in agreement with the data of Zhao Wen-rong (9), Sudár (10), Tao Zhenlan (11) and Irvine (12) experiments but the data of Clark (13) seem to be too high. The experimental data of Burgus (14) are not shown on the figure since they were presented only in graphical form and the presented data were given as a lower limit of the cross section.

$^{nat}Fe(d,x)^{57}Co$

For the formation of ^{57}Co on natural iron the $^{56}Fe(d,n)$, $^{57}Fe(d,2n)$ and $^{58}Fe(d,3n)$ reactions can contribute. The (d,3n) process on ^{58}Fe starts only above 14 MeV and its contribution comparing to the other two reaction channel is negligible. Our data are in good agreement with the only data measured by Tao Zhenlan earlier (11).

APPLICATIONS AND CONCLUSION

The use of monitor reactions is often the only possible way to determine beam current and/or beam energy in many application of charged particle activation technique. The $^{27}Al(d,x)^{24}Na$ reaction is frequently used for monitoring deuteron beam above 12 MeV but there is no recommended reaction in the lower energy region. Considering the decay data and the production cross sections the $^{nat}Cu(d,x)^{65}Zn$, $^{nat}Fe(d,x)^{56}Co$ and $^{nat}Fe(d,x)^{57}Co$ reactions appear to be promising for monitoring beam performances. These reactions allow to monitor both low energy (^{57}Co, from 3 MeV up to 15 MeV) and middle energy deuterons (^{56}Co and ^{65}Zn up to 25 MeV). However, more experimental data are needed, that cover the entire investigated energy region (from threshold up to 25 MeV) in order to reduce the discrepancies and to provide more experimental data for the evaluators to create a recommended database in the near future.

To investigate corrosion and/or wear processes and for quantifying the mass loss of Cu and Fe containing materials the TLA method is frequently used. Having two or more reaction channels open gives possibility to use multi profile activation, like in the case of ^{56}Co and ^{57}Co isotopes where the activity distribution ends at different depths. The first step is to determine the distribution of activity as the function of depth a so called calibration curve. Individual calibration curves are required for different matrix materials and for different nuclear reactions. There are two ways of creating calibration curves: by experiment and by calculation. One can construct the calibration curve experimentally, when irradiating a sample, then grinding away the irradiated surface of the test sample step by step and measuring precisely the thickness of the removed layers and removed and/or remaining radioactivity. The removed and remaining activity as a function of depth give the differential and integral calibration curves respectively. The procedure of determining experimental calibration curves requires special techniques and the process is very time consuming. To produce calibration curves by calculation is easier but requires well known excitation function and stopping power data. By combining the two methods (measurements and calculation) for not well known or for complex material, the calculated activity distribution can be fitted to a few experimentally determined points and this way the calibration process can be accelerated and shortened in time, as we described earlier in (15). In Fig. 3 we reproduced several integral calibration curves measured by Romanov (16), Asher (17), Ivanov (18) and Racolta (19) on natural copper and on natural iron, and also present calibration curves obtained by calculation based on our measured excitation functions. The direct measured and the calculated integral activity of ^{65}Zn and ^{57}Co as a function of depth are in good agreement for pure, homogenous copper and iron. But the experimental data of Romanov (16) have significant deviation from our calculated ones which indicates presence some systematic error in the data of Romanov.

The final conclusion is that for the above three processes the status of the existing data still seems to be not satisfactory to provide consistent and well defined (low error) recommended data base for the whole investigated energy region. In the case of ^{65}Zn and ^{56}Co the application of their excitation functions can be extended beyond 20 MeV where the values of the cross section are still high enough.

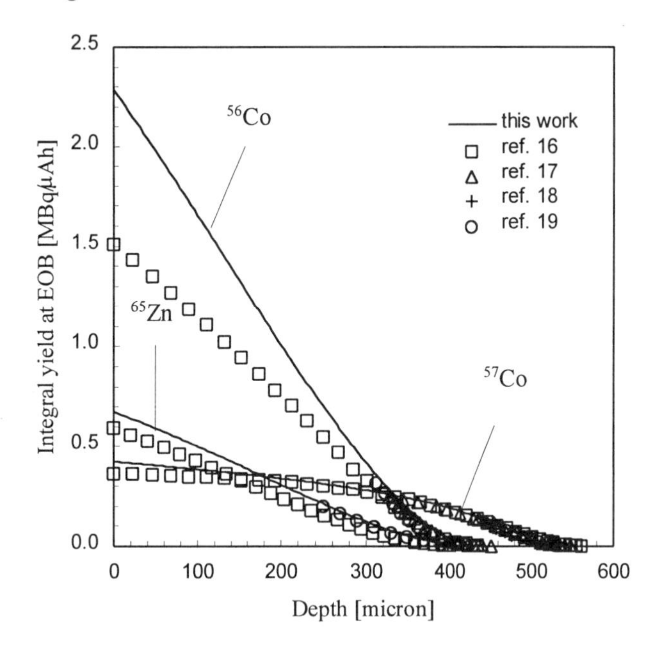

FIGURE 3: Comparison of TLA calibration curves measured directly and calculated from our experimental excitation function for $^{nat}Cu(d,x)^{65}Zn$, $^{nat}Fe(d,x)^{56}Co$ and $^{nat}Fe(d,x)^{57}Co$ processes.

ACKNOWLEDGEMENTS

This work was done in the frame of a Co-ordinated Research Project "Development of Reference Charged Particle Cross Section Data Base for Medical Radioisotope Production" organised and partly supported by the International Atomic Energy Agency, Vienna, Austria, and in collaboration between the National Fund for Scientific Research, Belgium and the Hungarian Academy of Sciences, Hungary.

REFERENCES

1. Schwerer O and Okomoto K, Status Report on Cross-sections of Monitor Reactions for Radioisotope Production, IAEA NDS, *Vienna, INDC(NDS)-218/GZ+,* 73-80, (1989)
2. Tárkányi F., Szelecsényi F. and Kopecky P, *Journal of Applied Radiation and Isotopes,* **42**, 513-517, (1991)
3. Sonck M., Van Hoyweghen J.and Hermanne A., *Journal of Applied Radiation and Isotopes*, **47**, 445-449, (1996)
4. Andersen H. H. and Ziegler J. F., *Hydrogen Stopping Powers and Ranges in All Elements*, New York, Pergamon Press, 1977
5. Browne E. and Firestone R. B., (Shirley S. ed.), *Table of Radioactive Isotopes,* London, Wiley, 1986
6. Pement F.W. and Wolke R. L., *Journal of Nuclear Physics* **86,** 429-442, (1966)
7. Okamura H. and Tamagawa S., *Journal of Nuclear Physics* **A169**, 401- 406, (1971)
8. Fulmer C. B. and Williams I. R., *Journal of Nuclear Physics* **A155,** 40-48, (1970)
9. Zhao Wen-rong, Lu Han-lin, Yu Wei-xiang and Cheng Jian-tao, *Chinese Journal of Nuclear Physics,* **17**, 163-166, (1995)
10. Sudár S. and Qaim S. M., *Physical Review* **C50,** 2408-2419, (1994)
11. Tao Zhenlan, Znu Fuying, Qui Huiyuan and Wang Gonging *Atomic Energy Sciences and Technology* **5,** 506-509, (1983)
12. Irvine J. W., (Ed. Goodman C.) *The Science and Engineering of Nuclear Power, Vol II. 223* ., Addison-Wesley Press Inc. Cambridge (Mass) 1949
13. Clark J. W, Fulmer C. B and Williams I. R, *Physical Review*, **179,** 1104-1108, (1969)
14. Burgus W. H., Cowan G. A., Hadley J. W.,Hess W., Shull T., Stevenson M. L. and York H. F., *Physical Review* **95,** 750-751, (1954)
15. Takács S., Vasváry L. and Tárkányi F., *Nuclear Instruments and Methods in Physics Research* **B 89**, 88-94, (1994)
16. Romanov V. A., *Final Report, The measurement of calibration curves used in thin layer activation techniques for chemical elements irradiated by charged particles,* private communication, 1996
17. Asher J. and Conlon T. W., *Nuclear Instruments and Methods* **179**, 201-203, (1981)
18. Ivanov E. A., Pascovici G. and Racolta P. M., *Nuclear Instruments and Methods in Physics Research* **B82**, 604-606, (1993)
19. Racolta P. M., Popa-Simil L.,Ivanov E. A. and Alexandreanu B., *Journal of Radiation Phys. Chem.* **47**, 677-680, (1996)

THE USE OF PROMPT GAMMA ACTIVATION ANALYSIS IN SEDIMENT SAMPLES FROM A POLLUTED MARINE ENVIRONMENT

A. M. YUSOF and M. MARPONGAHTUN

Department of Chemistry, Universiti Teknologi Malaysia, Locked Bag No. 791, 80990 Johor Bahru, Malaysia

R. M. LINDSTROM

Inorganic Analytical Research Division, National Institute of Standards and Technology, Gaithersburg, MD 20899, U.S.A.

Abstract

Marine sediments pollution could either be caused by point sources discharge or from natural processes such as coastal erosion. A study on the pollution trend of a fairly polluted marine environment was carried out using the cold neutron prompt gamma activation analysis (CNPGAA) facility to identify certain elements with radionuclides of relatively short half-lives. This technique serves as a complementary tool in elemental analysis. Some of the results obtained can be used to construct a depth profile in terms of elemental distribution over time.

1. Introduction

While the use of the conventional thermal neutron activation analysis has provided high sensitivity in trace elements analysis for most environmental samples, prompt gamma activation analysis (PGAA)[1] has offered an advantage in the non-destructive determination of a wider range of elements such as H, B, C and N apart from an array of elements present in the samples. The use of the 'cold' neutrons with an average energy of 0.005 eV has managed to alleviate the sensitivities for multielemental determination in environmental samples. The advantages of using this technique have been documented elsewhere[2]. The presence of about 2% epicadmium neutrons and the extensive neutron shielding material with high hydrogen contents renders this system unsuitable for the determination of hydrogen though high hydrogen count rates have been observed in the marine samples. This work describes the analysis of marine sediment samples using the Cold Neutron Research Facility (CNRF) at the National Institute of Standards and Technology (NIST) Research Reactor.

2. Experimental design

Sediment core samples from a polluted a marine environment[3] were taken using a mechanical box-coring device (30 cm x 20 cm Ø) with the central core cut into strata of between 2 - 3 cm intervals which were later treated accordingly to produce a homogenous dried powder of < 200 mesh size. The CNRF and PGAA instrument[4] at NIST were used in the analysis of the sediment core samples. The neutron beam free of epithermal and fast neutrons as well as gamma rays has a flux of 1.5×10^8 $cm^{-2}s^{-1}$ at the sample position placed 41 m from the cold source. Figure 1 shows the experimental setup at the new cold-neutron PGAA work station in the guide hall of the reactor with the neutron beam emerging from the guide through a 0.23 mm thick magnesium alloy window and collimated by a fused $^6Li_2CO_3$ plate prior to striking the sample. Figure 2 shows the PGAA spectrometer assembly with the sample in place.

CP392, *Application of Accelerators in Research and Industry,* edited by J. L. Duggan and I. L. Morgan
AIP Press, New York © 1997

Sediment samples were weighed (~ 1g) and the powder pressed into a cylindrical disk of 1 cm in diameter and sealed in Teflon bags cut out from the sheet and later suspended by Teflon strings between the prongs of an aluminium fork. The disk was then irradiated in the neutron flux for approximately two hours. Residual neutrons were eliminated by using a beam stop of ^{6}Li glass[5] placed just behind the sample. The shutter made of ^{6}Li glass regulates the flow of neutrons to the sample target.

The Ge gamma-ray detector with 27% efficiency and a resolution of 1.7 keV relative to NaI in the PGAA spectrometer assembly detects the prompt gamma-rays emitted from the samples. The spectra of the gamma-rays emitted are taken using a 16384-channel fixed conversion time analog-digital converter (Canberra Nuclear Data ND582) coupled to an ND556 AIM multichannel analyzer. The details of the data acquisiton procedure using the CNRF at NIST is described elsewhere.[6]

3. Results and Discussion

Absolute concentrations of the elements of interest may be difficult to determine but this was overcome by using the sensitivity ratios of the elements in the samples and the selective standards used. Since the CNRF at NIST is free of epithermal and fast neutrons, the background radiation levels due to H and B should be quite low. However significant amount of H was observed at 2223.23 keV and this could be due to the hydrogen present in the paint coating the guide as reported by Paul *et.al.*[6] and also from the samples with high organic contents. Apart from that, strong peaks of Si at 3538 keV and Cl at 1165.18 keV were observed suggesting that, as expected, the sediment samples were of marine origin although the Si could also be from the glass of the adjacent guide.[4] Nitrogen was first observed at a depth of between 4 - 6 cm at 480 keV and continued to be present in subsequent layers of the samples. This suggested that nitrogenous compunds, possibly due to anaerobic activities, may have been formed from wastes or organic materials deposited over the years.

Results on the quantitative analysis based on the sensitivities of some elements of interest in 12 different sediment strata are given in Table 1. The results are expressed in the count rates (c/s). The count rates obtained can also serve as an indication of the amount of elemental concentrations present in each sample and also the sensitivity of the peak energy. The whole series of elements could be difficult to determine quantitatively due to constraints in getting the individual standards. However a great deal of information could be obtained on the status of the samples analyzed with regard to the anticipated detection limits for some elements.

Scandium, being non-anthropogenic in origin, has a relatively high sensitivity of between 9 - 19 cps and this supported the fact that it can be used as a reference crustal element in the ratioing technique in the quantitative assessment of environmental samples.[3] Vanadium (6 - 15 cps) with good potentials in oil spills monitoring exhibited almost a similar sensitivity range shown by Sc. The same advantage is also offered by Cd with a sensitivity range of 8 - 22 cps. Most of the lanthanides (La, Ce, Nd, Sm, Yb and Lu) which are useful in chondrite-normalized plots have reasonable levels of sensitivities, thus allowing PGNAA to be used as a complementary tool in sedimentation pattern studies of coastal and offshore areas.

4. Conclusion

The ability of CNPGAA in detecting H (2223.23 keV) and N (480 keV) in the sediment samples of homogenous composition will offer new avenues in the field of agriculture related to plant growth studies. Waste disposal management system in the marine environment will also stand to benefit especially in the study of anaerobic decompositioin of wastes using nitrogen as the indicator. The use of elemental ratios in the concentration determination has made this technique very simple and yet reliable to use especially in the analysis of homogenous samples.

References

[1] R. E. Chiren, *Practical Uses of Neutron Gamma-Ray Spectroscopy*, in *Neutron Radiative Capture*, ed. R. E. Chiren, NEA-NDC (OECD) Series "Neutron Physics and Nuclear Data in Science and Technology", Vol.3, Pergamon Press, Oxford (1984) 187.

[2] R. M. Lindstrom and C. Yonezawa, *Prompt Gamma Activation Analysis with Guided Neutron Beams*, ed. Z. B. Alfassi and C. Chung in *Prompt Gamma Neutron Activation Analysis*, CRC Press, Boca Raton (1995) 93.

[3] A. M. Yusof, A. K. H. Wood and Z. Ahmad, Nucl.Instrum.Methods in Phys.Res., **B 99** (1995) 502.

[4] R. M. Lindstrom, R. Zeisler, D. H. Vincent, R. R. Greenberg, C. A. Stone, E. A. Mackey, D. L. Anderson and D. D. Clark, J.Radioanal.Nucl.Chem., Articles, **167(2)** (1993) 121.

[5] C. A. Stone, R. Zeisler, D. H. Blackbum, D. A. Kaufmann, D. C. Cranmer, in *NIST Techn.Note 1285*, U.S. Govt. Print. Off., Washington, D. C., (1990) 116.

[6] R. L. Paul, R. M. Lindstrom and D. H. Vincent, J.Radioanal.Nucl.Chem., **180(2)** (1994) 263.

*Work supported by the National Science Development Council under Contract No. IRPA 4-07-07-018, IRPA 4-06-05-013 and IAEA Regular TC Programme under Grant No. MAL/7/004.

Table 1
Sensitivities of elements in sediment strata samples at varying depths expressed in count rate (c/s)

Element	Sc	V	Cr	Cd	La	Ce	Nd	Sm	Yb	Lu
Energy (keV)	585.2	645.9	835.0	558.6	162.7	334.8	697.3	333.4	475.4	457.7
Depth (cm)										
0 - 2	9.00±0.10	7.00±0.01	9.00±0.08	8.00±0.07	5.00±0.01	18.00±0.42	15.00±0.43	7.00±0.02	33.00±1.67	7.00±0.06
2 - 4	6.00±0.04	8.00±0.01	10.00±0.20	9.00±0.16	-	15.00±0.35	-	6.00±0.11	22.00±0.70	6.00±0.07
4 - 6	11.00±0.17	8.00±0.03	35.00±0.60	8.00±0.08	-	17.00±0.07	18.00±0.07	9.00±0.38	28.00±0.70	5.00±0.04
6 - 8	11.00±0.15	6.00±0.01	34.00±0.27	10.00±0.10	5.00±0.01	17.00±0.68	18.00±0.05	9.00±0.26	29.00±0.86	6.00±0.05
8 - 10	11.00±0.11	15.00±0.04	34.00±0.38	16.00±0.26	4.00±0.01	18.00±0.34	-	-	-	7.00±0.06
10 - 12	11.00±0.12	9.00±0.02	34.00±0.59	17.00±0.27	4.00±0.01	18.00±0.59	17.00±0.06	10.00±0.23	31.00±1.00	7.00±0.05
12 - 14	10.00±0.09	14.00±0.03	33.00±0.35	11.00±0.15	5.00±0.01	15.00±0.17	19.00±0.08	10.00±0.19	30.00±0.57	7.00±0.05
14 - 16	10.00±0.10	13.00±0.08	33.00±0.53	18.00±0.23	5.00±0.01	15.00±0.18	18.00±0.07	11.00±0.28	30.00±0.66	7.00±0.06
16 - 18	15.00±0.23	10.00±0.17	35.00±0.56	16.00±0.23	-	19.00±0.11	-	11.00±0.33	34.00±0.57	11.00±0.14
18 - 20	15.00±0.19	9.00±0.01	36.00±0.49	20.00±0.25	7.00±0.02	18.00±0.25	-	24.00±0.88	-	11.00±0.10
20 - 23	16.00±0.21	10.00±0.01	36.00±0.58	19.00±0.25	7.00±0.02	23.00±0.48	38.00±1.18	24.00±0.95	-	10.00±0.13
23 - 26	14.00±0.18	11.00±0.05	35.00±0.72	22.00±0.33	-	16.00±0.24	38.00±1.56	24.00±0.82	37.00±1.08	10.00±0.13

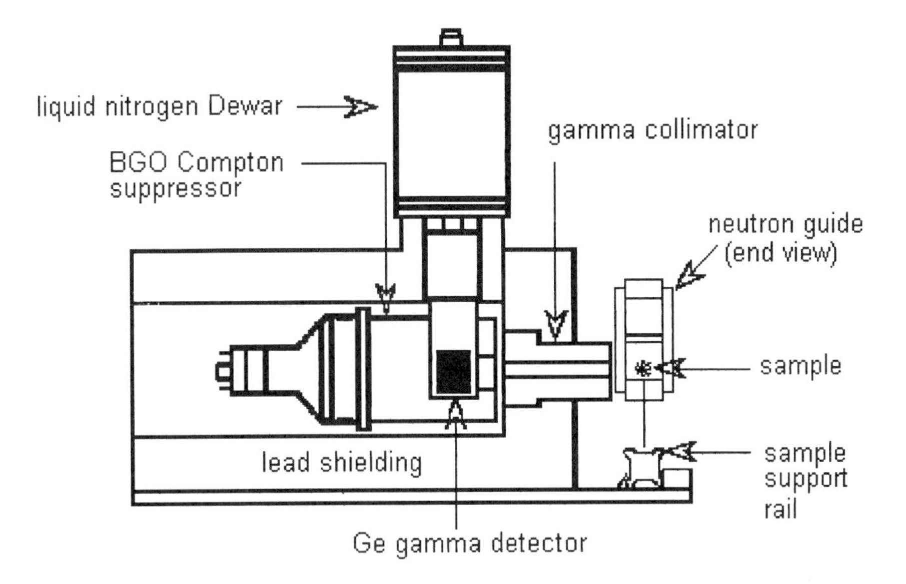

Fig. 1 Experimental setup of the new CNPGAA spectrometer at NIST (*Courtsey of NIST*)

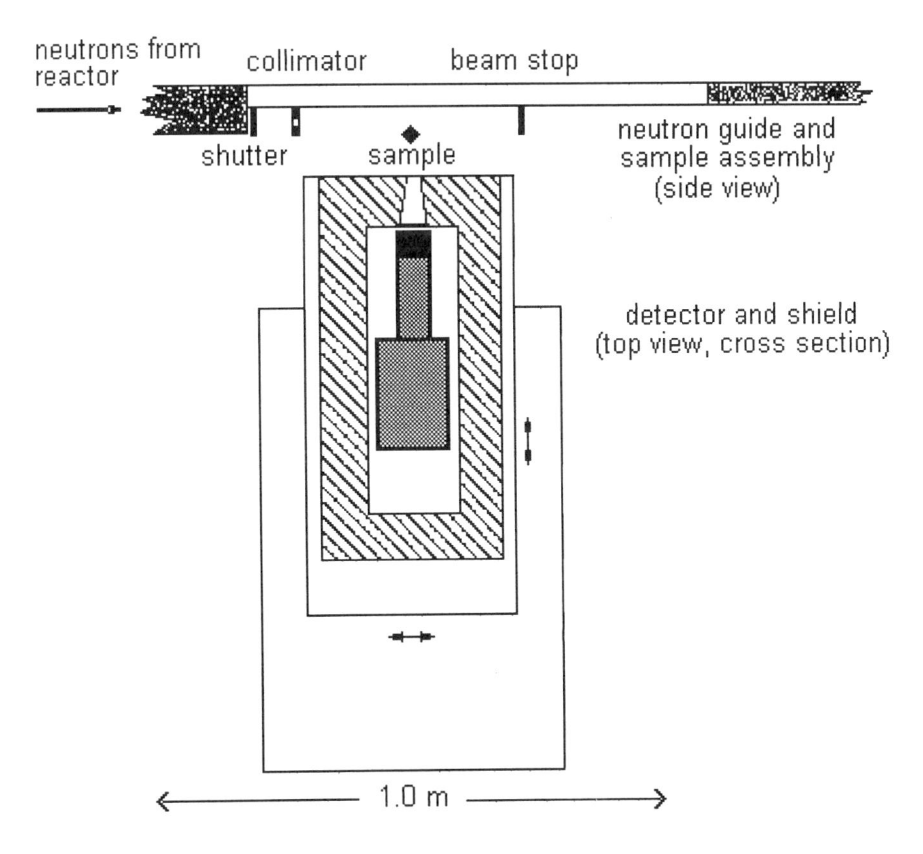

Fig 2. Scale drawing of the PGAA spectrometer with the cross section of the shield and detector (*Courtsey of NIST*)

WEAR RATE QUANTIFYING IN REAL-TIME USING THE CHARGED PARTICLE SURFACE ACTIVATION[1]

B. ALEXANDREANU[2], L. POPA–SIMIL, D. VOICULESCU, P.M. RACOLTA

Institute of Atomic Physics, Cyclotron Laboratory, P.O. Box MG–6, 76900 Bucharest, Romania

Surface activation, commonly known as Thin Layer Activation (TLA), is currently employed in over 30 accelerator laboratories around the world for wear and/or corrosion monitoring in industrial plants [1-6]. TLA was primarily designed and developed to meet requirements of potential industrial partners, in order to transfer this technique from research to industry.
The method consists of accelerated ion bombardment of a surface of interest e.g. a machine part subjected to wear. Loss of material owing to wear, erosive corrosion or abrasion is characterized by monitoring the resultant changes in radioactivity.
In principle, depending upon the case at hand, one may choose to measure either the remnant activity of the component of interest or to monitor the activity of the debris. For applications of the second type, especially when a lubricating agent is involved, dedicated installations have been constructed and adapted to an engine or a tribological testing stand in order to assure oil circulation around an externally placed detection gauge. This way, the wear particles suspended in the lubricant can be detected and the material loss rates quantified in real time. Moreover, in specific cases, such as the one presented in this paper, remnant activity measurements prove to be useful tools for complementary results. This paper provides a detailed presentation of such a case: in-situ resistance-to-wear testing of two types of piston rings.

INTRODUCTION

The piston ring and cylinder jacket are the two main tribologic components of any thermal engine. Hence, their matching has been the subject of various attempts at determining wear levels. Several of the TLA's intrinsic characteristics had been taken into account when our industrial partners decided to employ the TLA-based wear monitoring method [7, 8, 9]. Of those, we mention:

– It allows fast and precise material loss monitoring without it being necessary to stop and disassemble an engine;

– Conventional measuring methods, such as profilography, tri-dimensional analysis, etc., cannot be used when mass loss levels must be quantified in real time;

–The total cost of the TLA-based experiment usually does not exceed the cost of conventional methods.

EXPERIMENT

Two types of piston rings have been tested: Mo-coated and nitrited steel. The piston rings of a 1400 cm^3 car engine were labelled by the $^{95}Mo(d,2n)^{95m}Tc$ (E_d = 13.5 MeV) and $^{56}Fe(d,n)^{57}Co$ (E_d = 8.5 MeV) nuclear reactions utilising in-air irradiation. Rotating target holders were used. The width of circular labelling was 3 mm. Since the labelled zones must have constant specific radioactivities, autoradiographic films were used for checking the uniformity of the initial irradiation.

The above mentioned types of piston rings were tested simultaneously on the same engine. Therefore, several parameters such as irradiation time, beam intensity, and

[1] Work partially supported by the International Atomic Energy Agency, Vienna, Austria
[2] Corresponding author (Department of Physics, University of Miami, P.O. Box 248064, Coral Gables, FL 33124, Tel: (305) 284-3211, e-mail: boalex@phyvax.ir.miami.edu)

appropriate cooling time (allowing short-lived isotopes to disappear) were correlated so as to avoid the useful γ-rays resulting from ^{95m}Tc and ^{57}Co from overlapping with any undesired background. The in-laboratory γ–spectroscopy was performed with an adequate acquisition system based on a NaI(Tl) detector (Ø 2" × 2" area, 9% resolution at 661 keV) and a Canberra 35 MCA. The γ-spectra and the total and specific activity of the produced radionuclides were also measured with a HPGe Ortec detector (energy resolution: 1.8 keV at 1.332 MeV).

A schematic view of the experimental set-up is presented in Fig. 1. The γ-rays originating from both ^{95m}Tc (204 + 582 keV) and ^{57}Co (122 keV) were simultaneously recorded by two NaI(Tl)-based detection systems (two-channel analyzers). One NaI(Tl) detector was placed in a detection vessel - the so-called "flow chamber". The lubricant oil was pumped in a closed circuit through the flow chamber; this way, the wear particles suspended in the oil could be detected in a ~ 4π detection geometry. The wear rates were derived on-line by transforming the radioactivity variations into material loss data. The sensitivity threshold was 40 μg/cm^2 (~ 0.01 μm).

The calibration of the above described mechanical-spectrometric arrangement was made using standards irradiated at the same time as the machine components to be studied, and then dissolved chemically in the flow chamber's volume for activity measurements. The volume of the flow chamber and the NaI(Tl) detector's type were chosen after performing several tests of detection efficiency vs. γ-ray energy [10, 11].

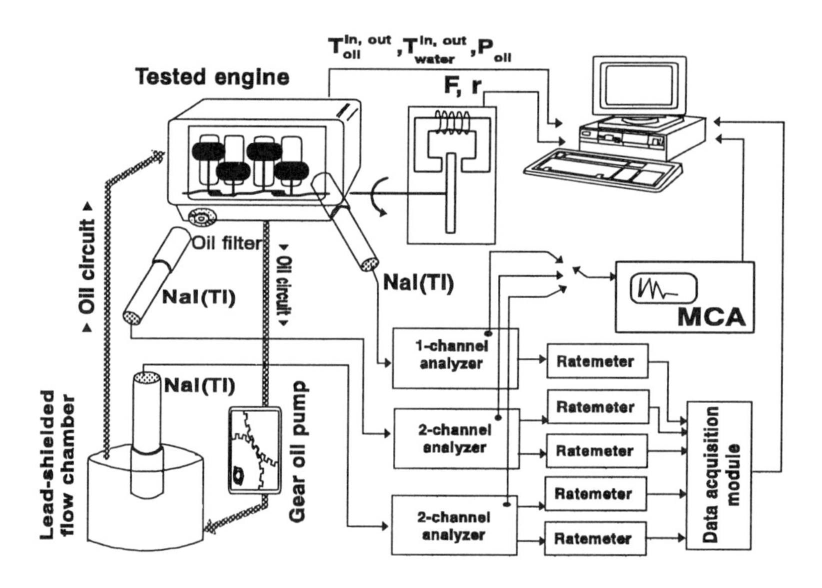

FIGURE 1. Schematic view of the experimental set-up.

The proportionality between the radioactivity level and the concentration of the labelled material particles removed by friction processes, and therefore suspended in the lubricating oil is given by:

$$M_t = K\,V\,C_i\,N_i$$

where: M_t (g) is the total wear value; K (1/l) is the specific calibration factor - a constant of each mechanical-spectrometric arrangement; V (l) is the total volume of the lubricant; C_i (g sec/pulse) is the irradiated standard's mass and counting rate ratio; N_i (pulse/sec) is the counting rate caused by radioactive wear particles spread in the lubricant.

For monitoring oil filter's retention capacity, a second two-channel detection system continuously recorded its activity. Filter's calibration was made in a way similar to that described above.

A third monochannel detection system was placed on the engine at the cylinders' level for remnant activity measurements. This approach was intended to provide a means of control only, as the sensitivity for the remnant method is ~ 1÷2 μm.

RESULTS AND DISCUSSION

The resultant wear diagram is presented in Fig. 2. Nitriding significantly improves piston ring performance compared to Mo-alloy coating. The uncertainties with which the experimental points in Fig. 2. are given are 10 to 15%. These values were estimated considering both filter retention and lubricant losses, which are usual processes for any thermal engine. However, it should be stressed that the above results are given for a working time of only 20 hours. That is why, service life predictions cannot be made at this stage.

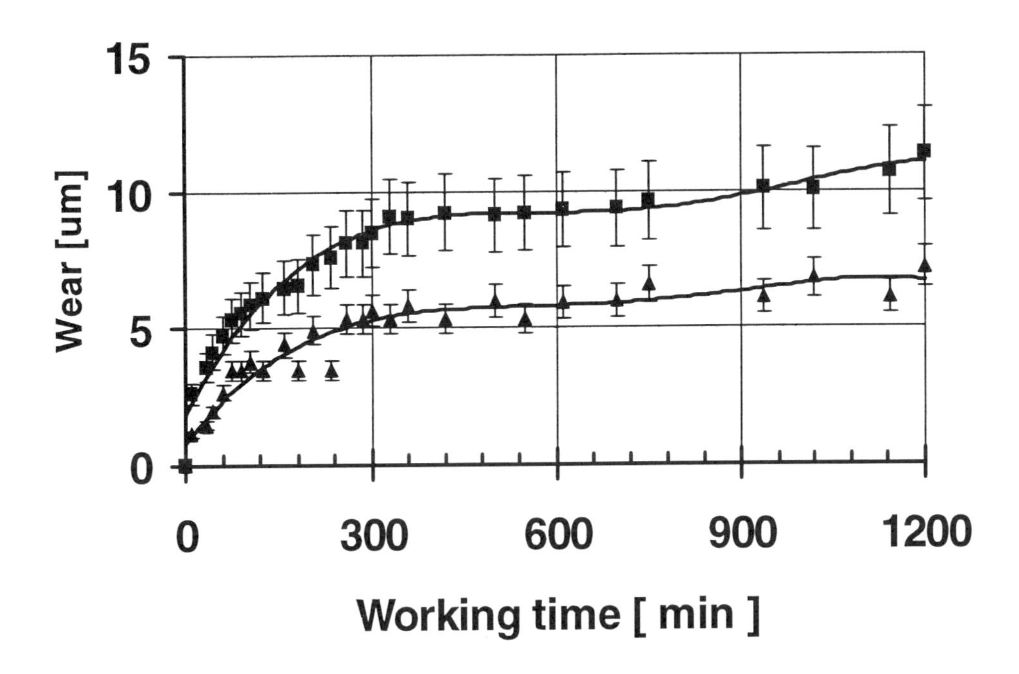

FIGURE 2. Wear diagram obtained for Mo-alloy coated (■) and nitrided (▲)piston rings.

CONCLUSIONS

The objective of this paper was to provide a detailed presentation of a key example where simultaneous comparative studies of engine components were made in-situ utilising combined TLA-based measuring methods.

ACKNOWLEDGEMENTS

The authors gratefully acknowledge Eng. I. Adamesteanu, Eng. V. Bezman and Eng. A. Costa of the Institute for Thermal Engines "MASTER SA" - Bucharest for their generous help and assistance received during the measurements on the engine testing stand.

REFERENCES

1. Conlon T. W., Contemp. Phys. 26, 521 (1985).
2. Fehsenfeld P., Kleinrahm A., Schweikert H., J. Radioanal. Nucl. Chem. 160, 141 (1992).
3. Leterrible P., Blondiaux G., Valladon M., Debrun J.L., Ducreux M. and Guerrand M., Nucl. Instr. and Meth. B 10/11, 1054 (1985).
4. Kaiser P., MTZ Motortechnische Zeit. 34, 1 (1973).
5. Konstantinov, N.N. Krasnov, J. Radioanal. Chem. 8, 357 (1971)
6. Vasvay S., Ditroi F., Takacs F., Szabo Z., Szucs J., Kundrak J., Mahunka I., Nucl. Instr. and Meth. B 85 (1994) 225
7. Racolta P.M., Popa-Simil L., Ivanov E., Pascovici G. and Alexandreanu B., Application of Particle and Laser Beams in Materials Technology, Dordrecht: Kluwer, 1995, p. 415
8. Racolta P.M., Appl. Rad. Isot. 46, 663 (1996).
9. Racolta P.M., Popa-Simil L., Ivanov E. and Alexandreanu B., Radiat. Phys. Chem. 47, 677 (1996).
10. Lacroix O., Blondiaux G., Sauvage T., Racolta P.M., Popa-Simil L. and Alexandreanu B., Nucl. Instr. and Meth. A 396, 426 (1996).
11. Racolta P.M., Popa-Simil L. and Alexandreanu B., Nucl. Instr. and Meth. B 113, 420 (1996).

ARIBA, AN All ROUND ION BEAM ACQUISITION PROGRAM

B. Brijs, J. Deleu, W. De Coster*, D. Wils and W. Vandervorst**

IMEC, Kapeldreef 75, B-3001 Leuven, Belgium
**Present address : Philips. Research Lab, P. Holstlaan 4, Nl-5656 Eindhoven, The Netherlands*
***Nucomat, Mercatorstraat 206, B-9100 Sint-Niklaas*

ARIBA is a computer program for data acquisition and experimental control of the IBA (Ion Beam Analysis) facility at IMEC The program is written in C and runs on an IBM PC under WIN/OS2 . Data acquisition is performed by two Acquisition Interface Modules (AIMS) with a built-in ethernet (LAN) interface . Peripheral devices (goniometer, current integrator, beam shutters) are IEEE controlled and are optically isolated from the computer . The package has been designed to handle all Pulse Height Analysis (PHA) acquisition mode experiments (RBS, angular channeling, autochanneling,....) and LIST mode acquisition experiments (TOF-ERD,....).

1. INTRODUCTION

In most of the ion beam analysis laboratories data acquisition is routinely performed by PC's containing a variety of interface boards : counters, MCA boards, I/O cards,.... Such an integrated system meets all requirements of a specific set-up, but is inconvenient if the PC must be replaced or if an expansion of the set-up has to be incorporated (too limited available slots) . Our aim therefore was to develop a generic acquisition system set-up independent from the PC and a data acquisition software package covering a wide range of ion beam based analysis methods such as the single parameter techniques RBS, ERD, NRA, ... and the multi parameter techniques such as Time-of-Flight measurements, position sensitive detection,...The approach consisted of using only commercially available electronic modules .

2. HARDWARE

ARIBA was designed for operation on an IBM or compatible PC and is written in C language for portability . WIN/OS2 was selected as the operating system for the whole set-up : the data acquisition runs under OS2, while the peripheral device control operates under WIN . This choice has mainly been driven by the availability of drivers for the different devices . The communication between both operating systems occurs via the internal memory of the PC . All data communication between PC and AIMS occurs via ethernet , whereas the peripheral modules control is organised through IEEE communication . PC plug-in boards are only installed for the communication . All data acquisition modules are installed in the direct neighbourhood of the endstation where the target and the detectors are mounted . This results in a better electrical isolation and less groundloop induced problems . The schematic diagram of the hardware set-up is shown in Fig. 1.

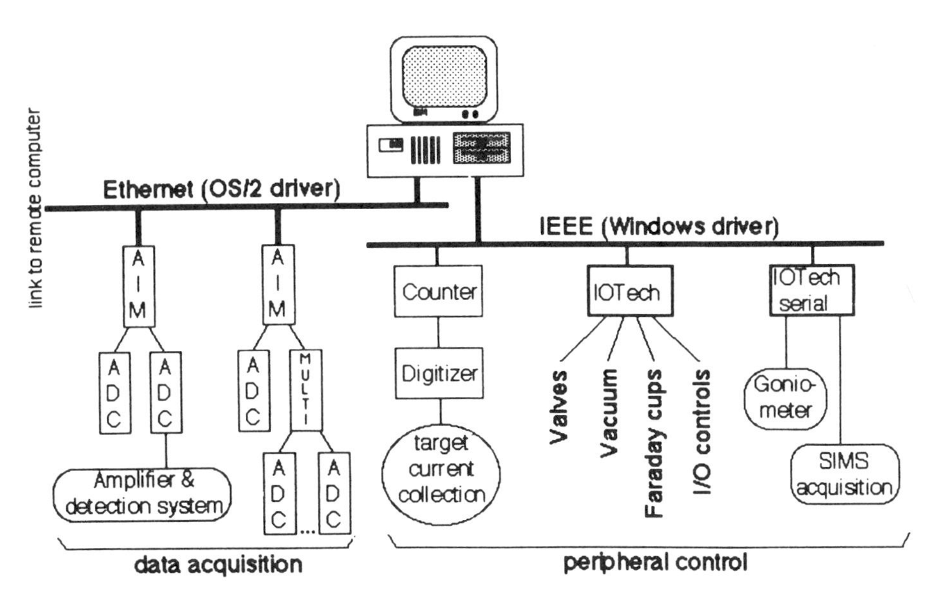

FIGURE 1. Schematic Overview of Acquisition System .

CP392, *Application of Accelerators in Research and Industry*, edited by J. L. Duggan and I. L. Morgan
AIP Press, New York © 1997

Data acquisition

The whole data acquisition system is built around the Acquisition Interface Modules (AIM Model 556 of Canberra) . This AIM is essentially a 64K channel two-input multichannel analyzer . The AIM has been designed with two high speed ADC ports and a built-in ethernet interface (LAN) compatible with IEEE 802.2/803.2 communication standard . Multiple AIMS can be placed in the same network . The AIM can acquire data as well in PHA as in LIST mode from either ADC port independently . The LIST mode however is only possible in combination with the Canberra multi parameter system (ND9900) . This module expands a single ADC port of the AIM into a multi parameter experiment configuration . The multi parameter module provides coincident gating control from 2 to 8 ADCs , allowing TOF-ERD or position sensitive detection experiments .

Peripheral Control

The peripheral devices are controlled via the IEEE protocol . Currently the IEEE board, which converts the PC in an IEEE 488.2 controller, is mounted inside the PC limiting the distance between peripheral devices and the computer to a range of roughly 10 meters . If necessary, this can be replaced by a converter board ethernet-IEEE making complete remote control of the experimental set-up possible

The core of the communication between the peripheral devices and the IEEE bus consists of 2 I/OTech 488 contollers . The first I/OTech controller (Digital 488HS/32) is a module which allows a high speed transfer between 32 I/O lines and the IEEE bus . This TTL level digital input and output capability in combination with OPTO22 solid state relays form an ideal way to manipulate Faraday cups, valves, beam shutters, alignment laser, BCD compatible equipment...etc.. The second I/OTech controller (Serial 488/11) controls up to 4 devices equiped with serial communication RS-232 interface from one controller . This kind of controller has been integrated to ensure a maximal flexibility .

The previous set-up is conceived in a generic and modular way and can be adapted to particular needs . As an example, Figure 2 shows our dedicated set-up [1] for the analysis by RBS of the artifacts induced by low energy ion sputtering during SIMS depth profiling measurements.

The RBS analysis chamber is for this purpose equipped with an Atomika quadrupole based SIMS system. An NEC microprobe has been integrated in the He beam line . To allow maximum flexibility of these experiments, the 2 AIMs are installed and equipped with 3 PIPS detectors . The 3-axes Panmure goniometer in the RBS chamber, as well as the SIMS data collection system is controlled via the I/OTech serial converter . The peripheral equipment including a current integrator to control the current collection on the target, several I/Os for valves, pumps. Faraday cups,...etc... are controlled via the other I/OTech box .

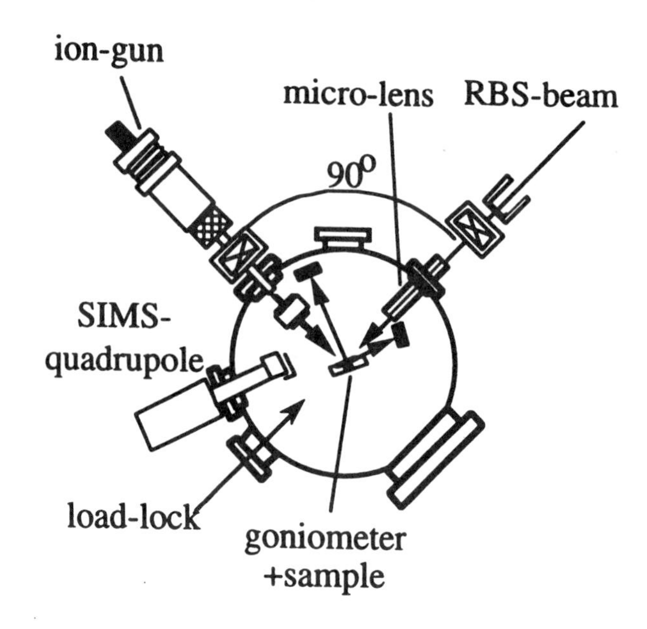

FIGURE 2. Schematic representation of the experimental sputter/RBS set-up . Standard UHV chamber with 3-axes goniometer, connected to a beam line of an NEC 6SDH accelerator (with micro-lens) and to a telefocus ion gun .

3. SOFTWARE

The main objective when developing this program was to guarantee the maximum flexibility and quality of the performed analysis measurements . This requirement stems from :

(1) the different character of the performed analysis measurements going from more repetitive routine control measurements (doping calibration, layer thickness calibration, ...) to research driven experiments (in-situ RBS analysis of sputtered layers,...);

(2) the multitude of available ion beam analysis methods (RBS, ERD, Sputter-RBS/SIMS,...);

(3) the need of future developments to switch to multi parameter acquisition measurements (TOF-ERD, position sensitive detector measurements,...) without separate acquisition system .

In this context a dedicated routine, called AUTOCOLLECT, has been developed to allow the user to define 10 different measurements or modes which can be executed automatically and unattended .

Figure 3 shows the acquisition screen with the different windows for acquisition selection, current counter, MCA

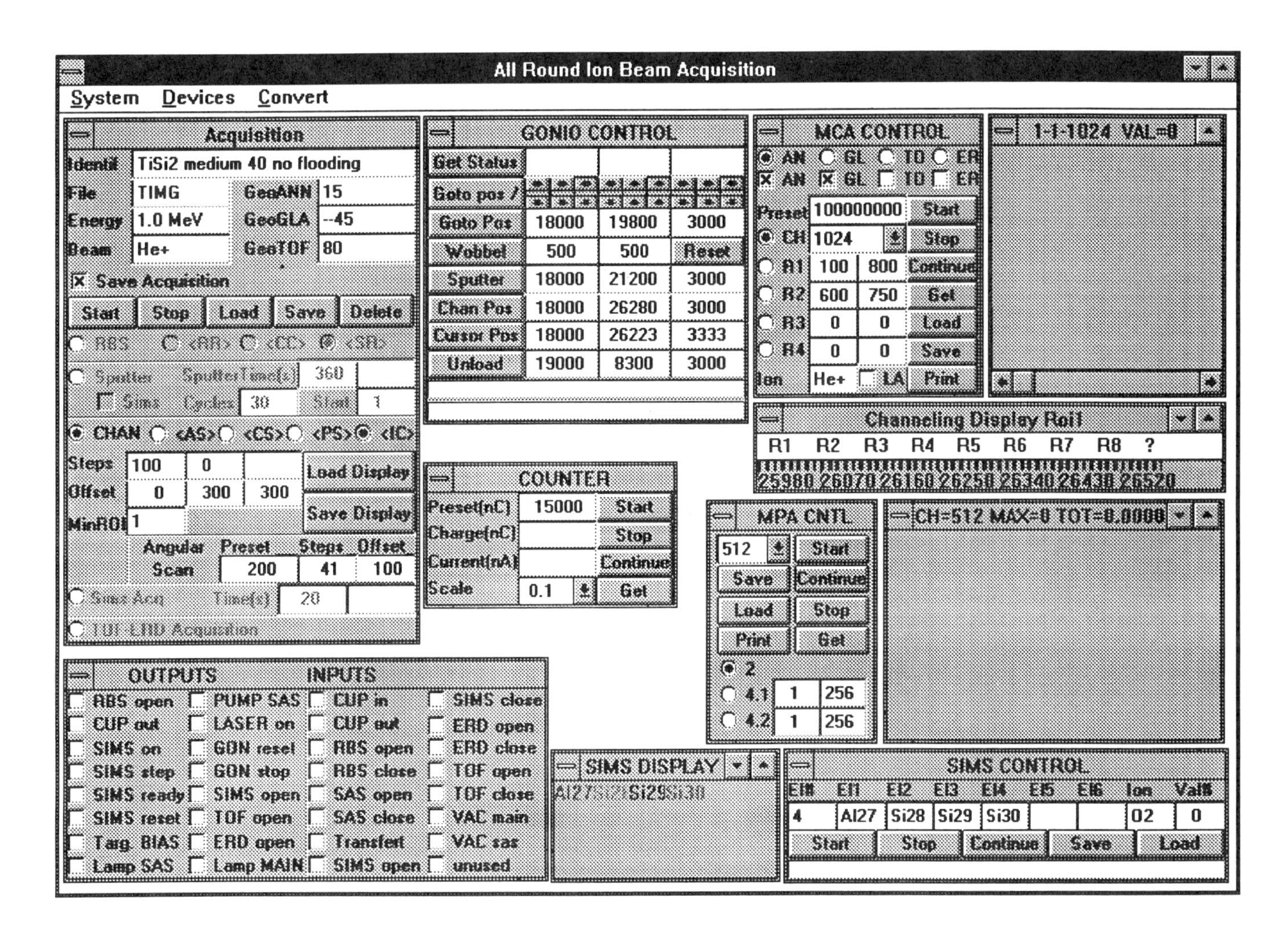

FIGURE 3. The acquisition screen of ARIBA to configure the set-up for a dedicated experiment . From left to right : acquisition menu, goniometer control, MCCA control, display for MCA, counter, channeling display menu, MPA control, display for MPA, display for SIMS, SIMS control, input/output monitoring.

control, display for collected spectrum, I/O monitoring, goniometer control, MPA control and MPA display .

When selecting the RBS acquisition mode , 4 ADC's can acquire data independently and simultaneously . As a consequence 4 detectors can be mounted permanently in the analysis end station of which only those relevant to the measurement can be activated . All different RBS modes : static random (SR), rotating random (RR) and aligned along (CC) are supported by the program . As the set-up contains a 3 axes goniomter the RR routine for instance steers a rotation of the target of minus/plus a number of degrees around a chosen angle . If however a multi sample holder is mounted, than the RR routine foresees the possibility of rotating only 10 degrees for each new sample . This procedure guarantees an undisturbed geometry, what is not the case in most industrial acquisition stations where a RR mode is simulated by changing the both tilt angles of a two-axes goniometer [2].

When selecting the channeling mode, different scans can be performed : angular scan, polar scan and crystal image scan . Moreover a complete automated channeling procedure has been integrated . Special attention has been paid to acquire high quality angular scans . As is well known, many automated systems restrict themselves to the measurement of the channeling dip around one of the 2 tilt angles of the goniometer, often resulting in non symmetrical angular scans and therefore an inaccurate determination of channeling width at half depth . Our procedure consists of localizing the major planes by collecting a crystal image scan or a polar scan and subsequently selecting coordinates of a non planar region . Doing so, the symmetry of the angular scans is guaranteed . The crystal scans are displayed during the data collection and not reconstructed after the measurement has been completed as is the case in industrial systems . Another advantage of this real time display of image scans is that misalignment can be noticed very rapidly. An image scan can be displayed for different energy windows of the RBS spectrum . Moreover, it is foreseen that cursors can indicate the coordinates of a certain region of the image scan. This allows the user to localize very easily planes, axes and even a static random position .

When selecting the multi parameter acquisition mode currently the choice between 2 (for TOF-ERD) or 4 (for position sensitive detector measurements, cfr. ref. [3]) input signals in LIST mode is foreseen . In both cases the dimension of the conversion matrix is optional (128x128, 256x256 or 512x512 pixels). During a position sensitive detection two energy levels can be acquired simultaneously .

When selecting the sputter-RBS-SIMS experimental mode automated alternating sputter cycles, SIMS analysis and RBS analysis cycles are performed . The collection of the secondary ion signal is performed through the the serial output of the quadrupole mass spectrometer . By comparison of the subsequent RBS spectra of sputtered layers to the RBS spectrum of the original target composition and by analysis of the SIMS spectra, different artifacts such as preferential sputtering, compound formation and segregation can be studied [4,5].

The software package ARIBA is designed for a variety of IBA-data acquisition . Acquired data can be dispayed or printed but no analysis routines have been incorporated.

Actually many good analysis programs exist each of which has its particular strengths and weaknesses . Unfortunately, up to now IBA data files have no standard format . ARIBA currently supports all conversion interfaces needed to use the most widespread RBS analysis packages such as RUMP [6], GISA [7], SCATT/HYPRA [8] .

4. CONCLUSION

Ion Beam Analysis plays an important role in the field of material analysis . Presently, commercially available accelerators deliver beams adequate for RBS, PIXE as well for ERD and TOF-ERD . The All Round Acquisition package ARIBA has been conceived in a generic and modular way . The hardware has been composed of commercially available electronics . The software was developed to sustain all IBA measurements . ARIBA is the first data acquisition package that integrates data acquisition of MCA and multiparameter based data, peripheral device control and data conversion to most widespread analysis software.

REFERENCES

[1] B. Brijs, H. Bender, W. De Coster, R. Moons and W. Vandervorst, *Nucl. Instr.&Meth.* B(79) (1993) 446. In situ RBS Analysis of Ion Beam Mixing during low Energy Sputtering.

[2] M.D. Strathman, *Nucl. Instr.&Meth.* B10/11 (1985) 600. Rutherford Backscattering in an Industrial Environment .

[3] J. De Wachter, P. Hendrickx, H. Pattyn, G. Langouche, B. Brijs and W. Vandervorst , *Nucl. Instr.&Meth.* B(79) (1993)446. Channeling Detection using Position Sensitive Detectors .

[4] B. Brijs, W. De Coster, H. Bender, J. Allay, P. Osceanu and W. Vandervorst, *Nucl. Instr.&Meth.* B(85) (1994) 306. Sputtering phenomena of CoSi2 under low energy oxygen bombardment .

[5] W. De Coster, B. Brijs, P. Osceanu, J. Allay, M. Caymax and W. Vandervorst, *Nucl. Instr.&Meth.* B(85) (1994) 911. Ion Beam Mixing and oxidation of a Si/Ge multilayer under oxygen bombardment .

[6] L.R. Doolittle, *Nucl. Instr.&Meth.* B(15) (1986) 227 . A Semi-Automatic Algorithm for Rutherford Backscattering Analysis .

[7] J. Saarilahti and E. Rauhala, *Nucl. Instr.&Meth.* B(64) (1992) 734. Interactive Personal Computer Data Analysis of In Backscattering Spectra .

[8] M..D. Strathman and S. Bauman, *Nucl. Instr.&Meth.* B(64) (1992) 840 . Angle-Resolved Imaging of Single Crystal Materials with MeV Helium Ions .

ACCELERATOR LIMITATIONS TO ION BEAM ANALYSIS

Ch. Klatt, B. Hartmann and S. Kalbitzer

Max-Planck-Institut für Kernphysik, Postfach 103980, D-69029 Heidelberg, Germany

Energy spread of ion beams is an important limitation of nuclear reaction analysis. While in single-stage electrostatic accelerators ion source or high voltage instabilities predominate, tandem type machines exhibit the stripping process as an additional and rather complex source of energy broadening. We have investigated the energy widths of the ions ^{1}H and ^{15}N for resonance reactions and ^{4}He from a 3 MV tandem accelerator operated with a gas stripping column. Major differences between the final energy spread of injected atomic and molecular negative ions have been observed. Molecular break-up and energy-loss straggling may easily amount to a few keV[1], whereas the contribution by high voltage ripple is negligible in most cases. The results will be discussed in view of their impact on different analytical techniques for high-resolution analysis of thin solid films.

INTRODUCTION

The demands upon precision of nuclear analysis have reached new levels as compared to former standards. Within this context the energy spread of the incident ion beam is of fundamental importance for most measurement techniques. Ion beam scattering techniques e.g. Rutherford backscattering spectrometry, RBS, will suffer from degradation of mass and depth resolution, as well as nuclear reaction spectrometry, e.g. nuclear reaction Doppler analysis, from disturbance of the respectiv spectra.

While a preliminary report (1) was given before and a full description of the energy spread mechanism (2) is under preparation, this paper emphasizes the consequences of these measurements on the most frequently used ion beams of ^{1}H, ^{4}He and ^{15}N.

A precise knowledge of all quantities contributing to beam energy spread, such as fluctuation of the accelerating voltage (ripple), charge exchange process in the gas-stripper, and energy-loss straggling will allow to correct for the energy broadening processes and therefore to optimize the analytical tools. Thus, new applications may come under consideration.

EXPERIMENTS AND RESULTS

A short description of the energy-spread measurements will be given here, since a detailed explanation can be found elsewhere (1,2).

Ion beams of H^-, He^- and CN^-, produced by either the 'SNICS' sputter source or the 'Alphatross' radiofrequency source, were injected into the 3 MV NEC 9SDH-2 Pelletron tandem accelerator. The charge-exchange stripper canal with a length of 60 cm was operated with N_2. The beam energy profiles were determined by using our automated energy scanning device (3) as an energy spectrometer, consisting of the 90°-magnet and the 'Amsel steerers' (electrostatic deflection plates). With slit settings of 0.05 mm the energy resolution of the analyzing spectrometer amounts to 0.01 %. During the measurements the accelerator potential was stabilized in the generating voltmeter mode and the ripple signals, derived from the capacitive charge pick-up, were analyzed with a multichannel analyzer.

Voltage Ripple and Energy-Loss Straggling

For applied reasons ^{1}H and ^{4}He beams are of particular interest: for nuclear reaction and scattering analysis. A plot of the normalized current density per energy interval as a function of the deflector voltage energy equivalent is shown in Fig.1 for $^{1}H^{+}$ and $^{4}He^{+}$. The line shape of the analyzed $^{1}H^{+}$ beam is a nearly perfect Gaussian curve with a width of 240 eV; this is also true for all other energies above 3 MeV independent of the stripper-gas pressures in the range of $10^{-2} \geq p(N_2) \geq 10^{-5}$ mbar in the exchange canal.

As Fig.2 shows, the terminal voltage ripple T dominates the energy width of the H^-/H^+-system: we obtain the relation $T=0.27E^{1/2}(MeV)^{½}$. Thus, a total spread of W=0.27 keV for a proton beam of 1 MeV is obtained. These results show that for protons the energy spread is fully explained by the terminal ripple and the straggling is negligible. Thus, a further reduction of the proton energy spread is

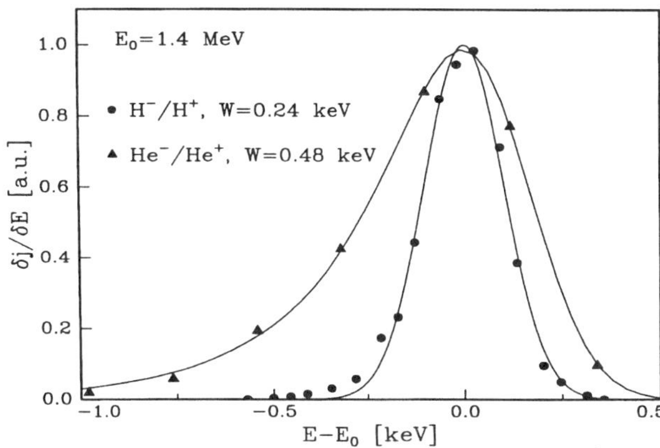

FIGURE 1. Beam energy profiles of 1.4 MeV H^-/H^+, $p=4.5\cdot10^{-5}$ mbar N_2, and He^-/He^+, $p=5\cdot10^{-5}$ mbar N_2. The symmetric Gaussian profile of $^{1}H^{+}$ is due to voltage ripple, the He^{+} profile to energy-loss straggling.

[1] All energy spreads will be quoted as FWHM unless stated otherwise.

CP392, *Application of Accelerators in Research and Industry*, edited by J. L. Duggan and I. L. Morgan
AIP Press, New York © 1997

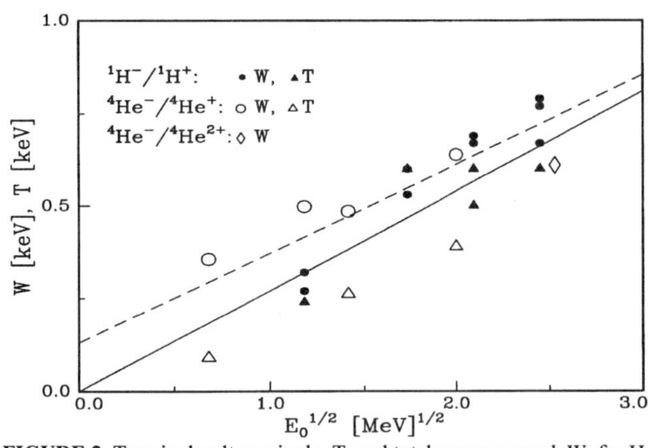

FIGURE 2. Terminal voltage ripple, T, and total energy spread, W, for H and He as a function of the beam energy. $p(N_2)$ was set to maximum yield, the lines are fits to W and T, respectively.

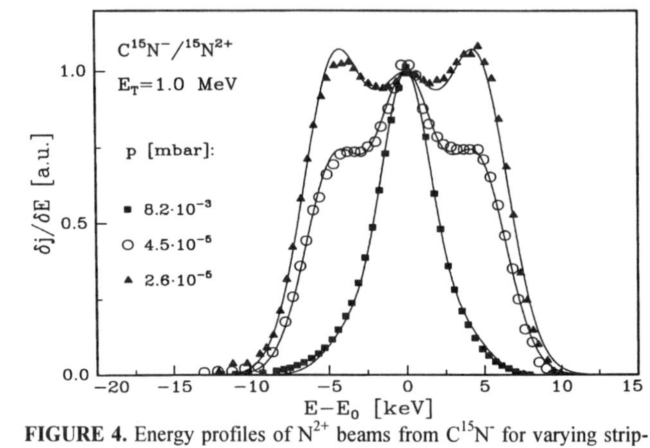

FIGURE 4. Energy profiles of N^{2+} beams from $C^{15}N^-$ for varying stripper-gas pressures. Central and two satellite peaks appear for low stripper-gas pressures with strong changes in the respective amplitudes.

possible by using an anti-ripple system reducing the fluctuations of the terminal voltage. Ripples of about 50 eV and less have been reported for MeV tandems (4). In case of $He^{+(+)}$, energ-loss straggling phenomena are already visible (Fig.1). The measured energy spread is mainly due to electronic energy straggling, Ω. Thus, for the Bohr regime the relation

$$\Omega^2 = 32\,\pi\,\ln 2\,(Z_1 e^2)^2\,Z_2 N\Delta x \qquad (1)$$

holds. The stripper-gas pressure was set to $8\cdot10^{-4}$ mbar, where the highest He^+ yield is obtained. Figure 2 shows that W for He is also rising with E_0. For energies below 2 MeV Ω is dominant, and above T:

$$W^2 = T^2 + \Omega^2. \qquad (2)$$

Extracting the energy-loss straggling contributions from eq.(2) the main contribution to energy broadening for higher Z_1 projectiles becomes apparent: For C^{2+} and F^{2+} Ω^2 as a function of stripper-gas pressure is shown in Fig.3. He^+ at 2 MeV is also presented for comparison.

For all three ion species straight lines could be fitted, confirming the proportionality of Ω^2 to the areal stripper-gas

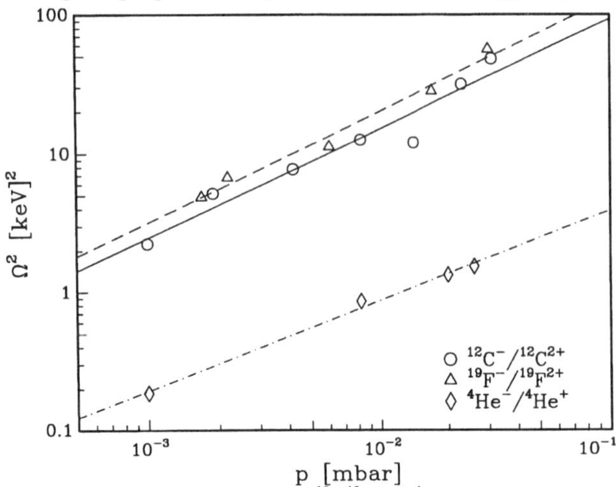

FIGURE 3. Energ-loss straggling of ^{19}F, ^{12}C and ^4He beams as a function of stripper-gas pressure, E_0=0.5 MeV/amu.

density NΔx of atoms according to eq.(1). Slopes of the C and F fits are comparable to the slope of He which confirms the energy independence of the straggling in the Bohr regime. The offsets of the fits are primarily controlled by the effective charge states as measured, e.g. $Z_1^*\sim2$ for C,F, $Z_1^*\sim1$ for He and the corresponding pathlengths in the stripper canal.

The two systems of F^-/F^{2+} and C^-/C^{2+} allow an interpolation to the case of N^-/N^{2+}, which cannot be realized in a tandem, but would follow a regression line inbetween. The N^-/N^{2+} system is of interest for a systematic comparison with corresponding molecular systems of the type NX^-/N^{2+}.

Molecular Break-up

For the analysis of ^1H via the $^1H(^{15}N,\alpha\gamma)^{12}C$ reaction ^{15}N ion beams are mainly produced from negatively charged molecules such as $C^{15}N^-$ or $^{15}NH_2^-$. In ref.(2) the break-up mechanism is described in detail. Figure 4 shows the current/energy profiles of CN^-/N^{2+} beams of 2.6 MeV (molecular terminal energy E_0= 1MeV) for a widely varied stripper-gas pressure. For standard analysis conditions a pressure of $\sim10^{-2}$ mbar is typical (boxes). The line width amounts to 4.6 keV, approximately twice the values of the F^-/F^{2+} and C^-/C^{2+}spectra and with large tails towards higher and lower energies.

At lower stripper-gas pressures the profiles consist of 3 Gaussian components of strongly varying amplitudes. At gas pressures of $p\geq10^{-4}$mbar, the Gaussian center peak is so dominant that the 2 tailing contributions can hardly be distinguished. At rather low pressures the two Gaussian satellite peaks are attributed to electrostatic energy release, or Coulomb explosion, of the charged fragments. The center peak represents a neutral break-up process without any excess energy of the fragments. Its width is primarily controlled by energy-loss straggling in the stripper canal. The satellite peaks correspond to an energy release parallel or antiparallel to the beam direction, causing a shift:

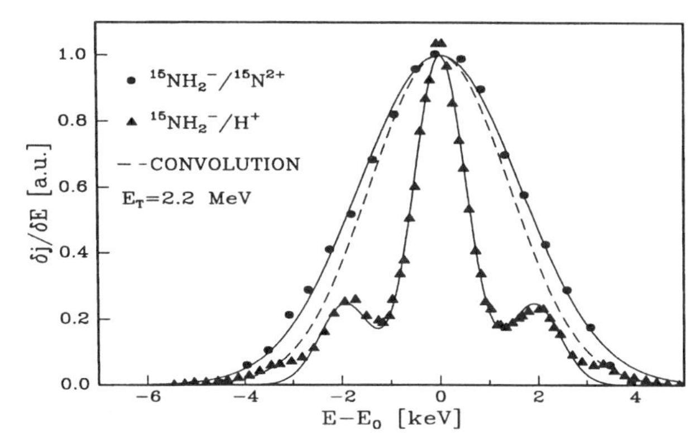

FIGURE 5. Energy profiles of N^{2+} and H^+ from NH_2^- for a stripper-gas pressure of $1 \cdot 10^{-3}$mbar. Whereas the H^+ profile shows a 5-peak structure, the N^{2+} profile (ΔE=3.8 keV) exhibits a single peak. A Gaussian fit to the H^+profile convoluted with a straggling width of 1.55 keV yields a width of 3.3 keV (dashed line).

$$E_S = \pm 2 \cdot [(m_1 m_2)/(m_1+m_2)^2 E_C E_T)]^{1/2}, \text{ with } E_C=Z_1 Z_2 e^2/d \quad (3)$$

where m, E_C, E_T denote mass, Coulomb energy and terminal energy, respectively. For CN, with a bond length d=1.16 Å, Z_1Z_2=2, E_C=25 eV and E_T=1 MeV we obtain E_S=5.0 keV. The experimental value is E_S~4.6 keV (Fig.4).

Experimentally well known is the smaller energy spread for $N^{+(+)}$ ions from NH^-_2 as compared to CN^-. The results of such measurements are shown in Fig.5. For a stripper-gas pressure of $p=1 \cdot 10^{-3}$mbar the system NH^-_2/H^+ (triangles) shows the characteristic structure due to Coulomb explosion of the both fragments with Z_1Z_2=1 and Z_1Z_2=2. A necessary condition for resolving these H^+ satellites peaks is a negligible energy-loss straggling of H^+ in the stripper-gas.

The comparison with the other molecular break-up partner N^{2+} confirms an interesting result of eq.(3): energy shift of the satellite and energy profiles are identical for both constituents. For H^+ at a terminal energy of 2.2 MeV(NH_2^-) straggling amounts to less than 0.1 keV for a stripper-gas setting of $p=1 \cdot 10^{-3}$ mbar. The corresponding straggling of N^{2+} interpolated from Fig.3 amounts to Ω=1.55 keV. A Gaussian fit to the convoluted H profile with the straggling contribution of 1.55 keV leads to the dashed line. This Gaussian approximation with an energy width of 3.3 keV is close to the experimental value of 3.8 keV. The remaining differences in the energy widths may result from different ionization probabilities of satellite and central peaks of H^+ and N^{2+}.

CONSEQUENCES FOR NUCLEAR REACTION ANALYSIS

For depth analysis by resonance reactions and elastic ion scattering, depth resolution is impaired by energy-loss straggling. With the knowledge of both beam energy and energy-loss widths now a full deconvolution of the measured spectra in a depth concentration profile can be carried through.

Resonance Reactions

Earlier publications on the nuclear reaction Doppler analysis could not quantitatively explain the observed line widths. The insufficient knowledge of the analyzing beam energy profile resulted in a Doppler deficit between theoretical estimates and experimental data (5-8). The measured yield curve, Y, is a convolution of the natural line width, Γ, with the energy widths of the analyzing ion beam, W, and the Doppler effect, R:

$$Y(E)=\Gamma(E)*R(E)*W(E). \quad (4)$$

This case will be demonstrated with an oriented ^{1}H δ-layer on a (111) Si surface (1). Since R~10 keV, the Lorentzian line width of the ^{15}N reaction of Γ=1.86 keV will momentarily be neglected.

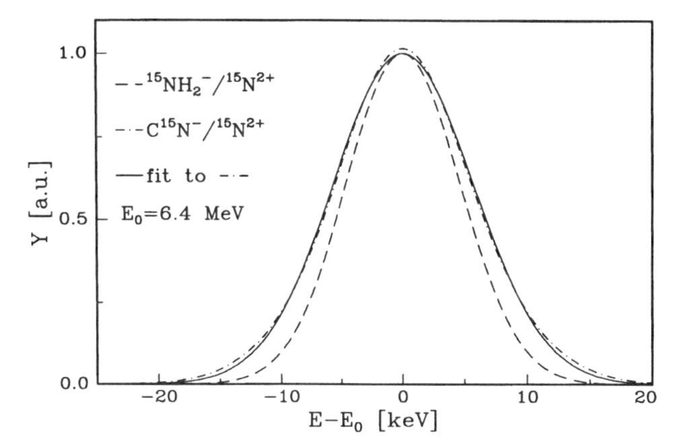

FIGURE 6. Convolution of a 10 keV wide Doppler line with energy parameters of NH_2^-/N^{2+} and CN^-/N^{2+} ion beams.

To this end, the Gaussian Doppler broadening R(E) was convoluted 1) with the Gaussian energy profile of a NH^-_2/N^{2+} ion beam with a single energy width of W=3.8 keV or for comparison 2) with the 3-fold Gaussian profile of a CN^-/N^{2+} beam. As Fig.6 shows, the first case leads to a purely Gaussian profile with a linewidth of Y=10.7 keV, whereas the second case results in a strongly increased non-Gaussian width of Y=13.5 keV corresponding to a CN^-/N^{2+} line width of 9.1 keV. This effect is due to the two satellite peaks produced by Coulomb explosion. Consequently, NH^-_2/N^{2+} is the preferred analyzing ion beam for Doppler analysis of vibrational states of ^{1}H in solids.

Measurements on oriented ^{1}H terminated Si surfaces are now well understood: All relevant widths are precisely known, so that R(E) can be derived reliably, as shown in Tab.1, where Γ was taken into account.

Analysis with both kinds of ^{15}N-beam systems lead now within experimental errors to the same Doppler broadening and therefore binding energy.The main error is due to the numerical fit of the measured γ-profile which can be reduced to 3% only, whereas the precision of the energy beam spread is <<3%. Therefore, other ^{1}H material systems with lower vibrational energies can now be considered, as Tab.2 demonstrates.

TABLE 1. Total Gaussian line width Y, beam energy spread W and Doppler width R of a δ-^{1}H layer on Si(111). The Lorentzian contribution has been corrected for by subtracting linearly Γ/2.

beam	Y/keV	W/keV	R/keV
NH^-_2/N^{2+}	11.8±0.6	4.8±0.1	10.2±0.5
CN^-/N^{2+}	12.5±0.8	7.7±0.2	9.8±0.8
theoretical (2)	-		10.2

Ground-state binding energies of the stretch vibration for H-X systems are shown in Tab.2. From these values, with the assumption of the coupling to the corresponding partner mass and a Gaussian energy profile of W=4 keV of the analyzing ion beam, the expected Doppler broadening is calculated. Doppler measurements with precisions of better 5% appear to be feasible even for the Pd-H system.

TABLE 2. Energies of bond-stretching vibrations E_V of H-X systems, with the corresponding Doppler widths R, yield curve widths Y, and precision of Doppler measurements ΔR/R.

System	E_V/eV	R/keV	Y/keV	ΔR/R
H-H	0.258	11.7	12.4	0.03
O-H	0.240	15.5	16.0	0.03
C-H	0.180	13.3	13.9	0.03
Ge-H	0.129	11.6	12.3	0.03
Ta-H	0.068	8.5	9.4	0.03
Ca-H	0.041	6.5	7.7	0.04
Pd-H	0.028	5.4	6.7	0.04

Ion Scattering Spectrometry

The requirements upon beam quality becomes even more challenging for surface and mass analysis. Therefore for RBS detection systems as electrostatic analyzers (ESA) and magnets with much better energy resolution are required (9, 10). These detection methods are mainly limited by the energy spread of the incident ion beam, whereas in elastic recoil detection analysis (ERDA) the analyzing detector with ΔE/E~1% already limits mass and depth resolution.

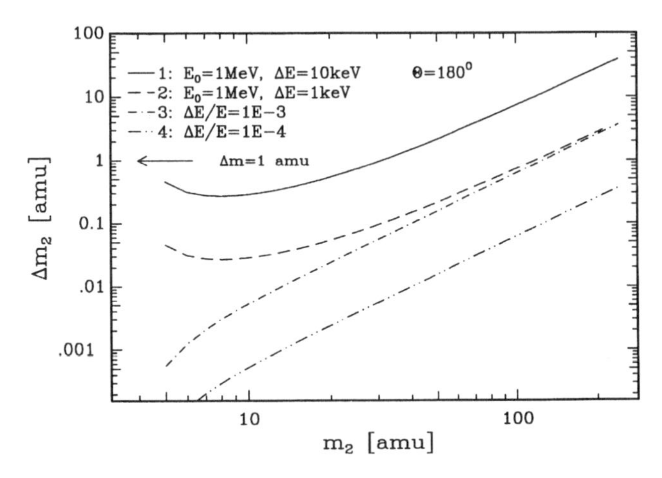

FIGURE 7. Mass resolution Δm_2 as a function of analyzed mass m_2 for various analysis conditions, m_1=4. 1) standard Si detector, 2) ESA with ΔE=const. , 3, 4) ESA with ΔE/E=const.

Figure 7 shows mass resolution Δm_2 as a function of the analyzed mass m_2 for various energy conditions, dominated either by beam energy-spread or detector resolution ΔE. Isotopic mass resolution, $\Delta m_2=\Delta E/E_0(dk/dm_2)^{-1}\leq 1$, k denoting the kinemtatic factor, is achieved in the standard case of E_0=1 MeV ^{4}He beam and a Si detector with ΔE~10 keV to m_2~35. With an ESA with ΔE~1 keV at 1 MeV He^+, isotopic resolution can be improved up to mass 120. Taking into account the ESA resolution of ΔE/E=const. over the whole energy range an improvement in mass resolution for lighter masses is obtained.

Curve 4 shows the best possible case for ^{4}He as analyzing beam: Although the beam energy profile still limits the mass resolution for an energy spread of $\Delta W/W=10^{-4}$ at 1 MeV $^4He^+$ isotopic resolution is provided for all elements with m_2>4: a powerful mass spectrometer with a mass resolution of $m_2/\Delta m_2$~1000 is provided.

CONCLUSIONS

The limitations of tandem type accelerators for ion beam analysis may be specified as:

- The energy spread of H^+ beams is dominated by terminal voltage ripple, which can be significantly reduced by use of an anti-ripple system.
- The width of all ion beams with Z_1>1, produced from atomic elemental species, is dominated by energy-loss straggling effects in the stripper canal.
- The beam energy spread of ions, produced from molecular species, is dominated by both Coulomb explosion in the stripper canal and energy-loss straggling.
- With a precise knowledge of the ^{15}N beam-energy profile Doppler analysis may be extended to other ^{1}H/material systems with smaller vibrational energies.
- Mass resolution in ion scattering analysis with electrostatic detection systems is limited by beam-energy spread at surface position and by straggling in the deeper bulk.

REFERENCES

1. B. Hartmann, S. Kalbitzer and Ch.Klatt, *Nucl. Instr. and Meth.* **B113** 50 (1996)
2. B. Hartmann, Ch. Klatt and S. Kalbitzer, to be published
3. M. Zinke-Allmang, V. Koessler and S. Kalbitzer, *Nucl. Instr. and Meth.* **B15** 563 (1986)
4. J. A. Ferry, *Nucl. Instr. and Meth.* **A328** 28 (1993)
5. M. Horn and W. A. Lanford, *Nucl. Instr. and Meth.* **B29** 609 (1988)
6. Y. Iwata, *Nucl. Instr. and Meth.* **B56/57** 469 (1991)
7. K. Izsak, S. Kalbitzer, M. Weiser and M. Zinke-Allmang, *Nucl.Instr. and Meth.* **B33** 578 (1988)
8. S. Jans, S. Kalbitzer, P. Oberschachtsiek and J. Sellschop, *Nucl. Instr. and Meth.* **B85** (1994) 321
9. D. O. Boerma, W. M. Arnoldbik and W. Wolfswinkel, "Use of an ultra-high resolution magnetic spectrograph for materials research", presented at the *1. Int. Sym. Mat. Sci. Appl. Ion Beam Techn.*, Seeheim, Germany, September 9.-12. 1996
10. A. Feuerstein, H. Grahmann, S. Kalbitzer and H. Oetzmann, *Proc. 2. Int. Conf. Ion Beam Surface Layer Analysis*, (1976) p.471

Convolution Fitting and Deconvolution Methods for Fluorine Depth Profiling by Resonant NRA

J. Jin, D. L. Weathers, J. P. Biscar, B.F. Hughes, J. L. Duggan, F. D. McDaniel, and S. Matteson

Ion Beam Modification and Analysis Laboratory, Department of Physics, University of North Texas, Denton, TX 76203

The $^{19}F(p,\alpha\gamma)^{16}O$ resonant nuclear reaction at a proton energy of 872 keV is used to study the fluorine penetration in archaeological flints and other materials. The experimentally acquired data presented as the γ-ray yield curve Y(E) is the convolution of the fluorine depth profile, C(x), with the excitation peak G(E, x). G(E, x) changes its centroid and width with depth x due to energy loss and energy straggling. G(E, x) is obtained by measuring the γ-ray yield curves of a thin layer of calcium fluoride covered by varying thicknesses of SiO_2, and fitted to a Pearson function. Two numerical techniques, namely, the convolution fitting and deconvolution method, have been developed to extract the depth profile C(x) from the experimental data. The performance and application of each method are discussed.

Introduction

Fluorine depth profiling using resonant Nuclear Reaction Analysis (NRA) is a potential dating technique for archaeological flints.[1,2] The $^{19}F(p, \alpha\gamma)^{16}O$ reaction has several resonant peaks in the energy range from several hundred keV to over 1 MeV.[3] The resonance at the proton energy of 872 keV is chosen for the present measurements because of its large cross section. This allows minimization of the background γ-rays from the reactions of protons with other nuclides such as ^{17}O, ^{29}Si, and ^{30}Si in the target. However, the trade-off of this choice is that the width of this peak is rather wide (4.7 keV), and the interference from the two adjacent resonance reactions at 902 keV and 935 keV limits the range of depth measurement (see Fig. 1).

For NRA depth profiling, the γ-ray yield curve Y(E) is measured by recording the reaction production of γ-ray counts at each scanned proton energy near the resonant peak. The impurity depth distribution information C(x) is convoluted in Y(E). If G(E, x) is the probability for an incoming proton of energy E to produce a detected γ-ray count for unit concentration in the vicinity dx of depth x, then the following relation is satisfied[4]:

$$Y(E)=\int_0^\infty C(x)G(E,x)dx.$$

The aim of this work is to extend the depth measurements by including all three resonance peaks shown in Fig. 1 in G(E,x). The actual depth distribution C(x) then will be extracted by using numerical techniques, namely, the convolution fitting method and the deconvolution method.

Modeling of the excitation function G(E,x)

Basically, G(E, x) is the reaction cross section $\sigma(E)$ convoluted with the proton energy distribution caused by the beam energy spread and energy straggling inside the target, multiplied the factors of detector efficiency and solid angle.[4] The cross section $\sigma(E)$ is closely approximated by the Breit-Wigner law,[5]

$$\sigma(E)=\sigma_0\frac{(\Gamma/2)^2}{(\Gamma/2)^2+(E-E_R)^2},$$

which is a Lorenzian function, while the energy spread from the energy straggling and the instruments are close to Gaussian. Since the Pearson function[6] is a good approximation for the convolution of a Lorenzian with a Gaussian, it is considered as the model for G(E,x). Then, a single peak of the excitation function has the following form:

$$G(E,x)=\frac{A_0\Gamma[\alpha(x)]\sqrt{2^{\frac{1}{\alpha(x)}}-1}}{\sqrt{\pi}D(x)\Gamma[\alpha(x)-1/2][1+4(\frac{E-E_c(E,x)}{D(x)})^2(2^{\frac{1}{\alpha(x)}}-1)]^{\alpha(x)}}.$$

In this expression, A_0 is the area of the peak. D(x) is the FWHM of the peak, which varies with x due to energy straggling. $\alpha(x)$ is the shape index of the peak: when $\alpha=1$, the peak becomes a Lorenzian; when α is very large, it is approximately a Gaussian. $E_c(E,x)$ is the center of the peak, which varies with depth x due to energy loss. It satisfies:

$$E_c(E,x)=E_R+\beta(E)x.$$

Here E_R is the resonance energy, $\beta(E)$ is the average stopping power from E to E_R, calculated by TRIM[7] and fitted to a linear function.

Experimental measurements have been performed to determine G(E, x) for our experimental system. 100Å of calcium fluoride was evaporated onto Si wafers and then covered by different thicknesses of SiO_2, which is the main component of flints. However, the SiO_2 only approximates flint, since the evaporated SiO_2 is expected to be amorphous while flint is "cryptocrystalline," i.e., very fine-grained polycrystalline. The measured γ-ray yield curves from these samples are proportional to G(E,x) at different depths x. Fig. 2 shows these measured G(E,x) peaks. It was found that all the peaks are well fitted to the Pearson functions, while they are not well fitted to either Lorenzian or Gaussian functions.

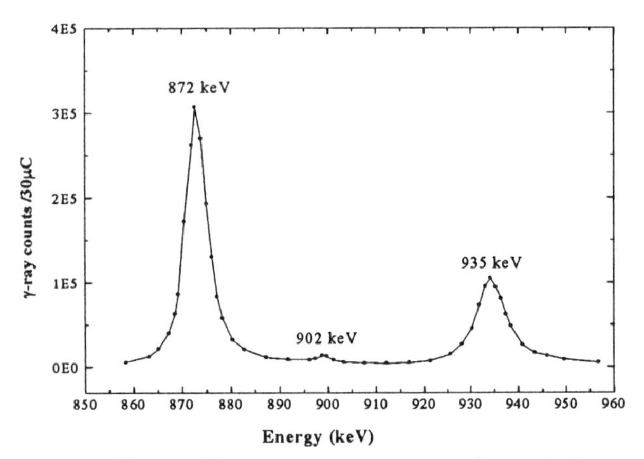

FIGURE 1. $^{19}F(p,\alpha\gamma)^{16}O$ Nuclear Resonance Reaction excitation curve. The three resonances at about 872keV, 902 keV, and 935 keV are indicated.

Fig. 3 shows $D(x_i)$ and $\alpha(x_i)$ for Pearson peaks fitted to the γ-ray yield curves at the discrete x_i for which measurements were made. The expressions of D(x) and α(x) are obtained by fitting these parameters as a function of depth x. The calculated FWHM by Bohr's theory[3] is also shown in the figure. It is clearly above the experimental data on the graph.

Finally, the excitation function G(E,x) includes all three Pearson peaks corresponding to the three resonances in Fig. 1:

$$G(E,x)=G_1(E,x)+G_2(E,x)+G_3(E,x).$$

The numerical techniques

The convolution fitting method

The convolution fitting method is similar to the standard least squares fitting method. It can be briefly described as follows:

(1) Assuming that the fluorine depth distribution is C(x, **p**), where **p** is the parameter vector, calculate the expected value $Y_{th}(E_k,\mathbf{p})$ for each energy point E_k:

$$Y_{th}(E_k,\mathbf{p})=\int_0^\infty G(E_k,x)C(x,\mathbf{p})dx.$$

(2) Define the function F(**p**) as the sum of the squares of the difference between the expected value and the experimental value for all the energy points:

$$F(\mathbf{p})=\sum_{k=1}^{m}(Y_{th}(E_k,\mathbf{p})-Y(E_k))^2.$$

(3) Find the optimal value $\mathbf{p}_{opt}$ to minimize the F(**p**).

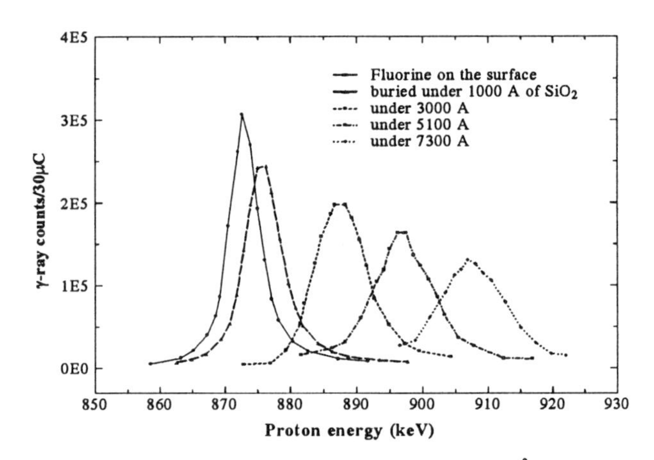

FIGURE 2. Resonance response peaks G(E,x) for 100Å of CaF_2 evaporated on Si and then covered by different thickness of SiO_2.

The Levenberg-Marquardt compromise algorithm is used to find the optimal parameter vector. The details of this algorithm can be found in reference.[8]

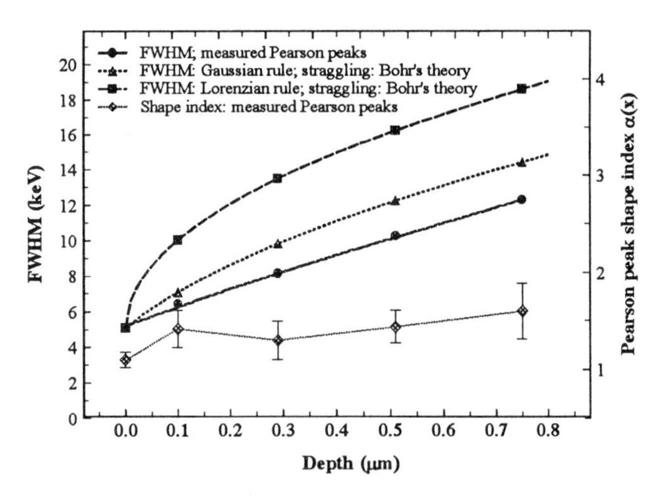

FIGURE 3. The resonance peak width versus depth in the sample shows the energy straggling effect.

A correct concentration distribution model C(x,**p**) is the key to the success of this fitting method. Since the mechanism of fluorine diffusion in the flints is still unknown, several trial models were used, including erfc(x/d) (the complementary error function) , exp(-x/d) (the exponential function), and a linear combination of the two. According to diffusion theory, the erfc(x/d) pattern corresponds to fluorine bulk diffusion in the material when it is exposed to an environment with constant fluorine concentration, while the exponential pattern corresponds to fluorine diffusion along grain boundaries.[9]

The Deconvolution technique

The deconvolution method is the direct recovery of the best estimation of C(x) from Y(E). The standard convolution relation is:

$$Y(E)=\int_{-\infty}^{\infty} g(E-x)C(x)dx,$$

where g(E) is the peak with area equal to 1.

Jansson has developed an effective iterative algorithm given in reference:[10]

$$C_n(x)=C_{n-1}(x)+\lambda(C_{n-1})[Y(E)-\int_{-\infty}^{\infty} g(E-x)C_{n-1}(x)dx]|_{E=x}.$$

Here the starting value is $C_0(x)$ =Y(x), and $\lambda(C)$ is a correction factor that ensures the iteration will be convergent.

Notice that G(E,x) for our study is not exactly like g(E-x) in the standard deconvolution problem. However, it meets the basic characteristics of a broadening peak. As depth x varies, the area under the peak remains constant, while the peak's center varies almost linearly. Jansson's iterative relation was modified as follows:

$$C_n(x)=C_{n-1}(x)+\lambda(C_{n-1})\frac{\beta(E)}{A_0}[Y(E)-\int_{0}^{\infty} G(E,x)C_{n-1}(x)dx]|_{E=E_R+\beta(E)x}.$$

Here A_0 is the area of the major resonance peak, $\beta(E)$ is the average stopping power. The starting value,

$$C_0(x)=\frac{\beta(E)}{A_0}Y(E)|_{E=E_R+\beta(E)x},$$

gives the zeroth-order of concentration C(x) when G(E,x) is considered as a δ function. Each iteration gives the correction by considering the contribution caused by the spread of the actual excitation peak G(E,x).

This iterative relation requires continuous Y(E) data, or at least a very narrow interval E_k between points, to ensure that the calculation of the integral is precise. The experimental data are sometimes very sparse. The Lagrange interpolation scheme is used to interpolate the value between two data points and smoothly connect them to other data points.

This deconvolution method has usually been found to be converge. After about 10 iterations, the standard deviation (or sum of squares) will decrease to a small value. If C(x) is reconvoluted with G(E, x), it fits to the experimental data Y(E) with little deviation.

Application

The two computer programs are written in C in the Borland C environment. Both of them have run times of less than 5 minutes on a Pentium 586 if fewer than 30 data points and 4 parameters are used. The programs have been used to analyze fluorine depth profiles of several archeological arrowhead and flint samples.

Fig. 4 show the γ-ray yield curves and their convolution fittings for two different samples. Both samples were coated with gold before resonance measurement to avoid surface charging effect. The yield curve of sample #1 (a) is quite well fitted by an exponential distribution rather than erfc(x/d), while the data from sample #2 (b) are not fitted well by either a single exponential distribution or erfc. However, the data are quite well fitted with the two-part distribution:

$$C(x)=C_1 erfc(x/d_1)+C_2\exp(-x/d_2).$$

These results may suggest that fluorine diffusion in the flints is mostly along grain boundaries (sample #1), or is the combination of bulk diffusion and grain boundary diffusion (sample #2).

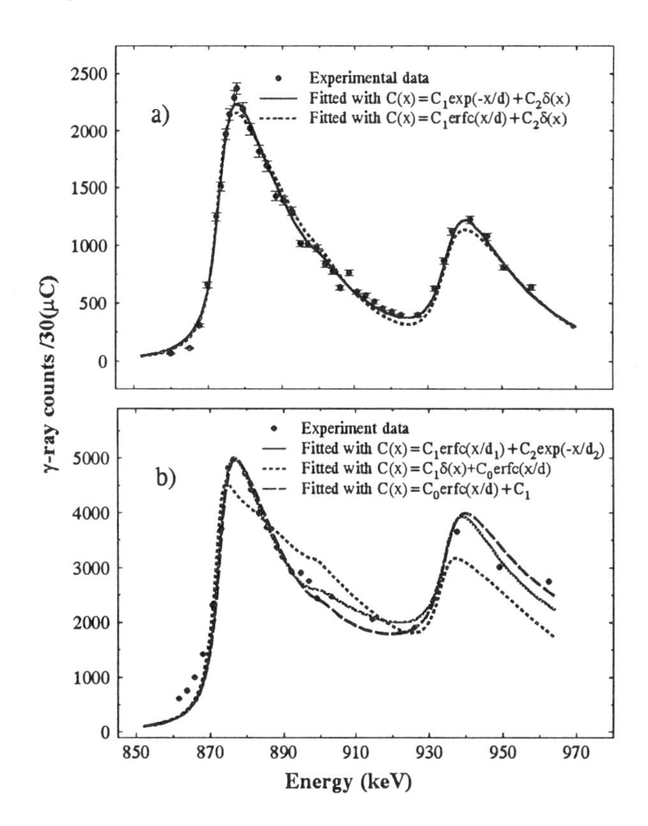

FIGURE 4. γ-ray yield curves and their convolution fittings. a): flints sample, better fitted by $C(x)=C_1\exp(-x/d)$; b): arrow head sample, better fitted by $C(x)=C_1 ercf(x/d_1)+C_2\exp(-x/d_2)$.

It should be noted that some unsmooth undulation is apparent in both γ-ray yield curves. We still have not determined whether these undulations were from experimental noise or from the actual fluorine distribution in the flints. We checked the studies by Walter *et al.*,[1,2] and found that the yield curves they measured in samples subjected to diffusion also showed irregular oscillations, while in the fluorine-implanted samples they were relatively smooth. Alternatively, the oscillations in the work of Walter *et al.* might be attributable to random statistical fluctuations in some of their data.

Fig. 5 shows the deconvoluted distribution patterns for the same two samples. The undulations in the original γ-ray curves correspond to peaks in the deconvoluted distribution. These peaks may be caused by the fast fluorine diffusion along the grain boundary network, or fluorine trapping at various depths when diffused into the flints. Further experiments are being performed to study the temperature controlled fluorine diffusion in flints, the flints microstructure, and the possible noise sources

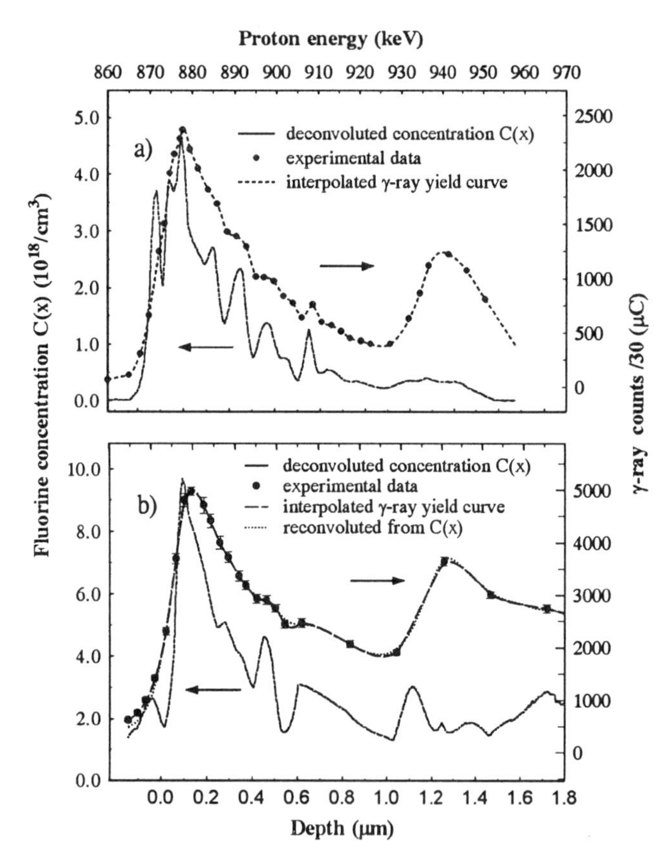

FIGURE 5. Fluorine concentration versus depth in the samples determined from deconvolution.

in the experiment. Nevertheless, if the experiment noise could be reduced, the deconvolution method would dramatically improve the resolution of the depth profiling.

Notice that a peak appears in each curve at depth about x=0. It is believed that this peak is from fluorine contamination on the gold surface, or pinholes in the very thin (<200Å) evaporated gold layer. However, the peak at depth about x=1.1 μm of sample #2 may be caused by poor interpolation because the experimental data at that range are very sparse. Therefore, a reasonable interval of experimental data is required to perform deconvolution.

Summary

A Pearson function has been used as the model for G(E,x), the excitation function of nuclear reaction $^{19}F(p,\alpha\gamma)^{16}O$ for impurity fluorine in SiO_2. The detailed expression of this function has been obtained by experimental measurement and TRIM calculation. Based on this, computer programs have been written to extract the actual fluorine depth profile C(x) from the experimental data. The convolution fitting program has been used to evaluate different models of fluorine diffusion in flints. The deconvolution program has been used to directly extract fluorine distribution in flint samples, the depth resolution was improved by using the deconvolution method.

Acknowledgments

This work supported in part by the Texas Advanced Research Program, the Office of Naval Research, the National Science Foundations, and the Robert A. Welch Foundation.

References

[1]P. Walter, M. Menu, I.C. Vickridge, *Nucl Instrum. and Methods* **B45** (1990) 119.
[2]P. Walter, M. Menu, J.-C. Dran , *Nucl Instrum. and Methods* **B64** (1992) 494.
[3]J. W. Mayer, E. Rimini, *Ion Beam Handbook for Material Analysis* (Academic Press, Inc, New York, 1977).
[4]B. Maurel, G. Amsel, J.P. Dadai, *Nucl Instrum. and Methods* **197** (1982) 1.
[5]E. Segre, *Experimental Nuclear Physics* (John Willy & Sons, New York, 1953).
[6]PeakFit User's Manual (Jandel scientific software, 1995).
[7]J. F. Ziegler, J. P. Biersack, *The stopping and range of ions in solid* (Pergamon press, New York, 1985).
[8]Jens G. Reich, *C Curve Fitting and Modeling for Scientists and Engineers* (McGraw-Hill , New York, 1992).
[9]L. A. Girifalco, *Atomic Migration in Crystals* (Blaisdell publishing Company, New York, 1964).
[10]W. E. Blass, G. W. Halsey, *Deconvolution of Absorption Spectra* (Academic Press, Inc, New York, 1981).

NRA OF HYDROGEN IN GLASSY POLYMERIC CARBON

R. L. Zimmerman[ab], D. Ila[a*], G. M. Jenkins[a], J. K. Hirvonen[c] and H. Maleki[d]

[a] Center for Irradiation of Materials, Department of Physics, Alabama A&M University, Normal AL 35762-1447
[b]University of São Paulo, DFM, FFCLRP, Ribeirão Preto SP, Brazil
[c] Materials Division, US Army Research Laboratory AMSR-MA-CC; APG, MD 21005-5069
[d] Illinois Institute of Technology, Department of Chemical and Environmental Engineering, Chicago, IL 60616

Glassy Polymeric Carbon (GPC) is prepared from a precursor resin by careful heat treatment. Heat Treatment Temperatures (HTT) above 1500°C are believed to expel all hydrogen and oxygen from the grafene structure of GPC. However, we have shown that significant amounts of oxygen remain sequestered in the pores of GPC even at HTT's above 1500°C. In the present study we report the detection of similar amounts of hydrogen for various heat treatment temperatures up to 2500°C. A Nuclear Reaction Analysis (NRA) method with the $^{1}H(^{15}N,\alpha\gamma)^{12}C$ reaction and a specifically designed coincidence array is used to detect the 4.43 MeV gamma ray whose yield is proportional to the hydrogen content in the GPC. The H:C atomic ratio decreases with increasing HTT and we show that it exceeds 1:100 even for a HTT of 2500°C.

INTRODUCTION

Glassy Polymeric Carbon (GPC) is formed when the precursor resol $C_7H_8O_2$, a liquid, converts to fully cured phenolic resin C_7H_6O of specific gravity 1.25 on heat treating at 150°C to 200°C. Heat treatment above 550°C produces a conducting hydrocarbon medium whose room temperature conductivity improves progressively as the heat treatment temperature is increased while water and hydrogen are released. On heat treating to 1000°C the material further transforms without disruption and with no change in shape into a porous but impermeable glass-like polymeric form of pure carbon (1,2) of specific gravity 1.45, significantly less than the X-ray density 2.25. High temperature pyrolysis removes impurities and leaves a pure, fully carbonized product (3). Decomposition gases must diffuse to the surface such that the rate of increase of the heat treatment temperature must be carefully controlled.

The final product consists of long ribbon-like molecules of sp^2 carbon atoms aggregated locally to form sub-crystalline domains which are arranged randomly in space. X-ray analysis shows that the interlaced ribbons are composed of randomly oriented grafene sheets (4) whose edges and faces enclose pores.

The pores have a mean separation of a few nm and occupy about 30% of the volume between the ribbons. The porosity and permeability of GPC may be controlled by a combination of heat treatment (5) and by ion bombardment (6) of the surface.

* Corresponding author: Tel. (205) 851-5866 fax (205 851-5868,
E-mail ILA@CIM.AAMU.EDU

TECHNICAL APPROACH

GPC Sample Preparation

For heat treatment temperatures (HTT) below 450°C hydrogen is still a bonded constituent of the resin. For HTT between 550°C and 750°C the hydrogen bonds are broken and the pores are connected such that hydrogen and other gases may escape or remain adsorbed on the available surface area. For HTT above 750°C the pore connectivity disappears and the material becomes impermeable. Besides the precursor, we prepared samples heat treated to 500°C, 650°C, 1000°C and 2500°C, representing HTT both below and well above the temperature required to break all C-H bonds.

Pair Production Photon Detector

We have developed a new photon detector array (7) that uses Nuclear Reaction Analysis (NRA) to determine the quantity of hydrogen that remains in GPC after heat treatment.

Ion beam accelerators have assumed an important role in the analysis of solid materials. The best example, Rutherford Backscattering Spectroscopy (RBS), identifies the atomic mass of most elements in a solid material, hydrogen being a notable exception. Only by detecting forward scattered protons can hydrogen be detected. Sensitive non-nuclear methods, such as Nuclear Magnetic Resonance (NMR), lack depth resolution. Secondary Ion Mass Spectrometry (SIMS) gives an adequate depth

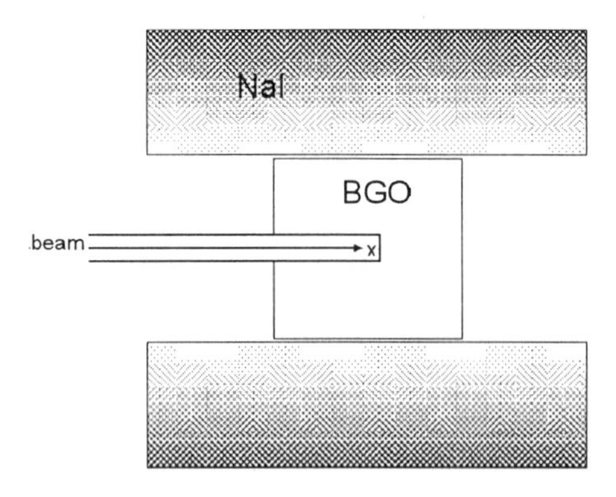

Figure 1. The detector array. A 7x7 inch annular NaI detector surrounds a 2.75x3 inch BGO well detector. The ion beam produces nuclear reactions in targets at the **x**.

resolution in materials for which the method may be calibrated, but it is a destructive analytical technique.

A powerful alternate technique has been developed (8-14) for detecting hydrogen in solid samples that makes use of NRA. W. A. Lanford (8) has published a definitive review of these methods. The technique involves accelerating an appropriate light atomic ion and bombarding the hydrogen-containing sample. Protons react relatively easily with light nuclei giving characteristic and easily identifiable reaction products, often with the liberation of an energy far higher than the incident ion energy. The nuclear reaction occurs, and energy is liberated, in proportion to the hydrogen concentration present in the sample. Moreover, when the nuclear reaction occurs only for a well determined incident energy, depth profiling of the hydrogen concentration is obtained.

The nuclear reaction employed here is $^{1}H(^{15}N,\alpha,\gamma)^{12}C$. This reaction has a resonance for 6.39 MeV ^{15}N ions impinging on a hydrogen target. The reaction product ^{12}C is left in an excited nuclear state that produces a single 4.43 MeV gamma-ray. Its yield is proportional to the hydrogen concentration at a depth in the target determined by the difference between the initial ^{15}N ion energy and 6.39 MeV and by the stopping power dE/dx for ^{15}N ions in the target material.

The sensitivity of this technique depends on the efficiency of the detector as well as its ability to discriminate the 4.43 MeV gamma ray from background radiation. Low energy ^{15}N ions cannot produce significant nuclear reactions in targets other than hydrogen. Thus, if the accelerator analyzer magnet keeps other energetic light ions away from the target chamber, no other background is expected from the accelerator. Natural radioactivity produces no gamma rays with energies as high as 4.43 MeV. The only significant background comes from cosmic rays which produce pulses in a gamma ray detector. This cosmic ray component must be subtracted from the detector yield, and ultimately establishes the lower limit for detection of hydrogen in samples bombarded with a ^{15}N ion beam.

A large detector is required to increase the probability that the 4.44 MeV gamma-ray will convert its energy to electrons. However, the cosmic ray counting rate increases with the projected area of the detector. Thus, a point of diminishing returns is reached for a detector size such that the efficiency is less than 100%. The large detectors reported in the literature (15-19) reduce the cosmic ray background by shielding and by the use of anticoincidence detectors. A detection limit of 150 atomic parts per million (ppma) has been reported (16) using a single 15x15 cm cylindrical NaI scintillation detector for the 4.43 MeV gamma-ray. This detector is reported to have an efficiency of 20%.

To maximize the efficiency for detection of the 4.43 MeV gamma-ray and minimize the cosmic ray counting we have designed a detector with less than 100% efficiency but with an ability to discriminate the 4.43 gamma-ray from cosmic rays. Pair production (positron and electron) is the most probable reaction for a 4.43 MeV gamma-ray in high-Z detector material. The presence of a positron in the detector permits the desired discrimination, since few positrons are produced by cosmic rays. We have designed a detector of 4.43 MeV gamma rays that reduces background by requiring a coincidence of the 0.511 MeV annihilation radiation in one detector with the remaining 3.92 MeV in another.

Figure 1 shows schematically the arrangement of the components of the photon detector array. An inner 4π BGO scintillator surrounds the photon source. An electron-positron pair produced in this inner detector will dissipate an energy E minus 2x0.511 MeV, where E is the primary photon energy and 0.511 MeV is the electron rest energy. The annihilation of the positron after it comes to rest produces two secondary photons each with 0.511 MeV. In a large detector there is a high probability that the energy of both these annihilation photons is captured, thus the name "total energy detector."

In a smaller detector, one or sometimes both of the annihilation photons may escape. Another detector, shown as an annular cylinder in Figure 1, is deployed around the inner scintillator to capture one, or both, of the annihilation photons. The requirement of coincidence pulses and appropriate energy discrimination, eliminates environmental background and most cosmic ray events.

The efficiency is equal to the product of three probabilities: 1. Electron-positron production in the inner detector. 2. Escape of an annihilation photon from the inner detector. 3. Detection of the escaped annihilation photon in the outer detector.

The first increases with the size of the inner detector, but the second decreases because the annihilation radiation is absorbed and cannot escape a big detector. There is an optimum size of the inner detector for maximum efficiency. The third probability is independent of the others and increases with the size of the outer detector.

Both the first and the second probabilities are dependent on where the positron comes to rest. One must integrate the

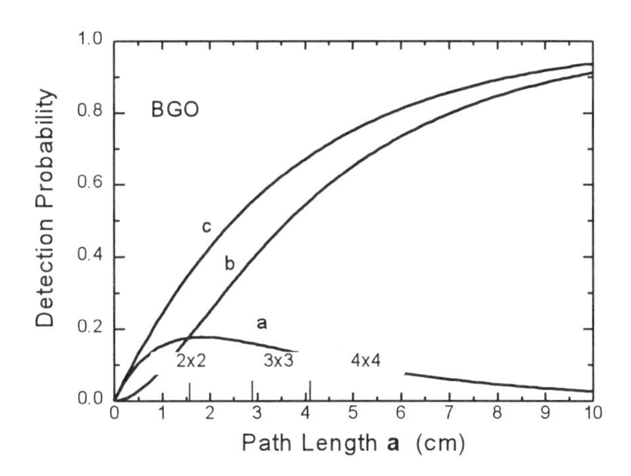

Figure 2. The relative efficiency of a BGO scintillation detector (a) with the escape of one annihilation photon , (b) requiring total energy capture and (c) accepting all events.

product of the two over the volume of the inner detector. With simplifying geometrical assumptions the integral is accomplished and multiplied by the third probability to give the efficiency **e**:

$$\mathbf{e} = F\frac{\mu_1}{\mu_2-\mu_1}(e^{-\mu_2 a}-e^{-\mu_1 a})(1-e^{-\mu_3 b}) \qquad (1)$$

where μ_1 and μ_2 are the total absorption coefficients of the primary photon and of the 0.511 MeV annihilation photon, respectively, in the material of the inner detector. a and b are radial dimensions of the inner and outer detectors, respectively. F is the fraction of total absorption events in the inner material that result in the production of an electron-positron pair. μ_3 is the total absorption coefficient for 0.511 MeV radiation in the material of the outer detector.

The maximum efficiency is obtained for a detector smaller than those used for total energy detectors, a consequence of the desire to detect annihilation radiation outside the central detector. The optimum radial dimension **a** for the inner detector is

$$\mathbf{a} = \frac{\ln(\mu_2/\mu_1)}{\mu_2-\mu_1} \qquad (2)$$

One can calculate the efficiency of a practical detector with the optimum dimensions. In BGO $1/\mu_1$ = 3.60 cm for a 4.43 MeV and $1/\mu_2$ = 1.06 cm for the 0.511 MeV annihilation radiation. Substituting in Equation (3), we find that the optimum internal radial dimension is 1.84 cm.

This optimization is independent of the material and size of the outer detector. The ideal outer detector would detect 100% of the 0.511 MeV photons, and Equation (1) would give the array 7.4% efficiency using a fraction F = 0.42 of 4.43 MeV photon events in BGO that produce an electron-positron pair. Extrapolating published values, this same inner detector, operated as a total energy detector, would be 15% efficient (Figure 2) comparable to large detectors described in the literature (15-19).

While the treatment above displays all the concepts needed to understand the design of the new detector, the final dimensions have been determined partly influenced by other factors: 1. No detector material can be placed in the way of the entry beam. 2. The angular distribution of the 4.43 photon produced by the ^{15}N reaction with ^{1}H is slightly peaked fore and aft (20) along the ion beam direction.

These factors together dictate an internal detector longer along the axis than the dimension given by Equation (2). The inner detector is 2.75 (diameter) x 3.50 inch cylindrical BGO, with a 0.75 inch diameter well to the center. Factor 1 and economic factors influenced the selection of an annular cylindrical outer detector 7x7 inches NaI.

The outer detector, the most costly component, reduces the efficiency of the array from the ideal 7.2% to 4.5%, mostly from the escape of the 0.511 MeV photon through the NaI (28% loss) and out the open ends (14% loss).

The detectors were furnished by BICRON Inc., Cleveland, Ohio. Single channel discriminators with 10% windows were used to select signals corresponding to 0.511 MeV in the outer detector and 3.93 MeV in the inner detector. If these signals were within 250 ns of coincidence, an event was counted.

RESULTS AND DISCUSSION

We have designed a gamma ray detector array that has a low cosmic ray background owing to the requirement that positron annihilation radiation is detected in an outer detector in coincidence with the remaining gamma ray energy in the central detector. The penalty in efficiency is offset by background reduction, such that low gamma ray yields may be detected in Nuclear Reaction Analysis. The array is used to measure the hydrogen content in samples of GPC using the $^{1}H(^{15}N,\alpha\gamma)^{12}C$ resonant nuclear reaction. We used 6.8 MeV ^{15}N ions to detect hydrogen a few microns within the precursor resin and the GPC samples. Low energy background from radioactive decay of ^{232}Th and ^{40}K is eliminated by pulse height discrimination. High energy cosmic background is reduced by the coincidence counting technique.

The gamma ray yield is proportional to the hydrogen content in the samples which is compared in Figure 3 with chemical measurements (1) for the same material.

For targets with stopping powers similar to carbon, we may expect to detect 25 ppma hydrogen with unit signal to noise ratio with 100 nA N^{15+} on the target.

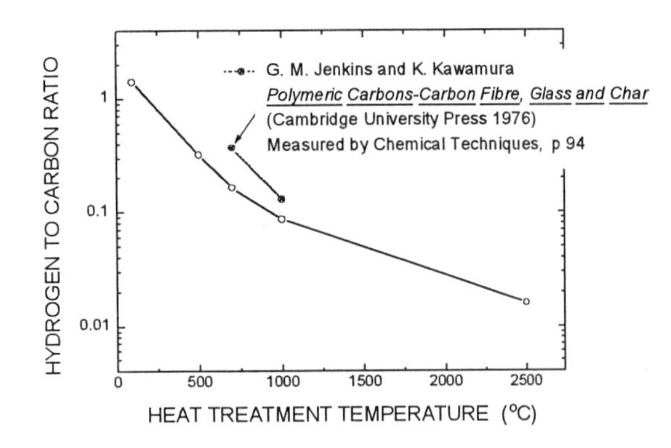

Figure 3 The hydrogen concentration as a function of heat treatment temperature of the material, including that of the precursor, resol $C_7H_8O_2$.

CONCLUSIONS

Nuclear reaction analysis of the precursor resin shows a hydrogen to carbon ratio appropriate for resol and its hydrocarbon solvents. Heat treatment to 500 °C reduces the hydrogen content dramatically owing to the removal of the solvents and H_2O from the polymer. The H_2 removal, which should occur between 550 °C and 750 °C, is observed to be incomplete. The atomic percent is almost 0.1 for samples heat treated to 1000 °C, and does not fall below 0.01 even for samples heat treated to 2500 °C.

GPC heat treated to temperatures for which only carbon bonds remain nevertheless retain hydrogen. Since the material is impermeable, it is thought that hydrogen is retained in the GPC pores. This unexpected result suggests more study of the diffusion of gasses in GPC during high temperature heat treatment, as was done for the critical heat treatment temperatures closely above 550 °C (21).

ACKNOWLEDGMENTS

This work partially supported by USArmy contract DAAH04-93-2-0003 and the Center for Irradiation of Materials at Alabama A&M University. RLZ is grateful for support from the Coodenação de Aperfeiçoamento de Pessoal de Ensino Superior.

REFERENCES

1. G. M. Jenkins and K. Kawamura, *Polymeric Carbon, Carbon Fiber Glass and Char,* Cambridge University Press, 1976, 140.
2. D. Ila, G. M. Jenkins, L. R. Holland, A. L. Evelyn and H. Jena, *Vacuum* **45,** 451 (1994).
3. D. Ila, A. L. Evelyn, H. Jena and G. M. Jenkins, *Carbon* **31**, 1211-1212 (1993).
4. J. H. Fisher, L. R. Holland, G. M. Jenkins and H. Maleki, *Carbon* **34**, 789-795 (1996).
5. D. Ila, G. M. Jenkins, R. L. Zimmerman and A. L.Evelyn, *Mat. Res. Soc Symp. Proc.* **331**, 281-285 (1994).
6. Zimmerman, R. L., Ila, D., Jenkins, G. M., Maleki, H. and Poker, D. B., *Nucl. Instr. and Meth.* **B106**, 550 (1995).
7. Zimmerman, R. L., and Ila, D., *Nucl. Instr. and Meth* **B, (in** print)
8. Lanford, W. A., *Nucl. Instr. and Meth.* **B66**, 65-82 (1992).
9. Leich, D. A., and Tombrello, T. A., *Nucl. Instr. and Meth.* **108**, 67 (1973).
10. Xiong, F., Rauch, F., Shi, C., Zhov, Z., Livi , R. S. and Tombrello, T. A., *Nucl. Instr. and Meth.* **B27**, 432 (1987).
11. Ryden, J., Hjorvarsson, B., Ericsson, T., Karkson, E., Krozer, A., and Kasemo, B., *Journal of Less Common Metals* **152**, 295 (1989).
12 Ziegler, J. F., *et al., Nucl. Instr.and Meth.* **149**, 19 (1989).
13 Iwata, Y., *et al., J. of Appl. Phys.* **26L**, 1026 (1987).
14. Forster, J. S., Leslie, J. R. and Laursen, T., *Nucl. Instr. and Meth.* **B66**, 215-220 (1992).
15. Damjantschitsch, H., Weiser, M., Heusser, H., Kalbitzer, S., and Mannsperger, H., *Nucl. Instr. and Meth.* **218**, 129-140 (1983).
16. Briere, M. A., Wulf, F., and Bräunig, D., *Nucl. Instr. and Meth.* **B45**, 45-48 (1990).
17. Kuhn, D., Rauch, F., Baumann, H., *Nucl. Instr. and Meth.* **B45**, 252-255 (1990).
18. Kuhn, D., and Rauch, F., *Nucl. Instr. and Meth.* **B58**, 113-115 (1991)
19. Endisch, D., Rauch, F., Götzelmann, A., Reiter, G., and Stamm, M., *Nucl. Instr. and Meth.* **B62**, 513-520 (1992).
20. Kraus Jr., A. A., French, A. P., Fowler, William A. and Lauritsen, C. C., *Phys. Rev.* **89**, 299-302 (1953).
21. Maleki, H., Holland, L.R., Jenkins, G. M., Zimmerman, R. L. and Porter, W., *J. of Mat. Res.* **11,** No. 9, 391 (1996).

HIGH-ENERGY ION-BEAM ANALYSIS IN COMBINATION WITH KEV SPUTTERING

J. Maldener, F. Rauch

Institut für Kernphysik, J. W. Goethe Universität, Frankfurt am Main, Germany

Potential advantages of combining high-energy ion-beam analysis with pre-analysis sputter treatment are discussed. Examples are shown which demonstrate improvements of sensitivity, quantification and depth range in different analytical problems. The sputter gun used is quite compact and may be easily incorporated into an analysis vacuum chamber.

INTRODUCTION

One of the main features of high-energy ion-beam analysis (HE-IBA) is its capability to yield depth-resolved information over a larger depth range without layer-by-layer removal. The latter is required for, e.g., secondary ion mass spectrometry (SIMS), Auger electron spectroscopy (AES) and electron spectroscopy for chemical analysis (ESCA) and is achieved by low-energy ion sputtering. However, there are certain situations in which sputter treatment may also be a good complement to HE-IBA. For example, it could be used for the removal of hydrocarbon contamination layers which persist on sample surfaces even under good vacuum conditions. (This is probably done in several laboratories, although we are not aware of published descriptions.) Furthermore, near-surface layers which differ chemically from the sample material can be removed in order to expose the sample interior. By prolonged sputter treatment one could take off thick layers in order to extend the depth range of HE-IBA methods.
Commercially available sputter guns provide currents in the μA range. Sputtering yields for several keV Ar ions are around 1 or higher. Thus, one can expect reasonably short sputter treatment times.
This paper reports on various experiments we have carried out to test the usefulness of the HE-IBA/sputtering combination, using an arrangement for in-situ sputter treatment. These experiments were initiated in the course of a program on H analysis of nominally anhydrous minerals with the $^{1}H(^{15}N,\alpha\gamma)^{12}C$ resonant nuclear reaction. The concentration range to be covered is about 10 to 500 at.ppm H. As discussed in (1), at these concentrations γ-ray background from reactions in the hydrocarbon layer may limit the analysis sensitivity. Furthermore, the mineral samples often have a H-rich subsurface layer arising from sample preparation by polishing. In both cases, sputter treatment should provide a remedy.

THE SPUTTER GUN

The sputter gun used by us is of simple construction and comparably inexpensive.* It is based on a saddle-field ion source and works without magnetic field and filaments. This sputter gun has been described before in an application to target production (2). Its small dimensions (diameter: 35 mm, length: 40 mm) allow convenient incorporation into a HE-IBA vacuum chamber. The accelerating-voltage range is 2 - 10 kV. In our experiments, Ar was used as the source gas, the accelerating voltage was 6 kV, and the sputter treatment was done at normal incidence. The beam produced by this sputter gun contains a certain fraction of neutrals; at 6 kV, employing an optional extended front cathode, the neutral fraction is about 60%. Even though neutral components are detrimental to dose monitoring using current integration, they have the advantage of improving the sputter treatment of insulators. The dose can alternatively be determined from the measured mass removal from known standards. A measurement of the beam intensity profile (see below) showed that a large part of the beam is contained in a narrow cone of about 4°.

EXPERIMENTAL RESULTS

Density profile of the sputter beam

To determine the beam intensity profile, we found a convenient method which may also be regarded as a first utilisation of the HE-IBA/sputtering combination. A gold film (200 nm, evaporated on a Si substrate) was bombarded with the Ar beam at a distance of about 14 cm for a certain time. The dimensions of the sputter crater were determined

* Supplier: Atom Tech Ltd., West Molesey, Surrey, England.

CP392, *Application of Accelerators in Research and Industry,* edited by J. L. Duggan and I. L. Morgan
AIP Press, New York © 1997

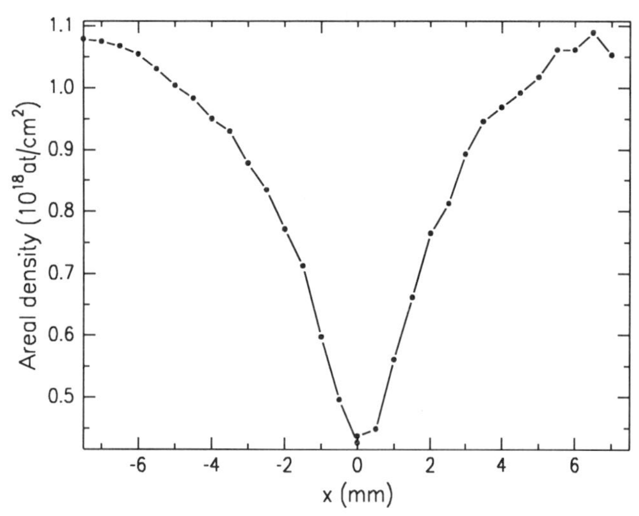

FIGURE 1. Lateral variation of the areal density of an Au film after 5 min sputter treatment determined by RBS (compare text).

by stepping the ^{4}He beam (diameter about 1 mm) laterally across the film and measuring the intensity of the Au signal at each point with Rutherford backscattering spectrometry (RBS). From the resulting profile, see fig. 1, one can infer that about 10% of the beam intensity falls within the central region of 2 mm diameter. In all measurements reported below, the HE-IBA analyzing beam was targeted on this central region of the sputter crater. Since the crater is not flat at the bottom, the depth resolution is degraded compared to measurements on flat surfaces. It is also degraded because of collisional damage in the near-surface layer. This fact was not important in the applications discussed here.

Using the total volume of the sputter crater and a sputtering yield of Y=5.0 for 6 keV Ar on Au (3), the sputter beam current delivered onto the target was determined to be approximately 18 µA. Though this value differs substantially from the manufacturer's quoted maximum output for the gun, the amount of current is sufficient for the sputter-treatment applications investigated.

Removal of hydrogen-bearing adsorption layers

Hydrogen-bearing adsorption layers are commonly thought to consist of hydrocarbon molecules with an overall composition of about CH_2 (4). We have made survey experiments on the effect of the Ar sputter beam with respect to changes in the H areal density, S(H). Complementary measurements on the C areal density will follow. S(H) was determined with the ^{15}N technique by integrating over the so-called surface peak (1,5). As an example, fig. 2 shows yield curves obtained for an Au film (200 nm, evaporated on a Si substrate) and a commercial Al foil (0.3 mm). The measurements were performed at a vacuum of 4×10^{-8} mbar and with the use of a LN_2 trap.

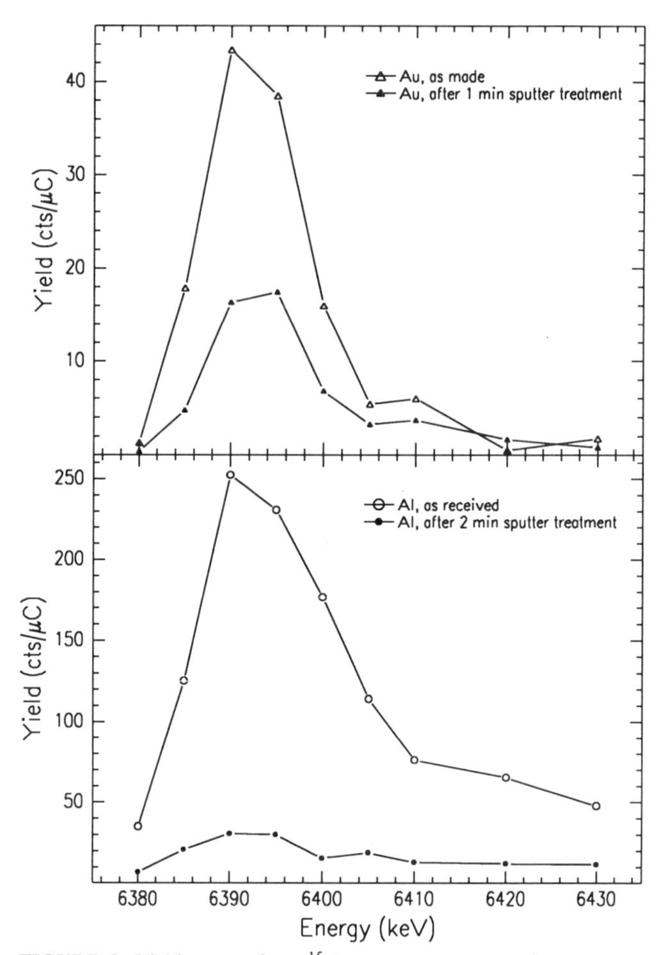

FIGURE 2. Yield curves from ^{15}N measurements near the resonance energy showing the effect of sputter treatment on Au (upper panel) and Al (lower panel) samples.

As can be seen, short sputter treatments result in a distinct reduction of the H areal density from 4.4 ML (1 ML $=10^{15}$/cm^2) to 1.8 ML for Au in 1 min and from 36 ML to 6.1 ML for Al in 2 min. Reduction of S(H) by high energy bombardment, which has previously been used (1,5), is more time-consuming and may lead to radiation effects inside the sample. It was found that the H areal density increased only slightly when the samples were left without further bombardment; for Al S(H) increased only to 7.6 ML after 6 hours. This means that under the vacuum conditions described, a single short sputter treatment is sufficient for ensuring an adequately low contamination level.

The importance of reducing the H areal density can be appreciated by considering the fact that the off-resonance yield from ^{15}N+^{1}H reactions in a surface layer of an Al sample containing 36 ML H corresponds to a H-concentration value of about 210 at.ppm (430 at.ppm) at 1 µm (2 µm) depth.

Similar considerations hold for hydrogen detection with elastic recoil detection (ERD), where the low energy tail behind the surface peak, caused by multiple scattering, limits the sensitivity.

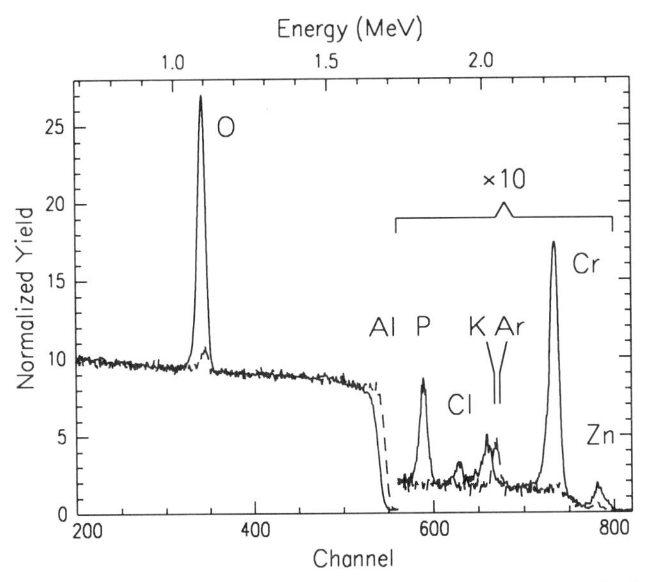

FIGURE 3. Backscattering spectra of an aluminium sample as received (solid line) and after sputter treatment for 15 minutes (dashed line). The spectra are scaled by a factor 10 above channel 560.

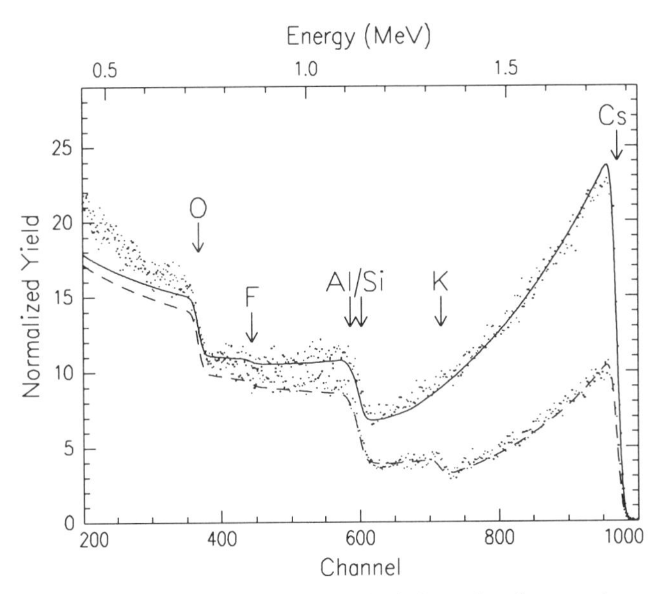

FIGURE 4. RBS spectra and RUMP simulations of a glass sample as received (solid line) and after sputter treatment for 2 hours (dashed line).

Removal of an unwanted surface layer

We had also acquired another piece of Al as a material having a very low H content, to be used in conjunction with high-sensitivity H analysis. Survey measurements showed the presence of a H-rich (up to 8 at.%) surface layer of about 100 nm thickness. Successful sputter removal was achieved, as verified by using the ^{15}N technique. In addition, backscattering measurements were performed in order to see whether the H enrichment was connected with an oxide or impurity layer. The beam selected was ^{4}He at an energy of 3.05 MeV. In this case the cross section for scattering on O is about 20 times higher than the Rutherford cross section (6). As seen in fig. 3, the as-received sample showed indeed an oxygen peak and, in addition, sitting on a somewhat indistinct background, several other peaks which indicate various contaminants.

Sputter treatment and intermittent backscattering analysis revealed that all these peaks decreased in parallel, proving that they belong to just one contamination layer and allowing unequivocal element identification (P, Cl, K, Cr, Zn) and determination of areal densities by use of the computer simulation program RUMP (7). The RBS-spectrum taken after 15 min sputtering shows that this layer has been removed completely and that the O signal is strongly reduced. The total areal density of the removed layer determined from the peak areas is about $2\times10^{17}/cm^2$ and is compatible with the observed shift of the Al edge of 24 keV. The remaining yield above the Al subspectrum is flat and can be attributed to Cr as a deep-reaching contamination layer or an alloying element. The peak occurring at an energy of 2.05 MeV arises from Ar implanted during sputter treatment. (The implantation of Ar did not cause a problem here; in cases where it may be troublesome, lower sputter beam energies have to be used.)

Extending the depth range of analysis

As an example for the feasibility of extending the depth range of analysis, fig. 4 shows measured RBS spectra of a glass sample before and after sputter treatment. The glass components are Si, O, Al, B, K and F. In this sample, the K component had been replaced in a layer of about 2 μm thickness by Cs through diffusional ion exchange in a salt bath (8). The Cs depth profile has the shape of an error-function, the K depth profile that of the error-function complement. RUMP simulations showed that the Cs surface concentration in the sputter-treated sample is reduced to 3.8 at.% compared to 9.0 at.% at the original surface. This can be converted to a removal of a layer with a thickness of approximately 500 nm. The K signal is now visible, allowing a more direct examination of the ion-exchange depth profile than possible with the original sample. From a comparison of the removal rate with that of the Au film, a sputtering yield of about 1.0 can be estimated. This is compatible with the value expected for sputtering of oxides (9).

As noted above, the depth resolution was degraded under the sputtering conditions used. It may be improved by producing a flat sputter crater (scanning of the Ar beam or the sample) and working with lower beam energies for reducing collisional mixing. Then, one can examine rapid concentration changes at depths where in "normal" HE-IBA measurements the depth resolution is limited by straggling.

CONCLUSION

In conclusion, it was shown how sputter treatment for controlled layer removal can be used to tailor the sample according to the analytical task and thus extend the possibilities of high-energy ion-beam analysis.

ACKNOWLEGMENT

We are indebted to K. Horn, Albuquerque, for helpful comments and H.J. Maier, Universität München, for drawing our attention to the usefulness of the type of ion gun used here.

This work was supported by the Deutsche Forschungsgemeinschaft.

REFERENCES

1. Endisch, D., Sturm, H., and Rauch, F., *Nuclear Instruments and Methods* **B84**, 380-392 (1994).
2. Maier, H.J., *Nuclear Instruments and Methods* **A303**, 172-181 (1991)
3. Matsunami, N., et. al., *Atomic Data and Nuclear Data Tables* **31**, 1-80 (1984).
4. Amsel, G., and Maurel, B., *Nuclear Instruments and Methods* **B218**, 183-196 (1983)
5. Thomas, J. P., Fallavier, M., Pijolat, C., and Tousset, J., *Radiation Effects* **61**, 207-214 (1982).
6. Leavitt, J. A., McIntyre, L. C., et. al., *Nuclear Instruments and Methods* **B44**, 260-265 (1990).
7. Doolittle, L. R., *Nuclear Instruments and Methods* **B9**, 344 (1985).
8. Gießler, K. H., Rauch, F., and Fabricius, N., *Nuclear Instruments and Methods* **B**, in press.
9. Kelly, R., "The Sputtering of Insulators", in *Ion Beam Modification of Insulators*, ed. Mazzoldi, P., and Arnold, G.W., Amsterdam, Elsevier, 1987, ch. 2, 57-113.

ION BEAM TECHNIQUES APPLIED TO THE STUDY OF OXYGEN DIFFUSION IN $YBa_2Cu_3O_{7-\delta}$ FILMS

Yupu Li, Jairui Liu, and Wei-Kan Chu

Texas Center for Superconductivity, University of Houston, Houston, Texas 77204-5932, USA

J. A. Kilner

Department of Materials, Imperial College of Science Technology and Medicine, London SW7 2BP, UK

T. J. Tate

Department of Electrical and Electronic Engineering, Imperial College of Science Technology and Medicine, London SW7 2BT, UK

Ion beam techniques applied to the study of oxygen diffusion in YBCO films can be summarized in three areas: (1) implanting ^{18}O into YBCO films either at trace levels or as a tracer; (2) SIMS depth profiling of the exchanged ^{18}O (without irradiation damage) and as-implanted ^{18}O concentration distributions and their evaluation during annealing; and (3) analyzing of relative ^{18}O content and depth distributions with the ^{18}O (p, α) ^{15}N nuclear reaction. This paper will only summarize and review the authors work in the above areas. It is observed that the apparent diffusivity in the c-direction is larger than that for bulk materials. Hence, the mechanism of diffusion in c-axis oriented YBCO films is linked with the presence of short circuit diffusion along planar defects. It also showed that a thin buried $LaAlO_3$ layer in epitaxial multi-layer of $YBa_2Cu_3O_{7-\delta}/LaAlO_3/YBa_2Cu_3O_{7-\delta}$/substrate can act as a barrier to diffusing oxygen. The ^{18}O (p, α) ^{15}N nuclear reaction analyses of the ^{18}O implanted samples and annealed samples showed that both (1) implantation at 500°C and (2) post-irradiation vacuum annealing at 500°C resulted in "in-situ" oxygen migration and release.

INTRODUCTION

It has been found that the superconducting properties of ceramic superconductors depend on the oxygen content and oxygen ordering of the material, which can be changed by the diffusion of oxygen either into or out of the materials. The motion of oxygen has been a major focus in the literature. An activation energy of 1.2±0.2eV for oxygen self-diffusion in the a-b plane and 1.6±0.3eV for oxygen self-diffusion in the c-direction, together with an anisotropy factor $D_{a-b}/D_c \sim 10^4$—10^6, can be generally summarized from the published results [1—7]. It should be noted that the above results were derived mainly from bulk single crystal or poly-crystalline YBCO samples. Epitaxial YBCO films hold great promise for practical applications, but there are few data about oxygen self-diffusion coefficient in epitaxial YBCO films [8] since the thin films usually have a thickness less than 1 micron and so it is more difficult to design suitable experimental conditions for direct measurement of oxygen diffusivity in the films. There are no publications reporting on oxygen diffusion in the epitaxial multi-layer of $YBa_2Cu_3O_{7-\delta}/LaAlO_3/YBa_2Cu_3O_{7-\delta}$ /substrate ($LaAlO_3$).

The present paper is essentially a summary and review of our experimental results about oxygen diffusion in YBCO films. By using as-implanted ^{18}O implantation and exchanged ^{18}O as trace, it is observed that the apparent diffusivity in the c-direction is larger than that for bulk materials. Hence, the mechanism of diffusion in c-axis oriented YBCO films is linked with the presence of short circuit diffusion along planar defects [8]. In fact, at present only sparse data are available and thus more research needs to be carried out to determine such parameters as the activation energy for oxygen self-diffusion and diffusion mechanisms in these YBCO films and the HTS multi-layer.

CP392, *Application of Accelerators in Research and Industry*, edited by J. L. Duggan and I. L. Morgan
AIP Press, New York © 1997

EXPERIMENTAL

Samples: c-axis oriented YBCO films on (100) $LaAlO_3$ single crystal were prepared by inverted cylindrical magnetron sputtering of a stoichiometric $YBa_2Cu_3O_{7-\delta}$ target [9], at the National Physical Laboratory, UK [8] and at Texas Center for Superconductivity, with a thickness range from 800 nm — 1100 nm. A laser ablated epitaxial multi-layer of $YBa_2Cu_3O_{7-\delta}$ /$LaAlO_3$/$YBa_2Cu_3O_{7-\delta}$/substrate ($LaAlO_3$) was provided by the Center for Superconductivity at the University of Maryland for studying oxygen diffusion in such sample.

Implanting ^{18}O into YBCO films as trace: irradiation was performed with 200keV $^{18}O^+$ to a dose of $5x10^{16}/cm^2$, using a 200kV WHICKHAM implanter at Imperial College. The energy and dose were designed after considering the simulated profile by TRIM [10], which showed that ^{18}O will be built up within a 400 nm top layer of the YBCO film, leaving the bottom of the film undamaged. During implantation, two target temperatures were chosen: room temperature and 500°C, since we aim to study oxygen migration and loss during the higher temperature implantation. In addition, post-irradiation annealing in flowing oxygen ambient and vacuum (~$5x10^{-6}$ Torr) were also performed for following evaluation of the ^{18}O distributions.

Another method to introduce ^{18}O into YBCO films as tracer is through ^{18}O gas/^{16}O solid exchange [7,8]. Simply, annealing of YBCO films in a quartz tube containing ^{18}O at a pressure of 800 mbar to 1 bar at the certain temperature will meet this aim.

SIMS depth profiling: SIMS analyses and stable isotopic tracers have been used to obtain self diffusion data in variety of materials. ^{18}O isotopic ratio profiles after implantation (or oxygen isotopic exchange) and post-irradiation annealing steps were obtained on an Atomika 6500 SIMS instrument (at Imperial College) by using a 10keV Cs^+ primary beam at normal incidence and analyzing the negative secondary ions. Details of the SIMS analysis on such samples can be found in our previous publication [8].

The ^{18}O (p, α) ^{15}N reaction: the relative ^{18}O contents and distributions in the 250 nm top layer of the YBCO films after implantation and after post-irradiation annealing was measured by the ^{18}O (p, α) ^{15}N reaction nuclear reaction resonance at 629keV, using a 2x1.7 MV accelerator at the Texas Center for Superconductivity at University of Houston. The initial energy was 579keV and it was continuously increased by steps of 2—5keV in order to follow the change of the distribution in a shorter time. In other words, if in some depth range the profile is flat, then the energy step is set at a larger increment, saving measuring time.

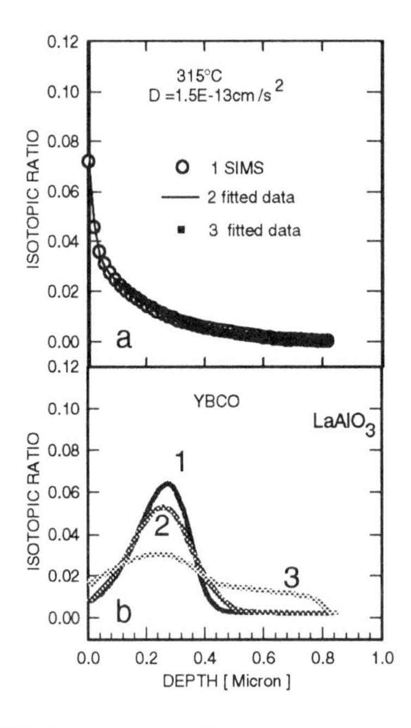

FIGURE 1. (a) Line 1, ^{18}O penetration profile along the c direction of the YBCO film after oxygen isotopic exchange annealing at 315°C for 1h. Line 2 and 3: fitted data (see the text); (b) ^{18}O isotopic ratio profiles for the ^{18}O implanted film before annealing (line 1), after annealing at 300°C for 1h (line 2), and after rapid thermal annealing at 450°C for 2 min (line 3).

RESULTS AND DISCUSSION

We have designed our experimental conditions to achieve direct measurement of the oxygen tracer diffusivity in YBCO films. In our experiments, the atomic motion of oxygen in a c-axis oriented YBCO films was studied with exchanged ^{18}O or implanted ^{18}O as a tracer. For example, line 1 in Fig. 1a shows the exchange/diffusion profile of ^{18}O for the YBCO film annealed at 315°C for 1h in a quartz tube containing ^{18}O gas at a pressure of 990 mbar. The ^{18}O isotopic fraction profile was obtained from the intensity ratio of $^{18}O^-/(^{18}O^-+^{16}O^-)$ after SIMS measurement. Line 2 in Fig. 1a, which is very close to the experimental points, is the curve fitted to the solution appropriate for a semi-infinite solid with both volume and fast planar (short-circuit) diffusion [8]. It is also found that it is possible to obtain a second "good" fit using two apparent volume component, i.e. the slow one fitted to the first 80 nm and the second (fast one, i.e. line 3 in Fig.1b) to the remainder of profile shown in line 1 in Fig. 1a, which gives

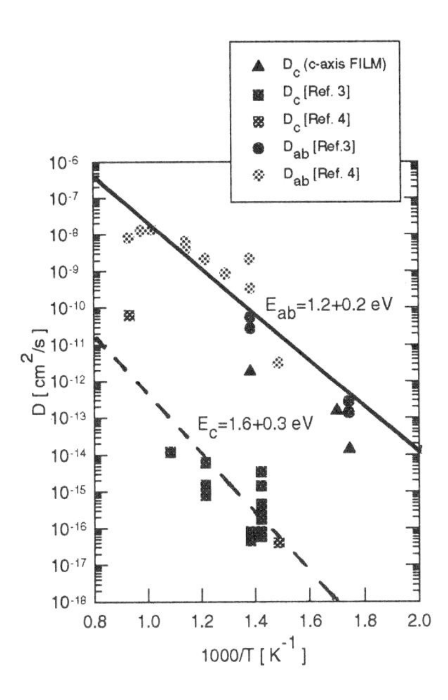

FIGURE 2. Arrhenius plot for oxygen diffusion in bulk and c-oriented thin film of $YBa_2Cu_3O_{7-\delta}$.

an apparent diffusivity in the deeper region D_A of $1.5x10^{-13}$ cm^2/s [8]. Fig.1b shows SIMS ^{18}O isotopic fraction depth profiles for the $^{18}O^+$ implanted film before and after annealing in flowing natural oxygen gas. Line 1 in Fig.1b shows the as-implanted ^{18}O isotopic ratio profile in a 830 nm thick YBCO film. lines 2 and 3 in Fig.1b show the ^{18}O isotopic ratio profiles after annealing at 300°C for 60 min (line 2) and after rapid thermal annealing at 450°C for 2 min (line 3). The area (under the ^{18}O isotopic ratio curve) which corresponds to the retained dose of ^{18}O for each case can be estimated to be $95.5\%Q_0$ and $\sim95.6\%Q_0$, respectively, where Q_0 is the retained dose after implantation. It is clear that the implanted ^{18}O plays a role in monitoring the oxygen migration within the film. However, it should be mentioned here that calculations show that a 10% oxygen enrichment of the film by the natural oxygen gas (99.8% ^{16}O) ambient will induce 10% decrement of the area and thus can only play a minor role in the change of the area under ^{18}O isotopic ratio curve. At 300°C the implanted ^{18}O is clearly mobile (see line 2 in Fig.1b). The profile becomes broader and some of the implanted ^{18}O migrates into the deeper layer of the film where the irradiation damage is less. Using $\Delta X^2=2D_At$ (i.e. via Brownian movement, see ref. [11]), the apparent diffusion coefficient

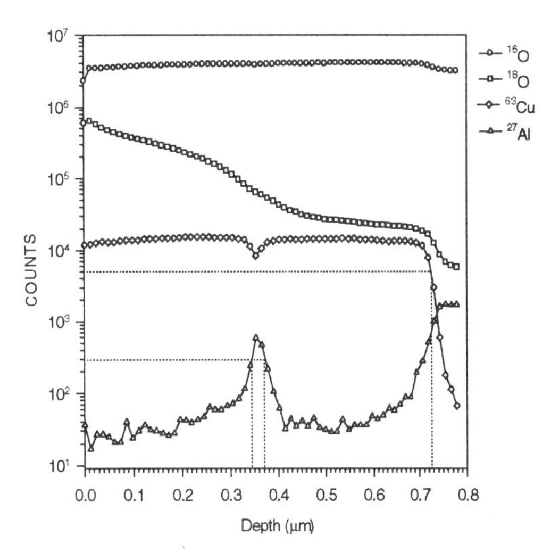

FIGURE 3. SIMS depth profile from a multi-layer sample after oxygen isotopic exchange annealing at 415°C for 1 h (899 mbar).

in the irradiated film, D_A, is $\sim1.4x10^{-14}$ cm^2/s at 300°C. Here, we have defined the mean square displacement of the diffusing ^{18}O, ΔX^2, as equal to the difference between the square of the right half-height widths for line 2 and line 1 in Fig.1b. From line 3 in Fig.1b, D_A can be simply estimated as $1.8x10^{-12}$ cm^2/s [12]. Diffusivity data obtained from our ^{18}O exchanged and $^{18}O^+$ implanted samples [8,12] have been included in Figure 2, which also shows a compilation of data for the diffusion of oxygen in bulk YBCO. The values obtained in the c-direction of the film are about one order of magnitude smaller than in the a-b plane but much higher than in the c-direction of bulk YBCO single crystals. These facts suggest that some fast diffusion paths exist. Short-circuit diffusion is thus playing an important role in determining the high mobility of oxygen in the c-axis oriented YBCO film.

Figure. 3 shows the SIMS depth profile obtained from a multi-layer of $YBa_2Cu_3O_{7-\delta}$ /$LaAlO_3$ /$YBa_2Cu_3O_{7-\delta}$/substrate ($LaAlO_3$). The $^{18}O^-$ signal (open squares in Fig.3) for the upper YBCO layer is seen to have a different gradient when compared to the lower YBCO layer, with the turning point being located at the depth for the buried $LaAlO_3$ layer (nominal about 11 nm thick). This result for such a sample showed that a thin buried $LaAlO_3$ layer can act as a barrier to diffusing oxygen. This means that any unsuitable annealing process during the deposition may result in the lower YBCO layer having a different oxygen content and oxygen ordering to that for the upper YBCO layer. This has clear implications for device fabrication. It is well known that the physical properties $YBa_2Cu_3O_{7-\delta}$ materials depend on both oxygen concentration and oxygen ordering. Thus, a study of oxygen diffusion in such a multi-layered sample may play an important role in optimizing the conditions for device fabrication.

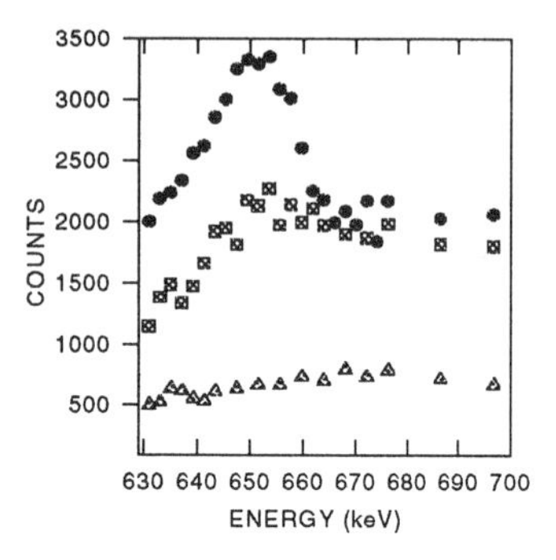

FIGURE 4. The $^{18}O(p, \alpha)^{15}N$ 629 keV resonance spectra from 800 nm thick YBCO films following (1) implantation with 200keV $^{18}O^+$ to a dose of $5x10^{16}/cm^2$ at room temperature (solid circle), (2) implantation with same energy ^{18}O to same dose but at 500°C (solid square), and (3) implantation at room temperature plus vacuum annealing at 500°C for 45 min (solid triangle).

Figure 4 shows some results where the $^{18}O(p, \alpha)^{15}N$ 629 keV resonance was used to determine relative ^{18}O content and difference of the ^{18}O distributions in the $^{18}O^+$ implanted and annealed c-oriented YBCO films. Each point in Fig. 4 was calculated from the α peak integral after each measurement at a given energy. If the energy step is 2keV, then the depth scale is about 2keV/100eV/nm=20nm, where an average stopping power of 100eV/nm was estimated by using TRIM [10]. Therefore, each depth profile shown in Fig.4 stops at a depth of ~710 nm. Because the resonance is 2.1keV wide, the depth resolution on the surface is approximately 21 nm. However, it should be noted that the resonance (at 629keV) sits on a continuous background and the nuclear reaction cross sections increase with increasing energy, therefore, the yield stays in a quite larger count level when the resonance happens in the inside of the film. In other words, the distributions within about 250 nm thick upper layer of the film can be inferred as the ^{18}O distributions. Behind this depth (e.g. 250 nm) the contribution from non-resonance reaction between the higher energy proton (>629keV) and the ^{18}O atoms located at the shallower depth cannot be ignored, which makes the main contribution for the high count level in the bottom of the film (as shown in Fig.4). It is possible to remove the contribution of such non-resonance by considering the cross sections measured by Amsel et al. [13] and doing special data processing. Here, we only focus on the upper distributions within the upper films. For the untreated YBCO film under the same measurement conditions (containing natural ^{18}O at an abundance of 0.2% of total oxygen atoms), in the surface region the yield is about 170 counts. Therefore, from Fig. 4 we can conclude that both (1) implantation at 500°C and post-irradiation vacuum annealing at 500°C resulted in a significant migration and release of ^{18}O. However, for the piece after room temperature implantation plus vacuum annealing, in the surface region the yield is still about 2-3 times higher than that for the unimplanted piece, which shows that some implanted ^{18}O still remain in the film.

ACKNOWLEDGEMENTS

For three of us (Y. Li, J.R. Liu, and Wei-Kan Chu), this work is currently supported by the State of Texas through the Texas Center For Superconductivity at the University of Houston. The authors would like to thank Dr. T. Venkatesan for supplying the multi-layer sample, and also thank Dr. P.G. Quincey for supplying a YBCO film for previous work [8].

REFERENCES

1. Rothman, S.J., Routbort, J.L., and Baker, J.E. *Phys. Rev.* **B40**, 8852 (1989).

2. Turrillas, X, Kilner, J.A., Kontoulis, I., and Steele, B.C. H., *J. Less-Common Metals*, **151**, 229 (1989).

3. Rothman, S.J., Routbort, J.L., Welp, U., and Baker, J..E., *Phys. Rev.* **B44** , 2326 (1991).

4. Bredikhin, S.I., Emel'chenko, G.A., Shechtman, V.S., Zhokhov, A.A., Carter, S., Chater, R.J., Kilner, J.A., and Steele, B.C.H., *Physica* **C179**, 286 (1991).

5. Sabras, J., Dolin, C., Ayache, J., Monty, C., Maury, R., and Fert, A., *Colloque De Physique, Colloque* **C1**, 51, 1035 (1990).

6. Ikuma, Y. and Akiyoshi, S., *J. Appl. Phys.* **64**, 3915 (1988).

7. Chater, R.J., Carter, S., Kilner, J. A.., and Steele, B.C.H., *Solid State Ionic* **53-56** , 859 (1992).

8. Li, Yupu, Kilner, J.A., Tate, T.J., Lee, M.J.., Chater, R.J., Fox, H. , Souza, R.A. De., and Quincey, P.G., *Phys. Rev.* **B51** 8498 (1995).

9 Xi, X.X., Linker, G., Meyer, O. , Nold, E., Obst, B, Ratzel, F., Smithey, R., Strehlau, B., Weschenfelder, F., and , Geerk, J., Z. , *Phys. B- Condensed Matter* **74**, 13 (1989).

10 Ziegler, J.F., Biersack, J.P., and Littmark, U. *The Stopping and Ranges of Ions in Solids* (Pergamon, New York, 1985),Ch.5.

11. see, for example, Jost, W., *Diffusion in Solids, Liquid, and gases* (Academic Press Inc., Publishers, New York, 1960).

12. Li, Yupu, Kilner, J.A., Tate, T.J., and Quincey, P..G. *Nucl. Instr. and Meth.* **B** (Proc. the 12th Int. Conf. on IBA) in press).

13. Amsel, G. and Samuel, D., *Anal. Chem.* **39** , 1689 (1967).

CHARACTERIZATION OF FLUORINATED SILICON DIOXIDE FILMS BY NUCLEAR REACTION ANALYSIS AND OPTICAL TECHNIQUES

A. Kumar, H. Bakhru and A. W. Haberl
Department of Physics, University at Albany, Albany, NY 12222

R. A. Carpio and A. Ricci
SEMATECH, Austin, TX 78741

Fluorinated silicon dioxide films have been characterized using nuclear reaction analysis (NRA) and optical techniques. The use of prompt gamma-ray emitting nuclear reaction for measuring the concentration profiles of fluorine in thin films (>200 Å)of SiO_2:F is discussed. Films of SiO_2:F were studied using the $^{19}F(p,\alpha\gamma)^{16}O$ reaction at 340 keV resonance, by observing the emitted gamma rays in a 3 inch x 3 inch BGO detector. The film thickness, fluorine distributions and fluorine concentrations were determined from the NRA data. Film thickness was also determined from spectroscopic ellipsometery measurement and fluorine concentrations were also obtained from Fourier transform infrared spectroscopy (FTIR) data. For NRA analysis, the system has been calibrated using several ion implanted fluorine standards and well known fluorine compounds. This technique is now intrinsically capable of determining the fluorine content of any sample with a sensitivity of 100 ppm (atomic) without the use of reference standard samples. FTIR and NRA techniques are shown as a useful combination to study the changes in structure of SiO_2:F films.

INTRODUCTION

Plasma deposited silicon dioxide (SiO_2) has for years been widely used as the insulating material for on-chip device interconnect structures in integrated circuit (IC) fabrication. Undoped SiO_2 films have a dielectric constant of approximately 4.0, which is too high for sub-quarter-micron technology. The recent level of interest in SiO_2:F thin films has been driven by the semiconductor industry's need for reduced capacitance between adjacent metal conductors for advanced multilayer interconnect applications. The addition of fluorine into the SiO_2 matrix has been shown to reduce the dielectric constant of this material, while maintaining the mechanical and insulating properties required for successful integration (1). Films of SiO_2:F have been deposited using a variety of chemistries and commercially available plasma enhanced CVD reactors (2-5), providing uniform and reproducible processes over large substrate areas. Stability of the resultant films is of primary concern (6-7), depending strongly on the amount of incorporated fluorine. Continuing development requires measurement of fluorine content down to 1 or 2 atomic %.

In the past, quantitative analysis of fluorine (8-12) has been difficult due to its light mass; methods based on X-Ray or Auger emission do not work well. Also other techniques like Rutherford Backscattering (RBS)(13), particle induced X-Ray (PIXE) and neutron activation are only useful for light atoms within a heavier substrate if the concentration is high (several atomic percent). Secondary ion mass spectrometry (SIMS) needs standards of similar composition which are difficult to obtain. Nuclear techniques, such as nuclear reaction analysis (NRA), and optical spectroscopy, particularly Fourier Transform Infrared Spectroscopy (FTIR), are useful combination. The NRA technique can be used to determine absolute concentrations and can be used to provide depth profiling. The NRA technique, for example, has been used for measurements of hydrogen incorporation in silicon dioxide and silicon nitride thin films. In addition, the technique of neutron depth profiling has been used for accurate determination of boron in borophosphosilicate glass films (14). The FTIR technique is in widespread use in the semiconductor industry due to (a) low cost, (b) nondestructive nature, (c) fairly good spatial resolution, (d) sensitivity, and (e) ability to provide information on species which are chemically bound. Previous studies of SiO_2:F films are mentioned in ref. 15-17.

In this report, NRA, FTIR, and spectroscopic ellipsometry are used together to characterize fluorinated silicon dioxide films. These techniques provide a means to study the amount of fluorine incorporation as well as the manner in which the fluorine is incorporated. Changes in the structure of these films with processing can also be

CP392, *Application of Accelerators in Research and Industry*, edited by J. L. Duggan and I. L. Morgan
AIP Press, New York © 1997

ascertained. It is the purpose of this paper to illustrate the principles and applications of these techniques. Detailed comparison of different deposition methods will be the subject of future publication.

EXPERIMENTAL METHOD

The SiO_2:F thin film samples were deposited in a commercially available plasma CVD reactor. A mixture of tetraethylorthosilicate (TEOS), O_2 and C_2F_6 is introduced through a showerhead immediately above the silicon substrate. RF energy at a frequency of 13.56 MHz is applied to the showerhead to promote the reaction. RF energy at a lower frequency (100-400 kHz) is applied independently to the substrate to enhance film stability and step coverage. The amount of incorporated fluorine is controlled by the C_2F_6 gas flow rate relative to the other reactants. Higher fluorine content in the film was produced by increasing the C_2F_6 to O_2 flow ratio, whereas lower fluorine content was achieved by decreasing this ratio. A deposited thickness of 8000 Å was used for all samples.

The fluorine was measured using the nuclear reaction $^{19}F(p,\alpha\gamma)^{16}O$ at 340 keV, which has two times better resolution and fewer background problems than the resonance used in reference 17. The above nuclear reaction produces gamma-rays of energy 6.3 MeV (3%), 6.72 MeV (0.5%) and 7.12 MeV (96.5%). These photo peaks are well separated from background radiation and are measured by a 3-inch x 3 inch BGO detector mounted outside the vacuum chamber about 2cm directly behind the target. The chamber is the same as used for hydrogen profiling (18). The characteristic gamma-rays penetrate the vacuum wall of the analysis chamber with negligible absorption. The analysis chamber containing the sample is electrically insulated from the beam line by an insulating coupling. The sample is surrounded by a properly biased faraday cup for best beam integration. A proton beam from the Dynamitron Accelerator at the University at Albany was used for the analysis. For each sample an excitation curve was measured in steps of 1 or 3 keV, taking care to advance only in one direction, in order to avoid hysteresis effects in the analyzing magnet. The accelerator energy and the counting system at the Albany Accelerator Laboratory are computer controlled. The program automatically records gamma-ray yield vs. beam energy for a fixed charge. Typically, with 500 pico-Coulombs, at a current of 40 nA, a profile takes about 20 minutes to complete. In this resonance reaction, the gamma-rays come from the surface when the proton energy is equal to the resonance energy. As the proton energy is increased, the gamma-rays come from deeper within the target.

FTIR measurements were performed with a Nicolet ECO-8SN spectrometer using a DTGS detector. Measurements were performed at a resolution of $4cm^{-1}$. An uncoated Si wafer was used as the reference in order to remove the substrate spectral features. Measurements were performed on the as-deposited films, and again after a high humidity treatment which was conducted by sealing the 200 mm coated wafer in a plastic bag with DI water and holding at 100 deg. C for at least one day. Spectroscopic ellipsometry measurements were performed with a Woollam Variable Angle Spectroscopic Ellipsometer. The films were modeled using the optical constants for undoped SiO_2 films.

ANALYSIS

To convert the raw data (γ -ray counts for a fixed charge versus proton energy) into a fluorine concentration profile, depth X is given by

$$X = \frac{E - E_{res}}{dE/dX}$$

where E and Eres are the beam energy and the resonance energy (340 keV) respectively. The fluorine concentration ρ_F at this depth is given by

$$\rho_F = K\frac{dE}{dX} N_\gamma$$

where K is an experimental constant reflecting the $^{19}F(p,\alpha\gamma)^{16}O$ reaction cross section and the γ-ray detector efficiency. dE/dx is the energy loss rate of the incident ions in the target being analyzed and N_γ is the γ-ray counts.

There are two ways by which we have measured the value of constant K. The first method utilizes a known amount of fluorine implanted in a clean wafer of Si. We have used $1E16/cm^2$ fluorine implant at 100 keV, $5E15/cm^2$ fluorine implant at 100 keV and $1E16/cm^2$ fluorine implant at 300 keV as our standards. Fig. 1 shows one of these implants of fluorine in Si. The implant profile determined by us is shown with a solid line and is in fair agreement with the implanted profile obtained by TRIM (19). K is determined by comparing the measured total fluorine per square centimeter with the known amount.

The second method to determine constant K is to measure γ-ray yields from samples with known fluorine stoichiometry. Figure 2 shows the NRA data for a CaF_2 crystal. The ratio of fluorine to total atoms is correctly obtained as 0.67.

The depth resolution and range of depths profiled depend on the dE/dX of protons at the resonance energy in the target material. For CaF_2, the value of dE/dX = 0.108 MeV/micron is used (19). The resonance width is 2.4 keV. Hence, the near surface depth resolution is ~200Å. Figure 3 shows NRA data for measurement of fluorine in one of the three thin films of SiO_2 where oxygen is replaced by one atom of fluorine.

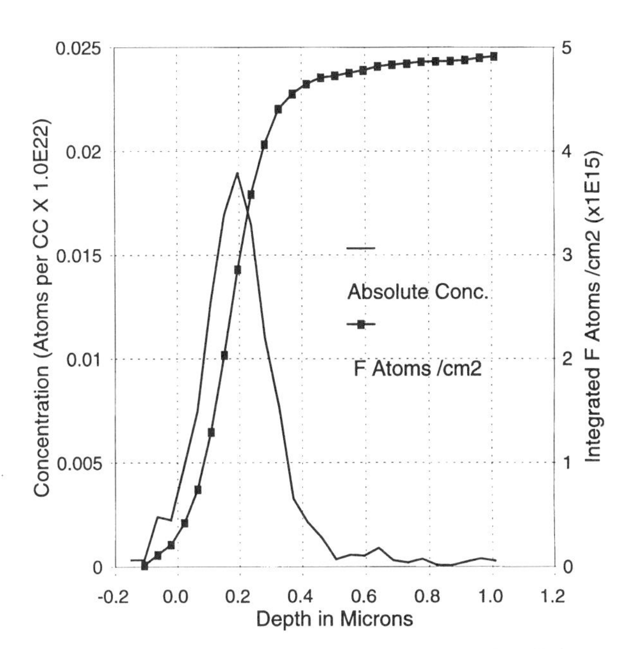

FIGURE 1. Fluorine concentration vs depth for 100keV 5E15 F implant in Si

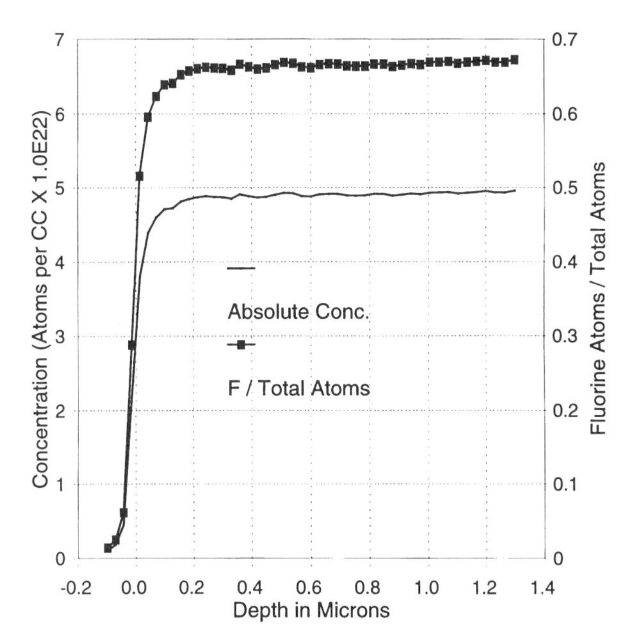

FIGURE 2. Fluorine concentration vs depth for thick CaF_2 crystal

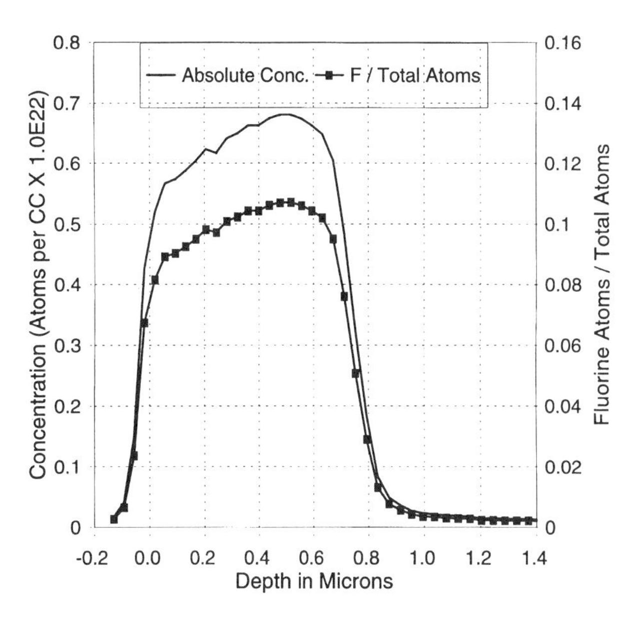

FIGURE 3. Fluorine concentration vs depth for SiO_2:F film

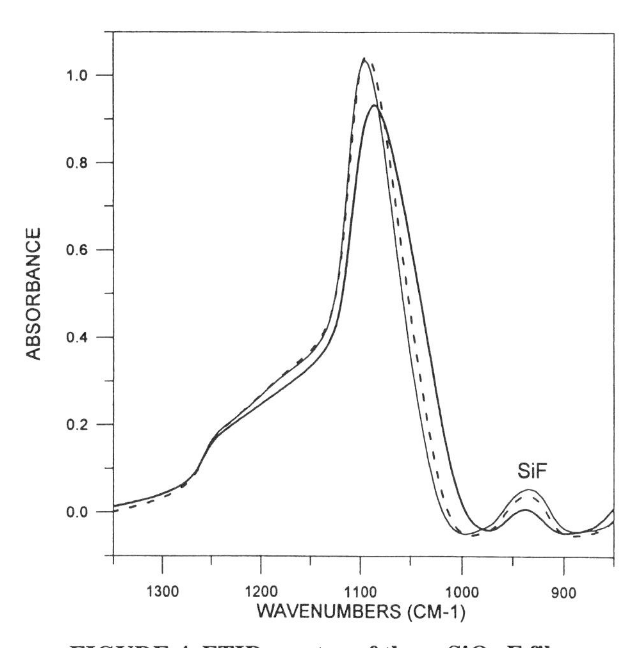

FIGURE 4. FTIR spectra of three SiO_2:F films

Similar plots were obtained for the other two samples showing various concentrations of fluorine.

An overlay of the FTIR spectra (determined at the wafer center) for the three films which were analyzed by NRA are shown in Figure 4. Figure 5 shows the relationship between the ratio of the integrated area of the Si-F band at 933 cm^{-1} to that of the asymmetric Si-O stretch vs. F concentration in atoms/cc. The use of such a ratio allows the normalization for thickness.

It was observed that the SiF/SiO ratio is also linearly related to the peak position of the asymmetric SiO stretch. The use of the peak position as a means to monitor the F content is obviously simpler than the use of the SiF ratio. This linear relation has been observed earlier (20).

Humidity treatments were performed on films with an SiF/SiO ratio of 0.033. Figures 6 and 7, respectively, reveal how the Si-F and SiO band area changes across the wafer as a result of humidification. The humidity treatment increases the SiOH band area and results in a decrease in the SiF content.

The mechanism for the loss of HF is thought to be

$$\text{-Si-F} + H_2O \text{ ------------------}> \text{-Si-OH} + HF$$

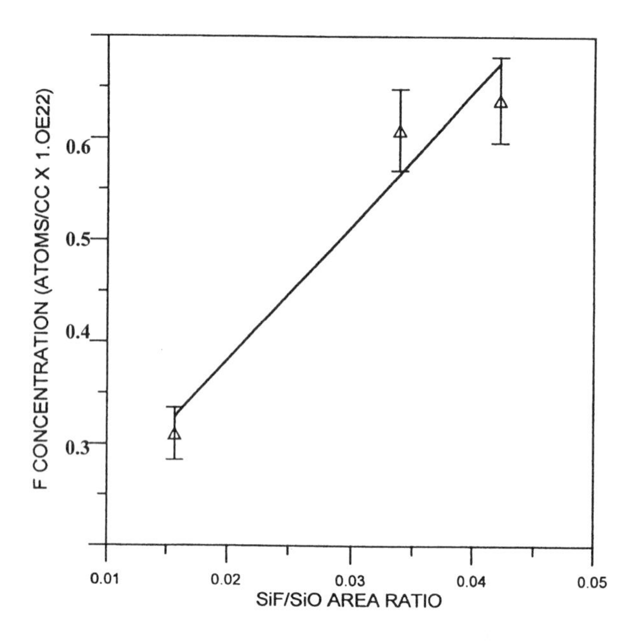

FIGURE 5. Fluorine concentration vs SiF/SiO area ratio

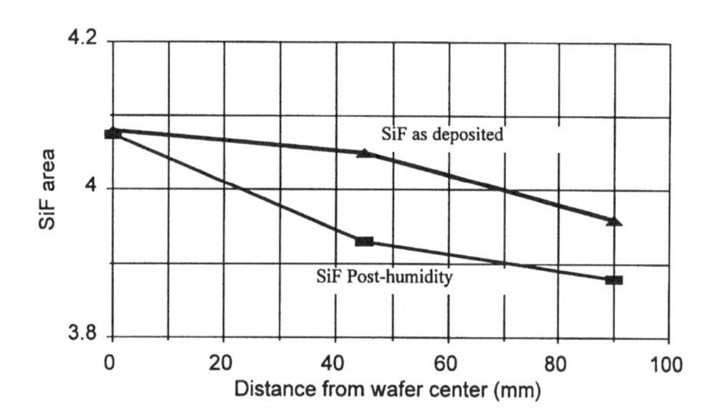

FIGURE 6. SiF spectral changes resulting from humidity treatment

These results emphasize the need for a capability to monitor the F as well as the -OH concentration in both process development and process monitoring applications.

ACCURACY OF THE MEASUREMENT

Concentration measurement by NRA depends on the stability of the experimental apparatus, the accuracy of the dE/dX used, and on the statistical error in counting. In the present work, the experimental apparatus is repeatable within 7%. The dE/dX values were obtained from STOP (19), and are estimated to be within 10% (21). Statistical variation is typically less than 1%. Total error is thus within about 12% for concentrations at or above 0.1 atomic percent. Base noise in the measurement is at about 0.01 atomic percent, which limits the sensitivity to 100 ppm. In contrast, base noise in comparable RBS measurements is about 3 atomic percent.

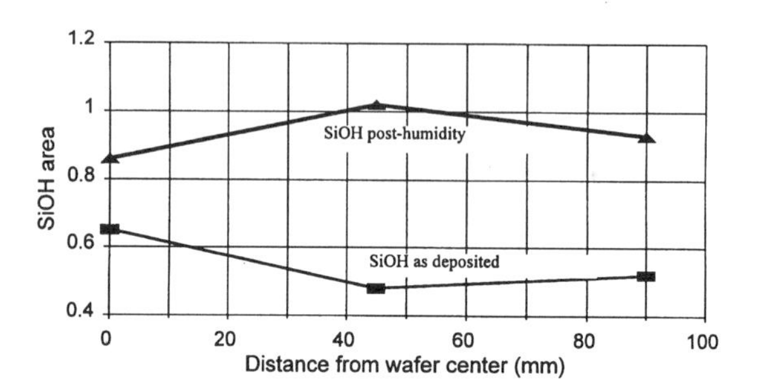

FIGURE 7. SiOH spectral changes resulting from humidity treatment

CONCLUSIONS

The NRA and FTIR measurements are shown to be very useful in characterizing SiO_2:F films. Future work will utilize this new NRA technique to develop a more complete FTIR quantitative procedure for SiOF films, based upon a partial least squares chemometric approach. The goal of this future work will be to utilize the entire FTIR spectrum to simultaneously derive fluorine concentrations, film thickness, and dielectric constant.

REFERENCES

1. Carl, D., et. al., *Proceedings of DUMIC* (1995), pp 234-240.
2. Fukada, T. and Akahori, T., *Proceedings of International Conference on Solid State Devices and Materials* (1993), pp. 158-160.
3. Matsuda, T., Shapiro, M., and Nguyen,, S., *Proceedings of DUMIC* (1995), pp. 22-28.
4. Hayasaka, N., et. al., *Proceeding of Dry Process Symposium* (1993), pp. 163-168.
5. Qian, L., et. al., *Proceedings of DUMIC* (1995), pp. 50-56.
6. Carl, D., et. al., *Proceedings of VMIC* (1995), pp. 97-103.
7. Pai, C. S., et. al., *Proceedings of VMIC* (1995), pp. 406-408.
8. Coote, G.E., *Nucl. Instr. And Meth.* **B66**, 191(1992).
9. Bird, J.R., *Nucl. Instr. And Meth.* **168**, 85(1980).
10. Zironi, E.P., Richards, J., Maldono, A., and Asomoza, R., *Nucl. Instr. And Meth.* **B45**, 115(1990).
11. Deconnick, G., and Oystayen, B. V., *Nucl. Instr. And Meth.* **218**, 165(1983).
12. Bodart, F., and Deconnick, G., *Nucl. Instr. And Meth.* **197**, 59(1982).
13. Bakhru, H., *Encyclopedia of Material Science and Engineering*, Edited by Michael B. ,Bever, Peragamon Press 1990, pp. 2402-2407.
14. Zeitzoff, P. M., et. al., *J. Electrochem. Soc.* **137**, 3917-3922(1990).
15. Homma, T., *J. Electrochem. Soc.* **143**, 707-711 (1996).
16. Yeh, C. F., Chen, C. L., Lur, W., and Yen, P. W., *Appl. Phys. Lett.* **66**(8), 938-940 (1995).
17. Sapro, A., Denison, D. R., and Lam, J., *Proceedings of DUMIC* (1996), pp. 239-246.
18. Lanford, W., *Handbook of Modern Ion Beam Material Analysis*, edited by Michael Natasi, Materials Research Society 1995, ch. 8.
19. Ziegler J.F., and Biersack, J.P., *TRIM, The Stopping and Range of Ions in Solids*, Pergamon Press, New York 1985.
20. Miyajima, H., Katsumata, R., Nakasaki, Y., Hayasaka, N., and Okano, H., *Proceedings of VMIC* (1995), pp. 391-393.
21. Alfassi, Z.B., Peisach, M., *Elemental Analysis by Particle Accelerators*, CRC Press, (1992), p. 386.

HYDROGEN DEPTH PROFILING AT IRMM

L. Persson, M. Hult, G. Giorginis and A. Crametz

CEC-JRC-IRMM, Retieseweg, B-2440 Geel, Belgium

At the Institute for Reference Materials and Measurements (IRMM) a facility for hydrogen profiling has been implemented. Two techniques are being used; Nuclear Resonant Reaction Analysis (NRRA) using the $^{1}H(^{15}N,\alpha\gamma)^{12}C$ reaction and Elastic Recoil Detection Analysis (ERDA) using a He^{+} ion beam. The ERDA chamber is a common type of scattering chamber with a load lock system for quick changing of the samples. NRRA is performed in a UHV chamber in which the samples can be cooled to liquid nitrogen temperature and sputter cleaned with a 5 kV ion gun. The 4.43 MeV γ rays are detected using a 4 in.×4 in. BGO detector. Using only passive shielding a detection limit of hydrogen in silicon of 5×10^{18} atoms/cm^3 has been achieved. Experimental results which exemplify the major features of the facility are presented.

INTRODUCTION

Hydrogen is one of the most common contaminant elements and has important effects on the chemical, physical and electrical properties of many materials. It can for example passify both donors and acceptors in semiconductors. The need for quick and reliable hydrogen analysis is an essential tool in materials science in order to optimize e.g. production processes.

Nuclear Resonant Reaction Analysis (NRRA) and Elastic Recoil Detection Analysis (ERDA) are two nuclear techniques using MeV ion beams that are well suited to profile hydrogen. Nuclear techniques have the important feature that they are insensitive to matrix effects, except through channeling and stopping cross section, which makes it possible to obtain quantitative results without having to rely on standards with a similar composition which is the case for secondary ion mass spectrometry. Excellent descriptions of the NRRA and ERDA techniques can be found in (1) and (2) and the references therein.

This paper describes the implementation of NRRA and ERDA at the Institute for Reference Materials and Measurements (IRMM) and presents experimental results which exemplify the major features that the facility offers.

EXPERIMENTAL SET-UPS

Accelerator

The accelerator used to produce both the He^{+} and the $^{15}N^{+}$ ion beams is a 7 MV vertical single ended CN Van de Graaff. For profiling measurements using the NRRA technique the $^{1}H(^{15}N,\alpha\gamma)^{12}C$ reaction with resonance energy 6.385 MeV is used. The fact that the accelerator produces singly charged ^{15}N ions means that the accessible depth is limited to approximately 200 nm in Si and 130 nm in InP. In order to extend the profiling depth, work is currently being undertaken to improve the ion source for the production of multiply charged ions.

The NRRA Set-up

A schematic drawing of the target chamber is shown in Fig. 1. The detector is located at 0° with respect to the beam line and a CCD-camera, a 5-kV ion gun, a mass spectrometer, a sorption pump, a movable Si particle detector and a movable Faraday cup are connected to the various side ports of the chamber. On the top flange is a high precision XYZ translator mounted onto which the sample holder system is connected. The ion gun is intended to be used for cleaning the sample surface and to produce implanted samples.

In order to produce a differentially pumped system a 12 mm aperture is located 1.4 m from the chamber and a second 10 mm aperture is located 0.4 m from the chamber. The pressure between the two apertures is better than 10^{-6} Pa and obtained using a 500 l/s turbo molecular pump. The total volume after the second aperture is approximately 6 dm^3 and in order to obtain UHV in this large volume the vacuum system consists of a 230 l/s turbo molecular pump backed by an oil free membrane pump, a 600 l/s ion pump, a Ti sublimation pump and a liquid nitrogen cryoshroud. A sorption pump is connected to the chamber via a leak valve which makes it possible to pump very slowly from atmospheric pressure

CP392, *Application of Accelerators in Research and Industry*, edited by J. L. Duggan and I. L. Morgan
AIP Press, New York © 1997

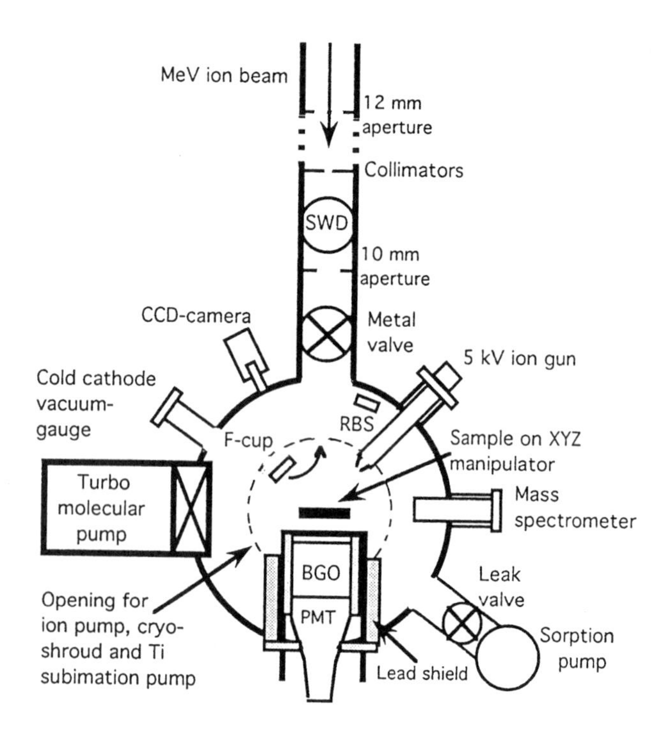

FIGURE 1: Schematic drawing of the UHV chamber used for ^{15}N profiling of hydrogen.

to pre-vacuum. This feature is expected to be useful when studying hydrogen terminated surfaces because a quick roughing of the system may destruct the formed hydrogen layers.

The system is normally under vacuum and after opening to insert new samples a vacuum of 10^{-6} Pa can be reached without bakeout and 10^{-8} Pa after bakeout.

Detector System

The detector employed to measure the 4.43 MeV γ rays emitted at the resonance energy is a 4 in.×4 in. BGO detector. It is positioned near the center of the chamber using a stainless steel holder which has the shape of a tube and a thickness of 4 mm. Placing the detector at 0° has the advantage of being in the optimum direction of γ-ray emission from the ^{1}H(^{15}N,$\alpha\gamma$)^{12}C reaction (3). Unwanted beam induced γ rays can be reduced by using heated tantalum collimators (4) or by locating a beam collimator far away from the sample. The second alternative has been chosen here and the distance between the beam collimator and the detector is 1200 mm. The collimator is made of tantalum with a backing of oxygen free copper and there are five choices regarding the size; 2, 3, 5, 8 and 10 mm.

The γ-ray energy region of interest is $\sim$ 3.7 - 4.7 MeV which encompasses the full energy and the single escape peaks. Without any shielding the background count rate in this energy region is 0.16/s. In order to reduce this background, which arises mainly from cosmic rays (5), passive shielding in the form of a two lead shields has been used. The two shields reduce the background level with approximately 36 %. A more efficient background reduction can be accomplished by combining passive shielding with an anticoincidence shielding unit (4),(6), consisting of plastic scintillator material surrounding the detector. This can reduce the background further by a factor four (6). With the present shielding it is possible to reach a detection limit of approximately 5×10^{18} atoms/cm^3 or 100 at.ppm in Si.

Sample Holder System

The sample holder system consists of a sample stick made from oxygen free copper, vacuum bonded to a stainless steel tube which serves as a liquid nitrogen container. It is connected to a differentially pumped rotary feedthrough via a ceramic break. This is all mounted on top of the XYZ translator. Samples up to 18 mm wide can be accommodated and depending on the length of the samples it is possible to mount several samples at one time since samples can be mounted on both sides of the sample holder which is 200 mm long. The minimum distance between the detector and sample is 15 mm and the total attenuation of 4.43 MeV γ rays on the way through the sample holder and detector holder is 14 %.

Dose Measurements

The incident beam dose is measured using a spinning wire dosimetry system similar to the system described by Musket *et al.* (7). It consists of two spinning W-wires that rotate with a frequency of 25 Hz and the backscattered particles are collected with a Si particle detector. The yield is calibrated with the help of a Faraday cup that can intercept the beam in front of the sample. The backscattered particles from the sample are also measured in order to control the accuracy of the spinning wire system which can be quite sensitive to a change in the beam profile or a movement of the beam. This sensitivity is believed to be due to the fact that the geometry of the two rotating wires is quite complex. A typical beam current is 100 nA.

The ERDA Set-up

The ERDA set-up consists of a movable surface barrier detector to measure the recoils located at an angle of 30° with respect to the beam line. In front of the detector sits an absorber foil to prevent forward scattered particles from reaching the detector. An RBS detector is positioned at 165°. The ^{4}He$^+$ beam with a typical energy of 2.7 MeV and a beam current of approximately

70 nA, hits the sample at 15° glancing incidence. The charge is monitored by counting the number of particles backscattered from a thin layer ($\sim$ 50 nm) of Au that sits on a Si backing mounted on a pendulum. The pendulum intercepts the beam with a frequency of about 1 Hz. The indirect charge measured with the pendulum is calibrated with a Faraday cup.

The system is made up of UHV compatible components and pumped by a 500 l/s turbo pump. A typical vacuum during measurement is 10^{-5} Pa. The detection limit of the system has been checked using hydrogen implanted Si standards and is better than 10^{15} atoms/cm^2. This corresponds to better than 7×10^{19} atoms/cm^3 or approximately 1400 at.ppm in Si.

MEASUREMENTS

Quantitative Measurements

In order to obtain absolute hydrogen concentrations with the NRRA system an amorphorized Si sample implanted with 1×10^{16} 10 keV H^+ ions cm^{-2} is used. Using this standard the hydrogen concentration can be established via the determination of the material independent calibration factor (8). The major uncertainties in this procedure stem from the uncertainty of the implanted dose and the uncertainty to which the implanted area is measured during the hydrogen profiling measurement.

The on-resonance to off-resonance cross section ratio is about 20800 (3) but even so it can be necessary to correct for off-resonance contributions. Off-resonance contributions are especially a concern when measuring low concentrations of hydrogen in the interior of a sample when the overlaying layers contain a significant amount of hydrogen. Off-resonance correction is carried out using the correction formula established by Horn and Lanford (3).

Although it is not necessary to use standards to determine absolute concentrations with the ERDA technique, the results obtained are always checked against one of two amorphous Si samples implanted with 35 keV H^+ ions at doses of 10^{16} and 10^{17} ions cm^{-2} respectively.

Results

Examples of hydrogen analysis at IRMM are shown in Figs. 2 to 4. Figure 2 shows a hydrogen profile of an a-Si/SiO_2 structure obtained using NRRA. The good depth resolution of the method, $\sim$ 10 nm, makes it possible to discern the surface peak from the high level of hydrogen in the amorphous Si and also to clearly see an enrichment at the interface between the Si and the underlying SiO_2. Three measurements were performed on this sample and as can be seen in Fig. 2 they all

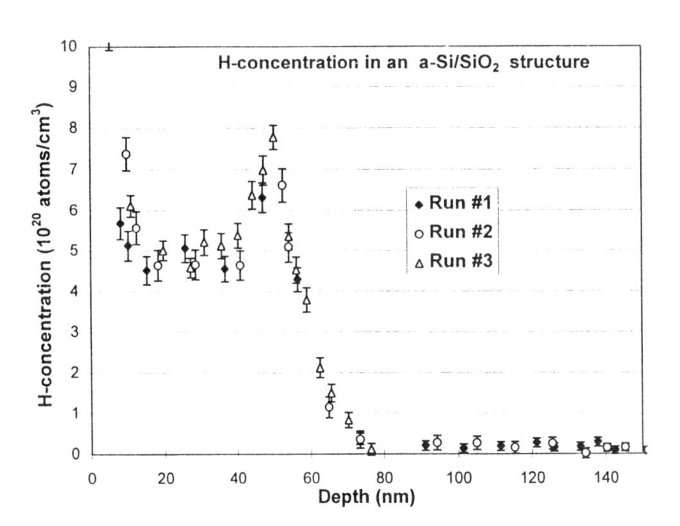

FIGURE 2: Hydrogen depth profiles from three different measurements of an a-Si/SiO2 structure obtained using NRRA. The surface peak, which is located at 0 nm and cut off in the figure in order to show the rest of the profile, is followed by a high concentration of hydrogen in the a-Si layer. At the interface there is a clear enrichment of hydrogen followed by a very low concentration in the oxide layer.

yielded the same result. This shows the robustness of the method and one can also conclude that there is no loss of hydrogen during the measurements.

ERDA profiles of blank InP, InP implanted with 100 keV 5×10^{15} atoms/cm^2 hydrogen and one sample that was annealed in an N_2 atmosphere at 300 °C for 20 min. after implantation are shown in Fig. 3. In order to reach a depth of 1600 nm the beam energy was set to 4.45 MeV. At this energy (α,p) reactions with phosphorus are rather pronounced and therefore the spectra quite complex making it necessary to run a blank sample. The implanted peak is clearly seen and after annealing the

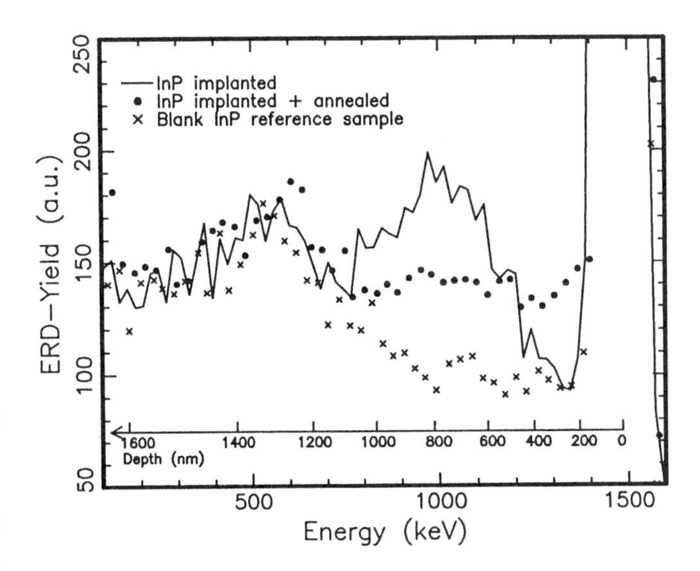

FIGURE 3: Hydrogen depth profiles obtained by ERDA using 4.45 MeV He+ ions on three different InP samples. Note that the yield scale has been cut off.

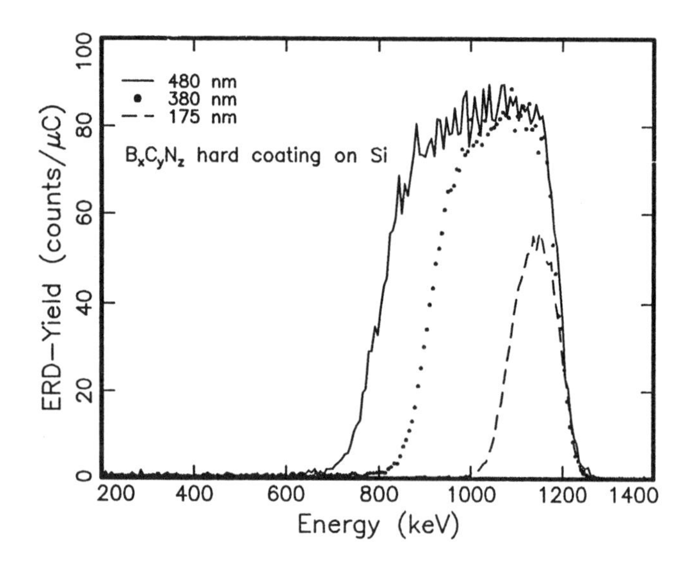

FIGURE 4: ERDA measurement using 2.7 MeV He^+ ions of three $B_xN_yC_z$ covered Si samples containing a high and homogeneous concentration of hydrogen.

hydrogen is smeared out with an enrichment near the surface. The near surface hydrogen profile ($\sim$ 130 nm), a problem for ERDA due to its limited depth resolution, was studied using NRRA.

The third example, Fig. 4, shows the energy spectra of recoil protons from three different Si samples covered with thin layers of $B_xN_yC_z$ that were analyzed for hydrogen using ERDA at 2.7 MeV. The hydrogen content is very high and homogeneous throughout the layer making ERDA a suitable method. No loss of hydrogen was registered during ERDA measurements. When performing NRRA, the ^{15}N beam caused an initial loss of hydrogen making accurate quantification difficult.

DISCUSSION AND CONCLUSIONS

The two set-ups installed at IRMM for hydrogen measurements are somewhat complementary in nature. Presently it is possible to reach a detection limit of about 5×10^{18} atoms/cm^3 or 100 at.ppm in Si using NRRA and better than 7×10^{19} atoms/cm^3 using ERDA. One advantage of the ERDA technique compared to NRRA is that the commonly used 2.7 MeV He beam usually does not induce any noticeable beam damage in electronic materials, however the depth resolution using ERDA is worse compared with NRRA.

A new target holder has been ordered for the NRRA system which will increase the maximum sample width to 55 mm. The movement in Z will be increased which will make it possible to change samples through an interlock system. In order to reduce the background further it is planned to install an anticoincidence shielding unit consisting of plastic scintillator material. When the active shielding unit is in place the background should be reduced considerably and the detection limit should approach that of other low-level detection systems currently in use, for example (4),(6), which have detection limits in the 10 at.ppm range.

A sample holder system similar to the one used for NRRA has been ordered for ERDA consisting of a motorized differentially pumped rotary feedthrough, a high precision XYZ translator and the same type of sample holder. These improvements will make it possible to cool the samples, facilitate sample handling, and lead to a more accurate determination of the target tilt angle compared with the present system.

ACKNOWLEDGEMENTS

The authors would like to thank Dr. Serge Mathot and Dr. Pascal Courel for their efforts in the starting-up phase of the project. The work of Victor Garcia Fernandez is also acknowledged.

REFERENCES

1. Tesmer, J. R., and Nastasi, M., editors, *Handbook of Modern Ion Beam Analysis,* Material Research Society, 9800 McKnight Road, Pittsburgh, PA 15237 USA, 1995, Chapters 5 and 8.
2. Lanford, W. A., *Nucl. Instr. and Meth.,* **B66,** 65-82 (1992).
3. Horn, K. M., and Lanford, W. A., *Nucl. Instr. and Meth.,* **B34**, 1-8 (1988).
4. Torri, P., Keinonen, J., and Nordlund, K., *Nucl. Instr. and Meth.*, **B84**, 105-110 (1994).
5. Damjantschitsch, H., Weiser, M., Heusser, G., Kalbitzer, S., and Mannsperger, H., *Nucl. Instr. and Meth.*, **218**, 129-140 (1983).
6. Endisch, D., Sturm, H., and Rauch, F., *Nucl. Instr. and Meth.*, **B84**, 380-392 (1994).
7. Musket, R. G., Daley, R. S., and Patterson, R. G., *Nucl. Instr. and Meth.*, **B83**, 425-429 (1993).
8. Lanford, W. A., *Solar Cells*, **2**, 351-363 (1980).

SCANNING ION BEAM MICROSCOPY: A NEW TOOL FOR MAPPING THE TRANSPORT PROPERTIES OF SEMICONDUCTORS AND INSULATORS

C. Manfredotti*, F. Fizzotti*, P. Polesello*, E. Vittone*, M. Jaksic‡, I. Bodganovic‡, V. Valkovic‡

*Università di Torino, Dip. Fisica Sperimentale, INFN Sez. Torino, INFM Unità Torino, Via P.Giuria 1 I-10125 Torino (I)
‡Ruder Boskovic Institute, Laboratory for Nuclear Microanalysis, P.O.Box 1016, 41001 Zagreb, (HR)

Proton microbeams of energy from 3 to 5 MeV have been used to investigate the cross section of Si, CdTe and CVD diamond samples by recording the charge pulses delivered at the electrodes by the single protons with a charge-sensitive electronic chain. The investigated depth varies from 50 to 140 μm depending on the proton energy and on the material. In the case of an homogeneous material (Si, CdTe), lifetime and mobility values can be obtained, together with the in-depth electrical field profiles. For polycrystalline materials, the maps of collection efficiency can be correlated with morphological maps putting in evidence a columnar-like structure due to the film growth mechanism. In these cases, it is relatively difficult to separate the effects due to the electric field from the carriers transport properties (mobility, lifetime). Anyway, maps of collection length are quite important in order to detect the electrical inhomogeneities of polycrystalline materials.

INTRODUCTION

Micro-IBIC (Ion Beam Induced Current) is a powerful method for a microscopic investigation of p-n junctions around drain and source of MOS structures, in order to locate particular electrical features under the metallization and passivation layers (1). The advantages of a long penetration depth, of the order of 100 μm in Si for 3 MeV protons, and of the low proton scattering, with respect to EBIC (Electron Beam Induced Current) (2) are quite clear. However, samples are not homogeneous in depth, and it is consequently not easy to separate the effects of the local electric field from the effects due to the local physical properties of carriers. In this paper we introduce a new kind of technique, which we call "lateral micro-IBIC", in which the sample is homogeneous, it is equipped with electrodes in a planar capacitor geometry and it investigated not frontally, through the electrodes, but laterally, along a cross section limited by the electrodes themselves. At high electric fields, generated carriers recombination can be excluded and charge pulse height can be related to the local value of what is called "charge collection length" (3,4). In such a way, a direct map of the electric field as a function of depth is possible, and values of (mobility)x(lifetime) product both for electrons and holes can be obtained (5).

If the sample is not homogeneous (i.e. is polycrystalline) and if the collection length is much lower than the sample thickness (i.e. the whole travel path per carriers), local values of the collection length can be obtained and displayed in a two-dimensional map. Of course, there are average values along the particle penetration depth and they can be influenced, for short penetration depths, by the surface properties. However, if energies are large enough and the surfaces are cleaved, these effects can be excluded and the collection length maps can give a good representation of the "electronic morphology" of the material. This is the particular case of CVD diamond.

EXPERIMENTAL

The micro-IBIC measurements were carried out at Ruder Boskovic Institute of Zagreb, by using the proton microbeam facility. The proton microbeam, of energy from 3 MeV to 5 MeV, was focused onto a spot of 2÷3 μm size, in order to scan a region of maximum 2 mm x 2 mm of a cross section of the sample, which was either finely polished (Si, CdTe) of cleaved (CVD diamond). In order to avoid fringing fields effects, electrodes (generally Au, differently obtained) were drawn close to the borders of the investigated surface. A standard sensitive electronic chain was used, connected to a PC: for mapping purposes, files containing the beam spot coordinates and the charge pulse height were recorded. By displaying collected charge spectra, noise level was detected and a threshold for pulse recording was set. Pulse height spectra could be recorded for each pixel, for a row extending from one electrode to the other one and for a whole (x,y) region. ROI's could be also placed in order to map regions of collection efficiency values in a certain range. Pulse counting rate was kept very low (about 100 Hz) in order to avoid not only radiation damage, but also the effect of the space-charge created by

CP392, *Application of Accelerators in Research and Industry*, edited by J. L. Duggan and I. L. Morgan
AIP Press, New York © 1997

the carriers locally trapped (at the surface of Si, CdTe, in the bulk for CVD diamond). Of course, fluctuations of the proton flux cannot be avoided for currents much lower than 1 fA.

RESULTS AND DISCUSSION

Silicon

Silicon is a very good material to start investigations on electric field distribution, since junction behavior is well known and junction devices fabrication is very

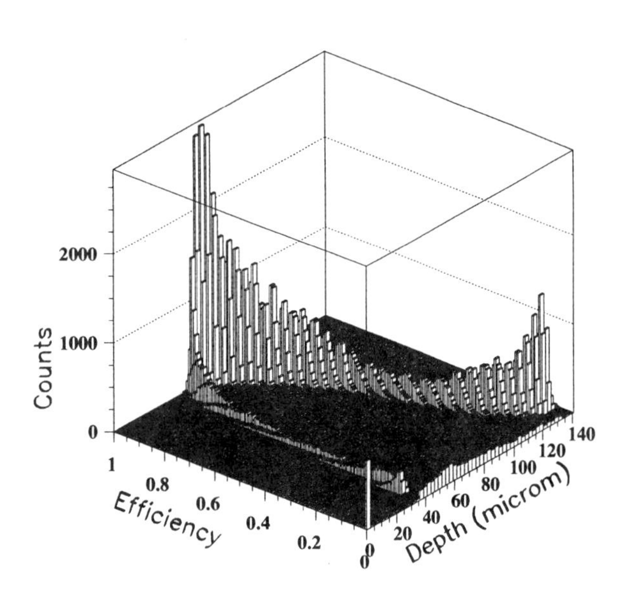

FIGURE 1. Charge collection efficiency distribution along the depletion region of a p^+-n-n^+ Si structure biased at 100 V. The three dimensional plot is given as a function of position (front axis, p^+ region on the left), while the lateral axis is used to give a measure of the broadening of charge collection efficiency in each point.

reproducible. We used a silicon power device, with a structure p^+-n-n^+ and with a very abrupt and quite deeply lying p^+-n junction at about 40 μm below upper metallization. The device was cut and the cross section was finely polished. The depletion region of the device was scanned along a line from the upper contact down to a depth of 200 μm, for reverse bias of 100, 200, 300 V and also at zero bias. It was very interesting to follow the expansion of the depletion region with increasing bias voltage, clearly indicated by the movement of the charge collection efficiency distribution towards the bottom of the device. At 100 V (see Fig. 1) this distribution is quite large, reaching the value of 100% at 60 μm and extending

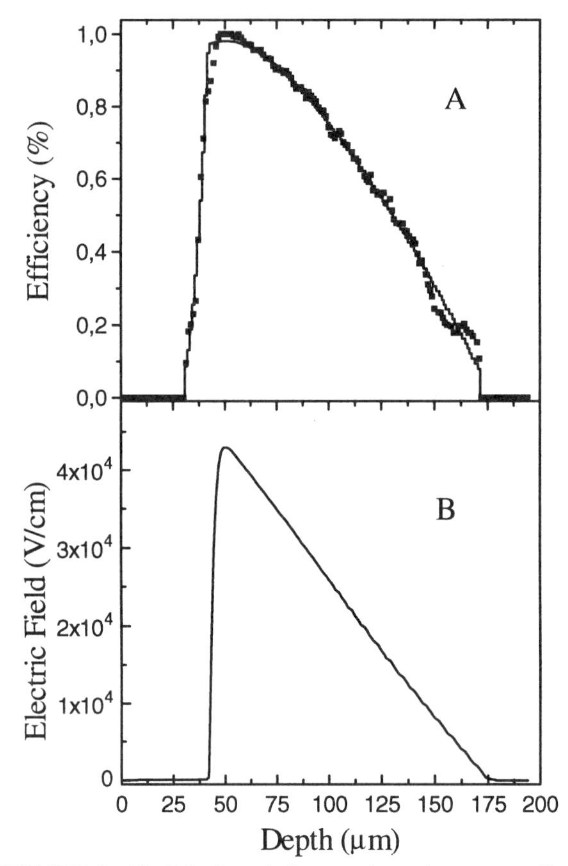

FIGURE 2. (A) Behavior of the experimental average collection efficiency in the depleted region of a Si p^+-n-n^+ structure, together with the result of a theoretical fit carried out with a triangle-shaped electric field distribution (see Fig. B). (B) Behavior of the electric field distribution which has been used to fit experimental collection efficiency data (see Fig.A)

down to 180 μm, and it is characterized by quite small fluctuations, with a clear indication of a good homogeneity of the material properties along the proton track.

The classical triangular distribution of the electric field is a very good approximation in this case (see Fig. 2B), since the fit with experimental data (Fig. 2A), after a deconvolution of the two contributions of electrons and holes to the collection efficiency, is excellent. The dopants profile was obtained by the spreading resistance method and it was used to calculate the mobility profile for electrons and holes and the electric field profile used in the fit. The free parameters were the electrical depth of the depletion region and the trapping times for electrons and holes: the former one followed the theoretical relationship with bias voltage, while the latter two turned out to be 20 ns and 0.6 ns respectively. The maximum value of charge collection length, given by the product among mobility μ, electric field E and trapping time τ was 3 mm for electrons and 30 μm for holes. The transit times across the depletion region were of the order of 1 ns for electrons and 2 ns for holes, accounting this last value for the relatively low collection efficiency for holes (about 18% maximum).

CdTe

The problem of CdTe as x-ray detector is given by the electrical contacts and by the non-homogeneity of the

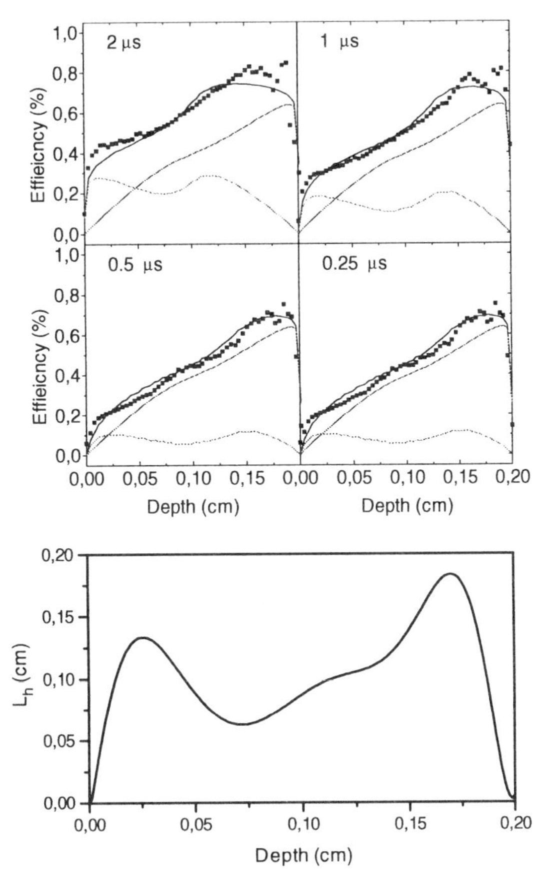

FIGURE 3. Experimental (points) and fitted (curves) distributions of charge collection efficiency along the depth of a CdTe sample 2 mm thick biased at 100 V. The four upper figures refer to data obtained for different shaping times (2 μs, 1 μs, 0.5 μs and 0.25 μs respectively). The contribution of electrons (right peaked curve) and holes (lowest curve) to the total collection efficiency is shown. At the bottom, the calculated behavior of the hole collection length (L_h, in cm) as a function of depth is shown. For constant values of μ and τ, the same behavior can be attributed to the electric field.

electric field, which may give rise to low collection efficiency for holes, characterized by low mobilities (80 $cm^2V^{-1}s^{-1}$). Since the transit time for holes across a detector may be relatively long (about 2.5 μs at V=200 V for a 2 mm thick detector), attention should be paid to electronic shaping time, in order not to have a ballistic deficit for holes. Figure 3 shows four experimental profiles of the large collection efficiency, as measured for different shaping times and deconvoluted according to the different contributions of holes and electrons, by assuming constant mobility and lifetime, which, according to the fit, turned out to be 80 $cm^2V^{-1}s^{-1}$ and 2.4 μs for holes, 800 $cm^2V^{-1}s^{-1}$ and 0.3 μs for electrons respectively. These data are in agreement with literature (6). The progressive decrease of the hole collection efficiency is quite evident, particularly for electrons.

CVD diamond

CVD diamond is known to be polycrystalline with a columnar structure along the direction of growth (3). In this case, both the electric field and the transport properties of carriers (mobility, trapping time) are certainly not homogeneous and maps of collection efficiency may show interesting details (see fig. 4).

The columnar structure is evident also in all the maps, confirming that single crystals (but not all of them) display much larger collection efficiencies with respect to the rest

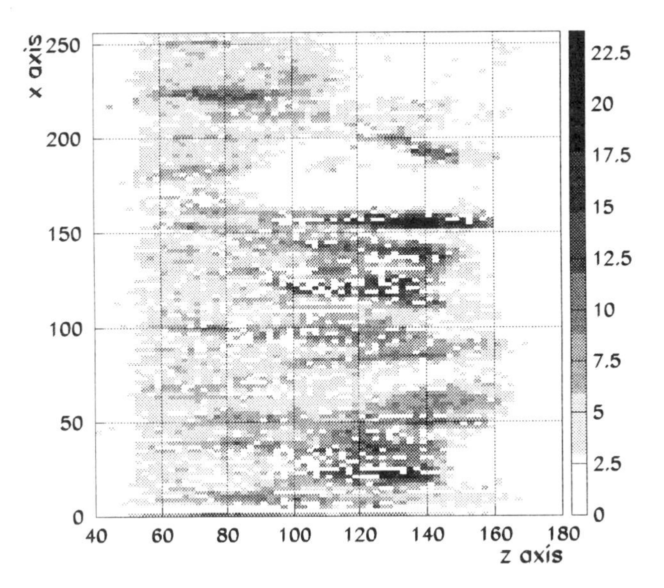

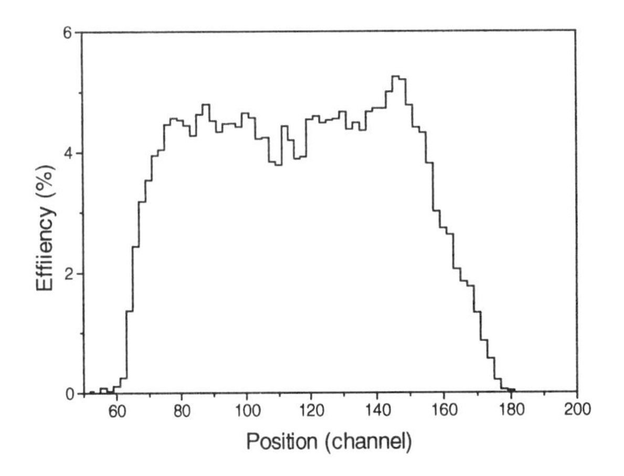

FIGURE 4. (Top) Map of the average charge collection efficiency across a section of a CVD diamond sample 400 μm thick. Electrodes are approximately at channel z=52 and z=160 (growth side). Bias is -750 V (on the left electrode). The average collection efficiency is displayed according to a gray scale (on the right, in percent). (Bottom) Distribution of the average collection efficiency as a function of depth for the same region (top). The average has been calculated from all the pixels with the same z coordinate.

of the sample, that these are larger in the growth side (on the right), and that there are also regions in which the collection efficiency is very low, because of a lower electric field and/or lower mobilities and lifetimes. For low collection efficiencies, the maps can be interpreted as collection length maps since

$$\eta = \frac{Q}{Q_0} \cong \frac{(\mu\tau\mathbf{E})_e + (\mu\tau\mathbf{E})_h}{d} \tag{1}$$

where η is the collection efficiency, Q and Q_0 are the collected and deposited charges respectively, **E** is the local value of electric field, d the sample thickness and the subscripts e and h refer to electrons and holes respectively. The mean efficiency in this region (Fig. 4, bottom) can be considered as an average behavior of the electric field, multiplied by the average of $(\mu\tau)_{e,h}$. The average electric field, therefore, is relatively constant, apart from a decay at the right-hand side, due to big granularity of the growth surface (the sample is not polished). It is also interesting to look at residual electric field behavior, when the bias is set to zero (Fig. 5): this is due to a polarization effect produced by trapped carriers and, also, to the relatively large $(\mu\tau)$ values at the growth surface of the film. In the "strips" of the collected maps (Fig. 6), there are likely regions displaying relatively large values of collection efficiency; for instance, in row 225, there is a region in which the collection efficiency is larger than 40%. Since the sample thickness is 400 µm, the corresponding collection length is about 160 µm, i.e. much larger than in natural IIa diamond. A fact like this could be easily attributed to the much lower nitrogen content in CVD diamond (N impurities give rise to trapping centers), and to the perfection of a part of these crystalline columns. It is also interesting to note that strips of this "polished" sample are now more central than in the unpolished one (Fig. 4). For polished samples, the electric field penetrates deeper into the sample, being the surface homogeneously flat, without dips and peaks. There are also much wider regions

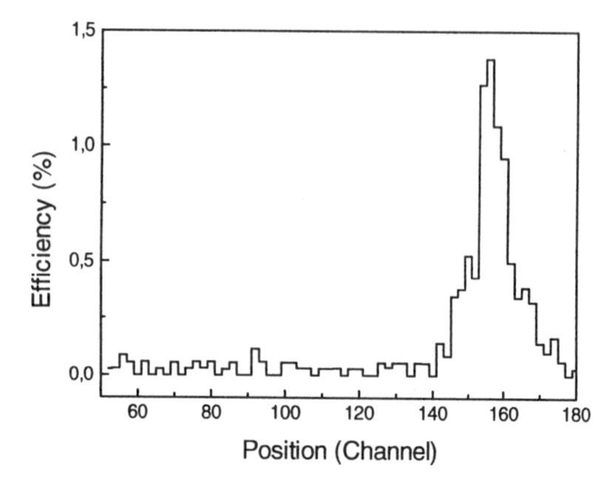

FIGURE 5. Distribution of the average charge collection efficiency as a function of depth for a CVD diamond sample, with the bias reset to zero.

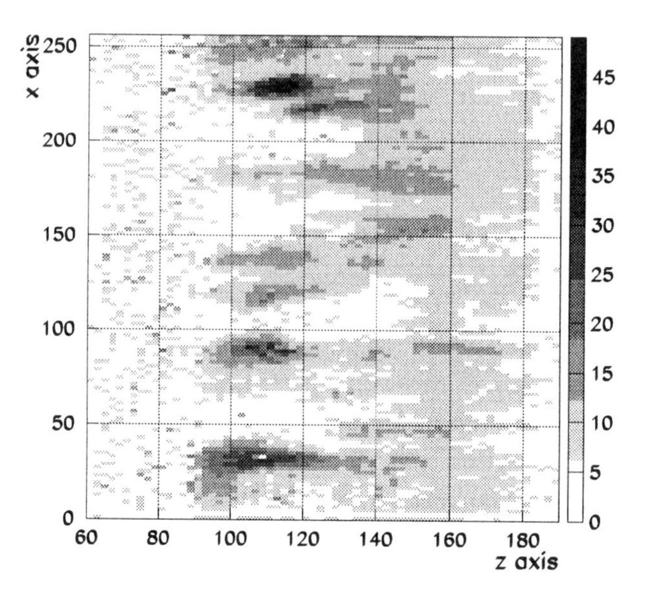

FIGURE 6. Map of the average charge collection efficiency across a section of a CVD diamond sample 400 µm thick biased at -600V at the left electrode. For details of representation, see Fig. 4.

of lower efficiency and/or of lower electric field: it could be interesting to explore these regions as a function of shaping time, by collecting charges coming also from sites with low electric field.

CONCLUSIONS

A particular version of micro-IBIC can be successfully used to map the charge collection length both in homogeneous and in non-homogeneous samples. A new kind of ion microscopy for mapping electric fields, mobility and lifetime of carriers inside bulk samples and devices can be easily envisaged.

REFERENCES

1. M.B.H.Breese,G.W.Grime,F. Watt, *Nucl. Instr. Meth. In Phys. Res.* **B77**, 301 (1993)
2. H.J. Leamy, *J. Appl. Phys.* **53**, R51 (1982)
3. C.Manfredotti, F.Fizzotti, P.Polesello, E.Vittone, F.Wang, *Phys. Stat. Solidi (a)* **154**, 327 (1996)
4. C.Manfredotti, F.Fizzotti, E.Vittone, M.Boero, P.Polesello, S.Galassini, M.Jaksic, S.Fazinic, *Nucl.Instr.Meth.Phys.Res.***B109** 555 (1996)
5. C.Manfredotti, F.Fizzotti, P.Polesello, P.P.Trapani, E.Vittone, M.Jaksic, S.Fazinic, to be published on *Nucl. Instr. Meth. In Phys. Res. A*
6. P. Siffert, *Nucl. Instr. Meth. In Phys. Res.* **150**, 1 (1978)

THE HIGH ENERGY ION NANOPROBE AT LEIPZIG*

T. Butz, R.-H. Flagmeyer, D. Lehmann, J. Vogt

Fakultät für Physik und Geowissenschaften, Universität Leipzig, Linnéstr. 5, D-04103 Leipzig, Germany

A new dedicated high energy ion nanoprobe will be installed at Leipzig University, Germany. We describe the concept and layout.

High energy ion microprobes have made their way into several disciplines like material science, geosciences, environmental sciences, art and archaeology, bio- and medical sciences, and several others already some years ago. The penetration and depth sensitivity of ions, the element specificity, the quantitation, the imaging and tomographic capabilities, and the versatility of ion microprobes are advantages, to mention just the few important ones. The future progress depends, at least in several disciplines e.g. semiconductor technology or biomedical applications, on the achievement of ultra-high resolution below the micrometre-regime. Therefore, a new dedicated high energy microprobe with true nanoprobe capabilities was planned and will be installed at Leipzig University, Germany, in early 1997.

The basic components of the system will be described below with first and preliminary numbers of performance. Figure 1 shows the layout.

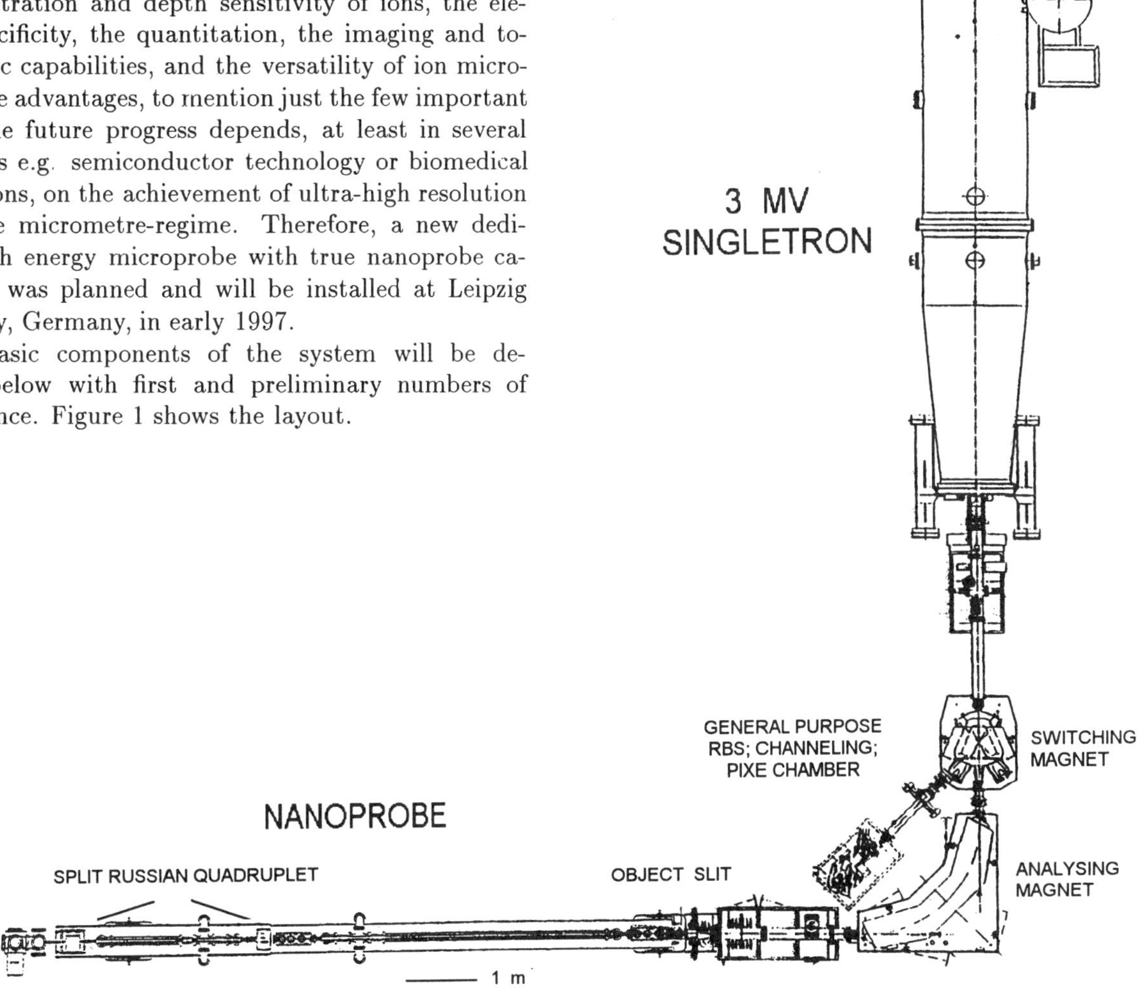

Figure 1. Layout of the Leipzig Nanoprobe.

* Supported by the German Federal Ministry of Education, Science, Research, and Technology (BMBF), the German Science Foundation (DFG), and the Saxonian Ministry of Science and Culture (SMWK).

CP392, *Application of Accelerators in Research and Industry,* edited by J. L. Duggan and I. L. Morgan
AIP Press, New York © 1997

FOUNDATIONS

A key issue for the submicrometre resolution is the minimization of vibrations. We shall have two separate foundations for the accelerator (including switching magnet, general purpose RBS/channeling/PIXE chamber, 90°-analyzing magnet) and for the microprobe. Both are build on pillars 5-6 m deep with a massive concrete table. Measured vibration amplitudes *without* the foundations are below 1 μm in the frequency range from 0-10 Hz. We expect to reduce this amplitude by one order of magnitude *with* foundations.

ACCELERATOR

The accelerator is a 3 MV SINGLETRON by HVEE with a Cockcroft-Walton solid-state power supply, designed to deliver more than 3.5 MV. Proton currents are in the 20 μA range. Ripple value are below 30 V_{pp} summed over all frequencies (50 Hz, 300 Hz, 36 kHz). Preliminary stability measurements indicate drifts below 10 V/h. With an RF-ion source reduced brightness measurements at 2.25 MeV with a H^+-beam yielded values well above 20 A/rad^2m^2eV. All these quantities are of crucial importance for ultra-high resolution work. There are steerers before the switching magnet, but we do not have quadrupoles before and after the 90°-magnet.

MICROPROBE

The microprobe will be supplied by MARC, Melbourne. Contrary to previous layouts, we shall have a *split* Russian quadruplet lens with a large demagnification factor and optional antiscattering slits. We aim at lateral resolutions well below 100 nm thus justifying the name "nanoprobe".

High priority will be given to the following techniques: proton induced X-ray emission (PIXE), proton induced X-ray emission tomography (PIXE-T), scanning transmission ion microscopy (STIM), and tomography (STIM-T), single event upset (SEU), ionoluminescence (IL), secondary electron imaging (SEI).

Research areas will be nanostructuration, main and trace elemental analysis in thin films and biological samples, *in-situ* monitoring of morphogenesis on surfaces and intercalation compounds, and single ion bombardment of living cells. This ambitions programme depends strongly on interdisciplinary collaborations and a strong external scientific input which will be invited in due time. We hope that we shall soon be able to join the high energy ion microprobe community in order to push this exciting field a little further.

ACKNOWLEDGMENTS

The fruitful collaboration with G. J. F. Legge, MARC Melbourne, and D. Mous, HVEE, Amersfoort is gratefully acknowledged.

ERD SPECTRUM TO DEPTH PROFILE CONVERSION PROGRAM FOR WINDOWS®

F. Schiettekatte and G.G. Ross

INRS-Énergie et Matériaux, 1650 boulevard Lionel Boulet, Varennes, Québec J3X 1S2, Canada

Alegria is a new PC-based program to convert ERD and some NRA spectra into depth profiles. The version 1.0 of the program is intended for one implant in one substrate, but will be improved for multi-element detection and multilayers. It is a user friendly Windows application that takes advantage of the Windows functionalities such as "drag and drop" for file managing, multitasking, full memory access, etc. The stopping power is evaluated trough fitting formulae. The iterative integration of the stopping power is made by the RUNGE-KUTTA adaptive step algorithm according to the atomic concentration found in the previous iteration. A demonstration, showing the progression of the solution with the iterations, and an application are presented.

INTRODUCTION

Elastic recoil detection (ERD) [1] has been developed as an easy-to-use alternative method to nuclear reaction analysis (NRA) [2] for light elements profiling in the near surface region of materials. These methods are complementary to the Rutherford backscattering spectrometry [3] because RBS has a low sensitivity to light elements in heavier substrate. Ref. [4] gives a recent review of these nuclear microanalysis techniques.

In the last two decades, while these techniques were evolving as methods for quantitative characterisations of materials, many computer programs have been developed for the ion beam material analysis. For the last 10 years, PC-based (MS-DOS) programs such as RUMP [5], GISA [6] and RBX [7] have become available especially for RBS, but also capable of managing ERD spectrum. Those programs perform simulation and analysis (i.e. iterative approaching) of a multi-element spectra, as obtained from RBS. They are generally intended for layered structures as frequently encountered in semiconductor technology (although GISA has depth profile capabilities). This is not suitable when irregular profile shapes are treated (e.g. ionic implantation, diffusion process, plasma-surface interaction in Tokamak devices).

SPECTRUM CONVERSION PROGRAMS

Since the spectra of the different masses are separated in ERD, they are often directly convertible to depth profiles. No assumption needs to be made about the shape of the atom distribution although it can be more or less affected by the resolution. The equations which have to be solved for spectrum-to-depth profile conversion are greatly simplified compared to simulations. Usually, the spectrum is converted in a depth profile by an iterative process. At each iteration, the stopping power is corrected, according to Bragg's rule, for the atomic fractions found in the previous iteration.

Actually, many authors have already their own codes to make the conversion. Unfortunately, because these programs are generally intended for their specific needs, they are often neither flexible nor user-friendly.

An exception to this is the program developed by Oxorn et al. [8]. According to the authors, it independently profiles all elements in a multi-layered structure from separated mass energy spectra. The package includes a menu-driven database and is intended for overnight batch processing of a large amount of spectra. Nevertheless, the package was developed on a VAX mainframe computer and uses many VAX-specific file and screen management routines.

PROGRAM DESCRIPTION

Alegria outlines

As a consequence to the improvement in computer technology, the micro-computer tends to replace the mainframe computer for data acquisition and analysis. Thus, we felt that it would be useful to develop a flexible and user-friendly PC-based program for spectrum to depth profile conversion. *Alegria* stands for ALternative Erd GRaphical Interface Analyser. It is developed to run in the Windows® 95 and Windows® 3.11 environment so:

- It runs on PC computers.
- The user interface is user-friendly.
- It runs in multi-task.
- It directly access the full computer memory (> 640 Kb).
- File management is simple. *Alegria* supports Windows drag & drop. The user cans pick a file with the mouse in the File Manager and drop it on the graph to open the file.
- Plotting can be made on every printer supported by Windows.

In spite of its name, Alegria can also manage some NRA spectra, such as (p,α) reactions. In that case, if the Q of reaction, the masses of the reaction products and the cross section are given to the program, the reaction's spectra can be converted to a depth profile.

CP392, *Application of Accelerators in Research and Industry*, edited by J. L. Duggan and I. L. Morgan
AIP Press, New York © 1997

TABLE 1. List of graphics in *Alegria*

Experimental setup schema	Cross section plot
Energy spectrum	Charge fraction data
Resulting depth profile	Count-to-concentration corresp.
Stopping power plot	Energy-to-depth correspondence

The user interface consists of eight windows, as listed in Table 1. The experimental setup window shows all the parameters implied in the conversion. The user can set them by pressing the corresponding button.

Actuality, the seven other windows are graphics. Four of them appear on Fig. 1. The user can manipulate a graph (axis, etc.) and the data contained in it (e.g. background noise subtraction). *Alegria* allows basic operations (+-×÷) with constants and addition/subtraction of two data sets. It has polynomial regression and spline fitting capabilities in linear, logarithmic and exponential modes. It is also implemented for zooming.

The program operation

Program inputs

The spectrum that has to be converted is loaded in the Spectrum window. *Alegria*, version 1.0, can make the conversion of the spectrum of only one atom in one substrate. It will be improved for multiple atom detection, as discussed in the conclusion. In the mean time, the energy spectrum needs to be clean (i.e. background and other parts of the spectrum subtracted).

As mentioned above, the user controls many physical and computational parameters, listed in Table 2. The beam and detection angles are given relatively to the surface. The energy calibration is set by a 2nd order polynomial. When the cross section is non-Rutherford, for instance with the NRA technique, the user must supply a file containing the cross section data in which the program can interpolate.

Stopping power evaluation and integration

In *Alegria*, the stopping power is evaluated by means of analytic formulae. The formulæ supplied with the program are those of Andersen and Ziegler [9] (AZ) and EPS model [10] for the proton stopping power and Ziegler [11] (JFZ) for helium energy loss. The parameters for AZ and JFZ, supplied with the program, come from an ICRU report [12] and are used to calculate the stopping power in units of eV/[at/cm^2]. When non of the above can be used, a spline fitting on the stopping power data has to be made, and its parameters saved before the conversion.

During each iteration, for every depth slice, the stopping power of a given ion with energy E becomes

$$S(x,E)_{total} = S(E)_{substrate} + f(x)\ S(E)_{atom} \qquad (1)$$

where $S(E)_{substrate}$ and $S(E)_{atom}$ are the stopping power of the ion in the substrate and the detected atom, respectively, and f(x) is the concentration of the detected atom in the substrate found at a depth x in the previous iteration. For a stopping cross section given in units of eV/[at/cm^2], the depth units will result in **at/cm^2 of substrate**. However, *Alegria* is programmed in a way to be eventually implemented for multi-element detection. In this case, the stopping power will be calculated as follows

$$S(x,E)_{total} = \sum f_i(x)\ S_i(E) \qquad (2)$$

with a depth scale in the depth units of the stopping power formulae.

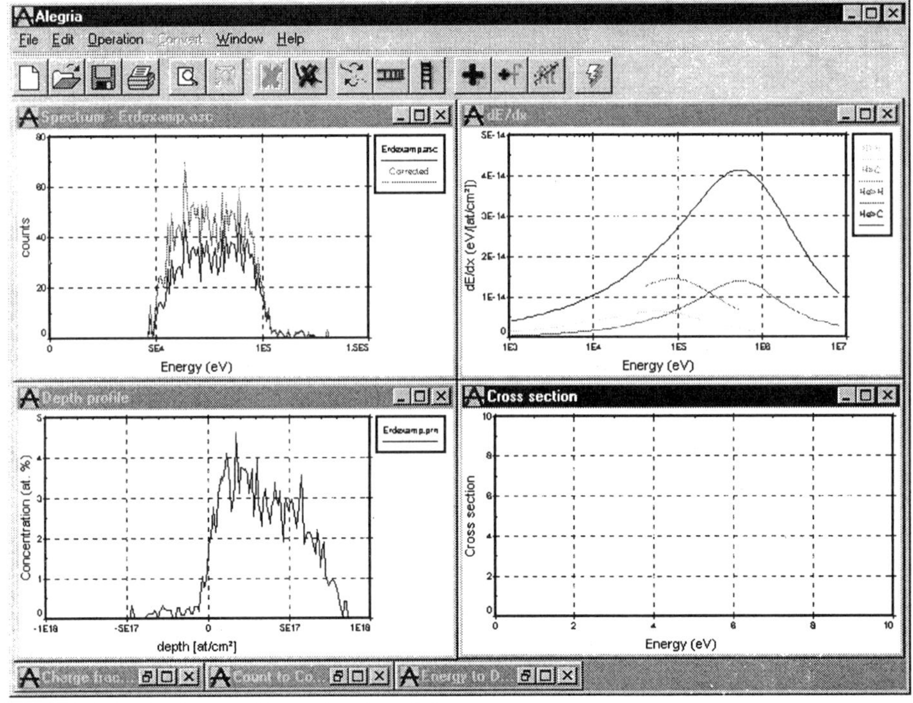

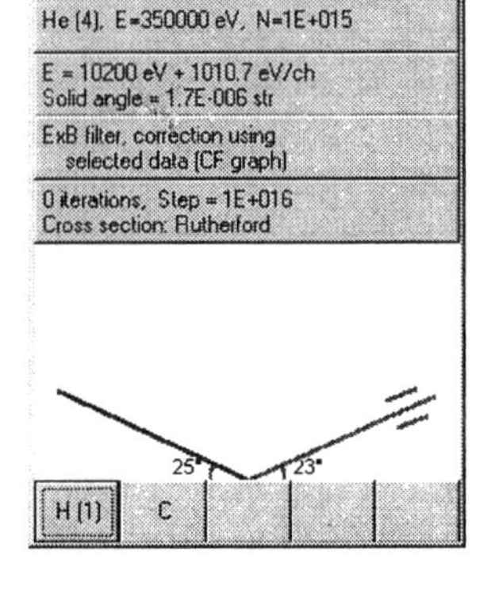

FIGURE 1. Example of the graphic user interface of *Alegria*

TABLE 2. Parameters controlled by the user in *Alegria*

Beam	Recoil	Substrate	Detector	Type of filter	Calculation parameters
Name, Z_1	Name, Z_2	Name	Solid angle, Ω	E×B filter: charge fraction file	Depth slices thickness, Δx
Mass, M_1	Mass, M_2		Energy calibration	Range foil: name & thickness	Calculation precision
Energy, E_0	Detection angle, β		(2nd order polynomial)	NRA: Q of reaction and M_4	Number of iterations
Incident atoms, N					Cross section, $d\sigma/d\Omega$
Beam angle, α					(Rutherford or data file)

Then, in an iterative process, *Alegria* calculates the beam and recoil energy loss by integrating these analytic stopping power formulae by means of a RUNGE-KUTTA adaptive step algorithm [13]. Thus, in the calculation of

$$\Delta E = \int_{\frac{x}{\sin\gamma}}^{\frac{x+\Delta x}{\sin\gamma}} dx\ S(x,E)_{total}\ , \tag{3}$$

the precision is independent of the thickness of the depth slice Δx, chosen by the user. It is rather dependent on the precision parameter of the algorithm. Here, γ is either the angle of the beam (α) or recoil (β) relative to the surface. An adaptive step algorithm will adjust the step length dx at each step, depending of the integrated function roughness. Thus there is no computation time loss with small steps when the integrated function is smooth (relatively to dx), but the step will be reduced to get the required precision when the function is more complicated. The purpose of using such an algorithm here is that the stopping power S(x,E) may have some discontinuities, for instance at layer interfaces in the sample. The precision is no longer influenced by the number of output points chosen by the user via Δx.

Calculation process

In the case of ERD with an E×B filter [14], the recoil energy spectrum has to be divided by the corresponding charge fraction data. A linear interpolation is made between the data given in the file.

At each step Δx, although the beam energy loss is calculated only for this step Δx, the recoil energy loss has to be calculated every time backward to the surface. This becomes necessary because the stopping power will change with the implant concentration found in the previous iteration. So the calculation time scales as $\sim 1/\Delta x^2$. When a range foil is used, the recoil energy loss needs also to be calculated through it at every step.

The implant concentration (implant atoms / substrate atom) is evaluated by the equation

$$\frac{\sin(\alpha)}{\Omega N \frac{d\sigma}{d\Omega} \Delta x} \sum_{E_i=E_a}^{E_b} P(E_i) \tag{4}$$

where $P(E_i)$ is the recoil energy spectrum, with discrete values of E_i. E_a and E_b are the detected energies of the recoils coming from the depth x+Δx and x, respectively. Table 2 identifies the other variables. However, E_a or E_b usually fall within the jth energy channel in the spectrum $P(E_j)$ so only a fraction of $P(E_j)$ has to be considered in the summation. Assuming that E_i is the energy at the begining of the ith channel, the first term of the summation is then $P(E_i)(E_{i+1}-E_a)/(E_{i+1}-E_i)$, $E_i<E_a<E_{i+1}$. The last term is then $P(E_i)(E_b-E_i)/(E_{i+1}-E_i), E_i<E_b<E_{i+1}$. The concentration found in each [x, x+Δx] interval is used in the next iteration to calculate the stopping power in equation (1).

Program outputs

Figure 1 shows an example of the graphic user interface of *Alegria* where the experimental setup, stopping power curves, energy spectrum and the resulting depth profile appear. The graphs that can be displayed are listed in Table 1. The contents of the screen can be printed on any printer supported by Windows. When the user saves the depth profile, *Alegria* appends to the file the first five moments of the distribution.

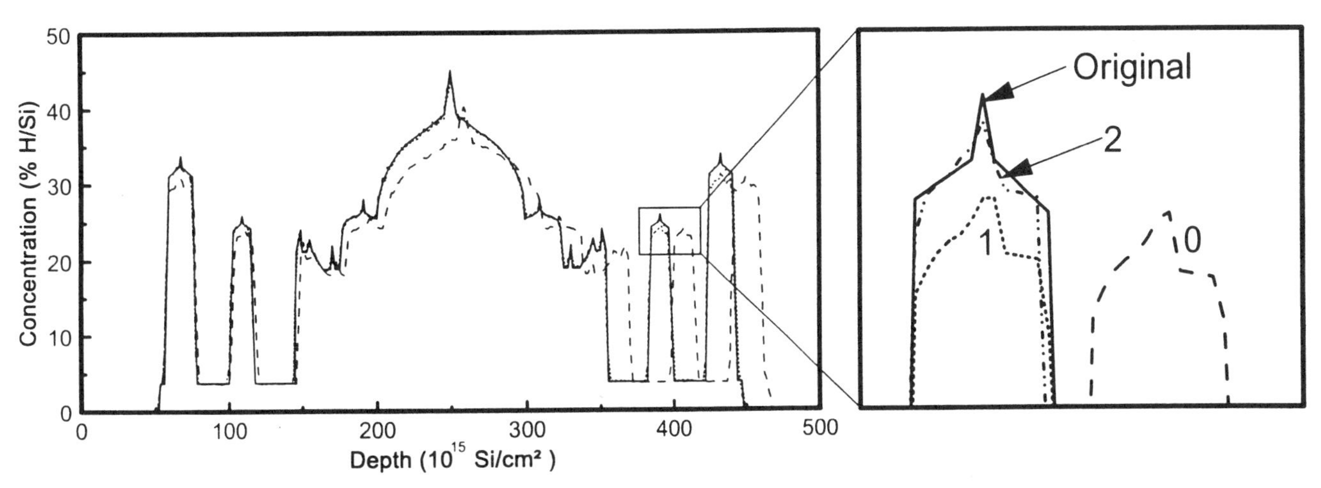

FIGURE 2. Conversion of an arbitrary profile. The original profile (——) is shown together with the progression of the conversion with the number of iterations 0 (- - - -), 1 (······), and 2 (- ·· - ··-).

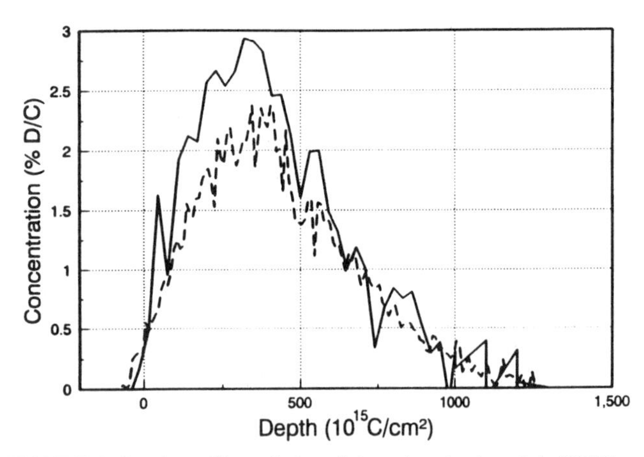

FIGURE 3. Depth profile evolution of deuterium implanted in HOPG graphite at 1.6 keV under a 2.54 MeV ^{15}N bombardement. Profiles obtained with fluences of $4.16 \cdot 10^{14}$ N/cm² (———) and $1.04 \cdot 10^{16}$ N/cm² (- - - -).

DEMONSTRATION AND APPLICATION

An arbitrary profile has been set up and a simulation has been made to get the corresponding energy spectrum. For this purpose, the profile was considered as a hydrogen distribution in silicon. The simulation was made for a 300 keV ^{4}He beam with the H recoils detected trough an E×B filter. Resolution effects have not been taken into account.

The energy spectrum has then been converted to a depth profile using Alegria. Figure 2 shows the original profile (actually the profile of the Tāj Mahal) and the three first iterations. The beginning of the profile is already adequately reproduced at the first iteration because no hydrogen is present from the surface to that depth. So no correction to the stopping power is to be considered. The inset of Fig. 2 shows clearly the progression of the depth profile with the iterations. The number of iterations necessary to have a precise depth profile depends on the concentration (correction to the stopping power) and on the maximum depth.

Figure 3 shows deuterium profiles (converted by means of *Alegria*) where a beam induced desorption process appears. Deuterium has been implanted in HOPG graphite at 1.6 keV. Then, the profile is obtained by means of ERD E×B technique using a 2.54 MeV ^{15}N beam at IPNL, France [15]. The desorption is observed only in the first $5 \cdot 10^{17}$ C/cm² of the surface. This suggests that a diffusion mechanism probably interferes in the desorption process. The desorption yield is also dependent of the concentration (2nd order process). An extensive investigation concerning beam induced profile modification will be published in a subsequent article. Profiles are still resolution broadened.

CONCLUSION

A spectrum obtained from ERD may be simulated or analysed in the same way as a RBS spectrum in order to extract quantitative information. However, this procedure can result in limited information on the shape of the detected particles distribution. It can be more appropriate to convert directly the spectra into depth profile without any assumption on the shape of the distribution.

A PC computer program named *Alegria* is proposed to convert such spectra. It is a user friendly Windows application that takes advantage of the Windows functionalities (e.g. full memory access, file managing, multitasking, etc.) The stopping power is evaluated by fitting formulae and integration of the stopping power is made by the RUNGE-KUTTA adaptive step algorithm. *Alegria*, version 1.0, can convert a single species in a single and uniform compound layer.

However *Alegria* will go through the following improvements:

- Multiple recoil detection capabilities
- Multiple substrate (or undetected species) layers
- Integration of the Ziegler stopping power formula for the heavy ion beam and recoils (TRIM/SR) [16]
- Replacement of the interpolation in cross section data by an adaptive step integration.
- Optional use of the current formulation of Bragg's rule: $S(x,E)_{total} = \sum f(x)\, S(E)$ for stopping power calculation.
- Eventually, integration in the program of the correction method for the resolution developed earlier [17].

ACKNOWLEDGEMENTS

The authors want to thank Dr. Luc Leblanc for fruitful discussion and suggestions, especially for stopping power data and models during the development of *Alegria*. This work has been supported by the Natural Sciences and Engineering Research Council of Canada.

REFERENCES

1. J. L'Écuyer, C. Brassard, C. Cardinal, J. Chabbal, L. Deschênes, J.P. Labrie, B. Terreault, J.G. Martel and R. St-Jacques. J. Appl. Phys. **47**, 381 (1976)
2. D.A. Leich and T.A. Tombrello. Nucl. Inst. Meth. **108** (1973) 67.
3. E. Rutherford. Philos. Mag. **21** (1911) 669
4. J.R. Tesmer and M. Nastasi, Handbook of modern ion beam material analysis (Materials Research Society, Pittsburgh, 1995)
5. L.R. Doolittle, Nucl. Inst. and Meth. **B9** (1985) 344.
6. J. Saarilahti and E. Rauhala. Nucl. Inst. and Meth. **B64** (1992) 734.
7. E. Kótai. Nucl. Inst. and Meth. **B85** (1994) 588.
8. K. Oxorn, S.C. Gujrathi, S. Bultena, L. Cliche and J. Miskin. Nucl.Inst. and Meth. **B45** (1990) 166
9. H.H. Andersen and J.F. Zeigler, Hydrogen stopping power and ranges in all elements. Pergamon Press, New York (1977)
10. C.Eppacher and D.Semrad, Nucl. Inst. and Meth. **B35** (1988) 109
11. J.F. Zeigler, Helium: stopping power and ranges in all elemental matter. Pergamon Press, New York (1977)
12. Stopping Power and Ranges for Protons and Alpha Particles. International commission on radiation units and measurements report 49. Bethesda, USA (1993)
13. W.H. Press, B.P. Flannery, S.A. Teulolsky, W.T. Vetterling. Numerical Recipes in C. ch 15, Cambridge University Press, Cambridge (1990)
14. G.G. Ross, B. Terreault, G. Gobeil, G. Abel, C. Boucher and G. Veilleux, J. Nucl. Mater. **128/129** (1984) 730; G.G. Ross and L. Leblanc, Nucl. Inst. and Meth. **B62** (1992) 484.
15. F.Schiettekatte, A. Chevarier, N. Chevarier, A. Plantier and G.G. Ross. Quantitative depth profiling of light elements by means of the ERD E×B technique. Nucl. Inst. and Meth. In Press.
16. J.F. Ziegler and J.P. Biersack. The stopping and Range of Ions in Solids. Pergamon Press, New York (1995)
17. F. Schiettekatte, R. Marchand and G.G. Ross. Nucl. Inst. and Meth. **B93** (1994) 334

SIMULATION OF TWO DIMENSIONAL TIME OF FLIGHT AND ENERGY RECOIL SPECTROMETRY DATA

P.N. Johnston, M. El Bouanani, W.B. Stannard and I.F. Bubb

Department of Applied Physics, Royal Melbourne Institute of Technology, GPO Box 2476V, Melbourne 3001, Australia.

P. Jönsson, Y. Zhang and H.J. Whitlow

Department of Nuclear Physics, Lund Institute of Technology, Box 118, S-221 00 Lund, Sweden.

A new analytical procedure for treatment of data from Mass and Energy Dispersive Heavy Ion Elastic Recoil Detection Analysis (HIERDA), involving simulation of the Time of Flight and Energy (ToF-E) data has been implemented. The analysis can be considered as two parts (i) the simulation of the physical processes which take place in the sample and timing foils, which yield the distribution of exiting ions arriving at the detector and (ii) the response behaviour of the detectors which determines the pulse height spectrum. The model takes into account straggling and stopping of the incident ions, scattering of the incident ions which gives rise to recoils and scattered ions as well as determines both energy and intensity of exiting ions, and stopping and straggling of the exiting ions in the target and the timing foils. This approach leads to intermediate data which is stored in linked lists of exiting ions. This data is then combined with detector response functions and energy and time calibrations. The simulated spectra can readily be compared with the raw experimental data.

INTRODUCTION

Heavy Ion Elastic Recoil Detection Analysis (HIERDA), otherwise known as Recoil Spectrometry (RS) is being developed as a widely applicable ion beam analysis tool by groups in Germany (1), Canada (2), France (3), Finland (4), Sweden (5) and Australia (6) using high energy (0.5-2 A MeV) heavy ion projectiles. These groups employ heavy ion tandem accelerators or cyclotrons with two basic detector types (i) Time of Flight and Energy (ToF-E) telescopes which are mass resolving and (ii) Energy Loss and Energy (ΔE-E) telescopes which are nuclear charge resolving.

Most of the research work undertaken to date has been dedicated to the optimisation of the RS technique and demonstration of the advantages of using heavy ions and high energies. The technique offers exceptional potential because it allows analysis of the entire periodic table. It has been demonstrated through many highly topical applications of technological importance such as metallisation of semiconductors (7-9), superconductors (3) and corrosion studies (10, 11). RS is a superior technique to Rutherford Backscattering Spectrometry (RBS) because of its capacity for profiling simultaneously and unambiguously a wide range of the elements present in many samples. RS also offers better sensitivity because of the dependence of the recoil cross section on Z_1^2.

The ToF-E (4, 12-14) and ΔE-E (1-3) telescopes are the basic detector types used in HIERDA. Both ToF-E and ΔE-E telescopes use Si detectors for energy measurement. The ΔE-E system allows the use of a large detection solid angle with the advantage of short beam exposure (less sample damage), but requires individual elemental calibrations. ToF-E systems allow a precise multi-elemental internal energy calibration which is an important requirement for accurate and quantitative depth profiling.

Despite the obvious advantages of RS, the analytic procedures for dealing with recoil data are not well developed. If all of the isotopes from a given sample are resolved and the scattered incident ions cause no interference then a simple slab analysis of each isotope (or small group of isotopes) in the spectrum is possible and this is the manner in which recoil data has been mainly treated. However when signals from various isotopes are overlapping, analysis is more difficult.

For ToF-E data, there is a scheme for the conversion of the raw data, i.e. data pairs of time and Si detector pulse height (T,X) into data pairs of pseudo-mass and energy

CP392, *Application of Accelerators in Research and Industry*, edited by J. L. Duggan and I. L. Morgan
AIP Press, New York © 1997

(M,E) (14). This allows separation of the signals from different isotopes by projection of energy slices onto the mass axis combined with fitting (6). This works well in some cases, but encounters problems when an element of interest has a range of isotope masses. In addition, the process of transforming and projecting the data results in the loss of information as M is derived from T and X which are strongly correlated quantities and because energy broadening is the dominant contribution to mass broadening.

In an attempt to improve the quantity and quality of information that we can extract from RS data, we have implemented 2-d simulation of the (T, X) data from a defined material structure. The simulation includes a semi-empirical description of the resolution of the Si detector. Straggling is also calculated on semi-empirical basis.

ALGORITHMS AND SOFTWARE DETAIL

Analysis using 2-d simulation of the spectral data consists of (i) calibration of the data for time and energy, (ii) simulation of the ToF-E data from physical principals and a model structure based on an estimated sample structure and (iii) refinement of the sample structure until features of the simulation match the experimental data. This allows a quantitative analysis of the depth profile of the sample. The first stages use the CERN analysis package PAW (15) and the macro package TASS (16). The second and third stages use RMIT software for analysis of ToF-E data in 2 dimensions - 2DTOFE.

Calibration

Time calibration is established from the ToF of recoiling and scattered ions from different sample surfaces as described by Stannard et al (17).

The energy calibration is based on pulse height distributions from narrow time slices and for known masses (i.e known energy) using the method of El Bouanani et al (14) as described by Stannard et al (17). The centroids and widths of the pulse height distributions are fitted to a model X(E,A) to provide the energy calibration. This is different to earlier analytical methodology (14). In this case we simulate the ToF-E data so it is necessary to convert energy, for a given nuclear mass, into Si detector pulse height, i.e. X(E,A), whereas previously E(X,A) was determined as well as a mass calibration.

Simulation

The program for simulating the two dimensional spectra coming from a TOF-E Recoil Spectrometer is derived from the program ANALNRA (18). It runs on a DEC ALPHA workstation under UNIX. The program has two main parts: (i) the simulation of the physical processes that take place in the sample and timing foils which yield the distribution of exiting ions arriving at the detector and (ii) the response behaviour of the detector which determines the pulse height spectrum.

The modelling of physical processes takes into account
a) stopping and straggling of the incident ion,
b) scattering of the incident ions which gives rise to recoiling ions and scattered ions, i.e. scattering determines the energy and intensity of exiting ions, and
c) stopping and straggling of the exiting ions in the target and timing foils.

Stopping powers are taken from the ZBL theory of Ziegler et al. (19). Straggling of heavy ions is poorly described by the Bohr model (20)

$$\Omega_B^2 = \frac{e^4}{4\pi\varepsilon_0} Z_1^2 Z_2 N x,$$

consequently the straggling is estimated using the semi-empirical model of Yang et al. (21) This semi-empirical model divides straggling into two components: (i) straggling based on an effective charge model analogous to that used in ZBL stopping and (ii) a correlation component due to high ionisation density of heavy ions. The first part (I) comes from a model by Chu et al. (22, 23). We have calculated effective charge (γ) using eqn. 3-16 from Ziegler et al. (20). The second part (ii) uses Yang, O'Connor and Wang's semi-empirical formula:

$$\frac{\Omega^2}{\Omega_B^2} = \gamma^2(Z_1, Z_2, v)\frac{\Omega_{CHU}^2}{\Omega_B^2} + \left[\frac{\Delta\Omega^2}{\Omega_B^2}\right]_{ion}$$

The straggling is calculated for each species of incident and exiting ion. The correlation component is only computed on entry to each layer.

In generating the spectrum each piece of exiting ion data is distributed into 10 energy components distributed according to the computed straggling which are handled separately to account for the energy spread due to straggling.

The simulation does not take into account multiple scattering which may explain the differences in the low energy tails which are commonly seen in HIERDA.

This leads to intermediate data which is stored in linked lists of exiting ions. This data is then combined with detector responses which include:-
a) the energy detector resolution of the Si detector extrapolated from Amsel's model (24, 25):

$$\delta E = a + bE^{1/3}$$

The parameters a and b are functions of A. Almost all the available data is for light ions with Z<10. This is extrapolated to heavy ions with the available data for Cl and Br (25).

b) experimentally determined energy calibration
c) experimentally determined time calibration
d) time resolution of the detector.

The simulated spectrum is stored in an array which can be compared with the raw experimental data.

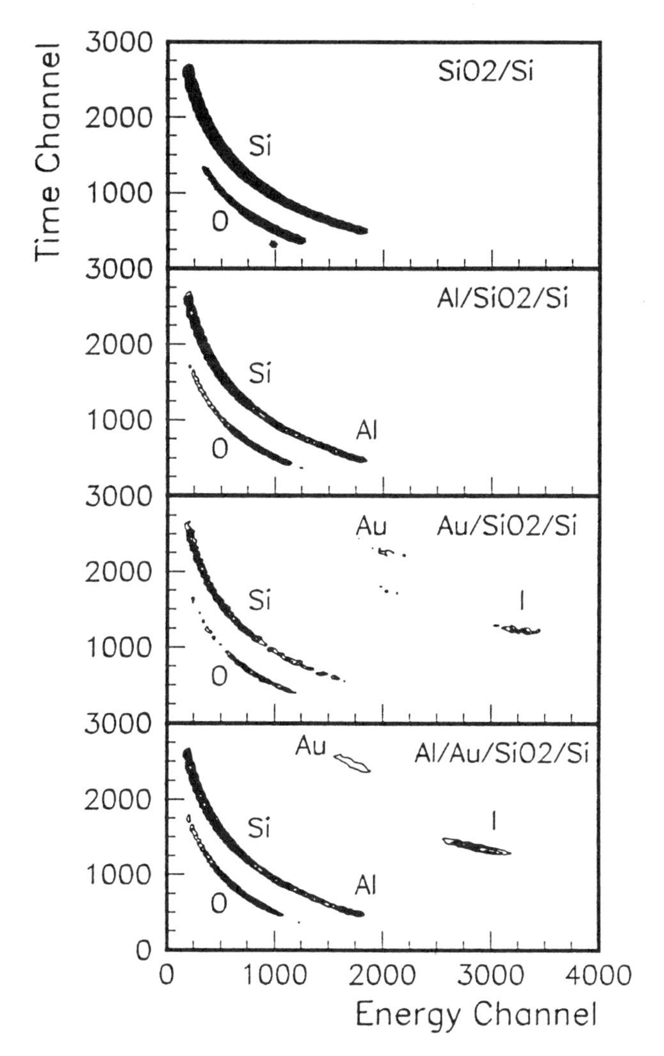

FIGURE 1. Raw Data from test structure measurements

EXAMPLES AND TESTING

Tests of the simulation were performed on real data as well as specially generated test data. The test measurements were performed at Uppsala using the 6 MV EN Tandem accelerator. 60 MeV $^{127}I^{11+}$ ions were used to analyse test structures of SiO_2/Si, $Al/SiO_2/Si$, $Au/SiO_2/Si$ and $Al/Au/SiO_2/Si$ (Figure 1). The test samples were manufactured by depositing 100 and 50 nm respectively of Al and Au on SiO_2/Si substrates.

After making the measurements, the spectra were simulated for the test structures (Figure 2). Although the two dimensional pictures appear alike, it is very difficult to assess the similarities in detail without comparing one dimensional projections on the raw data and simulations (Figures 3 & 4).

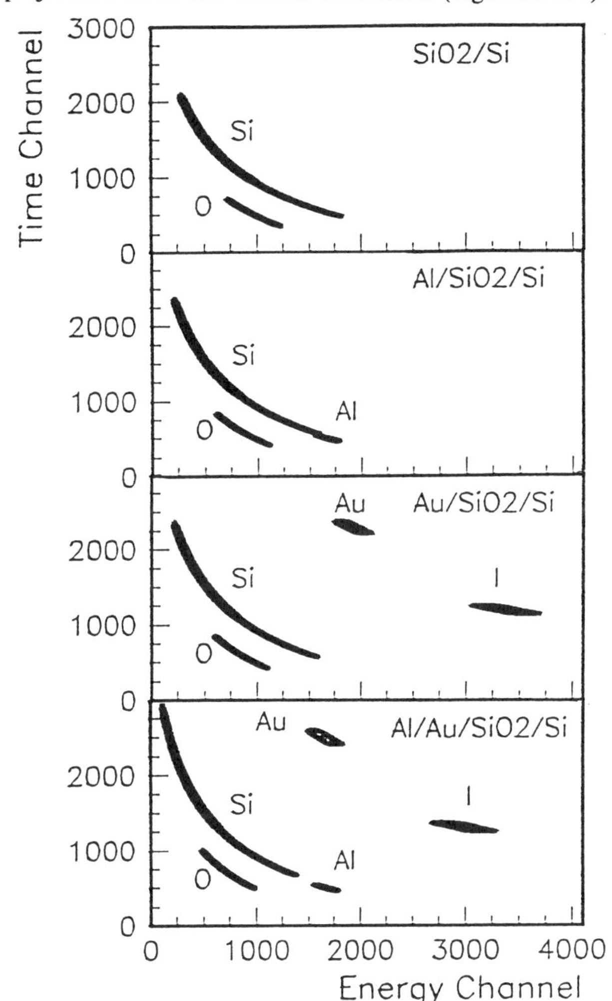

FIGURE 2. Simulated Data for test structures

CONCLUSION

The 2d-simulation of the ToF-E RS data has been implemented and provides a good description of the characteristics of the raw data..

Future development of the analytical procedure is planned to include improvements in the simulation software to include multiple scattering of ions to fully describe the long tails observed in real data, methods of comparing 2-d data and improved energy calibration.

ACKNOWLEDGEMENTS

This work is supported by the Australian Research Council, The Australian Department of Industry, Science and Technology, The Australian Institute of Nuclear Science and Engineering and Strängs Donationsfond. The authors wish to thank Ulma of the Department of Solid State Physics, Lund Institute of Technology for sample preparation.

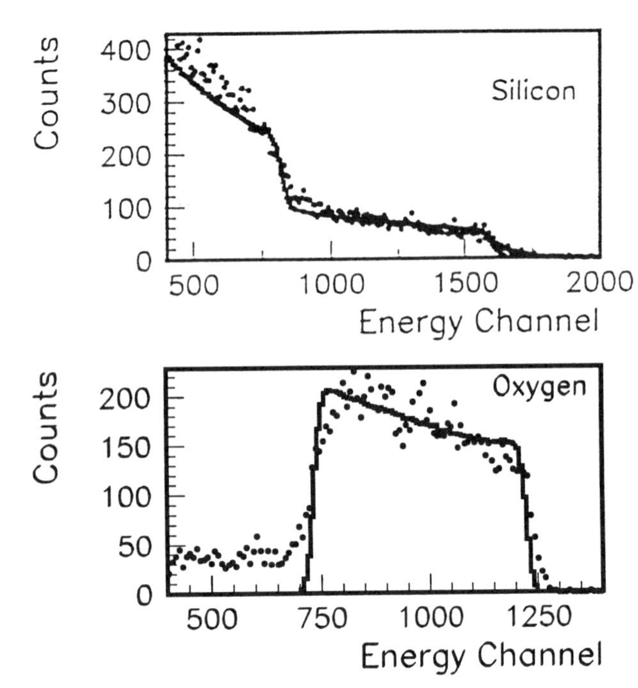

FIGURE 3. E projections of Si and O for SiO_2/Si sample. The dots are the raw data and the curve is the simulation.

REFERENCES

1. Assmann, W., Huber, H., Steinhausen, Ch., Dobler, M., Gluckler, H. and Weidinger, A., Nucl. Instr. and Meth. B **89**, 131 (1994).
2. Siegele, R., Haugen, H.K., Davies, J.A., Forster, J.S. and Andrews, H.R., J. Appl. Phys. **76**, 4524 (1994).
3. Stoquert, J.P., Guillaume, G., Hage-Ali, M., Grob, J.J., Ganter, C. and Siffert, P., Nucl. Instr. and Meth. B **44**, 184 (1989).
4. Jokkinen, J., Private communication.
5. Whitlow, H.J., Possnert, G. and Petersson, C.S., Nucl. Instr. and Meth. B **27**, 448 (1987).
6. Stannard, W.B., Johnston, P.N., Walker, S.R., Bubb, I.F., Scott, J.F., Cohen, D.D., Dytlewski, N., Martin, J.W., Integ. Ferroelectrics **9**, 243 (1995).
7. Hult, M., Whitlow, H.J., Ostling, M., Lundberg, N., Zaring, C., Cohen, D.D., Dytlewski, N., Johnston, P.N. and Walker, S.R., Nucl. Instr. and Meth. B **85**, 916 (1994).
8. Whitlow, H.J., Andersson, M., Hult, M., Persson, L., El Bouanani, M., Ostling, M., Zaring, C., Lundberg, N., Cohen, D.D., Dytlewski, N., Johnston, P.N., Bubb, I.F. and Walker, S.R., Vacuum **46**, 737 (1995).
9. Stannard, W.B., Johnston, P.N., Walker, S.R., Bubb, I.F., Scott, J.F., Cohen, D.D., Dytlewski, N., Martin, J.W., Nucl. Instr. and Meth. B **99**, 447 (1995).
10. Whitlow, H.J., Johansson, E., Ingemarsson, P.A. and Hogmark, S., Nucl. Instr. and Meth. B **63**, 445 (1992).
11. Forster, J.S., Tapping, R.L., Davis, J.A., Siegele, R. and Wallace, S.G., Unpublished report.
12. Walker, S.R., Johnston, P.N., Bubb, I.F., Stannard, W.B., Jamieson, D.N., Dooley, S.P., Cohen, D.D., Dytlewski, N. and Martin, J.W., Nucl. Instr. and Meth. B **113**, 312 (1996).
13. Gujrathi, S.C., Hetherington, D.W., Hinrichsen, P.F. and Bentourkia, M., Nucl. Instr. and Meth. B **45**, 260 (1990).
14. El Bouanani, M., Hult, M., Persson, L., Swietlicki, E., Andersson, M., Ostling, M., Lundberg, N., Zaring, C., Cohen, D.D., Dytlewski, N., Johnston, P.N., Walker, S.R., Bubb, I.F. and Whitlow, H.J.. Nucl. Instr. and Meth. B **94,** 530 (1994).
15. CERN, PAW Manual Version 1.14, (Application Software Group, Computing and Networks Division, CERN, Geneva, Switzerland, 1992).
16. Whitlow, H.J., TASS, Internal Report, (Dept.Nuclear Physics, Lund Inst. Technology, Sölvegatan 14 S-223 62 Lund, Sweden, 1993).
17. 'Experimental Studies of Interfacial Phenomena in Barium Strontium Titanate (BST) Devices', Stannard, W.B., Johnston, P.N., Walker, S.R., Bubb, I.F., Scott, J.F., Cohen, D.D., Dytlewski, N. and Martin, J.W., Integrated Ferroelectrics (in press).
18. Johnston, P.N., Nucl. Instr. and Meth. B **79**, 506 (1993).
19. Zeigler, J.F., Biersack, J.P. and Littmark, U., The Stopping and Ranges of Ions in Matter Volume 1 (Pergamon, New York, 1985).
20. Bohr, N., Mat. Fys. Medd. Dan. Vid. Selsk., **18**, No.8 (1948).
21. Yang, Q., O'Connor, D.J. and Wang, Z., Nucl. Instr. Meth. B **61**, 149 (1991).
22. Chu, W.K., Phys.Rev.A **13**, 2057 (1976).
23. Chu, W.K., In: Ion Beam handbook for Materials Analysis, eds. Mayer, J.W. and Rimini, E., (Academic Press, New York, 1977).
24. Amsel, G., Cohen, C. and L'Hoir, A., Ion Beam Surface Layer Analysis Volume 2, Meyer, O., Linker, G. and Kappeler, F., eds. (Plenum, New York, 1976).
25. Hinrichsen, P.F., Hetherington, D.W., Gujrathi, S.C. and Cliche, L., Nucl. Instr. and Meth. B **45**, 275 (1990).

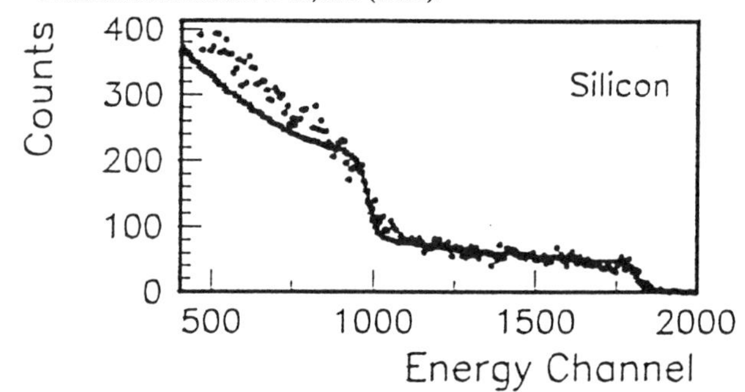

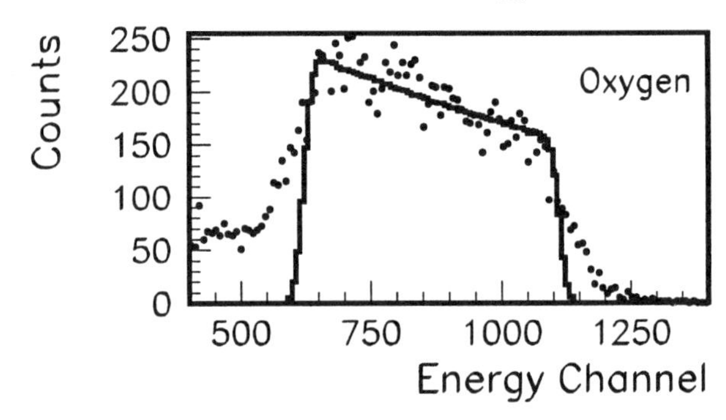

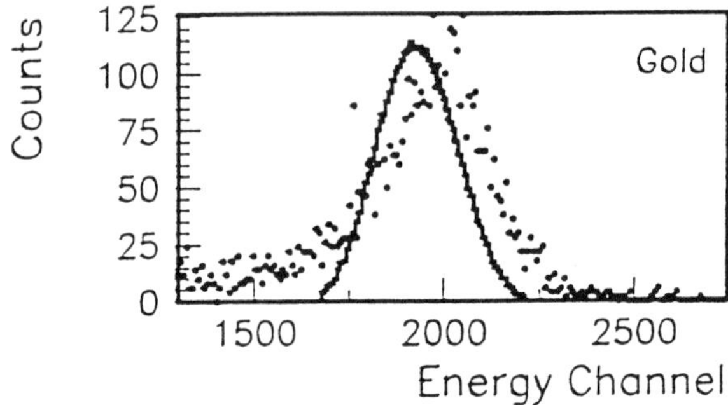

FIGURE 4. E projections of Si, O and Au for SiO_2/Si sample. The dots are the raw data and the curve is the simulation.

LATERAL AND DEPTH DISTRIBUTION OF HYDROGEN IN POLY-CRYSTALLINE SILICON FOR APPLICATION IN SOLAR CELLS

C.L. Churms, V.M. Prozesky, T.K. Marais, R. Pretorius, W.F. van der Weg* and W. Sinke#

Van de Graaff Group, National Accelerator Centre, P.O. Box 72, Faure 7131, South Africa
**Debye Institute, Utrecht University, The Netherlands*
#ECN, Petten, The Netherlands

The use of hydrogen as passivator in silicon solar cells is well known. The function of hydrogen is to occupy the dangling bonds in silicon that occur at defects, such as dislocations and grain boundaries. In this study we used the micro-ERDA (Elastic Recoil Detection Analysis) technique to determine the lateral and depth distribution of hydrogen in poly-crystalline silicon solar cell material. To discriminate against atmospheric hydrogen contamination, the solar cells were manufactured with deuterium as passivation. The ability of ERDA, using ^{4}He as primary ions, to discriminate between hydrogen and deuterium signals enabled us to study the location of both passivation deuterium and atmospheric hydrogen. Within the spatial resolution of the technique, the distribution of deuterium was found not to peak at the grain boundaries, but to be homogeneously distributed in the grains, although some grains did tend to have higher concentrations of deuterium than others.

INTRODUCTION

The use of solar cells as a clean energy source has gained popularity during the last few decades, due to both the severe environmental side-effects of fossil and nuclear fuels and the considerable reduction in the manufacturing cost of solar cells which has been achieved. For this reason many efforts around the world focus on research in the physics and technology of photo-voltaic conversion of sunlight into electricity. Still, the parameter of major importance in the competitiveness of solar power when compared to fossil or nuclear fuel, apart from environmental advantages, is the cost of power production --- the sum of the manufacturing, installation and running costs. A major factor in the determination of the cost of photo-voltaic electricity is the efficiency of conversion of sunlight into electricity. The maximum theoretical efficiency achievable in a solar cell is approximately 30%, whereas the current efficiency achieved by poly-crystalline silicon cells at module level is only about half of this value(1). Thus increase in conversion efficiency is still one of the main drives of solar cell research.

The most favoured semiconductor for mass production of solar cells is silicon. The main advantages of silicon are a lower production cost and the advanced state of silicon technology, when compared to the technology of other possible solar cell candidates. There are three candidates for silicon solar cells, namely single crystal, poly-crystalline and amorphous silicon. Of these, single crystal silicon is the most efficient, but unfortunately also the most costly to produce. Amorphous silicon is cheap to produce, but the conversion efficiency is low, and other effects, such as the efficiency degradation with time, are not yet well understood(2).

Polycrystalline silicon offers a compromise between production costs and efficiency of the resulting solar cells. Such cells normally consist of an array of small single crystal grains, with dimensions from a few μm to hundreds of μm. The

CP392, *Application of Accelerators in Research and Industry*, edited by J. L. Duggan and I. L. Morgan
AIP Press, New York © 1997

small single crystals offer the opportunity of high conversion efficiency, but the presence of grain boundaries which can act as charge traps is the price paid for the poly-crystallinity.

The operation of the photo-voltaic solar cell is based on the creation of free charge carriers by the incident photons of sunlight. These carriers must then move under an internal electric field to an area of collection, from where useful work can be done with this charge. The efficiency of the solar cell is determined to a large extent by the fraction of charge carriers excited into the conduction band, that eventually reach the charge accumulation contacts. Charge carriers can, *inter alia* be trapped and eventually recombine at crystal defects such as dislocations and grain boundaries. For this reason, hydrogen is added to the silicon solar cells to passivate these crystal defects. The silicon is normally heated to around 100-400 °C in a hydrogen atmosphere or plasma, and the quick diffusion of hydrogen into the bulk ensures efficient passivation.

As hydrogen is used for the passivation of these sites of trapping, a study of the distribution of hydrogen, both laterally over the face of the solar cell and as a function of depth, is an appropriate way of establishing the sites of passivation. These would be expected, in the case of poly-crystalline silicon, to be concentrated at the grain boundaries.

One of the problems of the detection of hydrogen in materials is the presence of atmospheric hydrogen, mostly due to water on the surface of the material. This interference is difficult to get rid of, and can lead to erroneous interpretation of results. For this reason the poly-crystalline silicon solar cells studied were manufactured using deuterium for passivation. This, coupled with the ability of the ERDA analysis technique(3) to discriminate between hydrogen and deuterium in the sample, enabled us to study both passivation and environmental hydrogen in the cells.

EXPERIMENTAL

Poly-crystalline silicon solar cells were manufactured at ECN, using conventional poly-crystalline silicon, with individual grains varying in dimension from 30 to 200 μm. The silicon was 0.25mm thick, and was etched with both NaOH and HF as surface treatment. Passivation was obtained by heating the material in a deuterium atmosphere. The estimated areal density of deuterium was 10^{15} - 10^{16} atoms per cm^2. Some of the resulting poly-crystalline silicon was converted into solar cells, whereas other specimens were transported to the National Accelerator Centre (NAC) for analysis.

The Nuclear Microprobe (NMP) of NAC was used for the analysis. The NMP has been described in detail elsewhere (4-6). For the analyses, a beam of 2.5MeV ^4He was accelerated using the 6MV Van de Graaff accelerator, and focused to small dimensions in the NMP chamber. Due to the nature of the recoil process, detection must be performed at forward angles, and in this case, the angles between the incident beam and target surface normal, and target surface normal and detector, were both selected as 75°. This yielded a recoil angle of 30° with respect to the incident beam direction. Detection was performed using a silicon surface barrier detector, with a nominal resolution of 16keV for 5.5MeV ^4He. The solid angle of the detector was 6.9 msterad and a 8.5 μm Al stopping foil was used to stop elastically scattered ^4He particles.

A square projection of the beam spot was obtained by focusing the beam to a rectangle of around 2x7μm. This projected to a beam spot size around 7x7μm on the tilted specimen. Currents of 2 to 5nA were used during analyses.

RESULTS

A typical ERDA spectrum obtained from the poly-crystalline silicon is shown in Fig. 1, with the contributions of both hydrogen and deuterium included.

The conversion from recoil energy to depth was performed using the SENRAS computer package(7). It is clear that the contributions of the two isotopes are fully separated, enabling use of the full depth measured for both isotopes. Also indicated in Fig. 1 are the energy windows H1..4 and D1..D4, selected for mapping of different depth regions for both isotopes. A region consisting of various grains was scanned, and data accumulation was done via list-mode (event-by-event counting).

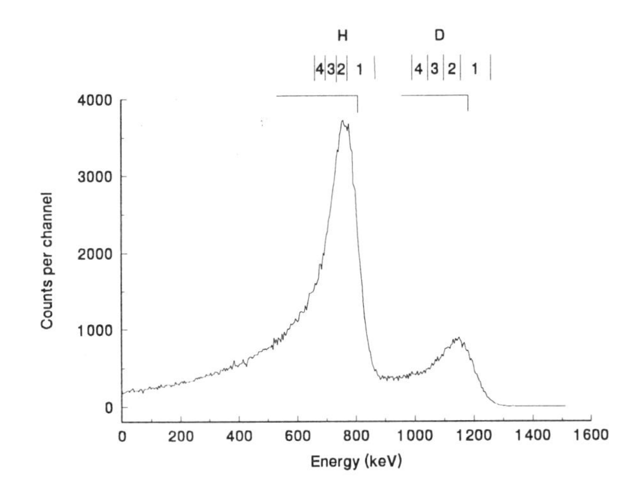

FIGURE 1 Total ERDA spectrum obtained, showing the setting of the gates D1..4 and H1..4 (in 50 nm depth steps), which gave rise to the corresponding maps of H and D concentration as a function of position in Fig. 2.

These scans over the surface, and the energy windows used, allowed 3D imaging of the hydrogen and deuterium concentration in the surface layers of the silicon.

The results of the imaging are shown in Fig. 2, with the contributions of hydrogen and deuterium indicated as a function of depth. The contrast of the maps was selected such that dark pixels represent low concentrations and light pixels indicate high concentrations. The tilt axis of the sample was not in the same plane as one of the scan axes, with the result that the area imaged was not a rectangle, but a parallelogram. A photograph with the scanned area exhibited as a dark parallelogram is included, on which the differing contrast for grains of differing orientation can also be seen. A schematic of the location of some of the larger grains, is also included to illustrate how hydrogen and deuterium concentrations are related to grain orientation.

CONCLUSIONS

Within the resolution of the micro-beam (7x7μm), the deuterium is not seen to be concentrated around the grain boundaries. i.e. Any increase in deuterium concentration in the region of the grain boundaries is, when averaged over 7μm, not large enough to be detected above the experimental noise. However, differing concentrations of deuterium are clearly identified in the different grains. These results would seem to imply that, contrary to expectation, passivation deuterium does not segregate at the grain boundaries.

Considering concentration as a function of depth, some of the grains show a depletion of deuterium with increasing depth, while other grains are enriched. Whcrcas a simplc variation in deuterium concentration with depth could be understood in the light of factors such as deuterium diffusion, the tendency for the variation to be inverted in some grains relative to the others is currently not understood.

The hydrogen results are disconcertingly similar to the deuterium and raise the question as to whether either atmospheric or radiation induced effects have an effect on the deuterium distribution in the solar cells. The effects of radiation induced movement of deuterium was studied by looking at the results of short (a few seconds) analyses on the deuterium concentration, but no difference was detected on this time scale. On a longer time scale (of the order of 5 hours) small effects were observed, in that some of the grain boundaries became slightly enriched with both hydrogen and deuterium.

The results of this preliminary investigation seem to indicate that, even though deuterium is expected to perform a larger role in the passivation of defects at grain boundaries, it is distributed roughly homogeneously throughout the grains. It is furthermore surprising to detect this homogeneous concentration as differing from grain to grain, and also with depth. The possibility of channeling in single crystal grains that might result in the seemingly different concentrations in different grains was also investigated by simultaneous RBS measurements. These measurements showed no channeling effects, and the variations of concentration with depth are real.

One viable explanation for these results is atmospheric influence, as the cells were analysed a few months after production. Furthermore, whereas changes in concentration profiles caused by radiation enhanced diffusion could not be detected under the current experimental conditions, this possibility can not be excluded. However, results of Fink et.al.(8) seem also to predict only limited mobility under these conditions. Diffusion of hydrogen into poly-crystalline silicon is in any case not well understood due to complex diffusion mechanisms(9).

The technique was thus found to be powerful in the separate detection of hydrogen and deuterium in poly-crystalline solar cells. However, various unexpected results give rise to the need for further experimental clarification.

REFERENCES

1. Green, M.A., *Solar Cells*, University of New South Wales, Kensington, 1992.
2. Street, R.A., *Hydrogenated Amorphous Silicon,* Cambridge University Press, Cambridge, 1991.
3. Tesmer, J.R. and Nastasi, M, eds, *Handbook of Modern Ion Beam Materials Analysis*, Materials Research Society, Pittsburgh, p.83, 1995.
4. Tapper, U.A.S, *et. al., Nucl. Instr. and Meth.*, **B77**:17, 1993.
5. Churms, C.L. *et.al. Nucl. Instr. and Meth.*, **B77**:56, 1993.
6. Prozesky, V.M. *et.al., Nucl. Instr. and Meth.*, **B104**:36, 1995.
7. Vizkelethy, G., *Nucl. Instr. and Meth.*, **B45**:1, 1990.
8. Fink, D, *et. al., Applied Physics*, **A61**:381, 1995.
9. Möller, H.J., *Semiconductors for Solar Cells,* Artech House, Norwood, p.259, 1993.

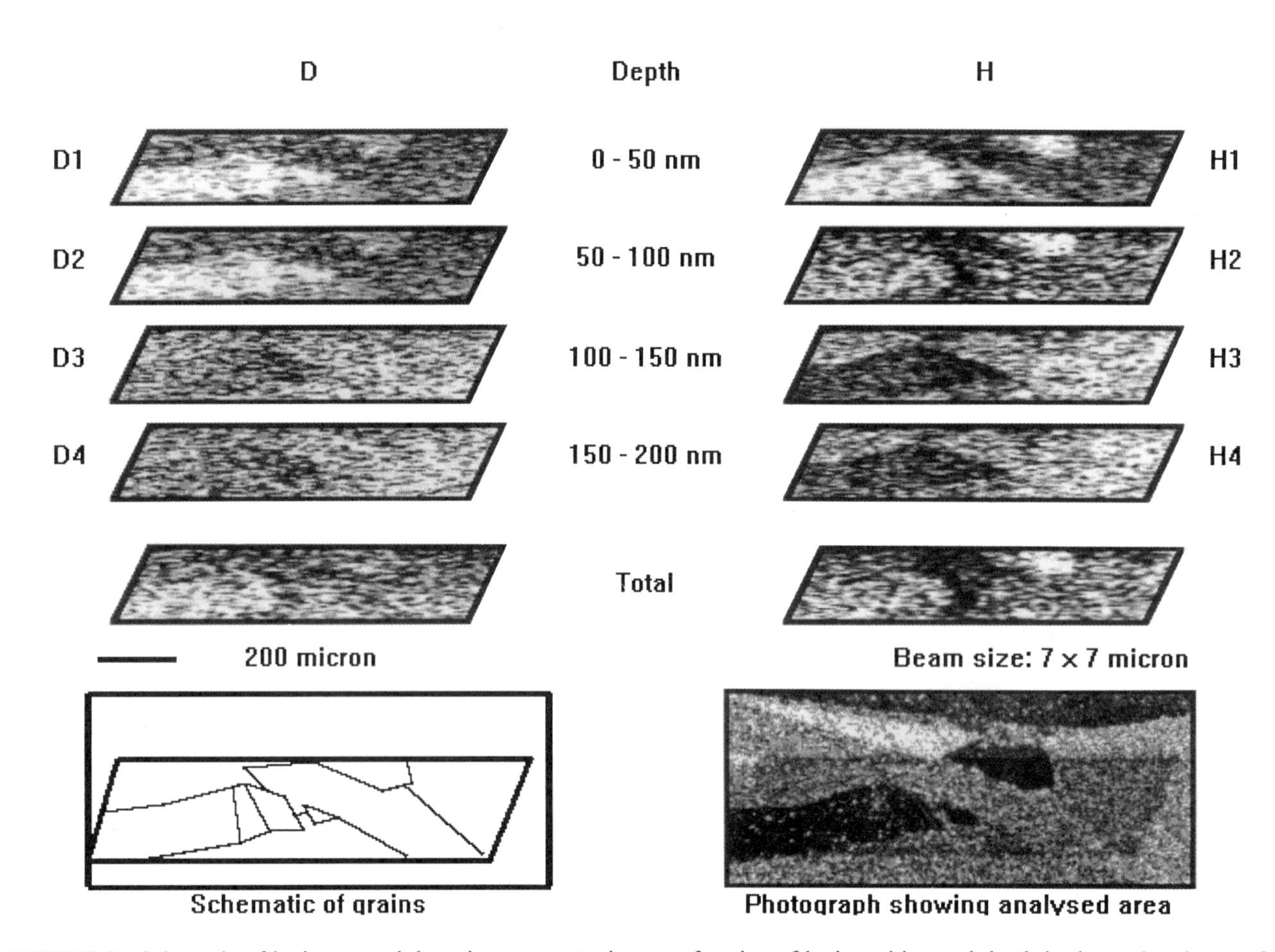

FIGURE 2. Schematic of hydrogen and deuterium concentration as a function of both position and depth in the analysed area of a poly-crystalline silicon specimen. Light areas imply high concentration, and dark areas lower concentrations. The H1..4 and D1..4 maps show detected particle count as a function of position, within the energy gates set as shown in Fig. 1. The dark parallelogram in the photograph demarcates the area scanned and the larger grains are indicated in the schematic.

High depth resolution ERD of light elements by means of an electrostatic spectrometer for MeV ions in combination with a time of flight technique

S. Jamecsny, H.D. Carstanjen

Max-Planck Institut für Metallphysik, Institut für Physik, 70569 Stuttgart, Germany

At the Max-Planck-Institut für Metallphysik in Stuttgart, Germany, a 6 MV Pelletron accelerator is being used for surface analysis using high resolution RBS and ERD of hydrogen by means of an electrostatic spectrometer. While for ERD analysis of hydrogen a thin foil in front of the detector is sufficient to eliminate the background of backscattered particles, for the ERD analysis of heavier elements, e.g. oxygen, the recording of a second parameter of the detected particles is necessary to separate particles of the same energy, but of different mass and charge state. For this purpose a set-up was installed that combines the high energy resolution of the spectrometer with a measurement of the time of flight of the ions through the spectrometer using a chopped ion beam. The measurement of depth profiles of elements lighter than argon, with a depth resolution in the few monolayer range, is now possible. The technique is demonstrated by the analysis of the oxidation of aluminum single crystal surfaces. The experiments revealed a strong anisotropy in the oxidation. The technique also allows to measure charge state dependent depth profiles. Measurements of backscattering of Ne and Ar ions on Au and of recoil O ions from Ta_2O_5 and Al_2O_3 have been performed. From the measuremants charge state distributions after the ion-atom collision and mean free path lengths for electron capture and loss are derived.

INTRODUCTION

In the analysis of near surface layers of solids by elastic recoil techniques one commonly encounters the problem of having to discriminate between ion species of the same energy but of different mass. For example in the analysis of targets containing hydrogen one has to discriminate between primary ions, e.g. He or Ne, that are elastically scattered by the matrix atoms, and the recoil hydrogen ions, whose number usually is much smaller. Here the discrimination is usually done by applying a thin foil in front of the detector which completely stops the scattered primary ions, but allows the recoil hydrogen ions to penetrate. With this technique even high depth resolution can be achieved, if the influence of the energy straggling in the foil can be limited. This is achieved for example by the use of an electrostatic spectrometer [1].

For the discrimination of heavier ions in the few MeV range as it is encountered in the analysis of oxygen with a 1.5 MeV argon beam, the foil technique fails, since the ranges of the scattered primary ions and the recoil ions in such a stopper foil are of about the same magnitude.

To overcome this problem various techniques have been used. Two commonly used methods are the use of detector systems which simultaneously measure the stopping power and the total particle energy, or of a time of flight technique in combination with a common silicon surface barrier detector.

Both techniques usually require high primary beam energies with the drawback of small recoil cross sections. Since in both cases the particles have to pass through one or two detector windows the attained energy and depth resolutions are rather limited.

In this paper we want to present a method that combines the high energy resolution of an electrostatic spectrometer with a time of flight technique, using a chopped argon or neon beam with energies in the few MeV range. The measurement of the time of flight of the ions allows us to discriminate between different ion masses and ions of equal masses, but of different charge states without spoiling the excellent energy resolution of the electrostatic analyser, see figure 1 [2].

Using this set-up, the oxidation behaviour of aluminum single crystals was studied. Aluminum is known to be highly reactive. In air it forms an oxide which covers the metal with a thin, thermally stable oxide layer with excellent adhesion. Surprisingly, when studying the oxidation behaviour of (110) Al surfaces we observe the dissolution of the oxide layer when the

CP392, *Application of Accelerators in Research and Industry*, edited by J. L. Duggan and I. L. Morgan
AIP Press, New York © 1997

sample is heated to temperatures above 400°C. This oxidation behaviour was compared to that of an (111) Al surface which shows the expected oxidation behaviour, i.e. forms a thin thermally stable oxide layer.

The set-up was also used to study depth dependent charge state distributions of backscattered and recoil ions from various targets. From these data mean free path lengths of electron capture and loss as well as the initial charge state distributions after the collision were derived.

EXPERIMENTAL SET-UP

The experiments were performed at the 6 MV Pelletron accelerator of the Max-Planck-Institut für Metallforschung in Stuttgart, Germany. A high frequency ion source produces H, D, He, Ne, and Ar ions. For the heavier ions the maximum beam energy is limited by the maximum magnetic field of the bending magnet, e.g. for singly charged argon ions the maximum energy amounts to 1.5 MeV.

Very high energy resolution is obtained with an electrostatic spectrometer that consists of an electrostatic lens system, a 700 mm electrostatic analyser, and a position sensitive detector [3], see also figure 1. With this system an energy resolution of 1.44 keV was obtained in He-RBS experiments with a 10 mm Si surface barrier detector at 1 MeV primary beam energy. For heavier ions it is slightly larger. Currently a position sensitive channel plate detector is being tested, and first experiments reveal a by far better energy resolution, especially for heavier ions, i.e. Ne and Ar. Energy resolutions of approx. 0.7 keV at incident beam energies of 1 MeV were obtained.

The time of flight measurement is based on a chopped beam. After a first 1 mm aperture the beam passes through a plate capacitor consisting of two parallel aluminum plates of 90 mm length at a distance of 4 mm. By applying a sinusoidal high voltage (up to 6 kV at 300 kHz) to this plate capacitor the beam is scanned over a second aperture with a typical diameter of 0.5 mm at a distance of 1 m from the centre of the capacitor. With this set-up pulse lengths of 10 ns have been realised. Start pulses are generated by the impact of an ion on the position sensitive detector, while stop pulses are derived from the high frequency generator. The length of the flight path through the spectrometer is 3.33 m, typical times of flight are in the order of 1 μs.

With this method three-dimensional yield versus energy and time of flight spectra were recorded. In these spectra the relevant areas are separated and summed over the time of flight. In this way one-dimensional spectra of only the relevant ion species are obtained.

The time of flight resolution is a few percent only, which however is enough to separate light elements. Nevertheless depth resolutions have been obtained which are in the order of 0.2 to 0.4 nm, depending on the scattering geometry and the target material.

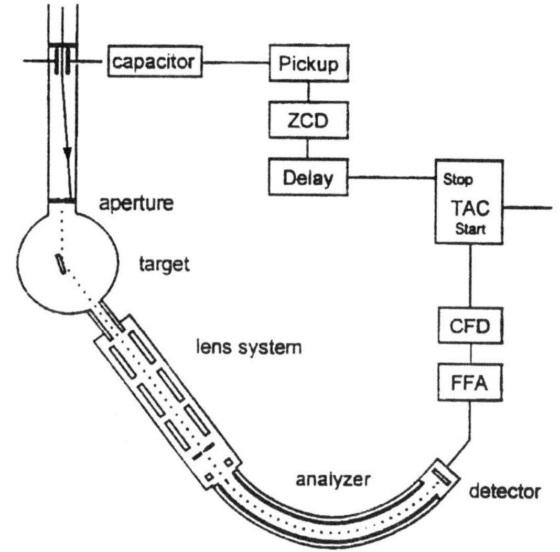

FIGURE 1. Experimental set-up, schematically

RESULTS

Charge State Distributions

Prior to starting an analysis experiment of thin surface layers with the technique described above, one needs to know which one of the charge states of the backscattered or recoiled ions is able to provide energy spectra which allow the straight forward derivation of depth profiles. For this purpose highly resolved energy spectra of backscattered and/or recoiled ions were recorded for various charge states.

In a first step argon ions that were backscattered from a solid gold target were analysed. Two different scattering geometries were used: in one case the incidence and exit angles amount to 19° each, in the other case the exit angle was 3°. In both cases the scattering angle was 38°. Figure 2 shows the recorded energy spectra of the experiment with symmetric geometry.

The spectra recorded for an exit angle of the back scattered ions of 3° essentially show the same structure, i.e. highly charged ions coming from the very surface only, ions of low charge state mainly arising from the bulk. The highest recorded charge state was 8+ in the latter case.

These results were compared to a Monte-Carlo simulation of the experiment. The comparison yields free mean path lengths for electron capture λ_c and loss λ_l (see table 1). The simulation was performed with two different dependencies of the stopping power on the charge state. A quadratic dependence on the charge state, as e.g. the model by Lindhard suggests, was not able to explain the recorded spectra, whereas a model in which the stopping power is independent or only weakly dependent of the charge state could easily reproduce the depth dependent charge state distributions observed in the experiment. One set of parameters in the simulation is the charge state distribution after the collision, which was fitted to the recorded data (see table 1). In a next step we want to eliminate this fitting by an experiment, where argon ions are backscattered on gold vapour.

q	f [%]	λ_c and λ_l [nm]	q	f [%]	λ_c and λ_l [nm]
0	0	- 1.05	5	11	1.45 30.00
1	0	λ_c = 4.20 λ_l = 1.80	6	30	1.31 30.00
2	0	3.00 3.00	7	50	0.99 30.00
3	0	1.86 3.60	8	7	0.66 30.00
4	0	1.68 30.00	9	2	0.48 -

TABLE 1. Parameters of the best fit of the simulation of the scattering of Ar ions on Gold. Charge state q, fraction after the collision f, mean free path for electron capture and loss λ_c and λ_l.

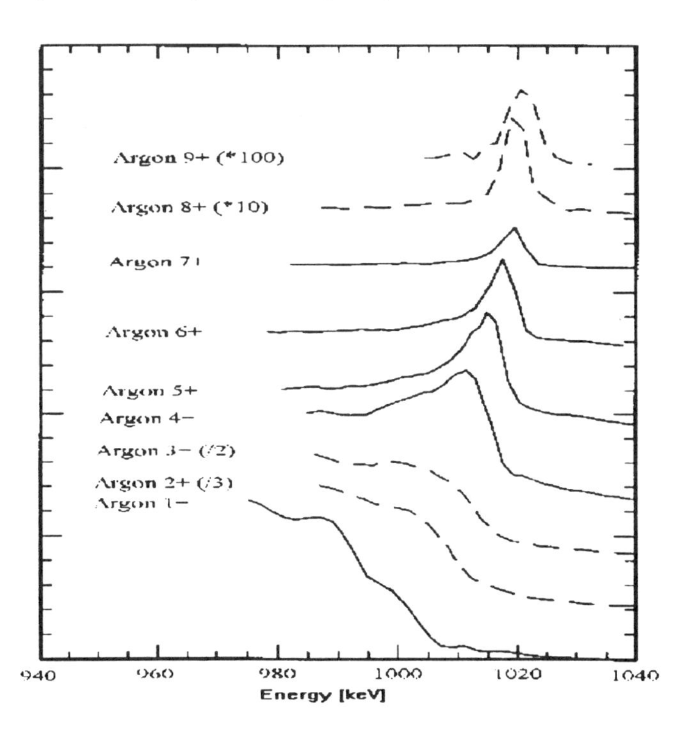

FIGURE 2. Yield versus energy spectra of backscattered argon ions of different charge states from a solid gold target. The yield is in arbitrary units, the zero lines of the spectra have different offsets for better comparison.

In a further experiment backscattered neon ions from a solid gold target were analysed and finally oxygen recoil ions from two different targets (Ta_2O_5 and Al_2O_3, fig. 3). The charge state distributions of the recoil oxygen ions as obtained for the two targets were independent of the target material. The charge state distribution of the backscattered neon ions was similar to that of the recoiled oxygen ions.

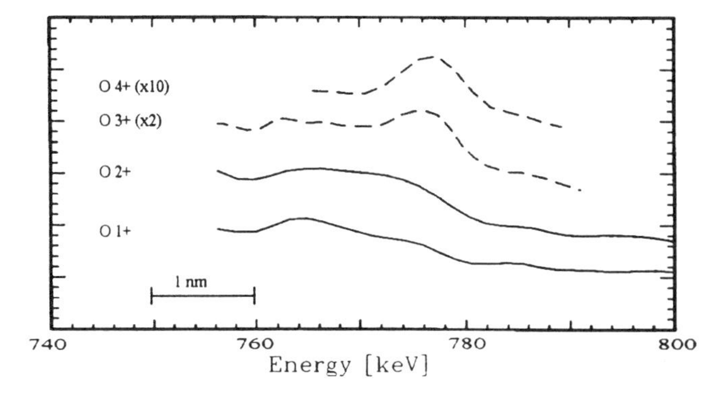

FIGURE 3. Yield versus energy spectra of recoileded oxygen ions of different charge states from a solid alumina target.

As figure 3 shows, doubly (and also triply) charged oxygen ions show a rather steep edge at the surface and a rather flat signal in the region of the sample bulk as one would expect from a solid target. Doubly charged oxygen ions are now being used for the analysis of oxidation.

Oxidation of Aluminum

Aluminum single crystals with (110) and (111) surface orientations were mechanically polished and cleaned in UHV (better than 10^{-9} mbar) by sputtering with 1 keV argon and annealed at 300°C for one hour prior to oxidation. With these samples essentially two types of oxidation experiments were performed:

- The samples, one with (111) and one with (110) surface orientation, were oxidised at room temperature by offering an amount of approx. 10 L of oxygen gas. At this temperature instantly a thin oxide layer is formed. The samples were then heated, in steps, up to 500°C and oxygen spectra recorded (figure 4). Up to about 400°C the thin oxide layers remain stable for both surface orientations. At temperatures above 400°C the two surface orientations show a significantly different behaviour: while on the (111) surface the oxygen layer is still stable up to high temperatures, it starts to dissolve on the (110) surface, with the oxygen diffusing into the bulk of the sample.
- In a set of experiments the sample temperature was set to approx. 490°C and an increasing amount of oxygen offered to the samples. The obtained oxygen recoil spectra, taken at different stages of the oxidation process are shown in figure 5 (for reason of comparison the oxygen spectrum obtained from an exposure at room temperature is also shown). The (111) surface forms again a thin oxide layer which

increases slightly with the amount of applied oxygen. In contrast the oxygen spectrum of the (110) surface shows a wide diffusion profile; already after an exposure to 10 L of oxygen the profile extends to a depth of approx. 3 nm. With increasing exposure time the oxygen layer gains in thickness with a diffusion profile-like tail into the bulk. After an exposure of 1000 L the total amount of absorbed oxygen seems to reach saturation. At this stage a plateau in the oxygen spectrum is seen close to the surface.

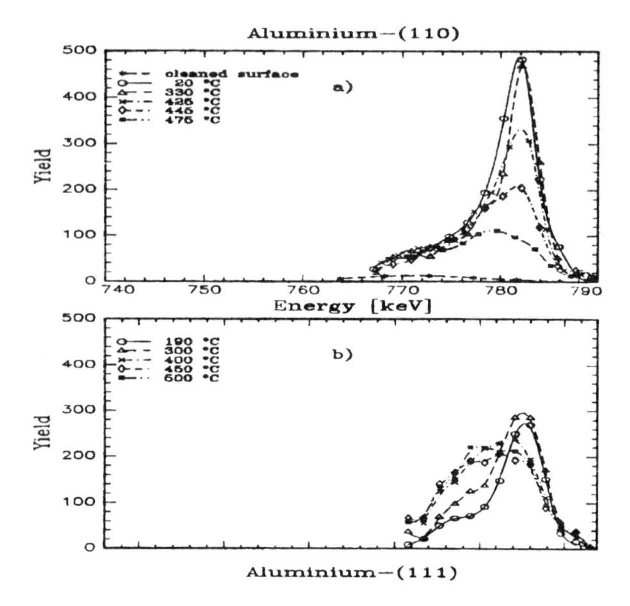

FIGURE 4. Oxygen depth profiles of oxidised aluminum surfaces (10 L at 20°C) during annealing treatments at the temperatures indicated in the figure. a) (110) surface, b) (111) surface. While on the (111) surface the amount of oxygen remains constant (only a minor change in the distribution is visible), the oxide on the (110) surface dissolves above 400°C, by oxygen diffusion into the bulk. For comparison the depth profile of a clean surface is shown in figure. 4 a). 10 keV energy loss represent a layer thickness of 1 nm.

The anomalous oxidation behaviour of the aluminum (110) surface found in this investigation is most probably due to the premelting of the surface that consists in the melting of a thin surface layer at temperatures well below the melting temperature of the bulk. It occurs on loosely bound surfaces and has been observed e.g. on (110) surfaces of aluminum and lead by as well ion channelling as X-ray diffraction. In aluminum surface melting has been observed down to approx. 190 K below the bulk melting temperature of 660°C. At this temperature approx. 1 monolayer of the material is disordered; but since the temperature dependence of the thickness of the melted layer shows a logarithmic behaviour, some disorder may be present at even lower temperatures. The effect of the premelting on the oxygen diffusion consists probably in the increased mobility of the oxygen due to the presence of defects in the partially disordered Al host lattice. In contrast to these findings the premelting of lead (110) surfaces is inhibited by the presence of oxygen.

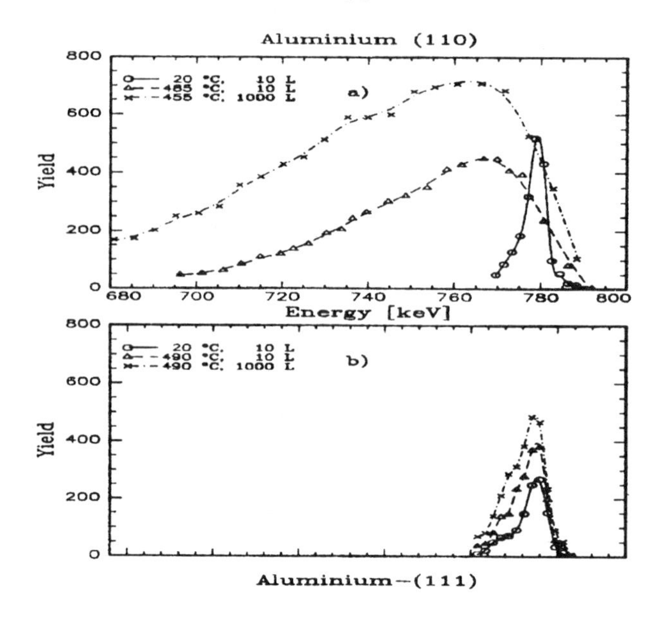

FIGURE 5. Oxygen depth profiles of aluminum surfaces after different exposures to oxygen at temperatures around 400°C. For comparison the depth profile after an exposure of 10 L at room temperature is also shown. While on the (111) surface a thin oxide layer is formed, which slightly increases with temperature and exposure (figure 6b), the oxygen diffuses into the bulk of the sample for the (110) surface (fig. 6a)). 10 keV energy loss represent a layer thickness of 1 nm.

AKNOWLEDGEMENTS

The authors would like to thank all members of the ion beam group and the staff of the Pelletron accelerator at the Max-Planck-Institut für Metallforschung in Stuttgart for kind cooperation.

REFERECES

[1] O. Kruse and H.D. Carstanjen, Nucl. Instr. Methods **B89**, 191 (1994)

[2] R. Plieninger and H.D. Carstanjen, Nucl. Instr. Methods **B**, in press

[3] Th. Enders, M. Rilli, and H.D. Carstanjen, Nucl. Instr. Methods **B64**, 817 (1992)

SINGLE EVENT TRIGGER IN THE FOCUSED MICROBEAM OF THE LABORATORI NAZIONALI DI LEGNARO

F.Cervellera[(1)], G. P. Egeni[(1)], G. Fortuna[(1)], M. Pegoraro[(2)], M. Poggi[(1)], P. Rossi[(2)], V. Rudello[(1)] and M. Viviani[(1)]

[(1)]*Laboratori Nazionali di Legnaro (INFN), Legnaro(PD), Italy.*
[(2)]*Dipartimento di Fisica dell'Università and INFN, via Marzolo 8, Padova, Italy*

The problems which arise in developing microbeam apparata for ions of a few MeV with single event capability are outlined and the solutions adopted for the focused microbeam of the Laboratori Nazionali di Legnaro are described. Possible single event triggers, suited for thin and thick samples, for light and heavy ions, are taken into account. We then pay special attention to scintillating plastic layers of a few dozens of microns, which look very promising as thick samples trigger. We evaluate the beam spatial straggling introduced by the layers with the aid of a simulation program, and propose optimal solutions for light guides and detectors. The challenging case of irradiation in air of cells in radiobiological experiments is considered.

INTRODUCTION

Ionic beams of a few MeV, when reduced to micrometric size and brought onto samples of various nature, are an important tool for research in several fields, and numerous apparata have been realized in different countries to this purpose. A great many of them are devoted to scanning elemental analysis and are based on focusing quadrupole magnetic lenses, which allow microbeams of considerable intensity. Among them a few have been upgraded in order to perform the so called single event (SE) physics, which consists in striking the sample with a single ion in a specific spot to generate a radiation damage, or some useful modification, that can afterwards be evaluated off-line. Specimens of this kind can be as diverse as living cells, solid state detectors and microchip devices.

A SE programme needs the addition of a fast deflector, either electrostatic or magnetic, and of a SE trigger, fast and capable of detecting a single particle with high efficiency, while introducing a very low beam degradation. It is moreover very useful to have a diverging quadrupole doublet upstream of the object slit, i.e. the slit that sizes the beam, in order to reduce the current to less than 0.1 pA. In fact the time for switching off the beam, including the trigger signal delay along the cables and the deflector turning off, can hardly be smaller than 200 ns. To be nearly sure no other particles can escape the deflection it is convenient to have a mean time between pulses of 1 or 2 μs, which is of various orders of magnitude larger than that of a typical microbeam.

It is difficult to compare in a quantitative way focused SE apparata with collimation systems, based on micro holes or adjustables slits. There have been studies to optimize the hole shape, which biases significantly the performance. What we can say is that collimation unavoidably brings with it some energy dispersion, due to energy loss of ions crossing the hole walls, and a prominent angular halo, that hardly could allow a spot size smaller than 5 μm FWHM. On the other hand, in spite of better quality, like monochromaticity and spot sizes up to 0.5 μm, the building of a focused microbeam only for single event physics is often discouraged by the high cost and complexity of operation. However the transformation of a preexisting microprobe into a SE apparatus requires limited changes. In this latter case a particular care has to be devoted to SE trigger, which should not spoil the excellent qualities of the microbeam, considerably better than those of collimation systems.

This article will describe the project of a SE trigger developed for the focused proton microbeam of the Laboratori Nazionali di Legnaro, fed by a 2.5 MV Van De Graaf. The following sections will handle a discussion of the various kind of possible triggers, a detailed trigger design for irradiation in air, a description of a kind of scintillating plastic layers suited for that and an evaluation of their efficiency, finally a discussion of possible light detectors with their electronic chain.

WHY A SCINTILLATING TRIGGER ?

The type of trigger and the difficulty of realizing it change according to whether the sample under study is thin or thick. In the first case the particles pass through the sample and their residual energy can be measured by a total absorbtion detector. The current output of a typical

CP392, *Application of Accelerators in Research and Industry*, edited by J. L. Duggan and I. L. Morgan
AIP Press, New York © 1997

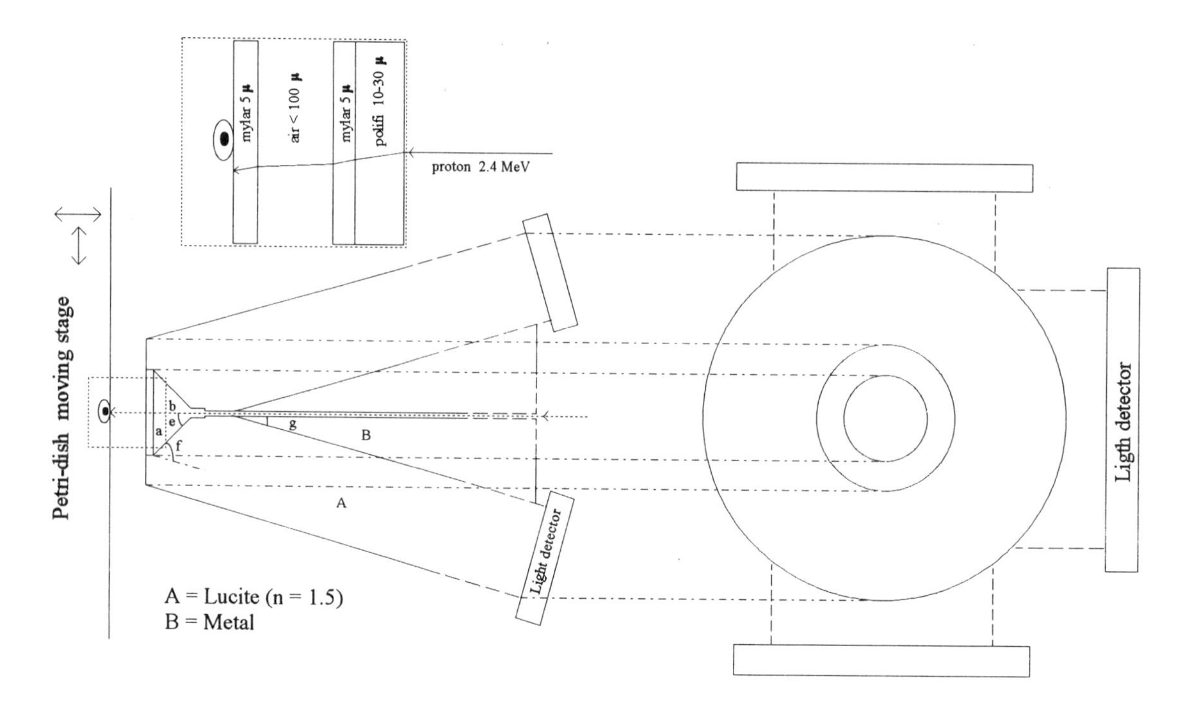

FIGURE 1. Layout of a Single Event Trigger based on scintillating foils.

silicon SBD detector can react in a dozen of ns, providing the required fast signal. In the second case the difficulty arises from the requirement that the beam not be degraded, i.e. slowed or deviated. But the very peculiar case in which the sample is a fast detector itself, the only totally non-degrading upstream triggers handle heavy ions (Z>18 [Ar]). In fact these cause the release of considerable numbers of electrons from the surface of the sample itself, which can then be amplified and detected, e.g. by a channeltron (1).

For protons or light ions it is necessary to interpose some material to provide signals. One should note that, in gaseous detectors, the walls containing the gas have to be considered as well. In all these cases the difficulty arises from the presence of multiple scattering, which causes a severe particle deflection. For example, for the limiting value of 10 $\mu g/cm^2$ of carbon, the smallest usable value to guarantee the mechanical integrity of a foil, protons of 2.4 MeV are deviated on average about 1 mrad FWHM. The multiple scattering is the basic factor that shapes the design of the proposed trigger. The material for the detection has thus to be placed not more than few dozens of microns from the sample. It cannot be based on release of charges, since that would require high voltages, and therefore the risk of discharge.

Among the traditional systems only the detection of scintillation light seems adequate.

SE TRIGGER FOR IRRADIATION IN AIR

The idea outlined in the previous section was developed in detail for the case of irradiation in air of living cells by protons of 2.4 MeV. Even if a living cell, of 5 - 10 μm water equivalent thickness, is in itself a thin sample, it is usually inserted into liquid cultures ('Petri dish') many mm thick and it would be surely not practical, if at all possible, to use more minute systems. Moreover the cells have to live in air and hence the proton, coming from a vacuum chamber, must cross a window.

The scheme of the trigger shown in Fig.1 was conceived to maximize light collection and hence detection efficiency, which must be very close to 1. One should note:

-The scintillating blade, from 10 to 30 μm thick, is glued to a 5 μm metallized mylar foil and placed at the vacuum side. In this way the external light is stopped, and the light going forward from the scintillating blade is reflected towards the beam pipe. The vacuum integrity can be guaranteed also by such a thin mylar layer provided the opening is small enough (2 mm in figure).

-The light pipe (lucite with reflection index n=1.5 and limiting angle ~ 42°) is shaped as an overturned cone, with base angle of 45° and base radius a ~ b ~1mm, surrounding the beam. The angles e, f and g are such to maximize the number of rays which, upon entering the pipe, will travel nearly parallel to the pipe and reach the detector with no reflections. Of course the pipe is metallized wherever necessary allowing a partial recovery of light. Finally the lucite is protected against radiation damage by a conic metallic support, pierced in the center for beam crossing.

-The light should be collected at more than one end to generate a coincidence and hence an effective background rejection. The four external pipes of the front section in Fig.1 allow an optimal light collection: any ray will touch their walls with an angle bigger than the limiting angle of lucite. A logic will chose the two biggest signals coming from the detectors and check for coincidence. One should note that the total system could be kept very small, provided the light detectors are of reduced size. This is the case if the APD detectors, with an external diameter od 10 mm, will be used (see last section). In this case the miniaturisation will be actually remarkable.

-The beam must be centered at the beginning of a measurement session and no longer displaced. Instead it will be the sample to be moved by a remotely controlled stage. To avoid damage the motion will take three steps: back-off from the window a few mm, lateral movement, move-in again to the final position, which must be as close as possible to the mylar window. It is of course necessary that this latter be the most external part of the beam line to avoid barriers to stage movement.

We worked out a computer simulation, based on the program TRIM (2), to study the passage of a 2.4 MeV proton through the various materials of the trigger. We took into account three different thicknesses of the scintillator, based on polystirene (see next section), i.e. 10, 30 and 50 μm, and four values of the air interspace, i.e. 10, 50, 100, 200 μm. The mylar vacuum window and the mylar cell support have been set at 5 μm. We evaluated the spot size (FWHM), i.e. 2.35·σ, where σ is the rms of the density distribution of deviated protons as function of the distance between the final particle position and the unperturbed beam direction. We note a strong dipendence of spot size from the air gap. This could be hardly lowered to less than 50 - 100 μm, because samples have to be displaced with no damage of the vacuum window and the Petri-dish support, whose planarity will not be perfect. We note also that, with an air gap of 100 μm, the spot size could vary from 1.5 to 5 μm according to the thickness of the scintillating blade.

These values show the importance of reducing as much as possible this thickness, which is possible only by maximizing the light collection efficiency and the detector sensitivity. A study of this latter subject will follow in the last section.

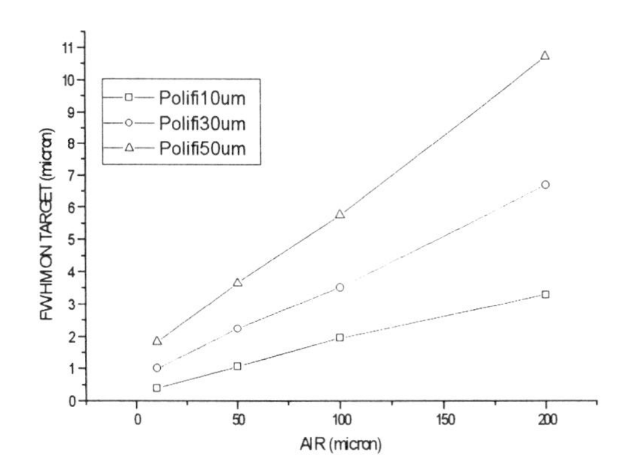

FIGURE 2. Proton lateral displacement due to multiple scattering

SCINTILLATING PLASTIC FOILS

The choice of scintillating material depends on the photon yield dL/dE and on the rise time of the light signal(τ). This time must be no bigger than few hundreds of ns in order to be usable as a trigger and to be matched for shaping and amplification by standard electronics. This requirement rules out most of the inorganic scintillators. Among the organic ones, anthracene in crystalline form ($C_{14}H_{10}$, ρ=1.25 g/cm^2, n=1.62, τ=30 ns, λ_{max}=425 nm), and plastics based on polystirene ($C_5H_8O_2$, ρ=1.1, n=1.58) stand out both in terms of a notable dL/dE and acceptable rise time. The plastics, although slighty lower regards photon yield, have the advantage of being easily workable into very thin foils.

We employed a doped polystirene, POLIFI type 0046-100, produced by the firm POL.HI.TECH (3) as material for scintillating fibers, with a τ=3 ns, λ_{max}=435 nm and $(dL/dE)_{POLIFI}$ = 0.55 · $(dL/dE)_{anth}$. The light yield per unit thickness, for a 2.4 MeV proton, can be calculated starting from that of anthracene and is $(dL/dx)_{POLIFI}$ ~ 450 ph/10μm.

The light pipe geometry of Fig.1 allows a good light collection efficiency. Although we still have to measure it, we can imagine it is more than 10% and very easily close to 50%. In fact the light directed backwards will be collected nearly totally with no reflections at metallized surfaces, while the forward light, being subject to various reflections and transmissions, will be partially absorbed. With the very gloomy estimate of a 10% collection efficiency, we still have a 45 ph, which will be shared in the worst case evenly among 4 detectors. This yield will already allow a coincidence between the two biggest signals, taking into account the background, if the 'right'solution for light detectors is adopted, as will be made clear in the next section.

Finally we describe the recipe for the preparation of plastic foils. We dissolved known weights of POLIFI powder, obtained by scraping a plastic block, into known volumes of toluene, and spread the solution over a level glass side (4). The thickness of solution deposit is determined by the surface tension and does not depend on the total quantity. The solution was then allowed to evaporate for a few hours in a dust-free environment until a film is left. Its thickness is determined by the concentration of POLIFI in toluene and can be tuned rather exactly. To remove the film from the glass surface it is often necessary to dip the glass in deionized water. The film is eventually glued on a transparent support, for example lucite, which will act as a light pipe.

The uniformity of the film thickness has been measured roughly by a mechanical micrometer at the level of 1 μm and found to be better than 10% for thin films, i.e. less than 30 μm, while the situation looks worse for thicker films, still requiring an improvement of the procedure.

LIGHT DETECTORS AND ELECTRONIC CHAIN

The photodetector and the associated electronic chain play a critical role in this apparatus having to satisfy several requirements. The detector should have a high quantum efficiency together with a high gain in order to be able to detect very few photons and to provide an electric signal strong enough to overwhelm the noise of the attached preamplifier; the intrinsic noise (mainly generated by detector capacitance and dark current) should be very low as well. For these reasons we have focused on two possible choices based on HPD (hybrid photodetector by DEP either electrostatically or proximity focused) (5) or on APD (avalanche photodiodes with UV enhanced sensitivity by Advanced Photonix or Hamamatsu) devices (6).These two approaches are presently under investigation because of the recent development and related continous upgrades they undergo. Other aspects, such as environment constraints (e.g. allowed space and magnetic field) and difficulties of construction (e.g. light guide couplings and high voltage biasing) are being considered.

The following table summarizes what we foresee for a proximity focusing HPD and an Hamamatsu APD when connected to a readout chain performing a gaussian shaping with peaking time of about 200 ns:

PARAMETER	HPD	APD
C_{det}	100 pF	15 pF
I_{dark}	1 nA	1 nA @ G = 50 (gain dependent)
Q.E. (440 nm)	20 %	60 %
Gain	2500	to be investigated
ENC	1700 e	750 e
<Nph> corresponding to 3•ENC	10	3750 / Gain
Dimensions	37 mm diam.	10 mm diam.
Draw backs	HV >10 kV	needs Peltier cooling to control I_{dark} and temperat. dependent biasing

As one can see, the APD option must be checked on the critical item of the gain which should be greater then 350 to be competitive. The gain can be raised up to values of thousands by increasing biasing voltage but this produces a higher dark current, and consequently higher electric noise, with a dependence very similar to that of the gain: some papers report that in Hamamatsu devices the ratio gain/dark-current is constant for values of gain ranging from 30 to 150 as expected from the multiplication mechanism of the APD. If this characteristic is valid for a broader range, one can think of working at gain of some hundreds at the expense of a higher noise current.

The electrostatically focused HPD on the other hand exhibits a much lower noise with respect to the proximity one, but is both larger and more sensitive to magnetic fields.

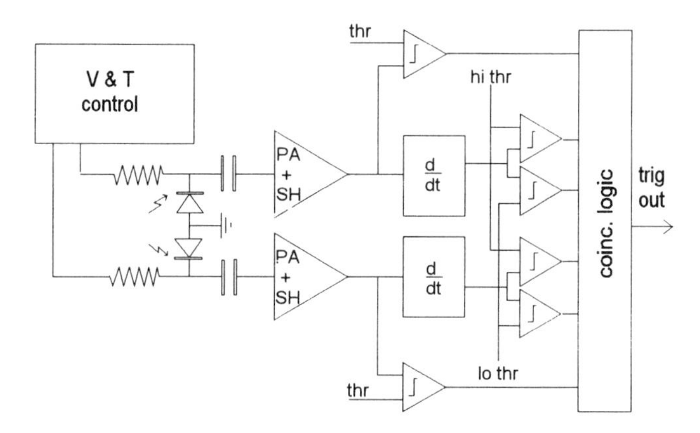

FIGURE 3. Ligth detection electronics.

The electronics of the proposed trigger is sketched in Fig.3 for a 2 photodetectors prototype; in case of 4 photodetectors the difference is in a majority logic that performs the coincidence of 2 out of 4 channels.
The first part of each channel is a FET preamplifier plus a 5 poles gaussian shaper whose output is discriminated to give a first timing information; the same output after differentiation is sent to a zero cross discriminator to obtain the time of the peak with a precision of dozens of ns in order to reduce the coincidence window. All the timing signals are finally compared by the logic, which decides on whether a coincidence has occurred.

This configuration should allow to gain in efficiency and in rejection of spurious signal in such a way to satisfy the requirements of the apparatus.

ACKNOWLEDGMENTS

We thank Dr. Jeff Wyss for critical revision of manuscript and the students Davide Bernardi and Fabio Parere for valuable help in preparing it.

REFERENCES

1. Fischer B.E., "Single - particle techniques", NIM B54 (1991) 401 - 406.
2. Ziegler J.F., "Trim instruction manual", IBM-Research, 28-0 Yorktown, NY 10598, USA.
3. POL.HI.TECH s.r.l., S.P. Turanese Km 44.400 - 67061 Carsoli (AQ) - Italy
4. Manduchi C., Russo-Manduchi M.T. and Segato G.F., "Response of ultra-thin scintillator foils to fission fragments", NIM A243 (1986) 453-458.
5. DEP, Delft Instruments, Dwazzie Weglu 2, Roden, the Netherlands
6. Karar A. et al., "Investigation of Avalanche Photodiodes for EM calorimeter at LHC", Ecole Polytecnique IN2P3 - CNRS, report X - LPNHE/95 - 10, or CMS TN/95-135 (30 october 1995).

CHARGED PARTICLE MICROPROBES WITH MINIMUM BEAM SPOT SIZE FOR A GIVEN BEAM CURRENT

Alexander D. Dymnikov and Genoveva Martínez

Dept. Física Aplicada III, Fac. de Física
Universidad Complutense, E-28040, Madrid, Spain

In a nuclear microprobe the beam (usually 1-3 MeV protons) is focused on the target to a lateral spot size of micrometer dimensions, using a focusing lens and at least two beam emittance-defining apertures. This type of beam has gained an increased interest during the last decade as an analytical instrument. In this paper the limits to high spatial resolution in ion microprobes and the requirements for its achievement are presented. Different focusing systems with quadrupole and rotational symmetry are studied and compared. The optimal geometry, the minimum spot size and the appropriate optimal radii of two apertures and the distance between them for a given emittance are found.

1 INTRODUCTION

The microprobe system consists of the object drift space, the focusing system, where the electromagnetic field acts on the beam motion, and the image drift space. We use the following notations: the length of the object space is the object distance a, the length of the focusing system is l, the length of the image space is the working distance g and the total length of the microprobe system is $l_t = a + l + g$. The electrostatic round lenses and quadrupole lenses are widely used for forming and transportation of beams themselves as well as a part of more complicated systems. We consider the focusing system which consists of two, three and four quadrupole lenses (1) and electrostatic coaxial multiple cylinder lenses (2), having equal diameter with thin walls. The beam propagating along the central axis z has initial energy E_0 and a given emittance em which is determined by the object diaphragm with radius r_1 and the aperture diaphragm with radius r_2 separated by a distance l_{12}. The following problem is solved: what are the values of lens excitations, r_1, l_{12} and the geometry of the system which provide, at the given emittance, the minimum beam spot size on the target situated at the image plane.

2 THE MATRIX METHOD FOR SOLVING THE NONLINEAR EQUATION OF MOTION

We consider the differential equation of motion of the charged particles accurate to terms of third order inclusive. It is convenient to choose a set of variables, for which the phase volume remains unchanged during the beam motion. For the static field these variables have the following form:

$$x_1 = x,\ x_2 = \frac{p(z)}{p(0)}x',\ y_1 = y,\ y_2 = \frac{p(z)}{p(0)}y',$$

where $p(z) = \sqrt{\gamma^2(z) - 1}$,

$$\gamma(z) = \gamma(0) + \varphi(0) - \varphi(z),\quad \gamma(0) = 1 + \frac{E_0}{W_0}.$$

Here $p(z)$ is the dimensionless momentum of the axial particle, $\gamma(z)$ is its relative total energy, W_0 is the rest energy of an axial particle, $\varphi(z) = V(z)q/W_0$ is the dimensionless axial potential and $V(z)$ is the axial potential. For the quadrupole lenses $p(z) = \text{const} = p(0)$ and $V(z) = \text{const} = V(0)$.

The analysis and calculation of the nonlinear systems of equations for monochromatic beam formation in the static field are considerably simplified by transforming from the nonlinear differential equations of motion in the phase space (x_1, x_2, y_1, y_2) to the system of linear equations in extended phase space - the phase-moment space. This is the essence of the method of embedding in phase-moment space (3). For the differential equation of motion of the particles accurate to terms of k-order we have the phase-moment space of k-order. If the motion of the monochromatic beam is described by the third order equation, we have two linear equations for two phase-moment vectors of the third order:

$$\tilde{x}[3] = \{x_1, x_2, x_1^3, x_1^2x_2, x_1x_2^2, x_2^3, x_1y_1^2, x_1y_1y_2, x_1y_2^2, x_2y_1^2, x_2y_1y_2, x_2y_2^2\}$$

and

$$\tilde{y}[3] = \{y_1, y_2, y_1^3, y_1^2y_2, y_1y_2^2, y_2^3, y_1x_1^2, y_1x_1x_2, y_1x_2^2, y_2x_1^2, y_2x_1x_2, y_2x_2^2\}.$$

The writing of the nonlinear equation in a linearized form makes it possible to construct its solution using a matrizant, which is independent of the initial vector $\{x_{10}, x_{20}, y_{10}, y_{20}\}$, whereas the solution of the nonlinear equation is sought for each value $\{x_{10}, x_{20}, y_{10}, y_{20}\}$.

CP392, *Application of Accelerators in Research and Industry*, edited by J. L. Duggan and I. L. Morgan
AIP Press, New York © 1997

To the nonlinear equation of the third order we can associate a linear equation for the phase moments: $dx[3]/dz = P_x(z)x[3]$, where $P_x(z)$ depends on the axial electromagnetic field and its derivatives. The solution of this equation is written in terms of the matrizant $R_x(z/z_0)$ in the form:

$$x[3] = R_x(z/z_0)x_0[3], \quad R_x(z_0/z_0) = I$$

The equations in y- plane are obtained from the equations in x -plane if $x \to y$, $y \to x$. For the cylinder lenses $P_x(z) = P_y(z)$ and $R_x(z/z_0) = R_y(z/z_0)$.

3 THE ANALYTICAL MODEL OF THE AXIAL ELECTROMAGNETIC FIELD

We choose a rectangular analytical model of the axial electromagnetic field for the quadrupole lenses and an analytical model of the axial electrostatic potential for the lenses with rotational symmetry in the form of piecewise-continuous function:

$V(z) = V_{2j-1}(z)$, if $z_{2j-2} \le z \le z_{2j-1}$, and $V(z) = V_{2j}(z)$, if $z_{2j-1} \le z \le z_{2j}$, where $V_{2j-1}(z) = \text{const} = U_{2j-1}$,

$$V_{2j}(z) = U_{2j-1} + (z - z_{2j-1})^5 \left(\frac{\Delta U_j}{\Delta z_j^5} + (z - z_{2j})(-5\frac{\Delta U_j}{\Delta z_j^6} + (z - z_{2j})(15\frac{\Delta U_j}{\Delta z_j^7} + 35(2z - 3\Delta z_j - 2z_{2j-1})(z - z_{2j})\frac{\Delta U_j}{\Delta z_j^9}))\right),$$

where $\Delta U_j = U_{2j+1} - U_{2j-1}$, $\Delta z_j = z_{2j} - z_{2j-1}$, $V_{2j}(z)$ is changed from U_{2j-1} to U_{2j}, while z is changed from z_{2j-1} to z_{2j} and the four first derivatives of $V_{2j}(z)$ are zero for $z = z_{2j-1}$ and for $z = z_{2j}$.

4 THE METHOD OF THE AVERAGE MOMENTS

It is known that the information about the averaged characteristics of a beam can be obtained by calculating the moments of the particle distribution function in phase space - the average moments (4). We consider the beam motion as a motion of the closed phase set. This allows us to introduce the matrix of the average moments M, where $M_x(z) = \int_\Omega f(x_1, x_2, y_1, y_2)x[3]\tilde{x}[3]dx_1dx_2dy_1dy_2$.

The integration is performed over the apertures of two diaphragms for the rectangular distribution of f. The averaged radius $r(z)$ of the beam is determined by the matrix element $M_{x11}(z)$. In this case we obtain $r_x(z) = \sqrt{M_{x11}(z)}$, $M_x(z) = R_x(z/z_0)M_x(0)\tilde{R}_x(z/z_0)$, $M_x(0) = \int_\Omega x_0[3]\tilde{x}_0[3]dx_{10}dx_{20}dy_{10}dy_{20}$, where $M_x(0)$ is a function of r_1, of emittance $em = r_1 r_2 / l_{12}$, and of l_{12}.

5 THE OPTIMIZATION PROCEDURE

At first, for a given geometry and initial energy E_0, we find from the stigmatic equations the values of the excitations (or the potential V) which provides the stigmatic property of the system. For these excitations the third order matrizant $R(z/0)$ is calculated. We choose the merit function as $\rho = \max[\rho_x, \rho_y]$, $\rho_x = r_x(l_t)$, $\rho_y = r_y(l_t)$.

The merit function for the chosen emittance depends on r_1 and l_{12}. All remaining parameters are fixed when we are seeking for the minimum value of ρ, and we find this minimum for different parameters. The radius of the object diaphragm has the strongest influence on the beam spot size for the given emittance. It is very important to use the optimal r_1 for obtaining the smallest beam spot size. The optimal value of l_{12} provides the best distribution of density of the beam in the spot on the target and allows to avoid tailing of the beam.

6 THE QUADRUPOLE DOUBLET

The doublet is the simplest system which is applied in ion microprobe. It consists of two quadrupole lenses, separated by the drift space with length s.

TABLE 1.

s (cm)	4.0	12.0	4.0
l_1 (cm)	5.0	5.0	20.0
κ_1	0.587	0.462	0.856
κ_2	0.693	0.610	1.154
D_1	-14.7	-10.9	-6.8
D_2	-56.8	-59.8	-49.0
r (mm)	1.23 1.01	1.63 1.49	2.50 2.40
r_1 (μm)	9.82 11.9	7.94 8.68	5.18 5.40
r_2 (μm)	16.1 16.1	10.9 25.9	7.70 14.5
l_{12} (cm)	15.8 19.1	8.70 22.5	4.00 7.80
ρ_x (μm)	0.65 0.73	0.69 0.75	0.73 0.74
ρ_y (μm)	0.37 0.73	0.58 0.75	0.45 0.74

The effective lengths and the excitations of quadrupole lenses are denoted respectively by l_1 or l_2 and by κ_1 or κ_2. The doublet has different demagnification D_1 and D_2 in the x- and in y- planes. In the doublet we can consider only one mode of the excitations which produce negative demagnification. Our calculations show that to obtain the doublet with a minimum spot size one needs to take l_1, l_2 and s as small as possible.

The results of such calculations for a chosen emittance (10^{-9} m) and for different geometries are given in Table 1 (g = 20 cm, $l_1=l_2$, l_t= 8 m). In this Table and also in Tables 2 and 3 the upper (lower) values have been obtained for magnetic (electrostatic) systems, r is the maximum radius of the beam inside of the focusing system.

7 THE QUADRUPOLE TRIPLET

The simplest quadrupole system which gives the same absolute value of demagnification is a triplet. For the first mode of excitations, the triplet T_1 has negative demagnification in two planes and has no intermediate crossover. For the triplet T_2 with the second mode of excitations, there will be a crossover in one plane only and therefore one demagnification is positive and one is negative. For these two types of systems it is possible to find the excitations which produce equal absolute value of demagnification in both planes. For the optimal system the distance s_2 between the second and the third lenses should be chosen as small as possible. The distance s_1 in the triplet T_1 has a very small influence on the value of the minimum spot size and on the value of the demagnification. The separated triplet T_2 has a demagnification approximately two times that of the triplet T_1. The results of calculations for the triplets T_1 and T_2 for chosen emittance (10^{-9} m) and for different geometries are given in Table 2 (g = 20 cm, l_1 = l_2= l_3 = 5 cm, l_t= 8 m).

8 THE RUSSIAN QUADRUPLET

The most popular quadruplet system is the Russian quadruplet (RQ). This is the system where the first and last quadrupole lenses as well as the middle two are electrically coupled. It is an orthomorphic system, which means that demagnification is the same horizontally and vertically. In this system $s_1 = s_3$, $l_1 = l_4$, $l_2 = l_3$, $\kappa_1 = \kappa_4$ and $\kappa_2 = \kappa_3$. The RQ as used on the first focused probe at Harwell is the most popular configuration, partly because of its symmetry and its orthomorphic character which permits the use of circular object diaphragms. Our calculations show that the Russian

TABLE 2.

s_1 (m)	0.04	3.0	3.0
κ_1	0.505	0.126	0.213
κ_2	0.717	0.589	0.596
κ_3	0.597	0.690	0.704
D_1	-24.3	-23.4	39.0
D_2	-24.3	-23.4	-39.0
r (mm)	0.98 0.77	1.23 0.93	1.32 0.93
r_1 (μm)	10.9 13.8	9.87 13.1	16.7 22.0
r_2 (μm)	30.0 33.9	24.7 24.7	37.1 38.0
l_{12} (cm)	32.0 46.8	24.3 22.4	62.1 83.7
ρ_x (μm)	0.43 0.55	0.40 0.55	0.41 0.56
ρ_y (μm)	0.47 0.55	0.48 0.54	0.43 0.56

TABLE 3.

em (nm)	1	1	10^{-3}
g (cm)	20.0	20.0	5.0
s_2 (m)	0.04	3.0	3.0
κ_1	0.407	0.703	1.057
κ_2	0.658	0.599	0.704
D	-20.5	190.9	1094
r (mm)	0.94 0.82	1.22 0.93	0.194 0.146
r_1 (μm)	10.0 11.5	82.1 108.0	1.87 2.49
r_2 (μm)	24.3 29.9	41.3 38.1	1.74 1.41
l_{12} (cm)	24.4 34.4	339 412	325 351
ρ_x (μm)	0.48 0.51	0.43 0.59	0.0017 0.0023
ρ_y (μm)	0.50 0.54	0.41 0.56	0.0017 0.0023

quadruplet for the first mode of excitations with no crossover before the target (RQ_1) has a minimum spot size on the target if all values of s_1, s_2, l_1 and l_2 are chosen as small as possible (1).

The Russian quadruplet for the second mode of excitations with one crossover in both planes before the target (RQ_2) has a minimum spot size on the target if the distance between the last and last but one lenses is as small as possible and the distance between the last but one and the preceding lenses is $\approx(0.25\text{-}0.5)\ l_t$. The optimal RQ_2 is the separated Russian quadruplet (SRQ). Major advantages to be found in the SRQ are the possibility to obtain the largest symmetric demagnification D up to 1000 and even more and the very small crossover in both planes of focusing. The choice of a larger value for D enables the use of larger object diaphragms, with consequent reduction in scattering. In the SRQ system the optimal object and aperture diaphragms are comparable in size and any scattering that occurs at either of these diaphragms can be virtually eliminated by slits placed at each of the subsequent crossovers. Using these systems, it is possible to obtain probes with nanometer resolution. SRQ systems are being installed at Melbourne, Huntsville, Denton, Krakow and Leipzig. Some results of our numerical investigation are presented in Table 3. The last column of this table gives the example of the nanoprobe.

9 THE ELECTROSTATIC COAXIAL MULTIPLE CYLINDER SYSTEM

We have investigated the possibility of the use of the axisymmetric electrostatic multiple cylinder lenses. For this purpose we have proposed and used a new method of the optimal synthesis of electrostatic focusing systems (2). This method allows to determine the real parameters of focusing electrostatic systems with rotational symmetry, which gives the minimum beam spot size on the target for a given emittance. For some systems the optimal synthesis has been performed and this minimum has been found together with the appropriate values of radii of two diaphragms . We studied the physical model of the beam focusing system which consists of two round diaphragms and $2n+1$ cylinders, having equal radius r_{cyl} with thin walls and rotational symmetry about the central axis z. The $2j+1-$th cylinders have equal length l_1 $(j=1,2,...n)$ and zero potential. The $2j-$th cylinders have the same potential V and the same length l_2. The system has equal gaps l_g between cylinders. We use the following notations for the geometry of the system: $\Delta z_1 = a$, $\Delta z_{2j} = \Delta z_{2j+2} = l_f$, $\Delta z_{2j+1} = l_c$, $\Delta z_{2n+3} = g$.

The results of our calculations for this optimal system (l_c =2 cm, l_t = 8 m, $l_1 = l_2$, g = 20 cm) are given in Table 4. From these results it follows that the use of many cylinders provides a smaller potential V but at the same time the demagnification is decreasing and the minimum spot size is growing.

TABLE 4.

n	1	1	5
l_f (cm)	8.0	20.0	8.0
eV/E_0	0.492	0.605	0.195
r_{cyl} (cm)	1.52	4.06	1.52
l_g (cm)	2.0	4.4	2.0
l_2 (cm)	8.0	18.6	8.0
D	-26.2	-17.8	-9.35
r (mm)	1.33	2.33	1.48
r_1 (μm)	14.0	7.98	10.5
r_2 (μm)	23.6	15.8	17.9
l_{12} (cm)	33.0	11.0	18.8
ρ (μm)	0.49	0.41	1.02

10 CONCLUSION

The general matrix method for the optimisation of the quadrupole configuration and for axisymmetric electrostatic systems accurate to terms of third order inclusive is proposed and developed. A new beam program, using this method and based on the software Mathematica, has been written and realized. A numerical optimization has been performed for the quadrupole doublet, triplet, Russian quadruplet and for the electrostatic multiple cylinder lenses. The best microprobe focusing system is the Russian quadruplet with the second mode of excitations which provides a very big range of demagnification, up to 1000 and more, and gives the smallest spot size. The minimum beam spot size is proportional to $em^{3/4}$ or to $I^{3/8}$ and for the electrostatic quadrupoles is 1.1÷1.4 times bigger, than for magnetic ones. The use of electrostatic multiple cylinder lenses allows to design the simplest and cheapest 1-MeV microprobe for 200 keV potential V.

REFERENCES

1. Dymnikov A.D., Jamieson D.N., Legge G.J.F., *Nucl. Instr. and Meth.*, **B 104**, 64-67 (1995).
2. Dymnikov A.D. and Martínez G., *Nucl. Instr. and Meth.*, (in press).
3. Dymnikov A.D. and Hellborg R., *Nucl. Instr . and Meth.*, **A 330**, 323-362, (1993).
4. Dymnikov A.D. and Perelshtein E. A., *Nucl. Instr. and Meth.*, **148** , 567-571, (1978).

*Work supported by DGICYT, Ref.: SAB95-0208

IONOLUMINESCENCE COMBINED WITH PIXE IN THE NUCLEAR MICROPROBE FOR THE STUDY OF INORGANIC MATERIALS

C. Yang[1], K. G. Malmqvist[1], J. M. Hanchar[2], R. J. Utui[1], M. Elfman[1] P. Kristiansson[1], J. Pallon[1], A. Sjöland[1]

[1]*Nuclear Physics Department, Lund University and Institute of Technology, Sölvegatan 14, S-223 62, Lund, Sweden*
[2]*Rensselaer Polytechnic Institute, Department of Earth and Environmental, Sciences, Troy, NY 12180-4389; present address: Argonne National Laboratory, Chemical Technology Division, 9700 South Cass Ave., Argonne, IL 60439-4837, USA.*

Ionoluminescence (IL) produced by MeV/amu particles interacting with materials in the nuclear microprobe and often observed as visible light, carries useful information on lattice properties and nature of luminescence centers, such as trace ion substituents and structural defects in inorganic materials. IL also allows chemical valence identification for some elements and the determination of optical/electrical properties for some materials. IL has been observed to be very sensitive to changes in lattice properties. Beam modification and damage effects using a nuclear microprobe have also been investigated. $ZrSiO_4$ and MgO crystals were used in the study, although this tool may also be applied to the other types of crystals. The combination of the IL technique with the well-established PIXE method in the nuclear microprobe provides a unique tool for both general luminescence studies and material analysis; it adds a new dimension to the study of the microstructure in various inorganic materials.

INTRODUCTION

Ionoluminescence (IL) is a luminescence phenomenon induced by energetic ion particles in solids. In contrast to processes leading to the production of X-rays or gamma-rays, in which the inner electron shell or the nucleus is involved, the process resulting in luminescence is usually related to the valence electrons in the outermost shell of the atom or in the sub-shell near the outermost shell. Thus, the luminescence process is chemically, optically and electronically sensitive to the chemical state of the ions and to the local conditions of chemical binding in the sample, or the conditions of the crystal lattice.

Publications concerning the investigation of ionoluminescence introduced by a beam of kinetic particles are to be found throughout the scientific literature published between the 1950's and today (1, 2, 3, 4, 5). A systematic investigation of ionoluminescence as an analytical tool for the characterization of materials was initiated by Yang et al. in Lund in 1992 in cooperation with Australian groups (6). The aim was to develop a new tool to obtain chemical/optical information for use in a nuclear microprobe and to enhance the application of traditional methods of ion beam analysis such as Particle Induced X-ray Emission (PIXE), Rutherford Back-Scattering (RBS) and Nuclear Reaction Analysis (NRA). The traditional ion beam methods are not sensitive to states of valence electrons. Introducing ionoluminescence into ion beam analysis, and in particular employing it in the nuclear microprobe, can enhance the strength of ion beam analysis considerably (7, 8, 9, 10). The simultaneous, quantitative data of multiple trace elements provided by PIXE helps interpret the IL spectroscopic and microscopic results. IL is very sensitive to the dynamic changes of the properties of the samples during ion beam analysis, making it useful as a diagnostic tool to study the processes involved in material modification by the ion beam. IL shows a great potential for the study of defects in crystals. It can reveal changes in lattice properties of a crystal that reflect the changes in geological conditions it has been subjected to during its growth. It can further identify lattice-damage induced internally over geological time scale, by radioactive-isotope impurities in some minerals. IL/PIXE studies of zircon grains provide valuable information for geochronology. The principles of the technique demonstrated in this paper on the two large-band-gap crystals $ZrSiO_4$ (zircon) and MgO can also be extended to investigations of other types of inorganic materials.

INSTRUMENTAION

The IL system employed here was partially described elsewhere (4, 5, 6). The operation range of the monochromator extends from UV to near-IR (1000 nm), with a grating of 1200 lines/mm. For spectroscopic analysis, the scan speed of the grating is usually set at a 0.5 nm/sec. Two photomultiplier tubes (PMTs) are used for IL

CP392, *Application of Accelerators in Research and Industry*, edited by J. L. Duggan and I. L. Morgan
AIP Press, New York © 1997

wavelength response from 160 nm to 650 nm; the PMT has a very low dark current and it is often used as an imaging device due to its high signal-to-noise-ratio. The other PMT is of the Hamamatsu R943-02 type, with a wide response region of wavelength from 160 nm to 950 nm; it is often applied for IL spectral analysis due to this wide response range.

A photodiode-array based system for spectral analysis, Hamamatsu PMA-50, is also applied for IL spectral analysis. It makes very high-speed spectral acquisition possible, in a wavelength range from 350 nm to 900 nm. The exposure time can be set to anywhere from 20 millisecond to the order of minutes. It is especially useful for beam modification studies. By contrast, a spectral analysis system, operated with a PMT and grating-scan control, usually takes about 10 to 20 minutes to acquire a full IL spectrum. However, since the PMT offers a higher signal-to-noise ratio than a photodiode-array based system in the blue region, the PMT is very useful for IL imaging application in blue-region.

An IL image can be recorded by employing a focused beam and a PMT for light detection, where the combination of the IL imaging with μ-PIXE imaging is straight forward. An IL image can also be recorded simply with an optical microscope and a video camera (or film).

ION BEAM MODIFICATION EFFECTS IN A LARGE BAND GAP CRYSTAL

The IL method can be combined with PIXE in the Nuclear Microprobe for characterizing crystal micro-structures and for studying the effect of beam-modification of solid samples. As an example, a synthetic zircon crystal ($ZrSiO_4$) doped with holmium was tested. The PIXE result gives the following contents of trace elements: Ho 2390 ppm, Hf 180 ppm, Pb less than 30 ppm, U less than 30 ppm, Fe less than 5 ppm, Mn less than 5 ppm.

As shown in Fig. 1, the location indicated by the letter "M" in the maps was modified by using a 2.55 MeV proton beam with a fixed-beam spot (about 10 μm beam size) and a beam dose of 6.0 10^{17} protons/cm^2. The intensity of the IL signal in the short wavelength region (centered at 360 nm) was quenched (shown on the lower and the left hand map), while the IL intensity in the long wavelength region (from 500 nm to 800 nm) is enhanced (shown on the lower and right hand map) due to the beam modification of the crystal.

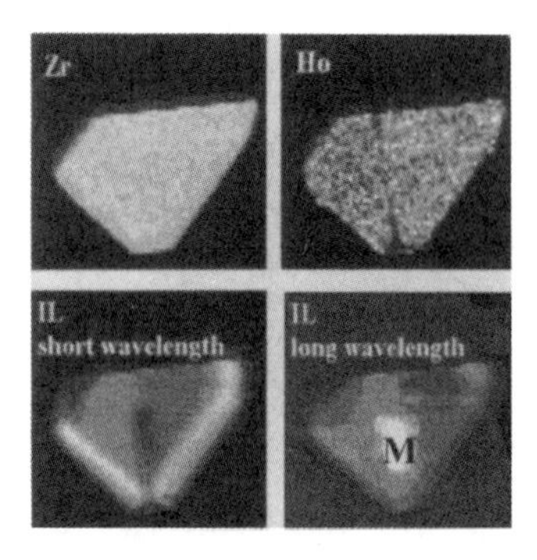

FIGURE 1. IL and PIXE imaging of a synthetic zircon ($ZrSiO_4$) crystal doped with Ho. The location indicated by the letter "M" was previously probed by IL/PIXE spectral analysis using a fixed-beam spot.

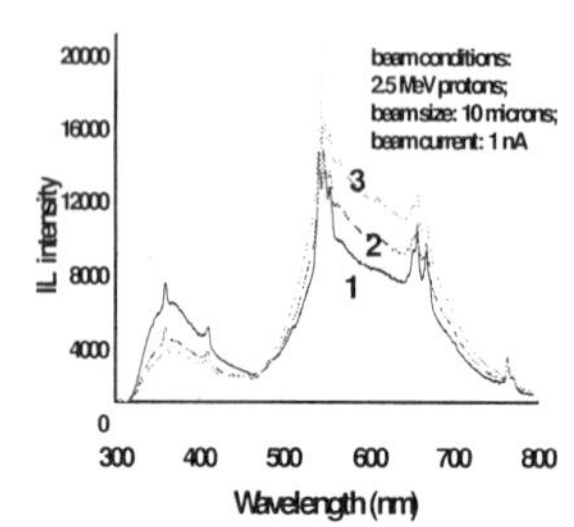

FIGURE 2. IL spectra of $ZrSiO_4$ (Ho doped). Exposure time 0.5 sec and interval time 1.0 sec. Beam conditions: 2.55 MeV protons, beam size 10 μm, beam current 1 nA. Exposure time 0.5 sec and interval time 1.0 sec.

The IL spectrum at the location probed by the static beam (indicated by letter "M" in Fig. 1) is shown in Fig. 2. The first spectral line was obtained with a fresh start for an exposure time of 0.5 sec. The beam was kept striking the same location of the sample thereafter. After an interval time of 1.0 second, the second spectrum was acquired with the same exposure time of 0.5 sec and in the same way the third spectrum was acquired. There are two broad, major distributions: one is from 300 to 470 nm, the other from 480 to 800 nm. The broad distributions are very sensitive to beam modifications. The narrow peaks superimposed on the two broad distributions, are contributed by REE^{3+} emissions. The heights of these narrow peaks are almost unchanged if the beam current variation (a few percent) during the measurement is corrected. For the broad distributions, we observed that the continuous accumulation of the beam dose at the same location leads to a maximized level of the IL emission at the region (480 to 800 nm). A further beam dose eventually reduces the IL emission in all regions.

The IL emission intensity at the region (480 to 800 nm) was observed to be enhanced by a factor of 5 to 10 at a maximum with a certain beam dose. The modification effects caused by the beam may be applied in manufacturing a light-emission-diode (LED) made from this type of large-band-gap crystal due to the fact that electroluminescence (EL) may also show the similar color features as IL. The broad distribution of the luminescence light, in this case, from 480 nm to 800 nm, can be greatly enhanced through the optimization of beam modification of the crystal. Some narrow emission lines can also be introduced into an optical system by doping crystals with REEs. The intensities of the narrow lines can be controlled by controlling the level of the doping.

DEFECTS IN CRYSTALS

A MgO crystal was investigated by IL panchromatic imaging (wavelength range from 350 nm to 650 nm) by employing a blue-sensitive PMT (Hamamatsu R585). The result is shown in Fig. 3. The imaging range is approximately 200 μm, and a beam of 2.55 MeV protons focused down to the beam size of 2 microns was applied. The dislocation defects are clearly revealed in the IL imaging. PIXE analysis in the imaged area yields the following trace elements contents: Fe 300 ppm, Mn less than 2 ppm. Datta et al. 1980 (11) made an observation of a MgO crystal by CL method; they studied the dislocation

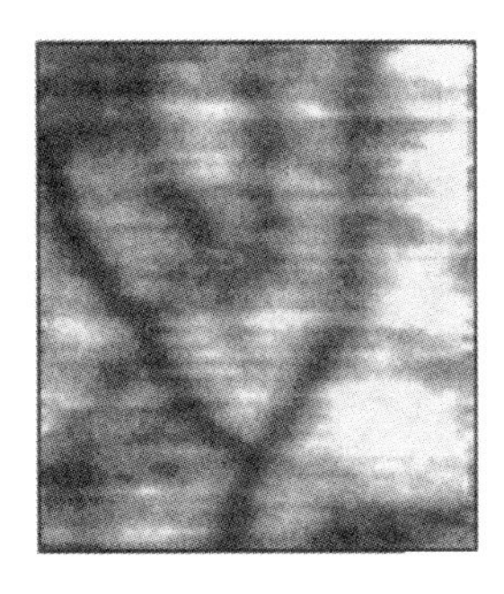

FIGURE 3. The dislocation defect in a MgO crystal is revealed by the panchromatic IL image. A blue-sensitive PMT (Hamamatsu R585) was applied. Beam conditions: 2.55 MeV protons, beam size 2 μm, beam current 10 pA, accumulation time 10 min.

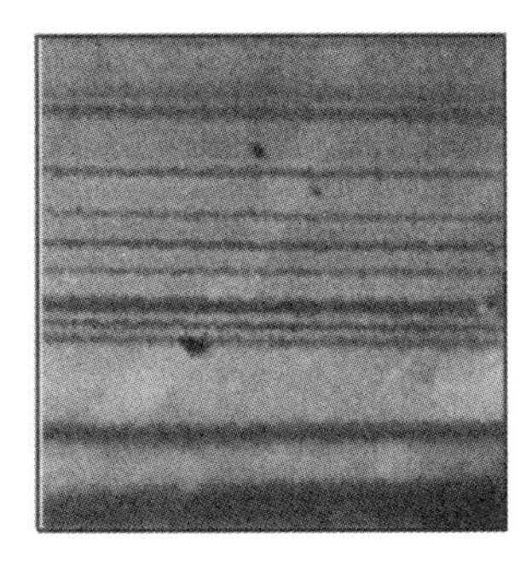

FIGURE 4. A zonation pattern in a zircon crystal (natural mineral) under a panchromatic-IL investigation. The zonation pattern is related to the dislocation defects formed over a long period of mineral growth through different geological periods. There is a strong emission with a broad distribution centered at 580 nm. The IL spectrum is shown in the Fig. 5.

by chromatic CL imaging and reported that the slip lines of the deformed sample yield a broad CL emission peak centered at around 466 nm.

A zircon crystal (a natural mineral) was investigated showing zonation pattern in a panchromatic IL investigation with an imaging range 300 μm, as displayed in Fig. 4. The zonation pattern can be related to the dislocation defects formed over a long period of mineral growth in different geological environments. The IL spectrum in Fig. 5, shows a strong emission with a broad distribution centered at 580 nm. PIXE analysis of the sample gives the following trace element contents: Dy 110 ppm, Er 150 ppm, Hf 9000 ppm, Fe 20 ppm, Ni 20 ppm, Cu 48 ppm. Efforts were made to look for a possible link between the IL pattern and the distribution of trace elements by the combination of the IL imaging method with μ-PIXE imaging method. There is no evidence that any luminescence activator can be related to the IL pattern. However, a correlation exists between the IL pattern and a partial distribution pattern of iron. This can be explained as the quenching effects of Fe^{2+} that is cooperated into the zircon lattice and therefore is optically activated. The partially non-optically active Fe distributed in the sample plays no role to influence the IL pattern.

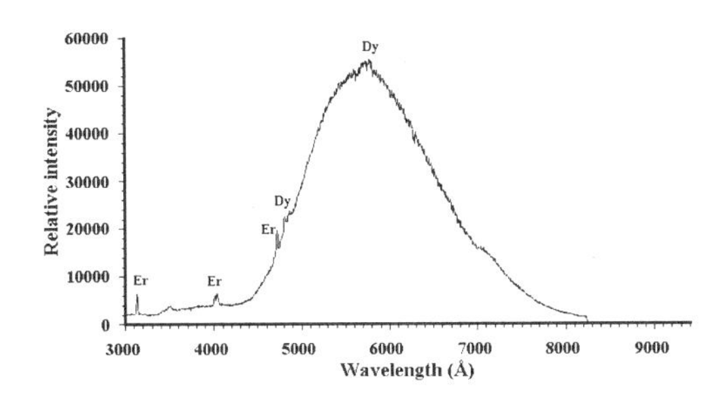

FIGURE 5. IL spectrum of the natural zircon. Trace element contents in the sample: Dy 110 ppm, Er 150 ppm, Hf 9000 ppm, Fe 20 ppm, Ni 20 ppm. Beam conditions for the spectral analysis: 2.55 MeV protons, beam size 50 μm, beam current 0.5 nA.

We noticed that the dominant IL emission at 580 nm can be enhanced by beam modification with a certain beam dose.

The growth pattern revealed by IL can be very useful in re-constructing the information on previous geological environments, during which a crystal was formed or to which a crystal was subjected. The IL/PIXE method may also be useful in the diagnostic analysis of crystal growth characteristics in the study of crystal synthesis.

ZIRCON SAMPLE USED FOR GEOCHRONOLOGY

A zircon grain used for geological age-dating was investigated by back-scattered electron (BSE) method in the scanning electron microscope (SEM) and IL/PIXE method in the nuclear microprobe. The zircon sample, AM-86-11a, is from a fayalite-bearing leucogranite from Ausable Falls in the Adirondack Mountains, of New York State (12). Zircons from this granite have been dated with the single crystal U/Pb method (13). The data plot on a chord with an upper intercept of 1045 +/- 3 Ma. This granite is part of a group of *leucogranitic granitoids* which occur throughout the *Adirondack* region.

The zircons from these rocks commonly contain high U rims (dark in IL or CL, and bright in BSE) that when not abraded to remove these high-U regions often yield discordant U/Pb analyses. When the rims are abraded the U/Pb data from the zircons plot within error of *concordia*. The rims do not yield statistically younger ages than the inner regions of the zircons.

A BSE image of the zircon grain from AM-86-11a is displayed in Fig. 6a, and elemental and IL images obtained with the nuclear microprobe are displayed in Fig. 6b. A beam of 2.55 MeV protons focused down to about 2 or 3 micrometers was used for the IL/PIXE imaging.

The IL/PIXE study of the zircon followed an electron microprobe investigation of the same grain where a 30 keV electron beam was used. The proton beam probes a deeper excitation volume than the electron beam, therefore, the IL signal yield is still high even after the considerable beam damage made by the electron beam in a relatively shallow surface of the sample. IL is usually

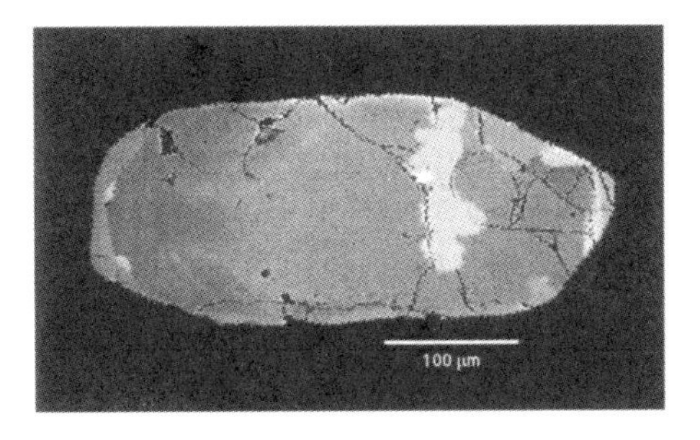

FIGURE 6a. A back-scattered electron image of zircon from sample AM-86-11a. The elemental and IL images of the same zircon are displayed in Fig. 6b.

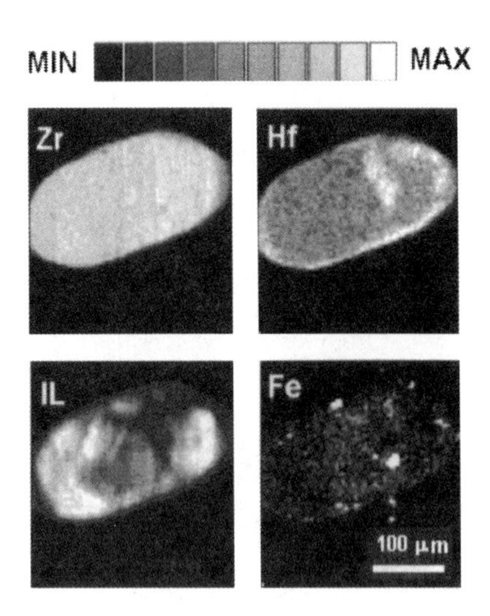

FIGURE 6b. The elemental and IL images of the zircon grain in Fig. 6a. The beam conditions for the panchromatic IL imaging: 2.55 MeV protons, beam size 2 µm, beam current 10 pA; for the PIXE imaging, the beam current was increased to approximately 100 pA.

volume sensitive if a high energy proton beam is used in the nuclear microprobe in which a smearing of the IL image can occur due to the summation of the signals from different depths. A variation of lattice properties in zircons, such as the structure of a core and overgrowth, or an area of large lattice-damage, sometimes can not be revealed by BSE and CL methods in the SEM if the structure is not uncovered by a polishing process of sample preparation, since the 30 keV electrons can only probe a depth in the order of 1 micrometer. IL/PIXE can probe much deeper, for an example, a 2.55 MeV proton beam can reach a depth of about 10 to 40 micrometers in zircons. Therefore, it is very practical to apply the PIXE/IL method for the investigation of homogeneity of samples for selection of zircon grains used for U-Pb age-dating (4).

For the IL imaging a beam current of a few to 10 pA was applied and the IL data were collected for approximately 10 minutes. This assures that the information of interest is obtained before significant beam modification/damage takes place. The subsequent PIXE imaging was performed over the same area as the IL imaging, however, with a much higher beam current (about 100 pA).

From the IL and CL imaging patterns, it is evident that there is considerable lattice damage in the dark IL and CL regions in the zircon. These areas are enriched in trace elements including the heavy rare earth elements (HREEs), Fe, Mn, U, and Pb (as determined with micro-PIXE). Obviously, the areas of the zircon with severe lattice damage cannot contain valuable geochemical information as is possible in a normal zircon crystal. Likewise a damaged region of the zircon cannot preserve the geological age information provided by U/Pb dating technique, since the amorphous condition make it possible for various impurities to migrate and diffuse through the damaged area from/to the external environment. This could yield a considerable error in the U-Pb dating.

CONCLUSIONS

IL/PIXE can be a unique tool for the study of beam modification and damage and also for the diagnostic analysis of crystal defects. IL/PIXE can provide useful information for the study of zircons used for geochronology. The combination of the IL technique with the well-established PIXE method in the Nuclear Microprobe adds a new dimension to the study of inorganic materials.

ACKNOWLEGEMENTS

The authors wish to thank the following Swedish foundations: The Swedish Natural Science Research Council, The Royal Physiographic Society, and C. Trygger for financial support received for the present study. One of the authors (Dr. Changyi Yang) wishes to acknowledge the support from Chinese Academy of Sciences.

REFERENCES

1. Leverenz HW (1950), *An introduction to luminescence of solids*, New York, John Wiley and sons, Inc., 1950, ch. 7, pp.399-471.
2. Derham CJ, Geake DJE and Walker G, Luminescence of Enstatite Achondrite Meteorites, *Nature*, **203**, 134-136 (1964).
3. Townsend PD (1987), Optical effects of ion implantation, *Rep. Prog. Phys.* **50**, 501-558 (1987).
4. C. Yang, N.P.-O. Homman, K.G. Malmqvist, L. Johansson, N.M. Halden, V. Barbin, Ionoluminescence: a new tool for nuclear microprobe in geology, *Scanning Microscopy*, **9-1**, 43-62 (1995).
5. Malmqvist KG, Elfman M, Remond G, and Yang C, PIXE and Ionoluminescence - a Synergetic Analytical Combination, *Nucl. Instr. and Meth.* **B109/110**, 227-233 (1996).
6. Yang C, Larsson NP-O, Swietlicki E, Malmqvist KG, Jamieson DN, Ryan CG, Imaging with Ionoluminescence (IL) in a Nuclear Microprobe, *Nucl. Instr. and Meth.* **B77**, 188-194 (1993).
7. N.P.-O. Homman, C. Yang, and K. G. Malmqvist, A highly sensitive method for rare-earth element analysis using Ionoluminescence combined with PIXE, *Nucl. Instr. and Meth.* **A353**, 610-614 (1994).
8. C. Yang, N.P.-O. Homman, L. Johansson, K.G. Malmqvist, Micro-characterising zircon mineral grain by Ionoluminescence combined with PIXE, *Nucl. Instr. and Meth.* **B85**, 808-814 (1994).
9. Bettiol AA, Jamieson DN, Prawer S and Allen MG (1994), Ion beam induced luminescence from diamond and other crystals from a nuclear microprobe, *Nucl. Instr. and Meth,* **B85**, 775-779 (1994).
10. Malmqvist, K.G., Analytical techniques in nuclear microprobes, *Nucl. Inst. and Meth.* **B 104**, 138-151 (1995).
11. Datta S., Aeberli K.E., Boswarva I.M., and Holt D.B., *J. Microscopy* **118**, 367-369 (1980).
12. Chiarenzelli, J.R. and McLelland, J.M., Age and regional relationships of granitoid rocks of the Adirondack highlands, *Journal of Geology* **99**, 571-590 (1990).
13. McLelland, J.M. and McLelland, J.M., New high precision U/Pb zircon ages for Lyon Mtn. Gneiss, Adirondacks, and tectonic implications, *Geological Society of America Abstracts with Programs*, **28**, 80-88 (1996).

RESONANCE IONIZATION OF SPUTTERED ATOMS: QUANTITATIVE ANALYSIS IN THE NEAR-SURFACE REGION OF SILICON WAFERS

W. F. Calaway,[1] D. R. Spiegel,[2] A. H. Marshall,[2] S. W. Downey,[3] and M. J. Pellin[1]

[1]*Material Science, Chemical Technology, and Chemistry Divisions, Argonne National Laboratory, Argonne, IL 60439*

[2]*Department of Physics, Trinity University, San Antonio, TX 78212*

[3]*Lucent Technologies, Orlando, FL 32819*

The unambiguous identification and quantification of low levels of metallic impurities on Si wafers are difficult problems due to the rapidly changing chemical activity near the surface. Air-exposed Si surfaces typically possess a native oxide layer several atoms thick plus a top monolayer of various silicon-containing molecules. Resonance ionization spectroscopy (RIS) used for postionization in secondary neutral mass spectrometry (SNMS) is uniquely suited to this task. The high sensitivity of this technique allows detection of metals at parts per billion levels with monolayer sensitivity. The high selectivity of RIS allows unambiguous identification of elements, while the reduced matrix effects of SNMS allow quantification of the photoionized element. Characterization of Si surfaces using RIS/SNMS has been explored by measuring the concentration profiles of Ca in the near-surface region of Si wafers of varying degrees of cleanliness. Calcium detection can be problematic due to the isobaric interference with SiC, particularly in the near-surface region during fabrication of devices due to the use of organic photoresist. Three different resonance ionization schemes for Ca have been examined and compared for effectiveness by calculating detection limits for Ca in Si in the chemically active near-surface region.

INTRODUCTION

Surface analysis and materials characterization have played key roles in many of the remarkable advances made by the electronics industry, and it is anticipated that this trend will continue in the future. Secondary neutral mass spectrometry (SNMS) combined with postionization techniques that employ resonance ionization spectroscopy (RIS) is an emerging method that could help meet this future need of the semiconductor industry.(1) RIS/SNMS instruments are very sensitive tools for the unambiguous identification of elemental impurities in bulk materials and on surfaces. Because of the high selectivity and sensitivity of the technique, concentrations of various contaminants at part per billion (ppb) levels and below can be obtained on or near the surface.(2-4) In addition, the technique has the potential for quantitative determinations(2-4) and imaging(5,6) at these low concentrations.

One key advantage of RIS is the unambiguous identification of a specific element in the sputter flux. Because atomic transitions are discrete and narrow, a laser tuned to an atomic absorption can ionize that element without ionizing the remainder of the sputtered flux. In traditional mass spectrometry, identification of elements can be compromised by isobaric interferences. A prime example is $^{28}Si^{12}C$ with ^{40}Ca. This is a difficult problem for semiconductor manufacturing where photoresist may be used for patterning. Typically, such interferences are circumvented by monitoring a minor isotope of an element that is free from isobaric interferences or by using a high resolution instrument.(7) In either case, suppression of the interference results in reduced sensitivity. RIS, on the other hand, can selectively photoionize an element from a large background of potentially interfering neutral molecules and atoms, eliminating mass analyses that reduce sensitivity.

The ability to distinguish elemental impurities in the presence of mass-interfering molecules makes RIS/SNMS instruments particularly well-suited for trace analysis in samples of rapidly changing special chemical composition such as a Si wafer. Metallic Si reacts with air at room temperature to form a thin self-limiting oxide. On top of the silicon oxide is a mixed layer of air contaminants sometimes called an aerosol layer. Determining trace metallic impurities in these thin layers is very difficult, because the amount of material available for analysis is small and backgrounds are generally large due to isobaric interferences. In addition to the benefit from selectivity, SNMS has a specific advantage because secondary neutral yields of ground state atoms remains relatively constant even during compositional changes of the matrix, improving quantitative measurements. Thus, due to the high selectivity and small matrix effects, elemental impurity concentrations in the aerosol, oxide, and metallic layers can be measured, permitting determinations of depth versus concentration profiles for these thin layers.

This paper describes experiments that examined the detection of Ca in Si by RIS/SNMS. The aim was to evaluate ionization schemes that could be used to detect Ca on the surface of Si wafers, where SiC concentrations are anticipated to be large enough to present a severe isobaric interference problem. To that end, a series of Si wafers in various states of cleanliness has been analyzed by three different RIS schemes, the sensitivity and selectivity of each scheme have been evaluated, and detection limits determined.

CP392, *Application of Accelerators in Research and Industry*, edited by J. L. Duggan and I. L. Morgan
AIP Press, New York © 1997

EXPERIMENTAL

All analyses were performed with a SNMS instrument especially designed and built for high detection sensitivity when employing laser postionization. The instrument, commonly called surface analysis by resonance ionization of sputtered atoms (SARISA), is composed of a pulsed ion source, a time-of-flight mass spectrometer, and tunable dye lasers. Data collection and control of the instrument are accomplished via a personal computer. Details of the instrument have been previously reported.(8)

For these experiments, the primary ion beam was composed of 4 keV Ar^+ ions having a peak current of 3 μA in a 250 μm diameter spot. During analyses the ion beam was typically pulsed on target for 600 ns prior to the triggering of the laser. The same ion beam was also used to clean and ion mill (sputter erode) samples for measuring depth profiles. Here the ion beam impinged on the sample continuously while being swept in a 2 × 2 mm square raster pattern, allowing rapid removal of material (~1 monolayer/s) and preventing crater wall effects from biasing results during depth profile measurements.

Postionization was accomplished by employing resonance ionization to detect sputtered neutral ground-state calcium atoms. Three different ionization schemes were tested to determine ionization efficiency and the photoionization background that they generate. These are shown in Fig. 1. Two excimer-pumped dye lasers were used to produce the desired frequencies of laser light. A single XeCl excimer laser (308 nm) operating at >100 mJ/pulse and 77 Hz was used as the pump laser. Each dye laser produced between 1 and 5 mJ/pulse of tunable light depending upon the efficiency of the dye. The 240-nm light was generated by frequency doubling 480-nm light and had an intensity of about 0.1 mJ/pulse. The cross sections of the laser beams were reduced to a 1-mm spot in front of the sample using a long focal length lens. Saturation of the photoionization was confirmed by plotting signal versus laser intensity for each wavelength. Only the ionization step for the 1-color (240 nm + 240 nm) scheme did not reach saturation. During

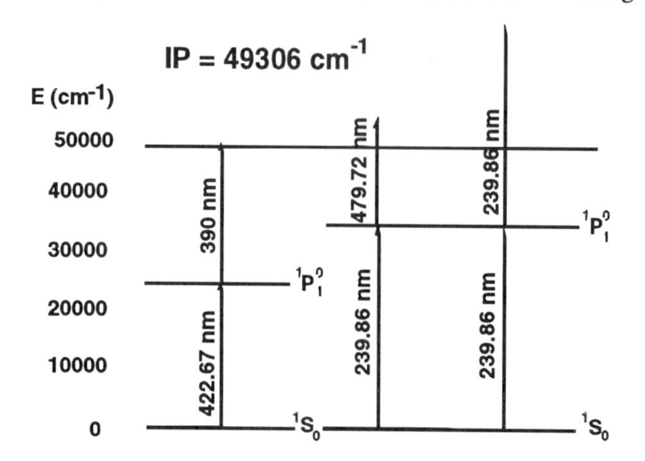

FIGURE 1. Schematic energy level diagram showing the resonance ionization schemes used to detect Ca.

experiments, the intensity of each laser beam was attenuated as much as possible while still maintaining saturation of each transition in order to minimize nonresonant ionization.

Samples were prepared at Bell Laboratories. One sample consisted of an uncleaned Si substrate topped by 100 nm of SiO_2. The other samples had only the typical native oxide layer. Photoresist was laid down on all wafers and ashed by plasma. The samples without the oxide cap were then cleaned by various procedures, before being shipped to Argonne for analysis. The Si wafers were cleaved into squares, 0.5 cm on a side, and attached to sample stubs with conductive epoxy in a class 10 desktop Cleansphere (Safetech, Ltd.) before being loaded into the SARISA instrument.

Alignment and calibration of the instrument were accomplished using a pure Ca metal target as described previously.(9) Measurements on the Ca target were performed before and after each day's analyses to determine whether the instrument calibration had drifted over the day. Significant memory effects were found to occur due to the sputtering of the pure Ca target. However, it was determined that sputtering a Si or Mo target for an hour between the Ca metal measurement and a sample measurement eliminated any discernible memory effect.

Data were collected as a function of depth by measuring and integrating the ^{40}Ca mass spectral peak as the Si surface was eroded by the ion beam. To probe changes in composition in the near-surface region, a number of spectra were collected sequentially with only the pulsed ion beam, as it probed the sample, used to erode the surface. Once a dose sufficient to remove approximately one monolayer was accumulated in this mode, additional data were collected by ion milling the surface between periods of data collection. Depth was calculated from the total ion fluence assuming a sputtering yield of 1.5.

RESULTS AND DISCUSSION

The potential of RIS for unambiguous elemental determinations is shown in Fig. 2, where two mass spectra from a Si sample in the mass region near Ca are presented. The data result from removing 1.5×10^{11} ions in a 250 μm spot (7×10^{14} ions/cm^2) for each spectrum, corresponding to 12,000 laser shots for our particular instrument. Thus, the data were collected with removal of approximately one monolayer. The two mass spectra are identical except that the wavelength of the dye laser used to access the Ca resonance transition was tuned off the atomic absorption by 0.2 nm for the off-resonance spectrum. As can be seen in Fig. 2, when the laser is tuned on resonance, Ca atoms are ionized and detected by the mass spectrometer. When the resonance transition is not accessed, no Ca atoms are detected. It therefore is straightforward to identify the actual Ca signal, even in the presence of large backgrounds, by subtracting the two spectra. The results shown in Fig. 2 indicate that Ca is present in the aerosol layer of this particular sample at a concentration of 3 ppm.

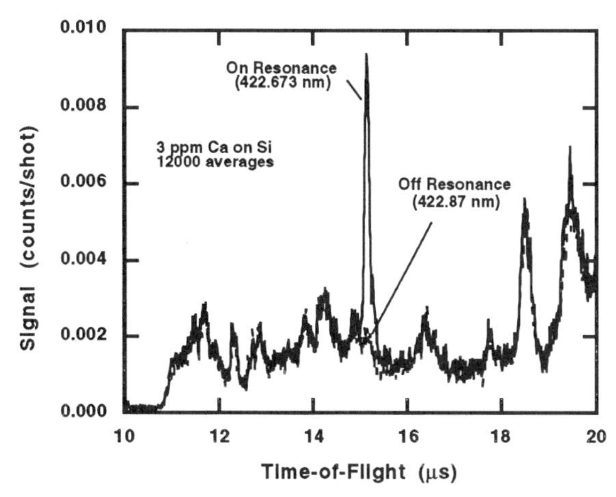

FIGURE 2. Mass spectra showing how the signal from Ca can be distinguished from a large background by tuning on (422.673 nm) and off (422.870 nm) a resonance frequency.

Besides the Ca signal, the mass spectra contain several peaks that correspond to major constituents sputtered from the surface as neutrals and then photoionized nonresonantly. Three of these peaks (i.e., Si, Si_2, and SiO) can be used to monitor the transitions between the aerosol, oxide, and metal layers.(10) For the oxide-capped wafer, a signal change is observed very near to the known oxide thickness of 100 nm, indicating that our dose to depth conversion is approximately correct. For the "cleaned" wafers, the nonresonant signals indicate that the metal/oxide interface is in the region between 1 and 10 nm, a reasonable thickness for a native oxide layer on Si. Changes in signal levels for Si, Si_2, and SiO much nearer the surface (<1 nm) are also observed in all depth profiles. These signal changes mark the location of the aerosol/oxide interface. However, because the sputtering yield from the aerosol layer is unknown, the calibration of the depth scale at these very small distances is highly uncertain.

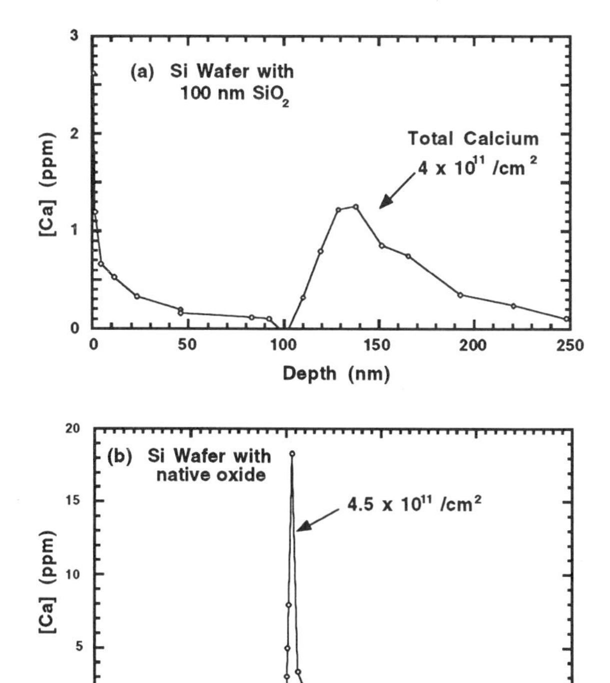

FIGURE 3. Depth profile of Ca concentration in an (a) oxide-capped and (b) cleaned Si wafer. Note that the depth axis in spectrum (b) has been shifted so that the oxide/metal interfaces for the two spectra are approximately aligned.

Shown in Fig. 3 are the Ca signals as a function of depth for two Si wafers. The upper panel (Fig. 3a) shows results for the capped oxide wafer. The lower panel (Fig. 3b) shows data from a wafer where the photoresist has been removed by cleaning. There are two features that are striking about the results. First, in both measurements, the Ca is found to be located at the SiO_2/Si interface. This was found to be true for all measurements. Second, the total amount of Ca detected in each measurement is the same within experimental uncertainty. Integrating the Ca versus depth curves in Fig. 3, the amount of Ca detected is found to be approximately $4 \times 10^{11}/cm^2$ for both samples.

As can be seen in Fig. 3, the Ca is spread over a much wider depth range (50-100 nm) in the oxide-capped wafer. By examining the nonresonant Si, SiO and Si_2 signals, it was found that the Ca had diffused from the SiO_2/Si interface into the Si metal phase. Presumably, the Ca was present on the wafer before the oxide cap was laid down and diffused into the Si upon heating during plasma etching.

It was anticipated that the cleaning procedures would remove most if not all of the Ca; and so, it was surprising that Ca was detected on all samples at such high and similar concentrations. The most reasonable explanation for this result is that the wafers were contaminated by Ca after cleaning rather than during the photoresist steps. If this is the case, then the Ca must be diffusing across the native oxide layer to the SiO_2/Si interface during the analysis and piling up there. Similar results have been reported for other elements in Si, particularly alkali metals.(11,12) To investigate this hypothesis, the nonresonant signal from potassium as a function of depth was examined. It was found that the potassium concentration tracked the Ca concentration and also piled up at the oxide/metal interface. It thus appears that the observation of Ca at the SiO_2/Si interface is due to charge- or ion-induced diffusion and that the source of Ca is contamination following wafer preparations.

Calcium detection limit for the three resonance ionization schemes used in this study have been determined. They were calculated assuming that the uncertainties in the measurements are dominated by random error associated with pulse counting (Poison statistics). For measurements where a background is subtracted from a signal and assuming a limiting signal-to-noise ratio of unity, the detection limit, D_L, is calculated from the background count

Table 1. Calcium detection limits (ppb) in the aerosol layer and bulk when analyzing a Si matrix. The two measurements correspond to removing 6×10^{11} and 2×10^{13} atoms, respectively.

RIS Scheme	Aerosol Layer	Bulk Si
240 + 240 nm	5000	110
240 + 480 nm	640	11
423 + 390 nm	80	5

rate, B, the number of averages, n, and the sensitivity of the instrument, S, from

$$D_L = S \cdot \frac{1+\sqrt{1+8nB}}{2n} \quad . \tag{1}$$

Calculations of D_L, have been made for two types of analyses and are presented in Table 1. Detection limits were calculated for measurements in the aerosol layer where the amount of material is limited. Here it is assumed that one monolayer (6×10^{11} atoms = 1.2×10^4 averages) would be removed. The second set of detection limits is for a bulk measurement and assumes that data collection time, which is controlled by the laser repetition rate, limits the analysis. For our present instrument, if the signal and background are both averaged for 5×10^5 shots (about 1 hour at 77 Hz), approximately 2×10^{13} atoms of Si are consumed (25 monolayers) during the analysis.

As can be seen from Table 1, there are wide variations in detection limits for the different ionization schemes and also between the aerosol layer and bulk for each scheme. The bulk detection limits are expected to be lower than the aerosol limits by about a factor of 6.5 because of the larger number of averages assumed for these analyses. The bulk detection limits are lower than the aerosol detection limits by more than 6.5 for all three ionization schemes. This is because the backgrounds in the aerosol layer are larger than the corresponding backgrounds in the bulk Si. This is found to be true independent of laser intensity or frequency, suggesting that the increase in background is due to secondary ions. It appears that the much larger secondary ion yield in the aerosol, compared to the bulk Si, overwhelms SARISA's secondary ion suppression capabilities. Methods for improving suppression of secondary ions are currently being examined.

The poorer detection limits for the 1-color scheme (240 nm + 240 nm) are mainly due to the fact that saturation of the ionization step was not achieved for this scheme. However, even if a more intense 240-nm laser beam was available, the benefit may not be as great as anticipated because the 240-nm light appears to photoionize the isobarically interfering SiC molecules more efficiently than the other wavelengths. In fact, the poorer detection limit for the 1-color scheme is at least partially attributable to nonresonant photoionization of SiC by 240-nm light.

The uncertainty in the detection limits listed in Table 1 is about a factor of two.(9) Thus, when analyzing Ca in bulk Si, it is unclear whether there is any difference between the two 2-color ionization schemes that were tested. However, when searching for Ca in the aerosol layer, the benefit of using the 423-nm resonance is clear. This difference is mainly due to background differences since both ionization schemes were saturated and thus produced similar sensitivities. As mentioned above, 240-nm light appears to photoionize SiC; however, only low-intensity 240-nm light (~10 μJ/pulse) was used in the 2-color scheme, since this resonance transition is strong and saturates at low laser intensities. While we were unable to find any spectroscopic information on SiC, it has been reported that SiC_2 absorbs near 480 nm.(13) It may be that the background at mass 40 is due to fragmenting SiC_2.

CONCLUSIONS

Resonance ionization schemes for the detection of Ca by postionized SNMS have been tested to determine sensitivity and detection limits in a Si matrix. The two-photon, two-color schemes that were tested both work well for clean bulk Si, producing detection limits in the low ppb range. For Ca detection at the surface, where SiC may be present, the ionization path using the resonance transition at 423 nm was clearly superior. The ubiquitous nature of Ca appears to have caused contamination of the wafers after preparation at a level of 4×10^{11} atoms/cm^2.

ACKNOWLEDGMENTS

Work supported by the U.S. Department of Energy, ER/LTT and ER/BES-Materials Sciences, under Contract W-31-109-ENG-38. DRS and AHM gratefully acknowledge support from the U.S. Department of Energy's Division of Education Programs.

REFERENCES

1. Calaway, W. F., Coon, S. R., Pellin, M. J., Gruen, D. M., Gordon, M., Diebold, A. C., Maillot, P., Banks, J. C., and Knapp, J. A., *Surf. Interface Anal.* **21**, 131-137 (1994).
2. Pellin, M. J., Young, C. E., Calaway, W. F., Whitten, J. E., Gruen, D. M., Blum, J. D., Hutcheon, I. D., and Wasserburg, G. J., *Phil. Trans. R. Soc. Lond. A* **333**, 133-146 (1990).
3. Arlinghaus, H. F., Spaar, M. T., and Thonnard, N., *J. Vac. Sci. Technol. A* **8**, 2318-2322 (1990).
4. Pappas, D. L., Hrubowchak, D. M., Erwin, M. H., and Winograd, N., *Science* **243**, 64-66 (1989).
5. Ma, Z., Thompson, R. N., Lykke, K. R., Pellin, M. J., and Davis, A. M., *Rev. Sci. Instrum.* **66**, 3168-3176 (1995).
6. Wood, M., Zhou, Y., Brummel, C. L., and Winograd, N., *Anal. Chem.* **66**, 2425-2432 (1994).
7. Niehuis, N., "Time-of-Flight SIMS," in *Proceedings of the Eighth International Conference on Secondary Ion Mass Spectrometry (SIMS VIII)*, 1991, pp. 269-280.
8. Pellin, M. J., Young, C. E., and Gruen, D. M., *Scanning Microsc.* **2**, 1353-1364 (1988).
9. Calaway, W. F., Coon, S. R., Pellin, M. J., Young, C. E., Whitten, J. E., Wiens, R. C., Gruen, D. M., Stingeder, G., Penka, V., Grasserbauer, M., and Burnett, D. S., *Inst. Phys. Conf. Ser.* **128**, 271-274 (1992).
10. Emerson, A. B., Ma, Y., Wise, M. L., Green, M. L., and Downey, S. W., *J. Vac. Sci. Technol. B* **14**, 301-304 (1996).
11. Magee, C. W., and Harrington, W. L., *Appl. Phys. Lett.* **33**, 193-196 (1978).
12. Morgan, A. E., and Maillot, P., *Appl. Phys. Lett.* **50**, 959-961 (1987).
13. Verma, R. D., and Nagaraj, S., *Can. J. Phys.* **52**, 1938-1948 (1974).

AIP Conference Proceedings

	Title	L.C. Number	ISBN
No. 286	Ordering Disorder: Prospect and Retrospect in Condensed Matter Physics: Proceedings of the Indo-U.S. Workshop (Hyderabad, India 1993)	93-072549	1-56396-255-1
No. 287	Production and Neutralization of Negative Ions and Beams: Sixth International Symposium (Upton, NY 1992)	93-72821	1-56396-103-2
No. 288	Laser Ablation: Mechanismas and Applications-II: Second International Conference (Knoxville, TN 1993)	93-73040	1-56396-226-8
No. 289	Radio Frequency Power in Plasmas: Tenth Topical Conference (Boston, MA 1993)	93-72964	1-56396-264-0
No. 290	Laser Spectroscopy: XIth International Conference (Hot Springs, VA 1993)	93-73050	1-56396-262-4
No. 291	Prairie View Summer Science Academy (Prairie View, TX 1992)	93-73081	1-56396-133-4
No. 292	Stability of Particle Motion in Storage Rings (Upton, NY 1992)	93-73534	1-56396-225-X
No. 293	Polarized Ion Sources and Polarized Gas Targets (Madison, WI 1993)	93-74102	1-56396-220-9
No. 294	High-Energy Solar Phenomena: A New Era of Spacecraft Measurements (Waterville Valley, NH 1993)	93-74147	1-56396-291-8
No. 295	The Physics of Electronic and Atomic Collisions: XVIII International Conference (Aarhus, Denmark, 1993)	93-74103	1-56396-290-X
No. 296	The Chaos Paradigm: Developments an Applications in Engineering and Science (Mystic, CT 1993)	93-74146	1-56396-254-3
No. 297	Computational Accelerator Physics (Los Alamos, NM 1993)	93-74205	1-56396-222-5
No. 298	Ultrafast Reaction Dynamics and Solvent Effects (Royaumont, France 1993)	93-074354	1-56396-280-2
No. 299	Dense Z-Pinches: Third International Conference (London, 1993)	93-074569	1-56396-297-7
No. 300	Discovery of Weak Neutral Currents: The Weak Interaction Before and After (Santa Monica, CA 1993)	94-70515	1-56396-306-X
No. 301	Eleventh Symposium Space Nuclear Power and Propulsion (3 Vols.) (Albuquerque, NM 1994)	92-75162	1-56396-305-1 (Set) 156396-301-9 (pbk. set)

	Title	L.C. Number	ISBN
No. 302	Lepton and Photon Interactions/ XVI International Symposium (Ithaca, NY 1993)	94-70079	1-56396-106-7
No. 303	Slow Positron Beam Techniques for Solids and Surfaces Fifth International Workshop (Jackson Hole, WY 1992)	94-71036	1-56396-267-5
No. 304	The Second Compton Symposium (College Park, MD 1993)	94-70742	1-56396-261-6
No. 305	Stress-Induced Phenomena in Metallization Second International Workshop (Austin, TX 1993)	94-70650	1-56396-251-9
No. 306	12th NREL Photovoltaic Program Review (Denver, CO 1993)	94-70748	1-56396-315-9
No. 307	Gamma-Ray Bursts Second Workshop (Huntsville, AL 1993)	94-71317	1-56396-336-1
No. 308	The Evolution of X-Ray Binaries (College Park, MD 1993)	94-76853	1-56396-329-9
No. 309	High-Pressure Science and Technology—1993 (Colorado Springs, CO 1993)	93-72821	1-56396-219-5 (Set)
No. 310	Analysis of Interplanetary Dust (Houston, TX 1993)	94-71292	1-56396-341-8
No. 311	Physics of High Energy Particles in Toroidal Systems (Irvine, CA 1993)	94-72098	1-56396-364-7
No. 312	Molecules and Grains in Space (Mont Sainte-Odile, France 1993)	94-72615	1-56396-355-8
No. 313	The Soft X-Ray Cosmos ROSAT Science Symposium (College Park, MD 1993)	94-72499	1-56396-327-2
No. 314	Advances in Plasma Physics Thomas H. Stix Symposium (Princeton, NJ 1992)	94-72721	1-56396-372-8
No. 315	Orbit Correction and Analysis in Circular Accelerators (Upton, NY 1993)	94-72257	1-56396-373-6
No. 316	Thirteenth International Conference on Thermoelectrics (Kansas City, Missouri 1994)	95-75634	1-56396-444-9
No. 317	Fifth Mexican School of Particles and Fields (Guanajuato, Mexico 1992)	94-72720	1-56396-378-7
No. 318	Laser Interaction and Related Plasma Phenomena 11th International Workshop (Monterey, CA 1993)	94-78097	1-56396-324-8
No. 319	Beam Instrumentation Workshop (Santa Fe, NM 1993)	94-78279	1-56396-389-2
No. 320	Basic Space Science (Lagos, Nigeria 1993)	94-79350	1-56396-328-0
No. 321	The First NREL Conference on Thermophotovoltaic Generation of Electricity (Copper Mountain, CO 1994)	94-72792	1-56396-353-1

	Title	L.C. Number	ISBN
No. 322	Atomic Processes in Plasmas Ninth APS Topical Conference (San Antonio, TX)	94-72923	1-56396-411-2
No. 323	Atomic Physics 14 Fourteenth International Conference on Atomic Physics (Boulder, CO 1994)	94-73219	1-56396-348-5
No. 324	Twelfth Symposium on Space Nuclear Power and Propulsion (Albuquerque, NM 1995)	94-73603	1-56396-427-9
No. 325	Conference on NASA Centers for Commercial Development of Space (Albuquerque, NM 1995)	94-73604	1-56396-431-7
No. 326	Accelerator Physics at the Superconducting Super Collider (Dallas, TX 1992-1993)	94-73609	1-56396-354-X
No. 327	Nuclei in the Cosmos III Third International Symposium on Nuclear Astrophysics (Assergi, Italy 1994)	95-75492	1-56396-436-8
No. 328	Spectral Line Shapes, Volume 8 12th ICSLS (Toronto, Canada 1994)	94-74309	1-56396-326-4
No. 329	Resonance Ionization Spectroscopy 1994 Seventh International Symposium (Bernkastel-Kues, Germany 1994)	95-75077	1-56396-437-6
No. 330	E.C.C.C. 1 Computational Chemistry F.E.C.S. Conference (Nancy, France 1994)	95-75843	1-56396-457-0
No. 331	Non-Neutral Plasma Physics II (Berkeley, CA 1994)	95-79630	1-56396-441-4
No. 332	X-Ray Lasers 1994 Fourth International Colloquium (Williamsburg, VA 1994)	95-76067	1-56396-375-2
No. 333	Beam Instrumentation Workshop (Vancouver, B. C., Canada 1994)	95-79635	1-56396-352-3
No. 334	Few-Body Problems in Physics (Williamsburg, VA 1994)	95-76481	1-56396-325-6
No. 335	Advanced Accelerator Concepts (Fontana, WI 1994)	95-78225	1-56396-476-7 (Set) 1-56396-474-0 (Book) 1-56396-475-9 (CD-Rom)
No. 336	Dark Matter (College Park, MD 1994)	95-76538	1-56396-438-4
No. 337	Pulsed RF Sources for Linear Colliders (Montauk, NY 1994)	95-76814	1-56396-408-2
No. 338	Intersections Between Particle and Nuclear Physics 5th Conference (St. Petersburg, FL 1994)	95-77076	1-56396-335-3

	Title	L.C. Number	ISBN
No. 339	Polarization Phenomena in Nuclear Physics Eighth International Symposium (Bloomington, IN 1994)	95-77216	1-56396-482-1
No. 340	Strangeness in Hadronic Matter (Tucson, AZ 1995)	95-77477	1-56396-489-9
No. 341	Volatiles in the Earth and Solar System (Pasadena, CA 1994)	95-77911	1-56396-409-0
No. 342	CAM -94 Physics Meeting (Cacun, Mexico 1994)	95-77851	1-56396-491-0
No. 343	High Energy Spin Physics Eleventh International Symposium (Bloomington, IN 1994)	95-78431	1-56396-374-4
No. 344	Nonlinear Dynamics in Particle Accelerators: Theory and Experiments (Arcidosso, Italy 1994)	95-78135	1-56396-446-5
No. 345	International Conference on Plasma Physics ICPP 1994 (Foz do Iguaçu, Brazil 1994)	95-78438	1-56396-496-1
No. 346	International Conference on Accelerator-Driven Transmutation Technologies and Applications (Las Vegas, NV 1994)	95-78691	1-56396-505-4
No. 347	Atomic Collisions: A Symposium in Honor of Christopher Bottcher (1945-1993) (Oak Ridge, TN 1994)	95-78689	1-56396-322-1
No. 348	Unveiling the Cosmic Infrared Background (College Park, MD, 1995)	95-83477	1-56396-508-9
No. 349	Workshop on the Tau/Charm Factory (Argonne, IL, 1995)	95-81467	1-56396-523-2
No. 350	International Symposium on Vector Boson Self-Interactions (Los Angeles, CA 1995)	95-79865	1-56396-520-8
No. 351	The Physics of Beams Andrew Sessler Symposium (Los Angeles, CA 1993)	95-80479	1-56396-376-0
No. 352	Physics Potential and Development of $\mu^+ \mu^-$ Colliders: Second Workshop (Sausalito, CA 1994)	95-81413	1-56396-506-2
No. 353	13th NREL Photovoltaic Program Review (Lakewood, CO 1995)	95-80662	1-56396-510-0
No. 354	Organic Coatings (Paris, France, 1995)	96-83019	1-56396-535-6
No. 355	Eleventh Topical Conference on Radio Frequency Power in Plasmas (Palm Springs, CA 1995)	95-80867	1-56396-536-4
No. 356	The Future of Accelerator Physics (Austin, TX 1994)	96-83292	1-56396-541-0
No. 357	10th Topical Workshop on Proton-Antiproton Collider Physics (Batavia, IL 1995)	95-83078	1-56396-543-7

	Title	L.C. Number	ISBN
No. 358	The Second NREL Conference on Thermophotovoltaic Generation of Electricity	95-83335	1-56396-509-7
No. 359	Workshops and Particles and Fields and Phenomenology of Fundamental Interactions (Puebla, Mexico 1995)	96-85996	1-56396-548-8
No. 360	The Physics of Electronic and Atomic Collisions XIX International Conference (Whistler, Canada, 1995)	95-83671	1-56396-440-6
No. 361	Space Technology and Applications International Forum (Albuquerque, NM 1996)	95-83440	1-56396-568-2
No. 362	Two-Center Effects in Ion-Atom Collisions (Lincoln, NE 1994)	96-83379	1-56396-342-6
No. 363	Phenomena in Ionized Gases XXII ICPIG (Hoboken, NJ, 1995)	96-83294	1-56396-550-X
No. 364	Fast Elementary Processes in Chemical and Biological Systems (Villeneuve d'Ascq, France, 1995)	96-83624	1-56396-564-X
No. 365	Latin-American School of Physics XXX ELAF Group Theory and Its Applications (México City, México, 1995)	96-83489	1-56396-567-4
No. 366	High Velocity Neutron Stars and Gamma-Ray Bursts (La Jolla, CA 1995)	96-84067	1-56396-593-3
No. 367	Micro Bunches Workshop (Upton, NY, 1995)	96-83482	1-56396-555-0
No. 368	Acoustic Particle Velocity Sensors: Design, Performance and Applications (Mystic, CT, 1995)	96-83548	1-56396-549-6
No. 369	Laser Interaction and Related Plasma Phenomena (Osaka, Japan 1995)	96-85009	1-56396-445-7
No. 370	Shock Compression of Condensed Matter-1995 (Seattle, WA 1995)	96-84595	1-56396-566-6
No. 371	Sixth Quantum 1/f Noise and Other Low Frequency Fluctuations in Electronic Devices Symposium (St. Louis, MO, 1994)	96-84200	1-56396-410-4
No. 372	Beam Dynamics and Technology Issues for + - Colliders 9th Advanced ICFA Beam Dynamics Workshop (Montauk, NY, 1995)	96-84189	1-56396-554-2
No. 373	Stress-Induced Phenomena in Metallization (Palo Alto, CA 1995)	96-84949	1-56396-439-2
No. 374	High Energy Solar Physics (Greenbelt, MD 1995)	96-84513	1-56396-542-9
No. 375	Chaotic, Fractal, and Nonlinear Signal Processing (Mystic, CT 1995)	96-85356	1-56396-443-0
No. 376	Chaos and the Changing Nature of Science and Medicine: An Introduction (Mobile, AL 1995)	96-85220	1-56396-442-2

	Title	L.C. Number	ISBN
No. 377	Space Charge Dominated Beams and Applications of High Brightness Beams (Bloomington, IN 1995)	96-85165	1-56396-625-7
No. 378	Surfaces, Vacuum, and Their Applications (Cancun, Mexico 1994)	96-85594	1-56396-418-X
No. 379	Physical Origin of Homochirality in Life (Santa Monica, CA 1995)	96-86631	1-56396-507-0
No. 380	Production and Neutralization of Negative Ions and Beams / Production and Application of Light Negative Ions (Upton, NY 1995)	96-86435	1-56396-565-8
No. 381	Atomic Processes in Plasmas (San Francisco, CA 1996)	96-86304	1-56396-552-6
No. 382	Solar Wind Eight (Dana Point, CA 1995)	96-86447	1-56396-551-8
No. 383	Workshop on the Earth's Trapped Particle Environment (Taos, NM 1994)	96-86619	1-56396-540-2
No. 384	Gamma-Ray Bursts (Huntsville, AL 1995)	96-79458	1-56396-685-9
No. 385	Robotic Exploration Close to the Sun: Scientific Basis (Marlboro, MA 1996)	96-79560	1-56396-618-2
No. 386	Spectral Line Shapes, Volume 9 13th ICSLS (Firenze, Italy 1996)		1-56396-656-5
No. 387	Space Technology and Applications International Forum (Albuquerque, NM 1997)	96-80254	1-56396-679-4 (Case set) 1-56396-691-3 (Paper set)
No. 388	Resonance Ionization Spectroscopy 1996 Eighth International Symposium (State College, PA 1996)	96-80324	1-56396-611-5
No. 389	X-Ray and Inner-Shell Processes 17th International Conference (Hamburg, Germany 1996)	96-80388	1-56396-563-1
No. 390	Beam Instrumentation Proceedings of the Seventh Workshop (Argonne, IL 1996)	97-70568	1-56396-612-3
No. 391	Computational Accelerator Physics (Williamsburg, VA 1996)	97-70181	1-56396-671-9
No. 392	Applications of Accelerators in Research and Industry: Proceedings of the Fourteenth International Conference (Denton, TX 1996)	97-71846	1-56396-652-2
No. 393	Star Formation Near and Far Seventh Astrophysics Conference (College Park, MD 1996)	97-71978	1-56396-678-6
No. 394	NREL/SNL Photovoltaics Program Review Proceedings of the 14th Conference—A Joint Meeting (Lakewood, CO 1996)	97-72645	1-56396-687-5